# HANDBOOK OF SOLVENTS

George Wypych, Editor

ChemTec Publishing

William Andrew publishing

Toronto – New York 2001

Published by ChemTec Publishing
38 Earswick Drive, Toronto, Ontario M1E 1C6, Canada

Co-published by William Andrew Inc.
13 Eaton Avenue, Norwich, NY 13815, USA


ISBN 1-895198-24-0



Canadian Cataloguing in Publication Data

Handbook of Solvents

Includes bibliographical references and index

ISBN 1-895198-24-0 (ChemTec Publishing)
ISBN 0-8155-1458-1 (William Andrew Inc.)
Library of Congress Catalog Card Number: 00-106798

1. Solvents--Handbooks, manuals, etc. I. Wypych, George

TP247.5.H35 2000 661'.807 C00-900997-3

Printed in Canada by Transcontinental Printing Inc., 505 Consumers Rd.
Toronto, Ontario M2J 4V8

# TABLE OF CONTENTS


| | | |
|---|---|---|
| | *Preface* | *xxvii* |
| | GEORGE WYPYCH | |
| 1 | **INTRODUCTION** | 1 |
| | CHRISTIAN REICHARDT | |
| 2 | **FUNDAMENTAL PRINCIPLES GOVERNING SOLVENTS USE** | 7 |
| 2.1 | Solvent effects on chemical systems | 7 |
| | ESTANISLAO SILLA, ARTURO ARNAU, IÑAKI TUÑÓN | |
| 2.1.1 | Historical outline | 7 |
| 2.1.2 | Classification of solute-solvent interactions | 10 |
| 2.1.2.1 | Electrostatic | 11 |
| 2.1.2.2 | Polarization | 12 |
| 2.1.2.3 | Dispersion | 13 |
| 2.1.2.4 | Repulsion | 14 |
| 2.1.2.5 | Specific interactions | 15 |
| 2.1.2.6 | Hydrophobic interactions | 16 |
| 2.1.3 | Modelling of solvent effects | 17 |
| 2.1.3.1 | Computer simulations | 18 |
| 2.1.3.2 | Continuum models | 20 |
| 2.1.3.3 | Cavity surfaces | 21 |
| 2.1.3.4 | Supermolecule models | 22 |
| 2.1.3.5 | Application example: glycine in solution | 23 |
| 2.1.4 | Thermodynamic and kinetic characteristics of chemical reactions in solution | 27 |
| 2.1.4.1 | Solvent effects on chemical equilibria | 27 |
| 2.1.4.2 | Solvent effects on the rate of chemical reactions | 28 |
| 2.1.4.3 | Example of application: addition of azide anion to tetrafuranosides | 30 |
| 2.1.5 | Solvent catalytic effects | 32 |
| 2.2 | Molecular design of solvents | 36 |
| | KOICHIRO NAKANISHI | |
| 2.2.1 | Molecular design and molecular ensemble design | 36 |
| 2.2.2 | From prediction to design | 37 |
| 2.2.3 | Improvement in prediction method | 38 |
| 2.2.4 | Role of molecular simulation | 39 |
| 2.2.5 | Model system and paradigm for design | 40 |
| | Appendix. Predictive equation for the diffusion coefficient in dilute solution | 41 |
| 2.3 | Basic physical and chemical properties of solvents | 42 |
| | GEORGE WYPYCH | |
| 2.3.1 | Molecular weight and molar volume | 43 |
| 2.3.2 | Boiling and freezing points | 44 |
| 2.3.3 | Specific gravity | 46 |
| 2.3.4 | Refractive index | 47 |
| 2.3.5 | Vapor density and pressure | 48 |
| 2.3.6 | Solvent volatility | 49 |
| 2.3.7 | Flash point | 50 |
| 2.3.8 | Flammability limits | 51 |
| 2.3.9 | Sources of ignition and autoignition temperature | 52 |
| 2.3.10 | Heat of combustion (calorific value) | 54 |
| 2.3.11 | Heat of fusion | 54 |
| 2.3.12 | Electric conductivity | 54 |
| 2.3.13 | Dielectric constant (relative permittivity) | 54 |
| 2.3.14 | Occupational exposure indicators | 56 |
| 2.3.15 | Odor threshold | 56 |

2.3.16 Toxicity indicators . . . 57
2.3.17 Ozone-depletion and creation potential . . . 58
2.3.18 Oxygen demand . . . 58
2.3.19 Solubility . . . 58
2.3.20 Other typical solvent properties and indicators . . . 60
**3 PRODUCTION METHODS, PROPERTIES, AND MAIN APPLICATIONS . . . 65**
3.1 Definitions and solvent classification . . . 65
GEORGE WYPYCH
3.2 Overview of methods of solvent manufacture . . . 69
GEORGE WYPYCH
3.3 Solvent properties . . . 74
GEORGE WYPYCH
3.3.1 Hydrocarbons . . . 75
3.3.1.1 Aliphatic hydrocarbons . . . 75
3.3.1.2 Aromatic hydrocarbons . . . 76
3.3.2 Halogenated hydrocarbons . . . 78
3.3.3 Nitrogen-containing compounds (nitrates, nitriles) . . . 79
3.3.4 Organic sulfur compounds . . . 80
3.3.5 Monohydric alcohols . . . 81
3.3.6 Polyhydric alcohols . . . 83
3.3.7 Phenols . . . 84
3.3.8 Aldehydes . . . 85
3.3.9 Ethers . . . 86
3.3.10 Glycol ethers . . . 87
3.3.11 Ketones . . . 88
3.3.11 Acids . . . 90
3.3.12 Amines . . . 91
3.3.13 Esters . . . 92
3.3.14 Comparative analysis of all solvents . . . 94
3.4 Terpenes . . . 96
TILMAN HAHN, KONRAD BOTZENHART, FRITZ SCHWEINSBERG
3.4.1 Definitions and nomenclature . . . 96
3.4.2 Occurrence . . . 96
3.4.3 General . . . 96
3.4.4 Toxicology . . . 97
3.4.5 Threshold limit values . . . 97
**4 GENERAL PRINCIPLES GOVERNING DISSOLUTION OF MATERIALS IN SOLVENTS . . . 101**
4.1 Simple solvent characteristics . . . 101
VALERY YU. SENICHEV, VASILIY V. TERESHATOV
4.1.1 Solvent power . . . 101
4.1.2 One-dimensional solubility parameter approach . . . 103
4.1.3 Multi-dimensional approaches . . . 110
4.1.4 Hansen's solubility . . . 112
4.1.5 Three-dimensional dualistic model . . . 116
4.1.6 Solubility criterion . . . 119
4.1.7 Solvent system design . . . 120
4.2 Effect of system variables on solubility . . . 124
VALERY YU. SENICHEV, VASILIY V. TERESHATOV
4.2.1 General considerations . . . 124
4.2.2 Chemical structure . . . 126
4.2.3 Flexibility of a polymer chain . . . 127
4.2.4 Crosslinking . . . 128
4.2.5 Temperature and pressure . . . 128
4.2.6 Methods of calculation of solubility based on thermodynamic principles . . . 130

4.3 Polar solvation dynamics: Theory and simulations . . . 132
ABRAHAM NITZAN
4.3.1 Introduction. . . . 132
4.3.2 Continuum dielectric theory of solvation dynamics . . . 133
4.3.3 Linear response theory of solvation dynamics . . . 136
4.3.4 Numerical simulations of solvation in simple polar solvents: The simulation model . . . 138
4.3.5 Numerical simulations of solvation in simple polar solvents: Results and discussion . . . 140
4.3.6 Solvation in complex solvents . . . 144
4.3.7 Conclusions. . . . 145
4.4 Methods for the measurement of solvent activity of polymer solutions . . . 146
CHRISTIAN WOHLFARTH
4.4.1 Introduction. . . . 146
4.4.2 Necessary thermodynamic equations. . . . 149
4.4.3 Experimental methods, equipment and data reduction. . . . 154
4.4.3.1 Vapor-liquid equilibrium (VLE) measurements . . . 154
4.4.3.1.1 Experimental equipment and procedures for VLE-measurements. . . . 155
4.4.3.1.2 Primary data reduction . . . 170
4.4.3.1.3 Comparison of experimental VLE-methods . . . 175
4.4.3.2 Other measurement methods . . . 178
4.4.3.2.1 Membrane osmometry . . . 178
4.4.3.2.2 Light scattering . . . 181
4.4.3.2.3 X-ray scattering. . . . 184
4.4.3.2.4 Neutron scattering . . . 185
4.4.3.2.5 Ultracentrifuge . . . 186
4.4.3.2.6 Cryoscopy (freezing point depression of the solvent) . . . 188
4.4.3.2.7 Liquid-liquid equilibrium (LLE) . . . 189
4.4.3.2.8 Swelling equilibrium . . . 193
4.4.4 Thermodynamic models for the calculation of solvent activities of polymer solutions. . . . 195
4.4.4.1 Models for residual chemical potential and activity coefficient in the liquid phase . . . 196
4.4.4.2 Fugacity coefficients from equations of state . . . 207
4.4.4.3 Comparison and conclusions . . . 214
Appendix 4.4A . . . 223
**5 SOLUBILITY OF SELECTED SYSTEMS AND INFLUENCE OF SOLUTES . . . 243**
5.1 Experimental methods of evaluation and calculation of solubility parameters of polymers and solvents. Solubility parameters data . . . 243
VALERY YU. SENICHEV, VASILIY V. TERESHATOV
5.1.1 Experimental evaluation of solubility parameters of liquids. . . . 243
5.1.1.1 Direct methods of evaluation of the evaporation enthalpy . . . 243
5.1.1.2 Indirect methods of evaluation of evaporation enthalpy . . . 244
5.1.1.3 Static and quasi-static methods of evaluation of pair pressure. . . . 245
5.1.1.4 Kinetic methods . . . 245
5.1.2 Methods of experimental evaluation and calculation of solubility parameters of polymers. . . . 246
5.2 Prediction of solubility parameter . . . 253
NOBUYUKI TANAKA
5.2.1 Solubility parameter of polymers . . . 253
5.2.2 Glass transition in polymers . . . 254
5.2.2.1 Glass transition enthalpy . . . 254
5.2.2.2 $C_p$ jump at the glass transition . . . 256
5.2.3 Prediction from thermal transition enthalpies . . . 258
5.3 Methods of calculation of solubility parameters of solvents and polymers . . . 261
VALERY YU. SENICHEV, VASILIY V. TERESHATOV

5.4 Mixed solvents, a way to change the polymer solubility . . . 267
LIGIA GARGALLO AND DEODATO RADIC
5.4.1 Introduction . . . 267
5.4.2 Solubility-cosolvency phenomenon . . . 268
5.4.3 New cosolvents effects. Solubility behavior . . . 273
5.4.4 Thermodynamical description of ternary systems. Association equilibria theory of preferential adsorption . . . 274
5.4.5 Polymer structure of the polymer dependence of preferential adsorption. Polymer molecular weight and tacticity dependence of preferential adsorption . 277
5.5 The phenomenological theory of solvent effects in mixed solvent systems . . . 281
KENNETH A. CONNORS
5.5.1 Introduction . . . 281
5.5.2 Theory . . . 281
5.5.2.1 Principle . . . 281
5.5.2.2 The intersolute effect: solute-solute interactions . . . 282
5.5.2.3 The solvation effect: solute-solvent interaction . . . 283
5.5.2.4 The general medium effect: solvent-solvent interactions . . . 284
5.5.2.5 The total solvent effect . . . 285
5.5.3 Applications . . . 285
5.5.3.1 Solubility . . . 285
5.5.3.2 Surface tension . . . 288
5.5.3.3 Electronic absorption spectra . . . 290
5.5.3.4 Complex formation . . . 291
5.5.3.5 Chemical kinetics . . . 295
5.5.3.6 Liquid chromatography . . . 298
5.5.4 Interpretations . . . 298
5.5.4.1 Ambiguities and anomalies . . . 298
5.5.4.2 A modified derivation . . . 299
5.5.4.3 Interpretation of parameter estimates . . . 300
5.5.4.4 Confounding effects . . . 301
Solute-solute interactions . . . 301
Coupling of general medium and solvation effects . . . 301
The cavity surface area . . . 301
The role of interfacial tension . . . 302
**6 SWELLING . . . 305**
6.1 Modern views on kinetics of swelling of crosslinked elastomers in solvents . . 305
E. YA. DENISYUK, V. V. TERESHATOV
6.1.1 Introduction . . . 305
6.1.2 Formulation of swelling for a plane elastomer layer . . . 306
6.1.3 Diffusion kinetics of plane layer swelling . . . 310
6.1.4 Experimental study of elastomer swelling kinetics . . . 314
6.1.5 Conclusions . . . 317
6.2 Equilibrium swelling in binary solvents . . . 318
VASILIY V. TERESHATOV, VALERY YU. SENICHEV
6.3 Swelling data on crosslinked polymers in solvents . . . 327
VASILIY V. TERESHATOV, VALERY YU. SENICHEV
6.4 Influence of structure on equilibrium swelling . . . 331
VASILIY V. TERESHATOV, VALERY YU. SENICHEV
**7 SOLVENT TRANSPORT PHENOMENA . . . 339**
7.1 Introduction to diffusion, swelling, and drying . . . 339
GEORGE WYPYCH
7.1.1 Diffusion . . . 339
7.1.2 Swelling . . . 344
7.1.3 Drying . . . 348
7.2 Bubbles dynamics and boiling of polymeric solutions . . . 356
SEMYON LEVITSKY, ZINOVIY SHULMAN
7.2.1 Rheology of polymeric solutions and bubble dynamics . . . 356

7.2.1.1 Rheological characterization of solutions of polymers . . . 356
7.2.1.2 Dynamic interaction of bubbles with polymeric liquid . . . 363
7.2.2 Thermal growth of bubbles in superheated solutions of polymers . . . 372
7.2.3 Boiling of macromolecular liquids . . . 377
7.3 Drying of coated film . . . 386
SEUNG SU KIM AND JAE CHUN HYUN
7.3.1 Introduction . . . 386
7.3.2 Theory for the drying . . . 388
7.3.2.1 Simultaneous heat and mass transfer . . . 388
7.3.2.2 Liquid-vapor equilibrium . . . 389
7.3.2.3 Heat and mass transfer coefficient . . . 390
7.3.2.4 Prediction of drying rate of coating . . . 392
7.3.2.5 Drying regimes: constant drying rate period (CDRP) and falling drying rate period (FDRP) . . . 394
7.3.3 Measurement of the drying rate of coated film . . . 396
7.3.3.1 Thermo-gravimetric analysis . . . 396
7.3.3.2 Rapid scanning FT-IR spectrometer analysis . . . 399
7.3.3.3 High-airflow drying experiment using flame ionization detector (FID) total hydrocarbon analyzer . . . 401
7.3.3.4 Measurement of drying rate in the production scale dryer . . . 404
7.3.4 Miscellaneous . . . 407
7.3.4.1 Drying of coated film with phase separation . . . 407
7.3.4.2 Drying defects . . . 409
7.3.4.2.1 Internal stress induced defects . . . 409
7.3.4.2.2 Surface tension driven defects . . . 412
7.3.4.2.3 Defects caused by air motion and others . . . 414
7.3.4.3 Control of lower explosive level (LEL) in a multiple zone dryer . . . 414
**8 INTERACTIONS IN SOLVENTS AND SOLUTIONS . . . 419**
JACOPO TOMASI, BENEDETTA MENNUCCI, CHIARA CAPPELLI
8.1 Solvents and solutions as assemblies of interacting molecules . . . 419
8.2 Basic simplifications of the quantum model . . . 420
8.3 Cluster expansion . . . 424
8.4 Two-body interaction energy: the dimer . . . 424
8.4.1 Decomposition of the interaction energy of a dimer: variational approach . . . 426
The electrostatic term . . . 426
The induction term . . . 428
The exchange term . . . 428
The charge transfer term . . . 429
The dispersion term . . . 430
The decomposition of the interaction energy through a variational approach: a summary . . . 432
8.4.2 Basis set superposition error and counterpoise corrections . . . 433
8.4.3 Perturbation theory approach . . . 436
8.4.4 Modeling of the separate components of ΔE . . . 441
The electrostatic term . . . 441
The induction term . . . 445
The dispersion term . . . 446
The exchange (or repulsion) term . . . 447
The other terms . . . 448
A conclusive view . . . 448
8.4.5 The relaxation of the rigid monomer constraint . . . 449
8.5 Three- and many-body interactions . . . 451
Screening many-body effects . . . 453
Effective interaction potentials . . . 454
8.6 The variety of interaction potentials . . . 456
8.7 Theoretical and computing modeling of pure liquids and solutions . . . 461
8.7.1 Physical models . . . 461

| | | |
|---|---|---|
| 8.7.1.1 | Integral equation methods | 465 |
| 8.7.1.2 | Perturbation theories | 467 |
| 8.7.2 | Computer simulations | 468 |
| 8.7.2.1 | Car-Parrinello direct QM simulation | 470 |
| 8.7.2.2 | Semi-classical simulations | 472 |
| | Molecular dynamics | 472 |
| | Monte Carlo | 473 |
| | QM/MM | 478 |
| 8.7.3 | Continuum models | 479 |
| 8.7.3.1 | QM-BE methods: the effective Hamiltonian | 482 |
| 8.8 | Practical applications of modeling | 487 |
| | Dielectric constant | 487 |
| | Thermodynamical properties | 490 |
| | Compressibilities | 490 |
| | Relaxation times and diffusion coefficients | 491 |
| | Shear viscosity | 492 |
| 8.9 | Liquid surfaces | 492 |
| 8.9.1 | The basic types of liquid surfaces | 493 |
| 8.9.2 | Systems with a large surface/bulk ratio | 495 |
| 8.9.3 | Studies on interfaces using interaction potentials | 497 |
| **9** | **MIXED SOLVENTS** | **505** |
| | Y. Y. FIALKOV, V. L. CHUMAK | |
| 9.1 | Introduction | 505 |
| 9.2 | Chemical interaction between components in mixed solvents | 505 |
| 9.2.1 | Processes of homomolecular association | 505 |
| 9.2.2 | Conformic and tautomeric equilibrium. Reactions of isomerization | 506 |
| 9.2.3 | Heteromolecular association | 507 |
| 9.2.4 | Heteromolecular associate ionization | 507 |
| 9.2.5 | Electrolytic dissociation (ionic association) | 508 |
| 9.2.6 | Reactions of composition | 508 |
| 9.2.7 | Exchange interaction | 509 |
| 9.2.8 | Amphoterism of mixed solvent components | 509 |
| 9.2.8.1 | Amphoterism of hydrogen acids | 509 |
| 9.2.8.2 | Amphoterism of L-acids | 509 |
| 9.2.8.3 | Amphoterism in systems H-acid-L-acid | 510 |
| 9.2.8.4 | Amphoterism in binary solutions amine-amine | 510 |
| 9.3 | Physical properties of mixed solvents | 511 |
| 9.3.1 | The methods of expression of mixed solvent compositions | 511 |
| 9.3.1.1 | Permittivity | 513 |
| 9.3.1.2 | Viscosity | 515 |
| 9.3.1.3 | Density, molar volume | 516 |
| 9.3.1.4 | Electrical conductivity | 516 |
| 9.3.2 | Physical characteristics of the mixed solvents with chemical interaction between components | 517 |
| 9.3.2.1 | Permittivity | 518 |
| 9.3.2.2 | Viscosity | 519 |
| 9.3.2.3 | Density, molar volume | 521 |
| 9.3.2.4 | Conductivity | 522 |
| 9.3.3 | Chemical properties of mixed solvents | 524 |
| 9.3.3.1 | Autoprotolysis constants | 524 |
| 9.3.3.2 | Solvating ability | 526 |
| 9.3.3.3 | Donor-acceptor properties | 527 |
| 9.4 | Mixed solvent influence on the chemical equilibrium | 527 |
| 9.4.1 | General considerations | 527 |
| 9.4.2 | Mixed solvent effect on the position of equilibrium of homomolecular association process | 529 |
| 9.4.3 | Mixed solvent influence on the conformer equilibrium | 530 |

9.4.4 Solvent effect on the process of heteromolecular association . . . 532
9.4.4.1 Selective solvation. Resolvation . . . 538
9.4.5 Mixed solvent effect on the ion association process . . . 546
9.4.6 Solvent effect on exchange interaction processes . . . 552
Systems with non-associated reagents . . . 552
Systems with one associated participant of equilibrium . . . 553
Systems with two associated participants of equilibrium . . . 553
9.4.7 Mixed solvent effect on processes of complex formation . . . 556
9.5 The mixed solvent effect on the chemical equilibrium thermodynamics . . . 557
**10 ACID-BASE INTERACTIONS . . . 565**
10.1 General concept of acid-base interactions . . . 565
GEORGE WYPYCH
10.2 Effect of polymer/solvent acid-base interactions: relevance to the aggregation of PMMA . . . 570
S. BISTAC, M. BROGLY
10.2.1 Recent concepts in acid-base interactions . . . 570
10.2.1.1 The nature of acid-base molecular interactions . . . 571
10.2.1.1.1 The original Lewis definitions . . . 571
10.2.1.1.2 Molecular Orbital (MO) approach to acid-base reactions . . . 571
10.2.1.1.3 The case of hydrogen bonding . . . 573
10.2.1.2 Quantitative determination of acid-base interaction strength . . . 574
10.2.1.2.1 Perturbation theory . . . 574
10.2.1.2.2 Hard-Soft Acid-Base (HSAB) principle . . . 574
10.2.1.2.3 Density functional theory. . . . 575
10.2.1.2.4 Effect of ionocity and covalency: Drago's concept . . . 576
10.2.1.2.5 Effect of amphotericity of acid-base interaction: Gutmann's numbers . . . 577
10.2.1.2.6 Spectroscopic measurements: Fowkes' approach . . . 578
10.2.2 Effect of polymer/solvent interactions on aggregation of stereoregular PMMA 578
10.2.2.1 Aggregation of stereoregular PMMA . . . 578
10.2.2.2 Relation between the complexing power of solvents and their acid-base properties . . . 579
10.2.3 Influence of the nature of the solvent on the $\alpha$ and $\beta$-relaxations of conventional PMMA . . . 581
10.2.3.1 Introduction. . . . 581
10.2.3.2 Dielectric spectroscopy results . . . 581
10.2.4 Concluding remarks . . . 582
10.3 Solvent effects based on pure solvent scales . . . 583
JAVIER CATALÁN
Introduction. . . . 583
10.3.1 The solvent effect and its dissection into general and specific contributions . . 584
10.3.2 Characterization of a molecular environment with the aid of the probe/homomorph model. . . . 585
10.3.3 Single-parameter solvent scales: the Y, G, $E_T(30)$, $P_y$, Z, $\chi_R$, $\Phi$, and S' scales. . 587
10.3.3.1 The solvent ionizing power scale or Y scale . . . 587
10.3.3.2 The G values of Allerhand and Schleyer. . . . 588
10.3.3.3 The $E_T(30)$ scale of Dimroth and Reichardt . . . 588
10.3.3.4 The $P_y$ scale of Dong and Winnick. . . . 589
10.3.3.5 The Z scale of Kosower . . . 589
10.3.3.6 The $\chi_R$ scale of Brooker . . . 590
10.3.3.7 The $\Phi$ scale of Dubois and Bienvenüe . . . 590
10.3.3.8 The S' scale of Drago. . . . 591
10.3.4 Solvent polarity: the SPP scale. . . . 591
10.3.5 Solvent basicity: the SB scale . . . 600
10.3.6 Solvent acidity: the SA scale. . . . 601
10.3.7 Applications of the pure SPP, SA and SB scales. . . . 605
10.3.7.1 Other reported solvents scales . . . 605
10.3.7.2 Treatment of the solvent effect in: . . . 608

10.3.7.2.1 Spectroscopy . . . 608
10.3.7.2.2 Kinetics . . . 611
10.3.7.2.3 Electrochemistry . . . 612
10.3.7.2.4 Thermodynamics . . . 612
10.3.7.3 Mixtures of solvents. Understanding the preferential solvation model . . . 612
10.4 Acid-base equilibria in ionic solvents (ionic melts) . . . 616
VICTOR CHERGINETS
10.4.1 Acid-base definitions used for the description of donor-acceptor interactions in ionic media . . . 617
10.4.1.1 The Lewis definition . . . 617
10.4.1.2 The Lux-Flood definition. . . . 618
10.4.2 The features of ionic melts as media for acid-base interactions . . . 618
10.4.2.1 Oxygen-less media . . . 619
10.4.2.2 Oxygen-containing melts. . . . 619
10.4.2.3 The effect of the ionic solvent composition on acid-base equilibria . . . 620
10.4.3 Methods for estimations of acidities of solutions based on ionic melts . . . 623
10.4.4 On studies of the homogeneous acid-base reactions in ionic melts . . . 625
10.4.4.1 Nitrate melts . . . 625
10.4.4.2 Sulphate melts . . . 627
10.4.4.3 Silicate melts . . . 628
10.4.4.4 The equimolar mixture KCl-NaCl . . . 629
10.4.4.5 Other alkaline halide melts . . . 631
10.4.5 Reactions of melts with gaseous acids and bases . . . 632
10.4.5.1 High-temperature hydrolysis of molten halides . . . 632
10.4.5.2 The processes of removal of oxide admixtures from melts . . . 633
**11 ELECTRONIC AND ELECTRICAL EFFECTS OF SOLVENTS . . . 639**
11.1 Theoretical treatment of solvent effects on electronic and vibrational spectra of compounds in condensed media. . . . 639
MATI KARELSON
11.1.1 Introduction. . . . 639
11.1.2 Theoretical treatment of solvent cavity effects on electronic-vibrational spectra of molecules . . . 647
11.1.3 Theoretical treatment of solvent electrostatic polarization on electronic-vibrational spectra of molecules . . . 649
11.1.4 Theoretical treatment of solvent dispersion effects on electronic-vibrational spectra of molecules . . . 671
11.1.5 Supermolecule approach to the intermolecular interactions in condensed media 674
11.2 Dielectric solvent effects on the intensity of light absorption and the radiative rate constant . . . 680
TAI-ICHI SHIBUYA
11.2.1 The Chako formula or the Lorentz-Lorenz correction . . . 680
11.2.2 The generalized local-field factor for the ellipsoidal cavity . . . 680
11.2.3 Dielectric solvent effect on the radiative rate constant. . . . 682
**12 OTHER PROPERTIES OF SOLVENTS, SOLUTIONS, AND PRODUCTS OBTAINED FROM SOLUTIONS. . . . 683**
12.1 Rheological properties, aggregation, permeability, molecular structure, crystallinity, and other properties affected by solvents . . . 683
GEORGE WYPYCH
12.1.1 Rheological properties . . . 683
12.1.2 Aggregation . . . 689
12.1.3 Permeability . . . 693
12.1.4 Molecular structure and crystallinity. . . . 697
12.1.5 Other properties affected by solvents . . . 700
12.2 Chain conformations of polysaccharides in different solvents. . . . 706
RANIERI URBANI AND ATTILIO CESÀRO
12.2.1 Introduction. . . . 706
12.2.2 Structure and conformation of polysaccharides in solution . . . 707

12.2.2.1 Chemical structure . . . 707
12.2.2.2 Solution chain conformation . . . 707
12.2.3 Experimental evidence of solvent effect on oligosaccharide conformational equilibria . . . 711
12.2.4 Theoretical evaluation of solvent effect on conformational equilibria of sugars 715
12.2.4.1 Classical molecular mechanics methods . . . 715
12.2.4.2 Molecular dynamic methods . . . 720
12.2.5 Solvent effect on chain dimensions and conformations of polysaccharides . . . 722
12.2.6 Solvent effect on charged polysaccharides and the polyelectrolyte model . . . 726
12.2.6.1 Experimental behavior of polysaccharides polyelectrolytes . . . 726
12.2.6.2 The Haug and Smidsrød parameter: description of the salt effect on the chain dimension . . . 727
12.2.6.3 The statistical thermodynamic counterion-condensation theory of Manning . . 729
12.2.6.4 Conformational calculations of charged polysaccharides . . . 731
12.2.7 Conclusions . . . 733
**13 EFFECT OF SOLVENT ON CHEMICAL REACTIONS AND REACTIVITY . . . 737**
13.1 Solvent effects on chemical reactivity . . . 737
ROLAND SCHMID
13.1.1 Introduction . . . 737
13.1.2 The dielectric approach . . . 737
13.1.3 The chemical approach . . . 738
13.1.4 Dielectric vs. chemical approach . . . 742
13.1.5 Conceptual problems with empirical solvent parameters . . . 744
13.1.6 The physical approach . . . 746
13.1.7 Some highlights of recent investigations . . . 753
The like dissolves like rule . . . 753
Water's anomalies . . . 755
The hydrophobic effect . . . 758
The structure of liquids . . . 762
Solvent reorganization energy in ET . . . 765
The solution ionic radius . . . 768
13.1.8 The future of the phenomenological approach . . . 772
13.2 Solvent effects on free radical polymerization . . . 777
MICHELLE L. COOTE AND THOMAS P. DAVIS
13.2.1 Introduction . . . 777
13.2.2 Homopolymerization . . . 777
13.2.2.1 Initiation . . . 777
13.2.2.2 Propagation . . . 778
13.2.2.3 Transfer . . . 779
13.2.2.4 Termination . . . 779
13.2.3 Copolymerization . . . 779
13.2.3.1 Polarity effect . . . 780
13.2.3.1.1 Basic mechanism . . . 780
13.2.3.1.2 Copolymerization model . . . 781
13.2.3.1.3 Evidence for polarity effects in propagation reactions . . . 781
13.2.3.2 Radical-solvent complexes . . . 782
13.2.3.2.1 Basic mechanism . . . 782
13.2.3.2.2 Copolymerization model . . . 782
13.2.3.2.3 Experimental evidence . . . 783
13.2.3.3 Monomer-solvent complexes . . . 785
13.2.3.3.1 Introduction . . . 785
13.2.3.3.2 Monomer-monomer complex participation model . . . 785
13.2.3.3.3 Monomer-monomer complex dissociation model . . . 790
13.2.3.3.4 Specific solvent effects . . . 791
13.2.3.4 Bootstrap model . . . 791
13.2.3.4.1 Basic mechanism . . . 791

13.2.3.4.2 Copolymerization model . . . 791
13.2.3.4.3 Experimental evidence . . . 793
13.2.4 Concluding remarks . . . 795
13.3 Effects of organic solvents on phase-transfer catalysis . . . 798
MAW-LING WANG
13.3.1 Two-phase phase-transfer catalytic reactions . . . 801
13.3.1.1 Theoretical analysis of the polarity of the organic solvents and the reactions . . 801
13.3.1.2 Effect of organic solvent on the reaction in various reaction systems . . . 805
13.3.1.3 Effects of the organic solvents on the reactions in other catalysts . . . 811
13.3.1.4 Effect of the volume of organic solvent and water on the reactions in various reaction systems . . . 822
13.3.1.5 Effects of organic solvents on other phase-transfer catalytic reactions . . . 825
13.3.1.6 Other effects on the phase-transfer catalytic reactions . . . 828
13.3.2 Three-phase reactions (triphase catalysis) . . . 830
13.3.2.1 The interaction between solid polymer (hydrophilicity) and the organic solvents . . . 830
13.3.2.2 Effect of solvents on the reaction in triphase catalysis . . . 833
13.3.2.3 Effect of volume of organic solvent and water on the reactions in triphase catalysis . . . 836
13.4 Effect of polymerization solvent on the chemical structure and curing of aromatic poly(amideimide) . . . 841
NORIO TSUBOKAWA
13.4.1 Introduction . . . 841
13.4.2 Effect of solvent on the chemical structure of PAI . . . 842
13.4.2.1 Imide and amide bond content of PAI . . . 842
13.4.2.2 Intrinsic viscosity and carboxyl group content . . . 844
13.4.3 Effect of solvent on the curing of PAI by heat treatment . . . 844
13.4.3.1 Chemical structure of PAI after heat treatment . . . 844
13.4.3.2 Curing PAI by post-heating . . . 845
13.4.4 Conclusions . . . 846
**14 SOLVENT USE IN VARIOUS INDUSTRIES . . . 847**
14.1 Adhesives and sealants . . . 847
GEORGE WYPYCH
14.2 Aerospace . . . 852
GEORGE WYPYCH
14.3 Asphalt compounding . . . 855
GEORGE WYPYCH
14.4 Biotechnology . . . 856
14.4.1 Organic solvents in microbial production processes . . . 856
MICHIAKI MATSUMOTO, SONJA ISKEN, JAN A. M. DE BONT
14.4.1.1 Introduction . . . 856
14.4.1.2 Toxicity of organic solvents . . . 859
14.4.1.3 Solvent-tolerant bacteria . . . 862
14.4.1.4 Biotransformation using solvent-tolerant microorganisms . . . 863
14.4.2 Solvent-resistant microorganisms . . . 865
TILMAN HAHN, KONRAD BOTZENHART
14.4.2.1 Introduction . . . 865
14.4.2.2 Toxicity of solvents for microorganisms . . . 865
14.4.2.2.1 Spectrum of microorganisms and solvents . . . 865
14.4.2.2.2 Mechanisms of solvent toxicity for microorganisms . . . 866
14.4.2.3 Adaption of microorganisms to solvents - solvent-resistant microorganisms . . 867
14.4.2.3.1 Spectrum of solvent-resistant microorganisms . . . 867
14.4.2.3.2 Adaption mechanisms of microorganisms to solvents . . . 868
14.4.2.4 Solvents and microorganisms in the environment and industry - examples . . . 869
14.4.2.4.1 Examples . . . 869
14.4.3 Choice of solvent for enzymatic reaction in organic solvent . . . 872
TSUNEO YAMANE

14.4.3.1 Introduction. . . . . 872
14.4.3.2 Classification of organic solvents . . . . . 872
14.4.3.3 Influence of solvent parameters on nature of enzymatic reactions in organic media. . . . . 873
14.4.3.4 Properties of enzymes affected by organic solvents . . . . . 875
14.4.3.5 Concluding remarks . . . . . 879
14.5 Coil coating. . . . . 880
GEORGE WYPYCH
14.6 Cosmetics and personal care products . . . . . 881
GEORGE WYPYCH
14.7 Dry cleaning - treatment of textiles in solvents . . . . . 883
KASPAR D. HASENCLEVER
14.7.1 Dry cleaning . . . . . 883
14.7.1.1 History of dry cleaning . . . . . 883
14.7.1.2 Basis of dry cleaning . . . . . 884
14.7.1.3 Behavior of textiles in solvents and water . . . . . 885
14.7.1.4 Removal of soiling in dry cleaning. . . . . 886
14.7.1.5 Activity of detergents in dry cleaning . . . . . 887
14.7.1.6 Dry cleaning processes . . . . . 888
14.7.1.7 Recycling of solvents in dry cleaning . . . . . 890
14.7.2 Spotting. . . . . 891
14.7.2.1 Spotting in dry cleaning . . . . . 891
14.7.2.2 Spotting agents . . . . . 891
14.7.2.3 Spotting procedure . . . . . 892
14.7.3 Textile finishing . . . . . 893
14.7.3.1 Waterproofing . . . . . 893
14.7.3.2 Milling . . . . . 893
14.7.3.3 Antistatic finishing . . . . . 893
14.8 Electronic industry - CFC-free alternatives for cleaning in electronic industry. 894
MARTIN HANEK, NORBERT LÖW, ANDREAS MÜHLBAUER
14.8.1 Cleaning requirements in the electronic industry . . . . . 894
14.8.2 Available alternatives. . . . . 896
14.8.2.1 Water based systems; advantages and disadvantages . . . . . 897
14.8.2.1.1 Cleaning with DI - water . . . . . 897
14.8.2.1.2 Cleaning with alkaline water-based media. . . . . 898
14.8.2.1.3 Aqueous-based cleaning agents containing water soluble organic components . 898
14.8.2.1.4 Water-based cleaning agents based on MPC® Technology (MPC = Micro Phase Cleaning) . . . . . 899
14.8.2.1.5 Advantages and disadvantages of aqueous cleaning media . . . . . 899
14.8.2.2 Semi-aqueous cleaners based on halogen-free solvents, advantages and disadvantages. . . . . 900
14.8.2.2.1 Water insoluble cleaning fluids . . . . . 901
14.8.2.2.2 Water-soluble, water-based cleaning agents . . . . . 901
14.8.2.2.3 Comparison of the advantages (+) and disadvantages (-) of semi-aqueous cleaning fluids . . . . . 901
14.8.2.3 Other solvent based cleaning systems . . . . . 902
14.8.3 Cleaning of tools and auxiliaries . . . . . 904
14.8.3.1 Cleaning substrates and contamination. . . . . 904
14.8.3.2 Compatibility of stencil and cleaning agent . . . . . 905
14.8.3.3 Different cleaning media . . . . . 906
14.8.3.4 Comparison of manual cleaning vs. automated cleaning. . . . . 908
14.8.3.5 Cleaning equipment for stencil cleaning applications . . . . . 909
14.8.3.6 Stencil cleaning in screen printing machines. . . . . 911
14.8.3.7 Summary . . . . . 911
14.8.4 Cleaning agents and process technology available for cleaning PCB . . . . . 911
14.8.4.1 Flux remove and aqueous process . . . . . 911
14.8.4.1.1 The limits of a no-clean process . . . . . 911

14.8.4.1.2 Different cleaning media and cleaning processes . . . 912
14.8.4.1.3 Semi-aqueous cleaning . . . 913
14.8.4.1.4 Aqueous cleaning in spray in air cleaning equipment . . . 913
14.8.4.2 Flux removal from printed circuit boards - water-free cleaning processes . . . 914
14.8.4.2.1 Water-free cleaning processes using HFE (hydrofluoroethers) in combination with a cosolvent . . . 915
14.8.4.2.2 Water-free cleaning processes in closed, one-chamber vapor defluxing systems 916
14.8.5 Criteria for assessment and evaluation of cleaning results . . . 917
14.8.6 Cost comparison of different cleaning processes. . . . 919
14.9 Fabricated metal products . . . 920
GEORGE WYPYCH
14.10 Food industry - solvents for extracting vegetable oils . . . 923
PHILLIP J. WAKELYN, PETER J. WAN
14.10.1 Introduction . . . 923
14.10.2 Regulatory concerns . . . 924
14.10.2.1 Workplace regulations . . . 925
14.10.2.1.1 Air Contaminants Standard (29 CFR 1910.1000) . . . 925
14.10.2.1.2 Hazard Communication Standard (HCS) (29 CFR 1910.1200) . . . 926
14.10.2.1.3 Process Safety Management (PSM) Standard (29 CFR 1910.119) . . . 927
14.10.2.2 Environmental regulations . . . 927
14.10.2.2.1 Clean Air Act (CAA; 42 U.S. Code 7401 et seq.) . . . 929
14.10.2.2.2 Clean Water Act (CWA; 33 U.S. Code 1251 et seq.) . . . 932
14.10.2.2.3 Resource Conservation and Recovery Act (RCRA; 42 U.S.Code 6901 et seq.). 932
14.10.2.2.4 Emergency Planning and Community Right-to-Know Act (EPCRA; 42 U.S. Code 11001 et seq.) . . . 933
14.10.2.2.5 Toxic Substances Control Act (TSCA; 15 U.S. Code 2601 et seq.) . . . 933
14.10.2.3 Food safety . . . 934
14.10.3 The solvent extraction process . . . 935
14.10.3.1 Preparation for extraction . . . 936
14.10.3.2 Oil extraction . . . 938
14.10.3.3 Processing crude oil . . . 938
14.10.4 Review of solvents studied for extraction efficiency. . . . 940
14.10.4.1 Hydrocarbon solvents . . . 941
14.10.4.1.1 Nomenclature, structure, composition and properties of hydrocarbons . . . 942
14.10.4.1.2 Performance of selected hydrocarbon solvents. . . . 942
14.10.5 Future trends . . . 946
14.11 Ground transportation . . . 950
GEORGE WYPYCH
14.12 Inorganic chemical industry . . . 950
GEORGE WYPYCH
14.13 Iron and steel industry . . . 951
GEORGE WYPYCH
14.14 Lumber and wood products - Wood preservation treatment: significance of solvents. . . . 953
TILMAN HAHN, KONRAD BOTZENHART, FRITZ SCHWEINSBERG, GERHARD VOLLAND
14.14.1 General aspects . . . 953
14.14.2 Role of solvents . . . 954
14.14.2.1 Occurrence . . . 954
14.14.2.2 Technical and environmental aspects . . . 955
14.15 Medical applications . . . 955
GEORGE WYPYCH
14.16 Metal casting . . . 957
GEORGE WYPYCH
14.17 Motor vehicle assembly . . . 958
GEORGE WYPYCH
14.18 Organic chemical industry . . . 962
GEORGE WYPYCH

14.19 Paints and coatings . . . . 963
14.19.1 Architectural surface coatings and solvents . . . . 963
TILMAN HAHN, KONRAD BOTZENHART, FRITZ SCHWEINSBERG, GERHARD VOLLAND
14.19.1.1 General aspects . . . . 963
14.19.1.2 Technical aspects and properties of coating materials . . . . 963
14.19.2 Recent advances in coalescing solvents for waterborne coatings . . . . 969
DAVID RANDALL
14.19.2.1 Introduction . . . . 969
14.19.2.2 Water based coatings . . . . 970
14.19.2.3 Emulsion polymers . . . . 970
14.19.2.4 Role of a coalescing solvent . . . . 971
14.19.2.5 Properties of coalescing agents . . . . 972
14.19.2.5.1 Hydrolytic stability . . . . 972
14.19.2.5.2 Water solubility . . . . 972
14.19.2.5.3 Freezing point . . . . 972
14.19.2.5.4 Evaporation rate . . . . 972
14.19.2.5.5 Odor . . . . 972
14.19.2.5.6 Color . . . . 973
14.19.2.5.7 Coalescing efficiency . . . . 973
14.19.2.5.8 Incorporation . . . . 973
14.19.2.5.9 Improvement of physical properties . . . . 973
14.19.2.5.10 Biodegradability . . . . 973
14.19.2.5.11 Safety . . . . 973
14.19.2.6 Comparison of coalescing solvents . . . . 973
14.19.2.7 Recent advances in diester coalescing solvents . . . . 974
14.19.2.8 Appendix - Classification of coalescing solvents . . . . 975
14.20 Petroleum refining industry . . . . 975
GEORGE WYPYCH
14.21 Pharmaceutical industry . . . . 977
14.21.1 Use of solvents in the manufacture of drug substances (DS) and drug products (DP) . . . . 977
MICHEL BAUER, CHRISTINE BARTHÉLÉMY
14.21.1.1 Introduction . . . . 977
14.21.1.2 Where are solvents used in the manufacture of pharmaceutical drugs? . . . . 979
14.21.1.2.1 Intermediates of synthesis, DS and excipients . . . . 979
14.21.1.2.2 Drug products . . . . 984
14.21.1.3 Impacts of the nature of solvents and their quality on the physicochemical characteristics of raw materials and DP . . . . 985
14.21.1.3.1 Raw materials (intermediates, DS, excipients) . . . . 985
14.21.1.3.2 Drug product . . . . 988
14.21.1.3.3 Conclusions . . . . 989
14.21.1.4 Setting specifications for solvents . . . . 990
14.21.1.4.1 Solvents used for the raw material manufacture . . . . 990
14.21.1.4.2 Solvents used for the DP manufacture . . . . 991
14.21.1.5 Quality of solvents and analysis . . . . 991
14.21.1.5.1 Quality of solvents used in spectroscopy . . . . 991
14.21.1.5.2 Quality of solvents used in chromatography . . . . 993
14.21.1.5.3 Quality of solvents used in titrimetry . . . . 996
14.21.1.6 Conclusions . . . . 996
14.21.2 Predicting cosolvency for pharmaceutical and environmental applications . . . . 997
AN LI
14.21.2.1 Introduction . . . . 997
14.21.2.2 Applications of cosolvency in pharmaceutical sciences and industry . . . . 998
14.21.2.3 Applications of cosolvency in environmental sciences and engineering . . . . 1000
14.21.2.4 Experimental observations . . . . 1001
14.21.2.5 Predicting cosolvency in homogeneous liquid systems . . . . 1003
14.21.2.6 Predicting cosolvency in non-ideal liquid mixtures . . . . 1007

14.21.2.7 Summary . . . . . 1013
14.22 Polymers and man-made fibers. . . . . 1016
GEORGE WYPYCH
14.23 Printing industry . . . . . 1020
GEORGE WYPYCH
14.24 Pulp and paper . . . . . 1023
GEORGE WYPYCH
14.25 Rubber and plastics. . . . . 1025
GEORGE WYPYCH
14.26 Use of solvents in the shipbuilding and ship repair industry. . . . . 1026
MOHAMED SERAGELDIN, DAVE REEVES
14.26.1 Introduction. . . . . 1026
14.26.2 Shipbuilding and ship repair operations . . . . . 1026
14.26.3 Coating operations . . . . . 1026
14.26.4 Cleaning operations using organic solvents . . . . . 1027
14.26.4.1 Surface preparation and initial corrosion protection . . . . . 1027
14.26.4.2 Cleaning operations after coatings are applied. . . . . 1028
14.26.4.3 Maintenance cleaning of equipment items and components . . . . . 1031
14.26.5 Marine coatings. . . . . 1031
14.26.6 Thinning of marine coatings . . . . . 1032
14.26.7 Solvent emissions . . . . . 1033
14.26.8 Solvent waste . . . . . 1035
14.26.9 Reducing solvent usage, emissions, and waste. . . . . 1036
14.26.10 Regulations and guidelines for cleaning solvents . . . . . 1037
14.27 Stone, clay, glass, and concrete . . . . . 1039
GEORGE WYPYCH
14.28 Textile industry. . . . . 1041
GEORGE WYPYCH
14.29 Transportation equipment cleaning. . . . . 1042
GEORGE WYPYCH
14.30 Water transportation . . . . . 1042
GEORGE WYPYCH
14.31 Wood furniture . . . . . 1043
GEORGE WYPYCH
14.32 Summary . . . . . 1045
**15 METHODS OF SOLVENT DETECTION AND TESTING. . . . . 1053**
15.1 Standard methods of solvent analysis . . . . . 1053
GEORGE WYPYCH
15.1.1 Alkalinity and acidity. . . . . 1053
15.1.2 Autoignition temperature. . . . . 1054
15.1.3 Biodegradation potential . . . . . 1054
15.1.4 Boiling point . . . . . 1055
15.1.5 Bromine index . . . . . 1055
15.1.6 Calorific value . . . . . 1056
15.1.7 Cleaning solvents. . . . . 1056
15.1.8 Color . . . . . 1056
15.1.9 Corrosion (effect of solvents) . . . . . 1057
15.1.10 Density . . . . . 1057
15.1.11 Dilution ratio . . . . . 1057
15.1.12 Dissolving and extraction . . . . . 1058
15.1.13 Electric properties . . . . . 1058
15.1.14 Environmental stress crazing. . . . . 1059
15.1.15 Evaporation rate . . . . . 1059
15.1.16 Flammability limits. . . . . 1059
15.1.17 Flash point . . . . . 1060
15.1.18 Freezing point . . . . . 1061
15.1.19 Free halogens in halogenated solvents . . . . . 1061

15.1.20 Gas chromatography . . . . . 1061
15.1.21 Labeling . . . . . 1062
15.1.22 Odor . . . . . 1062
15.1.23 Paints standards related to solvents . . . . . 1063
15.1.24 pH. . . . . 1063
15.1.25 Purity . . . . . 1063
15.1.26 Refractive index . . . . . 1066
15.1.27 Residual solvents . . . . . 1066
15.1.28 Solubility . . . . . 1066
15.1.29 Solvent partitioning in soils . . . . . 1066
15.1.30 Solvent extraction . . . . . 1067
15.1.31 Specifications. . . . . 1067
15.1.32 Sustained burning . . . . . 1067
15.1.33 Vapor pressure . . . . . 1068
15.1.34 Viscosity . . . . . 1068
15.1.35 Volatile organic compound content, VOC . . . . . 1069
15.2 Special methods of solvent analysis . . . . . 1078
15.2.1 Use of breath monitoring to assess exposures to volatile organic solvents . . . 1078
MYRTO PETREAS
15.2.1.1 Principles of breath monitoring . . . . . 1078
15.2.1.2 Types of samples used for biological monitoring . . . . . 1080
15.2.1.3 Fundamentals of respiratory physiology . . . . . 1080
15.2.1.3.1 Ventilation . . . . . 1081
15.2.1.3.2 Partition coefficients . . . . . 1081
15.2.1.3.3 Gas exchange. . . . . 1082
15.2.1.4 Types of exhaled air samples. . . . . 1083
15.2.1.5 Breath sampling methodology . . . . . 1084
15.2.1.6 When is breath monitoring appropriate? . . . . . 1087
15.2.1.7 Examples of breath monitoring. . . . . 1088
15.2.2 A simple test to determine toxicity using bacteria . . . . . 1095
JAMES L. BOTSFORD
15.2.2.1 Introduction. . . . . 1095
15.2.2.2 Toxicity defined . . . . . 1095
15.2.2.3 An alternative. . . . . 1097
15.2.2.4 Chemicals tested . . . . . 1099
15.2.2.5 Comparisons with other tests. . . . . 1103
15.2.2.6 Toxic herbicides . . . . . 1107
15.2.2.7 Toxicity of divalent cations . . . . . 1108
15.2.2.8 Toxicity of organics in the presence of EDTA. . . . . 1108
15.2.2.9 Mechanism for reduction of the dye . . . . . 1110
15.2.2.10 Summary . . . . . 1111
15.2.3 Description of an innovative GC method to assess the influence of crystal texture and drying conditions on residual solvent content in pharmaceutical products. . . . . 1113
CHRISTINE BARTHÉLÉMY, MICHEL BAUER
15.2.3.1 Description of the RS determination method . . . . . 1113
15.2.3.2 Application: Influence of crystal texture and drying conditions on RS content . 1114
15.2.3.2.1 First example: monocrystalline particles of paracetamol . . . . . 1116
15.2.3.2.2 Second example: polycrystalline particles of meprobamate and ibuprofen . . . 1119
15.2.3.2.3 Third example: polycrystalline particles of paracetamol. . . . . 1122
**16 RESIDUAL SOLVENTS IN PRODUCTS . . . . . 1125**
16.1 Residual solvents in various products . . . . . 1125
GEORGE WYPYCH
16.2 Residual solvents in pharmaceutical substances . . . . . 1129
MICHEL BAUER, CHRISTINE BARTHÉLÉMY
16.2.1 Introduction. . . . . 1129
16.2.2 Why should we look for RS? . . . . . 1129

16.2.2.1 Modifying the acceptability of the drug product . . . 1129
16.2.2.2 Modifying the physico-chemical properties of drug substances (DS) and drug products (DP) . . . 1130
16.2.2.3 Implications of possible drug/container interactions . . . 1131
16.2.2.4 As a tool for forensic applications . . . 1131
16.2.2.5 As a source of toxicity . . . 1131
16.2.2.5.1 General points . . . 1131
16.2.2.5.2 Brief overview of the toxicology of solvents . . . 1132
16.2.3 How to identify and control RS in pharmaceutical substances? . . . 1133
16.2.3.1 Loss of weight . . . 1133
16.2.3.2 Miscellaneous methods . . . 1133
16.2.3.3 Gas chromatography (GC) . . . 1134
16.2.3.3.1 General points . . . 1134
16.2.3.3.2 Review of methods . . . 1135
16.2.3.3.3 Official GC methods for RS determination . . . 1139
16.2.4 How to set specifications? Examination of the ICH guidelines for residual solvents . . . 1140
16.2.4.1 Introduction . . . 1143
16.2.4.2 Classification of residual solvents by risk assessment . . . 1143
16.2.4.3 Definition of PDE. Method for establishing exposure limits . . . 1143
16.2.4.4 Limits for residual solvents . . . 1143
16.2.4.5 Analytical procedures . . . 1145
16.2.4.6 Conclusions regarding the ICH guideline . . . 1145
16.2.5 Conclusions . . . 1146
**17 ENVIRONMENTAL IMPACT OF SOLVENTS . . . 1149**
17.1 The environmental fate and movement of organic solvents in water, soil, and air . . . 1149
WILLIAM R. ROY
17.1.1 Introduction . . . 1149
17.1.2 Water . . . 1150
17.1.2.1 Solubility . . . 1150
17.1.2.2 Volatilization . . . 1150
17.1.2.3 Degradation . . . 1151
17.1.2.4 Adsorption . . . 1151
17.1.3 Soil . . . 1151
17.1.3.1 Volatilization . . . 1151
17.1.3.2 Adsorption . . . 1152
17.1.3.3 Degradation . . . 1153
17.1.4 Air . . . 1153
17.1.4.1 Degradation . . . 1153
17.1.4.2 Atmospheric residence time . . . 1154
17.1.5 The 31 solvents in water . . . 1154
17.1.5.1 Solubility . . . 1154
17.1.5.2 Volatilization from water . . . 1155
17.1.5.3 Degradation in water . . . 1155
17.1.6 Soil . . . 1157
17.1.6.1 Volatilization . . . 1157
17.1.6.2 Adsorption . . . 1159
17.1.6.3 Degradation . . . 1160
17.1.7 Air . . . 1161
17.2 Fate-based management of organic solvent-containing wastes . . . 1162
WILLIAM R. ROY
17.2.1 Introduction . . . 1162
17.2.1.1 The waste disposal site . . . 1163
17.2.1.2 The advection-dispersion model and the required input . . . 1164
17.2.1.3 Maximum permissible concentrations . . . 1164
17.2.1.4 Distribution of organic compounds in leachate . . . 1164

17.2.2 Movement of solvents in groundwater . . . . . 1166
17.2.3 Mass limitations . . . . . 1167
17.3 Environmental fate and ecotoxicological effects of glycol ethers . . . . . 1169
JAMES DEVILLERS, AURÉLIE CHEZEAU, ANDRÉ CICOLELLA, ERIC THYBAUD
17.3.1 Introduction . . . . . 1169
17.3.2 Occurrence . . . . . 1170
17.3.3 Environmental behavior . . . . . 1171
17.3.4 Ecotoxicity . . . . . 1175
17.3.4.1 Survival and growth . . . . . 1175
17.3.4.2 Reproduction and development . . . . . 1185
17.3.5 Conclusion . . . . . 1187
17.4 Organic solvent impacts on tropospheric air pollution . . . . . 1188
MICHELLE BERGIN, ARMISTEAD RUSSELL
17.4.1 Sources and impacts of volatile solvents . . . . . 1188
17.4.2 Modes and scales of impact . . . . . 1189
17.4.2.1 Direct exposure . . . . . 1189
17.4.2.2 Formation of secondary compounds . . . . . 1190
17.4.2.3 Spatial scales of secondary effects . . . . . 1190
17.4.2.3.1 Global impacts . . . . . 1190
17.4.2.3.2 Stratospheric ozone depletion . . . . . 1191
17.4.2.3 Global climate forcing . . . . . 1191
17.4.2.4 Urban and regional scales . . . . . 1192
17.4.3 Tropospheric ozone . . . . . 1192
17.4.3.1 Effects . . . . . 1192
17.4.3.2 Tropospheric photochemistry and ozone formation . . . . . 1193
17.4.3.3 Assessing solvent impacts on ozone and VOC reactivity . . . . . 1195
17.4.3.3.1 Quantification of solvent emissions on ozone formation . . . . . 1196
17.4.4 Regulatory approaches to ozone control and solvents . . . . . 1198
17.4.5 Summary . . . . . 1299
**18 CONCENTRATION OF SOLVENTS IN VARIOUS INDUSTRIAL ENVIRONMENTS . . . . . 1201**
18.1 Measurement and estimation of solvents emission and odor. . . . . . 1201
MARGOT SCHEITHAUER
18.1.1 Definition “solvent” and “volatile organic compounds” (VOC) . . . . . 1201
18.1.2 Review of sources of solvent emissions . . . . . 1203
18.1.2.1 Causes for emissions . . . . . 1203
18.1.2.2 Emissions of VOCs from varnishes and paints . . . . . 1203
18.1.2.3 VOC emissions from emulsion paints . . . . . 1205
18.1.3 Measuring of VOC-content in paints and varnishes . . . . . 1205
18.1.3.1 Definition of low-emissive coating materials . . . . . 1205
18.1.3.2 Determination of the VOC content according to ASTM D 3960-1 . . . . . 1205
18.1.3.3 Determination of the VOC content according to ISO/DIS 11 890/1 and 2 . . . 1206
18.1.3.3.1 VOC content > 15% . . . . . 1206
18.1.3.3.2 VOC content > 0.1 and < 15 %. . . . . . 1208
18.1.3.4 Determination of VOC-content in water-thinnable emulsion paints (in-can VOC) . . . . . 1208
18.1.4 Measurement of solvent emissions in industrial plants . . . . . 1209
18.1.4.1 Plant requirements . . . . . 1209
18.1.4.2 The determination of the total carbon content in mg C/Nm³ . . . . . 1214
18.1.4.2.1 Flame ionization detector (FID) . . . . . 1214
18.1.4.2.2 Silica gel approach . . . . . 1214
18.1.4.3 Qualitative and quantitative assessment of individual components in the exhaust-gas . . . . . 1215
18.1.4.3.1 Indicator tubes . . . . . 1215
18.1.4.3.2 Quantitative solvent determination in exhaust gas of plants by means of gas-chromatography . . . . . 1215
18.1.5 “Odor” definition . . . . . 1219

18.1.6 Measurement of odor in materials and industrial plants . . . . . . . . . . . . . 1222
18.1.6.1 Introduction. . . . . . . . . . . . . . . . . . . . . . . . . . . . . . . . . . . . . . . 1222
18.1.6.2 Odor determination by means of the "electronic nose" . . . . . . . . . . . . . 1222
18.1.6.3 Odor determination by means of the olfactometer . . . . . . . . . . . . . . . . 1223
18.1.6.4 Example for odor determination for selected materials: Determination of odorant concentration in varnished furniture surfaces . . . . . . . . . . . . . 1223
18.1.6.5 Example of odor determination in industrial plants: Odor measurement in an industrial varnishing plant. . . . . . . . . . . . . . . . . . . . . . . . . . . . 1225
18.2 Prediction of organic solvents emission during technological processes . . . . 1227
KRZYSZTOF M. BENCZEK, JOANNA KURPIEWSKA
18.2.1 Introduction. . . . . . . . . . . . . . . . . . . . . . . . . . . . . . . . . . . . . . . 1227
18.2.2 Methods of degreasing . . . . . . . . . . . . . . . . . . . . . . . . . . . . . . . . 1227
18.2.3 Solvents. . . . . . . . . . . . . . . . . . . . . . . . . . . . . . . . . . . . . . . . . 1228
18.2.4 Identification of the emitted compounds . . . . . . . . . . . . . . . . . . . . . . 1228
18.2.5 Emission of organic solvents during technological processes . . . . . . . . . . 1228
18.2.6 Verification of the method . . . . . . . . . . . . . . . . . . . . . . . . . . . . . . 1230
18.2.7 Relationships between emission and technological parameters . . . . . . . . . 1231
18.2.7.1 Laboratory test stand . . . . . . . . . . . . . . . . . . . . . . . . . . . . . . . . . 1231
18.2.7.2 The influence of temperature on emission . . . . . . . . . . . . . . . . . . . . . 1231
18.2.7.3 The influence of air velocity on emission . . . . . . . . . . . . . . . . . . . . . 1232
18.2.7.4 The relationship between the mass of solvent on wet parts and emissions . . . 1232
18.2.8 Emission of solvents . . . . . . . . . . . . . . . . . . . . . . . . . . . . . . . . . 1232
18.2.9 Verification in industrial conditions . . . . . . . . . . . . . . . . . . . . . . . . 1232
18.3 Indoor air pollution by solvents contained in paints and varnishes . . . . . . . 1234
TILMAN HAHN, KONRAD BOTZENHART, FRITZ SCHWEINSBERG, GERHARD VOLLAND
18.3.1 Composition - solvents in paints and varnishes. Theoretical aspects . . . . . . 1234
18.3.2 Occurrence of solvents in paints and varnishes . . . . . . . . . . . . . . . . . . 1235
18.3.2.1 Solvents in products . . . . . . . . . . . . . . . . . . . . . . . . . . . . . . . . . 1235
18.3.2.2 Paints and varnishes . . . . . . . . . . . . . . . . . . . . . . . . . . . . . . . . . 1237
18.3.3 Emission of solvents . . . . . . . . . . . . . . . . . . . . . . . . . . . . . . . . . 1240
18.3.3.1 Emission . . . . . . . . . . . . . . . . . . . . . . . . . . . . . . . . . . . . . . . . 1240
18.3.3.2 Immission. . . . . . . . . . . . . . . . . . . . . . . . . . . . . . . . . . . . . . . . 1242
18.3.4 Effects on health of solvents from paints and varnishes . . . . . . . . . . . . . 1243
18.3.4.1 Exposure . . . . . . . . . . . . . . . . . . . . . . . . . . . . . . . . . . . . . . . . 1243
18.3.4.2 Health effects . . . . . . . . . . . . . . . . . . . . . . . . . . . . . . . . . . . . . 1243
18.3.4.2.1 Toxic responses of skin and mucose membranes . . . . . . . . . . . . . . . 1243
18.3.4.2.2 Neurological disorders . . . . . . . . . . . . . . . . . . . . . . . . . . . . . . 1244
18.3.4.2.3 Carcinogenic effects . . . . . . . . . . . . . . . . . . . . . . . . . . . . . . . 1245
18.3.4.2.4 Respiratory effects . . . . . . . . . . . . . . . . . . . . . . . . . . . . . . . . 1246
18.3.4.2.5 Toxic responses of blood . . . . . . . . . . . . . . . . . . . . . . . . . . . . . 1247
18.3.4.2.6 Toxic responses of the reproductive system . . . . . . . . . . . . . . . . . . 1247
18.3.4.2.7 Toxic responses of other organ systems . . . . . . . . . . . . . . . . . . . . 1247
18.3.5 Methods for the examination of solvents in paints and varnishes . . . . . . . . 1248
18.3.5.1 Environmental monitoring . . . . . . . . . . . . . . . . . . . . . . . . . . . . . 1248
18.3.5.1.1 Solvents in products . . . . . . . . . . . . . . . . . . . . . . . . . . . . . . . 1248
18.3.5.1.2 Emission of solvents . . . . . . . . . . . . . . . . . . . . . . . . . . . . . . . 1248
18.3.5.2 Biological monitoring of solvents in human body fluids . . . . . . . . . . . . 1248
18.3.5.2.1 Solvents and metabolites in human body fluids and tissues . . . . . . . . . 1248
18.3.5.2.2 Biomarkers . . . . . . . . . . . . . . . . . . . . . . . . . . . . . . . . . . . . . 1248
18.4 Solvent uses with exposure risks . . . . . . . . . . . . . . . . . . . . . . . . . . . 1251
PENTTI KALLIOKOSKI, KAI SAVOLINEN
18.4.1 Introduction. . . . . . . . . . . . . . . . . . . . . . . . . . . . . . . . . . . . . . . 1251
18.4.2 Exposure assessment . . . . . . . . . . . . . . . . . . . . . . . . . . . . . . . . . 1252
18.4.3 Production of paints and printing inks . . . . . . . . . . . . . . . . . . . . . . . 1255
18.4.4 Painting . . . . . . . . . . . . . . . . . . . . . . . . . . . . . . . . . . . . . . . . . 1256
18.4.5 Printing . . . . . . . . . . . . . . . . . . . . . . . . . . . . . . . . . . . . . . . . . 1257
18.4.6 Degreasing, press cleaning and paint removal . . . . . . . . . . . . . . . . . . . 1258

18.4.7 Dry cleaning . . . 1260
18.4.8 Reinforced plastics industry . . . 1261
18.4.9 Gluing . . . 1262
18.4.10 Other . . . 1262
18.4.11 Summary . . . 1263
**19 REGULATIONS . . . 1267**
CARLOS M. NUÑEZ
19.1 Introduction . . . 1267
19.2 Air laws and regulations . . . 1282
19.2.1 Clean Air Act Amendments of 1990 . . . 1282
19.2.1.1 Background . . . 1282
19.2.1.2 Title I - Provisions for Attainment and Maintenance of National Ambient Air Quality Standards . . . 1284
19.2.1.3 Title III - Hazardous Air Pollutants . . . 1288
19.2.1.4 Title V - Permits . . . 1292
19.2.1.5 Title VI - Stratospheric Ozone Protection . . . 1292
19.3 Water laws and regulations . . . 1293
19.3.1 Clean Water Act . . . 1293
19.3.1.1 Background . . . 1293
19.3.1.2 Effluent Limitations . . . 1293
19.3.1.3 Permit Program . . . 1294
19.3.2 Safe Drinking Water Act . . . 1294
19.3.2.1 Background . . . 1294
19.3.2.2 National Primary Drinking Water Regulations . . . 1295
19.4 Land laws & regulations . . . 1295
19.4.1 Resource Conservation and Recovery Act (RCRA) . . . 1295
19.4.1.1 Background . . . 1295
19.4.1.2 RCRA, Subtitle C - Hazardous Waste . . . 1296
19.5 Multimedia laws and regulations . . . 1297
19.5.1 Pollution Prevention Act of 1990 . . . 1297
19.5.1.1 Background . . . 1297
19.5.1.2 Source Reduction Provisions . . . 1298
19.5.2 Toxic Substances Control Act . . . 1300
19.5.2.1 Background . . . 1300
19.5.2.2 Controlling toxic substances . . . 1300
19.6 Occupational laws and regulations . . . 1301
19.6.1 Occupational Safety and Health Act . . . 1301
19.6.1.1 Background . . . 1301
19.6.1.2 Air contaminants exposure limits . . . 1301
19.6.1.3 Hazard Communication Standard . . . 1302
19.7 International perspective . . . 1302
19.7.1 Canada . . . 1303
19.7.2 European Union . . . 1303
19.8 Tools and resources for solvents . . . 1304
19.9 Summary . . . 1306
19.10 Regulations in Europe . . . 1311
TILMAN HAHN, KONRAD BOTZENHART, FRITZ SCHWEINSBERG
19.10.1 EEC regulations . . . 1311
19.10.2 German regulations . . . 1312
**20 TOXIC EFFECTS OF SOLVENT EXPOSURE . . . 1315**
20.1 Toxicokinetics, toxicodynamics, and toxicology . . . 1315
TILMAN HAHN, KONRAD BOTZENHART, FRITZ SCHWEINSBERG
20.1.1 Toxicokinetics and toxicodynamics . . . 1315
20.1.1.1 Exposure . . . 1315
20.1.1.2 Uptake . . . 1315
20.1.1.2.1 Inhalation . . . 1316
20.1.1.2.2 Dermal uptake . . . 1316

20.1.1.2 Metabolism, distribution, excretion . . . 1317
20.1.1.3 Modeling of toxicokinetics and modifying factors . . . 1317
20.1.2 Toxicology . . . 1318
20.1.2.1 General effects . . . 1318
20.1.2.2 Specific non-immunological effects . . . 1318
20.1.2.3 Immunological effects . . . 1319
20.1.2.4 Toxic effects of solvents on other organisms . . . 1320
20.1.2.5 Carcinogenicity . . . 1320
20.1.2.6 Risk assessment . . . 1323
20.1.3 Conclusions . . . 1323
20.2 Cognitive and psychosocial outcome of chronic occupational solvent neurotoxicity . . . 1326
JENNI A OGDEN
20.2.1 Introduction . . . 1326
20.2.2 Acute symptoms of solvent neurotoxicity . . . 1327
20.2.3 Categorization of OSN . . . 1327
20.2.4 Assessment of OSN . . . 1328
20.2.5 Do the symptoms of Type 2 OSN resolve? . . . 1330
20.2.6 Individual differences in susceptibility to OSN . . . 1331
20.2.7 Psychosocial consequences of OSN, and rehabilitation . . . 1331
20.3 Pregnancy outcome following maternal organic solvent exposure . . . 1333
KRISTEN I. MCMARTIN, GIDEON KOREN
20.3.1 Introduction . . . 1333
20.3.2 Animal studies . . . 1334
20.3.3 Pregnancy outcome following maternal organic solvent exposure: a meta-analysis of epidemiologic studies . . . 1338
20.3.4 Pregnancy outcome following gestational exposure to organic solvents: a prospective controlled study . . . 1345
20.3.5 A proactive approach for the evaluation of fetal safety in chemical industries . . . 1347
20.3.6 Overall conclusion . . . 1353

20.4 Industrial solvents and kidney disease . . . 1355
NACHMAN BRAUTBAR
20.4.1 Introduction . . . 1355
20.4.2 Experimental animal studies . . . 1356
20.4.3 Case reports . . . 1356
20.4.4 Case control studies . . . 1357
20.4.5 Epidemiological assessment . . . 1360
20.4.6 Mechanism . . . 1361
20.5 Lymphohematopoietic study of workers exposed to benzene including multiple myeloma, lymphoma and chronic lymphatic leukemia . . . 1363
NACHMAN BRAUTBAR
20.5.1 Introduction . . . 1363
20.5.2 Routes of exposure . . . 1363
20.5.3 Hematopoietic effects of benzene . . . 1365
20.5.4 Carcinogenic effects of benzene . . . 1365
20.5.5 Risk assessment estimates . . . 1367
20.5.6 Levels of exposure . . . 1367
20.5.7 Cell types: hematolymphoproliferative effects of benzene . . . 1369
20.5.8 Epidemiological studies . . . 1369
20.5.9 Solvents and benzene . . . 1370
20.5.10 Genetic fingerprint theory . . . 1372
20.6 Chromosomal aberrations and sister chromatoid exchanges . . . 1375
NACHMAN BRAUTBAR
20.7 Hepatotoxicity . . . 1379
NACHMAN BRAUTBAR

20.7.1 Introduction . . . 1379
20.7.2 Individual variability and hepatotoxicity of solvents . . . 1384
20.7.3 Non-halogenated solvents . . . 1385
20.7.4 Solvent mixtures . . . 1386
20.7.5 Trichloroethylene . . . 1387
20.7.6 Tetrachloroethylene . . . 1388
20.7.7 Toluene . . . 1388
20.7.8 Dichloromethane . . . 1389
20.7.9 Stoddard solvent . . . 1389
20.7.10 1,1,1-Trichloroethane . . . 1389
20.7.11 Summary . . . 1390
20.8 Solvents and the liver . . . 1393
DAVID K. BONAUTO, C. ANDREW BRODKIN, WILLIAM O. ROBERTSON
20.8.1 Normal anatomic and physiologic function of the liver . . . 1393
20.8.1.1 Factors influencing solvent hepatotoxicity . . . 1394
20.8.1.2 Microscopic, biochemical and clinical findings associated with liver injury due to solvents . . . 1394
20.8.2 Hepatotoxicity associated with specific solvents . . . 1395
20.8.2.1 Haloalkanes and haloalkenes . . . 1396
20.8.2.2 Carbon tetrachloride . . . 1396
20.8.2.3 Chloroform . . . 1397
20.8.2.4 Dichloromethane . . . 1398
20.8.2.5 Trichloroethanes . . . 1398
20.8.2.6 1,1,2,2-Tetrachloroethane . . . 1398
20.8.2.7 Tetrachloroethylene and trichloroethylene . . . 1399
20.8.2.8 Other halogenated hydrocarbons . . . 1399
20.8.2.9 Styrene and aromatic hydrocarbons . . . 1399
20.8.2.10 N-substituted amides . . . 1400
20.8.2.11 Nitroparaffins . . . 1400
20.8.2.12 Other solvents and mixed solvents . . . 1401
20.9 Toxicity of environmental solvent exposure for brain, lung and heart . . . 1404
KAYE H. KILBURN
**21 SUBSTITUTION OF SOLVENTS BY SAFER PRODUCTS AND PROCESSES . . . 1419**
21.1 Supercritical solvents . . . 1419
AYDIN K. SUNOL, SERMIN G. SUNOL
21.1.1 Introduction . . . 1419
21.1.1.1 A promising path to green chemistry . . . 1422
21.1.1.2 Unique and tunable physico-chemical properties . . . 1422
21.1.1.3 Sustainable applications in many different areas . . . 1422
21.1.2 Fundamentals . . . 1423
21.1.2.1 Phase behavior with supercritical solvents . . . 1423
21.1.2.1.1 Experimental methods . . . 1426
21.1.2.1.2 Computational aspects . . . 1428
21.1.2.1.3 Modeling . . . 1429
21.1.2.2 Transport properties of supercritical solvents . . . 1431
21.1.2.2.1 Viscosity . . . 1431
21.1.2.2.2 Diffusivity . . . 1432
21.1.2.2.3 Thermal conductivity . . . 1433
21.1.2.2.4 Surface tension . . . 1435
21.1.2.3 Entrainer (co-solvent effects) of supercritical solvents . . . 1435
21.1.2.4 Reaction rate implication in supercritical solvents . . . 1436
21.1.2.5 Sorption behavior of supercritical solvents . . . 1437
21.1.2.6 Swelling with supercritical solvents . . . 1437
21.1.2.7 Surfactants and micro-emulsions . . . 1438
21.1.3 Separation with supercritical solvents . . . 1438
21.1.3.1 Leaching - generic application . . . 1441

21.1.3.2 Extraction - generic applications . . . 1442
21.1.3.3 Crystallization - generic applications. . . . 1443
21.1.3.4 Sorption - generic applications . . . 1443
21.1.4 Reactions in supercritical solvents . . . 1444
21.1.4.1 Homogenous reactions in supercritical solvents - examples . . . 1445
21.1.4.1.1 Homogeneous reactions catalyzed by organometallic compounds . . . 1446
21.1.4.1.2 Homogeneous reactions of supercritical water. . . . 1447
21.1.4.1.3 Homogeneous non-catalytic reactions in supercritical solvents . . . 1448
21.1.4.2 Heterogeneous reactions in supercritical solvents - examples . . . 1448
21.1.4.2.1 Heterogeneous catalytic reactions in supercritical solvents . . . 1449
21.1.4.2.2 Heterogeneous non-catalytic reactions in supercritical solvents. . . . 1450
21.1.4.3 Biochemical reactions - examples . . . 1451
21.1.4.4 Polymerization reactions - examples. . . . 1451
21.1.4.5 Materials processing with supercritical solvents. . . . 1452
21.1.4.6 Particle synthesis - generic application. . . . 1453
21.1.4.7 Encapsulation - generic application . . . 1454
21.1.4.8 Spraying and coating - generic application. . . . 1454
21.1.4.9 Extrusion - generic application. . . . 1454
21.1.4.10 Perfusion (impregnation) - generic application . . . 1454
21.1.4.11 Parts cleaning - generic application . . . 1455
21.1.4.12 Drying - generic application . . . 1455
21.2 Ionic liquids . . . 1459
D.W. ROONEY, K.R. SEDDON
21.2.1 Introduction. . . . 1459
21.2.2 Fundamental principles of the formation of room temperature ionic liquids . . 1461
21.2.2.1 Development of ionic liquids. . . . 1461
21.2.2.2 Binary ionic liquid systems. . . . 1465
21.2.3 Catalysis in ionic liquids . . . 1466
21.2.3.1 Reactions involving first generation chloroaluminate(III) ionic liquids . . . 1467
21.2.3.2 Reactions in neutral or second generation ionic liquids . . . 1469
21.2.4 Electrochemical applications. . . . 1472
21.2.4.1 Electrosynthesis . . . 1473
21.2.5 Physical characterization . . . 1473
21.2.5.1 Viscosity . . . 1473
21.2.5.2 Density . . . 1478
21.2.6 Summary . . . 1480
21.3 Oxide solubilities in ionic melts . . . 1484
VICTOR CHERGINETS
21.3.1 Methods used for solubility estimations in ionic melts . . . 1484
21.3.1.1 Isothermal saturation method. . . . 1485
21.3.1.2 Potentiometric titration method . . . 1486
21.3.2 Oxygen-containing melts. . . . 1487
21.3.3 Halide melts . . . 1487
21.3.3.1 The eutectic mixture KCl-LiCl (0.41:0.59) . . . 1487
21.3.3.2 Molten KCl-NaCl (0.50:0.50) . . . 1488
21.3.3.3 Other chloride-based melts. . . . 1491
21.3.3.4 Other alkaline halides . . . 1493
21.3.4 On the possibility to predict oxide solubilities on the base of the existing data . 1494
21.3.4.1 The estimation of effect of anion. . . . 1494
21.3.4.2 The estimation of effect of melt acidity . . . 1494
21.3.4.3 The estimation of effect of temperature . . . 1495
21.3.5 Conclusions. . . . 1495
21.4 Alternative cleaning technologies/drycleaning installations . . . 1497
KASPAR D. HASENCLEVER
21.4.1 Drycleaning with liquid carbon dioxide (LCD) . . . 1497
21.4.1.1 Basics. . . . 1497
21.4.1.2 State of the art . . . 1498

21.4.1.3 Process technology . . . . . 1498
21.4.1.4 Risks . . . . . 1499
21.4.1.5 Competition . . . . . 1500
21.4.2 Wet cleaning . . . . . 1501
21.4.2.1 Kreussler textile cleaning system. . . . . 1501
21.4.2.2 Possibilities . . . . . 1503
21.4.2.3 Limitations . . . . . 1504
21.4.2.4 Adapting to working practices . . . . . 1504
21.4.3 Future . . . . . 1505
**22 SOLVENT RECYCLING, REMOVAL, AND DEGRADATION . . . . . 1507**
22.1 Absorptive solvent recovery . . . . . 1507
KLAUS-DIRK HENNING
22.1.1 Introduction . . . . . 1507
22.1.2 Basic principles . . . . . 1509
22.1.2.1 Fundamentals of adsorption . . . . . 1509
22.1.2.2 Adsorption capacity . . . . . 1510
22.1.2.3 Dynamic adsorption in adsorber beds . . . . . 1511
22.1.2.4 Regeneration of the loaded adsorbents . . . . . 1512
22.1.3 Commercially available adsorbents . . . . . 1513
22.1.3.1 Activated carbon . . . . . 1513
22.1.3.2 Molecular sieve zeolites . . . . . 1514
22.1.3.3 Polymeric adsorbents . . . . . 1515
22.1.4 Adsorptive solvent recovery systems . . . . . 1515
22.1.4.1 Basic arrangement of adsorptive solvent recovery with steam desorption . . . . 1515
22.1.4.2 Designing solvent recovery systems . . . . . 1518
22.1.4.2.1 Design basis . . . . . 1518
22.1.4.2.2 Adsorber types . . . . . 1519
22.1.4.2.3 Regeneration . . . . . 1521
22.1.4.2.4 Safety requirements . . . . . 1522
22.1.4.3 Special process conditions . . . . . 1523
22.1.4.3.1 Selection of the adsorbent . . . . . 1523
22.1.4.3.2 Air velocity and pressure drop . . . . . 1526
22.1.4.3.3 Effects of solvent-concentration, adsorption temperature and pressure . . . . . 1526
22.1.4.3.4 Influence of humidity . . . . . 1528
22.1.4.3.5 Interactions between solvents and activated carbon . . . . . 1529
22.1.4.3.6 Activated carbon service life . . . . . 1531
22.1.5 Examples from different industries . . . . . 1531
22.1.5.1 Rotogravure printing shops . . . . . 1531
22.1.5.2 Packaging printing industry . . . . . 1532
22.1.5.2.1 Fixed bed adsorption with circulating hot gas desorption . . . . . 1533
22.1.5.2.2 Solvent recovery with adsorption wheels . . . . . 1535
22.1.5.3 Viscose industry . . . . . 1535
22.1.5.4 Refrigerator recycling . . . . . 1539
22.1.5.5 Petrochemical industry and tank farms . . . . . 1539
22.1.5.6 Chemical industry . . . . . 1541
22.2 Solvent recovery . . . . . 1543
ISAO KIMURA
22.2.1 Activated carbon in fluidized bed adsorption method . . . . . 1543
22.2.2 Application of molecular sieves . . . . . 1544
22.2.3 Continuous process for air cleaning using macroporous particles as adsorption agents . . . . . 1546
22.2.4 Solvent recovery from hazardous wastes . . . . . 1548
22.2.5 Halogenated solvent recovery . . . . . 1549
22.2.5.1 Coating process . . . . . 1549
22.2.5.2 Tableting process of pharmaceutical products . . . . . 1552
22.2.6 Energy recovery from waste solvent . . . . . 1553
22.3 Solvent treatment in a paints and coating plant . . . . . 1555

DENIS KARGOL
22.4 Application of solar photocatalytic oxidation to VOC-containing airstreams . . 1559
K. A. MAGRINI, A. S. WATT, L. C. BOYD, E. J. WOLFRUM, S. A. LARSON, C. ROTH
G. C. GLATZMAIER
22.4.1 Solvent degradation by photocatalytic oxidation. . . . . . . . . . . . . . . . 1559
22.4.2 PCO pilot scale systems . . . . . . . . . . . . . . . . . . . . . . . . . . . 1560
22.4.2.1 Air stripper application . . . . . . . . . . . . . . . . . . . . . . . . . . 1560
22.4.2.2 Paint booth application . . . . . . . . . . . . . . . . . . . . . . . . . . 1562
22.4.3 Field test results . . . . . . . . . . . . . . . . . . . . . . . . . . . . . . 1564
22.4.3.1 Air stripper application . . . . . . . . . . . . . . . . . . . . . . . . . . 1564
22.4.3.2 Paint booth application . . . . . . . . . . . . . . . . . . . . . . . . . . 1566
22.4.4 Comparison with other treatment systems . . . . . . . . . . . . . . . . . . 1568
**23 CONTAMINATION CLEANUP: NATURAL ATTENUATION AND ADVANCED REMEDIATION TECHNOLOGIES . . . . . . . . . . . . . 1571**
23.1 Natural attenuation of chlorinated solvents in ground water. . . . . . . . . . . 1571
HANADI S. RIFAI, CHARLES J. NEWELL, TODD H. WIEDEMEIER
23.1.1 Introduction. . . . . . . . . . . . . . . . . . . . . . . . . . . . . . . . . 1571
23.1.2 Natural attenuation processes affecting chlorinated solvent plumes . . . . . . . 1572
23.1.2.1 Advection. . . . . . . . . . . . . . . . . . . . . . . . . . . . . . . . . 1572
23.1.2.2 Dispersion . . . . . . . . . . . . . . . . . . . . . . . . . . . . . . . . 1573
23.1.2.3 Sorption. . . . . . . . . . . . . . . . . . . . . . . . . . . . . . . . . . 1574
23.1.2.4 One-dimensional advection-dispersion equation with retardation . . . . . . . . 1577
23.1.2.5 Dilution (recharge) . . . . . . . . . . . . . . . . . . . . . . . . . . . . 1577
23.1.2.6 Volatilization . . . . . . . . . . . . . . . . . . . . . . . . . . . . . . . 1578
23.1.2.7 Hydrolysis and dehydrohalogenation . . . . . . . . . . . . . . . . . . . . 1579
23.1.2.8 Reduction reactions. . . . . . . . . . . . . . . . . . . . . . . . . . . . . 1581
23.1.3 Biodegradation of chlorinated solvents . . . . . . . . . . . . . . . . . . . . 1581
23.1.3.1 Halorespiration or reductive dechlorination using hydrogen. . . . . . . . . . . 1582
23.1.3.1.1 Stoichiometry of reductive dechlorination . . . . . . . . . . . . . . . . . . 1585
23.1.3.1.2 Chlorinated solvents that are amenable to halorespiration . . . . . . . . . . . 1585
23.1.3.2 Oxidation of chlorinated solvents . . . . . . . . . . . . . . . . . . . . . . 1586
23.1.3.2.1 Direct aerobic oxidation of chlorinated compounds . . . . . . . . . . . . . . 1586
23.1.3.2.2 Aerobic cometabolism of chlorinated compounds . . . . . . . . . . . . . . . 1587
23.1.3.2.3 Anaerobic oxidation of chlorinated compounds . . . . . . . . . . . . . . . . 1587
23.1.4 Biodegradation rates for chlorinated solvents . . . . . . . . . . . . . . . . . 1588
23.1.4.1 Michaelis-Menten rates. . . . . . . . . . . . . . . . . . . . . . . . . . . 1588
23.1.4.2 Zero-order rates. . . . . . . . . . . . . . . . . . . . . . . . . . . . . . . 1590
23.1.4.3 First-order rate constants . . . . . . . . . . . . . . . . . . . . . . . . . . 1591
23.1.5 Geochemical evidence of natural bioremediation at chlorinated solvent sites. . 1599
23.1.5.1 Assessing reductive dechlorination at field sites . . . . . . . . . . . . . . . 1599
23.1.5.2 Plume classification schemes. . . . . . . . . . . . . . . . . . . . . . . . . 1599
23.1.5.2.1 Type 1 . . . . . . . . . . . . . . . . . . . . . . . . . . . . . . . . . . . 1599
23.1.5.2.2 Type 2 . . . . . . . . . . . . . . . . . . . . . . . . . . . . . . . . . . . 1600
23.1.5.2.3 Type 3 . . . . . . . . . . . . . . . . . . . . . . . . . . . . . . . . . . . 1601
23.1.5.2.4 Mixed environments . . . . . . . . . . . . . . . . . . . . . . . . . . . . 1601
23.1.6 Chlorinated solvent plumes - case studies of natural attenuation . . . . . . . . 1602
23.1.6.1 Plume databases . . . . . . . . . . . . . . . . . . . . . . . . . . . . . . 1602
23.1.6.2 Modeling chlorinated solvent plumes . . . . . . . . . . . . . . . . . . . . 1605
23.1.6.2.1 BIOCHLOR natural attenuation model . . . . . . . . . . . . . . . . . . . . 1605
23.1.6.3 RT3D numerical model . . . . . . . . . . . . . . . . . . . . . . . . . . . 1609
23.1.6.4 CS case study - The Plattsburgh Air Force Base. . . . . . . . . . . . . . . . 1611
23.2 Remediation technologies and approaches for managing sites impacted by hydrocarbons . . . . . . . . . . . . . . . . . . . . . . . . . . . . . . . . . . 1617
BARRY J. SPARGO, JAMES G. MUELLER
23.2.1 Introduction. . . . . . . . . . . . . . . . . . . . . . . . . . . . . . . . . 1617
23.2.1.1 Understanding HC and CHC in the environment . . . . . . . . . . . . . . . 1617
23.2.1.2 Sources of HC in the environment . . . . . . . . . . . . . . . . . . . . . . 1617

23.2.1.3 Sources of CHC in the environment . . . . . . . . . . . . . . . . . . . . . 1618
23.2.2 In situ biotreatment . . . . . . . . . . . . . . . . . . . . . . . . . . . . . . 1618
23.2.2.1 Microbial-enhanced natural attenuation/bioremediation . . . . . . . . . . . . . 1618
23.2.2.1.1 Case study - Cooper River Watershed, Charleston, SC, USA . . . . . . . . . . 1620
23.2.2.2 Phytoremediation . . . . . . . . . . . . . . . . . . . . . . . . . . . . . . . 1622
23.2.2.2.1 Case study - phytoremediation for CHCs in groundwater at a chemical plant in Louisiana . . . . . . . . . . . . . . . . . . . . . . . . . . . . . . . 1622
23.2.3 In situ treatment technologies . . . . . . . . . . . . . . . . . . . . . . . . 1623
23.2.3.1 Product recovery via GCW technology . . . . . . . . . . . . . . . . . . . . . 1623
23.2.3.1.1 Case study - GCW recovery of creosote, Cabot/Kopper's Superfund Site, Gainesville, FL . . . . . . . . . . . . . . . . . . . . . . . . . . . . . . . . 1624
23.2.3.2 Surfactant enhanced product recovery . . . . . . . . . . . . . . . . . . . . . 1625
23.2.3.2.1 Case study - Surfactant-aided chlorinated HC DNAPL recovery, Hill Air Force Base, Ogden, Utah . . . . . . . . . . . . . . . . . . . . . . . . 1625
23.2.3.3 Foam-enhanced product recovery . . . . . . . . . . . . . . . . . . . . . . . . 1626
23.2.3.4 Thermal desorption - Six Phase Heating . . . . . . . . . . . . . . . . . . . . 1626
23.2.3.4.1 Case study - Six-Phase Heating removal of CHC at a manufacturing facility near Chicago, IL . . . . . . . . . . . . . . . . . . . . . . . . . . . . . . . 1627
23.2.3.5 In situ steam enhanced extraction (Dynamic Underground Stripping) . . . . . 1628
23.2.3.6 In situ permeable reactive barriers (funnel and gate) . . . . . . . . . . . . . 1628
23.2.3.6.1 Case study - CHC remediation using an in situ permeable reactive barrier at Naval Air Station Moffett Field, CA . . . . . . . . . . . . . . . . . . . . . 1628
23.2.4 Conclusions . . . . . . . . . . . . . . . . . . . . . . . . . . . . . . . . . . 1629
**24 PROTECTION . . . . . . . . . . . . . . . . . . . . . . . . . . . . . . . . . . 1631**
GEORGE WYPYCH
24.1 Gloves . . . . . . . . . . . . . . . . . . . . . . . . . . . . . . . . . . . . . 1631
24.2 Suit materials . . . . . . . . . . . . . . . . . . . . . . . . . . . . . . . . . 1633
24.3 Respiratory protection . . . . . . . . . . . . . . . . . . . . . . . . . . . . . 1633
**25 NEW TRENDS BASED ON PATENT LITERATURE . . . . . . . . . . . 1637**
GEORGE WYPYCH
25.1 New solvents . . . . . . . . . . . . . . . . . . . . . . . . . . . . . . . . . . 1637
25.2 Adhesives . . . . . . . . . . . . . . . . . . . . . . . . . . . . . . . . . . . . 1638
25.3 Aerospace . . . . . . . . . . . . . . . . . . . . . . . . . . . . . . . . . . . . 1640
25.4 Agriculture . . . . . . . . . . . . . . . . . . . . . . . . . . . . . . . . . . . 1640
25.5 Asphalt . . . . . . . . . . . . . . . . . . . . . . . . . . . . . . . . . . . . . 1640
25.6 Automotive applications . . . . . . . . . . . . . . . . . . . . . . . . . . . . 1641
25.7 Coil coating . . . . . . . . . . . . . . . . . . . . . . . . . . . . . . . . . . . 1641
25.8 Composites and laminates . . . . . . . . . . . . . . . . . . . . . . . . . . . 1642
25.9 Cosmetics . . . . . . . . . . . . . . . . . . . . . . . . . . . . . . . . . . . . 1643
25.10 Cleaning . . . . . . . . . . . . . . . . . . . . . . . . . . . . . . . . . . . . 1644
25.11 Fibers . . . . . . . . . . . . . . . . . . . . . . . . . . . . . . . . . . . . . . 1645
25.12 Furniture and wood coatings . . . . . . . . . . . . . . . . . . . . . . . . . . 1646
25.13 Paper . . . . . . . . . . . . . . . . . . . . . . . . . . . . . . . . . . . . . . 1647
25.14 Printing . . . . . . . . . . . . . . . . . . . . . . . . . . . . . . . . . . . . . 1647
25.15 Stone and concrete . . . . . . . . . . . . . . . . . . . . . . . . . . . . . . . 1648
25.16 Wax . . . . . . . . . . . . . . . . . . . . . . . . . . . . . . . . . . . . . . . 1648
25.17 Summary . . . . . . . . . . . . . . . . . . . . . . . . . . . . . . . . . . . . 1649
**ACKNOWLEDGMENTS . . . . . . . . . . . . . . . . . . . . . . . . . . . . 1653**
**INDEX . . . . . . . . . . . . . . . . . . . . . . . . . . . . . . . . . . . . . 1657**

# PREFACE

Although the chemical industry can trace its roots into antiquity, it was during the industrial revolution that it started to become an actual industry and began to use the increased knowledge of chemistry as a science and technology to produce products that were needed by companion industries and consumers. These commercial efforts resulted in the synthesis of many new chemicals. Quite quickly, in these early days, previously unknown materials or materials that had been present only in low concentrations, were now in contact with people in highly concentrated forms and in large quantities. The people had little or no knowledge of the effects of these materials on their bodies and the natural biological and physical processes in the rivers and oceans, the atmosphere, and in the ground.

Until the end of the nineteenth century these problems were not addressed by the chemical industry and it is only recently that the industry began to respond to public criticism and political efforts. Legal restrictions aimed at preserving the quality of life have been directed at health, safety and longevity issues and the environment. Solvents have always been mainstays of the chemical industry and because of their widespread use and their high volume of production they have been specifically targeted by legislators throughout the world. The restrictions range from total prohibition of production and use, to limits placed on vapor concentrations in the air. As with any arbitrary measures some solvents have been damned unfairly. However, there is no question that it is best to err on the side of safety if the risks are not fully understood. It is also true that solvents should be differentiated based on their individual properties.

This book is intended to provide a better understanding of the principles involved in solvent selection and use. It strives to provide information that will help to identify the risks and benefits associated with specific solvents and classes of solvents. The book is intended to help the formulator select the ideal solvent, the safety coordinator to safeguard his or her coworkers, the legislator to impose appropriate and technically correct restrictions and the student to appreciate the amazing variety of properties, applications and risks associated with the more than one thousand solvents that are available today.

By their very nature, handbooks are intended to provide exhaustive information on the subject. While we agree that this is the goal here, we have attempted to temper the impact of information, which may be too narrow to make decision.

Many excellent books on solvents have been published in the past and most of these are referenced in this book. But of all these books none has given a comprehensive overview of all aspects of solvent use. Access to comprehensive data is an essential part of solvent evaluation and it has been a hallmark of such books to provide tables filled with data to the point at which 50 to 95% of the book is data. This approach seems to neglect a fundamental requirement of a handbook - to provide the background, explanations and clarifications that are needed to convert data to information and assist the reader in gaining the knowledge to make a decision on selecting a process or a solvent. Unfortunately, to meet the goal of providing both the data and the fundamental explanations that are needed, a book of 4,000 to 5,000 pages might be required. Even if this was possible, much of the data would fall out of date quite quickly. For example, a factor that defines solvent safety such as threshold limit values (TLVs) for worker exposure or some single toxicity determinants may change frequently. This book would be huge and it would have to be updated frequently to continue to claim that it is current.

What we have attempted to do here is to give you a book with a comprehensive and extensive analysis of all current information on solvents then use other media to present the supporting data on individual solvents. These data are provided on a CD-ROM as a searchable database. Data are provided on more than 1140 solvents in 110 fields of data. The medium permits frequent updates. If the same data were presented in book form, more than 2,000 pages would be needed which exceeds the size of any data in handbook form offered to date.

The best approach in presenting an authoritative text for such a book is to have it written by experts in their fields. This book attracted well-known experts who have written jointly 47 books and authored or coauthored hundreds of papers on their areas of expertise. The authors have made their contributions to this book in late 1999 and early 2000 providing the most current picture of the technology. Their extreme familiarity with their subjects enables them to present information in depth and detail, which is essential to the reader's full understanding of the subject.

The authors were aware of the diversity of potential readers at the outset and one of their objectives was to provide information to various disciplines expressed in a way that all would understand and which would deal with all aspects of solvent applications. We expect professionals and students from a wide range of businesses, all levels of governments and academe to be interested readers. The list includes solvent manufacturers, formulators of solvent containing products, industrial engineers, analytical chemists, government legislators and their staffs, medical professionals involved in assessing the impact on health of solvents, biologists who are evaluating the interactions of solvents with soil and water, environmental engineers, industrial hygienists who are determining protective measures against solvent exposure, civil engineers who design waste disposal sites and remediation measures, people in industries where there are processes which use solvents and require their recovery and, perhaps most important, because understanding brings improvements, those who teach and learn in our universities, colleges and schools.

A growing spirit of cooperation is evident between these groups and this can be fostered by providing avenues of understanding based on sharing data and information on common problems. We hope to provide one such avenue with this book. We have tried to present a balanced picture of solvent performance by dealing not only with product performance and ease of processing but also by giving environmental and health issues full consideration.

Data and information on known products and processes should be cornerstones of the understanding of a technology but there is another aspect of technology, which can lead to advances and improvements in utility, safety and in safeguarding the environment. This must come from you, the reader. It is your ideas and creative thinking that will bring these improvements. The authors have crammed their ideas into the book and we hope these will stimulate responsible and effective applications of solvents. Francis Bacon wrote, "The end of our foundation is the knowledge of causes, and the secret motion of things, and the enlarging of the bound of human Empire, to the effecting of all things possible."

Today there are few technical activities that do not employ solvents. Almost all industries, almost all consumer products, almost everything we use can, if analyzed, be shown to contain or to have used in its processing, a solvent. Solvent elimination need never be a technical objective. Rather, we need to use our increasing understanding and knowledge to find the safest and the most effective means of meeting our goals.

I would like to thank the authors for their relentless efforts to explain the difficult in an interesting way. In advance, I would like to thank the reader for choosing this book and encourage her or him to apply the knowledge to make our world a better, more livable place.

George Wypych
Toronto, August 3, 2000

# 1

# INTRODUCTION

CHRISTIAN REICHARDT
**Department of Chemistry, Philipps University, Marburg, Germany**

Chemical transformations can be performed in a gas, liquid, or solid phase, but, with good reasons, the majority of such reactions is carried out in the liquid phase in solution. At the macroscopic level, a liquid is the ideal medium to transport heat to and from exo- and endothermic reactions. From the molecular-microscopic point of view, solvents break the crystal lattice of solid reactants, dissolve gaseous or liquid reactants, and they may exert a considerable influence over reaction rates and the positions of chemical equilibria. Because of nonspecific and specific intermolecular forces acting between the ions or molecules of dissolved reactants, activated complexes as well as products and solvent molecules (leading to differential solvation of all solutes), the rates, equilibria, and the selectivity of chemical reactions can be strongly influenced by the solvent. Other than the fact that the liquid medium should dissolve the reactants and should be easily separated from the reaction products afterwards, the solvent can have a decisive influence on the outcome (i.e., yield and product distribution) of the chemical reaction under study. Therefore, whenever a chemist wishes to perform a certain chemical reaction, she/he has to take into account not only suitable reaction partners and their concentrations, the proper reaction vessel, the appropriate reaction temperature, and, if necessary, the selection of the right reaction catalyst but also, if the planned reaction is to be successful, the selection of an appropriate solvent or solvent mixture.

Solvent effects on chemical reactivity have been studied for more than a century, beginning with the pioneering work of Berthelot and Saint Gilles[1] in Paris in 1862 on esterification reactions and with that of Menschutkin[2] in St. Petersburg in 1880 on the quaternization of tertiary amines by haloalkanes. At this time Menschutkin remarked that "a reaction cannot be separated from the medium in which it is performed... Experience shows that solvents exert considerable influence on reaction rates." Today, we can suggest a striking example to reinforce his remark, the rate of the unimolecular heterolysis of 2-chloro-2-methylpropane observed in water and benzene increases by a factor of approximately$10^{11}$ when the nonpolar benzene is replaced by water.[3,4] The influence of solvents on the position of chemical equilibria was discovered in 1896 by Claisen[5] in Aachen, Knorr[6] in Jena, Wislicenus[7] in Würzburg, and Hantzsch[8] in Würzburg. They investigated almost simultaneous but independent of one another the keto-enol tautomerism of 1,3-dicarbonyl compounds and the nitro-isonitro tautomerism of primary and secondary aliphatic nitro compounds. With this example, the enol content of acetylacetone increases from 62 to 95 % when acetonitrile is substituted with n-hexane.[3,9]

The proper solvent and solvent mixture selection is not only important for chemical but also for physical processes such as recrystallization, all extraction processes, partitioning, chromatographic separations, phase-transfer catalytic reactions, etc. Of particular interest in this context is the influence of solvents on all types of light absorption processes, e.g., on UV/Vis, IR, ESR, and NMR spectra, caused by differential solvation of the ground and excited states of the absorbing species.[3,12] In 1878, Kundt[10] in Zürich proposed the rule that increasing dispersion interactions between the absorbing solute and the solvent lead in general to a bathochromic shift of an UV/Vis absorption band. Later, in 1922, Hantzsch[11] termed the solvent-dependence of UV/Vis absorption spectra "solvatochromism". UV/Vis absorption of solute molecules can be influenced not only by the surrounding solvent sphere, but also by other entities in the surroundings such as solids, polymers, glasses, and surfaces. In order to emphasize this influence, the use of the more general term "perichromism" (from Greek peri = around) has been recommended.[12,13] A typical, more recent, example of extraordinary solvatochromism is the intramolecular charge-transfer Vis-absorption of 2,6-diphenyl-4-(2,4,6-triphenyl-l-pyridinio)phenolate, a zwitterionic betaine dye: its corresponding absorption band is shifted from $\lambda_{max}$ = 810 nm to $\lambda_{max}$ = 453 nm ($\Delta\lambda$ = 357 nm) when diphenyl ether is replaced by water as solvent.[3,12] Such solvatochromic dyes can be used as empirical solvent polarity indicators.[12]

The number of solvents generally available to chemists working in research and industrial laboratories is between 250 and 300[3,14] (there is an infinite number of solvent mixtures), and this number is increasing. More recently and for obvious reasons, the search for new solvents has been intensified: peroxide-forming solvents are being substituted by solvents which are more stable against oxidation (e.g., diethyl ether by t-butyl methyl ether or by formaldehyde dialkyl acetals), toxic solvents are being replaced by nontoxic ones (e.g., the cancerogenic hexamethylphosphoric triamide, HMPT, by N,N'-dimethylpropyleneurea, DMPU[15]) and environmentally dangerous solvents by benign ones (e.g., tetrachloromethane by perfluorohexane[16]). The development of modern solvents for organic syntheses is the subject of much current research.[17] Amongst these modern solvents, also called "neoteric solvents" (neoteric = recent, new, modern) in contrast to the classical ones, are ionic liquids (i.e., room-temperature liquid salts such as 1-ethyl-3-methylimidazolium tetra-chloroaluminates[18,19]), supercritical-fluid solvents, SCF, (such as SCF carbon dioxide[20,21]), and perfluorinated solvents (e.g., partially or perfluorinated hydrocarbons as used in so-called "fluorous biphase catalysis reactions", making possible mono-phase reactions and a two-phase separation of catalyst and reaction products[22-24]). Even plain water has found a magnificent renaissance as a solvent for organic reactions.[25,26] These efforts have also recently strengthened the search for completely solvent-free reactions, thus avoiding the use of expensive, toxic, and environmentally problematic solvents.[27,28]

With respect to the large and still increasing number of valuable solvents useful for organic syntheses, a chemist needs, in addition to his experience and intuition, to have general rules, objective criteria, and the latest information about the solvents' physical, chemical, and toxicological properties for the selection of the proper solvent or solvent mixture for a planned reaction or a technological process. To make this often cumbersome and time-consuming task easier, this "Handbook of Solvents" with its twenty-five chapters is designed to provide a comprehensive source of information on solvents over a broad range of applications. It is directed not only to chemists working in research laboratories, but also to all industries using solvents for various purposes. A particular advantage is that the printed handbook is accompanied by a compact-disc (CD-ROM) containing additional solvent databases with hundred ten fields for over eleven hundred solvents. This makes large data sets easily available for quick search and retrieval and frees the book text from bulky tables, thus giving more room for a thorough description of the underlying theoretical and practical fundamental subjects.

Fundamental principles governing the use of solvents (i.e., chemical structure, molecular design as well as physical and chemical properties of solvents) are given in Chapter 2. Solvent classification, methods of solvent manufacture together with properties and typical applications of various solvents are provided in Chapter 3. Chapters 4, 5 and 6 deal with all aspects of the dissolution of materials in solvents as well as with the solubility of selected systems (e.g., polymers and elastomers) and the influence of the solute's molecular structure on its solubility behavior. In particular, the valuable solubility-parameter concept is extensively treated in these chapters. All aspects of solvent transport within polymeric system and the drying of such polymeric systems, including coated films, are described in Chapter 7. The fundamentals of the interaction forces acting between ions or molecules of the solvents themselves and between solutes and solvents in solutions are presented in Chapter 8. Chapter 9 deals with the corresponding properties of solvent mixtures. Specific solute/solvent interactions, particularly Lewis acid/base interactions between electron-pair donors (EPD) and electron-pair acceptors (EPA), are reviewed in Chapter 10, together with the development of empirical scales of solvent polarity and Lewis acidity/basicity, based on suitable solvent-dependent reference processes, and their application for the treatment of solvent effects. The theory for solvent effects on electronic properties is provided in Chapter 11 and extended to solvent-dependent properties of solutes such as fluorescence spectra, ORD and CD spectra. Aggregation, swelling of polymers, their conformations, the viscosity of solutions and other solvent-related properties are treated in Chapter 12. A review concerning solvent effects on various types of chemical reactivity is given in Chapter 13, along with a discussion of the effects of solvent on free-radical polymerization and phase-transfer catalysis reactions.

The second part of this handbook (Chapters 14-25) is devoted more to the industrial use of solvents. Formulating with solvents applied in a broad range of industrial areas such as biotechnology, dry cleaning, electronic industry, food industry, paints and coatings, petroleum refining industry, pharmaceutical industry, textile industry, to mention only a few, is extensively described in Chapter 14. Standard and special methods of solvent detection and solvent analysis as well as the problem of residual solvents in various products, particularly in pharmaceutical ones, are the topics of Chapters 15 and 16.

At present, large-scale chemical manufacturing is facing serious solvent problems with respect to environmental concerns. National and international regulations for the proper use of hazardous solvents are becoming increasingly stringent and this requires the use of environmentally more benign but nevertheless economical liquid reaction media. This has enormously stimulated the search for such new solvent systems within the framework of so-called green chemistry. Supercritical fluids, SCF,[20,21] and ionic liquids (room temperature liquid salts)[18,19] have been known and have been the subject of scientific interest for a long time. It is only recently, however, that the potential benefits of these materials in solvent applications have been realized.[17] This handbook includes in Chapters 17-25 all the knowledge necessary for a safe handling of solvents in research laboratories and in large-scale manufacturing, beginning with the environmental impact of solvents on water, soil, and air in Chapter 17, followed by considerations about safe solvent concentrations and the risks of solvent exposure in various industrial environments in Chapter 18. Chapter 19 summarizes the corresponding legal regulations, valid for North America and Europe, and Chapter 20 describes in detail the toxic effects of solvent exposure to human beings. Authors specializing in different fields of solvent toxicity give the most current information on the effect of solvent exposure from the point of view of neurotoxicity, reproductive and maternal effects, nephrotoxicity, cancerogenicity, hepatotoxicity, chromosomal aberrations, and toxicity to brain, lungs, and heart. This information brings both the results of documented studies and an evaluation of risk in different industrial environments in a comprehensive but easy to understand form to engineers and decision-makers in industry. Chapter 21 is focused on the substitution of harmful solvents by safer ones and on the development of corresponding new technological processes. Chapter 22 describes modern methods of solvent recovery, solvent recycling. When recycling is not possible, then solvents have to be destroyed by incineration or other methods of oxidation, as outlined in Chapter 22. Chapter 23 describes natural attenuation of solvents in groundwater and advanced remediation technologies as well as management strategies for sites impacted by solvent contamination. Protection from contact with solvents and their vapors is discussed in Chapter 24. Finally, new trends in solvent chemistry and applications based on the recent patent literature are discussed in Chapter 25.

In most cases, the intelligent choice of the proper solvent or solvent mixture is essential for the realization of certain chemical transformations and physical processes. This handbook tries to cover all theoretical and practical information necessary for this often difficult task for both academic and industrial applications. It should be used not only by chemists, but also by physicists, chemical engineers, and technologists as well as environmental scientists in academic and industrial institutions. It is to be hoped that the present compilation of all relevant aspects connected with the use of solvents will also stimulate further basic and applied research in the still topical field of the physics and chemistry of liquid media.

## REFERENCES

1 M Berthelot, L Péan de Saint Gilles, *Ann. Chim. Phys.*, 3. Sér., **65**, 385 (1862); *ibid.* **66**, 5 (1862); *ibid.* **68**, 255 (1863).
2 N Menschutkin, *Z. Phys. Chem.*, **5**, 589 (1890); *ibid.* **6**, 41 (1890); *ibid.* **34**, 157 (1900).
3 C Reichardt, **Solvents and Solvent Effects in Organic Chemistry**, 2nd ed., *VCH*, Weinheim, 1988.
4 (a) G F Dvorko, E A Ponomareva, *Usp. Khim.*, **53**, 948 (1984); *Russ. Chem. Rev.*, **53**, 547 (1984); (b) M H Abraham, *Pure Appl. Chem.*, **57**, 1055 (1985); and references cited therein.
5 L Claisen, *Justus Liebigs Ann. Chem.*, **291**, 25 (1896).
6 L Knorr, *Justus Liebigs Ann. Chem.*, **293**, 70 (1896).
7 W Wislicenus, *Justus Liebigs Ann. Chem.*, **291**, 147 (1896).
8 A Hantzsch, O W Schultze, *Ber. Dtsch. Chem. Ges.*, **29**, 2251 (1896).
9 M T Rogers, J L Burdett, *Can. J Chem.*, **43**, 1516 (1965).
10 A Kundt, *Poggendorfs Ann. Phys. Chem. N. F.*, **4**, 34 (1878); *Chem. Zentralbl.*, 498 (1878).
11 A Hantzsch, *Ber. Dtsch. Chem. Ges.*, **55**, 953 (1922).
12 C Reichardt, *Chem. Rev.*, **94**, 2319 (1994).
13 Prof. E M Kosower, Tel Aviv, private communication to C.R.
14 Y Marcus, **The Properties of Solvents**, *Wiley*, Chichester, 1998.
15 (a) Editorial, *Chimia*, **39**, 147 (1985); (b) D Seebach, *Chemistry in Britain*, **21**, 632 (1985).
16 S M Pereira, G P Sauvage, G. W. Simpson, *Synth. Commun.*, **25**, 1023 (1995).

17 P Knochel (Ed.), **Modern Solvents in Organic Synthesis**, Topics in Current Chemistry, Vol. 206, *Springer*, Berlin, 1999.
18 Y Chauvin, H Olivier-Bourbigou, *CHEMTECH*, **25**(9), 26 (1995).
19 (a) K R Seddon, *Kinetika i Kataliz*, **37**, 743 (1996); *Kinetics and Catalysis*, **37**, 693 (1996); *Chem. Abstr.*, **125**, 285927s (1996);
(b) K R Seddon, *J. Chem. Technol. Biotechnol.*, **68**, 351 (1997); *Chem. Abstr.*, **126**, 306898w (1997).
20 R Noyori (Ed.), Supercritical Fluids, *Chem. Rev.*, **99**, 353-633 (1999).
21 W Leitner, *Top. Curr. Chem.*, **206**, 107 (1999).
22 B Cornils, *Angew. Chem.*, **109**, 2147 (1997); *Angew. Chem., Int. Ed. Engl.*, **36**, 2057 (1997).
23 B Betzemeier, P Knochel, *Top. Curr. Chem.*, **206**, 61 (1999).
24 J J Maul, P J Ostrowski, G A Ublacker, B Linclau, D P Curran, *Top. Curr. Chem.*, **206**, 79 (1999).
25 P A Grieco, **Organic Synthesis in Water**, *Blackie Academic and Professional*, Hampshire, 1998.
26 A Lubineau and J. Augé, *Top. Curr. Chem.*, **206**, 1 (1999).
27 J O Metzger, *Angew, Chem.*, **110**, 3145 (1998); *Angew. Chem., Int. Ed. Engl.*, **37**, 2975 (1998).
28 A Loupy, *Top. Curr. Chem.*, **206**, 153 (1999).

2

# Fundamental Principles Governing Solvents Use

## 2.1 SOLVENT EFFECTS ON CHEMICAL SYSTEMS

Estanislao Silla, Arturo Arnau and Iñaki Tuñón
**Department of Physical Chemistry, University of Valencia, Burjassot (Valencia), Spain**

### 2.1.1 HISTORICAL OUTLINE

According to a story, a little fish asked a big fish about the ocean, since he had heard it being talked about but did not know where it was. Whilst the little fish's eyes turned bright and shiny full of surprise, the old fish told him that all that surrounded him was the ocean. This story illustrates in an eloquent way how difficult it is to get away from every day life, something of which the chemistry of solvents is not unaware.

The chemistry of living beings and that which we practice in laboratories and factories is generally a chemistry in solution, a solution which is generally aqueous. A daily routine such as this explains the difficulty which, throughout the history of chemistry, has been encountered in getting to know the effects of the solvent in chemical transformations, something which was not achieved in a precise way until well into the XX century. It was necessary to wait for the development of experimental techniques in vacuo to be able to separate the solvent and to compare the chemical processes in the presence and in the absence of this, with the purpose of getting to know its role in the chemical transformations which occur in its midst. But we ought to start from the beginning.

Although essential for the later cultural development, Greek philosophy was basically a work of the imagination, removed from experimentation, and something more than meditation is needed to reach an approach on what happens in a process of dissolution. However, in those remote times, any chemically active liquid was included under the name of "divine water", bearing in mind that the term "water" was used to refer to anything liquid or dissolved.[1]

Parallel with the fanciful search for the philosopher's stone, the alchemists toiled away on another impossible search, that of a universal solvent which some called "alkahest" and others referred to as "menstruum universale", which term was used by the very Paracelsus (1493-1541), which gives an idea of the importance given to solvents during that dark and obscurantist period. Even though the "menstruum universale" proved just as elusive as the philosopher's stone, all the work carried out by the alchemists in search of these illusionary materials opened the way to improving the work in the laboratory, the development of new methods, the discovery of compounds and the utilization of novel solvents. One of the tangible results of all that alchemistry work was the discovery of one of the first experimental rules of chemistry: "similia similibus solvuntor", which reminds us of the compatibility in solution of those substances of similar nature.

Even so, the alchemistry only touched lightly on the subject of the role played by the solvent, with so many conceptual gulfs in those pre-scientific times in which the terms dissolution and solution referred to any process which led to a liquid product, without making any distinction between the fusion of a substance - such as the transformation of ice into liquid water -, mere physical dissolution - such as that of a sweetener in water - or the dissolution which takes place with a chemical transformation - such as could be the dissolution of a metal in an acid. This misdirected vision of the dissolution process led the alchemists down equally erroneous collateral paths which were prolonged in time. Some examples are worth quoting: Hermann Boerhaave (1688-1738) thought that dissolution and chemical reaction constituted the same reality; the solvent, (menstruum), habitually a liquid, he considered to be formed by diminutive particles moving around amongst those of the solute, leaving the interactions of these particles dependent on the mutual affinities of both substances.[2] This paved the way for Boerhaave to introduce the term affinity in a such a way as was conserved throughout the whole of the following century.[3] This approach also enabled Boerhaave to conclude that combustion was accompanied by an increase of weight due to the capturing of "particles" of fire, which he considered to be provided with weight by the substance which was burned. This explanation, supported by the well known Boyle, eased the way to considering that fire, heat and light were material substances until when, in the XIX century, the modern concept of energy put things in their place.[4]

Even Bertollet (1748-1822) saw no difference between a dissolution and a chemical reaction, which prevented him from reaching the law of definite proportions. It was Proust, an experimenter who was more exacting and capable of differentiating between chemical reaction and dissolution, who made his opinion prevail:

> "The dissolution of ammonia in water is not the same as that of hydrogen in azote (nitrogen), which gives rise to ammonia"[5]

There were also alchemists who defended the idea that the substances lost their nature when dissolved. Van Helmont (1577-1644) was one of the first to oppose this mistaken idea by defending that the substance dissolved remains in the solution in aqueous form, it being possible to recover it later. Later, the theories of osmotic pressure of van´t Hoff (1852-1911) and that of electrolytic dissociation of Arrhenius (1859-1927) took this approach even further.

Until almost the end of the XIX century the effects of the solvent on the different chemical processes did not become the object of systematic study by the experimenters. The effect of the solvent was assumed, without reaching the point of awakening the interest of the chemists. However, some chemists of the XIX century were soon capable of unraveling the role played by some solvents by carrying out experiments on different solvents, classified according to their physical properties; in this way the influence of the solvent both on chemical equilibrium and on the rate of reaction was brought to light. Thus, in 1862, Berthelot and Saint-Gilles, in their studies on the esterification of acetic acid with ethanol, discovered that some solvents, which do not participate in the chemical reaction, are capable of slowing down the process.[6] In 1890, Menschutkin, in a now classical study on the reaction of the trialkylamines with haloalcans in 23 solvents, made it clear how the choice of one or the other could substantially affect the reaction rate.[7] It was also Menschutkin who discovered that, in reactions between liquids, one of the reactants could constitute a solvent inadvisable for that reaction. Thus, in the reaction between aniline and acetic acid to produce acetanilide, it is better to use an excess of acetic acid than an excess of aniline, since the latter is a solvent which is not very favorable to this reaction.

The fruits of these experiments with series of solvents were the first rules regarding the participation of the solvent, such as those discovered by Hughes and Ingold for the rate of the nucleophilic reactions.[8] Utilizing a simple electrostatic model of the solute - solvent interactions, Hughes and Ingold concluded that the state of transition is more polar than the initial state, that an increase of the polarity of the solvent will stabilize the state of transition with respect to the initial state, thus leading to an increase in the reaction rate. If, on the contrary, the state of transition is less polar, then the increase of the polarity of the solvent will lead to a decrease of the velocity of the process. The rules of Hughes-Ingold for the nucleophilic aliphatic reactions are summarized in Table 2.1.1.

**Table 2.1.1. Rules of Hughes-Ingold on the effect of the increase of the polarity of the solvent on the rate of nucleophilic aliphatic reactions**

| Mechanism | Initial state | State of transition | Effect on the reaction rate |
|---|---|---|---|
| $S_N2$ | $Y^- + RX$ | $[Y\text{--}R\text{--}X]^-$ | slight decrease |
| | $Y + RX$ | $[Y\text{--}R\text{--}X]$ | large increase |
| | $Y^- + RX^+$ | $[Y\text{--}R\text{--}X]$ | large decrease |
| | $Y + RX^+$ | $[Y\text{--}R\text{--}X]^+$ | slight decrease |
| $S_N1$ | $RX$ | $[R\text{--}X]$ | large increase |
| | $RX^+$ | $[R\text{--}X]^+$ | slight decrease |

In 1896 the first results about the role of the solvent on chemical equilibria were obtained, coinciding with the discovery of the keto-enolic tautomerism.[9] Claisen identified the medium as one of the factors which, together with the temperature and the substituents, proved to be decisive in this equilibrium. Soon systematic studies began to be done on the effect of the solvent in the tautomeric equilibria. Wislicenus studied the keto-enolic equilibrium of ethylformylphenylacetate in eight solvents, concluding that the final proportion between the keto form and the enol form depended on the polarity of the solvent.[10] This effect of the solvent also revealed itself in other types of equilibria: acid-base, conformational, those of isomerization and of electronic transfer. The acid-base equilibrium is of particular interest. The relative scales of basicity and acidity of different organic compounds and homologous families were established on the basis of measurements carried out in solution, fundamentally aqueous. These scales permitted establishing hypotheses regarding the effect of the substituents on the acidic and basic centers, but without being capable of separating this from the effect of the solvent. Thus, the scale obtained in solution for the acidity of the α-substituted methyl alcohols [$(CH_3)_3COH > (CH_3)_2CHOH > CH_3CH_2OH > CH_3OH$][11] came into conflict with the conclusions extracted from the measurements of movements by NMR.[12] The irregular order in the basicity of the methyl amines in aqueous solution also proved to be confusing [$NH_3 < CH_3NH_2 < (CH_3)_2NH > (CH_3)_3N$],[13] since it did not match any of the existing models on the effects of the substituents. These conflicts were only resolved when the scales of acidity-basicity were established in the gas phase. On carrying out the abstraction of the solvent an exact understanding began to be had of the real role it played.

The great technological development which arrived with the XIX century has brought us a set of techniques capable of giving accurate values in the study of chemical processes in the gas phase. The methods most widely used for these studies are:

- The High Pressure Mass Spectrometry, which uses a beam of electron pulses[14]
- The Ion Cyclotron Resonance and its corresponding Fourier Transform (FT-ICR)[15]
- The Chemical Ionization Mass Spectrometry, in which the analysis is made of the kinetic energy of the ions, after generating them by collisions[16]
- The techniques of Flowing Afterglow, where the flow of gases is submitted to ionization by electron bombardment[17-19]

All of these techniques give absolute values with an accuracy of ±(2-4) Kcal/mol and of ±0.2 Kcal/mol for the relative values.[20]

During the last decades of XX century the importance has also been made clear of the effects of the solvent in the behavior of the biomacromolecules. To give an example, the influence of the solvent over the proteins is made evident not only by its effects on the structure and the thermodynamics, but also on the dynamics of these, both at local as well as at global level.[21] In the same way, the effect of the medium proves to be indispensable in explaining a large variety of biological processes, such us the rate of interchange of oxygen in myoglobin.[22] Therefore, the actual state of development of chemistry, as much in its experimental aspect as in its theoretical one, allows us to identify and analyze the influence of the solvent on processes increasingly more complex, leaving the subject open for new challenges and investigating the scientific necessity of creating models with which to interpret such a wide range of phenomena as this. The little fish became aware of the ocean and began explorations.

### 2.1.2 CLASSIFICATION OF SOLUTE-SOLVENT INTERACTIONS

Fixing the limits of the different interactions between the solute and the solvent which envelopes it is not a trivial task. In the first place, the liquid state, which is predominant in the majority of the solutions in use, is more difficult to comprehend than the solid state (which has its constitutive particles, atoms, molecules or ions, in fixed positions) or the gaseous state (in which the interactions between the constitutive particles are not so intense). Moreover, the solute-solvent interactions, which, as has already been pointed out, generally happen in the liquid phase, are half way between the predominant interactions in the solid phase and those which happen in the gas phase, too weak to be likened with the physics of the solid state but too strong to fit in with the kinetic theory of gases. In the second place, dissecting the solute-solvent interaction into different sub-interactions only serves to give us an approximate idea of the reality and we should not forget that, in the solute-solvent interaction, the all is not the sum of the parts. In the third place, the world of the solvents is very varied from those which have a very severe internal structure, as in the case of water, to those whose molecules interact superficially, as in the case of the hydrocarbons. At all events, there is no alternative to meeting the challenge face to face.

If we mix a solute and a solvent, both being constituted by chemically saturated molecules, their molecules attract one another as they approach one another. This interaction can only be electrical in its nature, given that other known interactions are much more intense and of much shorter range of action (such as those which can be explained by means of nuclear forces) or much lighter and of longer range of action (such as the gravitational force). These intermolecular forces usually also receive the name of van der Waals forces, from the fact that it was this Dutch physicist, Johannes D. van der Waals (1837-1923), who recognized them as being the cause of the non-perfect behavior of the real gases, in a period in which the modern concept of the molecule still had to be consolidated. The intermolecular forces not only permit the interactions between solutes and solvents to be explained but also determine the properties of gases, liquids and solids; they are essential in the chemical transformations and are responsible for organizing the structure of biological molecules.

The analysis of solute-solvent interactions is usually based on the following partition scheme:

$$\Delta E = \Delta E_i + \Delta E_{ij} + \Delta E_{jj} \qquad [2.1.1]$$

where i stands for the solute and j for the solvent.This approach can be maintained while the identities of the solute and solvent molecules are preserved. In some special cases (see below in specific interactions) it will be necessary to include some solvent molecules in the solute definition.

The first term in the above expression is the energy change of the solute due to the electronic and nuclear distortion induced by the solvent molecule and is usually given the name solute polarization. $\Delta E_{ij}$ is the interaction energy between the solute and solvent molecules. The last term is the energy difference between the solvent after and before the introduction of the solute. This term reflects the changes induced by the solute on the solvent structure. It is usually called cavitation energy in the framework of continuum solvent models and hydrophobic interaction when analyzing the solvation of nonpolar molecules.

The calculation of the three energy terms needs analytical expressions for the different energy contributions but also requires knowledge of solvent molecules distribution around the solute which in turn depends on the balance between the potential and the kinetic energy of the molecules. This distribution can be obtained from diffraction experiments or more usually is calculated by means of different solvent modelling. In this section we will comment on the expression for evaluating the energy contributions. The first two terms in equation [2.1.1] can be considered together by means of the following energy partition :

$$\Delta E_i + \Delta E_{ij} = \Delta E_{el} + \Delta E_{pol} + \Delta E_{d-r} \qquad [2.1.2]$$

Analytical expressions for the three terms (electrostatic, polarization and dispersion-repulsion energies) are obtained from the intermolecular interactions theory.

#### 2.1.2.1 Electrostatic

The electrostatic contribution arises from the interaction of the unpolarized charge distribution of the molecules. This interaction can be analyzed using a multipolar expansion of the charge distribution of the interacting subsystems which usually is cut off in the first term which is different from zero. If both the solute and the solvent are considered to be formed by neutral polar molecules (with a permanent dipolar moment different from zero), due to an asymmetric distribution of its charges, the electric interaction of the type dipole-dipole will normally be the most important term in the electrostatic interaction. The intensity of this interaction will depend on the relative orientation of the dipoles. If the molecular rotation is not restricted, we must consider the weighted average over different orientations

$$\langle E_{d-d} \rangle = -\frac{2}{3}\frac{\mu_1^2 \mu_2^2}{(4\pi\varepsilon)^2 kTr^6} \qquad [2.1.3]$$

where:

| | |
|---|---|
| $\mu_i$, $\mu_j$ | dipole moments |
| k | Boltzmann constant |
| $\varepsilon$ | dielectric constant |
| T | absolute temperature |
| r | intermolecular distance |

The most stable orientation is the antiparallel, except in the case that the molecules in play are very voluminous. Two dipoles in rapid thermal movement will be orientated sometimes in a way such that they are attracted and at other times in a way that they are repelled. On the average, the net energy turns out to be attractive. It also has to be borne in mind that the thermal energy of the molecules is a serious obstacle for the dipoles to be oriented in an optimum manner. The average potential energy of the dipole-dipole interaction, or of orientation, is, therefore, very dependent on the temperature.

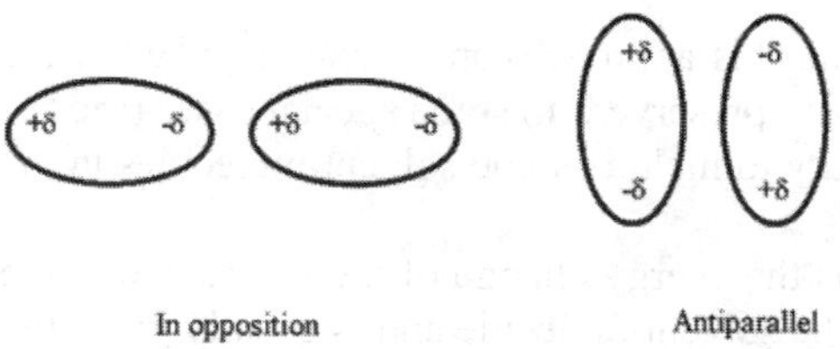


Figure 2.1.1. The dipoles of two molecules can approach one another under an infinite variety of attractive orientations, among which these two extreme orientations stand out.

In the event that one of the species involved were not neutral (for example an anionic or cationic solute) the predominant term in the series which gives the electrostatic interaction will be the ion-dipole which is given by the expression:

$$\langle E_{i-d} \rangle = -\frac{q_i^2 \mu_j^2}{6(4\pi\varepsilon)^2 kTr^4} \qquad [2.1.4]$$

## 2.1.2.2 Polarization

If we dissolve a polar substance in a nonpolar solvent, the molecular dipoles of the solute are capable of distorting the electronic clouds of the solvent molecules inducing the appearance in these of new dipoles. The dipoles of solute and those induced will line up and will be attracted and the energy of this interaction (also called interaction of polarization or induction) is:

$$\langle E_{d-id} \rangle = -\frac{\alpha_j \mu_i^2}{(4\pi\varepsilon)^2 r^6} \qquad [2.1.5]$$

where:

| | |
|---|---|
| $\mu_i$ | dipole moment |
| $\alpha_j$ | polarizability |
| r | intermolecular distance |

In a similar way, the dissolution of an ionic substance in a nonpolar solvent also will occur with the induction of the dipoles in the molecules of the solvent by the solute ions.

These equations make reference to the interactions between two molecules. Because the polarization energy (of the solute or of the solvent) is not pairwise additive magnitude, the consideration of a third molecule should be carried out simultaneously, it being impossible to decompose the interaction of the three bodies in a sum of the interactions of two bodies. The interactions between molecules in solution are different from those which take place between isolated molecules. For this reason, the dipolar moment of a molecule may vary considerably from the gas phase to the solution, and will depend in a complicated fashion on the interactions which may take place between the molecule of solute and its specific surroundings of molecules of solvent.

#### 2.1.2.3 Dispersion

Even when solvent and solute are constituted by nonpolar molecules, there is interaction between them. It was F. London who was first to face up to this problem, for which reason these forces are known as London's forces, but also as dispersion forces, charge-fluctuations forces or electrodynamic forces. Their origin is as follows: when we say that a substance is nonpolar we are indicating that the distribution of the charges of its molecules is symmetrical throughout a wide average time span. But, without doubt, in an interval of time sufficiently restricted the molecular movements generate displacements of their charges which break that symmetry giving birth to instantaneous dipoles. Since the orientation of the dipolar moment vector is varying constantly due to the molecular movement, the average dipolar moment is zero, which does not prevent the existence of these interactions between momentary dipoles. Starting with two instantaneous dipoles, these will be oriented to reach a disposition which will favor them energetically. The energy of this dispersion interaction can be given, to a first approximation, by:

$$E_{disp} = -\frac{3I_i I_j}{2(4\pi\varepsilon)^2 (I_i + I_j)} \frac{\alpha_i \alpha_j}{r^6} \qquad [2.1.6]$$

where:

| | |
|---|---|
| $I_i$, $I_j$ | ionization potentials |
| $\alpha_i$, $\alpha_j$ | polarizabilities |
| r | intermolecular distance |

From equation [2.1.6] it becomes evident that dispersion is an interaction which is more noticeable the greater the volume of molecules involved. The dispersion forces are often more intense than the electrostatic forces and, in any case, are universal for all the atoms and molecules, given that they are not seen to be subjected to the requirement that permanent dipoles should exist beforehand. These forces are responsible for the aggregation of the substances which possess neither free charges nor permanent dipoles, and are also the protagonists of phenomena such as surface tension, adhesion, flocculation, physical adsorption, etc. Although the origin of the dispersion forces may be understood intuitively, this is of a quantum mechanical nature.

### 2.1.2.4 Repulsion

Between two molecules where attractive forces are acting, which could cause them to be superimposed, it is evident that also repulsive forces exist which determine the distance to which the molecules (or the atoms) approach one another. These repulsive forces are a consequence of the overlapping of the electronic molecular clouds when these are nearing one another. These are also known as steric repulsion, hard core repulsion or exchange repulsion. They are forces of short range which grow rapidly when the molecules which interact approach one another, and which enter within the ambit of quantum mechanics. Throughout the years, different empirical potentials have been obtained with which the effect of these forces can be reproduced. In the model hard sphere potential, the molecules are supposed to be rigid spheres, such that the repulsive force becomes suddenly infinite, after a certain distance during the approach. Mathematically this potential is:

$$E_{rep} = \left(\frac{\sigma}{r}\right)^{\infty} \quad [2.1.7]$$

where:

| | |
|---|---|
| r | intermolecular distance |
| $\sigma$ | hard sphere diameter |

Other repulsion potentials are the power-law potential:

$$E_{rep} = \left(\frac{\sigma}{r}\right)^{n} \quad [2.1.8]$$

where:

| | |
|---|---|
| r | intermolecular distance |
| n | integer, usually between 9 and 16 |
| $\sigma$ | sphere diameter |

and the exponential potential:

$$E_{rep} = C\exp\left(-\frac{r}{\sigma_0}\right) \quad [2.1.9]$$

where:

| | |
|---|---|
| r | intermolecular distance |
| C | adjustable constant |
| $\sigma_o$ | adjustable constant |

These last two potentials allow a certain compressibility of the molecules, more in consonance with reality, and for this reason they are also known as soft repulsion.

If we represent the repulsion energy by a term proportional to $r^{-12}$, and given that the energy of attraction between molecules decreases in proportion to $r^{-6}$ at distances above the molecular diameter, we can obtain the total potential of interaction:

$$E = -Ar^{-6} + Br^{-12} \quad [2.1.10]$$

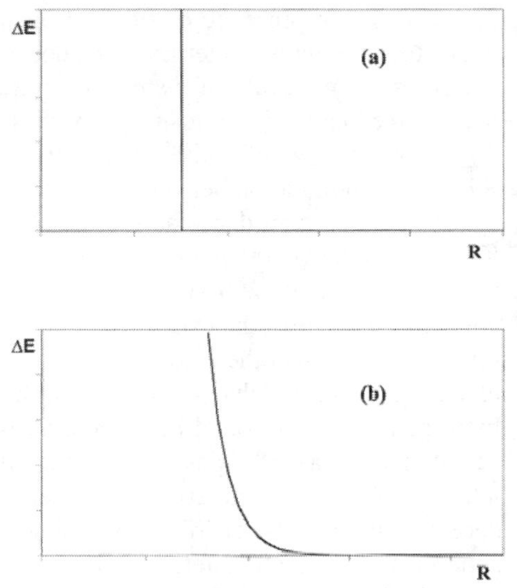


Figure 2.1.2. Hard-sphere repulsion (a) and soft repulsion (b) between two atoms.

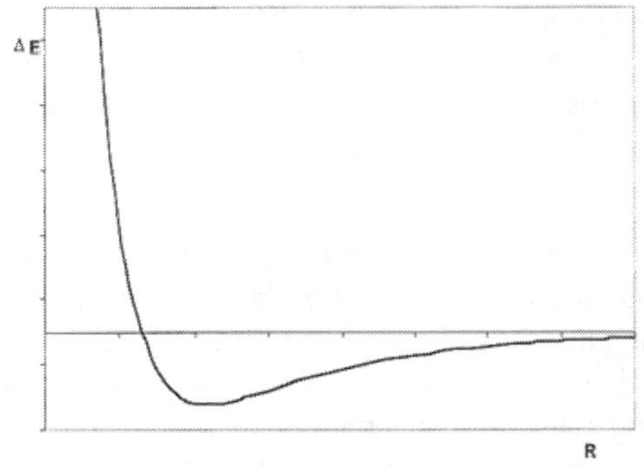


Figure 2.1.3. Lennard-Jones potential between two atoms.

where:

| | |
|---|---|
| r | intermolecular distance |
| A | constant |
| B | constant |

which receives the name of potential "6-12" or potential of Lennard-Jones,[23] widely used for its mathematical simplicity (Figure 2.1.3)

### 2.1.2.5 Specific interactions

Water, the most common liquid, the "universal solvent", is just a little "extraordinary", and this exceptional nature of the "liquid element" is essential for the world which has harbored us to keep on doing so. It is not normal that a substance in its solid state should be less dense than in the liquid, but if one ill-fated day a piece of ice spontaneously stopped floating on liquid water, all would be lost, the huge mass of ice which is floating in the colder seas could sink thus raising the level of water in the oceans.

For a liquid with such a small molecular mass, water has melting and boiling temperatures and a latent heat of vaporization which are unexpectedly high. Also unusual are its low compressibility, its high dipolar moment, its high dielectric constant and the fact that its density is maximum at 4 °C. All this proves that water is an extraordinarily complex liquid in which the intermolecular forces exhibit specific interactions, the so-called hydrogen bonds, about which it is necessary to know more.

Hydrogen bonds appear in substances where there is hydrogen united covalently to very electronegative elements (e.g., F, Cl, O and N), which is the case with water. The hydrogen bond can be either intermolecular (e.g., $H_2O$) or intramolecular (e.g., DNA). The protagonism of hydrogen is due to its small size and its tendency to become positively polarized, specifically to the elevated density of the charge which accumulates on the mentioned compounds. In this way, hydrogen is capable, such as in the case of water, of being doubly bonded: on the one hand it is united covalently to an atom of oxygen belonging to its molecule and, on the other, it electrostatically attracts another atom of oxygen belonging to another molecule, so strengthening the attractions between molecules. In this way, each atom of oxygen of a molecule of water can take part in four links with four more molecules of water, two of these links being through the hydrogen atoms covalently united to it and the other two links through hydrogen bonds thanks to the two pairs of solitary electrons which it

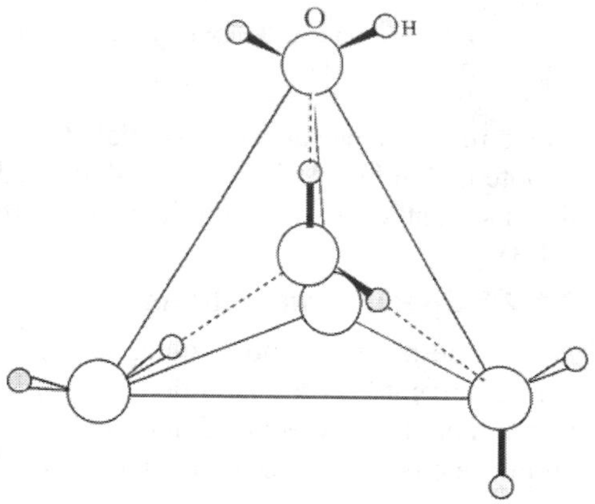


Figure 2.1.4. Tetrahedric structure of water in a crystal of ice. The dotted lines indicate the hydrogen bridges.

possesses. The presence of the hydrogen bonds together with this tetrahedric coordination of the molecule of water constitute the key to explaining its unusual properties.

The energy of this bond (10-40 KJ/mol) is found to be between that corresponding to the van der Waals forces (~1 KJ/mol) and that corresponding to the simple covalent bond (200-400 KJ/mol). An energetic analysis of the hydrogen bond interaction shows that the leading term is the electrostatic one which explains that strong hydrogen bonds are found between hydrogen atoms with a partial positive charge and a basic site. The second term in the energy decomposition of the hydrogen bond interaction is the charge transfer.[24]

The hydrogen bonds are crucial in explaining the form of the large biological molecules, such as the proteins and the nucleic acids, as well as how to begin to understand more particular chemical phenomena.[25]

Those solutes which are capable of forming hydrogen bonds have a well known affinity for the solvents with a similar characteristic, which is the case of water. The formation of hydrogen bonds between solute molecules and those of the solvent explains, for example, the good solubility in water of ammonia and of the short chain organic acids.

### 2.1.2.6 Hydrophobic interactions

On the other hand, those nonpolar solutes which are not capable of forming hydrogen bonds with water, such as the case of the hydrocarbons, interact with it in a particular way. Let us imagine a molecule of solute incapable of forming hydrogen bonds in the midst of the water. Those molecules of water which come close to that molecule of solute will lose some or all of the hydrogen bonds which they were sharing with the other molecules of water. This obliges the molecules of water which surround those of solute to arrange themselves in space so that there is a loss of the least number of hydrogen bonds with other molecules of water. Evidently, this rearrangement (solvation or hydration) of the water molecules around the nonpolar molecule of solute will be greatly conditioned by the form and the size of this latter. All this amounts to a low solubility of nonpolar substances in water, which is known as the hydrophobic effect. If we now imagine not one but two nonpolar molecules in the midst of the water, it emerges that the interaction between these two molecules is greater when they are interacting in a free space. This phenomenon, also related to the rearrangement of the molecules of water around those of the solute, receives the name of hydrophobic interaction.

The hydrophobic interaction term is used to describe the tendency of non-polar groups or molecules to aggregate in water solution.[26] Hydrophobic interactions are believed to play a very important role in a variety of processes, specially in the behavior of proteins in aqueous media. The origin of this solvent-induced interactions is still unclear. In 1945 Frank and Evans[27] proposed the so-called iceberg model where emphasis is made on the enhanced local structure of water around the non-polar solute. However, computational studies and experimental advances have yielded increasing evidence against the traditional interpretation,[28] and other alternative explanations, such as the reduced freedom of water molecules in the solvation shell,[29] have emerged.

To understand the hydrophobic interaction at the microscopic level molecular simulations of non-polar compounds in water have been carried out.[30] The potential of mean force between two non-polar molecules shows a contact minimum with an energy barrier. Computer simulations also usually predict the existence of a second solvent-separated minimum. Although molecular simulations provide valuable microscopic information on hydrophobic interactions they are computationally very expensive, specially for large systems, and normally make use of oversimplified potentials. The hydrophobic interaction can also be alternatively studied by means of continuum models.[31] Using this approach a different but complementary view of the problem has been obtained. In the partition energy scheme used in the continuum models (see below and Chapter 8) the cavitation free energy (due to the change in the solvent-solvent interactions) is the most important contribution to the potential of mean force between two non-polar solutes in aqueous solution, being responsible for the energy barrier that separates the contact minimum. The electrostatic contribution to the potential of mean force for two non-polar molecules in water is close to zero and the dispersion-repulsion term remains approximately constant. The cavitation free energy only depends on the surface of the cavity where the solute is embedded and on the solvent physical properties (such as the surface tension and density).

### 2.1.3 MODELLING OF SOLVENT EFFECTS

A useful way of understanding the interaction between the molecules of solvent and those of solute can be done by reproducing it by means of an adequate model. This task of imitation of the dissolution process usually goes beyond the use of simple and intuitive structural models, such as "stick" models, which prove to be very useful both in labors of teaching as in those of research, and frequently require the performance of a very high number of complex mathematical operations.

Even though, in the first instance, we could think that a solution could be considered as a group of molecules united by relatively weak interactions, the reality is more complex, especially if we analyze the chemical reactivity in the midst of a solvent. The prediction of reaction mechanisms, the calculation of reaction rates, the obtaining of the structures of minimum energy and other precise aspects of the chemical processes in solution require the support of models with a very elaborated formalism and also of powerful computers.

Traditionally, the models which permit the reproduction of the solute-solvent interactions are classified into three groups:[32]

i Those based on the simulation of liquids by means of computers.
ii Those of continuum.
iii Those of the supermolecule type.

In the models classifiable into the first group, the system analyzed is represented by means of a group of interacting particles and the statistical distribution of any property is calculated as the the average over the different configurations generated in the simulation. Especially notable among these models are those of Molecular Dynamics and those of the Monte Carlo type.

The continuum models center their attention on a microscopic description of the solute molecules, whilst the solvent is globally represented by means of its macroscopic properties, such as its density, its refractive index, or its dielectric constant.

Finally, the supermolecule type models restrict the analysis to the interaction among just a few molecules described at a quantum level which leads to a rigorous treatment of their interactions but does not allow us to have exact information about the global effect of the solvent on the solute molecules, which usually is a very long range effect.

The majority of these models have their origin in a physical analysis of the solutions but, with the passage of time, they have acquired a more chemical connotation, they have centered the analysis more on the molecular aspect. As well as this, recourse is more and more being made to combined strategies which use the best of each of the methods referred to in pursuit of a truer reproduction of the solute-disolvente interactions. Specially useful has been shown to be a combination of the supermolecular method, used to reproduce the specific interactions between the solute and one or two molecules of the solvent, with those of continuum or of simulation, used to reproduce the global properties of the medium.

### 2.1.3.1 Computer simulations

Obtaining the configuration or the conformation of minimum energy of a system provides us with a static view of this which may be sufficient to obtain many of its properties. However, direct comparison with experiments can be strictly be done only if average thermodynamic properties are obtained. Simulation methods are designed to calculate average properties of a system over many different configurations which are generated for being representative of the system behavior. These methods are based in the calculation of average properties as a sum over discrete events:

$$\langle F \rangle = \int dR_1 \ldots dR_n P(R) F(R) \approx \sum_{i=1}^{N} P_i F_i \qquad [2.1.11]$$

Two important difficulties arise in the computation of an average property as a sum. First, the number of molecules that can be handled in a computer is of the order of a few hundred. Secondly, the number of configurations needed to reach the convergency in the sum can be too great. The first problem can be solved by different computational strategies, such as the imposition of periodic boundary conditions.[33] The solution of the second problem differs among the main used techniques in computer simulations.

The two techniques most used in the dynamic study of the molecular systems are the Molecular Dynamics, whose origin dates back[34] to 1957, and the Monte Carlo methods, which came into being following the first simulation of fluids by computer,[35] which occurred in 1952.

**Molecular Dynamics**

In the Molecular Dynamic simulations, generation of new system configurations or sequence of events is made following the trajectory of the system which is dictated by the equations of motion. Thus, this methods leads to the computation of time averages and permits the calculation of not only equilibrium but also transport properties. Given a configuration of the system, a new configuration is obtained moving the molecules according to the total force exerted on them:

$$\frac{md^2R_j}{dt^2} = -\sum_{k=1}^{n} \nabla_j E(R_{jk}) \qquad [2.1.12]$$

If we are capable of integrating the equations of movement of all the particles which constitute a system, we can find their paths and velocities and we can evaluate the properties of the system in determined time intervals. Thus, we can find how the system being studied evolved as time moves forward. In the first simulations by Molecular Dynamics of a condensed phase,[34] use was made of potentials as simple as the hard sphere potential, under which the constituent particles move in a straight line until colliding elastically. The collisions happen when the separation between the centers of the spheres is equal to the diameter of the sphere. After each collision, the new velocities are obtained by making use of the principle of conservation of the linear moments. But a chemical system requires more elaborate potentials under which the force, which at every instant acts between two atoms or molecules, changes in relation to the variation of the distance between them. This obliges us to integrate the equations of movement of the system in very small time intervals, in which it is assumed that the force which acts on each atom or molecule is constant, generally lying between 1 and 10 femtoseconds. For each of these intervals, the positions and velocities of each of the atoms is calculated, after which they are placed in their new positions and, once again, the forces are evaluated to obtain the parameters of a new interval, and so on, successively. This evolution in time, which usually requires the evaluation of hundreds of thousands of intervals of approximately 1 femtosecond each, allows us to know the properties of the system submitted to study during the elapse of time. In fact, the task commences by fixing the atoms which make up the system being studied in starting positions, and later move them continuously whilst the molecules being analysed rotate, the bond angles bend, the bonds vibrate, etc., and during which the dispositions of the atoms which make up the system are tabulated at regular intervals of time, and the energies and other properties which depend on each of the conformations, through which the molecular system makes its way with the passage of time, are evaluated. Molecular Dynamics is Chemistry scrutinized each femtosecond.

**Monte Carlo methods**

The first simulation by computer of a molecular system was carried out using this method. It consists of generating configurations of a system introducing random changes in the position of its constituents.

In order to obtain a good convergence in the sequence of configurations, Metropolis *et al.*[35] suggested an interesting approach. This approach avoids the generation of a very long random configurations as follows: instead of choosing random configurations and then weighing them according to the Boltzmann factor, one generates configurations with a probability equal to the Boltzmann factor and afterwards weigh them evenly. For this purpose once a new configuration is generated the difference in the potential energy with respect to the previous one is computed ($\Delta U$) and a random number $0 \le r \le 1$ is selected. If the Boltzmann factor $\exp(-\Delta U/kT)>r$ then the new configuration is accepted, if $\exp(-\Delta U/kT)<r$ is rejected.

In the Monte Carlo method, every new configuration of the system being analysed can be generated starting from the random movement of one or more atoms or molecules, by rotation around a bond, etc. The energy of each new configuration is calculated starting from a potential energy function, and those configurations to which correspond the least energy are selected. Once a configuration has been accepted, the properties are calculated. At the end of the simulation, the mean values of these properties are also calculated over the ensemble of accepted configurations.

When the moment comes to face the task of modelling a dissolution process, certain doubts may arise about the convenience of using Molecular Dynamics or a Monte Carlo method. It must be accepted that a simulation with Molecular Dynamics is a succession of configurations which are linked together in time, in a similar way that a film is a collection of scenes which follow one after the other. When one of the configurations of the system analysed by Molecular Dynamics has been obtained, it is not possible to relate it to those which precede it or those which follow, which is something which does not occur in the Monte Carlo type of simulations. In these, the configurations are generated in a random way, without fixed timing, and each configuration is only related to that which immediately preceded it. It would therefore seem advisable to make use of Molecular Dynamics to study a system through a period of time. On the other hand, the Monte Carlo method is usually the most appropriate when we are able to do without the requirement of time. Even so, it is advisable to combine adequately the two techniques in the different parts of the simulation, using a hybrid arrangement. In this way, the evolution of the process of the dissolution of a macromolecule can be followed by Molecular Dynamics and make use of the Monte Carlo method to resolve some of the steps of the overall process. On the other hand, its is frequently made a distinction in the electronic description of solute and solvent. So, as the chemical attention is focused on the solute, a quantum treatment is used for it while the rest of the system is described at the molecular mechanics level.[36]

### 2.1.3.2 Continuum models

In many dissolution processes the solvent merely acts to provide an enveloping surrounding for the solute molecules, the specific interactions with the solvent molecules are not of note but the dielectric of the solvent does significantly affect the solute molecules. This sort of situation can be confronted considering the solvent as a continuum, without including explicitly each of its molecules and concentrating on the behavior of the solute. A large number of models have been designed with this approach, either using quantum mechanics or resorting to empirical models. They are the continuum models.[32] The continuum models have attached to their relative simplicity, and therefore a lesser degree of computational calculation, a favorable description of many chemical problems in dissolution, from the point of view of the chemical equilibrium, kinetics, thermochemistry, spectroscopy, etc.

Generally, the analysis of a chemical problem with a model of continuum begins by defining a cavity - in which a molecule of solute will be inserted - in the midst of the dielectric medium which represents the solvent. Differently to the simulation methods, where the solvent distribution function is obtained during the calculation, in the continuum models this distribution is usually assumed to be constant outside of the solute cavity and its value is taken to reproduce the macroscopic density of the solvent.

Although the continuum models were initially created with the aim of the calculation of the electrostatic contribution to the solvation free energy,[37] they have been nowadays extended for the consideration of other energy contributions. So, the solvation free energy is usually expressed in these models as the sum of three different terms:

$$\Delta G_{sol} = \Delta G_{ele} + \Delta G_{dis-rep} + \Delta G_{cav} \qquad [2.1.13]$$

where the first term is the electrostatic contribution, the second one the sum of dispersion and repulsion energies and the third one the energy needed to create the solute cavity in the solvent. Specific interactions can be also added if necessary.

For the calculation of the electrostatic term the system under study is characterized by two dielectric constants, in the interior of the cavity the constant will have a value of unity, and in the exterior the value of the dielectric constant of the solvent. From this point the total electrostatic potential is evaluated. Beyond the mere classical outlining of the problem, Quantum Mechanics allows us to examine more deeply the analysis of the solute inserted in the field of reaction of the solvent, making the relevant modifications in the quantum mechanical equations of the system under study with a view to introducing a term due to the solvent reaction field.[38] This permits a widening of the benefits which the use of the continuum methods grant to other facilities provided for a quantical treatment of the system, such as the optimization of the solute geometry,[39] the analysis of its wave function,[40] the obtaining of its harmonic frequencies,[41] etc.; all of which in the presence of the solvent. In this way a full analysis of the interaction solute-solvent can be reached at low computational cost.

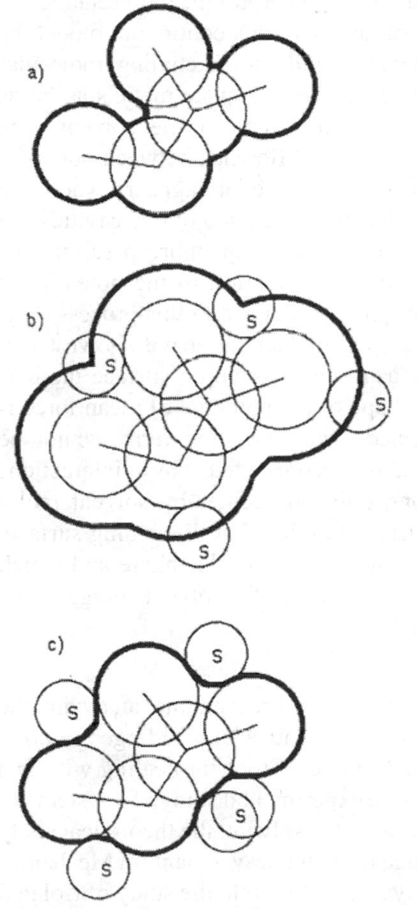


Figure 2.1.5. Molecular Surface models. (a) van der Waals Surface; (b) Accessible Surface and (c) Solvent Excluding Surface.

Continuum models essentially differ in:

- How the solute is described, either classically or quantally
- How the solute charge distribution and its interaction with the dielectric are obtained
- How the solute cavity and its surface are described.

The first two topics are thoroughly considered in Chapter 8 by Prof. Tomasi, here we will consider in detail the last topic.

### 2.1.3.3. Cavity surfaces

Earliest continuum models made use of oversimplified cavities for the insertion of the solute in the dielectric medium such as spheres or ellipsoids. In the last decades, the concept of molecular surface as become more common. Thus, the surface has been used in microscopic models of solution.[38,42] Linear relationships were also found between molecular surfaces and solvation free energies.[43] Moreover, given that molecular surfaces can help us in the calculation of the interaction of a solute molecule with surroundings of solvent molecules, they are one of the main tools in understanding the solution processes and solvent effects on chemical systems. Another popular application is the generation of graphic displays.[44]

One may use different types of molecular cavities and surfaces definitions (e.g. equipotential surfaces, equidensity surfaces, van der Waals surfaces). Among them there is a subset that shares a common trait: they consider that a molecule may be represented as a set of rigid interlocking spheres. There are three such surfaces: a) the van der Waals surface (WSURF), which is the external surface resulting from a set of spheres centered on the atoms or group of atoms forming the molecule (Figure 2.1.5a); b) the Accessible Surface (ASURF), defined by Richards and Lee[45] as the surface generated by the center of the solvent probe sphere, when it rolls around the van der Waals surface (Figure 2.1.5b); and c) the Solvent Excluding Surface (ESURF) which was defined by Richards[46] as the molecular surface and defined by him as composed of two parts: the contact surface and the reentrant surface. The contact surface is the part of the van der Waals surface of each atom which is accessible to a probe sphere of a given radius. The reentrant surface is defined as the inward-facing part of the probe sphere when this is simultaneously in contact with more than one atom (Figure 2.1.5c). We defined[47] ESURF as the surface envelope of the volume excluded to the solvent, considered as a rigid sphere, when it rolls around the van der Waals surface. This definition is equivalent to the definition given by Richards, but more concise and simple.

Each of these types of molecular surfaces is adequate for some applications. So, the van der Waals surface is widely used in graphic displays. However, for the representation of the solute cavity in a continuum model the Accessible and the Excluding molecular surfaces are the adequate models as far as they take into account the solvent. The main relative difference between both molecular surface models appears when one considers the separation of two cavities in a continuum model and more precisely the cavitation contribution to the potential of mean force associated to this process (Figure 2.1.6). In fact, we have shown[31] that only using the Excluding surface the correct shape of the potential of mean force is obtained. The cavitation term cannot be correctly represented by interactions among only one center by solvent molecule, such as the construction of the Accessible surface implies. The Excluding surface, which gives the area of the cavity not accessible to the solvent whole sphere and which should be close to the true envelope of the volume inaccessible to the solvent charge distribution, would be a more appropriate model (Figure 2.1.7).

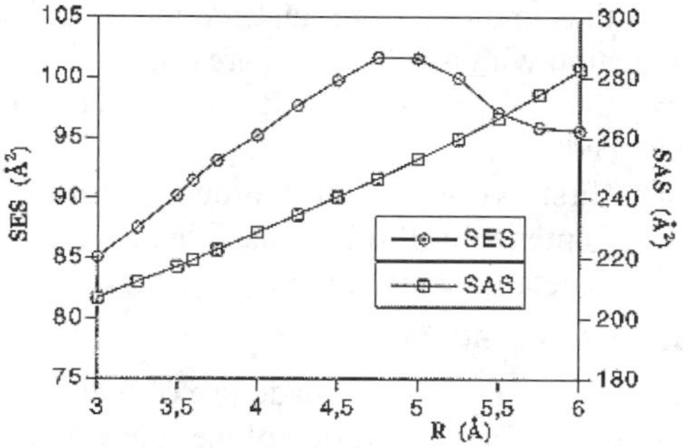

Figure 2.1.6. Variation of the area of the Solvent Accessible Surface and Solvent Excluding Surface of a methane dimer as a function of the intermolecular distance.

### 2.1.3.4 Supermolecule models

The study of the dissolution process can also be confronted in a direct manner analyzing the specific interactions between one or more molecules of solute with a large or larger group of solvent molecules. Quantum Mechanics is once again the ideal tool for dealing with this type of system. Paradoxically, whilst on the one hand the experimental study of a system becomes complicated when we try to make an abstraction of the solvent, the theoretical study becomes extraordinarily complicated when we include it. In this way, Quantum Mechanics has been, since its origins, a useful tool and relatively simple to use in the study of isolated molecules, with the behavior of a perfect gas. For this reason, Quantum Mechanics becomes so useful in the study of systems which are found in especially rarefied gaseous surroundings, such as the case of the study of the molecules present in the interstellar medium.[48] Nev-

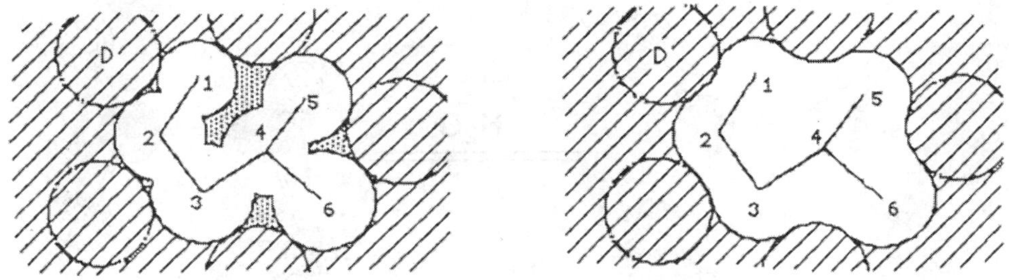


Figure 2.1.7. The Solvent Excluding Surface of a particular solute is the envelope of the volume excluded to the solvent considered as a whole sphere that represents its charge density.

ertheless, when studying systems in solution, the goodness of the results grows with the number of solvent molecules included in the calculation, but so also does the computational effort. Because of this, the main limitation of the supermolecule model is that it requires computational possibilities which are not always accessible, especially if it is desired to carry out the quantum mechanical calculations with a high level of quality. This problem is sometimes resolved by severely limiting the number of solvent molecules which are taken into account. However, this alternative will limit our hopes to know what is the global effect of the solvent over the molecules of solute, specifically the far reaching interactions solute-solvent. Where the supermolecule model is effective is in the analysis of short distance interactions between the molecules of the solute and those of the solvent, provided a sufficient number of solvent molecules are included in the system being studied to reproduce the effect being studied and provided that the calculation can be tackled computationally.[49] In this manner the supermolecule model takes the advantage over other models of modelling the solute-solvent interaction, which is the case of the continuum models. Recent advances are removing the boundaries between the different computational strategies that deal with solvent effects. A clear example is given by the Car-Parrinello approach where a quantum treatment of the solute and solvent molecules is used in a dynamic study of the system.[50]

### 2.1.3.5 Application example: glycine in solution

A practical case to compare the advantages obtained with the utilization of the different models can be found in the autoionization of the aminoacids in aqueous solution. The chemistry of the aminoacids in an aqueous medium has been at the forefront of numerous studies,[51-57] due to its self-evident biological interest. Thanks to these studies, it is well known that, whilst in the gas phase the aminoacids exist as non-autoionized structures, in aqueous solution it is the zwiterionic form which is prevailing (Figure 2.1.8).

This transformation suggests that when an aminoacid molecule is introduced from vacuum into the middle of a polar medium, which is the case of the physiological cellular media, a severe change is produced in its properties, its geometry, its energy, the charges which its atoms support, the dipolar molecular moment, etc., set off and conditioned by the presence of the solvent.

Utilizing Quantum Mechanics to analyze the geometry of the glycine in the absence of solvent, two energy minima are obtained (Figure 2.1.9), differentiated essentially by the rotation of the acid group. One of them is the absolute energy minimum in the gas phase (I), and the other is the conformation directly related with the ionic structure (II): The structure which is furthest from the zwiterionic form, I, is 0.7 Kcal/mol more stable than II, which

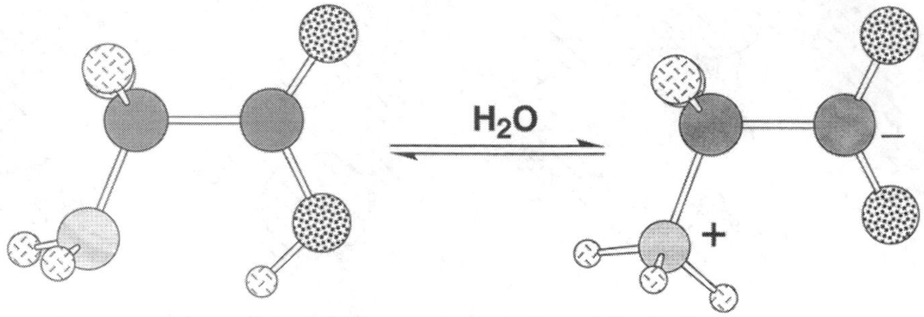


Figure 2.1.8. Neutral and zwitterionic form of glycine.

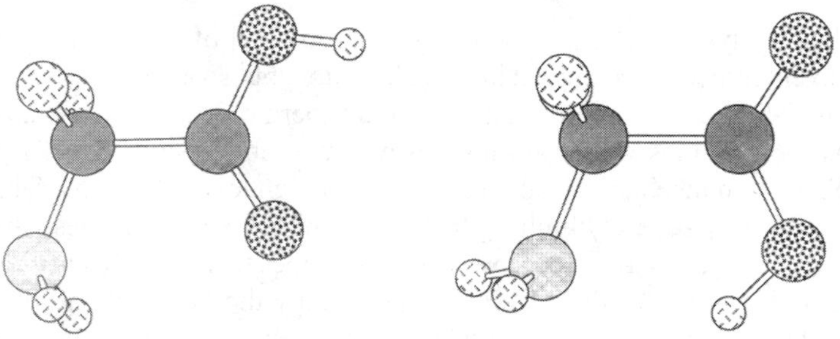

I II

Figure 2.1.9. Quantum Mechanics predicts the existence of two conformers of glycine in the gas phase whose main difference is the rotation of the acid group around the axis which joins the two carbons. The structure I, the furthest from the zwiterionic geometry, is revealed to be the most stable in gas phase.

highlights the importance of the solvent for the autoionized form of the aminoacids to prevail.

An analysis of the specific effects of the solvent in the formation and the stability of the zwitterion can be addressed carrying out the calculations of the reaction path with and without a molecule of water (Figure 2.1.10), this being within the philosophy of the supermolecule calculations. This discrete molecule of solvent has been located such that it actively participates in the migration of the proton from the oxygen to the nitrogen.[55]

That molecule of water forms two hydrogen bonds so much with the neutral glycine as zwitterionic, and when the glycine is transformed from the neutral configuration to the zwitterion the interchange of two atoms of hydrogen is produced between the aminoacid and the molecule of solvent. Thus, we reproduce from a theoretical point of view the process of protonic Table transfer of an aminoacid with the participation of the solvent (intermolecular mechanism). Table 2.1.2 shows the relative energies of the three solute-solvent structures analysed, as well as that of the system formed by the amino acid and the molecule of solvent individually.

The table shows that the neutral glycine - molecule of water complex is the most stable. Moreover, the energy barrier which drives the state of transition is much greater than that which is obtained when Quantum Mechanics is used to reproduce the autoionization of the glycine with the presence of the solvent by means of an intramolecular process.[55] This data suggests that even in the case where a larger number of water molecules are included in

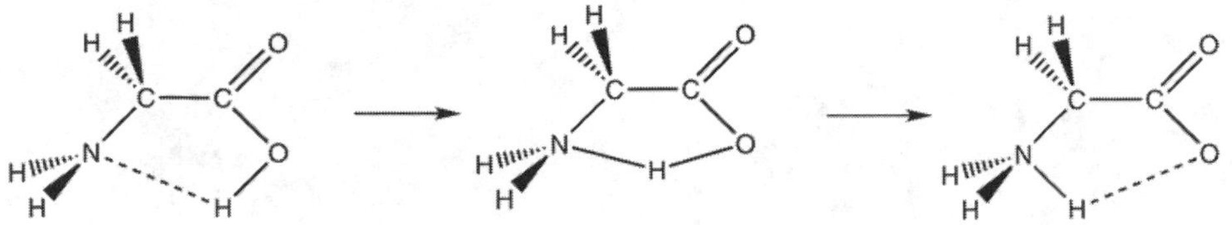

NE TS ZW

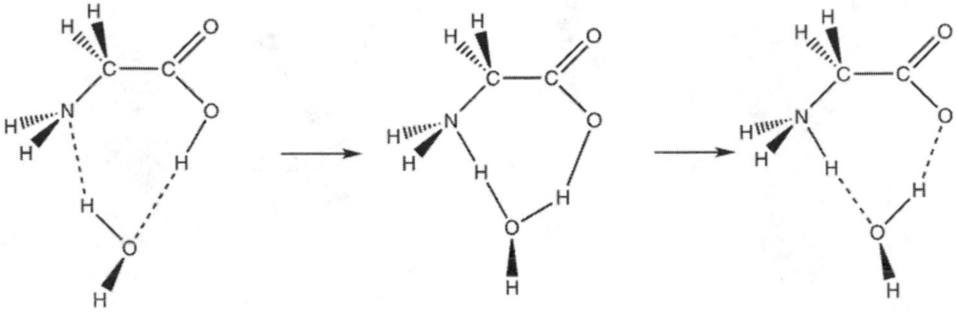

NE $xH_2O$ TS$xH_2O$ ZW$xH_2O$

Figure 2.1.10. Intramolecular and intermolecular proton transfer in glycine, leading from the neutral form to the zwitterionic one.

**Table 2.1.2. Relative energies (in Kcal/mol) for the complexes formed between a molecule of water with the neutral glycine (NE·$H_2O$), the zwitterion (ZW·$H_2O$), and the state of transition (TS·$H_2O$), as well as for the neutral glycine system and independent molecule of water (NE+$H_2O$), obtained with a base HF/6-31+G****

| NE·$H_2O$ | TS·$H_2O$ | ZW·$H_2O$ | NE+$H_2O$ |
|---|---|---|---|
| 0 | 29.04 | 16.40 | 4.32 |

the supermolecule calculation, the intramolecular mechanism will still continue as the preferred in the autoionization of the glycine. The model of supermolecule has been useful to us, along with the competition of other models,[54,55] to shed light on the mechanism by which an aminoacid is autoionized in aqueous solution.

If the calculations are done again but this time including the presence of the solvent by means of a continuum model (implemented by means of a dielectric constant = 78.4; for the water), the situation becomes different.[55,58-60] Now it is the conformation closest to the zwitterionic (II) which is the most stable, by some 2.7 Kcal/mol. This could be explained on the basis of the greater dipolar moment of the structure (6.3 Debyes for the conformation II compared to 1.3 of the conformation I).

The next step was to reproduce the formation of the zwitterion starting with the most stable initial structure in the midst of the solvent (II). Using the continuum model, we observe how the structure II evolves towards a state of transition which, once overcome, leads

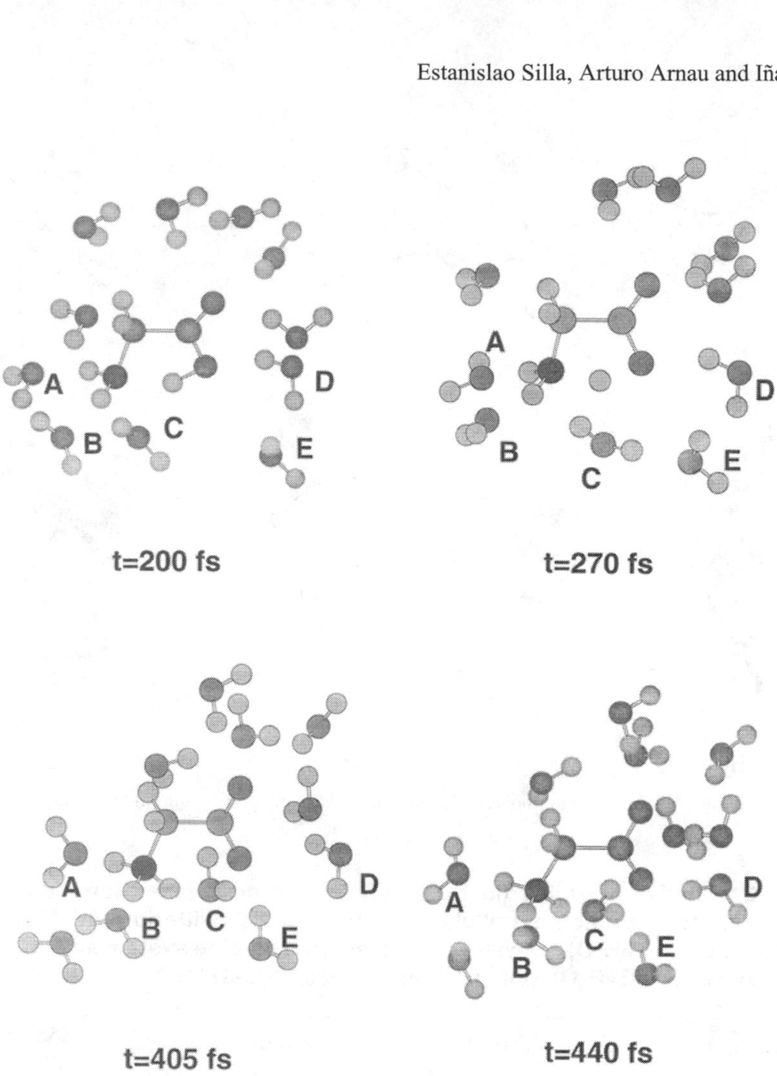


Figure 2.1.11. The microscopic environment in which the formation of zwitterion takes place is exhibited in these four "snapshots". It is possible to see the re-ordering of the structure of the molecule of glycine, along with the re-ordering of the shell of molecules of water which surround it, at 200, 270, 405 and 440 femtoseconds from the beginning of the simulation of the process by Molecular Dynamics.

to the zwitterion (Figure 2.1.7). This transition structure corresponds, then, to an intramolecular protonic transfer from the initial form to the zwiterionic form of the aminoacid. The calculations carried out reveal an activation energy of 2.39 Kcal/mol, and a reaction energy of -1.15 Kcal/mol at the MP2/6-31+G** level.

A more visual check of the process submitted to study is achieved by means of hybrid QM/MM Molecular Dynamics,[58] which permits "snapshots" to be obtained which reproduce the geometry of the aminoacid surrounded by the solvent. In Figure 2.1.11 are shown four of these snapshots , corresponding to times of 200, 270, 405 and 440 femtoseconds after commencement of the process.

In the first of these, the glycine has still not been autoionized. Two molecules of water (identified as A and B) form hydrogen bonds with the nitrogen of the amine group, whilst only one (D) joins by hydrogen bond with the $O_2$. Two more molecules of water (C and E) appear to stabilize electrostatically the hydrogen of the acid ($H_1$). The second snapshot has been chosen whilst the protonic transfer was happening. In this, the proton ($H_1$) appears jumping from the acid group to the amine group. However, the description of the first shell of hydratation around the nitrogen atoms, oxygen ($O_2$) and of the proton in transit remain essentially unaltered with respect to the first snapshot. In the third snapshot the aminoacid has now reached the zwiterionic form, although the solvent still has not relaxed to its surroundings. Now the molecules of water A and B have moved closer to the atom of nitrogen, and a third molecule of water appears imposed between them. At the same time, two molecules of water (D and E) are detected united clearly by bridges of hydrogen with the atom of oxygen ($O_2$). All of these changes could be attributed to the charges which have been placed on the atoms of nitrogen and oxygen. The molecule of water (C) has followed a proton in its transit and has fitted in between this and the atom of oxygen ($O_2$). In the fourth snapshot of the Figure 2.1.8 the relaxing of the solvent, after the protonic transfer, is now observed, permitting the molecule of water (C) to appear better orientated between the proton transferred ($H_1$) and the atom of oxygen ($O_2$), and this now makes this molecule of solvent strongly attached to the zwitterion inducing into it appreciable geometrical distortions.

From the theoretical analysis carried out it can be inferred that the neutral conformer of the glycine II, has a brief life in aqueous solution, rapidly evolving to the zwitterionic species. The process appears to happen through an intramolecular mechanism and comes accompanied by a soft energetic barrier. For its part, the solvent plays a role which is crucial both to the stabilization of the zwitterion as well as to the protonic transfer. This latter is favored by the fluctuations which take place in the surroundings.

## 2.1.4 THERMODYNAMIC AND KINETIC CHARACTERISTICS OF CHEMICAL REACTIONS IN SOLUTION

It is very difficult for a chemical equilibrium not to be altered when passing from gas phase to solution. The free energy standard of reaction, $\Delta G^o$, is usually different in the gas phase compared with the solution, because the solute-solvent interactions usually affect the reactants and the products with different results. This provokes a displacement of the equilibrium on passing from the gas phase to the midst of the solution.

In the same way, and as was foreseen in section 2.1.1, the process of dissolution may alter both the rate and the order of the chemical reaction. For this reason it is possible to use the solvent as a tool both to speed up and to slow down the development of a chemical process.

Unfortunately, little experimental information is available on how the chemical equilibria and the kinetics of the reactions become altered on passing from the gas phase to the solution, since as commented previously, the techniques which enable this kind of analysis are relatively recent. It is true that there is abundant experimental information, nevertheless, on how the chemical equilibrium and the velocity of the reaction are altered when one same process occurs in the midst of different solvents.

### 2.1.4.1 Solvent effects on chemical equilibria

The presence of the solvent is known to have proven influences in such a variety of chemical equilibria: acid-base, tautomerism, isomerization, association, dissociation, conformational, rotational, condensation reactions, phase-transfer processes, etc.,[1] that its detailed analysis is outside the reach of a text such as this. We will limit ourselves to analyzing superficially the influence that the solvent has on one of the equilibria of greatest relevance, the acid-base equilibrium.

The solvent can alter an acid-base equilibrium not only through the acid or basic character of the solvent itself, but also by its dielectric effect and its capacity to solvate the different species which participate in the process. Whilst the acid or basic force of a substance in the gas phase is an intrinsic characteristic of the substance, in solution this force is also a reflection of the acid or basic character of the solvent and of the actual process of solvatation. For this reason the scales of acidity or basicity in solution are clearly different from those corresponding to the gas phase. Thus, toluene is more acid than water in the gas phase but less acid in solution. These differences between the scales of intrinsic acidity-basicity and in solution have an evident repercussion on the order of acidity of some series of chemical substances. Thus, the order of acidity of the aliphatic alcohols becomes inverted on passing from the gas phase to solution:

$$\text{in gas phase: } R{-}CH_2{-}OH < R{-}\overset{R'}{\overset{|}{C}}H{-}OH < R{-}\underset{R''}{\underset{|}{\overset{R'}{\overset{|}{C}}}}{-}OH$$

$$\text{in solution: } R{-}CH_2{-}OH > R{-}\overset{R'}{\overset{|}{C}}H{-}OH > R{-}\underset{R''}{\underset{|}{\overset{R'}{\overset{|}{C}}}}{-}OH$$

The protonation free energies of MeOH to t-ButOH have been calculated in gas phase and with a continuum model of the solvent.[61] It has been shown that in this case continuum models gives solvation energies which are good enough to correctly predict the acidity ordering of alcohols in solution. Simple electrostatic arguments based on the charge delocalization concept, were used to rationalize the progressive acidity of the alcohols when hydrogen atoms are substituted by methyl groups in the gas phase, with the effect on the solution energies being just the opposite. Thus, both the methyl stabilizing effect and the electrostatic interaction with the solvent can explain the acid scale in solution. As both terms are related to the molecular size, this explanation could be generalized for acid and base equilibria of homologous series of organic compounds:

$$AH \Leftrightarrow A^- + H^+$$

$$B + H^+ \Leftrightarrow BH^+$$

In vacuo, as the size becomes greater by adding methyl groups, displacement of the equilibria takes place toward the charged species. In solution, the electrostatic stabilization is lower when the size increases, favoring the displacement of the equilibria toward the neutral species. The balance between these two tendencies gives the final acidity or basicity ordering in solution. Irregular ordering in homologous series are thus not unexpected taking into account the delicate balance between these factors in some cases.[62]

### 2.1.4.2 Solvent effects on the rate of chemical reactions

When a chemical reaction takes place in the midst of a solution this is because, prior to this, the molecules of the reactants have diffused throughout the medium until they have met. This prior step of the diffusion of the reactants can reach the point of conditioning the performance of the reaction, especially in particularly dense and/or viscous surroundings. This is the consequence of the liquid phase having a certain microscopic order which, although much less than that of the solid state, is not depreciable. Thus, in a solution, each molecule of solute finds itself surrounded by a certain number of molecules of solvent which envelope it forming what has been denominated as the solvent cage. Before being able to escape from the solvent cage each molecule of solute collides many times with the molecules of solvent which surround it.

In the case of a dilute solution of two reactants, A and B, their molecules remain for a certain time in a solvent cage. If the time needed to escape the solvent cage by the molecules A and B is larger than the time needed to suffer a bimolecular reaction, we can say that this will not find itself limited by the requirement to overcome an energetic barrier, but that the reaction is controlled by the diffusion of the reactants. The corresponding reaction rate will, therefore, have a maximum value, known as diffusion-controlled rate. It can be demonstrated that the diffusion-limited bimolecular rate constants are of the order of $10^{10}$-$10^{11}$ $M^{-1}s^{-1}$, when A and B are ions with opposite charges.[63] For this reason, if a rate constant is of this order of magnitude, we must wait for the reaction to be controlled by the diffusion of the reactants. But, if the rate constant of a reaction is clearly less than the diffusion-limited value, the corresponding reaction rate is said to be chemically controlled.

Focusing on the chemical aspects of the reactivity, the rupture of bonds which goes along with a chemical reaction usually occurs in a homolytic manner in the gas phase. For this reason, the reactions which tend to prevail in this phase are those which do not involve a separation of electric charge, such as those which take place with the production of radicals. In solution, the rupture of bonds tends to take place in a heterolytic manner, and the solvent is one of the factors which determines the velocity with which the process takes place. This explains that the reactions which involve a separation or a dispersion of the electric charge can take place in the condensed phase. The effects of the solvent on the reactions which involve a separation of charge will be very drawn to the polar nature of the state of transition of the reaction, whether this be a state of dipolar transition, isopolar or of the free-radical type. The influence of the solvent, based on the electric nature of the substances which are reacting, will also be essential, and reactions may occur between neutral nonpolar molecules, between neutral dipolar molecules, between ions and neutral nonpolar molecules, between ions and neutral polar molecules, ions with ions, etc. Moreover, we should bear in mind that alongside the non specific solute-solvent interactions (electrostatic, polarization, dispersion and repulsion), specific interactions may be present, such as the hydrogen bonds.

**Table 2.1.3. Relative rate constants of the Menschutkin reaction between triethylamine and iodoethane in twelve solvents at 50°C. In 1,1,1-trichloroethane the rate constant is $1.80\times10^{-5}$ l mol$^{-1}$ s$^{-1}$. Data taken from reference 40**

| Solvent | Relative rate constant | Solvent | Relative rate constant |
|---|---|---|---|
| 1,1,1-Trichloroethane | 1 | Acetone | 17.61 |
| Chlorocyclohexane | 1.72 | Cyclohexanone | 18.72 |
| Chlorobenzene | 5.17 | Propionitrile | 33.11 |
| Chloroform | 8.56 | Benzonitrile | 42.50 |
| 1,2-Dichlorobenzene | 10.06 | | |

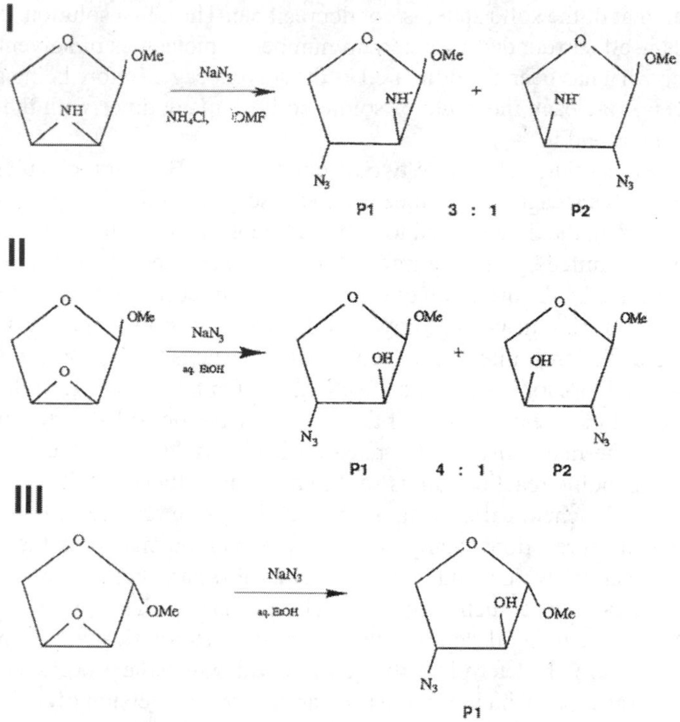


Figure 2.1.12. Models of the reaction studied for the addition of azide anion to tetrafuranosides. The experimental product ratio is also given.

The influence of the solvent on the rate at which a chemical reaction takes place was already made clear, in the final stages of the XIX century, with the reaction of Menschutkin between tertiary amines and primary haloalkanes to obtain quaternary ammonium salts.[64] The reaction of Menschutkin between triethylamine and iodoethane carried out in different media shows this effect (Table 2.1.3):

### 2.1.4.3 Example of application: addition of azide anion to tetrafuranosides

The capacity of the solvent to modify both the thermodynamic and also the kinetic aspects of a chemical reaction are observed in a transparent manner on studying the stationary structures of the addition of azide anion to tetrafuranosides, particularly: methyl 2,3-dideoxy-2,3-epimino-α-L-erythrofuranoside (I), methyl 2,3-anhydro-α-L-erythrofuranoside (II), and 2,3-anhydro-β-L-erythrofuranoside (III). An analysis with molecular orbital methods at the HF/3-21G level permits the potential energy surface in vacuo to be characterized, to locate the stationary points and the possible reaction pathways.[65] The effect of the solvent can be implemented with the aid of a polarizable continuum model. Figure 2.1.12 shows the three tetrafuranosides and the respective products obtained when azide anion attacks in $C_3$ (P1) or in $C_4$ (P2).

The first aim of a theoretical study of a chemical reaction is to determine the reaction mechanism that corresponds to the minimum energy path that connects the minima of reactant and products and passes through the transition state (TS) structures on the potential en-

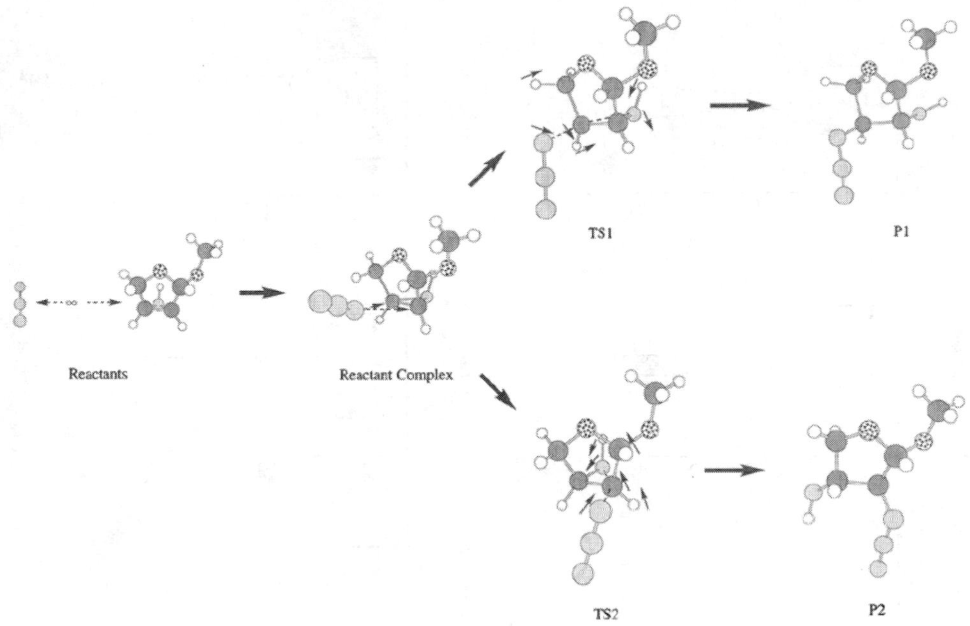


Figure 2.1.13. Representation of the stationary points (reactants, reactant complex, transition states, and products) for the molecular mechanism of compound I. For the TS´s the components of the transition vectors are depicted by arrows.

ergy surface. In this path the height of the barrier that exists between the reactant and TS is correlated to the rate of each different pathway (kinetic control), while the relative energy of reactants and products is correlated to equilibrium parameters (thermodynamic control). The second aim is how the solute-solvent interactions affect the different barrier heights and relative energies of products, mainly when charged or highly polar structures appear along the reaction path. In fact, the differential stabilization of the different stationary points in the reaction paths can treat one of them favorably, sometimes altering the relative energy order found in vacuo and, consequently, possibly changing the ratio of products of the reaction. An analysis of the potential energy surface for the molecular model I led to the location of the stationary points showed in Figure 2.1.13.

The results obtained for the addition of azide anion to tetrafuranosides with different molecular models and in different solvents can be summarized as follows:[65]

- For compound I, in vacuo, P1 corresponds to the path with the minimum activation energy, while P2 is the more stable product (Figures 2.1.14 and 2.1.15). When the solvent effect is included, P1 corresponds to the path with the minimum activation energy and it is also the more stable product. For compound II, in vacuo, P2 is the more stable product and also presents the smallest activation energy. The inclusion of the solvent effect in this case changes the order of products and transition states stability. For compound III, P1 is the more stable product and presents the smallest activation energy both in vacuo and in solution.
- A common solvent effect for the three reactions is obtained: as far as the dielectric constant of the solvent is augmented, the energy difference between P1 and P2

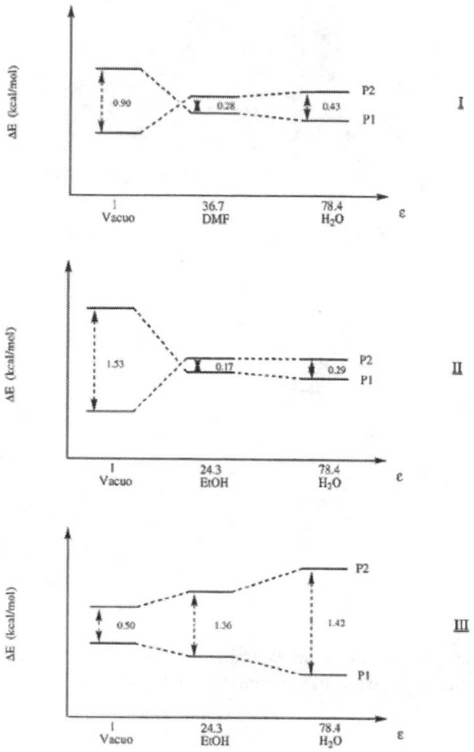


Figure 2.1.14. Schematic representation of the relative energies (in Kcal/mol) for the products P1 and P2, in vacuo (ε = 1), DMF (ε = 36.7), or EtOH (ε = 24.3), and in water (ε = 78.4).

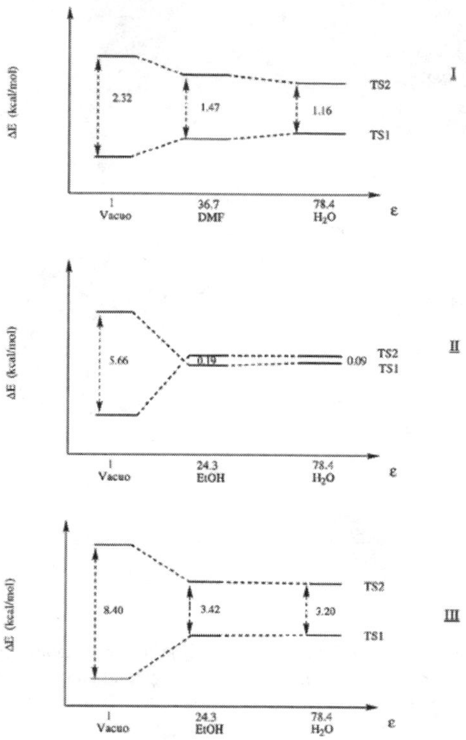


Figure 2.1.15. Schematic representation of the relative energies of activation (in Kcal/mol) for the transition states TS1 and TS2, in vacuo (ε = 1), DMF (ε = 36.7), or EtOH (ε = 24.3), and in water (ε = 78.4).

increases, favoring thermodynamically the path leading to P1. On the other hand, an opposite influence is evident in the case of the transition states, so an increase of the dielectric constant kinetically favors the path leading to P2.

All this data makes evident the crucial role which the solvent plays both in the thermodynamics and in the kinetics of the chemical reaction analysed.

## 2.1.5 SOLVENT CATALYTIC EFFECTS

Beyond the solvent as merely making possible an alternative scenery to the gas phase, beyond its capacity to alter the thermodynamics of a process, the solvent can also act as a catalyst of some reactions, and can reach the point of altering the mechanism by which the reaction comes about.

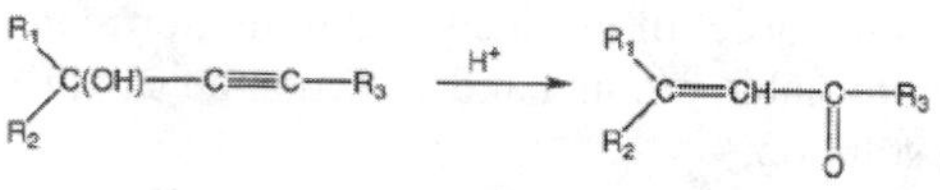


Figure 2.1.16. Reaction of Meyer-Schuster.

An example of a reaction in which the solvent is capable of altering the mechanism through which the reaction takes place is that of Meyer-Schuster, which is much utilized in organic synthesis.[66-71] This consists of the isomerization in an acid medium

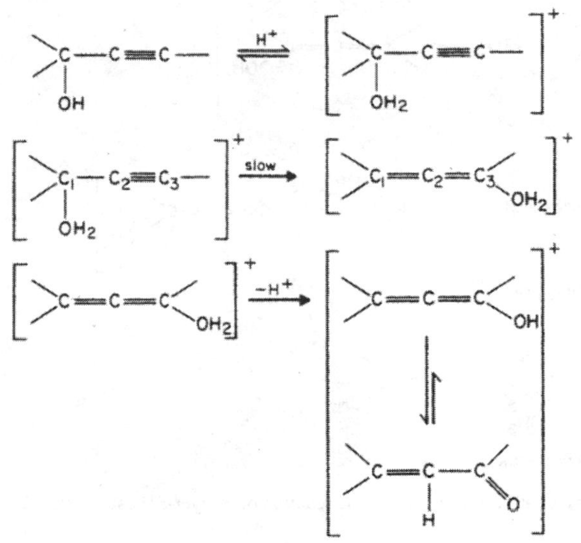


Figure 2.1.17. Steps of the reaction of Meyer-Schuster.

of secondary and tertiary α-acetylenic alcohols to carbonylic α,β-unsaturated compounds (Figure 2.1.16).

Its mechanism consists of three steps (Figure 2.1.17). The first one is the protonation of the oxygen atom. The second, which determines the reaction rate, is that in which the 1.3 shift from the protonated hydroxyl group is produced through the triple bond to give way to the structure of alenol. The last stage corresponds to the deprotonation of the alenol, producing a keto-enolic tautomerism which displaces towards the ketonic form.

For the step which limits the reaction rate (rate limiting step), three mechanisms have been proposed, two of which are intramolecular - denominated intramolecular, as such, and solvolytic - and the other intermolecular (Figure 2.1.18). The first of these implies a covalent bond between water and the atoms of carbon during the whole of the transposition. In the solvolytic mechanism there is an initial rupture from the O-$C_1$ bond, followed by a nucleophilic attack of the $H_2O$ on the $C_3$. Whilst the intermolecular mechanism corresponds to a nucleophilic attack of $H_2O$ on the terminal carbon $C_3$ and the loss of the hydroxyl group protonated of the $C_1$.

The analysis of the first two mechanisms showed[72] the solvolytic mechanism as the most favorable localizing itself during the reaction path to an alquinylic carbocation interacting electrostatically with a molecule of water. This fact has been supported by the experimental detection of alquinylic carbocations in solvolytic conditions. Things being like that, two alternatives remain for the slow stage of the Meyer-Schuster reaction, the solvolytic and the intermolecular mechanism, and it seems that the solvent has a lot to say in this.

Although both mechanisms evolve in two steps, these are notably different. In the intermolecular mechanism, the first transition state can be described as an almost pure electrostatic interaction of the entrant molecule of water with the $C_3$, whilst the $C_1$ remains united covalently to the protonated hydroxyl group. This first transition state leads to a intermediate in which the two molecules of water are covalently bonded to the $C_1$ and $C_3$ atoms. The step from the intermediate to the product takes place through a second transition state, in which the $C_3$ is covalently bonded to the molecule of entrant water and there is an electrostatic interaction of the other water molecule. In the mechanism which we call solvolytic, the first transition state corresponds to the pass from covalent to electrostatic interaction of the $H_2O$ united to the $C_1$, that is to say, to a process of solvolysis, so, the water molecule remains interacting electrostatically with the carbons $C_1$ and $C_3$.

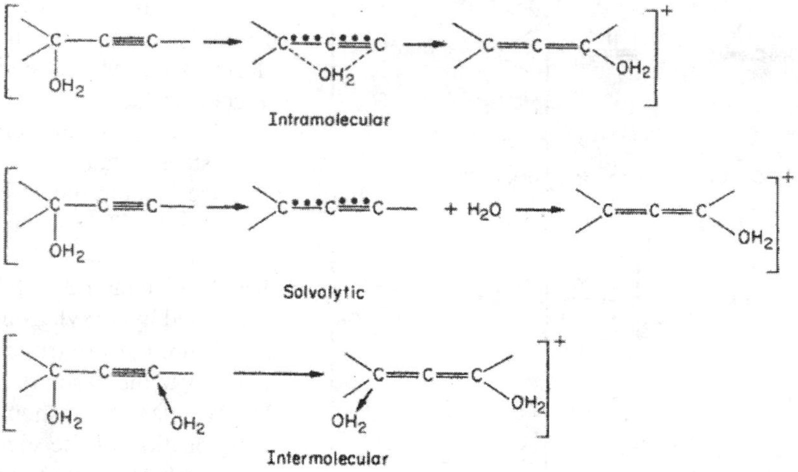


Figure 2.1.18. Three different mechanism for the rate-limiting step of the reaction of Meyer-Schuster.

On comparing the solvolytic and intermolecular processes a smaller potential barrier is observed for the latter, thus the solvent plays an active part in the Meyer-Schuster reaction, being capable of changing radically the mechanism through which this takes place. It seems clear that in the presence of aqueous solvents the nucleophilic attack on the $C_3$ precedes the loss of the water (solvolysis): a lesser activation energy corresponds to the intermolecular process than to the solvolytic. Moreover, if we analyze the intermolecular mechanism we can verify that the solvent stabilizes both the reactants as well as the products by the formation of hydrogen bridges.

**Epilogue**

As a fish in the midst of the ocean, the reactants are usually found in the midst of a solution in our laboratory tests. In the same way as in the ocean where there is both danger and a heaven for the fish, in the internal scenery of a solution the chemical reactions can be speeded up or slowed down, favored thermodynamically or prejudiced. In this way, on passing from vacuum to a solution, the molecules of the reactants can experience alterations in their geometry, the distribution of their charges, or their energy, which can have an effect on the outcome of the reaction. In the preceding pages we have attempted to make clear these solute-solvent influences, and to achieve this we have plunged, hand in hand with theoretical chemistry, into the microscopic and recondite environment of the solutions.

## REFERENCES

1 Ch. Reichardt, **Solvents and Solvent Effects in Organic Chemistry**, *VCH*, Weinheim, 1988, p. 1.
2 H. Metzger, Newton, Stahl, **Boerhaave et la doctrine chimique**, *Alcan*, Paris, 1930, pp. 280-289.
3 M.M. Pattison, **A History of Chemical Theories and Laws**, *John Wiley & Sons*, New York, 1907, p. 381.
4 H.M. Leicester, **Panorama histórico de la química**, *Alhambra*, Madrid, 1967, pp. 148, 149. (Transcription from: **The Historical Background of Chemistry**, *John Wiley & Sons*, New York.)
5 J.L. Proust, *J. Phys.*, **63**, 369, 1806.
6 M. Berthelot and L. Péan de Saint-Gilles, *Ann. Chim. et Phys., Ser. 3*, **65**, 385 (1862); **66**, 5 (1862); **68**, 255 (1863).
7 N. Menschutkin, *Z. Phys. Chem.*, **6**, 41 (1890).
8 E. D. Hughes, C. K. Ingold, *J. Chem. Soc.*, **244**, 252 (1935).
9 L. Claisen, *Liebigs Ann. Chem.*, **291**, 25 (1896).

10 W. Wislicenus. Liebigs Ann. Chem., 291, 147 (1896).
11 J. E. Bartmess, J. A. Scott, R. T. McIver Jr., *J. Am. Chem. Soc.*, **101**, 6056 (1979).
12 L. M. Jackman, D. P. Kelly, *J. Chem. Soc., (B)* 102 (1970).
13 H. C. Brown, H. Bartholomay, M. D. Taylor, *J. Am. Chem. Soc.*, **66**, 435 (1944).
14 J. P. Briggs, R. Yamdagni, P. Kebarle, *J. Am. Chem. Soc.*, **94**, 5128 (1972).
15 T. A. Leheman, M. M. Bursey, **Ion Cyclotron Resonance Spectrometry**, *Wiley*, New York, 1976.
16 S. A. McLukey, D. Cameron, R. G. Cooks, *J. Am. Chem. Soc.*, **103**, 1313 (1981).
17 B. K. Bohme, P. Fennelly, R. S: Hemsworth, H. J. Schiff, *J. Am. Chem. Soc. b*, 7512 (1973).
18 J. E. Bartmess, R. T. McIver, **Gas phase Ion Chemistry**, M. T. Bowe ed., Ed., *Academic Press*, New York, 1979.
19 J. W. Larson, T. B. McMahon, *J. Am. Chem. Soc.*, **104**, 6255 (1982).
20 M. Meotner, L. W: Sieck, *J. Am. Chem. Soc.*, **105**, 2956 (1983).
21 C. L. Brooks III, M. Karplus, *Methods Enzymol.*, **127**, 369 (1986).
22 D. Beece, L. Eisenstein, H. Frauenfelder, D. Good, M. C. Marden, L. Reinisch, A. H. Reynolds, L. B. Sorensen, K. T. Yue, *Biochemistry*, **19**, 5147 (1980).
23 J.E. Lennard-Jones, *Proc. Phys. Soc. London*, **43**, 461 (1931).
24 J. P. Daudey, *Int. J. Quantum Chem., b*, **29** (1974); J. L. Rivail, **Éléments de Chimie Quantique à l'usage des Chimistes**, *InterÉditions-CNRS Éditions*, Paris (1994).
25 C. Sandorfy, R. Buchet, L.S. Lussier, P. Ménassa and L. Wilson, *Pure Appl. Chem.*, **58**, 1115 (1986). J. N., Israelachvili, **Intermolecular and surface forces**, **Academic Press**, San Diego, 1989.
26 W. Kauzmann, *Adv. Protein Chem.*, **14**, 1 (1959).
27 H. S. Frank and M. W. Evans, *J. Chem. Phys.*, **13**, 507 (1945).
28 W. Blokzilj and J. B. F. N. Engberts, *Angew. Chem. Int. Ed. Engl.*, **32**, 1545 (1993).
29 B. Lee, *Biopolymers,* **24**, 813 (1985), B. Lee, *Biopolymers*, **31**, 993 (1991).
30 W. L. Jorgensen, J. K. Bukner, S. Boudon and J. Tirado-Rives, *Chem. Phys.*, **89**, 3742 (1988); C. Pangali, M. Rao and B. J. Berne, *J. Chem. Phys.*, **71**, 2975(1979); D. van Belle and S. J. Wodak, *J. Am. Chem. Soc.*, **115**, 647 (1993); M. H. New and B. J. Berne, *J. Am. Chem. Soc.*, **117**, 7172 (1995); D. E. Smith, L. Zhang and A. D. J. Haymet, *J. Am. Chem. Soc.*, **114**, 5875 (1992); ); D. E. Smith and A. D. J. Haymet, *J. Chem. Phys.*, **98**, 6445(1994); L. X. Dang, *J. Chem. Phys.*, **100**, 9032(1994); T. Head-Gordon, *Chem. Phys. Lett.*, **227**, 215 (1994).
31 J. Pitarch, V. Moliner, J. L. Pascual-Ahuir, E. Silla and I. Tuñçon , *J. Phys. Chem.*, **100**, 9955(1996).
32 J. Tomasi and M. Persico, *Chem. Rev.*, **94**, 2027-2094 (1994).
33 A. Rahman and F.M. Stillinger, *J. Chem. Phys.*, **57**, 3336 (1971). A. R. Leach, **Molecular Modelling**, *Addison Wesley Longman*, Singapore, 1997.
34 B.J. Alder and T.E. Wainwright, *J. Chem. Phys.*, **27**, 1208-1209 (1957).
35 N. Metropolis, A.W. Rosenbluth, M.N. Rosenbluth, A. Teller and E. Teller, *J. Chem. Phys.*, **21**, 1087 (1953)
36 A. Warshel, M. Levitt, *J. Mol. Biol.*, **103**, 227-249 (1973). P. A. Bash, M. J. Field, M. Karplus, *J. Am. Chem. Soc.*, **109**, 8092-8094 (1987).
37 M. Born, *Z. Physik,* **1**, 45 (1920); J. G. Kirkwood, *J. Chem. Phys.*, **2**, 351 (1934); J. G. Kirkwood, F. H. Westheimer, *J. Chem. Phys.*, **6**, 506 (1936); L. Onsager, *J. Am. Chem. Soc.*, **58**, 1486 (1936).
38 E. Scrocco, J. Tomasi, *Top. Curr. Chem.*, **42**, 97 (1973). J. L. Pascual-Ahuir, E. Silla, J. Tomasi, R. Bonaccorsi, *J. Comput. Chem.*, **8**, 778 (1987).
39 D. Rinaldi, J. L. Rivail, N. Rguini, *J. Comput. Chem., b*, 676 (1992).
40 I. Tuñón, E. Silla, J. Bertrán, *J. Chem. Soc. Fraday Trans.*, **90**, 1757 (1994).
41 X. Assfeld, D. Rinaldi, AIP Conference Proceedings ECCC1, F. Bernardi, J. L. Rivail, Eds., AIP, Woodbury, New York, 1995.
42 F. M. Floris, J. Tomasi, J. L. Pascual Ahuir, *J. Comput. Chem.*, **12**, 784 (1991); C. J. Cramer and D. G. Trhular, *Science*, **256**, 213 (1992); V. Dillet, D. Rinaldi, J. G. Angyan and J. L. Rivail, *Chem. Phys. Lett.*, **202**, 18 (1993).
43 I. Tuñón, E. Silla, J. L. Pascual-Ahuir, *Prot. Eng.*, **5**, 715 (1992); I. Tuñón, E. Silla, J. L. Pascual-Ahuir, *Chem. Phys. Lett.*, **203**, 289 (1993); R. B. Hermann, *J. Phys. Chem.*, **76**, 2754 (1972); K. A. Sharp, A. Nicholls, R. F: Fine, B. Honig, *Science*, **252**, 106 (1991).
44 L. H. Pearl, A. Honegger, *J. Mol. Graphics*, **1**, 9 (1983); R. Voorintholt, M. T. Kosters, G. Vegter, G. Vriend, W. G. H. Hol, *J. Mol. Graphics*, **7**, 243 (1989); J. B. Moon, W. J. Howe, *J. Mol. Graphics*, **7**, 109 (1989); E. Silla, F. Villar, O. Nilsson, J.L. Pascual-Ahuir, O. Tapia, *J. Mol. Graphics*, **8**, 168 (1990).
45 B. Lee, F. M. Richards, *J. Mol. Biol.*, **55**, 379 (1971).
46 F. M. Richards, *Ann. Rev. Biophys. Bioeng.*, **6**, 151 (1977).
47 J. L. Pascual-Ahuir, E. Silla, I. Tuñón, *J. Comput. Chem.*, **15**, 1127-1138 (1994).

48 V. Moliner, J. Andrés, A. Arnau, E. Silla and I. Tuñón, *Chem. Phys.*, **206**, 57-61 (1996); R. Moreno, E. Silla, I. Tuñón and A. Arnau, *Astrophysical J.*, **437**, 532-539 (1994); A. Arnau, E. Silla and I. Tuñón, *Astrophysical J. Suppl. Ser.*, **88**, 595-608 (1993); A. Arnau, E. Silla and I. Tuñón, *Astrophysical J.*, **415**, L151-L154 (1993).
49 M. D. Newton, S. J. Ehrenson, *J. Am. Chem. Soc.*, **93**, 4971 (1971). G. Alagona, R. Cimiraglia, U. Lamanna, *Theor. Chim. Acta*, **29**, 93 (1973). J. E: del Bene, M. J. Frisch, J. A. Pople, *J. Phys. Chem.*, **89**, 3669 (1985). I. Tuñón, E. Silla, J. Bertrán, *J. Phys. Chem.*, **97**, 5547-5552 (1993).
50 R. Car, M. Parrinello, *Phys. Rev. Lett.*, **55**, 2471 (1985).
51 P. G. Jonsson and A. Kvick, *Acta Crystallogr. B*, **28**, 1827 (1972).
52 A. G. Csázár, *Theochem.*, **346**, 141-152 (1995).
53 Y. Ding and K. Krogh-Jespersen, *Chem. Phys. Lett.*, **199**, 261-266 (1992).
54 J. H. Jensen and M.S.J. Gordon, *J. Am. Chem. Soc.*, **117**, 8159-8170 (1995).
55 F. R. Tortonda., J.L. Pascual-Ahuir, E. Silla, and I. Tuñón, *Chem. Phys. Lett.*, **260**, 21-26 (1996).
56 T.N. Truong and E.V. Stefanovich, *J. Chem. Phys.*, **103**, 3710-3717 (1995).
57 N. Okuyama-Yoshida et al., *J. Phys. Chem. A*, **102**, 285-292 (1998).
58 I. Tuñón, E. Silla, C. Millot, M. Martins-Costa and M.F. Ruiz-López, *J. Phys. Chem. A*, **102**, 8673-8678 (1998).
59 F. R. Tortonda, J.L. Pascual-Ahuir, E. Silla, I. Tuñón, and F.J. Ramírez, *J. Chem. Phys.*, **109**, 592-602 (1998).
60 F.J. Ramírez, I. Tuñón, and E. Silla, *J. Phys. Chem. B*, **102**, 6290-6298 (1998).
61 I. Tuñón, E. Silla and J.L. Pascual-Ahuir, *J. Am. Chem. Soc.*, **115**, 2226 (1993).
62 I. Tuñón, E. Silla and J. Tomasi, *J. Phys. Chem.*, **96**, 9043 (1992).
63 K.A. Connors, **Chemical Kinetics**, *VCH*, New York, 1990, pp. 134-135.
64 N. Menschutkin, *Z. Phys. Chem.*, **6**, 41 (1890); *ibid.*, **34**, 157 (1900).
65 J. Andrés, S. Bohm, V. Moliner, E. Silla, and I. Tuñón, *J. Phys. Chem.*, **98**, 6955-6960 (1994).
66 S. Swaminathan and K.V. Narayan, *Chem. Rev.*, **71**, 429 (1971).
67 N.K. Chaudhuri and M. Gut, *J. Am. Chem. Soc.*, **87**, 3737 (1965).
68 M. Apparu and R. Glenat, *Bull. Soc. Chim. Fr.*, 1113-1116 (1968).
69 S. A. Vartanyan and S.O. Babayan, *Russ. Chem. Rev.*, **36**, 670 (1967).
70 L.I. Olsson, A. Claeson and C. Bogentoft, *Acta Chem. Scand.*, **27**, 1629 (1973).
71 M. Edens, D. Boerner, C. R. Chase, D. Nass, and M.D. Schiavelli, *J. Org. Chem.*, **42**, 3403 (1977).
72 J. Andrés, A. Arnau, E. Silla, J. Bertrán, and O. Tapia, *J. Mol. Struct. Theochem.*, **105**, 49 (1983); J. Andrés, E. Silla, and O. Tapia, *J. Mol. Struct. Theochem.*, **105**, 307 (1983); J. Andrés, E. Silla, and O. Tapia, *Chem. Phys. Lett.*, **94**, 193 (1983); J. Andrés, R. Cárdenas, E. Silla, O. Tapia, *J. Am. Chem. Soc.*, **110**, 666 (1988).

## 2.2 MOLECULAR DESIGN OF SOLVENTS

KOICHIRO NAKANISHI
**Kurashiki Univ. Sci. & the Arts, Okayama, Japan**

### 2.2.1 MOLECULAR DESIGN AND MOLECULAR ENSEMBLE DESIGN

Many of chemists seem to conjecture that the success in developing so-called high-functional materials is the key to recover social responsibility. These materials are often composed of complex molecules, contain many functional groups and their structure is of complex nature. Before establishing the final target compound, we are forced to consider many factors, and we are expected to minimize the process of screening these factors effectively.

At present, such a screening is called "design". If the object of screening is each molecule, then it is called "molecular design". In similar contexts are available "material design", "solvent design", "chemical reaction design", etc. We hope that the term "molecular ensemble design" could have the citizenship in chemistry. The reason for this is as below.

Definition of "molecular design" may be expressed as to find out the molecule which has appropriate properties for a specific purpose and to predict accurately via theoretical approach the properties of the molecule. If the molecular system in question consists of an isolated free molecule, then it is "molecular design". If the properties are of complex macroscopic nature, then it is "material design". Problem remains in the intermediate between the above two. Because fundamental properties shown by the ensemble of molecules are not always covered properly by the above two types of design. This is because the molecular design is almost always based on quantum chemistry of free molecule and the material design relies too much on empirical factor at the present stage. When we proceed to molecular ensemble (mainly liquid phase), as the matter of fact, we must use statistical mechanics as the basis of theoretical approach.

Unfortunately, statistical mechanics is not familiar even for the large majority of chemists and chemical engineers. Moreover, fundamental equations in statistical mechanics cannot often be solved rigorously for complex systems and the introduction of approximation becomes necessary to obtain useful results for real systems. In any theoretical approach for molecular ensemble, we must confront with so-called many-body problems and two-body approximations must be applied. Even in the frameworks of this approximation, our knowledge on the intermolecular interaction, which is necessary in statistical mechanical treatment is still poor.

Under such a circumstance, numerical method should often be useful. In the case of statistical mechanics of fluids, we have Monte Carlo (MC) simulation based on the Metropolis scheme. All the static properties can be numerically calculated in principle by the MC method.

Another numerical method to supplement the MC method should be the numerical integration of the equations of motion. This kind of calculation for simple molecular systems is called molecular dynamics (MD) method where Newton or Newton-Euler equation of motion is solved numerically and some dynamic properties of the molecule involved can be obtained.

These two methods are invented, respectively, by the Metropolis group (MC, Metropolis et al., 1953)[1] and Alder's group (MD, Alder et al., 1957)[2] and they are the molecular versions of computer experiments and therefore called now molecular simulation.[3] Molecular simulation plays a central role in "molecular ensemble design". They can reproduce thermodynamics properties, structure and dynamics of a group of molecules by using high speed supercomputer. Certainly any reasonable calculations on molecular ensemble need long computer times, but the advance in computer makes it possible that this problem becomes gradually less serious.

Rather, the assignment is more serious with intermolecular interaction potential used. For simple molecules, empirical model potential such as those based on Lennard-Jones potential and even hard-sphere potential can be used. But, for complex molecules, potential function and related parameter value should be determined by some theoretical calculations. For example, contribution of hydrogen-bond interaction is highly large to the total interaction for such molecules as $H_2O$, alcohols etc., one can produce semi-empirical potential based on quantum-chemical molecular orbital calculation. Molecular ensemble design is now complex unified method, which contains both quantum chemical and statistical mechanical calculations.

## 2.2.2 FROM PREDICTION TO DESIGN

It is not new that the concept of "design" is brought into the field of chemistry. Moreover, essentially the same process as the above has been widely used earlier in chemical engineering. It is known as the prediction and correlation methods of physical properties, that is, the method to calculate empirically or semi-empirically the physical properties, which is to be used in chemical engineering process design. The objects in this calculation include thermodynamic functions, critical constant, phase equilibria (vapor pressure, etc.) for one-component systems as well as the transport properties and the equation of state. Also included are physical properties of two-components (solute + pure solvent) and even of three-components (solute + mixed solvent) systems. Standard reference, "The properties of Gases and Liquids; Their Estimation and Prediction",[4] is given by Sherwood and Reid. It was revised once about ten years interval by Reid and others. The latest 4th edition was published in 1987. This series of books contain excellent and useful compilation of "prediction" method. However, in order to establish the method for molecular ensemble design we need to follow three more stages.

(1) Calculate physical properties of any given substance. This is just the establishment and improvement of presently available "prediction" method.

(2) Calculate physical properties of model-substance. This is to calculate physical properties not for each real molecule but for "model". This can be done by computer simulation. On this stage, compilation of model"substance data base will be important.

(3) Predict real substance (or corresponding "model") to obtain required physical properties. This is just the reverse of the stage (1) or (2). But, an answer in this stage is not limited to one particular substance.

The scheme to execute these three stages for a large variety of physical properties and substances has been established only to a limited range. Especially, important is the establishment of the third stage, and after that, "molecular ensemble design" will be worth to discuss.

### 2.2.3 IMPROVEMENT IN PREDICTION METHOD

Thus, the development of "molecular ensemble design" is almost completely future assignment. In this section, we discuss some attempts to improve prediction at the level of stage (1). It is taken for the convenience's sake from our own effort. This is an example of repeated improvements of prediction method from empirical to molecular level.

The diffusion coefficient $D_1$ of solute 1 in solvent 2 at infinitely dilute solution is a fundamental property. This is different from the self-diffusion coefficient $D_0$ in pure liquid. Both $D_1$ and $D_0$ are important properties. The classical approach to $D_1$ can be done based on Stokes and Einstein relation to give the following equation

$$D_1\eta_2 = kT / 6\pi r_1 \quad [2.2.1]$$

where $D_1$ at constant temperature T can be determined by the radius $r_1$ of solute molecule and solvent viscosity $\eta_2$ (k is the Boltzmann constant). However, this equation is valid only when the size of solvent molecule is infinitely small, namely, for the diffusion in continuous medium. It is then clear that this equation is inappropriate for molecular mixtures. The well known Wilke-Chang equation,[5] which corrects comprehensively this point, can be used for practical purposes. However, average error of about 10% is inevitable in the comparison with experimental data.

One attempt[6] to improve the agreement with experimental data is to use Hammond-Stokes relation in which the product $D_1\eta_2$ is plotted against the molar volume ratio of solvent to solute Vr. The slope is influenced by the following few factors, namely,

(1) self-association of solvent,
(2) asymmetricity of solvent shape, and
(3) strong solute-solvent interactions.

If these factors can be taken properly into account, the following equation is obtained.

$$D_1\eta_2 / T = K_1 / V_0^{1/3} + K_2 Vr / T \qquad [2.2.2]$$

Here constants $K_1$ and $K_2$ contain the parameters coming from the above factors and $V_0$ is the molar volume of solute.

This type of equation [2.2.2] cannot be always the best in prediction, but physical image is clearer than with other purely empirical correlations. This is an example of the stage (1) procedure and in order to develop a stage (2) method, we need MD simulation data for appropriate model.

## 2.2.4 ROLE OF MOLECULAR SIMULATION

We have already pointed out that statistical mechanical method is indispensable in "molecular ensemble design". Full account of molecular simulation is given in some books[3] and will not be reproduced here. Two types of approaches can be classified in applying this method.

The first one makes every effort to establish and use as real as exact intermolecular interaction in MC and MD simulation. It may be limited to a specific type of compounds. The second is to use simple model, which is an example of so-called Occam's razor. We may obtain a wide bird-view from there.

In the first type of the method, intermolecular interaction potential is obtained based on quantum-chemical calculation. The method takes the following steps.

(1) Geometry (interatomic distances and angles) of molecules involved is determined. For fundamental molecules, it is often available from electron diffraction studies. Otherwise, the energy gradient method in molecular orbital calculation can be utilized.
(2) The electronic energy for monomer $E_1$ and those for dimers of various mutual configurations $E_2$ are calculated by the so-called *ab initio* molecular orbital method, and the intermolecular energy $E_2$-$2E_1$ is obtained.
(3) By assuming appropriate molecular model and semi-empirical equation, parameters are optimized to reproduce intermolecular potential energy function.

Representative example of preparation of such potential energy function, called now *ab initio* potential would be MCY potential for water-dimer by Clementi et al.[7] Later, similar potentials have been proposed for hetero-dimer such as water-methanol.[8] In the case of such hydrogen-bonded dimers, the intermolecular energy $E_2$-$2E_1$ can be fairly large value and determination of fitted parameters is successful. In the case of weak interaction, *ab initio* calculation needs long computer time and optimization of parameter becomes difficult. In spite of such a situation, some attempts are made for potential preparation, e.g., for benzene and carbon dioxide with limited success.

To avoid repeated use of long time MO calculation, Jorgensen et al.[9] has proposed MO-based transferable potential parameters called TIPS potential. This is a potential ver-

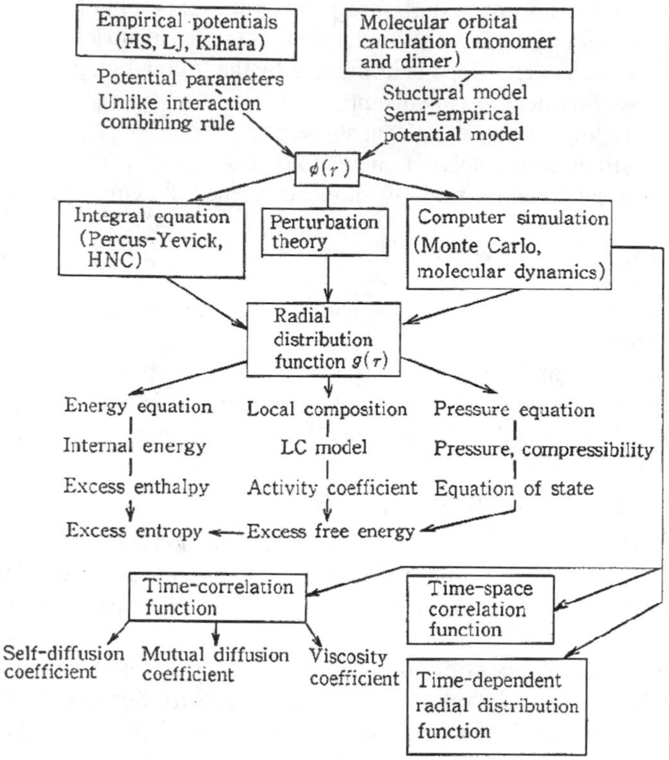


Figure 2.2.1. A scheme for the design of molecular ensembles.

sion of additivity rule, which has now an empirical character. It is however useful for practical purposes.

## 2.2.5 MODEL SYSTEM AND PARADIGM FOR DESIGN

The method described above is so to speak an orthodox approach and the ability of present-day's supercomputer is still a high wall in the application of molecular simulation. Then the role of the second method given in the last section is highly expected.

It is the method with empirical potential model. As the model, the approximation that any molecule can behave as if obeying Lennard-Jones potential seems to be satisfactory. This (one-center) LJ model is valid only for rare gases and simple spherical molecules. But this model may also be valid for other simple molecules as a zeroth approximation. We may also use two-center LJ model where interatomic interactions are concentrated to the major two atoms in the molecule. We expect that these one-center and two-center LJ models will play a role of Occam's razor.

We propose a paradigm for physical properties prediction as shown in Figure 2.2.1. This corresponds to the stage (2) and may be used to prepare the process of stage (3), namely, the molecular ensemble design for solvents.

Main procedures in this paradigm are as follows; we first adopt target molecule or mixture and determine their LJ parameters. At present stage, LJ parameters are available only for limited cases. Thus we must have method to predict effective LJ parameters.

For any kinds of mixtures, in addition to LJ parameters for each component, combining rule (or mixing rule) for unlike interaction should be prepared. Even for simple liquid mixtures, conventional Lorentz-Berthelot rule is not good answer.

Once potential parameters have been determined, we can start calculation downward following arrow in the figure. The first key quantity is radial distribution function g(r) which can be calculated by the use of theoretical relation such as Percus-Yevick (PY) or Hypernetted chain (HNC) integral equation. However, these equations are an approximations. Exact values can be obtained by molecular simulation. If g(r) is obtained accurately as functions of temperature and pressure, then all the equilibrium properties of fluids and fluid mixtures can be calculated. Moreover, information on fluid structure is contained in g(r) itself.

On the other hand, we have, for non-equilibrium dynamic property, the time correlation function TCF, which is dynamic counterpart to g(r). One can define various TCF's for each purpose. However, at the present stage, no extensive theoretical relation has been derived between TCF and $\phi(r)$. Therefore, direct determination of self-diffusion coefficient, viscosity coefficient by the molecular simulation gives significant contribution in dynamics studies.

**Concluding Remarks**

Of presently available methods for the prediction of solvent physical properties, the solubility parameter theory by Hildebrand[10] may still supply one of the most accurate and comprehensive results. However, the solubility parameter used there has no purely molecular character. Many other methods are more or less of empirical character.

We expect that the 21th century could see more computational results on solvent properties.

## REFERENCES


1 N. Metropolis, A. W. Rosenbluth, M. N. Rosenbluth, A. H. Teller and E. Teller, *J. Chem. Phys.*, **21**, 1087 (1953).

2 B. J. Alder and T. E. Wainwright, *J. Chem. Phys.*, **27**, 1208 (1957).

3 (a) M. P. Allen and D. J. Tildesley, **Computer Simulation of Liquids**, *Clarendon Press*, Oxford, 1987. (b) R. J. Sadus, **Molecular Simulation of Fluids**, *Elsevier*, Amsterdam, 1999.

4 R. C. Reid, J. M. Prausnitz and J. E. Poling, **The Properties of Gases and Liquids; Their Estimation and Prediction**, 4th Ed., *McGraw-Hill,* New York, 1987.

5 C. R. Wilke and P. Chang, *AIChE J.*, **1**, 264 (1955).

6 K. Nakanishi, *Ind. Eng. Chem. Fundam.*, **17**, 253 (1978).

7 O. Matsuoka, E. Clementi and M. Yoshimine, *J. Chem. Phys.*, **60**, 1351 (1976).

8 S. Okazaki, K. Nakanishi and H. Touhara, *J. Chem. Phys.*, **78**, 454 (1983).

9 W. L. Jorgensen, *J. Am. Chem. Soc.*, **103**, 345 (1981).

10 J. H. Hildebrand and R. L. Scott, **Solubility of Non-Electrolytes**, 3rd Ed., *Reinhold*, New York, 1950.


## APPENDIX

## PREDICTIVE EQUATION FOR THE DIFFUSION COEFFICIENT IN DILUTE SOLUTION

Experimental evidence is given in Figure 2.2.2 for the prediction based on equation [2.2.1]. The diffusion coefficient $D_0$ of solute A in solvent B at an infinite dilution can be calculated using the following equation:

$$D_0 = \left( \frac{9.97 \times 10^{-8}}{[I_A V_A]^{1/3}} + \frac{2.40 \times 10^{-8} A_B S_B V_B}{I_A S_A V_A} \right) \frac{T}{\eta_B} \qquad [2.2.3]$$

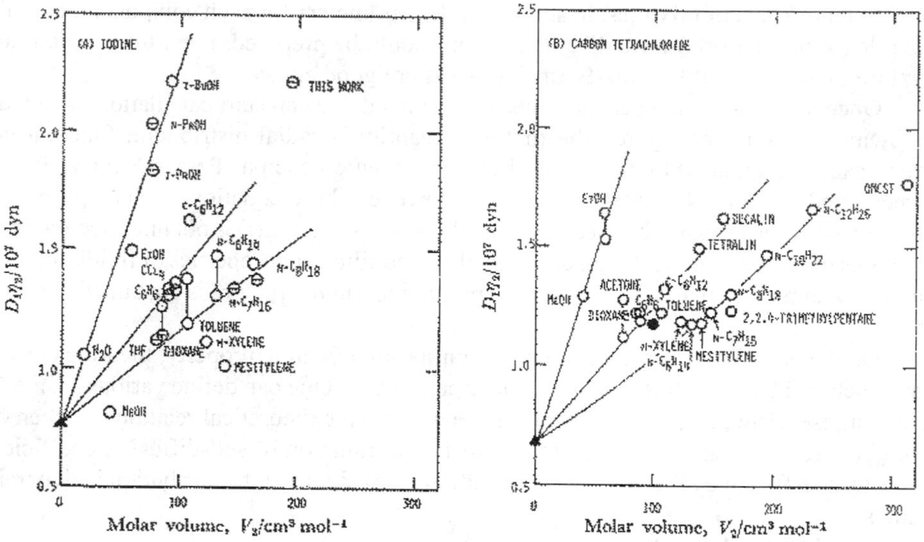


Figure 2.2.2. Hammond-Stokes plot for diffusion of iodine and carbon tetrachloride in various solvents at 298.15K. O, ⊖: $D_1\eta_2$, ●: $(D_1)_{self}\eta_2$, ▲: $D_1\eta_2$ at the Stokes-Einstein limit. Perpendicular lines connect two or more data for the same solvent from different sources. [Adapted, by permission, from K. Nakanishi, *Bull. Chem. Soc. Japan*, **51**, 713 (1978).

where $D_0$ is in $cm^2 s^{-1}$. $V_A$ and $V_B$ are the liquid molar volumes in $cm^3 mol^{-1}$ of A and B at the temperature T in K, and the factors I, S, and A are given in original publications,[4,6] and $\eta$ is the solvent viscosity, in cP.

Should the pure solute not be a liquid at 298 K, it is recommended that the liquid volume at the boiling point be obtained either from data or from correlations.[4] Values of $D_0$ were estimated for many (149) solute-solvent systems and average error was 9.1 %.

## 2.3 BASIC PHYSICAL AND CHEMICAL PROPERTIES OF SOLVENTS

George Wypych
**ChemTec Laboratories, Inc., Toronto, Canada**

This section contains information on the basic relationships characterizing the physical and chemical properties of solvents and some suggestions regarding their use in solvent evaluation and selection. The methods of testing which allow us to determine some of physical and chemical properties are found in Chapter 15. The differences between solvents of various chemical origin are discussed in Chapter 3, Section 3.3. The fundamental relationships in this chapter and the discussion of different groups of solvents are based on extensive CD-ROM database of solvents which can be obtained from ChemTec Publishing. The database has 110 fields which contain various data on solvent properties which are discussed below. The database can be searched by the chemical name, empirical formula, molecular weight, CAS number and property. In the first case, full information on a particular solvent is returned. In the second case, a list of solvents and their values for the selected property are given in tabular form in ascending order of the property in question.

## 2.3.1 MOLECULAR WEIGHT AND MOLAR VOLUME

The molecular weight of a solvent is a standard but underutilized component of the information on properties of solvents. Many solvent properties depend directly on their molecular weights. The hypothesis of Hildebrand-Scratchard states that solvent-solute interaction occurs when solvent and polymer segment have similar molecular weights. This is related to the hole theory according to which a solvent occupying a certain volume leaves the same volume free when it is displaced. This free volume should be sufficient to fit the polymer segment which takes over the position formerly occupied by the solvent molecule.

Based on this same principle, the diffusion coefficient of a solvent depends on its molecular mass (see equations [6.2] and [6.3]). As the molecular weight of a solvent increases its diffusion rate also increases. If there were no interactions between solvent and solute, the evaporation rate of the solvent would depend on the molecular weight of the solvent. Because of various interactions, this relationship is more complicated but solvent molecular weight does play an essential role in solvent diffusion. This is illustrated best by membranes which have pores sizes which limit the size of molecules which may pass through. The resistance of a material to solvents will be partially controlled by the molecular weight of the solvent since solvent molecules have to migrate to the location of the interactive material in order to interact with it.

The chemical potential of a solvent also depends on its molecular weight (see eq. [6.6]). If all other influences and properties are equal, the solvent having the lower molecular weight is more efficiently dissolving materials, readily forms gels, and swells materials. All this is controlled by the molecular interactions between solvent and solute. In other words, at least one molecule of solvent involved must be available to interact with a particular segment of solute, gel, or network. If solvent molecular weight is low more molecules per unit weight are available to affect such changes. Molecular surface area and molecular volume are part of various theoretical estimations of solvent properties and they are in part dependent on the molecular mass of the solvent.

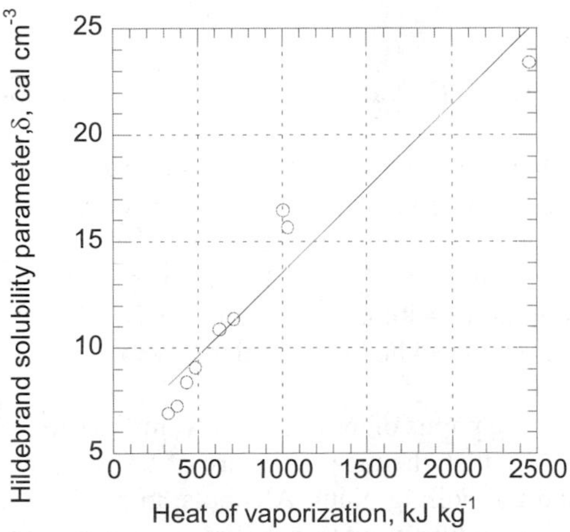

Figure 2.3.1. Hildebrand solubility parameter vs. heat of vaporization of selected solvents.

Many physical properties of solvents depend on their molecular weight, such as boiling and freezing points, density, heat of evaporation, flash point, and viscosity. The relationship between these properties and molecular weight for a large number of solvents of different chemical composition is affected by numerous other influences but within the same chemical group (or similar structure) molecular weight of solvent correlates well with its physical properties.

Figure 2.3.1 gives an example of interrelation of seemingly unrelated parameters: Hildebrand solubility parameter and heat of vaporization (see more on the subject in the Section 2.3.19). As heat of vaporization increases, the solubility parameter also increases.

Molar volume is a rather speculative, theoretical term. It can be calculated from Avogadro's number but it is temperature dependent. In addition, free volume is not taken into consideration. Molar volume can be expressed as molecular diameter but solvent molecules are rather non-spherical therefore diameter is often misrepresentation of the real dimension. It can be measured from the studies on interaction but results differ widely depending on the model used to interpret results.

### 2.3.2 BOILING AND FREEZING POINTS

Boiling and freezing points are two basic properties of solvents often included in specifications. Based on their boiling points, solvents can be divided to low (below 100°C), medium (100-150°C) and high boiling solvents (above 150°C).

The boiling point of liquid is frequently used to estimate the purity of the liquid. A similar approach is taken for solvents. Impurities cause the boiling point of solvents to increase but this increase is very small (in the order of 0.01°C per 0.01% impurity). Considering that the error of boiling point can be large, contaminated solvents may be undetected by boiling point measurement. If purity is important it should be evaluated by some other, more sensitive methods. The difference between boiling point and vapor condensation temperature is usually more sensitive to admixtures. If this difference is more than 0.1°C, the presence of admixtures can be suspected.

The boiling point can also be used to evaluate interactions due to the association among molecules of solvents. For solvents with low association, Trouton's rule, given by the following equation, is fulfilled:

$$\Delta S^o_{bp} = \frac{\Delta H^o_{bp}}{T_{bp}} = 88 J\,mol^{-1}\,K^{-1} \qquad [2.3.1]$$

where:

$\Delta S^o_{bp}$ molar change of enthalpy
$\Delta H^o_{bp}$ molar change of entropy
$T_{bp}$ boiling point

If the enthalpy change is high it suggests that the solvent has a strong tendency to form associations.

Boiling point depends on molecular weight but also on structure. It is generally lower for branched and cyclic solvents. Boiling and freezing points are important considerations for solvent storage. Solvents are frequently stored under nitrogen blanket and they contribute to substantial emissions during storage. Freezing point of some solvents is above temperatures encountered in temperate climatic conditions. Although, solvents are usually very stable in their undercooled state, they rapidly crystallize when subjected to any mechanical or sonar impact.

Figures 2.3.2 to 2.3.6 illustrate how the boiling points of individual solvents in a group are related to other properties. Figure 2.3.2 shows that chemical structure of a solvent affects the relationship between its viscosity and the boiling point. Alcohols, in particular, show a much larger change in viscosity relative to boiling point than do aromatic hydrocarbons, esters and ketones. This is caused by strong associations between molecules of alcohols, which contain hydroxyl groups. Figure 2.3.3 shows that alcohols are also less volatile

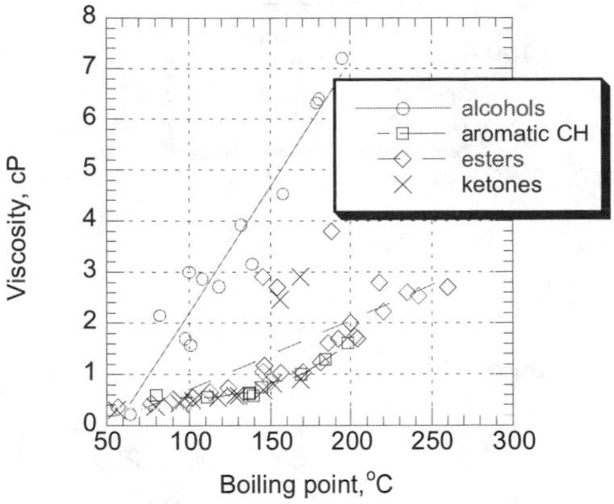


Figure 2.3.2. Effect of boiling point on solvent viscosity.

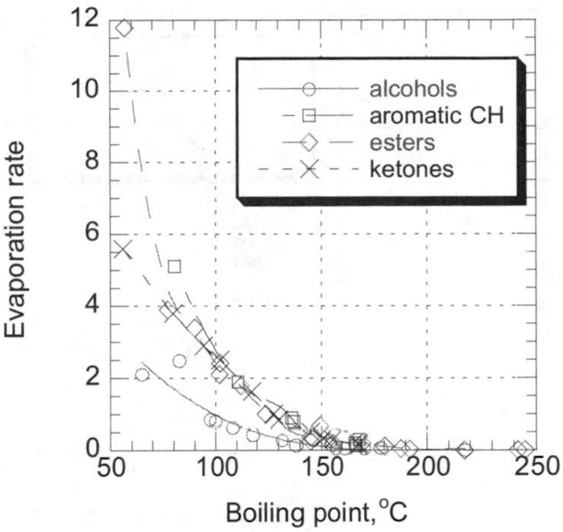


Figure 2.3.3. Effect of boiling point on solvent evaporation rate (relative to butyl acetate = 1).

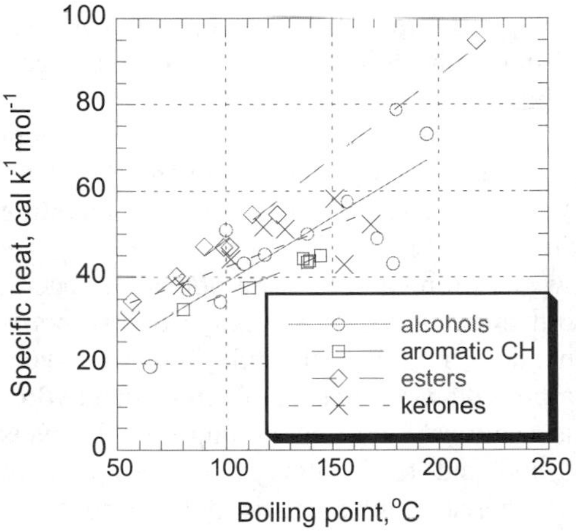


Figure 2.3.4. Specific heat of solvents vs. their boiling point.

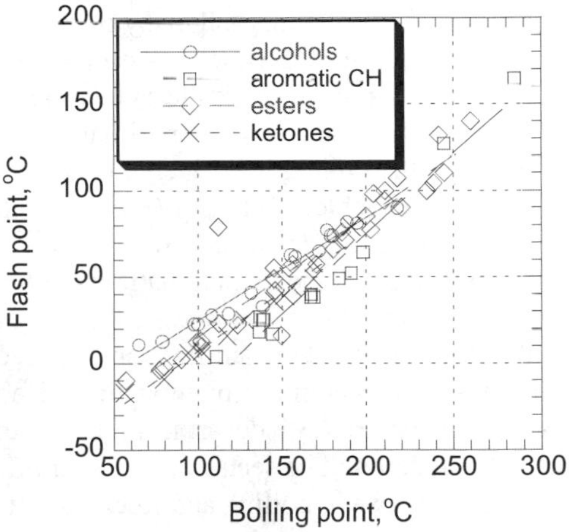


Figure 2.3.5. Flash point of solvents vs. their boiling point.

than other three groups of solvents and for the same reason. Viscosity and evaporation rate of aromatic hydrocarbons, esters, and ketones follow single relationship for all three groups of solvents, meaning that the boiling point has strong influence on these two properties. There are individual points on this set of graphs which do not fall close to the fitted curves. These discrepancies illustrate that chemical interactions influence viscosity and evaporation rate. However, for most members of the four groups of solvents, properties correlate most strongly with boiling point.

All linear relationships in Figure 2.3.4 indicate that specific heat is strongly related to the boiling point which is in agreement with the fact that boiling point is influenced by molecular weight. However, there are substantial differences in the relationships between dif-

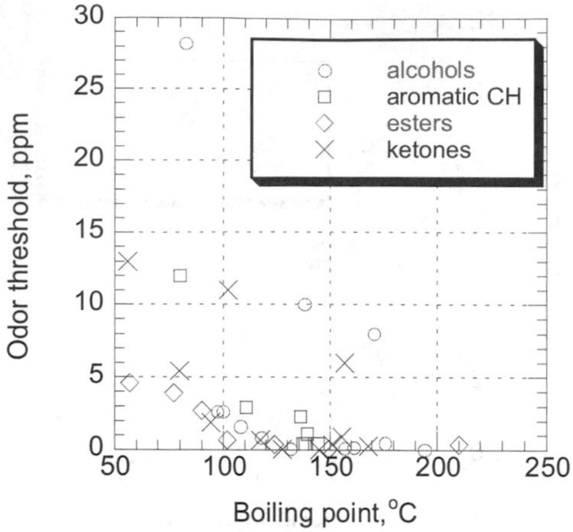


Figure 2.3.6. Odor threshold of solvent vs. its boiling point.

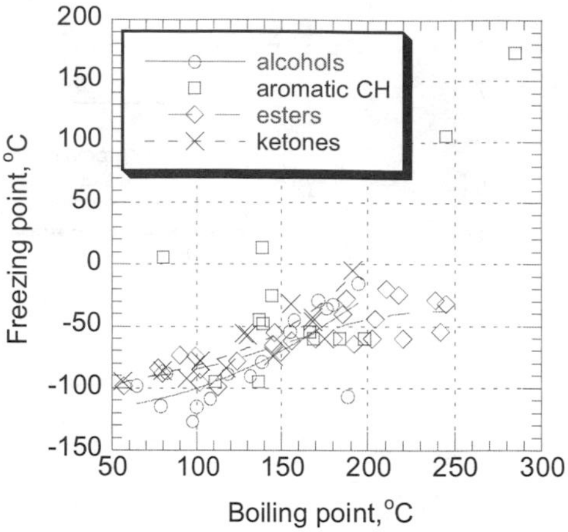


Figure 2.3.7. Relationship between boiling and freezing points of solvents.

ferent groups of solvents and many experimental points are scattered. Figure 2.3.5 verifies the origin of flash point which has strong correlation with boiling point. Here, all four chemical groups of solvents have the same relationship.

Odor threshold is an approximate but quite unreliable method of detection of solvent vapors. As the boiling point increases, the odor threshold (concentration in air when odor becomes detectable) decreases (Figure 2.3.6). This may suggest that slower evaporating solvents have longer residence time close to the source of contamination.

Figure 2.3.7 shows the relationship between boiling and freezing points. A general rule is that the difference between boiling and freezing points for analyzed solvents is 190±30°C. Relatively small fraction of solvents does not follow this rule. Natural solvent mixtures such as aromatic or aliphatic hydrocarbons deviate from the rule (note that hydrocarbons in Figure 2.3.7 depart from the general relationship). If more groups of solvents is investigated, it will be seen that CFCs, amines, and acids tend to have a lower temperature difference between boiling and freezing points whereas some aliphatic hydrocarbons and glycol ethers have a tendency towards a larger difference.

### 2.3.3 SPECIFIC GRAVITY

The specific gravity of most solvents is lower than that of water. When solvent is selected for extraction it is generally easy to find one which will float on the surface of water. Two groups of solvents: halogenated solvents and polyhydric alcohols have specific gravity greater than that of water. The specific gravity of alcohols and ketones increases with increasing molecular weight whereas the specific gravity of esters and glycol ethers decreases as their molecular weight decreases.

The specific gravity of solvents affects their industrial use in several ways. Solvents with a lower density are more economical to use because solvents are purchased by weight but many final products are sold by volume. The specific gravity of solvent should be considered in the designs for storage systems and packaging. When switching the solvent types in storage tanks one must determine the weight of new solvent which can be accommodated in the tank. A container of CFC with a specific gravity twice that of most solvents, may be

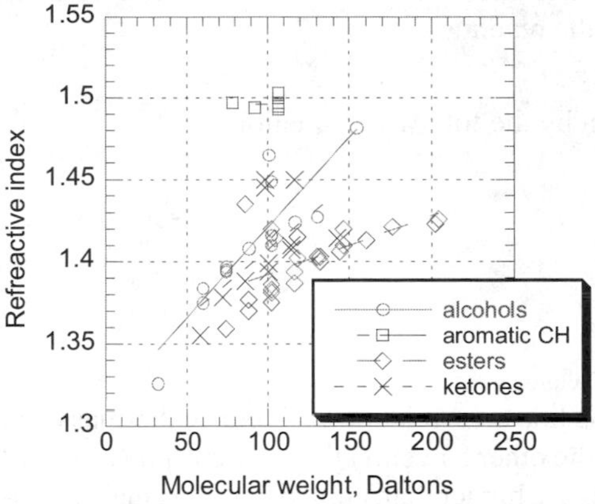


Figure 2.3.8. Refractive index for four groups of solvents as the function of their molecular weight.

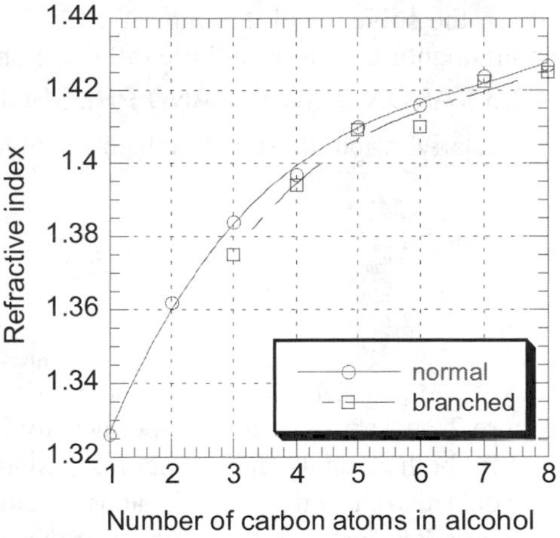


Figure 2.3.9. Refractive index of normal and branched alcohols as the function of number of carbon atoms in alcohol.

too heavy to handle. When metering by volume, temperature correction should always be used because solvent specific gravity changes substantially with temperature.

## 2.3.4 REFRACTIVE INDEX

Refractive index is the ratio of the velocity of light of a specified wavelength in air to its velocity in the examined substance. When the principle of measurement is used it may be defined as the sine of the angle of incidence divided by the the sine of the angle of refraction. The absolute angle of refraction (relative to vacuum) is obtained by dividing the refractive index relative to air by a factor of 1.00027 which is the absolute refractive index of air. The ratio of the sines of the incident and refractive angles of light in the tested liquid is equal to the ratio of light velocity to the velocity of light in vacuum (that is why both definitions are correct). This equality is also referred to as Snell's law.

Figures 2.3.8 and 2.3.9 show the relationship between the molecular weight of a solvent and its refractive index. Figure 2.3.8 shows that there is a general tendency for the refractive index to increase as the molecular weight of the solvent increases. The data also indicates that there must be an additional factor governing refractive index. The chemical structure of the molecule also influences refractive index (Figure 2.3.9). Normal alcohols have a slightly higher refractive index than do branched alcohols. Cyclic alcohols have higher refractive indices than the linear and branched alcohols. For example, 1-hexanol has refractive index of of 1.416, 4-methyl-2-pentanol 1.41, and cyclohexanol 1.465. Aromatic hydrocarbons are not dependent on molecular weight but rather on the position of substituent in the benzene ring (e.g., m-xylene has refractive index of 1.495, o-xylene 1.503, and p-xylene 1.493).

The data also show that the differences in refractive indices are rather small. This imposes restrictions on the precision of their determination. Major errors stem from poor instrument preparation and calibration and inadequate temperature control. The refractive index may change on average by 0.0005/°C. Refractive index is useful tool for determination of solvent purity but the precision of this estimation depends on relative difference between the solvent and the impurity. If this difference is small, the impurities, present in small quantities, will have little influence on the reading.

### 2.3.5 VAPOR DENSITY AND PRESSURE

The relative vapor density of solvents is given by the following equation:

$$d_{vp} = \frac{M_s}{M_{air}} \quad [2.3.2]$$

where:

| | |
|---|---|
| $M_s$ | the molecular mass of solvents |
| $M_{air}$ | the molecular mass of air (28.95 Daltons) |

Figure 2.3.10 shows that the vapor density has linear correlation with molecular mass and that for both alcohols and ketones (as well as the other solvents) the relationship is similar. The data also show that solvent vapor densities are higher than air density. This makes ventilation a key factor in the removal of these vapors in the case of spill or emissions from equipment. Otherwise, the heavier than air vapors will flow along floors and depressions filling pits and subfloor rooms and leading to toxic exposure and/or risk of ignition and subsequent explosions.

The Clausius-Clapeyron equation gives the relationship between molecular weight of solvent and its vapor pressure:

$$\frac{d\ln p}{dT} = \frac{M\Lambda}{RT^2} \quad [2.3.3]$$

where:

| | |
|---|---|
| p | vapor pressure |
| T | temperature |
| M | molecular mass of solvent |
| Λ | heat of vaporization |
| R | gas constant |

Figure 2.3.11 shows that the vapor pressure of alcohols increases as the number of carbon atoms in the molecules and the molecular mass increases. A small increase in vapor pressure is produced when branched alcohols replace normal alcohols.

Vapor pressure at any given temperature can be estimated by the use of Antoine (eq. [2.3.4]) or Cox (eq. [2.3.5]) equation (or chart). Both equations are derived from Clausius-Clapeyron equation:

$$\log p = A - \frac{B}{C+T} \quad [2.3.4]$$

$$\log p = A - \frac{B}{T+230} \quad [2.3.5]$$

where:

| | |
|---|---|
| A, B, C | constants. The constants A and B are different in each equation. The equations coincide when C = 230 in the Antoine equation. |

From the above equations it is obvious that vapor pressure increases with temperature.

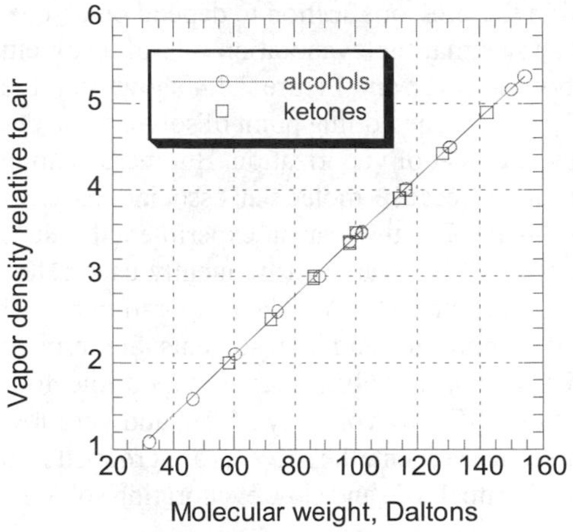


Figure 2.3.10. Vapor density relative to air of alcohols and ketones vs. their molecular weight.

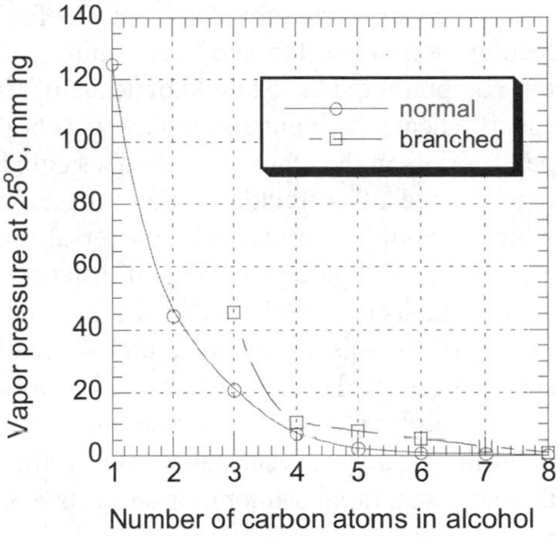


Figure 2.3.11. Vapor pressure at 25°C of normal and branched alcohols vs. number of carbon atoms in their molecule.

The vapor pressure at the boiling point of a pure solvent is equal to atmospheric pressure. When solvents are used in mixtures or solutions, the vapor pressure is affected by other components present in the mixture. For example, if a solvent is hydrophilic, the addition of a hydrophilic solute decreases the vapor pressure. The addition of a hydrophobic solute to a hydrophilic solvent increases the vapor pressure. Alcohols have hydrophobic chains, therefore addition of small quantities of alcohol to water increases vapor pressure of resulting solution. Because of these phenomena and other types of associations between solvents in their mixtures, theory cannot be used to accurately predict the resulting vapor pressure.

Raoult's Law has limited prediction capability of the vapor pressure of two miscible solvents:

$$p_{12} = m_1 p_1 + (1 - m_1) p_2 \quad [2.3.6]$$

where:

| | |
|---|---|
| $p_{12}$ | vapor pressure of the mixture |
| $m_1$ | molar fraction of the first component |
| $p_1$, $p_2$ | vapor pressures of the components |

If associations exist between molecules in the mixture, the vapor pressure of the mixture is lower than that predicted by the law.

## 2.3.6 SOLVENT VOLATILITY

The evaporation rate of solvents is important in many applications. This has resulted in attempts to model and predict solvent volatility. The evaporation rate of a solvent depends on its vapor pressure at the processing temperature, the boiling point, specific heat, enthalpy and heat of vaporization of the solvent, the rate of heat supply, the degree of association between solvent molecules and between solvent and solute molecules, the surface tension of the liquid, the rate of air movement above the liquid surface, and humidity of air surrounding the liquid surface.

The vapor pressure of solvent was found in the previous section to depend on its molecular weight and temperature. Figure 2.3.3 shows that the evaporation rate of a solvent may be predicted based on knowledge of its boiling point and Figure 2.3.4 shows that the specific heat of solvent also relates to its boiling point. The boiling point of solvent also depends on its molecular weight as does enthalpy and heat of vaporization. But there is not a high degree of correlation among these quantities because molecular associations exist which cannot be expressed by universal relationship. For this reason experimental values are used to compare properties of different solvents. The two most frequently used reference solvents are: diethyl ether (Europe) and butyl acetate (USA). The evaporation rate of other solvents is determined under identical conditions and the solvents are ranked accordingly. If diethyl ether is used as a reference point, solvents are grouped into four groups: high volatility < 10, moderate volatility 10-35, low volatility 35-50, and very low volatility > 50. If butyl acetate is used as the reference solvent, the solvents are grouped into three classes: rapid evaporation solvents > 3, moderate 0.8-3, and slow evaporating solvents < 0.8.

In some applications such as coatings, casting, etc., evaporation rate is not the only important parameter. The composition must be adjusted to control rheological properties, prevent shrinkage, precipitation, formation of haze, and to provide the required morphology. Solvents with different evaporation rates can address all existing requirements.

Both the surface tension of mixture and solvent diffusion affect the evaporation rate. This becomes a complex function dependent not only on the solvents present but also on influence of solutes on both surface tension and diffusion. These relationships affect the real evaporation rates of solvents from the complicated mixtures in the final products. In addition, solvent evaporation also depends on relative humidity and air movement.

## 2.3.7 FLASH POINT

Flash point is the lowest temperature, corrected to normal atmospheric pressure (101.3 kPa), at which the application of an ignition source causes the vapors of a specimen to ignite under the specific conditions of the test. Flash point determination methods are designed to be applied to a pure liquid but, in practice, mixtures are also evaluated. It is important to understand limitations of such data. The flash point of a solvent mixture can be changed by adding various quantities of other solvents. For example, the addition of water or halogenated hydrocarbons will generally increase the flash point temperature of mixture. The flash point can also be changed by forming an azeotropic mixture of solvents or by increasing the interaction between solvents. At the same time, the flash point of single component within the mixture is not changed. If conditions during production, application, or in a spill allow the separation or removal of a material added to increase the flash point, then the flash point will revert to that of the lowest boiling flammable component.

An approximate flash point can be estimated from the boiling point of solvent using the following equation:

$$Flash\ point = 0.74T_b \qquad [2.3.7]$$

Figure 2.3.5 shows that there is often good correlation between the two but there are instances where the relationship does not hold. The correlation for different groups of solvents varies between 0.89 to 0.96.

Flash point can also be estimated from vapor pressure using the following equation:

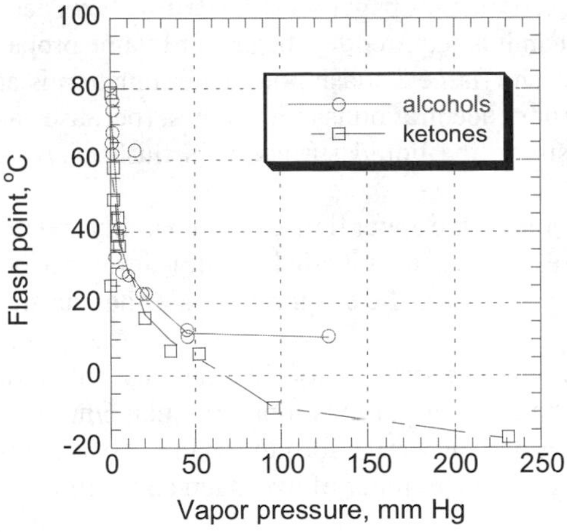


Figure 2.3.12. Flash point vs. vapor pressure of solvent.

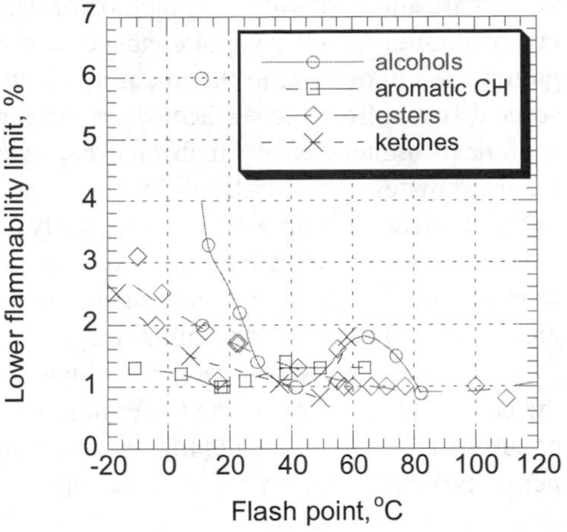


Figure 2.3.13. Lower flammability limit of solvents vs their flash points.

$$Flash\ point = a \log p + b \qquad [2.3.8]$$

The constants a and b are specific to each group of solvents. Figure 2.3.12 shows that estimation of flash point from vapor pressure of solvent is less accurate than its estimation from boiling point.

## 2.3.8 FLAMMABILITY LIMITS

Two limits of solvent flammability exist. The lower flammability limit is the minimum concentration of solvent vapor in oxidizing gas (air) that is capable of propagating a flame through a homogeneous mixture of the oxidizer and the solvent vapor. Below the lower flammability limit the mixture is too lean to burn or explode. The upper flammability limit is the maximum concentration of solvent vapor in an oxidizing gas (air) above which propagation of flame does not occur. Mixtures with solvent vapor concentrations above the upper flammability limit are too rich in solvent or too lean in oxidizer to burn or explode.

The flammable limits depend on oxygen concentration, concentration of gases other than oxygen, the inert gas type and concentration, the size of the equipment, the direction of flame propagation, and the pressure, temperature, turbulence and composition of the mixture. The addition of inert gases to the atmosphere containing solvent is frequently used to reduce the probability of an explosion. It is generally assumed that if the concentration of oxygen is below 3%, no ignition will occur. The type of inert gas is also important. Carbon dioxide is more efficient inert gas than nitrogen. The size of equipment matters because of the uniformity of vapor concentration. A larger head space tends to increase the risk of inhomogeneity. The cooling effect of the equipment walls influences the evaporation rate and the vapor temperature and should be used in risk assessment.

The flash point is not the temperature at which the vapor pressure in air equals the lower flammable limit. Although both parameters have some correspondence there are large differences between groups of solvents. There is a general tendency for solvents with a lower flammability limit to have a lower flash point. The flash point determination uses a downward and horizontal propagation of flame. Flame propagation in these directions generally requires a higher vapor concentration than it is required for the upward flame propagation used to determine flammability limits. The flame in flash point determination is at some distance from the surface where the vapor concentration is at its highest (because vapors have higher density than air) than exists on the liquid surface thus flush analysis underestimates concentration of vapor.

An increased vapor pressure typically increases the upper limit of flammability and reduces the lower limit of flammability. Pressures below atmospheric have little influence on flammability limits. An increase in temperature increases the evaporation rate and thus decreases the lower limit of flammability.

There are a few general rules which help in the estimation of flammability limits. In the case of hydrocarbons, the lower limit can be estimated from simple formula: 6/number of carbon atoms in molecule; for benzene and its derivatives the formula changes to: 8/number of carbon atoms. To calculate the upper limits, the number of hydrogen and carbon atoms is used in calculation.

The lower flammability limit of a mixture can be estimated from Le Chatelier's Law:

$$LFL_{mix} = \frac{100}{\frac{\phi_1}{LFL_1} + \frac{\phi_2}{LFL_2} + \cdots + \frac{\phi_n}{LFL_n}} \quad [2.3.9]$$

where:

$\phi_i$ fraction of components 1, 2, ..., n
$LFL_i$ lower flammability limit of component 1, 2, ..., n

### 2.3.9 SOURCES OF IGNITION AND AUTOIGNITION TEMPERATURE

Sources of ignition can be divided to mechanical sources (impact, abrasive friction, bearings, misaligned machine parts, choking or jamming of material, drilling and other maintenance operations, etc), electrical (broken light, cable break, electric motor, switch gear, liquid velocity, surface or personal charge, rubbing of different materials, liquid spraying or jetting, lightning, stray currents, radio frequency), thermal (hot surface, smoking, hot transfer lines, electric lamps, metal welding, oxidation and chemical reactions, pilot light, arson, change of pressure, etc.), and chemical (peroxides, polymerization, catalysts, lack of inhibitor, heat of crystallization, thermite reaction, unstable substances, decomposition reactions). This long list shows that when making efforts to eliminate ignition sources, it is also essential to operate at safe concentrations of volatile, flammable materials because of numerous and highly varied sources of ignition.

The energy required for ignition is determined by the chemical structure of the solvent, the composition of the flammable mixture, and temperature. The energy of ignition of hydrocarbons decreases in the order alkanes > alkenes > alkynes (the presence of double or triple bond decreases the energy energy required for ignition). The energy requirement increases with an increase in molecular mass and an increase branching. Conjugated structure generally requires less ignition energy. Substituents increase the required ignition energy in the following order: mercaptan < hydroxyl < chloride < amine. Ethers and ketones require higher ignition energy but an aromatic group has little influence. Peroxides require extremely little energy to ignite.

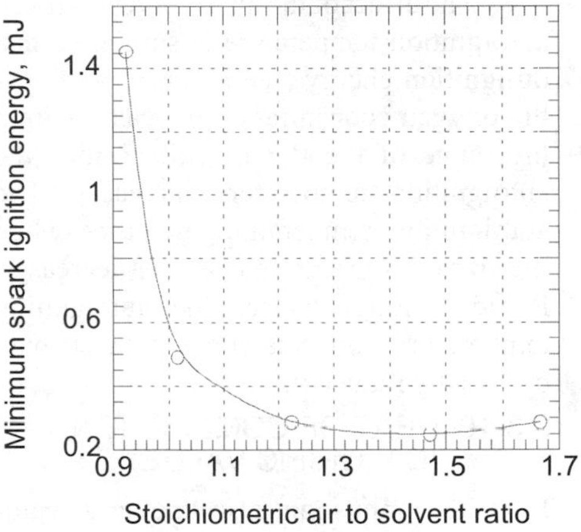


Figure 2.3.14. Minimum ignition energy vs. stoichiometric ratio of air to methyl ethyl ketone. [Data from H F Calcote, C A Gregory, C M Barnett, R B Gilmer, *Ind. Eng. Chem.*, **44**, 2656 (1952)].

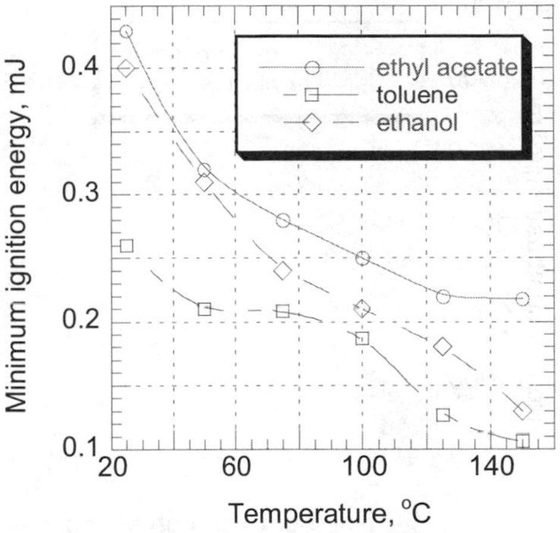


Figure 2.3.15. Minimum ignition energy vs. temperature for selected solvents. [Data from V S Kravchenko, V A Bondar, **Explosion Safety of electrical Discharges and Frictional Sparks**, *Khimia*, Moscow, 1976].

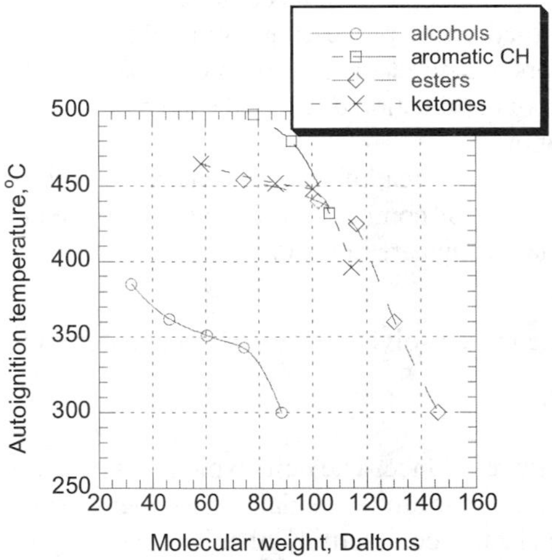


Figure 2.3.16. Autoignition temperature of selected solvents vs. their molecular weight.

Figure 2.3.14 shows the effect of changing the ratio of air to methyl ethyl ketone on the minimum spark ignition energy. The ignition energy decreases within the studied range as the amount of air increases (less flammable content). Figure 2.3.15 shows the effect of temperature on the minimum ignition energy of selected solvents. There are differences between solvents resulting from differences in chemical structure as discussed above but the trend is consistent – a decrease of required energy as temperature increases.

The autoignition temperature is the minimum temperature required to initiate combustion in the absence of a spark or flame. The autoignition temperature depends on the chemical structure of solvent, the composition of the vapor/ air mixture, the oxygen concentration, the shape and size of the combustion chamber, the rate and duration of heating, and on catalytic effects. Figure 2.3.16 shows the effect of chemical structure on autoignition temperature. The general trend for all groups of solvents is that the autoignition temperature decreases as molecular weight increases. Esters and ketones behave almost identically in this respect and aromatic hydrocarbons are very similar. The presence of a hydroxyl group substantially reduces autoignition temperature.

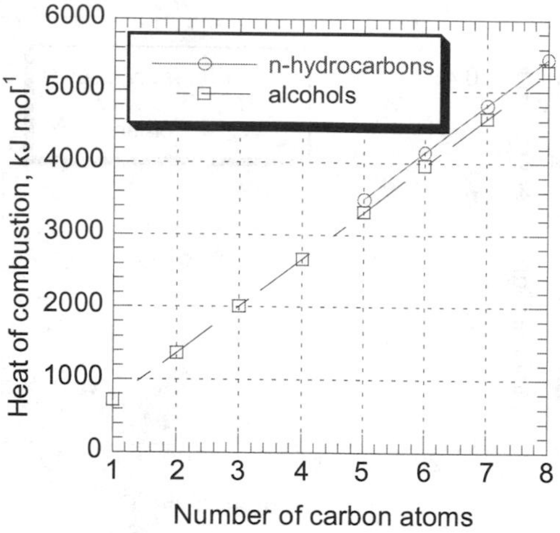


Figure 2.3.17. Heat of combustion vs. number of carbon atoms in molecule.

The effect of the air to solvent ratio on autoignition temperature is similar to that on ignition energy (see Figure 2.3.14). As the oxygen concentration increases within the range of the flammability limits, the autoignition temperature increases. The autoignition temperature increases when the size of combustion chamber decreases. Rapid heating reduces the autoignition temperature and catalytic substance may drastically reduce it.

## 2.3.10 HEAT OF COMBUSTION (CALORIFIC VALUE)

Heat of combustion, also known as calorific value, is the quantity of energy per mole released during combustion. It coincides with the heat of reaction. Solvents have higher heats of combustion than typical fuels such as natural gas, propane or butane. They can be very good source of energy in plants which process solutions. In addition to supplying energy, the combustion of solvents can be developed to be one of the cleanest method of processing from solutions. Two approaches are commonly used: solvent vapors are directed to a combustion chamber or spent solvents are burned in furnaces. The heat of combustion of a liquid solvent is less than 1% lower than the heat of combustion of a vapor.

Figure 2.3.17 shows the relationship between the heat of combustion and the number number of carbon atoms in the molecule. The heat of combustion increases as molecular weight increases and decreases when functional groups are present.

## 2.3.11 HEAT OF FUSION

Heat of fusion is the amount of heat to melt the frozen solvent. It can be used to determine the freezing point depression of solute.

## 2.3.12 ELECTRIC CONDUCTIVITY

Electric conductivity is the reciprocal of specific resistance. The units typically used are either $ohm^{-1}$ $m^{-1}$ or, because the conductivities of solvents are very small picosiemens per meter which is equivalent to$10^{-12}$ $ohm^{-1}$ $m^{-1}$. The electric conductivity of solvents is very low (typically between $10^{-3}$ - $10^{-9}$ $ohm^{-1}$ $m^{-1}$). The presence of acids, bases, salts, and dissolved carbon dioxide might contribute to increased conductivity. Free ions are solely responsible for the electric conductivity of solution. This can be conveniently determined by measuring the conductivity of the solvent or the conductivity of the water extract of solvent impurities. The electronic industry and aviation industry have the major interest in these determinations.

## 2.3.13 DIELECTRIC CONSTANT (RELATIVE PERMITTIVITY)

The dielectric constant (or relative permittivity) of a solvent reflects its molecular symmetry. The value of the dielectric constant is established from a measurement relative to vacuum. The effect is produced by the orientation of dipoles along an externally applied

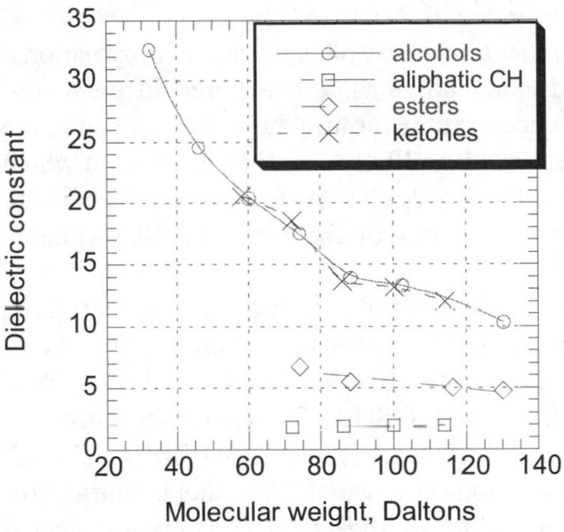


Figure 2.3.18. Dielectric constant vs. molecular weight of selected solvents.

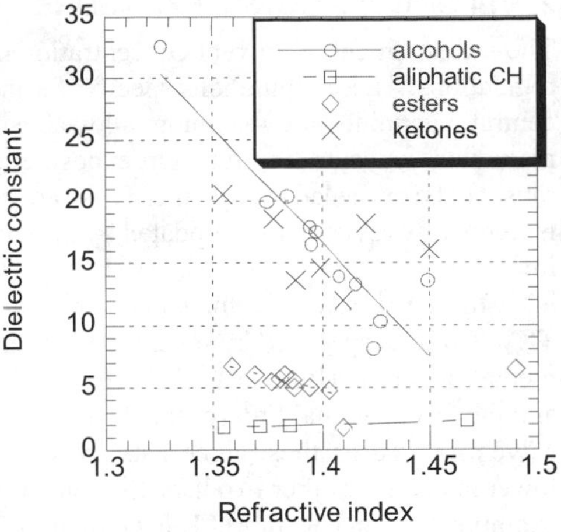


Figure 2.3.19. Dielectric constant of selected solvents vs. their refractive index.

electric field and from the separation of charges in apolar molecules. This orientation causes polarization of the molecules and a drop in electric field strength. Dielectric constant data may be used in many ways. In particular, it is the factor which permits the evaluation of electrostatic hazards. The rate of charge decay is a product of dielectric constant and resistivity.

In solvent research, dielectric constant has a special place as a parameter characteristic of solvent polarity. The dielectric constant, ε, is used to calculate dipole moment, μ:

$$\mu = \frac{\varepsilon - 1}{\varepsilon - 2} V_M \qquad [2.3.10]$$

where:

$V_M$ molar volume

The product of dipole moment and dielectric constant is called the electrostatic factor and it is a means of classifying solvents according to their polarity.

Figure 2.3.18 shows that the dielectric constant correlates with molecular weight. It is only with aliphatic hydrocarbons that the dielectric constant increases slightly as the molecular weight increases. The dielectric constant of alcohols, esters, and ketones decreases as their molecular weight increases, but only alcohols and ketones have the same relationship. The dielectric constants of esters are well below those of alcohols and ketones.

The dielectric constant also correlates with refractive index. In the case of aliphatic hydrocarbons, the dielectric constant increases slightly as refractive index increases. Both aromatic and aliphatic hydrocarbons have dielectric constants which follow the relationship: $\varepsilon \approx n_D^2$. The dielectric constants of alcohols, esters and ketones decrease as the refractive constants increase but only alcohols and ketones form a similar relationship. The dielectric constants of ketones poorly correlate with their refractive indices.

### 2.3.14 OCCUPATIONAL EXPOSURE INDICATORS

The measurement of solvent concentration in the workplace place is required by national regulations. These regulations specify, for individual solvents, at least three different concentrations points: the maximum allowable concentration for an 8 hour day exposure, the maximum concentration for short exposure (either 15 or 30 min), and concentration which must not be exceeded at any time. These are listed in the regulations for solvents. The listing is frequently reviewed and updated by the authorities based on the most currently available information.

In the USA, the threshold limit value, time-weighted average concentration, TLV-TWA, is specified by several bodies, including the American Conference of Governmental Industrial Hygienists, ACGIH, the National Institute of Safety and Health, NIOSH, and the Occupational Safety and Health Administration, OSHA. The values for individual solvents stated on these three lists are very similar. Usually the NIOSH TLV-TWA are lower than on the other two lists. Similar specifications are available in other countries (for example, OES in UK, or MAK in Germany). The values for individual solvents are selected based on the presumption that the maximum allowable concentration should not cause injury to a person working under these conditions for 8 hours a day.

For solvent mixtures, the following equation is used in Germany to calculate allowable limit:

$$I_{MAK} = \sum_{i=1}^{i=n} \frac{c_i}{MAK_i} \qquad [2.3.11]$$

where:

| | |
|---|---|
| $I_{MAK}$ | evaluation index |
| c | concentrations of components 1, 2, ..., n |
| MAK | maximum permissible concentrations for components 1, 2, ..., n |

The maximum concentrations for short exposure is the most frequently limit specified for an exposure of 15 min with a maximum of 4 such occurrences per day each occurring at least 60 min apart from each other. These values are 0-4 times larger than TLVs. They are selected based on the risks associated with an individual solvent.

Solvent concentrations which should not be exceeded at any time are seldom specified in regulations but, if they are, the values stated as limits are similar to those on the three lists.

In addition, to maintaining concentration below limiting values, adequate protection should be used to prevent the inhalation of vapors and contact with the skin (see Chapter 24).

### 2.3.15 ODOR THRESHOLD

The principal for odor threshold was developed to relate the human sense of smell to the concentration of the offending substances. If the substance is toxic, its detection may provide early warning to the danger. However, if the odor threshold is higher than the concentration at which harm may be caused it is not an effective warning system. Toxic substances may have very little or no odor (e.g., carbon monoxide) and an individual's sense of smell may vary widely in its detection capabilities. A knowledge of odor threshold is most useful in determining the relative nuisance factor for an air pollutant when designing a control system to avoid complaints from neighboring people surrounding a facility. Regulations often state (as they do for example in Ontario, Canada) that, even when the established concentration limits for air pollutants are met, if neighbors complain, penalties will be applied. Figure 2.3.6 shows that odor threshold is related to the boiling point (although odor threshold decreases with boiling point increasing). It is known from comparisons of TLV and odor detection that odor detection is not a reliable factor.

### 2.3.16 TOXICITY INDICATORS

Lethal dose, LD50, and lethal concentration, LC50, are commonly used indicators of substance toxicity. LD50 is reported in milligrams of substance per kilogram of body weight to cause death in 50% of tested animals (exception is LC50 which is given in ppm over usually the period of 4 hours to produce the same effect). It is customary to use three values: LD50-oral, LD50-dermal, and LC50-inhalation which determine the effect of a chemical substance on ingestion, contact with the skin and inhalation. The preferred test animal for LD50-oral and LC50-inhalation is the rat. The rabbit is commonly used for LD50-dermal determination but other test animals are also used.

There is no official guideline on how to use this data but the Hodge-Sterner table is frequently referred to in order to assign a particular substance to a group which falls within certain limits of toxicity. According to this table, dangerously toxic substances are those which have LD50 < 1 mg/kg, seriously toxic - 1-50, highly toxic - 50-500, moderately toxic - 500-5,000, slightly toxic - 5,000-15,000, and extremely low toxic - >15,000 mg/kg. Using this classification one may assess the degree of toxicity of solvents based on a lethal dose scale. No solvent is classified as a dangerously toxic material. Ethylenediaminetetraacetic acid and furfural are seriously toxic materials. Butoxyethanol, ethylene oxide, formaldehyde, metasulfonic acid, 3-methyl-2-butanone, N-nitrosodimethylamine, and triethylamine are classified as highly toxic material. The remaining solvents fall into the moderately, slightly, and extremely low toxic material classes.

The LD50-oral is usually assigned a lower value than LD50-dermal but there are many cases where the opposite applies. Toxicity information is usually further expanded by adding more details regarding test animals and target organs.

In addition to estimates of toxicity for individual solvents, there are lists which designate individual solvents as carcinogenic, mutagenic, and reproductively toxic. These lists contain the name of solvent with yes or no remark (or similar). If a solvent is not present on the list that does not endorse its benign nature because only materials that have been tested are included in the lists. To further elaborate, materials are usually divided into three categories: substance known to cause effect on humans, substance which has caused responses in animal testing and given reasons to believe that similar reactions can be expected with human exposures, and substance which is suspected to cause responses based on experimental evidence.

In the USA, four agencies generate lists of carcinogens. These are: the Environmental Protection Agency, EPA, the International Agency for Research on Cancer, IRAC, the National Toxicology Program, NTP, and the Occupational Safety and Health Administration, OSHA. Although, there is a good agreement between all four lists, each assessment differs in some responds. The following solvents made at least one of the lists (no distinction is given here to the category assignment but any known or suspected carcinogen found on any list is given (for more details see Chapter 3)): acetone, acrolein, benzene, carbon tetrachloride, dichloromethane, 1,4-dioxane, ethylene oxide, formaldehyde, furfural, d-limonene, N-nitrosodimethyl amine, propylene oxide, tetrachloroethylene, 2,4-toluenediisocyanate, 1,1,2-trichloroethylene, and trichloromethane.

Mutagenic substances have the ability to induce genetic changes in DNA. The mutagenicity list maintained in the USA includes the following solvents: all solvents listed above for carcinogenic properties with exception of dichloromethane, d-limonene, and tetrachloroethylene. In addition, the following long list solvents: 1-butanol, 2-butanol, γ-butyrolactone, 2-(2-n-butoxyethoxy)ethanol, chlororodifluoromethane, chloromethane, diacetone alcohol, dichloromethane, diethyl ether, dimethyl amine, dimethylene glycol dimethyl ether, dimethyl sulfoxide, ethanol, 2- ethoxyethanol, 2-ethoxyethanol acetate, ethyl acetate, ethyl propionate, ethylbenzene, ethylene glycol diethyl ether, ethylene glycol methyl ether acetate, ethylene glycol monophenyl ether, ethylenediaminetetraacetic acid, formic acid, furfuryl alcohol, heptane, hexane, methyl acetate, 3-methyl-2-butanol, methyl ester of butyric acid, methyl propionate, N-methylpyrrolidone, monomethylamine, 1-octanol, 1-pentanol, 1-propanol, propyl acetate, sulfolane, 1,1,1-trichloroethane, triethylene glycol, triethylene glycol dimethyl ether, trifluoromethane, trimethylene glycol, and xylene (mixture only). It is apparent that this much longer list includes commonly used solvents from the groups of alcohols, halogenated solvents, hydrocarbons, glycols, and esters.

The following solvents are reported to impair fertility: chloroform, ethylene glycol and its acetate, 2-methoxypropanol, 2-methoxypropyl acetate, dichloromethane, methylene glycol and its acetate, and N,N-dimethylformamide.

### 2.3.17 OZONE - DEPLETION AND CREATION POTENTIAL

Ozone depletion potential is measured relative to CFC-11 and it represents the amount of ozone destroyed by emission of a vapor over its entire atmospheric lifetime relative to that caused by the emission of the same mass of CFC-11.

Urban ozone formation potential is expressed relative to ethene. It represents the potential of an organic solvents vapor to form ozone relative to that of ethene (($g\ O_3$/g solvent)/($g\ O_3$/g ethene)). Several groups of solvents, including alcohols, aldehydes, amines, aliphatic and aromatic hydrocarbons, esters, ethers, and ketones are active in ozone formation. Aldehydes, xylenes, some unsaturated compounds, and some terpenes are the most active among those.

### 2.3.18 OXYGEN DEMAND

There are several indicators of solvent biodegradation. Most solvents have a biodegradation half-life of days to weeks and some biodegrade even faster.

The amount of oxygen required for its biodegradation is a measure of a solvent's impact on natural resources. Several factors are used to estimate this, such as biological oxygen demand, BOD, after 5-day and 20-day aerobic tests, chemical oxygen demand, COD, and theoretical oxygen demand, TOD. All results are given in grams of oxygen per gram of solvent. COD is the amount of oxygen removed during oxidation in the presence of permanganate or dichromate. TOD is the theoretically calculated amount of oxygen required to oxidize solvent to $CO_2$ and $H_2O$. Most alcohols and aromatic hydrocarbons have a highest BOD5. They consume twice their own weight in oxygen.

### 2.3.19 SOLUBILITY

The prediction of solubility of various solutes in various solvents is a major focus of research. An early theory has been that "like dissolves like". Regardless of the apparent merits of this theory it is not sufficiently rigorous and is overly simple.

A universal approach was developed by Hildebrand who assumed that the mutual solubility of components depends on the cohesive pressure, c. The square root of cohesive pressure is the Hildebrand's solubility parameter, δ:

$$\delta = \sqrt{c} = \sqrt{\frac{\Delta H_v - RT}{V_m}} \qquad [2.3.12]$$

where:

| | |
|---|---|
| $\Delta H_v$ | heat of vaporization |
| R | gas constant |
| T | temperature |
| $V_m$ | molar volume of solvent = M/d |
| M | molecular mass of solvent |
| d | density of solvent |

Frequently, the term RT is neglected because it accounts for only 5-10% of the heat of vaporization. This equation explains the reasons for the correlation between the Hildebrand solubility parameter and heat of vaporization as given in Figure 2.3.1.

The Hildebrand model takes into account only the dimensions of molecules or of the molecular segments participating in the process of solvation and dispersion interactions. The model is useful, therefore, in predicting the solubility of non-polar substrates. The solubility parameters of solvents and solutes are compared and if they are similar there is high probability (exceptions exist) that the solvents are miscible that a solute is soluble in a solvent. Two solvents having the same solubility parameters should have the same dissolving capabilities. If one solvent has solubility parameter slightly below the solubility parameter of solute and the second solvent has solubility parameter above the solute, the mixture of both solvents should give better results than either solvent alone. This model is an experimental and mathematical development of the simple rule of "like dissolves like".

Solvents and solutes also interact by donor-aceptor, electron pair, and hydrogen bonding interactions. It can be predicted that the above concept is not fully universal, especially in the case of solutes and solvents which may apply these interactions in their solubilizing action. Hansen developed a three-dimensional scale with parameters to expand theory in order to include these interactions. Hansen defined solubility parameter by the following equation:

$$\delta^2 = \delta_d^2 + \delta_p^2 + \delta_h^2 \qquad [2.3.13]$$

where:

| | |
|---|---|
| $\delta_d$ | dispersion contribution to solubility parameter |
| $\delta_p$ | polar contribution to solubility parameter |
| $\delta_h$ | hydrogen bonding contribution to solubility parameter |

Hansen defined solvent as a point in three-dimensional space and solutes as volumes (or spheres of solubility). If a solvent point is within the boundaries of a solute volume space then the solute can be dissolved by the solvent. If the point characterizing the solvent is outside the volume space of a solute (or resin) such a solvent does not dissolve the solute. The solubility model based on this concept is broadly applied today by modern computer techniques using data obtained for solvents (the three components of solubility parameters) and solutes (characteristic volumes). A triangular graph can be used to outline the limits of

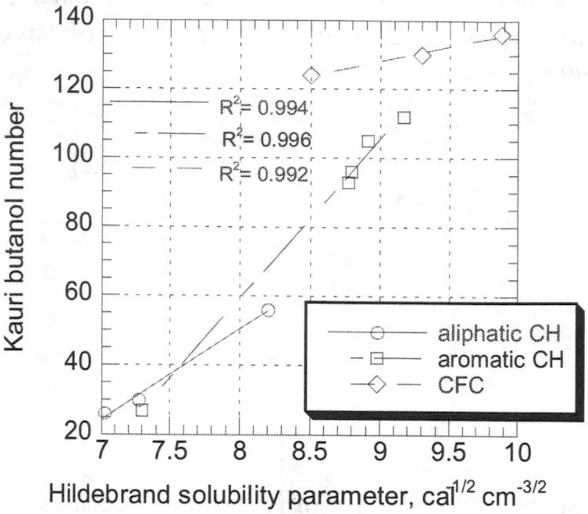


Figure 2.3.20. Kauri butanol number vs. Hildebrand solubility parameter.

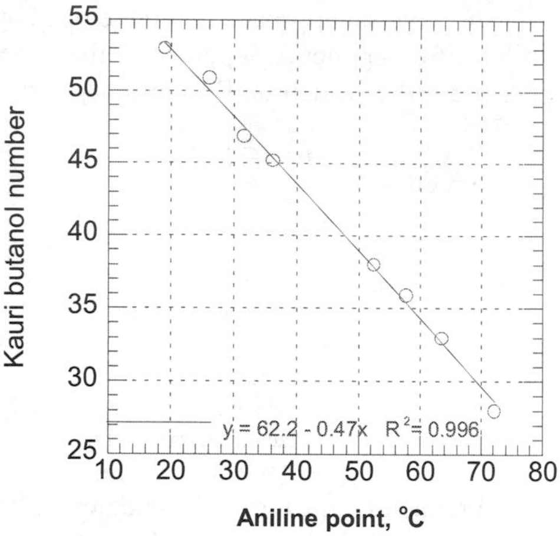


Figure 2.3.21. Kauri butanol number vs. aniline point.

solubility and place different solvents within the matrix to determine their potential dissolving capability for a particular resin.

Simpler methods are also used. In the paint industry, Kauri butanol values are determined by establishing the tolerance of a standard solution of Kauri resin in n-butanol to the addition of diluents. This method is applicable to hydrocarbons (both aromatic and aliphatic) and CFCs. Figure 2.3.20 shows that there is a good correlation between the Kauri butanol number and the Hildebrand solubility parameter. The Kauri butanol number can be as high as 1000 (amyl ester of lactic acid) or 500 (Freon solvent M-162).

The aniline point determination is another method of establishing the solubilizing power of a solvent by simple means. Here, the temperature is measured at which a solution just becomes cloudy. Figure 2.3.21 shows that there is a good correlation between the Kauri number and the aniline point. Also dilution ratio of cellulose solution is measured by standardized methods (see Chapter 15).

## 2.3.20 OTHER TYPICAL SOLVENT PROPERTIES AND INDICATORS

There are many other solvent properties and indices which assist in solvent identification and selection and help us to understand the performance characteristics of solvents. Most data characterizing the most important properties were discussed in the sections above. The solvent properties and classification indicators, which are discussed below, are included in the Solvent Database available on CD-ROM from ChemTec Publishing.

**Name**. A solvent may have several names such as common name, Chemical Abstracts name, and name according to IUPAC systematic nomenclature. Common names have been used throughout this book and in the CD-ROM database because they are well understood by potential users. Also, CAS numbers are given in the database to allow user of the database to use the information with Chemical Abstract searches. In the case of commercial solvents which are proprietary mixtures, the commercial name is used.

The molecular formula for each solvent is given in the database, followed by the molecular formula in Hills notation, and the molecular mass (if solvent is not a mixture).

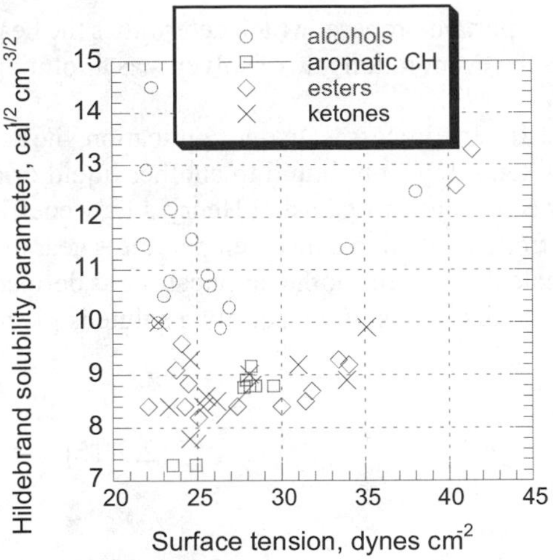


Figure 2.3.22. Hildebrand solubility parameter vs. surface tension for four groups of solvents.

**The CAS number** identifies the chemical compound or composition without ambiguity.

**RTECS number** is the symbol given by Registry of Toxic Effects of Chemical Substances (e.g. AH4025000) to identify toxic substances.

**Composition** is given for solvents which are manufactured under trade name and have a proprietary composition (if such information is available).

**Solvent purity** (impurities) is given as a percent and known impurities and their concentrations are provided.

**Hygroscopicity and water solubility** of solvents is an important characteristic in many applications. The data are given in the database either in mg of water per kilogram of solvent or in a generic statement (e.g. miscible, slight). Many solvents are hygroscopic, especially those which contain hydroxyl groups. These solvents will absorb water from their surrounding until equilibrium is reached. The equilibrium concentration depends on the relative humidity of air and the temperature. If solvents must maintain a low concentration of water, vents of storage tanks should be fitted with silica gel or molecular sieves cartridges or tanks should be sealed and equipped with pressure and vacuum relief vents which open only to relieve pressure or to admit a dry inert gas to replace the volume pumped out. Preferably prevention of water from contacting solvents or the selection of solvents with a low water content is more economical than the expensive operation of drying a wet solvent.

**Surface tension** and solubility parameter have been related in the following equation:

$$\delta = 2.1K\left[\frac{\gamma}{V^{1/3}}\right]^a \qquad [2.3.14]$$

where:

| | |
|---|---|
| $\delta$ | Hildebrand solubility parameter |
| K, a | constants |
| $\gamma$ | surface tension |
| V | molar volume of the solvent |

However, Figure 2.3.22 shows that the parameters correlate only for aromatic hydrocarbons. For three other groups the points are scattered. The equation has a very limited predictive value.

**Viscosity**. Figure 2.3.2 shows that viscosity of solvents correlates with their boiling point. There are substantial differences in the viscosity - boiling point relationship among alcohols and other groups of solvents. These are due to the influence of hydrogen bonding on the viscosity of alcohols.

**Thermal conductivity** of solvents is an important property which determines the heat transfer in a solvent or solution and influences the evaporation rate of solvents as a solution is being dried.

**Activity coefficients** may be applied to different processes. In one application, the activity coefficient is a measure of the escaping tendency from liquid to another liquid or a gaseous phase (in the liquid to gas phase they can be quantified using Henry's Law coefficient). These activity coefficients are derived from distillation data at temperatures near the boiling point or from liquid-liquid extraction calculations. In another application as defined by Hildebrand and Scratchard solvent activity to dissolve a non-electrolyte solute is given by equation:

$$\ln f = \frac{V_m\left(\delta_{solute} - \delta_{sovent}\right)^2}{RT} \qquad [2.3.15]$$

where:

| | |
|---|---|
| f | activity coefficient |
| $V_m$ | molar volume of solute |
| $\delta_{solute}$ | solubility parameter of solute |
| $\delta_{solvent}$ | solubility parameter of solvent |
| RT | gas constant x temperature |

This coefficient is used to express rate constants of bimolecular reactions.

**Azeotropes.** One solvent may form azeotropes with another solvent due to molecular association. This physical principle can be exploited in several ways. The most important in solvent applications is the possibility of reducing the boiling temperature (some azeotropes have lower boiling point) therefore an applied product such as a coating may lose its solvents and dry faster. The formation of such azeotrope also lowers flash point by which it increases hazards in product use. The formation of an azeotrope is frequently used to remove water from a material or a solvent. It affects the results of a distillation since azeotrope formation makes it difficult to obtain pure components from a mixture by distillation. Azeotrope formation can be suppressed by lowering the boiling point (distillation under vacuum). One benefit of azeotropic distillation is the reduction in the heat required to evaporate solvents.

**Henry's constant** is a measure of the escaping tendency of a solvent from a very dilute solution. It is given by a simple equation: Henry's constant = $p \times \phi$, where p is the pressure of pure solvent at the solution temperature and $\phi$ is the solvent concentration in the liquid phase. A high value of Henry's constant indicates that solvent can be easily stripped from dilute water solution. It can also be used to calculate TLV levels by knowing concentration of a solvent in a solution according to the equation: TLV (in ppm) = [18 H (concentration of solvent in water)]/ molecular weight of solvent.

**pH** and **corrosivity**. The pH of solvents is of limited value but it is sometimes useful if the solvent has strong basic or acidic properties which could cause corrosion problems.

**The acid dissociation constant** is the equilibrium constant for ionization of an acid and is expressed in negative log units.

**The color** of a solvent may influence the effect of solvent on the final product and allow the evaluation of solvent quality. The colorless solvents are most common but there are many examples of intrinsically colored solvents and solvents which are colored because of an admixture or inadequate storage conditions or too long storage.

**Odor**. Odor threshold (discussed above) is not a precise tool for estimation concentration of vapors. The description of odor has little relevance to the identification of solvent but description of odors in the database may be helpful in the selection of solvent to minimize odor or make it less intrusive.

**UV absorption** maxima for different solvents given in the database are useful to predict the potential effect of solvent on UV absorption from sun light. The collection of data is also useful for analytical purposes.

**Solvent partition - activated carbon and between octanol and water**. Solvents can be economically removed from dilute solutions by activated carbon or ion exchange resins. Activated carbon partition coefficient which helps to determine the amount of activated carbon needed to remove a contaminant can be obtained using the following equation:

$$m = \frac{w}{Pc} \qquad [2.3.16]$$

where:

| | |
|---|---|
| w | total weight of solvent in solution |
| P | partition coefficient |
| c | residual concentration of solvent remaining after treatment. |

The octanol/water partition coefficient is the log of solubility of the solvent in water relative to that in octanol. This coefficient is used to estimate biological effects of solvents. It can also be used to estimate the potential usefulness of a solvent extraction from water by any third solvent.

**Soil adsorption constant** is a log of the amount of a solvent absorbed per unit weight of organic carbon in soil or sediment.

**Atmospheric half-life** of solvents due to reaction in the atmosphere with hydroxyl radicals and ozone is a measure of the persistence of particular solvent and its effect on atmospheric pollution.

**Hydroxyl rate constant** is the reaction rate constant of the solvent with hydroxyl radicals in the atmosphere.

**Global warming potential** of a well-mixed gas is defined as the time-integrated commitment to radiative forcing from the instantaneous release of 1 kg of trace gas expressed relative to that from the release of 1 kg of $CO_2$.

**Biodegradation half-life** determines persistence of the solvent in soil. Commercial proprietary solvents mixtures are classified as biodegradable and solvents having known chemical compositions are classified according to the time required for biodegradation to cut their initial mass to half.

**Target organs** most likely affected organs by exposure to solvents. The database contains a list of organs targeted by individual solvents.

**Hazchem Code** was developed in the UK for use by emergency services to determine appropriate actions when dealing with transportation emergencies. It is also a useful to apply as a label on storage tanks. It consists a number and one or two letters. The number informs about the firefighting medium to be used. The first letter gives information on explosion risk, personal protection and action. A second letter (E) may be added if evacuation is required.

# 3

# Production Methods, Properties, and Main Applications

## 3.1 DEFINITIONS AND SOLVENT CLASSIFICATION

George Wypych
ChemTec Laboratories, Inc., Toronto, Canada

Several definitions are needed to classify solvents. These are included in Table 3.1.1

**Table 3.1.1 Definitions**

| Term | Definition |
|---|---|
| Solvent | A substance that dissolves other material(s) to form solution. Common solvents are liquid at room temperature but can be solid (ionic solvents) or gas (carbon dioxide). Solvents are differentiated from plasticizers by limiting their boiling point to a maximum of 250°C. To differentiate solvents from monomers and other reactive materials - a solvent is considered to be non-reactive. |
| Polarity | Polarity is the ability to form two opposite centers in the molecule. The concept is used in solvents to describe their dissolving capabilities or the interactive forces between solvent and solute. Because it depends on dipole moment, hydrogen bonding, entropy, and enthalpy, it is a composite property without a physical definition. The dipole moment has the greatest influence on polar properties of solvents. Highly symmetrical molecules (e.g. benzene) and aliphatic hydrocarbons (e.g. hexane) have no dipole moment and are considered non-polar. Dimethyl sulfoxide, ketones, esters, alcohol are examples of compounds having dipole moments (from high to medium, sequentially) and they are polar, medium polar, and dipolar liquids. |
| Polarizability | The molecules of some solvents are electrically neutral but dipoles can be induced by external electromagnetic field. |
| Normal | A normal solvent does not undergo chemical associations (e.g. the formation of complexes between its molecules). |

| Term | Definition |
|---|---|
| Aprotic/Protic | Aprotic solvents (also commonly called inert) have very little affinity for protons and are incapable to dissociating to give protons. Aprotic solvents are also called indifferent, non-dissociating, or non-ionizing. Protic solvents contain proton-donating groups. |
| Protogenic | An acidic solvent capable of donating protons. |
| Protophilic | A basic solvent able to combine with hydrogen ion or to act as a proton acceptor. |
| Acidic/Basic | Lewis acidity/basicity determines the solvent's ability to donate or accept a pair of electrons to form a coordinate bond with solute and/or between solvent molecules. A scale for this acid/base property was proposed by Gutman (DN and AN - donor and acceptor number, respectively) based on calorimetric determination. The complete proton transfer reaction with formation of protonated ions is determined by proton affinity, gas phase acidity, acid or base dissociation constants. Both concepts differ in terms of net chemical reaction. |
| Hydrogen-bonding | A bond involving a hydrogen atom, which is bound covalently with another atom, is referred to as hydrogen bonding. Two groups are involved: hydrogen donor (e.g., hydroxyl group) and hydrogen acceptor (e.g., carbonyl group). |
| Solvatochromism | Shift of UV/Vis absorption wavelength and intensity in the presence of solvents. A hypsochromic (blue) shift increases as solvent polarity increases. The shift in the red direction is called bathochromic. |
| Dielectric constant | A simple measure of solvent polarity (the electrostatic factor is a product of dielectric constant and dipole moment). The electrical conductivity of solvent indicates if there is a need to earth (or ground) the equipment which handles solvent to prevent static spark ignition. Admixtures affect solvent conductivity. These are most important in electronics industry. |
| Miscible | Solvents are usually miscible when their solubility parameters do not differ by more than 5 units. This general rule does not apply if one solvent is strongly polar. |
| Good solvent | Substances readily dissolve if the solubility parameters of solvent and solute are close (less than 6 units apart). This rule has some exceptions (for example, PVC is not soluble in toluene even though the difference of their solubility parameters is 2.5). |
| Θ solvent | The term relates to the temperature of any polymer/solvent pair at which chain expansion is exactly balanced by chain contraction. At this temperature, called Θ temperature chain dimensions are unperturbed by long-range interactions. |
| Reactivity | Solvent, according to this definition, should be a non-reactive medium but in some processes solvent will be consumed in the reaction to prevent its evaporation (and pollution). Solvents affect reactivity in two major ways: viscosity reduction and decreasing the barrier of Gibbs activation energy. |
| Hygroscopicity | Some solvents such as alcohols and glycols are hygroscopic and, as such, are unsuitable for certain applications which require a moisture-free environment or a predetermined freezing point. Solvents which are not hygroscopic may still contain moisture from dissolved water. |

| Term | Definition |
|---|---|
| Solvent strength | Solvent strength is used to establish required solvent concentration to form a clear solution and to estimate the diluting capabilities of pre-designed system. Two determined quantities are used for the purpose: Kauri butanol value and aniline point. |
| Solvent partition | Solvent partition is determined for three purposes: to estimate the potential for solvent removal from dilute solution by carbon black adsorption, to evaluate the partition of solute between water and solvent for the purpose of studying biological effects of solvents and solutes, and to design system for solvent extraction. |
| Volatility | Solvent volatility helps in estimation of the solvent evaporation rate at temperatures below its boiling point. The Knudsen, Henry, Cox, Antoine, and Clausius-Clapeyron equations are used to estimate the vapor pressure of a solvent over a liquid, its evaporation rate, and the composition of the atmosphere over the solvent. The boiling point of a solvent gives an indication of its evaporation rate but it is insufficient for its accurate estimation because of the influence of the molar enthalpy of evaporation. |
| Residue | This may refer to either the non-volatile residue or the potential for residual solvent left after processing. The former can be estimated from the solvent specification, the later is determined by system and technology design. |
| Carcinogenic | Solvents may belong to a group of carcinogenic substances. Several groups of solvents have representatives in this category (see listings in Section 3.3) |
| Mutagenic | Mutagenic substance causes genetic alterations, such as genetic mutation or a change to the structure and number of chromosomes (mutagens listed in Section 3.3). |
| Impairing reproduction | Several solvents in the glycols and formamides groups are considered to impair fertility. |
| Toxicity | LD50 and LC50 give toxicity in mg per kg of body weight or ppm, respectively. Threshold limit values place a limit on permissible concentration of solvent vapors in the work place. Also "immediate-danger-to-life" and "short-term-exposure-limits" are specified for solvents. Odor threshold values have limited use in evaluating the potential danger to solvent exposure. |
| Flammable | Several data are used to evaluate the dangers of solvent explosion and flammability. Flash point and autoignition temperature are used to determine a solvent's flammability and its potential for ignition. The flash points for hydrocarbons correlate with their initial boiling points. Lower and upper explosive limits determine the safe ranges of solvent concentration. |
| Combustible | The net heat of combustion and the calorific value help to estimate the potential energy which can be recovered from burning used solvents. In addition, the composition of the combustion products is considered to evaluate potential corrosiveness and the effect on the environment. |
| Ozone depleter | Ozone depletion potential is the value relative to that of CFC-11. It represents the amount of ozone destroyed by the emission of gas over its entire atmospheric life-time. Photochemical ozone creation potential is a relative value to that of ethene to form ozone in an urban environment. Numerous solvents belong to both groups. |

| Term | Definition |
|---|---|
| Biodegradability | Several methods are used to express biodegradability. These include biodegradation half-life, biological oxygen demand, chemical and theoretical oxygen demand. |
| Cost | Cost of solvent is a key factor in solvent selection. |

The above list of terms and parameters is not exhaustive. These and related subjects are discussed at length in various parts of the book. The table is presented to assist in an understanding solvent classification.

A review of the selected definitions suggests that there are many important determinants of solvent quality for specific application. Some solvent parameters are conflicting, some not well quantified, and each solvent application requires a unique set of solvent performance criteria. It can be thus anticipated, prior to any analysis, that the chemical structure can be used as the best means of solvent classification for any application. Such a classification is used in this book because of its broad application. Chemical names used are the common names because they are generally understood by all solvents users.

Other means of classification are briefly analyzed below because they are useful in some applications. For a classification to be useful, it must be based on a model and a method which permits its quantification.

In organic synthesis, the solvent's polarity plays an important role. Dimroth and Reichardt[1] developed a classification based on the normalized empirical parameter of solvent polarity, $E_T^N$, given by the following equation:

$$E_T^N = \frac{E_T(solvent) - E_T(TMS)}{E_T(water) - E_T(TMS)} = \frac{E_T(solvent) - 30.7}{32.4} \qquad [3.1.1]$$

where:

$E_T$ excitation energy
TMS tetramethylsilane

The values of $E_T$ are known for several hundred solvents based on measurements of solvent-induced shifts with betaine dye used as the solvatochromic indicator. Based on such data, solvents can be divided into 3 groups: protic ($E_T^N$ from 0.5 to 1), dipolar non-hydrogen donating ($E_T^N$ from 0.3 to 0.5) and apolar ($E_T^N$ from 0 to 0.3). The $E_T^N$ values have a good linear correlation with light absorption, reaction rates, and chemical equilibria. In addition, the $E_T^N$ values have a very good correlation with the Kosower's polarity parameter, Z, for which there is also large amount of data available. Both sets of data can be converted using the following equation:

$$E_T = 0.752Z - 7.87 \qquad [3.1.2]$$

Gutman[2,3] chose the reaction enthalpy of solvent with the reference acceptor (antimony pentachloride) to quantify Lewis-donor properties. The donor number, DN, is a dimensionless parameter obtained from negative values of reaction enthalpy. The data obtained from electrochemical and NMR studies were combined into one scale in which data are available for several hundred solvents. These data have a linear correlation with $E_T^N$ according to the following equation:[4]

$$AN = -59.9 + 1850E_T \quad [3.1.3]$$

The donor number is frequently used in various fields of polymer chemistry (see Chapter 10). Another classification based on acidity/basicity of solvents allows the division of solvents into six groups containing protic-neutral; protogenic; protophilic; aprotic-protophilic; aprotic-protophobic; and aprotic-inert.[4]

Snyder[5,6] developed classification of solvents for chromatography which arranges solvents according to their chromatographic strength. It is classification based on the solvent's ability to engage in hydrogen bonding or dipole interaction using the experimentally determined partition coefficients by Rohrschneider.[7] Eight groups of solvents were defined based on cluster analysis. In addition to the usefulness of this classification in chromatography, it was found recently that it is also useful in the design of coatings which do not affect undercoated paints.[8]

Numerous other classifications and sets of data are available, such as those included in various databases on solvent toxicity, their environmental fate, combustion properties, explosive limits, etc.

## REFERENCES

1 Ch Reichardt, **Solvents and Solvents Effects in Organic Chemistry**, *VCH*, Weinheim, 1988.
2 V Gutman, G Resch, **Lecture Notes on Solution Chemistry**, *Wold Scientific*, Singapore, 1995.
3 G Gritzner, *J. Mol. Liq.*, **73**,74, 487-500 (1997).
4 Y Marcus, **The Properties of Solvents**, *John Wiley & Sons*, Chichester, 1999.
5 L R Snyder, *J. Chromatographic Sci.*, **16**, 223 (1978).
6 S C Rutan, P W Carr, J Cheong, J H Park, L R Snyder, *J. Chromatography*, **463**, 21 (1989).
7 L Rohrschneider, *Anal. Chem.*, **45**, 1241 (1973).
8 I R Owen, **US Patent 5,464,888**, 3M, 1995.

# 3.2 OVERVIEW OF METHODS OF SOLVENT MANUFACTURE

George Wypych
**ChemTec Laboratories, Inc., Toronto, Canada**

Crude oil is the major raw material source for solvents. Aliphatic and aromatic hydrocarbons are produced by physical processes used in petrochemical industry. Other solvents are synthetic but their starting raw materials are usually products of the petrochemical industry. Figure 3.2.1 shows the main groups of materials produced by petrochemical industry from crude oil.

Two observations are pertinent: the main goal of petrochemical industry is to convert crude oil to fuels. Solvents are only a small fraction of materials produced. Figure 3.2.2 shows that solvents are not only used directly as solvents but are also the building blocks in the manufacture of a large number of materials produced by organic chemistry plants.

Desalting of crude oil is the first step in crude oil processing. It is designed to remove corrosive salts which may cause catalyst deactivation. After desalting, crude oil is subjected to atomospheric distillation. Figure 3.2.3 shows a schematic diagram of the crude oil distillation process. The raw material is heated to 400°C and separated into fractions on 30-50 fraction trays in distillation column. The diagram in Figure 3.2.3 shows the main fractions obtained from this distillation. The heavier fraction cannot be distilled under atmospheric

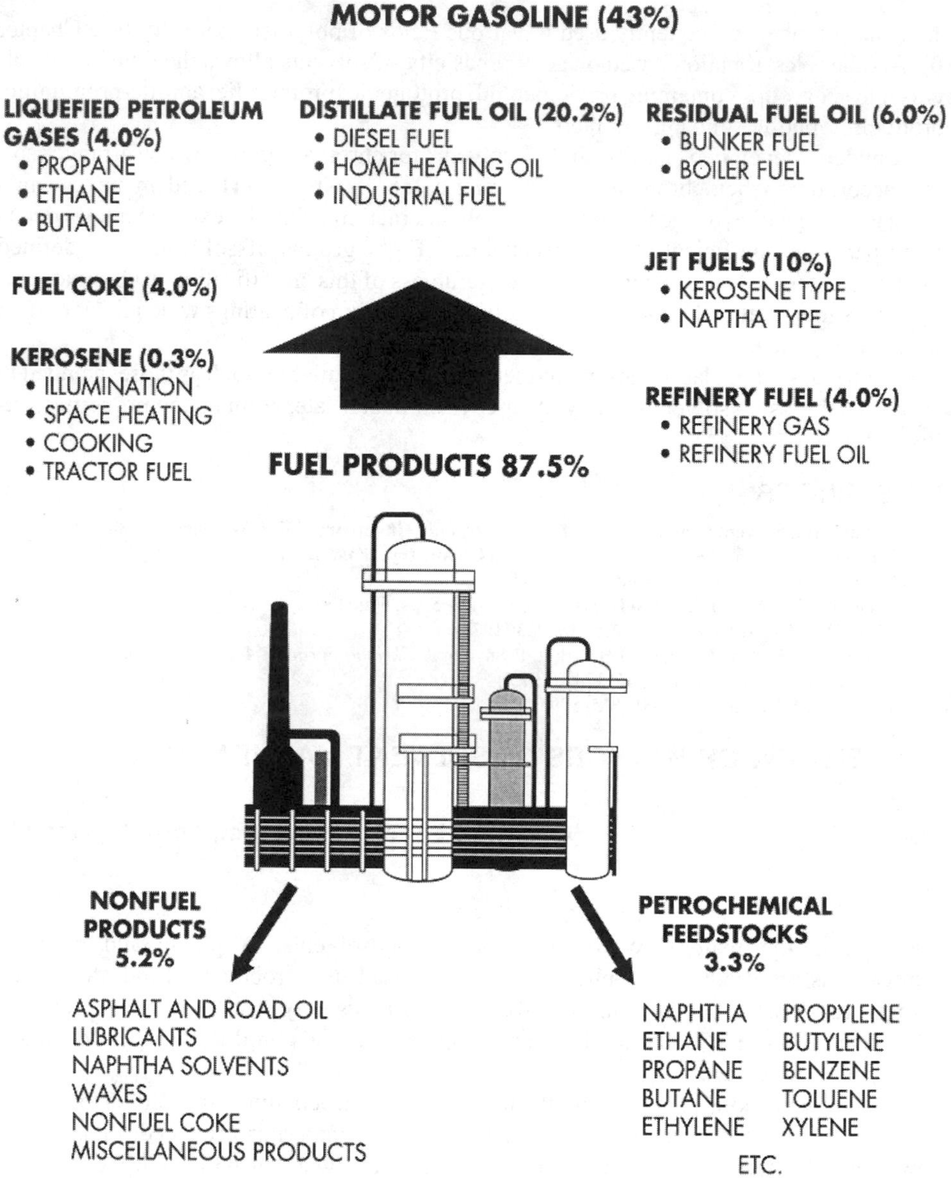


Figure 3.2.1. Schematic diagram of petroleum industry products and yields. [Reproduced from reference 1]

pressure therefore in the next step vacuum is applied to increase volatilization and separation.

Certain fractions from the distillation of crude oil are further refined in thermal cracking (visbreaking), coking, catalytic cracking, catalytic hydrocracking, alkylation,

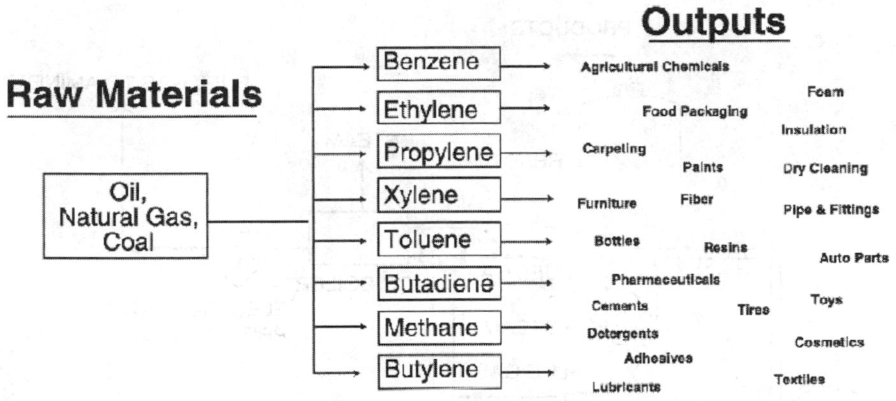


Figure 3.2.2. Organic chemicals and building block flow diagram. [Reproduced from reference 2]

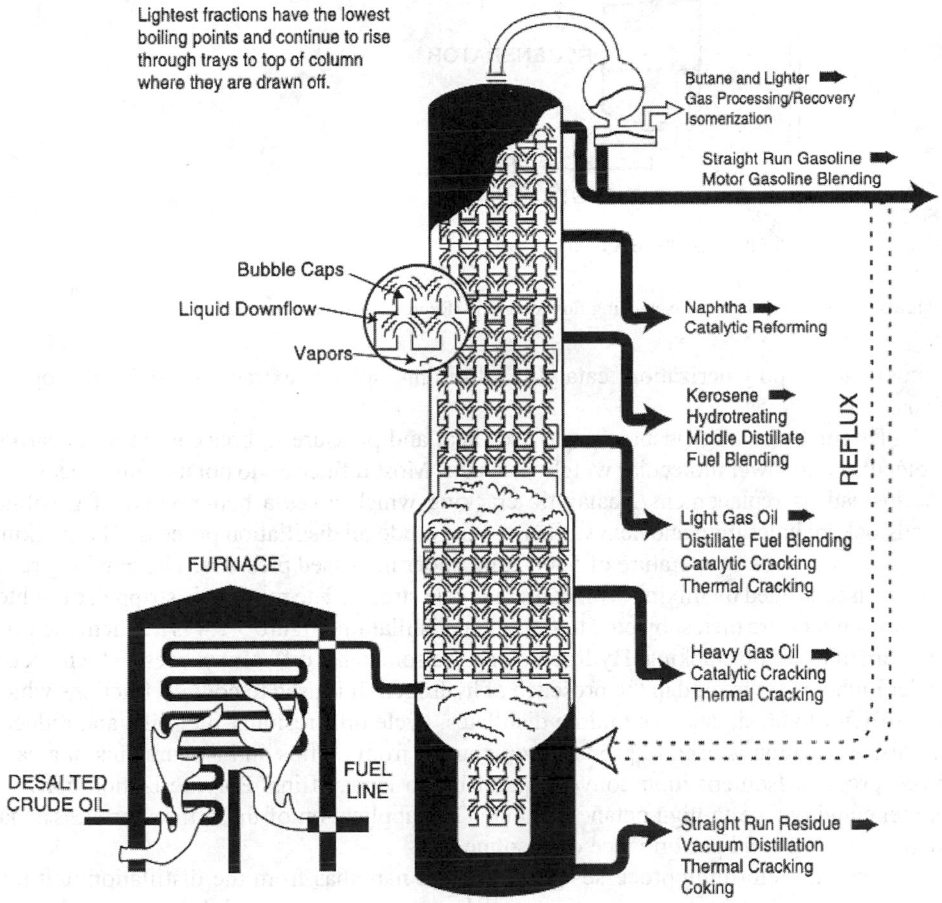


Figure 3.3.3. Crude oil distillation. [Reproduced from reference 1]

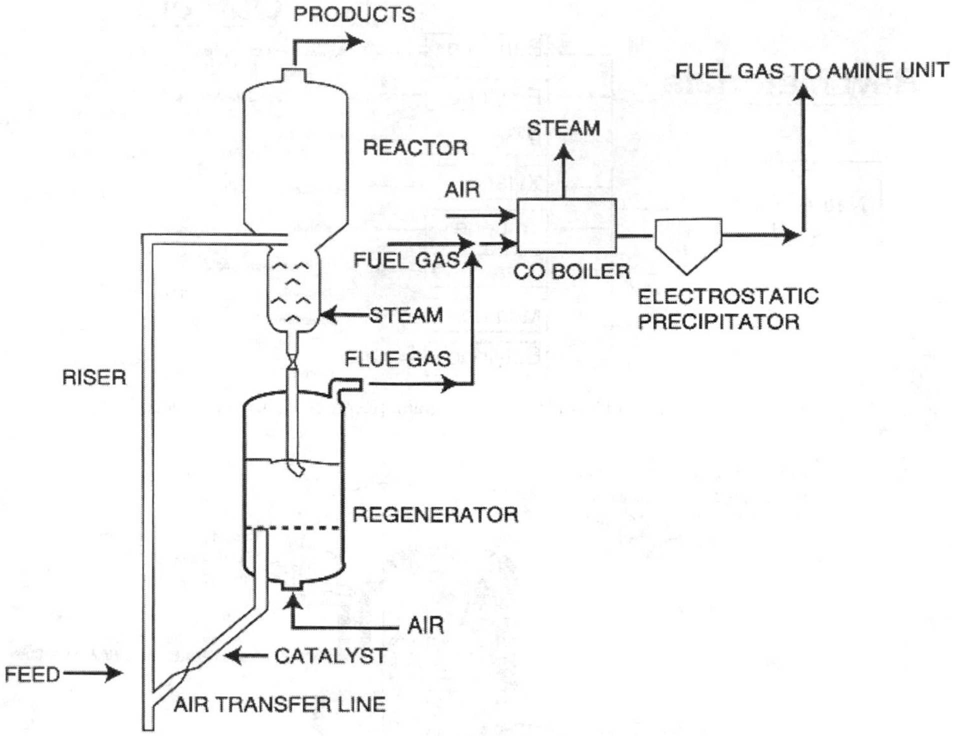


Figure 3.3.4. Simplified catalytic cracking flow diagram. [Reproduced from reference 1]

isomerization, polymerization, catalytic reforming, solvent extraction, and other operations.

Thermal cracking (visbreaking), uses heat and pressure to break large hydrocarbon molecules into lower molecular weight products. Most refineries do not use this process but use instead its replacement – catalytic cracking which gives a better yield of gasoline. Feedstock includes light and heavy oils from the crude oil distillation process. The cracking process occurs at a temperature of 550°C and under increased pressure. The cracking reaction is discontinued by mixing with cooler recycle stream. The mixture is stripped of lighter fractions which are then subjected to fractional distillation. Figure 3.2.4 is a schematic flow diagram of catalytic cracking. Hydrocracking is a somewhat different process which occurs under higher pressure and in the presence of hydrogen. It is used to convert fractions which are difficult to crack, such as middle distillates, cycle oils, residual fuel oils, and reduced crudes. Alkylation is used to produce compounds from olefins and isoparaffins in a catalyzed process. Isomerization converts paraffins to isoparaffins. Polymerization converts propene and butene to high octane gasoline. The application of these three processes has increased output and performance of gasoline.

Catalytic reforming processes gasolines and naphthas from the distillation unit into aromatics. Four major reactions occur: dehydrogenation of naphthenes to aromatics, dehydrocyclization of paraffins to aromatics, isomerization, and hydrocracking.

In some cases mixtures of solvents are required to meet a particular requirement. For example, linear paraffins have a very low viscosity. Branched paraffins have high viscosity but very good low temperature properties and low odor. A combination of the two carried out in the conversion process (not by mixing) results in a solvent which has the desirable properties of both solvents: low viscosity and good low temperature properties.[3]

The recovery of pure aromatics from hydrocarbon mixtures is not possible using distillation process because the boiling points of many non-aromatics are very close to benzene, toluene, etc. Also, azeotropes are formed between aromatics and aliphatics. Three principle methods are used for separation: azeotropic distillation, liquid-liquid extraction, and extractive distillation. Three major commercial processes have been developed for separation: Udex, Sulpholane, and Arosolvan. Over 90% plants now use one of these processes. Each use an addition of solvent such as a mixture of glycols, tetramethylene sulfone, or N-methyl-2-pyrrolidone to aid in the extraction of aromatics. This occurs with high precision and efficiency. Pure benzene, toluene, and xylene are produced by these processes.

These three are used for synthesis of several other important solvents. Benzene is used in the production of ethyl benzene (alkylation), cyclohexanone, cyclohexanol, cyclohexane, aniline (hydrogenation), acetone, nitrobenzene, and chlorobenzene. Toluene is used in the production of cresol and benzene. Xylene is the raw material for the production of ethyl benzene and the fractionation of the xylene mixture to isomers.

Lower boiling fractions from the primary distillation are also used in the production of solvents. Ethylene is used to produce ethylene dichloride, ethylene glycol, ethanol, and ethyl benzene. Propylene is used to produce isopropyl alcohol. Halogenation, hydrohalogenation, alkylation, and hydrolysis reactions are used in these conversions.

With different feedstock and methods of processing it is inevitable that there will be some differences between products coming from different feedstock sources and manufacturers. Over the years processes have been refined. Long practice, globalization of technology has occurred and restrictions have been imposed. Today, these differences are small but in some technological processes even these very small differences in solvent quality may require compensating actions.

There are many other unitary operations which are used by organic chemistry plants to manufacture synthetic solvents. These include: alkoxylation (ethylene glycol), halogenation (1,1,1-trichloethane), catalytic cracking (hexane), pyrolysis (acetone and xylene), hydrodealkylation (xylene), nitration (nitrobenzene), hydrogenation (n-butanol, 1,6-hexanediol), oxidation (1,6-hexanediol), esterification (1,6-hexanediol), and many more.

In the manufacture of oxygenated solvents, the typical chemical reactions are hydration, dehydration, hydrogenation, dehydrogenation, dimerization and esterification. For example methyl ethyl ketone is manufactured from 1-butene in a two step reaction. First, 1-butene is hydrated to 2-butanol then a dehydrogenation step converts it to methyl ethyl ketone. The production of methyl isobutyl ketone requires several steps. First acetone is dimerized producing diacetone alcohol which, after dehydration, gives mesityl oxide subjected in the next step to hydrogenation to result in the final product. Ethylene glycol is a product of the addition (ethylene oxide and ethanol) followed by esterification with acetic acid. 18% of all phenol production is converted to cyclohexanone.[4] Some solvents are obtained by fermentation processes (e.g., ethanol, methanol).

Synthetic routes are usually quite complex. For example, the manufacture of N-methyl-2-pyrrolidone involves the reaction of acetylene with formaldehyde. The resulting but-2-ine-1,4-diol is hydrated to butane-1,4-diol and then dehydrated to γ-butyrolactone then reacted with monomethyl amine to give the final product. Strict process control is essential to obtain very high purity.

New processes have been developed to produce solvents which are based on non-conventional materials (e.g., lactide and drying oil). The resultant solvent is non-volatile and useful in production of coatings, paints and inks.[5] These new technological processes are driven by the need to reduce VOCs.

## REFERENCES


1 EPA Office of Compliance Sector Notebook Project. Profile of the Petroleum Refining Industry. US Environmental Protection Agency, 1995.
2 EPA Office of Compliance Sector Notebook Project. Profile of the Organic Chemical Industry. US Environmental Protection Agency, 1995.
3 R J Wittenbrink, S E Silverberg, D F Ryan, **US Patent 5,906,727**, Exxon Research And Engineering Co., 1999.
4 A M Thayer, *Chem. Eng. News*, **76**, 16, 32-34 (1998).
5 D Westerhoff, **US Patent 5,506,294**, 1996.


## 3.3 SOLVENT PROPERTIES


George Wypych
**ChemTec Laboratories, Inc., Toronto, Canada**


The purpose of this section is to provide analysis of properties of major groups of solvents. The data on individual solvents are included in a separate publication on CD-ROM as a searchable database containing 1145 the most common solvents. The information on properties of solvents is included in 110 fields containing chemical identification of solvent, physical chemical properties, health and safety data, and its environmental fate. Here, the analysis of this data is provided in a form of tables to show the range of properties for different groups of solvents and their strengths and weaknesses. For each group of solvents a separate table is given below. No additional discussion is provided since data are self-explanatory. The data are analyzed in the final table to highlight the best performance of various groups of solvents in different properties.

By their nature, these data have a general meaning and for the exact information on particular solvent full data on CD-ROM should be consulted. For example, in the list of target organs, there are included all organs involved on exposure to solvents included in the group which does not necessary apply to a particular solvent. The data are given to highlight overall performance of entire group which may contain very diverse chemical materials.

One very obvious method of use of this information is to review the list of carcinogens and mutagens to and relate them to actually used in particular application. These solvents should be restricted from use and possibly eliminated or equipment engineered to prevent exposure of workers and release to environment. The comparative tables also provide suggestions as to where to look for suitable substitutes based on physical chemical characteristics and potential health and environmental problems.

Other application of this data is in selection of solvents for new products. The bulk of the data allows to analyze potential requirements critical for application and select group or groups which contain solvents having these properties. Further, based on their health and environmental characteristics suitable candidates can be selected.

The bulk data are also very useful in constructing set of requirements for specification. Many specifications for industrial solvents are very simplistic, which is partially caused by the lack of data provided by manufacturers of these solvents. This may cause potential fear of future problems with such solvents since solvent replacement in formulated product is not always a trivial substitution. It is also possible that solvents for which inadequate data exist at the present moment will be more studied in future for their environmental and health impacts and will then require to be replaced. Also, incomplete specification means that solvents may contain undesirable admixtures and contaminations and were used on premisses that manufacturer considered its application for a particular purpose which was, in fact, never intended by solvent manufacturer for this particular application. These details should be obtained from manufacturers and relevant parameters included in the specification. It is known from experience of creation of the database of solvents that there are still large gaps in information which should be eliminated by future efforts. The practice of buying solvents based on their boiling point and specific gravity does not serve the purpose of selecting reliable range of raw materials.

These and other aspects of created reference tables should be continuously updated in future to provide a reliable base of data which will be broadly used by industry. Application of full information allows to decrease quantity of solvents required for task and eliminate questionable materials and wastes due to solvent evaporation too rapid to make an impact on product properties at the time of its application. Also many sources of the problems with formulated products are due to various manifestations of incompatibility which can be eliminated (or predicted) based on solvent's characteristics.

## 3.3.1 HYDROCARBONS

### 3.3.1.1 Aliphatic hydrocarbons

| Property | Value | | |
|---|---|---|---|
| | minimum | maximum | median |
| boiling temperature, °C | -11.7 | 285 | 124 |
| freezing temperature, °C | -189 | 18 | -75 |
| flash point, °C | -104 | 129 | 46 |
| autoignition temperature, °C | 202 | 640 | 287 |
| refractive index | 1.29 | 1.46 | 1.41 |
| specific gravity, g/cm$^3$ | 0.51 | 0.84 | 0.74 |
| vapor density (air=1) | 1 | 5.90 | 4.5 |
| vapor pressure, kPa | 0.00 | 1976 | 4.42 |
| viscosity, mPa.s | 0.21 | 1.58 | 0.46 |
| surface tension, mN/m | 15 | 40 | 21 |

| Property | Value | | |
|---|---|---|---|
| | minimum | maximum | median |
| donor number, DN, kcal/mol | 0 | 0 | 0 |
| acceptor number, AN | 0 | 1.6 | 0 |
| polarity parameter, $E_T(30)$, kcal/mol | 30.9 | 31.1 | 31 |
| coefficient of cubic expansion, $10^{-4}$/°C | 8 | 10 | 9 |
| specific heat, cal/K mol | 41.24 | 93.41 | 60.33 |
| heat of vaporization, cal/g | 6.32 | 8.47 | 7.54 |
| heat of combustion, MJ/kg | 4.35 | 49.58 | 46.52 |
| dielectric constant | 1.8 | 2.15 | 2 |
| Kauri-butanol number | 22 | 56 | 32 |
| aniline point, °C | 21 | 165 | 81 |
| Hildebrand solubility parameter, $cal^{1/2}$ $cm^{-3/2}$ | 6.8 | 8.2 | 7.4 |
| Henry's Law constant, atm/$m^3$ mol | 2.59E-4 | 4.56E-4 | 3.43E-4 |
| evaporation rate (butyl acetate = 1) | 0.006 | 17.5 | 0.45 |
| threshold limiting value - 8h average, ppm | 0.1 | 1000 | 300 |
| maximum concentration (15 min exp), ppm | 375 | 1000 | 500 |
| LD50 oral, mg/kg | 218 | 29820 | 2140 |
| route of entry | absorption, contact. ingestion, inhalation | | |
| target organs | blood, bone marrow, central nervous system, eye, gastrointestinal tract, heart, kidney, lymphatic system, liver, lung, nervous system, peripheral nervous system, respiratory system, skin, spleen, stomach, testes, thyroid | | |
| carcinogenicity | - | | |
| mutagenic properties | n-hexane | | |
| theoretical oxygen demand, g/g | 3.46 | 3.56 | 3.53 |
| biodegradation probability | days-weeks | | |
| octanol/water partition coefficient | 2.3 | 5.98 | |
| urban ozone formation | 0.11 | 0.13 | 0.12 |

### 3.3.1.2 Aromatic hydrocarbons

| Property | Value | | |
|---|---|---|---|
| | minimum | maximum | median |
| boiling temperature, °C | 74 | 288 | 168 |

| Property | Value | | |
|---|---|---|---|
| | minimum | maximum | median |
| freezing temperature, °C | -96 | 5.5 | -31 |
| flash point, °C | -11 | 144 | 52 |
| autoignition temperature, °C | 204 | 550 | 480 |
| refractive index | 1.43 | 1.61 | 1.5 |
| specific gravity, $g/cm^3$ | 0.71 | 1.02 | 0.87 |
| vapor density (air=1) | 2.8 | 4.8 | 4.1 |
| vapor pressure, kPa | 0.00 | 21.3 | 0.43 |
| viscosity, mPa.s | 0.58 | 6.3 | 1.04 |
| surface tension, mN/m | 24.3 | 36.8 | 28 |
| donor number, DN, kcal/mol | 0.1 | 10 | 5 |
| acceptor number, AN | 6.8 | 8.2 | 7.3 |
| polarity parameter, $E_T(30)$, kcal/mol | 32.9 | 34.8 | 34.7 |
| coefficient of cubic expansion, $10^{-4}$/°C | 8 | 10.7 | 8 |
| specific heat, cal/K mol | 32.44 | 63.48 | 52.41 |
| heat of vaporization, cal/g | 8.09 | 10.38 | 10.13 |
| heat of combustion, MJ/kg | 41.03 | 43.5 | 41.49 |
| dielectric constant | 2.04 | 2.6 | 2.3 |
| Kauri-butanol number | 33 | 112 | 86 |
| aniline point, °C | 7 | 85 | 26 |
| Hildebrand solubility parameter, $cal^{1/2} cm^{-3/2}$ | 7.9 | 9.3 | 8.8 |
| Henry's Law constant, $atm/m^3$ mol | 5.19E-3 | 3.8E-1 | 7.6E-3 |
| evaporation rate (butyl acetate = 1) | 0.006 | 5.1 | 0.16 |
| threshold limiting value - 8h average, ppm | 0.3 | 100 | 50 |
| maximum concentration (15 min exp), ppm | 6 | 150 | 125 |
| LD50 oral, mg/kg | 636 | 6989 | 4300 |
| LD50 dermal, mg/kg | 4400 | 17800 | 12400 |
| route of entry | absorption, contact. ingestion, inhalation | | |
| target organs | blood, bone marrow, central nervous system, eye, gastrointestinal tract, kidney, liver, lung, respiratory system, skin | | |
| carcinogenicity | benzene, styrene | | |
| mutagenic properties | benzene, ethylbenzene, xylene | | |

| Property | Value | | |
|---|---|---|---|
| | minimum | maximum | median |
| biological oxygen demand, 5-day test, g/g | 0.92 | 2.53 | 1.65 |
| chemical oxygen demand, g/g | 2.15 | 2.91 | 2.56 |
| theoretical oxygen demand, g/g | 2.41 | 3.29 | 3.17 |
| biodegradation probability | days-to weeks, weeks | | |
| octanol/water partition coefficient | 2.13 | 4.83 | |
| urban ozone formation | 0.03 | 1.13 | 0.90 |

## 3.3.2 HALOGENATED HYDROCARBONS

| Property | Value | | |
|---|---|---|---|
| | minimum | maximum | median |
| boiling temperature, °C | -40.6 | 253 | 87 |
| freezing temperature, °C | -189 | 17 | -36 |
| flash point, °C | -50 | 350 | 45 |
| autoignition temperature, °C | 240 | 648 | 557 |
| refractive index | 1.20 | 1.63 | 1.43 |
| specific gravity, $g/cm^3$ | 0.9 | 3 | 1.6 |
| vapor density (air=1) | 1.8 | 33.4 | 6.5 |
| vapor pressure, kPa | 0.01 | 4000 | 10.2 |
| viscosity, mPa.s | 0.02 | 5.14 | 1.1 |
| surface tension, mN/m | 0.03 | 33.4 | 15.2 |
| pH | 4 | 8 | 7 |
| donor number, DN, kcal/mol | 0 | 4 | 3 |
| acceptor number, AN | 8.6 | 23.1 | 16.2 |
| polarity parameter, $E_T(30)$, kcal/mol | 32.1 | 41.3 | 36.7 |
| specific heat, cal/K mol | 12.32 | 205 | 41.5 |
| heat of vaporization, cal/g | 3.76 | 81.2 | 11.54 |
| heat of combustion, MJ/kg | 6.27 | 29.14 | 15.73 |
| dielectric constant | 1.0 | 8.93 | 2.32 |
| Kauri-butanol number | 31 | 500 | 90 |
| Hildebrand solubility parameter, $cal^{1/2}$ $cm^{-3/2}$ | 5.9 | 10.75 | 8.3 |
| Henry's Law constant, $atm/m^3$ mol | 3.6E-4 | 8.5E0 | 3E-2 |
| evaporation rate (butyl acetate = 1) | 0.9 | 14.5 | 1 |

| Property | Value | | |
|---|---|---|---|
| | minimum | maximum | median |
| threshold limiting value - 8h average, ppm | 0.5 | 1000 | 100 |
| maximum concentration (15 min exp), ppm | 20 | 1250 | 250 |
| maximum concentration any time, ppm | 5 | 500 | 400 |
| LD50 oral, mg/kg | 214 | 13000 | 1210 |
| LD50 dermal, mg/kg | 500 | 20000 | 8750 |
| route of entry | absorption, contact. ingestion, inhalation | | |
| target organs | central nervous system, eye, gastrointestinal tract, heart, kidney, liver, lung, respiratory system, skin | | |
| carcinogenicity | benzotrichloride, carbon tetrachloride,chloroform, 1,2-dibromomethane, 1,4-dichlorobenzene, 1,2-dichloroethane, Dowper, 1,1,2,2-tetrachlroethylene | | |
| mutagenic properties | benzotrichloride, carbon tetrachloride, chloroform, chloromethane, chlorodifluoromethane, dichloromethane, 1,2-dibromomethane, Freon MS-117 TE, Freon MS-178 TES, 1,1,2,2-tetrachloroethylene, 1,1,1-trichloroethane, 1,1,2-trichloroethylene, trifluoromethane | | |
| theoretical oxygen demand, g/g | 0 | 0.19 | 0.09 |
| biodegradation probability | weeks | | |
| octanol/water partition coefficient | 0.64 | 4.02 | |
| ozone depletion potential | 0.00 | 1.1 | 0.8 |
| global warming potential | 0.25 | 11700 | 4600 |
| urban ozone formation | 0.00 | 0.09 | 0.01 |

## 3.3.3 NITROGEN-CONTAINING COMPOUNDS (NITRATES, NITRILES)

| Property | Value | | |
|---|---|---|---|
| | minimum | maximum | median |
| boiling temperature, °C | 77 | 234 | 134 |
| freezing temperature, °C | -112 | 6 | -50 |
| flash point, °C | 2 | 101 | 36 |
| autoignition temperature, °C | 414 | 550 | 481 |
| refractive index | 1.34 | 1.55 | 1.39 |
| specific gravity, g/cm$^3$ | 0.78 | 1.21 | 0.87 |
| vapor density (air=1) | 1.4 | 4.2 | 3.1 |

| Property | Value | | |
|---|---|---|---|
| | minimum | maximum | median |
| vapor pressure, kPa | 0.01 | 11.0 | 2.62 |
| viscosity, mPa.s | 0.34 | 1.96 | 0.77 |
| dissociation constant, pKa | 7.67 | 10.21 | 8.98 |
| donor number, DN, kcal/mol | 4.4 | 16.6 | 11.0 |
| acceptor number, AN | 14.8 | 20.5 | 17.7 |
| polarity parameter, $E_T(30)$, kcal/mol | 41.5 | 46.7 | 43.6 |
| heat of combustion, MJ/kg | 22.46 | 30.38 | 27.25 |
| Hildebrand solubility parameter, $cal^{1/2}$ $cm^{-3/2}$ | 9.5 | 12.3 | 10.5 |
| evaporation rate (butyl acetate = 1) | 1.15 | 2.3 | 2.1 |
| threshold limiting value - 8h average, ppm | 1 | 100 | 25 |
| LD50 oral, mg/kg | 39 | 3800 | 510 |
| route of entry | absorption, contact. ingestion, inhalation | | |
| target organs | blood, central nervous system, eye, kidney, liver, respiratory system, skin | | |
| carcinogenicity | acrylonitrile, 2-nitropropane | | |
| mutagenic properties | acrylonitrile, 2-nitropropane | | |
| biodegradation probability | days-weeks | | |
| octanol/water partition coefficient | -0.3 | 1.86 | |

## 3.3.4 ORGANIC SULFUR COMPOUNDS

| Property | Value | | |
|---|---|---|---|
| | minimum | maximum | median |
| boiling temperature, °C | 37 | 287 | 142 |
| freezing temperature, °C | -116 | 32 | -38 |
| flash point, °C | -38 | 177 | 43 |
| autoignition temperature, °C | 206 | 470 | 395 |
| refractive index | 1.38 | 1.62 | 1.47 |
| specific gravity, $g/cm^3$ | 0.80 | 1.43 | 1.00 |
| vapor density (air=1) | 2.14 | 4.35 | 3.05 |
| vapor pressure, kPa | 0.00 | 19.00 | 1.05 |
| viscosity, mPa.s | 0.28 | 10.29 | 0.97 |
| surface tension, mN/m | 35.5 | 42.98 | 39.00 |

| Property | Value | | |
|---|---|---|---|
| | minimum | maximum | median |
| dissociation constant, pKa | -1.54 | 15.3 | 13.6 |
| donor number, DN, kcal/mol | 2 | 41 | 29.8 |
| acceptor number, AN | 7.5 | 19.3 | 19.2 |
| polarity parameter, $E_T(30)$, kcal/mol | 26.8 | 54.4 | 38.4 |
| specific heat, cal/K mol | 36.61 | 43 | 40.1 |
| dielectric constant | 43.26 | 46.45 | 44.30 |
| Hildebrand solubility parameter, $cal^{1/2}$ $cm^{-3/2}$ | 8.2 | 12.6 | 9.8 |
| Henry's Law constant, atm/$m^3$ mol | 4.96E-8 | 4.85E-6 | 1.25E-6 |
| evaporation rate (butyl acetate = 1) | 0.005 | 0.026 | 0.013 |
| threshold limiting value - 8h average, ppm | 0.1 | 10 | 0.5 |
| LD50 oral, mg/kg | 505 | 14500 | 1941 |
| LD50 dermal, mg/kg | 380 | 40000 | 20000 |
| route of entry | absorption, contact. ingestion, inhalation | | |
| target organs | central nervous system, eye, liver, lung | | |
| carcinogenicity | diethyl sulfate, dimethyl sulfate | | |
| mutagenic properties | diethyl sulfate, dimethyl sulfoxide, sulfolane | | |
| theoretical oxygen demand, g/g | 1.73 | 1.84 | 1.75 |
| biodegradation probability | days-weeks | | |
| octanol/water partition coefficient | -1.35 | 2.28 | |
| urban ozone formation | 0.07 | 0.23 | 0.15 |

## 3.3.5 MONOHYDRIC ALCOHOLS

| Property | Value | | |
|---|---|---|---|
| | minimum | maximum | median |
| boiling temperature, °C | 64.55 | 259 | 155 |
| freezing temperature, °C | -129 | 71 | -38.6 |
| flash point, °C | 11 | 156 | 67 |
| autoignition temperature, °C | 231 | 470 | 295 |
| refractive index | 1.277 | 1.539 | 1.42 |
| specific gravity, g/$cm^3$ | 0.79 | 1.51 | 0.81 |
| vapor density (air=1) | 1.10 | 5.50 | 3.0 |
| vapor pressure, kPa | 0.00 | 21.20 | 0.4 |

| Property | Value | | |
|---|---|---|---|
| | minimum | maximum | median |
| viscosity, mPa.s | 0.59 | 41.1 | 4.4 |
| surface tension, mN/m | 21.99 | 40.0 | 26.2 |
| dissociation constant, pKa | 9.3 | 19.0 | 15.4 |
| donor number, DN, kcal/mol | 5 | 44 | 30 |
| acceptor number, AN | 22.2 | 66.7 | 37.1 |
| polarity parameter, $E_T(30)$, kcal/mol | 41 | 65.3 | 48.8 |
| coefficient of cubic expansion, $10^{-4}$/°C | 9 | 12.2 | 10.3 |
| specific heat, cal/K mol | 19.47 | 78.03 | 43.03 |
| heat of vaporization, cal/g | 8.95 | 15.40 | 12.32 |
| heat of combustion, MJ/kg | 22.66 | 38.83 | 34.56 |
| dielectric constant | 8.17 | 32.66 | 17.51 |
| Hildebrand solubility parameter, $cal^{1/2}$ $cm^{-3/2}$ | 9.26 | 23 | 11.5 |
| Henry's Law constant, atm/$m^3$ mol | 4.1E-9 | 3.44E+1 | 3.1E-5 |
| evaporation rate (butyl acetate = 1) | 0.005 | 2.9 | 0.39 |
| threshold limiting value - 8h average, ppm | 1 | 1000 | 100 |
| maximum concentration (15 min exp), ppm | 4 | 500 | 125 |
| LD50 oral, mg/kg | 275 | 50000 | 2300 |
| LD50 dermal, mg/kg | 400 | 20000 | 3540 |
| route of entry | absorption, contact. ingestion, inhalation | | |
| target organs | central nervous system, eye, kidney, liver, lung, lymphatic system, respiratory system, skin | | |
| carcinogenicity | - | | |
| mutagenic properties | 1-butanol, 2-butanol, ethanol, 1-octanol, 1-pentanol, 1-propanol | | |
| biological oxygen demand, 5-day test, g/g | 0.41 | 2.37 | 1.5 |
| chemical oxygen demand, g/g | 1.5 | 2.97 | 2.46 |
| theoretical oxygen demand, g/g | 1.5 | 2.9 | 2.59 |
| biodegradation probability | days-weeks | | |
| octanol/water partition coefficient | -1.57 | 2.97 | |
| urban ozone formation | 0.04 | 0.45 | 0.16 |

## 3.3.6 POLYHYDRIC ALCOHOLS

| Property | Value | | |
|---|---|---|---|
| | minimum | maximum | median |
| boiling temperature, °C | 171 | 327.3 | 214 |
| freezing temperature, °C | -114 | 60 | -4 |
| flash point, °C | 85 | 274 | 152 |
| autoignition temperature, °C | 224 | 490 | 371 |
| refractive index | 1.43 | 1.48 | 1.44 |
| specific gravity, $g/cm^3$ | 0.92 | 1.22 | 1.12 |
| vapor density (air=1) | 2.14 | 6.70 | 3.10 |
| vapor pressure, kPa | 0.00 | 0.32 | 0.01 |
| viscosity, mPa.s | 21 | 114.6 | 54.65 |
| surface tension, mN/m | 33.1 | 48.49 | 44.13 |
| dissociation constant, pKa | 14.1 | 15.1 | 14.5 |
| donor number, DN, kcal/mol | 19 | 20 | 19 |
| acceptor number, AN | 34.5 | 46.6 | 36.2 |
| polarity parameter, $E_T(30)$, kcal/mol | 51.8 | 56.3 | 54.1 |
| specific heat, cal/K mol | 36.1 | 294 | 77.6 |
| heat of vaporization, cal/g | 13.0 | 18.7 | 16.2 |
| heat of combustion, MJ/kg | 19.16 | 29.86 | 23.69 |
| dielectric constant | 7.7 | 35.0 | 28.8 |
| Hildebrand solubility parameter, $cal^{1/2}$ $cm^{-3/2}$ | 10.7 | 16.18 | 12.81 |
| Henry's Law constant, $atm/m^3$ mol | 4.91E-13 | 2.3E-7 | 6E-8 |
| evaporation rate (butyl acetate = 1) | 0.001 | 0.01 | 0.01 |
| threshold limiting value - 8h average, ppm | 1 | 25 | 10 |
| maximum concentration any time, ppm | 25 | 50 | |
| LD50 oral, mg/kg | 105 | 50000 | 16000 |
| LD50 dermal, mg/kg | 2000 | 225000 | 20000 |
| route of entry | absorption, contact. ingestion, inhalation | | |
| target organs | blood, eye, gastrointestinal tract, kidney, lymphatic system, liver, lung, respiratory system, skin, spleen | | |
| carcinogenicity | - | | |

| Property | Value | | |
|---|---|---|---|
| | minimum | maximum | median |
| mutagenic properties | tetraethylene glycol, triethylene glycol, trimethylene glycol | | |
| biological oxygen demand, 5-day test, g/g | 0.03 | 1.08 | 0.18 |
| chemical oxygen demand, g/g | 1.29 | 1.64 | 1.57 |
| theoretical oxygen demand, g/g | 1.07 | 1.68 | 1.60 |
| biodegradation probability | days-weeks | | |
| octanol/water partition coefficient | -0.92 | -2.02 | |
| urban ozone formation | 0.16 | 0.47 | 0.38 |

## 3.3.7 PHENOLS

| Property | Value | | |
|---|---|---|---|
| | minimum | maximum | median |
| boiling temperature, °C | 182 | 245 | 202 |
| freezing temperature, °C | -18 | 105 | 25 |
| flash point, °C | 43 | 127 | 95 |
| autoignition temperature, °C | 558 | 715 | 599 |
| refractive index | 1.52 | 1.60 | 1.54 |
| specific gravity, g/cm$^3$ | 0.93 | 1.34 | 1.02 |
| vapor density (air=1) | 3.2 | 4.4 | 3.7 |
| vapor pressure, kPa | 0.00 | 0.23 | 0.02 |
| viscosity, mPa.s | 3.5 | 11.55 | 9.4 |
| dissociation constant, pKa | 9.1 | 10.85 | 10.3 |
| donor number, DN, kcal/mol | 11 | | |
| acceptor number, AN | 44.8 | 50.4 | |
| polarity parameter, $E_T(30)$, kcal/mol | 50.3 | 60.8 | 53.3 |
| Hildebrand solubility parameter, cal$^{1/2}$ cm$^{-3/2}$ | 8.7 | 12.1 | 10.6 |
| Henry's Law constant, atm/m$^3$ mol | 3.84E-11 | 3.14E-9 | |
| threshold limiting value - 8h average, ppm | 5 | 5 | 5 |
| LD50 oral, mg/kg | 40 | 320000 | 810 |
| LD50 dermal, mg/kg | 950 | 1040 | |
| route of entry | absorption, contact. ingestion, inhalation | | |

| Property | Value | | |
|---|---|---|---|
| | minimum | maximum | median |
| target organs | central nervous system, eye, respiratory system, skin | | |
| carcinogenicity | 3-chlorophenol, o-chlorophenol | | |
| mutagenic properties | 3-chlorophenol, o-chlorophenol | | |
| octanol/water partition coefficient | 0.59 | 2.47 | |

## 3.3.8 ALDEHYDES

| Property | Value | | |
|---|---|---|---|
| | minimum | maximum | median |
| boiling temperature, °C | -21 | 253 | 162 |
| freezing temperature, °C | -123 | 12.4 | -86 |
| flash point, °C | -39 | 102 | 13 |
| autoignition temperature, °C | 180 | 424 | 196 |
| refractive index | 1.33 | 1.62 | 1.44 |
| specific gravity, $g/cm^3$ | 0.70 | 1.25 | 0.85 |
| vapor density (air=1) | 1 | 4.5 | 2.5 |
| vapor pressure, kPa | 0.00 | 438 | 2.30 |
| viscosity, mPa.s | 0.32 | 5.4 | 1.32 |
| surface tension, mN/m | 23.14 | 41.1 | 32.00 |
| donor number, DN, kcal/mol | | | 16 |
| acceptor number, AN | | | 12.8 |
| heat of vaporization, cal/g | 5.53 | 10.33 | 6.77 |
| Hildebrand solubility parameter, $cal^{1/2}$ $cm^{-3/2}$ | 8.33 | 11.7 | 10.85 |
| evaporation rate (butyl acetate = 1) | | | 7.8 |
| threshold limiting value - 8h average, ppm | 0.1 | 100 | 2 |
| maximum concentration (15 min exp), ppm | 0.3 | 150 | |
| maximum concentration any time, ppm | 0.2 | 0.3 | |
| LD50 oral, mg/kg | 46 | 3078 | 100 |
| LD50 dermal, mg/kg | 270 | 16000 | 582 |
| route of entry | absorption, contact. ingestion, inhalation | | |
| target organs | eye, heart, liver, respiratory system, skin | | |
| carcinogenicity | acetaldehyde, formaldehyde, furfural | | |

| Property | Value | | |
|---|---|---|---|
| | minimum | maximum | median |
| mutagenic properties | acrolein, formaldehyde, furfural | | |
| biological oxygen demand, 5-day test, g/g | 0.00 | 0.77 | 0.74 |
| chemical oxygen demand, g/g | 1.72 | | |
| theoretical oxygen demand, g/g | 1.07 | 2.00 | 1.67 |
| biodegradation probability | days-weeks | | |
| octanol/water partition coefficient | 0.35 | 1.48 | |
| urban ozone formation | 0.94 | 1.55 | 1.23 |

## 3.3.9 ETHERS

| Property | Value | | |
|---|---|---|---|
| | minimum | maximum | median |
| boiling temperature, °C | 34.4 | 289 | 104 |
| freezing temperature, °C | -137 | 64 | -58 |
| flash point, °C | -46 | 135 | 25 |
| autoignition temperature, °C | 189 | 618 | 429 |
| refractive index | 1.35 | 1.57 | 1.42 |
| specific gravity, g/cm$^3$ | 0.71 | 1.21 | 0.89 |
| vapor density (air=1) | 1.5 | 6.4 | 4.0 |
| vapor pressure, kPa | 0.00 | 174.7 | 1.33 |
| viscosity, mPa.s | 0.24 | 1.1 | 0.42 |
| surface tension, mN/m | 17.4 | 38.8 | 24.8 |
| dissociation constant, pKa | -5.4 | -2.08 | -2.92 |
| donor number, DN, kcal/mol | 6 | 24 | 19 |
| acceptor number, AN | 3.3 | 10.8 | 8 |
| polarity parameter, $E_T(30)$, kcal/mol | 16 | 43.1 | 36 |
| specific heat, cal/K mol | 28.77 | 60.22 | 45.86 |
| heat of vaporization, cal/g | 5.99 | 13.1 | 10.57 |
| heat of combustion, MJ/kg | 34.69 | 38.07 | 36.58 |
| dielectric constant | 2.2 | 13.0 | 4.5 |
| Hildebrand solubility parameter, cal$^{1/2}$ cm$^{-3/2}$ | 7 | 10.5 | 9.2 |
| Henry's Law constant, atm/m$^3$ mol | 54E-9 | 8.32E0 | 3.19E-4 |
| evaporation rate (butyl acetate = 1) | 0.004 | 11.88 | 8.14 |

| Property | Value | | |
|---|---|---|---|
| | minimum | maximum | median |
| threshold limiting value - 8h average, ppm | 1 | 1000 | 200 |
| maximum concentration (15 min exp), ppm | 2 | 500 | 250 |
| LD50 oral, mg/kg | 72 | 30900 | 4570 |
| LD50 dermal, mg/kg | 250 | 2000 | 7600 |
| route of entry | absorption, contact. ingestion, inhalation | | |
| target organs | blood, central nervous system, eye, kidney, liver, respiratory system, skin | | |
| carcinogenicity | bis(chloromethyl) ether, chloromethyl methyl ether, 1,4-dioxane, epichlorohydrin, ethylene oxide, propylene oxide | | |
| mutagenic properties | diethyl ether, 1,4-dioxane, ethylene oxide, propylene oxide | | |
| biological oxygen demand, 5-day test, g/g | 0.06 | 0.48 | 0.19 |
| chemical oxygen demand, g/g | 1.74 | 1.75 | 1.75 |
| theoretical oxygen demand, g/g | 1.07 | 2.95 | 2.21 |
| biodegradation probability | days-weeks | | |
| octanol/water partition coefficient | -0.56 | 5.10 | |
| urban ozone formation | 0.02 | 0.49 | 0.31 |

## 3.3.10 GLYCOL ETHERS

| Property | Value | | |
|---|---|---|---|
| | minimum | maximum | median |
| boiling temperature, °C | 117 | 265 | 191 |
| freezing temperature, °C | -148 | 14 | -83 |
| flash point, °C | 27 | 143 | 85 |
| autoignition temperature, °C | 174 | 406 | 255 |
| refractive index | 1.39 | 1.53 | 1.43 |
| specific gravity, g/cm$^3$ | 0.83 | 1.11 | 0.95 |
| vapor density (air=1) | 3.00 | 8.01 | 5.25 |
| vapor pressure, kPa | 0.00 | 1.33 | 0.12 |
| viscosity, mPa.s | 0.7 | 20.34 | 3.3 |
| surface tension, mN/m | 25.6 | 42.0 | 28.5 |
| acceptor number, AN | 9 | 36.1 | |

| Property | Value | | |
|---|---|---|---|
| | minimum | maximum | median |
| polarity parameter, $E_T(30)$, kcal/mol | 38.6 | 52 | 51 |
| coefficient of cubic expansion, $10^{-4}$/°C | 9.7 | 11.5 | 11.2 |
| specific heat, cal/K mol | 24.85 | 108 | 65.27 |
| heat of vaporization, cal/g | 10.33 | 14.3 | 13.3 |
| heat of combustion, MJ/kg | 24.3 | 30.54 | 28.75 |
| dielectric constant | 5.1 | 29.6 | 10.5 |
| Hildebrand solubility parameter, $cal^{1/2}$ $cm^{-3/2}$ | 8.2 | 12.2 | 8.8 |
| Henry's Law constant, atm/$m^3$ mol | 6.5E-10 | 7.3E-5 | 7.3E-8 |
| evaporation rate (butyl acetate = 1) | 0.001 | 1.05 | 0.37 |
| threshold limiting value - 8h average, ppm | 5 | 100 | 25 |
| maximum concentration (15 min exp), ppm | 150 | | 150 |
| LD50 oral, mg/kg | 470 | 16500 | 6500 |
| route of entry | absorption, contact. ingestion, inhalation | | |
| target organs | blood, brain, central nervous system, eye, kidney, lymphatic system, liver, lung, respiratory system, skin, spleen, testes | | |
| carcinogenicity | - | | |
| mutagenic properties | diethylene glycol monobutyl ether, diethylene glycol dimethyl ether, 2-ethoxyethanol, ethylene glycol diethyl ether, ethylene glycol monophenyl ether, triethylene glycol dimethyl ether | | |
| biological oxygen demand, 5-day test, g/g | 0.12 | 1.18 | 0.71 |
| chemical oxygen demand, g/g | 1.69 | 2.20 | 1.85 |
| theoretical oxygen demand, g/g | 1.07 | 3.03 | 2.17 |
| biodegradation probability | days-weeks | | |
| octanol/water partition coefficient | -1.57 | 3.12 | |
| urban ozone formation | 0.27 | 0.58 | 0.44 |

## 3.3.11 KETONES

| Property | Value | | |
|---|---|---|---|
| | minimum | maximum | median |
| boiling temperature, °C | 56.1 | 306 | 147 |
| freezing temperature, °C | -92 | 28 | -55 |

| Property | Value | | |
|---|---|---|---|
| | minimum | maximum | median |
| flash point, °C | -18 | 143 | 44 |
| autoignition temperature, °C | 393 | 620 | 465 |
| refractive index | 1.35 | 1.55 | 1.41 |
| specific gravity, $g/cm^3$ | 0.74 | 1.19 | 0.82 |
| vapor density (air=1) | 2 | 4.9 | 3.5 |
| vapor pressure, kPa | 0.00 | 30.8 | 1.1 |
| viscosity, mPa.s | 0.30 | 2.63 | 0.68 |
| surface tension, mN/m | 22.68 | 35.05 | 25.50 |
| dissociation constant, pKa | -8.3 | 24.2 | 20.5 |
| donor number, DN, kcal/mol | 11 | 18 | 17 |
| acceptor number, AN | | | 12.5 |
| polarity parameter, $E_T(30)$, kcal/mol | 36.3 | 42.2 | 39.8 |
| coefficient of cubic expansion, $10^{-4}$/°C | 9.7 | 13 | 13 |
| specific heat, cal/K mol | 29.85 | 58.22 | 51.0 |
| heat of vaporization, cal/g | 7.48 | 12.17 | 9.94 |
| heat of combustion, MJ/kg | 26.82 | 40.11 | 36.35 |
| dielectric constant | 11.98 | 20.56 | 16.1 |
| Hildebrand solubility parameter, $cal^{1/2} cm^{-3/2}$ | 7.54 | 11.0 | 9.2 |
| Henry's Law constant, $atm/m^3$ mol | 4.4E-8 | 2.7E-4 | 8.7E-5 |
| evaporation rate (butyl acetate = 1) | 0.02 | 6.6 | 0.83 |
| threshold limiting value - 8h average, ppm | 5 | 750 | 50 |
| maximum concentration (15 min exp), ppm | 75 | 1000 | 300 |
| maximum concentration any time, ppm | 5 | | |
| LD50 oral, mg/kg | 148 | 5800 | 2590 |
| LD50 dermal, mg/kg | 200 | 20000 | 6500 |
| route of entry | absorption, contact. ingestion, inhalation | | |
| target organs | central nervous system, eye, kidney, liver, lung, peripheral nervous system, respiratory system, skin, stomach, testes | | |
| carcinogenicity | - | | |
| mutagenic properties | diacetone alcohol, methyl isopropyl ketone | | |
| biological oxygen demand, 5-day test, g/g | 0.68 | 2.03 | 1.37 |

| Property | Value | | |
|---|---|---|---|
| | minimum | maximum | median |
| chemical oxygen demand, g/g | 1.92 | 2.88 | 2.31 |
| theoretical oxygen demand, g/g | 1.67 | 2.93 | 2.44 |
| biodegradation probability | | days-weeks | |
| octanol/water partition coefficient | -1.34 | 2.65 | |
| urban ozone formation | 0.01 | 0.65 | 0.15 |

### 3.3.11 ACIDS

| Property | Value | | |
|---|---|---|---|
| | minimum | maximum | median |
| boiling temperature, °C | 20 | 337 | 164 |
| freezing temperature, °C | -83 | 137 | -3 |
| flash point, °C | 37 | 140 | 100 |
| autoignition temperature, °C | 298 | 539 | 380 |
| refractive index | 1.285 | 1.551 | 1.421 |
| specific gravity, g/cm$^3$ | 0.9 | 1.83 | 1.08 |
| vapor density (air=1) | 0.7 | 5.0 | 3.3 |
| vapor pressure, kPa | 0.00 | 410 | 0.08 |
| viscosity, mPa.s | 0.25 | 23.55 | 2.82 |
| surface tension, mN/m | 27.4 | 37.6 | 33 |
| dissociation constant, pKa | 0.23 | 4.88 | 4.25 |
| donor number, DN, kcal/mol | 2.3 | 20 | 10.5 |
| acceptor number, AN | 18.5 | 105 | 52.9 |
| polarity parameter, $E_T(30)$, kcal/mol | 43.9 | 57.7 | 54.4 |
| specific heat, cal/K mol | 2367 | 29.42 | 2.612 |
| heat of vaporization, cal/g | 4.8 | 5.58 | 4.80 |
| dielectric constant | 6.17 | 58.5 | 40.5 |
| Hildebrand solubility parameter, cal$^{1/2}$ cm$^{-3/2}$ | 9.79 | 15.84 | 12.29 |
| Henry's Law constant, atm/m$^3$ mol | 1.26E-8 | 4.4E-5 | 1.67E-7 |
| evaporation rate (butyl acetate = 1) | 0.00 | 1.34 | 0.3 |
| threshold limiting value - 8h average, ppm | 1 | 10 | 4 |
| maximum concentration (15 min exp), ppm | 10 | 15 | 10 |
| LD50 oral, mg/kg | 200 | 74000 | 3310 |

| Property | Value | | |
|---|---|---|---|
| | minimum | maximum | median |
| route of entry | absorption, contact. ingestion, inhalation | | |
| target organs | eye, kidney, liver, respiratory system, skin | | |
| carcinogenicity | - | | |
| mutagenic properties | formic acid | | |
| biological oxygen demand, 5-day test, g/g | 0.2 | 0.65 | |
| chemical oxygen demand, g/g | 0.36 | 1.09 | |
| theoretical oxygen demand, g/g | 0.35 | 1.07 | 0.67 |
| biodegradation probability | days-weeks | | |
| octanol/water partition coefficient | -0.17 | +1.88 | |
| urban ozone formation | 0-0.09 | | |

## 3.3.12 AMINES

| Property | Value | | |
|---|---|---|---|
| | minimum | maximum | median |
| boiling temperature, °C | -33 | 372 | 152 |
| freezing temperature, °C | -115 | 142 | -6 |
| flash point, °C | -37 | 198 | 55 |
| autoignition temperature, °C | 210 | 685 | 410 |
| refractive index | 1.32 | 1.62 | 1.48 |
| specific gravity, g/cm$^3$ | 0.7 | 1.66 | 1.02 |
| vapor density (air=1) | 0.54 | 10.09 | 3.2 |
| vapor pressure, kPa | 0.00 | 1.013 | 0.13 |
| viscosity, mPa.s | 0.13 | 4000 | 3.15 |
| surface tension, mN/m | 19.11 | 48.89 | 32.43 |
| dissociation constant, pKa | 8.96 | 11.07 | 10.78 |
| pH | 7.2 | 12.1 | 11 |
| donor number, DN, kcal/mol | 24 | 61 | 33.1 |
| acceptor number, AN | 1.4 | 39.8 | 18.8 |
| polarity parameter, $E_T(30)$, kcal/mol | 32.1 | 55.8 | 42.2 |
| specific heat, cal/K mol | 30.4 | 74.1 | 53.4 |
| heat of vaporization, cal/g | 5.65 | 16.13 | 8.26 |
| heat of combustion, MJ/kg | | | 30.22 |

| Property | Value | | |
|---|---|---|---|
| | minimum | maximum | median |
| dielectric constant | 2.42 | 37.78 | 29.36 |
| Hildebrand solubility parameter, $cal^{1/2} cm^{-3/2}$ | 7.4 | 15.5 | 10.5 |
| Henry's Law constant, $atm/m^3$ mol | 1.7E-23 | 3.38E+1 | 1.56E-8 |
| evaporation rate (butyl acetate = 1) | 0.001 | 3.59 | 0.06 |
| threshold limiting value - 8h average, ppm | 0.1 | 100 | 5 |
| maximum concentration (15 min exp), ppm | 6 | 35 | 15 |
| maximum concentration any time, ppm | 5 | | |
| LD50 oral, mg/kg | 100 | 12760 | 470 |
| LD50 dermal, mg/kg | 64 | 8000 | 660 |
| route of entry | absorption, contact. ingestion, inhalation | | |
| target organs | eye, kidney, lymphatic system, liver, lung, respiratory system, skin, testes | | |
| carcinogenicity | acetamide, p-chloroaniline, N,N-dimethylformamide, hydrazine, N-nitrosodimethylamine, o-toluidyne | | |
| mutagenic properties | dimethylamine, ethylene diamine tetracetic acid, methylamine, N-methylpyrrolidone, N-nitrosomethyl amine, tetraethylene pentamine | | |
| biological oxygen demand, 5-day test, g/g | 0.01 | 2.24 | 0.84 |
| chemical oxygen demand, g/g | 1.28 | 1.9 | 1.53 |
| theoretical oxygen demand, g/g | 0.65 | 2.85 | 1.8 |
| biodegradation probability | days-weeks | | |
| octanol/water partition coefficient | -1.66 | 1.92 | |
| urban ozone formation | 0.00 | 0.51 | 0.21 |

### 3.3.13 ESTERS

| Property | Value | | |
|---|---|---|---|
| | minimum | maximum | median |
| boiling temperature, °C | 32 | 343 | 165 |
| freezing temperature, °C | -148 | 27.5 | -54 |
| flash point, °C | -19 | 240 | 64 |
| autoignition temperature, °C | 252 | 505 | 400 |
| refractive index | 1.34 | 1.56 | 1.44 |
| specific gravity, $g/cm^3$ | 0.81 | 1.38 | 0.92 |

| Property | Value | | |
|---|---|---|---|
| | minimum | maximum | median |
| vapor density (air=1) | 2.5 | 9.60 | 5.2 |
| vapor pressure, kPa | 0.00 | 64.0 | 0.27 |
| viscosity, mPa.s | 0.42 | 32.7 | 1.07 |
| surface tension, mN/m | 23.75 | 41.39 | 28.6 |
| dissociation constant, pKa | 10.68 | 13.3 | 12.0 |
| pH | 5 | 7 | 7 |
| donor number, DN, kcal/mol | 11 | 23.7 | 16.3 |
| acceptor number, AN | 6.7 | 18.3 | 16.3 |
| polarity parameter, $E_T(30)$, kcal/mol | 36.7 | 48.6 | 40.9 |
| coefficient of cubic expansion, $10^{-4}/^oC$ | 8.76 | 10.3 | |
| specific heat, cal/K mol | 31.54 | 119 | 46.9 |
| heat of vaporization, cal/g | 7.72 | 21.8 | 10.04 |
| heat of combustion, MJ/kg | 18.5 | 36.35 | 28.19 |
| dielectric constant | 4.75 | 64.9 | 64.0 |
| Kauri-butanol number | 62 | 1000 | 1000 |
| Hildebrand solubility parameter, $cal^{1/2}$ $cm^{-3/2}$ | 7.34 | 12.6 | 8.8 |
| Henry's Law constant, $atm/m^3$ mol | 9.9E-8 | 1.9E-2 | 3.6E-4 |
| evaporation rate (butyl acetate = 1) | 0.001 | 11.8 | 0.22 |
| threshold limiting value - 8h average, ppm | 0.2 | 400 | 100 |
| maximum concentration (15 min exp), ppm | 2 | 310 | 150 |
| LD50 oral, mg/kg | 500 | 42000 | 5600 |
| LD50 dermal, mg/kg | 500 | 20000 | 5000 |
| route of entry | absorption, contact. ingestion, inhalation | | |
| target organs | blood, brain, central nervous system, eye, gastrointestinal tract, lung, respiratory system, skin, spleen | | |
| carcinogenicity | ethyl acrylate, vinyl acetate | | |
| mutagenic properties | methyl ester of butyric acid, γ-butyrlactone, dibutyl phthalate, 2-ethoxyethyl acetate, ethyl acetate, ethyl propionate, ethylene glycol methyl ether acetate, methyl propionate, n-propyl acetate | | |
| biological oxygen demand, 5-day test, g/g | 0.25 | 1.26 | 0.6 |
| chemical oxygen demand, g/g | 1.11 | 2.32 | 1.67 |
| theoretical oxygen demand, g/g | 1.09 | 2.44 | 1.67 |

| Property | Value | | |
|---|---|---|---|
| | minimum | maximum | median |
| biodegradation probability | | days-weeks | |
| octanol/water partition coefficient | -0.56 | +3.97 | |
| urban ozone formation | 0.02 | 0.42 | 0.08 |

### 3.3.14 COMPARATIVE ANALYSIS OF ALL SOLVENTS

| Property | Value | | |
|---|---|---|---|
| | minimum | maximum | range |
| boiling temperature, °C | CFCs | PHA | -40.6-372 |
| freezing temperature, °C | CFCs | amines | -189-142 |
| flash point, °C | aliphatic HC | CFCs (none) | -104-350 |
| autoignition temperature, °C | glycol ethers | phenols | 174-715 |
| refractive index | CFCs | halogenated | 1.20-1.63 |
| specific gravity, $g/cm^3$ | aliphatic HC | CFCs | 0.51-3 |
| vapor density (air=1) | aldehydes | CFCs | 1-33.4 |
| vapor pressure, kPa | many | CFCs | 0.00-4000 |
| viscosity, mPa.s | CFCs | PHA | 0.02-114.6 |
| surface tension, mN/m | CFCs | PHA | 0.03-48.49 |
| dissociation constant, pKa | ethers | alcohols | -8.3-19.00 |
| pH | acids | amines | 1-14 |
| donor number, DN, kcal/mol | hydrocarbons | amines | 0-61 |
| acceptor number, AN | hydrocarbons | acids | 0-105 |
| polarity parameter, $E_T(30)$, kcal/mol | ethers | alcohols | 16-65.3 |
| coefficient of cubic expansion, $10^{-4}$/°C | alcohols | ethers | 7-14.5 |
| specific heat, cal/K mol | CFCs | PHA | 12.32-294 |
| heat of vaporization, cal/g | CFCs | halogenated | 3.76-81.2 |
| heat of combustion, MJ/kg | CFCs | aliphatic HC | 6.57-44.58 |
| dielectric constant | CFCs | esters | 1.0-64.9 |
| Kauri-butanol number | aliphatic HC | esters | 22-1000 |
| aniline point, °C | aromatic HC | aliphatic HC | 7-165 |
| Hildebrand solubility parameter, $cal^{1/2}\ cm^{-3/2}$ | CFCs | alcohols | 5.9-23 |
| Henry's Law constant, $atm/m^3$ mol | amines | alcohols | 1.7E-23-34.4 |
| evaporation rate (butyl acetate = 1) | many | aliphatic HC | 0-17.5 |

| Property | Value | | |
|---|---|---|---|
| | minimum | maximum | range |
| threshold limiting value - 8h average, ppm | several | several | 0.1-1000 |
| maximum concentration (15 min exp), ppm | aldehydes | CFCs | 0.3-1250 |
| LD50 oral, mg/kg | aldehydes | phenols | 46-320000 |
| LD50 dermal, mg/kg | amines | alcohols | 64-225000 |
| route of entry | absorption, contact. ingestion, inhalation | | |
| target organs | blood, brain, bone marrow, central nervous system, eye, gastrointestinal tract, heart, kidney, lymphatic system, liver, lung, peripheral nervous system, respiratory system, skin, spleen, stomach, testes, thyroid | | |
| carcinogenicity | some in the following groups: aromatic hydrocarbons, halogenated hydrocarbons, nitrogen-containing compounds, organic sulfur compounds, phenols, aldehydes, ethers, amines, esters | | |
| mutagenic properties | each group contains some species | | |
| theoretical oxygen demand, g/g | CFCs | aliphatic HC | 0-3.56 |
| biodegradation probability | days-weeks in the most cases | | |
| ozone depletion potential | CFCs | | |
| global warming potential | CFCs | | |
| urban ozone formation | CFCs | aldehydes | 0-1.55 |

HC - hydrocarbons, PHA - polyhydric alcohols

The comparative chart allocates for each group the highest and the lowest position in relationship to their respective values of particular parameters. The chart allows to show that the fact of having many good solvent properties does not warrant that solvent is suitable for use. For example, CFCs have many characteristics of good solvents but they are still eliminated from use because they cause ozone depletion and are considered to be a reason for global warming. On the other hand, esters do not appear on this chart frequently but they are very common solvents.

The chart also shows that solvents offer very broad choice of properties, which can be selected to satisfy any practical application.

## 3.4 TERPENES

Tilman Hahn, Konrad Botzenhart, Fritz Schweinsberg
**Institut für Allgemeine Hygiene und Umwelthygiene**
**University of Tübingen, Tübingen, Germany**

### 3.4.1 DEFINITIONS AND NOMENCLATURE

Terpenes are natural products. Terpenoids enlarge this division to include natural or synthetic derivatives.

The structures of terpenes or terpenoids varies widely and are classified according to various aspects, e.g. number of isoprene units ($C_{10}$ monoterpenes, $C_{15}$ sesquiterpenes, $C_{20}$ diterpenes, $C_{25}$ sesterterpenes, $C_{30}$ triterpenes, $C_{40}$ tetraterpenes) or division in acyclic, mono-, bi-, tri-, tetra-, pentacyclic terpenes.[1,2] Often terpenes are named by their trivial names, e.g. d-limonene.

### 3.4.2 OCCURRENCE

The occurrence of terpenes is ubiquitous. Natural terpenes are found in plants and animals in minute amounts. Especially in higher plants, terpenes characterize the type of plant (chemotaxonomy): mono- and sesquiterpenes in essential oils, sesqui-, di-, triterpenes in balsams and resins, tetraterpenes in pigments and polyterpenes in latexes.[1,3,4,5] Therefore, terpenes are often emitted from natural products such as citrus fruits or trees, e.g. conifers.

Terpenes are components of various products: e.g. tobacco smoke, wax pastes (furniture and floor polishes etc.), liquid waxes (floor polishes etc.), cleansers (detergents etc.), polishes, dyes and varnishes, synthetic resins, so-called "natural" building products, deodorants, perfumes, softeners, air fresheners, foods, beverages, pharmaceutical products (e.g. camomile oil, eucalyptus oil).[1,3,4,5] In these products terpene compounds such as geraniol, myrcene (beta-myrcene), ocimene, menthol, alpha-pinene, beta-pinene, d-limonene, 3-carene, cineole, camphene or caryophyllene can be detected.[1,3,4,5]

Often terpenes may be included as additives, e.g. food additives licensed by the FDA. Terpenes detected in indoor air are mainly the monoterpenes alpha-, beta- pinenes, 3-carene and d-limonene which occur primarily in conifer products.[5] Some of the monoterpenes may be converted into well-known epoxides and peroxides with high allergic potential.[5]

Several products containing terpenes are more highly refined which influences quality and quantities of terpenes in these products: e.g. in oil of turpentine or in resin components. The quantity of monoterpenes is essentially influenced by the composition of the raw materials, e.g. d-limonene dominates in agrumen oils as citrus oil products.[5]

Terpene products are often associated with "natural positive" properties, e.g. attributes such as "biological, positive health effects and good biodegradability" which are often neither substantiated nor proven.

Indoor concentrations of some terpenes, e.g., d-limonene and pinene, are highest in the group of VOCs.[6]

### 3.4.3 GENERAL

Terpenes are synthesized in chloroplasts, mitochondria and microsomes of plants or in the liver of animals. Typical biosynthesis pathways of terpenes are well-known, e.g. via decarboxylation, isomerization and acetyl-CoA-processes.[1] Degradation of terpenes is possible by microorganisms, e.g. Pseudomonas and Aspergillus sp., in plants and in animals.[1]

In terpene products, terpenes exist as two enantiomers in different mixture ratios. Enantiomers are associated with characteristic odour (e.g. d-limonene in orange-oil).[5] Odors of terpenes are essential criteria in the classification of terpenes. Some terpenes can be smelt in extremely low concentrations.[5]

Threshold limit values are in the range of $\mu g/m^3$ in indoor air.[5] This applies primarily to monoterpenes pinenes, d-limonene, carenes and sesquiterpenes longifolenes and caryophyllenes.

## 3.4.4 TOXICOLOGY

Most terpenes show low acute oral toxicity and low dermal toxicity. Contact dermatitis is the most common symptom described as a result of exposure to terpenes. Other allergic reactions occur more rarely: e.g. allergic rhinitis or allergic bronchial asthma. The most common products with an allergic potential (contact dermatitis) are oils of turpentine.[5] Older turpentine products show higher allergic potential than freshly distilled products. Turpentines have now been replaced by other less toxic petrochemical products.

Many terpenes or metabolites are well-known contact allergens causing allergic dermatitis, e.g. d-limonene or oxygenated monocyclic terpenes which are produced by autoxidation of d-limonene.[7] Normally the highest allergic potential is associated with photo-oxidants (e.g. peroxides, epoxides) which are formed from terpenes. The symptoms of allergic dermatitis disappear if dermal contact to the causative terpene allergens is removed.

Exposure to the monoterpenes (alpha-pinene, beta-pinene and 3-carene) showed no major changes in lung function, but showed chronic reaction in the airways (reduced lung function values which persist between shifts) in workers of joinery shops.[8] In studies of dwellings, bronchial hyper-responsiveness could be related to indoor air concentrations of d-limonene.[9] Other studies did not find significant changes in the respiratory tract.[10,11] Nevertheless, these studies postulate effects of metabolites of terpenes (e.g. pinenes) as relevant causative agents. It is suggested in some cases that "Multiple Chemical Sensitivity" (MCS) may be attributed to increased values of terpenes and aromatic hydrocarbons.[12]

Often mixtures of terpene products or so-called "natural products" show allergic effects, e.g. fragrant mixtures containing d-limonene,[7] tea tree oil,[13] oilseed rape.[14] Consumer products such as deodorants or perfumes also contain terpenes with allergic potential.[15,16]

## 3.4.5 THRESHOLD LIMIT VALUES

Relevant threshold limit values for terpenes are rare because of a lack of basic information about specific terpene products and by-products on the one hand, and occupational and environmental exposures on the other hand.[5] The threshold limit values which have been documented the best concern oil of turpentine. A MAK-value of 100 ppm is defined in German regulations and noted to be dermally sensitive.[17] For other terpenes, such as d-limonene which is also classified as dermally sensitive, it has not yet been possible to establish a MAK-value because of a lack of information of their effects on animals or humans. With terpenes, as is often the case, aggregate concentration parameters are used as limit values such as the minimum level goals recommended by the former German Federal Health Authority.[18] These suggested minimum values bear in mind actual levels detected in indoor areas.

Although Mohr[5] quotes values of 30 μg/m³ for terpene aggregate and 15 μg/m³ for a single terpene compound from a scheme of Seifert, he considers values of 60 μg/m³ for terpene aggregate to be more appropriate according to his experience. He also discusses aggregate values of 200 μg/m³ (these exceed the olfactory threshold level) as posing possible health hazards in individual cases. These concentration limits only concern monoterpenes (pinenes, d-limonene, 3-carene) not other terpene products (e.g. sesquiterpenes).[5]

**Table 3.4.1 Selected examples of terpenes**

| Substance group | Examples | Common occurrence | Selected properties |
|---|---|---|---|
| Acyclic monoterpenes | geraniol | in essential oils, perfume products and luxury foods, production from other terpene products, e.g. beta-pinene | acyclic, double unsaturated alcohol, several possible reactions, occurrence as esters, typical rose odor |
| | myrcene | naturally occurring in plant oils and organisms, industrial production | very reactive, pleasant odor, part of many synthetic reactions (synthesis of other terpenes) |
| | ocimene | in essential oils and perfume products | pleasant odor, sensitive to oxidation |
| Monocyclic monoterpenes | p-menthane | in essential oils, e.g. eucalyptus fruits | typical odor (fennel) |
| | p-cimene | perfume and soap products (musk perfumes), in various production processes, e.g., sulfite leaching of wood | typical odor (aromatic hydrocarbons), inflammable |
| Bicyclic monoterpenes | pinenes | in essential oils and conifer products, industrial production and use, e.g. in fragrance and flavor industry | typical odor (turpentine), softening agent |
| Acyclic sesquiterpenes | farnesol | in essential oils, in perfume and soap products | typical odor, sensitive to oxidation, heat and light |
| Monocyclic sesquiterpenes | bisabolenes | in various oils, e.g. myrrh and limete oil, in perfume and fragrance products | balsamic odor |
| Bicyclic sesquiterpenes | caryophyllenes | in essential oils, as perfume and fragrance products, chewing gum, synthesis of other perfumes | typical odor (clove) |
| Tricyclic sesquiterpenes | longifolene | in essential oils, e.g. turpentine, solvent additive, production of perfumes | colorless, oily liquid |

## REFERENCES


1 **Ullmann's Encyclopedia of Industrial Chemistry**, 1998.
2 O.W. Thiele, Lipide, **Isoprenoide mit Steroiden**, *Thieme Verlag,* Stuttgart, 1979.
3 Römpp Lexikon, **Encyclopedia of Chemistry**. *Chemie-Lexikon*, Thieme, 1998.
4 Seifert, B., Esdorn, H., Fischer, M., Rüden, H., Wegner, J. (eds.), Indoor air '87. Proceeding of the 4th International Conference on Indoor Air Quality and Climate. Institute for Soil, Water and Air Hygiene, 1987.
5 Mohr, S., Neue Lasten durch ökologische Baustoffe ? Vorkommen und Bewertung von Terpenen in der Innenraumluft. In Behrends, B. (ed.), Gesundes Wohnen durch ökologisches Bauen, Hannoversche Ärzte-Verlags-Union, 61-101, 1998.
6 Lance, A., Wallace, Ph.D., **Volatile organic compounds**, In Samet, J.M., and Spengler, J.D. (eds.): Indoor air pollution: a health perspective. *The Johns Hopkins University Press*, 1991.
7 Karlberg, A.T., DoomsGoossens, A., Contact allergy to oxidized d-limonene among dermatitis patients. *Contact Dermatitis*, **36**, 201-206, 1997.
8 Eriksson, K.A., Levin, J.O., Sandstrom, T., Lindstrom-Espeling, K., Linden, G., Stjernberg, N.L., Terpene exposure and respiratory effects among workers in Swedish joinery shops, *Scan. J. Work. Environ. Health*, **23**, 114-120, 1997.
9 Norback, D., Bjornsson, E., Janson, C., Widstrom, J., Boman, G., Asthmatic symptoms and volatile organic compounds, formaldehyde, and carbon dioxide in dwellings, *Occup. Environ. Med.*, **52**, 388-395, 1995.
10 Falk, A.A., Hagberg, M.T., Lölf, A.E. , Wigaeus-Hjelm, E.M., Zhiping, W., Uptake, distribution and elimination of pinene in man after exposure by inhalation, *Scand. J. Work Environ. Health*, **16**, 372-378, 1990.
11 Eriksson, K., Levin, J.O., Identification of cis- and trans-verbenol in human urine after occupational exposure to terpenes. *Int. Arch. Occup. Environ. Health*, **62**, 379-383, 1990.
12 Ring, J., Eberlein-Konig, B., Behrendt, H., "Eco-syndrome"- "Multiple Chemical Sensitivity" (MCS), *Zbl. Hyg. Umweltmed.*, **202**, 207-218, 1999.
13 Rubel, D.M., Freeman, S., Southwel, I.A., Tea tree oil allergy: what is the offending agent ? Report of three cases of tea tree oil allergy and review of the literature. *Austral. J. Dermatol.*, **39**, 244-247, 1998.
14 McEwan, M., MacFarlane-Smith, W.H., Identification of volatile organic compounds emitted in the field by oilseed rape (Brassica napus ssp. Oleifera) over the growing season. *Clin. Exp. Allergy*, **28**, 332-338, 1998.
15 Rastogi, S.C., Johanson, J.D., Frosch, P., Menne, T., Bruze, M., Lepoittevin, J.P., Dreier, B., Andersen, K.E., White, I.R., Deodorants on the European market: quantitative chemical analysis of 21 fragrances, *Contact Dermatitis*, **38**, 29-35, 1998.
16 Johanson, J.D., Rastogi, S.C., Andersen, K.E., Menne, T., Content and reactivity to product perfumes in fragrance mix positive and negative eczema patients. A study of perfumes used in toileries and skin-care products, *Contact Dermatitis*, **36**, 291-296, 1997.
17 DFG (Deutsche Forschungsgemeinschaft), **MAK- und BAT-Werte-Liste** 1999, DFG. *WILEY-VCH*, 1999.
18 Anonymous, Bewertung der Luftqualität in Innenräumen. Evaluation of Indoor Air Quality, *Bundesgesundhbl.,* 3/93, 117-118, 1993.

4

# General Principles Governing Dissolution of Materials in Solvents

## 4.1 SIMPLE SOLVENT CHARACTERISTICS

Valery Yu. Senichev, Vasiliy V. Tereshatov
**Institute of Technical Chemistry**
**Ural Branch of Russian Academy of Sciences, Perm, Russia**

Polymer dissolution is a necessary step in many of the polymer processing methods, such as blending, separation, coating, casting, etc. The developments in physical chemistry of non-electrolyte solutions relate the capabilities of solvents to dissolve materials with their physical properties. The relationships were developed within the framework of the concept of solubility parameters.

### 4.1.1 SOLVENT POWER

The usual problem of polymer engineering is a selection of proper solvent(s) for a given polymer. This selection implies that the solvent must form with polymer a thermodynamically stable mixture in the whole range of concentrations and temperatures. Such choice is facilitated by use of numerical criterion of a solvent power. Solvent power might be taken from the thermodynamic treatment (for example, a change of Gibbs' free energy or chemical potentials of mixing of polymer with solvent) but these criteria depend not only on the solvent properties but also on polymer structure and its concentration. For this reason, various approaches were proposed to estimate solvent power.

Kauri-butanol value, KB is used for evaluation of dissolving ability of hydrocarbon solvents. It is obtained by titration of a standard Kauri resin solution (20 wt% in 1-butanol) with the solvent until a cloud point is reached (for example, when it becomes impossible to read a text through the solution). The amount of the solvent used for titration is taken as KB value. The relationship between KB and solubility parameter, $\delta$, fits the following empirical dependence:[1]

$$\delta = 12.9 + 0.06KB \qquad [4.1.1]$$

The KB value is primarily a measure of the aromaticity of solvents. Using KB value, it is possible to arrange solvents in sequence: aliphatic hydrocarbons < naphthenic hydrocarbons < aromatic hydrocarbons.

Dilution ratio, DR, is used to express the tolerance of solvents to diluents, most frequently, toluene. DR is the volume of a solvent added to a given solution that causes precipitation of the dissolved resin. This ratio can characterize the compatibility of a diluent with a resin solution in primary solvent. When compatibility is high, more diluent can be added. Only a multi-parameter approach provides a satisfactory correlation with solubility parameters.[2-3] DR depends on the polymer concentration. With polymer concentration increasing, DR increases as well. Temperature influences DR in a similar manner. Determination of DR must be performed at standard conditions. DR can be related to the solubility parameters but such correlation depends on concentration.

Aniline point, AP, is the temperature of a phase separation of aniline in a given solvent (the volume ratio of aniline : solvent = 1:1) in the temperature decreasing mode. AP is a critical temperature of the aniline - solvent system. AP can be related to KB value using the following equations:

At KB<50

$$KB = 99.6 - 0.806\rho - 0.177AP + 0.0755\left(358 - \frac{5}{9}T_b\right) \qquad [4.1.2]$$

At KB>50

$$KB = 177.7 - 10.6\rho - 0.249AP + 0.10\left(358 - \frac{5}{9}T_b\right) \qquad [4.1.3]$$

where:

$T_b$ a solvent boiling point.

AP depends on the number of carbon atoms in the hydrocarbon molecule. AP is useful for describing complex aromatic solvents.

The solvent power can also be presented as a sum of factors that promote solubility or decrease it:[4]

$$S = H + B - A - C - D \qquad [4.1.4]$$

where:

H a factor characterizing the presence of active sites of opposite nature in solvent and polymer that can lead to formation of hydrogen bond between polymer and solvent
B a factor related to the difference in sizes of solute and solvent molecules
A a factor characterizing solute "melting"
C a factor of the self-association between solvent molecules
D a factor characterizing the change of nonspecific cohesion forces in the course of transfer of the polymer molecule into solution.

The equations for calculation of the above-listed factors are as follows:

$$B = (\alpha + b)\frac{V_m}{V_s}(1 - \varphi_p) \qquad [4.1.5]$$

$$D = \frac{V_m}{RT}\left(\delta'_p - \delta'_s\right)^2 \left(1-\varphi_p\right)^2 \quad [4.1.6]$$

$$H = \ln\left[1 + \frac{K\left(1-\varphi_p\right)}{V_s}\right] \quad [4.1.7]$$

$$C = \ln\frac{1+\left(K_{pp}/V_m\right)}{1+\left(K_{pp}\varphi_p/V_m\right)} \quad [4.1.8]$$

where:

| | |
|---|---|
| $\alpha$ | constant depending on the choice of the equation for the entropy of mixing. Usually it equals 0.5. |
| b | constant depending on the structure of solvent. b=1 for unstructured solvents, b=-1 for solvents with single H-bond (e.g. alcohols) and b=-2 for solvents with double H-bonds chains such as water. |
| $V_m$ | the molar volume of a repeating segment |
| $V_s$ | the molar volume of solvent |
| $\varphi_p$ | the volume fraction of polymer |
| $\delta_p$, $\delta_s$ | modified solubility parameters without regard to H-bonds. |
| K | the stability constant of the corresponding solvent-polymer hydrogen bond |
| $K_{pp}$ | the constant of the self-association of polymer segments |

Several polymers such as polyethylmethacrylate, polyisobutylmethacrylate and polymethylmethacrylate were studied according to Huyskens-Haulait-Pirson approach. The main advantage of this approach is that it accounts for entropy factors and other essential parameters affecting solubility. The disadvantages are more numerous, such as lack of physical meaning of some parameters, great number of variables, and insufficient coordination between factors influencing solubility that have reduced this approach to an approximate empirical scheme.

### 4.1.2 ONE-DIMENSIONAL SOLUBILITY PARAMETER APPROACH

The thermodynamic affinity between components of a solution is important for quantitative estimation of mutual solubility. The concept of solubility parameters is based on enthalpy of the interaction between solvent and polymer. Solubility parameter is the square root of the cohesive energy density, CED:

$$\delta = (CED)^{1/2} = \left(\frac{\Delta E_i}{V_i}\right)^{1/2} \quad [4.1.9]$$

where:

| | |
|---|---|
| $\Delta E_i$ | cohesive energy |
| $V_i$ | molar volume |

Solubility parameters are measured in $(MJ/m^3)^{1/2}$ or $(cal/sm^3)^{1/2}$ (1 $(MJ/m^3)^{1/2}$=2.054 $(cal/sm^3)^{1/2}$). The cohesive energy is equal in magnitude and opposite in sign to the potential energy of a volume unit of a liquid. The molar cohesive energy is the energy associated with all molecular interactions in one mole of the material, i.e., it is the energy of a liquid relative to its ideal vapor at the same temperature (see Chapter 5).

$\delta$ is a parameter of intermolecular interaction of an individual liquid. The aim of many studies was to find relationship between energy of mixing of liquids and their $\delta$. The first attempt was made by Hildebrand and Scatchard[5,6] who proposed the following equation:

$$\Delta U^m = (x_1V_1 + x_2V_2)\left[\left(\frac{\Delta E_1}{V_1}\right)^{1/2} - \left(\frac{\Delta E_2}{V_2}\right)^{1/2}\right]^2 \varphi_1\varphi_2 =$$

$$= (x_1V_1 + x_2V_2)(\delta_1 - \delta_2)^2 \varphi_1\varphi_2 \quad [4.1.10]$$

where:

| | |
|---|---|
| $\Delta U^m$ | internal energy of mixing, that is a residual between energies of a solution and components, |
| $x_1, x_2$ | molar fractions of components |
| $V_1, V_2$ | molar volumes of components |
| $\varphi_1, \varphi_2$ | volume fractions of components |

The Hildebrand-Scatchard equation became the basis of the Hildebrand theory of regular solutions.[5] They interpreted a regular solution as a solution formed due to the ideal entropy of mixing and the change of an internal energy. The assumed lack of the volume change makes an enthalpy or heat of mixing equated with the right members of the equation. The equation permits calculation heat of mixing of two liquids. It is evident from equation that these heats can only be positive. Because of the equality of of components, $\Delta H^m$=0.

The free energy of mixing of solution can be calculated from the equation

$$\Delta G^m = (x_1V_1 + x_2V_2)\left[\left(\frac{\Delta E_1}{V_1}\right)^{1/2} - \left(\frac{\Delta E_2}{V_2}\right)^{1/2}\right]^2 \varphi_1\varphi_2 -$$

$$-T\Delta S_{id} = V(\delta_1 - \delta_2)^2 \varphi_1\varphi_2 - T\Delta S \quad [4.1.11]$$

The change of entropy, $\Delta S_{id}$, is calculated from the Gibbs equation for mixing of ideal gases. The calculated values are always positive.

$$\Delta S_{id} = -R(x_1 \ln x_1 + x_2 \ln x_2) \quad [4.1.12]$$

where:

| | |
|---|---|
| R | gas constant |

Considering the signs of the parameters $\Delta S_{id}$ and $\Delta H^m$ in Eq. [4.1.10], the ideal entropy of mixing promotes a negative value of $\Delta G^m$, i.e., the dissolution and the value of $\Delta H^m$ reduces the $\Delta G^m$ value. It is pertinent that the most negative $\Delta G^m$ value is when $\Delta H^m$=0, i.e., when $\delta$ of components are equal. With these general principles in mind, the components with solubility parameters close to each other have the best mutual solubility. The theory of regular solutions has essential assumptions and restrictions.[7] The Eq. [4.1.10] is deduced under assumption of the central role of dispersion forces of interaction between components of solution that is correct only for the dispersion forces. Only in this case it is possible to accept that the energy of contacts between heterogeneous molecules is a geometric mean value of energy of contacts between homogeneous molecules:

$$\varepsilon_{12}^* = \sqrt{\varepsilon_{11}^* \varepsilon_{22}^*} \quad [4.1.13]$$

where:

$\varepsilon_{ii}^*$ potential energy of a pair of molecules

This assumption is not justified in the presence of the dipole-dipole interaction and other more specific interactions. Therefore the theory of regular solutions poorly suits description of the behavior of solutions of polar substances. Inherent in this analysis is the assumption of molecular separation related to molecular diameters which neglects polar or specific interactions. The theory also neglects volume changes on dissolution. This leads to a disparity (sometimes very large) between internal energy of mixing used in the theory and the constant pressure enthalpy measured experimentally.

The correlation between these values is given by equation:

$$\left(\Delta H^m\right)_p = \left(\Delta U^m\right)_V + T(\partial p / \partial T)_V \left(\Delta V^m\right)_p \quad [4.1.14]$$

where:

$(\partial p / \partial T)_V$ thermal factor of pressure which has value of the order 10-14 atm/degree for solutions and liquids.

Therefore, even at small changes of volume, the second term remains very large and brings substantial contribution to the value of $(\Delta H^m)_p$. For example, for a system benzene (0.5 mol) - cyclohexane (0.5 mol):

$$\Delta V^m = 0.65\ cm^3, \quad \left(\Delta H^m\right)_p - 182\ cal, \quad \left(\Delta U^m\right)_V = 131\ cal$$

The theory also assumes that the ideal entropy is possible for systems when $\Delta H^m \neq 0$. But the change of energy of interactions occurs in the course of dissolution that determines the inevitable change of entropy of molecules. It is assumed that the interactive forces are additive and that the interactions between a pair of molecules are not influenced by the presence of other molecules. Certainly, such an assumption is simplistic, but at the same time it has allowed us to estimate solubility parameters using group contributions or molar attractive constants (see Subchapter 5.3).

The solubility parameter $\delta$ is relative to the cohesion energy and it is an effective characteristic of intermolecular interactions. It varies from a magnitude of 12 $(MJ/m^3)^{1/2}$ for nonpolar substances up to 23 $(MJ/m^3)^{1/2}$ for water. Knowing $\delta$ of solvent and solute, we can estimate solvents in which particular polymer cannot be dissolved. For example, polyisobutylene for which $\delta$ is in the range from 14 to 16 $(MJ/m^3)^{1/2}$ will not be dissolved in solvents with $\delta$=20-24 $(MJ/m^3)^{1/2}$. The polar polymer with $\delta$=18 $(MJ/m^3)^{1/2}$ will not dissolve in solvents with $\delta$=14 or $\delta$=26 $(MJ/m^3)^{1/2}$. These are important data because they help to narrow down a group of solvents potentially suitable for a given polymer. However, the opposite evaluation is not always valid because polymers and solvents with the identical solubility parameters are not always compatible. This limitation comes from integral character of the solubility parameter. The solubility depends on the presence of functional groups in molecules of solution components which are capable to interact with each other and this model does not address such interactions. The latter statement has become a premise for the development of the multi-dimensional approaches to solubility that will be the subject of the following subchapter. The values of solubility parameters are included in Table 4.1.1.

**Table 4.1.1 Solubility parameters and their components (according Hansen's approach) for different solvents**

| Solvent | $V_1$, Kmol/m$^3$ | $\delta$, (MJ/m$^3$)$^{1/2}$ | $\delta_d$, (MJ/m$^3$)$^{1/2}$ | $\delta_p$, (MJ/m$^3$)$^{1/2}$ | $\delta_h$, (MJ/m$^3$)$^{1/2}$ |
|---|---|---|---|---|---|
| Alkanes | | | | | |
| n-Butane | 101.4 | 14.1 | 14.1 | 0 | 0 |
| n-Pentane | 116.2 | 14.3 | 14.3 | 0 | 0 |
| n-Hexane | 131.6 | 14.8 | 14.8 | 0 | 0 |
| n-Heptane | 147.4 | 15.1 | 15.1 | 0 | 0 |
| n-Octane | 163.5 | 14.0 | 14.0 | 0 | 0 |
| n-Nonane | 178.3 | 15.4 | 15.4 | 0 | 0 |
| n-Decane | 195.9 | 15.8 | 15.8 | 0 | 0 |
| n-Dodecane | 228.5 | 16.0 | 16.0 | 0 | 0 |
| Cyclohexane | 108.7 | 16.7 | 16.7 | 0 | 0 |
| Methylcyclohexane | 128.3 | 16.0 | 16.0 | 0 | 0.5 |
| Aromatic hydrocarbons | | | | | |
| Benzene | 89.4 | 18.7 | 18.4 | 1.0 | 2.9 |
| Toluene | 106.8 | 18.2 | 18.0 | 1.4 | 2.0 |
| Naphthalene | 111.5 | 20.3 | 19.2 | 2.0 | 5.9 |
| Styrene | 115.6 | 19.0 | 17.8 | 1.0 | 3.1 |
| o-Xylene | 121.2 | 18.4 | 17.6 | 1.0 | 3.1 |
| Ethylbenzene | 123.1 | 18.0 | 17.8 | 0.6 | 1.4 |
| Mesitylene | 139.8 | 18.0 | 18.0 | 0 | 0.6 |
| Halo hydrocarbons | | | | | |
| Chloromethane | 55.4 | 19.8 | 15.3 | 6.1 | 3.9 |
| Dichloromethane | 63.9 | 20.3 | 18.2 | 6.3 | 7.8 |
| Trichloromethane | 80.7 | 18.8 | 17.6 | 3.0 | 4.2 |
| n-Propyl chloride | 88.1 | 17.4 | 16.0 | 7.8 | 2.0 |
| 1,1-Dichloroethene | 79.0 | 18.6 | 17.0 | 6.8 | 4.5 |
| 1-Chlorobutane | 104.5 | 17.2 | 16.2 | 5.5 | 2.0 |
| 1,2-Dichloroethane | 79.4 | 20.0 | 19.0 | 7.4 | 4.1 |
| Carbon tetrachloride | 97.1 | 17.6 | 17.6 | 0 | 0 |

| Solvent | $V_1$, Kmol/m$^3$ | $\delta$, (MJ/m$^3$)$^{1/2}$ | $\delta_d$, (MJ/m$^3$)$^{1/2}$ | $\delta_p$, (MJ/m$^3$)$^{1/2}$ | $\delta_h$, (MJ/m$^3$)$^{1/2}$ |
|---|---|---|---|---|---|
| Perchloroethylene | 101.1 | 19.0 | 19.0<br>18.8 | 6.5<br>0 | 2.9<br>1.4 |
| 1,1,2,2-Tetrachloroethane | 105.2 | 19.8 | 18.8 | 5.1 | 9.4<br>5.3 |
| Chloro-difluoromethane (Freon 21) | 72.9 | 17.0 | 12.3 | 6.3 | 5.7 |
| 1,1,2-Trichloro-trifluoro-ethane (Freon 113) | 119.2 | 14.8 | 14.5 | 1.6 | 0 |
| Dichloro-difluoro-methane (Freon 12) | 92.3 | 12.2 | 12.2 | 2.0 | 0 |
| Chlorobenzene | 102.1 | 19.5 | 18.9 | 4.3 | 2.0 |
| o-Dichlorobenzene | 112.8 | 20.4 | 19.1 | 6.3 | 3.3 |
| Bromoethane | 76.9 | 19.6 | 15.8 | 3.1 | 5.7 |
| Bromobenzene | 105.3 | 20.3 | 20.5<br>18.9 | 5.5<br>4.5 | 4.1<br>5.1 |
| Ethers | | | | | |
| Epichlorohydrin | 72.5 | 18.5 | 17.8 | 1.8 | 5.3 |
| Tetrahydrofuran | 79.9 | 22.5 | 19.0 | 10.2 | 3.7 |
| 1,4-Dioxane | 81.7 | 18.5 | 16.8 | 5.7 | 8.0 |
| Diethyl ether | 85.7 | 20.5 | 19.0 | 1.8 | 7.4 |
| Diisopropyl ether | 104.8 | 15.6 | 14.4 | 2.9 | 5.1 |
| Ketones | | | | | |
| Acetone | 74.0 | 19.9 | 15.5 | 10.4 | 6.9 |
| Methyl ethyl ketone | 90.1 | 18.9 | 15.9 | 9.0<br>8.4 | 5.1 |
| Cyclohexanone | 104.0 | 18.8 | 16.3 | 7.1 | 6.1 |
| Diethyl ketone | 106.4 | 18.1 | 15.8 | 7.6 | 4.7 |
| Mesityl oxide | 115.6 | 16.7 | 15.9<br>17.6 | 3.7 | 4.1 |
| Acetophenone | 117.4 | 19.8 | 16.5 | 8.2 | 7.3 |
| Methyl isobutyl ketone | 125.8 | 17.5 | 15.3 | 6.1 | 4.1 |
| Methyl isoamyl ketone | 142.8 | 17.4 | 15.9 | 5.7 | 4.1 |
| Isophorone | 150.5 | 18.6 | 16.6 | 8.2 | 7.4 |
| Diisobutyl ketone | 177.1 | 16.0 | 16.0 | 3.7 | 4.1 |

| Solvent | $V_1$, Kmol/m$^3$ | $\delta$, (MJ/m$^3$)$^{1/2}$ | $\delta_d$, (MJ/m$^3$)$^{1/2}$ | $\delta_p$, (MJ/m$^3$)$^{1/2}$ | $\delta_h$, (MJ/m$^3$)$^{1/2}$ |
|---|---|---|---|---|---|
| Aldehydes | | | | | |
| Acetaldehyde | 57.1 | 21.1 | 14.7 | 8.0 | 11.3 |
| Furfural | 83.2 | 22.9 | 18.6 | 14.9 | 5.1 |
| n-Butyraldehyde | 88.5 | 18.4 | 14.7<br>18.7 | 5.3<br>8.6 | 7.0 |
| Benzaldehyde | 101.5 | 19.2 | 19.4 | 7.4 | 5.3 |
| Esters | | | | | |
| Ethylene carbonate | 66.0 | 30.1 | 29.6 | 19.4 | 21.7 |
| Methyl acetate | 79.7 | 19.6 | 15.5 | 7.2 | 7.6 |
| Ethyl formate | 80.2 | 19.6 | 15.5 | 8.4 | 8.4 |
| Propylene-1,2-carbonate | 85.0 | 27.2 | 20.0 | 18.0 | 4.1 |
| n-Propyl formate | 97.2 | 19.5 | 15.0 | 5.3 | 11.2 |
| Propyl acetate | 115.1 | 17.8 | 15.5 | 4.5 | 7.6 |
| Ethyl acetate | 98.5 | 18.6 | 15.8 | 5.3 | 7.2 |
| n-Butyl acetate | 132.5 | 17.4 | 15.8 | 3.7 | 6.3 |
| n-Amyl acetate | 149.4 | 17.3 | 15.6 | 3.3 | 6.7 |
| Isobutyl acetate | 133.5 | 17.0 | 15.1 | 3.7 | 6.3 |
| Isopropyl acetate | 117.1 | 17.3 | 14.4 | 6.1 | 7.4 |
| Diethyl malonate | 151.8 | 19.5 | 15.5 | 4.7 | 10.8 |
| Diethyl oxalate | 135.4 | 22.5 | 15.5 | 5.1 | 15.5 |
| Isoamyl acetate | 148.8 | 16.0 | 15.3 | 3.1 | 7.0 |
| Dimethyl phthalate | 163 | 21.9 | 18.6 | 10.8 | 4.9 |
| Diethyl phthalate | 198 | 20.5 | 17.6 | 9.6 | 4.5 |
| Dibutyl phthalate | 266 | 19.0 | 17.8 | 8.6 | 4.1 |
| Dioctyl phthalate | 377 | 16.8 | 16.6 | 7.0 | 3.1 |
| Phosphorous compounds | | | | | |
| Trimethyl phosphate | 116.7 | 25.2 | 16.7 | 15.9 | 10.2 |
| Triethyl phosphate | 169.7 | 22.2 | 16.7 | 11.4 | 9.2 |
| Tricresyl phosphate | 316 | 23.1 | 19.0 | 12.3 | 4.5 |
| Nitrogen compounds | | | | | |
| Acetonitrile | 52.6 | 24.3 | 15.3 | 17.9 | 6.1 |
| n-Butyronitrile | 86.7 | 20.4 | 15.3 | 12.5 | 5.1 |

| Solvent | $V_1$, Kmol/m$^3$ | $\delta$, (MJ/m$^3$)$^{1/2}$ | $\delta_d$, (MJ/m$^3$)$^{1/2}$ | $\delta_p$, (MJ/m$^3$)$^{1/2}$ | $\delta_h$, (MJ/m$^3$)$^{1/2}$ |
|---|---|---|---|---|---|
| Propionitrile | 70.9 | 22.1 | 15.3 | 14.3 | 5.5 |
| Benzonitrile | 102.6 | 19.9 | 17.4 | 9.0 | 3.3 |
| Nitromethane | 54.3 | 25.1 | 15.7 | 18.8 | 5.1 |
| Nitroethane | 71.5 | 22.6 | 15.9 | 15.9 | 4.5 |
| 2-Nitropropane | 86.9 | 20.4 | 16.1 | 12.0 | 4.1 |
| Nitrobenzene | 102.7 | 20.5 | 20.0 | 8.6 | 4.1 |
| Ethylenediamine | 67.3 | 25.2 | 16.6 | 8.8 | 17.0 |
| 2-Pyrrolidinone | 76.4 | 30.1 | 19.4 | 17.4 | 11.3 |
| Pyridine | 80.9 | 21.9 | 19.0 | 8.8 | 5.9 |
| Morpholine | 87.1 | 22.1 | 18.8 | 4.9 | 9.2 |
| Aniline | 91.5 | 21.1 | 19.4 | 5.1 | 10.2 |
| n-Butylamine | 99.0 | 17.8 | 16.2 | 4.5 | 8.0 |
| 2-Aminoethanol | 59.7 | 31.3 | 17.1 | 15.5 | 21.2 |
| Di-n-propyl amine | 136.8 | 15.9 | 15.3 | 1.4 | 4.1 |
| Diethylamine | 103.2 | 16.4 | 14.9 | 2.3 | 6.1 |
| Quinoline | 118.0 | 22.1 | 19.4 | 7.0 | 7.6 |
| Formamide | 39.8 | 39.3 | 17.2 | 26.2 | 19.0 |
| N,N-Dimethylformamide | 77.0 | 24.8 | 17.4 | 13.7 | 11.2 |
| Sulfur compounds | | | | | |
| Carbon disulfide | 60.0 | 20.3 | 20.3 | 0 | 0 |
| Dimethyl sulfoxide | 71.3 | 26.4 | 18.4 | 16.3 | 10.2 |
| Diethyl sulfide | 107.6 | 17.2 | 16.8 | 3.1 | 2.0 |
| Dimethyl sulfone | 75 | 29.7 | 19.0 | 19.4 | 12.3 |
| Monohydric alcohols and phenols | | | | | |
| Methanol | 40.7 | 29.1 | 15.1 | 12.2 | 22.2 |
| Ethanol | 66.8 | 26.4 | 15.8 | 8.8 | 19.4 |
| Allyl alcohol | 68.4 | 24.1 | 16.2 | 10.8 | 16.8 |
| 1-Propanol | 75.2 | 24.4 | 15.8 | 6.7 | 17.3 |
| 2-Propanol | 76.8 | 23.5 | 15.8 | 6.1 | 16.4 |
| Furfuryl alcohol | 86.5 | 25.6 | 17.4 | 7.6 | 15.1 |
| 1-Butanol | 91.5 | 23.1 | 15.9 | 5.7 | 15.7 |
| 2-Butanol | 92.0 | 22.1 | 15.8 | 5.7 | 14.5 |

| Solvent | $V_1$, Kmol/m$^3$ | $\delta$, $(MJ/m^3)^{1/2}$ | $\delta_d$, $(MJ/m^3)^{1/2}$ | $\delta_p$, $(MJ/m^3)^{1/2}$ | $\delta_h$, $(MJ/m^3)^{1/2}$ |
|---|---|---|---|---|---|
| 1-Pentanol | 108.3 | 21.6 | 15.9 | 4.5 | 13.9 |
| Benzyl alcohol | 103.6 | 24.8 | 18.4 | 6.3 | 13.7 |
| Cyclohexanol | 106.0 | 23.3 | 17.3 | 4.1 | 13.5 |
| Ethylene glycol monomethyl ether | 79.1 | 23.3 | 16.2 | 9.2 | 16.4 |
| Ethylene glycol monoethyl ether | 97.8 | 21.5 | 16.2 | 9.2 | 14.3 |
| Ethylene glycol monobutyl ether | 142.1 | 20.8 | 15.9 | 4.5 | 12.7 |
| 1-Octanol | 157.7 | 21.1 | 17.0 | 3.3 | 11.9 |
| m-Cresol | 104.7 | 22.7 | 18.0 | 5.1 | 12.9 |
| Carboxylic acids | | | | | |
| Formic acid | 37.8 | 24.8 | 14.3 | 11.9 | 16.6 |
| Acetic acid | 57.1 | 20.7 | 14.5 | 8.0 | 13.5 |
| n-Butyric acid | 110 | 21.5 | 14.9 | 4.1 | 10.6 |
| Polyhydric alcohols | | | | | |
| Ethylene glycol | 55.8 | 33.2 | 16.8 | 11.0 | 25.9 |
| Glycerol | 73.3 | 43.8 | 17.3 | 12.0 | 29.2 |
| Diethylene glycol | 95.3 | 29.8 | 16.0 | 14.7 | 20.4 |
| Triethylene glycol | 114.0 | 21.9 | 16.0 | 12.5 | 18.6 |
| Water | 18.0 | 47.9 | 15.5 | 16.0 | 42.3 |

### 4.1.3 MULTI-DIMENSIONAL APPROACHES

These approaches can be divided into three types:

1 H-bonds are not considered. This approach can be applied only for nonpolar and weak polar liquids.

2 H-bonds taken into account by one parameter.

3 H-bonds taken into account by two parameters.

Blanks and Prausnitz[8,9] proposed two-component solubility parameters. They decomposed the cohesion energy into two contributions of polar and non-polar components:

$$-\frac{E}{V_1} = -\frac{E_{nonpolar}}{V_1} - \frac{E_{polar}}{V_1} = \lambda^2 + \tau^2 \qquad [4.1.15]$$

where:

λ non-polar contribution to solubility parameter

τ polar contribution to solubility parameter

This approach has become a constituent of the Hansen approach and has not received a separate development.

Polar interactions can themselves be divided into two types:

- Polar interactions where molecules having permanent dipole moments interact in solution with the dipole orientation in a symmetrical manner. It follows that the geometric mean rule is obeyed for orientation interactions and the contribution of dipole orientations to the cohesive energy and dispersion interactions.
- Polar interactions accompanied by the dipole induction. These interactions are asymmetrical.

Thus for a pure polar liquid without hydrogen bonds:[10]

$$\delta^2 = \delta_d^2 + \delta_{or}^2 + 2\delta_d\delta_{in} \qquad [4.1.16]$$

where:

$\delta_d$ dispersion contribution to the solubility parameter
$\delta_{or}$ orientation contribution to the solubility parameter
$\delta_{in}$ induction contribution to the solubility parameter.

A more traditional approach of contribution of the induction interaction was published elsewhere;[11] however, it was used only for the estimation of the common value of the $\delta$ parameter rather than for evaluation of solubility:

$$\delta^2 = \delta_d^2 + \delta_p^2 + \delta_i^2 \qquad [4.1.17]$$

The first method taking into account the hydrogen bonding was proposed by Beerbower et al.,[12] who expressed hydrogen bonding energy through the hydrogen bonding number $\Delta\nu$. The data for various solvents were plotted into a diagram with the solubility parameter along the horizontal axis and the hydrogen bonding number $\Delta\nu$ along the vertical axis. Data were obtained for a given polymer for suitable solvents. All solvents in which a given polymer was soluble got a certain regions. Lieberman also plotted two-dimensional graphs of solubility parameters versus hydrogen-bonding capabilities.[13]

On the base of work by Gordy and Stanford, the spectroscopic criterion, related to the extent of the shift to lower frequencies of the OD infrared absorption of deuterated methanol, was selected. It provides a measure of the hydrogen-bonding acceptor power of a solvent.[14,15] The spectrum of a deuterated methanol solution in the test solution was compared with that of a solution in benzene and the hydrogen-bonding parameter was defined as

$$\gamma = \Delta\nu / 10 \qquad [4.1.18]$$

where:

$\Delta\nu$ OD absorption shift (in wavenumber).

Crowley et al.[16] used an extension of this method by including the dipole moment of the solvents. One of the axis represented solubility parameter, the second the dipole moment, and the third hydrogen bonding expressed by spectroscopic parameter $\gamma$. Because this method involved an empirical comparison of a number of solvents it was impractical. Nelson et al.[17] utilized this approach to hydrogen bond solubility parameters. Hansen (see the next section) developed this method.

Chen introduced a quantity $\chi_H$[18]

$$\chi_H = \frac{V_S}{RT}\left[\left(\delta_{d.S} - \delta_{d.P}\right)^2 + \left(\delta_{p.S} - \delta_{p.P}\right)^2\right] \qquad [4.1.19]$$

where $\delta_{d.S}$, $\delta_{d.P}$, $\delta_{p.S}$, $\delta_{p.P}$ are Hansen's parameters of solvent and polymer (see the next section).

Chen implied that $\chi_H$ was the enthalpy contribution to the Flory-Huggins parameter $\chi_1$ and plotted the solubility data in a $\delta_h - \chi_H$ diagram where $\delta_h$ was the H-bond parameter in the Hansen approach. In these diagrams sphere-like volumes of Hansen's solubility have degenerated to circles.

The disadvantage of this method lies in the beforehand estimating characteristics of the polymer. Among other two-dimensional methods used for the representation of solubility data was the $\delta_p$-$\delta_h$ diagram proposed by Henry[19] and the $\delta$-$\delta_h$ diagram proposed by Hoernschemeyer,[20] but their representations of the solubility region were less correct. All these approaches involving hydrogen bond parameter ignored the fact that hydrogen bond interaction was the product of hydrogen bonding donating and accepting capability.[21-23]

On the basis of chemical approach to hydrogen bonding, Rider proposed a model of solubility for liquids in which the enthalpy limited the miscibility of polymers and solvents.[24,25] For substances capable to form hydrogen bonds, Rider proposed a new factor relating miscibility with an enthalpy of mixing which depends on an enthalpy of the hydrogen bond formation. He has introduced the quantity of a hydrogen bond potential (HBP). If the quantity of HBP is positive it promotes miscibility and if it is negative it decreases miscibility.

$$HBP = \left(b_1 - b_2\right)\left(C_1 - C_2\right) \qquad [4.1.20]$$

where:

$b_1$,$b_2$ donor parameters of solvent and solute, respectively
$C_1$, $C_2$ acceptor parameters of solvent and solute, respectively

For certain polymers Rider has drawn solubility maps. Thus the area of solubility was represented by a pair of symmetric quarters of a plane lying in coordinates b,C.[24] Values of parameters were defined from data for enthalpies of hydrogen bonds available from the earlier works. The model is a logical development of the Hansen method. A shortcoming of this model is in neglecting all other factors influencing solubility, namely dispersion and polar interactions, change of entropy, molecular mass of polymer and its phase condition. The model was developed as a three-dimensional dualistic model (see Section 4.1.5).

## 4.1.4 HANSEN'S SOLUBILITY

The Hansen approach[26-30] assumed that the cohesive energy can be divided into contributions of dispersion interactions, polar interactions, and hydrogen bonding.

$$E = E_d + E_p + E_h \qquad [4.1.21]$$

where:

E total cohesive energy
$E_d$, $E_p$, $E_h$ contributions of dispersion forces, permanent dipole-permanent dipole forces, and hydrogen bonds.

Dividing this equation by the molar volume of solvent, $V_1$, gives:

$$\frac{E}{V_1} = \frac{E_d}{V_1} + \frac{E_p}{V_1} + \frac{E_h}{V_1} \qquad [4.1.22]$$

or

$$\delta^2 = \delta_d^2 + \delta_p^2 + \delta_h^2 \qquad [4.1.23]$$

where:

$\delta$ total solubility parameter

$\delta_d, \delta_p, \delta_h$ components of the solubility parameter determined by the corresponding contributions to the cohesive energy.

Hansen gave a visual interpretation of his method by means of three-dimensional spheres of solubility, where the center of the sphere has coordinates corresponding to the values of components of solubility parameter of polymer. The sphere can be coupled with a radius to characterize a polymer. All good solvents for particular polymer (each solvent has been represented as a point in a three-dimensional space with coordinates) should be inside the sphere, whereas all non-solvents should be outside the solubility sphere. An example is given in Section 4.1.7.

In the original work these parameters were evaluated by experimental observations of solubility. It was assumed that if each of the solubility parameter components of one liquid is close to the corresponding values of another liquid, then the process of their mixing should readily occur with a more negative free energy. The solubility volume has dimensions $\delta_d$, $\delta_p$, $2\delta_h$. The factor 2 was proposed to account for the spherical form of solubility volumes and had no physical sense. However, it is necessary to notice that, for example, Lee and Lee[31] have evaluated spherical solubility volume of polyimide with good results without using the factor 2. Because of its simplicity, the method has become very popular.

Using the Hansen approach, the solubility of any polymer in solvents (with known Hansen's parameters of polymer and solvents) can be predicted. The determination of polymer parameters requires evaluation of solubility in a great number of solvents with known values of Hansen parameters. Arbitrary criteria of determination are used because Hansen made no attempts of precise calculations of thermodynamic parameters.

The separation of the cohesion energy into contributions of various forces implies that it is possible to substitute energy for parameter and sum contributions proportional to the second power of a difference of corresponding components. Hansen's treatment permits evaluation of the dispersion and polar contribution to cohesive energy. The fitting parameter of the approach (the solubility sphere radius) reflects on the supermolecular structure of polymer-solvent system. Its values should be higher for amorphous polymers and lower for glass or crystalline polymers.

The weak point of the approach is the incorrect assignment of the hydrogen bond contribution in the energy exchange that does not permit its use for polymers forming strong hydrogen bonds.

**Table 4.1.2. Solubility parameters and their components for solvents (after refs 37,40)**

| Polymer | $\delta$, $(MJ/m^3)^{1/2}$ | $\delta_d$, $(MJ/m^3)^{1/2}$ | $\delta_p$, $(MJ/m^3)^{1/2}$ | $\delta_h$, $(MJ/m^3)^{1/2}$ |
|---|---|---|---|---|
| Polyamide-66 | 22.77 | 18.5 | 5.1 | 12.2 |

| Polymer | $\delta$, $(MJ/m^3)^{1/2}$ | $\delta_d$, $(MJ/m^3)^{1/2}$ | $\delta_p$, $(MJ/m^3)^{1/2}$ | $\delta_h$, $(MJ/m^3)^{1/2}$ |
|---|---|---|---|---|
| Polyacrylonitrile | 25.10 | 18.19 | 15.93 | 6.74 |
| Polyvinylchloride | 21.41 | 18.68 | 10.01 | 3.06 |
| Polymethylmethacrylate | 20.18 | 17.72 | 5.72 | 7.76 |
| Polystyrene | 19.81 | 19.68 | 0.86 | 2.04 |
| Polytetrafluoroethylene | 13.97 | 13.97 | 0.00 | 0 |
| Polyethyleneterephthalate | 21.6 | 19.5 | 3.47 | 8.58 |

A large number of data were accumulated for different solvents and polymers (see Tables 4.1.1, 4.1.2). A variation of the Hansen method is the approach of Teas.[33] He showed for some polymer-solvent systems that it was possible to use fractional cohesive energy densities plotted on a triangular chart to represent solubility limits:

$$E_d = \frac{\delta_d^2}{\delta_0^2}, \quad E_p = \frac{\delta_p^2}{\delta_0^2}, \quad E_h = \frac{\delta_h^2}{\delta_0^2} \qquad [4.1.24]$$

where $\delta_0^2 = \delta_d^2 + \delta_p^2 + \delta_h^2$

Teas used fractional parameters defined as

$$f_d = \frac{100\delta_d}{\delta_d + \delta_p + \delta_h}, \quad f_p = \frac{100\delta_p}{\delta_d + \delta_p + \delta_h}, \quad f_h = \frac{100}{\delta_d + \delta_p + \delta_h} \qquad [4.1.25]$$

This representation was completely empirical without any theoretical justification.

Some correlations between components of solubility parameters and physical parameters of liquids (surface tension, dipole moment, the refraction index) were generalized elsewhere.[11]

$$\delta_d^2 + 0.632\delta_p^2 + 0.632\delta_h^2 = 13.9V_1^{-1/3}\gamma_l \qquad \text{non-alcohols} \qquad [4.1.26]$$

$$\delta_d^2 + \delta_p^2 + 0.06\delta_h^2 = 13.9V_1^{-1/3}\gamma_l \qquad \text{alcohols} \qquad [4.1.27]$$

$$\delta_d^2 + 2\delta_p^2 + 0.48\delta_h^2 = 13.9V_1^{-1/3}\gamma_l \qquad \text{acids, phenols} \qquad [4.1.28]$$

where:

$\gamma_l$ surface tension.

Koenhan and Smolder proposed the following equation applicable to the majority of solvents, except cyclic compounds, acetonitrile, carboxylic acids, and multi-functional alcohols.[34]

$$\delta_d^2 + \delta_p^2 = 13.8V_1^{-1/3}\gamma_l \qquad [4.1.29]$$

They also proposed a correlation between polar contribution to the solubility parameter and refractive index:

$$\delta_d = 9.55n_D - 5.55 \quad [4.1.30]$$

where:

$n_D$ refractive index

Alternatively Keller et. al.[35] estimated that for nonpolar and slightly polar liquids

$$\delta_d = 62.8x \qquad \text{for } x \le 0.28 \quad [4.1.31]$$

$$\delta_d = -4.58 + 108x - 119x^2 + 45x^3 \qquad \text{for } x > 0.28 \quad [4.1.32]$$

where:

$$x = \frac{n_D^2 - 1}{n_D^2 + 2}$$

Peiffer suggested the following expression:[35]

$$\delta_d^2 = K\left(4\pi\pi l^2 / 3d\right)\left(N / V_1\right)^3 \quad [4.1.33]$$

where:

K packing parameter
I ionization potential
$\alpha$ molecular polarizability
N number of molecules in the volume unit
$V_1$ $=Nr^{*3}/K$
r* the equilibrium distance between molecules.

For the estimation of nonpolar component of δ Brown et al.[36] proposed the homomorph concept. The homomorph of a polar molecule is the nonpolar molecule most closely resembling it in the size and the structure (e.g., n-butane is the homomorph of n-butyl alcohol). The nonpolar component of the cohesion energy of a polar solvent is taken as the experimentally determined total vaporization energy of the corresponding homomorph at the same reduced temperature (the actual temperature divided by the critical temperature in Kelvin's scale). For this comparison the molar volumes must also be equal. Blanks and Prausnitz proposed plots of dependencies of dispersion energy density on a molar volume for straight-chain, alicyclic and aromatic hydrocarbons. If the vaporization energies of appropriate hydrocarbons are not known they can be calculated by one of the methods of group contributions (See Chapter 5).

Hansen and Scaarup[28] calculated the polar component of solubility parameter using Bottcher's relation to estimating the contribution of the permanent dipoles to the cohesion energy:

$$\delta_p^2 = \frac{12108}{V_1^2}\frac{\varepsilon - 1}{2\varepsilon - n_D^2}\left(n_D^2 + 2\right)\mu^2 \quad [4.1.34]$$

where:

$\varepsilon$ dielectric constant,
$\mu$ dipole moment

$$\delta_p = 50.1\frac{\mu}{V_1^{3/4}} \qquad [4.1.35]$$

Peiffer[11] proposed the expressions which separates the contributions of polar forces and induction interactions to the solubility parameters:

$$\delta_p^2 = K\left(2\pi\mu^4 / 3kTp\right)\left(N / V_1\right)^3 \qquad [4.1.36]$$

$$\delta_i^2 = K\left(2\pi\alpha\mu / i\right)\left(N / V_1\right)^3 \qquad [4.1.37]$$

where:

| | |
|---|---|
| ε* | interaction energy between two molecules at the distance r* |
| N | number of hydroxyl groups in molecule |

$$p = 2\mu^4 / 3kT\varepsilon^* \qquad [4.1.38]$$

$$i = 2\alpha\mu^2 / \varepsilon^* \qquad [4.1.39]$$

It should be noted that in Hansen's approach these contributions are cumulative:

$$\delta_p^2 = \delta_p^2 + \delta_i^2 \qquad [4.1.40]$$

For the calculation of hydrogen-bonding component, $\delta_h$, Hansen and Scaarup[28] proposed an empirical expression based on OH-O bond energy (5000 cal/mol) applicable to alcohols only:

$$\delta_h = \left(20.9N / V_1\right)^{1/2} \qquad [4.1.41]$$

In Subchapter 5.3, the values of all the components of a solubility parameter are calculated using group contributions.

### 4.1.5 THREE-DIMENSIONAL DUALISTIC MODEL

The heat of mixing of two liquids is expressed by the classical theory of regular polymer solutions using Eq. [4.1.10]. This expression is not adequate for systems with specific interactions. Such interactions are expressed as a product of the donor parameter and the acceptor parameter. The contribution of H-bonding to the enthalpy of mixing can be written in terms of volume units as follows:[21]

$$\Delta H'_{mix} = \left(A_1 - A_2\right)\left(D_1 - D_2\right)\varphi_1\varphi_2 \qquad [4.1.42]$$

where:

| | |
|---|---|
| $A_1$, $A_2$ | effective acceptor parameters |
| $D_1$, $D_2$ | donor parameters, |
| $\varphi_1$, $\varphi_2$ | volume fractions, |

Hence enthalpy of mixing of two liquids per volume unit can be expressed by:[32]

$$\Delta H_{mix} = \left[\left(\delta'_1 - \delta'_2\right)^2 + \left(A_1 - A_2\right)\left(D_1 - D_2\right)\right]\varphi_1\varphi_2 = B\varphi_1\varphi_2 \qquad [4.1.43]$$

This equation is used for the calculation of the enthalpy contribution to the Huggins parameter (see Subchapter 4.2) for a polymer-solvent system:

$$\chi_H = V_1\left[(\delta'_1 - \delta'_2)^2 + (A_1 - A_2)(D_1 - D_2)\right]/RT \qquad [4.1.44]$$

where:

$\delta'_1, \delta'_2$ dispersion-polar components of solubility parameters (values of solubility parameters excluding H-bonds contributions).

Results of calculations using Eq. [4.1.44] of three-dimension dualistic model coincide with the experimental values of $\chi_H$ and the $\chi_H$ values calculated by other methods.[37] Values A, D, and δ′ can been obtained from IR-spectroscopy evaluations and Hansen's parameters.[24,25] Values of the TDM parameters are presented in Tables 4.1.3, 4.1.4. It should be noted that Hansen parameters are used for estimation of values of TDM parameters from equation $\delta^2_{di} + \delta^2_{pi} = \delta'^2_i$, $\delta^2_{hi} = A_i D_i$.

**Table 4.1.3. TDM parameters of some solvents. [Adapted, by permission, from V.Yu. Senichev, V.V. Tereshatov , *Vysokomol . Soed* ., B31 , 216 (1989).]**

| # | Solvent | D | A | δ′ | δ | $V_1 \times 10^6$ $m^3/mol$ |
|---|---|---|---|---|---|---|
| | | $(MJ/m^3)^{1/2}$ | | | | |
| 1 | Isopropanol | 11.8 | 13.3 | 20.0 | 23.6 | 76.8 |
| 2 | Pentanol | 9.9 | 11.2 | 19.7 | 22.3 | 108.2 |
| 3 | Acetone | 3.8 | 13.1 | 18.5 | 19.8 | 74.0 |
| 4 | Ethyl acetate | 4.9 | 10.4 | 17.0 | 18.3 | 98.5 |
| 5 | Butyl acetate | 4.8 | 9.0 | 16.4 | 17.6 | 132.5 |
| 6 | Isobutyl acetate | 4.8 | 8.4 | 15.7 | 16.9 | 133.3 |
| 7 | Amyl acetate | 4.7 | 7.8 | 16.2 | 17.3 | 148.9 |
| 8 | Isobutyl isobutyrate | 4.7 | 7.4 | 14.6 | 15.7 | 165.0 |
| 9 | Tetrahydrofuran | 5.2 | 12.2 | 17.7 | 19.4 | 81.7 |
| 10 | o-Xylene | 0.5 | 7.4 | 18.3 | 18.4 | 121.0 |
| 11 | Chlorobenzene | 0.6 | 7.3 | 19.3 | 19.4 | 102.1 |
| 12 | Acetonitrile | 2.7 | 14.0 | 24.4 | 24.5 | 52.6 |
| 13 | n-Hexane | 0 | 0 | 14.9 | 14.9 | 132.0 |
| 14 | Benzene | 0.6 | 8.5 | 18.6 | 18.8 | 89.4 |
| 15 | N,N-Dimethylformamide | 8.2 | 15.4 | 22.0 | 27.7 | 77.0 |
| 16 | Toluene | 0.6 | 7.8 | 18.1 | 18.2 | 106.8 |
| 17 | Methanol | 18.2 | 17.0 | 23.5 | 29.3 | 41.7 |
| 18 | Ethanol | 14.4 | 14.7 | 21.7 | 26.1 | 58.5 |

| # | Solvent | D | A | δ′ | δ | $V_1 \times 10^6$ $m^3/mol$ |
|---|---|---|---|---|---|---|
| | | $(MJ/m^3)^{1/2}$ | | | | |
| 19 | 1-Propanol | 11.9 | 13.4 | 21.0 | 24.5 | 75.2 |
| 20 | Methyl ethyl ketone | 2.3 | 11.6 | 18.3 | 19.0 | 90.1 |
| 21 | Cyclohexanone | 2.7 | 11.1 | 19.5 | 20.3 | 104.0 |
| 22 | Diethyl ether | 7.6 | 12.4 | 12.0 | 15.4 | 104.8 |
| 23 | Ethylbenzene | 0.3 | 7.3 | 17.9 | 18.0 | 123.1 |
| 24 | Pyridine | 1.9 | 19.6 | 21.0 | 21.8 | 80.9 |
| 25 | Propyl acetate | 4.8 | 9.6 | 16.4 | 17.8 | 115.2 |
| 26 | 1,4-Dioxane | 8.3 | 12.9 | 17.7 | 20.5 | 85.7 |
| 27 | Aniline | 6.2 | 16.8 | 20.0 | 22.5 | 91.5 |

**Table 4.1.4. TDM parameters of some polymers, $(MJ/m^3)^{1/2}$. [Adapted, by permission, from V.Yu. Senichev, V.V. Tereshatov , *Vysokomol . Soed .*, B31 , 216 (1989)]**

| Polymer | D | A | δ′ | δ |
|---|---|---|---|---|
| Polymethylmethacrylate | 2.5 | 6.5 | 18.6 | 19.8 |
| Polyvinylacetate | 4.9 | 10.4 | 17.8 | 19.2 |
| Polystyrene | 0.3 | 7.3 | 17.9 | 18.0 |
| Polyvinylchloride | 11.6 | 10.6 | 16.1 | 19.5 |

$\delta_h$ can be separated into donor and acceptor components using values of enthalpies of the hydrogen bond formation between proton-donors and proton-acceptors. In the absence of such data it is possible to evaluate TDM parameters by means of analysis of parameters of compounds similar in the chemical structure. For example, propyl acetate parameters can been calculated by the interpolation of corresponding parameters of butyl acetate and ethyl acetate,[24] parameters of benzene can be calculated by decomposition of $\delta_h$ into acceptor and donor components in the way used for toluene elsewhere.[25]

The solubility prediction can be made using the relationship between solubility and the $\chi_1$ parameter (see Subchapter 4.2). The total value of the $\chi_1$ parameter can be evaluated by adding the entropy contribution:

$$\chi_1 = \chi_S + \chi_H \qquad [4.1.45]$$

where:

$\chi_S$ an empirical value. Usually it is 0.2-0.4 for good solvents.

The value of the parameter is inversely proportional to coordination number that is number of molecules of a solvent interacting with a segment of polymer. The value of the entropy contribution to the parameter should be included in solubility calculations. The value $\chi_S = 0.34$ is then used in approximate calculations.

## 4.1.6 SOLUBILITY CRITERION

The polymer superstructure influences its solubility. Askadskii and Matveev proposed a new criterion of solubility for linear polymers based on interaction of forces of a surface tension on wetting.[38] The solubility parameter of polymer should be lower or equal to the work of rupture by solvent of a bond relative to a volume unit of the bond element. The condition of solubility can be expressed as follows:

$$\mu \le 2\rho\Phi\left(\frac{\gamma_p}{\gamma_s}\right)^{1/2} \qquad [4.1.46]$$

where:

| | |
|---|---|
| $\mu$ | $= \delta_p^2 / \delta_s^2$ |
| $\delta_p$, $\delta_s$ | solubility parameters for polymer and solvent accordingly. |

$$\rho = \frac{\varepsilon_{max}^p r_s}{\varepsilon_{max}^s r_p} \qquad [4.1.47]$$

$$\Phi = \frac{4\left(V_p V_s\right)^{1/2}}{\left(V_p^{1/2} + V_s^{1/2}\right)^2} \qquad [4.1.48]$$

where:

| | |
|---|---|
| $\varepsilon_{max}^p$, $\varepsilon_{max}^s$ | maximum deformations of polymer and solvent at rupture |
| $r_s$, $r_p$ | characteristic sizes of Frenkel's swarms for solvent and and small radius of globule of bond for polymer, respectively |
| $V_p$, $V_s$ | molar volumes of polymer and solvent (per unit) |
| $\Phi \approx 1$ | |
| $\rho \approx$ const | |

The above expression was obtained with neglecting the preliminary swelling. Consideration of swelling requires correction for surface tension of swelled surface layers:

$$\mu < 2\rho\Phi\left[\Phi - \left(\Phi^2 - 1 + a\right)^{1/2}\right] \qquad [4.1.49]$$

where:

$$a = \gamma_{ps} / \gamma_p \qquad [4.1.50]$$

$$\gamma_{ps} = \gamma_p + \gamma_s - 2\Phi\left(\gamma_p \gamma_s\right)^{1/2} \qquad [4.1.51]$$

For practical purposes, the magnitude of $\rho$ estimated graphically is 0.687. Thus

$$\mu < 1.374\beta \qquad [4.1.52]$$

$$\text{for } \beta = \Phi\left[\Phi - \left(\Phi^2 - 1 + a\right)^{1/2}\right]$$

For both polymers and solvents, the values of solubility parameters can be obtained experimentally (see Subchapters 5.1, 5.3). The surface tension of polymer can be calculated using parahor:

$$\gamma = (P/V)^4$$ [4.1.53]

where:

V molar volume of a repeated polymer unit

Then the value of $V_p$ is calculated:

$$V_p = \frac{N_A \sum_i \Delta V_i}{k_{av}}$$ [4.1.54]

where:

$k_{av}$ = 0.681

If the density of polymer $d_p$ is known, then $V_p = M/d_p$, where M is the molecular mass of a repeating unit. The values of parahors are given in Table 4.1.5.

**Table 4.1.5. Values of parahors**

| Atom | C | H | O | $O_2$ | N | S | F | Cl | Br | I |
|---|---|---|---|---|---|---|---|---|---|---|
| P | 4.8 | 17.1 | 20.0 | 60.0 | 12.5 | 48.2 | 27.5 | 54.3 | 68.0 | 91.0 |

| Increment | Double bond | Triple bond | 3-member ring | 4-member ring | 5-member ring | 6-member ring |
|---|---|---|---|---|---|---|
| P | 23.2 | 46.4 | 16.7 | 11.6 | 8.5 | 6.1 |

The value of Φ is calculated from Eq. [4.1.48]. $V_p$, $V_s$ are defined from ratios $V_p$=M/$d_p$ and $V_s$=M/$d_s$ where $d_p$, $d_s$ are the densities of polymer and solvent, respectively. Then μ is calculated from Eq. [4.1.49]. The obtained value of μ from Eq. [4.1.49] is compared with value of $\mu = \delta_p^2 / \delta_s^2$ if the last value is lower or equal to the value of μ calculated from Eq. [4.1.49], polymer should dissolve in a given solvent with probability of 85 %.

## 4.1.7 SOLVENT SYSTEM DESIGN

**One-component system**. Solvents can be arranged in accordance to their solubility parameter as shown in Figure 4.1.1. It is apparent that a set of compatible solvents can be selected for polymer, determining their range based on the properties of polymer.

The simplest case is expressed by the Gee's equation for equilibrium swelling:[41]

$$Q = q_{max} \exp\left[-V_s\left(\delta_s - \delta_p\right)^2\right]$$ [4.1.55]

where:

$\delta_s$, $\delta_p$ solubility parameters for solvent and polymer.

The value of solubility parameter of solvent mixture with components having similar molar volumes is relative to their volume fractions and solubility parameters:

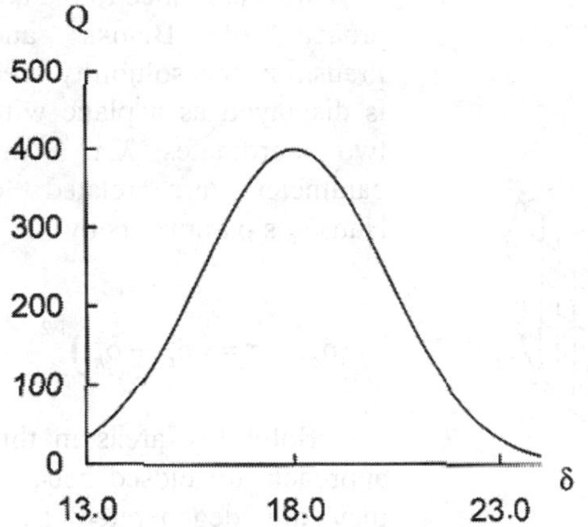


Figure 4.1.1. Representation of the one-component solubility parameter system. The curve represents Eq. [4.1.56] for polymer with $\delta = 18\ (MJ/m^3)^{1/2}$.

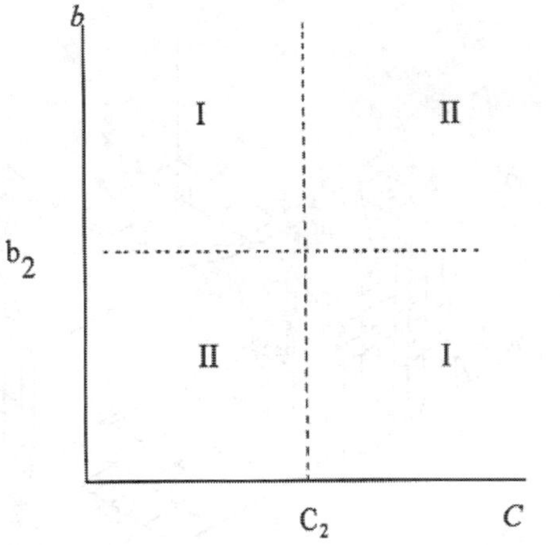


Figure 4.1.2. Representation of the two-component solubility parameter system in the Rider's approach. $b_2$, $C_2$ are the values of Rider's parameters for polymer.

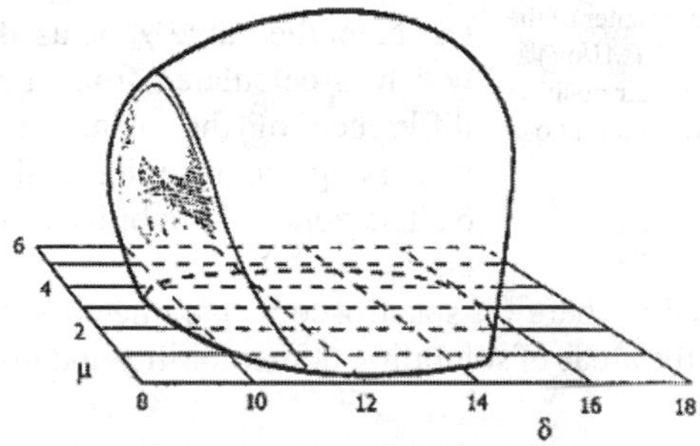


Figure 4.1.3. The solubility volume for cellulose acetate butyrate in terms of one-component solubility parameter, δ, dipole moment, μ, and (on vertical axis) the spectroscopic parameter, γ, from the approach developed by Crowley ct.al.[2] [Adapted, by permission, from J.D. Crowley, G.S. Teague and J.W. Lowe, *J. Paint Technol.*, **38**, 269 (1966)]

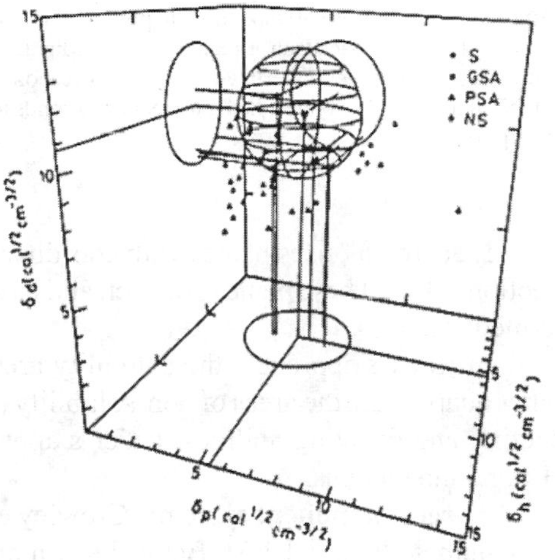


Figure 4.1.4. Hansen's solubility volume of polyimide synthesized from 3,3'4,4'-benzophenone tetracarboxylic dianhydride and 2,3,5,6-tetramethyl-p-phenylene diamine (after Lee[31]). [Adapted, by permission, from H.-R. Lee, Y.-D. Lee, *J. Appl. Polym. Sci.*, **40**, 2087 (1990)]

$$\bar{\delta} = \sum_i \delta_i \varphi_i \qquad [4.1.56]$$

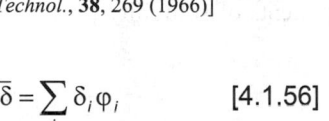

Solute is frequently soluble in a mixture of two non-solvents, for example, the mixture of diisopropyl ether ($\delta = 15.6\ (MJ/m^3)^{1/2}$) and ethanol ($\delta = 26.4\ (MJ/m^3)^{1/2}$) is a solvent for nitrocellulose ($\delta = 23\ (MJ/m^3)^{1/2}$).

**Two-component systems**. Two parametrical models of solubility use two-dimensional graphs of solubility area. Two-dimensional solubility areas may be closed or open.

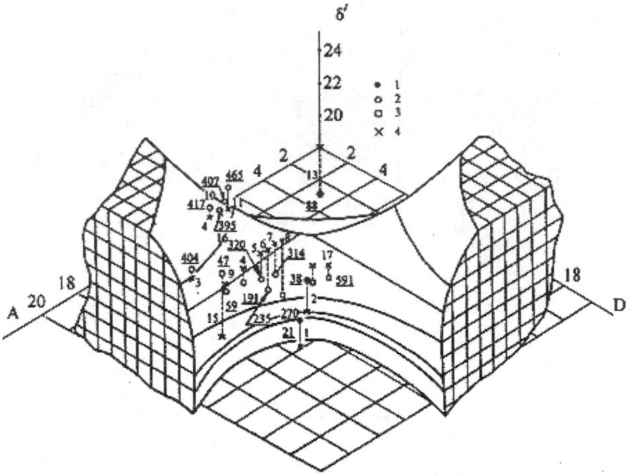

Figure 4.1.5. Volume of the increased swelling of crosslinked polybutadiene urethane elastomer, $\chi_H \le 0.85$. Labels of points: 1-outside volume, points with coordinates corresponding to solvents; 2-inside volume, points with coordinates corresponding to solvents; 3-the points with coordinates corresponding to polymer. This is the center of the volume. 4- points placed on a plane with coordinate $\delta = 18$ $(MJ/m^3)^{1/2}$. Number of solvent (not underlined number) corresponds to their position in the Table 4.1.3. The underlined number corresponds to swelling ratio at the equilibrium.

In accordance to the approach of Blanks and Prausnitz, the solubility area is displayed as a plane with two coordinates, $\lambda, \tau$.[8,9] The parameters are related to Hansen's parameters by

$$\lambda = \delta_d, \quad \tau = \left(\delta_p^2 + \delta_h^2\right)^{1/2}$$

Solubility areas in this approach are closed because they are degenerated from Hansen's spheres (see below). Another example of application of degenerate Hansen's spheres was given by Chen.[18] Instead of parameters $\delta_p$, $\delta_d$ the value $\chi_H$ is used which is calculated from the difference of the polar and dispersing contributions, $\delta_p$, $\delta_d$. The zone of solubility is a circle.

Lieberman[13] uses planes with coordinates $\delta$, $\gamma$ where $\gamma$ is spectroscopic parameter (see Section 4.1.3). These planes are open and have the areas of solubility, non-solubility and intermediate.

In Rider's approach, the solubility area is a system of two quarters on a plane; two other quarters are the areas of non-solubility (Figure 4.1.2). Coordinates of this plane are accepting and donating abilities. Rider's approach finds application for solvents with high H-bond interactions.

**Three-component systems**. Crowley et. al.[2] proposed the three-dimensional solubility volumes (Figure 4.1.3). Better known are Hansen's three-dimensional solubility volumes (Figure 4.1.4). In Hansen's approach, the components of solubility parameters for mixed solvents $\delta_j$ are calculated from Eq. [4.1.56]:

$$\overline{\delta}_j = \sum_i \delta_{ji} \varphi_i \qquad [4.1.57]$$

The choice of a solvent for polymer is based on coordinates of polymer in space of coordinates $\delta_d$, $\delta_p$, $\delta_h$ and the radius of a solubility volume. For mixed solvents, their coordinates can be derived by connecting coordinates of individual solvents. This can be used to determine synergism in solvents for a particular polymer but it cannot demonstrate antisynergism of solvent mixtures (a mixture is less compatible with polymer than the individual solvents).

In the TDM approach, the solubility volume has also three coordinates, but the H-bond properties are taken into account. For this reason solubility volume can be represented by hyperbolic paraboloid (Figure 4.1.5). The use of this model permits to evaluate potential of synergism and antisynergism of solvent mixtures. It can be demonstrated by position of a point of a solvent mixture moving from a zone of good compatibility into the similar zone through the zone of inferior compatibility or, on the contrary, from a zone of an incompatibility into a similar zone through the zone of the improved compatibility. In the case of polymers that have no hydrogen bond forming abilities, this approach is equivalent to the Hansen or Blanks-Prausnitz approaches.

The above review of the methods of solvent evaluation shows that there is a broad choice of various techniques. Depending on the complexity of solvent-polymer interactions, the suitable method can be selected. For example, if solvents and/or polymer do not have functional groups to interact with as simple method as one-dimensional model is adequate. If weak hydrogen bonding is present, Hansen's approach gives good results (see further applications in Subchapter 5.3). In even more complex interactions, TDM model is always the best choice.

## REFERENCES


1 W.W. Reynolds and E.C. Larson, *Off. Dig. Fed. Soc. Paint Technol.*, **34**, 677 (1972).
2 J.D. Crowley, G.S.Teague and J.W. Lowe, J. Paint Technol., **38**, 269 (1966).
3 H. Burrel, *Off. Dig. Fed. Paint. Varn. Prod. Clubs*, **27**, 726 (1955).
4 P.L. Huyskens, M.C.Haulait-Pirson, *Org. Coat. Sci. Technol.*, **8**, 155 (1986).
5 I.H. Hildebrand , *J. Amer. Chem. Soc.*, **38**, 1452, (1916).
6 G. Scatchard, *J. Amer. Chem. Soc.*, **56**, 995 (1934).
7 A.A. Tager, L.K. Kolmakova, *Vysokomol. Soed.*, **A22**, 483 (1980).
8 R.F. Blanks, J.M. Prausnitz, *Ind. Eng. Chem., Fundam.*, **3**, 1, (1964).
9 R.F. Weimer and J.M. Prausnitz, *Hydrocarbon Process.*, **44**, 237 (1965).
10 R.A Keller, B.L.Karger and L.R. Snyder, *Gas Chromatogr., Proc. Int Symp. (Eur)*, **8**, 125 (1971).
11 D.G. Peiffer, *J. Appl. Polym. Sci.*, **25**, 369 (1980).
12 A. Beerbower, L.A. Kaye and D.E. Pattison, *Chem. Eng.*, 118 (1967).
13 E.P. Lieberman, *Off. Dig. Fed. Soc. Paint Technol.*, **34**, 30 (1962).
14 W.Gordy and S.C. Stanford, *J. Chem. Phys.*, **8**, 170 (1940).
15 W.Gordy and S.C. Stanford, *J. Chem. Phys.*, **9**, 204 (1941).
16 J.D. Crowley, G.S. Teague and J.W. Lowe, *J. Paint Technol.*, **39**, 19 (1967).
17 R.C. Nelson, R.W. Hemwall and G.D Edwards, *J. Paint Technol.*, **42**, 636 (1970).
18 S.-A. Chen, *J. Appl. Polym. Sci.*, **15**, 1247 (1971).
19 L.F. Henry, *Polym. Eng. Sci.*, **14**, 167 (1974)
20 D. Hoernschemeyer, *J. Appl. Polym. Sci.*, **18**, 61 (1974).
21 P.A. Small, *J. Appl. Chem.*, **3**, 71 (1953).
22 H. Burrel, *Adv. Chem. Ser.*, No. 124, 1 (1973).
23 H. Renon, C.A Reckert and J.M. Prausnitz, *Ind. Eng. Chem.*, **3**, 1(1964).
24 P.M Rider, *J. Appl. Polym. Sci.*, **25**, 2975 (1980).
25 P.M Rider, *Polym. Eng. Sci.*, **23**, 180 (1983).
26 C.M. Hansen, *J. Paint. Technol.*, **39**, 104 (1967).
27 C.M. Hansen, *J. Paint. Technol.*, **39**, 505 (1967).
28 C.M. Hansen and K. Scaarup, *J. Paint. Technol.*, **39**, 511 (1967).
29 C.M. Hansen, **Three Dimensional Solubility Parameters and Solvent Diffusion Coefficient**, *Danish Technical Press*, Copenhagen, 1967.
30 C.M. Hansen and A. Beerbower in **Kirk-Othmer Encyclopedia of Chemical Technology**, Suppl. Vol., 2nd ed., A.Standen Ed., 889, 1971.
31 H.-R. Lee, Y.-D. Lee, *J. Appl. Polym.Sci.*, **40**, 2087 (1990).
32 V.Yu. Senichev, V.V. Tereshatov, *Vysokomol. Soed.*, **B31**, 216 (1989).
33 J.P.Teas, J. Paint Technol., **40**, 19 (1968).
34 D.M. Koenhen, C.A. Smolder, *J. Appl. Polym. Sci.*, **19**, 1163 (1975).

35 R.A Keller, B.L. Karger and L.R. Snyder, *Gas Chromtogr., Proc. Int Symp. (Eur)*, **8**, 125(1971).
36 H.C. Brown, G.K. Barbaras, H.L. Berneis, W.H. Bonner, R.B. Johannesen, M. Grayson and K.L. Nelson, *J. Amer. Chem. Soc.*, **75**, 1 (1953).
37 A.E. Nesterov, **Handbook on physical chemistry of polymers. V. 1. Properties of solutions**, *Naukova Dumka*, Kiev, 1984.
38 A.A. Askadskii, Yu.I. Matveev, M.S. Matevosyan, *Vysokomol. Soed.* **32**, 2157 (1990).
39 Yu.I. Matveev, A.A. Askadskii, *Vysokomol. Soed.*, **36**, 436 (1994).
40 S.A. Drinberg, E.F. Itsko, **Solvents for paint technology**, *Khimiya*, Leningrad, 1986.
41 A.F. Barton, *Chem Rev.*, **75**, 735 (1975).

## 4.2 EFFECT OF SYSTEM VARIABLES ON SOLUBILITY

VALERY YU. SENICHEV, VASILIY V. TERESHATOV
**Institute of Technical Chemistry**
**Ural Branch of Russian Academy of Sciences, Perm, Russia**

Solubility in solvents depends on various internal and external factors. Chemical structure, molecular mass of solute, and crosslinking of polymer fall into the first group of factors, in addition to temperature and pressure in the second group of factors involved.

### 4.2.1 GENERAL CONSIDERATIONS

The process of dissolution is determined by a combination of enthalpy and entropy factors. The dissolution description can be based on the Flory-Huggins equation. Flory[1-3] and Huggins[4] calculated the entropy of mixing of long-chain molecules under the assumption that polymer segments occupy sites of a "lattice" and solvent molecules occupy single sites.

The theory implies that the entropy of mixing is combinatorial, i.e., it is stipulated by permutations of molecules into solution in which the molecules of mixed components differ greatly in size. The next assumption is that $\Delta V_{mix} = 0$ and that the enthalpy of mixing does not influence the value of $\Delta S_{mix}$. The last assumptions are the same as in the Hildebrand theory of regular solutions.[5] The expression for the Gibbs energy of mixing is

$$\frac{\Delta G}{RT} = x_1 \ln \varphi_1 + x_2 \ln \varphi_2 + \chi_1 \varphi_1 \varphi_2 \left( x_1 + x_2 \frac{V_2}{V_1} \right) \qquad [4.2.1]$$

where:

$x_1, x_2$ molar fractions of solvent and polymer, respectively
$\chi_1$ Huggins interaction parameter

The first two terms result from the configurational entropy of mixing and are always negative. For $\Delta G$ to be negative, the $\chi_1$ value must be as small as possible. The theory assumes that the $\chi_1$ parameter does not depend on concentration without experimental confirmation.

$\chi_1$ is a dimensionless quantity characterizing the difference between the interaction energy of solvent molecule immersed in the pure polymer compared with interaction energy in the pure solvent. It is a semi-empirical constant. This parameter was introduced by Flory and Huggins in the equation for solvent activity to extend their theory for athermic processes to the non-athermic processes of mixing:

$$\ln a_1 = \frac{\Delta\mu_1}{RT} = \ln(1-\varphi_2) + \varphi_2 + \chi_1\varphi_2^2 \qquad [4.2.2]$$

where:

$$\chi_1 = z\Delta\varepsilon_{12}^* / kT \qquad [4.2.3]$$

$\Delta\varepsilon_{12}^* = 0.5(\varepsilon_{11}^* + \varepsilon_{22}^*) - \varepsilon_{12}^*$, $a_1$ - solvent activity, $\varepsilon_{11}$, $\varepsilon_{22}$ - energy of 1-1 and 2-2 contacts formation in pure components, $\varepsilon_{12}$- energy of 1-2 contacts formation in the mixture, $\mu_1$- chemical potential of solvent.

The critical value of $\chi_1$ sufficient for solubility of polymer having large molecular mass is 0.5. Good solvents have a low $\chi_1$ value. $\chi_1$ is a popular practical solubility criterion and comprehensive compilations of these values have been published.[6-9]

Temperature is another factor. It defines the difference between polymer and solvent. Solvent is more affected than polymer. This distinction in free volumes is stipulated by different sizes of molecules of polymer and solvent. The solution of polymer in chemically identical solvent should have unequal free volumes of components. It causes important thermodynamic consequences. The most principal among them is the deterioration of compatibility between polymer and solvent at high temperatures leading to phase separation.

The theory of regular solutions operates with solutions of spherical molecules. For the long-chain polymer molecules composed of segments, the number of modes of arrangement in a solution lattice differs from a solution of spherical molecules, and hence it follows the reduction in deviations from ideal entropy of mixing. It is clear that the polymer-solvent interactions differ qualitatively because of the presence of segments.

Some novel statistical theories of solutions of polymers use the $\chi_1$ parameter, too. They predict the dependence of the $\chi_1$ parameter on temperature and pressure. According to the Prigogine theory of deformable quasi-lattice, a mixture of a polymer with solvents of different chain length is described by the equation:[10]

$$R\chi_1 = A(r_A / T) + (BT / r_A) \qquad [4.2.4]$$

where:

| | |
|---|---|
| A, B | constants |
| $r_A$ | number of chain segments in homological series of solvents. |

These constants can be calculated from heats of mixing, values of parameter $\chi_1$, and from swelling ratios. The Prigogine theory was further developed by Patterson, who proposed the following expression:[11]

$$\chi_1 = \left(\frac{U_1}{RT}\right)\nu^2 + \left(\frac{C_{P1}}{2R}\right)\tau^2 \qquad [4.2.5]$$

where:

| | |
|---|---|
| $U_1$ | configuration energy ($-U_1$ - enthalpy energy) |
| $C_{P1}$ | solvent thermal capacity |
| $\nu$, $\tau$ | molecular parameters |

The first term of the equation characterizes distinctions in the fields of force of both sizes of segments of polymer and solvent. At high temperatures, in mixtures of chemically similar components, its value is close to zero. The second term describes the structural contribution to $\chi_1$ stipulated by the difference in free volumes of polymer and solvent. Both terms of the equation are larger than zero, and as temperature increases the first term decreases and the second term increases. The expression can be given in a reduced form (with some additional substitutions):[12]

$$\frac{\chi_1}{V_1^*} = \frac{P_1^*}{RT_1^*}\left[\frac{\tilde{V}_1^{1/3}}{\tilde{V}_1^{1/3}-1}\left(\frac{X_{12}}{P_1^*}\right) + \frac{\tilde{V}_1^{1/3}}{2\left(4/3-\tilde{V}_1^{1/3}\right)}\tau^2\right] \qquad [4.2.6]$$

where:

$\tilde{V}_1, P_1^*, T_1^*$ reduced molar volume of solvent, pressure and temperature consequently
$X_{12}$ contact interaction parameter.

These parameters can be calculated if factors of the volumetric expansion, isothermal compressibility, thermal capacity of a solvent and enthalpy of mixing of solution components are known.

With temperature decreasing, the first term of the right side of the expression [4.2.6] increases and the second term decreases. Such behavior implies the presence of the upper and lower critical temperatures of mixing. Later Flory developed another expression for $\chi_1$ that includes the parameter of contact interactions, $X_{12}$:[13,14]

$$\chi_1 = \frac{P_1^* V_1^*}{\tilde{V}_1 RT}\left[\left(\frac{s_2}{s_1}\right)^2 \frac{X_{12}}{P_1^*} + \frac{\alpha_1 T}{2}\left\{\left(\frac{P_2^*}{P_1^*}\right)\tau - \frac{s_2}{s_1}\frac{X_{12}}{P_1^*}\right\}^2\right] \qquad [4.2.7]$$

where:

$s_1, s_2$ ratios of surfaces of molecules to their volumes obtained from structural data.

The large amount of experimental data is then an essential advantage of the Flory's theory.[9] Simple expressions exist for parameter $X_{12}$ in the terms of $X_{ij}$ characteristic parameters for chemically different segments of molecules of components 1 and 2. Each segment or chemical group has an assigned value of characteristic length ($\alpha_i$, $\alpha_j$) or surface area as a fraction of the total surface of molecule:[15]

$$X_{12} = \sum_{i,j}\left(\alpha_{i,1} - \alpha_{i,2}\right)\left(\alpha_{j,1} - \alpha_{j,2}\right)X_{ij} \qquad [4.2.8]$$

Bondi's approach may be used to obtain surface areas of different segments or chemical groups.[16] To some extent Huggins' new theory[17-21] is similar to Flory's theory.

### 4.2.2 CHEMICAL STRUCTURE

Chemical structure and the polarity determine dissolution of polymers. If the bonds in polymer and solvent are similar, then the energy of interaction between homogeneous and heterogeneous molecules is nearly identical which facilitates solubility of polymer. If the chemical structure of polymer and solvent molecule differ greatly in polarity, then swelling and dissolution does not happen. It is reflected in an empirical rule that "like dissolves like".

Nonpolar polymers (polyisoprene, polybutadiene) mix infinitely with alkanes (hexane, octane, etc.) but do not mix with such polar liquids as water and alcohols. Polar polymers (cellulose, polyvinylalcohol, etc.) do not mix with alkanes and readily swell in water. Polymers of the average polarity dissolve only in liquids of average polarity. For example, polystyrene is not dissolved or swollen in water and alkanes but it is dissolved in aromatic hydrocarbons (toluene, benzene, xylene), methyl ethyl ketone and some ethers. Polymethylmethacrylate is not dissolved nor swollen in water nor in alkanes but it is dissolved in dichloroethane. Polychloroprene does not dissolve in water, restrictedly swells in gasoline and dissolves in 1,2-dichloroethane and benzene. Solubility of polyvinylchloride was considered in terms of relationship between the size of a solvent molecule and the distance between polar groups in polymer.[22]

The above examples are related to the concept of the one-dimensional solubility parameter. However the effects of specific interactions between some functional groups can change compatibility of the system. Chloroalkanes compared with esters are known to be better solvents for polymethylmethacrylate. Aromatic hydrocarbons although having solubility parameters much higher than those of alkanes, dissolve some rubbers at least as well as alkanes. Probably it is related to increase in entropy change of mixing that has a positive effect on solubility.

The molecular mass of polymer significantly influences its solubility. With molecular mass of polymer increasing, the energy of interaction between chains also increases. The separation of long chains requires more energy than with short chains.

### 4.2.3 FLEXIBILITY OF A POLYMER CHAIN

The dissolution of polymer is determined by chain flexibility. The mechanism of dissolution consists of separating chains from each other and their transfer into solution. If a chain is flexible, its segments can be separated without a large expenditure of energy. Thus functional groups in polymer chain may interact with solvent molecules.

Thermal movement facilitates swelling of polymers with flexible chains. The flexible chain separated from an adjacent chain penetrates easily into solvent and the diffusion occurs at the expense of sequential transition of links.

The spontaneous dissolution is accompanied by decrease in free energy ($\Delta G < 0$) and that is possible at some defined values of $\Delta H$ and $\Delta S$. At the dissolution of high-elasticity polymers $\Delta H \geq 0$, $\Delta S > 0$ then $\Delta G < 0$. Therefore high-elasticity polymers are dissolved in solvents completely.

The rigid chains cannot move gradually because separation of two rigid chains requires large energy. At usual temperatures the value of interaction energy of links between polymer chains and molecules of a solvent is insufficient for full separation of polymer chains. Amorphous linear polymers with rigid chains having polar groups swell in polar liquids but do not dissolve at normal temperatures. For dissolution of such polymers, the interaction between polymer and solvent (polyacrylonitrile in N,N-dimethylformamide) must be stronger.

Glassy polymers with a dense molecular structure swell in solvents with the heat absorption $\Delta H > 0$. The value of $\Delta S$ is very small. Therefore $\Delta G > 0$ and spontaneous dissolution is not observed and the limited swelling occurs. To a greater degree this concerns crystalline polymers which are dissolved if $\Delta H < 0$ and $|\Delta H| > |T\Delta S|$.

When molecular mass of elastic polymers is increased, $\Delta H$ does not change but $\Delta S$ decreases. The $\Delta G$ becomes less negative. In glassy polymers, the increase in molecular mass is accompanied by a decrease in ΔH and ΔS. The ΔS value changes faster than the ΔH value, therefore the ΔG value becomes more negative, which means that the dissolution of polymeric homologues of the higher molecular weight becomes less favorable.

Crystalline polymers dissolve usually less readily than amorphous polymers. Dissolution of crystalline polymers requires large expenditures of energy for chain separation. Polyethylene swells in hexane at the room temperature and dissolves at elevated temperature. Isotactic polystyrene does not dissolve at the room temperature in solvents capable to dissolve atactic polystyrene. To be dissolved, isotactic polystyrene must be brought to elevated temperature.

## 4.2.4 CROSSLINKING

The presence of even a small amount of crosslinks hinders chain separation and polymer diffusion into solution. Solvent can penetrate into polymer and cause swelling. The swelling degree depends on crosslink density and compatibility of polymer and solvent.

The correlation between thermodynamic parameters and the value of an equilibrium swelling is given by Flory-Rehner equation[23] used now in a modified form:[24]

$$\ln(1-\varphi_2)+\varphi_2+\chi_1\varphi_2^2=-\frac{\nu_2}{V}V_s\left(\varphi_2^{1/3}-\frac{2\varphi_2}{f}\right) \quad [4.2.9]$$

where:

| | |
|---|---|
| $\varphi_2$ | polymer volume fraction in a swollen sample |
| $\nu_2/V$ | volume concentration of elastically active chains |
| f | the functionality of polymer network |

The value of $\nu_2/V$ is determined by the concentration of network knots. These knots usually have a functionality of 3 or 4. This functionality depends on the type of curing agent. Crosslinked polyurethanes cured by polyols with three OH-groups are examples of the three-functional network. Rubbers cured through double bond addition are examples of four-functional networks.

Eq. [4.2.9] has different forms depending on the form of elasticity potential but for practical purposes (evaluation of crosslinking density of polymer networks) it is more convenient to use the above form. The equation can be used in a modified form if the concentration dependence of the parameter $\chi_1$ is known.

The value of equilibrium swelling can be a practical criterion of solubility. Good solubility of linear polymers is expected if the value of equilibrium swelling is of the order of 300-400%. The high resistance of polymers to solvents indicates that the equilibrium swelling does not exceed several percent.

In engineering data on swelling obtained at non-equilibrium conditions (for example, for any given time), swelling is frequently linked to the diffusion parameters of a system (see more on this subject in Subchapter 6.1).[25]

An interesting effect of swelling decrease occurs when swollen polymer is placed in a solution of linear polymer of the same structure as the crosslinked polymer. The decrease of solvent activity causes this effect. The quantitative description of these processes can be made by the scaling approach.[26]

## 4.2.5 TEMPERATURE AND PRESSURE

The temperature effect on solubility may have different characters depending on the molecular structure of solute. For systems of liquid-amorphous polymer or liquid-liquid, the tem-

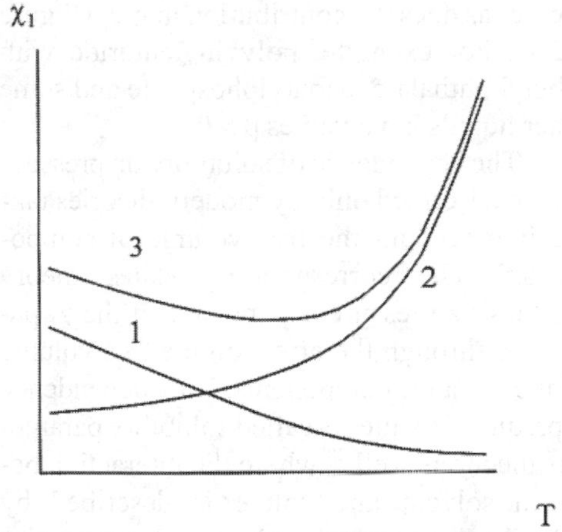


Figure 4.2.1 Two contributions to the $\chi_1$ parameter. 1-$\chi_H$, 2 -$\chi_S$, 3 - the total value of $\chi_1$.

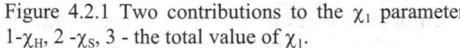

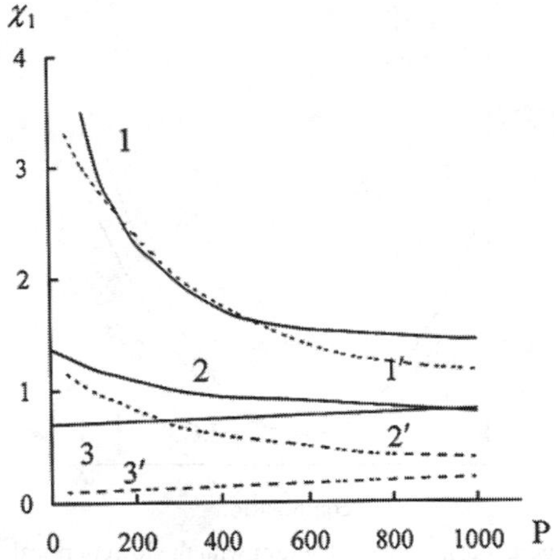


Figure 4.2.2 The $\chi_1$ parameter as a function of pressure at T = 300K, $r_1 = 3.5$. The curves are for the following values of $\varepsilon_{11}^* / \varepsilon_{22}^* = 0.85$; 1.0 and 1.3 (After refs.[9,28]).

perature raise can cause improvement of compatibility. Such systems are considered to have the upper critical solution temperature (UCST). If the system of two liquids becomes compatible at any ratio at the temperature below the defined critical point, the system is considered to have the lower critical solution temperature (LCST). Examples of a system with UCST are mixtures of methyl ethyl ketone-water (150°C) and phenol-water (65.8°C). An example of a system with LCST is the mixture of water-triethylamine (18°C). There are systems with both critical points, for example, nicotine-water and glycerol-benzyl-ethylamine.

Presence of UCST in some cases means a rupture of hydrogen bonds on heating; however, in many cases, UCST is not determined by specific interactions, especially at high temperatures, and it is close to critical temperature for the liquid-vapor system.

There are suppositions that the UCST is the more common case but one of the critical points for polymer-solvent system is observed only at high temperatures. For example, polystyrene ($M = 1.1 \times 10^5$) with methylcyclopentane has LCST 475K and UCST 370K. More complete experimental data on the phase diagrams of polymer-solvent systems are published elsewhere.[27]

The solubility of crystalline substances increases with temperature increasing. The higher the melting temperature of the polymer, the worse its solubility. Substances having higher melting heat are less soluble, with other characteristics being equal. Many crystalline polymers such as polyethylene or polyvinylchloride are insoluble at the room temperature (only some swelling occurs); however, at elevated temperature they dissolve in some solvents.

The experimental data on the temperature dependence of $\chi_1$ of polymer-solvent systems are described by the dependence $\chi_1 = \alpha + \beta / T$. Often in temperatures below 100°C, $\beta < 0$.[9] In a wide temperature range, this dependence decreases non-linearly. The negative contribution to $\Delta S$ and positive contribution to the $\chi_1$ parameter are connected with the difference of free volume. On heating the difference in free volumes of polymer and solvent in-

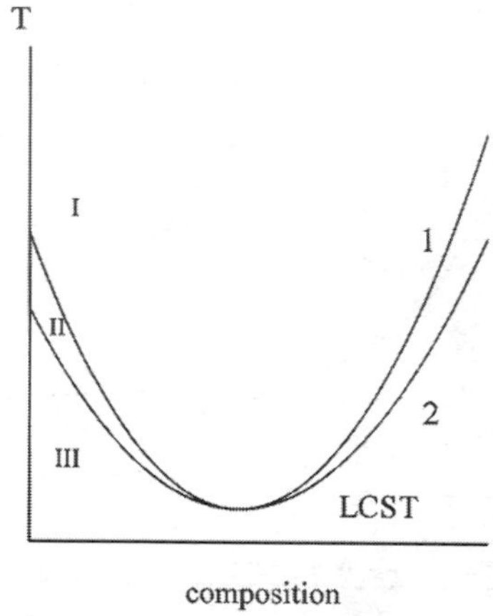


Figure 4.2.3. A phase diagram with the lower critical solution temperature (LCST).1-binodal, 2-spinodal; I- zone of non-stable conditions, II- zone of metastable conditions, III- zone of the one phase conditions.

creases as does the contribution into $\chi_S$ (Figure 4.2.1). For example, polyvinylchloride with dibutyl phthalate, tributylphosphate and some other liquids have values $\beta > 0$.

The dependence of solubility on pressure can be described only by modern theories taking into account the free volume of components.[28] The corresponding states theory predicts[12] a pressure dependence of the $\chi_1$ parameter through the effect on the free volume of the solution components. This dependence is predicted by the so-called solubility parameters theory as well,[28] where the interaction between solvent and solute is described by solubility parameters with their dependencies on temperature and pressure (Fig. 4.2.2). When $\varepsilon^*_{11} / \varepsilon^*_{22} \le 1$ then $\delta_1 < \delta_2$ and hence $\partial\chi_1 / \partial P < 0$ in the solubility parameters theory and in the corresponding states theory. When $\varepsilon^*_{11} / \varepsilon^*_{22}$ is greater than unity $\delta_1 > \delta_2$ and the solvent becomes less compressible than the polymer. Then pressure can increase the $(\delta_1 - \delta_2)$ value, giving $\partial\chi_1 / \partial P > 0$.

## 4.2.6 METHODS OF CALCULATION OF SOLUBILITY BASED ON THERMODYNAMIC PRINCIPLES

Within the framework of the general principles of thermodynamics of solutions, the evaluation of solubility implies the evaluation of value of the Gibbs energy of mixing in the whole range of concentrations of solution. However, such evaluation is difficult and for practical purposes frequently unnecessary. The phase diagrams indicate areas of stable solutions. But affinity of solvent to polymer in each of zone of phase diagram differs. It is more convenient to know the value of the interaction parameter, possibly with its concentration dependence. Practical experience from solvent selection for rubbers gives foundations for use of equilibrium swelling of a crosslinked elastomer in a given solvent as a criterion of solubility. The equilibrium swelling is related to $\chi_1$ parameter by Eq. [4.2.9]. As previously discussed in Subchapter 4.1, the value of the $\chi_1$ parameter can be determined as a sum of entropy and enthalpy contributions. In the one-dimensional solubility parameter approach, one may use the following equation:

$$\chi_1 = \chi_S + \frac{(\delta_1 - \delta_2)^2 V_1}{RT} \qquad [4.2.10]$$

where:

$\chi_S$ the entropy contribution

In TDM approach, Eq. [4.1.45] can be used. Similar equations can be derived for the Hansen approach. All existing systems of solubility imply some constancy of the entropy contribution or even constancy in some limits of a change of cohesion characteristics of polymers. Frequently $\chi_1 = 0.34$ is used in calculations.

Phase diagrams are characterized by critical temperatures, spinodals and binodals. A binodal is a curve connecting equilibrium structures of a stratified system. A spinodal is a curve defining boundary of metastables condition (Fig.4.2.3).

Binodals are evaluated experimentally by light scattering at cloud point,[29] by volume changes of coexisting phases,[30] or by the electron probe R-spectral analysis.[31]

It is possible to calculate phase behavior, considering that binodals correspond to a condition:

$$(\Delta\mu_i)' = (\Delta\mu_i)'' \quad [4.2.11]$$

where:

i a component of a solution

$(\Delta\mu_i)', (\Delta\mu_i)''$ changes of a chemical potential in phases of a stratified system

The equation of the spinodal corresponds to the condition

$$\frac{\partial^2(\Delta G)}{\partial\varphi_i^2} = \frac{\partial(\Delta\mu_i)}{\partial\varphi_i} = 0 \quad [4.2.12]$$

where:

$\Delta G$ the Gibbs free mixing energy

$\varphi_i$ volume fraction of a component of a solvent.

At a critical point, binodal and spinodal coincide

$$\frac{\partial^2(\Delta G)}{\partial\varphi_i^2} = \frac{\partial^3(\Delta G)}{\partial\varphi_i^3} \quad [4.2.13]$$

In the elementary case of a two-component system, the Flory-Huggins theory gives the following solution:[3]

$$\Delta\mu_i = RT\left[\ln\varphi_i + \left(1 - \frac{r_i}{r_j}\right)(1-\varphi_i) + r_i\chi_1\varphi_i^2\right] \quad [4.2.14]$$

where:

$r_i, r_j$ numbers of segments of corresponding component.

The last equation can be solved if one takes into account the equality of chemical potentials of a component in two co-existing phases of a stratified system.

$$\ln\varphi_i' + \left(1 - \frac{x_i}{x_j}\right)\varphi_j' + x_i\chi_{ij}(\varphi_i')^2 = \ln\varphi_i'' + \left(1 - \frac{x_i}{x_j}\right)\varphi_j'' + x_i\chi_{ij}(\varphi_i'')^2 \quad [4.2.15]$$

## REFERENCES


1 P.J. Flory, *J. Chem. Phys.*, **9**, 660, (1941).
2 P.J. Flory, *J. Chem. Phys.*, **10**, 51 (1942).
3 P.J. Flory, **Principles of polymer chemistry**, *Cornell University Press*, Ithaca, 1953.
4 M.L. Huggins, *J. Chem. Phys.*, **9**, 440 (1941).

5 J.H. Hildebrand and R.L. Scott, **Solubility of none-electrolytes**. 3rd ed., *Reinhold*, New-York, 1950.
6 R.M.Masegosa, M.G. Prolonga, A. Horta, *Macromolecules*, **19**, 1478 (1986)
7 G.M. Bristow, *J. Polym. Sci.*, **36**, 526 (1959).
8 J. Rehner, *J. Polym. Sci.*, **46**, 550 (1960).
9 A.E. Nesterov, **Handbook on physical chemistry of polymers. V. 1. Properties of solutions**, *Naukova Dumka*, Kiev, 1984.
10 I. Prigogine, **The molecular theory of solutions**, New York, **Interscience**, 1959.
11 D. Patterson, G. Delmas, T. Somsynsky, *J. Polym. Sci.*, **57**, 79 (1962).
12 D. Patterson, *Macromolecules*, **2**, 672 (1969).
13 P.J. Flory, R.A. Orwoll, A Vrij, *J. Amer. Chem. Sci.*, **86**, 3507 (1968).
14 P.J. Flory, *J. Amer. Chem. Sci.*, **87**, 1833 (1965).
15 A. Abe, P.J Flory, *J. Amer. Chem. Sci.*, **87**, 1838 (1965).
16 V. Crescenzi, G. Manzini, *J. Polym. Sci., C*, **54**, 315 (1976).
17 A. Bondi, **Physical properties of molecular crystals, liquids and glasses**. New York, *Wiley*, 1968
18 M.L. Huggins, *J. Phys. Chem.*, **74**, 371 (1970).
19 M.L. Huggins, *Polymer*, **12**, 389 (1971).
20 M.L. Huggins, *J. Phys. Chem.*, **75**, 1255 (1971).
21 M.L. Huggins, *Macromolecules*, **4**, 274 (1971).
22 K. Thinius, *Chem. Techn.*, **6**, 330 (1954).
23 P.J. Flory, J. Rehner, *J. Chem. Phys.*, **11**, 512 (1943).
24 A.E. Oberth, R.S. Bruenner, *J. Polym. Sci.*, **8**, 605 (1970).
25 U.S. Aithal, T.M. Aminabhavi, *Polymer*, **31**, 1757 (1990).
26 J. Bastide, S. Candau, L. Leibner, *Macromolecules*, **14**, 319 (1981).
27 A.E.Nesterov, Y.S. Lipatov, **Phase condition of polymer solutions and mixtures**, *Naukova dumka*, Kiev, 1987.
28 J. Biros, L. Zeman, D.Patterson, *Macromolecules*, **4**, 30 (1971).
29 J.W. Kennedy , M. Gordon, G.A. Alvarez, *J. Polym. Sci.*, **20**, 463 (1975).
30 R. Konningsveld, *Brit. Polym. J.*, **7**, 435 (1975).
31 N.N. Avdeev, A.E. Chalych, Y.N. Moysa, R.C. Barstein, *Vysokomol. soed.*, **A22**, 945 (1980).


## 4.3 POLAR SOLVATION DYNAMICS: THEORY AND SIMULATIONS


ABRAHAM NITZAN

**School of Chemistry,**

**The Sackler Faculty of Sciences, Tel Aviv University, Tel Aviv, Israel**


### 4.3.1 INTRODUCTION

When an ion or, more generally, a charge distribution associated with a given solute is placed in a dielectric solvent, the solvent responds to accommodate this solute and to minimize the free energy of the overall system. Equilibrium aspects of this phenomenon are related to macroscopic observables such as solvation free energy and microscopic properties such as the structure of the solvation 'shell'. Dynamical aspects of this process are manifested in the time evolution of this solvent response.[1] A direct way to observe this dynamics is via the time evolution of the spectral line-shifts following a pulse excitation of a solute molecule into an electronic state with a different charge distribution.[1] Indirectly, this dynamics can have a substantial effect on the course and the rate of chemical reactions that involve a redistribution of solute charges in the polar solvent environment.[2] Following a brief introduction to this subject within the framework of linear response and continuum dielectric theories, this chapter describes numerical simulation studies of this process, and contrasts the results obtained from such simulations with those obtained from linear response continuum models. In particular we focus on the following issues:

- How well can the solvation process be described by linear response theory?
- To what extent can the dynamics of the solvation process be described by continuum dielectric theory?
- What are the signatures of the solute and solvent structures in the deviation of the observed dynamics from that predicted by continuum dielectric theory?
- What are the relative roles played by different degrees of freedom of the solvent motion, in particular, rotation and translation, in the solvation process?
- How do inertial (as opposed to diffusive) solvent motions manifest themselves in the solvation process?

This chapter is not an exhaustive review of theoretical treatments of solvation dynamics. Rather, it provides, within a simple model, an exposition of the numerical approach to this problem. It should be mentioned that a substantial effort has been recently directed towards developing a theoretical understanding of this phenomenon. The starting point for such analytical efforts is linear response theory. Different approaches include the dynamical mean spherical approximation (MSA),[3,4] generalized transport equations,[5-8] and ad hoc models for the frequency and wavevector dependence of the dielectric response function $\varepsilon(k, \omega)$.[9] These linear response theories are very valuable in providing fundamental understanding. However, they cannot explore the limits of validity of the underlying linear response models. Numerical simulations can probe non-linear effects, but are very useful also for the direct visualization and examination of the interplay between solvent and solute properties and the different relaxation times associated with the solvation process. A substantial number of such simulations have been carried out in recent years.[10,11] The present account describes the methodology of this approach and the information it yields.

## 4.3.2 CONTINUUM DIELECTRIC THEORY OF SOLVATION DYNAMICS

The Born theory of solvation applies continuum dielectric theory to the calculation of the solvation energy of an ion of charge q and radius a in a solvent characterized by a static dielectric constant, $\varepsilon_s$. The well known result for the solvation free energy, i.e., the reversible work needed to transfer an ion from the interior of a dielectric solvent to vacuum, is

$$W = \frac{q^2}{2a}\left(\frac{1}{\varepsilon_s} - 1\right) \quad [4.3.1]$$

Eq. [4.3.1] corresponds only to the electrostatic contribution to the solvation energy. In experiments where the charge distribution on a solute molecule is suddenly changed (e.g. during photoionization of the solute) this is the most important contribution because short range solute-solvent interactions (i.e., solute size) are essentially unchanged in such processes. The origin of W is the induced polarization in the solvent under the solute electrostatic field.

The time evolution of this polarization can be computed from the dynamic dielectric properties of the solvent expressed by the dielectric response function $\varepsilon(\omega)$.[12] Within the usual linear response assumption, the electrostatic displacement and field are related to each other by

$$D(t) = \int_{-\infty}^{t} dt' \varepsilon(t-t')E(t') \qquad [4.3.2]$$

and their Fourier transforms (e.g. $E(\omega) = \int_{-\infty}^{\infty} dt e^{-i\omega t} E(t)$) satisfy

$$D(\omega) = \varepsilon(\omega)E(\omega) \qquad [4.3.3]$$

where

$$\varepsilon(\omega) \equiv \int_{0}^{\infty} dt e^{-i\omega t} \varepsilon(t) \qquad [4.3.4]$$

rewriting ε(t) in the form

$$\varepsilon(t) = 2\varepsilon_e \delta(t) + \tilde{\varepsilon}(t) \qquad [4.3.5]$$

we get

$$D(t) = \varepsilon_e E(t) + \int_{-\infty}^{t} dt' \tilde{\varepsilon}(t-t')E(t') \qquad [4.3.6]$$

$$D(\omega) = \varepsilon_e E(\omega) + \tilde{\varepsilon}(\omega)E(\omega) \qquad [4.3.7]$$

In Eq. [4.3.5] $\varepsilon_e$ is the "instantaneous" part of the solvent response, associated with its electronic polarizability. For simplicity we limit ourselves to the Debye model for dielectric relaxation in which the kernel $\tilde{\varepsilon}$ in [4.3.5] takes the form

$$\tilde{\varepsilon}(t) = \frac{\varepsilon_s - \varepsilon_e}{\tau_D} e^{-t/\tau_D} \qquad [4.3.8]$$

This function is characterized by three parameters: the electronic $\varepsilon_e$ and static $\varepsilon_s$ response constants, and the Debye relaxation time, $\tau_D$. In this case

$$\varepsilon(\omega) = \varepsilon_e + \int_{0}^{\infty} dt \frac{\varepsilon_s - \varepsilon_e}{\tau_D} e^{-t/\tau} e^{-i\omega t} = \varepsilon_e + \frac{\varepsilon_s - \varepsilon_e}{1 + i\omega\tau_D} \qquad [4.3.9]$$

In this model a step function change in the electrostatic field

$$E(t) = 0, t < 0; \quad E(t) = E, t \geq 0 \qquad [4.3.10]$$

leads to

$$D(t) = \varepsilon_e E(t) + \int_{0}^{t} \frac{\varepsilon_s - \varepsilon_e}{\tau_D} e^{-(t-t')/\tau_D} E(t')dt' = \left[\varepsilon_s\left(1 - e^{-t/\tau_D}\right) + \varepsilon_e e^{-t/\tau_D}\right]E \qquad [4.3.11]$$

For t→ 0, D(t) becomes $\varepsilon_e E$, and for t→ ∞ it is $D = \varepsilon_s E$. The relaxation process which carries the initial response to its final value is exponential, with the characteristic relaxation time, $\tau_D$.

The result [4.3.11] is relevant for an experiment in which a potential difference is suddenly switched on and held constant between two electrodes separated by a dielectric spacer. This means that the electrostatic field is held constant as the solvent polarization relaxes. For this to happen the surface charge density on the electrodes, i.e. the dielectric displacement **D**, has to change under the voltage source so as to keep the field constant.

The solvation dynamics experiment of interest here is different: Here at time t = 0 the charge distribution ρ(**r**) is switched on and is kept constant as the solvent relaxes. In other words, the dielectric displacement **D**, the solution of the Poisson equation $\nabla \mathbf{D} = 4\pi\rho$ that corresponds to the given ρ is kept constant while the solvent polarization and the electrostatic field relax to equilibrium. To see how the relaxation proceeds in this case we start again from

$$D(t) = \varepsilon_e E(t) + \int_{-\infty}^{t} dt' \tilde{\varepsilon}(t-t') E(t') \qquad [4.3.12]$$

take the time derivative of both sides with respect to t

$$\frac{dD}{dt} = \varepsilon_e \frac{dE}{dt} + E(t)\tilde{\varepsilon}(0) + \int_{-\infty}^{t} dt' \left(\frac{d\tilde{\varepsilon}}{dt}\right)_{t-t'} E(t') \qquad [4.3.13]$$

use the relations $\tilde{\varepsilon}(0) = (\varepsilon_s - \varepsilon_e)/\tau_D$ and

$$\int_{-\infty}^{t} dt' \left(\frac{d\tilde{\varepsilon}}{dt}\right)_{t-t'} E(t') = -\frac{1}{\tau_D} \int_{-\infty}^{t} \tilde{\varepsilon}(t-t') E(t') = -\frac{1}{\tau_D}\left(D(t) - \varepsilon_e E(t)\right)$$

(cf Eq. [4.3.8]), to get

$$\frac{d}{dt}(D - \varepsilon_e E) = -\frac{1}{\tau_D}(D - \varepsilon_s E) \qquad [4.3.14]$$

When D evolves under a constant E, Eq. [4.3.14] implies that $(d/dt)D = (-1/\tau_D)D$ + constant, so that D relaxes exponentially with the time constant $\tau_D$, as before. However if E relaxes under a constant D, the time evolution of E is given by

$$\frac{d}{dt}E = -\frac{\varepsilon_s}{\varepsilon_e \tau_D}\left(E - \frac{1}{\varepsilon_s}D\right) \qquad [4.3.15]$$

i.e.

$$E(t) = \frac{1}{\varepsilon_s}D + Ae^{-t/\tau_L} \qquad [4.3.16]$$

where A is an integration constant and $\tau_L$ is the longitudinal Debye relaxation time

$$\tau_L = \frac{\varepsilon_e}{\varepsilon_s}\tau_D \qquad [4.3.17]$$

The integration constant A is determined from the initial conditions: Immediately following the switch-on of the charge distribution, i.e. of D, E is given by E(t=0) = $D/\varepsilon_e$, so $A = \left(\varepsilon_e^{-1} - \varepsilon_s^{-1}\right)D$. Thus, finally,

$$E(t) = \frac{1}{\varepsilon_s}D + \left(\frac{1}{\varepsilon_e} - \frac{1}{\varepsilon_s}\right)De^{-t/\tau_L} \qquad [4.3.18]$$

We see that in this case the relaxation is characterized by the time $\tau_L$ which can be very different from $\tau_D$. For example, in water $\varepsilon_e / \varepsilon_s \cong 1/40$, and while $\tau_D \cong 10$ps, $\tau_L$ is of the order of 0.25ps.

### 4.3.3 LINEAR RESPONSE THEORY OF SOLVATION DYNAMICS

The continuum dielectric theory of solvation dynamics is a linear response theory, as expressed by the linear relation between the perturbation D and the response of E, Eq. [4.3.2]. Linear response theory of solvation dynamics may be cast in a general form that does not depend on the model used for the dielectric environment and can therefore be applied also in molecular theories.[13,14] Let

$$H = H_0 + H' \qquad [4.3.19]$$

where $H_0$ describes the unperturbed system that is characterized by a given potential surface on which the nuclei move, and where

$$H' = \sum_j X_j F_j(t) \qquad [4.3.20]$$

is some perturbation written as a sum of products of system variables $X_j$ and external time dependent perturbations $F_j(t)$. The nature of X and F depend on the particular experiment: If for example the perturbation is caused by a point charge q(t) at position $r_j$, $q(t)\delta(r-r_j)$, we may identify F(t) with this charge and the corresponding $X_j$ is the electrostatic potential operator at the charge position. For a continuous distribution $\rho(\mathbf{r},t)$ of such charge we may write $H' = \int d^3\mathbf{r}\Phi(\mathbf{r})\rho(\mathbf{r},t)$, and for $\rho(\mathbf{r},t) = \sum_j q_j(t)\delta(\boldsymbol{r}-\boldsymbol{r}_j)$ this becomes $\sum_j \Phi(\boldsymbol{r}_j)q_j(t)$. Alternatively we may find it convenient to express the charge distribution in terms of point moments (dipoles, quadrupoles, etc.) coupled to the corresponding local potential gradient tensors, e.g. H′ will contain terms of the form $\mu\nabla\Phi$ and $\mathbf{Q}:\nabla\nabla\Phi$ where $\mu$ and $\mathbf{Q}$ are point dipoles and quadrupoles respectively.

In linear response theory the corresponding solvation energies are proportional to the corresponding products $q\langle\Phi\rangle$, $\mu\langle\nabla\Phi\rangle$ and $\mathbf{Q}:\langle\nabla\nabla\Phi\rangle$ where $\langle\rangle$ denotes the usual observable average. For example, the average potential $\langle\Phi\rangle$ is proportional in linear response to the perturbation source q. The energy needed to create the charge q is therefore $\int_0^q dq'\langle\Phi\rangle \approx (1/2)q^2 \approx (1/2)q\langle\Phi\rangle$.

Going back to the general expressions [4.3.19] and [4.3.20], linear response theory relates non-equilibrium relaxation close to equilibrium to the dynamics of equilibrium fluctuations: The first fluctuation dissipation theorem states that following a step function change in F:

$$F_j(t) = 0, t < 0; \quad F_j(t) = F_j, t \geq 0 \qquad [4.3.21]$$

the corresponding averaged system's observable relaxes to its final equilibrium value as $t \to \infty$ according to

$$\langle X_j(t)\rangle - \langle X_j(\infty)\rangle = \frac{1}{k_B T}\sum_l F_l\left(\langle X_j X_l(t)\rangle - \langle X_j\rangle\langle X_l\rangle\right) \qquad [4.3.22]$$

where all averages are calculated with the equilibrium ensemble of $H_0$. Applying Eq. [4.3.22] to the case where a sudden switch of a point charge $q \to q + \Delta q$ takes place, we have

$$\langle \Phi(t)\rangle - \langle \Phi(\infty)\rangle = \frac{\Delta q}{k_B T}\left(\langle \Phi\Phi(t)\rangle - \langle \Phi\rangle^2\right) = \frac{\Delta q}{k_B T}\langle \delta\Phi\delta\Phi(t)\rangle \qquad [4.3.23]$$

The left hand side of [4.3.23], normalized to 1 at t = 0, is a linear approximation to the solvation function

$$S(t) - \frac{E_{solv}(t) - E_{solv}(\infty)}{E_{solv}(0) - E_{solv}(\infty)} \underset{=}{LR} \frac{\langle \Phi(t)\rangle - \langle \Phi(\infty)\rangle}{\langle \Phi(0)\rangle - \langle \Phi(\infty)\rangle} \qquad [4.3.24]$$

and Eq. [4.3.23] shows that in linear response theory this non equilibrium relaxation function is identical to the equilibrium correlation function

$$S(t) \underset{=}{LR} C(t) \equiv \frac{\langle \delta\Phi(0)\,\delta\Phi(t)\rangle}{\langle \delta\Phi^2\rangle} \qquad [4.3.25]$$

C(t) is the time correlation function of equilibrium fluctuations of the solvent response potential at the position of the solute ion. The electrostatic potential in C(t) will be replaced by the electric field or by higher gradients of the electrostatic potential when solvation of higher moments of the charge distribution is considered.

The time dependent solvation function S(t) is a directly observed quantity as well as a convenient tool for numerical simulation studies. The corresponding linear response approximation C(t) is also easily computed from numerical simulations, and can also be studied using suitable theoretical models. Computer simulations are very valuable both in exploring the validity of such theoretical calculations, as well as the validity of linear response theory itself (by comparing S(t) to C(t)). Furthermore they can be used for direct visualization of the solute and solvent motions that dominate the solvation process. Many such simulations were published in the past decade, using different models for solvents such as water, alcohols and acetonitrile. Two remarkable outcomes of these studies are first, the close qualitative similarity between the time evolution of solvation in different simple solvents, and second, the marked deviation from the simple exponential relaxation predicted by the Debye relaxation model (cf. Eq. [4.3.18]). At least two distinct relaxation modes are observed, a fast Gaussian-like component and a slower relaxation mode of an exponential character which may correspond to the expected Debye relaxation. In what follows we describe these and other features observed in computer simulations of solvation dynamics using simple generic model dielectric solvents.

### 4.3.4 NUMERICAL SIMULATIONS OF SOLVATION IN SIMPLE POLAR SOLVENTS: THE SIMULATION MODEL[11a]

The simplest simulated system is a Stockmayer fluid: structureless particles characterized by dipole-dipole and Lennard-Jones interactions, moving in a box (size L) with periodic boundary conditions. The results described below were obtained using 400 such particles and in addition a solute atom A which can become an ion of charge q embedded in this solvent. The long range nature of the electrostatic interactions is handled within the effective dielectric environment scheme.[15] In this approach the simulated system is taken to be surrounded by a continuum dielectric environment whose dielectric constant ε′ is to be chosen self consistently with that computed from the simulation. Accordingly, the electrostatic potential between any two particles is supplemented by the image interaction associated with a spherical dielectric boundary of radius $R_c$ (taken equal to L/2) placed so that one of these particles is at its center. The Lagrangian of the system is given by

$$L\left(R,\dot{R},\mu,\dot{\mu}\right)=\frac{1}{2}M_A\dot{R}_A^2+\frac{1}{2}M\sum_{i=1}^{N}\dot{R}_i^2+\frac{1}{2}M\sum_{i=1}^{N}\frac{I}{\mu^2}\dot{\mu}_i^2-\frac{1}{2}\sum_{i\neq j}^{N}V_{ij}^{lJ}\left(R_{ij}\right)-\sum_{i=1}^{N}V_{iA}^{lJ}\left(R_{iA}\right)-$$

$$-\frac{1}{2}\sum_{i\neq j}^{N}V^{DD}\left(R_i,R_j,\mu_i,\mu_j\right)-\sum_{i=1}^{N}V^{AD}\left(R_A,R_i,\mu_i\right)-\sum_{i=1}^{N}\lambda_i\left(\mu_i^2-\mu^2\right) \qquad [4.3.26]$$

where N is the number of solvent molecules of mass M, μ dipole moment, and I moment of inertia. $R_A$ and $R_i$ are positions of the impurity atom (that becomes an ion with charge q) and a solvent molecule, respectively, and $R_{ij}$ is $|R_i - R_j|$. $V^{LJ}$, $V^{DD}$ and $V^{AD}$ are, respectively, Lennard-Jones, dipole-dipole, and charge-dipole potentials, given by

$$V_{ij}^{LJ}(R)=e\varepsilon_D\left[\left(\sigma_D/R\right)^{12}-\left(\sigma_D/R\right)^{6}\right] \qquad [4.3.27]$$

($V_{ij}^{LJ}$ is of the same form with $\sigma_A$ and $\varepsilon_A$ replacing $\sigma_D$ and $\varepsilon_D$) and

$$V^{DD}\left(R_i,R_j,\mu_i,\mu_j\right)=\frac{\mu_i\mu_j-3\left(n\mu_i\right)\left(n\mu_j\right)}{R_{ij}^3}-\frac{2(\varepsilon'-1)}{(2\varepsilon'+1)R_C^3}\mu_i\mu_j \qquad [4.3.28]$$

where $n = (R_i - R_j)/R_{ij}$,

$$V^{AD}\left(R_i,\mu_i,R_A\right)=q\left(\frac{1}{R_{ij}^3}-\frac{2(\varepsilon'-1)}{(2\varepsilon'+1)R_C^3}\right)\mu_i\left(R_i-R_A\right) \qquad [4.3.29]$$

The terms containing ε′ in the electrostatic potentials $V^{DD}$ and $V^{AD}$ are the reaction field image terms.[15,16] The last term in Eq. [4.3.26] is included in the Lagrangian as a constraint, in order to preserve the magnitude of the dipole moments $\left(|\mu_i|=\mu\right)$ with a SHAKE like algorithm.[17] In this representation the mass, M, and the moment of inertia, I, of the solvent molecules are independent parameters, which makes it possible to study the relative importance of translational and rotational motions in the solvation process without affecting other potentially relevant parameters such as the molecular size. The time evolution is done using the velocity Verlet algorithm, with the value of $\lambda(t)$ determined as in the SHAKE algorithm, and with the Andersen[18] thermalization used to keep the system at constant temperature.

For the Stockmayer solvent, the initial molecular parameters are taken to approximate the $CH_3Cl$ molecule: $\sigma_D = 4.2$ Å, $\varepsilon_D = 195K$, M = 50 amu, I = 33.54 amu Å$^2$, and $\mu = 1.87$ D. The parameters taken for the solvated ion are $M_A = 25$ amu, $\sigma_A = 3.675$ Å, and $\varepsilon_A = 120K$, q is taken to be one electron charge e. These parameters can be changed so as to examine their effect on the solvation dynamics. Most of the results described below are from simulations done at 240K, and using L = 33.2Å for the edge length of the cubic simulation cell was, corresponding to the density $\rho = 1.09\times10^{-2}$ Å$^{-3}$, which is the density of $CH_3Cl$ at this temperature. In reduced units we have for this choice of parameters $\rho^* \equiv \rho\sigma_D^3 = 0.81$, $\mu^* \equiv \mu\left(\varepsilon_D\sigma_D^3\right)^{1/2} = 1.32$, $T^* \equiv k_BT/\varepsilon_D = 1.23$, and $I^* \equiv I(M\sigma^2) = 0.038$. A simple switching function

$$f(R) = \begin{cases} 1 & R < R_s \\ (R_c - R)/(R_c - R_s) & R_s < R < R_c \\ 0 & R > R_c \end{cases} \qquad [4.3.31]$$

is used to smoothly cut off this electrostatic potential. $R_c$ and $R_s$ are taken to be $R_c = L/2$ and $R_s = 0.95R_c$. Under these simulation conditions the pressure fluctuates in the range 500±100 At.The dielectric constant is computed from pure solvent simulations, using the expression[33]

$$\frac{(\varepsilon - 1)(2\varepsilon' + 1)}{2\varepsilon' + \varepsilon} = \frac{1}{k_BTR_c^3}\langle PP(R_c)\rangle \qquad [4.3.32]$$

where

$$P = \sum_{i=1}^{N} \mu_i \qquad [4.3.33]$$

and

$$P(R_c) = \frac{1}{N}\sum_{j=1}^{N}\sum_{k=1}^{N}{}' \mu_k \qquad [4.3.34]$$

where the prime on the inner summation indicates the restriction $R_{jk} < R_c$. The result for our solvent is $\varepsilon = 17$, compared with $\varepsilon(CH_3Cl) = 12.6$ at 253K. After evaluating $\varepsilon$ in this way the external dielectric constant $\varepsilon'$ is set to $\varepsilon$ and the computation is, in principle, repeated until convergence, i.e., until the evaluated $\varepsilon$ is equal to the environmental $\varepsilon'$. In fact, we have found that our dynamical results are not sensitive to the magnitude of $\varepsilon'$.

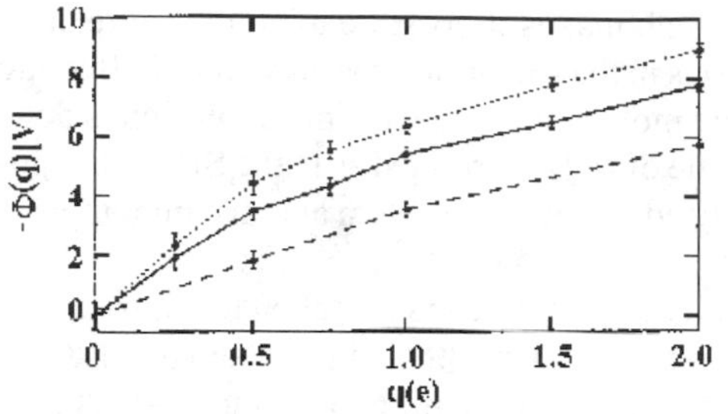


Figure 4.3.1. The electrostatic response potential <Φ> induced by the solvent at the position of the solute ion, as a function of the solute charge. Dashed line - the Stockmayer-$CH_3Cl$ model described in Section 4. Full and dotted lines, model polyether solvents described in the text. [From Ref. 11b].

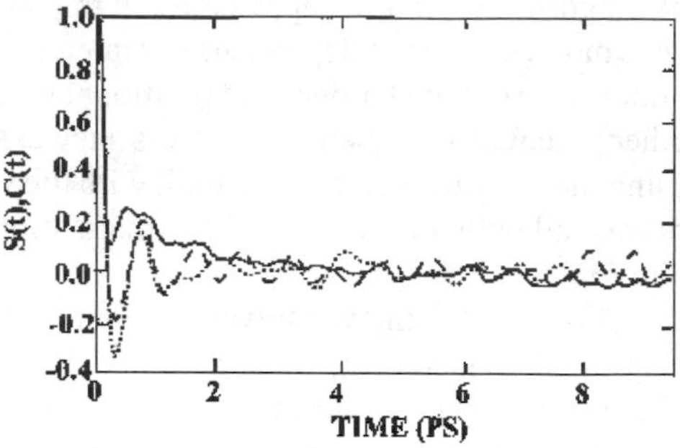


Figure 4.3.2. The linear response relaxation function C(t) (dashed and dotted lines] and the non-equilibrium solvation function S(t) (solid line) computed for the Stockmayer-$CH_3Cl$ model described in Section 4. In the nonequilibrium simulation the ion charge is switched on at t = 0. The dotted and dashed lines represent C(t) obtained from equilibrium simulations with uncharged and charged ion, respectively. [From Ref. 11a].

A typical timestep for these simulation is 3fs. In the absence of thermalization this gives energy conservation to within $10^{-4}$ over ~80,000 time steps. After equilibrating the system at 240K, the equilibrium correlation function C(t) is evaluated from equilibrium trajectories with both a charged (q = e) and an uncharged (q = 0) impurity atom. The non-equilibrium solvation function S(t) can also be computed from trajectories that follow a step function change in the ion charge from q = 0 to q = e. These calculations are done for the $CH_3Cl$ solvent model characterized by the above parameters and for similar models with different parameters. In particular, results are shown below for systems characterized by different values of the parameter[15]

$$p' = I / 2M\sigma^2 \qquad [4.3.35]$$

which measures the relative importance of rotational and translational solvent motions.

## 4.3.5 NUMERICAL SIMULATIONS OF SOLVATION IN SIMPLE POLAR SOLVENTS: RESULTS AND DISCUSSION

The dashed line of Figure 4.3.1 shows the equilibrium solvent induced electrostatic potential Φ at the position of the ion, as a function of the ion charge q obtained for the Stockmayer-$CH_3Cl$ model described in Section 4. Clearly the solvent response is linear with q all the way up to q = e, with slight deviations from linearity starting at q > e. The slope (~4) of the linear dependence of the dashed line in Figure 1 (for q < e) is considerably smaller from that obtained from $\Phi = q / a\varepsilon_s$ (taking $a = \sigma_A/2$ gives a slope of 7.4) that is used to get Eq. [4.3.1]. A more advanced theory of solvation based on the mean spherical approximation predicts (using $\sigma_A$ and $\sigma_D$ for the diameters of the ion and the solvent, respectively) a slope of 4.6.

The linearity of the response depends on the nature of the solvent. As examples Figure 4.3.1 also shows results obtained for models of more complex solvents, $H(CH_2OCH_2)_nCH_3$

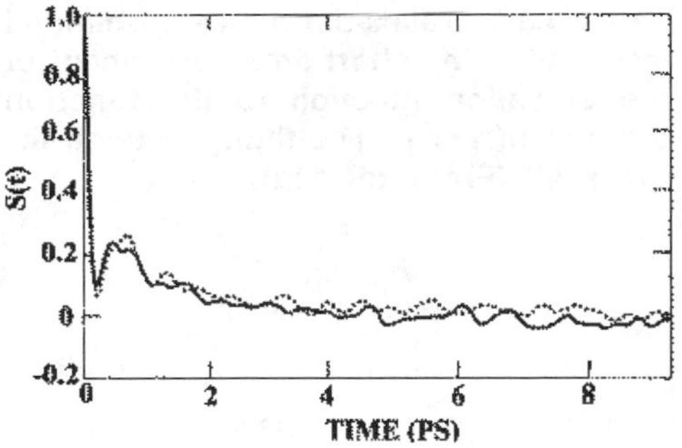


Figure 4.3.3. The function S(t) obtained with $\varepsilon' = 17.0$ (solid line, same as in Figure 4.3.2), together with the same function obtained with $\varepsilon' = 1.0$ (dashed line). $\varepsilon'$ is the continuum dielectric constant associated with the reaction field boundary conditions. [From Ref. 11a].

with n = 1 (ethyl methyl ether, full line) and n = 2 (1,2-methoxy ethoxy ethane, dotted line). In these solvents the main contribution to the solvation energy of a positive ion comes from its interaction with solvent oxygen atoms. Because of geometric restriction the number of such atoms in the ion's first solvation shell is limited, leading to a relatively early onset of dielectric saturation.

Figure 4.3.2 shows the time evolution of the solvation functions C(t), Eq. [4.3.25] and S(t) (Eq. [4.3.24]). C(t) is evaluated from an equilibrium trajectory of 220 ps for a system consisting of the solvent and a charged or uncharged atom. The nonequilibrium results for S(t) are averages over 25 different trajectories, each starting from an initial configuration taken from an equilibrium run of an all-neutral system following switching, at t = 0, of the charge on the impurity atom from q = 0 to q = e.

These results show a large degree of similarity between the linear response (equilibrium) and nonequilibrium results. Both consist of an initial fast relaxation mode that, at closer inspection is found to be represented well by a Gaussian, $\exp[-(t/\tau)^2]$, followed by a relatively slow residual relaxation. The initial fast part is more pronounced in C(t). The latter is also characterized by stronger oscillations in the residual part of the relaxation. The fact that the linear response results obtained for equilibrium simulations with an uncharged solute and with a solute of charge q are very similar give further evidence to the approximate validity of linear response theory for this systems.

The sensitivity of these results to the choice of boundary conditions is examined in Figure 4.3.3. We note that the use of reaction field boundary conditions as implemented here is strictly valid only for equilibrium simulations, since the dynamic response of the dielectric continuum at $R > R_c$ is not taken into account. One could argue that for the short-time phenomena considered here, $\varepsilon'$ should have been taken smaller than the static dielectric constant of the system. Figure 4.3.3 shows that on the relevant time scale our dynamical results do not change if we take $\varepsilon' = 1$ instead of $\varepsilon' = \varepsilon = 17$. (The absolute solvation energy does depend on $\varepsilon'$, and replacing $\varepsilon' = 17$ by $\varepsilon' = 1$ changes it by $\cong 5\%$.)

In the simulations described so far the solvent parameters are given by the aforementioned data. For these, the dimensionless parameter $p'$, Eq. [4.3.35], is 0.019. In order to separate between the effects of the solvent translational and rotational degrees of freedom, we can study systems characterized by other $p'$ values. Figure 4.3.4 shows results obtained for $p' = 0$ (dotted line), 0.019 (solid line), 0.25 (dashed line), and $\infty$ (dashed-dotted line). Except for $p' = 0$, these values were obtained by changing the moment of inertia I, keeping M=50 amu. The value $p' = 0$ was achieved by taking $M = M_A = \infty$ and I=33.54 amu Å$^2$. Note that the values $p' = 0$ and $p' = \infty$ correspond to models with frozen translations and frozen rotations, respectively. Figures 4.3.4(a) and 4.3.4(b) show, respectively, the solvation energy $E_{solv}(t)$ and the solvation function S(t) obtained for these different systems. The following points are noteworthy:

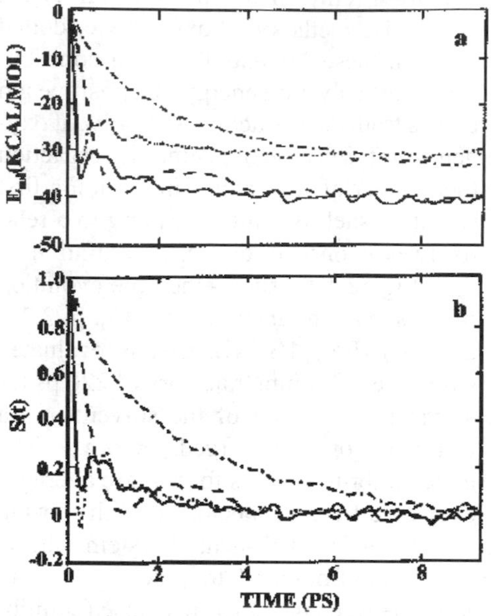


Figure 4.3.4. The solvation energy, $E_{solv}$ (a) and the non-equilibrium solvation function S(t) (b), plotted against time (after switching the ion charge from 0 to e at t = 0) for different solvent models characterized by the parameter p′ (Eq. [4.3.35]). Dotted line, p′= 0; solid line, p′=0.019; dashed line, p′=0.25; dashed-dotted line, p′=8. [From Ref. 11a].

**Table 4.3.1. Relaxation times τ obtained from fitting the short time component of the solvation function to the function $S(t) = \exp[-(t/\tau)^2]$. The fitting is done for S(t) > 0.3 [From Ref. 11a].**

| p′ | M amu | I amu Å$^2$ | τ ps |
|---|---|---|---|
| 0.0 | ∞ | 33.54 | 0.206 |
| 0.019 | 50 | 33.54 | 0.170 |
| 0.125 | 50 | 220.5 | 0.347 |
| 0.250 | 50 | 441.0 | 0.421 |

(1) The asymptotic ($t \rightarrow \infty$) values of $E_{solv}$ (Figure 4.3.4a) are different for p′ = 0 (M = ∞) and p′ = ∞ (I = ∞) then in the other cases because of the freezing of solvent translations and rotations, respectively. Note, however, that the I = ∞ curves converge very slowly (the solvent compensates for the lack of rotations by bringing into the neighborhood of the solute solvent molecules with the "correct" orientation) and probably did not reach its asymptotic value in Figure 3.4.4a.

(2) Except for the rotationless system (p′ = ∞) all the other systems exhibit a bimodal relaxation, with a fast relaxation component that accounts for most of the solvation energy. The relaxation of the rotationless solvent is exponential (a fit to $\exp(-t/\tau)$ yields $\tau = 2.2$ ps).

(3) A closer look at the fast component in the finite p′ systems shows a Gaussian behavior, a fit to $\exp[-(t/\tau)^2]$ yields the τ values summarized in Table 4.3.1. τ increases with increasing solvent moment of inertia (recall that this is how p′ is changed for p′ > 0, still for the range of p′ studied, it stays distinct from the long component.

(4) The oscillations and the thermal noise seen in the relatively small slow relaxation component make it difficult to estimate the long relaxation time. A fit of the long time component for the p′ = 0.019 case to an exponential relaxation $\exp(-t/\tau)$ yields $\tau \cong 2.7 \pm 0.7$ ps. The long-time components in the other systems relax on similar time scales.

The nature of this fast relaxation component of the solvation dynamics has been discussed extensively in the past decade.[19] Carter and Hynes[20] were the first to suggest that this initial relaxation is associated with inertial, as opposed to diffusive, solvent motion. In this mode of motion solvent molecules move under the suddenly created force-field without their mutual interactions and consequent energy exchange having a substantial influence on this fast time scale. Simulations and analytical calculations[19,21,22] have confirmed this assertion for simple polar fluids.

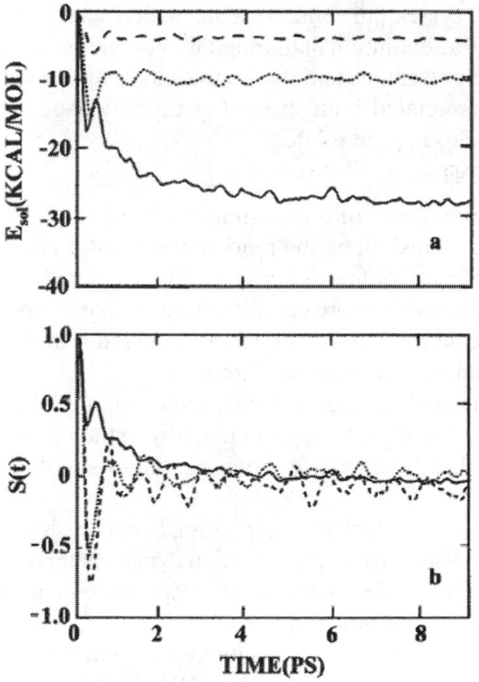


Figure 4.3.5.(a) The solvation energy E(t) and (b) the solvation function S(t) associated with the three solvation shells defined in the text, plotted against time after the ion charge is switched on, for the system with $\rho^*=0.019$. Solid line, nearest shell; dotted line, second shell, dashed line, outer shell. [From Ref. 11a].

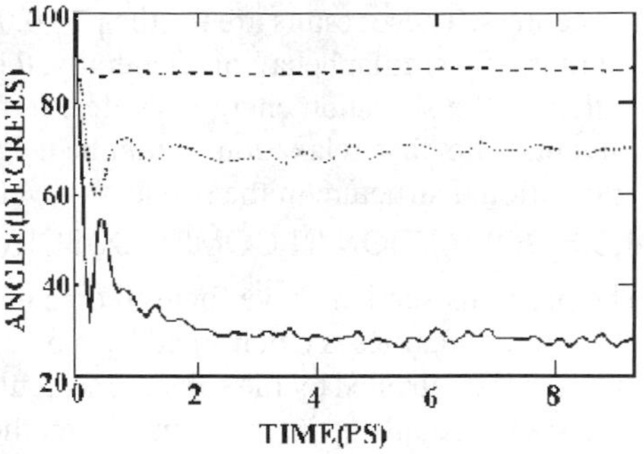


Figure 4.3.6. The time dependence of the average angle between the molecular dipoles and between the corresponding radius vectors to the ion center, associated with molecules in the three different solvation shells defined in the text, plotted against time following the switching on of the ionic charge. [From Ref. 11a].

Next we examine the relative contributions of different solvation shells to the solvation process. This issue is important for elucidating the solvation mechanism, and has been under discussion since an early remark made by Onsager[23] that the shorter time scales are associated mostly with solvent layers further away from the solute, and that the longer ~ $\tau_D$ times are associated with the individual response of solvent molecules nearest to the solute. From the structure of the solute-solvent radial distribution function of the simulated system one can estimate[11a] that the first solvation shell about the solute consists of the eight nearest neighbor solvent particles at distance closer than 5.5 Å from the solute center, and the second solvation shell encompasses the next nearest 26 solvent particles at distance smaller than ~10 Å from the solute center. Taking the rest of the solvent particles in the simulation box as the "third solvation shell", Figure 4.3.5 shows the contributions of these layers to the time evolution of the solvation energy and of the solvation function. It seems that the fast component in these time evolutions is faster for the contribution from the first solvation shell. The same shell also shows a distinct slow component which is much smaller or absent in the contribution from the further shells. Also note that the solvation energy is dominated by the first solvation shell: the first, second, and third shells contribute ~67%, 24%, and 9%, respectively, to the solvation energy. The fast relaxation component accounts for ~80% of the solvation energy. It should be kept in mind, however, that the contribution from outer shell molecules is suppressed by the finite size of the simulated system.

Finally, the nature of the motion that gives rise to the fast relaxation component is seen in Figure 6, which depicts as functions of time the average angles between the molecular dipoles in the different solvation shells and between the corresponding radius vectors to the ion centers. These results are for the $p' = 0.019$ system; the other systems with $p' < \infty$ show qualitatively similar behavior. Generally, the time evolution of the angular motion is similar to that of the solvation energy. Typical to the present system that represents simple polar solvents, the fast relaxation component is associated with the initial relaxation of the orientational structure in the solvation layers close to the solute.

### 4.3.6 SOLVATION IN COMPLEX SOLVENTS

The previous sections have focused on a generic model of a very simple solvent, in which solvation dynamics is determined by molecular translations and reorientations only. These in turn are controlled by the solvent molecular mass, moment of inertia, dipole moment and short-range repulsive interactions. When the solvent is more complex we may expect specific structures and interactions to play significant roles. Still, numerical simulations of solvation dynamics in more complex systems lead to some general observations:

(a) In large molecular solvents, solvation may be associated with binding of the solute to particular solvents sites. As seen in Figure 4.3.1, deviations from linearity in the solvent response potential are associated with the fact that the fraction of polar binding sites constitutes a relatively small fraction of the solvent molecule.

This deviation from linearity shows itself also in the solvation dynamics. Figure 4.3.7 shows the linear response functions and the non-equilibrium solvation function, C(t) and S(t), respectively, computed as before, for the di-ether $H(CH_2OCH_2)_2CH_3$ solvent. Details of this simulations are given in Ref.11b. If linear response was a valid approximation all the lines in Figure 4.3.7: The two lines for C(t) that correspond to q=0 and q=1, and the two lines for S(t) for the processes $q=0 \rightarrow q=1$ and the process $q=1 \rightarrow q=0$, would coalesce. The marked differences between these lines shows that linear response theory fails for this system.

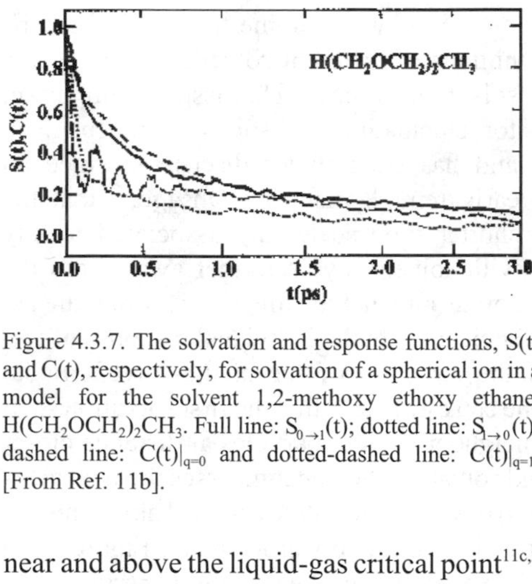


Figure 4.3.7. The solvation and response functions, S(t) and C(t), respectively, for solvation of a spherical ion in a model for the solvent 1,2-methoxy ethoxy ethane, $H(CH_2OCH_2)_2CH_3$. Full line: $S_{0\rightarrow 1}(t)$; dotted line: $S_{1\rightarrow 0}(t)$; dashed line: $C(t)|_{q=0}$ and dotted-dashed line: $C(t)|_{q=1}$. [From Ref. 11b].

(b) Linear response theory was also shown to fail for low-density solvents (e.g. near and above the liquid-gas critical point[11c,24]). In this case the origin of the non-linearity is the large rearrangement in the solvent structure near the solute during the solvation process. This rearrangement is associated with a local density change in such highly compressible low-density solvents.

(c) Similarly, solvation in mixed solvents usually involve large rearrangement of solvent structure near the solute because the latter usually have a higher affinity for one of the solvent components. Solvation in electrolyte solutions provides a special example.[25,26] In this case the solvent dynamics about the newly created charge distribution is not much different than in the pure dielectric solution, however in addition the mobile ions rearrange about this charge distribution, and on the timescale of this process linear response theory fails.[27]

(d) In the situations discussed in (b) and (c) above, new dynamical processes exist: While the dielectric response in normal simple polar solvents is dominated by molecular rotations, the motions that change the local structure about the solute are usually dominated by solvent translation. This gives rise to a new, usually slower, relaxation components. Solvation in electrolyte solutions clearly shows this effect: In addition to the dielectric response on the picosecond timescale, a much slower relaxation component is observed on the nanosecond scale.[25] Numerical simulations have identified the origin of this relaxation component as the exchange between a water molecule and an ion in the first solvation shell about the solute.[26]

Finally, it is intuitively clear that in large molecule complex solvents simple molecular rotation as seen in Figure 4.3.6 can not be the principal mode of solvation. Numerical simulations with polyether solvents show that instead, hindered intramolecular rotations that distort the molecular structure so as to bring more solvating sites into contact with the ion dominate the solvation dynamics.[11b] The bi-modal, and in fact multi-modal, character of the solvation is maintained also in such solvents, but it appears that the short time component of this solvation process is no longer inertial as in the simple small molecule solvents.[11]

## 4.3.7 CONCLUSIONS

Numerical simulations of solvation dynamics in polar molecular solvents have been carried out on many models of molecular systems during the last decade. The study described in sections 4.3.4-4.3.5 focused on a generic model for a simple polar solvent, a structureless Stockmayer fluid. It is found that solvation dynamics in this model solvent is qualitatively similar to that observed in more realistic models of more structured simple solvents, including solvents like water whose energetics is strongly influenced by the H-bond network. In particular, the bimodal nature of the dynamics and the existing of a prominent fast Gaussian relaxation component are common to all models studied.

Such numerical simulations have played an important role in the development of our understanding of solvation dynamics. For example, they have provided the first indication that simple dielectric continuum models based on Debye and Debey-like dielectric relaxation theories are inadequate on the fast timescales that are experimentally accessible today. It is important to keep in mind that this failure of simple theories is not a failure of linear response theory. Once revised to describe reliably response on short time and length scales, e.g. by using the full k and $\omega$ dependent dielectric response function $\varepsilon(k,\omega)$, and sufficiently taking into account the solvent structure about the solute, linear response theory accounts for most observations of solvation dynamics in simple polar solvents.

Numerical simulations have also been instrumental in elucidating the differences between simple and complex solvents in the way they dynamically respond to a newly created charge distribution. The importance of translational motions that change the composition or structure near the solute, the consequent early failure of linear response theory in such systems, and the possible involvement of solvent intramolecular motions in the solvation process were discovered in this way.

We conclude by pointing out that this report has focused on solvation in polar systems where the solvent molecule has a permanent dipole moment. Recently theoretical and experimental work has started on the dynamics of non-polar solvation.[28] This constitutes another issue in our ongoing effort to understand the dynamics of solvation processes.

## REFERENCES

1 For recent reviews see M. Maroncelli, *J. Mol. Liquids*, **57**, 1 (1993); G.R. Fleming and M. Cho, *Ann. Rev. Phys. Chem.*, **47**, 109 (1996).

2 See, e.g., J.T. Hynes, in **Ultrafast Dynamics of Chemical Systems**, edited by J.D. Simon, *Kluwer*, Dordrecht, 1994, pp. 345-381; L.D. Zusman, *Zeit. Phys. Chem.*, **186**, 1 (1994); S. Roy and B. Bagchi, *J. Chem. Phys.*, **102**, 6719; 7937 (1995).
3 P. G. Wolynes, *J. Chem. Phys.*, **86**, 5133 (1987).
4 I. Rips, J. Klafter and J. Jortner, *J. Chem. Phys.*, **88**, 3246 (1988); **89**, 4288 (1988).
5 D. F. Calef and P. G. Wolynes, *J. Chem. Phys.*, **78**, 4145 (1983).
6 B. Bagchi and A. Chandra, *J. Chem. Phys.*, **90**, 7338 (1989); **91**, 2594 (1989); **97**, 5126 (1992).
7 F.O Raineri, H. Resat, B-C Perng, F. Hirata and H.L. Friedman, *J. Chem. Phys.*, **100**, 1477 (1994)
8 R. F. Loring and S. Mukamel, *J. Chem. Phys.*, **87**, 1272 (1987); L. E. Fried and S. Mukamel, *J. Chem. Phys.*, **93**, 932 (1990).
9 A. A. Kornyshev. A. M. Kuznetsov, D. K. Phelps and M. J. Weaver, *J. Chem. Phys.*, **91**, 7159 (1989).
10 See, e.g. M. Maronelli and G.R. Fleming *JCP*, **89**, 5044 (1988); M. Maroncelli, *J. Chem. Phys.*, **94**, 2084 (1991); Perera and Berkowitz, *J. Chem. Phys.*, **96**, 3092 (1992); P.V. Kumar and M. Maroncelli, *J. Chem. Phys.*, **103**, 3038 (1995).
11 (a) E. Neria and A. Nitzan, *J. Chem. Phys.*, **96**, 5433 (1992).
(b) R. Olender and A. Nitzan, *J. Chem. Phys.*, **102**, 7180 (1995).
(c) P. Graf and A. Nitzan, *Chem. Phys.*, **235**, 297(1998).
12 A. Mozumder in **Electron-solvent and anion- solvent interactions**, L. Kevan and B. Webster, Editors, *Elsevier*, Amsterdam, 1976.
13 M. Maronelli and G.R. Fleming in Ref. 11.
14 E.A. Carter and J.T. Hynes, *JCP,* **94**, 2084 (1991).
15 J. W. de Leeuw, J. W. Perram and E. R. Smith, *Annu. Rev. Phys. Chem.*, **37**, 245 (1986).
16 C. J. F. Böttcher and P. Bordewijc, **Theory of Electric Polarization**, 2nd. ed. *Elsevier,* Amsterdam, 1978, Vol. 2, Chap. 10.
17 M. P. Allen and D. J. Tildesely, **Computer Simulation of Liquids**, *Oxford*, London, 1989).
18 H. C. Andersen, *J. Chem. Phys.*, **72**, 2384 (1980).
19 See, e.g. M. Maroncelli, in Ref. 11.
20 E. A. Carter and J.T. Hynes, *J. Chem. Phys.*, **94**, 5961 (1991).
21 M. Maroncelli, V.P. Kumar and A. Papazyan, *J. Phys. Chem.*, **97**, 13 (1993).
22 L. Perera and M.L. Berkowitz, *J. Chem. Phys.*, **97**, 5253 (1992).
23 L. Onsager, *Can. J. Chem.*, **55**, 1819 (1977).
24 R. Biswas and B. Bagchi, *Chem. Phys. Lett.*, 290 (1998).
25 V. Itah and D. Huppert, *Chem. Phys. Lett.*, **173**, 496 (1990); E. Bart and D. Huppert, *ibid.* **195**, 37 (1992).
26 C.F. Chapman and M. Maroncelli, *J. Phys. Chem.*, **95**, 9095 (1991).
27 E. Neria and A. Nitzan, *J. Chem. Phys.*, **100**, 3855 (1994).
28 B.M. Ladanyi and M. Maroncelli, *J. Chem. Phys.*, **109**, 3204 (1998); M. Berg, *J. Phys. Chem.*, **A102,** 17 (1998); *Chem. Phys.*, **233**, 257, (1998).

## 4.4 METHODS FOR THE MEASUREMENT OF SOLVENT ACTIVITY OF POLYMER SOLUTIONS

CHRISTIAN WOHLFARTH
**Martin-Luther-University Halle-Wittenberg, Institute of Physical Chemistry, Merseburg, Germany, e-mail: Wohlfarth@chemie.uni-halle.de**

### 4.4.1 INTRODUCTION

Knowledge of solvent activities in polymer solutions is a necessity for a large number of industrial and laboratory processes. Such data are an essential tool for understanding the thermodynamic behavior of polymer solutions, for studying their intermolecular interactions and for getting insights into their molecular nature. Besides, they are the necessary basis for any development of theoretical thermodynamic models. Scientists and engineers in academic and industrial research need such data.

Solvent activities of polymer solutions have been measured for about 60 years now. However, the database for polymer solutions is still modest, in comparison with the enormous amount of data available for mixtures of low-molecular substances. Three explicit databases have been published in the literature up to now.[1-3] The database prepared by Wen Hao et al.[1] is summarized in two parts of the DECHEMA Chemistry Data Series. Danner and High[2] provided a database and some calculation methods on a floppy disk with their book. Wohlfarth[3] prepared the most complete data collection regarding vapor-liquid equilibrium data of polymer solutions. His annually updated electronic database is not commercially available; however, personal requests can be made via his e-mail address given above.

Some implicit databases are provided within the Polymer Handbook[4] by Schuld and Wolf[5] or by Orwoll[6] and in two papers prepared earlier by Orwoll.[7,8] These four sources list tables of Flory's $\chi$-function and tables where enthalpy, entropy or volume changes, respectively, are given in the literature for a large number of polymer solutions. The tables of second virial coefficients of polymers in solution, which were prepared by Lechner and coworkers[9] (also provided in the Polymer Handbook), are a valuable source for estimating the solvent activity in the dilute polymer solution. Bonner reviewed vapor-liquid equilibria in concentrated polymer solutions and listed tables containing temperature and concentration ranges of a certain number of polymer solutions.[10] Two CRC-handbooks prepared by Barton list a larger number of thermodynamic data of polymer solutions in form of polymer-solvent interaction or solubility parameters.[11,12]

An up-to-date list of all polymer-solvent systems for which solvent activities or vapor pressures from vapor-liquid equilibrium measurements were published in the literature is provided in Appendix 4.4.A of this Subchapter 4.4 (please see below).

Solvent activities in polymer solutions can be determined by rather different techniques. However, no one is really a universal method but covers a certain concentration range of the polymer solution. Figure 4.4.1 explains in short the situation.

Corresponding to the different regions in the diagram, different experimental techniques were used for the measurement of the solvent activity in a homogeneous polymer solution:

(i) Solvent activities of highly diluted polymer solutions can be obtained from scattering methods such as light scattering, small angle X-ray scattering and small angle neutron scattering via the second osmotic virial coefficient, which is often related to investigations for polymer characterization. These methods are able to resolve the very small difference between the thermodynamic limit of 1.0 for the activity of the pure solvent and the actual value of perhaps 0.9999x at the given (very low) polymer concentration.

(ii) Solvent activities of polymer solutions with polymer concentrations up to about 30 wt% can be measured by osmometry (membrane as well as vapor-pressure osmometry), light scattering, ultracentrifuge (of course, all these methods can also be applied for polymer characterization and can be extrapolated to zero polymer concentration to obtain the second virial coefficient), and differential vapor pressure techniques. Cryoscopy and ebulliometry can also be used to measure solvent activities in dilute and semidilute polymer solutions, but with limited success only.

(iii) The concentrated polymer solution between 30 and 85 wt% is covered by vapor pressure measurements which were usually performed by various isopiestic sorption meth-

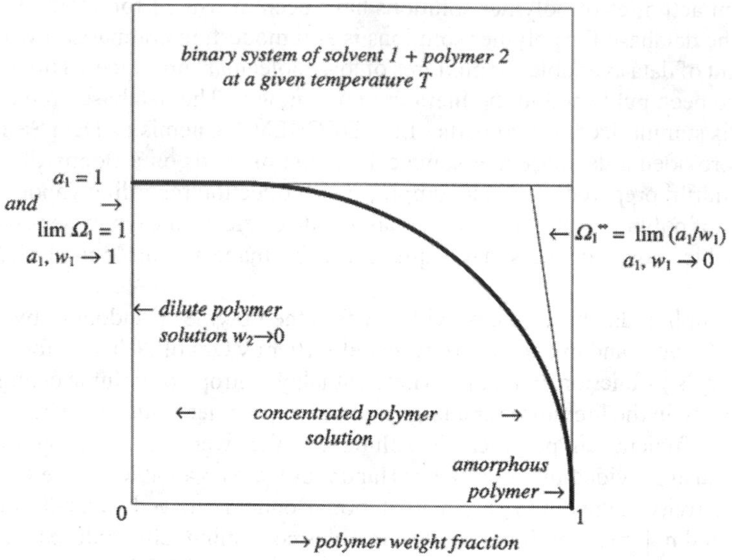


Figure 4.4.1. Typical isotherm for the solvent activity of a homogeneous binary polymer solution.

ods. The ultracentrifuge can also be applied for solutions up to 80 wt% polymer, but this was only scarcely done in the literature.

(iv) A special problem are polymer solutions with concentrations higher than 90 wt% up to the limit of the region of Henry's law. For this purpose, the inverse gas-liquid chromatography (IGC) is the most useful method. Measurements can be made at infinite dilution of the solvent for determining the activity coefficient at infinite dilution or Henry's constant, but IGC can also be performed at finite solvent concentrations up to 10-15 wt% of the solvent to get solvent activities for highly concentrated polymer solutions. Some sorption experiments in this concentration range were reported by piezoelectric quartz crystal technique; however, thermodynamic equilibrium absorption is difficult to obtain, as discussed below. At least, melting point depression can be applied in some cases for small amounts of solvents in semicrystalline polymers, but obtaining reliable results seems to be difficult.

(v) There is another possibility to measure solvent activities in polymer solutions if the state of the solution is inhomogeneous, i.e., for the region of liquid-liquid equilibrium. Binodal and/or spinodal curves can be reduced to solvent activity data by means of a thermodynamic ansatz for the Gibbs free energy of mixing in dependence on temperature, concentration (and pressure if necessary), which has to be solved according to thermodynamic equilibrium conditions. In the case of polymer networks, swelling equilibria can be measured instead. The solvent activity in a swollen network arises from two parts, a mixing part with the (virtually) infinite-molar-mass polymer, and a contribution from elastic network deformation. The second follows from statistical theory of rubber elasticity and also needs certain model approximations for data reduction.

In summary, investigations on vapor-liquid equilibrium of polymer solutions are the most important source for obtaining solvent activities in polymer solutions. Therefore, emphasis is laid in this subchapter on the experimental methods, which use this equilibrium.

Reviews on experimental methods, sometimes including tables with thermodynamic data were prepared more or less continuously during the last three decades. Especially methods and results of the application of IGC to polymers and polymer solutions are carefully reviewed.[13-25] Special reviews on determining solvent activities from various scattering techniques could not be found. However, there is a large number of reviews and books on scattering methods and their applications. Some references may give a starting point for the interested reader.[26-32] Experimental techniques for vapor-pressure measurements were reviewed in the paper by Bonner.[10] Ebulliometry, cryoscopy and vapor-pressure osmometry were reviewed by Cooper,[33] Glover,[34] Mays and Hadjichristidis,[40] and a recent summary can be found in a new book edited by Pethrick and Dawkins.[26] Reviews that account for the measurement of thermodynamic data from sedimentation equilibria using the ultracentrifuge are given by Fujita,[35] Harding et al.[36] or Munk.[37] An overview on membrane osmometry was given by Adams,[38] Tombs and Peacock[39] or Mays and Hadjichristidis,[40] and a recent summary can again be found in the book edited by Pethrick and Dawkins.[26] Reviews on liquid-liquid demixing of polymer solutions will not be summarized in detail here, some references should be enough for a well-based information.[41-45] A short summary on equipment and thermodynamic equations of most techniques was given in Danner's handbook.[2] Finally, the classical books on polymer solutions written by Flory,[46] by Huggins,[47] and by Tompa[48] must not be forgotten for the historical point of view on the topic of this subchapter.

### 4.4.2 NECESSARY THERMODYNAMIC EQUATIONS

Here, the thermodynamic relations are summarized which are necessary to understand the following text. No derivations will be made. Details can be found in good textbooks, e.g., Prausnitz et al.[49]

The activity of a component i at a given temperature, pressure, and composition can be defined as the ratio of the fugacity of the solvent at these conditions to the solvent fugacity in the standard state; that is, a state at the same temperature as that of the mixture and at specified conditions of pressure and composition:

$$a_i(T,P,x) \equiv f_i(T,P,x) / f_i\left(T,P^0,x^0\right) \qquad [4.4.1a]$$

where:

| | |
|---|---|
| $a_i$ | activity of component i |
| T | absolute temperature |
| P | pressure |
| x | mole fraction |
| $f_i$ | fugacity of component i |

In terms of chemical potential, the activity of component i can also be defined by:

$$a_i(T,P,x) \equiv \exp\left\{\frac{\mu_i(T,P,x) - \mu_i\left(T,P^0,x^0\right)}{RT}\right\} \qquad [4.4.1b]$$

where:

$\mu_i$ chemical potential of component i
R gas constant

$P^0$ and $x^0$ denote the standard state pressure and composition. For binary polymer solutions the standard state is usually the pure liquid solvent at its saturation vapor pressure at T. The standard state fugacity and the standard state chemical potential of any component i are abbreviated in the following text by their symbols $f_i^0$ and $\mu_i^0$, respectively.

Phase equilibrium conditions between two multi-component phases I and II require thermal equilibrium,

$$T^I = T^{II} \qquad [4.4.2a]$$

mechanical equilibrium,

$$P^I = P^{II} \qquad [4.4.2b]$$

and the chemical potential of each component i must be equal in both phases I and II.

$$\mu_i^I = \mu_i^{II} \qquad [4.4.3]$$

For Equation [4.4.3] to be satisfied, the fugacities of each component i must be equal in both phases.

$$f_i^I = f_i^{II} \qquad [4.4.4]$$

Applying fugacity coefficients, the isochemical potential expression leads to:

$$\phi_i^I x_i^I = \phi_i^{II} x_i^{II} \qquad [4.4.5]$$

where:

$\phi_i$ fugacity coefficient of component i.

Fugacity coefficients can be calculated from an equation of state by:

$$\ln \phi_i = \frac{1}{RT} \int_V^\infty \left[ \left( \frac{\partial P}{\partial n_i} \right)_{T,V,n_j} - \frac{RT}{V} \right] dV - \ln \frac{PV}{RT} \qquad [4.4.6]$$

where a pressure explicit equation of state is required to use Equation [4.4.6]. Not all equations of state for polymers and polymer solutions also are valid for the gaseous state (see section in Subchapter 4.4.4), however, and a mixed gamma-phi approach is used by applying Equation [4.4.7]. Applying activity coefficients in the liquid phase, the isochemical potential expression leads, in the case of the vapor-liquid equilibrium (superscript V for the vapor phase and superscript L for the liquid phase), to the following relation:

$$\phi_i^V y_i P = \gamma_i x_i^L f_i^0 \qquad [4.4.7]$$

where:

$y_i$ mole fraction of component i in the vapor phase with partial pressure $P_i = y_i P$
$\gamma_i$ activity coefficient of component i in the liquid phase with activity $a_i = x_i^L \gamma_i$

or in the case of liquid-liquid equilibrium to

$$\gamma_i^I x_i^I f_i^{0I} = \gamma_i^{II} x_i^{II} f_i^{0II} \qquad [4.4.8]$$

If the standard state in both phases is the same, the standard fugacities cancel out in Equation [4.4.8]. Equation [4.4.8] also holds for solid-liquid equilibria after choosing appropriate standard conditions for the solid state, but they are of minor interest here.

All expressions given above are exact and can be applied to small molecules as well as to macromolecules. The one difficulty is having accurate experiments to measure the necessary thermodynamic data and the other is finding correct and accurate equations of state and/or activity coefficient models to calculate them.

Since mole fractions are usually not the concentration variables chosen for polymer solutions, one has to specify them in each case. The following three quantities are most frequently used:

mass fractions $$w_i = m_i / \sum m_k \qquad [4.4.9a]$$

volume fractions $$\varphi_i = n_i V_i / \sum n_k V_k \qquad [4.4.9b]$$

segment (hard-core volume) fractions $$\psi_i = n_i V_i^* / \sum n_k V_k^* \qquad [4.4.9c]$$

where:

$m_i$ mass of component i
$n_i$ amount of substance (moles) of component i
$V_i$ molar volume of component i
$V_i^*$ molar hard-core (characteristic) volume of component i.

With the necessary care, all thermodynamic expressions given above can be formulated with mass or volume or segment fractions as concentration variables instead of mole fractions. This is the common practice within polymer solution thermodynamics. Applying characteristic/hard-core volumes is the usual approach within most thermodynamic models for polymer solutions. Mass fraction based activity coefficients are widely used in Equations [4.4.7 and 4.4.8] which are related to activity by:

$$\Omega_i = a_i / w_i \qquad [4.4.10]$$

where:

$\Omega_i$ mass fraction based activity coefficient of component i
$a_i$ activity of component i
$w_i$ mass fraction of component i

Classical polymer solution thermodynamics often did not consider solvent activities or solvent activity coefficients but usually a dimensionless quantity, the so-called Flory-Huggins interaction parameter $\chi$.[44,45] $\chi$ is not only a function of temperature (and pressure), as was evident from its foundation, but it is also a function of composition and polymer molecular mass.[5,7,8] As pointed out in many papers, it is more precise to call it $\chi$-function (what is in principle a residual solvent chemical potential function). Because of its widespread use and its possible sources of mistakes and misinterpretations, the necessary relations must be included here. Starting from Equation [4.4.1b], the difference between the chemical potentials of the solvent in the mixture and in the standard state belongs to the first derivative of the Gibbs free energy of mixing with respect to the amount of substance of the solvent:

$$\Delta\mu_1 = \mu_1 - \mu_1^0 = \left(\frac{\partial n\Delta_{mix}G}{\partial n_1}\right)_{T,p,n_{j\neq 1}} \quad [4.4.11]$$

where:

| | |
|---|---|
| $n_i$ | amount of substance (moles) of component i |
| $n$ | total amount of substance (moles) of the mixture: $n = \Sigma n_i$ |
| $\Delta_{mix}G$ | molar Gibbs free energy of mixing. |

For a truly binary polymer solutions, the classical Flory-Huggins theory leads to:[46,47]

$$\Delta_{mix}G / RT = x_1 \ln\varphi_1 + x_2 \ln\varphi_2 + gx_1\varphi_2 \quad [4.4.12a]$$

or

$$\Delta_{mix}G / RTV = (x_1 / V_1)\ln\varphi_1 + (x_2 / V_2)\ln\varphi_2 + BRT\varphi_1\varphi_2 \quad [4.4.12b]$$

where:

| | |
|---|---|
| $x_i$ | mole fraction of component i |
| $\varphi_i$ | volume fraction of component i |
| g | integral polymer-solvent interaction function that refers to the interaction of a solvent molecule with a polymer segment, the size of which is defined by the molar volume of the solvent $V_1$ |
| B | interaction energy-density parameter that does not depend on the definition of a segment but is related to g and the molar volume of a segment $V_{seg}$ by $B = RTg/V_{seg}$ |
| V | molar volume of the mixture, i.e. the binary polymer solution |
| $V_i$ | molar volume of component i |

The first two terms of Equation [4.4.12] are named combinatorial part of $\Delta_{mix}G$, the third one is then a residual Gibbs free energy of mixing. Applying Equation [4.4.11] to [4.4.12], one obtains:

$$\Delta\mu_1 / RT = \ln(1-\varphi_2) + \left(1 - \frac{1}{r}\right)\varphi_2 + \chi\varphi_2^2 \quad [4.4.13a]$$

or

$$\chi = \left[\Delta\mu_1 / RT - \ln(1-\varphi_2) - \left(1 - \frac{1}{r}\right)\varphi_2\right] / \varphi_2^2 \quad [4.4.13b]$$

or

$$\chi = \left[\ln a_1 - \ln(1-\varphi_2) - \left(1 - \frac{1}{r}\right)\varphi_2\right] / \varphi_2^2 \quad [4.4.13c]$$

where:

| | |
|---|---|
| r | ratio of molar volumes $V_2/V_1$, equal to the number of segments if $V_{seg} = V_1$ |

$\chi$ Flory-Huggins interaction function of the solvent

The segment number r is, in general, different from the degree of polymerization or from the number of repeating units of a polymer chain but proportional to it. One should note that Equations [4.4.12 and 4.4.13] can be used on any segmentation basis, i.e., also with $r = V_2^* / V_1^*$ on a hard-core volume segmented basis and segment fractions instead of volume fractions, or with $r = M_2/M_1$ on the basis of mass fractions. It is very important to keep in mind that the numerical values of the interaction functions g or $\chi$ depend on the chosen basis and are different for each different segmentation!

From the rules of phenomenological thermodynamics, one obtains the interrelations between both parameters at constant pressure and temperature:

$$\chi = g + \varphi_1 \frac{\partial g}{\partial \varphi_1} = g - (1-\varphi_2)\frac{\partial g}{\partial \varphi_2} \qquad [4.4.14a]$$

$$g = \frac{1}{\varphi_1}\int_0^{\varphi_1} \chi d\varphi_1 \qquad [4.4.14b]$$

A recent discussion of the g-function was made by Masegosa et al.[50] Unfortunately, g- and $\chi$-functions were not always treated in a thermodynamically clear manner in the literature. Sometimes they were considered to be equal, and this is only true in the rare case of composition independence. Sometimes, and this is more dangerous, neglect or misuse of the underlying segmentation basis is formed. Thus, numerical data from literature has to be handled with care (using the correct data from the reviews[5-8,11] is therefore recommended).

A useful form for their composition dependencies was deduced from lattice theory by Koningsveld and Kleintjens:[51]

$$g = \alpha + \frac{\beta}{(1-\gamma\varphi_2)} \quad and \quad \chi = \alpha + \frac{\beta(1-\gamma)}{(1-\gamma\varphi_2)^2} \qquad [4.4.15]$$

where:

$\alpha$ acts as constant within a certain temperature range

$\beta$ describes a temperature function like $\beta = \beta_0 + \beta_1 / T$

$\gamma$ is also a constant within a certain temperature range.

Quite often, simple power series are applied only:

$$\chi = \sum_{i=0}^{n} \chi_i \varphi_2^i \quad and \quad g = \sum_{i=0}^{n}\left(\frac{\chi_i}{i+1}\right)\left(\frac{1-\varphi_2^{i+1}}{1-\varphi_2}\right) \qquad [4.4.16]$$

where:

$\chi_i$ empirical fitting parameters to isothermal-isobaric data

Both interaction functions are also functions of temperature and pressure. An empirical form for these dependencies can be formulated according to the rules of phenomenological thermodynamics:

$$g = \beta_{00} + \beta_{01}/T + (\beta_{10} + \beta_{11}/T)P \text{ or } \chi = a + b/T + (c + d/T)P \qquad [4.4.17]$$

where:

| | |
|---|---|
| a,b,c,d | empirical fitting parameters for the $\chi$-function |
| $\beta_{00}, \beta_{01}, \beta_{10}, \beta_{11}$ | empirical fitting parameters for the g-function |
| T | absolute temperature |
| P | pressure |

All these fitting parameters may be concentration dependent and may be included into Equations [4.4.15 or 4.4.16]. Details are omitted here. More theoretical approaches will be discussed shortly in Subchapter 4.4.4.

The $\chi$-function can be divided into an enthalpic and an entropic parts:

$$\chi = \chi_H + \chi_S \text{ with } \chi_H = -T\left(\frac{\partial \chi}{\partial T}\right)_{p,\varphi} \text{ and } \chi_S = \left(\frac{\partial \chi T}{\partial T}\right)_{P,\varphi} \qquad [4.4.18]$$

where:

| | |
|---|---|
| $\chi_H$ | enthalpic part |
| $\chi_S$ | entropic part |

An extension of all these equations given above to multi-component mixtures is possible. Reviews of continuous thermodynamics which take into account the polydisperse character of polymers by distribution functions can be found elsewhere.[52-54]

## 4.4.3 EXPERIMENTAL METHODS, EQUIPMENT AND DATA REDUCTION

### 4.4.3.1 Vapor-liquid equilibrium (VLE) measurements

Investigations on vapor-liquid equilibrium of polymer solutions are the most important source for obtaining solvent activities in polymer solutions. Therefore, emphasis is laid to the experimental methods which use this equilibrium. These methods are:

(i) absolute vapor pressure measurement,
(ii) differential vapor pressure measurement,
(iii) isopiestic sorption/desorption methods, i.e. gravimetric sorption, piezoelectric sorption, or isothermal distillation,
(iv) inverse gas-liquid chromatography (IGC) at infinite dilution, IGC at finite concentrations, and head-space gas-chromatography (HSGC),
(v) ebulliometry and
(vi) the non-equilibrium steady-state method vapor-pressure osmometry (VPO).

The measurement of vapor pressures for polymer solutions is generally more difficult and more time-consuming than that of low-molecular mixtures. The main difficulties can be summarized as follows: Polymer solutions exhibit strong negative deviations from Raoult's law. These are mainly due to the large entropic contributions caused by the difference between the molar volumes of solvents and polymers, as was explained by the classical Flory-Huggins theory[46,47] about 60 years ago, Equation [4.4.12]. However, because of this large difference in molar mass, vapor pressures of dilute solutions do not differ markedly from the vapor pressure of the pure solvent at the same temperature, even at polymer concentrations of 10-20 wt%. This requires special techniques to measure very small differences in solvent activities. Concentrated polymer solutions are characterized by rapidly increasing viscosities with increasing polymer concentration. This leads to a strong increase in time required to obtain real thermodynamic equilibrium caused by a slow solvent diffusion effects (in or out of a non-equilibrium-state polymer solution). Furthermore, only the solvent coexists in both phases because polymers do not evaporate. The experimental techniques used for the measurement of vapor pressures of polymer solutions have to take into account all these effects.

Vapor pressures of polymer solutions are usually measured in the isothermal mode by static methods. Dynamic methods are seldom applied, see under ebulliometry below. At least, one can consider measurements by VPO to be dynamic methods, where a dynamic (steady-state) balance is obtained. However, limits for the applicable ranges of polymer concentration and polymer molar mass, limits for the solvent vapor pressure and the measuring temperature and some technical restrictions prevent its broader application, see below. Static techniques usually work at constant temperature. The three different methods (i)-(iii) were used to determine most of the vapor pressures of polymer solutions in the literature. All three methods have to solve the problems of establishing real thermodynamic equilibrium between liquid polymer solution and solvent vapor phase, long-time temperature constancy during the experiment, determination of the final polymer concentration and determination of pressure and/or activity. Methods (i) and (ii) were mostly used by early workers. The majority of recent measurements was done with the isopiestic sorption methods. Gas-liquid chromatography as IGC closes the gap at high polymer concentrations where vapor pressures cannot be measured with sufficient accuracy. HSGC can be considered as some combination of absolute vapor pressure measurement with GLC. The following text will now explain some details of experimental equipment and measuring procedures as well as of data reduction procedures to obtain solvent activities. A recent review by Williamson[55] provides corresponding information related to low-molecular mixtures.

#### *4.4.3.1.1 Experimental equipment and procedures for VLE-measurements*

**(i) Absolute vapor pressure measurement**

Absolute vapor pressure measurement may be considered to be the classical technique for our purposes, because one measures directly the vapor pressure above a solution of known polymer concentration. Refs. 56-65 provide a view of the variety of absolute vapor pressure apparatuses developed and used by different authors. The common principle of an absolute vapor pressure apparatus is shown in Figure 4.4.2.

Vapor pressure measurement and solution equilibration were made separately: A polymer sample is prepared by weighing, the sample flask is evacuated, degassed solvent is introduced into the flask and the flask is sealed thereafter. All samples are equilibrated at elevated temperature in a thermostat for some weeks (!). The flask with the equilibrated polymer solution is connected to the pressure measuring device (in Figure 4.4.2 a Hg-manometer) at the measuring temperature. The vapor pressure is measured after reaching equilibrium and the final polymer concentration is obtained after correcting for the amount of evaporated solvent. Modern equipment applies electronic pressure sensors and digital techniques to measure the vapor pressure, e.g. Schotsch and Wolf[57] or Killmann et al.[58] Data processing can be made online using computers. Figure 4.4.3 shows a schematic diagram of the equipment used by Killmann and coworkers.[58]

A number of problems have to be solved during the experiment. The solution is usually in an amount of some $cm^3$ and may contain about 1 g of polymer or even more. Degassing is absolutely necessary. For example, Killmann et al.[58] included special degassing units for each component of the entire equipment. All impurities in the pure solvent have to

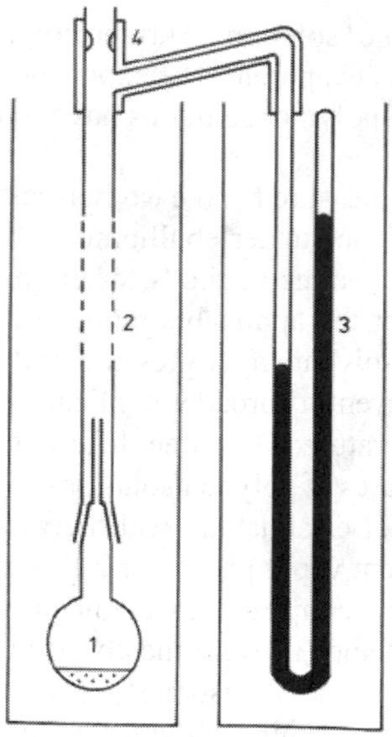


Figure 4.4.2. Schematic of the common principle of an absolute vapor pressure apparatus: 1 - polymer solution, 2 - connection to the manometer, 3 - Hg-manometer, 4 - heating coils. The whole construction is thermostated at the measuring temperature, the connection to the manometer is kept slightly above the measuring temperature to avoid condensation.

be eliminated. Equilibration of all prepared solutions is very time-consuming (liquid oligomers need not so much time, of course). Increasing viscosity makes the preparation of concentrated solutions more and more difficult with further increasing amount of polymer. Solutions above 50-60 wt% can hardly be prepared (depending on the solvent/polymer pair under investigation). The determination of the volume of solvent vaporized in the unoccupied space of the apparatus is difficult and can cause serious errors in the determination of the final solvent concentration. To circumvent the vapor phase correction, one can measure the concentration directly by means, for example, of a differential refractometer. The contact of solvent vapor with the Hg-surface in older equipment may cause further errors. Complete thermostating of the whole apparatus is necessary to avoid condensation of solvent vapors at colder spots. Since it is disadvantageous to thermostat Hg-manometers at higher temperatures, null measurement in-

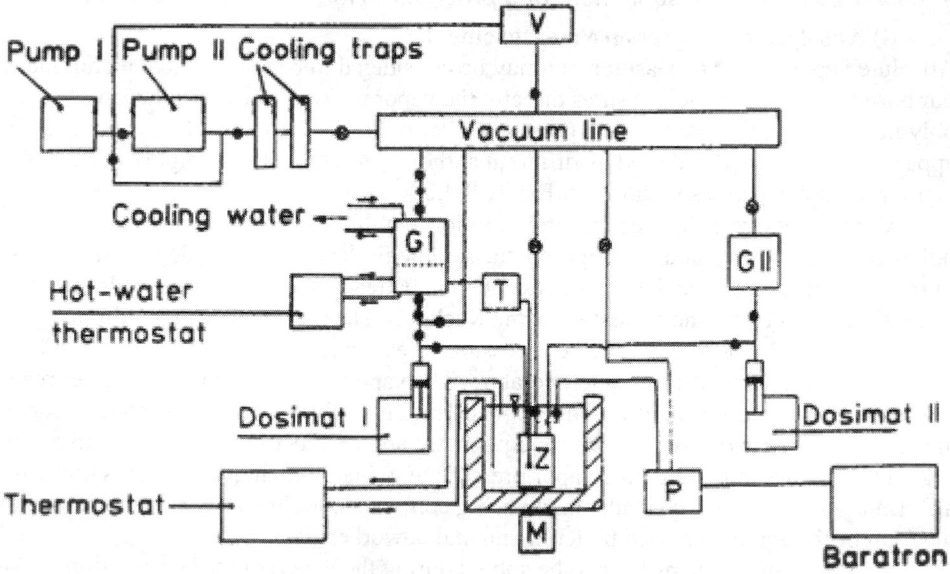


Figure 4.4.3. Schematic diagram of a modern absolute vapor pressure apparatus: T - temperature meter, P - vapor-pressure meter, V - vacuum meter, Z - measuring cell, M - magnetic stirrer, GI and GII - degassing units for the solvent and for the polymer. [Reprinted with permission from Ref. 58, Copyright 1990,Wiley-VCH].

struments with pressure compensation were sometimes used, e.g., Baxendale et al.[59] Modern electronic pressure sensors can be thermostated within certain temperature ranges. If pressure measurement is made outside the thermostated equilibrium cell, the connecting tubes must be heated slightly above the equilibrium temperature to avoid condensation.

The measurement of polymer solutions with lower polymer concentrations requires very precise pressure instruments, because the difference in the pure solvent vapor pressure becomes very small with decreasing amount of polymer. At least, no one can really answer the question if real thermodynamic equilibrium is obtained or only a frozen non-equilibrium state. Non-equilibrium data can be detected from unusual shifts of the $\chi$-function with some experience. Also, some kind of hysteresis in experimental data seems to point to non-equilibrium results. A common consistency test on the basis of the integrated Gibbs-Duhem equation does not work for vapor pressure data of binary polymer solutions because the vapor phase is pure solvent vapor. Thus, absolute vapor pressure measurements need very careful handling, plenty of time, and an experienced experimentator. They are not the method of choice for high-viscous polymer solutions.

**(ii) Differential vapor pressure measurement**

The differential method can be compared under some aspects with the absolute method, but there are some advantages. The measuring principle is to obtain the vapor pressure difference between the pure solvent and the polymer solution at the measuring temperature. Figure 4.4.4 explains the basic principle as to how it is used by several authors. References[66-75] provide a view of a variety of differential vapor pressure apparatuses developed and used by different authors.

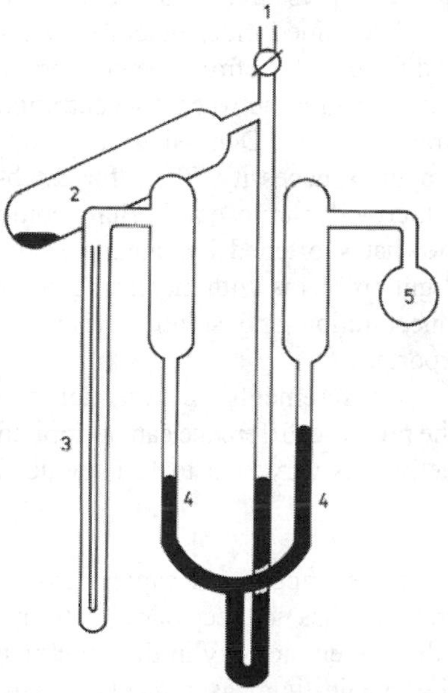

Figure 4.4.4. Schematic diagram of a differential vapor-pressure apparatus: 1 - connection to vacuum pump, 2 - Hg-storage bulb, 3 - burette, 4 - Hg-manometer, 5 - polymer solution. The whole apparatus is kept constant at the measuring temperature within a thermostat.

The polymer sample is put, after weighing, into the sample flask and the apparatus is evacuated. Degassed solvent is distilled into the measuring burette and from there a desired amount of solvent is distilled into the sample flask. The Hg-manometer is filled from the storage bulb and separates the polymer solution from the burette. Care must be taken to avoid leaving any solvent in the manometer. The apparatus is kept at constant measuring temperature, completely immersed in a thermostat for several days. After reaching equilibrium, the pressure is read from the manometer difference and the concentration is calculated from the calibrated burette meniscus difference corrected by the amount of vaporized solvent in the unoccupied space of the equipment. The pure solvent vapor pressure is usually precisely known from independent experiments.

Difference/differential manometers have some advantages from their construction: They are comparatively smaller and their resolution is much higher (modern pressure transducers can resolve differences of 0.1 Pa and less). However, there are the same disadvantages with sample/solution preparation (solutions of grams of polymer in some $cm^3$ vol-

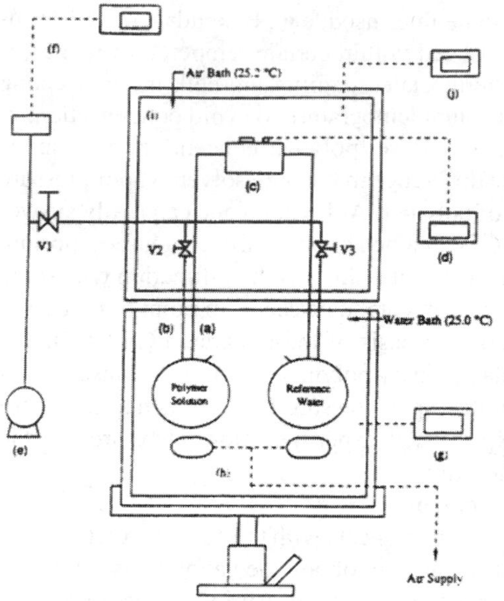


Figure 4.4.5. Differential vapor-pressure apparatus. 100 ml Pyrex flasks connected (a) to a differential pressure transducer (c) with digital readout (d) and (b) to vacuum pump (e) and absolute pressure vacuum thermocouple gauge (f). The constant temperature in the water bath is maintained by a temperature controller (g). The transducer and connecting glassware are housed in an insulated box (i) and kept at constant temperature slightly above the measuring temperature by controller (j). Polymer solution and pure solvent (here water) are stirred by underwater magnetic stirrers (h). [Reprinted with permission from Ref. 66, Copyright 1989, American Chemical Society].

ume, degassing, viscosity), long-time thermostating of the complete apparatus because of long equilibrium times (increasing with polymer molar mass and concentration/viscosity of the solution), correction of unoccupied vapor space, impurities of the solvent, connection to the Hg-surface in older equipment and there is again the problem of obtaining real thermodynamic equilibrium (or not) as explained above.

Modern equipment uses electronic pressure sensors instead of Hg-manometers and digital technique to measure the vapor pressure. Also thermostating is more precise in recent apparatuses. The apparatus developed by Haynes et al.[66] is shown in Figure 4.4.5 as example.

Problems caused by the determination of the unoccupied vapor space were avoided by Haynes et al., since they measure the pressure difference as well as the absolute vapor pressure. Also, the concentration is determined independently by using a differential refractometer and a normalized relation between concentration and refractive index. Degassing of the liquids remains a necessity. Time for establishing thermodynamic equilibrium could be somewhat shortened by intensive stirring (slight problems with increasing polymer concentration and solution viscosity were reported).

In comparison to absolute vapor-pressure measurements, differential vapor-pressure measurements with a high resolution for the pressure difference can be applied even for dilute polymer solutions where the solvent activity is very near to 1. They need more time than VPO-measurements, however.

**(iii) Isopiestic sorption/desorption methods**

Isopiestic measurements allow a direct determination of solvent activity or vapor pressure in polymer solutions by using a reference system (a manometer has not necessarily to be applied). There are two general principles for lowering the solvent activity in the reference system: concentration lowering or temperature lowering. Isopiestic measurements have to obey the condition that no polymer can vaporize (as it might be the case for lower-molecular oligomers at higher temperatures).

Concentration lowering under isothermal conditions is the classical isopiestic technique, sometimes also called isothermal distillation. A number of solutions (two as the minimum) are in contact with each other via their common solvent vapor phase and solvent

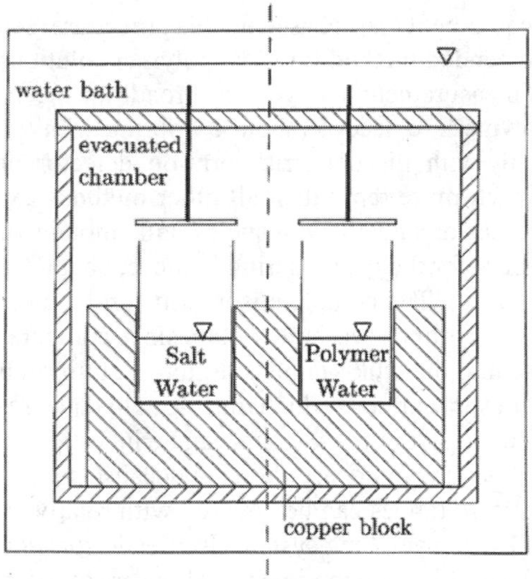


Figure 4.4.6. Schematic of the experimental arrangement for an isopiestic measurements. [Reprinted with permission from Ref. 77, Copyright 1995, Elsevier Science].

evaporates and condenses (this is the isothermal "distillation" process) between them as long as the chemical potential of the solvent is equal in all solutions. At least one solution serves as reference system, i.e., its solvent activity vs. solvent concentration dependence is precisely known. After an exact determination of the solvent concentration in all equilibrated solutions (usually by weighing), the solvent activity in all measured solutions is known from and equal to the activity of the reference solution, Equations [4.4.1 to 4.4.5]. This method is almost exclusively used for aqueous polymer solutions, where salt solutions can be applied as reference systems. It is a standard method for inorganic salt systems. Examples of this technique are given elsewhere.[76-80] Figure 4.4.6 provides a scheme of the experimental arrangement for isopiestic measurements as used by Grossmann et al.[77] to illustrate the common principle.

The complete apparatus consists of six removable stainless steel cells placed in hexagonal pattern in a copper block. The copper block is mounted in a chamber which is thermostated. Each cell has a volume of about 8 $cm^3$ and is closed by a removable lid. During the experiment, the cells are filled with about 2 $cm^3$ polymer solution (or reference solution) and placed into the copper block. The chamber is sealed , thermostated and evacuated. The lids are then opened and solvent is allowed to equilibrate between the cells as explained above. After equilibration, the cells are closed, removed from the chamber and weighed precisely. Equilibrium requires usually a couple of days up to some weeks. During this time, the temperature of the thermostat does not fluctuate by less than ±0.1 K, which is realized by the copper block that works as a thermal buffer.

Temperature lowering at specified isobaric or isochoric conditions is the most often used technique for the determination of solvent vapor pressures or activities in polymer solutions. The majority of all measurements is made using this kind of an isopiestic procedure where the pure solvent is used as reference system. The equilibrium condition of equal chemical potential of the solvent in the polymer solution as well as in the reference system is realized by keeping the pure solvent at a lower temperature ($T_1$) than the measuring temperature ($T_2$) of the solution. At equilibrium, the vapor pressure of the pure solvent at the lower temperature is then equal to the partial pressure of the solvent in the polymer solution, i.e., $P_1^s(T_1) = P_1(T_2)$. Equilibrium is again established via the common vapor phase for both subsystems. The vapor pressure of the pure solvent is either known from independent data or measured additionally in connection with the apparatus. The composition of the polymer solution can be altered by changing $T_1$ and a wide range of compositions can be studied (between 30-40 and 85-90 wt% polymer, depending on the solvent). Measurements above 85-90 wt% polymer are subject to increasing errors because of surface adsorption effects.

There is a broad variety of experimental equipment that is based on this procedure (see below). This isopiestic technique is the recommended method for most polymer solutions since it is advantageous in nearly all aspects of measurement: It covers the broadest concentration range. Only very small amounts of polymer are needed (about 30-50 mg with the classical quartz spring balance, about 100 μg with piezoelectric sorption detector or microbalance techniques - see below). It is much more rapid than all other methods explained above, because equilibrium time decreases drastically with such small amounts of polymer and polymer solution (about 12-24 hours for the quartz spring balance, about 3-4 hours for piezoelectric or microbalance techniques). The complete isotherm can be measured using a single loading of the apparatus. Equilibrium is easier to obtain since comparatively small amounts of solvent have to diffuse into the bulk sample solution. Equilibrium can better be tested by measuring sorption and desorption runs which must lead to equal results for thermodynamic absorption equilibrium. Supercritical solvents can be investigated if the piezoelectric detector is used (otherwise buoyancy in dense fluids may cause serious problems). Much broader pressure and temperature ranges can be covered with relatively simple equipment, what may again be limited by the weighing system. Isopiestic sorption measurements can be automated and will allow also kinetic experiments. There are two disadvantages: First, isopiestic sorption measurements below about 30 wt% polymer are subject to increasing error because very small temperature differences (vapor pressure changes) are connected with large changes in concentration. Second, problems may arise with precise thermostating of both the solvent and the solution at different constant temperatures over a longer period of time.

Because of their importance, several technical solutions will now be presented in some detail. The classical concept is the sorption method using a quartz spring balance. Refs. 81-90 provide some examples, where the concentration (mass) of the solution is measured by the extension of the quartz spring according to Hook's law (linear relationship, no hysteresis). It was not originally developed for polymer solutions but for gas-solid adsorption measurements by McBain.[91] The principle was introduced into the investigation of polymer solutions by van der Waals and Hermans[84] and became popular after the work of Bonner and Prausnitz.[85] In this method, a weighed quantity of the (non-volatile) polymer is placed on the pan of the quartz spring balance within a measuring cell. The determination of spring extension vs. mass has to be made in advance as a calibration procedure. Reading of the spring extension is usually made by means of a cathetometer. The cell is sealed, evacuated and thermostated to the measuring temperature ($T_2$), and the solvent is then introduced into the measuring cell as solvent vapor. The solvent vapor is absorbed by the polymer sample to form the polymer solution until thermodynamic equilibrium is reached. The solvent vapor is provided from a reservoir either of pure liquid solvent thermostated at a lower temperature ($T_1$) or of a reference liquid solution of known concentration/solvent partial pressure like in the case of the isothermal distillation procedure as described above. A compact version of such an apparatus was developed by Illig[82] and widely used within the author's own work (see Appendix 4.4A for the corresponding references). Figure 4.4.7a shows the details of the equilibrium cell, which has a vacuum double-walled jacket.

The following problems have to be solved during the experiment: The equilibrium cell has to be sealed carefully to avoid any air leakage over the complete duration of the measurements (to measure one isotherm lasts about 14 days). Specially developed thin Teflon sealing rings were preferred to grease. The polymer sample has to withstand the tempera-

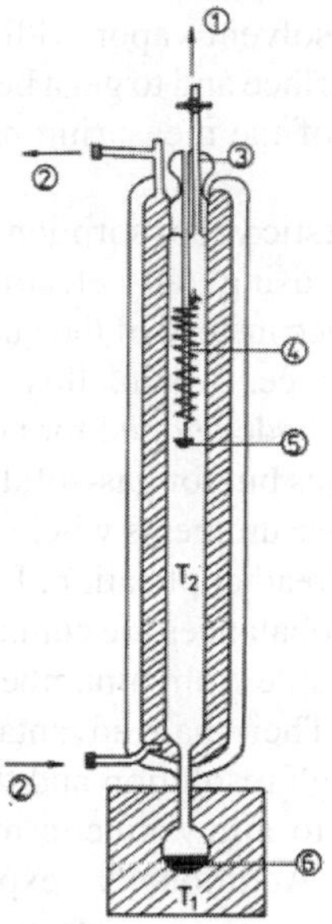


Figure 4.4.7a. Isopiestic vapor-sorption apparatus using a quartz spring: 1 - connection to the vacuum line, 2 - connection to the thermostating unit which realizes the constant measuring temperature $T_2$ (the correct value of $T_2$ is obtained by a Pt-100 resistance thermometer within the cell that is not shown), 3 - closing plug, 4 - quartz spring (reading of its extension is made by a cathetometer), 5 - sample pan with the polymer solution, 6 - pure solvent reservoir at temperature $T_1$. [Reprinted with permission from Ref. 82, Copyright 1982, Wiley-VCH].

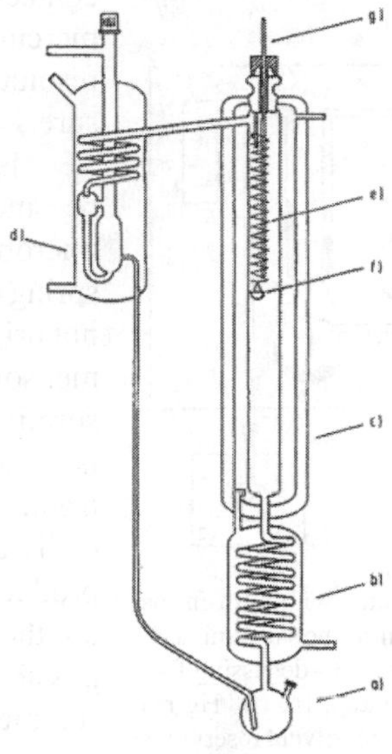


Figure 4.4.7b. Dynamic isopiestic vapor-sorption apparatus using a quartz spring (drawing provided by G. Sadowski): a) evaporator, b) superheater, c) measuring cell, d) condenser, e) quartz spring, f) polymer sample/solution, g) Pt-100 resistance thermometer. [Reprinted with permission from Ref. 87, Copyright 1995, Wiley-VCH].

ture. Changes by thermal ageing during the experiment must be avoided. The temperatures provided by the thermostats must not fluctuate more than ±0.1 K. Condensation of solvent vapor at points that become colder than $T_2$ has to be avoided by slight overheating (this problem may arise at the closing plug and at the connection between the pure solvent reservoir flask and the double-walled jacket). An intelligent improvement of this compact apparatus was made by Sadowski,[87] see Figure 4.4.7b, where this vapor-sorption apparatus is combined with technical solutions from ebulliometry (more about ebulliometers can be found below).

In comparison to the usual ebulliometric equipment where the polymer solution is placed into the evaporator, only pure solvent is evaporated. The vapor flows through the cell and is condensed at its head-condenser to flow back into the reservoir at the bottom. The vapor pressure is kept constant using a manostat and is measured additionally outside the apparatus after the condenser. Equilibrium times decrease somewhat, degassing of the solvent is not necessary, air leakage does not play any role.

As was stated by different authors, additional measurement of the vapor pressure inside the isopiestic sorption apparatus seems to be necessary if there is some doubt about the real pressure or if no reliable pure solvent vapor pressure data exist for the investigated temperature range. Figure 4.4.8 shows an apparatus used by the author for measurements between room temperature and 70°C and pressures up to 1.5 bar. It combines mercury float valves with Hg-manometers to avoid the use of any grease within the measuring system, a kind of equipment proposed earlier by Ashworth and Everett.[88] Up to four quartz springs can be inserted into the equilibrium cell (only one is shown). Reading of the manometer and of the extension of the quartz spring was made using a cathetometer.

The direct pressure measurement has the advantage that absolute pressures can be obtained and pressure fluctuations can be observed. More modern equipment applies electronic pressure sensors instead of Hg-manometers to avoid the problems caused by the

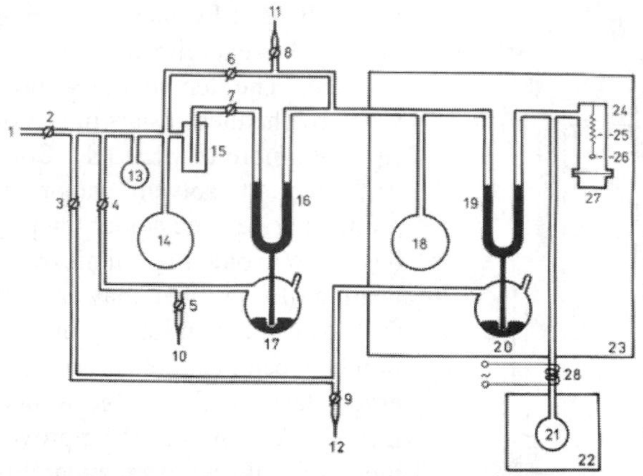


Figure 4.4.8. Isopiestic vapor-sorption apparatus with built-in manometer using a quartz spring: 1 - connection to the vacuum, 2-9 - stop corks, 10, 11, 12 - connections to nitrogen, 13 - degassing flask for the pure solvent, 14, 18 - buffers, 15 - cold trap, 16, 19 - Hg-manometers, 17, 20 - mercury float valves, 21 - pure solvent reservoir at temperature $T_1$ provided by 22 - thermostat, 23 - temperature controlled air box, 24 - measuring cell, 25 - quartz spring (four quartz springs can be inserted into the equilibrium cell, only one is shown), 26 - pan with the polymer solution, 27 - closing plug sealed with epoxy resin, 28 - heating to avoid solvent condensation.

contact of solvent vapor with the mercury surface and to get a better resolution of the measuring pressure.

Isopiestic vapor sorption can be made using an electronic microbalance instead of the quartz spring balance. Again, this was not originally developed for polymer solutions but for gas-solid adsorption measurements where this is a widespread application. Electronic microbalances are commercially available from a number of producers. Their main advantages are their high resolution and their possibility to allow kinetic measurements. Additionally, experiments using electronic microbalances can easily be automated and provide computing facilities. The major disadvantage with some of these microbalances is that they cannot be used at high solvent vapor pressures and so are limited to a relatively small concentration range. However, since thin polymer films can be applied, this reduces both the time necessary to attain equilibrium (some hours) and the amount of polymer required and equilibrium solvent absorption can be obtained also at polymer mass fractions approaching 1 (i.e., for small solvent concentrations). Depending on their construction, the balance head is situated inside or outside the measuring apparatus. Problems may arise when it is inside where the solvent vapor may come into contact with some electronic parts. Furthermore, all parts of the balance that are inside the apparatus have to be thermostated to the measuring temperature to enable the correct equilibration of the polymer solution or even slightly above to avoid condensation of solvent vapor in parts of the balance. The allowed temperature range of the balance and its sensitivity to solvent corrosion determine then the accessible measuring range of the complete apparatus. Yoo and coworkers[92,93] have recently measured various polymer solutions with such equipment and Figure 4.4.9 shows some details of their apparatus.

Two thermostats maintain the pure solvent temperature $T_1$ and the measuring temperature $T_2$ as described above for the spring balance technique, thermostat three protects the essential part of the balance for solvent vapor condensation and damage. A calibrated weight was loaded on the left side of the balance. A granular type of quartz was used as reference weight in order to prevent possible solvent vapor condensation. A dish-type quartz sorption cell was used to load the polymer sample. Platinum wire was used to link both arms to the balance to prevent possible oxidative corrosion of the arm by the solvent. The vapor pressure is measured directly by applying a W-tube Hg-Manometer. The manome- ter reading was made using a cathetometer.

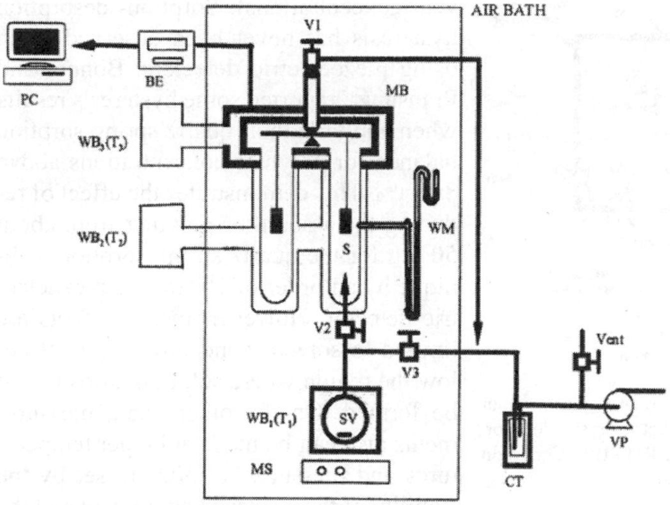


Figure 4.4.9. Schematic diagram of an isopiestic vapor sorption apparatus using an electronic microbalance: PC - personal computer, MB - microbalance, WB1-3 - water bath thermostats with $T_3>T_2>T_1$, V1-3 - valves, WM - W-tube mercury manometer, S - polymer sample/solution, SV - solvent reservoir, MS - magnetic stirrer, CT - cold trap, VP - vacuum pump. [Reprinted with permission from Ref. 92, Copyright 1998, American Chemical Society].

Comparable apparatuses were constructed and used, for example, by Bae et al.[94] or by Ashworth,[95] Ashworth and Price[96] applied a magnetic suspension balance instead of an electronic microbalance. The magnetic suspension technique has the advantage that all sensitive parts of the balance are located outside the measuring cell because the balance and the polymer solution measuring cell are in separate chambers and connected by magnetic coupling only. This allows its application even at very high temperatures of some hundred degrees as well as pressures up to hundreds of MPa.

The most sensitive solvent vapor sorption method is the piezoelectric sorption detector. The amount of solvent vapor absorbed by a polymer is detected by a corresponding change in frequency of a piezoelectric quartz crystal coated with a thin film of the polymer because a frequency change is the response of a mass change at the surface of such a crystal. The frequency of the crystal decreases as mass increases when the crystal is placed in a gas or vapor medium. The frequency decrease is fairly linear. The polymer must be coated onto the crystal from a solution with some care to obtain a fairly uniform film. Measurements can be made at dynamic (vapor flow) or static conditions. With reasonable assumptions for the stability of the crystal's base frequency and the precision of the frequency counter employed, the piezoelectric method allows the detection of as little as 10 nanograms of solvent using a 10 MHz crystal. This greatly reduces both the time necessary to attain equilibrium (3-4 hours) and the amount of polymer required. Saeki et al.[97-99] extensively applied this method to various polymer solutions in a concentration range between 60 and 100 wt% polymer. Recently, Wong et al.[100] and Mikkilineni et al.[101] presented some new investigations with this method. Figure 4.4.10 shows a schematic diagram of the general equipment.

A resolution of nanograms could be realized by Mikkilineni et al.[101] Measurements were also made as a function of time to obtain diffusion coefficients. Comparison with gravimetric sorption measurements demonstrated the accuracy of the experiment. Ref.[100] presents some details about the electronic circuit, the mounting arrangements for the quartz crystals and the sorption cell. Because very thin films are applied, equilibrium solvent absorption also can be obtained at polymer mass fractions approaching 1 (i.e., for small sol-

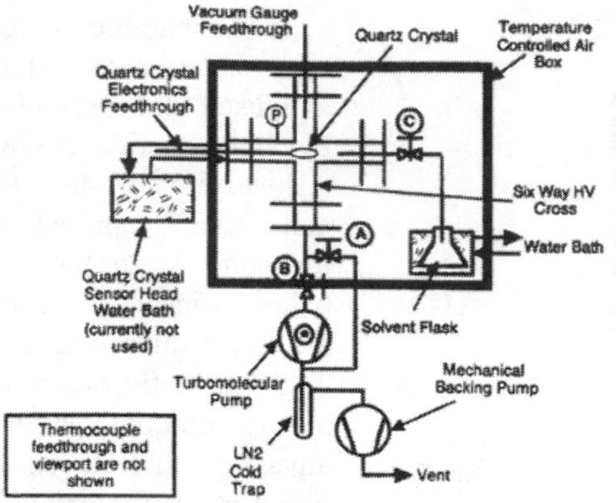


Figure 4.4.10. Schematic diagram of an isopiestic vapor sorption apparatus using a piezoelectric crystal detector. [Reprinted with permission from Ref. 101, Copyright 1995, American Chemical Society].

vent concentrations). Sorption- desorption hysteresis has never been observed when using piezoelectric detectors. Bonner and Prausnitz[85] reported some hysteresis results when applying their quartz spring sorption balance for polymer concentrations above 85 wt%. This demonstrates the effect of reducing the amount of polymer from about 50 mg for the quartz spring sorption technique by an order of $10^3$ for the piezoelectric detector. However, measurements are limited to solvent concentrations well below the region where solution drops would be formed. On the other hand, measurements also can be made at higher temperatures and pressures. Limits are set by the stability of the electrical equipment and the construction of the measuring cell.

**(iv) Gas-liquid chromatography (GLC)**

In 1969 Smidsrod and Guillet[102] demonstrated that GLC could be used to determine the activity coefficient of a solute in a (molten) polymer at essentially zero solute concentration. This type of activity coefficient is known as an infinite-dilution activity coefficient. Smidsrod and Guillet also introduced the term "inverse" gas-liquid chromatography (IGC) because in IGC the liquid polymer in the stationary phase acts as a solvent for the very small amount of an injected solvent sample like the solute in this case. Methods and results of the application of IGC to polymers and polymer solutions have been reviewed continuously[13-25] so that an extensive discussion is not required here. The equipment in principle does not differ very much from that used in analytical GLC. Figure 4.4.11 is a schematic of a simple IGC unit.

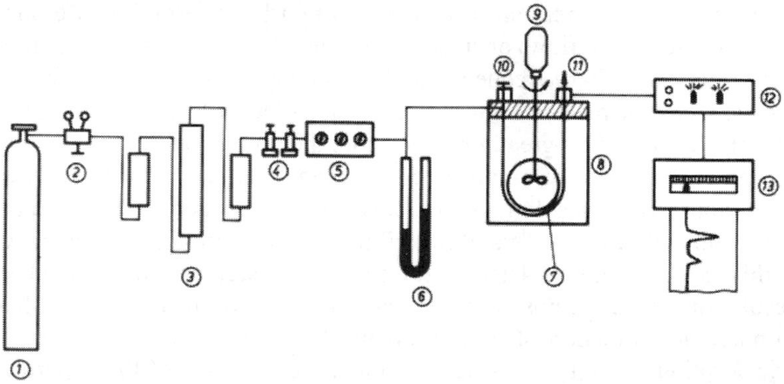


Figure 4.4.11. Schematic diagram of an IGC apparatus: 1 - carrier gas, 2 - pressure reducer, 3 - gas cleaning unit (if necessary) , 4+5 - gas-pressure regulation and control unit, 6 - manometer, 7 - column, 8 - thermostat, 9 - mechanical mixer, 10 - inlet syringe, 11 - detector (the gas flows after the detector through a bubble flow meter that is not shown here), 12 - electronics, 13 - recorder.

For infinite dilution operation the carrier gas flows directly to the column which is inserted into a thermostated oil bath (to get a more precise temperature control than in a conventional GLC oven). The output of the column is measured with a flame ionization detector or alternately with a thermal conductivity detector. Helium is used today as carrier gas (nitrogen in earlier work). From the difference between the retention time of the injected solvent sample and the retention time of a non-interacting gas (marker gas), the thermodynamic equilibrium behavior can be obtained (equations see below). Most experiments were made up to now with packed columns, but capillary columns were used, too. The experimental conditions must be chosen so that real thermodynamic data can be obtained, i.e., equilibrium bulk absorption conditions. Errors caused by unsuitable gas flow rates, unsuitable polymer loading percentages on the solid support material and support surface effects as well as any interactions between the injected sample and the solid support in packed columns, unsuitable sample size of the injected probes, carrier gas effects, and imprecise knowledge of the real amount of polymer in the column, can be sources of problems, whether data are nominally measured under real thermodynamic equilibrium conditions or not, and have to be eliminated. The sizeable pressure drop through the column must be measured and accounted for.

Column preparation is the most difficult task within the IGC-experiment. In the case of packed columns, the preparation technique developed by Munk and coworkers[103,104] is preferred, where the solid support is continuously soaked with a predetermined concentration of a polymer solution. In the case of capillary IGC, columns are made by filling a small silica capillary with a predetermined concentration of a degassed polymer solution. The one end is then sealed and vacuum is applied to the other end. As the solvent evaporates, a thin layer of the polymer is laid down on the walls. With carefully prepared capillary surfaces, the right solvent in terms of volatility and wetting characteristics, and an acceptable viscosity in the solution, a very uniform polymer film can be formed, typically 3 to 10 μm thick. Column preparation is the most time-consuming part of an IGC-experiment. In the case of packed columns, two, three or even more columns must be prepared to test the reproducibility of the experimental results and to check any dependence on polymer loading and sometimes to filter out effects caused by the solid support. Next to that, various tests regarding solvent sample size and carrier gas flow rate have to be done to find out correct experimental conditions.

There is an additional condition for obtaining real thermodynamic equilibrium data that is caused by the nature of the polymer sample. Synthetic polymers are usually amorphous or semi-crystalline products. VLE-based solvent activity coefficients require the polymer to be in a molten state, however. This means that IGC-measurements have to be performed for our purpose well above the glass transition temperature of the amorphous polymer or even above the melting temperature of the crystalline parts of a polymer sample. On the other hand IGC can be applied to determine these temperatures. The glass transition of a polymer does not take place at a fixed temperature but within a certain temperature range depending on the probing technique applied because it is a non-equilibrium effect. Figure 4.4.12 demonstrates the appearance of the glass transition region in an IGC-experiment.

The S-shaped part of the curves in Figure 4.4.12 is the glass transition region. Its minimum describes the glass transition temperature as obtained by IGC. Only data from the straight line on the left side at temperatures well above the glass transition temperature lead

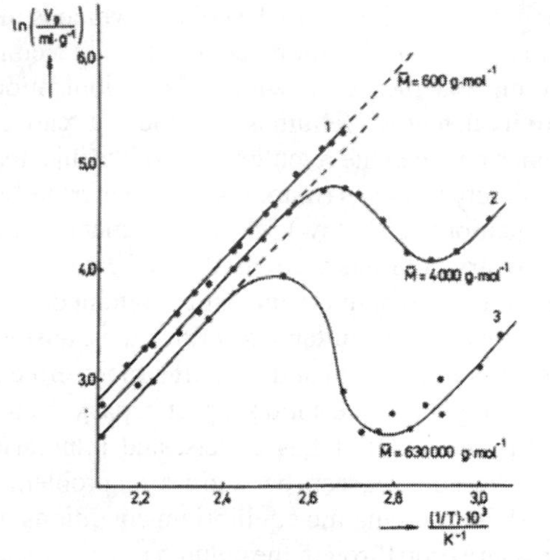

Figure 4.4.12. Temperature dependence of the specific retention volume $V_g$ of p-xylene in polystyrenes of varying molar masses, experimental data were measured by Glindemann.[105]

to real thermodynamic vapor-liquid equilibrium data. As a rule of thumb, the experimental temperature must exceed the glass transition temperature by about 50K. Form and width of the S-shaped region depend somewhat on the solvent used and, as can be seen from the picture, there is a certain dependence on molar mass of the polymer. Data on the right side at temperatures below the glass transition describe mainly surface adsorption effects.

Brockmeier et al.[106,107] showed that GLC can also be used to determine the partial pressure of a solute in a polymer solution at concentrations as great as 50 wt% solute. In this case of finite concentration IGC, a uniform background concentration of the solute is established in the carrier gas. The carrier gas is diverted to a saturator through a metering valve. In the saturator it passes through a diffuser in a well-stirred, temperature-controlled liquid bath. It leaves the separator with the solute equilibrium vapor pressure in the carrier gas. The solute concentration is varied by changing the saturator temperature. Precise control of the temperature bath is needed in order to obtain a constant plateau concentration. Upon leaving the saturator the gas flows to the injector block and then to the column. As in the infinite dilute case a small pulse of the solvent is then injected. This technique is known as elution on a plateau, Conder and Purnell.[108,109] Because finite concentration IGC is technically more complicated, only few workers have applied it. Price and Guillet[110] demonstrated that results for solvent activity, activity coefficient or $\chi$-function are in good agreement with those obtained by traditional isopiestic vapor sorption methods. Whereas the vapor sorption results are more accurate at higher concentrations, the reverse is true for finite concentration IGC since larger injection volumes have to be used, which strains the theory on which the calculations are based. Also, at large vapor concentrations the chromatographic peaks become more spread out, making the measurement of retention times less precise. Additionally, the concentration range is limited by the requirement that the saturator temperature must be below that of the column. Clearly, at higher measuring temperatures, higher solvent concentrations may be used. Finite concentration IGC can be extended to multi-component systems. Especially ternary polymer solutions were investigated to some extend with this technique, e.g., Bonner and coworkers[111,112] or Glover and coworkers.[113-115] Data reduction is somewhat complicated, however.

Danner et al.[116] tested the frontal analysis by characteristic point (FACP) technique to measure thermodynamic data for polymer-solvent systems at finite concentrations. In the FACP technique, a complete isotherm can be derived from the shape of one breakthrough profile. A point on an isotherm is obtained by measuring the retention volume of the characteristic point at the corresponding concentration. The methods to determine thermodynamic data by FACP technique were discussed in detail by Conder and Young.[117]

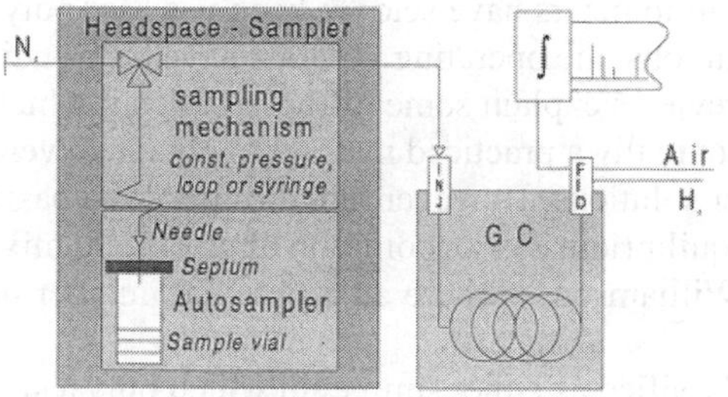


Figure 4.4.13. Schematic of applying head-space gas-chromatography (HSGC) to VLE-measurements in polymer solutions (drawing provided by B. A. Wolf, Univ. Mainz, Germany).

Wolf and coworkers[118,119] applied another GLC technique to VLE-measurements for polymer-solvent systems, the so-called head-space gas-chromatography (HSGC). This is practically a combination of static vapor pressure measurement with gas-chromatographic detection. HSGC experiments were carried out with an apparatus consisting of a head-space-sampler and a normal gas chromatograph. Figure 4.4.13 shows a schematic diagram of the equipment.

The pneumatically driven thermostated headspace-sampler samples a constant amount of gas phase (that must be in equilibrium with the liquid polymer solution, of course) and injects this mixture into the gas chromatograph. Helium is used as carrier gas. After separation of the components of the gaseous mixture in a capillary column they are detected individually by a thermal conductivity detector. The signals are sent to an integrator which calculates the peak areas, which are proportional to the amount of gas in the sample volume and consequently to the vapor pressure. Calibration can be made by measuring the pure solvent in dependence on temperature and to compare the data with the corresponding vapor pressure vs. temperature data. Measurements can be done between about 25 and 85 wt% polymer in the solution (again depending on temperature, solvent and polymer investigated). In order to guarantee thermodynamic equilibrium, solutions have to be conditioned for at least 24 h at constant temperature in the head-space-sampler before measurement. Degassing is not necessary and solvents have to be purified only to the extent necessary to prevent unfavorable interactions in the solution. The experimental error in the vapor pressures is typically of the order of 1-3%. Details about theory and practice of HSGC were discussed by Kolb and Ettre.[120] One great advantage of HSGC is its capability to measure VLE-data, not only for binary polymer solutions but also for polymer solutions in mixed solvents, since it provides a complete analysis of the vapor phase in equilibrium. This is usually not the case with the classical isopiestic sorption balances where PVT-data and a material-balance calculation must be included into the data reduction to calculate vapor phase concentrations, e.g., Refs.[121-123]

**(v) Ebulliometry (boiling point elevation of the solvent)**

As pointed out above, dynamic vapor-liquid equilibrium measurement methods are not very suitable for concentrated polymer solutions, especially due to their heavy foaming behavior. For dilute polymer solutions, however, there is continuing application of ebulliometry as an absolute method for the direct determination of the number-average molecular mass $M_n$. Dedicated differential ebulliometers allow the determination of values up to an order of 100,000 g/mol. Ebulliometry as a method for molar mass determination was recently reviewed by Cooper,[33] Glover,[34] and Mays and Hadjichristidis.[40]

The major requirements for a successful ebulliometry experiment are thermal stability, equilibration of both concentration and temperature, temperature measurement and control and pressure measurement and control. It is an advantage of ebulliometry to know very exactly the constant pressure applied since pressure constancy is a prerequisite of any successful experiment. Commercially sold ebulliometers have seldom been used for polymer solutions. For application to polymer solutions, the operating systems have been individually constructed. The above-mentioned reviews explain some of these in detail which will not be repeated here as ebulliometry is not really a practiced method to obtain solvent activities and thermodynamic data in polymer solutions. However, ebulliometry is a basic method for the investigation of vapor-liquid equilibrium data of common binary liquid mixtures, and we again point to the review by Williamson,[55] where an additional number of equilibrium stills is shown.

Ebulliometers have traditionally been classified as either simple, in which only a single temperature is measured, or differential, in which the boiling temperatures of the pure solvent and of the solution were measured simultaneously. In differential ebulliometers, two independent temperature sensors can work, or a single differential temperature measurement is done. Essentially, all ebulliometers for polymer solutions are of the differential type. The manner in which the reference boiling temperature of the pure solvent is provided differs, however. Establishment and maintenance of both temperature and concentration equilibrium are accomplished in a variety of ways. The common method is the use of a vapor lift pump (a Cottrell pump) where the boiling liquid is raised to a position from which it can flow in a thin film until superheat is dissipated and its true boiling temperature can be measured. This technique has one disadvantage: the pumping rate depends on the heat input. This is of particular importance with polymer solutions in which problems due to foaming occur. To overcome this problem mechanical pumps were sometimes applied. Other ebulliometer types have been reported that use the methods of surface volatilization, spray cooling, two-stage heating, or rotating ebulliometer; for more details please see Refs.[33,34,40] Methods of temperature measurement within ebulliometer experiments will not be discussed here, as they change rapidly with continuing progress of electronics and computerization. Pressure control is important for single temperature ebulliometers, as the boiling temperature depends on pressure. It is not so important in differential type ebulliometers, owing to the simultaneous and compensating change in reference temperatures. Therefore, direct changes in boiling temperatures present no serious problem if sufficient time is allowed for calibration. It is usually recommended that the ebulliometer be thoroughly cleaned and dried between experiments. Small amounts of polymer adsorbed on the surface must be avoided.

**(vi) Vapor-pressure osmometry (VPO)**

Vapor-pressure osmometry is, from its name, compared with membrane osmometry by considering the vapor phase to act like the semipermeable membrane, however, from its principles it is based on vapor pressure lowering or boiling temperature elevation. Since the direct measure of vapor pressure lowering of dilute polymer solutions is impractical because of the extreme sensitivity that is required, VPO is in widespread use for oligomer solutions ($M_n$ less than 20,000 g/mol) by employing the thermoelectric method as developed by Hill in 1930.[124] In the thermoelectric method, two matched temperature-sensitive thermistors are placed in a chamber that is thermostated to the measuring temperature and where the atmosphere is saturated with solvent vapor. If drops of pure solvent are placed on both thermistors, the thermistors will be at the same temperature (zero point calibration). If a solution drop is placed on one thermistor, a temperature difference $\Delta T$ occurs which is caused by condensation of solvent vapor onto the solution drop. From equilibrium thermodynamics follows that this temperature increase has its theoretical limit when the vapor pressure of

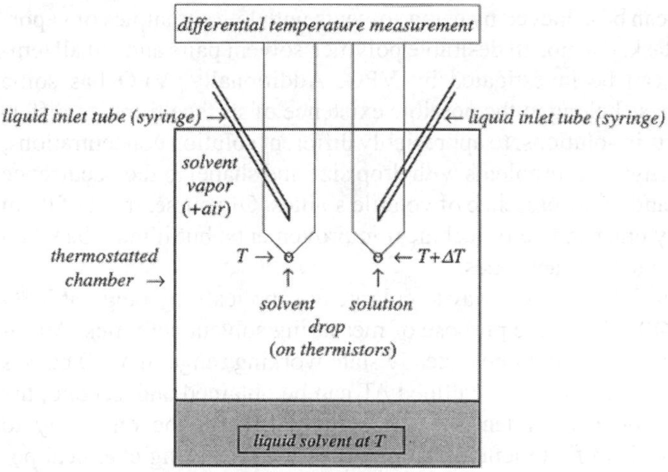


Figure 4.4.14. Principle scheme of a vapor-pressure osmometer of the hanging drop type. T - measuring temperature, ΔT - obtained temperature difference (time dependent), measurements are made at atmospheric pressure where T determines the partial vapor pressure of the solvent $P_1$ in air.

the solution is equal to that of the pure solvent, i.e., at infinite dilution. The obtained temperature difference is very small, about $10^{-5}$K. Because solvent transfer effects are measured, VPO is a dynamic method. This leads to a time-dependent measurement of ΔT. The principle scheme of a VPO apparatus is given in Figure 4.4.14.

Today, vapor-pressure osmometers are commercially available from a number of producers. They can be divided into two basic types: those that employ conventional hanging drop thermistors as in Figure 4.4.14 and those that use vertical thermistors. The vertical thermistors automatically control drop size to ensure more reproducible response. The hanging drop design requires the operator to manually monitor and control drop size. Furthermore, commercial instruments have been developed which utilize vertical thermistors having cups or pieces of platinum gauze to control drop size in a highly reproducible manner. More details about instrumentation and techniques can be found in the reviews given by Glover,[34] Mays and Hadjichristidis.[40] A very recent presentation can be found in a new book edited by Pethrick and Dawkin.[26]

Depending on technical details of the equipment, on the sensitivity of the temperature detector, on measuring temperature, solvent vapor pressure and polymer concentration in the solution drop, a steady state for ΔT can be obtained after some minutes. The value of $\Delta T^{st}$ is the basis for thermodynamic data reduction (see below). If measuring conditions do not allow a steady state, an extrapolation method to ΔT at zero measuring time can be employed for data reduction. Sometimes a value is used that is obtained after a predetermined time; however, this may lead to some problems with knowing the exact polymer concentration in the solution. The extrapolation method is somewhat more complicated and needs experience of the experimentator but gives an exact value of polymer concentration. Both methods are used within solvent activity measurements when polymer concentrations are higher and condensation is faster than in common polymer characterization experiments. A way to avoid these problems is discussed below.

Experience has shown that careful selection of solvent and temperature is critical to the success of the VPO experiment. Nearly all common solvents, including water (usually, there are different thermistor sensors for organic solvents and for water), can be used with VPO. The measuring temperature should be chosen so that the vapor pressure of the solvent will be greater than 6,000 Pa, but not so high as to lead to problems with evaporation from the chamber. Solvent purity is critical, especially volatile impurities, and water must be avoided. Greater sensitivity can be achieved by using solvents with low enthalpies of vaporization. This means, for our task, that not all desirable polymer-solvent pairs and not all temperature (pressure) ranges can be investigated by VPO. Additionally, VPO has some inherent sources of error. These belong to the possible existence of surface films, to differences in diffusion coefficients in solutions, to appreciably different solution concentrations, to differences in heat conductivity, to problems with drop size and shape, to the occurrence of reactions in the solution, and to the presence of volatile solutes. Of course, most of them can be avoided by laboratory practice and/or technical improvements, but it must be taken into account when measuring solvent activities.

Regener and Wohlfarth[125] developed a way to enlarge the applicability range of VPO to polymer concentrations ≤40wt% for the purpose of measuring solvent activities. An increase of polymer concentration over the linear steady state working range of VPO causes some problems. First, no thermodynamically defined $\Delta T$ can be obtained and, second, the calibration constant may become dependent on concentration. Thus, the only way to achieve higher concentrations is to find methods to minimize the increasing chemical potential difference of the solvent between the two drops. This can be achieved by using a reference solution of known solvent activity instead of the pure solvent. The instrument is then used as a zero-point detector comparing the solvent activity of the reference solution with solvent activity of the polymer solution. The reference concentration has to be varied until $\Delta T = 0$ is found. The only assumption involved in this method is equal solvent condensation and diffusion. The extrapolation method to $\Delta T$ at zero measuring time can be used to minimize these influences. It is not really necessary to find the reference solution at exactly $\Delta T=0$, but it is sufficient to measure a small $\Delta T < 0$ and small $\Delta T > 0$ and to interpolate between both known solvent activities. An example is shown in Figure 4.4.15, where benzene was used as solute for the reference solutions.

Since the polymer solution remains quasi unchanged in concentration, this modified VPO-method is faster than isopiestic isothermal distillation experiments with organic solvents and polymer solutions. Difficulties with the increasing viscosity of concentrated polymer solutions set limits to its applicability, because solutions should flow easily to form drops.

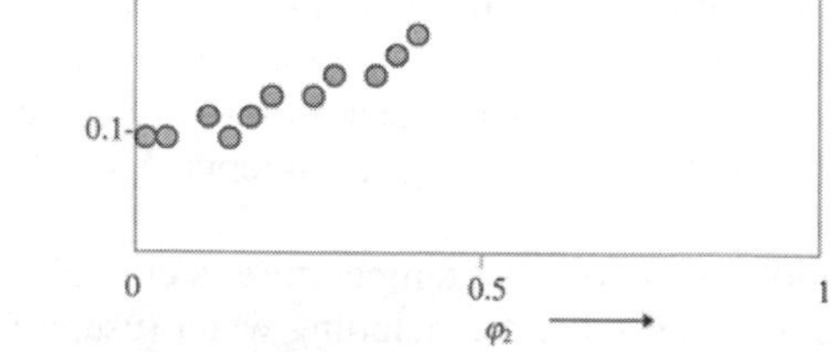

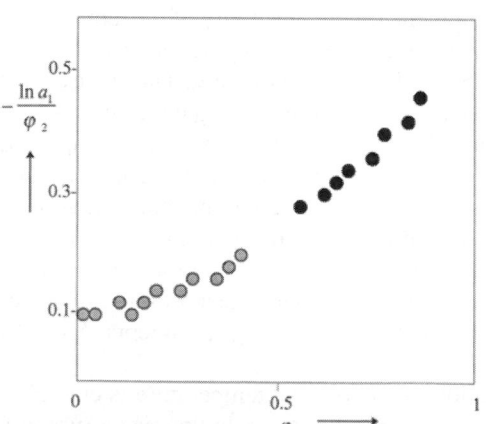

Figure 4.4.15. Experimental data of the system toluene + polystyrene, $M_n = 1380$ g/mol, at 323.15K, isopiestic vapor pressure/sorption measurement (full circles), VPO at higher concentrations (gray circles), data from authors own work.

Recently, Gaube et al.[126,127] or Eliassi et al.[128] measured water activities in aqueous solutions of poly(ethylene glycol) and showed that the conventional VPO method also can be used for higher polymer concentrations with good success.

#### *4.4.3.1.2 Primary data reduction*

Equation [4.4.7] is the starting relation for data from VLE-measurements. Two relations are necessary to obtain the solvent activity $a_1$: one for the fugacity coefficient of the solvent vapor and one for the standard state fugacity of the liquid solvent. In principle, every kind of equation of state can be applied to calculate the solvent vapor fugacity coefficient. This is done if dedicated equations of state are applied for further modeling. However, in most cases it is common practice to use the virial equation of state for the purpose of reducing primary VLE-data of polymer solutions. This procedure is sufficient for vapor pressures in the low or medium pressure region where most of the VLE-measurements are performed. The virial equation is truncated usually after the second virial coefficient, and one obtains from Equation [4.4.6]:

$$\ln \phi_i = \left( 2\sum_{j=1}^{m} y_i B_{ij} - \sum_{i=1}^{m}\sum_{j=1}^{m} y_i y_j B_{ij} \right) \frac{P}{RT} \qquad [4.4.19]$$

where:

| | |
|---|---|
| $B_{ii}$ | second virial coefficient of pure component i at temperature T |
| $B_{jj}$ | second virial coefficient of pure component j at temperature T |
| $B_{ij}$ | second virial coefficient corresponding to i-j interactions at temperature T. |

In the case of a strictly binary polymer solutions Equation [4.4.19] reduces simply to:

$$\ln \phi_1 = \frac{B_{11} P}{RT} \qquad [4.4.20]$$

To calculate the standard state fugacity, we consider the pure solvent at temperature T and saturation vapor pressure $P^s$ for being the standard conditions. The standard state fugacity is then calculated as:

$$f_1^0 = P_1^s \exp\left[ \frac{V_1^L \left( P - P_1^s \right) + B_{11} P_1^s}{RT} \right] \qquad [4.4.21]$$

where:

| | |
|---|---|
| $P_1^s$ | saturation vapor pressure of the pure liquid solvent 1 at temperature T |
| $V_1^L$ | molar volume of the pure liquid solvent 1 at temperature T |

The so-called Poynting correction takes into account the difference between the chemical potentials of the pure liquid solvent at pressure P and at saturation pressure $P_1^s$ assuming that the liquid molar volume does not vary with pressure. Combining Equations [4.4.7, 4.4.20 and 4.4.21] one obtains the following relations:

$$a_1 = \phi_1^V y_1 P / f_1^0 = \left( P_1 / P_1^s \right) \exp\left[ \frac{\left( B_{11} - V_1^L \right)\left( P - P_1^s \right)}{RT} \right] \qquad [4.4.22a]$$

$$\gamma_1 = a_1 / x_1^L = \left( P_1 / x_1^L P_1^s \right) \exp\left[ \frac{\left( B_{11} - V_1^L \right)\left( P - P_1^s \right)}{RT} \right] \qquad [4.4.22b]$$

$$\Omega_1 = a_1 / w_1^L = \left(P_1 / w_1^L P_1^s\right) \exp\left[\frac{\left(B_{11} - V_1^L\right)\left(P - P_1^s\right)}{RT}\right] \quad [4.4.22c]$$

These relations can be applied to VLE-data from all experimental methods.

The data reduction for infinite dilution IGC starts with the usually obtained terms of retention volume or net retention volume.

$$V_{net} = V_r - V_{dead} \quad [4.4.23]$$

where:

| | |
|---|---|
| $V_{net}$ | net retention volume |
| $V_r$ | retention volume |
| $V_{dead}$ | retention volume of the (inert) marker gas, dead retention, gas holdup |

These retention volumes are reduced to specific ones by division of Equation [4.4.23] with the mass of the liquid (here the liquid, molten polymer), corrected for the pressure difference between column inlet and outlet pressure and reduced to $T_0 = 273.15K$.

$$V_g^0 = \left(\frac{V_{net}}{m_2}\right)\left(\frac{T_0}{T}\right)\frac{3\left(P_{in} / P_{out}\right)^2 - 1}{2\left(P_{in} / P_{out}\right)^3 - 1} \quad [4.4.24]$$

where:

| | |
|---|---|
| $V_g^0$ | specific retention volume corrected to 0°C |
| $m_2$ | mass of the polymer in the liquid phase within the column |
| $P_{in}$ | column inlet pressure |
| $P_{out}$ | column outlet pressure |

Theory of GLC provides the relation between $V_g^0$ and thermodynamic data for the low-molecular component (solvent) 1 at infinite dilution:

$$\left(\frac{P_1}{x_1^L}\right)^{\infty} = \frac{T_0 R}{V_g^0 M_2} \quad or \quad \left(\frac{P_1}{w_1^L}\right)^{\infty} = \frac{T_0 R}{V_g^0 M_1} \quad [4.4.25]$$

where:

| | |
|---|---|
| $M_2$ | molar mass of the liquid (molten) polymer |
| $M_1$ | molar mass of the low-molecular component (solvent). |

The activity coefficients at infinite dilution follow immediately from Equation [4.4.22] by introducing the above result, if we neglect interactions to and between carrier gas molecules (which is normally helium):

$$\gamma_1^{\infty} = \left(\frac{T_0 R}{V_g^0 M_2 P_1^s}\right) \exp\left[\frac{\left(B_{11} - V_1^L\right)\left(P - P_1^s\right)}{RT}\right] \quad [4.4.26a]$$

$$\Omega_1^\infty = \left(\frac{T_0 R}{V_g^0 M_2 P_1^s}\right) \exp\left[\frac{P_1^s\left(V_1^L - B_{11}\right)}{RT}\right] \qquad [4.4.26b]$$

The standard state pressure P has to be specified. It is common practice by many authors to define here zero pressure as standard pressure since pressures are usually very low during GLC-measurements. Then, Equations [4.4.26a and b] change to:

$$\gamma_1^\infty = \left(\frac{T_0 R}{V_g^0 M_2 P_1^s}\right) \exp\left[\frac{P_1^s\left(V_1^L - B_{11}\right)}{RT}\right] \qquad [4.4.27a]$$

$$\Omega_1^\infty = \left(\frac{T_0 R}{V_g^0 M_1 P_1^s}\right) \exp\left[\frac{P_1^s\left(V_1^L - B_{11}\right)}{RT}\right] \qquad [4.4.27b]$$

One should keep in mind that mole fraction-based activity coefficients become very small values for common polymer solutions and reach the value of 0 for $M_2 \rightarrow \infty$, which means a limited applicability to at least oligomer solutions. Therefore, the common literature provides only mass fraction-based activity coefficients for (high-molecular) polymer/(low-molecular) solvent pairs. Furthermore, the molar mass $M_2$ of the polymeric liquid is an average value according to the usual molar-mass distribution of polymers. Additionally, it is a second average if mixed stationary liquid phases are applied.

Furthermore, thermodynamic VLE-data from GLC-measurements are provided in the literature as values for $(P_1/w_1)^\infty$, see Equation [4.4.25], i.e., classical mass fraction based Henry's constants (if assuming ideal gas phase behavior):

$$H_{1,2} = \left(\frac{P_1}{w_1^L}\right)^\infty = \frac{T_0 R}{V_g^0 M_1} \qquad [4.4.28]$$

Thus, Equation (4.4.27b) reduces to

$$\Omega_1^\infty = \frac{H_{1,2}}{P_1^s} \exp\left[\frac{P_1^s\left(V_1^L - B_{11}\right)}{RT}\right] \qquad [4.4.29]$$

The data reduction for finite concentration IGC by elution on a plateau is more complicated than for infinite dilution IGC via Equations [4.4.24 to 26] and will not be explained here. A detailed analysis of the elution on plateau mode was made by Conder and Purnell.[108,109] For the determination of thermodynamic properties of polymer solutions by finite-concentration IGC the reader is referred to the paper by Price and Guillet[110] who provide a comprehensive derivation of all necessary equations.

The data reduction of ebulliometric measurements can be made either by using Equations [4.4.22] or by applying the relation for the boiling point elevation of a binary mixture:

$$\Delta T^{ebull} = -\frac{RT^2}{\Delta_{vap}H_1^0}\ln a_1 \qquad [4.4.30]$$

where:

| | |
|---|---|
| $T$ | measuring temperature (= boiling point temperature of the pure solvent) |
| $\Delta T^{ebull}$ | temperature difference of boiling point elevation |
| $\Delta_{vap}H_1^0$ | molar enthalpy of vaporization of the pure solvent 1 at temperature T. |

The ratio $M_1RT^2/\Delta_{vap}H_1^0$ is called the ebulliometric constant. For the determination of solvent activities from ebulliometric data, tabulated ebulliometric constants should not be used, however. On the other side, it is sometimes recommended to use reference solutes to establish an experimental relationship for the equipment in use, i.e., unprecise data for the enthalpy of vaporization or perhaps some non-equilibrium effects cancel out of the calculation. Enthalpies of vaporization are provided by several data collections, e.g., by Majer and Svoboda,[129] or through the DIPPR database.[130]

The data reduction of vapor-pressure osmometry (VPO) follows to some extent the same relations as outlined above. However, from its basic principles, it is not an equilibrium method, since one measures the (very) small difference between the boiling point temperatures of the pure solvent drop and the polymer solution drop in a dynamic regime. This temperature difference is the starting point for determining solvent activities. There is an analogy to the boiling point elevation in thermodynamic equilibrium. Therefore, in the steady state period of the experiment, the following relation can be applied if one assumes that the steady state is sufficiently near the vapor-liquid equilibrium and linear non-equilibrium thermodynamics is valid:

$$\Delta T^{st} = -k_{VPO}\frac{RT^2}{\Delta_{vap}H_1^0}\ln a_1 \qquad [4.4.31]$$

where:

| | |
|---|---|
| $T$ | measuring temperature (= temperature of the pure solvent drop) |
| $\Delta T^{st}$ | temperature difference between solution and solvent drops in the steady state |
| $k_{VPO}$ | VPO-specific constant |
| $\Delta_{vap}H_1^0$ | molar enthalpy of vaporization of the pure solvent 1 at temperature T . |

Recent examples of solvent activity measurements by VPO in aqueous solutions of poly(ethylene glycol) by Eliassi et al.[128] and of poly(ethylene glycol) or dextran by Gaube et al.[126,127] demonstrate the obtainable high quality if precise experiments were made.

The so-called VPO-specific constant contains all deviations from equilibrium state and it is to be determined experimentally. It depends on certain technical details from the equipment used and also on the temperature and solvent applied. It is assumed not to depend on the special solute under investigation and can therefore be obtained by calibration. Equation [4.4.31] can also be used if not the steady state, but the temperature difference extrapolated to a measuring time of zero is determined by the experimentator. However, the values of $k_{VPO}$ are different for both methods. A more detailed discussion about calibration problems can be found in the papers of Bersted,[131,132] or Figini.[133-135]

Usually, VPO-data are reduced to virial coefficients and not to solvent activities. Power series expansion of Equation [4.4.31] leads to the following relations:

$$\frac{\Delta T^{st}}{c_2} = k_{VPO} \frac{RT^2 V_1}{\Delta_{LV} H_1^0} \left[ \frac{1}{M_2} + \Gamma_2 c_2 + \Gamma_3 c_2^2 + \ldots \right] \quad [4.4.32a]$$

or

$$\frac{\Delta T^{st}}{c'_2} = k_{VPO} \frac{RT^2 V_1}{\Delta_{LV} H_1^0} \left[ \frac{1}{M_2} + \Gamma'_2 c'_2 + \Gamma'_3 c'^2_2 + \ldots \right] \quad [4.4.32b]$$

where:

| | |
|---|---|
| $c_2$ | mass by volume concentration $c_2 = m_2/v$ |
| $c'_2$ | mass by mass concentration $c'_2 = m_2/m_1$ |
| v | volume of the polymer solution |
| $m_i$ | mass of component i |
| $V_1$ | molar volume of the solvent |
| $M_1$ | molar mass of the solvent |
| $M_2$ | molar mass of the polymer |
| $\Gamma_2, \Gamma_3$, ... | second, third, ... VPO-virial coefficients based on g/cm$^3$ concentrations |
| $\Gamma'_2, \Gamma'_3$, .. | second, third, ... VPO-virial coefficients based on g/g concentrations |

In the dilute concentration region, these virial equations are usually truncated after the second virial coefficient which leads to a linear relationship. These truncated forms of Equation [4.4.32] are the basis for applying VPO to polymer characterization, which will not be discussed here - please see Refs.[26,34,40] Solvent activities can be estimated from second virial coefficients with some care regarding the necessary accuracy of all numerical values included. The molar mass of the polymer, $M_2$, is the number-average, $M_n$, if polydisperse samples are investigated. Corresponding averages of the virial coefficients can be introduced, too. The estimation of higher virial coefficients than the second one is difficult and hardly leads to satisfying results, because measurements at high polymer concentrations cause a lot of problems. In some cases, however, as in the above-mentioned paper by Gaube et al.,[126,127] precise measurements were done for polymer concentrations up to 30-40 wt% and second and third virial coefficients were obtained in good quality.

As pointed out above, there is another way VPO can be applied to measure activity differences between two polymer solution drops that differ slightly in concentration (in the same solvent, of course). In this case, VPO is quasi an isopiestic experiment and the unknown activity can be determined by using reference solutions with known solvent activity values:[125]

$$a_1(T, w_{polymer}) = a_1(T, w_{reference}) \quad [4.4.33]$$

Reference solutions can be made with the same organic solutes that are used for calibration. In the case of water, NaCl or KCl solutions may be applied as it is done for many isopiestic (isothermal distillation) measurements with aqueous solutions.

#### *4.4.3.1.3 Comparison of experimental VLE-methods*

The general aim of all experiments is to measure solvent activities in polymer solutions over the complete concentration range and for all desired temperatures (and pressures). Additionally, the dependence on molar mass of the polymer has to be taken into account. As is clear from all explanations above, there is no really universal method to fulfill all purposes.

Vapor pressure/vapor sorption measurements cover nearly the complete concentration range if different apparatuses are used. Measurements can be made with good accuracy. Principal limits with respect to temperature or pressure do not exist, but most apparatuses in the literature are constructed only for temperatures between 20 and 100°C, sometimes up to 150°C, and pressures between 1 and 100 - 200 kPa. Vapor pressure/vapor sorption measurements are very time-consuming experiments. To obtain a complete isotherm one needs usually about a month with conventional techniques or, at least, some days with microbalances or piezoelectric sensors. This demands long-time stability for thermostating and precise temperature control. Furthermore, the equilibrium cell has to be sealed in such a way that air leakage is avoided for the complete duration of the measurement. Experimentators need quite a lot of experience until they observe really good data. Experiments can only partially be automated depending on the method and equipment applied. The accuracy of the final data depends on the method applied, the temperature or pressure investigated, and also the given concentration. Measurements above about 85 wt% polymer showed sometimes sorption-desorption hysteresis. Solvent degassing is absolutely necessary with the exception of the apparatus proposed by Sadowski where degassing takes place automatically during the experiment (see above). The solvent must be purified from all other impurities. This is true of course also for the polymer investigated. According to their capabilities, different apparatuses should be used: differential pressure measurements for 5-30 wt% polymer in the solution, isopiestic sorption techniques for 30-85 wt% polymer, piezoelectric or microbalance detection for 60-99 wt% polymer. These limits can change somewhat with molar mass of the polymer. Oligomer solutions are easier to handle and can be measured even with conventional VLE-technique as developed for low-molecular liquid mixtures. There may be limits in temperature and pressure that depend on the nature of the solvent/polymer pair. Usually, the solutions investigated should not show liquid-liquid demixing and solutions should not become solid. Thermodynamic equilibrium data can only be obtained if the polymer is investigated well above its glass transition temperature. There is a depression of the glass transition temperature with increasing solvent concentration, but there are polymers that can be investigated only at temperatures above 100°C and more, even in concentrated solutions.

VPO is more limited with respect to the measurement of solvent activities. It is designed only for dilute polymer solutions (in the maximum up to 40 wt% polymer), optimum temperature and pressure (well below normal pressure) ranges and molar masses up to about 20,000 g/mol for the polymer. Not all solvents can be applied. On the other hand, VPO is a well-established method, commercially available, possessing a high resolution for very small differences of solvent activities with respect to the pure solvent and does not need much time. Steady-state conditions are obtained within minutes and quite a lot of measurements can be made during a working day. There are no problems with external thermostating or long-time stability. Experimental results from VPO are in good agreement with measurements from scattering techniques. VPO measurements close the gap between 0 and 30 wt% polymer in the solution with respect to conventional vapor pressure/vapor sorption measurements (of course, only within its limits explained). Experimentators easily acquire the necessary experience with the measuring equipment.

The piezoelectric sorption technique is a method that is especially suitable for the low solvent concentration range. It is the most sensitive solvent vapor sorption method. A resolution of nanograms can be realized. Measurements can also be made as a function of time to obtain diffusion coefficients. Comparison with gravimetric sorption measurements demonstrated the accuracy of the experiment. Because very thin films are applied, equilibrium solvent absorption also can be obtained at polymer mass fractions approaching 1, as with the IGC experiment. Comparison to IGC-data gives good agreement. Sorption-desorption hysteresis has never been observed when using piezoelectric detectors. Measurements are limited to a concentration range where the swollen polymer film is still stable at the crystal surface. Equilibrium is rather quickly established, usually after 3-4 hours, i.e., an isotherm can be measured within some days. With the corresponding equipment, high pressures and high temperatures can be applied, too.

IGC is the most rapid method and it is the recommended technique for the infinite dilution range of the solvent in the (liquid, molten) polymer. Measurements can also be made to obtain diffusion coefficients. Column preparation and finding optimum experimental conditions are the most time-consuming tasks. These tasks require quite a lot of experience. The final measurements can be automated and provide quick, reliable and reproducible results. Temperature and solvent dependencies can easily be investigated. The common accuracy is 1-3% with respect to data of the $\chi$-function or Henry's constant. There is no need to degas the solvents or to purify them except from impurities which may react with the polymer. Limits are mainly given by the glass transition temperature of the polymer as explained above. Due to this problem, most IGC measurements are made at temperatures well above 100°C. On the other hand, temperatures well above 100°C can cause the problem of thermal ageing and degradation of the polymer sample if temperatures are too high. In comparison to IGC, vapor pressure measurements were made in most cases below 100°C. There were some special investigations in earlier literature to compare IGC-data at infinite dilution with those from vapor pressure measurements at concentrated solutions, e.g., Refs.[110,136-138] Differences between IGC-data and vapor pressure measurements reported in older papers are mainly caused by errors with the IGC technique. Temperatures were used too near or even within the glass transition region, unsuitable polymer loading was applied, non-equilibrium conditions were used. But, there are also errors from/within vapor pressure data, mainly sorption/desorption hysteresis at too high polymer concentrations because of non-equilibrium conditions. Today it is accepted that there are no differences between IGC-data and vapor pressure measurements if all thermodynamic equilibrium conditions are carefully obeyed. In contrast to vapor pressure measurements, IGC can also be applied with thermodynamically bad solvents. It is the only method to obtain limiting activity coefficients for strong non-solvents. Even mass fraction based activity coefficients above 25 or $\chi$-values of 2 or more can be measured.

Finite concentration IGC provides the possibility to connect advantages from IGC and vapor pressure measurements because it can be applied between 50 and 100 wt% polymer. However, the experimental technique is more sophisticated, data reduction is more complicated, and only few workers have applied it. On the other hand, much experimental time can be saved since finite concentration IGC is a rapid method. One isotherm can be observed within one day (or two). Price and Guillet[110] or Danner et al.[116] demonstrated that results for solvent activity coefficients and $\chi$-functions or sorption isotherms are in good agreement with those obtained by traditional isopiestic vapor sorption methods. The concentration range of finite concentration IGC is limited by the requirement that the saturator temperature must be below that of the column. Clearly, at higher measuring temperatures, higher solvent concentrations may be used. Finite concentration IGC can be extended to multi-component systems.

Head-space gas chromatography is a modern tool for the measurement of vapor pressures in polymer solutions that is highly automated. Solutions need time to equilibrate, as is the case for all vapor pressure measurements. After equilibration of the solutions, quite a lot of data can be measured continuously with reliable precision. Solvent degassing is not necessary. Measurements require some experience with the equipment to obtain really thermodynamic equilibrium data. Calibration of the equipment with pure solvent vapor pressures may be necessary. HSGC can easily be extended to multi-component mixtures because it determines all components in the vapor phase separately.

In summary, the decision for a special equipment depends to some extend on concentration, temperature and pressure ranges one is interested in. From the experience of the author, the combination of isopiestic vapor pressure/vapor sorption measurements for the determination of solvent activities with infinite dilution IGC for the determination of Henry's constants provides good experimental data and covers a temperature range that is broad enough to have a sufficient data basis for thermodynamic modeling. If one is interested in both solvent solubility and diffusion data, finite concentration IGC or piezoelectric sorption techniques should be applied.

### 4.4.3.2 Other measurement methods

This subchapter summarizes all other experimental methods mentioned in subchapter 4.4.1 in order of their special importance and use regarding the determination of solvent activities in polymer solutions.

#### *4.4.3.2.1 Membrane osmometry*

Apart from VLE-measurements, membrane osmometry is the next important method that has been used for measuring solvent activities in polymer solutions. This follows from the tables in Refs.[1,2,5,8] according to its occurrence in comparison to the other methods. Most of these measurements were made in the dilute solution regime; only a small number of papers dealt with high-pressure osmometry where one also can measure solvent activities for concentrated solutions with polymer concentrations up to about 50 wt%, e.g. Refs.[139-145]

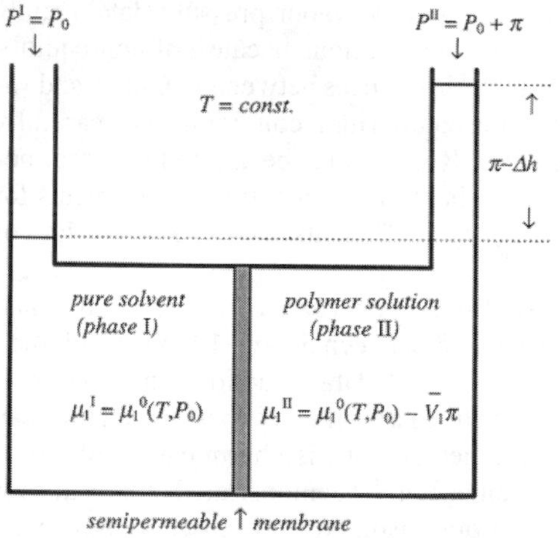


Figure 4.4.16. Principle scheme of a membrane osmometer: 1 - solvent, 2 - polymer, $\pi$ - osmotic pressure, $\Delta h$ - hydrostatic height difference, $P_0$ - ordinary pressure or measuring pressure, $\overline{V}_1$- partial molar volume of the solvent in the polymer solution.

Laboratory designed instruments were developed in the 40's and 50's, e.g. by Zimm[145] or by Flory.[144] Later on, high speed membrane osmometers are commercially available, e.g., from Knauer, Hewlett-Packard or Wescan Instruments. External pressures may be applied to balance the osmotic pressure if necessary, e.g., Vink.[140] The principle scheme of a membrane osmometer together with the corresponding

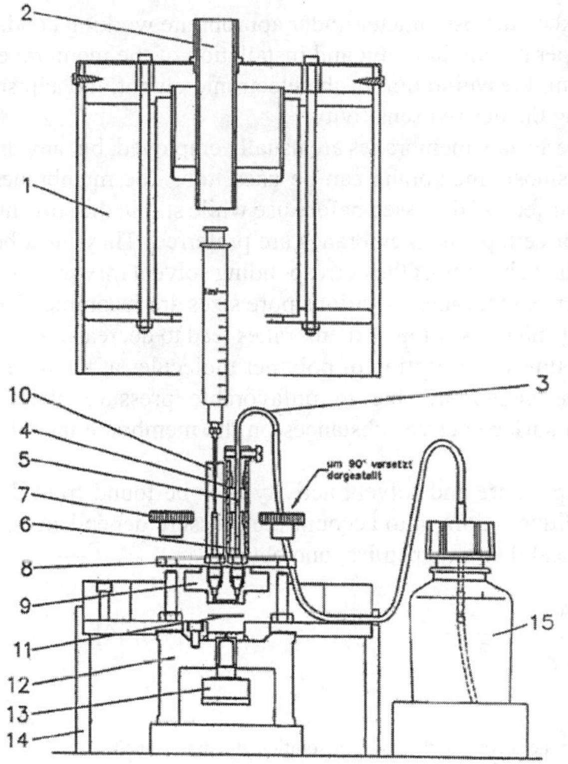


Figure 4.4.17. Description of the principal construction of a Knauer membrane osmometer A 300: 1 - head thermostat, 2 - channel for syringe, 3 - calibration device with suction tube, 4 - calibration glass, 5 - capillary position MEASUREMENT, 6 - capillary position CALIBRATION, 7 - tension screws, 8 - cell retaining disc, 9 - upper half of measuring cell, 10 - sample introduction system, 11 - semipermeable membrane, 12 - lower half of measuring cell, 13 - pressure measuring system, 14 - cell thermostat, 15 - suction of calibration bottle. [Reprinted from the operating manual with permission from Dr. H. Knauer GmbH (Germany)].

thermodynamic situation is illustrated in Figure 4.4.16.

Technical details of the different apparatuses will not be presented here, however, the principle construction of the measuring cell and the heating thermostat of the Knauer membrane osmometer A 300 is shown in Figure 4.4.17 for illustration and as example.

As a general feature of most osmometers, the membrane is clamped into a stainless steel thermostated chamber (the measuring cell and the pressure measuring system of modern osmometers are built into a high-grade electronically stabilized thermostat) and serves as barrier between the pure solvent and the polymer solution sides of the chamber. The solvent side (bottom) is in juxtaposition with a pressure sensor, e.g. the diaphragm of a capacitance strain gauge or a piezo-chip. The solvent transport is measured across the bottom side of the membrane in the direction of the solution which is topside the membrane. The amount of flowing solvent is in the range of $10^{-6}$ ml and equilibrium is established after some minutes when hydrostatic pressure prevents further solvent flow. This is indicated by the electronics of the equipment as well as any changes in equilibrium such as thermal drift of solute diffusion through the membrane. Other osmometers apply compensation methods where the increase of the hydrostatic height of the solution side is automatically compensated by changing the filling height. Due to this procedure, only very small amounts of solvent have to permeate through the membrane and equilibrium is reached within 10-20 minutes. The classical procedure was only used in older laboratory designed instruments where one started at zero and measured the hydrostatic height difference as a function of time until equilibrium is reached. More details about instrumentation and techniques can be found in the reviews by Adams,[38] Tombs and Peacock,[39] Mays and Hadjichristidis.[40] A very recent presentation can be found in a new book edited by Pethrick and Dawkins.[26]

Some efforts are necessary to keep the osmometer under appropriate working conditions. This relates mainly to the proper preconditioning and installation of the membrane, the attainment of thermal equilibrium, the calibration of the electronic output, the adjustment of solvent zero, and to choosing the desired sensitivity.

For aqueous solutions, cellulose acetate membranes are usually employed, but any dialysis or ultrafiltration or reverse osmosis membrane can be used, too. The membranes should be conditioned in solvent or buffer and degassed before use while still in the solvent. For organic solvents, gel cellulose or cellophane membranes are preferred. They must be conditioned to a new solvent by gradual changes of the corresponding solvent mixture. Details are usually given by the supplier. Membranes in various pore sizes are recommended for solutes of low molar mass. Aging and deswelling of membranes lead to decreasing permeability and increasing measuring times. Adsorption of polymer molecules at the membrane surface, "ballooning" of the membrane due to unfavorable pressure effects, membrane asymmetry and action of surface active substances on the membrane must be avoided.

The relation between osmotic pressure and solvent activity is to be found from the chemical potential equilibrium condition, taking into account the pressure dependence of $\mu_1$. From the rules of phenomenological thermodynamics, one obtains:

$$\left(\frac{\partial \mu_1^0}{\partial P}\right)_T = V_1 \quad and \quad \left(\frac{\partial \ln a_1}{\partial P}\right)_T = \frac{\overline{V}_1 - V_1}{RT} \qquad [4.4.34]$$

where:

$\overline{V}_1$ partial molar volume of the solvent in the polymer solution at temperature T
$V_1$ molar volume of the pure solvent at temperature T

Integration is performed between $P_0$, i.e., the ordinary pressure or measuring pressure, and $\pi$, the osmotic pressure, and results in Equation [4.4.35].

$$\Delta\mu_1 = RT \ln a_1 = -\overline{V}_1 \pi \qquad [4.4.35]$$

Usually, the experimental data are reduced to virial coefficients and not to solvent activities. Series expansion of Equation [4.4.35] leads to the following relation:

$$\frac{\pi}{c_2} = RT\left[\frac{1}{M_2} + A_2 c_2 + A_3 c_2^2 + \ldots\right] \qquad [4.4\text{-}36]$$

where:

$c_2$ mass by volume concentration $c_2 = m_2/v$
$A_2, A_3, \ldots$ second, third, ... osmotic virial coefficients

The molar mass of the polymer, $M_2$, is the number-average, $M_n$, if polydisperse samples are investigated. Corresponding averages of the virial coefficients can be introduced, too. In the dilute concentration region, the virial equation is usually truncated after the second virial coefficient which leads to a linear relationship. A linearized relation over a wider concentration range can be constructed, Equation [4.4.38], if the Stockmayer-Casassa relation,[146] Equation [4.4.37], between $A_2$ and $A_3$ is applied:

$$A_3M_n = (A_2M_n / 2)^2 \quad [4.4.37]$$

$$\left(\frac{\pi}{c_2}\right)^{0.5} = \left(\frac{RT}{M_n}\right)^{0.5}\left[1 + \frac{A_2M_n}{2}c_2\right] \quad [4.4.38]$$

Examples for experimentally determined virial coefficients can be found in the above mentioned papers[139-145] and in the tables prepared by Lechner et al.[9] Solvent activities can be calculated via Equations [4.4.35 to 38] from osmotic second virial coefficients with some care regarding the necessary accuracy of all numerical values included. The partial molar volume of the solvent can be approximated in most cases by the molar volume of the pure solvent. Noda et al.[139] published a combined investigation of the thermodynamic behavior of poly(α-methylstyrene)s having sharp molar mass distributions and covering a wide range of molar masses in toluene. They applied osmotic pressure, light scattering and vapor pressure measurements and demonstrated the capabilities of these methods in comprehensive and detailed form. Gaube et al.[126,127] could show that in the case of aqueous dextran solutions, water activity data and virial coefficients measured by VPO and by membrane osmometry are in good agreement.

#### *4.4.3.2.2 Light scattering*

Light scattering is one of the most widespread characterization techniques for polymers. Therefore, technical and methodical details will not be explained here - please see Refs.[26-32] for such information. The general set-up of a scattering principle is illustrated by Figure 4.4.18. The scattering vector **q** is the difference between the wave vectors $\mathbf{k}_i$ and $\mathbf{k}_s$ of the incident and the scattered plane waves, the scattering angle θ is the angle between both vectors. Both are related by $|\mathbf{q}| = (4\pi/\lambda_0)\sin(\theta/2)$, where $\lambda_0$ is the wavelength of light in vacuum. Laser light is used today for the light source.

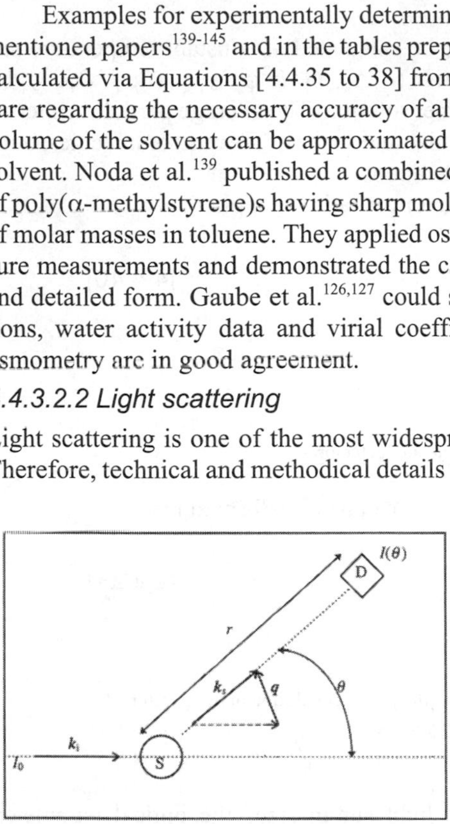


Figure 4.4.18. General set-up of a scattering experiment: $\mathbf{k}_i$, $\mathbf{k}_s$ - wave vectors of the incident and the scattered plane waves, **q** - scattering (or wave) vector, D- detector, S - sample, θ - scattering angle from the transmitted beam, $I_0$ - incident intensity of unpolarized light, r - the distance between sample and detector.

Light scattering in homogeneous fluids is caused by fluctuations in the dielectric constant. In pure liquids these are due to density fluctuations, in homogeneous solutions mainly to concentration fluctuations which generally lead to much larger fluctuations in dielectric constant than density variations. The difference between solution and pure solvent is called excess scattering. This excess scattering is of interest here, since it is related to the second derivative of Gibbs free energy of mixing with respect to concentration (and via this way to solvent activities):

$$I^{excess} \propto \frac{TP(\theta)}{\left|\dfrac{\partial^2 \Delta_{mix} G}{\partial\varphi_i \partial\varphi_j}\right|} \quad [4.4.39]$$

where:

| | |
|---|---|
| $I^{excess}$ | excess scattering intensity |
| T | absolute temperature |
| $\Delta_{mix}G$ | Gibbs free energy of mixing. |
| $\varphi_i$ | volume fractions of component i (= i[th] polymer species in the molecular distribution) |
| P(θ) | properly averaged particle scattering factor |
| θ | scattering angle from the transmitted beam |

The determinant in the denominator is to be calculated at constant temperature and pressure. It reduces to the single second derivative $(\partial^2\Delta_{mix}G/\partial^2\varphi_2)_{P,T} = (\partial\mu_1/\partial\varphi_2)_{P,T}$ for the case of a strictly binary monodisperse polymer solution. The average particle scattering factor is of primary importance in studies of the size and shape of the macromolecules, but it is merely a constant for thermodynamic considerations.

Conventionally, the so-called Rayleigh factor (or ratio) is applied:

$$R(\theta) \equiv \frac{I^{excess} r^2}{I_0 V_0 \left(1 + \cos^2\theta\right)} \qquad [4.4.40]$$

where:

| | |
|---|---|
| R(θ) | Rayleigh factor |
| $I_0$ | incident intensity of unpolarized light |
| $r^2$ | square of the distance between sample and detector |
| $V_0$ | detected scattering volume |

and, neglecting P(θ), Equations [4.4.39 and 4.4.40] can be transformed to:

$$R(\theta) \equiv \frac{RTK\varphi_2\overline{V}_1}{\left(\partial\mu_1/\partial\varphi_2\right)_{P,T}} \qquad [4.4.41]$$

where:

| | |
|---|---|
| $\overline{V}_1$ | partial molar volume of the solvent in the polymer solution at temperature T |
| $\varphi_2$ | volume fraction of the monodisperse polymer |
| K | optical constant |

The optical constant for unpolarized light summarizes the optical parameters of the experiment:

$$K \equiv 2\pi^2 n_o^2 \left(dn/dc_2\right)_{P,T}^2 / \left(N_{Av}\lambda_0^4\right) \qquad [4.4.42]$$

where:

| | |
|---|---|
| $n_o$ | refractive index of the pure solvent |
| n | refractive index of the solution |
| $c_2$ | mass by volume concentration $c_2 = m_2/v$ |
| $N_{Av}$ | Avogadro's number |
| $\lambda_0$ | wavelength of light in vacuum |

For dilute polymer solutions, the partial derivative in Equation [4.4.41] is a weak function of composition and the scattering intensity increases roughly proportional to the volume fraction of the polymer. While Equation [4.4.41] permits any light scattering data to be interpreted as partial derivatives of solvent chemical potential or activity, dilute solution measurements are conventionally again presented in terms of the osmotic virial expansion:

$$(Kc_2)/R(\theta=0)=(1/M_2)+2A_2c_2+3A_3c_2^2+\ldots \quad [4.4.43]$$

where:

$A_2, A_3, \ldots$ second, third, ... osmotic virial coefficients
$M_2$ molar mass of the polymer

According to the scattering theory of polydisperse polymers (please see, for example, the book written by Strobl),[147] the molar mass of the polymer, $M_2$, is equal to the mass average, $M_w$, of polydisperse polymers. The exact application of Equation [4.4.43] is at the scattering angle $\theta = 0$. Interpretation of scattering data at larger angles has to take into account the interparticle interference and the angular variation of the excess scattering intensity. Usually, such data have been analyzed with a Zimm plot,[148] where $(Kc_2)/R(\theta)$ is graphed as a function of $\sin^2(\theta/2)$ + const.*$c_2$ and measurements are extrapolated at constant angle to zero concentration values and at constant concentration to zero angle values. Connecting each of these two sets of points gives the curve specified by Equation [4.4.41]. The intercept gives $M_w$, the slope of the zero angle data yields the second virial coefficient. In many cases, non-linear Zimm plots were observed.[31]

An illustrative example for the variation of the second virial coefficient with molar mass and temperature from endothermic to exothermic conditions is given in the paper by Wolf and Adam,[149] or in the paper by Lechner and Schulz[150] for the variation of the second virial coefficient on pressure, both obtained by light scattering. A recent example for a light scattering investigation on the molar mass dependence of $A_2$ and $A_3$ was published by Nakamura et al.[151]

The virial expansion is inappropriate as the polymer concentration increases. In these cases, scattering data can be analyzed in terms of a thermodynamic ansatz for the Gibbs free energy of mixing. For example, Scholte[152] analyzed light scattering data of concentrated polystyrene solutions by means of the Flory-Huggins approach and determined $\chi$-data. But recently, Hasse, et al.[153] combined laser-light scattering with isopiestic measurements and obtained second and third virial coefficients of aqueous poly(ethylene glycol) solutions with high accuracy. Both virial coefficients could be correlated over a temperature range between 278 and 313K, including a derived theta-temperature of 375.7K in good agreement with results from liquid-liquid equilibrium. Additional measurements using membrane osmometry agreed well with the results of the simultaneous correlation of light scattering and isopiestic data. Corresponding measurements of aqueous dextran solutions by Kany et al.[154] showed again the resources inherent in such a combination of different methods.

Light scattering provides another interesting tool to determine thermodynamic data of polymer solutions. Starting from Equation [4.4.39], Scholte[155,156] developed the idea of measuring spinodal curves, i.e., the border between metastable and unstable liquid-liquid demixing behavior of polymer solutions. At this spinodal curve, the determinant in Equation [4.4.39] vanishes, i.e., it becomes equal to zero. At small enough scattering angles (near 30°) and in a temperature range of about $0.03 < \Delta T < 5$K around the critical temperature, a proportionality of $I^{excess} \propto 1/\Delta T$ can be obtained that leads to a simple linear behavior of $1/I_{30}$ against T for various concentrations around the critical demixing concentration. The extrapolated curves of $1/I_{30}$ vs. T to $1/I_{30} = 0$ lead to spinodal temperatures as function of the corresponding polymer concentrations. If, for example, one combines $\Delta_{mix}G$-ansatz by Koningsveld and Kleintjens,[51] Equation [4.4.15], with the spinodal condition:

$$\left|\frac{\partial^2 \Delta_{mix} G}{\partial \varphi_i \partial \varphi_j}\right| = 0 \qquad [4.4.44]$$

one obtains for a binary polymer solution:

$$\alpha + \frac{\beta(1-\gamma)}{(1-\gamma\varphi_2)^3} - \frac{1}{\varphi_2 r_w} - \frac{1}{1-\varphi_2} = 0 \qquad [4.4.45]$$

where:

| | |
|---|---|
| $\alpha$ | acts as constant within a certain temperature range |
| $\beta$ | describes a temperature function like $\beta = \beta_0 + \beta_1/T$ |
| $\gamma$ | is also a constant within a certain temperature range. |
| $r_w$ | mass average segment number, compare r in Equation (4.4.13) |
| $\varphi_2$ | total volume fraction of the polymer |

The adjustable parameters $\alpha$, $\beta$, $\gamma$ have to be fitted to spinodal data $\varphi_2^{spinodal}$ vs. $T^{spinodal}$ and solvent activities can be calculated from the following relation:

$$\ln a_1 = \ln(1-\varphi_2) + \left(1 - \frac{1}{r_n}\right)\varphi_2 + \alpha\varphi_2^2 + \frac{\beta(1-\gamma)}{(1-\gamma\varphi_2)^2}\varphi_2^2 \qquad [4.4.46]$$

where:

| | |
|---|---|
| $r_n$ | number average segment number, compare r in Equation [4.4.13] |

Gordon and coworkers[157-159] improved this method and developed the so-called PICS (pulse-induced critical scattering) apparatus - details and history were summarized by Galina et al.[160] PICS enables not only investigations within the metastable range, i.e., nearer to the spinodal, but also of high-viscous solutions and polymer blends for determining spinodal and binodal (cloud-point) curves. How to obtain solvent activities from demixing equilibrium is explained in the text below.

#### *4.4.3.2.3 X-ray scattering*

X-ray scattering can be measured by the classical Kratky camera or more modern synchrotron techniques. Technical details can be found in a number of books, e.g., those by Guinier and Fouret,[161] Chen and Yip,[162] or Glatter and Kratky.[163]

Small angle X-ray scattering (SAXS) can be used in analogy to light scattering to measure second virial coefficients of binary polymer solutions. Zimm-diagrams can be constructed following the same ways as in light scattering. This was demonstrated, for example, in papers by Kirste and coworkers.[164-166] In analogy to Equation [4.4.43], one can derive

$$(Kc_2)/I(\theta = 0) = (1/M_2) + 2A_2c_2 + 3A_3c_2^2 + \ldots \qquad [4.4.47]$$

where:

| | |
|---|---|
| $c_2$ | mass by volume concentration $c_2 = m_2/v$ |

| | |
|---|---|
| $A_2, A_3, \ldots$ | second, third, ... osmotic virial coefficients |
| $M_2$ | molar mass of the polymer |

According to the scattering theory of polydisperse polymers (please see, for example, the book written by Strobl),[147] the molar mass of the polymer, $M_2$, is equal to the mass average, $M_w$, of polydisperse polymers. The exact application of Equation [4.4.47] is to be made again at the scattering angle $\theta = 0$. The constant K is now given by:

$$K \equiv e^4 \Delta z^2 / \left(m^2 N_{Av} c^4\right) \qquad [4.4.48]$$

where:

| | |
|---|---|
| e | electron charge |
| m | electron mass |
| $\Delta z$ | excess number of electrons |
| c | speed of light in vacuum |

In principle, there is agreement between values of second virial coefficients from light scattering or X-ray scattering. Okano et al.[167,168] applied SAXS to semidilute solutions of polystyrene in cyclohexane in the poor solvent regime and obtained virial coefficients in good agreement with liquid-liquid data from a coexistence curve. Takada et al.[169] provided a more recent example for poly(vinyl methyl ether) in cyclohexane, Horkay et al.[170] for poly(vinyl acetate) in toluene and poly(dimethyl siloxane) in octane. In comparison to data from osmotic pressure and neutron scattering, they observed good agreement.

#### *4.4.3.2.4 Neutron scattering*

Neutron scattering is an important method for investigating conformation and dynamics of polymer molecules, Higgins,[171] or polymer mixtures, Hammouda.[172] A recent presentation of various techniques can be found in a new book edited by Pethrick and Dawkins.[26] Thermodynamics of polymer solutions is not the first task in neutron scattering experiments. The general set-up of the neutron scattering experiment is equivalent to the one used for light scattering, but applying a neutron source, and elastic neutron scattering at small angles (SANS) can be applied like light scattering or X-ray scattering to obtain second virial coefficients in dilute solutions. Similarly to the scattering of photons, it is the difference in scattering power between solvent molecules and polymer segments which determines the absolute scattering intensity. Formally, the virial equation has the same form as Equations [4.4.43 and 47], again neglecting $P(\theta)$:

$$(Kc_2) / \Sigma(\theta) = (1/M_2) + 2A_2 c_2 + 3A_3 c_2^2 + \ldots \qquad [4.4.49]$$

and

$$K = \left(b_2 - b_1 \rho_1 \nu_{2,spez}\right) / N_{Av} \qquad [4.4.50]$$

where:

| | |
|---|---|
| $c_2$ | mass by volume concentration $c_2 = m_2/v$ |
| $A_2, A_3, \ldots$ | second, third, ... osmotic virial coefficients |
| $M_2$ | molar mass of the polymer |
| $\Sigma(\theta)$ | differential scattering cross section per volume unit |
| K | contrast factor for neutron scattering |
| $b_1, b_2$ | densities of solvent and polymer scattering length |
| $\rho_1$ | density of the solvent |

$v_{2,spez}$ specific volume of the polymer (more exact, $\bar{v}^{\infty}_{2,spez}$, the partial specific volume at infinite dilution)

According to the scattering theory of polydisperse polymers, please see, for example, in the book written by Strobl,[147] the molar mass of the polymer, $M_2$, is equal to the mass average, $M_w$, of polydisperse polymers. The contrast factor for neutron scattering takes into account for the difference in scattering power of solvent molecules and polymer segments. Again, Zimm plots can be constructed, as was explained above, for light scattering measurements to take into account for angular and concentration dependence - for a demonstration see Vennemann et al.[173]

The transformation of the obtained second virial coefficients into solvent activities is as explained above, Equations [4.4.34 and 4.4.35]. A recent example for the determination of second virial coefficients from SANS is the investigation of aggregation phenomenon in associating polymer solutions by Pedley et al.,[174] where sodium sulfonated polystyrene ionomers in deuterated xylene were considered. Enthalpic and entropic contributions to $A_2$ were calculated (as in the paper by Wolf and Adams[149] for $A_2$ from light scattering) and an enthalpy of aggregation was estimated from these data. A high-pressure investigation on aqueous poly(ethylene oxide) solutions was made by Vennemann et al.[173] who measured second virial coefficients by a SANS experiment for pressures up to 200 MPa and combined these data with PVT-measurements to obtain also excess and partial excess volumes and gained information about the pressure dependence of the chemical potential.

#### *4.4.3.2.5 Ultracentrifuge*

The analytical ultracentrifuge is a powerful tool for polymer characterization. Technical details of ultracentrifugation will not be considered here - please see Refs.[26,35-37] for more information. In a typical ultracentrifuge experiment, the polymer solution is put in a sample tube and rotated at high speed. Thermodynamic data can be obtained either from the sedimentation velocity (sedimentation coefficient) or from the sedimentation-diffusion equilibrium since the centrifugal forces are balanced by the activity gradient. The concentration gradient is conventionally measured via the refractive index gradient along the axis of the tube using Schlieren photography or various optics.

The sedimentation coefficient is defined as the sedimentation velocity in a unit force field:

$$s = \frac{dh/dt}{\omega^2 h} \qquad [4.4.51]$$

where:

s sedimentation coefficient
h distance from the center of rotation
t time
ω angular velocity

For a given polymer-solvent system, the sedimentation coefficient is dependent on temperature, pressure and polymer concentration. For obtaining thermodynamic data from sedimentation coefficients, one additionally has to measure the diffusion coefficient. This can be made with an ultracentrifuge in special diffusion cells[35] or with dynamic light scattering[32] based on the theory of Pecora.[175] Nearly all diffusion coefficients have been measured by this method since it became available in 1970. The determination of sedimentation and diffusion coefficient yields virial coefficients of a polymer solution. The so-called Svedberg equation reads:

$$\left(\frac{D}{s}\right)\left(1-\bar{v}_{2,spez}\rho_1\right)=RT\left[\frac{1}{M_2}+2A_2c_2+3A_3c_2^2+\ldots\right] \qquad [4.4.52]$$

where:

| | |
|---|---|
| D | diffusion coefficient |
| $\bar{v}_{2,spez}$ | partial specific volume of the polymer |
| $\rho_1$ | density of the solvent |
| $c_2$ | mass by volume concentration $c_2 = m_2/v$ |
| $A_2, A_3, \ldots$ | second, third, ... osmotic virial coefficients |
| $M_2$ | molar mass of the polymer |

Equation [4.4.52] is strictly valid for monodisperse polymers, i.e., one single component 2. For polydisperse polymers, different averages were obtained for the sedimentation and the diffusion coefficient, which depends on the applied measuring mode and the subsequent calculations. The averages of $M_2$ correspond with averages of D and s and are mixed ones that have to be transformed into the desired common averages - for details please see Refs.[35-37]

Sedimentation-diffusion equilibrium in an ultracentrifuge gives also a virial series:[35]

$$\omega^2 h\left(1-\bar{v}_{2,spez}\rho_1\right)\left(\frac{\partial \ln c_2}{\partial h}\right)=RT\left[\frac{1}{M_2}+2A_2c_2+3A_3c_2^2+\ldots\right] \qquad [4.4.53]$$

where:

| | |
|---|---|
| h | distance from the center of rotation |
| $\omega$ | angular velocity |

Equation [4.4.53] is again valid for monodisperse polymers only. Polydisperse polymers lead to apparent molar mass averages and to averages of the virial coefficients which have to be transformed into the desired common averages by appropriate calculation methods.[35-37]

A somewhat different way of avoiding the virial expansion in Equation [4.4.53] was developed by Scholte.[176,177] Without going into details, his final relation was:

$$\omega^2 h\left(1-\bar{v}_{2,spez}\rho_1\right)\left(\frac{M_1 w_2}{w_2-1}\right)=\left[\frac{\partial \Delta\mu_1}{\partial w_2}+RT\left(\frac{M_1}{M_n}-\frac{M_1}{M_w}\right)\right]\left(\frac{dn}{dh}\right)/\left(\frac{dn}{dw_2}\right) \qquad [4.4.54]$$

where:

| | |
|---|---|
| $w_2$ | mass fraction of the polymer |
| $M_1$ | molar mass of the solvent |
| $M_n$ | number average molar mass of the polymer |
| $M_w$ | mass average molar mass of the polymer |
| n | refractive index of the solution |

Some assumptions were made for the derivation of Equation [4.4.54], especially the partial specific volume, the refractive index, and the derivative $dn/dw_2$ must not depend on the molar mass distribution of the polymer. If one further assumes that the Flory-Huggins

$\chi$-function depends only on temperature and concentration, but not on molar mass, the partial derivative of the chemical potential can be calculated by Equation [4.4.13a] to obtain values of the $\chi$-function. Scholte carried out experiments for solutions of polystyrene in cyclohexane or toluene at different temperatures and in a concentration range of 0-80 wt%.

Thus, the sedimentation method is able to cover nearly the total concentration range of a polymer solution; however, values obtained by this method were slightly higher than values determined by other methods. Since the measurement of thermodynamic data by sedimentation equilibrium is not very frequent in the literature this is certainly not a final statement. A combined determination of second osmotic virial coefficients of poly(ethylene glycol)s in methanol, water and N,N-dimethylformamide by Elias and Lys[178] using light scattering, VPO and sedimentation equilibrium showed good agreement between all methods. This was also confirmed in a recent investigation on poly(1-phenyl-1-propene) in toluene by Hirao et al.,[179] where second virial coefficients were determined by light scattering and by sedimentation equilibrium over a wide range of molar mass. Some further $A_2$ data from sedimentation measurements can be found in the tables by Lechner et al.[9] The transformation of the obtained second virial coefficients into solvent activities is as explained above.

#### *4.4.3.2.6 Cryoscopy (freezing point depression of the solvent)*

In the cryoscopic method, the freezing temperature of a solution is compared with that of the pure solvent. The polymer must be solvable in the solvent at the freezing temperature and must not react with the solvent either chemically or physically. Difficulties may arise from limited solubility and from the formation of solid solutions on freezing. Application of cryoscopy to polymer solutions is not widespread in literature despite the simplicity of the required equipment. Cryoscopy was reviewed by Glover,[34] who also discussed technical details and problems in concern with application to polymer solutions. A detailed review on cryometers and cryoscopic measurements for low-molar mass systems was recently made by Doucet.[180] Cryometers are sold commercially, e.g., Knauer. Measurements of thermodynamic data are infrequent. Applications usually determine molar masses. Accurate data require precise temperature measurement and control as well as caution with the initiation of the crystallization process and the subsequent establishment of equilibrium (or steady state) conditions. High purity is required for the solvent and also for the solute.

Data reduction of cryoscopic measurements is made by applying the relation for the freezing point depression of a binary mixture to obtain solvent activities:

$$\frac{1}{{}^{SL}T_1^0} - \frac{1}{{}^{SL}T_1} = \frac{R}{\Delta_{SL}H_1^0}\ln a_1 \qquad [4.4.55]$$

where:

| | |
|---|---|
| ${}^{SL}T_1^0$ | solid-liquid equilibrium melting temperature of the pure solvent |
| ${}^{SL}T_1$ | solid-liquid equilibrium melting temperature of the solvent in the polymer solution |
| $\Delta_{SL}H_1^0$ | molar enthalpy of fusion of the pure solvent. |

Kawai[181] determined some values of the $\chi$-function for benzene solutions of polystyrene or poly(vinyl acetate) and aqueous solutions of poly(vinyl alcohol). In comparison with various data from the tables given by Orwoll,[8] larger deviations with respect to other methods have to be stated. Just recently, Hoei et al.[182] made a more sophisticated analysis of solid-liquid coexistence data of benzene in natural rubber and found good agreement to other data.

Equation [4.4.55] could in principle also be used for the determination of thermodynamic data from the melting point depression of (semi)crystalline polymers when the subscripts changed from 1 to 2. This enables a second approach to data for the infinite dilution range of the solvent in the polymer. Such investigations have been made in the literature. However, these data are regarded as being less reliable by a number of reasons and no further discussion will be made here.

#### *4.4.3.2.7 Liquid-liquid equilibrium (LLE)*

There are two different situations for the liquid-liquid equilibrium in polymer-solvent systems:

(i) the equilibrium between a dilute polymer solution (sol) and a polymer-rich solution (gel) and

(ii) the equilibrium between the pure solvent and a swollen polymer network (gel).

Case (i) is considered now, case (ii) is specially considered below as swelling equilibrium.

LLE-measurements do not provide a direct result with respect to solvent activities. Equation (4.4.8) says that solvent activities at given temperature and pressure must be equal in both coexisting phases. Since the solvent activity of such a coexisting phase is a priori not known, one has to apply thermodynamic models to fit LLE-data as functions of temperature and concentration. Solvent activities can be obtained from the model in a subsequent step only.

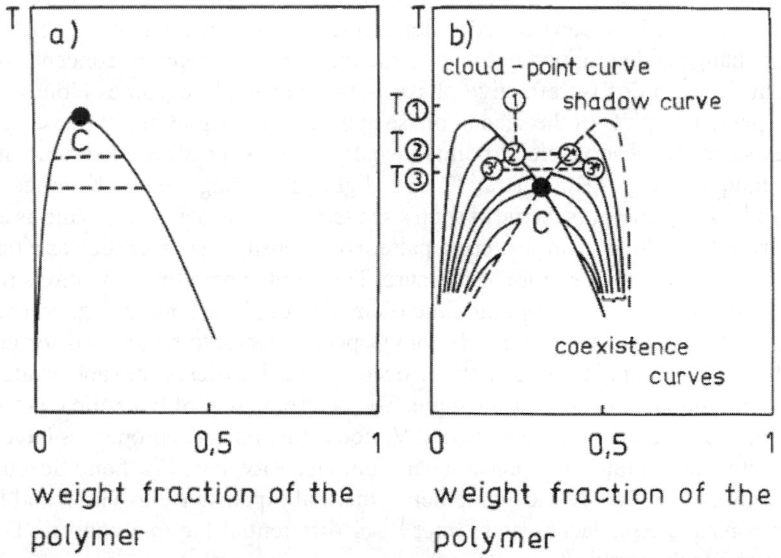


Figure 4.4.19. Principles of liquid-liquid demixing in polymer solutions, a) - strictly binary polymer solution of a monodisperse polymer, b) - quasi-binary polymer solution of a polydisperse polymer which is characterized by a distribution function: C - critical point, dashed lines - tie lines, T(1) - temperature/concentration in the homogeneous region, T(2) - temperature/concentrations of the cloud point (phase′) and the corresponding shadow point (phase′′), T(3) - temperature in the heterogeneous LLE region, coexistence concentrations of phase′ and phase′′ at T(3) are related to the starting concentration = cloud point concentration of (2).

Furthermore, there is another effect which causes serious problems with LLE-data of polymer solutions. This is the strong influence of distribution functions on LLE, because fractionation occurs during demixing - see, for example, Koningsveld.[44,183] Figure 4.4.19 illustrates the differences between the LLE-behavior of a strictly binary polymer solution of a monodisperse polymer and a quasi-binary polymer solution of a polydisperse polymer which is characterized by a distribution function.

One can see the very complicated behavior of quasi-binary solutions where the phase boundary is given by a cloud-point curve and where an infinite number of coexistence curves exists (one pair for each starting concentration, i.e., each cloud-point). The cloud-point is a point in the T-$w_2$- or the P-$w_2$-diagram where a homogeneous solution of concentration $w_{02}$ begins to demix (where the "first" droplet of the second phase occurs, T(2) in Figure 4.4.19). If $w_{02}$ is smaller than the critical concentration, the cloud-point belongs to the sol-phase, otherwise to the gel-phase.

As this subchapter is devoted to solvent activities, only the monodisperse case will be taken into account here. However, the user has to be aware of the fact that most LLE-data were measured with polydisperse polymers. How to handle LLE-results of polydisperse polymers is the task of continuous thermodynamics, Refs.[52-54] Nevertheless, also solutions of monodisperse polymers or copolymers show a strong dependence of LLE on molar mass of the polymer,[184] or on chemical composition of a copolymer.[185] The strong dependence on molar mass can be explained in principle within the simple Flory-Huggins $\chi$-function approach, please see Equation [4.4.61].

Experimental methods can be divided into measurements of cloud-point curves, of real coexistence data, of critical points and of spinodal curves:

Due to distinct changes in a number of physical properties at the phase transition border, quite a lot of methods can be used to determine cloud-points. In many cases, the refractive index change is determined because refractive indices depend on concentration (with the seldom exception of isorefractive phases) and the sample becomes cloudy when the highly dispersed droplets of the second phase appear at the beginning of phase separation. Simple experiments observe cloud-points visually. More sophisticated equipment applies laser techniques, e.g., Kuwahara,[186] and light scattering, e.g., Koningsveld and Staverman.[187] The principle scheme of such a scattering experiment is the same as explained with Figure 4.4.18. Changes in scattering pattern or intensity were recorded as a function of decreasing/increasing temperature or pressure. The point, where first deviations from a basic line are detected, is the cloud-point. Since demixing or phase homogenization need some time (especially for highly viscous solutions), special care is to be applied for good data. Around the critical point large fluctuations occur (critical opalescence) and scattering data have to be measured at 90° scattering angle. The determination of the critical point is to be made by independent methods (see below). Various other physical properties have been applied for detecting liquid-liquid phase separation: viscosity, e.g., Wolf and Sezen,[188] ultrasonic absorption, e.g., Alfrey and Schneider,[189] thermal expansion, e.g., Greer and Jacobs,[190] dielectric constant, e.g., Jacobs and Greer,[191] or differential thermal analysis DTA, e.g., Muessig and Wochnowski.[192]

There are only a small number of investigations where real coexistence data were measured. This is mainly due to very long equilibrium times (usually weeks) which are necessary for obtaining thermodynamically correct data. A common method is to cool homogeneous solutions in ampullae very slowly to the desired temperature in the LLE-region and equilibrium is reached after both phases are sharply separated and clear, Rehage et al.[193] After separating both phases, concentrations and distribution functions were measured. Acceptable results can be obtained for low polymer concentrations (up to 20 wt%). Scholte and Koningsveld[194] developed a method for highly viscous polymer solutions at higher concentrations by constructing a modified ultracentrifuge where the equilibrium is quickly established during cooling by action of gravitational forces. After some hours, concentrations, phase volume ratios and concentration differences can be determined. Rietfeld with his low-speed centrifuge[195] and Gordon with a centrifugal homogenizer[196] improved this technique and expanded its applicability up to polymer melts, e.g., Koningsveld et al.[197]

The methods for obtaining spinodal data have already been discussed above with the light scattering technique, please see Subchapter 4.4.3.2.2.

Special methods are necessary to measure the critical point. For solutions of monodisperse polymers, it is the maximum of the binodal. Binodals of polymer solutions can be rather broad and flat. The exact position of the critical point can be obtained by the method of the rectilinear diameter. Due to universality of critical behavior, a relation like Equation [4.4.56] is valid, deGennes:[198]

$$\left(\varphi_2^I - \varphi_2^{II}\right)/2 - \varphi_2^{crit} \propto \left(1 - T/T^{crit}\right)^{1-\alpha} \qquad [4.4.56]$$

where:

| | |
|---|---|
| $\varphi_2^I$ | volume fraction of the polymer in coexisting phase I |
| $\varphi_2^{II}$ | volume fraction of the polymer in coexisting phase II |
| $\varphi_2^{crit}$ | volume fraction of the polymer at the critical point |
| $T^{crit}$ | critical temperature |
| $\alpha$ | critical exponent |

and critical points can be obtained by using regression methods to fit LLE-data to Equation [4.4.56].

For solutions of polydisperse polymers, such a procedure cannot be used because the critical concentration must be known in advance to measure its corresponding coexistence curve. Additionally, the critical point is not the maximum in this case but a point at the right-hand side shoulder of the cloud-point curve. Two different methods were developed to solve this problem, the phase-volume-ratio method, e.g., Koningsveld,[199] where one uses the fact that this ratio is exactly equal to one only at the critical point, and the coexistence concentration plot, e.g. Wolf,[200] where an isoplethal diagram of values of $\varphi_2^I$ and $\varphi_2^{II}$ vs. $\varphi_{02}$ gives the critical point as the intersection point of cloud-point and shadow curves.

Since LLE-measurements do not provide a direct result with respect to solvent activities, Equation [4.4.8] and the stability conditions are the starting points of data reduction. As pointed out above, the following explanations are reduced to the strictly binary solution of a monodisperse polymer. The thermodynamic stability condition with respect to demixing is given for this case by (see Prausnitz et al. [49]):

$$\left(\partial^2 \Delta_{mix} G / \partial^2 \varphi_2\right)_{P,T} > 0 \qquad [4.4.57]$$

If this condition is not fulfilled between some concentrations $\varphi_2^I$ and $\varphi_2^{II}$, demixing is obtained and the minimum of the Gibbs free energy of mixing between both concentrations is given by the double tangent at the corresponding curve of $\Delta_{mix}G$ vs. $\varphi_2$, Equation [4.4.8].

The resulting curve in the T vs. $\varphi_2$ diagram, see Figure 4.4.19a, is the binodal curve. Applying Equations [4.4.3 to 4.4.5 and 4.4.13a], one gets two relations which have to be solved simultaneously to fit an empirical $\chi$(T)-function to experimental binodal (coexistence) data. For the most simple case of $\chi$ being only a function of T (or P) and not of $\varphi_2$ (the so-called one-parameter approach) these relations read:

$$\ln\frac{\left(1-\varphi_2^I\right)}{\left(1-\varphi_2^{II}\right)}+\left(1-\frac{1}{r}\right)\left(\varphi_2^I-\varphi_2^{II}\right)+\chi\left(\varphi_2^{I2}-\varphi_2^{II2}\right)=0 \quad [4.4.58a]$$

and

$$\frac{1}{r}\ln\frac{\varphi_2^I}{\varphi_2^{II}}-\left(1-\frac{1}{r}\right)\left(\varphi_2^I-\varphi_2^{II}\right)+\chi\left(\varphi_1^{I2}-\varphi_1^{II2}\right)=0 \quad [4.4.58b]$$

where:

| | |
|---|---|
| $\varphi_2^I$ | volume fraction of the polymer in coexisting phase I |
| $\varphi_2^{II}$ | volume fraction of the polymer in coexisting phase II |
| r | ratio of molar volumes $V_2/V_1$ what is the number of segments with $V_{seg} = V_1$ |
| $\chi$ | Flory-Huggins interaction function of the solvent |

and solvent activities result from Equation [4.4.13a]. However, this simple approach is of limited quality. More sophisticated models have to be applied to improve calculation results.

A special curve is obtained with the border line to the instability region, i.e., the spinodal curve, for which the second derivative in Equation [4.4.57] is equal to zero. If one applies again the one-parameter approach with an empirical $\chi$(T)-function, the following simple result can be derived:

$$2\chi\left(T^{spinodal}\right)-\frac{1}{r\varphi_2^{spinodal}}-\frac{1}{1-\varphi_2^{spinodal}}=0 \quad [4.4.59]$$

where:

| | |
|---|---|
| $T^{spinodal}$ | spinodal temperature |
| $\varphi_2^{spinodal}$ | volume fraction of the polymer at the spinodal curve |

which has to fit an empirical $\chi$(T)-function. An example for the spinodal relation of a polydisperse polymer was given above by Equations [4.4.44 and 4.4.45].

The common point of spinodal and binodal curve is the critical point. The critical point conditions are:

$$(\partial^2\Delta_{mix}G/\partial^2\varphi_2)_{P,T}=0 \quad (\partial^3\Delta_{mix}G/\partial^3\varphi_2)_{P,T}=0$$

and [4.4.60]

$$(\partial^4\Delta_{mix}G/\partial^4\varphi_2)_{P,T}>0$$

If one applies again the one-parameter approach with an empirical $\chi$(T)-function, two simple results can be derived:

$$\varphi_2^{crit} = 1/\left(1 + r^{0.5}\right) \quad and \quad \chi^{crit}\left(T^{crit}\right) = 0.5\left(1 + 1/r^{0.5}\right)^2 \qquad [4.4.61]$$

where:

$\varphi_2^{crit}$ volume fraction of the polymer at the critical point

$T^{crit}$ critical temperature

This means that the critical concentration is shifted to lower values with increasing segment number (molar mass) of the polymer and becomes zero for infinite molar mass. Equation [4.4.61] explains also why the $\chi$(T)-function becomes 0.5 for infinite molar mass. The critical temperature of these conditions is then called theta-temperature. Solvent activities can be calculated from critical $\chi$(T)-function data via Equation [4.4.13]. However, results are in most cases of approximate quality only.

#### 4.4.3.2.8 Swelling equilibrium

Polymers, crosslinked sufficiently to form a three-dimensional network, can absorb solvent up to a large extent. The maximum possible solvent concentration, the degree of swelling, are a function of solvent activity. If solvent is present in excess, this swelling equilibrium is reached when the chemical potential of the pure solvent outside the network is equal to the chemical potential inside the swollen sample. This means, there must be an additional contribution to the Gibbs free energy of mixing (as is the case with the osmotic equilibrium) besides the common terms caused by mixing the (virtually) infinite-molar-mass polymer and the solvent. This additional part follows from the elastic deformation of the network. The different aspects of chemical and physical networks will not be discussed here, for some details please see Refs.[201-205] The following text is restricted to the aspect of solvent activities only.

One method to obtain solvent activities in swollen polymer networks in equilibrium is to apply vapor pressure measurements. This is discussed in detail above in the Subchapter 4.4.3.1.1 and most methods can be used also for network systems, especially all sorption methods, and need no further explanation. The VPO-technique can be applied for this purpose, e.g., Arndt.[206,207] IGC-measurements are possible, too, if one realizes a definitely crosslinked polymer in the column, e.g., Refs.[208-210]

Besides vapor sorption/pressure measurements, definite swelling and/or deswelling experiments lead to information on solvent activities. Swelling experiments work with pure solvents, deswelling experiments use dilute solutions of macromolecules (which must not diffuse into or adsorb at the surface of the network) and allow measurements in dependence on concentration. Deswelling experiments can be made in dialysis cells to prevent diffusion into the network. The determination of the equilibrium swelling/deswelling concentration can be made by weighing, but, in most cases, by measuring the swelling degree. Some methods for measuring the swelling degree are: measuring the buoyancy difference of the sample in two different liquids of known density, e.g. Rijke and Prins,[211] measuring the volume change by cathetometer, e.g., Schwarz et al.,[212] measuring the volume change by electrical (inductive) measurements, e.g., Teitelbaum and Garwin.[213]

The swelling degree can be defined as the ratio of the masses (mass based degree) or of the volumes (volume based degree) of the swollen to the dry polymer network sample:

$$Q_m = 1 + m_1 / m_N \quad or \quad Q_v = 1 + v_1 / v_n = 1 + (Q_m - 1)\rho_n / \rho_1 \qquad [4.4.62]$$

where:

| | |
|---|---|
| $Q_m$ | mass based degree of swelling |
| $Q_v$ | volume based degree of swelling |
| $m_1$ | absorbed mass of the solvent |
| $m_N$ | mass of the dry polymer network |
| $v_1$ | absorbed volume of the solvent |
| $v_N$ | volume of the dry polymer network |
| $\rho_1$ | density of the solvent |
| $\rho_N$ | density of the dry polymer network |

Since $Q_v = 1/\varphi_2$, usually volume changes were measured. If the sample swells isotropically, the volume change can be measured by the length change in one dimension. Calculation of solvent activities from measurements of the swelling degree needs a statistical thermodynamic model. According to Flory,[46] a closed thermodynamic cycle can be constructed to calculate the Gibbs free energy of swelling from the differences between the swelling process, the solution process of the linear macromolecules, the elastic deformation and the crosslinking. The resulting equation can be understood in analogy to the Flory-Huggins relation, Equation [4.4.13a] with $r \to \infty$, and reads:

$$\Delta\mu_1 / RT = \ln(1-\varphi_2) + \varphi_2 + \chi\varphi_2^2 + \nu_c V_1\left(A\eta\varphi_2^{1/3} - B\varphi_2\right) \qquad [4.4.63]$$

where:

| | |
|---|---|
| $\nu_c$ | network density $\nu_c = \rho_N/M_c$ |
| $V_1$ | molar volume of the pure liquid solvent 1 at temperature T |
| $\eta$ | memory term |
| A | microstructure factor |
| B | volume factor |
| $M_c$ | molar mass of a network chain between two network knots |
| $\varphi_2$ | equilibrium swelling concentration = $1/Q_v$ |
| $\chi$ | Flory-Huggins $\chi$-function |

A numerical calculation needs knowledge of the solvent activity of the corresponding homopolymer solution at the same equilibrium concentration $\varphi_2$ (here characterized by the value of the Flory-Huggins $\chi$-function) and the assumption of a deformation model that provides values of the factors A and B. There is an extensive literature for statistical thermodynamic models which provide, for example, Flory:[46] A = 1 and B = 0.5; Hermans:[214] A = 1 and B = 1; James and Guth[215] or Edwards and Freed:[216] A = 0.5 and B = 0. A detailed explanation was given recently by Heinrich et al.[203]

The swelling equilibrium depends on temperature and pressure. Both are related to the corresponding dependencies of solvent activity via its corresponding derivative of the chemical potential:

$$\left(\frac{\partial T}{\partial \varphi_1}\right)_P = \frac{1}{\Delta S_1}\left(\frac{\partial \mu_1}{\partial \varphi_1}\right)_{P,T} = \frac{T}{\Delta H_1}\left(\frac{\partial \mu_1}{\partial \varphi_1}\right)_{P,T} \qquad [4.4.64]$$

where:

| | |
|---|---|
| $\varphi_1$ | equilibrium swelling concentration of the solvent |
| $\Delta S_1$ | differential entropy of dilution at equilibrium swelling |
| $\Delta H_1$ | differential enthalpy of dilution at equilibrium swelling where $\Delta H_1 = T\Delta S_1$ |

The first derivative in Equation [4.4.64] describes the slope of the swelling curve. Since the derivative of the chemical potential is always positive for stable gels, the positive or negative slope is determined by the differential enthalpy or entropy of dilution, Rehage.[217]

$$\left(\frac{\partial P}{\partial \varphi_1}\right)_T = \frac{1}{\left(V_1 - \overline{V}_1\right)}\left(\frac{\partial \mu_1}{\partial \varphi_1}\right)_{P,T} \qquad [4.4.65]$$

where:

$\overline{V}_1$ partial molar volume of the solvent in the polymer solution at temperature T

$V_1$ molar volume of the pure solvent at temperature T

In analogy to membrane osmometry, swelling pressure measurements can be made to obtain solvent activities. Pure solvents as well as dilute polymer solutions can be applied. In the case of solutions, the used macromolecules must not diffuse into or adsorb at the surface of the network. Two types of swelling pressure apparatuses have been developed. The anisochoric swelling pressure device measures swelling degrees in dependence on pressure and swelling volume, e.g., Borchard.[218] The isochoric swelling pressure device applies a compensation technique where the volume is kept constant by an external pressure which is measured, e.g., Borchard.[219] Swelling pressures can also be measured by sedimentation equilibrium using an ultracentrifuge for details, please see Borchard.[220,221]

The swelling pressure $\pi_{swell}$ is directly related to the solvent activity by:

$$\Delta\mu_1 = RT\ln a_1 = -\overline{V}_1\pi_{swell} \qquad [4.4.66]$$

and may be in the range of some MPa.

In comparison with all methods of determination of solvent activities from swelling equilibrium of network polymers, the gravimetric vapor sorption/pressure measurement is the easiest applicable procedure, gives the most reliable data, and should be preferred.

### 4.4.4 THERMODYNAMIC MODELS FOR THE CALCULATION OF SOLVENT ACTIVITIES OF POLYMER SOLUTIONS

Since measurements of solvent activities of polymer solutions are very time-consuming and can hardly be made to cover the whole temperature and pressure ranges, good thermodynamic theories, and models are necessary, which are able to calculate them with sufficient accuracy and which can interpolate between and even extrapolate from some measured basic data over the complete phase and state ranges of interest for a special application.

Many attempts have been made to find theoretical explanations for the non-ideal behavior of polymer solutions. There exist books and reviews on this topic, e.g., Refs.[2,41,42,46-49] Therefore, only a short summary of some of the most important thermodynamic approaches and models will be given here. The following explanations are restricted to concentrated polymer solutions only because one has to describe mainly concentrated polymer solutions when solvent activities have to be calculated. For dilute polymer solutions, with the second virial coefficient region, Yamakawa's book[222] provides a good survey.

There are two different approaches for the calculation of solvent activities of polymer solutions:

(i) the approach which uses activity coefficients, starting from Equation [4.4.11]

(ii) the approach which uses fugacity coefficients, starting from Equations [4.4.5 and 4.4.6].

From the historical point of view and also from the number of applications in the literature, the common method is to use activity coefficients for the liquid phase, i.e., the polymer solution, and a separate equation-of-state for the solvent vapor phase, in many cases the truncated virial equation of state as for the data reduction of experimental measurements explained above. To this group of theories and models also free-volume models and lattice-fluid models will be added in this paper because they are usually applied within this approach. The approach where fugacity coefficients are calculated from one equation of state for both phases was applied to polymer solutions more recently, but it is the more promising method if one has to extrapolate over larger temperature and pressure ranges.

Theories and models are presented below without going into details and without claiming completeness, since this text is not dedicated to theoretical problems but will only provide some help to calculate solvent activities.

### 4.4.4.1 Models for residual chemical potential and activity coefficient in the liquid phase

Since polymer solutions in principle do not fulfill the rules of the ideal mixture but show strong negative deviations from Raoult's law due to the difference in molecular size, the athermal Flory-Huggins mixture is usually applied as the reference mixture within polymer solution thermodynamics. Starting from Equation [4.4.11] or from

$$RT\ln a_1 = RT\ln x_1\gamma_1 = \Delta\mu_1 = \mu_1 - \mu_1^0 = \left(\frac{\partial n\Delta_{mix}G}{\partial n_1}\right)_{T,P,n_{j\neq 1}} \qquad [4.4.67]$$

where:

| | |
|---|---|
| $a_1$ | activity of the solvent |
| $x_1$ | mole fraction of the solvent |
| $\gamma_1$ | activity coefficient of the solvent in the liquid phase with activity $a_1 = x_1\gamma_1$ |
| $\mu_1$ | chemical potential of the solvent |
| $\mu_1^0$ | chemical potential of the solvent at standard state conditions |
| R | gas constant |
| T | absolute temperature |
| $n_1$ | amount of substance (moles) of the solvent |
| n | total amount of substance (moles) in the polymer solution |
| $\Delta_{mix}G$ | molar Gibbs free energy of mixing, |

the classical Flory-Huggins theory[46,47] leads, for a truly athermal binary polymer solution, to:

$$\ln a_1^{athermal} = \Delta\mu_1^{athermal} / RT = \ln(1-\varphi_2) + \left(1-\frac{1}{r}\right)\varphi_2 \qquad [4.4.68a]$$

or

$$\ln\gamma_1^{athermal} = \ln\left[1-\left(1-\frac{1}{r}\right)\varphi_2\right] + \left(1-\frac{1}{r}\right)\varphi_2 \qquad [4.4.68b]$$

where:

| | |
|---|---|
| $\varphi_2$ | volume fraction of the polymer |

which is also called the combinatorial contribution to solvent activity or chemical potential, arising from the different configurations assumed by polymer and solvent molecules in solution, ignoring energetic interactions between molecules and excess volume effects. The Flory-Huggins derivation of the athermal combinatorial contribution contains the implicit assumption that the r-mer chains placed on a lattice are perfectly flexible and that the flexibility of the chain is independent of the concentration and of the nature of the solvent. Generalized combinatorial arguments for molecules containing different kinds of energetic contact points and shapes were developed by Barker[223] and Tompa,[224] respectively. Lichtenthaler et al.[225] have used the generalized combinatorial arguments of Tompa to analyze VLE of polymer solutions. Other modifications have been presented by Guggenheim[226,227] or Staverman[228] (see below). The various combinatorial models are compared in a review by Sayegh and Vera.[229] Recently, Freed and coworkers[230-232] developed a lattice-field theory, that, in principle, provides an exact mathematical solution of the combinatorial Flory-Huggins problem. Although the simple Flory-Huggins expression does not always give the (presumably) correct, quantitative combinatorial entropy of mixing, it qualitatively describes many features of athermal polymer solutions. Therefore, for simplicity, it is used most in the further presentation of models for polymer solutions as reference state.

The total solvent activity/activity coefficient/chemical potential is simply the sum of the athermal part as given above, plus a residual contribution:

$$\ln a_1 = \ln a_1^{athermal} + \ln a_1^{residual} \qquad [4.4.69a]$$

or

$$\mu_1 = \mu_1^0 + \mu_1^{athermal} + \mu_1^{residual} \qquad [4.4.69b]$$

The residual part has to be explained by an additional model and a number of suitable models is now listed in the following text.

The Flory-Huggins interaction function of the solvent is the residual function used first and is given by Equations [4.4.12 and 4.4.13] with $\mu_1^{residual} / RT = \chi\varphi_2^2$. It was originally based on van Laar's concept of solutions where excess entropy and excess volume of mixing could be neglected and $\chi$ is represented only in terms of an interchange energy $\Delta\varepsilon/kT$. In this case, the interchange energy refers not to the exchange of solvent and solute molecules but rather to the exchange of solvent molecules and polymer segments. For athermal solutions, $\chi$ is zero, and for mixtures of components that are chemically similar, $\chi$ is small compared to unity. However, $\chi$ is not only a function of temperature (and pressure) as was evident from this foundation, but it is also a function of composition and polymer molecular mass, see e.g., Refs.[5,7,8] If we neglect these dependencies, then the Scatchard-Hildebrand theory,[233,234] i.e., their solubility parameter concept, could be applied:

$$\mu_1^{residual} / RT = (V_1 / RT)(\delta_1 - \delta_2)^2 \varphi_2^2 \qquad [4.4.70]$$

with

$$\delta_1 = \left(\Delta_{vap} U_1^0 / V_1\right)^{1/2} \qquad [4.4.71]$$

where:

| | |
|---|---|
| $V_1$ | molar volume of the pure liquid solvent 1 at temperature T |
| $\Delta_{vap}U_1^0$ | molar energy of vaporization of the pure liquid solvent 1 at temperature T |
| $\delta_1$ | solubility parameter of the pure liquid solvent |
| $\delta_2$ | solubility parameter of the polymer |

Solubility parameters of polymers cannot be calculated from energy of vaporization since polymers do not evaporate. Usually they have been measured according to Equation [4.4.70], but details need not be explained here. Equation [4.4.70] is not useful for an accurate quantitative description of polymer solutions but it provides a good guide for a qualitative consideration of polymer solubility. For good solubility, the difference between both solubility parameters should be small (the complete residual chemical potential term cannot be negative, which is one of the disadvantages of the solubility parameter approach). Several approximate generalizations have been suggested by different authors - a summary of all these models and many data can be found in the books by Barton.[11,12] Calculations applying additive group contributions to obtain solubility parameters, especially of polymers, are also explained in the book by Van Krevelen.[235]

Better-founded lattice models have been developed in the literature. The ideas of Koningsveld and Kleintjens, e.g., Ref.,[51] lead to useful and easy to handle expressions, as is given above with Equations [4.4.15, 4.4.17 and 4.4.46] that have been widely used, but mainly for liquid-liquid demixing and not so much for vapor-liquid equilibrium and solvent activity data. Comprehensive examples can be found in the books by Fujita[41] or Kamide.[42]

The simple Flory-Huggins approach and the solubility parameter concept are inadequate when tested against experimental data for polymer solutions. Even for mixtures of n-alkanes, the excess thermodynamic properties cannot be described satisfactorily - Flory et al.[236-239] In particular, changes of volume upon mixing are excluded and observed excess entropies of mixing often deviate from simple combinatorial contributions. To account for these effects, the PVT-behavior has to be included in theoretical considerations by an equation of state. Pure fluids have different free volumes, i.e., different degrees of thermal expansion depending on temperature and pressure. When liquids with different free volumes are mixed, that difference contributes to the excess functions. Differences in free volumes must be taken into account, especially for mixtures of liquids whose molecules differ greatly in size, i.e., the free volume dissimilarity is significant for polymer solutions and has important effects on their properties, such as solvent activities, excess volume, excess enthalpy and excess entropy. Additionally, the free volume effect is the main reason for liquid-liquid demixing with LCST behavior at high temperatures.[240,241]

Today, there are two principal ways to develop an equation of state for polymer solutions: first, to start with an expression for the canonical partition function utilizing concepts similar to those used by van der Waals (e.g., Prigogine,[242] Flory et al.,[236-239] Patterson,[243,244] Simha and Somcynsky,[245] Sanchez and Lacombe,[246-248] Dee and Walsh,[249] Donohue and Prausnitz,[250] Chien et al.[251]), and second, which is more sophisticated, to use statistical thermodynamics perturbation theory for freely-jointed tangent-sphere chain-like fluids (e.g., Hall and coworkers,[252-255] Chapman et al.,[256-258] Song et al.[259,260]). A comprehensive review about equations of state for molten polymers and polymer solutions was given by Lambert et al.[261] Here, only some resulting equations will be summarized under the aspect of calculating solvent activities in polymer solutions.

The theories that are usually applied within activity coefficient models are given now, the other theories are summarized in Subchapter 4.4.4.2.

The first successful theoretical approach of an equation of state model for polymer solutions was the Prigogine-Flory-Patterson theory.[236-242] It became popular in the version by Flory, Orwoll and Vrij[236] and is a van-der-Waals-like theory based on the corresponding-states principle. Details of its derivation can be found in numerous papers and books and need not be repeated here. The equation of state is usually expressed in reduced form and reads:

$$\frac{\tilde{P}\tilde{V}}{\tilde{T}} = \frac{\tilde{V}^{1/3}}{\tilde{V}^{1/3} - 1} - \frac{1}{\tilde{V}\tilde{T}} \qquad [4.4.72]$$

where the reduced PVT-variables are defined by

$$\tilde{P} = P / P^*, \tilde{T} = T / T^*, \tilde{V} = V / T^*, P^* V^* = rcRT^* \qquad [4.4.73]$$

and where a parameter c is used (per segment) such that 3rc is the number of effective external degrees of freedom per (macro)molecule. This effective number follows from Prigogine's approximation that external rotational and vibrational degrees of freedom can be considered as equivalent translational degrees of freedom. The equation of state is valid for liquid-like densities and qualitatively incorrect at low densities because it does not fulfill the ideal gas limit. To use the equation of state, one must know the reducing or characteristic parameters P*, V*, T*. These have to be fitted to experimental PVT-data. Parameter tables can be found in the literature - here we refer to the book by Prausnitz et al.,[49] a review by Rodgers,[262] and the contribution by Cho and Sanchez[263] to the new edition of the Polymer Handbook.

To extend the Flory-Orwoll-Vrij model to mixtures, one has to use two assumptions: (i) the hard-core volumes $\upsilon^*$ of the segments of all components are additive and (ii) the intermolecular energy depends in a simple way on the surface areas of contact between solvent molecules and/or polymer segments. Without any derivation, the final result for the residual solvent activity in a binary polymer solution reads:

$$\ln a_1^{residual} = \frac{P_1^* V_1^*}{RT}\left[3\tilde{T}_1 \ln\frac{\tilde{V}_1^{1/3} - 1}{\tilde{V}^{1/3} - 1} + \left(\frac{1}{\tilde{V}_1} - \frac{1}{\tilde{V}}\right)\right] +$$

$$+ \frac{V_1^*}{RT}\left(\frac{X_{12}}{\tilde{V}}\right)\theta_2^2 + \frac{PV_1^*}{RT}(\tilde{V} - \tilde{V}_1) \qquad [4.4.74]$$

where:

| | |
|---|---|
| $X_{12}$ | interaction parameter |
| $\theta_2$ | surface fraction of the polymer |

The last term in Equation [4.4.74] is negligible at normal pressures. The reduced volume of the solvent 1 and the reduced volume of the mixture are to be calculated from the same equation of state, Equation [4.4.72], but for the mixture the following mixing rules have to be used (if random mixing is assumed):

$$P^* = P_1^* \psi_1 + P_2^* \psi_2 - \psi_1 \theta_2 X_{12} \quad [4.4.75a]$$

$$T^* = P^* / \left[ \left( P_1^* \psi_1 / T_1^* \right) + \left( P_2^* \psi_2 / T_2^* \right) \right] \quad [4.4.75b]$$

where the segment fractions $\psi_i$ and the surface fractions $\theta_i$ have to be calculated according to:

$$\psi_i = n_i V_i^* / \sum n_k V_k^* = m_i V_{spez,i}^* / \sum m_k V_{spez,k}^* = x_i r_i / \sum x_k r_k \quad [4.4.76a]$$

$$\theta_i = \psi_i s_i / \sum \psi_k s_k \quad [4.4.76b]$$

where:

| | |
|---|---|
| $m_i$ | mass of component i |
| $x_i$ | mole fraction of component i |
| $r_i$ | number of segments of component i, here with $r_i/r_k = V_i^* / V_k^*$ and $r_1 = 1$ |
| $s_i$ | number of contact sites per segment (proportional to the surface area per segment) |

Now it becomes clear from Equation [4.4.74] that the classical Flory-Huggins $\chi$-function ($\chi\psi_2^2 = \ln a_1^{residual}$) varies with composition, as found experimentally. However, to calculate solvent activities by applying this model, a number of parameters have to be considered. The characteristic parameters of the pure substances have to be obtained by fitting to experimental PVT-data as explained above. The number of contact sites per segment can be calculated from Bondi's surface-to-volume parameter tables[264] but can also be used as fitting parameter. The $X_{12}$-interaction parameter has to be fitted to experimental data of the mixture. Fitting to solvent activities, e.g. Refs.,[265,266] does not always give satisfactorily results. Fitting to data for the enthalpies of mixing gives comparable results.[266] Fitting to excess volumes alone does not give acceptable results.[142] Therefore, a modification of Equation [4.4.74] was made by Eichinger and Flory[142] by appending the term $-(V_1^* / R)Q_{12}\theta_2^2$ where the parameter $Q_{12}$ represents the entropy of interaction between unlike segments and is an entropic contribution to the residual chemical potential of the solvent. By adjusting the parameter $Q_{12}$, a better representation of solvent activities can be obtained.

There are many papers in the literature that applied the Prigogine-Flory-Patterson theory to polymer solutions as well as to low-molecular mixtures. Various modifications and improvements were suggested by many authors. Sugamiya[267] introduced polar terms by adding dipole-dipole interactions. Brandani[268] discussed effects of non-random mixing on the calculation of solvent activities. Kammer et al.[269] added a parameter reflecting differences in segment size. Shiomi et al.[270,271] assumed non-additivity of the number of external degrees of freedom with respect to segment fraction for mixtures and assumed the sizes of hard-core segments in pure liquids and in solution to be different. Also Panayiotou[272] accounted for differences in segment size by an additional parameter. Cheng and Bonner[273] modified the concept to obtain an equation of state which provides the correct zero pressure limit of the ideal gas. An attractive feature of the theory is its straightforward extension to multi-component mixtures,[274] requiring only parameters of pure components and binary ones as explained above. A general limitation is its relatively poor description of the compressibility behavior of polymer melts, as well as its deficiencies regarding the description of the pressure dependence of thermodynamic data of mixtures.

Dee and Walsh[249] developed a modified version of Prigogine's cell model that provides an excellent description of the PVT-behavior of polymer melts:

$$\frac{\tilde{P}\tilde{V}}{\tilde{T}} = \frac{\tilde{V}^{1/3}}{\tilde{V}^{1/3} - 0.8909q} - \frac{2}{\tilde{T}}\left(\frac{1.2045}{\tilde{V}^{2}} - \frac{1.011}{\tilde{V}^{4}}\right) \qquad [4.4.77]$$

where the reduced variables and characteristic parameters have the same definitions as in the Flory model above. Equation [4.4.77] is formally identical with Prigogine's result, except for the additional constant parameter q, which can also be viewed as a correction to the hard-core cell volume. The value of q = 1.07 corresponds approximately to a 25% increase in the hard-core volume in comparison with the original Prigogine model. Characteristic parameters for this model are given in Refs.[249,262] The final result for the residual solvent activity in a binary polymer solution reads:

$$\ln a_1^{residual} = \frac{P_1^* V_1^*}{RT}\left[3\tilde{T}_1 \ln\frac{\tilde{V}_1^{1/3} - 0.8909q}{\tilde{V}^{1/3} - 0.8909q} + \left(\frac{1.2045}{\tilde{V}_1^{2}} - \frac{1.2045}{\tilde{V}^{2}}\right) - \left(\frac{0.5055}{\tilde{V}_1^{4}} - \frac{0.5055}{\tilde{V}^{4}}\right)\right]$$

$$+ \frac{V_1^* \theta_2^2}{RT}\left(X_{12} - TQ_{12}\tilde{V}\right)\left(\frac{1.2045}{\tilde{V}^{2}} - \frac{0.5055}{\tilde{V}^{4}}\right) + \frac{PV_1^*}{RT}\left(\tilde{V} - \tilde{V}_1\right) \qquad [4.4.78]$$

The last term in Equation [4.4.78] is again negligible at normal pressures, which is the case for the calculation of solvent activities of common polymer solutions. The reduced volume of the mixture is to be calculated from the equation of state where the same mixing rules are valid, as given by Equations [4.4.75, 4.4.76] if random mixing is assumed. Equation [4.4.78] is somewhat more flexible than Equation [4.4.74]. Again, entropic parameter $Q_{12}$ and interaction parameter $X_{12}$ have to be fitted to experimental data of the mixture. There is not much experience with the model regarding thermodynamic data of polymer solutions because it was mainly applied to polymer blends, where it provides much better results than the simple Flory model.

To improve on the cell model, two other classes of models were developed, namely, lattice-fluid and lattice-hole theories. In these theories, vacant cells or holes are introduced into the lattice to describe the extra entropy change in the system as a function of volume and temperature. The lattice size, or cell volume, is fixed so that the changes in volume can only occur by the appearance of new holes, or vacant sites, on the lattice. The most popular theories of such kind were developed by Simha and Somcynsky[245] or Sanchez and Lacombe.[246-248]

The Sanchez-Lacombe lattice-fluid equation of state reads:

$$\frac{\tilde{P}\tilde{V}}{\tilde{T}} = \frac{1}{r} - 1 - \tilde{V}\ln\frac{\tilde{V} - 1}{\tilde{V}} - \frac{1}{\tilde{V}\tilde{T}} \qquad [4.4.79]$$

where the reduced parameters are given in Equation [4.4.73], but no c-parameter is included, and the size parameter, r, and the characteristic parameters are related by

$$P^* V^* = (r/M)RT^* \qquad [4.4.80]$$

where:

r size parameter (segment number)

M molar mass

In comparison with Equation [4.4.72], the size parameter remains explicit in the reduced equation of state. Thus, a simple corresponding-states principle is not, in general, satisfied. But, in principle, this equation of state is suitable for describing thermodynamic properties of fluids over an extended range of external conditions from the ordinary liquid to the gaseous state (it gives the correct ideal gas limit) and also to conditions above the critical point where the fluid is supercritical. Equation of state parameters for many liquids and liquid/molten polymers have recently been reported by Sanchez and Panayiotou[275] and for polymers by Rodgers[262] and by Cho and Sanchez.[263] To extend the lattice fluid theory to mixtures, appropriate mixing rules are needed. There is a fundamental difficulty here, because the segment size of any component is not necessarily equal to that of another but the molecular hard-core volume of a component must not change upon mixing. Consequently, the segment number of a component in the mixture, $r_i$, may differ from that for the pure fluid, $r_i^0$. But, following the arguments given by Panayiotou,[276] the number of segments may remain constant in the pure state and in the mixture. This assumption leads to a simpler formalism without worsening the quantitative character of the model. Thus, the following mixing rules may be applied:

$$P^* = P_1^*\psi_1 + P_2^*\psi_2 - \psi_1\psi_2 X_{12} \qquad [4.4.81a]$$

$$V^* = \Sigma\Sigma\psi_i\psi_j V_{ij}^* \qquad [4.4.81b]$$

$$1/r = \Sigma\psi_i / r_i \qquad [4.4.81c]$$

where $V_{ii}^* = V_i^*$ and $V_{ij}^*$ provides an additional binary fitting parameter and Equation [4.4.80] provides the mixing rule for T*. The final result for the residual solvent activity in a binary polymer solution reads:

$$\ln a_1^{residual} = r_1\left(\tilde{V}-1\right)\ln\left(\frac{\tilde{V}-1}{\tilde{V}}\right) - r_1\left(\tilde{V}_1-1\right)\ln\left(\frac{\tilde{V}_1-1}{\tilde{V}_1}\right) - \ln\frac{\tilde{V}}{\tilde{V}_1}$$

$$-\frac{r_1}{\tilde{T}_1}\left(\frac{1}{\tilde{V}}-\frac{1}{\tilde{V}_1}\right) + r_1\left(\frac{X_{12}}{\tilde{V}}\right)\psi_2^2 + \frac{r_1\tilde{P}_1\left(\tilde{V}-\tilde{V}_1\right)}{\tilde{T}_1} \qquad [4.4.82]$$

The last term in Equation (4.4.82) is again negligible at normal pressures. Various other approximations were given in the literature. For example, one can assume random mixing of contact sites rather than random mixing of segments,[277,278] as well as non-random mixing.[277,279] The model is applicable to solutions of small molecules as well as to polymer solutions. Like the Prigogine-Flory-Patterson equation of state, the lattice-fluid model and its variations have been used to correlate the composition dependence of the residual solvent activity.[277,279] These studies show that again entropic parameter $Q_{12}$ and interaction parameter $X_{12}$ have to be fitted to experimental data of the mixture to provide better agreement with measured solvent activities. The model and its modifications have been successfully used to represent thermodynamic excess properties, VLE and LLE for a variety of mixtures, as summarized by Sanchez and Panayiotou.[275] In addition to mixtures of polymers with normal solvents, the model can also be applied to polymer-gas systems, Sanchez and Rodgers.[280]

In lattice-hole theories, vacant cells or holes are introduced into the lattice, which describe the major part of thermal expansion, but changes in cell volume are also allowed which influence excess thermodynamic properties as well. The hole-theory for polymeric liquids as developed by Simha and Somcynsky[245] provides a very accurate equation of state that works much better than the Prigogine-Flory-Patterson equation of state or the Sanchez-Lacombe lattice-fluid model with respect to the precision how experimental PVT-data can be reproduced. However, the Dee-Walsh equation of state, Equation [4.4.77], with its more simple structure, works equally well. The Simha-Somcynsky equation of state must be solved simultaneously with an expression that minimizes the partition function with respect to the fraction of occupied sites and the final resulting equations for the chemical potential are more complicated. Details of the model will not be provided here. Characteristic parameters for many polymers have recently been given by Rodgers[262] or Cho and Sanchez.[263] The model is applicable to solutions of small molecules as well as to polymer solutions. Binary parameters have to be fitted to experimental data as with the models explained above. Again, one can assume random mixing of contact sites rather than random mixing of segments as well as non-random mixing, as was discussed, for example, by Nies and Stroeks[281] or Xie et al.[282,283]

Whereas the models given above can be used to correlate solvent activities in polymer solutions, attempts also have been made in the literature to develop concepts to predict solvent activities. Based on the success of the UNIFAC concept for low-molecular liquid mixtures,[284] Oishi and Prausnitz[285] developed an analogous concept by combining the UNIFAC-model with the free-volume model of Flory, Orwoll and Vrij.[236] The mass fraction based activity coefficient of a solvent in a polymer solution is given by:

$$\ln\Omega_1 \ln(a_1 / w_1) = \ln\Omega_1^{comb} + \ln\Omega_1^{res} + \ln\Omega_1^{fv} \qquad [4.4.83]$$

where:

| | |
|---|---|
| $\Omega_1$ | mass fraction based activity coefficient of solvent 1 at temperature T |
| $\Omega_1^{comb}$ | combinatorial contribution to the activity coefficient |
| $\Omega_1^{res}$ | residual contribution to the activity coefficient |
| $\Omega_1^{fv}$ | free-volume contribution to the activity coefficient |

Instead of the Flory-Huggins combinatorial contribution, Equation [4.4.68], the Staverman relation[228] is used.

$$\ln\Omega_1^{comb} = \ln\frac{\psi_1}{w_1} + \frac{z}{2}q_1 \ln\frac{\theta_1}{\psi_1} + l_1 - \frac{\psi_1 M_1}{w_1}\sum_j \frac{\psi_j M_j}{w_j} \qquad [4.4.84]$$

where the segment fractions $\psi_i$ and the surface area fractions $\theta_i$ have to be calculated according to

$$\psi_i = (w_i r_i / M_i) / \Sigma(w_k r_k / M_k) \qquad [4.4.85a]$$

$$\theta_i = (w_i q_i / M_i) / \Sigma(w_k q_k / M_k) \qquad [4.4.85b]$$

and the $l_i$-parameter is given by

$$l_i = (z/2)(r_i - q_i) - (r_i - 1) \qquad [4.4.85c]$$

where:

| | |
|---|---|
| i,j | components in the solution |
| z | lattice coordination number (usually = 10) |
| $q_i$ | surface area of component i based on Bondi's van-der-Waals surfaces |
| $M_i$ | molar mass of component i (for polymers the number average is recommended) |
| $w_i$ | mass fraction of component i |
| $r_i$ | segment number of component i based on Bondi's van-der-Waals volumes |

The molecules must be divided into groups as defined by the UNIFAC method. The segment surface areas and the segment volumes are calculated from Bondi's tables[264] according to

$$r_i = \sum_k \nu_\kappa^{(i)} R_k \quad and \quad q_i = \sum_k \nu_k^{(i)} Q_k \qquad [4.4.86]$$

where:

| | |
|---|---|
| k | number of groups of type k in the solution |
| $\nu_k$ | number of groups of type k in the molecule i |
| $R_k$ | Van-der-Waals group volume parameter for group k |
| $Q_k$ | Van-der-Waals group surface parameter for group k |

The residual activity coefficient contribution for each component is given by

$$\ln\Omega_i^{res} = \sum_k \nu_k^{(i)} \left[\ln\Gamma_k - \ln\Gamma_k^{(i)}\right] \qquad [4.4.87]$$

where:

| | |
|---|---|
| $\Gamma_k$ | the residual activity coefficient of group k in the defined solution |
| $\Gamma_k^{(i)}$ | the residual activity coefficient of group k in a reference solution containing pure component i only |

The residual activity coefficient of group k in the given solution is defined by

$$\ln\Gamma_k = Q_k \left[ 1 - \ln\left( \sum_m \Lambda_m \exp\left\{-\frac{a_{mk}}{T}\right\}\right) - \sum_m \left( \frac{\Lambda_m \exp\left\{-\frac{a_{km}}{T}\right\}}{\sum_p \Lambda_p \exp\left\{-\frac{a_{pm}}{T}\right\}} \right) \right] \qquad [4.4.88]$$

where:

| | |
|---|---|
| $a_{mn}$ and $a_{nm}$ | group interaction parameter pair between groups m and n |
| $\Lambda_m$ | group surface area fraction for group m |

The group surface area fraction $\Lambda_m$ for group m in the solution is given by

$$\Lambda_m = \frac{Q_m X_m}{\sum_p Q_p X_p} \qquad [4.4.89a]$$

where:

$X_m$ group mole fraction for group m

The group mole fraction $X_m$ for group m in the solution is given by

$$X_m = \frac{\sum_j \nu_m^{(j)} w_j / M_j}{\sum_j \sum_p \nu_p^{(j)} w_j / M_j} \qquad [4.4.89b]$$

The residual activity coefficient of group k in reference solutions containing only component i, $\Gamma_k^{(i)}$ is similarly determined using Equations [4.4.88, 4.4.89], with the exception that the summation indices k, m, p refer only to the groups present in the pure component and the summations over each component j are calculated only for the single component present in the reference solution.

The group interaction parameter pairs $a_{mn}$ and $a_{nm}$ result from the interaction between the groups m and n. These parameter are unsymmetric values that have to be fitted to experimental VLE-data of low-molecular mixtures. They can be taken from UNIFAC tables, e.g., Refs.[2,284,286-289] and, additionally, they may be treated as temperature functions.

The free-volume contribution, which is essential for nonpolar polymer solutions, follows, in principle, from Equation [4.4.74] with parameter $X_{12} = 0$ as applied by Raetzsch and Glindemann,[290] or in a modified form from Equation [4.4.90] as introduced by Oishi and Prausnitz and used also in Danner's Handbook.[2]

$$\ln Q_1^{fv} = 3c_1 \ln\left(\frac{\tilde{V}_1^{1/3} - 1}{\tilde{V}^{1/3} - 1}\right) - c_1\left(\frac{\tilde{V}_1 - \tilde{V}}{\tilde{V}\left(1 - \tilde{V}_1^{1/3}\right)}\right) \qquad [4.4.90]$$

where:

$c_1$ external degree of freedom parameter of the solvent 1, usually fixed = 1.1

To get a predictive model, the reduced volumes and the external degree of freedom parameter are not calculated from Flory's equation of state, Equation [4.4.72], but from some simple approximations as given by the following relations:[2]

$$\tilde{V}_i = \frac{\upsilon_{spez,i} M_i}{0.01942 r_i} \qquad [4.4.91a]$$

$$\tilde{V} = \frac{\sum_i \upsilon_{spez,i} w_i}{0.01942 \sum_i r_i w_i / M_i} \qquad [4.4.91b]$$

where:

$\upsilon_{spez,i}$ specific volume of component i in $m^3/kg$

$r_i$ segment number of component i based on Bondi's van-der-Waals volumes

$M_i$ molar mass of component i (for polymers the number average is recommended)

$w_i$ mass fraction of component i

There has been a broad application of this group-contribution UNIFAC-fv concept to polymer solutions in the literature. Raetzsch and Glindemann[290] recommended the use of the real free-volume relation from the Flory-Orwoll-Vrij model to account for realistic PVT-data. Problems arise for mixtures composed from chemically different components that posses the same groups, e.g., mixtures with different isomers. Kikic et al.[291] discussed the influence of the combinatorial part on results obtained with UNIFAC-fv calculations. Gottlieb and Herskowitz[292,293] gave some polemic about the special use of the $c_1$-parameter within UNIFAC-fv calculations. Iwai et al.[294,295] demonstrated the possible use of UNIFAC-fv for the calculation of gas solubilities and Henry's constants using a somewhat different free-volume expression. Patwardhan and Belfiore[296] found quantitative discrepancies for some polymer solutions. In a number of cases UNIFAC-fv predicted the occurrence of a demixing region during the calculation of solvent activities where experimentally only a homogeneous solution exists. Price and Ashworth[297] found that the predicted variation of residual solvent activity with polymer molecular mass at high polymer concentrations is opposite to that measured for polydimethylsiloxane solutions. But, qualitative correct predictions were obtained for poly(ethylene glycol) solutions with varying polymer molecular mass.[298-300] However, UNIFAC-fv is not capable of representing thermodynamic data of strongly associating or solvating systems.

Many attempts have been made to improve the UNIFAC-fv model which cannot be listed here. A comprehensive review was given by Fried et al.[301] An innovative method to combine the free-volume contribution within a corrected Flory-Huggins combinatorial entropy and the UNIFAC concept was found by Elbro et al.[302] and improved by Kontogeorgis et al.[303] These authors take into account the free-volume dissimilarity by assuming different van der Waals hard-core volumes (again from Bondi's tables[264]) for the solvent and the polymer segments

$$V_i^{fv} = q_i\left(V_i^{1/3} - V_{i,vdW}^{1/3}\right)^3 \qquad [4.4.92a]$$

$$\varphi_i^{fv} = x_i V_i^{fv} / \sum_j x_j V_j^{fv} \qquad [4.4.92b]$$

where:

$q_i$ surface area of component i based on Bondi's van der Waals surfaces
$x_i$ mole fraction of component i
$V_i$ molar volume of component i
$V_{i,vdW}$ van der Waals hard-core molar volume of component i

and introduced these free-volume terms into Equation [4.4.68] to obtain a free-volume corrected Flory-Huggins combinatorial term:

$$\ln\gamma_i^{fv} = \ln\left(\varphi_i^{fv} / x_i\right) + 1 - \left(\varphi_i^{fv} / x_i\right) \qquad [4.4.93a]$$

or

$$\ln\Omega_i^{fv} = \ln\left(\varphi_i^{fv} / w_i\right) + 1 - \left(\varphi_i^{fv} / x_i\right) \qquad [4.4.93b]$$

To obtain the complete activity coefficient, only the residual term from the UNIFAC model, Equation [4.4.87], has to be added. An attempt to incorporate differences in shape between solvent molecules and polymer segments was made by Kontogeorgis et al.[302] by adding the Staverman correction term to obtain:

$$\ln\gamma_i^{fv} = \ln\frac{\varphi_i^{fv}}{x_i} + 1 - \frac{\varphi_i^{fv}}{x_i} - \frac{zq_i}{2}\left(\ln\frac{\psi_i}{\theta_i} + 1 - \frac{\psi_i}{\theta_i}\right) \qquad [4.4.93c]$$

where the segment fractions $\psi_i$ and the surface area fractions $\theta_i$ have to be calculated according to Equations [4.4.85a+b]. Using this correction, they get somewhat better results only when Equation [4.4.93] leads to predictions lower than the experimental data.

Different approaches utilizing group contribution methods to predict solvent activities in polymer solutions have been developed after the success of the UNIFAC-fv model. Misovich et al.[304] have applied the Analytical Solution of Groups (ASOG) model to polymer solutions. Recent improvements of polymer-ASOG have been reported by Tochigi et al.[305-307] Various other group-contribution methods including an equation-of-state were developed by Holten-Anderson et al.,[308,309] Chen et al.,[310] High and Danner,[311-313] Tochigi et al.,[314] Lee and Danner,[315] Bertucco and Mio,[316] or Wang et al.,[317] respectively. Some of them were presented again in Danner's Handbook.[2] Detail are not provided here.

### 4.4.4.2 Fugacity coefficients from equations of state

Total equation-of-state approaches usually apply equations for the fugacity coefficients instead of relations for chemical potentials to calculate thermodynamic equilibria and start from Equations [4.4.2 to 6]. Since the final relations for the fugacity coefficients are usually much more lengthy and depend, additionally, on the chosen mixing rules, only the equations of state are listed below. Fugacity coefficients have to be derived by solving Equation [4.4.6]. After obtaining the equilibrium fugacities of the liquid mixture at equilibrium temperature and pressure, the solvent activity can be calculated from Equation [4.4.1]. The standard state fugacity of the solvent can also be calculated from the same equation of state by solving the same equations but for the pure liquid. Details of this procedure can be found in textbooks, e.g., Refs.[318,319]

Equations of state for polymer systems that will be applied within such an approach have to be valid for the liquid as well as for the gaseous state like lattice-fluid models based on Sanchez-Lacombe theory, but not the free-volume equations based on Prigogine-Flory-Patterson theory, as stated above. However, most equations of state applied within such an approach have not been developed specially for polymer systems, but, first, for common non-electrolyte mixtures and gases. Today, one can distinguish between cubic and non-cubic equations of state for phase equilibrium calculations in polymer systems. Starting from the free-volume idea in polymeric systems, non-cubic equations of state should be applied to polymers. Thus, the following text presents first some examples of this class of equations of state. Cubic equations of state came later into consideration for polymer systems, mainly due to increasing demands from engineers and engineering software where three-volume-roots equations of state are easier to solve and more stable in computational cycles.

About ten years after Flory's development of an equation of state for polymer systems, one began to apply methods of thermodynamic perturbation theory to calculate the thermodynamic behavior of polymer systems. The main goal was first to overcome the restrictions of Flory's equation of state to the liquid state, to improve the calculation of the compressibility behavior with increasing pressure and to enable calculations of fluid phase equilibria at any densities and pressures from the dilute gas phase to the compressed liquid including molecules differing considerably in size, shape, or strength of intermolecular potential energy. More recently, when more sophisticated methods of statistical mechanics were developed, deeper insights into liquid structure and compressibility behavior of model polymer chains in comparison to Monte Carlo modelling results could be obtained applying thermodynamic perturbation theory. Quite a lot of different equations of state have been developed up to now following this procedure; however, only a limited number was applied to real systems. Therefore, only some summary and a phenomenological presentation of some equations of state which have been applied to real polymer fluids should be given here, following their historical order.

The perturbed-hard-chain (PHC) theory developed by Prausnitz and coworkers in the late 1970s[320-322] was the first successful application of thermodynamic perturbation theory to polymer systems. Since Wertheim's perturbation theory of polymerization[323] was formulated about 10 years later, PHC theory combines results from hard-sphere equations of simple liquids with the concept of density-dependent external degrees of freedom in the Prigogine-Flory-Patterson model for taking into account the chain character of real polymeric fluids. For the hard-sphere reference equation the result derived by Carnahan and Starling[324] was applied, as this expression is a good approximation for low-molecular hard-sphere fluids. For the attractive perturbation term, a modified Alder's[325] fourth-order perturbation result for square-well fluids was chosen. Its constants were refitted to the thermodynamic equilibrium data of pure methane. The final equation of state reads:

$$\frac{PV}{RT} = 1 + c\frac{4y^2 - 2y}{(y-1)^3} + c\sum_n \sum_m \frac{mA_{nm}}{\tilde{V}^m \tilde{T}^n} \quad [4.4.94]$$

where:

- y — packing fraction with $y = V/(V_0\tau)$ and $\tau = (\pi/6)2^{0.5} = 0.7405$ (please note that in a number of original papers in the literature the definition of y within this kind of equations is made by the reciprocal value, i.e., $\tau V_0/V$)
- c — degree of freedom parameter, related to one chain-molecule (not to one segment)
- $V_0$ — hard-sphere volume for closest packing
- $A_{nm}$ — empirical coefficients from the attractive perturbation term

The reduced volume is again defined by $\tilde{V} = V/V_0$ and the reduced temperature by $\tilde{T} = T/T^*$. The coefficients $A_{nm}$ are given in the original papers by Beret[320,321] and are considered to be universal constants that do not depend on the chemical nature of any special substance. The remaining three characteristic parameters, c, T* and $V_0$, have to be adjusted to experimental PVT-data of the polymers or to vapor-liquid equilibrium data of the pure solvents. Instead of fitting the c-parameter, one can also introduce a parameter P* by the relation $P^* = cRT^*/V_0$. In comparison with Flory's free-volume equation of state, PHC-equation of state is additionally applicable to gas and vapor phases. It fulfills the ideal gas limit, and it describes the PVT-behavior at higher pressures better and without the need of temperature and/or pressure-dependent characteristic parameters, such as with Flory's model. Values for characteristic parameters of polymers and solvents can be found in the original literature. A review for the PHC-model was given by Donohue and Vimalchand,[326] where a number of extensions and applications also are summarized. Application to mixtures and solutions needs mixing rules for the characteristic parameters and introduction of binary fitting parameters[322,327,328] (details are not given here). Examples for applying PHC to polymer solutions are given by Liu and Prausnitz[328] or Iwai, Arai and coworkers.[329-331]

The chain-of-rotators (COR) equation of state was developed by Chao and coworkers[332] as an improvement of the PHC theory. It introduces the non-spherical shape of molecules into the hard-body reference term and describes the chain molecule as a chain of rotators with the aim of an improved model for calculating fluid phase equilibria, PVT and derived thermodynamic properties, at first only for low-molecular substances. Instead of hard spheres, the COR-model uses hard dumbbells as reference fluid by combining the result of Boublik and Nezbeda[333] with the Carnahan-Starling equation for a separate consideration of rotational degrees of freedom; however, still in the sense of Prigogine-Flory-Patterson regarding the chain-character of the molecules. It neglects the effect of rotational motions on intermolecular attractions; however, the attractive portion of the final equation of state has an empirical dependence on rotational degrees of freedom given by the prefactor of the double sum. For the attractive perturbation term, a modified Alder's fourth-order perturbation result for square-well fluids was chosen, additionally improved by an empirical temperature-function for the rotational part. The final COR equation reads:

$$\frac{PV}{RT} = 1 + \frac{4y^2 - 2y}{(y-1)^3} + c\left(\frac{\alpha - 1}{2}\right)\frac{3y^2 + 3\alpha y - (\alpha + 1)}{(y-1)^3}$$

$$+\left(1 + \frac{c}{2}\left\{B_0 + B_1/\tilde{T} + B_2\tilde{T}\right\}\right)\sum_n\sum_m \frac{mA_{nm}}{\tilde{V}^m\tilde{T}^n} \qquad [4.4.95]$$

where:

| | |
|---|---|
| y | packing fraction with $y = V/(V_0\tau)$ and $\tau = (\pi/6)2^{0.5} = 0.7405$ (please note that in a number of original papers in the literature the definition of y within this kind of equations is made by its reciprocal value, i.e., $\tau V_0/V$) |
| c | degree of freedom parameter, related to one chain-molecule (not to one segment) |
| $V_0$ | hard-sphere volume for closest packing |
| $A_{nm}$ | empirical coefficients from the attractive perturbation term |
| $B_0,B_1,B_2$ | empirical coefficients for the temperature dependence of the rotational part |
| $\alpha$ | accounts for the deviations of the dumbbell geometry from a sphere |

As can be seen from the structure of the COR equation of state, the Carnahan-Starling term becomes very small with increasing chain length, i.e., with increasing c, and the rotational part is the dominant hard-body term for polymers. The value of c is here a measure of rotational degrees of freedom within the chain (and related to one chain-molecule and not to one segment). It is different from the meaning of the c-value in the PHC equation. Its exact value is not known a priori as chain molecules have a flexible structure. The value of $\alpha$ for the various rotational modes is likewise not precisely known. Since $\alpha$ and c occur together in the product $c(\alpha - 1)$, departure of real rotators from a fixed value of $\alpha$ is compensated for by the c-parameter after any fitting procedure. As usual, the value of $\alpha$ is assigned a constant value of 1.078 calculated according to the dumbbell for ethane as representative for the rotating segments of a hydrocarbon chain. The coefficients $A_{nm}$ and the three parameters $B_0$,

$B_1$, $B_2$ were refitted to the thermodynamic equilibrium data of pure methane, ethane, and propane.[332] Both $A_{nm}$ matrices for PHC and COR equation of state contain different numerical values. The remaining three characteristic parameters, c, T* and $V_0$, have to be adjusted to experimental equilibrium data. Instead of fitting the c-parameter, one can also introduce a parameter P* by the relation $P^* = cRT^*/V_0$. Characteristic parameters for many solvents and gases are given by Chien et al.[332] or Masuoka and Chao.[334] Characteristic parameters of more than 100 polymers and many solvents are given by Wohlfarth and coworkers,[335-348] who introduced segment-molar mixing rules and group-contribution interaction parameters into the model and applied it extensively to polymer solutions at ordinary pressures as well as at high temperatures und pressures, including gas solubility and supercritical solutions. They found that it may be necessary sometimes to refit the pure-component characteristic data of a polymer to some VLE-data of a binary polymer solution to calculate correct solvent activities, because otherwise demixing was calculated. Refitting is even more necessary when high-pressure fluid phase equilibria have to be calculated using this model.

A group-contribution COR equation of state was developed Pults et al.[349,350] and extended into a polymer COR equation of state by Sy-Siong-Kiao et al.[351] This equation of state is somewhat simplified by replacing the attractive perturbation term by the corresponding part of the Redlich-Kwong equation of state.

$$\frac{PV}{RT} = 1 + \frac{4y^2 - 2y}{(y-1)^3} + c\left(\frac{\alpha - 1}{2}\right)\frac{3y^2 + 3\alpha y - (\alpha + 1)}{(y-1)^3} - \frac{a(T)}{RT[V + b(T)]} \quad [4.4.96]$$

where:

| | |
|---|---|
| a | attractive van der Waals-like parameter |
| b | excluded volume van der Waals-like parameter |
| c | degree of freedom parameter, related to one chain-molecule (not to one segment) |
| y | packing fraction |
| $\alpha$ | accounts for the deviations of the dumbbell geometry from a sphere |

Exponential temperature functions for the excluded volume parameter b and the attractive parameter a were introduced by Novenario et al.[352-354] to apply this equation of state also to polar and associating fluids. Introducing a group-contribution concept leads to segment-molar values of all parameters a, b, c which can easily be fitted to specific volumes of polymers.[351,354]

The statistical associating fluid theory (SAFT) is the first and the most popular approach that uses real hard-chain reference fluids, including chain-bonding contributions. Its basic ideas have been developed by Chapman et al.[256-258] Without going into details, the final SAFT equation of state is constructed from four terms: a segment term that accounts for the non-ideality of the reference term of non-bonded chain segments/monomers as in the equations shown above, a chain term that accounts for covalent bonding, and an association term that accounts for hydrogen bonding. There may be an additional term that accounts for other polarity effects. A dispersion term is also added that accounts for the perturbing potential, as in the equations above. A comprehensive summary is given in Praunsitz's book.[49] Today, there are different working equations based on the SAFT approach. Their main differences stem from the way the segment and chain terms are estimated. The most common version is the one developed by Huang and Radosz,[355] applying the fourth-order perturbation approach as in COR or PHC above, but with new refitted parameters to argon, as given by Chen and Kreglewski,[356] and a hard-sphere pair-correlation function for the chain term as following the arguments of Wertheim. The Huang-Radosz-form of the SAFT-equation of state without an association term reads:[355]

$$\frac{PV}{RT} = 1 + r\frac{4y^2 - 2y}{(y-1)^3} + (1-r)\frac{5y-2}{(y-1)(2y-1)} + r\sum_n\sum_m \frac{mD_{nm}}{\tilde{V}^m}\left[\frac{u}{kT}\right]^n \quad [4.4.97]$$

where:

| | |
|---|---|
| $D_{nm}$ | empirical coefficients from the attractive perturbation term |
| k | Boltzmann's constant |
| r | chain segment number |

In comparison to the PHC equation of state, the new term between the Carnahan-Starling term and the double sum accounts for chain-bonding. The terms are proportional to the segment number, r , of the chain molecule. However, the hard-sphere volume, $V_0$, is now a slight function of temperature which is calculated according to the result of Chen and Kreglewski:[356]

$$V_0 = V^{00}\left[1 - 0.12\exp(-3u_0 / kT)\right]^3 \quad [4.4.98]$$

where:

| | |
|---|---|
| $V^{00}$ | hard-sphere volume at T = 0 K |
| $u_0$ | well-depth of a square-well potential $u/k = u_0/k\ (1 + 10K/T)$ with 10K being an average for all chain molecules. |

The ratio $u_0/k$ or u/k is analogous to the characteristic parameter T* in the equations above. There are two additional volume and energy parameters if association is taken into account. In its essence, the SAFT equation of state needs three pure component parameters which have to be fitted to equilibrium data: $V^{00}$, $u_0/k$ and r. Fitting of the segment number looks somewhat curious to a polymer scientist, but it is simply a model parameter, like the c-parameter in the equations above, which is also proportional to r. One may note additionally that fitting to specific volume PVT-data leads to a characteristic ratio r/M (which is a specific r-value), as in the equations above, with a specific c-parameter. Several modifications and approximations within the SAFT-framework have been developed in the literature. Banaszak et al.[357-359] or Shukla and Chapman[360] extended the concept to copolymers. Adidharma and Radosz[361] provides an engineering form for such a copolymer SAFT approach. SAFT has successfully applied to correlate thermodynamic properties and phase behavior of pure liquid polymers and polymer solutions, including gas solubility and supercritical solutions by Radosz and coworkers[355,357-359,361-368] Sadowski et al.[369] applied SAFT to calculate solvent activities of polycarbonate solutions in various solvents and found that it may be necessary to refit the pure-component characteristic data of the polymer to some VLE-data of one binary polymer solution to calculate correct solvent activities, because otherwise demixing was calculated. Groß and Sadowski[370] developed a "Perturbed-Chain SAFT" equation of state to improve for the chain behavior within the reference term to get better calculation results for the PVT- and VLE-behavior of polymer systems. McHugh and coworkers applied SAFT extensively to calculate the phase behavior of polymers in supercritical fluids, a comprehensive summary is given in the review by Kirby and McHugh.[371] They also state that characteristic SAFT parameters for polymers from PVT-data lead to wrong phase equilibrium calculations and, therefore, also to wrong solvent activities from such calculations. Some ways to overcome this situation and to obtain reliable parameters for phase equilibrium calculations are provided in Ref.,[371] together with examples from the literature that will not be repeated here.

The perturbed-hard-sphere-chain (PHSC) equation of state is a hard-sphere-chain theory that is somewhat different to SAFT. It is based on a hard-sphere chain reference system and a van der Waals-type perturbation term using a temperature-dependent attractive parameter $a(T)$ and a temperature-dependent co-volume parameter $b(T)$. Song et al.[259,260] applied it to polymer systems and extended the theory also to fluids consisting of heteronuclear hard chain molecules. The final equation for pure liquids or polymers as derived by Song et al. is constructed from three parts: the first term stems (as in PHC, COR or SAFT) from the Carnahan-Starling hard-sphere monomer fluid, the second is the term due to covalent chain-bonding of the hard-sphere reference chain and the third is a van der Waals-like attraction term (more details are given also in Prausnitz's book[49]):

$$\frac{PV}{RT} = 1 + r\frac{4\eta - 2\eta^2}{(1-\eta)^3} + (1-r)\left\{\frac{(1-\eta/2)}{(1-\eta)^3} - 1\right\} - \frac{r^2 a(T)}{RTV} \qquad [4.4.99]$$

where:

| | |
|---|---|
| $\eta$ | reduced density or packing fraction |
| a | attractive van der Waals-like parameter |
| r | chain segment number |

The reduced density or packing fraction $\eta$ is related to an effective and temperature-dependent co-volume $b(T)$ by $\eta = r\,b(T)\rho/4$, with $\rho$ being the number density, i.e., the number of molecules per volume. However, PHSC-theory does not use an analytical intermolecular potential to estimate the temperature dependence of $a(T)$ and $b(T)$. Instead, empirical temperature functions are fitted to experimental data of argon and methane (see also[49]).

We note that the PHSC equation of state is again an equation where three parameters have to be fitted to thermodynamic properties: $\sigma$, $\varepsilon/k$ and r. These may be transformed into macroscopic reducing parameters for the equation of state by the common relations $T^*=\varepsilon/k$, $P^* = 3\varepsilon/2\pi\sigma^3$ and $V^* = 2\pi r\sigma^3/3$. Parameter tables are given in Refs.[86,260,372-374] PHSC was successfully applied to calculate solvent activities in polymer solutions, Gupta and Prausnitz.[86] Lambert et al.[374] found that it is necessary to adjust the characteristic parameters of the polymers when liquid-liquid equilibria should correctly be calculated.

Even with simple cubic equations of state, a quantitative representation of solvent activities for real polymer solutions can be achieved, as was shown by Tassios and coworkers.[375,376] Using generalized van der Waals theory, Sako et al.[377] obtained a three-parameter cubic equation of state which was the first applied to polymer solutions:

$$\frac{PV}{RT} = \frac{V - b + bc}{V - b} - \frac{a(T)}{RT(V + b)} \qquad [4.4.100]$$

where:

| | |
|---|---|
| a | attractive van der Waals-like parameter |
| b | excluded volume van der Waals-like parameter |

c 3c is the total number of external degrees of freedom per molecule

When c = 1, Equation [4.4.100] reduces to the common Soave-Redlich-Kwong (SRK) equation of state.[378] Temperature functions and combining/mixing rules for parameters a,b,c are not discussed here because quite different approximations may be used. Problems, how to fit these parameters to experimental PVT-data for polymers, have been discussed by several authors.[375-380]

Orbey and Sandler[380] applied the Peng-Robinson equation of state as modified by Stryjek and Vera[381] (PRSV):

$$\frac{PV}{RT} = \frac{V}{V-b} - \frac{a(T)}{RT\left(V^2 + 2bV - b^2\right)} \qquad [4.4.101]$$

to calculate solvent activities in polymer solutions using Wong-Sandler mixing rules[382] that combine the equation of state with excess energy models (EOS/GE-mixing rules). They have shown that a two-parameter version can correlate the solvent partial pressure of various polymer solutions with good accuracy over a range of temperatures and pressures with temperature-independent parameters. Harrismiadis et al.[379] worked out some similarities between activity coefficients derived from van der Waals like equations-of-state and Equations (4.4.92 and 93), i.e., the Elbro-fv model. Zhong and Masuoka[383] combined SRK equation of state with EOS/GE-mixing rules and the UNIFAC model to calculate Henry's constants of solvents and gases in polymers. Additionally, they developed new mixing rules for van der Waals-type two-parameter equations of state (PRSV and SRK) which are particularly suitable for highly asymmetric systems, i.e., also polymer solutions, and demonstrated that only one adjustable temperature-independent parameter is necessary for calculations within a wide range of temperatures.[384] In a following paper,[385] some further modifications and improvements could be found. Orbey et al.[386] successfully proposed some empirical relations for PRSV-equation-of-state parameters with polymer molar mass and specific volume to avoid any special parameter fitting for polymers and introduced a NRTL-like local-composition term into the excess energy part of the mixing rules for taking into account of strong interactions, for example, in water + poly(propylene glycol)s. They found infinite-dilution activity coefficient data, i.e., Henry's constants, to be most suitable for fitting the necessary model parameter.[386]

Orbey et al.[387] summarized three basic conclusions for the application of cubic equations of state to polymer solutions:

(i) These models developed for conventional mixtures can be extended to quantitatively describe VLE of polymer solutions if carefully selected parameters are used for the pure polymer. On the other hand, pure-component parameters of many solvents are already available and VLE between them is well represented by these cubic equations of state.

(ii) EOS/GE-mixing rules represent an accurate way of describing phase equilibria. Activity coefficient expressions are more successful when they are used in this format than directly in the conventional gamma-phi approach.

(iii) It is not justifiable to use multi-parameter models, but it is better to limit the number of parameters to the number of physically meaningful boundary conditions and calculate them according to the relations dictated by these boundary conditions.

#### 4.4.4.3 Comparison and conclusions

The simple Flory-Huggins $\chi$-function, combined with the solubility parameter approach may be used for a first rough guess about solvent activities of polymer solutions, if no experimental data are available. Nothing more should be expected. This also holds true for any calculations with the UNIFAC-fv or other group-contribution models. For a quantitative representation of solvent activities of polymer solutions, more sophisticated models have to be applied. The choice of a dedicated model, however, may depend, even today, on the nature of the polymer-solvent system and its physical properties (polar or non-polar, association or donor-acceptor interactions, subcritical or supercritical solvents, etc.), on the ranges of temperature, pressure and concentration one is interested in, on the question whether a special solution, special mixture, special application is to be handled or a more universal application is to be found or a software tool is to be developed, on numerical simplicity or, on the other hand, on numerical stability and physically meaningful roots of the non-linear equation systems to be solved. Finally, it may depend on the experience of the user (and sometimes it still seems to be a matter of taste).

There are deficiencies in all of these theories given above. These theories fail to account for long-range correlations between polymer segments which are important in dilute solutions. They are valid for simple linear chains and do not account for effects like chain branching, rings, dentritic polymers. But, most seriously, all of these theories are of the mean-field type that fail to account for the contributions of fluctuations in density and composition. Therefore, when these theories are used in the critical region, poor results are often obtained. Usually, critical pressures are overestimated within VLE-calculations. Two other conceptually different mean-field approximations are invoked during the development of these theories. To derive the combinatorial entropic part correlations between segments of one chain that are not nearest neighbors are neglected (again, mean-field approximations are therefore not good for a dilute polymer solution) and, second, chain connectivity and correlation between segments are improperly ignored when calculating the potential energy, the attractive term.

Equation-of-state approaches are preferred concepts for a quantitative representation of polymer solution properties. They are able to correlate experimental VLE data over wide ranges of pressure and temperature and allow for physically meaningful extrapolation of experimental data into unmeasured regions of interest for application. Based on the experience of the author about the application of the COR equation-of-state model to many polymer-solvent systems, it is possible, for example, to measure some vapor pressures at temperatures between 50 and 100°C and concentrations between 50 and 80 wt% polymer by isopiestic sorption together with some infinite dilution data (limiting activity coefficients, Henry's constants) at temperatures between 100 and 200°C by IGC and then to calculate the complete vapor-liquid equilibrium region between room temperature and about 350°C, pressures between 0.1 mbar and 10 bar, and solvent concentration between the common polymer solution of about 75-95 wt% solvent and the ppm-region where the final solvent and/or monomer devolatilization process takes place. Equivalent results can be obtained with any other comparable equation of state model like PHC, SAFT, PHSC, etc.

The quality of all model calculations with respect to solvent activities depends essentially on the careful determination and selection of the parameters of the pure solvents, and also of the pure polymers. Pure solvent parameter must allow for the quantitative calculation of pure solvent vapor pressures and molar volumes, especially when equation-of-state approaches are used. Pure polymer parameters strongly influence the calculation of gas solubilities, Henry's constants, and limiting solvent activities at infinite dilution of the solvent in the liquid/molten polymer. Additionally, the polymer parameters mainly determine the occurrence of a demixing region in such model calculations. Generally, the quantitative representation of liquid-liquid equilibria is a much more stringent test for any model, what was not discussed here. To calculate such equilibria it is often necessary to use some mixture properties to obtain pure-component polymer parameters. This is necessary because, at present, no single theory is able to describe correctly the properties of a polymer in both the pure molten state and in the highly dilute solution state. Therefore, characteristic polymer parameters from PVT-data of the melt are not always meaningful for the dilute polymer solution. Additionally, characteristic polymer parameters from PVT-data also may lead to wrong results for concentrated polymer solutions because phase equilibrium calculations are much more sensitive to variations in pure component parameters than polymer densities.

All models need some binary interaction parameters that have to be adjusted to some thermodynamic equilibrium properties since these parameters are a priori not known (we will not discuss results from Monte Carlo simulations here). Binary parameters obtained from data of dilute polymer solutions as second virial coefficients are often different from those obtained from concentrated solutions. Distinguishing between intramolecular and intermolecular segment-segment interactions is not as important in concentrated solutions as it is in dilute solutions. Attempts to introduce local-composition and non-random-mixing approaches have been made for all the theories given above with more or less success. At least, they introduce additional parameters. More parameters may cause a higher flexibility of the model equations but leads often to physically senseless parameters that cause troubles when extrapolations may be necessary. Group-contribution concepts for binary interaction parameters in equation of state models can help to correlate parameter sets and also data of solutions within homologous series.

## 4.4.5 REFERENCES


1 Wen Hao, H. S. Elbro, P. Alessi, **Polymer Solution Data Collection, Pt.1: Vapor-Liquid Equilibrium, Pt.2: Solvent Activity Coefficients at Infinite Dilution, Pt. 3: Liquid-Liquid Equilibrium**, *DECHEMA Chemistry Data Series*, Vol. XIV, Pts. 1, 2+3, DECHEMA, Frankfurt/M., 1992.

2 R. P. Danner, M. S. High, **Handbook of Polymer Solution Thermodynamics**, DIPPR, *AIChE*, New York, 1993.

3 Ch. Wohlfarth, **Vapour-liquid equilibrium data of binary polymer solutions, Physical Science Data 44, Elsevier**, Amsterdam, 1994.

4 J. Brandrup, E. H. Immergut, E. A. Grulke (eds.), **Polymer Handbook**, 4th ed., *J. Wiley & Sons., Inc.*, New York, 1999.

5 N. Schuld, B. A. Wolf, in **Polymer Handbook**, J. Brandrup, E. H. Immergut, E. A. Grulke (eds.), 4th ed., *J. Wiley & Sons., Inc.*, New York, 1999, pp. VII/247-264.

6 R. A. Orwoll, in **Polymer Handbook**, J. Brandrup, E. H. Immergut, E. A. Grulke (eds.), 4th ed., *J. Wiley & Sons., Inc.*, New York, 1999, pp. VII/649-670.

7 R. A. Orwoll, P. A. Arnold, in **Physical Properties of Polymers Handbook**, J. E. Mark (Ed.), *AIP Press*, Woodbury, New York, 1996, pp. 177-198.

8 R. A. Orwoll, *Rubber Chem. Technol.*, **50**, 451 (1977).

9 M. D. Lechner, E. Nordmeier, D. G. Steinmeier, in **Polymer Handbook**, J. Brandrup, E. H. Immergut, E. A. Grulke (eds.), 4th ed., *J. Wiley & Sons.*, Inc., New York, 1999, pp. VII/85-213.

10 D. C. Bonner, *J. Macromol. Sci. - Revs. Macromol. Chem., C*, **13**, 263 (1975).

11 A. F. M. Barton, **CRC Handbook of Polymer-Liquid Interaction Parameters and Solubility Parameters**, *CRC Press*, Boca Raton, 1990.

12 A. F. M. Barton, **CRC Handbook of Solubility Parameters and Other Cohesion Parameters**, 2nd Ed., *CRC Press*, Boca Raton, 1991.
13 J.-M. Braun, J. E. Guillet, *Adv. Polym. Sci.*, **21**, 108 (1976).
14 A. E. Nesterov, J. S. Lipatov, **Obrashchennaya Gasovaya Khromatografiya v Termodinamike Polimerov**, *Naukova Dumka*, Kiev, 1976.
15 D. C. Locke, *Adv. Chromatogr.*, **14**, 87 (1976).
16 D. G. Gray, *Progr. Polym. Sci.*, **5**, 1 (1977).
17 J. E. G. Lipson, J. E. Guillet, *Developm. Polym. Character.*, **3**, 33 (1982).
18 J. S. Aspler, *Chromatogr. Sci.*, **29**, 399 (1985).
19 A. E. Nesterov, **Obrashchennaya Gasovaya Khromatografiya Polimerov**, *Naukova Dumka*, Kiev, 1988.
20 R. Vilcu, M. Leca, **Polymer Thermodynamics by Gas Chromatography**, *Elsevier*, Amsterdam, 1989.
21 G. J. Price, *Adv. Chromatogr.*, **28**, 113 (1989).
22 D. R. Lloyd, T. C. Ward, H. P. Schreiber, C. C. Pizana (eds.), **Inverse Gas Chromatography**, *ACS Symp. Ser., 391*, 1989.
23 P. Munk, in **Modern Methods of Polymer Characterization**, H. G. Barth, J. W. Mays (eds.), *J. Wiley & Sons, Inc.*, New York, 1991, pp.151-200.
24 Bincai Li, *Rubber Chem. Technol.*, **69**, 347 (1996).
25 Z. Y. Al-Saigh, *Int. J. Polym. Anal. Charact.* **3**, 249 (1997).
26 R. A. Pethrick, J. V. Dawkins (Eds.), **Modern Techniques for Polymer Characterization**, *J. Wiley & Sons, Inc.*, Chichester, 1999.
27 G. D. Wignall, in **Encyclopedia of Polymer Science and Engineering**, 2nd Ed., J. Kroschwitz (Ed.), *J. Wiley & Sons, Inc.*, New York, 1987, Vol. 10, pp. 112-184.
28 G. C. Berry, in **Encyclopedia of Polymer Science and Engineering**, 2nd Ed., J. Kroschwitz (Ed.), *J. Wiley & Sons, Inc.*, New York, 1987, Vol. 8, pp. 721-794.
29 M. B. Huglin (Ed.), **Light Scattering from Polymer Solutions**, *Academic Press*, Inc., New York, 1972.
30 E. F. Casassa, G. C. Berry, in **Polymer Molecular Weights**, Pt. 1, *Marcel Dekker*, Inc., New York, 1975, pp. 161-286.
31 P. Kratochvil, **Classical Light Scattering from Polymer Solutions**, *Elsevier*, Amsterdam, 1987.
32 B. Chu, **Laser Light Scattering**, *Academic Press*, Inc., New York, 1991.
33 A. R. Cooper, in **Encyclopedia of Polymer Science and Engineering**, 2nd Ed., J. Kroschwitz (Ed.), *J. Wiley & Sons, Inc.*, New York, 1987, Vol. 10, pp. 1-19.
34 C. A. Glover, in **Polymer Molecular Weights, Pt. 1**, *Marcel Dekker, Inc.*, New York, 1975, pp. 79-159.
35 H. Fujita, **Foundations of Ultracentrifugal Analysis**, *J.Wiley & Sons, Inc.*, New York, 1975.
36 S. E. Harding, A. J. Rowe, J. C. Horton, **Analytical Ultracentrifugation in Biochemistry and Polymer Science**, *Royal Society of Chemistry*, Cambridge, 1992.
37 P. Munk, in **Modern Methods of Polymer Characterization**, H. G. Barth, J. W. Mays (eds.), *J. Wiley & Sons, Inc.*, New York, 1991, pp. 271-312.
38 E.T.Adams, in **Encyclopedia of Polymer Science and Engineering**, 2nd Ed., J. Kroschwitz (Ed.), *J. Wiley & Sons, Inc.*, New York, 1987, Vol. 10, pp. 636-652.
39 M. P. Tombs, A. R. Peacock, **The Osmotic Pressure of Macromolecules**, *Oxford University Press*, London, 1974.
40 J. W. Mays, N. Hadjichristidis, in **Modern Methods of Polymer Characterization**, H. G. Barth, J. W. Mays (eds.), *J.Wiley & Sons, Inc.*, New York, 1991, pp. 201-226.
41 H. Fujita, **Polymer Solutions**, *Elsevier*, Amsterdam, 1990.
42 K. Kamide, **Thermodynamics of Polymer Solutions**, *Elsevier*, Amsterdam, 1990
43 Y. Einaga, *Progr. Polym. Sci.*, **19**, 1 (1994).
44 R. Koningsveld, *Adv. Colloid Interfacial Sci.*, **2**, 151 (1968).
45 B. A. Wolf, *Adv. Polym. Sci.*, **10**, 109 (1972).
46 P. J. Flory, **Principles of Polymer Chemistry**, *Cornell University Press*, Ithaca, 1953.
47 M. L. Huggins, **Physical Chemistry of High Polymers**, *J.Wiley & Sons., Inc.*, New York, 1958.
48 H. Tompa, **Polymer Solutions**, *Butterworth*, London, 1956.
49 J. M. Prausnitz, R. N. Lichtenthaler, E. G. de Azevedo, **Molecular Thermodynamics of Fluid Phase Equilibria**, 3rd Ed., *Prentice Hall*, Upper Saddle River, New Jersey, 1999.
50 R. M. Masegosa, M. G. Prolongo, A. Horta, *Macromolecules*, **19**, 1478 (1986).
51 R. Koningsveld, L. A. Kleintjens, *Macromolecules*, **4**, 637 (1971).
52 M. T. Raetzsch, H. Kehlen, in **Encyclopedia of Polymer Science and Engineering**, 2nd Ed., J. Kroschwitz (Ed.), *J. Wiley & Sons, Inc.*, New York, 1989, Vol. 15, pp. 481-491.

53 M. T. Raetzsch, H. Kehlen, *Progr. Polym. Sci.*, **14**, 1 (1989).
54 M. T. Raetzsch, Ch. Wohlfarth, *Adv. Polym. Sci.*, **98**, 49 (1990).
55 A. G. Williamson, in **Experimental Thermodynamics, Vol.II, Experimental Thermodynamics of non-reacting Fluids**, B. Le Neindre, B. Vodar (eds.), *Butterworth*, London, pp. 749-786.
56 K. Schmoll, E. Jenckel, *Ber. Bunsenges. Phys. Chem.*, **60**, 756 (1956).
57 K. Schotsch, B. A. Wolf, *Makromol. Chem.*, **185**, 2161 (1984).
58 E. Killmann, F. Cordt, F. Moeller, *Makromol. Chem.*, **191**, 2929 (1990).
59 J. H. Baxendale, E. V. Enustun, J. Stern, *Phil. Trans. Roy. Soc. London, A*, **243**, 169 (1951).
60 M. L. McGlashan, A. G. Williamson, *Trans. Faraday Soc.*, **57**, 588 (1961).
61 M. L. McGlashan, K. W. Morcom, A. G. Williamson, *Trans. Faraday Soc.*, **57**, 581, (1961).
62 H. Takenaka, *J. Polym. Sci.*, **24**, 321 (1957).
63 R. J. Kokes, A. R. Dipietro, F. A. Long, *J. Amer. Chem. Soc.*, **75**, 6319 (1953).
64 P. W. Allen, D. H. Everett, M. F. Penney, *Proc. Roy. Soc. London A*, **212**, 149 (1952).
65 D. H. Everett, M. F. Penney, *Proc. Roy. Soc. London A*, **212**, 164 (1952).
66 C. A. Haynes, R. A. Beynon, R. S. King, H. W. Blanch, J. M. Praunsitz, *J. Phys. Chem.*, **93**, 5612 (1989).
67 W. R. Krigbaum, D. O. Geymer, *J. Amer. Chem. Soc.*, **81**, 1859 (1959).
68 M. L. Lakhanpal, B. E. Conway, *J. Polym. Sci.*, **46**, 75 (1960).
69 K. Ueberreiter, W. Bruns, *Ber. Bunsenges. Phys. Chemie*, **68**, 541 (1964).
70 R. S. Jessup, *J. Res. Natl. Bur. Stand.*, **60**, 47 (1958).
71 Van Tam Bui, J. Leonard, *Polym. J.*, **21**, 185 (1989).
72 C. E. H. Bawn, R. F. J. Freeman, A. R. Kamaliddin, *Trans. Faraday Soc.*, **46**, 677 (1950).
73 G. Gee, L. R. J. Treloar, *Trans. Faraday Soc.*, **38**, 147 (1942).
74 C. Booth, G. Gee, G. Holden, G. R. Williamson, *Polymer*, **5**, 343 (1964).
75 G. N. Malcolm, C. E. Baird, G. R. Bruce, K. G. Cheyne, R. W. Kershaw, M. C. Pratt, *J. Polym. Sci., Pt. A-2*, **7**, 1495 (1969).
76 L.R. Ochs, M. Kabri-Badr, H. Cabezas, *AIChE-J.*, **36**, 1908 (1990).
77 C. Grossmann, R. Tintinger, G. Maurer, *Fluid Phase Equil.*, **106**, 111 (1995).
78 Z. Adamcova, *Sci. Pap. Prag. Inst. Chem. Technol.*, **N2**, 63 (1976).
79 Z. Adamcova, *Adv. Chem.*, **155**, 361 (1976).
80 K. Kubo, K. Ogino, *Polymer*, **16**, 629 (1975).
81 C. G. Panayiotou, J. H. Vera, *Polym. J.*, **16**, 89 (1984).
82 M. T. Raetzsch, G. Illig, Ch. Wohlfarth, *Acta Polymerica*, **33**, 89 (1982).
83 P. J. T. Tait, A. M. Abushihada, *Polymer*, **18**, 810 (1977).
84 J. H. Van der Waals, J. J. Hermans, *Rec. Trav. Chim. Pays-Bas*, **69**, 971 (1950).
85 D. C. Bonner, J. M. Prausnitz, *J. Polym. Sci., Polym. Phys. Ed.*, **12**, 51 (1974).
86 R. B. Gupta, J. M. Prausnitz, *J. Chem. Eng. Data*, **40**, 784 (1995).
87 S. Behme, G. Sadowski, W. Arlt, *Chem.-Ing.-Techn.*, **67**, 757 (1995).
88 A. J. Ashworth, D. H. Everett, *Trans. Faraday Soc.*, **56**, 1609 (1960)
89 J. S. Vrentas, J. L. Duda, S. T. Hsieh, *Ind. Eng. Chem., Process Des. Dev.*, **22**, 326 (1983).
90 Y. Iwai, Y. Arai, *J. Chem. Eng. Japan*, **22**, 155 (1989).
91 J. W. McBain, A. M. Bakr, *J. Amer. Chem. Soc.*, **48**, 690 (1926).
92 S. Hwang, J. Kim, K.-P. Yoo, *J. Chem. Eng. Data*, **43**, 614 (1998).
93 J. Kim, K. C. Joung, S. Hwang, W. Huh, C. S. Lee, K.-P. Yoo, *Korean J. Chem. Eng.*, **15**, 199 (1998).
94 Y. C. Bae, J. J. Shim, D. S. Soane, J. M. Prausnitz, *J. Appl. Polym. Sci.*, **47**, 1193 (1993).
95 A. J. Ashworth, *J. Chem. Soc., Faraday Trans. I*, **69**, 459 (1973).
96 A. J. Ashworth, G. J. Price, *Thermochim. Acta*, **82**, 161 (1984).
97 S. Saeki, J. C. Holste, D. C. Bonner, *J. Polym. Sci., Polym. Phys. Ed.*, **20**, 793, 805 (1982).
98 S. Saeki, J. C. Holste, D. C. Bonner, *J. Polym. Sci., Polym. Phys. Ed.*, **19**, 307 (1981).
99 S. Saeki, J. Holste, D. C. Bonner, *J. Polym. Sci., Polym. Phys. Ed.*, **21**, 2049 (1983)
100 H. C. Wong, S. W. Campbell, V. R. Bhetnanabotla, *Fluid Phase Equil.*, **139**, 371 (1997).
101 S. P. V. N. Mikkilineni, D.A. Tree, M. S. High, *J. Chem. Eng. Data*, **40**, 750 (1995).
102 O. Smidsrod, J. E. Guillet, *Macromolecules*, **2**, 272 (1969).
103 Z. Y. Al-Saigh, P. Munk, *Macromolecules*, **17**, 803 (1984).
104 P. Munk, P. Hattam, Q. Du, *J. Appl. Polym. Sci., Appl. Polym. Symp.*, **43**, 373 (1989).
105 D. Glindemann, Dissertation, TH Leuna-Merseburg, 1979.
106 N. F. Brockmeier, R. W. McCoy, J. A. Meyer, *Macromolecules*, **5**, 130, 464 (1972).
107 N. F. Brockmeier, R. E. Carlson, R. W. McCoy, *AIChE-J.*, **19**, 1133 (1973).
108 J. R. Conder, J. R. Purnell, *Trans. Faraday Soc.*, **64**, 1505, 3100 (1968).

109 J. R. Conder, J. R. Purnell, *Trans. Faraday Soc.*, **65**, 824, 839 (1969).
110 G. J. Price, J. E. Guillet, *J. Macromol. Sci.-Chem.*, **A23**, 1487 (1986).
111 D. C. Bonner, N. F. Brockmeier, *Ind. Eng. Chem., Process Des. Developm.*, **16**, 180 (1977).
112 S. Dincer, D. C. Bonner, R. A. Elefritz, *Ind. Eng. Chem., Fundam.*, **18**, 54 (1979).
113 W. A. Ruff, C. J. Glover, A. T. Watson, *AIChE-J.*, **32**, 1948, 1954 (1986).
114 L. L. Joffrion, C. J. Glover, *Macromolecules*, **19**, 1710 (1986).
115 T. K. Tsotsis, C. Turchi, C. J. Glover, *Macromolecules*, **20**, 2445 (1987).
116 R. P. Danner, F. Tihminlioglu, R. K. Surana, J. L. Duda, *Fluid Phase Equil.*, **148**, 171 (1998).
117 J. R. Conder, C. L. Young, **Physicochemical Measurements by Gas Chromatography,** *J.Wiley & Sons, Inc.*, New York, 1979.
118 C. Barth, R. Horst, B. A. Wolf, *J. Chem. Thermodyn.*, **30**, 641 (1998).
119 H.-M. Petri, N. Schuld, B. A. Wolf, *Macromolecules*, **28**, 4975 (1995).
120 B. Kolb, L. S. Ettre, **Static Headspace Gas Chromatography - Theory and Practice**, *Wiley-VCH*, Weinheim 1997.
121 H. Wang, K. Wang, H. Liu, Y. Hu, *J. East China Univ. Sci. Technol.*, **23**, 614 (1997).
122 T. Katayama, K. Matsumura, Y. Urahama, *Kagaku Kogaku*, **35**, 1012 (1971).
123 K. Matsumura, T. Katayama, *Kagaku Kogaku*, **38**, 388 (1974).
124 A. V. Hill, *Proc. Roy. Soc. London, Ser. A*, **127**, 9 (1930).
125 E. Regener, Ch. Wohlfarth, *Wiss. Zeitschr. TH Leuna-Merseburg*, **30**, 211 (1988).
126 J. Gaube, A. Pfennig, M. Stumpf, *J.Chem. Eng. Data,* **38**, 163 (1993).
127 J. Gaube, A. Pfennig, M. Stumpf, *Fluid Phase Equil.*, **83**, 365 (1993).
128 A. Eliassi, H. Modarress, G. A. Mansoori, *J. Chem. Eng. Data,* **35**, 52 (1999).
129 V. Majer, V. Svoboda, **Enthalpies of Vaporization of Organic Compounds**, *Blackwell Sci. Publ.*, Oxford, 1985.
130 T. E. Daubert, R. P. Danner, **Physical and Thermodynamic Properties of Pure Chemicals. Data Compilation**, *Hemisphere Publ.Corp.*, New York, 1989 and several Supplements up to 1999, data are online available from DIPPR database.
131 B. H. Bersted, *J. Appl. Polym. Sci.*, **17**, 1415 (1973).
132 B. H. Bersted, *J. Appl. Polym. Sci.*, **18**, 2399 (1974).
133 R. V. Figini, *Makromol. Chem.*, **181**, 2049 (1980).
134 R. V. Figini, M. Marx-Figini, *Makromol. Chem.*, **182**, 437 (1980).
135 M. Marx-Figini, R. V. Figini, *Makromol. Chem.*, **181**, 2401 (1980).
136 P. J. T. Tait, A. M. Abushihada, *Polymer*, **18**, 810 (1977).
137 A. J. Ashworth, C.-F. Chien, D. L. Furio, D. M. Hooker, M. M. Kopecni, R. J. Laub, G. J. Price, *Macromolecules*, **17**, 1090 (1984).
138 R. N. Lichtenthaler, D. D. Liu, J. M. Prausnitz, *Macromolecules*, **7**, 565 (1974).
139 I. Noda, N. Kato, T. Kitano, M. Nagasawa, *Macromolecules*, **14**, 668 (1981).
140 H. Vink, Europ. *Polym. J.*, **7**, 1411 (1971) and **10**, 149 (1974).
141 A. Nakajima, F. Hamada, K. Yasue, K. Fujisawa, T. Shiomi, *Makromol. Chem.*, **175**, 197 (1974).
142 B. E. Eichinger, P. J. Flory, *Trans. Faraday Soc.*, **64**, 2035, 2053, 2061, 2066 (1968).
143 H. Hoecker, P. J. Flory, *Trans. Faraday Soc.*, **67**, 2270 (1971).
144 P. J. Flory, H. Daoust, *J. Polym. Sci.*, **25**, 429 (1957).
145 B. H. Zimm, I. Myerson, *J. Amer. Chem. Soc.*, **68**, 911 (1946).
146 W. H. Stockmayer, E. F. Casassa, *J. Chem. Phys.*, **20**, 1560 (1952).
147 G. Strobl, **The Physics of Polymers**, *Springer Vlg*., Berlin, 1996.
148 B. H. Zimm, *J. Chem. Phys.*, **16**, 1093 (1948).
149 B. A. Wolf, H.-J. Adams, *J. Chem. Phys.*, **75**, 4121 (1981).
150 M. Lechner, G. V. Schulz, Europ. *Polym. J.*, **6**, 945 (1970).
151 Y. Nakamura, T. Norisuye, A. Teramoto, *J. Polym. Sci., Polym. Phys. Ed.*, **29**, 153 (1991).
152 Th. G. Scholte, *Eur. Polym. J.*, **6**, 1063 (1970).
153 H. Hasse, H.-P. Kany, R. Tintinger, G. Maurer, *Macromolecules*, **28**, 3540 (1995).
154 H.-P. Kany, H. Hasse, G. Maurer, *J. Chem. Eng. Data*, **35**, 230 (1999).
155 Th. G. Scholte, *J. Polym. Sci., Pt. A-2*, **9**, 1553 (1971).
156 Th. G. Scholte, *J. Polym. Sci., Pt. C*, **39**, 281 (1972).
157 K. W. Derham, J. Goldsborough, M. Gordon, *Pure Appl. Chem.*, **38**, 97 (1974).
158 M. Gordon, P. Irvine, J. W. Kennedy, *J. Polym. Sci., Pt. C*, **61**, 199 (1977).
159 M. Gordon, *NATO ASI Ser., Ser. E,* **89**, 429 (1985).

160 H. Galina, M. Gordon, B. W. Ready, L. A. Kleintjens, in **Polymer Solutions**, W. C. Forsman (Ed.), *Plenum Press*, New York, 1986, pp. 267-298.
161 A. Guinier, G. Fournet, **Small Angle Scattering of X-rays**, *J.Wiley & Sons, Inc.*, New York, 1955.
162 S. H. Chen, S. Yip (eds.), **Spectroscopy in Biology and Chemistry, Neutron, X-ray and Laser**, *Academic Press*, New York, 1974.
163 O. Glatter, O. Kratky (eds.), **Small-Angle X-ray Scattering**, *Academic Press*, London, 1982.
164 R. G. Kirste, W. Wunderlich, *Z. Phys. Chem. N.F.*, **58**, 133 (1968).
165 R. G. Kirste, G. Wild, *Makromol.Chem.*, **121**, 174 (1969).
166 G. Meyerhoff, U. Moritz, R. G. Kirste, W. Heitz, *Europ. Polym. J.*, **7**, 933 (1971).
167 K. Okano, T. Ichimura, K. Kurita, E. Wada, *Polymer*, **28**, 693 (1987).
168 T. Ichimura, K. Okano, K. Kurita, E. Wada, *Polymer*, **28**, 1573 (1987).
169 M. Takada, K. Okano, K. Kurita, *Polym. J.*, **26**, 113 (1994).
170 F. Horkay, A.-M. Hecht, H. B. Stanley, E. Geissler, *Eur. Polym. J.*, **30**, 215 (1994).
171 J. S. Higgins, A. Macconachie, in **Polymer Solutions**, W. C. Forsman (Ed.), *Plenum Press*, New York, 1986, pp. 183-238.
172 B. Hammouda, *Adv. Polym. Sci.*, **106**, 87 (1993).
173 N. Vennemann, M. D. Lechner, R. C. Oberthuer, *Polymer*, **28**, 1738 (1987).
174 A. M. Pedley, J. S. Higgins, D. G. Pfeiffer, A. R. Rennie, *Macromolecules*, **23**, 2492 (1990).
175 B. J. Berne, R. Pecora, **Dynamic Light Scattering**, *J.Wiley & Sons, Inc.*, New York, 1976.
176 Th. G. Scholte, *J. Polym. Sci., Pt. A-2*, **8**, 841 (1970).
177 B. J. Rietveld, Th. G. Scholte, J. P. L. Pijpers, *Brit. Polym. J.*, **4**, 109 (1972).
178 H.-G. Elias, H. Lys, *Makromol.Chem.*, **92**, 1 (1966).
179 T. Hirao, A. Teramoto, T. Sato, T. Norisuye, T. Masuda, T. Higashimura, *Polym. J.*, **23**, 925 (1991).
180 Y. Doucet, in **Experimental Thermodynamics, Vol.II, Experimental Thermodynamics of non-reacting Fluids,** B. Le Neindre, B. Vodar (eds.), *Butterworth*, London, pp. 835-900.
181 T. Kawai, *J. Polym. Sci.*, **32**, 425 (1958).
182 Y. Hoei, Y. Ikeda, M. Sasaki, *J. Phys. Chem. B*, **103**, 5353 (1999).
183 R. Koningsveld, in **Polymer Science**, E. D. Jenkins (Ed.), *North-Holland Publ.Comp.*, Amsterdam, 1972, pp. 1047-1134.
184 A. Imre, W. A. Van Hook, *J. Polym. Sci., Pt. B, Polym. Sci.*, **34**, 751 (1996) and *J. Phys. Chem. Ref. Data*, **25**, 637 (1996).
185 O. Pfohl, T. Hino, J. M. Prausnitz, *Polymer*, **36**, 2065 (1995).
186 N. Kuwahara, D. V. Fenby, M. Tamsuy, B. Chu: *J. Chem. Phys.*, **55**, 1140 (1971).
187 R. Koningsveld, A. J. Staverman, *J. Polym. Sci., Pt. A-2*, **6**, 349 (1968).
188 B. A. Wolf, M. C. Sezen, *Macromolecules*, **10**, 1010 (1977).
189 G. F. Alfrey, W. G. Schneider, *Discuss. Faraday Soc.*, **15**, 218 (1953).
190 S. C. Greer, D. T. Jacobs, *J. Phys. Chem.*, **84**, 2888 (1980).
191 D. T. Jacobs, S. C. Greer, *Phys. Rev. A*, **24**, 2075 (1981).
192 B. Muessig, H. Wochnowski, *Angew. Makromol. Chem.*, **104**, 203 (1982).
193 G. D. Rehage, D. Moeller, D. Ernst, *Makromol. Chem.*, **88**, 232 (1965).
194 Th. G. Scholte, R. Koningsveld, *Kolloid Z. Z. Polym.*, **218**, 58 (1968).
195 B. J. Rietfeld, *Brit. Polym. J.*, **6**, 181 (1974).
196 M. Gordon, L. A. Kleintjens, B. W. Ready, J. A. Torkington, *Brit. Polym. J.*, **10**, 170 (1978).
197 R. Koningsveld, L. A. Kleintjens, M. H. Onclin, *J. Macromol. Sci. Phys. B*, **18**, 357 (1980).
198 P. G. de Gennes, **Scaling Concepts in Polymer Physics**, *Cornell Univ. Press*, Ithaca, 1979.
199 R. Koningsveld, A. J. Staverman, *J. Polym. Sci., Pt. A-2,* **6**, 325 (1968).
200 B. A. Wolf, *Makromol. Chem.*, **128**, 284 (1969).
201 L. R. G. Treloar, **The Physics of Rubber Elasticity**, 3rd Ed., *Clarendon Press*, Oxford, 1975.
202 K. Dusek, W. Prins, *Adv. Polym. Sci.*, **6**, 1 (1969).
203 G. Heinrich, E. Straube, G. Helmis, *Adv. Polym. Sci.*, **84**, 33 (1988).
204 K. Dusek (Ed.), Responsive Gels I and II, *Adv. Polym. Sci.*, **109** and **110** (1996).
205 G. Rehage, *Ber. Bunsenges. Phys. Chem.*, **81**, 969 (1977).
206 K. F. Arndt, J. Schreck, *Acta Polymerica*, **36**, 56 (1985), **37**, 500 (1986).
207 K. F. Arndt, I. Hahn, *Acta Polymerica*, **39**, 560 (1988).
208 R. Sanetra, B. N. Kolarz, A. Wlochowicz, *Polimery*, **34**, 490 (1989).
209 D. G. Gray, J. E. Guillet, *Macromolecules*, **5**, 316 (1972).
210 M. Galin, *Macromolecules*, **10**, 1239 (1977).
211 A. M. Rijke, W. Prins, *J. Polym. Sci.*, **59**, 171 (1962).

212 J. Schwarz, W. Borchard, G. Rehage, *Kolloid-Z. Z. Polym.*, **244**, 193 (1971).
213 B. J. Teitelbaum, A. E. Garwin, *Vysokomol. Soedin., Ser. A*, **10**, 1684 (1968).
214 J. J. Hermans, *J. Polym. Sci.*, **59**, 191 (1962).
215 H. M. James, E. Guth, *J. Chem. Phys.*, **21**, 1039 (1953).
216 S. F. Edwards, E. K. Freed, *J. Phys. C*, **3**, 739, 750 (1970).
217 G. Rehage, *Kolloid-Z. Z. Polym.*, **199**, 1 (1964).
218 W. Borchard, Dissertation, RWTH Aachen, 1966.
219 W. Borchard, Progr. *Colloid Polym. Sci.*, **57**, 39 (1975).
220 W. Borchard, Progr. *Colloid Polym. Sci.*, **86**, 84 (1991).
221 W. Borchard, H. Coelfen, *Makromol. Chem., Macromol. Symp.*, **61**, 143 (1992).
222 H. Yamakawa, **Modern Theory of Polymer Solutions**, *Harper & Row*, New York, 1971.
223 J. A. Barker, *J. Chem. Phys.*, **20**, 1256 (1952).
224 H. Tompa, T*rans. Faraday Soc.*, **48**, 363 (1952).
225 R. N. Lichtenthaler, D. S. Abrams, J. M. Prausnitz, *Can. J. Chem.*, **53**, 3071 (1973).
226 E. A. Guggenheim, *Proc. Roy. Soc. London, A,* **183**, 203 (1944).
227 E. A. Guggenheim, **Mixtures**, *Clarendon Press*, Oxford, 1952.
228 A. J. Staverman, *Recl. Trav. Chim. Pays-Bas*, **69**, 163 (1950).
229 S. G. Sayegh, J. H. Vera, *Chem. Eng. J.*, **19**, 1 (1980).
230 K. F. Freed, *J. Phys. A, Math. Gen.*, **18**, 871 (1985).
231 M. G. Bawendi, K. F. Freed, *J. Chem. Phys.*, **87**, 5534 (1987), **88**, 2741 (1988).
232 W. G. Madden, A. I. Pesci, K. F. Freed, *Macromolecules*, **23**, 1181 (1990).
233 G. Scatchard, *Chem. Rev.*, **44**, 7 (1949).
234 J. H. Hildebrand, R. L. Scott, **The Solubility of Nonelectrolytes**, 3rd Ed., *Reinhold Publ. Corp.*, New York, 1950.
235 D. W. Van Krevelen, **Properties of Polymers**, 3rd Ed., *Elsevier*, Amsterdam, 1990.
236 P. J. Flory, R. A. Orwoll, A. Vrij, *J. Amer. Chem. Soc.*, **86**, 3507, 3515 (1964).
237 P. J. Flory, *J. Amer. Chem. Soc.*, **87**, 1833 (1965).
238 P. J. Flory, R. A. Orwoll, *J. Amer. Chem. Soc.*, **89**, 6814, 6822 (1964).
239 P. J. Flory, *Discuss. Faraday Soc.*, **49**, 7 (1970).
240 P. I. Freeman, J. S. Rowlinson, *Polymer*, **1**, 20 (1960).
241 A. H. Liddell, F. L. Swinton, *Discuss. Faraday Soc.*, **49**, 215 (1970).
242 I. Prigogine, **The Molecular Theory of Solutions**, *North-Holland Publ.*, Amsterdam, 1957.
243 D. Patterson, *Macromolecules*, **2**, 672 (1969).
244 D. Patterson, G. Delmas, *Discuss. Faraday Soc.*, **49**, 98 (1970).
245 R. Simha, T. Somcynsky, *Macromolecules*, **2**, 342 (1969).
246 I. C. Sanchez, R. H. Lacombe, *J. Phys. Chem.*, **80**, 2352, 2568 (1976).
247 I. C. Sanchez, R. H. Lacombe, *J. Polym. Sci., Polym. Lett. Ed.*, **15**, 71 (1977).
248 I. C. Sanchez, R. H. Lacombe, *Macromolecules*, **11**, 1145 (1978).
249 G. T. Dee, D. J. Walsh, *Macromolecules*, **21**, 811, 815 (1988).
250 M. D. Donohue, J. M. Prausnitz, *AIChE-J.*, **24**, 849 (1978).
251 C. H. Chien, R. A. Greenkorn, K. C. Chao, *AIChE-J.* **29**, 560 (1983).
252 K. G. Honnell, C. K. Hall, *J. Chem. Phys.*, **90**, 1841 (1989).
253 H. S. Gulati, J. M. Wichert, C. K. Hall, *J. Chem. Phys.*, **104**, 5220 (1996).
254 H. S. Gulati, C. K. Hall, *J. Chem. Phys.*, **108**, 5220 (1998).
255 A. Yethiraj, C. K. Hall, *J. Chem. Phys.*, **95**, 1999 (1991).
256 W. G. Chapman, G. Jackson, K. E. Gubbins, *Mol. Phys.*, **65**, 1057 (1988).
257 W. G. Chapman, K. E. Gubbins, G. Jackson, M. Radosz, *Fluid Phase Equil.*, **52**, 31 (1989).
258 W. G. Chapman, K. E. Gubbins, G. Jackson, M. Radosz, *Ind. Eng. Chem. Res.*, **29**, 1709 (1990).
259 Y. Song, S. M. Lambert, J. M. Prausnitz, *Macromolecules*, **27**, 441 (1994).
260 Y. Song, S. M. Lambert, J. M. Prausnitz, *Ind. Eng. Chem. Res.*, **33**, 1047 (1994).
261 S. M. Lambert, Y. Song, J. M. Prausnitz, in **Equations of State for Fluids and Fluid Mixtures**, J. V. Sengers et al. (eds.), *Elsevier*, Amsterdam, 2000, in press.
262 P. A. Rodgers, *J. Appl. Polym. Sci.*, **48**, 1061 (1993).
263 J. Cho, I. C. Sanchez, in **Polymer Handbook**, J. Brandrup, E. H. Immergut, E. A. Grulke (eds.), 4th ed., *J. Wiley & Sons*., Inc., New York, 1999, pp. VI/591-601.
264 A. Bondi, **Physical Properties of Molecular Crystals, Liquids and Glasses**, *John Wiley & Sons*, New York, 1968.
265 D. C. Bonner, J. M. Prausnitz, *AIChE-J.*, **19**, 943 (1973).

266 M. T. Raetzsch, M. Opel, Ch. Wohlfarth, *Acta Polymerica*, **31**, 217 (1980).
267 K. Sugamiya, *Makromol. Chem.*, **178**, 565 (1977).
268 U. Brandani, *Macromolecules*, **12**, 883 (1979).
269 H.-W. Kammer, T. Inoue, T. Ougizawa, *Polymer*, **30**, 888 (1989).
270 T. Shiomi, K. Fujisawa, F. Hamada, A. Nakajima, *J. Chem. Soc. Faraday Trans II*, **76**, 895 (1980).
271 F. Hamada, T. Shiomi, K. Fujisawa, A. Nakajima, *Macromolecules*, **13**, 279 (1980).
272 C. G. Panayiotou, *J. Chem. Soc. Faraday Trans. II*, **80**, 1435 (1984).
273 Y. L. Cheng, D. C. Bonner, *J. Polym. Sci., Polym. Phys. Ed.*, **15**, 593 (1977), **16**, 319 (1978).
274 J. Pouchly, D. Patterson, *Macromolecules*, **9**, 574 (1976).
275 I. C. Sanchez, C. G. Panayiotou, in **Models for Thermodynamics and Phase Equilibrium Calculations**, S. I. Sandler (Ed.), *Marcel Dekker*, New York, 1994, pp. 187-285.
276 C. G. Panayiotou, *Makromol. Chem.*, **187**, 2867 (1986).
277 C. G. Panayiotou, J. H. Vera, *Polym. J.*, **14**, 681 (1982).
278 C. G. Panayiotou, J. H. Vera, *Can. J. Chem. Eng.*, **59**, 501 (1981).
279 C. G. Panayiotou, *Macromolecules*, **20**, 861 (1987).
280 I. C. Sanchez, P. A. Rodgers, *Pure Appl. Chem.*, **62**, 2107 (1990).
281 E. Nies, A. Stroeks, *Macromolecules*, **23**, 4088, 4092 (1990).
282 H. Xie, E. Nies, *Macromolecules*, **26**, 1689 (1993).
283 H. Xie, R. Simha, *Polym. Int.*, **44**, 348 (1997).
284 A. Fredenslund, J. Gmehling, P. Rasmussen, **Vapor-Liquid Equilibria Using UNIFAC**, *Elsevier Sci. Publ.*, New York, 1977.
285 T. Oishi, J. M. Prausnitz, *Ind. Eng. Chem. Process Des. Dev.*, **17**, 333 (1978).
286 S. Skjold-Jorgensen, B. Kolbe, J. Gmehling, P. Rasmussen, *Ind. Eng. Chem. Process Des. Dev.*, **18**, 714 (1979).
287 J. Gmehling, P. Rasmussen, A. Fredenslund, *Ind. Eng. Chem. Process Des. Dev.*, **21**, 118 (1982).
288 E. A. Macedo, U. Weidlich, J. Gmehling, P. Rasmussen, *Ind. Eng. Chem. Process Des. Dev.*, **22**, 676 (1983).
289 D. Tiegs, J. Gmehling, P. Rasmussen, A. Fredenslund, *Ind. Eng. Chem. Res.*, **26**, 159 (1987).
290 M. T. Raetzsch, D. Glindemann, *Acta Polymerica*, **30**, 57 (1979).
291 I. Kikic, P. Alessi, P.Rasmussen, A. Fredenslund, *Can. J. Chem. Eng.*, **58**, 253 (1980).
292 M. Gottlieb, M. Herskowitz, *Ind. Eng. Chem. Process Des. Dev.*, **21**, 536 (1982).
293 M. Gottlieb, M. Herskowitz, *Macromolecules*, **14**, 1468 (1981).
294 Y. Iwai, M. Ohzono, Y. Arai, *Chem. Eng. Commun.*, **34**, 225 (1985).
295 Y. Iwai, Y. Arai, *J. Chem. Eng. Japan*, **22**, 155 (1989).
296 A. A. Patwardhan, L. A. Belfiore, *J. Polym. Sci. Pt. B Polym. Phys.*, **24**, 2473 (1986).
297 G. J. Price, A. J. Ashworth, *Polymer*, **28**, 2105 (1987).
298 M. T. Raetzsch, D. Glindemann, E. Hamann, *Acta Polymerica*, **31**, 377 (1980).
299 E. L. Sorensen, W. Hao, P. Alessi, *Fluid Phase Equil.*, **56**, 249 (1990).
300 M. Herskowitz, M. Gottlieb, *J. Chem. Eng. Data*, **30**, 233 (1985).
301 J. R. Fried, J. S. Jiang, E. Yeh, *Comput. Polym. Sci.*, **2**, 95 (1992).
302 H. S. Elbro, A. Fredenslund, P. A. Rasmussen, *Macromolecules*, **23**, 4707 (1990).
303 G. M. Kontogeorgis, A. Fredenslund, D. P. Tassios, *Ind. Eng. Chem. Res.*, **32**, 362 (1993).
304 M. J. Misovich, E. A. Grulke, R. F. Blanks, *Ind. Eng. Chem. Process Des. Dev.*, **24**, 1036 (1985).
305 K. Tochigi, S. Kurita, M. Ohashi, K. Kojima, *Kagaku Kogaku Ronbunshu*, **23**, 720 (1997).
306 J. S. Choi, K. Tochigi, K. Kojima, *Fluid Phase Equil.*, **111**, 143 (1995).
307 B. G. Choi, J. S. Choi, K. Tochigi, K. Kojima, *J. Chem. Eng. Japan*, **29**, 217 (1996).
308 J. Holten-Andersen, A. Fredenslund, P. Rasmussen, G. Carvoli, *Fluid Phase Equil.*, **29**, 357 (1986).
309 J. Holten-Andersen, A. Fredenslund, P. Rasmussen, *Ind. Eng. Chem. Res.*, **26**, 1382 (1987).
310 F. Chen, A. Fredenslund, P. Rasmussen, *Ind. Eng. Chem. Res.*, **29**, 875 (1990).
311 M. S. High, R. P. Danner, *Fluid Phase Equil.*, **53**, 323 (1989).
312 M. S. High, R. P. Danner, *AIChE-J.*, **36**, 1625 (1990).
313 M. S. High, R. P. Danner, *Fluid Phase Equil.*, **55**, 1 (1990).
314 K. Tochigi, K. Kojima, T. Sako, *Fluid Phase Equil.*, **117**, 55 (1996).
315 B.-C. Lee, R. P. Danner, *Fluid Phase Equil.*, **117**, 33 (1996).
316 A. Bertucco, C. Mio, *Fluid Phase Equil.*, **117**, 18 (1996).
317 W. Wang, X. Liu, C. Zhong, C. H. Twu, J. E. Coon, *Fluid Phase Equil.*, **144**, 23 (1998).
318 J. M. Smith, H. C. Van Ness, M. M. Abbott, **Introduction to Chemical Engineering Thermodynamics**, 5th Ed., *McGraw-Hill*, Singapore, 1996.
319 S. M. Walas, **Phase equilibria in chemical engineering**, *Butterworth*, Boston, 1985.

320 S. Beret, J. M. Prausnitz, *AIChE-J.*, **21**, 1123 (1975).
321 S. Beret, J. M. Prausnitz, *Macromolecules*, **8**, 878 (1975).
322 M. D. Donohue, J. M. Prausnitz, *AIChE-J.*, **24**, 849 (1978).
323 M. S. Wertheim, *J.Chem.Phys.*, **87**, 7323 (1987).
324 N. F. Carnahan, K. E. Starling, *J.Chem.Phys.*, **51**, 635 (1969).
325 B. J. Alder, D. A. Young, M. A. Mark, *J.Chem.Phys.*, **56**, 3013 (1972).
326 M. D. Donohue, P. Vimalchand, *Fluid Phase Equil.*, **40**, 185 (1988).
327 R. L. Cotterman, B. J. Schwarz, J. M. Prausnitz, *AIChE-J.*, **32**, 1787 (1986).
328 D. D. Liu, J. M. Prausnitz, *Ind. Eng. Chem. Process Des. Dev.*, **19**, 205 (1980).
329 M. Ohzono, Y. Iwai, Y. Arai, *J. Chem. Eng. Japan*, **17**, 550 (1984).
330 S. Itsuno, Y. Iwai, Y. Arai, *Kagaku Kogaku Ronbunshu*, **12**, 349 (1986).
331 Y. Iwai, Y. Arai, *J. Japan Petrol. Inst.*, **34**, 416 (1991).
332 C. H. Chien, R. A. Greenkorn, K. C. Chao, *AIChE-J.*, **29**, 560 (1983).
333 T. Boublik, I. Nezbeda, *Chem. Phys. Lett.*, **46**, 315 (1977).
334 H. Masuoka, K. C. Chao, *Ind. Eng. Chem. Fundam.*, **23**, 24 (1984).
335 M. T. Raetzsch, E. Regener, Ch. Wohlfarth, *Acta Polymerica*, **37**, 441 (1986).
336 E. Regener, Ch. Wohlfarth, M. T. Raetzsch, *Acta Polymerica*, **37**, 499, 618 (1986).
337 E. Regener, Ch. Wohlfarth, M. T. Raetzsch, S. Hoering, *Acta Polymerica*, **39**, 618 (1988).
338 Ch. Wohlfarth, E. Regener, *Plaste & Kautschuk*, **35**, 252 (1988).
339 Ch. Wohlfarth, *Plaste & Kautschuk*, **37**, 186 (1990).
340 Ch. Wohlfarth, B. Zech, *Makromol. Chem.*, **193**, 2433 (1992).
341 U. Finck, T. Heuer, Ch. Wohlfarth, *Ber. Bunsenges. Phys. Chem.*, **96**, 179 (1992).
342 Ch. Wohlfarth, U. Finck, R. Schultz, T. Heuer, *Angew. Makromol. Chem.*, **198**, 91 (1992).
343 Ch. Wohlfarth, *Acta Polymerica*, **43**, 295 (1992).
344 Ch. Wohlfarth, *Plaste & Kautschuk*, **39**, 367 (1992).
345 Ch. Wohlfarth, *J. Appl. Polym. Sci.*, **48**, 1923 (1993).
346 Ch. Wohlfarth, *Plaste & Kautschuk*, **40**, 272 (1993).
347 Ch. Wohlfarth, *Plaste & Kautschuk*, **41**, 163 (1994).
348 Ch. Wohlfarth, *Macromol. Chem. Phys.*, **198**, 2689 (1997).
349 J. C. Pults, R. A. Greenkorn, K. C. Chao, *Chem. Eng. Sci.*, **44**, 2553 (1989).
350 J. C. Pults, R. A. Greenkorn, K. C. Chao, *Fluid Phase Equil.*, **51**, 147 (1989).
351 R. Sy-Siong-Kiao, J. M. Caruthers, K. C. Chao, *Ind. Eng. Chem. Res.*, **35**, 1446 (1996).
352 C. R. Novenario, J. M. Caruthers, K. C. Chao, *Fluid Phase Equil.*, **142**, 83 (1998).
353 C. R. Novenario, J. M. Caruthers, K. C. Chao, *Ind. Eng. Chem. Res.*, **37**, 3142 (1998).
354 C. R. Novenario, J. M. Caruthers, K. C. Chao, *J. Appl. Polym. Sci.*, **67**, 841 (1998).
355 S. H. Huang, M. Radosz, *Ind. Eng. Chem.*, **29**, 2284 (1990).
356 S. S. Chen, A. Kreglewski, *Ber. Bunsenges. Phys. Chem.*, **81**, 1048 (1977).
357 M. Banaszak, Y. C. Chiew, M. Radosz, *Phys. Rev. E*, **48**, 3760 (1993).
358 M. Banaszak, Y. C. Chiew, R. O'Lenick, M. Radosz, *J. Chem. Phys.*, **100**, 3803 (1994).
359 M. Banaszak, C. K. Chen, M. Radosz, *Macromolecules*, **29**, 6481 (1996).
360 K. P. Shukla, W. G. Shapman, *Mol. Phys.*, **91**, 1075 (1997).
361 H. Adidharma, M. Radosz, *Ind. Eng. Chem. Res.*, **37**, 4453 (1998).
362 S.-J. Chen, M. Radosz, *Macromolecules*, **25**, 3089 (1992).
363 S.-J. Chen, I. G. Economou, M. Radosz, *Macromolecules*, **25**, 4987 (1992).
364 C. J. Gregg, S.-J. Chen, F. P. Stein, M. Radosz, *Fluid Phase Equil.*, **83**, 375 (1993).
365 S.-J. Chen, M. Banaszak, M. Radosz, *Macromolecules,* **28**, 1812 (1995).
366 P. Condo, M. Radosz, *Fluid Phase Equil.*, **117**, 1 (1996).
367 K. L. Albrecht, F. P. Stein, S. J. Han, C. J. Gregg, M. Radosz, *Fluid Phase Equil.*, **117**, 117 (1996).
368 C. Pan, M. Radosz, *Ind. Eng. Chem. Res.*, **37**, 3169 (1998) and **38**, 2842 (1999).
369 G. Sadowski, L. V. Mokrushina, W. Arlt, *Fluid Phase Equil.*, **139**, 391 (1997).
370 J. Groß, G. Sadowski, submitted to *Fluid Phase Equil.*
371 C. F. Kirby, M. A. McHugh, *Chem. Rev.*, **99**, 565 (1999).
372 R. B. Gupta, J. M. Prausnitz, *Ind. Eng. Chem. Res.*, **35**, 1225 (1996).
373 T. Hino, Y. Song, J. M. Prausnitz, *Macromolecules*, **28**, 5709, 5717, 5725 (1995).
374 S. M. Lambert, Song, J. M. Prausnitz, *Macromolecules*, **28**, 4866 (1995).
375 G. M. Kontogeorgis, V. I. Harismiadis, A. Fredenslund, D. P. Tassios, *Fluid Phase Equil.*, **96**, 65 (1994).
376 V. I. Harismiadis, G. M. Kontogeorgis, A. Fredenslund, D. P. Tassios, *Fluid Phase Equil.*, **96**, 93 (1994).
377 T. Sako, A. W. Hu, J. M. Prausnitz, *J. Appl. Polym. Sci.*, **38**, 1839 (1989).

378 G. Soave, *Chem. Eng. Sci.*, **27**, 1197 (1972).
379 V. I. Harismiadis, G. M. Kontogeorgis, A. Saraiva, A. Fredenslund, D. P. Tassios, *Fluid Phase Equil.*, **100**, 63 (1994).
380 N. Orbey, S. I. Sandler, *AIChE-J.*, **40**, 1203 (1994).
381 R. Stryjek, J. H. Vera, *Can. J. Chem. Eng.*, **64**, 323 (1986).
382 D. S. H. Wong, S. I. Sandler, *AIChE-J.*, **38**, 671 (1992).
383 C. Zhong, H. Masuoka, *Fluid Phase Equil.*, **126**, 1 (1996).
384 C. Zhong, H. Masuoka, *Fluid Phase Equil.*, **126**, 59 (1996).
385 C. Zhong, H. Masuoka, *Fluid Phase Equil.*, **144**, 49 (1998).
386 H. Orbey, C.-C. Chen, C. P. Bokis, *Fluid Phase Equil.*, **145**, 169 (1998).
387 H. Orbey, C. P. Bokis, C.-C. Chen, *Ind. Eng. Chem. Res.*, **37**, 1567 (1998).

## APPENDIX 4.4A

### Table of polymer-solvent systems for which experimental VLE-data have been reported in the literature (the references are given at the end of this table)

| Solvent | T, K | Ref. | Solvent | T, K | Ref. |
|---|---|---|---|---|---|
| **acrylonitrile/butadiene copolymer** | | | | | |
| acetonitrile | 333.15 | 121 | n-hexane | 333.15 | 121 |
| chloroform | 333.15 | 121 | n-octane | 333.15 | 121 |
| cyclohexane | 333.15 | 121 | n-pentane | 333.15 | 121 |
| **acrylonitrile/styrene copolymer** | | | | | |
| benzene | 343.15 | 130, 134 | o-xylene | 398.15 | 130, 134 |
| 1,2-dichloroethane | 343.15 | 121 | m-xylene | 398.15 | 130, 134 |
| 1,2-dichloroethane | 353.15 | 121 | p-xylene | 373.15 | 130, 134 |
| propylbenzene | 398.15 | 130, 134 | p-xylene | 398.15 | 130, 134 |
| toluene | 343.15 | 130, 134 | p-xylene | 423.15 | 130, 134 |
| toluene | 373.15 | 130, 134 | | | |
| **p-bromostyrene/p-methylstyrene copolymer** | | | | | |
| toluene | 293.20 | 72 | | | |
| **cellulose acetate** | | | | | |
| acetone | 303.15 | 81 | 1,4-dioxane | 308.15 | 81 |
| acetone | 308.15 | 81 | methyl acetate | 303.15 | 81 |
| N,N-dimethylformamide | 322.85 | 75 | methyl acetate | 308.15 | 81 |
| N,N-dimethylformamide | 342.55 | 75 | pyridine | 303.15 | 81 |
| 1,4-dioxane | 303.15 | 81 | pyridine | 308.15 | 81 |
| **cellulose triacetate** | | | | | |
| chloroform | 303.15 | 77 | dichloromethane | 293.15 | 77 |
| chloroform | 308.15 | 77 | dichloromethane | 298.15 | 77 |

| Solvent | T, K | Ref. | Solvent | T, K | Ref. |
|---|---|---|---|---|---|
| **dextran** | | | | | |
| water | 293.15 | 111, 112, 119, 154 | water | 313.15 | 111 |
| water | 298.15 | 76 | water | 333.15 | 111, 119 |
| **di(trimethylsilyl)-poly(propylene oxide)** | | | | | |
| toluene | 323.15 | 75 | n-decane | 342.45 | 75 |
| toluene | 342.65 | 75 | | | |
| **ethylene/vinyl acetate copolymer** | | | | | |
| benzene | 303.15 | 36 | n-propyl acetate | 343.15 | 37 |
| benzene | 323.15 | 36 | n-propyl acetate | 363.15 | 37 |
| benzene | 328.15 | 36 | toluene | 303.15 | 36 |
| benzene | 333.15 | 35 | toluene | 323.15 | 36 |
| benzene | 343.15 | 36 | toluene | 333.15 | 35 |
| benzene | 353.15 | 35 | toluene | 343.15 | 36 |
| benzene | 373.15 | 34, 35 | toluene | 353.15 | 35 |
| butyl acetate | 323.15 | 37 | toluene | 363.15 | 36 |
| butyl acetate | 343.15 | 37 | toluene | 373.15 | 34, 35 |
| butyl acetate | 363.15 | 37 | o-xylene | 323.15 | 36 |
| chloroform | 333.15 | 121 | o-xylene | 343.15 | 36 |
| cyclohexane | 353.15 | 121 | o-xylene | 363.15 | 36 |
| ethyl acetate | 303.25 | 37 | p-xylene | 323.15 | 36 |
| ethyl acetate | 323.15 | 37 | p-xylene | 333.15 | 35 |
| ethyl acetate | 343.15 | 37 | p-xylene | 343.15 | 36 |
| methyl acetate | 303.15 | 37 | p-xylene | 353.15 | 35 |
| methyl acetate | 323.15 | 37 | p-xylene | 363.15 | 36 |
| n-propyl acetate | 303.15 | 37 | p-xylene | 373.15 | 34, 35 |
| n-propyl acetate | 323.15 | 37 | | | |
| **hydroxypropyl cellulose** | | | | | |
| acetone | 298.15 | 54 | tetrahydrofuran | 298.15 | 54 |
| ethanol | 298.15 | 54 | water | 298.15 | 141 |
| **hydroxypropyl starch** | | | | | |
| water | 293.15 | 117 | water | 298.15 | 160 |

| Solvent | T, K | Ref. | Solvent | T, K | Ref. |
|---|---|---|---|---|---|
| **natural rubber** | | | | | |
| acetone | 273.15 | 25 | 2-butanone | 318.15 | 25 |
| acetone | 298.15 | 25 | ethyl acetate | 298.15 | 25 |
| benzene | 298.15 | 6 | ethyl acetate | 323.15 | 25 |
| 2-butanone | 298.15 | 25 | toluene | 303.00 | 82 |
| **nitrocellulose** | | | | | |
| acetone | 293.00 | 79 | ethyl propyl ether | 293.00 | 79 |
| acetone | 303.15 | 81 | methyl acetate | 303.15 | 81 |
| acetone | 308.15 | 81 | methyl acetate | 308.15 | 81 |
| acetonitrile | 293.00 | 79 | 3-methyl-2-butanone | 293.00 | 79 |
| ethyl formate | 293.15 | 78 | 3-methylbutyl acetate | 293.15 | 78 |
| cyclopentanone | 293.00 | 79 | nitromethane | 293.00 | 79 |
| 3,3-dimethyl-2-butanone | 293.00 | 79 | 2-pentanone | 293.00 | 79 |
| 2,4-dimethyl-3-pentanone | 293.00 | 79 | ethyl propionate | 293.15 | 78 |
| 1,4-dioxane | 293.00 | 79 | propyl acetate | 293.15 | 78 |
| ethyl acetate | 293.15 | 78 | | | |
| **nylon 6,6** | | | | | |
| water | 296.15 | 103 | | | |
| **nylon 6,10** | | | | | |
| water | 296.15 | 103 | | | |
| **poly(acrylic acid)** | | | | | |
| ethanol | 303.15 | 145 | water | 303.15 | 145 |
| **poly(acrylonitrile)** | | | | | |
| 1,2-dichloroethane | 353.15 | 121 | N,N-dimethylformamide | 343.55 | 75 |
| N,N-dimethylformamide | 323.25 | 75 | | | |
| **polyamidoamine dendrimers** | | | | | |
| acetone | 308.15 | 144 | methanol | 308.15 | 144 |
| acetonitrile | 313.15 | 144 | 1-propylamine | 308.15 | 144 |
| chloroform | 308.15 | 144 | | | |
| **poly(benzyl ether) dendrimers** | | | | | |
| acetone | 323.15 | 144 | tetrahydrofuran | 343.15 | 144 |
| chloroform | 323.15 | 144 | toluene | 343.15 | 144 |
| chloroform | 343.15 | 144 | n-pentane | 313.15 | 144 |

| Solvent | T, K | Ref. | Solvent | T, K | Ref. |
|---|---|---|---|---|---|
| cyclohexane | 333.15 | 144 | | | |
| **poly(γ-benzyl-L-glutamate)** | | | | | |
| chloroform | 303.15 | 51 | | | |
| **poly(p-bromostyrene)** | | | | | |
| toluene | 293.20 | 72 | | | |
| **polybutadiene** | | | | | |
| benzene | 296.65 | 7 | ethylbenzene | 403.15 | 106 |
| chloroform | 296.65 | 7 | n-hexane | 296.65 | 7 |
| chloroform | 298.15 | 69 | n-hexane | 333.15 | 121 |
| chloroform | 333.15 | 121 | n-nonane | 353.15 | 106 |
| cyclohexane | 296.65 | 7 | n-nonane | 373.15 | 106 |
| cyclohexane | 333.15 | 121 | n-nonane | 403.15 | 106 |
| dichloromethane | 296.65 | 7 | n-pentane | 333.15 | 121 |
| ethylbenzene | 353.15 | 106 | tetrachloromethane | 296.65 | 7 |
| ethylbenzene | 373.15 | 106 | toluene | 296.65 | 7 |
| **poly(n-butyl acrylate)** | | | | | |
| benzene | 296.65 | 61 | tetrachloromethane | 296.65 | 61 |
| chloroform | 296.65 | 61 | toluene | 296.65 | 61 |
| dichloromethane | 296.65 | 61 | | | |
| **poly(n-butyl methacrylate)** | | | | | |
| benzene | 323.65 | 75 | mesitylene | 373.15 | 75 |
| benzene | 343.45 | 75 | mesitylene | 403.15 | 75 |
| 2-butanone | 323.65 | 75 | 3-pentanone | 323.55 | 75 |
| 2-butanone | 344.45 | 75 | 3-pentanone | 343.95 | 75 |
| chloroform | 323.75 | 75 | propylbenzene | 344.35 | 75 |
| chloroform | 343.75 | 121 | tetrachloromethane | 323.65 | 75 |
| cumene | 373.15 | 75 | tetrachloromethane | 344.45 | 75 |
| cumene | 403.15 | 75 | toluene | 323.35 | 75 |
| cyclohexane | 308.15 | 125 | toluene | 343.75 | 75 |
| cyclohexane | 318.15 | 125 | toluene | 373.15 | 75 |
| cyclohexane | 328.15 | 125 | o-xylene | 344.45 | 75 |
| cyclohexane | 338.15 | 125 | o-xylene | 373.15 | 75 |
| diethyl ether | 298.15 | 159 | o-xylene | 403.15 | 75 |

| Solvent | T, K | Ref. | Solvent | T, K | Ref. |
|---|---|---|---|---|---|
| 1,2-dichloroethane | 323.95 | 75 | m-xylene | 343.95 | 75 |
| 1,2-dichloroethane | 343.15 | 75 | m-xylene | 373.15 | 75 |
| 3,3-dimethyl-2-butanone | 323.65 | 75 | m-xylene | 403.15 | 75 |
| 3,3-dimethyl-2-butanone | 344.45 | 75 | p-xylene | 344.45 | 75 |
| ethylbenzene | 343.75 | 75 | p-xylene | 373.15 | 75 |
| ethylbenzene | 373.15 | 75 | p-xylene | 403.15 | 75 |
| ethylbenzene | 403.15 | 75 | | | |
| **poly(tert-butyl methacrylate)** | | | | | |
| benzene | 323.15 | 75 | 3-pentanone | 342.65 | 75 |
| benzene | 342.65 | 75 | propylbenzene | 342.65 | 75 |
| 2-butanone | 323.15 | 75 | tetrachloromethane | 323.15 | 75 |
| 2-butanone | 342.65 | 75 | tetrachloromethane | 342.65 | 75 |
| chloroform | 323.15 | 75 | toluene | 323.15 | 75 |
| cumene | 373.15 | 75 | toluene | 342.75 | 75 |
| 1,2-dichloroethane | 323.15 | 75 | toluene | 373.15 | 75 |
| 1,2-dichloroethane | 342.65 | 75 | o-xylene | 342.65 | 75 |
| 3,3-dimethyl-2-butanone | 323.15 | 75 | o-xylene | 373.15 | 75 |
| 3,3-dimethyl-2-butanone | 342.65 | 75 | m-xylene | 342.65 | 75 |
| ethylbenzene | 342.65 | 75 | m-xylene | 373.15 | 75 |
| ethylbenzene | 373.35 | 75 | p-xylene | 342.65 | 75 |
| mesitylene | 373.15 | 75 | p-xylene | 373.15 | 75 |
| 3-pentanone | 323.15 | 75 | | | |
| **poly(ε-caprolacton)** | | | | | |
| tetrachloromethane | 338.15 | 65 | | | |
| **polycarbonate-bisphenol-A** | | | | | |
| chlorobenzene | 413.15 | 123, 139 | mesitylene | 453.15 | 139 |
| chlorobenzene | 433.15 | 139 | n-pentane | 303.15 | 145 |
| chlorobenzene | 453.15 | 139 | toluene | 413.15 | 139 |
| ethanol | 303.15 | 145 | water | 303.15 | 145 |
| ethylbenzene | 413.15 | 139 | m-xylene | 413.15 | 139 |
| ethylbenzene | 433.15 | 139 | m-xylene | 453.15 | 139 |
| mesitylene | 413.15 | 139 | p-xylene | 413.15 | 139 |
| mesitylene | 433.15 | 139 | | | |

| Solvent | T, K | Ref. | Solvent | T, K | Ref. |
|---|---|---|---|---|---|
| **poly(o-chlorostyrene)** | | | | | |
| benzene | 298.15 | 115 | 2-butanone | 313.15 | 115 |
| benzene | 313.15 | 115 | | | |
| **poly(p-chlorostyrene)** | | | | | |
| toluene | 293.20 | 72 | | | |
| **polydecene** | | | | | |
| toluene | 303.15 | 3 | | | |
| **poly(dimethyl siloxane)** | | | | | |
| benzene | 298.15 | 52, 62 | hexamethyl disiloxane | 298.15 | 62 |
| benzene | 303.00 | 63, 66, 101 | n-nonane | 313.15 | 157 |
| benzene | 313.15 | 62 | octamethyl cyclotetrasiloxane | 413.15 | 85 |
| 2-butanone | 303.15 | 53 | n-octane | 298.15 | 62 |
| chloroform | 303.00 | 99 | n-octane | 303.15 | 157 |
| cyclohexane | 293.15 | 161 | n-octane | 313.15 | 62 |
| cyclohexane | 303.15 | 63, 66, 161 | n-pentane | 303.15 | 63, 147 |
| dichloromethane | 303.00 | 99 | n-pentane | 313.15 | 145 |
| n-heptane | 298.15 | 62 | toluene | 298.15 | 62 |
| n-heptane | 303.15 | 63, 157 | toluene | 308.15 | 124 |
| n-heptane | 313.15 | 62 | toluene | 313.15 | 62 |
| n-hexane | 298.09 | 66 | toluene | 318.15 | 124 |
| n-hexane | 303.15 | 63, 66, 100, 147 | toluene | 328.15 | 124 |
| n-hexane | 308.08 | 66 | 2,2,4-trimethylpentane | 298.15 | 62 |
| n-hexane | 313.15 | 145 | 2,2,4-trimethylpentane | 313.15 | 62 |
| **poly(1,3-dioxolane)** | | | | | |
| benzene | 303.15 | 105 | benzene | 313.15 | 105 |
| **polydodecene** | | | | | |
| toluene | 303.15 | 3 | | | |
| **poly(ethyl acrylate)** | | | | | |
| benzene | 296.65 | 61 | tetrachloromethane | 296.65 | 61 |
| chloroform | 296.65 | 61 | toluene | 296.65 | 61 |
| dichloromethane | 296.65 | 61 | | | |

| Solvent | T, K | Ref. | Solvent | T, K | Ref. |
|---|---|---|---|---|---|
| **polyethylene** | | | | | |
| chlorobenzene | 393.15 | 109 | propyl acetate | 426.15 | 137 |
| chlorobenzene | 403.15 | 109 | propyl acetate | 474.15 | 137 |
| chlorobenzene | 413.15 | 109 | 2-propylamine | 427.15 | 137 |
| cyclopentane | 425.65 | 137 | 2-propylamine | 475.15 | 137 |
| cyclopentane | 474.15 | 137 | toluene | 393.15 | 109 |
| ethylbenzene | 413.15 | 116 | o-xylene | 413.15 | 116 |
| n-heptane | 382.05 | 2 | m-xylene | 413.15 | 116 |
| n-pentane | 423.65 | 137 | p-xylene | 353.15 | 1 |
| n-pentane | 474.15 | 137 | p-xylene | 363.15 | 1, 35 |
| 3-pentanol | 423.15 | 137 | p-xylene | 373.15 | 35 |
| 3-pentanol | 473.15 | 137 | p-xylene | 383.15 | 35 |
| 3-pentanone | 425.15 | 137 | p-xylene | 403.15 | 116 |
| 3-pentanone | 477.15 | 137 | p-xylene | 413.15 | 116 |
| 1-pentene | 423.65 | 137 | p-xylene | 423.15 | 116 |
| 1-pentene | 474.15 | 137 | | | |
| **poly(ethylene glycol)** | | | | | |
| benzene | 297.75 | 43 | 1-propanol | 323.15 | 45 |
| benzene | 307.75 | 43 | 1-propanol | 333.15 | 94 |
| benzene | 313.15 | 42 | 1-propanol | 343.15 | 45 |
| benzene | 323.15 | 42 | 1-propanol | 353.15 | 94 |
| benzene | 343.15 | 42 | 1-propanol | 373.15 | 45 |
| 1-butanol | 323.15 | 42 | tetrachloromethane | 303.15 | 95, 96 |
| 1-butanol | 343.15 | 42 | toluene | 323.15 | 42, 75 |
| 1-butanol | 373.15 | 42 | toluene | 343.15 | 42, 75 |
| 1-butanol | 403.15 | 42 | toluene | 373.15 | 42 |
| chloroform | 323.15 | 146 | water | 293.15 | 104, 111, 112, 117, 119 |
| chloroform | 333.15 | 146 | water | 298.15 | 39, 40, 41, 76, 128, 129, 158, 160 |
| ethanol | 303.15 | 45, 157 | water | 303.15 | 38 |
| ethanol | 313.15 | 45 | water | 308.15 | 40, 151 |
| ethanol | 323.15 | 45 | water | 313.15 | 104, 111 |

| Solvent | T, K | Ref. | Solvent | T, K | Ref. |
|---|---|---|---|---|---|
| ethylbenzene | 323.15 | 42, 44 | water | 318.15 | 151 |
| ethylbenzene | 343.15 | 42, 44 | water | 323.15 | 38, 108 |
| ethylbenzene | 343.75 | 75 | water | 328.15 | 38, 151 |
| ethylbenzene | 373.15 | 42, 44 | water | 333.15 | 38, 104, 108, 111, 119, 146 |
| ethylbenzene | 403.15 | 44 | water | 338.15 | 38, 151 |
| 1-hexanol | 323.15 | 46 | water | 343.15 | 108 |
| 1-hexanol | 373.15 | 46 | p-xylene | 323.15 | 45 |
| 1-hexanol | 403.15 | 46 | p-xylene | 343.15 | 45 |
| methanol | 303.15 | 157 | p-xylene | 373.15 | 45 |
| 1-propanol | 303.15 | 45 | p-xylene | 403.15 | 45 |
| **poly(ethylene glycol) dimethyl ether** | | | | | |
| chloroform | 278.68 | 80 | tetrachloromethane | 303.15 | 96 |
| tetrachloromethane | 278.68 | 80 | | | |
| **poly(ethylene glycol) monomethyl ether** | | | | | |
| tetrachloromethane | 303.15 | 96 | | | |
| **poly(ethylene oxide)** | | | | | |
| acetone | 323.15 | 152 | benzene | 423.55 | 47 |
| acetone | 353.15 | 136 | 2-butanone | 353.15 | 136 |
| benzene | 318.85 | 74 | chloroform | 298.15 | 48 |
| benzene | 323.45 | 74 | chloroform | 323.15 | 152 |
| benzene | 328.15 | 65 | chloroform | 343.15 | 152 |
| benzene | 343.15 | 65, 74 | chloroform | 333.15 | 121 |
| benzene | 348.25 | 47 | cyclohexane | 353.15 | 136 |
| benzene | 353.15 | 136 | toluene | 353.15 | 136 |
| benzene | 361.25 | 47 | toluene | 372.98 | 68 |
| benzene | 375.15 | 47 | p-xylene | 353.15 | 136 |
| benzene | 398.85 | 47 | | | |
| **poly(ethylene oxide)-b-poly(tert-butyl methacrylate) diblock copolymer** | | | | | |
| toluene | 323.15 | 75 | toluene | 343.75 | 75 |
| **poly(ethylene oxide)-b-poly(methyl methacrylate) diblock copolymer** | | | | | |
| toluene | 323.41 | 68 | toluene | 373.27 | 68 |

| Solvent | T, K | Ref. | Solvent | T, K | Ref. |
|---|---|---|---|---|---|
| toluene | 343.27 | 68 | | | |
| **poly(ethylene oxide)-b-poly(methyl methacrylate)-b-poly(ethylene oxide) triblock copolymer** | | | | | |
| toluene | 323.08 | 68 | toluene | 343.17 | 68 |
| toluene | 373.26 | 68 | | | |
| **poly(ethylene oxide)-b-poly(propylene oxide) diblock copolymer** | | | | | |
| tetrachloromethane | 303.15 | 96 | | | |
| **poly(ethylene oxide)-b-poly(propylene oxide)-b-poly(ethylene oxide) triblock copolymer** | | | | | |
| tetrachloromethane | 303.15 | 96 | toluene | 343.75 | 75 |
| toluene | 323.35 | 75 | | | |
| **poly(ethylene oxide)-b-polystyrene-b-poly(ethylene oxide) triblock copolymer** | | | | | |
| toluene | 323.35 | 75 | toluene | 343.75 | 75 |
| **poly(ethyl methacrylate)** | | | | | |
| benzene | 296.65 | 61 | tetrachloromethane | 296.65 | 61 |
| chloroform | 296.65 | 61 | toluene | 296.65 | 61 |
| dichloromethane | 296.65 | 61 | | | |
| **polyheptene** | | | | | |
| toluene | 303.15 | 3 | | | |
| **poly(4-hydroxystyrene)** | | | | | |
| acetone | 293.15 | 97 | acetone | 308.15 | 97 |
| acetone | 298.15 | 97 | acetone | 313.15 | 97 |
| acetone | 303.15 | 97 | acetone | 318.15 | 97 |
| **polyisobutylene** | | | | | |
| benzene | 298.15 | 57, 8, 107 | ethylbenzene | 338.15 | 107 |
| benzene | 300.05 | 9 | n-heptane | 296.65 | 60 |
| benzene | 313.15 | 8, 57, 70, 107 | n-heptane | 338.15 | 156 |
| benzene | 333.20 | 70 | n-hexane | 298.15 | 107 |
| benzene | 338.15 | 8, 107 | n-hexane | 313.15 | 107 |
| benzene | 353.20 | 70 | n-hexane | 338.15 | 107, 156 |
| n-butane | 298.15 | 4 | 2-methylbutane | 298.15 | 4 |
| n-butane | 308.15 | 4 | 2-methylbutane | 308.15 | 4 |
| n-butane | 319.65 | 4 | 2-methylbutane | 319.65 | 4 |
| chloroform | 296.65 | 60 | 2-methylpropane | 308.15 | 4 |

| Solvent | T, K | Ref. | Solvent | T, K | Ref. |
|---|---|---|---|---|---|
| cyclohexane | 298.15 | 8, 58, 60, 107, 147 | 2-methylpropane | 319.65 | 4 |
| cyclohexane | 308.15 | 125 | n-nonane | 338.15 | 156 |
| cyclohexane | 313.15 | 70, 107 | n-octane | 338.15 | 156 |
| cyclohexane | 315.15 | 8 | n-pentane | 298.15 | 4, 5, 59, 65 |
| cyclohexane | 318.15 | 125 | n-pentane | 308.15 | 4, 5 |
| cyclohexane | 328.15 | 125 | n-pentane | 318.15 | 5 |
| cyclohexane | 333.20 | 70 | n-pentane | 319.65 | 4 |
| cyclohexane | 338.15 | 8, 107, 125 | n-pentane | 328.15 | 5 |
| cyclopentane | 296.65 | 60 | propane | 308.15 | 4 |
| 2,2-dimethylbutane | 296.65 | 60 | tetrachloromethane | 296.65 | 60 |
| 2,2-dimethylpropane | 298.15 | 4 | toluene | 298.15 | 107 |
| 2,2-dimethylpropane | 308.15 | 4 | toluene | 313.15 | 107 |
| 2,2-dimethylpropane | 319.65 | 4 | toluene | 338.15 | 107 |
| ethylbenzene | 298.15 | 107 | 2,2,4-trimethylpentane | 296.65 | 60 |
| ethylbenzene | 313.15 | 107 | | | |
| **1,4-cis-polyisoprene** | | | | | |
| benzene | 296.65 | 7 | dichloromethane | 296.65 | 7 |
| benzene | 353.15 | 10 | tetrachloromethane | 296.65 | 7 |
| chloroform | 296.65 | 7 | toluene | 296.65 | 7 |
| cyclohexane | 296.65 | 7 | | | |
| **polyisoprene, hydrogenated** | | | | | |
| cyclohexane | 323.15 | 161 | | | |
| **poly(maleic anhydride)** | | | | | |
| acetone | 323.15 | 140 | methanol | 333.15 | 140 |
| **poly(methyl acrylate)** | | | | | |
| benzene | 296.65 | 61 | tetrachloromethane | 296.65 | 61 |
| chloroform | 296.65 | 61 | toluene | 296.65 | 61 |
| dichloromethane | 296.65 | 61 | | | |
| **poly(methyl methacrylate)** | | | | | |
| acetone | 308.15 | 131, 150 | 3,3-dimethyl-2-butanone | 343.45 | 56 |
| acetone | 323.15 | 133 | ethyl acetate | 308.15 | 131 |
| benzene | 296.65 | 61 | ethylbenzene | 398.15 | 153 |

| Solvent | T, K | Ref. | Solvent | T, K | Ref. |
|---|---|---|---|---|---|
| benzene | 323.15 | 56 | mesitylene | 403.15 | 67 |
| benzene | 343.15 | 56 | methyl acetate | 323.15 | 133 |
| 2-butanone | 308.15 | 132, 150 | tetrachloromethane | 323.15 | 56 |
| 2-butanone | 323.15 | 23, 56 | tetrachloromethane | 343.75 | 56 |
| 2-butanone | 343.55 | 56 | toluene | 296.65 | 61 |
| chloroform | 296.65 | 61 | toluene | 323.15 | 23, 56, 75 |
| chloroform | 303.15 | 146 | toluene | 343.15 | 56, 75 |
| chloroform | 308.15 | 132 | toluene | 373.97 | 68 |
| chloroform | 323.15 | 56, 133, 142, 146 | toluene | 433.15 | 19 |
| cyclohexanone | 323.15 | 146 | p-xylene | 323.15 | 56 |
| dichloromethane | 296.65 | 61 | p-xylene | 343.15 | 56 |
| 1,2-dichloroethane | 323.15 | 67 | p-xylene | 373.15 | 56 |
| 1,2-dichloroethane | 343.15 | 67 | p-xylene | 403.15 | 56 |
| 3,3-dimethyl-2-butanone | 323.35 | 56 | p-xylene | 409.35 | 56 |
| **poly(α-methylstyrene)** | | | | | |
| cumene | 338.15 | 83 | tetrahydrofuran | 298.15 | 92 |
| 1,4-dioxane | 313.15 | 89 | toluene | 298.15 | 28, 90 |
| α-methylstyrene | 303.15 | 90 | toluene | 303.15 | 90 |
| α-methylstyrene | 308.15 | 90 | toluene | 308.15 | 90 |
| α-methylstyrene | 313.15 | 90 | toluene | 313.15 | 90 |
| α-methylstyrene | 338.15 | 83 | | | |
| **poly(p-methylstyrene)** | | | | | |
| toluene | 293.20 | 72 | | | |
| **polyoctadecene** | | | | | |
| toluene | 303.15 | 3 | | | |
| **polypropylene** | | | | | |
| 2,4-dimethyl-3-pentanone | 298.15 | 11 | 3-pentanone | 318.15 | 11 |
| 2,4-dimethyl-3-pentanone | 318.15 | 11 | tetrachloromethane | 298.15 | 84 |
| 3-pentanone | 298.15 | 11 | | | |
| **poly(propylene glycol)** | | | | | |
| n-decane | 343.45 | 75 | methanol | 298.15 | 49 |

| Solvent | T, K | Ref. | Solvent | T, K | Ref. |
|---|---|---|---|---|---|
| ethylbenzene | 342.65 | 75 | tetrachloromethane | 303.15 | 96 |
| n-hexane | 298.10 | 87 | toluene | 323.15 | 75 |
| n-hexane | 312.65 | 87 | toluene | 343.75 | 75 |
| n-hexane | 323.15 | 87 | water | 298.15 | 41, 87 |
| methanol | 263.15 | 49 | water | 303.15 | 38 |
| methanol | 273.15 | 49 | water | 312.65 | 87 |
| methanol | 288.15 | 49 | water | 323.15 | 38 |
| **poly(propylene glycol) dimethyl ether** | | | | | |
| chloroform | 278.68 | 50 | tetrachloromethane | 278.68 | 50 |
| **poly(propylene imine) dendrimers** | | | | | |
| acetone | 323.15 | 155 | n-heptane | 348.15 | 155 |
| acetonitrile | 343.15 | 155 | n-hexane | 338.15 | 155 |
| acetonitrile | 348.15 | 155 | n-nonane | 338.15 | 155 |
| chloroform | 323.15 | 155 | n-octane | 338.15 | 155 |
| chloroform | 343.15 | 155 | tetrahydrofuran | 323.15 | 155 |
| n-heptane | 338.15 | 155 | toluene | 343.15 | 155 |
| n-heptane | 343.15 | 155 | triethylamine | 338.15 | 155 |
| **poly(propylene oxide)** | | | | | |
| benzene | 298.15 | 147 | methanol | 298.15 | 147 |
| benzene | 320.35 | 73 | methanol | 303.15 | 157 |
| benzene | 333.35 | 73 | methanol | 313.15 | 145 |
| benzene | 343.05 | 73 | propanol | 303.15 | 157 |
| benzene | 347.85 | 73 | water | 303.15 | 157 |
| ethanol | 303.15 | 157 | | | |
| **poly(propylene oxide)-b-poly(ethylene oxide) diblock copolymer** | | | | | |
| ethylbenzene | 343.75 | 75 | | | |
| **poly(propylene oxide)-b-poly(ethylene oxide)-b-poly(propylene oxide) triblock copolymer** | | | | | |
| toluene | 323.05 | 75 | toluene | 342.65 | 75 |
| **polystyrene** | | | | | |
| acetone | 298.15 | 27 | dichloromethane | 296.65 | 17 |
| acetone | 323.15 | 27, 152 | 1,4-dioxane | 293.15 | 16 |
| acetone | 393.15 | 122 | 1,4-dioxane | 323.15 | 23, 26 |
| acetone | 423.15 | 122 | dipropyl ether | 293.15 | 16 |

| Solvent | T, K | Ref. | Solvent | T, K | Ref. |
|---|---|---|---|---|---|
| acetonitrile | 393.15 | 122 | ethyl acetate | 313.15 | 146 |
| acetonitrile | 423.15 | 122 | ethyl acetate | 333.15 | 146 |
| anisole | 323.15 | 26 | ethylbenzene | 303.15 | 15 |
| benzene | 288.15 | 102 | ethylbenzene | 323.15 | 22 |
| benzene | 293.15 | 16 | ethylbenzene | 343.15 | 22 |
| benzene | 296.65 | 17 | ethylbenzene | 398.15 | 153 |
| benzene | 298.15 | 65 | ethylbenzene | 403.15 | 19 |
| benzene | 303.15 | 12, 71, 102, 149 | ethylbenzene | 413.15 | 19 |
| benzene | 313.15 | 149 | ethylbenzene | 433.15 | 19 |
| benzene | 318.15 | 102 | ethylbenzene | 443.15 | 19 |
| benzene | 323.15 | 12, 138 | ethylbenzene | 451.15 | 19 |
| benzene | 333.15 | 102, 71 | n-hexane | 393.15 | 122 |
| benzene | 343.15 | 12 | n-hexane | 423.15 | 122 |
| benzene | 353.20 | 71 | methyl acetate | 323.15 | 133 |
| benzene | 393.15 | 122 | n-nonane | 403.15 | 106 |
| benzene | 403.15 | 19 | n-nonane | 423.15 | 106 |
| benzene | 423.15 | 122 | n-nonane | 448.15 | 106 |
| benzene | 428.15 | 19 | 3-pentanone | 293.15 | 16 |
| 2-butanone | 298.15 | 24, 26 | propyl acetate | 298.15 | 27 |
| 2-butanone | 321.65 | 23 | propyl acetate | 343.14 | 27 |
| 2-butanone | 343.15 | 24 | tetrachloromethane | 293.15 | 16 |
| 2-butanone | 393.15 | 122 | tetrachloromethane | 296.65 | 17 |
| 2-butanone | 423.15 | 122 | toluene | 293.15 | 16 |
| butyl acetate | 308.15 | 159 | toluene | 296.65 | 17 |
| butyl acetate | 323.15 | 26 | toluene | 298.15 | 24, 102, 138 |
| tert-butyl acetate | 283.15 | 64 | toluene | 303.15 | 14, 15, 20 |
| tert-butyl acetate | 303.15 | 64 | toluene | 308.15 | 127 |
| tert-butyl acetate | 323.15 | 64 | toluene | 313.15 | 149 |
| tert-butyl acetate | 343.15 | 64 | toluene | 321.65 | 23 |
| tert-butyl acetate | 363.15 | 64 | toluene | 323.15 | 14, 20, 71, 138, 149 |
| chloroform | 296.65 | 17 | toluene | 333.15 | 24, 71 |
| chloroform | 298.15 | 27, 65, 138 | toluene | 343.15 | 20 |

| Solvent | T, K | Ref. | Solvent | T, K | Ref. |
|---|---|---|---|---|---|
| chloroform | 323.15 | 27, 138, 152 | toluene | 353.20 | 71 |
| cyclohexane | 293.15 | 16 | toluene | 373.15 | 20, 21, 123 |
| cyclohexane | 296.65 | 17 | toluene | 383.15 | 19 |
| cyclohexane | 297.15 | 13 | toluene | 393.15 | 21, 122, 123 |
| cyclohexane | 303.15 | 12, 14, 15, 108, 149 | toluene | 403.15 | 19 |
| cyclohexane | 308.15 | 13, 125 | toluene | 413.15 | 19 |
| cyclohexane | 313.15 | 71, 108, 149 | toluene | 423.15 | 122 |
| cyclohexane | 318.15 | 13, 125 | 1,2,4-trimethylbenzene | 443.15 | 19 |
| cyclohexane | 323.15 | 12, 14, 108, 152 | o-xylene | 323.15 | 26 |
| cyclohexane | 328.15 | 125 | o-xylene | 373.15 | 21 |
| cyclohexane | 333.15 | 71, 108 | o-xylene | 403.15 | 21 |
| cyclohexane | 343.15 | 12 | m-xylene | 323.15 | 146 |
| cyclohexane | 353.20 | 71 | m-xylene | 373.15 | 21 |
| cyclohexane | 338.15 | 125 | m-xylene | 403.15 | 106 |
| cyclohexane | 393.15 | 122 | m-xylene | 423.15 | 106 |
| cyclohexane | 423.15 | 122 | m-xylene | 448.15 | 106 |
| cyclohexanone | 313.15 | 146 | p-xylene | 373.15 | 21 |
| cyclohexanone | 333.15 | 146 | p-xylene | 393.15 | 122 |
| 1,2-dichloroethane | 343.15 | 121 | p-xylene | 403.15 | 21 |
| 1,2-dichloroethane | 353.15 | 121 | p-xylene | 423.15 | 122 |
| **polystyrene-b-polybutadiene-b-polystyrene triblock copolymer** | | | | | |
| cyclohexane | 323.15 | 161 | cyclohexane | 373.15 | 161 |
| cyclohexane | 348.15 | 161 | cyclohexane | 393.15 | 161 |
| **polystyrene-b-poly(ethylene oxide) diblock copolymer** | | | | | |
| toluene | 322.95 | 75 | toluene | 342.65 | 75 |
| **polystyrene-b-polyisoprene-b-polystyrene triblock copolymer** | | | | | |
| cyclohexane | 323.15 | 161 | | | |
| **polystyrene-b-poly(methyl methacrylate) diblock copolymer** | | | | | |
| benzene | 343.15 | 153 | ethylbenzene | 398.15 | 153 |
| 1,4-dimethylbenzene | 398.15 | 153 | toluene | 343.15 | 153 |
| ethylbenzene | 373.15 | 153 | 1,3,5-trimethylbenzene | 398.15 | 153 |

| Solvent | T, K | Ref. | Solvent | T, K | Ref. |
|---|---|---|---|---|---|
| | | **poly(tetramethylene oxide)** | | | |
| benzene | 318.15 | 91 | 1,4-dioxane | 303.15 | 55 |
| tetrahydrofuran | 318.15 | 91 | 1,4-dioxane | 313.15 | 55 |
| | | **poly(vinyl acetate)** | | | |
| acetone | 298.15 | 110 | 1-chloropropane | 313.15 | 30 |
| acetone | 303.15 | 29, 30 | 1,2-dichloroethane | 299.55 | 86 |
| acetone | 308.15 | 110 | ethyl acetate | 303.15 | 29 |
| acetone | 313.15 | 30 | methanol | 303.15 | 29 |
| acetone | 318.15 | 110 | methanol | 353.15 | 121 |
| acetone | 323.15 | 30 | 1-propanol | 323.15 | 30 |
| allyl chloride | 313.15 | 30 | 1-propylamine | 313.15 | 30 |
| benzene | 303.15 | 29, 30, 31 | 2-propylamine | 313.15 | 30 |
| benzene | 313.15 | 98 | toluene | 299.55 | 86 |
| benzene | 323.15 | 30 | toluene | 308.15 | 19 |
| benzene | 333.15 | 98 | toluene | 313.15 | 19, 98, 120 |
| 1-butanol | 353.15 | 121 | toluene | 333.15 | 98, 120 |
| chloroform | 308.15 | 19 | toluene | 353.15 | 120 |
| chloroform | 313.15 | 19 | vinyl acetate | 303.15 | 31 |
| chloroform | 333.15 | 121 | | | |
| | | **poly(vinyl alcohol)** | | | |
| water | 303.15 | 32, 145, 147 | | | |
| | | **poly(vinylcarbazol)** | | | |
| benzene | 279.15 | 33 | benzene | 308.15 | 33 |
| benzene | 288.15 | 33 | benzene | 318.15 | 33 |
| benzene | 298.15 | 33 | benzene | 328.15 | 33 |
| | | **poly(vinyl chloride)** | | | |
| 2-butanone | 333.15 | 146 | tetrachloromethane | 338.15 | 65 |
| cyclohexanone | 313.15 | 146 | tetrahydrofuran | 315.65 | 23 |
| cyclohexanone | 333.15 | 146 | toluene | 316.35 | 23 |
| dibutyl ether | 315.35 | 23 | vinyl chloride | 340.15 | 93 |
| 1,4-dioxane | 315.65 | 23 | | | |
| | | **poly(vinyl methyl ether)** | | | |
| benzene | 298.15 | 65 | ethylbenzene | 398.15 | 118 |

| Solvent | T, K | Ref. | Solvent | T, K | Ref. |
|---|---|---|---|---|---|
| benzene | 323.15 | 75 | propylbenzene | 373.15 | 75 |
| benzene | 343.15 | 75 | toluene | 323.15 | 75 |
| chlorobenzene | 343.15 | 75 | toluene | 343.15 | 75 |
| chlorobenzene | 373.15 | 75 | o-xylene | 363.15 | 75 |
| chloroform | 298.15 | 65 | o-xylene | 373.15 | 118 |
| cyclohexane | 308.15 | 125 | o-xylene | 398.15 | 118 |
| cyclohexane | 318.15 | 125 | m-xylene | 373.15 | 118 |
| cyclohexane | 328.15 | 125 | m-xylene | 398.15 | 118 |
| cyclohexane | 338.15 | 125 | p-xylene | 373.15 | 118 |
| ethylbenzene | 343.15 | 75 | p-xylene | 398.15 | 118 |
| ethylbenzene | 373.15 | 118 | | | |
| **starch** | | | | | |
| water | 353.15 | 143 | water | 383.15 | 143 |
| water | 363.15 | 143 | water | 393.15 | 143 |
| water | 373.15 | 143 | water | 403.15 | 143 |
| **styrene/butadiene copolymer** | | | | | |
| acetone | 323.15 | 121 | ethylbenzene | 403.15 | 113 |
| acetone | 333.15 | 121 | n-hexane | 343.15 | 121 |
| benzene | 343.15 | 126, 134 | mesitylene | 398.15 | 126, 134 |
| chloroform | 323.15 | 121 | n-nonane | 373.15 | 114 |
| cyclohexane | 296.65 | 121 | n-nonane | 403.15 | 114 |
| cyclohexane | 333.15 | 121 | n-pentane | 333.15 | 121 |
| cyclohexane | 343.15 | 126, 134 | toluene | 343.15 | 126, 134 |
| ethylbenzene | 373.15 | 113, 126, 134 | toluene | 373.15 | 126, 134 |
| ethylbenzene | 398.15 | 126, 134 | p-xylene | 398.15 | 126, 134 |
| **styrene/butyl methacrylate copolymer** | | | | | |
| acetone | 333.15 | 121 | chloroform | 343.15 | 121 |
| **styrene/docosyl maleate copolymer** | | | | | |
| acetone | 323.15 | 140 | cyclohexane | 333.15 | 140 |
| acetone | 343.15 | 140 | methanol | 333.15 | 140 |
| **styrene/dodecyl maleate copolymer** | | | | | |
| acetone | 323.15 | 140 | cyclohexane | 333.15 | 140 |
| acetone | 343.15 | 140 | methanol | 333.15 | 140 |

| Solvent | T, K | Ref. | Solvent | T, K | Ref. |
|---|---|---|---|---|---|
| **styrene/maleic anhydride copolymer** | | | | | |
| acetone | 323.15 | 140 | methanol | 333.15 | 140 |
| **styrene/methyl methacrylate copolymer** | | | | | |
| acetone | 323.15 | 133 | mesitylene | 398.15 | 135 |
| benzene | 343.15 | 135 | methyl acetate | 323.15 | 133 |
| chloroform | 323.15 | 133 | toluene | 343.15 | 135 |
| ethylbenzene | 373.15 | 135 | toluene | 373.15 | 135 |
| ethylbenzene | 398.15 | 135 | p-xylene | 398.15 | 135 |
| **styrene/pentyl maleate copolymer** | | | | | |
| acetone | 323.15 | 140 | cyclohexane | 333.15 | 140 |
| acetone | 343.15 | 140 | methanol | 333.15 | 140 |
| **vinyl acetate/vinyl chloride copolymer** | | | | | |
| benzene | 398.15 | 148 | ethylbenzene | 428.15 | 148 |
| benzene | 418.15 | 148 | n-octane | 398.15 | 148 |
| 1-butanol | 353.15 | 121 | n-octane | 418.15 | 148 |
| chlorobenzene | 398.15 | 148 | p-xylene | 398.15 | 148 |
| ethylbenzene | 398.15 | 148 | p-xylene | 418.15 | 148 |
| ethylbenzene | 418.15 | 148 | | | |

## REFERENCE LIST OF APPENDIX 4.4A


1 G. Krahn, Dissertation, TH Leuna-Merseburg, 1973*).
2 J. H. Van der Waals, J. J. Hermans, *Rec. Trav. Chim. Pays-Bas*, **69**, 971 (1950).
3 P. J. T. Tait, P. J. Livesey, *Polymer*, **11**, 359 (1970).
4 S. Prager, E. Bagley, F. A. Long, *J. Amer. Chem. Soc.*, **75**, 2742 (1953).
5 C. H. Baker, W. B. Brown, G. Gee, J. S. Rowlinson, D. Stubley, R. E. Yeadon, *Polymer*, **3**, 215 (1962).
6 P. E. Eichinger, P. J. Flory, *Trans. Faraday Soc.*, **64**, 2035 (1968).
7 S. Saeki, J. C. Holste, D. C. Bonner, *J. Polym. Sci., Polym. Phys. Ed.*, **20**, 793 (1982).
8 C. E. H. Bawn, R. D. Patel, *Trans. Faraday Soc.*, **52**, 1664 (1956).
9 R. S. Jessup, *J. Res. Natl. Bur. Stand.*, **60**, 47 (1958).
10 D. C. Bonner, J. M. Prausnitz, *J. Polym. Sci., Polym. Phys. Ed.*, **12**, 51 (1974).
11 W. B. Brown, G. Gee, W. D. Taylor, *Polymer*, **5**, 362 (1964).
12 M. T. Raetzsch, M. Opel, Ch. Wohlfarth, *Acta Polymerica*, **31**, 217 (1980).
13 W. R. Krigbaum, D. O. Geymer, *J. Amer. Chem. Soc.*, **81**, 1859 (1959).
14 K. Schmoll, E. Jenckel, *Ber. Bunsenges. Phys. Chem.*, **60**, 756 (1956).
15 T. Katayama, K. Matsumara, Y. Urahama, *Kagaku Kogaku*, **35**, 1012 (1971).
16 E. C. Baughan, *Trans. Faraday Soc.*, **44**, 495 (1948).
17 S. Saeki, J. C. Holste, D. C. Bonner, *J. Polym. Sci., Polym. Phys. Ed.*, **19**, 307 (1981).
18 M. Opel, Dissertation, TH Leuna-Merseburg, 1978*).
19 J. S. Vrentas, J. L. Duda, S. T. Hsieh, *Ind. Eng. Chem., Process Des. Dev.*, **22**, 326 (1983).
20 M. Braeuer, Dissertation, TH Leuna-Merseburg, 1983*).
21 G. Illig, Dissertation, TH Leuna-Merseburg, 1981*).
22 M. T. Raetzsch, G. Illig, Ch. Wohlfarth, *Acta Polymerica*, **33**, 89 (1982).
23 P. J. T. Tait, A. M. Abushihada, *Polymer*, **18**, 810 (1977).

24 C. E. H. Bawn, R. F. J. Freeman, A. R. Kamaliddin, *Trans. Faraday Soc.*, **46**, 677 (1950).
25 C. Booth, G. Gee, G. Holden, G. R. Williamson, *Polymer*, **5**, 343 (1964).
26 G. Illig, Diploma Paper, TH Leuna-Merseburg, 1973*).
27 C. E. H. Bawn, M. A. Wajid, *Trans. Faraday Soc.*, **52**, 1658 (1956).
28 I. Noda, N. Kato, T. Kitano, M. Nagasawa, *Macromolecules*, **14**, 668 (1981).
29 K. Matsumara, T. Katayama, *Kagaku Kogaku,* **38**, 388 (1974).
30 R. J. Kokes, A. R. Dipietro, F. A. Long, *J. Amer. Chem. Soc.*, **75**, 6319 (1953).
31 A. Nakajima, H. Yamakawa, I. Sakurada, *J. Polym. Sci.*, **35**, 489 (1959).
32 I. Sakurada, A. Nakajima, H. Fujiwara, *J. Polym. Sci.*, **35**, 497 (1959).
33 K. Ueberreiter, W. Bruns, *Ber. Bunsenges. Phys. Chemie*, **68**, 541 (1964).
34 J. Belorussow, Diploma Paper, TH Leuna-Merseburg, 1973*).
35 K. Peinze, Diploma Paper, TH Leuna-Merseburg, 1972*).
36 M. Boblenz, D. Glindemann, Diplom Paper, TH Leuna-Merseburg, 1975*).
37 D. Kiessling, Diploma Paper, TH Leuna-Merseburg, 1976*).
38 G. N. Malcolm, J. S. Rowlinson, *Trans. Faraday Soc.*, **53**, 921 (1957).
39 Z. Adamcova, *Sci. Pap. Prag. Inst. Chem. Technol.*, **N2**, 63 (1976).
40 M. L. Lakhanpal, K. S. Chhina, S. C. Sharma, *Indian J. Chem.*, **6**, 505 (1968).
41 Z. N. Medved, P. P. Petrova, O. G. Tarakanov, *Vysokomol. Soedin., Ser. B*, **24**, 674 (1982).
42 Ch. Wohlfarth, M. T. Raetzsch, *Acta Polymerica*, **37**, 86 (1986).
43 M. L. Lakhanpal, H. G. Singh, S. C. Sharma, *Indian J. Chem.*, **6**, 436 (1968).
44 Ch. Wohlfarth, W. Zschoch, M. T. Raetzsch, *Acta Polymerica*, **32**, 616 (1981).
45 Ch. Wohlfarth, H. Hahmann, M. T. Raetzsch, *Acta Polymerica*, **32**, 674 (1982).
46 E. Regener, Diploma Paper, TH Leuna-Merseburg, 1983*).
47 S. H. Chang, D. C. Bonner, *J. Appl. Polym. Sci.*, **19**, 2457 (1975).
48 G. Allen, C. Booth, G. Gee, M. N. Jones, *Polymer*, **5**, 367 (1964).
49 M. L. Lakhanpal, B. E. Conway, *J. Polym. Sci.,* **46**, 75 (1960).
50 R. W. Kershaw, G. N. Malcolm, *Trans. Faraday Soc.*, **64**, 323 (1968).
51 K. Kubo, K. Ogino, *Polymer*, **16**, 629 (1975).
52 M. J. Newing, *Trans. Faraday Soc.*, **46**, 613 (1950).
53 A. Muramoto, *Polymer*, **23**, 1311 (1982).
54 J. S. Aspler, D. G. Gray, *Polymer*, **23**, 43 (1982).
55 S. C. Sharma, M. L. Lakhanpal, *J. Polym. Sci., Polym. Phys. Ed.*, **21**, 353 (1983).
56 E. Regener, Ch. Wohlfarth, M. T. Raetzsch, S. Hoering, *Acta Polymerica*, **39**, 618 (1988).
57 P. E. Eichinger, P. J. Flory, *Trans. Faraday Soc.*, **64**, 2053 (1968).
58 P. E. Eichinger, P. J. Flory, *Trans. Faraday Soc.*, **64**, 2061 (1968).
59 P. E. Eichinger, P. J. Flory, *Trans. Faraday Soc.*, **64**, 2066 (1968).
60 S. Saeki, J. C. Holste, D. C. Bonner, *J. Polym. Sci., Polym. Phys. Ed.*, **20**, 805 (1982).
61 S. Saeki, J. Holste, D. C. Bonner, *J. Polym. Sci., Polym. Phys. Ed.*, **21**, 2049 (1983).
62 E. Dolch, M. Glaser, A. Heintz, H. Wagner, R. N. Lichtenthaler, *Ber. Bunsenges. Phys. Chem.*, **88**, 479 (1984).
63 A. J. Ashworth, C.-F. Chien, D. L. Furio, D. M. Hooker, M. M. Kopecni, R. J. Laub, G. J. Price, *Macromolecules*, **17**, 1090 (1984).
64 K. Schotsch, B. A. Wolf, *Makromol. Chem.*, **185**, 2161 (1984).
65 C. Panayiotou, J. H. Vera, *Polym. J.*, **16**, 89 (1984).
66 A. J. Ashworth, G. J. Price, *Thermochim. Acta*, **82**, 161 (1984).
67 D. Hailemariam, Diploma Paper, TH Leuna-Merseburg, 1985*).
68 M. Krcek, Diploma Paper, TH Leuna-Merseburg , 1986*).
69 C. Booth, G. Gee, M. N. Jones, W. D. Taylor, *Polymer*, **5**, 353 (1964).
70 N.-H. Wang, S. Takashima, H. Masuoka, *Kagaku Kogaku Ronbunshu*, **15**, 313 (1989).
71 N.-H. Wang, S. Takashima, H. Masuoka, *Kagaku Kogaku Ronbunshu*, **15**, 795 (1989).
72 R. Corneliussen, S. A. Rice, H. Yamakawa, *J. Chem. Phys.*, **38**, 1768 (1963).
73 C. Booth, C. J. Devoy, *Polymer*, **12**, 320 (1971).
74 C. Booth, C. J. Devoy, *Polymer*, **12**, 309 (1971).
75 Ch. Wohlfarth, unpublished data, TH Leuna-Merseburg, Institute of physical chemistry 1979 - 1993, Martin-Luther-University, Institute of physical chemistry 1993 - 1999*).
76 C. A. Haynes, R. A. Beynon, R. S. King, H. W. Blanch, J. M. Praunsitz, *J. Phys. Chem.*, **93**, 5612 (1989).
77 W. R. Moore, R. Shuttleworth, *J. Polym. Sci., Pt. A*, **1**, 1985 (1963).

78 A. L. Jones, *Trans. Faraday Soc.*, **52**, 1408 (1956).
79 E. C. Baughan, A. L. Jones, K. Stewart, *Proc. Roy. Soc., London, Ser. A*, **225**, 478 (1954).
80 G. N. Malcolm, C. E. Baird, G. R. Bruce, K. G. Cheyne, R. W. Kershaw, M. C. Pratt, *J. Polym. Sci., Pt. A-2*, **7**, 1495 (1969).
81 W. R. Moore, R. Shuttleworth, *J. Polym. Sci., Pt. A*, **1**, 733 (1963).
82 K. H. Meyer, E. Wolff, Ch. G. Boissonnas, *Helv. Chim. Acta*, **23**, 430 (1940).
83 S. G. Canagratna, D. Margerison, J. P. Newport, *Trans. Faraday Soc.*, **62**, 3058 (1966).
84 H. Ochiai, K. Gekko, H. Yamamura, *J. Polym. Sci., Pt. A-2*, **9**, 1629 (1971).
85 R. C. Osthoff, W. T. Grubb, *J. Amer. Chem. Soc.*, **76**, 399 (1954).
86 A. A. Tager, A. I. Suvorova, Yu.S. Bessonov, A. I. Podlesnyak, I.A. Koroleva, L. V. Adamova, *Vysokomol. Soedin., Ser. A*, **13**, 2454 (1971).
87 A. A. Tager, L. V. Adamova, Yu.S. Bessonov, V. N. Kuznetsov, T. A. Plyusnina, V. V. Soldatov, *Vysokomol. Soedin., Ser. A*, **14**, 1991 (1972).
88 T. V. Gatovskaya, V. A. Kargin, A. A. Tager, *Zh. Fiz. Khim.*, **29**, 883 (1955).
89 J. Leonard, Van Tam Bui, *Polymer*, **28**, 1041 (1987).
90 A. Hamdouni, J. Leonard, Van Tam Bui, *Polymer Commun.*, **31**, 258 (1990).
91 Van Tam Bui, J. Leonard, *J. Chem. Soc., Faraday Trans. I*, **82**, 899 (1986).
92 Van Tam Bui, J. Leonard, *J. Chem. Soc., Faraday Trans. I*, **81**, 1745 (1985).
93 A. H. Abdel-Alim, *J. Appl. Polym. Sci.*, **22**, 3597 (1978).
94 U. Messow, Inst. Phys. Chem., Univ. Leipzig, personal communication.
95 F. Cordt, Dissertation, TU Muenchen, 1985.
96 F. Moeller, Dissertation, TU Muenchen, 1989.
97 G. Luengo, G. Rojo, R. G. Rubio, M. G. Prolongo, R. M. Masegosa, *Macromolecules*, **24**, 1315 (1991).
98 N. H. Wang, K. Hattori, S. Takashima, H. Masuoka, *Kagaku Kogaku Ronbunshu*, **17**, 1138 (1991).
99 A. J. Ashworth, G. J. Price, *Macromolecules*, **19**, 362 (1986).
100 A. J. Ashworth, G. J. Price, *J. Chem. Soc., Faraday Trans. I*, **81**, 473 (1985).
101 A. J. Ashworth, G. J. Price, *Macromolecules*, **19**, 358 (1986).
102 I. Noda, Y. Higo, N. Ueno, T. Fujimoto, *Macromolecules*, **17**, 1055 (1984).
103 H. W. Starkweather, Jr., *J. Appl. Polym. Sci.*, **2**, 129 (1959).
104 M. Herskowitz, M. Gottlieb, *J. Chem. Eng. Data*, **30**, 233 (1985).
105 Van Tam Bui, J. Leonard, *Polym. J.*, **21**, 185 (1989).
106 Y. Iwai, Y. Arai, *J. Chem. Eng. Japan*, **22**, 155 (1989).
107 H. Masuoka, N. Murashige, M. Yorizane, *Fluid Phase Equil.*, **18**, 155 (1984).
108 Y. C. Bae, J. J. Shim, D. S. Soane, J. M. Prausnitz, *J. Appl. Polym. Sci.*, **47**, 1193 (1993).
109 Ch. Wohlfarth, *Plaste & Kautschuk*, **40**, 272 (1993).
110 G. Luengo, R. G. Rubio, I. C. Sanchez, C. G. Panayiotou, *Macromol. Chem. Phys.*, **195**, 1043 (1994).
111 J. Gaube, A. Pfennig, M. Stumpf, *J. Chem. Eng. Data*, **38**, 163 (1993).
112 J. Zhu, Dissertation, Univ. Kaiserslautern, 1991.
113 Y. Iwai, S. Miyamoto, K. Nakano, Y. Arai, *J. Chem. Eng. Japan*, **23**, 508 (1990).
114 Y. Iwai, T. Ishidao, S. Miyamoto, H. Ikeda, Y. Arai, *Fluid Phase Equil.*, **68**, 197 (1991).
115 K. Gekko, K. Matsumara, *Bull. Chem. Soc. Japan*, **46**, 1554 (1973).
116 Ch. Wohlfarth, *Plaste & Kautschuk*, **41**, 163 (1994).
117 M. Koester, Diploma Paper, TH Darmstadt, 1994.
118 Ch. Wohlfarth, ELDATA: *Int. Electron. J. Phys.-Chem. Data*, **1**, 113 (1995).
119 C. Grossmann, R. Tintinger, G. Maurer, *Fluid Phase Equil.*, **106**, 111 (1995).
120 S. P. V. N. Mikkilineni, D.A. Tree, M. S. High, *J. Chem. Eng. Data*, **40**, 750 (1995).
121 R. B. Gupta, J. M. Prausnitz, *J. Chem. Eng. Data*, **40**, 784 (1995).
122 J. S. Choi, K. Tochigi, K. Kojima, *Fluid Phase Equil.*, **111**, 143 (1995).
123 S. Behme, G. Sadowski, W. Arlt, *Chem.-Ing.-Techn.*, **67**, 757 (1995).
124 H.-M. Petri, N. Schuld, B. A. Wolf, *Macromolecules*, **28**, 4975 (1995).
125 H.-M. Petri, Dissertation, Univ. Mainz, 1994.
126 Ch. Wohlfarth, ELDATA: *Int. Electron. J. Phys.-Chem. Data*, **2**, 13 (1996).
127 K. Wang, Y. Chen, J. Fu, Y. Hu, Chin. *J. Chem. Eng.*, **1**, 65 (1993).
128 D.-Q. Lin, L.-H. Mei, Z.-Q. Zhu, Z.-X. Han, *Fluid Phase Equil.*, **118**, 241 (1996).
129 L.R. Ochs, M. Kabri-Badr, H. Cabezas, *AIChE-J.*, **36**, 1908 (1990).
130 Ch. Wohlfarth, *ELDATA: Int. Electron. J. Phys.-Chem. Data*, **2**, 163 (1996).
131 K. Wang, Q. Xue, J. Fu, Y. Hu, *Huadong Ligong Daxue Xuebao*, **22**, 330 (1996).

132 K. Wang, Q. Xue, Y. Hu, *Huadong Ligong Daxue Xuebao*, **23**, 109 (1997).
133 J. O. Tanbonliong, J. M. Prausnitz, *Polymer*, **38**, 5775 (1997).
134 Ch. Wohlfarth, *Macromol. Chem. Phys*., **198**, 2689 (1997).
135 Ch. Wohlfarth, ELDATA: *Int. Electron. J. Phys.-Chem. Data*, **3**, 47 (1997).
136 K. Tochigi, S. Kurita, M. Ohashi, K. Kojima, *Kagaku Kogaku Ronbunshu*, **23**, 720 (1997).
137 R. K. Surana, R. P. Danner, A. B. De Haan, N. Beckers, *Fluid Phase Equil*., **139**, 361 (1997).
138 H. C. Wong, S. W. Campbell, V. R. Bhetnanabotla, *Fluid Phase Equil*., **139**, 371 (1997).
139 G. Sadowski, L. V. Mokrushina, W. Arlt, *Fluid Phase Equil*., **139**, 391 (1997).
140 C. Mio, K. N. Jayachandran, J. M. Prausnitz, *Fluid Phase Equil*., **141**, 165 (1997).
141 J. S. Aspler, D. G. Gray, *Macromolecules*, **12**, 562 (1979).
142 K. N. Jayachandran, P. R. Chatterji, J. M. Prausnitz, *Macromolecules*, **31**, 2375 (1998).
143 Benczedi, D., Tomka, I., Escher, F., *Macromolecules*, **31**, 3062 (1998).
144 C. Mio, S. Kiritsov, Y. Thio, R. Brafman, J. M. Prausnitz, C. Hawker, E. E. Malmstroem, *J. Chem. Eng. Data*, **43**, 541 (1998).
145 S. Hwang, J. Kim, K.-P. Yoo, *J. Chem. Eng. Data*, **43**, 614 (1998).
146 F. Wie, W. Wenchuan, F. Zhihao, *J. Chem. Ind. Eng. China*, **49**, 217 (1998).
147 J. Kim, K. C. Joung, S. Hwang, W. Huh, C. S. Lee, K.-P. Yoo, *Korean J. Chem. Eng*., **15**, 199 (1998).
148 N. H. Kim, S. J. Kim, Y. S. Won, J. S. Choi, *Korean J. Chem. Eng*., **15**, 141 (1998).
149 Y. Dahong, S. Jibin, L. Xiaohui, H. Ying, *J. East China Univ. Sci. Technol*., **23**, 608 (1997).
150 W. Hiyan, W. Kun, L. Honglai, H. Ying, *J. East China Univ. Sci. Technol*., **23**, 614 (1997).
151 A. Eliassi, H. Modarress, G. A. Mansoori, *J. Chem. Eng. Data*, **44**, 52 (1999).
152 C. Mio, J. M. Prausnitz, *Polymer*, **39**, 6401 (1998).
153 Ch. Wohlfarth, *ELDATA: Int. Electron. J. Phys.-Chem. Data*, **4**, 83 (1998).
154 H.-P. Kany, H. Hasse, G. Maurer, *J. Chem. Eng. Data*, **44**, 230 (1999).
155 J. G. Lieu, M. Liu, J. M. J. Frechet, J. M. Prausnitz, *J. Chem. Eng. Data*, **44**, 613 (1999).
156 J. G. Lieu, J. M. Prausnitz, *Polymer*, **40**, 5865 (1999).
157 J. Kim, E. Choi, K.-P. Yoo, C. S. Lee, *Fluid Phase Equil*., **161**, 283 (1999).
158 L. Ninni, M. S. Camargo, A. J. A. Meirelles, *Thermochim. Acta*, **328**, 169 (1999).
159 K. Wang, Y. Hu, D. T. Wu, *J. Chem. Eng. Data*, **39**, 916 (1994).
160 D.-Q. Lin, Y.-T. Wu, Z.-Q. Zhu, L.-H. Mei, S.-J. Yao, *Fluid Phase Equil*., **162**, 159 (1999).
161 R. N. French, G. J. Koplos, *Fluid Phase Equil*., **158-160**, 879 (1999).

*) all data from the Merseburg group before 1994 have been published in Ref. Ch. Wohlfarth, Vapor-liquid equilibrium data of binary polymer solutions, Physical Science Data 44, Elsevier, Amsterdam, 1994

# 5

# Solubility of Selected Systems and Influence of Solutes

## 5.1 EXPERIMENTAL METHODS OF EVALUATION AND CALCULATION OF SOLUBILITY PARAMETERS OF POLYMERS AND SOLVENTS. SOLUBILITY PARAMETERS DATA.

Valery Yu. Senichev, Vasiliy V. Tereshatov
**Institute of Technical Chemistry**
**Ural Branch of Russian Academy of Sciences, Perm, Russia**

### 5.1.1 EXPERIMENTAL EVALUATION OF SOLUBILITY PARAMETERS OF LIQUIDS

The value of solubility parameter can be calculated from the evaporation enthalpy of liquid at given temperature:[1]

$$\delta = \left( \frac{\Delta H_p - RT}{V} \right)^{1/2} \qquad [5.1.1]$$

where:

$\Delta H_p$ latent heat of vaporization
$V$ molar volume

#### 5.1.1.1 Direct methods of evaluation of the evaporation enthalpy

For measurement of the evaporation enthalpy of volatile substances, adiabatic apparatuses were developed. They require significant quantities of highly purified substances. The accuracy is determined to a large degree by equipment design and precision of measurement.

Most calorimeters that measure a latent heat of vaporization work under isobaric conditions. The measurement of a latent heat of vaporization requires monitoring heat input into calorimeter and the amount of liquid evaporated during measurement time.[2-4]

In calorimeters of a flowing type,[5-6] a liquid evaporates from a separate vessel of a calorimeter. The vapors are directed into the second calorimeter where the thermal capacity of gas is measured. The design of such calorimeters ensures a precise measurement of the stream rate. Heaters and electrical controls permit control of heat flow with high precision and highly sensitive thermocouples measure temperature of gas. There are no excessive thermal losses, thus single-error corrections for the heat exchange can be used to increase precision of measurement.

The calorimeters used for the measurement of the heat of reaction can also be used for a measurement of the latent heat of vaporization. These are calorimeters for liquids, micro-calorimeters, mass calorimeters, and double calorimeters.[7]

The calorimeters with carrier gas are also used.[8-9] Evaporation of substance is accelerated by a stream of gas (for example, nitrogen) at reduced pressure. The heat loss by a calorimeter, due to evaporation, is compensated by an electrical current to keep temperature of calorimeter constant and equal to the temperature of the thermostating bath.

### 5.1.1.2 Indirect methods of evaluation of evaporation enthalpy

Because the calorimetric methods of measurement of enthalpy of vapor formation are very difficult, the indirect methods are used, especially for less volatile substances. The application of generalized expression of the first and second laws of thermodynamics to the heterogeneous equilibrium between a condensed phase in isobaric- thermal conditions is given in the Clausius-Clapeyron equation that relates enthalpy of a vapor formation at the vapor pressure, P, and temperature, T. For one component system, the Clausius-Clapeyron equation has the form:[7]

$$dP/dT = \Delta H_p / T\Delta V \quad [5.1.2]$$

where:

$\Delta V$ difference between molar volumes of vapor and liquid

The ratio that neglects volume of a condensed phase with assumption that vapor at low pressure is ideal can be derived from the above equation:

$$d\ln P / d(1/T) = -\Delta H_p / R \quad [5.1.3]$$

After integration:

$$\ln P = -\Delta H_p / RT + const \quad [5.1.4]$$

Introducing compressibility factors of gas and liquids, $\Delta Z$, the Clausius-Clapeyron equation can be written as:

$$d\ln P / d(1/T) = -\Delta H_p / R\Delta Z \quad [5.1.5]$$

where:

$\Delta Z$ difference between compressibility factors of gas and liquids

The value $\Delta Z$ includes corrections for volume of liquid and non-ideality of a vapor phase. The simplifying assumptions give the equation:

$$\ln P = A + B/T \quad [5.1.6]$$

Approximate dependence of a vapor pressure on inverse temperature is frequently linear but the dependence may also be non-linear because of changing ratio of $\Delta H_p/\Delta Z$ on heating. The mathematical expressions of the dependence lnP on 1/T of real substances in a wide range of temperatures should be taken into account. If $\Delta H_p/\Delta Z = a + bT$, it results in an equation with three constants:

$$\ln P = A + B / T + C \ln T \quad [5.1.7]$$

In more complicated dependancies, the number of constants may further increase.

Another convenient method is based on empirical relation of $\Delta H_p$ at 25°C with the normal boiling point, $T_b$, of non-polar liquids:[1]

$$\Delta H_p = T_b^2 + 23.7T_b - 2950 \quad [5.1.8]$$

Methods of evaluation of vapor pressure may be divided into static, quasi-static, and kinetic methods.

### 5.1.1.3 Static and quasi-static methods of evaluation of pair pressure

Manometric method[10] consists of thermostating with a high precision (0.01K) and vapor pressure measurement by a level of mercury with the help of a cathetometer or membrane zero-manometer. The accuracy of measurement is 0.1-0.2 mm Hg.

Ebulliometric method[11] is used for a simultaneous measurement of the boiling and condensation temperature that is required for evaluation of purity of a substance and its molecular mass.

### 5.1.1.4 Kinetic methods

These methods were developed based on the molecular kinetic theory of gases. The Langmuir method is based on the evaporation of substance from a free surface into a vacuum. The Knudsen method is based on the evaluation of the outflow rate of a vapor jet from a mesh.

The basic expression used in Langmuir method[12] is:

$$P = \left(\frac{m}{St\alpha}\right)\left(\frac{2\pi RT}{M}\right)^{1/2} \quad [5.1.9]$$

where:

| | |
|---|---|
| m | mass of evaporated substance |
| S | surface of evaporation |
| t | time of evaporation. |

The Knudsen method[13] is based on a measurement of the mass rate of the vapor outflow through a hole. Knudsen proposed the following expression:

$$P_k = \left(\frac{\Delta m}{S_h t\beta}\right)\left(\frac{2\pi RT}{M}\right)^{1/2} \quad [5.1.10]$$

where:

| | |
|---|---|
| $\Delta m$ | mass output of substance |
| $S_h$ | surface area of the hole |
| t | time of vaporization |
| $\beta$ | Clausing parameter |
| M | molecular mass |

The method uses special effusion cameras with holes of a definite form, maintaining high vacuum in the system. The method is widely applied to the measurements of a vapor pressure of low volatile substances.

The detailed comparative evaluation of experimental techniques and designs of equipment used for determination of enthalpy of evaporation can be found in the appropriate monographs.[7,14] Values of solubility parameters of solvents are presented in Subchapter 4.1.

## 5.1.2 METHODS OF EXPERIMENTAL EVALUATION AND CALCULATION OF SOLUBILITY PARAMETERS OF POLYMERS

It is not possible to determine solubility parameters of polymers by direct measurement of evaporation enthalpy. For this reason, all methods are indirect. The underlining principles of these methods are based on the theory of regular solutions that assumes that the best mutual dissolution of substances is observed at the equal values of solubility parameters (see Chapter 4).

Various properties of polymer solutions involving interaction of polymer with solvent are studied in a series of solvents having different solubility parameters. A value of a solubility parameter is related to the maximum value of an investigated property and is equated to a solubility parameter of polymer.

This subchapter is devoted to the evaluation of one-dimensional solubility parameters. The methods of the evaluation of components of solubility parameters in multi-dimensional approaches are given in the Subchapter 4.1.

According to Gee,[15] a dependence of an equilibrium swelling of polymers in solvents on their solubility parameters is expressed by a curve with a maximum where the abscissa is equal to the solubility parameter of the polymer. For exact evaluation of δ, a swelling degree is represented by an equation resembling the Gaussian function:

$$Q = Q_{max} \exp\left[-V_1(\delta_1 - \delta_2)^2\right] \qquad [5.1.11]$$

where:

| | |
|---|---|
| $Q_{max}$ | the degree of swelling at the maximum on the curve |
| $V_1$ | molar volume of solvent |
| $\delta_1, \delta_2$ | solvent and polymer solubility parameters. |

Then

$$\delta_2 = \delta_1 \pm \left(\frac{1}{V_1}\ln\frac{Q_{max}}{Q}\right)^{1/2} \qquad [5.1.12]$$

The dependence $[(1/V_1)\ln(Q_{max}/Q)]^{1/2} = f(\delta_1)$ is expressed by a direct line intersecting the abscissa at $\delta_1 = \delta_2$. This method is used for calculation of the parameters of many crosslinked elastomers.[16-19]

The Bristow-Watson method is based on the Huggins equation deduced from a refinement of the lattice approach:[20]

$$\chi = \beta + (V_1 / RT)(\delta_1 - \delta_2)^2 \qquad [5.1.13]$$

where:

| | |
|---|---|
| β | $=(1/z)(1-1/m)$ |
| z | a coordination number |
| m | the chain length. |

β may be rewritten as $\chi_S$ entropy contribution to χ (see Chapter 4).

Accepting that Eq. [5.1.12] represents a valid means of assignment of a constant $\delta_2$ to polymer, the rearrangement of this equation gives:

$$\frac{\delta_1^2}{RT} - \frac{\chi}{V_1} = \left(\frac{2\delta_2}{RT}\right)\delta_1 - \frac{\delta_2^2}{RT} - \frac{\chi_S}{V_1} \qquad [5.1.14]$$

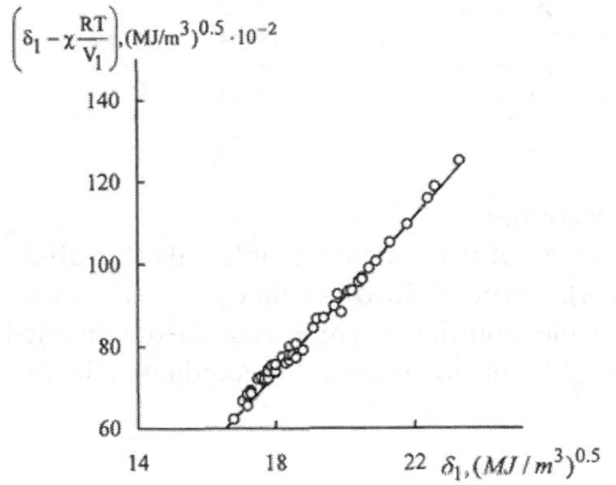


Figure 5.1.1. Dependence for equilibrium swelling of crosslinked elastomer on the base of polyether urethane. [Adapted, by permission, from V.Yu. Senichev in Synthesis and properties of cross-linked polymers and compositions on their basis. *Russian Academy of Sciences Publishing*, Sverdlovsk,1990, p.16]

Now it is assumed that $\chi_S$ is of the order of magnitude suggested above and that, in accordance with the Huggins equation, it is not a function of $\delta_2$. Therefore $\chi_S/V_1$ is only about 3% or less of $\delta_2^2/RT$ for reasonable values of $\delta_2$ of 10-20 $(MJ/m^3)^{1/2}$. Hence Eq. [5.1.14] gives $\delta_2$ from the slope and intercept on plot against $\delta_1$ (see Figure 5.1.1).

This method was improved[21] by using calculations that exclude strong deviations of $\chi$. When ($\chi_S RT/V_1 \approx$ const), Eq. [5.1.14] is close to linear (y = A + Bx), where $y = \delta_1^2 - \chi_S(RT/V_1)$, $A = -\chi_S(RT/V_1) - \delta_2^2$, $B = 2\delta_2$, $x = \delta_1$.

But $\delta_2$ enters into expression for a tangent of a slope angle and intercept which is cut off on the ordinates axes. This can be eliminated by introduction of a sequential approximation of $\chi_S(RT/V_1)$ and grouping of experimental points in areas characterized by a definite interval of values $\chi_S(RT/V_1)$. Inside each area $\chi_S(RT/V_1) \rightarrow$ const and Eq. [5.1.14] becomes more precise.

The intervals of values $\chi_S(RT/V_1)$ are reduced in the course of computations. For n experimental points, the files X ($x_1, x_2, \ldots x_n$) and Y ($y_1, y_2, \ldots y_n$) are gathered. Tangent of the slope angle is defined by the method of least squares and the current value (at the given stage) of a solubility parameter of a polymer is:

$$\delta_{2j} = \frac{\sum_{i=1}^{n} x \sum_{i=1}^{n} y - n\sum_{i=1}^{n} xy}{\left(\sum_{i=1}^{n} x\right)^2 - n\sum_{i=1}^{n} x^2} \qquad [5.1.15]$$

where:

j a stage of computation

$\chi_S(RT/V_1)$ is then calculated using the equation derived from Eqs. [5.1.13] and [5.1.14]:

$$\left(\chi_S \frac{RT}{V_1}\right)_j = -y_i - \delta_{2j}^2 + 2x_i\delta_{2j} \qquad [5.1.16]$$

**Table 5.1.1. Modification of $\bar{\delta}_{2j}$ values during stages of computation**

| Polymer | j = 1 | j = 2 | j = 3 | j = 4 |
|---|---|---|---|---|
| | $\bar{\delta}_{2j}$, $(MJ/m^3)^{1/2}$ | | | |
| polydiene urethane epoxide | 17.64 | 17.88 | 17.78 | 17.8 |
| polydiene urethane | 17.72 | 18.17 | 17.93 | 17.82 |
| poly(butylene glycol) urethane | 19.32 | 18.89 | 18.95 | 18.95 |
| poly(diethylene glycol adipate) urethane | 19.42 | 19.44 | 19.44 | - |

where:

$\delta_{2j}$ value of $\delta_2$ at the given stage of computation

By sorting of all experimental points into a defined amount of intervals (for 30-50 points it is more convenient to take 5-6 intervals), it is possible to calculate $\delta_2$ for each interval separately. The current average weighted value (contribution of $\delta_2$ is defined), obtained for each interval, is proportional to the amount of points in the interval according to the following formula:

$$\bar{\delta}_{2j} = \frac{1}{M}\sum_{k=1}^{k} \delta_{2k} m_k \qquad [5.1.17]$$

where:

k = 1,2,.... k, number of intervals
$m_k$ number of points in k-interval
M the total number of points
j stage of computation.

The shaping of subarrays of points is made in the following order, ensuring that casual points are excluded: 1) account is made in a common array of points of $\delta_2$ and $\chi_S(RT/V_1)_i$; 2) partition of a common array into a population of subarrays of $\chi_S(RT/V_1)$ in limits defined for elimination of points not included in intervals and points which do not influence consequent stages of computation, 3) reductions of intervals in each of the subarrays (this stage may be repeated in some cases).

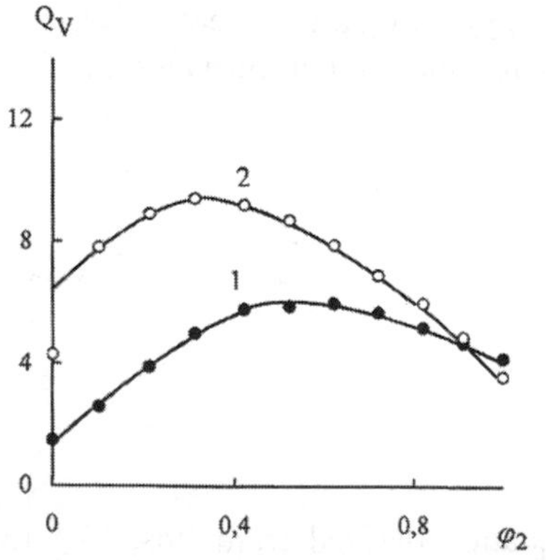

Figure 5.1.2. Dependence of equilibrium swelling of crosslinked elastomer of polyester urethane (1) and polybutadiene nitrile rubber (2) on the volume fraction of acetone in the toluene-acetone mixture. [Adapted, by permission, from V.V. Tereshatov, V.Yu. Senichev, A.I. Gemuev, *Vysokomol. soed.*, **B32**, 412 (1990)]

At the each stage the sequential approximation to constant value of $\chi_S(RT/V_1)$ is produced in a separate form, permitting one to take into account the maximum number of points. The procedure gives a sequence of values $\delta_{2j}$ as shown in Table 5.1.1.

In still other methods of evaluation,[22] the solvents are selected so that the solubility parameter of polymer occupies an intermediate position between solubility parameters of solvents. Assumption is made that the maximum polymer swelling occurs when the solubility parameters of the solvent mixture and polymer are equal. This is the case when the solubility parameter of polymer is lower than the solubility parameter of the primary solvent and the solubility parameter of the secondary solvent is higher. The dependence of the swelling ratio on the composition of solvent mixture has a maximum (see Figure 5.1.2). Such mixed solvents are called symmetric liquids. The reliability of the method is examined by a narrow interval of change of the solubility parameter of a binary solvent. The data obtained by this method in various mixtures differ by no more than 1.5% with the data obtained by other methods. Examples of results are given in Tables 5.1.2, 5.1.3.

**Table 5.1.2. Values of solubility parameters for crosslinked elastomers from swelling in symmetric liquids. [Adapted, by permission, from V.V. Tereshatov, V.Yu. Senichev, A.I. Gemuev, *Vysokomol. soed.*, B32, 412 (1990).]**

| Elastomer | Symmetric liquids | $\varphi_2^H$ at $Q_{max}$ | $\delta_p$, $(MJ/m^3)^{1/2}$ |
|---|---|---|---|
| Polyether urethane | Toluene( 1) - acetone (2)<br>Cyclohexane (1) -acetone (2) | 0.11<br>0.49 | 18.4<br>18.2 |
| Ethylene-propylene rubber | Hexane (1) - benzene (2)<br>Toluene (1) - acetone (2) | 0.38<br>0.32 | 16.2<br>18.9 |
| Butadiene-nitrile rubber | Ethyl acetate (1) - acetone (2)<br>Benzene (1) - acetone (1) | 0.22<br>0.27 | 19.1<br>19.1 |
| Polyester-urethane | Toluene (1) - acetone (2)<br>o-Xylene (1) - Butanol -1(2) | 0.62<br>0.21 | 19.6<br>19.5 |
| Butyl rubber | Octane (1) - benzene (2) | 0.26 | 16.1 |

**Table 5.1.3. Values of solubility parameters of crosslinked elastomers from swelling in individual solvents and symmetric liquids. [Adapted, by permission, from V.V. Tereshatov, V.Yu. Senichev, A.I. Gemuev, *Vysokomol. soed.*, B32, 412 (1990).]**

| Elastomer | Symmetric liquids | Individual solvents | |
|---|---|---|---|
| | | Gee method | Bristow-Watson method |
| | $(MJ/m^3)^{1/2}$ | | |
| Polyether-urethane | 18.3±0.1 | 17.8, 18.4, 19.4 | 19.2 |
| Ethylene-propylene rubber | 16.2 | 16.1-16.4 | - |
| Butadiene-nitrile rubber | 19.0±0.1 | 18.9-19.4 | 18.7 |
| Polyester-urethane | 19.5-19.6 | 19.3, 19.9 | 19.5 |
| Butyl rubber | 16.1 | 15.9-16.6 | 14.9 |

The calculations were made using equation:

$$\delta_p = \delta_{mix} = \left[\delta_1^2(1-\varphi_1'') + \delta_2^2\varphi_1'' - \Delta H_{mix} / V_{12}\right]^{1/2} \qquad [5.1.18]$$

where:

$\varphi_1''$ volume fraction of a solvent in a binary mixture of solvents causing a maximum of equilibrium swelling

$\Delta H_{mix}$ experimental value of the mixing enthalpy of components of binary solvent. It can be taken from literature.[23-24]

$V_{12}$ the molar volume of binary solvent

If there are no volume changes, $V_{12}$ can be calculated using the additivity method:

$$V_{12} = V_1\varphi_1 + V_2\varphi_2 \quad [5.1.19]$$

Attempts[25] were made to relate intrinsic viscosity [η] to solubility parameters of mixed solvents. $\delta_2$ of polymer was calculated from the equation: $[\eta] = f(\delta_1)$. The authors assumed that the maximum value of [η] is when $\delta_1 = \delta_2$ of polymer. However, studying [η] for polymethylmethacrylate in fourteen liquids, the authors found a large scatter of experimental points through which they have drawn a curve with a diffusion maximum. Thus the precision of $\delta_2$ values was affected by 10% scatter in experimental data.

This method was widely used by Mangaray et al.[26-28] The authors have presented [η] as the Gaussian function of $(\delta_1 - \delta_2)^2$. Therefore, dependence $\{(1/V_1)\ln[\eta]_{max}/[\eta]\}^{1/2} = f(\delta_1)$ can be expressed by a straight line intersecting the abscissa at a point for which $\delta_1 = \delta_2$. For natural rubber and polyisobutylene, the paraffin solvents and ethers containing alkyl chains of a large molecular mass were studied.[26] For polystyrene, aromatic hydrocarbons were used. For polyacrylates and polymethacrylates esters (acetates, propionates, butyrates) were used.[27,28] The method was used for determination of $\delta_2$ of many polymers.[29-31] In all cases, the authors observed extrema in the dependence of $[\eta] = f(\delta_1)$, and the obtained values of $\delta_2$ coincided well with the values determined by other methods. But for some polymers it was not possible to obtain extremum in dependence of $[\eta] = f(\delta_1)$.[32]

A method of the evaluation [η] in one solvent at different temperatures was used for polyisobutylene[33] and polyurethanes.[34]

For polymers soluble in a limited range of solvents, more complex methods utilizing [η] relationship are described.[35-37]

In addition to the above methods, $\delta_2$ of polymer can be determined from a threshold of sedimentation[38] and by critical opalescence.[39] In recent years the method of inverse gas-liquid chromatography has been used to evaluate $\delta_2$ of polymers.[40,41] One may also use some empirical ratios relating solubility parameters of polymers with some of their physical properties, such as, surface tension[42-44] and glass transition temperature.[45]

The solubility parameters for various polymers are given in Table 5.1.3.

**Table 5.1.4. Solubility parameters of some polymers[36,46]**

| Polymer | δ | | Polymer | δ | |
|---|---|---|---|---|---|
| | $(cal/cm^3)^{1/2}$ | $(MJ/m^3)^{1/2}$ | | $(cal/cm^3)^{1/2}$ | $(MJ/m^3)^{1/2}$ |
| Butyl rubber | 7.84 | 16.0 | Polydimethylsiloxane | 9.53 | 19.5 |
| Cellulose diacetate | 10.9 | 22.2 | Polydimethylphenyleneoxide | 8.6 | 17.6 |
| Cellulose dinitrate | 10.6 | 21.6 | Polyisobutylene | 7.95 | 16.2 |
| Polyamide-66 | 13.6 | 27.8 | Polymethylmethacrylate | 9.3 | 19.0 |

| Polymer | δ | | Polymer | δ | |
|---|---|---|---|---|---|
| | $(cal/cm^3)^{1/2}$ | $(MJ/m^3)^{1/2}$ | | $(cal/cm^3)^{1/2}$ | $(MJ/m^3)^{1/2}$ |
| Natural rubber | 8.1 | 16.5 | Polymethylacrylate | 9.7 | 19.8 |
| Neoprene | 8.85 | 18.1 | Polyoctylmethacrylate | 8.4 | 17.2 |
| Cellulose nitrate | 11.5 | 23.5 | Polypropylene | 8.1 | 16.5 |
| Polyacrylonitrile | 14.5 | 29.6 | Polypropylene oxide | 7.52 | 15.4 |
| Polybutadiene | 8.44 | 17.2 | Polypropylene sulphide | 9.6 | 19.6 |
| Poly-n-butylacrylate | 8.7 | 17.8 | Polypropylmethacrylate | 8.8 | 18.0 |
| Polybutylmethacrylate | 8.7 | 17.8 | Polystyrene | 8.83 | 18.0 |
| Polybutyl-tert-methacrylate | 8.3 | 16.9 | Polyethylene | 7.94 | 16.2 |
| Polyvinylacetate | 9.4 | 19.2 | Polyethyleneterephthalate | 10.7 | 21.8 |
| Polyvinylbromide | 9.55 | 19.5 | Polyethylmethacrylate | 9.1 | 18.6 |
| Polyvinylidenechloride | 12.4 | 25.3 | Polybutadienenitrile (82:18 w) | 8.7 | 17.8 |
| Polyvinylchloride | 9.57 | 19.5 | Polybutadienenitrile (75:25) | 9.38 | 19.2 |
| Polyhexyl methacrylate | 8.6 | 17.6 | Polybutadienenitrile (70:30) | 9.64 | 19.7 |
| Polyglycol terephthalate | 10.7 | 21.8 | Polybutadienenitrile (61:39) | 10.30 | 21.0 |
| Polydiamylitaconate | 8.65 | 17.7 | Polybutadienevinylpyridine (75:25 mas.) | 9.35 | 19.1 |
| Polydibutylitaconate | 8.9 | 18.2 | Polybutadienestyrene (96:4) | 8.1 | 16.5 |
| Polysulfone | 10.5 | 21.4 | Polybutadienestyrene (87.5:12.5) | 8.31 | 17.0 |
| Polytetrafluorethylene | 6.2 | 12.7 | Polybutadienestyrene (85:15) | 8.5 | 17.4 |
| Polychloroacrylate | 10.1 | 20.6 | Polybutadienestyrene (71.5:28.5) | 8.33 | 17.0 |
| Polycyanoacrylate | 14.0 | 28.6 | Polybutadienestyrene (60:40) | 8.67 | 17.7 |
| Polyethylacrylate | 9.3 | 19.0 | Chlorinated rubber | 9.4 | 19.2 |
| Polyethylenepropylene | 7.95 | 16.2 | Ethylcellulose | 10.3 | 21.0 |

## REFERENCES


1 J.H. Hildebrand and R.L. Scott, **Solubility of none-electrolytes**. 3rd ed., *Reinhold*, New-York, 1950.
2 J. Mathews, *J. Amer. Chem. Soc.*, **48**, 562 (1926).
3 A. Coolidge, *J. Amer. Chem. Soc.*, **52**, 1874 (1930).
4 N. Osborn and D. Ginnings, *J. Res. Nat. Bur. Stand.*, **49**, 453 (1947).
5 G. Waddington, S. Todd, H. Huffman, *J. Amer. Chem. Soc.*, **69**, 22 (1947).
6 J. Hales, J. Cox, E. Lees, *Trans. Faraday Soc.*, **59**, 1544 (1963).
7 Y. Lebedev and E. Miroshnichenko, **Thermochemistry of vaporization of organic substances**, *Nauka*, Moscow, 1981.
8 F. Coon and F. Daniel, *J. Phys. Chem.*, **37**, 1 (1933).
9 J. Hunter, H. Bliss, *Ind. Eng. Chem.*, **36**, 945 (1944).
10 J. Lekk, **Measurement of pressure in vacuum systems**, *Mir*, Moscow, 1966.

11 W. Swietoslawski, **Ebulliometric measurement**, *Reinhold*, N.-Y., 1945.
12 H. Jones, I. Langmuir, G. Mackay, *Phys. Rev*., **30**, 201 (1927).
13 M. Knudsen, *Ann. Phys*., **35**, 389 (1911).
14 **Experimental thermochemistry**, Ed. H. Skinner, *Intersci. Publ*., 1962.
15 G. Gee, *Trans. Faraday Soc*., **38**, 269 (1942).
16 R.F. Boyer, R.S. Spencer, *J. Polym. Sci*., **3**, 97 (1948).
17 R.L. Scott, M. Magat, *J. Polym. Sci*., **4**, 555(1949).
18 P.I. Flory, I. Rehner, *J. Chem. Phys*., **11**, 521 (1943).
19 N.P. Apuhtina, E.G. Erenburg, L.Ya. Rappoport, *Vysokomol. soed*., **A8**, 1057 (1966).
20 G.M. Bristow., W.F. Watson, *Trans. Faraday Soc*., **54**, 1731 (1959).
21 V.Yu. Senichev in **Synthesis and properties of crosslinked polymers and compositions on their basis**. *Russian Academy of Sciences Publishing*, Sverdlovsk,1990, pp.16-20.
22 V.V.Tereshatov, V.Yu. Senichev, A.I. Gemuev, *Vysokomol. soed*., **B32**, 412(1990).
23 V.P. Belousov, A.G. Morachevskiy, **Mixing heats of liquids**, *Chemistry*, Leningrad, 1970.
24 V.P. Belousov, A.G. Morachevskiy, **Heat properties of non-electrolytes solutions**. **Handbook, Chemistry**, Leningrad, 1981.
25 T. Alfrey, A.I.Goldberg, I.A. Price, *J. Colloid. Sci*., **5**, 251 (1950).
26 D. Mangaray, S.K. Bhatnagar, Rath S.B., *Macromol. Chem*., **67**, 75 (1963).
27 D. Mangarey, S. Patra, S.B. Rath, *Macromol. Chem*., **67**, 84 (1963).
28 D. Mangarey, S. Patra, P.C. Roy, *Macromol. Chem*., **81**, 173 (1965).
29 V.E. Eskin, U. Guravlev, T.N. Nekrasova, *Vysokomol. soed*., **A18**, 653 (1976).
30 E. G. Gubanov, S.V. Shulgun, V.Sh. Gurskay, N.A. Palihov, B.M. Zuev, B.E. Ivanov, *Vysokomol. soed*., **A18**, 653 (1976).
31 C.I. Scheehan, A.L. Bisio, *Rubber Chem. Technol*., **39**, 149 (1966).
32 A.A. Tager, L.K. Kolmakova, G. Ya. Shemaykina, Ya.S. Vigodskiy, S.N. Salazkin, *Vysokomol. soed*., **B18**, 569 (1976).
33 W.R. Song, D.W. Brownawell, *Polym. Eng. Sci*., **10**, 222 (1970).
34 Y.N. Hakimullin, Y.O. Averko-Antonovich, P.A. Kirpichnikov, M.A. Gasnikova *Vysokomol. soed*., **B17**, 287 (1975).
35 F.P. Price, S.G. Martin, I.P. Bianchi, *J. Polymer Sci*., **22**, 49 (1956).
36 G.M. Bristow, W.F. Watson, *Trans. Faraday Soc*., **54**, 1731, 1742 (1958).
37 T.G. Fox, *Polymer*, **3**, 11 (1962).
38 K.W. Suh, D.H. Clarke, *J. Polymer Sci., A-1*, **4**, 1671 (1967).
39 V.E. Eskin, A.E. Nesterov, *Vysokomol. soed*., **A8**, 1051 (1966).
40 S.K. Ghosh, *Macromol. Chem*., **143**, 181 (1971).
41 A.G. Grozdov, B.N. Stepanov, *Vysokomol. soed*., **B17**, 907(1975).
42 A.A. Berlin, V.E. Basin, **Fundamentals of adhesion of polymers**, *Chemistry*, Moscow, 1974.
43 S.M. Ygnaytskaya, S.S. Voutskiy, L. Y. Kapluynova, *Colloid J*., **34**, 132 (1972).
44 R.M. Koethen, C.A. Smolders, *J. Appl. Polym. Sci*., **19**, 1163 (1975)
45 R.A. Hayes, *J. Appl. Polym. Sci*., **5**, 318, (1961).
46 E.S. Lipatov, A.E. Nesterov, T.M. Gritsenko, R.A Veselovski, **Handbook on polymer chemistry**. *Naukova dumka*, Kiev, 1971.
47 Z.Grubisic-Gallot, M. Picot, Ph. Gramain, H. Benoit, *J. Appl. Polym. Sci*., **16**, 2931(1972).
48 H.-R. Lee, Y.-D. Lee, *J. Appl. Polym. Sci*, **40**, 2087(1990).

# 5.2 PREDICTION OF SOLUBILITY PARAMETER

NOBUYUKI TANAKA
**Department of Biological and Chemical Engineering**
**Gunma University, Kiryu, Japan**

## 5.2.1 SOLUBILITY PARAMETER OF POLYMERS

For the purpose of searching for the solvents for a polymer, the solubility parameter of polymers, $\delta_p$, is defined as:[1-11]

$$\delta_p = (h_0 / v)^{1/2} \quad [5.2.1]$$

where:

| | |
|---|---|
| $h_0$ | the cohesive enthalpy per molar structural unit for a polymer (cal/mol) |
| v | the volume per molar structural unit for a polymer ($cm^3$/mol) |

because $\delta_p$ is equivalent to the solubility parameter of solvents, $\delta_s$, that shows the minimum of the dissolution temperature[7] or the maximum of the degree of swelling[1,2] for the polymer. For $\delta_s$, $h_0$ is the molar energy of vaporization that is impossible to measure for polymers decomposed before the vaporization at elevated temperatures, and v is the molar volume of a solvent. The measurements of the dissolution temperature and the degree of swelling are only means to find the most suitable solvents for a polymer by trial and error.

In order to obtain easily the exact value of $\delta_p$ at a temperature, T, the possibility of $\delta_p$ prediction from the thermal transition behaviors such as the glass transition and the melting has been discussed.[9-11] Consequently, it was found that the sum of their transition enthalpies gave $h_0$ in equation [5.2.1] approximately:

for crystalline polymers,

$$h_0 \approx h_g + h_x + h_u \qquad T \le T_g \quad [5.2.2]$$

$$h_0 \approx h_x + h_u \qquad T_g < T < T_m \quad [5.2.3]$$

for amorphous polymers,

$$h_0 \approx h_g + h_x \qquad T \le T_g \quad [5.2.4]$$

where:

| | |
|---|---|
| $h_g$ | the glass transition enthalpy per molar structural unit for a polymer |
| $h_u$ | the heat of fusion per molar structural unit for a polymer |
| $h_x$ | the transition enthalpy per molar structural unit due to ordered parts in the amorphous regions |
| $T_g$ | the glass transition temperature; here the onset temperature of heat capacity jump at the glass transition |
| $T_m$ | the melting temperature |

In the following sections, the physical meanings of $h_g$ and $h_x$ are shown in the theoretical treatments of the glass transition, and for several polymers, $\delta_p$ is predicted using these thermodynamic quantities.

## 5.2.2 GLASS TRANSITION IN POLYMERS

The glass transition in polymers is the same kind of physical phenomenon as observed generally for amorphous materials.[12] At $T_g$ in the cooling process, polymers are frozen glasses and the molecular motions are restricted strictly. However, the actual states of glasses are dependent on the cooling rate; if the cooling rate is rapid, the glasses formed should be imperfect, such as liquid glasses or glassy liquids.[13,14] The annealing for imperfect glasses results in the enthalpy relaxation from imperfect glasses to perfect glasses. At $T_g$ in the heating process, the strong restriction of molecular motions by intermolecular interactions is removed and then the broad jump of heat capacity, $C_p$, is observed.[15] Annealing the glasses, the $C_p$ jump curve becomes to show a peak.[15,16]

### 5.2.2.1 Glass transition enthalpy

For polymer liquids, the partition function, Ω, normalized per unit volume is given by:[10,14,17,18]

$$\Omega = \left(Z^N / N!\right)\left(2\pi mkT / \mathbf{h}^2\right)^{3Nx/2} \left(q / v_f\right)^{Nx} \exp\left\{-Nxh^{\mathrm{int}} / (RT)\right\} \quad [5.2.5]$$

with $v_f = qv\exp\{-h^{int}/(RT)\}$

where:

| | |
|---|---|
| $h^{int}$ | the intermolecular cohesive enthalpy per molar structural unit for a polymer |
| $\mathbf{h}$ | Planck's constant |
| k | Boltzmann's constant |
| m | the mass of a structural unit for a polymer |
| N | the number of chains |
| q | the packing factor of structural units for a polymer |
| R | the gas constant |
| $v_f$ | the free volume per molar structural unit for a polymer |
| x | the degree of polymerization |
| Z | the conformational partition function per a chain |

From equation [5.2.5], the enthalpy and the entropy per molar chain for polymer liquids, $H_l$ and $S_l$, are derived:[10]

$$H_l = RT^2 d\ln Z / dT + (3/2)RxT - RxT^2 d\ln v_f / dT + xh^{\mathrm{int}} \quad [5.2.6]$$

$$S_l = \left(R\ln Z + RTd\ln Z/dT\right) + (3/2)Rx - x\left(R\ln v_f + RTd\ln v_f/dT\right) + xS_d \quad [5.2.7]$$

with $S_d = (3R/2)\ln(2\pi mkT/\mathbf{h}^2) - (1/x)(R/N)\ln N! + R\ln q$

The first terms on the right hand side of equations [5.2.6] and [5.2.7] are the conformational enthalpy and entropy per molar chain, $xh^{conf}$ and $xs^{conf}$, respectively.[19] Assuming that chains at $T_g$ are in quasi-equilibrium state, the criterions on $T_g$ are obtained:

$$f_{flow}\ (= h_{flow} - T_g s_{flow}) \approx 0 \quad [5.2.8]$$

$$\text{and } s_{flow} \approx 0 \text{ (hence } h_{flow} \approx 0) \quad [5.2.9]$$

with $h_{flow} = H_l/x - 3RT_g/2$ and $s_{flow} = S_l/x - 3R/2$

From equations [5.2.8] and [5.2.9], which show the conditions of thermodynamic quasi-equilibrium and freezing for polymer liquids, the conformational enthalpy and entropy per molar structural unit at $T_g$, $h_g^{conf}$ and $s_g^{conf}$, are derived, respectively:

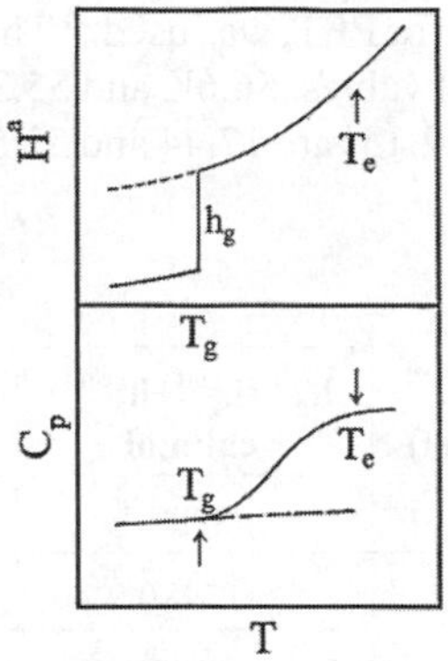


Figure 5.2.1. Schematic curves of $H^a$ and $C_p$ in the vicinity of $T_g$ for an amorphous polymer. Two lines of short and long dashes show $H^a$ for a supercooled liquid and $C_p$ for a superheated glass (hypothesized), respectively. $T_e$ is the end temperature of $C_p$ jump. [after reference 10]

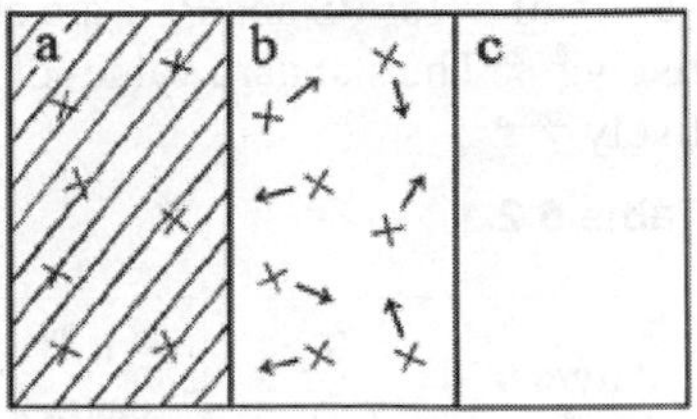


Figure 5.2.2. State models for an amorphous polymer in each temperature range of (a) $T \le T_g$, (b) $T_g < T \le T_e$, and (c) $T_e < T$, where hatching: glass parts, crosses: ordered parts, and blank: flow parts. The arrows show the mobility of ordered parts. [after reference 10]

$$h_g^{conf}\left\{=RT_g^2\left(d\ln Z/dT\right)/x)\right\} = RT_g^2 d\ln v_f/dT - h_g^{int} \qquad [5.2.10]$$

$$s_g^{conf}\left\{=\left(R\ln Z + RT_g d\ln Z/dT\right)/x\right\} = R\ln v_f + RT_g d\ln V_f/dT - S_d \qquad [5.2.11]$$

Rewriting equation [5.2.10], the glass transition enthalpy per molar structural unit, $h_g\ (= h_g^{int} + h_g^{conf})$, is obtained:

$$h_g = RT_g^2 d\ln v_f/dT \approx RT_g^2/c_2 \qquad [5.2.12]$$

with $c \approx \phi_g/\beta$

where:

| | |
|---|---|
| $h_g^{int}$ | $h^{int}$ at $T_g$ |
| $c_2$ | the constant in the WLF equation[20] |
| $\beta$ | the difference between volume expansion coefficients below and above $T_g$ |
| $\phi_g$ | the fractional free volume at $T_g$ |

The WLF equation on the time - temperature superposition of viscoelastic relaxation phenomena is given by:[20]

$$\log a_T = -c_1\left(T - T_g\right)/\left(c_2 + T - T_g\right) \qquad [5.2.13]$$

where:

| | |
|---|---|
| $a_T$ | the shift factor |
| $c_1$ | the constant |

Figure 5.2.1 shows the schematic curves of $H^a$ and $C_p$ for an amorphous polymer, where $H^a$ is the molar enthalpy for an amorphous polymer. Substituting $H_l/x$ for $H^a$, $H^a$ at $T_g$ corresponds to $3RT_g/2$, because of $h_{flow} = 0$. At $T_g$, the energy of $h_g$ is given off in the cooling process and absorbed in the heating process.

Table 5.2.1 shows the numerical values of $h_g$ (= $h_g^{int}$ + $h_g^{conf}$) and $c_2$ from equation [5.2.12], together with $T_g$,[12,16] $h_g^{int}$, $h_g^{conf}$, and $s_g^{conf}$, for several polymers. As the values of $h_g^{int}$, the molar cohesive energy of main residue in each polymer, e.g. -CONH- for N6 and

N66, -CH($CH_3$)- for iPP, -CH($C_6H_5$)- for iPS, and -$C_6H_4$- for PET, was used.[21] The predicted values of $c_2$ for PS and PET are close to the experimental values, 56.6K and 55.3K, respectively.[20,22] The standard values of $c_1$ and $c_2$ in equation [5.2.13] are 17.44 and 51.6K, respectively.[20,23]

**Table 5.2.1**

| Polymer | $T_g$ K | $h_g^{int}$ cal/mol | $h_g^{conf}$ cal/mol | $s_g^{conf}$ cal/(K mol) | $h_g$ (=$h_g^{int}$+$h_g^{conf}$) cal/mol | $c_2$ K |
|---|---|---|---|---|---|---|
| N6 | 313 | 8500 | 475 | 11.2 | 8980 | 21.7 |
| N66 | 323 | 17000 | 976 | 22.6 | 17980 | 11.5 |
| iPP | 270 | 1360 | 180 | 0.98 | 1540 | 94.1 |
| iPS | 359[16] | 4300 | 520 | 2.03 | 4820 (4520) | 53.1 (56.6) |
| PET | 342 | 3900 | 282 | 7.10 | 4180 (4200) | 55.6 (55.3) |

The numerical values in parentheses are the experimental values[20,22] of $c_2$ and $h_g$ (from $c_2$). N6: polycaproamide (nylon-6), N66: poly(hexamethylene adipamide) (nylon-6,6), iPP: isotactic polypropylene, iPS: isotactic polystyrene, PET: poly(ethylene terephthalate).

### 5.2.2.2 $C_p$ jump at the glass transition

The mechanism of $C_p$ jump at the glass transition could be illustrated by the melting of ordered parts released from the glassy states.[10] Figure 5.2.2 shows the state models for an amorphous polymer below and above $T_g$.

The ordered parts are generated near $T_g$ in the cooling process; in the glasses, the ordered parts are contained. In the heating process, right after the glassy state was removed at $T_g$, the melting of ordered parts starts and continues up to $T_e$, keeping an equilibrium state between ordered parts and flow parts. In this temperature range, the free energy per molar structural unit for polymer liquids contained ordered parts, $f_m$, is given by:[10]

$$f_m = f_x X_x + f_{flow}(1 - X_x) \qquad [5.2.14]$$

where:

| | |
|---|---|
| $f_x$ | the free energy per molar structural unit for ordered parts |
| $f_{flow}$ | the free energy per molar structural unit for flow parts |
| $X_x$ | the mole fraction of ordered parts |

From $(df_m/dX_x)_p = 0$, an equilibrium relation is derived:

$$f_m = f_x = f_{flow} \qquad [5.2.15]$$

Whereas, $C_p$ is defined as:

$$C_p = (dh_q / dT)_p \qquad [5.2.16]$$

with $h_q = f_q - T(df_q/dT)_p$, q = m, x or flow

From equations [5.2.15] and [5.2.16], it is shown that $C_p$ of ordered parts is equal to that of flow parts. Therefore, $h_x$ is given by:[10]

$$h_x \approx h_g + \Delta h \qquad [5.2.17]$$

with $\Delta h = \int_{T_g}^{T_e} \Delta C_p dT$ and $\{RT_g \ln(Z_g/Z_0)\}/x \le \Delta h \le T_g\{s_g^{conf} - (R\ln Z_0)/x\}$,

where:

| | |
|---|---|
| $\Delta C_p$ | the difference in the observed $C_p$ and the hypothesized super heated glass $C_p$ at the glass transition |
| $Z_g$ | the conformational partition function per a chain at $T_g$ |
| $Z_0$ | the component conformational partition function per chain regardless of the temperature in Z |
| $s_g^{conf}$ | the conformational entropy per molar structural unit at $T_g$ |

$h_x$ is also given by rewriting the modified Flory's equation, which expresses the melting point depression as a function of the mole fraction of major component, X, for binary random copolymers:[24-27]

$$h_x \approx 2h_u(1 - 1/a) \qquad [5.2.18]$$

with $a = -h_u(1/T_m(X) - 1/T_m^0)/(R\ln X)$

where:

| | |
|---|---|
| $T_m(X)$ | the melting temperature for a copolymer with X |
| $T_m^0$ | the melting temperature for a homopolymer of major component |

**Table 5.2.2**

| Polymer | $h_u$ cal/mol | $h_x$(eq.[5.2.17]) cal/mol | $h_x$(eq.[5.2.17]) - $h_u$ cal/mol | $h_x$ (eq.[5.2.18]) cal/mol | $h_x$ from $\delta_p$ cal/mol |
|---|---|---|---|---|---|
| N6 | 5100 | 9590 (10070) | 4490 (4970) | 4830 | - |
| N66 | 10300 | 19300 (20280) | 9000 (9980) | 9580 | 10070 |
| iPP | 1900 | 1600 (1780) | - | 1470 | 1420 |
| iPS | 2390[12] | 5030 (5550) | 2640 (3160) | - | 2410 - 5790 |
| PET | 5500 | 5380 (5670) | - | 6600 | 6790 |

The numerical values in parentheses were calculated using equation [5.2.17] with the second term of $T_g\{s_g^{conf} - (R\ln Z_0)/x\}$.

Table 5.2.2 shows the numerical values of $h_x$ from equations [5.2.17] and [5.2.18], and from the reference values[7,8] of $\delta_p$ using equations [5.2.1] ~ [5.2.4], together with the values[12,28,29] of $h_u$, for several polymers. The second term in the right hand side of equation [5.2.17] was calculated from $\{RT_g\ln(Z_g/Z_0)\}/x$. The numerical values in parentheses, which were calculated using equation [5.2.17] with the second term of $T_g\{s_g^{conf} - (R\ln Z_0)/x\}$, are a little more than in the case of $\{RT_g\ln(Z_g/Z_0)\}/x$. The relationship of $h_x$(eq.[5.2.17]) » $h_u$ found for N6 and N66 suggests two layer structure of ordered parts in the glasses, because, as shown in the fourth column, $h_x$(eq.[5.2.17]) - $h_u$ is almost equal to $h_x$ ($\approx h_u$) from equation [5.2.18] or $\delta_p$ (=13.6 $(cal/cm^3)^{1/2}$). For iPP, $h_x$(eq.[5.2.17]) is a little more than $h_x$ from equation [5.2.18] or $\delta_p$ (=8.2 $(cal/cm^3)^{1/2}$), suggesting that the ordered parts in glasses seem to be related closely to the helical structure. For iPS, $h_x$(eq.[5.2.17]) and $h_x$(eq.[5.2.17]) - $h_u$ are in the upper and lower ranges of $h_x$ from $\delta_p$ (=8.5~10.3 $(cal/cm^3)^{1/2}$), respectively. For PET, $h_x$(eq.[5.2.17]) is almost equal to $h_u$, but $h_x$ from equation [5.2.18] or $\delta_p$ (=10.7 $(cal/cm^3)^{1/2}$) is a little more than $h_u$, resulting from glycol bonds in bulk crystals distorted more than in ordered parts.[10,25,30]

## 5.2.3 $\delta_p$ PREDICTION FROM THERMAL TRANSITION ENTHALPIES

Table 5.2.3 shows the numerical values of $\delta_p$ predicted from equations [5.2.1] ~ [5.2.4] using the results of Tables 5.2.1 and 5.2.2, together with $h_x$, $h_u$, $h_g$, $h_0$, and $\delta_p^{\,r}$ (reference values)[7,8] for several polymers. The predicted values of $\delta_p$ for each polymer are close to $\delta_p^{\,r}$.

**Table 5.2.3**

| Polymer | $h_x$ cal/mol | $h_u$ cal/mol | $h_g$ cal/mol | $h_0$ cal/mol | $\delta_p$ $(cal/cm^3)^{1/2}$ | $\delta_p^{\,r}$ $(cal/cm^3)^{1/2}$ |
|---|---|---|---|---|---|---|
| N6 | 9590* | - | 8980 | 18570 | 13.4 | - |
| | (10070) | - | 8980 | (19050) | (13.6) | - |
| | 4830** | 5100 | 8980 | 18910 | 13.5 | - |
| N66 | 19300* | - | 17980 | 37280 | 13.4 | 13.6 |
| | (20280) | - | 17980 | (38260) | (13.6) | 13.6 |
| | 9580** | 10300 | 17980 | 37860 | 13.5 | 13.6 |
| iPP | 1600* | 1900 | - | 3500 | 8.4 | 8.2[7] |
| | (1780) | 1900 | - | (3680) | (8.6) | 8.2[7] |
| | 1470** | 1900 | - | 3370 | 8.3 | 8.2[7] |
| iPS | 5030* | - | 4820 | 9850 | 9.9 | 8.5 - 10.3 |
| | (5550) | - | 4820 | (10370) | (10.2) | 8.5 - 10.3 |
| PET | 5380* | 5500 | 4180 | 15060 | 10.2 | 10.7 |
| | (5670) | 5500 | 4180 | (15350) | (10.3) | 10.7 |
| | 6600** | 5500 | 4180 | 16280 | 10.6 | 10.7 |

The numerical values attached * and ** are $h_x$ from equations [5.2.17] and [5.2.18], respectively. The numerical values in parentheses were calculated using equation [5.2.17] with the second term of $T_g\{s_g^{conf} - (R\ln Z_0)/x\}$.

For atactic polypropylene (aPP) that could be treated as a binary random copolymer composed of meso and racemi dyads,[26,27,31] $\delta_p$ is predicted as follows.

For binary random copolymers, $h_g$ is given by:

$$h_g = h_g^{int} + h_g^{conf}(X_A) \qquad [5.2.19]$$

with $h_g^{int} = h_g^{int}(1) - (h_g^{int}(1) - h_g^{int}(0))(1 - X_A)$

where:

$h_g^{conf}(X_A)$ the conformational enthalpy per molar structural unit for a copolymer with $X_A$ at $T_g$

$h_g^{int}(1)$ the intermolecular cohesive enthalpy per molar structural unit for a homopolymer of component A ($X_A$=1) at $T_g$

$h_g^{int}(0)$ the intermolecular cohesive enthalpy per molar structural unit for a homopolymer of component B ($X_A$=0) at $T_g$
$X_A$ the mole fraction of component A

Further, $h_x$ is obtained from equation [5.2.17], but $Z_g$ is the conformational partition function for a copolymer with $X_A$ at $T_g$ and $Z_0$ is the component conformational partition function for a copolymer with $X_A$ regardless of the temperature in Z. Whereas for binary random copolymers, $T_g$ is given by:[32,33]

$$T_g = T_g(1)(h_g^{int} + h_g^{conf}(X_A)) / \{h_g^{int}(1) + h_g^{conf}(1) - T_g(1)(s_g^{conf}(1) - s_g^{conf}(X_A))\} \qquad [5.2.20]$$

where:

$h_g^{conf}(1)$ the conformational enthalpy per molar structural unit for a homopolymer ($X_A$=1) at $T_g$
$s_g^{conf}(1)$ the conformational entropy per molar structural unit for a homopolymer ($X_A$=1) at $T_g$
$s_g^{conf}(X_A)$ the conformational entropy per molar structural unit for a copolymer with $X_A$ at $T_g$
$T_g(1)$ the glass transition temperature for a homopolymer ($X_A$=1)

Thus, using equations [5.2.19] and [5.2.20], $\delta_p$ for binary random copolymers, including aPP, could be predicted.

**Table 5.2.4**

| $1 - X_A$ | $T_g$ K | $h_g$ cal/mol | $h_x$ cal/mol | $h_0$ cal/mol | | $\delta_p(cal/cm^3)^{1/2}$ | |
|---|---|---|---|---|---|---|---|
| | | | | $T>T_g$ | $T<T_g$ | $T>T_g$ | $T<T_g$ |
| 0 | 270 | 1540 | 1600 (1780) | 3500 (3680) | 5040 (5220) | 8.42 (8.62) | 10.10 (10.23) |
| 0.05 | 265 | 1480 | 1510 | 3410 | 4900 | 8.31 | 9.95 |
| 0.1 | 261 | 1450 | 1470 | 3370 | 4830 | 8.26 | 9.88 |
| 0.15 | 259 | 1440 | 1460 | 3360 | 4790 | 8.24 | 9.85 |
| 0.2 | 257 | 1430 | 1440 | 3340 | 4770 | 8.23 | 9.83 |
| 0.3 | 255 | 1420 | 1440 (1500) | 3340 (3400) | 4760 (4820) | 8.22 (8.29) | 9.82 (9.87) |
| 0.4 | 254 | 1420 | 1450 | 3350 | 4770 | 8.23 | 9.83 |
| 0.5 | 255 | 1440 | 1470 | 3370 | 4810 | 8.26 | 9.86 |
| 0.6 | 256 | 1450 | 1490 | 3390 | 4840 | 8.28 | 9.90 |
| 0.7 | 258 | 1470 | 1520 | 3420 | 4890 | 8.32 | 9.95 |
| 0.8 | 262 | 1500 | 1560 | 3460 | 4960 | 8.37 | 10.02 |
| 0.85 | 264 | 1510 | 1580 | 3480 | 4990 | 8.39 | 10.05 |
| 0.9 | 266 | 1520 | 1600 | 3500 | 5030 | 8.42 | 10.09 |
| 0.95 | 268 | 1540 | 1620 | 3520 | 5060 | 8.45 | 10.12 |
| 1 | 270 | 1550 | 1650 (1840) | 3550 (3740) | 5100 (5300) | 8.48 (8.69) | 10.16 (10.34) |

The numerical values in parentheses were calculated using equation [5.2.17] with the second term of $T_g\{s_g^{conf}(R\ln Z_g)/x\}$.

Table 5.2.4 shows the numerical values of $\delta_p$ for aPP, together with $T_g$, $h_g$, $h_x$, and $h_0$, which give the concave type curves against $1 - X_A$, respectively. Where $X_A$ is the mole fraction of meso dyads. As $h_u$ of crystals or quasi-crystals in aPP, 1900 cal/mol was used, because $h_u$(=1900 cal/mol) of iPP[28] should be almost equal to that of syndiotactic PP (sPP); $h^{conf} + h^{int} \approx 1900$ cal/mol at each $T_m$ for both PP, where $h^{conf} = 579.7$ cal/mol at $T_m = 457K$[28] for iPP, $h^{conf} = 536.8$ cal/mol at $T_m = 449K$[34] for sPP, and $h^{int} = 1360$ cal/mol for both PP. Here quasi-crystals are ordered parts in aPP with $1 - X_A \approx 0.30 \sim 0.75$, which do not satisfy the requirements of any crystal cell. As $T_g$ of iPP and sPP, 270K was used for both PP.[12,35] The experimental values of $T_g$ for aPP are less than 270K.[36-41] In this calculation, for aPP with $1 - X_A = 0.4$, the minimum of $T_g$, 254K, was obtained. $\delta_p$ in $T>T_g$ showed the minimum, 8.22 $(cal/cm^3)^{1/2}$, at $1 - X_A = 0.3$, which is 0.20 less than 8.42 $(cal/cm^3)^{1/2}$ of iPP, and $\delta_p$ in $T<T_g$ showed the minimum, 9.82 $(cal/cm^3)^{1/2}$, at $1 - X_A = 0.3$ and 10.10 $(cal/cm^3)^{1/2}$ for iPP. Substituting $T_g\{s_g^{conf} - (R\ln Z_0)/x\}$ for the second term in equation [5.2.17],[42,43] the numerical values of $h_x$, $h_0$, and $\delta_p$ become a little more than in the case of $\{RT_g\ln(Z_g/Z_0)\}/x$, as shown in Table 5.2.4. However, the increase of $h_x$ for iPP led $T_e$ (=375K) near the experimental values,[35] e.g. 362K and 376K, where $T_e$ was approximated by:

$$T_e \approx 2(h_x - h_g) / \Delta C_p^0 + T_g \qquad [5.2.21]$$

where:

$\Delta C_p^0$ the difference in liquid $C_p$ and glass $C_p$ before and after the glass transition; 4.59 cal/(K mol) for iPP[12]

Thus, we can browse the solvents of aPP which satisfy $\delta_p = \delta_s$.

Equations [5.2.2], [5.2.3] and [5.2.4] would be available as tools to predict $\delta_p$ from thermal transition behaviors.

## REFERENCES


1 M. Barton, *Chem. Rev.*, **75**, 731(1975).
2 G. M. Bristow and W. F. Watson, *Trans. Faraday Soc.*, **54**, 1742(1958).
3 W. A. Lee and J. H. Sewell, *J. Appl. Polym. Sci.*, **12**, 1397(1968).
4 P. A. Small, *J. Appl. Chem.*, **3**, 71(1953).
5 M. R. Moore, *J. Soc. Dyers Colourists*, **73**, 500(1957).
6 D. P. Maloney and J. M. Prausnity, *J. Appl. Polym. Sci.*, **18**, 2703(1974).
7 A. S. Michaels, W. R. Vieth, and H. H. Alcalay, *J. Appl. Polym. Sci.*, **12**, 1621(1968).
8 J. Brundrup and E. H. Immergut, **Polymer Handbook**, *Wiley*, New York, 1989.
9 N. Tanaka, *Sen-i Gakkaishi*, **44**, 541(1988).
10 N. Tanaka, *Polymer*, **33**, 623(1992).
11 N. Tanaka, *Current Trends in Polym. Sci.*, **1**, 163(1996).
12 B. Wunderlich, **Thermal Analysis**, *Academic Press*, New York, 1990.
13 N. Tanaka, *Polymer*, **34**, 4941(1993).
14 N. Tanaka, *Polymer*, **35**, 5748(1994).
15 T. Hatakeyama and Z. Liu, **Handbook of Thermal Analysis**, *Wiley*, New York, 1998.
16 H. Yoshida, *Netsusokutei*, **13**(4), 191(1986).
17 N. Tanaka, *Polymer*, **19**, 770(1978).
18 N. Tanaka, **Thermodynamics for Thermal Analysis in Chain Polymers**, *Kiryu Times*, Kiryu, 1997.
19 A. E. Tonelli, *J. Chem. Phys.*, **54**, 4637(1971).
20 J. D. Ferry, **Viscoelastic Properties of Polymers**, *Wiley*, New York, 1960.
21 C. W. Bunn, *J. Polym. Sci.*, **16**, 323(1955).
22 H. Sasabe, K. Sawamura, S. Sawamura, S. Saito, and K. Yoda, *Polymer J.*, **2**, 518(1971).
23 S. Iwayanagi, **Rheology**, *Asakurashoten*, Tokyo, 1971.
24 N. Tanaka and A. Nakajima, *Sen-i Gakkaishi*, **29**, 505(1973).
25 N. Tanaka and A. Nakajima, *Sen-i Gakkaishi*, **31**, 118(1975).
26 N. Tanaka, *Polymer*, **22**, 647(1981).

27 N. Tanaka, *Sen-i Gakkaishi*, **46**, 487(1990).
28 A. Nakajima, **Molecular Properties of Polymers**, *Kagakudojin*, Kyoto, 1969.
29 L. Mandelkern, **Crystallization of Polymers**, *McGraw-Hill*, New York, 1964.
30 N. Tanaka, Int. Conf. Appl. Physical Chem., Warsaw, Poland, November 13-16, 1996, Abstracts, Warsaw, 1996, p.60.
31 N. Tanaka, 36th IUPAC Int. Symp. Macromolecules, Seoul, Korea, August 4-9, 1996, Abstracts, Seoul, 1996, p.780.
32 N. Tanaka, *Polymer*, **21**, 645(1980).
33 N. Tanaka, 6th SPSJ Int. Polym. Conf., Kusatsu, Japan, October 20-24, 1997, Preprints, Kusatsu, 1997, p.263.
34 J. Schmidtke, G. Strobl, and T. Thurn-Albrecht, *Macromolecules*, **30**, 5804(1997).
35 D. R. Gee and T. P. Melia, *Makromol. Chem.*, **116**, 122(1968).
36 F. P. Reding, *J. Polym. Sci.*, **21**, 547(1956).
37 G. Natta, F. Danusso, and G. Moraglio, *J. Polym. Sci.*, **25**, 119(1957).
38 M. Dole and B. Wunderlich, *Makromol. Chem.*, **34**, 29(1959).
39 R. W. Wilkinson and M. Dole, *J. Polym. Sci.*, **58**, 1089(1962).
40 F. S. Dainton, D. M. Evans, F. E. Hoare, and T. P. Melia, *Polymer*, **3**, 286(1962).
41 E. Passaglia and H. K. Kevorkian, *J. Appl. Phys.*, **34**, 90(1963).
42 N. Tanaka, *AIDIC Conference Series*, **4**, 341 (1999).
43 N. Tanaka, 2nd Int. and 4th Japan-China Joint Symp. Calorimetry and Thermal Anal., Tsukuba, Japan, June 1-3, 1999, Preprints, Tsukuba, 1999, pp.92-93.


## 5.3 METHODS OF CALCULATION OF SOLUBILITY PARAMETERS OF SOLVENTS AND POLYMERS

Valery Yu. Senichev, Vasiliy V. Tereshatov
**Institute of Technical Chemistry**
**Ural Branch of Russian Academy of Sciences, Perm, Russia**

The methods of calculation of solubility parameters are based on the assumption that energy of intermolecular interactions is additive. Thus, the value of an intermolecular attraction can be calculated by addition of the contributions of cohesion energy of atoms or groups of atoms incorporated in the structure of a given molecule. Various authors use different physical parameters for contributions of individual atoms.

Dunkel proposed to use a molar latent heat of vaporization[1] as an additive value, describing intermolecular interactions. He represented it as a sum of the contributions of latent heat of vaporization of individual atoms or groups of atoms at room temperature.

Small's method[2] received the greatest interest. Small has used the data of Scatchard[3] showing that the square root of a product of cohesion energy of substance and its volume is a linear function of a number of carbon atoms in a molecule of substance. He also proposed additive constants for various groups of organic molecules that permit calculation of $(EV)^{1/2}$ value. He named these constants as molar attraction constants, $F_i$:

$$(EV)^{1/2} = \sum_i F_i \qquad [5.3.1]$$

Cohesion energy and solubility parameters could then be estimated for any molecule:

$$E = \frac{\left(\sum_i F_i\right)^2}{V}, \quad \delta = \frac{\sum_i F_i}{V} \qquad [5.3.2]$$

where:

V molar volume of solvent or a repeating unit of polymer

The molar attraction constants were calculated by Small based on literature data for a vapor pressure and latent heat of vaporization of liquids. The comprehensive values of these constants are given in Table 5.3.1.

**Table 5. 3.1. Molar attraction constants**

| Group | Small[2] | Van Krevelen[5] | Hoy[7] |
|---|---|---|---|
| | F, $(cal\ cm^3)^{1/2}\ mol^{-1}$ | | |
| >C< | -93 | 0 | 32.0 |
| >CH- | 28 | 68.5 | 86.0 |
| $-CH_2-$ | 133 | 137 | 131.5 |
| $-CH_3$ | 214 | 205.5 | 148.3 |
| $-CH(CH_3)-$ | 242 | 274 | (234.3) |
| $-C(CH_3)_2-$ | 335 | 411 | (328.6) |
| >C=CH- | 130 | 148.5 | 206.0 |
| -CH=CH- | 222 | 217 | 243.1 |
| $-C(CH_3)=CH-$ | (344) | 354 | (354.3) |
| Cyclopentyl | - | 676.5 | 633.0 |
| Cyclohexyl | - | 813.5 | 720.1 |
| Phenyl | 735 | 741.5 | 683.5 |
| 1,4-Phenylene | 658 | 673 | 704.9 |
| -O- | 70 | 125 | 115.0 |
| -OH | - | 368.5 | 225.8 |
| -CO- | 275 | 335 | 263.0 |
| -COO- | 310 | 250 | 326.6 |
| -COOH | - | 318.5 | (488.8) |
| -O-CO-O- | - | 375 | (441.6) |
| -CO-O-CO- | - | 375 | 567.3 |
| -CO-NH- | - | 600 | (443.0) |
| -O-CO-NH- | - | 725 | (506.6) |
| -S- | 225 | 225 | 209.4 |

| Group | Small[2] | Van Krevelen[5] | Hoy[7] |
|---|---|---|---|
| | F, (cal cm$^3$)$^{1/2}$ mol$^{-1}$ | | |
| -CN | 410 | 480 | 354.6 |
| -CHCN- | (438) | 548.5 | (440.6) |
| -F | (122) | 80 | 41.3 |
| -Cl | 270 | 230 | 205.1 |
| -Br | 340 | 300 | 257.9 |
| -I | 425 | - | - |

The Scatchard equation is correct only for nonpolar substances because they have only dispersive interactions between their molecules. Small eliminated from his consideration the substances containing hydroxyl, carboxyl and other groups able to form hydrogen bonds.

This method has received its further development due to Fedors' work,[4] who extended the method to polar substances and proposed to represent as an additive sum not only the attraction energy but also the molar volumes. The lists of such constants were published in several works.[5-8] The most comprehensive set of contributions to cohesion energy can be found elsewhere.[9]

Askadskii has shown[10] that Fedors' supposition concerning the additivity of contributions of volume of atoms or groups of atoms is not quite correct because the same atom in an environment of different atoms occupies different volume. In addition, atoms can interact with other atoms in different ways depending on their disposition and this should be taken into account for computation of cohesion energy. Therefore, a new scheme of the solubility parameters calculation was proposed that takes into account the nature of an environment of each atom in a molecule and the type of intermolecular interactions. This approach is similar to that described in the work of Rheineck and Lin.[6]

$$\delta = \left( \frac{\sum_i \Delta E_i}{N_A \sum_i \Delta V_i} \right)^{1/2} \qquad [5.3.3]$$

where:

$N_A$ Avogadro number
$\Delta E_i$ increment (contribution) to cohesion energy of atom or group of atoms
$\Delta V_i$ increment to the van der Waals volume of atom

The volume increment $\Delta V_i$ of an atom under consideration is calculated as volume of sphere of the atom minus volumes of spherical segments, which are cut off on this sphere by the adjacent covalently-bound atoms:

$$\Delta V_1 = \frac{4}{3}\pi R^3 - \sum_i \frac{1}{3}\pi h_i^3 (3R - h_i) \qquad [5.3.4]$$

where:

R van der Waals (intermolecular) radius of a considered atom
$h_i$ a height of segment calculated from the formula:

$$h_i = R - \frac{R^2 + d_i^2 - R_i^2}{2d_i} \quad [5.3.5]$$

where:

$d_i$ bond length between two atoms
$R_i$ van der Waals radius of the atom adjacent to the covalently-bonded atoms under consideration

The increments to the van der Waals volume for more than 200 atoms in various neighborhoods is available elsewhere.[11]

Using data from Tables 5.3.2 and 5.3.3, van der Waals volumes of various molecules can be calculated. The increments to the cohesion energy are given in Table 5.3.4. An advantage of this method is that the polymer density that is important for estimation of properties of polymers that have not yet been synthesized does not need to be known.

The calculation methods of the solubility parameters for polymers have an advantage over experimental methods that they do not have any prior assumptions regarding interactions of polymer with solvents. The numerous examples of good correlation between calculated and experimental parameters of solubility for various solvents support the assumed additivity of intermolecular interaction energy.

The method has further useful development in calculation of components of solubility parameters based on principles of Hansen's approach.[12] It may be expected that useful results will also come from analysis of donor and acceptor parameters used in TDM-approach (see Chapter 4).

In Table 5.3.5, the increments required to account for contributions to solubility parameters related to the dipole-dipole interactions and hydrogen bonds are presented.[13] Table 5.3.6 contains Hansen's parameters for some common functional groups.

**Table 5.3.2. Intermolecular radii of some atoms**

| Atom | R, Å | Atom | R, Å | Atom | R, Å | Atom | R, Å |
|---|---|---|---|---|---|---|---|
| C | 1.80 | F | 1.5 | Si | 2.10 | P | 1.90 |
| H | 1.17 | Cl | 1.78 | Sn | 2.10 | Pb | 2.20 |
| O | 1.36 | Br | 1.95 | As | 2.00 | B | 1.65 |
| N | 1.57 | I | 2.21 | S | 1.8 | | |

**Table 5.3.3. Lengths of bonds between atoms**

| Bond | $d_i$, Å | Bond | $d_i$, Å | Bond | $d_i$, Å | Bond | $d_i$, Å |
|---|---|---|---|---|---|---|---|
| C–C | 1.54 | C–F | 1.34 | C–S | 1.76 | N–P | 1.65 |
| C–C | 1.48 | C–F | 1.31 | C=S | 1.56 | N–P | 1.63 |
| C–C$^{arom}$ | 1,40 | C–Cl | 1.77 | H–O | 1.08 | S–S | 2.10 |
| C=C | 1.34 | C–Cl | 1.64 | H–S | 1.33 | S–Sn | 2.10 |

| Bond | $d_i$, Å | Bond | $d_i$, Å | Bond | $d_i$, Å | Bond | $d_i$, Å |
|---|---|---|---|---|---|---|---|
| C≡C | 1.19 | C–Br | 1.94 | H–N | 1.08 | S–As | 2.21 |
| C–H | 1.08 | C–Br | 1.85 | H–B | 1.08 | S=As | 2.08 |
| C–O | 1.50 | C–I | 2.21 | O–S | 1.76 | Si–Si | 2.32 |
| C–O | 1.37 | C–I | 2.05 | O–Si | 1.64 | P–F | 1.55 |
| C=O | 1.28 | C–P | 1.81 | O–P | 1.61 | P–Cl | 2.01 |
| C–N | 1.40 | C–B | 1.73 | N–O | 1.36 | P–S | 1.81 |
| C–N | 1.37 | C–Sn | 2.15 | N–N | 1.46 | B–B | 1.77 |
| C=N | 1.31 | C–As | 1.96 | O=N | 1.20 | Sn–Cl | 2.35 |
| C=N | 1.27 | C–Pb | 2.20 | O=S | 1.44 | As–Cl | 2.16 |
| $C–N^{arom}$ | 1.34 | C–Si | 1.88 | O=P | 1.45 | As–As | 2.42 |
| C≡N | 1.16 | C–Si | 1.68 | $N–P^{arom}$ | 1.58 | | |

**Table 5.3.4. Values of $\Delta E_i^*$ for various atoms and types of intermolecular interaction required to calculate solubility parameters according to equation [5.3.3] (Adapted from refs. 10,11)**

| Atom and type of intermolecular interaction | Label | $\Delta E_i^*$, cal/mol |
|---|---|---|
| C | $\Delta E_C^*$ | 550.7 |
| H | $\Delta E_H^*$ | 47.7 |
| O | $\Delta E_O^*$ | 142.6 |
| N | $\Delta E_N^*$ | 1205.0 |
| F | $\Delta E_F^*$ | 24.2 |
| S | $\Delta E_S^*$ | 1750.0 |
| Cl | $\Delta E_{Cl}^*$ | -222.7 |
| Br | $\Delta E_{Br}^*$ | 583 |
| I | $\Delta E_I^*$ | 1700 |
| Double bond | $\Delta E_{\neq}^*$ | -323 |
| Dipole-dipole interaction | $\Delta E_d^*$ | 1623 |
| Dipole-dipole interaction in nonpolar aprotic solvents of amide type | $\Delta E_{a,N}^*$ | 1623 |
| Dipole-dipole interaction in nonpolar aprotic solvents as in dimethylsulfoxide | $\Delta E_{a,S}^*$ | 2600 |
| Aromatic ring | $\Delta E_{ar}^*$ | 713 |
| Hydrogen bond | $\Delta E_h^*$ | 3929 |
| Specific interactions in the presence of $=CCl_2$ group | $\Delta E_{=CCl_2}^*$ | 2600 |
| Specific interactions in 3-5 member rings in the presence of O atom | $\Delta E_{O,r}^*$ | 2430 |
| Isomeric radicals | $\Delta E_i^*$ | -412 |

**Table 5.3.5 Increments of atoms or groups of atoms required in equation [5.3.3]**

| Atoms or their groups | $\Delta V_i$, Å$^3$ | Atoms or their groups | $\Delta V_i$, Å$^3$ |
|---|---|---|---|
| $-CH_3$ | 23.2 (22.9) | $>CH_2$ | 17.1 (16.8, 16.4) |
| >CH | 11.0 (10.7,10.4) | C | 5.0 (4.7,4.5) |
| $=CH_2$ | 21.1 | –CH= | 15.1 (14.7) |
| –O– | 3.4 (2.7, 2.1) | –OH | 10.3 (9.9) |
| >CO- | 18.65 (18.35,18.15) | –NH– | 8.8 (8.5) |
| $-NH_2$ | 16.1 | –CN | 25.9 |
| –F | 9.0 (8.9) | –Cl | 19.9 (19.5) |

The values in brackets correspond to one and two neighboring aromatic carbon atoms. In other cases values are given for the aliphatic neighboring carbon atoms.

**Table 5.3.6. Hansen's parameters[12]**

| Atom (group) | $\Delta V$, cm$^3$/mol | $V\delta_p$, (cal cm/mol)$^{1/2}$ | $V\delta_h^2$, cal/mol | |
|---|---|---|---|---|
| | | | aliphatic | aromatic |
| –F | 18.0 | 12.55±1.4 | ~0 | ~0 |
| –Cl | 24.0 | 12.5±4.2 | 100±200 | 100±20 |
| $>Cl_2$ | 26.0 | 6.7±1.0 | 165±10 | 180±10 |
| –Br | 30.0 | 10.0±0.8 | 500±10 | 500±100 |
| –I | 31.5 | 10.3±0.8 | 1000±200 | - |
| –O | 3.8 | 53±13 | 1150±300 | 1250±300 |
| >CO | 10.8 | 36±1 | 800±250 | 400±125 |
| >COO | 18.0 | 14±1 | 1250±150 | 800±150 |
| –CN | 24.0 | 22±2 | 500±200 | 550±200 |
| $-NO_2$ | 33.5 | 15±1.5 | 400±50 | 400±50 |
| $-NH_2$ | 19.2 | 16±5 | 1350±200 | 2250±200 |
| >NH | 4.5 | 22±3 | 750±200 | - |
| –OH | 10.0 | 25±3 | 4650±400 | 4650±500 |
| $(-OH)_n$ | n 10.0 | n (17±2.5) | n (4650±400) | n (4650±400) |
| –COOH | 28.5 | 8±0.4 | 2750±250 | 2250±250 |

## REFERENCES


1 M. Dunkel, *Z. Phys. Chem.*, **A138**, 42 (1928).
2 P.A. Small, *J. Appl. Chem.*, **3**, 71 (1953).
3 G. Scatchard, *J. Amer. Chem. Soc.*, **56**, 995 (1934).
4 R.F. Fedors, *Polym. Eng. Sci.*, **14**, 147(1974).

5 D.W. Van Krevelen, **Properties of polymers. Correlations with chemical structure**, *Elsevier Publishing Comp.*, Amsterdam-London-New York,1972.
6 A.E. Rheineck and K.F. Lin, *J. Paint Technol.*, **40**, 611 (1968).
7 K.L. Hoy, *J. Paint Technol.*, **42**, 76 (1970).
8 R.A. Hayes, *J. Appl. Polym. Sci.*, **5**, 318 (1961).
9 Y. Lebedev and E. Miroshnichenko, **Thermochemistry of evaporation of organic substances. Heats of evaporation, sublimation and pressure of saturated vapor**, *Nauka*, Moscow, 1981.
10 A.A. Askadskii, L.K. Kolmakova, A.A. Tager, et.al., *Vysokomol. soed.*, **A19**, 1004 (1977).
11 A.A. Askadskii, Yu.I. Matveev, **Chemical structure and physical properties of polymers**, *Chemistry*, Moscow, 1983.
12 C. Hansen, A. Beerbower, **Solubility parameters. Kirk-Othmer Encyclopedia of chemical technology**, 2nd ed., Supplement Vol., 1973, 889-910.
13 H.C. Brown, G.K. Barbaras, H.L. Berneis, W.H. Bonner, R.B. Johannesen, M. Grayson and K.L. Nelson, *J. Am. Chem. Soc.*, **75**, 1, (1953).


## 5.4 MIXED SOLVENTS, A WAY TO CHANGE THE POLYMER SOLUBILITY


Ligia Gargallo and Deodato Radic
**Facultad de Quimica**
**Pontificia Universidad Católica de Chile, Santiago, Chile**


### 5.4.1 INTRODUCTION

In general, a mixture is often found to be unexpectedly potent for a particular purpose. Examples are known in several branches of Science and Technology. In the field of solubility, the synergistic effect is, sometimes, spectacular. Innumerable cases of synergism in solvent extraction are known.[1-9]

Eucaliptus oil has been found to act as a cosolvent, special type of synergistic solvent in case of water-ethanol-gasoline.[10]

It has been frequently observed that certain polymers can be readily dissolved in mixtures of two or more solvents, whereas they are not soluble in the individual constituents of this mixture. This phenomenon, known as cosolvency is of great practical importance.

The scientific and technological importance of polymers has led to extensive study of their solution properties. Most techniques rely on dilute solution methods such as viscometry, light scattering, osmometry, gel swelling or dipole moments. However, these need solvents which completely dissolve the polymer, and many important polymers either are not very soluble, or completely insoluble. In general, polymer solubility normally increases with rising temperature, but negative temperature coefficients are also observed. Increase in polymer molecular weight reduces solubility. Certain combinations of two solvents may become nonsolvents. However, mixtures of non-solvents may sometimes become solvents. In this article, we will review major information for the evaluation of solubility behavior of polymers in binary solvents. Experimental results are compiled and discussed. The emphasis here is on solubility-cosolvency phenomenon, the thermodynamical description of ternary systems and the influence of the polymer structure on preferential adsorption. Finally, interrelationships between polymer structure and thermodynamic properties of the mixture will be discussed relative to the properties and conformation of the dissolved polymer.

Some new cosolvent effects will be also described particularly of polymer-supercritical $CO_2$ mixtures.

## 5.4.2 SOLUBILITY-COSOLVENCY PHENOMENON

The addition of a second liquid to a binary liquid-polymer system to produce a ternary system is used widely for a variety of purposes. If the second liquid is a poor solvent, or a precipitant for the polymer, the dissolving potential of the liquid medium can be reduced and eventually phase separation may even occur. This does not necessarily take place in every case and sometimes mixtures of two relatively poor solvents can even produce an enhancement of the solvent power.[11-16] The mixed solvent is then said to exhibit a synergistic effect.[17]

Cosolvency usually refers to a certain range of temperatures of practical use. Within that range, the polymer dissolves in the mixed solvent but not in the pure liquids. The term, "true cosolvency" has been coined to designate those more strict cases, in which the polymer does not dissolve in the pure liquids at any temperature, not just in a given range.[18]

An analysis of the phase behavior in terms of the free volume theory of polymer solutions has revealed that cosolvency has enthalpic origin.[19] In a classic cosolvent system, the single liquids are both very poor solvents for the polymer and the number of polymer-liquid contacts formed in each binary system are not enough to stimulate dissolution of the polymer, except at very short chain lengths. Mixing to produce a ternary system results in a combination of liquid(1)/polymer(3) and liquid(2)/polymer(3) contacts, which according to Cowie et al.[20] is sufficient to cause the polymer to dissolve if these contacts are of a different nature, i.e., if the two liquids tend to solvate the polymer at different sites along the chain and so lead to a reinforcement of solvation. If this occurs then, it is also likely that the expanded coil will allow further liquid-polymer contacts to develop even though these may be energetically weak.

This idea of a favorable solvation sheath is in agreement with observations made during preferential adsorption studies in ternary systems exhibiting cosolvency.[21] Maximum coil expansion is usually found to occur at a mixed solvent composition where there is not preferential adsorption. In other words, the driving force is an attempt to maintain the most favorable composition, thereby minimizing polymer-polymer contacts. However, the balance of interactions giving rise to cosolvency and to inversion in preferential sorption are different, so that both phenomena have to be studied separately. This has been exemplified by a series of systems in which the molecular sizes of the liquid solvents and the nature of their interactions are varied and each plays its role in determining maximum sorption and inversion in preferential sorption.[22-25]

The interesting phenomenon where a mixture of two poor solvents or nonsolvents for a polymer provides a medium that acts as a good solvent for the polymers[26,27] has been the objective of many studies, by light scattering,[21,29,30-34] viscometry,[35,36] sorption equilibrium,[37] and fluorescence.[38] From these techniques, it has been possible to appreciate how the second virial coefficient $A_2$[39,17,40-42] and the intrinsic viscosity $[\eta]$[13,43-45] preferential adsorption coefficient $\lambda$ and excimer and monomer emission ratio $I_E/I_M$ are involved by changing solvent composition. They present ($[\eta]$, $A_2$) a maximum or a variation at a certain solvent composition where the polymer behaves as through it were dissolved in a good solvent.

The quality of solvent or the cosolvent action has been established by determining the magnitude of the miscibility range between the two critical temperatures, UCST and LCST.[19,45-58] The application of pressure can widen the miscibility range.[59,60]

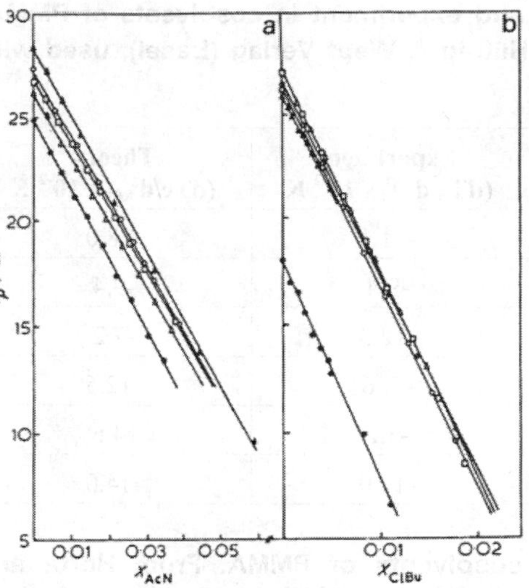


Figure 5.4.1. Phase separation temperature ($T_p$), as a function of solvent mixture composition, determined on the system PMMA-acetonitrile + chlorobutane, at several polymer concentrations, x(x $10^2$/g cm$^{-3}$): (a): c = Δ, 12.2; O, 9.17; □ 7.14; Δ, 4.36: ●, 2.13. (b): c = O, 9.47; ◘, 8.22; Δ, 5.89; □, 3.69; Δ, 2.17; ●, 0.442. From Fernandez-Pierola and Horta.[65] (Copyright by Hüthig & Wepf Verlag (Basel), used with permission).

The cosolvency phenomenon was discovered in 1920's experimentally for cellulose nitrate solution systems.[61] Thereafter cosolvency has been observed for numerous polymer/mixed solvent systems. Polystyrene (PS) and polymethylmethacrylate (PMMA) are undoubtedly the most studied polymeric solutes in mixed solvents.[62,63]

Horta et al.[64] have developed a theoretical expression to calculate a coefficient expressing quantitatively the cosolvent power of a mixture $(dT_c/dx)_0$, where $T_c$ is the critical temperature of the system and x is the mole fraction of liquid 2 in the solvent mixture, and subscript zero means $x \to 0$. This derivative expresses the initial slope of the critical line as a function of solvent composition (Figure 5.4.1).[65] Large negative values of $(dT_c/dx)$ are the characteristic feature of the powerful cosolvent systems reported.[65] The theoretical expression developed for $(dT_c/dx)_0$ has been written in terms of the interaction parameters $\chi_i$ for the binary systems:

$$(dT_c / dx)_0 = \frac{\left(\chi_{23} - \chi_{13} - \frac{V_2}{V_1}\chi_{12}\right)}{(-d\chi_{13} / dT)} \quad [5.4.1]$$

where:

$V_i$ the volume fraction of polymer and solvent, respectively

All the magnitudes on the right hand side of this equation are to be evaluated at the critical temperature corresponding to x = 0.

The expression provides a criterion to predict whether or not the mixed solvent is expected to be a cosolvent of the polymer. When $T_{c1}$, is a UCST (as is the case in these phase separation studies), $-d\chi_{13}/dT > 0$ and $(dT_c/dx)_0$ has the same sign as the numerator of the equation. Choosing solvent 1 such that $T_{c2} < T_{c1}$, then $(dT_c/dx)_0 < 0$ guarantees that the system will be a cosolvent one. Since $\chi_{23} - \chi_{13} > 0$, at $T_{c1}$, the numerator in the equation [5.4.1] is negative (cosolvent system) if the unfavorable interaction between the two liquids is large enough to compensate for their different affinity towards the polymer. The equation proposed gives a more detailed criterion for cosolvency than the simple criterion of $G^E > 0$. The information needed to predict $(dT_c/dx)_0$ from equation [5.4.1] includes the binary interaction parameters of the polymer in each one of the two solvents as a function of temperature, and $\chi_{12}(G^E)$ for the mixed solvent too. Table 5.4.1 summarizes results reported by Horta et al.[64] for some cosolvents of polymethylmethacrylate (PMMA).[65]

**Table 5.4.1. Comparison between theory and experiment in cosolvents of PMMA. From Fernandez-Pierola.[65] (Copyright by Hüthig & Wepf Verlag (Basel), used with permission)**

| Solvent 1 | Solvent 2 | Experiment $(dT_p/dx)_0 \times 10^{-2}$ K | Theory $(dTc/dx)_0 \times 10^{-2}$ K |
|---|---|---|---|
| Acetonitrile | Amylacetate | -11.1 | -18.0 |
| Acetonitrile | Chlorobutane | -9.4 | -9.4 |
| Acetonitrile | $CCl_4$ | -12.3 | -7.2 |
| Acetonitrile | BuOH | -7. 6 | -12.5 |
| 4-Chloro-n-butane | Acetonitrile | -3.2 | -4.6 |
| 1-Chloro-n-butane | BuOH | -17.0 | -14.0 |

**Table 5.4.2 Liquid mixtures which are cosolvents of PMMA. From Horta and Fernandez Pierola.[64] (Copyright by Butterworth-Heineman Ltd., used with permission)**

| | |
|---|---|
| Acetonitrile (AcN) + 4-Heptanone (Hna)[a)] | Formamide (FA) + Ethanol (EtOH)[a)] |
| Acetonitrile (AcN) + Isopentyl acetate (iPac)[a)] | 1-Chlorobutane (BuCl) + Isopentyl alcohol (iPOH)[a)] |
| Acetonitrile (AcN) + Pentylacetate (Pac)[a)] | 1-Chlorobutane (BuCl) + Pentyl alcohol (POH)[a)] |
| Acetonitrile (AcN) + 1-Chlorobutane (BuCl)[a)] | 1-Chlorobutane (BuCl) + 2-Butanol (sBuOH)[b)] |
| Acetonitrile (AcN) + Carbon tetrachloride ($CCl_4$)[a)] | 1-Chlorobutane (BuCl) + Isopropyl alcohol (iPrOH)[a)] |
| Acetonitrile (AcN) + Isopentyl alcohol (iPOH)[a)] | Carbon tetrachloride ($CCl_4$) + 1-Butanol (BuOH)[c)] |
| Acetonitrile (AcN) + Pentyl alcohol (POH)[a)] | Carbon tetrachloride ($CCl_4$) + Ethanol (EtOH)[c)] |
| Acetonitrile (AcN) + 1-Butanol (BuOH)[a)] | Carbon tetrachloride ($CCl_4$) + Methanol (MeOH)[c)] |
| Acetonitrile (AcN) + Isopropyl alcohol (iPrOH)[a)] | |

a) Ref. 65, b) Ref. 47, c) Ref. 6 6

In the case of PMMA, several powerful cosolvent mixtures have been reported.[65,46,47] In such systems, a small proportion of liquid 2, added to the polymer-liquid 1 solution, is enough to produce a large decrease in $T_p$, these cosolvents are accessible to phase separation determinations.

In Table 5.4.2, a number of liquid mixtures are listed to act as cosolvents of polymethylmethacrylate (PMMA).[65,47,64,67] Intrinsic viscosity [η] has been reported in cosolvent mixtures containing $CCl_4$: $CC1_4$/methanol,[42,66] ethanol,[66] 1-propanol,[66] 1-butanol,[66] 1-chlorobutane,[67] and acetonitrile,[68] and acetonitrile/methanol.[68] It was also reported the [η] of PMMA in the cosolvent mixture acetonitrile/1-chlorobutane.[42] The last system is a powerful cosolvent. Acetonitrile forms powerful cosolvents for PMMA too with other liquids having a wide variety of chemical groups.[65] [η] has been reported for PMMA in acetonitrile/pentylacetate as a powerful cosolvent.[44] On the contrary, the mixture of 1-chlorobutane/pentyl acetate is a co-nonsolvent of the polymer.[44] Mixing acetonitrile, with pentylacetate or mixing acetonitrile with 1-butanol greatly increases solubility and produces a large increase in [η]. This large increase of [η] in the cosolvents contrasts with the approximate constancy of [η] in the 1-chlorobutane/pentylacetate mixture.

The increase of [η] in a mixed solvent over its weighed mean value in the pure liquids is usually expressed as:

$$\Delta[\eta] = [\eta] - [\eta]_1 \phi_1 - [\eta]_2 \phi_2 \qquad [5.4.2]$$

where $[\eta]_1$ and $[\eta]_2$ refer to values in the pure liquids 1 and 2, and [η] refers to the value in the mixture. This increase in [η] has been attributed to the existence of unfavorable interactions between the two liquids.[39] For a given molecular weight, Δ[η] is usually taken to be proportional to the excess in Gibbs function of the mixture: $\Delta[\eta](\phi) \sim G^E(\phi)/RT$.[39]

The values of $G^E$ and $S^E$ for these cosolvent mixtures[69,70] at equimolecular composition, and 25°C, are given in Table 5.4.3.

**Table 5.4.3. Thermodynamic properties of the liquid mixtures used as cosolvents of PMMA. Excess Gibbs function $G^E$, and excess entropy $S^E$, of the binary mixtures at equimolecular composition (at 25°C). From Prolongo et al.[44] (Copyright by Butterworth-Heineman Ltd., used with permission)**

| Cosolvent mixture | $G^E$, J mol$^{-1}$ | $S^E$, J mol$^{-1}$ K$^{-1}$ |
|---|---|---|
| MeCN + BuOH | 1044 | 3.70 |
| MeCN + PAc | 646 | -0.58 |
| MeCN + ClBu | 1032 | - |

The values of $G^E$s have allowed for a qualitative interpretation of the relative values of [η] in these three cosolvent systems studied.[44]

Mixing cosolvents is much more effective in expanding the polymer coil than increasing temperature.[44] In fact, the same increase in [η] experienced by one sample in pure acetonitrile in going from 25°C to 45°C is reached at 25°C by adding just 9% pentyl acetate or 8% 1-butanol or 6% in the case of 1-chlorobutane,[71] for polymethylmethacrylate.

The sign of Δ[η] was in contradiction with the cononsolvent character attributed to this mixture by cloud point studies.[72] This apparent inconsistency could be due to the different range of concentrations in which [η] and cloud point temperature were determined.[44]

Systematic study of the cosolvency phenomenon has been practically limited to polymethylmethacrylate[42,46,64-68,72-81] and polystyrene.[45,82,83] The cosolvency is usually explained in terms of the molecular characteristics of the system, specially in terms of molecular interactions. In the powerful cosolvents of PMMA described in the literature[72,65,42,44,46,64] one of the liquid components is always either acetonitrile or an alcohol. These are non-random liquids with a certain degree of order in their structure. Two important characteristics seem to be present in these polymer cosolvent systems: the liquid order structure and the tendency of the polymer towards association.[65,14,66] The roles of these two factors were considered to interpret solvation of PMMA chains in cosolvent systems.[43,14] The mechanisms of cosolvent action have been discussed in terms of the competitive interactions between liquid components and one liquid component and the polymer.[44] The best example is the case of acetonitrile and a second liquid having a high proportion of methylene units in its molecule, the unfavorable nitrile-methylene interactions between acetonitrile and PMMA favor the nitrile ester group interaction and an extensive polymer solvation becomes possible. According to Prolongo et al.,[44] the number of methylene units or length of the n-alkyl chain is very important for reaching cosolvency when the second liquid is an ester (acetate). Another factor $G^E$ of the acetonitrile + acetate mixtures is larger for long alkyl chains such as in pentyl acetate, favors cosolvency.

In the majority of cases, the cosolvent mixtures for PMMA contain either $CCl_4$[73,77,80,81] or acetonitrile[13,42-44,67] as one of the liquid components. A study of the mixture formed by these two liquids and a comparison with the results obtained in the other cosolvents studied before has been also reported.[84] The total sorption of the coil (PMMA) was calculated from second virial coefficient and intrinsic viscosity data. According to these authors, acetonitrile can interact favorably with the ester group of PMMA and is unfavorable with its methylene backbone. The role of these opposing interactions and of liquid order in acetonitrile are taken into account to explain the dilute solution properties of PMMA in this cosolvent system.[84]

On the other hand, in the case of PMMA, in powerful cosolvents, a small proportion of liquid 2, added to the polymer/liquid 1 solution, is enough to produce a large decrease in the phase separation temperature ($T_p$).[64]

Katime et al.[85] have studied the influence of cosolvency on stereo-complex formation of isotactic and syndiotactic PMMA. The formation of PMMA stereo-complex has been attributed to the interactions between the ester group of the isotactic form and the $\alpha$-methyl group of the syndiotactic form.[86]

The stereo-complex was obtained at different compositions of the cosolvent mixtures acetonitrile/carbon tetrachloride, acetonitrile/butyl chloride and butyl chloride/carbon tetrachloride. The results show a high yield of complex formation in pure solvents and when approaches its solvency maximum a decrease of the yield of stereo-complex was observed, indicating that the interactions are impeded.[85]

The dilute solution viscosity of PMMA in the cosolvent mixture formed by acetonitrile (MeCN) and 1-chloro-n-butane (ClBu) at 25°C has been studied.[42,87] The cosolvent effect in this system is extremely large. It has been observed a large increase in the hydrodynamic volume of the macromolecule in solution,[42,17] and a step depression in the critical temperature of phase separation (UCST).[64,88] The quantitative determination of the magnitude of these effects has been reported.[42,64,88] Horta et al.,[71] have compared a relative capacities of temperature and of cosolvent mixing on expanding the macromolecular coils and the tendency of the polymer to associate in poor solvents. They have also shown that there is a connection which relates the dependencies of $[\eta]$ on temperature and solvent composition with the depression in critical temperature (UCST) caused by cosolvency.[71] The action of the cosolvent was much more effective in expanding the macromolecule than temperature was.[71] These authors have concluded that the comparison between the temperature and solvent composition variations of $[\eta]$ allows for a correct prediction of the cosolvent depression of the UCST. The comparison between cloud points and $[\eta]$ - T gives, in general, inconsistent results, but the combination of $[\eta]$ - T and $[\eta]$ - $\phi$ compensates such inconsistency and establishes a valid link between $T_c$ and $[\eta]$.

It was found that when the cosolvent power of the binary mixture increases, the complexing capacity decreases. These results were explained by taking into account the excess Gibbs free energy, $G^E$, and the order of the liquid.

## 5.4.3 NEW COSOLVENTS EFFECTS. SOLUBILITY BEHAVIOR

Cosolvent effect of alkyl acrylates on the phase behaviour of poly(alkyl acrylate)supercritical $CO_2$ mixtures has been reported.[89] Cloud-point data to 220 and 2000 bar are presented for ternary mixtures of poly(butyl acrylate)-$CO_2$-butyl acrylate (BA) and poly(ethylhexyl acrylate)-$CO_2$-ethylhexyl acrylate) (EHA). The addition of either BA or EHA to the respective polymer-solvent mixtures decreases the cloud-point pressures by as much as 1000 bar and changes the pressure-temperature slope of the cloud-point curves from negative to positive, which significantly increases the single-phase region.

The literature presents many studies on coil dimensions of synthetic polymers in mixed solvents. Most investigations involve liquid mixtures composed of a good and a poor solvents for the polymer. The action of mixed solvents has been reported to change coil dimensions, not only because of excluded volume effect or due to the interactions existing between the two liquids[90,91] but also due to the preferential adsorption of one of the solvent by the polymer.

Recently, the behavior of polysiloxanes with amino end-groups in toluene/nitromethane mixtures has been reported.[92] This mixture is solvent/non-solvent for the polymer.

The transition concentrations separating the concentration domain[93] chain flexibility aspects, excluded volume effects[94] and total and preferential adsorption coefficients[95] of the same system have been discussed.

The solubility curves, the cloud point curves and vitrification boundaries for several poly(lactide)-solvent-nonsolvent systems have been reported.[96] The liquid-liquid miscibility gap for the systems with the semicrystalline poly(L-lactide) (PLLA) were located in a similar composition range as the corresponding systems with the amorphous poly(DL-lactide) (PDLLA). The solvent-nonsolvent mixtures used for the experiments were: dioxane/water, N-methyl pyrrolidone (NMP)/water, chloroform/methanol and dioxane/methanol. For all PLLA solvent-nonsolvent systems studied solid-liquid demixing was preferred thermodynamically over liquid-liquid demixing. Attempts were made to correlate the experimental finding with predictions on the basis of the Flory-Huggins theory for ternary polymer solutions using interaction parameters derived from independent experiments. Qualitative agreement was found for the relative locations of the liquid-liquid miscibility gaps. The Flory-Huggins description of the solubility curves was less satisfactory.

The phase separation processes occurring in poly(L-lactide) (PLLA)-chloroform-methanol mixtures and poly(DL-lactide) (PDLLA)-chloroform-methanol mixtures have been also studied using differential scanning calorimetry, cloud point measurements and optical microscopy.[97] It was demonstrated that liquid-liquid demixing occurs in ternary solutions of PDLLA at sufficient high methanol concentrations. For PLLA-containing-solutions, both liquid-liquid demixing processes and soli-liquid demixing processes occur. Only a low cooling rates and high polymer concentration does solid-liquid demixing take place without the interference of liquid-liquid demixing.[97]

Another interesting effects are the changes of a polyelectrolyte in binary solvents. The complex inter and intramolecular interactions that take place due to the presence of hydrophilic and hydrophobic structural units in the macroion can modify the balance of the interactions and for this reason can change the solubility.[98]

## 5.4.4 THERMODYNAMICAL DESCRIPTION OF TERNARY SYSTEMS. ASSOCIATION EQUILIBRIA THEORY OF PREFERENTIAL ADSORPTION

Polymers dissolved in mixed solvents show the phenomenon of Preferential Adsorption. Experimentally, the preferential adsorption coefficient, $\lambda$, is determined. $\lambda$ is the volume of one of the liquids sorbed in excess by the polymer (per unit mass of polymer). In general, the Flory-Huggins model of polymer solutions is used to describe the Preferential Adsorption. More recently, equation of state theories have been applied.[13,43,99-101]

Description of experimental results of $\lambda$ (and another properties as intrinsic viscosities, second virial coefficients, etc.) necessitates the use of correcting terms in the form of ternary interaction parameters. Using equation of state theory it has been shown that such correcting terms can in part be explained by free volume and molecular surface effect.[100] Non-random interactions are important in many systems (hydrogen bonding, complex formation, etc.) (see Table 5.4.1). Strongly interacting species can be described taking into account the formation of associates in equilibrium with unassociated molecules (Association Equilibria Theory).

Experimental results[102] for polymethylmethacrylates in 1,4-dioxane/methanol have been reported , which indicate that the size of the substituent in the polymer ester group exerts an influence on the specific interaction between the methanol molecule and the carbonyl of the ester. In fact, the preferential adsorption of methanol is completely hindered when the lateral group is bulky enough. Similar results have been reported for substituted poly(phenyl methacrylate)s in the mixture tetrahydrofuran/water.

The description of these systems was not in agreement with predictions of classical thermodynamic theories.[39,103-105]

This behavior was analyzed in terms of specific interactions among the components of the ternary system. If the oxygen atoms of 1,4-dioxane can interact specifically with methanol by accepting a proton, then, methanol hydrogen bonds not only to poly(methacrylate)s but also to 1,4-dioxane. The new theoretical formation takes into account the case of solvent which is self-associated and interacts specifically with sites in the polymer chain and with sites in the other solvent molecule. Therefore, it must consider association constants of the solvent molecule and association constants of the self-associated-solvent (2) with (3) the polymer. It was assumed that the polymer molecule has one site for specific interactions with 2, that the constant for such specific interaction between one 2 molecule and one sites is $\eta_1$ and that the self-association of 2 over the 2 molecule attached to a site in 3 is characterized by constant $\sigma_2$, $\eta_2$ or $\sigma_2$ values, or both, should vary from polymer to polymer depending on the size of the substituent pending from the ester group. In addition, also association constants of 2 with the other solvent molecule (1) have been postulated. These are: $\eta_1$ and $\sigma_1$. The number of sites on molecule 1 was called t. The constant for the specific interaction of one 2 molecule with one site in 1 is $\eta_1$, and the self-association of 2 over the already attached 2 molecule is $\sigma_1$. With all these constants a quantitative description of the experimental results for the three poly(alkyl methacrylates) (alkyl + Me, Et, iBu) in 1,4-dioxane/methanol was reported.[106] In conclusion, the theory applied by Pouchly and Zivny to the simpler case in which one of the liquids is inert, was extended to more complex mixture.[106]

Table 5.4.4 summarizes the glossary of association constants and interaction parameters used in the theory in the case of poly(alkyl methacrylates). The results that are obtained for the minimum standard deviation in each case are shown on Table 5.4.4.

**Table 5.4.4. Parameter values giving the minimum deviation (δ) between theory and experiment, for the preferential adsorption coefficient, λ, calculated according to the Association Equilibria Theory. Reprinted with permission from Horta et al.[106] (Copyright (1989) American Chemical Society)**

| | Polymer | σ | gBA' | $\eta_a$ | $\sigma_c$ | gAC'-r rAgBC' | $\eta_c$ | δ |
|---|---|---|---|---|---|---|---|---|
| 1 | **PMMA** | 400 | 0.85 | 244 | 400 | 0.0 | 5.7 | 0.034 |
| 2 | **PEMA** | 400 | 0.85 | 244 | 400 | -1.2 | 59.3 | 0.028 |
| 3 | **PiBMA** | 400 | 0.85 | 244 | 400 | -2.3 | 129.5 | 0.018 |
| 4 | **PMMA** | 400 | 0.30 | 100 | 400 | -3.4 | 243.2 | 0.013 |
| 5 | **PEMA** | 400 | 0.30 | 100 | 400 | -2.75 | 169.2 | 0.014 |
| 6 | **PiBMA** | 400 | 0.30 | 100 | 400 | -1.6 | 188.2 | 0.009 |
| 7 | **PMMA** | 400 | 0.85 | 244 | 375 | -0.1 | 11.2 | 0.034 |
| 8 | **PEMA** | 400 | 0.85 | 244 | 375 | -2.2 | 145.4 | 0.028 |
| 9 | **PiBMA** | 400 | 0.85 | 244 | 375 | -1.6 | 100.7 | 0.035 |
| 10 | **PMMA** | 400 | 0.30 | 100 | 375 | -2.45 | 195.5 | 0.013 |
| 11 | **PEMA** | 400 | 0.30 | 100 | 375 | -1.75 | 125.0 | 0.016 |
| 12 | **PiBMA** | 400 | 0.30 | 100 | 375 | -0.5 | 45.6 | 0.016 |

The results obtained are a good description of the experimental data on these ternary systems. Effectively, the shape of the variation of λ with solvent composition was well reproduced by the association equilibria theory, as it is shown in Figure 5.4.2.

It has extended the same theoretical treatment to other closely related systems, a family of poly(dialkyl itaconates).[107] Preferential adsorption coefficient λ was determined and calculated according to the association equilibria theory, and using classical thermodynamic theories.[108,103,104,105]

The dependence of the preferential adsorption coefficient for poly(dimethyl itaconate) (PDMI), poly(diethyl itaconate) (PDEI), poly(dipropyl itaconate) (PDPI) and poly(dibutyl itaconate) (PDBI), in 1,4-dioxane/methanol mixtures, as a function of the methanol composition ($u_{BO}$) is shown in Figure 5.4.3. The results are very similar to those of poly(alkyl methacrylates) in the same solvent.[106]

Comparison between theory and experimental for λ, by using classical thermodynamic theorics[103-105,108] are shown in Figure 5.4.4.

The results obtained are a good description of the experimental data on these ternary systems. Effectively, the shape of the variation of λ with solvent composition was well reproduced by the association equilibria theory. It can be concluded that the association equilibria theory of preferential adsorption in systems with solvent-solvent and solvent-polymer interactions describe in a quantitative way the experimental results of λ. Although, the systems in which the theory has been applied are closely related. More recently, there was no other systems studied to apply this theory.

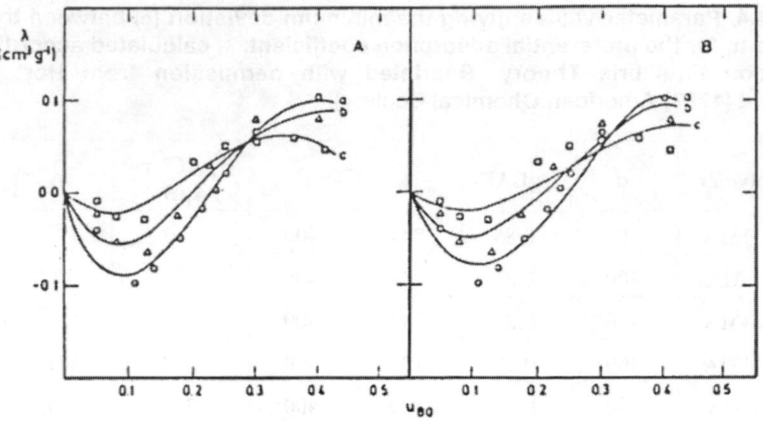


Figure 5.4.2. Comparison of theory and experiment for preferential adsorption coefficient, λ, of poly(alkyl methacrylate)s in 1,4-dioxane-methanol. ($u_{BO}$=methanol volume fraction). Points: Experimental results from ref.[6]. (O) PMMA (alkyl = Me); (Δ) PEMA (Et); (□) PiBMA (iBu). Association equilibria theory. (8A) Calculated with the parameter values shown in Table 5.4.4 and numbered as 4-6. Curves: (a) PMMA; (b) PEMA; c) PiBMA. (7B) Calculated with the parameter values shown in Table 5.4.4 and numbered as 10-12. Curves: (a) PMMA; (b) PEMA; c) PiBMa. Reprinted with permission from Horta et al.[106] (Copyright (1989) American Chemical Society).

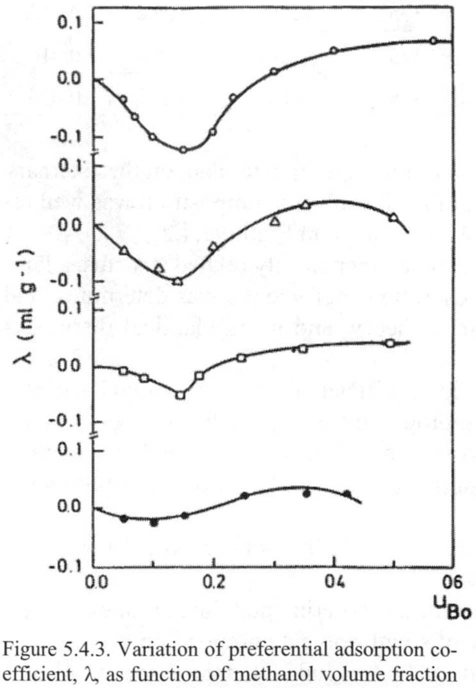


Figure 5.4.3. Variation of preferential adsorption coefficient, λ, as function of methanol volume fraction $u_{BO}$, for PDMI (O), PDEI (Δ), PDPI (□), and PDBI (●), at 298 K. Reprinted with permission from Horta et al.[107] (Copyright (1990) American Chemical Society).

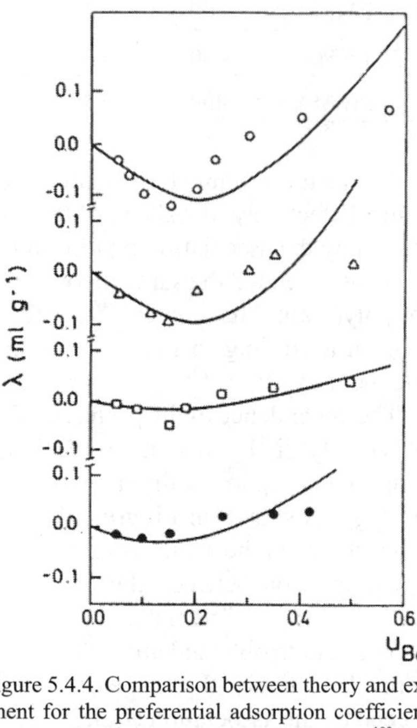


Figure 5.4.4. Comparison between theory and experiment for the preferential adsorption coefficient, λ, by using classical thermodynamic theories.[103] Points: experimental result of λ. Curves: classical thermodynamic theories.[108] Reprinted with permission from Horta et al.[107] (Copyright (1990) American Chemical Society).

## 5.4.5 POLYMER STRUCTURE OF THE POLYMER DEPENDENCE OF PREFERENTIAL ADSORPTION. POLYMER MOLECULAR WEIGHT AND TACTICITY DEPENDENCE OF PREFERENTIAL ADSORPTION

There are some important structural aspects of the polymer which are necessary to take into account in the analysis of the polymer behavior in mixture solvents, such as its polarity, chemical structure, microtacticity, molecular weight. The analysis of these properties shows that they are determinant factors in preferential adsorption phenomena involved. It has been pointed out[73] that the effect of tacticity, and particularly the molecular weight, is a complex problem. In the case of poly(2-vinylpyridine), when the polar solvent is preferentially adsorbed, preferential solvation is independent of molecular weight; but when the non-polar solvent is adsorbed, there is a dependence on the molecular weight.[62]

The reported experimental evidence[109-111,67,73,77] seems to show that the coefficient of preferential adsorption λ, for a given polymer in a mixed solvent of fixed composition depends on molecular weight of the polymer sample. It is important to remember, however, that this dependence of λ on M has not been always detected. Particularly for molecular weights lower than a certain value.[100,111,91]

According to Dondos and Benoit,[109] Read,[104] and Hertz and Strazielle,[112] the dependence of preferential adsorption coefficient λ, with molecular weight or with segment density can be expressed empirically as:

$$\lambda = \lambda_\infty + AM^{-1/2} \quad [5.4.4]$$

or

$$\lambda = \lambda_\infty + K[\eta]^{-1} \quad [5.4.5]$$

where:

| | |
|---|---|
| $\lambda_\infty$ | the value of λ extrapolated to $M \to \infty$ |
| $[\eta]$ | the intrinsic viscosity |
| A and K | constants |

It is interesting to note that the variation of λ with M is more pronounced. A in equation [5.4.4] is larger in mixtures which are poor solvents close to θ-conditions than in mixtures with excluded volume.[109,111,112]

Apparently, there will not be exhaustive results either with the chemical structure of the polymer on the preferential adsorption,[113] or the influence of the tacticity on the preferential adsorption. In the last years, investigations regarding the effect of ortho-substituents in polymers with aromatic bulky side groups on the preferential adsorption and viscometric behavior have been reported for poly(phenyl methacrylate) and its dimethyl and diisopropyl ortho derivatives in tetrahydrofuran/water.[114] Figure 5.4.5 from ref.[114] shows the λ values for three polymers in THF/water.

The λ values diminish when the volume of the side groups increases and there is a strong water adsorption. The behavior reported[114] indicates that the cosolvent effect decreases or disappears when the preferential adsorption is very small or is not observed.[114]

In another publication,[102] the systems studied were a series of poly(alkyl methacrylates) including the methyl (PMMA), ethyl (PEMA), isobutyl (PiBMA) and cyclohexyl (PCHMA) substituents in the mixture solvent 1,4-dioxane/methanol. The experimental

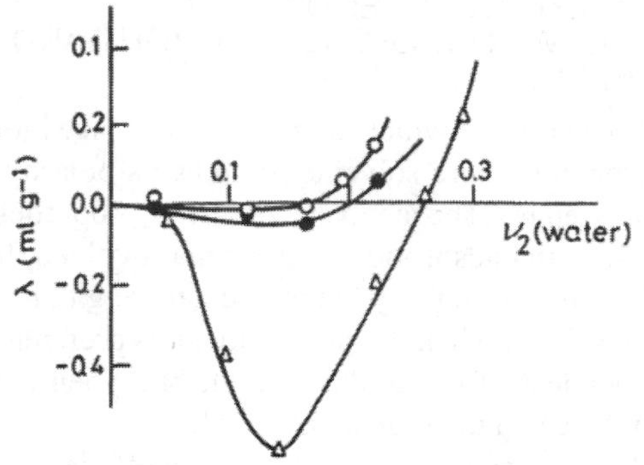


Figure 5.4.5. Variation of the preferential adsorption as a function of solvent composition (Δ) PPh (Gargallo et al., 1984): (●) PDMPh: (O) PDPPh. From Gargallo et al.[114] (Copyright by Springer Verlag, used with permission).

results[113] indicate that the size of the substituents on the polymer ester group exerts an influence on the specific interaction between the methanol molecule and the carbonyl of the ester. It has been shown that the preferential adsorption of methanol is completely hindered when the lateral groups are bulky enough.

Several attempts have been made to take into account the geometrical character of the polymer segment and the solvent molecule. None of them seems to give a unique explanation for the experimental results about the cosolvent effect.

The most important task in the field of research on ternary systems polymer in binary solvents is to examine the state of the macromolecular chain in solution and the analysis of the changes in solubility and then in composition involved in the total system itself.

The goal of this review was to present aspects of the preferential adsorption phenomena of solvents and polymers with a focus on their thermodynamic aspects. The idea behind this was to attract the attention of polymeric physico-chemists to this area, which is sufficiently related to a lot of different effects. In fact, preferential adsorption occupies a special place in the solubility of polymers in mixed solvents.

## REFERENCES


1 Y. A. Zolotov, and L. G. Gamilova, *J. Inorg. Nucl. Chem*., **31**, 3613 (1969).
2 V. Pandu Ranga Rao, and P.V.R. Bhaskara Sarma, *J. Inorg. Nucl. Chem.,* **31**, 2151 (1969).
3 T.V. Healy, *J. Inorg. Chem*., **31**, 49 (1969).
4 M.A. Carey and C.V. Banks, *J. Inorg. Nucl. Chem*., **31**, 533 (1969).
5 L. Newman and P. Klotz, *Inorg. Chem*., **5**, 461 (1966).
6 H.M.N.H. Irving and N.S. Al-Niaimi, *J. Inorg. Nucl. Chem*., **27**, 1671, 2231 (1965).
7 H.M.N.H. Irving and N.S. Al-Niaimi, *J. Inorg. Nucl. Chem*., **21**, 1671, 2231 (1965).
8 L. Schaefer, *Am. J. Pharm*., **85**, 439 (1913).
9 Y. Marcus, *Chem. Rev*., **63**, 161 (1963).
10 A.F.M. Barton and J. Tjandra, *Fuerl*, **68**, 11 (1989).
11 L. Gargallo, D. Radic', and E. Zeidan, *Bol. Soc. Chil. Qufm*., **27**, 15 (1982).
12 A.A. Abdel-Azim, S.S. Moustafa, M.M. Edl Dessouky, F. Abdel-Rehim, and S.A. Hassan, *Polymer*, **27**, 15 (1982).
13 R.M. Masegosa, M. G. Prolongo, I. Hernandez-Fuentes, A. Horta, *Macromolecules*, **17**, 1181 (1984).
14 J.R Ochoa, B. Caballero, R. Valenciano and I. Katime, *Mater. Chem. Phys*., **9**, 477 (1983).
15 J.M.G. Cowie and J.T. McCrindle, *Eur. Polym. J*., **8**, 1325 (1972).
16 T. Spychaj. *Angew, Makromol. Chem*., **149**, 127 (1987).
17 P. Munk, M.T. Abijaoude, M.E. Halbrook, *J. Polym. Sci. Polym. Phys*., **16**, 105 (1978).
18 a) S. Matsuda and K. Kamide. *Polym, J*., **19**, 2, 211 (1987); b) S. Matsuda and K. Kamide, *Polym. J*., **19**, 2, 203 (1987); c) B.A. Wolf, and R.J. Molinari, *Makromol. Chem*., **173**, 241 (1973).
19 C. Strazielle and H. Benoit, *J. Chem. Phys*., **38**, 675 (1961).
20 J.M.G. Cowie, *J. Polym. Sci*., **C23**, 267 (1968).
21 L. Gargallo, D. Radicand, I. Femandez-Pierola, *Makromol. Chem. Rapid. Commun*., **3**, 409 (1982).
22 C. Feyreisen, M. Morcellet and C. Loucheux, *Macromolecules*, **11**, 620 (1978).
23 B. Carton, V. Boltigliori, M. Morcellet, and C. Loucheux, *Makromol. Chem*., **179**, 2931 (1978).
24 R. Noel, D. Patterson and T. Somcynsky, *J. Polym. Sci*., **B42**, 561 (1960).

25 R. Gavara, B. Celda, A. Campos, *Eur. Polym. J.*, **22**, 373 (1986).
26 D.K. Sarkar, S.R Palit, *J. Polym. Sci.*, **C30**, 69 (1970).
27 K. Ishikawa, T. Kawai, *J. Chem. Soc. Jpn. Ind. Chem. Sci.*, **55**, 173 (1952).
28 H. Yamakawa, *J. Chem. Phys.*, **46**, 3, 973 (1967).
29 K. Takashima, K. Nake, M. Shibata and H. Yamakawa, *Macromolecules*, **7**, 641 (1974).
30 P. Kratochvil, *J. Polym. Sci. Polym. Symp.*, **50**, 487 (1975); **55**, 173 (1952).
31 J.M.G. Cowie, *Pure Appl. Chem.*, **23**, 355 (1970).
32 A. Campos, B. Celda, J. Mora and J.E. Figueruelo, *Polymer*, **25**, 1479 (1984).
33 W. Wunderlich, *Makromol. Chem.*, **182**, 2465 (1981).
34 M. Nakata andN. Numasawa, *Macromolecules*, **18**, 1736 (1985).
35 B. Friederich and K. Prochazka, *Eur. Polym. J.*, **15**, 873 (1979).
36 I. Katime, P.M. Sasiaand B. Eguia, *Eur. Polym. J.*, **24**, 12, 1159 (1988).
37 J. Pouchly and A. Zivny, *Makromol. Chem.*, **186**, 37 (1985).
38 R.M. Masegosa, I. Hernandez-Fuentes, I. Fernandez Pierola, A. Horta, *Polymer*, **28**, 231 (1987).
39 A. Dondos and D. Patterson, *J. Polym. Sci., A-2*, **7**, 209 (1969).
40 C. Tsitsilianis, E. Pierri and A. Dondos, *J. Polym. Sci. Polym. Lett. Ed.*, **21**, 685 (1983).
41 M. Nakata, Y. Nakano and K. Kawata, *Macromolecules*, **21**, 2509 (1988).
42 M. G. Prolongo, R.M. Masegosa, I. Hernandez-Fuentes and A. Horta, *Macromolecules*, **14**, 1526 (1981).
43 A. Horta and I. Fernandez-Pierola, *Macromolecules*, **14**, 1519 (1981).
44 M.G. Prolongo, R.M. Masegosa, I. Hernandez-Fuentes and A. Horta, *Polymer,* **25**, 1307 (1984).
45 F. Viras and K. Viral, *J. Polym. Sci. Phys. Ed.*, **26**, 2525 (1988).
46 J.M.G. Cowie and IT McEwen, *Macromolecules*, **7**, 291 (1974).
47 B.A. Wolf and G.J. Blaum, *Polym. Sci.*, **13**, 1115 (1975).
48 J.M.G. Cowie and IT McEwen, *Polymer*, **24**, 1449 (1983).
49 J.M.G. Cowie and IT McEwen, *Polymer*, **24**, 1445 (1983).
50 J.M.G. Cowie and IT McEwen, *Polymer*, **24**, 1453 (1983).
51 B.A. Wolf, J.W. Breitenbach and H. Senftl, *J. Polym. Sci.*, **C31**, 345 (1970).
52 J.M.G. Cowie, IT McEwen and M.T. Garay, *Polymer Commun.*, **27**, 122 (1986).
53 J.M.G. Cowie and IT McEwen, *J. Polym. Sci. B. Phys. Ed.*, **25**, 1501 (1987).
54 J.M.G. Cowie, M.A., Mohsin, I.J. McEwen, *Polymer*, **28**, 1569 (1987).
55 J.W. Breitenbach, B.A. Wolf, *Makromol.Chem.*, **117**, 163 (1968).
56 B.A. Wolf, J.W. Breitenbach and H. Senftl, *Monatsh, Chem.*, **101**, 57 (1970).
57 B.A. Wolf, J.W. Breitenbach and J.K. Rigler, *Ang. Makromol. Chem.*, **34**, 177 (1973).
58 J.M.G. Cowie, IT McEwen, *J. Polym. Sci. Phys. Ed.*, **25**, 1501 (1987).
59 G. Blaum and B.A. Wolf, *Macromolecules*, **9**, 579 (1976).
60 B.A. Wolf and G. Blaum, *Makromol. Chem.*, **177**, 1073 (1976).
61 F.D. Miles, **Cellulose Nitrate**, Oliver and Boyd Ed., Imperial Chem. Industries Ltda., London, 1955, Chapter V.
62 L. Gargallo and D. Radic', *Adv. Colloid. Interface Sci.*, **21**, 1-53 (1984).
63 A. Dondos and D. Patterson, *J. Polym. Sci., A-2*, **5**, 230 (1967).
64 A. Horta and I. Fernandez-Pierola, *Polymer*, **22**, 783 (1981).
65 I. Fernandez-Pierola and A. Horta, *Makromol. Chem.*, **182**, 1705 (1981).
66 P. Ch. Deb and S.R Palit, *Makromol. Chem.*, **166**, 227 (1973).
67 I. Katime, J.R Ochoa, L.C. Cesteros and J. Penafiel, *J. Polymer Bull.*, **6**, 429 (1982).
68 Z. Wenging and X. Chengwei, *Fenzi Kexu Xuebao*, **2**, 73 (1982).
69 R.M. Masegosa, M.G. Prolongo, I. Hernandez-Fuentes and A. Horta, *Ber. Bunseges. Phys. Chem.*, **58**, 103 (1984).
70 I. Fernandez-Pidrola and A. Horta, *J. Chim. Phys.*, **77**, 271 (1980).
71 A. Horta, M.G. Prolongo, R.M. Masegosa and I. Hernandez-Fuentes, *Polymer Commun.*, **22**, 1147 (1981).
72 I. Fernandez-Pierola and A. Horta, *Polym. Bull.*, **3**, 273 (1980).
73 I. Katime and C. Strazielle, *Makromol. Chem.*, **178**, 2295 (1977).
74 I. Katime, R. Valenciano and M. Otaduy, *An. Quim.*, **77**, 405 (1981).
75 I. Katime, M.B. Huglin and P. Sasia, *Eur. Polym. J.*, **24**, 561 (1988).
76 M.D. Blanco, J.M. Teij6n, J.A. Anrubia and I. Katime, *Polym. Bull.*, **20**, 563 (1988).
77 I. Katime, J. Tamarit and J.M. Teijon, *An. Quim.*, **75**, 7 (1979).
78 I. Fernandez-Pierola, and A. Horta, IUPAC Symposium Macromolecules, 1980, pp. 3, 47.
79 I. Fernandez-Pierola, Tesis Doctoral, Universidad Complutense, Madrid, 1981.

80 M.G. Prolongo, R.M. Masegosa, L. de Blas, I. Hernandez-Fuentes, and A. Horta, Commun. to the IUPAC-Macro'83, Bucarest, September, 1983.
81 N. Vidyarthi, M. Gupta and S.E. Palit, *J. Indian Chem. Soc.*, **61**, 697 (1979).
82 P.C. Read, and S.R. Palit, *Makromol. Chem.*, **128**, 123 (1972).
83 J.M.G. Cowie and J.T. McGrindle, *Eur. Polym. J.*, **8**, 1185 (1972).
84 J. Vasquez, L. de Blas, M. G. Prolongo, R.M. Masegosa, I. Hernandez-Fuentes, and A. Horta, *Makromol. Chem.*, **185**, 797 (1984).
85 I. Katime, J.R Quintana, J. Veguillas, *Polymer*, **24**, 903 (1983).
86 F. Bosscher, D. Keekstra and G. Challa, *Polymer*, **22**, 124 (1981).
87 M.G. Prolongo, R.M. Masegosa, I. Hernandez-Fuentes and A. Horta, IUPAC on Macromolecules, 1980, pp. 3, 51.
88 I. Fernandez-Pierola and A. Horta, Symp. on Macromolecules, Italy, 1980, pp. 3, 47.
89 M.A. McHugh, F. Rindfleisch, P.T. Kuntz, C. Schmaltz, M. Buback, *Polymer*, **39** (24), 6049 (1998).
90 A. Dondos, P. Rempp and H. Benoit, *J. Polym. Sci.*, **C30**, 9 (1970).
91 A. Dondos and H. Benoit, *Macromolecules*, **4**, 279 (1971).
92 G. Grigorescu, S. Ioan, V. Harabagin and B.C. Simionescu, *Eur. Polym. J.*, **34**, 827-832 (1998).
93 G. Grigorescu, S. Ioan, V. Harabagin and B.C. Simionescu, *Rev. Roum. Chim.*, **42**, 701 (1997).
94 G. Grigorescu, S. Ioan, V. Harabagin and B.C. Simionescu, *Macromol. Rep.*, **A33**, 163 (1966).
95 G. Grigorescu, S. Ioan, C. Cojocaru and B.C. Simionescu, *Pure Appl. Chem.* **A34**, 533 (1997).
96 P. van de Witte, P.J. Dijkstra, J.W.A. van den Berg and J. Feijen, *J. Polym. Sci., Part B: Polym. Phys.*, **34**, 2553 (1996).
97 P. van de Witte, A. Borsma, H. Esselbrugge, P.J. Dijkstra, J.W.A. van den Berg and J. Feijen, *Macromolecules*, **29**, 212 (1996).
98 V.O. Aseyev, S.I. Klenin and H. Tenhu, *J. Polym. Sci. Part B: Polym. Phys.*, **36**, 1107 (1998).
99 A. Horta, *Macromolecules*, **12**, 785 (1979).
100 A. Horta, *Macromolecules*, **18**, 2498 (1985).
101 A. Horta, and M. Criado-Sancho, *Polymer*, **23**, 1005 (1982).
102 I. Katime, L. Gargallo, D. Radic' and A. Horta, *Makromol. Chem.*, **186**, 2125 (1985).
103 T. Kawai, *Bull. Chem. Soc. Jpn.*, **26**, 6 (1953).
104 B.E. Read, *Trans. Faraday Soc.*, **56**, 382 (1960).
105 W.R. Krigbaum and D.K. Carpenter, *J. Phys. Chem.*, **59**, 1166 (1955).
106 A. Horta, D. Radic' and L. Gargallo, *Macromolecules*, **22**, 4267 (1989).
107 A. Horta, L. Gargallo and D. Radic', *Macromolecules*, **23**, 5320 (1990).
108 P.W. Morgan and S.L. Kwolek, *Macromolecules*, **8**, 104 (1975).
109 A. Dondos and H. Benoit, *Makromol. Chem.*, **133**, 119 (1970).
110 E.F. Casassa, *Polymer J.*, **3**, 517 (1972).
111 A. Dondos, and H. Benoit, in **Order in Polymer Solutions**, K. Solc. Ed., *Gordon and Breach*, London, 1976.
112 M. Hertz and C. Strazielle, in **Order in Polymer Solutions**, K. Solc Ed., *Gordon and Breach*, London, 1976, pp 195.
113 M. Yamamoto and J.L. White, *Macromolecules*, **5**, 58 (1972).
114 L. Gargallo, N. Hamidi, I. Katime and D. Radic, *Polym. Bull.*, **14**, 393 (1985).

# 5.5 THE PHENOMENOLOGICAL THEORY OF SOLVENT EFFECTS IN MIXED SOLVENT SYSTEMS

KENNETH A. CONNORS
**School of Pharmacy, University of Wisconsin, Madison, USA**

## 5.5.1 INTRODUCTION

We do not lack theories dealing with solvent effects on chemical and physical processes, as is made clear by other sections and authors in the present book. Some of these theories are fundamental in the sense that they invoke detailed physical descriptions of molecular phenomena (electrostatic interactions or the dispersion interaction, for example) whereas others are extensively empirical (such as the UNIQUAC and UNIFAC schemes for estimating activity coefficients, or extrathermodynamic correlations with model processes, exemplified by the Dimroth-Reichardt $E_T$ value). Given the abundance of theoretical and empirical approaches, it might seem that new attacks on the general problem of solvent effects would be superfluous. Yet when a solvent effect problem (in particular a solubility problem) arises, the extant theories often are in some measure inadequate. The empirical approaches tend to constitute special rather than general methods of attack, and the physical theories are either too complexly detailed or must be overly simplified to be usable. (It should be noted that our present concern is dominated by an interest in pharmaceutical systems, and therefore by aqueous and mixed aqueous solvents). The consequence is that a chemist or pharmacist confronted with a solubility problem (and such problems usually arise in situations allowing little commitment of time to the problem) often finds it more fruitful to solve the problem experimentally rather than theoretically. This is perfectly valid, but seldom does the solution lead to deeper understanding, and moreover the time required for the experimental effort may be excessive. Another factor is the frequent availability of only milligram quantities of material. It is these considerations that led to the development of this phenomenological theory of solvent effects in mixed aqueous-organic solvent systems. The theory is termed "phenomenological" because it includes elements of description such as equilibrium constants whose evaluation is carried out experimentally, yet it is based on physicochemical ideas.

## 5.5.2 THEORY

In the following development, the symbol x represents mole fraction, c is the molar concentration, component 1 is water, component 2 is a water-miscible organic cosolvent, and component 3 is the solute.

### 5.5.2.1 Principle

Before we can carry out any solution chemistry we must have a solution, and so we begin with the process of dissolution of a solid solute in a liquid solvent (which may itself be a mixture), the system being at equilibrium at constant temperature T.[1] The experimentally measured equilibrium solubility of the solute is $x_3$. Eq. [5.5.1] gives the free energy of solution per molecule, where k is the Boltzmann constant.

$$\Delta G^*_{soln} = kT \ln x_3 \qquad [5.5.1]$$

Conventionally the standard free energy of solution is given by eq. [5.5.2],

$$\Delta G^{0}_{soln} = kT \ln a_3 \qquad [5.5.2]$$

where $a_3$ is the mole fraction solute activity. Since we can write $a_3 = f_3x_3$, eq. [5.5.3] relates $\Delta G^{*}_{soln}$ and $\Delta G^{0}_{soln}$.

$$\Delta G^{*}_{soln} = \Delta G^{0}_{soln} + kT \ln f_3 \qquad [5.5.3]$$

The traditional approach would be to develop a theory for the activity coefficient $f_3$, which measures the extent of nonideal behavior. This seems to be a very indirect manner in which to proceed, so we will henceforth make no use of activity coefficients, but instead will develop an explicit model for $\Delta G^{*}_{soln}$.

The general principle is to treat $\Delta G^{*}_{soln}$ as the sum of contributions from the three types of pairwise interactions: solvent-solvent interactions, which give rise to the *general medium effect*; solvent-solute interactions, or the *solvation effect*; and solute-solute interactions (the *intersolute effect* in the present context). Thus we write.

$$\Delta G^{*}_{soln} = \Delta G_{gen\ med} + \Delta G_{solv} + \Delta G_{intersol} \qquad [5.5.4]$$

Our problem is to develop explicit expressions for the solvent-dependent quantities on the right-hand side of eq. [5.5.4].

#### 5.5.2.2 The intersolute effect: solute-solute interactions

There are two contributions to the intersolute effect. One of these comes from solute-solute interactions in the pure solute, which for solid solutes constitutes the crystal lattice energy. We will make the assumption that this contribution is independent of the nature (identity and composition) of the solvent. Usually this is a valid assumption, but exceptions are known in which the composition of the solid depends upon the composition of the solvent. Theophylline, for example, forms a hydrate in water-rich solvents, but exists as the anhydrous form in water-poor solvents; thus its crystal energy varies with the solvent composition.[2-4] Although the final theory is capable of empirically describing the solubility of such systems, it lacks a valid physical interpretation in these cases. Fortunately such solid solute behavior is not common, and our assumption that the pure solute interaction energy is solvent-independent is usually a good one.

The second contribution to the intersolute effect comes from solute-solute interactions in the solution phase. In pharmaceutics our motivation for incorporating organic solvents into an aqueous system commonly arises from an unacceptably low equilibrium solubility of a drug in pure water. This means that in water and in water-rich mixed solvents the extent of solution phase solute-solute interactions will be negligible because the solute concentration is in the extremely dilute range. At higher concentrations of the organic cosolvent it is true that the solute concentration may rise well above the dilute range, but in some degree this is offset by the diminished tendency for solute-solute interaction in such systems. Thus the hydrophobic interaction is sharply decreased by incorporating organic cosolvents. We will recognize these solution phase solute-solute interactions as a possible source of perturbation in our theory because of our assumption either that they are negligible, or that they do not vary with solvent composition.

### 5.5.2.3 The solvation effect: solute-solvent interaction

Our approach is to treat solvation as a stoichiometric equilibrium process. Let W symbolize water, M an organic cosolvent, and R the solute. Then we postulate the 2-step (3-state) system shown below.

$$RW_2 + M \underset{}{\overset{K_1}{\rightleftharpoons}} RWM + W \qquad [5.5.5]$$

$$RWM + M \underset{}{\overset{K_2}{\rightleftharpoons}} RM_2 + W \qquad [5.5.6]$$

In this scheme $K_1$ and $K_2$ are dimensionless solvation equilibrium constants, the concentrations of water and cosolvent being expressed in mole fractions. The symbols $RW_2$, RWM, $RM_2$ are not meant to imply that exactly two solvent molecules are associated with each solute molecule; rather $RW_2$ represents the fully hydrated species, $RM_2$ the fully cosolvated species, and RWM represents species including both water and cosolvent in the solvation shell. This description obviously could be extended, but experience has shown that a 3-state model is usually adequate, probably because the mixed solvate RWM cannot be algebraically (that is, functionally) differentiated into sub-states with data of ordinary precision.

Now we further postulate that the solvation free energy is a weighted average of contributions by the various states, or

$$\Delta G_{solv} = \Delta G_{WW} F_{WW} + \Delta G_{WM} F_{WM} + \Delta G_{MM} F_{MM} \qquad [5.5.7]$$

where $F_{WW}$, $F_{WM}$, and $F_{MM}$ are fractions of solute in the $RW_2$, RWM, and $RM_2$ forms, respectively. Eq. [5.5.7] can be written

$$\Delta G_{solv} = (\Delta G_{WM} - \Delta G_{WW})F_{WM} + (\Delta G_{MM} - \Delta G_{WW})F_{MM} + \Delta G_{WW} \qquad [5.5.8]$$

By combining definitions of $K_1$, $K_2$, $F_{WM}$, and $F_{MM}$ we get

$$F_{WM} = \frac{K_1 x_1 x_2}{x_1^2 + K_1 x_1 x_2 + K_1 K_2 x_2^2} \qquad F_{MM} = \frac{K_1 K_2 x_2^2}{x_1^2 + K_1 x_1 x_2 + K_1 K_2 x_2^2} \qquad [5.5.9]$$

for use in eq. [5.5.8]

Now observe this thermodynamic cycle [5.5.10]:

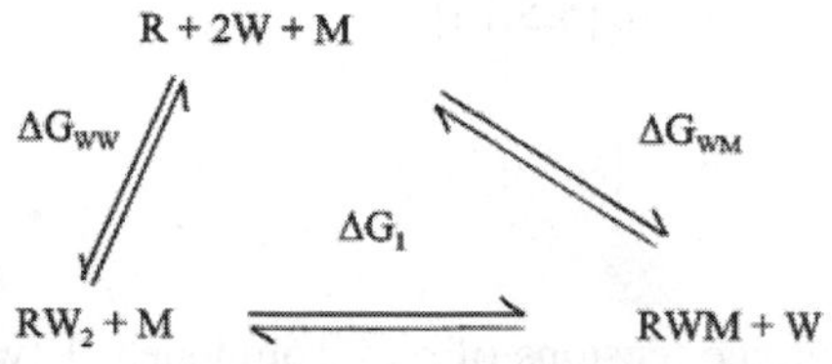


From this cycle we get

$$\Delta G_1 = \Delta G_{WM} - \Delta G_{WW} = -kT \ln K_1 \qquad [5.5.11]$$

A similar cycle yields eq. [5.5.12].

$$\Delta G_2 = \Delta G_{MM} - \Delta G_{WW} = -kT \ln K_1 K_2 \quad [5.5.12]$$

Combination of eqs. [5.5.8] - [5.5.12] then gives

$$\Delta G_{solv} = \frac{(-kT \ln K_1) K_1 x_1 x_2 + (-kT \ln K_1 K_2) K_1 K_2 x_2^2}{x_1^2 + K_1 x_1 x_2 + K_1 K_2 x_2^2} + \Delta G_{ww} \quad [5.5.13]$$

Obviously when $x_2 = 0$, $\Delta G_{solv} = \Delta G_{WW}$.[5] Eq. [5.5.13] is the desired expression relating the solvation energy to the solvent composition.

### 5.5.2.4 The general medium effect: solvent-solvent interactions

Here we make use of Uhlig's model,[6] writing eq. [5.5.14] as the energy required to create a molecular-sized cavity in the solvent.

$$\Delta G_{gen\,med} = gA\gamma \quad [5.5.14]$$

In eq. [5.5.14] g is a curvature correction factor, an empirical quantity that corrects the conventional surface tension $\gamma$ for the curvature of the cavity needed to contain a solute molecule. A is the surface area of this cavity in Å$^2$ molecule$^{-1}$; in Sections 5.5.3 and 5.5.4 we treat the meaning of A in more detail, but here we only make the assumption that it is essentially constant, that is, independent of $x_2$.

There is a subtlety in assigning the value of $\gamma$, for implicit in our model (but treated more fully in reference 1) is the condition that the $\gamma$ of eq. [5.5.14] is the surface tension of the cavity surface at its equilibrium composition. But this is the composition of the solvation shell immediately adjacent to the molecule, and this is in general different from the composition ($x_1$, $x_2$) of the bulk solvent mixture. Let $f_1$ and $f_2$ be the equilibrium mean fractional concentrations of water and cosolvent, respectively, in the solvation shell, so $f_1 + f_2 = 1$. These fractions are defined

$$f_1 = \frac{1}{2}(2F_{WW} + F_{WM}) \quad [5.5.15]$$

$$f_2 = \frac{1}{2}(F_{WM} + 2F_{MM}) \quad [5.5.16]$$

Now we define, for use in eq. [5.5.14],

$$\gamma = \gamma_1 f_1 + \gamma_2 f_2 \quad [5.5.17]$$

$$\gamma = \gamma_1 + (\gamma_2 - \gamma_1) f_2 \quad [5.5.18]$$

where $\gamma_1$ and $\gamma_2$ are the surface tensions of pure component 1 (water) and 2, respectively. Combining eqs. [5.5.14, 5.5.16, and 5.5.18] gives for the general medium effect

$$\Delta G_{gen\,med} = gA\gamma_1 + \frac{gA\gamma' K_1 x_1 x_2 + 2gA\gamma' K_1 K_2 x_2^2}{x_1^2 + K_1 x_1 x_2 + K_1 K_2 x_2^2} \quad [5.5.19]$$

where $\gamma' = (\gamma_2 - \gamma_1)/2$. Notice that the general medium and solvation effects are coupled through the solvation constants $K_1$ and $K_2$.

When $x_2 = 0$, eq. [5.5.19] yields $\Delta G_{gen\,med} = gA\gamma_1$. We interpret this as a quantitative expression for the *hydrophobic effect*. In general, eq. [5.5.19] describes the solvophobic effect. This is a phenomenological description, not a detailed structural description.

### 5.5.2.5 The total solvent effect

The solution free energy is now obtained by inserting eqs. [5.5.13] and [5.5.19] into eq. [5.5.4]. We obtain

$$\Delta G^*_{soln}(x_2) = gA\gamma_1 + \Delta G_{intersol} + \Delta G_{WW} + \frac{(gA\gamma' - kT\ln K_1)K_1x_1x_2 + (2gA\gamma' - kT\ln K_1K_2)K_1K_2x_2^2}{x_1^2 + K_1x_1x_2 + K_1K_2x_2^2} \qquad [5.5.20]$$

When $x_2 = 0$, eq. [5.5.20] gives

$$\Delta G^*_{soln}(x_2 = 0) = gA\gamma_1 + \Delta G_{intersol} + \Delta G_{WW} \qquad [5.5.21]$$

With the Leffler-Grunwald delta operator symbolism[7] we define

$$\delta_M \Delta G^* = \Delta G^*_{soln}(x_2) - \Delta G^*_{soln}(x_2 = 0) \qquad [5.5.22]$$

which, applied to eqs. [5.5.20] and [5.5.21], gives our final result:

$$\delta_M \Delta G^*_{soln} = \frac{(gA\gamma' - kT\ln K_1)K_1x_1x_2 + (2gA\gamma' - kT\ln K_1K_2)K_1K_2x_2^2}{x_1^2 + K_1x_1x_2 + K_1K_2x_2^2} \qquad [5.5.23]$$

The quantity $\delta_M\Delta G^*_{soln}$ can be read "the solvent effect on the solution free energy." Because of eq. [5.5.1], $\delta_M\Delta G^*$ is proportional to the "relative solubility," $\log[(x_3)_{x_2} / (x_3)_{x_2=0}]$, that is, the logarithm of the solubility in the mixed solvent of composition $x_2$ relative to the solubility in pure water. The subtraction that yields eq. [5.5.23], a workable equation with just three unknown parameters (gA, $K_1$, and $K_2$), has also prevented us from dealing with absolute solubilities.

## 5.5.3 APPLICATIONS

### 5.5.3.1 Solubility

It will be no surprise that the first use of eq. [5.5.23] was to describe the equilibrium solubility of solid nonelectrolytes in mixed aqueous-organic solvents.[1] Equilibrium solubility in mol $L^{-1}$, $c_3$, is converted to mole fraction, $x_3$, with eq. [5.5.24], where ρ is the saturated solution density, w is the wt/wt percentage of organic cosolvent, and $M_1$, $M_2$, $M_3$ are the molecular weights of water, cosolvent, and solute.[8]

$$x_3 = \frac{c_3}{c_3(1000\rho - c_3M_3)\left(\frac{w}{M_2} + \frac{(1-w)}{M_1}\right)} \qquad [5.5.24]$$

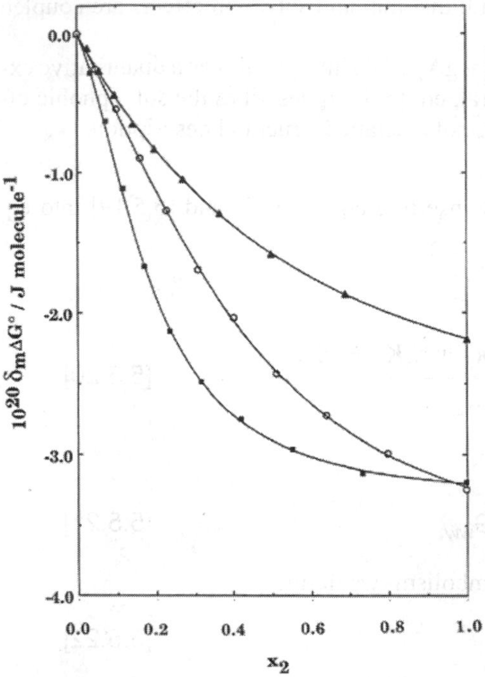


Figure 5.5.1. Solvent effect on the solubility of diphenylhydantoin. Cosolvents, top to bottom: glycerol, methanol, ethanol. The smooth lines were drawn with eq. 5.5.23. (Reproduced with permission from the *Journal of Pharmaceutical Sciences* reference 1.)

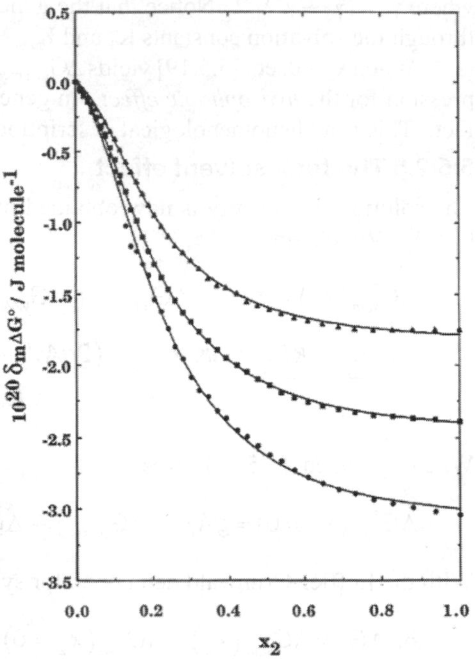


Figure 5.5.2. Solvent effect on the solubilities of barbituric acid derivatives in ethanol-water mixtures. Top to bottom: metharbital, butabarbital, amobarbital. The smooth lines were drawn with eq. 5.5.23. (Reproduced with permission from the *Journal of Pharmaceutical Sciences*, reference 1.)

The free energy of solution per molecule is then calculated with eq. [5.5.1], $\delta_M \Delta G^*$ is found with eq. [5.5.22], and $\delta_M \Delta G^*$ as a function of $x_2$ is fitted to eq. [5.5.23] by nonlinear regression, with gA, $K_1$, and $K_2$ being treated as adjustable parameters.[9] Figures 5.5.1 and 5.5.2 show some results.[1]

Clearly eq. [5.5.24] possesses the functional flexibility to describe the data. (In some systems a 1-step (2-state) equation is adequate. To transform eq. [5.5.24] to a 1-step version, set $K_2 = 0$ and let $\gamma' = \gamma_2 - \gamma_1$.) The next step is to examine the parameter values for their possible physical significance. It seems plausible that $K_1$ and $K_2$ should be larger than unity, but not "very large," on the basis that the solutes are organic and so are the cosolvents, but the cosolvents are water-miscible so they are in some degree "water-like." In fact, we find that nearly all $K_1$ and $K_2$ values fall between 1 and 15. Likewise the gA values seem, in the main, to be physically reasonable. Earlier estimates of g (reviewed in ref.[1]) put it in the range of 0.35-0.5. A itself can be estimated as the solvent-accessible surface area of the solute, and many of the gA values found were consistent with such estimates, though some were considerably smaller than expected. Since gA arises in the theory as a hydrophobicity parameter, it seemed possible that A in the equation represents only the nonpolar surface area of the molecule. An experiment whose results are summarized in Table 5.5.1 was designed to examine this possibility.[10]

**Table 5.5.1. Surface area estimates of biphenyls[10]**

| Compound | $A_{total}$ | $A_{nonpolar}$ | gA |
|---|---|---|---|
| Biphenyl | 179 (3) | 179 (3) | 74 (0.6) |
| 4-Hydroxybiphenyl | 185 (4) | 155 (7) | 69 (0.3) |
| 4,4'-Dihydroxybiphenyl | 203 (5) | 126 (7) | 53 (0.5) |
| 4-Bromobiphenyl | 217 (7) | 217 (7) | 87 (1.4) |

Areas in $Å^2$ molecule; standard deviations in parentheses. The cosolvent was methanol.

Evidently the experimental gA estimate is correlated with $A_{nonpolar}$ rather than with $A_{total}$, and the linear relationship yields the estimate g = 0.37. On the other hand, there was evidence[1] that g depends upon cosolvent identity (for a given solute), and LePree and Mulski[8,11] examined this possibility. Their findings led to an empirical but quite general correlation between gA and properties of the cosolvent and solute:

$$gA = -42 \log P_M + 11 \log P_R \qquad [5.5.25]$$

In eq. [5.5.25] $P_M$ is the 1-octanol/water partition coefficient of the pure organic cosolvent and $P_R$ is the partition coefficient of the solute. Table 5.5.2 gives examples of the application of eq. [5.5.25].

**Table 5.5.2 Experimental and calculated gA values[8,11]**

| Solute | Solvent | gA, $Å^2$ molecule$^{-1}$ | |
|---|---|---|---|
| | | Calculated | Observed |
| Naphthalene | Methanol | 66 | 63 |
| Naphthalene | Ethanol | 48 | 54 |
| Naphthalene | 2-Propanol | 35 | 43 |
| Naphthalene | 1,2-Propanediol | 92 | 71 |
| Naphthalene | 1,2-Ethanediol | 116 | 102 |
| Naphthalene | Acetone | 45 | 69 |
| Naphthalene | DMSO | 120 | 127 |
| 4-Nitroaniline | Methanol | 46 | 35 |
| 4-Nitroaniline | Ethanol | 29 | 21 |
| 4-Nitroaniline | 2-Propanol | 15 | 11 |
| 4-Nitroaniline | 1,2-Ethanediol | 96 | 84 |
| 4-Nitroaniline | Acetone | 25 | 37 |

| Solute | Solvent | gA, Å$^2$ molecule$^{-1}$ | |
|---|---|---|---|
| | | Calculated | Observed |
| 4-Nitroaniline | DMSO | 101 | 86 |
| 4-Nitroaniline | Acetonitrile | 30 | 29 |

We now encounter a curious observation. The parameter gA is constrained, in the nonlinear regression fitting program, to be constant as $x_2$ varies over its entire range from 0 to 1. We have identified A as the nonpolar surface area of the solute (though it may actually be the corresponding area of the solvent cavity, and so may show some cosolvent dependency, which we have ignored). The quantity g can then be estimated. For example, from Table 5.5.2 for naphthalene (A = 147 Å$^2$ molecule$^{-1}$), g varies from 0.29 (for 2-propanol) to 0.86 (for DMSO). Yet how can g possess these different values in different cosolvents, maintain its constancy as $x_2$ varies, and then collapse to the unique value it must possess in water, the reference solvent for all systems? An independent calculation gives g = 0.41±0.03 in water.[8] Some tentative explanations for this puzzle have been offered,[8] and we return to this issue in Section 5.5.4.

Turning to the $K_1$ and $K_2$ parameters, we have observed that these are relatively insensitive to the identity of the solute, but that they depend upon the cosolvent, whose polarity is a controlling factor. Table 5.5.3 gives some empirical correlations that provide routes to the prediction of $K_1$ and $K_1K_2$. In Table 5.5.3, $P_M$ is the 1-octanol/water partition coefficient of the pure cosolvent,[12] and $E_T$ is the Dimroth-Reichardt solvatochromic polarity parameter.[13] We thus have the capability of predicting gA, $K_1$, and $K_1K_2$, which extends the utility of eq. [5.5.23] from the merely descriptive to the predictive.

**Table 5.5.3. Empirical relationships for estimating solvation constants**

| Equation | n | r | Restrictions |
|---|---|---|---|
| $\log K_1 = -0.0316\,E_T + 2.24$ | 10 | 0.91 | - |
| $\log (K_1/K_2) = 0.0171\,E_T - 9.23$ | 4 | 0.98 | $E_T > 51$ |
| $\log (K_1/K_2) = -0.0959\,E_T + 4.60$ | 6 | 0.85 | $E_T < 52$ |
| $\log K_1K_2 = 1.31 \log P_M + 1.81$ | 6 | 0.91 | $\log P_M > -1.0$ |

### 5.5.3.2 Surface tension

In the development of the basic phenomenological model, eq. [5.5.23], we derived a relationship for the surface tension of the solvation shell. Combining eqs. [5.5.16] and [5.5.18] yields

$$\gamma = \gamma_1 + \gamma' \left[ \frac{K_1 x_1 x_2 + 2K_1 K_2 x_2^2}{x_1^2 + K_1 x_1 x_2 + K_1 K_2 x_2^2} \right] \qquad [5.5.26]$$

where $\gamma' = (\gamma_2 - \gamma_1)/2$. Now if we identify the solute-solvation shell system with the air-solvent interface, we are led to test eq. [5.5.26] as a description of the composition dependence

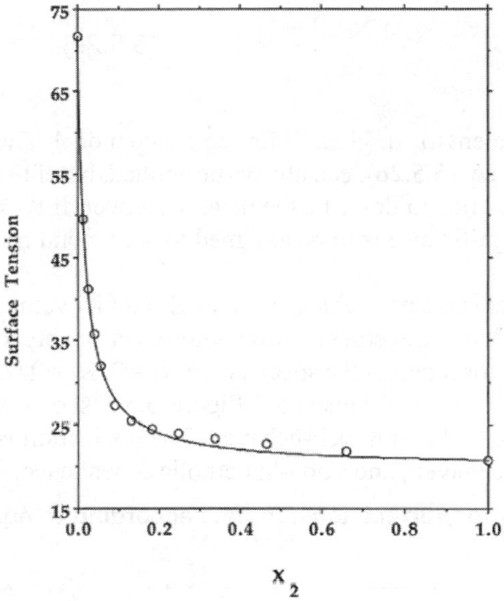


Figure 5.5.3. Surface tension of 2-propanol-water mixtures. The smooth line is drawn with eq. 5.5.26. (Reproduced with permission from the *Journal of Solution Chemistry*. reference 14.)

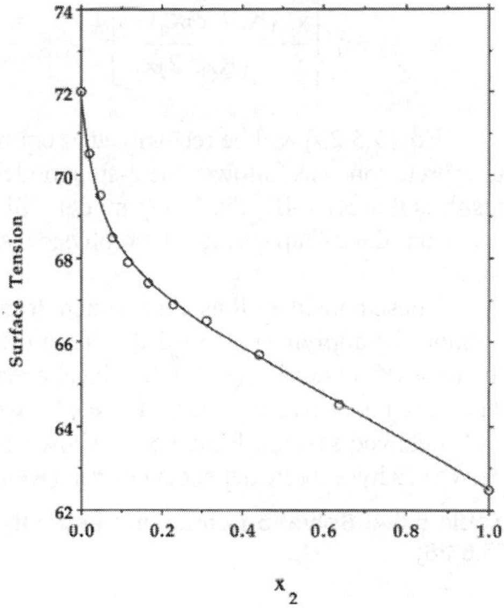


Figure 5.5.4. Surface tension of glycerol-water mixtures. The smooth line is drawn with eq. 5.5.26. (Reproduced with permission from the *Journal of Solution Chemistry*. reference 14.)

of surface tension in mixed solvent systems, air playing the role of the solute. Figures 5.5.3 and 5.5.4 show examples of these curve-fits.[14]

When $K_2 = 0$, eq [5.5.26] gives the 1-step model, eq. [5.5.27], where $\gamma' = \gamma_2 - \gamma_1$.

$$\gamma = \gamma_1 + \gamma' \left[ \frac{K_1 x_2}{x_1 + K_1 x_2} \right] \qquad [5.5.27]$$

We had earlier[15] published an equation describing the dependence of surface tension on composition, and a comparison of the two approaches has been given;[14] here we will restrict attention to eqs. [5.5.26] and [5.5.27].

Suppose we set $K_1 = 2$ and $K_2 = 1/2$ in eq. [5.5.26]. This special condition converts eq. [5.5.26] to

$$\gamma = \gamma_1 x_1 + \gamma_2 x_2 \qquad [5.5.28]$$

which corresponds to ideal behavior; the surface tension is a linear function of $x_2$. The restriction $K_1 = 2$, $K_2 = 1/2$ is, however, a unique member of a less limited special case in which $K_1 = 4K_2$. This important condition (except when it happens to occur fortuitously) implies the existence of two identical and independent binding sites.[16] Inserting $K_2 = K_1/4$ into eq. [5.5.26] yields, upon simplification, eq. [5.5.29], where $\gamma' = \gamma_2 - \gamma_1$.

$$\gamma = \gamma_1 + \gamma' \left[ \frac{(K_1 / 2)x_2}{x_1 + (K_1 / 2)x_2} \right] \qquad [5.5.29]$$

Eq. [5.5.29] will be recognized as equivalent to eq. [5.5.27] for the 1-step model. The interpretation is as follows: the 2-step model, eq. [5.5.26], can always be applied, but if the result is that $K_1 \approx 4K_2$ the 1-step model will suffice to describe the data. Moreover, if $K_1 \approx 4K_2$ from the 2-step treatment, no physical significance is to be assigned to the second parameter.

These considerations are pertinent to real systems. Table 5.5.4 lists $K_1$ and $K_2$ values obtained by applying eq. [5.5.26] to literature data.[15] Several systems conform reasonably to the $K_1 \approx 4K_2$ condition. Recall that ideal behavior requires the special case $K_1 = 2$, $K_2 = 1/2$. The less restrictive condition $K_1 \approx 4K_2$ we call "well-behaved." Figure 5.5.3 shows a well-behaved system; Figure 5.5.4 shows one that is not well-behaved. The distinction is between a hyperbolic dependence on $x_2$ (well-behaved) and a non-hyperbolic dependence.

**Table 5.5.4. Solvation parameter estimates for surface tension data according to eq. [5.5.26]**

| Cosolvent | $K_1$ | $K_2$ | $K_1/K_2$ |
|---|---|---|---|
| Methanol | 19.8 | 2.9 | 6.8 |
| 2-Propanol | 130 | 29.4 | 4.4 |
| 1-Propanol | 232 | 50 | 4.6 |
| t-Butanol | 233 | 65 | 3.6 |
| Acetic acid | 115 | 2.7 | 42.6 |
| Acetone | 138 | 7.1 | 19.4 |
| Acetonitrile | 33.4 | 14.5 | 2.3 |
| Dioxane | 62.1 | 7.4 | 8.4 |
| THF | 136 | 25.9 | 5.3 |
| Glycerol | 22.7 | 0.80 | 28.4 |
| DMSO | 12.3 | 1.43 | 8.6 |
| Formamide | 5.52 | 2.57 | 2.1 |
| Ethylene glycol | 9.4 | 2.62 | 3.6 |

A further observation from these results is that some of the $K_1$ values are much larger than those encountered in solubility studies. Correlations with log $P_M$ have been shown.[14]

### 5.5.3.3 Electronic absorption spectra

The energy of an electronic transition is calculated from the familiar equation

$$E_T = h\nu = \frac{hc}{\lambda} \qquad [5.5.30]$$

where h is Planck's constant, c is the velocity of light, v is frequency, and λ is wavelength. If λ is expressed in nm, eq. [5.5.31] yields $E_T$ in kcal $mol^{-1}$.

$$E_T = 2.859 \times 10^4 / \lambda \qquad [5.5.31]$$

The phenomenological theory has been applied by Skwierczynski to the $E_T$ values of the Dimroth-Reichardt betaine,[13] a quantity sensitive to the polarity of the medium.[17] The approach is analogous to the earlier development. We need only consider the solvation effect. The solute is already in solution at extremely low concentration, so solute-solute interactions need not be accounted for. The solvent cavity does not alter its size or shape during an electronic transition (the Franck-Condon principle), so the general medium effect does not come into play. We write $E_T$ of the mixed solvent as a weighted average of contributions from the three states:

$$E_T(x_2) = F_{WW}E_T(WW) + F_{WM}E_T(WM) + F_{MM}E_T(MM) \qquad [5.5.32]$$

where the symbolism is obvious. Although $E_T(WW)$ can be measured in pure water and $E_T(MM)$ in pure cosolvent, we do not know $F_T(WM)$, so provisionally we postulate that $E_T(WM) = [E_T(WW) + E_T(MM)]/2$. Defining a quantity Γ by

$$\Gamma = \frac{E_T(x_2) - E_T(WW)}{E_T(MM) - E_T(WW)} \qquad [5.5.33]$$

we find, by combining eqs. [5.5.9], [5.5.10], and [5.5.32],

$$\Gamma = \frac{K_1x_1x_1 / 2 + K_1K_2x_2^2}{x_1^2 + K_1x_1x_2 + K_1K_2x_2^2} \qquad [5.5.34]$$

The procedure is to fit Γ to $x_2$. As before, a 1-parameter version can be obtained by setting $K_2 = 0$:

$$\Gamma = \frac{K_1x_2}{x_1 + K_1x_2} \qquad [5.5.35]$$

Figure 5.5.5 shows a system that can be satisfactorily described by eq. [5.5.35], whereas the system in Figure 5.5.6 requires eq. [5.5.34]. The $K_1$ values are similar in magnitude to those observed from solubility systems, with a few larger values; $K_2$, for those systems requiring eq. [5.5.34], is always smaller than unity. Some correlations were obtained of $K_1$ and $K_2$ values with solvent properties. Figure 5.5.7 shows log $K_1$ as a function of log $P_M$, where $P_M$ is the partition coefficient of the pure organic solvent.

### 5.5.3.4 Complex formation.

We now inquire into the nature of solvent effects on chemical equilibria, taking noncovalent molecular complex formation as an example. Suppose species S (substrate) and L (ligand) interact in solution to form complex C, $K_{11}$ being the complex binding constant.

$$S + L \underset{}{\overset{K_{11}}{\rightleftharpoons}} C \qquad [5.5.36]$$

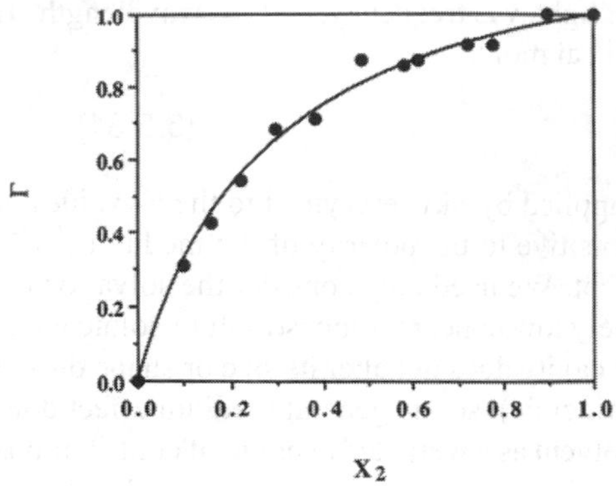


Figure 5.5.5. Dependence of $E_T$ on composition for the methanol-water system. The smooth line was drawn with eq. 5.5.35. (Reproduced with permission from the *Journal of the Chemical Society. Perkin Transactions 2*, reference 17.)

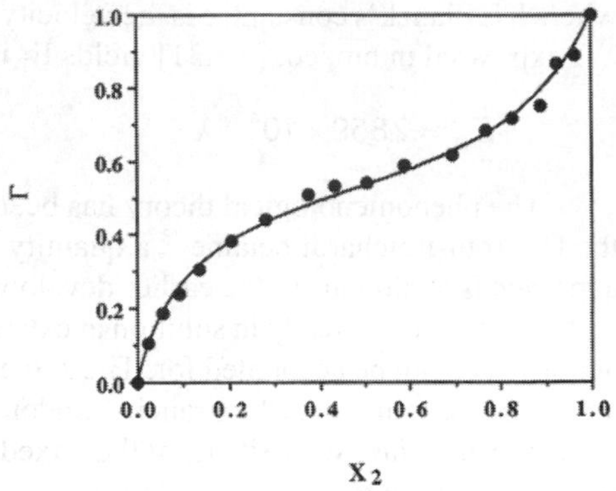


Figure 5.5.6. Dependence of $E_T$ on composition for the acetone-water system. The smooth line was drawn with eq. 5.5.34. (Reproduced with permission from the *Journal of the Chemical Society. Perkin Transactions 2*,. reference 17.)

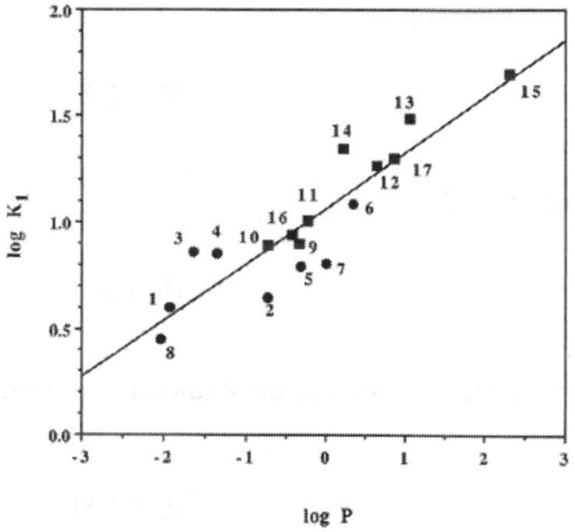


Figure 5.5.7. A plot of log $K_1$ from the $E_T$ data against log $P_M$; the circles represent 1-step solvents (eq. 5.5.35) and the squares, 2-step solvents (eq. 5.5.34). (Reproduced with permission from the *Journal of the Chemical Society, Perkin Transactions 2*, reference 17.)

It is at once evident that this constitutes a more complicated problem than those we have already considered inasmuch as here we have three solutes. We begin with the thermodynamic cycles shown as Figure 5.5.8; these cycles describe complex formation in the solid, solution, and gas phases horizontally, and the energy changes associated with the indicated processes. $\Delta G_{latt}$ corresponds to the crystal lattice energy (solute-solute interactions), $\Delta G_{cav}$ represents the energy of cavity formation (identical with the general medium effect of Section 5.5.2). $\Delta G_{comp}$ is the free energy of complex formation, which in the solution phase is given by eq. [5.5.37].

$$\Delta G_{comp}(l) = -kT \ln K_{11} \quad [5.5.37]$$

Eq. [5.5.37] gives the free energy with respect to a 1M standard state, because the unit of $K_{11}$ is $M^{-1}$. To calculate the unitary (mole fraction) free energy change we write, instead of eq. [5.5.37], eq. [5.5.38]:

$$\Delta G_{comp}(l) = -kT \ln\left(K_{11} M^{*} \rho\right) \quad [5.5.38]$$

where M* is the number of moles of solvent per kg of solvent and ρ is the solution density. The unitary free energy does not include the entropy of mixing.

From cycle gl in Figure 5.5.8 we obtain eq. [5.5.39].

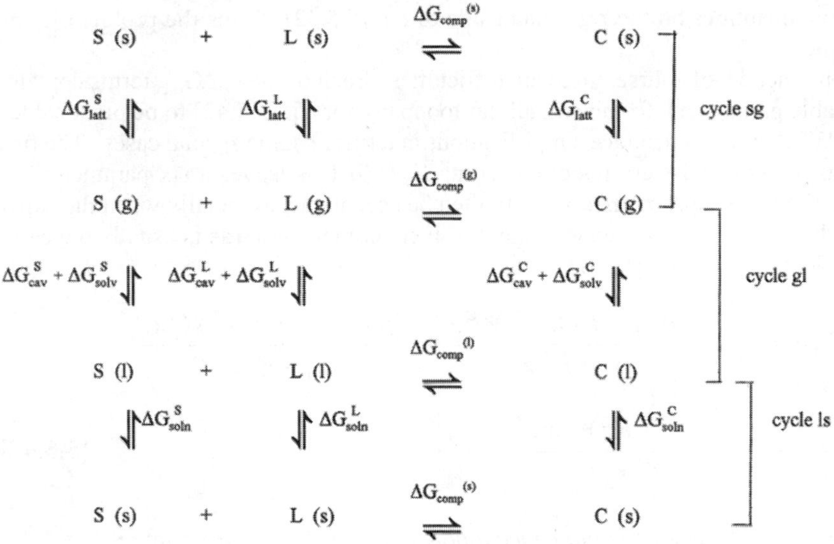


Figure 5.5.8. Thermodynamic cycles for bimolecular association. The symbols s, l, g represent solid, liquid, and gas phases; the superscripts refer to substrate S, ligand L, and complex C. (Reproduced with permission from the *Journal of Solution Chemistry*, reference 18.)

$$\Delta G_{comp}(g) + \left(\Delta G^{C}_{cav} + \Delta G^{C}_{solv}\right) - \Delta G_{comp}(l) - \left(\Delta G^{L}_{cav} + \Delta G^{L}_{solv}\right) - \left(\Delta G^{S}_{cav} + \Delta G^{S}_{solv}\right) = 0 \qquad [5.5.39]$$

We apply the $\delta_M$ operator to eq. [5.5.39]

$$\delta_M \Delta G^{C}_{cav} - \delta_M \Delta G^{S}_{cav} - \delta_M \Delta G^{L}_{cav} + \delta_M \Delta G^{C}_{solv} - \delta_M \Delta G^{S}_{solv} - \delta_M \Delta G^{L}_{solv} = \delta_M \Delta G_{comp}(l) \qquad [5.5.40]$$

where we have assumed $\delta_M \Delta G_{comp}(g) = 0$, which is equivalent to supposing that the structure of the complex (the spatial relationship of S and L) does not depend upon solvent composition, or that the intersolute effect is composition independent.

Also applying the $\delta_M$ treatment to eq. [5.5.4] gives

$$\delta_M \Delta G_{soln} = \delta_M \Delta G_{cav} + \delta_M \Delta G_{solv} \qquad [5.5.41]$$

for each species; recall that $\Delta G_{gen\ med}$ and $\Delta G_{cav}$ are identical. Use eq. [5.5.41] in [5.5.40]:

$$\delta_M \Delta G_{comp}(l) = \delta_M \Delta G^{C}_{soln} - \delta_M \Delta G^{L}_{soln} - \delta_M \Delta G^{S}_{soln} \qquad [5.5.42]$$

Eq. [5.5.42] says that the solvent effect on complex formation is a function solely of the solvent effects on the solubilities of reactants (negative signs) and product (positive sign). This is a powerful result, because we already have a detailed expression, eq. [5.5.23], for each of the three quantities on the right-hand side of eq. [5.5.42]. Thus the problem is solved in principle.[18]

In practice, of course, there are difficulties. Each of the $\delta_M\Delta G_{soln}$ terms contains three adjustable parameters, for nine in all, far too many for eq. [5.5.42] to be practicable in that form. We therefore introduce simplifications in terms of some special cases. The first thing to do is to adopt a 1-step model by setting $K_2 = 0$. This leaves a six-parameter equation, which, though an approximation, will often be acceptable, especially when the experimental study does not cover a wide range in solvent composition (as is usually the case). This simplification gives eq. [5.5.43].

$$\delta_M \Delta G^*_{comp} = \frac{(gA^C\gamma' - kT\ln K_1^C)K_1^C x_2}{x_1 + K_1^C x_2} - \frac{(gA^S\gamma' - kT\ln K_1^S)K_1^S x_2}{x_1 + K_1^S x_2} - \frac{(gA^L\gamma' - kT\ln K_1^L)K_1^L x_2}{x_1 + K_1^L x_2} \qquad [5.5.43]$$

Next, in what is labeled the *full cancellation approximation*, we assume $K_1^C = K_1^S = K_1^L = K_1$ and we write $\Delta gA = gA^C - gA^S - gA^L$. The result is

$$\delta_M \Delta G^*_{comp} = \frac{(kT\ln K_1 + \Delta gA\gamma')K_1 x_2}{x_1 + K_1 x_2} \qquad [5.5.44]$$

and we now have a 2-parameter model. The assumption of identical solvation constants is actually quite reasonable; recall from the solubility studies that $K_1$ is not markedly sensitive to the solute identity.

The particular example of cyclodextrin complexes led to the identification of another special case as the *partial cancellation approximation*; in this case we assume $K_1^C = K_1^S < K_1^L$, and the result is, approximately.[19]

$$\delta_M \Delta G^*_{comp} = \frac{(kT\ln K_1 - gA\gamma')K_1 x_2}{x_1 + K_1 x_2} \qquad [5.5.45]$$

Functionally eqs. [5.5.44] and [5.5.45] are identical; the distinction is made on the basis of the magnitudes of the parameters found. Note that gA in eq. [5.5.45] is a positive quantity whereas ΔgA in eq. [5.5.44] is a negative quantity. In eq. [5.5.45] it is understood that gA and $K_1$ refer to L. Eqs. [5.5.44] and [5.5.45] both have the form

$$\delta_M \Delta G^*_{comp} = \frac{(kT\ln K_1 + G\gamma')K_1 x_2}{x_1 + K_1 x_2} \qquad [5.5.46]$$

where G = ΔgA in eq. [5.5.44] and G = -gA in eq. [5.5.45]. Table 5.5.5 shows G and $K_1$ values obtained in studies of α-cyclodextrin complexes.[19,20] The assignments are made on the basis of the magnitude of $K_1$; those values substantially higher than typical solubility $K_1$ values suggest that the full cancellation condition is not satisfied. After the assignments are made, G can be interpreted as either ΔgA (full cancellation) or -gA (partial cancellation).

Notice in Table 5.5.5 that all full cancellation systems give substantial negative ΔgA values. If g is constant, ΔgA = gΔA, and the negative ΔA value leads to a solvophobic driving force of gΔAγ for complex formation. (The dioxane system in Table 5.5.5 is unassigned because its $K_1$ value suggests partial cancellation whereas its G value suggests full cancellation).

**Table 5.5.5. Parameter values of the 4-nitroaniline/α-cyclodextrin and methyl orange/ α-cyclodextrin systems[19,20]**

| Cosolvent | $K_1$ | G[a] | Cancellation assignment |
|---|---|---|---|
| 4-Nitroaniline | | | |
| Acetonitrile | 55 | +3 | Partial |
| 2-Propanol | 46 | -3 | Partial |
| Ethanol | 29 | -9 | Partial |
| Acetone | 10 | -57 | Full |
| Methanol | 3.1 | -68 | Full |
| Methyl orange | | | |
| Acetone | 46 | -3 | Partial |
| 2-Propanol | 43 | -11 | Partial |
| Acetonitrile | 40 | -13 | Partial |
| Dioxane | 31 | -38 | (Unassigned) |
| Ethylene glycol | 7.7 | -58 | Full |
| DMSO | 6.4 | -66 | Full |
| Methanol | 4.9 | -43 | Full |

[a]Units are $Å^2$ $molecule^{-1}$

### 5.5.3.5 Chemical kinetics

Treatment of the solvent effect on chemical reaction rates by means of the phenomenological theory is greatly facilitated by the transition state theory, which postulates that the initial and transition states are in (virtual) equilibrium. Thus the approach developed for complex formation is applicable also to chemical kinetics. Again we begin with a thermodynamic cycle, Figure 5.5.9, where R represents the reactant (initial state) in a unimolecular reaction, $R^{\ddagger}$ is the transition state, and P is the product. From Figure 5.5.9 we write eq. [5.5.47], where $\Delta G^{\ddagger}_{rxn(1)}$ subsequently writ-

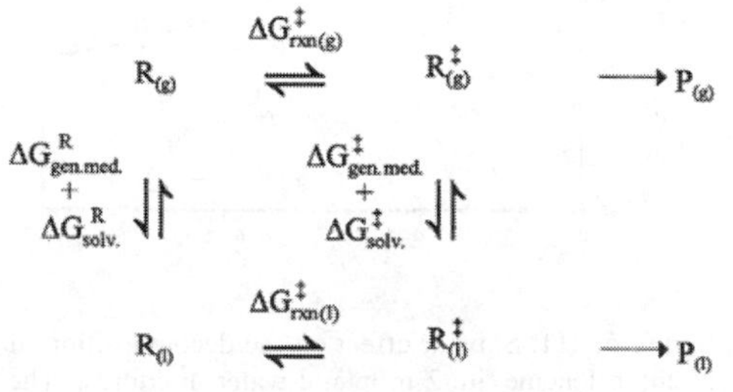


Figure 5.5.9. Thermodynamic cycle for a unimolecular reaction. (Reproduced with permission from the *Journal of Pharmaceutical Sciences*, reference 21.)

ten $\Delta G^\ddagger_{rxn}$, is the free energy of activation in the solution phase.

$$\Delta G^\ddagger_{rxn(1)} = \Delta G^\ddagger_{rxn(g)} + (\Delta G^\ddagger_{gen\ med} + \Delta G^\ddagger_{solv}) - (\Delta G^R_{gen\ med} + \Delta G^R_{solv}) \qquad [5.5.47]$$

Applying the $\delta_M$ operation gives eq. [5.5.48]:

$$\delta_M \Delta G^\ddagger_{rxn} = \Delta G^\ddagger_{rxn(x_2)} - \Delta G^\ddagger_{rxn(x_2=0)} \qquad [5.5.48]$$

The quantity $\Delta G^\ddagger_{rxn(g)}$, disappears in this subtraction, as do other composition-independent quantities. We make use of eq. [5.5.13] and eq. [5.5.19] to obtain a function having six parameters, namely $K_1^R$, $K_2^R$, $K^\ddagger_1$, $gA^R$, and $gA^\ddagger$. This function is made manageable by adopting the full cancellation approximation, setting $K_1^R = K^\ddagger_1 = K_1$ and $K_2^R = K^\ddagger_2 = K_2$. We then obtain

$$\delta_M \Delta G^\ddagger_{rxn} = (\Delta gA^\ddagger \gamma' K_1 x_1 x_2 + 2\Delta gA^\ddagger \gamma' K_1 K_2 x_2^2)/(x_1^2 + K_1 x_1 x_2 + K_1 K_2 x_2^2) \qquad [5.5.49]$$

where $\Delta gA^\ddagger = gA^\ddagger - gA^R$; this is the difference between the curvature-corrected molecular surface areas of the cavities containing the transition state and the reactant. This quantity may be positive or negative.

LePree[21] tested eq. [5.5.49] with the decarboxylative dechlorination of N-chloroamino acids in mixed solvents

$$RCH(NHCl)COOH + H_2O \rightarrow RCHO + NH_3 + HCl + CO_2 \qquad [5.5.50]$$

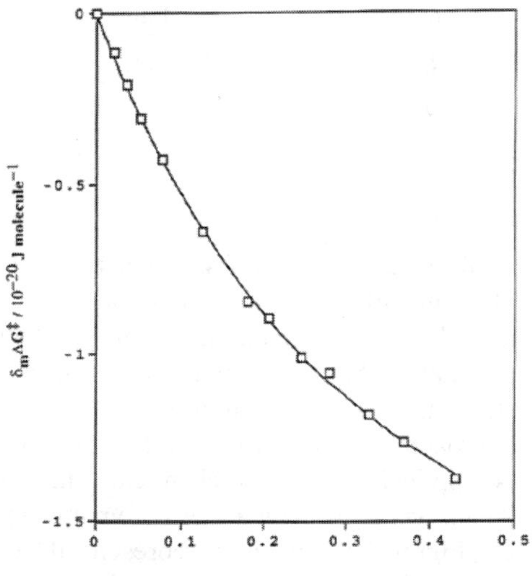


Figure 5.5.10. Solvent effect on the decomposition of N-chloroaniline in acetonitrile-water mixtures. The smooth curve is drawn with eq. 5.5.49. (Reproduced with permission from the *Journal of Pharmaceutical Sciences*, reference 21.)

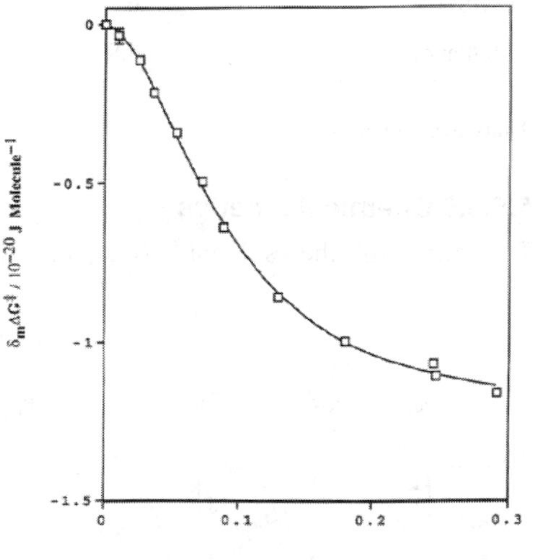


Figure 5.5.11. Solvent effect on the decomposition of N-chloroleucine in 2-propanol-water mixtures. The smooth curve is drawn with eq. 5.5.49. (Reproduced with permission from the *Journal of Pharmaceutical Sciences*, reference 21.)

For this test, the reaction possesses these very desirable features: (1) the kinetics are first order, and the rate-determining step is unimolecular; (2) the reaction rate is independent of pH over the approximate range 4-13; (3) the rate-determining step of the process is not a solvolysis, so the concentration of water does not appear in the rate equation; and (4) the reaction is known to display a sensitivity to solvent composition. Figures [5.5.10] and [5.5.11] show curve-fits, and Tables 5.5.6 and 5.5.7 give the parameter values obtained in the curve-fitting regression analysis. Observe that $\Delta gA^{\ddagger}$ is positive. This means that the transition state occupies a larger volume than does the reactant. This conclusion has been independently confirmed by studying the pressure dependence of the kinetics.[21,22]

**Table 5.5.6. Model parameters for solvent effects on the decomposition of N-chloroalanine[21]**

| Cosolvent | $K_1$ | $K_2$ | $\Delta gA^{\ddagger}$, $Å^2$ $molecule^{-1}$ |
|---|---|---|---|
| Methanol | 2.8 | 3.1 | 16.8 |
| Ethanol | 5.7 | 2.8 | 23.8 |
| 1-Propanol | 13.0 | 3.2 | 22.0 |
| 2-Propanol | 5.9 | 11 | 20.6 |
| Ethylene glycol | 4.4 | 1.7 | 34.8 |
| Propylene glycol | 5.2 | 6.0 | 27.0 |
| Acetonitrile | 7.0 | 2.0 | 42 |
| Dioxane | 8.9 | 4.3 | 55 |

**Table 5.5.7. Model parameters for solvent effects on the decomposition of N-chloroleucine[21]**

| Cosolvent | $K_1$ | $K_2$ | $\Delta gA^{\ddagger}$, $Å^2$ $molecule^{-1}$ |
|---|---|---|---|
| Methanol | 2.5 | 4.0 | 19.3 |
| 2-Propanol | 2.7 | 40 | 24.2 |
| Ethylene glycol | 4.4 | 3.2 | 38 |
| Acetonitrile | 7.5 | 3.9 | 43 |

The success shown by this kinetic study of a unimolecular reaction unaccompanied by complications arising from solvent effects on pH or water concentration (as a reactant) means that one can be confident in applying the theory to more complicated systems. Of course, an analysis must be carried out for such systems, deriving the appropriate functions and making chemically reasonable approximations. One of the goals is to achieve a practical level of predictive ability, as for example we have reached in dealing with solvent effects on solubility.

### 5.5.3.6 Liquid chromatography

In reverse phase high-pressure liquid chromatography (RP HPLC), the mobile phase is usually an aqueous-organic mixture, permitting the phenomenological theory to be applied. LePree and Cancino[23] carried out this analysis. The composition-dependent variable is the capacity factor k′, defined by eq. [5.5.51],

$$k' = \frac{V_R - V_M}{V_M} = \frac{t_R - t_M}{t_M} \qquad [5.5.51]$$

where $V_R$ is the retention volume of a solute, $t_R$ is the retention time, $V_M$ is the column dead volume (void volume), and $t_M$ is the dead time. It is seldom possible in these systems to use pure water as the mobile phase, so LePree reversed the usual calculational procedure, in which water is the reference solvent, by making the pure organic cosolvent the reference. This has the effect of converting the solvation constants $K_1$ and $K_2$ to their reciprocals, but the form of the equations is unchanged. For some solvent systems a 1-step model was adequate, but others required the 2-step model. Solvation constant values (remember that these are the reciprocals of the earlier parameters with these labels) were mostly in the range 0.1 to 0.9, and the gA values were found to be directly proportional to the nonpolar surface areas of the solutes. This approach appears to offer advantages over earlier theories in this application because of its physical significance and its potential for predicting retention behavior.

## 5.5.4 INTERPRETATIONS

The very general success of the phenomenological theory in quantitatively describing the composition dependence of many chemical and physical processes arises from the treatment of solvation effects by a stoichiometric equilibrium model. It is this model that provides the functional form of the theory, which also includes a general medium effect (interpreted as the solvophobic effect) that is functionally coupled to the solvation effect. The parameters of the theory appear to have physical significance, and on the basis of much experimental work they can be successfully generated or predicted by means of empirical correlations. The theory does not include molecular parameters (such as dipole moments or polarizabilities), and this circumstance deprives it of any fundamental status, yet at the same time enhances its applicability to the solution of practical laboratory problems. Notwithstanding the widespread quantitative success of the theory, however, some of the observed parameter values have elicited concern about their physical meaning, and it is to address these issues, one of which is mentioned in 5.5.3.1, that the present section is included.

### 5.5.4.1 Ambiguities and anomalies

Consider a study in which the solubility of a given solute (naphthalene is the example to be given later) is measured in numerous binary aqueous-organic mixed solvent systems, and eq. [5.5.23] is applied to each of the mixed solvent systems, the solvent effect $\delta_M \Delta G^*_{soln}$ being calculated relative to water (component 1) in each case. According to hypothesis, the parameter gA is independent of solvent composition. This presumably means that it has the same value in pure solvent component 1 and in pure solvent component 2, since it is supposed not to change as $x_2$ goes from 0 to 1. And in fact the nonlinear regression analyses support the conclusion that gA is a parameter of the system, independent of composition.

But in the preceding paragraph no restriction has been placed on the identity of solvent component 2, so the conclusion must apply to any cosolvent 2 combined with the common solvent 1, which is water. This means that all mixed solvent systems in this study as described should yield the same value of gA. But this is not observed. Indeed, the variation in gA can be extreme, in a chemical sense; see Table 5.5.2. This constitutes a logical difficulty.

There is another anomaly to be considered. In nearly all of the nonelectrolyte solubility data that have been subjected to analysis according to eq. [5.5.23] the solute solubility increases as $x_2$, the organic cosolvent concentration, increases, and gA is positive, the physically reasonable result. But in the sucrose-water-ethanol system, the sucrose solubility decreases as $x_2$ increases, and gA is negative. There appears to be no physically reasonable picture of a negative gA value.

A further discrepancy was noted in 5.5.3.2, where we saw that some of the solvation constants evaluated from surface tension data did not agree closely with the corresponding numbers found in solubility studies.

### 5.5.4.2 A modified derivation

Recognizing that the original condition that g and $\Lambda$ are independent of composition was unnecessarily restrictive, we replace eq. [5.5.18] with eq. [5.5.52], where the subscripts 1 and 2 indicate values in the pure solvents 1 and 2.

$$\hat{g}\hat{A}\gamma = g_1A_1\gamma_1 + (g_2A_2\gamma_2 - g_1A_1\gamma_1)f_2 \qquad [5.5.52]$$

It is important for the moment to maintain a distinction between gA in the original formulation, a composition-independent quantity, and $\hat{g}\hat{A}$ in eq. [5.5.52], a composition-dependent quantity. Eq. [5.5.52] combines the composition dependence of three entities into a single grouping, $\hat{g}\hat{A}\gamma$, which is probably an oversimplification, but it at least generates the correct values at the limits of $x_2 = 0$ and $x_2 = 1$; and it avoids the unmanageable algebraic complexity that would result from a detailed specification of the composition dependence of the three entities separately. Eq. [5.5.14] now is written $\Delta G_{gen\ med} = \hat{g}\hat{A}\gamma$ and development as before yields eq. [5.5.53] as the counterpart to eq. [5.5.23], where $\delta_M gA\gamma = g_2A_2\gamma_2 - g_1A_1\gamma_1$.

$$\delta_M \Delta G^{*}_{soln} = \frac{(\delta_M gA\gamma/2 - kT\ln K_1)K_1x_1x_2 + (\delta_M gA\gamma - kT\ln K_1K_2)K_1K_2x_2^2}{x_1^2 + K_1x_1x_2 + K_1K_2x_2^2} \qquad [5.5.53]$$

Comparison of eqs. [5.5.23] and [5.5.53] gives eq. [5.5.54], which constitutes a specification of the meaning of gA in the original formulation in terms of the modified theory.

$$gA(\gamma_2 - \gamma_1) = g_2\gamma_2A_2 - g_1\gamma_1A_1 \qquad [5.5.54]$$

Now, the right-hand side of eq. [5.5.54] is a constant for given solute and solvent system, so the left-hand side is a constant. This shows why gA in the original theory (eq. [5.5.23]), is a composition-independent parameter of the system. Of course, in the derivation of eq. [5.5.23] gA had been assumed constant, and in effect this assumption led to any composition dependence of gA being absorbed into $\gamma$. In the modified formulation we acknowledge that the composition dependence of the product $\hat{g}\hat{A}\gamma$ is being accounted for without claiming that we can independently assign composition dependencies to the separate factors in the product.

### 5.5.4.3 Interpretation of parameter estimates

Eq. [5.5.54] constitutes the basis for the resolution of the logical problem, described earlier, in which different cosolvents, with a given solute, yield different gA values, although gA had been assumed to be independent of composition. As eq. [5.5.54] shows, gA is determined by a difference of two fixed quantities, thus guaranteeing its composition independence, and at the same time permitting gA to vary with cosolvent identity.

Eq. [5.5.53] is therefore conceptually sounder and physically more detailed than is eq. [5.5.23]. Eq. [5.5.53] shows, however, that in the absence of independent additional information (that is, information beyond that available from the solubility study alone) it is not possible to dissect the quantity ($g_2\gamma_2A_2$ - $g_1\gamma_1A_1$) into its separate terms. In some cases such additional information may be available, and here we discuss the example of naphthalene solubility in mixed aqueous-organic binary mixtures. Table 5.5.8 lists the values of $gA(\gamma_2-\gamma_1)$ obtained by applying eq. [5.5.23] to solubility data in numerous mixed solvent systems.[8] In an independent calculation, the solubility of naphthalene in water was written as eq. [5.5.55],

$$\Delta G^{*}_{soln}(x_2 = 0) = \Delta G_{cryst} + g_1A_1\gamma_1 \qquad [5.5.55]$$

which is equivalent to eq. [5.5.4]. $\Delta G_{cryst}$ was estimated by conventional thermodynamic arguments and $\Delta G_{solv}$ was omitted as negligible,[24] yielding the estimate $g_1A_1\gamma_1 = 4.64 \times 10^{-20}$ J molecule$^{-1}$. With eq. [5.5.54] estimates of $g_2A_2\gamma_2$ could then be calculated, and these are listed in Table 5.5.8.

**Table 5.5.8. Parameter estimates and derived quantities for naphthalene solubility in water-cosolvent mixtures at 25°C[a]**

| Cosolvent | $\gamma_2$, erg cm$^{-2}$ | $10^{20}$ $gA(\gamma_2-\gamma_1)$, J molecule$^{-1}$ | $10^{20}$ $g_2A_2\gamma_2$, J molecule$^{-1}$ |
|---|---|---|---|
| Methanol | 22.4 | -3.11 | +1.53 |
| Ethanol | 21.8 | -2.70 | 1.94 |
| Isopropanol | 20.8 | -2.19 | 2.45 |
| Propylene glycol | 37.1 | -2.46 | 2.18 |
| Ethylene glycol | 48.1 | -2.24 | 2.22 |
| Acetone | 22.9 | -3.37 | 1.27 |
| Dimethylsulfoxide | 42.9 | -3.67 | 0.97 |

[a]Data from ref. (8); $\gamma_1 = 71.8$ erg cm$^{-2}$, $g_1A_1\gamma_1 = 4.64 \times 10^{-20}$ J molecule$^{-1}$.

Observe that $g_1A_1\gamma_1$ and $g_2A_2\gamma_2$ are positive quantities, as expected; $gA(\gamma_2-\gamma_1)$ is negative because of the surface tension difference. It is tempting to divide each of these quantities by its surface tension factor in order to obtain estimates of gA, $g_1A_1$, and $g_2A_2$, but this procedure may be unsound, as proposed subsequently.

### 5.5.4.4 Confounding effects

*Solute-solute interactions*

It is very commonly observed, in these mixed solvent systems, that the equilibrium solubility rises well above the dilute solution condition over some portion of the $x_2$ range. Thus solution phase solute-solute interactions must make a contribution to $\Delta G^*_{soln}$. To some extent these may be eliminated in the subtraction according to eq. [5.5.22], but this operation cannot be relied upon to overcome this problem. Parameter estimates may therefore be contaminated by this effect. On the other hand, Khossravi[25] has analyzed solubility data for biphenyl in methanol-water mixtures by applying eq. [5.5.23] over varying ranges of $x_2$; he found that $gA(\gamma_2-\gamma_1)$ was not markedly sensitive to the maximum value of $x_2$ chosen to define the data set. In this system the solubility varies widely, from $x_3 = 7.1 \times 10^{-7}$ ($3.9 \times 10^{-5}$ M) at $x_2 = 0$ to $x_3 = 0.018$ (0.43 M) at $x_2 = 1$.

*Coupling of general medium and solvation effects*

In this theory the general medium and solvation effects are coupled through the solvation exchange constants $K_1$ and $K_2$, which determine the composition of the solvation shell surrounding the solute, and thereby influence the surface tension in the solvation shell. But the situation is actually more complicated than this, for if surface tension-composition data are fitted to eq. [5.5.26] the resulting equilibrium constants are not numerically the same as the solvation constants $K_1$ and $K_2$ evaluated from a solubility study in the same mixed solvent. Labeling the surface tension-derived constants $K_1'$ and $K_2'$, it is usually found that $K_1'>K_1$ and $K_2'>K_2$. The result is that a number attached to $\gamma$ at some $x_2$ value as a consequence of a nonlinear regression analysis according to eq. [5.5.23] will be determined by $K_1$ and $K_2$, and this number will be different from the actual value of surface tension, which is described by $K_1'$ and $K_2'$. But of course the actual value of $\gamma$ is driving the general medium effect, so the discrepancy will be absorbed into gA. The actual surface tension (controlled by $K_1'$ and $K_2'$) is smaller (except when $x_2 = 0$ and $x_2 = 1$) than that calculated with $K_1$ and $K_2$. Thus $g_{apparent} = g_{true} \times \gamma(K_1',K_2')/\gamma(K_1,K_2)$. This effect will be superimposed on the curvature correction factor that g represents, as well as the direct coupling effect of solvation mentioned above.

*The cavity surface area*

In solubility studies of some substituted biphenyls, it was found (see 5.5.3.1) that gA evaluated via eq. [5.5.23] was linearly correlated with the nonpolar surface area of the solutes rather than with their total surface area; the correlation equation was $gA = 0.37\ A_{nonpolar}$. It was concluded that the A in the parameter gA is the nonpolar surface area of the solute. This conclusion, however, was based on the assumption that g is fixed. But the correlation equation can also be written $gA = 0.37\ F_{nonpolar}A_{total}$, where $F_{nonpolar} = A_{nonpolar}/A_{total}$ is the fraction of solute surface area that is nonpolar. Suppose it is admitted that g may depend upon the solute (more particularly, it may depend upon the solute's polarity); then the correlation is consistent with the identities $A = A_{total}$ and $g = 0.37\ F_{nonpolar}$.

Thus differences in gA may arise from differences in solute polarity, acting through g. But A may itself change, rather obviously as a result of solute size, but also as a consequence of change of solvent, for the solvent size and geometry will affect the shape and size of the cavity that houses the solute.

### *The role of interfacial tension*

In all the preceding discussion of terms having the gA$\gamma$ form, $\gamma$ has been interpreted as a surface tension, the factor g serving to correct for the molecular-scale curvature effect. But a surface tension is measured at the macroscopic air-liquid interface, and in the solution case we are actually interested in the tension at a molecular scale solute-solvent interface. This may be more closely related to an interfacial tension than to a surface tension. As a consequence, if we attempt to find (say) $g_2A_2$ by dividing $g_2A_2\gamma_2$ by $\gamma_2$, we may be dividing by the wrong number.

To estimate numbers approximating to interfacial tensions between a dissolved solute molecule and a solvent is conjectural, but some general observations may be helpful. Let $\gamma_X$ and $\gamma_Y$ be surface tensions (vs. air) of pure solvents X and Y, and $\gamma_{XY}$ the interfacial tension at the X-Y interface. Then in general,

$$\gamma_{XY} = \gamma_X + \gamma_Y - W_{XY} - W_{YX} \qquad [5.5.56]$$

where $W_{XY}$ is the energy of interaction (per unit area) of X acting on Y and $W_{YX}$ is the energy of Y acting on X. When dispersion forces alone are contributing to the interactions, this equation becomes[26]

$$\gamma_{XY} = \gamma_X + \gamma_Y - 2\left(\gamma_X^d \gamma_Y^d\right)^{1/2} \qquad [5.5.57]$$

where $\gamma_X^d$ and $\gamma_Y^d$ are the dispersion force components of $\gamma_X$ and $\gamma_Y$. In consequence, $\gamma_{XY}$ is always smaller than the larger of the two surface tensions, and it may be smaller than either of them.

Referring now to Table 5.5.8, if we innocently convert $g_2A_2\gamma_2$ values to estimates of $g_2A_2$ by dividing by $\gamma_2$, we find a range in $g_2A_2$ from 23 Å$^2$ molecule$^{-1}$ (for dimethylsulfoxide) to 118 Å$^2$ molecule$^{-1}$ (for isopropanol). But if the preceding argument is correct, in dividing by $\gamma_2$ we were dividing by the wrong value. Taking benzene ($\gamma$ = 28 erg cm$^{-2}$) as a model of supercooled liquid naphthalene, we might anticipate that those cosolvents in Table 5.5.8 whose $\gamma_2$ values are greater than this number will have interfacial tensions smaller than $\gamma_2$, hence should yield $g_2A_2$ estimates larger than those calculated with $\gamma_2$, and vice versa. Thus, the considerable variability observed in $g_2A_2$ will be reduced.

On the basis of the preceding arguments it is recommended that gA$\gamma$ terms (exemplified by $g_1A_1\gamma_1$, $g_2A_2\gamma_2$, and $gA(\gamma_2$-$\gamma_1)$) should not be factored into gA quantities through division by $\gamma$, the surface tension, (except perhaps to confirm that magnitudes are roughly as expected). This conclusion arises directly from the interfacial tension considerations.

Finally let us consider the possibility of negative gA values in eq. [5.5.23]. Eq. [5.5.54] shows that a negative gA is indeed a formal possibility, but how can it arise in practice? We take the water-ethanol-sucrose system as an example; gA was reported to be negative for this system. Water is solvent 1 and ethanol is solvent 2. This system is unusual because of the very high polarity of the solute. At the molecular level, the solute in contact with these solvents is reasonably regarded as supercooled liquid sucrose, whose surface tension is unknown, but might be modeled by that of glycerol ($\gamma$ = 63.4 erg cm$^{-2}$). In these very polar systems capable of hydrogen-bonding eq. [5.5.57] is not applicable, but we can anticipate that the sucrose-water interaction energies (the $W_{XY}$ and $W_{XY}$ terms in eq. [5.5.56] are greater than sucrose-ethanol energies. We may expect that the sucrose-water interfacial tension is very low.

Now, gA turned out to be negative because $gA(\gamma_2-\gamma_1)$, a positive quantity as generated by eq. [5.5.23], was divided by $(\gamma_2-\gamma_1)$, a difference of surface tensions that is negative. Inevitably gA was found to be negative. The interfacial tension argument, however, leads to the conclusion that division should have been by the difference in interfacial tensions. We have seen that the interfacial tension between sucrose and water may be unusually low. Thus the factor $(\gamma_2-\gamma_1)$, when replaced by a difference of interfacial tensions, namely [$\gamma$(sucrose/ethanol) - $\gamma$(sucrose/water)], is of uncertain magnitude and sign. We therefore do not know the sign of gA; we only know that the quantity we label $gA(\gamma_2-\gamma_1)$ is positive. This real example demonstrates the soundness of the advice that products of the form $gA\gamma$ not be separated into their factors.[27,28]

## 5.5.5 NOTES AND REFERENCES


1 D. Khossravi and K.A. Connors, *J. Pharm. Sci.*, **81**, 371 (1992).

2 R.R. Pfeiffer, K.S. Yang, and M.A. Tucker, *J. Pharm. Sci.*, **59**, 1809 (1970).

3 J.B. Bogardus, *J. Pharm. Sci.*, **72**, 837 (1983).

4 P.L. Gould, J.R. Howard, and G.A. Oldershaw, *Int. J. Pharm.*, **51**, 195 (1989).

5 Also, when $K_1 = 1$ and $K_2 = 1$, eq. [5.5.13] shows that $\Delta G_{solv} = \Delta G_{WW}$; in this special case the solvation energy is composition-independent.

6 H.H. Uhlig, *J. Phys. Chem.*, **41**, 1215 (1937).

7 J.E. Leffler and E. Grunwald, **Rates and Equilibria of Organic Reactions**, *J. Wiley & Sons*, New York, 1963, p. 22.

8 J.M. LePree, M.J. Mulski, and K.A. Connors, *J. Chem. Soc., Perkin Trans. 2*, 1491 (1994).

9 The curvature correction factor g is dimensionless, as are the solvation constants $K_1$ and $K_2$. The parameter gA is expressed in $Å^2$ $molecule^{-1}$ by giving the surface tension the units J $Å^{-2}$ (where 1 erg $cm^{-2} = 1 \times 10^{-23}$ J $Å^{-2}$).

10 D. Khossravi and K.A. Connors, *J. Pharm. Sci.*, **82**, 817 (1993).

11 J.M. LePree, Ph.D. Dissertation, University of Wisconsin-Madison, 1995, p. 29.

12 A. Leo, C. Hansch, and D. Elkins, *Chem. Revs.*, **71**, 525 (1971).

13 C. Reichardt, **Solvents and Solvent Effects in Organic Chemistry**, *VCH*, Weinheim, 1988.

14 D. Khossravi and K.A. Connors, *J. Solution Chem.*, **22**, 321 (1993).

15 K.A. Connors and J.L. Wright, *Anal. Chem.*, **61**, 194 (1989).

16 K.A. Connors, **Binding Constants**, *Wiley-Interscience*, New York, 1987, pp. 51, 78.

17 R.D. Skwierczynski and K.A. Connors, *J. Chem. Soc., Perkin Trans. 2*, 467 (1994).

18 K.A. Connors and D. Khossravi, *J. Solution Chem.*, **22**, 677 (1993).

19 M.J. Mulski and K.A. Connors, *Supramol, Chem.*, **4**, 271 (1995).

20 K.A. Connors, M.J. Mulski, and A. Paulson, *J. Org. Chem.*, **57**, 1794 (1992).

21 J.M. LePree and K.A. Connors, *J. Pharm. Sci.*, **85**, 560 (1996).

22 M.C. Brown, J.M. LePree, and K.A. Connors, *Int. J. Chem. Kinetics*, **28**, 791 (1996).

23 J.M. LePree and M.E. Cancino, *J. Chromatogr. A*, **829**, 41 (1998).

24 The validity of this approximation can be assessed. The free energy of hydration of benzene is given as -0.77 kJ $mol^{-1}$ (E. Grunwald, **Thermodynamics of Molecular Species**, *Wiley-Interscience*, New York, 1997, p. 290). Doubling this to -1.5 kJ $mol^{-1}$ because of the greater surface area of naphthalene and repeating the calculation gives $g_1A_1\gamma_1 = 4.88 \times 10^{-20}$ J $molecule^{-1}$, not sufficiently different from the value given in the text to change any conclusions.

25 D. Khossravi, Ph.D. Dissertation, University of Wisconsin-Madison, 1992, p. 141.

26 F.M. Fowkes, **Chemistry and Physics of Interfaces**; *American Chemical Society*: Washington, D.C., 1965, Chap. 1.

27 The introduction of the interfacial tension into the cavity term was first done by Yalkowsky et al.,[28] who also argue that a separate solute-solvent interaction term is unneeded, as the solute-solvent interaction is already embodied in the interfacial tension. In our theory we explicitly show the coupling between the solute-solvent and solvent-solvent interactions (eq. [5.5.19]), but this is in addition to the solute-solvent interaction (eq. [5.5.13]). This difference between the two theories is a subtle issue that requires clarification.

28 S.H. Yalkowsky, G.L. Amidon, G. Zografi, and G.L. Flynn, *J. Pharm. Sci.*, **64**, 48 (1975).

# 6

# SWELLING

## 6.1 MODERN VIEWS ON KINETICS OF SWELLING OF CROSSLINKED ELASTOMERS IN SOLVENTS

E. YA. DENISYUK
**Institute of Continuous Media Mechanics**
V. V. TERESHATOV
**Institute of Technical Chemistry**
**Ural Branch of Russian Academy of Sciences, Perm, Russia**

### 6.1.1 INTRODUCTION

Diffusion phenomena encountered in mass-transfer of low-molecular liquids play an important role in many technological processes of polymer manufacture, processing, and use of polymeric materials. Diffusion of organic solvents in crosslinked elastomers may cause considerable material swelling. In this case, the polymeric matrix experiences strains as large as several hundred percent, while a non-homogeneous distribution of a liquid due to diffusion results in establishing stress-strain state capable of affecting the diffusion kinetics. The processes of material deformation and liquid diffusion in such systems are interrelated and nonlinear in nature and are strongly dependent on physical and geometrical nonlinearities. Therefore, exact relations of nonlinear mechanics of elastic-deformable continuum are the mainstream of a sequential theory of mass-transfer processes of low-molecular liquids in elastomers.

The general principles of the development of nonlinear models of mass transfer in elastically deformed materials were developed in studies.[1,2] The general formulation of constitutive equations and the use of non-traditional thermodynamic parameters such as partial stress tensors and diffusion forces lead to significant difficulties in attempts to apply the theory to the description of specific objects.[3,4] Probably, because of this, the theory is little used for the solution of applied problems.

In the paper,[5] a theory for mechanical and diffusional processes in hyperelastic materials was formulated in terms of the global stress tensor and chemical potentials. The approach described in[1,2] was used as the basic principle and was generalized to the case of a multi-component mixture. An important feature of the work[5] is that, owing to the structure of constitutive equations, the general model can be used without difficulty to describe specific systems.

In the paper[6] the nonlinear theory[5] was applied to steady swelling processes of crosslinked elastomers in solvents. The analytical and numerical treatment reveals three possible mechano-diffusion modes which differ qualitatively. Self-similar solutions obtained for these modes describe asymptotic properties at the initial stage of swelling. These modes are related to thermodynamical material properties. The theoretical predictions have been verified in the experiments on real elastomers.

### 6.1.2 FORMULATION OF SWELLING FOR A PLANE ELASTOMER LAYER

Consider an infinite plane elastomer layer of thickness 2h embedded in a low-molecular liquid. Suppose that the elastomer initially does not contain liquid and is unstrained. This state is taken as a reference configuration. Let us introduce the Cartesian coordinates (x,y,z) with the origin placed in the layer center and relate them to a polymer matrix. In the examined problem, the Cartesian coordinates will be used as the material coordinates. With reference to the layer, the x axis has a transverse direction and the other axes have longitudinal directions. In our approach, we define the problem under consideration as a one-dimensional problem, in which all quantities characterizing the elastomer state depend only on the x-coordinate.

On swelling, the layer experiences transversal and longitudinal deformations which can be written as

$$X = X(x,t) \quad Y = \nu(t)y \quad Z = \nu(t)z \qquad [6.1.1]$$

where (X,Y,Z) are the spatial Cartesian coordinates specifying the actual configuration of the polymeric matrix. From this it follows that the relative longitudinal stretch of the layer is $\lambda_2 = \lambda_3 = \nu(t)$ and the relative transversal stretch is $\lambda_1 = \lambda(x,t) = \partial X / \partial x$. The quantity

$$J = \lambda_1 \lambda_2 \lambda_3 = \lambda \nu^2 \qquad [6.1.2]$$

characterizes a local relative change in the material volume due to liquid absorption.

The boundary conditions and the relations describing free swelling of the plane layer in the reference configuration are represented in[5] as

$$\frac{\partial N_1}{\partial t} = \frac{\partial}{\partial x}\left(D\frac{\partial N_1}{\partial x}\right), \quad N_1 = N_1(x,t) \qquad [6.1.3]$$

$$\partial N_2 / \partial t = 0 \qquad [6.1.4]$$

$$\partial \sigma_1 / \partial x = 0 \qquad [6.1.5]$$

$$N_1(x,0) = 0 \qquad [6.1.6]$$

$$\partial N_1(0,t) / \partial x = 0, \quad X(0,t) = 0 \qquad [6.1.7]$$

$$\mu(h,t) = 0, \quad \sigma_1(h,t) = 0 \qquad [6.1.8]$$

$$\langle \sigma_2(x,t) \rangle = \langle \sigma_3(x,t) \rangle = 0 \qquad [6.1.9]$$

where:

- $N_1, N_2$ — the molar concentrations of the liquid and the chains of polymeric network of elastomer, respectively,
- $\mu$ — the chemical potential of the liquid dissolved in material
- $\sigma_k$ — (k = 1,2,3) are the principal values of the Piola stress tensors.

The angular brackets denote integration with respect to coordinate x:

$$\langle \ldots \rangle = h^{-1} \int_0^h \ldots dx$$

Owing to the symmetry of the swelling process in the layer, the problem is solved for $0<x<h$.

The equation of the liquid transport [6.1.3] in a plane layer has the form of a general diffusion equation except for the diffusion coefficient of the liquid, which, in the general case, is defined by the function $D = D(N_1,\nu)$, implying that it depends on the liquid concentration and the relative longitudinal stretch of the layer.[5] Eq. [6.1.4] is the law of conservation of matter for the polymeric matrix, and Eq. [6.1.5] states that the process of elastomer swelling is in the state of mechanical equilibrium. The initial condition is explicitly defined by Eq. [6.1.6]. The constraint that the diffusion flux and the displacements of polymeric matrix along x-axis in the layer center are absent is given by Eq. [6.1.7]. Eq. [6.1.8] has the physical meaning that there exists a thermodynamical equilibrium at the elastomer-liquid interface and that elastomer is not subjected to transverse mechanical loading, while Eq. [6.1.9] means that the layer does not experience longitudinal stretch under the external force.

The assumption that the elastomer and the liquid are incompressible media can be mathematically represented by an incompressibility condition, which in the present case is written as[5]

$$J = \phi^{-1} \qquad [6.1.10]$$

where the volume fraction of the polymer is

$$\phi = N_2V_2 / (N_1V_1 + N_2V_2) \qquad [6.1.11]$$

where:

$V_1$ and $V_2$ the molar volumes of liquid and chains of the elastomer network, respectively

To make the definition of the examined problem complete, we need to add to the above model equations, the constitutive relations for mechanical stress tensor and chemical potential of a liquid. According to[5, 6] these equation are given by

$$\sigma_k = RTV_2^{-1}\left(\lambda_k - I_1\lambda_k^{-1}/3\right) - pJ\lambda_k^{-1} \qquad [6.1.12]$$

$$\mu = \mu_{mix}(\phi) + RTZ^{-1}\phi^{1/3}\Gamma_1/3 + V_1p \qquad [6.1.13]$$

$$\mu_{mix} = RT\left[\ln(1-\phi) + \phi + \chi\phi^2\right] \qquad [6.1.14]$$

where:

| | |
|---|---|
| R | the gas constant per mole |
| T | the absolute temperature |
| $\mu_{mix}$ | the chemical potential of mixing |
| p | pressure |
| $\chi$ | the Flory-Huggins interaction parameter |

Z $= V_2/V_1$
$I_1$ $= \lambda_1^2 + \lambda_2^2 + \lambda_3^2$
$\Gamma_1$ $= I_1/J^{2/3}$

The above equations follow from the classical high elasticity theory and the Flory theory of polymeric networks.[7]

From Eq. [6.1.5] and the second condition of Eq. [6.1.8], we find that $\sigma_1(x,t) = 0$. This equation together with Eq. [6.1.12] yields the expression for pressure. By substituting it in the formulas for chemical potential [6.1.13] and longitudinal stresses, we find, using Eqs. [6.1.2] and [6.1.10], that

$$\sigma_2 = \sigma_3 = RTV_2^{-1}\left(\nu - J^2 / \nu^5\right) \qquad [6.1.15]$$

$$\mu = \mu_{mix}(1/J) + RTZ^{-1}J / \nu^4 \qquad [6.1.16]$$

A substitution of Eq. [6.1.15] in Eq. [6.1.9] gives an expression for longitudinal stretch of the layer

$$\nu^6 = \left\langle J^2(x,t)\right\rangle \qquad [6.1.17]$$

With consideration of Eq. [6.1.16], the boundary condition at x = h is transformed to

$$\mu_{mix}(1/J)RT + Z^{-1}J / \nu^4 = 0 \qquad [6.1.18]$$

Thus, the initial swelling problem for a plane layer is reduced to a boundary value problem for diffusion equation [6.1.3] with boundary conditions of Eqs. [6.1.6], [6.1.7], [6.1.17] and [6.1.18]. The solution to this problem provides a full description of swelling processes in the plane layer. In other words, using Eqs. [6.1.1], [6.1.2], [6.1.10] and [6.1.15] we can define a current distribution of a liquid through the layer and calculate the stress-strain state of the material.

It should be noted that boundary conditions of Eq. [6.1.18] and Eq. [6.1.17] specify the existence of positive feedback in the system, which is responsible for the onset of unsteady boundary regime during material swelling. The nonlinear distributed systems with positive feedback are generally known as active media and are distinguished for their complex and multimode response.[8] In free swelling, the response of elastomers is, in a sense, similar to that of active media. Such behavior is most pronounced when the extent of material swelling is high, which makes this case worthwhile for detailed investigation.

For high-swelling elastomers, the volume fraction of polymer in equilibrium swelling state denoted in the following as $\varepsilon$ and the volume fraction of polymer at the elastomer-liquid interface $\phi = 1/J$ entering Eq. [6.1.18] are small quantities. The asymptotic behavior of the function $\mu_{mix}(\phi)$ at $\varphi \to 0$ is described by

$$\mu_{mix}(\phi) / RT = -b\phi^{\alpha} \qquad [6.1.19]$$

The constants b and $\alpha$ can be calculated using the Flory equation [6.1.14]. A second order expansion of $\ln(1-\phi)$ as a power series of $\phi$ gives $b = 1/2-\chi$ and $\alpha = 2$. The scaling approach gives a slightly different value of $\alpha$, which is found to be $\alpha = 9/4$ (des Cloizeaux law[9]).

A volume fraction of the polymer in equilibrium swelling state can be determined by substituting Eq. [6.1.19] in Eq. 6.1.18] and setting $\phi = J^{-1} = \varepsilon$ and $\nu = \varepsilon^{-1/3}$, yields $\varepsilon \approx (bZ)^{-3/(3\alpha-1)}$. Then, using Eqs. [6.1.18], [6.1.19] and the last relation, we arrive at the following expression for the volumetric swelling ratio of the layer at the elastomer-liquid interface:

$$J \approx \varepsilon^{-1}\left(\varepsilon^{1/3}\nu\right)^{6d} \qquad [6.1.20]$$

where

$$d = \frac{2}{3(\alpha + 1)} \qquad [6.1.21]$$

Note that approximate Eq. 6.1.20] defines the strain dependence of the equilibrium swell ratio of the elastomer in a liquid medium under conditions of biaxial symmetric material extension.

Substituting Eq. [6.1.17] in Eq. [6.1.20], we express the boundary swell ratio in terms of liquid distribution in the layer

$$J(h,t) = \varepsilon^{2d-1}\left\langle J^2(x,t)\right\rangle^{d} \qquad [6.1.22]$$

Then the problem is finally defined as

$$u_t = \left(k(u,l)u_x\right)_x; \quad x \in (0,1), \quad t > 0 \qquad [6.1.23]$$

$$u(x,0) = 0, \quad u_x(1t) = 0 \qquad [6.1.24]$$

$$(1-\varepsilon)u(0,t) + \varepsilon = \left\langle\left[(1-\varepsilon)u(x,t) + \varepsilon\right]^2\right\rangle^{d} \qquad [6.1.25]$$

Here we assign dimensions to the variables. The quantities h and $h^2/D_0$ (where $D_0$ is the value of diffusion coefficient in the state of ultimate elastomer swelling) are used as the units of distance and time. For the sake of convenience we transform, the coordinate to x→1-x. Integrating for x between the limits from 0 to 1 in Eq. [6.1.25] is designated by angular brackets. The function u(x,t) takes the value over the interval (0,1) and represents a dimensionless concentration of penetrating liquid. It is related to the liquid concentration and local material swelling by the following equations:

$$N_1 = V_1\left(\varepsilon^{-1} - 1\right)u(x,t), \quad J(x,t) = \varepsilon^{-1}\left[(1-\varepsilon)u(x,t) + \varepsilon\right] \qquad [6.1.26]$$

The quantity $l = \varepsilon^{1/3}\nu$ represents the longitudinal layer stretch normalized to unity. By virtue of [6.1.17] and [6.1.26] we may write

$$l^6(t) = \left\langle\left[(1-\varepsilon)u(x,t) + \varepsilon\right]^2\right\rangle \qquad [6.1.27]$$

Dimensionless diffusion coefficient is defined by the formula $k(u,l) = D(u,l)/D_0$.

The longitudinal stresses in the layer (15) are expressed in terms of dimensionless stresses q(x,t) by

$$\sigma_2 = \sigma_3 = RTV_2^{-1}q(x,t) \tag{6.1.28}$$

where according to Eqs. [6.1.15] and [6.1.27]

$$q(x,t) = \varepsilon^{-1/3} l(t)\left\{1 - \left[(1-\varepsilon)u(x,t) + \varepsilon\right]^2 / l^6(t)\right\} \tag{6.1.29}$$

Consider two functions

$$g_1(t) = \langle u(x,t)\rangle, \quad g_2(t) = \langle u^2(x,t)\rangle \tag{6.1.30}$$

which are integral characteristics of swelling kinetics for a plane layer and can be determined from experiments. The first function characterizes a relative amount of liquid absorbed by a polymer in time t and the second function according to Eq. [6.1.27] is related to longitudinal layer deformation. For high-swelling elastomers

$$g_2(t) \approx l^6(t) \tag{6.1.31}$$

The numerical results obtained by solving model problem of Eqs. [6.1.23] - [6.1.25] for a constant diffusion coefficient are plotted in Figure 6.1.1.[6] The obtained curves show the evolution of penetrating liquid concentration and longitudinal stresses. It is seen that the boundary liquid concentration during swelling monotonically increases.

### 6.1.3 DIFFUSION KINETICS OF PLANE LAYER SWELLING

Consider two stages of swelling process in a plane layer - the initial and final. In the initial stage, the influence of the opposite layer boundary on the swelling process is inessential and therefore diffusion in a layer of finite thickness at sufficiently small values of time can be considered as the diffusion in half-space.

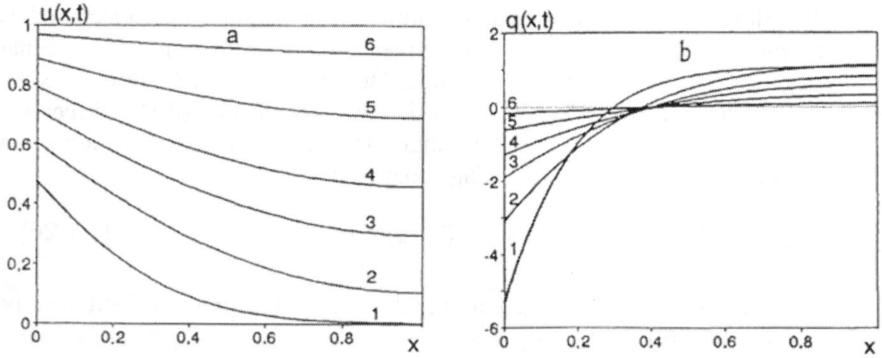


Figure 6.1.1. Distribution of penetrating liquid (a) and longitudinal stresses (b) during swelling of a plane layer with constant diffusion coefficient k(u,l) = 1 at ε = 0.1 and d = 2/9: 1 - t = 0.05; 2 - t = 0.2; 3 - t = 0.4; 4 - t = 0.6; 5 - t = 1; 6 - t = 1.8. [Adapted, by permission, from E. Ya. Denisyuk, V. V. Tereshatov, *Vysokomol. soed.*, **A42**, 74 (2000)].

At the very beginning of the swelling process the amount of absorbed liquid is rather small. Hence we may set $u(x,t) \approx 0$ in the right-hand parts of Eqs. [6.1.25] and [6.1.27] which results in

$$u(0,t) \approx \psi_0 = \left(\varepsilon^{2d} - \varepsilon\right) / (1-\varepsilon), \quad l(t) \approx \varepsilon^{1/3}$$

and Eq. [6.1.23] becomes an usual parabolic equation describing diffusion on a half-line with constant boundary concentration $\psi_0$. It has self-similar solution of the form $u(x,t)=\psi_0\theta(x/t^{1/2})$. The function $\theta(\xi)$ satisfies the equation $(k(\theta, \varepsilon^{1/3})\theta')' + \xi\theta'/2 = 0$ and the boundary conditions

$$\theta(0) = 1, \quad \theta(+\infty) = 0 \qquad [6.1.32]$$

From this follows the expression for the integral process characteristics

$$g_1(t) = \psi_0 M_1 t^{1/2}, \quad g_2(t) = \psi_0^2 M_2 t^{1/2}$$

where

$$M_p = \int_0^\infty \theta^p(\xi)d\xi \quad p = 1,2 \qquad [6.1.33]$$

These relations define the asymptotic properties of swelling at $t \to 0$.

As more and more amount of the liquid is absorbed, the longitudinal strains in the layer increase. By virtue of Eq. [6.1.25], this causes the growth of liquid concentration at the boundary. For high-swelling materials at sufficiently large values of time, all terms in Eq. [6.1.25] involving $\varepsilon$ as a multiplier factor can be neglected to the first approximation. The resulting expression is written as

$$u(0,t) = \left\langle u^2(x,t)\right\rangle^d \qquad [6.1.34]$$

where the angular brackets denote integrating for x in the limits from 0 to $+\infty$.

Since for arbitrary dependence of $k(u,l)$ a boundary-value solution of equation [6.1.23] on the half-line with boundary condition of Eq. [6.1.34] cannot be represented in a similar form, we restrict our consideration to a model problem with diffusion coefficients defined by

$$k(u,l) = u^s, \quad s \geq 0 \qquad [6.1.35]$$

$$k(u,l) = u^s l^{6p}, \quad s \geq 0, \quad p \geq 0 \qquad [6.1.36]$$

(Let us agree that $s = 0$ corresponds to a constant diffusion coefficient $k(u,l) = 1$). The analysis of this problem allows us to qualitatively explain many mechanisms of diffusion kinetics of elastomer swelling.

First, consider the diffusion coefficient defined by Eq. [6.1.35]. In this case, Eq. [6.1.23] on the half-line has a variety of self-similar solutions, which can be written as

$$u(x,t) = \psi(t)\theta(x / \varphi(t)) \qquad [6.1.37]$$

where the function $\theta(\xi)$ satisfies conditions of Eq. [6.1.32] and defines the profile of the diffusion wave, $\psi(t)$ describes boundary conditions and $\varphi(t)$ the function specifies the penetration depth of diffusion wave.

Eq. [6.1.37] satisfies the boundary condition of Eq. [6.1.34] and Eq. [6.1.23] on the half-line with the diffusion coefficient of Eq. [6.1.35] in the following cases: 1) $\psi(t)$, $\varphi(t)$ are the functions of power type (power swelling mode); 2) $\psi(t)$, $\varphi(t)$ are the function exponentially depending on time (exponential swelling mode); 3) $\psi(t) \sim (t_0 - t)^m$, $\varphi(t) \sim (t_0 - t)^n$, where m, $n < 0$ ( blow-up swelling mode).

Power swelling mode occurs at sufficiently small values of s. In this case, the amount of absorbed liquid, boundary concentration, the depth of diffusion wave penetration and the longitudinal layer deformation are the power function of time. If the parameter s approaches the critical value

$$s_c = 2/d - 4 \qquad [6.1.38]$$

Swelling process is governed by exponential law. And finally, if $s > s_c$, the swelling mode is of a blow-up nature. The solutions describing these modes are given below.

I. Power mode ($s < s_c$):

$$u(x,t) = M_2^{2m} t^m \theta(\xi), \quad \xi = x / M_2^{ms} t^n \qquad [6.1.39]$$

$$g_1(t) = M_1 M_2^{2q-1} t^q, \quad g_2(t) = M_2^{2r} t^r \qquad [6.1.40]$$

$$m = \frac{1}{s_c - s}, \quad n = \frac{1/d - 2}{s_c - s}, \quad q = \frac{1/d - 1}{s_c - s}, \quad r = \frac{1}{d(s_c - s)} \qquad [6.1.41]$$

II. Exponential mode ($s = s_c$):

$$u(x,t) = \exp\{M_2^2 (t - t_0)\}\theta(\xi), \; \xi = xM_2 / \exp\{s_c M_2^2 (t - t_0)/2\}$$

$$g_1(t) = (M_1 / M_2)\exp\{(s_c/2 + 1)M_2^2 (t - t_0)\}$$

$$g_2(t) = \exp\{(s_c/2 + 1)M_2^2 (t - t_0)\}$$

III. Blow-up mode ($s > s_c$):

$$u(x,t) = M_2^{2m}(t_0 - t)^m \theta(\xi), \quad \xi = x / M_2^{ms}(t_0 - t)^n$$

$$g_1(t) = M_1 M_2^{2q-1}(t_0 - t)^q, \quad g_2(t) = M_2^{2r}(t_0 - t)^r$$

where the exponents are defined by Eqs. [6.1.41] but if $s > s_c$, then m, n, q, $r<0$.

For power and blow-up swelling modes $\theta(\xi)$ are derived from the equation $(\theta^s \theta')' + |n| \xi\theta' - |m|\theta = 0$. For an exponential mode, this equation will be valid if we put

$n=s_c/2$ and $m=1$. The constants $M_1$ and $M_2$ are evaluated from Eq. [6.1.33] and the constant $t_0$ can be estimated by the order of magnitude from the condition $u(0,t_0)\sim\psi_0\sim\varepsilon^{2d}$. For the exponential mode, we have $t_0 \sim -2dM_2^{-2}\ln(\varepsilon)$, and for the blow-up mode, $t_0 \sim M_2^{-2}\,\varepsilon^{2d/m}$.

It is worth noting that at $m = 1/s$, which holds only if $s = 1/d - 2$, solution of Eq. [6.1.39] is expressed in terms of primary functions and describes the diffusion wave propagating with a constant velocity d:

$$u(x,t) = (1 - 2d)^{1/s}\,(dt - x)^{1/s}$$

The solution describing the blow-up mode turns into infinity at the finite time. In the global sense the boundary-value problems admitting such solutions are time unsolvable and are generally applied to modeling high rate physical-chemical processes.[10] It is quite evident that all solutions to the swelling problem for a layer of finite thick are limited and each of the self-similar solutions presented in this study describes asymptotic properties of diffusion modes at initial well-developed stage of swelling.

At the final swelling stage $u(x,t) \to 1$. In this case, an approximate solution to the problem can be obtained by its linearization in the vicinity of equilibrium $u = 1$. To this end, one needs to introduce a variable $v(x,t) = 1 - u(x,t)$. After transformation we get

$$v_t = v_{xx},\quad v_x(1,t) = 0,\quad v(0,t) = 2d\langle v(x,t)\rangle$$

By making use of the method of variable separation, we find

$$v(x,t) = \sum_{k=1}^{\infty} a_k \exp\left(-\alpha_k^2 t\right)\cos\left[\alpha_k(1-x)\right] \qquad [6.1.42]$$

where the values of $\alpha_k$ are determined from the equation

$$\alpha_k = 2d\tan(\alpha_k) \qquad [6.1.43]$$

Restricting ourselves to the first term of a series Eq. [6.1.42], we can write for the final swelling stage the equation of kinetic curve

$$g_1(t) = 1 - (a_1/\alpha_1)\exp\left(-\alpha_1^2 t\right)\sin(\alpha_k) \qquad [6.1.44]$$

The above expressions allow us to describe the shape of kinetic curves $g_1(t)$ in general terms. In particular, as it follows from Eqs. [6.1.39] - [6.1.41] at $q < 1$, the kinetic curves in the coordinates $(t,g_1)$ are upward convex and have the shape typical for pseudo-normal sorption. This takes place at $s<s_c/2-1$. If $s=s_c/2-1$, then $q = 1$, which corresponds to a linear mode of liquid absorption. At $s>s_c/2-1$ the lower part of the kinetic curve is convex in a downward direction and the whole curve becomes S-shaped. Note that in terms of coordinates $(t^{1/2},g_1)$ at $s \geq 0$ all the kinetic curves are S-shaped. Hence, the obtained solutions enable one to describe different anomalies of sorption kinetics observed in the experiments on elastomer swelling in low-molecular liquids.

Figure 6.1.2[6] gives the results of numerical solution to problems of Eq. [6.1.25] with diffusion coefficient defined by Eq. [6.1.35]. The kinetic curves of swelling at different values of s are depicted in Figure 6.1.3.[6]

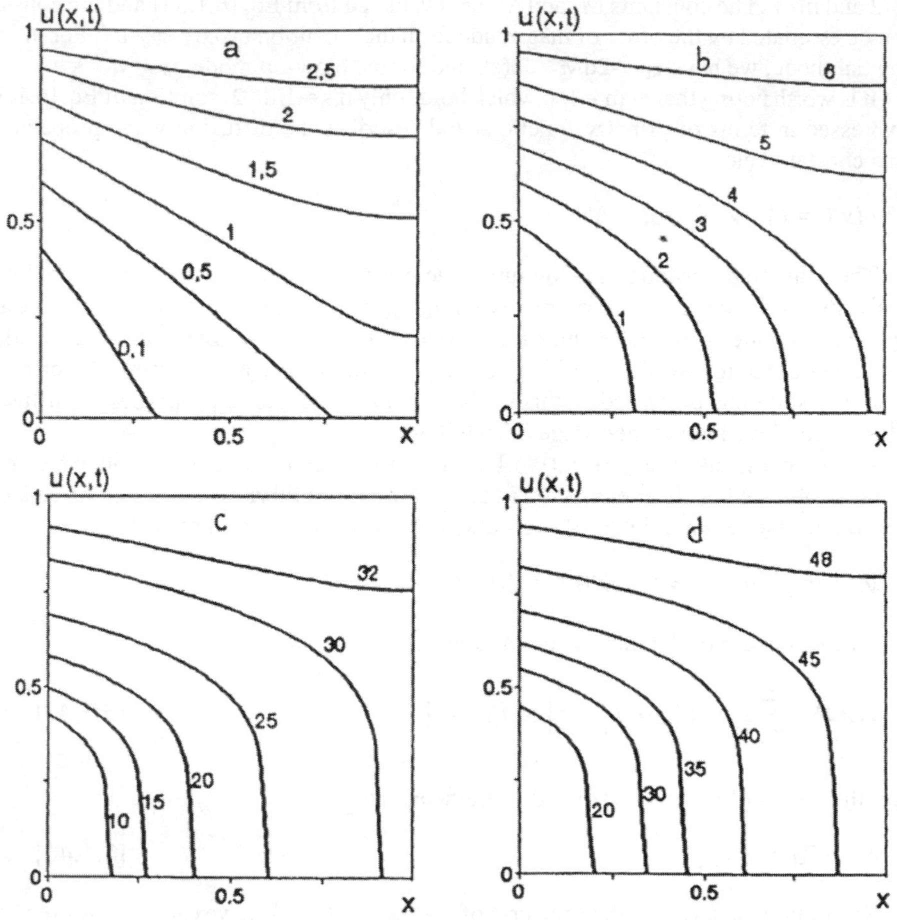


Figure 6.1.2. Diffusion kinetics of plane layer swelling for diffusion coefficient $k(u)=u^s$ at $\varepsilon = 0.1$ and $d = 2/9$; a, b are power swelling modes at s = 1 and s = 2.5, respectively; c - exponential swelling mode (s=5); d - blow-up swelling mode (s = 5.5). Numerals over curves denote correspond to instants of time. [Adapted, by permission, from E. Ya. Denisyuk, V. V. Tereshatov, *Vysokomol. soed.*, **A42**, 74 (2000)].

Here it is to be noted that strain dependence of the diffusion coefficient described by Eq. [6.1.36] does not initiate new diffusion modes. The obtained three self-similar solutions hold true. Only critical value $s_c$ is variable and is defined by expression $s_c = (2-p)/d-4$. This fact can be supported by a direct check of the solutions.

## 6.1.4 EXPERIMENTAL STUDY OF ELASTOMER SWELLING KINETICS

The obtained solutions can be applied to experimental study of the diffusive and thermodynamic properties of elastomers. In particular, with the relation

$$s = 2/d - 4 - (1/d - 1)/q \qquad [6.1.45]$$

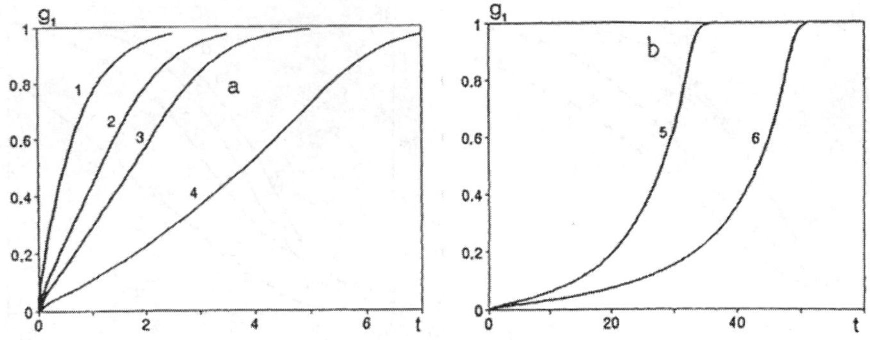


Figure 6.1.3. Kinetic curves of plane layer swelling at different values of concentration dependence of diffusion coefficient ($\varepsilon = 0.1$, $d = 2/9$): a - power law mode (1 - s = 0; 2 - s = 1; 3 - s = 1.5; 4 - s = 2.5); b - exponential (5 - s = 5) and blow-up mode (6 - s = 5.5). [Adapted, by permission, from E. Ya. Denisyuk, V. V. Tereshatov, *Vysokomol. soed.*, **A42**, 74 (2000)].

following from Eqs. [6.1.38] and [6.1.41] we can estimate the concentration dependence of the diffusion coefficient of a liquid fraction in elastomer. According to Eq. [6.1.40] the parameter q is determined from the initial section of the kinetic swelling curve.

Experimental estimates of the parameter r can be obtained from the strain curve l(t) using Eqs. [6.1.31] and [6.1.40]. Then by making use of the formula

$$d = 1 - q/r \qquad [6.1.46]$$

following from Eq. [6.1.41] we can evaluate the parameter d which characterizes the strain dependence of the equilibrium swelling ratio of elastomer under symmetric biaxial extension in Eq. [6.1.45]. Generally the estimation of this parameter in tests on equilibrium swelling of strained specimens proves to be a tedious experimental procedure.

The value of diffusion coefficient in an equilibrium swelling state can be determined from the final section of kinetic swelling curve using Eq. [6.1.44], which is expressed in terms of dimensional variables as

$$g_1(t) = 1 - C\exp(-\alpha_1^2 D_0 t / h^2) \qquad [6.1.47]$$

where $\alpha_1$ is calculated from Eq. [6.1.43]. For $d = 2/9$, $\alpha_1 \approx 1.2220$.

Note that all these relations are valid only for sufficiently high values of elastomer swelling ratio.

The obtained theoretical predictions have been verified in experiments on real elastomers. The elastomers tested in our experiments were amorphous polybutadiene urethanes (PBU) with polymer network of different density: 0.3 kmol/m$^3$ (PBU-1), 0.05 kmol/m$^3$ (PBU-2), 0.2 kmol/m$^3$ (PBU-3), 0.1 kmol/m$^3$ (PBU-4). Oligooxypropylene triol - Laprol 373 was used as a crosslinking agent at the curing of prepolymer of oligobutadiene diol. The elastomer specimens were manufactured in the form of disks, 35 mm in diameter and 2 mm thick. The kinetics of specimen swelling was determined in low-molecular liquids: toluene, dibutyl sebacate (DBS), dioctyl sebacate (DOS).

The typical kinetic and strain curves of free swelling are given in Figure 6.1.4.[6] The S-shape of the kinetic swelling curves in terms of coordinates ($t^{1/2}$, $g_1$) is indicative of

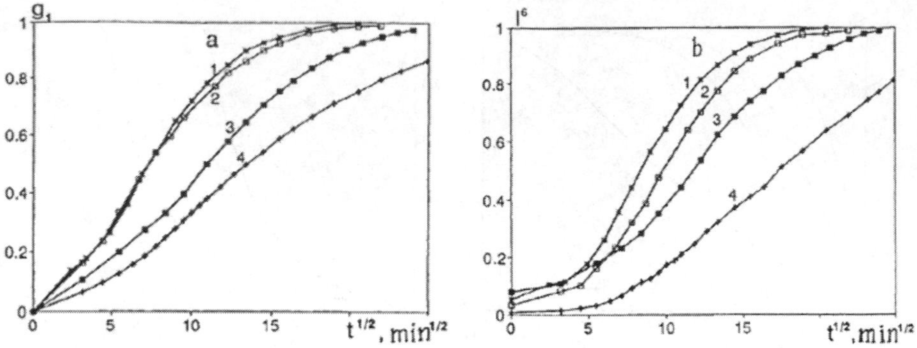


Figure 6.1.4. Kinetic (a) and strain (b) curves of elastomer swelling in toluene: 1 - PBU-3; 2 - PBU-4; 3 - PBU-1; 4 - PBU-2. [Adapted, by permission, from E. Ya. Denisyuk, V. V. Tereshatov, *Vysokomol. soed.*, **A42**, 74 (2000)].

abnormal sorption. The values of parameters q and r were obtained from kinetic and strain curves using the regression method. The values of correlation coefficient were 0.997 - 0.999 and 0.994 - 0.998 respectively. The obtained data and Eqs. [6.1.45], [6.1.46] were then used to calculate s and d. The diffusion coefficients were defined by the kinetic curves in terms of Eq. [6.1.47] under the assumption that d = 2/9. The obtained results were summarized in Table 6.1.1.[6] The analysis of these data shows that swelling of the examined elastomers is of power-mode type. The concentration dependence of the liquid diffusion coefficient defined by the parameter s is found to be rather weak. For elastomers under consideration no exponential or blow-up swelling modes have been observed.

**Table 6.1.1. Experimental characteristics of elastomer swelling kinetics[6]**

| Elastomer/Liquid | ε | q | r | s | d | $D_0$, cm²/s |
|---|---|---|---|---|---|---|
| PBU-1/toluene | 0.278 | 0.68 | 0.84 | 0 | 0.19 | $1.4\times10^{-6}$ |
| PBU-2/toluene | 0.093 | 0.83 | 1.09 | 0.8 | 0.24 | $4.9\times10^{-7}$ |
| PBU-3/toluene | 0.230 | 0.78 | 1.02 | 0.5 | 0.24 | $1.3\times10^{-6}$ |
| PBU-4/toluene | 0.179 | 0.71 | 0.97 | 0 | 0.27 | $1.4\times10^{-6}$ |
| PBU-1/DBS | 0.345 | 0.67 | 0.92 | 0 | 0.27 | $6.2\times10^{-8}$ |
| PBU-2/DBS | 0.128 | 0.78 | 1.04 | 0.5 | 0.25 | $2.5\times10^{-8}$ |
| PBU-3/DBS | 0.316 | 0.67 | 0.83 | 0 | 0.19 | $6.6\times10^{-8}$ |
| PBU-4/DBS | 0.267 | 0.69 | 0.88 | 0 | 0.21 | $7.4\times10^{-8}$ |
| PBU-1/DOS | 0.461 | 0.67 | 0.81 | 0 | 0.17 | $1.8\times10^{-8}$ |
| PBU-4/DOS | 0.318 | 0.63 | 0.81 | 0 | 0.22 | $3.4\times10^{-8}$ |

It is of interest to note that experimental values of the parameter d characterizing the strain dependence of equilibrium swelling ratio for elastomers subjected to uniform biaxial extension closely approximate the theoretical values. It will be recalled that this parameter is specified by Eq. [6.1.21]. Moreover, the Flory theory defines it as d = 2/9 = 0.22(2), whereas the des Cloizeaux law provides d ≈ 0,205, which suggests that the proposed model of elastomer swelling performs fairly well.

### 6.1.5 CONCLUSIONS

In this section, we have developed a geometrically and physically nonlinear model of swelling processes for an infinite plane elastomeric layer and obtained approximate solutions describing different stages of swelling at large deformations of a polymeric matrix. We have identified the strain-stress state of the material caused by diffusion processes and analyzed its influence on the swelling kinetics.

It has been found that a non-stationary boundary regime initiated by deformations arising in elastomer during swelling and increasing a thermodynamical compatibility of elastomer with a liquid is the main reason for swelling anomalies observed in the experiments. Anomalies of sorption kinetics turn out to be a typical phenomenon observable to one or another extent in elastic swelling materials.

The theory predicts the possibility for qualitatively different diffusion modes of free swelling. A particular mode is specified by a complex of mechanical, thermodynamical, and diffusion material properties. The results of analytical and numerical solutions for a plane elastomer layer show that the swelling process may be governed by three different laws resulting in the power, exponential, and blow-up swelling modes. Experimentally it has been determined that in the examined elastomers the swelling mode is governed by the power law. The existence of exponential and blow-up swelling modes in real materials is still an open question.

New methods have been proposed, which allow one to estimate the concentration dependence of liquid diffusion in elastomer and strain dependence of equilibrium swelling ratio under conditions of symmetric biaxial elastomer extension in terms of kinetic and strain curves of swelling.

### REFERENCES

1 A E Green, P M Naghdi, *Int. J. Eng. Sci.*, **3**, 231 (1965).
2 A E Green, T R Steel, *Int. J. Eng. Sci.*, **4**, 483 (1966).
3 K R Rajagopal, A S Wineman, MV Gandhi, *Int. J. Eng. Sci.*, **24**, 1453 (1986).
4 M V Gandhi, K R Rajagopal, AS Wineman, *Int. J. Eng. Sci.*, **25**, 1441 (1987).
5 E Ya Denisyuk, V V Tereshatov, *Appl. Mech. Tech. Phys.*, **38**, 913 (1997).
6 E Ya Denisyuk, V V Tereshatov, Vysokomol. soed., **A42**, 74 (2000) (in Russian).
7 P J Flory, **Principles of polymer chemistry**, *Cornell Univ. Press*, New York, 1953.
8 V A Vasilyev, Yu M Romanovskiy, V G Yahno, **Autowave Processes**, *Nauka,* Moscow, 1987 (in Russian).
9 P G De Gennes, **Scaling Concepts in Polymer Physics**, *Cornell Univ. Press*, Ithaca, 1980.
10 A A Samarskiy, V A Galaktionov, S P Kurdyumov, A P Mikhaylov, **Blow-up Modes in Problems for Quasilinear Parabolic Equations**, *Nauka*, Moscow, 1987 (in Russian).

## 6.2 EQUILIBRIUM SWELLING IN BINARY SOLVENTS

VASILIY V. TERESHATOV, VALERY YU. SENICHEV
**Institute of Technical Chemistry**
E.YA. DENISYUK
**Institute of Continuous Media Mechanics**
**Ural Branch of Russian Academy of Sciences, Perm, Russia**

Depending on the purposes and operating conditions of polymer material processing, the opposite demands to solubility of low-molecular-mass liquids in polymers exist. The products of polymer materials designed for use in contact with solvents should be stable against relative adsorption of these liquids. On the contrary, the well dissolving polymer solvents are necessary to produce the polymer films. The indispensable condition of creation of the plasticized polymer systems (for example, rubbers) is the high thermodynamic compatibility of plasticizers with the polymer basis of material.

Hence, it immediately follows the statement of the problem of regulation of thermodynamic compatibility of polymers with low-molecular-mass liquids in a wide range of its concentration in polymer material. The problem of compatibility of crosslinked elastomers with mixed plasticizers and volatile solvents is thus of special interest.

Depending on ratios between solubility parameters of the solvent 1, $\delta_1$, of the solvent 2, $\delta_2$, and polymer, $\delta_3$, solvents can be distinguished as "symmetric" liquids and "non-symmetric" ones. The non-symmetric liquids are defined as a mixture of two solvents of variable composition, solubility parameters $\delta_1$ and $\delta_2$ which are larger or smaller than solubility parameter of polymer ($\delta_2 > \delta_1 > \delta_3$, $\delta_3 > \delta_2 > \delta_1$). The symmetric liquid (SL) with relation to polymer is the mixture of two solvents, whose solubility parameter, $\delta_1$, is smaller, and parameter, $\delta_2$, is larger than the solubility parameter of polymer, $\delta_3$.

The dependence of equilibrium swelling on the non-symmetric liquid composition does not have a maximum, as a rule.[1] Research on swelling of crosslinked elastomers in SL is particularly interested in the regulation of thermodynamic compatibility of network polymers and binary liquids. Swelling in such liquids is characterized by the presence of maximum on the curve of dependence of network polymer equilibrium swelling and composition of a liquid phase.[2,3] The extreme swelling of crosslinked polymers of different polarity in SL was discussed elsewhere.[3] The following elastomers were used as samples: a crosslinked elastomer of ethylene-propylene rubber SCEPT-40 [$\delta_3 = 16\ (MJ/m^3)^{1/2}$, $(v_e/V_0)_x = 0.24\ kmol/m^3$], crosslinked polyester urethane, PEU, from copolymer of propylene oxide and trimethylol propane [$\delta_3 = 18.3\ (MJ/m^3)^{1/2}$, $(v_e/V_0)_x = 0.27\ kmol/m^3$], crosslinked polybutadiene urethane, PBU, from oligobutadiene diol [$\delta_3 = 17.8\ (MJ/m^3)^{1/2}$, $(v_e/V_0)_x = 0.07\ kmol/m^3$] and crosslinked elastomer of butadiene-nitrile rubber [$\delta_3 = 19\ (MJ/m^3)^{1/2}$, $(v_e/V_0)_x = 0.05\ kmol/m^3$].

The samples of crosslinked elastomers were swollen to equilibrium at 25°C in 11 SLs containing solvents of different polarity. The following regularities were established. With decrease in the solubility parameter value of component 1 (see Table 6.2.1) in SL ($\delta_1 < \delta_3$), the maximum value of equilibrium swelling, Q, shifts to the field of larger concentration of component 2 in the mixture (Figures 6.2.1 and 6.2.2). On the contrary, with decrease in the

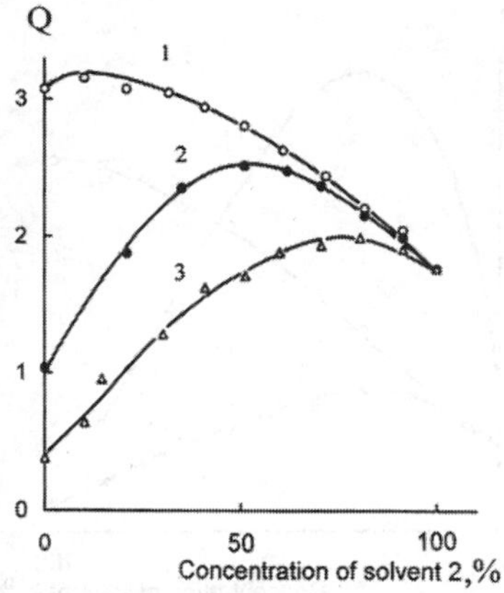


Figure 6.2.1. Dependence of equilibrium swelling of PEU on the acetone concentration in the mixtures: 1-toluene-acetone, 2-cyclohexane-acetone, 3-heptane-acetone. [Adapted, by permission, from V. V. Tereshatov, M. I. Balashova, A. I. Gemuev, **Prediction and regulating of properties of polymeric materials**, *Ural Branch of AS USSR Press*, Sverdlovsk, 1989, p. 3.]

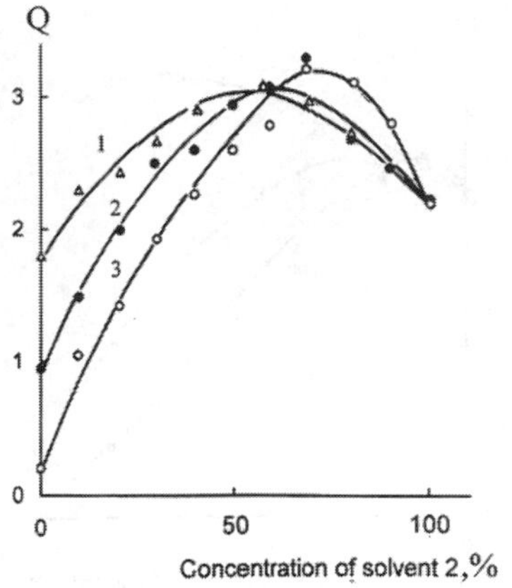


Figure 6.2.2. Dependence of equilibrium swelling of PBU on the DBP (component 2) concentration in the mixtures:1-DOS-DBP, 2-TO-DBP, 3- decane-DBP. [Adapted, by permission, from V. V. Tereshatov, M. I. Balashova, A. I. Gemuev, **Prediction and regulating of properties of polymeric materials**, *Ural Branch of AS USSR Press*, Sverdlovsk, 1989, p. 3.]

solubility parameter $\delta_2$ (see Table 6.2.1) of components 2 ($\delta_2 > \delta_3$), the maximum Q corresponds to composition of the liquid phase enriched by component 1 (Figure 6.2.3).

**Table 6.2.1. Characteristics of solvents and plasticizers at 298K. [Adapted, by permission, from V. V. Tereshatov, M. I. Balashova, A. I. Gemuev, Prediction and regulating of properties of polymeric materials, *Ural Branch of AS USSR Press*, Sverdlovsk, 1989, p. 3.]**

| Solvent/plasticizer | $\rho$, kg/m$^3$ | V×10$^6$, m$^3$ | $\delta$, (MJ/m$^3$)$^{1/2}$ |
|---|---|---|---|
| Cyclohexane | 779 | 109 | 16.8 |
| Heptane | 684 | 147 | 15.2 |
| Decane | 730 | 194 | 15.8 |
| Toluene | 862 | 106 | 18.2 |
| 1,4-Dioxane | 1034 | 86 | 20.5 |
| Acetone | 791 | 74 | 20.5 |
| Ethyl acetate | 901 | 98 | 18.6 |
| Amyl acetate | 938 | 148 | 17.3 |
| Dibutyl phthalate | 1045 | 266 | 19.0 |
| Dioctyl sebacate | 913 | 467 | 17.3 |
| Transformer oil | 890 | 296 | 16.0 |

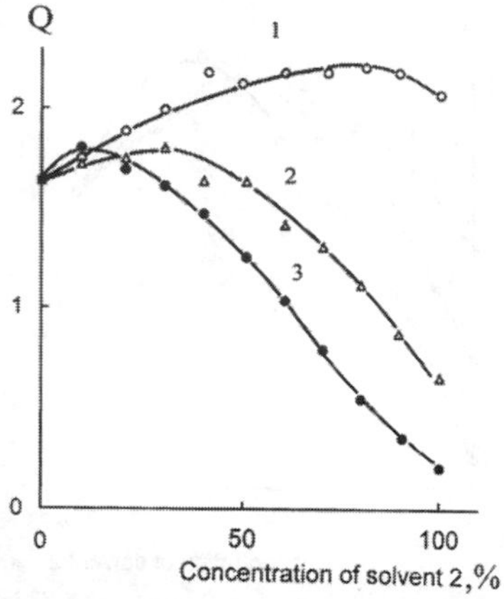


Figure 6.2.3. Dependence of equilibrium swelling of the crosslinked elastomer SCEPT-40 on the concentration of component 2 in the mixtures: 1-heptane-toluene, 2-heptane-amyl acetate, 3-heptane-ethyl acetate. [Adapted, by permission, from V. V. Tereshatov, M. I. Balashova, A. I. Gemuev, **Prediction and regulating of properties of polymeric materials**, *Ural Branch of AS USSR Press*, Sverdlovsk, 1989, p. 3.]

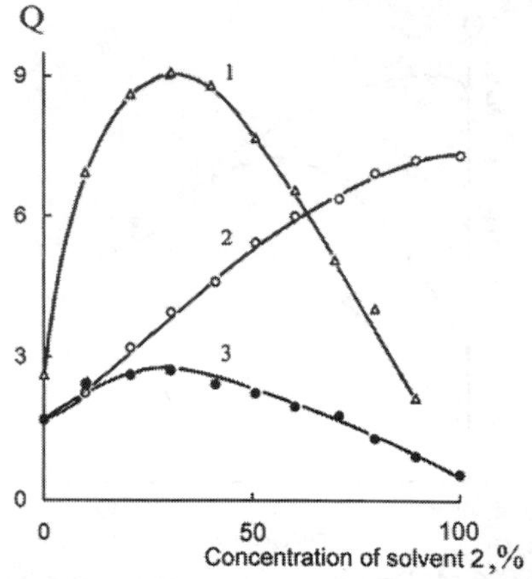


Figure 6.2.4. Dependence of equilibrium swelling of the crosslinked elastomer SCN-26 on the concentration of component 2 in the mixtures: 1-toluene-acetone, 2-ethyl acetate-dioxane, 3-ethyl acetate-acetone. [Adapted, by permission, from V. V. Tereshatov, M. I. Balashova, A. I. Gemuev, **Prediction and regulating of properties of polymeric materials**, *Ural Branch of AS USSR Press*, Sverdlovsk, 1989, p. 3.]

Neglecting the change of volume on mixing, the solubility parameter of the mixture of two liquids is represented by:

$$\delta_{12} = \delta_1 \varphi_1 + \delta_2 \varphi_2$$

where:

$\varphi_1$ and $\varphi_2$ volume fractions of components 1 and 2

More exact evaluation of the $\delta_{12}$ value is possible if the experimental data on enthalpy of mixing, $\Delta H$, of components of SL are taken into account:[4]

$$\delta_{12} = \left(\delta_1^2 \varphi_1 + \delta_2^2 \varphi_2 - \Delta H / V_{12}\right)^{1/2}$$

With a change in $\delta_1$ and $\delta_2$ parameters, the SL composition has the maximum equilibrium swelling which corresponds to shifts in the field of composition of the liquid phase. The $\delta_{12}$ parameter is close or equal to the value of the solubility parameter of polymer. Such a simplified approach to the extreme swelling of polymers in liquid mixtures frequently works very well in practice. If there is a maximum on the curve of swelling in SL, then the swelling has an extreme character (11 cases out of 12) (Figures 6.2.1-6.2.4).

The parameter of interaction, $\chi_{123}$, between polymer and a two-component liquid can be used as a co-solvency criterion for linear polymers (or criterion of extreme swelling), more general, than the equality ($\delta_{12} = \delta_3$):[5]

$$\chi_{123} = \chi_{13}\varphi_1 + \chi_{23}\varphi_2 - \varphi_1\varphi_2\chi_{12} \qquad [6.2.1]$$

where:

$\chi_{13}$ and $\chi_{23}$ parameters of interaction of components 1 and 2 with polymer, correspondingly
$\chi_{12}$ parameter of interaction of components 1and 2 of liquid mixture

In the equation obtained from the fundamental work by Scott,[5] mixed solvent is represented as "a uniform liquid" with the variable thermodynamic parameters depending on composition.

If the $\chi_{123}$ value is considered as a criterion of existence of a maximum of equilibrium swelling of polymer in the mixed solvent, a maximum of Q should correspond to the minimum of $\chi_{123}$. For practical use of Eq. [6.2.2] it is necessary to know parameters $\chi_{13}$, $\chi_{23}$, and $\chi_{12}$. The values $\chi_{13}$ and $\chi_{23}$ can be determined by the Flory-Rehner equation, with data on swelling of a crosslinked elastomer in individual solvents 1 and 2. The evaluation of the $\chi_{12}$ value can be carried out with use of results of the experimental evaluation of vapor pressure, viscosity and other characteristics of a binary mixture.[6,7]

To raise the forecasting efficiency of prediction force of such criterion as $\chi_{123}$ minimum, the amount of performance parameters determined experimentally must be reduced. For this purpose, the following expression for the quality criterion of the mixed solvent (the analogy with expression of the Flory-Huggins parameter for individual solvent-polymer system) is used:

$$\chi_{123} = \chi_{123}^{s} + \frac{V_{12}}{RT}(\delta_3 - \delta_{12})^2 \quad V_{12} = V_1 x_1 + V_2 x_2 \qquad [6.2.2]$$

where:

$V_{12}$ molar volume of the liquid mixture
$V_1$, $V_2$ molar volumes of components 1 and 2
$x_1$, $x_2$ molar fractions of components 1 and 2 in the solvent
R universal gas constant
T temperature, K.
$\chi_{123}^{s}$ constant, equal to 0.34

The values of entropy components of parameters $\chi_{13}^{s}$ and $\chi_{23}^{s}$ (essential for quality prediction)[8] were taken into account. Using the real values of entropy components of interaction parameters $\chi_{13}^{s}$ and $\chi_{23}^{s}$, we have:

$$\chi_{13}^{s} = \chi_{13} - \frac{V_1}{RT}(\delta_3 - \delta_1)^2 \qquad [6.2.3]$$

$$\chi_{23}^{s} = \chi_{23} - \frac{V_2}{RT}(\delta_3 - \delta_2)^2 \qquad [6.2.4]$$

quality criterion of mixed solvent is represented by:[3]

$$\chi_{123} = \chi_{13}^{s}\varphi_1 + \chi_{23}^{s}\varphi_2 + \frac{V_{12}}{RT}(\delta_3 - \delta_{12})^2 \qquad [6.2.5]$$

The values of parameters $\chi_{13}$ and $\chi_{23}$ were calculated from the Flory-Rehner equation, using data on swelling of crosslinked polymer in individual solvents, the values $\chi_{13}^{s}$, $\chi_{23}^{s}$, and $\chi_{123}$ were estimated from Eqs. [6.2.3-6.2.5]. Then equilibrium swelling of elastomer in SL was calculated from the equation similar to Flory-Rehner equation, considering the mixture as a uniform liquid with parameters $\chi_{123}$ and $V_{12}$ variable in composition. The calculated ratios of components of mixtures, at which the extreme swelling of elastomers is expected, are given in Table 6.2.2.

**Table 6.2.2. Position of a maximum on the swelling curve of crosslinked elastomers in binary mixtures. [Adapted, by permission, from V. V. Tereshatov, M. I. Balashova, A. I. Gemuev, Prediction and regulating of properties of polymeric materials, *Ural Branch of AS USSR Press*, Sverdlovsk, 1989, p. 3.]**

| Elastomer | Binary mixture | Components ratio, wt% | |
|---|---|---|---|
| | | Calculation | Experiment |
| PEU | toluene-acetone | 90/10 | 90/10 |
| | cyclohexanone-acetone | 40/60 | 50/50 |
| | heptane-acetone | 30/70 | 25/25 |
| PBU | decane-DBP | 30/70 | 30/70 |
| | TO-DBP | 40/60 | 40/60 |
| | DOS-DBP | 60/40 | 50/50 |
| SCN-26 | ethyl acetate-1,4-dioxane | - | - |
| | ethyl acetate-acetone | 70/30 | 70/30 |
| | toluene-acetone | 70/30 | 70/30 |

Results of calculation of the liquid phase composition at maximum swelling of elastomer correlate with the experimental data (see Table 6.2.2). The approach predicts the existence of an extremum on the swelling curve. This increases the forecasting efficiency of the prediction.

The application of "approximation of the uniform liquid" (AUL) for prediction of extreme swelling of crosslinked elastomers is proven under condition of coincidence of composition of a two-component solvent in a liquid phase and in the swollen elastomer. Results of study of total and selective sorption by crosslinked elastomers of components of SLs are given below.[3]

Crosslinked PBU [$(v_e/V_0)$ = 0.20 kmol/m$^3$] and crosslinked elastomer of butadiene-nitrile rubber SCN-26 [$(v_e/V_0)$ = 0.07 kmol/m$^3$] were used in this study. The tests were carried out at 25±0.10°C in the following mixtures: n-nonane-tributyl phosphate (TBP), n-hexane-dibutyl phthalate (DBP), n-hexane-dibutyl maleate (DBM), and dioctyl sebacate (DOS)-diethyl phthalate (DEP). A crosslinked elastomer SCN-26 was immersed to equilibrate in amyl acetate-dimethyl phthalate mixture. Liquid phase composition was in range of 5 to 10%. A sol-fraction of samples (plates of $0.9\times10^{-2}$ m diameter and in $0.3\times10^{-2}$ m thickness) was preliminary extracted with toluene.

For the high accuracy of the analysis of binary solvent composition, a volatile solvent, hexane, was used as a component of SL in most experiments. Following the attainment of equilibrium swelling, the samples were taken out of SL and held in air until the full evaporation of hexane, that was controlled by constant mass of sample. A content of a nonvolatile component of SL in elastomer was determined by the difference between the amount of liq-

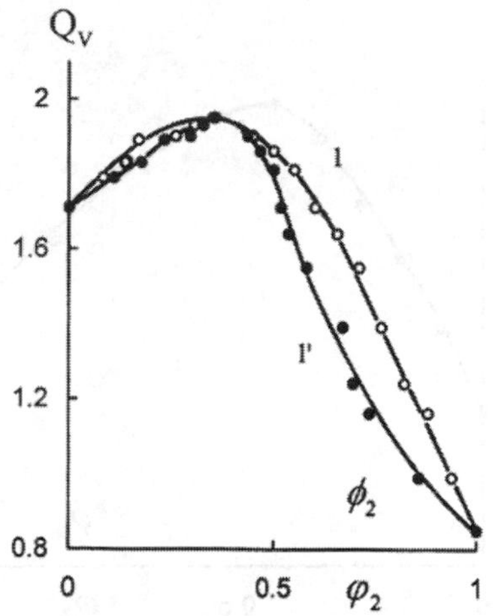


Figure 6.2.5. Dependence of equilibrium swelling of PBU on the $\varphi_2$ value for DEP in the liquid phase (1) and the $\phi_2$ value inside the gel (1') swollen in the mixture DOS-DEP.

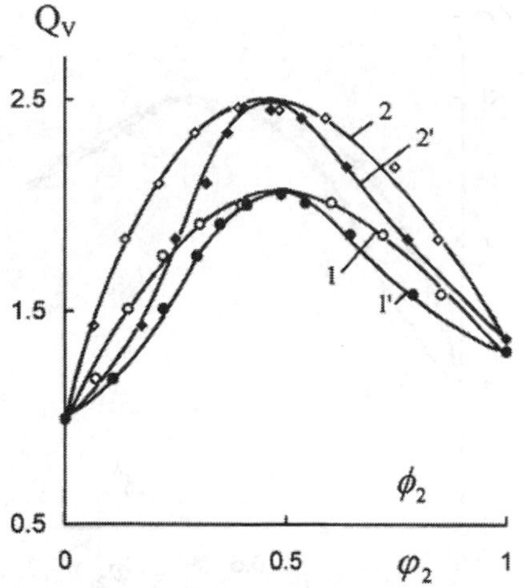


Figure 6.2.6. Dependence of equilibrium swelling of PBU on the $\varphi_2$ value for DEP in the liquid phase (curves 1 and 2) and the $\phi_2$ value inside the swollen gel (curves 1' and 2') in the mixtures: 1,1'-hexane(1)-DBM(2), 2,2'- hexane(1)-DBP(2).

uid in the swollen sample and the amount of hexane evaporated. In the case of SL with non-volatile components (such as, DBP or DOS) the ratio of SL components in the swollen PBU was determined by the gas-liquid chromatography, for which purpose toluene extract was used.

The volume fractions of components 1 and 2 of SL in a liquid phase $\varphi_1$ and $\varphi_2$ were calculated on the basis of their molar ratio and densities $\rho_1$ and $\rho_2$.

Volume fraction of polymer, $\upsilon_3$, in the swollen sample was calculated from the equation:

$$\upsilon_3 = \frac{1}{Q_V + 1}$$

where:

$Q_V$ volume equilibrium swelling of elastomer in SL.

Volume fractions $\varphi_1$ and $\varphi_2$ of components 1 and 2 of low-molecular-mass liquids inside the swollen gel were determined from the equation:

$$\phi_i = \frac{\upsilon_i}{1 - \upsilon_3}, \quad (i = 12)$$

where:

$\phi_i$ volume fraction of i-component related to the total volume of its low-molecular-mass part (not to the total volume of three-component system)

The results of study of sorption of two-component liquids in PBU and in crosslinked elastomer SCN-26 are shown in Figures 6.2.5-6.2.8, as dependencies of $Q_V$ on the volume

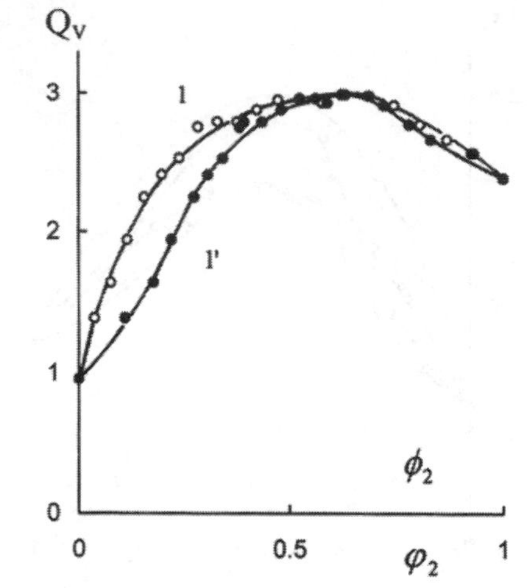


Figure 6.2.7. Dependence of equilibrium swelling of PBU on the $\varphi_2$ value for TBP in the liquid phase (1) and the $\phi_2$ value in the gel (1') swollen in the mixture: nonane-TBP.

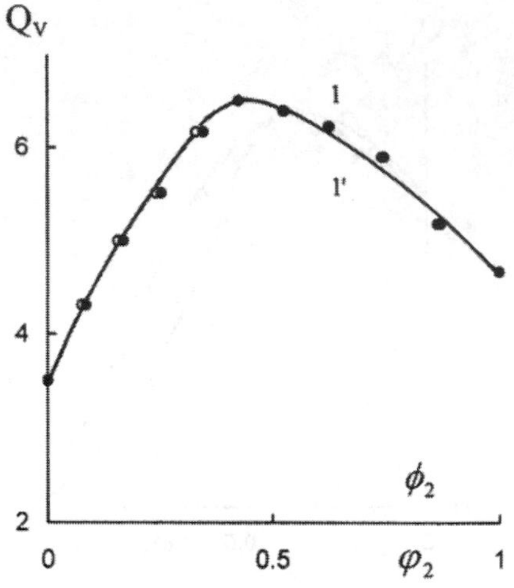


Figure 6.2.8. Dependence of equilibrium swelling of crosslinked elastomer SCN-26 on the $\varphi_2$ value for DMP in the liquid phase ( 1) and the $\phi_2$ value in the gel (1'), swollen in the mixture: amyl acetate-DMP.

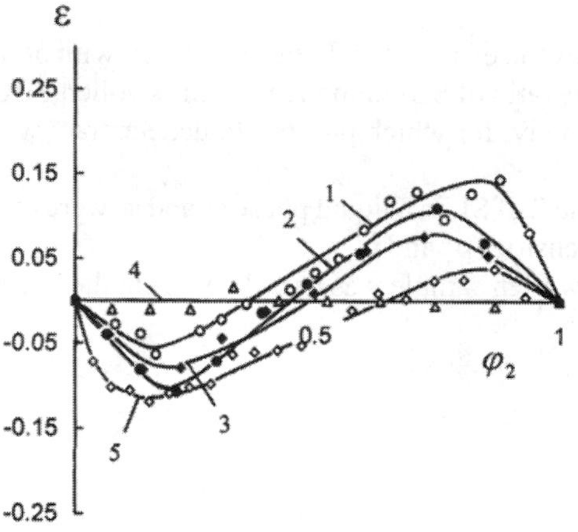


Figure 6.2.9. Experimental dependence of the preferential sorption ε on the volume fraction $\varphi_2$ of component 2 in the liquid phase: 1-DOS(1)-DEP(2)-PBU(3), 2-hexane(1)-DBP(2)-PBU(3), 3-hexane(1)-DBM(2)-PBU(3), 4-amyl acetate(1)-DMP(2)-SCN-26(3), 5-nonane(1)-TBP(2)-PBU(3).

fraction $\varphi_2$ of component 2 in a liquid phase and on the volume fraction $\phi_2$ of component 2 of SL that is a part of the swollen gel. The data vividly show that extremum swelling of crosslinked elastomers in SL can be observed in all investigated cases. At the maximum value of $Q_V$, the compositions of SL in the liquid phase and in the swollen elastomer practically coincide ($\varphi_2 \approx \phi_2$).

The total sorption can be determined by the total content of the mixed solvent in the swollen skin[9] or in the swollen elastic network (these results). The dependencies of $Q_V$ on $\varphi_2$ (Figures 6.2.5-6.2.8) reveal the influence of liquid phase composition on the total sorption. The total sorption of the binary solvent by polymer can also be measured by the value of the volume fraction of polymer in the swollen gel[9] because the value of $\upsilon_3$ is unequally related to the volume fraction of the absorbed liquid, $\upsilon_3 = 1-(\upsilon_1 + \upsilon_2)$. At the fixed total sorption ($Q_V$ or $\upsilon_3$ = const) the preferential sorption, ε, can be found from the following equation:

$$\varepsilon = \phi_1 - \varphi_1 = \varphi_2 - \phi_2$$

At the same values of $Q_v$ difference between coordinates on the abscissa axes of points of the curves 1-1', 2-2' (Figures 6.2.5-6.2.8) are equal $\varepsilon$ (Figure 6.2.9).

As expected,[10] preferential sorption was observed, with the essential distinction of molar volumes $V_1$ and $V_2$ of components of the mixed solvent (hexane-DBP, DOS-DEP). For swelling of the crosslinked elastomer SCN-26 in the mixture of components, having similar molar volumes $V_1$ and $V_2$ (e.g., amyl acetate-dimethyl phthalate) the preferential sorption of components of SL is practically absent. Influence of $V_1$ and $V_2$ and the influence of double interaction parameters on the sorption of binary liquids by crosslinked elastomers was examined by the method of mathematical experiment. Therewith the set of equations describing swelling of crosslinked elastomers in binary mixture, similar to the equations obtained by Bristow[6] from the Flory-Rehner theory[11] and from the work of Schulz and Flory,[12] were used:

$$\ln\varphi_1 + (1-l)\varphi_2 + \chi_{12}\varphi_2^2 = \ln\upsilon_1 + (1-l)\upsilon_2 + \upsilon_3 + \chi_{12}\upsilon_2^2 + \chi_{12}\upsilon_3^2 + $$

$$+(\chi_{12} + \chi_{13} - l\chi_{23})\upsilon_2\upsilon_3 + (\nu_e / V_3)V_1\left(\upsilon_3^{1/3} - \frac{2}{f}\upsilon_3\right) \qquad [6.2.6]$$

$$\ln\varphi_2 + \left(1-l^{-1}\right)\varphi_1 + l^{-1}\chi_{12}\varphi_1^2 = \ln\upsilon_2 + \left(1-l^{-1}\right)\upsilon_1 + \upsilon_3 + l^{-1}$$

$$\left[\chi_{12}\upsilon_1^2 + l\chi_{23}\upsilon_3^2 + (\chi_{12} + l\chi_{23} - \chi_{13})\upsilon_1\upsilon_3\right] + (\nu_e / V_3)V_2\left[\upsilon_3^{1/3} - \frac{2}{f}\upsilon_3\right] \qquad [6.2.7]$$

where:

| | |
|---|---|
| l | $=V_1/V_2$ |
| f | functionality of a network |

The analysis of calculations from Eqs. [6.2.6] and [6.2.7] has shown that if $V_1 < V_2$, the preferential sorption is promoted by difference in parameters of interaction, when $\chi_{12} > \chi_{13}$. The greater is $\chi_{12}$ value at $V_1 \neq V_2$ the more likely preferential sorption takes place with all other parameters being equal. The same applies to the diluted polymer solutions.

To improve calculations of the preferential sorption in the diluted solutions of polymers, the correction of Flory's theory is given in works[9,13-15] by introduction of the parameter of three-component interaction, $\chi_T$, into the expression for free energy of mixing, and substitution of $\chi_T$ by $q_T$.[14] This approach is an essential advancement in the analysis of a sorption of two-component liquids by polymer. On the other hand, the increase in the number of experimental parameters complicates the task of prediction of the preferential sorption.

In swelling crosslinked elastomers, the preferential sorption corresponds to the maximum $Q_V$ value and compositions in a swollen polymer and a liquid phase practically coincide (Figures 6.2.5-6.2.8). This observation can be successfully used for an approximate evaluation of the preferential sorption.

Assuming that $\varphi_1 = \phi_1$ and $\varphi_2 = \phi_2$ at extremum of the total sorption ($\phi_3 = \phi_3^{min}$), AUL[5] can be used to calculate the effective value of $\chi_{12}^e$ from the expression for $\chi_{123}$:[16]

$$\chi_{12}^e = \frac{\chi_{13}}{\varphi_2^m} + \frac{\chi_{23}V_1}{\varphi_1^m V_2} - \frac{\chi_{123}^m V_1}{V_{12}^m \varphi_1^m \varphi_2^m} \qquad [6.2.8]$$

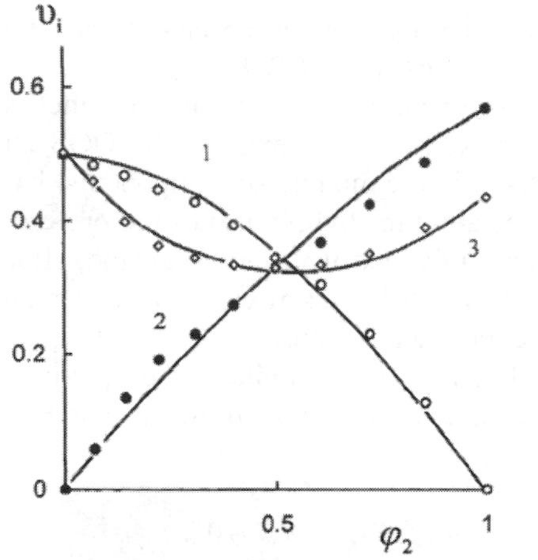


Figure 6.2.10. Experimental data and calculated dependence of equilibrium fraction of components $\upsilon_1$ (1), $\upsilon_2$ (2), and $\upsilon_3$ (3) in PBU sample swollen in the hexane-DBP mixture on the $\varphi_2$ value for DBP in the liquid phase: lines - calculation, points - experimental.

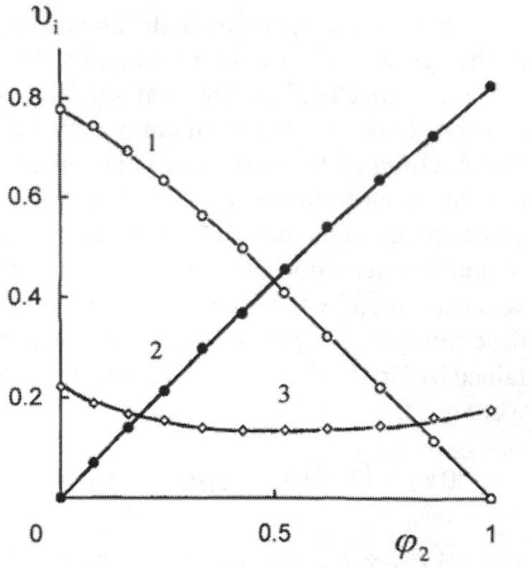


Figure 6.2.11. Dependence of equilibrium fraction of components $\upsilon_1$ (1), $\upsilon_2$ (2), and $\upsilon_3$ (3) in SCN-26 sample swollen in the amyl acetate-DMP mixture on the $\varphi_2$ value for DMP in the liquid phase: lines - calculation, points - experimental.

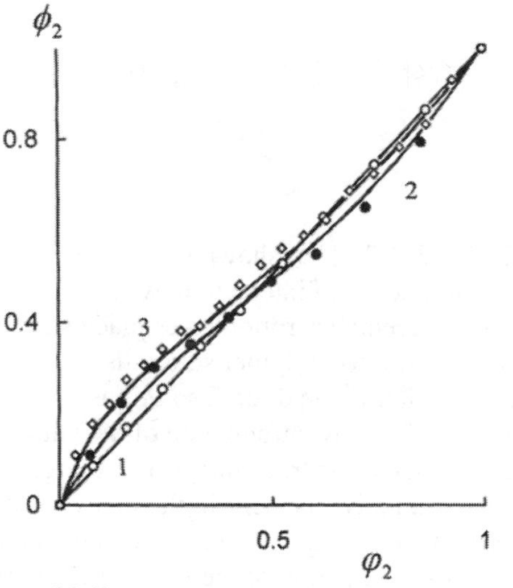


Figure 6.2.12. Dependence of equilibrium $\phi_2$ value on the $\varphi_2$ value in the liquid phase: 1-amyl acetate (1)-DMP (2)-SCN-26(3), 2-hexane(1)-DBM(2)-PBU (3), 3-nonane(1)-TBP(2)- PBU (3); lines - calculation, points - experimental.

where:

| | |
|---|---|
| m | the index for the maximum of sorption. |
| $\chi_{12}^{e}$ | the value, calculated from the equation [6.2.8], substitutes $\chi_{12}$ from the Eqs. [6.2.6] and [6.2.7] |

The results of calculations are presented as dependencies of volume fractions, $\upsilon_1, \upsilon_2$, and $\upsilon_3$ of the triple system components (swollen polymer) vs. the volume fraction $\varphi_2$ (or $\varphi_1$) of the corresponding components of the liquid phase (Figures 6.2.10 and 6.2.11). For practical purposes it is convenient to represent the preferential sorption as the dependence of equilibrium composition of a binary liquid (a part of the swollen gel) vs. composition of the liquid phase. These calculated dependencies (solid lines) and experimental (points) data for three systems are given in Figure 6.2.12. The results of calculations based on AUL, are in the satisfactory agreement with experimental data.

Hence the experimentally determined equality of concentrations of SL components in the swollen gel and in the liquid phase allows one to predict composition of liquid, whereby polymer swelling is at its maximum, and the preferential sorption of components of SL. To refine dependencies of $\phi_1$ on $\varphi_1$ or $\phi_2$ on $\varphi_2$, correction can be used, for example, minimization of the square-law deviation of calculated and experimental data on the total sorption of SL by polymer. Thus it is necessary to account for proximity of the compositions of solvent in the swollen gel and the liquid phase.

## REFERENCES


1 A Horta, *Makromol. Chem.*, **182**, 1705 (1981).
2 V V Tereshatov, V Yu Senichev, A I Gemuev, *Vysokomol. Soedin.*, **32B**, 422 (1990).
3 V V Tereshatov, M I Balashova, A I Gemuev, **Prediction and regulating of properties of polymeric materials**, *Ural Branch. of AS USSR Press*, Sverdlovsk, 1989, p. 3.
4 A A Askadsky, Yu I Matveev, **Chemical structure and physical properties of polymers**, *Chemistry*, Moscow, 1983.
5 R L Scott, *J. Chem. Phys.*, **17**, 268 (1949).
6 C M Bristow, *Trans. Faraday Soc.*, **55**, 1246 (1959).
7 A Dondos, D Patterson, *J. Polym. Sci.*, *A-2*, **5**, 230 (1967).
8 L N Mizerovsky, L N Vasnyatskaya, G M Smurova, *Vysokomol. Soedin.*, **29A**, 1512 (1987).
9 J Pouchly, A Zivny, *J. Polym. Sci.*, **10**, 1481 (1972).
10 W R Krigbaum, D K Carpenter, *J. Polym. Sci.*, **14**, 241 (1954).
11 P J Flory, J Rehner, *J. Chem. Phys.*, **11**, 521 (1943).
12 A R Shulz, P J Flory, *J. Polym. Sci.*, **15**, 231 (1955).
13 J Pouchly, A Zivny, *J. Polym. Sci.*, **23**, 245 (1968).
14 J Pouchly, A Zivny, *Makromol. Chem.*, **183**, 3019 (1982).
15 R M Msegosa, M R Comez-Anton, A Horta, *Makromol. Chem.*, **187**, 163 (1986).
16 J Scanlan, *J. Appl. Polym. Sci.*, **9**, 241 (1965).


## 6.3 SWELLING DATA ON CROSSLINKED POLYMERS IN SOLVENTS


Vasiliy V. Tereshatov, Valery Yu. Senichev

**Institute of Technical Chemistry**

**Ural Branch of Russian Academy of Sciences, Perm, Russia**


Data on equilibrium swelling of selected crosslinked elastomers in different solvents are presented in Table 6.3.2. All data were obtained in the author's laboratory during the last 5 years. The network density values for these elastomers are given in Table 6.3.1.

The data can be used for the quantitative evaluation of thermodynamic compatibility of solvents with elastomers and for calculation of $\Delta Z$, the thermodynamic interaction parameter. These data can be recalculated to the network density distinguished by the network density value, given in Table 6.3.1, using Eq. [4.2.9]. Network density values can be used for the calculation of the interaction parameter. It should be noted that each rubber has specific ratio for the equilibrium swelling values in various solvents. This can help to identify rubber in polymeric material.

Temperature influences swelling (see Subchapter 4.2). Equilibrium swelling data at four different temperatures are given in Table 6.3.3.

Polymer samples were made from industrial rubbers. The following polyols were used for preparation of polyurethanes: butadiene diol (M~2000) for PBU, polyoxybutylene glycol (M~1000) for SCU-PFL, polyoxypropylene triol (M~5000) for Laprol 5003, polydiethylene glycol adipate (M~2000) for P-9A and polydiethylene glycol adipate (M~800) for PDA. Polyurethanes from these polyols were synthesized by reaction with 2,4-toluylene diisocyanate with addition of crosslinking agent - trimetylol propane, PDUE, synthesized by reaction of oligobutadiene isoprene diol (M~4500) with double excess of 2,4-toluylene diisocyanate, followed by the reaction with glycidol.

**Table 6.3.1. Network density values for tested elastomers (data obtained from the elasticity modulus of samples swollen in good solvents)**

| Elastomer | Trade mark | Network density, $(kmol/m^3)\times 10^2$ |
|---|---|---|
| Silicone rubber | SCT | 6.0 |
| Butyl rubber | BR | 8.9 |
| Polybutadiene rubber | SCDL | 40.5 |
| Ethylene-propylene rubber | SCEPT | 2.7 |
| Isoprene rubber | SCI-NL | 8.9 |
| Butadiene-nitrile rubber | SCN-26 | 5.1 |
| Polydiene-urethane-epoxide | PDUE | 16.0 |
| Polybutadiene urethane | PBU | 20.8 |
| Polyoxybutylene glycol urethane | PFU | 10.0 |
| Polyoxypropylene glycol urethane | Laprol 5003 | 43.0 |
| Polydiethylene glycol adipate urethane | P-9A | 6.4 |
| Polydiethylene glycol adipate urethane | PDA | 9.6 |

**Table 6.3.2 Equilibrium swelling data (wt%) at 25°C**

| Solvent\elastomer | SCT | BR | SCDL | SCEPT | SCI -NL | SCN -26 | PDUE | PBU | SCU -PFL | Laprol | P-9A | PDA |
|---|---|---|---|---|---|---|---|---|---|---|---|---|
| Hydrocarbons | | | | | | | | | | | | |
| Pentane | 355 | | 75 | | 214 | | | | | | | |
| Hexane | 380 | 305 | 105 | 586 | 300 | | 122 | 78 | 16 | 44 | 4 | |
| Heptane | 367 | 357 | | 881 | | 10 | 137 | 84 | 16 | 42 | 0 | |
| Isooctane | 380 | 341 | | | | | 101 | | | 32 | 0 | |
| Decane | | 391 | | 1020 | | 7 | 139 | | | 28 | 1 | |
| Cyclohexane | | 677 | | | | 155 | 315 | | 39 | | | 3 |
| Transformer oil | 47 | | 230 | 1580 | 600 | 24 | 238 | | 3 | 53 | 4 | 0.5 |
| Toluene | | 384 | | 1294 | | 435 | 492 | 356 | 134 | 315 | 77 | 31 |
| Benzene | | 233 | | 511 | | 506 | 479 | 359 | 158 | 340 | 175 | 73 |
| o-Xylene | | 515 | | | | | 496 | 356 | 122 | 288 | 53 | 42 |
| Ethers | | | | | | | | | | | | |
| Diamyl | 310 | | 220 | 889 | 584 | 20 | 248 | | | 87 | 2 | |

| Solvent\elastomer | SCT | BR | SCDL | SCEPT | SCI -NL | SCN -26 | PDUE | PBU | SCU -PFL | Laprol | P-9A | PDA |
|---|---|---|---|---|---|---|---|---|---|---|---|---|
| Diisoamyl | 333 | | | | | | | | 33 | | | 3 |
| Didecyl | | | | 791 | | 7 | 168 | | | 22 | 2 | |
| Dioctyl | 120 | 428 | 180 | 896 | 702 | 11 | 203 | | 6 | 41 | 0.1 | 1 |
| Diethyl of diethylene glycol | 85 | | 156 | | 242 | | | | | 251 | | 54 |
| 1,4-Dioxane | | 23 | | | | | | 264 | 178 | | 782 | |
| Tetrahydrofuran | | | | | | | | 48 | 192 | | 848 | |
| Esters of monoacids | | | | | | | | | | | | |
| Heptyl propionate | 260 | 236 | 304 | 382 | | 287 | | | | | | |
| Ethyl acetate | | 20 | | | | | 170 | | 133 | | 213 | 88 |
| Butyl acetate | | 55 | | 55 | | 562 | 317 | 251 | 121 | | 64 | 40 |
| Isobutyl acetate | | 48 | | | | | 267 | 210 | | 238 | 48 | 35 |
| Amyl acetate | | 82 | | | | | 343 | | | 249 | 36 | |
| Isoamyl acetate | | 74 | | | | | | 236 | 100 | | | 22 |
| Heptyl acetate | | | | 213 | | | | 545 | 100 | | | 6 |
| Methyl capronate | | | | | | | 369 | 285 | 115 | 260 | 36 | 26 |
| Isobutyl isobuturate | | | | | | | 328 | 276 | 64 | | 6 | 6 |
| Esters of multifunctional acids | | | | | | | | | | | | |
| Dihexyl oxalate | | | | 28 | | 350 | | | 73 | 166 | 11 | 9 |
| Diethyl phthalate | | | | 6 | | 837 | 47 | 93 | 165 | 272 | 496 | 126 |
| Dibutyl phthalate | | | | 9 | | 767 | 123 | 154 | 121 | 228 | 47 | 20 |
| Dihexyl phthalate | | | | | | 570 | | | | 187 | 9 | |
| Diheptyl phthalate | | | | | | | | | 37 | | | 3 |
| Dioctyl phthalate | 8 | | | 28 | | | 189 | | 29 | | | 1 |
| Dinonyl phthalate | | | | | | | 211 | 181 | 33 | | | 1.4 |
| Didecyl phthalate | | | | | | | | | 23 | | | 0.5 |
| Diamyl maleate | | | | | | 425 | 197 | | 71 | 232 | 19 | 14 |
| Dimethyl adipate | | | | | | 560 | | | | 246 | 716 | 6 |
| Diamyl adipate | | | | | | | 294 | 203 | 88 | | | |
| Dihexyl adipate | | | | | | 280 | | | | | 8 | |
| Dioctyl adipate | | | | | | | 295 | 199 | 42 | | 2 | 2 |
| Dipropyl sebacate | | | | 20 | | | | | 126 | | 12 | 0.6 |

| Solvent\elastomer | SCT | BR | SCDL | SCEPT | SCI -NL | SCN -26 | PDUE | PBU | SCU -PFL | Laprol | P-9A | PDA |
|---|---|---|---|---|---|---|---|---|---|---|---|---|
| Diamyl sebacate | | | | | | | | | 70 | | | 2 |
| Diheptyl sebacate | | | | 60 | | | | | | 118 | 3 | |
| Dioctyl sebacate | 13 | | 160 | 127 | 445 | 39 | 261 | 162 | 5.5 | 74 | 0.4 | 0.7 |
| Triacetyne | | | | | | 34 | | 15 | 28 | | 489 | 129 |
| Tricresyl phosphate | | | | | | 800 | 45 | | 140 | | | 80 |
| Tributyl phosphate | | | | | | | 300 | 244 | 238 | 256 | 93 | 105 |
| Ketones | | | | | | | | | | | | |
| Cyclohexanone | | | | | | | 577 | | | 327 | 523 | |
| Acetone | | | | 7 | | | 44 | | 88 | | 218 | 99 |
| Alcohols | | | | | | | | | | | | |
| Ethanol | | | | | | | | | 49 | | | 16 |
| Butanol | | | | | | | | 25 | 72 | | 7 | 12 |
| Pentanol | | | | | | | 28 | 33 | 66 | | 10 | 5 |
| Hexanol | | | | | | | 46 | 34 | 62 | | 9 | 4 |
| Heptanol | | | | | | | | 36 | 67 | | 19 | |
| Nonanol | | | | | | | | | 57 | | | 3 |
| Halogen compounds | | | | | | | | | | | | |
| $CCl_4$ | | | | 2830 | | 284 | 999 | | | 584 | 66 | |
| Chlorobenzene | | | | | | | | | 234 | | 277 | 114 |
| Chloroform | | | | | | | | 1565 | 572 | | | 616 |
| Fluorobenzene | | | | | | | | | 205 | | | 113 |
| Nitrogen compounds | | | | | | | | | | | | |
| Diethyl aniline | | | | | | | | 336 | 120 | | | 34 |
| Dimethylformamide | | | | | | | | 52 | | | 918 | |
| Nitrobenzene | | | | | | | 277 | 298 | 242 | 377 | 799 | 241 |
| Capronitrile | | | | | | 769 | 142 | | 115 | | 84 | |
| Acetonitrile | | | | | | 47 | 69 | 16 | | | 323 | |
| Aniline | | | | | | | | | 317 | | | 423 |

**Table 6.3.3 Equilibrium swelling data (wt%) at -35,-10, 25, and 50°C**

| Solvent\Elastomer | Laprol* | SCU-PFL | P-9A | PBU | SCN-26 |
|---|---|---|---|---|---|
| Isopropanol | 8,12,37,305 | 8,19,72,- | 8,10,18,28 | 3,5,18,28 | 20,23,26,41 |

| Solvent\Elastomer | Laprol* | SCU-PFL | P-9A | PBU | SCN-26 |
|---|---|---|---|---|---|
| Pentanol | 27,34,80,119 | 16,25,78,104 | 7,9,10,18 | 8,9,33,52 | 31,37,49,60 |
| Acetone | 80,366,381,- | 80,122,132,- | 48,73,218,- | 56,321,347,- | 162,254,269,382 |
| Ethyl acetate | 214,230,288,314 | 113,114,133,138 | 209,210,210,213 | 165,171,187,257 | 324,343,395,407 |
| Butyl acetate | 294,304,309,469 | 119,122,123,125 | 41,44,64,80 | 288,293,307,473 | 404,426,518,536 |
| Isobutyl acetate | 259,259,288,- | 89,92,100,105 | 23,28,48,57 | 210,212,223,348 | 309,326,357,362 |
| Amyl acetate | 315,309,315,426 | 124,136,144,155 | 16,18,36,44 | 285,297,298,364 | 343,347,358,359 |
| Tetrahydrofuran | 91,96,97,119 | 119,177,192,- | 838,840,848,859 | 45,51,57,80 | 145,150,166,- |
| o-Xylene | 363,368,389,471 | 117,121,122,137 | 24,25,53,60 | 328,357,369,375 | 428,440,405,406 |
| Chlorobenzene | 522,537,562,- | 241,242,242,246 | 238,244,277,333 | 517,528,559,698 | 967,973,985,1256 |
| Acetonitrile | 19,21,28,33 | 23,26,39,41 | 254,263,323,362 | 6,10,16,27 | 47,47,47,82 |
| Hexane | 50,52,61,75 | 2,3,9,14 | 4,4,4,28 | 51,56,63,80 | - |
| Toluene | 360,366,380, 397 | 120,122,134,149 | 50,57,77,85 | 331,339,372,415 | - |

*$(v_e/V) = 0.149$ kmol/m$^3$. Network density values for other elastomers correspond to Table 6.3.2

## 6.4 INFLUENCE OF STRUCTURE ON EQUILIBRIUM SWELLING

Vasiliy V. Tereshatov, Valery Yu. Senichev
**Institute of Technical Chemistry**
**Ural Branch of Russian Academy of Sciences, Perm, Russia**

Swelling of single-phase elastomers is, other parameters being equal, limited by the chemical network. Microphase separation of hard and soft blocks can essentially influence swelling of block-copolymers. Hard domains formed in this process are knots of physical network, which can be resistant to action of solvents.[1-4] Swelling of polyurethane block-copolymers with urethane-urea hard segments is investigated. The maximum values of swelling, $Q^{max}$, of segmented polyurethane (prepolymer of oligopropylene diol with functional isocyanate groups) cured by 4,4'-methylene-bis-o-chloroaniline (MOCA) are given in Tables 6.4.1 and 6.4.2. The prepolymer Vibratane B 600 was obtained by the reaction of oligopropylene diol with 2,4-toluylene diisocyanate. The swelling experiments were carried out in four groups of solvents of different polarity and chemical structure.[5] The effective molecular mass of elastically active chains, $M_c$, between network crosslinks was estimated from the Flory-Rehner equation.[6] The interaction parameter of solvent with polymer was determined by calculation.

In the first variant of calculation, a classical method based on the solubility parameter concept, was used with application of Bristow and Watson's semi-empirical relationship for $\chi_1$:[7]

$$\chi_1 = \chi_1^S + \left(\frac{V_1}{RT}\right)(\delta_1 - \delta_p)^2 \qquad [6.4.1]$$

where:

| | |
|---|---|
| $\chi_1^S$ | a lattice constant whose value can be taken as 0.34 |
| $V_1$ | the molar volume of the solvent |
| $R_1$ | the gas constant |
| T | the absolute temperature |
| $\delta_1$ and $\delta_p$ | the solubility parameters of the solvent and the polymer, respectively |

However, to evaluate $\chi_1$ from Eq. [6.4.1], we need accurate data for $\delta_1$.

In some cases the negative $M_c$ values are obtained (Tables 6.4.1, 6.4.2). The reason is that the real values of the entropy component of the $\chi_1$ parameter can strongly differ from the 0.34 value.

**Table 6.4.1. Characteristics of polyurethane-solvent systems at 25°C. [Adapted, by permission, from U.S. Aithal, T.M. Aminabhavi, R.H. Balundgi, and S.S. Shukla, *JMS - Rev. Macromol. Chem. Phys.*, 30C (1), 43 (1990).]**

| Penetrant | $Q^{max}$ | $\chi_1$ | $M_C$ |
|---|---|---|---|
| Monocyclic aromatics | | | |
| Benzene | 0.71 | 0.378 | 630 |
| Toluene | 0.602 | 0.454 | 790 |
| p-Xylene | 0.497 | 0.510 | 815 |
| Mesitylene | 0.402 | 0.531 | 704 |
| Chlorobenzene | 1.055 | 0.347 | 905 |
| Bromobenzene | 1.475 | 0.347 | 1248 |
| o-Dichlorobenzene | 1.314 | 0.357 | 1106 |
| Anisole | 0.803 | 0.367 | 821 |
| Nitrobenzene | 1.063 | 0.356 | 835 |
| Aliphatic alcohols | | | |
| Methanol | 0.249 | 1.925 | -90 |
| Ethanol | 0.334 | 0.349 | 172 |
| n-Propanol | 0.380 | 0.421 | 484 |
| Isopropanol | 0.238 | 0.352 | 249 |
| n-Butanol | 0.473 | 2.702 | -131 |
| 2-Butanol | 0.333 | 1.652 | -307 |
| 2-Methyl-1-propanol | 0.389 | 0.440 | 535 |
| Isoamyl alcohol | 0.414 | 0.357 | 398 |

**Table 6.4.2. Characteristics of polyurethane-solvent systems at 25°C. [Adapted, by permission, from U.S. Aithal, T.M. Aminabhavi, R.H. Balundgi, and S.S. Shukla, *JMS - Rev. Macromol. Chem. Phys.*, 30C (1), 43 (1990).]**

| Penetrant | $Q^{max}$ | $\chi_1$ | $M_C$ |
|---|---|---|---|
| Halogenated aliphatics | | | |
| Chloroform | 4.206 | 0.362 | 3839 |
| Bromoform | 5.583 | 0.435 | 4694 |
| 1,2-Dibromoethane | 1.855 | 0.412 | 770 |
| 1,3-Dibromopropane | 1.552 | 0.444 | 870 |
| Dichloromethane | 2.104 | 0.34 | 1179 |
| Trichloroethylene | 1.696 | 0.364 | 1098 |
| Tetrachlorethylene | 0.832 | 0.368 | 429 |
| 1,2-Dichlorethane | 1.273 | 0.345 | 728 |
| Carbon tetrachloride | 1.058 | 0.538 | 860 |
| 1,4-Dichlorobutane | 0.775 | 0.348 | 626 |
| 1,1,2,2- Tetrachloroethane | 5.214 | 0.34 | 6179 |
| Miscellaneous liquids | | | |
| Methyl acetate | 0.494 | 0.341 | 370 |
| Ethyl acetate | 0.509 | 0.422 | 513 |
| Ethyl benzoate | 0.907 | 0.34 | 1111 |
| Methyl ethyl ketone | 1.261 | 0.364 | 2088 |
| Tetrahydrofuran | 2.915 | 0.39 | 2890 |
| 1,4-Dioxane | 2.267 | 0.376 | 2966 |
| DMF | 1.687 | 1.093 | -1184 |
| DMSO | 0.890 | 1.034 | -975 |
| Acetonitrile | 0.184 | 0.772 | 182 |
| Nitromethane | 0.302 | 1.160 | 5734 |
| Nitroethane | 0.463 | 0.578 | 382 |
| n-Hexane | 0.069 | 1.62 | 2192 |
| Cyclohexane | 0.176 | 0.753 | 287 |
| Benzyl alcohol | 4.221 | | |

In the second variant of evaluation of $M_c$, data on the temperature dependence of volume fraction of polymer, $\varphi_2$, in the swollen gel were used. Results of calculations of $M_c$ from the Flory-Rehner equation in some cases also gave negative values. The evaluation of

$M_c$ in the framework of this approach is not an independent way, and $M_c$ is an adjustment parameter, as is the parameter $\chi_1$ of interaction between solvent and polymer.

In the last decades of evaluation of the physical network density of SPU was done on samples swollen to equilibrium in two solvents.[2,9,10] Swelling of SPU in toluene practically does not affect hard domains.[9,11] Swelling of SPU (based on oligoethers diol) in a tributyl phosphate (a strong acceptor of protons) results in full destruction of the physical network with hard domains.[2] The effective network density was evaluated for samples swollen to equilibrium in toluene according to the Cluff-Gladding method.[12] Samples were swollen to equilibrium in TBP and the density of the physical network was determined from equation:[2,10]

$$(v_e / V_0)_{dx} - (v_e / V_0)_x = (v_e / V_0)_d \qquad [6.4.2]$$

As the result of unequal influence of solvents on the physical network of SPU, the values of effective density of networks calculated for the "dry" cut sample can essentially differ. The examples of such influence of solvents are given in the work.[2] SPU samples with oligodiene soft segments and various concentration of urethane-urea hard blocks were swollen to equilibrium in toluene, methyl ethyl ketone (MEK), tetrahydrofuran (THF), 1,4-dioxane and TBP (experiments 1-6, 9, 10, 12). SPU based on oligoether (experiment 7), with urethane-urea hard segments, and crosslinked single-phase polyurethanes (PU) on the base of oligodiene prepolymer with functional isocyanate groups, cured by oligoether triols (experiments 8, 11) were also used. The effective network densities of materials swollen in these solvents were evaluated by the Cluff-Gladding method through the elasticity equilibrium modulus. The data are given in Table 6.4.3 per unit of initial volume.

The data shows that the lowest values of the network density are obtained for samples swollen to equilibrium in TBP. Only TBP completely breaks down domains of hard blocks. If solvents which are acceptors of protons (MEK, THF, 1,4-dioxane) are used in swelling experiment an intermediate values are obtained between those for toluene and TBP.

The network densities of SPU (experiments 8 and 11) obtained for samples swollen in toluene, TBP, 1,4-dioxane and THF coincide (Table 6.4.3).[2]

**Table 6.4.3. Results of the evaluation of equilibrium swelling and network parameters of PUE. [Adapted, by permission, from E.N. Tereshatova, V.V. Tereshatov, V.P. Begishev, and M.A. Makarova, *Vysokomol. Soed.*, 34B, 22 (1992).]**

| # | ρ kg/m$^3$ | Qv | $(v_e/V_0)$ | Qv | $(v_e/V_0)$ | Qv | $(v_e/V_0)$ | Qv | $(v_e/V_0)$ | Qv | $(v_e/V_0)$ |
|---|---|---|---|---|---|---|---|---|---|---|---|
| | | Toluene | | THF | | MEK | | 1,4-Dioxane | | TBP | |
| 1 | 996 | 1.56 | 1.03 | 0.46 | 0.77 | 1.08 | 0.41 | 3.76 | 0.12 | 5.12 | 0.06 |
| 2 | 986 | 2.09 | 0.53 | 0.52 | 0.43 | 1.40 | 0.23 | 6.22 | 0.04 | 11.15 | 0.02 |
| 3 | 999 | 1.74 | 0.77 | 0.49 | 0.73 | 1.11 | 0.38 | 7.88 | 0.03 | 18.48 | 0.01 |
| 4 | 1001 | 2.41 | 0.44 | 0.49 | | 1.75 | 0.13 | 14.3 | 0.02 | ∞ | |
| 5 | 1003 | 4.46 | 0.15 | 0.47 | | 3.29 | | ∞ | | ∞ | |
| 6 | 984 | 2.63 | 0.40 | 0.57 | 0.39 | 1.45 | 0.19 | 3.44 | 0.14 | 4.21 | 0.08 |
| 7 | 1140 | 0.86 | 1.83 | 2.06 | | 1.40 | 0.73 | 1.86 | 0.31 | 7.22 | 0.04 |

| # | ρ kg/m³ | $Q_V$ | $(v_e/V_0)$ | $Q_V$ | $(v_e/V_0)$ | $Q_V$ | $(v_e/V_0)$ | $Q_V$ | $(v_e/V_0)$ | $Q_V$ | $(v_e/V_0)$ |
|---|---|---|---|---|---|---|---|---|---|---|---|
| | | Toluene | | THF | | MEK | | 1,4-Dioxane | | TBP | |
| 8 | 972 | 2.42 | 0.35 | 1.84 | 0.36 | 1.01 | 0.54 | 2.08 | 0.34 | 1.48 | 0.35 |
| 9 | 979 | 2.26 | 0.39 | 1.72 | 0.37 | 0.98 | 0.54 | 1.97 | 0.35 | 1.62 | 0.31 |
| 10 | 990 | 2.00 | 0.63 | 0.47 | | 0.99 | 0.55 | 2.42 | 0.29 | 2.19 | 0.24 |
| 11 | 991 | 4.11 | 0.18 | 2.79 | 0.18 | 3.98 | 0.17 | 3.42 | 0.18 | 2.54 | 0.17 |
| 12 | 997 | 2.42 | 0.50 | 1.00 | 0.32 | 2.68 | 0.12 | 4.06 | 0.13 | 4.32 | 0.08 |

$(v_e/V_0)$, kmol/m³

To understand the restrictions to swelling of SPU caused by the physical network containing hard domains, the following experiments were carried out. Segmented polybutadiene urethane urea (PBUU) on the base of oligobutadiene diol urethane prepolymer with functional NCO-groups (M ≈ 2400), cured with MOCA, and SPU-10 based on prepolymer cured with the mixture of MOCA and oligopropylene triol (M ≈ 5000) were used. The chemical network densities of PBUU and SPU were 0.05 and 0.08 kmol/m³, respectively. The physical network density of initial sample, $(v_e/V_0)_d$, of PBUU was 0.99 kmol/m³ and of SPU-10 was 0.43 kmol/m³.

Samples of PBUU and SPU-10 were swollen to equilibrium in solvents of different polarity: dioctyl sebacate (DOS), dioctyl adipate (DOA), dihexyl phthalate (DHP), transformer oil (TM), nitrile of oleic acid (NOA), dibutyl carbitol formal (DBCF), and tributyl phosphate (TBP).

The values of equilibrium swelling of elastomers in these solvents, $Q_1$, (ratio of the solvent mass to the mass of the unswollen sample) are given in Table 6.4.4. After swelling in a given solvent, samples were swollen in toluene to equilibrium. The obtained data for swelling in toluene, $Q_V^T$, indicate that the physical networks of PBUU and SPU-10 do not change on swelling in most solvents. Equilibrium swelling in toluene of initial sample and the sample previously swollen in other solvents is practically identical. Several other observations were made from swelling experiments, including sequential application of different solvents. If preliminary disruption of the physical network of PBUU and SPU-10 by TBP occurs, swelling of these materials in toluene strongly increases. Similarly, samples previously swollen in TBP have higher equilibrium swelling, $Q_2$, when swollen in other solvents. The value of $Q_2$ is likely higher than equilibrium swelling $Q_1$ of initial sample (Table 6.4.4). $Q_2$ for PBUU is closer to the value of equilibrium swelling of a single-phase crosslinked polybutadiene urethane, PBU, having chemical network density, $(v_e/V_0)_x = 0.04$ kmol/m³. Thus, the dense physical network of polyurethane essentially limits the extent of equilibrium swelling in solvents, which do not breakdown the domain structure of a material.

**Table 6.4.4. Equilibrium swelling of SPBUU and SPU-10 (the initial sample and sample after breakdown of hard domains by TBP) and swelling of amorphous PBU at 25°C**

| Solvent | SPBUU | | | SPU-10 | | | PBU |
|---|---|---|---|---|---|---|---|
| | Initial structure | | After breakdown of domains | Initial structure | | After breakdown of domains | |
| | $Q_1$ | $Q^T$ | $Q_2$ | $Q_1$ | $Q^T$ | $Q_2$ | $Q_3$ |
| - | | 1.35 | | | 2.10 | | |
| DOS | 0.62 | 1.34 | 3.97 | 0.66 | 2.09 | 3.04 | 5.56 |
| DOA | 0.68 | 1.34 | 4.18 | 0.84 | 2.05 | 3.26 | |
| DBP | 0.60 | 1.36 | 3.24 | 0.90 | 2.12 | 3.38 | |
| DHP | 0.63 | 1.34 | 3.71 | 0.85 | 2.10 | 3.22 | 4.47 |
| TO | 0.57 | 1.35 | 2.29 | 0.58 | 2.07 | 1.92 | 1.99 |
| NOA | 0.59 | 1.36 | 4.03 | 0.72 | 2.10 | 2.86 | 4.66 |
| DBCP | 0.69 | 1.37 | 4.60 | 0.155 | 2.96 | 3.45 | 5.08 |
| TBP | 5.01 | 7.92* | 5.01 | 4.22 | 5.43 | 4.22 | |

* The result refers to the elastomer, having physical network completely disrupted

Swelling of SPU can be influenced by changes in elastomer structure resulting from mechanical action. At higher tensile strains of SPU, a successive breakdown of hard domains as well as micro-segregation may come into play causing reorganization.[13-15] If the structural changes in segmented elastomers are accompanied by breakdown of hard domains and a concomitant transformation of a certain amount of hard segments into a soft polymeric matrix or pulling of some soft blocks (structural defects) out of the hard domains,[14] variation of network parameters, is inevitable. It is known[4] that high strains applied to SPU causes disruption of the physical network and a significant drop in its density.

The data of equilibrium swelling of SPU-1 with oligodiene soft segments obtained by curing prepolymer by the mixture of MOCA and oligobutadiene diol ($M \approx 2000$) with the molar ratio 1/1 are given in Table 6.4.5. The initial density of the SPU-1 network $(v_e/V_0)_{dx}$ = 0.206 kmol/m$^3$, the physical network density $(v_e/V_0)_d$ = 0.170 kmol/m$^3$. After stretching by 700 % and subsequent unloading, the value $(v_e/V_0)_{dx}$ = 0.115 kmol/m$^3$ and $(v_e/V_0)_d$ = 0.079 kmol/m$^3$. The density of the chemical network $(v_e/V_0)_x$ = 0.036 kmol/m$^3$ did not change. Samples were swollen to equilibrium in a set of solvents: toluene, n-octane, cyclohexane, p-xylene, butyl acetate, DBCF, DBP, DHP, dihexyl sebacate (DHS), DOA and DOS (the values of density, $\rho_1$, and molar volume of solvents, $V_1$, are given in Table 6.2.5).

The data in Table 6.4.5 shows that, after stretching, the volume equilibrium swelling of samples does not change in DBCF.[4] In other solvents equilibrium swelling is noticeably increased.

Swelling experiments of samples previously swollen in solvents and subsequently in toluene have shown that the value of equilibrium swelling in toluene varies only for samples previously swollen in DBCF. The effective network densities evaluated for SPU-1 samples, swollen in DBCF, and then in toluene, have appeared equal (0.036 and 0.037 kmol/m$^3$, respectively) and these values correspond to the value of the chemical network parameter,

$(v_e/V_0)_x$, of SPU-1. It means that DBCF breaks down the physical network of elastomer; thus, the values of $Q_v$ in DBCF do not depend on the tensile strain. In all other cases, swelling of SPU-1 in toluene differs only marginally from the initial swelling of respective samples in toluene alone. Therefore, these solvents are unable to cause the breakdown of the physical network.

**Table 6.4.5. Equilibrium swelling of SPU-1 samples in various solvents before and after stretching at 25°C (calculated and experimental data). [Adapted, by permission, from V.V. Tereshatov, Vysokomol. soed., 37A, 1529 (1995).]**

| Solvent | $\rho_1$, kg/m$^3$ | $V_1 \times 10^3$, m$^3$/kmol | $Q_V$ (ε=0) | $\chi_1$ (ε=0) | $Q_V$ (ε=700%) | | $Q_V^T$ (ε=700%) | $Q_V^T$ (ε=0) |
|---|---|---|---|---|---|---|---|---|
| | | | | | calc | exp | | |
| Toluene | 862 | 107 | 3.54 | 0.32 | 5.06 | 4.87 | | |
| DBP | 1043 | 266 | 1.12 | 0.56 | 1.47 | 1.53 | 4.85 | 3.49 |
| n-Octane | 698 | 164 | 0.79 | 0.75 | 0.90 | 0.93 | 4.94 | 3.60 |
| Cyclohexane | 774 | 109 | 1.46 | 0.60 | 1.78 | 1.95 | 4.99 | 3.63 |
| p-Xylene | 858 | 124 | 2.94 | 0.36 | 4.17 | 3.93 | 4.81 | 3.52 |
| Butyl acetate | 876 | 133 | 2.28 | 0.44 | 3.13 | 2.97 | 5.12 | 3.69 |
| DOS | 912 | 468 | 1.21 | 0.38 | 1.60 | 1.71 | 5.10 | 3.63 |
| DOA | 924 | 402 | 1.49 | 0.32 | 2.19 | 2.18 | 5.00 | 3.61 |
| DHS | 923 | 340 | 1.74 | 0.30 | 2.55 | 2.44 | 5.01 | 3.59 |
| DHP | 1001 | 334 | 1.25 | 0.47 | 1.71 | 1.74 | 4.79 | 3.47 |
| DBCP | 976 | 347 | 4.48 | | | 4.51 | 9.92 | 9.98 |

The calculation of equilibrium swelling of SPU-1 (after its deformation up to 700 % and subsequent unloading) in a given solvent was carried out using the Flory - Rehner equation. In our calculations we used values of the $\chi_1$ parameter of interaction between solvents and elastomer calculated from experimental data of equilibrium swelling of initial sample (ε=0) in these solvents.

The diference between calculated and experimental data on equilibrium swelling of SPU-1 samples after their deformation does not exceed 9%. Therefore, the increase in swelling of SPU-1 is not related to the change of the $\chi_1$ parameter but it is related to a decrease in the $(v_e/V_0)_{dx}$ value of its three-dimensional network, caused by restructure of material by effect of strain.

The change of the domain structure of SPU exposed to increased temperature (~200°C) and consequent storage of an elastomer at room temperature may also cause a change in equilibrium swelling of material.[16]

## REFERENCES


1 V.P. Begishev, V.V. Tereshatov, E.N. Tereshatova, Int. Conf. Polymers in extreme environments, Nottingham, July 9-10, 1991, Univ. Press, Nottingham, 1991, pp. 1-6.
2 E.N. Tereshatova, V.V. Tereshatov, V.P. Begishev, and M.A. Makarova, *Vysokomol. Soed.*, **34B**, 22 (1992).
3 V.V. Tereshatov, E.N. Tereshatova, V.P. Begishev, V.I. Karmanov, and I.V. Baranets, *Polym. Sci.*, **36A**, 1680 (1994).

4 V.V. Tereshatov, *Vysokomol. soed.*, **37A**, 1529 (1995).
5 U.S. Aithal, T.M. Aminabhavi, R.H. Balundgi, and S.S. Shukla, *JMS - Rev. Macromol. Chem. Phys.*, **30C** (1), 43 (1990).
6 P.J. Flory, **Principles of Polymer Chemistry**, *Cornell Univ. Press*, Ithaca, 1953.
7 G.M. Bristow, and W.F. Watson, *Trans. Faraday. Soc.*, **54**, 1731 (1958).
8 L.N. Mizerovsky, L.N. Vasnyatskaya, and G.M. Smurova, *Vysokomol. Soed.*, **29A**, 1512 (1987).
9 D. Cohen, A. Siegman, and M. Narcis, *Polym. Eng. Sci.*, **27**, 286 (1987).
10 E. Konton, G. Spathis, M. Niaounakis and V. Kefals, *Colloid Polym. Sci.*, **26B**, 636 (1990).
11 V.V. Tereshatov, E.N. Tereshatova, and E.R. Volkova, *Polym. Sci.*, **37A**, 1157 (1995).
12 E.E. Cluff, E.K. Gladding, and R. Pariser, *J. Polym. Sci.*, **45**, 341 (1960).
13 Yu.Yu. Kercha, Z.V. Onishchenko, I.S. Kutyanina, and L.A. Shelkovnikova, **Structural and Chemical Modification of Elastomers**, *Naukova Dumka*, Kiev, 1989.
14 Yu.Yu. Kercha, **Physical Chemistry of Polyurethanes**, **Naukova Dumka**, Kiev, 1979.
15 V.N. Vatulev, S.V. Laptii, and Yu.Yu. Kercha, **Infrared Spectra and Structure of Polyurethanes**, *Naukova Dumka*, Kiev, 1987.
16 S.V. Tereshatov, Yu.S. Klachkin, and E.N. Tereshatova, *Plastmassy*, **7**, 43 (1998).

7

# Solvent Transport Phenomena

Many industrial processes rely on dissolution of raw materials and subsequent removal of solvents by various drying process. The formation of a solution and the subsequent solvent removal depends on a solvent transport phenomena which are determined by the properties of the solute and the properties of the solvent. Knowledge of the solvent movement within the solid matrix by a diffusion process is essential to design the technological processes. Many of the final properties, such as tribological properties, mechanical toughness, optical clarity, protection against corrosion, adhesion to substrates and reinforcing fillers, protective properties of clothing, the quality of the coated surface, toxic residues, morphology and residual stress, ingress of toxic substances, chemical resistance, depend not only on the material chosen but also on the regimes of technological processes. For these reasons, solvent transport phenomena are of interest to modern industry.

## 7.1 INTRODUCTION TO DIFFUSION, SWELLING, AND DRYING

George Wypych
**ChemTec Laboratories, Inc., Toronto, Canada**

Small molecule diffusion is the driving force behind movement of small molecules in and out of the solid matrix. Although both swelling and drying rely on diffusion, these processes are affected by the surfaces of solids, the concentration of small molecules in the surface layers, the morphology of the surface, and the interface between phases in which diffusion gradient exists. For these reasons, swelling and drying are treated as specific phenomena.

### 7.1.1 DIFFUSION

The free-volume theory of diffusion was developed by Vrentas and Duda.[1] This theory is based on the assumption that movement of a small molecule (e.g., solvent) is accompanied by a movement in the solid matrix to fill the free volume (hole) left by a displaced solvent molecule. Several important conditions must be described to model the process. These include the time scales of solvent movement and the movement of solid matrix (e.g. polymer segments, called jumping units), the size of holes which may fit both solvent molecules and jumping units, and the energy required for the diffusion to occur.

The timescale of the diffusion process is determined by the use of the diffusion Deborah, number De, given by the following equation:

$$De = \frac{\tau_M}{\tau_D} \quad [7.1.1]$$

where:

| | |
|---|---|
| $\tau_M$ | the molecular relaxation time |
| $\tau_D$ | the characteristic diffusion time |

If the diffusion Deborah number is small (small molecular relaxation time or large diffusion time) molecular relaxation is much faster than diffusive transport (in fact, it is almost instantaneous).[2] In this case the diffusion process is similar to simple liquids. For example, diluted solutions and polymer solutions above glass transition temperature fall in this category.

If the Deborah number is large (large molecular relaxation time or small diffusion time), the diffusion process is described by Fickian kinetics and is denoted by an elastic diffusion process.[1] The polymeric structure in this process is essentially unaffected and coefficients of mutual and self-diffusion become identical. Elastic diffusion is observed at low solvent concentrations below the glass transition temperature.[2]

The relationships below give the energy required for the diffusion process and compare the sizes of holes required for the solvent and polymer jumping unit to move within the system. The free-volume coefficient of self-diffusion is given by the equation:[2]

$$D_1 = D_o \exp\left[-\frac{E}{RT}\right] \times \exp\left[-\frac{\gamma\left(\omega_1\hat{V}_1^* + \omega_2\xi\hat{V}_2^*\right)}{\hat{V}_{FH}}\right] \quad [7.1.2]$$

where:

| | |
|---|---|
| $D_1$ | self-diffusion coefficient |
| $D_o$ | pre-exponential factor |
| E | energy per molecule required by the molecule to overcome attractive forces |
| R | gas constant |
| T | temperature |
| $\gamma$ | overlap factor introduced to address the fact that the same free volume is available for more than one molecule |
| $\omega$ | mass fraction (index 1 for solvent, index 2 for polymer) |
| $V^*$ | specific free hole volume (indices the same as above) |
| $\xi$ | the ratio of the critical molar volume of the solvent jumping unit to the critical molar volume of the polymer jumping unit (see equation [7.1.3]) |
| $\hat{V}_{FH}$ | average hole free volume per gram of mixture. |

$$\xi = \hat{V}_1^* / \hat{V}_2^* = \hat{V}_1^* M_1 / \hat{V}_2^* M_2 \quad [7.1.3]$$

where:

| | |
|---|---|
| M | molecular weight (1 - solvent, 2 - polymer jumping unit) |

The first exponent in equation [7.1.2] is the energy term and the second exponent is the free-volume term. Figure 7.1.1 shows three regions of temperature dependence of free-volume: I - above glass transition temperature, II - close to transition temperature, and III - below the transition temperature. In the region I, the second term of the equation [7.1.2] is negligible and thus diffusion is energy-driven. In the region II both terms are significant. In the region III the diffusion is free volume-driven.[3]

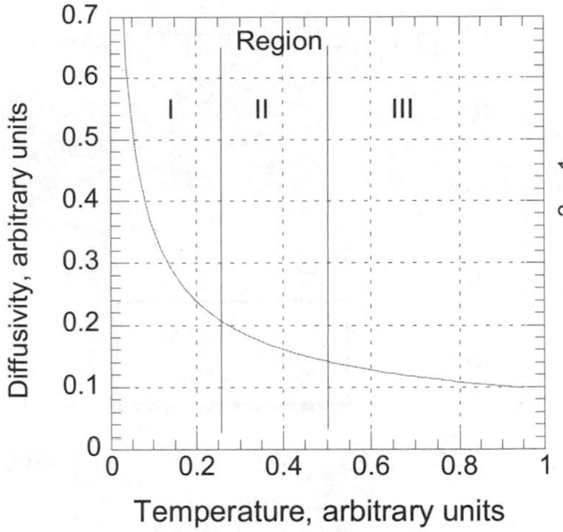


Figure 7.1.1. Temperature dependence of the solvent self-diffusion coefficient. [Adapted, by permission, from D Arnauld, R L Laurence, *Ind. Eng. Chem. Res.*, **31**(1), 218-28 (1992).]

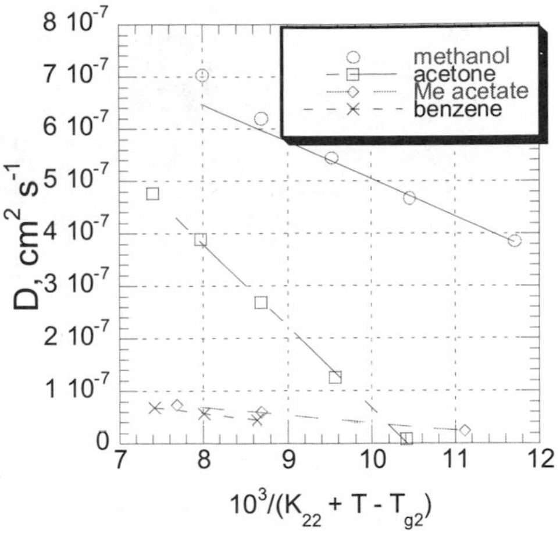


Figure 7.1.2. Free-volume correlation data for various solvents. [Data from D Arnauld, R L Laurence, *Ind. Eng. Chem. Res.*, **31**(1), 218-28 (1992).]

The mutual diffusion coefficient is given by the following equation:

$$D = \frac{D_1 \omega_1 \omega_2}{RT} \left( \frac{\partial \mu_1}{\partial \omega_1} \right)_{T,p} = D_1 Q \qquad [7.1.4]$$

where:

| | |
|---|---|
| D | mutual diffusion coefficient |
| $\mu_1$ | chemical potential of a solvent per mole |
| Q | thermodynamic factor. |

These equations are at the core of diffusion theory and are commonly used to predict various types of solvent behavior in polymeric and other systems. One important reason for their wide application is that all essential parameters of the equations can be quantified and then used for calculations and modelling. The examples of data given below illustrate the effect of important parameters on the diffusion processes.

Figure 7.1.2 shows the effect of temperature on the diffusivity of four solvents. The relationship between diffusivity and temperature is essentially linear. Only solvents having the smallest molecules (methanol and acetone) depart slightly from a linear relationship due to the contribution of the energy term. The diffusivity of the solvent decreases as temperature decreases. Several other solvents show a similar relationship.[3]

Figure 7.1.3 shows the relationship between the solvent's molar volume and its activation energy. The activation energy increases as the solvent's molar volume increases then levels off. The data show that the molar volume of a solvent is not the only parameter which affects activation energy. Flexibility and the geometry of solvent molecule also affect activation energy.[3] Branched aliphatic solvents (e.g., 2-methyl-pentane, 2,3-dimethyl-butane) and substituted aromatic solvents (e.g., toluene, ethylbenzene, and xylene) show large departures from free volume theory predictions.

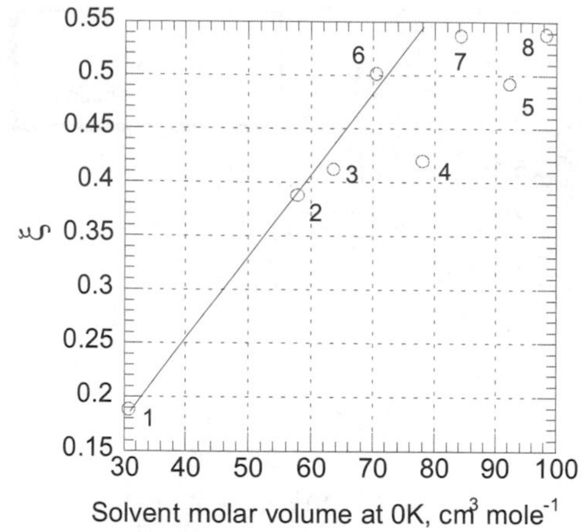


Figure 7.1.3. Parameter ξ vs. solvent molar volume. 1 - methanol, 2 - acetone, 3 - methyl acetate, 4 - ethyl acetate, 5 - propyl acetate, 6 - benzene, 7 - toluene, 8 - ethylbenzene. [Adapted, by permission, from D Arnauld, R L Laurence, *Ind. Eng. Chem. Res.*, **31**(1), 218-28 (1992).]

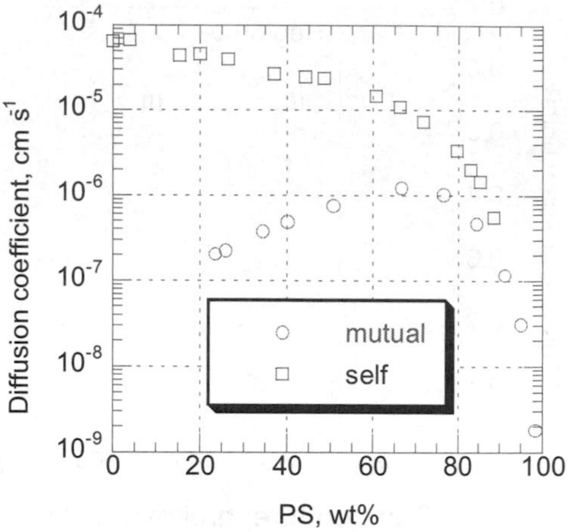


Figure 7.1.4. Mutual and self-diffusion coefficients for polystyrene/toluene system at 110°C. [Adapted, by permission, from F D Blum, S Pickup, R A Waggoner, *Polym. Prep.*, **31** (1), 125-6 (1990).]

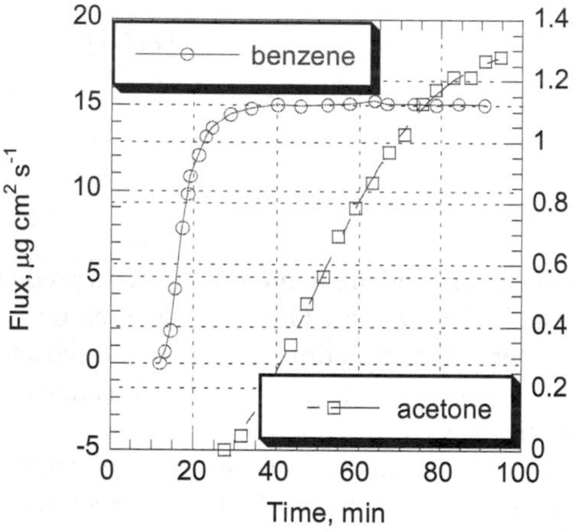


Figure 7.1.5. Flux vs. time for two solvents. [Adapted by permission from C F Fong, , Y Li, De Kee, J Bovenkamp, *Rubber Chem. Technol.*, **71**(2), 285-288 (1998).]

Many experimental methods such as fluorescence, reflection Fourier transform infrared, NMR, quartz resonators, and acoustic wave admittance analysis, are used to study diffusion of solvents.[4-11] Special models have been developed to study process kinetics based on experimental data.

Figure 7.1.4 shows the effect of concentration of polystyrene on mutual and self-diffusion coefficients measured by pulsed-gradient spin-echo NMR. The data show that the two coefficients approach each other at high concentrations of polymer as predicted by theory.[4]

Studies on solvent penetration through rubber membranes (Figure 7.1.5), show that in the beginning of the process, there is a lag time called break-through time. This is the time required for solvent to begin to penetrate the membrane. It depends on both the solvent and the membrane. Of solvents tested, acetone had the longest break-through time in natural rubber and toluene the longest in nitrile rubber. After penetration has started the flux of solvent increases rapidly and ultimately levels off.[7] This study is relevant in testing the permeability of protective clothing and protective layers of coatings.

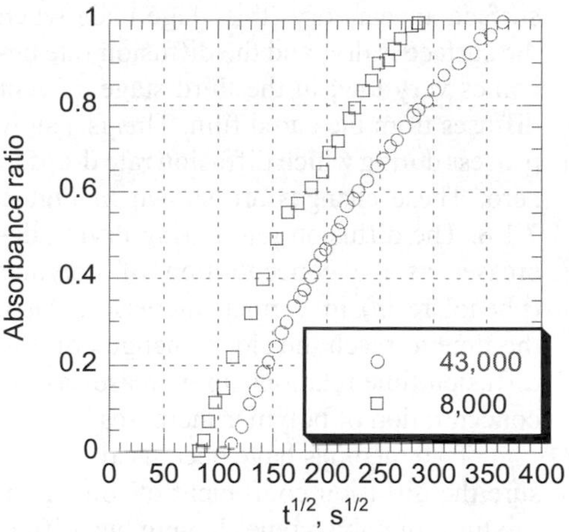


Figure 7.1.6. Absorbance ratio vs. exposure time to water for PMMA of different molecular weights. [Adapted, by permission, from I Linossier, F Gaillard, M Romand, J F Feller, *J. Appl. Polym. Sci.*, **66**, No.13, 2465-73 (1997).]

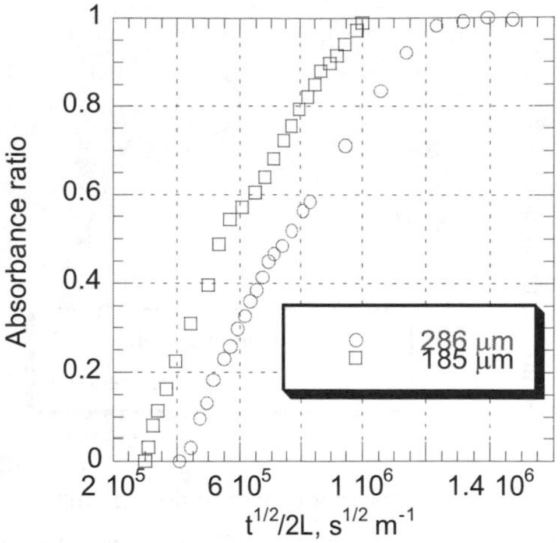


Figure 7.1.7. Absorbance ratio vs. exposure time to water for PMMA films of different thickness. [Adapted, by permission, from I Linossier, F Gaillard, M Romand, J F Feller, *J. Appl. Polym. Sci.*, **66**, No.13, 2465-73 (1997).]

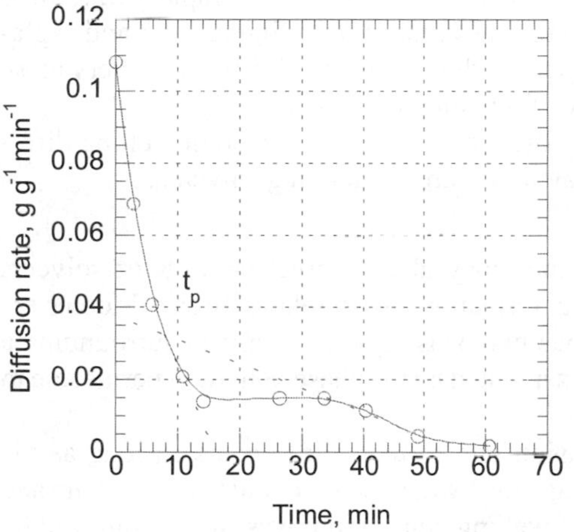


Figure 7.1.8. Relative diffusion rate vs. curing time. [Adapted, by permission, from Jinhua Chen, Kang Chen, Hong Zhang, Wanzhi Wei, Lihua Nie, Shouzhuo Yao, *J. Appl. Polym. Sci.*, **66**, No.3, 563-71 (1997).]

Similar observations were made using internal reflection Fourier transform infrared to measure water diffusion in polymer films.[9] Figure 7.1.6 shows that there is a time lag between the beginning of immersion and water detection in polymer film. This time lag increases as the molecular weight increases (Figure 7.1.6) and film thickness increases (Figure 7.1.7). After an initial increase in water concentration, the amount levels off. Typically, the effect of molecular weight on the diffusion of the penetrant does not occur. High molecular weight polymer has a shift in the absorption peak from 1730 to 1723 $cm^{-1}$ which is associated with the hydrogen bonding of the carbonyl group. Such a shift does not occur in low molecular weight PMMA. Water can move at a higher rate in low molecular weight PMMA. In some other polymers, this trend might be reversed if the lower molecular weight polymer has end groups which can hydrogen bond with water.

Bulk acoustic wave admittance analysis was used to study solvent evaporation during curing.[8] Three characteristic stages were identified: in the first stage viscosity increases accompanied by a rapid decrease in diffusion rate; in the second stage the film is formed, the

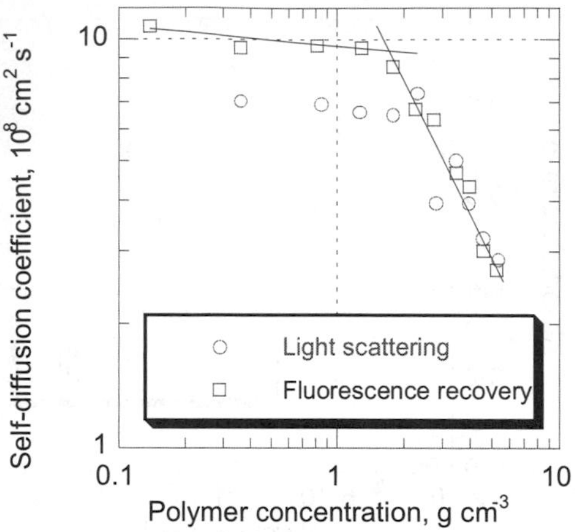


Figure 7.1.9. Self-diffusion coefficient vs. polystyrene concentration. [Adapted, by permission, from L Meistermann, M Duval, B Tinland, *Polym. Bull.*, **39**, No.1, 101-8 (1997).]

surface appears dry (this stage ends when the surface is dry) and the diffusion rate becomes very low; in the third stage solvent diffuses from the cured film. This is a slow process during which diffusion rate drops to zero. These changes are shown in Figure 7.1.8. The diffusion rate during drying decreases as the concentration of polymer (phenol resin) in varnish increases. Also, the time to reach the slope change point in diffusion/time relationship increases as the concentration of polymer increases.[8]

Two methods have been used to measure the diffusion coefficient of toluene in mixtures of polystyrenes having two different molecular weights: one was dynamic light scattering and the other, fluorescence recovery after bleaching.[10] The data show that the relationship between the diffusion coefficient and polymer concentration is not linear. The crossover point is shown in Figure 7.1.9. Below a certain concentration of polymer, the diffusion rate drops rapidly according to different kinetics. This is in agreement with the above theory (see Figure 7.1.1 and explanations for equation [7.1.2]). The slope exponent in this study was -1.5 which is very close to the slope exponent predicted by the theory of reptation (-1.75).

The above data show that theoretical predictions are accurate when modelling diffusion phenomena in both simple and complicated mixtures containing solvents.

## 7.1.2 SWELLING

Polymers differ from other solids because they may absorb large amounts of solvents without dissolving. They also undergo large deformations when relatively small forces are involved.[12] Swelling occurs in a heterogeneous two phase system a solvent surrounding a swollen body also called gel. Both phases are separated by the phase boundary permeable to solvent.[13]

The swelling process (or solvent diffusion into to the solid) occurs as long as the chemical potential of solvent is large. Swelling stops when the potentials are the same and this point is called the swelling equilibrium. Swelling equilibrium was first recognized by Frenkel[14] and the general theory of swelling was developed by Flory and Rehner.[15,16]

The general theory of swelling assumes that the free energy of mixing and the elastic free energy in a swollen network are additive. The chemical potential difference between gel and solvent is given by the equation:

$$(\mu_1 - \mu_1^0) = (\mu_1 - \mu_1^0)_{mix} + (\mu_1 - \mu_1^0)_{el} \qquad [7.1.5]$$

where:

$\mu_1$ chemical potential of gel
$\mu_1^0$ chemical potential of solvent

The chemical potential is the sum of the terms of free energy of mixing and the elastic free energy. At swelling equilibrium, $\mu_1 = \mu_1^0$, and thus the left hand term of the equation becomes zero. The equation [7.1.5] takes the following form:

$$(\mu_1 - \mu_1^0)_{mix} = -(\mu_1 - \mu_1^0)_{el} = RT[\ln(1 - v_2) - v_2 + \chi v_2^2] \quad [7.1.6]$$

where:

| | |
|---|---|
| $v_2$ | $= n_2V_2/(n_1V_1 + n_2V_2)$ volume fraction of polymer |
| $n_1$, $n_2$ | moles of solvent and polymer, respectively |
| $V_1$, $V_2$ | molar volumes of solvent and polymer, respectively |
| R | gas constant |
| T | absolute temperature |
| $\chi$ | Flory-Huggins, polymer-solvent interaction parameter. |

The interaction between the solvent and solid matrix depends on the strength of such intermolecular bonds as polymer-polymer, solvent-solvent, and polymer-solvent. If interaction between these bonds is similar, the solvent will easily interact with polymer and a relatively small amount of energy will be needed to form a gel.[12] The Hildebrand and Scatchard hypothesis assumes that interaction occurs between solvent and a segment of the chain which has a molar volume similar to that of solvent.[12] Following this line of reasoning the solvent and polymer differ only in potential energy and this is responsible for their interaction and for the solubility of polymer in the solvent. If the potential energies of solvents and polymeric segments are similar they are readily miscible. In crosslinked polymers, it is assumed that the distance between crosslinks is proportional to the molecular volume of the polymer segments. This assumption is the basis for determining molecular mass between crosslinks from results of swelling studies.

The result of swelling in a liquid solvent (water) is determined by equation:[13]

$$\left(\frac{\partial T}{\partial w_1}\right)_P^{g/l} = \frac{T\left(\dfrac{\partial \mu_1^g}{\partial w_1}\right)_{T,P}}{\Delta H_1^{g/l}} \quad [7.1.7]$$

where:

| | |
|---|---|
| T | thermodynamic (absolute) temperature |
| $w_1$ | mass fraction of solvent in gel at saturation concentration |
| g | phase symbol (for gel) |
| l | symbol for liquid |
| P | pressure |
| $\mu_1^g$ | chemical potential of solvent in gel phase dependent on temperature |
| $\Delta H_1^{g/l}$ | $= H_1^g - H_{01}^{l}$ is the difference between partial molar enthalpy of solvent (water) in gel and pure liquid solvent (water) in surrounding |

Contrast this with the equation for water in the solid state (ice):

$$\left(\frac{\partial T}{\partial w_1}\right)_P^{g/cr} = \frac{T\left(\dfrac{\partial \mu_1^g}{\partial w_1}\right)_{T,P}}{\Delta H_1^{g/cr}} \quad [7.1.8]$$

where: cr phase symbol for crystalline solvent (ice)

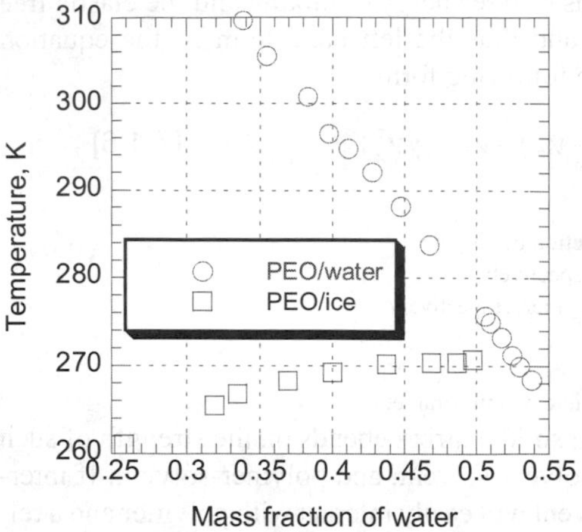


Figure 7.1.10. Swelling of crosslinked polyurethane in water and ice. [Adapted, by permission, from B Hladik, S Frahn, W Borchard, *Polym. Polym. Compos.*, **3**, No.1, 21-8 (1995).]

A comparison of equations [7.1.7] and [7.1.8] shows that the slope and sign of swelling curve are determined by the quantity $\Delta H_1^{g/l \text{ or } cr}$. Since the melting enthalpy of water is much larger than the transfer enthalpy of water, the swelling curves of gel in liquid water are very steep. The sign of the slope is determined by the heat transfer of the solvent which may be negative, positive or zero depending on the quality of solvent. The melting enthalpy is always positive and therefore the swelling curve in the presence of crystalline solvent is flat with a positive slope. A positive slope in temperatures below zero (for ice) means that gel has to deswell (release water to its surrounding, or dry out) as temperature decreases.[13] Figure 7.1.10 illustrates this.

For practical purposes, simple equations are used to study swelling kinetics. The degree of swelling, α, is calculated from the following equation:[17]

$$\alpha = \frac{V_1 - V_0}{V_0} \qquad [7.1.9]$$

where:

$V_1$ volume of swollen solid at time t=t
$V_0$ volume of unswollen solid at time t=0

The swelling constant, K, is defined by:

$$K = \frac{k_1}{k_2} = \frac{\alpha}{1-\alpha} \qquad [7.1.10]$$

where:

$k_1$ rate constant of swelling process
$k_2$ rate constant of deswelling process

This shows that the swelling process is reversible and in a dynamic equilibrium. The distance of diffusion is time-dependent:

$$distance \propto (time)^n \qquad [7.1.11]$$

The coefficient n is between 0.5 for Fickian diffusion and 1.0 for relaxation-controlled diffusion (diffusion of solvent is much faster than polymer segmental relaxation).[18] This relationship is frequently taken literally[19] to calculate diffusion distance from a measurement of the change of the linear dimensions of swollen material.

The following equation is used to model changes based on swelling pressure measurements:

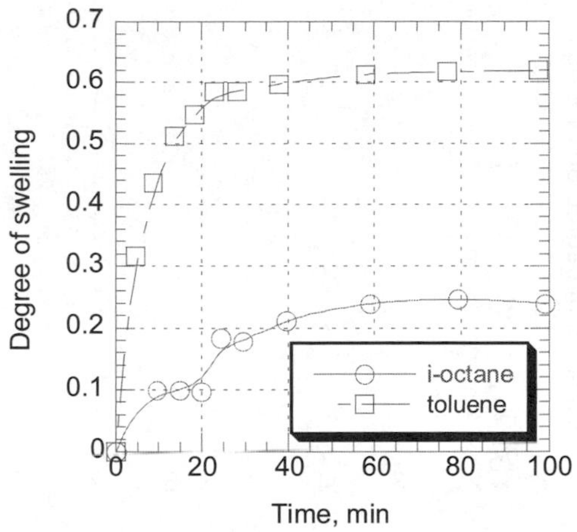


Figure 7.1.11. Swelling kinetics of EVA in toluene and i-octane. [Data from H J Mencer, Z Gomzi, *Eur. Polym. J.*, **30**, 1, 33-36, (1994).]

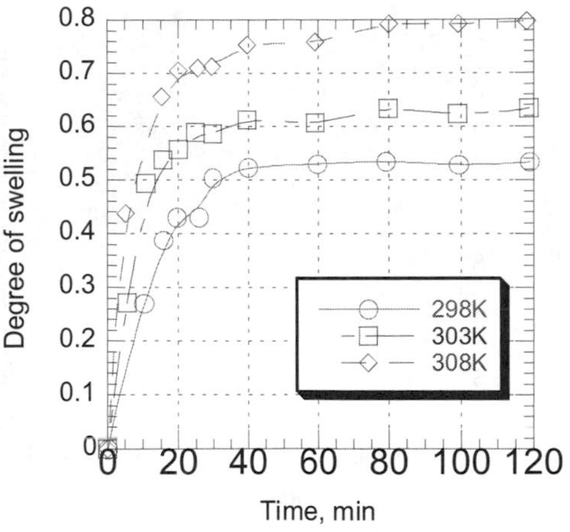


Figure 7.1.12. Swelling kinetics of EVA in tetrahydrofuran at different temperatures. [Adapted, by permission, from H J Mencer, Z Gomzi, *Eur. Polym. J.*, **30**, 1, 33-36, (1994).]

$$\Pi = A\varphi^n = -\frac{RT}{v_1}\left[\ln(1-\varphi) + \varphi + \chi\varphi^2\right] \qquad [7.1.12]$$

where:

| | |
|---|---|
| $\Pi$ | osmotic pressure |
| A | coefficient |
| $\varphi$ | volume fraction of polymer in solution |
| n | = 2.25 for good solvent and = 3 for $\Theta$ solvent |
| $v_1$ | molar volume of solvent |
| $\chi$ | Flory-Huggins, polymer-solvent interaction parameter |

The above relationship is used in to study swelling by measuring shear modulus.[20]

Figure 7.1.11 shows swelling kinetic curves for two solvents. Toluene has a solubility parameter of 18.2 and i-octane of 15.6. The degree to which a polymer swells is determined by many factors, including the chemical structures of polymer and solvent, the molecular mass and chain flexibility of the polymer, packing density, the presence and density of crosslinks, temperature, and pressure. In the example presented in Figure 7.1.11 the solubility parameter has a strong influence on swelling kinetics.[17] The effect of temperature on swelling kinetics is shown in Figure 7.1.12. Increasing temperature increases swelling rate. During the initial stages of the swelling process the rate of swelling grows very rapidly and then levels off towards the swelling equilibrium.

In Figure 7.1.13, the diffusion distance is almost linear with time as predicted by equation [7.1.11]. The coefficient n of the equation was 0.91 meaning that the swelling process was relaxation rate controlled.

Figure 7.1.14 shows the relationship between hydrogel swelling and pH. Hydrogels are particularly interesting because their swelling properties are controlled by the conditions around them (e.g. pH).[21-23] Because they undergo controllable volume changes, they find applications in separation processes, drug delivery systems, immobilized enzyme sys-

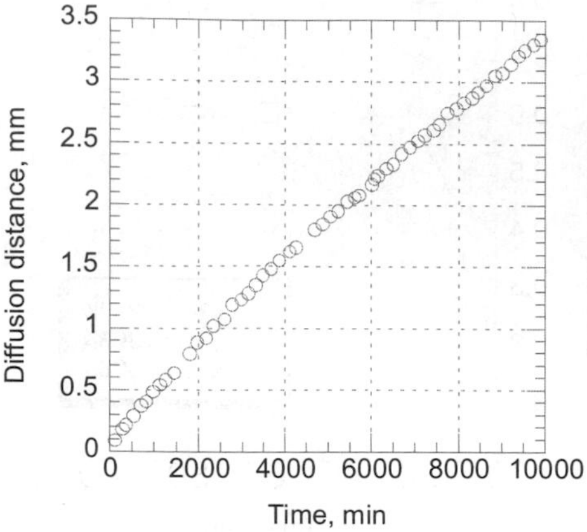


Figure 7.1.13. Diffusion distance of 1,4-dioxane into PVC vs. time. [Adapted, by permission, from M Ercken, P Adriensens, D Vanderzande, J Gelan, *Macromolecules*, **28**, 8541-8547, (1995).]

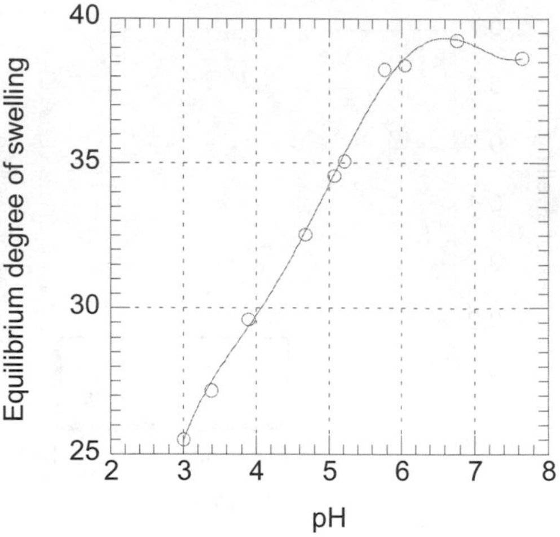


Figure 7.1.14. Effect of pH on the equilibrium swelling of hydrogel. [Data from M Sen, A Yakar, O Guven, *Polymer*, **40**, No.11, 2969-74 (1999).]

tems, etc. Maximum swelling is obtained at pH = 7. At this point there is complete dissociation of the acidic groups present in the hydrogel. The behavior of hydrogel can be modelled using Brannon-Peppas equation.[22]

Figure 7.1.15 shows the 1,1,2,2-tetrachoroethylene, TCE, uptake by amorphous poly(ether ether ketone), PEEK, as a function of time.[24,25] The swelling behavior of PEEK in this solvent is very unusual - the sample mass is increased by 165% which is about 3 times more than with any other solvent. In addition, the solvent uptake by PEEK results in a change in optical properties of the solution from clear to opaque. A clear solution is typical of amorphous PEEK and the opaque solution of crystalline PEEK. It was previously suggested by Fokes and Tischler[26] that polymethylmethacrylate forms weak complexes with various acid species in solution. This may also explain the unusual swelling caused by TCE. Because of the presence of C=O, C-O-C, and aromatic groups, PEEK acts as organic base. TCE is an electron acceptor due to electron-deficient atoms in the molecule.[24] This interaction may explain the strong affinity of solvent and polymer. Figure 7.1.16 illustrates the effect of crystallization. Below 250°C, the carbonyl frequency decreases. But a more rapid decrease begins below 140°C which is glass transition temperature. Above 250°C, the carbonyl frequency increases rapidly. Above the glass transition temperature there is rapid crystallization process which continues until the polymer starts to melt at 250°C then it gradually reverts to its original amorphous structure. The presence of solvent aids in the crystallization process.

## 7.1.3 DRYING

Solvent removal can be accomplished by one of three means: deswelling, drying or changes in the material's solubility. The deswelling process, which involves the crystallization of solvent in the surrounding gel, was discussed in the previous section. Here attention is focused on drying process. The changes due to material solubility are discussed in Chapter 12.

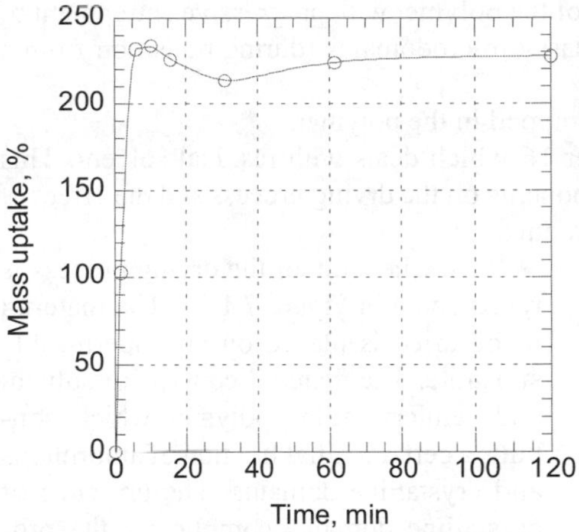


Figure 7.1.15. The mass uptake of TCE by PEEK vs. time. [Adapted, by permission, from B H Stuart, D R Williams, *Polymer*, **35**, No.6, 1326-8 (1994).]

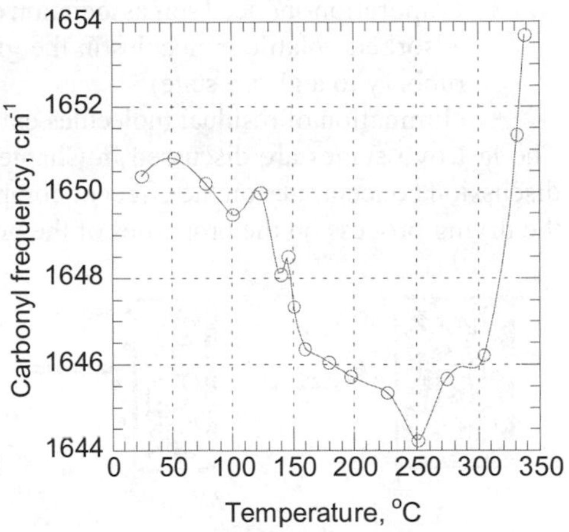


Figure 7.1.16. The frequency of carbonyl stretching mode of PEEK vs. temperature. [Adapted, by permission, from B H Stuart, D R Williams, *Polymer*, **36**, No.22, 4209-13 (1995).]

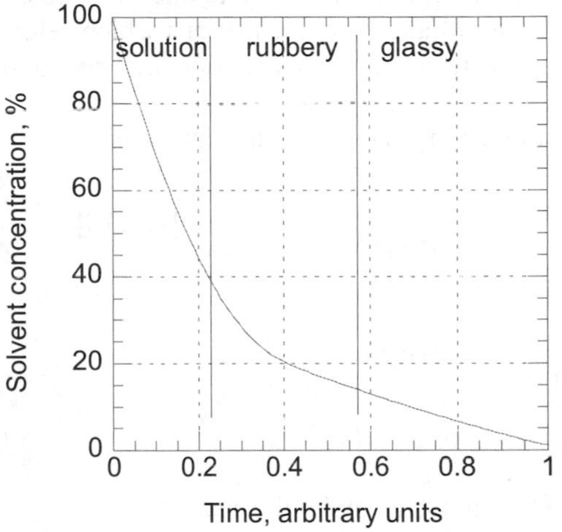


Figure 7.1.17. Solvent concentration vs. drying time.

Figure 7.1.1 can be discussed from a different perspective of results given in Figure 7.1.17. There are also three regions here: region 1 which has a low concentration of solid in which solvent evaporation is controlled by the energy supplied to the system, region 2 in which both the energy supplied to the system and the ability of polymer to take up the free volume vacated by solvent are important, and region 3 where the process is free volume controlled. Regions 2 and 3 are divided by the glass transition temperature. Drying processes in region 3 and to some extent in region 2 determine the physical properties of dried material and the amount of residual solvent remaining in the product. A sharp transition between region 2 and 3 (at glass transition temperature) might indicate that drying process is totally homogeneous but it is not and this oversimplifies the real conditions at the end of drying process. The most realistic course of events occurring close to the dryness point is presented by these four stages:[27,28]

- elimination of the volatile molecules not immobilized by the adsorption onto the polymer
- elimination of adsorbed molecules from the polymer in its rubbery state

- evaporation-induced self association of the polymer with progressive entrapment of adsorbed volatile molecules in the glassy microdomains (during transition from a rubbery to a glassy state)
- elimination of residual molecules entrapped in the polymer.

The last two stages are discussed in Chapter 16 which deals with residual solvent. This discussion concentrates on the effect of components on the drying process and the effect of the drying process on the properties of the product.

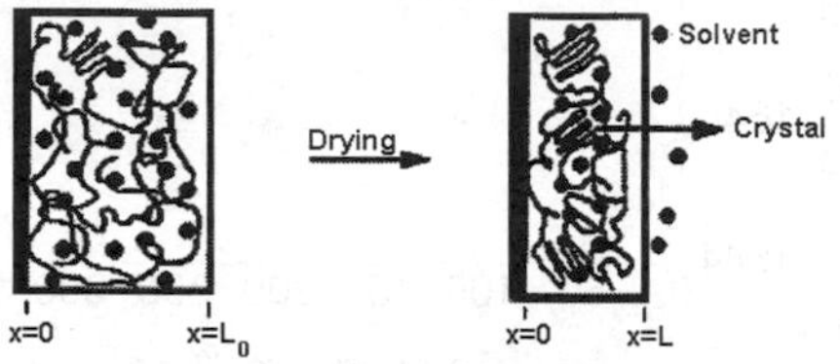

Figure 7.1.18. Schematic representation of drying a polymer slab. [Adapted, by permission, from M O Ngui, S K Mallapragada, *J. Polym. Sci.: Polym. Phys. Ed.*, **36**, No.15, 2771-80 (1998).]

A schematic of the drying process is represented in Figure 7.1.18. The material to be dried is placed on an impermeable substrate. The material consists of solvent and semicrystalline polymer which contains a certain initial fraction of amorphous and crystalline domains. The presence of crystalline domains complicates the process of drying because of the reduction in diffusion rate of the solvent. Evaporation of solvent causes an inward movement of material at the surface and the drying process may change the relative proportions of amorphous and crystalline domains.[29] Equations for the change in thickness of the material and kinetic equations which relate composition of amorphous and crystalline domains to solvent concentration are needed to quantify the rate of drying.

The thickness change of the material during drying is given by the equation:

$$v_1 \frac{dL}{dt} = \left( D \frac{\partial v_1}{\partial x} \right)_{x=L} \qquad [7.1.13]$$

where:

| | |
|---|---|
| $v_1$ | volume fraction of solvent |
| L | thickness of slab as in Figure 7.1.18 |
| t | time |
| D | diffusion coefficient |
| x | coordinate of thickness |

The rate of change of crystalline volume fraction is given by the equation:

$$\frac{\partial v_{2c}}{\partial t} = k_1 v_1 \qquad [7.1.14]$$

where:

| | |
|---|---|
| $v_{2c}$ | volume fraction of crystalline phase |
| $k_1$ | rate change of crystalline phase proportional to folding rate |

The rate of change of amorphous volume fraction is given by the equation:

$$\frac{\partial v_{2a}}{\partial t} = \frac{\partial}{\partial x}\left( D \frac{\partial v_1}{\partial x} \right) - k_1 v_1 \qquad [7.1.15]$$

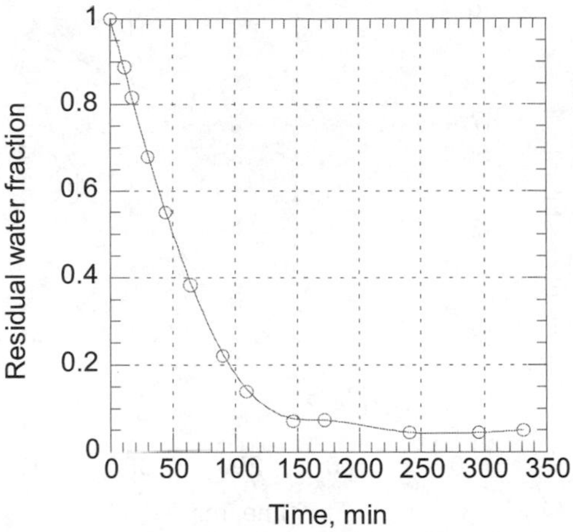


Figure 7.1.19. Fraction of water remaining in PVA as a function of drying time at 23°C. [Data from M O Ngui, S K Mallapragada, *J. Polym. Sci.: Polym. Phys. Ed.*, **36**, No.15, 2771-80 (1998).]

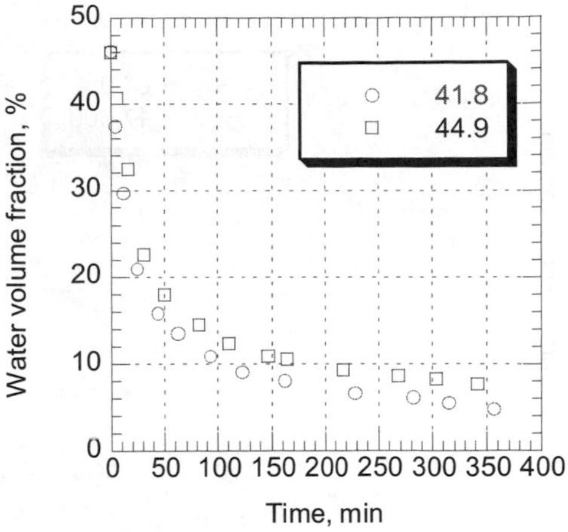


Figure 7.1.20. Water volume fraction vs. drying time for PVA of different crystallinity. [Adapted, by permission, from M O Ngui, S K Mallapragada, *J. Polym. Sci. Polym. Phys. Ed.*, **36**, No.15, 2771-80 (1998).]

where:

$v_{2a}$ volume fraction of amorphous phase

The rate of drying process is determined by the diffusion coefficient:

$$D = D_0 \left[\exp(\alpha_D v_1)\right](1 - v_{2c}) / \tau \qquad [7.1.16]$$

where:

$D_0$ initial diffusion coefficient dependent on temperature

$\alpha_D$ constant which can be determined experimentally for spin echo NMR studies[30]

$\tau$ constant equal to 1 for almost all amorphous polymers ($v_{2c} \le 0.05$) and 3 for semi-crystalline polymers

According to this equation, the coefficient of diffusion decreases as crystallinity increases because the last term decreases and $\tau$ increases.

Figure 7.1.19 shows that the fraction of solvent (water) decreases gradually as drying proceeds. Once the material reaches a glassy state, the rate of drying rapidly decreases. This is the reason for two different regimes of drying. Increasing temperature increases the rate of drying process.[29]

Figure 7.1.20 shows that even a small change in the crystallinity of the polymer significantly affects drying rate. An increase in molecular weight has similar effect (Figure 7.1.21). This effect is due both to lower mobility of entangled chains of the higher molecular weight and to the fact that higher molecular weight polymers are more crystalline.

Figure 7.1.22 shows that drying time increases as the degree of polymer crystallinity increases. Increased crystallinity slows down the diffusion rate and thus the drying process.

The physical properties of liquids, such as viscosity (Figure 7.1.23) and surface tension (Figure 7.1.24), also change during the evaporation process. The viscosity change for this system was a linear function of the amount of solvent evaporated.[31] This study on waterborne coatings showed that the use of a cosolvent (e.g. i-butanol) caused a reduction in

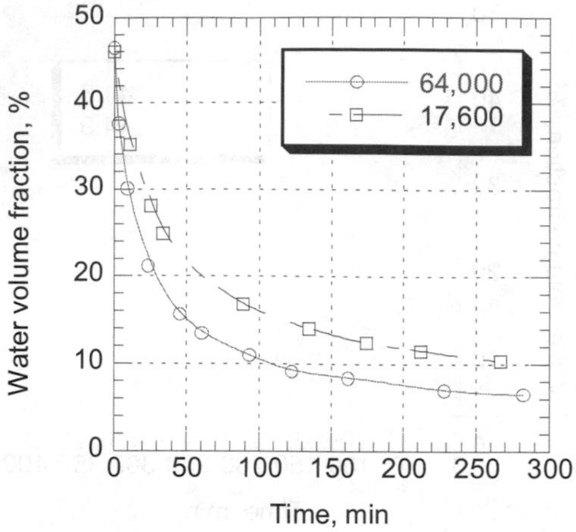


Figure 7.1.21. Water volume fraction vs. drying time for PVA of different molecular weight. [Adapted, by permission, from M O Ngui, S K Mallapragada, *J. Polym. Sci.: Polym. Phys. Ed.*, **36**, No.15, 2771-80 (1998).]

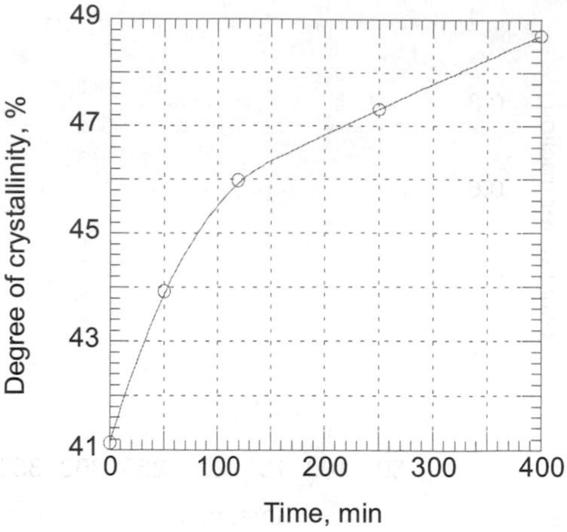


Figure 7.1.22. Degree of crystallinity vs. PVA drying time at 25°C. [Adapted, by permission, from M O Ngui, S K Mallapragada, *J. Polym. Sci.: Polym. Phys. Ed.*, **36**, No.15, 2771-80 (1998).]

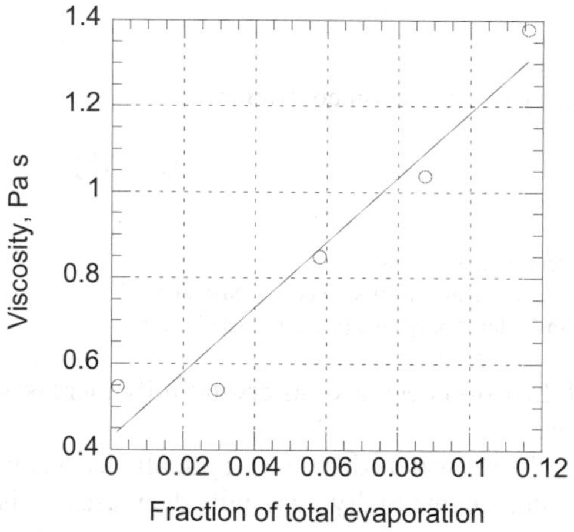


Figure 7.1.23. Viscosity vs. fraction of total evaporation of water from waterborne coating. [Data from S Kojima, T Moriga, *Polym. Eng. Sci.*, **35**, No.13, 1098-105 (1995).]

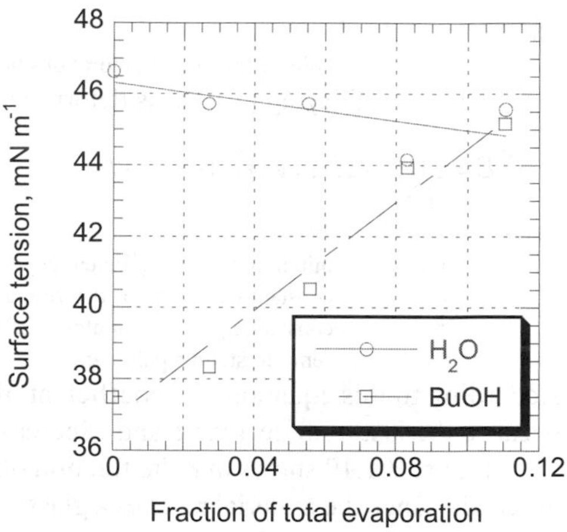


Figure 7.1.24. Surface tension vs. fraction of total evaporation of solvent from waterborne coating. [Data from S Kojima, T Moriga, *Polym. Eng. Sci.*, **35**, No.13, 1098-105 (1995).]

overall viscosity. Micelles of smaller size formed in the presence of the cosolvent explain lower viscosity.[31] Figure 7.1.24 shows that the surface tension of system containing a cosolvent (i-butanol) increases as the solvents evaporate whereas the surface tension of a system containing only water decreases. The increase of surface tension in the system containing cosolvent is due to the preferential evaporation of the cosolvent from the mixture.[31]

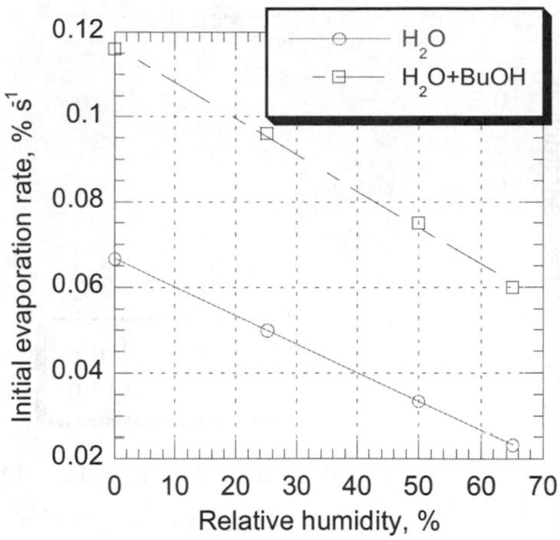


Figure 7.1.25. Initial evaporation rate from waterborne coating vs. relative humidity for two solvent systems. [Data from S Kojima, T Moriga, *Polym. Eng. Sci.*, **35**, No.13, 1098-105 (1995).]

**Table 7.1.1. Experimentally determined initial evaporation rates of waterborne coating containing a variety of solvents at 5% level (evaporation at 25°C and 50% RH)**

| Co-solvent | Initial evaporation rate, μg $cm^{-2}$ $s^{-1}$ |
|---|---|
| none | 3.33 |
| methyl alcohol | 4.44 |
| ethyl alcohol | 3.56 |
| n-propyl alcohol | 4.00 |
| n-butyl alcohol | 3.67 |
| i-butyl alcohol | 3.67 |
| n-amyl alcohol | 3.33 |
| n-hexyl alcohol | 3.22 |
| cthylene glycol mono-butyl ester | 3.22 |
| ethylene glycol mono-hexyl ester | 3.33 |
| butyl carbinol | 3.11 |
| methyl-i-butyl ketone | 4.89 |

Table 7.1.1 shows the effcct of cosolvent addition on the evaporation rate of solvent mixture.

The initial rate of evaporation of solvent depends on both relative humidity and cosolvent presence (Figure 7.1.25). As relative humidity increases the initial evaporation rate decreases. The addition of cosolvent doubles the initial evaporation rate.

In convection drying, the rate of solvent evaporation depends on airflow, solvent partial pressure, and temperature. By increasing airflow or temperature, higher process rates can be achieved but the risk of skin and bubble formation is increased. As discussed above, Vrentas-Duda free-volume theory is the basis for predicting solvent diffusion, using a small number of experimental data to select process conditions. The design of a process and a dryer which uses a combination of convection heat and radiant energy is a more complex process. Absorption of radiant energy is estimated from the Beer's Law, which, other than for the layers close to the substrate, predicts:[32]

$$Q_r(\xi)=I_0\alpha\exp[-\alpha(\beta h-\xi)] \quad [7.1.17]$$

where:

| | |
|---|---|
| $Q_r$ | radiant energy absorption |
| $\xi$ | distance from substrate |
| $I_0$ | intensity of incident radiation |
| $\alpha$ | volumetric absorption coefficient |
| $\beta$ | fractional thickness of the absorbing layer next to the substrate |
| h | thickness |

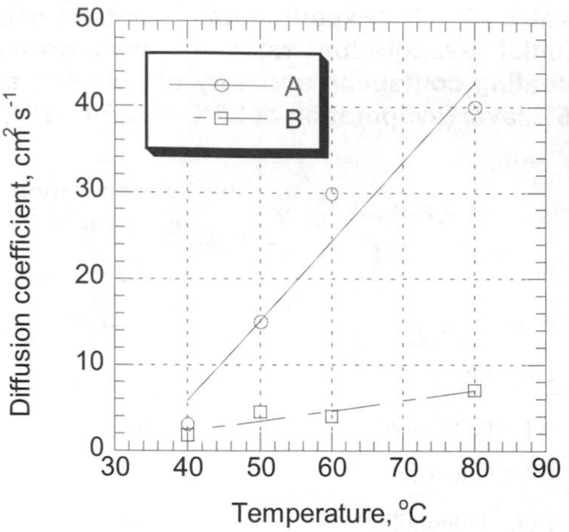


Figure 7.1.26. Diffusion coefficients for two polyurethanes (A & B) dissolved in N,N-dimethylacetamide vs. temperature. [Data from G A Abraham, T R Cuadrado, *International Polym. Process.*, **13**, No.4, Dec.1998, 369-78.]

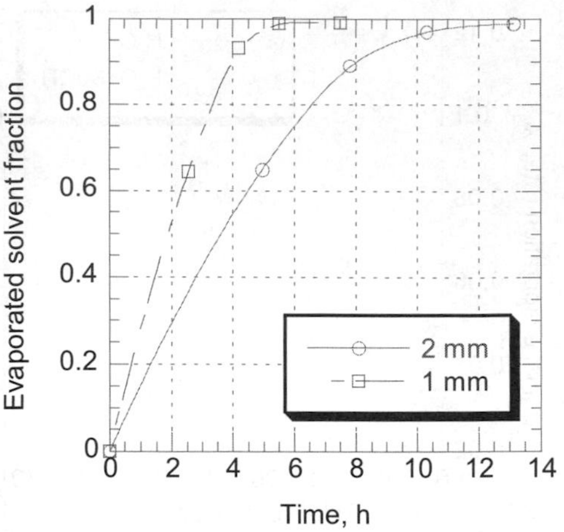


Figure 7.1.27. Effect of film thickness on evaporation rate. [Adapted, by permission from G A Abraham, T R Cuadrado, *International Polym. Process.*, **13**, No.4, Dec.1998, p.369-78.]

The radiant energy delivered to the material depends on the material's ability to absorb energy which may change as solvent evaporates. Radiant energy

- compensates for energy lost due to the evaporative cooling - this is most beneficial during the early stages of the process
- improves performance of the dryer when changes to either airflow rate or energy supplied are too costly to make.

Radiant energy can be used to improve process control. For example, in a multilayer coating (especially wet on wet), radiant energy can be used to regulate heat flow to each layer using the differences in their radiant energy absorption and coefficients of thermal conductivity and convective heat transfer. Experimental work by Cairncross *et al.*[32] shows how a combination dryer can be designed and regulated to increase the drying rate and eliminate bubble formation (more information on the conditions of bubble formation is included in Section 7.2 of this chapter). Shepard[33] shows how drying and curing rates in multilayer coating can be measured by dielectric analysis. Koenders *et al.*[34] gives information on the prediction and practice of evaporation of solvent mixtures.

Vrentas and Vrentas papers provide relevant modelling studies.[35,36] These studies were initiated to explain the earlier observations by Crank[37] which indicated that maintaining a slightly increased concentration of solvent in the air flowing over a drying material may actually increase the evaporation rate. Modelling of the process shows that although the diffusion of solvent cannot be increased by an increased concentration of solvent on the material's surface, an increased concentration of solvent in the air may be beneficial for the evaporation process because it prevents the formation of skin which slows down solvent diffusion.

Analysis of solvent evaporation from paint and subsequent shrinkage[38] and drying of small particles obtained by aerosolization[39] give further insight into industrial drying processes. The two basic parameters affecting drying rate are temperature and film thickness. Figure 7.1.26 shows that the diffusion coefficient increases as temperature increases. Figure 7.1.27 shows that by reducing film thickness, drying time can be considerably reduced.[40] Further information on the design and modelling of drying processes can be found in a review paper[41] which analyzes drying process in multi-component systems.

## REFERENCES


1 J S Vrentas and J.L. Duda, *J. Polym. Sci., Polym. Phys. Ed.*, **15**, 403 (1977); **17**, 1085, (1979).
2 J S Vrentas, C M Vrentas, *J. Polym. Sci., Part B: Polym Phys.*, **30**(9), 1005-11 (1992).
3 D Arnauld, R L Laurence, *Ind. Eng. Chem. Res.*, **31**(1), 218-28 (1992).
4 F D Blum, S Pickup, R A Waggoner, *Polym. Prep.*, **31** (1), 125-6 (1990).
5 C Bouchard, B Guerrier, C Allain, A Laschitsch, A C Saby, D Johannsmann, *J. Appl. Polym. Sci.*, **69**, No.11, 2235-46 (1998).
6 G Fleischer, J Karger, F Rittig, P Hoerner, G Riess, K Schmutzler, M Appel, *Polym. Adv. Technol.*, **9**, Nos.10-11, 700-8 (1998).
7 C F Fong, , Y Li, De Kee, J Bovenkamp, *Rubber Chem. Technol.*, **71**(2), 285-288 (1998).
8 Jinhua Chen, Kang Chen, Hong Zhang, Wanzhi Wei, Lihua Nie, Shouzhuo Yao, *J. Appl. Polym. Sci.*, **66**, No.3, 563-71 (1997).
9 I Linossier, F Gaillard, M Romand, J F Feller, *J. Appl. Polym. Sci.*, **66**, No.13, 2465-73 (1997).
10 L Meistermann, M Duval, B Tinland, *Polym. Bull.*, **39**, No.1, 101-8 (1997).
11 J S Qi, C Krishnan, J A Incavo, V Jain, W L Rueter, *Ind. Eng. Chem. Res.*, **35**, No.10, 3422-30 (1996).
12 Z Hrnjak-Murgic, J Jelencic, M Bravar, *Angew. Makromol. Chem.*, **242**, 85-96 (1996).
13 B Hladik, S Frahn, W Borchard, *Polym. Polym. Compos.*, **3**, No.1, 21-8 (1995).
14 J Frenkel, *Rubber Chem. Technol.*, **13**, 264 (1940).
15 P J Flory, J Rehner, *J. Chem. Phys.*, **11**, 521 (1943).
16 P J Flory, *J. Chem. Phys.*, **18**, 108 (1950).
17 H J Mencer, Z Gomzi, *Eur. Polym. J.*, **30**, 1, 33-36, (1994).
18 A G Webb, L D Hall, *Polymer*, **32**(16), 2926-38 (1991).
19 M Ercken, P Adriensens, D Vanderzande, J Gelan, *Macromolecules*, **28**, 8541-8547, (1995).
20 F Horkay, A M Hecht, E Geissler, *Macromolecules*, **31**, No.25, 8851-6 (1998).
21 M Sen, O Guven, *Polymer*, **39**, No.5, 1165-72 (1998).
22 M Sen, A Yakar, O Guven, *Polymer*, **40**, No.11, 2969-74 (1999).
23 Y D Zaroslov, O E Philippova, A R Khokhlov, *Macromolecules*, **32**, No.5, 1508-13 (1999).
24 B H Stuart, D R Williams, *Polymer*, **35**, No.6, 1326-8 (1994).
25 B H Stuart, D R Williams, *Polymer*, **36**, No.22, 4209-13 (1995).
26 F M Fowkes, D O Tischler, *J. Polym. Sci., Polym. Chem. Ed.*, **22**, 547 (1984).
27 L A Errede, *Macromol. Symp.*, **114**, 73-84 (1997).
28 L A Errede, P J Henrich, J N Schroepfer, *J. Appl. Polym. Sci.*, **54**, No.5, 649-67 (1994).
29 M O Ngui, S K Mallapragada, *J. Polym. Sci.: Polym. Phys. Ed.*, **36**, No.15, 2771-80 (1998).
30 N A Peppas, J C Wu, E D von Meerwall, *Macromolecules*, **27**, 5626 (1994).
31 S Kojima, T Moriga, *Polym. Eng. Sci.*, **35**, No.13, 1098-105 (1995).
32 R A Cairncross, S Jeyadev, R F Dunham, K Evans, L F Francis, L Scriven, *J. Appl. Polym. Sci.*, **58**, No.8, 1279-90 (1995).
33 D C Shepard, *J. Coat. Technol.*, **68**, No.857, 99-102 (1996).
34 L Khal, E Juergens, J Petzoldt, M Sonntag, *Urethanes Technology*, **14**, No.3, 23-6 (1997).
35 J S Vrentas, C M Vrentas, *J. Polym. Sci., Part B: Polym Phys.*, **30**(9), 1005-11 (1992).
36 J S Vrentas, C M Vrentas, *J. Appl. Polym. Sci.*, **60**, No.7, 1049-55 (1996).
37 J Crank, *Proc. Phys. Soc.*, **63**, 484 (1950).
38 L Ion, J M Vergnaud, *Polym. Testing*, **14**, No.5, 479-87 (1995).
39 S Norasetthekul, A M Gadalla, H J Ploehn, *J. Appl. Polym. Sci.*, **58**, No.11, 2101-10 (1995).
40 G A Abraham, T R Cuadrado, *International Polym. Process.*, **13**, No.4, Dec.1998, 369-78.
41 Z Pakowski, *Adv. Drying*, **5**, 145-202 (1992).

# 7.2 BUBBLES DYNAMICS AND BOILING OF POLYMERIC SOLUTIONS

SEMYON LEVITSKY
**Negev Academic College of Engineering, Israel**
ZINOVIY SHULMAN
**A.V. Luikov Heat and Mass Transfer Institute, Belarus**

## 7.2.1 RHEOLOGY OF POLYMERIC SOLUTIONS AND BUBBLE DYNAMICS

### 7.2.1.1 Rheological characterization of solutions of polymers

Solutions of polymers exhibit a number of unusual effects in flows.[1] Complex mechanical behavior of such liquids is governed by qualitatively different response of the medium to applied forces than low-molecular fluids. In hydrodynamics of polymers this response is described by rheological equation that relates the stress tensor, σ, to the velocity field. The latter is described by the rate-of-strain tensor, e

$$e_{ij} = \frac{1}{2}\left(\frac{\partial v_i}{\partial x_j} + \frac{\partial v_j}{\partial x_i}\right) \qquad [7.2.1]$$

where:

$v_i$ components of the velocity vector, $\vec{v}$
$x_i$ Cartesian coordinates (i, j = 1, 2, 3)

The tensors σ and e include isotropic, p, $e_{kk}$ and deviatoric, τ, s contributions:

$$\sigma = -pI + \tau, \quad p = -\frac{1}{3}\sigma_{kk}, \quad e = \frac{1}{3}e_{kk}I + s, \quad e_{kk} = \nabla \cdot \vec{v} \qquad [7.2.2]$$

where:

I unit tensor
∇ Hamiltonian operator

For incompressible fluid $\nabla \cdot \bar{v} = 0$, e = s and rheological equation can be formulated in the form of the τ- dependence from e. Compressibility of the liquid must be accounted for in fast dynamic processes such as acoustic waves propagation, etc. For compressible medium the dependence of pressure, p, on the density, ρ, and the temperature, T, should be specified by equation, p = p(ρ,T), that is usually called the equation of state.

The simplest rheological equation corresponds to incompressible viscous Newtonian liquid and has the form

$$\tau = 2\eta_0 e \qquad [7.2.3]$$

where:

$\eta_0$ viscosity coefficient

Generalizations of the Newton's flow law [7.2.3] for polymeric liquids are aimed to describe in more or less details the features of their rheological behavior. The most important among these features is the ability to accumulate elastic deformation during flow and thus to exhibit the memory effects. At first we restrict ourselves to the case of small deformation rates to discuss the basic principles of the general linear theory of viscoelasticity based upon thermodynamics of materials with memory.[2] The central idea of the theory is the postulate that instantaneous stresses in a medium depend on the deformation history. This suggestion leads to integral relationship between the stress and rate-of-strain tensors

$$\tau_{ij} = 2\int_{-\infty}^{t} G_1(t-t')s_{ij}(t')dt', \quad \sigma_{kk} = -3p_0 + 3\int_{-\infty}^{t} G_2(t-t')e_{kk}(t')dt' - 3\int_{-\infty}^{t} G_3(t-t')\frac{\partial\theta}{\partial t'}dt' \quad [7.2.4]$$

where:

$\theta$ = T - $T_0$ deviation of the temperature from its equilibrium value
$p_0$ equilibrium pressure
$G_1, G_2, G_3$ relaxation functions

The relaxation functions $G_1(t)$ and $G_2(t)$ satisfy restrictions, following from the entropy production inequality

$$\frac{\partial G_1}{\partial t} \le 0, \quad \frac{\partial G_2}{\partial t} \le 0 \quad [7.2.5]$$

Additional restriction on $G_{1,2}(t)$ is imposed by the decaying memory principle[3] that has clear physical meaning: the state of a medium at the present moment of time is more dependent on the stresses arising at $t = t_2$ than on that stresses arising at $t = t_1$ if $t_1 < t_2$. This principle implies that the inequality $\left(\partial^2 G_{1,2} / \partial t^2\right) \ge 0$ must be satisfied.

The necessary condition for the viscoelastic material to be a liquid means

$$\lim_{t\to\infty} G_1(t) = 0 \quad [7.2.6]$$

Unlike $G_1(t)$, the functions $G_2(t)$ and $G_3(t)$ contain non-zero equilibrium components, such as

$$\lim_{t\to\infty} G_2(t) = G_{20}, \quad \lim_{t\to\infty} G_3(t) = G_{30} \quad [7.2.7]$$

From the physical point of view this difference between $G_1$ and $G_{2,3}$ owes to the fact that liquid possesses a finite equilibrium bulk elasticity.

Spectral representations of relaxation functions, accounting for [7.2.5] - [7.2.7], have form ($G_{10} = 0$)

$$G_i(t) = G_{i0} + \int_0^{\infty} F_i(\lambda)\exp(-t/\lambda)d\lambda \quad [7.2.8]$$

where:

$\lambda$ relaxation time
$F_i(\lambda)$ spectrum of relaxation times (i = 1, 2, 3)

Equation [7.2.8] defines the functions $G_i(t)$ for continuous distribution of relaxation times. For a discrete spectrum, containing $n_i$ relaxation times, the distribution function takes the form

$$F_i(\lambda) = \sum_{k=1}^{n_i} G_{ik}\delta(\lambda - \lambda_{ik}) \quad [7.2.9]$$

where:

$G_{ik}$ partial modules, corresponding to $\lambda_{ik}$
$\delta(\lambda - \lambda_{ik})$ Dirac delta function

In this case the integration over the spectrum in Equation [7.2.8] is replaced by summation over all relaxation times, $\lambda_{ik}$.

For polymeric solutions, as distinct to melts, it is convenient to introduce in the right-hand sides of equations [7.2.4] additional terms, $2\eta_s s_{ij}$ and $3\rho_0\eta_v e_{kk}$, which represent contributions of shear, $\eta_s$, and bulk, $\eta_v$, viscosities of the solvent. The result has the form

$$\tau_{ij} = 2\int_{-\infty}^{t}\int_{0}^{\infty} F_1(\lambda)\exp\left(-\frac{t-t'}{\lambda}\right)s_{ij}(t')d\lambda dt' + 2\eta_s s_{ij} \qquad [7.2.10]$$

$$p = p_0 - G_{20}\int_{-\infty}^{t} e_{kk}(t')dt' - \int_{-\infty}^{t}\int_{0}^{\infty} F_2(\lambda)\exp\left(-\frac{t-t'}{\lambda}\right)e_{kk}(t')d\lambda dt' + G_{30}\theta +$$

$$+\int_{-\infty}^{t}\int_{0}^{\infty} F_3(\lambda)\exp\left(-\frac{t-t'}{\lambda}\right)\frac{\partial\theta(t')}{\partial t'}d\lambda dt' - \rho_0\eta_v e_{kk}, \quad p = -1/3\sigma_{kk} \qquad [7.2.11]$$

Equilibrium values of bulk and shear viscosity, $\eta_b$ and $\eta_p$, can be expressed in terms of the relaxation spectra, $F_1$ and $F_2$, as:[1]

$$\eta_p - \eta_s = \int_0^{\infty}\lambda F_1(\lambda)d\lambda, \quad \eta_b - \eta_v = \int_0^{\infty}\lambda F_2(\lambda)d\lambda \qquad [7.2.12]$$

In the special case when relaxation spectrum, $F_1(\lambda)$, contains only one relaxation time, $\lambda_{11}$, equation [7.2.10] yields

$$\tau_{ij} = 2G_{11}\int_{-\infty}^{t}\exp\left(-\frac{t-t'}{\lambda_{11}}\right)s_{ij}(t')dt' + 2\eta_s s_{ij}, \quad G_{11} = (\eta_p - \eta_s)/\lambda_{11} \qquad [7.2.13]$$

At $\eta_s = 0$ the integral equation [7.2.13] is equivalent to the linear differential Maxwell equation

$$\tau_{ij} + \lambda_{11}\dot{\tau}_{ij} = 2\eta_p s_{ij} \qquad [7.2.14]$$

Setting $\eta_s = \lambda_2\eta_0/\lambda_1$, $\lambda_1 = \lambda_{11}$, $\eta_p = \eta_0$, where $\lambda_2$ is the retardation time, one can rearrange equation [7.2.14] to receive the linear Oldroyd equation[3]

$$\tau_{ij} + \lambda_1\dot{\tau}_{ij} = 2\eta_0\left(s_{ij} + \lambda_{ij}\dot{s}_{ij}\right), \quad \lambda_1 \ge \lambda_2 \qquad [7.2.15]$$

Thus, the Oldroyd model represents a special case of the general hereditary model [7.2.10] with appropriate choice of parameters. Usually the maximum relaxation time in the spectrum is taken for $\lambda_1$ in equation [7.2.15] and therefore it can be used for quantitative description and estimates of relaxation effects in non-steady flows of polymeric systems.

Equation [7.2.11], written for quasi-equilibrium process, helps to clarify the meaning of the modules $G_{20}$ and $G_{30}$. In this case it gives

$$p = p_0 + G_{20}\rho_0^{-1}(\rho - \rho_0) + G_{30}\theta \quad [7.2.16]$$

Thermodynamic equation of state for non-relaxing liquid at small deviations from equilibrium can be written as follows

$$p = p_0 + \left(\frac{\partial p}{\partial \rho}\right)_{T=T_0}(\rho - \rho_0) + \left(\frac{\partial p}{\partial T}\right)_{\rho=\rho_0}\theta \quad [7.2.17]$$

The thermal expansion coefficient, $\alpha$, and the isothermal bulk modulus, $K_{is}$, are defined as[4]

$$\alpha = -\rho_0\left(\frac{\partial \rho}{\partial T}\right)_{\rho=\rho_0}, \quad K_{is} = \rho_0\left(\frac{\partial p}{\partial \rho}\right)_{T=T_0} \quad [7.2.18]$$

Therefore, equation [7.2.17] can be rewritten in the form

$$p = p_0 + K_{is}\rho_0^{-1}(\rho - \rho_0) + \alpha K_{is}\theta \quad [7.2.19]$$

From [7.2.16] and [7.2.18] it follows that $G_{20} = K_{is}$, $G_{30} = \alpha K_{is}$.

In rheology of polymers complex dynamic modulus, $G_k$*, is of special importance. It is introduced to describe periodic deformations with frequency, $\omega$, and defined according to:

$$G_k^* = \int_0^\infty \frac{F_k(\lambda)(\omega\lambda)(\omega\lambda + i)d\lambda}{1 + (\omega\lambda)^2} \quad [7.2.20]$$

Equations of motion of the liquid follow from momentum and mass conservation laws. In the absence of volume forces they mean:

$$\rho\frac{d\vec{v}}{dt} = \nabla \cdot \sigma \quad [7.2.21]$$

$$\frac{d\rho}{dt} + \rho\nabla \cdot \vec{v} = 0 \quad [7.2.22]$$

For polymeric solution the stress tensor, $\sigma$, is defined according to [7.2.10], [7.2.11]. To close the system, it is necessary to add the energy conservation law to equations [7.2.21], [7.2.22]. In the case of liquid with memory it has the form[2]

$$k\nabla^2\theta - T_0\frac{\partial}{\partial t}\int_{-\infty}^{t} G_3(t - t')e_{kk}(t')dt' = T_0\frac{\partial}{\partial t}\int_{-\infty}^{t} G_4(t - t')\frac{\partial\theta}{\partial t'}dt'$$

$$G_4(t - t') = G_{40} - \int_0^\infty F_4(\lambda)\exp\left(-\frac{t - t'}{\lambda}\right)d\lambda, \quad G_{40} = \rho_0 c_v T_0^{-1} \quad [7.2.23]$$

where:

k heat conductivity of the liquid

$c_v$ equilibrium isochoric specific heat capacity

Equations [7.2.10], [7.2.11], [7.2.21] - [7.2.23] constitute complete set of equations for linear thermohydrodynamics of polymeric solutions.

The non-linear generalization of equation [7.2.4] is not single. According to the Kohlemann-Noll theory of a "simple" liquid,[5] a general nonlinear rheological equation for a compressible material with memory may be represented in terms of the tensor functional that defines the relationship between the stress tensor, σ, and the deformation history. The form of this functional determines specific non-linear rheological model for a hereditary medium. Such models are numerous, the basic part of encountered ones can be found elsewhere.[1,6-8] It is important, nevertheless, that in most cases the integral rheological relationships of such a type for an incompressible liquid may be brought to the set of the first-order differential equations.[9] Important special case of this model represents the generalized Maxwell's model that includes the most general time-derivative of the symmetrical tensor

$$\tau = \sum_k \tau^{(k)}, \quad \tau^{(k)} + \lambda_k F_{abc} \tau^{(k)} = 2\eta_k e \qquad [7.2.24]$$

$$F_{abc}\tau = \frac{D\tau}{Dt} + a(\tau \cdot e + e \cdot \tau) + b I tr(\tau \cdot e) + ce\, tr(\tau), \quad \frac{D\tau}{Dt} = \frac{d\tau}{dt} - w \cdot \tau + \tau \cdot w, \quad w = \tfrac{1}{2}\left(\nabla\vec{v} - \nabla\vec{v}^T\right)$$

where:

D/Dt Jaumann's derivative[1]

d/dt ordinary total derivative

tr trace of the tensor, $tr\tau = \tau_{kk}$

w vorticity tensor

$\nabla\vec{v}^T$ transpose of the tensor $\nabla\vec{v}$

$\lambda_k$, $\eta_k$ parameters, corresponding to the Maxwell-type element with the number k

At a = -1, b = c = 0, equations [7.2.24] correspond to the Maxwell liquid with a discrete spectrum of relaxation times and the upper convective time derivative.[3] For solution of polymer in a pure viscous liquid, it is convenient to represent this model in such a form that the solvent contribution into total stress tensor will be explicit:

$$\tau = \sum_k \tau^{(k)} + 2\eta_s e, \quad \tau^{(k)} + \lambda_k \left[\frac{D\tau^{(k)}}{Dt} - \left(\tau^{(k)} \cdot e + e \cdot \tau^{(k)}\right)\right] = 2\eta_k e \qquad [7.2.25]$$

To select a particular nonlinear rheological model for hydrodynamic description of the fluid flow, it is necessary to account for kinematic type of the latter.[7] For example, the radial flows arising from the bubble growth, collapse or pulsations in liquid belong to the elongational type.[3] Therefore, the agreement between the experimental and theoretically predicted dependencies of elongational viscosity on the elongational deformation rate should be a basic guideline in choosing the model. According to data[10-13] the features of elongational viscosity in a number of cases can be described by equations [7.2.25]. More simple version of equation [7.2.25] includes single relaxation time and additional parameter $1/2 \le \alpha \le 1$, controlling the input of nonlinear terms:[7]

$$\tau = \tau^{(1)} + \tau^{(2)}, \quad \tau^{(1)} + \lambda\left[\frac{D\tau^{(1)}}{Dt} - \alpha\left(\tau^{(1)}\cdot e + e\cdot\tau^{(1)}\right)\right] = 2\eta\beta e, \quad \tau^{(2)} = 2\eta(1-\beta)e \qquad [7.2.26]$$

Parameter β governs the contribution of the Maxwell element to effective viscosity, η,(Newtonian viscosity of the solution). Equation [7.2.26] is similar to the Oldroyd-type equation [7.2.15] with the only difference that in the former the upper convective derivative is used to account for nonlinear effects instead of partial derivative, $\partial / \partial t$.

Phenomenological parameters appearing in theoretical models can be found from appropriate rheological experiments.[6] Certain parameters, the most important being relaxation times and viscosities, can be estimated from molecular theories. According to molecular theory, each relaxation time $\lambda_k$ is relative to mobility of some structural elements of a polymer. Therefore, the system as a whole is characterized by the spectrum of relaxation times. Relaxation phenomena, observed at a macroscopic level, owe their origin to the fact that response of macromolecules and macromolecular blocks to different-in-rate external actions is described by different parts of their relaxation spectrum. This response is significantly affected by temperature - its increase "triggers" the motion of more and more complex elements of the macromolecular hierarchy (groups of atoms, free segments, coupled segments, etc).

The most studied relaxation processes from the point of view of molecular theories are those governing relaxation function, $G_1(t)$, in equation [7.2.4]. According to the Rouse theory,[1] a macromolecule is modeled by a bead-spring chain. The beads are the centers of hydrodynamic interaction of a molecule with a solvent while the springs model elastic linkage between the beads. The polymer macromolecule is subdivided into a number of equal segments (submolecules or subchains) within which the equilibrium is supposed to be achieved; thus the model does not permit to describe small-scale motions that are smaller in size than the statistical segment. Maximal relaxation time in a spectrum is expressed in terms of macroscopic parameters of the system, which can be easily measured:

$$\lambda_{11} = \frac{6(\eta_p - \eta_s)M}{\pi^2 c R_G T} \qquad [7.2.27]$$

where:

| | |
|---|---|
| M | molecular mass of the polymer |
| c | concentration of polymer in solution |
| $R_G$ | universal gas constant |

The other relaxation times are defined as $\lambda_{1k} = \lambda_{11}/k^2$. In Rouse theory all the modules $G_{1k}$ are assumed to be the same and equal to $cR_GT/M$.

In the Kirkwood-Riseman-Zimm (KRZ) model, unlike Rouse theory, the hydrodynamic interaction between the segments of a macromolecular chain is accounted for. In the limiting case of a tight macromolecular globe, the KRZ theory gives the expression for $\lambda_{11}$ that is similar to [7.2.27]:

$$\lambda_{11} = \frac{0.422(\eta_p - \eta_s)M}{c R_G T} \qquad [7.2.28]$$

The differences between other relaxation times in the both spectra are more essential: the distribution, predicted by the KRZ model, is much narrower than that predicted by the Rouse theory.

The KSR and Rouse models were subjected to numerous experimental tests. A reasonably good agreement between the theoretical predictions and experimental data was demonstrated for a variety of dilute polymeric solutions.[14] Further advance in the molecular-kinetic approach to description of relaxation processes in polymeric systems have brought about more sophisticated models.[15,16] They improve the classical results by taking into account additional factors and/or considering diverse frequency, temperature, and concentration ranges, etc. For the aims of computer simulation of the polymeric liquid dynamics in hydrodynamic problems, either simple approximations of the spectrum, $F_1(\lambda)$, or the model of subchains are usually used. Spriggs law[17] is the most used approximation

$$\lambda_{1k} = \lambda_{11} / k^Z, \quad z \geq 2 \qquad [7.2.29]$$

The molecular theory predicts strong temperature dependence of the relaxation characteristics of polymeric systems that is described by the time-temperature superposition (TTS) principle.[18] This principle is based on numerous experimental data and states that with the change in temperature the relaxation spectrum as a whole shifts in a self-similar manner along t axis. Therefore, dynamic functions corresponding to different temperatures are similar to each other in shape but are shifted along the frequency axis by the value $a_T$; the latter is named the temperature-shift factor. With $\omega a_T$ for an argument it becomes possible to plot temperature-invariant curves $\mathrm{Re}\{G_1^*(\omega a_T)\}$ and $\mathrm{Im}\{G_1^*(\omega a_T)\}$. The temperature dependence of $a_T$ is defined by the formula

$$a_T = \frac{\rho(T_0)T_0(\eta_p(T) - \eta_s(T))}{\rho(T)T(\eta_p(T_0) - \eta_s(T_0))} \qquad [7.2.30]$$

The dependence of viscosity on the temperature can be described by the activation theory[19]

$$\eta_p = \eta_{p0} \exp\left[E_p(R_G T_0)^{-1}(T_0 / T - 1)\right], \quad \eta_s = \eta_{s0} \exp\left[E_s(R_G T_0)^{-1}(T_0 / T - 1)\right] \qquad [7.2.31]$$

where:

$E_p$, $E_s$ activation energies for the solution and solvent, respectively

The $E_s$ value is usually about 10 to 20 kJ/mol. For low-concentrated solutions of polymers with moderate molecular masses, the difference between these two activation energies, $\Delta E = E_p - E_s$, does not exceed usually 10 kJ/mol.[18,20] For low-concentrated solutions of certain polymers in thermodynamically bad solvents negative $\Delta E$ values were reported.[20]

The Newtonian viscosity of solution related to the polymer concentration can be evaluated, for example, using Martin equation[18]

$$\eta_p / \eta_s = 1 + \tilde{c} \exp(k_M \tilde{c}), \quad \tilde{c} = c[\eta] \qquad [7.2.32]$$

where:

$\tilde{c}$ reduced concentration of polymer in the solution

$[\eta]$ intrinsic viscosity

Martin equation is usually valid in the range of reduced concentrations, $\tilde{c} \le 10$. For evaluation of $[\eta]$, the Mark-Houwink relationship[21] is recommended

$$[\eta] = KM^{a} \qquad [7.2.33]$$

where K and a are constants for a given polymer-solvent pair at a given temperature over a certain range of the molecular mass variation. The parameter a (the Mark-Houwink exponent) lies in the range 0.5 to 0.6 for solutions of flexible chains polymers in thermodynamically bad solvents and in the range 0.7- 0.8 for good solvents. For the former ones the constant $K \approx 10^{-2}$ (if the intrinsic viscosity $[\eta]$ is measured in $cm^3/g$), while for the latter $K \approx 10^{-3}$.

Thus, the spectral functions, $F_1(\lambda)$, are comprehensively studied both experimentally and theoretically. The behavior of relaxation functions, $G_2(t)$ and $G_3(t)$, is still much less known. The properties of the function, $G_2(t)$, were mainly studied in experiments with longitudinal ultrasound waves.[22-24] It has been found that relaxation mechanisms manifested in shear and bulk deformations are of a similar nature. In particular, polymeric solutions are characterized by close values of the temperature-shift factors and similar relaxation behavior of both shear and bulk viscosity. The data on the function, $F_3(\lambda)$, indicate that relaxation behavior of isotropic deformation at thermal expansion can be neglected for temperatures well above the glass-transition temperature.[22]

### 7.2.1.2 Dynamic interaction of bubbles with polymeric liquid

Behavior of bubbles in liquid at varying external pressure or temperature is governed by cooperative action of a number of physical mechanisms, which are briefly discussed below. Sufficiently small bubbles execute radial motions (growth, collapse, pulsations) retaining their spherical shape and exchanging heat, mass and momentum with environment. Heat transfer between phases at free oscillations of gas bubbles is caused by gas heating during compression and its cooling when expanding. Due to the difference in thermal resistance of liquid and gas, the total heat flux from gas to liquid is positive for the oscillation period. This unbalanced heat exchange is the source of so-called heat dissipation. The magnitude of the latter depends on the relation between the natural time of the bubble (the Rayleigh time) $t_0 = R_0(\rho_{f0}/p_{f0})^{1/2}$, governed by the liquid inertia, and the time of temperature leveling in gas (characteristic time of heat transfer in a gas phase), $t_T = R_0^2/a_g$, that is from the thermal Peklet number, $Pe_T = t_T/t_0$,

where:

| | |
|---|---|
| R | radius of the bubble |
| $a_g$ | thermal diffusivity of gas, $a_g = k_g/(\rho_{g0}c_{gp})$ |
| $k_g$ | heat conductivity of gas |
| $c_{gp}$ | specific heat capacity of gas at constant pressure |
| 0 | index, referring to equilibrium state |
| f | index, referring to liquid |
| g | index, referring to gas |

In the limiting cases $Pe_T >> 1$ and $Pe_T << 1$, when behavior of gas inside the bubble is close to adiabatic and isothermal one, correspondingly, the heat dissipation is small. For a bubble oscillating in a sound field, the above conditions should be changed for $\bar{\omega} Pe_T >> 1$ and $\bar{\omega} Pe_T << 1$ with $\bar{\omega} = \omega t_0$ being the non-dimensional frequency.

At thermodynamic conditions, close to normal ones, the equilibrium vapor content of the bubble can be neglected, but it grows with temperature (or pressure reduction). In the vapor presence, the pressure and/or temperature variations inside the oscillating bubble cause evaporation-condensation processes that are accompanied by the heat exchange. In polymeric solutions, transition from liquid to vapor phase and conversely is possible only for a low-molecular solvent. The transport of the latter to the bubble-liquid interface from the bulk is controlled by the diffusion rate. In general case the equilibrium vapor pressure at the free surface of a polymeric solution is lower than that for a pure solvent. If a bubble contains a vapor-gas mixture, then the vapor supply to the interface from the bubble interior is controlled by diffusion rate in the vapor-gas phase. The concentration inhomogeneity within the bubble must be accounted for if $t_{Dg} > t_0$ or $Pe_{Dg} > 1$

where:

| | |
|---|---|
| $t_{Dg}$ | characteristic time of binary diffusion in vapor-gas phase, $t_{Dg} = R_0^2/D_g$ |
| $D_g$ | diffusion coefficient |
| $Pe_{Dg}$ | diffusion Peklet number for the vapor-gas phase, $Pe_{Dg} = t_{Dg}/t_0$ |

Fast motions of a bubble surface produce sound waves. Small (but non-zero) compressibility of the liquid is responsible for a finite velocity of acoustic signals propagation and leads to appearance of additional kind of the energy losses, called acoustic dissipation. When the bubble oscillates in a sound field, the acoustic losses entail an additional phase shift between the pressure in the incident wave and the interface motion. Since the bubbles are much more compressible than the surrounding liquid, the monopole sound scattering makes a major contribution to acoustic dissipation. The action of an incident wave on a bubble may be considered as spherically-symmetric for sound wavelengths in the liquid $l_f >> R_0$.

When the spherical bubble with radius $R_0$ is at rest in the liquid at ambient pressure, $p_{f0}$, the internal pressure, $p_{in}$, differs from $p_{f0}$ by the value of capillary pressure, that is

$$p_{in} = p_{f0} + 2\sigma / R_0 \qquad [7.2.34]$$

where:

| | |
|---|---|
| $\sigma$ | surface tension coefficient |

If the system temperature is below the boiling point at the given pressure, $p_{f0}$, the thermodynamic equilibrium of bubble in a liquid is possible only with a certain amount of inert gas inside the bubble. The pressure in vapor-gas mixture follows the Dalton law, that suggests that both the solvent vapor and the gas are perfect gases:

$$p_{in} = p_g + p_v = \left(\rho_g B_g + \rho_v B_v\right) T_m = \rho_m B_m T_m, \quad \rho_m = \rho_g + \rho_v \qquad [7.2.35]$$

$$B_m = (1 - k_0) B_g + k_0 B_v, \quad B_{g,v} = R_g / \mu_{g,v}$$

where:

| | |
|---|---|
| $k_0$ | equilibrium concentration of vapor inside the bubble |
| $\mu_{g,v}$ | molar masses of gas and solvent vapor |
| v | index, referring to vapor |

From [7.2.34], [7.2.35] follows the relation for $k_0$:

$$k_0 = \left[1 + B_v B_g^{-1} \left\{(1 + 2\bar{\sigma}) / \bar{p}_{v0} - 1\right\}\right]^{-1}, \quad \bar{p}_{v0} = p_{v0} / p_{f0}, \quad \bar{\sigma} = \sigma / (p_{v0} R_0) \qquad [7.2.36]$$

The equilibrium temperature enters equation [7.2.36] via the dependence of the saturated vapor pressure, $p_{v0}$, from $T_0$. Figure 7.2.1 illustrates the relation [7.2.36] for air-vapor bubbles in toluene.[25] The curves 1-3 correspond to temperatures $T_0 = 363, 378, 383.7$K (the

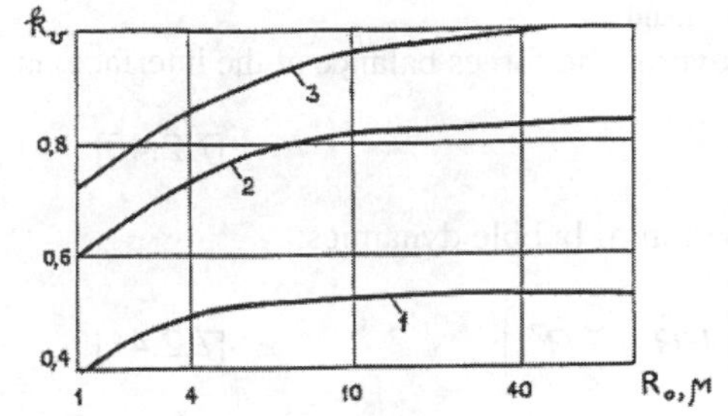


Figure 7.2.1. Dependence of the vapor content of air-vapor bubble in toluene from radius at different temperatures. [Adapted, by permission, from *Nauka i Tekhnica Press*, from the referencc 25]

latter value is equal to the saturation temperature $T_s$ for toluene at $p_{f0} = 10^5$ Pa). It follows from the calculated data that the vapor content dependence on the bubble radius is manifested only for minor bubbles as a result of capillary forces. The effect vanish for $R_0 >> 10$ mkm. For $T_0 << T_s$ the $k_0$ value is small and the bubble can be treated as a pure gas-filled.

The solvent vapor pressure above polymeric solution depends not only from the temperature but also on concentration of polymer. In most cases this dependence differs essentially from the linear Raul's law[21] and can be approximated by the Flory-Huggins equation, that must be used in evaluation of $k_0$ by the relation [7.2.36]

$$p_{v0} / p^0_{v0} = \phi_1 \exp\left[1 - \phi_1 + \chi(1 - \phi_1)^2\right], \quad \phi_1 = \kappa\left[\kappa + (1 - \kappa)K_p\right]^{-1}, \quad K_p = v_2 / v_1 \qquad [7.2.37]$$

where:

| | |
|---|---|
| $\phi_1, \kappa$ | volume and mass fraction of the volatile component |
| $p^0_{v0}, p_{v0}$ | vapor pressure above pure solvent and polymeric solution |
| $v_1, v_2$ | specific volumes of solvent and polymer |
| $\chi$ | parameter of thermodynamic interaction |

At $\phi_1 \to 0$ equation [7.2.37] gives a linear dependence of the relative vapor pressure, $p^0_{v0}/p_{v0}$ on the solvent volume concentration with the angle coefficient $\exp(1+\chi)$. At $\phi_1 \to 1$ solution obeys the Raul's law. Note that the value of $\phi_1$ in [7.2.37] is temperature-dependent due to difference in thermal expansion coefficients of components. The $\chi$ value for a given solvent depends on the concentration and molar mass of a polymer as well as on temperature. However, to a first approximation, these features may be ignored.[20] Usually $\chi$ varies within the range 0.2 - 0.5. For example, for solutions of polyethylene, natural rubber, and polystyrene in toluene $\chi = 0.28$, 0.393 and 0.456, correspondingly.

To describe the dynamic interaction of bubble with polymeric solution it is necessary to invoke equations of liquid motion, heat transfer and gas dynamics. General approach to description of bubble growth or collapse in a non-Newtonian liquid was formulated and developed.[26-31] The radial flow of incompressible liquid around growing or collapsing bubble is described by equations, following from [7.2.21], [7.2.22]:[27]

$$\rho_{f0}\left(\frac{\partial v_r}{\partial t} + v_r \frac{\partial v_r}{\partial r}\right) = -\frac{\partial p_f}{\partial r} + \frac{\partial \tau_{rr}}{\partial r} + \frac{2(\tau_{rr} - \tau_{\phi\phi})}{r} \qquad [7.2.38]$$

$$r\frac{\partial v_r}{\partial t} + 2v_r = 0 \qquad [7.2.39]$$

where:

| | |
|---|---|
| r, $\phi$ | radial and angular coordinates of the spherical coordinate system with the origin at the center of the bubble |

$v_r$ radial component of velocity in the liquid

The dynamic boundary condition that governs the forces balance at the interface, is:

$$p_g(R,t) = p_f(R,t) + 2\sigma R^{-1} - \tau_{rr}(R,t) \qquad [7.2.40]$$

From [7.2.38] - [7.2.40] follows the equation of bubble dynamics:

$$J + p_f(\infty) - p_g + 2\sigma R^{-1} = S, \quad J = \rho_{f0}\left(R\ddot{R} + \frac{3}{2}\dot{R}^2\right) \qquad [7.2.41]$$

$$S = 2\int_0^\infty (\tau_{rr} - \tau_{\phi\phi})(3y + R^3)^{-1} dy, \quad y = \frac{1}{3}(r^3 - R^3)$$

where:

$p_f(\infty)$ pressure in the liquid at infinity
$p_g$ pressure in the bubble
$\dot{R}, \ddot{R}$ derivatives of the bubble radius with respect to time, t

The equation [7.2.41] was investigated[25] for growing and collapsing cavity in a liquid, described by rheological model [7.2.26], at $p_f(\infty) - p_g$ = const. Similar analysis for another rheological model of the solution (the so-called "yo-yo" model of the polymer dynamics) was developed.[32] It was shown that viscoelastic properties of solution can be approximately accounted for only in the close vicinity of the interface, that is through the boundary condition [7.2.40]. In this case the integro-differential equation for R(t), following from [7.2.41], [7.2.26], can be reduced to a simple differential equation. The latter was analyzed accounting for the fact[7] that the effective viscosity $\eta_l$ of a polymeric solution in elongational flow around collapsing cavity can increase by the factor of $10^2$ to $10^3$. If the corresponding Reynolds number of the flow $Re = (\eta_p/\eta_l)Re_p$ ($Re_p = t_0/t_p$, $t_p = 4\eta_p(p_f(\infty) - p_g)^{-1}$) is small, the inertial terms in equation [7.2.41] can be neglected. For high-polymer solutions the inequalities $Re << Re_p$ and $Re < 1$ may be satisfied in elongational flow even in the case of $Re_p >> 1$. Under these assumptions the equation for the relative velocity of the bubble surface takes the form

$$\dot{z} + 2(z - z_1)(z - z_2) = 0, \quad z = \dot{x}/x, \quad x = R/R_0, \quad z_{1,2} = -A/4 \pm (A^2/16 + B/2)^{1/2}$$

$$A = \bar{\lambda}^{-1}(1-\beta)^{-1}(1 - 2k\bar{\lambda}Re_p), \quad B = (1-\beta)^{-1}k\bar{\lambda}^{-1}Re_p, \quad \bar{\lambda} = \lambda/t_0, \quad k = sign(p_g - p_f(\infty))$$

Here was adopted for simplicity that $\alpha = 1/2$ and $\bar{\sigma} << 1$ (the latter inequality is satisfied for bubbles with $R_0 >> 1$ mkm). Phase plot of this equation is presented in Figure 7.2.2. It is seen that for k = - 1 (collapsing cavity) $z \to z_1$ as $\bar{t} \to \infty$ if $z_0 > z_2$. The stationary point $z = z_2$ is unstable. The rate of the cavity collapse $z = z_1$ in the asymptotic regime satisfies inequality $z_p \le z_1 \le 0$, where $z_p = -Re_p$ is equal to the collapse rate of the cavity in a pure viscous fluid with viscosity of polymeric solution $\eta$. It means that the cavity closure in viscoelastic solution of polymer at asymptotic stage is slower than in a viscous liquid with the same equilibrium viscosity. On the contrary, the expansion under the same conditions is faster: at k = 1 $z_p \le z_1 \le z_x$, where $z_p = Re_p$ and $z_s = Re_s = (1 - \beta)^{-1}Re_p$ is the asymptotic rate of the cavity expansion in a pure solvent with the viscosity $(1 - \beta)\eta$. This result is explained by different behavior of the stress tensor component $\tau_{rr}$, controlling the fluid rheology effect on the cav-

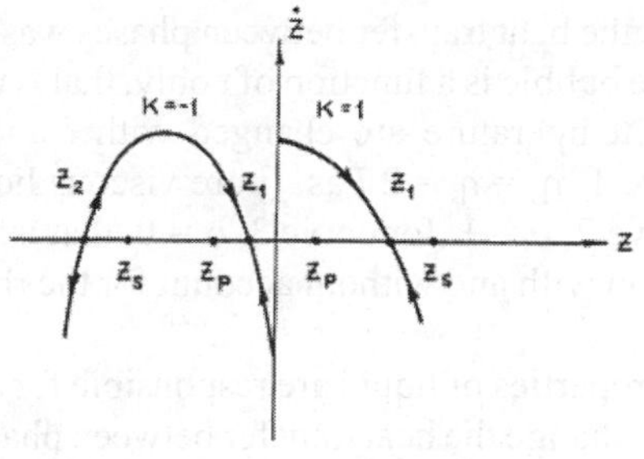

Figure 7.2.2. Phase plane for expanding and collapsing cavity in polymeric solution. [By permission of *Nauka i Tekhnica Press*, from the reference 25]

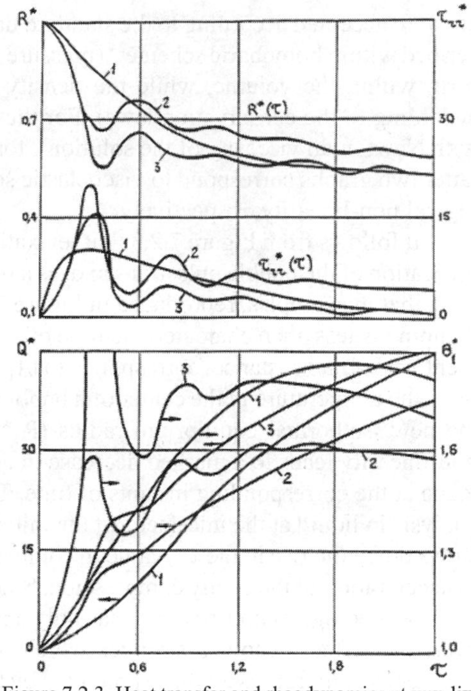

Figure 7.2.3. Heat transfer and rheodynamics at non-linear oscillations of a bubble in polymeric liquid. [By permission of *Nauka i Tekhnica Press*, from the reference 25]

ity dynamics, in extensional and compressional flows, respectively.[9] In the former case, the $\tau_{rr}$ value may be considerably greater than in the latter one.

Heat transfer between phases is a strong dissipative factor that in principle can mask the rheological features in bubble dynamics. Nevertheless, even with account for heat dissipation the theoretical dependencies of R(t) are sensitive to rheological properties of solution. Typical results of air bubble dynamics simulations at a sudden pressure change in the solution with $\eta_p >> \eta_s$ are presented on the Figure 7.2.3,

where:

| | |
|---|---|
| R* | dimensionless radius of the bubble, $R^* = R/R_0$ |
| $\tau$ | dimensionless time, $\tau = t/t_0$ |
| $\tau^*_{rr}$ | dimensionless radial component of the extra-stress tensor at the interface, $\tau^*_{rr} = \tau_{rr}(R,t)/p_{f0}$ |
| $\theta_1^0$ | dimensionless temperature at the center of the bubble, $\theta_1^0 = T_g(0,t)/T_0$ |
| Q* | dimensionless heat, transferred to liquid from the gas phase in a time $\tau$, $Q^* = Q/(R_0T_0k_gt_0)$ |
| $k_g$ | heat conductivity of gas |
| $\Delta p_f^*$ | dimensionless pressure change in the liquid at initial moment of time, $p_f^*(\infty) = 1 + \Delta p_f^* h(t)$, $\Delta p_f^* = \Delta p_f/p_{f0}$ |
| h(t) | unit step function |

Calculations have been done for the rheological model [7.2.25] with 20 relaxation elements in the spectrum, distributed according to the law [7.2.29] with z = 2. To illustrate the contribution of rheological non-linearity in equation [7.2.25] the numerical coefficient $\alpha$ ($\alpha$=1 or 0) was introduced in the term with $\lambda_k$, containing material derivative. The value $\alpha$=1 corresponds to non-linear model [7.2.25], while at $\alpha = 0$ equation [7.2.25] is equivalent to the linear hereditary model [7.2.10] with a discrete spectrum. Other parameters of the system were chosen as follows: $\eta_p = 2$ Pas, $\eta_s = 10^{-2}$ Pas, $\lambda_1 = 10^{-5}$ s, $R_0 = 50$ mkm, $\Delta p_f^* = 10$, $p_{f0} = 10^5$ Pa, $\rho_{f0} = 10^3$ kg/m$^3$, $T_0 = 293$K, $\sigma = 0.05$ N/m. Thermodynamic parameters of the air were accepted according to the standard data,[33] the heat transfer between phases was described within homobaric scheme[34] (pressure in the bubble is a function of t only, that is uniform within the volume, while the density and temperature are changed with r and t according to the conservation laws). For the curve 1, $\eta_p = \eta_s = 2$ Pas (pure viscous liquid with Newtonian viscosity of the solution), for curve 2, $\alpha = 1$, for curve 3, $\alpha = 0$, that is, the latter two graphs correspond to viscoelastic solution with and without account for the rheological non-linearity, respectively.

It follows from Figure 7.2.3 that relaxation properties of liquid are responsible for amplification of the bubble pulsations and, as a result, change the heat transfer between phases. Note that in examples, reproduced in Figure 7.2.3, the characteristic time of the pulsations' damping is less than characteristic time of the temperature leveling in gas, $t_T$ (the time moment $\tau = 2.4$, for instance, corresponds to $t/t_T \approx 0.1$). Therefore, after completion of oscillations the temperature in the center of a bubble is reasonably high, and the R* value exceeds the new isothermal equilibrium radius ($R_1^* = 0.453$). The manifestation of rheological non-linearity leads to a marked decrease in deviations of the bubble radius from the initial value at the corresponding instants of time. The explanation follows from stress dynamics analysis in liquid at the interface. At the initial stage, the relaxation of stresses in the liquid slows down the rise in the $\tau^*_{rr}$ value in comparison with similar Newtonian fluid. This leads to acceleration of the cavity compression. Since the stresses are small during this time interval, the rheological non-linearity has only a minor effect on the process. Further on, however, this effect becomes stronger which results in a considerable increase of normal stresses as compared with those predicted by the linear theory. It leads to deceleration of the cavity compression and, as a result, to decrease both in the maximum temperature of gas in a bubble and in the integral heat loss.

More detailed information about rheological features in gas bubble dynamics in polymeric solutions can be received within linear approach to the same problem that is valid for small pressure variations in the liquid. The equation describing gas bubble dynamics in a liquid with rheological equation [7.2.10] follows from [7.2.41], [7.2.10] and has the form[35]

$$\rho_{f0}\left(R\ddot{R} + \frac{3}{2}\dot{R}^2\right) = p_{g0}(R_0 / R)^{3\gamma} - p_f(\infty) - 2\sigma R^{-1} - 4\eta_s R / \dot{R} -$$

$$-4\int_0^\infty F_1(\lambda)\left\{\int_0^t \exp\left(-\frac{t-t'}{\lambda}\right)\dot{R}(t')R^{-1}(t')dt'\right\}d\lambda \qquad [7.2.42]$$

Here is supposed that gas in the bubble follows polytropic process with exponent λ. This equation was solved in linear approximation[35] by operational method with the aim to analyze small amplitude, natural oscillations of the constant mass bubble in relaxing liquid. It was taken $R = R_0 + \Delta R$, $\Delta R/R_0 << 1$, $\Delta R \sim \exp(ht)$ with h being the complex natural frequency. Logarithmic decrement, Λ, and dimensionless frequency, μ, of the oscillations are defined according to formulas

$$\Lambda = -2\pi \mathrm{Re}\{h\}\mathrm{Im}^{-1}\{h\}, \quad \mu = (2\pi)^{-1} t_0 \,\mathrm{Im}\{h\} \qquad [7.2.43]$$

Typical data for $\Lambda$ and $\mu$, calculated from the solution of linearized equation [7.2.42] for air bubble in the case of discrete spectrum with $n_1$ relaxation times, distributed by the Rouse law, are presented in Figures 7.2.4, 7.2.5. The maximum time, $\lambda_1$, was evaluated from equations [7.2.27], [7.2.32]. For curves 1- 6 and 1'- 6', $n_1$ = 1, 5, 10, 20, 50 and 100, correspondingly. Curve 7 refers to Newtonian fluid with $\eta = \eta_p$. For curves 1- 6, A = 200, for curves 1'- 6', A = 1000 with dimensionless parameter $A = p_{f0}M[\eta]/(R_G T_0)$. Other parameters were adopted as follows: $k_M = 0.4$, $\eta_s^* = 10^{-2}$, $\gamma = 1.36$. For atmospheric pressure and $\rho_{f0} = 10^3$ kg/m$^3$ these values correspond approximately to $\eta_s = 0.1$ Pas, $R_0 = 10^{-3}$ m. At these parameters, the relative effect of acoustic dissipation on damping of bubble pulsations is small.[31] The value of heat decrement is denoted in Figure 7.2.4 by $\Lambda_1$, the value of the rheological one - by $\Lambda_3$.

It is seen from the plots that viscoelasticity of the solution is responsible for reduction of rheological dissipative losses. The effect increases with reduced concentration of the polymer, $\tilde{c}$. When the number of relaxation times in the spectrum, accounted for in simulations, grows, the magnitude of $\Lambda_3$ also grows. Nevertheless, for all considered values the weakening of the damping, as compared with Newtonian fluid of the same viscosity, is essential. It follows from the analysis that in polymeric solution, pulsations of the bubble are possible even at relatively large c values (to the right of the dashed line in Figure 7.2.4). In this region of reduced concentrations the Newtonian viscosity of the solution is so high that only aperiodic solution for $\Delta R$ exists when the liquid follows the Newtonian model. The increase in molecular mass of the dissolved polymer at $\tilde{c}$ = const causes decrease in rheological losses.

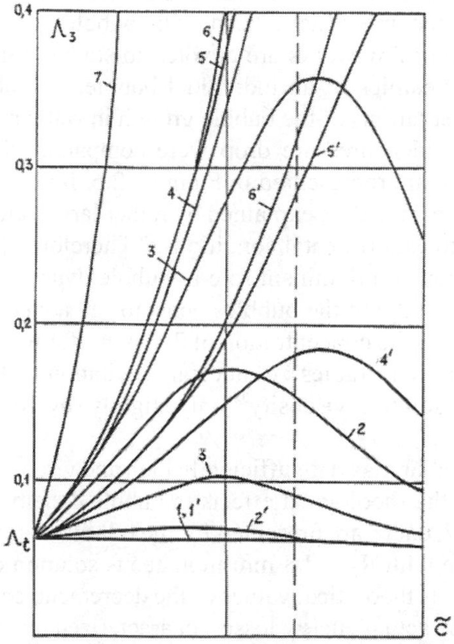

Figure 7.2.4. Rheological dissipative losses versus reduced concentration of a polymer. [By permission of *Nauka i Tekhnica Press*, from the reference 25]

The natural frequency of the bubble raises with $\tilde{c}$ and this behavior is qualita-

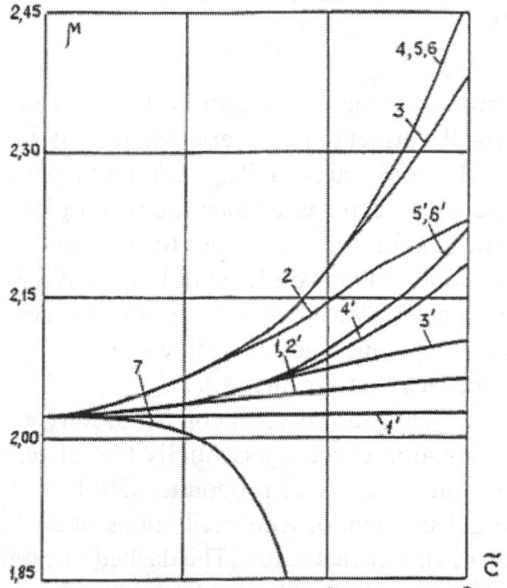

Figure 7.2.5. The effect of reduced polymer concentration on the natural frequency of bubbles. [By permission of *Nauka i Tekhnica Press*, from the reference 25]

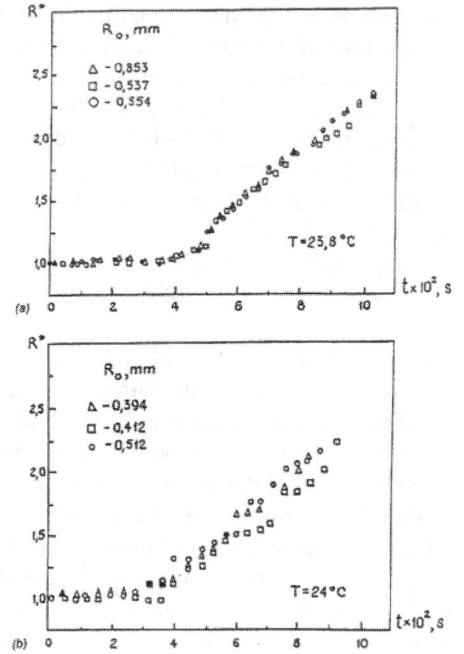


Figure 7.2.6. Expansion of bubbles in water and POE aqueous solution at a pressure drop. (a) - water, (b) - solution of POE with c = 420 ppm. [By permission of the American Institute of Physics from R.Y. Ting, and A.T. Ellis, *Phys. Fluids.*, **17**, 1461, 1974, the reference 36]

tively different from the concentration dependence of μ for a bubble in a pure viscous fluid with the same Newtonian viscosity as the solution (curve 7 in Figure 7.2.5).

The rheological properties of polymeric solution highly depend on temperature. Therefore, its variation affects the bubble pulsations. Thermorheological features in bubble dynamics have been studied[35] on the basis of temperature superposition principle, using relations [7.2.30], [7.2.31]. It was shown that the temperature rise leads to decrease of the decrement $\Lambda_3$, and this effect is enhanced with the increase of $n_1$. Note that the higher is the equilibrium temperature of the liquid, the less sensitive are the $\Lambda_3$ values to variations in $n_1$. It is explained by narrowing of the relaxation spectrum of the solution.[25]

Experimental results on bubble dynamics in solution of polymers are not numerous. Existing data characterize mainly the integral effects of polymeric additives on bulk phenomena associated with bubbles while only few works are devoted to studying the dynamics of an individual bubble. The observations of the bubble growth in water at a sudden pressure drop were compared with those for aqueous solution of POE.[36] The results are represented in Figure 7.2.6. It is seen that the effect of polymeric additives is small. This result is explained by rather large value of Reynolds number $Re_p = t_0/t_p$ corresponding to experimental conditions.[36] Therefore, the inertial effects rather than the rheological ones play a dominant role in bubble dynamics. The conclusions[36] were confirmed later[37] when studying the bubbles behavior in aqueous solution of POE (trade mark Polyox WS 301) with the concentration of 250 ppm (for water 1 ppm = 1 g/m$^3$). A similar result was received also in studies[38] of nucleate cavitation in the PAA aqueous solution with c = 0.1 kg/m$^3$. The solution viscosity[38] only slightly (by 10%) exceeded that of water.

When the polymer concentration (or molecular mass) is sufficiently high and viscosity of solution exceeds essentially that of solvent, the rheological effects in bubble dynamics become much more pronounced. In Figure 7.2.7, data[39] are presented for the relative damping decrement of free oscillations of air bubble with $R_0$ = 2.8 mm in aqueous solution of POE via concentration. The dashed line represents theoretical values of the decrement, corresponding to Newtonian liquid with $\eta = \eta_p$. The actual energy losses, characterized by experimental points, remained almost unchanged, despite the sharp rise in the "Newtonian" decrement, $\Lambda_p$, with c. This result correlates well with the above theoretical predictions and it is explained by viscoelastic properties of the solution. The same explanation has the phe-

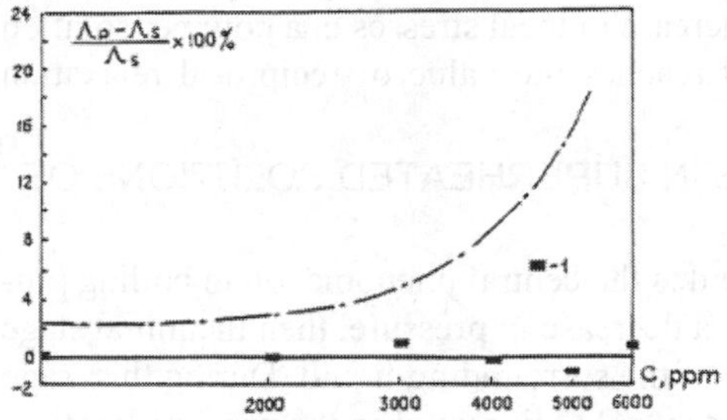


Figure 7.2.7. Relative decrement of free oscillations of air bubble versus concentration of POE in water. [By permission of IOP Publishing Limited from W.D. McComb, and S. Ayyash, *J. Phys. D. Appl. Phys.*, **13**, 773, 1980, the reference 39]

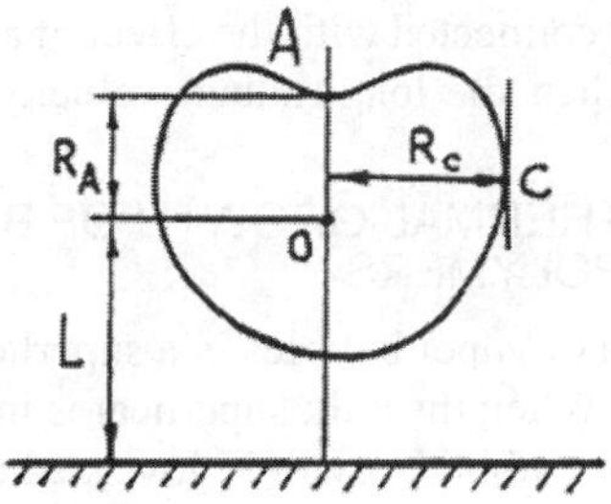


Figure 7.2.8. Geometrical parameters of the collapsing bubble. [By permission of the American Institute of Physics from G.L. Chahine, and D.H. Fruman, *Phys. Fluids.*, **22**, 1406, 1979, the reference 37]

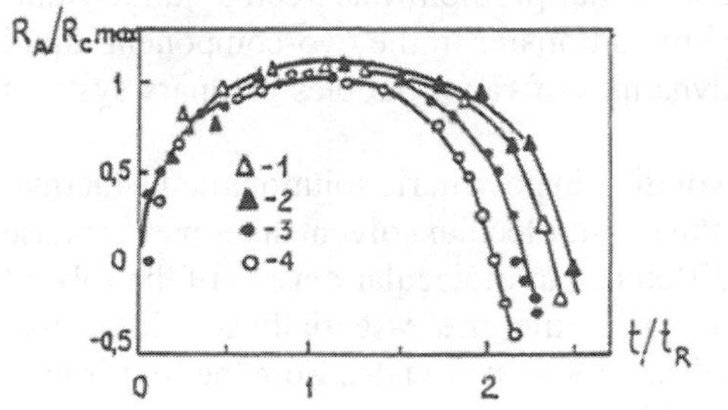


Figure 7.2.9. Collapse of a bubble near a solid wall in water (curves 1 and 4) and in dilute POE aqueous solution (curves 2, 3). [By permission of the American Institute of Physics from G.L. Chahine, and D.H. Fruman, *Phys. Fluids.*, **22**, 1406, 1979, the reference 37]

nomenon observed[40] in ultrasonic insonification of liquid polybutadiene with molecular mass $M \sim 10^5$ and $\eta_p \sim 10^6$ Pas. At ultrasound frequency f~18 kHz the acoustic cavitation and well-developed pulsations of bubbles were detected, in spite of for low-molecular liquids with so high viscosity it is completely impossible.[41]

For non-spherical bubbles the effect of polymeric additives becomes essential at lower concentrations, as compared to the spherical bubbles. For example, retardation of the bubble collapse near a solid wall was observed[37,42] in such concentration interval where dynamics of spherical bubbles has remained unchanged. In Figures 7.2.8 and 7.2.9 the data[37] are reproduced, where the Rayleigh time, $t_0$, was chosen for a scaling time, $t_R$, and for curves 1- 4 $R_{c,max}/L = 0.5, 0.56, 1.39, 1.25$, respectively. The bubble collapse is accompanied by generation of a microjet towards the wall and addition of polymer led to stabilization of the bubble shape and retardation of the jet formation. This effect is connected with the increase in the elongational viscosity of a polymeric solution in flow around collapsing bubble.

The effect of polymeric additives on collective phenomena, associated with the dynamics of bubbles, can be illustrated by the example of hydrodynamic cavitation, caused by abrupt decrease in local pressure (in flows around bodies, after stream contraction, in jets). It has been found that the use of polymers permits to decrease the cavitation noise, lower the cavitation erosion, and delay the cavitation inception. For example, adding a small amount of POE to a water jet issuing from the orifice caused the decrease of the critical cavitation number, $\kappa_{cr}$, by 35-40%.[43,44] In experiments with rotating disk[45] the value of $\kappa_{cr}$ was decreased by 65% with addition of 500 ppm POE. Note, however, that these features are linked not only to the changes in individual bubble dynamics, but to the influence of macromolecules on the total flow regime as well. In particular, phenomena listed above are closely connected with the effect of a strong increase in local stresses in a polymer solution flow when the longitudinal velocity gradient reaches the value of reciprocal relaxation time.[3]

## 7.2.2 THERMAL GROWTH OF BUBBLES IN SUPERHEATED SOLUTIONS OF POLYMERS

Growth of vapor bubbles in a superheated liquid is the central phenomenon in boiling processes. When the bulk superheat is induced by a decrease in pressure, then the initial stage of vapor bubble growth is governed by inertia of the surrounding liquid. During this stage the rheological properties of liquid play important role, discussed in the previous section. The basic features, characterizing this stage, are pressure changes within bubbles and their pulsations. After leveling of pressure in the phases, the process turns into the thermal stage when the cavity growth rate is controlled by ability of the liquid to supply the heat necessary for phase transitions. Expansion of vapor bubble in the thermal regime was examined[46] for the case of liquid representing a binary solution. Similar problem was treated[47] under additional assumption that the convective heat and mass transfer in the two-component liquid phase is insignificant. More recent works on dynamics of vapor bubbles in binary systems are reviewed elsewhere.[48-50]

The features, peculiar to vapor bubbles evolution in polymeric solutions at the thermal stage, owe mainly to the following. First, only the low-molecular solvent takes part in phase transitions at the interface because of a large difference in molecular masses of the solvent and polymer. The second, polymeric solutions, as a rule, are essentially non-ideal and, therefore, saturated vapor pressure of the volatile component deviates from the Raul's law. Finally, the diffusion coefficient in solution is highly concentration dependent that can greatly influence the rate of the solvent transport toward the interface. The role of the listed factors increases at boiling of systems that possess a lower critical solution temperature (LCST) and thus are subjected to phase separation in the temperature range $T < T_s$, where $T_s$ is the saturation temperature. In the latter case the rich-in-polymer phase which, as a rule, is more dense, accumulates near the heating surface (when a heater is placed at the bottom). As a consequence, the growth of bubbles proceed under limited supply of the volatile component.

Consider the expansion of a vapor cavity in a polymer solution with equilibrium mass concentration of the solvent, $k_0$, at the temperature $T_{f0} > T_s(k_0, p_{f0})$, assuming that both pressure and temperature in the vapor phase are constant

$$p_v = p_{f0}, \quad T_v = T_s(p_{f0}, k_R) = T_{fR}, \quad k_R = k(R, t), \quad T_{fR} = T_f(R, t) \qquad [7.2.44]$$

Parameters $k_0$, $T_{f0}$ characterize the state of solution far from the bubble (at $r = \infty$). Unlike a one-component liquid, the temperature $T_{fR}$ here is unknown. It is related to the surface concentration of solvent, $k_R$, by the equation of phase equilibrium at the interface.

Equations for heat transfer and diffusion in the solution have the form

$$\frac{\partial T_f}{\partial t} + v_{fR} \frac{R^2}{r^2} \frac{\partial T_f}{\partial r} = r^{-2} \frac{\partial}{\partial r}\left( a_f r^2 \frac{\partial T_f}{\partial r} \right) \qquad [7.2.45]$$

$$\frac{\partial k}{\partial t} + v_{fR} \frac{R^2}{r^2} \frac{\partial k}{\partial r} = r^{-2} \frac{\partial}{\partial r}\left( D r^2 \frac{\partial k}{\partial r} \right) \qquad [7.2.46]$$

Since the thermal diffusivity of solution, $a_f$, is less affected[20] by variations of temperature and concentration over the ranges $T_s(k_R) < T_f < T_{f0}$ and $k_R < k < k_0$, respectively, than the binary diffusion coefficient, D, it is assumed henceforward that $a_f$ = const. Furthermore, since the thermal boundary layer is much thicker than the diffusion layer, it is appropriate to assume that within the latter $D = D(k, T_{fR})$.

The boundary conditions for equations [7.2.45], [7.2.46] are as

$$T_f = T_{f0}, \qquad k = k_0 \qquad \text{at} \qquad r = \infty \qquad [7.2.47]$$

$$\dot{R} - v_{fR} = \rho_f^{-1} j, \quad \dot{R} = \rho_v^{-1} j \qquad [7.2.48]$$

$$j = \left(\dot{R} - v_{fR}\right)\rho_f k_R + \rho_f D \frac{\partial k}{\partial r}, \quad jl = k_f \frac{\partial T_f}{\partial r} \quad \text{at } r = R(t) \qquad [7.2.49]$$

where:

| | |
|---|---|
| j | phase transition rate per unit surface area of a bubble |
| $v_{fR}$ | radial velocity of the liquid at the interface |
| $k_f$ | heat conductivity of liquid |
| $\rho_f, \rho_v$ | densities of solution and solvent vapor |

Equations [7.2.48] and [7.2.49] yield

$$j = \frac{\rho_f}{1 - k_R} D \frac{\partial k}{\partial r}\Big|_{r=R} \qquad [7.2.50]$$

If thermodynamic state of the system is far from the critical one, $\varepsilon = \rho_v / \rho_f << 1$ and it is possible to assume that $v_{fR} = \dot{R}(1-\varepsilon) \approx \dot{R}$. The solution of equations [7.2.45], [7.2.46] is searched in the form $T_f = T(\eta)$, $k = k(\eta)$ with $\eta = r/R(t)$. The concentration dependence of the diffusion coefficient is represented as $D = D_0(1 + f(k))$. The self-similar solution of the problem exists if

$$h = R\dot{R}a_f^{-1} = const, \quad h_1 = R\dot{R}D_0^{-1} = const \qquad [7.2.51]$$

In this case the functions $T(\eta)$, $k(\eta)$ satisfy the following equations:

$$\frac{\partial^2 T}{\partial \eta^2} + \left[h\eta + 2\eta^{-1} - (1-\varepsilon)h\eta^{-2}\right]\frac{\partial T}{\partial \eta} = 0 \qquad [7.2.52]$$

$$\frac{\partial^2 k}{\partial \eta^2} + \left[\bar{D}^{-1}h_1\eta + 2\eta^{-1} - \bar{D}^{-1}(1-\varepsilon)h_1\eta^{-2}\right]\frac{\partial k}{\partial \eta} + \bar{D}^{-1}\frac{df}{dk}\left(\frac{dk}{d\eta}\right)^2 = 0, \bar{D} = D/D_0 \qquad [7.2.53]$$

Equation [7.2.53], as opposed to [7.2.52], is non-linear and cannot be solved analytically for arbitrary function $D = D(k(\eta))$. Note that in the case of the planar non-linear diffusion, if the self-similarity conditions are satisfied, the problem has analytical solution for particular forms of the dependencies $D = D(k)$ (e.g., linear, exponential, power-law, etc[51]). However, the resulting relationships are rather cumbersome. The approximate solution of the problem was derived in the case $J_a >> 1$, using the perturbation method:[52]

$$h = (6/\pi)Ja^2 = (6\pi)Le^{-1}Di^2(1+M_1)^2, \quad Di = \varepsilon^{-1}K_\alpha[1+f(k_R)], \quad K_\alpha = (k_0 - k_R)/(1-k_R) \quad [7.2.54]$$

where:

| | |
|---|---|
| Ja | Jacob number, $Ja = c_f\Delta T_f(\varepsilon l)^{-1}$ |
| $\Delta T_f$ | superheat of the solution with respect to the interface, $\Delta T_f = T_{f0} - T_{fR}$ |
| Le | Lewis number, $a_f/D_0$ |
| $K_\alpha$ | mass fraction of the evaporated liquid[46] |

Here $M_1$ follows certain cumbersome equation,[52] including f(k). The approximation Ja>>1 corresponds to the case of a thin thermal boundary layer around the growing bubble. Since, for polymeric solutions Le >> 1, the condition of small thickness of the diffusion boundary layer is satisfied in this situation as well.

We start the analysis of the solution [7.2.54] from the approximation f = 0 that corresponds to $D \approx D_0 = \text{const}$. Then from [7.2.54] it follows:

$$K_\alpha = \left(\sqrt{Le}\right)c_f l^{-1}\Delta T \quad [7.2.55]$$

Because of the diffusion resistance, the solvent concentration at the interface is less then in the bulk, $k_R < k_0$. Writing the equation of phase equilibrium in linear approximation with respect to $\Delta k = k_0 - k$, from [7.2.55] one can receive[49,53]

$$\Delta T/\Delta T^* = \left[1 - c_f l^{-1}(1-k_R)\sqrt{Le}\left(\frac{\partial T_s}{\partial k}\right)_{k=k_0}\right]^{-1}, \quad \Delta T^* = T_{f0} - T_s(k_0) \quad [7.2.56]$$

Here $\Delta T^*$ represents the superheat of the solution at infinity. For solutions of polymers $\partial T_s/\partial k < 0$ and, therefore, the actual superheat of the liquid $\Delta T < \Delta T^*$. Additional simplification can be achieved if $1 - k_R >> k_0 - k_R$. It permits to assume in [7.2.56] $k_R \approx k_0$ and, hence, to find easily the vapor temperature.

In the diffusion-equilibrium approximation (i.e. $Le \to 0$) $\Delta T = \Delta T^*$. When the diffusion resistance increases, the actual superheat $\Delta T$ lowers and, according to [7.2.56], at $Le \to \infty$ $\Delta T \to 0$. However, in the latter case the assumptions made while deriving [7.2.56], are no longer valid. Indeed, the Ja number, connected with the superheat of the solution with respect to the interface, is related to the $Ja_0$ value, corresponding to the bulk superheat, by $Ja = Ja_0(\Delta T/\Delta T^*)$. Since the ratio $\Delta T/\Delta T^*$ varies in the range (0, 1), then, at small diffusion coefficients, it may be that Ja << 1 even when $Ja_0$ >> 1. In this case, the asymptotic solution of the problem takes the form[46] h = Ja, and, for thin diffusion boundary layer, it can be received instead of [7.2.54]:

$$h = Ja = (6/\pi)Le^{-1}Di^2(1+M_1) \quad [7.2.57]$$

Finally, at Di << 1 and Ja << 1, the non-linear features in the diffusion transport can be neglected and the expressions for h and $k_R$ (or $T_{fR}$) take the form

$$h = Ja = Le^{-1}Di \quad [7.2.58]$$

The bubble growth in the thermal regime follows the law[54]

$$R = C\sqrt{t}, \quad C = \sqrt{2a_f h}$$

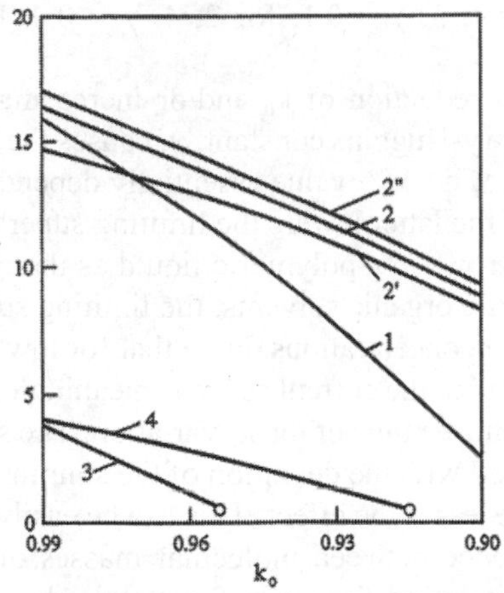


Figure 7.2.10. Limiting superheat at vapor bubble growth in polymeric solution. For all graphs $K_p = 0.7$, the symbol "o" corresponds to $J\hat{a}_0 = 1$. [Reprinted from Z.P. Shulman, and S.P. Levitsky, *Int. J. Heat Mass Transfer*, **39**, 631, Copyright 1996, the reference 52, with permission from Elsevier Science]

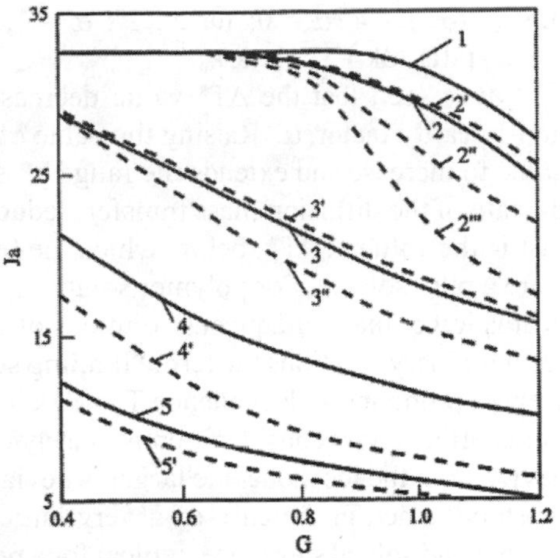


Figure 7.2.11. Dependence of the effective Jacob number for a vapor bubble, growing in a superheated aqueous solution of a polymer, on the parameter G. [Reprinted from Z.P. Shulman, and S.P. Levitsky, *Int. J. Heat Mass Transfer*, **39**, 631, Copyright 1996, the reference 52, with permission from Elsevier Science]

where the constant C can be evaluated through h from [7.2.54], [7.2.57] and [7.2.58]. Note that since $Ja < Ja_0$, the bubble growth rate in a polymer solution is always lower than that in a similar one-component liquid.

The set of equations, formulated above, is closed by the equation of phase equilibrium [7.2.37]. The temperature dependence of the pure solvent vapor pressure is described by equation[54] $p^0_{v0} = A\exp(-B/T)$.

Numerical simulations of vapor bubble growth in a superheated solution of polymer were performed,[52] using iterative algorithm to account for the diffusion coefficient dependence on concentration in the interval ($k_R$, $k_0$). The results are reproduced in Figures 7.2.10-7.2.12,

where:

| | |
|---|---|
| Sn | Scriven number, $Sn = \Delta T/\Delta T^*$ |
| G | dimensionless parameter, $G = \varepsilon Ja_0 Le^{1/2}$ |
| $\Delta\hat{T}^*$ | superheat of the solution at infinity, evaluated from the condition $Sn = 0.99$ |

A characteristic feature of the liquid-vapor phase equilibrium curves for polymeric solutions in the coordinates p, k or T, k is the existence of plateau-like domain in the region of small polymer concentrations ($k^* \le k_0 \le 1$). For this concentration range, the number $J\hat{a}_0$ can be defined so that at $1 < Ja_0 < J\hat{a}_0$ the diffusion-induced retardation of the vapor bubble growth does not manifest itself because of weak dependence of $T_s$ (or $p_s$) on $k_R$. The $J\hat{a}_0$ value or the corresponding limiting superheat, $\Delta\hat{T}^*$, can be estimated from the condition $Sn = 0.99$ (i.e. the deviation of effective superheat, $\Delta T$, from the bulk one, $\Delta T^*$, does not exceed 1%). The dependence of the so-defined parameter, $\Delta\hat{T}^*$, on $k_0$ is represented in Figure 7.2.10. For curves 1, 2, 2', 2": $l = 2.3\times10^6$ $Jkg^{-1}$, $c_f = 3\times10^3$ $Jkg^{-1}K^{-1}$, $a_f = 10^{-7}$ $m^2s^{-1}$, $D_0 = 5\times10^{-11}$ $m^2s^{-1}$; for 3, 4: $l = 3.6\times10^5$ $Jkg^{-1}$, $c_f = 2\times10^3$ $Jkg^{-1}K^{-1}$, $a_f = 8\times10^{-8}$ $m^2s^{-1}$, $D_0 = 5\times10^{-11}$

$m^2s^{-1}$; for 1 - 4: $\alpha = 0$; for 2', 2'': $\alpha = 1, -1$; for 1, 3: $\chi = 0.1$; for 2, 4: $\chi = 0.4$. Here $\alpha = k_R^{-1}\left(d\bar{D}/d\bar{k}\right)_{\bar{k}=\bar{k}_0}$, $\bar{k}=k/k_R$.

It is seen that the $\Delta\hat{T}^*$ value decreases with reduction of $k_0$ and/or increasing the non-linearity factor, $\alpha$. Raising the value of the Flory-Huggins constant, $\chi$, causes the $\Delta\hat{T}^*$ value to increase and extends the range $k^* \le k_0 \le 1$. The $\Delta T^*$ value essentially depends on the rate of the diffusion mass transfer; reduction of the latter lowers the limiting superheat, that is the value of $\Delta T^*$, below which the bubble grows in a polymeric liquid as though it were a pure solvent. For polymer solutions in volatile organic solvents, the limiting superheat is lower than for aqueous solutions of the same concentrations. Note that for low molecular binary solutions the term "limiting superheat" in the current sense is meaningless in view of pronounced dependence $T_s = T_s(k_0)$ in the entire range of the $k_0$ variation. The scale of the effect under consideration is closely connected with the deviation of the solution behavior from the ideal one: the larger is deviation the less is the effect. This can be easily understood, since in the case of a very large difference between molecular masses of the solvent and solved substance, typical for a polymer solution, the graph $T_s = T_s(k_0)$, plotted in accordance with the Raul law, nearly coincides with the coordinate axes.[20] For this reason, the bubble growth rate in a polymer solution that obeys the Flory-Huggings law, is always lower than in a similar ideal solution.

The reduction of the diffusion mass transfer rate ($G \sim (Le)^{1/2}$) at a fixed superheat, $\Delta T^*$, leads to a substantial decrease in the effective Jackob's number, Ja. The growth of the content of a polymer in a solution leads to the same result. This follows from Figure 7.2.11 where curves 1 - 5 correspond to $k_0 = 0.99, 0.95, 0.7, 0.5, 0.3$; 2' - 2''': $k_0 = 0.95$; 3' - 3'': $k_0=0.7$; 4', 5': $k_0 = 0.5, 0.3$; 1 - 5: $\alpha = 0$; 2' - 3': $\alpha = -0.5$; 2'': $\alpha = 0.8$; 2''', 3'', 4', 5': $\alpha = 2$. For all graphs, $\Delta T^*= 15$ K, $\chi = 0.1$, $K_p = 0.7$.

The influence of non-linearity of diffusional transport is higher for diluted solutions. This is explained by a decrease in the deviation of the surface concentration, $k_R$, from the bulk $k_0$ with lowering $k_0$. This takes place due to simultaneous increase in $|\partial T_s/\partial k|$ that is characteristic of polymeric liquids. The presence of a nearly horizontal domain on the curve $Ja = Ja(G)$ at $k_0 \ge 0.95$ is explained by the existence of the limiting superheat dependent on the Lewis number.

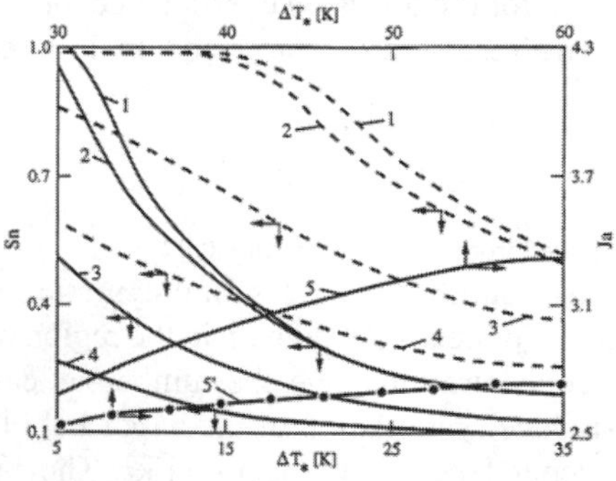


Figure 7.2.12. Effect of the solution bulk superheat on the Scriven and Jacob numbers. (–) - solution of polymer in toluene, (- - -) - aqueous solution. [Reprinted from Z.P. Shulman, and S.P. Levitsky, *Int. J. Heat Mass Transfer*, **39**, 631, Copyright 1996, the reference 52, with permission from Elsevier Science]

The role of diffusion-induced retardation increases with the bulk superheat. This reveals itself in reduction of the number Sn with a growth in $\Delta T^*$ (Figure 7.2.12). For solutions of polymers in volatile organic liquids, such as solvents, the effect is higher than in aqueous solutions. For concentrated solutions the difference between the effective $\Delta T$ and bulk $\Delta T^*$ superheats makes it practically impossible to increase substantially the rate of vapor bubble growth by increasing the bulk superheat. Curves 5 and 5' clearly demonstrate this. They are calculated for solution of polystyrene in toluene at $k_0 = 0.3$, therewith for the curve 5 the dependence of the diffusion coefficient from temperature and concentration D(T,k) was neglected ($\alpha = 0$), whereas for the curve 5' it was accounted for according to the experimental data.[55] Other curves were evaluated with the following parameter values: curves 1 - 5 correspond to $\alpha = 0$, $k_0 = 0.99, 0.95, 0.7, 0.5, 0.3$; 1 - 4: $\chi = 0.1$. Thermophysical parameters of the liquid and vapor are the same as in the Figure 7.2.10.

Thus, the rate of expansion of vapor bubbles in superheated solution of polymer is lower than in pure solvent due to diffusion resistance. But in diluted solution at rather small superheats the mechanism of diffusional retardation can be suppressed due to a weak dependence of $T_s$ on $k_0$ in this concentration range. Another important conclusion is that in concentrated solutions it is practically impossible to attain values Ja >> 1 by increasing the superheat because of low values of the corresponding Sn numbers.

### 7.2.3 BOILING OF MACROMOLECULAR LIQUIDS

Experimental investigations of heat transfer at boiling of polymeric liquids cover highly diluted (c = 15 to 500 ppm), low-concentrated (c ~ 1%), and concentrated solutions (c>10%). The data represent diversity of physical mechanisms that reveal themselves in boiling processes. The relative contribution of different physical factors can vary significantly with changes in concentration, temperature, external conditions, etc., even for polymers of the same type and approximately equal molecular mass. For dilute solutions this is clearly demonstrated by the experimentally detected both intensification of heat transfer at nucleate boiling and the opposite effect, viz. a decrease in the heat removal rate in comparison with a pure solvent.

Macroscopic effects at boiling are associated with changes in the intrinsic characteristics of the process (e.g., bubble shape and sizes, nucleation frequency, etc.). Let's discuss the existing experimental data in more detail.

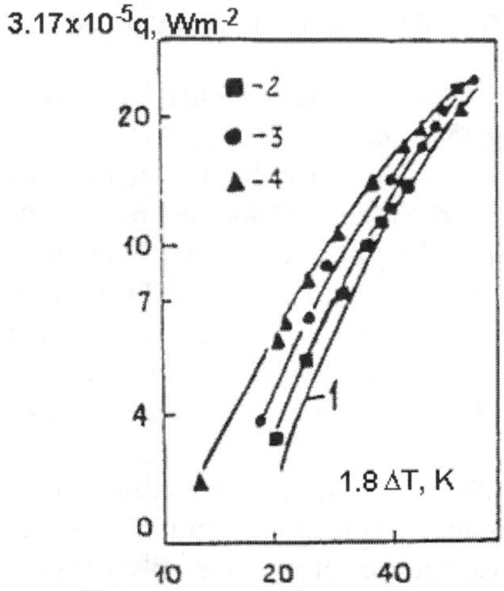


Figure 7.2.13. Effect of the HEC additives on the boiling curve. 1 - pure water; 2, 3 and 4 - HEC solution with c = 62.5, 125 and 250 ppm, correspondingly. [Reprinted from P. Kotchaphakdee, and M.C. Williams, *Int. J. Heat Mass Transfer*, **13**, 835, Copyright 1970, the reference 52, with permission from Elsevier Science]

One of the first studies on the effect of water-soluble polymeric additives on boiling was reported elsewhere.[56] For a plane heating element a significant increase in heat flux at fixed superheat, $\Delta T$ = 10-35K, was found in aqueous solutions of PAA Separan NP10 (M = $10^6$), NP20 (M = $2\times10^6$), and HEC (M ~ $7\times10^4$ to about $10^5$) at concentrations of 65 to 500 ppm (Figure 7.2.13). The experiments were performed at atmospheric pressure; the viscosity of the solutions did not exceed $3.57\times10^{-3}$ Pas. The following specific features of boiling of polymer solution were revealed by visual observations: (i) reduction in the departure diameter of bubbles, (ii) more uniform bubble-size distribution, (iii) decrease in the tendency to coalescence between bubbles. The addition of HEC led to faster covering of the heating surface by bubbles during the initial period of boiling and bubbles were

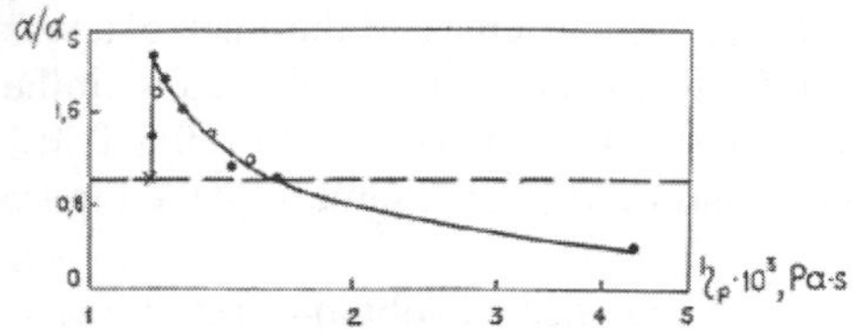


Figure 7.2.14. The relation between the relative heat transfer coefficient for boiling PIB solutions in cyclohexane and the Newtonian viscosity of the solutions measured at T=298 K. ΔT = 16.67 K; ● - PIB Vistanex L-100 in cyclohexane, o - PIB Vistanex L-80 in cyclohexane, x - pure cyclohexane. [Reprinted from H.J. Gannett, and M.C. Williams, *Int. J. Heat Mass Transfer*, **14**, 1001, Copyright 1971, the reference 57, with permission from Elsevier Science]

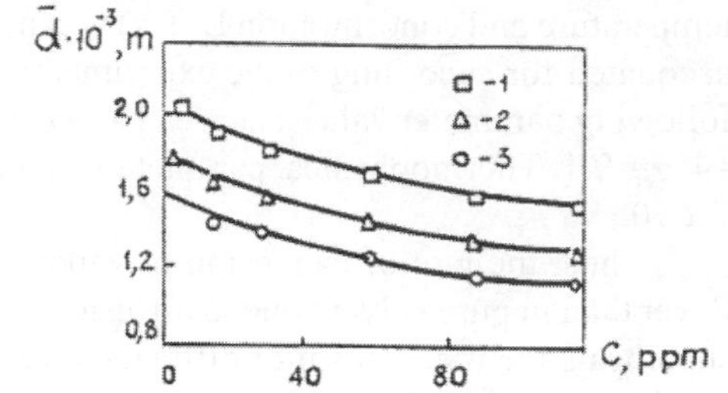


Figure 7.2.15. The average bubble detachment diameter in boiling dilute aqueous solutions of PEO.[59] ΔT = 15K. For curves 1-3 the flow velocity v = 0, $5\times10^{-2}$, and $10^{-1}$ m/s, respectively. [Adapted, from S.P. Levitsky, and Z.P. Shulman, **Bubbles in polymeric liquids**, *Technomic Publish. Co.*, Lancaster, 1995, with permission from Technomic Publishing Co., Inc., copyright 1995]

smaller in size than in water and aqueous solutions of PAA.

Non-monotonous change in the heat transfer coefficient, α, with increasing the concentration of PIB Vistanex L80 ($M = 7.2\times10^5$) or L100 ($M = 1.4\times10^6$) in boiling cyclohexane has been reported.[57] The results were received in a setup similar to that described earlier.[56] It was found that the value of α increases with c in the range 22 ppm < c < 300 ppm and decreases in the range 300 ppm < c < 5150 ppm. Viscosity of the solution, corresponding to $\alpha_{max}$ value, according to the data[57] only slightly exceeds that of the solvent (Figure 7.2.14). Within the entire range of concentrations at supercritical (with respect to pure solvent) superheats, the film boiling regime did not appear up to the maximum attainable value ΔT ~ 60K. The growth of polymer concentration in the region of "delayed " nucleate boiling led to a considerable decrease in heat transfer.

These findings[56,57] were confirmed[58] in a study of the nucleate boiling of aqueous solutions of HEC Natrosol 250HR ($M = 2\times10^5$), 250GR ($M = 7\times10^4$), and PEO ($M \sim (2\text{-}4)\times10^6$) at forced convection of the liquid in a tube. A decrease in the size of bubbles in the solution and reduction of coalescence intensity were recognized. Similar results were presented also in study,[59] where the increase in heat transfer at boiling of aqueous solutions of PEO WSR-301 ($M=2\times10^6$) and PAA Separan AP-30 (15 ppm < c < 150 ppm) on the surface of a conical heater was observed. In aqueous solutions of PAA with c > 60 ppm the α value began to decrease. With an increase in c the detachment diameter of bubbles decreased (Figure 7.2.15), the nucleation frequency increased, and the tendency to coalescence was suppressed.

Boiling of PEO solutions with c = 0.002 to 1.28% at atmospheric and sub-atmospheric pressures was examined[60] for subcoolings in the range 0 to 80K. It was demonstrated that at saturated boiling the dependence of the heat transfer coefficient α on the polymer concentration is non-monotonous: as c grows, α first increases, attaining the maximal value at c≈0.04% , whereas at c = 1.28% the value of α is smaller than in water ($\alpha < \alpha_s$) (Figure 7.2.10). With a decrease in pressure the effect of polymeric additives weakens and for solution with greatest PEO concentration (in the investigated range) the α value increases, approaching $\alpha_s$ from below. The critical heat flux densities in PEO solutions are smaller than those for water.

In view of the discussed results, the work[61] attracts special attention since it contains data on boiling of dilute solutions, opposite to those reported earlier.[56-60] The addition of PAA, PEO and HEC to water in concentrations, corresponding to the viscosity increase up to $\eta_p = 1.32\times10^{-3}$ Pas, has brought about reduction in heat transfer. The boiling curve in coordinates q (heat flux) vs. ΔT displaced almost congruently to the region of larger ΔT values with c (Figure 7.2.17a). It was demonstrated[61] that the observed decrease in α with addition of polymer to water can be both qualitatively and quantitatively (with the Rohsenow pool boiling correlation for the heat transfer coefficient[62]) associated with the increase in the solution viscosity (Figure 7.2.17, b). The experiments[61] were performed using a thin platinum wire with diameter 0.3 mm.

Explanation of experimental data needs more detailed discussion of physical factors that can reveal themselves in boiling of polymeric solutions. They include possible changes in capillary forces on interfaces in the presence of polymeric additives; absorption of macromolecules on the heating surface; increase in the number of weak points in the solution, which facilitates increase in the number of nuclei; thermodynamic peculiarities of the polymer-solvent system; the effect of macromolecules on the diffusion mass transfer in evaporation of solvent; hydrodynamics of convective flows in a boiling layer and the motion of bubbles; manifestation of rheological properties of solution.

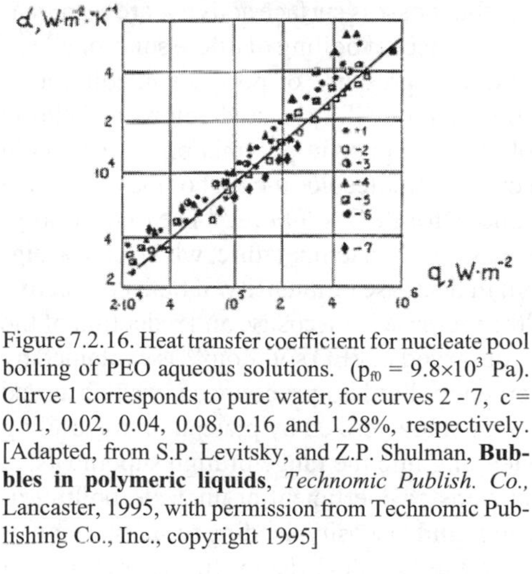


Figure 7.2.16. Heat transfer coefficient for nucleate pool boiling of PEO aqueous solutions. ($p_{f0} = 9.8\times10^3$ Pa). Curve 1 corresponds to pure water, for curves 2 - 7, c = 0.01, 0.02, 0.04, 0.08, 0.16 and 1.28%, respectively. [Adapted, from S.P. Levitsky, and Z.P. Shulman, **Bubbles in polymeric liquids**, *Technomic Publish. Co.*, Lancaster, 1995, with permission from Technomic Publishing Co., Inc., copyright 1995]

The capillary effects were indicated as one of the reasons for the intensification of heat transfer, since many polymers (in particular, HEC, PEO, etc.),[63] similar to low-molecular surfactants,[64] are capable of decreasing the surface tension. As a result, they decrease both the work of the nucleus formation, $W_{cr}$, and the critical size of bubble, $R_{cr}$:

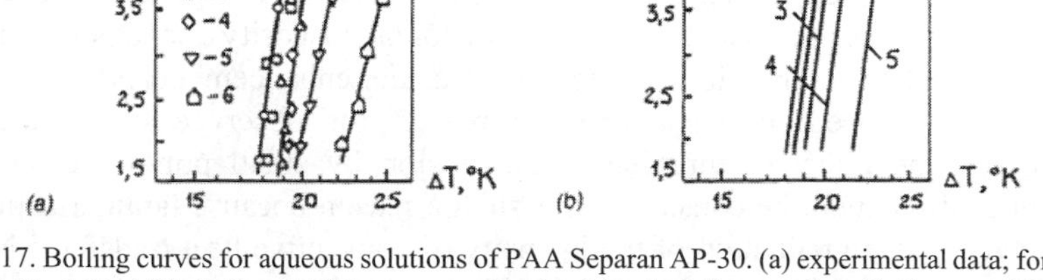


Figure 7.2.17. Boiling curves for aqueous solutions of PAA Separan AP-30. (a) experimental data; for curves 1 - 6 $\eta_r$ = 1.00, 1.01, 1.04, 1.08, 1.16 and 1.32, correspondingly; (b) calculations made with the use of the Rohsenow pool boiling correlation; for curves 1 - 5, $\eta_r$ = 1.00, 1.01, 1.04, 1.16 and 1.32, respectively ($\eta_r = \eta_p/\eta_s$). [By permission from D.D. Paul, and S.I. Abdel-Khalik, *J. Rheol.*, **27**, 59, 1983, reference 61]

$$W_{cr} = 16/3\pi\sigma^3\Phi(\theta)\left[(dp/dT)\Delta T(1-\rho_v/\rho_f)\right]^2, \quad \Phi(\theta) = 1/4\left(2+3\cos\theta-\cos^3\theta\right) \quad [7.2.59]$$

$$R_{cr} = 2\sigma\left[(dp/dT)\Delta T(1-\rho_v/\rho_f)\right]^{-1}$$

where:

$\theta$ wetting angle

$\sigma$ surface tension coefficient

However, it should be noted that the integral effect of the heat transfer enhancement, observed in highly diluted solutions, can not be attributed to the capillary phenomena alone, since the main change in $\sigma$ occurs in the range of low polymer concentrations[59] ($c < 50$ ppm) and further increase in c does not affect the value of $\sigma$, whereas the $\alpha$ value continues to grow. PAA, for example, does not behave like surfactants at all. It should be noted also that in the presence of polymer not only the value of $\sigma$ changes, but also the wetting angle, $\theta$, in the formula [7.2.59]. The latter may lead to manifestation of different behavior.

Absorption of macromolecules onto a heating surface favors the formation of new centers of nucleation. Together with an increase in nucleation sites in the boundary layer of a boiling liquid it explains the general growth in the number of bubbles. Both this factor and reduction in the $\sigma$ value for solutions of polymers that possess surface activity, are responsible for a certain decrease in superheat needed for the onset of boiling of dilute solutions.[57,60]

The decrease in the water vapor pressure due to presence of polymer in solution at c~1% can be neglected. However, if the solution has the LCST, located below the heating wall temperature, the separation into rich-in-polymer and poor-in-polymer phases occurs in the wall boundary layer. At low concentration of macromolecules the first of these exists in a fine-dispersed state that was observed, for example, for PEO solutions.[60] The rich-in-polymer phase manifests itself in a local buildup of the saturation temperature, which can be significant at high polymer content after separation; in decrease of intensity of both convective heat transfer and motion of bubbles because of the increase in viscosity; and reduction of the bubble growth rate. The so-called "slow" crisis, observed in PEO solutions[60] is explained by integral action of these reasons. Similar phenomenon, but less pronounced, was observed also at high enough polymer concentrations.[58] It is characterized by plateau on the boiling curves for solutions of PIB in cyclohexane, extending into the range of high superheats.

The main reason for the decrease in heat transfer coefficient at nucleate boiling of polymeric solutions with c ~ 1% is the increase in liquid viscosity, leading to suppression of microconvection and increasing the resistance to the bubbles' rising. In the presence of LCST, located below the boiling temperature, the role of this factor increases because appearance in the boiling layer of the rich-in-polymer phase in fine-dispersed state. Another reason for the decrease of $\alpha$ in the discussed concentration range is the decrease in the bubble growth rate at the thermal stage, when the superheat $\Delta T > \Delta T^*$ (Section 7.2.2).

In highly diluted solutions the change in Newtonian viscosity due to polymer is insignificant, and though the correlation between heat transfer enhancement and increase in viscosity has been noticed, it cannot be the reason for observed changes of $\alpha$. In hydrodynamics, the effect of turbulence suppression by small polymeric additives is known, but it also cannot be considered for such a reason because laminarization of the boundary layer leads to reduction of the intensity of convective heat transfer.[65] Nevertheless, the phenomenon of the decrease of hydrodynamic resistance and enhancement of heat transfer in boiling dilute solutions have a common nature. The latter effect was connected

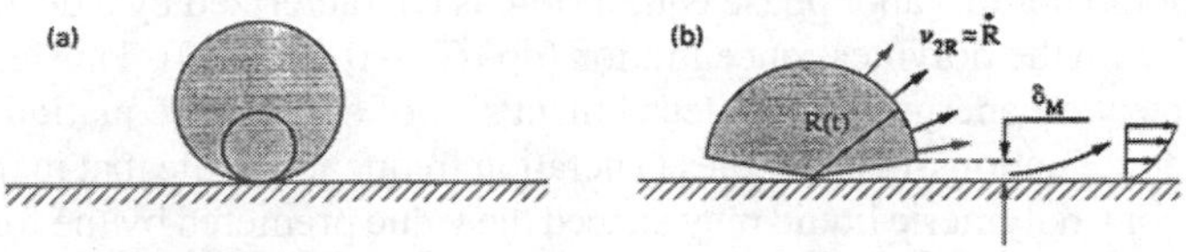


Figure 7.2.18. Growth of vapor bubbles on the heating surface at high (a) and low (b) pressures. [Reprinted from S.P. Levitsky, B.M. Khusid and Z.P. Shulman, *Int. J. Heat Mass Transfer*, **39**, 639, Copyright 1996, the reference 66, with permission from Elsevier Science]

with manifestation of elastic properties of the solution at vapor bubble growth on the heating surface.[66]

The general character of the bubbles evolution at boiling under atmospheric and subatmospheric pressures, respectively, is clarified schematically in Figure 7.2.18. In the first case (at high pressures) the base of a bubble does not "spread"[67] but stays at the place of its nucleation. Under such conditions the decrease in the curvature of the bubble surface with time, resulting from the increase in bubble radius, R, leads to liquid displacement from the zone between the lower part of the microbubble and the heating surface. This gives rise to the local shear in a thin layer of a polymer solution. A similar shear flow is developed also in the second case (at low pressures), when a microlayer of liquid is formed under a semi-spherical bubble. As known, at shear of a viscoelastic fluid appear not only tangential but also normal stresses, reflecting accumulation of elastic energy in the strained layer (the Weissenberg effect[3]). The appearance of these stresses and elastic return of the liquid to the bubble nucleation center is the reason for more early detachment of the bubble from the heating surface, reduction in its size and growth in the nucleation frequency. All this ultimately leads to enhancement of the heat transfer.

The above discussion permits to explain the experimental results.[61] Their reasons are associated with substantial differences in the conditions of boiling on a thin wire and a plate or a tube. Steam bubbles growing on a wire have a size commensurable with the wire diameter (the growing bubble enveloped the wire[61]). This results in sharp reduction of the boundary layer role, the same as the role of the normal stresses. Besides, the bubble growth rate on a wire is smaller than on a plane (for a wire $R \sim t^n$ where $n < 1/4$).[67]

The elastic properties of the solution are responsible also for stabilization of the spherical shape of bubbles observed in experiments on boiling and cavitation. Finally, the observed reduction in a coalescence tendency and an increase in the bubble sizes uniformity can also be attributed to the effects of normal stresses and longitudinal viscosity in thin films separating the drawing together bubbles.

The linkage between the enhancement of heat transfer at boiling of dilute polymer solutions and the elastic properties of the system is confirmed by the existence of the optimal concentration corresponding to $\alpha_{max}$ (Figure 7.2.14). Similar optimal concentration was established in addition of polymers to water to suppress turbulence - the phenomenon that also owes its origin to elasticity of macromolecules.[1,3,9] Therefore, it is possible to expect that the factors favoring the chain flexibility and increase in the molecular mass, should lead to strengthening of the effect.

The data on boiling of concentrated polymeric solutions[20] demonstrate that in such systems thermodynamic, diffusional, and rheological factors are of primary importance.

The diagram of the liquid-vapor phase equilibrium is characterized by a decrease in the derivative dp/dT with the polymer concentration (dp/dT $\to 0$ at k $\to 0$). This leads to increase in both the nucleation energy and the detachment size of a bubble (equation [7.2.59]) and, consequently, to reduction of the bubbles generation frequency. Note that in reality the critical work, $W_{cr}$, for a polymeric liquid may exceed the value predicted by the formula [7.2.59] because of manifestation of the elasticity of macromolecules.

As known,[62] the heat transfer coefficient in the case of developed nucleate boiling of low-molecular liquids is related to the heat flux, q, by the expression $\alpha = Aq^n$ where $n \approx 0.6$-$0.7$. For concentrated polymeric solutions the exponent n is close to zero. The decrease in heat transfer is explained by the increase in the viscosity of the solution near the heating surface, resulting from the evacuation of the solvent with vapor. Another reason for the decrease of $\alpha$ in such systems is the reduction of the bubble growth rate with lowering $k_0$ and the impossibility to achieve large Ja numbers by rising the solution superheat.

Since in boiling of concentrated polymer solutions the $\alpha$ value is small, the superheat of the wall at a fixed q increases. This can give rise to undesirable phenomena such as burning fast to the heating surface, structure formation, and thermal decomposition. Usually, in this case the heat transfer is intensified by mechanical agitation. Note that one of the promising trends in this field may become the use of ultrasound, the efficiency of which should be evaluated with account for considerable reduction in real losses at acoustically induced flows and pulsations of bubbles in viscoelastic media.[68,69]

Specific features of boiling of high-molecular solutions are important for a number of applications. One of examples is the heat treatment of metals, where polymeric liquids find expanding employment. The shortcomings of traditionally used quenching liquids, such as water and oil, are well known.[70] Quenching in oil, due to its large viscosity and high boiling temperature, does not permit to suppress the perlite transformation in steels. From the other hand, water as a quenching medium is characterized by high cooling rate over the temperature ranges of both perlite and beinite transformations. However, its maximum quenching ability lies in the temperature range of martensite formation that can lead to cracking and shape distortion of a steel article. Besides, the quenching oils, ensuring the so-called "soft" quenching, are fire-hazardous and have ecological limitations.

The polymeric solutions in a certain range of their physical properties combine good points of both oil and water as quenching liquids and permit to control the cooling process over wide ranges of the process parameters. For the aims of heat treatment a number of water-soluble polymers are used, e.g. PVA, PEO, PAA, polymethacrylic acids (PMAA, PAA) and their salts, cellulose compounds, etc.[71-73] The optimal concentration range is 1 to 40% depending on molecular mass, chemical composition, etc. Typical data[73] are presented in Figure 7.2.19, where curves 1-5 correspond to the solution viscosity $\eta_p$ = (1.25, 2.25, 3.25, 5, 11)$\times 10^{-3}$ Pas, measured at 40°C. The quenchants based on water-soluble polymers sustain high cooling

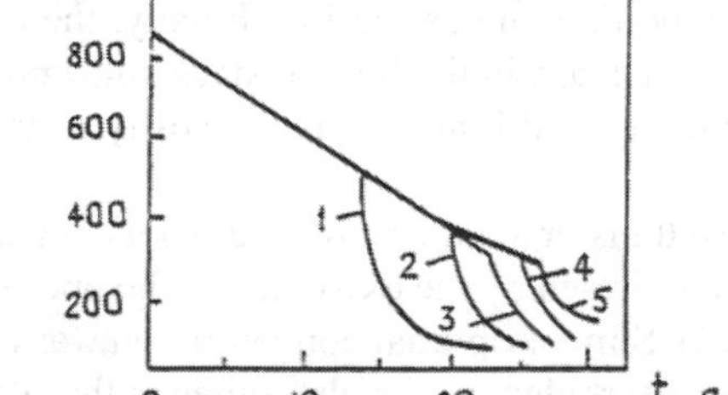

Figure 7.2.19. Cooling curves for a silver specimen quenched in a polymer aqueous solution at 20°C. [Adapted, from S.P. Levitsky, and Z.P. Shulman, **Bubbles in polymeric liquids**, *Technomic Publish. Co.*, Lancaster, 1995, with permission from Technomic Publishing Co., Inc., copyright 1995]

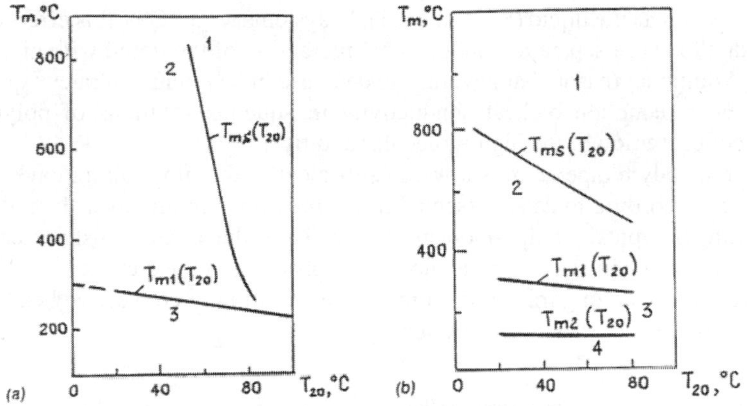


Figure 7.2.20. Stability diagrams for film boiling.[73] Quenching in water (a) and in aqueous polymer solution with $\eta_s = 3.27\times10^{-3}$ Pas at 40°C; (b): 1, 2 - stable and unstable film boiling, 3 - nucleate boiling, 4 - convection. $T_{ms}$, $T_{m1}$ and $T_{m2}$ are the surface temperatures of the specimen, corresponding to the destabilization of the regimes 1 - 3, respectively. [Adapted, from S.P. Levitsky, and Z.P. Shulman, **Bubbles in polymeric liquids**, *Technomic Publish. Co.*, Lancaster, 1995, with permission from Technomic Publishing Co., Inc., copyright 1995]

rates during the initial stage of quenching that permits to obtain fine-grained supercooled austenite, and relatively low intensity of heat removal at moderate temperatures, when martensite transformations take place.

The main feature of quenching in polymeric solutions is the prolongation of the cooling period as a whole in comparison with water that is explained by extended range of a stable film boiling (Figure 7.2.20). The increase in polymer concentration leads to reduction of $\alpha$ on the stage of nucleate boiling and growth of the temperature, corresponding to the onset of free convection regime.

The effect of polymeric additives on the initial stage of the process was the subject of a special investigation.[74] Experiments were performed with aqueous polymer solutions of Breox, PEO and some other polymers with $M = 6\times10^3$ to $6\times10^5$ at pressure of 0.1 MPa. The platinum heater with short time lag, submerged in solution, was heated in a pulsed regime with $\dot{T} \sim 10^5$ to $10^6$ K/s. The experimental results revealed the existence of a period with enhanced heat transfer (as compared to water) in solutions with $c \sim 1\%$, which lasted for 10 to 100 μs after the onset of ebullition. The sensitivity of heat transfer to the polymer concentration was sufficiently high. After formation of the vapor film the secondary ebullition was observed, which resulted from superheating of the liquid outside the region of concentration gradients near the interface. The mechanism of this phenomena was described.[75] It is associated with the fact that heat transfer has a shorter time lag than mass transfer, and thus the thermal boundary layer in a liquid grows faster than the diffusion one.

The experimental data and theoretical results on the growth of vapor bubbles and films in polymeric solutions explain the efficiency of quenchants, based on water-soluble polymers. The main reason is stabilization of the film-boiling regime at initial stage of quenching. Such stabilization is connected with elastic properties of the liquid skin layer, adjacent to the interface that is enriched by polymer due to solvent evaporation. Appearance of this layer leads to fast growth of longitudinal viscosity and normal stresses, when perturbations of the interface arise, thus increasing the vapor film stability. A similar mechanism is responsible for stabilization of jets of polymeric solutions[9] as well as for retardation of bubble collapse in a viscoelastic liquid (Section 7.2.1). In systems with LCST this effect can be still enhanced due to phase separation, induced by interaction of the liquid with high-temperature body. Additional reason that governs the decrease in heat removal rate at quenching is connected with reduction of heat conductivity in aqueous solutions of polymers with growth of concentration of the high-molecular additive.

After the body temperature is lowered sufficiently, the film boiling gave way to the nucleate one. According to data, presented in the previous section, over the polymer concentration ranges, typical for high-molecular quenchants, the α value must decrease in comparison with water. In fact, this is normally observed in experiments. Rise of the temperature, characterizing transition from nucleate boiling to convective heat transfer, is associated with the increase in liquid viscosity.

### *Abbreviations*

| | |
|---|---|
| HEC | hydroxyethylcellulose |
| LCST | lower critical solution temperature |
| PS | polystyrene |
| PIB | polyisobutylene |
| PAA | polyacrylamide |
| PEO | polymethyleneoxide |
| POE | polyoxyethylene |
| PVA | polyvinyl alcohol |
| TTS | time-temperature superposition |

## REFERENCES


1 A.B.Bird, R.C.Armstrong, and O.Hassager, **Dynamics of polymeric liquids**, *J.Wiley & Sons*, New York, 1977.
2 R. M.Christensen, **Theory of viscoelasticity. An introduction**, *Academic Press*, New York, 1982.
3 G. Astarita, and G. Marucci, **Principles of non-Newtonian fluid mechanics**, *McGraw-Hill Book Co.*, London, 1974.
4 L. D. Landau, and E. M. Lifshits, **Statistical physics**, *Pergamon Press*, Oxford, 1980.
5 C. Truesdell, **A first course in rational continuum mechanics**, *Johns Hopkins University*, Baltimore, 1972.
6 K. Walters, **Rheometry**, *Chapman and Hall*, London, 1975.
7 C.J.S. Petrie, **Elongational flows**, *Pitman*, London, 1979.
8 P. J. Carreau, and D. De Kee, *Can. J. Chem. Eng.*, **57**, 3 (1979).
9 Z.P. Shulman, and B. M. Khusid, **Non-stationary convective transfer processes in hereditary media**, *Nauka i Technika*, Minsk, 1983.
10 K.M. Baid, and A. B. Metzner, *Trans. Soc. Rheol.*, **21**, 237 (1977).
11 R.Y. Ting, *J. Appl. Polym. Sci.*, **20**, 1231 (1976).
12 J.M. Dealy, *Polym. Eng. Sci.*, **11**, 433 (1971).
13 M.M. Denn, and G. Marrucci, *AIChE J.*, **17**, 101 (1971).
14 J.D. Ferry, **Viscoelastic properties of polymers**, *J.Wiley & Sons*, New York, 1980.
15 M. Doi, and S. F. Edwards, **The theory of polymer dynamics**, *Clarendon Press*, Oxford, 1986.
16 P.-G. de Gennes, **Scaling concepts in polymer physics**, *Cornell University Press*, Ithaca and London, 1979.
17 T. W. Spriggs, *Chem. Eng. Sci.*, **20**, 931, (1965).
18 G. V. Vinogradov, and A. Ya. Malkin, **Rheology of polymers**, *MIR*, Moscow, 1980.
19 Ya.I. Frenkel, **Kinetic theory of liquids**, *Dover Publications*, New York, 1955.
20 V.P. Budtov, and V.V. Konsetov, **Heat and mass transfer in polymerization processes**, *Khimiya*, Leningrad, 1983.
21 D. W. Van Krevelen, **Properties of polymers: correlations with chemical structure**, *Elsevier Publishing Co.*, Amsterdam, 1972.
22 R. S. Marvin, and J. E. McKinney in **Physical Acoustics: Principles and Methods, vol.II**, p.B, W. P. Mason, Ed., *Academic Press*, New York, 1965, pp. 193-265.
23 B. Froelichb, C. Noelb, and L. Monnerie, *Polymer*, **20**, 529 (1979).
24 D. Pugh, and D.A. Jones, *Polymer*, **19**, 1008 (1978).

25 S.P. Levitsky, and Z.P. Shulman, **Dynamics and heat and mass transfer of bubbles in polymeric liquids**, *Nauka i Tekhnika*, Minsk, 1990.
26 W.J. Yang, and H.C. Yeh, *Phys. Fluids.*, **8**, 758 (1965).
27 H.S. Fogler, and J.D. Goddard, *Phys. Fluids.*, **13**, 1135 (1970).
28 J.R. Street, A.L. Fricke, and L.P. Reiss, *Ind. Eng. Chem. Fundam.*, **10**, 54 (1971).
29 S.P. Levitsky, and A.T. Listrov, *J. Appl. Mech. Techn. Phys.*, **15**, 111 (1974).
30 G. Pearson, and S. Middleman, *AIChE J.*, **23**, 714 (1977).
31 S.P. Levitsky, *J. Appl. Mech. Techn. Phys.*, **20**, 74 (1979).
32 G. Ryskin, *J. Fluid Mech.*, **218**, 239 (1990).
33 R.C. Reid, J.M.Prausnitz, and T.K. Sherwood, **The properties of gases and liquids**, *McGraw-Hill*, New York, 1977.
34 R.I. Nigmatulin, and N.S. Khabeev, *Fluid Dynamics*, **9**, 759 (1974).
35 S.P. Levitsky, and Z.P. Shulman, **Bubbles in polymeric liquids**, *Technomic Publish. Co.*, Lancaster, 1995.
36 R.Y. Ting, and A.T. Ellis, *Phys. Fluids*, **17**, 1461 (1974).
37 G.L. Chahine, and D.H. Fruman, *Phys. Fluids*, **22**, 1406 (1979).
38 A. Shima, Y. Tomito, and T. Ohno, *Phys. Fluids*, **27**, 539 (1984).
39 W.D. McComb, and S. Ayyash, *J. Phys. D: Appl. Phys.*, **13**, 773 (1980).
40 S.L. Peshkovsky, M.L. Fridman, V.I. Brizitsky et al., *Doklady Akad. Nauk SSSR*, **258**, 706 (1981).
41 R.T. Knapp, J.W. Daily, and F.G. Hammit, **Cavitation**, *McGraw-Hill*, New York, 1970.
42 P.R. Williams, P.M. Williams, and S.W. Brown, *J. Non-Newtonian Fluid Mech.*, **76**, 307 (1998).
43 J.W. Hoyt, Trans. *ASME. J. Fluids Eng.*, **98**, 106 (1976).
44 R.Y. Ting, *AIChE J.*, **20**, 827 (1974).
45 R.Y. Ting, *Phys. Fluids*, **21**, 898 (1978).
46 L.E. Scriven, *Chem. Eng. Sci.*, **10**, 1 (1959).
47 P.J. Bruijn, *Physica*, **26**, 326 (1960).
48 R.A. Shock, in **Multiphase science and technology**, *Hemisphere Publishing Corporation*, New York, 1981, pp. 281-386.
49 J.R. Thome, and R. A. W. Shock, *Adv. Heat Transfer*, **16**, 60 (1984).
50 S.G. Kandlikar, Trans. *ASME., J. of Heat Transfer*, **120**, 380 (1998).
51 J. Crank, **The mathematics of diffusion**, *Clarendon Press*, Oxford, 1975.
52 Z.P. Shulman, and S.P. Levitsky, *Int. J. Heat Mass Transfer*, **39**, 631 (1996).
53 L.W. Florshuets, and A. R. Khan, Heat Transfer-70, v.6, p. B7, 31970, Paris, 1970, pp.1-11.
54 R.I. Nigmatulin, **Dynamics of multiphase flow**, *Hemisphere Publishing Corporation*, New York, 1990.
55 J.L. Duda, J.S. Vrentas, S.T. Ju, and H.T. Liu, *AIChE J.*, **28**, 279 (1982).
56 P. Kotchaphakdee, and M.C. Williams, *Int. J. Heat Mass Transfer,* **13**, 835 (1970).
57 H.J. Gannett, and M.C. Williams, *Int. J. Heat Mass Transfer*, **14**, 1001 (1971).
58 H. Wei, and J.R. Maa, *Int. J. Heat Mass Transfer*, **25**, 431 (1982).
59 A.T. Papaioannou, and N.G. Koumoutsos, 7th Int. Heat Transfer Conf. Proc., 1982, Munchen, 1982, v.4, pp. 67-72.
60 B.P. Avksent'yuk, and Z.S. Mesarkishvili, Boiling of aqueous PEO solutions at reduced pressures under the conditions of natural convection, Institute of Thermophysics, RAS, Novosibirsk, 1983, preprint No. 108.
61 D.D. Paul, and S.I. Abdel-Khalik, *J. Rheol.*, **27**, 59 (1983).
62 **Heat exchanger design handbook**, v.1, contributors D.B. Spalding and J. Taborek, *Hemisphere Publishing Corporation*, New York, 1987.
63 R.Y.Z. Hu, A.T.A. Wang, and J.P. Hartnett, *Exp. Therm. Fluid Sci.*, **4**, 723 (1991).
64 Yan Yu Min, *Int. Commun. Heat and Mass Transfer*, **17**, 711 (1990).
65 Z.P. Shulman, B.M. Khusid, and S.P. Levitsky, *Heat Transfer Research*, **25**, 872 (1996).
66 S.P. Levitsky, B.M. Khusid and Z.P. Shulman, *Int. J. Heat Mass Transfer*, **39**, 639 (1996).
67 M.G. Cooper, and A.J.P. Lloyd, *Int. J. Heat & Mass Transfer*, **12**, 895 (1969).
68 Z.P. Shulman, and S.P. Levitsky, *Int. J. Heat & Mass Transfer*, **35**, 1077 (1992).
69 S.P. Levitsky, and Z.P. Shulman, *Soviet Phys. Acoust.*, **31**, 208 (1985).
70 N.R. Suttie, *Heat Treat. Metals*, **6**, 19 (1979).
71 K.Y. Mason, and T. Griffin, *Heat Treat. Metals*, **9**, 77 (1982).
72 F. Moreaux, and G. Beck, *Heat Transfer*, **4**, 2067 (1986).
73 F. Morou, and P. Arshambol, *Prom. Teplotekhn.*, **11**, 48 (1989).
74 G.B.Okonishnikov, N.V. Novikov, P.A. Pavlov, and S.L. Tsukrov, Heat and Mass Transfer - 92, v.6, Heat and Mass Transfer Institute, Minsk, 1992, pp. 3-6.
75 S.P. Levitsky, and Z.P.Shulman, *Thermophys. of High Temperatures*, **33**, 616 (1995).

# 7.3 DRYING OF COATED FILM

SEUNG SU KIM
**SKC Co., Ltd., Chon-an City, Korea**
JAE CHUN HYUN
**Department of Chemical Engineering, Korea University, Seoul, Korea**

## 7.3.1 INTRODUCTION

Thin film coating and drying technology are the key technologies for manufacturing diverse kinds of functional films, such as photographic films, adhesives, image media, magnetic media and recently lithium battery coating. Coating applied to a substrate as a liquid need some degree of solidification in order to be final products. The degree of solidification can be low in the case of pressure-sensitive adhesives (PSA) and it ranges to high in the case of dense metal-oxides.[48] The final structure and properties of coating are greatly influenced by the drying conditions.[21] Poorly chosen operating conditions of drying cause unwanted internal gradients,[6,7] phase separations,[29,35,36] colloidal transformations that lead to the wrong microstructure,[43,51,56] inappropriate non-uniformities, and stress-related defects.[12-15,21,22]

Here we address a subject of solvent removal, or drying, which is a part of solidification processes.

Typically a formulation of coating solution is consisted of pigments, binders (polymer resins) and solvents. The solvents are used to solubilize the coating formulation and to give the coating solution(or dispersion) the rheology necessary for the application. The coating solution is deposited onto a substrate or web at the coating station and is dried by passing through a series of separate ovens (zones). A substrate can be an impermeable material such as plastic film and permeable in the case of paper coating. The dryer is consisted of ovens

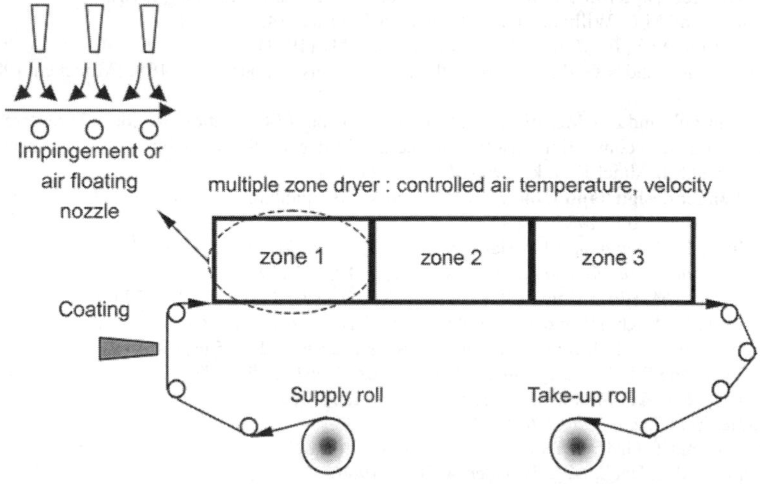


Figure 7.3.1. An example of industrial coating and drying apparatus. A coated liquid is deposited onto a substrate which is unwound from a supply roll at the coating station and passes through the three separate ovens(dryer) where the temperature and velocity of air is controlled independently. Finally dried coated substrate is taken up by a take-up roll.

(zones) in which the temperature and velocity of air are controlled independently. The air impinges the coated and back side surface of the substrate through the nozzles and sweeps away the solvent vapor from the coated surface. In a case of single sided impingement dryer, the air impinges only on the coated surface. The coated film must be dry before the rewind station. The residence time of coated substrate in the dryer is as short as several seconds in a case of high speed magnetic coating processes and it ranges to as long as several minutes in a case of lithium battery coating processes.[28] Finally the dried coated substrate is taken up by a take-up roll (Figure 7.3.1).

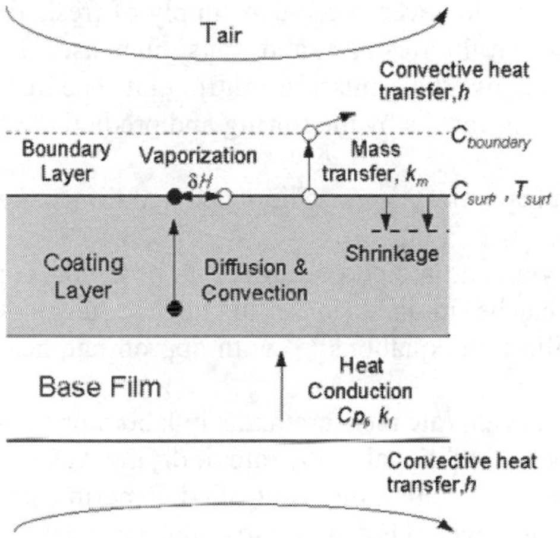


Figure 7.3.2. Elemental process of drying and typical parameters of drying. [After references 28,30].

The elemental process of drying is depicted in the Figure 7.3.2. Solvent is evaporated from the exposed surface of a liquid coating into the adjacent air. Diffusion of evaporating solvent into stagnant air is a comparatively slow process. Commonly the rate of diffusion of solvent vapor is greatly enhanced by the forced convective sweeping of the exposed surface. The rate of solvent evaporation per unit area is a product of the two factors : 1) the difference in partial pressure of the solvent at the surface of coating and in the bulk of nearby gas and 2) the mass transfer coefficient, which represents the combined action of convection and diffusion.[16]

The energy, which is needed to supply the latent heat of solvent evaporation, is delivered mainly by blowing hot air onto the coating surface. This convective heat transfer not only can deliver the needed energy to the coating, but also can enhance the transport of solvent vapor away from the surface of the coating. Commonly the conductive and radiative heating are accompanied with convective heating if they are necessary.

As long as the temperature and the concentration of solvent at the exposed surface of coating is constant, so does the evaporation rate of solvent. This is true during the initial stages of drying when the exposed surface of coating is fully wetted with the solvent. This period is commonly called as constant drying rate period (CDRP).[10] In CDRP, all the heat, which is supplied to the coating, is used to supply latent heat of vaporization. Thus the temperature of coating surface is nearly constant. In a case of aqueous coating, the temperature of coating surface is equivalent to the wet bulb temperature of a given air humidity.[9,10] Therefore in CDRP the external mass transfer resistance to drying limits the rate of drying.

As the solvents depart the coating, the rate controlling step for the drying steadily shifts from the external mass transfer to the mass transfer within the coating. The solvents within the coating can move to the exposed surface by diffusion along with diffusion-induced convection[1,8,17] and pressure gradient-driven flow in a porous coating.[43,44] In polymeric solutions, the diffusion coefficient of solvent dramatically drops as solvent concentration falls.[17] Thus the concentration of solvent at the exposed coating surface falls with drying proceeded, and so does the vapor pressure of solvent there. Therefore the drying rate steadily falls. This period is called as falling drying rate period (FDRP). Generally the diffusion coefficient of solvent in a polymeric solutions is a strong function of solvent concentration, temperature and molecular size.[53-55,60] The binary diffusion coefficient can be estimated by the free volume theory of Vrentas and Duda[17] and can be measured by the NMR. However, there are many difficulties to measure or estimate the ternary diffusion coefficient.

The evaporated solvents are swept away by the hot air and are exhausted to the outside of the dryer. The vapor concentration of solvent of a zone (oven) must be lower than the lower explosive level (LEL) of solvents to meet the safety. Thus enough fresh air should be supplied to each zone to meet this LEL safety.[10,28] However, too much supply of fresh air brings about too much energy consumption to heat the fresh air, and it also increases the VOC containing waste gas which should be treated by VOC emission control unit. The LEL of a zone can be controlled within appropriate LEL ranges by measuring and predicting of drying rate.[28]

We are going to deal with some important features of drying a liquid coated film in the subsequent sections.

First, we will introduce theory for the drying and method of modeling of the drying process. A drying process of liquid coated film can be simulated by setting up heat and mass transfer equations[8,9,40,59] and vapor liquid equilibrium equations[19,39] with appropriate heat and mass transfer coefficient.[37,45]

Secondly, we will try to give examples of drying rate measurements in laboratory experiments and in a pilot or production scale dryer. Traditionally, the rate of drying was obtained by measuring weight loss using a balance under the controlled experimental conditions of air velocity and temperature.[5,40,51] However, it is difficult to obtain such data in the high airflow rate experiments, for the balance is disturbed by the high airflow motion. Therefore the alternative way of measuring was proposed in which hydrocarbon analyzers[57] or FT-IR[47,50] was used instead of balance. Moreover, the weight changes of each solvent in a multi-solvent coating system could be found by applying rapid scanning FT-IR method.[50]

However, rare data has been given about measuring the actual drying rate and solvent concentration profile along the dryer while operating.[28] The drying rate of coated film in an industrial dryer is measured by analyzing dryer exit gas, and we will illustrate how the solvent concentration profiles of coating at the exit of each zone are estimated from the experimentally measured drying rate data.[28]

In third, we will try to illustrate the dependence of dried coating structure on the drying path by means of the basic phenomena, phase equilibrium.[29,35,38,41,42,56,58] We will also briefly introduce the mechanisms behind the formation of microstructure of various kind of functional films.[44,51]

Finally, we will categorize the drying related defects according to the origin of defects, and we will show some examples of defects and the cause and curing of them.[8,15,21,25]

## 7.3.2 THEORY FOR THE DRYING

### 7.3.2.1 Simultaneous heat and mass transfer

Because evaporation is an endothermic process, heat must be delivered to the system, either through convection, conduction, radiation, or a combination of these methods. The solvents are evaporated from the coating surface and at the same time the latent heat of solvent cool down the coating surface. Thus the heat and mass are transferred simultaneously through the surface of coating.

Assuming that the heat is supplied only by convection of hot air and the substrate is impermeable. Further if we neglect the internal resistance of solvent transport to the coating surface, the heat and mass balance consist a lumped parameter system.

A schematic diagram of the modeled system is shown in Figure 7.3.3. The corresponding mass and heat balance of the systems are as follows[28]

$$R_{mass,i} = \rho_i b_C \frac{dz_i}{dt} = k_m\left(C_i^{sat} - C_i^{\infty}\right) \qquad [7.3.1]$$

$$\left(\rho_f b_f C_{Pf} + \sum_{i=1}^{k} \rho_i z_i b_C C_{Pi}\right)\frac{dT}{dt} = h\left(T - T^{\infty}\right) - \sum_{i=1}^{k} R_{mass,i}\left(\delta H_{evap}\right)_i \qquad [7.3.2]$$

where:

| | |
|---|---|
| $R_{mass,i}$ | evaporation rate of component i |
| $z_i$ | volume fraction of component i |
| $\rho_i$ | density of pure component i |
| b | thickness |
| $C_i^{sat}$ | saturated solvent concentration of component i |
| $C_i^{\infty}$ | solvent concentration of component i in the bulk air |
| $k_m$ | mass transfer coefficient |
| h | heat transfer coefficient |
| T | temperature of coated film |
| $T^{\infty}$ | temperature of drying air |
| $\delta H$ | latent heat of solvent |
| $C_p$ | heat capacity |

Subscript f and C mean the substrate and coating layer respectively. Equation 7.3.1 and 7.3.2 apply to the each component of coating.

### 7.3.2.2 Liquid-vapor equilibrium

The equilibrium saturated solvent concentration is related to the concentration of solvent at the coating surface by thermodynamic equilibrium relations, such as Henry's law, Raoult's law and the Flory-Huggins equation.[8] The Raoult's law is

$$C_i^{sat} = \gamma_i z_i \frac{P_i^{sat}}{RT} \qquad [7.3.3]$$

where:

| | |
|---|---|
| $\gamma_i$ | activity coefficient of component i |
| $P_i^{sat}$ | saturated vapor pressure of component i |
| R | gas constant |

The saturated vapor pressure is calculated from the Antoine equation at the specific temperature.

$$\log_{10} P^{sat} = A - \frac{B}{T + C} \qquad [7.3.4]$$

The Antoine equation coefficients A, B, C for various organic solvents can be found elsewhere.[19] The UNIFAC group contribution method was used to calculate the activity coefficient of each solvent.[49]

### 7.3.2.3 Heat and mass transfer coefficient

The rate of heat transfer to the coating depends on the two factors as shown in Equation 7.3.2: the difference between the temperature of coating and ambient air (the driving force) and the geometry where the heat transfer occurs (heat transfer coefficient). The heat transfer coefficient is a function of the nozzle geometry and blowing air properties. Many variables affect the heat transfer coefficient of nozzles, such as nozzle geometry and size, nozzle to coating surface distance, nozzle to nozzle spacing, velocity of air at the nozzle exit and air motion above the coating surface.[37,45]

Therefore the average heat transfer coefficient of a zone can be expressed as follows,

$$\bar{h} = f(\text{Geometry of nozzle, Properties of air}) \times f(w) \qquad [7.3.5]$$

where:

$\bar{h}$ average heat transfer coefficient of a zone
w velocity of air at the nozzle exit.

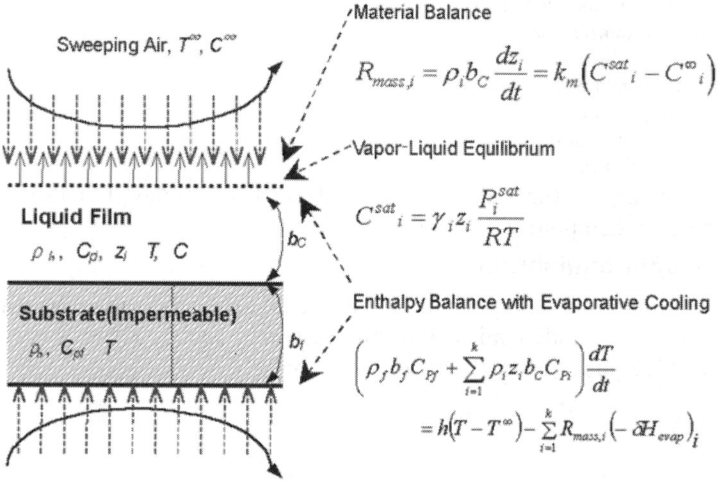


Figure 7.3.3. Schematic diagram of modeled drying of coated film. [After reference 28].

The accuracy of drying rate calculation greatly depends on the proper estimation of the heat and mass transfer coefficient of nozzles. Many researches have been done to find out the heat transfer coefficient of nozzles, among them the Martin's correlation is the notable one which correlates the geometry of impinging jet nozzle and air velocity to the heat and mass transfer coefficients.[37] For a multiple slot jet nozzles, which is depicted in Figure 7.3.4, Martin suggested following empirical correlation,

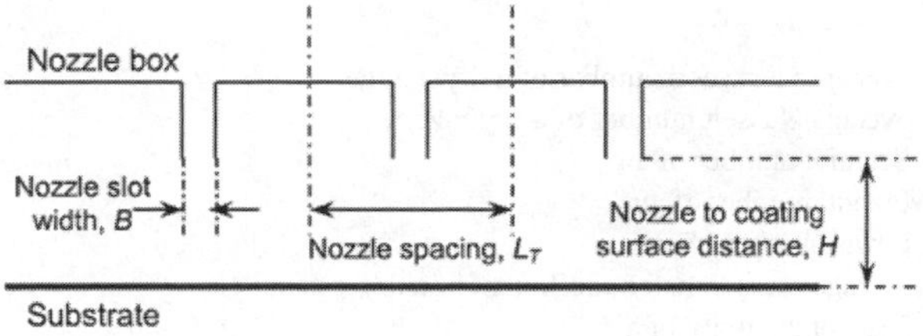


Figure 7.3.4. Geometry of slot nozzles for the Martin's correlation [After reference 37].

$$\frac{\overline{Nu}}{\Pr^{0.42}} = \frac{2}{6} f_O^{3/4} \left( \frac{4\mathrm{Re}}{\frac{f}{f_O} + \frac{f_O}{f}} \right)^{2/3} \qquad [7.3.6]$$

where:

| | |
|---|---|
| $\overline{Nu}$ | average Nusselt number over a zone ($2\bar{h}B/\kappa_a$) |
| Re | Reynolds number ($2wB/\nu_a$) |
| Pr | Prandtl number of air ($\nu_a/\alpha_a$) |
| f | Fraction open area ($B/L_T$) |
| $f_O$ | $[60 + 4(H/2B - 2)^2]^{-1/2}$ |
| B | Nozzle slot width (Figure 7.3.4) |
| $L_T$ | Nozzle spacing (Figure 7.3.4) |
| H | Nozzle to coating surface distance (Figure 7.3.4) |

Range of applicability is

$1{,}500 \le Re \le 40{,}000$

$0.008 \le f \le 2.5 f_O$

$2 \le H/B \le 80$

The heat transfer coefficient of the arrays of round jet nozzles and the other shapes of nozzles can be found elsewhere.[37,45] In general the heat transfer coefficient can be written in the following form[9,10]

$$\bar{h} = Kw^n \qquad [7.3.7]$$

where:

| | |
|---|---|
| K | a constant that depends on the physical properties of the air and geometric properties of the dryer |
| n | 0.6 ~ 0.8 (depending on the nozzle geometry) |

One can easily calculate the actual heat transfer coefficient of a zone by running a heavy gauge web through a dryer and measuring temperature rising of web using non-contacting infrared thermometer in the early part of the dryer where the web is heating up.[10] The procedures of measuring and accompanied calculation are illustrated in Table 7.3.1.

The mass transfer coefficient is related to the heat transfer coefficient through the Chilton-Colburn analogy.[16] Thus,

$$\frac{\overline{Sh}}{\overline{Nu}} = \left(\frac{Sc}{\Pr}\right)^{0.42} = (Le)^{0.42} \quad or \quad k_m = \frac{\bar{h}}{k_a} Le^{0.42} \qquad [7.3.8]$$

where:

| | |
|---|---|
| $\overline{Sh}$ | average Sherwood number of a dryer zone |
| $\overline{Nu}$ | average Nusselt number of a dryer zone |
| Sc | Schmidt number of air |
| Pr | Prandtl number of air |
| Le | Lewis number of air |
| $\overline{k}_m^{\cdot}$ | average mass transfer coefficient of a dryer zone |
| $k_a$ | heat conductivity of air |

**Table 7.3.1. Procedure for calculating h in dryer. [Adapted, by permission, from Cohen, E. D. and E. B. Gutoff, Modern Coating and Drying Technology, VCH Publishers, Inc., New York, 1992]**

1 Set dryer air conditions and measure air inlet temperature

2 Run heavy gauge polyester web through dryer

3 Install two infrared thermometers relatively close together at initial part of drying zone

4 Run web at varying line speeds and measure temperature rise in web

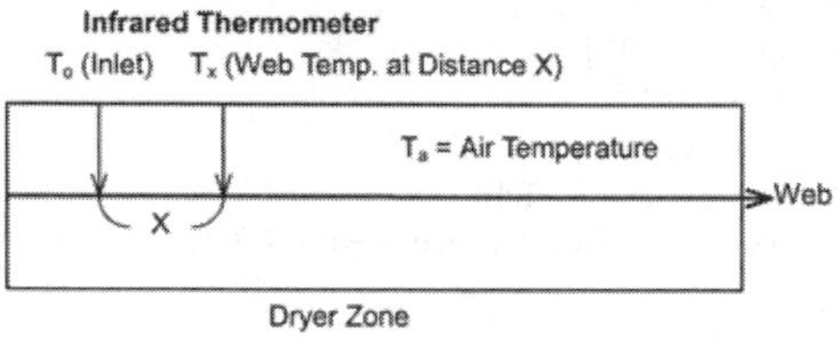

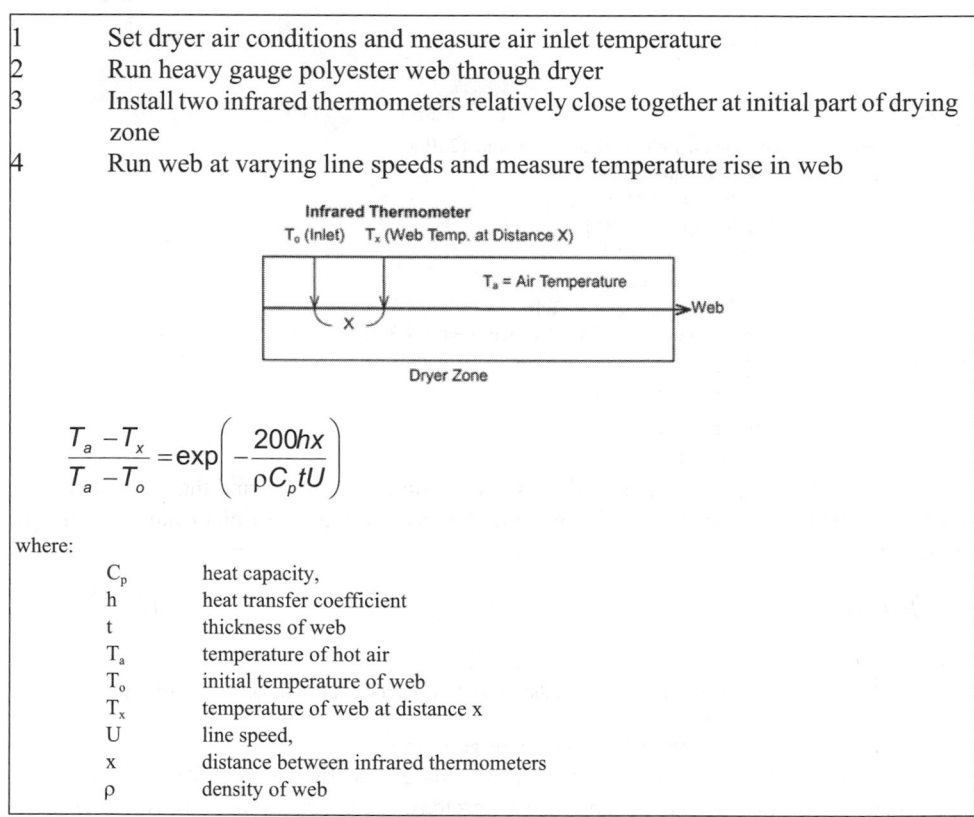

$$\frac{T_a - T_x}{T_a - T_o} = \exp\left(-\frac{200hx}{\rho C_p tU}\right)$$

where:

| | |
|---|---|
| $C_p$ | heat capacity, |
| h | heat transfer coefficient |
| t | thickness of web |
| $T_a$ | temperature of hot air |
| $T_o$ | initial temperature of web |
| $T_x$ | temperature of web at distance x |
| U | line speed, |
| x | distance between infrared thermometers |
| ρ | density of web |

The typical value of heat transfer coefficient of modern industrial dryer ranges from 50 to 150 J/m$^2$sec°C. Figure 7.3.5 shows the calculated heat transfer coefficient of an impingement dryer with varying nozzle exit velocity of air and fraction open area (nozzle spacing), calculation was done according to the Martin's correlation.

### 7.3.2.4 Prediction of drying rate of coating

The drying rate of coating and the subsequent residual solvent amount along with dryer length can be found by applying above equations. In magnetic media manufacturing pro-

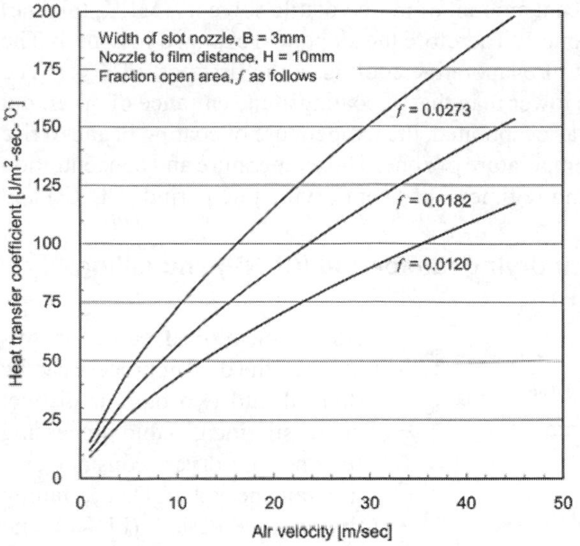


Figure 7.3.5. Calculated heat transfer coefficient of industrial dryer with varying air velocity.

cess, magnetic particulate dispersed solution is coated on the PET (polyethlylene terephthalate) film. The wet coating thickness is about 5 to 6 μm and the thickness of substrate is 14 to 15 μm. The line speed is normally 400 m/min to 1000 m/min. Usually several kinds of solvents are used for the coating solution, here we used three kinds of solvents - toluene, methyl ethyl ketonc (MEK) and cyclohexanone (CYC). Equations 7.3.1 and 7.3.3 are applied to each solvent component. The temperatures of dryer zones were 50 to 130°C, and the air velocities at the nozzle exit were 10 to 20 m/sec. The average heat transfer coefficient of a zone could be found by applying Martin's correlation for the slot nozzles, however in this case we had the empirical coefficient for the Equation 7.3.7 which was supplied by the dryer nozzle manufacturer. And we obtained average heat transfer coefficients which were ranging from 80 to 140 $J/m^2s°C$ according to the air velocities of the zones.

The coupled and non-linear set of equations is solved by standard numerical method (such as Runge-Kutta-Gill method). The temperature and concentration of coating at time 0 was given as the initial concentration of coating. The concentration of coating at the next time step was calculated by solving Equation 7.3.1 with assuming that there was no significant temperature change of coating. This is a plausible assumption if the time step is small enough. Then the result of concentration of coating was substituted to the Equation 7.3.2 and the temperature of coating at that time step was calculated.

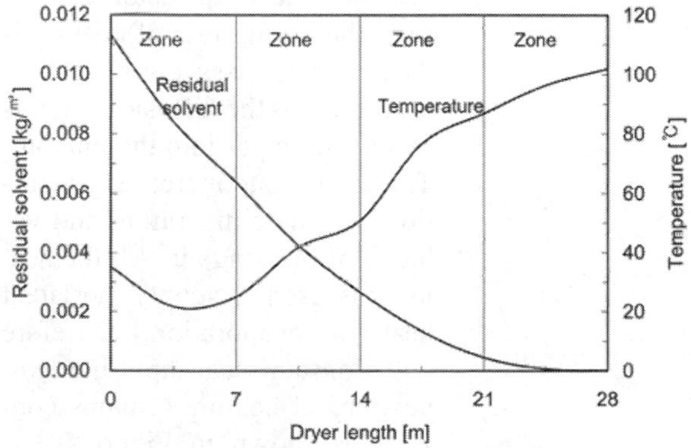


Figure 7.3.6. Predicted residual solvent and temperature profile along with the dryer length.

The calculated concentration and temperature profile of coating are given as in the Figure 7.3.6. The concentration of solvent gradually decreases along with the dryer length. The slope of solvent concentration profile at zone 1 was nearly constant, but the slop was changed within the zone 1 though there were no changes in the drying conditions. It was due to the fact that the less volatile solvent (cyclohexanone) began to evaporate only after significant amount of more volatile solvents (MEK, toluene) were evaporated in the first part of zone 1. Therefore the slope was changed in zone 1. The temperature profile shows the effect of evaporative cooling in the first part of dryer. The temperature of coating at zone 1 was lower than that of coating at the entrance of dryer, but after significant amount of solvent was evaporated, the temperature of coating began to rise and eventually it approached to the temperature of zone. The temperature and concentration profile successfully explain the drying regimes - constant drying rate period (CDRP) and falling drying rate period (FDRP).

### 7.3.2.5 Drying regimes: constant drying rate period (CDRP) and falling drying rate period (FDRP)

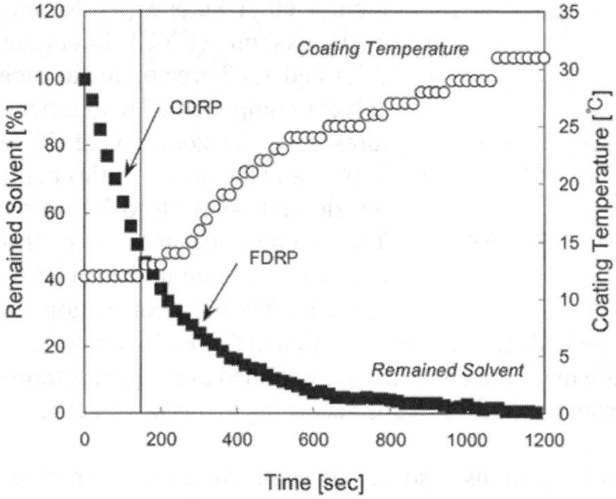


Figure 7.3.7. Characteristic drying curve of polymeric solution (methanol - ethyl acetate - acrylic resin solution).

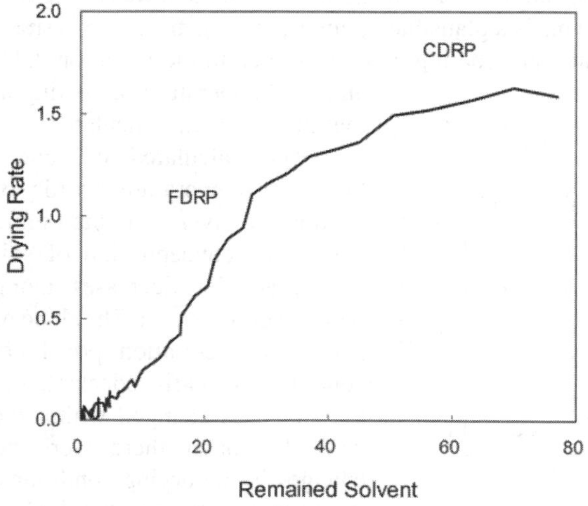


Figure 7.3.8. Typical drying rate curve (methanol - ethyl acetate - acrylic resin solution).

As we mentioned earlier in this chapter, the drying process can be divided into two or four distinct and easily identifiable processing regions: pre-dryer, constant drying rate period (CDRP), falling drying rate period (FDRP) and equilibration, or simply CDRP and FDRP.[10] The pre-dryer and equilibrium regions denote the part of dryer which is located between the coating station and the dryer, and between the dryer exit and the take up roll, respectively.

In the early stages of drying of polymeric solution coating, the surface of coating is fully wetted with solvent, thus the evaporation of solvent takes place as if there were no polymer or solute in it, as in the case of evaporation of pure solvent. The only factor that affects the drying rate of coating is the external mass transfer resistance such as the diffusion process of solvent vapor into the bulk air. The drying rate increases proportional to the temperature and velocity of blowing air. All the heat input is used to supply the latent heat of evaporation, therefore with constant heat input the temperature of coating remains constant as shown in Figure 7.3.7. Figures 7.3.7 and 7.3.8 represent typical drying curve of polymeric

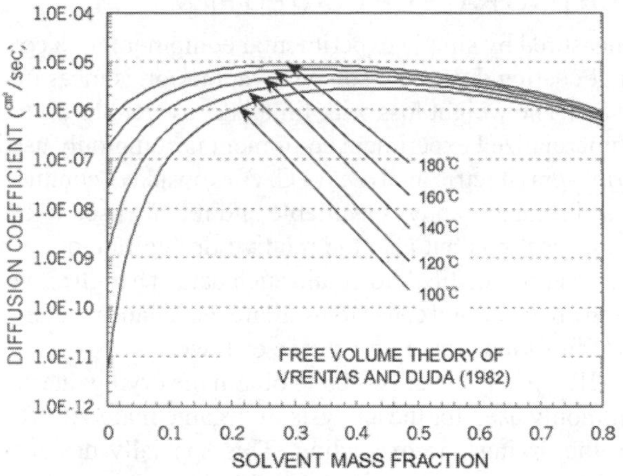


Figure 7.3.9. Concentration and temperature dependence of the binary diffusion coefficient of a polystyrene-toluene solution according to the free volume theory of Vrentas and Duda. [After References 17].

coating solution. In CDRP, as shown in the figures the slope of solvent content loss with time (drying rate) and the coating temperature are nearly constant. These drying rate curves were obtained by measuring weight changes of coating in an oven with blowing hot air above the coating samples - see Section 7.3.3.1.

However, as the solvent content decreases the drying rate of coating also decreases gradually as shown in Figure 7.3.8. In this falling rate regime, mass transfer within the coating becomes the limiting factor, and with constant heat input the temperature of coating rises and drying rate decreases. The solvent is transported to the exposed coating surface by diffusion and convection, where the solvent evaporates to the bulk air in a polymeric coating system. In most drying processes of thin film coatings the convection of solvent within the coating is negligible, thus the diffusion of solvent is the only method for solvent to reach the coating surface. The diffusivity of solvent in a polymeric solution falls dramatically when the solvent content is low as shown in Figure 7.3.9. Therefore the diffusion process of solvent within the coating controls the rate of drying in FDRP.

To understand and improve the drying process during the FDRP, it is important to estimate diffusion coefficient of solvent within the coating. Vrentas and Duda predicted the binary diffusion coefficient by using free volume theory.[17] Figure 7.3.9 shows the binary diffusion coefficient of toluene in the polystyrene-toluene solution that was found by applying free volume theory. The diffusion coefficient falls by a number of magnitudes in the low solvent concentration range, and it increases with increasing temperature. By the way the concentration dependency of diffusion coefficient at the low solvent concentration range declines at a high temperature. For example, at the temperature of 100°C the diffusion coefficient of toluene falls 5 orders of magnitude (about $10^{-5}$) from its highest value, but at the temperature of 180°C it falls only one order of magnitude. This shows us that it is highly required to eliminate residual solvent of coating at the low solvent concentration range the temperature of oven should be high enough. Therefore the temperatures of final zones keep high enough to achieve complete dryness in a multiple zone dryer. Normally the temperature of oven is restricted by the onset of deformation of substrate such as wrinkles and it is also restricted by the properties of coated material, such as melting point of binders.

The parameters for the free volume theory of binary solution systems can be found in the literatures,[17,53-55,60] and they have been effectively used in modeling drying process.[1,2,6-8] But there are many difficulties in estimating diffusion coefficient for the ternary systems. Until now, almost all the empirical and theoretical correlations are restricted to the binary solution systems.

### 7.3.3 MEASUREMENT OF THE DRYING RATE OF COATED FILM

The drying rate of coating is easily measured by simple experimental equipment. In a controlled air condition, the weight loss of coating due to the solvent evaporation is measured by electrical balance and filed at the PC. The weight loss with time is converted to drying rate of coating per unit area. The commercialized experiment equipment is commonly used to obtain the drying rate, such as thermo-gravimetric analyzer (TGA). Nowadays commercial TGA equipped with FT-IR or gas chromatography is available and readily used to obtain not only the overall drying rates of coatings but also the relative drying rate of each solvent in a multi-solvent system. However it is difficult to obtain such data at a high air velocity, because the air stream disturbs the balance and cause to oscillate the balance reading. Practically the available air velocity of this kind would be 1 m/sec or lower.

Fourier-Transform-Infrared (FT-IR) spectrometer is used to obtain the drying rate at a higher air velocities.[47,50] FT-IR is commonly used for the analysis of organic materials. Recently, FT-IR is applied to measure the drying rate of coating. This specially designed FT-IR with coating apparatus and air blowing system made it possible to measure solvent content without disturbances of airflow, and moreover it enabled us to find the content of each solvent with drying proceeded.[47,50] This can be used to study selective evaporation of solvent and phase separation phenomena in a multi-solvent system.[50]

The air velocity of industrial dryer is up to 20 m/s or more and the temperature of oven is normally up to 200°C. A specially designed drying chamber was suggested to measure solvent concentration in these drying conditions.[52,57] The chamber is equipped with flame ionization detector (FID) total hydrocarbon analyzer, and the oven exit gas which contains the evaporated solvent flows to the analyzer. The drying rate is calculated by multiplying the solvent concentration with exit gas flow rate. It provides access to a wider range of drying conditions that approximate industrial conditions.

However, it is much more difficult to find out the actual drying rate of coating in an industrial dryer. In a continuous industrial dryer it is impossible to measure the drying rate by any of the above methods, because the coated substrate is running through the dryer at the speed of several hundreds meter per minute. To measure the drying rate of coating in such a condition, the dryer exit gas of each zone is analyzed by gas chromatography. Then the drying rate of each zone is calculated by multiplying solvent concentration with exit gas flow rate. From the drying rate of each zone, the evaporated amount of solvent is calculated. Thus the solvent concentration of coating is found at the point of each zone end.[28,30]

#### 7.3.3.1 Thermogravimetric analysis

The drying rate of a coating could be easily found by measuring coating weight loss during drying in a laboratory. The set-up of experimental equipment is relatively easy, and the commercial equipment can be readily available such as thermogravimetric analyzer (TGA).[5,53] As drying proceeded, the weight of coating decreases due to the solvent evaporation. The amount of solvent loss with time is monitored by the balance. The schematic representation of the equipment is shown in Figure 7.3.10. The sample, such as coated films or a tray that contains the coating liquid, is mounted on the balance to be monitored. The temperature of coating is measured by non-contact infrared thermometer. A thin wire type thermocouple can be used to measure the temperature, and the thermocouple is attached to the coating or sample tray. The air is made up by conventional blower and is heated up to a certain temperature by electrical heater. The air is supplied to the coating surface through

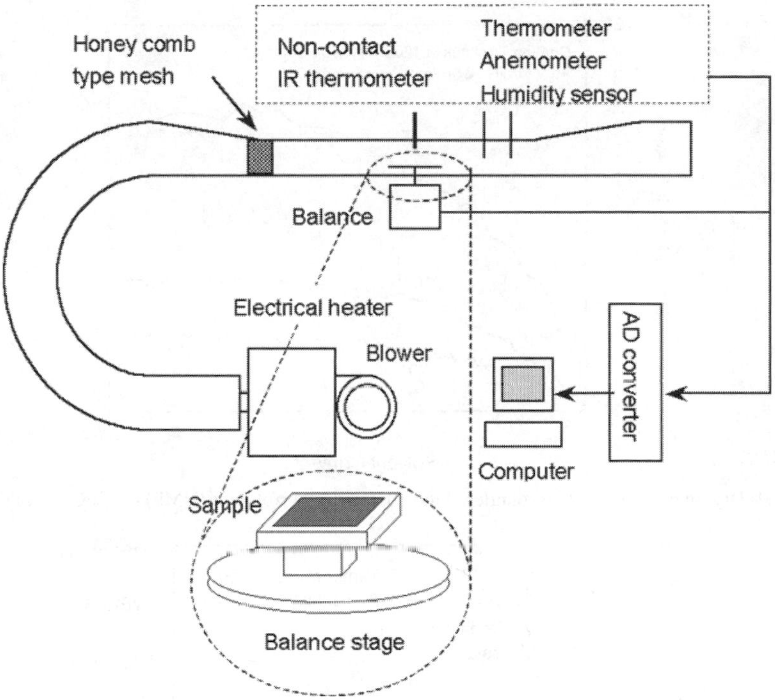


Figure 7.3.10. Schematic diagram of drying experiment apparatus.

ducts. To make the laminar airflow over the balance, the length of ducts and sizes are determined. The experiment is usually conducted in a low air velocity so as the air not to disturb the balance. Honeycomb style mesh is often helpful to filter the air and make the airflow a laminar one. It enabled us to do the drying experiment at a higher velocity of air. The air velocity can be as high as 1 ~ 2m/sec.

Actually the air velocity of industrial dryer is much higher than the experimental conditions and the directions of flow are normally perpendicular to the coating surface, not parallel with coating surface. Thus we can hardly expect to conduct a quantitative simulation from this experiment, but we can find the characteristic drying curve and mechanisms of drying of the given materials.

The drying rate of coating is the weight of solvent loss per time divided by the area of evaporation.

$$R_{mass} = -\frac{dW}{Adt} \approx -\frac{W_{t+\Delta t} - W_t}{A\Delta t} \qquad [7.3.9]$$

where:

| | |
|---|---|
| $R_{mass}$ | evaporation rate |
| W | weight of sample at a specific time |
| t | time |
| $\Delta t$ | time interval between measurements |
| A | evaporation area |

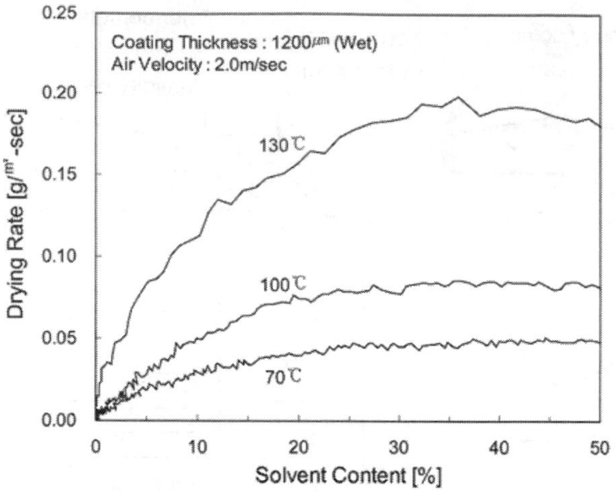


Figure 7.3.11. Drying rate curve of low volatile solution: N-methylpyrrolidone (NMP) - $LiCoO_2$ - PVDF solution.

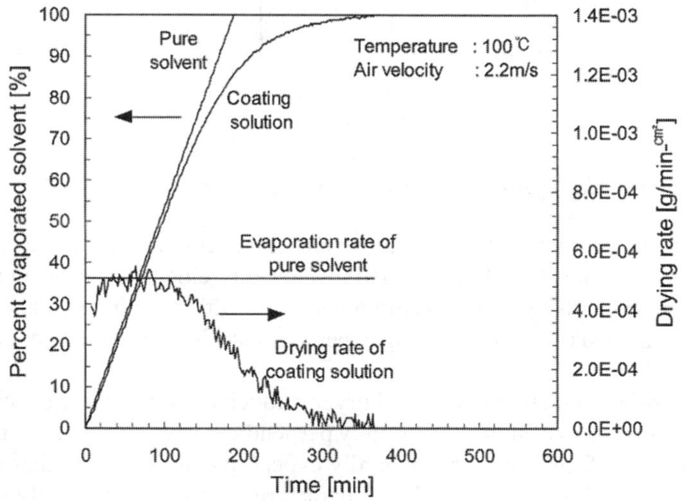


Figure 7.3.12. Comparison of drying rate between coating solution and pure solvent : N-methylpyrrolidone (NMP) - $LiCoO_2$ - PVDF solution and pure NMP.

The example of measurement is shown in Figure 7.3.7. The remained solvent means the percent of remained solvent to the total solvent load, and the drying rate of coating is readily calculated using Equation 7.3.9. The drying rate at the specific time equals the slope of the drying curve of Figure 7.3.7, and it is depicted in Figure 7.3.8. The drying rate of coating shows constant and falling rate period with solvent content decreases. The temperature of coating is nearly constant at the beginning of drying where the slope of weight loss remains constant. As shown in Figure 7.3.7, most of the solvent is evaporated during the constant rate period.

Figure 7.3.11 is the drying rate profile according to the solvent content at the various drying temperatures. The coating solution is consisted of $LiCoO_2$ and poly(vinylidene fluoride) (PVDF) solution. Polymer is dissolved in N-methylpyrrolidone (NMP), and this solu-

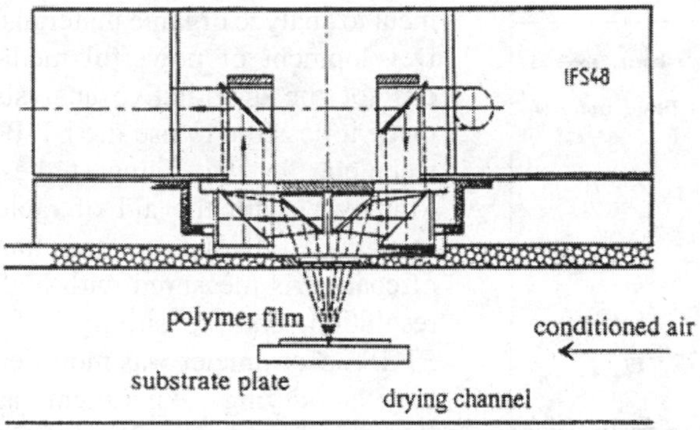


Figure 7.3.13. Schematic diagram of FT-IR spectrometer for the measurement of solvent content. [Adapted, by permission, from Saure, R. and V. Gnielinski, *Drying Technol.*, **12**, 1427 (1994)].

tion is originally prepared for the anode cell of the lithium ion battery. The solvent content represents the percent of remained solvent to the weight of non-volatile component, $(W-W_o)/W_o$, where W and $W_o$ represent the wet weight during drying and dry weight, respectively. NMP has very high boiling temperature (about 202°C) and low volatility, thus the drying of coating shows long period of constant rate. Figure 7.3.12 shows the comparison of drying rate between pure solvent and coating solution. The drying rate of coating solution is nearly equal to that of pure solvent during the CDRP. But the drying rate of coating gradually decreases as the solvent concentration falls, while the evaporation rate of pure solvent is constant throughout the evaporation.

### 7.3.3.2 Rapid scanning FT-IR spectrometer analysis

Recently Fourier-Transform-Infrared (FT-IR) spectrometer is applied to drying studies to obtain the drying rate at a higher air velocities.[47,50] FT-IR is a widely used analytical instru-

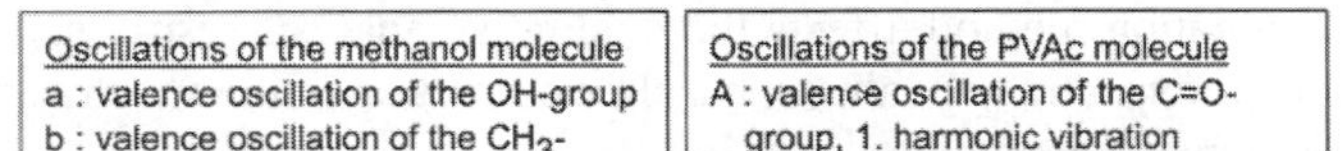

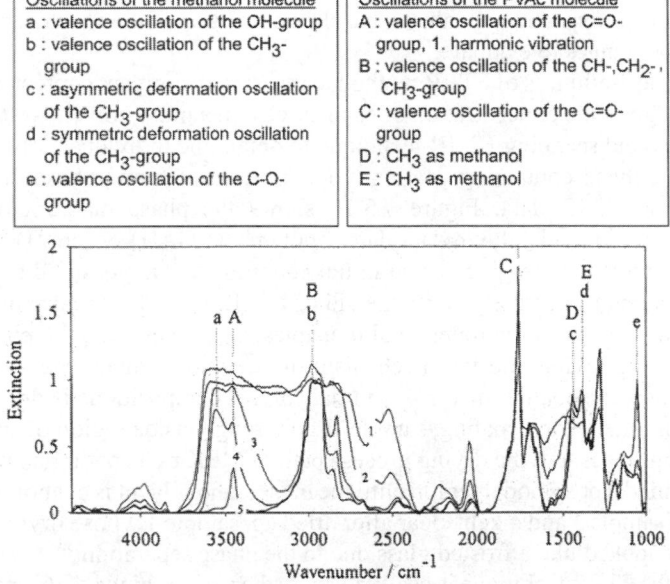


Figure 7.3.14. An example of FT-IR spectra of drying film. The methanol bands disappear slowly while the polymer bands remain constant. The ratio of the band height contains information on concentration. Methanol contents spectrum1 54 g/m², spectrum2 19 g/m², spectrum3 3.7 g/m², spectrum4 1.5 g/m², and spectrum5 0 g/m², PVAc content 57 g/m². [Adapted by permission, from Saure, R. and V. Gnielinski, *Drying Technol.*, **12**, 1427 (1994)].

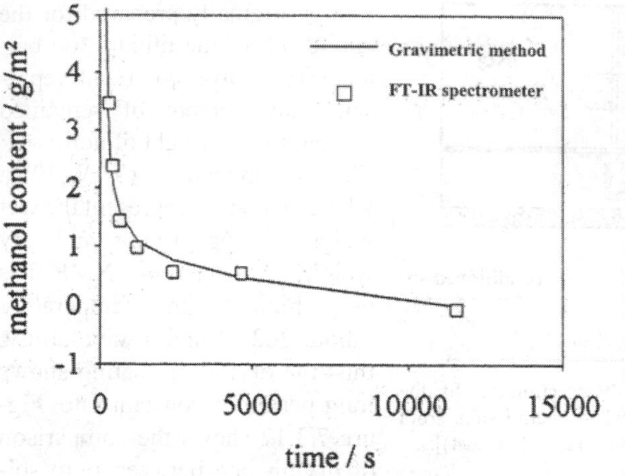


Figure 7.3.15. Comparison between gravimetric and FT-IR data. [Adapted, by permission, from Saure, R. and V. Gnielinski, *Drying Technol.*, **12**, 1427 (1994)].

ment to analyze organic materials. Development of powerful methods for the quantitative analysis made it possible to use the FT-IR technique for the drying studies. Moreover, with the aid of rapid scanning FT-IR the concentration of coating is measured with high resolution and sensitivity. The FT-IR spectrometer was modified for the drying experiment as shown in Figure 7.3.13.[47] A typical spectra of a coated film, which contains methanol and polyvinyl acetate (PVAc), is shown in Figure 7.3.14. The typical absorption bands of PVAc and methanol can be identified from the literature and is shown in the Figure 7.3.14 for a reference.[47] Basically the plot of FT-IR spectra and the ratio of band heights between the solvent and nonvolatile material give the information of concentration. The calculated methanol contents along with methanol bands are also seen in the Figure 7.3.14. The comparison between FT-IR and gravimetric data shows a good agreement as shown in Figure 7.3.15.[47] The calibration method affects the qualitative analysis result, so care should be given in selecting the spectra to evaluate, baseline correction of the spectra and selection of the wavenumber ranges to evaluate.[47]

Besides the usefulness of FT-IR method to measure the solvent content at the high airflow ranges, it gives the concentration of each solvent in a multi-solvent system. Suzuki et al. applied the rapid scanning FT-IR technique to obtain the individual solvent concentration of binary solvent containing coating.[50] The process path and phase diagram can be drawn from the FT-IR data. Figure 7.3.16 shows the phase diagram of MEK-toluene-polyvinylchloride and polyvinylacrylate copolymer (VGAH) system. The initial and final coating composition are given as an initial condition and a measured residual solvent content of a coating respectively. With the aid of FT-IR technique the drying process path between the two points can be found, and it enables us to investigate the phase separation phenomena during drying and the mechanism of structure formation of coating. Figure 7.3.16 shows how the coating of various initial solvent compositions is dried. The drying process path of toluene rich coating, sample 1, go through inside region of spinodal line in the early drying stages, but the drying process path of MEK rich coating do not go through the inside of binodal or spinodal region until the most of the solvent is evaporated. While the appearance of sample 3 and 4 kept clear after dried up, sample 1 whose drying process path was number 1, looked like a frosted glass due to the phase separation.[50]

Though the FT-IR technique is useful to study the various kinds of drying phenomena, the application of FT-IR spectrometer is restricted to a certain solvent system because the bands of spectrometer of each component of a solution must be separated. While the above

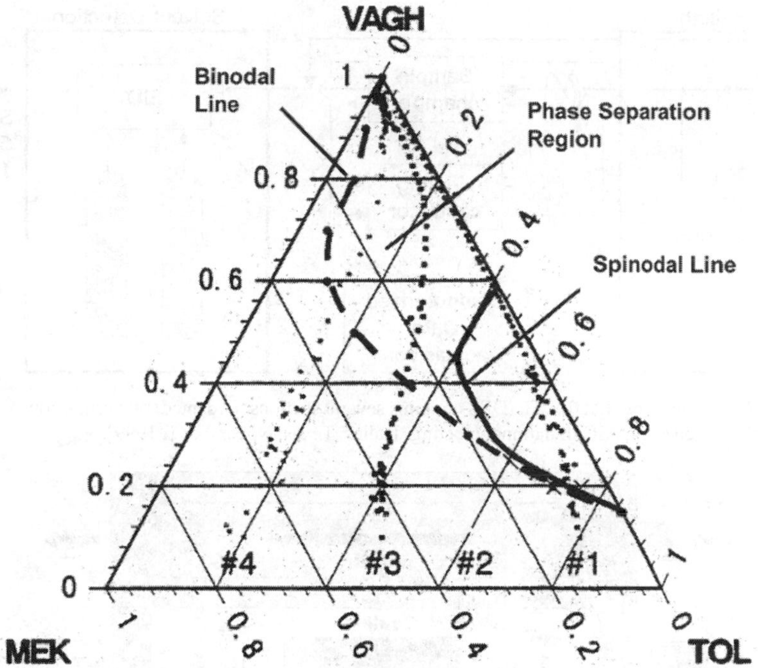


Figure 7.3.16. Phase diagram of MEK-toluene-VAGH system with the different initial solvent compositions. The concentration is measured by rapid scanning FT-IR method. [Adapted, by permission, Suzuki et al., Proceedings of 9th International Coating Science and Technology Symposium, Delaware, USA, May 17 - 20, 1998, pp. 21-24].

example of PVAc-methanol system is suitable for the FT-IR spectroscopy because of their separated bands, it is difficult to extend FT-IR technique to the other solvent systems.

#### 7.3.3.3 High-airflow drying experiment using flame ionization detector (FID) total hydrocarbon analyzer

The gravimetric method is limited to a certain air velocity level due to the oscillation of balance in the high airflow stream. However most of the industrial drying process accomplished by passing the coating under high air velocity jet of hot air. Thus to simulate the industrial drying conditions and according drying phenomena better, high airflow drying experiment setup (HADES) was suggested by Cairncross et al.[52,57]

Low air velocity results in low heat transfer coefficient, the heat transfer coefficient of conventional laboratory drying experiments ranges from 1 to 10 $J/m^2s^oC$, while that of industrial dryer is ranges about 20 to 200 $J/m^2s^oC$. The high heat and mass transfer at the evaporation surface may result in 'trapping skinning' because the surface evaporation is too high in compared with the rate of diffusion of solvent within the coating. The heat transfer coefficient of HADES is up to 26.4 $J/m^2s^oC$ which is equivalent with that of usual single-side impingement dryers,[52] and it was reported that the HADES successfully simulates the skinning phenomena with the solution of PMMA-toluene.[52]

HADES measures solvent concentration of exhaust gas from the sample chamber where the solvent is evaporated from the coating. Then the rate of evaporation is equal to the

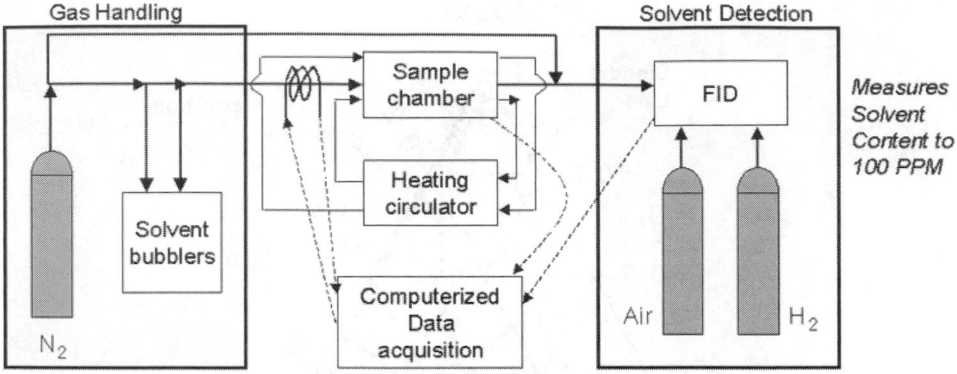


Figure 7.3.17. Schematic of HADES. HADES contains several sections. [Adapted, by permission, Vinjamur and Cairncross, Presented at the AIChE national meeting, Dallas, Texas, November 1, 1999].

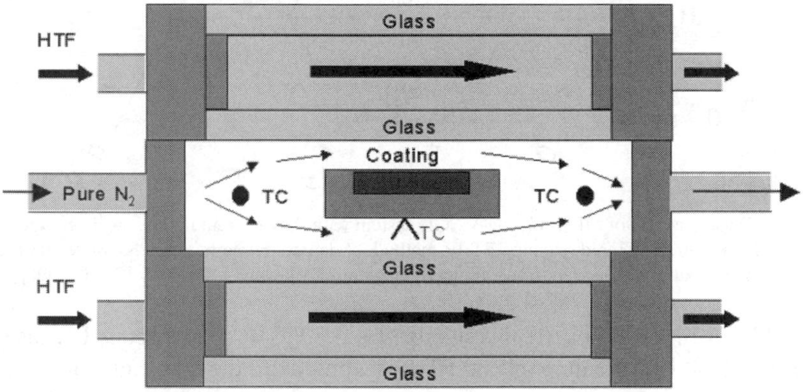


Figure 7.3.18. Schematic of sample chamber of HADES. Dry nitrogen flows in from the port on the left side and the exhaust flows out through the port on the right to a FID. [Adapted, by permission, Winward and Cairncross, 9th International Coating Science and Technology Symposium, Delaware, USA, May 17 - 20, 1998, pp.343 - 346].

solvent vapor concentration times the gas flow rate, and the solvent loss is found by integrating the evaporation rate.

The HADES is consisted of several sections as shown in Figure 7.3.17; gas handling system, a sample chamber and several process measurements.[52,57] The nitrogen gas flows into the sample chamber as shown in Figure 7.3.18. The temperature and rate of gas flow are controlled and the gas temperatures before and after the coating sample tray are measured with thermocouples. The coating temperature is measured with thermocouple which is installed at the coating sample tray, and the solvent laden gas flows into the total hydrocarbon analyzer which is equipped with flame ionization detector (FID). The concentration of solvent at the exhaust gas is measured by total hydrocarbon analyzer, which is calibrated with known concentration of solvent vapor (via solvent bubblers).

Figure 7.3.19 and Figure 7.3.20 show the examples of HADES running.[52] Figure 7.3.19 shows the measured solvent concentration of the PVAC-toluene system, as shown in the figure the residual solvent decreases according to the rate of airflow. However, over the 36 cm/s of airflow rate the drying rate wasn't changed, above this airflow rate, the residual

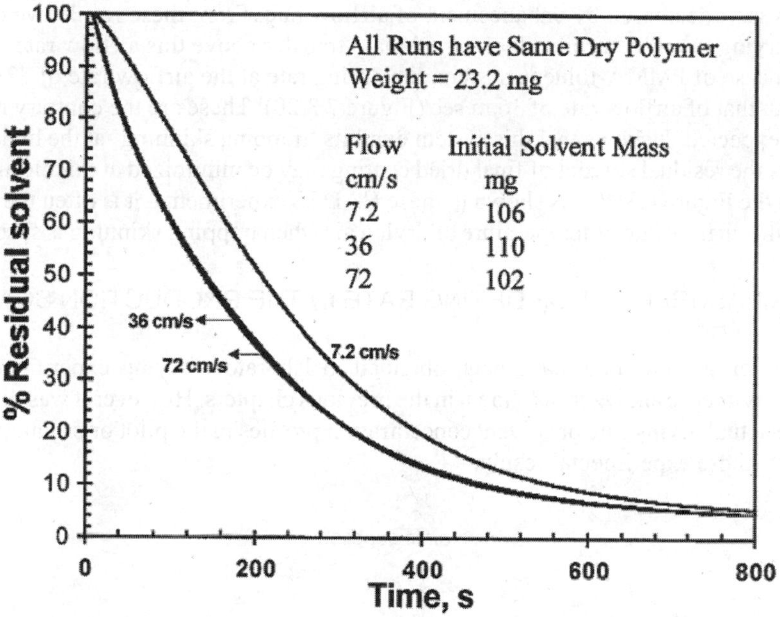


Figure 7.3.19. Measured solvent loss of PVAC-toluene system by HADES. [Adapted, by permission, Vinjamur and Cairncross, Presented at the AIChE national meeting, Dallas, Texas, November 1, 1999].

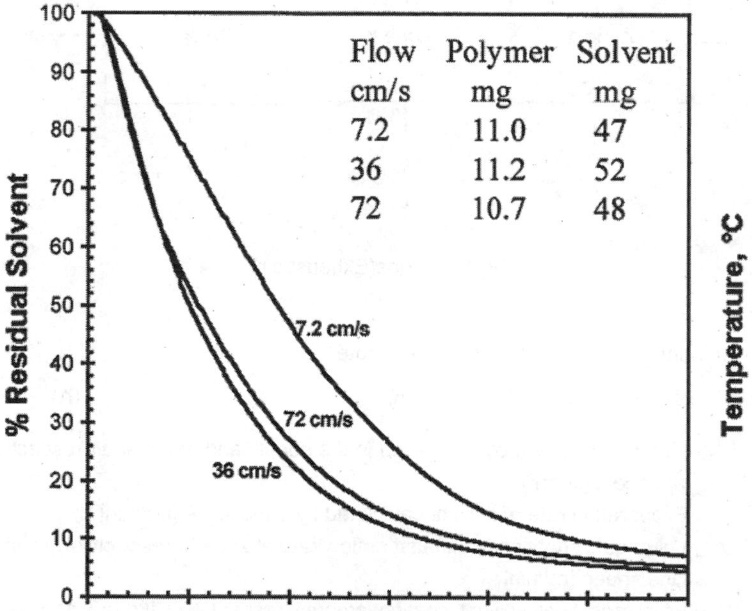


Figure 7.3.20. Measured solvent loss of PMMA-toluene system by HADES. [Adapted, by permission, Vinjamur and Cairncross, Presented at the AIChE national meeting, Dallas, Texas, November 1, 1999].

solvent wasn't decreased by enhancement of airflow rate. From these results we can infer that the drying rate is controlled by internal mass transfer above this airflow rate.

In a case of PMMA-toluene system, the drying rate at the airflow rate of 72 m/sec is lower than that of airflow rate of 36 m/sec (Figure 7.3.20). These are the contrary results to what we expected. It shows that this system exhibits 'trapping skinning' at the high airflow rate. Thus the residual solvent of final dried coating may be minimized at middle airflow as shown in the Figure 7.3.20. As shown in these HADES experiments, it is often the solution to lower the airflow rate or temperature of drying air when trapping skinning is suspected to occur.

### 7.3.3.4 MEASUREMENT OF DRYING RATE IN THE PRODUCTION SCALE DRYER

Numerous drying rate data have been obtained in laboratory drying experiments using thermo-gravimetric analyzers as shown in the previous chapters. However it was difficult to know the actual drying rate or solvent concentration profiles in the pilot or production scale dryer from these experimental results.

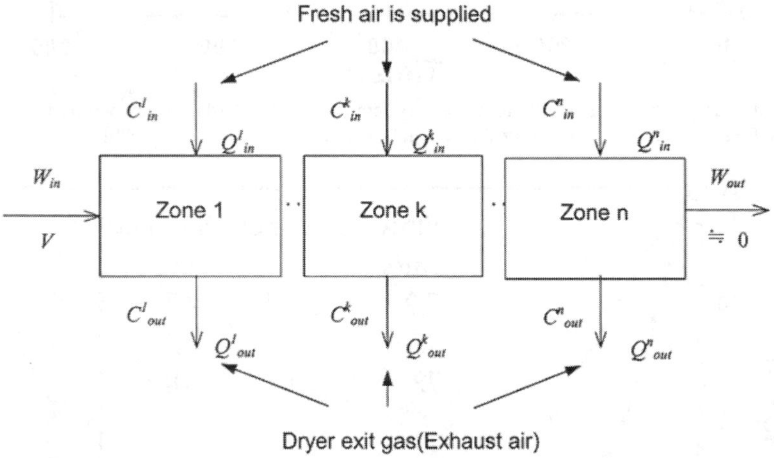


Total amount of evaporated solvent per minute

$$ER^{per} = (W_{in} - W_{out}) \times V \times WD \text{ [kg/min]} \qquad \text{(A)}$$

$C^k_{in}$, $C^k_{out}$ Measured solvent concentration in the supply and exhaust air respectively, at zone k [kg/m$^3$]

$ER^{per}$ Evaporation rate of solvent calculated by equation A [kg/min]

$Q^k_{in}$, $Q^k_{out}$ Volumetric supply and exhaust airflow rate at zone k, respectively[m$^3$/min]

$V$ Line speed [m/min]

$W_{in}$, $W_{out}$ Initial and final solvent coating amount respectively [kg/m$^2$] and $W_{out}$ is practically zero in this drying application (less than 1000 ppm)

$WD$ Coating width [m]

Figure 7.3.21. Construction of material balance to calculate the solvent concentration of the coating at each zone end from the measured solvent concentration of exit gas. [After reference 30].

However, in a multiple zone dryer the solvent concentration of coating at the end of each zone could be found by analyzing the dryer exit gas with measured airflow. This experimental method provides concentration profile of each solvent with drying as well as the total drying rate of coating. It gives precise measurement value at the early stages of drying in which the concentration of solvent in the dryer exit gas is high, and the length of each zone is long enough to ensure negligible intermixing of air between adjacent zones.

The solvent concentrations of the exit gas are measured by portable gas chromatography (GC), and the gas samples are taken at the exhausted air duct of each zone. From these concentration data and the airflow rate of each zone, the evaporation rate of solvent at each zone is calculated. (Figure 7.3.21)

*Evaporation rate of solvent at zone k,* $ER^k$ *[kg/min]*

$$= \text{Solvent concentration, } C^k_{out}\ [kg/m^3] \times \text{Airflow rate, } Q^k_{out}\ [m^3/min] \quad [7.3.10]$$

The example of solvent concentration measurement and the accompanying results of drying rate calculation are shown in Table 7.3.2., and the accompanying specification of dryer and the formulation of coating solution is given in Table 7.3.3 and 7.3.4. The solvent concentrations at the exhaust duct are measured by gas chromatography(MTI Analytical Instrument 200, Portable GC). The solvent concentrations were measured three times at the same point, and the deviations from the average value were less than ±5%.

**Table 7.3.2**

| Zone | Concentration, ppm | | | Evaporation rate, kg/h | | | |
|---|---|---|---|---|---|---|---|
| | MEK | Tol. | Cyc. | MEK | Tol. | Cyc. | Total |
| 1 | 1655 | 740 | 142 | 47.6 | 27.2 | 5.6 | 80.4 |
| 2 | 1150 | 988 | 359 | 40.5 | 44.4 | 17.2 | 102.0 |
| 3 | 485 | 1243 | 1866 | 14.0 | 46.0 | 73.5 | 133.6 |
| 4 | 308 | 157 | 2223 | 1.5 | 1.0 | 15.2 | 17.7 |
| 5, 6 | 269 | 99 | 278 | 6.6 | 3.1 | 9.4 | 19.2 |

**Table 7.3.3**

| | | |
|---|---|---|
| **Operating conditions** | Drying air temperature, °C | 50 ~ 120 |
| | Line speed, m/min | 900 |
| **Dryer specification** | Total number of zones | 6 |
| | Total dryer length, m | 45 |
| | Dry coating thickness, μm | 2.6 |
| | Dryer type | Air floating dryer |

**Table 7.3.4**

| Component | Weight fraction, % | Initial coating amount, $g/m^2$ |
|---|---|---|
| Pigment and binder | 35.5 | 6.23 |
| Methyl ethyl ketone | 21.5 | 3.77 |
| Toluene | 21.5 | 3.77 |
| Cyclohexanone | 21.5 | 3.77 |

The solvent concentration of coating at the entrance of dryer is known with coating formulation and the solvent concentration of final dried coating could be measured offline, thus the percent evaporated solvent at each zone can be calculated. In this example of drying of magnetic coated film, the final dried coating contains less than 1000 ppm solvent, so we assumed that all the coated solvent was evaporated within the dryer. Finally, we estimated how much percent of solvent that was evaporated at each zone. The fractional amount of the evaporated solvent at zone, k, is

$$F^k = \frac{ER^k}{\sum_{k=1}^{n} ER^k} \qquad [7.3.11]$$

where:

| | |
|---|---|
| $F^k$ | fractional evaporated solvent amount in zone k |
| $ER^k$ | evaporation rate at zone k |
| n | number of zone |

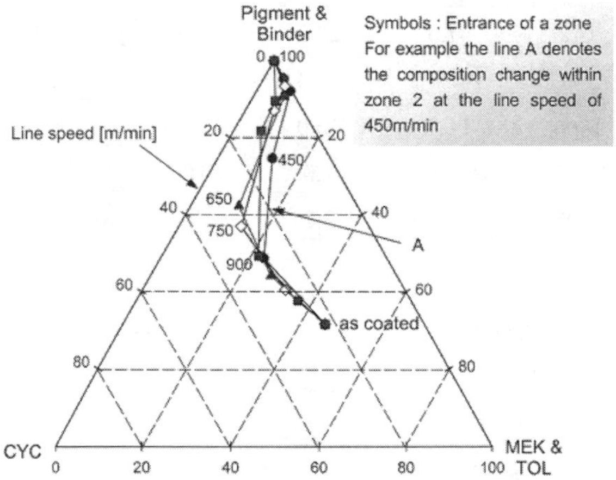

Figure 7.3.22. Measured coating composition changes with drying.

From these fractional solvent evaporation data we estimate the solvent concentration in the coated film at the end of each zone. The solvent content per unit area decreases according to the fractional solvent evaporation data at each zone, for example, if the fractional evaporation rate of solvent is 50% in zone 1 and the initial solvent content is given as 10 $g/m^2$, then the solvent content at the end of zone 1 should be 5 $g/m^2$, thus we can find the solvent concentration at the end of each zone. Moreover the measurements and calculations are applied to the each solvent in multiple solvent systems, therefore we can find the solvent content of each solvent at the end of zone. The resulting coating composition changes with drying are shown in Figure 7.3.22.

The theoretical evaporation rate to dry a given solvent load at the specified line speed can be calculated as in Figure 7.3.21. Theoretically and experimentally found evaporation rate should be equal in ideal case. We can check the measurement error by comparing measured and theoretical evaporation rate through the whole dryer. $ER^{per}$ is calculated by equation A in Figure 7.3.21.

$$\frac{ER^{per} - \sum_{k=1}^{n} ER^{k}}{ER^{per}} = \varepsilon \qquad [7.3.12]$$

In most cases the value of ε is about 0.05 ~ 0.10. $ER^{per}$ is the theoretical evaporation rate which assumes that all the solvent is evaporated only in the dryer. But in the real situation, some of the solvent is evaporated before the dryer, such as in the coating head and the pre-zone (between the coating head and the first dryer zone), and some of the solvent vapor leaks out of the dryer through the gaps where the substrate running in and out. Therefore measured ER was less than $ER^{per}$ by about 10%.

This experimental method gives precise results at the early stages of drying where the concentration of solvent at the exit gas is high. And it is a unique method to measure the solvent concentration of coating in the production and pilot scale dryer.

## 7.3.4 MISCELLANEOUS

### 7.3.4.1 Drying of coated film with phase separation

As a final process of coating process, drying plays an important role for the quality of products. The structure of coating is determined during the drying process. The structure formation of coating depends on the history of drying (drying process path) which represents the composition changes of coating during drying. The drying process path depends on the drying conditions such as temperature and velocity of hot air, residence time of coating in the dryer, humidity or solvent concentration of drying air, the initial composition of coating etc. Figure 7.3.23 shows the two different structures of coating according to the extremely different drying conditions (cellulose acetate solution of 10 wt% prepared in acetone 80 wt%

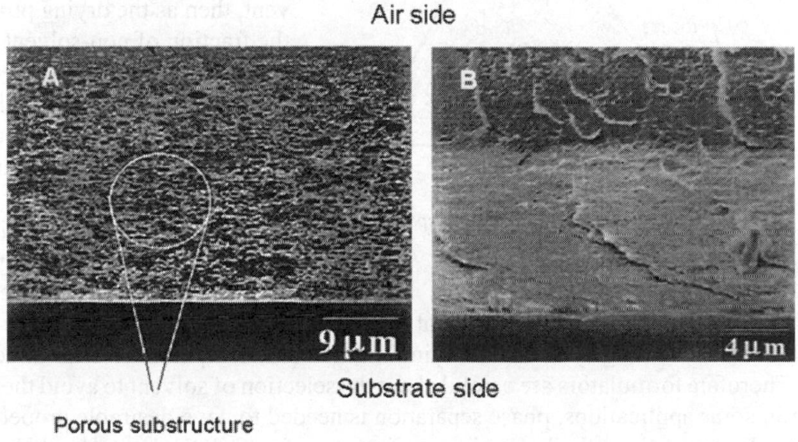


Figure 7.3.23. Cross-sectional SEM images of final coating microstructures prepared under slow and fast rates of external gas-phase mass transfer conditions. [Adapted, by permission, Mahendra et al., Proceedings of 9th International Coating Science and Technology Symposium, Delaware, USA, May 17 - 20, 1998, pp. 177 - 180].

and water 20 wt%).[35] Sample A was dried under the room conditions of free convection mass transfer, and sample B was dried under impinging high velocity air which corresponded to intense forced convection. Though the initial coating compositions were same for the two samples, the coating structures were totally different with each other after drying. As shown in the figure, sample A had a porous structure at the bottom (substrate side) and dense structure at the top (air side) while sample B had a dense structure trough the coating layer.[35]

The structure of coating is determined by the competition of phase separation and solidification phenomena. The final structure is determined by the onset of solidification.[35] The phase behavior of polymer solution is divided into three regions according to the thermodynamic stability; binodal (metastable region), spinodal (unstable region) and stable region. And the area of unstable and metastable region decreases with increasing temperature of solution.[48,49] The phase separation can also be induced by the saturation of polymer in the solution. When a polymer is saturated in a good solvent, we can easily observe the precipitation of polymer by adding some non-solvent to the system. The phase separation region can be found by thermodynamic relations such as Flory-Huggins equation,[35] and it is also be measured by adding non-solvent and observing the turbidity of solution (cloud point method).

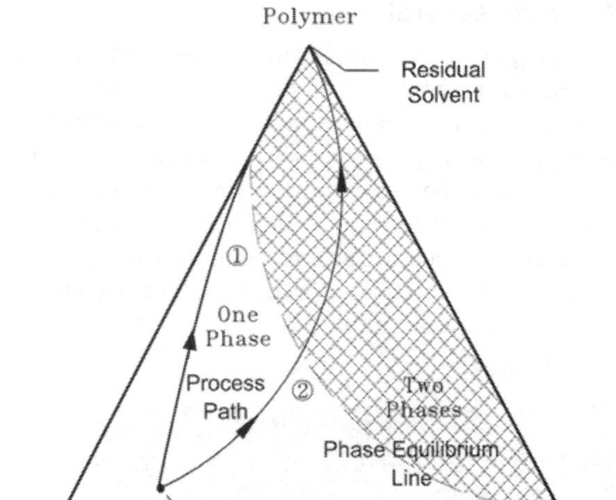

Figure 7.3.24. Conceptual representation of phase separation phenomena during drying.

In these polymer-solvent-solvent system, the drying induced phase separation can be explained by process path and phase equilibrium line. During the drying of coating the phase separation occurs when the isothermal drying path intersects two-phase region as shown in Figure 7.3.24. Let's say that the good solvent is more volatile than the non-solvent, then as the drying proceeded the fraction of non-solvent weight to the good solvent weight is steadily increased. Thus the drying process path intersects the phase equilibrium line as line ② in Figure 7.3.24. However, if the good solvent is less volatile than the non-solvent, then the process path doesn't intersect the phase equilibrium line until the significant amount of solvent is evaporated (line ①).

In most clear coating system, the drying induced phase separation is unwanted phenomena. Therefore formulators are careful about the selection of solvent to avoid the defect. However in some applications, phase separation is needed to have desirable properties for the coating. In the coating of adhesive layer of hot stamping foil, the layer should have discontinuous porous structure as shown in Figure 7.3.25.[29] Humidified air can be used to control the process path. In most polymer-solvent coating systems, the water acted as non-solvent. But it is difficult to handle or store the water laden coating solution, since the state of solution is very unstable. Therefore to supply water to the coating the highly humidified air is sometimes used at the early stages of drying (in most cases the first and the second zone of dryer). The water was condensed at the coating surface and it acted as non-solvent. By adjusting air humidity various kind of coating structures can be obtained as in Figure 7.3.25.

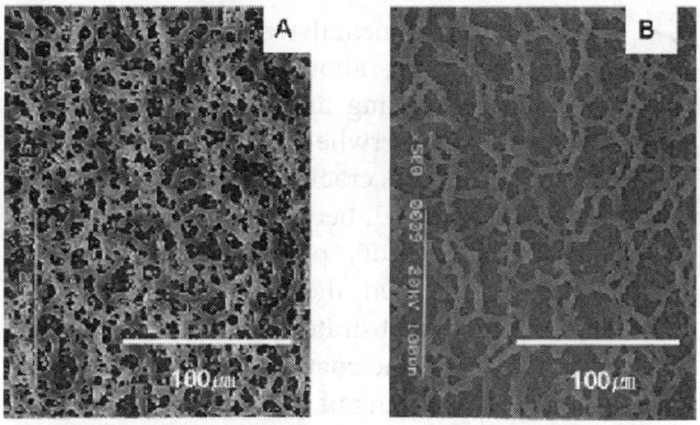

Figure 7.3.25. Surface structure of coatings with different drying conditions (SEM ×500). Humidity of air [kg water/kg dry air] : A - 0.062, B - 0.036, Initial solid content : 10.0 wt%.

The phase separation phenomena in polymer-polymer-solvent system can be called as polymer incompatibility.[41,42,56,58] The relative solubility of two polymer in the common solvent determines the surface structure of dried coating. Diverse kinds of surface structure can be obtained by adjusting drying rate, substrate surface properties, relative solubility of solvent.[56,58] The shapes of surface structure can be simulated by Cell Dynamic System.[41,42]

### 7.3.4.2 Drying defects

The maximum line speed of drying and operating conditions of dryer are often restricted by the onset of defects. The drying related defects could be classified according to the cause of defects - stress induced defects, surface tension driven defects, defects caused by air motion.

#### *7.3.4.2.1 Internal stress induced defects*

Evaporation of solvent is necessarily accompanied with the shrinkage of coating in volume as illustrated in Figure 7.3.26. This shrinkage can be only allowed in the coating thickness direction, because the adhesion of coating to the substrate prevents coating from shrinking in the plane of coating. Thus the stress is developed in the plane of coating. This stress brings about the defects which is known as curling, cracking and peeling.[10-15,21-23,34,46]

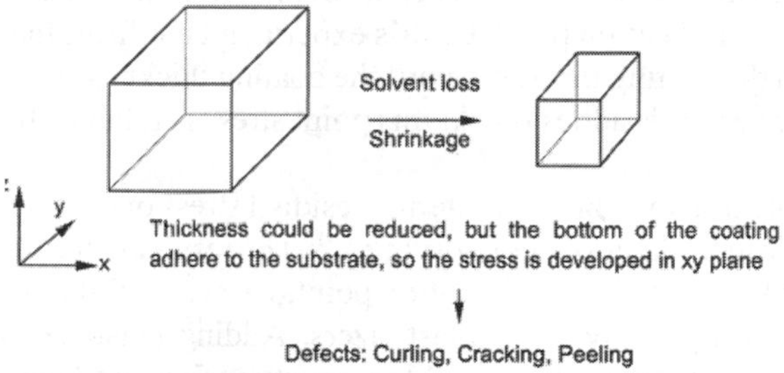

Figure 7.3.26. The origin of internal stress in a conventional polymer solution coating.

Curling and cracking

The internal stresses are build up with drying of wet coating. When the coating has enough mobility, the stresses that are developed during drying can be relieved by flow. After the coating solidified stresses will build up in the lateral dimensions. The solidification point equals the concentration at which the glass transition temperature of coating has risen

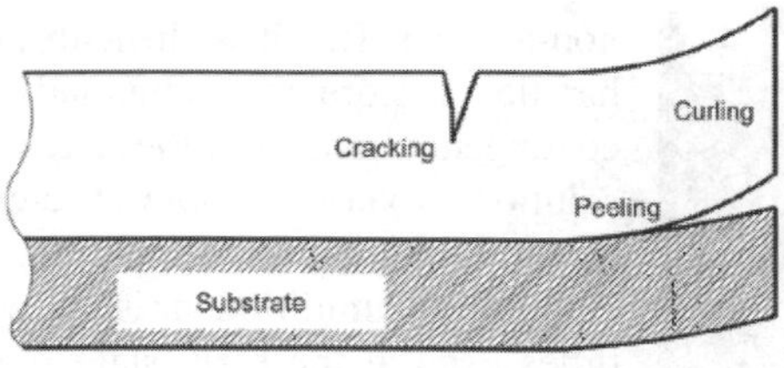

Figure 7.3.27. Internal stress related defects.

to the experimental temperature.[13-15] These stresses bring about the defects such as cracking, curling and peeling. If the local strength is overwhelmed by the local stress, the response is cracking. It is often called as 'mud cracking', because it looks like a field of mud in a dry period. And if the local stresses exceed the adhesion strength of coating to a substrate, the response is peeling or delamination as shown in Figure 7.3.27. At the edge of the coating the internal residual strength rises sharply, hence the stresses bring about the curling of coating at the edge if the adhesion strength of coating is sufficient to endure the peeling or delamination. The curling and peeling are also used to measure the internal stress of coating.[12,14,15,22]

With solvent evaporation, the coating becomes to be concentrated with polymers, and the coating becomes to have solid like nature (viscoelasticity). Thus the stresses depend not only on the strain rate, but also strain. Strain is deformation from the stress-free state.[49] After the solidification point, the solvent evaporation continues, so the stresses persist. Croll analyzed the origin of residual internal stress during drying and correlated the internal stress with the coating properties.[13-15] The residual internal stress, σ, for a coating is

$$\sigma = \frac{E}{1-\nu}\frac{\phi_s - \phi_r}{3} \qquad [7.3.13]$$

where:

| | |
|---|---|
| E | Young's modulus |
| $\nu$ | Poisson's ratio |
| $\phi_s$ | volume fraction of solvent at the point of solidification |
| $\phi_r$ | volume fraction of solvent in the dried coating |

As shown in the Equation 7.3.13, the internal stress depends on the difference of solvent volume fraction before and after solidification point. Croll's experiment confirms that the internal stress does not depend on the coating thickness until the coating thickness is so large that the net force on the interface exceeds adhesion. At this point stress is relieved by peeling.[12-15]

Adding plasticizer can often be helpful to reduce the internal residual stress of coating, because it makes coating more flexible to the later stages of drying.[13-15] The residual internal stress depends on the solvent volume loss from the solidification point, we can shift the solidification point to the later stages of drying by using plasticizers. Adding plasticizers makes the coating more flexible, but it is not always the desirable property of final products.

The residual internal stress is a result of combined action of stress and stress relaxation process. Therefore to give a sufficient relaxation time for the coating, we often dry a coated film in a mild operating conditions, e.g., dry at lower temperature and velocity of air.

Peeling

If the local stresses exceed the adhesive strength of coating to a substrate, then the coating is delaminated from the substrate. Peeling easily occurs when the coating is thick. With increasing thickness, one can find a critical thickness where the peeling occurs spontaneously. This critical thickness for the spontaneous peeling can be used as a method of measuring internal residual stress.[12]

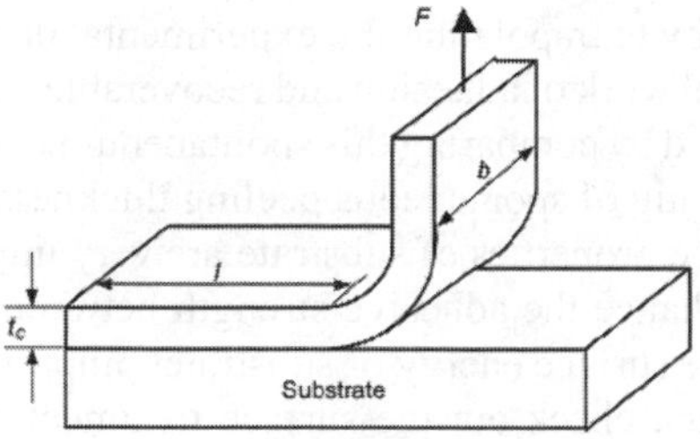


Figure 7.3.28. 90° Peel test configuration [After reference 12].

Croll illustrated that interfacial work of adhesion could be found by peeling test.[12] If the coating is peeled at a constant rate, the rate of change of total internal energy of coating is equal to zero. Thus the peeling strength is related to the interfacial work of adhesion as follows (Figure 7.3.28),

$$\frac{F}{b} = \gamma - t_C U_R \qquad [7.3.14]$$

where:

| | |
|---|---|
| γ | interfacial work of adhesion |
| b | sample width (Figure 7.3.28) |
| F | measured force of peeling |
| $t_C$ | thickness of sample (Figure 7.3.28) |
| $U_R$ | recoverable strain energy per unit volume stored in the coating |

In a case of spontaneous peeling,

$$\gamma = t_p U_R \qquad [7.3.15]$$

where:

$t_p$ critical thickness for the spontaneous peeling

Interfacial work of adhesion, γ, can be measured by spontaneous peeling test using above equation. Spontaneous peeling thickness, $t_p$, can be directly measured, and the recoverable strain energy, $U_R$, for an elastic material under a one-dimensional strain is,

$$U_R = \frac{1}{2} E \varepsilon_i^2 \qquad [7.3.16]$$

where:

| | |
|---|---|
| E | modulus of coating |
| $\varepsilon_i$ | internal strain in coating |

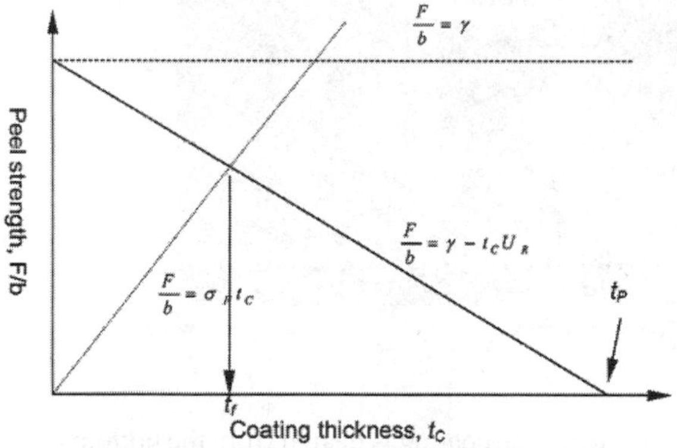


Figure 7.3.29. Graphical representation of the equations governing the 90° peel test. [Adapted, by permission, from Croll, S.G., *J. Coating Technol.*, **52**, 35 (1980)].

As shown in Figure 7.3.29 the interfacial work of adhesion and recoverable strain can also be found by measuring peeling strength at several different coating thickness. If the adhesive strength of the interface exceeds the cohesive strength of one of the coating component materials, there occurs cohesive failure - the coating layer loses the adhesive strength within the layer itself. At a small thickness, cohesive failure often occurs during the peeling test illustrated in Figure 7.3.28.

The spontaneous peeling thickness can be found by extrapolating the experimental data to the zero peeling strength. The validity of interfacial work of adhesion and recoverable strain which is obtained from this experiment can be tested by comparing this spontaneous peeling thickness with obtained from independent measuring of spontaneous peeling thickness.

To enhance the adhesive strength, the surface properties of substrate are very important. The surface treatments are often used to enhance the adhesive strength between the coating and the substrate, as well as to enhance the surface energy of substrate - improving wettability of coating. If adhesive problem occurs, check out the surface treatment processes such as flame, plasma, or corona.[21] Corona treatment does not persist permanently, therefore it should be done in-line, the corona is often applied directly before the coating station.

Sometimes a thin, high surface energy adhesive or subbing layer is coated on the substrate to improve the adhesion of coating.[21] As often the case, this layer is coated in-line during manufacturing of substrate - such as in-line coating of acrylic resin during bi-axial extension of PET film. If the adhesive failure occurred when one used these kind of treated substrate, one should check that the correct side was used where the subbing layer was coated and the status of coated subbing layer was perfect. Simple peel tests with adhesive tape can give a clue to the coating status of subbing layer. First laminating the adhesive tape with substrate where the coated material is applied, then pull off the tape rapidly. If there are problems with the subbing layer such as the partial un-coating of subbing layer, then the coating will be peeled off according to the un-coated pattern. By analyzing un-coated pattern, one can find the steps of process which have problems. If the substrate is suspected, change the substrate lots.

#### *7.3.4.2.2 Surface tension driven defects*

Defects can arise during drying of a coated film by building non-uniform surface tension gradient over the coating. They include convection or Benard cells, fat edges or picture framing, etc.

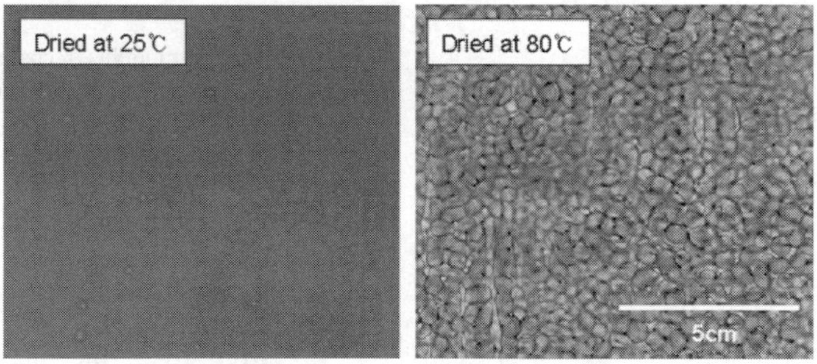


Figure 7.3.30. Surface tension driven defect: Convection cells.

<u>Convection cells</u>

Convection or Benard cells can arise when the coating is heated from the bottom of the coating, and the density or surface tension gradients are built across the coating thickness. These gradients lead to convection cells which look like regular close-packed hexagonal surface patterns (Figure 7.3.30).[10,21] The evaporative cooling can also be the cause of the temperature gradient. This gradient arises the density and surface tension gradient, which brings about the internal flows within the coating to form the convection or Benard cells.

The convection cells, which come from the surface tension gradient, can arise when the Marangoni number (Ma) exceeds 80.[10] The Marangoni number is

$$Ma = \frac{(d\sigma / dT)(dT / dy)h^2}{\mu\kappa} \quad [7.3.17]$$

where:

| | |
|---|---|
| σ | surface tension |
| μ | viscosity |
| κ | heat conductivity |
| T | temperature |
| h | thickness of coating |
| dT/dy | temperature gradient in the thickness direction |

Convection cells can arise at a lower Marangoni number when the coating thickness is above 2 mm.[21] Convection or Benard cells can be reduced or eliminated by adjusting operating conditions and formulation of coating solution. The possibility of having convection cells is reduced at the following conditions[21]

- Lower surface tension of coating liquid
- Reduce the thickness of the wet coating
- Increase the viscosity of coating liquid
- Reduce the drying rate by adding low volatile liquid or reducing the temperature of drying air

When the thickness of wet layer of coating is less than 1 mm, as is often the case with almost all coatings, convection cells are almost always due to the surface tension gradients.[10,21]

However, it is not desirable to lower the drying rate because it decreases the line speed of coating process. Frequently it is helpful to use surfactant to reduce the convection cells due to the surface tension gradients. Care should be given in selection of the surfactant and the amount of it not to deteriorate the final quality of products. The amount of surfactant should be minimized, excess surfactants can migrate to the coating surface and react with humidity at a coating station to form a haze coating surface. Moreover, the remaining excess surfactant, which is in the final products, can migrate to the coating surface while the products in use in a some environmental conditions to ruin the final quality of the products.

Fat edges

Fat edges can be built during drying of coating with non-uniform surface tension distribution at the edges of coating. Figure 7.3.31 illustrates the mechanisms of fat edge defects. The edges are usually thinner than the bulk of coating, thus as the evaporation proceeded the concentration of polymer is increased faster at edges than in the bulk of coating. Usually the surface tension of solid is much higher than that of polymer, the surface tension is higher at the edges than the bulk. The higher surface tension at the edges will

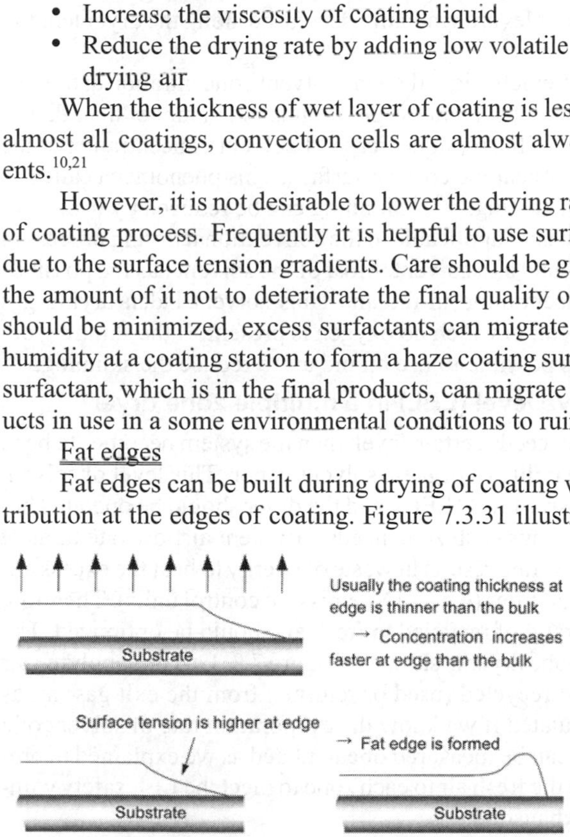

Figure 7.3.31. Formation of fat edge or picture framing.

cause to flow to the edges, giving fat edges. The surfactants are often helpful to eliminate the fat edges.

#### *7.3.4.2.3 Defects caused by air motion and others*

Dryer bands

At the early stages of drying, where the coating has enough mobility to flow, the coating layer is apt to being disturbed by the motion of drying air.[21] This defect is observed in the dryer which uses the arrays of round jet nozzles rather than the slot nozzles. Often it looks like a lane running in the machine direction and the width of bands is nearly equal to the nozzle diameter. When the air comes out of slots, bands are less likely if the flow across the slot is uniform.[10,21,25] The bands are more likely to be developed with low viscosity solution, thick coating and high air velocity. Thus we can reduce the bands by reducing air velocity (but this also decreases the drying rate) and by increasing the initial solid content so as to increase the viscosity of solution and to coat thinner layer with concentrated solution.

The geometry of arrays of round jet nozzles should be designed to avoid this defect by applying nozzles of larger diameters in the first part of zones. Large amount of fresh air is needed in the first part of zones of the multiple zone-dryer, because in the first zone the amount of evaporated solvent vapor is large, hence one should supply enough fresh air to ensure the dryer is operated below the lower explosive level (LEL) of a given solvent. Therefore the diameters of round jet nozzles are gradually decreased along the dryer length.

Skinning

If the drying rate of solvent is extremely high, then the solvent concentration at the surface of coating falls rapidly, for the rate of solvent evaporation is much higher than that of diffusion of solvent within the coating. This rapid decrease of solvent concentration at the surface causes to form a thin solid layer near the coating surface. This phenomena called as 'skinning', and it retards the drying of coating.[4,5,8-10] Skinning can be reduced by using solvent laden gas as a drying air,[4,8,10] but it isn't applicable to the conventional drying process in which uses the volatile organic compound as a solvent. Most of the solvent has the possibility of explosion above a certain concentration in the air. Thus solvent laden drying gas should be applied in a inert environment, in which no oxygen is present. If the skinning occurs in a aqueous coating system, it is possible to use humid air to reduce the skinning.

### 7.3.4.3 Control of lower explosive level (LEL) in a multiple zone dryer

If the solvent concentration of a zone exceeds certain level, then the system becomes to be in danger of explosion due to the flammability of organic solvent vapor. This level of solvent concentration is called as lower explosive level (LEL), and the dryer should be operated below the LEL prior to any constraints. Thus each zone needs sufficient airflow rate to meet these needs. However, too much airflow rate results in waste of energy to heat the excess air, and it increases the load of waste gas facilities (e.g., VOC emission control units). Therefore the airflow rate of each zone and the ratio of recycled to fresh air should be optimized. The typical airflow system for a multiple zone dryer is shown in Figure 7.3.32. The supplying air of each zone is consisted of fresh and recycled (used or returned from the exit gas) air as shown in the figure. LEL can be calculated if we know the evaporation rate of solvent of a zone. The evaporation rate of solvent can be measured or calculated as we explained in previous sections. Thus we can distribute the fresh air to each zone to meet the LEL safety without substantial increase in the total exhaust air.

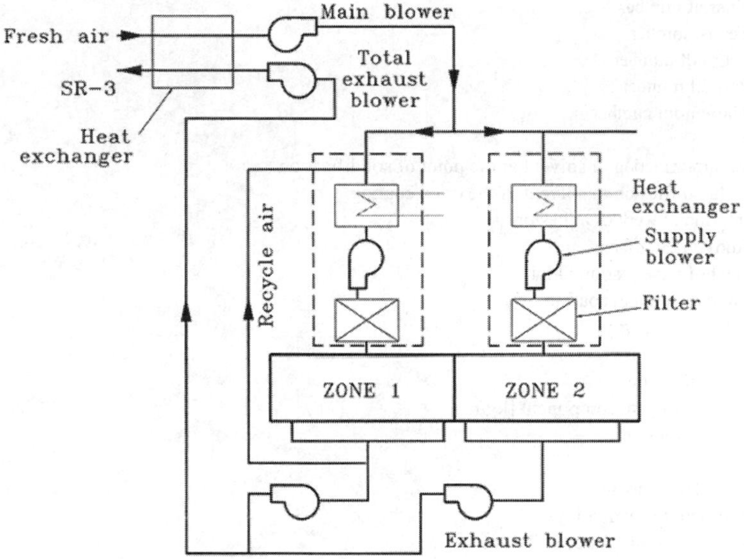


Figure 7.3.32. Schematic diagram of airflow system to a zone. [After reference 30].

## 7.3.5 NOMENCLATURE

| | |
|---|---|
| B | Nozzle slot width |
| b | Thickness [m] |
| C | Concentration of solvent at a gas phase [kg/m$^3$] |
| $C_P$ | Specific heat [J/kg] |
| E | Young's modulus |
| ER | Evaporation rate of solvent [kg/min] |
| F | measured force of peeling |
| f | Fraction open area ($B/L_T$) |
| $f_O$ | $[60 + 4(H/2B - 2)^2]^{-1/2}$ |
| G | Mass flow rate of dry air [kg/sec] |
| h | Heat transfer coefficient [J/m$^2$s$^o$C] |
| H | Nozzle to coating surface distance |
| k | Heat conductivity [J/m$^o$C] |
| $k_m$ | Mass transfer coefficient [m/sec] |
| $L_T$ | Nozzle spacing |
| P | Vapor pressure [atm] |
| Q | Volumetric flow rate of air [m$^3$/min] |
| R | Gas constant |
| T | Temperature [$^o$C] |
| t | Time [s] |
| $t_C$ | Thickness of sample |
| $U_R$ | Recoverable strain energy per unit volume stored in the coating |
| V | Line speed [m/min] |
| W | Solvent coating amount [kg/m$^2$] |
| w | Velocity of air at the nozzle exit [m/sec] |
| WD | Coating width [m] |
| z | Volume fraction of component [-] |

*Dimensionless Numbers*

| | |
|---|---|
| Sh | Sherwood number |

| | |
|---|---|
| Nu | Nusselt number |
| Le | Lewis number |
| Sc | Schmidt number |
| Pr | Prandtl number |
| Ma | Marangoni number |
| *Greek letters* | |
| $\phi_s$ | Volume fraction of solvent at the point of solidification |
| $\phi_r$ | Volume fraction of solvent in the dried coating |
| $\gamma$ | Interfacial work of adhesion |
| $\gamma$ | Activity coefficient |
| $\delta H$ | Heat of vaporization [J/kg] |
| $\varepsilon_i$ | Internal strain in coating |
| $\kappa$ | Heat conductivity |
| $\mu$ | Viscosity |
| $\nu$ | Poisson's ratio |
| $\rho$ | Density of pure component [kg/m$^3$] |
| $\sigma$ | Surface tension |
| *Subscripts* | |
| a | Pertinent to dry air |
| C | Pertinent to coating layer |
| f | Pertinent to substrate, PET film |
| fn | Outlet condition |
| i | Component |
| in | Inlet condition |
| *Superscripts* | |
| k | Zone number |
| $\infty$ | Bulk air condition |

## REFERENCES


1 Alsoy, S. and J. L. Duda, *Drying Technol.*, **16**, 15, (1998).

2 Aust, R., F. Dust and H. Raszillier, *Chem. Eng. Process.*, **36**, 469 (1997)

3 Bird, R. B., W. E. Stewart and E. N. Lightfoot. **Transport Phenomena**, *John Wiley & Sons*, 1960.

4 Bornside, D. E., C. W. Macosko, and L. E. Scriven, *J. Appl. Phys.*, **66**, 5185, (1989).

5 Cairncross, R. A., A. Limbert, L. F. Francis, and L. E. Scriven, 24th annual meeting of Fine Particle Society, 24-28 August 1993, Chicago, IL.

6 Cairncross, R. A., L. F.Francis, and L.E. Scriven, *AlChE J.*, **42**, 55 (1996).

7 Cairncross, R. A., S. Jeyadev, R.F. Dunham, K. Evans, L.F. Francis, and L.E. Scriven, *J. Appl. Polym. Sci.*, **58**,1279 (1995).

8 Cairncross, R. A., Ph.D. thesis, University of Minnesota, Minneapolis (1994).

9 Cary, J. D. and E. B. Gutoff, "Analyze the Drying of Aqueous Coatings" *Chem. Eng. Prog.*, **2**, 73 (1991).

10 Cohen, E. and E.Gutoff, **Modern Coating and Drying Technology**, *VCH Publishers, Inc.*, New York, 1992.

11 Cohen, E., E. J. Lightfoot and K. N. Christodoulou, *Ind. Coating Res.*, **3**, 45 (1995).

12 Croll, S.G. *J. Coating Technol.*, **52**, 35 (1980).

13 Croll, S.G. *J. Coating Technol.*, **51**, 64 (1979).

14 Croll, S.G. *J. Coating Technol.*, **50**, 33 (1978).

15 Croll, S.G. *J. Appl. Polym. Sci.*, **23**, 847 (1979).

16 Cussler, E. L. **Diffusion Mass Transfer in Fluid Systems**, 2nd Ed., *Cambridge University Press*, 1997.

17 Duda, J. L., J. S. Vrentas, S. T. Ju, and H. T. Lu, *AIChE J.*, **28**, 279 (1982).

18 Evans, K. J., Smith, W. R., Dunham, R. F., Leenhouts, T. J., Ceglinski; B. D., and Cairncross, R. A., **US Patent 5,394,622**.

19 Felder, R. M. and R. W. Rousseau. **Elementary Principles of Chemical Processes**, 2nd Ed., *John Wiley & Sons, Inc.*, 1986.

20 Gutoff, E. B., *Drying Technol.*, **14**, 1673 (1996).

21 Gutoff, E. and E. Cohen, **Coating and Drying Defects - Troubleshooting Operating Problems**, *John Wiley & Sons*, Inc., 1995.

22 Hoffman, R.W., *Surf. Interface Anal.*, **3**, 62 (1981).
23 Hu. S. M., *J. Appl. Phys.*, **50**, 4661 (1979).
24 Huelsman, G. L. and W. B. Kolb, **US Patent 5,694,701**.
25 Hunt, B. V., R. A. Yapel and R. K. Yonkoski, 9th International Coating Science and Technology Symposium, Delaware, USA, May 17 - 20, 1998, pp. 93 ~ 96.
26 Imakoma, H. and M. Okazaki, *Ind. Coating Res.*, **1**, 101 (1991).
27 Imakoma, H. and M. Okazaki, *Ind. Coating Res.*, **2**, 129 (1992).
28 Kim, S. S. and M. H. Kwon, 9th International Coating Science and Technology Symposium, Delaware, USA, May 17 - 20, 1998, pp. 353 - 356.
29 Kim, S. S., H. S. Park, I. C. Cheong and J. S. Hong, 3rd European Coating Symposium , Erlangen, Germany, Sep. 7 - 10, 1999.
30 Kim, S. S., M. H. Kwon, submitted to the *Drying Technol.* (1999).
31 Kolb, W.B. and G. L. Huelsman, 9th International Coating Science and Technology Symposium, Delaware, USA, May 17 - 20, 1998, pp.209 ~ 212.
32 Kolb, W.B. and M.S. Carvalho, 9th International Coating Science and Technology Symposium, Delaware, USA, May 17 - 20, 1998, pp.9 ~ 12.
33 Koschmieder, E.L., and M.I.Biggerstaff, *J. Fluid Mech.*, **167**, 49, (1986).
34 Lei, H, L. F. Francis, W. W. Gerberich and L. E. Scriven, 9th International Coating Science and Technology Symposium, Delaware, USA, May 17 - 20, 1998, pp.97 ~ 100.
35 Mahendra D., L. F. Francis and L. E. Scriven., 9th International Coating Science and Technology Symposium, Delaware, USA, May 17 - 20, 1998, pp 177 ~ 180.
36 Mahendra D., L.F.Francis and L.E. Scriven., 1st Asian and 7th Japan Coating Symposium, Fukuoka, Japan, Sept. 1997, pp 8 ~ 9.
37 Martin, H., **Advances in Heat Transfer**, 13, pp.1~60, *Academic Press*, New York, 1977.
38 Ohta, T., H. Nozaki and M. Doi, *J. Chem. Phys.*, **93**, 2664 (1990).
39 Oishi, T. and J. M. Prausnitz, *Ind. Eng. Chem. Process Des. Dev.*, **17**, 333 (1978).
40 Okazaki, M., K. Shioda, K. Masuda, and R. Toei, *J. Chem. Eng. Japan*, **7**, 99 (1974).
41 Oono, Y. and S. Puri, *Phys. Rev. A: At. Mol. Opt. Phys.*, **38**, 434, (1988).
42 Oono, Y. and S. Puri, *Phys. Rev. A: At. Mol. Opt. Phys.*, **38**, 1542, (1988).
43 Pan, S. X, H. T. Davis, and L. E. Scriven, *Tappi J.*, **78**, 127 (1995).
44 Pan, S. X., Ph.D. thesis, University of Minnesota, 1995.
45 Polat, S. *Drying Technol.*, **11**, 1147 (1993).
46 Sato, K., *Prog. Org. Coat.*, **8**, 143 (1980).
47 Saure, R. and V. Gnielinski, *Drying Technol.*, **12**, 1427 (1994).
48 Scriven, L.E. and W.J. Suszynski, Coating Process Fundamentals, Short course. June 19-21, 1996. University of Minnesota.
49 Smith, J. M and H. C. Van Ness, **Introduction to Chemical Engineering Thermodynamics**, *McGraw-Hill*, 1987.
50 Suzuki, I., Y. Yasui and A. Udagawa, 9th International Coating Science and Technology Symposium, Delaware, USA, May 17 - 20, 1998, pp. 21 - 24.
51 Takase, K., H. Miura, H. Tamon and M. Okazaki, *Drying Technol.*, **12**, 1279 (1994).
52 Vinjamur, M. and R. A. Cairncross, Presented at the AIChE national meeting, Dallas, Texas, November 1, 1999.
53 Vrentas, J. S. and C. M. Vrentas, *J. Polym. Sci. Part B : Polym. Phys.*, **32**, 187 (1994).
54 Vrentas, J. S. and J. L. Duda, *AIChE J.*, **25**, 1 (1979).
55 Vrentas, J. S. and J. L. Duda, *J. Appl. Polymer. Sci.*, **21**, 1715 (1977).
56 Walheim, S., Böltau, M., Mlynek, J., Krausch, G. and Steiner, U, *Macromol.*, **30**, 4995 (1997).
57 Winward, T. and R. A. Cairncross, 9th International Coating Science and Technology Symposium, Delaware, USA, May 17 - 20, 1998, pp.343 - 346.
58 Yamamura, M., K. Horiuchi, T. Kajiwara, and Adachi, K, 3rd European Coating Symposium, Erlangen, Germany, Sep. 7 - 10, 1999.
59 Yapel, R. A., M.S. Thesis, University of Minnesota, Minneapolis, 1988.
60 Zielinski J.M. and J. L. Duda, *AIChE J.*, **38**, 405 (1992).

# 8

# INTERACTIONS IN SOLVENTS AND SOLUTIONS

JACOPO TOMASI, BENEDETTA MENNUCCI, CHIARA CAPPELLI
**Dipartimento di Chimica e Chimica Industriale, Università di Pisa, Italy**

## 8.1 SOLVENTS AND SOLUTIONS AS ASSEMBLIES OF INTERACTING MOLECULES

A convenient starting point for the exposition of the topics collected in this chapter is given by a naïve representation of liquid systems.

According to this representation a liquid at equilibrium is considered as a large assembly of molecules undergoing incessant collisions and exchanging energy among colliding partners and among internal degrees of freedom. The particles are disordered at large scale, but often there is a local order that fades away. Solutions may be enclosed within this representation.

The collection of particles contains at least two types of molecules – those having a higher molar fraction are called the solvent – the others the solute. Collisions and exchange of energy proceed as in pure liquids, local ordering may be different, being actually dependent on the properties of the molecule on which attention is focused, and on those of the nearby molecules.

Liquids at a boundary surface require more specifications to be enclosed in the representation. There is another phase to consider which can be either a solid, another liquid, or a gas. Further specifications must be added to characterize a specific boundary system (see Section 8.8), but here it is sufficient to stress that the dynamic collision picture and the occurrence of local ordering is acceptable even for the liquid portion of a boundary.

This naïve description of liquid systems actually represents a model – the basic model to describe liquids at a local scale. As it has been here formulated, it is a classical model: use has been made of physical classical concepts, as energy, collisions (and, implicitly, classical moments), spatial ordering (i.e., distribution of elements in the space).

It is clear that the model, as formulated here, is severely incomplete. Nothing has been said about another aspect that surely has a remarkable importance even at the level of naïve representations: molecules exert mutual interactions that strongly depend on their chemical composition.

To say something more about molecular interactions, one has to pass to a quantum description.

Quantum mechanics (QM) is universally acknowledged as the appropriate theory to treat material systems at the level of phenomena of interest to chemistry. Therefore quantum mechanics is the legitimate theory level at which to treat liquids.

We recall here the opening statement of a famous book on quantum chemistry:[1] "In so far as quantum mechanics is correct, chemical questions are problems in applied mathematics." Actually, the mathematics to apply to liquid systems is a hard nut to crack. Fortunately a sequence of many, but reasonable, approximations can be introduced. We shall consider and exploit them in the following section of this chapter.

It is important to stress one point before beginning. The naïve picture we have summarized can be recovered without much difficulty in the quantum formulation. This will be a semiclassical model: the identity of the constituting particle is preserved; their motions, as well as those of nuclei within each molecule, are treated as in classical mechanics: the quantum methods add the necessary details to describe interactions. This semiclassical quantum description can be extended to treat problems going beyond the possibilities of purely classical models, as, for example, to describe chemical reactions and chemical equilibria in solution. The limits between the classical and quantum parts of the model are quite flexible, and one may shift them in favor of the quantum part of the model to treat some specific phenomenon, or in favor of the classical part, to make the description of larger classes of phenomena faster.

We shall enter into more details later. It is sufficient here to underline that the choice of using a quantum approach as reference is not a caprice of theoreticians: it makes descriptions (and predictions) safer and, eventually, simpler.

We shall start with the introduction of some basic simplifications in the quantum model.

## 8.2 BASIC SIMPLIFICATIONS OF THE QUANTUM MODEL

The quantum mechanical description of a material system is obtained as solution of the pertinent Schrödinger equations.

The first Schrödinger equation is the famous equation everybody knows:

$$H(x)\Psi(x) = E\Psi(x) \qquad [8.1]$$

The function Ψ(x) describes the "state of the system" (there are many states for each system). It explicitly depends on a set of variables, collectively called x, that are the coordinates of the particles constituting the system (electrons and nuclei). E is a number, obtained by solving the equation, which corresponds to the energy of the system in that state. H(x), called Hamiltonian, technically is an operator (i.e., a mathematical construct acting on the function placed at its right to give another function). Eq. [8.1] is an eigenvalue function: among the infinite number of possible functions depending on the variables x, only a few have the notable property of giving, when H(x) is applied to them, exactly the same function, multiplied by a number. To solve eq. [8.1] means to find such functions.

The first step in the sequence of operations necessary to reach a description of properties of the system is to give an explicit formulation of the Hamiltonian. This is not a difficult task; often its formulation is immediate. The problems of "applied mathematics" are related to the solution of the equation, not to the formulation of H.

The second Schrödinger equation adds more details. It reads:

$$i\hbar \frac{\partial}{\partial t}\Psi(x,t) = H\Psi(x,t) \qquad [8.2]$$

In practice, it expresses how the state $\Psi(x)$ of the system evolves in time (i.e., it gives $\partial\Psi(x)/\partial t$ when $\Psi(x)$ is known).

Note that in eq. [8.1] the time t was not included among the parameters defining H (and $\Psi$): with eq. [8.1] we are looking at stationary states, not depending on time. In spite of this, eq. [8.1] may be applied to liquids, which are characterized by a continuous dynamic exchange of energy through collisions, and by a continuous displacement of the constituting molecules. This is not a problem for the use of the time independent formulation of the Schrödinger equation for liquids: eq. [8.1] actually includes kinetic energy and all related dynamic aspects. A direct use of eq. [8.1] for liquids means to treat liquids at equilibrium. We shall not make explicit use of the second equation, even when the nonequilibrium problem is considered; the use of the semiclassical approximation permits us to treat time dependent phenomena at the classical level, simpler to use.

Let us come back to the expression of H(x). The number of parameters within (x) is exceedingly large if the system is a liquid. Fortunately, things may be simplified by using factorization techniques.

When a system contains two subsystems (say A and B) that do not interact, the Hamiltonian can be partitioned so:

$$H(x) = H_A(x_A) + H_B(x_B) \qquad [8.3]$$

This rigorously leads to a factorization of $\Psi$:

$$\Psi(x) = \Psi_A(x_A) + \Psi_B(x_B) \qquad [8.4]$$

and to a partition of the energy:

$$E = E_A + E_B \qquad [8.5]$$

Equation [8.1] is in consequence transformed, without simplifications, into two simpler equations:

$$H_A(x_A)\Psi_A(x_A) = E_A\Psi_A(x_A) \qquad [8.6a]$$

$$H_B(x_B)\Psi_B(x_B) = E_B\Psi_B(x_B) \qquad [8.6b]$$

In practice, there are always interactions between A and B. The Hamiltonian may always be written as:

$$H(x) = H_A(x_A) + H_B(x_B) + H_{AB}(x) \qquad [8.7]$$

The relative magnitude of the coupling term permits us to distinguish among three notable cases.

1) If the coupling term is very small, it may be neglected, coming back to equations [8.6].

2) If it is small but not negligible, it may be used to correct the solutions given by equations [8.6] with the use of appropriate mathematical tools: the factorized formulation with corrections continues to be simpler than the original one.

3) There are cases in which this factorization is profitably used, in spite of the fact that interactions between particles of A and B are of the same magnitude of interactions within both A and B.

We shall now briefly introduce two factorizations of the last type that are of paramount importance in molecular quantum chemistry, then consider other factorizations of direct interest for the description of liquid systems.

The first factorization we are introducing regards the electronic and nuclear coordinates (called r and R respectively) of the same system: $H(x) = H^e(r) + H^n(R)$. It is clear that the electrostatic interaction between two electrons (and between two nuclei) within a given molecule is of the same magnitude as that concerning a couple electron-nucleus. In spite of this, the factorization is performed, with a tremendous effect on the evolution of quantum chemistry. Its physical justification rests on the very different masses nuclei and electrons have, and therefore on their velocities. It is often called Born-Oppenheimer (BO) approximation and in the following we shall use this acronym. The BO approximation makes possible the introduction of the concept of potential energy surface (PES), a very useful model to describe the motion of nuclei (often in the semiclassical approximation) taking into account the interactions with the quantum description of the electrons.

Some more details can be useful in the following. To apply this factorization it is imperative to follow a given order. First, to solve the electronic equation with $H^e(R)$ at a fixed geometry of nuclei: the output is a wave function and an energy both parametrically dependent on the nuclear geometry R given as input: $\Psi^e(r;R)$ and $E^e(R)$ (actually there will be a set of electronic states, each with its energy). Second, to repeat the same calculations at different nuclear geometries R', many times, until a sufficiently detailed description of the function $E^e(R)$ is reached. $E^e(R)$ is the PES we have mentioned: it may be used to define the potential operator within the nuclear Hamiltonian $H^n(R)$ and then to compute vibrational and rotational states (another factorization is here introduced) or used in a semiclassical way to study the effect of nuclear motions.

The second factorization we are introducing regards the electronic part of the system (after the BO factorization). The main procedure in use leads to factorization into many separate parts, each regarding one electron only:

$$H^e(r) = \sum_i^{elec} h_i(r_i)$$

Here again the coupling terms are of the same order of magnitude as the interactions left within each one-electron Hamiltonian (that explicitly regards the electrostatic interaction of the electron with all the nuclei of the systems, placed at fixed positions). The trick allowing this factorization consists in introducing within each one-electron Hamiltonian (the symmetry of electrons makes them all equal) an averaged description of the couplings, based on the yet unknown wave functions of the other electrons. The calculation proceeds iteratively: starting from a first guess of the averaged interaction, the description is progressively refined using intermediate values of the one-electron wave functions. The final output is a $\Psi^e(r;R)$ expressed as an (anti)symmetrized product of the correct number of one-electron wave functions. Each electronic state will be described by its specific collection of such one-electron wave functions, called molecular orbitals (MO). The whole procedure is called SCF (self-consistent field), a name that reminds us of the technique in use, or HF (Hartree-Fock) to honor scientists working in its elaboration.

This partition is the second cornerstone of molecular quantum chemistry. Almost all QM molecular calculations use the SCF theory (passing then, when necessary, to a higher level of the molecular QM theory). The concepts of MOs and of orbital energies derive from this one-electron factorization.

In the following we shall make little explicit use of the two factorizations we have here examined, because we shall not enter into too technical details about how to obtain accurate molecular interaction potentials. They will be always in the background, however, and some concepts exposed here will be recalled when necessary.

The last factorization we have anticipated in the introduction regards systems composed by many molecules, as molecular crystals, clusters and liquids.

A factorization of a liquid into molecular subunits was implicit in the naïve model we have used in the introduction. The interactions within a molecule surely are larger than those among molecules; however, we cannot neglect these couplings, which are essential to describe a liquid. This is the main subject of this chapter and will be treated with due attention in the following sections.

To complete this preliminary overview, we stress that the factorization techniques are quite flexible and that they may be applied at different levels. One among them deserves mention, because it will be used in the following.

In the study of liquid systems there are several reasons to put more attention to a limited portion of the liquid, with the remainder of the liquid systems treated at a lower level of accuracy. We shall call the descriptions in which there is a portion of the whole system considered as the main component (called M) "focused models" while the remainder (called S) plays a supplementary or assisting role. In such an approximation, the Hamiltonian is partitioned in the following way:

$$H(x) = H_M(x_M) + H_S(x_S) + H_{MS}(x) \qquad [8.8]$$

Focused models are used to study local properties in pure liquids, solutions, and interfaces. The largest use is to study solvation effects and reactions in solutions. In these cases M is composed of a solute supplemented by a solvation cluster (also a single solute molecule may be used). In these models the more detailed description of M is ensured by using the BO approximation, followed by a MO-based description of the electronic structure (there are also methods that replace the QM description of the electronic structure with some simpler semiclassical model). The S components are generally described with the aid of the intermolecular potentials we shall examine in the following sections. The description of the coupling takes into account the nature of the description chosen for both M and S.

Focused models may also be used to get more detailed information on the structure of liquids, being in principle more accurate than descriptions solely based on intermolecular potentials. The computational cost is higher, of course, and this approach is now used only at the final stage of the assessment of models of the molecular interaction potential.

## 8.3 CLUSTER EXPANSION

The pure ab initio approach we have summarized has never been used to determine the molecular interaction potentials. Further simplifications are generally used.

We will start considering a model in which only the main component $H_M(x_M)$ of the Hamiltonian [8.8] is considered. This means to pass from a liquid to a cluster of molecules. The presence of the molecules within M is explicitly acknowledged. Adding now the further constraint of keeping each molecule at a fixed geometry, the dimensionality of the R space is severely reduced. In fact, six coordinates for each molecule will be sufficient to define the relative position of its center of mass an its orientation with respect to an arbitrary fixed reference frame. If M is composed of n molecules, the BO coordinate space will have 3n-6 coordinates.

It is convenient to consider now the whole PES $E_M(R_m)$, not limiting our attention to the minima (that correspond to equilibrium positions). For each point $R_m$ of the hyperspace we introduce the following cluster expansion of the energy

$$E_M(R_m) = \sum_A E_A + \frac{1}{2}\sum_A \sum_B E_{AB}(R_{AB}) + \frac{1}{6}\sum_A \sum_B \sum_C E_{ABC}(R_{ABC}) + \ldots \quad [8.9]$$

In this rather artificial (but exact) decomposition the whole energy is decomposed into the sum of the energies of the separate molecules, each at the geometry they have in the gas phase, followed by the sums of two-, three-, many-body terms. For each term of eq. [8.9] we have put in parentheses the indication of the pertinent nuclear coordinate subspace to emphasize that each dimeric interaction is defined in a 6-dimensional space ($R_{AB}$), each trimeric interaction over a 9-dimensional space, etc., while $E_M$ is defined over a 3n-6 space.

The total interaction energy is defined as the difference between [8.9] and the sum of the monomers's energies:

$$\Delta E(1,2,\ldots m; R_m) = \frac{1}{2}\sum_A \sum_B E_{AB}(R_{AB}) + \frac{1}{6}\sum_A \sum_B \sum_C E_{ABC}(R_{ABC}) + \ldots \quad [8.10]$$

The convergence of this expansion is relatively fast for clusters composed by neutral molecules, less fast when there are charged species. In any case, it is not possible to interrupt the expansion to the two-body terms. This contribution gives the additive terms of the interaction energy; the other terms describe non-additive effects that in principle cannot be neglected.

It is almost compulsory to proceed step by step and to study two-body interactions first. Each couple can be considered separately.

## 8.4 TWO-BODY INTERACTION ENERGY: THE DIMER

The definition of a two-body potential is given by:

$$\Delta E_{AB}(R) = E_{AB}(R) - [E_A + E_B] \quad [8.11]$$

We have kept the parameter $R=R_{AB}$ to underline that this potential depends both on the relative position of the two molecules and on their mutual orientation: it is a 6-dimensional function depending on these two sets of 3 parameters each, in the following indicated,

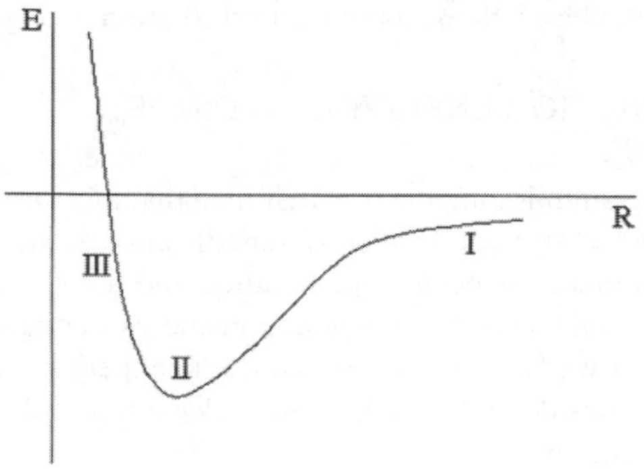


Figure 8.1. Interaction energy for a dimer with respect to the mutual approach distance at a fixed orientation.

where necessary, with $r_{ab}$ and $\Omega_{ab}$. $\Delta E_{AB}(R)$ has the status of a PES, with a shift in the reference energy, given here by the sum of the energies of the two monomers.

A function defined in a 6-dimensional space is hard to visualize. Many devices have been introduced to render in graphic form selected aspects of this function. Some will be used in the following: here we shall use the simplest graphical rendering, consisting in fixing an orientation ($\Omega_{ab}$) and two coordinates in the $r_{ab}$ set in such a way that the remaining coordinate corresponds to the mutual approach between molecule A and B, along a given straight trajectory and a fixed mutual orientation. A typical example is reported in Figure 8.1.

The energy curve may be roughly divided into three regions. Region I corresponds to large separation between interaction partners; the interaction is feeble and the curve is relatively flat. Region II corresponds to intermediate distances; the interactions are stronger compared to region I, and in the case shown in the figure, the energy (negative) reaches a minimum. This fact indicates that the interaction is binding the two molecules: we have here a dimer with stabilization energy given, in first approximation, by the minimum value of the curve. Passing at shorter distances we reach region III; here the interaction rapidly increases and there it gives origin to repulsion between the two partners. Before making more comments, a remark must be added. The shape of the interaction energy function in molecular systems is quite complex: by selecting another path of approach and/or another orientation, a completely different shape of the curve could be obtained (for example, a completely repulsive curve). This is quite easy to accept: an example will suffice. The curve of Figure 8.1 could correspond to the mutual approach of two water molecules, along a path leading to the formation of a hydrogen bond when their orientation is appropriate: by changing the orientation bringing the oxygen atoms pointing against each other, the same path will correspond to a continuously repulsive curve (see Figure 8.2 below for an even simpler example). We have to consider paths of different shape, all the paths actually, and within each path we have to consider all the three regions. To describe a liquid, we need to know weak long-range interactions as well as strong short-range repulsion at the same degree of accuracy as for the intermediate region. Studies limited to the stabilization energies of the dimers are of interest in other fields of chemical interest, such as the modeling of drugs.

The $\Delta E_{AB}(R)$ function numerically corresponds to a small fraction of the whole QM energy of the dimeric system. It would appear to be computationally safer to compute $\Delta E$ directly instead of obtaining it as a difference, as done in the formal definition [8.11]. This can be done, and indeed in some cases it is done, but experience teaches us that algorithms starting from the energies of the dimer (or of the cluster) and of the monomers are simpler and eventually more accurate. The approaches making use of this difference can be designed as variational approaches, the others directly aiming at $\Delta E$ are called perturbation approaches, because use is made of the QM perturbation theory. We shall pay more attention, here below on the variational approach, then adding the basic elements of the perturbation theory approach.

### 8.4.1 DECOMPOSITION OF THE INTERACTION ENERGY OF A DIMER: VARIATIONAL APPROACH

Coming back to Figure 8.1, its shape exhibiting a minimum makes it manifest that there are several types of interactions at work, with different signs and with different distance decays (at infinity all interactions go to zero). This subject has been amply studied and the main conclusions (there are details differing in the various schemes of interaction energy decomposition) are so widely known and intuitive that we feel authorized to introduce them here before giving a formal definition (that will be considered later). In Table 8.1 we report the names of these components of the interaction energy.

**Table 8.1. The main components of the bimolecular interactions $\Delta E_{AB}(R)$**

| Component name | Acronym | Physical Meaning |
|---|---|---|
| Electrostatic | ES | Coulomb interactions between rigid charge distributions |
| Induction | IND | Mutual electrostatic deformation of the two charge distributions |
| Exchange | EX | Quantum effect due to the Pauli exclusion principle |
| Dispersion | DIS | Interactions among fluctuations in the charge distributions of the two partners |
| Charge transfer | CT | Transfer of electrons between partners |

All components of $\Delta E_{AB}(R)$ are present in all points of the $R_{AB}$ 6-dimensional space we have introduced. It is convenient to examine them separately. We shall make references to a single decomposition scheme[2] that we consider more convenient. Reference to other schemes will be done when necessary. Another view of methods may be found in Tomasi et al.[3]

#### The electrostatic term

The ES term may be positive or negative: the shape of the ES(R) function strongly depends on the electric characteristics of the partners. If both molecules are reduced to dipoles, we have the two extreme situations (see Figure 8.2) which give rise to repulsive (i.e., positive) and attractive (i.e., negative) electrostatic contributions to $\Delta E_{AB}$, respectively. With different orientations of the two dipoles there will be different values of ES that in this simple case can be obtained with an analytical expression:

$$ES(R,\theta) = -2\mu_1\mu_2 \cos\theta / R^3 \qquad [8.12]$$

In more complex molecules, the ES(R) function has a complex shape that often determines the salient features of the whole $\Delta E_{AB}(R)$ function. This is the reason why in the studies of molecular recognition and molecular docking, great attention is paid to a proper representation of ES(R).

In the various methods for the decomposition of $\Delta E_{AB}(R)$ there are no differences in the definition of ES. This quantity can be easily computed with ab initio methods, and the following recipe is the speediest and more used way of doing it.

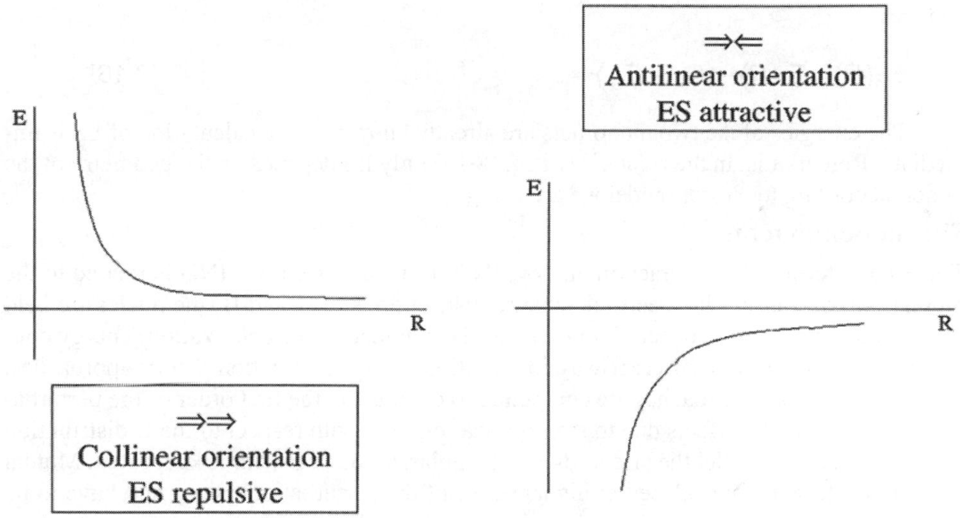


Figure 8.2. ES for two dipoles with collinear (left) and antilinear (right) orientation.

The electronic wave functions of A and B are separately computed, each with its own Hamiltonian $H_A$ and $H_B$. Ab initio methods give at every level of the formulation of the theory antisymmetric electronic wave functions, satisfying the Pauli exclusion principle, on which more details will be given later.

The two antisymmetric wave functions $\Psi_A$ and $\Psi_B$ are then used, without modifications, in connection with the Hamiltonian of the whole system $H_{AB}$ to get the expectation value of the energy; according to the standard notation of quantum chemistry we can write:

$$E_I = \langle \Psi_A \Psi_B | H_{AB} | \Psi_A \Psi_B \rangle \qquad [8.13]$$

It simply means the integral over the whole space of the complex conjugate of the function at the left (i.e., the function $(\Psi_A\Psi_B)^*$ ) multiplied by the function $H_{AB}(\Psi_A\Psi_B)$ (the application of an operator such as $H_{AB}$ to a function always gives a function). We can neglect complex conjugates (our functions are all real); the expression given above is a compact and clear indication of a set of operations ending with an integral, and we shall use it only to speed notations.

$H_{AB}$ differs from the sum of $H_A$ and $H_B$ according to the following expression

$$H_{AB} = H_A + H_B + V_{AB} \qquad [8.14]$$

$V_{AB}$ collects terms describing electrostatic interactions between the nuclei and electrons of A with electrons and nuclei of B. The wave function $\Psi_{AB}$ expressed as eigenfunction of $H_{AB}$ is of course antisymmetric with respect to all the electrons, of A as well as of B, but this has not yet been introduced in the model.

The energy obtained with this recipe (the calculations are a by-product of the calculation of the dimer energy) may be so decomposed as:

$$E_I(R) = E_A + E_B + ES(R) \qquad [8.15]$$

and so:

$$ES(R) = E_I(R) - (E_A + E_B) \quad [8.16]$$

The energies of the two monomers are already known, so the calculation of ES is immediate. Remark that in the right side of eq. [8.15] only ES depends on the geometry of the dimer, according to cluster model we are using.

### The induction term

The second term of the interaction energy, IND, is always negative. IND is related to the mutual polarization of the electronic charge distributions of A and B (the nuclei are held fixed), inducing additional stabilizing effects. This induction (or polarization) energy contribution is defined in a different way in variational and perturbation theory approaches. Perturbation theory approaches are compelled to compute at the first order of the perturbation scheme only the effects due to the polarization of A with respect to the B distribution kept fixed, and in parallel the effects due to the polarization of B with A kept fixed. Mutual induction effects are introduced at higher order of the perturbation theory and have to be separated in some way from dispersion effects computed at the same time.

In the variational approach use is made of an extension of the simple technique we have used for ES.

The separation between electrons of A and B is maintained but the product $\Psi_A\Psi_B$ is now subjected to a constrained variational optimization using the Hamiltonian $H_{AB}$. The two wave functions are so changed, allowing the effects of mutual polarization, because of the presence of the $V_{AB}$ term in the Hamiltonian: they will be so indicated as $\Psi_A^p$ and $\Psi_B^p$.

The resulting expectation value of the energy:

$$E_{II}(R) = \langle \Psi_A^p \Psi_B^p | H_{AB} | \Psi_A^p \Psi_B^p \rangle \quad [8.17]$$

may be so decomposed:

$$E_{II}(R) = E_I(R) + IND(R) \quad [8.18]$$

and so $IND(R) = E_{II}(R) - E_I(R)$. In this way, IND(R) contains all the mutual polarization effects.

### The exchange term

The next term, EX, is positive for all the molecular systems of interest for liquids. The name makes reference to the exchange of electrons between A and B. This contribution to ΔE is sometimes called repulsion (REP) to emphasize the main effect this contribution describes. It is a true quantum mechanical effect, related to the antisymmetry of the electronic wave function of the dimer, or, if one prefers, to the Pauli exclusion principle. Actually these are two ways of expressing the same concept. Particles with a half integer value of the spin, like electrons, are subjected to the Pauli exclusion principle, which states that two particles of this type cannot be described by the same set of values of the characterizing parameters. Such particles are subjected to a special quantum version of the statistics, the Fermi-Dirac statistics, and they are called fermions. Identical fermions have to be described with an antisymmetric wave function; the opposite also holds: identical particles described by an antisymmetric wave function are fermions and satisfy the Pauli exclusion principle. Introducing these concepts in the machinery of the quantum mechanical calculations, it turns out that at each coulomb interaction between two electrons described by MOs $\phi_\mu(1)$ and $\phi_\upsilon(2)$ (the standard expression of this integral, see eq.[8.13], is: $<\phi_\mu(1)\phi_\upsilon(2)|1/r_{12}|\phi_\mu(1)\phi_\upsilon(2)>$) one has to add a second term, in which there is an exchange of the two electrons in the conjugate function: $<\phi_\upsilon(1)\phi_\mu(2)|1/r_{12}|\phi_\mu(1)\phi_\upsilon(2)>$ with a minus sign (the exchange in the label of the two electrons is a permutation of order two, bearing a sign minus in the antisymmetric case).

There are other particles, called bosons, which satisfy other quantum statistics, the Bose-Einstein statistics, and that are described by wave functions symmetric with respect to the exchange, for which the Pauli principle is not valid. We may dispense with a further consideration of bosons in this chapter.

It is clear that to consider exchange contributions to the interaction energy means to introduce the proper antisymmetrization among all the electrons of the dimer. Each monomer is independently antisymmetrized, so we only need to apply to the simple product wave functions an antisymmetrizer restricted to permutations regarding electrons of A and B at the same time: it will be called $\mathbf{A}_{AB}$.

By applying this operator to $\Psi_A^p\Psi_B^p$ without other changes and computing the expectation value, one obtains:

$$E_{III}(R) = \langle \mathbf{A}_{AB}\Psi_A^p\Psi_B^p | H_{AB} | \mathbf{A}_{AB}\Psi_A^p\Psi_B^p \rangle \qquad [8.19]$$

with

$$E_{III}(R) = E_{II}(R) + EX(R) \qquad [8.20]$$

and

$$EX(R) = E_{III}(R) - E_{II}(R) \qquad [8.21]$$

Morokuma has done a somewhat different definition of EX: it is widely used, being inserted into the popular Kitaura-Morokuma decomposition scheme.[4] In the Morokuma definition $E_{III}(R)$ is computed as in eq. [8.19] using the original $\Psi_m$ monomer wave functions instead of the mutually polarized $\Psi_m^p$ ones. This means to lose, in the Morokuma definition, the coupling between polarization and antisymmetrization effects that have to be recovered later in the decomposition scheme. In addition, EX can be no longer computed as in eq. [8.20], but using $E_I$ energy: (i.e., $EX' = E_{III}(R) - E_I(R)$). The problem of this coupling also appears in the perturbation theory schemes that are naturally inclined to use unperturbed monomeric wave functions, not including exchange of electrons between A and B: we shall come back to this subject considering the perturbation theory approach.

### The charge transfer term

The charge transfer contribution CT may play an important role in some chemical processes. Intuitively, this term corresponds to the shift of some electronic charge from the occupied orbitals of a monomer to the empty orbitals of the other. In the variational decomposition schemes this effect can be separately computed by repeating the calculations on the dimer with deletion of some blocks in the Hamiltonian matrix of the system and taking then a difference of the energies (in other words: the same strategy adopted for the preceding terms, but changing the matrix blocks). In the Kitaura-Morokuma scheme CT also contains the couplings between induction and exchange effects.

We consider unnecessary to summarize the technical details; they can be found in the source paper[4] as well as in ref. [3] for the version we are resuming here.

In standard perturbation theory (PT) methods, the CT term is not considered; there are now PT methods able to evaluate it but they are rarely used in the modeling of interaction potentials for liquids.

### The dispersion term

The dispersion energy contribution DIS intuitively corresponds to electrostatic interactions involving instantaneous fluctuations in the electron charge distributions of the partners: these fluctuations cancel out on the average, but their contribution to the energy is different from zero and negative for all the cases of interest for liquids. The complete theory of these stabilizing forces (by tradition, the emphasis is put on forces and not on energies, but the two quantities are related) is rather complex and based on quantum electrodynamics concepts. There is no need of using it here.

The concept of dispersion was introduced by London (1930),[5] using by far simpler arguments based on the application of the perturbation theory, as will be shown in the following subsection. A different but related interpretation puts the emphasis on the correlation in the motions of electrons.

It is worth spending some words on electron correlation.

Interactions among electrons are governed by the Coulomb law: two electrons repel each other with an energy depending on the inverse of the mutual distance: $e^2/r_{ij}$, where e is the charge of the electron. This means that there is correlation in the motion of electrons, each trying to be as distant as possible from the others. Using QM language, where the electron distribution is described in terms of probability functions, this means that when one electron is at position $r_k$ in the physical space, there will be a decrease in the probability of finding a second electron near $r_k$, or in other words, its probability function presents a hole centered at $r_k$.

We have already considered similar concepts in discussing the Pauli exclusion principle and the antisymmetry of the electronic wave functions. Actually, the Pauli principle holds for particles bearing the same set of values for the characterizing quantum numbers, including spin. It says nothing about two electrons with different spin.

This fact has important consequences on the structure of the Hartree-Fock (HF) description of electrons in a molecule or in a dimer. The HF wave function and the corresponding electron distribution function take into account the correlation of motions of electrons with the same spin (there is a description of a hole in the probability, called a Fermi hole), but do not correlate motions of electrons of different spin (there is no the second component of the electron probability hole, called a Coulomb hole).

This remark is important because almost all the calculations thus far performed to get molecular interaction energies have been based on the HF procedure, which still remains the basic starting approach for all the ab initio calculations. The HF procedure gives the best definition of the molecular wave function in terms of a single antisymmetrized product of molecular orbitals (MO). To improve the HF description, one has to introduce in the calculations other antisymmetrized products obtained from the basic one by replacing one or more MOs with others (replacement of occupied MOs with virtual MOs). This is a procedure we shall see in action also in the context of the perturbation theory. At the variational level considered here, the procedure is called Configuration Interaction (CI): each configuration corresponds to one of the antisymmetrized products of MO we have introduced, including the HF one, and the coefficients in front of each component of this linear expansion of the exact (in principle) wave function are determined by applying the variational principle.

There are numerous alternative methods that introduce electron correlation in the molecular calculations at a more precise level that can be profitably used. We mention here the MC-SCF approach (the acronym means that this is a variant of HF (or SCF) procedure starting from the optimization no more of a single antisymmetric orbital product, but of many different products, or configurations), the Coupled-Cluster theory, etc., all methods based on a MO description of single-electron functions.

For readers wishing to reach a better appreciation of papers regarding the formulation of interaction potentials, we add that there is another way of introducing electron correlation effects in the calculations. It is based on the density functional theory (DFT).[6] There is a variety of DFT methods (detailed information can be found in the quoted monograph[6]); a family of these methods again makes use of MOs: they are called hybrid functional methods and give, on average, better results than other correlated methods at a lower computational cost.

The introduction of electron correlation in the description of the monomers produces changes in their charge distribution and in their propensity to be polarized. For this reason ES, IND, and EX computed with correlated wave functions are somewhat different with respect to the values obtained with the corresponding HF wave functions. The procedure sketched with equations [8.13]-[8.21] can be adapted, with some modifications, to correlated wave function using a MO basis.

The reader must be warned that here, as well as in other points of this chapter, we have simplified the discussion by omitting many details necessary for a proper handling and a fuller understanding the problem, but not essential to grasp the basic points.

What is of more practical interest here is that HF descriptions of the dimer cannot give a DIS term. This may be recovered by introducing CI descriptions of the system. The simpler CI description, now largely used in routine calculations on molecules and molecular aggregates, is called MP2. This acronym means that use has been made of a specialized version of the perturbation theory (called Møller-Plesset) limited to second order to determine the expansion coefficients.

MP2 wave functions contain elements able to give an appreciation of DIS, even if of limited accuracy. It is in fact possible to decompose MP2 values of $\Delta E_{AB}(R)$ using the strategy we have outlined for the HF case, adding to each term the appropriate MP2 correction. Each term of the decomposition is somewhat modified, because of the MP2 corrections: the remainder of the MP2 contribution can be taken as a first approximation to DIS.

There are other methods to get DIS values starting from HF wave functions, which may be more precise. Among them we quote the methods based on the response theory and on the use of dynamic polarizabilities. This powerful method has been developed during the years, and the outstanding contributions are due to McWeeny (1984) and Claverie (1986), to which reference is made for more details.[7]

### The decomposition of the interaction energy through a variational approach: a summary

We have so far given a description of the elements in which $\Delta E_{AB}(R)$ may be decomposed using variational approaches. We report here the final expression:

$$\Delta E_{AB} = ES + PL + EX + CT + COUP + DIS \quad [8.22]$$

At the HF level, all terms (but one: DIS) can be obtained with almost zero computational cost with respect to the numerical determination of $\Delta E_{AB}$; in the HF decomposition of the interaction energy there is a small additional term (called COUP) describing further couplings between the other components, which is often left as it is obtained (i.e., as the difference between the original $\Delta E_{AB}$ value and the sum of the other four components) or subjected to further decompositions.[8] The last term, DIS, may be calculated at this level using specialistic time-dependent formulations of the HF procedure. Going beyond the HF level, the decomposition can be obtained at the MP2 level essentially using the same techniques (DIS may be appreciated by separating some of the MP2 contributions that are strictly additive).

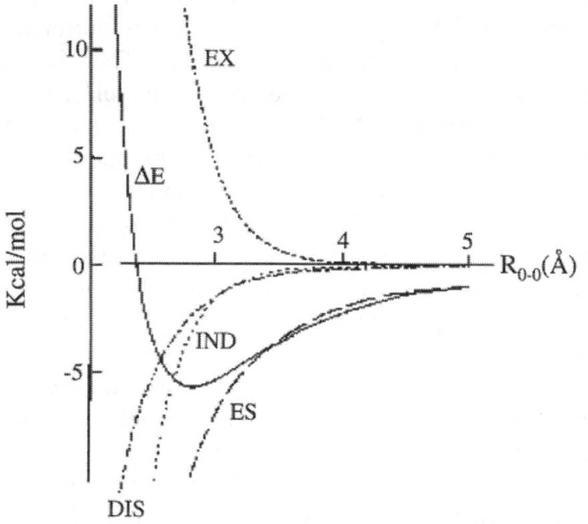

Figure 8.3. Decomposition of the interaction energy ΔE in $H_2O\cdots HOH$.

The variational approaches we have considered are able to describe, and to decompose, the interaction energy at the level of accuracy one wishes, once the necessary computational resources are available. In fact, as we shall see later, there is no need of reaching extreme precision in the preliminary calculations to model interaction potentials for liquids, because other approximations must be introduced that will drastically reduce the accuracy of the description.

We report in Figure 8.3 as an example the decomposition of the interaction energy of the water dimer, in the same orientation $\Omega$ used in Figure 8.1.

At large distances the interaction is dominated by ES; this contribution also gives a reasonable approximation to ΔE at the equilibrium distance. The IND decays with the distance more rapidly than ES; this contribution is particularly sensitive to the quality of the expansion basis set $\{\chi\}$. Old calculations of IND using restricted basis sets are not reliable. EX is a short ranged term: its contribution is however essential to fix the position of the minimum (and to describe portion III of the PES). CT and DIS terms have both a short-range character.

This schematic analysis is valid for almost all the dimeric interactions where at least one partner has a dipolar character. The presence of a hydrogen bond (as is the case for the example given in the figure) only introduces quantitative modifications, sufficient however to show by simple visual inspection if there is a hydrogen bond or not.

Analogous trends in the decomposition of $\Delta E$ are present in the interactions involving charged-neutral species (also in the case of apolar molecules). A special case is given by the interaction of a molecule with the bare proton: in this case there is no EX contribution.

If the partners have no permanent charge, or dipole, the interaction at large-medium distances is dominated by DIS, and by EX at small values of R.

### 8.4.2 BASIS SET SUPERPOSITION ERROR AND COUNTERPOISE CORRECTIONS

Calculations of the interaction energies are affected by a formal error that may have important consequences on the final value of the energy and on its decomposition. We shall consider here the case of variational calculations for dimers, but the basic considerations can be extended to larger clusters and to other computational methods. The origin of the error is a non-perfect balance in the quality of the calculation of dimer energy, $E_{AB}$ and of energies of the two monomers, $E_A$ and $E_B$. In fact, there are more computational degrees of freedom available for the dimer than for each monomer separately. The number of degrees of freedom corresponds to the number of basis functions available for the optimization of the electronic structure of the molecule, and hence for the minimization of the energy. Let us consider, to clarify the concept, the case of two water molecules giving origin to a dimer; each water molecule has ten electrons, while the quality and number of expansion functions is selected at the beginning of the calculation. This is called the expansion basis set (just basis set, or BS, for brevity) and it will be indicated for the molecule A with $\{\chi_A\}$. The number of these basis functions is fixed, for example, 30 functions. The second molecule will be described by a similar basis set $\{\chi_B\}$ containing in this example expansion functions of the same quality and number as for molecule A (the two molecules are in this example of the same chemical nature). The wave function of the dimer AB and its energy will be determined in terms of the union of the two basis sets, namely $\{\chi_{AB}\}=\{\chi_A \oplus \chi_B\}$, composed of 60 functions. It is evident that it is easier to describe 20 electrons with 60 parameters than 10 electrons with 30 only. The conclusion is that the dimer is better described than the two monomers, and so the dimer energy is relatively lower than the sum of the energies of the two monomers.

This is called the basis set superposition (BSS) error. Why superposition error? When the two components of the dimer are at large distance, $\{\chi_A\}$ and $\{\chi_B\}$ are well separated, i.e., they have small superposition (or overlap), and so the relative error we are considering is modest, zero at infinity. When the two monomers are at shorter distances the superposition of the two basis sets increases (the basis functions are always centered on the pertinent nuclei) as well as the error.

There is a simple recipe to correct this error: it consists of performing all the necessary calculations with the same basis set, the dimeric basis $\{\chi_{AB}\}$ which depends on the geometry of the dimer.[9] This means that the energy of the monomers must be repeated for each position in the $\{R^6\}$ configuration space (see Section 8.4 for its definition). We add a superscript CP (counterpoise) to denote quantities modified in such a way and we also add the specification of the basis set. We replace eq. [8.11] with the following one:

$$\Delta E_{AB}^{CP}(\chi_{AB};R) = E_{AB}(\chi_{AB};R) - \left[E_A^{CP}(\chi_{AB};R) + E_B^{CP}(\chi_{AB};R)\right] \qquad [8.23]$$

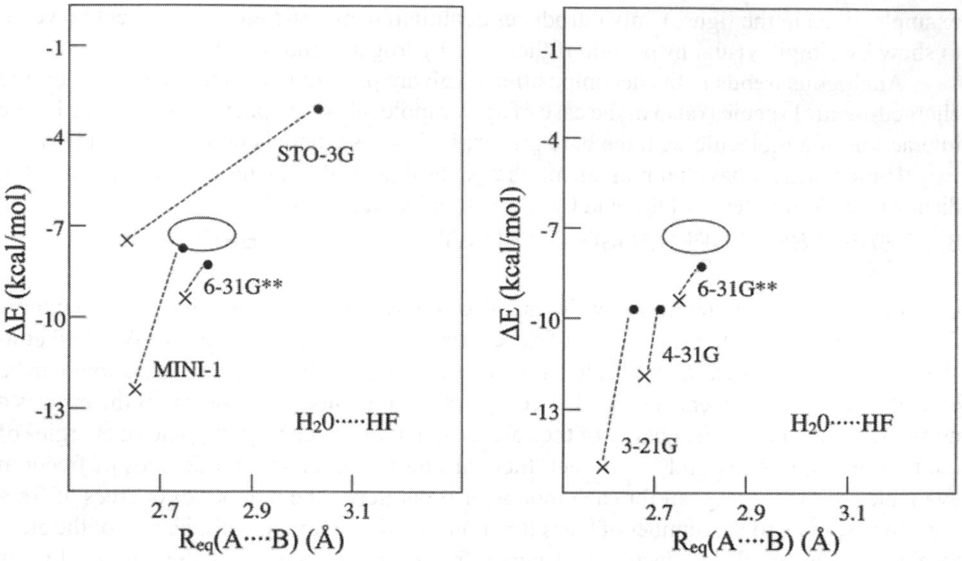


Figure 8.4. Location of the minimum energy for $H_2O\cdots HF$ dimer according to the various basis sets and the various methods. Crosses (×) refer to SCF calculations without CP corrections, full circles (●) to CP corrected calculations. The ovoidal area refer to an estimate of the location of the minimum at the HF limit (including an estimate of the error). Left side: minimal basis sets; right side: double valence-shell basis sets.

$\Delta E^{CP}$ is by far more corresponding to the exact potential energy functions in calculations performed at the HF level with a basis set of a small-medium size. Passing to calculations with larger basis sets, the BSS error is obviously smaller, but the CP correction is always beneficial.

We give in Figure 8.4 a graphical view of how CP corrections modify the equilibrium position of the dimer and, at the same time, its stabilization energy.

The same holds for calculations performed at higher levels of the QM theory, beyond the HF approximation, with decreasing effects of the CP corrections, however. We may leave this last subject to the attention of the specialists because for the determination of interaction potential for liquids, the HF approximation is in general sufficient; in some cases it may be supplemented by applying simple levels of description for electron correlation, as we have already said.

The CP corrected interaction energy may be decomposed into terms each having a definite physical meaning, in analogy with what we have exposed in the preceding subsection for ΔE without CP corrections. There are slightly different ways of doing it. We summarize here the strategy that more closely follows the physics of the problem.[10] When this correction is applied, the various terms better satisfy chemical intuition in passing from one dimer to another of different chemical composition. Each term of the ΔE decomposition is corrected with an additive term $\Delta^X$ (X stays for one of the components of the interaction energy) which is expressed as a difference in the monomers' energies computed with the opportune basis set.

Let us consider again the decomposition we have done in eq. [8.19].

The physical definition of ES (electrostatic interactions among rigid partners) clearly indicates that there is no room here for CP corrections, which would involve some shift of the monomer's charge on the ghost basis functions of the partner. This physical observation corresponds to the structure of the blocks of the Hamiltonian matrix used to compute ES: no elements regarding the BS of the partner are present in conclusion and thus no corrections to ES are possible.

In analogy, IND must be left unmodified. Physical analysis of the contribution and the structure of the blocks used for the calculation agree in suggesting it.

CP corrections must be performed on the other elements of ΔE, but here again physical considerations and the formal structure of the block partition suggest using different CP corrections for each term.

Let us introduce a partition into the BS space of each monomer and of the dimer. This partition can be introduced after the calculation of the wave function of the two monomers. At this point we know, for each monomer M (M stays for A or for B), how the complete monomer's BS is partitioned into occupied and virtual orbitals $\phi_M$:

$$\{\chi_M\} = \{\phi_M^0 \oplus \phi_M^v\} \qquad [8.24]$$

For the exchange term we proceed in the following way. The CP corrected term is expressed as the sum of the EX contribution determined as detailed above, plus a CP correction term called $\Delta^{EX}$:

$$EX^{CP} = EX + \Delta^{EX} \qquad [8.25]$$

$\Delta^{EX}$ in turn is decomposed into two contributions:

$$\Delta^{EX} = \Delta_A^{EX} + \Delta_B^{EX} \qquad [8.26]$$

with

$$\Delta_A^{EX} = \left[E_A(\chi_A) - E_A\left(\chi_A^{EX}\right)\right] \qquad [8.27]$$

and a similar expression for the other partner. The CP correction to EX is so related to the calculation of another energy for the monomers, performed on a basis set containing occupied MOs of A as well as of B:

$$\{\chi_A^{EX}\} = \{\phi_A^0 \oplus \phi_B^0\} \qquad [8.28]$$

We have used here as ghost basis the occupied orbitals of the second monomer, following the suggestions given by the physics of the interaction.

The other components of the interaction energy are changed in a similar way. The $\Delta^X$ corrections are all positive and computed with different extensions of the BS for the monomers, as detailed in Table 8.2.

**Table 8.2 Explicit expressions of the $\Delta^X$ counterpoise corrections to the interaction energy components**

$$\Delta_A^{EX} = \left[E_A(\chi_A) - E_A\left(\phi_A^0 \oplus \phi_B^0\right)\right]$$
$$\Delta_B^{EX} = \left[E_B(\chi_B) - E_B\left(\phi_A^0 \oplus \phi_B^0\right)\right]$$
$$\Delta_A^{CT} = \left[E_A(\chi_A) - E_A\left(\phi_A^0 \oplus \phi_B\right)\right]$$
$$\Delta_B^{CT} = \left[E_B(\chi_B) - E_B\left(\phi_B^0 \oplus \phi_A\right)\right]$$
$$\Delta_A^{TOT} = \left[E_A(\chi_A) - E_A\left(\chi_A \oplus \chi_B\right)\right]$$
$$\Delta_B^{TOT} = \left[E_B(\chi_B) - E_B\left(\chi_A \oplus \chi_B\right)\right]$$

The correction $\Delta^{TOT}$ permits to recover the full CP correction. It may be used to define the $\Delta^{COUP}$ correction:

$$\Delta^{COUP} = \Delta^{TOT} - \left(\Delta^{EX} + \Delta^{CT}\right) \quad [8.29]$$

or the correction to the remainder, if COUP is further decomposed.[8b] This last step is of little utility for solvent-solvent interactions, but useful for stronger chemical interactions. There is no need of analyzing here these refinements of the method.

The performances of the CP corrections to the interaction energy decomposition can be appreciated in a systematic study on representative hydrogen-bond dimers.[11] A general review on the CP theory has been done by van Dujneveldt et al.[12] (who prefer to use a different decomposition).

### 8.4.3 PERTURBATION THEORY APPROACH

The perturbation theory (PT) approach aims at exploiting a consideration we have already expressed, namely, that ΔE is by far smaller than the sum of the energies of the separate monomers. It would be safer (and hopefully easier) to compute ΔE directly instead of getting it as difference between two large numbers. The formulation of the perturbation theory for this problem that we shall give here below has all the basic premises to satisfy this program. In practice, things are different: the introduction of other aspects in this formulation of PT, made necessary by the physics of the problem, and the examination and correction of finer details, put in evidence by the analysis of the results, make the PT approach more costly than variational calculations of comparable accuracy. Modern PT methods are competitive in accuracy with variational procedures but are rarely used at accurate levels to model potentials for liquids, the main reason being the computational cost.

In spite of this we dedicate a subsection to PT methods, because this theoretical approach shows here, as in many other problems, its unique capability of giving an interpretation of the problem. To model interactions, we need, in fact, to have a clear vision of the physical elements giving origin to the interaction, and some information about the basic mathematical behavior of such elements (behavior at large distances, etc.).

The formal perturbation theory is simple to summarize.

The theory addresses the problem of giving an approximate solution for a target system hard to solve directly, by exploiting the knowledge of a simpler (but similar) system. The target system is represented by its Hamiltonian H and by the corresponding wave function Ψ defined as solution of the corresponding Schrödinger equation (see eq. [8.1])

$$H\Psi = E\Psi \quad [8.30]$$

The simpler unperturbed system is described in terms of a similar equation

$$H^0\Phi_0 = E_0\Phi_0 \quad [8.31]$$

The following partition of H is then introduced

$$H = H^0 + \lambda V \quad [8.32]$$

as well as the following expansion of the unknown wave function and energy as powers of the parameter λ:

$$\Psi = \Phi_0 + \lambda\Phi^{(1)} + \lambda^2\Phi^{(2)} + \lambda^3\Phi^{(3)} + \ldots \quad [8.33]$$

$$E = E_0 + \lambda E^{(1)} + \lambda^2 E^{(2)} + \lambda^3 E^{(3)} + \cdots \quad [8.34]$$

The corrections to the wave function and to the energy are obtained introducing the formal expressions [8.32]-[8.34] in the equation [8.30] and separating the terms according to their order in the power of λ. In this way one obtains a set of integro-differential equations to be separately solved. The first equation is merely the Schrödinger equation [8.31] of the simple system, supposed to be completely known. The others give, order by order, the corrections $\Phi^{(n)}$ to the wave function, and $E^{(n)}$ to the energy. These equations may be solved by exploiting the other solutions, $\Phi_1, \Phi_2, \Phi_3, \ldots, \Phi_K$, of the simpler problem [8.31], that constitute a complete basis set and are supposed to be completely known. With this approach every correction to Ψ is given as linear combination

$$\Phi^{(n)} = \sum_K C_K^n \Phi_k \quad [8.35]$$

The coefficients are immediately defined in terms of the integrals $V_{LK} = \langle\Phi_L|V|\Phi_K\rangle$ where the indexes L and K span the whole set of the eigenfunctions of the unperturbed system (including $\Phi_0$ where necessary). The corrections to the energy follow immediately, order by order. They only depend on the $V_{LK}$ integrals and on the energies $E_1$, $E_2$, ..., $E_K$, ... of the simple (unperturbed) system. The problem is so reduced to a simple summation of elements, all derived from the simpler system with the addition of a matrix containing the $V_{LK}$ integrals.

This formulation exactly corresponds to the original problem, provided that the expansions [8.33] and [8.34] of Ψ and E converge and that these expansions are computed until convergence.

We are not interested here in examining other aspects of this theory, such as the convergence criteria, the definitions to introduce in the case of interrupted (and so approximate) expansions, or the problems of practical implementation of the method.

The MP2 wave functions we have introduced in a preceding subsection are just the application of this method to another problem, that of the electronic correlation. In this case, the simpler unperturbed system is the HF approximation, the corrections are limited to the second order, and the corresponding contributions to the energy are expressed as a simple summation of elements. The MP2 method is currently used in the PT description of the intermolecular potential:[13] in such cases, two different applications of PT are used at the same time.

We pass now to apply PT to the calculation of $\Delta E_{AB}(R)$. The most reasonable choice consists of defining the unperturbed system as the sum of the two non-interacting monomers. We thus have:

$$H^0 = H_A + H_B \qquad [8.36]$$

$$\Phi_0 = |\Psi_0^A \Psi_0^B > \qquad [8.37]$$

$$E_0 = E_0^A + E_0^B \qquad [8.38]$$

$\Phi_0$ is the simple product of the two monomers' wave functions. The perturbation operator is the difference of the two Hamiltonians, that of the dimer and $H^0$. The perturbation parameter $\lambda$ may be set equal to 1:

$$V = H - H^0 = V_{AB} \qquad [8.39]$$

The set $\Phi_1, \Phi_2, \Phi_3, \ldots, \Phi_K$ of the solutions of the unperturbed system can be obtained by replacing within each monomer wave function, one, two or more occupied MO with vacant MOs belonging to the same monomer. The perturbation operator V only contains one- and two-body interactions, and so, being the MOs orthonormal, the only $V_{LK}$ integrals different from zero are those in which L and K differ at the maximum by two MOs.

The formulation is quite appealing: there is no need for repeated calculations, the decomposition of the interaction energy can be immediately obtained by separately collecting contributions corresponding to different ways of replacing occupied with virtual orbitals.

Interrupting the expansion at the second order, one obtains the following result:

$$\Delta E \approx E^{(1)} + E^{(2)} =$$

$$\langle \Psi_A \Psi_B | V \Psi_A \Psi_B \rangle + \qquad \text{I order: electrostatic term}$$

$$-\sum_K \frac{|\langle \Psi_0^A \Psi_0^B | V \Psi_K^A \Psi_0^B \rangle|^2}{E_K^A - E_0^A} + \qquad \text{II order: polarization of A}$$

$$-\sum_K \frac{|\langle \Psi_0^A \Psi_0^B | V \Psi_0^A \Psi_K^B \rangle|^2}{E_K^B - E_0^B} + \qquad \text{II order: polarization of B} \qquad [8.40]$$

$$-\sum_K \sum_L \frac{|\langle \Psi_0^A \Psi_0^B | V \Psi_K^A \Psi_L^B \rangle|^2}{(E_K^A + E_L^B) - (E_0^A + E_0^B)} + \qquad \text{II order: dispersion A-B + higher order terms}$$

We have here explicitly written the monomer wave functions, with a generic indication of their electronic definition: $\Psi_0^A$ is the starting wave function of A, already indicated with $\Psi_A$, $\Psi_K^A$ is another configuration for A with one or two occupied one-electron orbitals replaced by virtual MOs.

The first order contribution exactly corresponds to the definition we have done of ES with equation [8.16]: it is the Coulomb interaction between the charge distributions of A and B.

The second term of the expansion, i.e., the first element of the second order contributions, corresponds to the polarization of A, due to the fixed unperturbed charge distribution of B; the next term gives the polarization of B, due to the fixed charge distribution of A. The two terms, summed together, approximate IND. We have already commented that perturbation theory in a standard formulation cannot give IND with a unique term: further refinements regarding mutual polarization effects have to be searched at higher order of the PT expansion.

The last term of the second order contribution is interpreted as a dispersion energy contribution. The manifold of excited monomer states is limited here at single MO replacements within each monomer; the resulting energy contribution should correspond to a preliminary evaluation of DIS, with refinements coming from higher orders in the PT expansion.

Remark that in PT methods, the final value of $\Delta E$ is not available. It is not possible here to get a numerical appraisal of correction to the values obtained at a low level of the expansion. All contributions are computed separately and added together to give $\Delta E$.

In conclusion, this PT formulation, which has different names, among which "standard" PT and RS (Rayleigh-Schrödinger) PT, gives us the same ES as in the variational methods, a uncompleted value of IND and a uncompleted appraisal of DIS: higher order PT contributions should refine both terms. One advantage with respect to the variational approach is evident: DIS appears in PT as one of the leading terms, while in variational treatments one has to do ad hoc additional calculations.

Conversely, in the RS-PT formulation CT and EX terms are not present. The absence of CT contributions is quickly explained: RS-PT works on separated monomers and CT contributions should be described by replacements of an occupied MO of A with an empty MO of B (or by an occupied MO of B with an empty MO of A), and these electronic configurations do not belong to the set of state on which the theory is based. For many years the lack of CT terms has not been considered important. The attention focused on the examination of the interaction energy of very simple systems, such as two rare gas atoms, in which CT effects are in fact of very limited importance.

The absence of EX terms, on the contrary, indicates a serious deficiency of the RS formulation, to which we have to pay more attention.

The wave function $\Phi_K$ used in the RS formulation does not fully reflect the electron permutational symmetry of the dimer: the permutations among electrons of A and B are neglected. This leads to severe inconsistencies and large errors when RS PT is applied over the whole range of distances. One has to rework the perturbation theory in the search of other approaches. The simplest way would simply replace $\Phi_0 = |\Psi_A^0 \Psi_B^0>$ with $|\mathbf{A}_{AB}\Phi_0>$ where $\mathbf{A}_{AB}$ is the additional antisymmetry operator we have already introduced. Unfortunately $|\mathbf{A}_{AB}\Phi_0>$ is not an eigenfunction of $H^0$ as asked by the PT. Two ways of overcoming this difficulty are possible. One could abandon the natural partitioning [8.36] and define another unperturbed Hamiltonian having $|\mathbf{A}_{AB}\,\Phi_0>$ as eigenfunction. Changes in the definition of the unperturbed Hamiltonian are no rare events in the field of application of the perturbation theory and there are also other reasons suggesting a change in the definition of it in the study of dimeric interactions. This strategy has been explored with mixed success: some improvements are accompanied by the lack of well-defined meaning of the energy components when the expansion basis set grows toward completeness. A second way consists in keeping the original, and natural, definition of $H^0$ and in changing the PT. This is the way currently used at present: the various versions of the theory can be collected under the acronym SAPT (symmetry adapted perturbation theory)

It is worth remarking that in the first theoretical paper on the interaction between two atoms (or molecules) (Eisenschitz and London, 1930[14]) use was made of a SAPT; the problem was forgotten for about 40 years (rich, however, in activities for the PT study of weak interactions using the standard RS formulation). A renewed intense activity started at the end of the sixties, which eventually led to a unifying view of the different ways in which SAPT may be formulated. The first complete monograph is due to Arrighini[15] (1981), and now the theory can be found in many other monographs or review articles. We quote here our favorites: Claverie[16] (1976), a monumental monograph not yet giving a complete formal elaboration but rich in suggestions; Jeziorski and Kolos[17] (1982), short, clear and critical; Jeziorski et al.[13] (1994), clear and centered on the most used versions of SAPT.

SAPT theories are continuously refined and extended. Readers of this chapter surely are not interested to find here a synopsis of a very intricate subject that could be condensed into compact and elegant formulations, hard to decode, or expanded into long and complex sequences of formulas. This subject can be left to specialists, or to curious people, for which the above given references represent a good starting point.

The essential points can be summarized as follows. The introduction of the intra-monomer antisymmetry can be done at different levels of the theory. The simplest formulation is just to use $|\mathbf{A}_{AB}\Phi_0>$ as unperturbed wave function, introducing a truncated expansion of the antisymmetry operator. This means to pass from rigorous to approximate formulations.

One advantage is that the perturbation energies (see eq. [8.30]), at each order, may be written as the sum of the original RS value and of a second term related to the introduction of the exchange:

$$E^{(n)} = E^{(n)}_{RS} + E^{(n)}_{exc} \qquad [8.41]$$

The results at the lower orders can be so summarized:

$$\Delta E = \left\{E^{(1)}_{RS} + E^{(1)}_{exc}\right\} + \left\{E^{(2)}_{RS} + E^{(2)}_{exc}\right\} + \cdots \qquad [8.42]$$

$E^{(1)}_{exc}$ does not fully correspond to EX obtained with the variational approach. The main reason is that use has been made of an approximation of the antisymmetrizer: other contributions are shifted to higher order of the PT series. A second reason is that use has been here made of the original MOs $\varphi_i$ to be contrasted with the polarized ones $\varphi_i^p$, see Section 8.4.1: for this reason there will be in the next orders contributions mixing exchange and polarization effects.

At the second order we have:

$$E^{(2)}_{exc} = E^{(2)}_{exc-dis} + E^{(2)}_{exc-ind} \quad [8.43]$$

The partition of this second order contribution into two terms is based on the nature of the MO replacements occurring in each configuration $\Psi^{m}_{K}$ appearing in the sums of eq. [8.40]. These two terms give a mixing of exchange and dispersion (or induction) contributions.

Passing to higher orders $E^{(n)}$ the formulas are more complex. In the RS part it is possible to define pure terms, as $E^{(3)}_{RSdis}$, but in general they are of mixed nature. The same happens for the $E^{(n)}_{exc}$ contributions.

The examination of these high order contributions is addressed in the studies of the mathematical behavior of the separate components of the PT series. Little use has so far been made of them in the actual determination of molecular interaction potentials.

### 8.4.4 MODELING OF THE SEPARATE COMPONENTS OF ΔE

The numerical output of variational decompositions of ΔE (supplemented by some PT decompositions) nowadays represents the main source of information to model molecular interaction potentials. In the past, this modeling was largely based on experimental data (supported by PT arguments), but the difficulty of adding new experimental data, combined with the difficulty of giving an interpretation and a decoupling of them in the cases of complex molecules, has shifted the emphasis to theoretically computed values.

The recipes for the decomposition we have done in the preceding sections are too complex to be used to study liquid systems, where there is the need of repeating the calculation of $\Delta E(r,\Omega)$ for a very large set of the six variables and for a large number of dimers. There is thus the need of extracting simpler mathematical expressions from the data on model systems. We shall examine separately the different contributions to the dimer interaction energy. As will be shown in the following pages, the basic elements for the modeling are to a good extent drawn from the PT approach.

#### The electrostatic term

ES may be written in the following form, completely equivalent to eq. [8.16]

$$ES = \int\int \rho^{T}_{A}(r_1)\frac{1}{r_{12}}\rho^{T}_{B}(r_2)dr_1dr_2 \quad [8.44]$$

we have here introduced the total charge density function for the two separate monomers. The density function $\rho^{T}_{M}(r)$ is a one-electron function, which describes the distribution in the space of both electrons and nuclei. Formula [8.44] is symmetric both in A and B, as well as in $r_1$ and $r_2$.

It is often convenient to decompose the double integration given in eq. [8.44] in the following way

$$ES = \int \rho^{T}_{A}(r_1)V_B(r_1)dr_1 \quad [8.45]$$

where

$$V_B(r_1) = \int \rho^{T}_{B}(r_2)\frac{1}{r_{12}}dr_2 \quad [8.46]$$

$V_B(r_1)$ is the electrostatic potential of molecule B (often called MEP or MESP, according to the authors[3,18,19]). This a true molecular quantity, not depending on interactions, and it is often used to look at local details of the electrostatic interactions between molecules as required, for example, in chemical reactivity and molecular docking problems.

To model simplified expressions of the intermolecular potential the direct use of eq. [8.45] does not introduce significant improvements. The MEP, however, may be used in two ways. We consider here the first, consisting of defining, and using, a multipolar expansion of it. The theory of multipolar expansion is reported in all the textbooks on electrostatics and on molecular interactions, as well as in many papers, using widely different formalisms. It can be used to separately expand $\rho_A$, $\rho_B$, and $1/r_{12}$ of eq. [8.44] or $V_B$ and $\rho_A$ of eq. [8.45]. We adopt here the second choice. The multipolar expansion of $V_B(r_1)$ may be so expressed:

$$V_B(r_1) = \sum_{l=0}^{\infty} \sum_{m=-l}^{l} M_{l,m}^{B} R^{-(l+1)} Y_l^m(\theta, \varphi) \qquad [8.47]$$

It is a Taylor expansion in powers of the distance R from the expansion center. There are negative powers of R only, because this expansion is conceived for points lying outside a sphere containing all the elements of the charge distribution.

The other elements of [8.47] are the harmonic spherical functions $Y_l^m$ and the multipole elements $M_{l,m}^B$ which have values specific for the molecule:

$$M_{l,m}^{B} = \left(\frac{4\pi}{2l+1}\right) \int r^l Y_l^m(\theta, \varphi) \rho_B(r) d^3r = \left\langle \Psi_B | M_l^m \Psi_B \right\rangle \qquad [8.48]$$

The spherical harmonics are quite appropriate to express the explicit orientational dependence of the interaction, but in the chemical practice it is customary to introduce a linear transformation of the complex spherical functions $Y_l^m$ into real functions expressed over Cartesian coordinates, which are easier to visualize. In Table 8.3 we report the expressions of the multipole moments.

**Table 8.3. Multipole moments expressed with the aid of real Cartesian harmonics**

| | | | |
|---|---|---|---|
| $M_0$ | 1 element | Charge | Q |
| $M_1^m$ | 3 elements | Dipole | $\mu_x$, $\mu_y$, $\mu_z$ |
| $M_2^m$ | 5 elements | Quadrupole | $\theta_{xy}$, $\theta_{xz}$, $\theta_{yz}$, $\theta_{x^2-y^2}$, $\theta_{z^2}$ |
| $M_3^m$ | 7 elements | Octupole | $\omega_{xxy}$, etc. |

The leading parameter is l, which defines the $2^l$ poles. They are, in order, the monopole (l=0, a single element corresponding to the net molecular charge), the dipole (l=1, three elements, corresponding to the 3 components of this vector), the quadrupole (l=2, five components, corresponding to the 5 distinct elements of this first rank tensor), the octupole, etc. We have replaced the potential given by the diffuse and detailed charge distribution $\rho_B$ with that of a point charge, plus a point dipole, plus a point quadrupole, etc., placed all at the expansion center. Note that the potential of the $2^l$ pole is proportional to $r^{-(l+1)}$. This means that the potential of the dipole decreases faster than that of the monopole, the quadrupole faster than that of the dipole, and so on.

To get the electrostatic interaction energy, the multipolar expansion of the potential $V_B(r)$ is multiplied by a multipolar expansion of $\rho_A(r)$. The result is:

$$ES \approx \sum_{l=0}^{\infty}\sum_{l'=0}^{\infty} D_{ll'} R^{-(l+l'+1)} \qquad [8.49]$$

where

$$D_{ll'} = \sum_{m=-l}^{l} C_{ll'}^{m} M_{lm}^{A} M_{l'm}^{B} \qquad [8.50]$$

is the interaction energy of the permanent dipole l of A with the permanent dipole l' of B ($C_{ll'}^{m}$ is a numerical coefficient depending only on l, l', and m).

The calculation of ES via eq. [8.49] is much faster than through eq. [8.44]: the integrations are done once, to fix the $M_{lm}^{X}$ molecular multipole values and then used to define the whole ES(R) surface.

For small size and almost spherical molecules, the convergence is fast and the expansion may be interrupted at a low order: it is thus possible to use experimental values of the net charge and of the dipole moment (better if supplemented by the quadrupole, if available) to get a reasonable description of ES at low computational cost and without QM calculations.

We have, however, put a $\approx$ symbol instead of $=$ in eq. [8.49] to highlight a limitation of this expansion. To analyze this problem, it is convenient to go back to the MEP.

Expansion [8.47] has the correct asymptotic behavior: when the number of terms of a truncated expression is kept fixed, the description improves at large distances from the expansion center. The expansion is also convergent at large values of R: this remark is not a pleonasm, because for multipole expansions of other terms of the interaction energy (as for example the dispersion and induction terms), the convergence is not ensured.

Convergence and asymptoticity are not sufficient, because the expansion theorem holds (as we have already remarked) for points r lying outside a sphere containing all the elements of the charge distribution. At the QM level this condition is never fulfilled because each $\rho(r)$ fades exponentially to zero when $r \to \infty$. This is not a serious problem for the use of multipole expansions if the two molecules are far apart, and it simply reduces a little the quality of the results if almost spherical molecules are at close contact. More important are the expansion limitations when one, or both, molecules have a large and irregular shape. At strict contact a part of one molecule may be inserted into a crevice of the partner, within its nominal expansion sphere. For large molecules at close contact a systematic enlargement of the truncated expression may lead to use high value multipoles (apparent indication of slow convergence) with disastrously unphysical results (real demonstration of the lack of convergence).

The introduction of correction terms to the multipolar expansions of $V_B$ or of ES, acting at short distances and called "penetration terms", has been done for formal studies of this problem but it is not used in practical applications.

For larger molecules it is necessary to pass to many-center multipole expansions. The formalism is the same as above, but now the expansion regards a portion of the molecule, for which the radius of the encircling sphere is smaller than that of the whole molecule. Here it is no longer possible to use experimental multipoles. One has to pass to QM calculations supplemented by a suitable procedure of partition of the molecule into fragments (this operation is necessary to define the fragmental definition of multipoles with an analog of eq. [8.42]). There are many approaches to define such partitions of the molecular charge distribution and the ensuing multipole expansions: for a review see, e.g., Tomasi et al.[20]

It is important to remark here that such local expansions may have a charge term even if the molecule has no net charge (with these expansions the sum of the local charges must be equal to the total charge of the molecule).

There is freedom in selecting the centers of these local expansions as well as their number.

A formal solution to this problem is available for the molecular wave functions expressed in terms of Gaussian functions (as is the general rule). Each elementary electron distribution entering in the definition of $\Psi$, is described by a couple of basic functions $\chi_s^*\chi_t$ centered at positions $r_s$ and $r_t$. This distribution may be replaced by a single Gaussian function centered at the well defined position of the overlap center: $\chi_u$ at $r_u$. The new Gaussian function may be exactly decomposed into a finite local multipole expansion. It is possible to decompose the spherical function Y into a finite (but large) number of local multipole expansions each with a limited number of components. This method works (the penetration terms have been shown to be reasonably small) but the number of expansion centers is exceedingly large.

Some expedient approximations may be devised, introducing a balance between the number of expansion centers and the level of truncation of each expansion. This work has been done for years on empirical bases. One strategy is to keep each expansion at the lowest possible order (i.e., local charges) and to optimize number and location of such charges.

The modern use of this approach has been pioneered by Alagona et al.,[21] using a number of sites larger than the number of atoms, with values of the charges selected by minimization of the difference of the potential they generate with respect to the MEP function $V_B(r)$ (eq. [8.40]). This is the second application of the MEP function we have mentioned.

Alagona's approach has been reformulated by Momany in a simpler way, by reducing the number of sites to that of the nuclei present in the molecule.[22] This strategy has gained wide popularity: almost all the potentials in use for relative large molecules reduce the electrostatic contributions to Coulomb contributions between atomic charges. A relatively larger number of sites are in use for the intermolecular potentials of some simple molecules, as, for example, water, for which more accuracy is sought.

Momany's idea has led to a new definition of atomic charges. It would be possible to write volumes about atomic charges (AC), a concept that has no a precise definition in QM formalism, but is of extreme utility in practical applications. Many definitions of AC are based on manipulations of the molecular wave function, as, for example, the famous Mulliken charges.[23] Other definitions are based on different analyses of the QM definition of the charge density, as for example Bader's charges.[24] There are also charges derived from other theoretical approaches, such as the electronegativity equalization, or from experimental values, such as from the vibrational polar tensors.

The charges obtained using Momamy's idea of fitting the MEP with atomic charges are sometimes called PDAC (potential derived atomic charges). It has been realized that a fitting of MEP on the whole space was not convenient. It is better to reduce this fitting to the portion of space of close contact between molecules, i.e., near the van der Waals surface. This is the main technique now in use to define PDAC values.

### The induction term

It is possible to define a molecular index $P_A$ for the induction term to be used in combination with the MEP $V_A$ to get a detailed description of the spatial propensity of the molecule to develop electrostatic interactions of classical type. Both functions are used under the form of an interaction with a unit point charge q placed at position r. In the case of $V_A$ this means a simple multiplication; in the case of $P_A$ there is the need of making additional calculations (to polarize the charge distribution of A). There are fast methods to do it, both at the variational level[25a] and at the PT level.[25b] The analysis of $P_A$ has not yet extensively been used to model IND contributions to ΔE, and it shall not be used here. This remark has been added to signal that when one needs to develop interactions potentials for molecular not yet studied interactions including, e.g., complex solutes, the use of this approach could be of considerable help.

The multipole expansion of IND may be expressed in the following way:

$$IND = IND(A \leftarrow B) + IND(B \leftarrow A) \qquad [8.51]$$

with

$$IND(A \leftarrow B) \approx -\frac{1}{2}\sum_{ll'=1}^{\infty}\sum_{kk'=1}^{\infty} R^{-(l+l'+k+k'+2)} \sum_{m=-l<}^{l<} \sum_{m'=-k<}^{k<} C_{ll'}^{m} C_{kk'}^{m'} \times \pi_{lk}^{A}(m,m') M_{l',-m}^{B} M_{k',-m}^{B} \qquad [8.52]$$

and a similar expression for IND(B←A) with A and B labels exchanged.

We have here introduced the static polarizability tensor elements $\pi_{lk}^{X}(m,m')$: these tensors are response properties of the molecule with respect to external electric fields F (with components $F_\alpha$), electric field gradients ∇F (with elements $\nabla F_{\alpha\beta}$) and higher field derivatives, such as $\nabla^2 F$ (with elements $\nabla^2 F_{\alpha\beta\gamma}$), etc. It is possible to compute these quantities with the aid of PT techniques, but variational procedures are more powerful and more exact.

The polarizability tensors are defined as the coefficients of the expansion of one molecular property (the energy, the dipole, etc.) with respect to an external field and its derivatives.

Among these response properties the most known are the dipole polarizability tensors $\alpha, \beta$, and $\gamma$, that give the first three components of the expansion of a dipole moment $\mu$ subjected to an external homogeneous field F:

$$\mu_\alpha = -\frac{\partial E}{\partial F_\alpha} = \mu_\alpha^{(0)} + \alpha_{\alpha\beta} F_\beta + \beta_{\alpha\beta\gamma} F_\beta F_\gamma + \gamma_{\alpha\beta\gamma} F_\beta F_\gamma F_\delta + \ldots \qquad [8.53]$$

the convention of summation over the repeated indexes (which are the Cartesian coordinates) is applied here.

Other polarizabilities that may have practical importance are the quadrupole polarizability tensors A, B, etc. whose elements can be defined in terms of the expansion of a molecular quadrupole θ subjected to a uniform electric field:

$$\theta_{\alpha\beta} = -3\frac{\partial E}{\partial F_{\alpha\beta}} = \theta_{\alpha\beta}^{(0)} + A_{\gamma,\alpha\beta}F_{\gamma} + \frac{1}{2}B_{\gamma\delta,\alpha\beta}F_{\gamma}F_{\delta} + \ldots \qquad [8.54]$$

The A, B, ... terms are also present in the expansion series of the dipole when the molecule is subjected to higher derivatives of F. For example, the tensor A determines both the quadrupole induced by a uniform field and the dipole induced by a field gradient.

The electrical influence of a molecule on a second molecule cannot be reduced to a constant field F or to the combination of it with a gradient ∇F. So an appropriate handling of the formal expansion [8.52] may turn out to be a delicate task. In addition, the formal analysis of convergence of the series gives negative answers: there are only demonstrations that in special simple cases there is divergence. It is advisable not to push the higher limit of truncated expansions much with the hope of making the result better.

The use of several expansion centers surely improves the situation. The averaged value of dipole first polarizability α can be satisfactorily expressed as a sum of transferable group contributions, but the calculations of IND with distributed polarizabilities are rather unstable, probably divergent, and it is convenient to limit the calculation to the first term alone, as is actually done in almost all the practical implementations. Instability and lack of convergence are factors suggesting an accurate examination of the function to fit the opportune values for these distributed α polarizabilities.

### The dispersion term

The dispersion term DIS can be formally treated as IND. Asymptotically (at large R), DIS can be expressed in terms of the dynamic multipole polarizabilities of the monomers.

$$DIS \approx -\frac{1}{2\pi}\sum_{l,l'=1}^{\infty}\sum_{k,k'=1}^{\infty} R^{-(l+l'+k+k'+2)} \sum_{m=-l<}^{l<}\sum_{m'=-k<}^{k<} C_{ll'}^{m} C_{kk'}^{m'} \times \int_{0}^{\infty} \pi_{lk}^{A}(m,m';i\omega)\pi_{l'k'}^{B}(-m,-m';i\omega)d\omega \qquad [8.55]$$

The dynamic polarizabilities are similar to the static ones, but they are dependent on the frequency of the applied field, and they have to be computed at the imaginary frequency iω

Here again the standard PT techniques are not accurate enough, and one has to employ others techniques. It is worth remarking that at short or intermediate distances the dynamic multipole polarizabilities give an appreciation of DIS of poor quality, unless computed on the dimer basis set. This is a consequence of the BSS errors: the PT theory has thus to abandon the objective of computing everything solely relying on monomers properties.

At the variational level there are no formal problems to use the dynamic polarizabilities, for which there are efficient variational procedures.

The expansion is often truncated to the first term (dipole contributions only), giving origin to a single term with distance dependence equal to $R^{-6}$. From eq. [8.49] it turns out that all the members of this expansion have an even negative dependence on R. For accurate studies, especially for simple systems in the gas phase, the next terms, $C_8R^{-8}$, $C_{10}R^{-10}$, ... are often considered. To have an odd term, one has to consider third order elements in the PT theory, which generally are rarely used.

Limitations due to the non-convergence of these expansion series are reduced by the use of multicenter expansions: in such a way it is possible to reach a satisfactory representation even for polyatomic molecules at short distances using the first expansion term only.

In general, the expansion sites are the (heavy) atoms of the dimer, using a simpler formulation due to London. The London formulation uses the averaged (isotropic) value of the static dipole polarizabilities, $\overline{\alpha}_j$, of the atoms and their first ionization potentials $I_i$ and $I_j$:

$$DIS(6) \approx C_6(AB)R^{-6} = \sum_i^A \sum_j^B C_6(i,j)R_{ij}^{-6} \qquad [8.56]$$

with

$$C_6(i,j) = \frac{3}{2}\frac{\langle\alpha_i\rangle\langle\alpha_j\rangle I_i I_j}{I_i + I_j} \qquad [8.57]$$

This formula derives from the standard RS-PT with the expansion over the excited state truncated at the first term. For isolated atoms i and j, the coefficients $C_6(i,j)$ can be drawn from experimental data, while for atomic fragments of molecules only from computations.

**The exchange (or repulsion) term**

The exchange terms are related to the introduction of the intermonomer antisymmetrizer $\mathbf{A}_{AB}$ in the expression giving the electrostatic contribution. The correction factors introduced in the SAPT methods[15] may be related to a renormalization factor that may be written in the following form when $\mathbf{A}_{AB}$ is replaced with 1+P for simplicity:

$$\frac{1}{\langle\Phi_0|P\Phi_0\rangle} = \frac{1}{\langle\Psi_A\Psi_B|P\Psi_A\Psi_B\rangle} = \frac{1}{(1+\langle P\rangle)} \qquad [8.58]$$

This factor is different from 1, because <P> contains all the multiple overlap values between MOs of A and B. <P> may be then expanded into terms containing an increasing number of multiple overlaps (or exchanges of orbitals):

$$\langle P\rangle = \langle P_1\rangle + \langle P_2\rangle + \langle P_3\rangle + \langle P_4\rangle + \ldots \qquad [8.59]$$

The exchange energy may be expanded into increasing powers of MO overlap integrals S:

$$EX = EX(S_2) + EX(S_3) + EX(S_4) + \ldots \qquad [8.60]$$

with

$$EX(S_2) = \langle VP_2\rangle - \langle P_2\rangle \qquad [8.61]$$

This first term is sufficient for modeling intermolecular potentials. The overlaps depend on the monomer separation roughly as exp(-αR) so the following terms give small contribution at large-medium distances. At short distances the positive exchange term rapidly increases. Reasons of uniformity with the other terms of the interaction potential suggest replacing the exp(-αR) dependence with a highly negative power of R: $R^{-12}$ in most cases. The exchange terms describe what is often called the steric repulsion between molecules (actually there are other short range repulsive contributions, as the electrostatic penetration components not described by the multipole expansion).

### The other terms

Charge transfer contributions (CT) have been rarely introduced in the modeling of interaction potentials for liquids. Some attempts at modeling have been done for the study of interactions leading to chemical reactions, but in such cases, direct calculations of the interaction term are usually employed.

For curious readers, we add that the tentative modeling was based on an alternative formulation of the PT (rarely used for complete studies on the intermolecular interaction, because it presents several problems) in which the promotion of electrons of a partner on the virtual orbital of the second is admitted. The formal expressions are similar to those shown in eq. [8.40]. For the second order contributions we have

$$-\sum_K \frac{\left|\left\langle \Psi_0^a \Psi_0^b | V | \Psi_K^a \Psi_0^b \right\rangle\right|^2}{E_K^a - E_0^a}$$

The sum over K is strongly reduced (often to just one term, the HOMO-LUMO interaction), with K corresponding to the replacement of the occupied MO $\phi_r^0$ of A with the virtual MO $\phi_t^v$ of B (A is the donor, B the acceptor). The expression is then simplified: the numerator is reduced to a combination of two-electron Coulomb integrals multiplied by the opportune overlap. The CT contribution rapidly decays with increasing R.

COUP contributions, obtained as a remainder in the variational decomposition of ΔE, are not modeled. The contributions are small and of short-range character.

### A conclusive view

In this long analysis of the modeling of the separate components we have learned the following:

- All terms of the decomposition may be partitioned into short- and long-range contributions.
- The long-range contributions present problems of convergence, but these problems may be reduced by resorting to multicenter distributions based on suitable partitions of the molecular systems.
- The short-range contributions are in general of repulsive character and are dominated by the EX terms. The effect of the other short-range contributions is strongly reduced when multicenter expansions are used.

Adapting these remarks, one may write a tentative analytical expression for ΔE(R):

$$\Delta E \approx \sum_r \sum_t \sum_n C_{rt}^{(n)} R_{rt}^{-n} \qquad [8.62]$$

The expansion centers (called "sites") are indicated by the indexes r and t for the molecule A and B respectively. The index n indicates the behavior of the specific term with respect to the intersite distance $R_{rt}$; each couple r-t of sites has a specific set of allowed n values.

In such a way it is possible to combine sites corresponding only to a point charge with sites having more terms in the expansion. For two sites r-t both described by a point charge the only interaction term is $C_{rt}^{(1)}R_{rt}^{-1}$, while for a couple r-t described by a point charge and a dipole, the contribution to the interaction is limited to the $R^{-2}$ term, and, if it is described by two dipoles, to the $R^{-3}$ term. The last example shows that eq. [8.62] is cryptic or not well developed. In fact, if we examine eq. [8.12], which reports the interaction energy between two dipoles, it turns out that the dependence on the orientation angle $\theta$ apparently is missing in eq. [8.62]. The coefficients $C_{rt}^{(n)}$ must be specified in more detail as given in equations [8.50] and [8.52] by adding the opportune l, l' and m indexes (or the corresponding combination of Cartesian coordinates), or must be reduced to the isotropic form. This means to replace the three components of a dipole, and the five components of a quadrupole, and so on, by an average over all the orientations (i.e., over the m values). There is a loss of accuracy, very large in the case of two dipoles, and of decreasing importance in passing to higher multipoles. The loss in accuracy is greatly decreased when a multi-site development is employed. Many analytical expressions of interaction potentials use this choice. In such cases the $C_{rt}^{(n)}$ coefficients are just numbers.

The coefficients $C_{rt}^{(n)}$ can be drawn from experimental values (but this is limited to the cases in which there is one site only for both A and B), from ad hoc calculations, or by a fitting of $\Delta E$ values. In the past, large use has been made of "experimental" $\Delta E$ values, derived, e.g., from crystal packing energies, but now the main sources are the variationally computed QM values, followed by a fit.

The use of expression [8.62] for the fitting of QM values represents a remarkable improvement with respect to the past strategy of reaching a good fitting with complete freedom of the analytical form of the expression. There are in the literature interaction potentials using, e.g., non-integer values of n and/or other analytical expressions without physical meaning. This strategy leads to potentials that cannot be extended to similar systems, and that cannot be compared with the potential obtained by others for the same system.

Expression [8.62] has several merits, but also several defects. Among the latter we note that given powers of R may collect terms of different origin. For example, the term $R^{-6}$ describes both the induction dipole-induced dipole interaction and the dispersion dipole-dipole interaction: two contributions with a different sensitivity with respect to changes in the molecular system. Theory and computational methods both permit getting two separate $C_{rt}^{(6)}$ coefficients for the two contributions. This is done in a limited number of potentials (the most important cases are the NEMOn[26] and the ASP-Wn[27] families of models). The most popular choice of a unique term for contributions of different origin is motivated by the need of reducing computational efforts, both in the derivation and in the use of the potential. To fit a unique $C_{rt}^{(6)}$ coefficient means to use $\Delta E$ values only: for two coefficients there is the need of separately fitting IND and DIS contributions.

## 8.4.5 THE RELAXATION OF THE RIGID MONOMER CONSTRAINT

We have so far examined decompositions of two-body interaction potentials, keeping fixed the internal geometry of both partners. This constraint is clearly unphysical, and does not correspond to the naïve model we have considered in the introduction, because molecules always exhibit internal motions, even when isolated, and because molecular collisions in a liquid (as well as in a cluster) lead to exchanges of energy between internal as well as external degrees of freedom.

It may be convenient to introduce a loose classification of the origin of changes in the internal geometry of molecules in clusters and liquids, and to use it for the dimer case we are considering here:

1) permanent or semipermanent molecular interactions;
2) internal dynamism of the molecule;
3) molecular collisions.

In the variety of liquid systems there are quite abundant cases in which relatively strong interactions among partners induce changes in the internal geometry. The formation of hydrogen-bonded adducts is an example of general occurrence (e.g., in water solutions). The interactions of a metal cations and their first solvation shells is a second outstanding example but the variety of cases is very large. It is not easy to give general rules on the distinction between permanent and semipermanent interactions of this type. In general it depends on the time scale of the phenomenon being studied, but it is clear that the long residence time of water molecule (in some cases, on the order of years[28]) in the first solvation shell around a cation leads us to consider this effect as permanent.

In such cases, it is convenient to reconsider the definition of the cluster expansion and to introduce some extra variables in the nuclear coordinate subspace to span for the analysis. For example, in the case of a dimer $M^{n+}\cdot H_2O$, it is convenient to add three coordinates corresponding to the internal coordinates of the water molecule (the total number of degrees of freedom in such case is again 6, because of the spherical symmetry of the metal cation). There is no need of repeating, for this model, the variational decomposition of the energy (the PT approaches have more difficulties to treat changes in the internal geometry). The conclusions do not change qualitatively, but the quantitative results can be sensibly modified.

Another important case of permanent interactions is related to chemical equilibria with molecular components of the liquid. The outstanding example is the prototropic equilibrium, especially the case $AH + B \rightarrow A^- + HB^+$. There are water-water potentials, including the possibility of describing the ionic dissociation of $H_2O$.

The semipermanent interactions are generally neglected in the modeling of the potentials for liquids. Things are different when one looks at problems requiring a detailed local description of the interactions. Molecular docking problems are a typical example. In such cases, use is made of variational QM calculations, or especially for docking, where one molecule at least has a large size, use is made of molecular mechanics (MM) algorithms, allowing local modification of the systems.

The second category corresponds to molecular vibrations; rotations of a molecule as a whole have been already considered in, or definition of, the $\Delta E(R)$ potential. The third category, molecular collisions, gives rise to exchange of energy among molecules that can be expressed as changes in the translational and rotational energies of the rigid molecule and changes in the internal vibrational energy distributions. In conclusion, all the cases of non-rigidity we have to consider here can be limited to molecular vibrations considered in a broad sense (couplings among vibrations and rotations are generally neglected in intermolecular potentials for liquids).

The vibrations are generally treated at the classical level, in terms of local deformation coordinates. The local deformation functions are of the same type of those used in molecular mechanics (MM) methods. Nowadays, MM treats geometry changes for molecules of any dimension and chemical composition.

In the MM approach, the energy of the system is decomposed as follows:

$$E_{tot} = [E_{str} + E_{ben} + E_{tor} + E_{other}] + E_{elec} + E_{vdW} \qquad [8.63]$$

the terms within square brackets regard contributions due to the bond stretching, to the angle bending, and to torsional interactions, supplemented by other contributions due to more specific deformations and by couplings among different internal coordinates. The $E_{elec}$ terms regard Coulomb and inductions terms between fragments not chemically bound: among them there are the interactions between solute and solvent molecules. The same holds for the van der Waals interactions (i.e., repulsion and dispersion) collected in the last term of the equation.

It must be remarked that $E_{tot}$ is actually a difference of energy with respect to a statc of the system in which the internal geometry of each bonded component of the system is at equilibrium.

**Table 8.4**

| | |
|---|---|
| STRETCH | $E_{str} = k_s(1 - l_0)^2$ |
| BEND | $E_{ben} = k_b(\theta - \theta_0)^2$ |
| TORSION | $E_{tor} = k_t[1 \pm \cos(n\omega)]$ |
| $k_s$ | stretch force constant |
| $k_b$ | bend force constant |
| $k_t$ | torsion force constant |
| $l_0$ | reference bond length |
| $\theta_0$ | reference bond angle |

The versions of MM potentials (generally called force fields) for solvent molecules may be simpler, and in fact limited librarics of MM parameters are used to describe internal geometry change effects in liquid systems. We report in Table 8.4 the definitions of the basic, and simpler, expressions used for liquids.

To describe internal geometry effects in the dimeric interaction, these MM parameters are in general used in combination with a partitioning of the interaction potential into atomic sites. The coupling terms are neglected and the numerical values of the parameters of the site (e.g., the local charges) are left unchanged when there is a change of internal geometry produced by these local deformations. Only the relative position of the sites changes. We stress that interaction potential only regards interaction among sites of different molecules, while the local deformation affects the internal energy of the molecule.

## 8.5 THREE- AND MANY-BODY INTERACTIONS

The tree-body component of the interaction energy of a trimer ABC is defined as:

$$\Delta E(ABC;R_{ABC}) = E_{ABC}(R_{ABC}) - \frac{1}{2}\sum_A \sum_B E_{AB}(R_{AB}) - \sum_K E_K \qquad [8.64]$$

This function may be computed, point-by-point, over the appropriate $R_{ABC}$ space, either with variational and PT methods, as the dimeric interactions. The results are again affected by BSS errors, and they can be corrected with the appropriate extension of the CP procedure. A complete span of the surface is, of course, by far more demanding than for a dimer, and actually extensive scans of the decomposition of the $E_{ABC}$ potential energy surface have thus far been done for a very limited number of systems.

Analogous remarks hold for the four-body component $\Delta E(ABCD; R_{ABCD})$ of the cluster expansion energy, as well as for the five- and six- body components, the definition of which can be easily extracted from the eq. [8.9]. The computational task becomes more and more exacting in increasing the number of bodies, as well as the dimensions which rise to 18 with fixed geometries of the components, and there are no reasonable perspectives of having in the future such a systematic scan as that available for dimers.

The studies have been so far centered on four types of systems that represent four different (and typical) situations:

1) clusters composed of rare gas atoms or by small almost spherical molecules (e.g., methane);
2) clusters composed of polar and protic molecules (especially water);
3) clusters composed of a single non-polar molecule (e.g., an alkane) and a number of water molecules;
4) clusters composed of a single charged species (typically atomic ions) and a number of water molecules.

Inside the four categories, attention has mostly been paid to clusters with a uniform chemical composition (e.g., trimers AAA more than $A_2B$ or ABC for cases 1 and 2).

Clearly the types of many-body analyses are not sufficient to cover all cases of chemical relevance for liquids; for example, the mixed polar solvents (belonging to type 2) and the water solutions containing a cation in presence of some ligand (type 4) are poorly considered.

In spite of these deficiencies the available data are sufficient to draw some general trends that in general confirm what physical intuition suggests. We combine here below conclusions drawn from the formal analysis and from the examination of numerical cases.

ES contributions are strictly additive: there are no three- or many body terms for ES. The term "many-body correction" to ES introduced in some reviews actually regards two other effects. The first is the electron correlation effects which come out when the starting point is the HF description of the monomer. We have already considered this topic that does not belong, strictly speaking, to the many-body effects related to the cluster expansion [8.9]. The second regards a screening effect that we shall discuss later.

The other contributions are all non-additive. The cluster expansion [8.9] applies to these terms too:

$$TERM(ABCDE\ldots) = \frac{1}{2}\sum_A\sum_B TERM_{AB}(R_{AB}) + \frac{1}{6}\sum_A\sum_B\sum_C TERM_{ABC}(R_{ABC}) + \ldots \qquad [8.65]$$

The most important are the IND and DIS terms. CT is active for some special systems (long range electron transfer) but of scarce importance for normal liquids. EX many-body contributions rapidly fade away with the number of bodies: the three-body terms have been studied with some details,[29] but they are not yet used extensively to model interaction potentials for liquids.

The IND many-body contributions are dominated by the first dipole polarizability $\alpha$ contributions. The field F produced by the charge distribution of the other molecules and acting on the units for which $\alpha$ is defined, at a first (and good) approximation, as additive. The non-linearity derives from the fact that the contribution to the interaction energy is related to the square of the electric field. So in modeling many body effects for IND the attention is limited to molecular contributions of the type

$$E_{ind}(M) = -\frac{1}{2}\sum_{m=1}^{M} \mu_m^{ind} F(r_m) \quad [8.66]$$

the index m runs over all the sites of the molecule M. The dipole moment of site m is decomposed into a static and an induction dipole:

$$\mu_m = \mu_m^0 + \mu_m^{ind} \quad [8.67]$$

with the induction term $\alpha_m$ depending on the polarizability and on the total electric field $F^{tot}$. The latter comes from the other permanent charges and dipoles, as well as from the induced dipoles present in the system:

$$\mu_m^{ind} = \alpha_m F^{tot} = \alpha_m \sum_{s=1}^{S} F_s(r_m) \quad [8.68]$$

Generally, the calculation of F is not limited to three-body contributions but includes higher order terms; for example, in simulation methods the contributions coming from all the molecules included in the simulation box, which may be of the order of hundreds. These contributions are not computed separately, but just used to have a collective value of F.

The sum of three- and higher body contributions to IND thus may be written and computed in the following way:

$$IND(many-body) \approx \sum_m E_{ind}\left(m; F^{tot}\right) \quad [8.69]$$

When the system contains molecules of large size it is advisable to use a many-site description of the molecular polarizability $\alpha$. This of course means to increase the computational effort, which is not negligible, especially in computer simulations.

The many-body contribution to IND may be negative as well as positive. In general, it reinforces with a negative contribution minima on the PES.

The non-additive DIS contributions are described in terms of the first dipole dynamic polarizability or by the corresponding approximate expression we have introduced for the dimeric case.

DIS contributions are already present and are quite important for the liquid aggregation of spherical systems. Even spherical monomers exhibit dynamical dipoles and the related two- and many-body contributions to DIS.

The most important term is known as the Axilrod-Teller term and describes the interaction of three instantaneous dipoles. The sign of the Axilrod-Teller term strongly depends on the geometry. It is negative for almost linear geometries and positive in triangular arrangements. Four body contributions of the same nature are generally of opposite sign and thus they reduce the effect of these terms on the PES shape.

The potentials for polar liquids generally neglect many-body dispersion contributions.

### Screening many-body effects

When the material condensed system contains mobile charges, the electrostatic interactions, which are strong and with the slowest decay with the distance, are severely damped. Every charged component of the system tends to attract mobile components bearing the opposite charge. Typical examples are the counterion charge cloud surrounding each ion in salt solutions, and the counterion condensation effects acting on immobile charged species as DNA or other polyelectrolytes in salt solutions. In both cases, at large distance the electrostatic effect of the charge is screened by the similar electrostatic effect, but with an opposite sign, of the counterion distribution.

A similar screening effect is present for dipolar liquids: each molecule bearing a permanent dipole tends to organize around itself other dipolar molecules with the orientation which optimizes stabilization energy and screens the field produced by the singled out molecule.

The organization of the orientation of molecular dipoles around a singled-out charged component is even more effective. The interaction at long distance between two point charges screened by the solvent dipoles is $Q_1Q_2/\varepsilon R_{12}$ where $\varepsilon$ is the dielectric constant of the liquid. When the two charges are placed in a salt solution the value of the electrostatic energy is even lower, being screened by counterions as well as by oriented dipoles.

For the dispersion contributions there is no screening. We arrive to the apparently paradoxical situation that for relatively large bodies, even when bearing net charges, the interactions at large distances are governed by the dispersion forces, weak in comparison with others, and having a more rapid decrease with the distance.

### Effective interaction potentials

In the last sentences we have changed view with respect to the beginning of Section 8.4, where we introduced the cluster expansion of the interaction potentials. Still keeping a microscopic view, we passed from the examination of the properties of PES of increasing dimensions to a view that takes into consideration very limited portions of such surfaces selected on the basis of energetic effects, or, in other words, on the basis of probabilistic considerations. The screening effects we have mentioned are due to an averaged distribution of the other components around a single partner of the system, and its average is based on the relative energies of all the conformations of the system, spanning the whole pertinent PES.

This view based on averaged distributions is related to the "focused" model we introduced in Section 8.2.

If we suppose that the assisting part S of the system described by the Hamiltonian [8.8] with coordinates $X_S$ is in equilibrium with the main part M at every coordinate $X_M$, we may repeat the cluster decomposition analyses given in Section 8.3, starting again from the dimer and proceeding to the trimers, etc. The potential function $\Delta E_{AB}(R)$ we have called potential energy surface (PES) now assumes a different name, potential of mean force (MFP), emphasizing the fact that it regards no more interactions between A and B but also the mean effects of the other bodies present in the system. It must be remarked that the approach is not limited to equilibrium distributions of S: it may be extended to distributions of S displaced from the equilibrium. The non-equilibrium description has an important role, and the implementation of non-equilibrium methods today represents one of the frontiers of the theory of liquids.

An important aspect of this revisited analysis of the dissection of $\Delta E_{AB}$ into separate components is that all terms, including ES, are now described in terms of monomer distributions which feel the effects of S. The analyses we have done suggest that these changes can be expressed in terms of a monomer's MO basis $\{\phi_i^p\}$ "polarized" in a way similar to that we have used to describe IND. Actually, this "polarization" or deformation of the monomer is not due to classical polarization effects only, but to all the components of the interaction of S with A-B, namely dispersion, exchange, etc.

The results of the decomposition of the intermolecular interaction energy of cluster M are not qualitatively different from those found for the same cluster M in vacuo. There are quantitative changes, often of not negligible entity and that correspond, in general, to a damping of the effects.

The use of this approach requires us to consider the availability of efficient and accurate models and procedures to evaluate the effects of S on M, and at the same time, of M on S. The basic premises of this approach have been laid many years ago, essentially with the introduction of the concept of solvent reaction field made by Onsager in 1936,[30] but only recently have they been satisfactory formulated. There are good reasons to expect that their use on the formulation of "effective" intermolecular potentials will increase in the next few years.

Actually, the decomposition of the $\Delta E_M(R)$ interaction energy is preceded by a step not present for isolated systems.

There is the need of defining and analyzing the interactions of each monomer A, B, etc. with the solvent alone. The focused model is reduced to a single molecule alone, let us say A, which may be a solute molecule but also a component of the dominant mole fraction, the solvent, or a molecular component of the pure liquid.

This subject is treated in detail in other chapters of this book and something more will be added in the remainder of this chapter. Here, we shall be concise.

There are two main approaches, the first based on discrete (i.e., molecular) descriptions of S, the second on a continuous description of the assisting portion of the medium, via appropriate integral equations based on the density distribution of the medium and on appropriate integral kernels describing the various interactions (classical electrostatic, dispersion, exchange-repulsion). Both approaches aim at an equilibration of S with M: there is the need of repeating calculations until the desired convergence is reached. More details on both approaches will be given in the specific sections; here we shall limit ourselves to quoting some aspects related to interaction potentials, which constitute the main topic of this section.

The approach based on discrete descriptions of the solvent makes explicit use of the molecular interaction potentials which may be those defined without consideration of S; the calculations are rather demanding of computation time, and this is the main reason explaining why much effort has been spent to have simple analytical expressions of such potentials.

The second continuous approach is often called Effective Hamiltonian Approach (EHA) and more recently Implicit Solvation Method (ISM).[31] It is based on the use of continuous response functions not requiring explicit solvent-solvent and solute-solvent interaction potentials, and it is by far less computer-demanding than the simulation approach.

As we shall show in the section devoted to these methods, there are several possible versions; here it is worth quoting the Polarizable Continuum Model (PCM) for calculations at the ab initio QM level.[32]

PCM it is the only method used so far to get effective two body A-B potentials over the whole range of distances R (three-body corrections have been introduced in Ref. [33]).

Until now, the EHA approach has been mostly used to study solvent effects on a solute molecule. In such studies, M is composed of just a single solute molecule (M = A) or of a solute molecule accompanied by a few solvent molecules ($M = AS_n$). These alternative choices make the results indirectly relevant to the question of the difference between traditional and effective intermolecular potential energies. However, some hints can be extracted from the available data. The experience thus far collected shows that when both M and S have a polar nature, the classical electrostatic contributions are dominant with an exchange-repulsion term, giving an important contribution to the energy but a small effect on the charge distribution of M and, as a consequence, on the molecular properties on which the interaction potential depends (e.g., multipole distribution of the charge density, static and dynamic polarizabilities). Dispersion effects play a role when M is not polar and they become the dominant component of the solvation effects when both M and S are not polar.

The calculation of effective PCM interaction potentials has been so far limited to the cases in which the solvent effects are more sizeable, namely the interactions of metal cations with water. They have been used in computer simulations of models of very dilute ionic solutions (a single ion in water), noticeably improving the results with respect to similar studies using traditional potentials.[34]

## 8.6 THE VARIETY OF INTERACTION POTENTIALS

In the preceding sections we have examined the definition of the components of the intermolecular potential, paying attention mostly to rigorous methods and to some approximations based on these methods.

The beginning of qualitative and semi-quantitative computational studies on the properties of liquids dates back at least to the 1950s, and in the early years of these studies, the computational resources were not so powerful as they are today. For this reason, for many years, the potentials used have been simple, discarding what our understanding of the dimeric interactions suggested. There is no reason, however, to neglect here these simplified versions of the potential, because very simple and naïve expressions have given excellent results and are still in use. The struggle for better and more detailed potentials is justified, of course, because chemistry always tends to have more and more accurate estimates of the properties of interest, and there are many problems in which a semi-quantitative assessment is not sufficient.

We shall try to give a cursory view of the variety of intermolecular potentials in use for liquid systems, paying more attention to the simplest ones. We shall almost completely neglect potentials used in other fields, such as the scattering of two isolated molecules, the determination of spectroscopic properties of dimers and trimers, and the accurate study of local chemical interactions, which all require more sophisticated potentials.

The main criterion to judge the quality of a potential is *a posteriori*, namely, based on the examination of the performance of that potential in describing properties of the liquid system. In fact, the performances of a potential are to a good extent based on subtle factors expressing the equilibrium reached in the definition of its various components. The experience gained in using this criterion has unfortunately shown that there is no definite answer. Typically, a potential, or a family of potentials, better describe some properties of the liquid, while a second one is more proper for other properties. For this reason there are many potentials in use for the same liquid and there are in the literature continuous comparisons among different potentials. A second criterion is based on the computational cost, an aspect of direct interest for people planning to make numerical studies but of little interest for people only interested in the results.

Our exposition will not try to classify potentials according to such criteria, but simply show the variety of potentials in use. We shall pay attention to the description adopted for a single molecule, that must then be combined with that adopted for the interaction partners. In particular, we shall consider the number of sites and the shape of the molecule used in each description.

The number of sites reflects the possibility we have examined, and advocated, of using many-center expansions to improve the representation. Each expansion center will be a site. There are models with one, two, and more sites. This sequence of increasing complexity reaches the number of heavy atoms of the molecule and then the whole number of atoms, including hydrogens. It is not limited to the nuclei as expansion sites. There are potentials introducing other locations of sites, in substitution or in addition to the nuclei. For example potentials widely used in simulations adopt for water a four-site model; other potentials (rarely used in simulations) prefer to use the middle of the bonds instead of (or in addition to) nuclei. Each site of the molecule must be combined with the sites of the second (and other) molecule to give the potential.

The shape of the molecule reflects the effect of the exchange-repulsion interaction. For almost all many-site models the shape is not given, but it implicitly results to be that of the union of the spheres centered on the expansion sites provided by a source of exchange-repulsion potential. There are some simple models in which the shape is explicitly stated. There will be spheres, ellipsoids, cylinders and more complex shapes, as fused spheres, spherocylinders, etc. Some typical examples are reported in Table 8.5.

**Table 8.5. Single site-based potentials**

| # | Shape | Interaction | Name |
|---|---|---|---|
| 1 | Sphere | hard | HS |
| 2 | Sphere | soft | SS |
| 3 | Sphere | hard with charge | CHS |
| 4 | Sphere | soft with charge | CSS |
| 5 | Sphere | hard with rigid dipole | DHS |
| 6 | Sphere | soft with dipole | DSS |
| 7 | Disc | hard | HD |
| 8 | Disc | hard with dipole | SD |
| 9 | Ellipsoid | hard | |
| 10 | Ellipsoid | | Gay-Berne[35] |
| 11 | Quadrupolar shape | | Zewdie[36] |
| 12 | Sphere | repulsion dispersion | Lennard Jones (LJ) |
| 13 | Sphere | repulsion dispersion | Buckingham |
| 14 | Sphere | LJ + charge | LJ+q |
| 15 | Sphere | LJ + dipole | Stockmayer |
| 16 | Sphere | LJ + dipole+ soft sticky | SSD[37] |

Potential 1 is extremely simple: the only parameter is the radius of the sphere. In spite of this simplicity, an impressive number of physical results have been obtained using the HS potential on the whole range of densities and aggregations. Also, mixtures of liquids have been successfully treated, introducing in the computational machinery the desired number of spheres with appropriate radii. The HS model is at the basis of the Scaled Particle Theory (SPT),[38] which still constitutes a basic element of modern solvation methods (see later for more details).

Of course, HS cannot give many details. A step toward realism is given by potential 2, in which the hard potential is replaced by a steep but smoothly repulsive potential. It is worth reminding readers that the shift from HS to SS in computer simulations required the availability of a new generation of computers. Simulations had in the past, and still have at present, to face the problem that an increase in the complexity of the potential leads to a large increment of the computational demand.

Potentials 3 and 4 have been introduced to study ionic solutions and similar fluids containing mobile electric charges (as, for example, molten salts). In ionic solutions the appropriate mixture of charged and uncharged spheres is used. These potentials are the first examples of potentials in our list, in which the two-body characteristics become explicit. The interaction between two charged spheres is in fact described in terms of the charge values of a couple of spheres: $Q_kQ_l/r_{lk}$ (or $Q_kQ_l/\varepsilon r_{lk}$).

Potentials 5 and 6 are the first examples in our lists of anisotropic potentials. The interaction here depends on the mutual orientations of the two dipoles. Other versions of the dipole-into-a-sphere potentials (not reported in Table 8.5) include induced dipoles and actually belong to a higher level of complexity in the models, because for their use there is the need of an iterative loop to fix the local value of F. There are also other similar models simulating liquids in which the location of the dipoles is held fixed at nodes of a regular 3D grid, the hard sphere potential is discarded, and the optimization only regard orientation and strength of the local dipoles. This last type of model is used in combination with solutes M described in another, more detailed way.

Potentials 7-11 add more realism in the description of molecular shape. Among them, the Gay-Berne[35] potential (10) has gained a large popularity in the description of liquid crystals. The quite recent potential (11)[36] aims at replacing Gay-Berne potential. It has been added here to show that even in the field of simple potentials there is space for innovation.

Potentials 12 and 13 are very important for chemical studies on liquids and solutions.

The Lennard-Jones potential (12) includes dispersion and repulsion interactions in the form:

LJ: $$E^{LJ}_{AB} = \varepsilon\left[\left(\frac{\sigma}{R}\right)^{12} - \left(\frac{\sigma}{R}\right)^{6}\right] \qquad [8.70]$$

The dependence of the potential on the couple of molecules is here explicit, because the parameters ε and σ depend on the couple. In the original version, ε is defined as the depth of the attractive well and σ as the distance at which the steep repulsive wall begins.

More recently, the LJ expression has been adopted as an analytical template to fit numerical values of the interaction, using independently the two parameters.:

LJ with 2 parameters: $$E^{LJ}_{AB}(R) = A_{AB}\left(\frac{1}{R}\right)^{12} - C_{AB}\left(\frac{1}{R}\right)^{6} \quad [8.71]$$

LJ expressions are of extensive use in MM to model non-covalent interactions among molecular sites, often with the addition of local charges (potential 14)

LJ with charge: $$E_{ms}(r_{ms}) = A_{ms}\left(\frac{1}{r_{ms}}\right)^{12} - C_{ms}\left(\frac{1}{r_{ms}}\right)^{6} + \frac{q_m q_s}{r_{ms}} \quad [8.72]$$

Some potentials add an effective dielectric constant in the denominator of the Coulomb term to mimic polarization effects.

The Buckhingam potential is similar to LJ with a more physical description of the repulsion term given in terms of a decaying exponential function:

$$E^{(BU)}_{AB} = \varepsilon\left\{\frac{6}{\alpha - 6}\exp\left[-\alpha\left(\frac{r}{r^*}\right) - 1\right] - \frac{\alpha}{\alpha - 6}\left(\frac{r}{r^*}\right)^{-6}\right\} \quad [8.73]$$

where:

| | |
|---|---|
| $\varepsilon$ | depth of the attractive well |
| r | $R_{AB}$ |
| r* | distance of the minimum |
| $\alpha$ | numerical parameter |

The difference in the computational costs of LJ and Buckingham potentials is related to the difference between computing the square of a value already available ($R^{-12}$ from $R^{-6}$) and that of computing an exponential. Here again the increment in computational costs of simulations due to small changes in the potential plays a significant role. Also, the Buckingham + charge potential is used to describe liquids.

The last potential of Table 8.5 is specialized for water. It has been enclosed in the table to document the progress in one-site potentials; in this case, the LJ + dipole is supplemented by a short-range "sticky" tetrahedral interaction.

The listing of one-center potentials is not exhausted by the examples given in Table 8.5. We have, for example, neglected all the potentials in use for rare gases systems in which much attention is paid to using higher terms to describe the dispersion contribution. Our aim was just to show with a few examples how it is possible to define a large variety of potentials remaining within the constraints of using a single center.

The problem of giving a cursory but significant enough view of potentials becomes harder when one passes to many-site potentials. In Table 8.6 we report some analogues of HS with a more complex shape.

They consist of fused regular forms (spheres or combination of spheres with other regular solids). The hard version of these potentials is accompanied by soft modifications and by versions including dipole, charges, as in Table 8.5.

A step further along this way is given by a potential composed by spheres linked by 'spacers' with a constant length but allowing changes of conformation with appropriate torsion potentials. These potentials are used for polymers or for molecules having long hydrocarbon chains. This is not, however, the main trend in the evolution of potentials.

**Table 8.6. Many-site simple potentials**

| # | Shape | Interaction | Name |
|---|---|---|---|
| 1 | three fused spheres | hard | hard dumbbell HD |
| 2 | two half spheres + cylinder | hard | hard spherocylinder HSC |
| 3 | general convex shape | hard | hard convex core HCC |

The largest number of solvents are composed by polyatomic molecules of small-medium size, exhibiting a variable (but in general not excessive) degree of flexibility. In solutions, the interactions involve these solvents and molecules having the same characteristics as solvent or with more complex chemical composition. In all cases (pure liquids, solutions with solute of variable complexity), one has to take into account interactions having a remarkable degree of specificity. The chemical approach to the problem addresses this specificity on which the whole chemistry is based. This is the real field of application of the definition and analysis of molecular interactions on which we have spent the first sections of this chapter.

It is clear that we are entering here into a very complex realm, hard to summarize. A whole book would be necessary.

It is possible to make a rough distinction between potential of general applicability and potentials conceived for specific couples (or collections) of molecules.

In both cases, the many-site expansions are used, but for general potentials, in which transferability is asked, the description is obviously less detailed. The LJ+charge (potential 14 in Table 8.5) is a popular choice for these potentials; the sites are in general limited to the heavy atoms of the molecule. There are, however, many other versions with a variety of changes to this standard setting.

The potentials elaborated for specific cases are often more detailed and they adapt a larger number of devices to increase the accuracy. Many have been developed for pure liquids, but the number of potentials regarding specific combinations of solute and solvent is not negligible.

The number of solvents in chemistry is quite large. The number of available specific potentials is by far more limited, but too large to be summarized here. In addition, a simple list of names and references, not accompanied by critical remarks about the performances of such potentials in describing liquid systems would have little utility.

The numerous textbooks on intermolecular potentials are of little help, because outdated or paying little attention to the analytical exposition of computational models. More useful are monographs and reviews regarding computer simulations of liquids[39] or original papers making comparisons among different potentials.

These last are more abundant for water. This solvent, because of its importance and its molecular simplicity, has been the benchmark for many ideas about the description and modeling of intermolecular interactions. Almost all potentials that have been thus far proposed for not extremely flexible molecular component of liquid systems may be found in the literature regarding water. Alternative strategies in the defining number and position of sites, use diffuse instead of point charges, different approaches to model the many-body corrections, flexible potentials, dissociation of the molecule are some features that can be examined by looking at the water potential literature.

There is a large critical literature, but two recent reviews (Floris and Tani[40] and Wallqvist and Mountain,[41] both published in 1999) are here recommended as an excellent guide to this subject. The two reviews consider and analyze about 100 potentials for pure water. WM review starts from historical models, FM review pays attention to recent models (about 70 models, supplemented by ion-water potentials). The two reviews partly overlap, but they are to a good extent complementary, especially in the analysis of the performances of such models.

There are, to the best of our knowledge, no reviews of comparable accuracy for the potentials regarding other liquid systems.

# 8.7 THEORETICAL AND COMPUTING MODELING OF PURE LIQUIDS AND SOLUTIONS

## 8.7.1 PHYSICAL MODELS

In this chapter we shall present a necessarily partial review of the main theoretical approaches so far developed to treat liquid systems in terms of physical functions. We shall restrict ourselves to two basic theories, integral equation and perturbation theories, to keep the chapter within reasonable bounds. In addition, only the basic theoretical principles underlying the original methods will be discussed, because the progress has been less rapid for theory than for numerical applications. The latter are in fact developing so fast that it is an impossible task to try to give an exhaustive view in a few pages.

A fundamental approach to liquids is provided by the integral equation methods[42-44] (sometimes called distribution function methods), initiated by Kirkwood and Yvon in the 1930s. As we shall show below, one starts by writing down an exact equation for the molecular distribution function of interest, usually the pair function, and then introduces one or more approximations to solve the problem. These approximations are often motivated by considerations of mathematical simplicity, so that their validity depends on *a posteriori* agreement with computer simulation or experiment. The theories in question, called YBG (Yvon-Born-Green), PY (Percus-Yevick), and the HNC (hypernetted chain) approximation, provide the distribution functions directly, and are thus applicable to a wide variety of properties.

An alternative, and particularly successful, approach to liquids is provided by the thermodynamic perturbation theories.[42,44] In this approach, the properties of the fluid of interest are related to those of a reference fluid through a suitable expansion. One attempts to choose a reference system that is in some sense close to the real system, and whose properties are well known (e.g., through computer simulation studies or an integral equation theory).

As a last physical approach we mention, but do not further consider, the scaled-particle-theory (SPT)[38,45] which was developed about the same time as the Percus-Yevick theory. It gives good results for the thermodynamic properties of hard molecules (spheres or convex molecules). It is not a complete theory (in contrast to the integral equation and perturbation theories) since it does not yield the molecular distribution functions (although they can be obtained for some finite range of intermolecular separations).

Early work on the theory of dense fluids dealt almost exclusively with simple atomic fluids, in which the intermolecular forces are between the centers of spherical molecules and depend only on the separation distance r. However, in real fluids the intermolecular forces depend on the molecular orientations, vibrational coordinates, etc., in addition to r.

The complexity of the problem thus requires the assumption of some approximations; most of the formulated theories, for example, assume

(a) a rigid molecule approximation,

(b) a classical treatment of the traslational and rotational motions, and

(c) a pairwise additivity of the intermolecular forces.

In the rigid molecule approximation it is assumed that the intermolecular potential energy $V(\mathbf{r}^N,\omega^N)$ depends only on the positions of the centers of mass $\mathbf{r}^N = r_1 r_2 \ldots r_N$ for the N molecules and on their orientations $\omega^N = \omega_1\omega_2\ldots\omega_N$. This implies that the vibrational coordinates of the molecules are dynamically and statistically independent on the center of mass and orientation coordinates, and that the internal rotations are either absent, or independent of the $\mathbf{r}^N$ and $\omega^N$ coordinates. The molecules are also assumed to be in their ground electronic states.

At the bases of the second basic assumption made, e.g., that the fluids behave classically, there is the knowledge that the quantum effects in the thermodynamic properties are usually small, and can be calculated readily to the first approximation. For the structural properties (e.g., pair correlation function, structure factors), no detailed estimates are available for molecular liquids, while for atomic liquids the relevant theoretical expressions for the quantum corrections are available in the literature.

The third basic approximation usually introduced is that the total intermolecular potential energy $V(\mathbf{r}^N,\omega^N)$ is simply the sum of the intermolecular potentials for isolated pairs ij of molecules, i.e.,

$$V\left(\mathbf{r}^N, \omega^N\right) = \sum_{i<j} V\left(r_{ij}, \omega_i, \omega_j\right) \qquad [8.74]$$

In the sum i is kept less than j in order to avoid counting any pair interaction twice.

Eq. [8.74] is exact in the low-density gas limit, since interactions involving three or more molecules can be ignored. It is not exact for dense fluids or solids, however, because the presence of additional molecules nearby distorts the electron charge distributions in molecules i and j, and thus changes the intermolecular interaction between this pair from the isolated pair value. In order to get a more reliable description, three-body (and higher multi-body) correction terms should be introduced.

The influence of three-body terms on the physical properties has been studied in detail for atomic fluids,[46,47] while much less is known about molecular fluids. In the latter case, the accurate potentials are few and statistical mechanical calculations are usually done with model potentials. A particular model may be purely empirical (e.g., atom-atom), or semiempirical (e.g., generalized Stockmayer), where some of the terms have a theoretical basis. The so-called generalized Stockmayer model consists of central and non-central terms. For the central part, one assumes a two-parameter central form (the classical example is the Lennard-Jones, LJ, form introduced in the previous sections). The long-range, non-central part in general contains a truncated sum of multipolar, induction and dispersion terms. In addition, a short-range, angle-dependent overlap part, representing the shape or core of the potential, is usually introduced. The multipolar interactions are pair-wise additive, but the induction, dispersion and overlap interactions contain three-body (and higher multi-body) terms. Hence, three-body interactions are strongly suspected to be of large importance also for molecular liquids.

This review will be mainly restricted to a discussion of small rigid molecules in their ground electronic states. However, one can generalize in a number of directions by extending the pair potential $v(r_{ij},\omega_i,\omega_j)$ of eq. [8.74] so as to treat:

(a) non-rigid molecules,
(b) molecules with internal rotation,
(c) large (e.g., long chain) molecules which may be flexible,
(d) electronic excited state molecules.

In addition, the inclusion of covalence and charge transfer effects are also possible. The first two generalizations are straightforward in principle; one lets v depend on coordinates describing the additional degrees of freedom involved. As for (c), for very large molecules the site-site plus charge-charge model is the only viable one. Under (d) we can quote the so-called long-range 'resonance' interactions.

As the last note, we anticipate that in the following only equilibrium properties will be considered; however, it is fundamental to recall from the beginning that dynamical (e.g., non-equilibrium) analyses are an important and active field of research in the theory of physical models.[42,48,49]

Before entering into more details of each theory, it is worth introducing some basic definitions of statistical mechanics. All equilibrium properties of a system can be calculated if both the intermolecular potential energy and the distribution functions are known. In considering fluids in equilibrium, we can distinguish three principal cases:

(a) isotropic, homogeneous fluids, (e.g., liquid or compressed gas in the absence of an external field),
(b) anisotropic, homogeneous fluids (e.g., a polyatomic fluid in the presence of a uniform electric field, nematic liquid crystals), and
(c) inhomogeneous fluids (e.g., the interfacial region).

These fluid states have been listed in order of increasing complexity; thus, more independent variables are involved in cases (b) and (c), and consequently the evaluation of the necessary distribution functions is more difficult.

For molecular fluids, it is convenient to define different types of distribution functions, correlation functions and related quantities. In particular, in the pair-wise additive theory of homogeneous fluids (see eq. [8.74]), a central role is played by the angular pair correlation function $g(\mathbf{r}_{12}\omega_1\omega_2)$ proportional to the probability density of finding two molecules with position $\mathbf{r}_1$ and $\mathbf{r}_2$ and orientations $\omega_1$ and $\omega_2$ (a schematic representation of such function is reported in Figure 8.5).

In fact, one is frequently interested in some observable property <B>, which is (experimentally) a time average of a function of the phase variables $B(\mathbf{r}^N,\omega^N)$; the latter is often a sum of pair terms $b(\mathbf{r}_{ij},\omega_i,\omega_j)$ so that <B> is given by

$$\langle B\rangle = \int dr^N d\omega^N P\left(r^N\omega^N\right)B\left(r^N\omega^N\right) = \frac{1}{2}\rho N\int dr\langle g(r\omega_1\omega_2)b(r\omega_1\omega_2)\rangle_{\omega_1\omega_2} \qquad [8.75]$$

i.e., in terms of the pair correlation function g(12). Examples of such properties are the configurational contribution to energy, pressure, mean squared torque, and the mean squared force. In eq. [8.75] $\langle g(\mathbf{r}\omega_1\omega_2)b(\mathbf{r}\omega_1\omega_2)\rangle_{\omega_1\omega_2}$ means the unweighted average over orientations.

In addition to g(12), it is also useful to define:

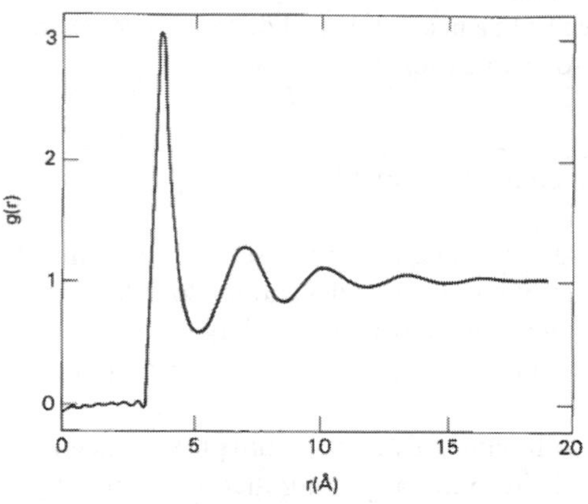


Figure 8.5. Example of the radial distribution function g(r) of a typical monoatomic liquid.

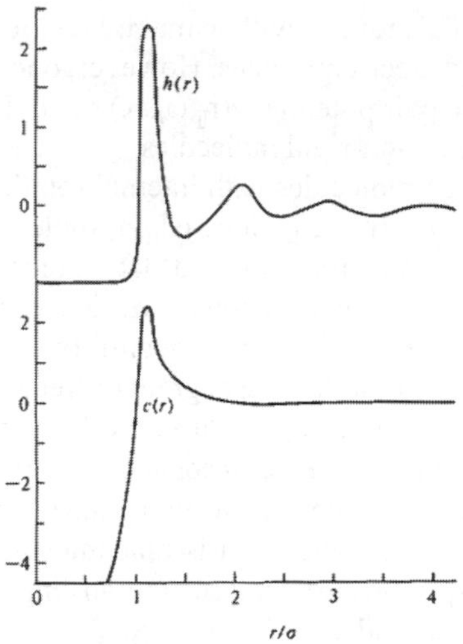


Figure 8.6. Examples of h(r) and c(r) functions for a typical monoatomic liquid.

a) the site-site correlation function $g_{\alpha\beta}(r_{\alpha\beta})$ proportional to the probability density that sites α and β on different molecules are separated by distance r, regardless of molecular orientations;
b) the total correlation function $h(r_{12}\omega_1\omega_2)= g(r_{12}\omega_1\omega_2) - 1$;
c) the direct correlation function $c(r_{12}\omega_1\omega_2)$.

Simple examples of h and c functions are reported in Figure 8.6.

In the list above, item (a) implies definite sites within molecules; these sites may be the nuclei themselves or sites at arbitrary locations within the molecules.

In addition, the total correlation, h, between molecules 1 and 2 can be separated into two parts: (i) a direct effect of 1 on 2 which is short-ranged and is characterized by c, and (ii) and indirect effect in which 1 influences other molecules, 3, 4, etc., which in turn affect 2. The indirect effect is the sum of all contributions from other molecules averaged over their configurations. For an isotropic and homogeneous fluid formed by non-spherical molecules, we have

$$h(\mathbf{r}_{12}\omega_1\omega_2) = c(\mathbf{r}_{12}\omega_1\omega_2) + \rho\int d\mathbf{r}_3 \left\langle c(\mathbf{r}_{13}\omega_1\omega_3)h(\mathbf{r}_{32}\omega_3\omega_2)\right\rangle_{\omega_3} \qquad [8.76]$$

which is the generalization of the Ornstein-Zernike (OZ) equation[50] to non-spherical molecules. The OZ equation is the starting point for many theories of the pair correlation function (PY, HNC, etc.); however, numerical solutions starting from [8.76] are complicated by the large number of variables involved. By expanding the direct and total correlation functions in spherical harmonics, one obtains a set of algebraically coupled equations relating the harmonic coefficients of h and c. These equations involve only one variable in the place of many in the original OZ equation. In addition, the theories we shall describe below truncate the infinite set of coupled equations into a finite set, thereby enabling a reasonably simple solution to be carried out.

### 8.7.1.1 Integral equation methods

The structure of the integral equation approach for calculating the angular pair correlation function $g(r_{12}\omega_1\omega_2)$ starts with the OZ integral equation [8.76] between the total (h) and the direct (c) correlation function, which is here schematically rewritten as h=h[c] where h[c] denotes a functional of c. Coupled to that a second relation, the so-called closure relation c=c[h], is introduced. While the former is exact, the latter relation is approximated; the form of this approximation is the main distinction among the various integral equation theories to be described below.

In the OZ equation, h depends on c, and in the closure relation, c depends on h; thus the unknown h depends on itself and must be determined self-consistently. This (self-consistency requiring, or integral equation) structure is characteristic of all many body problems.

Two of the classic integral equation approximations for atomic liquids are the PY (Percus-Yevick)[51] and the HNC (hypernetted chain)[52] approximations that use the following closures

$$h - c = y - 1 \qquad \text{(PY)} \qquad [8.77]$$

$$h - c = \ln y \qquad \text{(HNC)} \qquad [8.78]$$

where y is the direct correlation function defined by g(12)=exp(-β v(12))y(12) with v(12) the pair potential and β=1/kT.

The closures [8.77-8.78] can be also written in the form

$$c = g\left(1 - e^{\beta v(12)}\right) \qquad \text{(PY)} \qquad [8.79]$$

$$c = h - \beta v(12) - \ln g \qquad \text{(HNC)} \qquad [8.80]$$

For atomic liquids the PY theory is better for steep repulsive pair potentials, e.g., hard spheres, whereas the HNC theory is better when attractive forces are present, e.g., Lennard-Jones, and Coulomb potentials. No stated tests are available for molecular liquids; more details are given below.

As said before, in practice the integral equations for molecular liquids are almost always solved using spherical harmonic expansions. This is because the basic form [8.76] of the OZ relation contains too many variables to be handled efficiently. In addition, harmonic expansions are necessarily truncated after a finite number of terms. The validity of the truncations rests on the rate of convergence of the harmonic series that depends in turn on the degree of anisotropy in the intermolecular potential.

We recall that the solution to the PY approximation for the hard sphere atomic fluid is analytical and it also forms a basis for other theories, e.g., in molecular fluids the MSA and RISM theories to be discussed below.

The mean spherical approximation (MSA)[52] theory for fluids originated as the extension to continuum fluids of the spherical model for lattice gases. In practice it is usually applied to potentials with spherical hard cores, although extensions to soft core and non-spherical core potentials have been discussed.

The MSA is based on the OZ relation together with the closure

$$h = -1 \qquad r < \sigma \qquad [8.81]$$

$$c = -\beta v(12) \qquad r < \sigma \qquad [8.82]$$

where $\sigma$ is the diameter of the spherical hard core part of the pair potential v(12).

The condition [8.81] is exact for hard core potentials since g(12)=0 for $r < \sigma$, while the condition [8.82] is the approximation, being correct only asymptotically (large r). We thus expect the approximation to be worst near the hard core; we also note that the theory is not exact at low density (except for a pure hard sphere potential, where MSA=PY), but this is not too important since good theories exist for low densities, and we are therefore mainly interested in high densities. Once again the solution of the MSA problem is most easily accomplished using the spherical harmonic component of h and c.

In the list of the various correlation functions we have introduced at item (b) the so-called site-site pair correlation functions $g_{\alpha\beta}(r_{\alpha\beta})$; since they depend only on the radial variables $r_{\alpha\beta}$ (between sites), they are naturally simpler than $g(r_{12};\omega_1\omega_2)$ but, at the same time, they contain less information. The theories for $g_{\alpha\beta}$ fall into two categories based on site-site or particle-particle OZ equations, respectively. Various closures can be used with either category.

As concerns the site-site approach the most important theory is the 'reference interaction site model', or RISM.[53,54] This method applies to an intermolecular pair potential modeled by a site-site form, i.e., $V(r\omega_1\omega_2) = \Sigma_{\alpha\beta} v_{\alpha\beta}(r_{\alpha\beta})$ and its original intuitive derivation is based on exploring the possibility of decomposing $g(r;\omega_1\omega_2)$ also in the same form, i.e., as a sum of site-site $g_{\alpha\beta}(r)$s.

The RISM theory consists of (i) an OZ-like relation between the set $\{h_{\alpha\beta}\}$ (where $h_{\alpha\beta}(r)=g_{\alpha\beta}(r)-1$) and corresponding set $\{c_{\alpha\beta}\}$ of direct correlation functions, and (ii) a PY-type or other type of closure. The OZ-like relation is here a matrix one.

Although successful in many applications, the RISM approximation suffers from a number of major defects. First, it is not the best choice to calculate the equation of state, and the results that are obtained are thermodynamically inconsistent. Secondly, calculated structural properties show an unphysical dependence on the presence of 'auxiliary' sites, i.e., on sites that label points in a molecule but contribute nothing to the intermolecular potential. Thirdly, use of the RISM approximation leads to trivial and incorrect results for certain quantities descriptive of angular correlations in the fluid.[48]

Attempts to develop a more satisfactory theory within the interaction-site picture have developed along two different lines. The first relies on treating the RISM-OZ relation as providing the definition of the site-site direct correlation functions, $c_{\alpha\beta}$. In this way the usual RISM closures can be modified by adding terms to $c_{\alpha\beta}$ such that the asymptotic result is satisfied.

In the alternative approach, the basic observation is that the RISM-OZ relation, though plausible, does not provide an adequate basis for a complete theory of the structure of molecular fluids. Accordingly, it is there rather than in the closure relation that improvement must be sought.[55]

Despite the defects we have underlined, in the last years the RISM method has received much more attention than other integral equation methods; promising developments in this area are represented by the inclusion of polarizable media in RISM[56] and the coupling of the RISM equations to a quantum mechanical description.[57]

### 8.7.1.2 Perturbation theories

In perturbation theories[44,58] one relates the properties (e.g., the distribution functions or free energy) of the real system, for which the intermolecular potential energy is E, to those of a reference system where the potential is $E_0$, usually by an expansion in powers of the perturbation potential $E_1=E - E_0$.

The methods used are conveniently classified according to whether the reference system potential is spherically symmetric or anisotropic. Theories of the first type are most appropriate when the anisotropy in the full potential is weak and long ranged; those of the second type have a greater physical appeal and a wider range of possible application, but they are more difficult to implement because the calculation of the reference-system properties poses greater problems.

In considering perturbation theory for liquids it is convenient first to discuss the historical development for atomic liquids.

In many cases the intermolecular pair potential can be separated in a natural way into a sharp, short-range repulsion and a smoothly varying, long range attraction. A separation of this type is an explicit ingredient of many empirical representations of the intermolecular forces including, for example, the Lennard-Jones potential. It is now generally accepted that the structure of simple liquids, at least of high density, is largely determined by geometric factors associated with the packing of the molecular hard cores. By contrast, the attractive interactions may, in the first approximation, be regarded as giving rise to a uniform background potential that provides the cohesive energy of the liquid but has little effect on its structure. A further plausible approximation consists of modeling the short-range forces by the infinitely steep repulsion of the hard-sphere potential. In this way, the properties of a given liquid can be related to those of a hard-sphere reference system, the attractive part of the potential being treated as a perturbation. The choice of the hard-sphere fluid as a reference system is an obvious one, since its thermodynamic and structural properties are well known.

The idea of representing a liquid as a system of hard spheres moving in a uniform, attractive potential well is an old one; suffice here to recall the van der Waals equation. Roughly one can thus regard perturbation methods as attempts to improve the theory of van der Waals in a systematic fashion.

The basis of all the perturbation theories we shall consider is a division of the pair potential of the form

$$v(12) = v_0(12) + w(12) \qquad [8.83]$$

where $v_0$ is the pair potential of the reference system and w(12) is the perturbation. The following step is to compute the effect of the perturbation on the thermodynamic properties and pair distribution function of the reference system. This can be done systematically via an expansion in powers either of inverse temperature (the "$\lambda$ expansion") or of a parameter that measures the range of the perturbation (the "$\gamma$ expansion").

In spite of the fact that hard spheres are a natural choice of the reference system, for the reasons discussed above, realistic intermolecular potentials do not have an infinitely steep hard core, and there is no natural separation into a hard-sphere part and a weak perturbation. A possible improvement is that to take proper account of the "softness"' of the repulsive part of the intermolecular potential. A method of doing this was proposed by Rowlinson,[59] who used an expansion in powers of a softness parameter, $n^{-1}$, about $n^{-1}$=0 (corresponding to a hard sphere fluid). This method gives good results for the repulsive, but not for the attractive, part of the potential. Subsequently several researchers attempted to combine the advantages of both approaches. The first successful method was that of Barker and Henderson (BH),[60] who showed that a second-order theory gave quantitative results for the thermodynamic properties of a Lennard-Jones liquid. Even more rapidly convergent results were later obtained by Weeks, Chandler and Andersen (WCA)[61] using a somewhat different reference system.

Given the great success of perturbation theories in treating the properties of atomic liquids, a large effort has been devoted to extending the methods to deal with molecular systems. The first rigorous application of these theories to molecular fluids seems to have been made in 1951 by Barker,[62] who expanded the partition function for a polar fluid about that for a fluid of isotropic molecules.

Roughly, the basic problem is the same as in the atomic case, but the practical difficulties are much more severe. A possible approach is to choose a reference-system potential $v_0(r)$ spherically symmetric so that the integrations over the orientations (absent in atomic fluids) involve only the perturbation. The real system can be thus studied by a straightforward generalization of the λ-expansion developed for atomic fluids.

Perturbation theories based on spherically symmetric reference potentials, however, cannot be expected to work well when (as in the most real molecules) the short-range repulsive forces are strongly anisotropic. The natural approach in such cases is to include the strongly varying interactions in the specification of an isotropic reference potential to relate the properties of the reference system to those of hard molecules having the same shape.[63] Calculation of this type have been made by Tildesley,[64] exploiting a generalization of the WBA approach quoted above. In a similar way also, the BH perturbation theory has been generalized with good results.[65] The main disadvantage of these methods is the large computational effort that their implementation requires.

## 8.7.2 COMPUTER SIMULATIONS

In this section we will give a brief review of methodologies to perform computer simulations. These approaches nowadays have a major role in the study of liquid-state physics: they are developing so fast that it is impossible to give a complete view of all the different methodologies used and of the overwhelmingly large number of applications. Therefore we will only give a brief account of the basic principles of such methodologies, of how a computer simulation can be carried out, and we will briefly discuss limitations and advantages of the methods. The literature in the field is enormous: here we will refer interested readers to some of the basic textbooks on this subject.[42,66,67]

The microscopic state of a system may be specified in terms of the positions and momenta of the set of particles (atoms or molecules). Making the approximation that a classical description is adequate, the Hamiltonian H of a system of N particles can be written as a sum of a kinetic K and a potential V energy functions of the set of coordinates $q_i$ and momenta $p_i$ of each particle i, i.e.:

$$H(q,p) = K(p) + V(q) \qquad [8.84]$$

In the equation the coordinates q can simply be the set of Cartesian coordinates of each atom in the system, but in this case we treat the molecules as rigid bodies (as is usually done in most simulations); q will consist of the Cartesian coordinates of each molecular center of mass together with a set of other parameters specifying the molecular orientation. In any case p is the set of conjugate momenta.

Usually the kinetic energy K takes the form:

$$K = \sum_{i=1}^{N} \sum_{\mu} \frac{p_{i\mu}^2}{2m_i} \quad [8.85]$$

where the index μ runs over the components (x, y, z) of the momentum of molecule i and $m_i$ is the molecular mass.

The potential energy V contains the interesting information regarding intermolecular interactions. A detailed analysis on this subject has been given in the previous sections; here we shall recall only some basic aspects.

It is possible to construct from H an equation of motion that governs the time-evolution of the system, as well as to state the equilibrium distribution function for positions and momenta. Thus H (or better V) is the basic input to a computer simulation program. Almost universally, in computer simulation the potential energy is broken up into terms involving pairs, triplets, etc. of molecules, i.e.:

$$V = \sum_i v_1(r_i) + \sum_i \sum_{j>i} v_2(r_i, r_j) + \sum_i \sum_{j>i} \sum_{k>j>i} v_3(r_i, r_j, r_k) + \dots \quad [8.86]$$

The notation indicates that the summation runs over all distinct pairs i and j or triplets i, j and k, without counting any pair or triplet twice. In eq. [8.86], the first term represents the effect of an external field on the system, while the remaining terms represent interactions between the particles of the system. Among them, the second one, the pair potential, is the most important. As said in the previous section, in the case of molecular systems the interaction between nuclei and electronic charge clouds of a pair of molecules i and j is generally a complicated function of the relative positions $r_i$ and $r_j$, and orientations $\Omega_i$ and $\Omega_j$ (for atomic systems this term depends only on the pair separation, so that it may written as $v_2(r_{ij})$).

In the simplest approximation the total interaction is viewed as a sum of pair-wise contributions from distinct sites a in molecule i, at position $r_{ia}$, and b in molecule j, at position $r_{jb}$, i.e.:

$$v(r_{ij}, \Omega_i, \Omega_j) = \sum \sum v_{ab}(r_{ab}) \quad [8.87]$$

In the equation, $v_{ab}$ is the potential acting between sites a and b, whose inter-separation is $r_{ab}$.

The pair potential shows the typical features of intermolecular interactions as shown in Figure 8.1: there is an attractive tail at large separation, due to correlation between the electron clouds surrounding the atoms ('van der Waals' or 'London' dispersion), a negative well, responsible for cohesion in condensed phases, and a steeply rising repulsive wall at short distances, due to overlap between electron clouds.

Turning the attention to terms in eq. [8.86] involving triplets, they are usually significant at liquid densities, while four-body and higher are expected to be small in comparison with $v_2$ and $v_3$. Despite of their significance, only rarely triplet terms are included in computer simulations: that is due to the fact that the calculation of quantities related to a sum over triplets of molecules are very time-consuming. On the other hand, the average three-body effects can be partially included in a pair-wise approximation, leading it to give a good description of liquid properties. This can be achieved by defining an "effective" pair potential, able to represent all the many-body effects. To do this, eq. [8.86] can be re-written as:

$$V \approx \sum_i v_1(r_i) + \sum_i \sum_{j>i} v_2^{eff}(r_{ij}) \qquad [8.88]$$

The most widely used effective potential in computer simulations is the simple Lennard-Jones 12-6 potential (see eq. [8.70]), but other possibilities are also available; see Section 8.6 for discussion on these choices.

#### 8.7.2.1 Car-Parrinello direct QM simulation

Until now we have focused the attention to the most usual way of determining a potential interaction to be used in simulations. It is worth mentioning a different approach to the problem, in which the distribution of electrons is not treated by means of an "effective" interaction potential, but is treated *ab initio* by density functional theory (DFT). The most popular method is the Car-Parrinello (CP) approach,[68,69] in which the electronic degrees of freedom are explicitly included in the description and the electrons (to which a fictitious mass is assigned) are allowed to relax during the course of the simulation by a process called "simulated-annealing". In that way, any division of V into pair-wise and higher terms is avoided and the interatomic forces are generated in a consistent and accurate way as the simulation proceeds. This point constitutes the main difference between a CP simulation and a conventional MD simulation, which is preceded by the determination of the potential and in which the process leading to the potential is completely separated from the actual simulation. The forces are computed using electronic structure calculations based on DFT, so that the interatomic potential is parameter-free and derived from first principles, with no experimental input.

Let us consider a system for which the BO approximation holds and for which the motion of the nuclei can be described by classical mechanics. The interaction potential is given by:

$$V(R) = \langle \psi_0 | \hat{H} | \psi_0 \rangle \qquad [8.89]$$

where H is the Hamiltonian of the system at fixed R positions and $\psi_0$ is the corresponding instantaneous ground state. Eq. [8.89] permits to define the interaction potential from first principles.

In order to use eq. [8.89] in a MD simulation, calculations of $\psi_0$ for a number of configurations of the order of $10^4$ are needed. Obviously this is computationally very demanding, so that the use of certain very accurate QM methods (for example, the configuration interaction (CI)) is precluded. A practical alternative is the use of DFT. Following Kohn and Sham,[70] the electron density $\rho(r)$ can be written in terms of occupied single-particle orthonormal orbitals:

$$\rho(r) = \sum_i |\psi_i(r)|^2 \qquad [8.90]$$

A point on the BO potential energy surface (PES) is then given by the minimum with respect to the $\psi_i$ of the energy functional:

$$V[\{\psi_i\},\{R_I\},\{\alpha_v\}] = \sum_i \int_\Omega d^3r \psi_i^*(r)\left[-\frac{h^2}{2m}\nabla^2\right]\psi_i(r) + U[\rho(r),\{R_I\},\{\alpha_v\}] \quad [8.91]$$

where $\{\mathbf{R}_I\}$ are the nuclear coordinates, $\{\alpha_v\}$ are all the possible external constraints imposed on the system (e.g., the volume $\Omega$). The functional U contains the inter-nuclear Coulomb repulsion and the effective electronic potential energy, including external nuclear, Hartree, and exchange and correlation contributions.

In the conventional approach, the minimization of the energy functional (eq. [8.91]) with respect to the orbitals $\psi_i$ subject to the orthonormalization constraint leads to a set of self-consistent equations (the Kohn-Sham equations), i.e.:

$$\left\{-\frac{h^2}{2m}\nabla^2 + \frac{\partial U}{\partial \rho(r)}\right\}\psi_i(r) = \varepsilon_i \psi_i(r) \quad [8.92]$$

whose solution involves repeated matrix diagonalizations (and rapidly growing computational effort as the size of the system increases).

It is possible to use an alternative approach, regarding the minimization of the functional as an optimization problem, which can be solved by means of the simulated annealing procedure.[71] A simulated annealing technique based on MD can be efficiently applied to minimize the KS functional: the resulting technique, called "dynamical simulated annealing" allows the study of finite temperature properties.

In the "dynamical simulated annealing," the $\{R_i\}$, $\{\alpha_v\}$ and $\{\psi_i\}$ parameters can be considered as dependent on time; then the Lagrangian is introduced, i.e.:

$$L = \sum_i \frac{1}{2}\mu \int_\Omega d^3r |\dot{\psi}_i|^2 + \sum_I \frac{1}{2} M_I \dot{R}_I + \sum_v \frac{1}{2}\mu_v \dot{\alpha}_v^2 - V[\{\psi_i\},\{R_I\},\{\alpha_v\}] \quad [8.93]$$

with:

$$\int_\Omega d^3r \psi_i^*(r,t)\psi_j(r,t) = \delta_{ij} \quad [8.94]$$

In eq. [8.94] the dot indicates time derivative, $M_I$ are the nuclear masses and $\mu$ are arbitrary parameters having the dimension of mass.

Using eq. [8.94] it is possible to generate a dynamics for $\{R_i\}$, $\{\alpha_v\}$ and $\{\psi_i\}$ through the following equations:

$$M_I \ddot{R}_I = -\nabla_{R_I} V \quad [8.95]$$

$$\mu_v \ddot{\alpha}_v = -\left(\frac{\partial V}{\partial \alpha_v}\right) \quad [8.96]$$

$$\mu\ddot{\psi}(r,t) = -\frac{\partial V}{\partial \psi_i^*(r,t)} + \sum_k \Lambda_{ik}\psi_k(r,t) \qquad [8.97]$$

In eq. [8.97] the $\Lambda$ are the Lagrange multipliers introduced to satisfy eq. [8.94]. It is worth noticing that, while the nuclear dynamics [8.95] can have a physical meaning, that is not true for the dynamics associated with the $\{\alpha_v\}$ and $\{\psi_i\}$: this dynamics is fictitious, like the associated "masses" $\mu$.

If $\mu$ and the initial conditions $\{\psi_i\}_0$ and $\{d\psi_i/dt\}_0$ are chosen such that the two classical sets of degrees of freedom (nuclear and electronic) are only weakly coupled, the transfer of energy between them is small enough to allow the electrons to adiabatically follow the nuclear motion, then remaining close to their instantaneous BO surface. In such a metastable situation, meaningful temporal averages can be computed. The mentioned dynamics is meant to reproduce what actually occurs in real matter, that is, electrons adiabatically following the nuclear motion.

QM potentials have been widely used in molecular dynamics simulation of liquid water using the CP DFT algorithm. See, for example, refs. [72,73].

### 8.7.2.2 Semi-classical simulations

Computer simulations are methods addressed to perform "computer experimentation". The importance of computer simulations rests on the fact that they provide quasi-experimental data on well-defined models. As there is no uncertainty about the form of the interaction potential, theoretical results can be tested in a way that is usually impossible with results obtained by experiments on real liquids. In addition, it is possible to get information on quantities of no direct access to experimental measures.

There are basically two ways of simulating a many-body system: through a stochastic process, such as the Monte Carlo (MC) simulation,[74] or through a deterministic process, such as a Molecular Dynamics (MD) simulation.[75,76] Numerical simulations are also performed in a hybridized form, like the Langevin dynamics[42] which is similar to MD except for the presence of a random dissipative force, or the Brownian dynamics,[42] which is based on the condition that the acceleration is balanced out by drifting and random dissipative forces.

Both the MC and the MD methodologies are used to obtain information on the system via a classical statistical analysis but, whereas MC is limited to the treatment of static properties, MD is more general and can be used to take into account the time dependence of the system states, allowing one to calculate time fluctuations and dynamic properties.

In the following, we shall briefly describe the main features of MD and MC methodologies, focusing the attention to their use in the treatment of liquid systems.

#### Molecular dynamics

Molecular Dynamics is the term used to refer to a technique based on the solution of the classical equation of motion for a classical many-body system described by a many-body Hamiltonian H.

In a MD simulation, the system is placed within a cell of fixed volume, usually of cubic shape. A set of velocities is assigned, usually drawn from a Maxwell-Boltzmann distribution suitable for the temperature of interest and selected to make the linear momentum equal to zero. Then the trajectories of the particles are calculated by integration of the classical equation of motion. It is also assumed that the particles interact through some forces, whose calculation at each step of the simulation constitutes the bulk of the computational demand of the calculation. The first formulation of the method, due to Alder,[75] referred to a system of hard spheres: in this case, the particles move at constant velocity between perfectly elastic collisions, so that it is possible to solve the problem without making any approximations. More difficult is the solution of the equations of motion for a set of Lennard-Jones particles: in fact, in this case an approximated step-by-step procedure is needed, since the forces between the particles change continuously as they move.

Let us consider a point Γ in the phase space and suppose that it is possible to write the instantaneous value of some property A as a function A(Γ). As the system evolves in time, Γ and hence A(Γ) will change. It is reasonable to assume that the experimentally observable "macroscopic" property $A_{obs}$ is really the time average of A(Γ) over a long time interval:

$$A_{obs} = \langle A \rangle_{time} = \langle A(\Gamma(t)) \rangle_{time} = \lim_{t\to\infty} \frac{1}{t} \int_0^t A(\Gamma(t))dt \qquad [8.98]$$

The time evolution is governed by the well-known Newton equations, a system of differential equations whose solution is practical. Obviously it is not possible to extend the integration to infinite time, but the average can be reached by integrating over a long finite time, at least as long as possible as determined by the available computer resources. This is exactly what is done in a MD simulation, in which the equations of motion are solved step-by-step, taking a large finite number τ of steps, so that:

$$A_{obs} = \frac{1}{\tau} \sum_{\bar{\tau}} A(\Gamma(\bar{\tau})) \qquad [8.99]$$

It is worth stressing that a different choice in the time step is generally required to describe different properties, as molecular vibrations (when flexible potentials are used), translations and rotations.

A problem arising from the methodology outlined above is whether or not a suitable region of the phase space is explored by the trajectory to yield good time averages (in a relatively short computing time) and whether consistency can be obtained with simulations with identical macroscopic parameters but different initial conditions. Generally, thermodynamically consistent results for liquids can be obtained provided that careful attention is paid to the selection of initial conditions.

## Monte Carlo

As we have seen, apart from the choice of the initial conditions, a MD simulation is entirely deterministic. By contrast, a probabilistic (stochastic) element is an essential part of any Monte Carlo (MC) simulation.

The time-average approach outlined above is not the only possible: it is in fact practical to replace the time average by the ensemble average, being the ensemble a collection of points Γ distributed according to a probability density ρ(Γ) in the phase space. The density is determined by the macroscopic parameters, NPT, NVT, etc., and generally will evolve in time. Making the assumption that the system is "ergodic", the time average in eq. [8.98] can be replaced by an average taken over all the members of the ensemble at a particular time:

$$A_{obs} = \langle A \rangle_{ensemble} = \sum_{\Gamma} A(\Gamma)\rho_{ensemble}(\Gamma) \qquad [8.100]$$

Sometimes, instead of using the $\rho(\Gamma)$, a “weight” function $w(\Gamma)$ is used:

$$\rho(\Gamma) = \frac{w(\Gamma)}{Q} \qquad [8.101]$$

$$Q = \sum_{\Gamma} w(\Gamma) \qquad [8.102]$$

$$\langle A \rangle_{ensemble} = \frac{\sum_{\Gamma} w(\Gamma) A(\Gamma)}{\sum_{\Gamma} w(\Gamma)} \qquad [8.103]$$

Q is the partition function of the system.

From Eq. [8.102], it is possible to derive an approach to the calculation of thermodynamics properties by direct evaluation of Q for a particular ensemble. Q is not directly estimated, but the idea of generating a set of states in phase space sampled in accordance with the probability density $\rho(\Gamma)$ is the central idea of MC technique. Proceeding exactly as done for MD, replacing an ensemble average as in Eq. [8.103] with a trajectory average as in Eq. [8.99], a succession of states is generated in accordance with the distribution function $\rho_{NVE}$ for the microcanonical NVE ensemble. The basic aim of the MC method (so-called because of the role that random numbers play in the method), which is basically a technique for performing numerical integration, is to generate a trajectory in phase space that samples from a chosen statistical ensemble. It is possible to use ensembles different from the microcanonical: the only request is to have a way (physical, entirely deterministic or stochastic) of generating from a state $\Gamma(\tau)$ a next state $\Gamma(\tau + 1)$. The important point to be stressed is that some conditions have to be fulfilled: these are that the probability density $\rho(\Gamma)$ for the ensemble should not change as the system evolves, any starting distribution $\rho(\Gamma)$ should tend to a stationary solution as the simulation proceeds, and the ergodicity of the systems should hold. With these recommendations, we should be able to generate from an initial state a succession of points that in the long term are sampled with the desired probability density $\rho(\Gamma)$. In this case, the ensemble average will be the same as a “time average” (see Eq. [8.99]). In a practical simulation, $\tau$ runs over the succession of states generated following the previously mentioned rules, and is a large finite number. This approach is exactly what is done in an MC simulation, in which a trajectory is generated through phase space with different recipes for the different ensembles. In other words, in a MC simulation a system of particles interacting through some known potential is assigned a set of initial coordinates (arbitrarily chosen) and a sequence of configurations of the particles is then generated by successive random displacements (also called ‘moves’).

If f(R) is an arbitrary function of all the coordinates of the molecule, its average value in the canonical ensemble is:

$$\langle f \rangle = \frac{\int f(r) e^{-\beta V(R)} dR}{\int e^{-\beta V(R)} dR} \qquad [8.104]$$

where $\beta = 1/kT$ and V(R) is the potential energy.

In the MC methods the integration is done by re-writing <f> in terms of the probabilities $P_k$ of finding the configuration $R_k$:

$$\langle f \rangle = \sum_k f(R_k) P_k \qquad [8.105]$$

$$P_k = \frac{e^{-\beta V(R_k)}}{\sum_k e^{-\beta V(R_k)}} \qquad [8.106]$$

The search for the most probable configuration $R_k$ is done using Markov's chain theory, so that after setting the system at a given configuration $R_i$, another configuration $R_j$ is selected, making a random move in the coordinates. Not all configurations are accepted; the decision on whether or not to accept a particular configuration is made in such a way as to ensure that asymptotically the configuration space is sampled according to the probability density appropriate to a chosen ensemble. In particular:

if $$V(R_j) < V(R_i)$$

the move is accepted in the chain. Otherwise,

if $$V(R_j) > V(R_i)$$

the move is subjected to a second screening and accepted

if $$e^{-\beta(V_j - V_i)} < \gamma$$

where $\gamma$ is a random number between 0 and 1.

The original models used in MC were highly idealized representations of molecules, such as hard spheres and disks, but nowadays MC simulations are carried out on the basis of more reliable interaction potential. The use of realistic molecule-molecule interaction potential makes it possible to compare data obtained from experiments with the computer generated thermodynamic data derived from a model. The particle momenta do not enter the calculation, there is no scale time involved, and the order in which the configurations occur has no special significance.

It is worth noticing that, because only a finite number of states can be generated in a simulation, the results obtained from a MC simulation are affected by the choice of initial conditions, exactly as previously said for the MD results.

As we have already said, it is possible to extend the MC method, originally formulated in the microcanonical ensemble, to other ensembles. Particularly important in the study of liquid systems is the extension of the Monte Carlo method to the grand canonical $\mu VT$ ensemble:[76] that is due to the fact that such an extension permits to calculate the free energy of the system, a quantity of particular significance in the chemistry and physics of liquids. The grand canonical MC involves a two-stage process: the first stage is exactly identical to what is done in the conventional MC, with the particles moving and the move accepted or rejected as previously said. The second stage involves an attempt either to insert a particle at a randomly chosen point in the cell or to remove a particle that is already present. The decision whether to make an insertion or a removal is made randomly with equal probabilities. The trial configuration is accepted if the pseudo-Boltzmann factor:

$$W_N = e^{\left[N\beta\mu - \log N! - \beta V_N(r^N)\right]} \qquad [8.107]$$

increases. If it decreases, the change is accepted when equal to:

$$\frac{W_{N+1}}{W_N} = \frac{1}{N+1} e^{\left[\beta\mu - \beta(V_{N+1} - V_N\right]} \qquad [8.108]$$

for the insertion, or to:

$$\frac{W_{N-1}}{W_N} = N e^{\left[-\beta\mu - \beta(V_{N+1} - V_N\right]} \qquad [8.109]$$

for the removal.

This method works very well at low and intermediate densities. As the density increases, it becomes difficult to apply, because the probability of inserting or removing a particle is very small.

As we have already said, the grand canonical Monte Carlo provides a mean to determining the chemical potential, and hence, the free energy of the system. In other MC and MD calculations a numerical value for the free energy can always be obtained by means of an integration of thermodynamic relations along a path which links the state of interest to one for which the free energy is already known, for example, the dilute gas or the low-temperature solid. Such a procedure requires considerable computational effort, and it has a low numerical stability. Several methods have been proposed and tested.

The difficulties in using MC and MD arise from the heavy computational cost, due to the need of examining a large number of configurations of the system usually consisting of a large number of particles. The size of the system one can study is limited by the available storage on the computer and by the speed of execution of the program. The time taken to evaluate the forces or the potential energy is proportional to $N^2$.

Other difficulties in molecular simulations arise from the so-called quasi-ergodic problem,[77] i.e., the possibility that the system becomes trapped in a small region of the phase space. To avoid it, whatever the initial conditions, the system should be allowed to equilibrate before starting the simulation, and during the calculation, the bulk properties should be carefully monitored to detect any long-time drift.

As already said, apart from the initial conditions, the only input information in a computer simulation are the details of the inter-particle potential, almost always assumed to be pair-wise additive. Usually in practical simulations, in order to economize the computing time, the interaction potential is truncated at a separations $r_c$ (the cut-off radius), typically of the order of three molecular diameters. Obviously, the use of a cut-off sphere of small radius is not acceptable when the inter-particle forces are very long ranged.

The truncation of the potential differently affects the calculation of bulk properties, but the effect can be recovered by using appropriate "tail corrections". For energy and pressure for monoatomic fluids, for example, these "tail corrections" are obtained by evaluating the following integrals [8.110] and [8.111] between the limits $r=r_c$ and $r=\infty$, and assuming that $g(r)=1$ for $r>r_c$:

$$\frac{E}{N} = 2\pi\rho\int v(r)\,g(r)\,r^2dr \qquad [8.110]$$

$$\frac{\beta P}{\rho} = 1 - \frac{2\pi\beta\rho}{3}\int v(r)\,g(r)\,r^3dr \qquad [8.111]$$

As already mentioned above, the extent of "tail corrections" strongly depends on the property under study. In the case of pressure, this correction can be very large; for example, in the case of a Lennard-Jones potential truncated at $r_c=2.5\sigma$, at conditions close to the triple point, $E/N\varepsilon=-6.12$, to which the tail correction contributes -0.48, but $\beta P/\rho=0.22$, of which -1.24 comes to the tail. The neglecting of the tail correction would lead to a strongly positive value for pressure.[42]

Usually, when a cut-off radius is exploited, for the calculation of the interaction between a particle and the others, the "nearest-neighbor" convention is used. It means that, assuming the particles of interest lie in a cell and that this basic unit is surrounded by periodically repeated images of itself, a particle i lying on the central cell will interact only with the nearest image of any other particle j. The interaction is then set to zero if the distance between i and the nearest image is greater than $r_c$. The usual assumption $r_c<L/2$, with L the cell length, means it is assumed that at most one image of j lies in any sphere of radius $r<L/2$ centered on the particle i.

For the calculations of properties of a small liquid system, like a very small liquid drop, the cohesive forces between molecules may be sufficient to hold the system together during the course of a simulations: otherwise the system of molecules may be confined by a potential acting as a container. Such an arrangement is, however, not satisfactory for the simulation of bulk liquids, being the main obstacle the large fraction of molecules which lie on the surface. Molecules on the surface will experience quite different forces from molecules in the bulk.

To overcome the problem, in the simulation of bulk liquids a possible solution is to impose periodic boundary conditions. The system is in a cubic box, which is replicated throughout space to form an infinite lattice. During the simulation, as a molecule moves in the original box, its image in each of the neighboring boxes moves exactly in the same way. In particular as a molecule leaves the central box, one of its images will enter the central box through the opposite face. A common question to ask about the use of periodic boundary conditions is if the properties of a small, infinitely periodic system and the macroscopic system that it represents are the same. This will usually depend both on the range of the intermolecular potential and the phenomenon under investigation. In some cases, periodic boundary conditions have little effect on the equilibrium thermodynamic properties and structure of fluids away from phase transitions and where the interactions are short-ranged, but in other cases this is not true.

It is worth mentioning an alternative method to the standard periodic boundary conditions for simulating bulk liquids. A three dimensional system can be embedded in the surface of an hypersphere: the hypersphere is a finite system and cannot be considered as part of an infinitely repeating periodic system. Effects coming from the curved geometry (non-Euclidean geometry has to be considered) are expected to decrease as the system increases, so that in the case of solutes of large size, "spherical boundary conditions" are expected to be a noticeable method for simulating bulk liquids. In this case, boundary conditions can be obtained by inserting the spherical cell into a large sphere and then by performing a simulation of a lower computational level in the extended region. However, recent versions of the method tend to replace the second portion of the liquid, described at a molecular level, with a description by means of continuum models (see Section 8.7.3), which can be easily extended up to $r=\infty$.

Turning attention to the accuracy gained in the calculation of liquids properties by molecular simulations, it turns out that certain properties can be calculated more accurately than others. For example, the mean potential energy can be obtained with an uncertainty less than 1%; larger errors are associated with the calculation of thermodynamic properties, such as, the specific heat, that are linked to fluctuations in a microscopic variable. Typically, an uncertainty of the order of 10% is expected.[42]

The dynamical properties of liquids can be generally computed by calculating the time-correlation functions. They provide a quantitative description of the microscopic dynamics in liquids. Here, computer simulations play a key role, since they give access to a large variety of correlation functions, many of which are not measurable by laboratory experiments. In a MD simulation, the value of the correlation function $C_{AB}(t)$ at time t is calculated by averaging the product $A(t+s)B(s)$ of two dynamical variables, A(t) and B(t) over many choices of the time origin s. It has been shown that the uncertainty in time correlations between events separated by an interval $\tau$ increases as $\tau^{1/2}$. In addition, the correlated motion of large numbers of particles determines many dynamical properties. These collective properties are usually subjected to larger errors than in the case of single particle properties, and their calculation is very demanding from a computational point of view.[42]

**QM/MM**

As final remark of this section, we would like to briefly mention the so-called QM/MM method.[78,79]

In this approach, a part of the system (the solute M) is treated explicitly by a quantum mechanical (QM) method, while another part (the bulk solvent S) is approximated by a standard Molecular Mechanics (MM) force field. It is clear that such a method takes advantage of the accuracy and generality of the QM treatment, and of the computational efficiency of the MM calculation.

Using the BO approximation and assuming that no charge transfer occurs between S and M, the Hamiltonian of the system can be separated into three terms:

$$\hat{H} = \hat{H}_M + \hat{H}_{MS} + \hat{H}_{SS} \qquad [8.112]$$

where the first term is the Hamiltonian of the solute in vacuo, the second one represents the QM/MM solute-solvent interaction and couples solvent effects into QM calculations. It can be separated in an electrostatic term, a van der Waals terms and a polarization term, i.e.:

$$\hat{H}_{MS} = \hat{H}_{MS}^{el} + \hat{H}_{MS}^{vdW} + \hat{H}_{MS}^{pol} \qquad [8.113]$$

The third term in eq. [8.112] is the interaction energy between solvent molecules. It is usually approximated by the molecular mechanics force field, which in general contains bond stretching, angle bending, dihedral torsion and non-bonded terms (see Section 8.4.5).

Once the Hamiltonian of the system is defined, the total energy of an instantaneous configuration sampled during an MC or MD simulation is determined as:

$$E = \left\langle \Phi \middle| \hat{H} \middle| \Phi \right\rangle = E_M + E_{MS} + E_{SS} \qquad [8.114]$$

More details on the procedure can be found in literature.[78]

The forces used to integrate Newton's equation of motion in MD simulations are determined by differentiating eq. [8.114] with respect to nuclear coordinates.

In the practical QM/MM calculation, the solvent S is equilibrated first using simulations (or a RISM version of the integral equation approach, see Section 8.7.1.1), and then the electronic structure of the solute M is modified via an iterative QM procedure in the presence of a fixed potential of the S component. The procedure is repeated iteratively; at the next step the S distributions is determined again, now taking into account the modified description of M. This sequence of steps is repeated until convergence. A problem of such a methodology is that changes in the internal geometry of M must be treated apart, scanning point-by point the relevant PES, not being available analytical expressions for first and second derivatives of the energy with respect to nuclear coordinates, which would be necessary to take into account any modification in the solute geometry.

To overcome problems arising from the finite system size used in MC or MD simulation, boundary conditions are imposed using periodic-stochastic approximations or continuum models.[75] In particular, in stochastic boundary conditions the finite system is not duplicated but a boundary force is applied to interact with atoms of the system. This force is set as to reproduce the solvent regions that have been neglected. Anyway, in general any of the methods used to impose boundary conditions in MC or MD can be used in the QM/MM approach.

### 8.7.3 CONTINUUM MODELS

Originally, continuum models of solvent were formulated as dielectric models for electrostatic effects. In a dielectric model the solvent is modeled as a continuous medium, usually assumed homogeneous and isotropic, characterized by a scalar, static dielectric constant ε. This model of the solvent, that can be referred to the original work by Born, Onsager and Kirkwood 60-80 years ago, assumes linear response of the solvent to a perturbing electric field due to the presence of the molecular solute.

This simple definition then has been largely extended to treat more complex phenomena, including not only electrostatic effects; and nowadays continuum solvation models represent very articulate methodologies able to describe different systems of increasing complexity.

The history of, and the theory behind, continuum solvation models have been described exhaustively in many reviews[31a,80,81] and articles[82-84] in the past, so we prefer not to repeat them here. In addition, so large and continuously increasing is the amount of examples of theoretical developments on one hand, and of numerical applications on the other, that we shall limit our attention to a brief review of the basic characteristics of these models which have gained wide acceptance and are in use by various research groups.

Trying to give a more theoretically-stated definition of continuum models, but still keeping the same conciseness of the original one reported above, we have to introduce some basic notions of classical electrostatics.

The electrostatic problem of a charge distribution $\rho_M$ (representing the molecular solute) contained in a finite volume (the molecular cavity) within a continuum dielectric can be expressed in terms of the Poisson equation:

$$\vec{\nabla}\left[\varepsilon(\vec{r})\vec{\nabla}\phi(\vec{r})\right] = -4\pi\rho_M(\vec{r}) \qquad [8.115]$$

where $\phi$ is the total electrostatic potential.

Eq. [8.115] is subject to the constraint that the dielectric constant inside the cavity is $\varepsilon_{in}=1$, while at the asymptotic boundary we have:

$$\lim_{r\to\infty} r\phi(\vec{r}) = \alpha \qquad \lim_{r\to\infty} r^2\phi(\vec{r}) = \beta \qquad [8.116]$$

with $\alpha$ and $\beta$ finite quantities.

At the cavity boundary the following conditions hold:

$$\phi_{in}(\vec{r}) = \phi_{out}(\vec{r}) \qquad \frac{\partial\phi_{in}}{\partial\hat{n}} = \varepsilon_{out}(\vec{r})\frac{\partial\phi_{out}}{\partial\hat{n}} \qquad [8.117]$$

where subscripts in and out indicate functions defined inside and outside the cavity, and $\hat{n}$ is the unit vector perpendicular to the cavity surface and pointing outwards.

Further conditions can be added to the problem, and/or those shown above (e.g., the last one) can be modified; we shall consider some of these special cases later.

As said above, in several models an important simplification is usually introduced, i.e., the function $\varepsilon_{out}(\vec{r})$ is replaced by a constant $\varepsilon$ (from now on we skip the redundant subscript out). With this simplification we may rewrite the electrostatic problem by the following equations:

inside the cavity: $\nabla^2\phi(\vec{r}) = -4\pi\rho_M(\vec{r})$, outside the cavity: $\nabla^2\phi(\vec{r}) = 0$ [8.118]

Many alternative approaches have been formulated to solve this problem.

In biochemistry, methods that discretize the Poisson differential operator over finite elements (FE) have long been in use. In general, FEM approaches do not directly use the molecular cavity surface. Nevertheless, as the whole space filled by the continuous medium is partitioned into locally homogeneous regions, a careful consideration of the portion of space occupied by the molecular solute has still to be performed.

There is another family of methods, known as FD (finite difference) methods,[85] which exploit point grids covering the whole space, but conversely to the FEM point, in FD methods the points are used to replace differential equations by algebraic ones.

These methods can be applied both to the "simplified" model with constant $\varepsilon$, and to models with space dependent $\varepsilon(\vec{r})$ or real charges dispersed in the whole dielectric medium. They aim at solving the Poisson equation (1) expressed as a set of finite difference equations for each point of the grid. The linear system to be solved has elements depending both on $\varepsilon$

and $\rho_M$, and its solution is represented by a set of $\phi$ values in the grid points. These values have to be reached iteratively.

Many technical details of the methods have been changed in the years. We signal, among others, the strategy used to define the portion of the grid, especially near the molecular surface where there is a sharp change in $\varepsilon$, and the mathematical algorithms adopted to improve convergence in the iterative processes. Almost all FD methods have been conceived for, and applied to, classical descriptions of solutes, generally represented by atomic charges. The latter are usually drawn from previous QM calculations in vacuo or from Molecular Mechanics effective potentials, as in the study of large molecules, usually proteins. The deficiencies due to the use of rigid charges are partially alleviated by introducing a dielectric constant greater than one for the inner space ($\varepsilon$ between 2 and 4), and/or by including atomic polarization functions (akin to atomic dipole polarizabilities).

The DelPhi program[85a] is one of these proposals. This method, which has been extensively improved during the years, exists now in a QM version[86] exploiting a recent high level QM method, called PS-GVB,[87] based on a synthesis of grid point integration techniques and standard Gaussian orbital methods. From a PS-GVB calculation in vacuo, a set of atomic charges, the potential derived (PD) charges, is obtained by a fitting of the molecular electrostatic potential (MEP). These charges are used in DelPhi to solve the Poisson equation in the medium, and then to get, with the aid of a second application of DelPhi in vacuo, the reaction field (RF). The latter is represented by a set of point charges on the cavity surface, which define the solute-solvent interaction potential to be inserted in the PS-GVB Hamiltonian. Then a new computational cycle starts, and the whole process is repeated until convergence is reached.

In parallel, the classical FD procedure of Bashford and Karplus[88] has been coupled with the Amsterdam density fuctional theory (DFT)[89] code to give another version of QM-FD methods. Starting from a DFT calculation in vacuo, the RF potential is obtained as the difference between the solutions of the Poisson equation obtained with a FD method in the medium and in vacuo (in both cases, the electrostatic potential $\phi_M$ is expressed in terms of PD atomic charges). The RF potential is then re-computed to solve a modified Kohn-Sham equation. The calculations are repeated until convergence is achieved.

An alternative to the use of finite differences or finite elements to discretize the differential operator is to use boundary element methods (BEM).[90] One of the most popular of these is the polarizable continuum model (PCM) developed originally by the Pisa group of Tomasi and co-workers.[32] The main aspect of PCM is to reduce the electrostatic Poisson equation (1) into a boundary element problem with apparent charges (ASCs) on the solute cavity surface.

A method that is very similar to PCM is the conductor-like screening model (COSMO) developed by Klamt and coworker[91] and used later by Truong and coworkers,[92] and by Barone and Cossi.[93] This model assumes that the surrounding medium is well modeled as a conductor (thus simplifying the electrostatics computations) and corrections are made *a posteriori* for dielectric behavior.

Electrostatic BE methods can be supported by nonelectrostatic terms, such as dispersion and exchange, in different ways, and their basic theory can be extended to treat both classical and quantum mechanical solutes; in addition, many features, including analytical gradients with respect to various parameters, have been added to the original models so as to make BE methods one of the most powerful tools to describe solvated systems of increasing complexity. More details on their main aspects will be given in the following.

Further alternative methods for solving the Poisson equation are to represent the solute and the solvent reaction field by single-center multipole expansions (MPE), as first pioneered by the Nancy group of Rivail and coworkers.[94] Originally, this formalism was adopted because it simplified computation of solvation free energies and analytic derivatives for solute cavities having certain ideal shapes, e.g., spheres and ellipsoids. The formalism has been extended to arbitrary cavity shapes and multipole expansions at multiple centers. An alternative method still adopting the same MPE approach is that of Mikkelsen and coworkers;[95] this method has been implemented in the Dalton computational package.[96]

The idea of a distributed multipole expansion to represent the charge distribution is also employed by the so-called generalized Born (GB) approach to continuum solvation. In this case, however, only monopoles (i.e., atomic partial charges) are employed and instead of solving the Poisson equation with this charge distribution, one uses the generalized Born approximation. The most widely used solvation models adopting this procedure are the original generalized Born/surface area (GBSA) approach of Still and co-workers,[97] including the subsequent extensions and adaptations of several groups in which the GB electrostatics are combined with a classical MM treatment of the solute, and the SMx models of Cramer, Truhlar and coworkers[98] in which GB electrostatics are employed in a quantum mechanical treatment of the solute. In both of these approaches, semiempirical atomic surface tensions are used to account for nonelectrostatic solvation effects. For the SMx models, the value of x ranges from 1 to 5.42, corresponding to successively improved versions of the prescriptions for atomic charges and radii, the dielectric descreening algorithm, and the functional forms for the atomic surface tensions.

#### 8.7.3.1 QM-BE methods: the effective Hamiltonian

As anticipated above, in BE methods for solvation the basic feature is to solve the Poisson equation [8.115] with the related boundary conditions [8.117] (or in a simpler isotropic case the system [8.118]) by defining an apparent charge distribution $\sigma$ on the solute cavity surface (such charge is usually indicated as ASC).[31a] This continuous charge which depends on both the solvent dielectric constant and the solute charge and geometry, is then discretized in terms of point charges each placed on a different portion of surface identified by a collocation point $s_k$ and an area $a_k$. This partition of the cavity surface into patches, or tesserae, represents one of the main features of the method as well as the shape of the cavity itself, which here has not been limited to analytical simple shapes (as in the original MPE methods), but it can be well modeled according to the actual solute geometry.

The particular aspect we shall treat here is the generalization of the basic theory of these methods to QM descriptions of the solute. To do that, one has to introduce the concept of the Effective Hamiltonian (EH); for the isolated system, the fundamental equation to solve is the standard Schrödinger equation

$$H^0\Psi^0 = E^0\Psi^0 \qquad [8.119]$$

where the superscript 0 indicates the absence of any interactions with other molecules.

In the presence of a solvent treated in the ASC-BE framework, an external perturbation has to be taken into account: the reaction field due to the apparent charges placed on the cavity surface. Such perturbation can be represented in terms of an electrostatic potential $V^R$

to be added to the Hamiltonian $H^0$ so to obtain an effective Hamiltonian H which will satisfy a different Schrödinger equation with respect to [8.119], namely, we have

$$H\Psi = E\Psi \quad [8.120]$$

Both the wave function and the eigenvalue E will be modified with respect to [8.119], due to the presence of the solvent perturbation.

An important aspect, until now not introduced, is that the solvent apparent charges, and consequently the reaction operator $V^R$, depend on the solute charge, i.e., in the present QM framework, on the wave function they contribute to define. This mutual interactions between $\Psi$ and $V^R$ induces a complexity in the problem which can be solved through the standard iterative procedures characterizing the self-consistent (SC) methods. Only for an aspect the calculation in solution has to be distinguished from that in vacuo way; the energy functional to be minimized in a variational solution of [8.120] is not the standard functional E but the new free energy functional G

$$G = \left\langle \Psi \middle| H^0 + 1/2V^R \middle| \Psi \right\rangle \quad [8.121]$$

The 1/2 factor in front of $V^R$ accounts for the linear dependence of the operator on the solute charge (i.e., the quadratic dependence on $\Psi$). In a more physical description, the same factor is introduced when one considers that half of the interaction energy has been spent in polarizing the solvent and it has not been included in G.

In the standard original model the perturbation $V^R$ is limited to the electrostatic effects (i.e., the electrostatic interaction between the apparent point charges and the solute charge distribution); however, extensions to include dispersion and repulsion effects have been formulated. In this more general context the operator $V^R$ can be thus partitioned in three terms (electrostatic, repulsive and dispersive), which all together contribute to modify the solute wave function.

For clarity's sake in the following we shall limit the analysis to the electrostatic part only, referring the reader to refs. [31,83], and to the original papers quoted therein, for details on the other terms. For a more complete report we recall that nonelectrostatic effects can be also included *a posteriori*, i.e., independently of the QM calculation, through approximated or semiempirical models still exploiting the definition of the molecular cavity. In PCM and related methods, the term corresponding to the energy spent in forming the solute cavity in the bulk liquid, is usually computed using the SPT integral equation method (see Section 8.7.1.1).

Until now no reference to the specific form of the operator $V^R$ has been given, but here it becomes compulsory. Trying to remain at the most general level possible, we start to define the apparent charges $q(s_k)$ through a matrix equation, i.e.,

$$\mathbf{q} = -\mathbf{M}^{-1}\mathbf{f}(\rho) \quad [8.122]$$

where **M** is a square matrix depending only on the geometry of the system (e.g., the cavity and its partition in tesserae) and the solvent dielectric constant and **f**(ρ) is a column vector containing the values of a function f, which depends on the solute distribution charge ρ, computed on the cavity tesserae. The form of this function is given by the specific solvation model one uses; thus, for example, in the original version of PCM it coincides with the normal component of the electrostatic field to the cavity surface, while for both the COSMO model[91-93] and the revised PCM version, known as IEF (integral equation formalism),[99] it is represented by the electrostatic potential. In any case, however, this term induces a dependence of the solvent reaction to the solute charge, and finally to its wave function; in a standard SC procedure it has to be recomputed (and thus also the apparent charges **q**) at each iteration.

Also, the form of **M** depends on the solvation model; contrary to **f**, it can be computed once, and then stored, if the geometry of the system is not modified during the calculation.

Once the apparent charges have been defined in an analytical form through eq. [8.122], it is possible to define the perturbation $V^R$ as

$$V^R = \int \frac{q(sk)}{|r - sk|} \rho(\rho) dr \qquad [8.123]$$

In particular, by distinguishing the source of ρ, i.e., the solute nuclei and electrons, we can always define two equations [8.122], one for each source of f, and thus compute two sets of apparent charges, $\mathbf{q}^N$ and $\mathbf{q}^e$, depending on the solute nuclei and electrons, respectively. This allows one to partition the reaction operators in two terms, one depending on the geometry of the system and the solute nuclear charge, and the other still depending on the geometry and on the solute wave function.

Such partition is usually introduced to get an effective Hamiltonian whose form resembles that of the standard Hamiltonian for the isolated system. In fact, if we limit our exposition to the Hartree-Fock approximation in which the molecular orbitals are expressed as an expansion over a finite atomic orbital basis set, the Fock operators one defines for the two systems, the isolated and the solvated one, become (over the atomic basis set)

$$\mathbf{F}^0 = \mathbf{h} + \mathbf{G}(\mathbf{P}) \qquad \mathbf{F} = \mathbf{h} + \mathbf{j} + \mathbf{G}(\mathbf{P}) + \mathbf{X}(\mathbf{P}) \qquad [8.124]$$

where **h** and **G(P)** are the standard one and two-electron interaction matrices in vacuo (**P** represents the one-electron density matrix). In [8.124] the solvent terms are expressed by two matrices, **j** and **X(P)**, whose form follows directly from the nuclei and electron-induced equivalents of expression [8.122], respectively; for the electronic part, ρ has to be substituted by **P**.

The parallel form of $\mathbf{F}^0$ and **F** in [8.124] shows that the two calculations can be performed exactly in the same way, i.e., that solvent does not introduce any complication or basic modification to the procedure originally formulated for the isolated system. This is a very important characteristic of continuum BE solvation methods which can be generalized to almost any quantum mechanical level of theory. In other words, the definition of 'pseudo' one and two-electron solvent operators assures that all the theoretical bases and the formal issues of the quantum mechanical problem remain unchanged, thus allowing a stated solution of the new system exactly as in vacuo.

Above, we have limited the exposition to the basic features of single point HF calculations, and to the thermodynamic functions (the solvation free energy is immediately given by eqs. [8.121-8.124]). However, extensions to other QM procedures as well as to other types of analysis (the evaluation of molecular response functions, the study of chemical reactivity, the reproduction of spectra of various nature, just to quote the most common applications) have been already successfully performed.[83] Many of these extensions have been made possible by the formulation and the following implementation of efficient procedures to get analytical derivatives of the free energy in solution with respect to various parameters.[100] To understand the role of this methodological extension in the actual development of the solvation models we just recall the importance of an efficient computation of analytical gradients with respect to nuclear coordinates to get reliable geometry optimizations and dynamics.[101]

It is here important to recall that such improvements are not limited to BE solvation methods; for example, Rivail and the Nancy group[102] have recently extended their multipole-expansion formalism to permit the analytic computation of first and second derivatives of the solvation free energy for arbitrary cavity shapes, thereby facilitating the assignment of stationary points on a solvated potential energy surface. Analytic gradients for SMx models at ab initio theory have been recently described[103] (even if they have been available longer at the semiempirical level[104]), and they have been presented also for finite difference solutions of the Poisson equation and for finite element solutions.

Many other extensions of different importance have been added to the original BE models in the last few years. Here it may be worth recalling a specific generalization which also has made BE models preferable in particular applications until now almost completely restricted to FE and FD methods.

At the beginning of the present section we indicated the simplified system (4), valid only in the limit of an homogeneous and isotropic dielectric, the commonly studied problem; actually, this has been the only affordable system until recently, with some exceptions represented by systems constituted by two isotropic systems with a definite interface. The situation changed only a few years ago (1997-1998) when BE-ASC methods were extended to treat macroscopically anisotropic dielectrics (e.g., liquid crystals),[99] ionic solutions (e.g., isotropic solutions with nonzero ionic strength),[99] and supercritical liquids.[105]

Another important feature included in BE methods, as well in other continuum methods, are the dynamical effects.

In the previous section devoted to Physical Models, we have recalled that an important issue of all the solvation methods is their ability to treat nonequilibrium, or dynamical, aspects. There, for clarity's sake, we preferred to skip the analysis of this feature, limiting the exposition to static, or equilibrium problems; here, on the contrary, some comments will be added.

In continuum models no explicit reference to the discrete microscopical structure of the solvent is introduced; on one hand, this largely simplifies the problem, but on the other hand it represents a limit which can effect the quality of the calculations and their extensibility to more complex problems such as those involving dynamical phenomena. However, since the first formulations of continuum models, different attempts to overcome this limit have been proposed.

One of the theories which has gained more attention identifies the solvent response function with a polarization vector depending on time, P(t), which varies according to the variations of the field from which it is originated (i.e., a solute and/or an external field). This vector accounts for many phenomena related to different physical processes taking place inside and among the solvent molecules. In practice, strong approximations are necessary in order to formulate a feasible model to deal with such complex quantity. A very simple but also effective way of seeing this problem is to partition the polarization vector in terms accounting for the main degrees of freedom of the solvent molecules, each one determined by a proper relaxation time (or by the related frequency).

In most applications, this spectral decomposition is limited to two terms representing "fast" and "slow" phenomena, respectively[106,107]

$$P(t) = P_{fast} + P_{slow} \qquad [8.125]$$

In particular, the fast term is easily associated with the polarization due to the bound electrons of the solvent molecules which can instantaneously adjust themselves to any change of the inducing field. The slow term, even if it is often referred to as a general orientational polarization, has a less definite nature. Roughly speaking, it collects many different nuclear and molecular motions (vibrational relaxations, rotational and translational diffusion, etc.) related to generally much longer times. Only when the fast and the slow terms are adjusted to the actual description of the solute and/or the external field, on one hand, and to each other on the other hand, do we have solvation systems corresponding to full equilibrium. Nonequilibrium solvation systems, on the contrary, are characterized by only partial response of the solvent. A very explicative example of this condition, but not the only one possible, is represented by a vertical electronic excitation in the solute molecule. In this case, in fact, the immediate solvent response will be limited to its electronic polarization only, as the slower terms will not able to follow such fast change but will remain frozen in the equilibrium status existing immediately before the transition.

This kind of analysis is easily shifted to BE-ASC models in which the polarization vector is substituted by the apparent surface charge in the representation of the solvent reaction field.[107] In this framework, the previous partition of the polarization vector into fast and slow components leads to two corresponding surface apparent charges, $\sigma_f$ and $\sigma_s$, the sum of which gives the total apparent charge $\sigma$. The definition of these charges is in turn related to the static dielectric constant $\varepsilon(0)$ (for the full equilibrium total charge $\sigma$), to the frequency dependent dielectric constant $\varepsilon(\omega)$ (for the fast component $\sigma_f$), and to a combination of them (for the remaining slow component $\sigma_s$).

The analytical form of the frequency dependence of $\varepsilon(\omega)$ in general is not known, but different reliable approximations can be exploited. For example, if we assume that the solvent is polar and follows a Debye-like behavior, we have the general relation:

$$\varepsilon(\omega) = \varepsilon(\infty) + \frac{\varepsilon(0) - \varepsilon(\infty)}{1 - i\omega\tau_D} \qquad [8.126]$$

with $\varepsilon(0)$=78.5, $\varepsilon(\infty)$=1,7756, and $\tau_D$=0.85×10$^{-11}$s in the specific case of water.

It is clear that at the optical frequencies usually involved in the main physical processes, the value of $\varepsilon(\omega)$ is practically equal to $\varepsilon(\infty)$; variations with respect to this, in fact, become important only at $\omega\tau_D$ <<1, where $\varepsilon(\omega)$ is close to the static value. In practice, this means that for a polar solvent like water, where $\varepsilon(\infty)$ is much smaller than the static value, the dynamical solvent effects are small with respect to the static ones. Numerically, this means that all processes described in terms of a static or a dynamic solvent approach, as well as solute response properties computed in the full solute-solvent equilibrium or, instead in a condition of nonequilibrium, will present quite different values.

As a last note we underline that, while it is easy to see that nonequilibrium effects have to be taken into account in the presence of fast changes in the electronic distribution of the solute, or of an oscillating external field (as that exploited to measure molecular optical properties), nonequilibrium approaches for nuclear vibrational analyses are still open.

## 8.8 PRACTICAL APPLICATIONS OF MODELING

In this section we will present some examples of how to determine properties of pure liquids by means of the previously-shown computational procedures. In spite of the fact that most literature in this field is concerned with the study of the properties of water, we will rather focus the attention on the determination of the properties of a few organic solvents of common use (for a review of studies on water, we address interested readers to the already quoted paper of Floris and Tani).[40] Obviously, the following discussion cannot be considered as a complete review of all the literature in the field, but it is rather intended to give the reader a few suggestions on how the previously-shown computational methodologies can be used to obtain data about the most usual properties of pure liquids.

From what we said in the previous section, it should be clear to the reader that continuum models cannot properly be used in order to determine liquid properties: they are in fact concerned in the treatment of a solute in a solvent, and so not of direct use for the study of the solvent properties. For this reason, we will present in the following some examples concerning the application of integral equations and computer simulations.

The analysis will be limited to the examination of the most usual static and dynamic properties of liquids.

### Dielectric constant

Let us begin the discussion by considering a couple of examples regarding the calculation of the dielectric constant of pure liquids.

The critical role of the dielectric constant in determining the properties of a liquid and in particular in moderating the intermolecular interactions cannot be overemphasized. It is then extremely important that potential energy functions used, for example, in computer simulations are able to reproduce the experimental value of the dielectric constant and that such a calculation can be used to estimate the validity of any empirical potential to be used in simulations. The problem of the calculation of the dielectric response is complex:[108] many theories and methods are available, giving sometimes different and contrasting results.

It is accepted that for the calculation of the dielectric constant (as well as other properties dependent on intermolecular angular correlations) via computer simulations, it is compulsory to include long-range electrostatic interactions in the treatment. Such interactions can be taken into account by using either the Ewald summation approach or using the reaction field approach. Without going into details, we simply recall here that in the Ewald approach, electrostatic interactions are evaluated using an infinite lattice sum over all periodic images, i.e., by imposing a pseudo-crystalline order on the liquid. In the reaction field approach, however, electrostatic interaction beyond the spherical cutoff (see Section 8.7.2 for details) are approximated by treating the part of the system outside the cutoff radius as a polarizable continuum of a given dielectric constant $\varepsilon_{rf}$.

Several methods exist for the calculation of the dielectric constant by means of computer simulations. The most widely used is the calculation of the average of the square of the total dipole moment of the system, $<M^2>$ (fluctuation method). Using reaction field boundary conditions, the dielectric constant ε of the liquid within the cutoff radius is calculated by applying the following expression, where V is the volume of the system, k the Boltzmann constant and T the temperature:

$$\frac{4\pi}{9}\frac{\langle M^2\rangle}{kVT} = \left(\frac{\varepsilon - 1}{3}\right)\frac{2\varepsilon_{rf} + 1}{2\varepsilon_{rf} + \varepsilon} \quad [8.127]$$

In the application of eq. [8.127] careful attention has to be paid to the choice of $\varepsilon_{rf}$, the dielectric constant to use for the continuum part of the system. Usually, rapid convergence in calculations is obtained by choosing $\varepsilon_{rf} \approx \varepsilon$.

A different approach to the calculation of the dielectric constant via computer simulations is given by the polarization response method, i.e., to determine the polarization response of a liquid to an applied electric field $E_0$. If <P> is the average system dipole moment per unit volume along the direction of $E_0$, ε can be calculated using the following expression:

$$\frac{4\pi}{9}\frac{\langle P\rangle}{E^0} = \left(\frac{\varepsilon - 1}{3}\right)\frac{2\varepsilon_{rf} + 1}{2\varepsilon_{rf} + \varepsilon} \quad [8.128]$$

Even if this second approach is more efficient than the fluctuation method, it requires the perturbation of the liquid structure following the application of the electric field.

As an example, such a methodology has been applied by Essex and Jorgensen[109] to the calculation of the dielectric constant of formamide and dimethylformamide using Monte Carlo statistical mechanics simulations (see Section 8.7.2 for details). The simulation result for dimethylformamide, 32±2, is reasonably in agreement with respect to the experimental value, 37. However, in the case of formamide, the obtained value, 56±2, underestimates the experimental value of 109.3. The poor performance here addresses the fact that force field models with fixed charges underestimate the dielectric constant for hydrogen-bonded liquids.

Other methodologies exist for the calculation of the static dielectric constant of pure liquids by means of computer simulations. We would like to recall here the use of ion-ion potentials of mean force,[110] and an umbrella sampling approach whereby the complete probability distribution of the net dipole moment is calculated.[111]

Computer simulations are not the only methods which can be used to calculate the dielectric constant of pure liquids. Other approaches are given by the use of integral equations, in particular, the hypernetted chain (HNC) molecular integral equation and the molecular Ornstein-Zernike (OZ) theory (see Section 8.7.1 for details on such methodologies).

The molecular HNC theory gives the molecular pair distribution function g(12) from which all the equilibrium properties of a liquid can be evaluated. It is worth recalling here that the HNC theory consists for a pure liquid in solving a set of two equations, namely, the OZ equation [8.76] and the HNC closure relation [8.78] and [8.80].

In order to solve the HNC equations, the correlation functions are written in terms of a series of rotational invariants $\Phi^{mnl}_{\mu\nu}(12)$, which are defined in terms of cosines and sines of the angles defining the relative orientation of two interacting particles 1 and 2:

$$c(12) = \sum_{\substack{mnl \\ \mu\nu}} c_{\mu\nu}^{mnl}(R)\Phi_{\mu\nu}^{mnl}(12) \qquad [8.129]$$

where the coefficients $c_{\mu\nu}^{mnl}(12)$ depend only on the distance between the mass centers of the particles, and the indexes m, n, l, μ, and ν are integers satisfying the following inequalities:

$$|m-n| \le l \le m+n, \quad -m \le \mu \le m, \quad -n \le \nu \le n \qquad [8.130]$$

The dielectric constant can be obtained from the rotational invariant coefficients of the pair correlation function. For example, following Fries et al.,[112] it can be evaluated using the Kirkwood g factor:

$$g = 1 - \frac{4\pi}{\sqrt{3}}\rho \int_0^\infty R^2 h_{00}^{110}(R)dR \qquad [8.131]$$

$$\frac{4\pi}{9}\beta\rho m_e^2 g = \frac{(\varepsilon-1)(2\varepsilon+1)}{9\varepsilon} \qquad [8.132]$$

In the previous relations, ρ is the number density of molecules, β=1/kT, and $m_e$ is the effective dipole of molecules along their symmetry axes.

Such an approach has been used to evaluate the dielectric constant of liquid acetonitrile, acetone and chloroform by means of the generalized SCMF-HNC (self-consistent mean field hypernetted chain), and the values obtained are 30.8 for acetonitrile, 19.9 for acetone, and 5.66 for chloroform. The agreement with the experimental data, 35.9 for acetonitrile, 20.7 for acetone, and 4.8 for chloroform (under the same conditions), is good.

Another way of evaluating the dielectric constant is that used by Richardi et al.[113] who applied molecular Ornstein-Zernike theory and eq. [8.131] and [8.132]. The obtained values for both acetone and chloroform are low with respect to experimental data, but in excellent agreement with Monte Carlo simulation calculations. In this case, the dielectric constant is evaluated by using the following expressions:

$$\frac{(\varepsilon-1)(2\varepsilon_{rf}+1)}{3(2\varepsilon_{rf}+\varepsilon)} = \frac{\rho\mu^2}{9kT\varepsilon_0}\frac{\left\langle \sum_{i=1}^{N}\sum_{j=1}^{N}\mu_i\mu_j \right\rangle}{N\mu^2} \qquad [8.133]$$

where $\varepsilon_{rf}$ is the dielectric constant of the continuum part of the system, i.e., the part outside the cutoff sphere, N the number of molecules of the system and μ the dipole moment.

A further improvement in the mentioned approach has been achieved by Richardi et al.[114] by coupling the molecular Ornstein-Zernike theory with a self-consistent mean-field approximation in order to take the polarizability into account. For the previously-mentioned solvents (acetone, acetonitrile and chloroform), the calculated values are in excellent agreement with experimental data, showing the crucial role of taking into account polarizability contributions for polar polarizable aprotic solvents.

### Thermodynamical properties

Paying attention again to calculations for acetone and chloroform, we would like to mention two papers by the Regensburg group[113,114] exploiting molecular Ornstein-Zernike (MOZ) theory and site-site Ornstein-Zernike (SSOZ) theory for the calculation of excess internal energies. In the MOZ framework, the internal excess energy per molecule is computed by a configurational average of the pair potential V acting between molecules:

$$E^{ex} = \frac{\rho}{2\Omega^2}\int\int\int V(R_{12},\Omega_1,\Omega_2)\, g(R_{12},\Omega_1,\Omega_2)\, dR_{12}\, d\Omega_1 d\Omega_2 \qquad [8.134]$$

with ρ the density of molecules.

Using instead SSOZ, $E^{ex}$ can be directly calculated from the site-site distribution function, $g_{\alpha\beta}$, as:

$$E^{ex} = 2\pi\rho\int_0^{\infty}\sum_{\alpha=1}^{k}\sum_{\beta=1}^{k} g_{\alpha\beta}(r) V_{\alpha\beta}(r) r^2 dr \qquad [8.135]$$

where k is the number of sites per molecule.

In both cases the calculated values are in good agreement with experimental data and with calculations done exploiting the Monte Carlo computer simulation. In this case, $E^{ex}$ can be obtained as:

$$E^{ex} = \frac{1}{N}\sum_{i=1}^{N}\sum_{j>1}^{N}\sum_{\alpha=1}^{k}\sum_{\beta=1}^{k} V_{\alpha\beta}\left(r_{\alpha\beta,ij}\right) + \frac{\varepsilon_{rf}-1}{2\varepsilon_{rf}+1}\frac{z_\alpha z_\beta e_0^2 r_{\alpha\beta,ij}^2}{4\pi\varepsilon_0 r_{cutoff}^3} \qquad [8.136]$$

where k is the number of sites per molecule, $r_{\alpha\beta,ij}$ is the distance between site α of molecule i and site β of molecule j. Only pairs i,j of molecules for which $r_{ij}<r_{cutoff}$ are considered.

Further refinements of the mentioned approach have been proposed in the already quoted paper by Richardi et al.[114] by coupling the molecular Ornstein-Zernike theory with a self-consistent mean-field approximation. Here again, computed data are in excellent agreement with experiments.

As already said, the use of integral equation methodologies is not the only possible way of obtaining thermodynamical properties for liquids. Computer simulations are also widely used, and it is then possible to obtain equilibrium properties by mean of both Monte Carlo and Molecular Dynamics simulations.

As an example, we would like to mention some calculations on equilibrium properties for liquid ethanol at various temperatures.[115] The calculated values for the heat of vaporization are in overall agreement with experiments, showing an error of the order of 1-1.5%.

### Compressibilities

As an example of how to determine compressibilities by mean of integral equations, we would like to quote here the already cited paper of Richardi et al.[113] in which molecular Ornstein-Zernike (MOZ) theory and site-site Ornstein-Zernike (SSOZ) theory have been exploited. In particular, using the MOZ, the compressibility $\kappa_T$ can be evaluated using the equation:

$$\rho kT\kappa_T = 1 + 4\pi\rho\int_0^\infty \left(g_{00}^{000}(R_{12}) - 1\right)R_{12}^2 dR_{12} \qquad [8.137]$$

and exploiting SSOZ can be calculated from any site-site distribution function $g_{\alpha\beta}$ as:

$$\rho kT\kappa_T = 1 + 4\pi\rho\int_0^\infty \left(g_{\alpha\beta}(r) - 1\right)r^2 dr \qquad [8.138]$$

For acetone and chloroform, the obtained results are too large.

### Relaxation times and diffusion coefficients

Let us now consider an example of how to calculate dynamic properties of pure liquids by computational methodologies. As already said in Section 8.7.2, molecular dynamics simulations are able to take into account the time-dependence in the calculation of liquid properties.

Paying attention to pure acetone, Brodka and Zerda[116] have calculated rotational relaxation times and translational diffusion coefficients by molecular dynamics simulations. In particular, the calculated rotational times of the dipole moment can be compared with a molecular relaxation time $\tau_M$ obtained from the experimentally determined $\tau_D$ by using the following expression, which considers a local field factor:

$$\tau_M = \frac{2\varepsilon_s + \varepsilon_\infty}{3\varepsilon_s}\tau_D \qquad [8.139]$$

where $\varepsilon_s$ and $\varepsilon_\infty$ are the static and optical dielectric constants, respectively. The calculated data are in good agreement with experimental values.

Diffusion coefficients can be calculated directly from the velocity correlation functions or from mean square displacements, as:

$$D^{vcf} = \frac{1}{3}\int_0^\infty \langle v(t)v(0)\rangle dt \qquad [8.140]$$

$$D^{msd} = \lim_{i\to\infty}\frac{1}{6t}\left\langle \left|r(t)r(0)\right|^2\right\rangle \qquad [8.141]$$

The values obtained by using the [8.140] or the [8.141] are almost the same, and properly describe the experimental temperature and density dependencies of the diffusion coefficients, even if about 30% smaller than obtained using NMR spin-echo techniques.

Among the pure solvents we have treated so far, other available data regard the calculation of the self diffusion coefficient (D) for liquid ethanol at different temperatures.[115d] The D parameter was obtained from the long-time slope of the mean-square displacements of the center of mass: experimental changes of D over the 285-320K range of temperatures were acceptably reproduced by molecular dynamics simulations.

### Shear viscosity

To end this section, we would like to report an example of calculation of the shear viscosity again taking as an example pure acetone.

The viscosity can be computed by using molecular dynamics simulations from the virial form of the molecular pressure tensor P, which can be represented as a sum of four contributions:

$$PV = \sum_{i=1}^{N} \frac{p_i p_i}{m_i} + P_S V + P_R V + P_C V \qquad [8.142]$$

where V is the volume of the cell, N the number of molecules, $p_i$ and $m_i$ the momentum and mass of molecule i.

The first term in eq. [8.142] is the kinetic contribution, while $P_S$ is the short-range potential interaction contribution, $P_R$ the reciprocal-space portion of the pressure tensor, and $P_C$ a correction tensor term.

From the xy elements of the time-averaged molecular pressure tensor [8.142] and the applied shear (or strain) rate $\gamma = \partial v_x / \partial y$, the viscosity $\eta$ can be computed using the constitutive relation:

$$\eta = -\frac{\langle P_{xy} - P_{yx} \rangle}{2\gamma} \qquad [8.143]$$

the calculated values for acetone (extrapolated data) are in reasonable agreement with experimental data, as well as results for acetone/methanol and acetone/water mixtures.[117]

## 8.9 LIQUID SURFACES

Until now our task has been relatively easy, as we have only considered the general methodological aspects of the computational description of liquids.

To proceed further, we have to be more specific, entering into details of specific problems, all requiring the introduction and the characterization of different properties, and of the opportune methodological (computational) tools to treat them. We also note that, as additional complication, it can happen that things change considerably when the full range of macroscopic variables (temperature, pressure, volume) is considered, and when other components are added to the liquid system.

Not much effort is needed to convince readers of the overwhelming complexity of the task we are here considering. Let us take a simple pure liquid: to span the range of the thermodynamical variables means to consider systems in which the liquid is in presence of the solid, of the vapor, including supercritical as well as super-cooled states, and all the related phenomena, as ebullition, vaporization, condensation and freezing.

All states and all dynamical processes connecting states are of interest for chemistry, chemical engineering, and physical chemistry. Some properties can be defined for all the states (such as the thermodynamic basic quantities); others are specific for some situations, such as the surface tension. But also for properties of general definition, the techniques to use have often to be very different. It is not possible, for example, to use the same technique to study the temperature dependence of the Helmholtz free energy in a bulk liquid or in the same liquid under the form of a fog near condensation.

Things become more complex when a solute is added to the liquid, and more and more complex when the system contains three or more components.

We are not sure whether any case occurring in this field has been subjected to theoretical analysis and modeling; however, the number of problems considered by theoretical research is very large, and there are models, generally based on the use of molecular potentials, that can give hints, and often useful computational recipes.

These remarks, which are trivial for every reader, with even a minimal experience in working with solvents, have been added here to justify our choice of not trying to give a schematic overview, and also to justify the extremely schematic presentation of the unique specific more advanced subject we shall here consider, the interfacial properties of liquids.

### 8.9.1 THE BASIC TYPES OF LIQUID SURFACES

The surface of a liquid is often represented as a very small fraction of the liquid specimen, and it may be neglected. This statement is not of general validity, however, and often the surface plays a role; in some cases, the dominant role. To be more specific, it is convenient to consider a classification of liquid surfaces, starting from a crude classification, and introducing more details later.

The basic classification can be expressed as follows:

- Liquid/gas surfaces;
- Liquid/liquid surfaces;
- Liquid/solid surfaces.

Liquid/gas surfaces may regard a single substance present in both phases, liquid and vapor at equilibrium or out of equilibrium, but they may also regard a liquid in the presence of a different gas, for example, air (in principle, there is always a given amount of solvent in the gas phase). In the presence of a different gas, one has to consider it in the description of the liquid, both in the bulk and in the interface. In many cases, however, the role of the gas is marginal, and this is the reason why in many theoretical studies on this interface the gas is replaced by the vacuum. This is a very unrealistic description for liquids at equilibrium, but it simply reflects the fact that in simplifying the model, solvent-solute interactions may be sacrificed without losing too much in the description of the interfacial region.

The potentials used to describe these interfaces are those used to describe bulk liquids. The differences stay in the choice of the thermodynamical ensemble (grand canonical ensembles are often necessary), in the boundary conditions to be used in calculations, and in the explicit introduction in the model of some properties and concepts not used for bulk liquids, like the surface tension. Much could be said in this preliminary presentation of liquid/gas interfaces, but we postpone the few aspects we have decided to mention, because they may be treated in comparison with the other kind of surfaces.

Liquid/liquid surfaces regard, in general, a couple of liquids with a low miscibility coefficient, but there are also cases of one-component systems presenting two distinct liquid phases, for which there could be interest in examining the interfacial region at the coexistence point, or for situations out of the thermodynamical equilibrium.

If the second component of the liquid system has a low miscibility with the first, and its molar ratio is low, it is easy to find situations in which this second component is arranged as a thin layer on the surface of the main component with a depth reducing, in the opportune conditions, to the level of a single molecule. This is a situation of great practical importance to what we experience in our everyday life (petrol on the surface of water); however, it has to be related to a finer classification of liquid interfaces we shall consider later.

Once again the potentials used for liquid/liquid surfaces are in general those used for bulk liquids eventually with the introduction of a change in the numerical values of the parameters.[118] The difference from the bulk again regards the thermodynamical ensemble, the boundary conditions for the calculations and the use of some additional concepts.

The liquid/solid surfaces may be of very different types. They can be classified according to the electrical nature of the solid as:

- Liquid/conductor
- Liquid/dielectric

The presence of mobile electrons in metals and other solid substances (to which some liquids may be added, like mercury[119]), introduces in the modelling of the liquid interface new features not present in the previously-quoted interfaces nor in the bulk liquids. For many phenomena occurring at such interfaces there is the need for taking into explicit account electrons flowing from one phase to the other, and of related electron transfers occurring via redox mechanisms. The interaction potentials have to be modified and extended accordingly.[120]

The good electrical conductivity of the solid makes more sizeable and evident the occurrence of phenomena related to the presence of electric potentials at the interface (similar phenomena also occur at interfaces of different type, however[121]). A well-known example is the double layer at the liquid side of an electrolyte/electrode surface. For the double layer, actually there is no need of interaction potentials of special type: the changes in the modelling mainly regard the boundary conditions in the simulation or in the application of other models, of continuum or integral equation type.

It is clear that attention must be paid to the electrostatic part of the interaction. Models for the solid part similar to those used for bulk liquids play a marginal role. There are models of LJ type, but they are mainly used to introduce in the solid component the counterpart of the molecular motions of molecules, already present in the description of the liquid (vibrations of the nuclei: phonons, etc.).

The electrostatic problem may be treated with techniques similar to those we have shown for the continuum electrostatic model for solutions but with a new boundary condition on the surface, e.g., V=constant. In this case, a larger use is made (especially for planar surfaces) of the image method,[122] but more general and powerful methods (like the BEM) are gaining importance.

An alternative description of the metal that can be profitably coupled to the continuum electrostatic approach is given by the jellium model. In this model, the valence electrons of the conductor are treated as an interacting electron gas in the neutralizing background of the averaged distribution of the positive cores (some discreteness in this core distribution may be introduced to describe atomic motions). The jellium is described at the quantum level by the Density Functional Theory (DFT) and it represents a solid bridge between classical and full QM descriptions of such interfaces:[123] further improvements of the jellium which introduce a discrete nature of the lattice[124] are now often used.[125]

The liquid/dielectric surfaces exhibit an even larger variety of types. The dielectric may be a regular crystal, a microcrystalline solid, a glass, a polymer, an assembly of molecules held together by forces of different type.

The potentials in use for crystals and those used for covalently bound systems are cognate with those used for liquids, so it is not compulsory to open here a digression on this theme. We have already quoted the MM methods generally used for solids of molecular type; for more general classes of compounds, a variety of 'atom-atom' potentials have been given.[126] The difference with respect to the bulk is mainly due, again, to different ways of handling the opportune boundary conditions. It is evident that the techniques used to describe a water-protein interface greatly differ, in microscopic detail, time scale, etc., to those used to describe the surfaces between a liquid and a regular crystal surface.

For many problems, the solid surface in contact with the liquid may be held fixed: this characteristic has been amply exploited in simulation studies. We quote here two problems for which this simplification is not possible: problems in which the structure of the solid (or solid-like) component is subjected to conformational changes, and problems in which a progressive expansion (or reduction) of the solid phase plays the main role.

To the first case belong problems addressing the structure of the interfacial zone in the case of biopolymers (e.g., proteins) or similar covalently bound massive molecules, and ordered assemblies of molecules held together by strong non-covalent interactions (as membranes or similar structures). In both cases, the liquid and the solid component of the systems are generally treated on the same footing: interaction force fields are used for both components, introducing, where appropriate, different time scales in the molecular dynamics calculations.

The second case mainly regards the problem of crystal growth or dissolution in a liquid system (pure liquid or solution). A mobile surface of this type is not easy to treat with standard methods: it is better to pass to more specialistic approaches, for which there is the need of introducing a different mathematical treatment of the (mobile) boundary surface conditions. This is a subject that we cannot properly treat here.

### 8.9.2 SYSTEMS WITH A LARGE SURFACE/BULK RATIO

To complete this quick presentation of the three main types of liquid surfaces, it is convenient to consider systems in which the interfacial portion of the liquid is noticeably larger than in normal liquids.

Some among them are denoted in literature as constrained liquids, others as dispersed systems. This classification is not universal, and it does not take in account other cases.

A typical example of constrained liquid is that present in pores or capillaries. The surface is of liquid/solid type, and the same remarks about interaction potentials and computational methods we have already offered may be applied here. More attention must be paid in computations of interactions among separate portions of the liquid interface. In the slit pore case (i.e., in a system composed of two parallel plane solid surfaces, with a thin amount of liquid between them), there may be an interference between the two distinct surfaces, more evident when the thickness of the liquid is small, or when there is an electrical potential between the two faces. The slit pore model is extensively used to describe capillarity problems: in the most usual cylindrical capillary types there is a radial interaction similar in some sense to the interaction with a single solid surface.

Pores of different shape are present in many solid materials: zeolites and other aluminosilicates are outstanding examples. When the pore is very small there is no more distinction between superficial and bulk molecules of the liquid. Nature (and human ingenuity) offers examples over the complete range of situations. For all the systems we have here introduced (quite important in principle and for practical reasons), there is no need of introducing important changes in the standard theoretical approach with respect to that used for other liquid systems. Computer simulations are the methods more largely used to describe pores and micro-pores in molecular sieves of various nature.

A different type of confined liquid is given by a thin layer spread on a surface. The difference with respect to the micro-pore is that this is a three-component system: the surface may be a solid or a liquid, while the second boundary of the layer may regard another liquid or a gas. An example is a thin layer liquid on a metal, in the presence of air (the prodrome of many phenomena of corrosion); a second example is a layer on a second liquid (for example, oil on water). The standard approach may be used again, but other phenomena occurring in these systems require the introduction of other concepts and tools: we quote, for example, the wetting phenomena, which require a more accurate study of surface tension.

The thin layer in some case exhibits a considerable self-organization, giving rise to organized films, like the Blodgett-Langmuir films, which may be composed by several layers, and a variety of more complex structures ending with cellular membranes. We are here at the borderline between liquid films and covalently held laminar structures. According to the degree of rigidity, and to the inter-unit permeation interaction, the computational model may tune different potentials and different ways of performing the simulation.

Another type of confined liquid regards drops or micro-drops dispersed in a medium. We are here passing from confined liquids to the large realm of dispersed systems, some among which do not regard liquids.

For people interested in liquids, the next important cases are those of a liquid dispersed into a second liquid (emulsions), or a liquid dispersed in a gas (fogs), and the opposite cases of a solid dispersed in a liquid (colloids) or of a gas dispersed in a liquid (bubbles).

Surface effects on those dispersed systems can be described, using again the standard approach developed in the preceding sections: intermolecular potentials, computer simulations (accompanied where convenient by integral equation or continuum approaches). There is not much difference with respect to the other cases.

A remark of general validity but particular for these more complex liquid systems must be expressed, and strongly underlined.

We have examined here a given line of research on liquids and of calculations of pertinent physico-chemical properties. This line starts from QM formal description of the systems, then specializes the approach by defining intermolecular potentials (two- or many-body), and then it uses such potentials to describe the system (we have paid more attention to systems at equilibrium, with a moderate extension to non-equilibrium problems).

This is not the only research line conceivable and used, and, in addition, it is not sufficient to describe some processes, among them, those regarding the dispersed systems we have just considered.

We dedicate here a limited space to these aspects of theoretical and computational description of liquids because this chapter specifically addresses interaction potentials and because other approaches will be used and described in other chapters of the Handbook. Several other approaches have the QM formulation more in the background, often never mentioned. Such models are of a more classical nature, with a larger phenomenological character. We quote as examples the models to describe light diffraction in disordered systems, the classical models for evaporation, condensation and dissolution, the transport of the matter in the liquid. The number is fairly large, especially in passing to dynamical and out of equilibrium properties. Some models are old (dating before the introduction of QM), but still in use.

A more detailed description of molecular interactions may lead to refinements of such models (this is currently being done), but there is no reasonable prospect of replacing them with what we have here called the standard approach based on potentials. The standard approach, even when used, may be not sufficient to satisfactorily describe some phenomena. We quote two examples drawn from the presentation of liquid surfaces in dispersed systems: bubbles and solid small particles.

The description of the energetics of a single bubble, and of the density inside and around it, is not sufficient to describe collective phenomena, such as ebullition or separation of a bubble layer. The description of a single bubble is at a low level in a ladder of models, each introducing new concepts and enlarging the space (and time) scale of the model.

The same holds for small solid particles in a liquid. There is a whole section of chemistry, colloidal chemistry, in which our standard approach gives more basic elements, but there is here, again, a ladder of models, as in the preceding example.

Our standard model extends its range of applicability towards these more complex levels of theory. This is a general trend in physics and chemistry: microscopic approaches, starting from the study of the interaction of few elementary particles, are progressively gaining more and more importance over the whole range of sciences, from biology to cosmology, passing through engineering and, in particular, chemistry. In the case of liquids, this extension of microscopical models is far from covering all the phenomena of interest.

### 8.9.3 STUDIES ON INTERFACES USING INTERACTION POTENTIALS

To close this section on liquid surfaces, limited to a rapid examination of the several types of surface of more frequent occurrence, we report some general comments about the use of the standard approach on liquid surfaces.

The first information coming from the application of the method regards the density profile across the interface. Density may be assumed to be constant in bulk liquids at the equilibrium, with local deviations around some solutes (these deviations belong to the family of cybotactic effects, on which something will be said later). At each type of liquid surface there will be some deviations in the density, of extent and nature depending on the system.

A particular case is that of mobile surfaces, i.e., liquid/gas and liquid/liquid surfaces. Having liquids a molecularly grained structure, there will always be at a high level of spatial deviation a local deviation of the surface from planarity. If there are no other effects, this deviation (sometimes called corrugation, but the term is more convenient for other cases, like liquid/solid surfaces, where corrugation has a more permanent status) averages to zero.

Entropic forces are responsible for other effects on the density profile, giving origin at a large scale to a smooth behavior of the density profile, connecting the constant values of the two bulk densities. The most natural length unit in liquids is the bulk correlation length, $\xi$, i.e., the distance at which the pair correlation $g_{AB}(r)$ has decayed to 1. For bulk water $\xi$ is around 5-6 Å. This large scale behavior hides different local behaviors, that range among two extremes, from capillary waves to van der Waals regime.[127] Capillary waves correspond to a sharp boundary (corrugation apart) with "fingers" of a liquid protruding out of the bulk. Capillary waves may happen both at liquid and gas surfaces, in the former, in a more stable status (i.e., with a large mean life) than in the latter.

The van der Waals regime locally corresponds to the large scale energy profile, with a smooth shape.

The behavior of the density profile largely depends on the chemical nature of the composite system. Some phenomenological parameters, like mutual solubility, may give a good guess about the probable energy profile, but "surprises" are not rare events. It may be added that the density profile is sensitive to the quality of the potential and the method using the potentials. It is not clear, in our opinion, if some claims of the occurrence of a thin void (i.e., at very low density) layer separating two liquid surfaces is a real finding or an artifact due to the potentials. Old results must be considered with caution, because computations done with old and less powerful computers tend to reduce the size of the specimen, thus reducing the possibility of finding capillary waves. The dimension of the simulation box, or other similar limiting constraints, is a critical parameter for all phenomena with a large length scale (this is true in all condensed systems, not only for surfaces).

The density profile for liquid/solid surfaces has been firstly studied for hard spheres on hard walls, and then using similar but a bit more detailed models. It is easy to modify the potential parameters to have an increment of the liquid density near the wall (adhesion) or a decrement. Both cases are physically possible, and this effect plays an important role in capillarity studies.

A corrugation at the solid surface can be achieved by using potentials for the solid phase based on the atomic and molecular constitution. This is almost compulsory for studies on systems having a large internal mobility (biomolecules are the outstanding example), but it is also used for more compact solid surfaces, such as metals. In the last quoted example the discreteness of the potential is not exploited to address corrugation effects but rather to introduce mobility (i.e., vibrations. or phonons) in the solid part of the system, in parallel to the mobility explicitly considered by the models for the liquid phase.

Another type of mobility is that of electrons in the solid phase: this is partly described by the continuum electrostatic approach for dielectrics and conductors, but for metals (the most important conductors) it is better to resort to the jellium model. A QM treatment of jellium models permits us to describe electric polarization waves (polarons, solitons) and their mutual interplay with the liquid across the surface.

Jellium, and the other continuum descriptions of the solid, have the problem of exactly defining the boundary surface. This is the analog of the problem of the definition of the cavity boundary for solutes in bulk solvents, occurring in continuum solvation methods (see Section 8.7.3). The only difference is that there are more experimental data for solutes than for liquid/solid surfaces to have hints about the most convenient modeling. The few accurate ab-initio calculations on liquid/metal systems are of little help, because in order to reach an acceptable accuracy, one is compelled to reduce the solid to a small cluster, too small to describe effects with a large length scale.

The disturbances to the bulk density profile are often limited to a few solvent diameters (typically 2-3), but the effect may be larger, especially when external electric fields are applied.

The second information derived from the application of our methods to surfaces regards the preferred orientation of liquid molecules near the surface. To treat this point there is the need of a short digression.

In bulk homogeneous liquids at the equilibrium orientation, the location of a specific molecule is immaterial. We need to know the orientational modes (often reduced to librations) of the components of the liquid system to write the appropriate statistical partition function. In the case of solvation energies, the rotational component of the factorized solute partition function gives a not negligible contribution to $\Delta G_{solv}$. The translational component of the solute partition function can be factorized apart, as first suggested by Ben-Naim,[128] and included in the cratic component of $\Delta G_{solv}$ with a small correction (the cratic component is the numerical factor taking into account the relationships between standard states in a change of phase).

In the homogeneous bulk liquid we have to explicitly consider orientations for dynamical relaxation effects. The same holds for translation in other dynamical problems (diffusion, transport, etc.).

Coming back to the static problems, it turns out that near the limiting surface, the free energy of the components of the liquid also depends on positions and orientations. A quantity like $\Delta G_{solv}$ for the transfer of M in the solvent S in the bulk must be replaced by $\Delta G_{solv}$ (r, $\Omega$) specifying position and orientation of M with respect to the surface of S.

This effect must not be confused with the cybotactic effects we have mentioned, nor with the hole in the solute-solvent correlation function $g_{MS}(r)$ (see Figure 8.5). The hole in the radial correlation function is a consequence of its definition, corresponding to a conditional property, namely that it gives the radial probability distribution of the solvent S, when the solute M is kept at the origin of the coordinate system. Cybotactic effects are related to changes in the correlation function $g_{MS}(r)$ (or better $g_{MS}(r,\Omega)$) with respect to a reference situation. Surface proximity effects can be derived by the analysis of the $g_{MS}(r,\Omega)$ functions, or directly computed with continuum solvation methods. It must be remarked that the obtention of $g_{MS}(r)$ functions near the surface is more difficult than for bulk homogeneous liquids. Reliable descriptions of $g_{MS}(r, \Omega)$ are even harder to reach.

Orientational preferences are the origin of many phenomena, especially when at the surface there are electric charges, mobile or fixed. Outstanding examples occur in biological systems, but there is a large literature covering other fields. A relatively simple case that has been abundantly studied is the electric potential difference across the surface.[129]

We have not considered yet the behavior of solutes near the interface. The special, but quite important, case of the charged components of a salt solution near a charged boundary is widely known, and it is simply mentioned here.

More generally, the behavior of solutes largely depends on the chemical composition of the system. Small ions are repelled at the air/gas interface, hydrophobic molecules are attracted to the interface. Even the presence of a simple $CH_3$ group has some effect: the air interface of dilute solutions of methanol is saturated in $CH_3OH$ over a sizeable molar fraction range.[130] At the liquid/liquid interface between water and hydrocarbon solvents, solutes with a polar head and an alkylic chain tend to find an equilibrium position, with the polar head in water, at a distance from the surface depending on the chain length.[131]

All the quoted results have been obtained at the static level, employing a mean force potential, obtained in some cases with inexpensive continuum models; in other cases, with computer demanding simulations of the free energy profile.

Simulations can give more details than continuum methods, which exploit a simplified and averaged model of the medium. Among the simulation results we quote the recognition that small cations, exhibiting strong interactions with water molecules, tend to keep bound the first solvation layer when forced to pass from water to another solvent. Analogous effects are under examination for neutral polar solutes at water/liquid interfaces in case the bulk solubility of the two partners is not symmetric (a typical example is the water/octanol system).[132]

We have here mentioned the process of transport of a solute from a liquid phase to another. This is a typical dynamic process for which static descriptions based on mean forces simply represent the starting-point, to be completed with an explicit dynamical treatment. Molecular dynamics is the chosen tool for these studies, which are quite active and promise illuminating results. MD simulations also are the main approach to study another important topic: chemical reactions at the surface.[133]

The study of the various aspects of chemical reactions in bulk liquids, from the thermodynamical balance of the reaction, to the mechanistic aspects, to the reaction rates, represents nowadays a well explored field, in which several approaches may be used. The complexity of chemical systems does not permit us to consider this problem completely settled under the methodological and computational point of view, but important advances have been achieved.

Reactions at the surface are a frontier field, and the same remarks hold for reactions in dispersed liquid systems. Interaction potentials, simulations and other methods will find here renewed stimuli for more precise and powerful formulations.

## REFERENCES

1 M. Eyring, J. Walter, and G.E. Kimball, **Quantum Chemistry**, *Wiley*, New York, 1944.
2 R. Bonaccorsi, R. Cimiraglia, P. Palla, and J. Tomasi, *Int. J. Quant. Chem.*, **29**, 307 (1983).
3 J. Tomasi, B. Mennucci, and R. Cammi in **Molecular Electrostatic Potentials. Concepts and Applications**, J.S. Murray, K. Sen, Eds., *Elsevier*, Amsterdam, 1996, pp.1-103.
4 K. Kitaura, K. Morokuma, *Int. J. Quant. Chem.*, **10**, 325 (1976).
5 F. London, *Z. Phys.*, **60**, 245 (1930).
6 R. G. Parr, and W. Yang, **Density Functional Theory of Atoms and Molecules**, *Oxford University Press*, Oxford, 1989.
7 (a) R. Mc Weeny, *Croat. Chem. Acta*, **57**, 865 (1984); (b) P. Claverie, in **Molecules in Physics, Chemistry and Biology**, J. Maruani Ed., *Kluwer*, Dordrecht, 1988, vol.2, pp.393.
8 (a) R. Cammi, and J. Tomasi, *Theor. Chim. Acta*, **69**, 11 (1986); (b) S. Nagase, T. Fueno, S. Yamabe, and K. Kitaura, *Theor. Chim. Acta*, **49**, 309 (1978).
9 S.F. Boys, and F. Bernardi, *Mol. Phys.*, **19**, 953 (1970).
10 R. Cammi, R. Bonaccorsi, and J. Tomasi, *Theor. Chim. Acta*, **68**, 271 (1985).
11 G. Alagona, C. Ghio, R. Cammi, and J. Tomasi, in **Molecules in Physics, Chemistry and Biology**, J. Maruani Ed., *Kluwer*, Dordrecht, 1988, vol.2, pp.507.
12 F.B. van Duijneveldt, J.G.C.M. van Duijneveldt-van de Rijdt, and J.H. van Lenthe, *Chem. Rev.*, **94**, 1873 (1994).
13 (a) B. J. Jeziorski, R. Moszynski, and K. Szalewicz, *Chem. Rev.*, **94**, 1887 (1994); (b) K. Szalewicz, B. Jeziorski, *Mol. Phys.*, **38**, 191 (1979).
14 R. Eisenschitz, and F. London, *Z. Phys.*, **60**, 491 (1930)
15 G.P. Arrighini, **Intermolecular Forces and Their Evaluation by Perturbation Theory**, *Springer*, Berlin 1981.
16 P. Claverie in **Intermolecular Interactions: from Diatomics to Biopolimers**, B. Pullmann, Ed., *Wiley*, New York, 1978.
17 B. J. Jeziorski, and K. Kolos in **Molecular Interactions**, H. Rateiczak, and W. J. Orville-Thomas, Eds., *Wiley*, Chichester, 1978, pp. 29.
18 (a) E. Scrocco, and J. Tomasi, *Topics Curr. Chem.*, **42**, 97 (1973); (b) E. Scrocco, and J. Tomasi, *Adv. Quant. Chem.*, **11**, 115 (1978).
19 J. Tomasi in **Chemical Applications of Atomic and Molecular Electrostatic Potentials**, P. Politzer, and D. G. Truhlar, Eds., *Plenum*, New York, 1981.
20 J. Tomasi, R. Bonaccorsi, and R. Cammi in **Theoretical Models of Chemical Bonding**, Z. Maksic, Ed., *Springer*, Berlin, 1991, part IV.
21 G. Alagona, R. Cimiraglia, E. Scrocco, and J. Tomasi, *Theor Chim. Acta*, **25**, 103 (1972).

22 F. A. Momany, *J. Phys. Chem.*, **82**, 598 (1978).
23 J. Mulliken, *J. Chem. Phys.*, **23**, 1833 (1955).
24 F.W. Bader, **Atoms in Molecules : A Quantum Theory**, *Oxford University Press*, Oxford, 1994.
25 (a) R. Bonaccorsi, E. Scrocco, and J. Tomasi, *J. Am. Chem. Soc.*, **98**, 4049 (1976); (b) M.M. Francl, R.F. Hout, and W.J. Hehre, *J. Am. Chem. Soc.*, **106**, 563 (1984).
26 (a) A. Wallqvist, P. Ahlström, and G. Karlström, *J. Phys. Chem.*, **94**, 1649 (1990); (b) P. -O Åstrand, A. Wallqvist, and G. Karlström, *J. Chem. Phys.*, **100**, 1262 (1994); (c) P. -O Åstrand, P. Linse, and G. Karlström, *Chem. Phys.*, **191**, 195 (1995).
27 (a) M.P. Hodges, A.J. Stone, and S.S. Xantheas, *J. Phys. Chem. A*, **101**, 9163 (1997); (b) C. Millot, J-C. Soetens, M.T.C. Martins Costa, M.P. Hodges, and A.J. Stone, *J. Phys. Chem. A*, **102**, 754 (1998).
28 (a) L. Helm, and A.E. Merbach, *Coord. Chem. Rev.*, **187**, 151 (1999); (b) H. Ontaki, and T. Radnai, *Chem. Rev.*, **93**, 1157 (1993).
29 G. Chalasinski, J. Rak, M.M. Sczesniak, S.M. Cybulski, *J. Chem. Phys.*, **106**, 3301 (1997).
30 L. Onsager, *J. Am. Chem. Soc.*, **58**, 1486 (1936).
31 (a) J. Tomasi, and M. Persico, *Chem. Rev.*, **94**, 2027 (1994); (b) C. J. Cramer, and D. G. Truhlar, *Chem. Rev.*, **99**, 2161 (1999).
32 (a) S. Miertuš, E. Scrocco, and J. Tomasi, *Chem. Phys.*, **55**, 117 (1981); (b) R. Cammi, and J. Tomasi, *J. Comp. Chem.*, **16**, 1449 (1995).
33 (a) F.M. Floris, M. Persico, A. Tani, and J. Tomasi, *Chem. Phys.*, **195**, 207 (1995); (b) F.M. Floris, A. Tani, and J. Tomasi, *Chem. Phys.*, **169**, 11 (1993).
34 F.M. Floris, M. Persico, A. Tani, and J. Tomasi, *Chem. Phys. Lett.*, (a) **199**, 518 (1992); (b) **227**, 126 (1994).
35 J.G. Gay, and B.J. Berne, *J. Chem. Phys.*, **74**, 3316 (1981).
36 H. Zewdie, *J. Chem. Phys.*, **108**, 2117 (1998).
37 Y. Liu, and T. Ichiye *J. Phys. Chem.*, **100**, 2723 (1996).
38 (a) H. Reiss, H. L. Frisch, J.L. Lebowitz, *J. Chem. Phys.*, **31**, 369 (1959); (b) H. Reiss, *Adv. Chem. Phys.*, **9**, 1 (1965).
39 M. Sprik in **Computer Simulation in Chemical Physics**, M.P. Allen and D.J. Tildesley, Eds., *Kluwer*, Dordrecht, 1993.
40 F.M. Floris, and A. Tani in **Molecular Dynamics: from Classical to Quantum Methods**, P.B. Balbuena, and J.M. Seminario, Eds., *Elsevier*, Amsterdam, 1999; pp. 363-429.
41 A. Wallqvist, and R.D. Mountain, *Rev. Comp. Chem.*, **13**, 183 (1999).
42 J. -P. Hansen, and I. R. McDonald, **Theory of Simple Liquids**, *Academic Press*, London, 1986.
43 C. G. Gray, and K. E. Gubbins, **Theory of Molecular Fluids. Volume 1: Fundamentals**, *Clarendon Press*, Oxford, 1984.
44 J. A. Barker, and D. Henderson, *Rev. Mod. Phys.*, **48**, 587 (1976).
45 R.M. Gibbons, *Mol. Phys.*, **17**, 81 (1969); **18**, 809 (1970).
46 J.O. Hirschfelder (Ed.), Intermolecular Forces, *Adv. Chem. Phys.*, 12 (1967).
47 H. Margenau, and N. R. Kestner, **Theory of Intermolecular Forces**, *Pergamon*, Oxford, 1971; 2nd Edition, Chapt. 5.
48 D.A. McQuarrie, **Statistical Mechanics**, *Harper and Row*, New York, 1976.
49 (a) B.J. Berne, and, R. Pecora, **Dynamic Light Scattering**, *Wiley*, New York, 1976; (b) J.T. Hynes, *Ann. Rev. Phys. Chem.*, **28**, 301 (1977); (c)D. R. Bauer, J.I. Brauman, and R. Pecora, *Ann. Rev. Phys. Chem.*, **27**, 443 (1976).
50 H. L. Frisch, and J.L. Lebowitz, Eds., **The equilibrium theory of classical fluids**, *Benjamin*, New York, 1964.
51 J.K. Percus, and G. J. Yevick, *Phys. Rev.*, **110**, 1 (1958).
52 J.L. Lebovitz, and J.K. Percus, *Phys. Rev.*, **144**, 251 (1966).
53 (a) D. Chandler, and H.C. Andersen, *J. Chem. Phys.*, **57**, 1930 (1972); (b) L. R. Pratt, and D. C. Chandler, *J. Chem. Phys.*, **67**, 3683 (1977).
54 F. Hirata, P.J. Rossky, *J. Chem. Phys.*, **78**, 4133 (1983).
55 D. Chandler, R. Silbey, and B.M. Ladanyi, *Mol. Phys.*, **46**, 1335, (1982).
56 B.C. Perng, M.D. Newton, F.O. Raineri, H.L. Friedman, *J. Chem. Phys.*, **104**, 7153 (1996); 7177 (1996).
57 S. Ten-no, F. Hirata, and S. Kato, *Chem. Phys. Lett.*, **214**, 391 (1993); (b) H. Sato, F. Hirata, and S. Kato, *J. Chem. Phys.*, **105**, 1546 (1996); (c) L. Shao, H.A. Yu, J.L. Gao, J. *Phys. Chem. A*, **102**, 10366 (1998).
58 (a) H.C. Andersen, D. Chandler, and J.D. Weeks, *Adv. Chem. Phys.*, **34**, 105 (1976); (b) T. Boublik, *Fluid Phase Equilibria*, **1**, 37 (1977); (c) M.S. Wertheim, *Ann. Rev. Phys. Chem.*, **30**, 471 (1979).
59 J.S. Rowlinson, *Mol. Phys.*, **8**, 107 (1964).
60 J. A. Barker, and D. Henderson, *J. Chem. Phys.*, **47**, 4714 (1967).

61 J.D. Weeks, D. Chandler, and H.C. Andersen, *J. Chem. Phys.*, **54**, 5237 (1971).
62 J.A. Barker, *J. Chem. Phys.*, **19**, 1430 (1951).
63 (a) H.C. Andersen, and D. Chandler, *J. Chem. Phys.*, **57**, 1918 (1972); (b) B.M. Ladanyi, and D. Chandler, *J. Chem. Phys.*, **62**, 4308 (1975).
64 D.J. Tildesley, *Mol. Phys.*, **41**, 341 (1980)
65 M. Lombardero, J.L.F. Abascal, and S. Lago, *Mol. Phys.*, **42**, 999 (1981).
66 M. P. Allen, and D. J. Tildesley, **Computer Simulation of Liquids**, *Clarendon Press*, Oxford, 1987.
67 D. Frenkel, and B. Smit, **Understanding Molecular Simulation**, *Academic Press*, San Diego, 1996.
68 R. Car, and M. Parrinello, *Phys. Rev. Lett.*, **55**, 2471 (1985).
69 G. Galli, and M. Parrinello in **Computer Simulation in Materials Science**, M. Meyer, and V. Pontikis, Eds., *Kluwer*, Dordrecht, 1991.
70 W. Kohn, and L. J. Sham, *Phys. Rev.*, **140**, A1133 (1965).
71 S. Kirkpatrik, C. D. Gelatt, Jr., and M. P. Vecchi, *Science*, **220**, 671 (1983).
72 K. Laasonen, M. Sprik, M. Parrinello, and R. Car, *J. Chem. Phys.*, **99**, 9080 (1993).
73 E. S. Fois, M. Sprik, and M. Parrinello, *Chem. Phys. Lett.*, **223**, 411 (1994).
74 M. Metropolis, A. W. Rosenbluth, M. N. Rosenbluth, A. N. Teller, and E. Teller, *J. Chem. Phys.*, **21**, 1087 (1953).
75 B. J. Alder, and T. E. Wainwright, *J. Chem. Phys.*, (a) **27**, 1208 (1957); (b) **31**, 459 (1959).
76 D. Nicholson, and N. G. Parsonage, **Computer Simulation and the Statistical Mechanics of Adsorption**, *Academic Press*, New York, 1982.
77 W. W. Wood, and F. R. Parker, *J. Chem. Phys.*, **27**, 720 (1957).
78 J. Gao, *Rev. Comp. Chem.*, **7**, 119 (1996).
79 C. S. Pomelli, and J. Tomasi in **The Encyclopedia of Computational Chemistry**, P. V. R. Schleyer, N. L. Allinger, T. Clark, J. Gasteiger, P. A. Kollman, H. F. Schaefer III, and P. R. Schreiner, Eds., *Wiley*, Chichester, 1998, vol. 4, p. 2343-2350.
80 J.-L. Rivail, and D. Rinaldi in **Computational Chemistry, Review of Current Trends**, J. Leszczynski, Ed., *World Scientific*, New York, 1995; pp. 139.
81 O. Tapia, in **Quantum Theory of Chemical Reactions**, R. Daudel, A. Pullman, L. Salem, and A. Viellard, Eds., *Reidel*, Dordrecht, 1980; Vol. 2, pp. 25.
82 M. Orozco, C. Alhambra, X. Barril, J.M. Lopez, M. Busquest, and F.J. Luque, *J. Mol. Model.*, **2**, 1 (1996);
83 C.Amovilli, V. Barone, R. Cammi, E. Cancès, M. Cossi, B. Mennucci, C.S. Pomelli, J. Tomasi, *Adv. Quant. Chem.*, **32**, 227 (1998).
84 J.G. Angyan, *J. Math. Chem.*, **10**, 93 (1992).
85 (a) B. Honig, K. Sharp, and A.S. Yang, *J. Phys. Chem.*, **97**, 1101 (1993); (b) V.K. Misra, K.A. Sharp, R.A. Friedman, and B. Honig, *J. Mol. Biol.*, **238**, 245 (1994); (c) S.W. Chen, and B. Honig, *J. Phys. Chem. B*, **101**, 9113 (1997).
86 R.A. Friesner, *Ann. Rev. Phys. Chem.*, **42**, 341 (1991)
87 M. Ringnalda, J.M. Langlois, B. Greeley, R. Murphy, T. Russo, C. Cortis, R. Muller, B. Marten, R. Donnelly, D. Mainz, J. Wright, W.T. Polland, Y. Cao, Y. Won, G. Miller, W.A. Goddard III, and R.A. Friesner, PS-GVB Schrodinger Inc. (1994).
88 D. Bashford, and M. Karplus, *Biochemistry*, **29**, 10219 (1990).
89 G. Te Velde, and E.J. Baerends, *J. Comp. Phys.*, **99**, 84 (1992).
90 D. E. Beskos, **Boundary Element Methods in Mechanics**, *North Holland*, Amsterdam, 1997.
91 (a) A. Klamt, and G. Schuurmann, *J. Chem. Soc., Perkin Trans.*, **2**, 799 (1993); (b) J. Andzelm, C. Kolmel, and A. Klamt, *J. Chem. Phys.*, **103**, 9312 (1995).
92 (a) T.N. Truong and E.V. Stefanovich, *J. Chem. Phys.*, **103**, 3709 (1995); (b) E.V. Stefanovich and T.N. Truong, *J. Chem. Phys.*, **105**, 2961 (1996).
93 V. Barone, and M. Cossi, *J. Phys. Chem. A*, **102**, 1995 (1998).
94 D. Rinaldi, and J.-L. Rivail, *Theor. Chim. Acta*, **32**, 57 (1973); (b) J.-L. Rivail, and D. Rinaldi, *Chem. Phys.*, **18**, 233 (1976); (c) J.-L. Rivail, B. Terryn, and M.F. Ruiz-Lopez, *J. Mol. Struct. (Theochem)*, **120**, 387 (1985).
95 (a) K.V. Mikkelsen, E. Dalgaard and P. Swanstrøm, *J. Phys. Chem.*, **91**, 3081 (1987); (b) K.V. Mikkelsen, H. Ågren, H.J.Aa. Jensen and T. Helgaker, *J. Chem. Phys.*, **89**, 3086 (1988); (c) K.V. Mikkelsen, P. Jørgensen and H.J.Aa. Jensen, *J. Chem. Phys.*, **100**, 6597 (1994); (d) K.V. Mikkelsen, Y. Luo, H. Ågren and P. Jørgensen, *J. Chem. Phys.*, **100**, 8240 (1994).
96 DALTON, Release 1.0 (1997), T.Helgaker, H.J.Aa. Jensen, P. Jørgensen, J. Olsen, K. Ruud, H. Ågren, T. Andersen, K.L. Bak, V. Bakken, O. Christiansen, P. Dahle, E.K. Dalskov, T. Enevoldsen, B. Fernandez,

H. Heiberg, H. Hettema, D. Jonsson, S. Kirpekar, R. Kobayashi, H. Koch, K.V. Mikkelsen, P. Norman, M.J.Packer, T. Saue, P.R. Taylor, and O. Vahtras.

97 W.C. Still, A. Tempczyk, R.C. Hawley, and P. Hendrickson, *J. Am. Chem. Soc.*, **102**, 6127 (1990); (b) D. Qiu, P.S. Shemkin, F.P. Hollinger, and W.C. Still, *J. Phys. Chem. A*, **101**, 3005 (1997); (c) M.R. Reddy, M.D. Erion, A. Agarwal, V.N. Viswanadhan, D.Q. McDonald, and W.C. Still, *J. Comput. Chem.*, **19**, 769 (1998).

98 C.J. Cramer and D.G. Truhlar, (a) *J. Am. Chem. Soc.*, **113**, 8305 (1991); (b) *Science*, **256**, 213 (1992); (c) *J. Comp. Chem.*, **13**, 1089 (1992); (d) *J. Comput.-Aid. Mol. Des.*, **6**, 629 (1992); and references in (29b).

99 (a) E. Cancès, B. Mennucci and J. Tomasi, *J. Chem. Phys.*, **107**, 3032 (1997); (b) B. Mennucci, E Cancès, and J. Tomasi, *J. Phys. Chem. B*, **101**, 10506 (1997); (c) E. Cancès, and B. Mennucci, *J. Math. Chem.*, **23**, 309 (1998).

100 (a) R. Cammi, and J. Tomasi, *J. Chem. Phys.*, **100**, 7495 (1994); (b) R. Cammi, and J. Tomasi, *J. Chem. Phys.*, **101**, 3888 (1994).

101 (a) E. Cancès, and B. Mennucci, *J. Chem. Phys.*, **109**, 249 (1998); (b) E. Cancès, B. Mennucci, and J. Tomasi, *J. Chem. Phys.*, **109**, 260 (1998); (d) B. Mennucci, R. Cammi, J. Tomasi, *J. Chem. Phys.*, **110**, 6858 (1999).

102 V. Dillet, D. Rinaldi, J. Bertran, and J.-L. Rivail, *J. Chem. Phys.*, **104**, 9437 (1996).

103 T. Zhu, J. Li, D.A. Liotard, C.J. Cramer, and D.G. Truhlar, *J. Chem. Phys.*, **110**, 5503 (1999).

104 Y.-Y. Chuang, C.J. Cramer, and D.G. Truhlar, *Int. J. Quantum Chem.*, **70**, 887 (1998).

105 C.S. Pomelli, and J. Tomasi, *J. Phys. Chem. A*, **101**, 3561 (1997).

106 (a) Y. Ooshika, *J. Phys. Soc. Jpn.*, **9**, 594 (1954); (b) R. Marcus, *J. Chem. Phys.*, **24**, 966 (1956); (c) E.G. McRae, *J. Phys. Chem.*, **61**, 562 (1957); (d) S. Basu, *Adv. Quantum. Chem.*, **1**, 145 (1964); (e) V.G. Levich, *Adv. Electrochem. Eng.*, **4**, 249 (1966); (f) H.J. Kim, and J.T. Hynes, *J. Chem. Phys.*, **93**, 5194 and 5211 (1990); (g) A.M. Berezhkosvskii, *Chem. Phys.*, **164**, 331 (1992); (h) M.V. Basilevski, and G.E. Chudinov, *Chem. Phys.*, **144**, 155 (1990); (i) K.V. Mikkelsen, A. Cesar, H. Ågren, and H. J. Aa. Jensen, *J. Chem. Phys.*, **103**, 9010 (1995).

107 (a) M.A. Aguilar, F.J. Olivares del Valle, and J. Tomasi, *J. Chem. Phys.*, **98**, 7375 (1993); (b) R. Cammi, J. Tomasi, *Int. J. Quantum Chem: Quantum Chem. Symp.*, **29**, 465 (1995); (d) M.L. Sanchez, M.A. Aguilar, and F.J. Olivares del Valle, *J. Phys. Chem.*, **99**, 15758 (1995); (e) B. Mennucci, R. Cammi, and J. Tomasi, *J. Chem. Phys.*, **109**, 2798 (1998); (f) B. Mennucci, A. Toniolo, and C. Cappelli, *J. Chem. Phys.*, **110**, 6858 (1999).

108 S. H. Glarum, *Mol. Phys.*, **24**, 1327 (1972).

109 J. W. Essex, and W. L. Jorgensen, *J. Phys. Chem.*, **99**, 17956 (1995).

110 L. X. Dang, and B. M. Pettitt, *J. Phys. Chem.*, **94**, 4303 (1990).

111 Z. Kurtovic, M. Marchi, and D. Chandler, *Mol. Phys.*, **78**, 1155 (1993).

112 P. H. Fries, J. Richardi, and H. Krienke, *Mol. Phys.*, **90**, 841 (1997).

113 J. Richardi, P. H. Fries, R. Fischer, S. Rast, and H. Krienke, *Mol. Phys.*, **93**, 925 (1998).

114 J. Richardi, P. H. Fries, and H. Krienke, *Mol. Phys.*, **96**, 1411 (1999).

115 (a) W. L. Jorgensen, *J. Phys. Chem.*, **90**, 1276 (1986); (b) J. Gao, D. Habibollazadeh, L. Shao, *J. Phys. Chem.*, **99**, 16460 (1995); (c) M. E. van Leeuwen, *Mol. Phys.*, **87**, 87 (1996); (d) L. Saiz, J. A. Padro, and E. Guardia, *J. Phys. Chem. B*, **101**, 78 (1997).

116 A. Brodka, and T. W. Zerda, *J. Chem. Phys.*, **104**, 6313 (1996).

117 D. R. Wheeler, and R. L. Rowley, *Mol. Phys.*, **94**, 555 (1998).

118 A.R. Vanbuuren, S.J. Marriuk, and H.J.C. Berendsen, *J. Phys. Chem.*, **97**, 9206 (1993).

119 J.C. Shelley, G.N. Patey, D.R. Berard, and G.M. Torrie, *J. Chem. Phys.*, **107**, 2122 (1997).

120 J.C. Shelley, and D.R. Berard, *Rev. Comp. Chem.*, **12**, 137 (1998).

121 (a) M. Wintherhalter, and W. Helfrich, *J. Phys. Chem.*, **96**, 327 (1992); (b) W. Lorenz, *J. Phys. Chem.*, **95**, 10566 (1991); (c) S.J. Miklavic, *J. Chem. Phys.*, **103**, 4795 (1995).

122 (a) A. Krämer, M. Vassen, and F. Foistman, *J. Chem. Phys.*, **106**, 2792 (1996); (b) R.D. Berard, G.N. Patey, *J. Chem. Phys.*, **97**, 4372 (1992); (c) R.J. Clay, N.R. Goel, and P.F. Buff, *J. Chem. Phys.*, **56**, 4245 (1972).

123 (a) P. Gies, and R.R. Gerhardts, *Phys. Rev. B*, **33**, 982 (1986); (b) R.D. Berard, M. Kinoshita, X. Ye, and G.N. Patey, *J. Chem. Phys.*, **107**, 4719 (1997).

124 J.R. Mannier, and J.P. Perdew, *Phys. Rev. B*, **17**, 2595 (1978).

125 R. Saradha, M.V. Sangaranarayanan, *J. Phys. Chem. B*, **102**, 5099 (1998).

126 D.W. Brenner, O.A. Shenderova, and D.A. Areshkin, *Rev. Comp. Chem.*, **12**, 207 (1998).

127 L.-J. Chen, M. Knackstedt, and M. Robert, *J. Chem. Phys.*, **93**, 6800 (1990).

128 A. Ben-Naim, J. Phys. Chem., 82, 792 (1978); **Water and Aqueous Solutions**, *Plenum Press*, New York, 1987.

129 (a) A.J. Bard, and L.R. Faulkner, **Electrochemical Methods: Fundamentals and Applications**, *Wiley*, New York, 1980; (b) A.W. Adamson, **Physical Chemistry of Surfaces**, *Wiley*, New York, 1990; (c) L.R. Pratt, *J. Phys. Chem.*, **96**, 25 (1992).
130 M. Matsumoto, Y. Takaota, and Y. Kataoka, *J. Chem. Phys.*, **98**, 1464 (1993).
131 R. Bonaccorsi, E. Ojalvo, and J. Tomasi, *Coll. Czech. Chem. Comm.*, **53**, 2320 (1988).
132 D. Michael, and I. Benjamin, *J. Phys. Chem.*, **99**, 16810 (1994).
133 I. Benjamin, (a) in **Molecular Dynamics: from Classical to Quantum Methods**, P.B. Balbuena, and J.M. Seminario, Eds., *Elsevier*, Amsterdam, 1999; pp. 661-701; (b) *Chem. Rev.*, **96**, 1449 (1996).

# 9

# Mixed Solvents

Y. Y. Fialkov, V. L. Chumak
**Department of Chemistry**
**National Technical University of Ukraine, Kiev, Ukraine**

## 9.1 INTRODUCTION

One of the basic problems in conducting a chemical process in solution is control of the process parameters. The most important is the yield of the reaction products and process rates. The equilibrium constants and the rate constants of processes in solutions are multi-factor dependencies, that is, they depend on temperature and many solvent properties. The realization of the process in individual solvent often complicates such controls, and in some cases makes it completely impossible. At the same time, it is possible to choose properties determining the process characteristics in mixed solvents directly. Such properties relate, first of all, to the density, ρ, viscosity, η, permittivity, ε, and specific solvation energy.

In this chapter, the term "mixed solvent" refers to binary solvents - regardless of the relation of their components.

The use of mixed solvents in scientific and industrial practice began at the beginning of the 20th century. But systematic theoretical-experimental research began in the sixties, and its intensity has been growing ever since. There are two basic reports on physical properties of mixed solvents. One is the published in four-volume reference edition by Timmermans[1] in the fifties and published at that period by Krestov and co-author's monograph.[2] The detailed substantiation of information as well as the corresponding bibliography are included in the monograph.[3]

## 9.2 CHEMICAL INTERACTION BETWEEN COMPONENTS IN MIXED SOLVENTS

The developments in physical chemistry and chemical physics of liquid state substantiated common viewpoint that liquids are associated. Assuming this approach as a limit, we may consider the individual solvent as divided into non-associated molecules having smaller energy of intermolecular interaction than the energy of molecules in thermal motion, and to molecules, having energy higher than kT.

### 9.2.1 PROCESSES OF HOMOMOLECULAR ASSOCIATION

In mixed solvents formed of associated component A and non-associated component B (does not interact with component A), the following equilibrium takes place:

$$mA \xleftrightarrow{B} 2A_{m/2} \xleftrightarrow{B} \cdots \xleftrightarrow{B} A_m \qquad [9.1]$$

where:

m number of molecules A forming the homomolecular associate (in subscript - association degree)

The systems alcohol-carbon tetrachloride or carbon acids-cyclohexane are examples of mixed solvents of this type. Interesting examples of such mixed solvents are the systems formed of liquid tetraammonium salts, $R_4NA$, and various liquids that are solvate-inert towards such compounds.

The values of homomolecular association constants of alcohols and acids are high, typically in the range of $10^3$-$10^4$. Therefore concentration of the component A is rather low, even at high content of solvate-inert component B, and may be neglected.

The relative degree of the decomposition of heteromolecular associates into aggregates of a smaller association degree increases with increased permittivity of the non-associated solvate-inert component. In solutions of equal concentration of the two mixed solvents, acetic acid-cyclohexane ($\varepsilon$=1,88) and acetic acid-chlorobenzene ($\varepsilon$=5,6), the relative degree of acid heteromolecular association is higher in the first system than in the second system.

In mixed solvents, formed of two associated components in individual states $A_m$ and $B_n$ that do not interact with each other into specific solvation, the chemical equilibrium of mixed heteromolecular associates is established:

$$A_m + B_n \leftrightarrow x\left(A_{m/x}B_{n/x}\right) \qquad [9.2]$$

where:

x number of molecules of heteromolecular associate

Whether to consider these heteromolecular associates as real chemical adducts is more terminology than a chemical problem. In most cases investigators truly assume that in systems formed of two alcohols or of two carboxylic acids, specific interaction does not exist. In the case of mixed solvents formed of two carbon acids, that is true only when components have similar proton affinity, as is the case of a system such as acetic acid-propionic acid (see further paragraph 9.2.8).

### 9.2.2 CONFORMIC AND TAUTOMERIC EQUILIBRIUM. REACTIONS OF ISOMERIZATION

Various conformers of the same compound differ in their dipole moments. Thus, in the mixed solvents A-B, where A is a liquid whose molecules give conformer equilibrium, and B is solvate-inert towards A, the A conformer ratio changes on addition of B component.

In binary solvents A-B where component A can exist in two tautomeric forms $A_1$ and $A_2$ the ratio of concentrations of these forms changes on addition of component B. Common theory of this influence has been worked out by Kabachnic[4,5] who proposed the common scheme of equilibrium in systems of such type:

$$A_1H + S \xleftrightarrow{I} A_1HS \xleftrightarrow{II} A_1^- \cdot HS^+ \xleftrightarrow{III} A_2HS \xleftrightarrow{IY} A_2H + S$$

$$\uparrow\downarrow \text{III a} \qquad [9.3]$$

$$A_1^- + HS^+$$

The scheme [9.3] includes the following stages occurring in sequence:

I interaction of the first tautomeric form, $A_1H$, with the solvent that leads to formation of a addition product, $A_1HS$, (the solvate composition may be more complicated)

II ionization of solvate with formation of ionic couple, $A_1^-HS^+$

III ionic couple may decompose into ions (stage III a), transfer of ionized complex into a product of solvent addition already in the second tautomeric form

IV formation of the free tautomeric form, $A_2H$ $A_2H$.

The nature of the chemical type of influences on keto-enol equilibrium is obvious: the more basic the solvent, the higher the degree of keto-enol transformation into enol form. If solvent is indifferent with enough degrees of approximation, the well-defined dependence of the constants of keto-enol equilibrium on solvent permittivity is encountered. The difference in dipole moments of tautomeric forms is the basis of these relationships.

In mixed solvents such as ACR-B, the first component may undergo rearrangement reactions:

$$ACR \leftrightarrow ARC \qquad [9.4]$$

These occur under the solvent influence or even with its direct participation:

$$ACR \xleftrightarrow{+B} \cdots \xleftrightarrow{+B} ARC \qquad [9.4a]$$

### 9.2.3 HETEROMOLECULAR ASSOCIATION

When components of the binary solvent A-B are solvent-active to one another, the sum of chemical equilibrium is established:

$$mA + nB \leftrightarrow A_mB_n \qquad [9.5]$$

In most scenarios, the interaction proceeds in steps, i.e., equilibrium solution represents the mixture of heteroassociates of different stoichiometry: AB, $AB_2$, $A_2B$, etc. Systems formed by O-, S-, N-, P-bases with various H-donors (e.g., amines-carboxylic acids, esters-carboxylic acids (phenols), dimethylsulfoxide-carboxylic acid) refer to this type of interaction.

The systems having components which interact by means of donor-acceptor bond (without proton transfer) belong to the same type of solvents (e.g., pyridine-chloracetyl, dimethylsulfoxide-tetrachloroethylene, etc.). Components of mixed solvents of such type are more or less associated in their individual states. Therefore, processes of heteromolecular association in such solvents occur along with processes of homomolecular association, which tend to decrease heteromolecular associations.

### 9.2.4 HETEROMOLECULAR ASSOCIATE IONIZATION

In some cases, rearrangement of bonds leads to formation of electro-neutral ionic associate in binary solvents where heteromolecular associates are formed:

$$A_mB_n \leftrightarrow K_p^{q+}L_q^{p-} \qquad [9.6]$$

In many cases, ionic associate undergoes the process of ionic dissociation (see paragraph 9.2.5). If permittivity of mixture is not high, interaction can limit itself to formation of non-ionogenic ionic associate, as exemplified by the binary solvent anisol-m-cresol, where the following equilibrium is established:

$$C_6H_5OCH_3 + CH_3C_6H_4OH \leftrightarrow C_6H_5OCH_3 \bullet CH_3C_6H_4OH \leftrightarrow [C_6H_5OCH_3 \bullet H]^+ \bullet CH_3C_6H_4O^-$$

## 9.2.5 ELECTROLYTIC DISSOCIATION (IONIC ASSOCIATION)

In solvents having high permittivity, the ionic associate decomposes into ions (mostly associated) to a variable degree:

$$K_p^{q+}L_q^{p-} \leftrightarrow K_{p-1}^{q+}L_{q-1}^{p-} + K^{q+} + L^{p-} \leftrightarrow \cdots \leftrightarrow pK^{q+} + qL^{p-} \qquad [9.7]$$

Electrolyte solution is formed according to eq. [9.7]. Conductivity is a distinctive feature of any material. The practice of application of individual and mixed solvents refers to electrolyte solutions, having conductivities > $10^{-3}$Cm/m. Binary systems, such as carbonic acids-amines, cresol-amines, DMSO-carbonic acids, hexamethyl phosphorous triamide acids, and all the systems acids-water, are the examples of such solvents.

The common scheme of equilibrium in binary solvents, where components interaction proceeds until ions formation, may be represented by the common scheme:

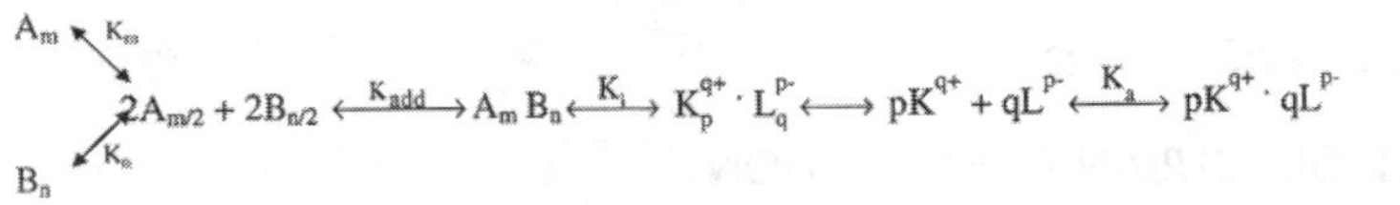


[9.8]

where:

| | |
|---|---|
| $K_m$, $K_n$ | the constants of homomolecular association; respectively |
| $K_{add}$ | the constant of the process formation of heteromolecular adduct |
| $K_i$ | the ionization constant |
| $K_a$ | the association constant. |

The system acetic acid-pyridine[4] may serve as an example of binary solvent whose equilibrium constants of all stages of the scheme [9.8] have been estimated.

## 9.2.6 REACTIONS OF COMPOSITION

As a result of interaction between components of mixed solvents, profound rearrangement of bonds takes place. The process is determined by such high equilibrium constant that it is possible to consider the process as practically irreversible:

$$mA + nB \rightarrow pC \qquad [9.9]$$

where:

| | |
|---|---|
| m, n, p | stoichiometric coefficients |
| C | compound formed from A and B |

Carbonic acid anhydrides-water (e.g., $(CH_3CO)_2O + H_2O \rightarrow 2CH_3COOH$), systems isothiocyanates-amines (e.g., $C_3H_5NCS + C_2H_5NH_2 \rightarrow C_2H_5NHCNSC_3H_5$) are examples of such binary solvents.

It is pertinent that if solvents are mixed at stoichiometric ratio (m/n), they form individual liquid.

## 9.2.7 EXCHANGE INTERACTION

Exchange interaction

$$mA + nB \leftrightarrow pC + qD \quad [9.10]$$

is not rare in the practice of application of mixed solvents. Carbonic acid-alcohol is an example of the etherification reaction. Carbonic acid anhydride-amine exemplifies acylation process $(CH_3CO)_2O + C_6H_5NH_2 \leftrightarrow C_6H_5NHCOCH_3 + CH_3COOH$; carbonic acid-another anhydride participate in acylic exchange: $RCOOH + (R_1CO)_2O \leftrightarrow RCOOCOR_1 + R_1COOH$. The study of the exchange interaction mechanism gives forcible argument to believe that in most cases the process goes through formation of intermediate - "associate":

$$mA + nB \leftrightarrow [A_mB_n] \leftrightarrow pC + qD \quad [9.11]$$

## 9.2.8 AMPHOTERISM OF MIXED SOLVENT COMPONENTS

The monograph of one of the authors of this chapter[7] contains a detailed bibliography of works related to this paragraph.

### 9.2.8.1 Amphoterism of hydrogen acids

In the first decades of the 20$^{th}$ century, Hanch ascertained that compounds which are typical acids show distinctly manifested amphoterism in water solutions while they interact with one another in the absence of solvent (water in this case). Thus, in binary solvents $H_2SO_4$-$CH_3COOH$ and $HClO_4$-$CH_3COOH$-acetic acid accepts proton forming acylonium, $CH_3COOH^{2+}$, cation. At present time, many systems formed by two hydrogen acids are being studied. In respect to one of the strongest mineral acids, $CF_3SO_3H$, all mineral acids are bases. Thus, trifluoromethane sulfuric acid imposes its proton even on sulfuric acid, forcing it to act as a base:[8]

$$CF_3SO_3H + H_2SO_4 \leftrightarrow CF_3SO_3^- \bullet H_3SO_4^+$$

Sulphuric acid, with the exception of the mentioned case, and in the mixtures with perchloric acid (no proton transfer), shows acidic function towards all other hydrogen acids.

The well-known behavior of acidic function of absolute nitric acid towards absolute sulphuric acid is a circumstance widely used for the nitration of organic compounds.

Trifluoroacetic acid reveals proton-acceptor function towards all strong mineral acids, but, at the same time, it acts as an acid towards acetic and monochloroacetic acids.

In mixed solvents formed by two aliphatic carbonic acids, proton transfer does not take place and interaction is usually limited to a mixed associate formation, according to the equilibrium [9.2].

### 9.2.8.2 Amphoterism of L-acids

In binary liquid mixtures of the so-called aprotic acids (Lewis acids or L-acids) formed by halogenides of metals of III-IV groups of the periodic system, there are often cases when one of the components is an anion-donor (base) and the second component is an anion-acceptor (acid):

$$M_IX_m + M_{II}X_n \leftrightarrow M_IX_{m-1}^+ + M_{II}X_{n+1}^- \quad [9.12]$$

The following is an example of eq. [9.12]: In the mixture $AsCl_3$ - $SnCl_4$, stannous tetrachloride is a base and arsenic trichloride is an acid:

$$2AsCl_3 + SnCl_4 \leftrightarrow (AsCl_2^+)_2 \bullet SnCl_6^{-2}$$

### 9.2.8.3 Amphoterism in systems H-acid-L-acid

Amphoterism phenomenon is the mechanism of acid-base interaction in systems formed by H-acids or two L-acids in binary systems. For example, in stannous tetrachloride-carboxylic acid system, acid-base interaction occurs with $SnCl_4$ being an acid:

$$SnCl_4 + 2RCOOH \leftrightarrow [SnCl_4(RCOO)_2]H_2$$

The product of this interaction is such a strong hydrogen acid that it is neutralized by the excess of RCOOH:

$$[SnCl_4(RCOO)_2]H_2 + RCOOH \leftrightarrow [SnCl_4(RCOO)_2]H^- + RCOOH_2^+$$

The mechanism of acid-base interaction in binary solvents of the mentioned type depends often on component relation. In systems such as stannous pentachloride-acetic acid, diluted by $SbCl_5$ solutions, interaction proceeds according to the scheme:

$$2SbCl_5 + 4HAc \leftrightarrow SbCl_3Ac_2 + SbCl_6^- + Cl^- + 2H_2Ac^+$$

In solutions of moderate concentrations, the following scheme is correct:

$$2SbCl_5 + 2HAc \leftrightarrow SbCl_4Ac + Cl^- + H_2Ac^+$$

### 9.2.8.4 Amphoterism in binary solutions amine-amine

If amines differ substantially in their energies of proton affinity, one of them has a proton-donor function in binary mixtures of these amines, so that it acts as an acid:

$$\begin{array}{l} | \quad\; | \qquad | \qquad\quad | \quad\; | \qquad\quad | \\ NH + N - \leftrightarrow N^- \bullet H^+ N \leftrightarrow N^- + H^+ N - \\ | \quad\; | \qquad | \qquad\quad | \quad\; | \qquad\quad | \end{array} \qquad [9.13]$$

In the overwhelming majority of cases, this acid-base interaction limits itself only to the first stage - formation of heteromolecular adduct. Interactions occurring in all stages of the scheme [9.13] are typical of binary solvents such as triethylene amine-pyridine or N-diethylaniline. Proton-donor function towards the amine component reveals itself distinctly in the case of diphenylamine.

To conclude the section discussing amphoterism in binary solvents, we have constructed a series of chemical equilibria that help to visualize the meaning of amphoterism as a common property of chemical compounds:

| base | | acid | | |
|---|---|---|---|---|
| $Et_3N$ | + | $PhNH_2$ | ↔ | $Et_3N \cdot PhNH_2$ |
| $PhNH_2$ | + | $Ph_2NH$ | ↔ | $PhNH_2 \cdot Ph_2NH$ |
| $Ph_2NH$ | + | $H_2O$ | ↔ | $Ph_2NH \cdot H_2O$ |
| $H_2O$ | + | HAc | ↔···↔ | $H_3O^+ + Ac^-$ |
| HAc | + | $CF_3COOH$ | ↔···↔ | $H_2Ac^+ + CF_3COO^-$ |
| $CF_3COOH$ | + | $HNO_3$ | ↔ | $CF_3COOH \cdot HNO_3$ |
| $HNO_3$ | + | $H_2SO_4$ | ↔ | $H_2NO_3^+ + HSO_4^-$ * |
| $H_2SO_4$ | + | $CF_3SO_3H$ | ↔ | $H_3SO_4^+ + CF_3SO_3^-$ |

*The first stage of absolute sulfuric and nitric acids

# 9.3 PHYSICAL PROPERTIES OF MIXED SOLVENTS

## 9.3.1 THE METHODS OF EXPRESSION OF MIXED SOLVENT COMPOSITIONS

In the practice of study and application of mixed solvents the fractional (percentage) method of composition expression, according to which the fraction, Y, or percentage, Y*100, of i-component in mixed solvent are equal, is the most common:

$$Y_i = c_i / \sum_{i=1}^{i=i} c_i \quad [9.14]$$

$$Y_i = \left( c_i / \sum_{i=1}^{i=i} c_i \right) \times 100\% \quad [9.14a]$$

where:

c a method of composition expression.

The volume fraction, V, of a given component in the mixture equals to:

$$V_i = v_i / \sum_{i=1}^{i=i} v_i \quad [9.15]$$

where:

v initial volume of component

Molar fraction, X, equals

$$X = m_i / \sum_{i=1}^{i=i} m_i \quad [9.16]$$

where:

m the number of component moles.

Mass fraction, P, equals to:

$$P = g_i / \sum_{i=1}^{i=i} g_i \qquad [9.17]$$

where:

g mass of a component.

It is evident that for a solvent mixture, whose components do not interact:

$$\sum V = \sum X = \sum P = 1 \qquad [9.18]$$

and expressed in percent

$$\sum V_{100} = \sum X_{100} = \sum P_{100} = 100 \qquad [9.19]$$

Fractions (percentage) [9.19] of mixed solvents, since they are calculated for initial quantities (volumes) of components, are called analytical. In the overwhelming majority of cases, analytical fractions (percentages) are used in the practice of mixed solvents application.

However, in mixed solvents A-B, whose components interact with formation of adducts, AB, $AB_2$, it is possible to use veritable fractions to express their composition. For example, if the equilibrium A+B $\leftrightarrow$ AB establishes, veritable mole fraction of A component equals to:

$$N_A = m_A / (m_A + m_B + m_{AB}) \qquad [9.20]$$

where:

$m_i$ number of moles of equilibrium participants

In the practice of mixed solvents application, though more seldom, molar, $c_M$, and molal, $c_m$, concentrations are used. Correlations between these various methods of concentration expression are shown in Table 9.1.

**Table 9.1. Correlation between various methods of expression of binary solvent A-B concentration (everywhere - concentration of component A)**

| Function | Argument | | | | |
|---|---|---|---|---|---|
| | **x** | **V** | **P** | **$c_M$** | **$c_m$** |
| x | | $V\theta_B / (\theta_A - \sigma_\theta V)$ | $PM_B / (M_A - \sigma_M P)$ | $c_M\theta_B / (10^3\rho - c_M\sigma_M)$ | $c_m M_B / (10^3 + c_m M_B)$ |
| V | $x\theta_A / (\sigma_\theta x + \theta_B)$ | | $P\rho_B / (\rho_A - \sigma_p P)$ | $c_M\theta_A / 10^3$ | $c_m M_B \theta_A / (10^3 \theta_B + c_m M_B \theta_A)$ |
| P | $xM_A / (\sigma_M x + M_B)$ | $V\rho_A / (\sigma_p V + \rho_B)$ | | $c_M M_A / 10^3\rho$ | $c_m M_A / (10^3 + c_m M_A)$ |
| $c_M$ | $10^3\rho x / (\sigma_M x + M_B)$ | $10^3 V / \theta_A$ | $10^3\rho P / M_A$ | | $10^3\rho c_m / (10^3 + c_m M_A)$ |
| $c_m$ | $10^3 x / M_B(1 - x)$ | $10^3\theta_B V / M_B\theta_A(1 - V)$ | $10^3 P / M_A(1 - P)$ | $10^3 c_M / (10^3\rho - M_A c_M)$ | |

$\sigma$ is the difference between the corresponding properties of components, i.e., $\sigma_y = y_A - y_B$, $\rho$ is density, and $\theta$ is molar volume.

### 9.3.1.1 Permittivity

In liquid systems with chemically non-interacting components, molar polarization, $P_M$, according to definition, is additive value; expressing composition in molar fractions:

$$P_M = \sum_{i=1}^{i=i} P_{M_i} x_i \qquad [9.21]$$

where:

$x_i$ molar fraction of corresponding mixture component

In the case of binary solvent system:

$$P_M = P_{M_1} x + P_{M_2}(1-x) = (P_{M_1} - P_{M_2})x + P_{M_2} \qquad [9.21a]$$

where:

x molar fraction of the first component.

Here and further definitions indices without digits relate to additive values of mixture properties and symbols with digits to properties of component solvents. As

$$P_M = f(\varepsilon)M/\rho = f(\varepsilon)\theta$$

then $$f(\varepsilon)\theta = \sum_{1}^{i} f(\varepsilon_i)\theta_i x_i \qquad [9.22]$$

or for binary solvent

$$f(\varepsilon)\theta = f(\varepsilon_1)\theta_1 x + f(\varepsilon_2)\theta_2(1-x) \qquad [9.22a]$$

where:

M an additive molecular mass
ρ density
θ molar volume
f(ε) one of the permittivity functions, e.g., 1/ε or (ε - 1)/(ε + 2)

In accordance with the rules of transfer of additive properties from one way of composition expression to another[3] based on [9.22], it follows that

$$f(\varepsilon) = \sum_{1}^{i} f(\varepsilon_i)V_i \qquad [9.23]$$

and for a binary system

$$f(\varepsilon) = f(\varepsilon_1)V + f(\varepsilon_2)(1-V) \qquad [9.23a]$$

where:

V the volume fraction.

Substituting f(ε) function into [9.23], we obtain volume-fractional additivity of permittivity

$$\varepsilon = \sum_{1}^{i} \varepsilon_i V_i \quad [9.24]$$

or for binary solvent

$$\varepsilon = \varepsilon_1 V + \varepsilon_2 (1 - V) \quad [9.24a]$$

Vast experimental material on binary liquid systems[1,2,3] shows that the equation [9.24] gives satisfactory accuracy for systems formed of components non-associated in individual states. For example, dielectric permittivity describes exactly systems n-hexane-pyridine, chlorobenzene-pyridine by equation [9.24a]. For systems with (one or some) associated components, it is necessary to consider fluctuations, which depend on concentration, density, orientation, etc. Consideration of ε fluctuations shows that calculation of dielectric permittivity for mixture of two chemically non-interacting liquids requires Δε-values determined from equation [9.24a] and substituted to equation:

$$\Delta\varepsilon = (\varepsilon_1 - \varepsilon_1)^2 (\Delta\bar{\rho}) / [\varepsilon(2 + \delta_\varepsilon)] \quad [9.25]$$

where:

| | |
|---|---|
| $\Delta\bar{\rho}$ | fluctuation of density |
| $\delta_\varepsilon$ | $= (\partial\varepsilon_{Add}/\partial V)/(\partial\varepsilon_{exp}/\partial V)$ |
| $\varepsilon_1$, $\varepsilon_2$ | dielectric permitivities of individual components |
| $\varepsilon_{Add}$ | additive value |
| $\varepsilon_{exp}$ | experimental value |

The Δε value may be calculated from a simplified equation. For equal volume mixtures (V=0.5) of two solvents:

$$\Delta\varepsilon_{V=0.5} = 0.043(\varepsilon_1 - \varepsilon_2)^2 / \varepsilon_{V=0.5} \quad [9.25a]$$

Thus, permittivity of equivolume mixture of two chemically non-interacting solvents can be calculated from the equation

$$\varepsilon_{V=0.5} = 0.5(\varepsilon_1 - \varepsilon_2) + \varepsilon_2 - \Delta\varepsilon_{V=0.5} \quad [9.26]$$

Having ε values of both initial components and the ε value for the mixture, it is not difficult to interpolate these values for mixtures for any given composition.

For systems formed of two associated liquids, there are no reliable methods of ε isotherms calculation. An empirical method has been proposed, which is based on the assumption that each associated component mixed with the other associated component that does not interact chemically with the first one introduces the rigorously defined contribution into ε deviation from the volume-fractional additivity. The extent of these deviations for the first representatives of the series of aliphatic carbonic acids and alcohols of normal structure, as well as for some phenols, are given elsewhere.[3]

Permittivity of mixtures at different temperatures can be calculated with high accuracy from absolute temperature coefficients of permittivity and their values using equations identical in form to [9.24]- [9.26]. For example, the calculation of temperature coefficients of permittivity, $\alpha_{\varepsilon}$, of the equivolume mixture of two chemically non-interacting associated solvents is carried out using the equation identical in its form to [9.26].

$$\Delta\alpha_{\varepsilon,V=0.5} = 0.5(\alpha_{\varepsilon_1} - \alpha_{\varepsilon_2}) + \alpha_{\varepsilon_2} - \Delta\alpha_{\varepsilon,V=0.5} \qquad [9.27]$$

where

$$\Delta\alpha_{\varepsilon,V=0.5} = 0.043(\alpha_{\varepsilon_1} - \alpha_{\varepsilon_2}) / \alpha_{\varepsilon,V=0.5} \qquad [9.28]$$

#### 9.3.1.2 Viscosity

There are a great number of equations available in literature intended to describe viscosity of mixtures of chemically non-interacting components. All these equations, regardless of whether they have been derived theoretically or established empirically, are divided into two basic groups. The first group includes equations relating mixture viscosity, $\eta$, to viscosity of initial components, $\eta_i$, and their content in mixture, c, (c in this case is an arbitrary method of concentration expression):

$$\eta = f(\eta_1, \eta_2, \cdots \eta_i; \quad c_1, c_2, \cdots c_{i-1}) \qquad [9.29]$$

The equations of the second group include various constants $k_1$, $k_2$, etc., found from experiment or calculated theoretically

$$\eta = f(\eta_1, \eta_2, \cdots \eta_i; \quad c_1, c_2, \cdots c_{i-1}; \quad k_1, k_2, \cdots) \qquad [9.30]$$

The accuracy of calculation of solvents mixture viscosity from [9.30] type of equations is not higher than from equations of [9.29] type. We, thus, limit discussion to equations [9.29].

The comparison of equations of type [9.29] for binary liquid systems[3] has shown that, in most cases, viscosity of systems with chemically non-interacting components is described by the exponential function of molar-fractional composition

$$\eta = \prod_{1}^{i} \eta_i^{x_i} \qquad [9.31]$$

or for binary solvent:

$$\eta = \eta_1^{x} \eta_2^{1-x} \qquad [9.31a]$$

where:

$x_i$ molar fractions of components of the binary system

Empirical expressions for calculation of viscosity of the system of a given type, which permit more precise calculations, are given in monograph.[3]

From [9.31], it follows that the relative temperature coefficient of viscosity $\beta_\eta = \partial\eta / \eta\partial T = \partial \ln\eta / \partial T$ for mixtures of chemically non-interacting liquids is described by equation:

$$\beta_\eta = x\beta_{\eta_1} + (1-x)\beta_{\eta_2} \qquad [9.32]$$

### 9.3.1.3 Density, molar volume

According to the definition, a density, ρ, of system is the volume-additive function of composition:

$$\rho = \sum_1^i \rho_i V_i \qquad [9.33]$$

or for binary system:

$$\rho = \rho_1 V_1 + \rho_2 (1 - V_1) \qquad [9.33a]$$

From [9.33], it follows that molar volume of ideal system is a molar-additive function of composition:

$$\theta = \sum_1^i \theta_i x_i \qquad [9.34]$$

Molar mass of mixture, according to the definition, is a molar-additive function too:

$$M = \sum_1^i M_i x_i \qquad [9.35]$$

For two-component mixture of solvents, these relations are as follows:

$$\theta = \theta_1 x + \theta_2 (1 - x) \qquad [9.36]$$

$$M = M_1 x + M_2 (1 - x) \qquad [9.37]$$

Non-ideality of the system leads to relative deviations of density and molar volume, as well as of other volumetric properties, having magnitude of about 1%. Here, contraction and expansion of solutions are approximately equiprobable. In systems in conditions close to separation, considerable contraction is possible. However, two component systems formed from the most common solvents, used in practical applications, form solutions far from separation at usual temperature range.

Taking into account the linear density dependence on temperature in equations [9.33] and [9.34], it is possible to calculate values of density and molar volume at any given temperature.

### 9.3.1.4 Electrical conductivity

High conductivity of solvent mixture formed from chemically non-interacting components may be related to the properties of only one or both components. Very high conductivity of mineral acids, carboxylic acids, some complexes of acids with amines, stannous chloride, and some tetraalkylammonium salts increases conductivity of their mixtures with other solvents.[3]

Normal practice does not often deal with mixed solvents of such high conductivity. Therefore, the theory of concentration dependence of conductivity of binary solvents is briefly discussed here.[3,10]

Two basic factors which influence conductivity of binary solvent mixture are viscosity and permittivity. The influence of these factors on specific conductivity is quantitatively considered in empirical equation:

$$\kappa = 1/\eta \exp[\ln \kappa_1 \eta_1 - L(1/\varepsilon - 1/\varepsilon_1)] \qquad [9.38]$$

where:

| | |
|---|---|
| $\kappa$ | conductivity (symbols without index belong to the mixture properties, index 1 belongs to electrolyte component) |
| L | $= a\varepsilon_1\varepsilon_2 / (\varepsilon_1 - \varepsilon_2)$ |
| a | proportionality coefficient |

and in non-empirically equation:

$$\ln \kappa = const_1 + 1/\varepsilon(const_2 + const_3 + const_4) + 1/2 \ln c - \ln \eta \qquad [9.39]$$

where:

| | |
|---|---|
| c | concentration, mol $l^{-1}$ |
| consts | calculated from the equations[11] based on crystallographic radii of the ions, dipole moments of molecules of mixed solution, and $\varepsilon$ of mixture components |

From equations [9.38] and [9.39], it follows that isotherms of the logarithm of conductivity corrected for viscosity and concentration $\ln\kappa\eta c^{1/2}$ have to be a linear function of reciprocal $\varepsilon$. Validity of this assumption is confirmed in Figure 9.1. Control of properties of multi-component mixture of solvents is achieved by methods of optimization based on analytical dependencies of their properties in composition:

$$y^I = f(y_1^I, y_2^I, \cdots c_1, c_2, \cdots c_{i-1}), \quad y^{II} = f(y_1^{II}, y_2^{II}, \cdots c_1, c_2 \cdots c_{i-1}) \qquad [9.40]$$

where:

| | |
|---|---|
| $y^I, y^{II}$ | any property of solvent |
| $y_1^I, y_2^{II}$ | initial property of solvent |

Concentration dependencies of permittivity, viscosity, density (molar volume) and conductivity described here permit to select with certainty the composition of mixed solvent, characterized by any value of mentioned properties.

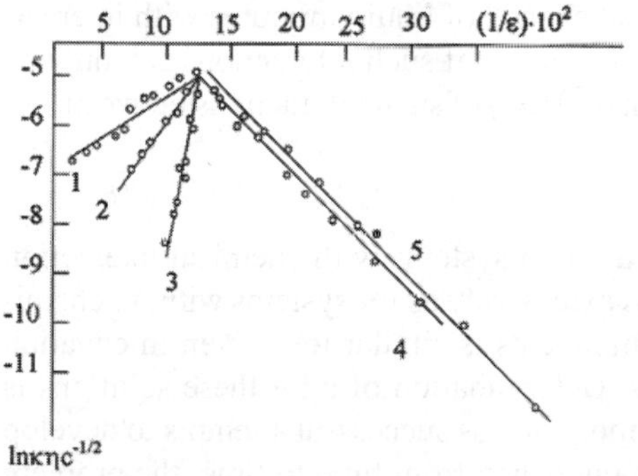


Figure 9.1. Dependencies $\ln\kappa\eta c^{-1/2}$ on $1/\varepsilon$ at 393.15K for systems containing tetrabutylammoniumbromide and aprotic solvent (1-nitromethane, 2-acetonitrile, 3-pyridine, 4-chlorobenzene, 5-benzene).

## 9.3.2 PHYSICAL CHARACTERISTICS OF THE MIXED SOLVENTS WITH CHEMICAL INTERACTION BETWEEN COMPONENTS

In this section, we discuss physical characteristic change due to the changes of binary mixed solvent composition in the systems with chemical interactions between components (these systems are the most commonly used in science and technology). Comprehensive discussion of the physical and chemical analysis of such systems, including stoichiometry and stability constants, determination of the formation of heteromolecular

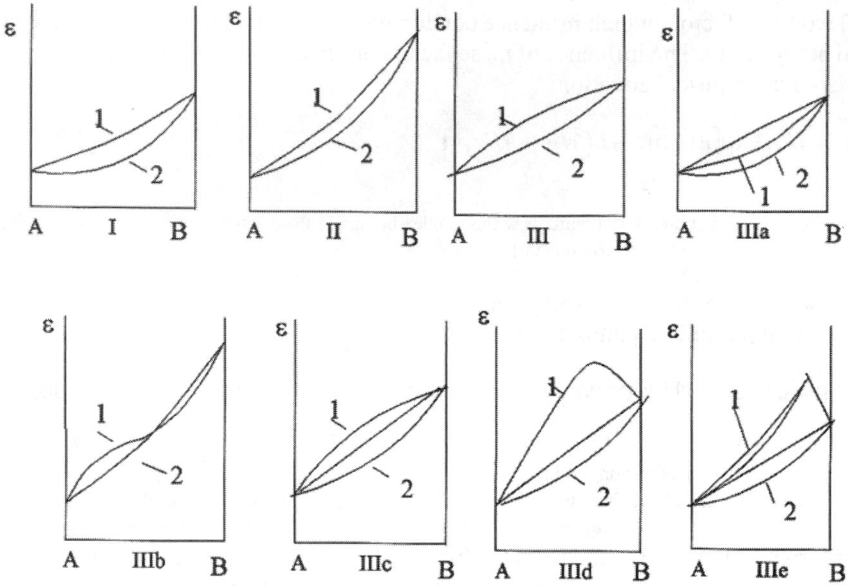


Figure 9.2. Classification of permittivity isotherms for binary solvents: 1 and 2 - isotherms for systems with interaction and without interaction, respectively.

associates can be found in a specialized monograph.[3]

In theory, if composition of the mixture is known, then calculation of any characteristic property of the multi-component mixture can be performed by the classical physical and chemical analysis methods.[3] For binary liquid system, it is possible only if all chemical forms (including all possible associates), their stoichiometry, stability constants, and their individual physical and chemical properties are well determined. A large volume of correct quantitative thermodynamic data required for these calculations is not available. Due to these obstacles, data on permittivity, viscosity and other macro-properties of mixed solvents with interacting components are obtained by empirical means. Data on empirical physical properties of liquid systems can be found in published handbooks.[1,2,10] Principles of characteristic changes due to the compositional change of liquid mixtures with interacting components are discussed here. Assessment of nature of such interactions can only be made after evaluation of the equilibrium constant (energy) of such interactions between solvents.

### 9.3.2.1 Permittivity

Permittivity, ε, is the only property of the mixed liquid systems with chemical interaction between components that has not been studied as extensively as for systems without chemical interaction.[10] When interaction between components is similar to a given in equation [9.8], formation of conductive solutions occurs. Determination of ε for these solutions is difficult, and sometimes impossible. Although, more or less successful attempts to develop such methods of permittivity determination are published from time to time, the problem until now has not been solved.

Classification of the isotherms ε vs. composition of liquid systems[3] is based on deviations of the experimental isotherms, which are then compared with isotherms of the system without chemical interaction between components. The latter isotherms are calculated from equations [9.24] - [9.26], (Figure 9.2).

Type I isotherm corresponds to the experimental isotherm with isotherms [9.24] or [9.26] of the liquid systems without interaction. The additive ε increase is compensated by ε decrease, due to homomolecular association process. The system diethyl ether - m-cresol illustrates this type of isotherm.

Type II is determined by the negative deviations from isotherm [9.26]. This type corresponds to systems with weak heteromolecular interactions between components, but with strong homomolecular association of one of the components. The system formic acid-anisole is an example of this kind of isotherm. Furthermore, these isotherms are characteristic when non-associating components in pure state form heteromolecular associates with lower dipole moment, DM, then DM of both components. The average DM for such kind of interaction in mixed systems is lower than correspondent additive value for non-interacting system. The system 1,2-dichloroethane - n-butylbromide can be referenced as an example of this kind of mixed binary solvent.

Type III combines ε isotherms, which are above the isotherm calculated from equation [9.26]. This kind of isotherm suggests interaction between components of mixed solvent. The variety of such systems allows us to distinguish between five isotherm subtypes.

Subtype IIIa is represented by isotherms lying between the curve obtained from eq. [9.26] and the additive line. This isotherm subtype occurs in systems with low value of heteromolecular association constant. The system n-butyric acid - water can be mentioned as a typical example of the subtype.

Subtype IIIb combines S-shaped ε isotherms. This shape is the result of coexistence of homo- and heteromolecular association processes. System pyridine-water is a typical example of this subtype.

Subtype IIIc combines most common case of ε isotherms - curves monotonically convex from the composition axis ($\partial\varepsilon / \partial V \neq 0, \partial^2\varepsilon / \partial V^2 < 0$). The system water-glycerol can serve as typical example of the subtype.

Subtype IIId is represented by ε isotherms with a maximum. Typically, this kind of isotherm corresponds to systems with high heteromolecular association constant. Component interaction results in associates with greater values of DM than expected from individual components. Carboxylic acids-amines have this type of isotherms.

Finally, subtype IIIe includes rare kind of ε isotherms with a singular maximum, which indicates equilibrium constant with high heteromolecular association. This kind of isotherm is represented by system $SnCl_4$-ethyl acetate.

### 9.3.2.2 Viscosity

Shapes of binary liquid system viscosity - composition isotherms vary significantly. The basic types of viscosity - composition isotherms for systems with interacting components are given in Figure 9.3.

Type I - viscosity isotherms are monotonically convex in direction to the composition axis ($\partial\eta / \partial x \neq 0, \partial^2\eta / \partial x^2 > 0$). Chemical interaction influences the shape of viscosity isotherm that is typical when experimental isotherm is situated above the curve calculated under assumption of absence of any interaction (i.e., from equation [9.31a]). The means of increasing heteromolecular association and determination of stoichiometry of associates for this type of isotherms was discussed elsewhere.[3] Piperidine-aniline system is an example of this kind of interacting system.

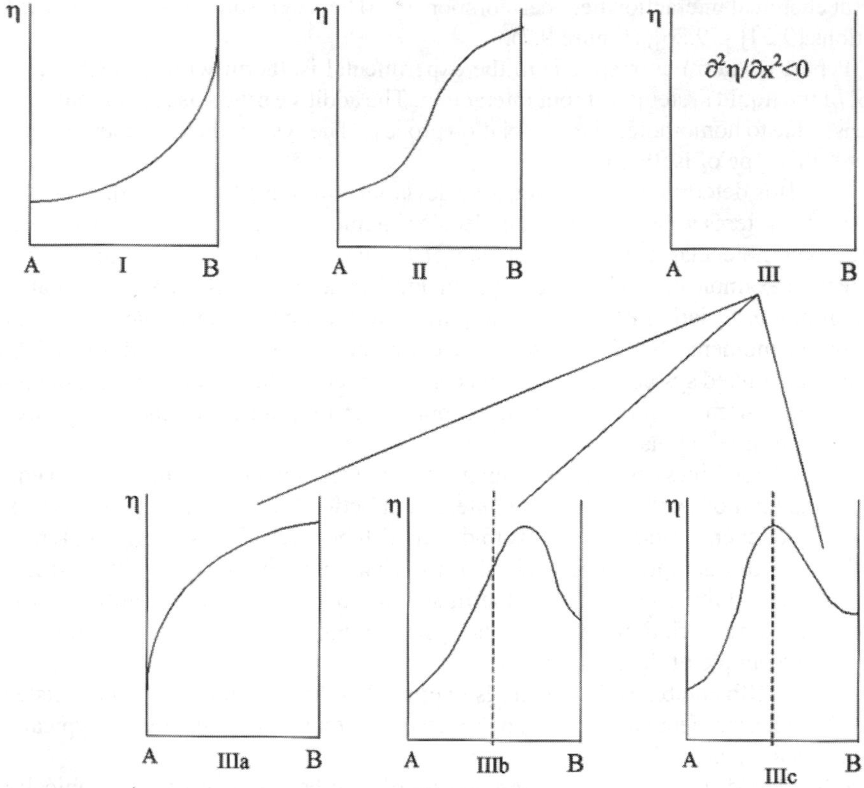


Figure 9.3. Classification of viscosity isotherms for binary solvents.

Type II - S-shaped viscosity curves ($\partial\eta / \partial x \neq 0, \partial^2\eta / \partial x^2 = 0$ at the inflection point). This type of isotherm is attributed to systems which have components differing substantially in viscosity, but have low yield of heteromolecular associates to form local maximum in isotherms. Systems such as sulfuric acid-pyrosulfuric acid and diphenylamine - pyridine are examples of this kind of viscosity isotherm.

Type III combines viscosity isotherms, which are convex-shaped from composition axis as its common attribute, i.e., $\partial^2\eta / \partial x^2 < 0$. In the whole concentration range, we have positive values of excessive viscosity $\eta^E = \eta_{exp} - [(\eta_1 - \eta_2)x + \eta_2$, where $\eta_{exp}$ is experimental value of viscosity. It shall be mentioned that, although the term $\eta^E$ is used in literature, unlike the excessive logarithm of viscosity $(\ln\eta)^E = \ln\eta_{exp} - (\ln\eta_1/\eta_2 + \ln\eta_2)$, it has no physical meaning. Viscosity isotherms of this type for interacting liquid systems are most commonly encountered and have shape diversity, which can be divided into three subtypes.

Subtype IIIa - isotherms monotonically convex shaped from composition axis, but without maximum (in the whole composition range $\partial\eta / \partial x \neq 0$). Dependence viscosity-composition of water-monochloroacetic acid can serve as an example.

Subtype IIIb - isotherms with "irrational" maximum (i.e., with maximum which does not correspond to any rational stoichiometric correlation of components in mixed solvent). This subtype is the most common case of binary mixed solvents with interacting components. System trifluoroacetic acid-acetic acid is common example. The true stoichiometry determination method for this case was described elsewhere.[3]

Subtype IIIc - isotherms with maximum at rational stoichiometry, which corresponds to composition of heteromolecular associate. This case can be exhibited by the system pyrosulfuric acid-monochloracetic acid.

Occasionally, isotherms with singular maximum can also occur. This behavior is characteristic of the system mustard oil-amine.

This classification covers all basic types of viscosity isotherms for binary mixed systems. Although the classification is based on geometrical properties of isotherm, heteromolecular associations determine specific isotherm shape and its extent. The relative level of interactions in binary mixed systems increases from systems with isotherms of type I to systems with type III isotherms.

### 9.3.2.3 Density, molar volume

Unlike the dependencies of density deviation (from additive values) on system composition, correspondent dependencies of specific (molar) volume deviation from the additive values:

$$\Delta\theta = \theta_{exp} - \left[(\theta_1 - \theta_2)x + \theta_2\right] \quad [9.41]$$

where:

| | |
|---|---|
| $\Delta\theta$ | departure from of molar volume |
| $\theta_{exp}$ | experimental value of molar volume |
| $\theta_i$ | molar volumes of components |
| x | molar fraction of the second component |

allow us to make a meaningful assessment of the heteromolecular association stoichiometry.[3] Classification of volume-dependent properties of binary mixed solvents are based on dependencies of $\Delta\theta$ value (often called "excessive molar volume", $\theta^E$).

It is evident both from "ideal system" definition and from equation [9.36], that for non-interacting systems $\Delta\theta \approx 0$.

Interaction is accompanied by formation of the heteromolecular associates. It can be demonstrated by analysis of volumetric equations for the liquid mixed systems, data on volume compression, i.e., positive density deviation from additivity rule, and hence negative deviations of experimental specific molar volume from partial molar volume additivity rule.

There are three basic geometrical types of $\Delta\theta$ isotherms for binary liquid systems with chemically interacting components[12] (Figure 9.4.).

Type I - rational $\Delta\theta$ isotherms indicating that only one association product is formed in the system (or that stability constants for other complexes in system are substantially lower than for main adduct). Most often, adduct of equimolecular composition is formed, and the maximum on $\Delta\theta$ isotherm occurs at x = 0.5. Fluorosulfonic acid-trifluoroacetic acid system is an example of the type.

Type II - irrational $\Delta\theta$ isotherms with maximum not corresponding to the stoichiometry of the forming associate. This irrationality of $\Delta\theta$ isotherms indicates formation of more than one adduct in the system. In the case of formation of two stable adducts in the system, the $\Delta\theta$ maximum position is between two compositions which are stoichiometric for each associate. If in the system, two adducts AB and $AB_2$ are formed,

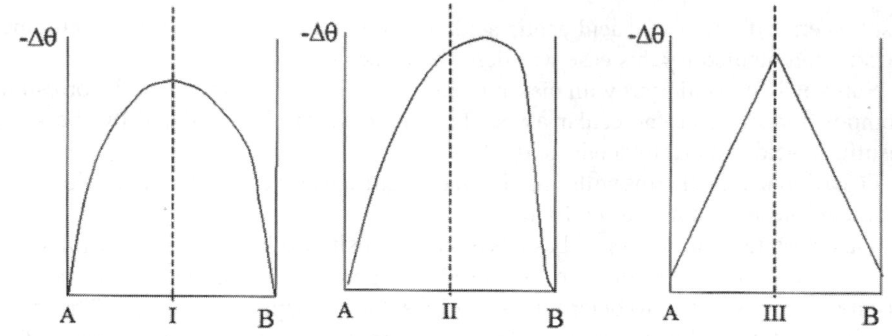


Figure 9.4. Classification of isotherms of molar volume deviation from additivity for binary solvents.

then Δθ position of maximum is between $x_A$ values of 0.5 and 0.33 molar fractions. This type of Δθ isotherm is most common for interacting binary liquid systems, for example, diethyl aniline-halogen acetic acids.

Type III includes Δθ isotherms with a singular maximum. According to metrical analysis,[3] dependencies of this type are typical of systems where adduct formation constant, $K_f \rightarrow \infty$. Mustard oil-amine is an example of this behavior.

Relative extent of the interaction increases from systems with Δθ isotherms of type I to systems of type III isotherms. This conclusion comes from analysis of metrical molar volume equations and comparison of binary systems composed of one fixed component and a number of components, having reactivity towards fixed component changing in well-defined direction.

### 9.3.2.4 Conductivity

The shape of specific conductivity relation on composition of the binary mixed solvent depends on conductivity of the components and some other parameters discussed below. Because of molar conductivity calculation of binary mixed solvent system, the ion associate concentration should be taken into account (eq. [9.7]). This quantity is known in very rare instances,[13,14] therefore, the conductivity of interacting mixed solvent is most often expressed in specific conductivity terms.

Classification of conductivity shapes for binary liquid systems may be based on the initial conductivity of components. Principal geometrical types of the conductivity isotherms are presented in Figure 9.5.

The type I isotherm includes the concentration dependence of conductivity of the systems (Figure 9.5).

These systems are not often encountered in research and technology practice. The type I can be subdivided into subtype Ia - isotherms with a minimum (for example, selenic acid-orthophosphoric acid) and subtype Ib - isotherms with a maximum (for instance, orthophosphoric acid-nitric acid).

The type II isotherms describe the concentration dependence of conductivity for binary systems containing only one conductive component. In this case also, two geometric subtypes may be distinguished. Subtype IIa - has monotonically convex shape (i.e., without extreme) in the direction of the composition axis. The binary system, selenic acid-acetic acid, can serve as an example of this kind of dependency. Subtype IIb includes isotherms with a maximum (minimum) - most widespread kind of isotherms (for example: perchloric

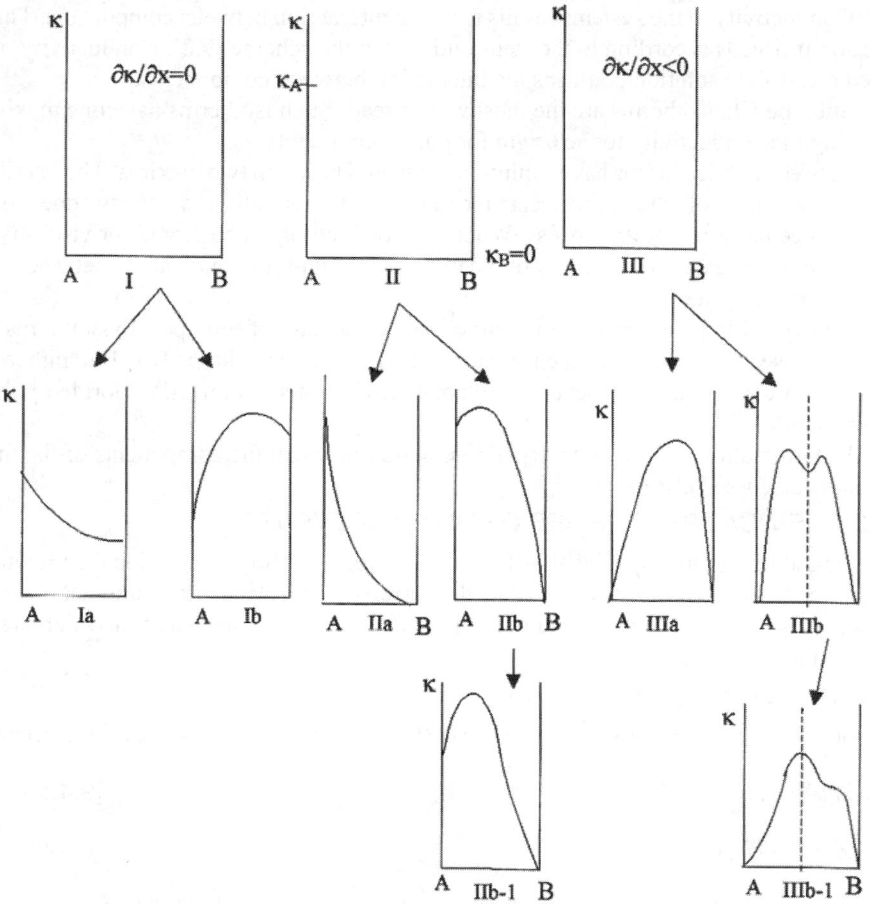


Figure 9.5.Classification of specific conductivity isotherms for binary solvents.

acid-trifluoroacetic acid). The next frequently observed case is separated as subtype IIb-1; these are isotherms with $\chi=0$ in the middle of concentration range. The system sulfuric acid-ethyl acetate illustrates the subtype IIb-1.

Geometrically, the conductivity isotherms of I and II types are similar to the $\chi$ isotherms for binary liquid systems without interaction (Section 9.3.1.4). But conductivity dependencies corrected for viscosity vs. the concentration of interacted system differ from the corresponding dependencies for non-interacting systems by the presence of maximum. Also, interacting and non-interacting systems differ based on analysis of effect of the absolute and relative temperature conductivity coefficients on concentration.[3] The relative temperature electric conductivity coefficient, $\beta_\chi$, differs from the electric conductivity activation energy, $E_{act}$, by a constant multiplier.[15]

Systems formed by non-electrolyte components are the most common types of electrolyte systems.

Conductivity of the systems results from the interaction between components. This interaction proceeds according to the steps outlined in the scheme [9.8]. Conductivity of the mixed electrolyte solution confirms the interaction between components.

Subtype IIIa isotherms are the most widespread. Such isotherms have one maximum and bring the conductivity to the origin for pure components.

Subtype IIIb isotherms have a minimum situated between two maxima. The maximum appears because of the significant increase of the solution viscosity due to the heteromolecular association process. When the conductivity is corrected for viscosity, the maximum disappears. Conductivity of the mixed solvent pyrosulfuric acid-acetic acid is an example of the system.

Subtype IIIb-1 isotherms with a curve is a special case of subtype IIIb isotherms. The curve is caused by viscosity influence. Isotherm IIIb turns to subtype IIIb-1 at higher temperatures. A concentration dependence of conductivity for stybium (III) chloride-methanol is an example of the systems.

Determination of stoichiometry of interaction between the components of the mixed solvent is discussed elsewhere.[3,12]

### 9.3.3 CHEMICAL PROPERTIES OF MIXED SOLVENTS

Solvating ability of mixed solvent differs from solvating ability of individual components. In addition to the permittivity change and the correspondent electrostatic interaction energy change, this is also caused by a number of reasons, the most important of which are discussed in the chapter.

#### 9.3.3.1 Autoprotolysis constants

Let both components of mixed solvent, AH-BH, to be capable of autoprotolysis process.

$$2HA \leftrightarrow H_2A^+ + A^- \qquad [9.42a]$$

$$2BH \leftrightarrow H_2B^+ + B^- \qquad [9.42b]$$

As Aleksandrov demonstrated,[15] the product of the activities of lionium ions sum ($a_{H_2A^+} + a_{H_2B^+}$) on the liate activities sum ($a_{A^-} + a_{B^-}$) is a constant value in the whole concentration range for a chosen pair of cosolvents. This value is named as ion product of binary mixed solvent.

$$K_{ap}^{HA-HB} = \left(a_{H_2A^+} + a_{H_2B^+}\right)\left(a_{A^-} + a_{B^-}\right) \qquad [9.43]$$

If components of binary mixed solvent are chemically interacting, then the equilibrium of protolysis processes [9.42] is significantly shifting to one or another side. In most such instances, concentrations of lionium ions for one of components and liate ions for another become so low that correspondent activities can be neglected, that is autoprotolysis constant can be expressed as:

$$K_{ap}^{HA-HB} = a_{H_2A^+} + a_{B^-} \qquad [9.44a]$$

$$K_{ap}^{HA-HB} = a_{H_2B^+} + a_{A^-} \qquad [9.44b]$$

As an example, let us consider mixed solvent water-formic acid. Both components of the system are subject to autoprotolysis:

$$2H_2O \leftrightarrow H_3O^+ + OH^- \text{ and } 2HCOOH \leftrightarrow HCOOH_2^+ + HCOO^-$$

It is obvious, that in the mixed solvent, concentrations of the $HCOOH_2^+$ ions (strongest acid from possible lionium ions in the system) and the $OH^-$ ions (strongest base from two possible liate ions) are neglected, because these two ions are mutually neutralized. Equilibrium constant of the direct reaction is very high:

$$HCOOH_2^+ + OH^- \leftrightarrow HCOOH + H_2O$$

Therefore, in accordance with [9.44a] autoprotolysis constant for this mixed solvent can be expressed as:

$$K_{ap}^{HA-HB} = a_{H_3A^+} + a_{HCOO^-}$$

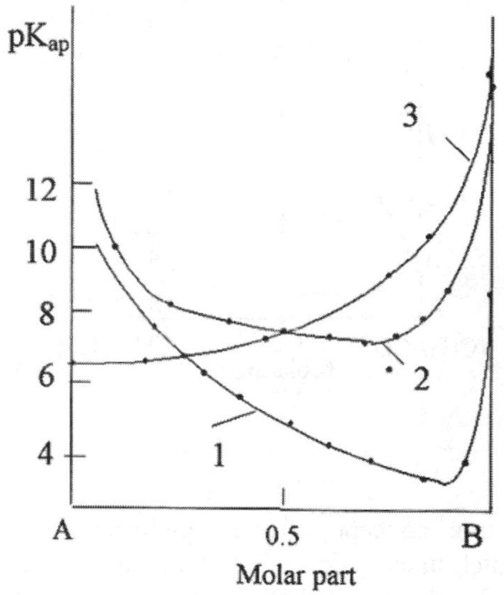


Figure 9.6. Dependence of $pK_{ap}$ ($-logK_{ap}$) on the mixed solvent composition: 1-DMSO-formic acid; 2-DMSO-acetic acid; 3- formic acid-acetic acid.

Because of lionium and liate ions activities, both mixed solvent components depend on solvent composition, as does autoprotolysis constant value for mixed solvent.

Figure 9.6 demonstrates dependencies of $pK_{ap}$ values on composition for some mixed solvents. Using the dependencies, one can change the mixed solvent neutrality condition in a wide range. That is, for mixed solvent composed of DMSO and formic acid, neutrality condition corresponds to concentration of $H^+$ (naturally solvated) of about $3\times10^{-2}$ g-ion*$l^{-1}$ at 80 mol% formic acid, but only $10^{-4}$ g-ion*$l^{-1}$ at 10 mol% acid component.

The method of autoprotolysis constant value determination based on electromotive force measurement in galvanic element composed from Pt,$H_2$(1 atm)|KOH(m), KBr(m), solvent|AgBr,Ag was proposed elsewhere.[15a] The method was used for polythermal study of autoprotolysis constant of binary mixed solvent 2-methoxyetanol - water. On the basis of ionization constants, polytherms data, and the autoprotolysis process, thermodynamic data were calculated.

A large data set on the electrophility parameter $E_T$ (Reichardt parameter[65]) for binary water-non-water mixed solvents was compiled.[66]

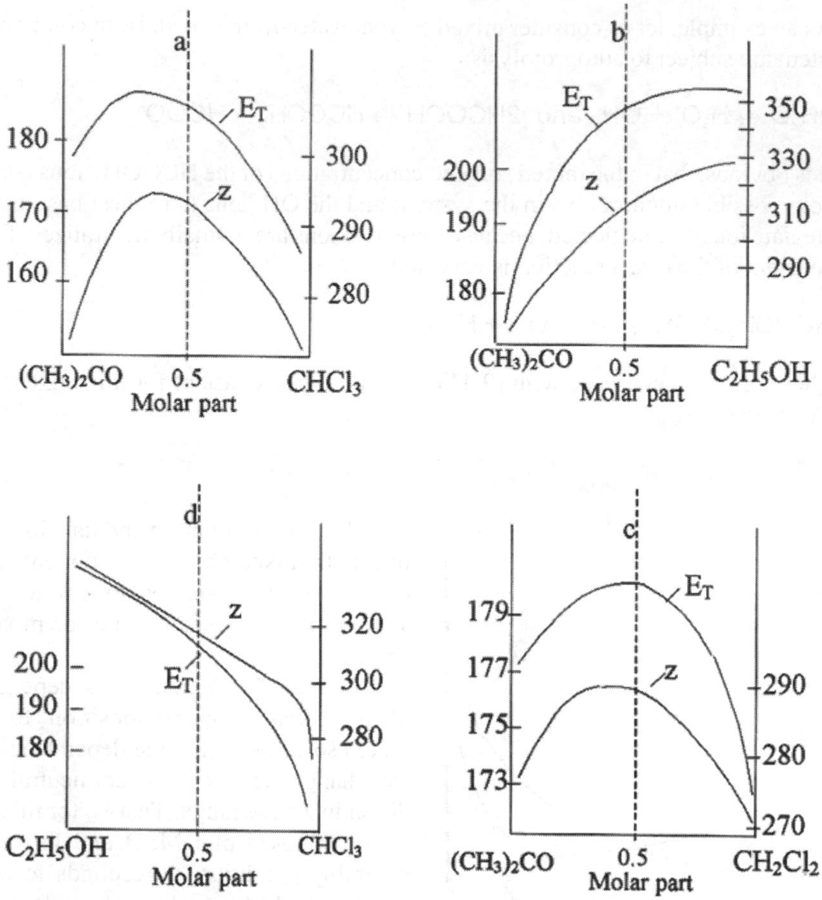

Figure 9.7. $E_T$ and Z parameters for some mixed solvents.

### 9.3.3.2 Solvating ability

Applying the widely-used chemical thermodynamics concept of free energy linearity, it is easy to demonstrate that any characteristic y which linearly depends on free energy or on free activation energy of process in the mixed solvent composed of specifically non-interacting components, shall be linear function of the components with partial molar concentration x.

$$y = \sum_{1}^{i} y_i x_i \qquad [9.44]$$

The formula was developed analytically by Palm.[16] In the mixed solvents with chemically interacting components, the function y=f(x) can substantially deflect from the linear form including appearance of extrema.

Selection of the mixed solvent components allows us in most cases to provide controlled solvation of all substances, participating in the chemical process performed in solution. Often, it can be achieved by combination of solvate active and solvate inert components. For example, it is obvious that in all compositional range of mixture DMSO-$CCl_4$ (except 100% $CCl_4$) specific solvation of acid dissolved in this mixture is realized by DMSO. Similarly in mixed solvent formic acid-chlorobenzene, solvation of the dissolved donor substance is performed exclusively by formic acid.

Selection of the second (indifferent) component's $\varepsilon$ also provides means to control the universal solvation ability of mixed solvent. In the above mentioned examples, increasing the solvate inert component concentration results in the decrease of mixed solvent $\varepsilon$. On the contrary, addition of indifferent component (propylene carbonate) into the systems such as acetic acid-propylene carbonate or propylene carbonate-aniline causes $\varepsilon$ to rise. Because in the last two systems acetic acid and aniline were chosen as solvate active components, it was obviously intended to use these mixtures for specific solvation of the dissolved donor and acceptor compounds respectively.

#### 9.3.3.3 Donor-acceptor properties

As was demonstrated,[17] the parameters $E_T$ and Z of the binary mixed solvent 1,2-dibromoethane - 1,2-dibromopropane are strictly additive function of molar composition. For the mixed solvents, having components engaged into specific interaction, dependencies $E_T$=f(x) and Z=F(x) are non-linear and even extremal, as can be seen from the examples in Figure 9.7.

The method was proposed[18] to linearize polarity index of mixed solvent by introducing a parameter, which connects the $E_T$ and Z values with fractional concentrations of components.

## 9.4 MIXED SOLVENT INFLUENCE ON THE CHEMICAL EQUILIBRIUM

### 9.4.1 GENERAL CONSIDERATIONS[7]

The chemical process established in a solvent can be represented in a general form:

$$E \leftrightarrow F \quad [9.45]$$

where:

E all chemical forms of reaction reagents
F all chemical forms of reaction products

Considering the traditional thermodynamic cycle, we can use the general equation of Gibbs' energy variation because of the process [9.45]:

$$\Delta G = \Delta G_{solv,E} - \Delta G_{solv,F} - \Delta G^{(v)} \quad [9.46]$$

where:

$\Delta G_{solv,E}$, $\Delta G_{solv,F}$ Gibbs' solvation energy of [9.45] equilibrium members
$\Delta G^{(v)}$ Gibbs' energy of process [9.45] in vacuum

The process [9.45] takes place in mixed solvent A-B. Solvation energies of equilibrium members are the algebraic sum of those for each of the mixed solvent components (this sum also takes into account the energy of mixed solvates such as $EA_xB_y$ and $FA_zB_t$):

$$\Delta G_{solv,i} = \Delta G_{solv,A} + \Delta G_{solv,B} = \sigma_{solv,i} \quad [9.47]$$

Therefore the equation [9.46] can be represented in the form

$$\Delta G = \sigma_{solv,E} - \sigma_{solv,F} - \Delta G^{(v)} \quad [9.48]$$

The variation of free energy due to any chemical process consists of both covalent and electrostatic components:

$$\Delta G = \Delta G^{(cov)} - \Delta G^{(el)} \quad [9.49]$$

Substituting [9.49] in [9.48], we come to an equation that in general describes the mixed solvent effect on the equilibrium chemical process:

$$\Delta G = -\Delta G^{(v)} + \left(\sigma^{(cov)}_{solv,E} - \sigma^{(cov)}_{solv,F}\right) + \left(\sigma^{(el)}_{solv,E} - \sigma^{(el)}_{solv,F}\right) \quad [9.50]$$

For universal (chemically indifferent) solvents where $\sigma^{(el)} >> \sigma^{(cov)}$, it may by assumed that

$$\Delta G = -\Delta G^{(v)} + \sigma^{(el)}_{solv,E} - \sigma^{(el)}_{solv,F} \quad [9.51]$$

Thus the mixed solvent effect on the equilibrium of the chemical process [9.45] is determined not only by the vacuum component but also by the solvation energy of each of the chemical forms of equilibrium members.

From free energy to equilibrium constants, one can obtain the equation describing mixed solvent effect on the equilibrium constant of the [9.45] process:

$$\ln K = \left[\Delta G^{(v)} + \left(\sigma^{(cov)}_{solv,F} - \sigma^{(cov)}_{solv,E}\right) + \left(\sigma^{(el)}_{solv,F} - \sigma^{(el)}_{solv,E}\right)\right] / RT \quad [9.52]$$

In the special case of universal media, i.e., mixed solvent formed by both solvation-inert components, this equation can be presented as:

$$\ln K^{(univ)} = \left[\Delta G^{(v)} + \sigma^{(el)}_{solv,f} - \sigma^{(el)}_{solv,E}\right] / RT \quad [9.53]$$

Because the energy of all types of electrostatic interaction is inversely proportional to permittivity, these equations can be rewritten in the form:

$$\ln K = \left[\Delta G^{(v)} + \left(\sigma^{(cov)}_{solv,F} - \sigma^{(cov)}_{solv,E}\right) + \left(\beta_{solv,F} - \beta_{solv,E}\right) / \varepsilon\right] / RT \quad [9.52a]$$

and

$$\ln K^{(univ)} = \left[\Delta G^{(v)} + \left(\beta_{solv,F} - \beta_{solv,E}\right) / \varepsilon\right] / RT \quad [9.53a]$$

where:

$\beta$ the multipliers of the magnitudes of reciprocal permittivity in equations of energy of the main types of electrostatic interactions such as dipole-dipole, ion-dipole, and ion-ion interactions

As follows from [9.52a], in binary mixed solvents formed from solvation-indifferent components (i.e., universal media), equilibrium constants of the [9.45] process are exponent dependent on the 1/ε values (i.e., it is a linear correlation between $\ln K^{(univ)}$ and 1/ε magnitudes).

The vacuum component of energy of the [9.45] process can be obtained by the assumption of hypothetical media with $\varepsilon \to \infty$ and $1/\varepsilon \to 0$:

$$\Delta G^{(v)} = RT \ln K^{(univ)}_{1/\varepsilon \to 0} \quad [9.54]$$

For binary solvents formed by solvation (active component A and indifferent component B) analysis of equation [9.52a] demonstrates that there is also a linear correlation between lnK and 1/ε. Such mixed solvents are proposed to be called as conventionally universal.

To analyze the solvent effect on the process [9.45], it is often convenient to represent the temperature and permittivity dependencies of lnK in approximated form:

$$\ln K = a_{00} + a_{01}/T + a_{02}/T^2 + \dots + \left(a_{10} + a_{11}/T + a_{12}/T^2 + \dots\right)/\varepsilon + + \left(a_{20} + a_{21}/T + a_{22}/T^2 + \dots\right)/\varepsilon^2 + \dots = \sum_{\substack{i=0 \\ j=o}}^{\substack{i=m \\ j=n}} \frac{a_{ij}}{\varepsilon^j T^i} \quad [9.55]$$

For the universal and conventionally-universal media, this dependence described by the equation can be represented in the form

$$\ln K^{(univ)} = a_{00} + a_{01}/T + a_{10}/\varepsilon + a_{11}/\varepsilon T \quad [9.56]$$

## 9.4.2 MIXED SOLVENT EFFECT ON THE POSITION OF EQUILIBRIUM OF HOMOMOLECULAR ASSOCIATION PROCESS

All questions this part deals with are considered based on the example of a special and wide-studied type of homomolecular association process, namely, monomer-dimer equilibrium:

$$2E \leftrightarrow E_2 \quad [9.57]$$

For this process, equation [9.52] can be presented in the form:

$$\ln K_{dim} = \left[\Delta G_{dim} + \Delta G^{(cov)}_{solv,E_2} - 2\Delta G^{(cov)}_{solv,E} + \left(\beta_{solv,E_2} - 2\beta_{solv,E}\right)/\varepsilon\right]/RT \quad [9.58]$$

where:

β value at the 1/ε is calculated in accordance with the following equation $\Delta G^{cl}_{d\text{-}d} = -120.6\mu^2 / r^3_{d\text{-}d}$ kJ / mol

μ dipole moment

Because of the electrostatic component of process [9.57], the free energy is conditioned by dipole-dipole interactions such as dimer-solvent and monomer-solvent.

In mixed solvent, $CCl_4$-$C_6H_5Cl$, universal relation to acetic acid (because the mixed solvent components do not enter into specific solvation with the acid), the dimerization constant dependence on the temperature and permittivity in accordance with [9.52.a] and [9.56] is described by equation:[19]

$$\ln K_{dim} = -1.72 + 1817.1/T + 0.92/\varepsilon + 1553.4/\varepsilon T$$

The dimer form concentration, $c_M$ (mol/l), is related to initial analytical concentration of dissolved compound and dimerization constant in equation:

$$c_M = \left[4K_{dim}c_M^o + 1 - \left(8K_{dim}c_M^o + 1\right)^{1/2}\right] / 8K_{dim} \qquad [9.59]$$

One can calculate, with the help of this equation, that in hexane at $c_M^o = 0.1$, half of the dissolved acetic acid is in dimer form ($K_{dim} \approx 1.5\times10^3$). In nitrobenzene solution ($K_{dim} \approx 10^2$), only a third of analytical concentration of acid is in dimer form. For phenol these values are 12% and 2%, respectively. Therefore the composition variation of universal mixed solvents, for instance, using hexane-nitrobenzene binary solvent, is an effective method to control the molecular composition of dissolved compounds able to undergo homomolecular association, in particular, through H-bonding.

Typically, in specific solvents, the process of monomer formation [9.57] is characterized by considerably lower dimerization constants, as compared with those in universal media. In fact, acetic acid dimerization constant in water-dioxane binary solvent, the components of which are solvation-active in respect to the acid, vary in the 0.05-1.2 range. Replacing the solvent is often the only method to vary the molecular association state of dissolved compound. To achieve dimer concentration in 0.1 M solution of phenol in n-hexane equal to dimer concentration in nitrobenzene solution (50% at 25°C), it would be necessary to heat the solution to 480°C, but it is impossible under ordinary experimental conditions.

### 9.4.3 MIXED SOLVENT INFLUENCE ON THE CONFORMER EQUILIBRIUM

Equilibrium took place in solutions

$$Conformer\ I \leftrightarrow Conformer\ II \qquad [9.60]$$

and the special case, such as

$$Cis\text{-}isomer \leftrightarrow Trans\text{-}isomer \qquad [9.61]$$

when investigated in detail. But the concentration-determining methods of equilibrium [9.60, 9.61] members are often not very precise. Dielcometry, or in our measurement the mean dipole moment, can be used only in the low-polar solvents. The NMR methods allowed us to obtain the equilibrium constants of processes by the measurement of chemical dislocation and constants of spin-spin interactions of nucleuses, which are quite precise but only at low temperatures (that is why the numerous data, on conformer and isomerization equilibrium, in comparison with other types of equilibria, are in solutions at low temperatures).

It is evident that the accuracy of the experimental definition of equilibrium constants of the investigated process is insufficient. Thus, the summarization of experimental data on solvent effect on [9.60, 9.61] equilibrium encounters several difficulties.

As a rule, the differences in conformer energies are not very high and change from one tenth to 10-12 kJ/mol. This is one order value with dipole-dipole interaction and specific solvation energies, even in low-active solvents. Moreover, the dipole moments of conformer highly differ one from another, so ε is one of the main factors which affects equilibrium, such as in equations [9.60, 9.61].

If equations of processes are identical to that of scheme [9.45] in the form and maintenance, it is possible to apply the equations from Section 9.4.1 to calculate the equilibrium constants of the investigated process.

Abrahem[20] worked out details of the solvent effect theory on changing equilibrium. This theory also accounts for the quadruple interactions with media dipoles. It demonstrates that the accuracy of this theory equations is not better than that obtained from an ordinary electrostatic model.

Because of low energy of [9.60, 9.61] processes, it is not possible to assign any solvents strictly to a universal or specific group. That is why only a limited number of binary solvents can be used for the analysis of universal media influcnce on conformer equilibrium constants. It is interesting to note that over the years, the real benzene basicity in liquid phase has not been sufficiently investigated. There were available conformer equilibrium constants in benzene, toluene, etc., highly differing from those in other low-polar media. Such phenomenon is proposed to be called the "benzene effect".[21]

Sometimes rather than equilibrium constants, the differences in rotamer energies - for example, gosh- and trans-isomers - were calculated from the experiment. It is evident that these values are linearly proportional to the equilibrium constant logarithm.

Different conformers or different intermediate states are characterized by highly distinguished values of dipole moments.[22] Indeed, the media permittivity, ε, change highly influences the energy of dipole-dipole interaction. Therefore, according to [9.53a], it is expected that conformer transformation energies and energies of intermediate processes in universal solvents are inversely proportional to permittivity. But equilibrium constants of reactive processes are exponent dependent on $1/\varepsilon$ value, i.e., there is a linear correlation between $\ln K_{conf}$ and reciprocal permittivity:

$$E_{conf\,(turning)} = A + B / \varepsilon \qquad [9.62]$$

$$\ln K_{conf\,(turning)} = a + b / \varepsilon \qquad [9.63]$$

The analysis of experimental data on conformer and intermediate equilibrium in universal media demonstrates that they can be described by these equations with sufficient accuracy not worse than the accuracy of experiment.

The differences in rotamer energies of 1-fluoro-2-chloroetane in mixed solvents such as alkane-chloroalkane can be described by equation:[20]

$$\Delta E = E_{gosh} - E_{trans} = -2.86 + 8.69/\varepsilon,\ \text{kJ/mol}$$

According to equation [9.51], the permittivity increase leads to decreasing absolute value of electrostatic components of conformer transformations free energies in universal solvents. For instance, conformer transformation free energy of α-bromocyclohexanone in cyclohexane (ε=2) is 5.2 kJ/mol, but in acetonitrile (ε=36) it is -0.3 kJ/mol.

Specific solvation is the effective factor which controls conformer stability. The correlation between equilibrium constants of the investigated processes [9.60] and [9.61] and polarity of solvation-inert solvents is very indefinite. For instance, diaxial conformer of 4-methoxycyclohexanone in acetone (as solvent) (ε=20.7) is sufficiently more stable than in methanol ($\varepsilon$=32.6). But the maximum stability of that isomer is reached in low-polar trifluoroacetic acid as solvent ($\varepsilon$=8.3).

Formation of the internal-molecular H-bonding is the cause of the increasing conformer stability, often of gosh-type. Therefore solvents able to form sufficiently strong external H-bonds destroy the internal-molecular H-bonding. It leads to change of conformer occupation and also to decrease of the internal rotation barrier.

Due to sufficiently high donority, the benzene has a comparatively high degree of conformer specific solvation ("benzene effect"). This leads to often stronger stabilization of conformer with higher dipole moment in benzene than in substantially more polar acetonitrile.

In a number of cases, the well-fulfilled linear correlation between conformer transformation constants and the parameter $E_T$ of mixed solvent exists (see Section 9.3.4.3).

It is often the solvent effect that is the only method of radical change of relative contents of different conformer forms. Thus, with the help of the isochore equation of chemical reaction, the data on equilibrium constants and enthalpies of dichloroacetaldehyde conformer transformation allow us to calculate that, to reach the equilibrium constant of axial rotamer formation in cyclohexane as solvent (it is equal to 0.79) to magnitude K=0.075 (as it is reached in DMSO as solvent), it is necessary to cool the cyclohexane solution to 64K (-209°C). At the same time, it is not possible because cyclohexane freezing point is +6.5°C. By analogy, to reach the "dimethylsulfoxide" constant to value of "cyclohexane", DMSO solution must be heated to 435K (162°C).

## 9.4.4 SOLVENT EFFECT ON THE PROCESS OF HETEROMOLECULAR ASSOCIATION

Solvent effect on the process of heteromolecular association

$$mE + nF \leftrightarrow E_mF_n \quad [9.64]$$

has been studied in detail.[22,23] In spite of this, monographs are mainly devoted to individual solvents.

In universal media formed by two solvate-inert solvents according to equation [9.53a], equilibrium constants of the process of heteromolecular adduct formation depend exponentially on reciprocal permittivity; thus

$$\ln K_{add} = a + b / \varepsilon \quad [9.65]$$

The process of adduct formation of acetic acid (HAc) with tributyl phosphate (TBP) nHAc•TBP in the mixed solvent n-hexane-nitrobenzene[7] can serve as an example of the [9.65] dependence validity. Equilibrium constants of the mentioned above adducts in individual and some binary mixed solvents are presented in Table 9.2.

The concentration of adduct EF $c_M$ is related to constant $K_{add}$ and initial concentration of the components $c_M^o$ by equation

$$c_M = \left[ 2K_{add}c_M^o + 1 - \left(4K_{add}c_M^o + 1\right)^{1/2} \right] / 2K_{add} \quad [9.66]$$

**Table 9.2. Equilibrium constants of adduct (nHAc•TBP) formation in the mixed solvent n-hexane - nitrobenzene (298.15K) and coefficients of equation [9.65]**

| Solvent | $K_{add}$ | | | |
|---|---|---|---|---|
| | HAc•TBP | 2HAc•TBP | 3HAc•TBP | 4HAc•TBP |
| Hexane (ε=2.23) | 327 | 8002 | $6.19\times10^5$ | $1.46\times10^{-7}$ |
| H+NB (ε=9.0) | 28.1 | 87.5 | 275 | 1935 |
| H+NB (ε=20.4) | 19.6 | 45.0 | 88.1 | 518 |
| Nitrobenzene (ε=34.8) | 17.4 | 36.2 | 60.8 | 337 |
| Coefficients of equation [9.65]: | | | | |
| a | 2.69 | 3.28 | 3.58 | 5.21 |
| b | 5.82 | 10.73 | 18.34 | 21.22 |

When stoichiometric coefficients are equal to m and n (equation [9.64]), $K_{add}$ expression is an equation of higher degree relative to $c_M$

$$K_{E_mF_n} = c_M / \left(c^o_{M,E} - nc_M\right)^n \left(c^o_{M,F} - mc_M\right)^m \qquad [9.67]$$

Let us consider an example of the interaction between acetic acid and tributyl phosphate (the change of permittivity of universal solvent permits us to change essentially the output of the reaction product). When initial concentration of the components equals to $c^o_M$=0.1 mol/l, the output (in %) of complexes of different composition is listed in table below:

| | 1:1 | 2:1 | 3:1 | 4:1 |
|---|---|---|---|---|
| In hexane | 84 | 46 | 32 | 22 |
| In nitrobenzene | 43 | 13 | 4.5 | 2 |

Relative concentration changes the more essentially, the larger the value of electrostatic component of the process free energy, $\Delta G^{el}_{\varepsilon=0}$.

Just as in the previous cases, the solvent use in this case is an effective means of process adjustment. To reduce the output of adduct HAc•TBF in nitrobenzene to the same value as in hexane solution, nitrobenzene solution must be cooled down to -78°C (taking into account that enthalpy of adduct formation equals 15 kJ/mol).[24] Naturally, the process is not possible because the nitrobenzene melting point is +5.8°C.

Let us consider the effect of specific solvation on equilibrium constant of the heteromolecular association process as an example of associate formation with a simplest stoichiometry:

$$E + F \leftrightarrow EF \qquad [9.68]$$

Only when A is a solvate-active component in the mixed solvent A-B, in the general case, both A and B initial components undergo specific solvation:

$$E + A \leftrightarrow EA \qquad [9.69]$$

$$F + A \leftrightarrow FA \qquad [9.70]$$

where:

EA, FA solvated molecules by the solvent (it is not necessary to take into account solvation number for further reasoning)

Since specific solvation of the adduct EF is negligible in comparison with specific solvation of initial components, A-B interaction in the solvent can be presented by the scheme:

$$EA + FA \leftrightarrow EF + 2A \qquad [9.71]$$

i.e., the heteromolecular association process in the specific medium is a resolvation process, since both initial components change their solvative surrounding. That is why equilibrium constants of the heteromolecular association process, calculated without consideration of this circumstance, belong indeed to the [9.71] process but not to the process [9.68].

Let us develop a quantitative relation between equilibrium constants for the process [9.68]

$$K_{EF} = [EF]/[E][F] \qquad [9.72]$$

and for the resolvation process [9.71]

$$K_{us} = [EF]/[EA][FA] \qquad [9.73]$$

Concentration of the solvate active solvent or the solvate active component A of mixed solvent in dilute solution is higher than the initial concentration of equilibrium components $[E]_0$ and $[F]_0$. Then activity of equilibrium components is equal to their concentrations. Then equations of material balance for components A and B can be set down as

$$[E]_0 = [E] + [EF] + [EA] \qquad [9.74]$$

$$[F]_0 = [F] + [EF] + [FA] \qquad [9.75]$$

Hence equilibrium constants of the process [9.69, 9.70] are presented by expressions

$$K_{EA} = [EA]/[E] = [EA]/([E]_0 - [EF] - [EA]) \qquad [9.76]$$

$$K_{FA} = [FA]/[F] = [FA]/([F]_0 - [EF] - [FA]) \qquad [9.77]$$

Definition [9.72] is presented in the form

$$K_{EF} = [EF]/([E]_0 - [EF] - [EA])([F]_0 - [EF] - [FA]) \qquad [9.78]$$

and re-solvation constant

$$K_{us} = [EF] / ([E]_0 - [EF])([F]_0 - [EF]) \quad [9.79]$$

Inserting [9.76] in [9.77] and [9.79] in [9.78], we come to the equation relating the equilibrium constants of all considered processes:

$$K_{EF} = K_{us}(1 + E_A)(1 + F_A) \quad [9.80]$$

A similar form of equation [9.80] has been proposed elsewhere.[25]

If it is impossible to neglect the solvent concentration (or the concentration of solvate component of the mixed solvent) in comparison with $[E]_0$ and $[F]_0$, equation [9.80] is written in the form:

$$K_{EF} = K_{us}(1 + K_{EA}^{A})(1 + K_{FA}^{S}) \quad [9.81]$$

Let E be acidic (acceptor) reagent in reaction [9.68], F basic (donor) reagent. Then, if A is acidic solvent, one can neglect specific solvation of the reagent E. Thus equations for equilibrium constant of the process [9.68] can be written in the form

$$K_{EF} = K_{us}(1 + K_{FS}) \quad K_{EF} = K_{us}(1 + K_{FS}^{A}) \quad [9.82]$$

If A is a basic component, equation [9.82] can be re-written to the form

$$K_{EF} = K_{us}(1 + K_{EA}) \quad K_{EF} = K_{us}(1 + K_{EA}^{A}) \quad [9.83]$$

Equations [9.82] and [9.83] were developed[26-28] for the case of specific solvation.

Thus for $K_{EF}$ calculation one must obtain the equilibrium constant of processes: [9.71] - $K_{us}$, [9.69] - $K_{EA}$ and [9.70] - $K_{FS}$ from conductance measurements. The constant $K_{EF}$ is identified in literature as "calculated by taking into account the specific solvation".[29] The value $K_{EF}$ characterizes only the universal solvation effect on the process of heteromolecular associate formation. The approach cited above can be illustrated by equilibrium:

$$(CH_3)_2SO + o\text{-}CH_3C_6H_4OH \leftrightarrow (CH_3)_2SO \bullet CH_3C_6H_4OH \quad [9.84]$$

studied in binary mixed solvents:

A) $CCl_4$ - hepthylchloride[30] formed by two solvate inert components [9.84]

B) $CCl_4$ - nitromethane[31] formed by solvate inert component ($CCl_4$) and acceptor (nitromethane) component

C) $CCl_4$ - ethyl acetate[32] formed by solvate inert component ($CCl_4$) and donor (ester) component;

The isotherms lnK vs. 1/ε (298.15K) are presented in Figure 9.8. These dependencies (right lines 1,2,4,5) are required for calculation of equilibrium constants of the heteromolecular association process free from specific solvation effect. It can be seen from Figure 9.8 that the values lnK, regardless of solvent nature, lie on the same line 3, which describes the change of equilibrium constants of the process [9.84] in the universal solution $CCl_4$-heptylchloride.

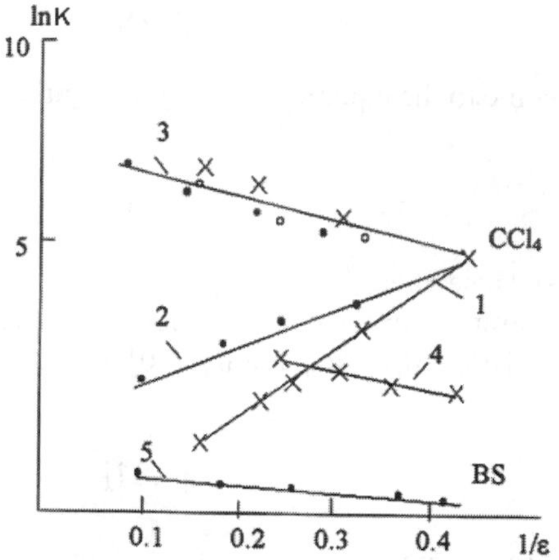

Figure 9.8. Dependence of equilibrium constant for the formation process of addition product DMSO to o-cresol on permittivity in solution based on heptylchloride (o), ethyl acetate (x) and nitromethane (•):1 - resolvation of o-cresol solvated in ethyl acetate in DMSO in mixed solvent $CCl_4$-ethyl acetate; 2 - resolvation of DMSO solvated in nitromethane in o-cresol in mixed solvent $CCl_4$-nitromethane; 3 - process [9.84] in mixed solvent $CCl_4$-heptylchloride; 4 - solvation of o-cresol in ethyl acetate; 5 - solvation of DMSO in nitromethane.

The data obtained for equilibrium [9.84] in binary mixed solvents of different physical and chemical nature are in need of some explanatory notes.

First of all, if solvents are universal or conditionally universal ones, in accordance with general rules of equilibrium constants, the dependence on permittivity and the dependence of K= f(1/ε) in all cases is rectilinear.

As appears from the above, specific solvation of one component decreases equilibrium constants in comparison with equilibrium constants in isodielectric universal solvent.

The increase of $K_{EF}$ with permittivity increase in DMSO - o-cresol system required supplementary study.[33]

The method of study of specific solvation effect on the process of heteromolecular association has been described in work,[34] devoted to study of the following interaction:

$$C_6H_3(NO_2)_3 + C_6H_5N(CH_3)_2 \leftrightarrow C_6H_3(NO_2)_3{\bullet}C_6H_5N(CH_3)_2 \qquad [9.85]$$

in the binary liquid solvent formed from a solvate-inert component (heptane) and one of the following solvate-active component: trifluoromethylbenzene, acetophenone or p-chlorotoluene.

Therefore trinitrobenzene, TNB, is solvated in initial mixed solvents, so its interaction with donor represents the resolvation process.

$$C_6H_3(NO_2)_3{\bullet}A + C_6H_5N(CH_3)_2 \leftrightarrow C_6H_3(NO_2)_3{\bullet}C_6H_5N(CH_3)_2 + (A) \qquad [9.85a]$$

Equation [9.62] applied to process of the heteromolecular association is written in the form:

$$\ln K = \left[\Delta G^{(v)} + \left(\sigma G^{(cov)}_{solv,EF} - \sigma G^{(cov)}_{solv,E} - \sigma G^{(cov)}_{solv,F}\right) + \sigma\beta_{solv} / \varepsilon\right] / RT \qquad [9.86]$$

But for equilibrium [9.85a] in mixed solvents one can assume $\sigma G^{(cov)}_{solv,EF} \approx 0$. That is why equation [9.52] can be converted for the process in conditionally universal media:

$$\ln K_{us} = \left[\Delta G^{(v)} - \sigma G^{(cov)}_{solv,E} + \sigma\beta_{solv} / \varepsilon\right] / RT \qquad [9.86a]$$

It follows from equation [9.86a] that dependence of $\ln K_{us}$ vs. $1/\varepsilon$ will be rectilinear. Experimental data of equilibrium constants for process [9.85a] are in agreement with the conclusion. Approximation of data by equation [9.65] is presented in Table 9.3.

It also follows from equation [9.86a] that RT, unlike in the cases of chemical equilibrium in universal solvents, describes, not the vacuum component of the process free energy, but the remainder $\Delta G^{(v)} - \sigma G^{(cov)}_{solv,TNB}$ and $\sigma G^{(el)}_{solv,TNB}$. The values of the remainder as well as electrostatic component of the free energy for the described chemical equilibrium are summarized in Table 9.4.

Although the solvate-active components are weak bases, their differences influence the values $\Delta G^{(v)} - \sigma G^{(cov)}_{solv,TNB}$ and $\sigma G^{(el)}_{\varepsilon=1}$. The first value characterizes the difference between the covalent components of solvation energy for trinitrobenzene in any solvate-active solvent. It follows from the table, that contributions of the covalent and electrostatic components are comparable in the whole concentration range of the mixed solvents. This indicates the same influence of both donor property and permittivity of the solvent on the [9.85a] process equilibrium.

**Table 9.3. Coefficients of equation [9.65] equilibrium constants of the [9.85a] process at 298.15K (r-correlation coefficient)**

| Solvent | a | b | r |
|---|---|---|---|
| Heptane-trifluoromethylbenzenc | -0.80 | 5.50 | 0.997 |
| Heptane-acctophenone | -1.49 | 6.74 | 0.998 |
| Heptane-chlorotoluene | -2.14 | 7.93 | 0.998 |

**Table 9.4. Free energy components (kJ/mol) of resolvation process [9.85a] in mixed solvents at 298.15K**

| $\varepsilon$ | -ΔG in system heptane - S, where S: | | |
|---|---|---|---|
| | trifluoromethylbenzene | acetophenone | chlorotoluene |
| 2 | 6.8 | 8.3 | 9.8 |
| 4 | 3.4 | 4.1 | 4.9 |
| 6 | 2.3 | 2.8 | 3.3 |
| 8 | 1.7 | 2.1 | - |
| $\Delta G^{(v)} - \sigma G^{(cov)}_{solv,TNB}$ | -2.0 | -3.7 | -5.3 |
| $\sigma G^{(el)}_{\varepsilon=1}$ | 13.6 | 16.7 | 19.6 |

According to the above energy characteristics of the heteromolecular association process (resolvation) in specific media, the solvent exchange affects the products' output (the relationship of output $c_M$ and $K_{us}$ is estimated from the equation [9.66]). This shows that the product output (with initial concentration of reagents 0.1M) can be changed from 34% (pure heptane) to 4 % (pure n-chlorotoluene) by changing the binary mixed solvent composition. The processes [9.85a] and [9.85] can be eliminated completely when the solvate active component (more basic then chlorotoluene) is used.

Possibility of management of the products` output may by illustrated by the data from reaction equilibrium [9.84]. As follows from Figure 9.8, maximum equilibrium constant (in investigated solvents) is for pure chloroheptane; the adducts output in this solvent equals to 86% (at initial concentration of reagent 0.1 M). Minimum equilibrium constant of the process is in pure ethyl acetate, where the products` output equals to 18 % (at initial concentration of reagent 0.1 M). Thus the products` output of the reaction [9.84] can be changed directionally in the range 90 up to 20% by means of choice of corresponding binary or individual solvent. If selected solvent is more basic than ethyl acetate or more acidic than nitromethane, the process is not possible.

Heteromolecular association process of o-nitrophenol and triethylamine can be an example of the management of products` output. Association constant of this process has been determined for some solvents.[35] Use of hexane instead of 1,2-dichloroethane, DHLE, increases products' output from 6% to 93%.

When both components of the mixed solvent are solvate-active towards the reagents of equilibrium [9.64], the following interactions take place:

$$E + A \leftrightarrow EA; \quad F + A \leftrightarrow FA; \quad E + B \leftrightarrow EB; \quad F + B \leftrightarrow FB \qquad [9.87]$$

The process of heteromolecular association [9.64] is due to displacement of the solvent components and formation of completely or partially desolvated adduct:

$$xEA + yFA + (m - x)EB \leftrightarrow E_mF_n + (A) + (B) \qquad [9.88]$$

It is easy to develop the equation for binary solvent formed from two solvate-active components similar to [9.90] by using the above scheme for the binary solvent with one solvate active component (equations [9.76 - 9.80]) and introducing equilibrium constant of the process [9.88] such as $K_{us}$:

$$K_{EF} = K_{us}(1 + K_{EA})(1 + K_{FA})(1 + K_{EB})(1 + K_{FB}) \qquad [9.89]$$

where:

$K_i$ equilibrium constant of the processes [9.87]

### 9.4.4.1 Selective solvation. Resolvation

When component B is added to solution E in solvent A (E is neutral molecule or ion), resolvation process takes place:

$$EA_p + qB = EB_q + pA \qquad [9.90]$$

Equilibrium constant of this process in ideal solution (in molar parts of the components) equals to:

$$K_{us} = m_{EB_q} m_A^p / m_{EA_p} m_B^q \qquad [9.91]$$

where:

$m_i$ a number of moles

p, q stoichiometric coefficients of reaction

On the other hand, resolvation constant $K_{us}$ equals to the ratio of equilibrium constants of solvation processes $E + pA \leftrightarrow EA_p$(I); $E + qB \leftrightarrow EB_q$(II):

$$K_{us} = K_{II} / K_{I} \qquad [9.92]$$

The method of $K_{us}$ determination, based on the differences between free energy values of electrolyte transfer from some standard solvent to A and B, respectively, leads to a high error.

If the concentration ratio of different solvative forms is expressed as α, the concentrations A and B are expressed as $1\text{-}x_B$ and $x_B$, respectively, ($x_B$ - molar part B), equation [9.91] may be presented in the form:

$$K_{us} = \alpha x_B^p / (1 - x_B)^q \qquad [9.91a]$$

or

$$\ln K_{us} = \ln a + p \ln x_B - q \ln(1 - x_B) \qquad [9.93]$$

The last equation permits to calculate value for the solvent of fixed composition and $x_B$ determination at certain composition of the solvate complexes p and q and resolvation constant $K_{us}$.

It follows from the equations [9.91] and [9.91a]

$$K_{us} = \left[m_E^{(q)} q / \left(m_E^o - m_E^{(q)}\right) p\right]\left[x_0^{(q)}\left(1 - x_B^{(q)}\right)^q\right] \qquad [9.91b]$$

where:

$m_E^o$ a number of moles E in solution
$m_E^{(q)}$ a number of moles E solvated by the solvent A

Hence

$$K_{us} = x'_B x_B^p / \left[(1 - x'_B)(1 - x_B)^q\right] \qquad [9.91c]$$

where:

$x'_B$ molar fraction of B in solvate shell

Equation [9.91c] developed for the ideal solution E-A-B permits us to establish the relationship between the composition of mixed solvent $x_B$ and the solvate shell composition $x'_B$. For the special case of equimolar solvates, the expression of resolvation constant is written in the form:

$$K_{us} = \alpha x_B / (1 - x_B) \qquad [9.91d]$$

It follows that even in ideal solution of the simplest stoichiometry, α is not a linear function of the mixed solvent composition $x_B$. Dependence of isotherm α on the solvent composition at $K_{us} = 1$ is presented in Figure 9.9a.

Analytical correlation between the composition of solvate shell $x'_B$ and the mixed solvent composition can be developed:

$$K_{us} = \left[x'_B / (1 - x'_B)\right]\left[x_B^p / (1 - x_B)^q\right] \qquad [9.94]$$

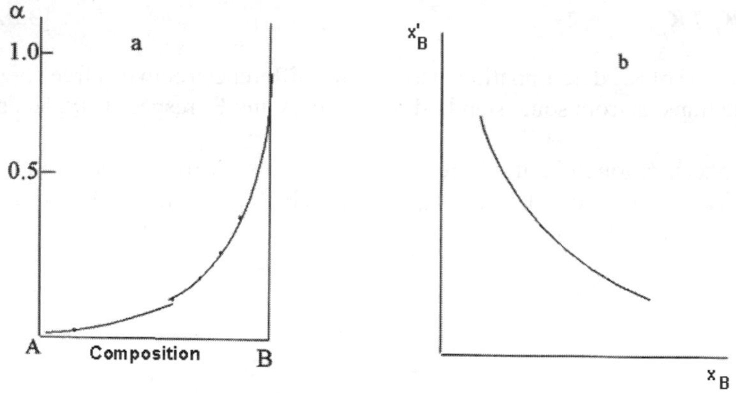


Figure 9.9. Characteristics of solvate shells in the mixed solvent A-B: a - dependence α = EA/EB on composition of the mixed solvent; b - dependence of the solvate shell composition, $x'_B$, on composition $x_B$ of the mixed solvent E-A-B.

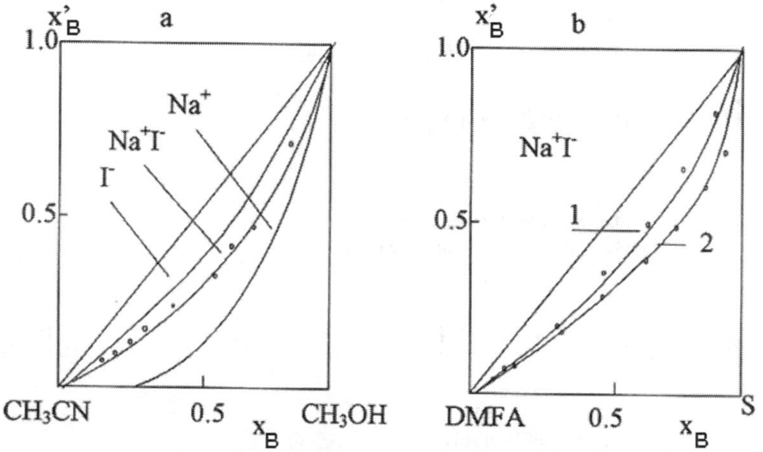


Figure 9.10. Selective solvation of NaI in mixed solvent cyanomethane-methanol (a) and DMFA-methanol (b, curve1), DMFA - cyanomethane (b, curve 2): $x_B$ - composition of mixed solvent; $x'_B$ - composition of solvate shell in molar parts of the second component.

The dependence for $K_{us} = 1$ is given in Figure 9.9b. As it follows from the analysis of equation [9.94] and Figure 9.9, the compositions of solvate shell and mixed solvent are different.

A similar approach has been developed for calculation of solvate shell composition $Na^+$ and $I^-$ in the mixed solvent formed by the components with close values of permittivity. Such solvent selection permits us to eliminate the permittivity effect on solvation equilibrium. Resolvation constants have been determined from the calorimetric study. The composition of anions solvate complex has been determined from experimental data of electrolyte $Bu_4NI$ assuming lack of the cation specific solvation. Experimental data are presented in Figure 9.10.

Padova[36] has developed this approach to non-ideal solutions. He has proposed an equation based on electrostatic interaction which relates molar fractions of the components ($x_B$ - in the mixed solvent and $x'_B$ - in the solvate shell) to the activity coefficient of components of the binary solvent:

$$\alpha = \ln\left[(1 - x'_B)/(1 - x_B)\right] = \ln \gamma_B^2 \quad [9.95]$$

Strengthening or weakening interaction (ion-dipole interaction or dipole-dipole interaction) of universal solvation leads to re-distribution of molecules in the mixed solvate and to the change of the composition of solvate shell in contrast to the composition of mixed solvent.

The method for determination of average filling of molecules` coordination sphere of dissolved substance by molecules of the mixed solvent (with one solvate-inert component) has been proposed.[37] The local permittivity is related to average filling of molecules' coordination sphere expressed by the equation:

$$\bar{\varepsilon}_p = \varepsilon_A x'_A + \varepsilon_B x'_B \quad [9.96]$$

where:

$$x'_A = z_A / (z_a + z_B); \quad x'_B = z_B / (z_A + z_B) \quad [9.97]$$

where:

$z_A$, $z_B$ average numbers of A and B molecules in the first solvate shell

The last equations can be used for development of the next expression permitting to calculate the relative content of B molecules in the solvate shell

$$x'_B = (\varepsilon_p - \varepsilon_A)/(\varepsilon_B - \varepsilon_A) \quad [9.98]$$

where:

$\varepsilon_p$ permittivity of binary solvent

$\varepsilon_A$, $\varepsilon_B$ permittivities of components

Value $x'_B$ can be found from the equation linking the location of maximum of absorption band of IR spectrum with refraction index and ε of the solution.

The data on selective solvation of 3-aminophthalimid by butanol from butanol-hexane mixture are presented in Figure 9.11. The data have been calculated from the equations presented above. Alcohol content in solvate shell has higher concentration than in solvent composition even at low concentration of alcohol in the solvent. For example, when molar fraction of n-butanol in the mixture was 7%, the relative molar fraction of n-butanol in the solvate shell of

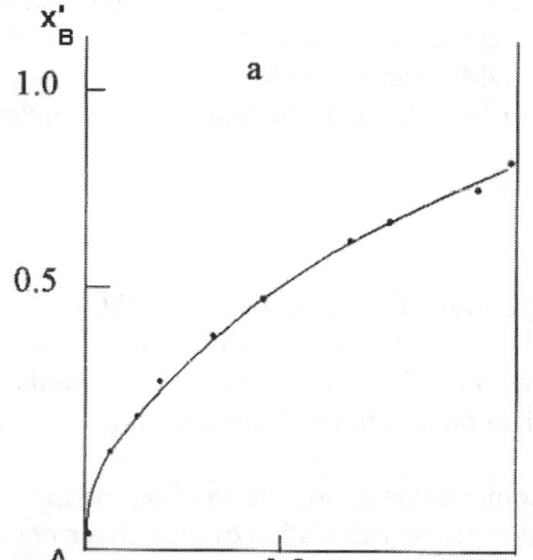


Figure 9.11. Selective solvation of 3-aminophthalimid (A) by n-butanol (B) from the mixed solvent hexane-n-butanol.

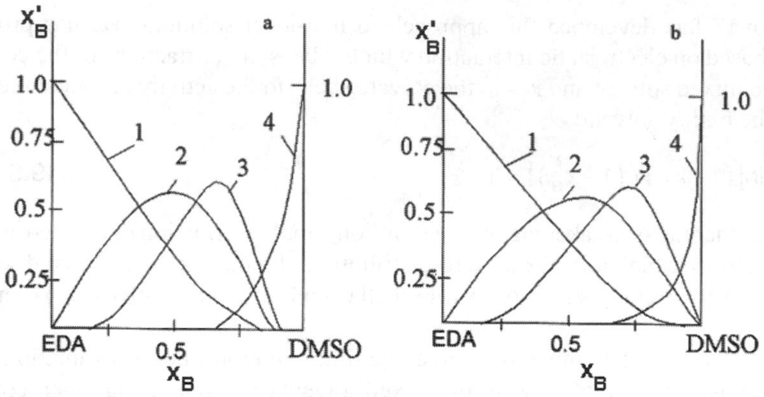


Figure 9.12. Dependence of concentration (in molar parts) of solvate forms $Li(EDA)_mS_n$ on molar fraction, $x_B$, of the second component in the binary solvents ethylenediamine-DMSO (a) and ethylenediamine-DMFA (b): 1-m=4, n=0;2- m=n=2;4-m=0, n=4.

aminophthalimide was 30%. Resolvation process completes at n-butanol concentration in solution ≈90%.

Mishustin[37] proposed a strict and accurate method for selective solvation study. The method is based on data of free energy transfer of electrolyte from individual solvent A to mixed solvent A - B. The method takes into account non-ideality of the system, and allows calculation of the concentration of different solvate forms and their dependence on the mixed solvent composition.

An example of application of this method is in the work.[39] Authors have calculated relative concentration of different solvate forms of $Li^+$ in the mixed solvent ethylenediamine - DMSO and ethylenediamine-DMFA (Figure 9.12). Free energy of lithium transfer from DMSO (DMFA) in the mixed solvent has been calculated from the time of spin-lattice relaxation of kernel $^7Li$. The curves presented in Figure 9.12 depict quantitatively the selectivity of $Li^+$ relative to ethylenediamine, which is more basic component in contrast to the second components of the mixed solvent, namely DMSO and DMFA.

The following systems can serve as examples of the effect of composition of the mixed solvent on the solvate shell composition:

$[Cr(NH)_S(H_2O)_m(DMSO)_n]^{3+}$ - $H_2O$ - DMSO[40]

$[Be(H_2O)_m(EG)_n]SO_4$ - $H_2O$ - ethylene glycol[41]

$[Be(H_2O)_m(HMPTA)_n]SO_4$ - $H_2O$ - HMPTA[42]

$[Ni(H_2O)_mS_n]ClO_4$ - $H_2O$ - S (where S is methanol, ethanol, propanol, DMSO)[43]

Data presented in Figure 9.13 contain information on the composition of solvate shell as a function of molar fraction of water in the mixed solvent $H_2O$-other solvents.[43] Monograph[44] contains collection of data on resolvation constants of the ions in the mixed solvents.

The above presented dependencies of the composition of solvate shell on the mixed solvent composition as well as resolvation constants permit calculation of the solvate composition by varying solvent composition. The dependence of resolvation constants on the permittivity of the solvent is discussed in the example of the proton resolvation process.

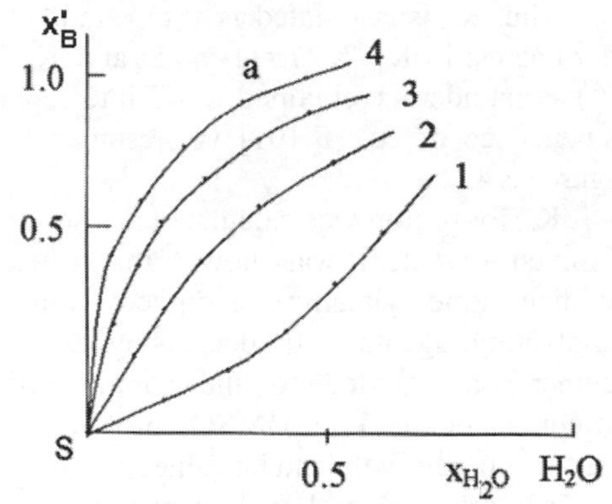


Figure 9.13. The composition of solvate shell $Ni^{2+}$ (in molar parts of water) in the mixtures of solvents formed from water and propanol (1), ethanol (2), methanol (3) and DMSO (4). Data from Ref. 43.

In the mixed solvents water - non-aqueous solvent, in spite of its donor and polar properties, water is a preferred solvating agent. This generalization has some exceptions (solvation in systems $Ag^+$ - $H_2O$ - acetonitrile, $Cr^{3+}$ - $H_2O$ - DMSO, $F^-$ - $H_2O$-ethylene glycol).[45]

Solvation energy of proton by donor solvents is very high. The regularities of the proton selective solvation and re-solvation processes were studied in more detail in comparison with other ions.

Let us consider the changes in the system, when donor component is added to protic acid HA in solvent A. Anion solvation can be neglected, if both solvents have donor character. The interaction influences the proton re-solvation process.

$$HA_p^+ + B \longleftrightarrow \left[HA_{p-1}B\right]^+ \xleftrightarrow{+B} \cdots \longleftrightarrow HB_p^+ + A \qquad [9.99]$$

or

$$HA_p^+ + qB \longleftrightarrow \cdots \longleftrightarrow HB_q^+ + pA \qquad [9.99a]$$

The model related to eq. [9.99a] was evaluated,[46] resulting in the supposition that the equilibrium of two forms of solvated proton: $HA_p^+$ and $HB_q^+$ is important. Solvation stoichiometry was not considered. Both proton solvated forms are denoted as $HA^+$ and $HB^+$.

If B is the better donor component (it is a necessary requirement for equilibrium [9.99] shift to the right hand side of equation), the equation from work[46] can be simplified to the form:[47]

$$K_a = K_a^A + \left(K_a^B - K_a^A\right)\left\{\left[K_{us}x_B/(1-x_B)\right]/\left[1+K_{us}x_B/(1-x_D)\right]\right\} \qquad [9.100]$$

where:

| | |
|---|---|
| $K_a$ | ionic association constant |
| $K_a^A, K_a^B$ | the ionic association constants of acid in individual solvents A and B |
| $K_{us}$ | a constant of resolvation process |
| $x_B$ | molar fraction of B |

For the calculation of resolvation constant, one must determine the experimental constant of HA association in the mixed solvent and determine independently $K_a^A$ and $K_a^B$.

When the resolvation process is completed at low concentration of the second component, the change of permittivity of mixed solvent A-B may be ignored. Thus, one may assume that $K_a^A$ and $K_a^B$ are constant and calculate $K_{us}$ from the equation [9.100] in the form:

$$1/\left(K_a - K_a^A\right) = 1/\left(K_a^B - K_a^A\right) + \left\{\left[(1-x_B)/x_B\right]\left[1/K_{us}\left(K_a^B - K_a^A\right)\right]\right\} \qquad [9.101]$$

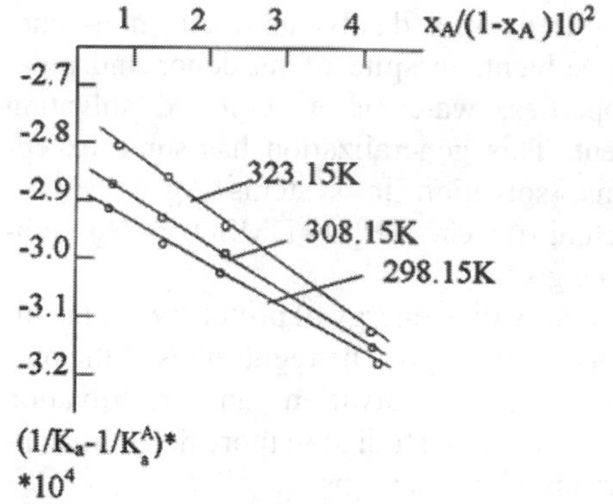


Figure 9.14. Dependence of $1/(K_a - K^A_a)$ for $CF_3COOH$ on the mixed solvent composition DMSO (A) - dimethylalanine at different temperatures.

Thus $K_{us}$ is calculated as a slope ratio of coordinates: $1/(K_a - K^A_a) = (1-x_B)/x_B$ and $(K_a - K^A_a)$ remainder is obtained as Y-intercept. Dependence of eq. [9.101] is presented in Figure 9.14.

$K_{us}$ for proton was calculated in a series of mixed solvents. It was shown[47] that, when pyridine, dimethylalanine or diphenylamine (resolvating agents with decreasing donor numbers) are added to the solution of trifluoroacetic acid in DMSO, proton decreases consecutively and its values are equal to $2.7x10^4$; $4.2x10^3$ and 35.4, respectively.

Consideration of permittivity of the mixed solvent has allowed calculation of proton in whole concentration range of the mixed solvent DMSO - diphenylamine. The data have been approximated using equation [9.55].

$$\ln K_{us} = -12.2 + 3400/T + 453.7/\varepsilon - 75043/\varepsilon T$$

Equilibrium constants for exchange process of alcohol shell of solvates to water shell were calculated:[48]

$$ROH_2^+ + H_2O \leftrightarrow H_3O^+ + ROH \qquad [9.102]$$

The solutions of HCl and $HOSO_3CH_3$ in aliphatic alcohol (i.e., $C_nH_{2n+1}OH$) - normal alcohol $C_1$-$C_5$ and isomeric alcohol $C_3$-$C_5$ have been studied. If the components taking part in resolvation process are capable of H-bonding, the anion solvation by these components cannot be neglected. The differences in $K_{us}$ values for both acids in different solvents may be explained as follows.

The dependence of $K_{us}$ on permittivity and temperate is described by the equation:

$$\ln K = a_{00} + a_{01}/T + a_{02}/T^2 + (a_{10} + a_{11}/T)/\varepsilon \qquad [9.103]$$

The coefficients of equation [9.103] are presented in Table 9.5.

**Table 9.5. Coefficients of the equation [9.103] for constants of resolvation process**

| System | $a_{00}$ | $a_{01}\times10^{-2}$ | $a_{02}\times10^{-4}$ | $a_{10}$ | $a_{11}\times10^{-3}$ | $\pm\delta$ |
|---|---|---|---|---|---|---|
| HCl - n-alcohol | 12.4 | 46.0 | 56.3 | 186.8 | 71.1 | 0.3 |
| HCl - isomeric alcohol | 13.1 | 5.9 | -36.9 | 318.4 | 94.7 | 0.2 |
| $HOSO_3CH_3$ - n-alcohol | 6.3 | 4.4 | -5.4 | 177.7 | 60.9 | 0.3 |

The solvent effect in the following resolvation process was studied:[49]

$$ROH_2^+ + Py \leftrightarrow HPy^+ + ROH \qquad [9.104]$$

The solutions of trifluoroacetic acid in ethanol and methanol in the temperature range 273.15-323.15K were investigated. The dependence of $K_{us}$ on permittivity and temperature is described by the equation:

$$\ln K_{us} = 9.44 + 1768/T + 77.8/\varepsilon - 11076/\varepsilon T$$

Unlike the processes considered above, in the case of process [9.104] permittivity decrease leads to decreasing $K_{us}$. The explanation of the results based on covalent and electrostatic components of resolvation process enthalpy is given.[50]

The reaction [9.104] also has been studied for isodielectric mixtures of alcohol-chlorobenzene with ε=20,2 (permittivity of pure n-propanol) and ε=17.1 (permittivity of pure n-butanol) to investigate the relative effect of universal and specific solvation on the resolvation process. The mixtures were prepared by adding chlorobenzene to methanol, ethanol, and $C_1$-$C_3$ alcohol. Alcohol is a solvate-active component in these isodielectric solvents. $K_{us}$ data are given in Table 9.6.

**Table 9.6. Equilibrium constants of the process [9.104] in isodielectric solvents**

| Solvents with ε=20.2 | $K_{us}\times10^{-5}$ | Solvents with ε=17.1 | $K_{us}\times10^{-5}$ |
|---|---|---|---|
| Methanol +23.5% chlorobenzene | 3.9 | Methanol + 32.8% chlorobenzene | 1.4 |
| Ethanol + 13% chlorobenzene | 4.9 | Ethanol + 24% chlorobenzene | 2.0 |
| n-Propanol | 7.3 | n-Butanol | 4.9 |

Insignificant increase of $K_{us}$ in n-butanol (or n-propanol) solution in comparison to methanol is due to relaxation of the proton-alcohol bond, when the distance of ion-dipole interaction increases.

The change of donor property of the solvate-active component is not significant. The equations relating $K_{us}$ to ε permit to divide free energy of resolvation process into the components. Corresponding data are presented in Table 9.7.

**Table 9.7. The components of free energy (kJ mol⁻¹) of proton resolvation process at 298.15K**

| Solvent | ε | $\sigma\Delta G^{el}$ in process | | |
|---|---|---|---|---|
| | | [9.102] | | [9.104] |
| | | HCl | $HOSO_3CH_3$ | $CF_3COOH$ |
| Methanol | 32.6 | 3.9 | 2.0 | -3.1 |
| Ethanol | 24.3 | 5.3 | 2.7 | -4.1 |
| n-Propanol | 20.1 | 6.4 | 3.3 | -5.0 |
| n-Butanol | 17.1 | 7.5 | 3.9 | -5.9 |
| n-Pentanol | 14.4 | 8.9 | 4.6 | - |

| Solvent | ε | σΔG$^{el}$ in process | | |
|---|---|---|---|---|
| | | [9.102] | | [9.104] |
| | | HCl | $HOSO_3CH_3$ | $CF_3COOH$ |
| $-\delta\Delta G^{cov}$ | | 8.2 | 17.7 | -38.1 |
| $-\delta\Delta G^{el}_{\varepsilon=1}$ | | 128 | 66.0 | -100.6 |

In contrast to the processes considered earlier, the vacuum electrostatic component of resolvation process has high value whereas $-\delta\Delta G^{el}$ values are comparable with the covalent component, $\delta\Delta G^{cov}$.

$\delta\Delta G$ values according to [9.51] are equal to:

$$\delta\Delta G = \left(\Delta G_{HB^+} + \Delta G_A - \Delta G_{HA^+} + \Delta G_B\right)_{sol}$$

For small additions of B (to component A), $(\Delta G_A)_{sol}=0$ and $(\Delta G_{HA})_{sol}=0$ , then

$$\delta\Delta G = (\Delta G_{HB+} + \Delta G_B)_{sol}$$

Solvation energy of complex $HB^+$ by solvent A is small because coordination vacancies of the proton are saturated to a considerable extent. Therefore the interaction energy between A and B influences significantly the value of σΔG. That is why, the mixed solvents (alcohol-water and alcohol-pyridine, for instance) are different because of the proton resolvation process. This can be explained in terms of higher energy of heteromolecular association for the alcohol-water in comparison with alcohol pyridine.

The concept of solvent effect on the proton resolvation process was confirmed by quantum chemical calculations.[51] Above phenomena determine the dependence of resolvation constant on physical and chemical properties.

Let the resolvation process proceeds at substantial abundance of the component A in mixed solvent and initial concentrations HA ($HA^+$) and B to be equal. The output of the process can be calculated from the equation similar to equation [9.66]. The large value of $K_{us}$ in all considered processes of proton resolvation indicates the effect of permittivity change on the yield of complex $HB^+$ formation. The output of resolvated proton in process [9.104] proceeding in methanol equals 100%, whereas in the same process in low polarity solvent (e.g., methanol-hexane), with abundance of the second component, the equilibrium is shifted to the left, resulting in solvate output of less than 0.1%. $K_{us}$ values in single alcohol solvents are large, thus the output of reaction does not depend on solvent exchange.

The process $H{\bullet}DMSO^+ + B \leftrightarrow HB^+ + DMSO$ may be considered as an example of the effect of chemical properties of B on the output of the reaction [9.99]. The output of $HB^+$ equals to 98%, if pyridine is included in the process at initial concentration of 0.1M. Use of diphenylamine, having lower donor properties, decreases the output to 60%. The output differs even more at smaller concentrations of component, such as $10^{-3}$ M, which gives yields of 83 and 33%, respectively.

### 9.4.5 MIXED SOLVENT EFFECT ON THE ION ASSOCIATION PROCESS

The ion association process (or opposite to it - ion dissociation process) has been studied in detail in comparison with other types of chemical equilibrium in solutions. The modern state of the theory of individual solvent effects on equilibrium constant of ion association process ($K_a$) is described in monographs by Izmailov[53] and Barthel.[54]

The formation of free (preferably solvated) ions is due to a successive equilibrium states proceeding in solution:

$$mE + nF \xleftrightarrow{K_{add}} E_mF_n \xleftrightarrow{K_i} Cat_p^{q+} \bullet An_q^{p-} \longleftrightarrow \cdots \xleftrightarrow{K_a} pCat^{q+} + qAn^{p-} \quad [9.105]$$

where:

| | |
|---|---|
| $K_{add}$ | a constant of adduct formation |
| $K_i$ | ionization constant |
| $K_a$ | association constant |

The true constant of ion association is a ratio of ionic associate concentration $Cat_p^{q+} \bullet An_q^{p-}$ to ion concentration product. If ionic associate concentration is unknown (as is true in many cases), ion association constant is calculated from the analytical concentration of dissolved substance:

$$K_a = \left\{c_0 - p\left[Cat^{q+}\right] - q\left[An^{p-}\right]\right\} / \left[Cat^{q+}\right]^p \left[An^{p-}\right]^q$$

where:

| | |
|---|---|
| $c_0$ | initial concentration of electrolyte |

The general theory of ionic equilibrium[53,54] leads to the conclusion that ion association constant in universal or conditionally universal media, in accordance to the equation [9.53a], depends exponentially on reciprocal value of permittivity:

$$\ln K_a = a_{00} + a_{01} / \varepsilon \quad [9.106]$$

where:

| | |
|---|---|
| $a_{00}$, $a_{01}$ | approximation coefficients |

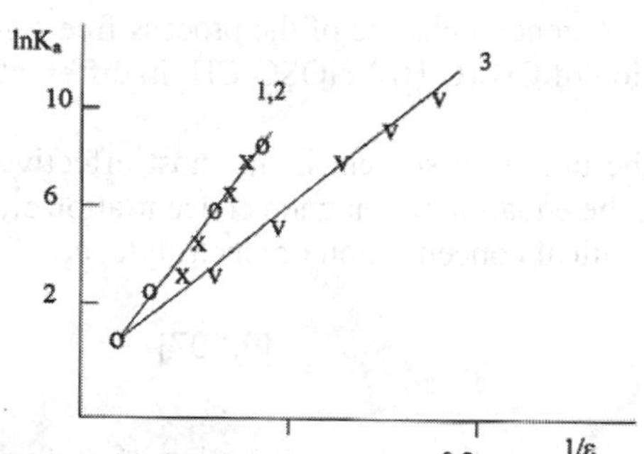


Figure 9.15. Dependence $K_a$ of $(C_2H_5)_4NBr$ on permittivity in mixed solvents formed by propylene carbonate with o-dichlorobenzene (1,2-o), pyridine (1,2-x) and acetic acid (3) at 298.15K.

If ion association process has high energy, the solvents are solvate-inert because of large ions, such as $(C_nH_{2n+1})_4N^+$, $(C_nH_{2n+1})_4P^+$, etc. Solution of $(C_2H_5)_4NBr$ in mixed solvents propylene carbonate - o-dichlorobenzene and propylene carbonate - pyridine[55] serve as an example. The components of these binary solvents (with the exception of inert dichlorobenzene) possess donor properties, though they do not solvate the large anion $R_4N^+$ because it is solvate-inert component in relation to anion. That is why $\ln K_a$ - $1/\varepsilon$ isotherms follow the same line (see Figure 9.15). But the $\ln K_a$ - $1/\varepsilon$ isotherm for solution of propylene carbonate - acetic acid differs from the other two because acetic acid is a solvate-active component in relation to anion $Br^-$.

Coefficients of equation 9.56 for the various systems are presented in Table 9.8.

**Table 9.8. Coefficients of equation [9.56] for ionic association of tetraethyl ammonium bromide at 298-323K**

| Solvent | $-a_{00}$ | $a_{10}$ | $a_{01}$ | $a_{11}$ |
|---|---|---|---|---|
| Propylene carbonate - o-dichlorobenzene | 5.2 | 1440 | 78.8 | 9523 |
| Propylene carbonate - pyridine | 6.3 | 1850 | 114 | 0 |
| Propylene carbonate - acetic acid | 7.7 | 2400 | 110.5 | -11500 |

The $\ln K_a$ - $1/\varepsilon$ dependence for universal media is linear in full range of permittivity.

Validity of the following equation was evaluated[7] for the mixed solvent propylene carbonate - 1,4-dioxane in the range of permittivity $\varepsilon$=65±3 for the solutions $Et_4NBr$

$$\ln K_a = 1.32 + 80.05/\varepsilon$$

**Table 9.9. Components of free energy (kJ mol$^{-1}$) of ion association of methyl octyl ammonium methyl sulfate in universal solvents at 298.15K**

| Solvent | $\varepsilon$ | $\sigma^{el}_{solv}$ |
|---|---|---|
| Pyridine-DMFA | 25.0 | 7.6 |
| Pyridine-acetonitrile | 19.2 | 9.9 |
| Pyridine-propylene carbonate | 16.0 | 11.9 |
| Nitrobenzene-acetic acid | 11.9 | 16.0 |
| Propylene carbonate-acetic acid | 10.6 | 17.9 |

Comparison of the free energy of components for ion association process in different solvents permits us to estimate the energy of specific solvation interaction.

Electrical component of the change of free energy, depending on ion-ion interaction in accordance with the nature of ion association process, must exceed a sum of vacuum and covalent components in equation [9.52]. $\sigma^{el}_{solv}$ for all above considered systems is higher than 85 - 90% because of the general change of the process free energy. The components of free energy of association of $CH_3(C_8H_{17})_3NOSO_3CH_3$ in different solvents[7] are presented in Table 9.9.

Evidently, the change of permittivity of the universal solvent is the most effective method of affecting the ion concentration. Using the equation relating ion concentration $c^*_M$ for every type of 1-1 electrolyte with $K_a$ and analytical concentration of electrolyte, $c_M$:

$$c_M = \left[\left(4K_a c^0_M + 1\right)^{1/2} - 1\right] / 2K_a \qquad [9.107]$$

one can ascertain, that $\varepsilon$ decrease from 185 to 6 changes the ion concentration of methyl octyl ammonium methyl sulfate in solution from 71 to 0.3%. In mixed solvent containing propylene carbonate, the ionic concentration of $Et_4NBr$ is also changed from 80% (in propylene carbonate) to 3% (in pyridine), and even to 0.8% (in o-dichlorobenzene or acetic acid). Thus, affecting electrolyte strength by solvent choice permits to change a strong electrolyte into non-electrolyte.

A similar phenomenon is observed in conditionally-universal media A-B (A is solvate-active component, B is solvate-inert component). Association constant can be calculated from equations [9.56] or [9.104] for LiBr and KCNS solutions.[55,57]

**Table 9.10. Coefficients of equations [9.56] and [9.106] for association constants of salts**

| Solvent | [9.56] | | | [9.106] (296K) | | |
|---|---|---|---|---|---|---|
| | $a_{00}$ | $a_{10}$ | $a_{01}$ | $a_{11}$ | $a_{00}$ | $a_{01}$ |
| Lithium bromide (298.15-323.15K) | | | | | | |
| Propylene carbonate-o-dichlorobenzene | 3.38 | -808 | -361 | $1.53\times10^5$ | 0.67 | 152.4 |
| Propylene carbonate-pyridine | -5.56 | 2040 | 135 | -6923 | 1.29 | 111.8 |
| Propylene carbonate-acetic acid | -3.55 | 1617 | 70.1 | -6160 | 1.87 | 49.4 |
| Potassium thiocyanate (298.15-348.15K) | | | | | | |
| Acetonitrile-chlorobenzene | 3.80 | -521 | -25.4 | $3.26\times10^4$ | 2.05 | 84.0 |
| Propylene carbonate-chlorobenzene | 1.88 | -229 | -1.53 | $2.60\times10^4$ | 1.11 | 85.7 |
| DMSO-chlorobenzene | 6.18 | -1580 | -16.7 | $2.73\times10^4$ | 0.88 | 74.9 |

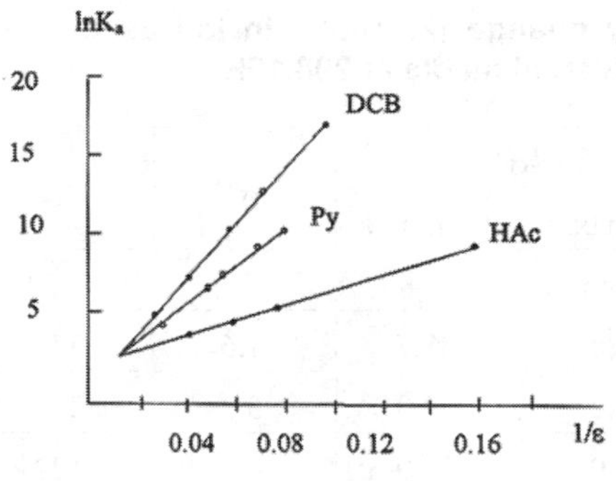


Figure 9.16. Dependence of $\ln K_a$ for lithium bromide on $1/\varepsilon$ in the mixed solvent based on propylene carbonate at 298.15K.

The dependence of $\ln K_a$ and $1/\varepsilon$ for LiBr in mixed solvent [9.53] shows an influence of solvate-active component on $K_a$. The difference of LiBr solvation energy in the presence of pyridine and acetic acid may be evaluated in accordance to [9.53], assuming low solvation energy of $Li^+$ by propylene carbonate in comparison with pyridine and negligible value of solvation energy of anion in solvents containing dichlorobenzene (DCB) and pyridine.

The data presented in Table 9.10 permit us to calculate the components of free energy for ion association processes. To interpret the data, one must take into account:

$$\sigma^{cov} = \Delta\sigma^{v}_{Cat\cdot An} + \left(\sigma_{sol,Cat\cdot An} - \sigma_{sol,Cat^+} - \sigma_{sol,An^-}\right)^{cov}$$

correspondingly:

$$\sigma^{el} = \left(\beta_{Cat\cdot An} - \beta_{Cat^+} - \beta_{An^-}\right)/\varepsilon$$

A small value $\sigma^{cov}$ in all considered systems can be explained by low energy of interaction between the filled orbital of cation and anion. On the contrary, contribution of ion-ion interaction to the process energy leads to high value $\sigma^{el}$ in comparison with the covalent component even in media of high permittivity (not to mention media of low permittivities).

From [9.106], isodielectric solution formed by two mixed solvents I and II follows the relationship:

$$\ln K_{a,I} = (a_{01,I} / a_{01,II}) \ln K_{a,II} - (a_{01,I} / a_{01,II}) a_{00,II} + a_{00,I} \qquad [9.108]$$

where:

00, 01 subscripts of coefficients of approximation for dependence on temperature and permittivity

I, II subscripts for solvents I and II

The leveling (leveling solvent reduces differences between constants of ion association) or differentiating (differentiating solvent enhances these differences) effect of the first mixed solvent is defined by the ratio of coefficients of equation [9.106] or by the ratio of vacuum electrostatic component of free energy of association process, $\delta\Delta G^{el}_{\varepsilon=1}$.

One can calculate ratio $a_{01,\ PC\text{-}DCB}/a_{01,\ PC\text{-}Py} = 1.36$ using the data from Tables 9.10 and 9.11 and assess the leveling effect of the solvent propylene carbonate-pyridine on electrolyte strength. The ratio $a_{PC\text{-}DCB}/a_{PC\text{-}HAc} = 3.1$ indicates that propylene carbonate-acetic acid has more pronounced leveling effect on electrolyte strength than propylene carbonate-pyridine.

**Table 9.11. The components of free energy change (kJ mol$^{-1}$) in ion association process of different salts in conditionally-universal media at 298.15K**

| Solvent | $-\sigma^{el}$ | | $-\sigma^{cov}$ | $\sigma^{el}_{\varepsilon=1}$ |
|---|---|---|---|---|
| | At max. ε | At min. ε | | |
| Lithium bromide | | | | |
| Propylene carbonate-o-dichlorobenzene | 5.8 | 37.7 | 1.6 | 377 |
| Propylene carbonate-pyridine | 4.2 | 22.9 | 3.2 | 277 |
| Propylene carbonate-acetic acid | 1.9 | 20.0 | 4.6 | 122 |
| Potassium thiocyanate | | | | |
| Acetonitrile-chlorobenzene | 5.7 | 37.1 | 5.1 | 208 |
| Propylene carbonate-chlorobenzene | 3.2 | 37.9 | 2.7 | 212 |
| DMSO-chlorobenzene | 4.0 | 33.1 | 2.2 | 185 |

ε change of the solvent and universal media affect the ion concentration in solution. The ion concentration in 0.1M solution of LiBr in the mixed solvent propylene carbonate-o-dichlorobenzene varies relative to its analytical concentration from 50% in propylene carbonate to 0.1% in dichlorobenzene. The ion concentration in 0.1M solution of KSCN in the solvent propylene carbonate-chlorobenzene varies from 60% in propylene carbonate to 0.06% in chlorobenzene.

The same effect can be obtained by means of solvent heating. For instance, LiBr solution in o-dichlorobenzene must be heated to ≈215°C to reach the same value of association constant as in propylene carbonate at room temperature. For this reason, solvent may be used as an effective means for tailoring electrolyte strength in conditionally universal media as well.

The influence of solvate active property of components of the mixed solvent is revealed in acid solution. The data for some acids in different mixed solvents are presented in Table 9.12.[7]

**Table 9.12. The properties of ion association process of H-acids in conditionally universal media at 298.15K**

| Acid | Solvent | Coefficients of equation [9.106] | | Components of energy of ion association process, kJ mol$^{-1}$ | | | |
|---|---|---|---|---|---|---|---|
| | | | | $-\sigma^{cov}$ | $\sigma^{el}_{\varepsilon=1}$ | $\sigma^{el}$ | |
| | | $a_{00}$ | $a_{01}$ | | | At max. ε | At min. ε |
| $CH_3SO_3H$ | Methanol - n-hexanol | 2.49 | 80.6 | 6.2 | 200 | 6.1 | 15.0 |
| $HSO_3F$ | Methanol-n-butanol | -1.50 | 165.0 | 3.7 | 408 | 12.5 | 23.9 |
| $CF_3COOH$ | | 4.86 | 190.9 | 12.0 | 473 | 15.5 | 26.7 |
| $H_2SO_4$* | | -1.82 | 165 | -4.5 | 408 | 12.6 | 23.1 |
| | DMSO - $CCl_4$ | -1.55 | 215 | -3.8 | 532 | 11.4 | 133 |
| | DMSO-o-dichlorobenzene | -1.34 | 183 | -3.3 | 453 | 9.7 | 75.6 |
| | DMSO-benzene | 0.55 | 145 | 1.4 | 359 | 7.7 | 120 |
| | DMSO-1,4-dioxane | 0.55 | 129 | 1.4 | 319 | 6.9 | 80 |
| | DMSO-pyridine | 1.50 | 90 | 3.7 | 223 | 4.8 | 15 |
| | Propylene carbonate -DMSO | 0.38 | 146 | 0.9 | 361 | 5.6 | 7.8 |
| $H_2SeO_4$* | Methanol-n-butanol | -2.06 | 162 | -5.1 | 401 | 12.3 | 22.7 |

*data for the first constant of ion association

The degree of ion association of acids as well as ionophores depends significantly on solvents` permittivity. The influence of the specific solvation by the solvate active component is pertinent from the comparison $\ln K_a$ - 1/ε isotherms for acid solutions in conditionally universal media $S^{(1)}$ - $S^{(2)}$, $S^{(1)}$ - $S^{(3)}$...

Figure 9.17[58] gives $\ln K_a$ - 1/ε relationships for solutions of sulfuric acid in binary solvent mixture containing DMSO and $CCl_4$, benzene, dioxane, pyridine. The increase in the second component basic capacity according to [9.50] leads to decrease in $\ln K_a$.

General analysis of the binary solvent mixtures formed by two solvate active components (these solvents are often used in analytical and electrochemistry) was conducted to evaluate their effect on H-acids.[59] The analysis was based on an equation which relates the constant of ion association, $K_a$, of the solvent mixture and constants of ion association of the acid $K_a^A$ and $K_a^B$ of each component of the mixed solvent, using equilibrium constants of scheme [9.105] - heteromolecular association constant, $K_{add}$; ionization constant of the

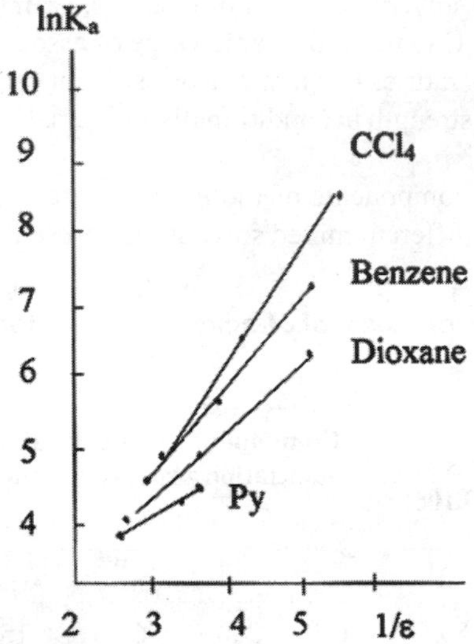


Figure 9.17. Dependence of $K_a$ of sulfuric acid on $1/\varepsilon$ in mixed solvents based on DMSO at 298.15K.

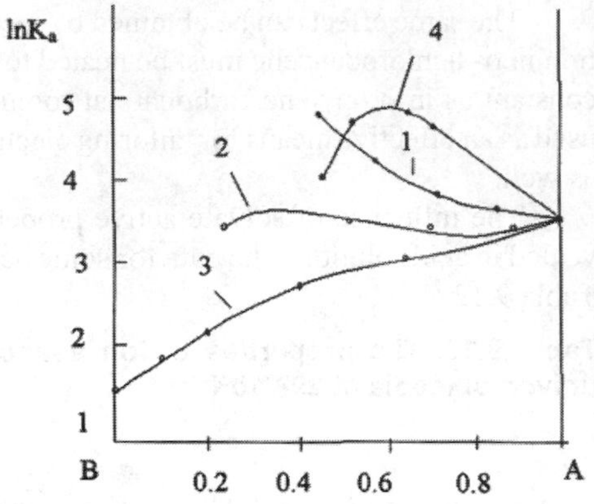


Figure 9.18. Dependence of $K_a$ for $HSO_3F$ on solvent composition in mixed solvents: A - n-butanol at 298.15K.[7] B : 1 - n-hexane; 2 - nitrobenzene; 3 - methanol; 4 - N,N- dimethyl aniline.

adduct, $K_i$; and resolvation constant, $K_{us}$, (e.g., equilibrium constant of process $HA_n \cdot A + B \leftrightarrow HA_n \cdot B + A$), and with molar fractions of components of the mixed solvent $x_A$ and $x_B$:

$$K_a = K_a^B + \left(K_a^B - K_a^A\right)\left[1 - \left(K_a^B / \left(1 + K_{add} x_B \left(1 + K_i\right)\right)\right)\right] \qquad [9.109]$$

The equation [9.109] changes into equation [9.100] when both components of the mixed solvent have donor character, e.i., $K_{us} = 0$. The analysis of equation [9.100][59] shows that the solvate active property of components of the mixed solvent A-B have to be changed to influence the dependence of $K_a$ on composition of the mixed solvent. Some examples of such influence are presented in Figure 9.18.

## 9.4.6 SOLVENT EFFECT ON EXCHANGE INTERACTION PROCESSES

Chemical processes of exchange interaction, given by equation:

$$EF + HL \leftrightarrow EL + HF \qquad [9.110]$$

are common in research and technology, but the solvent effect on their equilibrium was not frequently studied. Let us systematize the existing experimental data on these processes.[60]

### Systems with non-associated reagents

The process of exchange interaction between bis-(carbomethoxymethyl)mercury and mercury cyanide

$$(CH_3OCOCH_2)_2Hg + Hg(CN)_2 \leftrightarrow 2CH_3OCOCH_2HgCN \qquad [9.111]$$

has been studied in the mixed solvent DMSO-pyridine. The components of solvent possess almost equal ability to specific solvation of mercury-organic compounds. This ability is confirmed by equality of the constants of spin - spin interaction $^{2}I(^{1}H - ^{199}Hg)$ for different organic compounds in these solvents. This was also confirmed[61] for participants of the equilibrium [9.111].

### Systems with one associated participant of equilibrium

Methanolysis of triphenylchloromethane may serve as an example of such reaction:

$$Ph_3CCl + CH_3OH \leftrightarrow Ph_3COCH_3 + HCl \qquad [9.112]$$

This process has been studied in benzene, 1,2-dichloroethane, chloroform, trifluoromethylbenzene and in mixed solvents hexane - nitrobenzene, toluene - nitromethane, toluene - acetonitrile.

Low constant of the ion associate formation process of triphenylchloromethane, high constant of the ion association process, and low constant of the heteromolecular association process of HCl (HCl solutions in listed solvents obey the Henry's Law) show that only methanol is an associative participant of the equilibrium.

### Systems with two associated participants of equilibrium

These systems have been studied be means of reaction of acidic exchange between acids and anhydrides with different acidic groups:

$$CH_2ClCOOH + (CH_3CO)_2O \leftrightarrow CH_2ClCOOCOCH_3 + CH_3COOH \qquad [9.113a]$$

$$CH_3COOH + (CH_2ClCO)_2O \leftrightarrow CH_2ClCOOCOCH_3 + CH_2ClCOOH \qquad [9.113b]$$

in binary mixed solvents formed from tetrachloromethane with chlorobenzene, trifluoromethylbenzene and nitrobenzene.

The system trifluoroacetic acid and methanol, undergoing esterification reaction:

$$CF_3COOH + CH_3OH \leftrightarrow CF_3COOCH_3 + H_2O \qquad [9.114]$$

also belongs to the group.

The equilibrium process has been studied in binary mixed solvents such as hexane-chloroform, hexane-chlorobenzene, chloroform-chlorobenzene. Also, re-esterification

$$CH_3COOC_3H_7 + CX_3COOH \leftrightarrow CX_3COOC_3H_7 + CH_3COOH \qquad [9.115]$$

(where X is F or Cl) has been studied for the full concentration range of propyl acetate - trifluoro- and trichloroacetic acids.[62]

Etherification process of ethanol by caproic or lactic acids has been studied by these authors.[63] The reaction of acetal formation was studied[64]

$$RCHO + R'OH \leftrightarrow RCH(OR')_2 + H_2O \qquad [9.116]$$

in systems of isoamyl alcohol-benzaldehyde (a) and ethanol - butyric aldehyde (b).

The relationship between the maximum output of the reaction of exchange interaction and associative state of equilibrium [9.110] participants follows the Gibbs-Dugem-Margulis equation:

$$x_s d\ln a_s + \Sigma x_i d\ln a_i = 0 \quad [9.117]$$

where:

$x_S$, $x_i$ molar fractions

$a_s$, $a_i$ activity of the solvent and participants of the chemical process, correspondingly

Izmaylov[65] showed that maximum output of reaction depends on initial composition of reagents. Associative and dissociative processes of the chemical system components are accompanied by the change of their chemical potentials if the solvent potential depends on total activity of solution (it depends on concentration in dilute solutions):

$$\mu_s = \mu_s^o - RT\sum_{i=1}^{i} \ln c_i \quad [9.118]$$

Hence the condition of chemical equilibrium in non-ideal (regular) systems is presented in the following form:

$$(\partial G)_{P,T} = (\partial G_S)_{P,T} + \sum (\partial G_i)_{P,T} \quad [9.119]$$

$$\frac{\partial \ln K_{ch}}{\partial V} = n_s \frac{\partial \sum_{i=1}^{i} c_i}{\partial V} - \frac{\partial}{\partial V}\prod_{i=1}^{\iota} \gamma_i^{v_i} - \frac{\partial}{\partial V}\frac{\sum_{i=1}^{i} \mu_i^o v_i + n_s \Delta\mu_s^o}{RT} \quad [9.120]$$

Inserting values of chemical potentials for equilibrium participants [9.110] and solvent in equation [9.119], one can obtain the equation relating the equilibrium constant, $K_{ch}$, change relative to the solvent composition (in volume parts, V), mol number, $n_S$, activity coefficients, $\gamma_i$, stoichiometric coefficients, $v_i$, of process [9.110] and chemical potential, $\mu_i^o$, for every participant of equilibrium at concentration 1M.

Equation [9.120] changes to [9.121], if only EF and EL undergo homomolecular association:

$$\frac{\partial \ln K_{ch}}{\partial V} \approx 2n_s \sqrt{c_{EF}^o}\,\frac{\partial}{\partial V}\left(\frac{1}{\sqrt{K_{dim}^{EL}}} - \frac{1}{\sqrt{K_{dim}^{EF}}}\right) + \frac{dK_{ch}}{d\varepsilon} \quad [9.121]$$

where:

$K_{dim}$ equilibrium constant of homomolecular association

Equations [9.120] and [9.121] describe the solvent effect on $K_{ch}$ for a non-ideal liquid system. The dependence of $\ln K_{ch}$ and reciprocal ε is linear:

$$\ln K_{ch} = a_{00} + a_{01} / \varepsilon \quad [9.122]$$

Coefficients of the equation and some thermodynamic properties of exchange interaction processes are presented in Table 9.13.

The degree of heteromolecular association of esters in reaction [9.121] is higher than for acids when $K_{dim}$ values are equal. Thus the change of equilibrium constant according to [9.121] depends on electrostatic component. $K_{ch}$ for these reactions also depends on exponentially reciprocal ε.

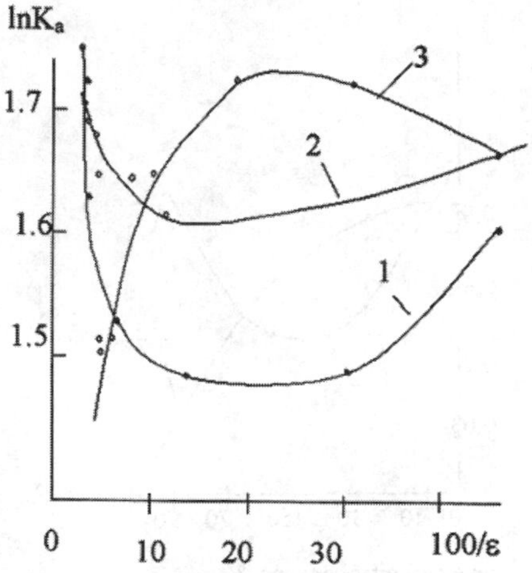


Figure 9.19. Dependence of equilibrium constants for triphenylchloromethane methanolysis on 1/ε of mixed solvent at 298.15K: 1 - hexane-nitrobenzene; 2 - toluene-nitromethane; 3 - toluene-acetonitrile.

The dependence of equilibrium constant for reaction [9.116] on ε is similar in a concentration range from 0.5 to 1.0 molar fraction of aldehyde, if the degrees of homomolecular association for aldehydes and acetals are smaller than for alcohol and water.

$K_{ch}$ increase with increasing ε of media is a general property of all processes of exchange interaction presented in the Table 9.13. Dependence of $\ln K_{ch}$ - 1/ε for process [9.111] is not linear in all mixed solvents (Figure 9.19).

Extreme $K_{ch}$ dependence on 1/ε is explained by chemical effect of solvation (it is explained in detail elsewhere[60]). Correlations with different empirical parameters of individual solvent parameters are held true for this process. The best of them is correlation (r=0.977) with parameter $E_T$.

**Table 9.13. Characteristics of some processes of exchange interaction**

| Reaction | Coefficients of equation [9.122] | | Free energy components, kJ/mol | | | |
|---|---|---|---|---|---|---|
| | $a_{00}$ | $-a_{01}$ | $-\sigma^{cov}$ | $\sigma^{el}_{\varepsilon=1}$ | σG at max. ε | σG at min. ε |
| [9.111] | 2.40 | 97.6 | 5.9 | 241 | 5.2 | 19.0 |
| [9.115], a, (F) | 1.72 | 19.6 | 4.3 | 48.5 | 3.1 | 6.3 |
| [9.115], b, (Cl) | 2.55 | 17.1 | 6.3 | 42.3 | 5.1 | 7.0 |
| [9.116], a | 1.24 | 47 | 3.1 | 116.4 | 7.5 | 9.9 |
| [9.116], b | 2.47 | 16.7 | 6.1 | 41.3 | 3.4 | 4.1 |

The influence of molecular state of participants of exchange interaction on the process equilibrium is reflected in $K_{ch}$ dependence (processes [9.113]) on the mixed solvent composition (Figure 9.20). The change of $K_{dim}$ for acetic acid is less pronounced than for monochloroacetic acid, when $CCl_4$ is replaced by chlorobenzene. Thus $K_{ch}$ should increase for reaction [9.113a] and it should decrease for reaction [9.113b], with ε increasing in accordance to equation [9.121].

Correlation of $K_{dim}$ for acetic and monochloroacetic acid is obtained when chlorobenzene is replaced by nitrobenzene. This leads to decrease of $K_{dim}$ when ε increases in the case of reaction [9.113a], and the opposite is the case for reaction [9.113b]. That is why a maximum is observed in the first process (see Figure 9.20 on the left hand side), and a minimum is observed for the second process.

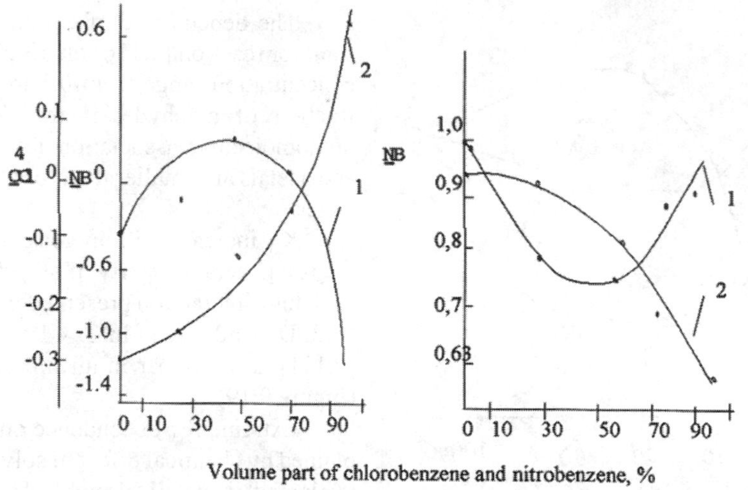


Figure 9.20. Dependence of $K_{ch}$ on the mixed solvent composition for processes [9.113a] and [9.113b] at 298.15K: 1-tetrachloromethane-nitrobenzene; 2- tetrachloromethane-chlorobenzene.

At equilibrium [9.114] (with three associated participants), association degree of associated participants decreases with media polarity increasing. It leads to $K_{ch}$ increasing and to $K_{add}$ decreasing in the process of heteromolecular association of acid and alcohol.

Thus, $K_{ch}$ depends to a low extent on the composition of studied mixed solvents. Vacuum component of free energy of processes of exchange interaction (Table 9.13) is low; $\sigma G^{el}$ values are low, too. This indicates, that $\varepsilon$ change does not lead to the change of the reaction output. Solvent replacement leads to essential change of reaction output in process [9.111], characterized by the high value of $\sigma^{el}$ (see Table 9.13), from 6% in pyridine to 60% in DMSO (in 0.1M solutions). $K_{ch}$ increase in pyridine solution to its value in DMSO requires cooling to -140°C.

## 9.4.7 MIXED SOLVENT EFFECT ON PROCESSES OF COMPLEX FORMATION

Non-aqueous solvent effect on equilibrium and thermodynamic of complexation processes are summarized in monographs.[67-69]

The solvent effect on complex formation processes has two aspects. The first is a change of coordination sphere composition, changing participation of solvent as a chemical regent:

$$[ML_m]^{p+} \xleftrightarrow{+A} [ML_{m-1}A]^{p+} + L \longleftrightarrow \cdots \longleftrightarrow [MA_n]^{p+} + mL \qquad [9.123]$$

where:

| | |
|---|---|
| $L_m$ | a ligand in initial complex |
| A, B | components of binary solvent |

If both components of the mixed solvent are solvate-active, equilibrium [9.123] may be presented as:

$$[ML_m]^{p+} \xleftrightarrow{+A+B} [ML_{m-2}AB]^{p+} + 2L \leftrightarrow \cdots \leftrightarrow [MA_{n-q}B_q]^{p+} + mL \quad [9.123a]$$

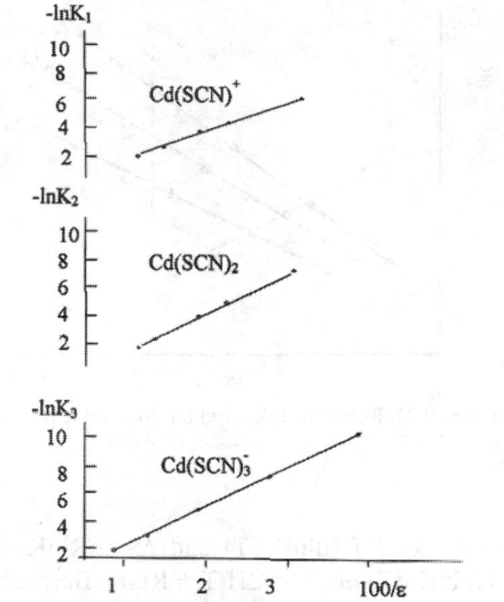


Figure 9.21. Dependence of stability constants for cadmium thiocyanate complexes on 1/ε in mixed solvent water-methanol at 298.15K.

Coordination solvents A and B taking part in the processes are the same ligands as L.

The second aspect of the solvent effect is related to the thermodynamic properties of the complex, such as formation constant. Discussing the problem of solvent effect on complex stability, authors often neglect the change of a complex nature, due to the change in solvent as follows from equations [9.123] and [9.123a]. These processes may also be regarded as resolvation processes.

When the components of mixed solvate are solvate-inert in relation to the complex, equation [9.53a] takes the following form:

$$\ln K_k = a_{00} + a_{01} / \varepsilon \qquad [9.124]$$

Coefficient $a_{00}$ has a physical meaning of logarithm of formation constant at $\varepsilon \to \infty$ related to the standard ionic strength I = 0. In Figure 9.21, this dependence is presented.

**Table 9.14. Complexation process characteristics for system $Cd^{2+}$-thiourea in mixed solvent water-methanol at 298.15K**

| Constant | Coefficient of equation | | Correlation coefficient r | Component σG, kJ/mol | | | |
|---|---|---|---|---|---|---|---|
| | $a_{00}$ | $a_{01}$ | | $\sigma G^{cov}$ | $\sigma^{el}_{\varepsilon=1}$ | $-\sigma G^{cov}$ in $H_2O$ | $-\sigma G^{cov}$ in MeOH |
| $K_2$ | -0.42 | 367.7 | 0.996 | -1.0 | 910 | 11.6 | 27.9 |
| $K_3$ | 1.07 | 561.9 | 0.987 | 2.6 | 1390 | 17.7 | 42.7 |
| $K_4$ | 7.79 | 631.3 | 0.991 | 19.3 | 1563 | 19.9 | 47.9 |
| $K_5$ | 2.72 | 837.0 | 0.988 | 6.7 | 2072 | 26.4 | 63.6 |

The data for cadmium thiourea complexes in water-methanol mixed solvent[70] are presented in Table 9.14. Dependence of stability constant on the solvent composition is very complex in the mixed solvents formed by two solvate-active components as defined by the relative activity, L, of A and B. Dependencies of stability constant for some complexes in water-B solution (where B is methyl acetate, methanol, etc.[69]) are presented in Figures 9.21 and 9.22 for complex $NiEn^{2+}$ (En–ethylenediamine) in water-non-aqueous solution (the constants are presented relative to stability constant in water).

## 9.5 THE MIXED SOLVENT EFFECT ON THE CHEMICAL EQUILIBRIUM THERMODYNAMICS

In most cases, thermodynamic characteristics of equilibrium processes are determined by temperature dependence of equilibrium constant K = f(T), according to the classical equa-

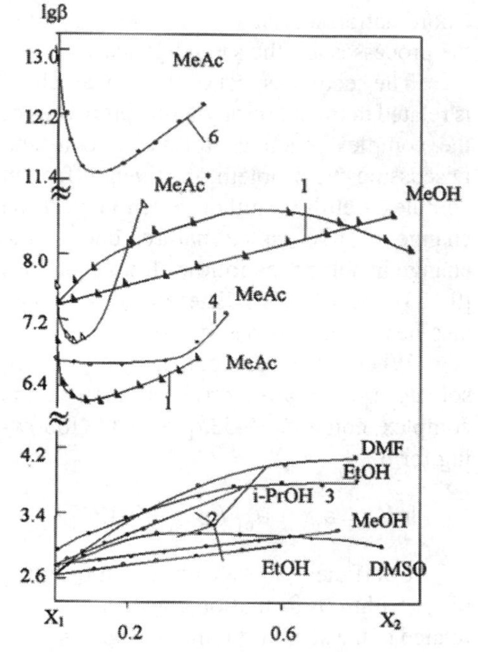


Figure 9.22. Stability constant of amine complexes of some $Ni(NH_3)^+$ in water-organic solvents: 1-$NiEn^{2+}$; 2-[N; $NH_3]^{2+}$; 3-$[CdCH_3NH_2]^{2+}$; 4-$[AgCH_3NH_2]^{2+}$; 5-$[AgEn_2]^{2+}$; 6-$[NiEn_2]^{2+}$.

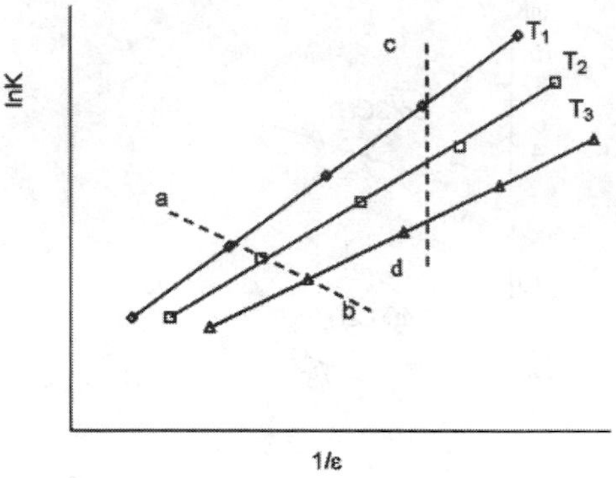


Figure 9.23. Isotherms lnK - f(ε) for different temperatures ($T_1<T_2<T_3$).

tion: $\Delta H = RT^2(d\ln K/dT)$ and $\Delta S = R\ln K + T(d\ln K/dT)$ or $\Delta S = \Delta H/T + R\ln K$. Because equilibrium constant is a function of both temperature and permittivity [9.52a], i.e.,

$$K = f(T, \varepsilon) \quad [9.125]$$

A change of constant during temperature change is a consequence of enthalpy of process (not equal to zero) and temperature dependence of permittivity. Thus, the change of equilibrium constant with temperature changing depends on both self-chemical equilibrium characteristics (here ΔH) and solvent characteristics (namely, solvent permittivity influence on equilibrium constant). This is pertinent from Figure 9.23, which shows dependence of equilibrium constant logarithm on reciprocal permittivity (isotherms). Equilibrium constants, obtained experimentally (dotted line a - b) correspond to various values of permittivity.

The thermodynamic characteristics of a chemical process, calculated from equilibrium constant polytherms, are integral characteristics ($\Delta H_i$ and $\Delta S_i$), i.e., they consist of terms related to process itself ($\Delta H_T$ and $\Delta S_T$) and terms depended on permittivity change with temperature changing ($\Delta H_\varepsilon$ and $\Delta S_\varepsilon$):

$$\Delta H_i = \Delta H_T + \Delta H_\varepsilon \quad [9.126a]$$

$$\Delta S_i = \Delta S_T + \Delta S_\varepsilon \quad [9.126b]$$

The thermodynamic description of a process in a solvent has to lead to defined values of $\Delta H_T$ and $\Delta S_T$, which are called van't Hoff or original parameters.[7] It follows from Figure 9.23 that one has to differentiate the equilibrium constant of polytherms corresponding to isodielectric values of the solvent on temperature (e.g., to equilibrium constant values of c-d dotted line).

Division of equation [9.126] into the constituents must be done with full assurance to maintain influence of different solvents. For this reason, it is necessary to have a few isotherms of equilibrium constant dependencies on permittivity (K=f(ε)T). We have developed an equation of a process depicted by scheme [9.45] after approximating each isotherm as lnK vs. 1/ε function and by following approximation of these equations for ε-1, ε-2, ε-3...ε-j conditions. We then can calculate the integral value of process entropy by differentiating relationship [9.55] versus T, e.g., ΔG= -RTlnK:

$$\Delta S_i = R\left[\sum_{\substack{i=0\\j=0}}^{\substack{i=m\\j=n}}(1-i)a_{ji}\left(T^i\varepsilon^i\right) - j(d\ln\varepsilon/dT)/\varepsilon^{j-1}\sum_{\substack{i=0\\j=0}}^{\substack{i=m\\j=n}}a_{ji}/T^{i-1}\right]/\varepsilon^i \quad [9.127]$$

Here, the first term of the sum in brackets corresponds to $\Delta S_T/R$ value and the second term to ΔS/R. We can determine "van't Hoff's" (original) constituents of the process entropy if all the terms containing dlnε/dT are equal to zero:

$$\Delta S_T = R\sum_{\substack{i=0\\j=0}}^{\substack{i=m\\j=n}}(1-i)a_{ji}/\left(T^i\varepsilon^j\right) \quad [9.128]$$

Van't Hoff's constituent of enthalpy is determined in analogous manner as ΔH=ΔG+TΔS ($\Delta H_T$):

$$\Delta H_T = -R\sum_{\substack{i=0\\j=0}}^{\substack{i=m\\j=n}}a_{ji}/\left(T^{i-1}\varepsilon^j\right) \quad [9.129]$$

Such approach may by illustrated by dependence of formic acid association constant on temperature and permittivity in mixed solvents: water-ethylene glycol,[72] which is approximated from equation

$$\ln K = a_{00} + a_{01}/T + a_{02}/T^2 + a_{10}/\varepsilon + a_{11}/(\varepsilon T)$$

where:

| | |
|---|---|
| $a_{00}$ | = 40.90 |
| $a_{01}$ | = $-2.17\times10^4$ |
| $a_{02}$ | = $3.11\times10^6$ |
| $a_{10}$ | = -425.4 |
| $a_{11}$ | = $2.52\times10^5$ |

Hence

$$\Delta H_i = -R\left[a_{01} + 2a_{02}/T + a_{11}/\varepsilon + T(d\ln\varepsilon/dT)(a_{10}T + a_{11})/\varepsilon\right]$$

$$\Delta S_i = R\left[a_{00} - a_{02}/T + a_{10}/\varepsilon - (d\ln\varepsilon/dT)(a_{10}T + a_{11})/\varepsilon\right]$$

and consequently,

$$\Delta H_T = -R(a_{01} + 2a_{02} / T + a_{11} / \varepsilon) \quad \Delta S_T = R(a_{00} - a_{02} / T + a_{10} / \varepsilon)$$

Comparison of $\Delta H_i$ with $\Delta H_T$ and $\Delta S_i$ with $\Delta S_T$ shows that integral values of thermodynamic characteristics of the ionic association process for HCOOH are not only less informative than van't Hoff's characteristics, but they contradict the physical model of the process in this case. Indeed, corresponding dependencies of compensative effect are described by equations:

$$\Delta H_i = -0.704\Delta S_i + 22.87\ kJ/mol$$

$$\Delta H_T = 0.592\Delta S_T - 22.22\ kJ/mol$$

Thus, if the compensative effect for integral thermodynamic functions is interpreted, one should conclude that characteristic temperature (this is $\tan\alpha = \Delta H/\Delta S$) is a negative value, but that is, of course, devoid of physical sense.

In common cases, when the dependence [9.125] is approximated using equation [9.56], "van't Hoff's" (original) thermodynamic constituents of equilibrium process are equal to:

$$\Delta H_T = -R(a_{01} + a_{11} / \varepsilon) \quad [9.130]$$

$$\Delta S_T = R(a_{00} + a_{10} / \varepsilon) \quad [9.131]$$

As it is evident from [9.130] and [9.131], the original thermodynamic characteristics of the process are summarized by the terms where the first of them does not depend, and the second one depends, on permittivity. These terms accordingly are called covalent and electrostatic constituents of the process in solution, i.e.,

$$\Delta H^{cov} = -Ra_{01} \quad and \quad \Delta H^{el} = -Ra_{11} / \varepsilon \quad [9.132]$$

$$\Delta S^{cov} = Ra_{00} \quad and \quad \Delta S^{el} = Ra_{10} / \varepsilon \quad [9.133]$$

It should be emphasized that $\Delta G_T$ values are substantially more informative than integral thermodynamic values for thermodynamic analysis of chemical equilibrium in solutions of mixed solvents. Indeed, in frequent cases of equilibrium constant polytherm characterized by an extreme (i.e., sign of $\Delta H_i$ changes), one may arrive at the wrong conclusion that the process nature changes greatly at temperature of extreme value. As has been shown,[71] appearance of extreme value of equilibrium constant polytherms is caused in the majority of cases by temperature change of solvent permittivity. The condition of extreme appearance for the process described by equation [9.55] is:

$$a_{01} \exp(\alpha T_{extr} + \gamma) + \alpha a_{11} T_{extr} (a_{10} T_{extr} + a_{11}) = 0 \quad [9.134]$$

where:

| | |
|---|---|
| $\alpha, \gamma$ | the coefficients of $\ln\varepsilon = \alpha T + \gamma$ equation |
| $\gamma$ | activity coefficient |
| $T_{extr}$ | temperature corresponding to extremum of isotherm of equilibrium constant |

Let us illustrate the principle of integral thermodynamic characteristics of chemical equilibrium in mixed solvent based on the example of ionic association process of methylsulfuric acid $HSO_3CH_3$ and its tetraalkylammonium salt $CH_3(C_8H_{17})_3N + OSO_3CH_3$ in mixed solvent - methanol - n-butanol.[73] The dependencies of ionic association constants for these electrolytes on temperature and permittivity are described by equations based [9.56] the relationship:

$$K_{a,acid} = 14.53 - 3436/T - 130.28/\varepsilon + 5.84\times10/\varepsilon T$$

$$K_{a,salt} = 11.8 - 2514/T - 198.70/\varepsilon + 7.79\times10/\varepsilon T$$

Thermodynamic characteristics for these systems, calculated based on [9.130]-[9.133] equations, are given in Tables 9.15 and 9.16.

**Table 9.15. Thermodynamic characteristics of ionic association process of $HSO_3CH_3$ in mixed solvent: methanol - n-butanol (298.15 K) ($\Delta H^{cov}$ = 28.56 kJ/mol, $\Delta S^{cov}$ = 120.8 J/(mol*K))**

| BuOH mol% | ε | ΔH, kJ/mol | | | ΔS, J/(mol* K) | | |
|---|---|---|---|---|---|---|---|
| | | $\Delta H_i$ | $\Delta H_T$ | $\Delta H_{el}$ | $\Delta S_i$ | $\Delta S_T$ | $\Delta S_{el}$ |
| 60.39 | 20 | 20.62 | 4.29 | -24.28 | 121.4 | 66.6 | -54.1 |
| 51.60 | 22 | 20.90 | 6.50 | -22.07 | 119.9 | 71.6 | -49.2 |
| 37.24 | 24 | 21.19 | 9.89 | -20.23 | 118.8 | 75.7 | -45.1 |
| 25.32 | 26 | 21.49 | 8.34 | -18.67 | 118.0 | 79.1 | -41.7 |
| 15.82 | 28 | 21.79 | 11.22 | -17.34 | 117.6 | 82.1 | -38.7 |
| 8.76 | 30 | 22.09 | 12.38 | -16.18 | 117.3 | 84.7 | -36.1 |

**Table 9.16. Thermodynamic characteristics of ionic association process of $CH_3(C_8H_{17})_3N + OSO_3CH_3$ in mixed solvent: methanol - n-butanol (298.15 K) ($\Delta H^{cov}$=20.90 kJ/mol, $\Delta S^{cov}$= 98.1 J/(mol K))**

| BuOH mol% | ε | ΔH, kJ/mol | | | ΔS, J/(mol K) | | |
|---|---|---|---|---|---|---|---|
| | | $\Delta H_i$ | $\Delta H_T$ | $\Delta H_{el}$ | $\Delta S_i$ | $\Delta S_T$ | $\Delta S_{el}$ |
| 60.39 | 20 | 3.85 | -11.23 | -32.13 | 66.1 | 15.5 | -82.6 |
| 51.60 | 22 | 5.00 | - 8.31 | -29.21 | 67.7 | 23.0 | -75.1 |
| 37.24 | 24 | 6.00 | - 5.88 | -26.78 | 69.1 | 29.3 | -68.8 |
| 25.32 | 26 | 6.90 | - 3.82 | -24.72 | 70.5 | 34.6 | -63.5 |
| 15.82 | 28 | 7.70 | -2.05 | -22.95 | 71.8 | 39.1 | -59.0 |
| 8.76 | 30 | 8.45 | - 0.52 | -21.42 | 73.1 | 43.0 | -55.1 |

The analysis of data leads to conclusion that $\Delta H_i$ depends weakly on permittivity for both acids and salts. This contradicts the nature of the process. In contrast, the rate of $\Delta H_T$

change is large enough. The analogous observations can be made regarding the character of $\Delta S_i$ and $\Delta S_T$ values changes along with permittivity change.

If considerable endothermic effect of the ionic association for acid can be explained by desolvation contribution of ionic pair formation (i.e., solvated molecule of acid), the endothermic effect of ionic pair formation by such voluminous cations contradicts the physical model of the process. Estimation of desolvation energy of ionic pair formation of salt ions according to the equation[74] shows that this energy is two orders of magnitude lower than the energy of heat movement of solvent molecules. For this reason, the process of ionic association of salt ions is exothermal, as seen from $\Delta H_T$ values. The exothermal character increases with permittivity decreasing, i.e., with increase of ion-ion interaction energy.

Dependencies $\Delta H_i$ on $\Delta S_i$ for salt and acid are not linear, i.e., the compensative effect is not fulfilled. Consequently, the substantial change of the ionic association process is due to change of solvent composition. Considering the same chemical characteristics of mixed solvent components, this conclusion is wrong because the compensative effect if fulfilled for van't Hoff's components is:

$$\Delta H_T = 0.448\ \Delta S_T - 28.5\ kJ/mol$$

$$\Delta H_T = 0.389\ \Delta S_T - 17.3\ kJ/mol$$

The smaller the magnitude of the characteristic temperature for salt (389K), in comparison to acid, the smaller the barrier of the ionic pairs formation process. That is true because the association process in the case of acid is accompanied by energy consumption for desolvation.

## REFERENCES


1 Timmermans J., **The Physico-Chemical Constants of Binary Systems in Concentrated Solutions** V1-4, N-Y-London, *Intern. Publ.* ING,1959.
2 Krestov G., Afanas'ev V., Efimova L., **Physico-chemical Properties of Binary Solvents**, Leningrad, *Khimiya*,1988.
3 Fialkov Yu., **Physical-Chemical Analysis of Liquid Systems and Solutions**, Kiev, *Naukova dumka*,1992.
4 Kabachnik M., *Doklady Akademii Nauk USSR*, **83**, 6, 1952 859-862
5 Kabachnik M., *Izvestiya Akademii Nauk USSR*, **1**, 1955, 98-103
6 Fialkov Yu., Diner L., *J. Gener. Chem.* (russ.), **48**, 2, 1978, 253-256.
7 Fialkov Yu., **Solvent as an Agent of Chemical Process Control**, Leningrad, *Khimiya*, 1990.
8 Fialkov Yu., Ligus V., *Doklady Akademii Nauk USSR*, **197**, 6,1971, 1353-1354.
9 Akhadov Ya., **Dielectric Properties of Binary Solutions,** Moskow, *Nauka*, 1977.
10 Fialkov Yu., Grishenko V., **Electrodeposition of Metals from Non-Aqueous Solutions**, Kiev, *Naukova dumka,* 1985.
11 Fialkov Yu., Chumak V., Kulinich N., *Electrochem.*(russ.), **18**, 8, 1024-1027.
12 Anosov V., Ozerova M., Fialkov Yu., **Bases of Physical-Chemical Analysis**, Moskow, *Nauka*, 1976, 1978.
13 Rudnickaya A., Maiorov V., Librovich N., Fialkov Yu., *Izvestiya Akademii Nauk USSR,* **5**, 1981, 960-966.
14 Rudnickaya A., Maiorov V., Librovich N., Fialkov Yu., *Izvestiya Akademii Nauk USSR*, **11**, 1981, 2478-2484.
15 Alexsandrov V., **Acidity of Non-Aqueous Solutions**, Kharkov, *Visha skola*,1981.
15a Prabir K.Guna, Kiron K.Kundu, *Can. J. Chem.*,1985, **63**, 804-808.
16 Palm V., **The Basis of Quantitative Theory of Organic Reactions**, Leningrad, *Khimiya*,1977
17 Balakrishnan S., Eastael A., *Austr. J. Chem.*, 1981, **34**, 5, 933-941.
18 Langhals M., *Chem. Ber.*, 1981, **114**, 8, 2907-2913.
19 Fialkov Yu., Barbash V., Bondarenko E., *Sovyet Progress in Chemistry*, **53**, 5, 490-491.
20 Abrachem R. J., Bretchnaider E., **Internal Rotation of Molecules**, 1977, 405-409.
21 Simoshin V., Zefirov N., *J. Mendeleev-Soc.*, 1984, **29**, 5, 521-530.

22 Minkin V., Osipov O., Zhdanov Yu., **Dipole Moments in Organic Chemistry**, Leningrad, *Khimiya,*1968.
23 Christian S., Lane E., in **Solutions and Solubilities**, V1, 1975, *John Wiley & Sons*, N-Y, 327-377.
24 Komarov E., Shumkov V., *Chemistry Chem. Thermodyn. Solution*, Leningrad, Univers,1973, **3**, 151-186.
25 Bishop R., Sutton L., *J. Chem. Soc.*, 1964, 6100-6105.
26 Bhownik B., Srimani P., *Spectrochim. acta,* 1973, **A29**, 6, 935—942.
27 Ewall R., Sonnesa A., *J. Amer. Chem. Soc.*, 1970, **92**, 9, 3845-3848.
28 Drago R., Bolles T., Niedzielski R., *J. Amer. Chem. Soc.*, 1966, **88**, 2717-2721.
29 Fialkov Yu., Barbash V., *Theoret. Exper. Chem.*(russ.), 1986, **22**, 248-252.
30 Barbash V., Fialkov Yu., Shepet`ko N., *J. Gener. Chem.*(russ.), 1983, **53**, 10, 2178-2180.
31 Fialkov Yu., Barbash V., *Sovyet Progress in Chemistry*, 1985, **51**, 4, 354-367.
32 Barbash V., Fialkov Yu., *Sovyet Progress in Chemistry*, 1983, **49**, 11, 1157-1160.
33 Barbash V., Golubev N., Fialkov Yu., *Doklady Akademii Nauk USSR*, **278**, 2, 1984, 390-391.
34 Borovikov A., Fialkov Yu., *J. Gener. Chem.*(russ.), 1978, **48**, 2, 250-253.
35 Pavelka Z., Sobczyk L., *Roczn. chem.*, 1975, **49**, 7/8, 1383-1394.
36 Padova J., *J. Phys. Chem.,* 1968, **72**, 796-801.
37 Bachshiev, **Spectroscopy of Intermolecular Interactions**, Leningrad, *Nauka*, 1972.
38 Mishustin A., *J. Phys. Chem.*(russ.), 1987, **61**, 2, 404-408.
39 Mishustin A., Kruglyak A., *J. Gener. Chem.*(russ.), 1987, **57**, 8, 1806-1810.
40 Reynolds W., *Inorg.Chem.*, 1985, **24**, 25, 4273-4278; *J. Chem. Soc. Comm.*, 1985, **8**, 526-529.
41 Busse J., *Ber. Buns. phys. Chem.*, 1985, **89**, 9, 982-985.
42 Fuldner H., *Ber. Buns. phys. Chem.*, 1982, **86**, 1, 68-71.
43 Kobayashi K., *Bull. Chem. Soc. Japan,* 1987, **60**, 467-470.
44 Marcus Y., **Ion Solvation**, N-Y, *Wiley*, Chichester,1985.
45 Gordon J., **The Organic Chemistry of Electrolyte Solutions**, 1977, N-Y-London-Sydney-Toronto, *John Wiley&Sons*.
46 Buleishvili M., Fialkov Yu., Chumak V., *Sovyet Progress in Chemistry*, 1983, **49**, 6, 599-602.
47 Fialkov Yu., Chumak V., Rudneva S., *Doklady Akademii Nauk Ukraine,*"B", 1984, **4**, 49-53.
48 Dorofeeva N., Kovalskaya V., Fialkov Yu., Shlyachanova V., *Sovyet Progress in Chemistry*, 1984, **50**, 5, 476-479.
49 Lukyunova S., Fialkov Yu., Chumak V., *Sovyet Progress in Chemistry*, 1987, **53**, 3, 252-255.
50 Levinskas A., Shmauskaite S., Pavidite Z., *Electrochemistry*(russ.), 1978, **14**, 12, 1885-1889.
51 Fialkov Yu., Pinchuk V., Rudneva S., Chumak V., *Doklady Akademii Nauk USSR*, 1984, **277**, 1, 136-138.
52 Fialkov Yu., Dorofeeva N., Chumak V., **Thermodynamic Properties of Solutions**, Ivanovo, 1984, 95-100.
53 Izmailov N., **Electrochemistry of Solutions**, *Khimiya*, Moskow, 1966.
54 Busse J., *Ber. Buns. phys. Chem.*, 1985, **89**, 9, 982-985.
55 Fialkov Yu., Chumak V., Borchashvili A., *Sovyet Progress in Chemistry*, 1984, **50**, 7, 10, 705-708, 1023-1027.
56 Fialkov Yu., Tarasenko Yu., Kryukov B., *Doklady Akademii Nauk Ukraine*,"B", 1985, **2**, 59-61.
57 Kryukov B., Tarasenko Yu., Fialkov Yu., Chumak V., *Sovyet Progress in Chemistry*, 1985, **51**, 1, 34-38.
58 Fialkov Yu., Chumak V., *Chem. Chem. Technol.(russ.)*, 1976, **19**, 2, 265-268.
59 Buleishvili M., Fialkov Yu., Chumak V., *Sovyet Progress in Chemistry*, 1983, **50**, 2, 214-216.
60 Fialkov Yu., Zhadaev B., **Thermodynamic Properties of Solutions** , Ivanovo, 1979, pp.90-104.
61 Zhadaev B., Fialkov Yu., *Doklady Akademii Nauk USSR*,1978, **242**, 6,1350-1351.
62 Fialkov Yu., Fenerli G., *J. Gener. Chem.(russ.)*, 1964, **34**, 10, 3146-3153.
63 Fialkov Yu., Fenerli G., *J. Gener. Chem.(russ.)*, 1966, **36**, 6, 967-973.
64 Fialkov Yu., Fenerli G., *J. Gener. Chem.(russ.)*, 1966, **36**, 6, 973-981.
65 Reichardt Ch., **Solvents and Solvent Effects in Organic Chemistry**, *VCH*, Weinheim, 1988.
66 Krigovski T., Wrona P., Zielikowska U., Reichardt Ch., *Tetrahedron*, 1985, **41**, 4519-4527.
67 Gutmann V., **Coordination Chemistry in Non-Aqueous Solvents**, 1968, *Springer-Verlag*,Wien-N-Y.
68 Burger K. **Solvatation, Ionic and Complex Formation Reactions in Non-Aqueous Solvents**, 1983, *Akademia Kiado,* Budapest.
69 **Complexing in Non-Aqueous Solution**, Ed. G. Krestov, Moscow, *Nauka*, 1989.
70 Migal P., Ciplakova V., *J. Neorg. Chem.*(russ.), 1964, **9**, 3, 601-605.
71 Fialkov Yu., *J. Chem. Thermodynamic Thermochemistry (russ.),* 1993, **2**, 2, 113-119.

72 Alexsandrov V., Mamina E., Bondarev, *Problems of Solvation and Complexing in Solutions*, Ivanovo, 1984, 1, 174-176.
73 Fialkov Yu., Koval`skaya V., *Chem. Phys. Reports*, 1996, **15**, 11, 1685-1692.
74 Fialkov Yu., Gorbachev V., *J. Neorg. Chem.(russ.)*, 1977, **42**, 8, 1401-1404.

# 10

# Acid-base Interactions

## 10.1 GENERAL CONCEPT OF ACID-BASE INTERACTIONS

George Wypych
ChemTec Laboratories, Inc., Toronto, Canada

Acid-base interactions have found numerous applications in research dealing with adsorption of molecules of liquids on the surfaces of solids. The main focus of this research is to estimate the thermodynamic work of adhesion, determine mechanism of interactions, analyze the morphology of interfaces and various surface coatings, develop surface modifiers, study the aggregation of macromolecular materials, explain the kinetics of swelling and drying, understand the absorption of low molecular weight compounds in polymeric matrices, and determine the properties of solid surfaces. In addition to these, there are many other applications.

Several techniques are used to determine and interpret acid-base interactions. These include: contact angle, inverse gas chromatography, IGC, Fourier transform infrared, FTIR, and X-ray photoelectron spectroscopy, XPS. These methods, as they are applied to solvents are discussed below.

Contact angle measurements have long been used because of common availability of instruments. In recent years, they have been developed from simple optical devices to the present day precise, sophisticated, computer-controlled instruments with sufficient precision. Van Oss and Good[1,2] developed the basic theory for this method. Their expression for surface free energy is used in the following form:

$$\gamma = \gamma^{LW} + \gamma^{AB} = \gamma^{LW} + 2\sqrt{\gamma^+\gamma^-} \qquad [10.1.1]$$

where:

| | |
|---|---|
| $\gamma^{LW}$ | Lifshitz-van der Waals interaction |
| $\gamma^{AB}$ | acid-base interaction |
| $\gamma^+$ | Lewis acid parameter of surface free energy |
| $\gamma^-$ | Lewis base parameter of surface free energy. |

The following relationship is pertinent from the equation [10.1.1]:

$$\gamma^{AB} = 2\sqrt{\gamma^+\gamma^-} \qquad [10.1.2]$$

A three-liquid procedure was developed[1-3] which permits the determination of the acid-base interaction from three measurements of contact angle:

$$\gamma_{L1}(1+\cos\theta_1)=2\left(\sqrt{\gamma_S^{LW}\gamma_{L1}^{LW}}+\sqrt{\gamma_S^{+}\gamma_{L1}^{-}}+\sqrt{\gamma_S^{-}\gamma_{L1}^{+}}\right)$$

$$\gamma_{L2}(1+\cos\theta_2)=2\left(\sqrt{\gamma_S^{LW}\gamma_{L2}^{LW}}+\sqrt{\gamma_S^{+}\gamma_{L2}^{-}}+\sqrt{\gamma_S^{-}\gamma_{L2}^{+}}\right) \qquad [10.1.3]$$

$$\gamma_{L3}(1+\cos\theta_3)=2\left(\sqrt{\gamma_S^{LW}\gamma_{L3}^{LW}}+\sqrt{\gamma_S^{+}\gamma_{L3}^{-}}+\sqrt{\gamma_S^{-}\gamma_{L3}^{+}}\right)$$

Solving this set of equations permits the calculation of solid parameters from equation [10.1.1]. The work of adhesion, Wa, between a solid and a liquid can be calculated using the Helmholtz free energy change per unit area:

$$-\Delta G_{SL}=Wa_{SL}=\gamma_S+\gamma_L-\gamma_{SL}=2\left(\sqrt{\gamma_S^{LW}\gamma_L^{LW}}+\sqrt{\gamma_S^{+}\gamma_L^{-}}+\sqrt{\gamma_S^{-}\gamma_L^{+}}\right) \qquad [10.1.4]$$

Determination is simple. However, several measurements (usually 10) should be taken to obtain a reliable averages (error is due mostly to surface inhomogeneity). There is a choice between measuring the advancing or the receding contact angle. Advancing contact angles are more representative of equilibrium contact angles.[3]

Inverse gas chromatography data are interpreted based on Papirer's equation:[4,5]

$$RT\ln V_N=2N(\gamma_S^D)^{1/2}a(\gamma_L^D)^{1/2}+c \qquad [10.1.5]$$

where:

| | |
|---|---|
| R | gas constant |
| T | temperature |
| $V_N$ | net retention volume |
| N | Avogadro's number |
| $\gamma^D$ | dispersion component of surface energy |
| S, L | indices for solid and liquid, respectively |
| a | molecular area of adsorbed molecule |
| c | integration constant relative to a given column. |

Several probes are used to obtain the relationship between $RT\ln V_N$ and $a\gamma_L^D$. From this relationship the reference retention volume, $V_N^{REF}$, is calculated and used to calculate the acid-base interaction's contribution to the free energy of desorption:

$$\Delta G_{AB}=RT\ln\frac{V_N}{V_N^{REF}} \qquad [10.1.6]$$

If data from a suitable temperature range for $\Delta G_{AB}$ can be obtained, the acid-base enthalpy, $\Delta H_{AB}$ can be calculated using the following equation:

$$\Delta H_{AB}=K_aDN+K_bAN \qquad [10.1.7]$$

where:

| | |
|---|---|
| $K_a$ | acid interaction constant |
| $K_b$ | base interaction constant |
| DN | donor number |
| AN | acceptor number |

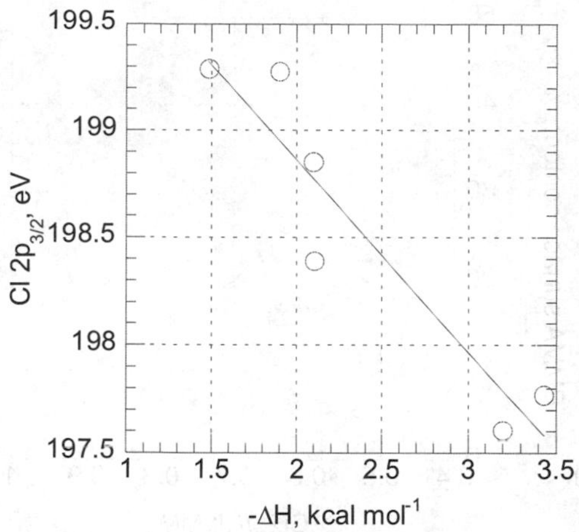


Figure 10.1.1. Correlation of binding energy and $\Delta H_{AB}$ for several polymers. [Adapted, by permission, from J F Watts, M M Chehimi, *International J. Adhesion Adhesives*, **15**, No.2, 91-4 (1995).]

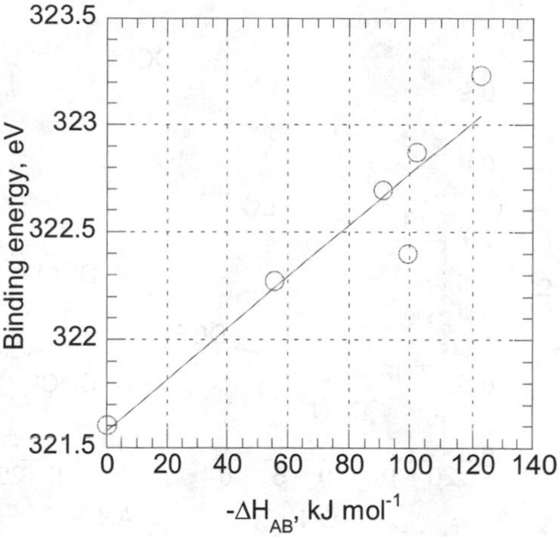


Figure 10.1.2. Correlation of binding energy and $\Delta H_{AB}$ for several solvents. [Adapted, by permission, from J F Watts, M M Chehimi, *International J. Adhesion Adhesives*, **15**, No.2, 91-4 (1995).]

Finally the plot of $\Delta H_{AB}/AN$ vs. DN/AN gives $K_a$ and $K_b$. Further details of this method are described elsewhere.[4-9] It can be seen that the procedure is complicated. The various conditions of experiments conducted in various laboratories were sufficiently different to prevent correlation of data between laboratories.[10] To rectify this situation, a large body of data was obtained for 45 solvents and 19 polymers tested under uniform conditions.[10]

Fowkes[11] showed that the carbonyl stretching frequency shifts to lower values as the dispersion component of surface tension increases. The following empirical relationship was proposed:

$$\Delta H_{AB} = 0.236 \Delta v^{AB} \qquad [10.1.8]$$

where:

| | |
|---|---|
| $\Delta H_{AB}$ | the enthalpy change on acid-base adduct formation |
| $\Delta v^{AB}$ | projection of carbonyl stretching frequency on dispersive line. |

It should be noted that, as Figure 7.1.16 shows, the change in frequency of carbonyl stretching mode is related to the process of crystallization.[12]

XPS is emerging as very precise method for evaluating acid-base interactions based on the works by Chehimi *et al.*[6-9] There is very good correlation between XPS chemical shift and the change in exothermic enthalpy of acid-base interaction. Drago's equation is used for data interpretation:[13]

$$-\Delta H_{AB} = E_A E_B + C_A C_B \qquad [10.1.9]$$

where:

| | |
|---|---|
| A, B | subscripts for acid and base, respectively |
| $E_A$, $E_B$ | susceptibility of acid or base species to undergo an electrostatic interaction |
| $C_A$, $C_B$ | susceptibility of acid or base species to undergo a covalent interaction. |

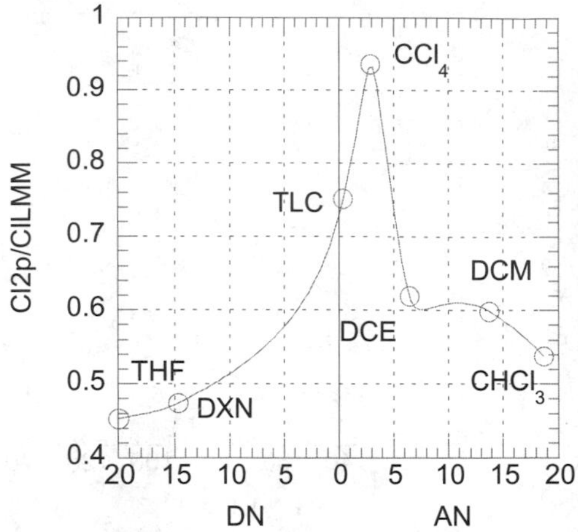


Figure 10.1.3. Intensity ratio vs DN and AN. [Adapted, by permission, from M L Abel, M M Chehimi, *Synthetic Metals*, **66**, No.3, 225-33 (1994).]

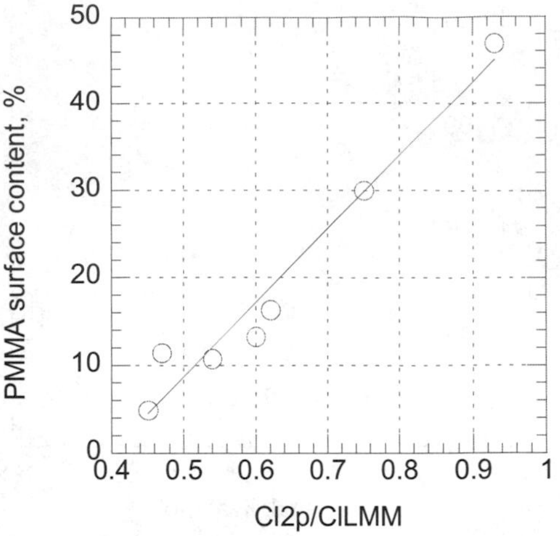


Figure 10.1.4. PMMA surface content vs. intensity ratio. [Data from M L Abel, M M Chehimi, *Synthetic Metals*, **66**, No.3, 225-33 (1994).]

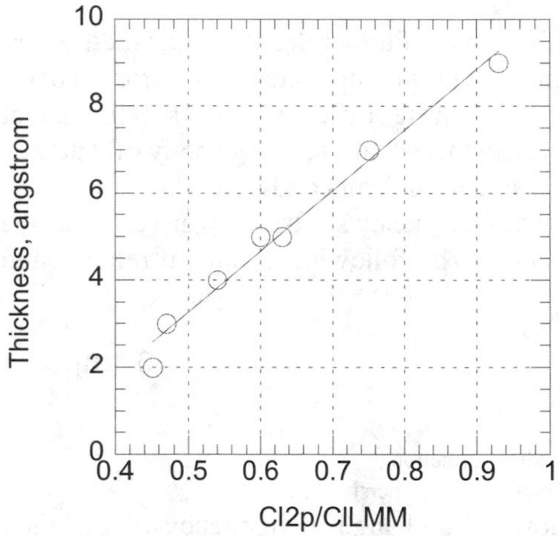


Figure 10.1.5. Thickness of PMMA overlayers vs. intensity ratio. [Data from M L Abel, M M Chehimi, *Synthetic Metals*, **66**, No.3, 225-33 (1994).]

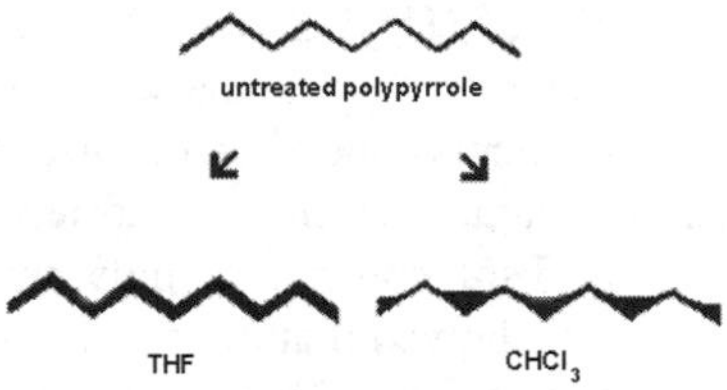


Figure 10.1.6. Schematic diagram of surface roughness of polypyrrole and deposition of PMMA from two different solvents. [Adapted, by permission, from M L Abel, J L Camalet, M M Chehimi, J F Watts, P A Zhdan, *Synthetic Metals*, **81**, No.1, 23-31 (1996).]

The determination of properties of unknown system requires that master curve be constructed. This master curve is determined by testing a series of different polymers exposed to a selected solvent. An example of such a relationship is given in Figure 10.1.1. The enthalpy change caused by the acid-base adduct formation is obtained from a study of the same solid with different solvent probes (see Figure 10.1.2). Having this data, coefficients E and C can be calculated from the chemical shifts in any system.

Figure 10.1.3 shows that there is a correspondence between DN and AN values of different solvents and Cl(2p)/Cl(LMM) intensity ratios.[7] Figure 10.1.4 shows that the type of solvent (measured by its Cl(2p)/Cl(LMM) intensity ratio) determines adsorption of basic PMMA on acidic polypyrrole. Figure 10.1.5 shows that also the thickness of PMMA overlayer corresponds to Cl(2p)/Cl(LMM) intensity ratio of solvent.

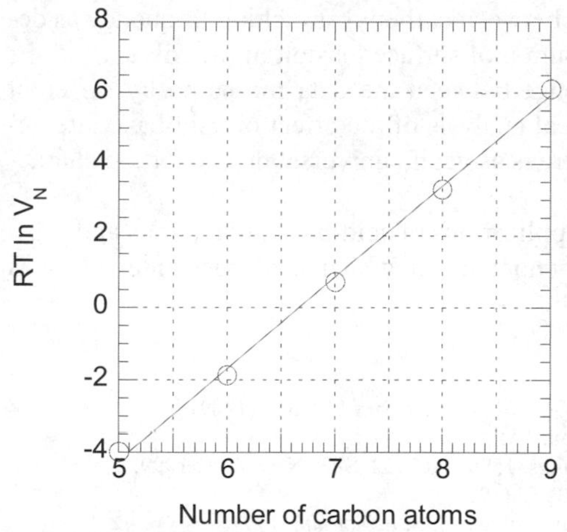


Figure 10.1.7. Relationship between RT ln $V_N$ and number of carbon atoms in n-alkanes. [Adapted, by permission from M M Chehimi, E Pigois-Landureau, M Delamar, J F Watts, S N Jenkins, E M Gibson, *Bull. Soc. Chim. Fr.*, **9**(2) 137-44 (1992).]

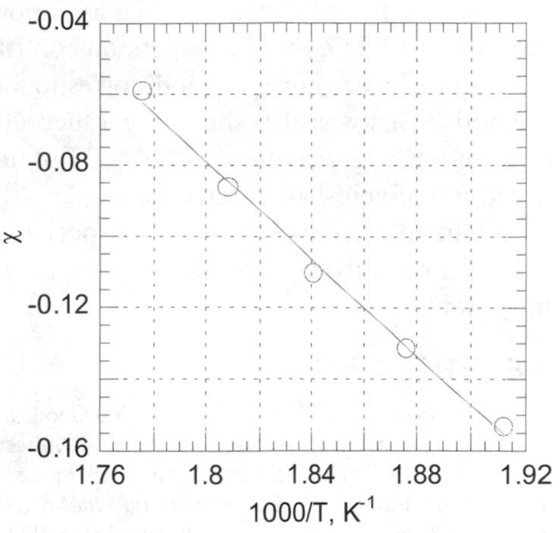


Figure 10.1.8. Flory interaction parameter vs. inverse temperature for polyamide. [Adapted, by permission, from L Bonifaci, G Cavalca, D Frezzotti, E Malaguti, G P Ravanetti, *Polymer*, **33**(20), 4343-6 (1992).]

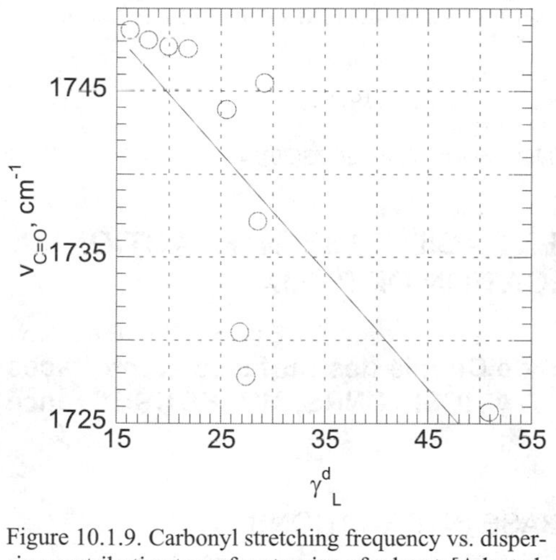


Figure 10.1.9. Carbonyl stretching frequency vs. dispersion contribution to surface tension of solvent. [Adapted, by permission from M M Chehimi, E Pigois-Landureau, M Delamar, J F Watts, S N Jenkins, E M Gibson, *Bull. Soc. Chim. Fr.*, **129**(2) 137-44 (1992).]

XPS analysis combined with analysis by atomic force microscopy, AFM, determines differences in the surface distribution of deposited PMMA on the surface of polypyrrole depending on the type of solvent used (Figure 10.1.6).[6] AFM, in this experiment, permitted the estimation of the surface roughness of polypyrrole. When PMMA was deposited from tetrahydrofuran, a poor solvent, it assumed a surface roughness equivalent to that of polypyrrole. Whereas PMMA, when deposited from a good solvent - $CHCl_3$,[6] developed a smooth surface. The above examples show the importance of solvents in coating morphology.

Figure 10.1.7 shows the correlation between the number of carbon atoms in n-alkanes and the net retention volume of solvent using IGC measurements.[8] Such a correlation must be established to calculate the acid-base interaction's contribution to the free energy of desorption, $\Delta G_{AB}$, as pointed out in discussion of equations [10.1.5] and [10.1.6]. Figure 10.1.8 shows that the Flory interaction parameter (measured by IGC) increases as the temperature increases.

Figure 10.1.9 shows a good correlation between carbonyl stretching frequency as determined by FTIR and the dispersion contribution of surface tension of the solvent.

The above shows a good correspondence between the data measured by different methods. This was also shown by a theoretical analysis of the effect of acid-base interactions on the aggregation of PMMA.[14] But there is as yet no universal theory for the characterization of acid-base interaction.

The following Sections show specific applications of acid-base interactions and present data on solvents applied to various phenomena in which acid base interaction is important.

## REFERENCES


1 C J van Oss, L Ju, M K Chaudhury, R J Good, *J. Colloid Interface Sci.*, **128**, 313 (1989).
2 R J Good, **Contact Angle, Wetting, and Adhesion**, *VSP*, 1993.
3 K-X Ma, C H Ho, T-S Chung, Antec '99 Proceedings, 1590 and 2212, SPE, New York, 1999.
4 C Saint Flour, E. Papirer, *Colloid Interface Sci.*, **91**, 63 (1983).
5 K C Xing, W. Wang, H P Schreiber, Antec '97 Proceedings 53, SPE, Toronto, 1997.
6 M L Abel, J L Camalet, M M Chehimi, J F Watts, P A Zhdan, *Synthetic Metals*, **81**, No.1, 23-31 (1996).
7 M L Abel, M M Chehimi, *Synthetic Metals*, **66**, No.3, 225-33 (1994).
8 M M Chehimi, E Pigois-Landureau, M Delamar, J F Watts, S N Jenkins, E M Gibson, *Bull. Soc. Chim. Fr.*, **129**(2) 137-44 (1992).
9 J F Watts, M M Chehimi, *International J. Adhesion Adhesives*, **15**, No.2, 91-4 (1995).
10 P Munk, P Hattam, Q Du, A Abdel-Azim, *J. Appl. Polym. Sci.: Appl. Polym. Symp.*, **45**, 289-316 (1990).
11 F M Fowkes, D O TIschler, J A Wolfe, L A Lannigan, C M Ademu-John, M J Halliwell, *J. Polym. Sci., Polym. Chem. Ed.*, **22**, 547 (1984).
12 B H Stuart, D R Williams, *Polymer*, **36**, No.22, 4209-13 (1995).
13 R S Drago, G C Vogel, T E Needham, *J. Am. Chem. Soc.*, **93**, 6014 (1971).
14 S J Schultz, *Macromol. Chem. Phys.*, **198**, No. 2, 531-5 (1997).
15 L Bonifaci, G Cavalca, D Frezzotti, E Malaguti, G P Ravanetti, *Polymer*, **33**(20), 4343-6 (1992).


## 10.2 EFFECT OF POLYMER/SOLVENT ACID-BASE INTERACTIONS: RELEVANCE TO THE AGGREGATION OF PMMA

S. Bistac, M. Brogly
**Institut de Chimie des Surfaces et Interfaces**
**ICSI - CNRS, MULHOUSE France**

### 10.2.1 RECENT CONCEPTS IN ACID-BASE INTERACTIONS

Polymer solvent interactions determine several properties, such as, solubility, solvent retention, plasticizer action, wettability, adsorption and adhesion. The solubility parameter is an important criterion for the choice of solvents. However, acid-basic characters of both solvent and polymer are also determinant parameters, which can affect the solution and final film properties. This part, devoted to the influence of acid-base interactions on the aggregation of poly(methyl methacrylate) will first present some recent concepts in acid-base interactions, followed by two practical examples based on experimental results obtained for PMMA/solvent systems.

### 10.2.1.1 The nature of acid-base molecular interactions

#### *10.2.1.1.1 The original Lewis definitions*

The Lewis definitions of acid-base interactions are now over a half a century old. Nevertheless they are always useful and have broadened their meaning and applications, covering concepts such as bond-formation, central atom-ligand interactions, electrophilic-nucleophilic reagents, cationic-anionic reagents, charge transfer complex formation, donor-acceptor reactions, etc. In 1923 Lewis reviewed and extensively elaborated the theory of the electron-pair bond,[1] which he had first proposed in 1916.[2] In this small volume which had since become a classic, Lewis independently proposed both the proton and generalized solvent-system definitions of acids and bases. He wrote:

> "An acid is a substance which gives off the cation or combines with the anion of the solvent; a base is a substance which gives off the anion or combines with the cation of the solvent".

The important point that Lewis revealed is that though the acid-base properties of species are obviously modified by the presence or absence of a given solvent, their ultimate cause should reside in the molecular structure of the acid or base itself, and in light of the electronic theory of matter, not in a common constituent such as $H^+$ or $OH^-$, but in an analogous electronic structure. He states that a basic substance is one which has a lone pair of electrons which may be used to complete the stable group of another atom (the acid) and that an acid substance is one which can employ a lone pair from another molecule (the base) in completing the stable group of one of its own atoms. Moreover by tying his definitions to the concept of chemical bond, Lewis linked their usefulness to contemporary views on the nature of the chemical bond itself. Hence, the shared electron-pair bond model explains the existence of both non-polar bond and polar link from the same premises. As the electrochemical natures of two atoms sharing an electron pair began to differ more and more, the pair should become more and more unequally shared, eventually becoming the sole property of the more electronegative atom and resulting in the formation of ions. Ionic and non-polar bonds appear as logical extremes of a continuum of intermediate bond type. Differences in the continuum are only attributable to variations in the electron-pair donation, i.e., in the way in which the charges are localized within the molecule. As a consequence the distinctions between salts, acids and bases, coordination compounds and organic compounds are not of fundamental nature.

#### *10.2.1.1.2 Molecular Orbital (MO) approach to acid-base reactions*

Translated into the idiom of molecular orbital theory,[3] the acid-base definitions should be read as follows:

- A base is a species, which employs a doubly occupied orbital in initiating a reaction.
- An acid is a species, which employs an empty orbital in initiating a reaction.

The term species may mean a discrete molecule, a simple or complex ion or even a solid exhibiting non-molecularity in one or more dimensions (graphite as an example). Free atoms seldom act as Lewis acids and bases. They usually have one or more unpaired electrons and their reactions are more accurately classified as free radical. The donor orbital is usually the highest occupied molecular orbital HOMO, and the acceptor orbital is usually the lowest unoccupied molecular orbital or LUMO. The molecular orbital definitions have a number of important consequences:

First, it is not necessary that the donor and acceptor orbitals be localizable on a single atom or between two atoms, as implied by Lewis dot structures. That is, the orbitals may be multi-centered even in a relatively localized representation. Thus donor-acceptor interactions involving delocalized electron systems (π-ring)[4-5] are naturally subsumed by the definitions.

Second, the HOMO or donor orbital on a base is likely to be either bonding or non-bonding in character, the latter always being the case for mono-atomic species. The LUMO or acceptor orbital on an acid is likely to be either anti-bonding or non-bonding in character, the latter always being the case for mono-atomic species.

Third, all degrees of electron donation are possible, ranging from essentially zero in the case of weak intermolecular forces and idealized ion associations to the complete transfer of one or more electrons from the donor to the acceptor. This continuity can be represented [10.2.1] by wave functions,[6] were the degree of donation increases as the ratio (a/b).[2]

$$\psi_{AB} = a\psi_A + b\psi_B \qquad [10.2.1]$$

where:

| | |
|---|---|
| $\psi_{AB}$ | wave function of the acid-base one-to-one adduct |
| $\psi_A$ | ground state wave function of the acid |
| $\psi_B$ | ground state wave function of the base |
| a, b | weighting coefficients |

We have reported in Table 10.2.1 the possible adducts classified in terms of the bonding properties of the donor and acceptor orbitals of the acid and base. Complete description of the mechanisms involved during the following adducts formation can be found in Jensen[7] complete review.

**Table 10.2.1. Possible acid-base adducts in terms of orbitals properties**

| | | | **Acceptor orbital of the acid** | | |
|---|---|---|---|---|---|
| | | | Non-bonding | Anti-bonding | |
| | | | **n** | σ* | π* |
| **Donor orbital of the base** | Non-bonding | **n** | n-n | n-σ* | n-π* |
| | Bonding | σ | σn | σσ* | σπ* |
| | | π | π-n | π-σ* | π-π* |

Even if the nature of an acid or a base is highly relative, a brief classification based on bonds properties and orbitals symmetry can be proposed:

| | |
|---|---|
| n donor: | Lewis bases, complex and simple anions, carbanions, amines, oxides, sulfurs, phosphines, sulfoxides, ketones, ethers, alcohols .... |
| π donor: | unsaturated and aromatic hydrocarbons having electron donor substituents..... |
| σ donor: | saturated hydrocarbons, CO, single links like C-C, C-H, polar links like NaCl, BaO, silanes.... |
| n acceptor: | Lewis acids, simple cations.... |
| π acceptor: | $N_2$, $SO_2$, $CO_2$, $BF_3$, dienic and unsaturated and aromatic hydrocarbons having electron acceptor substituents.... |

σ acceptor: Brönsted acids, boranes and alkanes having strong acceptor substituents, such as, $CHCl_3$, halogens....

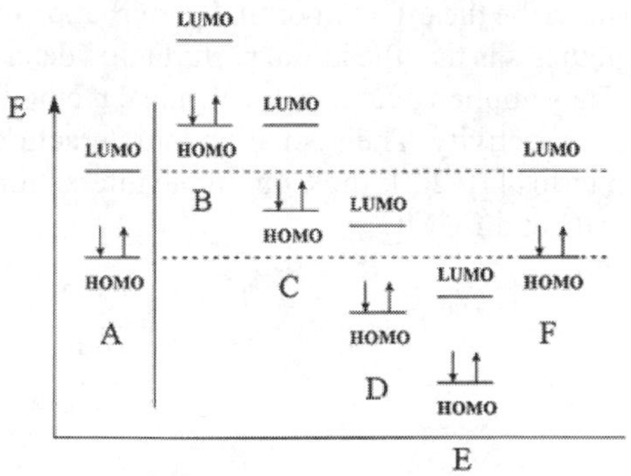


Figure 10.2.1. How do donor-acceptor molecular orbital interactions work?

To summarize how acid-base reactions do work on the basis of molecular orbitals perturbation theory, we have reported on Figure 10.2.1, the relative energies (as perturbed by the field of the other reactant) of the frontier orbitals HOMO and LUMO of a hypothetical species A and of the frontier orbitals of several hypothetical reaction partners B, C, D, E and F. This figure is intended to represent possible variations of donor-acceptor properties in the broadest possible context i.e. not only those species encountered in aqueous solution but also those stabilized by non-aqueous environments.

According to Figure 10.2.1, with respect to B, complete electron transfer from B to A will be favorable and A will act as an oxidizing agent. With respect to C, the A(LUMO) - C(HOMO) perturbation will be favorable and A will act as an acid. With respect to D, the A(HOMO) - D(LUMO) perturbation will be favorable and A will act as a base. Lastly, with respect to E, complete electron transfer from A to E will be favorable and A will act as a reducing agent. For F species, the frontiers orbitals are quite degenerated with those of A. Here neither species is clearly the donor nor acceptor and species may display both behavior simultaneously (case of multisite interactions encountered in concerted organic cycloaddition reactions).

#### *10.2.1.1.3 The case of hydrogen bonding*

In addition to the general discussion of Lewis acid-base interactions, hydrogen bonding represents a special case. According to Pauling, hydrogen bonding is partly covalent and partly ionic (polar). Nevertheless, it is obvious that electrostatic and charge transfer interactions are predominant for hydrogen bonds. In most cases the principal charge transfer contribution is derived from the proton acceptor - proton donor charge transfer complex through the σ-type interactions. Moreover hydrogen bonds complexes are linear and the angle between the H bond and the molecular axis of the proton acceptor are mainly 0° or 60°.

To conclude this first theoretical part, the chemical phenomena subsumed by the category of acid-base reaction, are the following:

- Systems defined by the Arrhenius description, solvent system, Lux-Flood and proton acid-base definitions
- Traditional coordination chemistry and "non-classical" complexes
- Solvation, solvolysis and ionic dissociation phenomena in both aqueous and non-aqueous solutions
- Electrophilic and nucleophilic reactions in organic and organometallic chemistry
- Charge transfer complexes, molecular addition compounds, weak intermolecular forces, hydrogen bonds
- Molten salt phenomena, salt formation

### 10.2.1.2 Quantitative determination of acid-base interaction strength

#### *10.2.1.2.1 Perturbation theory*

Hudson and Klopman[8] proposed an equation to describe the effect of orbital perturbation of two molecules on chemical reactivity. Their hypothesis is that the initial perturbation determines the course of a reaction or an interaction. They applied quantum-mechanical method[9] to treat the encounter of two interacting systems as reactivity. Their equation for interaction energy can be simplified by including only two terms [10.2.2]: the Coulombic interaction and the frontier orbital interaction between HOMO and LUMO.

$$\Delta E_{int} = -\frac{Q_N Q_E}{\varepsilon R} + \frac{2(C_N C_E \beta)^2}{E_{HOMO} - E_{LUMO}} \qquad [10.2.2]$$

where:

| | |
|---|---|
| $Q_N$ | total charges of nucleophile N |
| $Q_E$ | total charges of electrophile E |
| $C_N$ | coefficient of the atomic orbital N |
| $C_E$ | coefficient of the atomic orbital E |
| $\beta$ | resonance integral |
| $\varepsilon$ | permittivity |
| R | interatomic distance |
| $E_{HOMO}$ | energy level of the HOMO orbital |
| $E_{LUMO}$ | energy level of the LUMO orbital |

The relative magnitudes of the numerator and denominator of the second term determine the extent of perturbation and the type of reactivity. For electrostatic interactions, the first term dominates: this is the case of highly polar acceptors of low electron affinity and donors highly electronegative, while for electron donor - electron acceptor interaction, the second term dominates. Thus, a molecular interaction encompasses chiefly both acid-base and donor-acceptor interactions. On the other hand if the second frontier orbital term dominates, the extent of perturbation is large and the charge transfer is large, leading to a large gain in stability. The frontier orbital controlled processes are favored by the presence of weakly polar species lacking high charge density but possessing reactive atoms with large orbital radii. This second term is also benefitted by bases of low electronegativity and acids of high electron affinity.

#### *10.2.1.2.2 Hard-Soft Acid-Base (HSAB) principle*

The energy gap between HOMO and LUMO has been equated to the absolute hardness of the HSAB principle. This principle[10] describes some basic rules about kinetics and equilibrium of the acid-base interactions in solutions. The HSAB principle will be described as it has evolved in recent years on the basis of the density-functional theory.[11] For organic interactions, the following statements were proposed:

- A hard acceptor has a high energy LUMO and usually a positive charge
- A soft acceptor has a low energy LUMO but does not necessarily have a positive charge
- A hard donor has a low energy HOMO and usually a negative charge
- A soft donor has a high energy HOMO but does not necessarily have a negative charge

The principle states that hard acid prefers to interact with a hard base, and vice versa, a soft acid with a soft base. A hard-hard interaction is fast because of large Coulombic attraction, a soft-soft interaction is fast because of large orbital overlap between HOMO and LUMO. However the problem is, what is the physical meaning of hardness?

*10.2.1.2.3 Density functional theory*

Parr and al.[12] gave in 1988 a theoretical support to the absolute hardness. In the density functional theory two basic parameters were introduced. Any chemical system can be characterized by its chemical potential, μ, and its absolute hardness, η. The chemical potential measures the escaping tendency of an electronic cloud, while absolute hardness determines the resistance of the species to lose electrons. The exact definitions of these quantities are:

$$\mu = \left(\frac{\partial E}{\partial N}\right)_v \quad \text{and} \quad \eta = \frac{1}{2}\left(\frac{\partial^2 E}{\partial N^2}\right)_v \qquad [10.2.3]$$

where :

μ chemical potential
η absolute hardness
E electronic energy
N number of electrons
v potential due to the nuclei plus any external potential.

However, according to frontier orbital method,[13] the relationship between η and the HOMO and LUMO energies is reduced to:

$$\eta \approx -\frac{1}{2}\left(E_{HOMO} - E_{LUMO}\right) \qquad [10.2.4]$$

where :

η absolute hardness
$E_{HOMO}$ energy level of the HOMO orbital
$E_{LUMO}$ energy level of the LUMO orbital.

Of course the absolute softness is the reciprocal of the absolute hardness. The apparent success of the density-functional theory is to provide two parameters from which we can calculate the number of electrons transferred, resulting mainly from the charge transfer between two molecules, i.e., from electrons flow until chemical potential reaches an equilibrium. As a first approximation, the number of electron transferred is given[14] by:

$$N_{Trans} = \frac{\mu_B - \mu_A}{2(\eta_A - \eta_B)} \qquad [10.2.5]$$

where:

$N_{trans}$ number of electrons transferred
$\mu_A$ chemical potential of the acid
$\mu_B$ chemical potential of the base
$\eta_A$ absolute hardness of the acid
$\eta_B$ absolute hardness of the base.

This number varies from 0 to 1 and it is in most cases a fractional number. As an example[15] for the interaction between $Cl_2$ and substituted aromatic compounds, $N_{trans}$ varies from -0.013 to 0.20. Calculation for various solid polymers could be found in a paper of Lee.[16]

#### *10.2.1.2.4 Effect of ionocity and covalency: Drago's concept*

Similar to the perturbation theory, Drago and Wayland[17] proposed a four-parameter equation for predicting reactions enthalpies between acid and base species. Both species are each characterized by two independent parameters: an E value which measures their ability to participate in electrostatic bonding, and a C value which measures their ability to participate in covalent bond. Both E and C values are derived empirically to give the best curve fit of calculated to experimental heats of formation for the largest possible number of adducts, leading to:

$$-\Delta H^{AB} = E_A E_B + C_A C_B \qquad [10.2.6]$$

where:

| | |
|---|---|
| $\Delta H^{AB}$ | enthalpy of acid-base adduct formation |
| $E_A$ | ability of the acid to participate in electrostatic bonding |
| $E_B$ | ability of the base to participate in electrostatic bonding |
| $C_A$ | ability of the acid to participate in covalent bonding |
| $C_B$ | ability of the base to participate in covalent bonding. |

A self-consistent set of E and C values is now available[18] for 33 acids and 48 bases, allowing ΔH prediction for over 1584 adducts. It is assumed that the conditions under which measurements are made (gas phase or poorly coordinating solvents) give rather constant entropy contribution and that most of adducts are of one-to-one stoichiometry. Table 10.2.2 gathers Drago's parameters, given in $(kcal/mol)^{0.5}$ of some common solvents.

**Table 10.2.2. Drago's parameters of some common solvents [after reference 18]**

| Acid | $C_A$ | $E_A$ | Base | $C_B$ | $E_B$ |
|---|---|---|---|---|---|
| Phenol | 0.442 | 4.330 | Acetone | 2.330 | 0.987 |
| Chloroform | 0.159 | 3.020 | Benzene | 0.681 | 0.525 |
| Water | 2.450 | 0.330 | Ethyl acetate | 1.740 | 0.975 |
| tert-Butyl alcohol | 0.300 | 2.040 | Pyridine | 6.400 | 1.170 |
| Iodine | 1.000 | 1.000 | Methylamine | 5.880 | 1.300 |
| Pyrrole | 0.295 | 2.540 | Tetrahydrofuran | 4.270 | 0.978 |
| Trifluoroethanol | 0.451 | 3.88 | Dimethylformamide | 2.480 | 1.230 |

The major importance of the above four parameters is their relationship with the HSAB principle. Actually, through the plot of $E_A$ versus $C_A$ for several liquids on a solid, one can obtain indirectly the values of the chemical softness 1/η from the slope of C/E, as represented by the following equation :

$$E_A = -\frac{\Delta H^{AB}}{E_B} - C_A \frac{C_B}{E_B} \qquad [10.2.7]$$

With the chemical hardness, η, and the electronegativity (or negative chemical potential), one can easily apply the HSAB principle to acid-base interaction and calculate the number of electron transferred.

*10.2.1.2.5 Effect of amphotericity of acid-base interaction: Gutmann's numbers*

Solvatation, solvolysis and ionic dissociation phenomena, in both aqueous and nonaqueous solutions are subsumed by the Lewis definitions. In addition to the previous discussion of the dual polarity character of Lewis acids and bases, it should be noted that many of them are amphoteric, by definition. Donor number, DN, was developed[19] in order to correlate the behavior of a solute in a variety of donor solvents with a given basicity or donicity. A relative measurement of the basicity of a solvent D is given by the enthalpy of its reaction with an arbitrarily chosen reference acid ($SbCl_5$ in the Gutmann's scale). Latter Mayer[20] introduced an acceptor number, AN, as the relative $^{31}P$ NMR shift induced by triethylphosphine, and relative to acidic strength (AN=0 for hexane and 100 for $SbCl_5$). In 1989, Riddle and Fowkes[21] modify these AN numbers, to express them, AN*, in correct enthalpic unit (kcal/mol). Table 10.2.3 gathers electron acceptor number AN and AN* and electron donor number DN for amphoteric solvents.

**Table 10.2.3. Acceptor number, AN, and donor number, DN, for common solvents [After references 19, 21]**

| Amphoteric solvent | AN | AN*, kcal/mol | DN, kcal/mol |
|---|---|---|---|
| Acetone | 12.5 | 2.5 | 17.0 |
| Diethyl ether | 3.9 | 1.4 | 19.2 |
| Formamide | 39.8 | 9.3 | 26.6 |
| Tetrahydrofuran | 8.0 | 0.5 | 20.0 |
| Pyridine | 14.2 | 0.14 | 33.1 |
| Ethyl acetate | | 1.5 | 17.1 |
| Benzonitrile | 15.5 | 0.06 | 11.9 |
| Nitromethane | 20.5 | 4.3 | 2.7 |
| Water | 54.8 | 15.1 | 18.0 |

The most important assumption of Gutmann's approach is that the order of base strengths established remains constant for all other acids (solutes), the value of the enthalpy of formation of a given adduct is linearly related to the donor number of the base (solvent) through the equation [10.2.8]:

$$-\Delta H_{AB} = a_A DN_B + b_B \quad [10.2.8]$$

where:

$\Delta H_{AB}$ enthalpy of acid-base adduct formation
$DN_B$ donor number of the base
$a_A$, $b_A$ constants characteristic of the acid

Graphically this means that a plot of the DN for a series of donor solvents versus -ΔH of their adducts formation with a given acid gives a straight line, allowing the determination of $a_A$ and $b_A$. By experimentally measuring the enthalpy of formation of only two adducts for a given acid, one can predict, through the resulting $a_A$ an $b_A$ values, the enthalpy of adduct formation of this acid with any other donor solvent for which DN is known. Gutmann also proposes that the enthalpy of acid-base interaction could be approximated by a two-parameters equation of the form:

$$-\Delta H_{AB} = \frac{AN_A DN_B}{100} \qquad [10.2.9]$$

where:

$\Delta H_{AB}$ enthalpy of acid-base adduct formation
$DN_B$ donor number of the base
$AN_A$ acceptor number of the acid

The factor of 100 converts the AN value from a percentage of the $SbCl_5$ value to a decimal fraction. But one had to remind that on the 171 DN values reported in the literature[22] only 50 were determined precisely, i.e., calorimetrically.

#### *10.2.1.2.6 Spectroscopic measurements: Fowkes' approach*

Fowkes[23] has proposed that for specific functional groups involved in acid-base interaction, the enthalpy of acid-base adduct formation is related to the infrared frequency shift, Δν, of its absorption band according to the following equation:

$$\Delta H_{AB} = k_{AB} \Delta \nu_{AB} \qquad [10.2.10]$$

where:

$\Delta H_{AB}$ enthalpy of acid-base adduct formation
$\Delta \nu_{AB}$ infrared frequency shift
$k_{AB}$ characteristic correlation constant between IR wavenumber shift and enthalpy

$k_{AB}$ is a characteristic constant of the functional group determined on the basis of compared infrared and microcalorimetrical results of adduct formation. As an example, the latter is equal to -0.99 kJ/mol/cm for the carbonyl group C=O. The stretching frequency of the C=O vibration band is decreased by an amount $\Delta \nu_{AB}$ proportional to the enthalpy of acid-base bonding $\Delta H_{AB}$ according to $k_{AB}$. Such a methodology has recently been nicely confirmed not only for polymer-solvent adduction, but also for polymer/polymer[24] and polymer/metal[25] adduction.

## 10.2.2 EFFECT OF POLYMER/SOLVENT INTERACTIONS ON AGGREGATION OF STEREOREGULAR PMMA

### 10.2.2.1 Aggregation of stereoregular PMMA

PMMA chains are able to form aggregates in the presence of solvent[26] (in diluted solution, concentrated solution, gel or solid). Aggregation between isotactic and syndiotactic chains, after mixing in some solvents, leads to stereocomplexes, and self-aggregation corresponds to the aggregation of isotactic or syndiotactic chains together. These two kinds of aggregates result from the development of physical interactions between polymer chains. The formation of aggregates depends mainly on the PMMA degree of stereoregularity and on the nature of the solvent, but also on the temperature, the mixing time, and the isotactic/syndiotactic stoichiometric ratio for stereocomplexes.[26,27] Complexing solvents favor the formation of aggregates and non-complexing solvents hinder the formation of aggregates. Table 10.2.4 presents the solvents classification as a function of complexing power for PMMA stereocomplexes formation.

**Table 10.2.4. Solvents classification as a function of complexing power for PMMA stereocomplexes formation (after reference 26)**

| | |
|---|---|
| Strongly complexing solvents | $CCl_4$, acetonitrile, DMF, DMSO, THF, toluene, acetone |
| Weakly complexing solvents | benzene, o-dichlorobenzene, dioxane |
| Non-complexing solvents | chloroform, dichloromethane |

Several works have shown that the aggregation of isotactic and syndiotactic chains leads to the formation of stereocomplexes for which the iso/syndio stoichiometry is found equal to 1/2,[28,29] probably with a structure composed of a double-stranded helix of a 30/4 helicoidal isotactic chain surrounded by a 60/4 helicoidal syndiotactic chain.[30] Syndiotactic PMMA self-aggregates exhibit similar structures, with conformations close to extended chains.[31] Experimental data indicate that, in self-aggregated syndiotactic PMMA in solution, some of the ester groups are close in contact, probably in a double helix structure[32] with solvent molecules included in the cavities of inner- and inter-helices.[33] Isotactic PMMA self-aggregates also exhibit conformational helix structures.

These results prove that the presence of a complexing solvent leads to the formation of well-ordered structure made of paired PMMA chains.

A high degree of stereoregularity is needed to allow the formation of aggregates. Syndiotactic and isotactic sequences lengths have to be larger than a critical length in order to be involved in paired association. As the aggregation power of the solvent increases, the critical sequence length decreases.

Many authors explain aggregation as resulting from interactions between PMMA chains, probably through their ester functional groups.[34] The incidence of the nature of the solvent in the aggregate formation is not yet totally clarified in the literature. Apparently, there is no obvious relationship between the solvent polarity and the ability to induce aggregation of PMMA chains.

#### 10.2.2.2 Relation between the complexing power of solvents and their acid-base properties

A tentative explanation of the effect of solvent has been proposed recently by using the Lewis acid-base concept.[35] The acid or basic character of a given solvent was studied simultaneous with its strength of complexation in order to establish a relationship between both parameters.

PMMA stereocomplexes and self-aggregates exhibit some similar aspects, particularly the complexing power of the solvent or the helical structure of paired chains.

PMMA is classified as a basic polymer (electron donor, according to the Lewis concept), due to the presence of ester functional groups, where the carbonyl oxygen atom is the basic site.[36] PMMA is therefore able to exchange strong acid-base interactions with an acidic solvent, such as, chloroform, which is also a non-complexing solvent. According to Fowkes works, PMMA and chloroform can form acid-base complexes, resulting from the interaction between the ester basic group of PMMA and the hydrogen acid atom of chloroform.

According to Drago's classification, it appears therefore in Table 10.2.2 that chloroform exhibits an acid character.[37] PMMA presents a basic character, with $E_b$=0.68 and $C_b$=0.96.

The fact that an acidic solvent, such as, chloroform is able to exchange strong acid-base interactions with PMMA ester groups is able to explain its non-complexing behavior. Strong acid-base interactions between PMMA and chloroform molecules hinder the aggregation of PMMA chains, reflecting greater PMMA/solvent interactions compared to PMMA/PMMA interactions. In the presence of basic solvents, the lower PMMA/solvent acid-base interactions lead to inter-chains associations (and consequently aggregation), energetically favorable. Moreover, solvents, such as, acetone or THF possess a high degree of self-association (31%, 27% respectively) and Fowkes has shown that the contribution of solvent/solvent acid-base interactions to the heat of vaporization is proportional to the degree of self-association.[38] Acetone and THF are therefore strongly acid-base self-associated solvents. As a consequence, they would develop preferentially self-associations (between solvent molecules) rather than PMMA/solvent interactions. Hence the PMMA chains aggregation would be also favored.

**Table 10.2.5. AN and DN numbers for complexing and non-complexing solvents (after reference 19)**

| | AN | DN, kcal/mol |
|---|---|---|
| DMF | 16.0 | 26.6 |
| Acetone | 12.5 | 17.0 |
| THF | 8.0 | 20 |
| DMSO | 20.4 | 29.8 |
| Benzene | 8.2 | 0.1 |
| Chloroform | 23.1 | 0 |

Gutmann's numbers of complexing and non-complexing solvents are listed in Table 10.2.5.The results indicate also that chloroform is an acidic solvent, with a high Acceptor Number (acidic character), and a Donor Number (basic character) equal to zero. The advantage of Gutmann's approach is to consider the potential amphoteric character. As an example, benzene, which is classified as a basic solvent by Drago, exhibits an amphoteric character with a low AN, but also a DN close to zero. This result could explain that, in some cases, benzene (which is generally a complexing solvent) is described as a non-complexing solvent for syndiotactic PMMA.[31] In this case, amphoteric character of benzene favors acid-base interactions with PMMA.

To resume this part, it is possible to correlate the complexing power of solvents towards stereoregular PMMA with their acid-base character. It appears that complexing solvents exhibit basic character, like PMMA. Moreover, some of them are strongly self-associated. Solvent/solvent and chain/chain interactions are consequently favored, leading to the formation of PMMA aggregates.

On the contrary, chloroform is an acidic solvent, and the development of acid-base interactions with PMMA ester groups helps to hinder the formation of chains aggregates. However, this original view on the relationship between complexing power and solvent acid-base character does not take into account the way of action of the solvent in terms of steric effect and macromolecular conformations of aggregates.

## 10.2.3 INFLUENCE OF THE NATURE OF THE SOLVENT ON THE α AND β-RELAXATIONS OF CONVENTIONAL PMMA

### 10.2.3.1 Introduction

Thin polymer films are generally obtained by solution casting, in many applications, such as paints, varnishes, or adhesives. The properties of polymer films obtained from a solution differ from the original bulk properties, and this effect can have some significant consequences on the expected behavior of the final film.

Recent studies have analyzed the influence of the nature of the solvent on the relaxation temperatures of conventional PMMA solid films.[39,40] PMMA was dissolved in various solvents and the solid films (after solvent evaporation) were analyzed by dielectric spectroscopy. The purpose of the study was to investigate the influence of the nature of the solvent on the α and β-relaxations of PMMA solid films. The α-relaxation is related to the glass transition of PMMA and corresponds to the rotation of lateral groups around the main chain axis.[41] The β-relaxation, which occurs at a lower temperature, is induced by the rotation of the acrylate groups around the C-C bonding which links them to the main chain. Different solutions of conventional (atactic) PMMA in good solvents (chloroform, acetone, toluene and tetrahydrofuran) are cast on a metallic substrate (aluminum). After solvent evaporation at room temperature, solid films of PMMA are analyzed by dielectric spectroscopy (DETA) at 100 Hz and 10 KHz from -40 to 135°C with a scanning temperature of 2°C/min. The variation of the loss factor tan δ is studied as function of temperature. The temperatures of the tan δ peaks are related to the relaxations temperatures of the polymer.[42] The reference DETA spectrum is obtained by analyzing a bulk PMMA film obtained from heating press (without solvent).

At 10 KHz, only one peak is detectable (whatever the sample) which is attributed to the α-relaxation. However, the peak is not symmetrical, with a broadening towards the lower temperatures, due to the contribution of the β-relaxation. At 100 Hz, two peaks are present, the major at lower temperature, corresponding to the β-relaxation and the minor at higher temperature, attributed to the α-relaxation.

### 10.2.3.2 Dielectric spectroscopy results

Table 10.2.6 reports the temperatures of the α-relaxation (measured at 10 KHz) and the β-relaxation (measured at 100 Hz) for the reference sample (bulk PMMA) and the different solution-cast films.

**Table 10.2.6. Temperature of α and β-relaxations of bulk PMMA and solution-cast films (after reference 40)**

| | PMMA bulk | PMMA/ chloroform | PMMA/ toluene | PMMA/ acetone | PMMA/ THF |
|---|---|---|---|---|---|
| T° of α peak, °C at 10KHz | 108 | 75 | 96 | 85 | 82 |
| T° of β peak, °C at 100 Hz | 40 | 34 | 49 | 45 | 47 |

The results show that the temperature of the α-transition is significantly reduced for cast films compared to the bulk polymer, with a lower value for chloroform samples. The decrease of the α-transition temperature can be explained by the presence of residual solvent, inducing a plasticizing effect. On the contrary, an increase of the β-transition temperature is generally observed for the cast films, except for the 'chloroform' sample. An increase of the β-transition temperature indicates a lower mobility of the acrylate groups compared to the bulk polymer. However, for chloroform sample, this mobility is increased.

Two different effects appear and two groups of solvents can then be distinguished: the first group, with chloroform, which induces a higher mobility of both the main chain and the acrylate group (compared to the bulk polymer), and the second group, including acetone, toluene and THF which induce also an increase of the mobility of the main chain, but a decrease of the mobility of the lateral acrylate groups.

The influence of the nature of the solvent on the relaxation temperature can be explained by analyzing the acid-base properties of the polymer and the solvents. PMMA, which is a basic polymer can exchange strong acid-base interactions with an acidic solvent, such as, chloroform. In the solid cast films, acid-base interactions between acrylate groups and residual chloroform occur: some acrylate groups interact with chloroform molecules and PMMA/PMMA self-associations partially disappear, leading to a lower value of the β-transition temperature. The mobility of the lateral acrylate groups interacting with chloroform molecules is then higher than the mobility of acrylate groups interacting with other acrylate groups as in bulk PMMA.

Toluene, THF and acetone are described as basic solvents. The residual solvent molecules can therefore only weakly interact with the acrylate groups. Interactions between chains are then favored and self-associations between PMMA chains appear. The rotation of the acrylate groups becomes more difficult compared to the bulk PMMA, probably due to self-aggregation of some PMMA chains (even if the studied polymer is not stereoregular).

To resume, residual solvent molecules present in solid conventional PMMA films are able to significantly modify the polymer relaxation properties. The effect of residual solvent depends strongly on the nature of the solvent, specially its acid-base character.

## 10.2.4 CONCLUDING REMARKS

These works have shown that acid-basic character of solvent is able to have a major influence on polymer film properties. Acid-base interactions between stereoregular PMMA and some solvents can lead to the formation of aggregates, which modify the solution properties. Elsewhere, residual solvent molecules trapped in solid conventional PMMA films have an effect on the polymer chains mobility, depending on the acid-base character of the solvent. Both solvent effects, due to acid-base interactions, are able to modify the wetting and the adhesion properties of the films, but also the mechanical and the durability behavior of the final film. It is then necessary to take into account, in the choice of a solvent, its acid-base properties, in addition to its solubility parameter, especially for PMMA.

## REFERENCES

1 G.N. Lewis, **Valence and the structure of atoms and molecules**, The Chemical Catalog Co., N-Y, 1923.
2 G.N. Lewis, *J. Am. Chem. Soc.*, **38**, 762 (1916).
3 R.S. Mulliken, W.B. Pearson, **Molecular complexes : a lecture and reprint volume**, *Wiley-Interscience*, N-Y, 1969.
4 M. Brogly, M. Nardin, J. Schultz, *J. Adhesion*, **58**, 263 (1996).
5 M.F. Hawthorne, G.B. Dunks, *Science*, **178**, 462 (1972).
6 G. Klopman, **Chemical reactivity and Reaction paths**, *Wiley-Interscience*s, New-York, 1974.
7 W.B. Jensen, **The Lewis acid-base concepts: an overview**, *J.Wiley & Sons*, New-York, 1979.
8 R.F. Hudson, G. Klopman, *Tetrahedron Lett.*, **12**, 1103 (1967).
9 S.R. Cain, **Acid-base interactions : relevance to adhesion science and technology**, K.L. Mittal and H. Anderson, Jr (Eds), *VSP*, Zeist, The Netherlands, 1991.
10 R.G. Pearson, *J. Am. Chem. Soc.*, **85**, 3533 (1963).

11 R.G. Parr, W. Yang in **Density functional theory of atoms and molecules**, *Oxford University Press*, New-York, 1989.
12 M. Berkowitz and R.G. Parr, *J. Chem. Phys.*, **88**, 2554 (1988).
13 I. Fleming, **Frontier orbitals and organic chemical reactions**, *John Wiley*, London, 1976.
14 S. Shankar, R.G. Parr, *Proc. Natl. Acad. Sci., USA*, **82**, 264 (1985).
15 R.G. Pearson, *J. Org. Chem.*, **54**, 1423 (1989).
16 L.H. Lee, *J. Adhesion Sci. Technol.*, **5**, 71 (1991).
17 R.S. Drago, B. Wayland, *J. Am. Chem. Soc.*, **87**, 3571 (1965).
18 R.S. Drago, *Struct. Bonding*, **15**, 73 (1973).
19 V. Gutmann, **The donor-acceptor approach to molecular interaction**, *Plenum Press*, New-York, 1977.
20 U. Mayer, V. Gutmann, W. Greger, *Montatsh. Chem.*, **106**, 1235 (1975).
21 F.L. Riddle, Jr, F.M. Fowkes, *J. Am. Chem. Soc.*, **112**, 3259 (1990).
22 Y. Marcus, *J. Solution Chem.*, **13**, 599, (1984).
23 F.M. Fowkes et al., *J. Polym. Sci. Polym. Chem. Ed.*, **22**, 547 (1984).
24 M. Brogly, M. Nardin, J. Schultz, *Polymer*, **39**, 2185 (1998).
25 M.Brogly, S. Bistac, J. Schultz, *Macromolecules*, **31**, 3967 (1998).
26 J. Spevacek, B. Schneider, *Adv. Coll. Interf. Sci.*, **27**, 81 (1987).
27 M. Berghmans, S. Thijs, M. Cornette, H. Berghmans, F.C. De Schryver, P. Moldenaers, J. Mewis, *Macromolecules*, **27**, 7669 (1994).
28 K. Ohara, *Coll. Polym. Sci.*, **259**, 981 (1981).
29 K. Ohara, G. Rehage, *Coll. Polym. Sci.*, **259**, 318 (1981).
30 F. Bosscher, G. Ten Brinke, G. Challa, *Macromolecules*, **15**, 1442 (1982).
31 J. Spevacek, B. Schneider, J. Dybal, J. Stokr, J. Baldrian, Z. Pelzbauer, *J. Polym. Sci. : Polym. Phys. Ed.*, **22**, 617 (1984).
32 J. Dybal, J. Spevacek, B. Schneider, *J. Polym. Sci. : Polym. Phys. Ed.*, **24**, 657 (1986).
33 H. Kusuyama, N. Miyamoto, Y. Chatani, H. Tadokoro, *Polym. Comm.*, **24**, 119 (1983).
34 J. Spevacek, B. Schneider, *Colloid Polym. Sci.*, **258**, 621 (1980).
35 S. Bistac, J. Schultz, *Macromol. Chem. Phys.*, **198**, 531 (1997).
36 F.M. Fowkes, *J. Adhesion Sci. Tech.*, **1**, 7 (1987).
37 R.S. Drago, G.C. Vogel, T.E. Needham, *J. Amer. Chem. Soc.*, **93**, 6014 (1971).
38 F.M. Fowkes in **Acid-Base Interactions**, K.L. Mittal & H.R. Anderson, Ed, *VSP*, Utrecht 1991, pp. 93-115.
39 S. Bistac, J. Schultz, *Prog. in Org. Coat.*, **31**, 347, (1997).
40 S. Bistac, J. Schultz, *Int. J. Adhesion and Adhesives*, **17**, 197 (1997).
41 R.F. Boyer, *Rubb. Chem. Techn.*, **36**, 1303 (1982).
42 N.G. Mc Crum, B.E. Read, G. Williams, **Anelastic and Dielectric Effect in Polymeric Solids**, *Dover Publications*, New York, 1991.

## 10.3 SOLVENT EFFECTS BASED ON PURE SOLVENT SCALES

Javier Catalán
**Departamento de Química Fisíca Aplicada**
**Universidad Autónoma de Madrid, Madrid, Spain**

### INTRODUCTION

The solvent where a physico-chemical process takes place is a non-inert medium that plays prominent roles in chemistry. It is well-known[1] that, for example, a small change in the nature of the solvent can alter the rate of a reaction, shift the position of a chemical equilibrium, modify the energy and intensity of transitions induced by electromagnetic radiation or cause protein denaturation. As a result, the possibility of describing the properties of solvents in terms of accurate models has aroused the interest of chemists for a long time.

The interest of chemists in this topic originated in two findings reported more than a century ago that exposed the influence of solvents on the rate of esterification of acetic acid by ethanol, established in 1862 by Berthelot and Saint-Gilles,[2] and on the rate of quaternization of tertiary amines by alkyl halides, discovered in 1890 by Menschutkin.[3] In his study, Menschutkin found that even so-called "inert solvents" had strong effects on the reaction rate and that the rate increased by a factor about 700 from hexane to acetophenone. Subsequent kinetic studies have revealed even higher sensitivity of the reaction rate to the solvent. Thus, the solvolysis rate of tert-butyl chloride increases 340 000 times from pure ethanol to a 50:50 v/v mixture of this alcohol and water,[4,5] and by a factor of $2.88\times10^{14}$ from pentane to water.[6] Also, the decarboxylation rate of 6-nitrobenzisoxazol 3-carboxylate increases by a factor of $9.5\times10^{7}$ from water to HMPT.[7]

## 10.3.1 THE SOLVENT EFFECT AND ITS DISSECTION INTO GENERAL AND SPECIFIC CONTRIBUTIONS

Rationalizing the behavior of a solvent in a global manner, i.e., in terms of a single empirical parameter derived from a single environmental probe, appears to be inappropriate because the magnitude of such a parameter would be so sensitive to the nature of the probe that the parameter would lack predictive ability. One must therefore avoid any descriptions based on a single term encompassing every potential interaction of the solvent, often concealed under a global concept called "solvent polarity". One immediate way of dissecting the solvent effect is by splitting it into general (non-specific) interactions and specific interactions.

In relation to general interactions, the solvent is assumed to be a dielectric continuum. The earliest models for this type of interaction were developed by Kirkwood[8] and Onsager,[9] and were later modified with corrections for the effect of electrostatic saturation.[10,11] The intrinsic difficulty of these models in accurately determining the dimensions of the cybotactic region (viz. the solvent region where solvent molecules are directly perturbed by the presence of solute molecule) that surrounds each solute molecule in the bulk solvent, have usually raised a need for empirical approximations to the determination of a parameter encompassing solvent polarity and polarizability. An alternative approach to the general effect was recently reported that was derived from liquid-state theories. One case in point is the recent paper by Matyushov et al.,[12] who performed a theoretical thermodynamic analysis of the solvent-induced shifts in the UV-Vis spectra for chromophores; specifically, they studied p-nitroanisole and the pyridinium-N-phenoxide betaine dye using molecular theories based on long-range solute-solvent interactions due to inductive, dispersive and dipole-dipole forces.

A number of general empirical solvent scales have been reported;[1,13-15] according to Drago,[14] their marked mutual divergences are good proof that they do not reflect general effects alone but also specific effects of variable nature depending on the particular probe used to construct each scale. In any case, there have been two attempts at establishing a pure solvent general scale over the last decade. Thus, in 1992, Drago[14] developed the "unified solvent polarity scale", also called the "S' scale" by using a least-squares minimization program[16] to fit a series of physico-chemical properties ($\chi$) for systems where specific interactions with the solvents were excluded to the equation $\Delta\chi = PS' + W$. In 1995, our group reported the solvent polarity-polarizability (SPP) scale, based on UV-Vis measurements of the 2-N,N-dimethyl-7-nitrofluorene/2-fluoro-7-nitrofluorene probe/homomorph pair.[15]

According to Drago,[17] specific interactions can be described as localized donor-acceptor interactions involving specific orbitals in terms of two parameters, viz. E for electrostatic interactions and C for covalent interactions; according to Kamlet and Taft,[18a] such interactions can be described in terms of hydrogen-bonding acid-base interactions. In fact, ever since Lewis unified the acidity and basicity concepts in 1923,[19] it has been a constant challenge for chemists to find a single quantifiable property of solvents that could serve as a general basicity indicator. Specially significant among the attempts at finding one are the donor number (DN) of Gutmann et al.,[20] the B(MeOD) parameter of Koppel and Palm,[21] the pure base calorimetric data of Arnet et al.[22] and parameter β of Kamlet and Taft.[18b]

## 10.3.2 CHARACTERIZATION OF A MOLECULAR ENVIRONMENT WITH THE AID OF THE PROBE/HOMOMORPH MODEL

As a rule, a good solvent probe must possess two energy states such that the energy or intensity of the transition between them will be highly sensitive to the nature of the environment. The transition concerned must thus take place between two states affected in a different manner by the molecular environment within the measurement time scale so that the transition will be strongly modified by a change in the nature of the solvent. For easier quantification, the transition should not overlap with any others of the probe, nor should its spectral profile change over the solvent range of interest. Special care should also be exercised so that the probe chosen will be subject to no structural changes dependent on the nature of the solvent; otherwise, the transition will include this perturbation, which is external to the pure solvent effect to be quantified.

All probes that meet the previous requirements are not necessarily good solvent probes, however; in fact, the sensitivity of the probe may be the result of various types of interaction with the solvent and the results difficult to generalize as a consequence. A probe suitable for determining a solvent effect such as polarity, acidity or basicity should be highly sensitive to the interaction concerned but scarcely responsive to other interactions so that any unwanted contributions will be negligible. However, constructing a pure solvent scale also entails offsetting these side effects, which, as shown later on, raises the need to use a homomorph of the probe.

The use of a spectroscopic technique (specifically, UV-Vis absorption spectrophotometry) to quantify the solvent effect provides doubtless advantages. Thus, the solute is in its electronic ground state, in thermodynamic equilibrium with its environment, so the transition is vertical and the solvation sphere remains unchanged throughout. These advantages make designing a good environmental probe for any of the previous three effects quite easy.

A suitable probe for the general solvent effect must therefore be a polar compound. Around its dipole moment, the solvent molecules will arrange themselves as effectively as possible -in thermal equilibrium- and the interaction between the probe dipole and the solvent molecules that form the cybotactic region must be strongly altered by electronic excitation, which will result in an appropriate shift in charge; the charge will then create a new dipole moment enclosed by the same cybotactic cavity as in the initial state of the transition.

The change in the dipole moment of the probe will cause a shift in the electronic transition and reflect the sensitivity of the probe to the solvent polarity and polarizability. A high sensitivity in the solvent is thus usually associated with a large change in the dipole moment by effect of the electronic transition. However, the magnitude of the spectral shift depends not only on the modulus of the dipole moment but also on the potential orientation change. In fact, inappropriate orientation changes are among the sources of contamination of polarity probes with specific effects. For simplicity, orientation changes can be reduced to the following: (a) the dipole moment for the excited state is at a small angle to that for the ground state of the probe, so the orientation change induced by the electronic excitation can

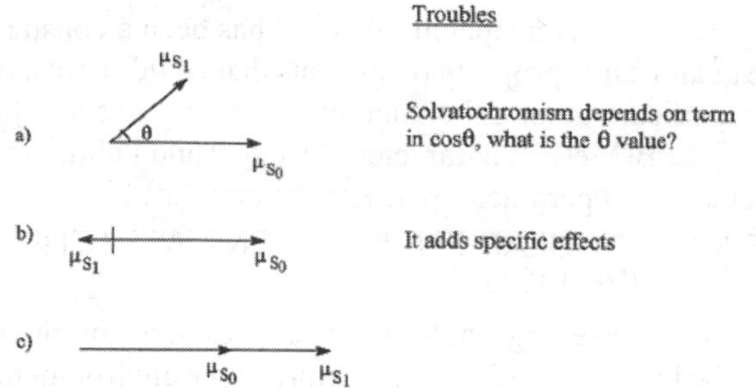


Scheme I.

be neglected; (b) the two dipole moments are at an angle near 180° to each other; and (c) the two dipole moments form an angle in between the previous two.

Any transitions where the dipole moments for the states involved in the transition are at angles markedly different from 0 or 180° should be avoided as the spectral shift includes contributions that depend on the cosine of the angle, which hinders its use in practice (see Scheme Ia). This difficulty vanishes at 0 and 180°, which are thus the preferred choices. A transition giving rise to an orientation change by 180° is usually one involving a molecule where charge is strongly localized at a given site and is discharged by delocalization upon electronic excitation; as a result, specific solvating effects in the initial state are strongly altered in the final state, so the transition concerned is strongly contaminated with specific interactions (see Scheme Ib). This situation is also inadvisable with a view to constructing a scale for the general solvent effect. The third possibility, where the transition causes no orientation change in the dipole moment, is the most suitable with a view to establishing the general effect of a solvent; the solvent box, which was the most suitable for solvating the probe, will continue to be so now with an increased dipole moment (see Scheme Ic). In summary, a suitable probe for estimating the general solvent effect must not only meet the above-described requirements but also be able to undergo a transition the dipole moment of which will increase markedly as a result while preserving its orientation.

A suitable basicity probe will be one possessing an acid group the acidity of which changes markedly with the electronic transition, so that the transition is affected mostly by solvent basicity. For its acidity to change to an appropriate extent upon excitation, the group concerned should resound with an appropriate electron acceptor and the resonance should increase during the transition by which the effect is probed. Obviously, the presence of this type of group endows the probe with a polar character; in addition, the charge transfer induced by the excitation also produces side effects that must be subtracted in order to obtain measurements contaminated with no extraneous effects such as those due to polarity changes or the acidity of the medium described above. Finding a non-polar probe with a site capable of inducing a specific orientation in the solvent appears to be a difficult task. The best choice in order to be able to subtract the above-mentioned effects appears to be an identical molecular structure exhibiting the same transition but having the acid group blocked. One interesting alternative in this context is the use of a pyrrole N-H group as probe and the corresponding N-Me derivative as homomorph.[23]

The same approach can be used to design a suitable acidity probe. The previous comments on the basicity probe also apply here; however, there is the added difficulty that the basic site used to probe solvent basicity cannot be blocked so easily. The situation is made much more complex by the fact that, for example, if the basic site used is a ketone group or a pyridine-like nitrogen, then no homomorph can be obtained by blocking the site. The solution, as shown below, involves sterically hindering the solvent approach by protecting the basic site of the probe with bulky alkyl (e.g., tert-butyl) groups at adjacent positions.[24]

One can therefore conclude that constructing an appropriate general or specific solvent scale entails finding a suitable molecular probe to preferentially assess the effect considered and a suitable homomorph to subtract any side effects extraneous to the interaction of interest. Any other type of semi-empirical evaluation will inevitably lead to scales contaminated with other solvent effects, as shown later on. Accordingly, one will hardly be able to provide a global description of solvent effects on the basis of a single-parameter scale.

### 10.3.3 SINGLE-PARAMETER SOLVENT SCALES: THE Y, G, $E_T(30)$, $P_y$, Z, $\chi_R$, $\Phi$, AND S' SCALES

Below are briefly described the most widely used single-parameter solvent scales, with special emphasis on their foundation and use.

#### 10.3.3.1 The solvent ionizing power scale or Y scale

$$(CH_3)_3CCl \xrightarrow{k_1} (CH_3)_3C^+ + Cl^-$$

Scheme II.

In 1948, Grunwald and Winstein[25] introduced the concept of "ionizing power of the solvent", Y, based on the strong influence of the solvent on the solvolysis rate of alkyl halides in general and tert-butyl chloride in particular (see Scheme II). Y is calculated from the following equation:

$$Y = \log k^{tBuCl} - \log k_0^{tBuCl} \qquad [10.3.1]$$

where $k^{tBuCl}$ and $k_0^{tBuCl}$ are the solvolysis rate constants at 25°C for tert-butyl chloride in the solvent concerned and in an 80% v/v ethanol/water mixture - the latter constant is used as reference for the process.

The strong solvent dependence of the solvolysis rate of tert-butyl chloride was examined by Grunwald and Winstein[25,26] in the light of the Brönsted equation. They found the logarithmic coefficient of activity for the reactant and transition state to vary linearly in a series of mixtures and the variation to be largely the result of changes in coefficient of activity for the reactant. By contrast, in the more poorly ionizing solvents, changes in k were found to be primarily due to changes in coefficient of activity for the transition state.

In a series of papers,[25-32] Grunwald and Winstein showed that the solvolysis rate constants for organic halides which exhibit values differing by more than 6 orders of magnitude in this parameter can generally be accurately described by the following equation:

$$\log k = mY + \log k_0 \qquad [10.3.2]$$

where k and $k_0$ are the rate constants in the solvent concerned and in the 80:20 v/v ethanol/water mixture, and coefficient m denotes the ease of solvolysis of the halide concerned relative to tert-butyl chloride. By grouping logarithms in eq. [10.3.2], one obtains

$$\log(k / k_0) = mY \qquad [10.3.3]$$

which is analogous to the Hammett equation:[33]

$$\log(k / k_0) = \rho\sigma \qquad [10.3.4]$$

There is thus correspondence between coefficient m and the Hammett reaction constant, ρ, and also between the ionizing power of the solvent, Y, and the Hammett substituent constant, σ.

In fact, the solvolysis of tert-butyl chloride is one of the cornerstones of physical organic chemistry.[34-36] Thus, some quantitative approaches to the kinetics of spontaneous reactions in various solvents -and, more interesting, solvent mixtures- are based on linear free-energy relations such as that of Grunwald and Winstein[25] or its extensions.[34,37-41] These equations allow one to interpolate or extrapolate rate constants that cannot be readily measured, and also to derive mechanistically significant information in the process.

#### 10.3.3.2 The G values of Allerhand and Schleyer

Allerhand and Schleyer[42] found that a proportionality exists between the stretching frequencies of vibrators of the type X=O (with X = C, N, P or S) and the corresponding stretching frequencies in situations involving hydrogen bonds of the X–H··B type in a variety of solvents. They thought these results to be indicative that the solvents studied interacted in a non-specific manner with both types of vibrator, in contradiction with the specific interaction-only approach advocated by Bellamy et al.[43-45] They used this information to construct a solvent polarity scale that they called "the G scale". Allerhand and Schleyer[42] used an empirical linear free-energy equation to define G:

$$G = \left(v^0 - v^s\right) / av^0 \quad [10.3.5]$$

where $v^0$ and $v^s$ are the corresponding stretching frequencies for one such vibrator in the gas phase and in solution, respectively; a is a function of the particular vibrator in a given probe and also a measure of its sensitivity to the solvent; and G is a function of the solvent alone. The scale was initially constructed from 21 solvents; zero was assigned to the gas phase and 100 to dichloromethane. Subsequently, other authors established the G values for additional solvents.[46,47]

#### 10.3.3.3 The $E_T(30)$ scale of Dimroth and Reichardt

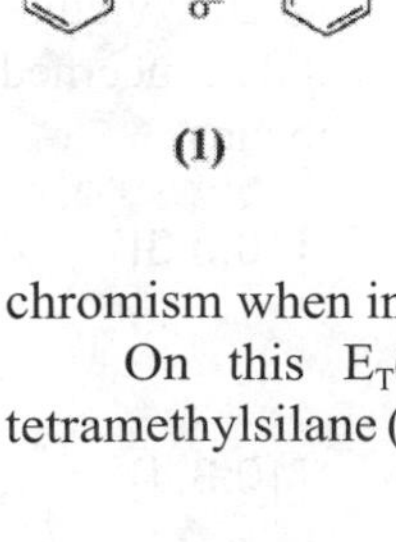

(1)

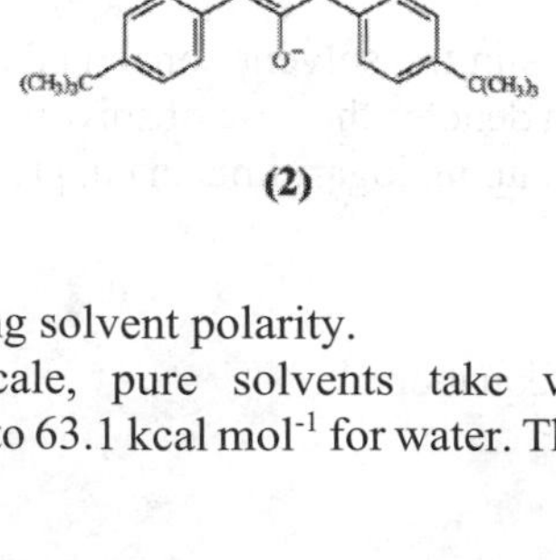


(2)

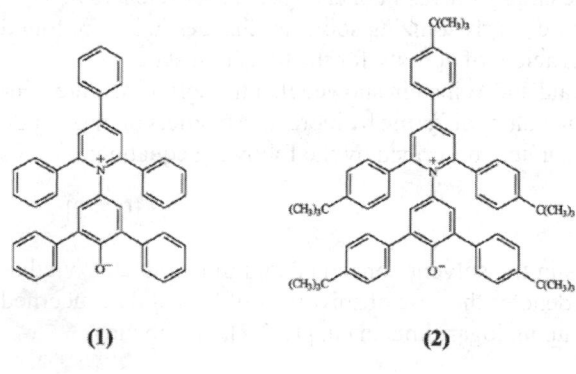

The $E_T(30)$ scale[48] is based on the extremely solvatochromic character of 2,6-diphenyl-4-(2,4,6-triphenyl-1-pyrido)phenoxide (**1**) and is defined by the position (in kcal $mol^{-1}$) of the maximum of the first absorption band for this dye, which has marked intramolecular charge-transfer connotations and gives rise to an excited electronic state that is much less dipolar than the ground state(see scheme III). This results in strong hypsochromism when increasing solvent polarity.

On this $E_T(30)$ scale, pure solvents take values from 30.7 kcal $mol^{-1}$ for tetramethylsilane (TMS) to 63.1 kcal $mol^{-1}$ for water. These data allow one to easily normal-

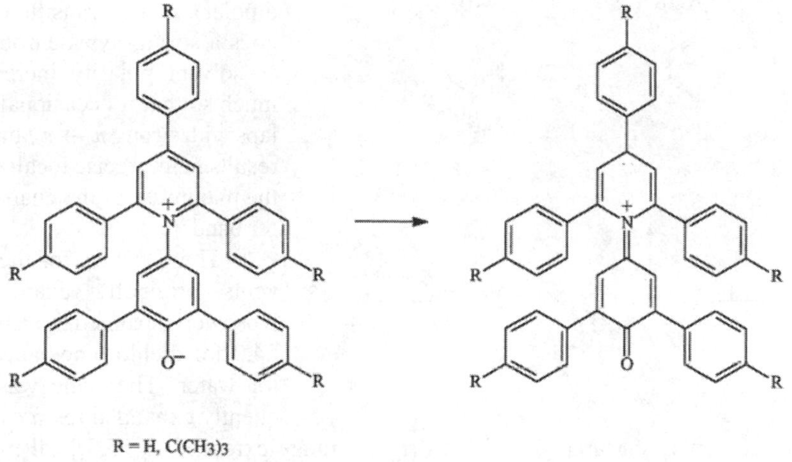


Scheme III.

ize the scale by assigning a zero value to tetramethylsilane and a unity value to water, using the following expression:

$$E_T^N = (E_T(solvent) - E_T(TMS)) / (E_T(water) - E_T(TMS)) = (E_T(solvent) - 30.7) / 32.4 \quad [10.3.6]$$

The corresponding $E_T^N$ values (between 0 and 1) allow one to rank all solvents studied. This is no doubt the most comprehensive solvent scale (it encompasses more than 300 solvents) and also the most widely used at present.[1,13] The probe (**1**) exhibits solubility problems in non-polar solvents that the authors have overcome by using a tert-butyl derivative (**2**). One serious hindrance to the use of this type of probe is its high basicity ($pK_a = 8.64$[49]), which raises problems with acid solvents.

### 10.3.3.4 The $P_y$ scale of Dong and Winnick

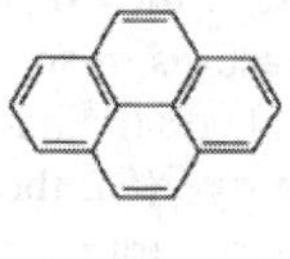

(**3**)

The $P_y$ scale[50] is based on the ratio between the intensity of components (0,0) $I_1$ and (0,2) $I_3$ of the fluorescence of monomeric pyrene (**3**) in various solvents. It was initially established from 95 solvents[50] and spans values from 0.41 for the gas phase to 1.95 for DMSO. This scale is primarily used in biochemical studies, which usually involve fluorescent probes. However, it poses problems arising largely from the difficulty of obtaining precise values of the above-mentioned intensity ratio; this has resulted in divergences among $P_y$ values determined by different laboratories.[51] One further hindrance is that the mechanism via which low-polar solvents enhance the intensity of symmetry-forbidden vibronic transitions through a reduction in local symmetry is poorly understood.[52]

### 10.3.3.5 The Z scale of Kosower

The Z scale[53,54] is based on the strong solvatochromism of the 1-ethyl-4-(methoxycarbonyl)pyridinium iodide ion-pair (**4**) and defined by the position (in kcal $mol^{-1}$) of the maximum of its first absorption band, which has marked intermolecular charge-transfer connotations according to Scheme IV (the excited state of the chromophore is much less

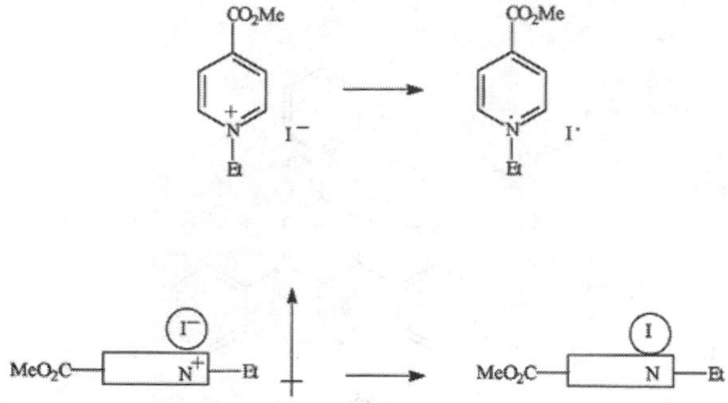


Scheme IV.

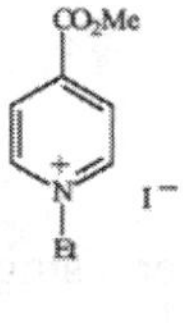


**(4)**

dipolar). This transition undergoes a strong hypsochromic shift as solvent polarity increases, so much so that it occasionally overlaps with strong $\pi \rightarrow \pi^*$ bands and results in imprecise localization of the maximum of the charge-transfer band.

The Z values for the 20 solvents originally examined by Kosower[53] spanned the range from 64.2 for dichloromethane to 94.6 for water. The scale was subsequently expanded to an overall 61 solvents by Marcus[56] and the original range extended to 55.3 kcal mol$^{-1}$ (for 2-methyltetrahydrofuran). Further expansion of this scale was precluded by the fact that high-polar solvents shift the charge-transfer band at the shortest wavelength to such an extent that it appears above the strong first $\pi \rightarrow \pi^*$ transition of the compound, thus hindering measurement; in addition, the probe (**4**) is scarcely soluble in non-polar solvents. One should also bear in mind that many solvents require using a high concentration of the probe in order to obtain a measurable charge-transfer band; as a result, the position of the band often depends on the probe concentration.

One other fact to be considered is that the interest initially aroused by this scale promoted attempts at overcoming the above-mentioned measurement problems by using correlations with other solvent-sensitive processes; as a result, many of the Z values currently in use are not actually measured values but extrapolated values derived from previously established ratios.

### 10.3.3.6 The $\chi_R$ scale of Brooker

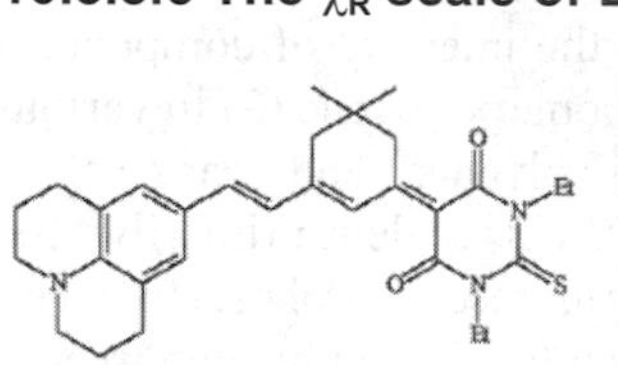


**(5)**

In 1951, Brooker[57] suggested for the first time that solvatochromic dyes could be used to obtain measures of solvent polarity. This author[58] constructed the $\chi_R$ scale on the basis of the solvatochromism of the merocyanine dye (**5**), the electronic transition of which gives rise to a charge-transfer from the amine nitrogen to a carboxamide group at the other end of the molecule. Hence, the excited status is more dipolar than the ground state, and the resulting band is shifted bathochromically as solvent polarity increases. $\chi_R$ values reflect the position of the maximum of the first band for the chromophore in kcal mol$^{-1}$.

The original scale encompassed 58 solvents spanning $\chi_R$ values from 33.6 kcal mol$^{-1}$ for m-cresol to 50.9 kcal mol$^{-1}$ for n-heptane.

### 10.3.3.7 The $\Phi$ scale of Dubois and Bienvenüe

Dubois and Bienvenüe[59] developed the $\Phi$ polarity solvent scale on the basis of the position of the $n \rightarrow \pi^*$ transition for eight selected aliphatic ketones that were studied in 23 solvents, using n-hexane as reference and the following equation for calculation:

$$\Delta v_H^S = v^S - v^H = \Phi\left(v^H - 32637\right) - 174 \qquad [10.3.7]$$

where $v^S$ is the absorption wavenumber in solvent S and $v^H$ in n-hexane (the reference solvent). The absorption maxima of these ketones depend both on the solvent and on their own structure.

The Φ values spanned by the 23 solvents studied range from -0.01 for carbon tetrachloride to 0.65 for formic acid. The values for DMSO and water are 0.115 and 0.545, respectively.

#### 10.3.3.8 The S' scale of Drago

Drago[60] developed a "universal polarity scale" (the S' scale) from more than three hundred spectral data (electronic transitions, $^{19}F$ and $^{15}N$ chemical shifts and RSE coupling constants) for 30 solutes in 31 non-protic solvents from cyclohexane to propylene carbonate. He used the equation

$$\Delta\chi = S'P + W \qquad [10.3.8]$$

where $\Delta\chi$ is the measured physico-chemical property, S' the solvent polarity, P the solvating susceptibility of the solute and W the value of $\Delta\chi$ at S' = 0. Drago assigned an arbitrary value of 3.00 to DMSO in order to anchor the scale.

The S' scale was constructed from carefully selected data. Thus, it excluded data for (a) all systems involving any contribution from donor-acceptor specific interactions (donor molecules where only measured in donor solvents and data for π solutes were excluded); (b) concentrated solutions of polar molecules in non-polar solvents (which might result in clustering); and (c) polar solvents occurring as rotamers (each rotamer would be solvated in a different way).

In order to extend the S' scale to acid solvents, Drago[61] devised an experiment involving separating general interactions from specific interactions using the $E_T(30)$ scale. First, the probe [$E_T(30)$] was dissolved in a non-coordinating solvent of weakly basic character -so that it would not compete with the solute for the specific interaction- and slightly polar nature -so that the probe would not cluster at low concentrations of the acceptor solvent-; then, the acceptor solvent was added and the variation of the maximum on the absorption band for the $E_T(30)$ probe was plotted as a function of the solvent concentration. By extrapolation to dilute solutions, the non-specific solvation component for the solvent was estimated. An interesting discussion of the solvent scales developed by Drago can be found in his book.[62]

### 10.3.4 SOLVENT POLARITY: THE SPP SCALE[15]

As noted earlier, a suitable probe for assessing solvent polarity should meet various requirements including the following: (a) the modulus of its dipole moment should increase markedly but its orientation remain unaltered in response to electronic excitation: (b) it should undergo no structural changes by effect of electronic excitation or the nature of the solvent; (c) its basicity or acidity should change as little as possible upon electronic excitation so that any changes will be negligible compared to those caused by polarity; (d) the spectral envelope of the electronic transition band used should not change with the nature of the solvent; and (e) its molecular structure should facilitate the construction of a homomorph allowing

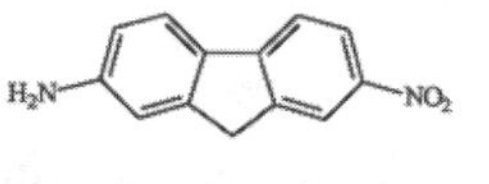


**(6)**

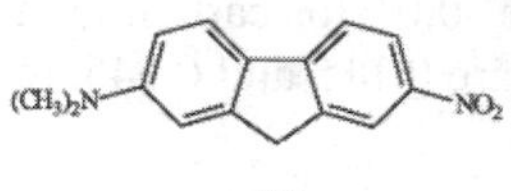


**(7)**

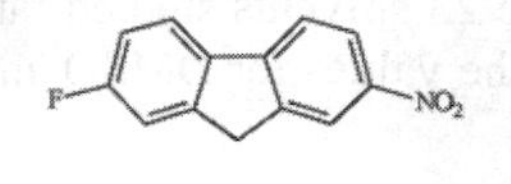


**(8)**

one to offset spurious contributions to the measurements arising from the causes cited in (b) to (d) above.

A comprehensive analysis of the literature on the choice of probes for constructing polarity scales led our group to consider the molecular structure of 2-amino-7-nitrofluorene (ANF) (**6**) for this purpose. This compound had previously been used by Lippert[63] to define his well-known equation, which relates the Stokes shift of the chromophore with the change in its dipole moment on passing from the ground state to the excited state. He concluded that, in the first excited state, the dipole moment increased by 18 D on the 5.8 D value in the ground state. This data, together with the dipole moment for the first excited electronic state (23 D), which was obtained by Czekalla et al.[64] from electric dichroism measurements, allow one to conclude that the direction of the dipole moment changes very little upon electronic excitation of this chromophore. Baliah and Pillay[65] analyzed the dipole moments of a series of fluorene derivatives at positions 2 and 7, and concluded that both positions were strongly resonant and hence an electron-releasing substituent at one and an electron-withdrawing substituent at the other would adopt coplanar positions relative to the fluorene skeleton. In summary, a change by 18 D in dipole moment of a system of these structural features reflects a substantial charge transfer from the donor group at position 2 to the acceptor group at 7 upon electronic excitation.

Although ANF seemingly fulfills requirements (a) and (b) above, it appears not to meet requirement (c) (i.e., that its acidity and basicity should not change upon electronic excitation). In fact, electronic excitation will induce a charge transfer from the amino group, so the protons in it will increase in acidity and the transition will be contaminated with specific contributions arising from solvent basicity. In order to avoid this contribution, one may in principle replace the amino group with a dimethylamino group (DMANF, 7), which will exhibit appropriate charge transfer with no significant change in its negligible acidity. The increase in basicity of the nitro group upon electronic excitation (a result of charge transfer from the N,N-dimethyl group), should result in little contamination as this group is scarcely basic and its basicity is bound to hardly change with the amount of charge transferred from the N,N-dimethylamino group at position 2 to the fluorene structure.

The analysis of the absorption spectra for DMANF in a broad range of solvents suggests that this probe possesses several interesting spectroscopic properties as regards its first absorption band, which is used to assess the polar properties of solvents. Thus,[15] (a) its first absorption band is well resolved from the other electronic bands (an increase in solvent polarity results in no overlap with the other electronic bands in the UV-Vis spectrum for this probe); (b) the position of this band is highly sensitive to solvent polarity and is bathochromically shifted with increase in it (the bathochromic shift in the absorption maximum between perfluorohexane and DMSO is 4130 $cm^{-1}$); (c) in less polar solvents, where the band appears at lower wavelengths, it is observed at ca. 376 nm (i.e., shifted to the visible region to an extent ensuring that no problems derived from the cut-off of the solvent concerned will be encountered); and (d) its first band becomes structured in non-polar solvents (see Figure 1 in ref. 15).

In summary, DMANF is a firm candidate for use as a solvent dipolarity/polarizability probe since its absorption is extremely sensitive to changes in the nature of the solvent, largely as a result of the marked increase in its dipole moment on passing from the electronic ground state to the first excited state. In addition, the change does not affect the dipole moment direction, which is of great interest if the compound is to be used as a probe. Because it possesses a large, rigid aromatic structure, DMANF is highly polarizable; consequently, its first electronic transition occurs at energies where no appreciable interferences with the cut-offs of ordinary solvents are to be expected.

However, the change in structure of the first absorption band for DMANF in passing from non-polar solvents to polar solvents and the potential contaminating effect of solvent acidity on the position of this band entails introducing a homomorph for the probe in order to offset the detrimental effects of these factors on the estimation of solvent polarities.

The homomorph to be used should essentially possess the same structure as the probe, viz. a nitro group at position 7 ensuring the occurrence of the same type of interaction with the solvents and an electron-releasing group at position 2 ensuring similar, through weaker, interactions with the nitro function at 7 in order to obtain a lower dipole moment relative to DMANF). The most suitable replacement for the $-NMe_2$ function in this context is a fluorine atom, which poses no structural problems and is inert to solvents. The homomorph chosen was thus 2-fluoro-7-nitrofluorene (FNF) (**9**). The analysis of the absorption spectra for FNF in a broad range of solvents clearly revealed that its first absorption band behaves identically with that for DMANF (its structure changes in passing from non-polar solvents to polar ones). However, the bathochromic shift in this band with increase in solvent polarity is much smaller than that in DMANF (see Figure 1 in ref. 15).

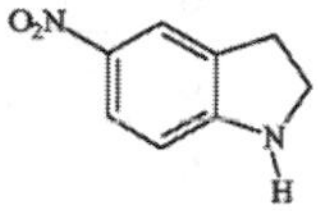


**(9)**

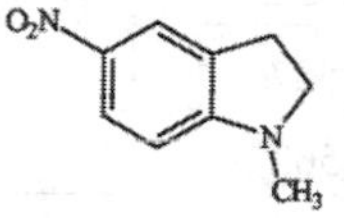


**(10)**

Obviously, the difference between the solvatochromism of DMANF and FNF will cancel many of the spurious effects involved in measurements of solvent polarity. Because the envelopes of the first absorption bands for FNF and DMANF are identical (see Figure 1 in ref. 15), one of the most common sources of error in polarity scales is thus avoided. The polarity of a solvent on the SPP scale is given by the difference between the solvatochromism of the probe DMANF and its homomorph FNF [$\Delta\nu(\text{solvent}) = \nu_{FNF} - \nu_{DMANF}$] and can be evaluated on a fixed scale from 0 for the gas phase (i.e., the absence of solvent) to 1 for DMSO, using the following equation:

$$SPP(solvent) = [\Delta v(solvent) - \Delta v(gas)] / [\Delta v(DMSO) - \Delta v(gas)] \quad [10.3.9]$$

Table 10.3.1 gives the SPP values for a broad range of solvents, ranked in increasing order of polarity. Data were all obtained from measurements made at our laboratory and have largely been reported elsewhere[15,66-70] -some, however, are published here for the first time.

**Table 10.3.1. The property parameters of solvents: Polarity/Polarizability SPP, Basicity SB, Acidity SA**

| Solvents | SPP | SB | SA |
|---|---|---|---|
| gas phase | 0 | 0 | 0[a] |
| perfluoro-n-hexane | 0.214 | 0.057 | 0[a] |
| 2-methylbutane | 0.479 | 0.053 | 0[a] |
| petroleum ether | 0.493 | 0.043 | 0[a] |
| n-pentane | 0.507 | 0.073 | 0[a] |
| n-hexane | 0.519 | 0.056 | 0[a] |
| n-heptane | 0.526 | 0.083 | 0[a] |
| isooctane | 0.533 | 0.044 | 0[a] |
| cyclopentane | 0.535 | 0.063 | 0[a] |
| n-octane | 0.542 | 0.079 | 0[a] |
| ethylcyclohexane | 0.548 | 0.074 | 0[a] |
| n-nonane | 0.552 | 0.053 | 0[a] |
| cyclohexane | 0.557 | 0.073 | 0[a] |
| n-decane | 0.562 | 0.066 | 0[a] |
| n-undecane | 0.563 | 0.080 | 0[a] |
| methylcyclohexane | 0.563 | 0.078 | 0[a] |
| butylcyclohexane | 0.570 | 0.073 | 0[a] |
| propylcyclohexane | 0.571 | 0.074 | 0[a] |
| n-dodecane | 0.571 | 0.086 | 0[a] |
| decahydronaphthalene | 0.574 | 0.056 | 0[a] |
| mesitylene | 0.576 | 0.190 | 0[a] |
| n-pentadecane | 0.578 | 0.068 | 0[a] |
| n-hexadecane | 0.578 | 0.086 | 0[a] |
| cycloheptane | 0.582 | 0.069 | 0[a] |
| tert-butylcyclohexane | 0.585 | 0.074 | 0[a] |
| cyclooctane | 0.590 | 0.077 | 0[a] |
| 1,2,3,5-tetramethylbenzene | 0.592 | 0.186 | 0[a] |
| cis-decahydronaphthalene | 0.601 | 0.056 | 0[a] |
| tripropylamine | 0.612 | 0.844 | 0[s] |
| m-xylene | 0.616 | 0.162 | 0[a] |
| triethylamine | 0.617 | 0.885 | 0[s] |

| Solvents | SPP | SB | SA |
|---|---|---|---|
| p-xylene | 0.617 | 0.160 | 0[a] |
| 1-methylpiperidine | 0.622 | 0.836 | 0[s] |
| tributylamine | 0.624 | 0.854 | 0[s] |
| 1,4-dimethylpiperazine | 0.627 | 0.832 | 0[s] |
| hexafluorobenzene | 0.629 | 0.119 | 0[a] |
| trimethylacetic acid | 0.630 | 0.130 | 0.471 |
| dibutylamine | 0.630 | 0.991 | 0[s] |
| di-n-hexyl ether | 0.630 | 0.618 | 0[a] |
| 1-methylpyrrolidine | 0.631 | 0.918 | 0[s] |
| tetrachloromethane | 0.632 | 0.044 | 0[a] |
| di-n-pentyl ether | 0.636 | 0.629 | 0[a] |
| butylbenzene | 0.639 | 0.149 | 0[a] |
| o-xylene | 0.641 | 0.157 | 0[a] |
| isobutyric acid | 0.643 | 0.281 | 0.515 |
| isovaleric acid | 0.647 | 0.405 | 0.538 |
| ethylbenzene | 0.650 | 0.138 | 0[a] |
| di-n-butyl ether | 0.652 | 0.637 | 0[a] |
| hexanoic acid | 0.656 | 0.304 | 0.456 |
| toluene | 0.655 | 0.128 | 0[a] |
| propylbenzene | 0.655 | 0.144 | 0[a] |
| tert-butylbenzene | 0.657 | 0.171 | 0[a] |
| N-methylbutylamine | 0.661 | 0.960 | 0[s] |
| heptanoic acid | 0.662 | 0.328 | 0.445 |
| di-isopropyl ether | 0.663 | 0.657 | 0[a] |
| N-methylcyclohexylamine | 0.664 | 0.925 | 0[s] |
| N,N-dimethylcyclohexylamine | 0.667 | 0.998 | 0[s] |
| benzene | 0.667 | 0.124 | 0[a] |
| 1,2,3,4-tetrahydronaphthlene | 0.668 | 0.180 | 0[a] |
| ethylenediamine | 0.674 | 0.843 | 0.047 |
| di-n-propyl ether | 0.676 | 0.666 | 0[a] |
| propionic acid | 0.690 | 0.377 | 0.608 |
| tert-butyl methyl ether | 0.687 | 0.567 | 0[a] |
| diethyl ether | 0.694 | 0.562 | 0[a] |

| Solvents | SPP | SB | SA |
|---|---|---|---|
| 2-methylbutyric acid | 0.695 | 0.250 | 0.439 |
| butyl methyl ether | 0.695 | 0.505 | 0[a] |
| pentafluoropyridine | 0.697 | 0.144 | 0[a] |
| 1,4-dioxane | 0.701 | 0.444 | 0.0 |
| 2-methyltetrahydrofuran | 0.717 | 0.584 | 0.0 |
| 1-methylnaphthalene | 0.726 | 0.156 | 0[a] |
| butylamine | 0.730 | 0.944 | 0.0 |
| ethoxybenzene | 0.739 | 0.295 | 0[a] |
| piperidine | 0.740 | 0.933 | 0.0 |
| 1-undecanol | 0.748 | 0.909 | 0.257 |
| isoamyl acetate | 0.752 | 0.481 | 0.0 |
| 1-decanol | 0.765 | 0.912 | 0.259 |
| fluorobenzene | 0.769 | 0.113 | 0[a] |
| 1-nonanol | 0.770 | 0.906 | 0.270 |
| tetrahydropyran | 0.778 | 0.591 | 0.0 |
| acetic acid | 0.781 | 0.390 | 0.689 |
| n-propyl acetate | 0.782 | 0.548 | 0.0 |
| n-butyl acetate | 0.784 | 0.525 | 0.0 |
| methyl acetate | 0.785 | 0.527 | 0.0 |
| 1-octanol | 0.785 | 0.923 | 0.299 |
| chloroform | 0.786 | 0.071 | 0.047 |
| 2-octanol | 0.786 | 0.963 | 0.088 |
| cyclohexylamine | 0.787 | 0.959 | 0.0 |
| trimethyl orthoformate | 0.787 | 0.528 | 0.0 |
| 1,2-dimethoxyethane | 0.788 | 0.636 | 0.0 |
| pyrrolidine | 0.794 | 0.990 | 0.0 |
| ethyl acetate | 0.795 | 0.542 | 0.0 |
| 1-heptanol | 0.795 | 0.912 | 0.302 |
| N,N,-dimethylaniline | 0.797 | 0.308 | 0.0 |
| methyl formate | 0.804 | 0.422 | 0.0 |
| 2-buthoxyethanol | 0.807 | 0.714 | 0.292 |
| 1-hexanol | 0.810 | 0.879 | 0.315 |
| 3-methyl-1-butanol | 0.814 | 0.858 | 0.315 |

| Solvents | SPP | SB | SA |
|---|---|---|---|
| propyl formate | 0.815 | 0.549 | 0.0 |
| 1-pentanol | 0.817 | 0.860 | 0.319 |
| dibenzyl ether | 0.819 | 0.330 | 0[a] |
| methoxybenzene | 0.823 | 0.299 | 0.084 |
| chlorobenzene | 0.824 | 0.182 | 0[a] |
| bromobenzene | 0.824 | 0.191 | 0[a] |
| cyclooctanol | 0.827 | 0.919 | 0.137 |
| 2,6-dimethylpyridine | 0.829 | 0.708 | 0.0 |
| 2-methyl-2-propanol | 0.829 | 0.928 | 0.145 |
| 2-pentanol | 0.830 | 0.916 | 0.204 |
| 2-methyl-2-butanol | 0.831 | 0.941 | 0.096 |
| 2-methyl-1-propanol | 0.832 | 0.828 | 0.311 |
| formamide | 0.833 | 0.414 | 0.674 |
| 2,4,6-trimethylpyridine | 0.833 | 0.748 | 0.0 |
| 3-hexanol | 0.833 | 0.979 | 0.140 |
| iodobenzene | 0.835 | 0.158 | 0[a] |
| ethyl benzoate | 0.835 | 0.417 | 0.0 |
| dibutyl oxalate | 0.835 | 0.549 | 0.0 |
| methyl benzoate | 0.836 | 0.378 | 0.0 |
| 1-chlorobutane | 0.837 | 0.138 | 0[a] |
| 1-butanol | 0.837 | 0.809 | 0.341 |
| 2-methyl-1-butanol | 0.838 | 0.900 | 0.289 |
| tetrahydrofuran | 0.838 | 0.591 | 0.0 |
| pyrrole | 0.838 | 0.179 | 0.387 |
| cycloheptanol | 0.841 | 0.911 | 0.183 |
| 2-butanol | 0.842 | 0.888 | 0.221 |
| 1,3-dioxolane | 0.843 | 0.398 | 0.0 |
| 3-methyl-2-butanol | 0.843 | 0.893 | 0.196 |
| 2-hexanol | 0.847 | 0.966 | 0.140 |
| cyclohexanol | 0.847 | 0.854 | 0.258 |
| 1-propanol | 0.847 | 0.727 | 0.367 |
| 2-propanol | 0.848 | 0.762 | 0.283 |
| ethanol | 0.853 | 0.658 | 0.400 |

| Solvents | SPP | SB | SA |
|---|---|---|---|
| 1,2-dimethoxybenzene | 0.854 | 0.340 | 0.0 |
| 2-methoxyethyl ether | 0.855 | 0.623 | 0.0 |
| methanol | 0.857 | 0.545 | 0.605 |
| methyl methoxyacetate | 0.858 | 0.484 | 0.0 |
| 3-pentanol | 0.863 | 0.950 | 0.100 |
| cyclopentanol | 0.865 | 0.836 | 0.258 |
| cyclohexanone | 0.874 | 0.482 | 0.0 |
| propionitrile | 0.875 | 0.365 | 0.030 |
| allyl alcohol | 0.875 | 0.585 | 0.415 |
| propiophenone | 0.875 | 0.382 | 0[a] |
| triacetin | 0.875 | 0.416 | 0.023 |
| dichloromethane | 0.876 | 0.178 | 0.040 |
| 2-methylpyridine | 0.880 | 0.629 | 0.0 |
| 2-butanone | 0.881 | 0.520 | 0.0 |
| acetone | 0.881 | 0.475 | 0.0 |
| 2-methoxyethanol | 0.882 | 0.560 | 0.355 |
| 3-pentanone | 0.883 | 0.557 | 0.0 |
| benzyl alcohol | 0.886 | 0.461 | 0.409 |
| 4-methyl-2-pentanone | 0.887 | 0.540 | 0.0 |
| trimethyl phosphate | 0.889 | 0.522 | 0.0 |
| 1-methylpyrrole | 0.890 | 0.244 | 0.0 |
| 1,2-dichloroethane | 0.890 | 0.126 | 0.030 |
| 2-phenylethanol | 0.890 | 0.523 | 0.376 |
| 2-chloroethanol | 0.893 | 0.377 | 0.563 |
| nitroethane | 0.894 | 0.234 | 0.0 |
| acetonitrile | 0.895 | 0.286 | 0.044 |
| chloroacetonitrile | 0.896 | 0.184 | 0.445 |
| N-methylacetamide | 0.897 | 0.735 | 0.328 |
| 1,2-butanediol | 0.899 | 0.668 | 0.466 |
| valeronitrile | 0.900 | 0.408 | 0.0 |
| N-methylaniline | 0.902 | 0.212 | 0.073 |
| 2,3-butanediol | 0.904 | 0.652 | 0.461 |
| acetophenone | 0.904 | 0.365 | 0.044 |

| Solvents | SPP | SB | SA |
|---|---|---|---|
| nitromethane | 0.907 | 0.236 | 0.078 |
| cyclopentanone | 0.908 | 0.465 | 0.0 |
| triethyl phosphate | 0.908 | 0.614 | 0.0 |
| 1,2-dichlorobenzene | 0.911 | 0.144 | 0.033 |
| tetrahydrothiophene | 0.912 | 0.436 | 0[a] |
| 2,2,2-trifluoroethanol | 0.912 | 0.107 | 0.893 |
| butyronitrile | 0.915 | 0.384 | 0.0 |
| N-methylformamide | 0.920 | 0.590 | 0.444 |
| isobutyronitrile | 0.920 | 0.430 | 0.0 |
| pyridine | 0.922 | 0.581 | 0.033 |
| 2-methyl-1,3-propanediol | 0.924 | 0.615 | 0.451 |
| 1,2-propanediol | 0.926 | 0.598 | 0.475 |
| propylen carbonate | 0.930 | 0.341 | 0.106 |
| N,N-diethylacetamide | 0.930 | 0.660 | 0.0 |
| hexamethylphosphoric acid triamide | 0.932 | 0.813 | 0[s] |
| 1,2-ethanediol | 0.932 | 0.534 | 0.565 |
| N,N-diethylformamide | 0.939 | 0.614 | 0.0 |
| 1,3-butanediol | 0.944 | 0.610 | 0.424 |
| 1,2,3-propanetriol | 0.948 | 0.309 | 0.618 |
| 1-methylimidazole | 0.950 | 0.668 | 0.069 |
| 1,4-butanediol | 0.950 | 0.598 | 0.424 |
| 1,3-propanediol | 0.951 | 0.514 | 0.486 |
| tetramethylurea | 0.952 | 0.624 | 0.0 |
| N,N-dimethylformamide | 0.954 | 0.613 | 0.031 |
| 2-pyrrolidinone | 0.956 | 0.597 | 0.347 |
| benzonitrile | 0.960 | 0.281 | 0.047 |
| 2,2,2-trichloroethanol | 0.968 | 0.186 | 0.588 |
| nitrobenzene | 0.968 | 0.240 | 0.056 |
| 1-methyl-2-pyrrolidinone | 0.970 | 0.613 | 0.024 |
| N,N-dimethylacetamide | 0.970 | 0.650 | 0.028 |
| water | 0.962 | 0.025 | 1.062 |
| γ-butyrolactone | 0.987 | 0.399 | 0.057 |
| dimethyl sulfoxide | 1.000 | 0.647 | 0.072 |

| Solvents | SPP | SB | SA |
|---|---|---|---|
| sulfolane | 1.003 | 0.365 | 0.052 |
| m-cresol | 1.000 | 0.192 | 0.697 |
| 1,1,1,3,3,3-hexafluoro-2-propanol | 1.014 | 0.014 | 1.00 |
| α α α-trifluoro-m-cresol | 1.085 | 0.051 | 0.763 |

[a]Assumed value because it is considered non acid solvent.
[a]Assumed value because the first band of the TBSB spectra exhibits vibronic structure.

A brief analysis of these SPP data allows one to draw several interesting conclusions from structural effects on solvent polarity, namely:

- Cyclohexane is used as the non-polar reference in many scales. In the SPP scale, the polarity gap between cyclohexane and the gas phase (0.557 SPP units) is as wide as that between cyclohexane and the highest polarity (0.443 SPP units).
- Alkanes span a wide range of SPP values (e.g., 0.214 for perfluoro-n-hexane, 0.479 for 2-methylbutane and 0.601 for decalin).
- Unsaturation increases polarity in alcohols. Thus, the SPP values for n-propanol, allyl alcohol and propargyl alcohol are 0.847, 0.875 and 0.915, respectively.

### 10.3.5 SOLVENT BASICITY: THE SB SCALE[72]

As stated above, for a probe of solvent basicity to be usable in UV-Vis spectroscopy, it should meet a series of requirements. One is that it should be acidic enough in its electronic ground state to allow characterization of the basicity of its environment. In addition, its acidity should increase upon electronic excitation such that its electronic transitions will be sensitive to the basicity of the medium. This behavior will result in a bathochromic shift in the absorption band the magnitude of which will increase with increasing basicity of the environment. The probe should also be free of potential conformational changes that might influence the electronic transition to be evaluated. Finally, its molecular structure should be readily converted into a homomorph lacking the acid site without any side effects that might affect the resulting solvent basicity.

5-Nitroindoline (NI, **9**) possesses the above-described electronic and structural features. It is an N-H acid with a single acid site borne by a donor group whose free rotation is hindered by an ethylene bridge on the ring. However, if charge transfer endows the compound with appropriate acid properties that increase with electronic excitation, the basicity of the acceptor (nitro) group will also increase and the polarity of the compound will be altered as a result. Both effects will influence the electronic transition of the probe that is to be used to evaluate basicity.

Replacement of the acid proton (N-H) by a methyl group in this molecular structure has the same side effects; as a result, the compound will be similarly sensitive to the polarity and acidity of the medium, but not to its basicity owing to the absence of an acid site. Consequently, 1-methyl-5-nitroindoline (MNI, **10**) possesses the required properties for use as a homomorph of 5-nitroindoline in order to construct our solvent basicity scale (SB). The suitability of this probe/homomorph couple is consistent with theoretical MP2/6-31G** data. Thus, both the probe and its homomorph exhibit the same sensitivity to solvent polarity/polarizability because of their similar dipole moments ($\mu_{NI}$ = 7.13, $\mu_{MNI}$ = 7.31 D) and polarizabilities (20.38 and 22.78 $\alpha/J^{-1}C^2m^2$ for NI and MNI, respectively). Both also have the same sensitivity to solvent acidity as their surface electrostatic potential minima are virtually coincident [$V_{S,min}$(NI) = -47.49 kcal $mol^{-1}$, $V_{S,min}$(MNI) = -47.58 kcal $mol^{-1}$]; according to Politzer et al.,[71] this means that they exhibit the same hydrogen-bonding basicity. In addition, the electrostatic potential surface in both molecules suggests that sole electrophilic sites are those on the oxygen atoms of the nitro group.

Obviously, the difference in solvatochromism between NI and MNI [$\Delta\nu(\text{solvent}) = \nu_{NI} - \nu_{MNI}$] will cancel many spurious effects accompanying the basicity effect of the solvent. In addition, NI and MNI possess several advantageous spectral features; thus, they exhibit a sharp first absorption band that overlaps with no higher-energy band in any solvent, and both have the same spectral envelope in such a band (see Figure 1 in ref. 72), which facilitates comparison between the two spectra and the precise establishment of the basicity parameter, SB. The basicity of a solvent (SB) on a scale encompassing values between zero for the gas phase -the absence of solvent- and unity for tetramethylguanidine (TMG) can be directly obtained from the following equation:

$$SB(\text{solvent}) = [\Delta\nu(\text{solvent}) - \Delta\nu(\text{gas})] / [\Delta\nu(\text{TMG}) - \Delta\nu(\text{gas})] \quad [10.3.10]$$

Table 10.3.1 lists the SB values for a wide range of solvents. All data were obtained from measurements made in our laboratory; most have been reported elsewhere[70,72] but some are published here for the first time.

A brief analysis of these SB data allows one to draw several interesting conclusions as regards structural effects on solvent basicity. Thus:

- Appropriate substitution in compound families such as amines and alcohols allows the entire range of the solvent basicity scale to be spanned with substances from such families. Thus, perfluorotriethylamine can be considered non-basic (SB=0.082), whereas N,N-dimethylcyclohexylamine is at the top of the scale (SB=0.998). Similarly, hexafluoro-2-propanol is non-basic (SB=0.014), whereas 2-octanol is very near the top (SB=0.963).
- The basicity of n-alkanols increases significantly with increase in chain length and levels off beyond octanol.
- Cyclization hardly influences solvent basicity. Thus, there is little difference in basicity between n-pentane (SB=0.073) and cyclopentane (SB=0.063) or between n-pentanol (SB=0.869) and cyclopentanol (SB=0.836).
- Aromatization decreases solvent basicity by a factor of 3.5-5.5, as illustrated by the following couples: pyrrolidine/pyrrole (0.99/0.18), N-methylpyrrolidine/N-ethylpyrrole (0.92/0.22), tetrahydrofuran/furan (0.59/0.11), 2-methyltetrahydrofuran/2-methylfuran (0.56/0.16), cyclohexylamine/aniline (0.96/0.26), N-methylcyclohexylamine/N-methylaniline (0.92/0.21), N,N-dimethylcyclohexylamine/N,N-dimethylaniline (0.99/0.30) and a structurally less similar couple such as piperidine/pyridine (0.93/0.58).

The potential family-dependence of our SB scale was examined and discarded elsewhere.[73]

### 10.3.6 SOLVENT ACIDITY: THE SA SCALE[74]

As stated above, a probe for solvent acidity should possess a number of features to be usable in UV-Vis spectroscopy. One is that it should be basic enough in its electronic ground state

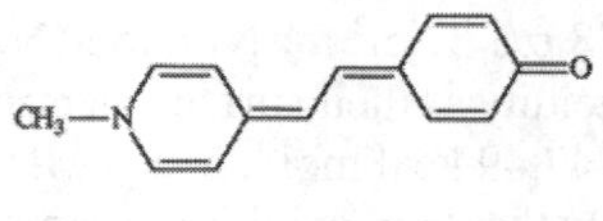

**(11)**

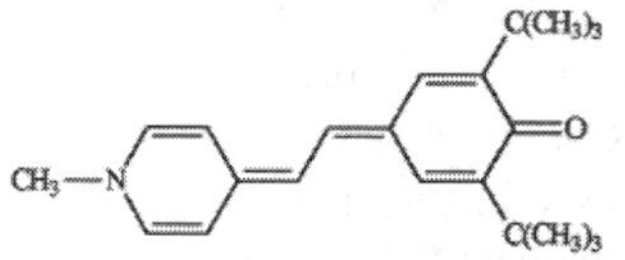

**(12)**

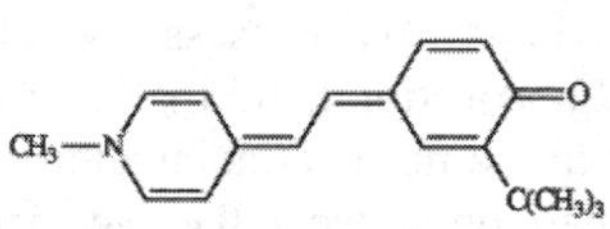

**(13)**

to be able to characterize the acidity of its environment. In addition, its basicity should increase upon electronic excitation such that its electronic transitions will be sensitive to the acidity of the medium. This behavior will result in a bathochromic shift in the absorption band the magnitude of which will increase with increasing acidity of the environment. The probe should also be free of potential conformational changes that might influence the electronic transition to be evaluated. Finally, its molecular structure should be easily converted into a homomorph lacking the basic site without any side effects potentially affecting the resulting solvent acidity.

Our group has shown[24,75] that the extremely strong negative solvatochromism of the chromophore stilbazolium betaine dye (**11**), about 6500 $cm^{-1}$, is not a result of a change in the non-specific effect of the solvent but rather of change in acidity. This was confirmed[75] by a study of the solvatochromic effect of a derivative of (**11**), o,o'-di-tert-butylstilbazolium betaine dye (DTBSB) (**12**) in the same series of solvents. In this compound, the basic site in the betaine dye (its oxygen atom) is protected on both sides by bulky tert-butyl groups; the specific effect of the solvent is hindered and the compound exhibits only a small solvatochromic effect that can be ascribed to non-specific solvent effects.[75]

Since the carbonyl group of stilbazolium betaine possesses two lone pairs in the plane of the quinoid ring, one can have two (**11**), one (TBSB, **13**) or no channels (DTBSB) to approach hydrogen bond-donor solvents, depending on the number of o-tert-butyl groups present in the molecular structure of the (**11**) derivative concerned. Our group[24] has also shown that, for at least 20 alkanols, the wavenumber difference between the maximum of the first absorption bands for (**11**) and TBSB, and for TBSB and DTBSB, is virtually identical.

These results show that an oxygen lone pair in (**11**) is basic enough, as a result of hydrogen bonding, for (**11**) to be used as an acidity probe in UV-Vis spectroscopy; thus, the absorption maximum for TBSB shifts by about 1000 $cm^{-1}$ from 1-decanol to ethanol,[24] whereas that for DTBSB shifts by only 300 $cm^{-1}$. This sensitivity to acidity, and the structural likeness of the probe and its homomorph -which must endow them with a similar sensitivity to solvent basicity and dipolarity/polarizability-, suggest that the two compounds make an appropriate probe-homomorph couple for developing a pure solvent acidity scale. For this purpose, the first visible absorption band for these compounds exhibits quite an interesting spectroscopic behavior. Thus, in a non-HBD solvent (i.e., one that cannot interact directly with the oxygen lone pair), the band is structured; on the other hand, the band loses its structure upon interaction with an HBD solvent.[75]

The SA acidity scale was established by comparing the solvatochromism of the probes TBSB and DTBSB using the method of Kamlet and Taft.[18a] In this method, the solvatochromism of DTBSB in a solvent is used as the reference for zero acidity. Consequently, non-acidic solvents obey the equation

$$\nu_{TBSB} = 1.409\nu_{DTBSB} - 6288.7 \qquad [10.3.11]$$

with n = 50, r = 0.961 and sd = 43.34 $cm^{-1}$.

Based on eq. [10.3.11] for non-acidic solvents, the SA value for a given solvent can be obtained from the following expression, where a value of 0.4 is assigned to ethanol, the acid solvatochromic behavior of which is exhibited at 1299.8 $cm^{-1}$:

$$SA = [[\nu_{TBSB} - (1.409\nu_{DTBSB} - 6288.7)] / 1299.8]0.4 \qquad [10.3.12]$$

Table 10.3.1 gives the SA values for a wide variety of C-H, N-H and O-H acid solvents. Data were all obtained from measurements made in our laboratory.[74] The homomorph was insoluble in some solvents, which therefore could not be measured, so they where assumed not to interact specifically and assigned a zero SA value if the probe exhibited a structured spectrum in them.

Evaluating the acidity of weakly acidic solvents entails using a probe basic enough to afford measurement of such an acidity; as a result, the probe is usually protonated by strongly acidic solvents and useless for the intended purpose. Stilbazolium betaines are subject to this constraint as they have pK values of about 10 -(**11**) and DTBSB have a $pK_a$ of 8.57[76] and ca. 10,[77] respectively, which makes them unsuitable for the evaluation of solvents more acidic than methanol. This forced us to find a suitable probe with a view to expanding the SA scale to more acidic solvents. This problem is not exclusive to our scale; in fact, it affects all acidity scales, which usually provide little information about strongly acidic solvents.

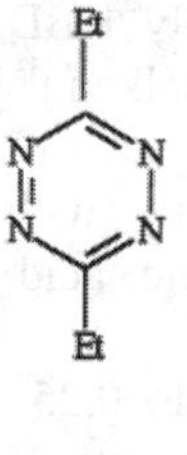

**(14)**

The problem was solved thanks to the exceptional behavior of 3,6-diethyltetrazine[70] (DETZ, **14**). This compound possesses two lone electron pairs located on opposite sites of its hexagonal ring; the strong interaction between the two pairs results in one antibonding n orbital in the compound lying at an anomalously high energy level; this, as shown below, has special spectroscopic implications. The π-electron system of DETZ is similar to that of benzene. Thus, the first $\pi \rightarrow \pi^*$ transition in the two systems appears at $\lambda_{max} = 252$ nm in DETZ and at 260 nm in benzene. On the other hand, the $\pi \rightarrow \pi^*$ transition in DETZ is strongly shifted to the visible region ($\lambda_{max} \cong 550$ nm), which is the origin of its deep color. This unique feature makes this compound a firm candidate for use as an environmental probe as the wide energy gap between the two transitions excludes potential overlap. Problems such as protonation of the DETZ probe by acid solvents can be ruled out since, according to Mason,[78] this type of substance exhibits pK values below zero.

The sensitivity of the $n \rightarrow \pi^*$ transition in DETZ to the solvent is clearly reflected in its UV/Vis spectra in methylcyclohexane, methanol, trifluoromethanol, hexafluoro-2-propanol, acetic acid and trifluoroacetic acid, with $\lambda_{max}$ values of 550, 536, 524, 517, 534 and 507 nm, respectively. In addition, the spectral envelope undergoes no significant change in the solvents studied, so a potential protonation of the probe -even a partial one- can be discarded.

The shift in the n $\rightarrow \pi^*$ band is largely caused by solvent acidity; however, an n $\rightarrow \pi^*$ transition is also obviously sensitive to the polarity of the medium, the contribution of which must be subtracted if acidity is to be accurately determined. To this end, our group used the solvatochromic method of Kamlet and Taft[18a] to plot the frequency of the absorption maximum for the n $\rightarrow \pi^*$ transition against solvent polarity (the SPP value). As expected, non-acidic solvents exhibited a linear dependence [eq. (10.3.13)], whereas acidic solvents departed from this behavior -the more acidic the greater the divergence. This departure from the linear behavior described by eq. [10.3.13] is quantified by $\Delta\nu_{DETZ}$,

$$\nu_{DETZ} = 1.015SPP + 17.51 \quad [10.3.13]$$

with n = 13, r = 0.983 and sd = 23 $cm^{-1}$.

The SA value of an acid solvent as determined using the DETZ probe is calculated using the expression established from the information provided by a series of solvents of increased acidity measurable by the TBSB/DTBSB probe/homomorph couple (viz. 1,2-butanediol, 1,3,-butanediol, ethanol, methanol and hexafluoro-2-propanol); this gives rise to eq. [10.3.14], which, in principle, should only be used to determine SA for solvents with $\Delta\nu_{DETZ}$ values above 100 $cm^{-1}$:

$$SA = 0.833\Delta\nu_{DETZ} + 0.339 \quad [10.3.14]$$

with n = 6 and r = 0.987.

Table 10.3.1 gives the SA values for a wide range of highly acidic solvents. Data were largely obtained from measurements made in our laboratory;[70] some, however, are reported here for the first time.

A brief analysis of these SA data allow one to draw several interesting conclusions from structural effects on solvent acidity. Thus:

- Substitution into alcoholic compounds allows the entire range of the solvent acidity scale to be spanned. Thus, 2-octanol can be considered to be scarcely HBD (SA=0.088) and ethylene glycol to be highly HBD (SA=0.717). Perfluoroalkanols are even more highly HBD.
- The acidity of 1-alkanols and carboxylic acids decreases significantly with increasing chain length but levels off beyond nonanol and octanoic acid, respectively.
- Cyclization decreases solvent acidity. Thus, SA decreases from 0.318 to 0.257 between 1-pentanol and cyclopentanol, and from 0.298 to 0.136 between 1-octanol and cyclooctanol.
- Aromatic rings make aniline (SA=0.131) and pyrrole (SA=0.386) more HBD than their saturated homologs: cyclohexylamine and pyrrolidine, respectively, both of which are non-HBD (SA=0).
- The acidity of alkanols decreases dramatically with increasing alkylation at the atom that bears the hydroxyl group [e.g., from 1-butanol (SA=0.340) to 2-butanol (SA=0.221) to tert-butyl alcohol (SA=0.146)].

### 10.3.7 APPLICATIONS OF THE PURE SPP, SA AND SB SCALES

The SPP general solvent scale, and the SA and SB specific solvent scales, are orthogonal to one another, as can be inferred from the small correlation coefficients obtained in mutual fittings involving the 200 solvents listed in Table 10.3.1 [$r^2$(SPP vs. SA) = 0.13, $r^2$(SPP vs. SB) = 0.10 and $r^2$(SA vs. SB) = 0.01]. These results support the use of these scales for the multi-parameter analysis of other solvent scales or data sets sensitive to the solvent effect on the basis of the following equation:

$$P = aSPP + bSA + c\,SB + P_0 \qquad [10.3.15]$$

where P is the quantity to be described in a given solvent; SPP, SA and SB are the corresponding polarity/polarizability, acidity and basicity values for such a solvent; coefficients a, b and c denote the sensitivity of P to such effects; and $P_0$ is the P value in the absence of solvent (i.e., the gas phase, which is given a zero value in our scales).

Each of these scales (SPP, SB and SA) has previously been compared with other reported solvent scales and found to be pure scales for the respective effects. However, eq. [10.3.15] is used below to perform a multi-parameter analysis of various reported scales and experimental data sets of interest.

#### 10.3.7.1 Other reported solvents scales

Reported scales for describing solvent polarity [$f(\varepsilon,n)$,[21,79] $\pi^*$[80] and S'[60]], basicity (DN[20] and $\beta$[18]) and acidity (AN[81] and $\alpha$[18]) were previously analyzed against our SPP, SA and SB scales in the originating references, so no further comment is made here.

Rather, this section analyses the behavior of reported single-parameter scales developed to describe the global behavior of solvents [viz. the Z, $\chi_R$, $P_y$, $\Phi$, G and $E_T(30)$ scales] against the SPP, SB and SA scales on the basis of the 200 solvents listed in Table 10.3.1. The Y scale, established to describe the behavior of solvolysis-like kinetics, is dealt with in the section devoted to the description of kinetic data. The data used in this analysis were taken from the following sources: those for the Z scale from the recent review paper by Marcus;[56] those for the $\chi_R$, $\Phi$ and G scales from the compilation in Table 7.2 of Reichardt's book;[1] those for parameter $P_y$ from the paper by Dong and Winnik;[50] and those for $E_T(30)$ from the recent review by Reichardt[13] or Table 7.3 in his book.[1]

The Z value for the 51 solvents in Table 10.3.1 reported by Marcus allow one to establish the following equation:

$$Z = 22.37(\pm 4.85)SPP + 7.68(\pm 1.64)SB + 31.27(\pm 1.81)SA + 41.03(\pm 4.15) \qquad [10.3.16]$$

with r = 0.944 and sd = 3.03 kcal $mol^{-1}$.

This equation reveals that Z is largely the result of solvent acidity and polarity, and also, to a lesser extent, of solvent basicity. The strong delocalization of charge in the structure of the probe (**4**) upon electronic excitation, see scheme IV, accounts for the fact that the electronic transition of this probe is hypsochromically shifted by solvent polarity and acidity. The shift is a result of the increased stabilizing effect of the solvent -exerted via general and specific interactions- in the electronic ground state being partly lost upon electronic excitation through delocalization of the charge in the electronic excited state, which gives rise to decreased polarity and specific solvation. On the other hand, the small contribution of solvent basicity appears to have no immediate explanation.

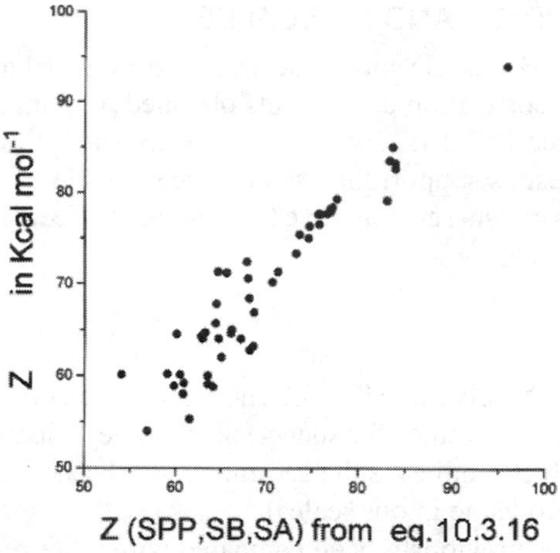


Figure 10.3.1. Plot of Kosower's Z values vs. the predicted Z values according to eq. [10.3.16].

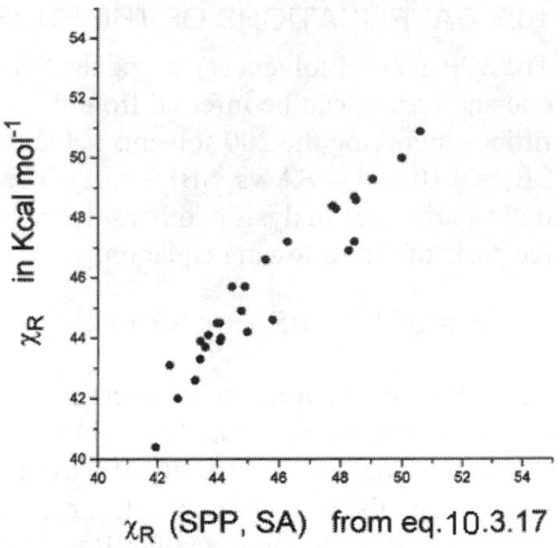


Figure 10.3.2. Plot of Brooker's $\chi_R$ values vs. the predicted $\chi_R$ values according to [eq. 10.3.17].

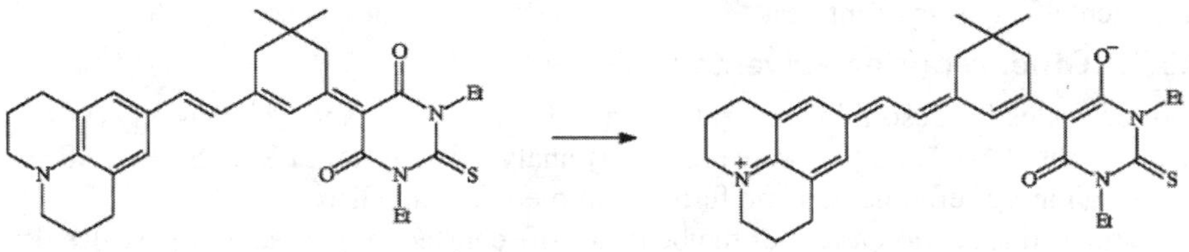

Scheme V.

It should be noted that the three solvent effects (polarity, acidity and basicity) increase Z; also, as can be seen from Figure 10.3.1, the greatest imprecision in this parameter corresponds to solvents with Z values below 70 kcal mol$^{-1}$.

The 28 solvents in Table 10.3.1 with known values of $\chi_R$ reveal that this parameter reflects solvent polarity and, to a lesser extent, also solvent acidity:

$$\chi_R = -15.81(\pm 1.19)SPP - 4.74(\pm 0.90)SA + 58.84(\pm 0.94) \qquad [10.3.17]$$

with r = 0.958 and sd = 0.78 kcal mol$^{-1}$.

It should be noted that the two solvent effects involved (polarity and acidity) contribute to decreasing $\chi_R$. As can be seen from Figure 10.3.2, the precision in $\chi_R$ is similarly good throughout the studied range. The structure of the probe used to construct this scale (**5**) reveals that its electron-donor site (an amino group) is not accessible for the solvent, so the charge transfer involved in the electronic excitation does not alter its solvation; rather, it increases its polarity and localization of the charge on the acceptor sites of the compound (carbonyl groups), see scheme V, so increasing solvent polarity and acidity will result in a bathochromically shifted electronic transition by effect of the increased stabilization of the excited state through general and specific interactions.

The results of the multi-parameter analysis of $P_y$ suggests that its values are highly imprecise; even if the value for the gas phase is removed, this parameter only reflects -in an im-

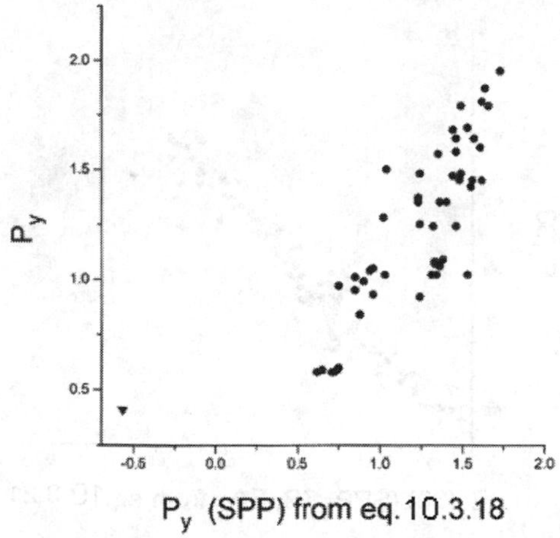


Figure 10.3.3. Plot of Dong and Winnck's $P_y$ values vs. the predicted $P_y$ values according to eq. [10.3.18] (Point for gas phase, ▼, is included in the plot, but not in the regression equation [10.3.18]).

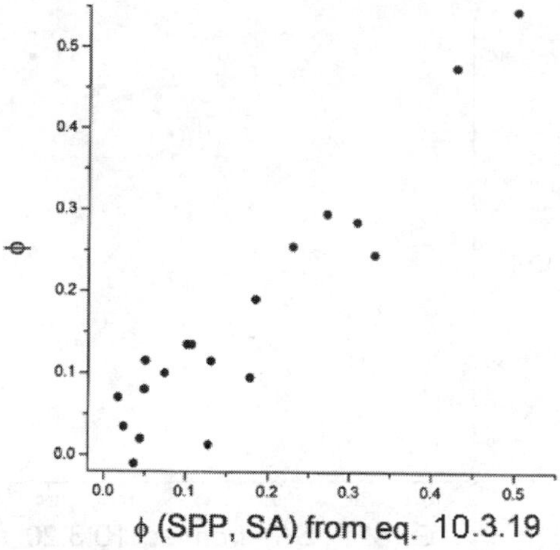


Figure 10.3.4. Plot of Dubois and Bienvenüe's Φ values vs. the predicted Φ values according to eq.[10.3.19].

precise manner- solvent polarity, as is clearly apparent from eq. [10.3.18] (see Figure 10.3.3):

$$P_y = 2.30(\pm 0.20)SPP - 0.57(\pm 0.16) \quad [10.3.18]$$

with r = 0.839 and sd = 0.20. The difficulty of understanding the performance of the probe precludes rationalizing its spectral behavior.

The 19 Φ values examined indicate that, based on eq. [10.3.19] and Figure 10.3.4, this parameter reflects mostly solvent acidity and, to a lesser extent, solvent polarity:

$$\Phi = 0.18(\pm 0.11)SPP + 0.38(\pm 0.04)SA - 0.07(\pm 0.08) \quad [10.3.19]$$

with r = 0.941 and sd = 0.05.

The Φ scale is based on the solvatochromic behavior of an n → π* electronic transition, so it reflects a high hypsochromic sensitivity to solvent acidity (specific solvation between the lone pair involved in the electronic transition and solvent acidity is lost when one electron in the pair is electronically excited). An n → π* electronic transition is also known to decrease the polarity of the chromophore concerned,[82] so an increase in SPP for the solvent will cause a hypsochromic shift in the electronic transition.

Based on Figure 10.3.5 and the following equation, the values of G parameters are also dictated by solvent polarity and, to a lesser extent, solvent acidity:

$$G = 102.96(\pm 9.02)SPP + 78.20(\pm 27.78)SA - 0.35(\pm 6.28) \quad [10.3.20]$$

with r = 0.965 and sd = 7.6.

It should be noted that the independent term for the gas phase is very small and highly imprecise in both the G scale and the Φ scale.

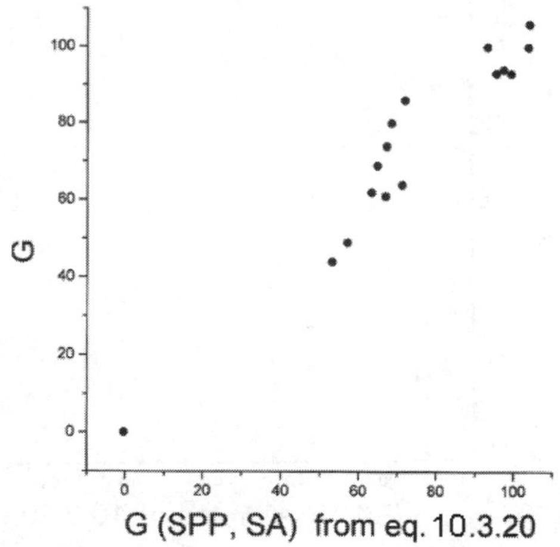


Figure 10.3.5. Plot of Allerhand and Schleyer's G values vs. the predicted G values according to eq. [10.3.20].

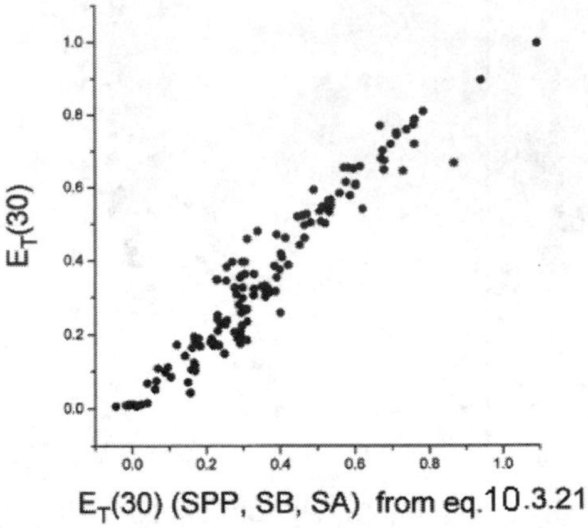


Figure 10.3.6. Plot of Reichardt's $E_T(30)$ values vs. the predicted $E_T(30)$ values according to eq. [10.3.21].

For the 138 solvents with a known $E_T(30)$ value listed in Table 10.3.1, this parameter reflects mostly solvent acidity and polarity, and, to a lesser degree, also solvent basicity (see Figure 10.3.6):

$$E_T(30) = 0.62(\pm 0.04)SPP + 0.12(\pm 0.02)SB + 0.77(\pm 0.02)SA - 0.31(\pm 0.03) \quad [10.3.21]$$

with r = 0.965 and sd = 0.06.

It should be noted that the three solvent effects (polarity, acidity and basicity) contribute to increasing $E_T(30)$, i.e., to a hypsochromic shift in the transition of the probe (see Figure 10.3.6).

The spectroscopic behavior of the probe used to construct this scale (**1**) is typical of a structure exhibiting highly localized charge on its carbonyl group in the electronic ground state, see scheme III, and hence strong stabilization by effect of increased solvent polarity and acidity. The electronic transition delocalizes the charge and results in an excited state that is much less markedly stabilized by increased polarity or acidity in the solvent. The decreased dipole moment associated to the electronic transition in this probe also contributes to the hypsochromic shift. The small contribution of solvent basicity to the transition of the probe (**1**) is not so clear, however.

### 10.3.7.2 Treatment of the solvent effect in:

#### *10.3.7.2.1 Spectroscopy*

Analyzing the SPP, SB and SA scales in the light of spectroscopic data that are sensitive to the nature of the solvent poses no special problem thanks to the vertical nature of the transitions, where the cybotactic region surrounding the chromophore is hardly altered. The papers where the SPP, SA and SB scales were reported discuss large sets of spectroscopic data in terms of the nature of the solvent. Some additional comments are made below.

Recently, Fawcet and Kloss,[83] analyzed the S=O stretching frequencies of dimethyl sulfoxide (DMSO) with a view to elucidating the behavior of 20 solvents and the gas phase.

They found the frequency of this vibration in DMSO to be shifted by more than 150 $cm^{-1}$ and to be correlated to the Gutmann acceptor number (AN) for the solvents. Based on our analysis, the frequency shift reflects the acidity and, to a lesser extent, the polarity of the solvent, according to the following equation:

$$\nu^{S=O} = 38.90(\pm 7.50)SPP - 89.67(\pm 4.60)SA + 1099.1 \qquad [10.3.22]$$

with n = 21, r = 0.979 and sd = 6.8 $cm^{-1}$.

This fit is very good, taking into account that it encompasses highly polar solvents such as DMSO itself, highly acidic solvents such as trifluoroacetic acid (SPP=1.016, SA=1.307), and highly non-polar and non-acidic solvents such as the gas phase.

Giam and Lyle[84] determined the solvent sensitivity of the $^{19}F$ NMR shifts on 4-fluoropyridine relative to benzene as an internal reference in 31 of the solvents listed in Table 10.3.1. If DMF is excluded on the grounds of its odd value, the chemical shifts for the remaining 30 solvents are accurately described by the following function of solvent acidity and polarity:

$$\Delta^{4F} = -1.66(\pm 0.57)SPP - 6.84(\pm 0.36)SA - 6.50(\pm 0.45) \qquad [10.3.23]$$

with r = 0.971 and sd = 0.34 ppm.

Clearly, the acidity of the medium is the dominant factor in the chemical shift measured for this compound, which reveals the central role played by the lone electron pair in pyridine.

Both absorption and emission electronic transitions are acceptably described by our scales, as shown by a recent study[85] on the solvation of a series of probes containing an intramolecular hydrogen bond, so no further comment is made here other than the following: even if one is only interested in evaluating the change in dipole moment upon electronic excitation via solvatochromic analysis, a multi-parameter analysis must be conducted in order to isolate the shifts corresponding to the pure dipolar effect of the solvent.[85]

Lagalante et al.[86] proposed the use of 4-nitropyridine N-oxide as a suitable solvatochromic indicator of solvent acidity. The hypsochromic shifts determined by these authors for 43 of the solvents in Table 10.3.1 are due largely to the acidity of solvent and, to a lesser extent, also to its basicity; based on the following equation, however, solvent polarity induces a bathochromic shift in the band:

$$\nu_{max} = -0.92(\pm 0.44)SPP + 0.68(\pm 0.17)SB + 3.63(\pm 0.23)SA + 29.10(\pm 0.31) \qquad [10.3.24]$$

with r = 0.943 and sd = 0.31 kK.

As can be seen from Figure 10.3.7, this probe classifies solvents in groups encompassing non-acidic solvents (below 29 kK), moderately acidic solvents (at about 30 kK) and highly acidic solvents (between 31 and 32 kK). Because the solvatochromism does not change gradually with increase in solvent acidity, the probe appears to be unsuitable for quantifying this effect.

Davis[87] determined the solvatochromism of a charge-transfer complex formed by tetra-n-hexylammonium iodide-nitrobenzene in 23 different solvents and, using the Z and $E_T(30)$ scales, observed a bilinear behavior in scarcely polar and highly polar solvents. A

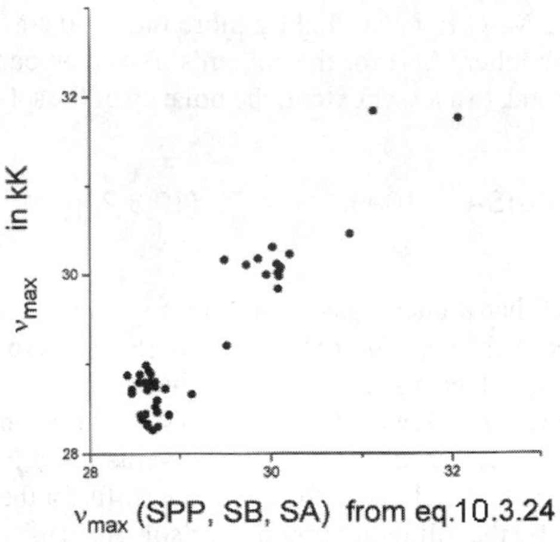


Figure 10.3.7. Plot of the experimental UV/Vis absorption maxima, $\nu_{max}$, of 4-nitropyridine N-oxide vs. the predicted $\nu_{max}$ values according to eq. [10.3.24].

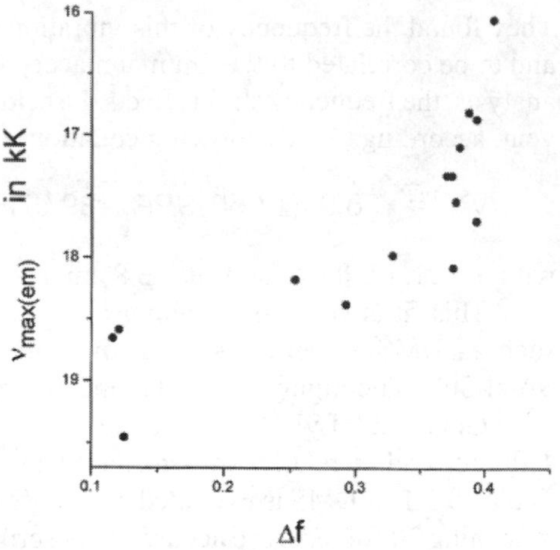


Figure 10.3.8. Plot of the fluorescence maxima, $\nu_{max(em)}$, for the Neutral Red in different solvents vs. the solvent polarity function Δf [$\Delta f = (\varepsilon -1)/(2\varepsilon +1)-(n^2 -1)/(2n^2 +1)$].

multi-parameter fit of the data for 22 of the solvents revealed that the solvatochromism is seemingly sensitive to solvent acidity only:

$$\nu_{c.t.} = 11.23(\pm 0.75)SA + 21.41(\pm 0.184) \qquad [10.3.25]$$

with n = 22, r = 0.959 and sd = 0.66 kK.

An interesting situation is encountered in the analysis of electronic transitions that results from chromophores which, assisted by solvent polarity, can undergo a change in the electronic state responsible for the fluorescence emission; this is a frequent occurrence in systems involving a TICT mechanism.[68,88-90] In this situation, a plot of Stokes shift against solvent polarity is a bilinear curve depending on whether the solvents are non-polar (where the fluorescence is emitted from the normal excited state of the chromophore) or polar enough for the transition to take place from the more polar state. Unless the solvents used are carefully selected, it is venturesome to assume that the electronic states will be inverted simply because the variation of solvatochromism with a function of solvent polarity is bilinear.

Recently, Sapre et al.[91] showed that a plot of the maximum fluorescence of Neutral Red (NR) against the solvent polarity function Δf is clearly bilinear (see Figure 10.3.8). Accordingly, they concluded that, in solvents with Δf > 0.37, the emitting state changes to a much more polar, ICT state. The analysis of this spectroscopic data in the light of our scales reveals that, in fact, the solvatochromism of NR is normal, albeit dependent not only on the polarity of the solvent (SPP), but also on its acidity (SA) (see Figure 10.3.9):

$$\nu_{max}^{NR} = -4.31(\pm 0.57)SPP - 1.78(\pm 0.19)SA + 21.72(\pm 0.47) \qquad [10.3.26]$$

with r = 0.974 and sd = 0.21 kK.

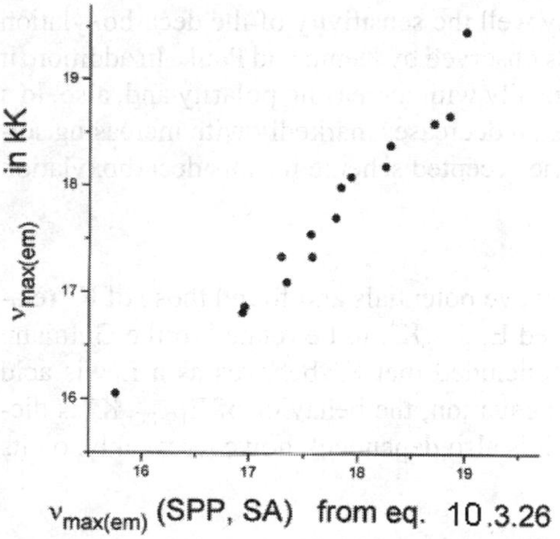


Figure 10.3.9. Plot of the fluorescence maxima, $\nu_{max(em)}$, for the Neutral Red in different solvents vs. the predicted, $\nu_{max(em)}$, values according to eq. [10.3.26].

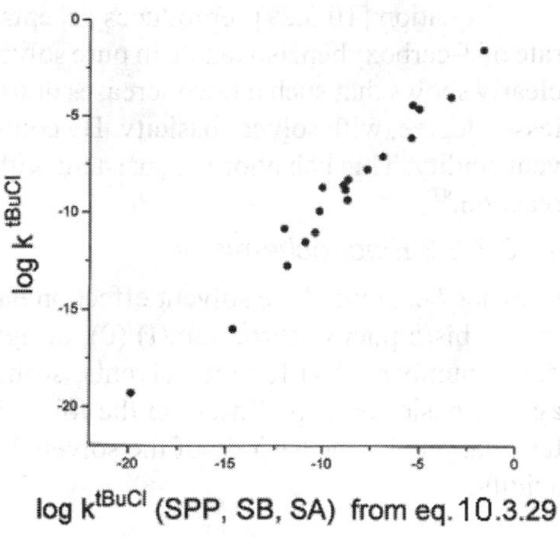


Figure 10.3.10. Plot of Grunwald and Wistein's log $k^{tBuCl}$ values vs. the predicted log $k^{tBuCl}$ values according to eq. [10.3.29].

### *10.3.7.2.2 Kinetics*

Kinetics so closely related to the solvent effect as those of the Menschutkin reaction between triethylamine and ethyl iodide [eq. (10.3.27)], the solvolysis of tert-butyl chloride [eq. (10.3.28)] or the decarboxylation of 3-carboxybenzisoxazole [eq. (10.3.29)], are acceptably described by our scales.[15,92,93]

The equation for the Menschutkin kinetics is

$$\log k_s/k_{Hex} = 8.84(\pm 0.66)SPP + 1.90(\pm 1.37)SA - 4.07(\pm 0.53) \quad [10.3.27]$$

with n = 27, r = 0.947 and sd = 0.41.

The rate of this reaction between triethylamine and ethyl iodide, which varies by five orders of magnitude from n-hexane ($1.35\times10^{-8}$ l mol$^{-1}$ s$^{-1}$) to DMSO ($8.78\times10^{-4}$ l mol$^{-1}$ s$^{-1}$), is accurately described by solvent polarity and acidity -the sensitivity to the latter is somewhat imprecise. The equation for the kinetics of solvolysis of tert-butyl chloride is:

$$\log k = 10.02(\pm 1.14)SPP + 1.84(\pm 0.99)SB + 8.03(\pm 0.69)SA - 19.85(\pm 0.70) \quad [10.3.28]$$

with n = 19, r = 0.985 and sd = 0.80.

Based on eq. [10.3.28], all solvent effects increase the rate of solvolysis. However, the strongest contribution is that of polarity and the weakest one that of acidity. Although much less significant, the contribution of solvent basicity is especially interesting as it confirms that nucleophilicity also assists in the solvolytic process. Taking into account that it encompasses data spanning 18 orders of magnitude (from log k = -1.54 for water to log k = -19.3 for the gas phase), the fit is very good (see Figure 10.3.10). The equation for the kinetics of decarboxylation of 3-carboxybenzisoxazole is:

$$\log k = 10.37(\pm 1.47)SPP + 2.59(\pm 0.74)SB - 5.93(\pm 0.58)SA - 9.74(\pm 1.17) \quad [10.3.29]$$

with n = 24, r = 0.951 and sd = 0.73.

Equation [10.3.29] reproduces acceptably well the sensitivity of the decarboxylation rate of 3-carboxybenzisoxazole in pure solvents observed by Kemp and Paul.[7] In addition, it clearly shows that such a rate increases dramatically with increasing polarity and, also, to a lesser degree, with solvent basicity. By contrast, it decreases markedly with increasing solvent acidity. This behavior is consistent with the accepted scheme for this decarboxylation reaction.[93]

#### *10.3.7.2.3 Electrochemistry*

Gritzner[94] examined the solvent effect on half-wave potentials and found those of $K^+$ relative to bis(biphenyl)chromium(I)/(0), designated $E_{1/2(BCr)}K^+$, to be related to the Gutmann donor number (DN) for the solvents, so he concluded that $K^+$ behaves as a Lewis acid against basic solvents. Based on the following equation, the behavior of $E_{1/2(BCr)}K^+$ is dictated largely by the basicity of the solvent but it is also dependent, however weakly, on its acidity:

$$E_{1/2(BCr)}K^+ = -0.51(\pm 0.05)SB + 0.12(\pm 0.04)SA - 1.06(\pm 0.03) \quad [10.3.30]$$

with n = 17, r = 0.941 and sd = 0.03 V.

It should be noted that solvent basicity increases the half-wave potentials of $K^+$ whereas solvent acidity decreases it.

#### *10.3.7.2.4 Thermodynamics*

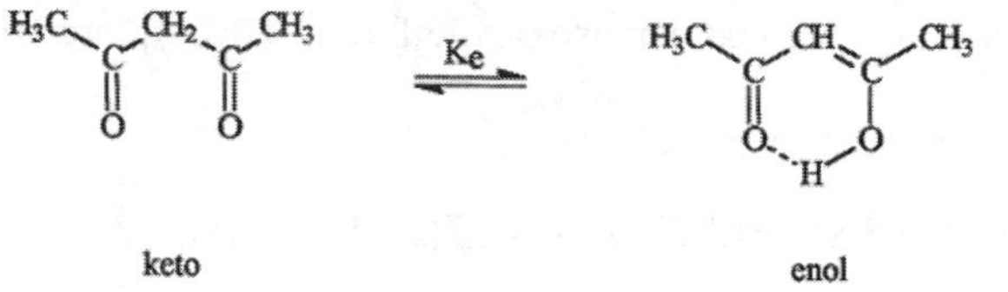


Scheme VI.

In order to compare the shifts of the conformational equilibrium position with solvent effects it is advisable to select species exhibiting negligible cavity effects on the equilibrium position. Two firm candidates in this respect are the conformational equilibria of 1,2,2-trichloroethane and the equilibrium between the equatorial and axial forms of 2-chlorocyclohexanone; both are accurately described by our scales.[15]

Of special interest is also the equilibrium between the keto and enol tautomers of pentane-2,4-dione (see the scheme VI).

The ΔG values for this equilibrium in 21 solvents reported by Emsley and Freeman[95] are accurately reproduced by solvent polarity and acidity according to

$$\Delta G = 18.45(\pm 2.41)SPP + 4.22(\pm 0.86)SA - 18.67(\pm 1.96) \quad [10.3.31]$$

with n = 21, r = 0.924 and sd = 1.34 kJ $mol^{-1}$.

The equilibrium position is mainly dictated by solvent polarity; however, there is clearly a specific contribution of solvent acidity from the carbonyl groups of the dione.

### 10.3.7.3 Mixtures of solvents. Understanding the preferential solvation model

The solvation of a solute, whether ionic or neutral, in a mixture of solvents is even more complex than in a pure solvent.[1] This is assumed to be so largely because a solvent mixture involves interactions not only between the solute and solvent but also among different molecules present in the mixture; the latter type of contribution also plays a central role in the solvation process. Among others, it results in significant deviations of the vapor pressure of a mixture with respect to the ideal behavior established by Raoult's law.

Solvation studies of solutes in mixed solvents have led to the conclusion that the above-mentioned divergences may arise from the fact that the proportion of solvent components may be significantly different around the solute and in the bulk solution. This would be the case if the solute were preferentially surrounded by one of the mixture components, which would lead to a more negative Gibbs energy of solvation.[1,96] Consequently, the solvent shell around the solute would have a composition other than the macroscopic ratio. This phenomenon is known as "preferential solvation", a term that indicates that the solute induces a change with respect to the bulk solvent in its environment; however, such a change takes place via either non-specific solute-solvent interactions called "dielectric enrichment" or specific solute-solvent association (e.g. hydrogen bonding).

Preferential solvation has been studied in the light of various methods, most of which are based on conductance and transference measurements[96], NMR measurements of the chemical shift of a nucleus in the solute[97] or measurements of the solvatochromism of a solute in the IR[98] or UV-Vis spectral region.[99] Plots of the data obtained from such measurements against the composition of the bulk solvent (usually as a mole fraction) depart clearly from the ideal behavior and the deviation is ascribed to the presence of preferential solvation.

Several reported methods aim to quantify preferential solvation;[96-100] none, however, provides an acceptable characterization facilitating a clear understanding of the phenomenon.

If preferential solvation is so strongly dictated by the polar or ionic character of the solute, then characterizing a mixture of solvents by using a molecular probe will be utterly impossible since any conclusions reached could only be extrapolated to solutes of identical nature as regards not only polarity and charge, but also molecular size and shape.

However, the experimental evidence presented below allows one to conclude that this is not the case and that solvent mixtures can in fact be characterized in as simple and precise terms as can a pure a solvent.

In the light of the previous reasoning, describing the solvolysis of tert-butyl chloride or the decarboxylation kinetics of 3-carboxybenzisoxazole in mixed solvents in terms of SPP, SB and SA for the mixtures appeared to be rather difficult owing to the differences between the processes concerned and the solvatochromism upon which the scales were constructed. However, the results are categorical as judged by the following facts:

(a) The solvolysis rate of tert-butyl chloride in 27 pure solvents and 120 binary mixtures of water with methanol (31 mixtures), ethanol (31), isopropyl alcohol (1), trifluoroethanol (8), dioxane (13), acetone (27) and acetic acid (9), in addition to the datum for the gas phase, all conform to the following equation:[92]

$$\log k = 10.62(\pm 0.44)SPP + 1.71(\pm 0.22)SB + 7.89(\pm 0.17)SA - 20.07(\pm 0.34) \quad [10.3.32]$$

with n = 148, r = 0.99 and sd = 0.40.

(b) The decarboxylation rate of 3-carboxybenzisoxazole in 24 pure solvents and 36 mixtures of DMSO with diglyme (4 mixtures), acetonitrile (4), benzene (7), dichloromethane (6), chloroform (6) and methanol (9) are accurately described by the following expression:

$$\log k = 10.03(\pm1.05)SPP + 2.41(\pm0.49)SB - 5.73(\pm0.40)SA - 9.58(\pm0.83) \quad [10.3.33]$$

with n = 60, r = 0.990 and sd = 0.60.

Other evidence obtained in our laboratory using solvent mixtures and probes as disparate in size and properties as the cation $Na^+$ and Reichardt's $E_T(30)$, also confirm that solvent mixtures are no more difficult to characterize than pure solvents.

## REFERENCES


1 C. Reichardt, **Solvents and Solvent Effects in Organic Chemistry**, 2nd edn., *VCH* Publishers, Weinheim, 1988.
2 M. Berthelot, L. Péan de Saint-Gilles, *Ann. Chim. Phys.*, **65**, 385 (1862); **66**, 5 (1862); **68**, 225 (1863).
3 N.A. Menschutkin, *Z. Phys. Chem.*, **5**, 589 (1890).
4 E. Grunwald, S. Wistein, *J. Am. Chem. Soc.*, **70**, 846 (1948).
5 S. Wistein, A.H. Fainberg, *J. Am. Chem. Soc.*, **79**, 5937 (1957).
6 M.H. Abraham, R.M. Dogerty, M.J. Kamlet, J.M. Harris, R.W. Taft, *J. Chem. Soc., Perkin Trans. 2*, 913 (1987).
7 D.S. Kemp, K.G. Paul, *J. Am. Chem. Soc.*, **97**, 7305 (1975).
8 J.G. Kirkwood, *J. Chem. Phys.*, **2**, 351 (1934); **7**, 911 (1939).
9 L. Onsager, *J. Am. Chem. Soc.*, **58**, 1486 (1936).
10 H. Block, S.M. Walker, *Chem. Phys. Lett.*, **19**, 363 (1973).
11 J.E. Brady, P.W. Carr, *J. Phys. Chem.*, **89**, 5759 (1985).
12 D.V. Matyushov, R. Schmid, B.M. Landansyi, *J. Phys. Chem. B*, **101**, 1035 (1997).
13 C. Reichardt, *Chem. Rev.*, **94**, 2319 (1994).
14 R.S. Drago, *J. Chem. Soc., Perkin Trans. 2*, 1827 (1992).
15 J. Catalán, V. López, P. Pérez, R. Martin-Villamil, J.G. Rodriguez, *Liebigs Ann.*, 241 (1995).
16 G. Taddei, E. Castelluci, F.D. Verderame, *J. Chem. Phys.*, **53**, 2407 (1970).
17. T.D. Epley, R.S. Drago, *J. Am. Chem. Soc.*, **89**, 5770 (1967); R.S. Drago, T.D. Epley, *J. Am. Chem. Soc.*, **91**, 2883 (1969); G.C. Vogel, R.S. Drago, *J. Am. Chem. Soc.*, **92**, 3924 (1970); R.S. Drago, L.B. Parr, C.S- Chamberlain, *J. Am. Chem. Soc.*, **99**, 3202 (1977); R.S. Drago, K.F. Purcell, *Prog. Inorg. Chem.*, **6**, 271 (1964); R.S. Drago, *Coord. Chem. Rev.*, **33**, 251 (1980); R.S. Drago, G.C. Vogel, *J. Am. Chem. Soc.*, **114**, 9527 (1992); R.S. Drago, *Inorg. Chem.*, **32**, 2473 (1993).
18 a) M.J. Kamlet, R.W. Taft, *J. Am. Chem. Soc.*, **98**, 377 (1976); b) R.W. Taft, M.J. Kamlet, *J. Am. Chem. Soc.*, **98**, 2886 (1976).
19 G.N. Lewis, **Valence and Structure of Atoms and Molecules**, *The Chemical Catalog Co.*, 1923, p 142.
20 V. Gutmann, E. Vychera, *Inorg. Nucl. Chem. Lett.*, **2**, 257 (1966).
21 I.A. Koppel, V.A. Palm in **Advances in Linear Free Energy Relationships** (Eds. N.B. Chapman, J. Shorter) Chapter 5, *Plenum Press*, London, 1972, p 204.
22 E.M. Arnett, L. Joris, E. Michell, T.S.S.R. Murty, T.M. Gorrie, P. v R. Scheyer, *J. Am. Chem. Soc.*, **96**, 3875 (1974).
23 J. Catalán, J. Gómez, A. Couto, J. Laynez, *J. Am. Chem. Soc.*, **112**, 1678 (1990).
24 J. Catalán, P. Pérez, J. Elguero, W. Meutermans, *Chem. Ber.*, **126**, 2445 (1993).
25 E. Grunwald, S. Winstein, *J. Am. Chem. Soc.*, **70**, 846 (1948).
26 S. Winstein, A.H. Fainberg, *J. Am. Chem. Soc.*, **79**, 5937 (1957).
27 S. Winstein, E. Grunwald, H.W. Jones, *J. Am. Chem. Soc.*, **73**, 2700 (1951)
28 A.H. Fainberg, S. Winstein, *J. Am. Chem. Soc.*, **78**, 2770 (1956).
29 A.H. Fainberg, S. Winstein, *J. Am. Chem. Soc.*, **79**, 1597 (1957).
30 A.H. Fainberg, S. Winstein, *J. Am. Chem. Soc.*, **79**, 1602 (1957).
31 A.H. Fainberg, S. Winstein, *J. Am. Chem. Soc.*, **79**, 1608 (1957).
32 S. Winstein, A.H. Fainberg, E. Grunwald, *J. Am. Chem. Soc.*, **79**, 4146 (1957).
33 L.P. Hammett, *J. Am. Chem. Soc.*, **59**, 96(1937); and in **Physical Organic Chemistry**, 2nd Edition, *McGraw-Hill,* New York 1970.
34 A. Streitwieser Jr., *Chem. Rev.*, **56**, 617 (1956); **Solvolytic Displacement Reactions**, *McGraw-Hill,* New York 1962.

35 T.W. Bentley, P.v R. Schleyer, *Adv. Phys. Org. Chem.*, **14**, 32 (1977).
36 T.W. Bentley, G. Llewellyn, *Prog. Phys. Org. Chem.*, **17**, 121 (1990).
37 C.G. Swain, R.B. Mosely, D.E. Bown, *J. Am. Chem. Soc.*, **77**, 3731 (1955).
38 T.W. Bentley, F.L. Schad, P.v R. Schleyer, *J. Am. Chem. Soc.*, **94**, 992 (1972).
39 F.L. Schad, T.W. Bentley, P.v R. Schleyer, *J. Am. Chem. Soc.*, **98**, 7667 (1976).
40 T.W. Bentley, G.E. Carter, *J. Am. Chem. Soc.*, **104**, 5741 (1982).
41 I.A. Koppel, V.A. Palm in **Advances in linear Free Energy Relationships**, N.B. Chapman, J. Shorter Eds., *Plenum Press*, London, 1972, p 203 and references therein.
42 G.A. Allerhand, P.V. Schleyer, *J. Am. Chem. Soc.*, **85**, 374 (1963).
43 L.J. Bellamy, H.E. Hallam, *Trans. Faraday Soc.*, **55**, 220 (1959).
44 L.J. Bellamy, H.E. Halam, R.L. Williams, *Trans. Faraday Soc.*, **54**, 1120 (1958).
45 L.J. Bellamy, R.L. Williams, *Proc. Roy. Soc.(London)*, **A255**, 22 (1960).
46 C. Walling, P.J. Wagner, *J. Am. Chem. Soc.*, **86**, 3368 (1964).
47 C. Somolinos, I. Rodriguez, M.I. Redondo, M.V. Garcia, *J. Mol. Struct.*, **143**, 301 (1986).
48 K. Dimroth, C. Reichardt, T. Siepmann, F. Bohlmann, *Liebigs Ann. Chem.*, **661**, 1 (1963).
49 M.A. Kessler, O.S. Wolfbeis, *Chem. Phys. Lipids*, **50**, 51 (1989); C.J. Drummond, F. Grieser, T.W. Healy, *Faraday Discuss. Chem. Soc.*, **81**, 95 (1986).
50 D.C. Dong , M.A. Winnick, *Can. J. Chem.*, **62**, 2560 (1984).
51 K.W. Street, W E. Acree, *Analyst*, **111**, 1197 (1986).
52 I. Kristjánsson, J. Ulstrup, *Chem. Scripta*, **25**, 49 (1985).
53 E.M. Kosower, *J. Am. Chem. Soc.*, **80**, 3253 (1958).
54 E.M. Kosower, J.A. Skorez, *J. Am. Chem. Soc.*, **82**, 2188 (1960).
55 T.R. Griffiths, D.C. Pugh, *Coord. Chem. Rev.*, **29**, 129 (1979).
56 Y. Marcus, *Chem. Soc. Rev.*, 409 (1993).
57 L.G.S. Brooker, G.H. Keyes, D.W. Heseltine, *J. Am. Chem. Soc.*, **73**, 5350-54 (1951).
58 L.G.S. Brooker, A.C. Craig, D.W. Heseltine, P.W. Jenkins, L.L. Lincoln, *J.Am. Chem. Soc.*, **87**, 2443 (1965).
59 J.E. Dubois, A. Bienvenue, *J.Chim. Phys.*, **65**, 1259 (1968).
60 R.S. Drago, *J. Chem. Soc., Perkin Trans II*, 1827 (1992).
61 R.S. Drago, M.S. Hirsch, D.C. Ferris, C.W. Chronister, *J. Chem. Soc., Perkin Trans II*, 219 (1994).
62 R.S. Drago, **Apllications of Electrostatic-Covalent Models in Chemistry**, *Surface Scientific Publishers*, Gainesville, 1994.
63 E. Lippert, *Z. Elektrochem.*, **61**, 962 (1957).
64 J. Czekalla, W. Liptay, K.O. Meyer, *Z. Elektrochem.*, **67**, 465 (1963).
65 V. Baliah, M.K. Pillay, *Indian J. Chem.*, **9**, 845 (1971).
66 J. Catalán, V. López, P. Pérez, *Liebigs Ann.*, 793 (1993).
67 J. Catalán, *J. Org. Chem.*, **60**, 8315 (1995).
68 J. Catalán. *New. J. Chem.*, **19**, 1233 (1995).
69 J. Catalán, C. Díaz, V. López, P. Pérez, R.M. Claramunt, *J. Phys..Chem.*, **100**, 18392 (1996).
70 J. Catalán, C. Díaz, *Eur. J. Org. Chem.*, 885 (1999).
71 J.S. Murray, S. Rauganathan, P. Politzer, *J. Org. Chem.*, **56**, 3734 (1991).
72 J. Catalán, C. Díaz, V. López, P. Pérez, J.L.G. de Paz, J.G. Rodríguez, *Liebigs Ann.*, 1785 (1996).
73 J. Catalán, J. Palomar, C. Díaz, J.L.G. de Paz, *J. Phys. Chem.*, **101**, 5183 (1997).
74 J. Catalán, C. Díaz, *Liebigs Ann.*, 1942 (1997).
75 J. Catalán, E. Mena, W. Meutermans, J. Elguero, *J. Phys. Chem.*, **96**, 3615 (1992).
76 J.E. Kuder, D. Wychick, *Chem. Phys. Lett.*, **24**, 69 (1974).
77 I. Guda, F. Bolduc, *J. Org. Chem.*, **49**, 3300 (1984).
78 S.F. Mason, *J.Chem. Soc.*, 1240 (1959).
79 J.A. Paéz, N. Campillo, J. Elguero, *Gazz. Chim. Ital.*, **126**, 307 (1996).
80 M.J. Kamlet, J.L.M. Aboud, R.W. Taft, *J.Am. Chem. Soc.*, **99**, 6027 (1977); *ibid*, **99**, 8325 (1977).
81 U. Mayer, V. Gutman, W. Gerger, *Monasth. Chem.*, **106**, 1235 (1975).
82 V.I. Minkin, O.A. Osipov, Yu A. Zhadanov, **Dipole Moments in Organic Chemistry**, *Plenum Press*, New York, 1970.
83 W.R. Fawcet, A.A. Kloss, *J. Phys. Chem.*, **100**, 2019 (1996).
84 C.S. Giam, J.L. Lyle, *J. Am. Chem. Soc.*, **95**, 3235 (1973).
85 J. Catalán, J.C. del Valle, C. Díaz, J. Palomar, J.L.G. de Paz, M. Kasha, *Internat.J. Quam. Chem.*, **72**, 421 (1999).
86 A.F. Lagalante, R.J. Jacobson, T.J. Bruno, *J. Org. Chem.*, **61**, 6404 (1996).
87 K.M.C. Davis, *J. Chem. Soc. (B)*, 1128 (1967).

88 F. Schneider, E. Lippert, *Ber.Bunsen-Ges. Phys. Chem.*, **72**, 1155 (1968).
89 J. Catalán, V. López, P. Pérez, *J. Fluorescence*, **6**, 15 (1996).
90 J. Catalán, C. Díaz, V. López, P. Pérez, R.M. Claramunt, *Eur. J. Org. Chem.*, 1697 (1998).
91 M.K. Singh, H. Pal, A.C. Bhasikuttan , A.V. Sapre, *Photochem & Photobiol.*, **68**, 32 (1998).
92 J. Catalán, C. Díaz, F. Garcia-Blanco, *J. Org. Chem.*, **64**, 6512 (1999).
93 J. Catalán, C. Díaz, F. Garcia-Blanco, *J. Org. Chem.*, in press.
94 G. Gritzner, *J. Phys. Chem.*, **90**, 5478 (1986).
95 J. Emsley, N.J. Freeman, *J. Mol. Struc.*, **161**, 193 (1987).
96 H. Scheneider, in **Solute-Solvent Interactions**, J.F. Coetzee and C.D. Ritchie (Eds), *Dekker*, New York, Vol 1 p 301 (1969).
97 L. S. Frankel, C.H. Langford, T.R. Stengle, *J. Phys. Chem.*, **74**, 1376 (1970). J.F. Hinton, E.S. Amis, *Chem. Rev:*, **67**, 367 (1967).
98 A.I. Popov, in **Soluto-Solvent Interactions**, J.F. Coetzee and C.D.Ritchie(eds) *Dekker*, New York, 1976, Vol 2 pg 271.
99 K. Dimroth , C. Reichardt, *Z. Anal. Chem.*, **215**, 344 (1966) J. G. Dawber, J.Ward, R. A. Williams, *J. Chem. Soc., Faraday Trans. 1*, **84**, 713 (1988).
100 J. Midwinter, P. Suppan, *Spectrochim. Acta*, **25A**, 953 (1969); P. Suppan, *J. Chem. Soc., Faraday Trans.,1*, **83**, 495 (1987); M. W. Muanda, J.B. Nagy, O.B. Nagy, *Tetrahedron Lett.*, **38**, 3424 (1974); O.B. Nagy, M.W. Muanda, J.B. Nagy, *J. Phys. Chem.*, **83**, 1961 (1979): H. Strehlow, H. Schneider, *Pure Appl. Chem.*, **25**, 327 (1971); M.S. Greenberg, A.I. Popov, *Spectrochim. Acta*, **31A**, 697 (1975); H. Langhals, *Angew. Chem. Int. Ed. Engl.*, **21**, 724 (1982); A. Ben-Nain, *J. Phys. Chem.*, **93**, 3809 (1989); P. Chatterjee, S. Bagchi, *J. Chem. Soc., Faraday Trans,* **87**, 587 (1991); W.E. Acree, Jr., S. A Tucker, D. C. Wilkins, *J. Phys. Chem.*, **97**, 11199 (1993); R.D. Skwierczynski, K.A. Connors, *J. Chem. Soc. Perkin Trans 2*, 467(1994) M. Rosés, C. Ráfols, J. Ortega, E. Bosch, *J. Chem. Soc. Perkin Trans 2*, 1607 (1995); W. E. Acree Jr., J. R. Powell, S. A. Tucker, *J. Chem. Soc. Perkin Trans 2*, 529 (1995).

## 10.4 ACID-BASE EQUILIBRIA IN IONIC SOLVENTS (IONIC MELTS)

VICTOR CHERGINETS
**Institute for Single Crystals, Kharkov, Ukraine**

Ionic melts are widely used in the science and engineering as media for performing different processes such as electrolysis, electrochemical synthesis, single crystals growing, etc. Practically complete dissociation of ionic media to the constituent ions creates high current densities at electrolysis. The absence of oxidants, similar to $H^+$, makes it possible to obtain products, which cannot be obtained from aqueous solvents (i.e., alkaline and alkaline earth metals, sub-ions, etc.). From the ecological standpoint, molten ionic media are especially available as technological solvents since their employment does not cause the accumulation of liquid wastes because cooling to the room temperature transforms ionic liquids into a solid state.

Processes taking place in ionic melt-solvents are considerably affected by impurities contained in the initial components of the melt or formed during preparation (mainly, melting) of solvents due to the high-temperature hydrolysis of melts or their interactions with container materials ($Al_2O_3$, $SiO_2$, etc.) or active components of atmosphere ($O_2$, $CO_2$, etc.). The list of these impurities is wide enough and includes multivalent cations of transition metals, different complex anions (oxo- or halide anions). The effect of the mentioned admixtures on the processes in ionic melts depends mainly on the degree of their donor-acceptor interactions with constituent parts of the melt.

## 10.4.1 ACID-BASE DEFINITIONS USED FOR THE DESCRIPTION OF DONOR-ACCEPTOR INTERACTIONS IN IONIC MEDIA

Donor-acceptor interactions in ionic media are often described as acid-base interactions according to Lewis[1] and Lux-Flood[2-5] definitions. The classic variant of the former definition considers acids as acceptors of electron pairs and bases as their donors. In modern variant of this definition, acids are the electron pair (or anion) acceptors or cation (proton) donors, bases are the electron pair (anion) donors or cation (proton) acceptors.

Lux-Flood definition considers bases as donors of oxide-ion, $O^{2-}$, its acceptors are Lux acids.

### 10.4.1.1 The Lewis definition

The Lewis acid-base process may be described by the following scheme:

$$A + :B = A:B, K \qquad [10.4.1]$$

where:

| | |
|---|---|
| A | an acid, |
| B | a base, |
| K | the equilibrium constant (used below pK ≡ -log K). |

This definition can be used for the description of interactions in ionic melts, containing complex anions undergoing heterolytic dissociation. Processes reverse to [10.4.1], i.e., the acid-base dissociation of A:B adducts (ions) in some solvents are considered as acid-base equilibria of these solvents.

Alkaline chloroaluminate melts containing the excess of $AlCl_3$ vs. the stoichiometry are promising solvents for preparing sub-ions, e.g., $Cd^+(Cd_2^{2+})$, $Bi^+$, $Bi_5^{3+}$,[6,7] the excess of acid ($AlCl_3$) favors their formation. Tremillon and Letisse,[8] Torsi and Mamantov[9,10] studied the acid-base properties of molten mixtures $AlCl_3$-MCl (M=Li, Na, K, Cs) with $AlCl_3$ concentration exceeding 50 mol% in the temperature range 175-400°C. Equilibrium molarities of $Cl^-$ were determined by a potentiometric method with the use of a chloride-reversible electrode. The solvents undergo the acid-base dissociation according to the following equation:

$$2AlCl_4^- = Al_2Cl_7^-\left(AlCl_3 \cdot AlCl_4^-\right) + Cl^-, \quad pK \qquad [10.4.2]$$

In the melts, chloride-ion donors were bases and substances increasing $AlCl_3$ concentration were acids. The pK values decrease with the temperature (from 7.1 to 5.0 at temperatures 175 and 400°C, respectively, Na-based melt[9]) elevation and from Cs to Li (at 400°C pK were 3.8, 5.0, 5.8, 7.4 for Li-, Na-, K- and Cs-based melts, respectively[10]). The latter effect may be explained from the point of view of "hard" and "soft" acids and bases[11-13] - Li-Cl complexes should be more stable than Cs-Cl ones as formed by "hard" base ($Cl^-$) and "more hard" acid ($Li^+$).

Dioum, Vedel and Tremillon[14] investigated molten $KGaX_4$ (X=Cl, I) as background for acid-base processes. The following acid-base equilibria exist in the pure solvents:

$$2GaCl_4^- = Ga_2Cl_7^-\left(GaCl_3 \cdot GaCl_4^-\right) + Cl^-, \quad pK=4.25\pm0.05 \qquad [10.4.3]$$

$$GaI_4^- = GaI_3 + I^-, \quad pK=2.6\pm0.05 \qquad [10.4.4]$$

Wagner[15] reported an electrochemical study of molten $NaBF_4$ at 420°C. The range of acid-base properties varying according to reaction:

$$BF_4^- = BF_3 + F^-, \qquad pK=1.8 \qquad [10.4.5]$$

has been estimated as ~2pF units. The acid-base ranges in ionic melts containing halide complexes are relatively narrow, it means that varying their halide basicity (i.e., equilibrium concentration of the halide ions) may be performed in 2-4 logarithmic units, although such changes are enough to obtain some unstable compounds with the intermediate oxidation degrees.[6,7]

#### 10.4.1.2 The Lux-Flood definition

Oxygen-containing impurities in molten salts are most usual, their effect on technological processes is mainly negative and consists of bonding acidic reagents with the formation of insoluble (suspensions) or slightly dissociated products. Similar reactions result in retarding the main processes caused by the decrease of the equilibrium concentrations of initial reagents, forming oxide inclusions in metals obtained by the electrolysis of melts, inclusions of insoluble particles in single crystals, considerable corrosion of container for crystal growth, etc. Therefore, quantitative studies of reactions with participation of oxide ions in ionic melts are of importance for scientific and engineering purposes.

Reactions with the transfer of oxide ions in ionic media are mainly considered as oxoacidity or acid-base equilibria by Lux-Flood:[2-5]

$$A + O^{-2} = B, K \qquad [10.4.6]$$

The addition of acids (bases) in a melt as it follows from [10.4.6] leads to changes of the oxide ion activity, for the characterization of melt acidities (basicities) Lux[2] proposed the oxygen index, or pO, which was similar to pOH (i.e., basicity index) in aqueous solutions:

$$pO \equiv -\log a_{O^{2-}} \left(\equiv -\log m_{O^{2-}}\right) \qquad [10.4.7]$$

where:

$a_{O^{2-}}$, $m_{O^{2-}}$ activity and molarity of oxide ions in the melt, respectively

The measurements of oxygen indices during various reactions in molten media define their thermodynamic characteristics - dissociation constants and solubility products. Most oxoacidity studies were made in nitrate melts, while there were much less communications dealing with the similar studies in higher-melting alkaline chlorides, molten bromides and iodides. There is a considerable scatter of obtained experimental results, which are often in conflict one with another. There were no serious attempts to generalize these data or consider them from the common point of view.

### 10.4.2 THE FEATURES OF IONIC MELTS AS MEDIA FOR ACID-BASE INTERACTIONS

Ionic melts as media for Lux-Flood acid-base reactions may be divided into two types on the base of constitutional (i.e., being a part of main components of the melt) oxygen ions: oxygen-less and oxygen-containing ones. Let us consider some features of acid-base interactions in the mentioned melts.[16]

### 10.4.2.1 Oxygen-less media

There are no oxide ions in the composition of these melts, therefore the pure melts cannot possess oxide-donor properties. Real oxygen-less melts contain small amounts of $O^{2-}$ owing to inevitable ingress of oxygen-containing impurities into the melt, but oxide ion concentration in the "pure" melts is variable, depending on the concentration and acid-base character of impurities. For example, even considerable amounts of sulfates in melts do not create appreciable $O^{2-}$ concentration, carbonate ion dissociation is substantially stronger, hydroxide ions may be referred to most strongly dissociated Lux bases. In "pure" oxygen-less melts oxygen index pO is usually in the range 3 to 4.5. The employment of strongest purifying agents (HCl, $CCl_4$, etc.) does not allow to decrease this concentration essentially, the latter being in "an unavoidable harm" causing errors at the quantitative investigations, especially when small amounts of acid or bases are studied. But, in some cases, oxygen index for such a "pure" melt has been used as the internal standard for construction of acidity scales.[17-20]

Quantitative studies of different Lux-Flood acids and bases in ionic melts were performed by two main ways: the construction of empirical acidity scales to estimate relative acidic strength of the substances and the determination of acid-base equilibria constants using potentiometric titration techniques. The first approach has been proposed in the classic work of Lux[2] who obtained the acidity scale for the equimolar mixture of potassium and sodium sulfates. Although this work was the basis for series of later studies, the results cannot be considered as undoubted ones. Addition of acids (bases) to the mentioned melt led to increasing (decreasing) e.m.f. of cell with the oxide-selective electrode vs. the corresponding magnitude in the neutral melt. Then, the e.m.f. value shifted progressively to that for the neutral melt because of $SO_3$ ($Na_2O$) evaporation from the acidic (basic) melt. Therefore, Lux extrapolated values of e.m.f. to the point of acid (base) addition to the melt. This resulted in the decrease of the data[2] accuracy.

The empirical acidity scales[17-20] give some information about strength of the acids and bases in melts studied. The principal error related to the term "acidity scale" in the case of oxygen-less melts is because it is not connected with the melt properties. Therefore, the values obtained could not be considered as the quantitative characteristics of the oxygen-less melts. The acidity scale length in a solvent is believed to be the interval (measured in acidity index units) between standard solutions of strong acid and base. If the solvent possesses its own acid-base autodissociation equilibrium then a substance creating unit concentration of acid (base) of solvent in the standard solution should be considered as the strongest acid (base). Addition of stronger acids (bases) should not result in extension of the acidity scale because of the known phenomenon of leveling acidic and basic properties by the solvent.[21] Oxygen-less melts do not possess acid-base equilibrium, therefore, values of the basicity index, in the "pure" melts may have any reasonable value.

### 10.4.2.2 Oxygen-containing melts

Acid-base processes in oxygen-containing melts are more complex than those in oxygen-less ones, since they are accompanied by competitive equilibria of own acid-base autodissociation of the melt-solvent. The coexistence of acidity and basicity "carriers" into melt is the characteristic feature of oxygen-containing melts making them similar to low-temperature molecular solvents with own acid-base equilibrium.

But, there exist some principal features due to relatively high temperatures of the liquid state. Own acids of the melts are often unstable or volatile,[2,22] therefore, acidic solutions, as a rule, lose acids because of their evaporation[2] or decomposition,[22] the oxygen index of such melts is shifted progressively to that of the neutral melt. Hence, in oxygen-containing melts it is possible to observe not only the leveling of acidic properties but also an upper limit of acidity.

$CrO_3$ and $MoO_3$ have been found[23] to breakdown the nitrate melt $KNO_3$ with the evaporation of $NO_2$. It should be noted, that the temperature elevation leads to the efficient shrinkage of the acidic region in the melt. The interactions between the oxides and the melt have been assumed to result in the formation of nitrogen (V) oxide, $N_2O_5$, and corresponding oxoanions. Since the acids studied ($CrO_3$, $MoO_3$) may be referred to as very strong acids, it may be supposed that reaction:

$$2CrO_3 + NO_3^- = Cr_2O_7^{2-} + NO_2^+ \qquad [10.4.8]$$

should be completely shifted to the right and the strength of the acid is determined by acidic properties of nitronium cation, $NO_2^+$. The stability of the latter is low and after achieving a certain concentration its reaction with the melt anions becomes intensive enough:

$$NO_2^+ + NO_3^- = 2NO_2 \uparrow + \frac{1}{2}O_2 \qquad [10.4.9]$$

The so-called kinetic methods of melt acidity determination, which will be considered below, are based just on this reaction. The reaction [10.4.9] leads to the decrease of nitronium concentration and the rate of $NO_2$ emission decreases until essentially constant acid concentration is determined by a sequence of consecutive measurements. This concentration is the "upper limit" of acidity of the melt. The thermal dependence of upper limit of acidity[23] in nitrate melts can be easily explained on the basis of the increase of the melt temperature, which leads not only to the reduction of nitronium stability but also to the elevation of process [10.4.9] rate. Sulphate melts have the upper limit of acidity too, it seems connected with limited and low solubility of $SO_3$ at elevated temperatures, such assumption may be confirmed by results.[2]

### 10.4.2.3 The effect of the ionic solvent composition on acid-base equilibria

The equilibrium parameters of Lux acid-base reactions in ionic media (solubility products of oxides and acid-base equilibrium constants) are essentially affected by the acidic properties of the molten alkaline halide mixtures, i.e., they are dependent on the constituent cation acidities. Therefore, one should consider the reverse problem - the estimation of the basicity indices of ionic melts on the basis of the calculated equilibrium constants.

The default acid-base processes are actually a superimposed effect of interactions between oxide-ions formed and the most acidic cations of the melt:

$$iMe^{m+} + O^{2-} \leftrightarrow Me_iO^{i \cdot m-2}, \qquad K_{l,i} \qquad [10.4.10]$$

where:

$Me^{m+}$ the most acidic cation of the solvent

The increase of melt acidity ($K_{l,i}$) leads the shift of interactions [10.4.10] to the right. The distribution of oxide ions added to the melt between different complexes with the melt cations may be presented by the following equation (N is mole fraction, m - molarity):

$$N^0_{O^{2-}} = N_{O^{2-}}\left(1 + \sum_{i=1}^{n} K_{l,i} N^i_{Me^+}\right)$$ [10.4.11]

It may be seen that the ratio "free oxide-ion/total oxide-ion" is the constant which may be designated as $I_l$:[24]

$$I_l = N_{O^{2-}} / N^0_{O^{2-}} \left(= m_{O^{2-}} / m^0_{O^{2-}}\right) = 1 / \left(1 + \sum_{i=1}^{n} K_{l,i} N^i_{Me^+}\right)$$ [10.4.12]

Now let us to estimate the equilibrium molarity of the constituent acidic cations in the melt, e.g., the eutectic KCl-LiCl melt (0.4:0.6) contains ~8.5 mole of $Li^+$ per 1 kg. Usually the ionic complexes in melts are characterized by the coordination number ~ 4-6.[21] For the solution of $O^{2-}$ of the 0.1 mole/kg concentration, the maximum possible quantity of fixed $Li^+$ concentration may be estimated as 0.4-0.6 mole/kg, i.e., efficiently lower than 8.5. In this case the change of actual $Li^+$ concentration is approximately equal to 5-7% and $m_{Me+}$ in this case may be suggested as constant. Therefore, for each melt the sum in the denominator of [10.4.12] is the constant reflecting its acidic properties. So, $pI_l$= $-\log I_l$ is a measure of melt acidities and may be denoted as the "oxobasicity index" of the melt. Since the determination of the "absolute" concentration of free $O^{2-}$ is practically impossible one should choose the "standard melt", for which $I_l$ is conditionally equal to 1 and $pI_l$=0. It is reasonable to choose the equimolar mixture KCl-NaCl as the "standard melt", since this melt is most frequently investigated. Further, one should choose "standard equilibria" and formulate the non-thermodynamic assumptions which usually postulate that the constant of the "standard equilibrium" calculated using "absolute" oxide ion concentrations remains the same for all other melts.

Now let us consider possible variants of such "standard equilibria" and conditions of their use.

One of the first attempts to define and estimate the oxoacidity parameters of ionic (chloride-based) melts was connected with the studies of the equilibrium

$$H_2O_{gas} + 2Cl^- \leftrightarrow 2HCl_{gas} + O^{2-}$$ [10.4.13]

in different chloride mixtures.[25,26] Here the partial pressures of $H_2O$ and HCl are known and the oxide ion concentration is calculated on the base of the calibration data as $m^0_{O^{2-}}$ ($N^0_{O^{2-}}$). The equilibrium constant of [10.4.13] may be represented in terms of the "absolute" mole fractions of $O^{2-}$ and $I_l$ by:

$$K_1 = \frac{p^2_{HCl}\left(N_{O^{2-}} I_l^{-1}\right)}{p_{H_2O} N^2_{Cl^-}} = K_{KCl-NaCl} I_l^{-1}$$ [10.4.14]

and, consequently:

$$pK_l = pK_{KCl-NaCl} - pI_l$$ [10.4.15]

Combes et al.[25] introduced the oxo-acidity function, Ω:

$$\Omega = 14 - pK_l + pO^{2-} \quad [10.4.16]$$

taking into account [10.4.15] eq. [10.4.16]

$$\Omega = (14 - pK_{KCl-NaCl}) + pI_l + pO^{2-} \quad [10.4.17]$$

where (14 - $pK_{KCl-NaCl}$) is the Ω value in KCl-NaCl standard solution (pO=0) and $pO^{2-}$ is the instrumental pO in the solvent studied.

Therefore, the shift of pO scales vs. KCl-NaCl depends only on the $I_l$ value. On the basis of [10.4.13] it has been estimated as +7-8 log units (Figure 3[25]) for KCl-LiCl at 1000K.

Homogeneous Lux acid-base equilibria of type of [10.4.6] have not been earlier considered as available for estimations of oxoacidity indices. However, pK for these reactions may be written as

$$pK_1 = -\log \frac{N_B}{N_A \left(N_{O^{2-}} I_l^{-1}\right)} = pK_{KCl-NaCl} + pI_l \quad [10.4.18]$$

For the use of this type of equilibrium it is necessary to make the following non-thermodynamic assumption:

$$\left(\frac{\gamma_B}{\gamma_A}\right)_l = \left(\frac{\gamma_B}{\gamma_A}\right)_{KCl-NaCl} \quad [10.4.19]$$

where:

γ corresponding activity coefficients of the acid and the base in solvent "l" and KCl-NaCl

The use of [10.4.16] where A and B are anion acid and base for estimations of the oxobasicity indices may be justified because the acid and the conjugated base are negatively charged (in the pair $Cr2O_7^{-2}/CrO_4^{-2}$ they are of the same charge).

Finally, let us consider the usability of the oxide solubility data

$$MeO_s = Me^{2+} + O^{2-} \quad [10.4.20]$$

for $pI_l$ estimations. The solubility product value, P, may be presented as

$$P_{MeO,l} = N_{Me^{2+}} \left(N_{O^{2-}} I_l^{-1}\right) = P_{MeO,KCl-NaCl} I_l^{-1} \quad [10.4.21]$$

or

$$pP_{MeO,l} = pP_{MeO,KCl-NaCl} - pI_l \quad [10.4.22]$$

Metal cations in molten halides form halide complexes. The reaction [10.4.20] for chloride melts ($N_{Cl^-}$ =1) suggests that the distribution of $Me^{2+}$ between different complexes $MeCl_n^{2-n}$ remains unchanged (similarly to eq. [10.4.10]-[10.4.12] for the oxide ion distribution). It is clear that the solubility data may be used for estimations of the oxoacidity indices only in melts with the same anion composition. The anion changes cause errors since $Me^{2+}$ in eq. [10.4.21] is referred to essentially different acids, e.g., halide complexes $MeCl_4^{2-}$ and

$MeBr_4^{2-}$ in the case of chloride and bromide melts. Therefore, regardless of the cation composition, the oxide solubilities are essentially affected by anion composition of a melt.[27]

Equations [10.4.14], [10.4.18] and [10.4.22] estimate $pI_{KCl-LiCl}$ vs. KCl-NaCl as 7-8,[25] 3.7[24] and 3.4[24], respectively. This estimation is considerably larger than the estimation based on eq. [10.4.18] and [10.4.22]. For $Ca^{2+}$ based chloride melts, the similar estimation made on the basis of [10.4.14] gave pI~$10^{25}$ vs. ~4 determined on the basis of [10.4.22].[28] So, the oxobasicity indices calculated according to $HCl/H_2O$ equilibrium constant are essentially different from those obtained using reactions without reactive gas atmosphere.

Probably, this discrepancy may be explained by the features of the water behavior in aprotic ionic melts. It is known that, similar to basic properties, water possesses oxoacidic properties according to the reaction:

$$H_2O + O^{2-} \leftrightarrow 2OH^- \left(or\,(OH)_2^{-2}\right) \qquad [10.4.23]$$

Such polynuclear complexes are stable according to spectral data[29] because of the H-bonds formation. Three acids ($HCl + Me^+ + H_2O$) exist in ionic melts saturated with water and HCl and consequently reaction [10.4.23] depends on the partial pressure of water. In Li- and Ca-based chloride melts, retention of water[30] and the solubility of $H_2O$ is appreciably higher than in e.g. KCl-NaCl.

Apparently, even the simplest reaction [10.4.13] is actually complicated by additional interactions favoring the fixation of “free” oxide ions. Values of oxobasicity indices calculated from eq. [10.4.13] are thus somewhat overestimated.

The oxobasicity index for KCl-LiCl (0.4:0.6) compared with KCl-NaCl at 700°C lies within the range 3.4-3.7 log units.

### 10.4.3 METHODS FOR ESTIMATIONS OF ACIDITIES OF SOLUTIONS BASED ON IONIC MELTS

The literature data show that many methods were used for oxoacidity studies and estimation of the acidic properties of melts. One simplest method[31] involves indicator. Acid-base indicators usually employed in aqueous solutions for protic acidity measurements have been used for acidity studies in molten $KNO_3$-$LiNO_3$ at 210°C and KSCN at 200°C. The color of indicator solution, relative to acidity, changes during titration of bases (sodium hydroxide or peroxide) by potassium pyrosulfate, $K_2S_2O_7$:

$$2HInd + O^{2-} = 2Ind^- + H_2O \qquad [10.4.24]$$

where:

| | |
|---|---|
| HInd | the protonized form of the acid-base indicator |
| Ind- | the anionic form of the said indicator |

In molten nitrates such transition was observed only for phenolphthalein (yellow-purple), other indicators seemingly were oxidized by the melt (the conclusion made[31] was “insoluble in the melt”). In molten KSCN not possessing oxidizing properties color transitions were observed for all indicators used (methyl red, thymolphthalein, etc.). The employment of indicators for acidity estimation is limited mainly by their thermal instability and tendency to oxidize at high temperatures. Since ionic melts, as a rule, have no constitutional water, the reverse transition of indicator into the protonized form [10.4.24] is hardly possible (the solution of phenolphthalein became yellow at the reverse transition “base-acid”[31]).

Therefore, the use of indicator method is possible only at low-temperatures (250°C) and in non-oxidizing ionic melts.

Spectral methods were also used for the melt basicity estimation.[32,33] The scale of "optical" basicity for metallurgical slugs with respect to CaO (standard) has been constructed.[32] The "optical" basicity, Λ, was estimated from spectral line shifts (the transition $^1S_0 \rightarrow {}^3P_1$) for $Tl^+$, $Pb^{2+}$, $Bi^{3+}$ cations in the basic medium. Ionic melts $Na_2O$-$B_2O_3$-$Al_2O_3$, $Na_2O$-$B_2O_3$-$SiO_2$, $Na_2O$-$SiO_2$-$Al_2O_3$ have been investigated by X-ray fluorescent spectroscopy,[32] relative oxide acidities decrease in the sequence $B_2O_3 > SiO_2 > Al_2O_3$.

The use of spectral methods is based on studies of cooled (quenched) samples, such a routine may distort results because of inconsistency between the solution temperature and that of acidity estimations.

Data on the melt basicity estimations by the determination of acidic gases ($SO_3$, $CO_2$, $H_2O$) solubilities are presented elsewhere.[34,35] Sulphur trioxide solubility in molten sodium phosphate was determined by thermogravimetric analysis,[34] the correlation between the melt basicity and the $SO_3$ solubility was found. The use of $CO_2$ and $H_2O$ for basicity estimation was described elsewhere.[35] The similar methods may be used only for basic melts since acids displace the acidic gas from the melt. Furthermore, the interaction products of the acidic gas with the melt may be relatively stable, especially in basic solutions, and their formation leads to irreversible changes of the melt properties.

A "kinetic method" was used for Lux acidity studies in molten nitrates. The interaction between the nitrate melts $KNO_3$-$NaNO_3$ and potassium pyrosulfate was studied.[36,37]

$$S_2O_7^{2-} + NO_3^- = 2SO_4^{2-} + NO_2^+ \qquad [10.4.25]$$

The formed nitronium cation, $NO_2^+$, reacted with $NO_3^-$ according to [10.4.9]. The latter stage is considerably slower than [10.4.25], hence, its rate allowed to estimate the melt acidity, which can be presented by the following sum: $T_A = [S_2O_7^{2-}] + [NO_2^+]$.[36]

The acid-base interactions between $Cr_2O_7^{2-}$ and chlorate ions, $ClO_3^-$, in molten $KNO_3$-$NaNO_3$ are described by the following equations.[38,39]

$$Cr_2O_7^{2-} + ClO_3^- = 2CrO_4^{2-} + ClO_2^+ \qquad [10.4.26]$$

$$ClO_2^+ + Cl^- \rightarrow Cl_2\uparrow + O_2\uparrow \qquad [10.4.27]$$

From two above processes, reaction [10.4.27] is the limiting stage, the rate of the chlorine evolution is proportional to the total acidity of the melt: $T_A = [Cr_2O_7^{2-}] + [ClO_2^+]$.

The reaction of bromate ions, $BrO_3^-$, with potassium dichromate in the nitrate melt has been studied.[38] Several studies[36-40] gave estimate of relative acidities of the oxo-cations. The acidities increased in the sequence $BrO_2^+ < ClO_2^+ < NO_2^+$.

Slama[41,42] used the kinetic method to investigate metal cation ($Cu^{2+}$, $Co^{2+}$) acidities in molten $KNO_3$-$NaNO_3$ in the temperature range 325-375°C. The cation acidities were estimated on the base of the $NO_2$ evolution rate [10.4.9]. Nitronium cation was formed as a product of the following reaction:

$$Me^{2+} + NO_3^- = MeO\downarrow + NO_2^+ \qquad [10.4.28]$$

where:

$Me^{2+}$ $Cu^{2+}$ or $Co^{2+}$

Metal cation acidities are comparable with acidic properties of dichromate-ion and mentioned above[36-40] cations.

A potentiometric method, or more accurately, method of e.m.f. measurements with liquid junction in cells with oxide-reversible electrodes was most advantageous in oxoacidity studies of various melts. The principal construction of electrochemical cells for oxoacidity studies may be presented by the following scheme:

*Reference electrode || Melt+O$^{2-}$| Oxide-reversible electrode* [10.4.29]

In all known studies, the potential of liquid junction was zero. The known reference electrodes were metallic (i.e., metal immersed into the solution with the definite concentration of the corresponding cation) or gas oxygen electrodes immersed into solutions with definite oxide ion concentrations (concentration cells).

### 10.4.4 ON STUDIES OF THE HOMOGENEOUS ACID-BASE REACTIONS IN IONIC MELTS

#### 10.4.4.1 Nitrate melts

Traditionally, nitrate melts are considered among most comprehensively studied ones. Relatively low melting temperatures both of individual nitrates and their mixtures allowed to use simpler experimental techniques. The potentiometric studies gave the acid-base equilibrium constants. Most studies were on oxoacidity reactions in molten $KNO_3$ at 350°C.[23,43-67] Constants of acid-base reactions of Group V highest oxides ($P_2O_5$, $As_2O_5$, $V_2O_5$) with Lux bases were also investigated.[43-48] All oxides breakdown the melt, the first stage is the formation of corresponding meta- acid salts and nitronium cation:

$$R_2O_5 + NO_3^- = 2RO_3^- + NO_2^+ \quad [10.4.30]$$

where:

R the designation of Group V element,

the second step is the redox interaction of $NO_2^+$ with $NO_3^-$ according to [10.4.9], which is shifted to the right, $NO_2$ evolution took place until e.m.f. values reached down to ~0.5 V, this magnitude corresponded to the upper limit of oxo-acidity in molten $KNO_3$. The oxygen index value for this limit can be estimated as ~$10^{-16}$.[43-48] Addition of $Na_2O_2$ as a Lux base led to the following neutralization steps

$$2RO_3^- + O^{2-} = R_2O_7^{4-} \quad [10.4.31]$$

$$R_2O_7^{4-} + O^{2-} = 2RO_4^{3-} \quad [10.4.32]$$

The acidic properties of oxides increased in the sequence $V_2O_5$<$P_2O_5$<$As_2O_5$. Besides of reactions [10.4.30]-[10.4.32],[43-45] the constants of the following assumed equilibria, with participation of hydroanions $HPO_4^{2-}$, $H_2PO_4^-$, $HAsO_4^{2-}$, $H_2AsO_4^{2-}$, were determined:

$$2H_2RO_4^- + O^{2-} = 2HRO_4^{2-} + H_2O\uparrow \quad [10.4.33]$$

$$2HRO_4^{2-} + O^{2-} = 2RO_4^{3-} + H_2O\uparrow \quad [10.4.34]$$

Since even the pure salts containing mentioned hydroanions undergo the decomposition at temperatures much lower then 350°C with the formation of corresponding meta- or pyro- salts,[68] equilibria [10.4.33] and [10.4.34] hardly take place in the absence of water vapor in atmosphere over the melt.

The acidic properties of highest oxides of VI Group elements (Cr, Mo, W) were investigated.[23,48,49] Similarly to Group V oxides[43-48] $CrO_3$ and $MoO_3$ destructed the nitrate melt with formation of $Cr_2O_7^{2-}$ and $Mo_3O_{10}^{2-}$, respectively. After the complete evolution of $NO_2$, e.m.f. values were close to 0.5 V, what was in a good agreement other studies.[43-48] The potentiometric titration curves of the formed products contained only one drop of e.m.f. (or pO). Chromate and molybdate ions were the final products of the neutralization reactions. $WO_3$ was weaker acid and did not destruct the nitrate melt. Its titration was the one-stage process resulting in the tungstate formation. The acidities of oxides increased in the order: $WO_3<MoO_3<CrO_3$.

The equilibrium constants[23,43-58] have incorrect values because calculated magnitudes had shift by order of a few pK units. Such phenomenon may take place either because of incorrect determination of reaction participants (stoichiometry) or slope changes of the calibration plot for the used oxygen electrode $Pt(O_2)$ in the acidic region.

Relative acidities of different metal cations in molten $KNO_3$ have been investigated[52-54] using potentiometric titration of corresponding carbonates by dichromate[52] and metaphosphate ions:[53]

$$Me_{n/2}CO_3 + A = n/2Me^{n+} + CO_2\uparrow + B \qquad [10.4.35]$$

The cations studied were arranged in the following sequence: K~Na>Li~Ba>Sr>Ca>Pb of basicity decrease. The cation sequence was in a good correlation with the electronegativity of metals. The proposed method of investigation had an essential problem since the acids used for titration fixed oxide ions with the simultaneous destruction of carbonate and formed insoluble chromates and phosphates with the majority of cations studied. As it has been shown,[36,37] the addition of $Ca^{2+}$ into nitrate melts leads to the shift of the interaction $Cr_2O_7^{2-}$ - $NO_3^-$ to formation of nitronium cation and the melt decomposition although the solution without $Ca^{2+}$ is stable. Such shift is explained by formation of slightly insoluble $CaCrO_4$ in nitrate melt and has no relation to $Ca^{2+}$ acidity. It is referred to the barium position in the sequence that gives evidence of the said complexes or precipitates formation, since their acidic properties are substantially weaker than those of $Li^+$. The formation of the $BaCrO_4$ precipitate leads to approaching of $Ba^{2+}$ acidity to $Li^+$ acidity. Probably, the same reason effected arrangement of others alkaline earth metals in this sequence.

The relative strength of different Lux bases in nitrate melts was investigated.[54,55] The bases studied were divided into two groups: the oxide ion group ($OH^-$, $O^{2-}$, $O_2^{2-}$) and the carbonate ion group ($CO_3^{2-}$, $HCO_3^-$, $(COO)_2^{2-}$, $CH_3COO^-$, $HCOO^-$). The bases belonging to the first group have been assumed[54,55] to be completely transformed into $O^{2-}$.The members of the first group of bases were weaker than the second group bases. From the thermal data it may be seen that at experimental temperature (350°C) all the second group bases are completely broken down to carbonate. The differences between single bases may be attributed to reduction properties - organic salts reduce nitrate ions to more basic nitrite ones and favor the accumulation of oxide and carbonate ions in the melt over the stoichiometric ratio.

Hence, the use of a majority of proposed bases for quantitative studies of oxoacidity is not possible, especially for acids having apparent oxidizing properties (chromates, vanadates, etc.).

Oxide ions in molten nitrates exist as the so-called pyronitrate ions $N_2O_7^{4-}$ which was a strong base with the basicity approximately equal to those for $CO_3^{2-}$.[55] The modern point of view excludes the existence of nitrogen compounds with the coordination number exceeding 3 by oxygen, hence pyronitrate was hardly existing in the mentioned conditions. Indeed, following oxoanion studies[51] have demonstrated the absence of the mentioned ion in basic nitrate solutions. A possible way for the stabilization of $O^{2-}$ in nitrate melts is the formation of silicates due to reactions with walls of the container (usually it is pyrex).

The anomalous behavior of the electrode $Pt(O_2)$ has been noted and explained by "peroxide" function of the gas oxygen electrodes (of the first type).[57,58] On the contrary to gas electrodes, metal-oxide electrodes (of the second type) were reversible with the slope corresponding to the reaction:

$$1/2O_2 + 2\bar{e} = O^{2-} \qquad [10.4.36]$$

The acid-base equilibria in complex solutions based on nitrates containing phosphates and molybdates were attributed[59,60] to the formation of heteropolyacids in molten nitrates. This is confirmed by cryoscopic studies.[61]

Some features of potentiometric titration techniques in molten salts have been discussed.[62,63] The automatic potentiometric titration described in details[62] consists of the following. A rod, prepared as frozen $K_2Cr_2O_7$ solution in the melt, was immersed into the melt containing a base with the constant rate. e.m.f. data were recorded by the recording potentiometer. Such a routine may be used only for the detection of equivalence points in melts (with accuracy of ~3.5 %).

The effect of acidic oxides on the thermal stability of molten $NaNO_3$ has been investigated.[64] The increase of oxide acidity results in the decrease of the melt decomposition temperature. This is explained by the reduction of the activation energy of the decomposition owing to fixing oxide ions by acids. But there was also another explanation: decreasing temperature of melt decomposition is caused by the equilibrium shift towards the oxide-ions and nitrogen oxide formation under the action of the acid. Acidic properties of oxides decrease in the sequence $SiO_2>TiO_2>ZrO_2>Al_2O_3>MgO$.

Potentiometric method[65-67] used metal-oxide electrodes ($Nb|Nb_2O_5$, $Ta|Ta_2O_5$, $Zr|ZrO_2$) for oxoacidity studies in the nitrate melt. The empirical acidity scale in molten potassium nitrate at 350°C was constructed: $NH_4VO_3>NaPO_3>NaH_2PO_4>K_2Cr_2O_7>K_2HPO_4>Na_4P_2O_7>NaHAsO_4>K_2CO_3>Na_2O_2$. This sequence includes some acids seldom existing in molten salts. $NH_4VO_3$ at 350°C should be completely transformed into $V_2O_5$ with a partial reduction of the latter by ammonia. Acidic salts mentioned were decomposed to corresponding pyro- and meta-salts. Therefore the scale obtained contains a number of errors.

#### 10.4.4.2 Sulphate melts

Sulphate melts are referred to melts with the own acid-base dissociation equilibrium:

$$SO_4^{2-} = SO_{3,l}\uparrow + O_l^{2-} \qquad [10.4.37]$$

There are rather few contributions devoted to the Lux acidity studies in the similar melts. This may be explained by relatively high melting points both of single sulfates and their eutectic mixtures. The weak stability of acidic solutions in sulphate melts can also be considered as a reason why these melts are studied insufficiently. Indeed, reaction [10.4.37] should result in $SO_3$ formation in the melt. Reference data,[68,69] however, show that decomposition temperatures of pyrosulfates (i.e., complexes $SO_3(SO_4^{2-})$ do not exceed 460°C, in dissolved state stability of solvated $SO_3$ should be even lower. The absence of $SO_3$ partial pressure over the melt gives rise to removing $SO_3$ from the latter and, since the equilibrium state of [10.4.37] is not attained, $SO_3$ should be removed completely. Among sulphate melts the ternary eutectic mixture $K_2SO_4$-$Li_2SO_4$-$Na_2SO_4$ (0.135:0.78:0.085) with the lowest melting point 512°C[70] may be most promising for the acidity studies since it is possible to perform studies at the temperatures significantly lower than in the classic work of Lux.[2] Under these conditions stability of acidic and basic solutions increases.

Lux[2] reported the oxygen electrode reversibility investigation and construction of empirical acidity scale in the molten eutectic mixture $K_2SO_4$-$Na_2SO_4$ at 800°C. The thermal stability of pyrosulfates has been shown by Flood[3] to decrease with melt cation acidity increasing. Kaneko and Kojima[71,72] investigated acid-base properties of different solutions in the molten $K_2SO_4$-$Li_2SO_4$-$Na_2SO_4$ at 550°C; the empirical acidity scale length was approximately 10 pO units between 0.01 mole/kg solutions of $S_2O_7^{2-}$ and $O_2^{2-}$ ions. From the latter data, the acid-base interval of this solvent can be easily estimated as the length of acidity scale between standard solution - ~14 pO units. The pO range of $VO^{2+}$ stability has been determined to be 8.5 to 10.6. To complete the consideration of investigations in sulphate melts the work[73] should be mentioned, it is concerned with potentiometric studies of the acidic properties of $MoO_3$, $Cr_2O_7^{2-}$, $PO_3^-$, $P_2O_7^{4-}$, $V_2O_5$ in molten $K_2SO_4$-$Li_2SO_4$-$Na_2SO_4$ at 625°C. Acids mentioned were neutralized by sodium carbonate. The potential (pO) drop in equivalence points was observed at "acid-base" ratio 1:3 for vanadium oxide and 1:1 for all other Lux acids. Values of equilibrium constants were not presented.

#### 10.4.4.3 Silicate melts

Studies performed in molten silicates are of importance for applied purposes, since systems under consideration are widely used in various industries (slugs, glasses, etc.). Flood et al.[74-76] studied the acidity dependence of silicate melts upon their composition. A system PbO-$SiO_2$ have been studied in the range of $SiO_2$ concentrations of $N_{SiO_2}$ from 0 to 0.6 at 1100-1200°C. The coexistence of $(SiO_3)_3^{6-}$ and $(SiO_{2.5})_6^{6-}$ polyanions together with the ordinary orthosilicate ions, $SiO_4^{4-}$ has been shown. The basicities of molten glasses vs. $Na_2O/SiO_2$ ratios were studied[75] by the potentiometric method with the use of the gas oxygen electrode $Ag(O_2)$ in concentration cells. The decrease of the mentioned ratio has been stated to cause a reduction of the basicity, while magnesium oxide has been found to have no effect on the melt basicity. The gas oxygen electrode reversibility in molten PbO-$SiO_2$, $Na_2O$-CaO-$SiO_2$, MeO-PbO-$SiO_2$, where $Me^{2+}$ - alkaline earth cation, has been investigated.[77] The electrode $Pt(O_2)$ has been shown to be reversible to $O^{2-}$.

The use of a membrane oxygen electrode $Pt(O_2)|ZrO_2$ for molten silicate acidity control during glass making has been described.[78] Studies of basicities in molten $Na_2O$-$Al_2O_3$-$SiO_2$ using a potentiometric method[79] showed that the acidic properties of $Al_2O_3$ were weaker, then $SiO_2$ ones. Processes of glass corrosion in molten alkaline earth nitrates have been studied for glass compositions $Na_2O{\cdot}xAl_2O_3{\cdot}2SiO_2$, where x was in the range from 0 to 0.4.[80] The degree of interaction between the melt and immersed glass increased with the reduction of the alkaline earth cation radius. The effect of the cation charge and the radius on the basicity of lead meta-silicate melt at 800, 850, 900°C by additions of $Tl_2O$, PbO, CdO, ZnO and $Bi_2O_3$ has been studied by potentiometry.[81]

#### 10.4.4.4 The equimolar mixture KCl-NaCl

This melt is the most investigated among chloride based melts. Some oxoacidity studies were performed by Shapoval et al.[82-87] Main purposes of these works were to investigate oxygen electrode reversibility and to obtain equilibria constants for acid-base reactions including oxo-compounds of $Cr^{VI}$, $Mo^{VI}$ and $W^{VI}$ at 700°C. Equilibrium constants of acid-base interactions for $CrO_3$ and $PO_3^-$ have been determined.[82-84] The titration of the former substance proceeds in two stages:

$$CrO_3 \xrightarrow{2.88\times10^3} Cr_2O_7^{2-} \xrightarrow{1.4\times10^{-1}} CrO_4^{2-} \qquad [10.4.38]$$

Sodium metaphosphate is two-basic acid too:

$$PO_3^- \xrightarrow{(2.88\pm1.2)\times10^3} P_2O_7^{4-} \xrightarrow{(2.5\pm1.4)\times10^1} PO_4^{3-} \qquad [10.4.39]$$

The excess of titrant, $O^{2-}$, has been found to cause the formation of basic phosphates with assumed composition $PO_3^-{\cdot}2O^{2-}$. When $Na_2CO_3$ was used as titrant, there was no formation of the mentioned basic products. $Ba^{2+}$ and $Li^+$ cations have been shown to possess appreciable acidic properties, corresponding constant values were estimated as $8.1\times10^1$ and $3.53\times10^2$.

An investigation of acidic properties of $MoO_3$ was performed:[85]

$$MoO_3 \xrightarrow{5\times10^4} Mo_2O_7^{2-} \xrightarrow{880\pm880} MoO_4^{2-} \qquad [10.4.40]$$

Acidic properties of oxo-compounds of $W^{VI}$ have been studied.[86,87] The titration of $WO_3$ in molten chlorides was a two-stage process:

$$WO_3 \xrightarrow{1460\pm390} W_2O_7^{2-} \xrightarrow{880\pm880} WO_4^{2-} \qquad [10.4.41]$$

Results formed the base for the following electrochemical studies of electroreduction processes of Group VI metals.[88-90] These studies developed theoretical bases and principles to control electrochemical processes of metals and their compounds (carbides, borides, silicides) deposition from ionic melts.[81]

The acid-base equilibria in the scheelite ($CaWO_4$) solutions in the said chloride mixture at 1000K were investigated.[92] CaO solubility in molten KCl-NaCl has been determined to be about 0.084 mol%, that of scheelite was $10^{-3.5}$ mole/kg. Equilibrium constants for acid-base reactions with $WO_3$ participation have been determined, too:

$$WO_3 \xrightarrow{10^{10}} WO_4^{2-} \qquad [10.4.42]$$

The reverse titration has been assumed by authors[92] to result in the formation of polyanion $W_3O_{10}^{2-}$, pK=12.7. The titration curve was irreversible. This result can be easily explained taking into account that $WO_3$ is unstable in strong acidic chloride melts and removed from as $WO_2Cl_2$. Therefore, additions of $WO_3$ into the acidic melt should lead to uncontrolled titrant loss affecting experimental results. Ditto referred to all results obtained by the reverse titration by $MoO_3$ and $CrO_3$ as the equivalence point.

Acidic properties of $P^V$ oxocompounds in the chloride melt at 700°C were investigated using a membrane oxygen electrode Ni,NiO|$ZrO_2$.[93] Polyphosphates of compositions with $Na_2O$-$P_2O_5$ ratio from 1.67 to 3 formed in the acidic region:

$$P_nO_{3n+1}^{-2(n+2)} + PO_4^{3-} = P_{n+1}O_{3n+5}^{-(n+5)} \quad [10.4.43]$$

The corresponding constants were estimated. Polyphosphate solubilities were determined by a cryoscopic method.

The titration of $V_2O_5$ proceed in two stages:[94]

$$V_2O_5 \xrightarrow{pK=-9.3\pm0.3} VO_3^- \xrightarrow{pK=-8.3\pm0.3} V_2O_7^{4-} \quad [10.4.44]$$

Some results of oxoacidity studies in molten KCl-NaCl were obtained in our works,[95-99] they are presented in Table 10.4.1.

**Table 10.4.1. The Lux acid-base equilibrium constants in molten KCl-NaCl and NaI (at the confidence level 0.95) [After references 95,96,98]**

| Equilibrium | -pK | | |
|---|---|---|---|
| | m | N | mol% |
| | KCl-NaCl | | |
| $2PO_3^- + O^{2-} = P_2O_7^{4-}$ | 8.01±0.1 | 10.67 | 6.67 |
| $PO_3^- + O^{2-} = PO_4^{3-}$ | 5.93±0.1 | 7.26 | 5.26 |
| $PO_3^- + 2O^{2-} = [PO_4^{3-} \cdot O^{2-}]$ | 7.24±0.1 | - | - |
| $CrO_3 + O^{2-} = CrO_4^{2-}$ | 8.41±0.1 | 9.77 | 7.77 |
| $Cr_2O_7^{2-} + O^{2-} = 2CrO_4^{2-}$ | 7.18±0.1 | 7.18 | 7.18 |
| $2CrO_4^{2-} + O^{2-} = [2CrO_4^{2-} \cdot O^{2-}]$ | 1.60±0.2 | 4.26 | 0.26 |
| $MoO_3 + O^{2-} = MoO_4^{2-}$ | 8.32±0.2 | 9.65 | 7.65 |
| $MoO_3 + 2O^{2-} = [MoO_4^{2-} \cdot O^{2-}]$ | 9.71±0.3 | 11.37 | 7.37 |
| $WO_3 + O^{2-} = WO_4^{2-}$ | 9.31±0.2 | 10.64 | 8.64 |
| $WO_3 + 2O^{2-} = [WO_4^{2-} \cdot O^{2-}]$ | 10.67±0.5 | 13.33 | 9.33 |
| $B_4O_7^{2-} + O^{2-} = 4BO_2^-$ | 4.82±0.1 | 2.16 | 4.16 |
| $BO_2^- + O^{2-} = BO_3^{3-}$ | 2.37±0.2 | 3.17 | 1.70 |
| $V_2O_5 + O^{2-} = 2VO_3^-$ | 6.95±0.2 | 6.95 | 6.95 |
| $V_2O_5 + 2O^{2-} = V_2O_7^{4-}$ | 12.23±0.1 | 14.89 | 10.89 |
| $V_2O_5 + 3O^{2-} = 2VO_4^{3-}$ | 12.30±0.1 | 14.96 | 10.96 |
| $V_2O_5 + 5O^{2-} = 2[VO_4^{3-} \cdot O^{2-}]$ | 13.88±0.5 | 19.20 | 11.20 |
| $2GeO_2 + O^{2-} = Ge_2O_5^{2-}$ | 4.18±0.4 | 6.84 | 2.84 |

| Equilibrium | -pK | | |
|---|---|---|---|
| | m | N | mol% |
| | NaI | | |
| $2VO_3^- + O^{2-} = V_2O_7^{4-}$ | 5.40±0.3 | 7.10 | 4.10 |
| $V_2O_7^{4-} + O^{2-} = 2VO_4^{3-}$ | 1.68±0.3 | 1.68 | 1.68 |
| $B_4O_7^{2-} + O^{2-} = 4BO_2^-$ | 5.02±0.8 | 3.37 | 5.37 |

The results included in the Table 10.4.1 are in good agreement with the similar studies, performed using a membrane oxygen electrode Ni,NiO|$ZrO_2$. The use of the gas platinum-oxygen electrode resulted in significant underrating of constant values (of order of 3-4 pK units).

### 10.4.4.5 Other alkaline halide melts

Some studies of acid-base equilibria in molten alkaline halides have been performed by Rybkin et al.[17-20] The empirical acidity scales have been constructed on the base of potentiometric studies in molten KCl[17,18] and CsI.[19] The practical significance of the mentioned works consists in the proposition to use some buffer solutions - $SO_4^{2-}/S_2O_8^{2-}$, $WO_4^{2-}/W_2O_7^{2-}$, $PO_4^{3-}/P_2O_7^{4-}$ as reference standards for the indicator electrode calibration and determinations of pO in molten alkaline halides.

Rybkin and Banik[20] obtained the empirical acidity scale in molten NaI at 700°C. The basicities of $CO_3^{2-}$ and $OH^-$ were close which is incorrect since 0.01 mole/kg solutions were used for the scale constructing while equations

$$OH^- = \frac{1}{2}H_2O + \frac{1}{2}O^{2-} \qquad [10.4.45]$$

$$CO_3^{2-} = CO_2\uparrow + O^{2-} \qquad [10.4.46]$$

showed that these additions were not equivalent. $CO_3^{2-}$ concentration recalculated to oxide ions was twice as high as OH concentration. Indeed, in the later work these authors[100,101] showed that hydroxide ion was stronger base than carbonate ion.

We have performed a study of acidic properties of boron (III) and vanadium (V) oxo-compounds in molten NaI at 700°C[98] (see Table 10.4.1). A comparative study of strength of Lux bases $OH^-$, $CO_3^{2-}$, $SO_4^{2-}$ was determined[100,101] by potentiometric titration using sodium pyrophosphate as acid. Two moles of the first base may be neutralized by 1 mole of $P_4O_7^{2-}$, while two other moles react with the acid in ratio 1:1. On the base of e.m.f. drop magnitude at the equivalence point, the bases have been arranged in sequence $OH^- > CO_3^{2-} > SO_4^{2-}$ of basicity decrease. The equilibrium constants were not estimated.[100,101]

There is no correlation[17,19] between substances entered into melt and those really existing in it. E.g., potassium nitrate and nitrite[17,18] are decomposed to $K_2O$ at temperatures considerably lower than the temperature of the experiment (700°C). Ditto[17,18] referred to $K_2S_2O_7$ and $Na_2S_2O_8$ as acidic phosphates.[19] In the latter work it has been found that pyrophosphate acidity is larger than that for metaphosphate. However, titration of phosphorus (V) oxo-compounds in melts runs according to the scheme [10.4.39]. The acidity decreases, and, hence, pyrophosphate acidity should be lower, since in pair "metaphosphate-pyrophosphate" the latter is the conjugated base.

The absence of correlation between pH in aqueous solutions and pO in melts for the same substance is often noted, this conclusion is made mainly from data for oxocompounds of $P^V$. This fact is due to the following reason. The stages of phosphoric acid neutralisation in aqueous solutions are:

$$H_3PO_4 \xrightarrow{OH^-} H_2PO_4^- \xrightarrow{OH^-} HPO_4^{2-} \xrightarrow{OH^-} PO_4^{3-} \qquad [10.4.47]$$

In melts, a similar sequence can be presented as

$$\frac{1}{2}P_2O_5 \xrightarrow{1/2O^{2-}} PO_3^- \xrightarrow{1/2O^{2-}} \frac{1}{2}P_2O_7^{4-} \xrightarrow{1/2O^{2-}} PO_4^{3-} \qquad [10.4.48]$$

Hence, correlations should be found between the anion in aqueous solution and corresponding anhydro-acid in melt.

$NaPO_3$ and $Na_4P_2O_7$ may be easily obtained by the calcination of the corresponding acidic salts, but reverse processes do not take place in aqueous media and pyro- and metaphosphate exist in water as salts of stable acids. For arsenates which are more prone to hydrolysis similar correlation should take place. The absence of the correlation is caused mainly by kinetic limitations.

## 10.4.5 REACTIONS OF MELTS WITH GASEOUS ACIDS AND BASES

The gases present in the atmosphere over molten salts can react both with their principal components and with impurities. In reactions of the first kind, which include, in particular, hydrolysis, the gas (water) behaves as a Lux base since its action results in the increase of $O^{2-}$ concentration. In reactions of the second type, which are usually employed for the purification of melts, the gas has acidic properties.

### 10.4.5.1 High-temperature hydrolysis of molten halides

These reactions were investigated mainly for alkaline halides and their mixtures. Hanf and Sole[102] studied the hydrolysis of solid and molten NaCl in the temperature range 600-950°C by so-called "dynamic method" which consisted of passing inert gas ($N_2$) containing water vapor through a layer of solid or fused NaCl. The developed routine determined the equilibrium constant of the following reaction:

$$NaCl_{s,1} + H_2O_{gas} = NaOH_{s,1} + HCl_{gas} \qquad [10.4.49]$$

The values of the equilibrium constants (log K) of [10.4.49] for solid and liquid sodium chloride were in the ranges -8 to -6 (650-800°C) and -6 to -5 (800-900°C).

The hydrolysis of NaCl occurs to a slight extent, although the rate of HCl evolution exceeded the expected value because of dissolution of NaOH formed in NaCl.

An evident disadvantage of the above method[102] is assumption that the partial pressure of HCl in the gas phase is equal to the equilibrium pressure. The same method was used to study the high-temperature hydrolysis and oxidation of sodium iodide.[103] According to the estimates, the heats of the reactions:

$$NaI + H_2O = NaOH + HI \qquad [10.4.50]$$

$$NaI + 1/2O_2 = 1/2Na_2O_2 + 1/2I_2 \quad [10.4.51]$$

are 32.8±2 and 13.9±1 kJ/mole for solid NaI and 30.6±2 and 9.3±2 kJ/mole for liquid one. These values suggest that NaI has a greater tendency to oxidation than to hydrolysis. This conclusions cannot be adopted without stipulation, because it is based on values ΔH and not ΔG, although only the latter values determine the direction of a reversible reactions at constant p and T.

A simple but unquestionable method has been used to investigate high-temperature hydrolysis of molten KCl-NaCl.[104] A mixture of HCl and $H_2O$ obtained by passing an inert gas through aqueous solutions at a definite concentration was passed into the melt. Measurements of the equilibrium $O^{2-}$ concentration by potentiometric method allowed to calculate the equilibrium constants of [10.4.13] as $pK=55.3\times10^3T^{-1}-40.2$. At 1000, pK=15.1, it means that equilibrium [10.4.13] in molten KCl-NaCl is displaced to the left.

A similar method has been used to study the hydrolysis of the eutectic KCl-LiCl at 500°C (pK=9.77±0.4).[105] The hydrolysis of the chloride melt is thermodynamically unfavorable and is completely suppressed in the presence of bases (even at a concentration of $\sim10^{-3}$ mole/kg).

The potentiometric cells:

$$(H_2O + H_2)Au|NMeOH + (1-N)MeX||MeX|C(X_2) \quad [10.4.52]$$

were used to investigate the hydrolysis of individual alkaline halide melts (with the exception of the lithium salts).[106]

The e.m.f. of cell [10.4.52] is related to the concentration of the reactants by the expression:

$$E = E^0 + \frac{RT}{F}\ln\frac{P_{H_2}^{1/2}P_{X_2}^{1/2}}{P_{H_2O}}a_{MeOH} \quad [10.4.53]$$

where $P_{H_2}^{1/2}$ and $P_{X_2}^{1/2}$ are partial pressures of hydrogen and halogen over the melt, respectively.

The logarithms of the equilibrium constants

$$\ln K = \frac{F}{RT}E - \frac{\Delta G_{HX}^0}{RT} + \ln\frac{P_{H_2}^{1/2}P_{X_2}^{1/2}}{P_{H_2O}}a_{MeOH} \quad [10.4.54]$$

are negative for all the melts studied, which indicates that the hydrolysis of alkali metal halide melts is thermodynamically unfavorable. The trend towards hydrolysis of melts of individual alkaline halides diminishes with the increase in the radii of both cation and anion. According to the results of Smirnov et al.[106] the solutions of alkali metal hydroxides in the corresponding individual halides are close to ideality.

### 10.4.5.2 The processes of removal of oxide admixtures from melts

The studies of the elimination of oxygen containing impurities from ionic melts are of an applied character and are performed in melts having industrial applications. Various halogenating agents, usually hydrogen halides or the halogens themselves, are used to eliminate the impurities:[107-111]

$$2HX\uparrow + O^{2-} = H_2O\uparrow + 2X^- \quad [10.4.55]$$

$$X_2\uparrow + O^{2-} = 1/2O_2\uparrow + 2X^- \quad [10.4.56]$$

However, the water formed in reaction [10.4.55] exhibits the properties of Lux acid owing to process opposite to [10.4.45] and this retards the purification process. The cessation of the HX effect may lead to partial hydrolysis, because reactions of type of [10.4.55] are reversible. The shift of [10.4.55] to the left is also promoted by the thermal instability of HBr and HI.

The effectiveness of the purification of melts based on lithium salts by hydrogen halides is much lower. Water dissolves in the KCl-LiCl melt in appreciable amounts and is firmly retained at temperatures up to 400°C.[30] When dry HCl is passed for 1 h, removal of $H_2O$ is incomplete.

The effectiveness of the purification by the halogens depends on their oxidation-reduction potentials and decreases from chlorine to iodine. In addition, the latter can disproportionate:

$$3I_2\uparrow + O^{2-} = 5I^- + IO_3^- \quad [10.4.57]$$

However, the addition of $NH_4X$ is recommended only for the drying and heat treatment of the initial components and its use for industrial purification of fused salts is undesirable because it is much easier to ensure the continuous supply of the corresponding hydrogen halide in the melt being purified. An additional lack of this purifying agent results in evolution of gaseous ammonia as one of the finishing products.

It is believed that the most convenient method of the elimination of oxygen-containing impurities from alkali metal halide melts is carbohalogenation.[112-118]

The reduction of $O^{2-}$ concentration under action of $CCl_4$, $Cl_2$, C (C is acetylene carbon black)+$Cl_2$ and $COCl_2$ may be explained by the occurrence of the following reactions:

$$CCl_4\uparrow + 2O^{2-} = 4Cl^- + CO_2\uparrow \quad [10.4.58]$$

$$C + Cl_2\uparrow + O^{2-} = 2Cl^- + CO\uparrow \quad [10.4.59]$$

$$COCl_2\uparrow + O^{2-} 2Cl^- + CO_2\uparrow \quad [10.4.60]$$

Thermodynamic analysis of processes [10.4.58]-[10.4.60] showed that the effectiveness of the purification is in all cases approximately the same,[113] but reaction [10.4.58] has the advantage that highly toxic reagents ($Cl_2$, $COCl_2$) are not used in it. For this reason, $CCl_4$ has been used in virtually all subsequent studies involving the carbochlorination of chloride melts.[114-118]

It has been suggested that $CHBr_3$, $CBr_4$, $C_2H_5Br$ and $CHI_3$ may be used for the removal of oxygen-containing impurities from bromide and iodide melts.[119] A significant advantage of carbohalogenation is that it leads to an appreciable decrease in the concentration of transition metal cations in the melt being purified. However, in the purification of melts by halogen-substituted compounds with a relatively low halogen content, carbon (carbon black) accumulates in the melt and is displaced by the crystallization front when single crystals are grown.[112]

The presence of a suspension of carbon in alkali metal halide melts is not always desirable and this method cannot therefore be recognized as universal.

The carbohalogenation of melts results in formation of $CO_2$ as one of products of [10.4.58]-[10.4.60] processes, carbon dioxide reacts with remaining oxide ions and induces the accumulation of fairly stable carbonate-ions. The latter is the main form in which oxygen-containing impurities exist in melts purified in this way. The purification threshold of carbohalogenation is estimated as $1\text{-}2\times10^{-4}$ mole/kg of $O^{2-}$.[24]

The use of halo-derivatives of silicon for the purification of iodide melts[120] is based on the reaction:

$$SiX_4\uparrow + 2O^{2-} = 4X^- + SiO_2\downarrow \qquad [10.4.61]$$

the product of [10.4.61] is silicon dioxide, which is separated from the melt as a consequence of the difference between the densities. Thermodynamic analysis showed[121] that the effectiveness of the purification of iodide melts diminishes in the sequence $SiI_4$>HI>$I_2$.

The use of silicon halides for the purification of the melts used in the growing single crystals does not lead to the appearance of additional impurities, because the processes are carried out in quartz (i.e., $SiO_2$) containers.

Although the methods of purification described are fairly effective in most cases, some kind of oxygen-containing and cationic impurities, characteristic of each method, always remains in the melt. The method of purification must therefore be selected depending on the aims of the subsequent application of the melt.

## REFERENCES


1 G.N. Lewis, **Valence and the structure of atoms and molecules**, *S.n.*, New York, 1923.
2 H. Lux, *Z. Elektrochem.*, **45**, 303(1939).
3 H. Flood and T. Forland, *Acta Chem.Scand.*, **1**, 592 (1947).
4 H. Flood and T. Forland, *Acta Chem.Scand.*, **1**, 781 (1947).
5 H. Flood, T. Forland and B. Roald, *Acta Chem.Scand.*, **1**, 790 (1947).
6 N.J. Bjerrum, C.R. Roston and G.P. Smith, *Inorg.Chem.*, **6**, 1162 (1967).
7 N.J. Bjerrum and G.P. Smith, *Inorg.Chem.*, **7**, 2528 (1968).
8 B. Tremillon and G. Letisse, *J. Electroanal. Chem.*, **17**, 371 (1968).
9 G. Torsi and G. Mamantov, *J. Electrocanal. Chem.*, **30**, 193 (1971).
10 G. Torsi and G. Mamantov, *Inorg. Chem.*, **11**, 1439 (1972).
11 R. G. Pearson, *J. Amer. Chem. Soc.*, **85**, 3533 (1963).
12 R. G. Pearson, *J. Chem. Educ.*, **45**, 581 (1968).
13 R. G. Pearson, *J. Chem. Educ.*, **45**, 643 (1968).
14 I. G. Dioum, J. Vedel, and B. Tremillon, *J. Electroanal. Chem.*, **137**, 219 (1982).
15 J. F. Wagner, Electrochemical study in molten sodium fluoroborate at 420°C, Rep. CEA-N-2350, 15(2), Abstr. No15:005221 (1983).
16 V. L. Cherginets, *Rus. Chem. Rev.*, **66**, 597 (1997).
17 Yu. F. Rybkin and A. S. Seredenko, *Ukr. Khim. Zhurn.*, **36**, 133 (1970).
18 Yu. F. Rybkin and A. S. Seredenko, *Ukr. Khim. Zhurn.*, **40**, 137 (1974).
19 Yu. F. Rybkin and V. V. Banik, In Single Crystals and Engineering , No. 2(9) issue, Inst. Single Crystals, Kharkov, 152 (1973).
20 N. N. Ovsyannikova and Yu. F. Rybkin, *Ukr. Khim. Zhurn.*, **42**, 151 (1976).
21 J. E. Huheey, **Inorganic chemistry. Principles of structure and reactivity**, *Harper and Row Publishers*, New York, 1983.
22 M. Dratovsky and D. Havlichek, *Electrochim. Acta*, **28**, 1761 (1983).

23 A. M. Shams El Din and A. A. El Hosary, *J. Electroanal. Chem.*, **9**, 349 (1965).
24 V. L. Cherginets and T. P. Rebrova, *Electrochim. Acta*, **45**, 469 (1999).
25 R. Combes, M. N. Levelut and B. Tremillon, *Electrochim. Acta*, **23**, 1291 (1978).
26 R. Combes, B. Tremillon, F. De Andrade, M. Lopes and H. Ferreira, *Anal. Lett.*, **15A**, 1585 (1982).
27 V. L. Cherginets, *Electrochim. Acta*, **42**, 3619 (1997).
28 Y. Castrillejo, A. M. Martinez, G. M. Haarberg, B. Borresen, K. S. Osen and R. Tunold, *Electrochim. Acta*, **42**, 1489 (1997).
29 A. I. Novozhilov and E. I. Ptchelina, *Zhurn. neorg. Khim.*, **22**, 2057 (1977).
30 W. J. Burkhardt and J. D. Corbett, *J. Amer. Chem. Soc.*, **79**, 6361 (1957).
31 B. J. Brough, D. H. Kerridge and M. Mosley, *J. Chem. Soc. (A)*, **N11**, 1556 (1966).
32 J. A. Duffy, M. D. Ingram and I. D. Sommerville, *J. Chem. Soc. Faraday Trans.*, **74**, 1410 (1978).
33 T. Maekawa and T. Yokokawa, *Nippon Kagaku*, **N6**, 900 (1982).
34 A. Kato R. Nishibashi, M. Hagano and I. Mochida, *J. Amer. Ceram. Soc.*, **55**, 183 (1972).
35 N. Iwamoto, *Youen*, **N21**(3), 287 (1978).
36 F. R. Duke and S. Yamamoto, *J. Amer. Chem. Soc.*, **80**, 5061 (1958).
37 F. R. Duke and S. Yamamoto, *J. Amer. Chem. Soc.*, **81**, 6378 (1959).
38 J. M. Shlegel, *J. Phys. Chem.*, **69**, 3638 (1965).
39 J. M. Shlegel, *J. Phys. Chem.*, **71**, 1520 (1967).
40 F. R. Duke, J. M. Shlegel, *J. Phys. Chem.*, **67**, 2487 (1963).
41 I. Slama, *Coll. Czechoslov. Chem. Commun.*, **28**, 985 (1963).
42 I. Slama, *Coll. Czechoslov. Chem. Commun.*, **28**, 1069 (1963).
43 A. M. Shams El Din and A. A. El Hosary, *J. Electroanal. Chem.*, **8**, 312 (1964).
44 A. M. Shams El Din and A. A. A. Gerges, *J. Inorg. Nucl. Chem.*, **26**, 1537 (1963).
45 A. M. Shams El Din and A. A. A. Gerges, *Electrochim. Acta*, **9**, 123 (1964).
46 A. M. Shams El Din and A. A. El Hosary, *J. Electroanal. Chem.*, **7**, 464 (1964).
47 A. M. Shams El Din and A. A. El Hosary, *J. Electroanal. Chem.*, **17**, 238 (1968).
48 A. M. Shams El Din, A. A. El Hosary and A. A. A. Gerges, *J. Electroanal. Chem.*, **6**, 131 (1963).
49 A. M. Shams El Din, *Electrochim. Acta*, **7**, 285 (1962).
50 A. M. Shams El Din and A. A. A. Gerges, *J. Electroanal. Chem.*, **4**, 309 (1962).
51 A. M. Shams El Din and, A. A. El Hosary, *J. Electroanal. Chem.*, **16**, 551 (1968).
52 A. M. Shams El Din, A. A. El Hosary and H. D. Taki El Din, *Electrochim. Acta*, **13**, 407 (1968).
53 A. A. El Hosary, M. E. Ibrahim and A. M. Shams El Din, *Electrochim. Acta*, **24**, 645 (1979).
54 A. M. Shams El Din and A. A. El Hosary, *Electrochim. Acta*, **13**, 135 (1968).
55 A. M. Shams El Din and A. A. El Hosary, *Electrochim. Acta*, **12**, 1665 (1967).
56 J. D. Burke and D. H. Kerridge, *Electrochim. Acta*, **19**, 251 (1974).
57 A. M. Shams El Din and A. A. A. Gerges, In **Electrochemistry**, *Pergamon Press*, 1964 pp. 562-577.
58 A. M. Shams El Din and A. A. A. Gerges, *Electrochim. Acta.*, **9**, 613 (1964).
59 N. Coumert and M. Porthault, J. -C. Merlin, *Bull. Soc. Chim. France*, **33**, 910 (1965).
60 N. Coumert and M. Porthault, J. -C. Merlin, *Bull. Soc. Chim. France*, **35**, 332 (1967).
61 M. Hassanein and N. S. Youssep, *Indian J. Chem.*, **A21**, 72 (1982).
62 C. Liteanu, E. Cordos and L. Margineanu, *Rev. Roumaine Chim.*, **15**, 583 (1970).
63 J. M. Schlegel, *J. Chem. Educ.*, **43**, 362 (1966).
64 Y. Hoshino, T. Utsunomiya and O. Abe, *Bull. Chem. Soc. Jpn.*, **54**, 135 (1981).
65 A. Baraka, A. Abdel-Razik and A. J. Abdel-Rohman, *Surface Technol.*, **25**, 31 (1985).
66 A. Baraka, A. J. Abdel-Rohman and E. A. El-Taher, *Mater. Chem. Phys.*, **9**, 447 (1983).
67 A. Baraka, A. J. Abdel-Rohman and E. A. El-Taher, *Mater. Chem. Phys.*, **9**, 583 (1983).
68 A. I. Efimov, **Properties of inorganic compounds. Handbook**, *Khimiya*, Leningrad, 1983.
69 I. T. Goronovskiy, Yu. P. Nazarenko and E. F. Nekryach, **Short handbook on chemistry**, *Naukova Dumka*, Kiev, 1987.
70 G. J. Janz, **Molten salts handbook**, *Academic Press*, New York, 1967.
71 Y. Kaneko and H. Kojima, Int. Symp. Molten Salts Chem. Technol., No. 1, 441 (1983) cited Chem. Abstr. 102, Abstr. No. 122309c (1988. )
72 Y. Kaneko and H. Kojima, *Youen*, **28**, 109 (1985).
73 A. Rahmel, *J. Electroanal. Chem.*, **61**, 333 (1975).
74 H. Flood and W. Knapp, *J. Amer. Ceram. Soc.*, **46**, 61 (1963).
75 T. Forland and M. Tashiro, *Glass Ind.*, **37**, 381 (1956).
76 T. Forland, *Glastekn. Tidskr.*, **17**, 35 (1962).

77 V. I. Minenko, C. M. Petrov and N. S. Ivanova, *Izv. vysshikh utchebnykh zavedeniy. Chernaya metallurgiya*, N7, 10 (1960).
78 US Patent 4,313,799; *Chem. Abstr.*, **96**, 115179g (1982).
79 H. Itoh and T. Yokokawa, *Trans. Jpn. Inst. Met.*, **25**, 879 (1984).
80 T. T. Bobrova, V. V. Moiseev, T. V. Permyakova and G. E. Sheshukova, *Izv. AN SSSR. Neorg. mater*, **9**, 1416 (1973).
81 R. Didtschenko and E. G. Rochow, *J. Amer. Chem. Soc.*, **76**, 3291 (1954).
82 V. I. Shapoval and O. G. Tsiklauri, Conf. "Phys. chemistry and electrochemistry of molten salts and solid electrolytes", Sverdlovsk, June 5-7, 1973, Part II, Sverdlovsk, 1973 pp. 32-33.
83 V. I. Shapoval, A. S. Avaliani and O. G. Tsiklauri, *Soobsch. AN Gruz. SSR*, **72**, 585 (1973).
84 Yu. K. Delimarsky, V. I. Shapoval, O. G. Tsiklauri and V. A. Vasilenko, *Ukr. Khim. Zhurn.*, **40**, 8 (1974).
85 V. I. Shapoval, A. S. Avaliani and N. A. Gasviani, *Soobsch. AN Gruz. SSR*, **72**, 105 (1973).
86 V. I. Shapoval, V. F. Grischenko and L. I. Zarubitskaya, *Ukr. Khim. Zhurn.*, **39**, 867 (1973).
87 V. I. Shapoval, V. F. Grischenko and L. I. Zarubitskaya, *Ukr. Khim. Zhurn.*, **38**, 1088 (1972).
88 V. I. Shapoval, Yu. K. Delimarsky and V. F. Grischenko, in **Ionic melts**, *Naukova Dumka*, Kiev, 1974, pp. 222-241.
89 O. G. Tsiklauri and N. A. Gasviani, In Materials of Conference of Young Scientists. Inst. of chemistry and electrochemistry of Academy of Sciences of Georgian SSR, Tbilisi, 1976, p. 63.
90 V. I. Shapoval, O. G. Tsiklauri and N. A. Gasviani, *Soobsch. AN Gruz. SSR*, **88**, 609 (1977).
91 Kh. B. Kushkhov and V. I. Shapoval, X All-Union Conference on Physical Chemistry and Electrochemistry of Ionic Melts and Solid Electrolytes, Ekaterinbourg, October 27-29, 1992, Ekaterinbourg, 1992, p. 3.
92 R. Combes and B. Tremillon, *J. Electroanal. Chem.*, **83**, 297 (1977).
93 M. Tazika, S. Mizoe, M. Nagano and A. Kato, *Denki kagaku.*, **46**, 37 (1978).
94 R. Combes, F. De Andrade and L. Carvalho, *C. r. Acad. Sci.*, **C285**, 137 (1977).
95 V. L. Cherginets and V. V. Banik, *Rasplavy*, N6, 92 (1990).
96 V. L. Cherginets and V. V. Banik, *Rasplavy*, N2, 118 (1991).
97 T. P. Boyarchuk, E. G. Khailova and V. L. Cherginets, *Ukr. Khim. Zhurn.*, **58**, 758 (1992).
98 V. L. Cherginets and V. V. Banik, *Zhurn. Fiz. Khim.*, **68**, 145 (1994).
99 V. L. Cherginets, Potentiometric studies of acidic properties of niobium (V) and germanium (IV) oxides in chloride melts, Kharkov, 1991; Dep in ONIITEKhim (Cherkassy), N 241 (1991).
100 Yu. F. Rybkin, V. V. Banik, In Methods of obtaining and investigations of single crystals and scintillators, Inst. Single Crystals, Kharkov, 1980, p. 121-125 (1980).
101 Yu. F. Rybkin, V. V. Banik, In Single crystals and engineering, Inst. Single Crystals, Kharkov, 1974, p. 111-114.
102 N. V. Hanf and M. J. Sole, *Trans. Faraday Soc.*, **66**, 3065 (1970).
103 Y. F. Rybkin and Y. A. Nesterenko, *Zhurn. Fiz. Khim.*, **50**, 781 (1976).
104 R. Combes, J. Vedel and B. Tremillon, *Electrochim. Acta*, **20**, 191 (1975).
105 V. L. Cherginets and V. V. Banik, *Rasplavy*, (4), 98 (1991).
106 M. V. Smirnov, I. V. Korzun and V. A. Oleynikova, *Electrochim. Acta*, **33**, 781 (1988).
107 C. Butler, J. R. Russel, R. B. Quincy, *J. Chem. Phys.*, **45**, 968 (1966).
108 F. Rosenberger, in **Ultrapurity methods and Techniques**, *Dekker Inc*, New York, 1972, pp. 3-70.
109 D. Ecklin, *Helv. Chim. Acta*, **50**, 1107 (1967).
110 J. M. Peech, D. A. Bower, R. O. Rohl, *J. Appl. Phys.*, **38**, 2166 (1967).
111 U. Gross, *Mat. Res. Bull.*, **5**, 117 (1970).
112 M. Lebl and J. Trnka, *Z. Phys.*, **186**, 128 (1965).
113 O. V. Demirskaya and Y. A. Nesterenko, In Physics and Chemistry of Crystals, Inst. Single Crystals, Kharkov, 1977, pp. 155-159.
114 Y. F. Rybkin and O. V. Demirskaya, In Single Crystals and Engineering, Inst. Single Crystals, Kharkov, 1974, No. 1(10), pp. 115-118.
115 R. C. Pastor and A. C. Pastor, *Mat. Res. Bull.*, **10**, 117 (1975).
116 R. C. Pastor and A. C. Pastor, *Mat. Res. Bull.*, **11**, 1043 (1976).
117 A. I. Agulyanskii and P. T. Stangrit, *Zhurn. Prikl. Khim.*, **10**, 1201 (1977).
118 V. V. Banik and V. L. Cherginets, VII All-Union Conf. Chem. and Technology of Rare Alkali Elements, Apatity, June 5-10, 1988, Cola Division of Acad. Sci. of USSR, Apatity, 1988, p. 116.
119 V. V. Banik, N. I. Davidenko and Y. A. Nesterenko, In Physics and Chemistry of Solids, Inst. Single Crystals, Kharkov, 1983, No. 10, pp. 139-141.
120 J. Ekstein, U. Gross and G. Rubinova, *Krist. Tech.*, **3**, 583 (1968).

121 N. N. Smirnov and V. R. Lyubinskii, In Single Crystals and Engineering, Inst. Single Crystals, Kharkov, 1971, No. 5, pp. 95-101.
122 O. B. Babushkina and S. V. Volkov, XIV Ukraine Conf. Inorg. Chem., Kiev, September 10-12, 1996, Kiev National University, Kiev, 1996, p. 33.

# 11

# ELECTRONIC AND ELECTRICAL EFFECTS OF SOLVENTS

## 11.1 THEORETICAL TREATMENT OF SOLVENT EFFECTS ON ELECTRONIC AND VIBRATIONAL SPECTRA OF COMPOUNDS IN CONDENSED MEDIA

MATI KARELSON
**Department of Chemistry, University of Tartu, Tartu, Estonia**

### 11.1.1 INTRODUCTION

The electronic-vibrational spectra of molecules can be substantially influenced by the surrounding condensed medium. The resultant effects arise from a variety of intermolecular interactions between the chromophoric solute and the solvent molecules in such media. Experimentally, these effects can be observed as

- the shifts of the spectral maxima (solvatochromic shifts)
- the change in the intensity of the spectral line or band
- the change of the shape and width of the spectral band

Each of those, so-called solvent effects can be described theoretically using different model approaches.

The solvatochromic spectral shifts are expected to arise from the difference in the solvation of the ground and the excited states of the molecule. As a result of the spectroscopic excitation, the charge distribution of the molecule changes and thus the interaction will be different in the ground state and in the excited state of the molecule. The direction and size of the respective spectral shift depends directly on the difference in the solvation energy of the molecule in those two states. The larger solvation energy of the ground state ($S_0$), as compared to that of the excited state ($S_1$), results in the negative solvatochromic shift (blue shift) of the spectral maximum (cf. Figure 11.1.1a). Alternatively, the stronger solvation of the excited state, as compared to the solvation of the ground state, leads to the decrease of the excitation energy and is reflected by the positive solvatochromic shift (red shift) in the spectrum of the compound (Figure 11.1.1b).

In each case, the direction and the size of the shift depends on the nature and electronic structure of the ground and excited state. For example, in the case of the n $\rightarrow \pi^*$ transition in acetone (Scheme 11.1.1) an electron from the n-orbital (lone pair) is transferred to the antibonding $\pi^*$ orbital localized on the C=O double bond. In

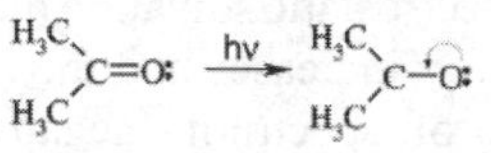

Scheme 11.1.1.

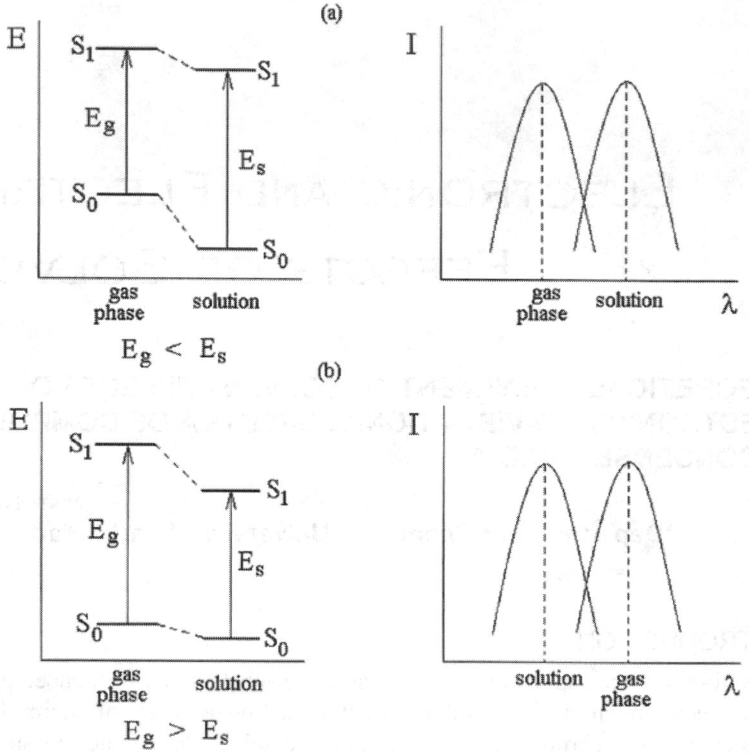


Figure 11.1.1. The origin of two types of solvatochromic shifts in the spectra of chromophoric compounds.

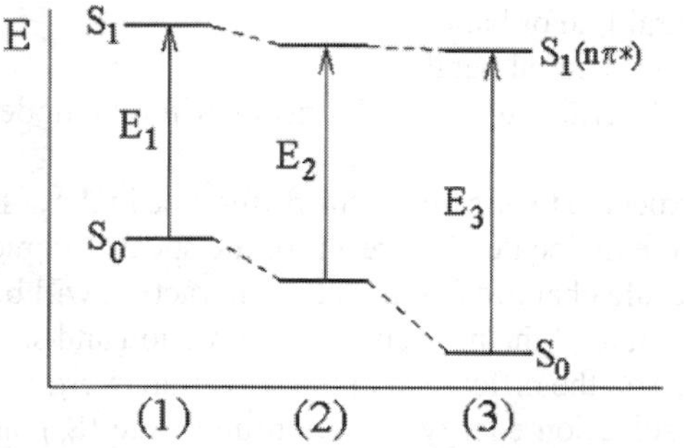


Figure 11.1.2. The schematic representation of the change of the solvation energy of the ground ($S_0$) and excited state ($S_1$) of acetone moving from a non-polar solvent (1) to a polar non-hydrogen bonded solvent (2) to a polar hydrogen-bonded solvent (3).

the ground state, the acetone molecule has a significant dipole moment (2.7 D) arising from the polarity of the C=O bond. Because of the difference in the electronegativity of bonded atoms, the electron distribution is shifted towards the oxygen that could be characterized by a negative partial charge on this atom. However, in the excited state the electron cloud is shifted from the oxygen atom to the bond and, correspondingly, the dipole moment of the molecule is substantially reduced. In result, the interaction of the dipole of the solute (acetone) with the surrounding medium in more polar solvents is larger in the ground state as compared to the excited state (Figure 11.1.2).

Thus, the energy of the ground state is lowered more by the electrostatic solvation than the energy of the excited state. Consequently, the excitation energy increases and the respective spectral maximum is shifted towards the blue end of spectrum (negative solvatochromic shift). In the hydrogen-bonding solvents, the ground state of acetone is additionally stabilized by the hydrogen bonding of the oxygen lone pair by the solvent that leads to further increase of the excitation energy (Figure 11.1.2) and the respective blue shift of the spectral maximum. Notably, the formation of this hydrogen bonding is impossible in the $S_1(n\pi^*)$ excited state of the acetone because of the electron transfer from the oxygen lone pair to the antibonding $\pi^*$ orbital.

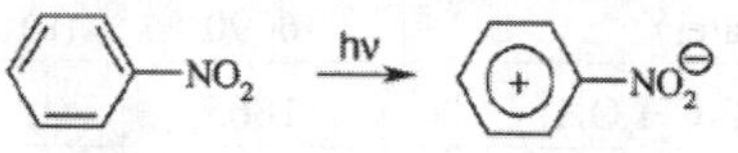


Scheme 11.1.2.

In many cases, the dipole moment increases in the excited state. For instance, in the nitrobenzene (Scheme 11.1.2) the $\pi \rightarrow \pi^*$ transition leads to the substantial redistribution of the electronic charge reflected by the shift of negative charge on the nitro-group. The real excited state is given by a combination of the $L_a$ and $L_b$ states. Nevertheless, the dipole moment of nitrobenzene is substantially increased in the first excited state ($S_1$). Because of the substantial charge redistribution in such excited states, they are often called the charge-transfer (CT) states. The much larger dipole of the nitrobenzene in the $S_1(\pi\pi^*)$ state is additionally stabilized by a more polar solvents that leads to the decrease in the excitation energy (Figure 3) and to the corresponding solvatochromic red shift of the spectral maximum.

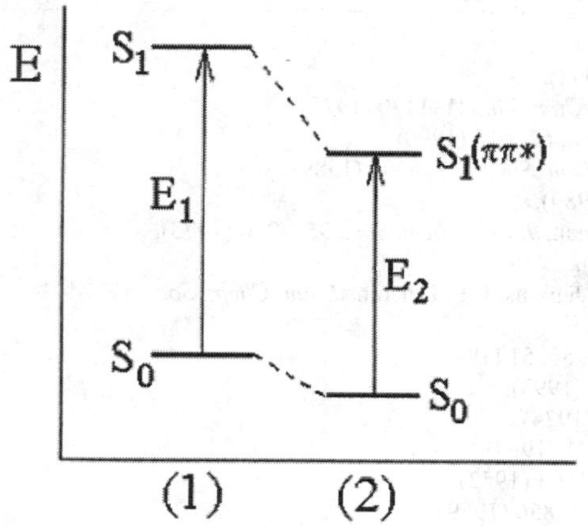


Figure 11.1.3. The relative energies of the ground ($S_0$) and the first excited state ($S_1(\pi\pi^*)$) of nitrobenzene in non-polar (1) and polar (2) solvents.

In Table 11.1.1, the solvatochromic shifts characterizing various positively and negatively solvatochromic compounds are listed. In most cases, the theoretical treatment of the solvatochromic shifts has been based on the calculation of the solvation energies of the chromophoric molecule in the ground and excited states, respectively.

**Table 11.1.1. The solvatochromic shifts for various positively and negatively solvatochromic compounds**

| Compound | $\nu_{max}$ (non-polar solvent), $cm^{-1}$ | $\nu_{max}$ (polar solvent), $cm^{-1}$ | $\Delta\nu_{max}$, $cm^{-1}$ | Ref. |
|---|---|---|---|---|
| (Scheme 3) | 30000 (hexane) | 25760 ($CF_3CH_2OH$) | +4240 | (a) |
| (Scheme 4) | 30170 (hexane) | 26140 (water) | +4030 | (b) |
| (Scheme 5) | 20640 (hexane) | 16860 (water) | +3780 | (c) |
| (Scheme 6) | 22620 (hexane) | 19920 (DMSO) | +2700 | (d) |
| (Scheme 7) | 20730 (hexane) | 18410 (methanol) | +2320 | (e) |
| (Scheme 8) | 27400 (cyclohexane) | 23230 (water) | +4170 | (f) |
| (Scheme 9) | 43370 (isooctane) | 41220 (water) | +2150 | (g) |
| (Scheme 10) | 14600 (toluene) | 24100 (water) | -9500 | (h) |

| Compound | $\nu_{max}$ (non-polar solvent), $cm^{-1}$ | $\nu_{max}$ (polar solvent), $cm^{-1}$ | $\Delta\nu_{max}$, $cm^{-1}$ | Ref. |
|---|---|---|---|---|
| (Scheme 11) | 15480 (tetrahydrofuran) | 24450 (water) | -8970 | (i) |
| (Scheme 12) | 20410 (chloroform) | 22080 (water) | -1670 | (j) |
| (Scheme 13) | 20160 (chloroform) | 22370 (water) | -2210 | (j) |
| (Scheme 14) | 16080 (chloroform) | 22170 (water) | -6090 | (k) |
| (Scheme 15) | 35870 ($CCl_4$) | 37735 ($CF_3CH_2OH$) | -1865 | (l) |
| (Scheme 16) | 16390 (pyridine) | 21280 (water) | -4890 | (m) |
| (Scheme 17) | 19560 (chloroform) | 22060 (water) | -2500 | (n) |


(a) S. Spange, D. Keutel, *Liebig's Ann. Chim.*, 423 (1992).
(b) S. Dähne, F. Shob, K.-D. Nolte, R. Radeglia, *Ukr. Khim. Zh.*, **41**, 1170 (1975).
(c) J.F. Deye, T.A. Berger, A.G. Anderson, *Anal. Chem.*, **62**, 615 (1990).
(d) D.-M. Shin, K.S. Schanze, D.G. Whitten, *J. Am. Chem. Soc.*, **111**, 8494 (1989).
(e) E. Buncel, S. Rajagopal, *J. Org. Chem.*, **54**, 798 (1989).
(f) M.J. Kamlet, E.G. Kayser, J.W. Eastes, W.H. Gilligan, *J. Am. Chem. Soc.*, **95**, 5210 (1973).
(g) E.M. Kosower, *J. Am. Chem. Soc.*, **80**, 3261 (1958).
(h) L.S.G. Brooker, A.C. Craig, D.W. Heseltine, P.W. Jenkins, L.L. Lincoln, *J. Am. Chem. Soc.*, **87**, 2443 (1965).
(i) M.A. Kessler, O.S. Wolfbeis, *Chem. Phys. Liquids*, **50**, 51 (1989).
(j) H. Ephardt, P. Fromherz, *J. Phys. Chem.*, **97**, 4540 (1993).
(k) H.W. Gibson, F.C. Bailey, *Tetrahedron*, **30**, 2043 (1974).
(l) G.E. Bennett, K.P. Johnston, *J. Phys. Chem.*, **98**, 441 (1994).
(m) N.S. Bayliss, E.G. McRae, *J. Am. Chem. Soc.*, **74**, 5803 (1952).
(n) E.M. Kosower, B.G. Ramsay, *J. Am. Chem. Soc.*, **81**, 856 (1959).


Scheme 3

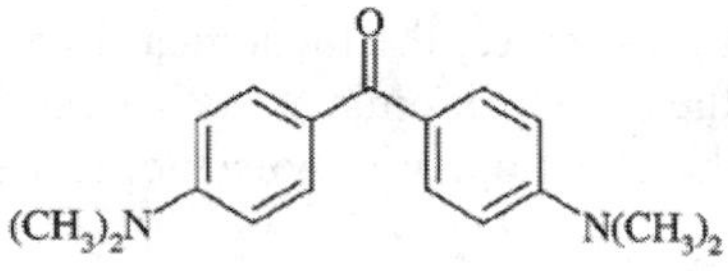


Scheme 4

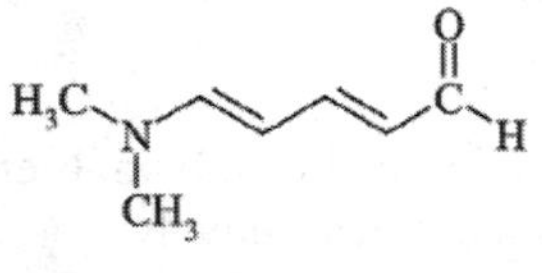


Scheme 5

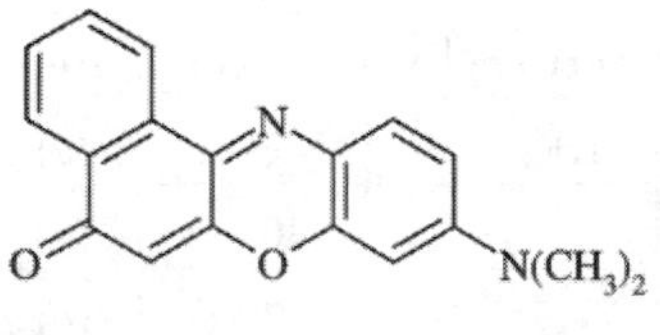


Scheme 6

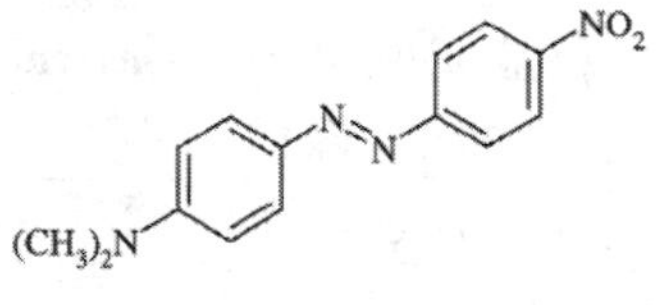


Scheme 7

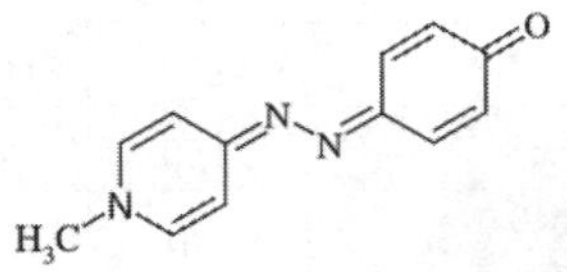


Scheme 8

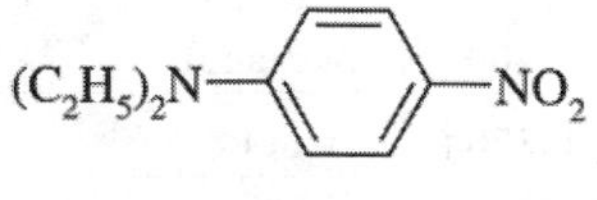

Scheme 9

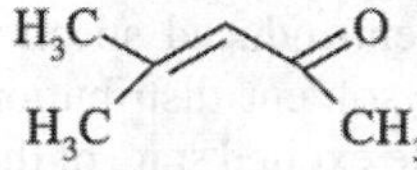

Scheme 10

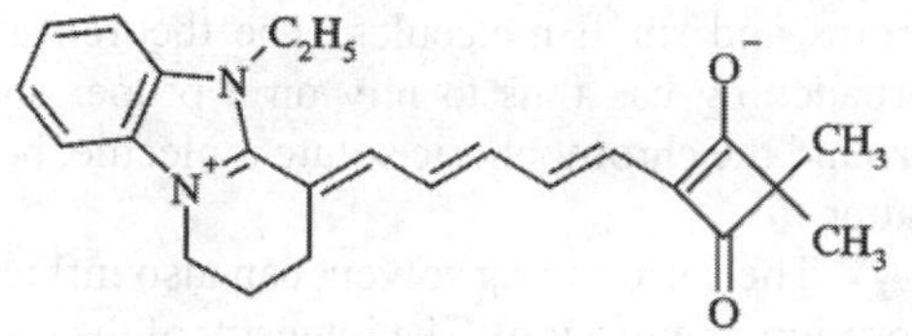

Scheme 11

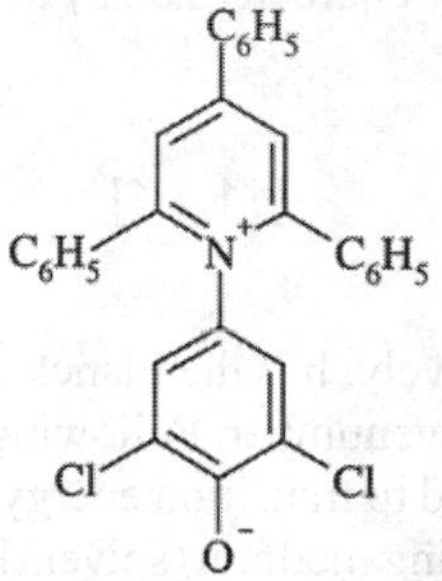

Scheme 12

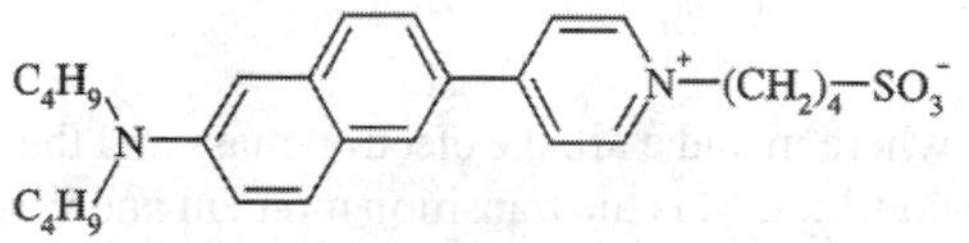

Scheme 13

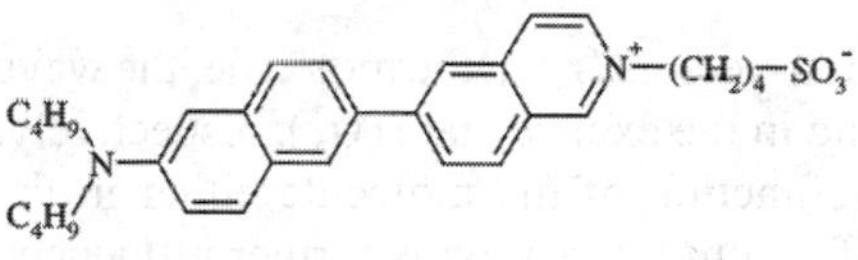

Scheme 14

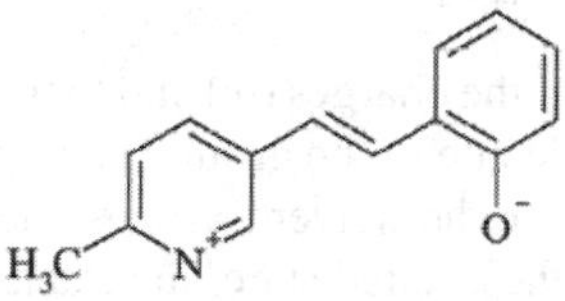

Scheme 15

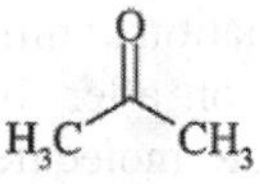

Scheme 16

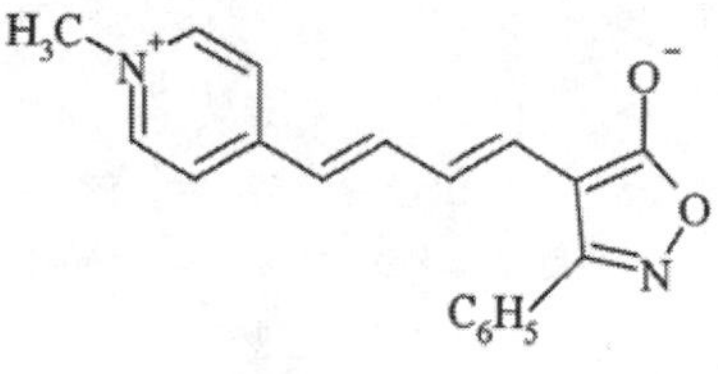

Scheme 17

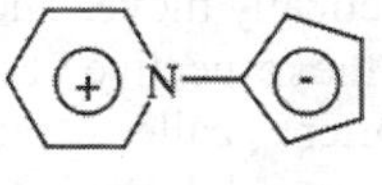

The solvent-induced broadening of the spectral lines and bands arises primarily from the variation of the local environment of the chromophoric solute molecule in the condensed medium caused by the thermal motion of the surrounding solvent molecules. At any given instant of time, there is a distribution of differently solvated solute molecules, each of which has characteristic transition energy to the excited state. The respective distribution of the transition energies leads to the broadening of the spectral band. It has to be kept in mind, however, that the broadening of spectral lines and bands can be also originated from adjoining the rotational and vibrational energy levels in the polyatomic molecule or from the Doppler and natural broadening of spectral lines. Those are more significant in the case of atoms and small molecules. The theoretical assessment of the solvent-induced spectral broadening has thus to rely on a proper statistical treatment of the solvent distribution around the chromophoric solute molecule, both in the ground and in the excited state of the latter.

The surrounding solvent can also influence the intensity of the spectral transition (absorption or emission). The intensity of the spectral transition is usually characterized by the oscillator strength f defined as follows

$$f = \left( \frac{8\pi m \bar{\nu}}{3he^2} \right) |\mathbf{M}|^2 \qquad [11.1.1]$$

where m and e are the electron mass and the electron charge, respectively, h is the Planck's constant, **M** is the transition moment and $\bar{\nu}$ is the mean absorption wavenumber. Following the last equation, the intensity of the spectrum is proportionally related to transition energy, provided that the transition moment **M** is independent of the surrounding medium (solvent). This may, however, be not the case. The definition of the transition moment[1]

$$\mathbf{M} = \sum_i \psi_0 | q_i \mathbf{r}_i | \psi_1^* \qquad [11.1.2]$$

includes, apart from the charges ($q_i$) and their position-vectors ($\mathbf{r}_i$) in the molecule, the wave function of the molecule in the ground state ($\psi_0$) and in the excited state ($\psi_1^*$), respectively. Therefore, whenever the solvent affects the wavefunction of the molecule either in the ground state or in the excited state, the intensity of spectral transition is further influenced by the change of the respective transition moment.

The analysis of the solvatochromic effects on molecular absorption and emission (fluorescence and phosphorescence) spectra is further complicated by the variation of time scales for the solvent relaxation after the spectral excitation of the solute molecule. The spectral transition is a very fast process that takes place within approximately $10^{-16}$ s. Thus, during this short period of time the atomic nuclei do not practically move. The excited state reached by the respective vertical transition is often called the Franck-Condon state (Figure 11.1.4).

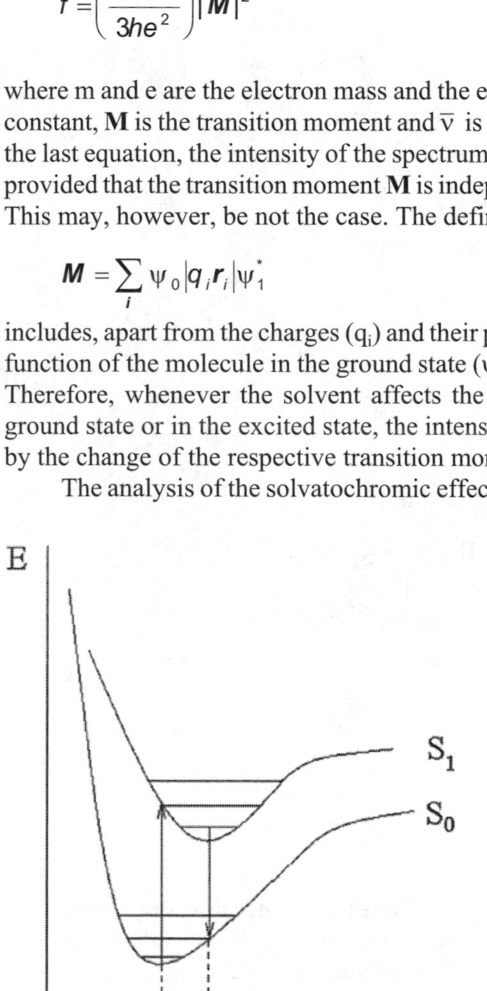

Figure 11.1.4. The Franck-Condon transitions during the excitation and the de-excitation of the molecule.

The lifetime of the fluorescent excited state may be long enough ($10^{-7}$ - $10^{-9}$ s) to allow in addition to the intramolecular nuclear relaxation ($10^{-12}$ s), also the solvent orientational relaxation. The latter, which is characterized by the relaxation times ranging from $10^{-10}$ s up to infinity (in the case of solids) may bring up the additional, solvent-induced stabilization of the relaxed excited state as compared to the Franck-Condon state. Thus, as a rule, the solvatochromic shifts in the absorption and fluorescence spectra are not equal.

The theoretical treatment of the time-dependent effects on molecular spectra in condensed phases is extremely complicated.[2] In most cases, it is assumed that only the electronic polarization of the solvent contributes to the solvation energy of the Franck-Condon state ($S_1$ in the case of absorption and $S_0$ in the case of emission). In the case of long-living states, i.e., the ground state and the relaxed excited state, a full relaxation of the solvent is assumed in the field of the solute molecule. The solvation energy of different states at different degrees of relaxation will thus be rather different that may result in rather different dependence of the absorption and emission transition energies on the polarity of the solvent. Some examples of solvatofluorochromical compounds are given in Table 11.1.2.

**Table 11.1.2. The solvatofluorochromic shifts for various positively and negatively solvatochromic compounds**

| Compound | $\nu_{max}$ (non-polar solvent), $cm^{-1}$ | $\nu_{max}$ (polar solvent), $cm^{-1}$ | $\Delta\nu_{max}$, $cm^{-1}$ | Ref. |
|---|---|---|---|---|
| (Scheme 18) | 24400 (hexane) | 16500 (water) | +7900 | (a) |
| (Scheme 19) | 21980 (cyclohexane) | 18210 (water) | +3770 | (b) |
| (Scheme 20) | ~20000 (cyclohexane) | ~17000 ($CH_3CN$) | +3000 | (c) |
| (Scheme 21) | 23530 (pentane) | 22730 (water) | +800 | (d) |
| (Scheme 22) | 24150 (hexane) | 21850 (methanol) | +2300 | (e) |

(a) I.A. Zhmyreva, V.V. Zelinskii, V.P. Kolobkov, N.D. Krasnitskaya, *Dokl. Akad. Nauk SSSR, Ser. Khim.*, **129**, 1089 (1959).
(b) M. Maroncelli, G.R. Fleming, *J. Chem. Phys.*, **86**, 6221 (1987).
(c) A. Safarzadeh-Amini, M. Thompson, U.J. Krall, *J. Photochem. Photobiol., Part A.*, **49**, 151 (1989).
(d) M.S.A. Abdel-Mottaleb, F.M. El-Zawawi, M.S. Antonious, M.M. Abo-Aly, M. El-Feky, *J. Photochem. Photobiol., Part A.*, **46**, 99 (1989).
(e) J. Catalán, C. Díaz, V. López, P. Pérez, *J. Phys. Chem.*, **100**, 18392 (1996).

Scheme 18

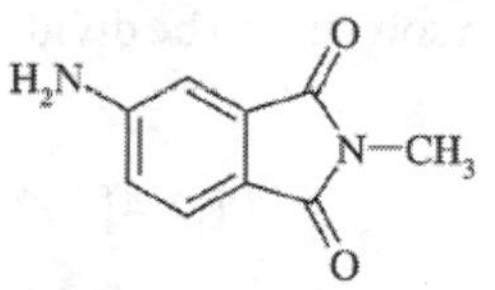


Scheme 19

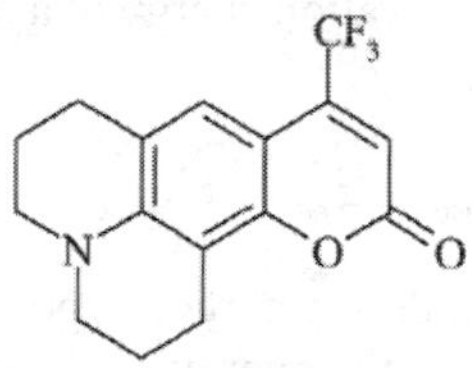


Scheme 20

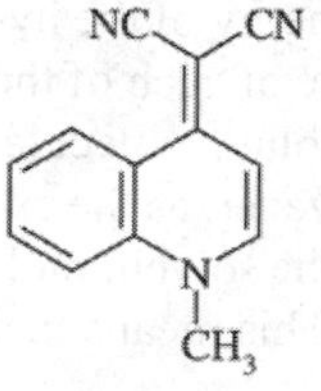

Scheme 21

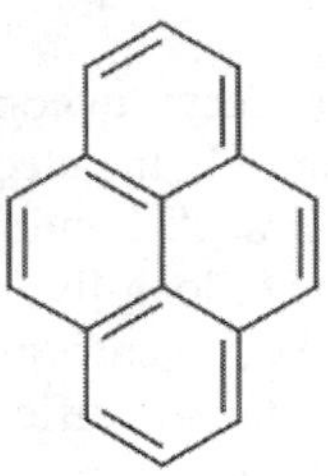

Scheme 22

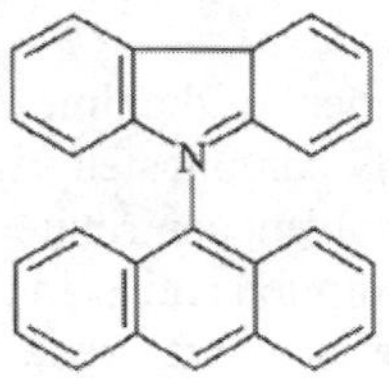


All theoretical treatments of solvatochromic shifts proceed from modelling the solvational interactions in the liquids and solutions. Theoretically, the interaction potential between a solute molecule and the surrounding solvent molecules Φ is given by the following integral

$$\Phi = C\int_0^\infty \varphi(R) g(R) R^2 dR \qquad [11.1.3]$$

where φ (R) and g(R) are the pair interaction potential between the solute and the solvent molecule and the solvent radial distribution function around the solute molecule, respectively, and C is a constant depending on the density of the system. The integration in the last formula is carried out over the distance between the solute and the solvent molecule, R. The equation [11.1.3] is derived proceeding from the assumption that the intermolecular forces in the condensed medium are additive. This assumption may be, however, violated because of possible three- and many-body interactions between the molecules in the solution. For most of the real systems, the application of Eq. [11.1.3] directly is rather impractical because of the low precision of φ (R) and g(R), particularly in the case of many-atomic molecules. Moreover, this equation will be strictly valid only for the solute-solvent systems in thermodynamic equilibrium and thus not applicable for the Franck-Condon states. Thus, almost all theoretical calculations of solvatochromic effects proceed from different physical models describing the intermolecular interactions in liquids and solutions.

Traditionally, the solvation energy of a molecule $E_{solv}$ in a given solvent can be divided into the following terms[3]

$$E_{solv} = E_{cav} + E_{disp} + E_{elst} + E_{H-bond} \qquad [11.1.4]$$

each of which corresponds to a certain type of intermolecular interaction in the condensed media. Thus, $E_{cav}$ denotes the energy of the cavity formation for the solute in the solvent, $E_{disp}$ is the dispersion energy and $E_{elst}$ the electrostatic energy of the interaction of the solute with the surrounding solvent molecules. The term $E_{H\text{-}bond}$ accounts for the energy of the hydrogen bond formation between the solute and solvent molecules. The value of each of the above terms will change as a result of the Franck-Condon excitation of the solute molecule. First, the size of the molecule increases, as a rule, during the excitation. However, as the excitation process is practically instantaneous, the position and orientation of the solvent molecules in the solvation sheath of the chromophoric solute will not change. This means that the average distance between the surface of the solute and the solvent molecules will decrease in the Franck-Condon excited state of the former that normally causes the enhanced solute-solvent repulsion in that state. At the same time, the dispersion energy that stabilizes the solute-solvent system will also increase in the absolute value, but to the opposite direction. In consequence, both effects may cancel each other and the net effect will be close to zero. For the polar solutes, both in the ground and in the excited state, the electrostatic solvation energy is therefore often considered as the most important term in Eq. [11.1.4].

During the excitation or de-excitation of the molecule, the molecular electronic wavefunction and the electron distribution may change significantly. In result, substantial differences are expected in the electrostatic and dispersion solvation energies of the ground and the excited state, respectively. In addition, the hydrogen bonding between the solute and solvent molecules may be affected by the excitation of the solute molecule that will be reflected as another contribution to the difference in the solvation energy of solute in the ground and in the excited state, respectively. In the following, we proceed with the systematic presentation of the theoretical methods developed for the description of the solvatochromic effects on molecular electronic and vibrational spectra in condensed disordered media (liquids, solutions, glasses etc.).

### 11.1.2 THEORETICAL TREATMENT OF SOLVENT CAVITY EFFECTS ON ELECTRONIC-VIBRATIONAL SPECTRA OF MOLECULES

As described above, the change (increase) in the size of the molecule during the excitation will result in increased van-der-Waals repulsion between the electron clouds of the chromophoric solute and the solvent molecules. Alternatively, the size of the molecule is expected to shrink as a result of the de-excitation of the molecule back to the ground state. In such case, the repulsion between the solute and solvent molecules will be reduced correspondingly. The respective energetic effect may be modeled as the difference in the cavity formation energies for the solute molecule in two states. The dependence of the cavity formation energy on the cavity size has been derived using several different model concepts.

The simplest approach is based on the concept of microscopic surface tension on the boundary between the solute cavity and the solvent. Within this approach, the free energy of cavity formation is assumed simply proportional to the surface of the solute cavity, $S_M$:

$$\Delta G_{cav} = \sigma S_M \qquad [11.1.5]$$

where $\sigma$ is the surface tension of the solvent. This formula has been applied for the evaluation of the free energy of transfer of electroneutral solutes between different solvents.[4] It has been extended to account for the size of the solvent molecule as follows:

$$\Delta G_{cav} = \sigma S_M - RT\ln(1 - V_s n_s) \qquad [11.1.6]$$

where $V_S$ is the intrinsic volume of a solvent molecule and $n_S$ is the number density of the solvent. In order to account for the chemical individuality of constituent atoms, it has been suggested to use different surface tension values $\sigma_i$ for different atomic types in the solute molecule.[5] Thus,

$$\Delta G_{cav} = C + \sum_i \sigma_i A_i \qquad [11.1.7]$$

where $A_i$ are the solvent-accessible surface areas of different atoms in the solute molecule and C is an empirically adjustable parameter. The quality of the description of experimental data has been, however, not significantly improved by the introduction of individual atomic surface tension characteristics.

Another theoretical approach for the calculation of the free energy of cavity formation proceeds from the theory of microscopic curved surfaces. According to this theory,[6]

$$\Delta G_{cav} = k_s^g\left(\frac{V_S}{V_M}\right)\sigma S \qquad [11.1.8]$$

where S is the area of the cavity and $k_s^g\,(V_S / V_M)$ is a correction factor, characteristic of a given solvent and depending on the ratio of molecular volumes of the solvent and solute. This factor has been approximated by the following formula

$$k_s^g\left(\frac{V_S}{V_M}\right) = 1 + \left(\frac{V_S}{V_M}\right)^{2/3}\left[k_s^g(1) - 1\right] \qquad [11.1.9]$$

where $k_s^g\,(1)$ is estimated from the solubility of a given solute in a given solvent. The main deficiency of this approach is connected with the introduction of additional empirical information, often not readily available.

The free energy of cavity formation has been also estimated from the data on isothermal compressibility, $\beta_T$, as follows[7]

$$\Delta G_{cav} = \frac{V_{cav}}{\beta_T} + C \qquad [11.1.10]$$

where $V_{cav}$ is the volume of the cavity and C is a constant term. However, the microscopic isothermal compressibility of water, calculated from the slope of Eq. [11.1.10], was found to be about an order higher than the respective experimental value for water ($\beta_T$(calc) = 23.5 vs. $\beta_T$(exp) = 3.14). Therefore, the use of the macroscopic surface tension or compressibility of the solvent for the respective microscopic model quantities is questionable.

An entropic approach to the calculation of the free energy of cavity formation proceeds from the scaled particle theory (SPT).[8,9] The free energy of the formation of a spherical cavity in a solvent, $\Delta G_{cav}$, can be calculated proceeding within the framework of SPT as follows

$$\Delta G_{cav} = RT\left\{1 - \ln(1-y) + \left(\frac{3y}{1-y}\right)\frac{a_M}{a_S} + \left[\frac{3y}{1-y} + \frac{9}{2}\left(\frac{y}{1-y}\right)^2\right]\left(\frac{a_M}{a_S}\right)^2\right\} \qquad [11.1.11]$$

where

$$y = \frac{4\pi\rho a_s^2}{3} \qquad [11.1.12]$$

is the reduced number density of the solvent. In the two last equations, $a_M$ and $a_S$ denote the intrinsic radii of the solute and solvent molecules, respectively, and ρ is the number density of the solvent. In the case of an ellipsoidal solute cavity, the SPT cavity formation energy has been given by the following equation[10]

$$\Delta G_{cav} = RT\left\{1 - \ln(1-y) + \left(\frac{\alpha y}{1-y}\right)\frac{a}{a_S} + \left[\frac{\beta y}{1-y} + \gamma\left(\frac{y}{1-y}\right)^2\right]\left(\frac{a}{a_S}\right)^2\right\} \quad [11.1.13]$$

where α, β and γ denote the geometrical coefficients and a is the characteristic length of the ellipsoid (the major semi-axis). The scaled particle theory has been extended to dilute solutions of arbitrary shaped solutes and has been successfully applied for the calculation of the solvation free energy of hydrocarbons in aqueous solutions.[11]

For most practical applications that involve the lowest excited states of the molecules, the increase in the cavity size during the excitation of the solute molecule would not be accompanied with a significant energetic effect. However, it may be important to account for the so-called Pauli repulsion between the solute electronic system and the surrounding medium. This interaction will force the solute electrons to stay inside the cavity and not to penetrate into the dielectric continuum (consisted of electrons, too) that surrounds it. The Pauli repulsion has been modeled by the respective model potentials, e.g., by expanding the potential in spherical Gaussian shells as follows:[12]

$$V_{PR} = \sum_i b_i \exp\left[-\beta_i (r - r_{o,i})^2\right] \quad [11.1.14]$$

where $b_i$ are the weight factors, $\beta_i$ the exponents and $r_{0,i}$ the radii of spherical shell functions. In general, the electrons in the solvent cavity could be treated as confined many-electron systems.[13]

### 11.1.3 THEORETICAL TREATMENT OF SOLVENT ELECTROSTATIC POLARIZATION ON ELECTRONIC-VIBRATIONAL SPECTRA OF MOLECULES

The origin of the solvatochromic shifts in the electronic spectra is related to the change in the electrostatic and dispersion forces between the solvent and the chromophoric solute molecule in the ground and in the excited state, respectively. The semiclassical approach to the treatment of the respective effects is based on the assumption that the solute and the solvent molecules are sufficiently separated to neglect the overlap between the electron distribution of these two molecular systems. The wave function for the whole system can then be approximated as the product of the wavefunctions of each individual system, i.e., the solute and individual solvent molecules:

$$\Psi = \psi^0_{s(1)} \psi^0_{s(2)} \cdots \psi^0_{s(n)} \psi^{(v)}_a \quad [11.1.15]$$

where $\psi^0_{s(1)}, \psi^0_{s(2)}, \cdots \psi^0_{s(n)}$, etc. are the wavefunctions of the respective solvent molecules in the ground state and $\psi^{(v)}_a$ is the wavefunction of the solute molecule in the v-th state. The antisymmetry of the total electronic wavefunction is ignored as the individual molecules are assumed separated enough not to allow the electron exchange. This approximation may not be valid in the case of strong semichemical interactions between the solute and the solvent molecules such as the hydrogen bonding or the formation of charge-transfer complexes. In such cases, the system consisting of the central solute molecule and the adjacent solvent molecules has to be treated as a supermolecule.

In the absence of strong semichemical interactions between the solute and solvent molecules, the interaction energy between them can be derived using the perturbation theory.[14] In the first approximation, the interaction between the nonionic molecules can be reduced to the dipole-dipole interaction between the molecules. The following perturbation operator can describe this interaction

$$\hat{\mathbf{H}}' = \sum_i \frac{\hat{\mu}_\alpha^{(v)} \hat{\Theta}_{ai} \hat{\mu}_{s(i)}}{R_{ai}^3} \qquad [11.1.16]$$

where $\hat{\mu}_\alpha^{(v)}$ and $\hat{\mu}_{s(i)}$ are the dipole moment operators for the solute a in the ν-th state and for the i-th solvent molecule in the ground state, respectively, $R_{ai}$ is the distance between the charge centroids of the interacting molecules and

$$\hat{\Theta}_{ai} = \hat{\mathbf{1}} - 3\vec{R}_{ai}\vec{R}_{ai} \qquad [11.1.17]$$

is the angular term describing the relative orientation of these two molecules in the space. The subsequent application of the perturbation theory to derive the energy of interaction between a pair of a solute and a set of N solvent molecules gives the following result[15]

$$\Delta E_a = \sum_{i=1}^{N} \frac{\mu_a^{(v)} \Theta_{ai} \mu_{s(i)}}{R_{ai}^3} - \frac{1}{2} \sum_{i=1}^{N} \frac{\mu_a^{(v)} \Theta_{ai} \alpha_s \Theta_{ai} \mu_a^{(v)}}{R_{ai}^6} -$$

$$- \frac{1}{2} \sum_{i=1}^{N} \sum_{j=1}^{N} \frac{\mu_{s(i)} \Theta_{ai} \alpha_a^{(v)} \Theta_{aj} \mu_{s(j)}}{R_{ai}^3 R_{aj}^3} - \sum_{i=1}^{N} \sum_{p \neq 0} \sum_{\lambda \neq v} \frac{\mu_{s(i)}^{(0p)} \Theta_{ai} \mu_a^{(\lambda v)} \mu_a^{(\lambda v)} \Theta_{aj} \mu_{s(i)}^{(0p)}}{R_{ai}^6} \qquad [11.1.18]$$

with the following notations:

$$\mu_a^{(v)} = \left\langle \psi_a^{(v)} \middle| \hat{\mu}_a \middle| \psi_a^{(v)} \right\rangle \qquad [11.1.19]$$

is the dipole moment of the solute in the ν-th state,

$$\mu_{s(i)} = \left\langle \psi_{s(i)}^0 \middle| \hat{\mu}_s \middle| \psi_{s(i)}^0 \right\rangle \qquad [11.1.20]$$

is the dipole moment of the i-th solvent molecule in the ground state,

$$\alpha_a^{(v)} = 2 \sum_{\lambda \neq v} \frac{\left\langle \psi_a^{(v)} \middle| \hat{\mu}_a \middle| \psi_a^{(\lambda)} \right\rangle \left\langle \psi_a^{(\lambda)} \middle| \hat{\mu}_a \middle| \psi_a^{(v)} \right\rangle}{E_v - E_\lambda} \qquad [11.1.21]$$

is the polarizability tensor of the solute molecule in the ν-th state,

$$\alpha_s = 2\sum_{p\neq 0} \frac{\langle \psi_s^0 | \hat{\mu}_s | \psi_s^{(p)} \rangle \langle \psi_s^{(p)} | \hat{\mu}_s | \psi_s^0 \rangle}{E_{0s} - E_{ps}} \quad [11.1.22]$$

is the polarizability tensor of the solvent molecule in the ground state, respectively, and

$$\mu_{s(i)}^{(0p)} = \langle \psi_{s(i)}^0 | \hat{\mu}_a | \psi_{s(i)}^{(p)} \rangle \quad [11.1.23]$$

and

$$\mu_a^{(\lambda\nu)} = \langle \psi_a^{(\lambda)} | \hat{\mu}_a | \psi_a^{(\nu)} \rangle \quad [11.1.24]$$

are the transition dipoles between the two states $(0 \rightarrow p)$ in the solvent and in the solute $(\lambda \rightarrow \nu)$ molecules, respectively. In the last formulae, $E_\nu$ and $E_\lambda$ denote the energy of the solute molecule in the respective ($\nu$-th and $\lambda$-th) states, and $E_{ps}$ and $E_{0s}$ - the energy of a solvent molecule in the p-th and in the ground state, respectively. The first term in equation [11.1.18] represents therefore the electrostatic interaction of the unperturbed charge distribution of the two molecules, given as the interaction between the respective permanent point dipoles. The second term in this equation corresponds to the interaction of the permanent dipole of the solute with the dipole induced in the solvent whereas the third term reflects the interaction of the permanent dipole of the solvent with the induced dipole of the solute. The last term represents the second-order interaction of both molecules in excited states and quantifies thus effectively the dispersion interaction in the solute-solvent system.

The equation [11.1.18] refers, of course, to a single fixed configuration of the solute and the solvent molecules. In order to find the effective interaction energy in the liquid medium, an appropriate statistical averaging over all configurations has to be carried out. In most practical applications, this procedure is very complicated and thus the semiclassical continuum approaches are employed to describe the solvent. The description of the electrostatic interactions between the solute and the solvent has been based on the Onsager's reaction field model. According to this model, the energy of electrostatic interaction between an arbitrary charge distribution inside the solute molecule and the surrounding polarizable dielectric medium is given by the following equation[16]

$$E_{el} = \frac{1}{8\pi} \int_V \mathbf{E}_s \mathbf{E}_o (\varepsilon - 1) dV \quad [11.1.25]$$

where $\mathbf{E}_o$ is the electrostatic field of the charges in the molecule in vacuo and $\mathbf{E}_s$ is the modified field in the presence of dielectric medium. Notably, within the formalism of the last [11.1.25], the dielectric constant $\varepsilon$ of the medium is still a function of the space coordinates, i.e., both the interior of the molecule and the surrounding medium are treated by the same equation. However, the integral in the last equation cannot be found analytically and even the numerical integration over the space presents a difficult mathematical task. Therefore, the electrostatic equation is usually simplified by the application of the Gauss divergence theorem. According to this theorem, the volume integral in [11.1.25] is transformed into a surface integral over some boundary

$$E_{el} = \frac{\varepsilon - 1}{8\pi} \int_S \Phi_S E_o \vec{\mathbf{n}} dS \qquad [11.1.26]$$

where S is the boundary surface, $\vec{\mathbf{n}}$ the outward normal unit vector on S and the reaction potential $\Phi_s$ is defined as follows: $\mathbf{E}_s$ = -grad $\Phi_S$. Depending on the shape of solute molecular cavity, different approaches have been applied for the calculation of the electrostatic solvation energies of compounds in liquids. Within the classical reaction field theory of Kirkwood and Onsager,[17,18] the solute molecule is represented by a set of point charges fixed inside of sphere of a radius $a_0$ and the electrostatic equation [11.1.26] is solved by applying the appropriate boundary conditions inside and outside the sphere. It is also assumed that the dielectric constant inside the cavity (sphere) is equal to unity (vacuum) and outside the cavity has a constant value, corresponding to the macroscopic dielectric constant of the medium studied. In that case, the energy of the electrostatic interaction between the solute charge distribution and the surrounding dielectric medium is given by the following infinite expansion

$$E_{el} = \frac{1}{2} \sum_{i,j} \mathbf{e}_i \mathbf{e}_j \sum_{l=0}^{\infty} \left[ \frac{(l+1)(1-\varepsilon)}{\varepsilon(l+1)+1} \right] \frac{\mathbf{r}_i^l \mathbf{r}_j^l}{a_0^{2l+1}} P_l\left(\mathbf{cos}\theta_{ij}\right) \qquad [11.1.27]$$

where $\mathbf{e}_i$ and $\mathbf{e}_j$ are the charges inside the sphere at positions $\mathbf{r}_i$ and $\mathbf{r}_j$, respectively, and $\theta_{ij}$ is the angle at the center of the sphere between the vectors $\mathbf{r}_i$ and $\mathbf{r}_j$. In the last equation, the summation proceeds over all charged particles (nuclei and electrons) of the solute and $P_l(\mathbf{cos}\theta_{ij})$ are the Legendre polynomials of l-th order. By expressing the Legendre polynomials as the products of the respective spherical harmonics of order m ($-l \le m \le l$), equation [11.1.27] can be rewritten as

$$E_{el} = -\frac{1}{2} \sum_{l=0}^{\infty} \sum_{m=-1}^{l} \mathbf{R}_l^m \mathbf{M}_l^m \qquad [11.1.28]$$

where

$$\mathbf{R}_l^m = f_l \mathbf{M}_l^m \qquad [11.1.29]$$

and

$$f_l = \frac{(l+1)(\varepsilon - 1)}{(l+1)\varepsilon + 1} \frac{1}{a_0^{2l+1}} \qquad [11.1.30]$$

In these equations, $\mathbf{M}_l^m$ and $\mathbf{R}_l^m$ represent the electrical momentum and the respective reaction field component. The first term ($l = 0$) in the expansion [11.1.27] gives the interaction of the excess (ionic) charge of the solute with the respective reaction field created in the dielectric medium (Born term)

$$E_{Born} = \frac{1-\varepsilon}{2\varepsilon} \frac{Q^2}{a_0} \qquad [11.1.31]$$

where Q is the numerical value of the ionic charge. The next term (l = 0) corresponds to the total dipole interaction with the corresponding reaction field (Onsager dipolar term)

$$E_{Onsager} = \frac{(1-\varepsilon)}{(2\varepsilon+1)}\frac{\vec{\mu}^2}{a_0^3} \qquad [11.1.32]$$

where $\vec{\mu}$ is the dipole moment of the solute. In most applications, only these two terms that are the largest by size are considered in the calculation of the electrostatic interaction energy. However, depending on system studied, the interaction of higher electrical moments with the corresponding reaction field may become also significant and the terms corresponding to higher moments of order $2^l$ (quadruple, octuple, hexadecuple, etc.) should be taken into account.

In many cases, the shape of the solute molecule may be very different from the sphere and therefore, it is necessary to develop the methods of calculation of the electrostatic solvation energy for more complex cavities. In the case of the ellipsoidal cavity with main semiaxes **a**, **b**, and **c**, the analytical formulas are still available for the calculation of the charge and dipolar terms of the electrostatic interaction with the reaction field. The charge term is simply

$$E^{ell}_{Born} = \frac{(1-\varepsilon)}{2\varepsilon}\frac{Q^2}{abc} \qquad [11.1.33]$$

whereas in the respective dipolar term[19]

$$E^{ell}_{Onsager} = \mathbf{R}\vec{\mu} \qquad [11.1.34]$$

the reaction field **R** is presented using a special tensor as follows

$$\mathbf{R} = \begin{pmatrix} \frac{3A_a(1-A_a)(1-\varepsilon)\mu_a}{abc[\varepsilon+(1-\varepsilon)A_a]} & 0 & 0 \\ 0 & \frac{3A_b(1-A_b)(1-\varepsilon)\mu_b}{abc[\varepsilon+(1-\varepsilon)A_b]} & 0 \\ 0 & 0 & \frac{3A_c(1-A_c)(1-\varepsilon)\mu_c}{abc[\varepsilon+(1-\varepsilon)A_c]} \end{pmatrix}\vec{\mu} \qquad [11.1.35]$$

where $A_a$, $A_b$ and $A_c$ are the standard ellipsoidal shape factor integrals, and $\mu_a$, $\mu_b$ and $\mu_c$ are the dipole moment components along the main semiaxes of the ellipsoid. Several methods have been developed to define the semiaxes of the ellipsoidal cavity. For instance, these have been taken collinear with the axes of the solute dipole polarizability tensor, and their lengths proportional to the respective eigenvalues.[20] Another definition proceeds from the inertia tensor of the van-der-Waals solid, i.e., a solid or uniform density composed of interlocking van-der-Waals spheres.[21] Also, the ellipsoidal surface has been defined in terms of the best fitting of a given molecular electrostatic isopotential surface.[22]

The above-discussed theoretical formulation of the electrostatic solute-solvent interaction is applicable for the fixed charge distribution inside the solute molecule. However, the solvent reaction field may cause a redistribution of the charge inside the solute. The magnitude of this redistribution depends on both the dielectric constant of the solvent and the polarizability of the solute molecule. Within the approximations of the spherical solute cavity and the point dipole interactions between the solute and solvent, the dynamically changed Onsager reaction field can be expressed by the following formula[14]

$$\boldsymbol{R}_l = \frac{\mu_a^{(v)} + \alpha_a^{(v)} \boldsymbol{R}_l}{a_0^3} \frac{2(\varepsilon - 1)}{(2\varepsilon + 1)} \qquad [11.1.36]$$

Notably, the use of the macroscopic dielectric constant $\varepsilon = \varepsilon_0$ in the last formula is justified only when the lifetime of the solute molecule in a given (ν-th) state is much longer than the rotational-vibrational relaxation time of the solvent at given temperature. This is not a valid assumption in the case of the Franck-Condon states, which have the lifetime much shorter than the rotational-vibrational relaxation time of the solvent. Therefore, the solvent is only partially relaxed for these states and the corresponding reaction field is characterized by the dielectric constant at infinite frequency of external electric field, $\varepsilon_\infty$. By inserting the expression for the reaction field [11.1.36] into the equation [11.1.18] and assuming that the static polarizability of the solute molecule is approximately equal to the one third of the cube of Onsager's cavity radius

$$\alpha_a^{(v)} \approx \frac{a_0^3}{3} \qquad [11.1.37]$$

the following semiclassical equation can be obtained for the solvation energy of the ν-th (Franck-Condon) state of the solute molecule[14]

$$E_s = -\left(\frac{\varepsilon_0 - 1}{\varepsilon_0 + 2} - \frac{\varepsilon_\infty - 1}{\varepsilon_\infty + 2}\right)\left[\frac{2\mu_a^0 \mu_a^{(v)}}{a_0^3} + \frac{2\mu_a^0 \mu_a^{(v)}}{a_0^3}\left(\frac{\varepsilon_0 - 1}{\varepsilon_0 + 2}\right)\right] - \frac{2\left(\mu_a^{(v)}\right)^2}{a_0^3}\left(\frac{\varepsilon_\infty - 1}{\varepsilon_\infty + 2}\right) +$$

$$+\left[\left(\frac{3}{2\varepsilon_0 + 1}\right)\frac{(2\varepsilon_0 + 1)^2(\varepsilon_0 - \varepsilon_\infty)^2}{(2\varepsilon_0 - \varepsilon_\infty)\varepsilon_0}\frac{kT}{3a_0^3} + \left(\frac{2\varepsilon_0 - 2}{2\varepsilon_0 + 1}\right)^2\right]\sum_{\lambda \neq v}\frac{\left(\mu_a^{(\lambda v)}\right)^2}{E_\lambda - E_v} +$$

$$+\left(\frac{\varepsilon_\infty - 1}{\varepsilon_\infty + 2}\right)\frac{2}{a_0^3}\sum_{\lambda \neq v}\left(\mu_a^{(\lambda v)}\right)^2\left(1 - \frac{E_v - E_\lambda}{E_{0s} - E_{ps}}\right) \qquad [11.1.38]$$

The solvatochromic shift due to the difference in the electrostatic solvation energy of the ground state and the excited state of the solute, respectively, is thus given as follows:

$$\Delta E_s = \frac{2}{a_0^3}\left\{\left(\frac{\varepsilon_0 - 1}{\varepsilon_0 + 2} - \frac{\varepsilon_\infty - 1}{\varepsilon_\infty + 2}\right)\left(\frac{2\varepsilon_0 + 1}{\varepsilon_0 + 2}\right)\left[(\mu_a^0)^2 - \mu_a^0 \mu_a^{(v)}\right] + \left(\frac{\varepsilon_\infty - 1}{\varepsilon_\infty + 2}\right)\left[(\mu_a^0)^2 - (\mu_a^{(v)})^2\right]\right\}$$

$$+\left\{\left(\frac{3}{2\varepsilon_0 + 1}\right)\frac{(2\varepsilon_0 + 1)^2(\varepsilon_0 - \varepsilon_\infty)}{(2\varepsilon_0 - \varepsilon_\infty)\varepsilon_0}\frac{kT}{3a_0^3}\left[\sum_{\lambda \neq 0}\frac{(\mu_a^{(0\lambda)})^2}{E_0 - E_\lambda} - \sum_{\lambda \neq v}\frac{(\mu_a^{(v\lambda)})^2}{E_v - E_\lambda}\right]\right\} +$$

$$+\left(\frac{2\varepsilon_0-2}{\varepsilon_0+2}\right)^2\left[\frac{\left(\mu_a^{(v)}\right)^2}{a_0^6}\sum_{\lambda\neq v}\frac{\left(\mu_a^{(v\lambda)}\right)^2}{E_v-E_\lambda}-\frac{\left(\mu_a^{(0)}\right)^2}{a_0^6}\sum_{\lambda\neq 0}\frac{\left(\mu_a^{(0\lambda)}\right)^2}{E_0-E_\lambda}\right]+$$

$$+\left(\frac{\varepsilon_\infty-1}{\varepsilon_\infty+2}\right)\frac{2}{a_0^3}\left[\sum_{\lambda\neq 0}\left(\mu_a^{(0\lambda)}\right)^2\left(1-\frac{E_\lambda-E_0}{E_{0s}-E_{ps}}\right)-\sum_{\lambda\neq v}\left(\mu_a^{(v\lambda)}\right)^2\left(1-\frac{E_\lambda-E_v}{E_{ps}-E_{0s}}\right)\right] \quad [11.1.39]$$

The last expression represents the solvent effect on the transition energy of the 0-0 band of the solute molecule.

McRae[15,23] has given a different derivation of the electrostatic solvation energy based on semiclassical reaction field approach. The final result is however similar to the above equation for the solvatochromic shift in the electronic transition:[15]

$$\Delta E_s=\frac{2}{a_0^3}\left\{\left(\frac{\varepsilon_0-1}{\varepsilon_0+2}-\frac{\varepsilon_\infty-1}{\varepsilon_\infty+2}\right)\left(\frac{2\varepsilon_0+1}{\varepsilon_0+2}\right)\left[\left(\mu_a^0\right)^2-\mu_a^0\mu_a^{(v)}\right]\right\}+$$

$$+\frac{1}{a_0^3}\left(\frac{\varepsilon_\infty-1}{\varepsilon_\infty+2}\right)\left[\left(\mu_a^{(v)}\right)^2-\left(\mu_a^0\right)^2\right]-\frac{1}{2}\alpha_a^{(v)}\left|\mathbf{E}_a^{(v)}\right|^2+\frac{1}{2}\alpha_a^0\left|\mathbf{E}_a^0\right|^2+D_a^{(v)}-D_a^0 \quad [11.1.40]$$

where $\mathbf{E}_a^0$ and $\mathbf{E}_a^{(v)}$ are the solvent fields due to the permanent dipole moments of the solvent molecules applying to the ground state and to the excited state of the solute molecule, respectively. The terms $D_a^0$ and $D_a^{(v)}$ denote the solute-solvent intermolecular dispersion energies in the corresponding states.

Abe[24] has developed an alternative semiclassical theory of the solvent effects on electronic spectra. This theory is based on the averaging of the intermolecular interaction energy over all solute-solvent configurations within the approximation of pair interactions. The theory involves the dipole moments and polarizabilities of the solute molecule and takes into account the temperature dependence arising from the Boltzmann factor.

In all above-listed theoretical approaches, the response of the solute charge distribution to the solvent field is expressed by using the static polarizability of the solute molecule. However, it would be plausible to account for this response directly within the quantum mechanical theoretical framework. The quantum-chemical approaches to the calculation of the solvation effects on the ground and excited states of the molecules in the solution can be classified using two possible ways. First, it can be based on the traditional division of the quantum chemistry into the non-empirical (ab initio) and the semiempirical methods. Within both those classes of methods, the Hartree-Fock method based on the independent particle model and the methods accounting for the static and dynamic electron correlation are usually distinguished. The second way of classification of methods can be based on the differences of the models used for the description of solute-solvent interactions. In general, these interactions can be taken into account in the framework of continuum representation of the solvent or using the resolution of solute-solvent interactions at molecular level. In the following, we first proceed with the review of models used for the solute-solvent interactions, with the subsequent elaboration of the quantum-chemical methodology for the calculation of the solvent effects on spectra.

The simplest continuum model is based on the classical Onsager reaction field theory assuming the spherical or ellipsoidal form of cavities for the solute molecules in dielectric media. The respective interaction energy is accounted for as a perturbation $\hat{\mathbf{V}}(a_0, \varepsilon)$ of the Hamiltonian of the isolated solute molecule, $\hat{\mathbf{H}}^0$.

$$\hat{\mathbf{H}} = \hat{\mathbf{H}}^0 + \hat{\mathbf{V}}(a_0, \varepsilon) \qquad [11.1.41]$$

Within the approximation of electrostatic interaction between the solute dipole and the respective reaction field, the perturbation term is simply

$$\hat{\mathbf{V}}(a_0, \varepsilon) = \Gamma \hat{\mu}_a^2 \qquad [11.1.42]$$

where

$$\Gamma = \frac{2(1-\varepsilon)}{(2\varepsilon + 1)a_0^3} \qquad [11.1.43]$$

In the case of ellipsoidal cavities, the last coefficient has to be substituted by the tensor given in equation [11.1.35].

A self-consistent reaction field method (SCRF) has been developed at the level of Hartree-Fock theory to solve the respective Schrödinger equation[25]

$$\hat{\mathbf{H}}\Psi = E\Psi \qquad [11.1.44]$$

Proceeding from the classical expression for the electrostatic solvation energy of a solute molecule in a dielectric medium in the dipole-dipole interaction approximation, the total energy of the solute is presented as follows[26]

$$E = E^o - \frac{1}{2}\Gamma\left(\langle\psi|\hat{\mu}|\psi\rangle\langle\psi|\hat{\mu}|\psi\rangle + 2\vec{\mu}_{nuc}\langle\psi|\hat{\mu}|\psi\rangle + \vec{\mu}_{nuc}^2\right) \qquad [11.1.45]$$

where $E^o = \langle\psi|\hat{\mathbf{H}}^0|\psi\rangle$, $\hat{\mathbf{H}}^0$ is the Hamiltonian for the reaction field unperturbed solute molecule and $\psi$ is the molecular electronic wave function. From the last equation, one can construct the variational functional

$$L = E^o - \frac{1}{2}\Gamma\left(\langle\psi|\hat{\mu}|\psi\rangle\langle\psi|\hat{\mu}|\psi\rangle + 2\vec{\mu}_{nuc}\langle\psi|\hat{\mu}|\psi\rangle + \vec{\mu}_{nuc}^2\right) - W\left(\langle\psi|\psi\rangle - 1\right) \qquad [11.1.46]$$

where W is the Lagrange multiplier ensuring the normalization of the variational wave function. The variation of the last equation with respect to the parameters of the wave function yields

$$\delta L = \delta E^o - \Gamma\left(\langle\delta\psi|\hat{\mu}|\psi\rangle\langle\psi|\hat{\mu}|\psi\rangle + \vec{\mu}_{nuc}\langle\psi|\hat{\mu}|\psi\rangle\right) - W\delta\left(\langle\psi|\psi\rangle\right)$$

$$= \langle\delta\psi|\hat{\mathbf{H}}^0|\psi\rangle - \Gamma\langle\delta\psi|\hat{\mu}|\psi\rangle\vec{\mu}_{tot} - W\langle\delta\psi|\psi\rangle + c.c. = 0 \qquad [11.1.47]$$

where $\hat{\mu}_{tot} = \hat{\mu}_{nuc} + \langle\psi|\vec{\mu}_{el}|\psi\rangle$ is the total dipole moment of the solute molecule. The latter is calculated during the SCRF procedure simultaneously with the total energy of the system.

From equation [11.1.47], the following Schrödinger equation is obtained for the electronic state $|\psi>$ of the solute molecule

$$\hat{\mathbf{H}}^0 - \Gamma\langle\psi|\hat{\mu}_{tot}|\psi\rangle\hat{\mu}_{el} = W|\psi\rangle \qquad [11.1.48]$$

W plays the role of the quantum mechanical motif that is directly obtained from the Schrödinger equation as follows

$$W = \left\langle\psi\left|\hat{\mathbf{H}}^0 - \Gamma\langle\psi|\hat{\mu}_{tot}|\psi\rangle\hat{\mu}_{el}\right|\psi\right\rangle = \left\langle\psi\left|\hat{\mathbf{H}}^0\right|\psi\right\rangle - \Gamma\langle\psi|\hat{\mu}_{tot}|\psi\rangle\langle\psi|\hat{\mu}_{el}|\psi\rangle \qquad [11.1.49]$$

By adding the part for the interaction of the nuclear component of the solute dipole with the total reaction field and assuming

$$\left\langle\psi\left|\hat{\mathbf{H}}^0\right|\psi\right\rangle \approx \left\langle\psi^0\left|\hat{\mathbf{H}}^0\right|\psi^0\right\rangle \qquad [11.1.50]$$

where $\psi^0$ is the wavefunction of the solute molecule, unperturbed by the reaction field, one obtains that

$$E_{el} = E_{rf} - E_0 = \left\langle\psi\left|\hat{\mathbf{H}}^0\right|\psi\right\rangle - \Gamma\left(\langle\psi|\hat{\mu}_{el}|\psi\rangle + \vec{\mu}_{nuc}\right)^2 - \left\langle\psi^0\left|\hat{\mathbf{H}}^0\right|\psi^0\right\rangle \approx -\Gamma\langle\vec{\mu}_{tot}\rangle^2 \qquad [11.1.51]$$

The comparison of the last equation with the starting equation [11.1.44] reveals a difference by the factor of two in the final result. Of course, the approximation [11.1.50] brings up a certain error and it has been therefore proposed[25,27] to correct the last formula by the addition of the "solvent cost", i.e., the additional work required to reorganize the solvent due to the electrostatic field of the solute

$$E_{el} = E_{rf} - E_0 + \frac{1}{2}\Gamma\langle\vec{\mu}_{tot}\rangle^2 \qquad [11.1.52]$$

Alternatively, the electrostatic solvation energy can be derived proceeding from the following variational functional[27]

$$L = E^0 - \frac{1}{4}\Gamma\left(\langle\psi|\hat{\mu}|\psi\rangle\langle\psi|\hat{\mu}|\psi\rangle + 2\vec{\mu}_{nuc}\langle\psi|\hat{\mu}|\psi\rangle + \vec{\mu}^2_{nuc}\right) - W\left(\langle\psi|\psi\rangle - 1\right) \qquad [11.1.53]$$

This leads to a Schrödinger equation which eigenvalue is directly related to the total electrostatic (dipolar) solvation energy, $E_{el}$,

$$\left(\hat{\mathbf{H}}^0 - \frac{1}{2}\Gamma\langle\psi|\hat{\mu}_{tot}|\psi\rangle\hat{\mu}_{el}\right)|\psi\rangle = E^{(e)}_{el}|\psi\rangle \qquad [11.1.54]$$

$$E_{el} = E^{(e)}_{rf} - \frac{1}{2}\Gamma\vec{\mu}_{nuc}\vec{\mu}_{tot} - E_0 \qquad [11.1.55]$$

For the solution of equation [11.1.54], the molecular wavefunction can be presented as a proper spin-projected antisymmetrized product of molecular (or atomic) orbitals[27]

$$|\psi\rangle = O_s A[\phi_1, \cdots, \phi_n] \qquad [11.1.56]$$

Recalling that the dipole moment operator is a one-electron operator (as are all electric moment operators), the following orbital equations are obtained

$$\mathbf{f}(k)\phi_i(k) = \varepsilon_i \phi_i(k) \qquad [11.1.57]$$

with

$$\mathbf{f}(k) = \mathbf{f}_0(k) - \Gamma\langle\psi|\hat{\mu}_{tot}|\psi\rangle\hat{\mu}_{el}(k) \qquad [11.1.58]$$

or

$$\mathbf{f}(k) = \mathbf{f}_0(k) - \frac{1}{2}\Gamma\langle\psi|\hat{\mu}_{tot}|\psi\rangle\hat{\mu}_{el}(k) \qquad [11.1.59]$$

where $\mathbf{f}_0(k)$ is the usual Fock operator for the isolated molecule, $\varepsilon_i$ is the molecular orbital energy for $|\phi_i>$ and $\hat{\mu}_{el}(k)$ is the electronic part of the dipole moment operator. Both equations are solved iteratively, using the usual SCF procedure and the expectation value of the total dipole moment from the previous SCF cycle.

A scheme for the treatment of the solvent effects on the electronic absorption spectra in solution had been proposed in the framework of the electrostatic SCRF model and quantum chemical configuration interaction (CI) method.[27] Within this approach, the absorption of the light by chromophoric molecules was considered as an instantaneous process. Therefore, during the photon absorption no change in the solvent orientational polarization was expected. Only the electronic polarization of solvent would respond to the changed electron density of the solute molecule in its excited (Franck-Condon) state. Consequently, the solvent orientation for the excited state remains the same as it was for the ground state, the solvent electronic polarization, however, must reflect the excited state dipole and other electric moments of the molecule. Considering the SCRF Hamiltonian

$$\hat{\mathbf{H}} = \hat{\mathbf{H}}^0 + \Gamma\langle\phi|\hat{\mu}_{tot}|\phi\rangle\hat{\mu} \qquad [11.1.60]$$

it is possible to write for the state $|\psi_I\rangle$ the following expression

$$H_{II} = \langle\psi_I|\hat{\mathbf{H}}^0 + \Gamma\langle\psi|\hat{\mu}|\psi\rangle\hat{\mu}|\psi_I\rangle = \langle\psi_I|\hat{\mathbf{H}}^0|\psi_I\rangle - \Gamma\langle\psi_0|\hat{\mu}|\psi_0\rangle\langle\psi_I|\hat{\mu}|\psi_I\rangle \qquad [11.1.61]$$

that is the zeroth order estimate of the energy of the state $|\psi_I\rangle$. Then, for a single excitation, $I \rightarrow A$, the excitation energy is given as follows

$$\Delta E_{ia} = \varepsilon_a - \varepsilon_i - J_{ia} + \begin{Bmatrix}2\\0\end{Bmatrix} K_{ia} = \varepsilon_a^0 - \varepsilon_i^0 - J_{ia} + \begin{Bmatrix}2\\0\end{Bmatrix} K_{ia} - -\Gamma\langle\psi_0|\hat{\mu}|\psi_0\rangle\left[\langle\psi_i^a|\hat{\mu}|\psi_i^a\rangle - \langle\psi_0|\hat{\mu}|\psi_0\rangle\right] \quad [11.1.62]$$

where $J_{ia}$ and $K_{ia}$ are the respective Coulomb' and exchange matrix elements and $\Gamma$ is the reaction field tensor at the dipole level. The terms $\varepsilon_i^0$ are the eigenvalues of the Fock operator for the k-th electron in the isolated solute molecule. The off-diagonal CI matrix elements are given by

$$H_{IJ} = \left\langle \psi_I \left| \hat{\mathbf{H}}^0 + \Gamma\langle\psi|\hat{\mu}|\psi\rangle\hat{\mu} \right| \psi_J \right\rangle = \left\langle \psi_I \left| \hat{\mathbf{H}}^0 \right| \psi_J \right\rangle - \Gamma\langle\psi_0|\hat{\mu}|\psi_0\rangle\langle\psi_I|\hat{\mu}|\psi_J\rangle \quad [11.1.63]$$

Equations [11.1.60] - [11.1.63] demonstrate that some part of the solvent effect is already included in thc ordinary CI treatment when proceeding from the SCRF Fock matrix. It has to be noticed that the terms $\langle\psi_0|\hat{\mu}|\psi_0\rangle$ should represent the ground-state dipole moment after CI, and therefore, an iterative procedure would be required to obtain a proper solution. However, at the CIS (CI single excitations) level, commonly used for the spectroscopic calculations, this is no concern because of Brillouin's theorem, which implies that the CI does not change the dipole moment of the molecule. Even at higher levels of excitation in CI, this effect should not be large and might be estimated from the respective perturbation operator.[27]

There are two approaches to address the instantaneous electronic polarization of the solvent during the excitation of the solute molecule. In the first case, the following correction term has to be added to the CI excitation energy

$$\Delta E_I = \frac{1}{2}\Gamma(\varepsilon_\infty)\left[\langle\psi_0|\hat{\mu}|\psi_0\rangle\langle\psi_I|\hat{\mu}|\psi_I\rangle - \left|\langle\psi_I|\hat{\mu}|\psi_I\rangle\right|^2\right] \quad [11.1.64]$$

where $\Gamma(\varepsilon_\infty)$ is the reaction field tensor for the optical relative dielectric permittivity of the solvent, $\varepsilon_\infty$. In the last equation, the first term removes the incorrect term arising from the SCRF orbitals and energies in forming the CI matrix whereas the second term adds the response of the electronic polarization of the solvent to the dipole of the excited state. Equation [11.1.64] is first order in electron relaxation. Higher orders can be examined by defining the perturbation

$$X^{(I)} = \lambda\Gamma(\varepsilon_\infty)\frac{\left[\langle\psi_0|\hat{\mu}|\psi_0\rangle - \langle\psi_I|\hat{\mu}|\psi_I\rangle\right]}{2} \quad [11.1.65]$$

which is clearly different for each excited state and would, if pursued, lead to a set of excited states that were nonorthogonal. In principle, these corrections need not to be small.

Depending on the Fock operator used (equation [11.1.58] or [11.1.59]), the excitation energy from the ground state $|\psi_0\rangle$ to the excited state $|\psi_I\rangle$ of a solute molecule in a dielectric medium is given as follows

$$W_I^A - W_0^A = \left\langle\psi_I\left|\hat{\mathbf{H}}\psi_I\right.\right\rangle - \left\langle\psi_0\left|\hat{\mathbf{H}}\psi_0\right.\right\rangle + \frac{1}{2}\Gamma\langle\psi_0|\hat{\mu}|\psi_0\rangle\left[\langle\psi_I|\hat{\mu}|\psi_I\rangle - \langle\psi_0|\hat{\mu}|\psi_0\rangle\right] -$$

$$-\frac{1}{2}\Gamma(\varepsilon_\infty)\langle\psi_I|\hat{\mu}|\psi_I\rangle\left[\langle\psi_I|\hat{\mu}|\psi_I\rangle-\langle\psi_0|\hat{\mu}|\psi_0\rangle\right] \qquad [11.1.66]$$

in the first case, and

$$W_I^B-W_0^B=\langle\psi_I|\hat{\mathbf{H}}\psi_I\rangle-\langle\psi_0|\hat{\mathbf{H}}\psi_0\rangle-\frac{1}{2}\Gamma(\varepsilon_\infty)\langle\psi_I|\hat{\mu}|\psi_I\rangle\left[\langle\psi_I|\hat{\mu}|\psi_I\rangle-\langle\psi_0|\hat{\mu}|\psi_0\rangle\right] \qquad [11.1.67]$$

in the second case. The last two equations are first order in electron polarization of the solvent.

The second approach to the calculation of spectra in solutions is based on the assumption that the ground and excited states are intimately coupled in an instantaneous absorption process.[28,29] In this model, the solute ground state electron distribution responds to the electron distribution in the excited state through the instantaneous polarization of the solvent. In such a case, the energy of the absorbing (ground) state is shifted by the following amount

$$\frac{1}{2}\Gamma(\varepsilon_\infty)\langle\psi_0|\hat{\mu}|\psi_0\rangle-\langle\psi_0|\hat{\mu}|\psi_0\rangle-\frac{1}{2}\Gamma(\varepsilon_\infty)\langle\psi_0|\hat{\mu}|\psi_0\rangle\left[\frac{1}{2}\langle\psi_I|\hat{\mu}|\psi_I\rangle-\langle\psi_0|\hat{\mu}|\psi_0\rangle\right] \qquad [11.1.68]$$

In the last equation, the first term removes the first order in the electron polarization part of the dielectric relaxation included in the SCRF of the ground state, and the second term adds back the appropriate interaction of the ground state with the "mean" reaction field, created by the excited state, $|\psi_I\rangle$. This leads to the following equation for the excitation energy

$$W_I^A-W_0^A=\langle\psi_I|\hat{\mathbf{H}}\psi_I\rangle-\langle\psi_0|\hat{\mathbf{H}}\psi_0\rangle+\frac{1}{2}\Gamma\langle\psi_0|\hat{\mu}|\psi_0\rangle\left[\langle\psi_I|\hat{\mu}|\psi_I\rangle-\langle\psi_0|\hat{\mu}|\psi_0\rangle\right]+$$

$$-\frac{1}{4}\Gamma(\varepsilon_\infty)\langle\psi_I|\hat{\mu}|\psi_I\rangle\left[\langle\psi_I|\hat{\mu}|\psi_I\rangle-\langle\psi_0|\hat{\mu}|\psi_0\rangle\right]^2 \qquad [11.1.69]$$

for the Fock operator [11.1.58] and

$$W_I^B-W_0^B=\langle\psi_I|\hat{\mathbf{H}}\psi_I\rangle-\langle\psi_0|\hat{\mathbf{H}}\psi_0\rangle+\frac{1}{4}\Gamma(\varepsilon_\infty)\left[\langle\psi_I|\hat{\mu}|\psi_I\rangle-\langle\psi_0|\hat{\mu}|\psi_0\rangle\right]^2 \qquad [11.1.70]$$

for the Fock operator [11.1.59]. However, it should be noticed that all four equations for spectral transition energies [[11.1.66], [11.1.67], [11.1.69] and [11.1.70]] yield very similar results when $\pi \rightarrow \pi^*$ and $n \rightarrow \pi^*$ transition solvatochromic shifts have been compared between the nonpolar and polar solvents.[27] In Table 11.1.3, the results of the INDO/S (ZINDO)[30,31] SCRF CIS calculated spectroscopic transition energies are given for some solvatochromic dyes. The relative shifts due to the solvent are reproduced theoretically in most cases. Even the absolute values of the spectroscopic transition energies are in satisfactory agreement with the respective experimental values, which demonstrates the applicability of the spectroscopic INDO/S parameterization for the spectra in solutions.

**Table 11.1.3. INDO/S SCRF CI calculated and experimental spectroscopic transition energies of some dyes in different solvents[27]**

| Molecule | Solvent | $\nu_{calc}$, cm$^{-1}$ | $\nu_{exp}$, cm$^{-1}$ |
|---|---|---|---|
| (Scheme 23) | Gas phase<br>Cyclohexane<br>Water | 29,700<br>26,300<br>22,500 | -<br>27,400<br>23,300 |
| (Scheme 24) | Gas phase<br>n-Hexane<br>Water | 36,900<br>34,800<br>31,100 | -<br>30,200<br>26,100 |
| (Scheme 25) | Gas phase<br>Chloroform<br>Water | 20,200<br>21,800<br>24,600 | -<br>19,600<br>22,100 |

Scheme 23

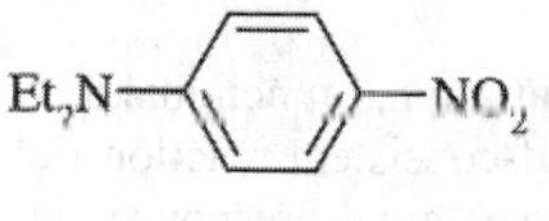


Scheme 24

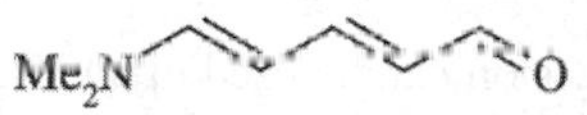


Scheme 25

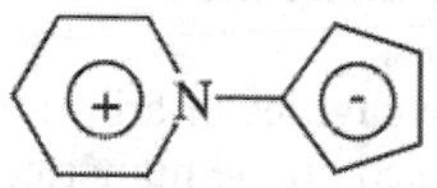


The INDO/S SCRF CI method has been also successfully applied for the prediction of the solvatochromic shifts in various nitro-substituted porphyrins.[32]

The SCRF methodology has been employed also for the prediction of the solvatochromic shifts on emission spectra.[33,34] A satisfactory agreement was obtained between the calculated and experimental fluorescence energies of p-N,N-dimethylaminobenzonitrile in different solvents. Finally, the solvent-induced shifts in the vibrational spectra of molecules have been also calculated using the SCRF theory.[35]

The SCRF approach has been also implemented for the treatment of solute-continuum solvent systems at the ab initio Hartree-Fock level of theory.[36,37] In addition, a general SCRF (GSCRF) approach has been proposed to account for the interaction of the solvent reaction field with the arbitrary charge distribution of the solute molecule. According to this theory,[38,39] the effective Hamiltonian of the solute in the solvent has the following form

$$\hat{\mathbf{H}}_s = \hat{\mathbf{H}}_s^0 + \int d\mathbf{r}\,\Omega_s(\mathbf{r})\left[V_m^0(\mathbf{r}) + \int d\mathbf{r}'\,G(\mathbf{r},\mathbf{r}')\Omega_s(\mathbf{r}')\right] \qquad [11.1.71]$$

where $\Omega_s(\mathbf{r})$ is the solute charge density operator given by

$$\Omega_s(\mathbf{r}) = -\sum_i \delta(\mathbf{r}-\mathbf{r}_i) + \sum_a Z_a \delta(\mathbf{r}-\mathbf{R}_a) \qquad [11.1.72]$$

where $\mathbf{r}_i$ stands for the i-th electron position vector operator, $\mathbf{R}_a$ is the position vector of the a-th nucleus with the charge $Z_a$ in the solute and $\delta(\mathbf{r})$ is the Dirac's delta function. The first term in square brackets in Eq. [11.1.71], $V_m^0(\mathbf{r})$, represents the electrostatic potential created by the solvent in the absence of the solute and the second, integral term corresponds to the reaction potential response function of the polarizable solvent. Together these terms produce the reaction field potential applying to the solvent molecule in the polarizable dielectric medium. Notably, a principal part of the Hamiltonian [11.1.71] is the solute charge density that can be represented using different approximations of which the multipolar expansion has been mostly applied. By using the distributed multipole model, it is possible to obtain the GSCRF equations for the molecules of a complex shape. However, it has been mentioned that the use of multipole expansions of the solvent electrostatic and reaction potentials in Eq. [11.1.71] may cause this Hamiltonian to become unbound and special damping procedures have been invented to overcome this difficulty.

The GSCRF total energy of the solute is given by the following equation

$$E_{GSCRF} = \langle\psi|\hat{\mathbf{H}}^0|\psi\rangle + \langle\psi|\int d\mathbf{r}\Omega_s(\mathbf{r})V_m^0(\mathbf{r})|\psi\rangle + \frac{1}{2}\langle\psi|\int d\mathbf{r}\Omega_s(\mathbf{r})|\psi\rangle\int d\mathbf{r}'G(\mathbf{r},\mathbf{r}')\Omega_s(\mathbf{r}') \quad [11.1.73]$$

This energy expression can be used to build up the respective variational functional to get the molecular orbitals [above]. A crucial step in the general self-consistent reaction field procedure is the estimation of the solvent charge density needed to obtain the response function $G(\mathbf{r},\mathbf{r}')$ and the reaction potential. The use of Monte Carlo or molecular dynamics simulations of the system consisting the solute and surrounding solvent molecules has been proposed to find the respective solvent static and polarization densities.

Several methods have been developed to account for the solute cavities of arbitrary shape in the solution. The polarizable continuum model (PCM) is based on the numerical integration of the relevant electrostatic equations describing the electrostatic interaction between the molecular charge distribution and the charge created on the boundary surface between the solute molecule and surrounding dielectric continuum.[40-45] Within this method, the solute cavity is usually constructed from the overlapping van der Waals spheres of constituent atoms in the solute molecule and the solvent reaction field arising from the solute charge distribution is calculated numerically. Alternatively, the cavity can be defined as constructed from the electron isodensity surface around the solute molecule (IPCM).[46] According to the classical electrostatics, the electrostatic potential at any point in the space can be described in terms of the apparent charge distribution, $\sigma$, on the cavity surface. It consists of two terms

$$\Phi_s = \Phi_M + \Phi_\sigma \quad [11.1.74]$$

the first of which ($\Phi_M$) corresponds to the electrostatic potential created by the charge distribution of the solute and the second ($\Phi_\sigma$) is due to the reaction potential by the solvent. The latter is directly connected with the apparent charge distribution on the surface of the cavity as follows:

$$\Phi_\sigma(\mathbf{r}) = \int_\Sigma \frac{\sigma(\mathbf{s})}{|\mathbf{r}-\mathbf{s}|}d^2\mathbf{s} \quad [11.1.75]$$

where Σ is the cavity surface and **s** vector defines a point on Σ. The Σ surface is usually divided into appropriate number of triangular small areas (tesserea), each of which has an area $\Delta S_k$ and contains the charge $q_k$ in some internal point $\mathbf{s}_k$. Thus, according to this, so-called boundary element method, the reaction potential is found as the following sum over all tesserea

$$\Phi_\sigma(\mathbf{r}) = \sum_k \frac{q_k}{|\mathbf{r} - \mathbf{s}_k|} \qquad [11.1.76]$$

with

$$q_k = \Delta S_k \sigma(\mathbf{s}_k) \qquad [11.1.77]$$

In the application of the boundary element method, it is crucial to select appropriate boundary surface for the solute cavity and to proceed as accurate as possible tessellation (triangulation) of this surface. For instance, it has been proposed that in the case of the cavity formation from overlapping van-der-Waals spheres, the atomic van-der-Waals radii should be multiplied by a coefficient equal to 1.2. Other possibilities of the surface definition include the closed envelope obtained by rolling a spherical probe of adequate diameter on the van-der-Waals surface of the solute molecule and the surface obtained from the positions of the center of such spherical probe around the solute.

Within the quantum-mechanical theory, the PCM model proceeds from the following Schrödinger equation for a solute molecule in the dielectric continuum

$$\left(\hat{\mathbf{H}}^0 + \hat{V}_{PCM}\right)\Psi = E\Psi \qquad [11.1.78]$$

where the reaction field potential is given by equation [11.1.76] as follows

$$\hat{V}_{PCM} = \sum_k \frac{q_k}{|\mathbf{r} - \mathbf{s}_k|} \qquad [11.1.79]$$

The charges on the boundary are found from the electrostatic polarization of the dielectric medium on the surface of the cavity due to the potential derived from the charge distribution of the solute and from other (induced) charges on the surface. The induced surface charge is evaluated iteratively at each step of the SCF procedure to solve the Schrödinger equation [11.1.78]. It has been reported that a simultaneous iteration of the surface charge with the Fock procedure reduces substantially the computation time without the loss in the precision of calculations.

Also, it has been shown that the expressions which determine the charges $q_k$ may be given as a set of linear equations. In the matrix form[42]

$$\mathbf{D}\mathbf{q}^{0f} = \mathbf{E}_{in} \qquad [11.1.80]$$

where **D** is a square nonsymmetric and nondiagonal matrix with the dimension equal to the number of surface elements, derived from the curvature of the surface and $\mathbf{q}^{0f}$ is a column vector, containing the unknown surface charges. $\mathbf{E}_{in}$ is also a column vector collecting the effective components of the solute electric field multiplied by the surface elements

$$E^{0}_{in,k} = -\Delta S_k \vec{\nabla}\Phi_{M,in}(\mathbf{s}_k)\mathbf{n}_k \quad [11.1.81]$$

The **D** matrix depends only on the shape of the cavity and the dielectric constant of the medium. Therefore, when the system of linear equations [11.1.80] has to be solved several times with different $\mathbf{E}_{in}$, as in the case of the polarizable solute, it may be convenient to work with the single inversion matrix $\mathbf{D}^{-1}$.

The PCM model has been implemented for the calculation of the electronic excitation energies of solvated molecules within the quantum-mechanical configuration interaction method.[47,48] The respective final expression for the excitation energy from the ground state (0) to the I-th state has the following form

$$\begin{aligned}\Delta W^{(0I)} &= \Delta E^{(0I)}_{CI} - \frac{1}{2}\left[\mathbf{J}_{20}(\mathbf{P}_I - \mathbf{P}_0) + \mathbf{P}_0\mathbf{T}_0(\mathbf{P}_I - \mathbf{P}_0)\right] + \\ &+ \frac{1}{2}\left[\mathbf{J}_{2\infty}(\mathbf{P}_I - \mathbf{P}_0) + \mathbf{P}_I\mathbf{T}_\infty(\mathbf{P}_I - \mathbf{P}_0)\right]\end{aligned} \quad [11.1.82]$$

where $\mathbf{P}_I$ and $\mathbf{P}_0$ are the electronic density matrices of the solute in the excited state and in the ground state, respectively,

$$\Delta E^{(0I)}_{CI} = E_I - E_0 + \frac{1}{2}\left[2(\mathbf{P}_I - \mathbf{P}_0)\mathbf{T}_0\mathbf{P}_0 + (\mathbf{P}_I - \mathbf{P}_0)\mathbf{J}_{10} + \mathbf{J}_{20}(\mathbf{P}_I - \mathbf{P}_0)\right] \quad [11.1.83]$$

are the diagonal elements of CI matrix, and

$$\mathbf{T} = C^t\mathbf{W}^{-1}\partial\hat{C} \quad [11.1.84]$$

$$\mathbf{J}_1 = C\mathbf{W}^{-1}\partial\hat{\mathbf{M}}\mathbf{Z} \quad [11.1.85]$$

$$\mathbf{J}_2 = \mathbf{Z}^t\mathbf{M}^t\mathbf{W}^{-1}\partial\hat{C} \quad [11.1.86]$$

$$\mathbf{B} = \mathbf{Z}^t\mathbf{M}^t\mathbf{W}^{-1}\partial\hat{\mathbf{M}}\mathbf{Z} \quad [11.1.87]$$

In the last equations, C and $\partial\hat{C}$ are the matrices representing the electrostatic potential and the electric field generated from the electron distribution in the solute molecule, respectively. The matrices **M** and $\partial\hat{\mathbf{M}}$ are the matrices representing the electrostatic potential and the electric field generated from the nuclear charges in the solute molecule, respectively. The diagonal elements of the matrix **W** are defined as the following function of the dielectric constant of the solvent

$$W_{ii} = \frac{\varepsilon + 1}{\varepsilon - 1} \quad [11.1.88]$$

The subscripts 0 and ∞ in the $\mathbf{J}_1$, $\mathbf{J}_2$, and **T** matrices in equation [11.1.82] correspond to the static and optical dielectric constant of the solvent. Equation [11.1.82] can be considered as an analog of equation [11.1.66] for the case of arbitrary cavity shape. The PCM-CI method has been applied for the calculation of solvatochromic shifts in the spectrum 4-[(4'-hydroxyphenyl)azo]-N-methylpyridine in a variety of solvents.[48]

An integral equation formalism (IEF) has been developed as particularly suitable for the description of solvent effects on spectral transition energies within the PCM model.[49] The respective theoretical equations have been applied for the calculation of solvatochromic shifts of several carbonyl-group containing molecules at the self-consistent field (SCF), configuration interaction (CI) and multiconfiguration self-consistent (MC SCF) field level of theory. The calculated spectral shifts accompanying the transfer of a solvatochromic compound from the gas phase to water were comparable with the experimental data. In Table 11.1.4, the results of calculations are presented for three carbonyl compounds, formaldehyde, acetaldehyde and acetone.

**Table 11.1.4. The calculated and experimental solvatochromic shifts (from the gas phase to water) in the spectra of some carbonyl compounds ($cm^{-1}$)[49]**

| Compound | ΔSCF | CI(SDT) | CAS SCF | Exp. |
|---|---|---|---|---|
| Formaldehyde | 1889 | 839 | 944 | 1700-1900[a] |
| Acetaldehyde | 1854 | 979 | 1049 | 1700-1900[a] |
| Acetone | 2273 | 1574 | 1259 | 1539-1889 |

[a]an estimate from other compounds

The advantage of the PCM method is in that it is applicable to the solute cavity of practically any shape in the solution. However, it is not clear how precisely should the molecular cavity be defined bearing in mind the classical (quasi-macroscopic) representation of the solvent. It is difficult to perceive that the solvent, e.g., the water molecules, can produce the electrical polarization corresponding to the statistically average distribution in the macroscopic liquid at infinitely small regions on the cavity surface. However, it is conceivable that larger chemical groups in the molecules may possess their own reaction field created by their charge distribution and the reaction fields of other groups in the solute molecule. A multi-cavity self-consistent reaction field (MCa SCRF) has been proposed[50] for the description of rotationally flexible molecules in condensed dielectric media. It proceeds from the observation that the interaction of the charge and higher electrical moments of a charge distribution in a spherical cavity with the corresponding reaction fields localized in the center of the cavity does not depend on the position of charge or (point) multipole centers in this cavity. Therefore, it is possible to divide a rotationally flexible solute molecule or a hydrogen-bonded molecular complex between two or more spherical cavities that embed the rotationally separated fragments of the solute or solute and solvent molecules, respectively. Assuming the classical Born-Kirkwood-Onsager charge density expansion (Eq. 11.1.27) for each of these fragments, the total energy of the solute in a dielectric medium can be expressed as a sum of terms that correspond to the energies arising from the interaction of the partial charge and the electric moments of a given molecular fragment with the reaction field of its own and the reaction fields of other fragments, as well as from the interaction between the reaction fields of different fragments. The Hartree-Fock-type equations derived from the variational functional for the total energy E can then be solved iteratively using the

SCRF procedure. The PCM approach has been further refined to account for the curvature of surface elements.[4] Also, this approach has been applied within different quantum-chemical frameworks.[47]

An alternative method for the description of solute-continuum electrostatic interaction has been developed as based on the notion that the electrostatic equations referring to the boundary surface between the solute and dielectric medium can be substantially simplified if to assume that the solvent is a homogeneous ideally conducting medium. Within this method (called the COSMO method), the electrostatic screening energy of a solute is given by the following equation (in matrix form)[51]

$$\Delta E = -\frac{1}{2}\mathbf{QBA}^{-1}\mathbf{BQ} \qquad [11.1.89]$$

with the following matrix elements

$$b_{ik} \approx \|\mathbf{t}_k - \mathbf{r}_i\|^{-1} \qquad [11.1.90]$$

for the point charges and

$$b_{ik} \approx \int_{\mu,\nu \in i} \frac{\chi_\mu(\mathbf{r})\chi_\nu(\mathbf{r})}{\|\mathbf{t}_k - \mathbf{r}\|} d^3\mathbf{r} \qquad [11.1.91]$$

for the continuous charge distribution, and

$$a_{kl} \approx \|\mathbf{t}_k - \mathbf{t}_l\|^{-1}, \quad k \neq l, \quad a_{kk} \approx 3.8|S_k|^{-1/2} \qquad [11.1.92]$$

In the last equations, $\mathbf{t}_k$ denotes the position vectors of the centers of small surface elements k on the arbitrary cavity surface; $\mathbf{r}_i$ are the position vectors of the point charges in the solute molecule; $\mathbf{r}$ is the vector for electronic charge position described on the atomic basis $\{\chi_\mu(\mathbf{r})\}$ and $S_k$ are the areas of the surface elements. In equation [11.1.82], $\mathbf{Q}$ is the matrix of source charges in the solute.

The COSMO model has been extended to account for the solvents with any dielectric constant.[52-54] Within the respective GCOSMO method,[53] the surface charges $\sigma(\mathbf{r})$ on the boundary between the solute and continuum solvent are first determined for the medium with the infinite dielectric constant under the assumption that the electrostatic potential on the surface S is zero. For a dielectric medium specified by the dielectric constant $\varepsilon$, the actual surface charges are then calculated by scaling the screening conductor surface charge $\sigma(\mathbf{r})$ by a factor of $f(\varepsilon) = (\varepsilon - 1)/\varepsilon$. This scaling preserves the validity of the Gauss theorem for the total surface charge. The boundary element method is applied for the calculation of the surface charge density, with the boundary divided into small areas and the surface charge approximated as the point charge in the center of this area. The charges are calculated either using the charge distribution in the molecule or by minimizing variationally of the total electrostatic solvation energy. The total free energy of the system of solute and surface charges is then calculated within the Hartree-Fock theory as

$$E_{tot} = \left[ P_{\mu\nu}\left(H^0_{\mu\nu} + H^s_{\mu\nu}\right) + \frac{1}{2}\left(G^0_{\mu\nu} + G^s_{\mu\nu}\right)\right] - \frac{1}{2} f(\varepsilon)\mathbf{Z}^+\mathbf{B}^+\mathbf{A}^{-1}\mathbf{B}\mathbf{Z} + E_{nn} + E_{nes} \quad [11.1.93]$$

where $E_{nn}$ is solute nuclear-nuclear repulsion and $E_{nes}$ is the solvation energy related to the dispersion and repulsion between the solute and solvent, and cavity formation; $H^0_{\mu\nu}$ and $G^0_{\mu\nu}$ are the one-electron and two-electron parts of the Fock matrix for the isolated solute, respectively, and $P_{\mu\nu}$ are the density matrix elements. The solvent perturbations to the corresponding operators have been expressed as

$$H^s_{\mu\nu} = -f(\varepsilon)\mathbf{Z}^+\mathbf{B}^+\mathbf{A}^{-1}\mathbf{B}\mathbf{L}_{\mu\nu} \quad [11.1.94]$$

and

$$G^s_{\mu\nu} = -f(\varepsilon)\left(\sum_{\lambda,\sigma} P_{\lambda\sigma}\mathbf{L}^+_{\lambda\sigma}\right)\mathbf{A}^{-1}\mathbf{L}_{\mu\nu} \quad [11.1.95]$$

where **A** and **B** are the N x N square matrices (N - number of atomic nuclei in the solute molecule) with the elements defined by equations [11.1.90] and [11.1.92]. The matrices $\mathbf{L}_{\mu\nu}$ consist of the one-electron integrals [11.1.91]. The first and second derivatives needed for the calculation of the molecular potential surfaces and the respective solvent effects on vibrational spectra have been also supplied within the framework of GCOSMO approach.[53] A semi-quantitative agreement between the computational and experimental results has been obtained for the vibrational frequencies of acetone in water.

Several approaches have been developed to account for the electron correlation effects on the solvation energy of both the ground and the excited states of the molecule in the solution. A multiconfigurational self-consistent reaction field (MC SCRF) theory has been proposed as based on the classical Onsager's reaction field model.[55] Notably, the higher order electrical moments of the solute molecule and the respective reaction field in the solvent were taken into account within this method. Thus, the dielectric solvation energy of a solute in a given state embedded into a linear isotropic medium has been calculated as the product of the expectation values for the reaction field $\langle \mathbf{R}^m_l \rangle$ and the respective multipole charge moments $\langle \mathbf{M}^m_l \rangle$ of the solute[55-58]

$$E_{el} = -\frac{1}{2}\sum_{l,m}\left\langle \mathbf{R}^{m^+}_l \right\rangle\left\langle \mathbf{M}^m_l \right\rangle \quad [11.1.96]$$

where

$$\left\langle \mathbf{M}^m_l \right\rangle = \sum_k Z_k S^m_l(R_k) - \left\langle S^m_l \right\rangle \quad [11.1.97]$$

$$\left\langle \mathbf{R}^m_l \right\rangle = f\left\langle \mathbf{M}^m_l \right\rangle \quad [11.1.98]$$

and

$$f_l = \frac{(l+1)(\varepsilon-1)}{1+\varepsilon(l+1)} \frac{1}{a^{-(2l+1)}} \quad [11.1.99]$$

$$S_l^m(\mathbf{r}) = \left[\frac{4\pi}{2l+1}\right]^{1/2} \mathbf{r}^l Y_{lm}(\theta, \varphi) \quad [11.1.100]$$

where $Y_{lm}(\theta, \varphi)$ are the Legendre' polynomials.

Interestingly, it had been suggested that the dispersion interaction energy between the solute and the solvent could be accounted for, in principle, in the framework of this approach as related to the full distribution of dielectric relaxation frequencies of the solvent. Thus, the formula for the MC SCRF solvation energy has been expressed as follows

$$E_{el} = -\frac{1}{2}\sum_{l,m} f_l (T_{lm})^2 \quad [11.1.101]$$

with the terms $T_{lm}$ obtained from the expectation values of the nuclear and electronic solvent operators

$$T_{lm} = T_{lm}^n - T_{lm}^e \quad [11.1.102]$$

$$T_{lm}^n = \sum_a Z_a R^{lm}(\mathbf{R}_a) \quad [11.1.103]$$

$$T_{lm}^e = R^{lm}(\mathbf{r}) \quad [11.1.104]$$

where $R^{lm}$ are the special solvent effect integrals.[55]

The solvent contributions have been developed also for the multiconfigurational energy gradient and Hessian, necessary for the solution of the MC SCRF equations. Notably, the results of the model calculations on water molecule implied that the higher multipole terms might play a significant role in the total electrostatic solvation energy of the molecule. Thus, the quadrupole term consisted approximately 20% of the dipolar term whereas the 4-th order term was even more significant (~ 30% of the dipolar term). The MC SCRF method has been applied for the calculation of the solvent effects on the spectral transitions of water and formaldehyde in different media.[59] The MC SCRF methodology has been further refined in the framework of the response theory approach.[60] This development describes the response of the solute or the solute-solvent complex to a time-independent or time-dependent high-frequency perturbation such as the spectral excitation.

In order to take into account the electron correlation effects, another combination of the self-consistent reaction field theory with the configuration interaction formalism has been introduced as follows.[61] Within this approach, the usual CI wavefunction has been constructed as follows

$$\Psi_{CI} = \sum_{a=1}^{N} C_a D_a \quad [11.1.105]$$

where $C_a$ are the CI expansion coefficients and $D_a$ are the basis functions (Slater determinants or their linear combinations). In the case of orthogonal basis functions, the normalization condition of the function $\Psi_{CI}$ is given as

$$\sum_{a=1}^{N} C_a^2 = 1 \qquad [11.1.106]$$

and the coefficients $C_a$ are determined from the following equation

$$\left(\hat{\mathbf{H}}^0 + \hat{V}_{rf}\right)\Psi = E\Psi \qquad [11.1.107]$$

where $\hat{\mathbf{H}}^0$ is the Hamiltonian for the molecule, unperturbed by the reaction field and $\hat{V}_{rf}$ is the reaction field perturbation. The latter can be presented, for example, in the framework of the boundary element method as follows (cf. Eq. [11.1.75])

$$\hat{V}_{rf} = \int d^2\mathbf{r}\frac{\sigma(\mathbf{r}')}{|\mathbf{r}-\mathbf{r}'|} \qquad [11.1.108]$$

where $\sigma(\mathbf{r}')$ is the charge density on the surface of the cavity.

For the spectroscopic applications, it would be again instructive to separate the noninertial and inertial components of the electrostatic polarization of the dielectric medium. The first of them corresponds to the electrostatic polarization of the electron charge distribution in the solvent that is supposedly instantaneous as compared to any electronic or conformational transition of the solute. The second component arises from the orientational polarization of the solvent molecules in the electrostatic field of the solute. The noninertial polarization can be described by the optical dielectric permittivity of the solvent that corresponds to the infinite frequency of external electromagnetic field ($\varepsilon_\infty \approx n_D^2$) whereas the inertial polarization represents the slow, orientational part of the total dielectric constant of the solvent, ε. In order to separate the noninertial polarization, it is helpful to determine the solute charge density as the sum of the respective nuclear and electronic parts

$$\rho = \rho_n + \rho_e = \sum_A \delta(\mathbf{r}-\mathbf{r}_A)Z_A + \sum_{a,b} C_a C_b (ab|\mathbf{r},\mathbf{r}') = \sum_{a,b} C_a C_b \rho_{ab} \qquad [11.1.109]$$

where $\delta(\mathbf{r} - \mathbf{r}_A)$ is the Dirac's delta function and $(ab|\mathbf{r},\mathbf{r}')$ are the elements of the single-determinant matrices of transitions between the configurations. Notably, the values of

$$\rho_{ab} = \delta_{ab}\rho_n - \rho(ab|\mathbf{r},\mathbf{r}') \qquad [11.1.110]$$

do not depend on the coefficients $C_a$ and $C_b$. The noninertial component of the polarization field, $\Phi_\infty(\mathbf{r})$, is always in equilibrium and thus it can be represented as follows

$$\Phi_\infty(\mathbf{r}) = \sum_{a,b} C_a C_b \Phi_{ab}^{(\infty)}(\mathbf{r}) \qquad [11.1.111]$$

where $\Phi_{ab}^{(\infty)}(\mathbf{r})$ is the solution of equation [11.1.107] for the basic charge distribution $\rho_{ab}$. Since the latter does not depend on the coefficients $C_a$ and $C_b$, the values of $\Phi_{ab}^{(\infty)}(\mathbf{r})$ can be determined before the calculation of the wavefunction $\Psi_{CI}$ [11.1.105]. The coefficients $C_a$ can be calculated, as usual, from the matrix equation

$$\mathbf{H}|C\rangle = E|C\rangle \qquad [11.1.112]$$

where $|C\rangle$ denotes the column-vector of the coefficients and the elements of the matrix **H** are given as follows[61]

$$H_{ab} = \left\langle D_a \left| \hat{\mathbf{H}}^0 \right| D_b \right\rangle + \sum_{a,b} C_a C_b \int d^3\mathbf{r}\rho_{ab}\Phi_{ab}^{(\infty)} + \int d^3\mathbf{r}\rho_{ab}\Phi_0 \qquad [11.1.113]$$

where $\Phi_0$ is inertial (nuclear) part of the polarization field. For a given $\Phi_0$, the set of equations [11.1.98) can be solved iteratively. Implicitly, the last equations describe both the electron subsystems of the solute and solvent, the latter being taken into account as the field of noninertial (electron) polarization of the solvent, $\Phi^{(\infty)}$.

The electron correlation effects on the solvation energy of a solute have been also accounted for within the framework of the perturbation theory.[62,63] By starting from the Hamiltonian of the solute molecule as follows

$$\hat{\mathbf{H}} = \hat{\mathbf{H}}^0 - M_l^m f_l^m \left\langle \psi \left| M_{l'}^{m'} \right| \psi \right\rangle \qquad [11.1.114]$$

where $\psi$ denotes the exact (correlated) wavefunction, the Hartree-Fock operator may be written as

$$\mathbf{F} = \mathbf{F}^0 - M_l^m f_l^m \left\langle \psi_0 \left| M_{l'}^{m'} \right| \psi_0 \right\rangle \qquad [11.1.115]$$

where $\psi_0$ denotes the electronic wavefunction at the Hartree-Fock level. The Hamiltonian may be then written as a perturbed expression of the Hartree-Fock operator

$$\hat{\mathbf{H}} = \mathbf{F} + \left(\hat{\mathbf{H}}^0 - \mathbf{F}^0\right) + M_l^m f_l^m \left( \left\langle \psi_0 \left| M_{l'}^{m'} \right| \psi_0 \right\rangle - \left\langle \psi \left| M_{l'}^{m'} \right| \psi \right\rangle \right) \qquad [11.1.116]$$

The perturbation has two contributions, the standard Møller-Plesset perturbation and the non-linear perturbation due to the solute-solvent interaction. If $C_j^{(i)}$ denotes the coefficient of the eigenstate $|J\rangle$ in the corrections of $\psi$ to the i-th order, then the perturbation operators $H^{(i)}$ of the i-th order are given by the following formulae

$$\hat{\mathbf{H}}^{(1)} = \hat{\mathbf{H}}^0 - \mathbf{F}^0 \qquad [11.1.117]$$

$$\hat{\mathbf{H}}^{(2)} = -2\sum_{I \neq 0} C_I^{(1)} M_l^m f_l^m \left\langle 0 \left| M_{l'}^{m'} \right| I \right\rangle \qquad [11.1.118]$$

$$\hat{\mathbf{H}}^{(3)} = -2\sum_{J \neq 0} C_J^{(2)} M_l^m f_l^m \left\langle 0 \left| M_{l'}^{m'} \right| J \right\rangle - \sum_{J \neq 0} \sum_{P \neq 0} C_J^{(1)} C_P^{(1)} M_l^m f_l^m \left\langle I \left| M_{l'}^{m'} \right| P \right\rangle \qquad [11.1.119]$$

The calculation is performed, as usual, by comparing the coefficients of the Schrödinger equation to successive orders.

The first order energy is the same as given by the usual Møller-Plesset treatment, $< 0|\hat{\mathbf{H}}^0|0 >$ and the first order electrostatic contribution to the free energy of solvation is identical to the result obtained at the Hartree-Fock level of theory. The second order correction to the free energy is given as

$$\Delta G_s^{(2)} = 2\sum_{S\neq 0} C_S^{(2)} \langle 0|M_{l'}^{m'}|S\rangle \boldsymbol{f}_l^m \langle 0|M_{l'}^{m'}|0\rangle + \sum_{D\neq 0}\sum_{D'\neq 0} C_D^{(1)} C_{D'}^{(1)} \langle D|M_{l'}^{m'}|D'\rangle \boldsymbol{f}_l^m \langle 0|M_{l'}^{m'}|0\rangle \quad [11.1.120]$$

where $|S\rangle$ stands for the singly excited states, and $|D\rangle$ and $|D'\rangle$ for a pair of doubly excited states different by just one orbital. Without excessive difficulty, it is possible to derive the correction terms to the electrostatic free energy of solvation of higher orders.

A many-body perturbation theory (MBPT) approach has been combined with the polarizable continuum model (PCM) of the electrostatic solvation.[64-66] The first approximation called by authors the perturbation theory at energy level (PTE) consists of the solution of the PCM problem at the Hartree-Fock level to find the solvent reaction potential and the wavefunction for the calculation of the MBPT correction to the energy. In the second approximation, called the perturbation theory at the density matrix level only (PTD), the calculation of the reaction potential and electrostatic free energy is based on the MBPT corrected wavefunction for the isolated molecule. At the next approximation (perturbation theory at the energy and density matrix level, PTED), both the energy and the wave function are solvent reaction field and MBPT corrected. The self-consistent reaction field model has been also applied within the complete active space self-consistent field (CAS SCF) theory[12,67] and the complete active space second-order perturbation theory.[12,67,68]

Several groups[69-73] have also proposed the quantum mechanical density functional theory (DFT) based methods for the calculation of the electrostatic solvation energy in dielectric media. However, the application of this theory for excited states is not straightforward.[74,75]

### 11.1.4 THEORETICAL TREATMENT OF SOLVENT DISPERSION EFFECTS ON ELECTRONIC-VIBRATIONAL SPECTRA OF MOLECULES

The dispersion interaction between two atomic or molecular systems can be theoretically presented at different levels of theory.[76-78] The modelling of the dispersion interactions in condensed media is more complicated and proceeds either from the discrete molecular description of the liquid or from the continuum model. According to a contemporary classification,[4] the theoretical approaches to the dispersion effect in solutions can be divided into following classes:

- pair-potential approaches
- reaction field based approaches
- cavity surface-dispersion energy relationship approaches

The pair-potential approach is based on the discrete representation of the pairs of solvent and solute molecules or some fragments of them. The respective dispersion potentials are expressed as truncated asymptotic expansions in powers of 1/r, the reciprocal of the distance between the interacting entities[4]

$$U_{ms}(disp) = \sum_{k=6,8,10} d_{ms}^{k} r_{ms}^{-k} \quad [11.1.121]$$

where the indexes m and s denote the structural entities (atoms, bonds, chemical groups) belonging to the solute and solvent molecules, respectively. The powers in expansion [11.1.121] are based on the formal theory of two-body interactions. In the practical calculations, only the first term of the expansion (k = 6) is frequently applied. The expansion coefficient $d_{ms}^{(6)}$ can be calculated using the London formula

$$d_{ms}^{(6)} = -\frac{3}{2}\alpha_m \alpha_s \frac{\bar{I}_m \bar{I}_s}{\bar{I}_m + \bar{I}_s} \qquad [11.1.122]$$

where $\alpha_m$ and $\alpha_s$ are the isotropic polarizabilities for interacting systems and $\bar{I}_m$ and $\bar{I}_s$ are the mean excitation energies of these systems. This approximate formula is, in principle, valid only for interacting atoms. In the case of molecular systems, the atomic or group polarizabilities and local excitation energies are, as a rule, not isotropic and require the use of the respective tensor quantities. The absence of information about of accurate solute-solvent atom-atom distribution functions in dense media complicates further the accurate treatment of the dispersion interaction. These distribution functions can be calculated either using the molecular dynamics or Monte Carlo computer simulations or from the experimental scattering data on the respective systems. However, almost all these methods give only the averaged distribution functions and lack, therefore, the information about the local anisotropy of the atom-atom distributions.

Similarly to the treatment of electrostatic effects, the dispersion potential can be limited to the dipole-dipole term and the mean excitation energies are approximated by the respective ionization potentials for the solute and solvent molecules. Thus, when a small cluster of solvent molecules surrounds the solute molecule, the first approximation of the dispersion energy can be presented by the following formula:[79]

$$U_{MS}(disp) = -\frac{x}{4}\frac{\bar{I}_M \bar{I}_S}{\bar{I}_M + \bar{I}_S}\sum_{u=1}^{B_M}\sum_{v=1}^{B_S}\left\{r_{uv}^{-6} Tr\left[\mathbf{T}_{uv}\mathbf{A}_u\mathbf{T}_{uv}\mathbf{A}_v\right]\right\} \qquad [11.1.123]$$

where $B_M$ and $B_S$ are the number of bonds in the solute and in the solvent molecules, respectively, $\bar{I}_m$ and $\bar{I}_s$ are the corresponding mean excitation energies, $\mathbf{T}_{uv}$ is the tensor

$$\mathbf{T}_{uv} = 3\frac{\mathbf{r}_{uv}}{r_{uv}} \otimes \frac{\mathbf{r}_{uv}}{r_{uv}} - 1 \qquad [11.1.124]$$

where $r_{uv}$ and $\mathbf{r}_{uv}$ are the distance and the radius-vector between the bonds u and v, respectively, and $\mathbf{A}_u$ is the polarizability tensor for bond u. The factor x in equation [11.1.123] is introduced to achieve the agreement between the molecule-molecule pair dispersion potential and a simpler expression derived on the basis of assumption that the dispersion energy between two molecules may be reduced to the sum of independent atom-atom contributions[80]

$$U_{MS}(disp) = \sum_m \sum_s d_{ms}^{(6)} r_{ms}^{-6} \qquad [11.1.125]$$

A scheme has been developed that reduces the spatial representation of the dispersion interaction into a surface representation of this interaction.[4] According to this approach, the average dispersion-repulsion energy of a solute-solvent system has been written as follows:

$$\left\langle E_{disp-rep} \right\rangle = \int \cdots \int U(\Omega)\, g(\Omega) d\Omega \qquad [11.1.126]$$

where Ω stands for the set of all coordinates of the molecules involved, g(Ω) is the solute-solvent pair distribution function and U(Ω) is expressed as a sum of two-body dispersion-repulsion potentials. In the case of the fixed geometry of the solute molecule

$$\left\langle E_{disp-rep} \right\rangle = n_S \sum_{s \in S} N_s \sum_{m \in M} \sum_{k} d_{ms}^{(k)} \int r_{ms}^{(k)} g_{ms}\left(\mathbf{r}_{ms}\right) dr_{ms}^3 \qquad [11.1.127]$$

The integrals in the last formula can be limited only to a certain minimum distance defined, for instance, by the van-der-Waals envelopes of interacting molecules. By introducing the auxiliary vector functions $A_{ms}^{(k)}(\mathbf{r}_{ms})$ such that

$$\vec{\nabla} A_{ms}^{(k)}\left(r_{ms}\right) = d_{ms}^{(k)} \mathbf{r}_{ms}^{-k} g_{ms}\left(\mathbf{r}_{ms}\right) \qquad [11.1.128]$$

the average dispersion-repulsion energy between thc solute and solvent molecules in solution may be written as follows

$$\left\langle E_{disp-rep} \right\rangle = n_S \sum_{s \in S} N_s \sum_{m \in M} \sum_{k} \int_{\Sigma_s} \mathbf{A}_{ms}^{(k)} \mathbf{n}_\sigma d\sigma \qquad [11.1.129]$$

where $\mathbf{n}_\sigma$ is the outer normal to the surface $\Sigma_s$ at the position σ. The integral in the last equation may be calculated numerically using an appropriate partitioning (tessellation) of the surface.

A quantum-mechanical method of calculation of the dispersion energy has been developed on the basis of the above-cited semiclassical Abe's theory.[81] According to this method, the dispersion energy, $E_{disp}$, for a solute molecule in a spherical cavity is given as follows

$$E_{disp} = -\frac{2}{3} \frac{1}{a_S^3 a_M^3} \sum_{J \neq I} \sum_{K \neq O} \frac{\left(\mu_{IJ}^M\right)^2 \left(\mu_{KO}^S\right)^2}{E_K^S - E_O^S + E_J^M - E_I^M} \qquad [11.1.130]$$

where the superscript S refers to the solvent molecule and the superscript M to the solute molecule. Thus, $\mu_{IJ}^M$ and $\mu_{KO}^S$ are the transition dipoles between the respective states of the solute (I and J) and the solvent (K and O) molecules. In equation [11.1.130], $E_K^S$, $E_O^S$ and $E_J^M$, $E_I^M$ denote the energies of the K-th and O-th state of the solvent and of thc J-th and I-th state of the solute molecule, respectively. The cavity radii for the solvent and solute molecules are denoted as $a_S$ and $a_M$, respectively.

**Table 11.1.5. The INDO/CI calculated solvatochromic shifts $\Delta\nu$ (from the gas phase to cyclohexane) of some aromatic compounds and the respective experimental data in low polarity solvents ($cm^{-1}$)[81]**

| Compound | Transition | $\Delta\nu$ (calc) | $\Delta\nu$ (exp) |
|---|---|---|---|
| Benzene | $^1B_{2u}$ | -316 | -209[a] |
| Naphthalene | $^1L_b(x)$<br>$^1L_b(y)$ | -332<br>-879 | -300[b]; -275[a]<br>-950[b]; -902[a] |
| Chrysene | $^1L_b(^1B_u)$<br>$^1L_a(^1B_u)$<br>$^1B_b(^1B_u)$ | -243<br>-733<br>-1666 | -252[a]<br>-1030[a]<br>-1620[a] |
| Azulene | $^1L_b(y)$<br>$^1L_a(x)$<br>$^1K_b(y)$<br>$^1B_b(x)$ | +162<br>-288<br>-446<br>-1475 | +164[c]<br>-333[c]<br>-285[c]<br>-1650[c] |

[a]in n-pentane, [b]in cyclohexane, [c]in 2-chloropropane

The equation [11.1.130] has been used within the semiempirical quantum-chemical INDO/CI formalism to calculate the solvent shifts of some aromatic compounds in cyclohexane.[81] The results compare favorably with the experimental data for some nonpolar solvents (cf. Table 11.1.5).

## 11.1.5 SUPERMOLECULE APPROACH TO THE INTERMOLECULAR INTERACTIONS IN CONDENSED MEDIA

The supermolecule approach to the calculation of solute-solvent interaction energies is based on the discrete molecular representation of the solvent. The supermolecule can be treated quantum-mechanically as a complex consisting of the central solute molecule and the surrounding closest solvent molecules. This supermolecule complex can be treated individually or as submerged into the dielectric continuum.[82] In the last case, some continuum theory (SCRF, PCM) is applied to the supermolecule complex consisting of the solute molecule and the solvent molecules in its first coordination sphere.[83-86] Therefore, the short-range solute-solvent electron correlation, dispersion and exchange-repulsion interactions are taken into account explicitly at the quantum level of theory as the electrons and nuclei both from the solute and solvent are included explicitly in the respective Schrödinger equation. The long-range electrostatic polarization of the solvent outside the first coordination sphere is, however, treated according to the dielectric continuum theory. Thus, the energy of solvation of a solute molecule can be expressed as follows:

$$E_{sol} = \left\langle \Psi_{SM} \left| \hat{\mathbf{H}}_{SM}^{(S)} \right| \Psi_{SM} \right\rangle - \left\langle \Psi_{SM} \left| \hat{\mathbf{H}}_{M}^{(0)} \right| \Psi_{SM} \right\rangle - n \left\langle \Psi_{SM} \left| \hat{\mathbf{H}}_{S}^{(0)} \right| \Psi_{SM} \right\rangle \qquad [11.1.131]$$

where $\hat{\mathbf{H}}_{SM}^{(S)}$ is the Hamiltonian for the supermolecule in the solution, and $\hat{\mathbf{H}}_{M}^{(0)}$ and $\hat{\mathbf{H}}_{S}^{(0)}$ are the Hamiltonians for the isolated solute and the solvent molecules, respectively. In the last equation, n denotes the number of the solvent molecules applied in the supermolecule, $\Psi_{SM}$ is the total wavefunction of the supermolecule immersed into dielectric medium, and $\Psi_M$ and $\Psi_S$ are the wavefunctions for isolated solute and solvent molecules, respectively.

For instance, the n $\rightarrow \pi^*$ electronic transition ($1^1A_1 \rightarrow 1^1A_2$) of formaldehyde solvated by varying number of water molecules has been investigated using multi-reference CI calculations.[87] This simple supermolecule approach has given already satisfactory results as compared to experimental shifts in liquid water. However, it has been shown that in general, both the short-range quantum mechanical effects and the long-range solvent polarization play important role in determining the spectral shifts in liquid media. Thus, the INDO/S SCRF CIS theory alone has explained the solvatochromic shifts of azoles in different solvents, except those observed in water (Table 11.1.6).[85] In both the water and acetonitrile, the compounds are predicted to have practically the same shift that is not the case in experiment. The explicit bonding of two water molecules to the nitrogen lone pairs leads in the cases of pyrimidine and pyridazine to the calculated large red shift instead of the experimentally observed solvatochromic blue shift. However, by treating the complex of an azole and two water molecules quantum-mechanically in the surrounding reaction field leads to quantitatively correct blue shifts.

**Table 11.1.6. The INDO/S SCRF/CI calculated and experimental spectral transition energies in different solvents for azoles ($cm^{-1}$)[85]**

| Molecule | Solvent | $\nu_{calc}$, $cm^{-1}$ | $\nu_{exp}$, $cm^{-1}$ |
|---|---|---|---|
| Pyrimidine | gas phase | 32966 | - |
| | isooctane | 33559 | 34200 |
| | diethyl ether | 34127 | 34400 |
| | acetonitrile | 34697 | 34800 |
| | water | 34743 | 36900 |
| | $2H_2O$ | 30982 | 36900 |
| | water + $2H_2O$ | 36572 | 36900 |
| Pyridazine | gas phase | 28329 | - |
| | isooctane | 29460 | 29740 |
| | diethyl ether | 30382 | 30150 |
| | acetonitrile | 31296 | 31080 |
| | water | 31368 | 33570 |
| | $2H_2O$ | 26490 | 33570 |
| | water + $2H_2O$ | 33927 | 33570 |
| Pyrazine | gas phase | 30387 | - |
| | isooctane | 30387 | 31610 |
| | diethyl ether | 30387 | 31610 |
| | acetonitrile | 30387 | 31740 |
| | water | 30387 | 33160 |
| | $2H_2O$ | 32900 | 33160 |
| | water + $2H_2O$ | 33301 | 33160 |

As a general remark, in the calculations of the intermolecular interactions using the supermolecule approach, the "size-extensivity"[88] of the methods applied is of crucial importance. Furthermore, the interaction energies calculated in the supermolecule approach usually suffer from what is called the basis set superposition error (BSSE),[89] a spurious energy improvement resulting from the use of truncated basis sets. This error seems to be unavoidable in most practical calculations except for very small systems.[90]

The intermolecular interactions that correspond to the fixed geometry of the solute-solvent complex can be also studied by a perturbation approach.[91,92] It has been suggested that the perturbation theory has some advantages over the supermolecule approach and may therefore be considered conceptually more appropriate for the calculation of intermolecular interaction energies. In this case, the interaction energy is calculated directly and it may be separated into components of well-defined physical meaning.

Within the direct reaction field (DRF) method,[93-96] the classical part of the solute-solvent system (solvent) is treated as a distribution of the polarizable point dipoles, interacting with each other. The DRF Hamiltonian of the solute-solvent system is thus given by the following formula:

$$\hat{\mathbf{H}}_{DRF} = \hat{\mathbf{H}}^0 - \frac{1}{2}\sum_{i,j}\sum_{p,q}\mathbf{F}_{ip}^{+}\alpha_{pq}\mathbf{F}_{jq} \qquad [11.1.132]$$

where indices i an j correspond to the solute particles (electrons and nuclei) and p and q run over the external polarizable points. $\mathbf{F}_{ip}$ is the field of the particle i at the position p, and $\alpha_{pq}$ gives the induced dipole at point q by a field applied at point p. The respective Schrödinger equation can be solved directly, without the iterative adjustment of the solvent charge distribution and the respective reaction field potential. The DRF method proceeds from the direct reaction field obtained as the linear solute-solvent interaction operator, proportional to the square of the electric field operator while the GSCRF approach uses the average reaction field model. It has been suggested that the additional energy contributions can be interpreted as due to the dispersion interaction between the solute and solvent molecules.[97] More recently, the DRF approach has been combined with the continuum approach by dividing the space around the solute into a closer surrounding treated by direct reaction field method and to more distant space represented by the macroscopic dielectric properties of the solvent.[98,99] A good quantitative agreement has been obtained between the experimental and DRF calculated solvatochromic shifts of the $n \rightarrow \pi^*$ transition of acetone in different solvents.[100,101]

Nevertheless, even at the highest theoretical level accessible for practical calculations, the static approach is strictly valid only for the description of the molecular clusters of fixed geometry. However, in the cases of strong and weak intermolecular interactions, the energy of interaction in the molecular cluster in the gas phase or on the inert-gas matrix is substantially different from the total solvation energy in the condensed phase.[102] A direct solution of a time-dependent Schrödinger equation for the condensed low-order bulk matter, needed to overcome this problem, is premature. Therefore, the molecular dynamics method (MD)[103-105] based on the computer modelling of a system of molecules which interact by the known model potential to each other and undergoes the rotational and translational movement in the field caused by this interaction according to the classical (Newtonian) mechanics is widely applied for this purpose. By applying various boundary conditions and performing the calculation of the potential energies and forces for hundreds of thousands configurations obtained by step-by-step time evolution the time - averages such as internal energy (or enthalpy) can be obtained. An alternative is the stochastic Monte Carlo method that is based on the ergodic theorem and provides the ensemble averages calculated from randomly generated and weighted configurations. These methods suffer from several shortcomings, of which the problem of the applicability of the ergodic theorem (i.e., the identity of the time-averaged and ensemble-averaged thermodynamic and dynamic observables), the path sampling and difficulties to obtain precise intermolecular interaction potentials are most serious. Also, the real dense systems, i.e., liquids and solutions are intrinsically quantified systems and therefore a quantum molecular dynamics should be developed which accounts for the quantum effects in the microscopic system from the first principles. The combined quantum-mechanical/molecular dynamics (QM/MD) or quantum-mechanical/molecular mechanics (QM/MM) approaches have been used for the calculation of solvatochromic shifts in different media.[106-108]

Numerous computational schemes have been developed to calculate the total molecular solvation energy or free energy using the combination of different theoretical solute-solvent interaction models. From these, one of the most popular is the SMx methodology.[109,110] This methodology proceeds from the division of the total molecular solvation energy into the solute-solvent electrostatic and inductive polarization terms, standard-state free energy of cavity creation in the solvent plus the solute-solvent dispersion interaction, and an empirical part of the nuclear motion free energy. The solvent polarization term is presented using the generalized Born formula:

$$G_p = -\frac{1}{2}\left(1-\frac{1}{\varepsilon}\right)\sum_{k=1}^{N}\sum_{k'=1}^{N} q_k q_{k'} \gamma_{kk'} \qquad [11.1.133]$$

where the double summation is performed over all atomic partial charges $q_k$ in the solute molecule, ε is the relative dielectric permittivity of the solvent and $\gamma_{kk'}$ - the Coulomb' integral between two centers k and k', parameterized for the interactions with the solvent. The cavity creation plus dispersion term is calculated as

$$G^0_{CD} = \sum_{k=1}^{N} \sigma_k A_k \qquad [11.1.134]$$

where N is the number of atoms in the solute, $A_k$ is the solvent accessible surface area of a given atom and $\sigma_k$ is the parameter for this atom that is called the accessible surface tension. The latter is obtained from the fit with the experimental data and is, thus, essentially an empirical parameter for a given type of atom. Thus, in essence the SMx methodology represents a semiempirical approach to the calculation of solvent effects.

## REFERENCES


1 P. Suppan, N. Ghoneim, **Solvatochromism**, *The Royal Society of Chemistry*, London, 1997.
2 N.G. Bakhshiev (Ed.), **Solvatochromism - Problems and Methods** (in Russian), *Leningrad Univ. Publ. House*, Leningrad, 1989.
3 M. Karelson, *Adv. Quant. Chem.*, **28**, 142 (1997).
4 J. Tomasi, M. Persico, *Chem. Rev.*, **94**, 2027 (1994).
5 T. Simonson, A.T. Brünger, *J. Phys. Chem.*, **98**, 4683 (1994).
6 O. Sinanoglu, *Chem. Phys. Lett.*, **1**, 283 (1967).
7 V. Gogonea, E. Osawa, *J. Mol. Struct. (THEOCHEM)*, **311**, 305 (1994).
8 H. Reiss, *Adv. Chem. Phys.*, **9**, 1 (1966).
9 R.A. Pierotti, *Chem. Rev.*, **76**, 717 (1976).
10 R.M. Gibbons, *Mol. Phys.*, **17**, 81 (1969).
11 M. Irisa, K. Nagayama, F. Hirata, *Chem. Phys. Lett.*, **207**, 430 (1993).
12 A. Bernhardsson, R. Lindh, G. Karlström, B.O. Roos, *Chem. Phys. Lett.*, **251**, 141 (1996).
13 W. Jaskólski, *Phys. Rep.*, **271**, 1 (1996).
14 S. Basu, *Adv. Quant. Chem.*, **1**, 145 (1964).

15 A.T. Amos, B.L. Burrows, *Adv,. Quant. Chem.*, **7**, 289 (1973).
16 R.J. Abraham, B.D. Hudson, M.W. Kermode, J.R. Mines, *J. Chem. Soc. Farad. Trans.*, **84**, 1911 (1988).
17 J.G. Kirkwood, *J. Chem. Phys.*, **7**, 911 (1939).
18 L. Onsager, *J. Am. Chem. Soc.*, **58**, 1486 (1936).
19 C.J.F. Böttcher and P. Bordewijk, **Theory of Electric Polarization**, 2nd ed., vol. 2, *Elsevier Co*, Amsterdam, 1978.
20 D. Rinaldi, M.F. Ruiz-Lopez, J.-L. Rivail, *J. Chem. Phys.*, **78**, **834** (1983).
21 D. Rinaldi, J.-L. Rivail, N. Rguini, *J. Comput. Chem.*, **13**, 675 (1992).
22 J.-L. Rivail, B. Terryn, D. Rinaldi, M.F. Ruiz-Lopez, *J. Mol. Struct. (THEOCHEM)*, **166**, 319 (1988).
23 E.G. McRae, *J. Phys. Chem.*, **61**, 562 (1957).
24 T. Abe, *Bull Chem. Soc. Japan*, **38**, 1314 (1965).
25 O. Tapia, O. Goscinski, *Mol. Phys.*, **29**, 1653 (1975).
26 M. Karelson, G.H.F. Diercksen, "Models for Simulating Molecular Properties in Condensed Systems, in: **Problem Solving in Computational Molecular Science. Molecules in Different Environments**, S. Wilson and G.H.F. Diercksen (Eds.), *Kluwer Academic Publishers*, Dordrecht, 1997, pp. 215- 248.
27 M. Karelson, M.C. Zerner, *J. Phys. Chem.*, **92**, 6949 (1992).
28 R. Marcus, *J. Chem. Phys.*, **24**, 979 (1956).
29 J. Jortner, *Mol. Phys.*, **5**, 257 (1962).
30 J. Ridley, M.C. Zerner, *Theor. Chim. Acta*, **42**, 223 (1976).
31 A.D. Bacon, M.C. Zerner, *Theor. Chim. Acta*, **53**, 21 (1979).
32 M. Karelson, K. Pihlaja, T. Tamm, A.Uri, and M.C. Zerner, *J. Photochem. and Photobiol. A Chemistry*, **85**, 119 (1995).
33 A. Broo, *Chem. Phys.*, **183**, 85 (1994).
34 A. Broo, M.C. Zerner, *Chem. Phys. Lett.*, **227**, 551 (1994).
35 J.-L. Rivail, D. Rinaldi, V. Dillet, *Mol. Phys.*, **89**, 1521 (1996).
36 M.M. Karelson, *Org. React.*, **17**, 357 (1980).
37 M.W. Wong, K.B. Wiberg, M.J. Frisch, *J. Am. Chem. Soc.*, **113**, 4776 (1991).
38 O. Tapia, F. Colonna, J. Ángyán, *J. Chim. Phys.*, **87**, 875 (1990).
39 G. Jansen, F. Colonna, J. Ángyán, *Int. J. Quant. Chem.*, **58**, 251 (1996).
40 S. Miertuš, E. Scrocco, J. Tomasi, *Chem. Phys.*, **55**, 117 (1981).
41 H. Hoshi, M. Sakurai, Y. Inoue, R. Chûjô, *J. Chem. Phys.*, **87**, 1107 (1987).
42 M.L. Drummond, *J. Chem. Phys.*, **88**, 5014 (1988).
43 B. Wang, G.P. Ford, *J. Chem. Phys.*, **97**, 4162 (1992).
44 T. Fox, N, Rösch, R.J. Zauhar, *J. Comput. Chem.*, **14**, 253 (1993).
45 F.J. Luque, M. Orozco, P.K. Bhadane, S.R. Gadre, *J. Chem. Phys.*, **100**, 6718 (1994).
46 C. Gonzalez, A. Restrepo-Cossio, M. Márquez, K.B. Wiberg, M. De Rosa, *J. Phys. Chem. A*, **102**, 2732 (1998).
47 M.A. Aguilar, J. Olivares del Valle, J. Tomasi, *J. Chem. Phys.*, **98**, 7375 (1993).
48 H. Houjou, M. Sakurai, Y. Inoue, *J. Chem. Phys.*, **107**, 5652 (1997).
49 B. Mennucci, R. Cammi, J. Tomasi, *J. Chem. Phys.*, **109**, 2798 (1998).
50 M. Karelson, T. Tamm, M.C. Zerner, *J. Phys Chem.*, **97**, 11901 (1993).
51 A. Klamt, G. Schüürmann, *J. Chem. Soc. Perkin Trans. 2*, 799 (1993).
52 A. Klamt, *J. Phys. Chem.*, **99**, 2224 (1995).
53 E.V. Stefanovich, T.N. Truong, *J. Chem. Phys.*, **105**, 2961 (1996).
54 A. Klamt, *J. Phys. Chem.,* **100**, 3349 (1996).
55 K.V. Mikkelsen, H. Ågren, H.J. Aa Jensen, T. Helgaker, *J. Chem. Phys.*, **89**, 3086 (1988).
56 J.L. Rivail, D. Rinaldi, *Chem. Phys.*, **18**, 233 (1976).
57 M.D. Newton, *J. Phys. Chem.*, **79**, 2795 (1975)
58 K.V. Mikkelsen, E. Dalgaard, P. Swanstrom, *J. Phys. Chem.*, **91**, 3081 (1987).
59 K.V. Mikkelsen, A. Cesar, H. Ågren, H.J. Aa. Jensen, *J. Chem. Phys.*, **103**, 9010 (1995).
60 K.V. Mikkelsen, P. Jørgensen, H.J. Aa. Jensen, *J. Chem. Phys.*, **100**, 6597 (1994).
61 M.V. Basilevsky, G.E. Chudinov, *Chem. Phys.*, **157**, 245 (1991).
62 J.-L. Rivail, *C. R. Acad. Sci. Paris*, **311 II**, 307 (1990).
63 J.-L . Rivail, D. Rinaldi, M.F. Ruiz-Lopez, in: **Theoretical and Computational Models for Organic Chemistry**; S.J. Formosinho, I.G. Csizmadia, and L.G. Arnaut, (Eds.), *Kluwer*, Dordrecht. 1991, pp. 77-92.
64 F.J. Olivares del Valle, J. Tomasi, *Chem. Phys.*, **150**, 139 (1991).
65 F.J. Olivares del Valle, M.A. Aguilar, *J. Comput. Chem.*, **13**, 115 (1992).
66 F.J. Olivares del Valle, M.A. Aguilar, *J. Mol. Struct. (THEOCHEM)*, **280**, 25 (1993).

67 N.A. Besley, J.D. Hirst, *J. Phys. Chem. A*, **102**, 10791 (1998).
68 L. Serrano-Andrés, M.P. Fülscher, G. Karlström, *Int. J. Quant. Chem.*, **65**, 167 (1997).
69 R.R. Contreras, A.J. Aizman, *Int. J. Quant. Chem. Quantum Chem. Symp.*, **25**, 281 (1991).
70 R.J.Hall, M.M. Davidson, N.A. Burton, I.H. Hillier, *J. Phys. Chem.*, **99**, 921 (1995).
71 R.R. Contreras, P. Pérez, A.J. Aizman, *Int. J. Quant. Chem.*, **56**, 433 (1995).
72 G.J. Tawa, R.L. Martin, L.R. Pratt, T.V. Russo, *J. Phys. Chem.*, **100**, 1515 (1996).
73 T.N. Truong, E.V. Stefanovich, *Chem. Phys. Lett.*, **240**, 253 (1995).
74 E.K.U. Gross, W. Kohn, *Phys. Rev. Lett.*, **55**, 2850 (1985).
75 C.A. Ullrich, E.K.U. Gross, *Phys. Rev. Lett.*, **74**, 872 (1995).
76 H. Margenau, N.R. Kestner, **Theory of Intermolecular Forces**, *Pergamon Press*, Oxford, 1971.
77 J. Mahanty, W.B. Ninham, **Dispersion Forces**, **Academic Press**, New York, 1976.
78 P. Hobza, P. Zahradnik, **Intermolecular Complexes**, *Elsevier*, Amsterdam, 1988.
79 M.J. Huron, P. Claverie, *Chem. Phys. Lett.*, **4**, 429 (1969).
80 A.J. Pertsin, A.I. Kitaigorodsky, **The Atom-Atom Pair Potential Method**, *Springer*, Berlin, 1986.
81 N. Rösch, M.C. Zerner, *J. Phys. Chem.*, **98**, 5817 (1994).
82 G.H.F. Diercksen, M. Karelson, T. Tamm and M.C. Zerner, *Int. J. Quant. Chem.*, **S28**, 339 (1994).
83 M.M. Karelson, *Org. React.*, **20**, 127 (1983).
84 O. Tapia, J.M. Lluch, R. Cardenas, J. Andres, *J. Am. Chem. Soc.*, **111**, 829 (1989).
85 M.M. Karelson, M.C. Zerner, *J. Am. Chem. Soc.*, **112**, 9405 (1990).
86 L.C.G. Freitas, R.L. Longo, A.M. Simas, *J. Chem. Soc. Faraday Trans.*, **88**, 189 (1992).
87 I. Frank, S. Grimme, M. von Arnim, S.D. Peyerimhoff, *Chem. Phys.*, **199**, 145 (1995).
88 E. Clementi, G. Corongiu, (Eds.) **Computational Chemistry**, METECC-95, STEF, Cagliari, 1995.
89 S.F. Boys, F. Bernardi, F. *Mol. Phys.*, **19**, 553 (1970).
90 Z. Latajka, S. Scheiner, *J. Comput. Chem.*, **8**, 674 (1987).
91 P. Arrighini, P. **Intermolecular Forces and Their Evaluation by Perturbation Theory**, Lecture Notes in Chemistry, vol. 25, *Springer-Verlag,* Berlin, 1981.
92 B.J. Jeziorski, R. Moszynski, A. Ratkiewicz, S. Rybak, K. Szalewicz, H.L. Williams, Chapter 3 in: **Methods and Techniques in Computational Chemistry**, METECC-94, vol. B. STEF, Cagliari, 1994.
93 B.T. Thole, P.Th. van Duijnen, *Chem. Phys.*, **71**, 211 (1982).
94 B.T. Thole, P.Th. van Duijnen, *Theor. Chim. Acta*, **63**, 209 (1983).
95 B.T. Thole, P.Th. van Duijnen, *Biophys. Chem.*, **18**, 53 (1983).
96 P.Th. van Duijnen, A.H. Juffer, H. Dijkman, H. *J. Mol. Struct. (THEOCHEM)*, **260**, 195 (1992).
97 A.H. deVries, P.Th. van Duijnen, *Int. J. Quant. Chem.*, **57**, 1067 (1996).
98 H. Dijkman, P. Th. van Duijnen, *Int. J. Quant. Chem. Quantum Biol. Symp.*, **18**, 49 (1991).
99 P.Th. van Duijnen, A.H. de Vries, *Int. J. Quant. Chem. Quantum Chem. Symp.*, **29**, 531 (1995).
100 A.H. de Vries, P.Th. van Duijnen, *Int. J. Quant. Chem.*, **57**, 1067 (1996).
101 F.C. Grozema, P.Th. van Duijnen, *J. Phys. Chem. A*, **102**, 7984 (1998).
102 F.H Stillinger, *Adv. Chem. Phys.*, **31**, 1 (1975).
103 S. Engström, B. Jönsson, R.W. Impey, *J. Chem. Phys.*, **80**, 5481 (1984).
104 W.F. van Gunsteren, *Current Opinion in Structural Biology*, **3**, 277 (1993).
105 C. Corongiu, V. Martorana, Chapter 3 in: **Methods and Techniques in Computational Chemistry**, METECC-94; vol. C. STEF, Cagliari, 1994.; W.R.P. Scott, W.F. van Gunsteren, Chapter 9 in: Methods and Techniques in Computational Chemistry, METECC-95; Ed. by E. Clementi and G. Corongiu, STEF, Cagliari, 1995.
106 J.T. Blair, K. Krogh-Jespersen, R.M. Levy, *J. Am. Chem. Soc.*, **111**, 6948 (1989).
107 V. Luzhkov, A. Warshel, *J. Am. Chem. Soc.*, **113**, 4491 (1989).
108 J.S. Bader, C.M. Cortis, B.J. Berne, *J. Chem. Phys.*, **106**, 2372 (1997).
109 C.J. Cramer, D.G. Truhlar, *J. Am. Chem. Soc.*, **113**, 8305 (1991).
110 C.J. Cramer, D.G. Truhlar, *J. Am. Chem. Soc.*, **113**, 8552 (1991).

# 11.2 DIELECTRIC SOLVENT EFFECTS ON THE INTENSITY OF LIGHT ABSORPTION AND THE RADIATIVE RATE CONSTANT

Tai-ichi Shibuya
**Faculty of Textile Science and Technology**
**Shinshu University, Ueda, Japan**

## 11.2.1 THE CHAKO FORMULA OR THE LORENTZ-LORENZ CORRECTION

The intensity of light absorption by a molecule is generally altered when the molecule is immersed in a solvent or transferred from one solvent to another. The change may be small if the solvents are inert and non-polar, but often a significant increase or decrease is observed. The first attempt to correlate such effects with the nature of the solvent was made by Chako[1] in 1934. Chako's formula reads as

$$\frac{f''}{f} = \frac{\left(n^2 + 2\right)^2}{9n} \qquad [11.2.1]$$

where:

| | |
|---|---|
| f | oscillator strength of an absorption band of a molecule |
| f″ | apparent oscillator strength of the molecule in solution |
| n | refractive index of the solution at the absorbing frequency |

The apparent oscillator strength is proportional to the integrated intensity under the molar absorption curve. To derive the formula, Chako followed the classical dispersion theory with the Lorentz-Lorenz relation (also known as the Clausius-Mosotti relation), assuming that the solute molecule is located at the center of the spherical cavity in the continuous dielectric medium of the solvent. Hence, the factor derived by Chako is also called the Lorentz-Lorenz correction. Similar derivation was also presented by Kortüm.[2] The same formula was also derived by Polo and Wilson[3] from a viewpoint different from Chako.

Chako's formula always predicts an increase of the absorption intensity with the refractive index. This does not hold, for instance, for the allowed $\pi \rightarrow \pi^*$ electronic transitions of cyclohexadiene and cyclopentadiene,[4] and monomethyl substituted butadienes.[5]

## 11.2.2 THE GENERALIZED LOCAL-FIELD FACTOR FOR THE ELLIPSOIDAL CAVITY

A natural generalization of the Chako formula was made by generalizing the spherical cavity to an ellipsoidal cavity. Such a generalization was shown by Shibuya[6] in 1983. The generalized formula derived by him reads as

$$\frac{f''}{f} = \frac{\left[s\left(n^2 - 1\right) + 1\right]^2}{n} \qquad [11.2.2]$$

where:

| | |
|---|---|
| s | shape parameter which takes a value between 0 and 1 |

This parameter s is more generally known as the depolarization factor, whose values are listed for special cases in general textbooks.[7] For the spherical cavity, s = 1/3 in any axis; for

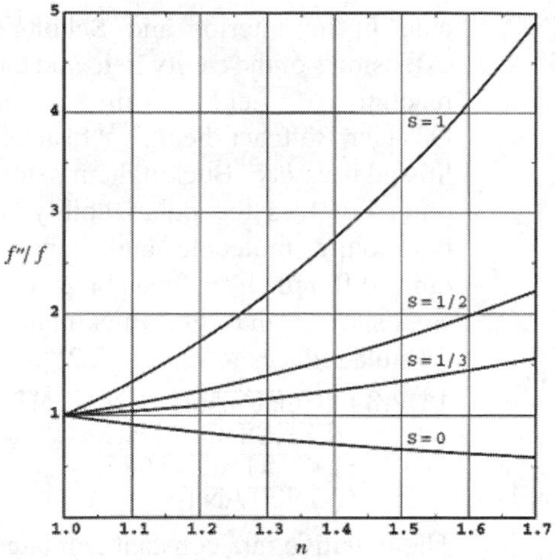


Figure 11.2.1. Dependence of f''/ f on the refractive index n for different values of s. [After reference 6]

a thin slab cavity, s = 1 in the normal direction and s = 0 in plane; and for a long cylindrical cavity, s = 0 in the longitudinal axis and s = 1/2 in the transverse direction. The shape of the ellipsoidal cavity is supposed to be primarily determined by the shape of the solute molecule. Typical cases are long polyenes and large planar aromatic hydrocarbons. One can assume s = 0 for the strong $\pi \rightarrow \pi^*$ absorption bands of these molecules. For smaller molecules, however, one should assume s ≈ 1/3 regardless of the shape of the solute molecule, as the cavity shape then may be primarily determined by the solvent molecules rather than the solute molecule. Note that Eq. [11.2.2] gives the Chako formula for s = 1/3, i.e., for the spherical cavity.

For transitions whose moments are in the longitudinal axis of a long cylindrical cavity or in the plain of a thin slab cavity, Eq. [11.2.2] with s = 0 leads to $f''/f = 1/n$, so that the absorption intensity always decreases with the refractive index. If the transition moment is normal to a thin slab cavity, Eq. [11.2.2] with s = 1 leads to $f''/f = n^3$. The dependence of the ratio f''/f on the refractive index n according to Eq. [11.2.2] is illustrated for different values of s in Figure 11.2.1. The slope of the ratio is always positive for s > 1/4. For 0 < s < 1/4, it is negative in the region $1 \le n \le \sqrt{(1-s)/3s}$ and positive in the other region.

Eq. [11.2.2] can be also written as the following form:

$$\sqrt{(nf'')} = \sqrt{f}s\left(n^2 - 1\right) + \sqrt{f} \qquad [11.2.3]$$

This equation shows a linear relationship between $\sqrt{(nf'')}$ and $(n^2 - 1)$. If a set of measured values of f'' vs. n are provided for a solute, the least-squares fitting to Eq. [11.2.3] of $\sqrt{(nf'')}$ against $(n^2 - 1)$ gives the values of f and s for the solute molecule. Note that f'' and f in Eq. [11.2.3] can be replaced by any quantities proportional to the oscillator strengths. Thus, they can be replaced by the integrated intensities or by their relative quantities.

Figure 11.2.2 shows such plots for the $\pi \rightarrow \pi^*$ absorption bands of β-carotene and the $n \rightarrow \pi^*$ absorption bands of pyrazine measured[8] in various organic solvents. Here, the relative intensities $f''/f''_c$, where $f''_c$ is the absorption intensity measured in cyclohexane as the reference solvent, are considered, and $y = \sqrt{nf''/f''_c}$ is plotted against $x = n^2 - 1$. The least-squares fittings give s = 0 for the allowed $\pi \rightarrow \pi^*$ transition of β-carotene and s = 0.29 for the vibronic $n \rightarrow \pi^*$ transition of pyrazine. Note that in this case the least-squares fitted line gives $\sqrt{f/f''_c}$ as its intercept and $s\sqrt{f/f''_c}$ as its slope so that s is given as the ratio of the slope divided by the intercept. A similar study was made[9] on the $n \rightarrow \pi^*$ absorption bands of acetone and cyclopentanone, giving the results s = 0.88 and $f = 1.8\times10^{-4}$ for acetone and s=0.72 and $f = 2.2\times10^{-4}$ for cyclopentanone.

A similar generalization was also made by Buckingham.[10] He followed Kirkwood's idea[11] in deriving the electric moment of a dielectric specimen produced by a fixed mole-

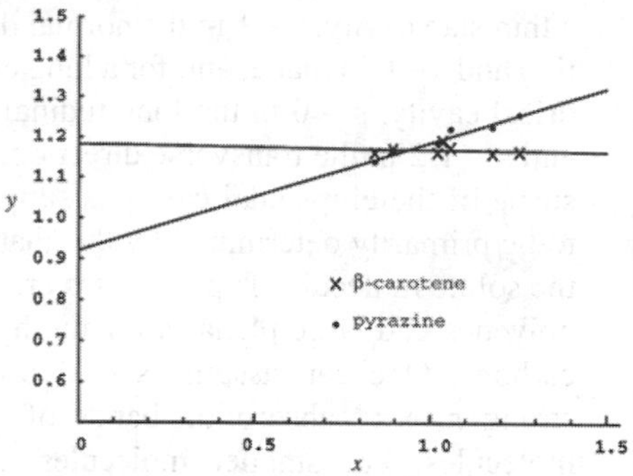


Figure 11.2.2. Plots of $y = \sqrt{nf''/f_c''}$ vs. $x = n^2 - 1$ for the $\pi \to \pi^*$ absorption bands of β-carotene (crosses) and the $n \to \pi^*$ absorption bands of pyrazine (solid circles). [After reference 6]

cule in its interior and Scholte's extension[12] of the cavity field and the reaction field in the Onsager-Böttcher theory[13,14] to an ellipsoidal cavity. Buckingham's formula involves the polarizability of the solute molecule and appears quite different from Eq. [11.2.2]. It was shown[6] that the Buckingham formula reduces to Eq. [11.2.2].

### 11.2.3 DIELECTRIC SOLVENT EFFECT ON THE RADIATIVE RATE CONSTANT

The radiative rate constant is related to the absorption intensity of the transition from the ground state to the excited state under consideration. The application of Eq. [11.2.2] leads[15] to

$$k_r'' / k_r = n\left[s\left(n^2 - 1\right) + 1\right]^2 \qquad [11.2.4]$$

where:

$k_r''$ apparent radiative rate constant of the solute molecule measured in a solvent of the refractive index n

$k_r$ radiative rate constant of the molecule in its isolated state

Note that the local-field correction factor $n[s(n^2 - 1) + 1]^2$ varies from n to $n^5$ as s varies from 0 to 1. For 9,10-diphenylanthracene (DPA), the correction factor was given[15] as $n[(0.128)(n^2 - 1) + 1]^2$, which lies between n and $n^2$. This agrees with the observed data[16] of fluorescence lifetimes of DPA in various solvents.

## REFERENCES


1 N. Q. Chako, *J. Chem. Phys.*, **2**, 644 (1934).
2 G. Kortüm, *Z. Phys. Chem.*, **B33**, 243 (1936).
3 (a) V. Henri and L. W. Pickett, *J. Chem. Phys.*, **7**, 439 (1939); (b) L. W. Pickett, E. Paddock, and E. Sackter, *J. Am. Chem. Soc.*, **63**, 1073 (1941).
4 L. E. Jacobs and J. R. Platt, *J. Chem. Phys.*, **16**, 1137 (1948).
5 S. R. Polo and M. K. Wilson, *J. Chem. Phys.*, **23**, 2376 (1955).
6 T. Shibuya, *J. Chem. Phys.*, **78**, 5176 (1983).
7 C. Kittel, **Introduction to Solid State Physics**, 4th Ed., *Wiley*, New York, 1971, Chap. 13.
8 A. B. Myers and R. R. Birge, *J. Chem. Phys.*, **73**, 5314 (1980).
9 T. Shibuya, *Bull. Chem. Soc. Jpn.* , **57**, 2991 (1984).
10 A. D. Buckingham, *Proc. Roy. Soc. (London)*, **A248**, 169 (1958); **A255**, 32 (1960).
11 J. G. Kirkwood, *J. Chem. Phys.*, **7**, 911 (1939).
12 T. G. Scholte, *Physica (Utrecht)*, **15**, 437 (1949).
13 L. Onsager, *J. Am. Chem. Soc.*, **58**, 1486 (1936).
14 C. J. F. Böttcher, (a) *Physica (Utrecht)*, **9**, 937, 945 (1942); (b) **Theory of Electric Polarization**, *Elsevier*, New York, 1952; 2nd Ed., 1973, Vol. I.
15 T. Shibuya, *Chem. Phys. Lett.*, **103**, 46 (1983).
16 R. A. Lampert, S. R. Meech, J. Metcalfe, D. Phillips, A. P. Schaap, *Chem. Phys. Lett.*, **94**, 137 (1983).

# 12

# Other Properties of Solvents, Solutions, and Products Obtained from Solutions

## 12.1 RHEOLOGICAL PROPERTIES, AGGREGATION, PERMEABILITY, MOLECULAR STRUCTURE, CRYSTALLINITY, AND OTHER PROPERTIES AFFECTED BY SOLVENTS

George Wypych
**ChemTec Laboratories, Inc., Toronto, Canada**

### 12.1.1 RHEOLOGICAL PROPERTIES

The modification of rheological properties is one of the main reasons for adding solvents to various formulations. Rheology is also a separate complex subject which requires an in-depth understanding that can only be accomplished by consulting specialized sources such as monographic books on rheology fundamentals.[1-3] Rheology is such a vast subject that the following discussion will only outline some of the important effects of solvents.

When considering the viscosity of solvent mixtures, solvents can be divided into two groups: interacting and non-interacting solvents. The viscosity of a mixture of non-interacting solvents can be predicted with good approximation by a simple additive rule rule:

$$\log\eta = \sum_{i=1}^{i=n} \phi_i \log\eta_i \qquad [12.1.1]$$

where:

| | |
|---|---|
| $\eta$ | viscosity of solvent mixture |
| i | iteration subscript for mixture components (i = 1, 2, 3, ..., n) |
| $\phi$ | fraction of component i |
| $\eta_i$ | viscosity of component i. |

Interacting solvents contain either strong polar solvents or solvents which have the ability to form hydrogen bonds or influence each other on the basis of acid-base interaction. Solvent mixtures are complicated because of the changes in interaction that occurs with changes in the concentration of the components. Some general relationships describe viscosity of such mixtures but none is sufficiently universal to replace measurement. Further details on solvent mixtures are included in Chapter 9.

The addition of solute(s) further complicates rheology because in such mixtures solvents may not only interact among themselves but also with the solute(s). There are also interactions between solutes and the effect of ionized species with and without solvent participation. Only very dilute solutions of low molecular weight substances exhibit Newtonian viscosity. In these solutions, viscosity is a constant, proportionality factor of shear rate and shear stress. The viscosity of these solutions is usually well described by the classical, Einstein's equation:

$$\eta = \eta_s (1 + 2.5\phi) \qquad [12.1.2]$$

where:

| | |
|---|---|
| $\eta_s$ | solvent viscosity |
| $\phi$ | volume fraction of spheres (e.g. suspended filler) or polymer fraction |

If $\phi$ is expressed in solute mass concentration, the following relationship is used:

$$\phi = \frac{NVc}{M} \qquad [12.1.3]$$

where:

| | |
|---|---|
| N | Avogadro's number |
| V | molecular volume of solute $((4/3)\pi R^3)$ with R - radius |
| c | solute mass concentration |
| M | molecular weight |

Combination of equations [12.1.2] and [12.1.3] gives:

$$\frac{\eta - \eta_s}{\eta_s c} = \frac{2.5NV}{M} \qquad [12.1.4]$$

The results of studies of polymer solutions are most frequently expressed in terms of intrinsic, specific, and relative viscosities and radius of gyration; the mathematical meaning of these and the relationships between them are given below:

$$[\eta] = \lim_{c \to 0} \left( \frac{\eta - \eta_s}{\eta_s c} \right) \qquad [12.1.5]$$

$$\frac{\eta_{sp}}{c} = [\eta] + k_1 [\eta]^2 c + \cdots \qquad [12.1.6]$$

$$\frac{\ln \eta_r}{c} = [\eta] - k_1' [\eta]^2 c + \cdots \qquad [12.1.7]$$

$$\eta_r = \eta_{sp} + 1 = \frac{\eta}{\eta_s} \qquad [12.1.8]$$

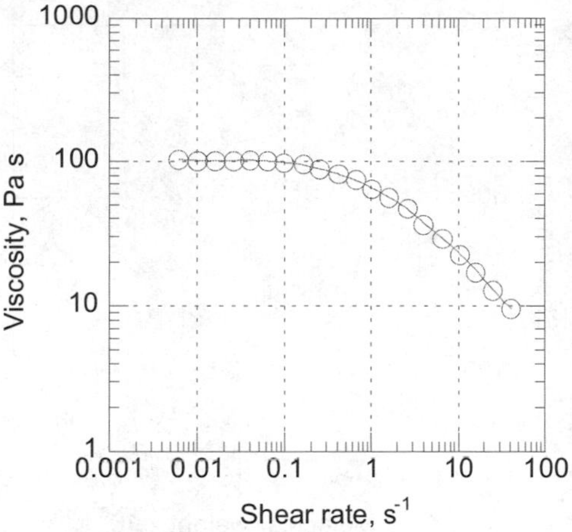


Figure 12.1.1. Viscosity vs. shear rate for 10% solution of polyisobutylene in pristane. [Data from C R Schultheisz, G B McKenna, Antec '99, SPE, New York, 1999, p 1125.]

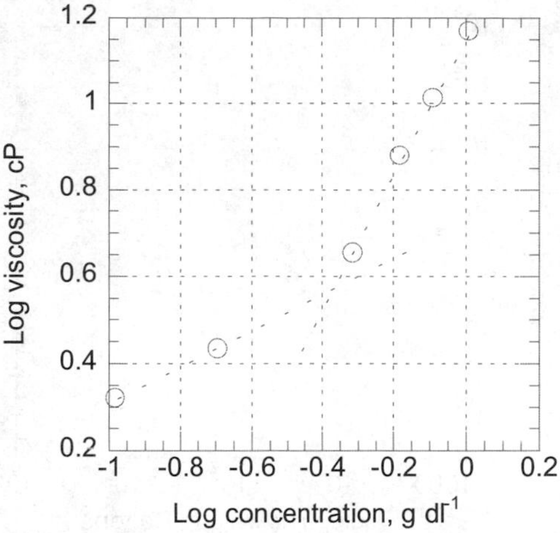


Figure 12.1.2. Viscosity of polyphenylene solution in pyrilidinone. [Data from F. Motamedi, M Isomaki, M S Trimmer, Antec '98, SPE, Atlanta, 1998, p. 1772.]

where:

| | |
|---|---|
| $[\eta]$ | intrinsic viscosity |
| $\eta_{sp}$ | specific viscosity |
| $\eta_r$ | relative viscosity |
| $k_1$ | coefficient of direct interactions between pairs of molecules |
| $k_1'$ | coefficient of indirect (hydrodynamic) interactions between pairs of molecules |

In Θ solvents, the radius of gyration of unperturbed Gaussian chain enters the following relationship:

$$[\eta]_0 = \frac{\Phi_0 R_{g,0}^3}{M} \qquad [12.1.9]$$

where:

| | |
|---|---|
| $\Phi_0$ | coefficient of intramolecular hydrodynamic interactions = $3.16 \pm 0.5 \times 10^{24}$ |
| $R_{g,0}$ | radius of gyration of unperturbed Gaussian chain |

In good solvents, the expansion of chains causes an increase of viscosity as described by the following equation:

$$[\eta] = \frac{\Phi_0 \alpha_\eta^3 R_{g,0}^3}{M} \qquad [12.1.10]$$

where:

| | |
|---|---|
| $\alpha_\eta$ | $= [\eta]^{1/3} / [\eta]_0^{1/3}$ is and effective chain expansion factor. |

Existing theories are far from being universal and precise in prediction of experimental data. A more complex treatment of measurement data is needed to obtain characteristics of these "rheological" liquids.

Figure 12.1.1 shows that the viscosity of a solution depends on shear rate. These data comes from the development of a standard for instrument calibration by NIST to improve

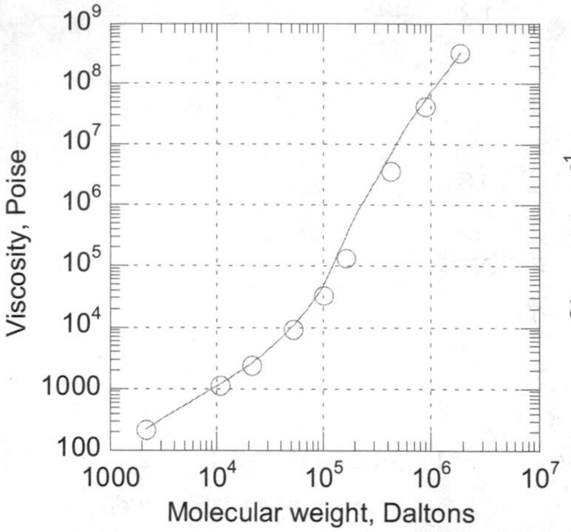


Figure 12.1.3. Viscosity of 40% polystyrene in di-2-ethyl hexyl phthalate. [Data from G D J Phillies, *Macromolecules*, **28**, No.24, 8198-208 (1995).]

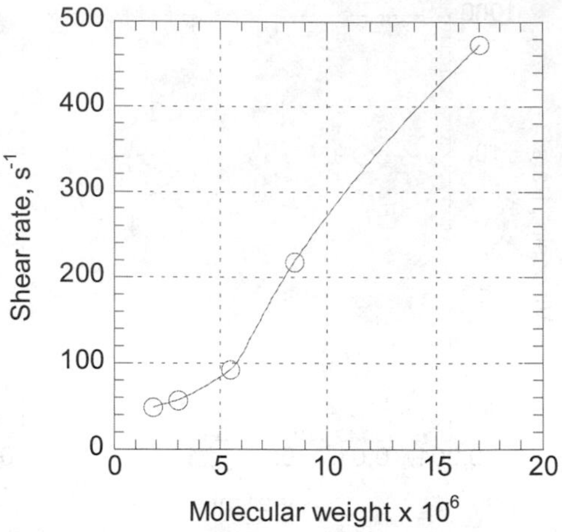


Figure 12.1.4. Shear rate of polystyrene in DOP vs. molecular weight. [Data from M Ponitsch, T Hollfelder, J Springer, *Polym. Bull.*, **40**, No.2-3, 345-52 (1998).]

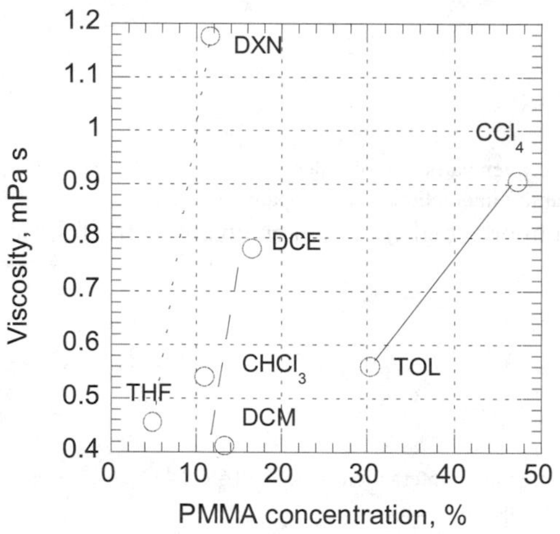


Figure 12.1.5. Viscosity of PMMA solutions in different solvents vs. PMMA concentration. Basic solvents: tetrahydrofuran, THF, and dioxane, DXN; neutral: toluene, TOL and $CCl_4$; acidic: 1,2-dichloroethane, DCE, $CHCl_3$, and dichloromethane, DCM. [Adapted, by permission, from M L Abel, M M Chehimi, *Synthetic Metals*, **66**, No.3, 225-33 (1994).]

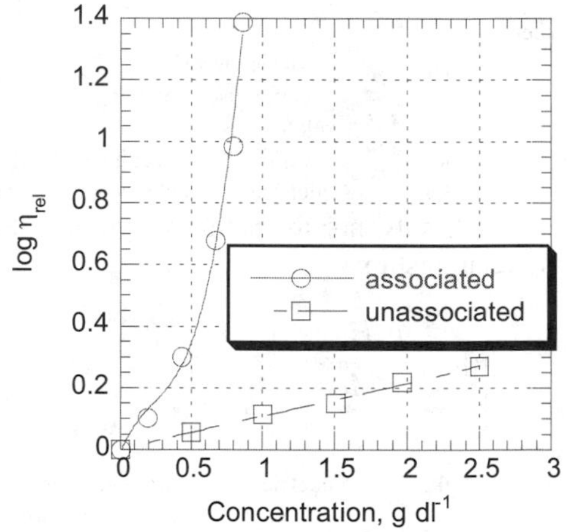


Figure 12.1.6. Relative viscosity of block copolymers with and and without segments capable of forming complexes vs. concentration. [Data from I C De Witte, B G Bogdanov, E J Goethals, *Macromol. Symp.*, **118**, 237-46 (1997).]

the accuracy of measurements by application of nonlinear liquid standards.[4] Figure 12.1.2 shows the effect of polymer concentration on the viscosity of a solution of polyphenylene in N-methyl pyrilidinone.[5] Two regimes are clearly visible. The regimes are divided by a critical concentration above which viscosity increases more rapidly due to the interaction of chains leading to aggregate formation. These two sets of data show that there

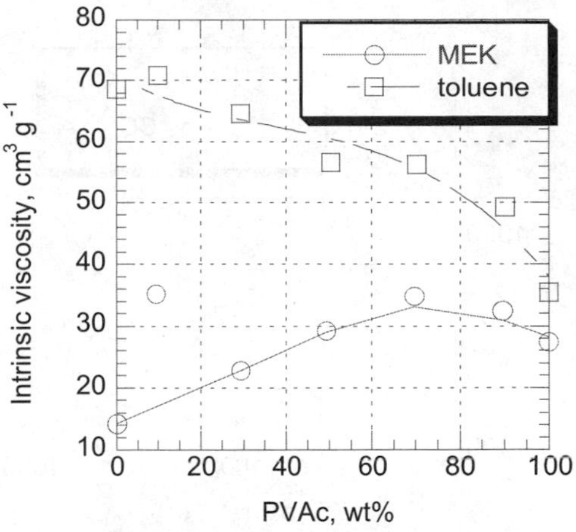


Figure 12.1.7. Intrinsic viscosity of PS/PVAc mixtures in methyl-ethyl-ketone, MEK, and toluene. [Data from H Raval, S Devi, *Angew. Makromol. Chem.*, **227**, 27-34 (1995).]

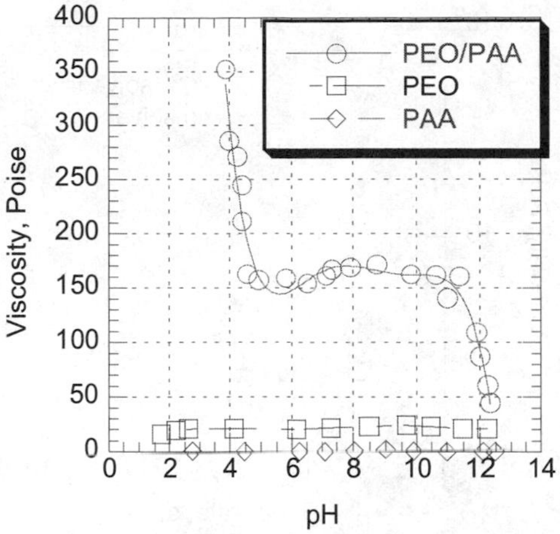


Figure 12.1.8. Viscosity of poly(ethylene oxide), PEO, poly(acrylic acid), PAA, and their 1:1 mixture in aqueous solution vs. pH. [Adapted, by permission, from I C De Witte, B G Bogdanov, E J Goethals, *Macromol. Symp.*, **118**, 237-46 (1997).

are considerable departures from the simple predictions of the above equations because, based on them, viscosity should be a simple function of molecular weight. Figure 12.1.3 shows, in addition, that the relationship between viscosity of the solution and molecular weight is nonlinear.[6] Also, the critical shear rate, at which aggregates are formed, is a non-linear function of molecular weight (Figure 12.1.4).[7]

These departures from simple relationships are representative of simple solutions. The relationships for viscosities of solution become even more complex if stronger interactions are included, such as the presence of different solvents, the presence of interacting groups within polymer, combinations of polymers, or the presence of electrostatic interactions between ionized structures within the same or different chains. Figure 12.1.5 gives one example of complex behavior of a polymer in solution. The viscosity of PMMA dissolved in different solvents depends on concentration but there is not one consistent relationship (Figure 12.1.5). Instead, three separate relationships exist each for basic, neutral, and acid solvents, respectively. This shows that solvent acid-base properties have a very strong influence on viscosity.

Figure 12.1.6 shows two different behaviors for unassociated and associated block copolymers. The first type has a linear relationship between viscosity and concentration whereas with the second there is a rapid increase in viscosity as concentration increases. This is the best described as a power law function.[8] Two polymers in combination have different reactions when dissolved in different solvents (Figure 12.1.7). In MEK, intrinsic viscosity increases as polymer concentration increases. In toluene, intrinsic viscosity decreases as polymer concentration increases.[9] The polymer-solvent interaction term for MEK is very small (0.13) indicating a stable compatible system. The interaction term for toluene is much larger (0.58) which indicates a decreased compatibility of polymers in toluene and lowers viscosity of the mixture. Figure 12.1.8 explicitly shows that the behavior of

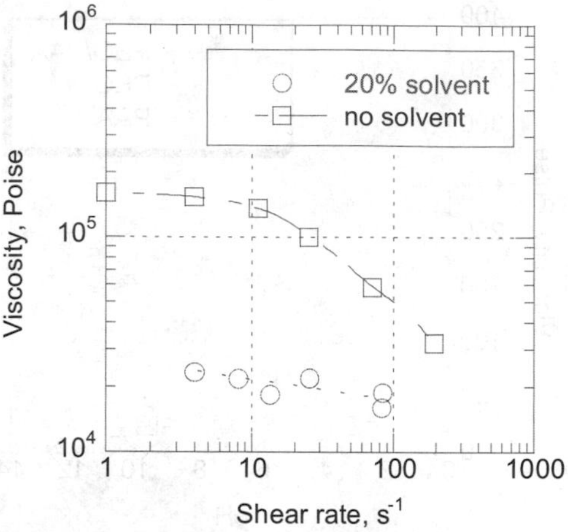


Figure 12.1.9. Apparent melt viscosity of original PET and PET containing 20% 1-methyl naphthalene vs. shear rate. [Adapted, by permission, from S Tate, S Chiba, K Tani, *Polymer*, **37**, No.19, 4421-4 (1996).]

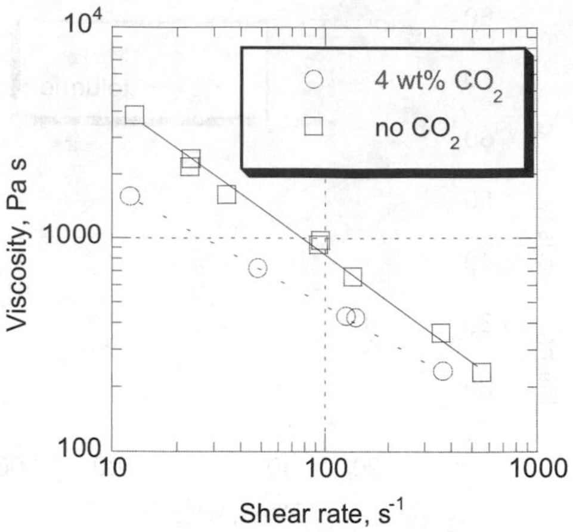


Figure 12.1.10. Viscosity behavior of PS with and without $CO_2$. [Data from M Lee, C Tzoganikis, C B Park, Antec '99, SPE, New York, 1999, p 2806.]

individual polymers does not necessarily have a bearing on the viscosity of their solutions. Both poly(ethylene oxide) and poly(propylene oxide) are not affected by solution pH but, when used in combination, they become sensitive to solution pH. A rapid increase of viscosity at a lower pH is ascribed to intermolecular complex formation. This behavior can be used for thickening of formulations.[8]

Figures 12.1.9 and 12.1.10 show one potential application in which a small quantity of solvents can be used to lower melt viscosity during polymer processing. Figure 12.1.9 shows that not only can melt viscosity be reduced but also that the viscosity is almost independent of shear rate.[10] In environmentally friendly process supercritical fluids can be used to reduce melt viscosity.

The above data illustrate that the real behavior of solutions is much more complex than it is intuitively predicted based on simple models and relationships. The proper selection of solvent can be used to tailor the properties of formulation to the processing and application needs. Solution viscosity can be either increased or decreased to meet process technology requirements or to give the desired material properties.

## REFERENCES

1 A Ya Malkin, **Rheology Fundamentals**, *ChemTec Publishing*, Toronto, 1994.
2 Ch W Macosko, **Rheology. Principles, Measurements, and Applications**, *VCH Publishers*, New York, 1994.
3 R I Tanner, K. Walters, **Rheology: an Historical Perspective**, *Elsevier*, Amsterdam, 1998.
4 C R Schultheisz, G B McKenna, Antec '99, SPE, New York, 1999, p 1125.
5 F Motamedi, M Isomaki, M S Trimmer, Antec '98, SPE, Atlanta, 1998, p. 1772.
6 G D J Phillies, *Macromolecules*, **28**, No.24, 8198-208 (1995).
7 M Ponitsch, T Hollfelder, J Springer, *Polym. Bull.*, **40**, No.2-3, 345-52 (1998).
8 I C De Witte, B G Bogdanov, E J Goethals, *Macromol. Symp.*, **118**, 237-46 (1997).
9 H Raval, S Devi, *Angew. Makromol. Chem.*, **227**, 27-34 (1995).

10 S Tate, S Chiba, K Tani, *Polymer*, 37, No.19, 4421-4 (1996).
11 M L Abel, M M Chehimi, *Synthetic Metals*, **66**, No.3, 225-33 (1994).
12 M Lee, C Tzoganikis, C B Park, Antec '99, SPE, New York, 1999, p 2806.

## 12.1.2 AGGREGATION

The development of materials with an engineered morphological structure, such as selective membranes and nanostructures, employs principles of aggregation in these interesting technical solutions. Here, we consider some basic principles of aggregation, methods of studies, and outcomes. The discipline is relatively new therefore for the most part, only exploratory findings are available now. The theoretical understanding is still to be developed and this development is essential for the control of industrial processes and development of new materials.

Methods of study and data interpretation still require further work and refinement. Several experimental techniques are used, including: microscopy (TEM, SEM);[1,2] dynamic light scattering[3-6] using laser sources, goniometers, and digital correlators; spectroscopic methods (UV, CD, fluorescence);[7,8] fractionation; solubility and viscosity measurements;[9] and acid-base interaction.[10]

Dynamic light scattering is the most popular method. Results are usually expressed by the radius of gyration, $R_g$, the second virial coefficient, $A_2$, the association number, p, and the number of arms, f, for starlike micelles.

Zimm's plot and equation permits $R_g$ and $A_2$ to be estimated:

$$\frac{KC}{R_\theta} = \frac{1}{\overline{M}_w}\left[1 + \frac{16}{3}\pi^2 \frac{R_g^2}{\lambda_0^2}\sin^2\theta + \cdots\right] + 2A_2C + \cdots \qquad [12.1.11]$$

where:

| | |
|---|---|
| $K$ | optical constant |
| $C$ | polymer concentration |
| $R_\theta$ | Rayleigh ratio for the solution |
| $\overline{M}_w$ | weight average molecular weight |
| $\lambda_0$ | wavelength of light |
| $\theta$ | scattering angle. |

The following equations are used to calculate p and f:

$$p = \frac{M_{agg}}{M_1}; M_{agg} = KI_{q\to 0} / C \qquad [12.1.12]$$

$$\frac{R_g}{R_{garm}} = f^{(1-v)/2} \qquad [12.1.13]$$

where:

| | |
|---|---|
| $M_{agg}$ | mass of aggregates |
| $M_1$ | mass of free copolymers |
| $I$ | scattered intensity |
| $R_{garm}$ | radius of gyration of linear polymer |
| $v$ | excluded volume exponent. |

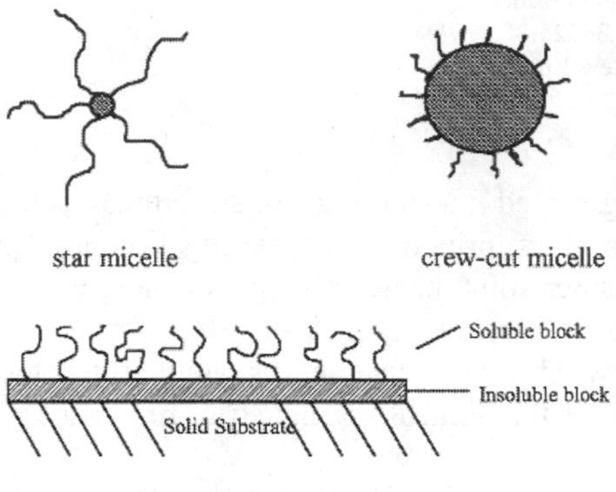


Figure 12.1.11. Morphological features: starlike and crew-cut. The bottom drawing illustrates polymer brush. [Adapted, by permission from G Liu, *Macromol. Symp.*, **113**, 233 (1997).]

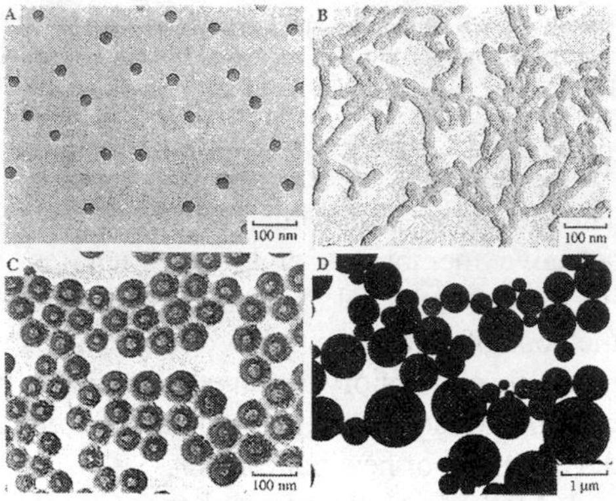


Figure 12.1.12. Morphological features of PS-PAA copolymer having different ratios of PS and PAA block lengths: A - 8.3, B- 12, C - 20.5, and D - 50. [Adapted, by permission, from L Zhang, A Eisenberg, *Macromol. Symp.*, **113**, 221-32 (1997).]

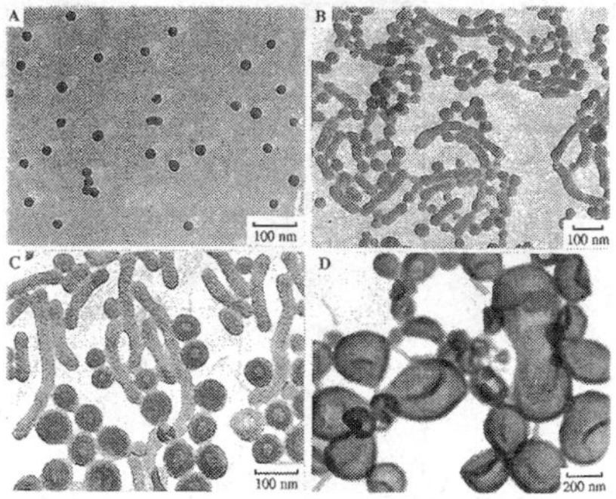


Figure 12.1.13. PS-PAA aggregates of different morphologies depending on its concentration in DMF: A - 2, B - 2.6, C - 3, D - 4 wt%. [Adapted, by permission, from L Zhang, A Eisenberg, *Macromol. Symp.*, **113**, 221-32 (1997).]

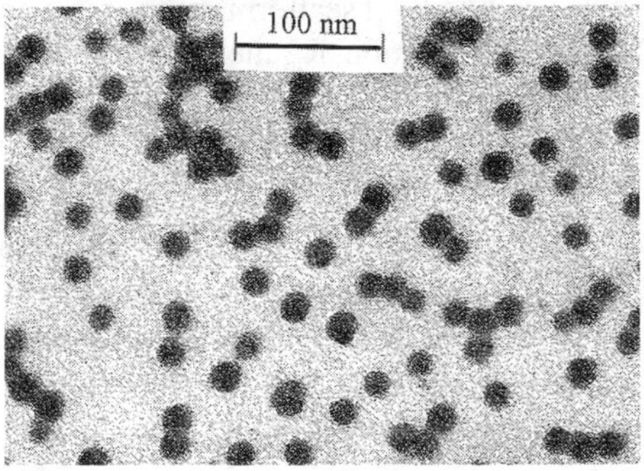


Figure 12.1.14. TEM micrograph of nanospheres. [Adapted, by permission, from G Liu, *Macromol. Symp.*, **113**, 233 (1997).]

Three major morphological features are under investigation: starlike, crew-cut, and polymer brushes (Figure 12.1.11). Morphological features have been given nick-names characterizing the observed shapes, such as "animals" or "flowers" to distinguish between various observed images.[6] Figure 12.1.12 shows four morphologies of aggregates formed by polystyrene, PS,-poly(acrylic acid), PAA, diblock copolymers.[1] The morphology produced was a direct result of the ratio between lengths of blocks of PS and PAA (see Figure 12.1.12). When this ratio is low (8.3), spherical micelles are formed. With a slightly higher ratio (12), rod-like micelles result which have narrow distribution of diameter but variable length. Increasing ratio even further (20.5) causes vesicular aggregates to form. With the highest ratio (50), large spherical micelles are formed. One of the reasons for the differences in these formations is that the surface tension between the core and the solvent varies widely. In order to decrease interfacial tension between the core and the solvent, the aggregate increases the core size.[1] Polymer concentration in solution also affects aggregate for-

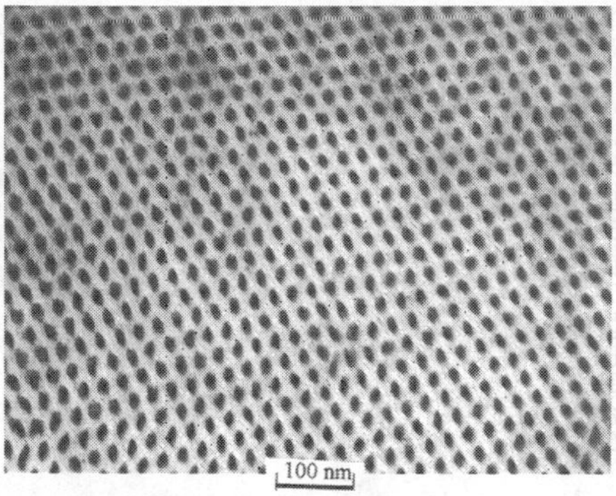


Figure 12.1.15. TEM micrograph of assemblies of cylindrical aggregates. [Adapted, by permission, from G Liu, *Macromol. Symp.*, **113**, 233 (1997).]

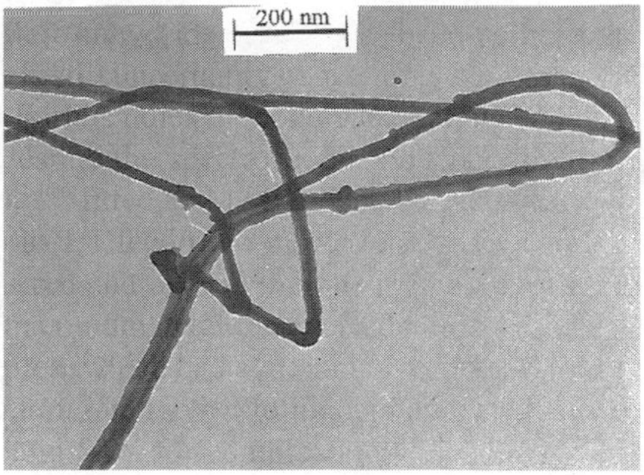


Figure 12.1.16. TEM micrograph of nanofibers. [Adapted, by permission, from G Liu, *Macromol. Symp.*, **113**, 233 (1997).]

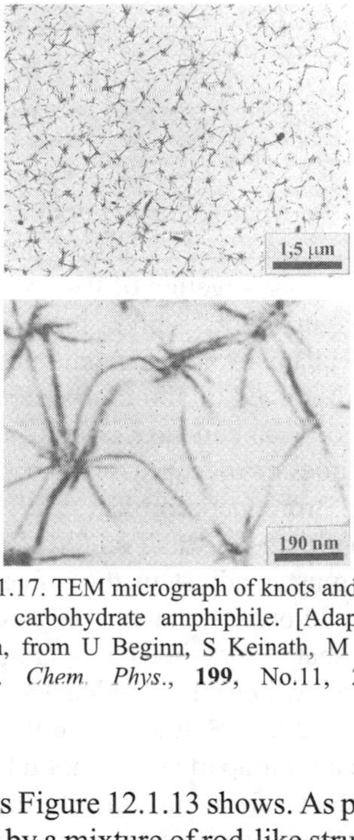


Figure 12.1.17. TEM micrograph of knots and strands formed in carbohydrate amphiphile. [Adapted, by permission, from U Beginn, S Keinath, M Moller, *Macromol. Chem. Phys.*, **199**, No.11, 2379-84 (1998).]

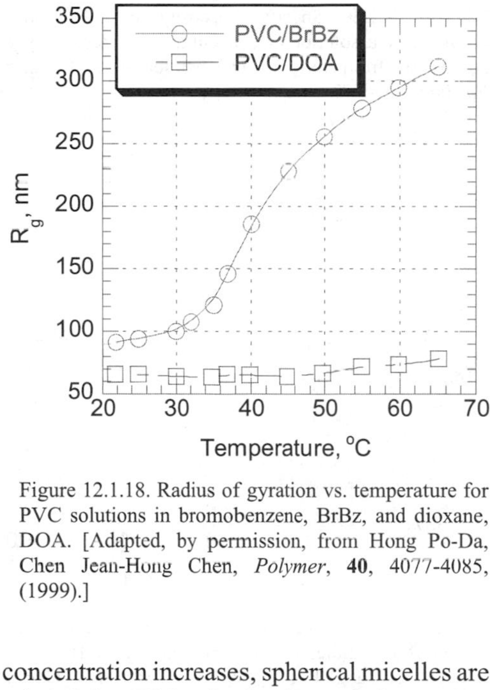


Figure 12.1.18. Radius of gyration vs. temperature for PVC solutions in bromobenzene, BrBz, and dioxane, DOA. [Adapted, by permission, from Hong Po-Da, Chen Jean-Hong Chen, *Polymer*, **40**, 4077-4085, (1999).]

mation as Figure 12.1.13 shows. As polymer concentration increases, spherical micelles are replaced by a mixture of rod-like structures and vesicles. With a further increase in concentration, the rod-like shapes disappear and only vesicles remain.

Figures 12.1.14-12.1.16 show a variety of shapes which have been observed. These include nanospheres, assemblies of cylindrical aggregates, and nanofibers.[11] Nanospheres were obtained by the gradual removal of solvent by dialysis, fibers were produced by a series of processes involving dissolution, crosslinking, and annealing. Figure 12.1.17 sheds some light on the mechanism of aggregate formation. Two elements are clearly visible from micrographs: knots and strands. Based on studies of carbohydrate amphiphiles, it is concluded that knots are formed early in the process by spinodal decomposition. Formation of

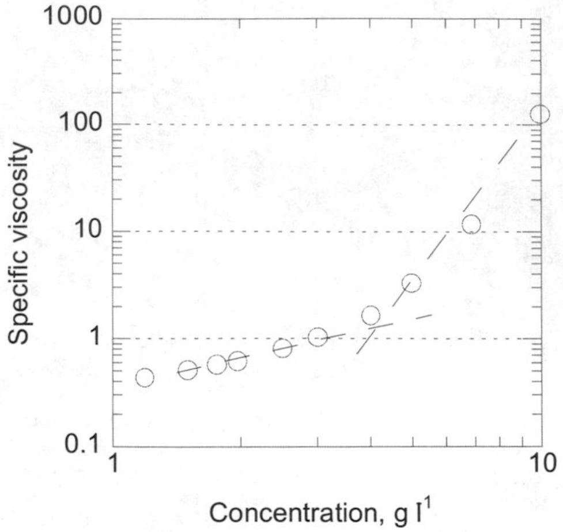


Figure 12.1.19. Specific viscosity of PVC in bromobenzene solution vs. concentration. [Adapted, by permission, from Hong Po-Da, Chen Jean-Hong Chen, *Polymer*, **40**, 4077-4085, (1999).]

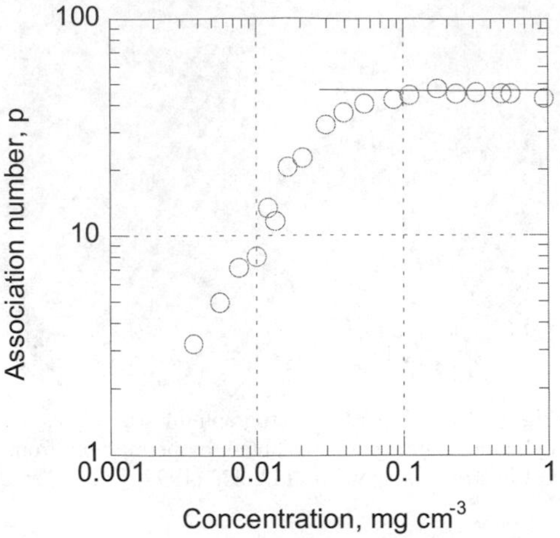


Figure 12.1.20. Diblock copolymer association number vs. its concentration. [Adapted, by permission, from D Lairez, M Adam, J-P Carton, E Raspaud, *Macromolecules*, **30**, No.22, 6798-809 (1997).]

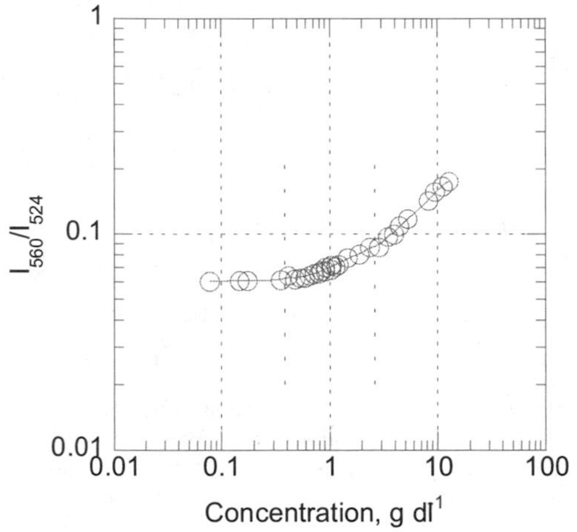


Figure 12.1.21. Fluorescence intensity ratio vs. concentration of polyimide in chloroform. [Data from H Luo, L Dong, H Tang, F Teng, Z Feng, *Macromol. Chem. Phys.*, **200**, No.3, 629-34 (1999).]

strands between knots occurs much later.[2] A similar course of events occurs with the molecular aggregation of PVC solutions.[5]

Figure 12.1.18 gives more details on the mechanism of aggregate formation.[5] PVC dissolved in bromobenzene, BrBz, undergoes a coil to globule transition which does not occur in dioxane, DOA, solution. Bromobenzene has a larger molar volume than dioxane and the polymer chain must readjust to the solvent molar volume in order to interact. Figure 12.1.19 shows that two regimes of aggregation are involved which are divided by a certain critical value. Below the critical value chains are far apart and do not interact due to contact self-avoidance. At the critical point, knots, similar to shown in Figure 12.1.17, begin to form and the resultant aggregation increases viscosity. As the polymer concentration increases, the association number, p, also increases, but then levels off (Figure 12.1.20).

Fluorescence studies (Figure 12.1.21) show that there are two characteristic points relative to concentration. Below the first point, at 0.13 g/dl, in very dilute solution, molecules are highly expanded and distant from one another and fluorescence does not change. Between the two points, individual chain coils begin to sense each other and become affected by the presence of neighbors forming intermolecular associations. Above the second point, coils begin to overlap leading to dense packing.[8] These data are essentially similar to those presented in Figure 12.1.20 but cover a broader concentration range.

The above studies show that there are many means of regulating aggregate size and shape which is likely to become an essential method of modifying materials by morphological engineering.

## REFERENCES


1 L Zhang, A Eisenberg, *Macromol. Symp.*, **113**, 221-32 (1997).
2 U Beginn, S Keinath, M Moller, *Macromol. Chem. Phys.*, **199**, No.11, 2379-84 (1998).
3 P A Cirkel, T Okada, S Kinugasa, *Macromolecules*, **32**, No.2, 531-3 (1999).
4 K Chakrabarty, R A Weiss, A Sehgal, T A P Seery, *Macromolecules*, **31**, No.21, 7390-7 (1998).
5 Hong Po-Da, Chen Jean-Hong Chen, *Polymer*, **40**, 4077-4085, (1999).
6 D Laircz, M Adam, J-P Carton, E Raspaud, *Macromolecules*, **30**, No.22, 6798-809 (1997).
7 R Fiesel, C E Halkyard, M E Rampey, L Kloppenburg, S L Studer-Martinez, U Scherf, U H F Bunz, *Macromol. Rapid Commun.*, **20**, No.3, 107-11 (1999).
8 H Luo, L Dong, H Tang, F Teng, Z Feng, *Macromol. Chem. Phys.*, **200**, No.3, 629-34 (1999).
9 A Leiva, L Gargallo, D Radic, *J. Macromol. Sci. B*, **37**, No.1, 45-57 (1998).
10 S Bistac, J Schultz, *Macromol. Chem. Phys.*, **198**, No. 2, 531-5 (1997).
11 G Liu, *Macromol. Symp.*, **113**, 233 (1997).


## 12.1.3 PERMEABILITY

The phenomenon of solvent transport through solid barriers has three aspects which discussed under the heading of permeability. These are the permeation of solvent through materials (films, containers, etc.); the use of pervaporation membranes to separate organic solvents from water or water from solvents; the manufacture of permeate selective membranes.

The permeability of different polymers and plastics to various solvents can be found in an extensive, specialized database published as a book and as a CD-ROM.[1] The intrinsic properties of polymers can be modified in several ways to increase their resistance and reduce their permeability to solvents. The development of plastic gas tanks was a major driving force behind these developments and now various plastic containers are manufactured using similar processes.

Fluorination of plastics, usually polypropylene or polyethylene, is by far the most common modification. Containers are typically manufactured by a blow molding process where they are protected by a fluorination process applied on line or, more frequently, off line. The first patent[2] for this process was issued in 1975 and numerous other patents, some of them issued quite recently have made further improvements to the process.[3] The latest processes use a reactive gas containing 0.1 to 1% fluorine to treat parison within the mold after it was expanded.[3] Containers treated by this process are barrier to polar liquids, hydrocarbon fuels, and carbon fuels containing polar liquids such as alcohols, ethers, amines, carboxylic acids, ketones, etc. Superior performance is achieved when a hydrogen purge precedes the exposure of the container to oxygen.

In another recent development,[4] containers are being produced from a blend of polyethylene and poly(vinylidene fluoride) with aluminum stearate as compatibilizer. This process eliminates the use of the toxic gas fluorine and by-products which reduces environmental pollution and disposal problems. It is argued that fluorinated containers may not perform effectively when the protective layer becomes damaged due to stress cracking.

Such a blend is intended as a replacement for previous blends of polyamide and polyethylene used in gas tanks and other containers. These older blends were exposed to a mixture of hydrocarbons but now oxygenated solvents (e.g., methanol) have been added which cause an unacceptable reduction of barrier properties. A replacement technology is needed. Improved barrier properties in this design[4] comes from increased crystallinity of the material brought about by controlling laminar thickness of polyethylene crystals and processing material under conditions which favor the crystallization of poly(vinylidene fluoride).

Another approach involves the use of difunctional telechelic polyolefins with ester, hydroxyl, and amine terminal groups.[5] Telechelic polyolefins can be used to make polyesters, polyamides, and polyurethanes with low permeability to solvents and gases.

Separation processes such as ultrafiltration and microfiltration use porous membranes which allow the passage of molecules smaller than the membrane pore size. Ultrafiltration membranes have pore sizes from 0.001 to 0.1 μm while microfiltration membranes have pore sizes in the range of 0.02 to 10 μm. The production of these membranes is almost exclusively based on non-solvent inversion method which has two essential steps: the polymer is dissolved in a solvent, cast to form a film then the film is exposed to a non-solvent. Two factors determine the quality of the membrane: pore size and selectivity. Selectivity is determined by how narrow the distribution of pore size is.[6,7] In order to obtain membranes with good selectivity, one must control the non-solvent inversion process so that it inverts slowly. If it occurs too fast, it causes the formation of pores of different sizes which will be non-uniformly distributed. This can be prevented either by an introduction of a large number of nuclei, which are uniformly distributed in the polymer membrane or by the use of a solvent combination which regulates the rate of solvent replacement.

Typical solvents used in membrane production include: N-methylpyrrolidinone, N,N-dimethylacetamide, N,N-dimethylformamide, dimethylsulfoxide, tetrahydrofuran, dioxane, dichloromethane, methyl acetate, ethyl acetate, and chloroform. They are used alone or in mixtures.[6] These are used most frequently as non-solvents: methanol, ethanol, 1-propanol, 2-propanol, 1-butanol, 2-butanol, and t-butanol. Polymers involved include: polysulfone, polyethersulfone, polyamide, polyimide, polyetherimide, polyolefins, polycarbonate, polyphenyleneoxide, poly(vinylidene fluoride), polyacrylonitrile, and cellulose and its derivatives.

A second method of membrane preparation is based on the thermally-induced phase separation process.[8] The goal of this process is to produce a membrane which has an ultrathin separation layer (to improve permeation flux) and uniform pores.[9] In this process, polymer is usually dissolved in a mixture of solvents which allow the mixture to be processed either by spinning or by coating. The desired membrane morphology is obtained through cooling to induce phase separation of

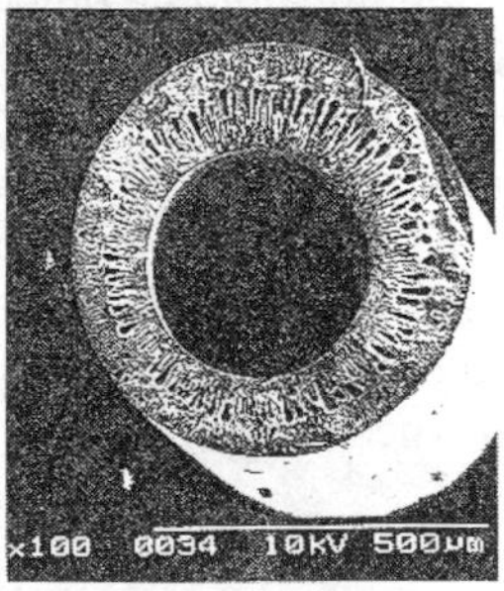


Figure 12.1.22. PEI fiber from NMP. [Adapted, by permission from D Wang, K Li, W K Teo, *J. Appl. Polym. Sci.*, **71**, No.11, 1789-96 (1999).]

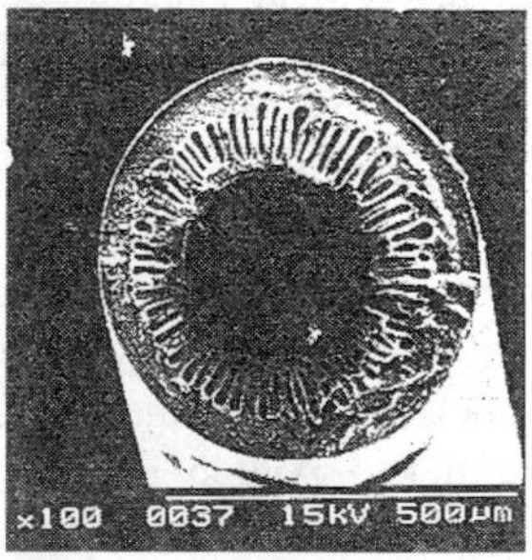


Figure 12.1.23. PEI fiber from DMF. [Adapted, by permission from D Wang, K Li, W K Teo, *J. Appl. Polym. Sci.*, **71**, No.11, 1789-96 (1999).]

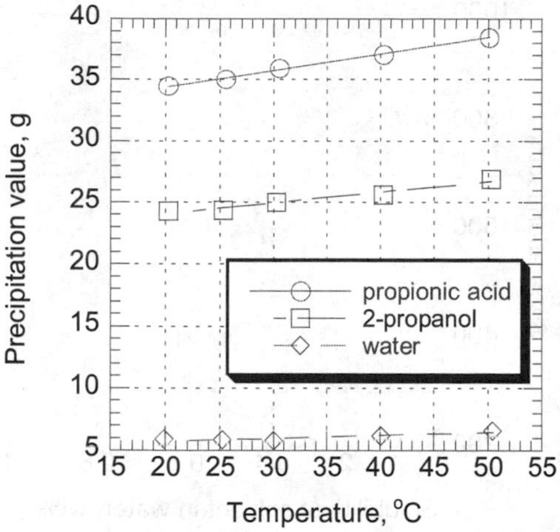


Figure 12.1.24. Precipitation value for PEI/NMP system with different non-solvents vs. temperature. [Data from D Wang, Li K, W K Teo, *J. Appl. Polym. Sci.*, **71**, No.11, 1789-96 (1999).]

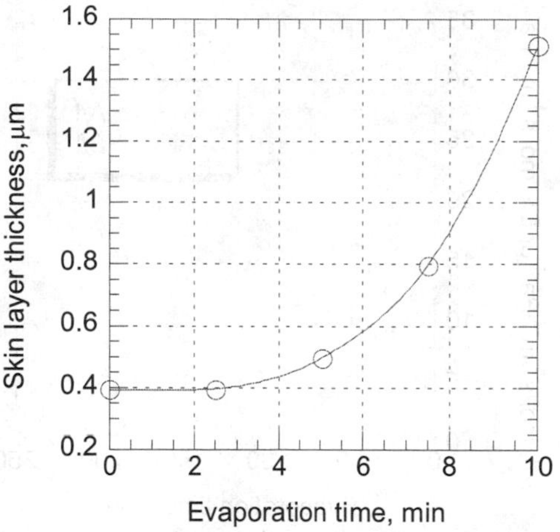


Figure 12.1.25. Skin thickness vs. evaporation time of formation of asymmetric membrane from polysulfone. [Data from A Yamasaki, R K Tyagi, A E Fouda, T Matsura, K Jonasson, *J. Appl. Polym. Sci.*, **71**, No.9, 1367-74 (1999).]

polymer/solvent/non-solvent system. One practices in use is the addition of non-solvents to the casting solution.[10] Thc mixture should be designed such that homogeneous casting is still possible but the thermodynamic condition approaches phase separation. Because of cooling and/or evaporation of one of the solvents, demixing occurs before material enters coagulation bath.[10]

Figures 12.1.22 and 12.1.23 explain technical principles behind formation of efficient and selective membrane. Figure 12.1.22 shows a micrograph of hollow PEI fiber produced from N-methyl-2-pyrrolidone, NMP, which has thin surface layer and uniform pores and Figure 12.1.23 shows the same fiber obtained from a solution in dimethylformamide, DMF, which has a thick surface layer and less uniform pores.[9] The effect depends on the interaction of polar and non-polar components. The compatibility of components was estimated based on their Hansen's solubility parameter difference. The compatibility increases as the solubility parameter difference decreases.[9] Adjusting temperature is another method of control because the Hansen's solubility parameter decreases as the temperature increases. A procedure was developed to determine precipitation values by titration with non-solvent to a cloud point.[9] Use of this procedure aids in selecting a suitable non-solvent for a given polymer/solvent system. Figure 12.1.24 shows the results from this method.[9] Successful in membrane production by either non-solvent inversion or thermally-induced phase separation requires careful analysis of the compatibilities between polymer and solvent, polymer and non-solvent, and solvent and non-solvent. Also the processing regime, which includes temperature control, removal of volatile components, uniformity of solvent replacement must be carefully controlled.

Efforts must be made to select solvents of low toxicity and to minimize solvent consumption. An older method of producing polyetherimide membranes involved the use of a mixture of two solvents: dichloromethane (very volatile) and 1,1,2-trichloroethane together

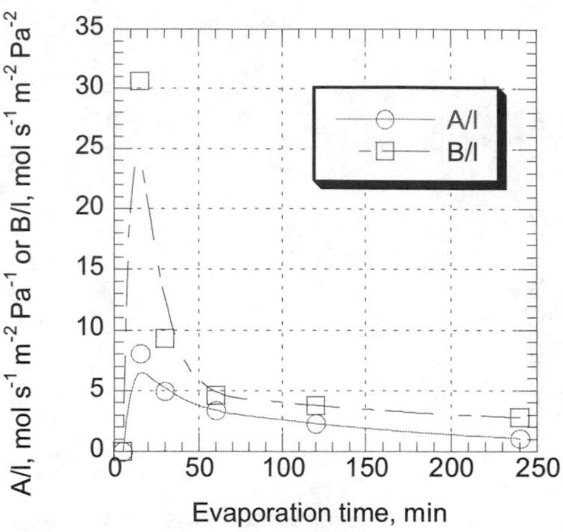


Figure 12.1.26. Pervaporation transport parameters: A/l (liquid transport) and B/l (vapor transport) vs. evaporation time during membrane preparation from aromatic polyamide. [Data from A Yamasaki, R K Tyagi, A Fouda, T Matsuura, *J. Appl. Polym. Sci.*, **57**, No.12, 1473-81 (1995).]

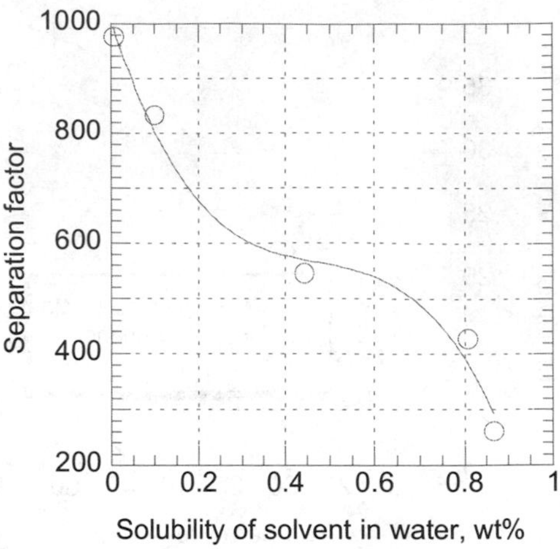


Figure 12.1.27. Separation factor vs. solubility of solvents in water. [Adapted, by permission, from M Hoshi, M Kobayashi, T Saitoh, A Higuchi, N Tsutomu, *J. Appl. Polym. Sci.*, **69**, No.8, 1483-94 (1998).]

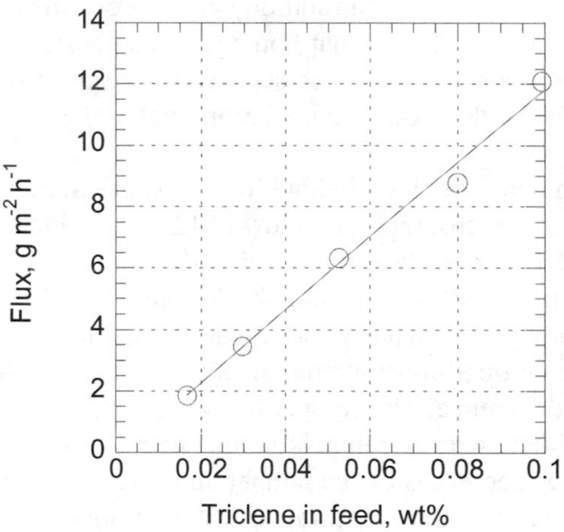


Figure 12.1.28. Effect of triclene concentration in feed solution on its pervaporation flux through acrylic membrane. [Data from M Hoshi, M Kobayashi, T Saitoh, A Higuchi, N Tsutomu, *J. Appl. Polym. Sci.*, **69**, No.8, 1483-94 (1998).]

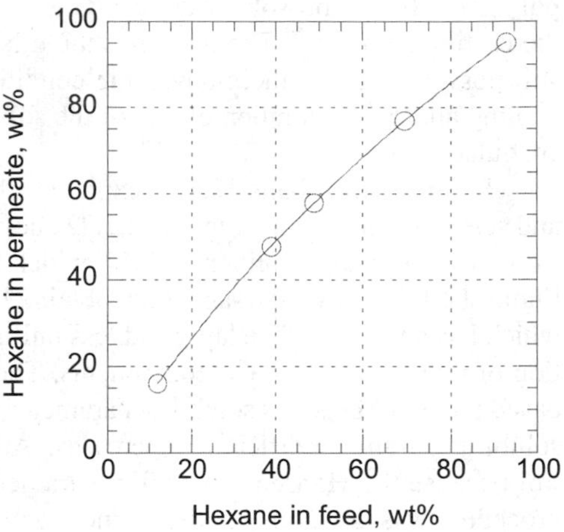


Figure 12.1.29. Effect of hexane concentration in feed solution on its concentration in permeate through acrylic membrane. [Data from M Hoshi, M Kobayashi, T Saitoh, A Higuchi, N Tsutomu, *J. Appl. Polym. Sci.*, **69**, No.8, 1483-94 (1998).]

with a mixture of non-solvents: xylene and acetic acid. In addition, large amounts of acetone were used in the coagulation bath. A more recent process avoided environmental problems by using tetrahydrofuran and γ-butyrolactone in a process in which water only is used in the coagulating bath (membrane quality remained the same).[10]

Thermally-induced phase separation has been applied to the production of polysilane foams.[11] Variation of polymer concentration, solvent type, and cooling rate have been used to refine the macrostructure. Both these membrane production methods rely on the formation of aggregates of controlled size and shape as discussed in the previous section.

Figure 12.1.25 shows the effect of of solvent evaporation time on the skin layer thickness. It is only in the beginning of the process that skin thickness does not increase. Thereafter there is rapid skin thickness growth. Polymer concentration is the other important determinant of skin thickness. Figure 12.1.26 shows that during very short evaporation times (5 min), even though skin is thin the pores diameter is so small that the pervaporation parameters A/l and B/l, which characterize liquid and vapor transport, respectively, do not increase. Longer evaporation times bring about a gradual decrease in transport properties.[12,13]

Figure 12.1.27 shows the relationship between the separation factor and the solubility of solvents in water. The separation of solvent by a pervaporation membrane occurs less efficiently as solvent solubility increases.The more concentrated the solution of solvent, the faster is the separation (Figure 12.1.28). Separation of hexane from a mixture with heptane is similar (Figure 12.1.29). The acrylic membrane shows good selectivity. These examples demonstrate the usefulness of pervaporation membranes in solvent recovery processes.

## REFERENCES


1 Permeability and Other Film Properties, Plastics Design Library, Norwich, 1996.
2 **US Patent 3,862,284**, 1975.
3 J P Hobbs, J F DeiTos, M Anand, **US Patent 5,770,135**, Air Products and Chemicals, Inc., 1998 and **US Patent 5,244,615**, 1992.
4 R T Robichaud, **US Patent 5,702,786**, Greif Bros. Corporation, 1997.
5 P O Nubel, H B Yokelson, **US Patent 5,731,383**, Amoco Corporation, 1998 and **US Patent 5,589,548**, 1996.
6 J M Hong, S R Ha, H C Park, Y S Kang, K H Ahn, **US Patent 5,708,040**, Korea Institute of Science and Technology, 1998.
7 K-H Lee, J-G Jegal, Y-I Park, **US Patent 5,868,975**, Korea Institute of Science and Technology, 1999.
8 J M Radovich, M Rothberg, G Washington, **US Patent 5,645,778**, Althin Medical, Inc., 1997.
9 D Wang, Li K, W K Teo, *J. Appl. Polym. Sci.*, **71**, No.11, 1789-96 (1999).
10 K-V Peinemann, J F Maggioni, S P Nunes, *Polymer*, **39**, No.15, 3411-6 (1998).
11 L L Whinnery, W R Even, J V Beach, D A Loy, *J. Polym. Sci.: Polym. Chem. Ed.*, **34**, No.8, 1623-7 (1996).
12 A Yamasaki, R K Tyagi, A Fouda, T Matsuura, *J. Appl. Polym. Sci.*, **57**, No.12, 1473-81 (1995).
13 A Yamasaki, R K Tyagi, A E Fouda, T Matsura, K Jonasson, *J. Appl. Polym. Sci.*, **71**, No.9, 1367-74 (1999).
14 M Hoshi, M Kobayashi, T Saitoh, A Higuchi, N Tsutomu, *J. Appl. Polym. Sci.*, **69**, No.8, 1483-94 (1998).


## 12.1.4 MOLECULAR STRUCTURE AND CRYSTALLINITY

The gelation of polymer-solvent systems was initially thought of as a process which occurs only in poor solvents which promote chain-chain aggregation. Further studies have revealed that many polymers also form gels in good solvents. This has prompted research which attempts to understand the mechanisms of gelation. These efforts have contributed the current understanding of the association between molecules of solvents and polymers.[1] Studies on isotactic, iPS, and syndiotactic, sPS, polystyrenes[2-6] have confirmed that, although both polymers have the same monomeric units, they are significantly different in terms of their solubility, gelation, and crystallization. In systems where sPS has been dissolved in benzene and carbon tetrachloride, gelation is accomplished in a few minutes

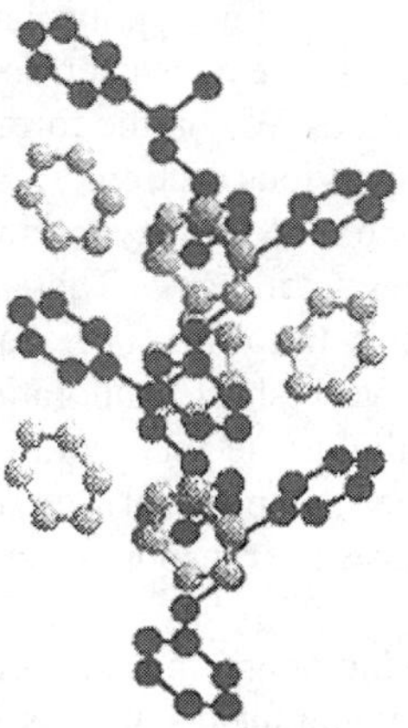

Figure 12.1.30. Molecular model of polystyrene and benzene rings alignment. [Adapted by permission, from J M Guenet, *Macromol. Symp.*, **114**, 97 (1997).]

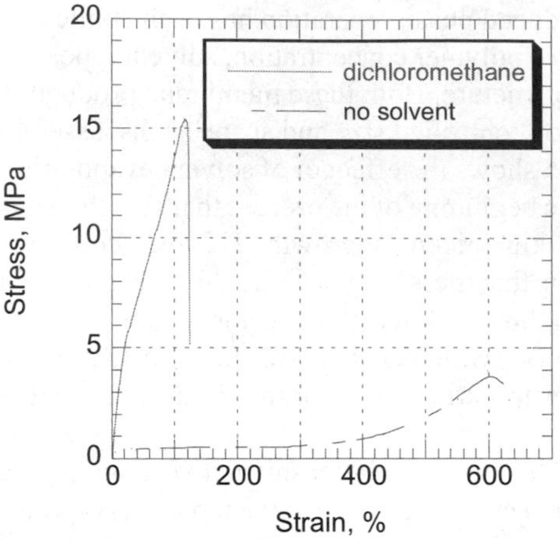


Figure 12.1.31. Stress-strain behavior at 130°C of glassy sPS with and without immersion in dichloromethane. [Data from C Daniel, L Guadagno, V Vittoria, *Macromol. Symp.*, **114**, 217 (1997).]

whereas in chloroform it takes tens of hours. Adding a small amount of benzene to the chloroform accelerates the gelation process. There is also evidence from IR which shows that the orientation of the ring plane is perpendicular to the chain axis. This suggests that there is a strong interaction between benzene and sPS molecules. The other interesting observation came from studies on decalin sPS and iPS systems. iPS forms a transparent gel and then becomes turbid, gradually forming trigonal crystallites of iPS. sPS does not convert to a gel but the fine crystalline precipitate particles instead. This shows that the molecular arrangement depends on both the solvent type and on the molecular structure of the polymer.[2]

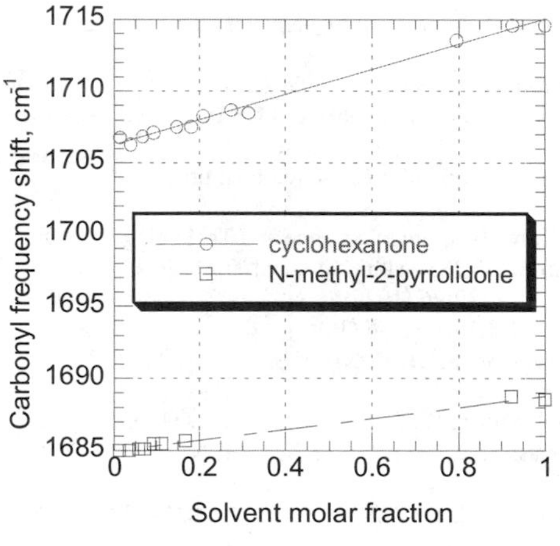


Figure 12.1.32. Carbonyl frequency vs. molar fraction of solvent. [Adapted, by permission, from J M Gomez-Elvira, P Tiemblo, G Martinez, J Milan, *Macromol. Symp.*, **114**, 151 (1997).]

Figure 12.1.30 shows how benzene molecules align themselves parallel to the phenyl rings of polystyrene and how they are housed within the helical form of the polymer structure which is stabilized by the presence of solvent.[1]

Work on poly(ethylene oxide) gels[7] indicates that the presence of solvents such as chloroform and carbon disulfide contributes to the formation of a uniform helical conformation. The gelation behavior and the gel structure depend on the solvent type which, in turn, is determined by solvent-polymer interaction. In a good solvent, polythiophene molecules exist in coiled conformation. In a poor solvent, the molecules form aggregates through the short substituents.[8]

Polyvinylchloride, PVC, which has a low crystallinity, gives strong gels. Neutron diffraction and scattering studies show that these strong gels result from the formation of

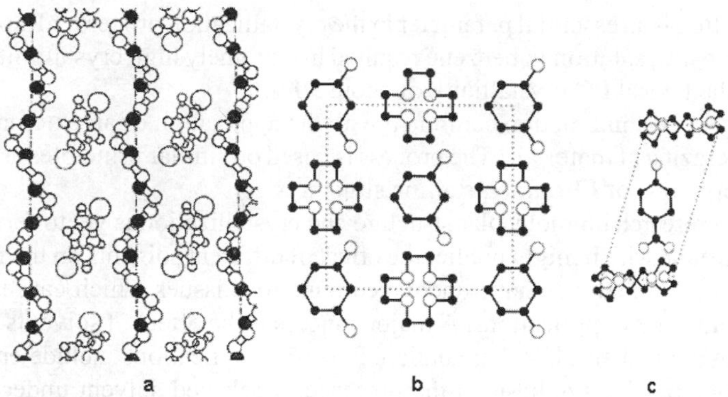


Figure 12.1.33. Crystal structure of poly(ethylene oxide) molecular complex with: (a) p-dichlorobenzene, (b) resorcinol, (c) p-nitrophenol. [Adapted, by permission, from M Dosiere, *Macromol. Symp.*, **114**, 51 (1997).]

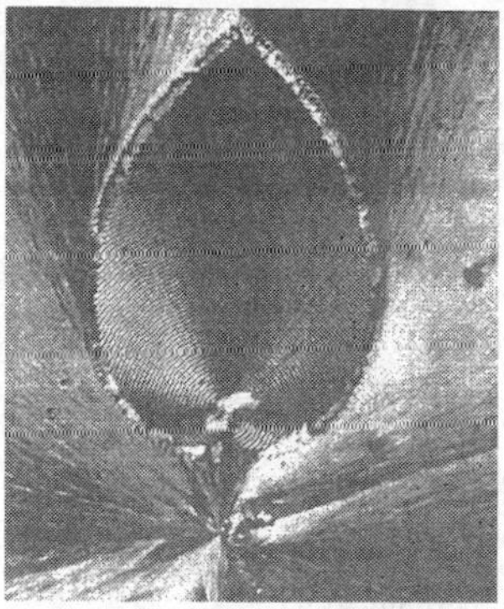

Figure 12.1.34. Optical micrograph of PEO-resorcinol complex. [Adapted, by permission, from M Dosiere, *Macromol. Symp.*, **114**, 51 (1997).]

PVC-solvent complexes.[9] This indicates that the presence of solvent may affect the mechanical properties of such a system. Figure 12.1.31 shows that the presence of dichloromethane in sPS changes the mechanical characteristic of the material. A solvent-free polymer has a high elongation and yield value. An oriented polymer containing dichloromethane has lower elongation, no yield value, and approximately four times greater tensile strength.[10]

The interaction of polymer-solvent affects rheological properties. Studies on a divinyl ether-maleic anhydride copolymer show that molecular structure of the copolymer can be altered by the solvent selected for synthesis.[11] Studies have shown that the thermal and UV stability of PVC is affected by the presence of solvents.[12-14]

Figure 12.1.32 shows that solvent type and its molar fraction affect the value of the carbonyl frequency shift. This frequency shift occurs at a very low solvent concentration, The slope angles for each solvent are noticeably different.[15] Similar frequency shifts were reported for various solvents in polyetheretherketone, PEEK, solutions.[16]

These associations are precursors of further ordering by crystallization. Figure 12.1.33 shows different crystalline structures of poly(ethylene oxide) with various solvents.[17] Depending on the solvent type, a unit cell contains a variable number of monomers and molecules of solvent stacked along the crystallographic axis. In the case of dichlorobenzene, the molecular complex is orthorhombic with 10 monomers and 3 solvent molecules. Figure 12.1.34 shows optical micrograph of spherulite formed from unit cells having 8 monomers and 4 molecules of solvent. The unit cell of the triclinic crystal, formed with nitrophenol, is composed of 6 monomers and 4 molecules of solvent. This underlines the importance of solvent selection in achieving the desired structure in the formed material. The technique is also advantageous in polymer synthesis, in improving polymer stability, etc.

Temperature is an essential parameter in the crystallization process.[18] Rapid cooling of a polycarbonate, PC, solution in benzene resulted in extremely high crystallinity (46.4%) as compared to the typical PC crystallinity of about 30%.

Polymer-solvent interaction combined with the application of an external force leads to the surface crazing of materials. The process is based on similar principles as discussed in this section formation of fibrilar crystalline structures.

Although research on molecular structure and crystallization is yet to formulate a theoretical background which might predict the effect of different solvents on the fine structure of different polymers, studies have uncovered numerous issues which cause concern but which also point to new applications. A major concern is the effect of solvents on craze formation, and on thermal and UV degradation. Potential applications include engineering of polymer morphology by synthesis in the presence of selected solvent under a controlled thermal regime, polymer reinforcement by interaction with smaller molecules, better retention of additives, and modification of surface properties to change adhesion or to improve surface uniformity, etc.

## REFERENCES


1 J M Guenet, *Macromol. Symp.*, **114**, 97 (1997).
2 M Kobayashi, *Macromol. Symp.*, **114**, 1-12 (1997).
3 T Nakaoki, M Kobayashi, *J. Mol. Struct.*, **242**, 315 (1991).
4 M Kobayashi, T Nakoaki, *Macromolecules*, **23**, 78 (1990).
5 M Kobayashi, T. Kozasa, *Appl. Spectrosc.*, **47**, 1417 (1993).
6 M. Kobayashi, T Yoshika, M Imai, Y Itoh, *Macromolecules,* **28**, 7376 (1995).
7 M Kobayashi, K Kitagawa, *Macromol. Symp.*, **114**, 291 (1997).
8 P V Shibaev, K Schaumburg, T Bjornholm, K Norgaard, *Synthetic Metals*, **97**, No.2, 97-104 (1998).
9 H Reinecke, J M Guenet, C. Mijangos, *Macromol. Symp.*, **114**, 309 (1997).
10 C Daniel, L Guadagno, V Vittoria, *Macromol. Symp.*, **114**, 217 (1997).
11 M Y Gorshkova, T L Lebedeva, L L Stotskaya, I Y Slonim, *Polym. Sci. Ser. A*, **38**, No.10, 1094-6 (1996).
12 G Wypych, **Handbook of Material Weathering**, *ChemTec Publishing*, Toronto, 1995.
13 G Wypych, **Poly(vinyl chloride) degradation**, *Elsevier*, Amsterdam, 1985.
14 G Wypych, **Poly(vinyl chloride) stabilization**, *Elsevier*, Amsterdam, 1986.
15 J M Gomez-Elvira, P Tiemblo, G Martinez, J Milan, *Macromol. Symp.*, **114**, 151 (1997).
16 B H Stuart, D R Williams, *Polymer*, **36**, No.22, 4209-13 (1995).
17 M Dosiere, *Macromol. Symp.*, **114**, 51 (1997).
18 Gending Ji, Fengting Li, Wei Zhu, Qingping Dai, Gi Xue, Xinhong Gu, *J. Macromol. Sci. A*, **A34**, No.2, 369-76 (1997).
19 I V Bykova, E A Sinevich, S N Chvalun, N F Bakeev, *Polym. Sci. Ser. A*, **39**, No.1, 105-12 (1997).


## 12.1.5 OTHER PROPERTIES AFFECTED BY SOLVENTS

Many other properties of solutes and solutions are affected by solvents. Here, we will discuss material stability and stabilization, some aspects of reactivity (more information on this subject appears in Chapter 13), physical properties, some aspects of electrical and electrochemical properties (more information on this subject appears in Chapter 11), surface properties, and polarity and donor properties of solvents.

Degradative processes have long been known to be promoted by the products of solvent degradation. Tetrahydrofuran is oxidized to form a peroxide which then dissociates to form two radicals initiating a chain of photo-oxidation reactions. Figure 12.1.35 shows the kinetics of hydroperoxide formation.[1] Similar observations, but in polymer system, were made in xylene by direct determination of the radicals formed using ESR.[2] An increased concentration in trace quantities of xylene contributed to the formation of n-octane radicals

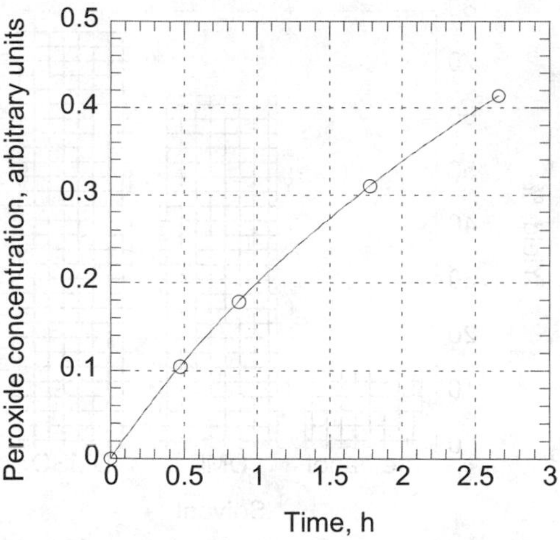


Figure 12.1.35. Peroxide formation from tetrahydrofuran during irradiation at 254 nm. [Data from J F Rabek, T A Skowronski, B Ranby, *Polymer*, **21**, 226 (1980).]

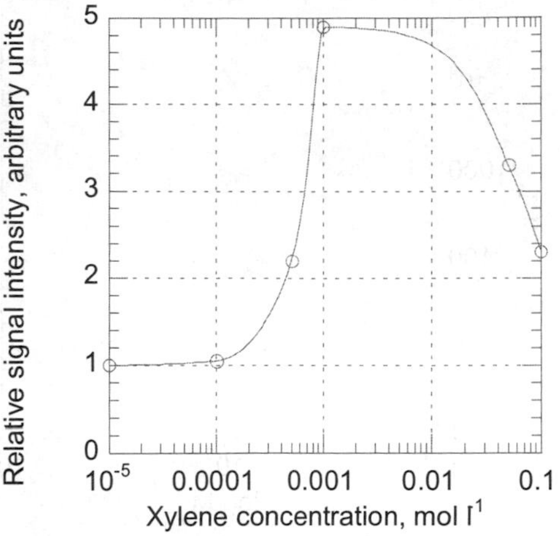


Figure 12.1.36. Spectral intensity of n-octane radical formed during irradiation of PE at -196°C for 30 min vs. concentration of p-xylene. [Adapted, by permission, from H Kubota, M Kimura, *Polym. Deg. Stab.*, **38**, 1 (1992).]

by abstracting hydrogen from polyethylene chain in an α-position to the double bond (Figure 12.1.36).

3-hydroperoxyhexane cleaves with formation of carboxylic acid and hydrocarbon radical (ethylene or propylene).[3] These known examples show that the presence of even traces of solvents may change the chemistry and the rate of photo-oxidative processes because of formation of radicals.

In another aspect of photo-oxidative processes, solvents influence the reactivity of small molecules and chain segments by facilitating the mobility of molecules and changing the absorption of light, the wavelength of emitted fluorescent radiation, and the lifetime of radicals. Work on anthraquinone derivatives, which are common photosensitizers, has shown that when photosensitizer is dissolved in isopropanol (hydrogen-donating solvent) the half-life of radicals is increased by a factor of seven compared to acetonitrile (a non-hydrogen-donating solvent).[4] This shows that the ability of a photosensitizer to act in this manner depends on the presence of a hydrogen donor (frequently the solvent). In studies on another group of photosensitizers – 1,2-diketones, the solvent cyclohexane increased the absorption wavelength of the sensitizer in the UV range by more than 10 nm compared with solutions in ethanol and chloroform.[5] Such a change in absorption may benefit some systems because the energy of absorbed radiation will become lower than the energy required to disrupt existing chemical bonds. But it may also increase the potential for degradation by shifting radiation wavelength to the range absorbed by a particular material. Presence of water may have a plasticizing action which increases the mobility of chains and their potential for interactions and reactions.[6]

Studies on photoresists, based on methacryloyethyl- phenylglyoxylate, show that, in aprotic solvents, the main reaction mechanism is a Norrish type II photolysis leading to chain scission.[7] In aprotic solvents, the polymer is photoreduced and crosslinks are formed.

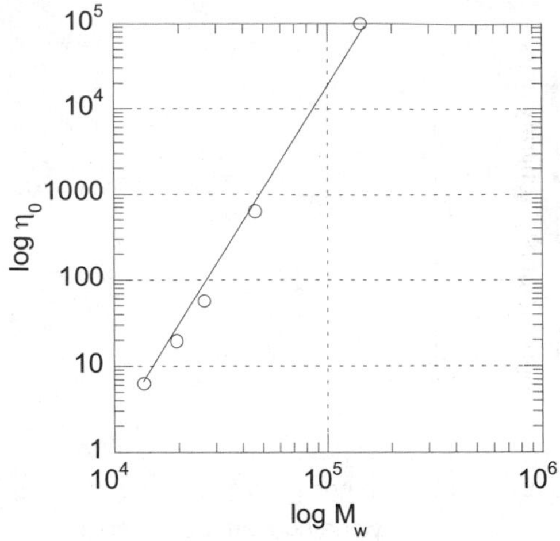


Figure 12.1.37. Effect of viscosity on molecular weight of polyethylene wax obtained by thermolysis in the presence of phenylether at 370°C. [Adapted, by permission, from L Guy, B Fixari, *Polymer*, **40**, No.10, 2845-57 (1999).]

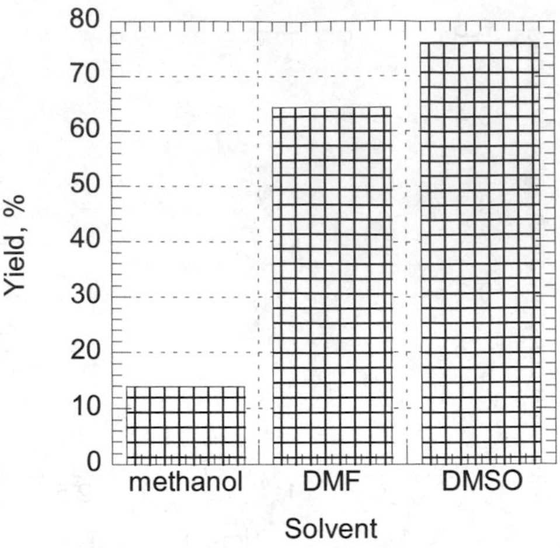


Figure 12.1.38. Yield of condensation products depending on solvent. [Adapted, by permission, from J Jeczalik, *J. Polym. Sci.: Polym. Chem. Ed.*, **34**, No.6, 1083-5 (1996).]

Polarity of the solvent determines quantum yields in polyimides.[8] The most efficient photocleavage was in medium-polar solvents. The selection of solvent may change the chemical mechanism of degradation and the associated products of such reactions.

Singlet oxygen is known to affect material stability by its ability to react directly with hydrocarbon chains to form peroxides. The photosensitizers discussed above are capable of generating singlet oxygen. Solvents, in addition to their ability to promote photosensitizer action, affect the lifetime of singlet oxygen (the time it has to react with molecules and form peroxides). Singlet oxygen has very short lifetime in water (2 μs) but much longer times in various solvents (e.g., 24 μs in benzene, 200 μs in carbon disulfide, and 700 μs in carbon tetrachloride).[9] Solvents also increase the oxygen diffusion coefficient in a polymer. It was calculated[10] that in solid polystyrene only less than 2% of singlet oxygen is quenched compared with more than 50% in solution. Also, the quantum yield of singlet oxygen was only 0.56 in polystyrene and 0.83 in its benzene solution.[10]

The photostabilizer must be durable. It was found that salicylic stabilizers are efficiently degraded by singlet oxygen in polar alkaline media but in a less polar, non-alkaline solvents these stabilizers are durable.[11] In hydrogen-bonding solvents, the absorption spectrum of UV absorbers is changed.[12]

Thermal decomposition yield and composition of polystyrene wastes is affected by addition of solvents.[13] Solvents play two roles: First, they stabilize radicals and, second, by lowering solution viscosity they contribute to homogeneity, increase the mobility of components and increase reactivity. 85% benzyl and phenoxy radicals were stabilized by hydrogens from tetralin.[13] In polyethylene thermolysis, the use of solvent to reduce viscosity resulted in obtaining lower molecular weight, narrow molecular weight distribution and high crystallinity in resultant polyethylene wax.[14] Figure 12.1.37 illustrates the relationship

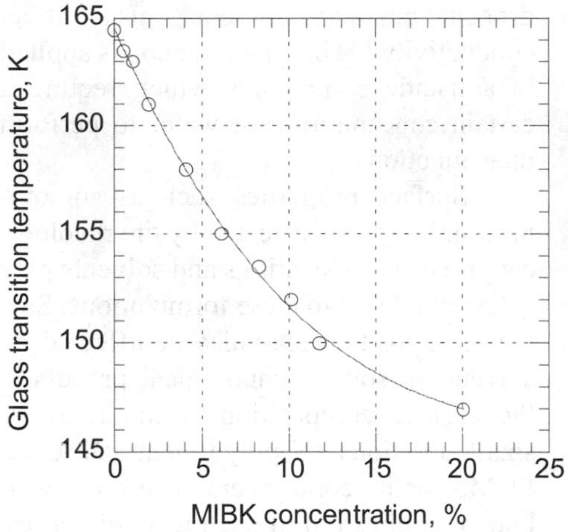


Figure 12.1.39. Glass transition temperature of perfluoro polyoxyalkylene oligomers vs. MIBK concentration. [Data from S Turri, M Scicchitano, G Gianotti, C Tonelli, *Eur. Polym. J.*, **31**, No.12, 1227-33 (1995).]

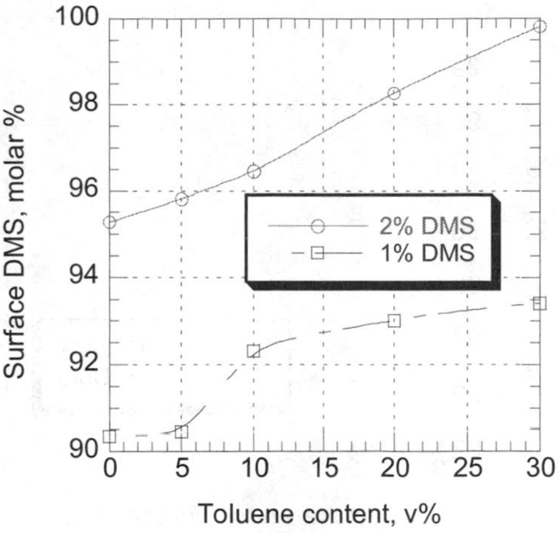


Figure 12.11.40. Surface DMS vs. toluene concentration for two bulk contents of DMS in polystyrene. [Data from Jiaxing Chen, J A Gardella, *Macromolecules*, **31**, No.26, 9328-36 (1998).]

between molecular weight and viscosity of solutions of polyethylene in phenylether as a solvent.[14]

Figure 12.1.38 shows that polymerization yield depends on the solvent selected for the polymerization reaction. The polymerization yield increases as solvent polarity increases.[15] Work on copolymerization of methyl methacrylate and N-vinylpyrrolidone shows that the solvent selected regulates the composition of copolymer.[16] The smaller the polarity of the solvent and the lower the difference between the resonance factors of the two monomers the more readily they can copolymerize. Work on electropolymerization[17] has shown an extreme case of solvent effect in the electrografting of polymer on metal. If the donocity of the monomer is too high compared with that of the solvent, no electrografting occurs. If the donocities are low, a high dielectric constant of solvent decreases grafting efficiency. This work, which may be very important in corrosion protection, illustrates that a variety of influences may affect the polymerization reaction.

The addition of solvent to polymer has a plasticizing effect. The increase in free volume has a further influence on the glass transition temperature of the polymer. Figure 12.1.39 shows the effect of methyl-isobutyl-ketone, MIBK, at various concentrations on glass transition temperature.[18] Several solvents were used in this study[18] to determine if the additivity rule can be useful to predict the glass transition temperature of a polymer-solvent system. The results, as a rule, depart from the linear relationship between glass transition temperature and solvent concentration. Generally, the better the solvent is for a particular polymer the higher is the departure.

Hydrogen bond formation as a result of the interaction of polymer with solvent was found to contribute to changes in the electric properties of polyaniline.[19] Hydrogen bonding causes changes in conformal structure of polymer chains. This increases the electrical conductivity of polyaniline. Water is especially effective in causing such changes but other hy-

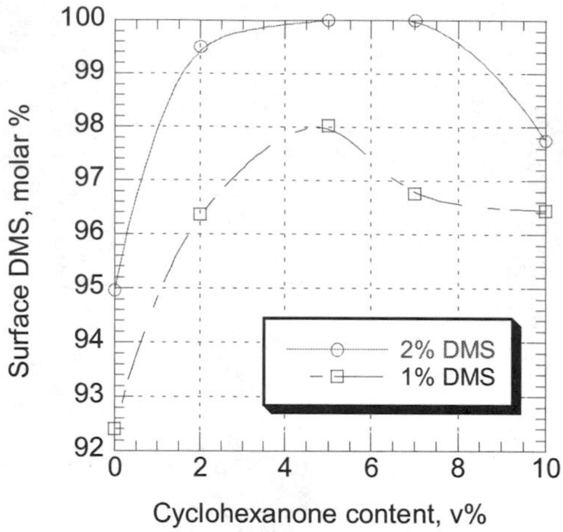


Figure 12.1.41. Surface DMS vs. cyclohexanone concentration for two bulk contents of DMS in polystyrene. [Data from Jiaxing Chen, J A Gardella, *Macromolecules*, **31**, No.26, 9328-36 (1998).]

drogen bonding solvents also affect conductivity.[19] This phenomenon is applied in antistatic compounds which require a certain concentration of water to perform their function.

Surface properties such as smoothness and gloss are affected by the rheological properties of coatings and solvents play an essential role in these formulations. Surface composition may also be affected by solvent. In some technological processes, the surface composition is modified by small additions of polydimethylsiloxane, PDMS, or its copolymers. The most well known application of such technology has been the protection of external surface of the space shuttle against degradation using PDMS as the durable polymer. Other applications use PDMS to lower the coefficient of friction . In both cases, it is important that PDMS forms a very high concentrations on the surface (possibly 100%). Because of its low surface energy, it has the intrinsic tendency to migrate to the surface consequently, surface concentrations as high as 90% can be obtained without special effort. It is more difficult to further increase this surface concentration. Figure 12.1.40 shows the effect of a chloroform/toluene mixture on the surface segregation of PDMS. When the poor solvent for PDMS (toluene) is mixed with chloroform, the decrease in the mobility of the PDMS segments makes the cohesive migration of the polystyrene, PS, segments more efficient and the concentration of PDMS on the surface increases as the concentration of toluene increases. Increase in PDMS concentration further improves its concentration on the surface to almost 100%. This shows the influence of polymer-solvent interaction on surface segregation and demonstrates a method of increasing the additive polymer concentration on the surface. Solvents with higher boiling points also increase surface concentration of PDMS because they extend the time of the segregation process.[20]

Figure 12.1.41 shows the effect of addition of cyclohexanone to chloroform.[20] This exemplifies yet another phenomenon which can help to increase the surface concentration of PDMS. Cyclohexanone is more polar solvent than chloroform (Hansen parameter, $\delta_p$, is 3.1 for chloroform and 6.3 for cyclohexanone). The solvation of cyclohexanone molecules will selectively occur around PDMS segments which helps in the segregation of PDMS and in the cohesive migration of PS. But the figure shows that PDMS concentration initially increases then decreases with further additions of cyclohexanone. The explanation for this behavior is in the amount of cyclohexanone required to better solvate PDMS segments. When cyclohexanone is in excess, it also increases the solvation of PS and the effect which produces an increased surface segregation is gradually lost.

The surface smoothness of materials is important in solvent welding. It was determined that a smooth surface on both mating components substantially increases the strength of the weld.[21] This finding may not be surprising since weld strength depends on close surface contact but it is interesting to note that the submicrostructure may have an influence on phenomena normally considered to be influenced by features which have a larger dimensional scale. Non-blistering primer is another example of a modification which tailors surface properties.[22] The combination of resins and solvents used in this invention allows solvent to escape through the surface of the cured primer before the material undergoes transitions that occur at high temperatures. Additional information on this subject can be found in Chapter 7.

A review paper[23] examines the nucleophilic properties of solvents. It is based on accumulated data derived from calorimetric measurements, equilibrium constants, Gibbs free energy, nuclear magnetic resonance, and vibrational and electronic spectra. Parameters characterizing Lewis-donor properties are critically evaluated and tabulated for a large number of solvents. The explanation of the physical meaning of polarity and discussion of solvatochromic dyes as the empirical indicators of solvent polarity are discussed (see more on this subject in Chapter 10).[23]

## REFERENCES


1 J F Rabek, T A Skowronski, B Ranby, *Polymer*, **21**, 226 (1980).
2 H Kubota, M Kimura, *Polym. Deg. Stab.*, **38**, 1 (1992).
3 G Teissedre, J F Pilichowski, J Lacoste, *Polym. Degrad. Stability*, **45**, No.1, 145-53 (1994).
4 M Shah, N S Allen, M Edge, S Navaratnam, F Catalina, *J. Appl. Polym. Sci.*, **62**, No.2, 319-40 (1996).
5 P Hrdlovic, I Lukac, *Polym. Degrad. Stability*, **43**, No.2, 195-201 (1994).
6 M L Jackson, B J Love, S R Hebner, *J. Mater. Sci. Materials in Electronics*, **10**, No.1, 71-9 (1999).
7 Hu Shengkui, A Mejiritski, D C Neckers, *Chem. Mater.*, **9**, No.12, 3171-5 (1997).
8 H Ohkita, A Tsuchida, M Yamamoto, J A Moore, D R Gamble, *Macromol. Chem. Phys.*, **197**, No.8, 2493-9 (1996).
9 G Wypych, **Handbook of Material Weathering**, *ChemTec Publishing*, Toronto, 1995.
10 R D Scurlock, D O Martire, P R Ogilby, V L Taylor, R L Clough, *Macromolecules*, **27**, No.17, 4787-94 (1994).
11 A T Soltermann, D de la Pena, S Nonell, F Amat-Guerri, N A Garcia, *Polym. Degrad. Stability*, **49**, No.3, 371-8 (1995).
12 K P Ghiggino, *J. Macromol. Sci. A*, **33**, No.10, 1541-53 (1996).
13 M Swistek, G Nguyen, D Nicole, *J. Appl. Polym. Sci.*, **60**, No.10, 1637-44 (1996).
14 L Guy, B Fixari, *Polymer*, **40**, No.10, 2845-57 (1999).
15 J Jeczalik, *J. Polym. Sci.: Polym. Chem. Ed.*, **34**, No.6, 1083-5 (1996).
16 W K Czerwinski, *Macromolecules*, **28**, No.16, 5411-8 (1995).
17 N Baute, C Calberg, P Dubois, C Jerome, R Jerome, L Martinot, M Mertens, P Teyssie, *Macromol. Symp.*, **134**, 157-66 (1998).
18 S Turri, M Scicchitano, G Gianotti, C Tonelli, *Eur. Polym. J.*, **31**, No.12, 1227-33 (1995).
19 E S Matveeva, *Synthetic Metals*, **79**, No.2, 127-39 (1996).
20 Jiaxing Chen, J A Gardella, *Macromolecules*, **31**, No.26, 9328-36 (1998).
21 F Beaume, N Brown, *J. Adhesion*, **47**, No.4, 217-30 (1994).
22 M T Keck, R J Lewarchik, J C Allman, **US Patent 5,688,598**, Morton International, Inc., 1996.
23 Ch Reichardt, *Chimia*, **45**(10), 322-4 (1991).

# 12.2 CHAIN CONFORMATIONS OF POLYSACCHARIDES IN DIFFERENT SOLVENTS

RANIERI URBANI AND ATTILIO CESÀRO
**Department of Biochemistry, Biophysics and Macromolecular Chemistry, University of Trieste, Italy**

## 12.2.1 INTRODUCTION

Carbohydrate monomers and polymers are present in all living organisms and are widely used in industrial applications. In life forms they show and express very diverse biological functions: as structural, storage and energy materials, as specific molecules in the immunochemistry of blood, as important polymers of cell walls determining cell-cell recognition,[1] antigenicity and viral infection, etc..

The variability of primary structure and conformation makes the carbohydrate molecule extremely versatile, for example, for specific recognition signals on the cell surface.[2] In any biological system whatsoever the shape and size adopted by carbohydrates in different solvent environments have been widely demonstrated to be responsible for the biological function of these molecules.

In recent years, significant progress has been made in the improvement of both the experimental and theoretical research tools needed to study the conformational complexity of carbohydrates in solution, such as X-ray and neutron scattering techniques (SAXS and SANS), atomic force microscopy (AFM), high-resolution NMR spectroscopy and relaxation techniques, and computational methods. All the experimental and computational methods unequivocally indicate the relevance of the environment (e.g., solvent composition, pH and salt conditions, temperature) on the topological shape and the properties of the carbohydrate solutes. The general problem of solvent effect on the conformational states and the preferential solvation of oligo- and polysaccharides has been tackled mainly to validate detailed molecular models which were developed for relating the structural characteristics of these macromolecules to their chemical, physical, and biological properties in solution.[3-5] Semi-empirical methods (unrefined in the sense that molecular parameters are not adjusted for the specific case studied) have been proven useful and generally applicable to different chain linkages and monomer composition. However, it must be clear that all these methods, unless specifically stated otherwise, refer to calculations of the unperturbed chain dimension and therefore do not take into account the excluded volume effects (which arise typically from long-range interactions). In most cases, solvation effects (which are short range) are not explicitly taken into consideration, e.g., molecular parameters of the solvent do not enter in the calculations, although some exceptions are found in literature.

The rationale for the correct setting of current knowledge about the shape of polysaccharides in solution is based on three factors: the correlation between primary structure (i.e., the chemical identity of the carbohydrates polymerized in the chain), intrinsic conformational features dictated by the rotational equilibria (often the major contributions are due to the rotation about the glycosidic linkages) and the interaction with the other molecular species in the system (mainly the solvent which determines, therefore, the solubility of the chains).

In this chapter the description of the solvent effect is given within the framework of some specific experimental results and computational methods for studying and predicting oligo- and polysaccharide conformations in solution. It is not the authors' intention to make an in depth investigation into the general methodologies which have been widely reported over recent years (and in this book) but rather to provide a step-wise presentation of some conformational features which have upheld theoretical predictions with experimental observations. The number of examples and approaches is necessarily limited and the choice undoubtedly reflects the authors' preferences. Nonetheless, the aim is to be as informative as possible about the conceptual difficulties and conceivable results.

## 12.2.2 STRUCTURE AND CONFORMATION OF POLYSACCHARIDES IN SOLUTION

### 12.2.2.1 Chemical structure

The primary structure of polysaccharides (glycans) is often complicated by different kinds of linkages in homopolymers and different kinds of monomeric units, which give rise to a huge number of different polymers. Glucans (see a general formula in Figure 12.2.1) are those composed exclusively of glucose, while glucuronans are polymers of glucuronic acid. Similarly mannans and galactans as well as mannuronans and galacturonans, are homopolymers of mannose, galactose, mannuronic acid and galacturonic acid, respectively. Although all the polysaccharides discussed here show a structural regularity, they may not be simple homopolymers. Their chemical structure can sometimes be fairly complicated.

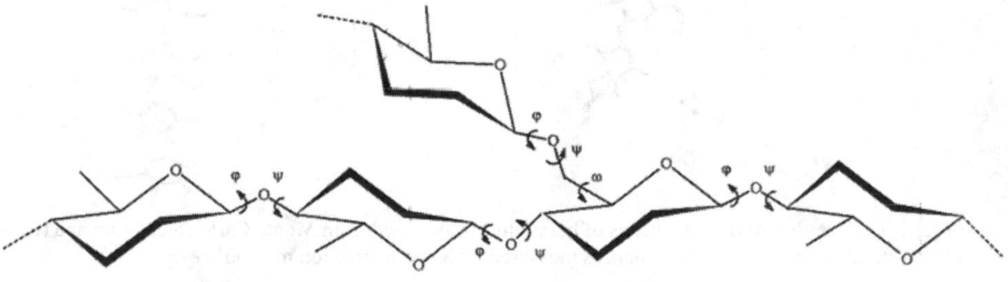

Figure 12.2.1. Example of the chemical structure of a polysaccharidic chain as a sequence of β-(1-4)- linked D-glucose units having a side chain of β-glucose linked (1-6) to the backbone. The glycosidic dihedral angles are also indicated.

### 12.2.2.2 Solution chain conformation

A regularity of primary structure could imply that the chains may assume ordered helical conformations, either of single or multiple strand type, both in the solid state and in solution. A knowledge of both polysaccharidic chain structure, up to the three-dimensional molecular shape, and the interaction of the polymers with other molecular components, is essential, in order to understand their capability to form supramolecular structures, including physical gels, with specific rheological properties, which have important implications for controlling and upgrading properties in industrial applications. The rationale is that the physico-chemi-

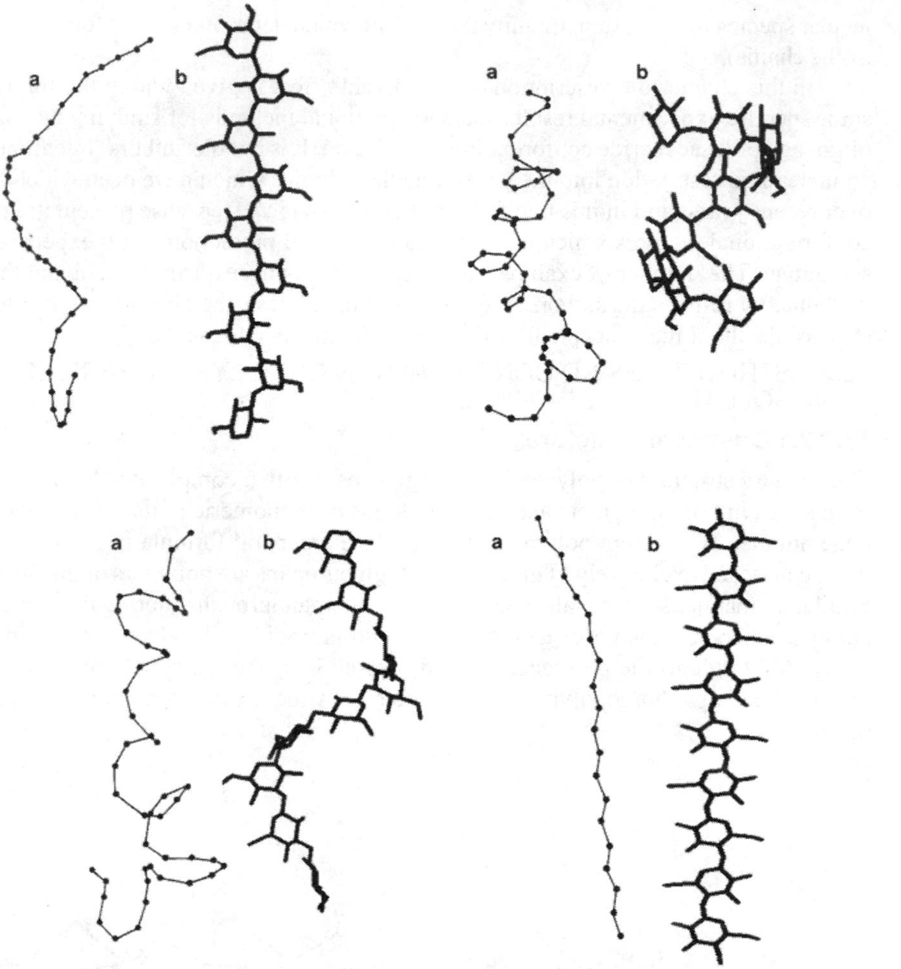


Figure 12.2.2. (a) Snapshot of random chains of homoglucans obtained from Monte Carlo calculation and (b) ordered helical structures of the same polymers as measured by X-ray diffraction method.

cal properties, which form the basis of the industrial applications of polysaccharides are directly related to the structure/conformation of the chain.

Many biological macromolecules in the solid state assume regular helices, which can be represented by means of a few geometrical parameters and symmetry relations. These helical structures originate from the stereo regularity of backbone monomers and are easily described, conceptually simple, and therefore used (all too often) as an idealized model for all the actual shapes. Elements of helical regularity are essential in the description of the structure of nucleic acids and polypeptides. Some biopolymers, e.g., globular proteins, almost completely preserve their structural regularity in solution. Such globular structures are not known in polysaccharides. Nonetheless, helical conformations have been proposed to represent the structure of many polysaccharides, microbial glycans in particular. An excellent overview of the stereo-regular helical conformations of polysaccharides deduced from X-ray fiber diffraction studies has recently been published.[6]

However, thermodynamic arguments suggest that a partially disordered state is an essential prerequisite for the stability of polymeric systems in solution. Under these circumstances, realistic chain pictures of polysaccharides will not necessarily be generated from the repetition of a single conformational state, which is usually identified by the minimum energy state found in the internal conformational energy calculations. Statistical approaches to these conformational energy surfaces[7,8] suggest a more disordered solution conformation (Figure 12.2.2, snapshots marked with a) than the chain structures proposed so as to fit the helical regularity deduced from x-ray fiber diffraction studies (Figure 12.2.2, structures marked with b). Thermal fluctuations are in general sufficient to generate delocalized disorder, unless diffuse interactions, cooperative in nature, ensure long-range order.

From the experimental point of view, several polysaccharides with different chain linkage and anomeric configuration have been studied to determine to what extent the polymeric linkage structure and the nature of the monomeric unit are responsible for the preferred solvation and for the chain topology and dimensions.[9] Conversely, since it is generally understood that the structure and topology of many macromolecules are affected by solvation, theoretical models must include these solvent effects in addition to the internal flexibility, in order to estimate changes in the accessible conformations as a result of the presence of the solvent molecules.

The concept of chain conformational disordering and dynamics in solution is associated with the existence of a multiplicity of different conformations with accessible energy and, moreover, exhibiting their topological differentiation. The above conformational variability of polymeric chains is implicitly recognizable by the great difficulty in crystallization and by the typical phenomenon of polymorphism. This discussion is relevant in particular to ionic polysaccharides (see below) because, from the polyelectrolytic point of view, the transition from a more compact conformation (also an "ordered chain") to an extended coil conformation, is usually associated with a net variation in ionic charge density along the chain.

Polysaccharides generally dissolve only in strongly solvating media. Water displays a complicated behavior: it is a good solvent for monomers and oligosaccharides inasmuch as it is able to compete with the specific inter- and intra-molecular hydrogen bond network (Figure 12.2.3). In many cases it is the thermodynamic stability of the solid-state form which protects the solute molecules from being solubilized.[10] Nevertheless, some other strong solvents, such as, dimethylsulfoxide, DMSO, and 1,4-dioxane are known to be good solvents for carbohydrate polymers. Many commercial applications of polysaccharides require compatibility with different solvents and solutes (organic solvents, salts, emulsifiers, plasticizers, enzymes, etc.), for example, in pharmaceutical matrices, paints and foods. In this field, solvent compatibility of some glycans has been improved and controlled by functionalization and derivatization in order to obtain a proper degree of substitution, which determines a wide range of compatibility properties.

At the molecular level, various specific and non-specific solvent-solute interactions may occur in polysaccharide solutions that may result in a change in the conformational shape, solubility, viscosity and other hydrodynamic and thermodynamic properties. Hydrophilic interactions such as hydrogen bonding and electrostatic interactions are believed to be factors that influence the conformation of polysaccharides in solution, although the question is being raised (more and more) as to the implication of patches of hydrophobic intermolecular interactions, especially for chain aggregations. One important feature is the

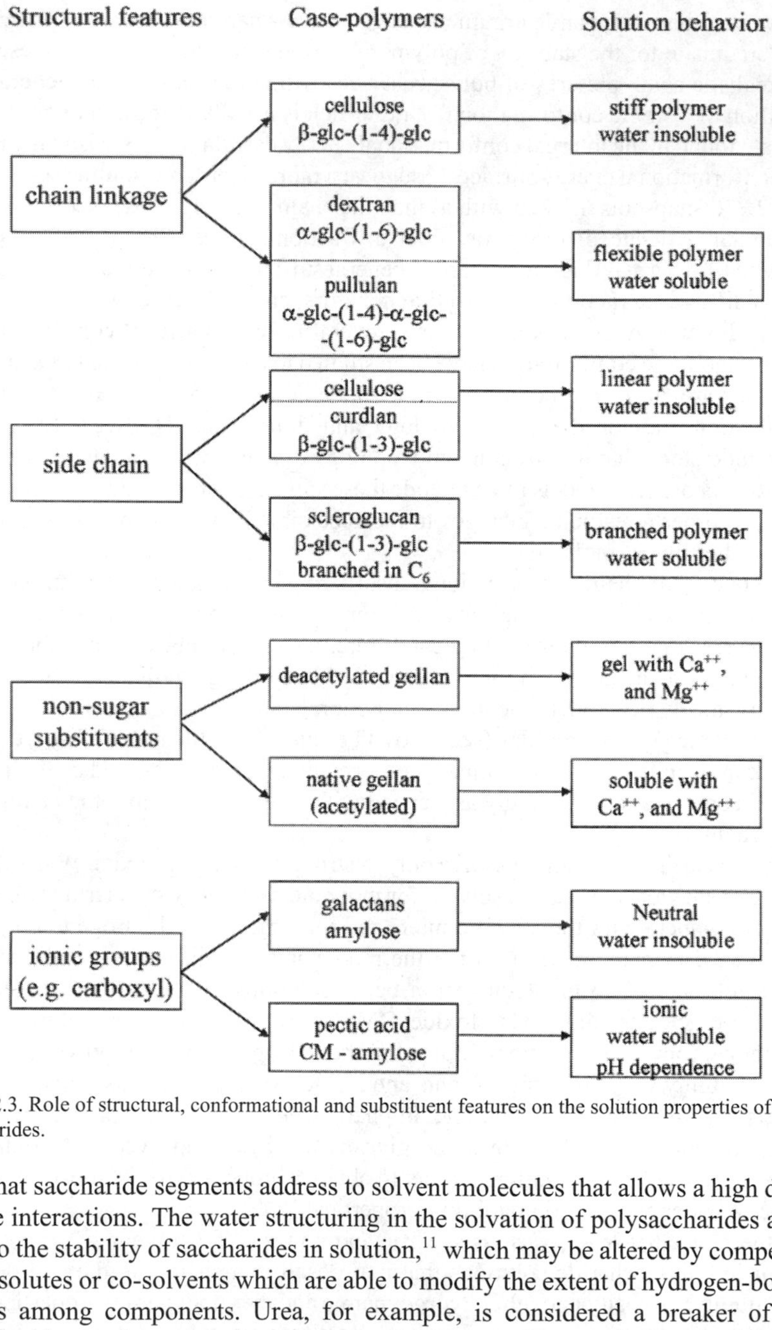


Figure 12.2.3. Role of structural, conformational and substituent features on the solution properties of oligo- and polysaccharides.

surface that saccharide segments address to solvent molecules that allows a high degree of favorable interactions. The water structuring in the solvation of polysaccharides also contributes to the stability of saccharides in solution,[11] which may be altered by competition of other co-solutes or co-solvents which are able to modify the extent of hydrogen-bonded interactions among components. Urea, for example, is considered a breaker of ordered polysaccharide conformation in gel networks, such as those obtained with agar and carrageenan as well as with microbial polysaccharides.

Over the last twenty years or so, several approaches have been made to determine the three-dimensional structure of oligosaccharides and theoretical calculations are becoming an increasingly important tool for understanding the structure and solution behavior of saccharides.[11-15] Although computer facilities and calculation speeds have grown exponentially over the years some of these techniques, like quantum mechanic methods (ab initio methods), have been shown not to be useful in dealing with the complexity of systems which involve many atoms such as in macromolecular systems. On the other hand, these techniques have been successfully applied to small molecules, e.g., mono- and disaccharides, in predicting charge distribution on the atoms, conformations and transitions among accessible conformations,[16] thus providing a background knowledge for the more complicated systems.

### 12.2.3 EXPERIMENTAL EVIDENCE OF SOLVENT EFFECT ON OLIGOSACCHARIDE CONFORMATIONAL EQUILIBRIA

The problem of sugar conformation and dynamics in solution is related to the question of to what extent are oligo- and polysaccharides intrinsically flexible under the different experimental conditions. The answer to this question, which requires a complete knowledge of the time-space dependence of the chain topology,[17] is often "rounded-off" by the use of empirical terms like "flexibility". A further problem is to what extent does the solvent contribute to stabilizing some conformational states rather than others. Solution properties are functions of the distribution of conformations of the molecules in the solvated states, in the sense that the experimental data are statistical thermodynamic averages of the properties over all the accessible conformational states of the molecule, taking each state with a proper statistical weight. This aspect can be better illustrated by taking into consideration the accessible conformational states of a simple sugar unit and the conformational perturbation arising from the changes in the interactions with the surrounding solvent medium.

For an α-pyranose ring, for example, the rotation about carbon-carbon bonds and the fluctuations of all ring torsional angles give rise to a great number of possible conformers with different energies (or probabilities). Some of these are identified as the preferred rotational isomeric states in various environments. The boat and boat-skew conformers of a pyranose ring, which are higher in energy than the preferred chair form, correspond to a major departure from the lowest energy chair conformation as illustrated by the globular conformational surface of Figure 12.2.4. Additional conformational mobility in a monosaccharide is due to the rotations of exocyclic groups, namely, OH and especially $CH_2OH$. Let us point out that in a polymer these ring deformations do not normally occur, but, when they do, they may determine to a great extent the equilibrium mean properties and the overall mean chain dimensions.[18,19] However, one very recent theory is that the elastic properties of amorphous polysaccharides are related to the glycosidic ring deformation.[20]

D-ribose is probably the best example of sugar that reaches a complex conformational equilibrium, giving rise to the mixture composition shown in Figure 12.2.5. The percentage of each form is taken from Angyal's data[21,22] with integration of the sub-splitting between the $^4C_1$ and the $^1C_4$ forms. The prime purpose of this analysis is to point out that the stability of each conformer is not determined solely by the intrinsic internal energy, which can be evaluated by means of e.g., ab-initio quantum mechanics calculations, but is strongly influenced by all the solvation contributions. Therefore, the conformer population may be shifted by changing temperature, solvent composition, or any other external variable such as, for example, adding divalent cations (see Table 12.2.1). The evaluation of the actual concentration of the several conformers involved in the equilibrium immediately leads to

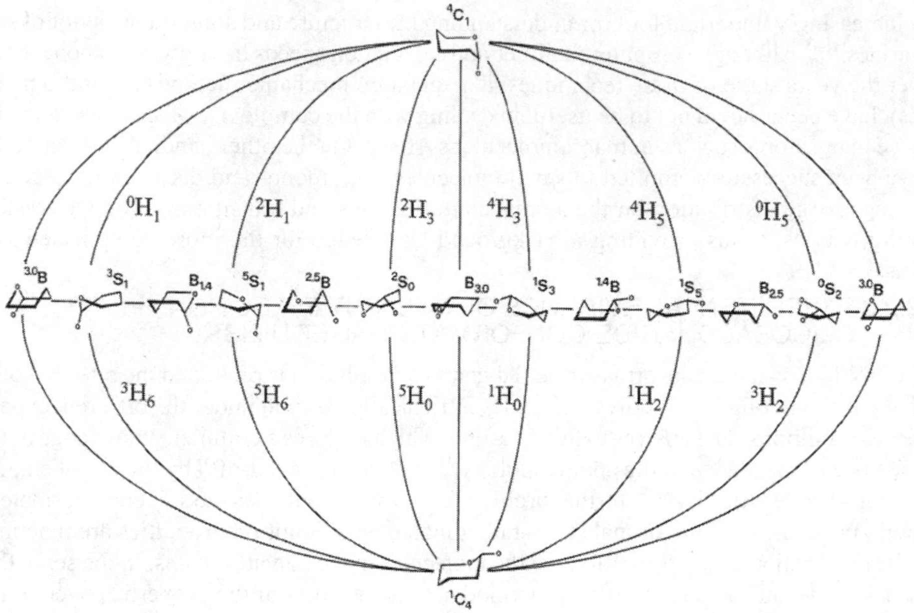


Figure 12.2.4. Representation of all possible conformations of the pyranosidic ring and respective inter-conversion paths.

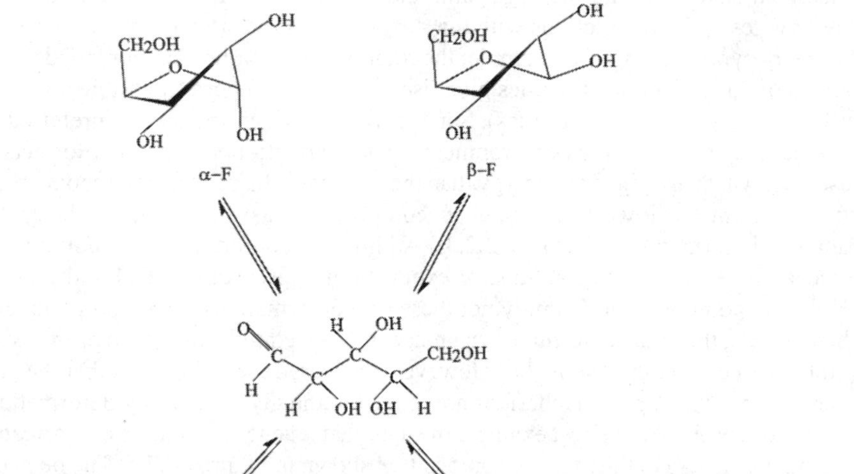

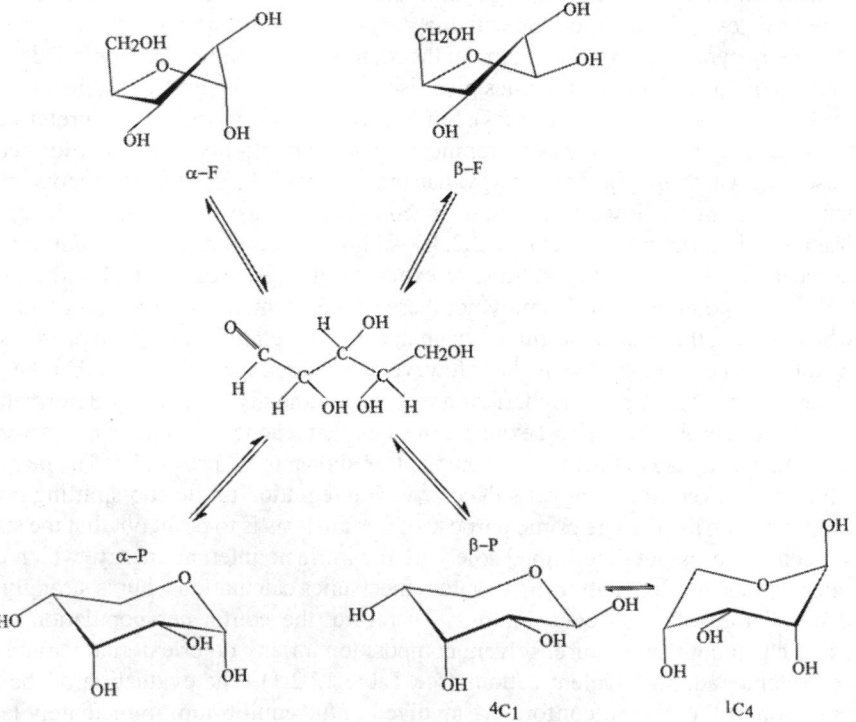


Figure 12.2.5. D-ribose conformational equilibria.

the partition function of the system (i.e., the free-energy change or the equilibrium constant for a transition between conformers).

**Table 12.2.1. Percentage composition of furanose and pyranose forms of D-ribose and glucose in different solvent systems**

| Sugar | Solvent | α-F | β-F | α-P | ($^4C_1$) β-P | ($^1C_4$) β-P |
|---|---|---|---|---|---|---|
| D-ribose | $H_2O$, 31°C | 13.5 | 6.5 | 21.5 | 44.0 | 14.5 |
| | $D_2O$, 30°C | 8.0 | 14.0 | 23.0 | 41.0 | 14.0 |
| | DMSO, 30°C | 6.0 | 22.0 | 16.2 | 31.0 | 24.8 |
| | $Ca^{2+}$, 1.27 M | 13.0 | 5.0 | 40.0 | 14.0 | 28.0 |
| Glucose | DMSO, 17°C | | | 45.0 | 55.0 | |
| | pyridine, 25°C | 0.6 | 1.0 | 45.0 | 53.0 | |

Passing from monomer units to oligomers (disaccharides and higher oligosaccharides), the dominant features of molecular flexibility become those due to rotations about the glycosidic linkages. Although other conformational fluctuations may contribute to the local dynamics of the atoms or group of atoms, only the glycosidic linkage rotations are able to dramatically change the conformational topology of oligomers at ambient temperature. The aim of a conformational analysis of oligosaccharides is thus to evaluate the probability (that is the energy) of all mutual orientations of the two monosaccharidic units, as a function of rotations about the glycosidic linkages, defined by the dihedral angles:

$$\varphi = [H1\text{-}C1\text{-}O1\text{-}Cm]$$

$$\psi = [C1\text{-}O1\text{-}Cm\text{-}Hm]$$

where m is the aglycon carbon number in the reducing ring. The important region of φ and ψ rotations is the one with energy variations in the order of kT, the thermal motion energy, because this may produce a large ensemble of accessible conformational states for the oligosaccharide. Even when the rotational motion is restricted to only a few angles, the fluctuations of many such glycosidic bonds is amplified along the chain backbone, as the molecular weight increases. The accumulation of even limited local rotations may produce very large topological variations in the case of polymeric chains and consequently relevant changes in thermodynamic, hydrodynamic and rheological properties of these systems. Other internal motions often make only small contributions to the observable properties on the macromolecular scale.[23]

Experimentally, NMR techniques are among the most valuable tools for studying conformations and dynamics of oligosaccharides in solution by determining chemical shifts, coupling constants, NOE's and relaxation time.[24-30] In the study of saccharide conformations the potential of coupling constants evaluation, specially the (hetero-nuclear) carbon-proton spin-spin coupling constants, $^nJ_{C,H}$, and their dependence on solvent medium is well known. Several empirical correlations between $^1J_{C,H}$ and structural parameters like dihedral angles have been reported,[31] although the low natural abundance of the $^{13}C$ isotope often made the measurements technically difficult and $^{13}C$ enrichment was required for the generation of data of sufficient quality for the quantitative analysis. Nowadays, due to the progress of FT-NMR spectrometers, measurements of the three-bond (vicinal) carbon-proton coupling constant, $^3J_{C,H}$ have indeed been made possible on the natural-abundance $^{13}C$ spectra. Much endeavor has been invested in establishing a general Karplus-like relation for the angular dependence of $^nJ_{C,H}$, especially on the dihedral angles in the glycosidic region because the overall three-dimensional structure of oligo- and polysaccharides are related to the glycosidic features in terms of dihedral (rotational) angles.

From comparison of experimental data and theoretical results (using quantum-mechanical and semi-empirical methods) on model compounds,[32-34] the angular dependence of the coupling constants, $^nJ_{C,H}$, on the anomeric and aglycon torsional angles φ and ψ,[31,34] and on the dielectric constant of the solvent[35] can be written in a general form such as:

$$^1J_{C,H} = A_1 \cos 2\chi + A_2 \cos\chi + A_3 \sin 2\chi + A_4 \sin\chi + A_5 + A_6\varepsilon$$

$$^3J_{C,H} = B_1 \cos^2\chi + B_2 \cos\chi + B_3 \sin\chi + B_4 \sin 2\chi + B_5 \qquad [12.2.1]$$

where:

| | |
|---|---|
| $A_i$, $B_i$ | constants different for α and β anomers |
| χ | φ or ψ |
| ε | solvent dielectric constant |

Experimental $^nJ_{C,H}$ values for conformationally rigid carbohydrate derivatives allow to calculate the constant values of $A_i$ and $B_i$ in equations [12.2.1]. The major practical use of these equations is their ability to estimate the glycosidic dihedral angles from experimental $^nJ_{C,H}$ data, in combination with other complementary results, for example, from NOESY,[36] X-ray and chiro-optical experiments. Since the experimental values are averaged over all the accessible conformational states in solution, they do not necessary reflect the property of only the most probable conformer[23] but they nevertheless include contributions of all the conformers, each one taken with its proper statistical weight. Thus, the quantitative interpretation of experimental data in terms of accessible conformational states of flexible molecules requires the additional theoretical evaluation of the energy of the molecule as a function of internal coordinates. Since the dependence of the observed coupling constants on the conformation is non-linear, it derives that:

$$\langle {}^nJ_{C,H}(\chi)\rangle \neq {}^nJ_{C,H}(\langle\chi\rangle)$$

Only for a linear dependence of $^nJ_{C,H}$ on χ, the equation could hold the equals sign. Because of the simultaneous dependence of the property on χ and on the complexity of potential energy function, E(χ), the ensemble average of $\langle ^nJ_{C,H}\rangle$, as well as of any property of interest, can be calculated only by taking into consideration the conformational energy surface.

Apart from NMR and the methods suitable for characterization of overall chain dimensions (which we will touch on below), there are not many simple experimental techniques that can be used to study sugar conformation and that are directly correlated to theoretical results based on calculated potential energy surfaces or force fields. An exception is given by chiro-optical techniques, which provide important (although empirical) structural information as optical rotation experiments have been shown to be very useful and informative in the study of saccharides in solution.[37,38] Literature gives the experimental evidence for the effects of the external conditions (solvent and temperature) on the optical rotation. The additivity methods, proposed in the Fifties by Whiffen[39] and Brewster,[40] were extended by Rees and co-workers,[37,41] who derived expressions for the contribution to the optical activity of changes in conformational states of oligo- and polysaccharide chains. More recently, Stevens and co-workers[42,43] developed a computational model for optical activity and circular dichroism based on the Kirkwood theory in which the calculations of the lowest energy component of a molecular $\sigma-\sigma^*$ transition, which derives from the mutual interaction of all $\sigma-\sigma^*$ transitions on C-C, C-H and C-O bonds, are carried out. The calculated optical properties are therefore geometry-dependent and the optical properties have been calculated as statistically averaged properties over an ensemble of all possible conformations, theoretically obtained as a function of $\varphi$ and $\psi$. This model also provides a useful tool for testing the quality of force field parametrizations and the possibility of some refinement of force constants for a non-explicit inclusion of solvent environment.

## 12.2.4 THEORETICAL EVALUATION OF SOLVENT EFFECT ON CONFORMATIONAL EQUILIBRIA OF SUGARS

### 12.2.4.1 Classical molecular mechanics methods

Theoretical approaches to the conformational analysis of oligosaccharides in solution become inherently more complicated than those for monomeric sugars, at least for the computer time required to optimize the structures. Depending on the degrees of freedom taken into consideration and on the level of sophistication of the method used, the conformational analysis may give different results. It has been the custom in the past to compute the conformational energy as a function of the dihedral angles $\varphi$ and $\psi$ only, by assuming the sugar ring to be rigid (rigid-residue method). More recently, all the internal coordinates are allowed to adjust at each increment of $\varphi$ and $\psi$, relaxing the structure toward a local minimum (relaxed-residue method). The rigid-residue approach is still considered suitable as a starting point, although some warning must be alerted to the use of the same set of structural coordinates for all sugars, as was initially done. The approach may however be useful as a starting point in the conformational analysis of polysaccharides, provided that the coordinates of all the atoms in the monomeric unit have been calculated by a suitable independent method.[13, 44]

Figure 12.2.6 reports, as a general example, the rigid-residue energy surfaces of two representative disaccharides, namely α-(1-4)-D-glucopyranosil-D-glucopyranose (D-maltose, Figure 12.2.6a) and the β-(1-4)-D-glucopyranosil-D-glucopyranose (D-cellobiose, Figure 12.26b). Both maps present multiple minima separated by barriers that are only a few kilocalories high. For many disaccharides the barriers between minima can be very high and they are often overestimated in the rigid-residue approximation.

In general, the conformational energy map shows that only a limited portion of the total $\varphi$, $\psi$ conformation space is actually accessible to the dissaccharide at room temperature. The dissaccharide is not frozen into its lowest energy conformation, nevertheless the steep walls of the allowed region of conformational energy may dramatically limit the multiplicity of thermally accessible conformational states.

One useful method, for taking into account the effect of solvent media upon the conformational properties of glycosidic structures, is the continuum reaction field method.[45, 46] This method is based on the Scaled Particle Theory (SPT) equations[47] and on Onsager's theory of the reaction field, as applied by Abraham[48] by considering the solvent as a dielectric continuum. In this approach the total conformational energy, $G_{tot}$, is given by:

$$G_{tot} = G_{conf} + G_{solv}$$

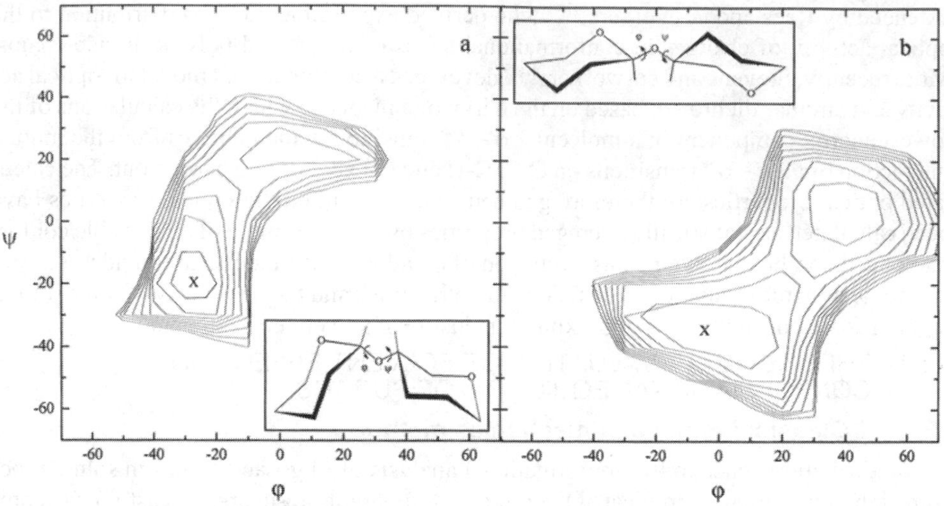


Figure 12.2.6. Conformational energy maps calculated for (a) D-maltose and (b) D-cellobiose. Contours are in kcal·mol$^{-1}$ with increments above the minimum (x) of 0.5 kcal·mol$^{-1}$.

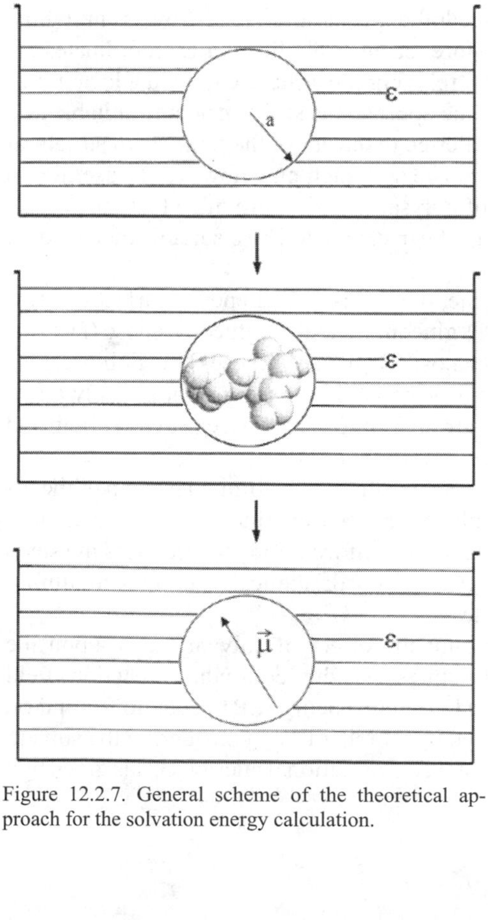


Figure 12.2.7. General scheme of the theoretical approach for the solvation energy calculation.

where the total conformational free energy, $G_{tot}$, is given by the sum of the contribution due to the in vacuo conformational state energy, $G_{conf}$, and the solvation contribution, $G_{solv}$. The latter term is a sum of contributions due to the energy required to create, in a given solvent, a cavity of suitable size to accommodate the solute molecule in a given conformational state and the interaction energy between the solute and the surrounding solvent molecules, as schematically shown in Figure 12.2.7:[49]

$$G_{solv} = G_{cav} + G_{el} + G_{disp} \quad [12.2.2]$$

The expression for $G_{cav}$, that is the free energy required for the formation of a cavity (first step in Figure 12.2.7), at a temperature T and a pressure P, is taken from the Scaled Particle Theory (SPT) which has been successfully applied in the study of thermodynamic properties of aqueous and non-aqueous solutions:[47,50]

$$\frac{G_{cav}}{RT} = -\ln(1-y) + \left(\frac{3y}{1-y}\right)\hat{R} + \left[\frac{3y}{1-y} + \frac{9}{2}\left(\frac{y}{1-y}\right)^2\right]\hat{R}^2 + \frac{yP}{\rho kT}\hat{R}^3 \quad [12.2.3]$$

where:

| | |
|---|---|
| y | $= 4\pi\rho a_v^3 / 3$, the reduced number density |
| $a_v$, $a_u$ | radii of hard sphere equivalent solute (u) and solvent (v) molecules |
| $\hat{R}$ | $= a_u / a_v$ |
| ρ | solvent density |
| R | gas constant |
| k | Boltzmann constant |

It is noteworthy that $G_{cav}$ is a function solely of solvent density and solvent and solute dimensions and represents a measure of the cohesive forces among solvent molecules. In a given solvent, this contribution is a function of the radius of solute molecule which may differ remarkably as a function of dihedral angles φ and ψ only.

The electrostatic interaction term $G_{el}$ between solute and solvent is based on the continuum reaction field[48] which takes into account a reaction potential induced by the solute dipole and quadrupole (third step in Figure 12.2.7) in a continuum medium of dielectric constant ε:

$$G_{el} = \frac{KX}{1-lX} + \frac{3HX}{5-X} bF\left[1-\exp\left(-\frac{bF}{16RT}\right)\right] \quad [12.2.4]$$

with:

$$K = \frac{\mu_u^2}{a^3}, \quad H = \frac{Q_u^2}{a^5}, \quad l = \frac{2(n_u^2-1)}{n_u^2+2}, \quad X = \frac{\varepsilon-1}{2\varepsilon+1}$$

$$F = \begin{cases} 0 \quad for \quad \varepsilon \le 2 \\ \left[\frac{(\varepsilon-2)(\varepsilon+1)}{\varepsilon}\right]^{1/2} \quad for \quad \varepsilon > 2 \end{cases}$$

$$b = 4.35\left(\frac{T}{300}\right)^{1/2}\left(a^{3/2}r_{uv}^3\right)^3\left(\frac{K+Ha^2}{r_{uv}^2}\right)^{1/2}$$

where:

| | |
|---|---|
| $\mu_u$ | solute dipole moment |
| $Q_u$ | solute quadrupole moment |
| $n_u$ | solute refractive index |
| a | $= a_v + a_u$ |
| $r_{uv}$ | $= a/2^{1/2}$ |

$r_{uv}$ is the average distance between the solvent and the solute molecule defined in terms of the radius of cavity a.

The free energy of dispersion, $G_{disp}$, in equation [12.2.2] takes into account both attractive and repulsive non-bonding interactions and is expressed as a combination of the London dispersion equation and Born-type repulsion:[51]

$$G_{disp} = -0.327 N_v \alpha_u \alpha_v \frac{I_u I_v}{I_u + I_v} r_{uv}^{-6} \qquad [12.2.5]$$

where:

| | |
|---|---|
| $\alpha$ | molecular polarizability |
| I | ionization potential |
| $N_v$ | nearest-neighbor solvent molecules |

$N_v$ is the number of molecules surrounding the solute molecule in a given conformation and is calculated from the following equation:[52]

$$N_V = \left[ \left( a_u + 2a_V \right)^3 - a_u^3 \right] \left( \frac{4\pi N_A}{3V_v} \right)^3$$

where:

| | |
|---|---|
| $V_v$ | solvent molar volume |
| $N_A$ | Avogadro number |

In order to elucidate the effect of different solvents on the conformation, the energy differences between conformers in a given solvent are more relevant than the absolute solvation energies in each solvent. In particular, it is important that the perturbation effect on the detailed shape in the low energy regions of the conformational map. Figure 12.2.8 shows the free-energy of solvation and energy contributions for the maltose dimer as a function of $\psi$ calculated at $\varphi$=-30° and refers to the energy of the (-30°, 180°) conformer for two solvents, water and DMSO (Figure 12.2.8 a and b, respectively). In this section, the cavity term is a complex function of the size of the maltose molecule, as determined by the spatial orientation of the two glucose residues. The cavity energy is, by definition, always unfavorable; the more expanded the conformers, which are usually located in the low energy region of the map (around $\psi$=0° for maltose), the larger the $G_{cav}$ due to the size of larger cavities to be created in the solvent.

The electrostatic free-energy, $G_{el}$, increases with the dipole and quadrupole moments of solute molecule and decreases with the radius of the cavity. On the other hand, the dipole moment $\mu$ is a function of dihedrals $\varphi$ and $\psi$ and for maltose has two large maxima around ($\varphi$, $\psi$)=(-40°,0°) and (180°, 140°) which is in agreement with the major contributions of $G_{el}$ shown in Figure 12.2.8. The electrostatic free-energy (eq. [12.2.4]) is also dependent on the solvent dielectric constant and increases in passing from DMSO ($\varepsilon$=46.68 debye) to water ($\varepsilon$=78.30 debye) at 25°C. The dispersion term (eq. [12.2.5]) makes a significant contribution (-20 to -70 $kJ \cdot mol^{-1}$) to the absolute value of solvation free-energy but the angular dependence is very small giving an almost equal contribution to the energy of conformers.

In an early paper[53] it was shown that, in the comparative cases of cellobiose and maltose, the probability distribution of conformers (Figure 12.2.9) is affected by the presence of the solvent, changing the shape of the function from one solvent to another. The minimum of the maltose map goes from (-20°,-30°) to (-10°,-20°) in water and DMSO, while that of cellobiose map goes from (0°, 50°) to (-30°, -20°) in both solvents.[53] These solvent perturbations (apparently small) on the conformational energies have a great effect on the proba-

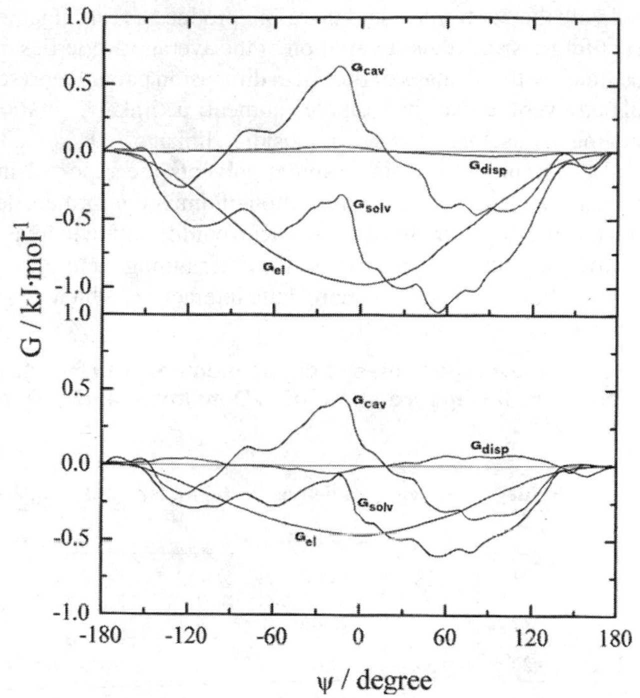


Figure 12.2.8. Section of the maltose energy map at φ=-30° showing the dependence of contributions to the $G_{solv}$ as a function of angle ψ in water (a) and DMSO (b).

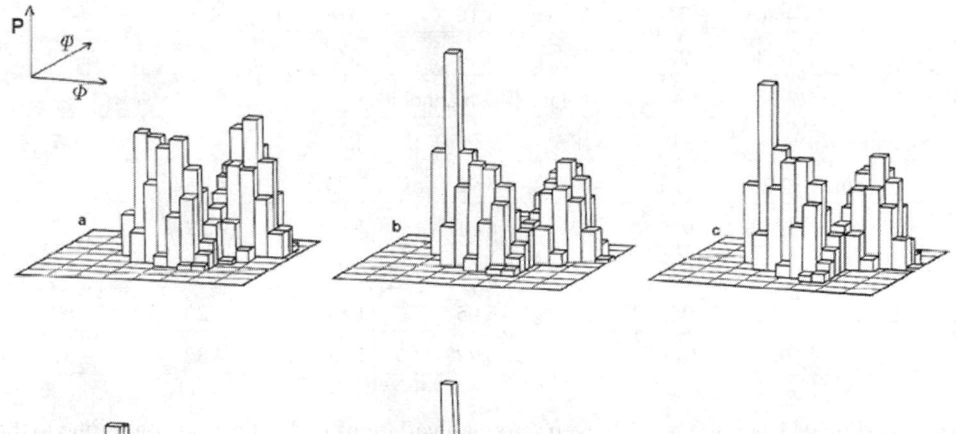

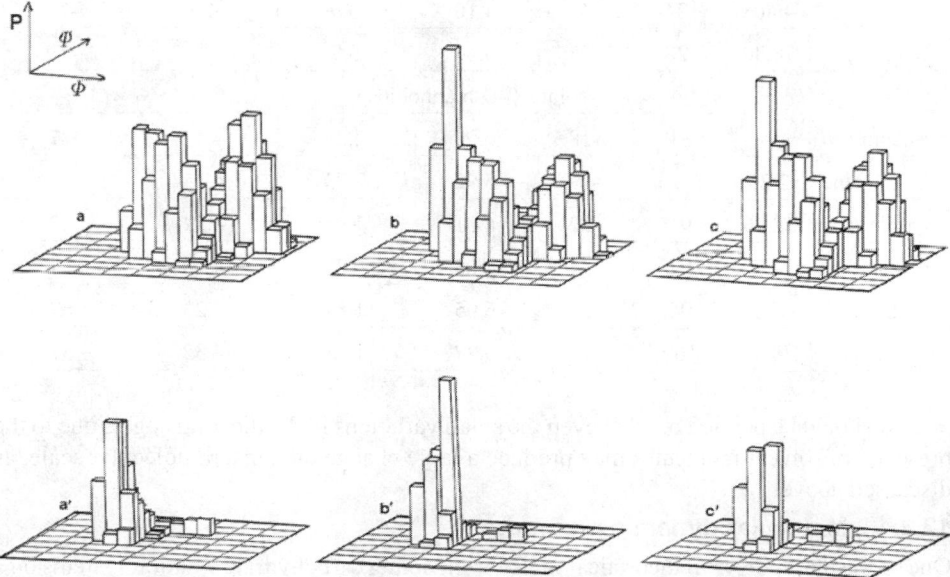


Figure 12.2.9. Histograms of probabilities of conformational states for cellobiose in vacuo (a), water (b) and DMSO (c), and for maltose (a', b', and c').

bilities associated with the conformers and therefore produce a rearrangement of statistical weights among conformers and, thus, a variation of the average properties of the system, as for example illustrated by the changes of the chain dimension parameters (see below Figure 12.2.11). Calculated average data like dipole moment, μ, linkage rotation, Λ, and proton-carbon coupling constants across glycosidic linkage, $^3J_{C1\text{-}H4}$ and $^3J_{C4'\text{-}H1}$, for β-D-maltose[14] and β-D-mannobiose[45] in different solvents are reported in Table 12.2.2. These results show that the equilibrium composition of dimer conformers depends strongly on the solvent and that the departure from the in vacuo conformation increases with increasing solvent dielectric constant. For mannobiose the determining factor of the solvent effect on the conformation is the intra-residue electrostatic interaction, which depends on φ and ψ in the same manner as the dipole moment.

**Table 12.2.2. Calculated average values of dipole moment, four-bonds proton-carbon coupling constant and linkage rotation of β-D-maltose and β-D-mannobiose in different solvents at 25°C**

| Solvent | ε | φ, deg | ψ, deg | μ, debye | $^3J_{C1\text{-}H4}$, Hz | $^3J_{C4'\text{-}H1}$, Hz | Λ, deg |
|---|---|---|---|---|---|---|---|
| solute: β-D-maltose | | | | | | | |
| vacuum | - | -21 | -28 | 3.80 | 4.2 | 4.5 | -31 |
| 1,4-dioxane | 2.21 | -22 | -22 | 3.89 | 4.1 | 4.5 | -30 |
| pyridine | 12.40 | -24 | -39 | 4.02 | 4.0 | 4.5 | -27 |
| ethanol | 24.55 | -24 | -41 | 4.09 | 3.9 | 4.5 | -27 |
| methanol | 32.70 | -25 | -46 | 4.17 | 3.9 | 4.5 | -25 |
| DMSO | 46.68 | -24 | -41 | 4.10 | 3.9 | 4.5 | -28 |
| water | 78.30 | -28 | -65 | 4.47 | 3.7 | 4.7 | -19 |
| solute: β-D-mannobiose | | | | | | | |
| vacuum | - | 81 | -18 | 6.00 | 2.29 | 4.14 | 77 |
| 1,4-dioxane | 2.21 | 82 | -19 | 5.92 | 2.23 | 4.15 | 76 |
| pyridine | 12.40 | 91 | -22 | 6.01 | 1.95 | 4.22 | 69 |
| methanol | 32.70 | 97 | -24 | 6.08 | 1.72 | 4.25 | 54 |
| DMSO | 46.68 | 94 | 23 | 6.05 | 1.84 | 4.23 | 67 |
| water | 78.30 | 100 | -18 | 6.27 | 1.53 | 4.32 | 57 |

It should be pointed out that even the small variations in the dihedral angles due to the presence of solvent molecules may produce a large change on a macromolecular scale, as discussed above.

### 12.2.4.2 Molecular dynamic methods

One of the most powerful theoretical tools for modeling carbohydrate solution systems on a microscopic scale and evaluating the degree of flexibility of these molecules is the molecular dynamics technique (MD) which has become popular over the last two decades. The first reported works of MD carbohydrate simulation appeared in 1986[54,55] and since then an increasing number of MD simulations on sugars has been carried out.[11,13,56,57] The explicit representation of solvent molecules is required especially in biological systems, where solvent structuring plays an important role. Starting with an appropriate potential energy function for sugar-water interactions,[13,58] the common solution simulations are carried out by placing the solute molecule in the center of a cubic box of finite dimensions containing a given number of solute molecules which usually corresponds to at least three solvation shells.[11] The macroscopic system is then simulated by using the approximation known as "periodic boundary conditions",[59] in which the entire box is replicated in every direction, leaving solute and solvent molecules to interact with each other both in central and in replica boxes and setting the long-range interactions to smoothly decrease to zero by using the appropriate switching functions. A pair distribution function, g(r), defined as:

$$g(r) = \frac{1}{4\pi\rho r^2}\frac{dN(r)}{dr}$$

where:

| | |
|---|---|
| r | interatomic distance |
| ρ | the bulk number density |
| N(r) | number of atoms of given type at distance r |

has been used to evaluate the normalized probability of finding a water oxygen atom at a distance r from a given atom on the carbohydrate molecule.[11] In this way the anisotropic distribution of solvent molecules around carbohydrate solutes was reported[11,60,61] showing an exceptional structuring of water molecules which extends to greater distances around the solute as compared to the pure solvent. One of the most interesting results of these simulations is the identification of the spatial distribution of water molecules on the van der Waals' surface of the carbohydrate molecules. Figure 12.2.10 shows a probability density excess of the water molecules in a channel, which is effectively a bisector of the two closest sites for hydrogen bonding.

When the energy maps have been obtained by molecular mechanics calculations, dynamics simulations of disaccharides in various conformations are carried out to analyze the

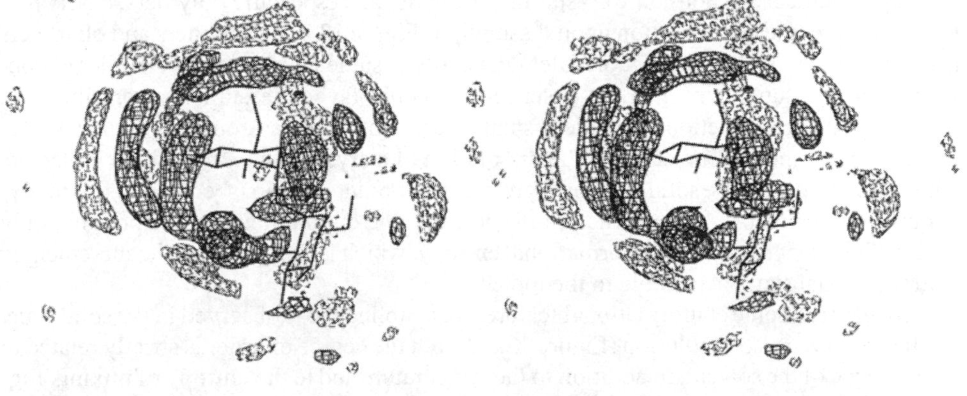

Figure 12.2.10. Contours of solvent anisotropic density around the α,α-trehalose disaccharide obtained from MD simulation [unpublished figure kindly provided by J.W. Brady and Q. Liu].

typical motions of the molecules along the conformational space at moderate temperatures. In general, initial conformations in the simulations are taken from one of those of minimized geometry as identified in the calculation of the energy map. Trajectories are then computed by assigning to the atoms velocity components randomly selected from a thermal distribution at a given temperature. By superimposing the trajectories of fluctuations of (φ, ψ) on the energy map, it is possible to observe a variety of motions, involving both the structures of the molecular rings and in some cases rotations about the glycosidic bonds and exocyclic torsions.[62] In the case of multiple minima, conformational transitions between low energy regions may also be revealed.

### 12.2.5 SOLVENT EFFECT ON CHAIN DIMENSIONS AND CONFORMATIONS OF POLYSACCHARIDES

In addition to the change of entropy, minimization of free energy in a binary dilute solution containing solvent and polymer occurs as a result of the favorable interactions between the chain segments and the solvent, which replace the homotactic interactions between solvent molecules and those between chain segments. In a good solvent and a very dilute solution, it is likely that only solute-solvent and solvent-solvent interactions prevail. However, in a bad solvent (and, in general, in a more concentrated solution) persistence of segmental interactions among chains is the major contribution in the macroscopic properties of the system. The theoretical evaluation of the entropic contributions arising from the configurational nature of the chain molecule is possible on the basis of thermodynamic-statistical models.[63] More troublesome is the contribution of enthalpy change of mixing, $\Delta_{mix}H$, of non-ionic polysaccharides and water which cannot be predicted, not even in sign. Furthermore, the scarcity of literature data does not allow any empirical rationalization although, in most cases, contributions significantly smaller than the related monomers are reported.

Notwithstanding the above limitations, a general picture can be drawn showing that the average dimension of a chain, experimentally obtained with light scattering or viscometric measurements, depends upon solvent interactions and behavior, in addition to the intrinsic features of the polysaccharide (chemical nature and linkage of monomers, conformational equilibria, etc.). The basic axiom is that solution properties are strictly related to the conformation of the molecules in the solvated state and this state, in turn, is only statistically defined from the primary (chemical) structure. In first instance, one can generalize the statement that non-ionic crystalline molecules and especially polymers, barely preserve their ordered conformation upon dissolution (Figure 12.2.2). Polymers and oligomers (including most carbohydrate molecules) generally assume a statistically disordered conformation in solution since, in the absence of specific favorable enthalpy contributions, polymer-solvent interactions provide a small increment in the entropy of mixing. Under these very common circumstances, the dissolution of a crystalline carbohydrate molecule, which is stabilized in the solid state by a great number of interactions, becomes a thermodynamically unfavorable event. Structurally speaking, the dissolution is made possible only by a sufficient increment in conformational entropy, which is described by the increment in conformational states accessible to the molecule.

In general, some conformational features, resembling those observed in the solid state, can be preserved also in solution (Figure 12.2.2), but the degree of order is strictly related to the presence of the solvent, in addition to the temperature and to the entropy of mixing. Figure 12.2.2 shows glucan chains with different types of linkages, and different pictorial trajectories which give different and sometimes surprising values of configurational entropy are also observed. For example, those of (1-4)-linked α-D-glucan and of (1-3)-linked β-D-glucan seem fairly restricted to some pseudo-helical character compared to the more disordered set of trajectories that would be obtained if rotations about the glycosidic bonds were completely unrestricted. Possible interactions between residues of the polysaccharide chain that are not nearest-neighbors in the primary sequence of the polymer can sometimes be ignored. In this case, a computer-based polysaccharide chain can be constructed from the conformational energy map of the dimeric units. The Monte Carlo method[64] and the Flory matrix methods[65] are commonly used in the so-called "nearest-neighbors approximation" to mimic the polymer chains in the pure amorphous state or in dilute solution.

The Monte Carlo sample of chains reflects the range of conformations experienced by any single chain as a function of time or, equivalently, the range of conformations in a large sample of chemically identical polymer molecules at any instant in time. In either sense, the sample can be analyzed to deduce both the characteristics of individual chain conformations and the mean properties of the sample as a whole, which correspond to those in the equilibrium state of the chain. Results refer, however, to an "unperturbed" chain model that ignores the consequences of the long range excluded volume effect, because only nearest-neighbor interactions are accounted for.

Given a sufficient Monte Carlo sample of chains in equally probable representative conformations, it is possible to assess many mean properties of the polymer in question simply by computing numerical (unweighted arithmetic) averages over the number of chains in the sample. For example, the mean square end-to-end distance, the mean square radius of gyration, or the angular dependence of scattered light (particle scattering factor) are all mean geometric properties readily computed from a knowledge of the coordinates of the atoms or atomic groups which are generated in the Monte Carlo sample. The average topological properties are described through the chain-length dependence of quantities such as the characteristic ratio, $C_n$, the persistence length, $P_n$, and the correlation function, $F_n$, defined as:

$$C_n = \frac{\langle r^2 \rangle_0}{nL^2}, \quad P_n = \left\langle \left( \frac{\vec{L}_1}{L_1} \right) \sum_{i=1}^{n} \vec{L}_i \right\rangle_0, \quad F_n = \langle \vec{u}\vec{u}_n \rangle_0 \qquad [12.2.6]$$

where:

| | |
|---|---|
| n | number of saccharide units |
| L | average virtual bond length |
| $\langle r^2 \rangle_0$ | mean square end-to-end distance |

The virtual bond vector is often defined for each monosaccharidic unit as connecting oxygen atoms involved in glycosidic linkages. Figure 12.2.11 shows the calculated properties of equations [12.2.6] for two homoglucan chains which differ only in the configuration of the anomeric carbon, i.e., (a) the $[\alpha\text{-D-(1-3)-glc}]_n$ (pseudonigeran) and (b) the $[\beta\text{-D-(1-3)-glc}]_n$ (curdlan). Whatever the chemical features, provided that the molecular weight is very large (that is for a degree of polymerization n which approaches infinity), the distribution of unperturbed polymer end-to-end length is Gaussian and $C_n$ equals an asymptotic characteristic ratio $C_\infty$, i.e.:

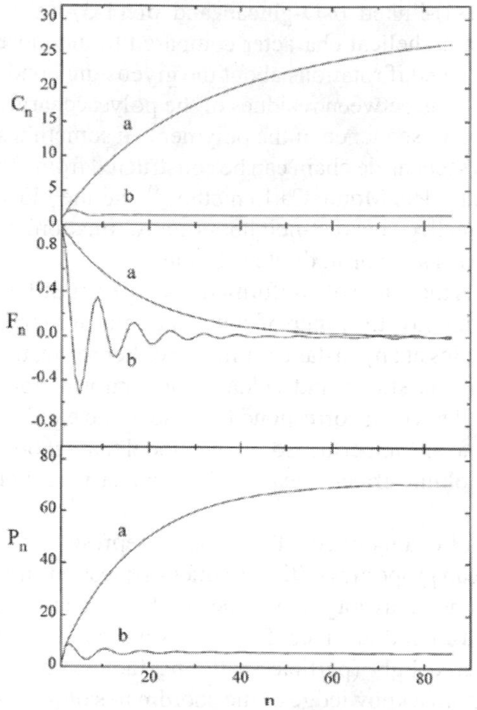


Figure 12.2.11. Characteristic ratio, correlation function and persistent length as a function of the degree of polymerization, n, for [α-D-(1-3)-glc]$_n$ (pseudonigeran) and (b) the [β-D-(1-3)-glc]$_n$ (curdlan) calculated on the in vacuo energy maps.

$$C_\infty = \lim_{n\to\infty} \frac{\langle r^2 \rangle}{nL^2} = \frac{6\langle R_g^2 \rangle}{nL^2}$$

where $R_g$ is the average square radius of gyration experimentally accessible. This relation is considered extremely important in the sense that any conformational perturbation is amplified by a factor n in the final value of $C_\infty$ (or of $<R_g>$), and therefore represents a highly demanding test for the appropriateness of the conformational calculations and at the same time a discriminating factor for conformation-dependent solution properties (e.g., viscosity).

An alternative measure of chain extension is the persistent length, $P_n$, which behaves in a similar asymptotic dependence on n as observed for $C_n$, in Figure 12.2.11. It is meant as the capacity of the chain to preserve the direction of the first residue (vector) of the chain. Therefore, as the directional persistence dissipates with increasing chain length and the direction of the terminal residue vector $L_n$ loses correlation with that of the initial vector $L_1$, $P_n$ approaches an asymptotic limit for sufficiently long chains. Both the $C_n$ and $P_n$ functions in Figure 12.2.11 reveal the [α-D-(1-3)-glc]$_n$ chain to be much more extended than that of [β-D-(1-3)-glc]$_n$.

The directional correlations are also well characterized by the correlation function $F_n$ of equations [12.2.6], which measures the average projection of a unit vector aligned with each virtual bond of the chain onto the unit vector relative to the first residue. The strongly oscillating character of $F_n$ for [β-D-(1-3)-glc]$_n$, which is also observed at low n in both the $C_n$ and $P_n$ functions, reflects the pseudo-helical persistence of the backbone trajectory which becomes uncorrelated (i.e., $F_n$ declines to zero) as the molecular weight increases. It is noteworthy that the oscillations in $F_n$ retain approximately the pseudo-helical periodicity present in the crystalline forms of that polysaccharide (Figure 12.2.2). The monotonic decline of $F_n$ (a) for [α-D-(1-3)-glc]$_n$, on the other hand, shows that this polymer possess a stronger directional correlation due to the different glycosidic linkage which has a dramatic effect on the character of the chain trajectory.

The chain properties illustrated above are often described with terms such as "stiffness" and "flexibility". It is important to clarify that stiffness and structural rigidity may not necessarily be alternative to flexibility and structural disorder, since two different concepts enter into the above definitions: one is concerned with the number of different accessible conformations, the other with the (average) direction of the sequential bonds, i.e., with the chain topology.

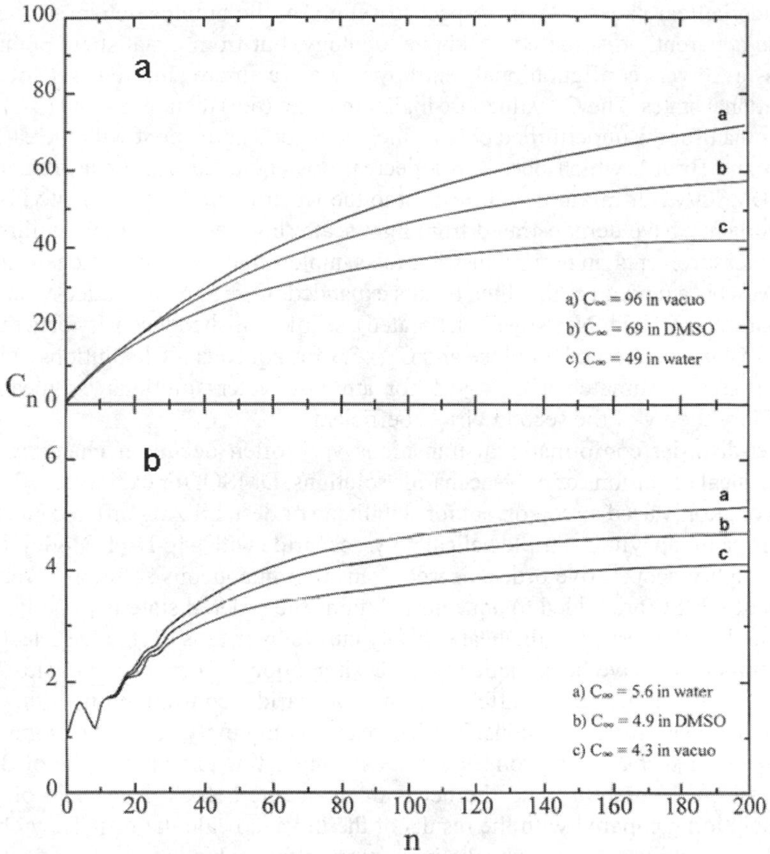


Figure 12.2.12. Characteristic ratio, $C_n$, as a function of degree of polymerization, n, for amylose (a) and cellulose (b) in vacuo, water and DMSO.

By calculating of the conformational energy surface of a dimer as a starting point for the prediction of mean chain properties, the effect of the solvation on the chain conformation is taken into account by evaluation of the perturbations of such surface due to the presence of the solvent. The characteristic ratio (eq. [12.2.6]) has been computed for (a) [β-D-(1-4)-glc]$_n$ (cellulose) and (b) [α-D-(1-4)-glc]$_n$ (amylose) in different solvents, namely, water and DMSO.[53] The results are reported in Figure 12.2.12 where, for comparison, the in vacuo $C_n$ for both polymers is shown. Figure 12.2.12a shows that the chain extension of cellulose decreases significantly in DMSO and much more so in water with a reduction of about 50% when compared to that in vacuo data ($C_\infty$=96). The solvation contribution of water seems to favor those conformations, which have smaller cavity volumes and thus force the chain to minor extensions. Tanner and Berry[66] reported, for cellulose derivatives in solution, a value for the limiting characteristic ratio of between 30 and 60, which is in good agreement with the theoretical data in Figure 12.2.12a.

Figure 12.2.12b shows the characteristic ratio of amylose chains computed in the same solvents as in Figure 12.2.12a. The amylose chain extension is considerably lower than that of cellulose and, as in the case of [β-D-(1-3)-glc]$_n$ shown above, presents at low n a remarkable pseudo-helical pattern. With respect to cellulose, the amylose chain shows a more coiled and apparently disordered backbone topology, but from a statistical point of view possesses a lower configurational entropy,[67] i.e., a more limited set of allowed conformational states. The $C_\infty$ values are higher in water (5.6) than in DMSO (4.9) and both are higher than for the unperturbed chain which is in good agreement with the earlier work of Jordan and Brant[67] which observed a decrease of about 20% in chain dimensions of amylose DMSO/water mixture with respect to the water alone. More recently, Nakanishi and co-workers[68] have demonstrated from light scattering, sedimentation equilibrium and viscosity measurements on narrow distribution samples, that the amylose chain conformation in DMSO is a random coil, which results expanded ($C_\infty = 5$) by excluded-volume effect at high molecular weight. Norisuye[69] elaborated a set of published viscosity data reporting a $C_\infty = 4.2\text{-}4.5$ for unperturbed amylose and $C_\infty = 5.3$ for aqueous KCl solutions, while Ring and co-workers[70] estimated a $C_\infty \cong 4.5$ for amylose/water solutions by means of the Orofino-Flory theory of the second virial coefficient.

Order-disorder conformational transitions very often occur on changing physical and/or chemical conditions of polysaccharide solutions. DMSO, for example, is the solvent, which is commonly used as co-solvent for stabilizing or destabilizing ordered solution conformations. Schizophyllan, a triple helical polysaccharide with a $[\beta\text{-D-(1-3)-glc}]_n$ backbone exhibits a highly cooperative order-disorder transition in aqueous solution.[71] When small quantities of DMSO are added to aqueous solutions the ordered state is remarkably stabilized, as has been observed in the heat capacity curves by means of the DSC technique.[71]

Several efforts have been made with MD simulations in order to explicitly take into account the solvent molecule effect on the saccharide conformation, although only oligomeric segments have been considered, given the complexity in terms of computational time required for such a multi-atoms system. One interesting example is that of Brady and co-workers[56] on the stability and the behavior of double-helix carrageenan oligomer in aqueous solution compared with the results of the in vacuo calculations. They observed a higher relative stability of the double helix in vacuo, a fact, which is consistent with experimental results under anhydrous conditions, as in the fiber diffraction studies. However, in aqueous solution, the interchain hydrogen bonds that stabilize the double-helix structure appear much less stable, as the glycosidic hydroxyl groups make more favorable interactions with water molecules. They concluded that in the solvation step the double-helix would seem to be unstable and an unwinding process is theoretically predicted, at least for the oligomers.

## 12.2.6 SOLVENT EFFECT ON CHARGED POLYSACCHARIDES AND THE POLYELECTROLYTE MODEL

### 12.2.6.1 Experimental behavior of polysaccharides polyelectrolytes

Based on the experimental evidence of polyelectrolyte solutions, whenever the degree of polymerization is sufficiently high, all ionic macromolecules are characterized by a peculiar behavior, which sets them apart from all other ionic low molecular weight molecules as well as from non-ionic macromolecules. A general consequence of the presence of charged groups in a chain is a favorable contribution to the solubility of polymer in water. A strongly attractive potential is generated between the charge density on the polymer and the opposite charges in solution. For example, the value of the activity coefficient of the counterions is strongly reduced with respect to that of the same ions in the presence of the univalent opposite charged species. If the charge density of the polyelectrolyte is sufficiently high, such a phenomenon is justified through a 'condensation process' of counterions and it has also been interpreted theoretically.

On the polymer side, among the dramatic changes that the presence of charged groups imparts to solution properties, there are the enhanced chain dimensions, the increased hydrodynamic volume (i.e., viscosity), and, in general, a strong influence on all conformational properties. Subject to the constraints imposed by the chemical structure of the chain, the distribution of charged groups and their degree of ionization contribute to determining the equilibrium chain conformation; both the Coulombic interaction among the charged groups and the distribution and concentration of the screening counterions are important. Most of the physico-chemical properties of the system result from a non-linear combination of these parameters.

However, one has not to forget that the variability of conformation alters the distances between charged groups on the polymeric chain and that the equilibrium is statistically defined by the Gibbs energy minimum of the system. As an important consequence of this energy balance, changes in temperature, ionic strength, pH, etc., can provoke changes in polyelectrolyte conformation, often cooperatively in the case of biopolymers, between states with different values of the charge density. These states may be characterized by different structural orders (e.g. helix $\rightarrow$ extended chain transition), by different degrees of flexibility of the chain (globular coil $\rightarrow$ expanded chain) or by different extent of aggregation (monomeric $\rightarrow$ dimeric or multimeric chains).

Theoretical calculations based on molecular grounds are still extremely complicated and incomplete[72] and other routes must be more empirically used in order to interpret the experimental data and to understand the correlation between conformational properties and structure. The central problem is to quantify the interactions among charges on the polymer and among these same charges and their respective counterions.

As far as it concerns the short-range interactions, the introduction of charged groups modifies the equilibrium geometry of the monomeric units and the contribution of the electrostatic nature on the nearest-neighbor conformational energy. These conclusions also derive from the already demonstrated effect of the solvent interactions on the unperturbed dimensions of amylose and cellulose,[53] and from the evidence of the perturbation on the conformational energy surface of several charged saccharidic units.[73]

There at least two approaches that may be relevant for this review; one is that described by Haug and Smidsrød[74] for the rationalization of the dimensional properties of polyelectrolytes as a function of salt concentration, the other is the formulation of a statistical thermodynamic theory for the "physical" framing of the ion-polyelectrolyte interactions. Both these theoretical formulations deal with the conformation of the polymer and predict that the conformational features must be function of ionic strength (see for example refs. 75 and 76).

#### 12.2.6.2 The Haug and Smidsrød parameter: description of the salt effect on the chain dimension

A peculiarity of the correlation between the viscometric parameters and the dimensions of the macromolecular chain has long been recognized and theoretical approaches have been developed for several chain models.[77,78] The behavior of polyelectrolytes adds some complications especially in the low ionic strength regime. It has however been understood that the intrinsic viscosity, $[\eta]$, of a polyion (i.e., its hydrodynamic volume) decreases with increasing ionic strength, I, as a consequence of the screening of the fixed charges on the

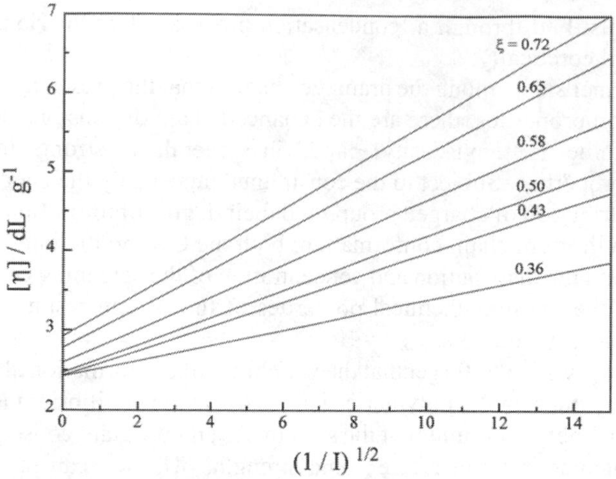


Figure 12.2.13. Dependence of the intrinsic viscosity [η] of hyaluronic acid (ξ=0.72) and its benzyl derivatives with decreasing linear charge density, ξ, on the inverse square root of ionic strength, I, at 25°C.

polyion. At infinite ionic strength, the chain dimensions may eventually correspond to the completely uncharged macromolecule. This is a sort of "ideal state" of the polyelectrolyte; "ideal" with respect to the long-range electrostatic repulsive interactions only, without relation to the Θ-conditions.

In the absence of a cooperative conformational transition, for many polysaccharide polyelectrolytes a linear dependence of [η] upon $I^{1/2}$ is observed, with the slope diminishing with the charge density associated to the polysaccharide chain (Figure 12.2.13). A theory has been presented for an estimation of the relative stiffness of the molecular chains by Smidsrød and Haug,[74] which is based on the Fixman's theory and Mark-Houwink equation. The chain stiffness parameter is estimated from the normalized slope B of [η] vs. the inverse square root of the ionic strength:

$$\frac{\partial[\eta]}{\partial I^{-1/2}} = slope = B\left([\eta]_{0.1}\right)^{\gamma} \quad [12.2.7]$$

where:

γ has a value between 1.2 and 1.4

The dependence of viscosity on the ionic strength, as given in equation above, has been increasingly popular in the field of polysaccharides with the purpose of comparing the chain stiffness of different macromolecules. The derivation is based on the Fixman theory, which defines the dependence of [η] on the molecular weight through an expansion coefficient which effectively takes into account the electrostatic interactions in the Debye-Hückel approximation. The semi-empirical treatment of the hydrodynamic properties of statistical polyelectrolytes (at sufficiently high values of the ionic strength) is built upon a straightforward extension of the theory of intermolecular interactions for uncharged polymers, for which a linear relation can be written between the expansion coefficient, $\alpha_\eta^3$, and the square root of the molecular weight, M. It should also be mentioned[78] that the various theoretical treatments of the salt dependence of the excluded volume and of the expansion coefficients led to a linearity of $\alpha_\eta^3$ on $(M/C_s)^{1/2}$ only over a limited range of salt concentration. On varying the salt concentration, one effectively deals with a set of binary solvents with a variety of interaction parameters. The approach proposed by Smidsrød overcomes the indetermination of some parameters by using the slope of [η] as a function of the inverse square-root of the ionic strength of the medium. The ultimate relationship is obtained between the constant B and the effective bond length $b_\theta$ ($B = const \cdot b_\theta^{-2}$), which, in the absence of any reasonable knowledge of the constant, can only be used in an empirical way.

There is a compelling although intuitive limit to the use of the Smidsrød-Haug approach for those macroions that do not counterbalance the effective electrostatic field exerted by the ionic strength through the conformational elasticity. Among these polymers, those characterized by a low value of fixed charges, of flexibility, and/or of molecular weight fall behind the limits of the correct applicability of the Smidsrød-Haug approach, which should maintain its validity only for gaussian chains with prominent electrostatic interactions. Most likely, the Smidsrød-Haug parameter has been abused in the field of polysaccharides without the authors' intention, in the sense that the original treatment was aware about the intrinsic limitations of the approach,[74] while the extensions have thereafter been considered as permitted. We wish to point out that the above comment does not imply the failure of the linearity of [η] with $I^{-1/2}$, but only a meaningless result for the values of $b_\theta$ obtained for low-charged polysaccharide polyelectrolytes. Reference can be made to two series of polysaccharides (chitosans and hyaluronans), which have been investigated in some detail for the specific application of the Smidsrød-Haug approach.[79,80]

#### 12.2.6.3 The statistical thermodynamic counterion-condensation theory of Manning

Linear polyelectrolytes bear a charge distribution along the chain, properly neutralized by small ions of opposite sign. In the absence of added salt, it is reasonably assumed that the charge density (when sufficiently high) will increase the local stiffness of the chain because of the electrostatic repulsion. For this reason the linear polyelectrolyte is often regarded as a charged rod. Osawa has introduced the concept of a critical charge density on the polymer and described the ion- pairing of polyion and counterions as condensation. The model described here has been extensively used in previous papers of the authors[81] and described in the original papers by Manning.[82] Let here simply give a few comments on the physical basis of this model relevant to the present case.

The counterion condensation (CC) theory, largely developed by Manning, gives analytical solutions to evaluate the electrostatic potential around a linear polyelectrolyte, provided its conformation is regular and fixed. The rigidity of the polymer seems therefore to be both a prerequisite and a result of the molecular polyelectrolyte theory, and it has been thought to be not too far from reality in many cases.[82] The application of the above theory to experimental results has been carried out with the assumption that the charge distribution is structurally defined by the monomer repeat as derived from the solid state fiber diffractograms, although sometimes the fully stretched chain conformation has been taken (Figure 12.2.14). The original work of Manning's counterion condensation theory has provided an elegant tool for describing several properties of polyelectrolytes in terms of the structural parameter, ξ, which is unequivocally defined as $e^2/\varepsilon kTb$, where e is the value of the elementary charge, ε is the dielectric constant of the medium, k is the Boltzmann constant, T is the Kelvin temperature, and b is the distance between the projections of the fixed charges of the polyelectrolyte on its contour axis. In the case of monovalent ions, for all

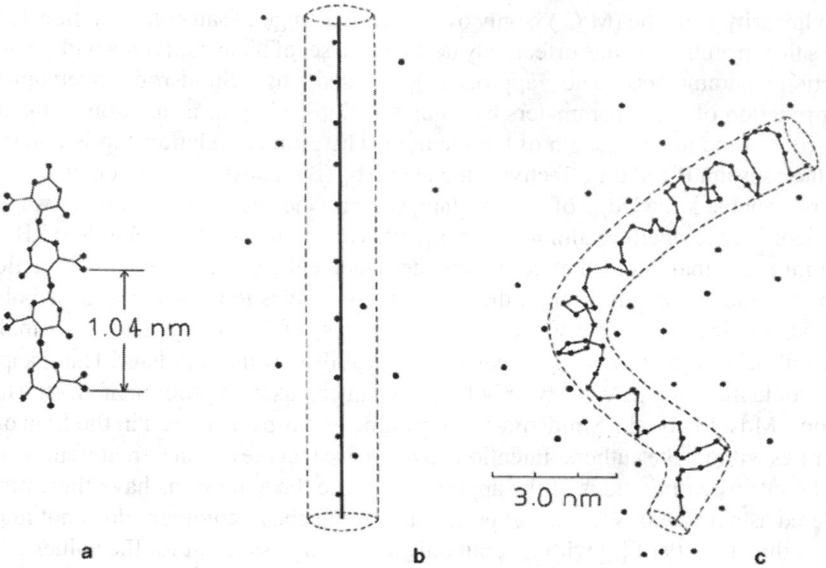


Figure 12.2.14. The rigid chain (a), the corresponding polyelectrolyte model (b) and the realistic flexible model of a polysaccharidic chain (c).

structural values of $\xi > 1$, a defined amount of counterions will "condense" from the solution into the domain of the polymer chain so as to reduce the "effective" value of $\xi$ to unity. For water, $\xi = 0.714/b$ (with b expressed in nm), and is practically independent of T, being the electrostatic-excess Gibbs free energy of the solution given by:

$$G_{el} = -\xi \ln\left[1 - \exp(-Kb)\right]$$

where K is the Debye-Hückel screening parameter. Application of this theory has been made to many experimental cases, and in particular an extensive correlation has been made between the theoretical predictions and the thermodynamic data on the processes of protonation, dilution and mixing with ions.[76]

Besides these nice applications of the theory to problems with a strong "academic" character, there is another very striking example of prediction of solvent-induced conformational changes for the effects of salts on the conformational stability of ordered polyelectrolytes. In fact, in addition to the condensation phenomena predicted by the polyelectrolitic theory, other physical responses may also occur (also simultaneously), which may mask this central statement of the Manning theory "that the onset of the critical value $\xi$ (for univalent ions $\xi > 1$) constitutes a thermodynamic instability which must be compensated by counterion condensation". In fact, chain extension and/or disaggregation of aggregated chains may occur or change upon the variation of charge density, and the energetic instability effectively becomes a function of the thermodynamic state of the polyelectrolytic chain.

The range of theoretical and experimental approaches has been, in particular, addressed to the problem of conformational transitions between two different states, provided they have different charge densities. For thermally induced, conformational transitions between states i and f of a polyelectrolyte, characterized by a set of $\xi_i$ and $\xi_f$(i.e., $b_i$ and $b_f$) values, polyelectrolyte theory predicts a simple relationship[82] between the values of the melting temperatures ($T_M$, the temperature of transition midpoint) and the logarithm of the ionic strength, I:

$$\frac{d\left(T_M^{-1}\right)}{d(\log I)} = -\frac{9.575F(\xi)}{\Delta_M H}$$

where $\Delta_M H$ is the value of the enthalpy of transition (in J per mole of charged groups) determined calorimetrically. This linearity implies, indeed, that the enthalpy change is essentially due to non-ionic contributions and largely independent of I. The function F(ξ) depends on the charge density of both the final state (subscript f) and the initial state (subscript i), within the common condition that $\xi_f < \xi_i$, that is the final state is characterized by a smaller value of the charge density. The value of F(ξ) is given in the literature.

This relation has been successfully applied first to the transition processes of DNA,[83] polynucleotides,[84] but also to many ionic polysaccharides (carrageenans,[85] xanthan,[86] succinoglycan,[87]) of great industrial interest. Accurate determination of the $T_M$ values of the polysaccharide as a function of the ionic strength is necessary.

### 12.2.6.4 Conformational calculations of charged polysaccharides

The major problem for conformational calculations of ionic polysaccharides arises from the correct evaluation of the electrostatic potential energy due to the charged groups along the chain and to the all other ions in solutions. The interaction between the polyion charges and the counterions is formally non-conformational but it largely affects the distribution of the conformational states. Ionic polymers are often simplistically treated either in the approximation of full screening of the charged groups or in the approximation of rigid conformational states (regular rod-like polyelectrolyte models).

A combination of the molecular polyelectrolyte theory[82,83] with the methods of statistical mechanics can be used at least for the description of the chain expansion due to charges along the polysaccharide chain. The physical process of the proton dissociation of a (weak) polyacid is a good way to assess the conformational role of the polyelectrolytic interactions, since it is possible of tuning polyelectrolyte charge density on an otherwise constant chemical structure. An amylose chain, selectively oxidized on carbon 6 to produce a carboxylic (uronic) group, has proved to be a good example to test theoretical results.[81]

If the real semi-flexible chain of infinite length is replaced by a sequence of segments, the average end-to-end distance <r> of each segment defines the average distance <b> between charges:

$$\langle b \rangle = \frac{\langle r \rangle}{N} \qquad [12.2.8]$$

where N is the number of charges in the segment. The distance between charges fluctuates within the limits of the conformational flexibility of the chain, as calculated by the proper non-bonding inter-residue interactions.

The probability function W'(r) of the end-to-end displacement r of a charged segment can be obtained by multiplying its a priori (non-ionic) probability W(r) with the Boltzmann

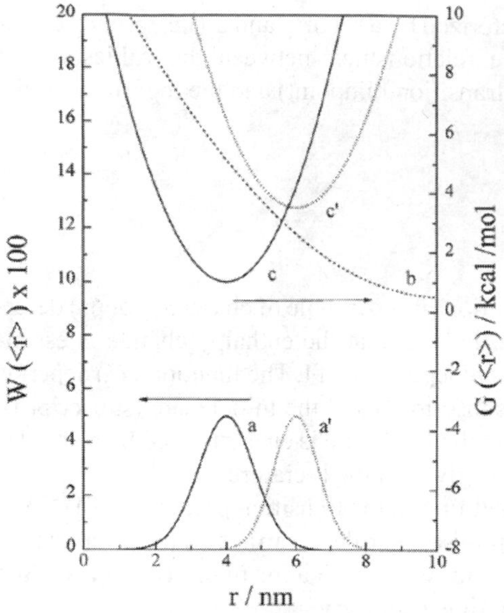


Figure 12.2.15. Dependence of the probability distribution function of a model semi-flexible chain, (a) uncharged and charged (a'), on the end-to-end distance and the respective total free energies (c and c'). The electrostatic contribution (b) is also reported.

term involving the excess electrostatic free energy (Figure 12.2.15). The probability theory guarantees both that the components (repeating units) of the segment vectors be distributed in a Gaussian way along the chain segment, and that high molecular weight polymers be composed by a statistical sequence of those segments. Consequence of the above approximation is that the distance r between any two points of the chain (separated by a sufficiently large number of residues, n) does not depend on the specific sequence and values of conformational angles and energies, but only upon the average potential summed over the number of residues n.

The calculation of the averaged (electrostatic) functions is reached in two steps. At the first, the proper flexibility of the polymer is evaluated either from conformational calculation or from suitable models, then the mean value of each property is calculated through the averaging procedure described below.

The computational procedure is the following:

- the conformational energy surface of the uncharged polymer is evaluated by the standard methods the conformational analysis;[65]
- the end-to-end distribution distance $W_n(r)$ for the (uncharged) polymer segments is determined by numerical Monte Carlo methods;[64]
- the dependence of the total (conformational) energy G(r) upon chain extension r is therefore estimated from the distribution of segment lengths; a Boltzmannian distribution is assumed.

In most cases the distribution function is Gaussian (or approximately so) and the corresponding free energy function can be approximated by a simple parabolic equation (Figure 12.2.15). In this case, we assume a Hookean energy (which is correct at least for the region around the maximum of the distribution curve), so we have:

$$W(r) = A\exp\left(-\frac{G}{RT}\right) \quad G(r) = k\left(r - r^0\right) \qquad [12.2.9]$$

where:

$r^0$ average segment length
k a constant which determines the flexibility of the chain

The ionic energy, that results from the process of charging the polymer groups, changes the probability of the end-to-end distance for the i-th segment, W'(r), to the probability of the average inter-charge separation distance <b>, W(b), following the definition of equation [12.2.8] and [12.2.9].

The conformational (non-ionic) free energy, obtained from the radial distribution function for non-ionic chains by Monte Carlo calculations, was used in conjunction with the electrostatic free energy to calculate the actual distribution function of the charged chain segments. The resulting expansion justifies almost quantitatively in many cases the experimental thermodynamic properties (such as $pK_a$, $H_{dil}$, etc.) and the dimensional properties (viscosity) of the ionic polysaccharides to which the approach has been applied.

## 12.2.7 CONCLUSIONS

Only some aspects of the solvent perturbation on the conformational properties of carbohydrate polymers have been covered in this chapter. One of the major concerns has been to develop a description of these "solvent effects" starting with the complex conformational equilibria of simple sugars. In fact, only recently it has been fully appreciated the quantitative relationship between conformational population and physical properties, e.g. optical rotation.

The chapter, however, does not give extensive references to the experimental determination of the polysaccharide shape and size in different solvents, but rather it attempts to focus on the molecular reasons of these perturbations. A digression is also made to include the electrostatic charges in polyelectrolytic polysaccharides, because of their diffusion and use and because of interesting variations occurring in these systems. Thus, provided that all the interactions are taken into account, the calculation of the energetic state of each conformation provides the quantitative definition of the chain dimensions.

## REFERENCES

1 J.R. Brisson and J.P. Carver, *Biochemistry*, **22**, 3671 (1983).
2 R.C. Hughes and N. Sharon, *Nature*, **274**, 637 (1978).
3 D.A. Brant, *Q. Rev. Biophys.*, **9**, 527 (1976).
4 B.A. Burton and D.A. Brant, *Biopolymers*, **22**, 1769 (1983).
5 G.S. Buliga and D.A. Brant, *Int. J. Biol. Macromol.*, **9**, 71 (1987).
6 V. S. R. Rao, P. K. Qasba, P. V. Balaji and R. Chandrasekaran, *Conformation of Carbohydrates, Harwood Academic Publ.*, Amsterdam, 1998, and references therein.
7 R. H. Marchessault and Y. Deslandes, *Carbohydr. Polymers,* **1**, 31 (1981).
8 D. A. Brant, *Carbohydr. Polymers,* **2**, 232 (1982).
9 P.R. Straub and D.A. Brant, *Biopolymers*, **19**, 639 (1980).
10 A. Cesàro in **Thermodynamic Data for Biochemistry and Biotechnology**, H.J. Hinz (Ed.), *Springer-Verlag*, Berlin, 1986, pp. 177-207.
11 Q. Liu and J.W. Brady, *J. Phys.Chem. B*, **101**, 1317 (1997).
12 J.W. Brady, *Curr. Opin. Struct.Biol.*, **1**, 711 (1991).
13 Q. Liu and J.W. Brady, *J. Am. Chem. Soc.*, **118**, 12276 (1996).
14 I. Tvaroška, *Biopolymers*, **21**, 188 (1982).
15 I. Tvaroška, *Curr. Opin. Struct.Biol.*, **2**, 661 (1991).
16 K. Mazeau and I. Tvaroška, *Carbohydr. Res.*, **225**, 27 (1992).
17 Perico, A., Mormino, M., Urbani, R., Cesàro, A., Tylianakis, E., Dais, P. and Brant, D. A., *Phys. Chem. B*, **103**, 8162-8171(1999).
18 K.D. Goebel, C.E. Harvie and D.A. Brant, *Appl. Polym. Symp.*, **28**, 671 (1976).
19 M. Ragazzi, D.R. Ferro, B. Perly, G. Torri, B. Casu, P. Sinay, M. Petitou and J. Choay, *Carbohydr. Res.*, **165**, C1 (1987).
20 P.E. Marszalek, A.F. Oberhauser, Y.-P. Pang and J.M. Fernandez, *Nature*, **396**, 661 (1998).
21 S.J. Angyal, *Aust. J.Chem.,* **21**, 2737 (1968).
22 S.J. Angyal, *Advan. Carbohyd. Chem. Biochem.*, **49**, 35 (1991).
23 D.A. Brant and M.D. Christ in **Computer Modeling of Carbohydrate Molecules**, A.D. French and J.W. Brady, Eds., ACS Symposium Series 430, *ACS,* Washington, DC, 1990, pp. 42-68.
24 R. Harris, T.J. Rutherford, M.J. Milton and S.W. Homans, *J. Biomol. NMR*, **9**, 47 (1997).
25 M. Kadkhodaei and D.A. Brant, *Macromolecules*, **31**, 1581 (1991).

26 I. Tvaroška and J. Gajdos, *Carbohydr. Res.,* **271**, 151 (1995).
27 F.R. Taravel , K. Mazeau and I. Tvaroška, *Biol. Res*., **28**, 723 (1995).
28 J.L. Asensio and J. Jimenez-Barbero, *Biopolymers*, **35**, 55 (1995).
29 P. Dais, Advan. Carbohyd. *Chem. Biochem*., **51**, 63 (1995).
30 F. Cavalieri, E. Chiessi, M. Paci, G. Paradossi, A. Flaibani and A. Cesàro, *Macromolecules*, (submitted).
31 I. Tvaroška and F.R. Taravel, *J. Biomol. NMR*, **2**, 421 (1992).
32 R.E. Wasylishen and T. Shaefer, *Can. J. Chem*., **51**, 961 (1973).
33 A.A. van Beuzekom, F.A.A.M. de Leeuw and C. Altona, *Magn. Reson. Chem*., **28**, 888 (1990).
34 I. Tvaroška, *Carbohydr. Res*., **206**, 55 (1990).
35 I. Tvaroška and F.R. Taravel, *Carbohydr. Res*., **221**, 83 (1991).
36 J.P. Carver, D. Mandel, S.W. Michnick, A. Imberty, J.W. Brady in **Computer Modeling of Carbohydrate Molecules**, A.D. French and J.W. Brady, Eds., ACS Symposium Series 430, *ACS*, Washington, DC, 1990, pp. 267-280.
37 D. Rees and D. Thom, *J. Chem. Soc. Perkin Trans. II*, 191 (1977).
38 E.S. Stevens, *Carbohydr. Res*., **244**, 191 (1993).
39 D.H. Whiffen, *Chem. Ind*., 964 (1956).
40 J.H. Brewster, *J. Am. Chem. Soc*., **81**, 5483 (1959).
41 D. A. Rees, *J. Chem. Soc. B*, 877 (1970).
42 E.S. Stevens and B.K. Sathyanarayana, *J.Am.Chem.Soc*., **111**, 4149 (1989).
43 C.A. Duda and E.S. Stevens, *Carbohydr. Res*., **206**, 347 (1990).
44 R. Urbani, A. Di Blas and A. Cesàro, *Int. J. Biol. Macromol*., **15**, 24 (1993).
45. I. Tvaroška, S. Perez, O. Noble and F. Taravel, *Biopolymers*, **26**, 1499 (1987).
46 C. Gouvion, K. Mazeau, A. Heyraud, F. Taravel, *Carbohydr. Res*., **261**, 261 (1994).
47 R.A. Pierotti, *Chem. Rev*., **76**, 717 (1976).
48 R.J. Abraham and E. Bretschneider in **Internal Rotation in Molecules**, W.J. Orville-Thomas, Ed., *Academic Press*, London, 1974, ch. 13.
49 D.L. Beveridge, M.M. Kelly and R.J. Radna, *J. Am. Chem. Soc*., **96**, 3769 (1974).
50 M. Irisa, K. Nagayama and F. Hirata, *Chem. Phys. Letters*, **207**, 430 (1993).
51 R.R. Birge, M.J. Sullivan and B.J. Kohler, *J. Am. Chem.Soc*., **98**, 358 (1976).
52 I. Tvaroška and T. Kozar, *J. Am. Chem. Soc*., **102**, 6929 (1980).
53 R. Urbani and A. Cesàro, *Polymers*, **32**, 3013 (1991).
54 J.W. Brady, *J. Am. Chem. Soc*., **108**, 8153 (1986).
55 C.B. Post, B.R. Brooks, M. Karplus, C.M. Dobson, P.J. Artymiuk, J.C. Cheetham and D.C. Phillips, *J. Mol. Biol*., **190**, 455 (1986).
56 K. Ueda, A. Imamura and J.W. Brady, *J. Phys. Chem*., **102**, 2749 (1998).
57 R. Grigera, *Advan. Comp. Biol*., **1**, 203 (1994).
58 S.N. Ha, A. Giammona, M. Field and J.W. Brady, *Carbohydr. Res*., **180**, 207 (1988).
59 C.L. Brooks, M. Karplus, B.M. Pettitt, **Proteins: A Theoretical Perspective of Dynamics, Structure and Thermodynamics**, *Wiley Interscience*, New York, 1988, vol. LXXI.
60 R. Schmidt, B. Teo, M. Karplus and J.W. Brady, *J. Am. Chem. Soc*., **118**, 541 (1996).
61 Q. Liu, R. Schmidt, B. Teo, P.A. Karplus and J.W. Brady, *J. Am. Chem. Soc*., **119**, 7851 (1997).
62 V.H. Tran and J.W. Brady, *Biopolymers*, **29**, 977 (1990).
63 M. Karplus and J.N. Kushick, *Macromolecules*, **31**, 1581 (1981).
64 R.C. Jordan, D.A. Brant and A. Cesàro, *Biopolymers*, **17**, 2617 (1978).
65 P.J. Flory, **Statistical Mechanics of Chain Molecules,** *Wiley-Interscience*, New York, 1969.
66 D.W. Tanner and G.C. Berry, *J. Polym. Sci., Polym. Phys. Edn*., **12**, 441 (1974).
67 R.C. Jordan and D.A. Brant, *Macromolecules*, **13**, 491 (1980).
68 Y. Nakanishi, T. Norisuye, A. Teramoto and S. Kitamura, *Macromolecules*, **26**, 4220 (1993).
69 T. Norisuye, *Food Hydrocoll*., **10**, 109 (1996).
70 S.G. Ring, K.J. Anson and V.J. Morris, *Macromolecules*, **18**, 182 (1985).
71 T. Hirao, T. Sato, A. Teramoto, T. Matsuo and H. Suga, *Biopolymers*, **29**, 1867 (1990).
72 C. Sagui and T.A. Darden, *Ann. Rev. Biophys. Biomol. Struct*., **28**, 155 (1999).
73 J.R. Ruggiero, R. Urbani and A. Cesàro, *Int. J. Biol. Macromol*., **17**, 205 (1995).
74 O. Smidsrød and A. Haug, *Biopolymers*, **10**, 1213 (1971).
75 G.S. Manning and S. Paoletti, in **Industrial Polysaccharides**, V. Crescenzi, I.C.M. Dea and S.S. Stivala, Eds., *Gordon & Breach*, N Y, 1987, pp. 305-324.
76 S. Paoletti, A. Cesàro, F. Delben, V. Crescenzi and R. Rizzo, in **Microdomains in Polymer Solutions**, P. Dubin Ed., *Plenum Press*, New York, (1985) pp 159-189.

77 H. Morawetz, **Macromolecules in Solution**, *Interscience*, New York, 1975, Ch. 7.
78 M. Bohdanecký and J. Kovár, **Viscosity of Polymer Solutions**, *Elsevier*, Amsterdam, 1982, p. 108.
79 M.W. Anthonsen, K.M. Vårum and O. Smidsrød, Carbohydr. *Polymers*, **22**, 193 (1993).
80 R. Geciova, A. Flaibani, F. Delben, G. Liut, R. Urbani and A. Cesàro, *Macromol. Chem. Phys*., **196**, 2891 (1995).
81 A. Cesàro, S. Paoletti, R. Urbani and J.C. Benegas, *Int. J. Biol. Macromol*., **11**, 66 (1989).
82 G.S. Manning, *Acc. Chem. Res*., **12**, 443 (1979).
83 G.S. Manning, Quart. *Rev. Biophys*., **11**, 179 (1978).
84 M.T. Record, C.F. Anderson and T.M. Lohman, *Quart. Rev. Biophys*., **11**, 103 (1978).
85 S. Paoletti, F. Delben, A. Cesàro and H. Grasdalen, *Macromolecules*, **18**, 1834 (1985).
86 S. Paoletti, A. Cesàro and F. Delben, *Carbohydr. Res*., **123**, 173 (1983).
87 T.V. Burova, I.A. Golubeva, N.V. Grinberg, A.Ya. Mashkevich, V.Ya. Grinberg, A.I.Usov, L. Navarini and A. Cesàro, *Biopolymers*, **39**, 517 (1996).

13

# Effect of Solvent on Chemical Reactions and Reactivity

## 13.1 SOLVENT EFFECTS ON CHEMICAL REACTIVITY

Roland Schmid
**Technical University of Vienna**
**Institute of Inorganic Chemistry, Vienna, Austria**

### 13.1.1 INTRODUCTION

About a century ago, it was discovered that the solvent can dramatically change the rate of chemical reactions.[1] Since then, the generality and importance of solvent effects on chemical reactivity (rate constants or equilibrium constants) has been widely acknowledged. It can be said without much exaggeration that studying solvent effects is one of the most central topics of chemistry and remains ever-increasingly active. In the course of development, there are few topics in chemistry in which so many controversies and changes in interpretation have arisen as in the issue of characterizing solute-solvent interactions. In a historical context, two basic approaches to treating solvent effects may be distinguished: a phenomenological approach and a physical approach. The former may be subdivided further into the dielectric approach and the chemical approach.

- Phenomenological approach
  - Dielectric
  - Chemical
- Physical approach

That what follows is not intended just to give an overview of existing ideas, but instead to filter seminal conceptions and to take up more fundamental ideas. It should be mentioned that solvent relaxation phenomena, i.e., dynamic solvent effects, are omitted.

### 13.1.2 THE DIELECTRIC APPROACH

It has soon been found that solvent effects are particularly large for reactions in which charge is either developed or localized or vice versa, that is, disappearance of charge or spreading out of charge. In the framework of electrostatic considerations, which have been around since Berzelius, these observations led to the concept of solvation. Weak electrostatic interactions simply created a loose solvation shell around a solute molecule. It was in this climate of opinion that Hughes and Ingold[2] presented the first satisfactory qualitative account of solvent effects on reactivity by the concept of activated complex solvation.

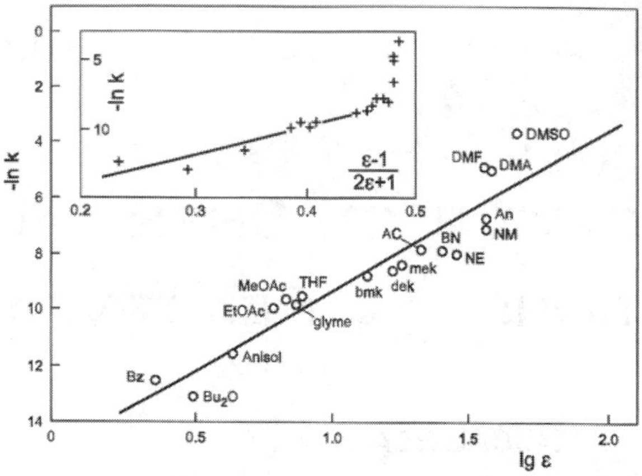


Figure 13.1.1. Relationship between second-order rate constants at 50°C of the reaction of p-nitrofluorobenzene and piperidine and the solvent dielectric properties [from ref. 21].

The first solvent property applied to correlate reactivity data was the static dielectric constant $\varepsilon$ (also termed $\varepsilon_s$) in the form of dielectric functions as suggested from elementary electrostatic theories as those by Born ($1/\varepsilon$), Kirkwood $(\varepsilon-1)/(2\varepsilon+1)$, Clausius-Mosotti $(\varepsilon-1)/(\varepsilon+2)$, and $(\varepsilon-1)/(\varepsilon+1)$. A successful correlation is shown in Figure 13.1.1 for the rate of the $S_N2$ reaction of p-nitrofluorobenzene with piperidine.[3] The classical dielectric functions predict that reactivity changes level out for dielectric constants say above 30. For instance, the Kirkwood function has an upper limiting value of 0.5, with the value of 0.47 reached at $\varepsilon = 25$. The insert in Figure 13.1.1 illustrates this point. Therefore, since it has no limiting value, the log $\varepsilon$ function may be preferred. A theoretical justification can be given in the framework of the dielectric saturation model of Block and Walker.[4]

Picturing the solvent as a homogeneous dielectric continuum means in essence that the solvent molecules have zero size and that the molecules cannot move. The most adequate physical realization would be a lattice of permanent point dipoles that can rotate but cannot translate.

### 13.1.3 THE CHEMICAL APPROACH

Because of the often-observed inadequacies of the dielectric approach, that is, using the dielectric constant to order reactivity changes, the problem of correlating solvent effects was next tackled by the use of empirical solvent parameters measuring some solvent-sensitive physical property of a solute chosen as the model compound. Of these, spectral properties such as solvatochromic and NMR shifts have made a spectacular contribution. Other important scales are based on enthalpy data, with the best-known example being the donor number (DN) measuring solvent's Lewis basicity.

In the intervening years there is a proliferation of solvent scales that is really alarming. It was the merit particularly of Gutmann and his group to disentangle the great body of empirical parameters on the basis of the famous donor-acceptor concept or the coordination-chemical approach.[5] This concept has its roots in the ideas of Lewis going back to 1923, with the terms donor and acceptor introduced by Sidgwick.[6] In this framework, the two outstanding properties of a solvent are its donor (nucleophilic, basic, cation-solvating) and acceptor (electrophilic, acidic, anion-solvating) abilities, and solute-solvent interactions are considered as acid-base reactions in the Lewis' sense.

Actually, many empirical parameters can be lumped into two broad classes, as judged from the rough interrelationships found between various scales.[7] The one class is more concerned with cation (or positive dipole's end) solvation, with the most popular solvent basicity scales being the Gutmann DN, the Kamlet and Taft β, and the Koppel and Palm B. The other class is said to reflect anion (or negative dipole's end) solvation. This latter class includes the famous scales π*, α, $E_T(30)$, Z, and last but not least, the acceptor number AN. Summed up:

Cation (or positive dipole's end) solvation

- Gutmann DN
- Kamlet and Taft β
- Koppel and Palm B (B*)

Anion (and negative dipole's end) solvation

- Gutmann AN
- Dimroth and Reichardt $E_T(30)$
- Kosower Z
- Kamlet and Taft α, π*

These two sets of scales agree in their general trend, but are often at variance when values for any two particular solvents are taken. Some intercorrelations have been presented by Taft et al., e.g., the parameters $E_T$, AN and Z can be written as linear functions of both α and π*.[8] Originally, the values of $E_T$ and π* were conceived as microscopic polarity scales reflecting the "local" polarity of the solvent in the neighborhood of solutes ("effective" dielectric constant in contrast to the macroscopic one). In the framework of the donor-acceptor concept, however, they obtained an alternative meaning, based on the interrelationships found between various scales. Along these lines, the common solvents may be separated into six classes as follows.

1. nonpolar aliphatic solvents
2. protics or protogenetic solvents (at least one hydrogen atom is bonded to oxygen)
3. aromatic solvents
4. (poly)halogenated solvents
5. (perhaps) amines
6. select (or "normal" according to Abraham) solvents defined as non-protonic, non-chlorinated, aliphatic solvents with a single dominant bond dipole.

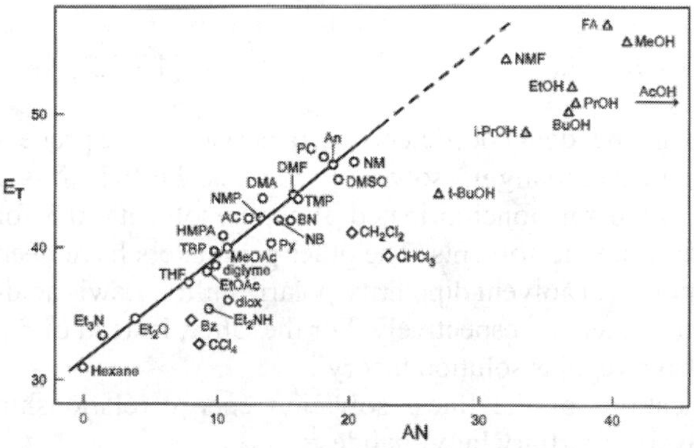


Figure 13.1.2. Relationship between the $E_T(30)$ values and the acceptor number [from ref. 21]. Triangles: protic solvents, squares: aromatic and chlorinated solvents.

A case study is the plot of AN versus $E_T$ shown in Figure 13.1.2. While there is a quite good correspondence for the select solvents (and likely for the nonpolar aliphatic solvents), the other classes are considerably off-line.[9] This behavior may be interpreted in terms of the operation of different solvation mechanisms such as electronic polarizability, dipole density, and/or hydrogen-bonding (HB) ability. For instance, the main physical difference between π* and $E_T(30)$, in the absence of

HB interactions, is claimed to lie in different responses to solvent polarizability effects. Likewise, in the relationship between the $\pi^*$ scale and the reaction field functions of the refractive index (whose square is called the optical dielectric constant $e_\infty$) and the dielectric constant, the aromatic and the halogenated solvents were found to constitute special cases.[10] This feature is also reflected by the polarizability correction term in eq. [13.1.2] below. For the select solvents, the various "polarity" scales are more or less equivalent. A recent account of the various scales has been given by Marcus,[11] and in particular of $\pi^*$ by Laurence et al.,[12] and of $E_T$ by Reichardt.[13]

However, solvation is not the only mode of action taken by the solvent on chemical reactivity. Since chemical reactions typically are accompanied by changes in volume, even reactions with no alteration of charge distribution are sensitive to the solvent. The solvent dependence of a reaction where both reactants and products are neutral species ("neutral" pathway) is often treated in terms of either of two solvent properties. The one is the cohesive energy density $\varepsilon_c$ or cohesive pressure measuring the total molecular cohesion per unit volume,

$$\varepsilon_c = (\Delta H_v - RT) / V \qquad [13.1.1]$$

where:

$\Delta H_v$ molar enthalpy of vaporization
$V$ molar liquid volume

The square root of $\varepsilon_c$ is termed the Hildebrand solubility parameter $\delta_H$, which is the solvent property that measures the work necessary to separate the solvent molecules (disrupt and reorganize solvent/solvent interactions) to create a suitably sized cavity for the solute. The other quantity in use is the internal pressure $P_i$ which is a measure of the change in internal energy U of the solvent during a small isothermal expansion, $P_i = (\partial U/\partial V)_T$. Interesting, and long-known, is the fact that for the highly dipolar and particular for the protic solvents, values of $\varepsilon_c$ are far in excess of $P_i$.[14] This is interpreted to mean that a small expansion does not disrupt all of the intermolecular interactions associated with the liquid state. It has been suggested that $P_i$ does not detect hydrogen bonding but only weaker interactions.

At first, solvent effects on reactivity were studied in terms of some particular solvent parameter. Later on, more sophisticated methods via multiparameter equations were applied such as[15]

$$XYZ = XYZ_0 + s(\pi^* + d\delta) + a\alpha + b\beta + h\delta_H \qquad [13.1.2]$$

where $XYZ_0$, s, a, b, and h are solvent-independent coefficients characteristic of the process and indicative of its sensitivity to the accompanying solvent properties. Further, $\delta$ is a polarizability correction term equal to 0.0 for nonchlorinated aliphatic solvents, 0.5 for polychlorinated aliphatics, and 1.0 for aromatic solvents. The other parameters have been given above, viz. $\pi^*$, $\alpha$, $\beta$, and $\delta_H$ are indices of solvent dipolarity/polarizability, Lewis acidity, Lewis basicity, and cavity formation energy, respectively. For the latter, instead of $\delta_H$, $\delta_H^2$ should be preferred as suggested from regular solution theory.[16]

Let us just mention two applications of the linear solvation energy relationship (LSER). The one concerns the solvolysis of tertiary butyl-halides[17]

$$\log k(Bu^tCl) = -14.60 + 5.10\pi^* + 4.17\alpha + 0.73\beta + 0.0048\delta_H^2$$

$$n = 21, r = 0.9973, s = 0.242$$

and the other deals with the transfer of tetramethylammonium iodide through solvents with methanol as the reference solvent,[16]

$$\Delta G^0_{tr} = 10.9 - 15.6\pi * - 6.2\alpha + 0.022\delta_H$$

$$n = 18, r = 0.997, s = 0.3$$

where:

| | |
|---|---|
| n | number of solvents |
| r | correlation coefficient |
| s | standard deviation |

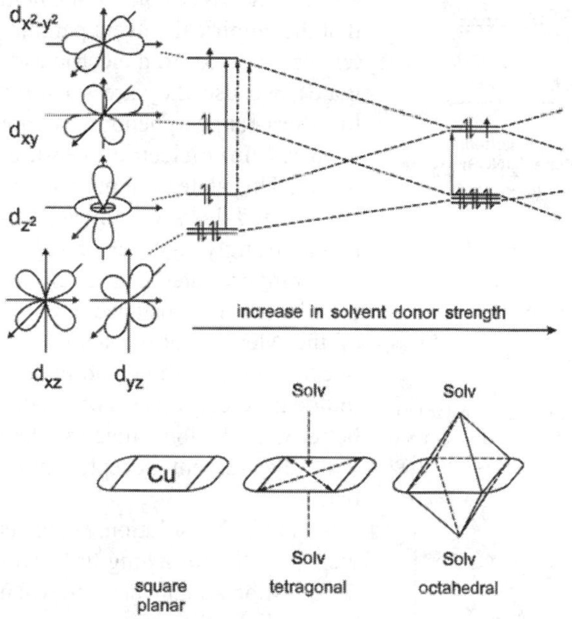


Figure 13.1.3. Relative orbital energy levels for $Cu^{2+}$ in square planar, tetragonal, and octahedral environments [adapted from ref. 18].

We will not finish this section without noting that there are also metal complexes available functioning as color indicators of the coordination properties of solvents.[18] Thus, Cu(tmen)(acac) $ClO_4$, where tmen = N,N,N',N'-tetramethylethylenediamine and acac = acetylacetonate, can be used as a Lewis-basicity indicator, and $Fe(phen)_2(CN)_2$, where phen = 1,10-phenanthroline, as a Lewis-acidity indicator. The physical origin of the underlying color changes is sketched in the Figures 13.1.3 and 13.1.4, as modified from ref. 18. These color indicators can be used as a quick method for assessing the coordination properties of solvents, solvent mixtures, and solutes not yet measured. This is very expedient since some classical parameters, particularly the donor numbers, are arduously amenable. The following equation

$$DN = 195.5 - 0.0102\nu_0 \quad [13.1.3]$$

$$n = 12, r = 0.990, s = 1.37$$

correlates the wave numbers $\nu_0$ (in $cm^{-1}$) of the visible band of $Cu(tmen)(acac)^+$ and the solvent donor numbers. Similarly, the acceptor numbers are expressed as a function of the wave numbers of the long wavelength absorption of $Fe(phen)_2(CN)_2$,

$$AN = -133.8 - 0.00933\nu_0 \quad [13.1.4]$$

$$n = 12, r = 0.980, s = 4.58$$

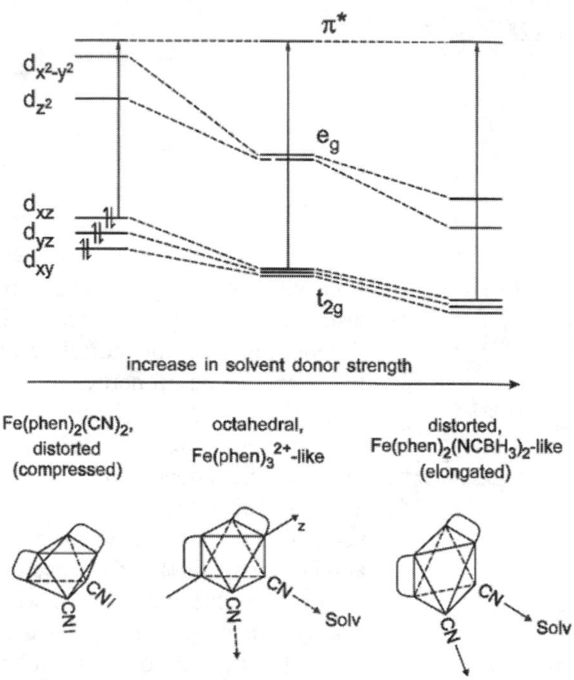


Figure 13.1.4. Simplified orbital scheme for the charge transfer transition in $Fe(phen)_2(CN)_2$ varying with solvation. The diagram, not drawn to scale, is adjusted so that π* is constant [adapted from ref. 18].

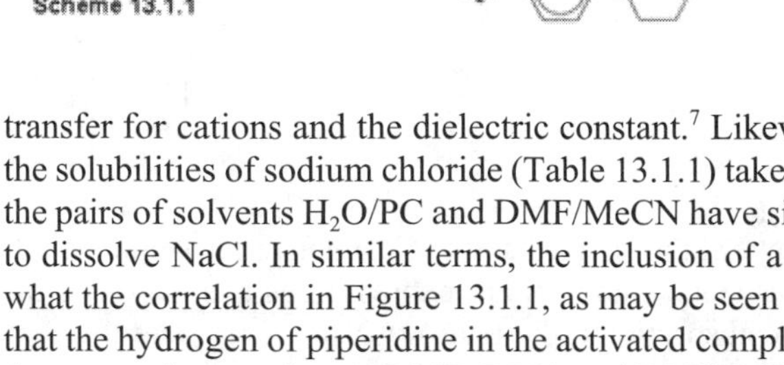


Scheme 13.1.1

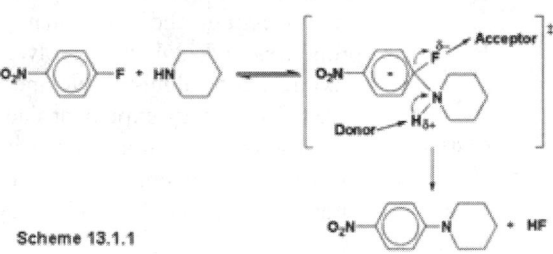

## 13.1.4 DIELECTRIC VS. CHEMICAL APPROACH

Although the success of the empirical solvent parameters has tended to downgrade the usefulness of the dielectric approach, there are correlations that have succeeded as exemplified by Figure 13.1.1. It is commonly held that the empirical solvent parameters are superior to dielectric estimates because they are sensitive to short-range phenomena not captured in dielectric measurements. This statement may not be generalized, however, since it depends strongly on the chemical reaction investigated and the choice of solvents. For instance, the rate of the Menschutkin reaction between tripropylamine and methyl iodide in select solvents correlates better with the log ε function than with the solvent acceptor number.[19]

Thus the solution chemists were puzzled for a long time over the question about when and when not the dielectric approach is adequate. In the meantime, this issue has been unraveled, in that dielectric estimates have no relevance to the solvation of positive (partial) charge. Thus, there is no relationship between the free energies of transfer for cations and the dielectric constant.[7] Likewise, note the solvent-dependence of the solubilities of sodium chloride (Table 13.1.1) taken from Mayer's work.[19] For instance, the pairs of solvents $H_2O$/PC and DMF/MeCN have similar ε's but vastly different abilities to dissolve NaCl. In similar terms, the inclusion of a donor number term improves somewhat the correlation in Figure 13.1.1, as may be seen in Figure 13.1.5. This would suggest that the hydrogen of piperidine in the activated complex becomes acidic and is attacked by the strong donor solvents DMF, DMA and DMSO (Scheme 13.1.1).

**Table 13.1.1. Standard free energies of solution of sodium chloride in various solvents at 25°C. Data of $\Delta G^0_{solv}$ are from reference 19**

| Solvent | $\Delta G^0_{solv}$, kJ mol$^{-1}$ | $\varepsilon_s$ | DN | AN |
|---|---|---|---|---|
| $H_2O$ | -9.0 | 78.4 | 18 | 55 |
| FA | -0.4 | 109 | 24 | 40 |
| NMF | +3.8 | 182 | 27 | 32 |
| MeOH | +14.1 | 32.6 | 19 | 41 |
| DMSO | +14.9 | 46.7 | 30 | 19 |
| DMF | +26.8 | 36.7 | 26 | 16 |
| PC | +44.7 | 65 | 15 | 18 |
| MeCN | +46.8 | 36 | 14 | 19 |

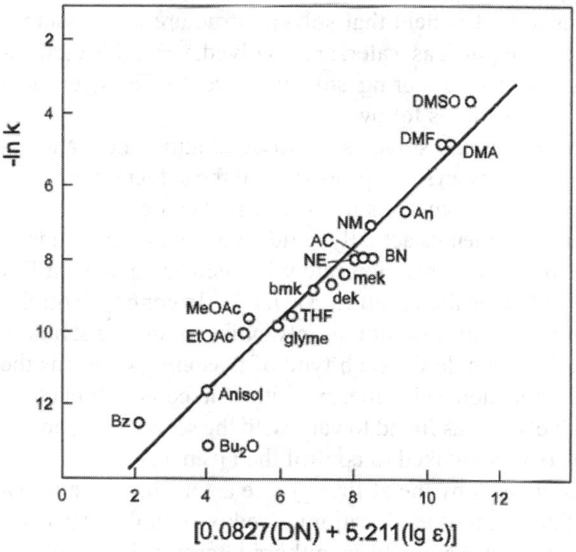

Figure 13.1.5. Correlation diagram for the same reaction as in Figure 13.1.1 [from ref. 21].

On the other hand, if negative charge is solvated in the absence of positive charge capable of solvation, the dielectric constant is often a pretty good guide to ranking changes in reactivity. As a consequence, the dielectric approach has still its place in organic chemistry while it is doomed to complete failure in inorganic reactions where typically cation solvation is involved. For select solvents, ultimately, the dielectric constant is related to the anion-solvating properties of solvents according to the regression equation[4]

$$\log \varepsilon = 0.32 + 0.073 (AN_E) \quad [13.1.5]$$

$$n = 31, r = 0.950, s = 0.129$$

where:

$AN_E$ — $E_T$-based acceptor numbers, $AN_E = -40.52 + 1.29 E_T$

This equation works also quite well for the aromatics and the halogenated solvents, but it does not hold for the protic solvents. For these, the predicted values of the dielectric constants are orders of magnitude too large, revealing how poorly the associates are dissociated by the macroscopically attainable fields. A correlation similar to [13.1.5] has been proposed[20] between the gas phase dipole moment and $\pi^*$

$$\mu(D) = 4.3\pi^* - 0.1 \qquad [13.1.6]$$

$n = 28$, $r = 0.972$, $s = 0.3$

Along these lines the dielectric and the chemical approach are brought under one roof.[4,21] The statement, however, that the terms "good acceptor solvent" and "highly polar solvent" may be used synonymously would seem, though true, to be provocative.

### 13.1.5 CONCEPTUAL PROBLEMS WITH EMPIRICAL SOLVENT PARAMETERS

A highly suspect feature behind the concept of empirical solvent parameters lies in the interpretation of the results in that condensed phase matters are considered from the narrow viewpoint of the solute only with the solvent's viewpoint notoriously neglected. However, the solute is actually probing the overall action of the solvent, comprising two modes of interactions: solute-solvent (solvation) and solvent-solvent (restructuring) effects of unknown relative contribution. Traditionally, it is held that solvent structure only assumes importance when highly structured solvents, such as water, are involved.[22] But this view increasingly turns out to be erroneous. In fact, ignoring solvent-solvent effects, even in aprotic solvents, can lead to wrong conclusions as follows.

In the donor-acceptor approach, solutes and solvents are divided into donors and acceptors. Accordingly, correlations found between some property and the solvent donor (acceptor) ability are commonly thought to indicate that positive (negative) charge is involved. In the case of solvent donor effects this statement is actually valid. We are unaware, in fact, of any exception to the rule saying: "Increase in reaction rate with increasing solvent DN implies that positive charge is developed or localized and *vice versa*".[21] In contrast, correlations with the acceptor number or related scales do not simply point to anion solvation, though this view is commonly held. An example for such type of reasoning concerns the medium effect on the intervalence transition (IT) energy within a certain binuclear, mixed-valence, 5+ cation.[23] As the salt effect was found to vary with the solvent AN, anion, that is counterion, solvation in ion pairs was invoked to control the IT energy.

A conceptual problem becomes obvious by the at first glance astonishing result that the reduction entropies of essentially non-donor cationic redox couples such as $Ru(NH_3)_6^{3+/2+}$ are correlated with the solvent AN.[24] These authors interpreted this solvent dependence as reflecting changes in solvent-solvent rather than solvent-ligand interactions. That the acceptor number might be related to solvent structure is easy to understand since all solvents of high AN always are good donors (but not *vice versa*!) and therefore tend to be increasingly self-associated.[21] There is since growing evidence that the solvent's AN and related scales represent ambiguous solvent properties including solvent structural effects instead of measuring anion solvation in an isolated manner. Thus, correlations between Gibbs energies of cation transfer from water to organic solvents and the solvent DN are improved by the inclusion of a term in $E_T$ (or a combination of $\alpha$ and $\pi^*$).[25] Consequently Marcus et al. rightly recognized that "$E_T$ does not account exclusively for the electron pair acceptance capacity of solvents".[26] In more recent work[27] a direct relationship has been found between the solvent reorganizational energy accompanying the excitation of ruthenium(II) cyano complexes and the solvent acceptor number.

In the basicity scales, on the other hand, complications by solvent structure are not as obvious. If restriction is to aprotic solvents, as is usual, various scales though obtained under different conditions, are roughly equivalent.[21,4] There is for instance a remarkably good relationship between the DN scale (obtained in dilute dichloromethane solution, i.e., with medium effects largely excluded) and the B scale (derived from measurements performed with 0.4 M solutions of MeOD in the various solvents[4]). The relationship between β and B, on the other hand, separates out into families of solvents.[20] Donor measures for protic solvents eventually are hard to assess and often are at considerable variance from one scale to another.[28,29] To rationalize the discrepancies, the concept of "bulk donicity" was introduced[7] but with little success. Instead, the consideration of structure changes accompanying solvation might better help tackle the problem.

Another suspect feature of the common method of interpreting solvent-reactivity correlations is that it is notoriously done in enthalpic (electronic, bond-strength etc.) terms. This way of thinking goes back to the Hughes-Ingold theory. However, many reactions in solution are not controlled by enthalpy changes but instead by entropy. Famous examples are the class of Menschutkin reactions and the solvolysis of t-butyl halides. Both these reaction types are characterized by the development of halide ions in the transition state, which can be considered as ion-pair like. In view of this, rate acceleration observed in good acceptor (or, alternatively, highly polar) solvents seems readily explainable in terms of solvation of the developing halide ion with concomitant carbon-halogen bond weakening. If this is true, most positive activation entropies and highest activation enthalpies should be expected to occur for the poor acceptor solvents. However, a temperature dependence study of the t-butyl halide solvolysis revealed just the opposite.[17] This intriguing feature points to changes in solvent structure as a major determinant of the reaction rate with the ionic transition state acting as a structure maker in poor acceptor solvents, and as a structure maker in the protic solvents.

It is rather ironic that the expected increase in rate with increasing solvent acceptor strength is a result of the coincidence of two, from the traditional point of view, unorthodox facts: (i) The intrinsic solvation of the developing halide ion disfavors the reaction via the entropy term. However, (ii), the extent of that solvation is greater in the poorly coordinating solvents (providing they are polarizable such as the aromatic solvents and the polyhalogenated hydrocarbons). In keeping with this interpretation, the Menschutkin reaction between benzyl bromide and pyridine is characterized by more negative activation volumina (i.e., stronger contraction of the reacting system in going to the activated complex) in poor acceptor (but polarizable) solvents.[30] The importance is evident of studying temperature or pressure dependencies of solvent effects on rate in order to arrive at a physically meaningful interpretation of the correlations.

Another problem with the interpretation of multiparameter equations such as [13.1.2] arises since some of the parameters used are not fully independent of one another. As to this, the trend between $\pi^*$ and $\alpha$ has already been mentioned. Similarly, the $\delta_H$ parameter displays some connection to the polarity indices.[31,32] Virtually, the various parameters feature just different blends of more fundamental intermolecular forces (see below). Because of this, the interpretations of empirical solvent-reactivity correlations are often based more on intuition or preconceived opinion than on physically defined interaction mechanisms. As it turns out, polar solvation has traditionally been overemphasized relative to nonpolar solvation (dispersion and induction), which is appreciable even in polar solvents.

The conceptual problems of the empirical solvent parameters summarized:

- The solvent acceptor number and other "polarity" scales include appreciable, perhaps predominant, contributions from solvent structure changes rather than merely measuring anion solvation.
- Care is urged in a rash interpretation of solvent-reactivity correlations in enthalpic terms, instead of entropic, before temperature-dependence data are available. Actually, free energy alone masks the underlying physics and fails to provide predictive power for more complex situations.
- Unfortunately, the parameters used in LSER's sometimes tend to be roughly related to one another, featuring just different blends of more fundamental intermolecular forces. Not seldom, fortuitous cancellations make molecular behavior in liquids seemingly simple (see below).

Further progress would be gained if the various interaction modes could be separated by means of molecular models. This scheme is in fact taking shape in current years giving rise to a new era of tackling solvent effects as follows.

## 13.1.6 THE PHYSICAL APPROACH

There was a saying that the nineteenth century was the era of the gaseous state, the twentieth century of the solid state, and that perhaps by the twenty-first century we may understand something about liquids.[33] Fortunately, this view is unduly pessimistic, since theories of the liquid state have actively been making breath-taking progress. In the meantime, not only equations of state of simple liquids, that is in the absence of specific solvent-solvent interactions,[34-36] but also calculations of simple forms of intermolecular interactions are becoming available. On this basis, a novel approach to treating solvent effects is emerging, which we may call the physical approach. This way of description is capable of significantly changing the traditionally accepted methods of research in chemistry and ultimately will lay the foundations of the understanding of chemical events from first principles.

A guiding principle of these theories is recognition of the importance of packing effects in liquids. It is now well-established that short-ranged repulsive forces implicit in the packing of hard objects, such as spheres or dumbbells, largely determine the structural and dynamic properties of liquids.[37] It may be noted in this context that the roots of the idea of repulsive forces reach back to Newton who argued that an elastic fluid must be constituted of small particles or atoms of matter, which repel each other by a force increasing in proportion as their distance diminishes. Since this idea stimulated Dalton, we can say that the very existence of liquids helped to pave the way for formulating modern atomic theory with Newton granting the position of its "grandfather".[38]

Since the venerable view of van der Waals, an intermolecular potential composed of repulsive and attractive contributions is a fundamental ingredient of modern theories of the liquid state. While the attractive interaction potential is not precisely known, the repulsive part, because of changing sharply with distance, is treatable by a common formalism in terms of the packing density $\eta$, that is the fraction of space occupied by the liquid molecules. The packing fraction is a key parameter in liquid state theories and is in turn related in a simple way to the hard sphere (HS) diameter $\sigma$ in a spherical representation of the molecules comprising the fluid:

$$\eta = \pi\rho\sigma^3 / 6 = \rho V_{HS} \qquad [13.1.7]$$

where:

| | |
|---|---|
| $\eta$ | packing density |
| $\rho$ | number density N/V= number of particles per unit volume |
| $\sigma$ | HS diameter |
| $V_{HS}$ | HS volume |

For the determination of $\sigma$ (and hence $\eta$), the most direct method is arguably that based on inert gas solubility data.[39,40] However, in view of the arduousness involved and the uncertainties in both the extrapolation procedure and the experimental solubilities, it is natural to look out for alternatives. From the various suggestions,[41,42] a convenient way is to adjust $\sigma$ such that the computed value of some selected thermodynamic quantity, related to $\sigma$, is consistent with experiment. The hitherto likely best method[43] is the following: To diminish effects of attraction, the property chosen should probe primarily repulsive forces rather than attractions. Since the low compressibility of the condensed phase is due to short-range repulsive forces, the isothermal compressibility $\beta_T = -(1/V)(\partial V/\partial P)_T$ might be a suitable candidate, in the framework of the generalized van der Waals (vdW) equation of state

$$\beta_T (RT / V)Q_r = 1 \qquad [13.1.8]$$

where Qr is the density derivative of the compressibility factor of a suitable reference system. In the work referred to, the reference system adopted is that of polar-polarizable spheres in a mean field,

$$Q_r = \left[ 2\frac{5\eta^2 - 2\eta^3}{(1-\eta)^4} - 1 - 2Z_\mu \right] \qquad [13.1.9]$$

where $Z_\mu$ − compressibility factor due to dipole-dipole forces,[43] which is important only for a few solvents such as MeCN and $MeNO_2$. The HS diameters so determined are found to be in excellent agreement with those derived from inert gas solubilities. It may be noted that the method of Ben-Amotz and Willis,[44] also based on $\beta_T$, uses the nonpolar HS liquid as the reference and, therefore, is applicable only to liquids of weak dipole-dipole forces. Of course, as the reference potential approaches that of the real liquid, the HS diameter of the reference liquid should more closely approximate the actual hard-core length. Finally, because of its popularity, an older method should be mentioned that relies on the isobaric expansibility $\alpha_p$ as the probe, but this method is inadequate for polar liquids. It turns out that solvent expansibility is appreciably determined by attractions.

Some values of $\eta$ and $\sigma$ are shown in Table 13.1.2 including the two extreme cases. Actually, water and n-hexadecane have the lowest and highest packing density, respectively, of the common solvents. As is seen, there is an appreciable free volume, which may be expressed by the volume fraction $\eta - \eta_0$, where $\eta_0$ is the maximum value of $\eta$ calculated for the face-centered cubic packing of HS molecules where all molecules are in contact with each other is $\eta_0 = \pi\sqrt{2}/6 = 0.74$. Thus, $1 - \eta_0$ corresponds to the minimum of unoccupied volume. Since $\eta$ typically is around 0.5, about a quarter of the total liquid volume is empty enabling solvent molecules to change their coordinates and hence local density fluctuations to occur.

| | Packing density η | |
|---|---|---|
| minimum | | maximum |
| ≈ 0 | 0.4 - 0.6 | 0.74 |
| perfect gas | liquids | cubic close packed |

**Table 13.1.2. Packing densities in some liquids**

| Liquid | η | free volume, % |
|---|---|---|
| $H_2O$ | 0.41 | 59 |
| n-$C_6$ | 0.50 | 50 |
| Benzene | 0.51 | 49 |
| MeOH | 0.41 | 59 |
| $Et_2O$ | 0.47 | 53 |
| n-$C_{16}$ | 0.62 | 38 |

These considerations ultimately offer the basis of a genuinely molecular theory of solvent effects, as compared to a mean-field theory. Thus, packing and repacking effects accompanying chemical reactions have to be taken into account for any realistic view of the solvent's role played in chemical reactions to be attained. The well-known cavity formation energy is the work done against intermolecular repulsions. At present, this energy is calculated for spherical cavities by the Boublik-Mansoori-Carnahan-Starling-Leland (BMCSL) mixed HS equation of state[45,46]

$$\frac{\Delta G_{rep}}{RT} = 2\frac{\eta d^3}{(1-\eta)^3} + 3\frac{\eta d^2}{(1-\eta)^2} + 3\frac{\eta d(-d^2 + d + 1)}{(1-\eta)} + (-2d^3 + 3d^2 - 1)\ln(1-\eta) \quad [13.1.10]$$

where $d = \sigma_0 / \sigma$ is the relative solute size ($\sigma_0$ is the solute HS diameter, and $\sigma$ is the solvent diameter). Quite recently, a modification of this equation has been suggested for high liquid densities and large solute sizes.[47] Notice that under isochoric conditions the free energy of cavity formation is a totally entropic quantity. Ravi et al[48] have carried out an analysis of a model dissociation reaction ($Br_2 \rightarrow 2Br$) dissolved in a Lennard-Jones solvent (Ne, Ar, and Xe). That and the previous work[49] demonstrated that solvent structure contributes significantly to both chemical reaction volumes (which are defined as the pressure derivatives of reaction free energies) and free energies, even in systems containing no electrostatic or dispersion long-ranged solvent-solute interactions.

Let us now turn to the more difficult case of intermolecular attractive forces. These may be subdivided into:

Long-ranged or unspecific
- dispersion
- induction
- dipole-dipole
- higher multipole

Short-ranged or specific
- electron overlap (charge transfer)
- H-bonding

For the first three ones (dispersion, induction, dipole-dipole forces) adequate calculations are just around the corner. Let us give some definitions.

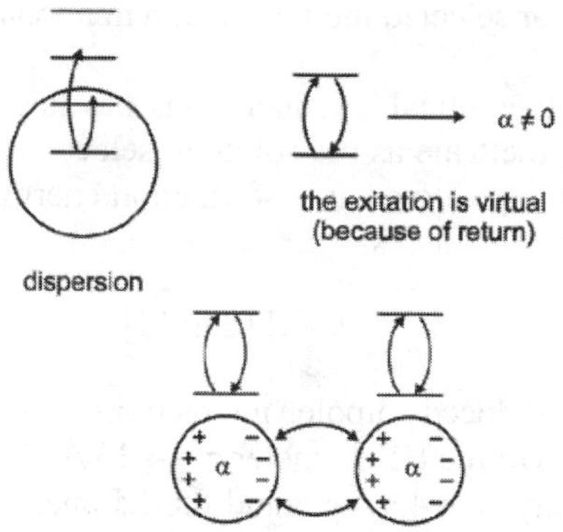

Figure 13.1.6

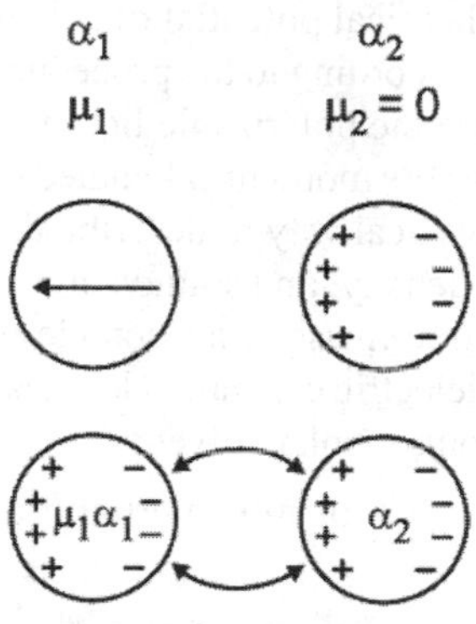

Figure 13.1.7

**Dispersion forces** are the result of the dipolar interactions between the virtually excited dipole moments of the solute and the solvent, resulting in a nonzero molecular polarizability. Although the average of every induced dipole is zero, the average of the product of two induced dipoles is nonzero (Figure 13.1.6).

**Induction forces** are caused by the interaction of the permanent solvent dipole with the solvent dipoles induced by the solute and solvent field (Figure 13.1.7).

Sometimes it is stated that dispersion is a quantum mechanical effect and induction is not. Thus, some clarifying comments are at place here. From the general viewpoint, all effects including polarizability are quantum mechanical in their origin because the polarizability of atoms and molecules is a quantum mechanical quantity and can be assessed only in the framework of quantum mechanics. However, once calculated, one can think of polarizability in classical terms representing a quantum molecular object as a classical oscillator with the mass equal to the polarizability, which is not specified in the classical framework. This is definitely wrong from a fundamental viewpoint, but, as it usually appears with harmonic models, a quantum mechanical calculation and such a primitive classical model give basically the same results about the induction matter. Now, if we implement this classical model, we would easily come up with the induction potential. However, the dispersion interaction will be absent. The point is that to get dispersions, one needs to switch back to the quantum mechanical description where both inductions and dispersions naturally appear. Thus the quantum oscillator may be used resulting in both types of potentials.[50] If in the same procedure one switches to the classical limit (which is equivalent to putting the Plank constant zero) one would get only inductions.

The calculation of the dispersive solvation energy is based on perturbation theories following the Chandler-Andersen-Weeks[51] or Barker-Henderson[52] formalisms, in which long-range attractive interactions are treated as perturbations to the properties of a hard body reference system. Essentially, perturbative theories of fluids are a modern version of van der Waals theory.[53] In the papers reviewed here, the Barker-Henderson approach was utilized with the following input parameters: Lennard-Jones (LJ) energies for the solvent, for which reliable values are now available, the HS diameters of solvent and solute, the solvent polarizability, and the ionization potentials of solute and solvent. A weak point is that in order to get the solute-solvent LJ parameters from the solute and solvent components, some combining rule has to be utilized. However, the commonly applied combining rules appear to be adequate only if solute and solvent molecules are similar in size. For the case of particles appreciably different both in LJ energy and size, the suggestion has been made to use an empirical scaling by introducing empirical coefficients so as to obtain agreement between calculated and experimental solvation energies for selected inert gases and nonpolar large solutes.[54]

In the paper referred to,[54] the relevance of the theoretical considerations has been tested on experimental solvation free energies of nitromethane as the solute in select solvents. The total solvation energy is a competition of the positive cavity formation energy and the negative solvation energy of dispersion and dipolar forces,

$$\Delta G = \Delta G_{cav} + \Delta G_{disp} + \Delta G_{dipolar} \qquad [13.1.11]$$

where the dipolar term includes permanent and induced dipole interactions. The nitromethane molecule is represented by the parameters of the HS diameter $\sigma = 4.36$ Å, the gas-phase dipole moment $\mu = 3.57$ D, the polarizability $\alpha = 4.95$ Å , and the LJ energy $\varepsilon_{LJ}/k$=391K. Further, the solvent is modeled by spherical hard molecules of spherical polarizability, centered dipole moment, and central dispersion potential. To calculate the dipolar response, the Padé approximation was applied for the chemical potential of solvation in the dipolar liquid and then extended to a polarizable fluid according to the procedure of Wertheim. The basic idea of the Wertheim theory is to replace the polarizable liquid of coupled induced dipoles with a fictitious fluid with an effective dipole moment calculated in a self-consistent manner. Further, the Padé form is a simple analytical way to describe the dependence of the dipolar response on solvent polarity, solvent density, and solute/solvent size ratio. The theory/experiment agreement of the net solvation free energy is acceptable as seen in Table 13.1.3 where solvent ordering is according to the dielectric constant. Note that the contribution of dispersion forces is considerable even in strongly polar solvents.

**Table 13.1.3. Thermodynamic potentials (kJ/mol) of dissolution of nitromethane at 25°C. Data are from reference 54**

| Solvent | $\varepsilon_s$ | $\Delta G_{cav}$ | $\Delta G_{disp}$ | $\Delta G_{dipolar}$ | ΔG(calc) | ΔG(exp) |
|---|---|---|---|---|---|---|
| n-$C_6$ | 1.9 | 23.0 | -32.9 | -2.2 | -12.1 | -12.1 |
| c-$C_6$ | 2.0 | 28.1 | -38.1 | -2.8 | -12.7 | -12.0 |
| $Et_3N$ | 2.4 | 24.2 | -33.5 | -2.9 | -12.2 | -15.2 |
| $Et_2O$ | 4.2 | 22.6 | -33.4 | -5.8 | -16.5 | -17.5 |
| EtOAc | 6.0 | 28.0 | -38.5 | -9.7 | -20.1 | -21.2 |
| THF | 7.5 | 32.5 | -41.5 | -12.2 | -21.1 | -21.3 |
| c-hexanone | 15.5 | 35.1 | -39.8 | -17.9 | -22.6 | -21.8 |
| 2-butanone | 17.9 | 28.3 | -33.9 | -18.9 | -24.5 | -21.9 |
| Acetone | 20.7 | 27.8 | -31.2 | -22.1 | -25.5 | -22.5 |
| DMF | 36.7 | 38.1 | -32.1 | -28.5 | -22.5 | -23.7 |
| DMSO | 46.7 | 41.9 | -31.1 | -31.0 | -20.2 | -23.6 |

With an adequate treating of simple forms of intermolecular attractions becoming available, there is currently great interest to making a connection between the empirical scales and solvation theory. Of course, the large, and reliable, experimental databases on

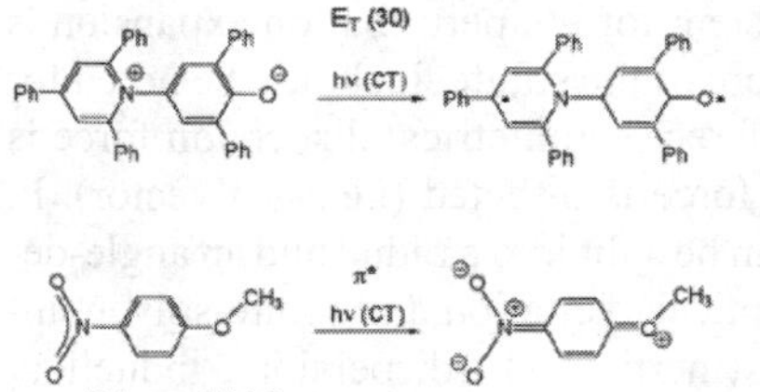

Scheme 13.1.2

empirical parameters are highly attractive for theoreticians for testing their computational models and improving their predictive power. At present, the solvatochromic scales are under considerable scrutiny. Thus, in a recent thermodynamic analysis, Matyushov et al.[55] analyzed the two very popular polarity scales, $E_T(30)$ and $\pi^*$, based on the solvent-induced shift of electronic absorption transitions (Scheme 13.1.2)

Solvatochromism has its origin in changes in both dipole moment and polarizability of the dye upon electronic excitation provoking differential solvation of the ground and excited states. The dipole moment, $\mu_e$, of the excited state can be either smaller or larger than the ground state value $\mu_g$. In the former case one speaks about a negatively solvatochromic dye such as betaine-30, whereas the $\pi^*$ dye 4-nitroanisole is positively solvatochromic. Thus, polar solvent molecules produce a red shift (lower energy) in the former and a blue shift (higher energy) in the latter. On the other hand, polarizability arguably always increases upon excitation. Dispersion interactions, therefore, would produce a red shift proportional to $\Delta\alpha = \alpha_e - \alpha_g$ of the dye. In other words, the excited state is stabilized through strengthening of dispersive coupling. Finally, the relative contributions of dispersion and dipolar interactions will depend on the size of the dye molecules with dispersive forces becoming increasingly important the larger the solute. Along these lines, the dye properties entering the calculations are given in Table 13.1.4. The purpose of the analysis was to determine how well the description in terms of "trivial" dipolar and dispersion forces can reproduce the solvent dependence of the absorption energies (and thereby, by difference to experiment, expose the magnitude of specific forces),

**Table 13.1.4. Dye properties used in the calculations. Data are from reference 55**

| Molecular parameter | Betaine-30 | 4-Nitroanisole |
|---|---|---|
| Vacuum energy gap (eV) | 1.62 | 4.49 |
| $R_0$ (Å) | 6.4 | 4.5 |
| $\alpha_g$ (Å) | 68 | 15 |
| $\Delta\alpha$ (Å) | 61 | 6 |
| $m_g$ (D) | 14.8 | 4.7 |
| $m_e$ (D) | 6.2 | 12.9 |
| $\Delta\mu$ (D) | -8.6 | +8.2 |

$$\hbar\omega_{abs} = \Delta + \Delta E_{rep} + \Delta E_{disp} + \Delta E_{dipolar} + \Delta E_{ss} \qquad [13.1.12]$$

where

| | |
|---|---|
| $\hbar\omega_{abs}$ | absorption energy |
| $\Delta$ | vacuum energy gap |
| $\Delta E_{rep}$ | shift due to repulsion solute-solvent interactions (taken to be zero) |
| $\Delta E_{disp}$ | shift due to dispersion interactions |
| $\Delta E_{dipolar}$ | shift due to dipolar forces of permanent and induced dipoles |
| $\Delta E_{ss}$ | solvent reorganization energy |

For the detailed and arduous calculation procedure, the reader may consult the paper cited. Here, let us just make a few general comments. The solvent influence on intramolecular optical excitation is treated by implementing the perturbation expansion over the solute-solvent attractions. The reference system for the perturbation expansion is chosen to be the HS liquid with the imbedded hard core of the solute. It should be noted for clarity that $\Delta E_{disp}$ and $\Delta E_{dipolar}$ are additive due to different symmetries: dispersion force is non-directed (i.e., is a scalar quantity), and dipolar force is directed (i.e., is a vector). In other words, the attractive intermolecular potential can be split into a radial and an angle-dependent part. In modeling the solvent action on the optical excitation, the solute-solvent interactions have to be dissected into electronic (inertialess) (dispersion, induction, charge-transfer) and molecular (inertial) (molecular orientations, molecular packing) modes. The idea is that the inertial modes are frozen on the time scale of the electronic transition. This is the Franck-Condon principle with such types of transitions called vertical transitions.

Thus the excited solute is to be considered as a Frank-Condon state, which is equilibrated only to the electronic modes, whereas the inertial modes remain equilibrated to the ground state. According to the frozen solvent configuration, the dipolar contribution is represented as the sum of two terms corresponding to the two separate time scales of the solvent, (i) the variation in the solvation potential due to the fast electronic degrees of freedom, and (ii) the work needed to change the solute permanent dipole moment to the excited state value in a frozen solvent field. The latter is calculated for accommodating the solute ground state in the solvent given by orientations and local packing of the permanent solvent dipoles. Finally, the solvent reorganization energy, which is the difference of the average solvent-solvent interaction energy in going from the ground state to the excited state, is extracted by treating the variation with temperature of the absorption energy. Unfortunately, experimental thermochromic coefficients are available for a few solvents only.

The following results of the calculations are relevant. While the contributions of dispersions and inductions are comparable in the $\pi^*$ scale, inductions are overshadowed in the $E_T(30)$ values. Both effects reinforce each other in $\pi^*$, producing the well-known red shift. For the $E_T(30)$ scale, the effects due to dispersion and dipolar solvation have opposite signs making the red shift for nonpolar solvents switch to the blue for polar solvents. Furthermore, there is overall reasonable agreement between theory and experiment for both dyes, as far as the nonpolar and select solvents are concerned, but there are also discrepant solvent classes pointing to other kinds of solute-solvent interactions not accounted for in the model. Thus, the predicted $E_T(30)$ values for protic solvents are uniformly too low, revealing a decrease in H-bonding interactions of the excited state with lowered dipole moment.

Another intriguing observation is that the calculated $\pi^*$ values of the aromatic and chlorinated solvents are throughout too high (in contrast to the $E_T(30)$ case). Clearly, these deviations, reminiscent of the shape of the plots such as Figure 13.1.2, may not be explained in terms of polarizability as traditionally done (see above), since this solvent property has been adequately accommodated in the present model via the induction potential. Instead, the theory/experiment discord may be rationalized in either of two ways. One reason for the additional solvating force can be sought in terms of solute-solvent $\pi$ overlap resulting in exciplex formation. Charge-transfer (CT) interactions are increased between the solvent and the more delocalized excited state.[55] The alternative, and arguably more reasonable, view considers the quadrupole moment which

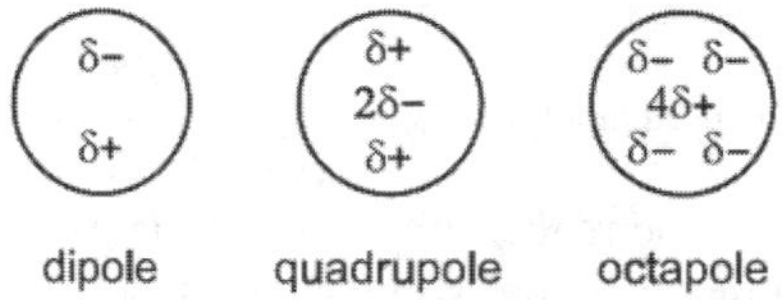


Figure 13.1.8

is substantial for both solvent classes.[56] Recently, this latter explanation in terms of dipole-quadrupole interactions is favored (Fig 13.1.8).

It is well-known that the interaction energy falls off more rapidly the higher the order of the multipole. Thus, for the interaction of an n-pole with an m-pole, the potential energy varies with distance as $E \propto 1/(r^{n+m+1})$. The reason for the faster decrease is that the array of charges seems to blend into neutrality more rapidly with distance the higher the number of individual charges contributing to the multipole. Consequently, quadrupolar forces die off faster than dipolar forces.

It has been calculated that small solute dipoles are even more effectively solvated by solvent quadrupoles than by solvent dipoles.[57] In these terms it is understandable that quadrupolar contributions are more important in the $\pi^*$ than in the $E_T(30)$ scale. Similarly, triethylphosphine oxide, the probe solute of the acceptor number scale, is much smaller than betaine(30) and thus might be more sensitive to quadrupolar solvation. Thus, at long last, the shape of Figure 13.1.2 and similar ones seems rationalized. Note by the way that the quadrupole and CT mechanisms reflect, respectively, inertial and inertialess solvation pathways, and hence could be distinguished by a comparative analysis of absorption and fluorescence shifts (Stokes shift analysis). However, for 4-nitroanisole fluorescence data are not available.

Reverting once more to the thermodynamic analysis of the $\pi^*$ and $E_T(30)$ scales referred to above, it should be mentioned that there are also other theoretical treatments of the solvatochromism of betaine(30). Actually, in a very recent computer simulation,[58] the large polarizability change $\Delta\alpha$ (nearly 2-fold, see Table 13.1.4) upon the excitation of betaine(30) has been (correctly) questioned. (According to a rule of thumb, the increase in polarizability upon excitation is proportional to the ground state polarizability, on the order $\Delta\alpha \approx 0.25\alpha_g$.[50]) Unfortunately, Matyushov et al.[55] derived this high value of $\Delta\alpha$ = 61 Å$^3$ from an analysis of experimental absorption energies based on aromatic, instead of alkane, solvents as nonpolar reference solvents. A lower value of $\Delta\alpha$ would diminish the importance of dispersion interactions.

Further theoretical and computational studies of betaine(30) of the $E_T(30)$ scale are reviewed by Mente and Maroncelli.[58] Despite several differences in opinion obvious in these papers, an adequate treatment of at least the nonspecific components of solvatochromism would seem to be "just around the corner". Finally, a suggestion should be mentioned on using the calculated $\pi^*$ values taken from ref. 55 as a descriptor of nonspecific solvent effects.[59] However, this is not meaningful since these values are just a particular blend of inductive, dispersive, and dipole-dipole forces.

## 13.1.7 SOME HIGHLIGHTS OF RECENT INVESTIGATIONS

### *The like dissolves like rule*

The buzzword "polarity", derived from the dielectric approach, is certainly the most popular word dealing with solvent effects. It is the basis for the famous rule of thumb "similia similibus solvuntur" ("like dissolves like") applied for discussing solubility and miscibility. Unfortunately, this rule has many exceptions. For instance, methanol and toluene, with dielectric constants of 32.6 and 2.4, respectively, are miscible, as are water (78.4) and isopropanol (18.3). The problem lies in exactly what is meant by a "like" solvent. Originally, the term "polarity" was meant to be an abbreviation of "static dipolarity" and was thus associated with solely the dielectric properties of the solvent. Later on, with the advent of the empirical solvent parameters, it has assumed a broader meaning, sometimes even that of the overall solvating power.[13] With this definition, however, the term "polarity" is virtually superfluous.

Clearly, neither the dielectric constant nor the dipole moment is an adequate means to define polarity. The reason is that there are liquids whose constituent molecules have no net dipole moment, for symmetry reasons, but nevertheless have local polar bonds. This class of solvents, already mentioned above, comprises just the notorious troublemakers in solvent reactivity correlations, namely the aromatic and chlorinated solvents. These solvents, called "nondipolar" in the literature,[60] stabilize charge due to higher solvent multipoles (in addition to dispersive forces) like benzene ("quadrupolar") and carbon tetrachloride ("octupolar"). Of this class, the quadrupolar solvents are of primary importance.

Thus, the gas-phase binding energy between of $K^+$ and benzene is even slightly greater than that of $K^+$-water. The interaction between the cation and the benzene molecule is primarily electrostatic in nature, with the ion-quadrupole interaction accounting for 60% of the binding energy.[61] This effect is size-dependent: Whereas at $K^+$ benzene will displace some water molecules from direct contact with the ion, $Na^+_{aq}$ is resistant towards dehydration in an aromatic environment, giving rise to selectivity in some $K^+$ channel proteins.[62]

For the polarity of the C-H bonds, it should be remembered that electronegativity is not an intrinsic property of an atom, but instead varies with hybridization. Only the $C(sp^3)$-H bond can be considered as truly nonpolar, but not so the $C(sp^2)$-H bond.[63] Finally, ethine has hydrogen atoms that are definitely acidic. It should further be mentioned that higher moments or local polarities cannot produce a macroscopic polarization and thus be detected in infinite wavelength dielectric experiments yielding a static dielectric constant close to the squared refractive index. Because of their short range, quadrupolar interactions do not directly contribute to the dielectric constant, but are reflected only in the Kirkwood $g_K$ factor that decreases due to breaking the angular dipole-dipole correlations with increasing quadrupolar strength.

In these terms it is strongly recommended to redefine the term polarity. Instead of meaning solely dipolarity, it should also include higher multipolar properties,

$$\textit{polarity} = \textit{dipolarity} + \textit{quadrupolarity} + \textit{octupolarity} \qquad [13.1.13]$$

This appears to be a better scheme than distinguishing between truly nonpolar and nondipolar solvents.[56] A polar molecule can be defined as having a strongly polar bond, but need not necessarily be a dipole. In this framework, the solvating power of the "nondipolar" solvents need no longer be viewed as anomalous or as essentially dependent on specific solvation effects.[10] Beyond this it should be emphasized that many liquids have both a dipole moment and a quadrupole moment, water for example. However, for dipolar solvents such as acetonitrile, acetone, and dimethyl sulfoxide, the dipolar solvation mechanism will be prevailing. For less dipolar solvents, like tetrahydrofuran, quadrupoles and dipoles might equally contribute to the solvation energetics.[57]

Notwithstanding this modified definition, the problem with polarity remains in that positive and negative charge solvation is not distinguished. As already pointed out above, there is no general relationship between polarity and the cation solvation tendency. For example, although nitromethane ($MeNO_2$) and DMF have the same dielectric constant, the extent of ion pairing in $MeNO_2$ is much greater than that in DMF. This observation is attributed to the weak basicity of $MeNO_2$ which poorly solvates cations. As a result, ion pairing is stronger in $MeNO_2$ in spite of the fact that long-range ion-ion interactions in the two solvents are equal.

Finally, a potential problem with polarity rests in the fact that this term is typically associated with enthalpy. But caution is urged in interpreting the like-dissolves-like rule in terms of enthalpy. It is often stated for example that nonpolar liquids such as octane and carbon tetrachloride are miscible because the molecules are held together by weak dispersion forces. However, spontaneous mixing of the two phases is driven not by enthalpy, but by entropy.

### *Water's anomalies*

The outstanding properties and anomalies of water have fascinated and likewise intrigued physicists and physical chemists for a long time. During the past decades much effort has been devoted to finding phenomenological models that explain the (roughly ten) anomalous thermodynamic and kinetic properties, including the density maximum at 4°C, the expansion upon freezing, the isothermal compressibility minimum at 46°C, the high heat capacity, the decrease of viscosity with pressure, and the remarkable variety of crystalline structures. Furthermore, isotope effects on the densities and transport properties do not possess the ordinary mass or square-root-mass behavior.

Some of these properties are known from long ago, but their origin has been controversial. From the increasingly unmanageable number of papers that have been published on the topic, let us quote only a few that appear to be essential. Above all, it seems to be clear that the exceptional behavior of water is not simply due to hydrogen bonding, but instead due to additional "trivial" vdW forces as present in any liquid. A hydrogen bond occurs when a hydrogen atom is shared between generally two electronegative atoms; vdW attractions arise from interactions among fixed or induced dipoles. The superimposition and competition of both is satisfactorily accommodated in the framework of a "mixture model".

The mixture model for liquid water, promoted in an embryonic form by Röntgen[64] over a century ago, but later discredited by Kauzmann[65] and others,[66] is increasingly gaining ground. Accordingly there are supposed to be two major types of intermolecular bonding configurations, an open bonding form, with a low density, such as occurs in ice-Ih, plus a dense bonding form, such as occurs in the most thermodynamically stable dense forms of ice, e.g., ice-II, -III, -V, and -VI.[67] In these terms, water has many properties of the glassy states associated with multiple hydrogen-bond network structures.[68] Clearly, for fluid properties, discrete units, $(H_2O)_n$, which can move independently of each other are required. The clusters could well be octamers dissociating into tetramers, or decamers dissociating into pentamers.[69,70] (Note by the way that the unit cell of ice contains eight water molecules.) However, this mixture is not conceived to be a mixture of ices, but rather is a dynamic (rapidly fluctuating) mixture of intermolecular bonding types found in the polymorphs of ice. A theoretical study of the dynamics of liquid water has shown that there exist local collective motions of water molecules and fluctuation associated with hydrogen bond rearrangement dynamics.[68] The half-life of a single H bond estimated from transition theory is about $2\times10^{-10}$ s at 300K.[71] In view of this tiny lifetime it seems more relevant to identify the two mixtures not in terms of different cluster sizes, but rather in terms of two different bonding modes. Thus, there is a competition between dispersion interactions that favor random dense states and hydrogen bonding that favors ordered open states. Experimental verification of the two types of bonding has been reviewed by Cho et al.[72]

From X-ray and neutron scattering, the open structure is characterized by an inner tetrahedral cage of four water molecules surrounding a central molecule, with the nearest-neighbor O···O distance of about 2.8 Å. This distance, as well as the nearest-neighbor count of four, remains essentially intact in both in all crystalline ice polymorphs and in the liquid water up to near the boiling point. In this open tetrahedral network the second-neighbor O···O distance is found at 4.5 Å. However, and most intriguing, another peak in the O···O radial distribution function (RDF), derived from a nonstandard structural approach (ITD), is found near 3.4 Å,[73] signaling a more compact packing than in an ordinary H-bonding structure. This dense bonding form is affected through dispersive O···O interactions supplanting H-bonding. Note, however, that in this array the H bonds may not be envisaged as being really broken but instead as being only bent. This claim is substantiated by a sophisticated analysis of vibrational Raman spectra[74] and mid-IR spectra[70] pointing to the existence of essentially two types of H bonds differing in strength, with bent H bonds being weaker than normal (i.e., linear) H bonds. It should be mentioned that the 3.4 Å feature is hidden by the ordinary minimum of open tetrahedral contributions to the RDF. Because of this, the ordinary integration procedure yields coordination numbers greater than four,[75] which confuses the actual situation. Instead, it is the outer structure that is changing whereas the inner coordination sphere remains largely invariant. Even liquid water has much of the tetrahedral H-bonding network of ice I.

As temperature, or pressure, is raised, the open tetrahedral hydrogen bonding structure becomes relatively less stable and begins to break down, creating more of the dense structure. Actually all the anomalous properties of water can be rationalized on the basis of this open → dense transformation. An extremum occurs if two opposing effects are superimposed. The density maximum, for instance, arises from the increase in density due to the thermal open → dense transformation and the decrease in density due to a normal thermal expansion.[72] As early as 1978 Benson postulated that the abnormal heat capacity of water is due to an isomerization reaction.[76] A clear explanation of the density anomaly is given by Silverstein et al.[77]: "The relatively low density of ice is due to the fact that H-bonding is stronger than the vdW interactions. Optimal H-bonding is incommensurate with the tighter packing that would be favored by vdW interactions. Ice melts when the thermal energy is sufficient to disrupt and disorder the H-bonds, broadening the distribution of H-bond angles and lengths. Now among this broadened H-bond distribution, the vdW interactions favor those conformations of the system that have higher density. Hence liquid water is denser than ice. Heating liquid water continues to further deform hydrogen bonds and increase the density up to the density anomaly temperature. Further increase of temperature beyond the density anomaly weakens both H bonds and vdW bonds, thus reducing the density, as in simpler liquids."

The same authors commented on the high heat capacity of water as follows: "Since the heat capacity is defined as $C_P = (\partial H/\partial T)_P$ the heat capacity describes the extent to which some kind of bonds are broken (increasing enthalpy) with increasing temperature. Breaking bonds is an energy storage mechanism. The heat capacity is low in the ice phase because thermal energy at those temperatures is too small to disrupt the H bonds. The heat capacity peaks at the melting temperature where the solid-like H bonds of ice are weakened to become the liquid-like H bonds of liquid water. The reason liquid water has a higher heat capacity than vdW liquids have is because water has an additional energy storage mechanism, namely the H bonds, that can also be disrupted by thermal energies."

Summed up, it appears that any concept to be used in a realistic study of water should have as a fundamental ingredient the competition between expanded, less dense structures, and compressed, more dense ones. Thus, the outer structure in the total potential of water should be characterized by a double minimum: open tetrahedral structure with a second-neighbor O···O distance of 4.5 Å and a bent H-bond structure with an O···O non-H-bonded distance of about 3.4 Å. Actually, according to a quite recent theoretical study, all of the anomalous properties of water are qualitatively explainable by the existence of two competing equilibrium values for the interparticle distance.[78] Along these lines the traditional point of view as to the structure of water is dramatically upset. Beyond that, also the classical description of the hydrogen bond needs revision. In contrast to a purely electrostatic bonding, quite recent Compton X-ray scattering studies have demonstrated that the hydrogen bonds in ice have substantial covalent character,[79] as already suggested by Pauling in the 1930s.[80] In overall terms, a hydrogen bond is comprised of electrostatic, dispersion, charge-transfer, and steric repulsion interactions. Similarly, there are charge-transfer interactions between biological complexes and water[81] that could have a significant impact on the understanding of biomolecules in aqueous solution.

Finally, we return to the physical meaning of the large difference, for the protic solvents, between the cohesive energy density $\varepsilon_c$ and the internal pressure $P_i$, quoted in section 13.1.3. For water this difference is highest with the factor $\varepsilon_c/P_i$ equal to 15.3. At first glance this would seem explainable in the framework of the mixture model if H bonding is insensitive to a small volume expansion. However, one should have in mind the whole pattern of the relationship between the two quantities. Thus, $\varepsilon_c - P_i$ is negative for nonpolar liquids, relatively small (positive or negative) for polar non-associated liquids, and strongly positive for H-bonded liquids. A more rigorous treatment[41] using the relations, $P_i = (\partial U/\partial V)_T = T(\partial P/\partial T)_V - P$ and the thermodynamic identity $(\partial S/\partial V)_T = (\partial P/\partial T)_V$ reveals that the relationship is not as simple and may be represented by the following equation with dispersion detached from the other types of association,

$$\varepsilon_c - P_i = P - \frac{RT}{V}\left[Z_0 + \frac{U_{disp}}{RT}\right] - \rho U_{ass} + \rho^2 T(\partial S_{ass} / \partial \rho)_T \qquad [13.1.14]$$

where:

| | |
|---|---|
| P | external pressure |
| V | liquid volume |
| $Z_0$ | compressibility factor due to intermolecular repulsion |
| $U_{disp}$ | potential of dispersion |
| $U_{ass}$ | potential of association excluding dispersion |
| ρ | liquid number density |
| $S_{ass}$ | entropy of association excluding dispersion |

With the aid of this equation we readily understand the different ranges of $\varepsilon_c - P_i$ found for the different solvent classes. Thus, for the nonpolar liquids, the last two terms are negligible, and for the usual values, $Z_o \approx 10$, $-U_{disp}/RT \approx 8$, and $V \approx 150\ cm^3$, we obtain the typical order of $\varepsilon_c - P_i \approx -(300 - 400)$ atm (equal to -(30 - 40) J cm$^{-3}$, since 1 J cm$^{-3} \equiv 9.875$ atm). For moderately polar liquids, only the last term remains small, while the internal energy of dipolar forces is already appreciable $-\rho U_{polar} \approx (200\text{-}500)$ atm giving the usual magnitude of $\varepsilon_c - P_i$. For H-bonded liquids, ultimately, the last term turns out to dominate reflecting the large increase in entropy of a net of H-bonds upon a small decrease in liquid density.

**Table 13.1.5. Solution thermodynamics of methane in water and carbon tetrachloride at 25°C. [Data from T. Lazaridis and M. E. Paulaitis, *J. Phys. Chem.*, 96, 3847 (1992) and ref. 108]**

| | Water | $CCl_4$ |
|---|---|---|
| $\Delta S^*$, J/mol K | - 64.4 | - 7.1 |
| $300\Delta S^*$, kJ/mol | - 19.3 | - 2.1 |
| $\Delta H^\circ$, kJ/mol | - 10.9 | - 1.2 |
| $\Delta G^*$, kJ/mol | + 8.4 | + 0.9 |
| $\Delta C_p$, J/mol K | 217.5 | 0 to 42 |

### *The hydrophobic effect*

In some respects the hydrophobic effect may be considered as the converse of the like-dissolves-like rule. The term hydrophobic effect refers to the unusual behavior of water towards nonpolar solutes. Unlike simple organic solvents, the insertion of nonpolar solutes into water is (1) strongly unfavorable though slightly favored by enthalpy, but (2) strongly opposed by a large, negative change in entropy at room temperature, and (3) accompanied by a large positive heat capacity. An example is given in Table 13.1.5 for the thermodynamic properties of methane dissolved in water and in carbon tetrachloride. In dealing with the entropy (and free energy) of hydration, a brief remark on the choice of standard states is in order. The standard molar entropy of dissolution, $\Delta_{solv}S^\circ$ pertains to the transfer from a 1 atm gas state to a 1 mol $L^{-1}$ solution and hence includes compression of the gas phase from 1 mol contained in 24.61 L (at 300 K) to 1 mol present in 1 L. Since theoretical calculations disregard volume contributions, it is proper to exclude the entropy of compression equal to -Rln24.61 = -26.63 J $K^{-1}$ $mol^{-1}$, and instead to deal with $\Delta_{solv}S^*$.[82] Thus,

$$\Delta_{solv}S^* = \Delta_{solv}S^o + 26.63JK^{-1}mol^{-1} \quad [13.1.15]$$

and

$$\Delta_{solv}G^*_{300} = \Delta_{solv}H^o - 300\Delta_{solv}S^* \quad [13.1.16]$$

Hydrophobicity forms the basis for many important chemical phenomena including the cleaning action of soaps and detergents, the influence of surfactants on surface tension, the immiscibility of nonpolar substances in water,[83] the formation of biological membranes and micelles,[84,85] the folding of biological macromolecules in water,[86] clathrate hydrate formation,[87] and the binding of a drug to its receptor.[88] Of these, particularly intriguing is the stabilization of protein structure due to the hydrophobicity of nonpolar groups. Hydrophobic interactions are considerably involved in self-assembly, leading to the aggregation of nonpolar solutes, or equivalently, to the tendency of nonpolar oligomers to adopt chain conformations in water relative to a nonpolar solvent.[89]

Ever-increasing theoretical work within the last years is being lifting the veil of secrecy about the molecular details of the hydrophobic effect, a subject of vigorous debate. Specifically, the scientific community would eagerly like to decide whether the loss in entropy stems from the water-water or the water-solute correlations. There are two concepts. The older one is the clathrate cage model reaching back to the "iceberg" hypothesis of Frank and Evans,[90] and the other, newer one, is the cavity-based model. It should be stressed here that the vast literature on the topic is virtually impossible to survey comprehensively. In the following we will cite only a few papers (and references therein) that paved the way to the present state of the art.

The clathrate cage model states that the structure of water is strengthened around a hydrophobic solute, thus causing a large unfavorable entropic effect. The surrounding water molecules adopt only a few orientations (low entropy) to avoid "wasting" hydrogen bonds, with all water configurations fully H-bonded (low energy). There is experimental evidence of structure strengthening, such as NMR and FT-IR studies,[91] NMR relaxation,[92] dielectric relaxation,[93] and HPLC.[94] A very common conclusion is that the small solubility of nonpolar solutes in water is due to this structuring process.

In the cavity-based model the hard core of water molecules is more important to the hydrophobic effect than H-bonding of water. The process of solvation is dissected into two components, the formation of a cavity in the water to accommodate the solute and the interaction of the solute with the water molecules. The creation of a cavity reduces the volume of the translational motion of the solvent particles. This causes an unfavorable entropic effect. The total entropy of cavity formation at constant pressure[54]

$$\Delta S_{cav,P} = \rho\alpha_p\left(\partial\Delta G_{cav} / \partial\rho\right)_T - \Delta G_{cav} / T \qquad [13.1.17]$$

where

$\Delta S_{cav,P}$ cavity formation entropy at constant pressure
$\rho$ liquid number density
$\Delta G_{cav}$ free energy of cavity formation

is the result of the opposing nature of the (positive) liquid expansibility term and the (negative) chemical potential summand. Along these lines the large and negative entropy of cavity formation in water is traced to two particular properties of water: the small molecular size ($\sigma - 2.87$ Å) and the low expansibility ($\alpha_p = 0.26\text{x}10^{-3}$ $K^{-1}$), with the latter having the greater impact. It is interesting to note that in both aspects water is extraordinary. Water's low expansibility reflects the fact that chemical bonds cannot be stretched by temperature. There is also a recent perturbation approach showing that it is more costly to accommodate a cavity of molecular size in water than in hexane as example.[95] Considering the high fractional free volume for water (Table 13.1.2), it is concluded that the holes in water are distributed in smaller packets.[96] Compared to a H-bonding network, a hard-sphere liquid finds more ways to configure its free volume in order to make a cavity.

In the cavity-based model, large perturbations in water structure are not required to explain hydrophobic behavior. This conclusion arose out of the surprising success of the scaled particle theory (SPT),[39] which is a hard-sphere fluid theory, to account for the free energy of hydrophobic transfers. Since the theory only uses the molecular size, density, and pressure of water as inputs and does not explicitly include any special features of H-bonding of water, the structure of water is arguably not directly implicated in the hydration thermodynamics. (However, the effect of H-bonds of water is implicitly taken into account through the size and density of water.) The proponents of this hypothesis argue that the entropic and enthalpic contributions arising from the structuring of water molecules largely compensate each other. In fact, there is thermodynamic evidence of enthalpy-entropy compensation of solvent reorganization.[97-100] Furthermore, recent simulations[101,102] and neutron scattering data[103-105] suggest that solvent structuring might be of much lower extent than previously believed. Also a recent MD study report[106] stated that the structure of water is preserved, rather than enhanced, around hydrophobic groups.

Finally, the contribution of solute-water correlations to the hydrophobic effect may be displayed, for example, in the framework of the equation

$$\Delta G_{sol} = \Delta G_{cav} + \Delta G_{att} \qquad [13.1.18]$$

where:

$\Delta G_{sol}$ free energy of dissolution
$\Delta G_{cav}$ free energy of cavity formation
$\Delta G_{att}$ free energy of attractive interactions

This equation has been used by de Souza and Ben-Amotz[107] to calculate values of $\Delta G_{att}$ from the difference between experimental solubilities of rare gases, corresponding to $\Delta G_{sol}$, and $\Delta G_{cav}$ assessed from eqn. [13.1.10], i.e., using a hard-sphere fluid (HF) model. The values of $\Delta G_{att}$ so obtained have been found to correlate with the solute polarizabilities suggesting a dispersive mechanism for attractive solvation. It is interesting to note that, in water, the solubility of the noble gases increases with increasing size, in contrast to the aliphatic hydrocarbons whose solubility decreases with size. This differential behavior is straightforwardly explained in terms of the high polarizability of the heavy noble gases having a large number of weakly bound electrons, which strengthens the vdW interactions with water. It can be shown that for noble gases, on increasing their size, the vdW interactions increase more rapidly than the work of cavity creation, enhancing solubility. On the contrary, for the hydrocarbons, on increasing the size, the vdW interactions increase less rapidly than the work of cavity creation, lowering the solubility.[108]

We have seen that there is evidence of either model, the clathrate cage model and the cavity-based model. Hence the importance of water structure enhancement in the hydrophobic effect is equivocal. The reason for this may be twofold. First, theoretical models have many adjustable parameters, so their physical bases are not always clear. Second, the free energy alone masks the underlying physics in the absence of a temperature dependence study, because of, amongst other things, the entropy-enthalpy compensation noted above. In place of the free energy, other thermodynamic derivatives are more revealing. Of these, the study of heat capacity changes arguably provides a better insight into the role of changes in water structure upon hydration than analysis of entropy or enthalpy changes alone. Note that heat capacity is the most complex of the four principal thermodynamic parameters describing solvation ($\Delta G$, $\Delta H$, $\Delta S$, $\Delta C_p$), with the following connections,

$$\Delta C_p = \frac{\partial \Delta H}{\partial T} = T\frac{\partial \Delta S}{\partial T} = -T^2\frac{\partial^2 \Delta G}{\partial T^2} \qquad [13.1.19]$$

It should be stressed that the negative entropy of hydration is virtually not the main characteristic feature of hydrophobicity, since the hydration of any solute, polar, nonpolar, or ionic, is accompanied by a decrease in entropy.[109] The qualitative similarity in hydration entropy behavior of polar and nonpolar groups contrasts sharply with the opposite sign of the heat capacity change in polar and nonpolar group hydration. Nonpolar solutes have a large positive heat capacity of hydration, while polar groups have a smaller, negative one. Thus, the large heat capacity increase might be what truly distinguishes the hydrophobic effect from other solvation effects.[110]

Recently, this behavioral difference of nonpolar and polar solutes could be reproduced by heat capacity calculations using a combination of Monte Carlo simulations and the random network model (RNM) of water.[110-112] It was found that the hydrogen bonds between the water molecules in the first hydration shell of a nonpolar solute are shorter and less bent (i.e., are more ice-like) compared to those in pure water. The opposite effect occurs around polar solutes (the waters become less ice-like). The increase in H-bond length and angle has been found to decrease the water heat capacity contribution, while decreases in length and angle have been found to cause the opposite effect.

Note further that a large heat capacity implies that the enthalpy and entropy are strong functions of temperature, and the free energy vs. temperature is a curved function, increasing at low temperatures and decreasing at higher temperatures. Hence there will be a temperature at which the solubility of nonpolar in water is a minimum. The low solubility of nonpolar species in water at higher temperatures is caused by unfavorable enthalpic interactions, not unfavorable entropy changes. Some light on these features has been shed by using a "simple" statistical mechanical MB model of water in which the water molecules are represented as Lennard-Jones disks with hydrogen bonding arms.[113] (the MB model is called this because of the resemblance of each model water to the Mercedes-Benz logo.) As an important result, the insertion of a nonpolar solute into cold water causes ordering and strengthening of the H bonds in the first shell, but the reverse applies in hot water. This provides a physical interpretation for the crossover temperatures $T_H$ and $T_S$, where the enthalpy and entropy of transfer equal zero. $T_H$ is the temperature at which H-bond reorganizations are balanced by solute-solvent interactions. On the other hand, $T_S$ is the temperature at which the relative H-bonding strengths and numbers of shell and bulk molecules reverse roles.

Although the large positive free energy of mixing of hydrocarbons with water is dominated by entropy at 25°C, it is dominated by enthalpy at higher temperatures (112°C from Baldwin's extrapolation for hydrocarbons, or 150°C from the measurements of Crovetto for argon)[113] where the disaffinity of oil for water is maximal. Ironically so, where hydrophobicity is strongest, entropy plays no role. For this reason, models and simulations of solutes that focus on cold water, around or below 25°C, miss much of the thermodynamics of the oil/water solvation process. Also, a clathrate-like solvation shell emerged from a recent computer simulation study of the temperature dependence of the structural and dynamical properties of dilute $O_2$ aqueous solutions.[114] In the first hydration shell around $O_2$, water-water interactions are stronger and water diffusional and rotational dynamics slower than in the bulk. This calls to one's mind an older paper by Hildebrand[115] showing that at 25°C, methane's diffusion coefficient in water is 40% less than it is in carbon tetrachloride ($D(H_2O) = 1.42x10^{-5}$ $cm^2/s$ vs $D(CCl_4) = 2.89x10^{-5}$ $cm^2/s$). Presumably the loose clathrate water cages serve to inhibit free diffusion of the nonpolar solute. From these data it seems that both the nonpolar solute and the aqueous solvent experience a decrease in entropy upon dissolving in water. It should also be mentioned in this context that pressure increases the solubility. The effect of pressure on the entropy was examined and it was found that increase in the pressure causes a reduction of orientational correlations, in agreement with the idea of pressure as a "structure breaker" in water.[116] Actually, frozen clathrate hydrates trapped beneath oceans and arctic permafrost may contain more than 50% of the world's organic carbon reserves.[117,118] Likewise, the solubility of aromatics is increased at high pressure and temperature, with $\pi$ bond interactions involved.[119]

Only at first glance, the two approaches, the clathrate cage model and the cavity-based model, looked very different, the former based on the hydrogen bonding of water, and the later on the hard core of water. But taken all results together it would appear that both are just different perspectives on the same physics with different diagnostics reporting consequences of the same shifted balance between H bonds and vdW interactions. Actually, in a very recent paper, a unified physical picture of hydrophobicity based on both the hydrogen bonding of water and the hard-core effect has been put forward.[120] Hydrophobicity features an interplay of several factors.

### *The structure of liquids*

A topic of abiding interest is the issue of characterizing the order in liquids which may be defined as the entropy deficit due to preferential orientations of molecular multipoles relative to random orientations (orientational order) and nonuniformly directed intermolecular forces (positional order). Phenomenologically, two criteria are often claimed to be relevant for deciding whether or not a liquid is to be viewed as ordered: the Trouton entropy of vaporization or Trouton quotient and the Kirkwood correlation factor $g_K$.[121] Strictly speaking, however, both are of limited relevance to the issue.

The Trouton quotient is related to structure of the liquids only at their respective boiling points, of course, which may markedly differ from their structures at room temperature. This should be realized especially for the high-boiling liquids. As these include the highly dipolar liquids such as HMPA, DMSO, and PC, the effect of dipole orientation to produce order in the neat liquids remains obscure. All that can be gleaned from the approximate constancy of the Trouton quotient for all sorts of aprotic solvents is that at the boiling point the entropy of attractions becomes unimportant relative to the entropy of unpacking liquid molecules, that is repulsions.[122] In terms of the general concept of separating the interaction potential into additive contributions of repulsion and attraction, the vaporization entropy can be expressed by

$$\Delta_v H / T_b = \Delta_v S = \Delta S_o - S_{att} \qquad [13.1.20]$$

where:

$\Delta_v H$ vaporization enthalpy at the boiling point $T_b$
$\Delta S_o$ entropy of depacking hard spheres = entropy of repulsion
$S_{att}$ entropy of attraction

$\Delta S_o$ can be separated into the entropy $f_o(\eta)$ of depacking hard spheres to an ideal gas at the liquid volume V, and the entropy of volume expansion to $V_g = RT/P$,

$$\Delta S_o / R = f_o(\eta) + \ln(V_g / V) \qquad [13.1.21]$$

Further, $f_o(\eta)$ can be derived from the famous Charnahan-Starling equation as[43]

$$f_o(\eta) = (4\eta - 3\eta^2) / (1-\eta)^2 \qquad [13.1.22]$$

It can in fact be shown that, for the nonpolar liquids, $\Delta_v S/R$ is approximately equal to $f_o(\eta)$.[122] Along these lines, Trouton's rule is traced to two facts: (i) The entropy of depacking is essentially constant, a typical value being $\Delta S_o/R \approx 9.65$, due to the small range of packing densities encompassed. In addition, the entropies of HS depacking and of volume change vary in a roughly compensatory manner. (ii) The entropy of attraction is insignificant for all the aprotics. Only for the protics the contributions of $S_{att}$ may not be neglected. Actually the differences between $\Delta_v S$ and $\Delta S_o$ reflect largely the entropy of hydrogen bonding. However, the application of the eqns [13.1.20] to [13.1.22] to room temperature data reveals, in contrast to boiling point conditions, not unimportant contributions of $S_{att}$ even for some aprotic liquids (see below).

On the other hand, the $g_K$ factor is, loosely speaking, a measure of the deviation of the relative dielectric constant of the solvent with the same dipole moment and polarizability would have if its dipoles were not correlated by its structure. However, the $g_K$ factor is only the average cosine of the angles between the dipole moments of neighboring molecules. There may thus be orientational order in the vicinity of a molecule despite a $g_K$ of unity if there are equal head-to-tail and antiparallel alignments. Furthermore, the $g_K$ factor is not related to positional order.

The better starting point for assessing order would be experimental room temperature entropies of vaporization upon applying the same method as described above for the boiling point conditions. (Note however, that the packing density, and hence the molecular HS diameter, varies with temperature. Therefore, in the paper[122] the packing densities have been calculated for near boiling point conditions.) Thus we choose the simplest fluid as the reference system. This is a liquid composed of spherical, nonpolar molecules, approximated to a HS gas moving in a uniform background or mean field potential provided by the attractive forces.[43] Since the mean field potential affects neither structure nor entropy, the excess entropy $S_{ex}$

$$S_{ex} / R = \ln\left(V_g / V\right) + f_o(\eta) - \Delta_v H / RT \qquad [13.1.23]$$

may be viewed as an index of orientational and positional order in liquids. It represents the entropy of attractions plus the contributions arising from molecular nonsphericity. This latter effect can be estimated by comparing the entropy deficits for spherical and hard convex body repulsions in the reference system. Computations available for three n-alkanes suggest that only $\approx$20% of $S_{ex}$ are due to nonsphericity effects.

**Table 13.1.6. Some liquid properties concerning structure. Data are from ref. 55 and 128 ($g_K$)**

| Solvent | $\Delta_v H/RT$ | $\Delta_v S_o/R$ | $-S_{ex}/R$ | $g_K$ |
|---|---|---|---|---|
| c-$C_6$ | 18.21 | 17.89 | 0.32 | |
| THF | 12.83 | 12.48 | 0.35 | |
| $CCl_4$ | 13.08 | 12.72 | 0.36 | |
| n-$C_5$ | 10.65 | 10.21 | 0.44 | |
| $CHCl_3$ | 12.62 | 12.16 | 0.46 | |
| $CH_2Cl_2$ | 11.62 | 11.14 | 0.48 | |
| Ph-H | 13.65 | 13.08 | 0.57 | |
| Ph-Me | 15.32 | 14.62 | 0.70 | |
| n-$C_6$ | 12.70 | 11.89 | 0.81 | |
| $Et_2O$ | 10.96 | 10.13 | 0.83 | |
| c-hexanone | 18.20 | 17.30 | 0.90 | |

| Solvent | $\Delta_v H/RT$ | $\Delta_v S_o/R$ | $-S_{ex}/R$ | $g_K$ |
|---|---|---|---|---|
| Py | 16.20 | 15.22 | 0.98 | |
| MeCN | 13.40 | 12.37 | 1.03 | 1.18 |
| $Me_2CO$ | 12.50 | 11.40 | 1.10 | 1.49 |
| MeOAc | 13.03 | 11.79 | 1.24 | |
| EtCN | 14.53 | 13.25 | 1.28 | 1.15 |
| $PhNO_2$ | 22.19 | 20.89 | 1.30 | 1.56 |
| $MeNO_2$ | 15.58 | 14.10 | 1.48 | 1.38 |
| NMP | 21.77 | 20.25 | 1.52 | 1.52 |
| EtOAc | 14.36 | 12.76 | 1.60 | |
| DMSO | 21.33 | 19.30 | 2.03 | 1.67 |
| DMF | 19.19 | 17.04 | 2.15 | 1.60 |
| DMA | 20.26 | 18.04 | 2.22 | 1.89 |
| HMPA | 24.65 | 22.16 | 2.49 | 1.44 |
| PhCN | 21.97 | 19.41 | 2.56 | |
| NMF | 22.69 | 19.75 | 2.94 | 4.52 |
| n-$C_{11}$ | 22.76 | 19.66 | 3.10 | |
| FA | 24.43 | 21.26 | 3.17 | 2.04 |
| $H_2O$ | 17.71 | 14.49 | 3.22 | 2.79 |
| MeOH | 15.10 | 11.72 | 3.38 | 2.99 |
| n-$C_{13}$ | 26.72 | 22.51 | 4.21 | |
| EtOH | 17.07 | 12.85 | 4.22 | 3.08 |
| PC | 26.33 | 22.06 | 4.27 | 1.86 |
| n-PrOH | 19.14 | 14.29 | 4.85 | 3.23 |
| n-BuOH | 21.12 | 15.90 | 5.22 | 3.26 |

The calculations for some common liquids are given in Table 13.1.6 ordered according to decreasing $S_{ex}$. An important result is the appreciable order produced by the hydrocarbon chain relative to polar groups and hydrogen-bonding effects. For instance, $S_{ex}$ would project for water the same degree of order as for undecane. In like terms, ethanol is comparable to tridecane. However, the same magnitude of the excess entropy does by no means imply that ordering is similar. Much of the orientational ordering in liquids composed of elongated molecules is a consequence of efficient packing such as the intertwining of chains. In contrast, the structure of water is largely determined by strong electrostatic interactions leading to sharply-defined directional correlations characteristic of H-bonding. Although contrary to chemical tradition, there are other indications that the longer-chain hydrocarbon liquids are to be classified as highly structured as judged from thermodynamic[123,124] and depolarized Rayleigh scattering data,[125,126] and vibrational spectra.[127]

For nonprotic fluids, as Table 13.1.6 further shows, the vaporization entropy is strongly dominated by the entropy of HS depacking. This is an at least qualitative representation of the longstanding claim that repulsions play the major role in the structure of dense fluids.[37] This circumstance is ultimately responsible for the striking success of the description of neutral reactions in the framework of a purely HS liquid, as discussed in Section 13.1.6.

Also included in the Table are values of $g_K$ as determined in the framework of a generic mean spherical approximation.[128] Since these values differ from those from other sources,[129,121] because of differences in theory, we refrain from including the latter. It is seen that the $g_K$ parameter is unsuited to scale order, since positional order is not accounted for. On the other hand, values of $g_K$ exceed unity for the highly dipolar liquids and thus both $S_{ex}$ and $g_K$ attest to some degree of order present in them.

### *Solvent reorganization energy in ET*

Electron transfer (ET) reactions in condensed matter continue to be of considerable interest to a wide range of scientists. The reasons are twofold. Firstly, ET plays a fundamental role in a broad class of biological and chemical processes. Secondly, ET is rather simple and very suitable to be used as a model for studying solvent effects and to relate the kinetics of ET reactions to thermodynamics. Two circumstances make ET reactions particularly appealing to theoreticians:

- Outer-sphere reactions and ET within rigid complexes of well-defined geometry proceed without changes in the chemical structure, since bonds are neither formed nor broken.
- The long-ranged character of interactions of the transferred electron with the solvent's permanent dipoles.

As a consequence of the second condition, a qualitative (and even quantitative) description can be achieved upon disregarding (or reducing through averaging) the local liquid structure changes arising on the length of molecular diameter dimensions relative to the charge-dipole interaction length. Because of this, it becomes feasible to use for outer-sphere reactions in strongly polar solvents the formalism first developed in the theory of polarons in dielectric crystals.[130] In the treatment, the polar liquid is considered as a dielectric continuum characterized by the high-frequency $\varepsilon_\infty$ and static $\varepsilon_s$ dielectric constants, in which the reactants occupy spherical cavities of radii $R_a$ and $R_d$, respectively. Electron transitions in this model are supposed to be activated by inertial polarization of the medium attributed to the reorientation of permanent dipoles. Along these lines Marcus[131] obtained his well-known expression for the free energy $\Delta F$ of ET activation

$$\Delta F = \frac{\left(\Delta F_o + E_r\right)^2}{4E_r} \qquad [13.1.24]$$

where $\Delta F_o$ is the equilibrium free energy gap between products and reactants and $E_r$ is the reorganization energy equal to the work applied to reorganize inertial degrees of freedom changing in going from the initial to the final charge distribution and can be dissected into inner-sphere and solvent contributions:

$$E_r = E_i + E_s \qquad [13.1.25]$$

where:

$E_i$ inner-sphere reorganization energy
$E_s$ solvent reorganization energy

For outer-sphere ET the solvent component $E_s$ of the reorganization energy

$$E_s = e^2 c_o g \qquad [13.1.26]$$

is the product of the medium-dependent Pekar factor $c_o = 1/\varepsilon_\infty - 1/\varepsilon_s$ and a reactant-dependent (but solvent independent) geometrical factor

$$g = 1/2R_a + 1/2R_d - 1/R \qquad [13.1.27]$$

where:

R the donor-acceptor separation
e the electron charge

Further advancements included calculations of the rate constant preexponent for nonadiabatic ET,[132] an account of inner-sphere[133] and quantum intramolecular[134-136] vibrations of reactants and quantum solvent modes.[137,138] The main results were the formulation of the dependencies of the activation energy on the solvent dielectric properties and reactant sizes, as well as the bell-shaped relationship between $\Delta F$ and $\Delta F_o$. The predicted activation energy dependence on both the solvent dielectric properties[139,140] and the donor-acceptor distance[141] has, at least qualitatively, been supported by experiment. A bell-shaped plot of $\Delta F$ vs $\Delta F_o$ was obtained for ET in exciplexes,[142] ion pairs,[143] intramolecular[144] and outer-sphere[145] charge shift reactions. However, the symmetric dependence predicted by eq. [13.1.24] has not yet been detected experimentally. Instead, always asymmetric plots of $\Delta F$ against $\Delta F_o$ are obtained or else, in the inverted region ($\Delta F_o < -E_r$), $\Delta F$ was found to be nearly invariant with $\Delta F_o$.[146] A couple of explanations for the asymmetric behavior are circulating in the literature (see, e.g., the review by Suppan[147]). The first one[148,149] considered vibrational excitations of high-frequency quantum vibrational modes of the donor and acceptor centers. Another suggestion[150] was that the frequencies of the solvent orientational mode are significantly different around the charged and the neutral reactants. This difference was supposed to be brought about by dielectric saturation of the polar solvent. This model is rightly questioned[151] since dielectric saturation cannot affect curvatures of the energy surface at the equilibrium point. Instead, dielectric saturation displays itself in a nonlinear deviation of the free energy surface from the parabolic form far from equilibrium. Hence, other sources of this behavior should be sought. Nevertheless, both concepts tend to go beyond the structureless description advocating a molecular nature of either the donor-acceptor complex or the solvent.

Nowadays, theories of ET are intimately related to the theories of optical transitions. While formerly both issues have developed largely independently, there is now growing desire to get a rigorous description in terms of intermolecular forces shifting the research of ET reactions toward model systems amendable to spectroscopic methods. It is the combination of steady state and transient optical spectroscopy that becomes a powerful method of studying elementary mechanisms of ET and testing theoretical concepts. The classical treatments of ET and optical transition have been facing a serious problem when extended to weakly polar and eventually nonpolar solvents. Values of $E_{op}$ (equal to $E_r$ in eq [13.1.25]) as extracted from band-shape analyses of absorption spectra were found to fall in the range 0.2 - 0.4 eV. Upon partitioning these values into internal vibrations and solvent degrees of freedom, although this matter is still ambiguous, the contribution of the solvent could well be on the order of 0.2 - 0.3 eV.[152-154] Unfortunately, all continuum theories predict zero solvent reorganization energies for ET in nonpolar liquids.

It is evident that some new mechanisms of ET, alternatively to permanent dipoles' reorientation, are to be sought. It should be emphasized that the problem cannot be resolved by treatments of fixed positions of the liquid molecules, as their electronic polarization follows adiabatically the transferred electron and thus cannot induce electronic transitions. On the other hand, the displacement of molecules with induced dipoles are capable of activating ET. In real liquids, as we have stated above, the appreciable free volume enables the solvent molecules to change their coordinates. As a result, variations in charge distribution in the reactants concomitantly alter the packing of liquid molecules. This point is corroborated by computer simulations.[155] Charging a solute in a Stockmayer fluid alters the inner coordination number from 11.8 for the neutral entity to 9.5 for the positively charged state, with the process accompanied by a compression of the solvation shell. It is therefore apparent that solvent reorganization involves reorganization of liquid density, in addition to the reorientational contribution. This concept has been introduced by Matyushov,[156] who dissected the overall solvent reorganization energy $E_s$ into a dipole reorganization component $E_p$ and a density reorganization component $E_d$,

$$E_s = E_p + E_d \qquad [13.1.28]$$

It should be mentioned that the two contributions can be completely separated because they have different symmetries, i.e., there are no density/orientation cross terms in the perturbation expansion involved in the calculations. The density component comprises three mechanisms of ET activation: (i) translations of permanent dipoles, (ii) translations of dipoles induced by the electric field of the donor-acceptor complex (or the chromophore), and (iii) dispersion solute-solvent forces. On the other hand, it appears that in the orientational part only the permanent dipoles (without inductions) are involved.

With this novel molecular treatment of ET in liquids the corundum of the temperature dependence of the solvent reorganization energy is straightforwardly resolved. Dielectric continuum theories predict an increase of $E_s$ with temperature paralleling the decrease in the dielectric constants. In contrast, experimental results becoming available quite recently show that $E_s$ decreases with temperature. Also curved Arrhenius plots eventually featuring a maximum are being reported, in weakly polar[157] and nonpolar[158] solvents. The bell-shaped temperature dependence in endergonic and moderately exergonic regions found for ET quenching reactions in acetonitrile[159] was attributed to a complex reaction mechanism. Analogously, the maximum in the Arrhenius coordinates, peculiar to the fluorescence of exciplexes formed in the intramolecular[160] and bimolecular[161] pathways, is commonly attributed to a temperature dependent competition of exciplex formation and deactivation rates. A more reasonable explanation can be given in terms of the new theory as follows.

A maximum in the Arrhenius coordinates follows from the fact that the two terms in eq. [13.1.28] depend differently on temperature. Density fluctuation around the reacting pair is determined mainly by the entropy of repacking hard spheres representing the repulsive part of the intermolecular interaction. Mathematically, the entropy of activation arises from the explicit inverse temperature dependence $E_d \propto 1/T$. Since the liquid is less packed at higher temperature, less energy is needed for reorganization. Repacking of the solvent should lead to larger entropy changes than those of dipoles' reorientation that is enthalpic in nature due to the long-range character of dipole-dipole forces. The orientational component increases with temperature essentially as predicted by continuum theories. In these ways the two solvent modes play complementary roles in the solvent's total response. This feature would lead to curved Arrhenius plots of ET rates with slight curvatures in the normal region of ET ($-\Delta F_o < E_r$), but even a maximum in the inverted region ($-\Delta F_o > E_r$).[162,163] The maximum may however be suppressed by intramolecular reorganization and should therefore be discovered particularly for rigid donor-acceptor pairs.

Photoinduced ET in binuclear complexes with localized electronic states provides at the moment the best test of theory predictions for the solvent dependent ET barrier. This type of reaction is also called metal-metal charge-transfer (MMCT) or intervalence transfer (IT). The application of the theory to IT energies for valence localized biruthenium complexes and the acetylene-bridged biferricenium monocation[164] revealed its superiority to continuum theories. The plots of $E_s$ vs. $E_{op}$ are less scattered, and the slopes of the best-fit lines are closer to unity. As a major merit, the anomalous behavior of some solvents in the continuum description - in particular HMPA and occasionally water - becomes resolved in terms of the extreme sizes, as they appear at the opposite ends of the solvent diameter scale.

Recently, it became feasible for the first time, to measure experimentally for a single chemical system, viz. a rigid, triply linked mixed-valence binuclear iron polypyridyl complex, $[Fe(440)_3Fe]^{5+}$, the temperature dependencies of both the rate of thermal ET and the optical IT energy (in acetonitrile-$d_3$).[165] The net $E_r$ associated with the intramolecular electron exchange in this complex is governed exclusively by low frequency solvent modes, providing an unprecedented opportunity to compare the parameters of the theories of thermal and optical ET in the absence of the usual complications and ambiguities. Acceptable agreement was obtained only if solvent density fluctuations around the reacting system were taken into account. In these ways the idea of density fluctuations is achieving experimental support. The two latest reports on negative temperature coefficients of the solvent reorganization energy (decrease in $E_s$ with temperature) should also be mentioned.[166,167]

Thus, two physically important properties of molecular liquids are absent in the continuum picture: the finite size of the solvent molecules and thermal translational modes resulting in density fluctuations. Although the limitations of the continuum model are long known, the necessity for a molecular description of the solvent, curiously, was first recognized in connection with solvent dynamic effects in ET. Solvent dynamics, however, affects the preexponent of the ET rate constant and, therefore, influences the reaction rate much less than does the activation energy. From this viewpoint it is suspicious that the ET activation energy has so long been treated in the framework of continuum theories. The reason of this affection is the otherwise relative success of the latter, traceable to two main features. First, the solvents usually used are similar in molecular size. Second, there is a compensation because altering the size affects the orientational and translational parts of the solvent barrier in opposite directions.

### *The solution ionic radius*

The solution ionic radius is arguably one of the most important microscopic parameters. Although detailed atomic models are needed for a full understanding of solvation, simpler phenomenological models are useful to interpret the results for more complex systems. The most famous model in this respect is that of Born, originally proposed in 1920,[168] representing the simplest continuum theory of ionic solvation. For a spherical ion, the Born excess free energy $\Delta G_B$ of solvation was derived by considering the free energy change resulting from the transfer of an ion from vacuum to solvent. The equation has a very simple dependence on the ionic charge z, the radius $r_B$, and the solvent dielectric constant ε (for the prime see eq. [13.1.16]):

$$\Delta G_B^* = \frac{-e^2 z^2}{2r_B}\left(1 - \frac{1}{\varepsilon}\right) \qquad [13.1.29]$$

While Born assumes that the dielectric response of the solvent is linear, nonlinear effects such as dielectric saturation and electrostriction should occur due to the high electric field near the ion.[169] Dielectric saturation is the effect that the dipoles are completely aligned in the direction of the field so that any further increase in the field cannot change the degree of alignment. Electrostriction, on the other hand, is defined as the volume change or compression of the solvent caused by an electric field, which tends to concentrate dipoles in the first solvation shell of an ion. Dielectric saturation is calculated to occur at field intensities exceeding $10^4$ V/cm while the actual fields around monovalent ions are on the order of $10^8$ V/cm.[170]

In the following we concentrate on ionic hydration that is generally the focus of attention. Unaware of nonlinear effects, Latimer et al.[171] showed that the experimental hydration free energies of alkali cations and halide anions were consistent with the simple Born equation when using the Pauling crystal radii $r_P$ increased by an empirical constant Δ equal to 0.85 Å for the cations and 0.1 Å to the anions. In fact three years earlier a similar relationship was described by Voet.[172] The distance $r_P + \Delta$ was interpreted as the radius of the cavity formed by the water dipoles around the ion. For cations, it is the ion-oxygen distance while for anions it is the ion-hydrogen distance of the neighboring water molecules. From those days onwards, the microscopic interpretation of the parameters of the Born equation has continued to be a corundum because of the ambiguity of using either an effective radius (that is a modification of the crystal radii) or an effective dielectric constant.

Indeed, the number of modifications of the Born equation is hardly countable. Rashin and Honig,[173] as example, used the covalent radii for cations and the crystal radii for anions as the cavity radii, on the basis of electron density distributions in ionic crystals. On the other hand, Stokes[174] put forward that the ion's radius in the gas-phase might be appreciably larger than that in solution (or in a crystal lattice of the salt of the ion). Therefore, the loss in self-energy of the ion in the gas-phase should be the dominant contributor. He could show indeed that the Born equation works well if the vdW radius of the ion is used, as calculated by a quantum mechanical scaling principle applied to an isoelectronic series centering around the crystal radii of the noble gases. More recent accounts of the subject are available.[175,176]

Irrespective of these ambiguities, the desired scheme of relating the Born radius with some other radius is facing an awkward situation: Any ionic radius depends on arbitrary divisions of the lattice spacings into anion and cation components, on the one hand, and on the other, the properties of individual ions in condensed matter are derived by means of some extra-thermodynamic principle. In other words, both properties, values of r and ΔG*, to be compared with one another, involve uncertain apportionments of observed quantities. Consequently, there are so many different sets of ionic radii and hydration free energies available that it is very difficult to decide which to prefer.

In a most recent paper,[82] a new table of absolute single-ion thermodynamic quantities of hydration at 298 K has been presented, based on conventional enthalpies and entropies upon implication of the thermodynamics of water dissociation. From the values of $\Delta_{hyd}G^*$ the Born radii were calculated from

$$r_B\,(\text{Å}) = -695\, z^2/\Delta_{hyd}G^*(kJ) \qquad [13.1.30]$$

as given in Table 13.1.7. This is at first a formal definition whose significance may be tested in the framework of the position of the first maximum of the radial distribution function (RDF) measured by solution X-ray and neutron diffraction.[177] However, the procedure is not unambiguous as is already reflected by the names given to this quantity, viz. (for the case of a cation) ion-water[178] or ion-oxygen distance. The ambiguity of the underlying interpretation resides in the circumstance that the same value of 1.40 Å is assigned in the literature to the radius of the oxide anion, the water molecule and the vdW radius of the oxygen atom.

**Table 13.1.7. Some radii (Å). Data are from ref. 82**

| Atom | $r_B$ [a] | $r_{aq}$ [b] | $r_{metal}$ |
|---|---|---|---|
| Li | 1.46 | 1.50 | 1.52 |
| Na | 1.87 | 1.87 | 1.86 |
| K | 2.33 | 2.32 | 2.27 |
| Rb | 2.52 | | 2.48 |
| Cs | 2.75 | 2.58 | 2.65 |
| Be | 1.18 | 1.06 | 1.12 |
| Mg | 1.53 | 1.52 | 1.60 |
| Ca | 1.86 | 1.86 | 1.97 |
| Sr | 2.03 | 2.02 | 2.15 |
| Ba | 2.24 | 2.27 | 2.17 |
| F | 1.39 | 1.29 | |
| Cl | 1.86 | 1.85 | |
| Br | 2.00 | 2.00 | |
| I | 2.23 | 2.30 | |

[a]From eqn. [13.1.30], [b]eqns. [13.1.31] and [13.1.32].

It seems that many workers would tend to equate the distance (d) corresponding to the first RDF peak with the average distance between the center of the ion and the centers of the nearest water molecules, $d = r_{ion} + r_{water}$. Actually, Marcus[179,180] presented a nice relationship between d, averaged over diffraction and simulation data, and the Pauling crystal radius in the form $d = 1.38 + 1.102\, r_p$. Notwithstanding this success, it is preferable to implicate not the water radius but instead the oxygen radius. This follows from the close correspondence between d and the metal-oxygen bond lengths in crystalline metal hydrates.[82]

The gross coincidence of the solid and solution state distances is strong evidence that the value of d measures the distance between the nuclei of the cation ad the oxygen rather than the center of the electron cloud of the whole ligand molecule. Actually, first RDF peaks for ion solvation in water and in nonaqueous oxygen donor solvents are very similar despite the different ligand sizes. Examples include methanol, formamide and dimethyl sulfoxide.[180]

Nevertheless, the division of d into ion and ligand components is still not unequivocal. Since the traditional ionic radius is often considered as a literal measure of size, it is usual to interpret crystallographic metal-oxygen distances in terms of the sum of the vdW radius of oxygen and the ionic radius of the metal. It should b emphasized, however, that the division of bond lengths into "cation" and "anion" components is entirely arbitrary. If the ionic radius is retained, the task remains to seek a connection to the Born radius, an issue that has a long-standing history beginning with the work of Voet.[172] Of course, any addition to the ionic radius necessary to obtain good results from the Born equation needs a physical explanation. This is typically done in terms of the water radius, in addition to other correction terms such as a dipolar correlation length in the MSA (mean spherical approximation).[181-183] In this case, however, proceeding from the first RDF peak, the size of the water moiety is implicated twice.

It has been shown[82] that the puzzle is unraveled if the covalent (atomic) radius of oxygen is subtracted from the experimental first peak position of the cation-oxygen radial distribution curve (strictly, the upper limits instead of the averages). The values of $r_{aq}$ so obtained are very close to the Born radius,

$$d(cation\text{-}O) - r_{cov}(O) = r_{aq} \approx r_B \qquad [13.1.31]$$

Similarly, for the case of the anions, the water radius, taken as 1.40 Å, is implicated,

$$d(anion\text{-}O) - r(water) = r_{aq} \approx r_B \qquad [13.1.32]$$

Furthermore, also the metallic radii (Table 13.1.7) are similar to values of $r_{aq}$. This correspondence suggests that the positive ion core dimension in a metal tends to coincide with that of the corresponding rare gas cation. The (minor) differences between $r_{aq}$ and $r_{metal}$ for the alkaline earth metals may be attributed, among other things, to the different coordination numbers (CN) in the metallic state and the solution state. The involvement of the CN is apparent in the similarity of the metallic radii of strontium and barium which is obviously a result of cancellation of the increase in the intrinsic size in going from Sr to Ba and the decrease in CN from 12 to 8.

Along these lines a variety of radii are brought under one umbrella, noting however a wide discrepancy to the traditional ionic radii. Cation radii larger than the traditional ionic radii would imply smaller anion radii so as to meet the (approximate) additivity rule. In fact, the large anion radii of the traditional sets give rise to at least two severe inconsistencies: (i) The dramatic differences on the order of 1 Å between the covalent radii and the anion radii are hardly conceivable in view of the otherwise complete parallelism displayed between ionic and covalent bonds.[184] (ii) It is implausible that non-bonded radii[185] should be smaller than ionic radii. For example, the ionic radius of oxygen of 1.40 Å implies that oxygen ions should not approach each other closer than 2.80 Å. However, non-bonding oxygen-oxygen distances as short as 2.15 Å have been observed in a variety of crystalline environments. "(Traditional) ionic radii most likely do not correspond to any physical reality," Baur notes.[186] It should be remarked that the scheme of reducing the size of the anion at the expense of that of the cation has been initiated by Gourary and Adrian, based on the electron density contours in crystals.[187]

The close correspondence seen between $r_B$ and $r_{aq}$ supports the idea that the Born radius (in aqueous solution) is predominantly a distance parameter without containing dielectric, i.e., solvent structure, contributions. This result could well be the outcome of a cancellation of dielectric saturation and electrostriction effects as suggested recently from simulations.[188-191] It should be emphasized, however, that the present discussion might be confined to water as the solvent. Recent theoretical treatments advise the cavity radius not to be considered as an intrinsic property of the solute, but instead to vary with solvent polarity, with orientational saturation prevailing at low polarity and electrostriction at high polarity.[191,192] It would appear that the whole area of nonaqueous ion solvation deserves more methodical attention. It should in addition be emphasized that the cavity radius is sensitive to temperature. Combining eq. [13.1.29] with

$$\Delta H^{\circ} = \Delta G^{*} + T\left(\frac{\partial \Delta G^{*}}{\partial T}\right) \qquad [13.1.33]$$

one obtains[177]

$$\Delta H^{\circ} = \Delta G^{*}\left[1 + \frac{T}{(\varepsilon - 1)\varepsilon}\left(\frac{\partial \varepsilon}{\partial T}\right)_P - \frac{T}{r}\left(\frac{\partial r}{\partial T}\right)_P\right] \qquad [13.1.34]$$

The derivation of $(\partial r/\partial T)_P$ from reliable values of ΔH° and ΔG* is interesting, in that nominally the dielectric effect (-0.018 for water) is smaller than the size effect (-0.069 for chloride), a result that has not given previously the attention due to it. Consequently, as Roux et al.[177] stated, unless a precise procedure for evaluating the dependence of the radius on the temperature is available, the Born model should be restricted to the free energy of solvation. Notwithstanding this, beginning with Voet,[172] the Born model has usually been tested by considering the enthalpies of hydration.[193] The reason for the relative success lies in the fact that ion hydration is strongly enthalpy controlled, i.e., ΔH° ~ ΔG*.

The discussion of radii given here should have implications to all calculations involving aqueous ionic radii, for instance the solvent reorganization energy in connection with eq. [13.1.26]. Thus, treatments using the crystal radii as an input parameter[194-197] may be revisited.

## 13.1.8 THE FUTURE OF THE PHENOMENOLOGICAL APPROACH

Originally, the empirical solvent parameters have been introduced to provide guidelines for the comparison of different solvent qualities and for an orientation in the search for an understanding of the complex phenomena in solution chemistry. Indeed, the choice of the right solvent for a particular application is an everyday decision for the chemist: which solvent should be the best to dissolve certain products, and what solvent should lead to increased reaction yields and/or rates of a reaction?

In the course of time, however, a rather sophisticated scheme has developed of quantitative treatments of solute-solvent interactions in the framework of LSERs.[198] The individual parameters employed were imagined to correspond to a particular solute-solvent interaction mechanism. Unfortunately, as it turned out, the various empirical polarity scales feature just different blends of fundamental intermolecular forces. As a consequence, we note at the door to the twenty-first century, alas with melancholy, that the era of combining empirical solvent parameters in multiparameter equations, in a scientific context, is beginning to fade away. As a matter of fact, solution chemistry research is increasingly being occupied by theoretical physics in terms of molecular dynamics (MD) and Monte Carlo (MC) simulations, the integral equation approach, etc.

In the author's opinion, it seems that further usage of the empirical parameters should more return to the originally intended purpose, emphasizing more the qualitative aspects rather than to devote too much effort to multilinear regression analyses based on parameters quoted to two decimal places. Admittedly, such a scheme may nevertheless still be used to get some insight concerning the nature of some individual solvent effect as in the recent case of an unprecedented positive wavelength shift in the solvatochromism of an aminobenzodifuranone.[199]

The physical approach, though still in its infancy, has been helping us to see the success of the phenomenological approach in a new light. Accordingly, the reason for this well documented and appreciated success can be traced back to the following features

- The molecular structure and the molecular size of many common solvents are relatively similar. The majority belongs to the so-called select solvents having a single dominant bond dipole, which, in addition, is typically hard, viz. an O- or N-donor. For example, if also soft donors (e.g., sulfur) had been employed to a larger extent, no general donor strength scale could have been devised. Likewise, we have seen that solvents other than the select ones complicate the issue.
- As it runs like a thread through the present treatment, various cancellations and competitions (enthalpy/entropy, repulsion/attraction, etc.) appear to be conspiring to make molecular behavior in complex fluids seemingly simple.

Notwithstanding this, the phenomenological approach will remain a venerable cornerstone in the development of unraveling solvent effects. Only time will tell whether a new generation of solvent indices will arise from the physical approach.

## REFERENCES

1 The reaction studied was the quaternarization of triethylamine by ethyl iodide at 100 °C [N. Menschutkin, *Z. Phys. Chem.*, **6**, 41 (1890)]. Menschutkin's first discussion on solvent effects dealt with the reactions between acetic anhydride and alcohols [*Z. Phys. Chem.*, **1**, 611 (1887)]. The catalytic role of solvents was already recognized in 1862 by Berthelot and Péan de Saint Gilles in their **Recherches sur les Affinités** [see, e. g., H. G. Grimm, H. Ruf, and Wolff, *Z. Phys. Chem.*, **B13**, 301 (1931)].
2 E. D. Hughes and C. K. Ingold, *J. Chem. Soc.*, 244 (1935).
3 H. Suhr, *Ber. Bunsenges. Phys. Chem.*, **67**, 893 (1963).
4 R. Schmid, *J. Solution Chem.*, **12**, 135 (1983).
5 V. Gutmann, **Donor-Acceptor Approach to Molecular Interactions**, *Plenum Press*, New York, 1978.
6 N. V. Sidgwick, **The Electronic Theory of Valency**, *Clarenton Press*, Oxford, 1927.
7 R. Schmid, *Rev. Inorg. Chem.*, **11**, 255 (1991).
8 R. W. Taft, N. J. Pienta, M. J. Kamlet, and E. M. Arnett, *J. Org. Chem.*, **46**, 661 (1981).
9 W. Linert and R. F. Jameson, *J. Chem. Soc. Perkin Trans. 2*, 1993, 1415.
10 V. Bekárek, *J. Phys. Chem.*, **85**, 722 (1981).
11 Y. Marcus, *Chem. Soc. Rev.*, 409 (1993).
12 C. Laurence, P. Nicolet, M. T. Dalati, J. M. Abboud, and R. Notario, *J. Phys. Chem.*, **98**, 5807 (1994).
13 C. Reichardt, *Chem. Rev.*, **94**, 2319 (1994).
14 M. R. J. Dack, *Aust. J. Chem.*, **28**, 1643 (1975).
15 M. J. Kamlet, J. M. Abboud, M. H. Abraham, and R. W. Taft, *J. Org. Chem.*, **48**, 2877 (1983).
16 R. W. Taft, M. H. Abraham, R. M. Doherty, and M. J. Kamlet, *J. Am. Chem. Soc.*, **107**, 3105 (1985).
17 M. H. Abraham, P. L. Grellier, A. Nasehzadeh, and R. A. C. Walker, *J. Chem. Soc. Perkin Trans. II*, 1988, 1717.
18 R. W. Soukup and R. Schmid, *J. Chem. Educ.*, **62**, 459 (1985).
19 U. Mayer, *Pure Appl. Chem.*, **51**, 1697 (1979).
20 M. J. Kamlet, J. L. M. Abboud, and R. W. Taft, *Prog. Phys. Org. Chem.*, **13**, 485 (1981).
21 R. Schmid and V. N. Sapunov, **Non-Formal Kinetics in Search for Chemical Reaction Pathways**, *Verlag Chemie*, Weinheim, 1982.
22 M. H. Abraham, R. M. Doherty, M. J. Kamlet, and R. W. Taft, *Chem. Br.*, **22**, 551 (1986).

23 N. A. Lewis, Y. S. Obeng, and W. L. Purcell, *Inorg. Chem.*, **28**, 3796 (1989).
24 J. T. Hupp and M. J. Weaver, *Inorg. Chem.*, **23**, 3639 (1984).
25 S. Glikberg and Y. Marcus, *J. Solution Chem.*, **12**, 255 (1983).
26 Y. Marcus, M. J. Kamlet, and R. W. Taft, *J. Phys. Chem.*, **92**, 3613 (1988).
27 C. J. Timpson, C. A. Bignozzi, B. P. Sullivan, E. M. Kober, and T. J. Meyer, *J. Phys. Chem.*, **100**, 2915 (1996).
28 U. Mayer and V. Gutmann, *Structure and Bonding*, **12**, 113 (1972).
29 Y. Marcus, *J. Solution Chem.*, **13**, 599 (1984).
30 Y. Kondo, M. Ohnishi, and N. Tokura, *Bull. Chem. Soc. Japan*, **45**, 3579 (1972).
31 H. F. Herbrandson and F. R. Neufeld, *J. Org. Chem.*, **31**, 1140 (1966).
32 M. J. Kamlet, P. W. Carr, R. W. Taft, and M. H. Abraham, *J. Am. Chem. Soc.*, **103**, 6062 (1981).
33 B. Widom, *Science*, **157**, 375 (1967).
34 J. A. Barker and D. Henderson, *J. Chem. Educ.*, **45**, 2 (1968).
35 Y. Song, R. M. Stratt, and E. A. Mason, *J. Chem. Phys.*, **88**, 1126 (1988).
36 S. Phan, E. Kierlik, M. L. Rosinberg, H. Yu, and G. Stell, *J. Chem. Phys.*, **99**, 5326 (1993).
37 C. G. Gray and K. E. Gubbins, **Theory of Molecular Fluids, Vol.1: Fundamentals**, *Clarendon Press*, Oxford, 1984.
38 L. K. Nash in **Harvard Case Histories in Experimental Science**, J. B. Conant, Ed., *Harvard University Press*, Harvard, 1950, p. 17.
39 R. A. Pierotti, J. Phys. Chem., 67, 1840 (1963); *Chem. Rev.*, **76**, 717 (1976).
40 E. Wilhelm and R. Battino, *J. Chem. Phys.*, **55**, 4021 (1971); *J. Chem. Thermodyn.*, **3**, 761 (1971); *J. Chem. Phys.*, **58**, 3558 (1973).
41 D. V. Matyushov and R. Schmid, *J. Chem. Phys.*, **104**, 8627 (1996).
42 D. Ben-Amotz and D. R. Herschbach, *J. Phys. Chem.*, **94**, 1038 (1990).
43 R. Schmid and D. V. Matyushov, *J. Phys. Chem.*, **99**, 2393 (1995).
44 D. Ben-Amotz and K. G. Willis, *J. Phys. Chem.*, **97**, 7736 (1993).
45 T. Boublik, *J. Chem. Phys.*, **53**, 471 (1970).
46 G. A. Mansoori, N. F. Carnahan, K. E. Starling, and T. W. Leland, Jr., *J. Chem. Phys.*, **54**, 1523 (1971).
47 D. V. Matyushov and B. M. Ladanyi, *J. Chem. Phys.*, **107**, 5851 (1997).
48 R. Ravi, L. E. S. Souza, and D. Ben-Amotz, *J. Phys. Chem.*, **97**, 11835 (1993).
49 D. Ben-Amotz, *J. Phys. Chem.*, **97**, 2314 (1993).
50 D. V. Matyushov and R. Schmid, *J. Chem. Phys.*, **103**, 2034 (1995).
51 J. D. Weeks, D. Chandler, and H. C. Andersen, *J. Chem. Phys.*, **54**, 5237 (1971).
52 J. A. Barker and D. Henderson, *Rev. Mod. Phys.*, **48**, 587 (1976).
53 F. Cuadros, A. Mulero, and P. Rubio, *J. Chem. Educ.*, **71**, 956 (1994).
54 D. V. Matyushov and R. Schmid, *J. Chem. Phys.*, **105**, 4729 (1996).
55 D. V. Matyushov, R. Schmid, and B. M. Ladanyi, *J. Phys. Chem. B*, **101**, 1035 (1997).
56 L. Reynolds, J. A. Gardecki, S. J. V. Frankland, M. L. Horng, and M. Maroncelli, *J. Phys. Chem.*, **100**, 10337 (1996).
57 D. V. Matyushov and G. A. Voth, *J. Chem. Phys.*, **111**, 3630 (1999).
58 S. R. Mente and M. Maroncelli, *J. Phys. Chem. B*, **103**, 7704 (1999).
59 J. Catalán, *J. Org. Chem.*, **62**, 8231 (1997).
60 B. C. Perng, M. D. Newton, F. O. Raineri, and H. L. Friedman, *J. Chem. Phys.*, **104**, 7177 (1996).
61 O. M. Carbacos, C. J. Weinheimer, and J. M. Lisy, *J. Chem. Phys.*, **108**, 5151 (1998).
62 O. M. Carbacos, C. J. Weinheimer, and J. M. Lisy, *J. Chem. Phys.*, **110**, 8429 (1999).
63 S. G. Bratsch, *J. Chem. Educ.*, **65**, 34 (1988).
64 W. C. Röntgen, *Ann. Phys. u. Chem.*, **45**, 91 (1892).
65 W. Kauzmann, *L'Eau Syst. Biol., Colloq. Int. C.N.R.S.*, **246**, 63 (1975).
66 See, for example: H. Endo, *J. Chem. Phys.*, **72**, 4324 (1980).
67 M. Vedamuthu, S. Singh, and G. W. Robinson, *J. Phys. Chem.*, **98**, 2222 (1994).
68 I. Ohmine, *J. Chem. Phys.*, **99**, 6767 (1995).
69 S. W. Benson and E. D. Siebert, *J. Am. Chem. Soc.*, **114**, 4269 (1992).
70 F. O. Libnau, J. Toft, A. A. Christy, and O. M. Kvalheim, *J. Am. Chem. Soc.*, **116**, 8311 (1994).
71 H. Kistenmacher, G. C. Lie, H. Pople, and E. Clementi, *J. Chem. Phys.*, **61**, 546 (1974).
72 C. H. Cho, S. Singh, and G. W. Robinson, *J. Chem. Phys.*, **107**, 7979 (1997).
73 L. Bosio, S. H. Chen, and J. Teixeira, *Phys. Rev.*, **A27**, 1468 (1983).
74 G. D'Arrigo, G. Maisano, F. Mallamace, P. Migliardo, and F. Wanderlingh, *J. Chem. Phys.*, **75**, 4264 (1981).
75 A. K. Soper and M. G. Phillips, *Chem. Phys.*, **107**, 47 (1986).

76 S. W. Benson, *J. Am. Chem. Soc.*, **100**, 5640 (1978).
77 K. A. T. Silverstein, A. D. J. Haymet, and K. A. Dill, *J. Am. Chem. Soc.*, **120**, 3166 (1998).
78 E. A. Jagla, *J. Chem. Phys.*, **111**, 8980 (1999).
79 E. D. Isaacs, A. Shukla, P. M. Platzman, D. R. Hamann, B. Barbiellini, and C. A. Tulk, *Phys. Rev. Lett.*, **82**, 600 (1999).
80 L. Pauling, *J. Am. Chem. Soc.*, **57**, 2680 (1935).
81 G. Nadig, L. C. Van Zant, S. L. Dixon, and K. M. Merz, Jr., *J. Am. Chem. Soc.*, **120**, 5593 (1998).
82 R. Schmid, A. M. Miah, and V. N. Sapunov, *Phys. Chem. Chem. Phys.*, **2**, 97 (2000).
83 A. Ben-Naim, **Hydrophobic Interactions**, Plenum, New York, 1980.
84 D. F. Evans and B. W. Ninham, *J. Phys. Chem.*, **90**, 226 (1986).
85 C. Tanford, **The Hydrophobic Effect: Formation of Micelles and Biological Membranes**, *Wiley*, New York, 1973.
86 K. A. Dill, *Biochemistry*, **29**, 7133 (1990).
87 O. K. Forrisdahl, B. Kvamme, and A. D. J. Haymet, *Mol. Phys.*, **89**, 819 (1996).
88 H. Wang and A. Ben-Naim, *J. Med. Chem.*, **39**, 1531 (1996).
89 H. S. Ashbaugh, E. W. Kaler, and M. E. Paulaitis, *J. Am. Chem. Soc.*, **121**, 9243 (1999).
90 H. S. Frank and M. W. Evans, *J. Chem. Phys.*, **13**, 507 (1945).
91 K. Mizuno, Y. Miyashita, Y. Shindo, and H. Ogawa, *J. Phys. Chem.*, **99**, 3225 (1995).
92 E. V. Goldammer and H. G. Hertz, *J. Phys. Chem.*, **74**, 3734 (1970).
93 R. Pottel and V. Kaatze, *Ber. Bunsenges. Phys. Chem.*, **73**, 437 (1969).
94 R. Silveston and B. Kronberg, *J. Phys. Chem.*, **93**, 6241 (1989).
95 M. Prévost, I. T. Oliveira, J. P. Kocher, and S. J. Wodak, *J. Phys. Chem.*, **100**, 2738 (1996).
96 A. Pohorille and L. R. Pratt, *J. Am. Chem. Soc.*, **112**, 5066 (1990).
97 E. Grunwald and C. Steel, *J. Am. Chem. Soc.*, **117**, 5687 (1995).
98 H. A. Yu, and M. Karplus, *J. Chem. Phys.*, **89**, 2366 (1988).
99 H. Qian and J. J. Hopfield, *J. Chem. Phys.*, **105**, 9292 (1996).
100 B. Lee, *J. Chem. Phys.*, **83**, 2421 (1985).
101 I. I. Vaisman, F. K. Brown, and A. Tropsha, *J. Phys. Chem.*, **98**, 5559 (1994).
102 M. Re, D. Laria, and R. Fernández-Prini, *Chem. Phys. Lett.*, **250**, 25 (1996).
103 A. K. Soper and J. L. Finney, *Phys. Rev. Lett.*, **71**, 4346 (1993).
104 J. Turner and A. K. Soper, *J. Chem. Phys.*, **101**, 6116 (1994).
105 A. K. Soper and A. Luzar, *J. Phys. Chem.*, **100**, 1357 (1996).
106 E. C. Meng and P. A. Kollman, *J. Phys. Chem.*, **100**, 11460 (1996).
107 L. E. S. de Souza and D. Ben-Amotz, *J. Chem. Phys.*, **101**, 9858 (1994).
108 G. Graziano, *J. Chem. Soc., Faraday Trans.*, **94**, 3345 (1998).
109 A. Ben-Naim and Y. Marcus, *J. Chem. Phys.*, **81**, 2016 (1984).
110 B. Madan and K. Sharp, *J. Phys. Chem.*, **100**, 7713 (1996).
111 K. Sharp and B. Madan, *J. Phys. Chem. B*, **101**, 4343 (1997).
112 B. Madan and K. Sharp, *J. Phys. Chem. B*, **101**, 11237 (1997).
113 K. A. T. Silverstein, A. D. J. Haymet, and K. A. Dill, , *J. Chem. Phys.*, **111**, 8000 (1999).
114 E. Fois, A. Gamba, and C. Redaelli, *J. Chem. Phys.*, **110**, 1025 (1999).
115 J. H. Hildebrand, *Proc. Natl. Acad, Sci. USA*, **89**, 2995 (1979).
116 T. Lazaridis and M. Karplus, *J. Chem. Phys.*, **105**, 4294 (1996).
117 T. S. Collett, *Chem. Eng. News*, **75**, 60 (1997).
118 T. P. Silverstein, *J. Chem. Educ.*, **75**, 116 (1998).
119 S. Furutaka and S. Ikawa, *J. Chem. Phys.*, **108**, 5159 (1998).
120 M. Ikeguchi, S. Seishi, S. Nakamura, and K. Shimizu, *J. Phys. Chem. B*, **102**, 5891 (1998).
121 Y. Marcus, *J. Solution Chem.*, **21**, 1217 (1992).
122 D. V. Matyushov and R. Schmid, *Ber. Bunsenges. Phys. Chem.*, **98**, 1590 (1994).
123 V. T. Lam, P. Picker, D. Patterson, and P. Tancrede, *J. Chem. Soc., Faraday Trans. 2*, **70**, 1465, 1479 (1974).
124 D. Patterson, *J. Solution Chem.*, **23**, 105 (1994).
125 P. Borothorel, *J. Colloid Sci.*, **27**, 529 (1968).
126 H. Quinones and P. Borothorel, *Compt. Rend.*, **277**, 133 (1973).
127 R. G. Snyder, *J. Chem. Phys.*, **47**, 1316 (1967).
128 L. Blum and W. R. Fawcett, *J. Phys. Chem.*, **100**, 10423 (1996).
129 Y. Marcus, **Ion Solvation**, *Wiley*, New York, 1985; Chapter 6.
130 S. I. Pekar, **Investigations in the Electronic Theory of Crystals**, *GIFML*, Moscow (in Russian).
131 R. A. Marcus, *J. Chem. Phys.*, **24**, 966 (1956).

132 R. R. Dogonadze and A. M. Kuznetsov, **Physical Chemistry. Kinetics**, *VINITI*, Moscow, 1973.
133 R. A. Marcus, *Disc. Farad. Soc.*, **29**, 21 (1960).
134 N. R. Kestner, J. Logan, and J. Jortner, *J. Phys. Chem.*, **78**, 2148 (1974).
135 T. Holstein, *Phil. Mag.*, **B37**, 499 (1978).
136 P. Siders and R. A. Marcus, *J. Am. Chem. Soc.*, **103**, 748 (1981).
137 A. A. Ovchinnimov and M. Ya. Ovchinnikova, *Sov. Phys. JETF*, **56**, 1278 (1969).
138 G. E. McManis, A. Gochev, and M. J. Weaver, *Chem. Phys.*, **152**, 107 (1991).
139 G. M. Tom, C. Creutz, and H. Taube, *J. Am. Chem. Soc.*, **96**, 7828 (1974).
140 H. Heitele, F. Pollinger, S. Weeren, and M. E. Michel-Beyerle, *Chem. Phys.*, **143**, 325 (1990).
141 M. J. Powers and T. J. Meyer, *J. Am. Chem. Soc.*, **98**, 6731 (1976).
142 P. P. Levin and P. K. N. Raghavan, *Chem. Phys. Lett.*, **171**, 309 (1990).
143 I. R. Gould and S. Farid, *J. Am. Chem. Soc.*, **110**, 7833 (1988).
144 G. L. Closs, L. T. Calcaterra, N. J. Green, K. W. Penfield, and J. R. Miller, *J. Phys. Chem.*, **90**, 3673 (1986).
145 J. R. Miller, J. V. Beitz, and R. K. Henderson, *J. Am. Chem. Soc.*, **106**, 5057 (1984).
146 D. Rehm and A. Weller, *Isr. J. Chem.*, **8**, 259 (1970).
147 P. Suppan, **Topics in Current Chemistry**, Vol. 163, *Springer*, Berlin, 1992.
148 S. Efrima and M. Bixon, *Chem. Phys. Lett.*, **25**, 34 (1974).
149 N. R. Kestner, J. Logan, and J. Jortner, *J. Phys. Chem.*, **78**, 2148 (1974).
150 T. Kakitani and N. Mataga, *Chem. Phys. Lett.*, **124**, 437 (1986); *J. Phys. Chem.*, **91**, 6277 (1987).
151 M. Tachiya, *Chem. Phys. Lett.*, **159**, 505 (1989); *J. Phys. Chem.*, **93**, 7050 (1989).
152 F. Markel, N. S. Ferris, I. R. Gould, and A. B. Myers, *J. Am. Chem. Soc.*, **114**, 6208 (1992).
153 K. Kulinowski, I. R. Gould, and A. B. Myers, *J. Phys. Chem.*, **99**, 9017 (1995).
154 B. M. Britt, J. L. McHale, and D. M. Friedrich, *J. Phys. Chem.*, **99**, 6347 (1995).
155 L. Perera and M. L. Berkowitz, *J. Chem. Phys.*, **96**, 3092 (1992).
156 D. V. Matyushov, *Mol. Phys.*, **79**, 795 (1993).
157 H. Heitele, P. Finckh, S. Weeren, F. Pöllinger, and M. E. Michel-Beyerle, *J. Phys. Chem.*, **93**, 5173 (1989).
158 E. Vauthey and D. Phillips, *Chem. Phys.*, **147**, 421 (1990).
159 H. B. Kim, N. Kitamura, Y. Kawanishi, and S. Tazuke, *J. Am. Chem. Soc.*, **109**, 2506 (1987).
160 F. D. Lewis and B. E. Cohen, *J. Phys. Chem.*, **98**, 10591 (1994).
161 A. M. Swinnen, M. Van der Auweraer, F. C. De Schryver, K. Nakatani, T. Okada, and N. Mataga, *J. Am. Chem. Soc.*, **109**, 321 (1987).
162 D. V. Matyushov and R. Schmid, *Chem. Phys. Lett.*, **220**, 359 (1994).
163 D. V. Matyushov and R. Schmid, *Mol. Phys.*, **84**, 533 (1995).
164 D. V. Matyushov and R. Schmid, *J. Phys. Chem.*, **98**, 5152 (1994).
165 C. M. Elliott, D. L. Derr, D. V. Matyushov, and M. D. Newton, *J. Am. Chem. Soc.*, **120**, 11714 (1998).
166 D. L. Derr and C. M. Elliott, *J. Phys. Chem. A*, **103**, 7888 (1999).
167 P. Vath, M. B. Zimmt, D. V. Matyushov, and G. A. Voth, *J. Phys. Chem. B*, **103**, 9130 (1999).
168 M. Born, *Z. Phys.*, **1**, 45 (1920).
169 B. E. Conway, **Ionic Hydration in Chemistry and Biophysics**, *Elsevier*, Amsterdam, 1981.
170 J. F. Böttcher, **Theory of Electric Polarization**, *Elsevier*, Amsterdam, 1973, Vol. 1.
171 W. M. Latimer, K. S. Pitzer, and C. M. Slansky, *J. Chem. Phys.*, **7**, 108 (1939).
172 A. Voet, *Trans. Faraday Soc.*, **32**, 1301 (1936).
173 A. A. Rashin and B. Honig, *J. Phys. Chem.*, **89**, 5588 (1985).
174 R. H. Stokes, *J. Am. Chem. Soc.*, **86**, 979 (1964); *J. Am. Chem. Soc.*, **86**, 982 (1964).
175 K. J. Laidler and C. Pegis, *Proc. R. Soc.*, **A241**, 80 (1957).
176 B. E. Conway and E. Ayranci, *J. Solution Chem.*, **28**, 163 (1999).
177 B. Roux, H. A. Yu, and M. Karplus, *J. Phys. Chem.*, **94**, 4683 (1990).
178 M. Mezei and D. L. Beveridge, *J. Chem. Phys.*, **74**, 6902 (1981).
179 Y. Marcus, *J. Solution Chem.*, **12**, 271 (1983).
180 Y. Marcus, *Chem. Rev.*, **88**, 1475 (1988).
181 D. Y. C. Chan, D. J. Mitchell and B. W. Ninham, *J. Chem. Phys.*, **70**, 2946 (1979).
182 L. Blum and W. R. Fawcett, *J. Phys. Chem.*, **96**, 408 (1992).
183 W. R. Fawcett, *J. Phys. Chem.*, **97**, 9540 (1993).
184 A. M. Pendás, A. Costales, and V. Luana, *J. Phys. Chem. B*, **102**, 6937 (1998).
185 M. O'Keeffe and B. G. Hyde, **Structure and Bonding in Crystals**, *Academic Press*, New York, 1981, vol. 1, p. 227.
186 W. H. Baur, *Cryst. Rev.*, **1**, 59 (1987).
187 B. S. Gourary and F. J. Adrian, *Solid State Phys.*, **10**, 127 (1960).

188 S. W. Rick and B. J. Berne, *J. Am. Chem. Soc.*, **116**, 3949 (1994).
189 J. K. Hyun and T. Ichiye, *J. Chem. Phys.*, **109**, 1074 (1998).
190 J. K. Hyun and T. Ichiye, *J. Phys. Chem. B*, **101**, 3596 (1997).
191 D. V. Matyushov and B. M. Ladanyi, *J. Chem. Phys.*, **110**, 994 (1999).
192 A. Papazyan and A. Warshel, *J. Chem. Phys.*, **107**, 7975 (1997).
193 D. W. Smith, *J. Chem. Educ.*, **54**, 540 (1977).
194 G. E. McManis, A. Gochev, R. M. Nielson, and M. J. Weaver, *J. Phys. Chem.*, **93**, 7733 (1989).
195 I. Rips, J. Klafter, and J. Jortner, *J. Chem. Phys.*, **88**, 3246 (1988).
196 W. R. Fawcett and L. Blum, *Chem. Phys. Lett.*, **187**, 173 (1991).
197 H. Heitele, P. Finckh, S. Weeren, F. Pollinger, and M. E. Michel-Beyerle, *J. Phys. Chem.*, **93**, 325 (1990).
198 **Quantitative Treatments of Solute/Solvent Interactions**, P. Politzer and J. S. Murray, Eds., *Elsevier*, Amsterdam, 1994.
199 A. A. Gorman, M. G. Hutchings, and P. D. Wood, *J. Am. Chem. Soc.*, **118**, 8497 (1996).

## 13.2 SOLVENT EFFECTS ON FREE RADICAL POLYMERIZATION

Michelle L. Coote and Thomas P. Davis
**Centre for Advanced Macromolecular Design, School of Chemical Engineering & Industrial Chemistry, The University of New South Wales, Sydney, Australia**

### 13.2.1 INTRODUCTION

Free radical polymerization is one of the most useful and lucrative fields of chemistry ever discovered - recent years have seen a tremendous increase in research into this area once considered a mature technological field. Free radical synthetic polymer chemistry is tolerant of diverse functionality and can be performed in a wide range of media. Emulsion and suspension polymerizations have been established as important industrial processes for many years. More recently, the 'green' synthesis of polymers has diversified from aqueous media to supercritical fluids and the fluorous biphase. An enduring feature of the research literature on free radical polymerization has been studies into specific solvent effects. In many cases the influence of solvent is small, however, it is becoming increasingly evident that solvent effects can be used to assist in controlling the polymerization reaction, both at the macroscopic and at the molecular levels. The purpose of this chapter is to give a brief introduction to the types of specific solvent effect that can be achieved in both free radical homo- and co-polymerizations.

### 13.2.2 HOMOPOLYMERIZATION

Free radical polymerization can be conveniently codified according to the classical chain reaction steps of initiation, propagation, transfer and termination. In cases where a significant solvent effect is operative then the effect is normally exerted in all of these steps. However, for the purpose of facilitating discussion this chapter is broken down into these specific reaction steps.

#### 13.2.2.1 Initiation

Solvent effects on the initiation reaction are primarily on the rate of decomposition of initiator molecules into radicals and in the efficiency factor, f, for polymerization. However, in some instances the solvent plays a significant role in the initiation process, for example, in initiation reactions with t-butoxy radical where the primary radical rarely initiates a chain but instead abstracts a hydrogen atom from the solvent medium, which subsequently initiates the chain.[1] The consequence of this is that the polymer chains contain fragments of solvent. As the stability of the chains to thermal and photochemical degradation is governed, in part, by the nature of the chain ends then the solvent moieties within the chain can have a substantial impact on the material performance of the polymer. The efficiency factor, f, decreases as the viscosity of the reaction medium increases.[2] This is caused by an increase in the radical lifetime within the solvent cage, leading to an increased possibility of radical-radical termination. In this regard the diffusion rates of the small radicals becomes an important consideration and Terazima and co-workers[3,4] have published results indicating that many small radicals diffuse slower than expected. They have attributed this to specific interactions between radical and solvent molecules.

#### 13.2.2.2 Propagation

The ability of solvents to affect the homopropagation rate of many common monomers has been widely documented. For example, Bamford and Brumby[5] showed that the propagation rate ($k_p$) of methyl methacrylate (MMA) at 25°C was sensitive to a range of aromatic solvents. Burnett et al.[6] found that the $k_p$ of styrene (STY) was depressed by increasing concentrations of benzonitrile, bromobenzene, diethyl phthalate, dinonyl phthalate and diethyl malonate, while in other studies[7,8] they found that the $k_p$ for MMA was enhanced by halobenzenes and naphthalene. More recent work by Zammit et al.[9] has shown that solvents capable of hydrogen-bonding, such as, benzyl alcohol and N-methyl pyrrolidone have a small influence on both the activation energy ($E_a$) and pre-exponential factor (A) in STY and MMA homopropagation reactions. These are but a few of the many instances of solvent effects in the homopolymerization reactions of two typical monomers, STY and MMA. For these monomers, solvent effects are relatively small, and this is indicative of the majority of homopropagation reactions. However, in some instances much larger effects are observed, especially in cases where specific interactions such as H-bonding or ionization occur. Examples of this type include the polymerization of N-vinyl-2-pyrrolidone (where water has been found to dramatically increase $k_p$)[10] and the polymerization of acrylamide (where pH plays a strong role).[11] There is only limited data on the Arrhenius parameters for homopropagation reactions in different solvents and this indicates that both the activation energy and pre-exponential factor are affected.[9,12] In some cases the solvent effect is not on the elementary rate constant $k_p$ but on the local monomer concentration (sometimes referred to as the 'Bootstrap" effect). This effect can originate in the preferential solvation of either the monomer (which is always present as a solvent) or the added solvent. It has also been suggested that in some instances the growing polymer coil can 'shield' the radical chain-end resulting in a low monomer concentration. This shielding effect would be expected to be greatest in poor solvents (hence a tighter coil).[13] For methyl methacrylate and styrene the largest solvent effects on propagation seem to be in the order of a 40% change in $k_p$.[14,15] In some solvents there seems to be reasonably strong evidence that the solvent does cause changes to the geometry of the transition state (e.g., dimethyl formamide and acetonitrile in styrene polymerization)[14] and in liquid carbon dioxide it appears that the 40% change in $k_p$ for methyl methacrylate can be ascribed to the poor solvent medium.[16] Recent work has found that some fluoro-alcohols[17] can influence the tacticity of free radical polymerization lending further credence to the concept of solvent-induced changes to the transition state of the radical addition reaction. The largest solvent effects observed on $k_p$ for homopropagations have been for vinyl acetate[18] and for α-(hydroxymethyl) ethyl acrylate.[12] In the former case the radical is highly unstable and some form of π-complexation between the vinyl acetate radical and aromatic solvents seems plausible. However, the large solvent effect cannot be explained by a simple radical stabilization argument (because of the early transition state for free radical propagation reactions)[19] and again the evidence points towards a change in the geometry of the transition state. The solvent effects on α-(hydroxymethyl) ethyl acrylate are in the order of 300% on $k_p$ and there are large changes in both the Arrhenius parameters as the solvent medium is changed.[12] In monomers exhibiting a strong solvent effect on propagation it is plausible that some control of the stereochemistry of the chains is possible by manipulating the solvent and possibly utilizing Lewis acids as additives. This approach is already being successfully applied to the control of radical reactions in conventional organic chemistry.[20]

#### 13.2.2.3 Transfer

Solvent effects on transfer reactions have not received too much attention. It would be expected (owing to the similarities between the transition states for radical addition and abstraction reactions) that these solvent effects should emulate those found in propagation reactions. However, there is potential for significant polar interactions in transfer reactions. Odian[21] has suggested that polar interactions play a significant role in the transfer reactions between styrene and carbon tetrachloride. More recent work supports this idea.[22] Significant solvent effects have been observed in catalytic chain transfer reactions using cobaloximes where the transfer reaction appears (in some cases at least) to be diffusion controlled and therefore the speed of the reaction is governed, in part, by the viscosity of the polymerizing medium.[23] In transfer reactions involving organometallic reagents then solvent effects may become important where ligand displacement may occur. This is thought to happen in catalytic chain transfer when pyridine is utilized as a solvent.[24]

#### 13.2.2.4 Termination

The solvent effects on the termination reaction have been extensively studied. In early work, it was established that the radical-radical termination reaction is diffusion controlled and the efficacy of termination was found to have a strong relationship with the solvent viscosity.[25] Subsequently, more complex models have been developed accounting for the quality of the solvent (hence the size of the polymer coil).[26] The current debate centers on the relative roles played by segmental and translational diffusion at different stages of conversion for a variety monomers. Clearly in both cases the nature of the solvent becomes important. Solvent effects are known to play a significant role in determining the strength and onset conversion of the gel effect. This work originated in the classical paper by Norrish and Smith[27] who reported that poor solvents cause an earlier gel effect in methyl methacrylate polymerization. Recent careful studies of the gel effect by Torkelson and co-workers[28] have reinforced observations made by Cameron and Cameron[29] over two decades ago concluding that termination is hindered in poor solvents due to formation of more tightly coiled polymer radicals.

### 13.2.3 COPOLYMERIZATION

When solvent effects on the propagation step occur in free-radical copolymerization reactions, they result not only in deviations from the expected overall propagation rate, but also in deviations from the expected copolymer composition and microstructure. This may be true even in bulk copolymerization, if either of the monomers exerts a direct effect or if strong cosolvency behavior causes preferential solvation. A number of models have been proposed to describe the effect of solvents on the composition, microstructure and propagation rate of copolymerization. In deriving each of these models, an appropriate base model for copolymerization kinetics is selected (such as the terminal model or the implicit or explicit penultimate models), and a mechanism by which the solvent influences the propagation step is assumed. The main mechanisms by which the solvent (which may be one or both of the comonomers) can affect the propagation kinetics of free-radical copolymerization reactions are as follows:

(1) Polarity effect
(2) Radical-solvent complexes
(3) Monomer-solvent complexes
(4) Bootstrap effect

In this chapter we explain the origin of these effects, show how copolymerization models for these different effects may be derived, and review the main experimental evidence for and against these models. Throughout this review the baseline model for copolymerization is taken as the terminal or Mayo-Lewis model.[30] This model can be used to derive well-known expressions for copolymer composition and copolymerization propagation kinetics. Deviations from this model have often been interpreted in terms of either solvent effects or penultimate unit effects, although the two are by no means mutually exclusive. Deviations which affect both the copolymer composition and propagation kinetics have been termed explicit effects by Fukuda[31] in deriving penultimate unit models, whereas deviations from the kinetics without influencing the copolymer composition have been termed implicit effects. In this review we use the same terminology with respect to solvent effects: that is, a solvent effect on $k_p$ only is termed an implicit solvent effect, while a solvent effect on composition, microstructure and $k_p$ is termed explicit. The relatively recent discovery by Fukuda and co-workers[32] of the seemingly general failure of the terminal model to predict $k_p$, even for bulk copolymerizations that follow the terminal model composition equation, led them to propose an implicit penultimate unit effect as a general phenomenon in free-radical copolymerization kinetics. We conclude this review with a brief examination of the possibility that a implicit solvent effect, and not an implicit penultimate unit effect, may instead be responsible for this failure of the terminal model $k_p$ equation.

### 13.2.3.1 Polarity effect

#### *13.2.3.1.1 Basic mechanism*

One type of solvent effect on free-radical addition reactions such as the propagation step of free-radical polymerization is the so-called 'polarity effect'. This type of solvent effect is distinguished from other solvent effects, such as complexation, in that the solvent affects the reactivity of the different types of propagation steps without directly participating in the reaction. The mechanism by which this could occur may be explained as follows. The transition states of the different types of propagation steps in a free-radical copolymerization may be stabilized by charge transfer between the reacting species. The amount of charge transfer, and hence the amount of stabilization, is inversely proportional to the energy difference between the charge transfer configuration, and the product and reactant configurations that combine to make up the wave function at the transition state.[33] Clearly, the stability of the charge transfer configuration would differ between the cross- and homopropagation reactions, especially in copolymerization of highly electrophilic and nucleophilic monomer pairs. Hence, when it is significant, charge transfer stabilization of the transition state occurs to different extent in the cross- and homopropagation reactions, and thus exerts some net effect on the monomer reactivity ratios. Now, it is known that polar solvents can stabilize charged species, as seen in the favorable effect of polar solvents on both the thermodynamics and kinetics of reactions in which charge is generated.[34] Therefore, when charge transfer in the transition state is significant, the stability of the charge transfer species and thus the transition state would be affected by the polarity of the solvent, and thus a solvent effect on reactivity ratios would result.

#### 13.2.3.1.2 Copolymerization model

There are two cases to consider when predicting the effect of solvent polarity on copolymerization propagation kinetics: (1) the solvent polarity is dominated by an added solvent and polarity is thus independent of the comonomer feed ratio, or (2) the solvent polarity does depend on the comonomer feed ratio, as it would in a bulk copolymerization. In the first case, the effect on copolymerization kinetics is simple. The monomer reactivity ratios (and additional reactivity ratios, depending on which copolymerization model is appropriate for that system) would vary from solvent to solvent, but, for a given copolymerization system they would be constant as a function of the monomer feed ratios. Assuming of course that there were no additional types of solvent effect present, these copolymerization systems could be described by their appropriate base model (such as the terminal model or the explicit or implicit penultimate models), depending on the chemical structure of the monomers.

In the second case, the effect of the solvent on copolymerization kinetics is much more complicated. Since the polarity of the reacting medium would vary as a function of the comonomer feed ratios, the monomer reactivity ratios would no longer be constant for a given copolymerization system. To model such behavior, it would be first necessary to select an appropriate base model for the copolymerization, depending on the chemical structure of the monomers. It would then be necessary to replace the constant reactivity ratios in this model by functions of the composition of the comonomer mixture. These functions would need to relate the reactivity ratios to the solvent polarity, and then the solvent polarity to the comonomer feed composition. The overall copolymerization kinetics would therefore be very complicated, and it is difficult to suggest a general kinetic model to describe these systems. However, it is obvious that such solvent effects would cause deviations from the behavior predicted by their appropriate base model and might therefore account for the deviation of some copolymerization systems from the terminal model composition equation.

#### 13.2.3.1.3 Evidence for polarity effects in propagation reactions

The idea of charge separation in the transition state of the propagation step of free radical polymerization reactions, as suggested by Price,[35] was discounted by Mayo and Walling[36] and many subsequent workers.[37] Their rejection of this idea was based upon the absence of any unambiguous correlation between the reactivity ratios of a system and the dielectric constant of the solvent. For instance, in the copolymerization of STY with MMA, it was reported that the reactivity ratios were independent of small quantities of water, ethyl benzene, dodecylmercaptans or hydroquinone, or the presence or absence of air[30,38,39] and were thus unaffected by the dielectric constant of the system. In contrast, other studies have found a relationship between dielectric constant and the reactivity ratios in specific systems.[40,41]

The apparent lack of a general relationship between the dielectric constant of the system and the monomer reactivity ratios does not necessarily discount a polarity effect on reactivity ratios. A polarity effect is only expected to occur if the charge transfer configurations of the transition state are sufficiently low in energy to contribute to the ground state wave function. Since this is not likely to occur generally, a comprehensive correlation between reactivity ratios and the solvent dielectric constant is unlikely. Furthermore, even in systems for which a polarity effect is operating, a correlation between solvent dielectric constant and monomer reactivity ratios may be obscured by any of the following causes.

- The operation of additional types of solvent effect, such as a Bootstrap effect, that would complicate the relationship between solvent polarity and reactivity ratios.
- Errors in the experimental data base from which the correlation was sought.
- The recognized inadequacy of simple reactivity - dielectric constant correlations, that take no account of specific interactions between the solvent and solute molecules.[34]

In fact, recent theoretical[33] and experimental studies[42] of small radical addition reactions indicate that charge separation does occur in the transition state when highly electrophilic and nucleophilic species are involved. It is also known that copolymerization of electron donor-acceptor monomer pairs are solvent sensitive, although this solvent effect has in the past been attributed to other causes, such as a Bootstrap effect (see Section 13.2.3.4). Examples of this type include the copolymerization of styrene with maleic anhydride[43] and with acrylonitrile.[44] Hence, in these systems, the variation in reactivity ratios with the solvent may (at least in part) be caused by the variation of the polarity of the solvent. In any case, this type of solvent effect cannot be discounted, and should thus be considered when analyzing the copolymerization data of systems involving strongly electrophilic and nucleophilic monomer pairs.

### 13.2.3.2 Radical-solvent complexes

#### *13.2.3.2.1 Basic mechanism*

Solvents can also interfere in the propagation step via the formation of radical-solvent complexes. When complexation occurs, the complexed radicals are more stable than their corresponding uncomplexed-radicals, as it is this stabilization that drives the complexation reaction. Thus, in general, one might expect complexed radicals to propagate more slowly than their corresponding free-radicals, if indeed they propagate at all. However, in the special case that one of the comonomers is the complexing agent, the propagation rate of the complexed radical may instead be enhanced, if propagation through the complex offers an alternative less-energetic reaction pathway. In any case, the complexed radicals would be expected to propagate at a rate different to their corresponding free-radicals, and thus the formation of radical-solvent complexes would affect the copolymerization propagation kinetics.

#### *13.2.3.2.2 Copolymerization model*

A terminal radical-complex model for copolymerization was formulated by Kamachi.[37] He proposed that a complex is formed between the propagating radical chain and the solvent (which may be the monomer) and that this complexed radical has a different propagation rate constant to the equivalent uncomplexed radical. Under these conditions there are eight different propagation reactions in a binary copolymerization, assuming that the terminal unit is the only unit of the chain affecting the radical reactivity. These are as follows.

$$RMi\cdot + Mj \xrightarrow{k_{ij}} RMiMj\cdot \text{ or } RMiMj\cdot S \text{ where: } i,j = 1 \text{ or } 2$$

$$RMi\cdot S + Mj \xrightarrow{k_{cij}} RMiMj\cdot \text{ or } RMiMj\cdot S \text{ where: } i,j = 1 \text{ or } 2$$

There are also two equilibrium reactions for the formation of the complex:

$$RMi\cdot + S \xleftrightarrow{K_i} RMi\cdot S \text{ where: } i,j = 1 \text{ or } 2$$

Applying the quasi-steady-state and long-chain assumptions to the above reactions, Kamachi derived expressions for $\bar{r}_i$ and $\bar{k}_{ii}$, which are used in place of $r_i$ and $k_{ii}$ in the terminal model equations for composition and $k_p$:

$$\bar{k}_{ii} = k_{ii}\frac{1+\bar{s}_{ci}K_i[C_i]}{1+K_i[C_i]} \quad and \quad \bar{r}_i = r_i\frac{1+\bar{s}_{ci}K_i[C_i]}{1+(r_i/\bar{r}_{ci})\bar{s}_{ci}K_i[C_i]}$$

where: $r_i = k_{ii}/k_{ij}$; $\bar{r}_{ic} = k_{cii}/k_{cij}$; $\bar{s}_{ci} = k_{cii}/k_{ii}$; $i, j = 1$ or $2$ and $i \neq j$

Variants of this model may be derived by assuming an alternative basis model (such as the implicit or explicit penultimate models) or by making further assumptions as to nature of the complexation reaction or the behavior of the complexed radical. For instance, in the special case that the complexed radicals do not propagate (that is, $\bar{s}_{ci} = 0$ for all i), the reactivity ratios are not affected (that is, $r_i = r_i$ for all i) and the complex formation serves only removal of radicals (and monomer, if monomer is the complexing agent) from the reaction, resulting in a solvent effect that is analogous to a Bootstrap effect (see Section 13.2.3.4).

#### *13.2.3.2.3 Experimental evidence*

There is certainly strong experimental evidence for the existence of radical-solvent complexes. For instance, Russell[45-47] and co-workers collected experimental evidence for radical-complex formation in studies of the photochlorination of 2,3-dimethylbutane in various solvents. In this work, different products were obtained in aliphatic and aromatic solvents, and this was attributed to formation of a π-complex between the Cl atom and the aromatic solvent. Complex formation was confirmed by flash photolysis.[48-51] Complex formation was also proposed to explain experimental results for the addition of trichloromethane radical to 3-phenylpropene and to 4-phenyl-1-butene[52] and for hydrogen abstraction of the t-butoxy radical from 2,3-dimethylbutane.[53] Furthermore, complexes between nitroxide radicals and a large number of aromatic solvents have been detected.[54-57] Evidence for complexes between polymer radicals and solvent molecules was collected by Hatada et al.,[58] in an analysis of initiator fragments from the polymerization of MMA-d with AIBN and BPO initiators. They discovered that the ratio of disproportionation to combination depended on the solvent, and interpreted this as evidence for the formation of a polymer radical-solvent complex that suppresses the disproportionation reaction.

There is also experimental evidence for the influence of radical-solvent complexes in small radical addition reactions. For instance, Busfield and co-workers[59-61] used radical-solvent to explain solvent effects in reactions involving small radicals, such as t-butoxyl radicals towards various electron donor-electron acceptor monomer pairs. The observed solvent effects were interpreted in terms of complex formation between the t-butoxyl radical and the electron-acceptor monomer, possibly via a sharing of the lone pair on the t-butoxyl oxygen with the π-system of the acceptor monomer. Several workers have invoked frontier orbital theory to rationalize such solvent effects in terms of radical-solvent complex formation, and thus provide a theoretical base.[37,62]

Many workers have suggested radical-solvent complexes as an explanation for the influence of aromatic compounds on the homopolymerization of vinyl monomers. For instance, Mayo[63] found that bromobenzene acts as a chain transfer agent in the polymerization of STY but is not incorporated into the polymer. He concluded that a complex is formed between the solvent molecule and either the propagating polystyryl radical or a proton derived from it. The influence of halobenzenes on the rate of polymerization of MMA was detected by Burnett et al.[8,64,65] They proposed that the efficiency of a number of different initiators increased in various halogenated aromatic solvents and, since enhanced initiator or solvent incorporation into the polymer was not observed, they concluded that initiator-solvent-monomer complex participation affected the initiator efficiency. Henrici-Olive and Olive[66-71] suggested that this mechanism was inadequate when the degree of polymerization was taken into account and they proposed instead a charge transfer complex between the polymer radical and aromatic solvent. The polymer radical can form a complex with either the monomer or solvent molecule, but only the former can propagate. Bamford and Brumby,[5] and later Burnett et al.,[72,73] interpreted their solvent-effects data for $k_p$ in terms of this donor-acceptor complex formation between aromatic solvents and propagating radicals.

Radical-solvent complexes are expected to be favored in systems containing unstable radical intermediates (such as vinyl acetate) where complexation may lead to stabilization. In this regard Kamachi et al[18] have noted that solvent effects on vinyl acetate homopolymerization result in a reduced $k_p$. Kamachi et al.[74] also measured the absolute rate constants of vinyl benzoate in various aromatic solvents and found that $k_p$ increased in the order:

benzonitrile < ethyl benzoate < anisole < chlorobenzene < benzene < fluorobenzene < ethyl acetate

They argued that this trend could not be explained by copolymerization through the solvent or transfer to the solvent because there was no correlation with the solvent dielectric constant or polarity, or with the rate constants for transfer to solvent. However, there was a correlation with the calculated delocalization stabilization energy for complexes between the radical and the solvent, which suggested that the propagating radical was stabilized by the solvent or monomer, but the solvent did not actually participate in the reaction.

As noted in the introduction to this section, radical-solvent complexes may enhance the propagation rate if propagation through the complex offers an alternative, less-energetic pathway for propagation. An example of this behavior is found in the homopolymerization of acrylamide. The homopropagation rate coefficient for this monomer shows a negative temperature dependence, which has been explained in terms of radical-complex formation. Pascal et al.[11,75] suggested that propagation proceeds via a complex that enhances the propagation rate, and this complex dissociates as temperature increases, thus explaining the normal temperature dependence of the propagation rate at high temperatures. This interpretation was supported by the observation that acrylamide behaves normally in the presence of reagents such as propionamide, which would be expected to inhibit complex formation.

Given the experimental evidence for the existence of radical-solvent complexes and their influence on free-radical addition reactions such as homopropagation, it is likely that radical-solvent complexes will affect the copolymerization kinetics for certain copolymerization systems, and indeed many workers have invoked the radical-complex model in order to explain solvent effects in copolymerization. For instance, Heublein and Heublein[76] have invoked a radical complex model in combination with a partitioning idea (see Section 13.2.3.4) to explain solvent effects on the copolymerization of vinyl acetate with acrylic acid. More recently, O'Driscoll and Monteiro[77] suggested that the effect of benzyl alcohol on the copolymerization of STY-MMA was best described by an RC-type model. This was supported by pulsed-laser studies[9] on the homopropagation reactions where $E_a$ values were found to be increased slightly by the presence of benzyl alcohol. Czerwinski (see for example reference[78] and references cited therein) has also published a variant of the RC model and has applied his model to a range of copolymerization experimental data. In conclusion, there is a strong experimental evidence for the importance of radical-solvent complexes in a number of specific copolymerization systems, especially when there is a large disparity in the relative stabilities of the different propagating radicals.

### 13.2.3.3 Monomer-solvent complexes

#### *13.2.3.3.1 Introduction*

A solvent may also interfere in the propagation step via complexation with the monomer. As was the case with radical-solvent complexes, complexed monomer might be expected to propagate at a different rate to free monomer, since complexation might stabilize the monomer, alter its steric properties and/or provide an alternative pathway for propagation. In examining the effect of such complexation on copolymerization kinetics, there are a number of different mechanisms to consider. In the case that the complex is formed between the comonomers, there are three alternatives: (1) the monomer-monomer complex propagates as a single unit, competing with the propagation of free monomer; (2) the monomer-monomer complex propagates as a single unit, competing with the propagation of free monomer, but the complex dissociates during the propagation step and only one of the monomers is incorporated into the growing polymer radical; (3) the monomer-monomer complex does not propagate, and complexation serves only to alter the free monomer concentrations. In the case that the complex is formed between one of the monomers and an added solvent, there are two further mechanisms to consider: (4) the complexed monomer propagates, but at a different rate to the free monomer; (5) the complexed monomer does not propagate. Models based on mechanisms (1) and (2) are known as the monomer-monomer complex participation (MCP) and dissociation (MCD) models, respectively. Mechanisms (3) and (5) would result in a solvent effect analogous to a Bootstrap effect, and will be discussed in Section 13.2.3.4. In this section, we review the MCP and MCD models, and conclude with a brief discussion of specific monomer-solvent interactions.

#### *13.2.3.3.2 Monomer-monomer complex participation model*

The use of monomer-monomer charge transfer complexes to explain deviations from the terminal model was first suggested by Bartlet and Nozaki,[79] later developed by Seiner and Litt,[80] and refined by Cais et al.[81] It was proposed that two monomers can form a 1:1 donor complex and add to the propagating chain as a single unit in either direction. The complex would be more reactive because it would have a higher polarizability due to its larger $\pi$-electron system that can interact more readily with the incoming radical. The complex would also have a higher pre-exponential factor, as a successful attack may be achieved over a wider solid angle.[80] The heavier mass of the complex would also serve to increase the pre-exponential factor.

In addition to the four terminal model reactions, four complex addition reactions and an equilibrium constant are required to describe the system.

$$RMi\cdot + Mj^{\circ} \xrightarrow{k_{ij}} RMiMj\cdot \quad where: i, j = 1 \, or \, 2$$

$$RMi\cdot + MiMj \xrightarrow{k_{iij}} RMiMiMj\cdot \quad where: i, j = 1 \, or \, 2 \, and \, i \neq j$$

$$RMi\cdot + MiMj \xrightarrow{k_{iji}} RMiMjMi\cdot \quad where: i, j = 1 \, or \, 2 \, and \, i \neq j$$

$$M1^{\circ} + M2^{\circ} \xleftrightarrow{K} M1M2$$

The composition and propagation rate can be expressed in terms of the following parameters.[31]

$$\frac{F_1}{F_2} = \frac{f_1^{\circ}}{f_2^{\circ}} \frac{(A_2B_1)r_1f_1^{\circ} + (A_1C_2)f_2^{\circ}}{(A_1B_2)r_2f_2^{\circ} + (A_2C_1)f_1^{\circ}}$$

$$\bar{k}_p = \frac{(A_2B_1)r_1\left(f_2^{\circ}\right)^2 + (A_1B_2)r_2\left(f_2^{\circ}\right)^2 + (A_1C_2 + A_2C_1)f_1^{\circ}f_2^{\circ}}{\left(A_2r_1f_1^{\circ} / k_{11}\right) + \left(A_1r_2f_2^{\circ} / k_{22}\right)}$$

where:

$A_1 = 1 + r_1s_1cQf_1^{\circ}$ and $A_2 = 1 + r_2s_2cQf_2^{\circ}$

$B_1 = 1 + s_1c(1 + r_1c^{-1})Qf_2^{\circ}$ and $B_2 = 1 + s_2c(1 + r_2c^{-1})Qf_1^{\circ}$

$C_1 = 1 + r_1s_1c(1 + r_1c^{-1})Qf_1^{\circ}$ and $C_2 = 1+ r_2s_2c(1 + r_2c^{-1})Qf_2^{\circ}$

$2Qf_i^{\circ} = \{[Q(f_j - f_i) +1]^2 +4Qf_i\}^{1/2} - [Q(f_i - f_j) +1]$ and $Q = k[M]$

$f_i$ is feed composition of Mi

$f_i^{\circ} = [Mi^{\circ}]/[M]$

$r_i = k_{ii}/k_{ij}$; $r_{ic} = k_{iij}/k_{iji}$; $s_{ic} = k_{iij}/k_{ii}$ where: i,j = 1 or 2 and $i \neq j$

The applicability of the MCP model to strongly alternating copolymerization has been a long standing point of contention. In essence, there are two opposing accounts of the strongly alternating behavior observed in copolymers of electron-donor-acceptor (EDA) monomer pairs. In the first account, this behavior has been attributed to the fact that the transition state is stabilized in cross-propagation reaction and destabilized in the homopropagation. Deviations from the terminal model are caused merely by penultimate unit effects. In the second account -the MCP model- the strongly alternating behavior is a result of propagation of a 1:1 comonomer complex which, as seen above, also leads to deviations from the terminal model. An intermediate mechanism, which will be discussed shortly, is the MCD model in which the complex dissociates during the propagation step. The main approach to discriminating between these models has been to compare their ability to describe the copolymerization data of various explicit systems, and to study the effect of added solvents on their behavior. Unfortunately, both approaches have led to inconclusive results.

As an example, the system STY with maleic anhydride (MAH) has been perhaps the most widely studied EDA system and yet there is still uncertainty concerning the role of the EDA complex in its propagation mechanism. Early studies[82,83] concluded that its behavior was best modelled by a penultimate model, despite the spectroscopic evidence for EDA complexes in this system. Later Tsuchida et al.[84,85] fitted an MCP model, based on the evidence that the rate went through a maximum at 1:1 feed ratio in benzene or $CCl_4$ but in strong donor solvents no such maximum occurred and instead the rate increased with the content of MAH in the feed. They argued that maximum in rate at 1:1 feed ratios was due to the fact that propagation occurred via the complex, which had a maximum concentration at this point. In strong donor solvents, the maximum rate moved to higher concentrations of MAH due to competition between the donor and STY for complexation with the MAH. However, a few years later, Dodgson and Ebdon[86,87] conducted an extensive study of STY-MAH in various solvents and discounted the MCP model on the basis of an absence of a dilution effect with the inert solvent MEK. In an MCP model a dilution effect would be expected due to the decrease in the relative concentration of the comonomer complex and the enhanced participation of the free monomer.[88] Later, Farmer et al.[89] reanalyzed this data and concluded that the composition data was consistent with both models and suggested sequence distribution may provide the answer. They also pointed out that there was a small dilution effect in MEK -greater than that predicted by the penultimate model and less than that predicted by the MCP model. Hill et al.[90] has suggested that interpretation of the effect of solvents is complicated by the fact that no solvent is truly inert, hence such results such be treated with caution. More recently Sanayei et al.[91] have performed a pulsed-laser polymerization study on STY-MAH copolymerization in butanone and acetonitrile. They concluded that whilst the complex participation model described the copolymer composition it failed to predict the average $k_p$ data. Consequently the best description of this copolymerization was given by the penultimate unit model.

There have been many other systems for which the MCP model has been proposed as an alternative to the penultimate unit model. For instance, Litt and Seiner used the MCP model to describe the composition of a number of systems, including MAH with 1-diphenylethylene, β-cyanacrolein with styrene,[92] and vinyl acetate with dichlorotetrafluoroacetone and with hexafluoroacetone.[80] An MCP model has also been suggested for the system STY-$SO_2$.[39,83,93-95] In this system, the composition changes with dilution or with solvent changes, strongly alternating behavior is observed across a range of feed ratios, and one of the comonomers ($SO_2$) does not undergo homopolymerization. However, while the MCP model appears to be appropriate for some systems, in other strongly alternating copolymerizations it is clearly not appropriate. For instance, there are many strongly alternating copolymerizations for which there is no evidence of complex formation.[36,39,88,96] Even when complex formation is known to occur, results cannot always be explained by the MCP model. For instance, measurements of sequence distribution data revealed that, while both the MCP and penultimate model could provide an adequate description of the composition of STY with acrylonitrile (AN), only the penultimate model could account for the sequence distribution data for this system.[97] As will be seen shortly, there is evidence that in some systems the heat of propagation would be sufficient to dissociate the EDA complex and hence it could not add to the monomer as unit. In this case an MCD model would be more appropriate. Thus, it might be concluded that the MCP, MCD and the penultimate models are needed to describe the behavior of strongly alternating systems, and the selection of each model should be on a case-by-case basis. There have been many more systems for which the MCP model has been evaluated against the penultimate model on the basis of kinetic behavior. These studies have been extensively reviewed by Hill et al.[90] and Cowie[98] will therefore not be reviewed here. Instead a few additional sources of evidence for the participation of the EDA complex will be highlighted.

**UV and NMR evidence for the existence of EDA complexes**

There is certainly a demonstrable existence of comonomer complexes in solutions of electron donor acceptor monomer pairs. These complexes can be detected, and their equilibrium constants measured, using UV or NMR spectroscopy. Techniques for this are described in detail in reviews of comonomer complexes by Cowie[98] and Hill et al.[90] The latter review[90] also includes a listing of the equilibrium constants for the numerous EDA complexes that have been experimentally detected. The existence of comonomer complexes is not sufficient evidence for their participation in the propagation step of copolymerization, but the fact that they exist in solutions from which strongly alternating copolymers are produced suggests that they play some role in the mechanism. Furthermore, the ability to measure their strengths and quantify the effects of solvents on their observed equilibrium constants without performing kinetic experiments, may provide the key to establishing their role in the propagation mechanism. Since the alternative models for copolymerization include (or in some cases omit) the equilibrium constant for these complexes in different ways; if the equilibrium constant was to be measured separately and not treated as an adjustable parameter in the kinetic analysis, more sensitive model discrimination would be possible. To date, such an analysis does not appear to have been performed but it should be included in subsequent kinetic analyses of these explicit systems.

**Temperature effects**

The study of the temperature dependence of copolymerization behavior may also provide evidence for the role of comonomer complexes. As was seen previously in the study of acrylamide, complexes dissociate at high temperatures and hence, if the complex is involved in controlling an aspect of the polymerization behavior, a change in this behavior should be observed at the temperature corresponding to the complete dissociation of the complex. Such evidence has been obtained by Seymour and Garner[99,100] for the copolymerization of MAH with a variety of vinyl monomers, including STY, VA, AN, and α-MSTY. They observed that the copolymers undergo a change from strongly alternating to random at high temperatures, and these temperatures are also the temperatures at which the concentration of the EDA complex becomes vanishingly small. It is true that, since reactivity ratios have an enthalpy component, they approach unity as temperature increases. Hence, most models predict that the tendency of copolymers to form a random microstructure increases as temperature increased. Indeed, more recent work by Klumperman[101] has shown that for STY-MAH copolymerization, the reactivity ratios do follow an Arrhenius type of temperature dependence. However, further work is required to verify this for the other copolymerization listed above. Based upon the existing copolymerization data, it appears that for many systems there are sudden transition temperatures that correspond to the dissociation of the complex, which does suggests that the complex is in some way responsible for the alternating behavior.[99,100]

**Stereochemical evidence for the participation of the complex**

Stereochemical data may provide evidence for participation of the EDA complex. The EDA complex will prefer a certain geometry -that conformation in which there is maximum overlap between the highest occupied molecular orbital (HOMO) of the donor and the lowest unoccupied molecular orbital (LUMO) of the acceptor.[102] If the complex adds to the propagating polymer chain as a unit, then this stereochemistry would be preserved in the polymer chain. If, however, only free monomer addition occurs, then the stereochemistry of the chain should be completely random (assuming of course that there are no penultimate unit effects operating). Hence, it is possible to test for the participation of the monomer complexes in the addition reaction by examination of the stereochemistry of the resulting polymer.

Such stereochemical evidence has been collected by a number of workers. For instance, Iwatsuki and Yamashia[103] observed an unusually high percentage of cis units in MAH/butadiene copolymers. Olson and Butler[104] studied the EDA system N-phenylmaleimide (NPM)/2-chloroethyl vinyl ether (CEVE) and found that the stereochemistry at succinimide units in NPM-CEVE copolymers is predominantly cis, and random elsewhere. Furthermore, they noted that the proportion of cis units was correlated with those variables with which the concentration of the EDA complex was also correlated. In these examples, the cis geometry is that which is most stable for the complex. However, Rätzsch and Steinert[105] have argued that this preference for cis geometry may also be explained by propagation occurring via a complex between the reacting free monomer and the chain end, as in an RC model. Thus this evidence should be used in conjunction with other evidence for model discrimination.

Further stereochemical evidence for the MCP model has been obtained by Butler et al.[106] They predicted that the usual preference for head-tail addition in free-radical polymerization would be overcome if propagation occurred via the EDA complex, and its favored geometry was a head-head conformation. They noted that for most EDA pairs head-tail geometry was favored and hence the predominance of head-tail linkages in these copolymers could not discriminate between free monomer addition and complex participation. To solve this problem, they designed and synthesized two monomer pairs for which a head-head conformation would be expected in their EDA complexes. These were the systems dimethyl cyanoethylene dicarboxylate (DMCE) with CEVE, and dimethyl cyanoethylene dicarboxylate (DMCE) with CEVE. They then showed that there were significant head-head linkages in the resulting copolymer and the proportion of these linkages was correlated with same types of variables that had previously affected the cis content of NPM/CEVE copolymers -that is, those variables which affected the concentration of the EDA complex. Thus they concluded that there was strong stereochemical evidence for the participation of the EDA complex in the propagation step.

**ESR evidence for the participation of the complex**

ESR studies have also been suggested as a means for providing information about the participation of the EDA complex. Since the addition of the complex is likely to occur more readily in one direction, if propagation occurs as the repeated addition of the complex then the propagating radical should be predominantly of one type. However, if free monomer addition predominantly occurs, both types of radical are likely to be present at any time. Thus ESR can be used to distinguish between the two mechanisms. This approach was used by Smirnov et al.[107] to show that, in the system phenyl vinyl ether/MAH, alternating addition of the free monomer predominates, but participation of EDA complexes is important for the system butyl vinyl ether/MAH. They argued that the difference in the behavior of the two EDA systems was a result of the different strengths of their EDA complexes. In another study, Golubev et al.,[108] used ESR to show that for dimethylbutadiene/MAH the cross-propagation of the free monomers dominated. However, Barton et al.[109] has questioned the assignments of ESR signals in the previous studies and suggested that the ESR evidence was inconclusive. Furthermore, the predominance of one type of ESR signal may also be explained without invoking the MCP model. Assuming that cross-propagation is the dominant reaction, and that one of the radicals is much less stable than the other, it might reasonably be expected that the less stable radical would undergo fast cross-propagation into the more stable radical, resulting in an ESR signal dominated by the more stable radical. Hence it appears that ESR is not able to discriminate between this and the MCP mechanism.

#### *13.2.3.3.3 Monomer-monomer complex dissociation model*

Tsuchida and Tomono[84] suggested that the monomer-monomer complex described in the MCP model may dissociate upon addition to the chain, with only one unit adding. The concept was developed by Karad and Schneider[110] who argued that the dissociation of the complex is likely since its heat of formation is typically less than the heat of propagation. As an example, they measured the heat of formation for a STY/fumaronitrile complex, and found that it was only 1.6 kcal/mol, significantly less than the heat of propagation (15-20 kcal/mol). Under a complex-dissociation mechanism, the role of the complex is merely to modify the reactivity of the reactant monomers.

A model based on the complex-dissociation mechanism was first formulated by Karad and Schneider[110] and generalized by Hill et al.[111] Again, eight rate constants and two equilibrium constants are required to describe the system.

$$RMi\cdot + Mj \xrightarrow{k_{ij}} RMiMj\cdot \quad where: i, j = 1 or 2$$

$$RMi\cdot + MjC \xrightarrow{k_{ijc}} RMiMj\cdot \quad where: i, j = 1 or 2$$

$$Mi + C \xleftrightarrow{K_i} MiC \quad where: i, j = 1 or 2$$

As for the previous models, expressions for $k_p$ and composition can be derived in terms of these parameters by first calculating $\bar{k}_{ii}$ and $\bar{r}_i$ and then using them in place of $k_{ii}$ and $r_i$ in the terminal model equations. The relevant formulae are:

$$\bar{k}_{ii} = k_{ii}\frac{1 + s_{ic}K_i[C_i]}{1 + K_i[C_i]} \text{ and } \bar{r}_i = r_i\frac{1 + s_{ic}K_i[C_i]}{1 + (r_i / r_{ci})s_{ic}K_i[C_i]}$$

where: $r_i = k_{ii}/k_{ij}$; $r_{ic} = k_{iic}/k_{ijc}$; $s_{ic} = k_{iic}/k_{ii}$; i,j = 1 or 2 and $i \neq j$

Efforts to compare this model with the MCP model have been hindered by the fact that similar composition curves for a given system are predicted by both models. Hill et al.[111] showed that the composition data of Dodgson and Ebdon[87] for STY/MAH at 60°C could be equally well described by the MCP, MCD or penultimate unit models. They suggested that sequence distribution would be a more sensitive tool for discriminating between these models. One study which lends some support to this model over the MCP model for describing this system was published by Rätzsch and Steinert.[105] Using Giese's[112] 'mercury method' to study the addition of monomers to primary radicals, they found that in mixtures of MAH and STY, only reaction products from the addition of free monomers, and not the EDA complex, to primary cyclohexyl radicals were found. Thus they concluded that the STY/MAH complex in the monomer solution is disrupted during the propagation step. It is likely that both the MCP and MCD mechanisms are valid and their validity in a specific system will depend on the relative strength of the EDA complex concerned. The MCD model may be useful for accounting for those systems, in which EDA complexes are known to be present but the MCP model has been shown not to hold.

##### *13.2.3.3.4 Specific solvent effects*

Several monomers are particularly susceptible to strong solvent effects via specific interactions such as hydrogen bonding, ionic strength and pH. The kinetic consequences of these specific interactions will vary from system to system. In some cases the radical and/or monomer reactivity will be altered and in other cases a Bootstrap effect will be evident. It is worth noting that monomers which are susceptible to strong medium effects will not have reliable Q-e values, a good example of this is 2-hydroxyethyl methacrylate (HEMA) where there is a large variation in reported values. The reactivity ratios of HEMA with STY have been reported to be strongly dependent on the medium,[113] similarly the copolymerization of HEMA with lauryl methacrylate is solvent sensitive;[114] behavior which has been attributed to non-ideal solution thermodynamics (cf Semchikov's work in Section 13.2.3.4). Chapiro[115] has published extensively on the formation of molecular associates in copolymerization involving polar monomers. Other common monomers which show strong solvent effects are N-vinyl-2-pyrrolidone, (meth)acrylic acids and vinyl pyridines.

### 13.2.3.4 Bootstrap model

#### *13.2.3.4.1 Basic mechanism*

In the Bootstrap model, solvent effects on $k_p$ are attributed to solvent partitioning and the resulting difference between bulk and local monomer concentrations. In this way, a solvent could affect the measured $k_p$ without changing the reactivity of the propagation step. Bootstrap effects may arise from a number of different causes. As noted previously, when radical-solvent and monomer-solvent complexes form and the complexes do not propagate, the effect of complexation is to alter the effective radical or monomer concentrations, thereby causing a Bootstrap effect. Alternatively, a Bootstrap effect may arise from some bulk preferential sorption of one of the comonomers around the growing (and dead) polymer chains. This might be expected to occur if one of monomers is a poor solvent for its resulting polymer. A Bootstrap effect may also arise from a more localized from of preferential sorption, in which one of the monomers preferentially solvates the active chain end, rather than the entire polymer chain. In all cases, the result is the same: the effective free monomer and/or radical concentrations differ from those calculated from the monomer feed ratios, leading to a discrepancy between the predicted and actual propagation rates.

#### *13.2.3.4.2 Copolymerization model*

Copolymerization models based upon a Bootstrap effect were first proposed by Harwood[116] and Semchikov[117] (see references cited therein). Harwood suggested that the terminal model could be extended by the incorporation of an additional equilibrium constant relating the effective and 'bulk' monomer feed ratios. Different versions of this so-called Bootstrap model may be derived depending upon the baseline model assumed (such as the terminal model or the implicit or explicit penultimate models) and the form of equilibrium expression used to represent the Bootstrap effect. In the simplest case, it is assumed that the magnitude of the Bootstrap effect is independent of the comonomer feed ratios. Hence in a bulk copolymerization, the monomer partitioning may be represented by the following equilibrium expression:

$$\frac{f_1}{f_2} = K\left(\frac{f_{1bulk}}{f_{2bulk}}\right)$$

The equilibrium constant K may be considered as a measure of the Bootstrap effect. Using this expression to eliminate the effective monomer fractions ($f_1$ and $f_2$) from the terminal model equations, replacing them with the measurable 'bulk' fractions ($f_{1bulk}$ and $f_{2bulk}$), the following equations for composition[43] and $k_p$[118] may be derived.

$$\frac{F_1}{F_2} = \frac{Kf_{1bulk}}{f_{2bulk}} \frac{r_1 Kf_{1bulk} + f_{2bulk}}{r_2 f_{2bulk} + Kf_{1bulk}}$$

$$k_p = \frac{1}{Kf_{1bulk} + f_{2bulk}} \frac{r_1 K^2 f_{1bulk}^2 + 2Kf_{1bulk} f_{2bulk} + r_2 f_{2bulk}^2}{r_1 Kf_{1bulk} / k_{11} + r_2 f_{2bulk} / k_{22}}$$

Examining the composition and $k_p$ equations above, it is seen that the Bootstrap effect K is always aliased with one of the monomer feed ratios (that is, both equations may be expressed in terms of $Kf_1$ and $f_2$). It is also seen that once $Kf_1$ is taken as a single variable, the composition equation has the same functional form as the terminal model composition equation, but the $k_p$ equation does not. Hence it may seen that, for this version of the Bootstrap effect, the effect is an implicit effect - causing deviation from the terminal model $k_p$ equation only. It may also be noted that, if K is allowed to vary as a function of the monomer feed ratios, the composition equation also will deviate from terminal model behavior - and an explicit effect will result. Hence it may be seen that it is possible to formulate an implicit Bootstrap model (that mimics the implicit penultimate model) but in order to do this, it must be assumed that the Bootstrap effect K is constant as a function of monomer feed ratios.

It should be noted that the above equations are applicable to a bulk copolymerization. When modelling solution copolymerization under the same conditions, the equations may be used for predicting copolymer composition since it is only the relationship between bulk and local monomer feed ratios that determines the effect on the composition and microstructure of the resulting polymer. However, some additional information about the net partitioning of monomer and solvent between the bulk and local phases is required before $k_p$ can be modelled. It should be observed that in a low-conversion bulk copolymerization, knowledge of the monomer feed ratios automatically implies knowledge of the individual monomer concentrations since, as there are no other components in the system, the sum of the monomer fractions is unity. However, in a solution copolymerization there is a third component - the solvent - and the monomer concentrations depend not only upon their feed ratio but also upon the solvent concentration. Modelling $k_p$ in a solution copolymerization could be achieved by re-writing the above equilibrium expression in terms of molar concentrations (rather than comonomer feed ratios), and including the solvent concentration in this expression.

The Bootstrap model may also be extended by assuming an alternative model (such as the explicit penultimate model) as the baseline model, and also by allowing the Bootstrap effect to vary as a function of monomer feed ratios. Closed expressions for composition and sequence distribution under some of these extended Bootstrap models may be found in papers by Klumperman and co-workers.[43,44]

#### *13.2.3.4.3 Experimental evidence*

The Bootstrap model was introduced by Harwood,[116] who studied three solvent sensitive copolymerizations (styrene/methacrylic acid, styrene/acrylic acid and styrene/acrylamide) and found that the copolymers of the same composition had the same sequence distribution irrespective of the solvent used. This meant that the conditional probabilities governing radical propagation were independent of the solvent. On this basis, he argued that composition and sequence distribution were deviating from their expected values because there was a difference between the monomer feed ratios in the vicinity of the active chain end, and those calculated on the basis of the bulk feed. In other words, the solvent was altering the rates of the individual propagation steps by affecting the reactant concentrations and not, as in the other solvent effects models, their reactivities. However, Fukuda et al.[31] have argued that the NMR evidence provided by Harwood is not conclusive evidence for the Bootstrap model, since Harwood's observations could also be described by variation of the reactivity ratios in such a way that their product ($r_1r_2$) remains constant. This has also been raised as an issue by Klumperman and O'Driscoll.[43] They showed mathematically that a variation in the local comonomer ratio is not reflected in the monomer sequence distribution versus copolymer composition -this relationship being governed by the $r_1r_2$ product only. An alternative explanation for Harwood's experimental data may be the stabilization or destabilization of the radicals by the solvent, an interpretation that would be analogous to the MCD model. Simple energy stabilization considerations, as used by Fukuda et al.[119] to derive the penultimate unit effect, also suggest the constancy of $r_1r_2$.

Prior to Harwood's work, the existence of a Bootstrap effect in copolymerization was considered but rejected after the failure of efforts to correlate polymer-solvent interaction parameters with observed solvent effects. Kamachi,[37] for instance, estimated the interaction between polymer and solvent by calculating the difference between their solubility parameters. He found that while there was some correlation between polymer-solvent interaction parameters and observed solvent effects for methyl methacrylate, for vinyl acetate there was none. However, it should be noted that evidence for radical-solvent complexes in vinyl acetate systems is fairly strong (see Section 3), so a rejection of a generalized Bootstrap model on the basis of evidence from vinyl acetate polymerization is perhaps unwise. Kratochvil et al.[120] investigated the possible influence of preferential solvation in copolymerizations and concluded that, for systems with weak non-specific interactions, such as STY-MMA, the effect of preferential solvation on kinetics was probably comparable to the experimental error in determining the rate of polymerization (±5%). Later, Maxwell et al.[121] also concluded that the origin of the Bootstrap effect was not likely to be bulk monomer-polymer thermodynamics since, for a variety of monomers, Flory-Huggins theory predicts that the monomer ratios in the monomer-polymer phase would be equal to that in the bulk phase.[122]

Nevertheless, there are many copolymerization systems for which there is strong evidence for preferential solvation, in particular, polymer solutions exhibiting cosolvency or where one of the solvents is a non-solvent for the polymer. Preferential adsorption and desorption are manifest where the polymer adjusts its environment towards maximum solvation. With this knowledge it may be expected that Bootstrap effects based on preferential solvation will be strongest where one of the monomers (or solvents) is a poor or non-solvent for the polymer (such as copolymerization of acrylonitrile or N-vinyl carbazole). Indeed, early experimental evidence for a partitioning mechanism in copolymerization was provided by Ledwith et al.[123] for the copolymerization of N-vinyl carbazole with MMA in the presence of a range of solvents.

Direct evidence for preferential solvation was obtained by Semchikov et al.,[124] who suggested that it could be detected by calculating, from measurements of the solution thermodynamics, the total and excess thermodynamic functions of mixing. Six monomer pairs were selected -Vac-NVP, AN-STY, STY-MA, Vac-STY, STY-BMA and MMA-STY. The first four of these monomer pairs were known to deviate from the terminal model composition equation, while the latter two were not. They found that these first four copolymerizations had positive $\Delta G^E$ values over the temperature range measured, and thus also formed non-ideal polymer solutions (that is, they deviated from Raoult's law). Furthermore, the extent of deviation from the terminal model composition equation could be correlated with the size of the $\Delta G^E$ value, as calculated from the area between the two most different composition curves obtained for the same monomer pair under differing reaction conditions (for example, initiator concentration; or type and concentration of transfer agent). For STY-MMA they obtained negative $\Delta G^E$ values over the temperature range considered, but for STY-BMA negative values were obtained only at 318 and 343K, and not 298K. They argued that the negative $\Delta G^E$ values for STY-MMA confirmed the absence of preferential solvation in this system, and hence its adherence to the terminal model composition equation. For STY-BMA they suggested that non-classical behavior might be expected at low temperatures. This they confirmed by polymerizing STY-BMA at 303K and demonstrating a change in reactivity ratios of STY-BMA with the addition of a transfer agent.

Based upon the above studies, it may be concluded that there is strong evidence to suggest that Bootstrap effects arising from preferential solvation of the polymer chain operate in many copolymerization systems, although the effect is by no means general and is not likely to be significant in systems such as STY-MMA. However, this does not necessarily discount a Bootstrap effect in such systems. As noted above, a Bootstrap effect may arise from a number of different phenomena, of which preferential solvation is but one example. Other causes of a Bootstrap effect include preferential solvation of the chain end, rather than the entire polymer chain,[121,125] or the formation of non-reactive radical-solvent or monomer-solvent complexes. In fact, the Bootstrap model has been successfully adopted in systems, such as solution copolymerization of STY-MMA, for which bulk preferential solvation of the polymer chain is unlikely. For instance, both Davis[125] and Klumperman and O'Driscoll[43] adopted the terminal Bootstrap model in a reanalysis of the microstructure data of San Roman et al.[126] for the effects of benzene, chlorobenzene and benzonitrile on the copolymerization of MMA-STY.

Versions of the Bootstrap model have also been fitted to systems in which monomer-monomer complexes are known to be present, demonstrating that the Bootstrap model may provide an alternative to the MCP and MCD models in these systems. For instance, Klumperman and co-workers have successfully fitted versions of the penultimate Bootstrap model to the systems styrene with maleic anhydride in butanone and toluene,[43] and styrene with acrylonitrile in various solvents.[44] This latter work confirmed the earlier observations of Hill et al.[127] for the behavior of styrene with acrylonitrile in bulk, acetonitrile and toluene. They had concluded that, based on sequence distribution data, penultimate unit effects were operating but, in addition, a Bootstrap effect was evident in the coexistent curves obtained when triad distribution was plotted against copolymer composition for each system. In the copolymerization of styrene with acrylonitrile Klumperman et al.[44] a variable Bootstrap effect was required to model the data. Given the strong polarity effects expected in this system (see Section 13.2.2), part of this variation may in fact be caused by the variation of the solvent polarity and its affect on the reactivity ratios. In any case, as this work indicates, it may be necessary to simultaneously consider a number of different influences (such as, for instance, penultimate unit effects, Bootstrap effects, and polarity effects) in order to model some copolymerization systems.

### 13.2.4 CONCLUDING REMARKS

Solvents affect free-radical polymerization reactions in a number of different ways. Solvent can influence any of the elementary steps in the chain reaction process either chemically or physically. Some of these solvent effects are substantial, for instance, the influence of solvents on the gel effect and on the polymerization of acidic or basic monomers. In the specific case of copolymerization then solvents can influence transfer and propagation reactions via a number of different mechanisms. For some systems, such as styrene-acrylonitrile or styrene-maleic anhydride, the selection of an appropriate copolymerization model is still a matter of contention and it is likely that complicated copolymerization models, incorporating a number of different phenomena, are required to explain all experimental data. In any case, it does not appear that a single solvent effects model is capable of explaining the effect of solvents in all copolymerization systems, and model discrimination should thus be performed on a case-by-case basis.

### REFERENCES


1 Bednarek, D., Moad, G., Rizzardo, E. and Solomon, D. H., *Macromolecules*, 1988, **21**, 1522.
2 Kiefer, H. and Traylor, T., *J. Am. Chem. Soc.*, 1967, **89**, 6667.
3 Terazima, M., Tenma, S., Watanabe, H. and Tominaga, T., *J. Chem. Soc.-Farad. Trans.*, 1996, **92**, 3057.
4 Terazima, M. and Hamaguchi, H., *J. Phys. Chem.*, 1995, **99**, 7891.
5 Bamford, C. H. and Brumby, S., *Makromol. Chem.*, 1967, **105**, 122.
6 Burnett, G. M., Cameron, G. G. and Joiner, S. N., *J. Chem. Soc., Faraday Trans.*, 1972, **69**, 322.
7 Burnett, G. M., Dailey, W. S. and Pearson, J. M., *Trans. Faraday Soc.*, 1972, **61**, 1216.
8 Anderson, D. B., Burnett, G. M. and Gowan, A. C., *J. Polym. Sci. Part A: Polym. Chem.*, 1963, **1**, 1456.
9 Zammit, M. D., Davis, T. P., Willett, G. D. and O'Driscoll, K. F., *J. Polym. Sci: Part A: Polym. Chem.*, 1997, **35**, 2311.
10 Shtamm, E. V., Skurlatov, Y. I., Karaputadse, I. M., Kirsh, Y. E. and Purmal, A. P., *Vysokomol. Soedin.B*, 1980, **22**, 420.
11 Pascal, P., Napper, D. H., Gilbert, R. G., Piton, M. C. and Winnik, M. A., *Macromolecules*, 1990, **23**, 5161.
12 Morrison, D. A. and Davis, T. P., *Macromol. Chem. Phys.*, 2000 - submitted.
13 Davis, T. P., O'Driscoll, K. F., Piton, M. C. and Winnik, M. A., *Macromolecules*, 1989, **22**, 2785.
14 Coote, M. L. and Davis, T. P., *Eur. Polym. J.*, 2000-in press.
15 Olaj, O. F. and Schnoll-Bitai, I., *Monatsh. Chem.*, 1999, **130**, 731.
16 Beuermann, S., Buback, M., Schmaltz, C. and Kuchta, F.-D., *Macromol. Chem. Phys.*, 1998, **199**, 1209.
17 Yamada, K., Nakano, T. and Okamoto, Y., *Macromolecules*, 1998, **31**, 7598.
18 Kamachi, M., Liaw, D. J. and Nozakura, S., *Polym. J.*, 1979, **12**, 921.
19 Coote, M. L. and Davis, T. P., *Progr. Polym. Sci.*, 2000, **24**, 1217.
20 Renaud, P. and Gerster, M., *Angew. Chem. Int. Ed.*, 1998, **37**, 2562.
21 Odian, G., **Principles of Polymerization**; 2nd edn. *Wiley*, New York, 1981.
22 Harrisson, S., Kapfenstein, H. M. and Davis, T. P., *Macromolecules*, 2000- submitted.

23 Heuts, J. P. A., Forster, D. J. and Davis, T. P., *Macromolecules*, 1999, **32**, 3907.
24 Davis, T. P., Haddleton, D. M. and Richards, S. N., *J. Macromol. Sci., Rev. Macromol. Chem. Phys.*, 1994, **C34**, 234.
25 Benson, S. W. and North, A. M., *J. Am. Chem. Soc.*, 1962, **84**, 935.
26 Mahabadi, H. K. and O'Driscoll, K. F., *J. Polym. Sci., Polym. Chem. Ed.*, 1977, **15**, 283.
27 Norrish, R. G. W. and Smith, R. R., *Nature* (London), 1942, **150**, 336.
28 O'Neil, G. A., Wisnudel, M. B. and Torkelson, J. M., *Macromolecules*, 1998, **31**, 4537.
29 Cameron, G. G. and Cameron, J., *Polymer*, 1973, **14**, 107.
30 Mayo, F. R. and Lewis, F. M., *J. Am. Chem. Soc.*, 1944, **66**, 1594.
31 Fukuda, T., Kubo, K. and Ma, Y.-D., *Prog. Polym. Sci.*, 1992, **17**, 875.
32 Fukuda, T., Ma, Y.-D. and Inagaki, H., *Macromolecules*, 1985, **18**, 17.
33 see for example(a) Wong, M. W., Pross, A., Radom, L., *J. Am. Chem. Soc.*, 1994, **116**, 6284; (b) Wong, M., W., Pross, A.; Radom, L., *J. Am. Chem. Soc.*, 1994, **116**, 11938.
34 For an account of solvent effects in chemical reactions see for example: Pross, A. **Theoretical and Physical Principles of Organic Reactivity**; *John Wiley & Sons*, Inc.: New York, 1995.
35 Price, C. C., *J. Polym. Sci.*, 1946, **1**, 83.
36 Mayo, F. R. and Walling, C., *Chem. Rev.*, 1950, **46**, 191.
37 Kamachi, M., *Adv. Polym. Sci.*, 1981, **38**, 56.
38 Lewis, F., Walling, C., Cummings, W., Briggs, E. R. and Mayo, F. R., *J. Am. Chem. Soc.*, 1948, **70**, 1519.
39 Nozaki, K., *J. Polym. Sci.*, 1946, **1**, 455.
40 Sandner, B. and Loth, E., *Faserforsch Textiltechnology*, 1976, **27**, 571.
41 Sandner, B. and Loth, E., *Faserforsch. Textiltechnology*, 1976, **27**, 633.
42 See for example: (a) Zytowski, T., Fischer, H., *J. Am. Chem. Soc.*, 1996, **118**, 437; (b) Heberger, K., Fischer, H., Int. *J. Chem. Kinetics*, 1993, **25**, 249.
43 Klumperman, B. and O'Driscoll, K. F., *Polymer*, 1993, **34**, 1032.
44 Klumperman, B. and Kraeger, I. R., *Macromolecules*, 1994, **27**, 1529.
45 Russell, G. A., *J. Am. Chem. Soc.*, 1958, **80**, 4897.
46 Russell, G. A., *Tetrahedron*, 1960, **8**, 101.
47 Russell, G. A., Ito, K. and Hendry, D. G., *J. Am. Chem. Soc.*, 1961, **83**, 2843.
48 Sadhir, R. K., Smith, J. D. B. and Castle, P. M., *J. Polym. Sci: Part A: Polym. Chem.*, 1985, **23**, 411.
49 Strong, R. L., Rand, S. J. and Britl, A. J., *J. Am. Chem. Soc.*, 1960, **82**, 5053.
50 Strong, R. L. and Perano, J., *J. Am. Chem. Soc.*, 1961, **83**, 2843.
51 Strong, R. L., *J. Phys. Chem.*, 1962, **66**, 2423.
52 Martin, M. M. and Gleicher, G. J., *J. Am. Chem. Soc.*, 1964, **86**, 238.
53 Russell, G. A., *J. Org. Chem.*, 1959, **24**, 300.
54 Burnett, G. M., Cameron, G. G. and Cameron, J., *Trans. Faraday Soc.*, 1973, **69**, 864.
55 Buchachenko, A. L., Sukhanova, O. P., Kalashnikova, L. A. and Neiman, M. B., *Kinetika i Kataliz*, 1965, **6**, 601.
56 Kalashnikova, L. A., Neiman, M. B. and Buchachenko, A. L., *Zh. Fiz. Khim.*, 1968, **42**, 598.
57 Kalashnikova, L. A., Buchachenko, A. L., Neiman, M. B. and Romantsev, E. G., *Zh. Fiz. Khim.*, 1969, **43**, 31.
58 Hatada, K., Kitayama, T. and Yuki, H., *Makromol. Chem., Rapid Commun.*, 1980, **1**, 51.
59 Busfield, W. K., Jenkins, I. D. and Monteiro, M. J., *Aust. J. Chem.*, 1997, **50**, 1.
60 Busfield, W. K., Jenkins, I. D. and Monteiro, M. J., *J. Polym. Sci.,;Part A, Polym. Sci.*, 1997, **35**, 263.
61 Busfield, W. K., Jenkins, I. D. and Monteiro, M. J., *Polymer*, 1996, **38**, 165.
62 Ratzsch, M. and Vogl, O., *Progr. Polym. Sci.*, 1991, **16**, 279.
63 Mayo, F., *J. Am. Chem. Soc.*, 1958, **80**, 4987.
64 Burnett, G. M., Dailey, W. S. and Pearson, J. M., *Trans. Faraday Soc.*, 1965, **61**, 1216.
65 Burnett, G. M., Dailey, W. S. and Pearson, J. M., *Eur. Polym. J.*, 1969, **5**, 231.
66 Hall Jr., H. K. and Daly, R. C., *Macromolecules*, 1975, **8**, 23.
67 Henrici-Olive, G. and Olive, S., *Makromol. Chem.*, 1963, **68**, 219.
68 Henrici-Olive, G. and Olive, S., *Z. Phys. Chem.*, 1965, **47**, 286.
69 Henrici-Olive, G. and Olive, S., *Z. Phys. Chem.*, 1966, **48**, 35.
70 Henrici-Olive, G. and Olive, S., *Z. Phys. Chem.*, 1966, **48**, 51.
71 Henrici-Olive, G. and Olive, S., *Makromol. Chem.*, 1966, **96**, 221.
72 Burnett, G. M., Cameron, G. G. and Zafar, M. M., *Eur. Polym. J.*, 1970, **6**, 823.
73 Burnett, G. M., Cameron, G. G. and Joiner, S. N., *J. Chem. Soc., Faraday Trans.*, 1973, **69**, 322.
74 Kamachi, M., Satoh, J. and Nozakura, S.-I., *J. Polym. Sci. Polym. Chem. Ed.*, 1978, **16**, 1789.

75 Pascal, P., Napper, D. H., Gilbert, R. G., Piton, M. C. and Winnik, M. A., *Macromolecules*, 1993, **26**, 4572.
76 Heublein, B. and Heublein, G., *Acta Polym.*, 1988, **39**, 324.
77 O'Driscoll, K. F. and Monteiro, M. J., 1996.
78 Czerwinski, W. K., *Macromolecules*, 1995, **28**, 5411.
79 Bartlett, P. D. and Nozaki, K., *J. Am. Chem. Soc.*, 1946, **68**, 1495.
80 Seiner, J. A. and Litt, M., *Macromolecules*, 1971, **4**, 308.
81 Cais, R. E., Farmer, R. G., Hill, D. J. T. and O'Donnell, J. H., *Macromolecules*, 1979, **12**, 835.
82 Bamford, C. H. and Barb, W. G., *Discuss. Faraday Soc.*, 1953, 208.
83 Barb, W. G., *J. Polym. Sci.*, 1953, **11**, 117.
84 Tsuchida, E. and Tomono, H., *Makromol. Chem.*, 1971, **141**, 265.
85 Tsuchida, E., Tomono, T. and Sano, H., *Makromol. Chem.*, 1972, **151**, 245.
86 Dodgson, K. and Ebdon, J. R., *Makromol. Chem.*, 1979, **180**, 1251.
87 Dodgson, K. and Ebdon, J. R., *Eur. Polym. J.*, 1977, **13**, 791.
88 Walling, C., Briggs, E. R., Wolfstern, K. B. and Mayo, F. R., *J. Am. Chem. Soc.*, 1948, **70**, 1537.
89 Farmer, R. G., Hill, D. J. T. and O'Donnell, J. H., *J. Macromol. Sci., Chem.*, 1980, **A14**, 51.
90 Hill, D. J. T., O'Donnell, J. J. and O'Sullivan, P. W., *Prog. Polym. Sci.*, 1982, **8**, 215.
91 Sanayei, R. A., O'Driscoll, K. F. and Klumperman, B., *Macromolecules*, 1994, **27**, 5577.
92 Litt, M., *Macromolecules*, 1971, **4**, 312.
93 Barb, W. G., *Proc. Roy. Soc., Ser. A*, 1952, **212**, 66.
94 Barb, W. G., *J. Polym. Sci.*, 1952, **10**, 49.
95 Booth, D., Dainton, F. S. and Ivin, K. J., *Trans. Faraday Soc.*, 1959, **55**, 1293.
96 Lewis, F. M., Walling, C., Cummings, W., Briggs, E. R. and Wenisch, W. J., *J. Am. Chem. Soc.*, 1948, **70**, 1527.
97 Hill, D. J. T., O'Donnell, J. H. and O'Sullivan, P. W., *Macromolecules*, 1982, **15**, 960.
98 Cowie, J. M. G., in **Alternating Copolymers**; Vol. , ed. Cowie, J. M. G., *Plenum*, New York, 1985, .
99 Seymour, R. B. and Garner, D. P., *Polymer*, 1976, **17**, 21.
100 Seymour, R. B. and Garner, D. P., *Polym. News*, 1978, **4**, 209.
101 Klumperman, B. "Free Radical Copolymerization of Styrene and Maleic Anhydride," PhD Thesis, Technische Universiteit Eindhoven, 1994.
102 Arnaud, R., Caze, C. and Fossey, J., *J. Macromol. Sci. - Chem.*, 1980, **A14**, 1269.
103 Iwatsuki, S. and Yamashita, Y., *Makromol. Chem.*, 1967, **104**, 263.
104 Olson, K. G. and Butler, G. B., *Macromolecules*, 1983, **16**, 710.
105 Rätzsch, M. and Seinert, V., *Makromol. Chem.*, 1984, **185**, 2411.
106 Butler, G. B., Olson, K. G. and Tu, C.-L., *Macromolecules*, 1984, **17**, 1884.
107 Smirnov, A. I., Deryabina, G. L., Kalabina, A. L., Petrova, T. L., Stoyachenko, I. L., Golubev, V. B. and Zubov, V. P., *Polym. Sci. USSR* (Engl. Transl.), 1978, **20**, 2014.
108 Golubev, V. B., Zubov, V. P., Georgiev, G. S., Stoyachenko, I. L. and Kabanov, V. A., *J. Polym. Sci., Polym. Chem. Ed.*, 1973, **11**, 2463.
109 Barton, J., Capek, I. and Tino, J., *Makromol. Chem.*, 1980, **181**, 255.
110 Karad, P. and Schneider, C., *J. Polym. Sci. Part A: Polym. Chem.*, 1983, **16**, 1295.
111 Hill, D. J. T., O'Donnell, J. H. and O'Sullivan, P. W., *Macromolecules*, 1983, **16**, 1295.
112 Giese, B. and Meister, J., *Chem. Ber.*, 1977, **110**, 2558.
113 Lebduska, J., Snuparek, J., Kaspar, K. and Cermak, V., *J. Polym. Sci: Part A: Polym. Chem.*, 1986, **24**, 777.
114 Ito, K., Uchida, K., Kitano, T., Yamada, E. and Matsumoto, T., *Polym. J.*, 1985, **17**, 761.
115 Chapiro, A. and Perec-Spitzer, L., *Eur. Polym. J.*, 1975, **25**, 713.
116 Harwood, H. J., *Makromol. Chem. Makromol. Symp.*, 1987, **10/11**, 331.
117 Semchikov, Y. D., *Macromol. Symp.*, 1996, **111**, 317.
118 Coote, M. L., Johnston, L. P. M. and Davis, T. P., *Macromolecules*, **30**, 8191 (1997).
119 Fukuda, T., Ma, Y.-D. and Inagaki, H., Makromol. Chem., *Rapid Commun.*, 1987, **8**, 495.
120 Kratochvil, P., Strakova, D., Stejskal, J. and Tuzar, Z., *Macromolecules*, 1983, **16**, 1136.
121 Maxwell, I. A., Aerdts, A. M. and German, A. L., *Macromolecules*, 1993, **26**, 1956.
122 Maxwell, I. A., Kurja, J., Doremaele, G. H. J. v. and German, A. L., *Makromol Chem.*, 1992, **193**, 2065.
123 Ledwith, A., Galli, G., Chiellini, E. and Solaro, R., *Polym. Bull.*, 1979, **1**, 491.
124 Egorochkin, G. A., Semchikov, Y., D., Smirnova, L. A., Karayakin, N. V. and Kut'in, A. M., *Eur. Polym. J.*, 1992, **28**, 681.
125 Davis, T. P., *Polym. Commun.*, 1990, **31**, 442.
126 San Roman, J., Madruga, E. L. and Puerto, M. A., *Angew Makromol. Chem.*, 1980, **86**, 1.
127 Hill, D. J. T., Lang, A. P., Munro, P. D. and O'Donnell, J. H., *Eur. Polym. J.*, 1992, **28**, 391.
128 Coote, M. L., Davis, T. P., Klumperman, B. and Monteiro, M. J., *J.M.S.-Rev. Macromol. Chem. Phys.*, 1998, **C38**, 567.

# 13.3 EFFECTS OF ORGANIC SOLVENTS ON PHASE-TRANSFER CATALYSIS

MAW-LING WANG
**Department of Chemical Engineering**
**National Chung Cheng University, Taiwan, ROC**

The reaction of two immiscible reactants is slow due to their low solubilities and limited contact surface area. The conventional way to improve the reaction rate or to elevate the conversion of reactants is to increase the agitation speed, temperature, or use the protic or aprotic solvent to dissolve the reactants. The increase in agitation speed can increase the contact surface area between two phases only to a certain value. Thus, the reaction rate or the conversion is limited by the increase in the agitation speed. Usually, the rate of reaction is increased by raising the temperature. However, byproducts are accompanied by elevating the solution temperature. The separation of product from byproducts or catalyst makes the cost to increase. Although protic solvent ($CH_3OH$, or $CH_3COOH$) can dissolve reactants, solvation and hydrogen bonding make the activity of the nucleophilic anion decrease significantly. Thus, the reaction rate using protic solvent is retarded. For the other case, the reaction rate is largely increased using aprotic solvent. The application of aprotic solvent is also limited because of cost and recovery difficulty. For this, the problem of two-phase reaction is not overcome until the development of phase-transfer catalysis (PTC). Phase-transfer catalytic reactions provide an effective method in organic synthesis from two immiscible reactants in recent development.[93,103,111,113]

In 1951, Jarrouse[47] found that the reaction of aqueous-soluble sodium cyanide (NaCN) and organic-soluble 1-chlorooctane (1-$C_8H_{17}Cl$) is dramatically enhanced by adding a small amount of quaternary ammonium salt ($R_4N^+X^-$, or $Q^+X^-$, $Q^+$: $R_4N^+$). The reaction is almost complete and a 95% conversion is obtained within two hours when a catalytic amount of tetra-n-butylammonium chloride ($(C_4H_9)_4N^+Cl^-$, or $Q^+Cl^-$, $Q^+$: $(C_4H_9)_4N^+$) is added. The mechanism of the reaction of sodium cyanide and 1-chlorooctane in organic solvent/water two-phase medium is expressed as

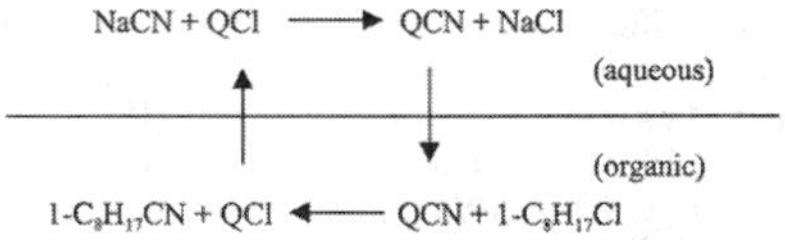


[13.3.1]

As shown in Equation [13.3.1], sodium cyanide (NaCN) and 1-chlorooctane (1-$C_8H_{17}Cl$) are soluble in aqueous phase and organic phase, respectively. In the aqueous phase, NaCN first reacts with tetra-n-butylammonium chloride ($(C_4H_9)_4N^+Cl^-$, $Q^+Cl^-$) to produce organic-soluble tetra-n-butylammonium cyanide ($(C_4H_9)_4N^+CN^-$, $Q^+CN^-$). Then, this tetra-n-butylammonium cyanide (QCN) further reacts with 1-chlorooctane (1-$C_8H_{17}Cl$) to produce 1-cyanooctane ($C_8H_{17}CN$) in the organic phase. Tetra-n-butylammonium chloride ($(C_4H_9)_4N^+Cl^-$), which is also produced from the organic-phase reaction, transfers to the aqueous phase, prepared for further regeneration. It is obvious that PTC reaction[107] involves a two-phase reaction (aqueous-phase and organic-phase reaction), transfer of QCN from aqueous phase to organic phase and transfer of QCl from organic phase to aqueous phase, and equilibrium partition of QCN and QCl between organic and aqueous phases, respectively. The overall reaction rate highly depends on the intrinsic rate constants in aqueous phase and organic phase, the mass transfer rate of QCN and QBr, and the equilibrium partition coefficient of QCN and QBr, which are all affected by the interaction of components and their environments. The organic solvent provides the environment for the interaction of reactants. Therefore, the organic solvent plays an important role in influencing the reaction rate and the conversion of reactant.

Since then, Makosza used an interfacial mechanism[65-67] to describe the behavior in the two-phase reaction. Later, Starks[107] used the extraction mechanism to explain the behavior in the two-phase reaction and selected phase-transfer catalysis (PTC) to describe this special chemical process.[14,161] The most important advantage of using PTC technique is in synthesizing specialty chemicals with almost no byproducts and moderate reaction conditions. Today, PTC is widely applied to various reactions via substitution, displacement, condensation, oxidation and reduction, polymer modification and polymerization to synthesize specialty chemicals. Based on the reaction mechanism, phase-transfer catalysis can be classified as: (1) normal phase-transfer catalysis (NPTC), (2) reverse phase-transfer catalysis (RPTC), and (3) inverse phase-transfer catalysis (IPTC). Equation [13.3.1] illustrates the typical reaction for NPTC. The phase-transfer catalyst ($Q^+$) brings the nucleophilic reagent ($CN^-$) from aqueous phase to organic phase. Quaternary ammonium salts, quaternary phosphonium salts, crown ethers, polyethylene glycols (PEGs) and tertiary amines are the common normal phase-transfer catalyst (NPTC).[17,29,94,108,109,110,128,130,152]

In general, the cation transfers from aqueous phase to organic phase in the RPTC. The principle of reverse phase-transfer catalysis (RPTC)[24,42-44,50] is that an ion pair is formed from catalyst and cation in the aqueous phase. This ion-paired compound then transfers to the organic phase reacting with an organic-phase reactant. Alkyl-aryl sulfonate ($RSO_3Na$), such as sodium 4-dodecylbenzene sulfonate (NaDBS) and tetraarylboronate such as sodium tetra(diperfluoromethyl)phenyl-boronate (TFPB) are the common reverse phase-transfer catalysts. However, few results were reported using reverse phase-transfer catalysis in synthesizing specialty chemicals.[24,42-44,50] A typical reaction mechanism in a liquid-liquid two-phase solution is given by Equation [13.3.2]

In the NPTC and RPTC, the function of the catalyst is that it first reacts with aqueous-phase reactant to produce an organic-soluble ion-pair compound. Mathias and Vaidya[69] found that an aqueous-soluble ion pair was produced in the organic phase from the reaction of alanine and benzoyl chloride catalyzed by 4-dimethylaminopyridine (DMAP). This discovery initiated the research of the field in inverse phase-transfer catalysis (IPTC), in which the catalyst first reacts with organic-phase reactant in the organic phase to produce an aqueous-soluble ion-paired intermediate. Then, this aqueous-soluble ion-paired intermediate transfers to the aqueous phase, prepared for reacting with aqueous-phase reactant to produce the desired product. Catalyst is released in the aqueous phase and transferred to the organic phase for further regeneration. A typical IPTC mechanism of the reaction of benzoyl chloride and sodium acetate to synthesize ester compound in the liquid-liquid two-phase reaction is expressed by Equation [13.3.3]

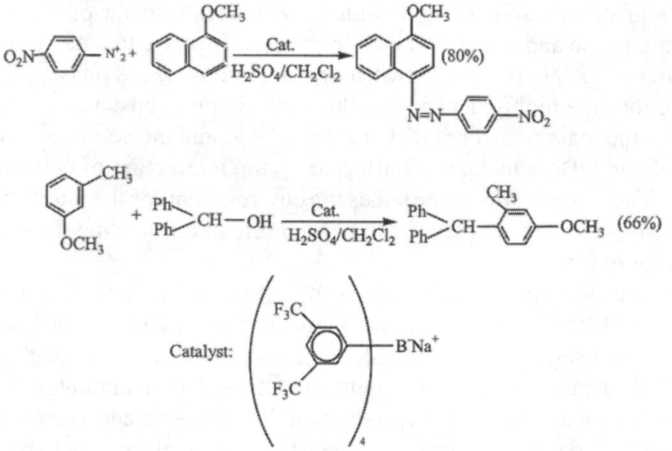


[13.3.2]

Inverse phase-transfer catalysis (IPTC) can be applied to synthesize symmetric and antisymmetric acid anhydride in organic synthesis.[26,54,69,102,148-150,153] Pyridine 1-oxide (PNO), 4-dimethylaminopyridine (DMAP), 4-pyrrolidinopyridine (PPY) and 1-methyl-2(1H)-pyridothione are usually used as the inverse phase-transfer catalysts (IPTC).[159]

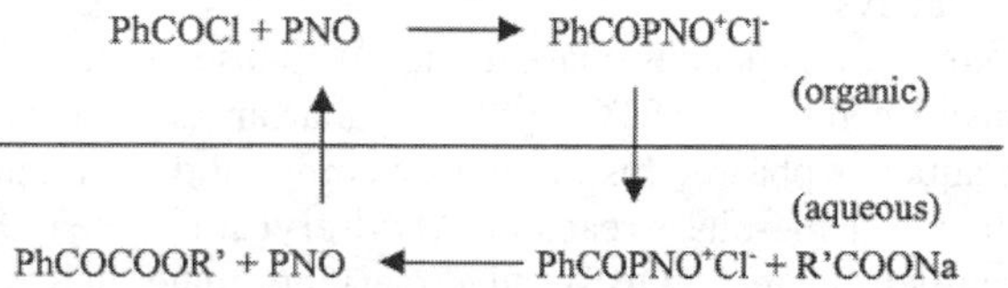


[13.3.3]

The characteristics of two-phase phase-transfer catalytic (PTC) systems are: the presence of at least two phases and at least one interfacial region separating the phases.[24,26,42,43,50,69,107,109,110,152] The reactions involve: (1) transfer of an ion or compound from its normal phase into the reaction phase or interfacial region, (2) reaction of the transferred ion or compound with the non transferred reactant located in the reaction phase or interfacial region, and (3) transfer of the product from the reaction phase or interfacial region into its normal phase. For example, a successful NPTC process involves (1) the maximization of the rate of transfer of reactant anions from the aqueous or solid phase to the organic phase, (2) the maximization of the rate of transfer of product anions from the organic phase to the aqueous phase or solid phase, and (3) the related equilibrium partitioning of the reactant and product anions between organic and aqueous or solid phases. The anion must not only transfer to the organic phase, but once there the anion must be in a highly reactive form. Some organic-phase reactions are so fast that the transferred anion requires little or no activation beyond just being delivered to the organic phase. Other reactions require substantial anion activation before useful and practical reaction rates can be achieved. It is obvious that the polarity of the organic solvent affects the activation of the anion as well as the difference in the cation-anion interionic distance for the two ion pairs. In principle, anions do not have a great affinity for nonpolar solvent and prefer to reside in an aqueous phase.

Ease or difficulty of transfer of most anions into organic-phase solution is also highly affected by the organic solvent, i.e., interaction of the organic solvent and the reactant. In general, a polar solvent may be necessary to obtain an appropriate rate of the anion transfer to the organic phase for a NPTC process. Solvent may be necessary to increase the rate of the organic-phase reaction. The most common solvent, dichloromethane ($CH_2Cl_2$), has been extensively used as a polar solvent in the PTC work because it readily dissolves most quaternary salts and other phase-transfer agents, and because it is polar to speed both the transfer step and the organic-phase reaction step. Although the hydrocarbons suffer from lack of polarity, they have also been extensively used as solvents for PTC systems. The main reasons are that they are reasonably safe, inexpensive and easy to recover in a high purity. One strategy for selecting organic solvent is that a high boiling point solvent is selected for a reaction in which the product has a low-boiling point. In other cases, a solvent might be chosen to minimize solubility of phase-transfer agent in the organic phase to force formation of third phase (catalyst) from which the phase-transfer catalyst may be more easily separated or extracted.

The phase-transfer catalytic reactions (NPTC, RPTC or IPTC) are usually carried out in a liquid-liquid two-phase medium. They have been extensively applied to liquid-gas, liquid-solid two-phase media.[18,21,63,128-130] However, purification of product from catalyst in the liquid phase of a final solution is difficult to produce a product of high purity. In 1975, Regen and coworkers[88-92] proposed triphase catalysis (TC) in which the catalyst is immobilized on a porous solid support (usually polymer). The solid catalyst is easily separated from the final products after reaction by mechanical separation processes, such as centrifugation or filtration. The organophilicity and the hydrophilicity of the solid polymer support greatly influence the content and the imbibed composition of the organic phase and the aqueous phase within the solid porous polymer support. Hence, the reaction rates are determined by the concentrations of reactants in both the organic phase and aqueous phase, they are controlled by the organic solvents. Therefore, it is important to understand the characteristics of the organic phase in the triphase catalyst as well as the characteristics of the organic solvent in the liquid-liquid two-phase PTC reaction.

## 13.3.1 TWO-PHASE PHASE-TRANSFER CATALYTIC REACTIONS

### 13.3.1.1 Theoretical analysis of the polarity of the organic solvents and the reactions

The transfer of anions from an aqueous phase to an organic phase may be achieved by choosing a phase-transfer cation that is not strongly solvated by water, that has organic-like characteristics, and is compatible with the organic phase for NPTC. The factors that affect the mass transfer and the distribution of the phase-transfer catalyst cation-anion pair between the organic and aqueous phases include:

(1) the charge-to-volume ratio of the anion, the polarizability, and the organic structure of the cation associated with the anion,
(2) the hydrophilic-organophilic balance of the associated cation;
(3) the polarity of the organic phase;
(4) the hydration of the anion;
(5) the presence of aqueous salts and/or aqueous hydroxide ions.

Both cation and anion of phase-transfer catalyst can affect the distribution of the PTC between two phases, and hence the reaction rate. The partitioning equilibrium of the anion between organic and aqueous phases can be qualitatively estimated from the free energies of the anion transfer from water to organic phase. A large positive free energy of transfer from the aqueous phase to the organic phase clearly indicates that the anion prefers to reside in the aqueous phase. For example, the free energies of transfer of $Cl^-$, $Br^-$, and $I^-$ from water to acetonitrile are +11.6, +8.1 and +4.8 Kcal/mol, respectively.[110] Thus, the transfer from aqueous to organic phase becomes less unfavorable as one proceeds from chloride to bromide to iodide. These trends may be understood in terms of the change in charge-to-volume ratios of the halide ions. Because chloride has the largest charge-to-volume ratio, it is the least polarizable and the most strongly hydrated. In contrast, iodide has a relatively diffuse charge and is less strongly hydrated.

A successful phase-transfer catalytic reaction occurs when the process is able to transfer the anions from the aqueous phase to the organic phase or *vice versa* for the reaction to proceed, and the transferred anions are active and prepared for reaction. An active catalyst needs to be sufficiently distributed in the organic phase for the reaction to occur. The distribution of catalysts and the associated anion in the organic phase strongly depends on the structure of the quaternary cation and the hydration of anion being transferred into the aqueous phase. Therefore, the following results are used for the reference in selective NPTC catalyst.

(1) Tetramethylammonium cation with a simple anion $(CH_3)_4N^+Y^-$ ($Y=Cl^-$, $Br^-$, $CN^-$, etc.) is not easily distributed in most organic solutions. Therefore, $(CH_3)_4N^+Y^-$ are usually not good PTCs. The only ways to increase the distribution of $(CH_3)_4N^+Y^-$ is to couple the cation with a large organic anion[13,49,78,95,155,156] or to use an organic solvent of high purity.[17,107,109,110,152]

(2) Tetraethylammonium ($(C_2H_5)_4N^+Y^-$) and tetrapropylammonium ($(C_3H_7)_4N^+Y^-$) salts are also poor catalysts for transferring small anions into most organic solutions.[51,105]

(3) Tetrabutylammonium salts show high efficiencies as phase-transfer catalysts. They are readily available in high purity on a commercial scale.

(4) Quaternary ammonium cations, $R_4N^+$, $R=C_5H_{11}$ to $C_{10}H_{21}$ easily extract anions into organic phase and exhibit higher catalytic activities.

(5) Higher tetraalkylammonium salts, $R_4N^+X$, R: $(C_{12}H_{25})_4N^+$ and higher groups, can easily extract anion into an organic phase. However, the interchange of anions between organic and aqueous phases is slow and the reaction rate decreases compared with quaternary salts where $R = C_5H_{11}$ to $C_{10}H_{21}$.

Table 13.3.1 shows the effect of catalyst structure on the rate of PTC reaction of thiophenoxide with 1-bromooctane.

In addition to the preference of anion to reside in the aqueous or organic phase, a distribution ratio (or partition coefficient), α, of phase-transfer catalyst (QX) cation between aqueous and organic phase is defined as

$$\alpha = [QX]_{org}/[QX]_{aq} \qquad [13.3.4]$$

Use of solvents having higher polarity facilitates distribution of quaternary salts into organic solvents. Hence, it also allows use of smaller quaternary salts as catalysts. With di-

**Table 13.3.1. Effect of catalyst structure on the rate of the reaction of thiophenoxide and 1-bromooctane in benzene/water solution**

| Catalyst | $k \times 10^3$, $Lmol^{-1}s^{-1}$ |
|---|---|
| $(CH_3)_4N^+Br^-$ | <0.0016 |
| $(C_3H_7)_4N^+Br^-$ | 0.0056 |
| $(C_4H_9)_4N^+Br^-$ | 5.2 |
| $(C_8H_{17})_3MeN^+Cl^-$ | 31 |
| $(C_8H_{17})_3EtP^+Br^-$ | 37 |
| $(C_6H_{13})Et_3N^+Br^-$ | 0.015 |
| $(C_8H_{17})Et_3N^+Br^-$ | 0.16 |
| $C_{10}H_{21}Et_3N^+Br^-$ | 0.24 |
| $C_{12}H_{25}Et_3N^+Br^-$ | 0.28 |
| $C_{16}H_{33}Et_3N^+Br^-$ | 0.15 |
| $C_{16}H_{33}Et_3N^+Br^-$ | 0.48 |

Data obtained from the work of Herriott and Picker[36]

chloromethane or solvent of a similar polarity, it is possible to use tetramethylammonium cation, $(CH_3)_4N^+$, or tetrapropylammonium cation, $(C_3H_7)_4N^+$, or cation salts as catalysts.

Wu et al.[158] measured the concentration distribution of the quaternary salt between dichloromethane (or chlorobenzene) and alkaline solution and determined the thermodynamic characteristics (the true extraction constant, the distribution coefficient, and the dissociation constant). The distribution coefficient, highly dependent on the organic solvent, increased with increasing NaOH concentration. However, the real dissociation constant decreased with increasing NaOH concentration. Konstantinova and Bojinov[52] synthesized several unsaturated 9-phenylxanthene dyes under phase transfer catalysis conditions. They determined the most favorable solvent.

The extraction constant of QX between two phases is given by equation:

$$E_{QX} = \frac{[Q^+]_{aq}[X^-]_{aq}}{[QX]_{org}} \qquad [13.3.5]$$

In addition to synthesizing specialty chemicals, the PTC technique can be used to analyze many lipid-rich samples.[106] Hydrolyzed samples were treated with phenylisothiocyanate and the phenylthiocarbamyl (PTC). Derivatives obtained were separated by reverse phase HPLC. The PTC/reverse HPLC method was used for analysis of chloroform/methanol extracts of spinal cord, lung and bile after chromatography on Lipidex 5000 in methanol/ethylene chloride, 4:1 (v/v).[106]

The extraction constant of hydroxide is about $10^4$ times smaller than that of chloride.[19,20] Table 13.3.2 shows the effect of solvents on the distribution of $(C_4H_9)_4N^+Br^-$ between aqueous and organic phases. In addition to improved transfer of anion to an organic phase, more polar solvents are commonly recognized in physical organic chemistry for enhancing the rate of organic-phase reactions by providing a more ion-compatible reaction mechanism.

The polarity of the organic phase in conjunction with the structure of the anion and the catalyst cation affects the selectivity of the phase-transfer catalyst partitioning into the organic phase. Increasing the polarity and the hydrogen-bonding ability of the organic phase has a strong favorable effect on the extraction of small ions (large charge-to-volume ratios) and on anion with substantial organic structure.

The polarity of the organic phase is thus an important factor influencing the reaction rate. The polarity of the organic phase depends on the polarity of solvent and organic-phase reactant. However, there are many PTC reactions without organic phase (near organic reactant is used as the organic phase). Thus, a substantial change in the polarity of the organic

**Table 13.3.2. Effect of organic solvents on the distribution of $(C_4H_9)_4N^+Br^-$ between organic and aqueous phases**

| Solvent | Extraction constant, $E_{QBr}$* |
|---|---|
| $C_6H_6$ | >10.0 |
| $C_6H_5Cl$ | >10.0 |
| o-$C_6H_4Cl_2$ | >10.0 |
| n-$C_4H_9Cl$ | >10.0 |
| $Cl(CH_2)_4Cl$ | 3.33 |
| $Cl(CH_2)_3Cl$ | 0.34 |
| $Cl(CH_2)_2Cl$ | 0.16 |
| $ClCH_2Cl$ | 0.028 |
| $CHCl_3$ | 0.021 |
| $C_2H_5COC_2H_5$ | 0.91 |
| $CH_3COC_2H_5$ | 0.071 |
| n-$C_4H_9OH$ | 0.014 |

*$E_{QBr} = [Q^+]_{aq}[Br^-]_{aq}/[QBr]_{org}$
Data adopted from Brandstrom[4]

phase may occur as the reaction proceeds. This may have some effect on the reaction rate of the organic-phase (may raise or lower the value of the rate constant). The changes in the polarity of the organic phase may increase or decrease, causing almost all catalyst cation-anion pairs to be partitioned into the organic phase. This behavior is evidenced in the cyanide displacement on 1-bromooctane catalyzed by tetra-n-butylphosphonium bromide ($(C_4H_9)_4P^+Br^-$, or $Q^+Br^-$).[107,109,110,111,114] The catalyst is only sparingly soluble in 1-bromooctane, but is substantially more soluble in aqueous sodium cyanide solution, so that initially little $Q^+CN^-$ is in the organic phase and the displacement reaction is slow. However, tetra-n-butylphosphonium salts ($(C_4H_9)_4P^+CN^-$, $Q^+CN^-$) are more soluble in the product 1-cyanooctane; therefore, as the conversion of 1-bromooctane to 1-cyanooctane continues, increasing quantities of the catalyst are taken into the organic phase, and the reaction rate accelerates. This behavior signals the autocatalytic character of reaction.

Not only does the solvent affect the reaction rate, but it also determines the reaction mechanism. In Starks' extraction mechanism of PTC, most reacting compound transfers to the bulk phase. However, reaction may occur at the interface of the two phases. For example: hexachlorocyclotriphosphazene has been reported to react very slowly with 2,2,2-trifluoroethanol in an alkaline solution of $NaOH/C_6H_5Cl$ two-phase system in the absence of phase-transfer catalyst.[136-140] Since sodium 2,2,2-trifluoroethanoxide is not soluble in chlorobenzene, the process probably proceeds at the interface region of the system. Similar is the reaction of benzylation of isobutyraldehyde in the presence of tetra-n-butylammonium iodide in an alkaline solution of NaOH/toluene, which is a two-phase system.[37] Makosza interfacial mechanism[65-67] was employed to rationalize the experimental results. The main reason is that the ammonium salt of the nucleophilic reagent is not soluble in toluene.

Usually, the nucleophilic substitutions under NPTC condition are described by an $S_N2$-type reaction both in solid-liquid and liquid-liquid systems in which they can proceed at the interface through the formation of cyclic adsorption complexes.[160] The activity of the nucleophilic reagent in the organic phase is determined by the polarity of the organic solution and the hydration in liquid-liquid system. In the solid-liquid system, the reaction is highly affected by the organic solvent.

## 13.3.1.2 Effect of organic solvent on the reaction in various reaction systems

(A) <u>Synthesis of ether compound catalyzed by quaternary ammonium salts (NPTC)</u>

One of the most useful synthesis applications of phase transfer catalysis (PTC) is in the preparation of ether according to the following general equation

$$R'X + ROX + OH^- \xrightarrow{\phantom{aa}PTC\phantom{aa}} ROR' + H_2O + X^- \qquad [13.3.6]$$

where R and R' are the primary or secondary alkyl or aryl groups, X is a halide and the caustic base is usually sodium or potassium hydroxide in the aqueous solution. The generally accepted reaction mechanism is

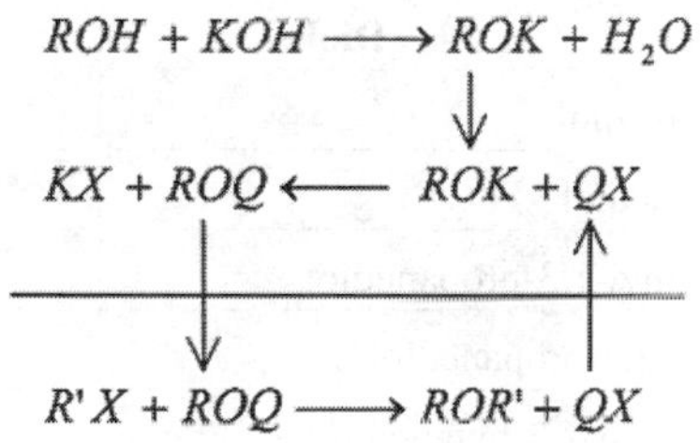

$$
\begin{array}{c}
ROH + KOH \longrightarrow ROK + H_2O \\
\downarrow \\
KX + ROQ \longleftarrow ROK + QX \\
\text{-----------------------} \\
\downarrow \qquad\qquad \uparrow \\
R'X + ROQ \longrightarrow ROR' + QX
\end{array}
\qquad [13.3.7]
$$

It is important to consider that the alkoxide ion (RO⁻) is a reactive nucleophilic but also a strong base. It was shown that $10^{-3}$ M $C_6H_5(CH_3)_3N^+OC_4H_9$ is 1000 times more basic than $KOC_4H_9$ (both in $C_4H_9OH$). Extracted alkoxide bases can be applied in principle to numerous base-catalyzed reactions, e.g., oxidations, eliminations and isomerization. Better quantitative understanding of the extraction of alkoxide into organic phase is important. Dehmlow et al.[20] investigated the extraction of aqueous sodium hydroxide solution with organic solvent containing various quaternary ammonium salts by mixing sodium hydroxide with organic solvent containing $R_4NX$. After phase separation, titration of the organic phase showed only traces of base presence if concentrated NaOH solution was employed and if Cl⁻ was the counter ion.

The Cl-OH exchange was found to be of the order 1-2% for all quaternary ammonium chloride with chlorobenzene as solvent; i.e., 98% of the salts remained in the $R_4NCl$ form. However, upon addition of trace amounts of various alcohols, a dramatic change in the behavior of the system was observed and significant amounts of base could be detected in the organic phase. Table 13.3.3 shows the experimental results where 50% aqueous caustic solutions were extracted by equal volumes of 0.1 M $(C_8H_{17})_4NBr$ in chlorobenzene containing 0.1 M of various alcohols.

**Table 13.3.3. Extraction of base by chlorobenzene solution of tetra-n-octyl-ammonium bromide and alcohols (0.1 M) from an equal volume of 50% NaOH (percent of the maximum possible basicity)**

| | % | | % |
|---|---|---|---|
| **Primary alcohols** | | 2-tert-Butylcyclohexanol | 2.0 |
| Ethanol | 4.5 | 3- Methylcyclohexanol | 2.0 |
| 1-Propanol | 5.0 | **tert-Alcohols** | |
| 2-Methyl-1-propanol | 4.4 | tert-Butanol | 0.3 |
| 1-Pentanol | 4.3 | 2-Methyl-2-butanol | 0.2 |
| 1- Hexanol | 4.3 | **Diols** | |
| 1-Heptanol | 4.8 | 1,5-Pentanediol | < 0.02 |
| 1-Octanol | 2.0 | 2,5-Hexanediol | 5.2 |
| 1-Dodecanol | 0.8 | 2,2-dimethyl-1,3-propanediol | 18.4 |
| **Secondary alcohols** | | 2-Methyl-2,4-propanediol | 28.0 |
| 2- Propanol | 1.9 | 2,3-Dimethyl-2,3-butanediol | 25.8 |
| 2-Pentanol | 1.2 | 2,5-Dimethyl-2,5-hexanediol | 32.0 |
| 2-Hexanol | 1.1 | **Diol monoethers** | |
| 2-Octanol | 0.7 | Ethylene glycol monoethylether | 8.9 |
| Cyclohexanol | 0.5 | Diethylene glycol monobutylether | 8.7 |
| 4-tert-Butylcyclohexanol | 1.5 | Glycerol isopropylideneacetal | 13.0 |

Data obtained from Dehmlow et al.[20]

It is apparent that the order of decreasing alkoxide extraction with monohydric alcohols is primary > secondary > tertiary. The better extractivity of diol anions can be attributed to the relatively high acidity of these alcohols in part, but it seems that the main factors are the distance between the two hydroxyl groups and the skeletal structure. In general, the concentration of the extracted base depends on the amount of alcohol added.[20] The concentration of aqueous sodium hydroxide is also an important factor in the extraction processes.

Herriott and Picker[36] carried out the reaction of thiophenoxide ion with 1-bromooctane in a two-phase system. They found that an increase in the ionic strength of the aqueous phase or change to a more polar organic solvent increased the reaction rate. The effect of organic solvent on the reaction rate under NPTC is given in Table 13.3.4. Correlations between the rate constants and the partition coefficients indicate that the major function of the catalyst is simply the solubilization of the nucleophilic in the organic phase. Conventional methods of synthesizing ethers, i.e., Williamson synthesis and alkoxymercuration have been well developed in organic chemistry.[76,96] The synthesis of formaldehyde acetal were carried out from the reaction of alcohol and dichloromethane in a 50% sodium hydroxide solution applying Tixoget VP clay as a catalyst. However, completing the reaction for such a low reaction rate takes long time. Dehmlow and Schmidt[15] first

**Table 13.3.4 Effect of solvent on the rate of reaction of thiophenoxide and bromooctane**

| Catalyst | Solvent | $k \times 10^3$ $M^{-1}s^{-1}$ |
|---|---|---|
| $(C_4H_9)_4N^+I^-$ | $C_7H_{16}$ | 0.02 |
| $(C_4H_9)_4N^+I^-$ | $C_6H_4Cl_2$ | 88 |
| $(C_3H_7)_4N^+Br^-$ | $C_6H_4Cl_2$ | 0.45 |
| $C_6H_5CH_2(C_2H_5)_3N^+Br^-$ | $C_6H_4Cl_2$ | 0.04 |
| $C_8H_{17}(C_2H_5)_3N^+Br^-$ | $C_6H_4Cl_2$ | 28 |
| $(C_6H_5)_4P^+Br^-$ | $C_7H_{16}$ | 0.0093 |
| $(C_6H_5)_4P^+Br^-$ | $C_6H_4Cl_2$ | 47 |
| $(C_6H_4)_4P^+Cl^-$ | $C_6H_4Cl_2$ | 180 |

Data obtained from Herriott and Picker[36]

used PTC technique to synthesize formaldehyde acetals from alcohol and dichloromethane in the aqueous phase. Wang and Chang[144-146] employed the PTC technique to synthesize formaldehyde acetals from the reaction of alcohol and dibromomethane in an alkaline solution of KOH/organic solvent. Alcohol (ROH) first reacted with KOH so as to form potassium alkoxide (ROK) in the aqueous phase. ROK further reacted with quaternary ammonium salt (QBr) in the aqueous phase to produce quaternary ammonium alkoxide (QOR) which is more soluble in the organic solvent. Dibromomethane reacted with QOR in the next step to form the desired product dialkoxymethane $CH_2(OR)_2$ in the organic phase, as shown in Equation [13.3.7].

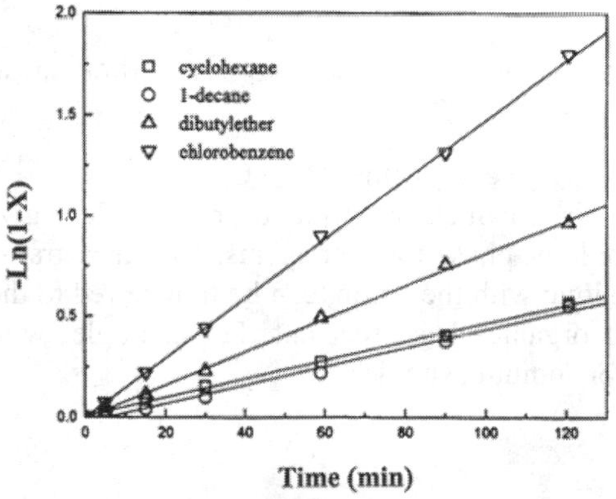


Figure 13.3.1 Effects of the organic solvents on the conversion of $CH_2Br_2$; $6.88\times10^{-2}$ mol of 1-butanol, 1.5 molar ratio of 1-butanol/1-octanol, 30 g of KOH, $2.76\times10^{-2}$ mol of $CH_2Br_2$, $3.11\times10^{-3}$ mol of TBAB catalyst, 10 mL of water, 50 mL of organic solvent, 1020 rpm, 50°C (Adapted from Ref. [145], by permission.)

Dibromomethane, which possesses weak dipole moment, may form a weak dipole-dipole bond with the organic solvent. However, this bond does not significantly affect the reaction rate. QOR solvates in a polar organic solvent. This solvation results in less energy in the nucleophilic agent than that in the transition state compound. The activation energy therefore becomes high due to the solvation of QOR with a highly polar solvent which is unfavorable in the present reaction system. The low polarity solvent neither solvates QOR, nor separates tetra-n-butylammonium ion ($Q^+$) from the alkoxide ion ($^-OR$). Thus, the reactivity in low polar solvent is low. Organic solvents of appropriate polarity, such as chlorobenzene or dibutyl ether, are the best solvents to obtain a high yields of various alcohols, as shown in Table 13.3.5. As shown in Figure 13.3.1, similar results were obtained in synthesis of unsymmetric acetals under PTC conditions.[144-146] The values of $k_{app}$, in which the reaction follows a pseudo-first-order kinetic rate law, are $4.59\times10^{-3}$, $4.58\times10^{-3}$, $8.17\times10^{-3}$, and $1.47\times10^{-2}$ $min^{-1}$ for reaction of $CH_2Br_2$ with butanol and octanol in cyclohexane, n-decane, dibutyl ether and chlorobenzene, respectively.

**Table 13.3.5. Effects of the organic solvents on the conversion of alcohols**

| Reactant | Organic solvent | | | |
|---|---|---|---|---|
| | chlorobenzene | dibutyl ether | xylene | benzene |
| | Conversion, X, % | | | |
| 1-Butanol | 83.22 | 58.94 | 49.65 | 55.48 |
| 1-Heptanol | 81.44 | 50.16 | 41.94 | 44.90 |
| 1-Octanol | 82.17 | 52.00 | 46.49 | 49.35 |
| Cyclohexanol (2 h) | 81.37 | 53.60 | 48.94 | 47.89 |
| 2-Ethoxyethanol (0.5 h) | 92.82 | 87.86 | 79.61 | 73.86 |
| 2-(2-Ethoxyethoxy)-ethanol (0.5 h) | 98.58 | 98.87 | 96.21 | 92.21 |
| Dielectric constant | 5.62 | 3.08 | 2.27 | 2.28 |

Data obtained from Wang and Chang;[145] $9.17\times10^{-2}$ mol of alcohols, 10 mL of $H_2O$, 0.028 mol of $CH_2Br_2$, 30 g of KOH, 1 g of TBAB catalyst, 50 mL of organic solvent, 1020 rpm, 50°C

(B) Synthesis of ether compound catalyzed by crown ether (NPTC)

The other type of phase transfer catalyst is crown ether, cryptands, polyethylene glycol (PEG) and their derivatives, and other nonionic phase-transfer agents. The phase transfer agent complexes with inorganic cation, along with the anion, can be transferred to the organic phase, preparing for reaction with organic-phase reactant. For example: with 18-crown-6 ether as a phase transfer agent for sodium cyanide:

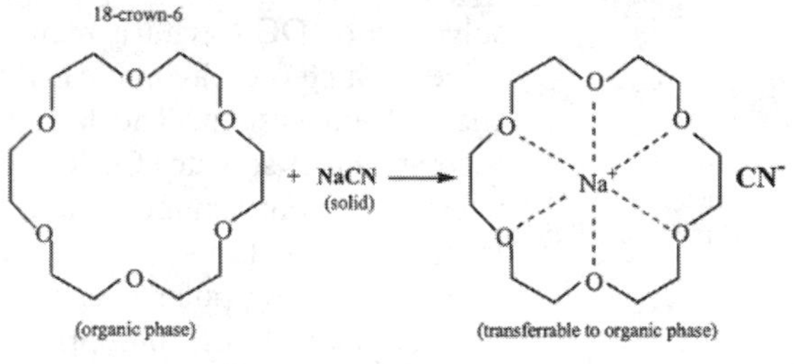


[13.3.8]

The function of crown ether is that it can chelate with metal ion, such as: lithium, sodium or potassium. Czech et al.[12] noted that crown ethers are a better phase-transfer catalyst for solid-liquid reactions, whereas quaternary salts are better for a liquid-liquid system. Table 13.3.6 shows the solubilities of potassium salts in acetonitrile at 25°C in the presence and absence of 18-crown-6 ether. The solubility of potassium salts in $CH_3CN$ highly depends on the addition of 18-crown-6 ether.

**Table 13.3.6. Solubilities of potassium salts in $CH_3CN$ at 25°C in the presence and absence of 18-crown-6 ether**

| Potassium salt | Solubility of potassium salt | | |
|---|---|---|---|
| | in 0.15 M crown in $CH_3CN$ (A) | in $CH_3OH$ (B) | Enhancement factor (A/B) |
| KF | $4.30\times10^{-3}$ | $3.18\times10^{-4}$ | 13.52 |
| KCl | $5.55\times10^{-2}$ | $2.43\times10^{-4}$ | 228.40 |
| KBr | $1.35\times10^{-1}$ | $2.08\times10^{-1}$ | 64.90 |
| KI | $2.02\times10^{-1}$ | $1.05\times10^{-1}$ | 1.92 |
| KCN | $1.29\times10^{-1}$ | $1.19\times10^{-3}$ | 108.40 |
| KOAc | $1.02\times10^{-1}$ | $5.00\times10^{-4}$ | 204 |
| KSCN | $8.50\times10^{-1}$ | $7.55\times10^{-1}$ | 1.13 |

Data adopted from the work of Liotta[60]

**Table 13.3.7. Rates of reaction of benzyl chloride with potassium cyanide at 85°C in the presence and absence of 18-crown-6 ether as a function of added water**

| Water | $k\times10^5$ $s^{-1}$ (crown) | $k\times10^5$ $s^{-1}$ (no crown) |
|---|---|---|
| 0.0 | 3.2 | 0.0 |
| 0.36 | 9.2 | 0.0 |
| 0.50 | 9.4 | 0.0 |
| 1.00 | 11.6 | 0.0 |
| 2.00 | 14.7 | 0.0 |
| 10.0 | 10.2 | 0.0 |
| 20.0 | 5.8 | 1.3 |
| 40.0 | 3.9 | 1.9 |
| 75.0 | 4.8 | 3.2 |

Data obtained from the work of Liotta;[61] 0.05 mol of benzyl chloride, 0.01 mol of 18-crown-6 ether, 0.15 mol of KBr, and 0.015 mol of KCN

The rates of reaction of benzyl bromide and benzyl chloride with potassium cyanide were studied as a function of added water in the presence and absence of crown ether in toluene at 85°C,[61] as shown in Table 13.3.7. The reaction is highly affected by the addition of 18-crown-6 ether. In addition to enhancing the reaction rate, it is important to note that in the absence of added water, the rates followed zero-order kinetics, while in the presence of added water, the rates followed first order kinetics.

(C) Synthesis of ether compound catalyzed by polyethylene glycols (NPTC)

Similar to quaternary ammonium salts, polyethylene glycols (PEGs) act as the phase transfer catalyst. There are two majors effects of PEG on the two-phase reactions. First, part of the PEG, existing in the organic phase, forms a complex with metal cation. The formation of a complex leads to an increase in the solubility of sodium alkoxide (RONa) or sodium phenoxide (PhONa) for the synthesis of ether in the organic phase. Hence, the reaction rate in the organic phase is promoted. Second, PEG acts as an excellent organic solvent, but it can also dissolve in water. Thus, part of the alkyl halide that is dissolved by PEG is brought into the aqueous phase from the organic phase. The dissolved alkyl halide directly reacts with phenoxide ($PhO^-$) or alkoxide ($RO^-$) ion in the aqueous phase, as shown in Figure 13.3.2.[143] The reaction rate in the aqueous phase is also enhanced. The mechanism of the reaction rate of alkyl halide (or allyl halide, RX) and phenoxide ($PhO^-$), both existing in the aqueous phase with PEG help, is different from that in presence of quaternary ammonium salt.

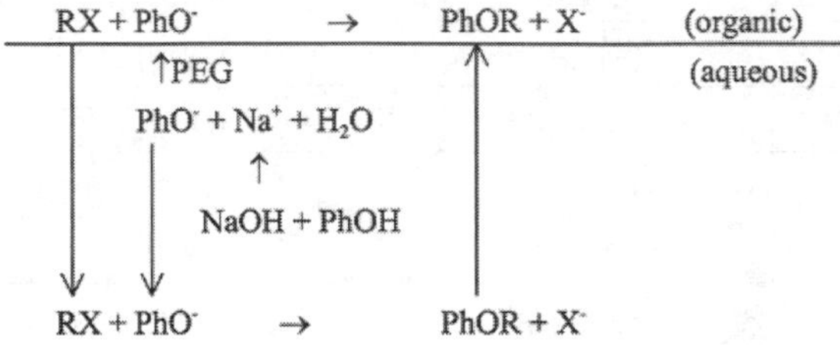


[13.3.9]

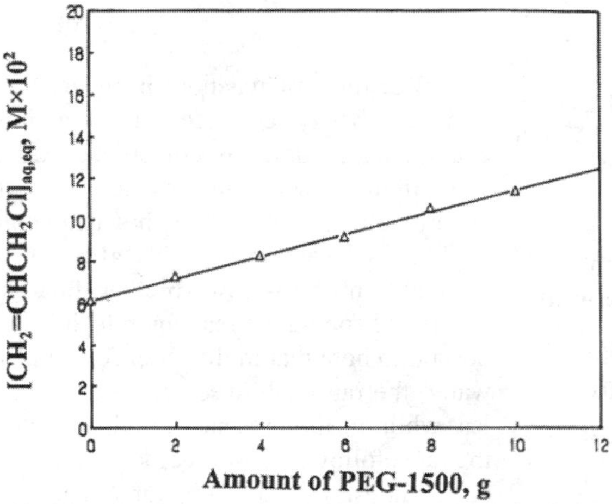


Figure 13.3.2 Dependence of the concentration of allyl chloride in the aqueous phase on the amount of PEG-1500 added, $V_{org}=V_{aq}=50$ mL, 30°C (Adapted from Ref. [143], by permission.)

The reaction catalyzed by PEG can be carried out either in a homogeneous phase or in a two-phase solution. The alkyl halide usually serves as the reactant as well as the solvent. The reaction proceeds because organic-phase reactant dissolves in an organic solvent in the presence of PEGs. Dichloromethane, chlorobenzene, ethyl ether, cyclohexane and n-decane are frequently used solvents.

The reaction mechanism of two-phase catalytic reaction by PEG includes formation of a complex of PEG with cation.[141,142] This is different than the reaction catalyzed by quaternary salts. Table 13.3.8 shows the initial reaction rate using PEG, ($(-r)_{i,PEG}$) and the initial reaction rate without using PEG, ($(-r)_{i,B}$) in various organic solvents. Both $(-r)_{i,B}$ and $(-r)_{i,PEG}$ decrease when the polarity of the organic solvent increases. The maximum reaction rate is obtained with n-decane, which has the lowest polarity, as the protic solvent. Same results were obtained from the work of Landini et al.[55] on the reaction of n-octylmethylene sulfonate and bromide ion in a homogeneous phase with $C_{16}H_{33}P^+(C_4H_9)_3Y^-$ as PTC. Wang and Chang[142] made a reasonable explanation for this peculiar phenomena, i.e., the transition state possesses a higher degree of dispersity of electric charge than does the ground state. Increasing the polarity of the solvent increases the relative activation energy between the transition state and the reactants. Hence the reaction rate is decreased.

**Table 13.3.8. Effect of the aprotic solvent on the initial reaction rate of the allylation of phenoxide**

| Solvents | Dielectric constant (20°C) | Initial reaction rate×$10^3$ $Mh^{-1}$ | | $(-r)_{i,PEG}/(-r)_{i,B}$ |
|---|---|---|---|---|
| | | $(-r)_{i,B}$ | $(-r)_{i,PEG}$ | |
| n-decane | 1.991 | 171.77 | 355.62 | 2.07 |
| cyclohexane | 2.023 | 127.5 | 277.68 | 2.16 |
| ethyl ether | 4.335 | 98 | 205.83 | 2.10 |
| cyclobenzene | 5.708 | 75.25 | 174.56 | 2.32 |
| dichloromethane | 9.080 | 45.61 | 86.31 | 1.89 |

Data obtained from the work of Wang and Chang[142]

### 13.3.1.3 Effects of the organic solvents on the reactions in other catalysts

(A) Quaternary ammonium salts as NPTC

The effects of the organic solvents on the reaction rate are given in Table 13.3.9.[121] The rates of the reactions of the tetra-n-butylammonium and potassium salts of phenoxide with 1-chlorobutane and 1-bromobutane in pure solvents and solvent mixtures varying in dielectric constant from 2.2 to 39 were obtained by Uglestad et al.[121]

**Table 13.3.9. Effect of organic solvent on the reaction of tetra-n-butylammonium salts of phenoxide and potassium salts of phenoxide with halobutane**

| Reactants Solvents | Rate constant×$10^5$, $Lmol^{-1}s^{-1}$ | | | | Dielectric constant (ε) |
|---|---|---|---|---|---|
| | 1-$C_4H_9Cl$ | 1-$C_4H_9Cl$ | 1-$C_4H_9Br$ | 1-$C_4H_9Br$ | |
| | $K^{+-}OC_6H_5$ | $Bu_4N^{+-}OC_6H_5$ | $K^{+-}OC_6H_5$ | $Bu_4N^{+-}OC_6H_5$ | |
| Dioxane 10% | | | 0.01 | 330 | 2.2 |
| $CH_3CN$ 50% | 0.0025 | 2.8 | 0.22 | 400 | 6 |
| $CH_3CN$ | 0.084 | 4.0 | 12 | 600 | |
| Acetonitrile | 0.33 | 2.2 | 40 | 300 | 39 |

Data obtained from Uglestad et al.,[121] 0.2 M $C_6H_5O^-$, 0.05 or 0.1 M $C_4H_9X$, 25°C

The rates of reaction of potassium phenoxide vary over three orders of magnitude with the changes in the dielectric constant of solvent, whereas the corresponding rates with the tetra-n-butylammonium salt vary by approximately a factor of 6.

(B) Tertiary amines as NPTC

Quaternary ammonium salts, PEGs and crown ethers are the common compounds, employed as PTC. The inexpensive tertiary amines have also been used as the phase transfer catalysts (PTC) in recent years. The synthetic process for producing 2-mercaptobenzimidazole (MBI) is a reaction of o-phenylene diamine ($C_6H_4(NH_2)_2$) and carbon disulfide ($CS_2$) in a two-phase medium affected by appropriate choice of solvent.[128-130]

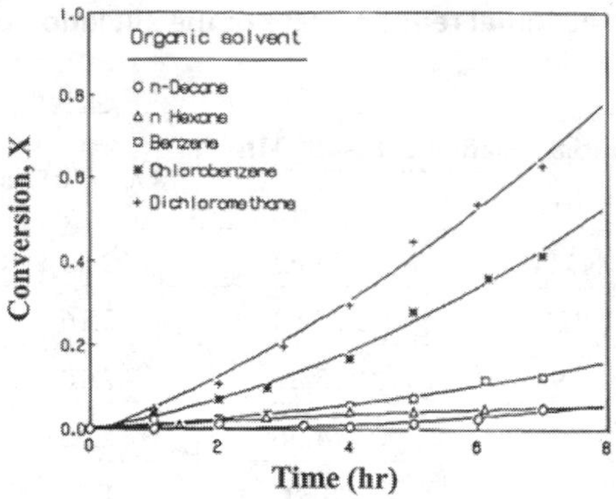


Figure 13.3.3. Effect of solvents on the conversion of o-phenylene diamine in the two-phase catalyzed reaction; $3.17 \times 10^{-3}$ mol of $C_6H_4(NH_2)_2$, $2.50 \times 10^{-2}$ mol of $CS_2$, $1.679 \times 10^{-3}$ mol of tributylamine (TBA), 1000 rpm, 30°C. (Adapted from Ref. [128], by permission.)

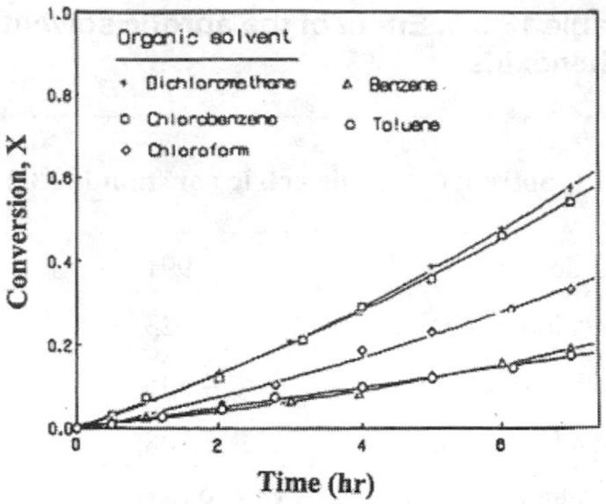


Figure 13.3.4. Effect of solvents on the conversion of o-phenylene diamine; 0.4 g of $C_6H_4(NH_2)_2$, 4.003 g of $CS_2$, 0.4 mL of tributylamine (TBA), 50 mL of organic solvent, 600 rpm, 30°C. (Adapted from Ref. [129], by permission.)

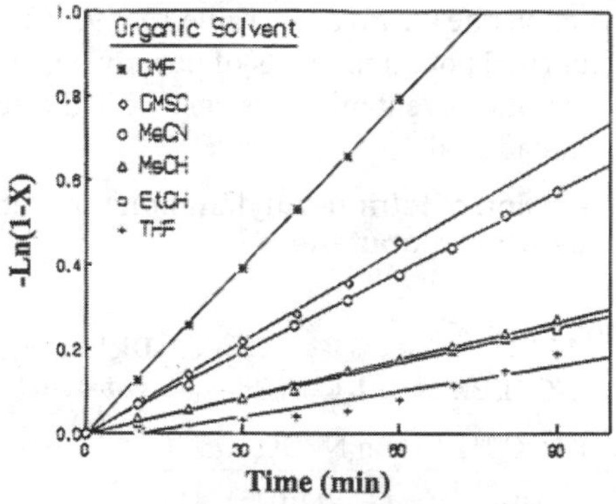


Figure 13.3.5. Effect of solvents on the conversion of o-phenylene diamine; $3.18 \times 10^{-3}$ mol of o-phenylene diamine, 8 molar ratio of $CS_2/C_6H_4(NH_2)_2$, 0.01 M of triethylamine (TEA), and 50 mL of $CH_3CN$, 600 rpm, 40°C. (Adapted from Ref. [130], by permission.)

Several solvents, such as: n-decane, n-hexane, benzene, chlorobenzene and dichloromethane, which are immiscible with water, were used. The effect of solvent on conversion is shown in Figure 13.3.3.[128] The order of the conversion of o-phenylene diamine $C_6H_4(NH_2)_2$ in various organic solvents is: dichloromethane > chlorobenzene > benzene > n-hexane > n-decane, which is consistent with the order of solvent polarity, i.e., the greater the polarity of solvent the higher the conversion of o-phenylene diamine.

Results for the reaction of o-phenylene diamine and carbon disulfide in a homogeneous phase (organic solvent) are given in Figure 13.3.4.[129] The order of the reactivities is: dichloromethane (8.91) > chlorobenzene (5.6) > chloroform (4.8) > toluene (2.4) > benzene (2.3). The reaction rate is related to the dielectric constant of the organic solvent. A larger conversion of o-phenylene diamine was obtained using solvent with a higher dielectric constant.

In choosing a polar organic solvent, such as: MeCN, MeOH, EtOH, DMSO, DMF and THF, a homogeneous solution was used for the reaction.[130] Figure 13.3.5 shows the effects of organic solvents (protic or aprotic) on the conversion of o-phenylene diamine. The order of the reactivities for these six organic solvents is: DMF > DMSO > > MeCN > MeOH > EtOH >THF. The corresponding dielectric constants of solvents are: DMF (37.71), DMSO (46.45), MeCN (35.94), MeOH (32.66), EtOH (24.55) and THF (7.58), respectively. The protic solvents, such as MeOH and EtOH, containing hydroxyl group possess acidic properties. The unpaired electrons on the oxygen atom associate with the anions. A relatively lower conversion is obtained in MeOH or EtOH solvent. This result indicates that the acidic hydrogen bond does not have a strong catalytic capability.

The aprotic solvents, which do not possess hydrogen bond, are highly polar. Therefore, the aprotic solvents possess high alkalinity and nucleophilicity required to obtain a high conversion of o-phenylene diamine in the synthesis of mercaptobenzimidazole (MBI). A larger conversion is obtained when using a protic solvent or aprotic solvent of high polarity. However, the structure of DMF, which is an amide, is similar to that the tertiary amine. It possesses similar catalytic property to dimethylaminopyridine (DMAP). The effect of DMF on the conversion of o-phenylene diamine is more pronounced than that of DMSO. The Arrhenius rate equations in various solvents for the reaction of o-phenylene diamine and carbon disulfide catalyzed by tributylamine are as follows:

$$\begin{aligned} &\text{DMF:} && k_{app} = 1.06\text{x}10^{15}\exp(-1.20\text{x}10^{4}/T) \\ &\text{DMSO:} && k_{app} = 7.82\text{x}10^{8}\exp(-7.78\text{x}10^{3}/T) \\ &\text{MeCN:} && k_{app} = 1.39\text{x}10^{13}\exp(-1.09\text{x}10^{4}/T) \\ &\text{MeOH:} && k_{app} = 9.62\text{x}10^{14}\exp(-1.24\text{x}10^{4}/T) \\ &\text{EtOH:} && k_{app} = 3.84\text{x}10^{10}\exp(-9.29\text{x}10^{3}/T) \\ &\text{THF:} && k_{app} = 3.25\text{x}10^{38}\exp(-2.99\text{x}10^{4}/T) \end{aligned} \qquad [13.3.10]$$

$k_{app}$ is the apparent rate constant in which the reaction follows pseudo-first-order rate law.

In two-phase phase-transfer catalytic reactions, the solvents significantly affect the reaction rate. The main reason is that the distribution of regenerating catalyst QX and the active catalyst QY between two-phases is highly dependent upon the polarity of the solvent. It is desirable for most of the intermediate products to stay in the organic phase and react with the organic-phase reactant. Therefore, a solvent with high polarity will be preferred for the reaction.

(C) Pyridine 1-oxide (PNO) as IPTC

The substitution reaction of benzoyl chloride (PhCOCl) and sodium acetate ($CH_3COONa$) using pyridine 1-oxide (PNO) as the inverse phase-transfer catalyst (IPTC) in a two-phase system of organic solvent and water was investigated by Wang, Ou and Jwo.[148-150] They found that the polarity of the organic solvent strongly affected conversion of benzoyl chloride, the yield of the main product (acetic benzoic anhydride ($PhCOOCOCH_3$)), and the reaction rate. The reaction follows a pseudo-first-order kinetic rate law. Dichloromethane, chloroform, tetrachloromethane and cyclohexanone ($C_6H_{10}O$) were used as the organic modifier in the two-phase reaction system. The results are given in Table 13.3.10. A linear reaction rate was observed for a more polar organic solvent. The order of relative reactivities in these solvents is cyclohexanone > dichloromethane > chloroform > tetrachloromethane, consistent with their polarities. Kuo and Jwo[54] obtained similar results. Wang, Ou and Jwo[148] also found that the conversion was substantially increased with initial concentration of PNO increasing in the aqueous phase with $CH_2Cl_2$ present as the organic solvent. The reason is that the concentration of carboxylate ion[54] influences the concentration of PNO in the organic phase.

**Table 13.3.10. Effect of the composition of organic solvent on the PNO-catalyzed $PhCOCl-CH_3COONa$ reaction in a two-phase medium**

| Organic phase | $k_{app}\times10^{-3}$ $min^{-1}$ at T= | | | | |
|---|---|---|---|---|---|
| | 5°C | 10°C | 18°C | 25°C | 33°C |
| $CH_2Cl_2$ | 24.7 | 32.3 | 48.5(5.73)$^b$ | 65.8(12.0)$^b$ | |
| $CH_2Cl_2+CCl_4$ | | | | | |
| $[CCl_4]$=1.00 M | 18.8 | 26.6 | 38.3 | 51.5 | 72.8 |
| $[CCl_4]$=3.00 M | | | 17.9 | 26.8 | |
| $[CCl_4]$=5.00 M | | 5.13 | 9.06 | 18.1 | |
| $CH_2Cl_2+C_6H_5NO_2$ | | | | | |
| $[C_6H_5NO_2]$=1.00 M | 28.0(2.93)$^b$ | 36.4(6.42)$^b$ | 62.7(14.6)$^b$ | 83.0 | 124 |
| $CH_2Cl_2+CHCl_3$ | | | | | |
| $[CHCl_3]$=5.00 M | | 26.9 | 37.6 | 49.0 | |
| $CH_2Cl_2+C_6H_{10}O$ | | | | | |
| $[C_6H_{10}O]$=3.00 M | | 35.0(4.49)$^b$ | 57.2(9.38)$^b$ | 91.4(20.4)$^b$ | |
| $CHCl_3$ | | 14.6 | 19.5 | 26.7 | |
| $CCl_4$ | | | 16.2 | | |
| $C_6H_{10}O$ | | 68.7 | 106 | 170 | |

Data obtained from Wang et al.[148] $2.00\times10^{-4}$ M of PNO, $1.00\times10^{-2}$ M of PhCOCl, 0.500 M of $CH_3COONa$, 18°C, 1200 rpm, 50 mL of $H_2O$, 50 mL of organic solvent, $^b$No PNO added; $C_6H_{10}O$, cyclohexanone

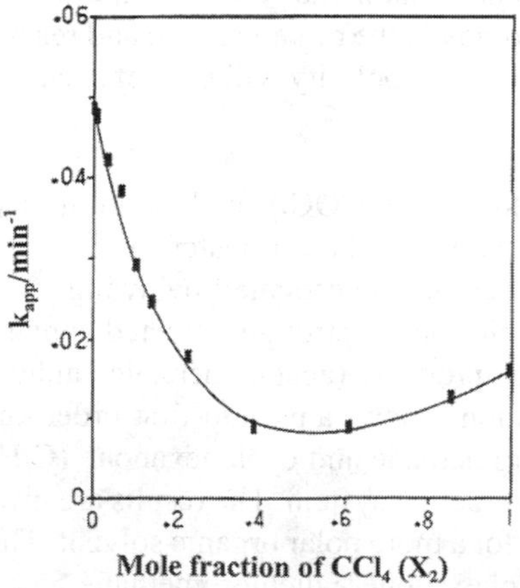


Figure 13.3.6. Effect of the mole fraction of $CCl_4$ in the mixed organic solvent on the $k_{app}$ value in the two-phase $H_2O/(CH_2Cl_2+CCl_4)$ medium; $1.00\times10^{-2}$ M of PhCOCl 0.500 M of $CH_3COONa$, $2.00\times10^{-4}$ M of PNO, 50 mL of $H_2O$, 50 mL of organic solvent ($CH_2Cl_2+CCl_4$), 18°C. (Adapted from Ref. [148], by permission.)

In the studies on the inverse phase-transfer catalytic reaction, Wang, Ou and Jwo[148] conducted two independent experiments in order to evaluate the effect of polarity of the organic phase on the reaction. In the first experiment, a relatively inert organic substances such as $C_6H_5CH_2CN$, $C_6H_5CN$, $C_6H_5N(Et)_2$, $C_6H_5NO_2$, $CH_3COOC_2H_5$ or $C_3H_7COOC_2H_5$ were individually added to the organic phase ($CH_2Cl_2$) as the mixed organic solvent in the two-phase reaction system. The reactions of these compounds with PhCOCl or PNO were negligibly slow compared to the reaction of PhCOCl and PNO. The results are given in Tables 13.3.11 and 13.3.12, respectively. It is shown that $k_{app}$ increased with added inert substance of high polarity,

**Table 13.3.11. Effect of the inert organic substance on the PNO-catalyzed $CH_3COONa$-PhCOCl reaction in a two-phase $H_2O/CH_2Cl_2$ medium**

| Organic substance, R | $k_{app}\times10^3$ $min^{-1}$ | Dipole moment, D |
|---|---|---|
| $C_6H_5CH_2CN$ | 61.2 | |
| $C_6H_5CN$ | 60.5 | 4.18 |
| $C_6H_5N(Et)_2$ | 56.8 | |
| $C_6H_5NO_2$ | 57.0 | 4.22 |
| $CH_3COOC_2H_5$ | 53.1 | 1.78 |
| $C_3H_7COOC_2H_5$ | 48.5 | |
| $CH_2Cl_2$ | 48.5 | 1.60 |
| $CCl_4$ | 42.5 | 0 |

Data obtained from Wang et al.[148] $1.00\times10^{-2}$ M of PhCOCl, 0.500 M of $CH_3COONa$, 0.500 M of R, $2.00\times10^{-4}$ M of PNO, 50 mL of $H_2O$, 50 mL of $CH_2Cl_2$, 18°C

such as nitrobenzene and ethyl acetate or basic organic substance, such as diethylaniline. The $k_{app}$-value increased to a greater extent with added highly polar and basic organic substance, such as benzyl cyanide and benzonitrile. In the second set of experiments, reactions were carried out with nonpolar $CCl_4$ added to the $CH_2Cl_2$ as the mixed organic solvent. As shown in Figure 13.3.6, that due to decreased polarity, the value of $k_{app}$ also decreased with increased amount of added $CCl_4$ to a minimum.[148] Then, it increased slightly on further addition of $CCl_4$ due to the increased rate of PNO-catalyzed hydrolysis of PhCOCl. Since the distribution of PhCOCl in the $CH_2Cl_2$ decreases with increased amount of $CCl_4$, the reaction rate of PhCOCl with PNO in the aqueous phase leads to the hydrolysis of PhCOCl. Wang, Ou and Jwo[148] observed that the yields of $PhCOOCOCH_3$ decrease with increased content of $CCl_4$.

**Table 13.3.12. Effect of the amount of inert organic substance on the PNO-catalyzed $CH_3COONa$-PhCOCl reaction in a two-phase $H_2O/CH_2Cl_2$ medium**

| Inert organic Substance $[R]_{org}$ | $k_{app}\times10^3$ $min^{-1}$ with $[R]_{org}$, M | | | | | | |
|---|---|---|---|---|---|---|---|
| | 0.100 | 0.300 | 0.500 | 0.800 | 1.00 | 1.50 | 2.00 |
| $C_6H_5CH_2CN$ | 57.3 | | 61.3 | | 62.4 | 71.8 | |
| $C_6H_5NO_2$ | 49.0 | | 57.0 | | 62.7 | 60.7 | 60.9 |
| $C_6H_5N(Et)_2$ | 55.1 | 61.1 | 56.8 | 54.9 | | | |
| $CCl_4$ | 47.6 | | 42.1 | | 38.3 | 28.5 | 24.7 |

Data obtained from Wang et al.[148] $1.00\times10^{-2}$ M of PhCOCl, 0.500 M of $CH_3COONa$, $2.00\times10^{-4}$ M of PNO, 50 mL of $H_2O$, 50 mL of $CH_2Cl_2$, 18°C

In Table 13.3.12, $k_{app}$ approached a constant value when nitrobenzene (1.0 M) was added. This result indicates that solvation of the transition structure for the reaction of benzoyl chloride with sodium nitrate reached an upper limit. Benzyl cyanide is a polar solvent. The value of $k_{app}$ increased with increased content of benzyl cyanide. Further, highly basic diethylaniline ($C_6H_5N(Et)_2$) could increase the concentration of free PNO and also the reaction rate. However, this compound is less polar than dichloromethane and the polarity decreased with increased proportion of diethylaniline, which caused the value of $k_{app}$ to decrease. Therefore, the value of $k_{app}$ reached a maximum, as shown in Table 13.3.12. In general, the value of $k_{app}$ increased with an increased proportion of highly polar inert organic substance, such as $C_6H_5CH_2CN$ and $C_6H_5NO_2$, and decreased with increased proportion of slightly polar inert organic substance, such as $C_6H_5N(Et)_2$ and $CCl_4$. In the case of $CCl_4$, the greater the proportion of nonpolar or less polar compound, the smaller the reaction rate. The results are due to a combination of the hydrolysis of benzoyl chloride, the distribution of PNO between two phases, and the mass transfer of PNO from organic phase to aqueous phase.

**Table 13.3.13. Initial rates of coupling between 4-nitrobenzene diazonium chloride and N-ethylcarbazole at 0°C in various solvent mixtures in the presence and absence of additives**

| Solvent[a] | Additve[b] | Rate[c]$\times 10^9$, $mol^{-1}dm^{-3}$ |
|---|---|---|
| $CH_2Cl_2$-$H_2O$ | none | 1.25 |
| $CH_2Cl_2$-$H_2O$ | NaDBS | 11.1 |
| $CH_2Cl_2$-$H_2O$ | 18-crown-C-6 ether | 0.28 |
| $CH_2Cl_2$-$H_2O$ | Lissapol NX[d] | 0.56 |
| Toluene-$H_2O$ | none | < 0.1 |
| Toluene-$H_2O$ | NaDBS | < 0.1 |
| EtOAc-$H_2O$ | none | 0.89 |
| EtOAc-$H_2O$ | NaDBS | 1.03 |
| $PhNO_2$-$H_2O$ | none | 1.39 |
| $PhNO_2$-$H_2O$ | NaDBS | 10.5 |
| AcOH-$H_2O$[e] | none | 0.47[f] |
| DMF-$H_2O$[g] | none | 0.53[f] |
| 1,4-Dioxane-$H_2O$[h] | none | 1.33 |

Data obtained form the work of Ellwood and Griffths;[24] [a]v/v=1/1 except where stated, [b]0.05 mmol except where stated, [c]initial rate of formation of azo dye, [d]commercial (ICI) non-ionic detergent, 0.05 mmol, [e]95% $H_2O$ v/v, containing $NaOAc\cdot 3H_2O$, [f]suspension of N-ethylcarbazole, [g]60% $H_2O$ v/v, [h]60% $H_2O$ v/v homogeneous solution containing 1.0 mmol diazonium ion and 1.0 mmol N-ethyl carbazole 100 mL solvent

(D) Electrophile reaction by NaDBS

Although the electrophile transferred to the organic phase from the aqueous phase by phase transfer catalysis (PTC), the role of organic solvent was still important. Ellwood and Griffiths[24] carried out the coupling reactions between 4-nitrobenzendiazonium chloride and N-ethylearbazole or N,N-diphenylamine in aqueous media. The coupling reactions were accelerated by using a two-phase water-dichloromethane containing sodium 4-dodecylbenzene sulfonate (NaDBS) as a transfer catalyst for the diazonium ion. Effects of solvents and catalyst (NaDBS) on the rate constants are given in Table 13.3.13. The NaDBS (0.05 molar proportions) increases the rate coupling in dichloromethane-water by a factor of at least 20 relative to the reaction in water-acetic acid. A part of this increase is attributed to incomplete solubility of N-ethyl carbazole in the latter solvent. Also, the polarity of the organic phase is important (cf. $CH_2Cl_2$, $C_6H_5NO_2$ have much higher dielectric constants than toluene and ethyl acetate). This may be attributed to the covalent character of diazonium arylsulfonates. Its ionization is greater in the former solvents. Crown ethers act as transfer agents for the diazonium ion, but the resultant complexes have low coupling reactivity.

(E) Oxidation by dimethyl polyethylene glycol and oxidant

Dimethyl polyethylene glycol solubilizes potassium permanganate in benzene or dichloromethane and can thus be used as a phase-transfer agent for permanganate oxidation. The reaction is highly dependent on the organic solvent. If benzene is used as solvent, dimethyl polyethylene glycol does not efficiently extract $KMnO_4$ from an aqueous solution, but it solubilizes the solid reagent when $CH_2Cl_2$ is used as the solvent, $KMnO_4$ may be transferred from either aqueous solution or from the solid phase. The effect of organic solvent on the distribution of products is given in Table 13.3.14.[58]

**Table 13.3.14. Oxidation of cyclododecene**

| Solvent system | Oxidation ratio[a] | Phase transfer agent | Products (%) |
|---|---|---|---|
| Benzene+17% acetic acid | 3.3 | Polyether[b] | 1,2-Cyclododecanedione (16), dodecanedioic acid (59), cyclododecane (23) |
| Benzene+17% acetic acid | 3.3 | Crown ether[c] | 1,2-Cyclododecanedione (22), dodecanedioic acid (56), cyclododecane (12) |
| Benzene+17% acetic acid | 3.3 | Adogen 464 | 1,2-Cyclododecanedione (8), dodecanedioic acid (58), cyclododecane (9) |
| Dichloromethane+17% acetic acid | 3.3 | Polyether | 1,2-Cyclododecanedione (8), dodecanedioic acid (77), cyclododecane (1) |
| Dichloromethane+17% acetic acid | 3.3 | Crown ether | 1,2-Cyclododecanedione (7), dodecanedioic acid (83), cyclododecene |
| Dichloromethane+17% acetic acid | 3.3 | Adogen 464 | 1,2-Cyclododecanedione (7), dodecanedioic acid (83), cyclododecene (2) |
| Dichloromethane+17% acetic acid | 2.2 | Adogen | 1,2-Cyclododecanedione (18), dodecanedioic acid (63), 2-hydroxycyclododecanone (6) |
| Dichloromethane+17% acetic acid | 1.6 | Agogen 464 | 1,2-Cyclododecanedione (14), dodecanedioic acid (40), 2-hydroxycyclododecanone (6), cyclododecene (23) |
| Dichloromethane+17% acetic acid | 2.2 | Adogen 464 | 1,2-Cyclododecanedione (19), dodecanedioic acid (27), 2-hydroxycyclododecone (7), cyclododecene (23) |
| Dichloromethane+10% acetic acid | 2.2 | Adogen 464 | 1,2-Cyclododecanedione (69), dodecanedioic acid (13), 2-hydroxycyclododecane (3), cyclododecene (9) |
| Dichloromethane/water +10% acetic acid | 2.2 | Polyether | 1,2-Cyclododecanedione (16), dodecanedioic acid (82) |
| Dichloromethane/aqueous NaOH | 1.0 | Benzyl triethyl ammonium chloride | 1,2-Cyclododecanediol (50) |

Data obtained from Lee and Chang;[58] [a]number of moles of potassium permanganate per mole of alkene, [b]dimethylpolyethylene glycol, [c]dicyclohexano-18-crown-6 ether

(F) Polymerization by PTC

Poly(ethylene glycol)-block-poly(butylacrylate), synthesized by radical polymerization,[56] were obtained by PTC in the Williamson reaction. The morphology and the crystallinity of the cast film of the block polymer were significantly affected by the organic solvent.

The technique of phase-transfer catalysis has been extensively applied to the two-phase polycondensation using various phase-transfer catalysts, such as quaternary ammonium and phosphonium salts, crown ethers and poly(ethylene glycol)s.[8,11,30,46,53,75,87,119,151] Various types of condensation polymers such as aromatic polysulfonates and polysulfides, aromatic polyethers, aliphatic and aromatic polysulfides, and carbon-carbon chain polymers of high molecular weights by the phase-transfer catalyzed polycondensation from combinations of aromatic disulfonyl chlorides, phosphonic dichlorides, activated aromatic dichlorides, and aliphatic dihalides, with bisphenol, aliphatic and aromatic dithiols, and active ethylene compounds. The two-phase polycondensation was generally carried out in a water-immiscible organic solvent-aqueous alkaline solution system at room temperature. The method of polycondensation offers a highly versatile and convenient synthetic method for a variety of condensation polymers.

Aromatic polysulfonates of high molecular weights can be prepared from aromatic disulfonyl chlorides and alkaline salts of bisphenols by interfacial polycondensation technique using onium salt accelerators. In the absence of the catalyst, only low molecular polysulfonate **III** was obtained, even though the reaction was continued for 254 hours, whereas the addition of these quaternary ammonium salts and crown ethers increased the average molecular weight of the polymer remarkably.

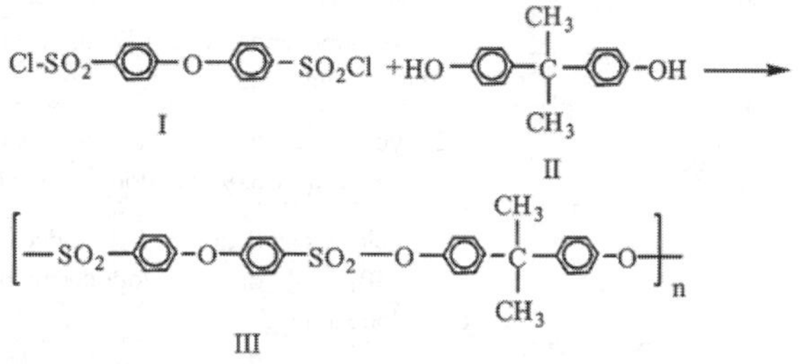


[13.3.11]

Among the catalysts, TBAC and DC-18-C-6 were found to be highly efficient, leading to the formation of the polysulfonate with an inherent viscosity [η] of as high as 1.4 $dLg^{-1}$.

Similarly, aromatic polysulfonates and aromatic polyether were synthesized from the polycondensation of phenylphosphonic dichloride (**IV**) with bisphenol A (**II**) leading to a polyphosphonate (**V**), aromatic dihalides (**VI**) with alkaline salts of bisphenols (**VII**), under various phase transfer catalysis in two-phase system, i.e.,

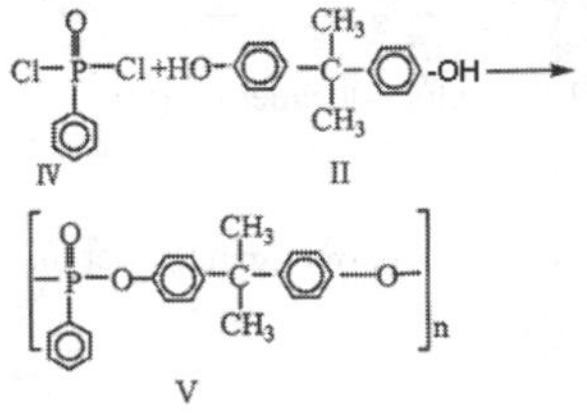


[13.3.12]

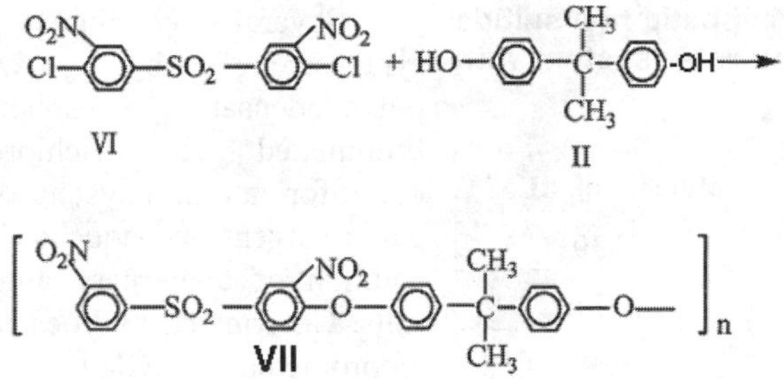

[13.3.13]

The effect of organic solvent on the two-phase polycondensation is shown in Table 13.3.15. Chloroform, 1,2-dichloroethane, nitrobenzene, acetophenone and anisole were all effective as the polymerization media to produce moderate molecular weight polymer (VII).

**Table 13.3.15. Synthesis of aromatic polyether VII in various organic solvent/water system with DC-18-C-6 catalyst**

| Solvent | Reaction temp., °C | Reaction time, h | Polymer [η], dLg$^{-1}$* |
|---|---|---|---|
| $CH_2Cl_2$ | 20 | 24 | 0.84 |
| $CHCl_3$ | 20 | 24 | 0.53 |
| $CH_2ClCH_2Cl$ | 20 | 24 | 0.42 |
| $C_6H_5NO_2$ | 20 | 24 | 0.47 |
| $C_6H_5NO_2$ | 80 | 2 | 0.51 |
| $C_6H_5NO_2$ | 100 | 1 | 0.42 |

Data obtained from the work of Imai;[41] Reaction conditions: 2.5 mmol of II, 2.5 mmol of VI, 0.05 mmol of DC-18-C-6 in 3.5 mL of solvent, and 5 mL of KOH (1.01 M) solution, *Measured at a concentration of 0.5 dLg$^{-1}$ in DMF at 30°C

A convenient method for the preparation of polysulfides by the two-phase polycondensation in a KOH solution is known.[41] Polycondensation of 1,4-dibromobutane (**VIII**) and 1,6-hexanedithiol (**IX**) leading to polysulfide (**X**) was carried out in various organic solvent/$H_2O$ system with DC-18-C-6 catalyst.

$$\underset{VIII}{Br(CH_2)_4Br} + \underset{IX}{HS(CH_2)_6SH} \rightarrow \underset{X}{\left[-(CH_2)_4-S-(CH_2)_6-S-\right]_n} \qquad [13.3.14]$$

The results of the polycondensation are given in Table 13.3.16. All polymerization media employed produced polysulfide with moderately high inherent viscosities; whereas the polymer with the highest viscosity was produced in the absence of organic solvents. Polycondensation conducted in the presence of any catalyst in this system led to the formation of a polymer with moderately high molecular weight.

**Table 13.3.16. Synthesis of aliphatic polysulfide X in various organic solvent-water system with DC-18-C-6 catalyst**

| Solvent | Reaction temp. °C | Polymer [η], $dLg^{-1}$* |
|---|---|---|
| $CH_2Cl_2$ | 20 | 0.30 |
| $CHCl_3$ | 80 | 0.31 |
| $C_6H_6$ | 80 | 0.30 |
| $C_6H_5NO_2$ | 80 | 0.58 |
| $CH_3CN$ | 80 | 0.58 |
| None | 80 | 0.73 |

Data obtained from Imai;[41] Polymerization conditions: 2.5 mmol of VIII and IX, 0.05 mmol of DC-18-C-6 in 2.5 mL of solvent and 5 mL of 1.01 M KOH solution for 48 h, *measured at a concentration of 0.5 $gdL^{-1}$ in chloroform at 30°C

Several polycarbonates[59] were synthesized by two-phase polycondensation of bisphenols and brominated with trichloromethyl chloroformate in a system of an organic solvent and aqueous alkaline solution of quaternary ammonium salts. Chlorinated hydrocarbons, dichloromethane (DCM), tetrachloromethane (TCM), tetrachloromethane (TCM) and nitrobenzene (NB) served as organic solvents. The effects of solvents on the reaction yields are given in Table 13.3.17.[59] Although polymers with a high yield were obtained using nitrobenzene (NB) as an organic solvent, the inherent viscosities were low. Polycarbonates were prepared by a two-phase condensation of TCF with bisphenol S. They precipitate from chlorinated hydrocarbon solvents such as DCM, TCM and DCE. According to both the yield and the inherent viscosity of these polymers, the use of BTEAC as a phase-transfer catalyst, sodium hydroxide as a base and DCE as an organic solvent was suitable to prepare a polycondensate having a large molar mass and a high yield.

**Table 13.3.17 Synthesis of bisphenol S-based homopolycarbonate by two-phase polycondensation catalyzed by PTC[a]**

| Reaction conditions | | Polymer yield | | |
|---|---|---|---|---|
| Solvent[b] | Catalyst | % | [η], $dLg^{-1}$* | State |
| DCM | TBAB | 76.7 | 0.21 | ppt. |
| DCM | TBAC | 76.6 | 0.10 | ppt. |
| DCM | BTEAC | 78.7 | 0.13 | ppt. |
| DCM | BTEAB | 79.3 | 0.20 | ppt. |
| TCM | TBAB | 60.0 | 0.11 | ppt. |
| TCM | BTEAC | 83.4 | 0.12 | ppt. |
| DCE | TBAB | 88.9 | 0.28 | ppt. |
| DCE | TBAC | 76.4 | 0.12 | ppt. |
| DCE | BTEAC | 86.5 | 0.32 | ppt. |
| DCE | BTEAB | 89.3 | 0.21 | ppt. |
| NB | TBAB | 85.0 | 0.17 | solution |
| NB | TBAC | 90.6 | 0.11 | solution |

| Reaction conditions | | Polymer yield | | |
|---|---|---|---|---|
| Solvent[b] | Catalyst | % | [η], $dLg^{-1}$* | State |
| NB | BTEAC | 92.2 | 0.19 | solution |

Data obtained from Liaw and Chang;[59] [a]Polymerization was carried out with bisphenol S (5.00 mmol) and TCF (7.50 mmol) in the organic solvent (37.5 mL) and water (30 mL) in the presence of catalyst (3.15 mmol) and sodium hydroxide (28.5 mmol) at room temperature for 2 h. [b]Abbreviations: DCM, dichloromethane; TCM, tetrachloromethane; DCE, 1,2-dichloroethane; NB, nitrobenzene. *Measured at a concentration of 0.5 $gdL^{-1}$ in DMF at 25°C

Another type of polysulfide (**XIII**) was synthesized by the two-phase polycondensation of bis-(3-chloroacryloy)benzenes (**XI**a and **XI**b) with 4,4'-oxybisbenzenethiol (**XII**). The polycondensation was carried out in a chloroform-water system at room temperature with some phase transfer catalysts.

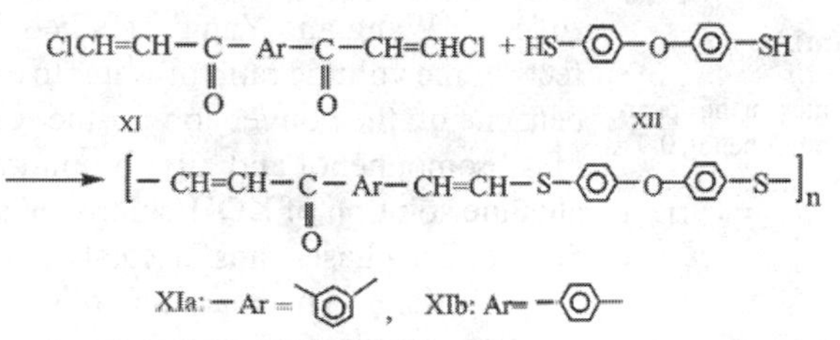


[13.3.15]

Table 13.3.18 shows the results of polycondensation. The polysulfides having inherent viscosities above 0.5 $dLg^{-1}$ were readily obtained from two bis(2-chloroacryloyl)benzene with or without use of phase transfer catalysts. These activated dichlorides are highly reactive, almost comparable to ordinary dicarboxylic acid chlorides. The use of catalysts, such as DC-18-C-6, was not essential to this type of polycondensation for producing high molecular weight of polysulfides **XIII**.

**Table 13.3.18. Synthesis of polysulfides XIII in organic solvent-water system[a]**

| Dichloride | Solvent | Catalyst | Reaction time, min | Polymer, [η], $dLg^{-1}$* |
|---|---|---|---|---|
| XIa | chloroform | none | 15 | 0.21 |
| XIa | chloroform | none | 60 | 0.61 |
| XIa | chloroform | DC-18-C-6 | 10 | 0.62 |
| XIa | chloroform | TBAC | 60 | 0.72 |
| XIa | dichloromethane | DC-18-C-6 | 60 | 0.55 |
| XIb | chloroform | none | 60 | 0.55 |
| XIb | dichloromethane | none | 15 | 0.42 |
| XIb | dichloromethane | DC-18-C-6 | 15 | 0.51 |

Data obtained from Imai;[41] [a]Reaction conditions: 2.5 mmol of XI, 2.5 mmol of XII, 0.05 mmol of catalyst, 5 mL of solvent, 5 mL of 1.01 M KOH at 15°C under nitrogen, *Measured at a concentration of 0.5 $gdL^{-1}$ in concentrate sulfuric acid at 30°C

### 13.3.1.4 Effect of the volume of organic solvent and water on the reactions in various reaction systems

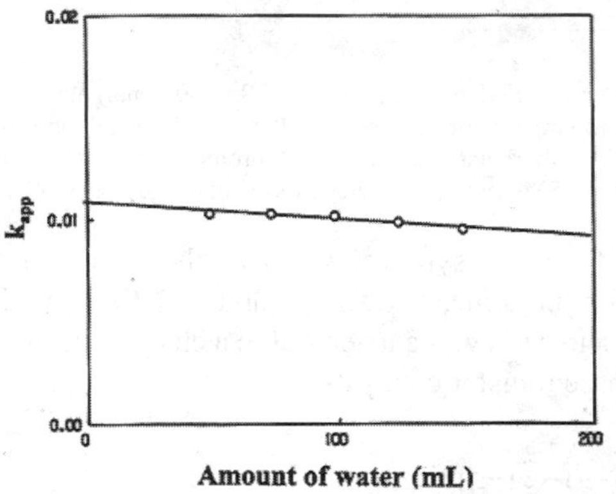


Figure 13.3.7. Effect of the amount of water on the apparent rate constant ($k_{app}$); 4 g of 2,4,6-tribromophenol, 0.9 g of KOH, 0.6 mL of benzyl bromide, 50 mL of $H_2O$, 50 mL of chlorobenzene, 40°C. (Adapted from Ref. [147], by permission.)

In general, the concentration of reactants in the aqueous phase is decreased by increased amount of water. The addition of water probably also decreases the concentration of the intermediate product in the organic phase. Hence, both the mass transfer rate and the degree of hydration with the anion are decreased, which also decreases the reaction rate. However, this argument is not necessarily correct. Figure 13.3.7 indicates that the conversion in the two-phase reaction is not affected by the amount of water added.[147] Wang and Yang[135] studied the effects of the volume ratio of water to chlorobenzene on the conversion for the reaction of 4-bromophenol and allyl bromide in an alkaline solution of KOH/chlorobenzene at 50°C under phase-transfer catalytic conditions. The reaction followed the pseudo-first-order rate law and the corresponding apparent rate constant decreased gradually when the water content was increased, as shown in Figure 13.3.8.[135] The reason was that the concentration of the intermediate product tetra-n-butylammonium phenoxide (ArOQ, or the active catalyst) in the aqueous phase decreased with the increase in the amount of water. The mass transfer rate of the intermediate product (or the active catalyst) from the aqueous phase to the organic phase decreased when a large amount of water was used. In addition, the dilution effect led to reduction of the reaction rate in the aqueous phase.

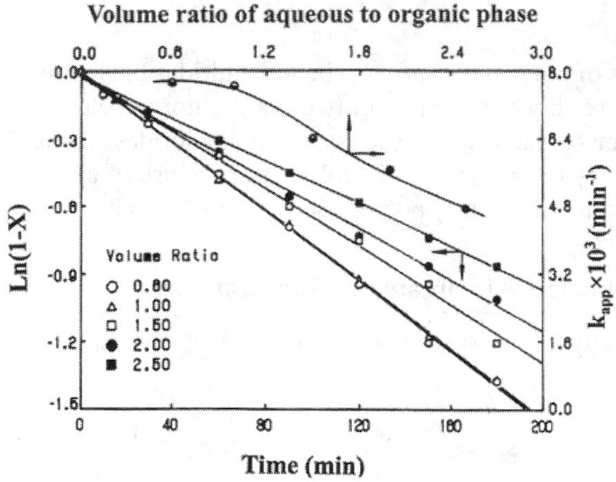


Figure 13.3.8. Effect of the volume ratio of water to chlorobenzene on the conversion; 1.568 g of 4-bromophenol, 1.0 g of KOH, 0.7 g of allyl bromide, 0.2 g of TBAB catalyst, 50 mL of chlorobenzene, 50°C. (Adapted from Ref. [135], by permission.)

In general, a higher concentration of the intermediate product (tetra-n-butylammonium alkoxide, or the active catalyst, ArOQ) in the aqueous phase enhances the reaction rate. This is due to a large concentration gradient across the interface in transferring the species from the aqueous phase to the organic phase. For the reaction of allyl bromide and 2,4-dibromophenol in synthesizing 2,4-dibromophenyl allyl ether in an

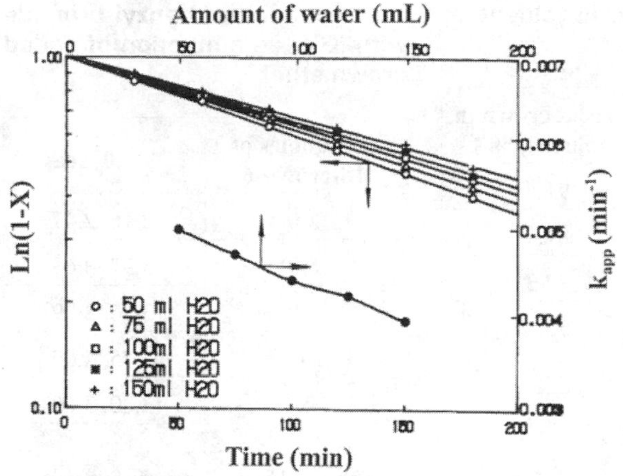


Figure 13.3.9. Effect of the water content on the conversion; 2.28 g of 2,4-dibromophenol, 0.2 g of TBAB catalyst, 0.7 g of allyl bromide, 1.0 g of KOH 50 mL of chlorobenzene, 50°C. (Adapted from Ref. [132], by permission.)

alkaline solution of KOH/chlorobenzene two-phase medium under PTC conditions,[132] the conversion increases with the increase in the concentration of ArOQ in the aqueous phase (or decreasing content of water). However, this change is small, reflecting a small mass transfer resistance, as shown in Figure 13.3.9.[132] The influence of the amount of water on the conversion in the reaction of carbon disulfide and o-phenylene diamine catalyzed by tertiary amine in a two-phase medium was studied. The conversion decreased with the increase in the amount of water. Therefore, the value of apparent rate constant ($k_{app}$), in which the reaction follows pseudo-first-order-rate law, decreases with the increase in the volume of water.[128] Wang and Chang[144-146] found that the conversion increases with the increase in the volume ratio of water to organic solution up to 1/5. The conversion is independent of the volume ratio of water to organic phase (chlorobenzene), greater than 1/5. The reason is that the reaction is carried out in a large amount of KOH (solid form). Probably, the omega phase is generated for the volume ratio of water to chlorobenzene less than 1/5. However, this change in the conversion vs. the volume ratio of water to chlorobenzene is not significant.

(A) Omega phase reaction

It is found that 92% of the 18-crown-6 ether added to a salt (KCN and KCl) and toluene system resided in the organic phase. However, all but approximately 1-2% of the crown ether was translocated onto the surface of the salt upon addition of small quantities of water. The results of Liotta et al.[62] are given in Table 13.3.19. The initial water added to the system coats the surface of the salt particles and it was this aqueous salt coating that extracted the crown from the organic phase. Liotta et al.[61,62] called this new region of the reaction system the omega phase. The 8% of the crown located on the surface of the salt particles prior to the addition of water was probably due to the presence of water already present in the salt. The distribution of 18-crown-6 ether between the organic phase and the omega phase was determined.[62] The amount of crown ether in the organic phase remained low and relatively constant (0.06-0.07 milimoles of 18-crown-6 ether in organic phase). The omega phase adsorbed most of added crown ether. For the accompanying pseudo first-order kinetics reaction of benzyl bromide with potassium cyanide, the results are given in Table 13.3.20.[122] There is a slight increase in the rate as the number of millimoles of 18-crown-6 ether increases, but the rate remains essentially constant with the increase of crown ether.

**Table 13.3.19 Effect of added water on the concentration of 18-crown-6 ether in toluene at room temperatures**

| Water, μL | Equiv. of water, mole of $H_2O$/mole of crown | Percent crown in toluene, % |
|---|---|---|
| 0 | 0.00 | 91.5 |
| 10 | 0.14 | 81.4 |
| 15 | 0.21 | 77.3 |
| 22 | 0.31 | 50.0 |
| 25 | 0.35 | 34.6 |
| 30 | 0.42 | 17.7 |
| 45 | 1.25 | 2.5 |
| 50 | 1.39 | 2.0 |
| 80 | 2.22 | 1.0 |

Data obtained from Liotta et al.[62] 0.0040 mole of 18-crown-6, 0.027 mol of KCN, 10 mL of toluene

**Table 13.3.20 18-Crown-6 catalyzed reactions of benzyl bromide with KCN as a function of added crown ether**

| Millimoles of 18-crown-6 | $k \times 10^5$ $sec^{-1}$ |
|---|---|
| 3.0 | 2.16, 2.47 |
| 5.0 | 3.97, 3.63 |
| 7.0 | 3.86, 3.99 |
| 10.0 | 3.75, 4.00 |
| 12.0 | 3.80, 3.60 |

Data obtained from Vladea and Simandan;[122] 1.0 mL of $H_2O$, 0.15 mole KBr, 0.15 mol of KCN, 50 mL of toluene, 25°C

(B) Reaction catalyzed by PEGs

The structure of polyethylene glycol ($HO(CH_2CH_2O)_nH$, PEG) is similar to that of crown ether. Polyethylene oxide chains ($CH_2CH_2O$) form complexes with cations, much like crown ethers, and these complexes cause the anion to be transferred into the organic phase and to be activated.[122] Table 13.3.21 shows the binding constant, K, for PEG complexes with sodium cation depend on both the value of n (i.e., average molecular weight of PEG or number of ($CH_2CH_2O$) unit) and on the end-group substituents.[112,114,115] Gokel and coworkers[31,32] determined the binding strength for $Na^+$ in anhydrous methanol solution with PEGs and obtained the binding constant K=1.4. They concluded that the strength of complexation is a function of the total number of binding sites present and not the number of polymer chains, suggesting that a long PEG chain may be involved in binding more than one cation.

PEGs and their derivatives have been extensively investigated as phase transfer catalysts and are used in many commercial processes. In the absence of strong acids, PEGs are nontoxic, inexpensive, and thermally stable. For some reactions such as with hydroxide transfer, PEGs are excellent catalysts, sometimes better than crown ethers, especially when used in liquid-solid PTC reactions with potassium salts, and with little or no added water, and with at least moderately polar organic solutions. PEGs are water soluble and if the organic phase is not sufficiently polar the PEG will reside almost completely in the aqueous phase; or with concentrated aqueous solutions of organic salts, the PEG may form a third catalyst-rich phase, a change that normally leads to a high level of catalytic activity.

PEGs are themselves soluble in water. To obtain partitioning of PEG into an organic solution may require use of a mono- or diether derivative. Harris and Case[34] found that with exception of dichloromethane as an organic-phase solvent, most PEGs are themselves partitioned into the aqueous phase, depending also on the concentration of dissolved salts in the aqueous phase.[35] To improve organic-phase solubility of PEG, several dialkyl ethers of PEGs as permanganate-PTC catalysts are suggested in Table 13.3.22.

**Table 13.3.21 Binding constants for complexation of PEGs and some symmetrical derivatives with sodium cation $Na^+ + PEG = [Na^+ PEG]$ complex**

| PEG Avg. MW | Log K (binding constant) with various sodium salts | | | | | |
|---|---|---|---|---|---|---|
| | Avg. n | HO- | $CH_3O$- | $C_2H_5O$- | PhO- | $cyC_5H_{10}N$- |
| 200 | 4.1 | 1.64 | | | 0.5 | |
| 300 | 6.4 | 2.02 | 1.55 | 1.25 | 1.05 | 1.16 |
| 400 | 8.7 | 2.26 | | | 1.49 | 1.51 |
| 600 | 13.2 | 2.59 | 2.09 | 1.99 | 1.87 | |
| 1000 | 22.3 | 2.88 | 2.55 | 2.48 | 2.37 | 2.46 |
| 1500 | 33.7 | 3.09 | 2.86 | 2.80 | 2.68 | |
| 2000 | 45.0 | 3.28 | 3.08 | 3.05 | 2.81 | 3.08 |

Data obtained from Szabo et al.[112,114,115]

**Table 13.3.22 Partition of PEG-dialkyl ethers between $C_6H_6$ and $H_2O$**

| PEG Ether | Partitioning, % in $C_6H_6$/% in $H_2O$ |
|---|---|
| $C_4$-PEG1500-$C_4$ | 14 |
| $C_6$-PEG1500-$C_6$ | 84 |
| $C_{18}$-PEG6000-$C_{18}$ | Emulsion |
| $C_{18}$-PEG750-Me | 108 |
| $C_{18}$-PEG1900-Me | 39 |
| $C_{18}$-PEG5000-Me | 37 |
| $C_8$-PEG5000-Me | 12 |
| $C_4$-PEG5000-Me | 13 |
| PEG6000 | <1 |

Data obtained from Harris et al.[34,35] $C_4$-PEG1500-$C_4$ represents a PEG of MW 1500, capped by a butyl group at both ends

Aliphatic hydrocarbons are immiscible with PEGs. Therefore, it is important to select a good organic-phase solvent such as aromatic hydrocarbons, chlorinated hydrocarbons, or acetonitrile.[80] In toluene, PEGs are more effective catalysts than crown ethers for the reaction of benzyl chloride and solid potassium acetate. In butanol, the effectiveness of PEGs and crown ethers as phase-transfer catalysts were the same for the reaction of benzyl chloride and solid potassium acetate.[27]

#### 13.3.1.5 Effects of organic solvents on other phase-transfer catalytic reactions

(A) Liquid-liquid-liquid three phase reaction

Not only will the organic solvent influence the NPTC reaction, but it also affects the character of NPTC. Another type of transfer mechanism takes place when the quaternary salt catalyst is not highly soluble in either the aqueous or the organic phase, but instead forms a third phase.[154] For example, tetrabutylammonium salts in the presence of highly concentrated aqueous solutions, and with toluene as solvent for the organic phase, form a third (quaternary compound) phase.[126] Jin et al.[48] investigated the relationship between the properties and the catalytic activity of the liquid-liquid-liquid three-phase phase-transfer catalytic system. The condition of formation of the third phase are investigated by changing the kind of PTC and organic solvent. With tetra-n-butylammonium bromide $Bu_4N^+Br^-$ is used as the phase transfer catalyst, the third phase is formed with both dodecane and toluene as the organic solvents. On the other hand, when tetra-n-hexylammonium bromide $(Hex)_4N^+Br^-$ is used as the phase transfer catalyst, the third phase forms with dodecane but not with toluene. In such situations, most of the reaction actually occurs in the third phase with both aqueous and organic reagent transferring to this phase for conversion. Third-phase reactions of this type may be faster than simple PTC reactions. Because formation of the third phase offers simplified catalyst removal and recovery procedures, third-phase catalyst is highly attractive for commercial operations.

In the conversion of benzyl chloride to benzyl bromide using tetra-n-butylammonium bromide as the catalyst, a third phase was also observed.[126] More rapid reaction rates were obtained in the presence of this additional phase. The kinetics associated with the base-catalyzed isomerization of p-allylanisole in the presence of a variety of polyethylene glycols (PEGs) has been reported.[79] The reaction followed first-order kinetics in p-allylanisole and the reaction system was characterized by three phases consisting of an organic solvent phase, an aqueous base phase, and a complex liquid phase consisting of PEG and potassium hydroxide. It was suggested that the isomerization reaction took place in the complex third phase.

PEG forms a third phase between toluene and aqueous KOH when some methanol is added.[39] Some methanol must be added to reduce the amount of PEG that otherwise would dissolve in the toluene phase. Thus, dehydrobromination of 2-bromooctane in toluene, using PEG as catalyst, could be accomplished by removal and replacement of the organic and aqueous phases after completion of reaction, and recycle of the catalyst phase. After four cycles no catalyst was lost. Reaction rates for KOH dehydrohalogenation of 2-bromooctane in toluene with PEG catalysts were increased by a maximum factor of 126 by addition of methanol.[38,40] The base efficiency (moles base per mole of catalyst) PEGs with molecular weights 3000 and 20,000 exceeded unity and reached a maximum of 12 on addition of methanol. Hydrogenative dehalogenation of polychlorinated aromatic halides can be accomplished by hypophosphite reduction using a quaternary ammonium salt as a phase-transfer catalyst in conjunction with a palladium-on-carbon co-catalyst.[68] The quaternary salt, being insoluble in both reactant phases, coats the Pd/C catalyst forming a third phase. The strongly alkaline medium and the phase-transfer agent are synergistic.

In principle, formation of a third catalyst could be accomplished either by (1) use of a phase-transfer catalyst that has limited solubility in both the organic and aqueous phase, or (2) by use of a special solvent, perhaps a fluorocarbon, having great affinity for the catalyst but only modest affinity for both of the reactants. Tetra-n-butyl-ammonium salts frequently form third layer when used in conjunction with an organic phase that has little polarity (e.g., neat 1-chlorooctane, toluene, or cyclohexane as solvent, but only dichloromethane) and with a concentrated aqueous solution of inorganic salts.

The catalyst may form a third layer during the PTC reaction, when using high salt concentrations and nonpolar organic solvents. Wang and Weng[123-126,154] found that a third layer phase was built when using a low polarity organic solvent in the PTC reaction system. This has been observed for PEGs, quaternary onium salts and crown ether. In such cases, catalyst recovery involves a simple phase separation. In industrial kettles (with limited visibility in the reactor) or when a "rag" layer distorts an otherwise sharp phase boundary, a phase separation operation may not be simple. Nevertheless, choosing conditions in which the catalyst separates as a third layer is usually advantageous. A dehydrohalgenation was performed four times consecutively, with no loss in catalytic activity by simply replacing the organic phase and replenishing the aqueous base phase after each use, leaving the third phase containing PEG in the reactor.

(B) Electrochemical and PTC reaction

The combination of phase transfer catalysis (PTC) with other processes, such as supercritical fluid extraction[22] and electrochemical process[5-7,23,120] have been investigated in detail. Do and Chou[23] carried out the anodic oxidation of benzyl alcohol in the two-phase system containing both the redox mediator, $ClO^-/Cl^-$, and a phase-transfer catalyst (PTC). The reaction mechanism and the factors which affect the efficiency of benzaldehyde production were explored. The current efficiency is mainly governed by the pH value and the nature of the organic solvent as well as the types and the concentration of phase transfer catalyst (PTC). When ethyl acetate and chlorinated hydrocarbons were used as solvents, the current efficiencies were between 64 and 86%, respectively as shown in Table 13.3.23.[23] With ethyl acetate as solvent, the current efficiency was less than for chlorinated hydrocarbons. The results indicate that the chlorinated hydrocarbons have a higher extraction capacity for $Bu_4N^+ClO^-$ from the aqueous phase than has ethyl acetate. Similar results were obtained by Dehmlow and Dehmlow.[16] The current efficiencics change slightly when different chlorinated hydrocarbons were used as organic solvents.

**Table 13.3.23 Effect of organic solvent on current efficiency**

| Organic solvent | Current efficiency, % |
|---|---|
| $CH_2Cl_2$ | 80 |
| $CH_2ClCH_2Cl$ | 75 |
| $CHCl_3$ | 86 |
| $CH_3COOC_2H_5$ | 64 |

Data obtained from the work of Do and Chou;[23] 0.5 M of $C_6H_5CH_2OH$ in the organic phase; 25°C, 600 rpm, 20 $mAcm^{-2}$ of current density, 70 mL of organic phase, 70 mL of aqueous phase, 0.005 M of $Bu_4NHSO_4$, 1.0 M of NaCl, pH=6.9, electricity passed 2 $Fmol^{-1}$ of benzyl alcohol

Tsai and Chou[120] carried out the indirect electrooxidation of cyclohexanol by using a double mediator consisting of ruthenium and chlorine redoxes in the multiphase system. Table 13.3.24 shows that the current efficiency had the highest value at 83% using carbon tetrachloride as organic solvent.[120] The current density decreased in order, carbon tetrachloride > chloroform > toluene > cyclohexane. The selectivity was 100% except when toluene was used as organic solvent. For this case, the concentration of cyclohexanol in carbon tetrachloride is higher than that of the other solvents.

**Table 13.3.24. Effect of organic solvent on the conversion and selectivity**

| Solvent | Conversion, % | Selectivity, % | C.E., % | Energy consumption, KWH/mole |
|---|---|---|---|---|
| $CCl_4$ | 42 | 100 | 83 | 0.226 |
| $CHCl_3$ | 38 | 100 | 70 | 0.268 |
| Toluene | 17 | 81 | 34 | 3.154 |
| Cyclohexane | 12 | 100 | 23 | 4.662 |

Data obtained from Tasi and Chou;[120] Reaction conditions: 0.8 M of cyclohexanol, 15 $mAcm^{-2}$ of current density, graphite cathode and graphite anode, 0.007 M of $RuO_2$, pH=4, 0.99 $Fmol^{-1}$ electricity passed, NaCl saturated solution as electrolyte, 1275 rpm, 5°C

### 13.3.1.6 Other effects on the phase-transfer catalytic reactions

Simple mechanical separation such as filtration, centrifugation or phase separation can be used to separate the product and the phase-transfer catalyst by use of insoluble catalysts. However, the more frequently encountered technical problems in use of PTC for industrial applications is the need to separate the product and the phase-transfer catalyst by chemical equilibrium separation method in the liquid-liquid two-phase phase transfer catalytic reaction. The most commonly used methods for separation of products and PTC catalysts on an industrial scale are extraction and distillation. Other separation methods include sorption[3,33,57] and reaction.[45]

The principle of extraction method used to separate PTC and product is based on solubility of quaternary ammonium salt in alkaline aqueous solution.[2,25,104] For example, tetrabutylammonium bromide is soluble to the extent of 27% in dilute (1% NaOH) aqueous solutions, but when the solution is made more concentrated (15% NaOH), the solubility of $Bu_4N^+Br^-$ decreases to 0.07%. When the products are obtained in PTC system, they can be usually separated from PTC by distillation method. PTC catalyst in the distillation residue may sometimes be reusable. With quaternary ammonium salts as catalysts, temperatures above 100-120°C usually result in partial or total decomposition of the quaternary salts to trialkylamines and other products. Mieczynska et al.[70] and Monflier et al.[72] investigated the hydrogenation and hydroformylation under phase transfer catalytic conditions. They found that the yield of aldehydes obtained in hydroformylation of 1-hexene strongly depends on solvent: 24% in toluene, 53-86% in toluene-water-ethanol mixture and 77-94% in water-ethanol solution. The mixture of water-ethanol as a solvent was also found to be the best for hydrogenation of 1-hexene (96% of hexane). Conversion of $Ph_2PCH(CH_3)(COOH)$ phosphine into sodium salt $Ph_2PCH(CH_3)(COONa)$ yields aldehyde in toluene, 92% in toluene-water and 94% in toluene-water-ethanol mixture.

In principle, hydroxide anion is very difficult to transfer from aqueous to organic phases, yet it is one of the most valuable and most commonly used anions in the PTC systems. Addition of small amounts of alcohols to PTC systems requiring hydroxide transfer causes a dramatic increase in rates. Therefore, addition of alcohol enhances the PTC reaction as the cocatalytic effect. For example: formation of alkoxide anions, $RO^-$, which are more readily transferred than the highly hydrated hydroxide anion, and which can serve as a strong base just as well as $OH^-$, and solvation of the hydroxide with alcohol rather than with water, making the hydroxide anion more organophilic and more easily transferred.[99,100]

Cyanide displacements catalyzed by quaternary ammonium salts usually do not proceed without the presence of water to facilitate exchange and transfer of anions. However, PTC displacement depends on alcohol structure. Benzyl alcohol is about 1.5-2 times as effective as either methanol or ethanol.

In the synthesis of BTPPC (benzyltriphenylphosphonium chloride) from benzyl chloride and triphenylphosphine, second-order rate constants and activation parameters for the reaction of benzyl chloride and triphenylphosphine were measured in several protic and aprotic solvents covering a wide range of dielectric constant were obtained by Maccarone et al.[64] Wang, Liu and Jwo[127] also used eight solvents in studying their effect on the reaction of triphenylphosphine and benzyl chloride. They classified these solvents into two categories depending on the solubility of benzyltriphenylphosphonium chloride (BTPPC). Solvents that dissolve BTPPC are acetic acid, dichloromethane, methanol and water. Solvents that do not dissolve BTPPC are acetone, benzene, toluene and ether. In general, triphenylphosphine (TP) does not dissolve in methanol or water. The effect of solvents on the reaction rate was measured by the apparent rate constant in which the reaction follows pseudo-first-order rate law. The order of relative activities of solvents is methanol (0.34 $h^{-1}$) > acetic acid (0.176 $h^{-1}$) > dichloromethane (0.0468 $h^{-1}$) > acetone (0.0114 $h^{-1}$) > diethyl ether (0.0043 $h^{-1}$) > benzene (0.0018 $h^{-1}$) > toluene (0.0008 $h^{-1}$).

**Table 13.3.25. Second-order rate constants and activation parameters for the reaction of benzyl chloride with triphenylphosphine in various solvents**

| Solvent | Dielectric constant (20°C) | $k \times 10^4$, $Lmol^{-1}s^{-1}$ | | | | |
|---|---|---|---|---|---|---|
| | | 60°C | 70°C | 80°C | 90°C | 100°C |
| Decalin | 2.26 | 0.00134 | | | | 0.0355 |
| Toluene | 2.38 | 0.0169 | | 0.0843 | 0.181 | 0.353 |
| Anisole | 4.33 | 0.0569 | | 0.260 | 0.466 | 1.15 |
| Bromobenzene | 5.40 | 0.0933 | | 0.276 | 0.457 | 0.909 |
| Chlorobenzene | 5.62 | 0.0512 | | 0.243 | | 1.13 |
| Benzyl alcohol | 13.1 | 8.94 | | 38.7 | | 107 |
| 1-Butanol | 17.1 | 4.93 | | 19.5 | | 43.7 |
| Acetopnenone | 17.39 | 0.300 | | 1.32 | | 9.19 |
| 1-Propanol | 20.1 | 4.41 | 9.86 | | | |
| Acetone | 20.3 | 0.166 | 0.417 | | | |
| Ethanol | 24.3 | 3.60 | 7.73 | | | |
| Benzonitrile | 25.2 | 0.545 | | 1.71 | | 6.38 |
| Nitroethane | 28.06 | 0.470 | | 3.05 | | 12.9 |
| Methanol | 32.65 | 8.86 | | | | |
| N,N-dimethylformanide | 36.7 | 0.460 | | 1.88 | | 6.59 |
| Acetonitrile | 37.5 | 1.20 | 2.79 | | | |

| Solvent | Dielectric constant (20°C) | k×10$^4$, Lmol$^{-1}$s$^{-1}$ | | | | |
|---|---|---|---|---|---|---|
| | | 60°C | 70°C | 80°C | 90°C | 100°C |
| N,N-Dimethylacetamide | 37.8 | 0.29 | | 1.06 | | 4.29 |
| N-Methylformamide | 189.5 | 6.08 | | 28.7 | | 86.5 |

Data obtained from the work of Maccarone et al.[64]

The BC-TP reaction shows better reactivity in protic or polar solvent since the activated complex is more polar than both reactant molecules.

### 13.3.2 THREE-PHASE REACTIONS (TRIPHASE CATALYSIS)

As stated, the solid PTC is suitable for the industrial processes concerning the removal of the catalyst from the reaction mixture and its economic recycle. The real mechanism of reaction in a triphase catalysis is not completely understood. However, the reaction rate and the conversion of reactant in a triphase catalysis (TC) is highly dependent on the organiphilicity (hydrophilicity or hydroprobicity) of the polymer support of the catalyst and the polarity of the organic solvent. Not only the partition of the organic to the aqueous solutions is affected by the organophilicity of the polymer-supported catalyst, but also the concentration distribution of the catalyst between two phases is influenced by the organophilicity of the polymer-supported catalyst.

Ohtani et al.[82-86] used polystyrene-supported ammonium fluoride as a phase transfer catalyst (triphase catalysis) for several base-catalyzed reactions, such as cyanoethylation, Knoevenage reaction, Claisen condensation and Michael addition. The catalytic activity of the polystyrene-supported ammonium fluid was comparable to that of tetrabutylammonium fluoride (TBAF). The ionic loading and the ammonium structure of the fluoride polymers hardly affected the catalytic efficiency. The reaction was fast in a non-polar solvent (e.g., octane or toluene) from which the rate-determining step of the base-catalyzed reaction is very similar to that of the $S_N^2$ nucleophilic substitution reactions.

The solvent may affect the catalytic activity in several ways. The greater its swelling power, the larger the volume fraction of catalytic occupied by the more mobile liquid, and the swollen volume fraction of the more rigid polymer network. The degree of swelling and the viscosity within the polymer matrix affect intraparticle diffusion rates. The solvent may also affect intrinsic reactivity at the active sites. Experimentally, it is difficult to distinguish solvent effect on diffusivity from solvent effects on reactivity. Tomoi and Ford[116] found that the triphase catalysis followed pseudo-first-order rate law. The corresponding apparent rate constant $k_{app}$ decreases with solvent in the order: chlorobenzene > toluene > decane over wide ranges of particle sizes and polymer crosslinking. The ability of the solvent to swell the catalysts decreases in the same order.

#### 13.3.2.1 The interaction between solid polymer (hydrophilicity) and the organic solvents

In triphase catalysis, solvated resin supports are important carriers for solid-phase organic synthesis in combinatorial chemistry. The physical properties of resin, resin swelling, dynamic solvation, and solvated supports are important factors in affecting the synthesis.[160] However, these factors are also affected by solvent. Selective solvation of resin alters the local reactivity and accessibility of the bound substrate and the mobility of the entrapped reagent. Resin solvation changes during the course of the reaction when the attached substrate changes its polarity or other physicochemical properties.

The basic steps involved in reactions with resin-supported PTC catalysts differ from ordinary two-phase PTC reactions in one important respect: ordinary PTC reactions require only one reagent to be transferred from their normal phase to the phase of the second reactant. Use of resin-supported catalysts requires that both reagents diffuse to active PTC sites on the catalyst surface, or for reactions with slow intrinsic rates, both reagents must also diffuse to the active sites inside the resin bulk phase. The need for diffusion processes with solid catalysts also means that both reagents are required to diffuse to and penetrate the stagnant outer layer of liquid(s) (the Nernst layer), coating the catalyst particle.

Ford and Tomoi[28] carried out the reaction of 1-bromooctane with aqueous sodium cyanide, catalyzed by tributylphosphonium groups bound into beads of an insoluble styrene-divinylbenzene resin,

$$1-C_8H_{17}Br_{(org)} + NaCN_{(aq)} \xrightarrow{polymer-C_8H_5PBu_3Br} 1-C_8H_{17}CH_{(org)} + NaBr_{(aq)} \quad [13.3.16]$$

The reaction includes the following steps:

(a) Diffusion of aqueous sodium cyanide through the bulk phase and through the resin bulk to active sites

(b) Equilibrium exchange of $CN^-$ for $Br^-$ at the active sites

(c) Diffusion of $Br^-$ out of the catalyst particle and into the aqueous bulk phase

(d) Diffusion of RBr ($1\text{-}C_8H_{17}Br_{(org)}$) through the organic bulk phase and through the bulk resin phase to active sites. Some reactions may occur at sites on the catalyst surface, but since the number of surface sites is small compared to the number of sites within the bulk of the catalyst, most of the reaction occurs inside the catalyst bulk.

(e) Chemical reaction (intrinsic reaction) between RBr and Resin-$PR_3^+CN^-$ at active sites to produce RCN and $Br^-$

(f) Diffusion of RCN out of the catalyst particle and into the organic phase.

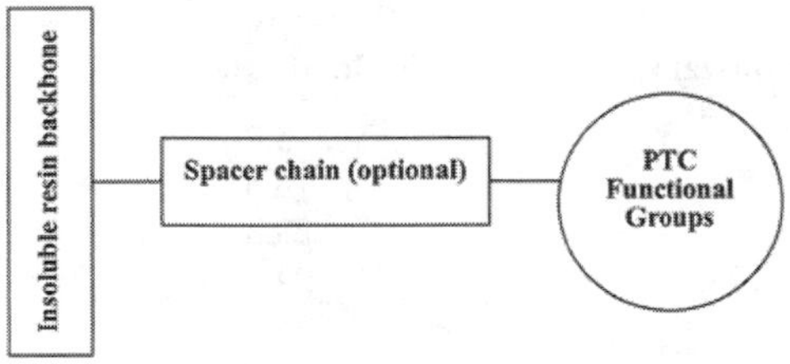

Figure 13.3.10. Schematic diagram of the general resin-bound PTC catalysts.

A schematic diagram of the general resin-bound PTC catalyst is given in Figure 13.3.10. As indicated in Figure 13.3.10, spacer chains can increase some reactions by removing the active site away from the polymer chain, and from other active sites. When active sites, particularly quaternary onium salts, are located close to one another, they join to form doublets, triplets, and higher aggregates that are less active catalyst centers, and that tend to present an "aqueous" face to the reactants. Thus, the use of spacer chains increases the rates of some reactions, such as nucleophilic displacements, two-to-four-fold.[1,9,73] When the spacer chain also contains complexable ether oxygen atoms, using 15-crown-5 ether as the PTC functional group, catalyst activity is even greater, as observed in halide exchange of KI with 1-bromooctane.[10]

Preparation of phase transfer catalyst (PTC) functional groups bound to insoluble resins and their activity for catalyzing two-phase reactions has been extensively stud-

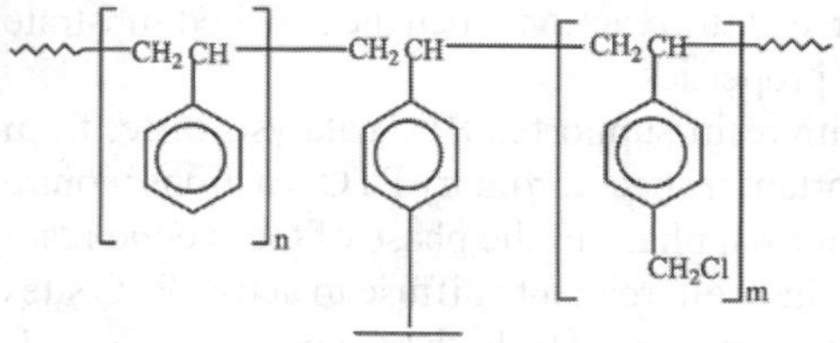

Figure 13.3.11. Crosslink to other polymer chains.

ied.[82,83,97,101] Much work has been done on the preparation and testing of phase-transfer catalysts supported on resins including extensive work by Montanari and co-workers[71,74] and by Ford and Tomoi[28] and their co-workers. Most published works on resin-bound phase-transfer catalysts use a styrene-divinylbenzene resin (SDV) and related resins, taking advantage of the huge amount of technology available on these resins due to their use as ion-exchange resin supports (Figure 13.3.11).

Tomoi and co-workers[117] suggest that solvents may affect rates of triphase-catalyzed reaction in three ways: intrinsic chemical reactivity; solvent effect on ion-exchange rate; and overall activity, including diffusion effects due to swelling of polymer-supported phosphonium salts under three-phase conditions. First, the intrinsic activity of the catalysts, as well as of soluble phosphonium salts, depended slightly on organic solvents for cyanide displacement reactions. Second, the exchange rate of chloride ion in the catalysts vs. that of acetate depends on the solvents when the degree of ring substitution is less than 16%. With 30% ring-substituted catalysts, the rate increases and hardly depends on the solvents. Third, the overall catalyst reactivity for the reaction of organic halides with NaCN depends on the substrate and organic solvents. For 1-bromooctane, the catalysts were more reactive in good solvents (e.g., chlorobenzene) than in poor solvent (e.g., octane). Shan and co-workers,[98] Wang and Wu[136,137] examined the effects of solvents and other resin-bound catalysts parameters (macroporosity, microporosity, crosslink density and size of catalyst pellet). They all show that the swelling in organic solvents is an important factor affecting the conversion of the reactant, as shown in Table 13.3.26.[98]

**Table 13.3.26. Effect of the organic solvent on the yield of ester from benzyl bromide and aqueous KOAc under standardized conditions**

| Catalyst[a] | Solvent (dielectric constant, ε)[b] | Yield of ester, % |
|---|---|---|
| Macro (6%)-400 | Cyclohexane (2.02) | 13.2 |
| | Toluene (2.38) | 16.5 |
| | Chlorobenzene (5.62) | 23.3 |
| | Benzyl ethyl ketone (17.4) | 94.5 |
| | Nitrobenzene | 100.0 |
| Micro (6%)-400 | Cyclohexane (2.02) | 9.2 |
| | Toluene (2.38) | 13.5 |
| | Chlorobenzene (5.62) | 18.2 |
| | Benzyl ethyl ketone (17.4) | 86.0 |
| | Nitrobenzene | 100.0 |

Data obtained from Shan, Kang and Li;[98] [a]6% crosslinking; [b]PEG-400 used for active sites on catalyst

### 13.3.2.2 Effect of solvents on the reaction in triphase catalysis

The disadvantage of using liquid-liquid phase-transfer catalysis (ll-PTC) is in the separation of catalyst from product after reaction. This problem can be overcome using the immobilized catalyst on a solid support (e.g., porous polymer pellet). Simple mechanical separation processes, such as filtration or centrifugation, can be employed to separate the solid catalyst from the product in liquid form. A detailed investigation of the effect of polymer particle on the reaction rate was conducted by Wang and coworkers.[131,133-137] Wang and Wu[136] studied the reaction of substitution of hexachlorocyclotriphosphazene and 2,2,2-trifluoroethanol in an organic solvent/alkaline solution by triphase catalysis. The polymer, which was prepared from the polymerization of styrene monomer and chloromethylstyrene monomer, served as the support for the immobilization of the catalyst.

The effects of the structure of the polymer support, which can be related to the factors of the degree of crosslinking, ring substitution (RS), lipophilicity of the polymer, the chloride density and solvents on the imbibed compositions[84,118] on the reaction rate or conversion were investigated. This imbibed composition, influenced by the internal structures of the triphase catalyst particles, affected the reactivities. The interaction of the polymer support pellet and the organic solvents play an important role in determining the reaction rate and the conversion of the reactant. The reaction could be improved to obtain a high reaction rate by using a polar solvent.

For investigating the degree of crosslinking of the polymer, the resistance of mass transfer within the catalyst pellet is small. When a smaller degree of crosslinking of the polymer support is used. This is due to the fact that a larger value of the swell of the polymer was obtained when a small degree of crosslinking of the polymer was used. Wang and Yu[131] have similar observations for the reaction of allyl bromide and 2,4-dibromophenol under triphase catalysis. A maximum value exists for the degree of swell and the imbibed composition, as shown in Table 13.3.28 for the degree of crosslinking.[136]

In Table 13.3.27,[136] the degree of swell for the polymer support with a 6% crosslinking is larger than that for the two other degrees of crosslinking. This implies that greater amounts of $NaOCH_2CF_3$ were imbibed into the catalyst pellet with a 6% crosslinking. The reaction rate is directly related to the amount of the imbibed composition. Also, in Table 13.3.27, the imbibed compositions are affected by the structure of the polymer support. The reactivity of the triphase catalysis can also be determined from the composition imbibed by the particles. It can be observed that the reactivities were highly affected by the lipophilicity of the catalyst pellet for the substitution reaction in the organic phase and the hydrophilicity of the catalyst pellet of the ion exchange in the aqueous phase. For example, the reaction rate in the organic phase was promoted by using a lipophilic polymer support catalyst when the substitution reaction rate was slow. In Table 13.3.27, the amount of chlorobenzene and water imbibed in the macroporous pellet was greater than that in the microporous pellets for most cases.[136] However, the macroporous pellet with 10% degree of crosslinking had the least lipophilicity and degree of swelling; therefore, the reactivity was the lowest for the macroporous pellet with a 10% crosslinking among the polymer-support catalysts. The reactivity environments which were created by the lipophilicity and the hydrophilicity of the polymer support plays an important role in determining the reactivity.

It is known that the distribution of organic phase and aqueous phase existing in the porous pellet is affected by a change of the ring substitution (RS) of the polymer support.[84] Wang and Wu[136] prepared three kinds of polymer supports with different numbers of ring substitution, such as 10%RS, 20%RS and 49%RS, to analyze the lipophilicity of the polymer support. The order of the lipophilicity was 10%RS > 20%RS> 49%RS, which is the same as the order of swelling. However, a maximum value of the apparent rate constant was obtained for using a 20%RS pellet catalyst among the three kinds of ring substitution polymer pellet. Therefore, it is concluded that the lipophilicity of the polymer cannot be too large to enhance the reaction rate. This is due to the fact that the ion exchange rate is retarded to lower the reaction rate because of using a high lipophilic polymer support. It is concluded that the lipophilicity and the hydrophilicity highly influence the reactivity in triphase catalysis.

For a two-phase PTC, it is recognized that the polarity of the organic solvent affects the reaction rate. In general, the reaction rate increases with the augmentation of the polarity of the solvents. Table 13.3.28 shows the effects of the organic solvents on the apparent rate constant, $k_{o,app}$ and $k_{a,app}$.[136] A higher value of the apparent rate constant was obtained using solvent of high polarity. This result is consistent with the swelling and the imbibed compositions that are given in Table 13.3.29.[136]

**Table 13.3.27. Compositions of the imbibed solvents and swelling volume of the triphase catalyst pellet with various polymer structures**

| Triphase catalyst | Conditions | $ClC_6H_5$ g | $H_2O$ g | $NaOCH_2CF_3$, g (calcd value, g) | Swelling volume ratio |
|---|---|---|---|---|---|
| microporous 2% | $ClC_6H_5$ | 1.31 | | | 2.4 |
| | $H_2O/ClC_6H_5$ | 1.23 | 0.33 | | 2.7 |
| | 2.8M $NaOCH_2CF_3/ClC_6H_5$ | 1.92 | 0.67 | 0.40 (0.29) | 3.6 |
| microporous 6% | $ClC_6H_5$ | 1.19 | | | 2.2 |
| | $H_2O/ClC_6H_5$ | 1.17 | 0.62 | | 2.8 |
| | 2.8M $NaOCH_2CF_3/ClC_6H_5$ | 1.90 | 0.50 | 0.60 (0.22) | 3.4 |
| microporous 10% | $ClC_6H_5$ | 1.06 | | | 2.0 |
| | $H_2O/ClC_6H_5$ | 0.96 | 0.29 | | 2.1 |
| | 2.8M $NaOCH_2CF_3/ClC_6H_5$ | 1.40 | 0.50 | 0.19 (0.22) | 2.9 |
| macroporous 2% | $ClC_6H_5$ | 1.28 | | | |
| | $H_2O/ClC_6H_5$ | 1.33 | 0.73 | | 3.1 |
| | 2.8M $NaOCH_2CF_3/ClC_6H_5$ | 2.29 | 0.50 | 0.34 (0.22) | 3.8 |
| macroporous 6% | $ClC_6H_5$ | 1.54 | | | 2.5 |
| | $H_2O/ClC_6H_5$ | 1.25 | 0.82 | | 3.1 |
| | 2.8M $NaOCH_2CF_3/ClC_6H_5$ | 2.2 | 0.59 | 0.42 (0.25) | 3.8 |
| macroporous 10% | $ClC_6H_5$ | 0.76 | | | |
| | $H_2O/ClC_6H_5$ | 1.05 | 0.63 | | 2.7 |
| | 2.8M $NaOCH_2CF_3/ClC_6H_5$ | 1.28 | 0.38 | 0.17 (0.16) | 2.6 |

Data obtained from Wang and Wu;[136] 50 mL of chlorobenzene, 20 mL of water, 0.059 mole of $(NPCl_2)_3$, 0.7 meq of catalyst was used, 20°C

**Table 13.3.28. Effects of the concentrations of $NaOCH_2CF_3$ and kind of solvent on the apparent intrinsic rate constants, $k_{o,app}$ and $k_{a,app}$**

| Solvent | $k_{o,app}$ for $[NaOCH_2CF_3]$ (M), $(min.meq)^{-1}$ | | | $k_{a,app}$ for $[NaOCH_2CF_3]$ (M), $(min.meq)^{-1}$ | | |
|---|---|---|---|---|---|---|
| | 1.6 M | 2.2 M | 2.8 M | 1.6 M | 2.2 M | 2.8 M |
| $CH_2Cl_2$ | 0.25 | 0.33 | 0.58 | 0.036 | 0.028 | 0.036 |
| $C_6H_5Cl$ | 0.063 | 0.12 | 0.19 | 0.017 | 0.015 | 0.014 |
| $C_6H_5CH_3$ | 0.027 | 0.056 | 0.15 | 0.0055 | 0.006 | 0.011 |
| n-$C_6H_{14}$ | 0.015 | 0.031 | 0.092 | 0.0008 | 0.0008 | 0.018 |

Data obtained from Wang and Wu;[136] 50 mL of solvent, 20 mL of water, 0.0059 mol of $(NPCl_2)_3$, 0.18 meq of macroporous catalyst, 20°C, 40-80 mesh of particle

**Table 13.3.29. Effects of solvents on the composition of the imbibed solvents and swelling volume of the triphase catalyst pellet**

| Solvent | Conditions | Solvent g | $H_2O$ g | $NaOCH_2CF_3$, g (calcd value, g) | Volume ratio |
|---|---|---|---|---|---|
| $CH_2Cl_2$ | $CH_2Cl_2$ | 2.75 | | | 3.2 |
| | $H_2O/CH_2Cl_2$ | 2.24 | 0.96 | | 3.8 |
| $ClC_6H_5$ | $ClC_6H_5$ | 1.28 | | | 2.2 |
| | $H_2O/ClC_6H_5$ | 1.33 | 0.73 | | 3.1 |
| | 2.8M $NaOCH_2CF_3/ClC_6H_5$ | 2.29 | 0.50 | 0.34 (0.18) | 3.8 |
| $CH_3C_6H_5$ | $CH_3C_6H_5$ | 0.57 | | | 1.7 |
| | $H_2O/CH_3C_6H_5$ | 0.61 | 0.29 | | 2.1 |
| | 2.8M $NaOCH_2CF_3/CH_3C_6H_5$ | 0.37 | 0.30 | 0.01 (0.10) | 1.8 |
| n-$C_6H_{14}$ | n-$C_6H_5$ | 0 | | | 1 |
| | $H_2O$/n-$C_6H_5$ | 0.10 | 0.30 | | 1.4 |
| | 2.8M $NaOCH_2CF_3$/n-$C_6H_5$ | 0.10 | 0.15 | 0.01 (0.05) | 1.24 |

Data obtained from Wang and Wu;[136] 30 mL of solvent, 0.80 meq of catalyst (1 g), 40-80 mesh of macroporous particle, 20°C

The overall kinetics can be divided into two steps by virtue of the presence of the two practically immiscible liquid phases, i.e.,

(1) a chemical conversion step in which the active catalyst sites (resin with 2,2,2-trifluoroethanoxide ions) react with hexachlorocyclotriphosphazene in the organic solvent, i.e.,

$$yResin^{+-}OCH_2CF_{3(s)} + (NPCl_2)_{3(org)} \rightarrow$$

$$yResin^{+}Cl^{-}_{(s)} + N_3P_3Cl_{6-y}(OCH_2CF_3)_{y(org)},\ y=1\text{-}6 \qquad [13.3.17]$$

(2) the ion exchange step in which the attached catalyst sites are in contact with the aqueous phase, i.e.,

$$Resin^{+}Cl^{-}_{(s)} + NaOCH_2CF_{3(aq)} \rightarrow$$

$$Resin^{+}OCH_2CF^{-}_{3(s)} + Na^{+}Cl^{-}_{(aq)} \qquad [13.3.18]$$

The total moles of the catalyst active sites are S; thus

$$S = [Resin^{+-}OCH_2CF_{3(s)} + Resin^{+}Cl^{-}_{(s)}] \qquad [13.3.19]$$

The reaction rates for $(NPCl_2)_3$ in the organic phase and for $NaOCH_2CF_3$ in the aqueous phase follow pseudo-first-order kinetics and can be written as

$$-\frac{d[(NPCl_2)_3]_0}{dt} = k_{0,app}S[(NPCl_2)_3]_{(org)} \qquad [13.3.20]$$

$$-\frac{d[NaOCH_2CF_3]_a}{dt} = k_{a,app}S[NaOCH_2CF_3]_{(org)} \qquad [13.3.21]$$

where $k_{0,app}$ and $k_{a,app}$ are the apparent rate constants of $(NPCl_2)_3$ per unit amount of catalyst (molar equivalent) in the organic phase for triphase catalysis and the apparent rate constant of $NaOCH_2CF_3$ per unit amount of catalyst (molar equivalent) in the aqueous phase for triphase catalysis, respectively.

Wang and Yang[134] carried out the reaction of 2,4,6-tribromophenol and allyl bromide catalyzed with tributylamine immobilized on the solid styrene-chloromethylstyrene polymer support in an alkaline solution of KOH/chlorobenzene. The experimental results indicate that the swelling power is enhanced in an organic solvent of high polarity. Thus, the reactivity of the reaction is increased with the increase in the polarity of the organic solvents.

### 13.3.2.3 Effect of volume of organic solvent and water on the reactions in triphase catalysis

In investigating the effect of the amount of water, the contents of other components are fixed. Changing the amount of water affects the volume ratio of organic phase to aqueous phase and the concentration of nucleophile in the aqueous phase. For the reaction of hexacyclotriphosphazene and sodium 2,2,2-trifluoroethanoxide catalyzed by tributylamine immobilized on the solid styrene-chloromethylstyrene polymer support catalyst.[138-140,157] As shown in Figure 13.3.12,[140] the reaction rate is decreased with the increase in the amount of water up to a concentration of sodium 2,2,2-trifluoroethanoxide at 2.8 M. However, the reaction rate is then increased with further increase in the amount of water larger than 2.8 M sodium 2,2,2-trifluoroethanoxide. This result indicates that a high concentration of sodium 2,2,2-trifluoroethanoxide reaction will decrease the reaction rate. The main reason is that the intraparticle diffusion is also affected by the concentration of sodium 2,2,2-trifluoroethanoxide due to changing the amount of water (or the volume ratio of organic phase to aqueous phase).

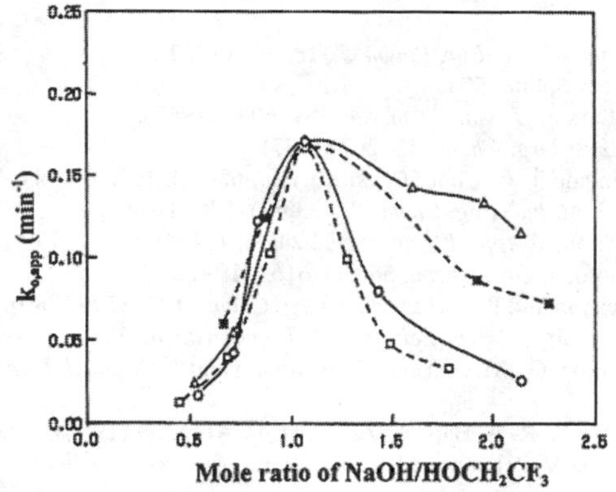


Figure 13.3.12. The apparent rate constants vs. the molar ratios of NaOH/$HOCH_2CF_3$; 0.059 mol of $(NPCl_2)_3$, $9.6\times10^{-5}$ mol of TBAB catalyst, 50 mL of chlorobenzene, 20 mL of $H_2O$, 20°C; and (Δ) 0.07 mol of $HOCH_2CF_3$, (*) 0.058 mol of $HOCH_2CF_3$, (O) 0.075 mol of NaOH, (□) 0.063 mol of NaOH (Adapted from Ref. [138], by permission.)

The other experiments, in which the concentrations of the components in the aqueous phase are fixed, were carried out by Wu[157]. The volume of organic phase is also fixed at 50 mL, in which the amount of catalyst and phosphazene are also fixed. A generalized apparent rate constant (pseudo-first-order rate law) $k'_{i,app}$ is defined as

$$k'_{i,app} = k_{i,app} / \left[ \text{Resin}_{(s)} / V_a \right]; \quad i = r, f \qquad [13.3.21]$$

where $Resin_{(s)}$ and $V_a$ indicate the total molar equivalent active sites and the volume of water. The results are given in Table 13.3.30.[157] The apparent rate constants $k'_{r,app}$ and $k'_{f,app}$ are increased with the increase in the volume of aqueous phase. These results are explained by low concentration of sodium 2,2,2-trifluoroethanoxide in the organic phase because $HOCH_2CF_3$ and $NaOCH_2CF_3$ are all insoluble in chlorobenzene. After 4 hours of reaction, only 8% of hexacyclotriphosphasene is reacted. The increase in reaction rate by increasing the volume of aqueous solution is not due to the increase in the concentration of $NaOCH_2CF_3$ in the organic phase. However, the mass transfer rate within the particles is obviously affected by increasing the concentration of $NaOCH_2CF_3$ in the aqueous phase.

**Table 13.3.30. Apparent rate constants in various $NaOCH_2CF_3$ concentration under constant amount of water or constant amount of $NaOCH_2CF_3$**

| $NaOCH_2CF_3$ (M) | Variation of $NaOCH_2CF_3$ | | | | Variation of water | | | |
|---|---|---|---|---|---|---|---|---|
| | $NaOCH_2CF_3$ (mole) | $k_{r,app}$ [a] | $k_{f,app}$ [a] | $k'_{f,app}$ [b] | Volume of water (mL) | $k_{r,app}$ [a] | $k_{f,app}$ [a] | $k'_{f,app}$ [b] |
| 1.3 | | | | | 50 | 0.033 | 0.0031 | 8.61 |
| 1.6 | 0.035 | 0.025 | 0.005 | 5.56 | 40 | 0.038 | 0.0033 | 7.33 |
| 2.2 | 0.0525 | 0.041 | 0.005 | 5.56 | 30 | 0.044 | 0.0041 | 6.83 |
| 2.8 | 0.07 | 0.055 | 0.005 | 5.56 | 20 | 0.055 | 0.0050 | 5.56 |

Data obtained from Wang and Wu;157 [a]$min^{-1}$, [b]$Lmin^{-1}meq\times10^{-4}$; Reaction conditions: 50 mL of chlorobenzene, 0.0059 moles of $(NPCl_2)_3$, 0.175 meq of catalyst, 20°C

## REFERENCES


1 A. Akelah and D. C. Sherrington, *Eur. Polym. J.*, **18**, 301 (1982).
2 B. C. Berris, **U. S. Patent 5,030,757** (1991).
3 G. E. Boyd and Q. V. Larson, *J. Am. Chem. Soc.*, **89**, 6038 (1967).
4 A. Brandstrom, *Adv. Phys. Org. Chem.*, **15**, 267 (1977).
5 L. M. Chen, Y. L. Chen and T. C. Chou, *Denki Kagaku*, **62**(12), 1173 (1994).
6 Y. L. Chen and T. C. Chou, *Ind. Eng. Chem. Res.*, **33**(3), 676 (1994).
7 Y. L. Chen and T. C. Chou, *J. Appl. Electrochem.*, **24**(5), 434 (1994).
8 T. Chiba and M. Okimoto, *J. Org. Chem.*, **56**(21), 6163 (1991).
9 M. S. Chiles, D. D. Jackson and P. C. Reeves, *J. Org. Chem.*, **187**, 357 (1980).
10 K. B. Chung and M. Tomoi, *J. Polym. Sci., Part A: Polym. Chem.*, **30**(6), 1089 (1992).
11 E. Cordoncillo, E., Monros, G., M. A. Tena, P. Escribano and J. Corda, *J. Non-Crystalline Solids*, **171**(2), 105 (1994).
12 B. P. Czech, M. J. Pugia and R. A. Bartsch, *Tetrahedron*, **41**, 5439 (1985).
13 W. H. Daly, J. D. Caldwell, V. P. Kien and R. Tang, *Polym. Prepr., Am. Chem. Soc., Div. Polym. Chem.*, **23**, 145 (1982).
14 O. I. Danilova and S. S. Yufit, *Mendeleev Commun.*, **4**, 165 (1993).
15 E. V. Dehmlow and J. Schmidt, *Tetrahedron Letters,* **2**, 95 (1976).
16 E. V. Dehmlow and S. S. Dehmlow, **Phase Transfer Catalysis**, *Verlag Chemie GmbH*, Weinheim (1980).
17 E. V. Dehmlow and S. S. Dehmlow, *Phase Transfer Catalysis*, Weinheim, Deerfield Beach, Florida; Basel, *Verlag Chemie*, (1983).
18 E. V. Dehmlow and T. Remmler, *J. Chem. Res.(S)*, 72 (1977).
19 E. V. Dehmlow, M. Slopianka and J. Heider, *Tetrahedron Letters*, **27**, 2361 (1977).
20 E. V. Dehmlow, R. Tiesen, Y. Sasson and E. Pross, *Tetrahedron*, **41**(14), 2927 (1985).
21 E. V. Dehmlow, *Tetrahedron Letters*, **9**, 1 (1976).
22 A. K. Dillow, S. L.J. Yun, D. Suleiman, D. L. Boatright, C. L. Liotta, and C. A. Eckert, *Ind. Eng. Chem. Res.*, **35**(6), 1801 (1996).
23 J. S. Do and T. C. Chou, *J. Appl. Electrochem.*, **19**, 922 (1989).
24 M. Ellwood and J. Griffiths, *J. Chem. Soc. Chem. Comm.*, 181 (1980).
25 T. L. Evans, **U. S. Patent 4,520,204** (1985).
26 W. K. Fife and X. Yue, *Bull. Chem. Soc. Jpn.*, **109**, 1278 (1987).
27 O. E. Filippova, I. N. Topchieva, V. V. Lutsenko and V. P. Zubov, *Vysolomol Soedin., Ser. A*, **26**, 402 (1984).
28 W. T. Ford and M. Tomoi, *Adv. Polym. Sci.*, **55**, 49 (1984).
29 H. H. Freedman, *Pure & Appl. Chem.*, **58**(6), 857 (1986).
30 A. Gobbi, D. Landini, A. Maia and S. Petricci, *J. Org. Chem.*, **63** (16), 5356 (1998).
31 G. W. Gokel, *ACS Ser.*, **326**, 24 (1987).
32 G. W. Gokel, D. M. Goli and R. A. Schultz, *J. Org. Chem.*, **48**, 2837 (1983).
33 H. P. Gregor and J. I. Bergaman, *J. Colloid. Sci.*, **6**, 323 (1951).
34 J. M. Harris and M. G. Case, *J. Org. Chem.*, **48**, 5390 (1983).
35 J. M. Harris, N. H. Hundley, T. G. Shannon and E. C. Struck, *J. Org. Chem.*, **47**, 4789 (1982).
36 A. W. Herriott, and D. Picker, *J. Am. Chem. Soc.*, **97**(9), 2345 (1975).
37 T. C. Huang and S. C. Lin, *J. Chin. Inst.Chem. Eng.*, **19**, 193 (1988).
38 T. Ido, M. Saiki and S. Goto, *Kagaku, Kagaku Ronbunshu*, **14**, 539 (1988).
39 T. Ido, M. Saiki and S. Goto, *Kagaku, Kagaku Ronbunshu*, **15**, 403 (1989).
40 T. Ido, Y. Kitamura and S. Goto, *Kagaku, Kogaku Ronbunshu*, **16**, 388 (1990).
41 Y. Imai, *J. Macromol. Sci., Chem.*, **A15**(5), 833 (1981).
42 H. Iwamoto, *Tetrahedron Letters*, **24**, 4703 (1983).
43 H. Iwamoto, H. Kobayashi, P. Murer, T. Sonoda and H. Zollinger, *Bull. Chem. Soc. Jpn.*, **66**(9), 2590 (1993).
44 H. Iwamoto, M. Yoshimara and T. Sonodo, *Bull. Chem. Soc. Jpn.*, **56**, 791 (1983).
45 D. Jaeger, M. D. Ward and A. K. Dutta, *J. Org. Chem.*, **53**, 1577 (1988).
46 Y. S. Jane and J. S. Shih, *J. Mol. Catal.*, **89**(1-2), 29 (1994).
47 J. Jarrouse, C. R. Hebd. *Seances Acad. Sci.*, Ser. **C232**, 1424 (1951).
48 G. Jin, T. Ido and S. Goto, *J. Chem. Eng. Jpn.*, **31**(5), 741 (1998).
49 T. Kawai, **British Patent 2,219,292** (1989).
50 H. Kobayashi, and T. Sonoda, *Chem. Letters,* 1185 (1982).
51 Kohjin Co., Ltd., **Japanese Kokai Tokkyo Koho JP 58/147,402** (1983).
52 T. N. Konstantinova and V. B. Bojinov, *Dyes and Pigments*, **39**(2), 69 (1998).

53 J. Kuji and T. Okano, *J. Syn. Org. Chem. Jpn.*, **52**(4), 276 (1994).
54 C. S. Kuo and J. J. Jwo, *J. Org. Chem.*, **57**, 1991 (1992).
55 D. Landini, A. Maia and F. Montanari, *J. Am. Chem. Soc.*, **100**, 2796 (1978).
56 M. D. Lang, G. Zhang, X. F. Chen, N. Feng and S. H. Li, *J. Appl. Sci.*, **70**(8), 1427 (1998).
57 H. Ledon, *Synthesis*, 347 (1974).
58 D. G. Lee and V. S. Chang, *J. Org. Chem.*, **43**(8), 1532 (1978).
59 D. J. Liaw and P. Chang, *J. Appl. Polym. Sci.*, **63**(2), 195 (1997).
60 C. L. Liotta, **Application of Macrocyclic Polydentate Ligands to Synthetic Transformations, in Synthetic Multidentate Compounds**, R. M. Izatt, J. J. Christensen, eds., *Academic Press,* New York, 111 (1978).
61 C. L. Liotta, E. M. Burgess, C. C. Ray, E. D. Black and B. E. Fair, *ACS Symp. Ser.*, **326**, 15 (1987).
62 C. L. Liotta,, E. M. Burgess and E. D. Black, *Polym. Prepr., Am. Chem. Soc., Div. Polym. Chem.*, **31**, 65 (1990).
63 B. L. Liu, A Study of Catalyzed Reaction of Synthesizing of 2-Mercaptobenzimidazole and Its Derivatives", Ph.D. Thesis, Dept. of Chem. Eng., Nat'l Tsing Hua Univ., Hsinchu, Taiwan (1995).
64 E. Maccarone, G. Perrini and M. Torre, *Gazzetta Chimica Italiana*, 112 (1982).
65 M. Makosza and E. Bialecka, *Tetrahedron Letters*, **2**, 1983 (1977).
66 M. Makosza, *Pure Appl. Chem.*, **43**, 439 (1975).
67 M. Makosza, *Russ. Chem. Rev.*, **46**, 1151 (1977).
68 C. A. Marques, M. Selva and P. Tundo, *J. Chem. Soc., Perkin Trans. I*, 529 (1993).
69 K. Mathias and R. A. Vaidya, *Bull. Chem. Soc. Jpn.*, **108**, 1093 (1986).
70 E. Mieczynski, A. M. Trzeciak, R. Grzybek and J. J. Ziolkowski, *J. Mol. Catal., A: Chemical*, **132** (2-3), 203 (1998).
71 H. Molinari, F. Montanari, S. Quici and P. Tundo, *J. Am. Chem. Soc.*, **101**, 3920 (1979).
72 E. Monflier, S. Tilloy, G. Fremy, Y. Castanet and A. Mortreux, *Tetrahedron Letters*, **36**(52), 9481 (1995).
73 F. Montanari, S. Quici and P. Tundo, *J. Org. Chem.*, **48**, 199 (1983).
74 F. Montanari, D. Landini, A. Maia, S. Quici and P. L. Anelli, *ACS Symp. Ser.*, **326**, 54 (1987).
75 P. W. Morgan, **Condensation Polymers by Interfacial and Solution Methods**, p. 349, *Interscience*, New York, (1965).
76 R. T. Morrison and R. N. Boyd, **Organic Chemistry**, p. 702, 5th ed., *Allynand Bacon, Inc.*, Mass, USA (1987).
77 K. Nakamura, S., Nishiyama, S. Tsuruya and M. Masai, *J. Mol. Catal.*, **93**(2), 195 (1994).
78 H. Namba, N. Takahashi, K. Abe and M. Saito, Japanese Patent, **Jpn. Kokai Tokkyo Koho 63/196547** (1988).
79 R. Neumann and Y. Sasson, *J. Org. Chem.*, **49**, 3448 (1984).
80 R. Neumann, S. Dermeik and Y. Sasson, *Israel J. Chem.*, **26**, 239 (1985).
81 T. Nishikubo and T. Iziawa, *J. Syn. Org. Chem. Jpn.*, **51**(2), 157 (1993).
82 N. Ohtani, *Yuki Gosei Kagaku Kyokaishi*, **43**, 313 (1985).
83 N. Ohtani, *Kagaku Kogyo*, **39**, 331 (1988).
84 N. Ohtani, C. A. Wilkie, A. Nigam and S. C. Regen, *Macromolecules*, **14**, 516 (1981).
85 N. Ohtani, Y. Inoue, A. Nomoto and S. Ohta, *React. Polym.*, **24**(1), 73 (1994).
86 N. Ohtani, Y. Inoue, H. Mizuoka and K. Itoh, *J. Polymer. Sci., Part A- Polym. Chem.*, **32**(13), 2589 (1994).
87 J. Pielichowski and P. Czub, *Angewandte Makromolekulare Chemie*, **251**, 1 (1997).
88 S. L. Regen, *J. Org. Chem.*, **42**, 875 (1977).
89 S. L. Regen, *J. Am. Chem. Soc.*, **97**, 5956 (1975).
90 S. L. Regen, J. C. K. Heh and J. McLick, *J. Org. Chem.*, **44**, 1961(1979).
91 S. L. Regen, J. J. Bese and J. McLick, *J. Am. Chem. Soc.*, **101**, 116 (1979).
92 S. L. Regen and J. J. Besse, *J. Am. Chem. Soc.*, **101**, 4059 (1979).
93 G. Rothenberg, L. Feldberg, H. Wiener and Y. Sasson, *J. Chem. Soc., Perkin Trans. 2*, **11**, 2429 (1998).
94 Y. Sasson and R. Neumann, **Handbook of Phase Transfer Catalysis**, *Blackie Academic & Professional, Chapman & Hall*, New York (1997).
95 Y. Sasson and S. Zbaida, **Ger. Patent DE 3,307,164** (1983).
96 E. G. See, C. A. Buehler and D. E. Pearson, **Survey of Organic Synthesis**, p. 285, *Wiley Interscience*, New York (1970).
97 F. Sevc, J. Kahovec and J. Hradil, *Chem. Listy*, **81**, 183 (1987).
98 Y. Shan, R. Kang and W. Li, *Ind. Eng. Chem. Res.*, **28**, 1289 (1989).
99 S. S. Shavanon, G. A. Tolstikov and G. A. Viktorov, *Zh. Obsch. Khim.*, **59**, 1615 (1989).
100 S. S. Shavanov, G. A. Tolstikov, T. V. Shutenkova, *Zh. Obshch. Khim.*, **57**, 1587 (1987).

101 D. C. Sherrington, K. J. Kelly, *Org. Coat. Appl. Polym. Sci., Proc.*, **46**, 278 (1981).
102 S. Shimizu, Y. Sasaki and C. Hirai, *Bull. Chem. Soc. Jpn.*, **63**, 176 (1990).
103 K. Shinoda, and K. Yasuda, *J. Org. Chem.*, **56**(12), 4081 (1991).
104 K. Sjoberg, *Aldrichimica Acta*, **13**, 55 (1980).
105 M. B. Smith, *Synth. Commun.*, **16**, 85 (1986).
106 M. Stark, Y. Q. Wang, O. Danielsson, H. Jornvall and J. Johansson, *Anal. Biochem.*, **265**(1), 97 (1998).
107 C. M. Starks and C. L. Liotta, **Phase Transfer Catalysis, Principles and Techniques**, *Academic Press*, New York (1978).
108 C. M. Starks, *J. Am. Chem. Soc.*, **93**, 195 (1971).
109 C. M. Starks, *ACS Symposium Series*, No. **326**, 1 (1987).
110 C. M. Starks, C. L. Liotta and M. Halpern, **Phase Transfer Catalysis, Fundamentals, Applications and Indutrial Perspectives**, *Chapman and Hall*, New York (1994).
111 C. M. Starks and R. M. Owens, *J.Am. Chem. Soc.,* **93**, 3615 (1973).
112 G. T. Szabo, K. Aranyosi and L. Toke, *Acta Chim. Acad. Sci., Hung.*, **110**, 215 (1982).
113 J. Y. Tarng and J. S. Shih, *J. Chin. Chem. Soc.*, **41**(1), 81 (1994).
114 L. Toke and G. T. Sazbo, *Acta chem. Acad. Sci., Hung.*, **93**, 421 (1977), as summarized by G. E. Totten and N. A. Clinton, *Rev. Macromol. Chem. Phys.*, **C28**, 293 (1988).
115 L. Toke, G. T. Szabo and K. Somogyi-Werner, *Acta Chim. Acad. Sci., Hung.*, **101**, 47 (1979).
116 M. Tomoi and W. T. Ford, *J. Am. Chem. Soc.*, **103**, 3821 (1981).
117 M. Tomoi, E. Nakamura, Y. Hosokawa and H. Kakuuichi, *J. Polym. Sci., Polym. Chem. Ed.*, **23**, 49 (1985).
118 M. Tomoi, E. Ogawa, Y. Hosokama and H. Kakiuchi, *J. Polym. Chem., Polym. Chem.*, **20**, 3421 (1982).
119 A. W. Trochimczuk, A. W. and B. N. Kolarz, *Eur. Polym. J.*, **28**(12), 1593 (1992).
120 M. L. Tsai and T. C. Chou, *J. Chin. Inst. Chem. Eng.*, **27**(6), 411 (1996).
121 J. Uglestad, T. Ellingsen and A. Beige, *Acta Chem. Scand.*, **20**, 1593 (1966).
122 R. Vladea and T. Simandan, *Rev. Chim.* (Bucharest), **41**, 421 (1990).
123 D. H. Wang and H. S. Weng, *J. Chin. Inst. Chem. Eng.*, **27**(3), 129 (1996).
124 D. H. Wang and H. S. Weng, *J. Chin. Inst. Chem. Eng.*, **26**(3), 147 (1995).
125 D. H. Wang and H. S. Weng, *Chem. Eng. Sci.*, **50**(21), 3477 (1995).
126 D. H. Wang and H. S. Weng, *Chem. Eng. Sci.*, **43**, 2019 (1988).
127 M. L. Wang and A. H. Liu and J. J. Jwo, *Ind. Eng. Chem. Res.*, **27**, 555 (1988).
128 M. L. Wang and B. L. Liu, *J.Mol. Catal. A: Chemical*, **105**, 49 (1996).
129 M. L. Wang, M. L. and B. L. Liu, *Int. J. Chem. Kinetics*, **28**, 885 (1996).
130 M. L. Wang and B. L. Liu, *Ind. Eng. Chem. Res.*, **34**(11), 3688 (1995).
131 M. L. Wang, M and C. C. Yu, *J. Polymer Sc., Part A: Polym. Chem.*, **31**, 1171 (1993).
132 M. L. Wang and C. C. Yu, *J. Chin. Inst. Eng.*, **23**(3), 153 (1992).
133 M. L. Wang, M. L. and C. Z. Peng, *J. Appl. Polym. Sci.*, **52**, 701 (1994).
134 M. L. Wang and H. M. Yang, *Ind. Eng. Chem. Res.*, **31**(8), 1868 (1992).
135 M. L. Wang and H. M. Yang, *J. Mol. Catal.*, **62**, 135 (1990).
136 M. L. Wang and H. S. Wu, *Ind. Eng. Chem. Res.*, **31**, 490 (1992).
137 M. L. Wang and H. S. Wu, *Ind. Eng. Chem. Res.*, **31**(9), 2238 (1992).
138 M. L. Wang and H. S. Wu, *Chem. Eng. Sci.*, **46**, 509 (1991).
139 M. L. Wang and H. S. Wu, J. *Chem. Soc., Perkin Trans.,* **2**, 841 (1991).
140 M. L. Wang and H. S. Wu, *J. Org. Chem.*, **55**, 2344 (1990).
141 M. L. Wang and K. R. Chang, *Ind. Eng. Chem. Res.*, **29**, 40 (1990).
142 M. L. Wang and K. R. Chang, *J. Chin. Inst. Eng.*, **16**(5), 675 (1993).
143 M. L. Wang and K. R. Chang, *J. Mol. Catal.*, **67**, 147 (1991).
144 M. L. Wang, M. L. and S. W. Chang, *Chem. Eng. Commum.*, **146**, 85 (1996).
145 M. L. Wang and S. W. Chang, *Bull. Chem. Soc. Jpn.*, **66**, 2149 (1993).
146 M. L. Wang, M. L. and S. W. Chang, *Ind. Eng. Chem. Res.*, **34**, 3696 (1995).
147 M. L. Wang and Y. M. Hsieh, *Develop. In Chem. Eng. and Min. Proc.*, **1**(4), 225 (1993).
148 M. L. Wang, M. L., C. C. Ou and J. J. Jwo, *Ind. Eng. Chem. Res.*, **33**, 2034 (1994).
149 M. L. Wang, M. L., C. C. Ou and J. J. Jwo, *J. Mol. Catal., A: Chemical*, **99**, 153 (1995).
150 M. L. Wang, M. L., C. C. Ou and J. J. Jwo, *Bull. Chem. Soc. Jpn.*, **67**, 2249 (1994).
151 Y. M. Wang, Z. P. Zhang, Z. Wang, J. B. Meng and P. Hodge, *Chin. J. Polym. Sci.*, **16**(4), 356 (1998).
152 W. P. Weber and G. W. Gokel, **Phase Transfer Catalysis in Organic Synthesis**, *Springer Verlag*: Berlin, Heideberg, New York (1977).
153 Y. L. Wen, M. Y. Yeh, Y. S. Lee and Y. P. Shih, *J. Chin. Inst. of Chem. Eng.*, **27**(6), 427 (1996).
154 H. S. Weng and D. H. Wang, *J. Chin. Inst. of Chem. Eng.*, **27**(6), 419 (1996).

155 C. R. White **U. S. Patent 4,642,399** (1987).
156 J. Wild and N. Goetz, **Ger. Patent DE 3,820,979** (1989).
157 H. S. Wu, Study on the Displacement Reaction of Phosphazene with Trifluoroethanol by Phase-Transfer Catalysis, Ph.D. Thesis, Department of Chemical Engineering, National Tsing Hua University, Hsinchu, Taiwan (1990).
158 H. S. Wu, Fang, T. R., S. S. Meng and K. H. Hu, J. of Mol. Cata., A: *Chemical*, **136**(2), 135 (1998).
159 M. Yamada, Y. Watabe, T. Sakakibara and R. Sudoh, *J. Chem. Soc. Chem. Commun*., 179 (1979).
160 B. Yan, *Combinatorial Chemistry & High Throughput Screening*, **1**(4), 215 (1998).
161 S. S. Yufit, *Russian Chemical Bulletin*, **44**(11), 1989 (1995).

## 13.4 EFFECT OF POLYMERIZATION SOLVENT ON THE CHEMICAL STRUCTURE AND CURING OF AROMATIC POLY(AMIDEIMIDE)

Norio Tsubokawa
**Faculty of Engineering, Niigata University, Niigata, Japan**

### 13.4.1 INTRODUCTION

Aromatic poly(amide-imide) (PAI) has an outstanding resistance not only to the thermal operations but also mechanical, electrical, and chemical operations. Although the properties of PAI are inferior to those of aromatic polyimide (PI), which is one of the most heat-resistant polymers, PAI has been widely utilized as a high performance heat resistant polymer as well as PI, because PAI is superior to PI in its workability in industry.[1,2]

In general, PAI is prepared by the following two processes: diamine process and diisocyanate process. Diisocyanate process is achieved by the direct polycondensation of trimellitic anhydride (TMAH) with aromatic diisocyanate, such as 4,4'-diphenylmethane diisocyanate, in a polar solvent such as N-methyl-2-pyrrolidone (NMP) as shown in Eq. [13.4.1].[1-3] On the other hand, the diamine process is achieved by a two-step reaction:[1]

(1) the polymerization (polycondensation and polyaddition) of trimellitic anhydride chloride (TMAH-CI) with aromatic diamine, such as 4,4'diaminodiphenylether (DDE) or 4,4'diaminodiphenylmethane (DDM) in a polar solvent such as N,N-dimethylacetamide (DMAc), to give poly(amic acid-amide) (PAAA) and

(2) the imidation of PAAA by heating as shown in Eq. [13.4.2].

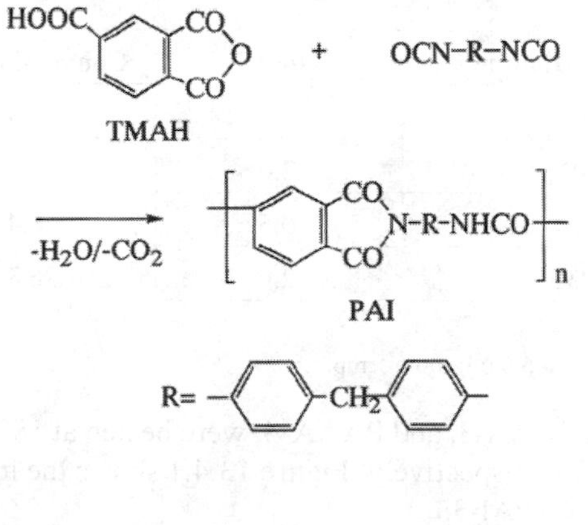


[13.4.1]

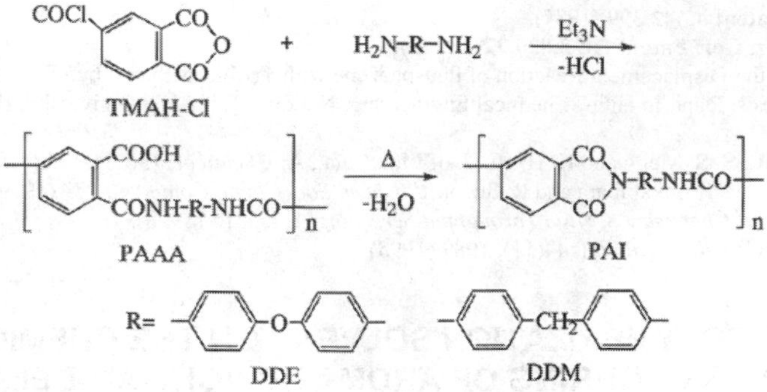


[13.4.2]

PAI is supposed to be a linear polymer containing equivalent amounts of amide and imide bonds. Therefore, PAI is soluble in polar solvents such as NMP, DMAc, and N,N-dimethylformamide (DMF). However, it is well known that PAI becomes insoluble and infusible when it is heated over 200°C. We have pointed out that some remaining carboxyl groups of PAI play an important role in the curing of PAI by heating.[4,5] Because of such a thermosetting property, PAI solution is used as a heat resistant temperature coating or an insulating enamel of magnet wire.

In the following section, the effect of polymerization solvent, DMAc and mixed solvent of methyl ethyl ketone (MEK) with water,[6] on the chemical structure and curing of PAIs prepared by diamine process will be summarized.[7]

## 13.4.2 EFFECT OF SOLVENT ON THE CHEMICAL STRUCTURE OF PAI

### 13.4.2.1 Imide and amide bond content of PAI

Four kinds of PAAA, as precursor of PAI, were prepared by the polycondensation of TMAH-Cl with DDE or DDM in DMAc and MEK/water mixed solvent (MEK containing 30 vol% of water) at room temperature. The results are shown in Table 13.4.1. It was found that the conversions reached 78-96% within 2 h at room temperature and the rate of the polycondensation of TMAHCl with diamine in MEK/water mixed solvent was much larger than that in DMAc.

**Table 13.4.1. Polymerization conditions and conversion of samples [Data from reference 7]**

| Sample No. | Solvent | Diamine | Time, h | Conversion, % |
|---|---|---|---|---|
| PAAA-1 | MEK/$H_2O$ | DDE | 0.5 | 90.3 |
| PAAA-2 | DMAc | DDE | 2.0 | 89.6 |
| PAAA-3 | MEK/$H_2O$ | DDM | 0.5 | 78.4 |
| PAAA-4 | DMAc | DDM | 2.0 | 96.8 |

TMAH-Cl=diamine=0.06 mol ; solvent, 172 ml; TEA, 5.0 ml; room temp.

These PAAAs, PAAA-1, PAAA-2, PAAA-3, and PAAA-4, were heated at 180°C for 2 h to give PAI-1, PAI-2, PAI-3, and PAI-4, respectively. Figure 13.4.1 shows the infrared spectra of (A) PAAA-1, (B) PAI-1, and (C) PAI-3.

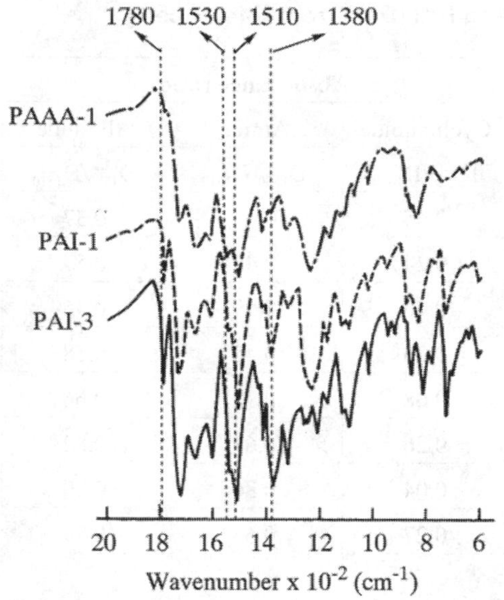

Figure 13.4.1 IR spectra of PAAA-1, PAI-1, and PAI-3 [Data from reference 7]

IR spectra of PAAA-1 show absorptions at 1530 and 1660 $cm^{-1}$, which are characteristic of amide bond, 1510 and 1590 $cm^{-1}$ which are characteristic of benzene nuclei, and 1230 $cm^{-1}$, which is characteristic of ether bond, but the absorption at 1380 and 1780 $cm^{-1}$, which are characteristic of imide bond, are hardly observed. On the other hand, IR spectra of PAI-1 and PAI-3, obtained by heating of PAAA-1 and PAAA-3, respectively, show new absorptions at 1380 and 1780 $cm^{-1}$.

Among the adsorptions of imide bond, the absorption at 1380 $cm^{-1}$ is assigned to C-N stretching vibrations of all imide bond (cyclic and acyclic imide bond as shown in Eq. [ 13.4.3 ] and 1780 $cm^{-1}$ is assigned to C=O stretching vibrations of five-member imide rings (cyclic imide bond). The absorption at 1530 $cm^{-1}$ is assigned to N-H stretching vibrations of amide bond. Furthermore, the absorption at 1510 $cm^{-1}$ is assigned to benzene nuclei that is stable to heat treatment.

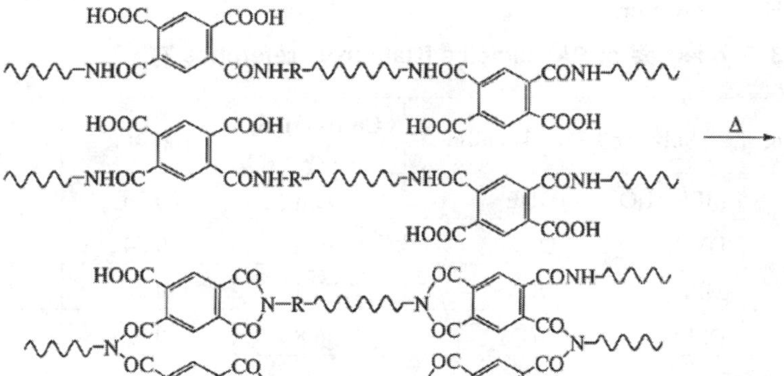

[13.4.3]

Therefore, the content of cyclic imide, amide and all imide bond of PAI was estimated by the absorbance ratio, $D_{1380}/D_{1510}$ (the absorbance ratio of absorbance at 1380 $cm^{-1}$ to that of benzene nuclei), $D_{1780}/D_{1510}$, and $D_{1530}/D_{1510}$, respectively.

Table 13.4.2 shows the absorbance ratio of cyclic imide, amide, and all imide bond before and after heat treatment of PAAAs. The considerable increase of cyclic and all imide bond content of PAAAs and the decrease of amide bond were observed by heating. But the effect of solvent on the imidation of PAAAs was hardly observed.

**Table 13.4.2. IR absorbance ratio of PAAA and PAI [Data from reference 7]**

| Sample No. | Heat treatment | | Absorbance ratio | | |
|---|---|---|---|---|---|
| | Temperature °C | Time h | Cyclic imide $D_{1780}/D_{1510}$ | Amide $D_{1530}/D_{1510}$ | All imide $D_{1380}/D_{1510}$ |
| PAAA-1 | | | 0.05 | 0.68 | 0.37 |
| PAI-1 | 180 | 2 | 0.24 | 0.43 | 0.57 |
| PAAA-2 | | | 0.10 | 0.59 | 0.36 |
| PAI-2 | 180 | 2 | 0.26 | 0.45 | 0.58 |
| PAAA-3 | | | 0.08 | 0.98 | 0.66 |
| PAI-3 | 180 | 2 | 0.26 | 0.69 | 0.82 |
| PAAA-4 | | | 0.04 | 0.86 | 0.29 |
| PAI-4 | 180 | 2 | 0.27 | 0.63 | 0.79 |

### 13.4.2.2 Intrinsic viscosity and carboxyl group content

Table 13.4.3 shows the effect of solvent on the intrinsic viscosity, [η], and carboxyl group content of PAIs. The carboxyl content was determined by potentiometric titration. It is interesting to note that intrinsic viscosity of PAI from PAAA prepared in DMAc is larger than that in MEK/water mixed solvent. This suggests that the polymerization degree decreases with decreasing activity of TMAH-Cl in MEK/water mixed solvent because of the hydrolysis of TMAH-Cl by water.

**Table 13.4.3. Properties of PAI samples [Data from reference 7]**

| Sample No. | Solvent | Diamine | Conversion, % | [η], dl/g[a] | COOH, eq/g[b] |
|---|---|---|---|---|---|
| PAI-1 | MEK/$H_2O$ | DDE | 90.3 | 0.36 | 214 |
| PAI-2 | DMAc | DDE | 89.6 | 0.54 | 44 |
| PAI-3 | MEK/$H_2O$ | DDM | 78.4 | 0.26 | 328 |
| PAI-4 | DMAc | DDM | 96.8 | 0.32 | 73 |

[a]Solvent, NMP ; 30.0°C. [b]Determined by potentiometric titration

In addition, the content of carboxyl groups in PAIs prepared in MEK/water mixed solvent was considerably larger than that prepared in DMAc. This also suggests the hydrolysis of TMAH-Cl in MEK/water mixed solvent.

## 13.4.3 EFFECT OF SOLVENT ON THE CURING OF PAI BY HEAT TREATMENT

### 13.4.3.1 Chemical structure of PAI after heat treatment

Table 13.4.4 shows the effect of solvent on the change of chemical structure and formation of insoluble part of PAI-1 and PAI-2 after heat treatment. It became apparent that the content of cyclic imide bond and all imide bonds of PAI-1 and PAI-2 further increased by heating of the corresponding PAAAs at 180°C for 2 h followed by heat treatment at 260°C for 2 h. The content of amide bond decreased by heating of PAAA at 180°C for 2 h, but increased by post-heating at 260°C for 2 h.

**Table 13.4.4. IR absorbance ratio and insoluble part of samples after heat treatment in air [Data from reference 7]**

| Sample No. | Heat treatment | | Absorbance ratio | | | Insoluble part % |
|---|---|---|---|---|---|---|
| | Temperature °C | Time h | Cyclic imide | Amide | All imide | |
| | | | $D_{1780}/D_{1510}$ | $D_{1530}/D_{1510}$ | $D_{1380}/D_{1510}$ | |
| PAAA-1 | | | 0.05 | 0.68 | 0.37 | 0 |
| PAI-1 | 180 | 2 | 0.24 | 0.43 | 0.57 | 0 |
| PAI-1 | 260 | 2 | 0.33 | 0.56 | 0.71 | 76.7 |
| PAAA-2 | | | 0.10 | 0.59 | 0.36 | 0 |
| PAI-2 | 180 | 2 | 0.26 | 0.45 | 0.58 | 0 |
| PAI-2 | 260 | 2 | 0.31 | 0.50 | 0.64 | 16.8 |

### 13.4.3.2 Curing PAI by post-heating

The formation of the insoluble part in NMP was observed and the amount of insoluble part formed by heating of PAI-1 (obtained in MEK/water mixed solvent) was larger than that by heating of PAI-2 (obtained in DMAc).

The increase of imide and amide bond content by post-heating at 260°C is considered as follows: the imidation of amic acid structure may be proceeded by both intermolecular and intramolecular imidation. The latter produces cyclic imide bond, but the former produces acyclic imide bond to give crosslinking material as shown in Eq. [13.4.3].

Since PAIs obtained by heating of PAAAs at 180°C are completely soluble in NMP, the intramolecular imidation preferentially proceeds at 180°C, but intermolecular imidation scarcely proceeds.

On the other hand, by post-heating at 260°C, the crosslinking reaction proceeds by the intermolecular imidation of terminal carboxyl groups of PAI with remaining amic-acid structure (Eq. [13.4.4]) and the amide bond in main chain of PAI (Eq. [13.4.5]) to give the insoluble part in NMP. The reaction induced the increase of imide bond content of PAI after post-heating. The increase of amide bond after post-heating may be due to the formation of crosslinking structure by the amidation of terminal amino groups of PAI with carboxyl groups of PAI in main chain (Eq. [13.4.5]-[13.4.8]).

Therefore, PAI obtained from the heating of PAAA obtained in MEK/water mixed solvent, which has many carboxyl groups, produces more insoluble part by post-heating.

Figure 13.4.2 shows the effect of heating temperature on the curing of PAI obtained by heating of PAAA at 180°C in MEK/water mixed solvent and DMAc. As shown in Figure 13.4.2, by post-heating at 280°C for 2 h in air, insoluble part in NMP reached 100%, indicating the almost complete curing of PAIs. It is interesting to note that PAIs formed by heating of PAAA obtained in MEK/water mixed solvent (PAI-1 and PAI-3) were found to be cured

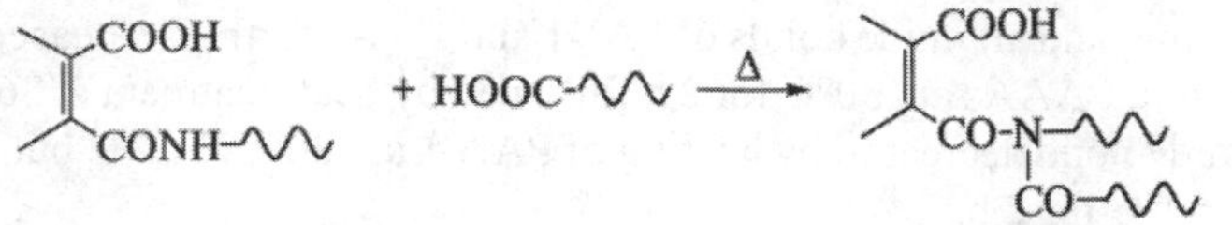

[13.4.4]

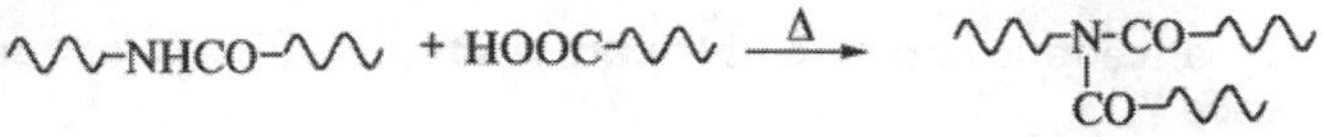

[13.4.5]

[13.4.6]

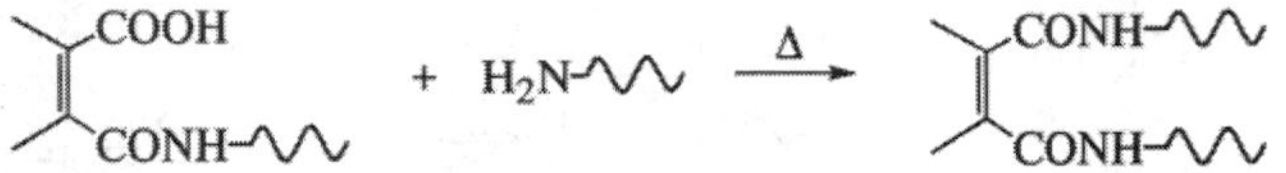

[13.4.7]

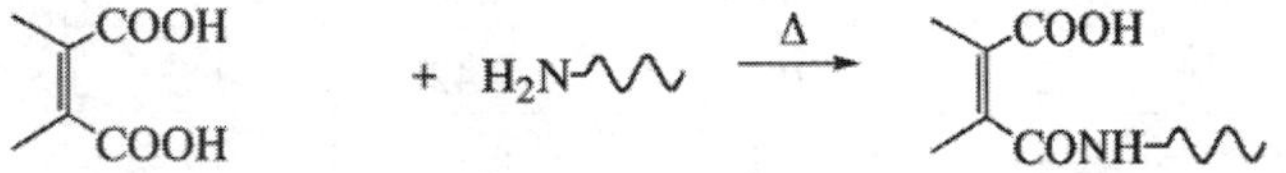

[13.4.8]

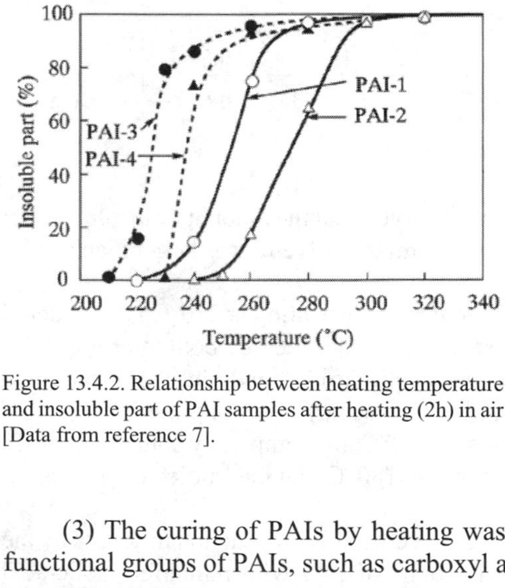

Figure 13.4.2. Relationship between heating temperature and insoluble part of PAI samples after heating (2h) in air [Data from reference 7].

more easily than in DMAc (PAI-2 and PAI-4). This is due to the fact that PAI-1 and PAI-3 contains more carboxyl groups than PAI-2 and PAI-4.

## 13.4.4 CONCLUSIONS

The effects of polymerization solvent on the curing of PAIs were investigated and the following results were obtained:

(1) The carboxyl group content of PAIs from PAAAs prepared in MEK/$H_2O$ mixed solvent was larger than that in DMAc.

(2) The curing of PAIs prepared in MEK/$H_2O$ proceeded easier than those in DMAc.

(3) The curing of PAIs by heating was due to the intermolecular reaction between functional groups of PAIs, such as carboxyl and amino groups.

## REFERENCES


1 M.Fujita and T.Fujita, *Rubber Digest*, **18** (1), 21 (1966).
2 S.Terney, J.Keating, J.Zielinski, J.Hakala, and H.Sheffer, *J. Polym. Sci., A-1*, **8**, 683 (1970).
3 Y.Imai and T.Takahasi, *Kobunshi Kagaku*, **29**, 182 (1972).
4 N.Tsubokawa, I.Yamamoto, and Y.Sone, *Kobunshi Ronbunshu*, **44**, 389 (1987).
5 N.Tsubokawa, I.Yamamoto, and Y.Sone, *Kobunshi Ronbunshu*, **44**, 831 (1987).
6 H.Uchiyama and Y.Imai, *Kobunshi Kagaku*, **28**, 73 (1971).
7 N.Tsubokawa, M.Murata, and Y.Sone, *Netsu Kokasei Jushi*, **12**, 12 (1991).

# Solvent Use in Various Industries

Attempts to reduce solvent use in the production of various materials require background information on the current inventory, the reasons for selecting certain solvents, the effect of various solvents on the properties of final products, future trends and the possibilities for solvent replacement.

Information on solvent use investigates these areas. This chapter is possible only because of a thorough evaluation by large groups of scientists and engineers assembled by US Environmental Protection Agency. This has produced Compliance Sector Notebooks which contain invaluable information on solvent use. Full documents can be found on the EPA website at http://es.epa.gov/oeca/sector/index.html. These are useful in the investigation of a particular industry. Similar data from other countries were not available but open literature and patents have been included to help the reader to understand changes occurring in other countries.

## 14.1 ADHESIVES AND SEALANTS

George Wypych
**ChemTec Laboratories, Inc., Toronto, Canada**

Adhesives and sealants are manufactured from a variety of polymers. Their selection and their combinations used impact solvent selection. Most solvent systems are designed to optimize the solubility of the primary polymer. Adhesives can be divided into ones which bond by chemical reaction and ones which bond due to physical processes.[1] Chemically reactive adhesives are further divided into three more categories for those that bond through polymerization, polyaddition, or polycondensation. Physically bonding adhesives include pressure sensitive and contact adhesives, melt, or solution adhesives, and plastisols. Polymerization adhesives are composed of cyanoacrylates (no solvents), anaerobic adhesives (do not contain solvents but require primers for plastics and some metals which are solutions of copper naphthenate),[2] UV-curable adhesives (solvent-free compositions of polyurethanes and epoxy), rubber modified adhesives (variety solvents discussed below).

Polyaddition adhesives include epoxy and polyurethane polymers which can either be 100% solids, water-based, reactive or non-reactive hot melts or contain solvents mostly to regulate viscosity. Typical solvents include methyl ethyl ketone, acetone, mineral spirits, toluene, and xylene.[3] Polycondensation adhesives include phenol-formaldehyde resin, polyamides, polyesters, silicones and polyimides. With the exception of polyesters (which require ethanol and N-methylpyrrolidone as solvents) and polyimides (which require methyl amyl ketone, butyl acetate, methyl ethyl ketone, 2-ethoxyhexyl acetate as solvents), these adhesives can be made without solvents.

Pressure sensitive and contact adhesives are made from a variety of polymers including acrylic acid esters, polyisobutylene, polyesters, polychloroprene, polyurethane, silicone, styrene-butadiene copolymer and natural rubber. With the exception of acrylic acid ester adhesives which can be processed as solutions, emulsions, UV curable 100% solids and silicones (which may contain only traces of solvents), all remaining rubbers are primarily formulated with substantial amounts of solvents such as hydrocarbon solvents (mainly heptane, hexane, naphtha), ketones (mainly acetone and methyl ethyl ketone), and aromatic solvents (mainly toluene and xylene).

Melt adhesives and plastisols do not contain solvents. The solution adhesives group includes products made from the following polymer-solvent systems: nitrocellulose (typical solvents include solvent combinations usually of a ketone or an ester, an alcohol and a hydrocarbon selected from isopropanol, 2-butylhexanol, amyl acetate, acetone, methyl ethyl ketone), nitrile rubber (main solvent - methyl ethyl ketone), polychloroprene (which is usually dissolved in a mixture of solvents including a ketone or an ester, an aromatic and aliphatic hydrocarbon selected from naphtha, hexane, acetone, methyl ethyl ketone, benzene, toluene), and polyvinyl acetate (water).

In addition to the solvents used in adhesives, solvents are needed for surface preparation[4] and primers. Their composition may vary and is usually designed for a particular substrate, often using fast evaporating solvents and environmentally unfriendly materials with significant adverse health effects.

Detailed data on the total amount of solvents used by adhesive industry could not be found. The adhesive manufacturing industry continues to grow at a very fast pace. Total adhesive production, according to Frost & Sullivan, was $18.25 billion in 1996 and this is expected to grow to $26.2 billion in 2003.[5] Solvent-based materials in 1995 constituted 13% of total production in North America, 14% in Europe, 15% in Japan, and 25% in the Far East.[6] Many industries which use solvent-based adhesives have moved to South America and Asia where regulations restricting emissions are less severe.[5] The shoe industry is now concentrated in South America. There are many initiatives to decrease solvent emissions. For example, World Bank's assistance program for developing countries focuses on this issue.[5] But in spite of the fact that solvent-based adhesives lost some of their markets (3.3% during the period of 1994-1996),[5] they still hold 14-15% of the European market.[6] It is estimated that the use of solvents contributes 24% of all VOC emissions. According to one source adhesives were responsible for a 6% share in these emissions in 1993.[7] Another source[8] blames adhesives for 7% of total VOC emissions in Germany in 1995.

Sealants are divided into groups according to the generic names of polymer base. The main groups include: polyurethanes, silicones, acrylics, polysulfide and others (PVC, polybutylene, styrene-butadiene-styrene copolymers, polychloroprene, and several others). The amount of solvent used in sealants is controlled by the standards which previously divided sealants into two groups: these below 10% VOC and those above. Recently, a provision was made to include water-based acrylics and the limit of VOC for class A sealants was increased to 20%. Polyurethane sealants and structural adhesives can be made without solvent (the first solvent-free polyurethane sealant was made in 1994).[9] Solvents are added to reduce sealant viscosity and to aid in the manufacture of polymer. Typical solvents used are mineral spirits, toluene, and xylene. A small amount of solvent is emitted from curatives which contain methyl ethyl ketone. Most formulations of silicone sealants do not contain solvents. In some sealants, traces of benzene and toluene can be found.

Acrylic sealants are water-based but they may also contain ethylene and propylene glycols, mineral spirits and mineral oil. There are also solvent-based acrylic sealants which contain substantial amounts of solvents such as mineral spirits, toluene and xylene. Polysulfide sealants usually contain toluene but methyl ethyl ketone is also used. The group of class B sealants contains substantially more solvents (up to 40% by volume) but there are some exceptions. PVC sealants are based on plastisols and they can be made without solvents. Butyl rubber based sealants usually contain hydrocarbons ($C_6$-$C_{12}$). Styrene-butadiene-styrene based sealants usually have a large amount of solvents selected from a group including toluene, heptane, hexane, methyl ethyl ketone, isobutyl isobutyrate, n-amyl acetate, n-amyl ketone. They are usually processed in solvent mixtures. Polychloroprene is usually dissolved in a mixture of solvents including ketones or esters, and aromatic and aliphatic hydrocarbons. The list includes naphtha, hexane, acetone, methyl ethyl ketone, benzene, and toluene.

The world market of sealants was estimated in 1996 at $2 billion and was expected to grow in 2003 to $2.75 billion with an annual growth rate of 4.5% which is slightly lower than that expected for adhesives (5.3%).[5]

The changing trends are clearly visible when developments in technology are studied but many barriers to reductions in solvent use exist such as the high investment required, longer processing time, frequently higher material cost of adhesives, and the psychological barrier of changing established adhesive practices. In many instances, adhesive performance is predicted by its superficial characteristics such as strong smell which might suggests that the material has superior properties, its initial green strength which for many indicates good bonding properties, and high viscosity often related to good processing characteristic.[10] Since the alternative materials may not have much odor, or require of longer time to reach strength and have a low viscosity, users are suspicious that their potential performance may be inferior. The following information reviews some recent findings which may contribute to future changes.

In the shoe industry, a major breakthrough occurred in 1928 when polychloroprene was first introduced.[1,10] The first, simple formulation is still manufactured and is used worldwide because the glue can be easily prepared by simply making a solution of the polymer. This gives a product with good adhesion to various substrates. Many new products are available today as potential replacements. Hot melt adhesives can be used in some applications but they still require solvents for cleaning, degreasing, and swelling. Also, their bond strength is frequently inadequate. Reactive systems are not yet used in the shoe industry but reactive hot melts are finding applications. Their broader use is hampered by their sensitivity to moisture which requires special equipment and special care.[10] Water-based adhesives are the most likely replacement product. They also need special equipment for processing because of the high heat of evaporation of water (although water based adhesives contain 50% polymer compared with 15-20% in solvent based adhesives).[10] Two sport shoe manufacturers, Nike and Reebok, already use this technology.

Traditional polychloroprene adhesives can be modified in several ways to be useful in water-based systems. Figure 14.1.1 shows peel strength of several adhesives. The solvent based adhesive (A) has excellent properties both in terms of green strength and bond strength. A simple emulsion of polychloroprene (B) has relatively good ultimate strength

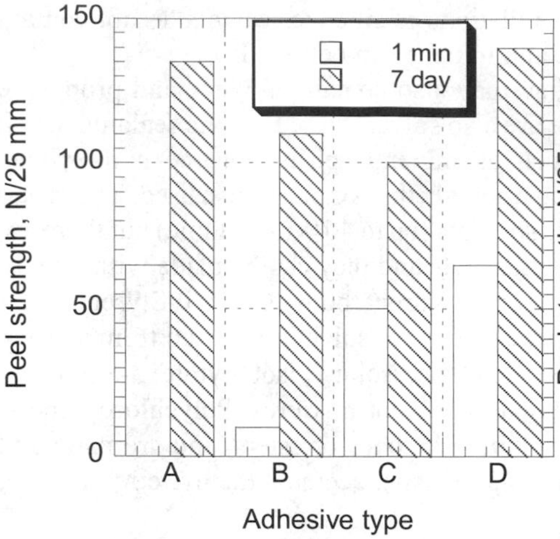


Figure 14.1.1. Green strength and adhesion of several adhesives. Symbols are explained in the text. [Data from B Archer, *International J. Adhesion & Adhesives*, **18**, No.1, 15-8 (1998).]

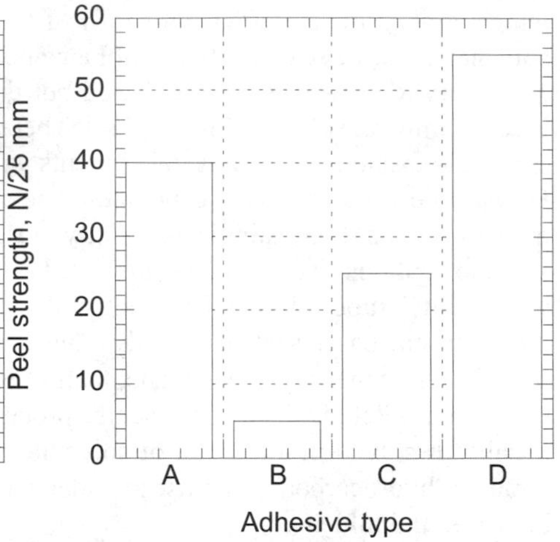


Figure 14.1.2. Contactability of various adhesives. Symbols are explained in the text. [Data from B Archer, *International J. Adhesion & Adhesives*, **18**, No.1, 15-8 (1998).

but lacks green strength and therefore does not meet the performance requirements of shoe manufacturers and other industries.[11] Adhesive (C) is a blend of polychloroprene with polyurethane in a water-based system. This modification gives both green strength and peel strength but Figure 14.1.2 shows that peel strength of a freshly applied adhesive is lower than that of solvent based polychloroprene which may cause problems in holding both adhering surfaces together. The adhesive (D) was developed in an interesting new process which involves the emulsification of a solvent based adhesive obtained from styrene-isoprene-styrene rubber.[11] After emulsification, the solvent is stripped under vacuum to produce a solvent-free adhesive. The reason for emulsification of the complete adhesive as opposed to emulsification of the rubber alone is to produce a homogeneous adhesive system which would otherwise suffer from separation of polymer particles surrounded by a layer of emulsifying agents. Figures 14.1.1 and 14.1.2 show that this system is superior to solvent-based adhesives. Adhesives can be further improved by polymer blending and by adhesive foaming. A foamed adhesive layer requires less material (approximately 4 times less than a conventional adhesive), it requires less drying time (less water and faster evaporation), and provides an improved bond strength. A foamed adhesive has a larger surface area which increases the surface area of contact with the substrate. The compressed rubber foam has a higher tear strength than unfoamed film of the same thickness.[11]

Also, regulations are helping to reduce the solvent content of adhesives.[12,13] The use of chlorinated solvents, frequently used in primers and for cleaning, has been discontinued based on the Montreal Protocol. From June 1998, the production of a pair of shoes in Europe should not involve the use of more than 20 g of solvent. This is only partially successful solution since shoe production is expected to move out of developed countries to less restrictive jurisdictions. Solvent Emission Directive will continue to restrict solvent use in Europe.

Many changes have occurred and more are expected in adhesives based on thermoplastic polyurethanes (TPU).[3,14,15]In the last 30 years, TPU based adhesives were manufactured with solvents for shoes, food packaging, and textile and plastic film lamination. Current technologies use TPU in the form of a hot melt, as a reactive PUR and as a thermoplastic laminating film.[3] Reactive hot-melts were first introduced in the early 1980s and since than have grown very rapidly. After application the adhesive is cured by moisture.[14] These adhesives are already in use by the automotive industry (bonding carpet to door panels, tray assembly, lenses to headlamp housing, and lamination of foam to fabric) and in furniture and building products (moldings, picture frames, decorative foil, edgebanding), in bookbinding, and in the footwear industry. Polyurethane water dispersions are expected to grow 8-10%/year from the current 5,000-6,000 tones/year market in Europe.[14] Applications are similar to those of hot melts.

UV-curable pressure-sensitive adhesives are the most recent application of the advancing radiation curing technology.[16] Low viscosity formulations allow the use of standard application techniques with several advantages such as improved production rate, energy efficiency, improved properties of the final products, and new potential applications for pressure-sensitive additives in thicker films with mechanical performance. It is expected that radiation cured materials will expand at a rate of 10%/year.[17] Adhesives constitute 16% by value and 13% by volume of radiation cured products (two major applications for radiation cured materials are coatings and inks). Henkel introduced a series of water-based laminating adhesives.[18] Hot melt systems, high-solids solvent systems with a 3 times higher solids content, and water based adhesives have been introduced to textile lamination to replace traditional low-solids solvent-based adhesives.[19]

Odor elimination is the additional benefit which has helped to drive the replacement of solvent-based systems.[20] In packaging materials, most odors are related to the solvents used in inks, coatings and adhesives. Also, coalescing solvents from water-based systems caused odors. Elimination of solvent is a priority but solvent replacement may also change the response to the odor because solvents such as toluene and xylene smell like lubricating oils or turpentine whereas isopropanol smells more like a disinfectant. Odors stem not only from solvents but also from products of the thermal and UV degradation of other components and solvents.[20]

In view of the above efforts, it is surprising that the majority of recent patents on adhesives are for solvent-based systems.[21-26] The new inventions include a universal primer,[21] an adhesive composition in which solvents have been selected based on Snyder's polarity (only solvents which belong to group III are useful in adhesive for automotive applications to avoid a deleterious effect on paint),[22] a low VOC adhesive for pipes and fittings,[23] a solvent-containing heat-resistant adhesive based on siloxane polyimide,[24] a water-based polyimide adhesive,[25] and two-component solvent-free polyurethane adhesive system for use in automotive door paneling.[26]

## REFERENCES


1 R Vabrik, G Lepenye, I Tury, I Rusznak, A Vig, *International Polym. Sci. Technol.*, **25**, No.3, T/1-9 (1998).
2 D Raftery, M R Smyth, R G Leonard, *International J. Adhesion Adhesives*, **17**, No.4, 349-52 (1997).
3 J B Samms, L Johnson, *J. Adhesive Sealant Coun.*, Spring 1998. Conference proceedings, *Adhesive & Sealant Council*, Orlando, Fl., 22nd-25th March 1998, p. 87.
4 A Stevenson, D Del Vechio, N Heiburg, Simpson R, *Eur. Rubber J.*, **181**, 1, 2829, (1999).
5 K Menzefricke, *Adhesive Technol.*, **15**, No.1, 6-7 (1998).
6 L White, *Eur. Rubber J.*, **179**, No.4, 24-5 (1997).
7 J Baker, *Eur. Chem. News*, **69**, No.1830, 20-2 (1998).

8 P Enenkel, H Bankowsky, M Lokai, K Menzel, W Reich, *Pitture e Vernici*, **75**, No.2, 23-31 (1999).
9 J. van Heumen, H Khalil, W Majewski, G. Nickel, G Wypych, **US Patent 5,288,797**, Tremco, Ltd., 1994.
10 S Albus, *Adhesive Technol.*, **16**, No.1, 30-1 (1999).
11 B Archer, *International J. Adhesion & Adhesives*, **18**, No.1, 15-8 (1998).
12 J C Cardinal, *Pitture e Vernici*, **74**, No.6, 38-9 (1998).
13 *Pitture e Vernici*, **73**, No.20, 39-40 (1997).
14 M Moss, *Pigment Resin Technol.*, **26**, No.5, 296-9 (1997).
15 Hughes F, TAPPI 1997 Hot Melt Symposium. Conference Proceedings. *TAPPI*. Hilton Head, SC, 15th-18th June 1997, p.15-21.
16 D Skinner, *Adhesive Technol.*, **15**, No.3, 22-4 (1998).
17 B Gain, *Chem. Week*, **160**, No.14, 28-30 (1998).
18 G Henke, *Eur. Adhesives & Sealants*, **14**, No.1, 18-9 (1997).
19 G Bolte, *J. Coated Fabrics*, **27**, 282-8 (1998).
20 R M Podhajny, *Paper, Film & Foil Converter*, **72**, No.12, 24 (1998).
21 M Levy, **US Patent 5,284,510**, Paris Laque Service, 1994.
22 I R Owen, **US Patent 5,464,888**, 3M, 1995.
23 C D Congelio, A M Olah, **US Patent 5,859,103**, BFGoodrich, 1999.
24 D Zhao, H Sakuyama, T Tomoko, L-C Chang, J-T Lin, **US Patent 5,859,181**, Nippon Mektron Ltd., 1999.
25 H Ariga, N Futaesaku, H Baba, **US Patent 5,663,265**, Maruzen Petrochemical Co. Ltd., 1997.
26 E Konig, U F Gronemeier, D Wegener, **US Patent 5,672,229**, Bayer AG, 1997.

## 14.2 AEROSPACE

GEORGE WYPYCH
**ChemTec Laboratories, Inc., Toronto, Canada**

Figure 14.2.1 shows a schematic diagram of the aerospace manufacturing process.[1] Metal finishing is the process in which most solvents and solvent containing materials are used. The main function of the metal finishing process is corrosion protection which requires proper cleaning, surface preparation, and the selection of suitable coatings.

The functions of coatings used in aircrafts are different from those used in ordinary coating applications therefore an extrapolation of the progress made with solvent replacement in other coating types is not justified. The typical flight conditions of operating altitude (about 10,000 m above the earth), speed (most frequently 900 km/h), temperature (very low in space at about -60°C and substantially higher after landing up to 80°C), humidity (low in space and high at earth level are combined with condensation due to the tempera-

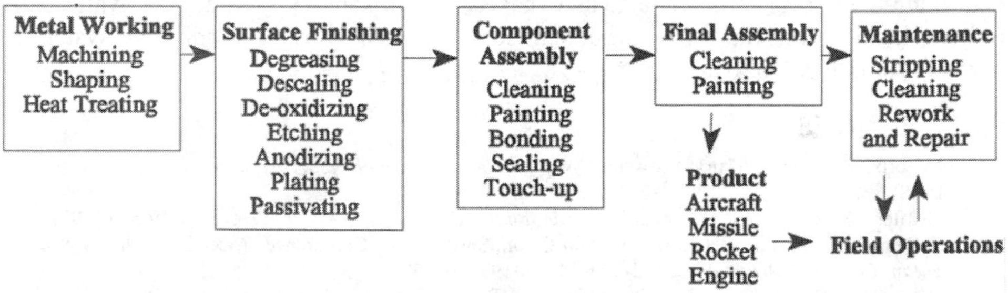


Figure 14.2.1. Schematic diagram of aerospace manufacturing process. [Reproduced from Profile of the Aerospace Industry. EPA Office of Compliance Sector Notebook Project. US Environmental Protection Agency. November 1998.]

**Table 14.2.1. Total releases of solvents by the aerospace industry. [Data from ref. 1]**

| Solvent | Amount, kg/year | Solvent | Amount, kg/year |
|---|---|---|---|
| benzene | 62,000 | methyl ethyl ketone | 995,000 |
| bromotrifluoromethane | 750 | methyl isobutyl ketone | 47,000 |
| n-butyl alcohol | 7,000 | methyl tetr-butyl ether | 550 |
| sec-butyl alcohol | 10,000 | tetrachloroethylene | 285,000 |
| cyclohexane | 400 | 1,1,1-trichloroethane | 781,000 |
| 1,2-dichlorobenzene | 600 | trichloroethylene | 429,000 |
| 1,1-dichloro-1-fluoroethanc | 10,000 | trichlorofluoromethane | 1,800 |
| dichloromethane | 314,000 | toluene | 414,000 |
| isopropyl alcohol | 1,000 | xylene | 103,000 |
| methanol | 21,000 | | |

ture difference), UV radiation (substantially higher during flight), mechanical abrasion due to the high speed of travel, exposure to salt in the atmosphere, exposure to higher level of acids and sulfur dioxide, and exposure to de-icing fluids during winter.[2,3] These unusual conditions should be considered in conjunction with the mechanical movement of the coating caused by rapid changes in temperature and the flexing of aircraft elements because of changes in pressure and severe load variations on wings.[2] In addition, because of their size, aircrafts must often be painted at low temperatures which requires a coating that will cure at these temperatures without leaving entrapped volatiles. These could evaporate in thc low pressure conditions at high altitude and cause the formation of voids whcrc corrosion could start. These factors make the design of an effective coating system a severe technological challenge.

Coatings are used by the aerospace industry both for OEM and maintenance purposes. In each case surface cleaning and preparation is required. A paint stripping operation is added to the task in maintenance repainting. Coatings are applied by spraying, brushing, rolling, flow coating, and dipping. Depending on the method of application, the rheological properties of coatings must be adjusted with solvents and, in some cases, with water. An alternative method of viscosity adjustment involves heating the coatings to lower its viscosity by increasing its temperature. This reduces solvent usage. Solvents are also used for equipment cleaning.

In addition to paints, sealants are also used. Sealants are mostly based on polysulfides, containing solvents as discussed in the previous section. Also, non-structural adhesives containing solvents are used as gaskets around windows and for carpeting.

Paint removal is accomplished by either chemical or blast depainting. Dichloromethane is the most common solvent used for this application. Aerospace industry estimates that 15,000 to 30,000 different materials are used for manufacturing some of which are potentially toxic, volatile, flammable, and contain chlorofluorocarbons. Some of these substances may result in air emissions, waste-waters, and solid waste.

**Table 14.2.2. Total transfers of solvents by the aerospace industry. [Data from ref. 1]**

| Solvent | Amount, kg/year | Solvent | Amount, kg/year |
|---|---|---|---|
| n-butyl alcohol | 2,600 | methyl ethyl ketone | 500,000 |
| cyclohexane | 18,000 | methyl isobutyl ketone | 17,000 |
| 1,2-dichlorobenzene | 4,000 | tetrachloroethylene | 110,000 |
| 1,1-dichloro-1-fluoroethane | 230 | 1,1,1-trichloroethane | 133,000 |
| dichloromethane | 68,000 | trichloroethylene | 98,000 |
| N,N-dimethylformamide | 500 | trichlorofluoromethane | 3,800 |
| ethylene glycol | 14,000 | toluene | 87,000 |
| methanol | 12,000 | xylene | 12,000 |

Air emissions result from sealing, painting, depainting, bonding, as well as from leakage in storage, mixing, drying, and cleaning. The most common solvents involved in coatings are trichloroethylene, 1,1,1-trichloroethane, toluene, xylene, methyl ethyl ketone, and methyl isobutyl ketone. Wastewater is generated through contamination by paints and solvents used for cleaning operations. Solid waste containing solvents comes from paint overspray intercepted by emission control devices, depainting, cleanup, and disposal of unused paint. Solvents used for cleaning are usually a mixture of dimethyl-benzene, acetone, 4-methyl-2-pentanone, butyl ester of acetic acid, naphtha, ethyl benzene, 2-butanone, toluene and 1-butanol. Some solvents used for painting and cleaning are either recycled are burned to recover energy.

In 1996, 199 aerospace facilities (out of 1885 analyzed in the report) released and transferred off-site or discharged to sewers about 12,000 kg of 65 toxic chemicals (solvents in these releases are reported in Table 14.2.1 and transfers in Table 14.2.2).

Methyl ethyl ketone, 1,1,1-trichloroethane, trichloroethylene, and toluene accounted for 66% of all releases. 70% of all transfers was for recycling purposes. The aerospace industry released 10,804 tons of VOC in 1997 which constituted 0.61% of the total releases from 29 industries which were analyzed. Thirteen other industries release more VOC than the aerospace industry.

Recycling and disposal of solvents in the aerospace industry equals the purchase cost of the solvents. Therefore reduction of solvent use is very cost effective. Some chemical stripping operations are now being replaced by cryogenic stripping with liquid nitrogen. Also, supercritical carbon dioxide has been used in Hughes Aircraft Company in some cleaning applications. Solvent emissions can be reduced through control of evaporation (lids, chillers), by dedicating process equipment (reduces cleaning frequency), production scheduling, immediate cleaning of equipment, better operating procedures, reuse of solvent waste, and use of optimized equipment for paint application. There are plans to evaluate powder coatings and water-based paints.[1,2] There are trials to use water as the paint thinner and to lower viscosity of paints by application of resins which have lower viscosity.[2] Work is under the way to replace dichloromethane/phenol stripper with benzyl alcohol.[2] The introduction of an intermediate layer between the primer and the top coat has been proposed.

This will aid the stripping action of the proposed stripping solvent, benzyl alcohol. VOC have already been reduced in several components: bonding primer (from 1030 to 850 g/l), undercoats (from 670 to 350), top coats (from 700-900 to 250-800), clear coats (from 700-800 to 250-520), surface cleaners (from 850 to 250) as well as other materials.[3]

## REFERENCES


1 Profile of the Aerospace Industry. EPA Office of Compliance Sector Notebook Project. US Environmental Protection Agency. November 1998.
2 R W Blackford, *Surface Coatings International*, **80**, No.12, 564-7 (1997).
3 R Blackford, *Polym. Paint Colour J.*, **186**, No.4377, 22-4 (1996).


## 14.3 ASPHALT COMPOUNDING

George Wypych
**ChemTec Laboratories, Inc., Toronto, Canada**

Numerous construction products are formulated from asphalt and coal tar for such applications as driveway sealers, cutback asphalts, flashing cements, concrete primers, concrete cold mixes, roof cements, expansion joint fillers, patch liquids, waterproofing liquid-applied membranes, and pipeline coatings. All these products are likely to contain solvents.

The simplest formulations are mixtures of asphalt and (usually) mineral spirits used for sealing , priming, and coating of concrete. These are usually very low performance products which are used in large quantities because of their low price. They release about 40% of their weight to atmosphere during and after application. Since they do not perform well they have to be re-applied at frequent intervals. Driveway sealer is an example of a product which is used every spring, in spite of the fact that, in addition to the pollution it causes, it also produces a gradual degradation and cracking of the driveway. The only solution for elimination of this unnecessary pollution seems to be banning the product by regulation. Some of these products can be replaced by asphalt emulsions which contain water in place of organic solvents.

Several products are used for patching and joint filling purposes. These materials (flashing cement, roof cement, patch liquid, and expansion joint filler) also use solvents to regulate viscosity. The solvents are usually mineral spirits, fuel oil, or polycyclic aromatic hydrocarbons. In addition to the base components, inexpensive fillers such as calcium carbonate or limestone but also still asbestos are added. These products harden on evaporation of the solvent and fill the joints, adhere to surfaces, and provide some waterproofing. These are again, low technology materials, traditionally used because of their very low cost. Most of these products can be replaced by modern sealants which will result in higher initial cost but longer service.

The most technologically advanced products are used for waterproofing and pipeline coatings. These products are also based on dispersion of asphalt in the above mentioned solvents but reinforced with addition of polymer. The addition of polymer modifies the plastic behavior of asphalt and renders it elastomeric. Additional solvents are usually added to improve the solubility of polymeric components. Reactive polyurethanes are the most frequently used modifiers for waterproofing liquid membranes. Toluene and xylene are the most frequently used additional solvents. These materials partially solidify because of evaporation of the solvent. Their elastomeric properties are derived from chain extension and crosslinking reactions which form an internal polymeric network which reinforces asphalt.

There is no data available on the solvent emissions from these materials but their scale of production suggests that their emissions are probably comparable with the entire rubber industry. This is one industry which should be closely monitored not only because of the emission of the above listed solvents but because some of the low grade solvents used contain large quantities of benzene and hexane. It is also cause for concern that asphalt and tar have carcinogenic components.

Recent inventions[1-5] are driven by product improvement needs and environmental aspects of application of these products. Janoski's patent[1] describes a product which is an anhydrous blend of polymer and asphalt and is substantially solvent-free. This technology shows that it is possible for an ingenious designer to produce low viscosity materials without using solvents but by selecting the appropriate type and concentration of bituminous materials, polyurethane components, and plasticizers.

In another invention,[2] a modifier is introduced to increase the adhesion of asphalt/water emulsions to aggregates. Emulsified asphalt is not so deleterious to the environment but its performance suffers from aggregate delamination. In yet another recent invention,[3] terpene solvent, which is a naturally occurring (but never in this high concentration), biodegradable material, was used to replace the mineral spirits, xylene, trichloroethane, toluene, or methyl ethyl ketone normally used in cutback formulations (cutback asphalt is a dispersion of asphalt in a suitable solvent to reduce viscosity and allow for cold application). The two other patents[4,5] discuss inventions leading to an improvement of high and low temperature properties of asphalt with no special impact on reduction of solvents used.

## REFERENCES

1 R J Janoski, **US Patent 5,319,008**, Tremco, Inc., 1994.
2 P Schilling, E Crews, **US Patent 5,772,749**, Westaco Corporation, 1998.
3 R W Paradise, **US Patent 5,362,316**, Imperbel America Corporation, 1994.
4 M P Doyle, J L Stevens, **US Patent 5,496,400**, Vinzoyl Petroleum Co., 1996.
5 M P Doyle, **US Patent 5,749,953**, Vinzoyl Technical Services, LLC, 1998.

# 14.4 BIOTECHNOLOGY

## 14.4.1 ORGANIC SOLVENTS IN MICROBIAL PRODUCTION PROCESSES

MICHIAKI MATSUMOTO, SONJA ISKEN, JAN A. M. DE BONT
**Division of Industrial Microbiology**
**Department of Food Technology and Nutritional Sciences**
**Wageningen University, Wageningen, The Netherlands**

### 14.4.1.1 Introduction

Solvents are not dominating compounds in the biosphere of our planet. Under natural conditions, their presence in appreciable amounts is restricted to specific areas. Only a very limited number of solvents is of biological origin and some may reach higher concentrations in nature. The best known example is ethanol. However, also butanol and acetone can be formed readily by microbes and locally high concentrations may occur. In fact, in the beginning of the 20th century, very large production facilities were in operation for the microbial production of butanol and acetone. Furthermore, terpenes are natural solvents that are produced mainly by plants and locally they can reach high concentrations. For instance, limonene is present in tiny droplets in the peel of oranges. All these solvents are toxic to microbial cells. Some others as higher hydrocarbons that are present for instance in olive oil, are not toxic to microbes as will be discussed later.

With the advent of the chemical industry, this picture has changed dramatically. In polluted locations, microorganisms may be confronted with a large number of solvents at high concentrations. With a few exceptions only, it has turned out that microbes can be found that are able to degrade these compounds if their concentration is low. This degradative potential is not unexpected in view of trace amounts that may be present locally in the natural biosphere. But the exposure of cells to unnatural high concentrations of these solvents usually leads to irreversible inactivation and finally to their death.

The chemical industry is largely based on solvent-based processes. But in biotechnological processes, the microbes usually are exploited in a water-based system. This approach is quite understandable in view of the preference of microbes for water and the problems solvents pose to whole cells. Solvents often are used to extract products from the aqueous phase but only after the production process has been completed. At this stage, damage to whole cells is obviously no longer relevant. In both, chemical industry and biotechnology, organic solvents have many advantages over water because of the nature of either product or substrate. Consequently, during the last decades many possibilities have been investigated to use solvents in biocatalytic processes.[1,2] The more simple the biocatalytic system, the less complex it is to use solvents.

Free or immobilized enzymes have been exploited already in a number of systems. Here, biocatalysis may take place in reversed micelles or in an aqueous phase in contact with an organic solvent.[3] In a powdered state some enzymes are able to function in pure organic solvents.[4] Furthermore, modified enzymes such as polymer bound enzymes[5] or surfactant-coated enzymes[6] have been developed so that they can solubilize in organic solvents to overcome diffusion limitation. The advantages of enzymatic reactions using organic solvents can be briefly summarized as follows:[1,3,4]

1) hydrophobic substances can be used;
2) synthetic reactions can take place;
3) substrate or production inhibition can be diminished and
4) bioproducts and biocatalysts can easily be recovered from the systems containing organic solvents.

Although organic solvents have often been used in enzymatic reactions, the application of organic solvents for biotransformation with whole-cell systems is still limited. Cells might be continuously in direct contact with the organic phase in a two-phase water-solvent system during the whole production cycle (Figure 14.4.1.1).[7] Alternatively, cells may remain separated from the bulk organic phase by using membrane bioreactors (Figure14.4.1.2).[8] In these instances, cells encounter phase toxicity[9] or molecular toxicity, respectively.

Because whole bacterial cells are more complex than enzymes, they pose by far greater problems in operating bioproduction processes when organic solvents are present. The most critical problem is the inherent toxicity of solvents to living organisms.[1,2,10,11] As

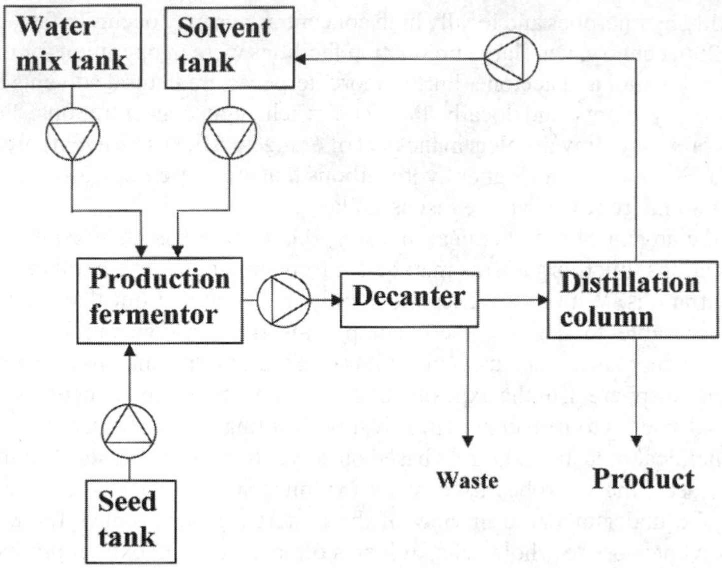


Fig.14.4.1.1. Schematic diagram of a two-phase bioreactor system for continuous 1-octanol production. [After reference 7]. 1-Octanol is produced from n-octane in hexadecene by *Pseudomonas oleovorans* or recombinant strains containing the alkane oxidation genes.

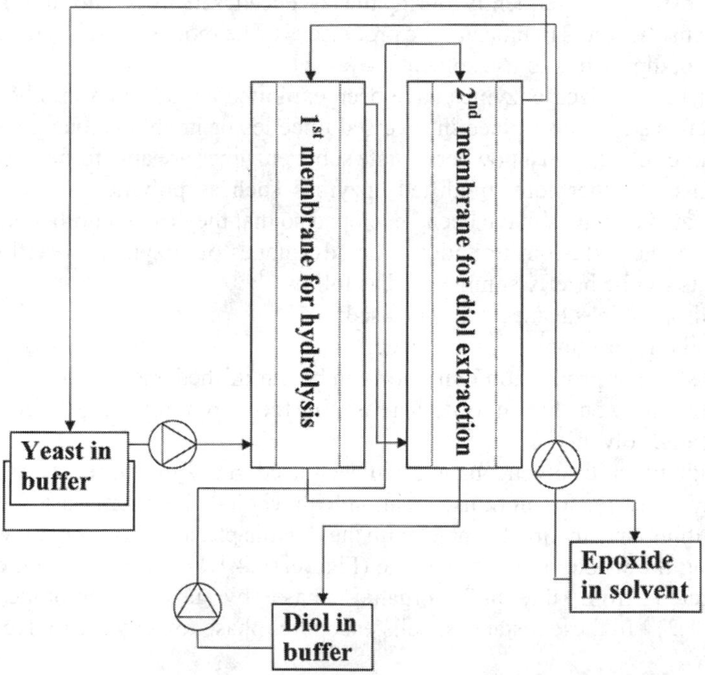


Fig.14.4.1.2. Schematic diagram of a two-phase hollow-fiber membrane bioreactor system for hydrolytic epoxide resolution. [After reference 8]. The yeast cells contain an epoxide hydrolase that enantioselectively hydrolyzes racemic epoxide resulting in enantiopure epoxide that partitions to the organic phase. Diol produced partitions to the water phase.

some solutions have been already found in this area, further progress is expected in the near future. A key problem is the selection of useful solvents in combination with a suitable microbe. Many solvents may be considered with their specific process properties, and many microorganisms may be considered also with their specific advantages or disadvantages. Only on the basis of a detailed understanding of the functioning of a microbial cell in the presence of solvents will it be possible to make a rational selection for a good combination solvent/organism. In the following section we will discuss these key questions. We will also describe recently found solvent-tolerant bacteria and will provide some examples of biotransformation using solvent-tolerant bacteria.

### 14.4.1.2 Toxicity of organic solvents

The toxicity of an organic solvent is closely related to its hydrophobicity, as expressed by the log $P_{O/W}$ values,[12-16] the logarithm of the partition coefficient of the solvent between 1-octanol and water (Table 14.4.1). In general, the Gram-negative bacteria show relatively higher solvent tolerances than the Gram-positive bacteria.[12,14] This may be caused by the difference in the composition of the cell envelope. Solvents with a high log $P_{O/W}$ resulted in the highest activities, but the range of log $P_{O/W}$ tolerated by an organism is dependent on the type of microorganism. In the following section we will describe what happens in the cell in the presence of solvents.

**Table 14.4.1. Relationship between the growth of a cell exposed to an organic solvent and the value of log $P_{O/W}$ of the solvent. [After reference 14]**

| Solvent | log $P_{O/W}$ | *Pseudomonas putida* IH-2000 | *Pseudomonas putida* IFO3738 | *Pseudomonas fluorescens* IFO3507 | *Escherichia coli* IFO3806 | *Achromobacter delicatulus* LAM1433 | *Alcaligenes faecalis* JCM1474 | *Agrobacterium tumefaciens* IFO3058 | *Bacillus subtilis* AHU1219 | *Saccharomyces uvarum* ATCC26602 |
|---|---|---|---|---|---|---|---|---|---|---|
| dodecane | 7.0 | + | + | + | + | + | + | + | + | + |
| decane | 6.0 | + | + | + | + | + | + | + | + | - |
| nonane | 5.5 | + | + | + | + | + | + | + | + | - |
| *n*-hexyl ether | 5.1 | + | + | + | + | + | + | + | + | - |
| octane | 4.9 | + | + | + | + | + | + | + | + | - |
| isooctane | 4.8 | + | + | + | + | + | + | + | - | - |
| cyclooctane | 4.5 | + | + | + | + | + | + | - | - | - |
| diphenyl ether | 4.2 | + | + | + | + | + | - | - | - | - |
| *n*-hexane | 3.9 | + | + | + | + | + | - | - | - | - |
| propylbenzene | 3.7 | + | + | + | + | - | - | - | - | - |
| *o*-dichlorobenzene | 3.5 | + | + | + | - | - | - | - | - | - |
| cyclohexane | 3.4 | + | + | + | - | - | - | - | - | - |
| ethylbenzene | 3.2 | + | + | - | - | - | - | - | - | - |

| Solvent | log $P_{O/W}$ | *Pseudomonas putida* IH-2000 | *Pseudomonas putida* IFO3738 | *Pseudomonas fluorescens* IFO3507 | *Escherichia coli* IFO3806 | *Achromobacter delicatulus* LAM1433 | *Alcaligenes faecalis* JCM1474 | *Agrobacterium tumefaciens* IFO3058 | *Bacillus subtilis* AHU1219 | *Saccharomyces uvarum* ATCC26602 |
|---|---|---|---|---|---|---|---|---|---|---|
| *p*-xylene | 3.1 | + | + | - | - | - | - | - | - | - |
| styrene | 2.9 | + | - | - | - | - | - | - | - | - |
| toluene | 2.6 | + | - | - | - | - | - | - | - | - |
| benzene | 2.1 | - | - | - | - | - | - | - | - | - |

+, growth; -, no growth

Bar[9] suggested that the toxicity in two-phase systems was caused by both the presence of a second phase (phase toxicity) and solvent molecules which dissolved in the aqueous phase (molecular toxicity). Basically, both mechanisms are governed by the same principle in that the solvent accumulates in the microbial membrane. In case of the direct contact between cells and pure solvent, the rate of entry of solvents in a membrane will be very high. If the solvent has to diffuse via the water phase, then the accumulation in membranes will be slower. This latter mechanism on the molecular toxicity has been investigated in more detail.[17] In experiments with liposomes from *E. coli*, and ten representative organic solvents labeled by $^{14}C$ under aqueous-saturating levels, it was observed that the solvents accumulate preferentially in the cell membrane. The partition coefficients (log $P_{M/B}$) of the solvents between the model liposome membrane and buffer correlate with those (log $P_{O/W}$) in a standard 1-octanol-water system:

$$\log P_{M/B} = 0.97 \times \log P_{O/W} - 0.64 \qquad [14.4.1.1]^{17}$$

The accumulation of an organic solvent in the membrane causes changes in the membrane structure. Organic solvents residing in the hydrophobic part of the membrane disturb the interactions between the acyl chains of the phospholipids. This leads to a modification of membrane fluidity which eventually results in the swelling of the bilayer.[10] In addition to this, conformations of the membrane-embedded proteins may be altered.[10] These changes in the integrity of the membrane also affect the membrane function.

The principal functions of the cytoplasmic membrane involve:
1) barrier function,
2) energy transduction and
3) formation of a matrix for proteins.

The disruption of lipid-lipid and lipid-protein interactions by the accumulation of organic solvents has a strong effect on the membrane's function as a selective barrier for ions and hydrophilic molecules. Permeability is of particular importance for protons because the leakage of protons directly affects the primary energy transducing properties of the membrane. The initial rates of proton influx in the absence and presence of different amounts of hydrocarbon were measured.[17] The permeability for protons increases with increasing amounts of hydrocarbon. Hence, leakage of protons occurs in the presence of organic sol-

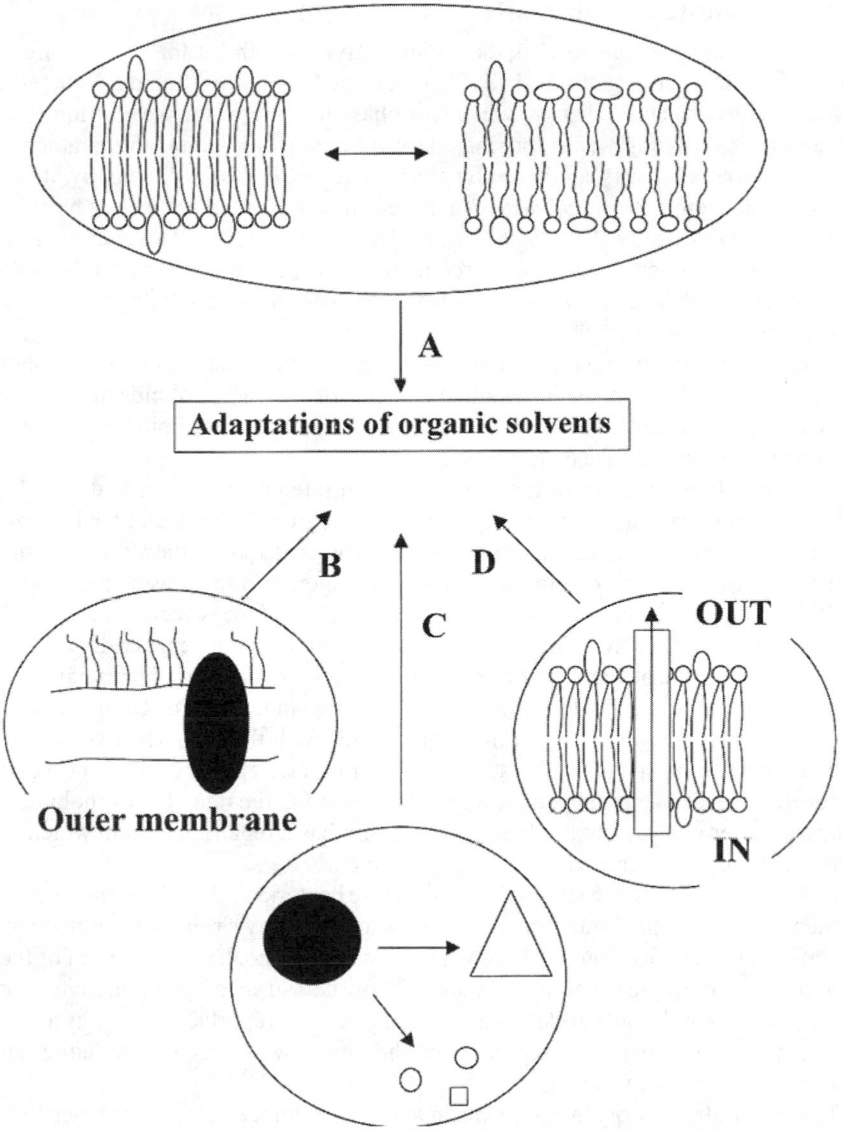


Figure 14.4.1.3 Schematic presentation of possible mechanisms of solvent tolerance. **A** Changes in the structure of cytoplasmic membrane. **B** Changes in the structure of outer membrane. **C** Transformation of the solvent. **D** Active export of the solvents [After reference 32]

vents.[17] Not only the impairment of the barrier function is caused by the alterations that occur in the membrane structure when it interacts with organic solvents. It is well known that the activities of the proteins embedded in the membrane are regulated by the membrane thickness, head group hydration, fluidity and fatty acid composition.[18,19] All these parameters are also known to be affected by the accumulation of organic solvents. The effects of solvents on these parameters were reviewed by Sikkema et al.[10]

### 14.4.1.3 Solvent-tolerant bacteria

As described in the previous section, the organic solvents with $1 < \log P_{O/W} < 4$ are considered to be toxic to microorganisms. In 1989, Inoue and Horikoshi[21] found a toluene-tolerant *Pseudomonas putida* strain that grew in a two-phase toluene-water system ($\log P_{O/W} = 2.5$ for toluene). This finding was surprising and went against the dominant paradigm at that time. Solvent tolerance was confirmed by other strains of *P. putida*[22-26] and by other representatives of the genus *Pseudomonas*.[27-30] Furthermore, solvent tolerance has been found in the strains of Gram-positive bacteria *Bacillus*[31,32] and *Rhodococcus*.[33] The key question now is: How do solvent-tolerant bacteria overcome the toxic effects of organic solvents? Some of the possible mechanisms involved in solvent tolerance according to various researchers are shown in Fig.14.4.1.3.[32]

Current research on changes in the structure of the cytoplasmic membrane shows the involvement of: 1) the composition of the fatty acids of the phospholipids like the *cis/trans* isomerization of unsaturated fatty acids; 2) composition of phospholipid headgroups and 3) rate of turnover of membrane components.

Organic solvents cause a shift in the ratio of saturated to unsaturated fatty acids.[34,35] In a solvent-tolerant strain, an increase in the saturation degree has been observed during adaptation to the presence of toluene. Solvent-tolerant strains also have the ability to synthesize *trans*-unsaturated fatty acids from the *cis*-form in response to the presence of organic solvents.[34,36-38] Increases in the saturation degree and the ratio of *trans*-form change the fluidity of the membrane and the swelling effects caused by solvents are depressed.

Alterations in the headgroups of lipids during the adaptation to solvents have also been observed in some solvent-tolerant strains.[37,39] The changes in the composition of the headgroups cause changes in the affinity of the lipids with the organic solvents and in the stability of membrane due to an alteration of bilayer surface charge density. These changes compensate the effect caused by the solvents. In one strain, the rate of phospholipid synthesis increases after exposure to a solvent.[40] This strain has a repairing system which is faster than the rate of damage caused by the organic solvent.

Unlike Gram-positive bacteria, Gram-negative bacteria such as *Pseudomonas* have an outer membrane. The outer membrane has been shown to play a role in the protection of the cell from solvent toxicity. Ions such as $Mg^{2+}$ or $Ca^{2+}$ stabilize the organization of the outer membrane and contribute to solvent tolerance.[38] Low cell surface hydrophobicity caused by changes in the lipopolysaccharide (LPS) content has been reported to serve as a defensive mechanism.[41,42] It has also been reported that the porins which are embedded in the outer membrane are relevant to solvent tolerance.[37,42-44]

The metabolism of organic solvents in solvent-tolerant strains contributes to solvent tolerance by degradation of the toxic compounds. This contribution, however, is considered to be limited[33,45] because many solvent-tolerant strains show non-specific tolerance against various organic compounds.

Non-specific tolerance to toxic compounds is well known in the field of antibiotic resistance. A wide range of structurally dissimilar antibiotics can be exported out of the cell by multidrug-efflux pumps. Could the export of organic solvents contribute to solvent tolerance?

Isken and de Bont[46] conducted experiments to determine whether the solvent tolerant *Pseudomonas putida* S12 was able to export toluene by monitoring the accumulation of $^{14}C$ labeled toluene in the cells. Toluene-adapted cells were able to export toluene from their

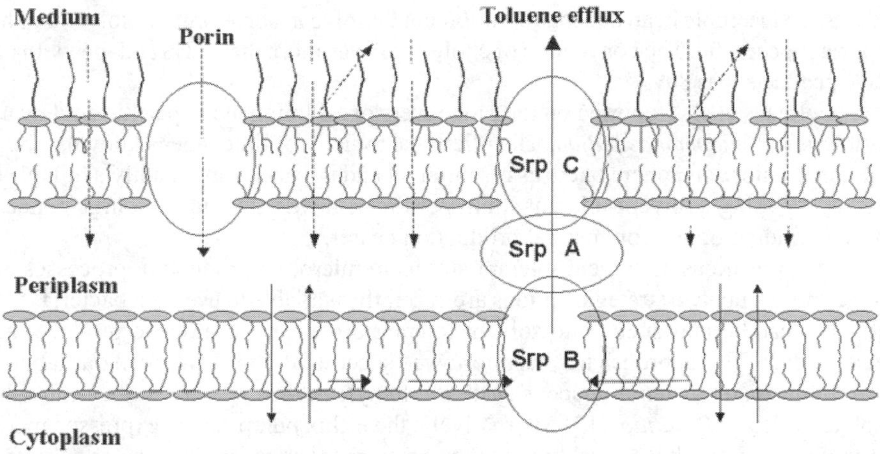


Figure 14.4.1.4. Schematic picture of toluene penetration and efflux in the solvent-tolerant *Pseudomonas putida* S12 [After reference 20].

membranes whereas the non-adapted cells were not. Furthermore, it was observed that in the presence of energy coupling inhibitors, toluene accumulation was the same as in the non-adapted cells. The amount of toluene in the cell was concluded to be kept at a low level by an active efflux system. The presence of a toluene-efflux system is supported by genetic research.[47,48] The pump has a striking resemblance to multidrug-efflux systems. Active efflux pumps for solvents have also been detected in other *Pseudomonas* strains.[25,26,30,37]

It is obvious that solvent tolerance is caused by a combination of the mechanisms described above. Figure 14.4.1.4 shows a schematic picture of toluene penetration and efflux in the solvent tolerant strain *P. putida* S12.[20] Toluene enters the cell through the outer membrane. At present, it is unclear whether toluene passes through porins or through the phospholipid part of the cell. The efflux pump recognizes and interacts with toluene in the cytoplasmic membrane. Toluene is then pumped into the extracellular medium.

### 14.4.1.4 Biotransformation using solvent-tolerant microorganisms

Many important fine chemicals, including catechols, phenols, aldehydes and ketones, low molecular epoxides and diepoxides, medium-chain alcohols, and terpenoids fall within the range of $1 < \log P_{O/W} < 4$. The discovery of solvent-tolerant bacteria leads to the new possibility of biocatalytic reaction systems containing organic solvents. By using solvent-tolerant bacteria, a variety of fine chemicals can be formed in microbial production processes.

The organic solvents used so far in published research had to be very hydrophobic ($\log P_{O/W} > 5$) in order to prevent microbial inactivation.[1] Consequently, many fine chemicals can not beneficially produced in the presence of such solvents because they simply would remain in the water phase and would not partition to the organic phase. The use of solvent-tolerant microorganisms enables the use of less hydrophobic solvents ($2.5 < \log P_{O/W} < 4$) and in such a system, chemicals with a $1 < \log P_{O/W} < 4$ preferentially will go into the organic phase. Up to now, however, only a few applications using solvent-tolerant microorganisms have been reported.

Aono et al.[49] reported the oxidative bioconversion of cholesterol as a model biocatalytic reaction using a solvent-tolerant *Pseudomonas* species. Cholesterol and its products are insoluble in an aqueous solution but dissolve in some organic solvents. The attempt was successful. The conversion of cholesterol was more than 98% and the yield of oxidative products was 80%.

Speelmans et al.[50] reported on the bioconversion of limonene to perillic acid by a solvent-tolerant *Pseudomonas putida*. The microbial toxicity of limonene is known to be very high. It is a major component of citrus essential oil and is a cheap and readily available base material. By using a solvent-tolerant strain perillic acid was obtained at a high concentration. This finding brings commercial production nearer.

The applications of solvent-tolerant strains in microbial production processes are at present limited, but two strategic options are currently available to use such bacteria.[20] Relevant genes can be introduced into solvent-tolerant organisms in order to produce the required product. This approach has been followed successfully by J. Wery in our laboratory who employed an 1-octanol-aqueous system. Methylcatechol was produced from toluene by solvent tolerant *P. putida* S12. Alternatively, the efflux pump can be expressed in a suitable solvent-sensitive host which would then be more tolerant for a particular solvent.

Other benefits may arrive from solvent-resistant bacteria. Ogino et al.[26] isolated *Pseudomonas aeruginosa* LST-03 which can grow in organic solvents with $\log P_{O/W} > 2.4$ and secrets organic solvent-stable lipolytic enzymes. They were able to purify an organic solvent-stable protease which was more stable than the commercially available proteases.[51] Hence, solvent-tolerant strains have become a source for new enzymes.[52]

In the near future, the use of solvent-tolerant strains will make the application of organic solvents in biotransformations by whole cells a more realistic option.

## References


1 R. León, P. Fernandes, H.M. Pinheiro, and J.M.S. Cabral, *Enz. Microb. Technol.*, **23**, 483 (1998).
2 B. Angelova and H.S. Schmauder, *J. Biotechnol.*, **67**, 13 (1999).
3 C. Laane, S. Boeren, K. Vos, and C. Veeger, *Biotechnol. Bioeng.*, **30**, 81 (1987).
4 A.M. Klibanov, *Trends Biotechnol.*, **15**, 97 (1997).
5 Y. Inada, A. Matsushima, M. Hiroto, H. Nishimura, and Y. Kodera, *Methods Enzymol.*, **242**, 65 (1994).
6 Y. Okahata, Y. Fujimoto, and K. Iijiro, *J. Org. Chem.*, **60**, 2244 (1995).
7 R.G. Mathys, A. Schmid, and B. Witholt, *Biotechnol. Bioeng.*, **64**, 459 (1999).
8 W.Y. Choi, C.Y. Choi, J.A.M. de Bont, and C.A.G.M. Weijers, *Appl. Microbiol. Biotechnol.*, **53**, 7 (1999).
9 R. Bar, *J. Chem. Technol. Biotechnol.*, **43**, 49 (1988).
10 J. Sikkema, J.A.M. de Bont, and B. Poolman, *Microbiol. Rev.*, **59**, 201 (1995).
11 M.D. Lilly and J.M. Woodley, in **Biocatalysis in organic synthesis**, J. Tramper, H.C. van der Plas, and P. Linko, Ed., *Elsevier*, Amsterdam, 1985, pp.179-192.
12 M. Vermuë, J. Sikkema, A. Verheul, and J. Tramper, *Biotechnol. Bioeng.*, **42**, 747 (1993).
13 S.D. Doig, A.T. Boam, D.J. Leak, A.G. Livingston, and D.C. Stuckey, *Biocatal. Biotransform.*, **16**, 27 (1998).
14 A. Inoue and K. Horikoshi, *J. Ferment. Bioeng.*, **71**, 194 (1991).
15 A.J. Harrop, M.D. Hocknull, and M.D. Lilly, *Biotechnol. Lett.*, **11**, 807 (1989).
16 A.N. Rajagopal, *Enz. Microb. Technol.*, **19**, 606 (1996).
17 J. Sikkema, J.A.M. de Bont, and B. Poolman, *J. Biol. Chem.*, **269**, 8022 (1994).
18 H. Sandermann, Jr., *Biochim. Biophys. Acta*, **515**, 209 (1978).
19 P.L. Yeagle, *FASEB J.*, **3**, 1833 (1989).
20 J.A.M. de Bont, *Trends Biotechnol.*, **16**, 493 (1998).
21 A. Inoue and K. Horikoshi, *Nature*, **338**, 264 (1989).
22 D.L. Cruden, J.H. Wolfram, R.D. Rogers, and D.T. Gibson, *Appl. Environ. Microbiol.*, **58**, 2723 (1992).
23 F.J. Weber, L.P. Ooykaas, R.M.W. Schemen, S. Hartmans, and J.A.M. de Bont, *Appl. Environ. Microbiol.*, **59**, 3502 (1993).
24 J.L. Ramos, E. Deque, M.J. Huertas, and A. Haïdour, *J. Bacteriol.*, **177**, 3911 (1995).
25 K. Kim, S.J. Lee, K.H. Lee, and D.B. Lim, *J. Bacteriol.*, **180**, 3692 (1998).

26 F. Fukumori, H. Hirayama, H. Takami, A. Inoue, and K. Horikoshi, *Extremophiles*, **2**, 395 (1998).
27 H. Nakajima, H. Kobayashi, R. Aono, and K. Horikoshi, *Biosci. Biotechnol. Biochem.*, **56**, 1872 (1992).
28 H. Ogino, K. Miyamoto, and H. Ishikawa, *Appl. Environ. Microbiol.*, **60**, 3884 (1994).
29 Y. Yoshida, Y. Ikura, and T. Kudo, *Biosci. Biotechnol. Biochem.*, **61**, 46 (1997).
30 X.Z. Li, L. Zhang, and K. Poole, *J. Bacteriol.*, **180**, 2987 (1998).
31 K. Moriya and K. Horikoshi, *J. Ferment. Bioeng.*, **76**, 397 (1993).
32 S. Isken and J.A.M. de Bont, *Extremophiles*, **2**, 229 (1998).
33 M.L. Paje, B.A. Neilan, and I. Couperwhite, *Microbiology*, **143**, 2975 (1997).
34 F.J. Weber, S. Isken, and J.A.M. de Bont, *Microbiology*, **140**, 2013 (1994).
35 H.C. Pinkart, J.W. Wolfram, R. Rogers, and D.C. White, *Appl. Environ. Microbiol.*, **62**, 1129 (1996).
36 H.J. Heipieper, G. Meulenbeld, Q. van Oirschot, and J.A.M. de Bont, *Chemosphere*, **30**, 1041 (1995).
37 J.L. Ramos, E. Duque, J.J. Rodriguez-Herva, P. Godoy, A. Haidour, F. Reyes, and A. Fernandez-Barrero, *J. Biol. Chem.*, **272**, 3887 (1997).
38 F.J. Weber and J.A.M. de Bont, *Biochim. Biophys. Acta*, **1286**, 225 (1996).
39 V. Pedrotta and B. Witholt, *J. Bacteriol.*, **181**, 3256 (1999).
40 H.C. Pinkart and D.C. White, *J. Bacteriol.*, **179**, 4219 (1997).
41 R. Aono and H. Kobayashi, *Appl. Environ. Microbiol.*, **63**, 3637 (1997).
42 H. Kobayashi, H. Takami, H. Hirayama, K. Kobata, R. Usami, and K. Horikoshi, *J. Bacteriol.*, **181**, 4493 (1999).
43 L. Li, T. Komatsu, A. Inoue, and K. Horikoshi, *Biosci. Biotechnol. Biochem.*, **59**, 2358 (1995).
44 H. Asano, K. Kobayashi, and R. Aono, *Appl. Environ. Microbiol.*, **65**, 294 (1999).
45 G. Mosqueda, M.-S. Ramos-González, and J.L. Ramos, *Gene*, **232**, 69 (1999).
46 S. Isken and J.A.M. de Bont, *J. Bacteriol.*, **178**, 6056 (1996).
47 J. Kieboom, J.J. Dennis, G.J. Zylstra, and J.A.M. de Bont, *J. Biol. Chem.*, **273**, 85 (1998).
48 J. Kieboom, J.J. Dennis, G.J. Zylstra, and J.A.M. de Bont, *J. Bacteriol.*, **180**, 6769 (1998).
49 R. Aono, N. Doukyu, H. Kobayashi, H. Nakajima, and K. Horikoshi, *Appl. Environ. Microbiol.*, **60**, 2518 (1994).
50 G. Speelmans, A. Bijlsma, and G. Eggink, *Appl. Microbiol. Biotechnol.*, **50**, 538 (1998).
51 H. Ogino, F. Watanabe, M. Yamada, S. Nakagawa, T. Hirose, A. Noguchi, M. Yasuda, and H. Ishikawa, *J. Biosci. Bioeng.*, **87**, 61 (1999).
52 N. Doukyu and R. Aono, *Appl. Environ. Microbiol.*, **64**, 1929 (1998).


## 14.4.2 SOLVENT-RESISTANT MICROORGANISMS


Tilman Hahn, Konrad Botzenhart

**Institut für Allgemeine Hygiene und Umwelthygiene**
**Universität Tübingen, Tübingen, Germany**


### 14.4.2.1 Introduction

Several main properties of microorganisms in relation to solvents can be considered:

- Toxic or antimicrobial effects of solvents
- Solvent resistance or adaptation of microorganisms
- Metabolic activities of microorganisms

Antimicrobial effects or solvent-resistant microorganisms are the main topic of this section.

### 14.4.2.2 Toxicity of solvents for microorganisms

#### *14.4.2.2.1 Spectrum of microorganisms and solvents*

The growth-inhibiting effects of several solvents on microorganisms are described.[1,2] Organic solvents have toxic effect on microorganisms. Table 14.4.2.1 summarizes the relevant organic solvents and their toxicity concerning selected microorganisms.

**Table 14.4.2.1. Toxicity of organic solvents - examples**

| Solvent | Microorganisms | References |
|---|---|---|
| Toluene, benzene, ethylbenzene, propylbenzene, xylene, hexane, cyclohexane | *Pseudomonas putida* | Isken et al. (1999)[3]<br>Gibson et al. (1970)[4] |
| Terpenes, e.g., alpha-pinene, limonene, β-pinene, terpinolene | *Bacillus sp., Saccharomyces cerevisiae*, isolated mitochondria | Andrews et al. (1980)[5]<br>Uribe et al. (1985)[6] |
| Styrene | soil microorganisms | Hartmans et al. (1990)[7] |
| Cyclohexane | yeast cells, isolated mitochondria | Uribe et al. (1990)[8] |
| Aromatic hydrocarbons | isolated bacterial and liposomal membranes | Sikkema et al. (1992)[9]<br>Sikkema et al. (1994)[1] |
| Ethanol | yeasts | Cartwright et al. (1986)[10]<br>Leao and van Uden (1984)[11] |

### *14.4.2.2.2 Mechanisms of solvent toxicity for microorganisms*

The toxicity of organic solvents or hydrophobic substances for microorganisms depends mainly on their effects on biological membranes[1,9,12-14] - similar to membrane effects of several anesthetics. This concerns especially effects on cytoplasmatic membranes. The following main changes of membrane structures and functions have been observed:

- Accumulation of hydrophobic substances such as organic solvents in cytoplasmatic membranes. This accumulation causes structural and functional changes in the cytoplasmatic membranes and microbial cells.
- Structural changes in cytoplasmatic membranes, e.g., swelling of membrane bilayers, increase of surface and thickness of the membranes, changes in the composition of the membrane (e.g., changes in the fatty acid composition), modification of the microviscosity, damage of membrane structures (see below).
- Loss of membrane integrity, especially disruption of cytoplasmatic membranes, less damage of outer membranes. Because of these damages often complex cellular structures (e.g., vesicula) or cell functions (decrease of respiratory activities of mitochondria) are destroyed or inhibited.
- Interactions of the accumulated lipophilic substances with the cytoplasmatic membranes and especially hydrophobic parts of the cell or cell membranes. Lipid-lipid interactions and interactions between proteins and lipids of the membrane structure (lipid bilayers, membrane-embedded proteins) are discussed.
- Effects on passive and active membrane transport systems, e.g., increase of passive efflux and flux of ions such as protons, cations $Mg^{++}$ and $Ca^{++}$ or small molecules, stimulation of the leakage of protons and potassium, changes in the uptake of compounds (e.g., solvents) and excretion (e.g., metabolic products), inhibition of active transport systems (e.g., ATP depletion).
- Damage of cellular homeostasis and cell physiology, e.g., reduction of transmembrane electrical potentials and proton chemical potentials or proton motive forces as a result of membrane changes (efflux of ions), changes of pH gradients.

- Changes in enzyme activities, e.g., inhibition of oxidases and depletion of ATP. Of special relevance are various interactions with enzymes (proteins) in the membrane, e.g., lipid-protein interactions.
- Loss of particular cellular functions, e.g., respiratory system of mitochondria or active transport systems (see above).
- Loss of complex cell functions, e.g., reduced growth rates and activities of microorganisms.

The extent of solvent toxicity to microorganisms is determined by various factors:

(a) Hydrophobic or lipophilic properties of solvents. The toxicity of solvents to microorganisms can be described by a partition coefficient (log $P_{o/w}$) between organic compounds (solvents) and water which is specific for the applied substance. This partition coefficient is based on a standard octanol-water system model.[1,15,16]

The toxicity and the affinity of solvents to cell structures increase with hydrophobic properties of solvents, e.g., high toxicities with $P_{o/w}$ values of 1-5.[1,14] The partition coefficient correlates with the membrane-buffer partition coefficient between membrane and aqueous system.[1,14,16] They also depend on membrane characteristics.[17,18]

(b) Accumulation, partitioning and concentrations of solvents in cell structures (membranes). Dissolution and partitioning of solvents depend essentially on solvent properties, e.g., polarity (specific partition coefficients), or membrane characteristics (influence on partition coefficients). Both dissolution and partitioning can be influenced by additional factors, e.g., cosolvents. The effects on microorganisms can depend typically on solvent concentrations, e.g., dose response effects.

(c) Biomass, ratio of concentrations of solvents and biomass. Effect of solvents depends on this ratio.

(d) Surrounding conditions, e.g., temperature which influences the proton leakage and microbial activities.[19]

The toxicity of solvents for microorganisms shows positive and negative consequences, e.g.:

- Positive aspects such as antibacterial effects[20] which are found in several products.
- Negative aspects such as reduced stability of biotransformation and bioremediation processes because of the inactivation of microorganisms.

### 14.4.2.3 Adaption of microorganisms to solvents - solvent-resistant microorganisms

#### *14.4.2.3.1 Spectrum of solvent-resistant microorganisms*

Different microorganisms are able to adapt and even to grow in the presence of solvents. Some relevant examples are given in Table 14.4.2.2.

**Table 14.4.2.2. Solvent-tolerant microorganisms and their resistance to organic solvents**

| Solvent-tolerant microorganisms | Solvents | References |
|---|---|---|
| bacteria from deep sea (1.168 m, Japan) | benzene, toluene, p-xylene, biphenyl, naphthalene | Abe et al. (1995)[21] |

| Solvent-tolerant microorganisms | Solvents | References |
|---|---|---|
| deep sea isolates, *Flavobacterium* strain DS-711, *Bacillus* strain DS-994 | hydrocarbons | Moriya and Horikoshi (1993)[22] |
| marine yeasts | n-alkanes | Fukamaki et al. (1994)[23] |
| *Pseudomonas putida* | toluene, m-, p-xylene, 1,2,4-trimethylbenzene, 3-ethyltoluene | Inoue and Horikoshi (1989)[24] Cruden et al. (1992)[25] |
| *Pseudomonas sp.* | alpha-pinene | Sikkema et al. (1995)[14] |
| *E. coli* K-12 | p-xylene | Aono et al. (1991)[26] |

The concentration of organic solvents which can be tolerated by the microorganisms varies extensively, e.g., growth of *Pseudomonas putida* in the presence of more than 50 % toluene[27] but tolerance of *E. coli* K-12 in the presence of only up to 10 % p-xylene.[26]

The growth characteristics of different species can differ to a great extent, e.g., no growth, growth with or without metabolizing solvents. Generally higher solvent tolerance of Gram-negative bacteria compared to Gram-positive bacteria is observed.[28] Some solvent-tolerant microorganisms cannot use organic solvents as substrate for growth and need other substrates in complex media.[26,27] Other microorganisms can use organic solvents in minimal media as a source of energy or carbon, e.g., *Pseudomonas putida* in the presence of xylene or toluene.[25]

#### *14.4.2.3.2 Adaption mechanisms of microorganisms to solvents*

The mechanisms of solvent tolerance are only partly known.[29-31] Relevant microbial adaptation mechanisms are:

- *Changes in the composition of the cytoplasmatic membrane*. The compounds of the membranes such as lipids or proteins can influence the membrane characteristics and, therefore, the adaption to solvents. Mainly phospholipids in the membrane bilayer determine the partitioning of solutes and especially the resistance to solvents[17,18] A reduction of the partition coefficient has been observed when the fatty acid composition was changed.[32] An increase of monounsaturated fatty acids and a decrease of saturated fatty acids correlated with a higher ethanol tolerance of *S. cerevisiae, E. coli*, and *Lactobacillus* strains.[33-35] An increase of unsaturated fatty acids is induced by polar solvents and low temperatures, an increase of saturated fatty acids is connected with more apolar solvents and high temperatures.[36] Even changes in the configuration of fatty acids, which are provoked by solvents, can lead to adaption mechanisms, e.g., *cis-trans* conversions.[29]

- *Changes in the microbial structure*. Various structural changes can cause a reduction of toxic solvent effects, e.g., the increase of membrane fluidity which is connected with an increase of unsaturated fatty acids (see above) and results in a decrease of the membrane permeability.

- *Specific structural characteristics of microorganisms*. Typical structures of microorganisms vary according to the microbial species, e.g., outer membrane characteristics of bacteria. For instance, Gram-negative bacteria such as *Pseudomonas sp.* tolerate higher concentrations of hydrophobic compounds compared to Gram-positive bacteria. The resistance of the outer membrane correlates with the solvent-tolerance.[28]

- *Alterations of the cell envelope structure (cell wall).* Mechanical alterations and chemical modifications of the cell wall can reduce the microbial resistance to solvents. The most interesting chemical modification concerns hydrophobic or hydrophilic abilities of the cell wall.[37] Decreasing hydrophobicity of the cell wall enhances the adaption of microorganisms to solvents.[38]

- *Suppression of the effects of solvents on membrane stability.*

- *Limitation of solvent diffusion into the cell* (see above).

- *Repairing mechanisms*, e.g. enhanced phospholipid biosynthesis.

- *Transport or export systems.* The excretion of compounds out of the microbial cell and cytoplasmatic membrane is well known but only documented for some substances, e.g., for drugs.[39] Passive and active transport systems are relevant, e.g., ATP driven systems. Export systems for the several solvents must be assumed.

- *Immobilization and mobilization of microorganisms and solvents.* The adsorption of solvents to microorganisms can be reduced if the contact is decreased. For instance, immobilization of microorganisms or solvents minimizes the contact. An immobilization and reduction of toxicity was shown if adsorption materials were added.[40]

- *Surrounding conditions*, e.g., low temperature, which can induce higher solvent resistance of microorganisms (see above).

## 14.4.2.4 Solvents and microorganisms in the environment and industry - examples

Microorganisms are frequently observed in organic-aqueous systems containing solvents are essential in natural and in industrial processes. The occurrence and role of microorganisms and organic compounds in these two-phase organic-aqueous systems are similar to the effects described above (see Section 14.4.2.2). Although toxic effects on microorganisms in these natural and industrial processes are well known, reliable data concerning solvent-resistant microorganisms are not available.

### *14.4.2.4.1 Examples*

14.4.2.4.1.1 Biofilms, biofouling, biocorrosion

Important examples for organic-aqueous systems are surface-associated biofilms which are a form of existence of microorganisms. Microorganisms, mostly bacteria, are embedded into a glycocalyx matrix of these biofilms.[41,42] This biofilm matrix mediates the adhesion of microorganisms to surfaces, concentrates substances and protects microorganisms from antimicrobial agents.[41,42] Several organic-aqueous systems can be observed, especially surface of surrounding materials (pipes, etc.) in relation to water or ingredients (e.g., oil in pipes) related to water between ingredients and surrounding materials (pipes, etc.).

Some aspects of solvents in these organic-aqueous biofilm systems are studied. Solvents can occur in water systems emitted from surrounding organic materials.[43] It was shown that solvents are important concerning microbial biocorrosion and biofouling processes, e.g., by swelling and hydrolysis of materials.[44]

Despite these well-known aspects, reliable data and studies concerning solvent-resistant microorganisms in biofilm, biofouling, or biocorrossion processes are not shown. Nevertheless similar mechanisms in biofilms must be assumed as described above (see Section 14.4.2.3) because similar conditions occur (organic-aqueous systems).

14.4.2.4.1.2 Antimicrobial effects, microbial test systems

The toxic mechanisms of solvents to microorganisms described above (see Section 14.4.2.2) are frequently used in effects of antimicrobial agents.

The damage of microbial biomembranes is fundamentally connected with the antimicrobial effects of several solvents on bacteria.[45]

Biomembranes and other microbial structures can be affected by solvents via similar processes as described. Relevant examples are naked viruses which are generally more resistant to viruzi agents because the envelopes of viruses are damaged by viruzi substances such as some solvents. Another example of solvent-like interactions are effects of antimicrobial agents to capsules of bacterial spores, e.g., Bacillus species.

Bacterial or enzymatic toxicity tests are used to assay the activity of organic compounds including solvents. A survey of environmental bacterial or enzymatic test systems is given by Bitton and Koopman.[46] The principles of these test systems are based on bacterial properties (growth, viability, bioluminescence, etc.) or enzymatic activities and biosynthesis. The toxicity of several solvents were tested in bacterial or enzymatic systems, e.g., pure solvents such as phenol in growth inhibition assays (Aeromonas sp.),[46] solvents in complex compounds such as oil derivates,[46,47] solvents in environmental samples such as sediments or solvents used in the test systems.[46,48,49] The efficiency of several test systems, e.g., Microtox tests or ATP assays, vary, e.g., looking at the effects of solvents.[46]

14.4.2.4.1.3 Industrial processes

In industrial processes the main microbial activities connected with solvents are:

- Processes in biotechnology, biotransformation and biocatalysis,[50,51] e.g., production of chemicals from hydrophobic substrates or use of solvents as starting materials for microbiological reactions.
- Bioremediation: degradation of environmental pollutants, e.g., wastewater treatment or bioremediation in biofilm reactors.[52-55]

Various microorganisms and microbial mechanisms are relevant in these industrial processes. Examples are conversion processes of organic substances, e.g., by bacterial oxygenases.[56,57] Normally low-molecular-weight aromatic hydrocarbons including solvents are converted in these biotransformation processes.[58]

## References

1 J. Sikkema, J.A.M. deBont, B.Poolman, *J. Biol. Chem.*, **269**, 8022 (1994).
2 G.J. Salter, D.B. Kell, *Crit. Rev. Biotechnol.*, **15**, 139 (1995).
3 S. Isken, A. Derks, P.F.G. Wolffs, J.A.M. deBont, *Appl. Environ. Microbiol.*, **65**, 2631 (1999).
4 D.T. Gibson, G.E. Cardini, F.C. Maseles, R.E. Kallio, *Biochemistry*, **9**, 1631 (1970).
5 R.E. Andrews, L.W. Parks, K.D. Spence, *Appl. Environ. Microbiol.*, **40**, 301 (1980).
6 S. Uribe, J. Ramirez, A. Pena, *J. Bacteriol.*, **161**, 1195 (1985).
7 S. Hartmans, M.J. van der Werf, J.A.M. de Bont, *Appl. Environ. Microbiol.*, **56**, 1347 (1990).
8 S. Uribe, P. Rangel, G. Espinola, G. Aguirre, *Appl. Environ. Microbiol.*, **56**, 2114 (1990).
9 J. Sikkema, B. Poolman, W.N. Konigs, J.A.M. deBont, *J. Bacteriol.*, **174**, 2986 (1992).
10 C.P. Cartwright, J.R. Juroszek, M.J. Beavan, F.M.S. Ruby, S.M.F. DeMorais, A.H. Rose, *J. gen. Microbiol.*, **132**, 369 (1986).
11 C. Leao, N. vanUden, *Biochim. Biophys. Acta*, **774**, 43 (1984).
12 M.J. DeSmet, J. Kingma, B. Witholt, *Biochem. Biophys. Acta*, **506**, 64 (1978).
13 M.R. Smith in **Biochemistryof Microbial Degradation**, C. Ratledge, Ed, *Kluwer Academic Press*, Dordrecht, 1993, pp. 347-378.
14 J. Sikkema, J.A.M. deBont, B. Poolman, B., *FEMS Microbiol. Rev.*, **59**, 201 (1995).
15 A. Leo, C. Hansch, D. Elkins, *Chem. Rev.*, **71**, 525 (1971).

16 W.R. Lieb and W.D. Stein in **Transport and Diffusion Across Cell Membranes**, W.D. Stein, Ed., *Academic Press*, N.Y., 1986, pp. 69-112.
17 M.C. Antunes-Madeira, V.M.C. Madeira, *Biochim. Biophys. Acta*, **861**, 159 (1986).
18 M.C. Antunes-Madeira, V.M.C. Madeira, *Biochim. Biophys. Acta*, **901**, 61 (1987).
19 W. DeVrij, A. Bulthuis, W.N. Konings, *J. Bacteriol.*, **170**, 2359 (1988).
20 F.M. Harold, *Adv. Microb. Physiol.*, **4**, 45 (1970).
21 A. Abe, A. Inoue, R. Usami, K. Moriya, K. Horikoshi, *Biosci. Biotechnol. Biochem.*, **59**, 1154 (1995).
22 K. Moriya, K. Horikoshi, *J. Ferment. Bioeng.*, **76**, 168 (1993).
23 T. Fukamaki, A. Inoue, K. Moriya, K. Horikoshi, *Biosci. Biotechnol. Biochem.*, **58**, 1784 (1994).
24 A. Inoue, K. Horikoshi, *Nature*, **338**, 264 (1989).
25 D.L. Cruden, J.H. Wolfram, R. Rogers, D.T. Gibson, *Appl. Environ. Microbiol.*, **58**, 2723 (1992).
26 R. Aono, K. Albe, A. Inoue, K. Horikoshi, *Agric. Biol. Chem.*, **55**, 1935 (1991).
27 A. Inoue, M. Yamamoto, K. Horikoshi, K., *Appl. Environ. Microbiol.*, **57**, 1560 (1991).
28 A.J. Harrop, M.D. Hocknull, M.D. Lilly, *Biotechnol. Lett.*, **11**, 807 (1989).
29 H.J. Heipieper, R. Diefenbach, H. Keweloh, *Appl. Environ. Microbiol.*, **58**, 1847 (1992).
30 H.C. Pinkart, D.C. White, *J. Bacteriol.*, **179**, 4219 (1997).
31 J.L. Ramos, E. Duque, J. Rodriguez-Herva, P. Godoy, A. Haidour, F. Reyes, A. Fernandez-Barrero, *J. Biol. Chem.*, **272**, 3887 (1997).
32 M.C. Antunes-Madeira, V.M.C., Madeira, *Biochim. Biophys. Acta*, **778**, 49 (1984).
33 P. Mishra, S. Kaur, *Appl. Microbiol. Biotechnol.*, **34**, 697 (1991).
34 L.O. Ingram, *J. Bacteriol.*, **125**, 670 (1976).
35 K. Uchida, *Agric. Biol. Chem.*, **39**, 1515 (1975).
36 L.O. Ingram, *Appl. Environ. Microbiol.*, **33**, 1233 (1977).
37 V. Jarlier, H. Nikaido, *FEMS Microbiol. Lett.*, **123**, 11 (1994).
38 Y.S. Park, H.N. Chang, B.H. Kim, *Biotechnol. Lett.*, **10**, 261 (1988).
39 D. Molenaar, T. Bolhuis, T. Abee, B. Poolman, W.N. Konings, *J. Bacteriol.*, **174**, 3118 (1992).
40 H. Bettmann, H.J. Rehm, *Appl. Microbiol. Biotechnol.*, **20**, 285 (1984).
41 J.W. Costerton, *J. Ind. Microbiol. Biotechnol.*, **22**, 551 (1999).
42 J.W. Costerton, *Int. J. Antimicrobial Agents*, **11**, 217 (1999).
43 K. Frensch, H.F. Scholer, D. Schoenen, *Zbl. Hyg. Umweltmed.*, **190**, 72 (1990).
44 W. Sand, *Int. Biodeterioration Biodegradation*, **40**, 183 (1997).
45 K. Wallhäußer, **Praxis der Sterilisation, Desinfektion, Konservierung**, *Thieme Verlag*, Stuttgart, 1998.
46 G. Bitton, B. Koopman, *Rev. Environ. Contamin. Toxicol.*, **125**, 1 (1992).
47 K.L.E. Kaiser, J.M. Ribo, *Tox. Assess*, **3**, 195 (1988).
48 B.J. Dutka, K.K. Kwan, *Tox. Assess*, **3**, 303 (1988).
49 M.H. Schiewe, E.G. Hawk, D.I. Actor, M.M. Krahn, *Can. J. Fish Aquat.*, **42**, 1244 (1985).
50 O. Favre-Bulle, J. Schouten, J. Kingma, B. Witholt, *Biotechnology*, **9**, 367, 1991.
51 C. Laane, R. Boeren, R. Hilhorst, C. Veeger, in C. Laane, J. Tramper, M.D. Lilly, Eds., **Biocatalysis in organic media**, *Elsevier Publishers,* B.V., Amsterdam, 1987.
52 C. Adami, R. Kummel, *Gefahrstoffe Reinhaltung Luft*, **57**, 365 (1997).
53 A.R. Pedersen, S. Moller, S. Molin, E. Arvin, *Biotechnol. Bioengn.*, **54**, 131 (1997).
54 P.J. Hirl, R.L. Irvine, *Wat. Sci. Technol.*, **35**, 49 (1997).
55 M.W. Fitch, D. Weissman, P. Phelps, G. Georgiou, *Wat. Res.*, **30**, 2655 (1996).
56 S.V. Ley, F. Sternfeld, S.C. Taylor, *Tetrahedron Lett.*, **28**, 225 (1987).
57 G.M. Whited, W.R. McComble, L.D. Kwart, D.T. Gibson, *J. Bacteriol.*, **166**, 1028 (1986).
58 M.R. Smith, *J. Bacteriol.*, **170**, 2891 (1990)

## 14.4.3 CHOICE OF SOLVENT FOR ENZYMATIC REACTION IN ORGANIC SOLVENT

TSUNEO YAMANE
**Graduate School of Bio- and Agro-Sciences, Nagoya University, Nagoya, Japan**

### 14.4.3.1 Introduction

The ability of enzymes to catalyze useful synthetic biotransformations in organic media is now beyond doubt. There are some advantages in using enzymes in organic media as opposed to aqueous medium, including

1) shifting thermodynamic equilibrium to favor synthesis over hydrolysis,
2) reduction in water-dependent side reaction,
3) immobilization of the enzyme is often unnecessary (even if it is desired, merely physical deposition onto solid surfaces is enough),
4) elimination of microbial contamination,
5) suitable for reaction of substrates insoluble and/or unstable in water, etc.

Here organic media as the reaction system are classified into two categories: substrates dissolved in neat organic solvents and solvent-free liquid substrates. Although the latter seems preferable to the former, if it works, there are a number of cases where the former is the system of choice: for example, when the substrate is solid at the temperature of the reaction, when high concentration of the substrate is inhibitory for the reaction, when the solvent used gives better environment (accelerating effect) for the enzyme, and so forth.

Prior to carrying out an enzymatic reaction in an organic solvent, one faces choice of a suitable solvent in the vast kinds of organic solvents. From active basic researches having been carried out in the past two decades, there has been a remarkable progress in our understanding of properties of enzymes in organic media, and in how organic solvents affect them. Some researchers call the achievement ‘medium engineering’. A comprehensive monograph was published in 1996 reviewing the progress.[1]

In this article organic solvents often used for enzymatic reactions are roughly classified, followed by influence of solvent properties on enzymatic reactions and then properties of the enzymes affected by the nature of the organic solvents are briefly summarized.

### 14.4.3.2 Classification of organic solvents

Among numerous kinds of organic solvents, the ones often used for enzymatic reactions are not so many, and may be classified into three categories (Table 14.4.3.1),[2] in view of the importance of water content of the organic solvent concerned (see Section 14.4.3.3).

1) water-miscible organic solvents

These organic solvents are miscible with water at the temperature of the reaction. Any cosolvent system having 0 - 100% ratio of the solvent/water can be prepared from this kind of solvent. Note that some organic solvents having limited water solubility at ambient temperature, and hence are not regarded as water-miscible, become miscible at elevated temperature.

2) Water-immiscible organic solvents

**Table 14.4.3.1. Classification of solvents commonly used for enzymatic reactions in organic media. [Adapted, by permission, from T. Yamane, Nippon Nogeikagaku Kaishi, 65, 1104(1991)]**

| |
|---|
| 1) Water-miscible organic solvents |
| Methanol, ethanol, ethylene glycol, glycerol, N,N'-dimethylformamide, dimethylsulfoxide, acetone, formaldehyde, dioxane, etc. |
| 2) Water-immiscible organic solvents (water solubility [g/l] at the temperature indicated) |
| • alcohols |
| (n-, iso-) propyl alcohol, (n-, s-, t-) butyl alcohol, (n-, s-, t-) amyl alcohol, n-octanol, etc. |
| • esters |
| methyl acetate, ethyl acetate (37.8, 40°C), n-butyl acetate, hexyl acetate, etc. |
| • alkyl halides |
| dichloromethane (2, 30°C), chloroform, carbon tetrachloride, trichloroethane (0.4, 40°C), etc. |
| • ethers |
| diethyl ether (12, 20°C), dipropyl ether, diisopropyl ether, dibutyl ether, dipentyl ether, etc. |
| 3) Water-insoluble organic solvents (water solubility [ppm] at the temperature indicated) |
| • aliphatic hydrocarbons |
| n-hexane (320, 40°C), n-heptane (310, 30°C), isooctane (180, 30°C), etc. |
| • aromatic hydrocarbons |
| benzene (1200, 40°C), toluene (880, 30°C), etc. |
| • allicyclic hydrocarbons |
| cyclohexane (160, 30°C), etc. |

These organic solvents have noticeable but limited solubility of water, ranging roughly 0.1 - 10 % its solubility. The water solubility is of course increased as the temperature is raised.

3) Water-insoluble organic solvents

These solvents are also water-immiscible and have very low water solubility so that they are regarded as water-insoluble, i.e., water is practically insoluble in the organic solvents. Most aliphatic and aromatic hydrocarbons belong to this category.

In Table 14.4.3.1 organic solvents often used for enzymatic reactions are listed together with their water solubilities (although not for all of them).

### 14.4.3.3 Influence of solvent parameters on nature of enzymatic reactions in organic media

Factors that influence the activity and stability of enzymes in organic media have been mostly elucidated. Several of them are mentioned below.

1) Water activity, $a_w$

Among factors, the amount of water existing in the reaction system is no doubt the most influential. To emphasize the effect of water, the author once proposed to say 'enzymatic reaction in microaqueous organic solvent', instead of merely say 'enzymatic reaction in organic solvent'.[3-5]

Very trace amount of water or nearly anhydrous state renders practically no enzymatic reaction. In this context, it should be reminded that commercially available enzyme preparations, or the enzymes even after lyophilization or other drying procedures, contain some water bound to the enzyme proteins. Whereas, excess water in the reaction system results in hydrolysis of the substrate, which is often unfavorable side reaction, giving rise to lower yield of product. Thus, there exist usually the optimal water content for each enzymatic reaction of concern.

Water molecules in the microaqueous system exist in three different states: 1) water bound to the enzyme protein, 2) water dissolved freely in the solvent (plus dissolved substrate), and 3) water bound to impurities existing in the enzyme preparation or bound to the support materials if immobilized enzyme particles are used. Therefore, the following equation with respect to water holds:

Total water = (Water bound to the enzyme) + (Water dissolved in the solvent) [14.4.3.1]

+ (Water bound to the immobilization support or to impurities of the enzyme preparation)

Water affecting most of the catalytic activity of the enzyme is the one bound to the enzyme protein.[6] From the above equation, it can be well understood that the effect of water varies depending on the amount of enzyme used and/or its purity, kind of solvent, and nature of immobilization support, etc. as far as the total water content is used as the sole variable. Also, it is often asked what is the minimal water content sufficient for enzymatic activity? It should be recognized that a relation between the degree of hydration of the enzyme and its catalytic activity changes continuously. There exist a thermodynamic isotherm-type equilibrium between the protein-bound water and freely dissolved water, and its relationship is quite different between water-miscible and water-insoluble solvents.[4]

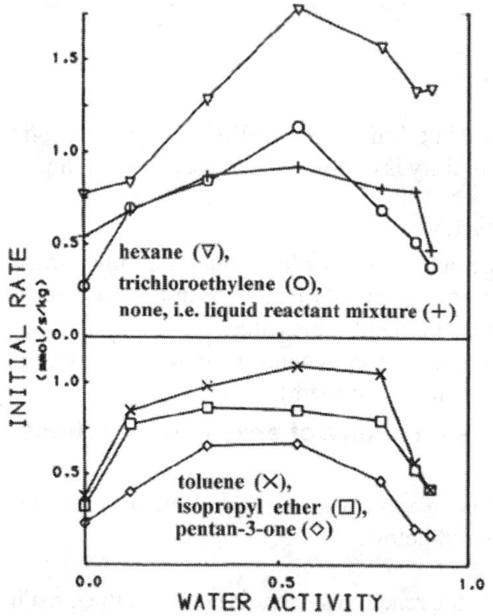


Figure 14.4.3.1 Activity of Lipozyme catalyst as a function of water activity in a range of solvents. [Adapted, by permission, from R.H. Valivety, P.J. Halling and A.R. Macrae, *Biochim. Biophys. Acta*, **1118**, 221 (1992)].

A parameter better than the water content, water activity, $a_w$, was proposed to generalize the degree of hydration of a biocatalyst in organic media.[7] $a_w$ is a thermodynamic parameter which determines how much water is bound to the enzyme and in turn decides the catalytic activity to a large extent among different kinds of the organic solvents. $a_w$ is especially useful when water-insoluble organic solvent is used because the precise water content is hard to be measured due to its low solubility. It was shown that profiles between $a_w$ and the reaction rate were similar when the same reaction was carried out in different solvents at varying water contents.[8] In different solvents, maximum reaction rate was observed at widely different water content, but if water content was qualified in terms of $a_w$, the optimum was observed at almost the same $a_w$ (Figure 14.4.3.1). However, as seen from Figure 14.4.3.1, the profile does not lay on a single curve, and the absolute optimal reaction rate varies depending on the kind of the solvent, implying that $a_w$ is not almighty.

2) Hydrophobicity (or polarity), logP

A hydrophobicity parameter, logP, was first proposed for microbial epoxidation of propene and 1-butene.[9,10] logP is the logarithm of P, where P is defined as the partition coefficient of a given compound in the standard n-octanol/water two phase:

$$P = \frac{\text{solubility of a given compound in } n-\text{octanol phase}}{\text{solubility of a given compound in water phase}} \quad [14.4.3.2]$$

Laane et al.[10] concluded as a general rule that biocatalysis in organic solvents is low in polar solvents having a logP < 2, is moderate in solvents having a logP between 2 and 4, and is high in apolar solvents having a logP > 4. They also stated that this correlation between polarity and activity paralleled the ability of organic solvents to distort the essential water layer bound to the enzyme that stabilized the enzyme. Since logP can easily be determined experimentally, or be estimated from hydrophobic fragmental constants, many biotechnologists have tried since then to correlate effects of organic solvents on biocatalysts they studied with logP approach. Their results have been successful, not completely but only partially. A number of exceptions to the 'logP rule' have been in fact reported.

3) Dielectric constant (or dipole moment), ε (or D)

Interactions between an enzyme and a solvent in which the enzyme is suspended are mostly non-covalent ones as opposed to interactions in water. These strong non-covalent interactions are essentially of electrostatic origin, and thus according to Coulomb's law, their strength is imposed dependent on the dielectric constant, ε, (which is higher for water than for almost all organic solvents). It is likely that enzymes are more rigid in anhydrous solvents of low ε than in those of high ε. Thus, ε of a solvent can be used as a criterion of rigidity of the enzyme molecule. For the enzyme to exhibit its activity, it must be dynamically flexible during its whole catalytic action so that its activity in a solvent of lower ε should be less than in a solvent of higher ε. On the other hand, its selectivity or specificity becomes higher when its flexibility decreases so that the selectivity in a solvent of lower ε should be higher than in a solvent of higher ε.

### 14.4.3.4 Properties of enzymes affected by organic solvents

1) Thermal stability (half-life), $t_{1/2}$

Stability of an enzyme in an organic solvent is estimated by its half-life, $t_{1/2}$, when its activity is plotted as a function of incubation time. Although, $t_{1/2}$ during the enzymatic reaction is more informative for practical purposes, $t_{1/2}$ under no substrate is often reported because of its easiness of measurement. Inactivation of an enzyme is caused mostly by change in its native conformation, or irreversible unfolding of its native structure. Water, especially enzyme-bound water makes a major contribution to the protein folding through van der Waals interaction, salt-bridges, hydrogen bonds, hydrophobic interaction, etc. When the enzyme molecule is put into organic solvent, water molecules bound to the enzyme molecule are more or less re-equilibrated, depending on the free water content. Therefore, both the nature of organic solvent and the free water content have profound effects on its stability.

It has been shown that a number of enzymes suspended in anhydrous (dry) organic solvents exhibit thermal stability far superior to that in aqueous solutions (Table 14.4.3.2).[11] This is because most of chemical processes that occur in the thermal inactivation involve water, and therefore do not take place in a water-free environment. Furthermore, increased rigidity in dry organic solvents hinders any unfolding process. The increased thermal stability in the dry organic solvent drops down to the stability in aqueous solution by adding small amount of water as demonstrated by Zaks and Klibanov[12] and others. Thus, thermal stability of the enzyme in organic solvent strongly depends on its free water content.

**Table 14.4.3.2. Stability of enzymes in non-aqueous vs. aqueous media. [Adapted, by permission, from 'Enzymatic Reactions in organic Media', A. M. P. Koskinen and A. M. Klibanov, Blackie Academic & Professional (An Imprint of Chapman & Hall), Glasgow, 1996, p. 84]**

| Enzyme | Conditions | Thermal property | References |
|---|---|---|---|
| PLL | tributyrin<br>aqueous, pH 7.0 | $t_{1/2} < 26$ h<br>$t_{1/2} < 2$ min | Zaks & Klibanov (1984) |
| *Candida* lipase | tributyrin/heptanol<br>aqueous, pH 7.0 | $t_{1/2} = 1.5$ h<br>$t_{1/2} < 2$ min | Zaks & Klibanov (1984) |
| Chymotrypsin | octane, 100°C<br>aqueous, pH 8.0, 55°C | $t_{1/2} = 80$ min<br>$t_{1/2} =15$ min | Zaks & Klibanov (1988)<br>Martinek et al. (1977) |
| Subtilisin | octane, 110°C | $t_{1/2} = 80$ min | Russell & Klibanov (1988) |
| Lysozyme | cyclohexane, 110°C<br>aqueous | $t_{1/2} = 140$ h<br>$t_{1/2} < 10$ min | Ahen & Klibanov (1986) |
| Ribonuclease | nonane, 110°C, 6 h<br>aqueous, pH 8.0, 90°C | 95% activity remains<br>$t_{1/2} < 10$ min | Volkin & Klibanov (1990) |
| $F_1$-ATPase | toluene, 70°C<br>aqueous, 70°C | $t_{1/2} > 24$ h<br>$t_{1/2} < 10$ min | Garza-Ramos et al. (1989) |
| Alcohol dehydrogenase | heptane, 55°C | $t_{1/2} > 50$ days | Kaul &Mattiasson (1993) |
| HindIII | heptane, 55°C, 30 days | no loss of activity | Kaul & Mattiasson (1993) |
| Lipoprotein lipase | toluene, 90°C, 400 h | 40% activity remains | Ottoline et al. (1992) |
| β-Glucosidase | 2-propanol, 50°C, 30 h | 80% activity remains | Tsitsimpikou et al. (1994) |
| Tyrosinase | chloroform, 50°C<br>aqueous solution, 50°C | $t_{1/2} = 90$ min<br>$t_{1/2} = 10$ min | Yang & Robb (1993) |
| Acid phosphatase | hexadecane, 80°C<br>aqueous, 70°C | $t_{1/2} = 8$ min<br>$t_{1/2} = 1$ min | Toscano et al. (1990) |
| Cytochrome oxidase | toluene, 0.3% water<br>toluene, 1.3% water | $t_{1/2} = 4.0$ h<br>$t_{1/2} = 1.7$ min | Ayala et al. (1986) |

For References, refer to Ref. 11.

2) Specificity and selectivity, $k_{cat}/K_m$

It is the most exciting and significant feature that the substrate specificity, enantioselectivity and regioselectivity can be profoundly affected by nature of solvents in which the enzyme molecule exists. This phenomenon has opened an alternative approach for changing specificity and selectivity of an enzyme other than both screening from nature and protein engineering in the field of synthetic organic chemistry. The ability of enzymes to discriminate substrate specificity among different, but structurally similar substrates, enantioselectivity among enantiomers, eantiofaces or identical functional groups linked to a prochiral center, and regioselectivity among identical functional groups on the same molecule, is expressed quantitatively on E value, which is the ratio of the specificity constants, $k_{cat}/K_m$, for the two kinds of substrate (or entiomers), i.e., $(k_{cat}/K_m)_1/(k_{cat}/K_m)_2$. For kinetic resolution of racemic mixture by the enzyme, E is called enantiomeric ratio. The higher E, the higher the enantiomeric excess (i.e., the optical purity), ee, of the product (or remaining substrate). It is said that an E value higher than 100 is preferable for pharmaceutical or biotechnological applications. For overview of this topic, see Refs. 13 and 14.

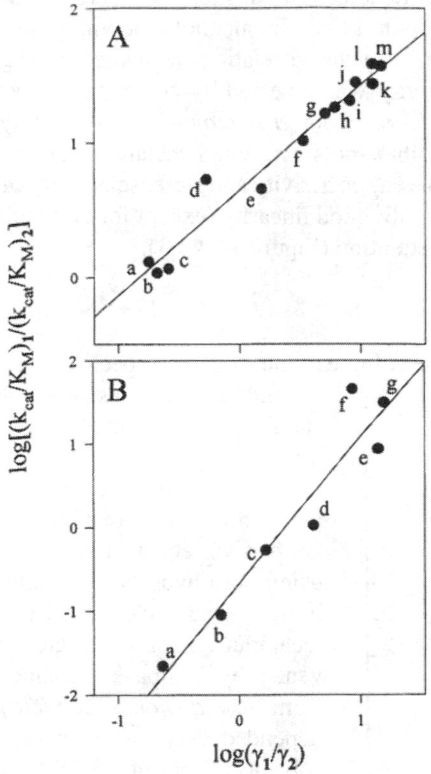


Figure 14.4.3.2. The dependence of (A) subtilisin Carlsberg and (B) a-chymotrypsin substrate specificity for substrates 1 and 2 on the ratio of their Raoult's law activity coefficients. For the structures of the substrates 1 and 2, and the solvents a through m in (A) and a to g in (B), refer to Ref. 16. [Adapted, by permission, from C.R. Wescott and A.M. Klibanov, *Biotechnol. Bioeng.*, **56**, 343(1997)].

(2a) Substrate specificity

Zaks and Klibanov reported that the substrate specificity of α-chymotrypsin, subtilisin, and esterase changed with an organic solvent.[15] The substrate specificity in octane was reversed compared to that in water. A thermodynamical model that predicted the substrate specificity of subtilisin Carlsberg and α-chymotrypsin in organic media on the basis of specificity of the enzyme in water and physicochemical characteristics of the solvents was developed by Wescott and Klibanov.[16] They determined $k_{cat}/K_m$ for the transesterification of N-acetyl-L-phenylalanine and N-acetyl-L-serine with propanol in 20 anhydrous solvents, and correlated the data of $(k_{cat}/K_m)_{Ser}/(k_{cat}/K_m)_{Phe}$, first with the solvent to water partition coefficients for the substrate, $P_{Phe}/P_{Ser}$. Later they examined the selectivity of subtilisin toward two different substrates with the Raoult's law activity coefficients, γ, by the following equation:[15]

$$\log\left[(k_{cat}/K_m)_1/(k_{cat}/K_m)_2\right]=\log(\gamma_1/\gamma_2)+constant \qquad [14.4.3.3]$$

The correlation was unexpectedly high as seen in Figure 14.4.3.2,[15] implying that the change of substrate specificity of enzyme in organic solvent stems to a large extent from the energy of desolvation of the substrate.

(2b) Enantioselectivity

Changes in enantioselectivity in various organic solvents was first discovered by Sakurai et al.[17] Later Fitzpatrick and Klibanov studied enantioselectivity of subtilisin, Carsberg in the transesterification between the sec-phenethyl alcohol (a chiral alcohol) and vinyl butyrate to find that it was greatly affected by the solvent. Only the correlations with ε or with D gave good agreements.[18] The enzyme enantioselectivity was inversed by changing solvents.[19] Nakamura et al. studied lipase (Amano AK from *Pseudomonas sp.*)-catalyzed transesterification of *cis*- and *trans*-methylcycolhexanols with vinyl acetate in various organic solvents, and investigated the effect of solvent on activity and stereoselectivity of the lipase.[20] They correlated their stereoselectivity with good linearity (except for dioxane and dibutyl ether) by the following two-parameter equation (Figure 14.4.3.3):

$$E=a(\varepsilon-1)/(2\varepsilon+1)+bV_m+c \qquad [14.4.3.4]$$

where ε and $V_m$ are dielectric constant and molar volume of the solvent, respectively, and a, b, and c are constants which should be experimentally determined.

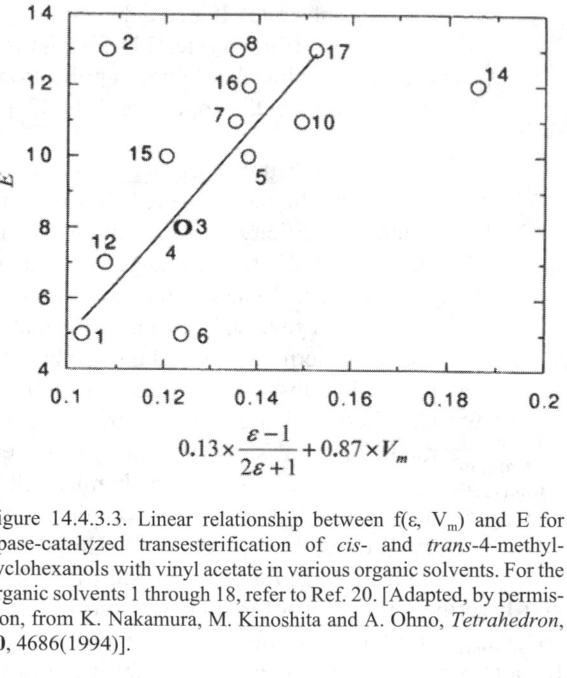


Figure 14.4.3.3. Linear relationship between f(ε, $V_m$) and E for lipase-catalyzed transesterification of *cis*- and *trans*-4-methylcyclohexanols with vinyl acetate in various organic solvents. For the organic solvents 1 through 18, refer to Ref. 20. [Adapted, by permission, from K. Nakamura, M. Kinoshita and A. Ohno, *Tetrahedron*, **50**, 4686(1994)].

Bianchi et al. also reported that for the resolution of antitussive agent, Dropropizine, using both hydrolysis in aqueous buffer and transesterification techniques in various organic solvents, by a lipase (Amano PS from *Pseudomonas cepacia*), E depended very much on organic solvents (Table 14.4.3.3), with the highest E value (589) in n-amyl alcohol and the lowest one (17) in water.[21] In this case, however, there was no correlation between the enantioselectivity and the physico-chemical properties of the solvents such as logP or ε.

**Table 14.4.3.3. Effect of the solvent on enantioselectivity of lipase PS. [Adapted, by permission, from D. Bianchi, A. Bosetti, P. Cesti and P. Golini, Tetrahedron Lett., 33, 3233(1992)].**

| Solvent | logP | ε | Nucleophile | E |
|---|---|---|---|---|
| $H_2O$ | | 78.54 | $H_2O$ | 17 |
| hexane | 3.5 | 1.89 | n-propanol | 146 |
| $CCl_4$ | 3.0 | 2.24 | n-propanol | 502 |
| toluene | 2.5 | 2.37 | n-propanol | 120 |
| iso-propyl ether | 1.9 | 3.88 | n-propanol | 152 |
| 2-methyl-2-butanol | 1.45 | 5.82 | n-propanol | 589 |
| 2-methyl-2-butanol | 1.45 | 5.82 | $H_2O$ | 63 |
| n-propanol | 0.28 | 20.1 | n-propanol | 181 |
| acetonitrile | -0.33 | 36.2 | n-propanol | 82 |
| 1,4-dioxane | -1.1 | 2.2 | n-propanol | 164 |

As mentioned above, no correlation was reported in a large number of articles on the effects of solvents on the enzyme specificity/selectivity, although correlations between the specificity/selectivity and physico-chemical properties of the solvents were successful in some combinations of an enzymatic reaction for a set of solvents. Therefore, attempts to rationalize the phenomena based on either physico-chemical properties of the solvents or on their structure, are at present clearly unsatisfactory from the point of view of predictable value. Further experiments carried out under more strictly defined condition are necessary to reach the quantitative explanation of the whole phenomena.

### 14.4.3.5 Concluding remarks

Activity, stability, and selectivity of an enzyme are affected considerably by nature of organic solvents as well as free water content in the enzyme-catalyzed reaction in the organic solvents. However, rational criteria for the selection of a proper solvent among vast variety of the organic solvents are very limited so far. Researchers are obliged to resort to empirical approach by examining some kinds of solvent for the enzyme-catalyzed reaction concerned at the present state of art.

### References


1 A.M.P. Koskinen and A.M. Klibasnov, Ed., **Enzymatic Reactions in Organic Media,** *Blackie Academic & Professional* (An Imprint of Chapman & Hall), London, 1996.
2 T. Yamane, *Nippon Nogeikagaku Kaishi*, **65**, 1103(1991).
3 T. Yamane, *J. Am. Oil Chem. Soc.*, **64**, 1657(1987).
4 T. Yamane, Y. Kozima, T. Ichiryu and S. Shimizu, *Ann. New York Acad. Sci.*, **542**, 282(1988).
5 T. Yamane, *Biocatalysis*, **2**, 1(1988).
6 J.A. Rapley, E. Gratton and G. Careri, *Trends Biol. Sci.*, Jan., 18(1983).
7 P.J. Halling, *Biochim. Biophys. Acta*, **1040**, 225(1990).
8 R.H. Valivety, P.J. Halling, A.R. Macrae, *Biochim. Biophys. Acta*, **1118**, 218(1992).
9 C. Laane, S. Boeren and K. Vos, *Trends Biotechnol.*, **3**, 251(1985).
10 C. Laane, S. Boeren, K. Vos and C. Veeger, *Biotechnol. Bioeng.*, **30**, 81(1987).

11 A. Zaks, in *Enzymatic Reactions in Organic Media* ed. by A.M.P. Koskinen and A.M. Klibanov, *Blackie Academic & Professional* (An Imprint of Chapman & Hall), London, 1996, p. 84.
12 A. Zaks and A.M. Klibanov, *Science*, **224**, 1249(1984).
13 C.R. Wescott and A.M. Klibanov, *Biocheim. Biophys. Acta*, 1206, 1(1994).
14 G. Carrea, G. Ottolina and S. Riva, *Trends Biotechnol.*, **13**, 63(1995).
15 A. Zaks and A.M. Klibanov, *J. Am. Chem. Soc.*, **108**, 2767(1986).
16 C.R. Wescott and A.M. Klibanov, *J. Am. Chem. Soc.*, **108**, 2767(1986).
17 T. Sakurai, A.L. Margolin, A.J. Russell and A.M. Klibanov, *J. Am. Chem. Soc.*, **110**, 7236(1988).
18 P.A. Fitzpatrick and A.M. Klibanov, *J. Am. Chem. Soc.*, **113**, 3166(1991).
19 S. Tawaki and A.M. Klibanov, *J. Am. Chem. Soc.*, **114**, 1882(1992).
20 K. Nakamura, M. Kinoshita and A. Ohno, *Terehedron*, **50**, 4681(1994).
21 D. Bianchi, A. Bosetti, P. Cesti and P. Golini, *Tetrahedron Lett.*, **33**, 3231(1992).

## 14.5 COIL COATING

George Wypych
**ChemTec Laboratories, Inc., Toronto, Canada**

The coil coating industry is under pressure to eliminate the use of solvents. Polyester coil coatings contain up to 40% of solvents such as glycol esters, aromatic hydrocarbons (e.g., Solvesso 150), alcohols, ketones, and butyl glycol.[1,2] A recent book[1] predicted that the solvent-based technology will not change during the next decade because the industry heavily invested in equipment to deal with solvents. Such changes in technology require long testing before they can be implemented. The coil coating industry normally recovers energy from evaporated solvents either by at-source incineration or by a recycling process which lowers emissions. Because of the large amount of solvents used, the use of PVC and fluorocarbon resins in some formulations, and the use of chromates in pretreatments the pressure remains on the industry to make improvements.[3] The coil coating industry is estimated to be consuming about 50,000 tons of solvents both in Europe and in the USA.[1] About half of these solvents are hydrocarbons.

According to the published studies,[3,4] efforts to change this situation did start in the early 1990s and by mid nineties research data were available to show that the technology can be changed. Two directions will most likely challenge the current technology: radiation curing and powder coating.

Coil coats are thin (about 30 μm wet thickness) but contain a high pigment loading. Consequently, UV curing is less suitable than electron beam curing. The application of this technology requires a change to the polymer system and acrylic oligomers are the most suitable for this application. This system can be processed without solvents. If a reduction of viscosity is required, it can be accomplished by the use of plasticizers (the best candidates are branched phthalates and linear adipates) and/or reactive diluents such as multifunctional monomers. Results[3] show that the UV stability of the system needs to be improved by using a polyester top coat or fluoropolymer. With top coat, material performs very well as learned from laboratory exposures and exposures in industrial environment.[3] At the time of the study (about 6 years ago), process of coating was less efficient than solvent-based system because production speed was about 6 times slower than the highest production rates in the industry (120 m/min). At the same time, it is known[2] that the quality of solvent-based coatings suffers from excessive production rates as well. Radiation curing has a disadvantage because of its high capital investment but it does have an economical advantage because the process is very energy efficient. Previous experiences with radiation curing technology show that the process has been successfully implemented in several industries such as paper, plastic processing, and wood coating where long term economic gains made the changes viable.

Comparison of solvent-based fluoropolymer and fluorocarbon powder coating developed in Japan[4] shows that elimination of solvent is not only good for environment but also improves performance (UV stability especially is improved). The study was carried out with a very well designed testing program to evaluate the weathering performance of the material.

These two technologies show that there is extensive activity to improve coil coatings with simultaneous elimination of solvents. Two recent patents contribute more information on the developments in the coil coating industry.[5,6] One problem in the industry is with the poor adhesion of the coating to steel.[2,3] A primer developed contains dipropylene glycol methyl ether and PM acetate which allows the deposition of relatively thick layers (20-40 μm) without blistering and at suitable rate of processing. However, the primer has a low solids content (30-45%).[5] A new retroreflective coating was also developed[6] which is based on ethyl acrylate-styrene copolymer and contains a mixture of xylene with another aromatic hydrocarbon (Solvesso 150) at relatively low concentration (11-12%).

## REFERENCES


1 B P Whim, P G Johnson, **Directory of Solvents**, Blackie Academic & Professional, London, 1996.
2 A L Perou, J M Vergnaud, *Polym. Testing*, **16**, No.1, 19-31 (1997).
3 G M Miasik, *Surface Coatings International*, **79**, No.6, 258-67 (1996).
4 C Sagawa, T Suzuki, T Tsujita, K Maeda, S Okamoto, *Surface Coatings International*, **78**, No.3, 94-8 (1995).
5 M T Keck, R J Lewarchik, J C Allman, **US Patent 5,688,598**, Morton International, Inc., 1997.
6 G L Crocker, R L Beam, **US Patent 5,736,602**, 1998.


## 14.6 COSMETICS AND PERSONAL CARE PRODUCTS

GEORGE WYPYCH
**ChemTec Laboratories, Inc., Toronto, Canada**

Several cosmetic products contain solvents. These include nail polish, nail polish remover, fragrances, hair dyes, general cleaners, hair sprays and setting lotions. In most cases, ethanol is the only solvent. Nail polish and nail polish remover contain a large variety of solvents. Several recent patents[3-6] give information on current developmental work in this area.

Nitrocellulose, polyester, acrylic and methacrylic ester copolymer, formaldehyde resin, rosin, cellulose acetate butyrate are the most frequently used polymers in nail polish formulations. Solvents were selected to suit the polymer used. These include acetone, methyl acetate, ethyl acetate, butyl acetate, methyl glycol acetate, methyl ethyl ketone, methyl isobutyl ketone, toluene, xylene, isopropyl alcohol, methyl chloroform, and naphtha. Solvents constitute a substantial fraction of the composition usually around 70%. Reformulation is ongoing to improve the flexibility and durability of the nail polish.[3] Other efforts are directed to improve antifungal properties,[4] to eliminate ketones and formaldehyde resin (ketones because of their toxicity and irritating smell and formaldehyde resins because they contribute to dermatitis),[5] and elimination of yellowing.[6] All efforts are directed towards improvements in drying properties and adhesion to the nails. These properties are partially influenced by solvent selection. The current trend is to greater use of ethyl acetate, butyl acetate, naphtha, and isopropanol which are preferable combinations to the solvents listed above.

Acetone used to be the sole component of many nail polish removers. It is still in use but there is a current effort to eliminate the use of ketones in nail polish removers. The combinations used most frequently are isopropanol/ethyl acetate and ethyl acetate/isopropanol/1,3-butanediol.

General cleaners used in hairdressing salons contain isopropanol and ethanol. Hair spray contains ethanol and propellants which are mixtures of ethane, propane, isobutane, and butane. The reported study[2] of chemical exposure in hairdresser salons found that although there were high concentrations of ethanol the detected levels were still below the NIOSH limit. The concentrations were substantially higher in non-ventilated salons (about 3 times higher) than those measured in well ventilated salons. Small concentrations of toluene were found as well, probably coming from dye components.

Recent patents[7,8] show that solvents may enter cosmetic products from other ingredients, such as components of powders and thickening agents. Some solvents such as dichloromethane and benzene, even though they are present is smaller quantities are reason for concern.

## REFERENCES

1 B P Whim, P G Johnson, **Directory of Solvents**, *Blackie Academic & Professional*, London, 1996.
2 B E Hollund, B E Moen, *Ann. Occup. Hyg.*, **42** (4), 277-281 (1998).
3 M F Sojka, **US Patent 5,374,674**, Dow Corning Corporation. 1994.
4 M Nimni, **US Patent 5,487,776**, 1996.
5 F L Martin, **US Patent 5,662,891**, Almell Ltd., 1997.
6 S J Sirdesai, G Schaeffer, **US Patent 5,785,958**, OPI Products, Inc., 1998.
7 R S Rebre, C Collete, T. Guertin, **US Patent 5,563,218**, Elf Atochem S. A., 1996.
8 A Bresciani, **US Patent 5,342,911**, 3V Inc., 1994.

# 14.7 DRY CLEANING - TREATMENT OF TEXTILES IN SOLVENTS

Kaspar D. Hasenclever
**Kreussler & Co.GmbH, Wiesbaden, Germany**

Most processes in manufacturing and finishing of textiles are aqueous. In order to prevent water pollution, some years ago developments were made to transform dyeing-, cleaning- and finishing-processes from water to solvents. During 1970-1980 solvent processes for degreasing, milling, dyeing and waterproofing of textiles could get limited economic importance. All these processes were done with tetrachloroethylene (TCE). After getting knowledge about the quality of TCE penetrating solid floors, stone and ground, contaminating groundwater, this technology was stopped.

Today the main importance of solvents in connection with textiles is given to dry cleaning, spotting and some special textile finishing processes.

## 14.7.1 DRY CLEANING

### 14.7.1.1 History of dry cleaning

Development of dry cleaning solvents

The exact date of discovery of dry cleaning is not known. An anecdote tells us that in about 1820 in Paris, a lamp filled with turpentine fell down by accident and wetted a textile. After the turpentine was vaporized, the wetted areas of the textile were clean, because the turpentine dissolved oily and greasy stains from it.

In 1825, Jolly Belin founded the first commercial "dry laundry" in Paris. He soaked textile apparel in a wooden tub filled with turpentine, cleaned them by manual mechanical action, and dried them by evaporating the turpentine in the air.

After getting the know-how to distill benzene from tar of hard coal in 1849, this was used as a solvent for dry cleaning because of its far better cleaning power. But benzene is a strong poison, so it was changed some decades later to petrol, which is explosive. In order to reduce this risk, petrol as dry cleaning solvent was changed to white spirit (USA: Stoddard solvent) with a flash point of 40 - 60°C (100 - 140°F) in 1925.

The flammability of the hydrocarbon solvents in dry cleaning plants was judged to be risky because of fire accidents. After finding the technology for producing inflammable chlorinated hydrocarbons, trichloroethane and tetrachloroethylene (TCE) were introduced in dry cleaning since about 1925. These solvents gave the opportunity for good cleaning results and economic handling. Up to 1980, TCE was the most important solvent for dry cleaning worldwide.

Compared to TCE, fluorinated chlorinated hydrocarbons (CFC) offer benefits to dry cleaning because of their lower boiling points and their more gentle action to dyestuffs and fabrics. So since 1960 these solvents have had some importance in North America, Western Europe, and the Far East. They were banned because of their influence on the ozone layer in the stratosphere by the UNESCO's Montreal Protocol in 1985.

At the same time, TCE was classified as a contaminant to groundwater and as a dangerous chemical to human health with the possible potential of cancerogenic properties. As a result of this, hydrocarbon solvents on the basis of isoparaffins with a flash point higher than 55°C (130°F) were at first used in Japanese and German dry cleaning operations since 1990.

Also, alternatives to conventional dry cleaning were developed. Wet cleaning was introduced by Kreussler in 1991 (Miele System Kreussler) and textile cleaning in liquid carbon dioxide was exhibited by Global Technologies at the Clean Show Las Vegas in 1997.

Development of dry cleaning machines

In the beginning, dry cleaning was done manually in wooden tubs filled with turpentine or benzene. About 1860, the Frenchman Petitdidier developed a wooden cylindrical cage, which was rotated in a tub filled with solvent. The apparel to be dry cleaned was brought into the cage and moved through the solvent by the rotation of the cylinder. This machine got the name "La Turbulente".

The next step was the addition of a centrifuge to the wooden machine. The dry cleaned apparel was transferred from the machine into the centrifuge and then dried by vaporizing the solvent in the open air.

About 1920, a tumble dryer was used for drying the dry cleaned textiles. Fresh air was heated up, blown through the dryer, where the air was saturated with solvent vapor and then blown out to be exhausted into the environment. The used solvent was cleaned to be recycled by separation of solid matter by centrifugal power with a separator and to be cleaned from dissolved contamination by distilling.

In 1950, dry to dry machines were developed by Wacker in Germany for use with TCE. The principle of working is as follows: The cage, filled with textiles, is rotating in a closed steel cylinder. The dry cleaning solvent is pumped from the storage tank into the cylinder so that the textiles in the cage are swimming in the solvent. After ending the cleaning process, the solvent is pumped back into the storage tank and the cage rotates with high speed (spinning) in order to separate the rest of the solvent from the textile. Then air is circulated through a heat exchanger to be heated up. This hot air is blown into the rotating cage in order to vaporize the remaining solvent from the textiles. The air saturated with solvent vapor is cleaned in a condenser, where the solvent is condensed and separated from the air. The air now goes back into the heat exchanger to be heated up again. This circulation continues until textiles are dry. The separated solvent is collected in a tank to be reused.

In order to reduce solvent losses and solvent emissions, since 1970 charcoal filters have been used in the drying cycle of dry cleaning machines, so that modern dry cleaning systems are separated from the surrounding air.

### 14.7.1.2 Basis of dry cleaning

Dry cleaning means a cleaning process for textiles, which is done in apolar solvents instead of water. If water is used, such cleaning process is called "washing" or "laundering".

Natural textile fibers, such as wool, cotton, silk and linen, swell in water because of their tendency to absorb water molecules in themselves. This causes an increase of their diameter and a change of the surface of yarns and fabrics. The result is shrinkage, felting and creasing.

Apolar solvents, such as hydrocarbons, are not absorbed by natural textile fibers because of the high polarity of the fibers. So there is no swelling, no shrinkage, no felting and no creasing. From the solvent activity, dry cleaning is very gentle to textiles, with the result that the risk of damaging garments is very low.

Because of the apolar character of dry cleaning solvents cleaning activity also deals with apolar contamination. Oils, fats, grease and other similar substances are dissolved in the dry cleaning operation. Polar contamination, such as, salt, sugar, and most nutrition and body excrements are not dissolved.

In the same way that water has no cleaning activity with regard to oils, fat, or grease, in dry cleaning there is no activity with regard to salts, sugar, nutrition and body excrements. And in the same way that water dissolves these polar substances, cleaning of textiles from these contaminants does not pose any problems in a washing process.

To make washing active to clean oils, fat and grease from textiles, soap (detergent) has to be added to water. To make dry cleaning active to clean salts, sugar and the like from textiles, dry cleaning detergent has to be added to the solvent.

In washing, fresh water is used for the process. After being used, the dirty washing liquid is drained off. In dry cleaning the solvent is stored in a tank. To be used for cleaning it is pumped into the dry cleaning machine. After being used, the solvent is pumped back into the storage tank. To keep the solvent clean it is constantly filtered during the cleaning time. In addition to this, a part of the solvent is pumped into a distilling vessel after each batch to be cleaned by distilling.

Dry cleaning solvents are recycled. The solvent consumption in modern machines is in the range of about 1 - 2 % per weight of the dry cleaned textiles.

### 14.7.1.3 Behavior of textiles in solvents and water

Fibers used for manufacturing textiles can be classified into three main groups:

- Cellulosic fibers: cotton, linen, rayon, acetate.
- Albumin fibers: wool, silk, mohair, camelhair, cashmere.
- Synthetic fibers: polyamide, polyester, acrylic.

Textiles made from cellulosic fibers and synthetics can be washed without problems. Apparel and higher class garments are made from wool and silk. Washing very often bears a high risk. So these kinds of textiles are typically dry cleaned.

Dependent to the relative moisture of the surrounding air fabrics absorb different quantities of water. The higher the polarity of the the fiber, the higher is their moisture content. The higher the swelling (% increase of fiber diameter) under moisture influence, the higher is the tendency of shrinkage in a washing or dry cleaning process.

**Table 14.7.1. Water content (%) in textile fibers dependent on relative humidity**

| Fiber | Relative humidity, % | | | Swelling, % |
|---|---|---|---|---|
| | 70 | 90 | max. | |
| Viscose | 14.1 | 23.5 | 24.8 | 115 |
| Wool | 15.6 | 22.2 | 28.7 | 39 |
| Silk | 11.2 | 16.2 | 17.7 | 31 |
| Cotton | 8.1 | 11.8 | 12.9 | 43 |
| Acetate | 5.4 | 8.5 | 9.3 | 62 |
| Polyamide | 5.1 | 7.5 | 8.5 | 11 |
| Acrylic | 2.1 | 4.0 | 4.8 | 9 |
| Polyester | 0.5 | 0.6 | 0.7 | 0 |

The same absorption of water occurs if textiles are immersed in solvent in a dry cleaning machine. If the relative humidity in the air space of the cylinder of a dry cleaning machine is higher, textiles absorb more water. The water content of textiles in solvent is equal to the water content of textiles in the open air if the relative humidity is the same.

If water additions are given to the dry cleaning solvent in order to intensify the cleaning effect with regard to polar contamination, then the relative humidity in the air space of the cylinder of the dry cleaning machine increases. As a result of this, the water content in the dry cleaned textiles increases, too, so that swelling begins and shrinkage may occur.

Woolen fabrics are particularly sensitive to shrinkage and felting because of the scales on the surface of the wool fiber. Not only in washing, but also in dry cleaning there is a risk of shrinkage on woolen garments. This risk is higher when the dry cleaning process is influenced by water addition. If the relative humidity in the air space of a dry cleaning machine is more than 70%, shrinkage and felting may occur, if the dry cleaning solvent is tetrachloroethylene (TCE). In hydrocarbon solvents wool is safe up to 80% relative humidity, because of the lower density of hydrocarbons compared to TCE, which reduces mechanical action.

#### 14.7.1.4 Removal of soiling in dry cleaning

"Soiling" means all the contamination on textiles during their use. This contamination is of very different sources. For cleaning purposes, the easiest way of classification of "soiling" is by solubility of soiling components. The classification can be done by definition of four groups:

- Pigments: Substances are insoluble in water and in solvents. Examples are: dust, particles of stone, metal, rubber; soot, scale of skin.
- Water soluble material: Examples are: salts, sugar, body excrements, sap and juice.
- Polymers: Substances are insoluble in solvents but can be soaked and swell in water. Examples are: starch, albumin and those containing material such as blood, milk, eggs, sauce.
- Solvent soluble material: Examples are: oil, fat, grease, wax, resins.

**Table 14.7.2 Average soiling of garments (apparel) in Europe**

| Soil type | Proportion, % | Solubility | Components |
|---|---|---|---|
| Pigments | 50 | not | dust, soot, metal oxides, rub-off, pollen, aerosols |
| Water soluble | 30 | water | sugar, salt, drinks, body excretions |
| Polymers | 10 | water | starch, albumin, milk, food |
| Solvent soluble | 10 | solvents | skin grease, resin, wax, oils, fats |

In practice, the situation is not so simple as it seems to appear after this classification. That is because soiling on garments almost always contains a mixture of different substances. For example, a spot of motor oil on a pair of trousers consists of solvent soluble oil, but also pigments of soot, metal oxides and other particles. The oil works as an adhesive for the pigments, binding them to fabric. In order to remove this spot, first the oil has to be dissolved then pigments can be removed.

The removal of solvent soluble "soiling" in dry cleaning is very simple. It will be just dissolved by physical action. Polymers can be removed by the combined activity of detergents, water and mechanical action. The efficiency of this process depends on the quality and concentration of detergent, the amount of water added into the system, and the operating time. Higher water additions and longer operating time increase the risk of shrinkage and felting of the textiles.

Water soluble "soiling" can be removed by water, emulsified in the solvent. The efficiency here depends on the emulsifying character of the detergent and the amount of water addition. The more water is emulsified in the solvent, the higher is the efficiency of the process and the higher is the risk of shrinkage and felting of the textiles.

Pigments can be removed by mechanical action and by dispersing activity of detergents. The higher the intensity of mechanical action, influenced by cage diameter, rotation, gravity of the solvent and operating time and the better the dispersing activity of the detergent, the better is the removal of pigments. The same parameters influence the care of the textiles. The better the cleaning efficiency, the higher the risk for textile damage.

#### 14.7.1.5 Activity of detergents in dry cleaning

The main component of detergents is surfactant. The eldest known surfactant is soap. Chemically soap is an alkaline salt of fatty acid. Characteristic of soap (and surfactant) is the molecular structure consisting of apolar - hydrophobic - part (fatty acid) and a polar - hydrophilic - part (-COONa), causing surface activity in aqueous solution.

Surface activity has its function in the insolubility of the hydrophobic part of molecule in water and the hydrophilic part of molecule influences water solubility. This gives a tension within soap molecules in water forming layers on every available surface and forming micelles if there is a surplus of soap molecules compared to the available surface. This soap behavior stands as an example of mechanism of action of surfactants in general.

Micelles of surfactants in water are formed by molecular aggregates of surfactants oriented in such a way that the hydrophobic parts are directed internally, so that the hydrophilic parts are directed outwards. In this way the aggregates form spheres, cylinders, or laminar layers, dependent on its concentration.

Because of this behavior, it is possible to remove oil, fat or grease from substrates in aqueous solutions, if surfactants are present. The surfactants act to disperse the oil into small particles and build up micelles around these particles, so that oil, fat or grease incorporated inside the micelle (Figure 14.7.1). If aggregates are small, the solution is clear. If aggregates are larger, the solution (emulsion - type oil in water) becomes milky.

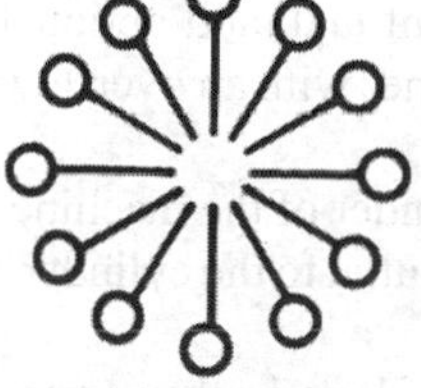

Figure 14.7.1. Schematic diagram of surfactants aggregated in micelle in aqueous solution.

In the same way, but in the opposite direction, surfactants form micelles in solvents (Figure 14.7.2). In this case, not the hydrophobic, but the hydrophilic part is directed internally and the hydrophobic outwards. Emulsions in this case are not formed by oil in water, but by water in oil (solvent).

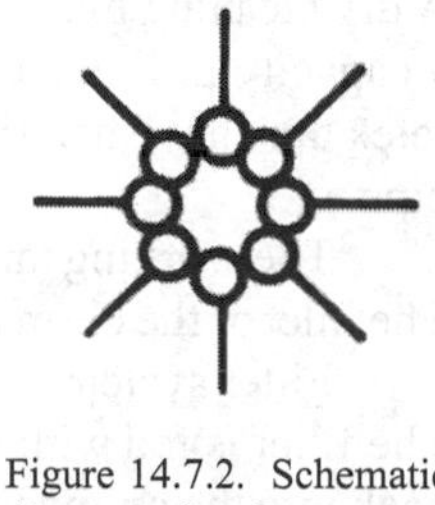

Figure 14.7.2. Schematic diagram of surfactants aggregated in micelle in solvent solution.

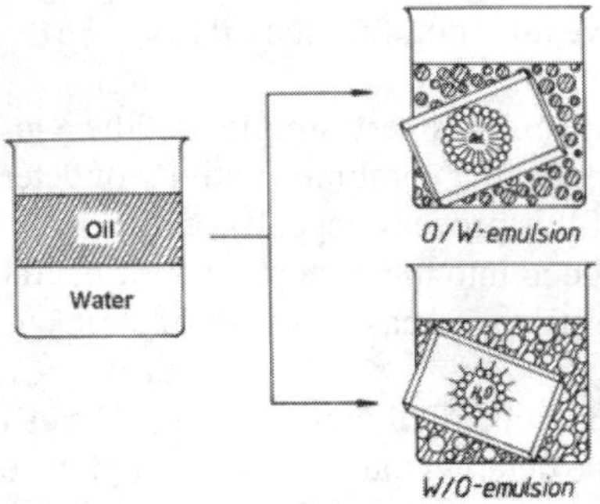


Figure 14.7.3. Aggregation of surfactants in emulsions of type oil in water (O/W) and water in oil (W/O).

Because of this behavior, it is possible to remove water soluble (polar) material from substrates in solvents, if surfactants are at present. The surfactants disperse the polar material into very small particles and build up micelles around the particles, so that the polar material is totally incorporated inside micelle. If the aggregates are small, the solution is clear. If they are large, the solution (emulsion - type water in oil) becomes milky.

Drycleaning detergents consist of surfactants, cosolvents, lubricants, antistatic compounds and water. They are used in order to increase the cleaning efficiency, to improve the handle of dry cleaned textiles, and to prevent electrostatic charge on textiles. They are formulated as liquids in order to be easily added to the solvent by automatic dosing equipment. In dry cleaning the drycleaning detergents play the same role as soap or laundry detergents in textile washing.

### 14.7.1.6 Dry cleaning processes

Process technology in dry cleaning has the target to clean garments as good as possible without damaging them, at lowest possible costs and with highest possible safety. Because of the environmental risks dependent on the use of TCE or hydrocarbon solvents, safe operation is the most important. Modern dry cleaning machines are hermetically enclosed, preventing infiltration of surrounding air, operating with electronic sensor systems, and they are computer controlled. The main part is the cylinder with the cleaning cage, the solvent storage tanks, the solvent recovering system, the solvent filter system and the distillation equipment (Figure 4.9.4).

Cylinder/cage:

A cage capacity is 20 liters per kg load capacity of the machine (about 3 gal per lb). That means, an average sized dry cleaning machine with load capacity of 25 kg (50 lbs) has a cage capacity of 500 liters (130 gal). This size is necessary because the same cage is used for cleaning and drying.

During the cleaning cycle, 3 - 5 liters of solvent per kg load (0.5 - 0.8 gal/lb) are pumped into the cylinder. In a 25 kg machine, it is 80 - 125 liters (25-40 gal). This solvent is filtered during the cleaning time.

Solvent storage tanks:

A dry cleaning machine has 2 - 3 solvent storage tanks. The biggest - the working tank - has a capacity in liters ten times the load capacity in kg. The clean solvent tank and optional retex tank have half the capacity. All the tanks are connected to each other with an overflow pipe.

The working tank with its inlet and outlet is connected to the cylinder of the machine. The inlet of the clean tank is connected to the distilling equipment, the outlet to the cylinder.

Filter system:

The filter is fed with solvent from the cylinder by pump pressure. The filtered solvent goes back into the cylinder. The filter has a further connection to the distilling equipment used as drain for the residue.

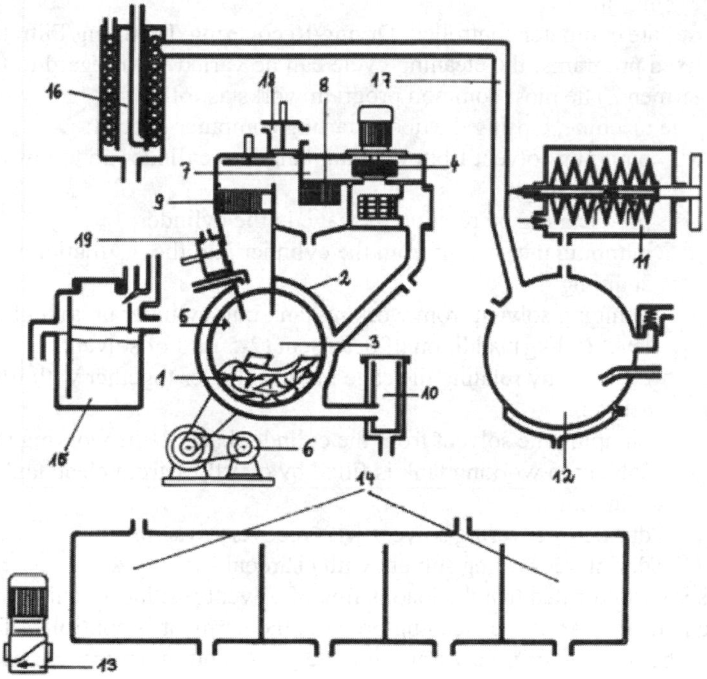

Figure 14.7.4. Scheme of a dry cleaning machine. 1 textiles to be cleaned, 2 cylinder, 3 cage, 4 fan, 5 heater, 6 drive for cage, 7 air canal, 8 cooler, 9 lint filter, 10 button trap, 11 filter, 12 distilling vessel, 13 pump, 14 solvent storage tanks, 15 water separator, 16 condenser, 17 steam pipe, 18 outlet to charcoal filter, 19 dosing unit.

Distilling equipment:

The distilling vessel has the same capacity as the biggest storage tank. Solvent to be distilled is pumped into the vessel from the cylinder or from the filter. The vessel is steam or electrical heated. The solvent vapor is directed into a condenser (water cooling) and then to a water separator, from where it flows into the clean solvent tank. The residue after distillation remains in the vessel and is pumped into special residue drums.

Recovering system:

In order to dry the dry cleaned garments from the solvent residue which remains after spinning, a fan extracts air out of the cylinder into a condenser, where solvent vapor is condensed out of the air. From there the air is blown into a heater and directed by fan back into the cylinder. In this way the solvent is vaporized away from the garment and after condensing, reused for cleaning.

Dosing equipment:

In order to get the right additions of dry cleaning detergent into the solvent, a dry cleaning machine has dosing equipment working on the basis of a piston pump, which doses the right amount of detergent at the right time into the system.

Computer control:

All the processes are computer controlled. Drying/Recovering, Distilling/Filtration operate according to fixed programs; the cleaning cycle can be varied with regard to the requirements of the garment. The most common program works as follows:

Loading the machine, closing the door, starting computer control:

| | |
|---|---|
| 1 min | pumping solvent from working tank into cylinder up to low dip level (3 l/kg); |
| 3 min | pre-cleaning by rotating the cage in the cylinder; |
| 1 min | pumping the solvent from the cylinder into the distillation vessel; |
| 1 min | spinning; |
| 2 min | pumping solvent from working tank into cylinder up to high dip level (5 l/kg); addition of detergent (2-5 ml/l of solvent); |
| 8 min | cleaning by rotating the cage in the cylinder together with filter action; |
| 1 min | pumping the solvent from the cylinder back to the working tank (the solvent in working tank is filled by overflow from clean tank); |
| 3 min | spinning |
| 8 min | drying /recovering solvent with recovery system |
| 5 min | drying/recovering solvent with charcoal filter. |

After the sensor has indicated that the load is free of solvent residue, opening the door and unloading the machine. After the cleaning process, the garment is controlled for cleaning quality and can be finished or if necessary, it undergoes spotting/recleaning before finishing.

#### 14.7.1.7 Recycling of solvents in dry cleaning

The recycling of solvents in dry cleaning is very important, because solvents are too expensive for single use. Three different systems are used for the cleaning of the solvent in order for it to be recycled:

- Filtration
- Adsorption
- Distillation

Filtration is a simple physical process, separating insoluble parts from the solvent. It is done during the cleaning cycle.

Adsorption is mainly used together with hydrocarbon solvents, because their high boiling temperature is insufficient to separate lower boiling contaminants from the solvent. Adsorption systems use charcoal or bentonites. The solvent is pumped to filters where contaminants with higher polarity than solvent are adsorbed by the adsorbing material. The adsorbing material can adsorb contaminants in a quantity of about 20% of its own weight. After being saturated, the adsorbing material must be replaced and changed to fresh material. The charged adsorbing material is disposed according to regulations, which is cost intensive.

Distillation is the best cleaning method if the boiling point of solvent is significantly lower than the boiling point of possible contaminants. With dry cleaning machines using TCE, distilling is the normal recycling method. The boiling point of TCE is 122°C, which makes steam heating possible, so that the process can be done safely and cost effectively. The distillation residue consists of removed soil and detergents. Its quantity is much lower than in the adsorption systems, because there is no waste from adsorbing material. Disposal costs are much lower than with adsorption.

## 14.7.2 SPOTTING

### 14.7.2.1 Spotting in dry cleaning

Spotting is the removal of stains from textiles during professional textile cleaning. Correct cleaning, good lighting conditions, appropriate equipment, effective spotting agents, and expert knowledge are indispensable.

Lighting conditions

Lighting of the spotting table with composite artificial light consisting of a bluish and a yellowish fluorescent tube attached above the table approx. 80 cm from the standing point is better suited than daylight. This ensures high-contrast, shadow-free lighting of the work surface and allows for working without fatigue.

Equipment

Basic equipment should comprise a spotting table with vacuum facility, sleeve board, steam and compressed air guns, and a spray gun for water. Spotting brushes should have soft bristles for gentle treatment of textiles. Use brushes with bright bristles for bright textiles and brushes with dark bristles for dark textiles. Use spatulas with rounded edges for removing substantial staining.

### 14.7.2.2 Spotting agents

In dry cleaning, normally three groups of spotting agents are used:

Brushing agents

Brushing agents containing surfactants and glycol ethers dissolved in low viscosity mineral oil and water. Brushing agents are used for pre-spotting to remove large stains from textiles. They are applied undiluted with a soft brush or sprayed onto the heavily stained areas before dry cleaning.

Special spotting agents

Special spotting agents are used for removing particular stains from textiles. The range consists of three different products in order to cover a wide range of different stains. The products are applied as drops directly from special spotting bottles onto the stain and are allowed to react. The three products are:

- Acidic solution of citric acid, glycerol surfactants, alcohol and water for removing stains originating from tannin, tanning agent, and fruit dye.
- Basic solution of ammonia, enzymes, surfactants, glycol ethers and water for removing stains originating from blood, albumin, starch, and pigments.
- Neutral solution of esters, glycol ethers, hydrocarbon solvents and surfactants for removing stains originating from paint, lacquer, resin, and adhesives.

Post-spotting agents

Stains that could not be removed during basic cleaning must be treated with post-spotting agents. Most common is a range of six products, which are used in the same way, as the special spotting agents:

- Alkaline spotting agent for stains originating from starch, albumin, blood, pigments
- Neutral spotting agent for stains originating from paint, lacquer, grease, and make-up.
- Acidic spotting agent for stains originating from tannin, fruits, beverages, and rust.
- Acidic rust remover without hydrofluoric acid.

- Solvent combination for removing oily and greasy stains from textiles.
- Bleaching percarbonate as color and ink remover.

### 14.7.2.3 Spotting procedure

Correct procedure for spotting

It is recommended to integrate the three stages of spotting

- Brushing
- Special spotting
- Post-spotting

into the process of textile cleaning. Perform brushing and special spotting when examining and sorting the textiles to be cleaned.

Brushing

Examine textiles for excessive dirt, particularly at collars, pockets, sleeves, and trouser legs. If textiles are to be dry cleaned check them particularly for stains originating from food or body secretions.

If the textiles are to be wet-cleaned check them particularly for greasy stains. Apply a small quantity of brushing agent onto the stained areas and allow to react for 10-20 minutes before loading the cleaning equipment.

Special spotting

Intensive staining found during the examination of the textiles can be treated with special spotting agents. The stain substance must be identified and related to one of the following categories:

- coffee, tea, fruit, red wine, grass, urine
- blood, food, pigments, sweat
- wax, paint, lacquer, make-up, pen ink, adhesives

Depending on the category of the stain substance, apply the special product with the dripping spouts in the work bottle onto the stain and tamp it gently with a soft spotting brush. Allow to react for 10-20 minutes before loading the cleaning equipment.

Post-spotting

Stains that could not be removed during dry-cleaning or wet cleaning with machines are subject to post-spotting. Proceed as follows:

- Place the garment with the stain area onto the perforated vacuum surface of the spotting table.
- Identify stain, drip appropriate product undiluted onto the stain and tamp it gently with the brush.
- For stubborn stains, allow product to react for up to 3 minutes.
- Use vacuum to remove product and rinse spotted area with steam gun.
- Dry with compressed air, moving the air gun from the edges to the center of the spotted area.

Hidden spot test

If it is suspected that a stain cannot be removed safely due to the textile material, the compatibility of the spotting agent should be checked by applying a small quantity of the agent at a hidden part of the garment. If the garment passes the test, the spotting agent is expected to be successful without damaging the garment.

### 14.7.3 TEXTILE FINISHING

The use of solvents for textile finishing has only some importance for the treatment of fully fashioned articles. The processes are done in industrial dry cleaning machines. The advantages are the same as in dry cleaning compared to washing: lower risk against shrinkage and damage of sensitive garments.

#### 14.7.3.1 Waterproofing

In Northern Europe, North America and Japan, some kinds of sportsware need to be waterproofed. The treatment is done in dry cleaning machines with a load capacity larger than 30 kg (> 60 lbs). The machines need to be equipped with a special spraying unit, which allows one to spray a solution of waterproofing agent into the cage of the machine.

The waterproofing agents consists of fluorocarbon resins dissolved in a mixture of glycol ethers, hydrocarbon solvents and TCE. This solution is sprayed onto the garments, which are brought into the cage of the machine. The spraying process needs about 5 - 10 min. After spraying, the solvent is vaporized in the same way as drying in dry cleaning, so that the fluorocarbon resin will stay on the fibers of the garments. In order to get good results and the highest possible permanence of water resistance, the resin needs to be thermally fixed. In order to meet these requirements, a drying temperature of > 80°C (= 175°F) is necessary.

#### 14.7.3.2 Milling

Solvent milling has some importance for the treatment of fully fashioned woolen knitwear. The process is very similar to normal dry cleaning. The specific difference is the addition of water together with the detergent, in order to force an exact degree of shrinkage and/or felting.

Milling agents are similar to dry cleaning detergent. They have specific emulsifying behavior, but no cleaning efficiency. The process runs like this: textiles are loaded into the machine, then solvent (TCE) is filled in before the milling agent diluted with water is added. After this addition, the cage rotates for 10 - 20 min. The higher the water addition, the higher the shrinkage; the longer the process time, the higher the felting. After this treatment, the solvent is distilled and textile load is dried. High drying temperature causes a rather stiff handle, low drying temperatures give more elastic handle.

#### 14.7.3.3 Antistatic finishing

Antistatic finishing is used for fully fashioned knitwear - pullovers made from wool or mixtures of wool and acrylic. The process is equal to dry cleaning. Instead of dry cleaning detergent, the antistatic agent is added.

Antistatic agents for the treatment of wool consist of cationic surfactants such as dialkyl-dimethylammonium chloride, imidazolidione or etherquats. Antistatic agents for the treatment of acrylic fibers are based on phosphoric acid esters.

# 14.8 ELECTRONIC INDUSTRY - CFC-FREE ALTERNATIVES FOR CLEANING IN ELECTRONIC INDUSTRY

MARTIN HANEK, NORBERT LÖW
**Dr. O. K. Wack Chemie, Ingolstadt, Germany**
ANDREAS MÜHLBAUER
**Zestron Corporation, Ashburn, VA, USA**

## 14.8.1 CLEANING REQUIREMENTS IN THE ELECTRONIC INDUSTRY

The global ban of the CFCs has fundamentally changed cleaning in the electronic industry. Manufacturing processes were developed with the goal of avoiding cleaning. However, this objective has only be partially realized.

Furthermore, there are still a lot of different areas in the electronic industry where cleaning of assemblies is highly recommended and necessary. Examples of such applications can be found in the aviation and space industries and in addition in the rapidly growing telecommunication industry. When investigating the production process of printed circuit assemblies, there are several important cleaning applications that have varying degrees of impact on the quality of the manufactured assemblies (see Figure 14.8.1).

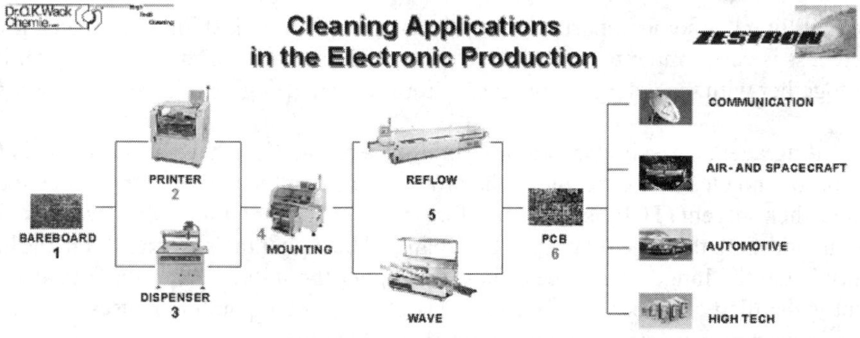

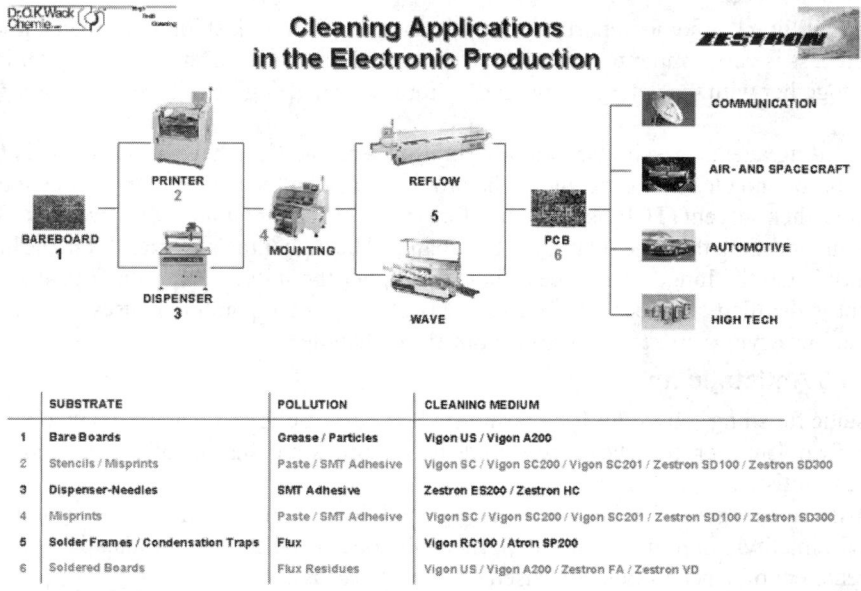

| | SUBSTRATE | POLLUTION | CLEANING MEDIUM |
|---|---|---|---|
| 1 | Bare Boards | Grease / Particles | Vigon US / Vigon A200 |
| 2 | Stencils / Misprints | Paste / SMT Adhesive | Vigon SC / Vigon SC200 / Vigon SC201 / Zestron SD100 / Zestron SD300 |
| 3 | Dispenser-Needles | SMT Adhesive | Zestron ES200 / Zestron HC |
| 4 | Misprints | Paste / SMT Adhesive | Vigon SC / Vigon SC200 / Vigon SC201 / Zestron SD100 / Zestron SD300 |
| 5 | Solder Frames / Condensation Traps | Flux | Vigon RC100 / Atron SP200 |
| 6 | Soldered Boards | Flux Residues | Vigon US / Vigon A200 / Zestron FA / Zestron VD |

Figure 14.8.1. Cleaning applications in the electronic production.

Cleaning applications in the SMT assembly line typically start with the cleaning of the bare boards (1). The preliminary substrates have to be cleaned in order to remove a variety of residues (particles, chemicals used in the manufacturing process of the bare boards). These residues may have a negative impact on the subsequent production steps. A detailed analysis[1] of different kinds of contamination which might occur on the bare boards revealed that these residues might cause failures in production steps far beyond the initial printing process. Therefore it is important to assure clean surfaces in each production step.

The cleaning applications (2-4) outlined in the SMT process line, shown in Figure 14.8.1, refer to the soldering process. Within this production step the components are placed on the bare board using solder paste that is printed on the solder pads through a stencil. The stencil can be cleaned continuously during the printing process by using a stencil printer underside wiping system. However, it has definitely to be cleaned after removing the stencil from the printer. The solder paste is removed from the apertures of the stencils in order to assure an accurate printing image (3).[2,3] If the assemblies are to be mounted on both sides, the components are placed on one side by using an epoxy adhesive (SMT adhesive), in order to prevent them from being accidentally removed during the reflow process (3).[3] The soldering process is carried out in a specially designed reflow oven. The cooling coils which contain condensed flux residue of such reflow ovens also have to be cleaned (5). In the case of wave soldering processes, the flux contaminated solder frames in addition must be cleaned.

The contamination that must be removed after the soldering process is predominantly from flux residues. The removal of flux residues from the soldered assemblies (6) is generally the most critical application. Thus it is the important that the assemblies are cleaned. A subsequent coating process demands a very clean and residue-free surface to assure long term stability of the coating against environmental stresses such as humidity.[1]

Asked for the reasons for cleaning most of the process engineers give the following answers:

1) Stencils and other tools, such as squeegees, are cleaned in order to assure a reproductive and qualitative satisfying printing process.[2]
2) Contaminations from printed circuit assemblies are removed to achieve long-term reliability.
3) Cleaning is an important manufacturing step within the SMT process line if subsequent processes such as coating and bonding are required.

In recent times an additional process step, which cleaning definitely is, was regarded to be time and cost inexpedient. Therefore a lot off optimization techniques have been applied to the SMT production process to try to avoid cleaning. However, considering new technologies in the electronic industry like fine pitch, flip chip and micro-BGA applications, the above statement is no longer true. The question if the removal of flux residues from printed circuit boards is necessary can only be answered by carrying out a detailed process analysis including the costs for cleaning and the increased reliability of the products. A large number of global players in the electronic industry prefer the advantages of better field reliability of their products. This aspect is even more important taking into consideration the outstanding competition in these industries.

Whether a product has to be cleaned, and, especially, when cleaning is necessary in the SMT production depends on the following factors.

- Process costs: cleaning is an additional process step that demands additional investment and resources. However, failures and downtime in other process steps can be diminished.
- Process reliability: due to a larger process window, the manufacturing process as a whole is more stable.
- Product reliability: the functionality and long term reliability of the products has to be assured.

After the decision has been made in order to choose the most suitable cleaning process, the following two questions have to be answered in advance:

- Which contamination has to be removed?
- What are the requirements regarding cleanliness of the surface, especially taking the subsequent process steps into account.

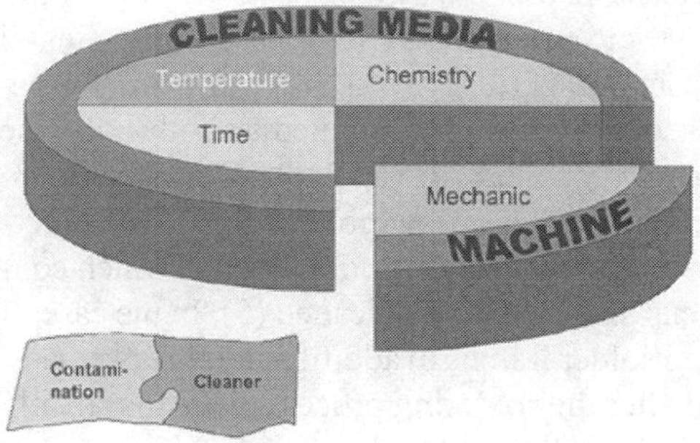

Figure 14.8.2. Factors influencing the cleaning process.

The most important consideration in achieving an optimal cleaning result, is the interaction between the chemistry of the cleaning agent and the type of contamination. This is explicitly shown in the below presented lock and key principle.

However, the cleaning result is not only depended on the nature of the chemistry used for cleaning. Other important factors in the cleaning process are the type of cleaning equipment used and the applied process parameters such as temperature, time and method of application of the cleaning agent (see Figure 14.8.2).

In order to optimize a cleaning process for the specific requirements, detailed information on the nature and composition of the kind of contamination that has to be removed is necessary. This tailoring of the chemistry is achieved by carrying out time consuming screenings matching typical contamination with suitable solvents.

Another very important aspect is the material compatibility of the cleaning agent and the substrates that need to be cleaned. The cleaning chemistry should not attack the different materials of the components in any manner. Consequently, a number of long-term compatibility tests have to be carried out to assure this very important process requirement.

Also, more environmental and worker safety issues are ever increasing when evaluating a new cleaning process. Cleaning processes approved by the industry have to pass even higher standards than are demanded from the government. They have to show, under the worst case scenario, a large process window and a large process margin of safety with respect to personal and environmental exposure.

If all of the above mentioned process requirements such as

- excellent cleaning result,
- long-term functional reliability of the substrates after cleaning,
- material compatibility between cleaning chemistry and the substrates, and
- no drawbacks due to environmental and safety issues

should be fulfilled, a lot of time and know-how has to be invested.

## 14.8.2 AVAILABLE ALTERNATIVES

Due to the global ban of the CFCs, a large number of different cleaning processes[4] have been developed. Each of them has advantages and disadvantages regarding the above mentioned process requirements. The available cleaning processes can be divided into the following main groups:

1) aqueous processes,
2) semi-aqueous processes,
3) water-free processes based on solvents, and

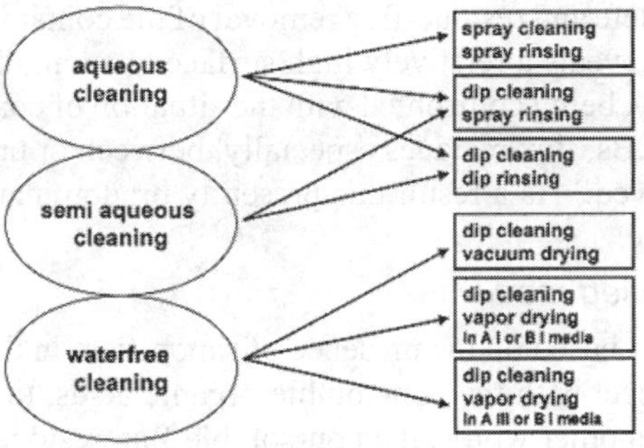

Figure 14.8.3. PCB cleaning.

4) special cleaning processes.

In an aqueous cleaning process,[5] cleaning is achieved by using water or a water-based cleaning agent. In order to avoid drying spots, rinsing is carried out with deionized water. By soldering with water-soluble solder pastes only hot water is necessary for cleaning. However, the use of an alkaline water-based cleaning agent is effective to remove more critical flux residues with low or high pressure spray in cleaning equipment.

In semi-aqueous processes,[4] cleaning is carried out using a high boiling mixture of organic solvents. Rinsing is achieved by using deionized water. The organic solvents used for these cleaners are predominantly based on alkoxypropanols, terpenes, high-boiling alcohols, or mixtures of different chemistries.

In applications based on solvents such as isopropanol or alkoxypropanols[6] a rinse with deionized water is not imperative, since these chemistries do not leave any residues if used for rinsing. Especially when using an uniformly boiling solvent-based cleaning agent, vapor rinsing with solvent vapor is possible. All there mentioned cleaning process will be discussed more detailed in the following sections.

Less widespread applications in the electronic industry include:

- Cleaning using supercritical carbon dioxide (SCF).[7]
- Cleaning using plasma.[7]

SCF cleaning alternative is especially useful for extremely sensitive and complex assemblies. At temperatures above 31°C and pressures above 73 bar, carbon dioxide transforms into a supercritical phase. Supercritical carbon dioxide reveals an extremely low surface tension. Consequently, the wetting of small gaps and complex assembly architectures can be achieved. However, the contamination that should be removed has to be nonpolar, and the compounds should be resistant to high pressures. Due to these basic limitations and the high costs of this cleaning process, this process does not play an important part in the electronic industry.

The advantage of plasma cleaning is in almost waste-free removal of contaminations. The contaminations are transformed into carbon dioxide and water. However, this cleaning process is only suitable for removal of the small amounts of residue from the substrate surface. Combinations with aqueous cleaning processes are feasible and common in the field.

## 14.8.2.1 Water based systems; advantages and disadvantages

### *14.8.2.1.1 Cleaning with DI - water*

Cleaning without using any kind of cleaning agent in the electronic assembly process is possible when water-based fluxes are used for the soldering process. Since these aqueous fluxes are highly activated, aggressive residues occur after the soldering process. These residues have to be removed immediately to avoid severe corrosion of solder joints. The use of water-based fluxes is very common in the US electronic market due to the main advantage of a large soldering process window for soldering. DI-water is used for this purpose since it is able to solubilize the polar residues. However, physical energy, such as high-spray-pressure, is in most cases necessary to assure complete and residue-free removal of the contamination. However, since pure deionized water reveals a relatively high surface tension, the wetting ability is extremely low.[8] Consequently, being confronted with the situation of constant increase of the packaging densities on PCBs, flux residues especially, between or under components, are very difficult to be removed.[8] As a result this presently predominant cleaning process will be in decline in the future.

#### *14.8.2.1.2 Cleaning with alkaline water-based media*

The principle of this kind of cleaning process is based on the presence of saponifiers in the cleaning fluid. These saponifiers are able to react with the non-soluble organic acids, that are the main ingredients of the flux residues. In other words, the non-soluble flux residues are transformed into water-soluble soaps due to a chemical reaction between the saponifier and the flux. This reaction is called saponification. For such a cleaning process an effective rinse with deionized water is imperative to minimize the level of ionic contamination.[1,5]

However, the presence of alkaline components in the aqueous cleaning fluid may cause the phenomenon of corrosion on different metals (Al, Cu, Sn) resulting in the visible dulling or discoloration of solder joints and pads.[9,10]

The chemical nature of the saponifiers is either organic or inorganic. Organic saponifiers are for example monoethanolamine or morpholine.[5] These soft bases transform the insoluble organic carboxylic acids to soluble soaps acting, in addition, as a kind of buffer to stabilize the pH of the cleaning solution. However, the above mentioned organic amines have an associated odor, and more important, are classified as volatile organic compounds (VOCs). Inorganic saponifiers based on the system sodium carbonate/sodium hydrogen carbonate are also very attractive alternatives exhibiting sufficient cleaning ability at pH-values below 11. By using these chemistries the VOC and odor issue can be avoided.[5]

Besides organic or inorganic saponifiers a large scale of different anionic, cationic or nonionic surfactants are present in water-based alkaline cleaning solutions. The main reason for adding surfactants is the requirement for lower surface tension in order to increase the wetting of the substrates. However, using surfactants in cleaning median in high-pressure spray in air cleaning applications, the issue of foaming has to be taken into consideration.

#### *14.8.2.1.3 Aqueous-based cleaning agents containing water soluble organic components*

There is a large number of products on the market that are mixtures of water and water-soluble organic solvents. These media are recommended for the removal of solder paste, SMT adhesives and flux residues from stencils, misprints and populated reflowed PCBs.[11]

The water-soluble organic solvents are predominantly natural long chain alcohols, glycol ether derivatives or furfuryl alcohol. The basic principle of the cleaning process using such fluids is based on the removal of lipophilic contamination such as oil, grease, flux or adhesive through the organic components of the mixtures. After the removal from the surface of the substrates, the lipophilic contamination precipitates out of the water-based cleaning agent and can be removed by using a filtration devices. In order to guarantee a sufficient cleaning performance, these mixtures contain organic components up to 50%. Consequently, these formulations tend to be classified as cleaning agents with high content of volatile organic compounds (VOCs).

### 14.8.2.1.4 Water-based cleaning agents based on MPC® Technology (MPC = Micro Phase Cleaning)

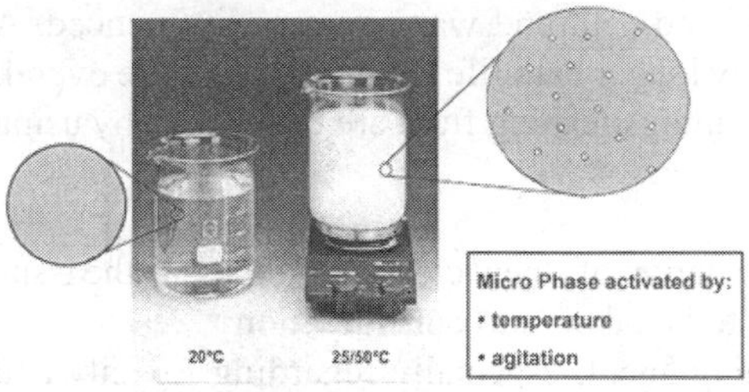


Figure 14.8.4. MPC® cleaning technology.

Cleaning agents based on MPC Technology can be formulated either neutral or alkaline. The media combine the advantages of solvent and aqueous-based formulations revealing an extremely large process window.[2,3] The principle of MPC systems is shown in Figure 14.8.4.

The MPC media are based on formulations consisting of alkoxypropanols and water. Their phase behavior is dependent on the temperature, agitation, and the dragged in contamination. The microphase, which is generated through temperature or media agitation, is responsible for the removal of the lipophilic components of the contamination. The ionics are removed by the aqueous phase.[12] Since the ability to keep the organic contamination in the microphase is limited through the degree of loading, the lipophilic substances are expedited into the aqueous phase. Due to their lipophilic character, they precipitate out of the fluid and can be removed by a simple filtration device. As a result the loading of the cleaning fluid with contamination is minimal thereby minimizing the amount of organic components that finally pollute the subsequent rinsing stages.[3]

Alkaline pH, that is essential for the saponification process, can be realized through biochemical buffer systems, predominately aminoalkanoles.

Due to the described cleaning principle, extremely long bath life times can be guaranteed with excellent cleaning results.[2,3,12] The MPC® media are tailor made for all kinds of different cleaning equipments such as high and low pressure spray systems, spray under immersion, and ultrasonic applications.[3] These cleaning agents based on MPC® Technology also contain low quantities of volatile organic compounds. However, due to this state of the art technology the percentage of VOCs are reduced to a minimal level.

### 14.8.2.1.5 Advantages and disadvantages of aqueous cleaning media

Prior to the implementation of a cleaning process in a production line, the following criteria need to be taken into consideration:

1) Cleaning performance

The substrates that have to be cleaned should be free of all kinds of residues that might have some negative influence on the functionality of the substrates. The contamination consist predominantly of flux residues, solder paste or SMT adhesive. Another important aspect with respect to the cleaning result is the compatibility between substrates and the cleaning chemistry. It is painless to remove the contamination while destroying the substrates.

2) Process reliability and process costs

Every engineer tries to implement a process with a wide process window, and minimizing potential problems during production. While evaluating a cleaning process, the process parameters should be stable over a long period of time. As a result the equipment down time can be minimized, due to maintenance thereby optimizing the overall process cost.

3) Environmental aspects

A cleaning agent should not only be considered based on its content of volatile organic compounds. A medium that does not contain any VOCs may on the one hand have no influence on air pollution. However, on the other hand, soil and water may be influenced. A VOC containing cleaning fluid has an inherently longer bath life. Consequently, the overall environmental calculation for some VOC containing cleaning fluid are better than by using VOC-free media.

4) Personal safety

Another very important aspect for the evaluation of a new cleaning process is the issue on personal safety. The following topics have to be taken into consideration:

- No risks during handling of the cleaning agent, especially regarding toxicity and flash point.
- Low odor of the cleaning fluid.

If the evaluated cleaning agent fulfils these important requirements, the chemistry will be accepted by the operator.

Taking the discussed issues (1-4) into account, the advantages (+) and disadvantages (-) for the different aqueous cleaning fluids are as follows:

a) Aqueous fluids using amines for saponification:

+ excellent removal of flux residues and unsoldered solder paste
+ predominantly VOC-free formulations
- low solid flux residues are difficult to remove
- limited removal of adhesives due to curing of the adhesive
- short bath life times resulting in high costs for media disposal

b) Aqueous fluids using inorganic saponifiers and buffer systems:

+ satisfying cleaning results on flux residues and solder paste
+ VOC-free formulations
+ longer bath life due to constant pH level caused by the buffering systems
- low solid flux residues and SMT adhesives are difficult to remove

c) Mixtures of organic solvents and water:

- small process window regarding flux residues and SMT adhesives
- VOC containing mixtures
- short bath life

d) pH neutral and alkaline fluids based on MPC® Technology:

+ extremely large process window regarding flux residues, solder paste and adhesive removal
+ extremely long bath life-times due to previous described cleaning principle
+ economical cleaning process
- small amounts of VOCs.

### 14.8.2.2 Semi-aqueous cleaners based on halogen-free solvents, advantages and disadvantages

A semi-aqueous cleaning process consists of a wash cycle using a mixture of different organic solvents followed by a rinse with deionized water. Organic contaminations, predominantly rosin flux residues, are removed through the lipophilic solvents, whereas the ionics are minimized by the polar rinse media. The different kinds of solvent formulations that are available on the market reveal an extremely wide process window. Almost all different fluxes that occur in the electronic assembly process can be removed efficiently.

In principle, the media that can be used for semi-aqueous cleaning processes can be classified into two main groups:[4]

- solvent-based mixtures that are not soluble in water, and
- water-soluble cleaning fluids.

The water solubility of the solvent-based cleaning agent is very important for the treatment of the subsequent water rinse.

#### *14.8.2.2.1 Water insoluble cleaning fluids*

The first media designed for this application were predominantly based on terpenes[13] and different mixtures of hydrocarbons and esters. In order to increase the rinse ability of the solvent-based mixtures with deionized water, different surfactants were added.[14] The principle of using a water insoluble cleaning fluid was based on the idea to avoid contaminating the rinse sections through the cleaning fluid. The lipophilic contamination should be kept in the organic cleaning media that can be easily separated from the rinse water by a skimming device. However, experience in the field revealed that this kind of media have some critical drawbacks.

Since the surface tension of the organic cleaning formulations are lower than the surface tension of deionized water (approx. 72 mN/m), the cleaning fluid stays on the surface and will not rinse off. Consequently, the cleaning result is not acceptable with regard to ionic contamination and surface resistivities.[1] Acceptable cleaning results can only be realized through complicated multi-stage rinse processes. Another important disadvantage is the potential risk of re-contamination of the substrate when lifting the substrate out of the rinse section, since the top layer of the rinse section consists of polluted cleaning media. Consequently, effective agitation and skimming of the rinse water is imperative to avoid this phenomenon.[4]

#### *14.8.2.2.2 Water-soluble, water-based cleaning agents*

The solvents used for this application are based on water soluble alcohols, alkoxypropanols and aminoalcohols.[4,6] The chemical structure of these compounds combines a hydrophobic and hydrophilic groups. Due to this ambivalent structure it is possible to optimize the water solubility and lipophilic character for an optimal cleaning performance. Consequently, cleaning fluids based on this principle reveal the largest process window. All different kinds of flux residues, especially low rosin fluxes, are easily rinseable with water resulting in an residue-free cleaning process. The quality and long-term stability of these processes are proven through many applications in the field.[8,15,16] Applications and processes will be discussed more detailed later on.

#### *14.8.2.2.3 Comparison of the advantages (+) and disadvantages (-) of semi-aqueous cleaning fluids*

In order to quantify the effectiveness of the two main groups of solvent based cleaning agents that can be used for semi-aqueous processes, the following key aspects has to be taken into consideration:

- cleaning performance,
- process reliability,
- environmental aspects, and

- personal safety.

a) Non water soluble cleaning agents:
- \+ wide process window regarding fluxes, solder pastes and adhesives
- \- low rosin flux residues are critical to remove
- \- possibility of re-contamination of the substrate in the rinsing stage
- \- cleaning fluids contain volatile organic compounds.

b) Water soluble solvent based cleaning agents:
- \+ extremely wide process window regarding all kinds of fluxes, solder pastes and adhesives
- \+ residue-free removal of the cleaning fluid due to excellent rinse ability
- \- VOC containing formulations

### 14.8.2.3 Other solvent based cleaning systems

Besides the already mentioned solvent based cleaners, there are some special solvents on the market. However, due to their physical and toxicological character, most of them are used for special applications. The advantages and disadvantages of the different products are summarized below.

14.8.2.3.1 <u>Isopropanol</u>

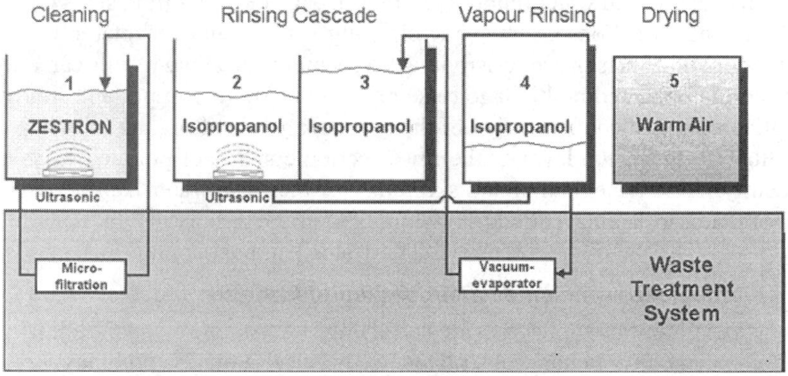


Figure. 14.8.5. Zestron FA/IPA.

Isopropanol is sometimes used for the cleaning of assemblies. More common, however, is the use of this solvent in stencil cleaning applications. Isopropanol is often also used as a rinsing media when cleaning with a solvent for hybrid or ceramic applications prior to bonding (see Figure 14.8.5). However, when using isopropanol for any kind of cleaning application there are some essential disadvantages that have to be taken into consideration:

- Flammability due to low flash point,
- Requires explosion-proof cleaning equipment,
- Small process window with respect to the ability to flux residues,
- Limited removal of adhesives due to curing of the adhesives, and
- White residues formation with some flux residues.

14.8.2.3.2 <u>Volatile siloxanes</u>

Due to their extremely low surface tensions combined with excellent wetting character, siloxanes were recommended for electronic cleaning applications. While siloxanes can be used for the removal of non-polar contaminations, ionic contaminations cannot be readily removed. Consequently, for an effective cleaning process, formulations containing siloxanes and alkoxypropanols are needed. However, the use of siloxanes containing cleaning fluids does not play an important role in electronic assembly cleaning applications with respect to the above mentioned disadvantages.

14.8.2.3.3 Chlorinated solvents

Chlorinated solvents such as trichloroethylene or 1.1.1-trichloroethane are nonflammable and excellent cleaning results can be achieved on especially non-polar residues. However, ionic contamination cannot be removed sufficiently.

There are predominant environmental issues, in particular, their ozone depletion potential, which makes their use in electronic cleaning applications obsolete.

14.8.2.3.4 n-Propylbromide (nPB)

Cleaning fluids based on formulations using n-propylbromide (nPB) have to be regarded very critically. Since these products also contain halogens, they cannot be considered to be real alternatives for chlorinated or fluorinated hydrocarbon mixtures. Due to these open environmental questions, cleaning applications using n-propylbromide formulations are still not very common, especially in Europe.

14.8.2.3.5 Alkoxypropanols

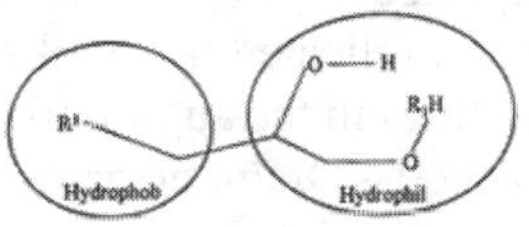


Cleaning agents based on alkoxypropanols show extremely satisfying cleaning results due to their chemical structure. The lipophilic part of the molecule is capable of removing organic soils such as greases or oils from the substrate surface, whereas the hydrophilic part is able to remove the ionic contaminations.

Furthermore, products based on this kind of chemistry show a large process window on all different types of flux residues. Different applications and processes will be discussed in detail in the following sections.

14.8.2.3.6 Hydrofluorinated ethers

Hydroflourinated ethers are a new generation of cleaning fluids revealing no ozone depletion potential. Since they have no flash point there is no need to use explosion-proof cleaning equipment, thereby the cost of cleaning equipment is dramatically reduced. In addition, with their extremely low surface tension and viscosity, they are able to penetrate the tightest spaces of assemblies with high packing densities. Physical properties and different applications and processes will be discussed in more detail below.

14.8.2.3.7 Advantages and disadvantages of solvent-based cleaning agents

*Advantages*:

- Solvents exhibit a large process window on all different kinds of fluxes, adhesives and solder pastes
- Most solvents can be regenerated through distillation,
- There is no need of extensive treatment systems for the rinse water,

*Disadvantages*:

- Higher proven costs due to investment in explosion-proof cleaning equipment,
- VOC containing solvents,

*Process*:

- Handling risks for the operator due to flammability, toxicity and odor, and

- Good material compatibility with respect to materials commonly found on assemblies.

### 14.8.3 CLEANING OF TOOLS AND AUXILIARIES

The prohibition of CFCs in 1993 in Europe found the electronic industry scrambling to undertake appropriate actions. Various alternative technologies have asserted themselves during the conversion to environmentally friendly processes. Basically, two courses have evolved as alternative to the use of CFCs for changing applications:

- New soldering techniques that avoid cleaning as a process have been developed and introduced in the electronic industry.
- New environmentally friendly cleaning processes have been introduced.

A large number of European companies decided in favor of no-clean processes, in spite of the associated disadvantages. In the USA cleaning has not be abandoned; it is still an important step in the SMT production line. However, with the increase of packing density and increased quality demands, more and more European companies are now returning to cleaning processes. In general, the purpose of changing is for the removal of contamination which might otherwise harm the operation of the electronic circuit assembly.

Ensuring high reliability for a circuit assembly by effective cleaning is extremely important in such domains as the military, aviation, telecommunications, and in the automotive industry. The need for change to high reliability can be expected to grow appreciably because of different factors. These factors range from more complicated assemblies with higher density to the greatly accelerating use of conformal coatings in the automotive, niche computer, consumer electronics, and telecommunications markets. Dramatic growth is expected to occur in the USA, Europe and Asia. The applications where cleaning is necessary can be divided into three main applications:

- the removal of flux residues from printed circuit assemblies (PCBs) after reflow or wave soldering;
- the removal of solder paste and/or SMT-adhesive from stencils screens or misprints, and
- the tool cleaning (dispensing nozzles, solder frames, cooling coils from reflow or wave soldering ovens, and squeegees).

#### 14.8.3.1 Cleaning substrates and contamination

In recent years stencil and misprint cleaning has become a significant issue for many users. Normally, in a typical PCB assembly line, the solder paste is printed through screens or stencils. As an alternative to glue dispensing, specially formulated SMT adhesives can now be printed through a stencil. Consequently, the stencils must be cleaned periodically to maintain the quality and yield of the process. However, the removal of SMT adhesive gives rise to completely new demands on the cleaning process and the condition of the stencils. The following specifies the points that must be observed in connection with stencil cleaning. Great importance has therefore been given to comparison of different cleaning chemistries, cleaning equipment, and comparison of manual and automated cleaning.

Ever-increasing component density and the result expected from the printed image give rise to stringent requirements regarding stencil cleanliness. A flawless printed image is very important because misprinted boards can no longer be used without cleaning. This can be very expensive, particularly if components have already been mounted onto one side of a board. As a consequence, stencil printing for the application of solder paste on SMT boards has become a standard. Stainless steel stencils with apertures that are either etched or cut with a laser are used for this purpose. They can also be used for fine-pitch apertures. Studies have revealed that more than 60% of the defects arising in the SMT process are caused by inadequate paste printing. This, in return, can be traced to insufficient cleaning that is often done manually. Consequently, a well working stencil cleaning process is an imperative.

With regard to the cleaning process it is very important that no paste or cleaner residues remain in a stencil aperture, as this will inevitably cause misprints (see Figure 14.8.6, right). The printing of adhesives, on the other hand, requires some points to be taken into consideration since this method has to produce different dot heights. The ratio between the stencil thickness and a hole opening is the decisive factor with this technology. Stencils for adhesive application are made of either stainless steel or plastic and are usually thicker than those used for paste printing. Cleaning is complicated by the thickness of the stencils (often exceeding 1 mm). Consequently, small holes require a cleaner with intense dissolving properties to ensure that the adhesives are completely removed (see Figure 14.8.6, left).

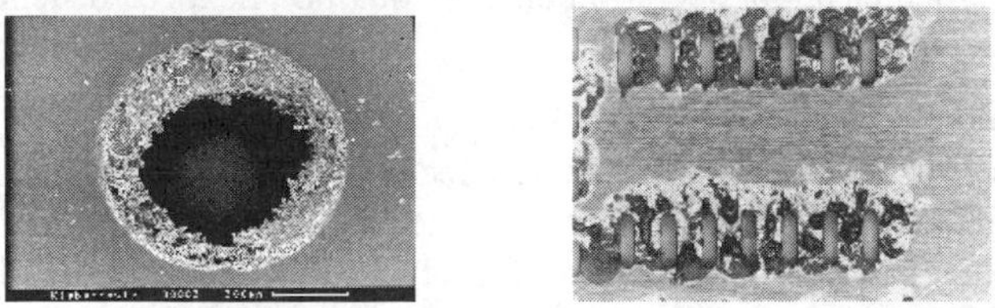

Figure 14.8.6. Left: Residues of SMT adhesive cured in the aperture of a stainless steel stencil. Right: Stainless steel stencil encrusted with solder paste.

Although cleaning of stencils improves the printing results drastically, the production of misprints during the printing process cannot be avoided completely. However, the cleaning of misprinted assemblies is an application that is still frequently ignored. This involves the removal of misprinted or smeared solder paste or SMT adhesive. The removal of solder paste as the most widespread application gives rise to the following demands:

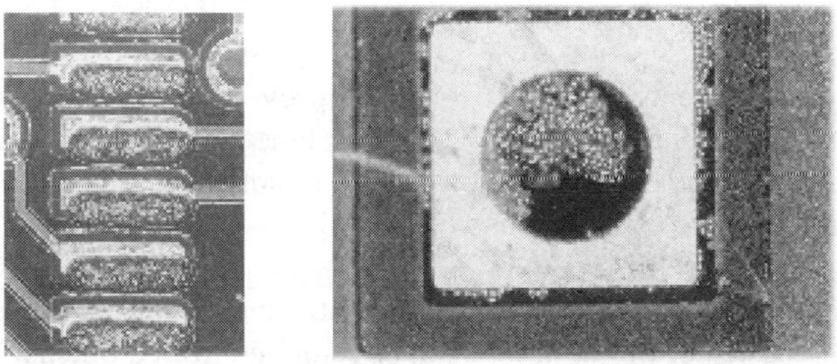

Figure 14.8.7. Left: Misprinted PCB, Right: Via filled with solder paste.

- No solder paste must remain on the surface, or the vias (see Figure 14.8.7).
- The surface must be perfectly clean and dry to ensure immaculate renewed printing.
- The solder flux on double-sided PCBs must also be removed from the side where soldering has been finished.

### 14.8.3.2 Compatibility of stencil and cleaning agent

In some cases, repeated washing can damage stencils cleaned in automated washing stations. This damage usually occurs through degradation of the frame-to-screen glue. This degradation occurs at the junction of the stencil and the mesh and also at the junction of the mesh and the frame. The causes for this degradation lay in two factors. First, the high temperatures that are needed to clean stencils with aqueous cleaners cause thermal stresses along the glue boundaries. Secondly, the glue itself, which in most cases is a cured epoxy material, may dissolve in the cleaning medium. The combination of these two factors can lead to severe stencil defects. A protective coating can be applied to the glue in cases where it is known that the glue is not stable towards the cleaning medium. Advanced cleaning agent manufacturers work closely with stencil manufacturers to avoid this problem. However, testing in advance of installed cleaning operation ensures that no unexpected stencil defects will affect the production process.

#### 14.8.3.3 Different cleaning media

When selecting a cleaning process, the chemistry should be selected first, because the cleaning chemistry has to be adopted to the chemical ingredients of the contamination. Only after the evaluation of the proper chemistry, can the method of applying the cleaning agent be evaluated.

Solvent-based cleaning agents such as hydrocarbons, alcohols, terpenes, and esters offer properties that make them effective cleaning agents for the electronics industry. Low surface tension, high solubility, and ease of drying allow organic solvents to clean effectively where water-based cleaning agents are less effective. These benefits are very similar to the physical properties of the chlorofluorocarbons (CFC) that were once in widespread use in electronics manufacturing operations. Solvents, however, have several significant drawbacks that have limited their use including:

- Flame and explosion hazards,
- Higher equipment costs, due to required explosion protection,
- Personal exposure concerns,
- Material costs, and
- Disposal of used solvents.

And while water-based cleaning agents have been much improved in recent times, their performance does not always match that of their solvent forerunners. In fact, an ideal water-based cleaning agent would combine the physical properties of solvent-based cleaning agents with the safety and convenience of water-based materials.

Solvent-based cleaning agents

Cleaning with solvents such as isopropanol or acetone still remains the most widely applied method because any type of solder paste or SMT adhesive can be removed from the stencils by specially formulated mixtures. Consequently, the cleaning process window for such modern solvents is sufficiently wide to ensure the required results when changing over to another paste or adhesive. Normally, such cleaning processes are performed in machines designed specifically for solvent cleaning. The use of cleaners with flash points in excess of 104°F (40°C) means that the machines can be set up without having to maintain an explosion protected area. Moreover installation in a special room may be necessary. However, the handling requirements of highly flammable cleaners, for such as isopropanol (IPA, flash point of 54°F, 12°C), are far more critical, and cleaning is not so effective, particularly with fine-pitch stencils and SMT adhesives. Besides, modern solvent cleaners have been formulated in such a manner that health hazards are minimized - provided, they are properly used.

Water-based cleaning agents

The benefits of cleaning with water-based agents are straightforward. They include:

- Ease of use,
- Absence of fire or explosive hazards,
- Simple disposal,

- Environmental and personal safety rules.

In some cases even pure water can be used as the cleaning medium (presupposing that the contaminants are fully water-soluble). In most cases, however, water requires the addition of chemical or mechanical energy to obtain good cleaning performance. This can come in the form of:

- A cosolvent,
- A chemical activator,
- Elevated temperature,
- Mechanical action, or
- High-pressure spray or ultrasonics.

Contaminants such as oils, greases and SMT adhesives that feature high organic matter content are not readily soluble in water. Adding chemical activators such as saponifiers to the formulation can increase their solubility. Unfortunately, saponifiers require in most cases elevated temperatures (38-60°C) in order to react with the contamination. However, temperatures in this range will most likely damage the stencil. Saponifiers improve cleaning by reducing the surface tension of the cleaning solution, permitting the agent to penetrate void spaces and stand-off gaps that pure water cannot reach. The saponifiers in water-based cleaning agents are either organic or inorganic. Organic saponifiers are, in most cases, fully miscible with water and can easily be rinsed from the stencil. Any residues remaining after rinsing evaporate during drying. Inorganic saponifiers are usually water-soluble materials applied as a solution in water. They are nonvolatile, and residues remaining after rinsing might leave contamination on the substrate. The alkaline saponifier is consumed during the cleaning process and requires constant replenishment. In addition, the high pH value and the elevated application temperatures can also cause an oxidative attack of the aluminum frame of the stencil.

Elevated temperatures are also commonly used to increase cleaning performance. An increase in temperature will lead to a corresponding increase in the solubilizing properties of the cleaning medium. There is of course a point at which the thermal stress induced by high temperature cleaning can damage the stencil adhesion.

Lastly, high pressures or ultrasonics are used as a way to remove contaminants by transferring mechanical energy to particulate contamination. Pressure helps to force the cleaning agent into void spaces and stand-off gaps. Pressure can only marginally improve cleaning performance when the contaminant is not in particulate form or when solubility is the limiting factor.

Water-based cleaners based on MPC® technology

Satisfactory results in stencil cleaning with aqueous systems can only be achieved for the removal of solder paste. Most aqueous systems used to remove adhesives failed in the past because the water caused the adhesives to set.

Micro Phase Cleaning (MPC)[2,3,12] refers to the use of a cleaning agent formulated to undergo a phase change at elevated temperature. When heated above a threshold temperature (typically 100-120°F, 40-50°C) a microphase cleaner changes from a clear colorless solution to a turbid milky mixture.

The phase transition produces a cleaning mixture that exhibits the properties of both solvent-based and water-based cleaners. "Water-like" properties allow the cleaning agent to effectively dissolve ionic contaminants and remove them from the substrate's surface.

Non-polar and organic residues are removed by contact with the hydrophobic or "solvent-like" phase.[12]

No specialized equipment is needed to handle microphase cleaning agents. In most cases a microphase cleaning agent can be used as a drop-in replacement in existing equipment. Microphase cleaners are effective in all types of cleaning equipment capable of applying the cleaning agent in a liquid form. This includes spray and ultrasonic equipment.

**Table 14.8.1. Comparison of different cleaner types**

| Cleaning Agent | Pros | Cons |
|---|---|---|
| Modern solvents | Removal of adhesives and solder paste<br>Wide process window<br>Cleaning at room temperature<br>Do not oxidize/corrode stencil | Flammable<br>Emits solvent vapors into work area<br>Slow drying<br>VOC (volatile organic compound)<br>Ozone depletion potential<br>Require explosion proof equipment |
| Aqueous alkaline cleaner | Cheap<br>Non-flammable<br>Non ozone depleting<br>No VOCs<br>Mild odor | Do not remove adhesives<br>Short bath life times<br>Narrow process window<br>Water rinse necessary |
| Aqueous based cleaners of MPC Technology | Removal of adhesives and solder paste<br>Wide process window<br>Non-flammable<br>Rapid drying<br>Residue-free drying<br>Long bath life | Agitation of the cleaner (spray, ultrasonic, spray under immersion, overflow) necessary |

Table 14.8.2 presents the typical technical characteristics of different cleaning agents developed by Dr. O.K. Wack Chemie for stencil and misprint cleaning applications.

### 14.8.3.4 Comparison of manual cleaning vs. automated cleaning

Printing screens are often cleaned manually (Figure 14.8.8) as users still shy away from the purchase of a cleaning machine because of the associated investment costs ($6,500 to $60,000). However, a more accurate assessment must take the following points into account:

**Table 14.8.2. Technical data of typical cleaning agents (water-based and solvent)**

| Cleaning Agent | Vigon® SC200 | Zestron® SD300 |
|---|---|---|
| Chemistry | water based | solvent based |
| Flash point | none | 106°F |
| Appl. Temp. | 77°F | 77°F |
| Process | spray in air ultrasonic manual | spray in air (ex proofed) manual |

(1) The mechanical rubbing action of the repeated manual cleaning impairs the stencil surface. This is particularly apparent with plastic stencils. These changes to the stencil surface can result in misprints and shorten the service life of the stencil, making it very difficult to track the source of the problem.

(2) Precise repeatability of the cleaning results is not guaranteed with manual

cleaning because manual applications are not based on precisely defined cleaning processes. This can also result in fluctuations of the printing result.

(3) Manual cleaning must be completed with the utmost care. Such care costs time and money, thus making it very expensive. The investments for a cleaning machine can be recovered within a very short period, independent of the number of substrates that are to be cleaned.

Example: A worker requires 1 hour per day for stencil cleaning. On the basis of \$40.00/h and 220 workdays, a small machine (\$8,700) can be a worthwhile investment within a year.

Some users clean misprinted assemblies manually. This can prove to be very labor and cost intensive when larger quantities are involved. The cleaning results and their reproducibility will be rather inconsistent. Manual cleaning of PCBs with vias and blind holes with a brush or cloth can mean that solder beads are actually rubbed into the holes. Such holes can only be satisfactorily freed by ultrasonic or other mechanical means. It is imperative that the via are free of solder particles, otherwise the consequential damage can be very expensive.

#### 14.8.3.5 Cleaning equipment for stencil cleaning applications

As previously mentioned, it may well prove to be an economical proposition to invest in a specially designed cleaning machine if this can be justified by the number of substrates that have to be cleaned. In general, PCB assemblers use the following different types of cleaning equipment for stencil and misprint cleaning:

- Spray in air,
- Spray under immersion, and
- Ultrasonic.

Spray in air cleaning equipment

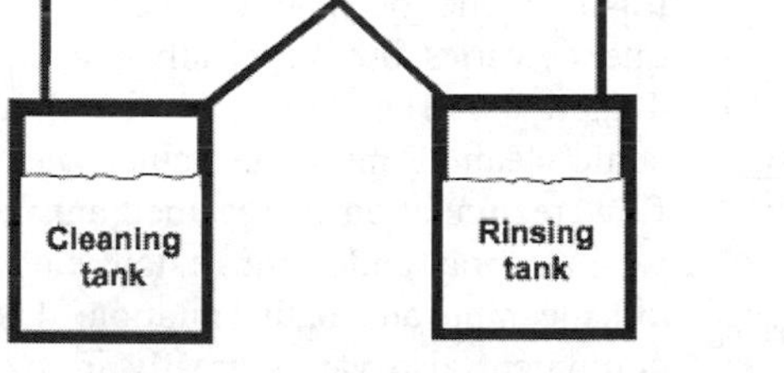

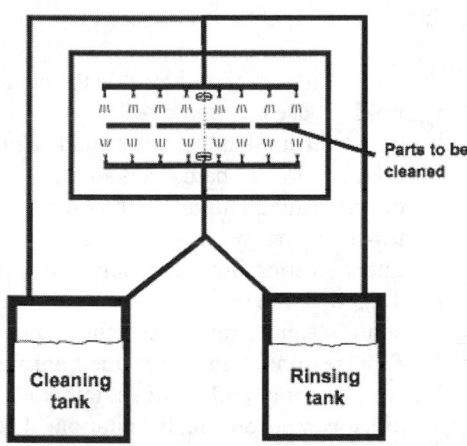

Figure 14.8.8. Scheme of typical spray in air type of cleaning equipment.

Spray in-air machines consist of either a single or multiple reservoirs containing wash and rinse fluids (detergents, solvents, saponifiers, water, etc.). The stencils are placed in a separate chamber, and the wash/rinse fluids are then pumped from the reservoirs into the stencil chamber, and delivered onto the stencil via low pressure (<70 psi) spray wands. These arms may be either co- or counter-rotating. When designed for using a solvent as a wash medium, these machines must be designed to be explosion protected (intrinsically safe electronics, pneumatic or TEFC pumps, sealed conduits, etc.). These machines may also be outfitted to use an inert atmosphere, such as nitrogen, in the stencil chamber.

When used with aqueous-based media, they may often be equipped with a wash and or rinse tank heater. Advantages of this type of machine are:

- Relatively inexpensive (although explosion protecting oftentimes significantly increases the cost of the machine),
- Easy to use,
- A large number of options (add-ons) available (in-line filtration, closed loop water recirculation, programming options, etc.),
- Do not require as much media as other types of stencil cleaners, because they do not require total immersion of the stencil,
- The stencil is also washed, rinsed, and dried in the same chamber, thus eliminating the need to move the stencil around manually.

Disadvantages include the sometimes large footprint that results from having to provide up to three separate chambers (wash, rinse, and stencil), and the sometimes inadequate agitation (spray pressure) generated by the machine. This can make it difficult to clean fine-pitch apertures.

Spray under immersion cleaning equipment

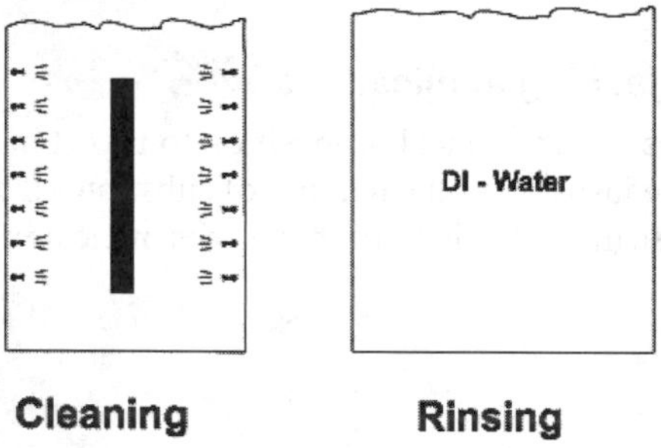


Figure 14.8.9. Scheme of typical spray under immersion batch cleaner.

In spray under immersion equipment, the stencil is placed into a bath (immersed) containing the cleaning medium. The medium is then recirculated using submerged spargers to provide agitation to the stencil surface. A separate chamber is then used for rinsing and drying. Advantages of this type of machine are that it is:

- Most gentle on the stencil, and
- The capital expense is relatively low.

However, because of the low pressure, cleaning cycles may require longer cleaning times. Also, the size of the machine and the quantity of cleaner required are dictated by the size of the stencil to be cleaned.

Ultrasonic cleaning equipment

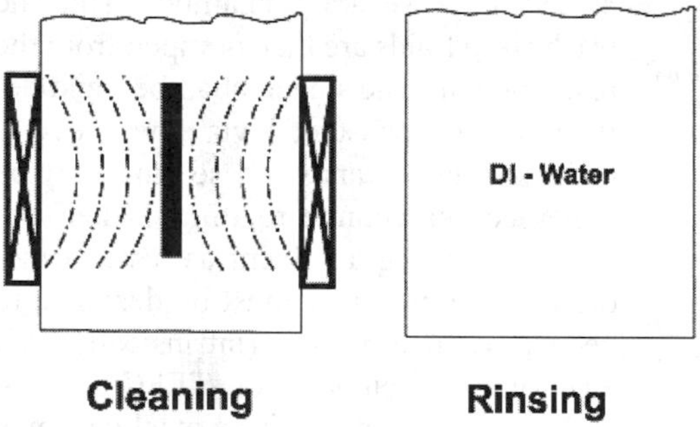


Figure 14.8.10. Scheme of typical ultrasonic batch cleaning equipment.

Ultrasonic machines, as the name implies, use ultrasonic "vibratory" energy to clean stencils. The stencil is immersed in a bath containing the cleaning medium, and exposed to ultrasonic energy. The ultrasonic frequency varies, but is typically between 40 and 100 kHz. Options for the ultrasonic cleaning machine include overflow recirculation of the medium, as well as spray under immersion jets to aid cleaning and bath agitation. The equipment also varies greatly in size (from tabletop models to large footprint models that include rinsing baths, air dryers, and robot arms). Advantages of this unit are that:

- The ultrasonic energy is much better suited to cleaning fine pitch openings, and

- External heaters often may not be needed when using aqueous medium, because of the bath heating that the ultrasonic energy accomplishes.

However, because of this heating, these units are not recommended for use with solvent cleaners. Also, precise control of the ultrasonic energy is required to prevent possible damage to the stencil and apertures. Comparatively, also, ultrasonic machines are more expensive than either type of spray equipment. The size, and thus the cost, is directly proportional to the size of the stencil to be cleaned.

#### 14.8.3.6 Stencil cleaning in screen printing machines

Today, many screen printers incorporate an automatic cleaning system for the undersides of stencils. During the process a wiper is moistened with a cleaner that wipes the stencil. On high-end printers this unit is programmable to permit subsequent dry wiping and vacuum drying. Cleaning the underside of stencils may not replace machine stencil cleaning, but it does extend the time, in some cases significantly, until machine cleaning becomes necessary.

A disadvantage of these automatic-cleaning units is the fact that only solder paste removal is possible. The main problem associated with cleaning the stencil underside of adhesive stencils is that it does not remove the adhesive from the stencil holes. This means that only the surface is cleaned, while the holes are untouched. Cleaning agents suitable for this application already exist.

#### 14.8.3.7 Summary

The cleaning process, which a user finally selects, depends to a major extent on the operating conditions. Each process has its own specific advantages and disadvantages, so users are supported by advice and testing. These can be accomplished under tightly simulated process conditions by a technical center.

### 14.8.4 CLEANING AGENTS AND PROCESS TECHNOLOGY AVAILABLE FOR CLEANING PCBs

#### 14.8.4.1 Flux remove and aqueous process

With the banning of CFCs - enforced in Germany since January 1993 - users have been confronted with the problem of ensuring cost effective elimination of ozone depleting chemicals from the production process while having to satisfy constantly rising quality requirements. In the electronic industry it is customary to set certain quality standards in the manufacturing process of electronic assemblies. Nevertheless, it is often not enough to check the quality of a part immediately after it has been manufactured as long-term quality must also be taken into account. However, technically proficient and efficient processes are available that can exclude long-term risks and even reduce overall costs.

A large number of companies have decided in favor of no-clean processes. Though, with the expansion of packing density and increased quality demands, more and more companies are now returning to cleaning processes.

##### *14.8.4.1.1 The limits of a no-clean process*

The no-clean concept (i.e., no cleaning after soldering) proceeds from the following assumptions:

(1) The used soldering fluxes have a low solid content of approx. 2 to 3% and contain strong activators. These activators, however, are critical as they can easily form highly conductive electrolytes under humid climatic conditions.

Figure 14.8.11. Soldering flux residues on a component.

Figure 14.8.12. Dentride growth between connecting contacts.

(2) Reduction of the resin content and quantity of activators simultaneously diminishes the risk of leakage current, electro-migration and corrosion.

After soldering, cleaning removes all residues from the pc-boards, whereas the no-clean soldering process leaves residues on the pc-boards surface, thereby exposing them to the influence of humidity. Consequently, the presence of soldering flux residues on assemblies (Figure 14.8.11) can cause significant functional disruptions. To guarantee quality assurance with no-clean fluxes, the standard tests previously employed in the days of CFC cleaning where transferred accordingly to no-clean pc-boards, including:

- Testing for impurity,
- Testing for surface resistance (SIR test),
- Testing for electro-migration,
- Testing for bondability, and
- Testing under climatic stresses.

Testing for ionic contamination and establishing the surface resistance usually produces good results with pc-boards soldered in a no-clean process. Nevertheless, the values are still significantly poorer than with cleaned assemblies. The migration and precipitation of metal ions can cause the dangerous growth of dendrites (Figure 14.8.12). These metal dendrites are outstanding electric conductors that diminish the surface resistance and, in the long-term, result in short-circuits. Coating can significantly reduce the risk of electro-migration. However, this demands a clean surface, otherwise homogeneous adhesion of the coating is not assured.

#### *14.8.4.1.2 Different cleaning media and cleaning processes*

When selecting a cleaning process, the cleaning chemistry is the most sensitive parameter, because the cleaning agent has to be adapted to the chemical ingredients of the contamination. Only after determining the most suitable chemistry, should the method of application be evaluated. In principle cleaning processes can be split into three separate categories:

(1) Semi-aqueous cleaning,
(2) Aqueous cleaning, and
(3) Solvent cleaning.

### 14.8.4.1.3 Semi-aqueous cleaning

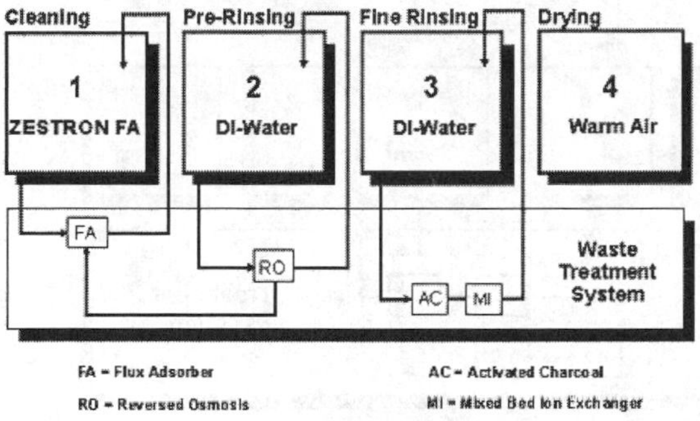


Figure 14.8.13. Diagram of a semi-aqueous cleaning process with ZESTRON® FA.

A process that has found widespread use in practice is semi-aqueous cleaning with ZESTRON® FA. This cleaner was especially developed for removal of no-clean soldering fluxes in ultrasonic batch or spray under immersion cleaning systems. It is used in the cleaning bath, rinsed with demineralized water, and then dried with hot air (see scheme in Figure 14.8.13.).

The carried-in flux constituents can be removed by a flux absorber system to ensure the cleaning medium maintains a constant quality. Cartridge filters also remove precipitated flux residues and particles. Two cascaded rinsing units - a pre-rinse and a fine rinse - follow the wash stage. The rinsing water contaminated with carried-over cleaning medium is passed through different sized particle filters, mixed-bed ion exchangers, and activated carbon filters. The quality of the rinsing water can be continuously monitored by conductivity measurement to establish its residual ion content. The conductivity value should be below 1 μS/cm.

The complete process in based on a closed-loop principle. Consequently, the cleaning process merely liberates those substances that are to be removed from the pc-boards (i.e., soldering flux residues). Closed-loop operation is not only advisable for ecological reasons, but is also a very sound economic proposition as it significantly minimizes process costs.

### 14.8.4.1.4 Aqueous cleaning in spray in air cleaning equipment

Low pressure spray in air batch cleaning systems (see Figure 14.8.14) can be attractive propositions for users who only have a low volume of pc-boards to clean and are required to furnish proof of a specified cleaning process to their customers. The water-based mildly alkaline cleaning agent VIGON® A200 was specifically developed for this application.

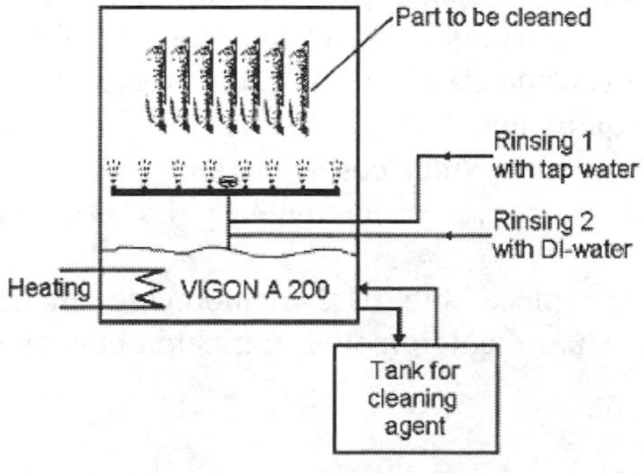


Figure 14.8.14. Diagram of an aqueous in-line cleaning process with VIGON® A200.

High-pressure spray in air in-line cleaning processes that remove flux residues from PCBs, are often employed by manufactures to clean large quantities of boards. This still represents the most economical way to clean large quantities of PCBs. Especially for contract manufactures that are required to clean different types of PCBs, often with several different flux residues.

The parts to be cleaned are fed, one after the other, into a continuous spray cleaner where they pass through the different process stages indicated in Figure 14.8.15. The

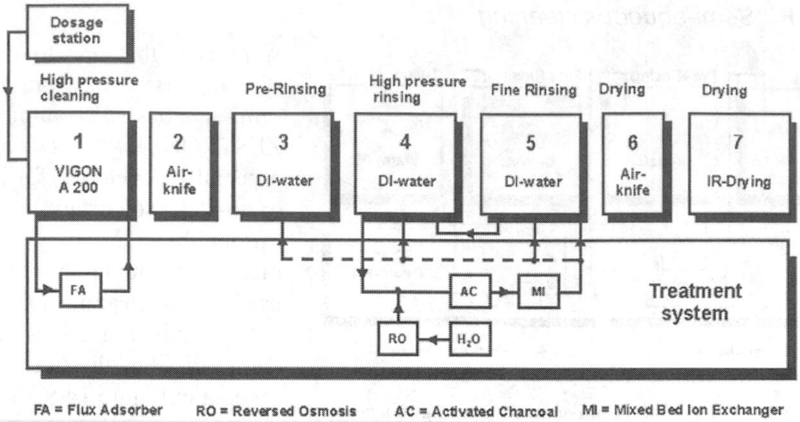


Figure 14.8.15. Scheme of a low-pressure batch cleaning system.

high-pressure cleaning unit is supplied with VIGON® A200 by a chemical dosing station. Cleaning medium quality can be maintained by installing a flux absorber system. The cleaned parts are blown-out with an air knife to minimize carry-over losses from the high-pressure cleaning section. Pre-rinsing the parts with deionized water follows the cleaning section. A fine rinse unit ensures minimum ionic contamination on the PCBs. Finally, the cleaned parts are then blown off with hot air. Complete drying, particularly of parts that retain a great deal of liquid, can be performed with an infrared dryer.[20,21]

#### 14.8.4.2 Flux removal from printed circuit boards - water-free cleaning processes

In the European Community, the production and sales of chlorinated fluorocarbons (CFCs) have been banned since 1995. Due to their ozone depletion potential (ODP) hydrochlorinated fluorocarbons (HCFCs) are also subject to be phased-out. The revision of EC Direction 3094/95 proposed in Article 5 the ban of HCFCs as solvents for parts cleaning. HCFCs are not allowed in open solvent applications and their use in contained cleaning systems, such as industrial equipment with ultrasonic agitation, will be banned by January 1, 2002. However, the aerospace and aeronautic industries are allowed to use HCFC until January 1, 2002. Furthermore, in some countries, including Germany and Switzerland, national ordinances have prohibited the use of HCFCs to replace CFC solvents for several years. Consequently, the search for alternative solutions to avoid cleaning (no clean technology) or to use environmentally acceptable solvents leads to new cleaning processes including new cleaning agents and new cleaning equipment.

- Novec™ hydrofluoroether (HFE) in combination with a cosolvent, and
- High boiling cleaning agents that can be recycled continuously by vacuum distillation.

These two new processes can permanently replace both 1,1,1-trichlorochlorethane and CFC-113, as well as, chlorinated solvents such as dichloromethane, trichloroethylene (TCE) , perchloroethylene and HCFC-141b.

### *14.8.4.2.1 Water-free cleaning processes using HFE (hydrofluoroethers) in combination with a cosolvent*

The cleaning process:

A cosolvent cleaning process is one, in which the cleaning and rinsing solvents are of significantly different composition. For example, in semi-aqueous systems (see Cleaning Technologies on the PCB Assembly Shop Floor (Part II) in EPP Europe, October 1999) cleaning is done with organic solvents and water is used for rinsing. In the cosolvent cleaning process, cleaning is accomplished primarily by the organic cosolvent, and rinsing is effected by a fluorochemical. Typically, in cosolvent systems, the boil (wash) sump contains a mixture of cosolvent and fluorochemical rinsing agent. The rinse sump normally contains essentially 100% fluorochemical rinsing agent.

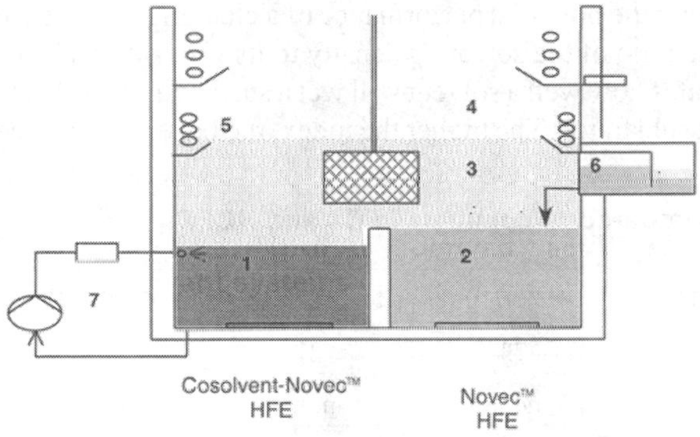


Figure 14.8.16. Typical vapor degreasing equipment.

Referring to Figurc 14.8.16, a cosolvent process operation can be summarized as follows. Printed circuit boards containing flux residues are immersed into the boil sump, which contains a mixture of cosolvent and fluorochemical rinsing agent, typically about equal volumes of each. After being cleaned in the boil sump, the substitutes are immersed in the rinse sump, which contains nearly 100% fluorochemical.

The primary function of the rinse sump is to remove the fluorochemical/cosolvent mixture and the soil dissolvcd in it. Following immersion in the rinse liquid, the parts are moved into the vapor phase for a final rinse with pure fluorochemical vapor. Finally, the parts are lifted into the freeboard zone, where any remaining fluorochemical rinsing agent evaporates from the parts and returns to the sump by condensation from cooling coils. At this point, the cleaning cycle is complete and the parts are clean and dry.

Furthermore, pure Novec™ HFEs can be used for the removal of light soils, halogenated compounds like fluorinated greases or oils, and other particles.

The increasing popularity of the cosolvent process is mainly due to its flexibility in allowing independent selection of both the solvating and rinsing agents that best meet the needs of a particular cleaning application. The fluorochemical cosolvent process is even capable of defluxing components with very complex geometry due to the advanced physical properties of the cosolvent.

The cleaning chemistry:

Novec™ hydrofluoroethers (Novec™ HFEs) are rapidly becoming recognized as the solvents of choice for superior cleaning in the electronics and precision engineering industries. The combined properties of these new materials have proven to be very effective alternatives to ozone depleting substances (ODSs) like the previously mentioned CFCs and HCFCs. The new developed chemical compounds methoxyfluorobuthylether (HFE 7100) and ethoxyfluorobuthylether (HFE 7200). HFE rinsing agents can be used in combination with respective cosolvents. The most effective cosolvents for defluxing applications in the market are based on alcoxypropanols which are specially designed for applications in combination with HFEs.

Since the boiling points of these two HFEs lie at the upper range covered by the traditional ODSs such as HCFC-141b, CFC-113 and 1,1,1-trichloroethane, a better cleaning efficiency is provided due to the increase in solubility of most soils with temperature.

Like ODS solvents, the HFEs are non-flammable, which leads to very simple and thus inexpensive equipment designs.

The low surface tension of HFEs helps these fluids to easily penetrate tight spaces, especially between and under components of populated printed circuit boards. In addition, this is supported by the low viscosity of the HFEs that also allow it to quickly drain from the parts that have been cleaned.

One useful parameter for assessing the potential performance of a cleaning agent is the wetting index, which is defined as the ratio of the solvent's density to its viscosity and surface tension. The wetting index indicates how well a solvent will wet a surface and penetrate into tight spaces of complex cleaning substrates. The higher the index, the better the surface penetration.

**Table 14.8.3. Properties of HFEs compared to common CFCs and HCFCs**

| Property | HFE-7100 | HFE-7200 | HCFC-141b | CFC-113 |
|---|---|---|---|---|
| Boiling point, °C | 60 | 73 | 32 | 48 |
| Flash point, °C | none | none | none | none |
| Surface tension at 25°C | 13.6 | 13.6 | 19.3 | 17.3 |
| Viscosity, Pa s | 0.61 | 0.61 | 0.43 | 0.7 |
| Wetting index | 181 | 172 | 149 | 133 |

Since HFEs contain no chlorine nor bromine, they do not have an ozone depletion potential. In addition, their atmospheric lifetime is short compared to CFCs. Furthermore, the Global Warming Potential of HFE fluids is significantly lower than other proposed fluorinated solvent replacements. Practical studies as well as laboratory tests have found that HFE solvent emissions are 5-10 times lower than the emissions of CFCs or HCFCs. Although, HFEs are volatile organic compounds (VOC), they are not controlled by related US directive regulations. Finally, toxicological tests have shown, that these products have extremely low toxicity which allows for higher worker exposure guidelines, unlike many of the chlorinated and brominated solvents offered as ODS alternatives.

#### *14.8.4.2.2 Water-free cleaning processes in closed, one-chamber vapor defluxing systems*

Liquid vapor degreasing using solvents has been an accepted method of precision cleaning for over 50 years. This cleaning process incorporates washing, rinsing, drying and solvent reclamation in a compact, cost-effective unit. Consequently, this makes it a very attractive process for many production cleaning applications. Environmental concerns raised in the last decade have changed both the chemistry and equipment technology for degreasing. Solvents have been modified or replaced to eliminate ozone depletion potential and other haz-

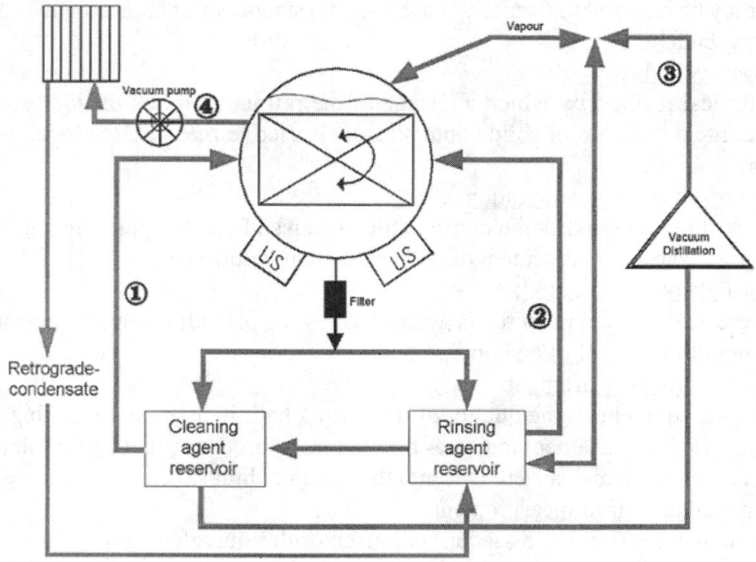


Figure 14.8.17. Closed-loop batch vapor degreasing process.

ards. Equipment designs and operating processes have been totally re-engineered to reduce solvent losses to near zero.

Referring to Figure 14.8.17, a modern vapor defluxing process can be described as follows. The cleaning equipment contains a cylindrical process chamber within which the complete cleaning, rinsing and drying process takes place. The process chamber may include spray manifolds and/or ultrasonic transducers (frequency of 40kHz) to provide the mechanical energy necessary for advanced precision cleaning. Following the final vapor rinse, remaining vapors are passed through a heat exchanger to condense the vapors to liquid, which is returned to the solvent recovering device. The vacuum then extracts any remaining solvent molecules, which are either exhausted from the machines directly or passed through carbon absorption system in order to limit solvent emissions to extremely low levels. The use of vacuum distillation for solvent recovery and vacuum drying assures minimal solvent consumption. Distillable solvents with a flash point above 55°C like Zestron® VD by Dr.O.K. Wack Chemie are suitable for such a process.

Many benefits provided by solvent cleaning are evident. Among these are:

- no process chemistry to mix or maintain,
- solvent is reclaimed automatically in the cleaning equipment,
- only few process variables to be managed,
- very easy and cost-effective to operate, and
- typically small footprint of the cleaning equipment.

## 14.8.5 CRITERIA FOR ASSESSMENT AND EVALUATION OF CLEANING RESULTS

There are a number of different processes that can be customized to monitor the cleaning process and cleaning results. Depending upon the existing requirements, simple "qualitative tests" can be applied to produce a "Good" or "Poor" rating, or more elaborate quantitative tests may be required to furnish a more in-depth statement of regarding the condition of the cleaning bath.

Refractive index:

An optical method by which a change of the refractive index of liquids, depending upon the content of liquid or solid contaminants, is used to measure the level of bath contamination.

Solid residue from evaporation:

This method establishes the non-volatile residues in cleaning and rinsing baths. This produces a very accurate statement of the bath contamination level.

Electrochemical potential:

The electrochemical potential is determined by the pH value, which furnishes an indirect statement of the level of contamination.

Conductivity measurement:

This is used to check the quality of the rinsing bath in assembly cleaning processes. The conductivity measurement indicates the amount of process-related ionic, and therefore conductive, contaminants carried over into the rinse medium.

Ionic contamination measurement:

Ionic contamination-expressed in sodium chloride equivalents is primarily used to assess the climatic resistance of electronic assemblies. Ionic measurement is suitable for statistical process control, but it does not furnish an absolute statement concerning the climatic resistance of an assembly.

Ink test:

Most organic residues on assemblies are not detected by ionic contamination meters. However, such residues do change the surface tension of assemblies, thereby influencing the adhesion of subsequent coatings. This method determines the surface tension by iterative use of corresponding testing inks.

Furthermore, the remaining protective ability of organic solder (OSP-coatings) on copper can be estimated. This test also provides an indication of the adhesion of lacquers and conformal coatings to the assembly.

Water-immersion test:

This economical test exposes the climatically unstable points of electronic components. Due to the nature of the test, the entire board is evaluated. This test accelerates the mechanisms of electrochemical migration. Consequently, faults that previously would appear after months or even years can be detected during the development process. To identify potential weak points, the assembly is operated in standby mode and immersed in deionized water. Testing while the assembly is in full operation is even more effective. The sensitivity of the circuit to moisture exposure is assessed on the basis of the recorded test current, combined with a subsequent examination of the assembly. Through weak point analysis, a Yes/No decision can be determined concerning the expected service life, of the assembly.

The growing diversity of cleaning media, cleaning equipment, and cleaning processes, together with more stringent legal requirements and quality expectations, are making it increasingly difficult for companies to develop their own cost-effective and result-optimized cleaning processes. Consequently, a lot of well known manufacturing companies opted for turnkey solutions for the entire development and installation of complete cleaning processes.

## 14.8.6 COST COMPARISON OF DIFFERENT CLEANING PROCESSES

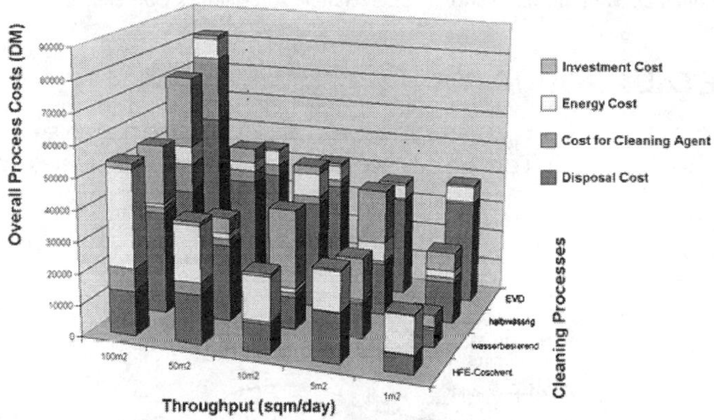

Figure 14.8.18. Comparison of overall process costs between different cleaning processes.

Long-term cost analysis experience on different cleaning processes have been summarized in Figure 14.8.18 which describes the overall process costs of the different cleaning applications discussed. Whereby the overall process costs are divided into four main sections:

- Investment costs,
- Costs for the cleaning agent,
- Costs for the disposal of used media/ recycling cost, and
- Energy costs.

By analyzing the numbers, the following key conclusions on the different cleaning processes are substantial:

- The total annual process costs are always connected with the throughput, but not necessarily in a linear function.
- The lowest investment costs can be achieved by using an aqueous-based or Novec™ HFE-cosolvent cleaning process.
- Closed vapor degreasing and Novec™ HFE-cosolvent processes are more competitive regarding waste disposal.
- A water-based cleaning process is the most competitive regarding media costs.

## REFERENCES


1 C. J. Tautscher, **Effects on Electronic Products**, *Marcel Dekker Inc.*, New York, 1991.
2 M. Hanek, System integration in Micro Electronics, Tutorial 4, Nürnberg, May 4-6, 1999.
3 N. Loew, *EPP Europe*, **10** ,44 (1999).
4 M. Hanek, R. Schreinert, *Galvanotechnik*, **83**, 1370 (1992).
5 F. Cala, A. E. Winston, **Handbook of Aqueous Cleaning Technology for Electronic Assemblies**, *Elektrochemical Publications Ltd.*, Ashai House, Church Road, Port Erin, Isle of Man, British Isles, 1996.
6 N. Loew, *EPP Europe*, **1/2**, 24 (2000).
7 S. J. Sackinger, F. G. Yost, *Material Technology*, **9**, 58 (1994).
8 J.M. Price, Proceedings Nepcon East`94, 365 (1994).
9 B. N. Ellis, **Cleaning and Contamination of Electronics Components and Assemblies**, *Electrochemical Publications Ltd*, Port Erin, Isle of Man, British Isles, 1986.
10 B. N. Ellis, *Circuit World*, **19**, 1, (1992).
11 http:\\www.Kyzen.com.
12 M. Hanek, N. Loew, Unveröffentliche Ergebnisse, Dr. Wack Chemie, Ingolstadt, Germany, 1999.
13 M. E. Hayes, R. E. Miller, **US Patent No. 4,640,719**, 1987.
14 L.Futch, K. R. Hrebenar, M. E. Hayes, **US Patent No. 4,934,391**, 1990.
15 M. Hanek, O. K. Wack, **Patent No. 0587917**, 1995.
16 K. Lessmann, Proceedings, Nepcon West `94, 2219 (1994).
17 H. Gruebel, *Galvanotechnik*, **87**, (1996).
18 M. Hanek, *Productronic*, **11**, 39, 1997.
19 P.Schlossarek, *Surface Mount Technology*, **7**, (1994).

20 N. Löw, *EPP Europe*, **8**, 42 (1999)..
21 N. Löw, P. Bucher, *Surface Mount Technology*, **6**, 36 (1998).
22 J. Owens, K. Warren, H. Yanome, D. Mibrath, International CFC Halone Alternatives Conference, Washington DC, October 1995.

## 14.9 FABRICATED METAL PRODUCTS

GEORGE WYPYCH
**ChemTec Laboratories, Inc., Toronto, Canada**

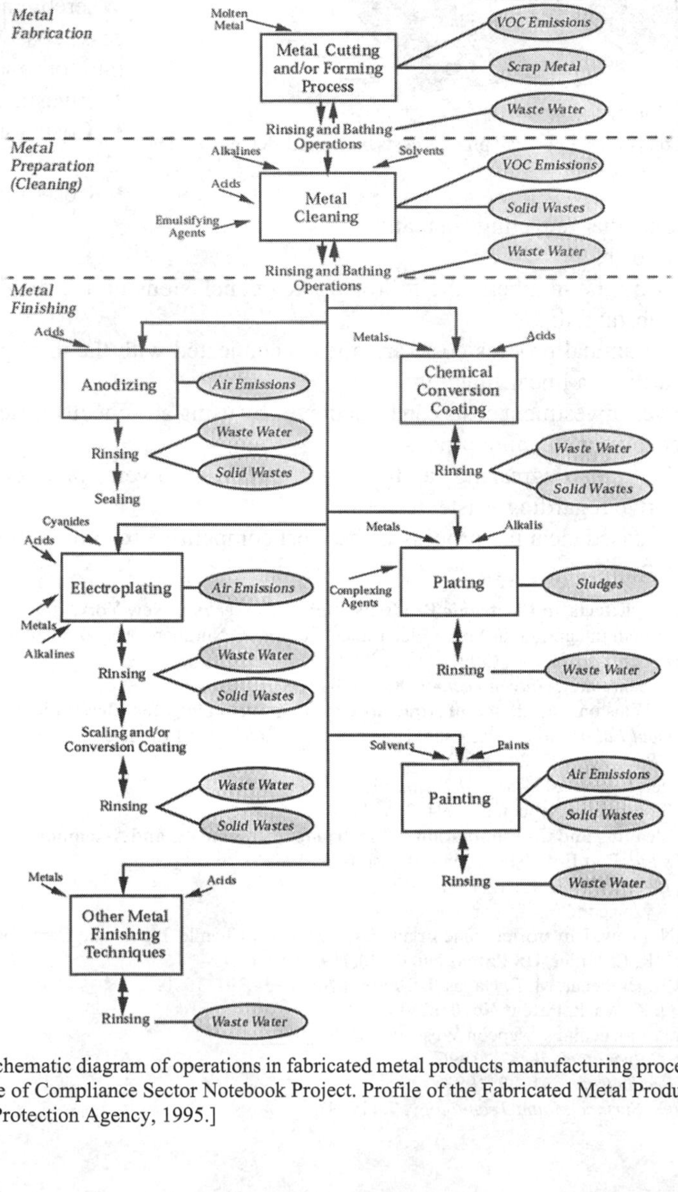


Figure 14.9.1. Schematic diagram of operations in fabricated metal products manufacturing process. [Reproduced from EPA Office of Compliance Sector Notebook Project. Profile of the Fabricated Metal Products Industry. US Environmental Protection Agency, 1995.]

**Table 14.9.1. Reported solvent releases from the metal fabricating and finishing facilities in 1993 [Data from Ref. 1]**

| Solvent | Amount, kg/year | Solvent | Amount, kg/year |
|---|---|---|---|
| acetone | 870,400 | 2-methoxyethanol | 22,800 |
| benzene | 1,800 | methyl ethyl ketone | 4,540,000 |
| n-butyl alcohol | 495,000 | methyl isobutyl ketone | 934,000 |
| sec-butyl alcohol | 13,600 | tetrachloroethylene | 844,000 |
| cyclohexanone | 303,000 | toluene | 3,015,000 |
| 1,2-dichlorobenzene | 10,900 | 1,1,1-trichloroethane | 2,889,000 |
| dichlorofluoromethane | 10,800 | trichloroethylene | 3,187,000 |
| dichloromethane | 1,350,000 | trichlorofluoromethane | 82,000 |
| ethylbenzene | 299,000 | 1,2,4-trimethylbenzene | 262,000 |
| ethylene glycol | 99,000 | xylene (mixture) | 4,814,000 |
| 2-ethoxy ethanol | 18,600 | m-xylene | 8,800 |
| isopropyl alcohol | 23,400 | o-xylene | 17,000 |
| methanol | 199,000 | p-xylene | 10 |

Figure 14.9.1 shows a schematic diagram of the operations in a fabricated metal products manufacturing process.[1] The diagram shows that solvent emissions, waste waters, and solid wastes are produced from three operations: the metal cutting and forming process, metal cleaning, and painting. In metal cutting and forming the major solvents used are 1,1,1-trichloethane, acetone, toluene, and xylene. In surface cleaning, the straight solvents are used for cleaning or some solvents such as kerosene or glycols are emulsified in water and the emulsion is used for cleaning. When emulsions are used the amount of solvent is decreased. Solvents are released due to evaporation, volatilization during storage, and direct ventilation of fumes. Waste waters are generated from rinse waters. These waste waters are typically cleaned on-site by conventional hydroxide precipitation. Solid wastes are generated from waste water cleaning sludge, still bottoms, cleaning tank residues, and machining fluid residues. In painting, two methods are used: spray painting and electrodeposition. Painting operations release benzene, methyl ethyl ketone, methyl isobutyl ketone, toluene, and xylene. Paint cleanup operations also contribute to emissions and waste generation. The predominant solvents used for equipment cleaning are tetrachloride, dichloromethane, 1,1,1-trichloroethane, and perchloroethylene. Sources of solid and liquid waste include components of emission controlling devices (e.g., paint booth collection system and ventilation filters), equipment washing, paint disposal, overspray, and excess paint discarded after expiration of paint's shelf-life.

Table 14.9.1 shows reported solvent releases from metal fabricating and finishing facilities and Table 14.9.2 shows reported solvent transfers from the same industry.

Xylene, methyl ethyl ketone, trichloroethylene, 1,1,1-trichloroethane, and dichloromethane are released in the largest quantities. Methyl ethyl ketone, xylene, toluene,

**Table 14.9.2. Reported solvent transfers from the metal fabricating and finishing facilities in 1993 [Data from Ref. 1]**

| Solvent | Amount, kg/year | Solvent | Amount, kg/year |
|---|---|---|---|
| acetone | 767,000 | methyl ethyl ketone | 5,130,000 |
| n-butyl alcohol | 282,000 | methyl isobutyl ketone | 846,000 |
| sec-butyl alcohol | 495 | tetrachloroethylene | 415,000 |
| cyclohexanone | 1,025 | toluene | 1,935,000 |
| dichloromethane | 237,000 | 1,1,1-trichloroethane | 920,000 |
| ethylbenzene | 263,000 | trichloroethylene | 707,000 |
| ethylene glycol | 68,000 | 1,2,4-trimethylbenzene | 42,000 |
| 2-ethoxy ethanol | 1,800 | xylene (mixture) | 2,335,000 |
| isopropyl alcohol | 41,000 | m-xylene | 27,000 |
| methanol | 138,000 | o-xylene | 37 |
| 2-methoxyethanol | 7,800 | p-xylene | 23 |

trichloroethylene, and methyl isobutyl ketone are transferred in largest quantities outside industry for treatment and disposal. The metal fabrication and finishing industry's contribution to reported releases and transfers in 1995 was 9.3% of the total for 21 analyzed industries. This industry is the fourth largest contributor to total reported releases and transfers.

Among the potential actions to reduce pollution, these are the most effective: recycling of solvents, employing better waste control techniques, and substituting raw materials. For solvent cleaning, several techniques are suggested. Employees should be required to obtain solvent from their supervisor. It is estimated that this will reduce waste by 49%. Pre-cleaning (wipe, blow part with air, etc.) should be adopted or the parts should be first washed with used solvent. Equipment for vapor degreasing should be modified by increasing its head space, covering degreasing units, installing refrigerator coils, rotating parts before removing, installing thermostatic controls, and adding filters. Trichloroethylene and other halogenated solvents should be replaced with liquid aqueous alkali cleaning compounds.

Other waste reduction can be realized by the better production scheduling, use of dry filters in the spray booth, prevention of leakage from the spray gun, separation of wastes, and recycling of solvents by distillation.

## REFERENCES

1 EPA Office of Compliance Sector Notebook Project. Profile of the Fabricated Metal Products Industry. US Environmental Protection Agency, 1995.
2 EPA Office of Compliance Sector Notebook Project. Sector Notebook Data Refresh - 1997. US Environmental Protection Agency, 1998.

# 14.10 FOOD INDUSTRY - SOLVENTS FOR EXTRACTING VEGETABLE OILS

PHILLIP J. WAKELYN
National Cotton Council, Washington, DC, USA
PETER J. WAN
USDA, ARS, SRRC, New Orleans, LA, USA

## 14.10.1 INTRODUCTION

Many materials including oils, fats, and proteins, for both food and nonfood use, are recovered from diverse biological sources by solvent extraction.[1,2] These materials include animal tissues (e.g., beef, chicken, pork and fish); crops specifically produced for oil or protein (e.g., soy, sunflower, safflower, rape/canola, palm, and olive); by-products of crops grown for fiber (e.g., cottonseed and flax); food (e.g., corn germ, wheat germ, rice bran, and coconut); confections (e.g., peanuts, sesame, walnuts, and almonds); nonedible oils and fats (castor, tung, jojoba); and other oil sources (oils and fats from microbial products, algae, and seaweed). There are many physical and chemical differences between these diverse biological materials. However, the similarities are that oils (edible and industrial) and other useful materials (e.g., vitamins, nutriceuticals, fatty acids, phytosterols, etc.) can be extracted from these materials by mechanical pressing, solvent extracting, or a combination of pressing and solvent extraction. The preparation of the various materials to be extracted varies. Some need extensive cleaning, drying, fiber removal (cottonseed), dehulling, flaking, extruding, etc., all of which affect the solvent-substrate interaction and, therefore, the yield, composition, and quality of the oils and other materials obtained.

Historically, the advancement of processing technology for recovering oils and other useful materials has been primarily driven by economics. For thousands of years stone mills, and for several centuries simple hydraulic or lever presses were used as batch systems. The continuous mechanical presses only became reality during the early 1900s. It was not until the 1930s, that extraction solvents were used more widely, which greatly enhanced the recovery of oil from oilseeds or other oil bearing materials.

In solvent extraction crude vegetable oil and other useful materials are dissolved in a solvent to separate them from the insoluble meal. Many solvents have been evaluated for commercial extraction. Commercial hexane has been the main solvent for the oilseed processing industry since the 1940s[1,2] because of its availability at reasonable cost and its suitable functional characteristics for oil extraction. However, the interest in alternative solvents to hexane has continued and is motivated by one of a combination of factors: desire for a nonflammable solvent, more efficient solvent, more energy efficient solvent, less hazardous and environmentally friendly solvent, solvent with improved product quality, and solvents for niche/specialty markets. Today commercial isohexane (hexane isomers) is replacing commercial hexane in a few oilseed extraction operations and other solvents (e.g., isopropanol, ethanol, acetone, etc.) are also being used for various extraction processes or have been evaluated for use as extracting solvents.[3-6]

With greater flexibility of operating hardware and availability of various solvents with tailored composition, the oilseed industry does have expanded options to choose the unique composition of solvents to obtain the desired final products. While the availability, cost, ease-of-use (i.e., no specialty processing hardware required and with low retrofitting costs), and acceptable product quality continue to be the principle factors used to choose a solvent for oil extraction, in recent years environmental concerns and health risk have become increasingly important criteria to be considered in selecting an oil extraction solvent.

This chapter presents information on solvents for extracting oilseeds and other biological materials for oils, fats, and other materials.

## 14.10.2 REGULATORY CONCERNS

Many workplace, environmental, food safety, and other regulations (see Tables 14.10.1 and 14.10.2 for summary information on U.S. laws and regulations) apply to the use of solvents for the extraction of oilseeds and other diverse biological materials. Some solvent may have fewer workplace and environmental regulatory requirements; offer benefits, such as an improved product quality and energy efficiency;[7] as well as other advantages, such as the ability to remove undesirable (e.g., gossypol from cottonseed) and desirable (e.g., vitamin E, tocopherols, lecithins, phytosterols, and long chain polyunsaturated fatty acids) constituents of oilseeds and other biological materials more selectively. Some of the regulations required in the United States are discussed. Many other countries have similar requirements, but if they do not, it would be prudent to consider meeting these regulations in solvent extraction and to have environmental, health and safety, and quality management programs.[19,44]

**Table 14.10.1. U.S. Worker health and safety laws and regulations**

**Laws**:

Occupational Safety and Health Act of 1970 (OSHA) (PL 91-596 as amended by PL 101 552; 29 U.S. Code 651 et. seq.)

**OSHA Health Standards**:

Air Contaminants Rule, 29 CFR 1910.1000
Hazard Communication Standard, 29 CFR 1910.1200

**OSHA Safety Standards**:

Process Safety Management, 29 CFR 1910.119
Emergency Action Plan, 29 CFR 1910.38(a)(1)
Fire Prevention Plan, 29 CFR 1910.38(b)(1)
Fire Brigades, 29 CFR 1910.156
Personal Protection Equipment (General Requirements 29 CFR 1910.132; eye and face protection, .133; respiratory protection, .134)

**Table 14.10.2. U.S. Environmental laws and regulations**

| Law or Regulation | Purpose |
|---|---|
| Environmental Protection Agency (EPA) (established 1970) | To protect human health and welfare and the environment |
| Clean Air Act (CAA) (42 U.S. Code 7401 et seq.) | To protect the public health and welfare. Provides EPA with the authority to set NAAQS, to control emissions from new stationary sources, and to control HAP. |
| Federal Water Pollution Control Act (known as the Clean Water Act) (CWA) (33 U.S. Code 1251 et seq.) | The major law protecting the "chemical, physical and biological integrity of the nation's waters." Allows the EPA to establish federal Limits on the amounts of specific Pollutants that can be released by municipal and industrial facilities. |

| | |
|---|---|
| Toxic Substances Control Act (TSCA) (15 U.S. Code 2601 et seq.) | Provides a system for identifying and evaluating the environmental and health effects of new chemicals and chemicals already in commerce. |
| Resource Conservation and Recovery Act (RCRA) (42 U.S. Code 6901 et seq.) | A system for handling and disposal of non-hazardous and hazardous waste. |
| Comprehensive Environmental Response, Compensation and Liability Act (CERCLA) (42 U.S. Code 9601 et seq.) | Known as "Superfund", gives the EPA power to recover costs for containment, other response actions, and cleanup of hazardous waste disposal sites and other hazardous substance releases. |
| Emergency Planning and Community Right-to-Know Act (EPCRA; also "SARA 313") (42 U.S. Code 1101 et seq.) | (Part of Superfund) Provides authority for communities to devise plans for preventing and responding to chemical spills and release into the environment; requires public notification of the types of hazardous substances handled or release by facilities; requires state and local emergency plans. |

### 14.10.2.1 Workplace regulations

Workplace regulations (see Table 14.10.1) are promulgated and enforced in the U.S. by the Occupational Safety and Health Administration (OSHA), which is in the Department of Labor. The purpose of OSHA is to ensure that the employers maintain a safe and healthful workplace. Several workplace standards that affect extraction solvents are discussed.

#### *14.10.2.1.1 Air Contaminants Standard (29 CFR 1910.1000)*

The purpose of the air contaminants standards are to reduce risk of occupational illness for workers by reducing permissible exposure limits (PEL) for chemicals. Table 14.10.3 lists the PELs [8-hr time-weighted average (TWA) exposure] for the solvents discussed. To achieve compliance with the PEL, administrative or engineering controls must first be determined and implemented, whenever feasible. When such controls are not feasible to achieve full compliance, personal protective equipment, work practices, or any other protective measures are to be used to keep employee exposure below the PEL.

**Table 14.10.3. U.S. Workplace regulations,[a] air contaminants**

| Chemical Name (CAS No.) | Permissible Exposure Limit (PEL) [Health Risk: Basis for the PEL] |
|---|---|
| n-Hexane (110-54-3) | 500 ppm/1800 mg/m$^3$; new PEL was 50 ppm/180 mg/m$^3$ same as ACGIH (TLV); [neuropathy] |
| Commercial hexane[b] (none) | (Same as n-hexane) |
| n-Heptane (142-82-5) | 500 ppm/200 mg/m$^3$; new PEL was 400 ppm/1640 mg/m$^3$, (500 ppm STEL) same as ACGIH (TLV); [narcosis] |
| Cyclohexane (110-82-7) | 300 ppm/1050 mg/m$^3$; ACGIH (TLV) 300 ppm/1030 mg/m$^3$; [sensory irritation] |
| Cyclopentane (287-92-3) | None; new PEL was 600 ppm, same as ACGIH (TLV); [narcosis] |
| Hexane isomers | None; new PEL was 500 ppm/1760 mg/m$^3$ (1000 ppm STEL) same as ACGIH (TLV); [narosis] |
| Commercial isohexane[c] (none) | (Same as hexane isomer) |

| Chemical Name (CAS No.) | Permissible Exposure Limit (PEL) [Health Risk: Basis for the PEL] |
|---|---|
| 2-Methyl pentane (isohexane) (2-MP) (107-83-5) | (Same as hexane isomer) |
| 3-Methyl pentane (3-MP) (96-14-0) | (Same as hexane isomer) |
| Methyl cyclopentane (MCP) (96-37-7) | (Same as hexane isomer) |
| 2,2 Dimethyl butane (neohexane) (2,2-DMB) (75-83-2) | (Same as hexane isomer) |
| 2,3 Dimethyl butane (2,3-DMB) (79-29-8) | (Same as hexane isomer) |
| Methyl cyclohexane (108-87-2) | 500 ppm; new PEL was 400 ppm/1610 mg/m$^3$, same as ACGIH (TLV); [narcosis] |
| Isopropyl alcohol (2-propanol) (IPA) (67-17-5) | 400 ppm/980 mg/m$^3$; ACGIH (TLV) same plus 500 ppm/1230 mg/m$^3$ STEL; [sensory irritation] |
| Ethyl alcohol (ethanol) (64-17-5) | 1000 ppm/1880 mg/m$^3$; ACGIH (TLV) same; [narcosis, irritation] |
| Acetone (67-64-1) | 1000 ppm/2400 mg/m$^3$; ACGIH (TLV) 750 ppm (1000 ppm STEL); [sensory irritation] |

[a]CAS No. is the Chemical Abstracts Service Registry Number; PEL is from 29 CFR 1910.1000, Table Z-1; American Conference on Governmental Industrial Hygienists (ACGIH), threshold limit value (TLV); under the HCS, a MSDS is required for all of the compounds (physical and/or chemical hazard); all of the solvents are flammable liquids or gasses, under the OSHA definition, and are regulated under the PSM Standard.
[b]Commercial hexane as used in the U.S. is usually about 65% n-hexane, and the rest is hexane isomers (e.g., methyl cyclopentane (MCP), 2-methyl pentane (2-MP), and 3-methyl pentane (3-MP)), and it contains less than 10 ppm benzene.
[c]Mixture of 2-MP (45-50%), 3-MP, 2,2-DMB, and 2,3-DMB. (ref 7)

In the case of a mixture of contaminants, an employer has to compute the equivalent exposure when the components in the mixture pose a synergistic threat (toxic effect on the same target organ) to worker health.[8,9]

#### *14.10.2.1.2 Hazard Communication Standard (HCS) (29 CFR 1910.1200)*

The HCS requires information on hazardous chemicals to be transmitted to employees through labels, material safety data sheets (MSDS), and training programs. A written hazard communications program and record keeping are also required.

A substance is a "hazardous chemical" if it is a "physical hazard" or a "health hazard". A flammable or explosive liquid is a "physical hazard". A flammable liquid means "any liquid having a flash point below 110°F (37.8°C), except any mixture having components with flash points of 100°F (37.8°C) or higher, the total of which make up 99% or more of the total volume of the mixture". "Health hazard" means "a chemical for which there is statistically significant evidence based on at least one valid study that acute or chronic health effects may occur in exposed employees". Hexane and all the solvents listed in Table 14.10.3 would require a MSDS, since all are flammable liquids (physical hazards) as defined by OSHA and/or possible health hazards because all, except hexane isomers, have an U.S. OSHA PEL. However, hexane isomers have an American Conference of Industrial Hygienist (ACGIH) threshold limit value (TLV),[10] which many states and countries enforce as a mandatory standard.

Chemical manufacturers and importers are required to review the available scientific evidence concerning the hazards of chemicals they produce or import, and to report the information to manufacturing employers who use their products. If a chemical mixture has not been tested as a whole to determine whether the mixture is a hazardous chemical, the mixture is assumed to present the same hazards as do the components that comprise 1% or greater of the mixture or a carcinogenic hazard if it contains a component in concentration of 0.1% or greater that is a carcinogen. Commercial hexane containing 52% n-hexane has been tested and found not to be neurotoxic unlike pure n-hexane.[11-13] So mixtures with less than 52% n-hexane should not be considered to be a neurotoxin, although n-hexane would have to be listed on the MSDS, if in greater quantity than 1% of the mixture.

##### *14.10.2.1.3 Process Safety Management (PSM) Standard (29 CFR 1910.119)*

PSM is for the prevention or minimization of the consequences of catastrophic releases of toxic, reactive, flammable, or explosive chemicals. This regulation applies to all processes that involve one or more of 137 listed chemicals (29 CFR 1910.119, Appendix A) above their threshold quantities or have 10,000 lbs. or more of a flammable liquid or gas, as defined by the U.S. OSHA HCS [29 CFR 1910.1200(c)]. This includes n-hexane, hexane isomers, and all solvents listed in Table 14.10.3.

In addition to the PSM standard, U.S. OSHA has been enforcing two other regulations for operations/processes with flammable liquids. First, under Personal Protective Equipment - General Requirements (29 CFR 1910.132), OSHA has cited or obtained voluntary agreement from organizations relative to flame-resistant (FR) clothing. Operators and other employees working in the area of a flammable process are being required to wear flame-resistant work clothing. For facilities that use hexane and other flammable solvents, it would be prudent to require FR clothing for all personnel working in areas where there is an exposure to a flammable liquid. Second, OSHA has cited organizations for failure to meet related safety regulations under Fire Brigades (29 CFR 1910.156), specifically for standards such as: training, both initial and annual refresher training; protective equipment availability and testing; and fitness for duty including periodic physicals. If an on-site fire brigade is part of the site's Emergency Response Plan (29 CFR 1910.38), then these requirements must also be met. In addition, the requirement of the PSM standard for an Emergency Response Plan triggers the requirements of Emergency Action Plan [29 CFR 1910.38(a)].

#### 14.10.2.2 Environmental regulations

The role of the U.S. Environmental Protection Agency (EPA) is to protect human health and welfare and the environment. The U.S. EPA administers all regulations affecting the environment and chemicals in commerce. The individual states and state environmental regulatory control boards implement and enforce most of the regulations. The legislation that serves as the basis for the regulations can be divided into: statutes that are media-specific [Clean Air Act (CAA) and Clean Water Act (CWA)]; statutes that manage solid and hazardous waste [Resources Conservation and Recovery Act (RCRA) and Comprehensive Environmental Response, Compensation and Liability Act (CERCLA; "Superfund")]; and, statutes that directly limit the production rather than the release of chemical substance [Toxic Substances Control Act (TSCA) and Federal Insecticide, Fungicide and Rodenicide Act (FIFRA)]. See Table 14.10.2 for a summary of the information on environmental laws and regulations, and Tables 14.10.4 and 14.10.5 for an overview of environmental requirements that apply to each solvent.

**Table 14.10.4. U.S. Environmental regulations, air and water**

| Chemical Name (CAS No.) | VOC | HAP | CWA[a] | Sol. in $H_2O$ |
|---|---|---|---|---|
| n-Hexane (110-54-3) | Yes | Yes | Yes | I |
| Commercial hexane(none) | Yes | | | |
| n-Heptane (148-82-5) | Yes | No | Yes | I |
| Cyclohexane (110-82-7) | Yes | No | Yes | 0.01 |
| Cyclopentane (287-92-3) | Yes | No | Yes | I |
| Hexane isomers (none) | Yes | No | Yes | |
| Commercial isohexane (none) | (same as hexane isomers) | No | Yes | |
| 2-Methyl pentane (isohexane) (107-83-5) | (a hexane isomer) Yes | No | Yes | 0.0014 |
| 3-Methyl pentane (96-14-0) | (a hexane isomer) Yes | No | Yes | 0.0013 |
| Methyl cyclopentane (96-37-7) | (a hexane isomer) Yes | No | Yes | 0.0013 |
| 2,2 dimethyl butane (neohexane) (75-83-2) | (a hexane isomer) Yes | No | Yes | 0.0018 |
| 2,3 dimethyl butane (79-29-8) | (a hexane isomer) Yes | No | Yes | 0.0011 |
| Methyl cyclohexane(108-87-2) | Yes | No | Yes | 0.0014 |
| Isopropyl alcohol (2-propanol) (67-17-5) | Yes | No | Yes | Misc. |
| Ethyl alcohol (ethanol) (64-17-5) | Yes | No | Yes | Misc. |
| Acetone (67-64-1) | No[b] | No | Yes | Misc. |

[a]Under the Clean Water Act there could be stormwater and NPDES permit requirements; none of the solvents are listed as priority toxic pollutants in 40 CFR 401.15.
[b]Acetone is considered by the U.S. EPA not be a VOC (60 FR 31643; June 16, 1995)
Abbreviations: CAS No., Chemical Abstracts Service Registry number; VOC, volatile organic chemical; HAP, hazardous air pollutant; CWA, Clean Water Act; I, insoluble in $H_2O$; MISC., miscible in $H_2O$; and 0.01, 0.01 parts soluble in 100 parts $H_2O$. Source for water solubility: Ref 47.

**Table 14.10.5 U.S. Environmental regulations, waste**

| Chemical Name (CAS No.) | (RCRA) | | (EPCRA/SARA Title III)[a] | |
|---|---|---|---|---|
| | RCRA Code[b] | Sec. 304 CERCLA RQ | Sec.311/312 | Sec.313 (TRI) |
| n-hexane (110-54-3) | | 5000[c] | Yes | Yes |
| Commercial hexane (none) | | | | Yes[d] |
| n-heptane (142-82-5) | | | Yes | No |

| Chemical Name (CAS No.) | (RCRA) | | (EPCRA/SARA Title III)[a] | |
|---|---|---|---|---|
| | RCRA Code[b] | Sec. 304 CERCLA RQ | Sec.311/312 | Sec.313 (TRI) |
| Cyclohexane (110-87-7) | U056 | 1000 | Yes | Yes |
| Cyclopentane (287-92-3) | | | Yes | No |
| Hexane isomers (none) | | | Yes | No[e] |
| Commercial isohexane (none) | | | | No[e] |
| 2-Methyl pentane (also called isohexane) (107-83-5) | | | Yes | No[e] |
| 3-Methyl pentane (96-14-0) | | | Yes | No[e] |
| Methyl cyclopentane (96-37-7) | | | Yes | No[e] |
| 2,2 Dimethyl butane (neohexane) (75-83-2) | | | Yes | No[e] |
| 2,3 Dimethyl butane (79-29-8) | | | Yes | No[e] |
| Methyl cyclohexane (108-87-2) | | | Yes | No |
| Isopropyl alcohol (2-propanol) (67-63-0) | | | Yes | No[f] |
| Ethyl alcohol (64-17-5) | | | Yes | No |
| Acetone (67-64-1) | U002 | 5000 | Yes | No |

[a]From Title III Lists of Lists, U.S. EPA, EPA 740-R-95-001 (April 1995); 40 CFR 52-99; (59 FR 4478; January 31, 1994) hexane added to TRI list; (60 FR 31633; June 16, 1995) acetone removed from TRI list.
[b]40 CFR 261.33, listed hazardous waste - EPA RCRA Hazardous Waste Number. All the solvents that are on the RCRA list are listed because of Section 3001 of RCRA (part for identification and listing of hazardous waste) except hexane which is on because of CAA Section 112 (HAP).
[c]RQ for hexane finalized June 12, 1995 (60 FR 30939).
[d]Only the amount of commercial hexane that is n-hexane has to be reported (e.g., if the commercial hexane is 62% n-hexane, only 62% of the emissions have to be reported for TRI).
[e]The EPA clarrified that the listing for hexane was only for n-hexane, other isomers of hexane are not included. (59 FR 61457; Nov. 30, 1994).
[f]The EPA has indicated (62 FR 22318, April 25, 1997) that IPA itself does not meet the criteria for listing on the TRI list. The EPA will remove IPA from the TRI list. Abbreviations: RQ, reportable quantity in pounds/24 h.

##### *14.10.2.2.1 Clean Air Act (CAA; 42 U.S. Code 7401 et seq.)*

To satisfy the CAA requirements, states and state air control boards are required to implement regulations and develop state implementation plans (SIP).[14,15] Criteria pollutants (e.g., ozone, oxides of nitrogen, carbon monoxide) are regulated with National Ambient Air Quality Standards (NAAQS) and hazardous air pollutants (HAP), such as hexane, with National Emissions Standards for Hazardous Air Pollutants (NESHAP).

The NAAQS are set at levels sufficient to protect public health (primary air quality standards) and welfare (secondary air quality standards; "welfare effects" include wildlife, visibility, climate, damage to and deterioration of property and effects on economic value and on personal comfort and well being) from any known or anticipated adverse effect of the pollutant with an adequate (appropriate) margin of safety. The 1990 CAA expanded the list of HAP to 188, including hexane, and more strictly regulates nonattainment areas for criteria pollutants such as ozone, particulate matter, carbon monoxide and oxides of nitrogen.

*NAAQS*: Volatile organic compounds (VOC) are essentially considered the same as the criteria pollutant ozone.[16,17] VOCs are very broadly defined by the U.S. EPA (40 CFR 51.100): any compound of carbon, excluding carbon monoxide, carbon dioxide, carbonic acid, metallic carbides or carbonates, and ammonium carbonate, that participates in atmospheric photochemical reactions. This includes any organic compound other than those specifically listed as having been determined to have negligible photochemical reactivity. Reactive VOCs are essentially all those judged to be clearly more reactive than ethane - the most reactive member of the "negligibly reactive" class. $C_4$ - $C_6$ paraffins are of relatively low kinetic reactivity but produce $NO_2$, and potentially ozone.[18] Hexane and all of the solvents discussed, except acetone, would be considered VOCs (see Table 14.10.4) that can undergo photochemical oxidation in the atmosphere to form ozone. In the U.S., acetone was added to the list of compounds excluded from the definition of VOCs in 1995, because it was determined to have negligible photochemical reactivity.[19]

Most U.S. vegetable-oil extracting facilities would be major sources of VOCs and would be covered by the requirements for ozone emissions and attainment, unless they used a solvent that was not classified as a VOC. (The definition of "major source" changes as the severity of the ozone nonattainment area increases. Plants in marginal and moderate areas are major if they emit 100 tons VOC/yr; in serious areas, 50 tons/yr; in severe areas, 25 tons/yr; and in extreme areas 10 tons/yr). All facilities in ozone non-attainment areas could be required to reduce emissions through implementing Reasonable Available Control Measures (RACM) standards or Best Available Control Measures (BACM). Any new or significantly modified facility would have to comply with the new source review (NSR) requirements and prevention of significant deterioration (PSD) requirements.

*Hazardous air pollutants (HAP) or air toxics*: If a facility is a major emitter of any of the chemicals on the CAA list of HAPs, EPA requires sources to meet emissions standards.[14,16,17] n-Hexane is on the HAP list but isohexane, acetone and other solvents listed in Table 14.10.4 are not.

The air toxic requirements of the CAA for establishing control measures for source categories are technology-based emission standards (not health based) established for major sources (10 tons/yr of one HAP or 25 tons/yr of total HAP) that require the maximum degree of reduction emissions, taking costs, other health and environmental impacts, and energy requirements into account. Standards are set based on known or anticipated effects of pollutants on the public health and the environment, the quantity emitted, and the location of emissions. Compliance involves the installation of Maximum Achievable Control Technology (MACT) - MACT essentially is maximum achievable emission reduction. For new sources, MACT standards must be no less stringent than the emission control achieved in practice by the best controlled similar source. The MACT standards for vegetable oil processing using n-hexane are expected to be issued in 2001. Once a standard has been promulgated for a source category, a source will have three years after the due date to comply. The requirements cover normal operations and startup, shutdown, and malfunction (SSM). The allowable emissions for solvent extraction for vegetable oil production in the U.S., as a 12-month rolling average based on a 64% n-hexane content, will vary from 0.2 gal/ton to greater than 0.7 gal/ton depending on the oilseed (65FR34252; May 26, 2000) (see Table

12.14.6). There are also variable emission requirements depending on the oilseed for allowable emissions for solvent extraction for vegetable oil production in Europe.

**Table 14.10.6. Oilseed solvent loss factors for allowable HAP loss (12-mo. rolling ave.)**

| Type of Oilseed Process | A source that... | Oilseed Solvent Loss Factor (gal/ton) | |
|---|---|---|---|
| | | Existing Sources | New Sources |
| Corn Germ, Wet Milling | processes corn germ that has been separated from other corn components using a wet process of centrifuging a slurry steeped in a dilute sulfurous acid solution | 0.4 | 0.3 |
| Corn Germ, Dry Milling | processes corn germ that has been separated from the other corn components using a dry process of mechanical chafing and air sifting | 0.7 | 0.7 |
| Cottonseed, Large | processes 120,000 tons or more of a combination of cottonseed and other listed oilseeds during all normal operating periods in a 12 operating month period | 0.5 | 0.4 |
| Cottonseed, Small | processes less than 120,000 tons of a combination of cottonseed and other listed oilseeds during all normal operating periods in a 12 operating month period | 0.7 | 0.4 |
| Flax | processes flax | 0.6 | 0.6 |
| Peanuts | processes peanuts | 1.2 | 0.7 |
| Rapeseed | processes rapeseed | 0.7 | 0.3 |
| Safflower | processes safflower | 0.7 | 0.7 |
| Soybean, Conventional | uses a conventional style desolventizer to produce crude soybean oil products and soybean animal feed products | 0.2 | 0.2 |
| Soybean, Specialty | uses a special style desolventizer to produce soybean meal products for human and animal consumption | 1.7 | 1.5 |
| Soybean, Small Combination Plant | processes soybeans in both specialty and conventional desolventizers and the quantity of soybeans processed in specialty desolventizers during normal operating periods is less than 3.3 percent of total soybeans processed during all normal operating periods in a 12 operating month period. The corresponding solvent loss factor is an overall value and applies to the total quantity of soybeans processed | 0.25 | 0.25 |
| Sunflower | processes sunflower | 0.4 | 0.3 |

A health-based standard would be for a boundary line level of a solvent (e.g., n-hexane) based on the inhalation reference concentration (RfC).[20] The current RfC for n-hexane is 200μg/$m^3$. Recent research suggests that the RfC for hexane should be at least 10 times higher (> 2000 μg/$m^3$).

*Federal permits*: All major sources of regulated solvents are required to have federally enforceable operating permits (FOP)[14,15] (also referred to as Title V permits).

*State permits:* Most states require state permits for facilities that emit listed air pollutants.[14,15] In some states federal permits and state permits are combined, while in other states facilities are required to have both a state or county (air district) permit and a federal permit. As part of annual emission inventory reporting requirements, many states already require reporting of HAP and VOC because of their state implementation plan (SIP).

#### *14.10.2.2.2 Clean Water Act (CWA; 33 U.S. Code 1251 et seq.)*

The CWA is the major law protecting the "chemical, physical and biological integrity of the nation's waters." Under it, the U.S. EPA establishes water-quality criteria used to develop water quality standards, technology-based effluent limitation guidelines, and pretreatment standards and has established a national permit program [National Pollution Discharge Elimination System (NPDES) permits; 40 CFR 122] to regulate the discharge of pollutants. The states have responsibility to develop water-quality management programs.

For extraction solvents vegetable oil extracting facilities are covered by basic discharge effluent limitations [direct discharges to receiving waters or indirect discharges to publicly owned treatment works (POTW)], and stormwater regulations.[15] The amount of solvent in effluent discharges and in stormwater (for those covered) needs to be determined and possibly monitored as part of an NPDES permit and as part of the visual examination or testing of stormwater quality.

#### *14.10.2.2.3 Resource Conservation and Recovery Act (RCRA; 42 U.S.Code 6901 et seq.)*

RCRA subtitle C (40 CFR 261) is a federal "cradle-to-grave" system to manage hazardous waste (including provisions for cleaning up releases and setting statutory and regulatory requirements). Subtitle D covers nonhazardous wastes. Materials or items are hazardous wastes if and when they are discarded or intended to be discarded. The act requires generators, transporters, and disposers to maintain written records of waste transfers, and requires the U.S. EPA to establish standards, procedures, and permit requirements for disposal. The act also requires states to have solid waste management plans, prohibits open dumping, and requires the EPA to establish criteria for sanitary landfills. EPA under RCRA also regulates underground storage tanks that store or have stored petroleum or hazardous substances.

Hazardous wastes are either listed wastes (40 CFR 261.30-.33) or characteristic wastes (40 CFR 261.21-.24). The U.S. EPA defines four characteristics for hazardous waste: ignitability (40 CFR 260.21); corrosivity (40 CFR 260.22); reactivity (40 CFR 260.23); and toxicity (40 CFR 260.24). Any waste that exhibits one or more of these characteristics is classified as hazardous under RCRA. The ignitability definition includes a liquid that has a flash point less than 60°C (140°F); the EPA included ignitability to identify wastes that could cause fires during transport, storage, or disposal (e.g., used solvents). All of the solvents in Table 14.10.5 have flash points less than 60°C, so all could be a RCRA ignitability waste.

#### 14.10.2.2.4 Emergency Planning and Community Right-to-Know Act (EPCRA; 42 U.S. Code 11001 et seq.)

Enacted as Title III of the 1986 Superfund Amendments and Reauthorization Act ("SARA"), the Act mandates the EPA to monitor and protect communities regarding releases of chemicals into the environment. It requires states to establish emergency planning districts with local committees to devise plans for preventing and responding to chemical spills and releases. ["Superfund" is the Comprehensive Environmental Response, Compensation and Liability Act (CERCLA) of 1980 that gives the U.S. EPA authority to force those responsible for hazardous waste sites or other releases of hazardous substances, pollutants, and contaminants to conduct cleanup or other effective response actions.]

*Section 304 (40 CFR 355.40)*: Facilities are subject to state and local reporting for accidental releases, in quantities equal to or greater than their reportable quantities (RQ), of extremely hazardous substances (EHS) or CERCLA hazardous substances (40 CFR 302, Table 302.4) under Section 304. n-Hexane, cyclohexane, acetone, and some of the other solvents discussed are CERCLA hazardous substances and have CERCLA RQ for spills (Table 14.10.5).

*Section 311, 312 (40 CFR 370.20-.21)*: Business must make MSDSs, for chemicals that are required to have an MSDS, available to state and local officials. Since all of the solvents discussed require MSDSs under the OSHA HCS, all are covered by these requirements.

*Section 313 (40 CFR 372), Toxic Release Inventory (TRI)*: Businesses are required to file annual reports with federal and state authorities of releases to air, water, and land above a certain threshold for chemicals on the TRI/Section 313 list (40 CFR 372.65) by July 1 each year for the previous year's releases.[21] TRI requirements are triggered if a facility is involved in manufacturing with 10 or more full-time employees, manufactures, processes, or otherwise uses with one or more listed substance(s) in a quantity above the statutory reporting threshold of 25,000 lbs./yr (manufactured or processed) or 10,000 lbs./yr (otherwise used). Beginning with the 1991 reporting year, such facilities also must report pollution prevention and recycling data for such chemicals pursuant to Section 6607 of the Pollution Prevention Act (42 U.S. Code 13106).

n-Hexane was added to the TRI list in 1994 with reporting for 1995 emissions.[19] The other solvents discussed are not on the TRI list. The EPA can add new chemicals to or delete chemicals from the TRI list as it deemed necessary and any person may petition the EPA to add chemicals or delete chemicals from the list.

#### 14.10.2.2.5 Toxic Substances Control Act (TSCA; 15 U.S. Code 2601 et seq.)

If a chemical's manufacture, processing, distribution, use, or disposal would create unreasonable risks, the U.S. EPA, under the TSCA, can regulate it, ban it, or require additional testing. TSCA mandates the U.S. EPA to monitor and control the use of toxic substances by requiring the Agency to review the health and environmental effects of new chemicals [referred to as "Premanufacturing Notice" or "PMN"; Section 5(a)(1) of TSCA] and chemicals already in commerce. The U.S. EPA also has Significant New Use Rules (SNUR) under Section 5(a)(2) of TSCA which provides a way for the U.S. EPA to restrict uses of a chemical substance already in commerce that are proposed for new uses. All of the solvents discussed are already commercially available, so a PMN would not apply; some could be subjected to SNUR (40 CFR 721, subpart A), since some are not presently being used as extraction solvents in large quantities.

Under Section 4(a) of TSCA, the U.S. EPA can require testing of a chemical substance or mixture to develop data relevant for assessing the risks to health and the environment. Section 8(d) of TSCA requires that lists of health and safety studies conducted or initiated with respect to a substance or mixture be submitted to the U.S. EPA. All new toxicological data of the effects of a chemical not previously mentioned must be reported immediately if the data reasonably supports the conclusion that such substance or mixture presents a substantial risk of injury to health or the environment [Section 8(e) of TSCA]. Testing (Section 4 test rule) was required for several of the solvents earlier (e.g., commercial hexane for which new toxicological information was reported to the U.S. EPA since 1992),[22] and any new toxicological information will have to be reported to the U.S. EPA under Section 8(e) and 8(d).

### 14.10.2.3 Food safety

In the U.S. the use of a solvent to extract oil, that is a human food product or used in a food product, from oilseeds and biological materials falls under the rules and regulatory jurisdiction of the U.S. Food and Drug Administration (FDA), which regulates all aspects of food, including food ingredients and labeling in the U.S. In order to be legally used as an oilseed extraction solvent in the U.S., a substance must have been subject to an approval by the U.S. FDA or the U.S. Department of Agriculture (USDA) during 1938-1958 for this use ("prior sanction"); be generally recognized as safe (GRAS) for this use; or be used in accordance with food additive regulations promulgated by the U.S. FDA.

Many prior sanctions and GRAS determinations are not codified in the U.S. FDA regulations. However, extracting solvents used in food manufacturing, such as n-hexane, have been labeled as a food additive, solvent, defoaming agent, component of a secondary food and color additives, minor constituent, or incidental additives (i.e., "additives that are present in a food at significant levels and do not have any technical or functional effect in that food") depending on the application. Incidental additives can be "processing aids," (i.e., "substances that are added to a food during processing but removed from the food before it is packaged"). Most food-processing substances, including solvents, can be regarded as "incidental additives" and thus are exempt from label declaration in the finished food product. Even if exempt from label declaration, all extraction solvents must be used in accordance with the U.S. FDA good manufacturing practices (GMP; 21 CFR 100).

In the U.S., the Flavor and Extract Manufacturers' Association (FEMA) has conducted a program since 1958 using a panel of expert pharmacologists and toxicologists to determine substances that are GRAS. This panel uses all available data, including experience based on common uses in food. This safety assessment program ("FEMA GRAS") is widely accepted and considered an industry/government partnership with the U.S. FDA.[23] A number of papers published in Food Technology since 1961[24,25] list the substances that the panel has determined to be GRAS and the average maximum levels in parts per million (ppm) at which each has been reported to be GRAS for different categories of food. The U.S. FDA has not incorporated these substances in their regulations but does recognize the findings of the Expert Panel of FEMA as GRAS substances.

Since vegetable oil and other human food grade oils undergo deodorization (steam distillation) and other purification processes (i.e., refining and bleaching) as part of the manufacturing process prior to being used as a food product, they should not contain any of

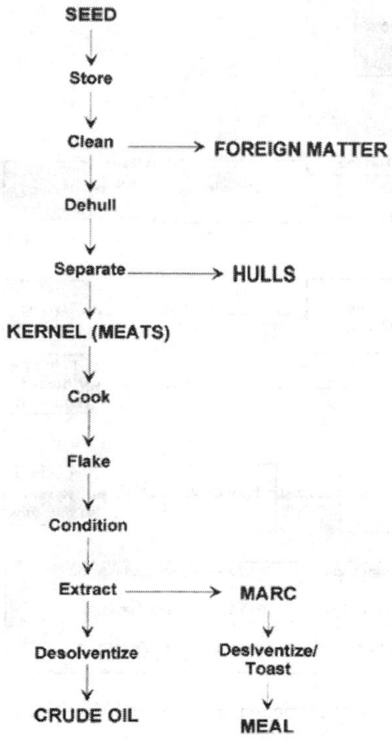


Figure 14.10.1a. Flow diagram of oilseed extraction process from seed to crude oil and meal.

the extraction solvent, if proper manufacturing practices are followed. (see Section 14.10.3.3 Processing crude oil, for more details.) Refining removes free fatty acids and other non-oil compounds (e.g., phospholipids, color, and trace metals); bleaching with acid-activated bleaching earth or clay (e.g., bentonite), removes color-producing substances and residual soaps; and deodorization, the last major processing step in edible oils refining removes volatile compounds (undesirable ingredients occurring in natural oils and those that may be imparted by prior unit processes or even storage, many of which are associated with undesirable flavors and odors).[26,27] Most commercial deodorizers operate at a temperature of 245-275°C (475-525°F) under a negative pressure of 2-10 mm Hg.[26,27] It has been reported that no hexane residue remains in the finished oil after processing due to its high volatility.[28] In addition, animal-feeding studies with expeller and solvent-extracted meals have not indicated any adverse health affects related to the extraction solvent.[29]

Hexane has been used since the 1940's as an oilseed-extraction solvent on the determination that it is GRAS and it may also be subject to a prior sanction. However, like many other food-processing substances, there is no U.S. FDA regulation specifically listing hexane as GRAS or prior sanctioned.

GRAS status may be determined by a company ("GRAS self-determination"), an industry, an independent professional scientific organization (e.g., FEMA GRAS), or the U.S. FDA. The Federal Food, Drug and Cosmetic Act (FFDCA; 21 U.S. Code 321 et seq.) does not provide for the U.S. FDA to approve all ingredients used in food, and the U.S. FDA explicitly recognizes that its published GRAS list is not meant to be a complete listing of all substances that are in fact GRAS food substances. Although there is no requirement to inform the U.S. FDA of a GRAS self-determination or to request FDA review or approval on the matter, the U.S. FDA has established a voluntary GRAS affirmation program under which such advice will be provided by the agency. Solvents that do not have prior sanction, a GRAS determination, or a tolerance set, probably should be evaluated for compliance under food safety requirements, if a facility is considering changing its extracting solvent or using a solvent for the extraction of the various biological materials for specialty markets.

## 14.10.3 THE SOLVENT EXTRACTION PROCESS

Three types of processing systems are used to extract oil from oil-bearing materials: expeller pressing, prepress solvent extraction, and direct solvent extraction. Only prepress solvent extraction and direct solvent extraction, which remove the oil from the conditioned, prepared seed with an organic solvent, will be discussed here[1,27] (see Figure 14.10.1). Oil-bearing materials have to be prepared for extraction to separate the crude oil from the

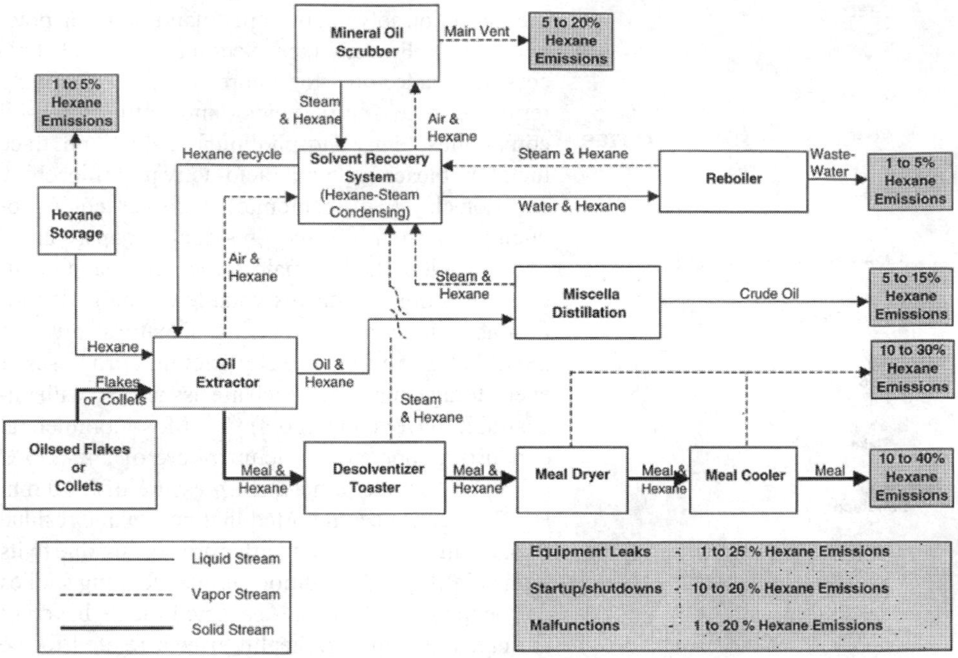


Figure 14.10.1b. Flow diagram of oilseed extraction process. Overview of extraction operation and identification of emission sources.

meal. Careful control of moisture and temperature during processing must be exercised to maintain the quality of the protein in the meal and to minimize the damage to the oil. Crude oils are refined by conditioning with phosphoric acid and treating with sodium hydroxide (alkali-refining) (see Figure 14.10.2). Refined oil is bleached with activated clay to remove color pigments. Bleached oils are then deodorized by steam distillation. The refined, bleached, and deodorized oil (RBD oil) is used to produce finished products, e.g., salad and cooking oils, shortening and margarine. Some of the finished products also require the oil to be hydrogenated, which changes the consistency of the oil, and increases stability to oxidation, which extends the shelf life of the finished products. Also some of the oils are winterized to remove the higher melting constituents, which can be used in confectionary products; the winterized oil is less likely to become cloudy in refrigerated storage.

### 14.10.3.1 Preparation for extraction

*Storage*: For optimum extraction and quality of oil, the oil-bearing material should be stored so that it remains dry and at relatively low temperature. If it is wet, it should be processed as soon as possible after harvest. Oils in the presence of water can deteriorate rapidly, forming free fatty acids and causing greater refining loss.

*Seed cleaning*: The first step in the commercial processing of oilseeds is "cleaning", to remove foreign materials, such as sticks, stems, leaves, other seeds, sand, and dirt using dry screeners and a combination of screens and aspiration. Permanent electromagnets are also used for the removal of trash iron objects. Final cleaning of the seed usually is done at the extraction plant just prior to processing.

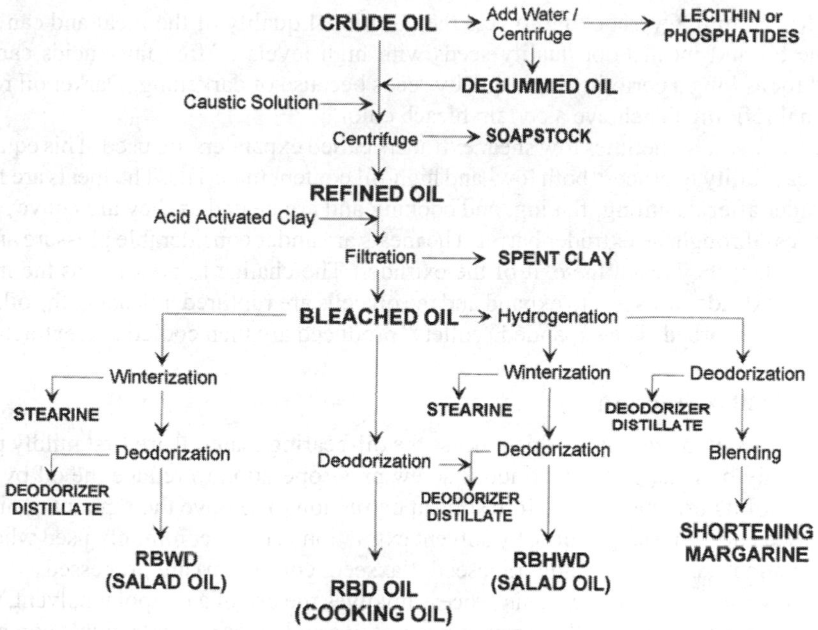


Figure 14.10.2. Flow diagram of edible oil processing.

*Dehulling*: After cleaning, it may be necessary to remove the seed's outer seed- coat (hull). The seedcoat contains little or no oil, so its inclusion makes the extraction less efficient. Also, the next processing step is grinding to reduce particle size, and any tough seedcoats would interfere with this process. If the hulls are not removed prior to extraction, they will reduce the total yield of oil by absorbing and retaining oil in thc press cake. An acceptable level of hull removal must be determined, depending on the desired protein level of the final meal. Hulls are removed by aspirator and undehulled seeds are removed from the kernels by screening and returned to the huller. Some meats still adhere to the hulls, which are beaten, then screened again to obtain the meat.

*Grinding, rolling, or flaking*: After dehulling, the meats are reduced in size, or "flaked," to facilitate oil removal. Proper moisture content of the seeds is essential for flaking, and if the moisture level is too low, the seeds are "conditioned," with water or steam, to raise the moisture to about 11%. For solvent extraction, flakes are commonly not less than 0.203-0.254 mm (0.008-0.010 inch), which can be solvent extracted efficiently with less than 1% residual oil. Thinner flakes tend to disintegrate during the solvent extraction process and reduce the miscella percolation rate.

*Cooking*: Prior to extraction, the flakes are heated. The purpose of cooking the flakes is: (1) cell walls are broken down, allowing the oil to escape; (2) oil viscosity is reduced; (3) moisture content is controlled; (4) protein is coagulated; (5) enzymes are inactivated and microorganisms are killed; and (6) certain phosphatides are fixed in the cake, which helps to minimize subsequent refining losses. Flakes are cooked in stack cookers to over 87.8°C (190°F) in the upper kettle. Flakes with high phosphatide content may benefit from being cooked at slightly lower temperatures to avoid elevating refining losses. The temperature of the flakes is raised to 110-132.2°C (230-270°F) in the lower kettles. The seeds are cooked for up to 120 min. Overcooking lowers the nutritional quality of the meal and can darken both the oil and meal. Poor-quality seeds with high levels of free fatty acids cannot be cooked for as long a period as high-quality seeds because of darkening. Darker oil requires additional refining to achieve a certain bleach color.

*Expanders*: Sometimes low shear extruders called expanders are used. This equipment has the capability to process both low- and high-oil content materials. The meats are fed into an extruder after dehulling, flaking, and cooking and are heated as they are conveyed by a screw press through the extruder barrel. The meats are under considerable pressure and temperature when they reach the exit of the extruder. The change in pressure as the material leaves the extruder causes it to expand and the oil cells are ruptured, releasing the oil, which is rapidly reabsorbed. The expanded "collets" produced are then cooled and extracted with solvent.

### 14.10.3.2 Oil extraction

*Prepress solvent extraction*: In this process the oil-bearing material are first mildly pressed mechanically by means of a continuous screw press operation to reduce the oil by half to two-thirds of its original level before solvent extraction to remove the remaining oil in the pre-pressed cake. Pressing follow by solvent extraction is more commonly used when high oil content materials (e.g., canola/rapeseed, flaxseed, corn germ) are processed.

*Direct solvent extraction*: This process involves the use of a nonpolar solvent, usually hexane, to dissolve the oil without removing proteins and other non-oil soluble compounds. Solvent extraction yields about 11.5% more oil than does the screw press method, and less oil remains in the meal. The cooked flakes or collets (if expanders are used) are mixed with hexane in a batch or continuous operation. The hexane vapor pressure limits the practical operating temperature of the extraction and its contents to about 50-55°C. The resulting miscella (oil-solvent mixture) and the marc (solvent laiden collets) are heated to evaporate the solvent, which is collected and reused. The oil is freed from the miscella, by using a series of stills, stripping columns, and associated condensers. The hexane-free oil (i.e., crude oil) is cooled and filtered before leaving the solvent-extraction plant for storage or further treatment. This is the crude oil normally traded in the commodity market. Occasional overheating of the oil-solvent miscella will cause irreversible color changes in the oil.

### 14.10.3.3 Processing crude oil

Most crude edible oils, obtained from oil-bearing materials, consist primarily of triglycerides (triacylglycerols). The triglycerides (approximately 95% of the crude oil) are the constituents recovered for use as neutral oil in the manufacture of finished products. The remaining nontriglyceride portion contains variable amounts of other lipophilic compounds, such as free fatty acids (FFA), nonfatty materials generally classified as "gums," phospholipids (phosphatides), tocopherols, color pigments, trace metals, sterols, meal, oxidized materials, waxes, moisture, and dirt. Most of these minor lipid components are detrimental to finished product color, flavor, and smoking stability, and so must be removed from the neutral oil by a processing/purification process. The object of the processing/purification steps is to remove the objectionable impurities while minimizing possible damage to the neutral oil and tocopherols and loss of oil during such processing.

Lecithin and cephalin are common phosphatides found in edible oils. Soybean, canola/rapeseed, corn, and cottonseed are the major oils that contain significant quantities of phosphatides. Alkaline treatment used for FFA reduction is also capable of removing most of the phosphatides from these crude oils. Tocopherols are important minor constituents of vegetable oils, which are natural antioxidant that retard the development of rancidity.

Refining, bleaching, and deodorization are the steps that are necessary if the oil is to be used in food applications. Oil that has only gone through these three steps is called "RBD" oil. Figure 14.10.2 illustrates the processing pathways.

*Refining*: Refining involves the removal of nonglyceride materials (phospholipids, color, and trace materials) and FFA. The goal is to produce a high-quality refined oil with the highest yield of purified triglycerides. Refining is by far the most important step in processing. An improperly refined oil will present problems in bleaching and deodorization and reduce quality.

Some solvent-extracted crude oils, including soybean or canola/rapeseed, contain approximately 2-3% gums, which are mainly phosphatides and require degumming. The principal phosphatides are lecithin and cephalin. Gums can cause problems through higher then necessary refining losses, or by settling out in storage tanks. The degumming operation exploits the affinity of most phosphatides for water, converting them to hydrated gums that are insoluble in oil and readily separated by centrifugal action. Lecithin can be recovered and concentrated from the gums in a separate solvent extraction process, usually with acetone.

Either water-degummed oil or crude oil can be treated with sodium hydroxide solution to saponify free fatty acids that are subsequently removed as soapstock by a primary refining centrifuge. Conventional alkali refining is by far the most widespread method of edible oil refining. The success of the alkali refining operation is the coordination of five prime factors: (1) use of the proper amount of reagent (sodium hydroxide), (2) proper mixing, (3) proper temperature control, (4) proper residual contact time, and (5) efficient separation.

Oil is alkali-refined by the addition of sodium hydroxide solution at a level sufficient to neutralize the FFA content of the oil. An excess of sodium hydroxide is required to reduce the color of the refined oil and to ensure the completion of the saponification reaction and to remove other trace elements. The amount and strength of the sodium hydroxide solution needed to neutralize the FFA is dependent on the amount of both FFAs and phosphatides present in the crude oil. Water-soluble soaps are formed in the primary reaction between the sodium hydroxide and FFAs. The hydratable phosphatides react with the caustic forming oil-insoluble hydrates. The caustic used in alkali refining is normally diluted to about 8-14% NaOH, although higher concentrations are occasionally used to reduce color. The proper amount of NaOH solution added to the oil will produce an adequately refined oil with the minimum of triglyceride oil loss. The amount of NaOH solution (neutralizing dose plus excess) is determined by experience and adjusted according to laboratory results.

After the NaOH solution is injected, it is mixed for 6-10 minutes to ensure thorough contact. The treated oil is then heated to assist in breaking of the emulsion prior to separation of the soapstock from oil in continuous centrifuges.

Any soap remaining, after the primary soapstock separation, is removed through continuous hot water washings. In this step, water is added at 10-15% at a temperature sufficient to prevent emulsification, generally 82-90.5°C (180-195°F). The oil is again separated from the soapy phase in water wash separators and drier prior to bleaching.

*Bleaching*: The oil is further purified by "bleaching", which removes color bodies and trace metals as well as entrained soaps, and products of oxidation that are adsorbed onto the surface of bleaching agents or adsorbents. Types of adsorbents most commonly used include neutral clay, acid activated clay, and activated carbon. The choice of adsorbent will depend on a balance between activity of the adsorbent, oil retention loss, and adsorbent cost.

The process is generally carried out via batch or continuous bleaching. Adsorbent is mixed with the refined oil creating a slurry that is agitated to enhance contact between the oil and the adsorbent. This is generally carried out under a vacuum at 90-95°C (194-203°F) for 15-30 minutes. Vacuum bleaching offers the advantages of an oil with improved oxidative and flavor stability. Finally the adsorbent is filtered from the oil using pressure leaf filters precoated with diatomaceous earth. Spent clay is steamed for efficient oil recovery.

*Deodorization*: Deodorization, which removes the volatile compounds along with residual FFA, is a critical step in ensuring the purity of any vegetable oil and improves flavor, odor, color, and oxidative stability. Many of the volatile compounds removed are formed by the auto-oxidation of fat, which produces aldehydes, ketones, alcohols, and hydrocarbons that are associated with undesirable flavors and odors. The process also is effective in removing any remaining pesticide residues or metabolites that may be in the oil.

Deodorization, which can be conducted as a batch operation in smaller plants or as a continuous or semicontinuous process by larger deodorizing facilities, consists of a steam distillation process in which the oil is heated to 230°C (446°F) under a vacuum of 2-10 mm Hg. Steam is sparged through the oil to carry away the volatiles and provide agitation. The odor and flavor compounds, which are more volatile than the triglycerides, are preferentially removed. After deodorization and during the cooling stage, 0.005-0.01% citric acid is generally added to chelate trace metals, which can promote oxidation. Deodorized oils preferably are stored in an inert atmosphere of nitrogen to prevent oxidation. Tocopherols and sterols are also partially removed in the deodorization process. Tocopherols can be recovered from the deodorizer distillate in a separate operation.

### 14.10.4 REVIEW OF SOLVENTS STUDIED FOR EXTRACTION EFFICIENCY

Research on solvents for extraction has been carried out for more than 150 years and has intensified since the first patent was issued to Deiss of France in 1855.[1,3,48] In the early effort of selecting an extraction solvent, the availability, operation safety, extraction efficiency, product quality and cost were the major concerns. In recent decades, toxicity, bio-renewability, environmental friendliness have been added to the solvent selection criteria. Among the solvents tested, a majority of the candidate solvents were excluded on the ground of toxicity and safety. Only a handful of solvents are used to any degree. These are acetone, alcohol, hexanes, heptane, and water.[1,4-7,49] Water is used in rendering of fat from animal tissues and fish and in coconut processing,[49] alcohol for spice and flavorants extraction,[5,49] acetone for lecithin separation and purification.[4] For commodity oils derived from vegetable sources, only hydrocarbon solvents have been used since 1930's. Acetone was used by an Italian cottonseed oil mill during the 1970's.[4] Aqueous acetone and acetone-hexane-water azeotrope were studied by the scientists at the Southern Regional Research Center of Agricultural Research Service, USDA during the 1960's and 1970's.[4] The effort was stopped due to the cost of retrofit required, the difficulties in managing the mixture of solvents with the presence of water and product quality concerns - a strong undesirable odor associated with the acetone extracted meals.[4] Ethanol and isopropanol were studied in the 1980's as a potential replacement of hexane for oil extraction. Both were proven technically feasible but economically unacceptable.[5,6] n-Hexane is listed as a HAP under the CAA[14,15] (See Section 14.10.2.2.1 CAA) and there are other regulatory requirements. As a way to meet environmental regulations, a short term option to commercial hexane appears to be hydrocarbons with significantly reduced n-hexane content.

#### 14.10.4.1 Hydrocarbon solvents

Extraction of oils has largely relied on mechanical or heat rendering process for centuries.[50] Increased demand of productivity to separate oils from oilseeds has been the principal factor driving the changes of oilseed processing from the ancient hydraulic press to a continuous screw press or expeller in early 1900's.[51,52] This operation still left more than 4-5% residual oil in the pressed cake.[51] More complete recovery of oil can only be effectively accomplished by solvent extraction.[53,54]

Solvent extraction of oils had an early beginning. Deiss in France received a patent to extract fat from bone and wool with carbon bisulfide in 1855.[53] A year later, Deiss received additional patents covering the extraction of oil bearing seeds. Large scale solvent extraction already was established in Europe in 1870.[55] The earliest extractors were unagitated single-unit batch extractors of small capacity and not very efficient.[56,57] These extractors were gradually modified by the addition of agitation. They were organized in a battery of ten batch extractors which can be operated in a countercurrent principle. Extractors of this type operated in European plants during the last three decades of the 19th century.[57]

Further development in solvent extraction technology was relatively slow until early twentieth century. Solvent extraction spread from Europe to various parts of the world including the United States and South America.[57] The first extraction plant in the United States was used to recover grease from garbage, bones, cracklings, and other packing-house wastes and to recover residual oil from castor pomace.[57] Wesson[58] reported his efforts applying solvent extraction to recover cottonseed oil from 1889 till the close of World War I. During the 1930's solvent extraction was introduced in the United States for the recovery of oil from soybeans and the German equipment of the continuous type was used almost exclusively.[57-61] Just prior to World War II the installation of continuous solvent extraction equipment was greatly accelerated and throughout the period of the War new plants were erected in an effort to keep pace with the constantly increasing production of soybeans. All of the later installations have been of American manufacture and in a number of cases of American design.[57-61]

Solvents used in the early effort to extract grease and oils were diverse. Besides carbon bisulfide used by Deiss,[53] chlorinated hydrocarbons, benzene, and alcohols were all being tried. Extracting oil from corn and cottonseed with both aviation gasoline and petroleum distillate was performed in the United States in 1915 and 1917 respectively.[62] The hydrocarbon paraffins became the preferred solvents for oilseed extraction during 1930's through the process of elimination.[63-71] Due to the prominent defects of early solvent extraction: dark crude oil, strong solvent odor in meal and high cost associated with solvent loss, low boiling hydrocarbons such as propane and butane were recommended as oil extraction media.[72] The flammability of hydrocarbons also prompted much research in 1940's using chlorinated hydrocarbons as the extraction solvents[73,75] before its meal was found unsafe as feed.[75-77] For the purpose of improved protein and oil quality[75,78] and of processing safety and biorenewable solvents,[75,79,80] both ethanol and isopropanol were investigated as the oil extracting solvents. While these alcohols offer various advantages in product quality and process safety and are renewable, they are still not economically feasible to replace hydrocarbons as oilseed extraction solvents.[81] Hexane rich solvent became popular for the oilseed industry,[54,57,82,83] because it is the most efficient solvent, extracts minimum non-oil material and is easy to separate from the crude oil and marc.[63] A thorough comparison of various hydrocarbon solvents for cottonseed oil extraction on a lab scale basis was reported by Ayers and Dooley in 1948.[84] A more recent study by Wan et al.[85] used a laboratory scale dynamic percolation extractor which operates at the conditions similar to those applied in the oil mill practice. Plant trials of isohexane and heptane solvents versus hexane in a 300 tons/day cottonseed factory revealed some interesting findings.[7]

#### *14.10.4.1.1 Nomenclature, structure, composition and properties of hydrocarbons*

Petroleum and natural gas are the most abundant and affordable sources of hydrocarbon. Sometimes naphtha is used to describe the low boiling liquid petroleum and liquid products of natural gas with a boiling range from 15.6°C (60°F) to 221°C (430°F). This large group of compounds can be structurally classified as aliphatic and aromatic. Aliphatic hydrocarbons include saturated alkanes (paraffins), unsaturated alkenes (olefins) and alkynes (acetylenes), and cycloparaffins (naphthenes). Paraffins can be linear such as n-butane, n-pentane, and n-hexane, and branched such as isobutane, isopentane, isohexane, etc. Example of olefins is ethylene; of cycloparaffins, cyclopentane and cyclohexane; and of an aromatic, benzene.[8,86,87] These compounds are derived from natural gas and petroleum oils which normally contain thousands of hydrocarbons with molecular weight ranging from methane to about 50,000 - 100,000 Daltons. Upon refining, the crude petroleum is divided into hydrocarbon gases (methane, ethane, propane and butane), light distillates (naphthas and refined oils), intermediate distillates (gas oil and absorber oil), heavy distillates (technical oils, paraffin wax and lubricating oils), residues (petroleum grease, residual fuel oil still wax, asphalts and coke), and refinery sludges (acid coke, sulphonic acid, heavy fuel oils and sulfuric acid).[88] Historically, various fractions of petroleum naphthas, pentane and hexane from the light naphthas, aviation gasoline and benzol from the intermediate naphthas, and aromatic hydrocarbon, benzene, have been tested for oil extraction.[62-71]

#### *14.10.4.1.2 Performance of selected hydrocarbon solvents*

*Factors affecting extraction*: There is little theoretical basis to be followed for the extraction of oilseeds.[66,89-91] The study of the extraction of oilseeds is complicated by the fact that the total extractible material is variable in quantity and composition.[66,89] Composition of the early extracted material is nearly pure triglycerides. As the extraction progresses, increasing amount of non-glyceride material will be extracted.[66,89] IT is believed that the majority of the oil from oilseed flakes is easily and readily extracted.[66,90] While the thickness of flakes affects extraction rate, the concentration of miscella below 20% does not greatly increase the amount of time to reduce the residual oil in flakes to 1%.[89] Good[91] summarized much of the early effort in soybean extraction: (1) The first oil extracted is superior in quality to the last small fraction; (2) While other solvents have been used in the past, hexane has become the primary solvent due to a combination of properties; (3) Flake thickness is the most important factor in achieving good extraction results; (4) Higher extractor temperatures up to nearly the boiling point, improve extraction results; (5) Moisture control is important throughout the extraction process; (6) Heat treatment affects the total extractibles; and (7) The soaking theory of extraction indicates that weak miscellas are very effective in helping to achieve good extraction results.

Particle size which relates to the surface area available for extraction and is obviously one of the most important factors for extraction study. Coats and Wingard[92] noticed that particle size was more influential when the seed grit was being extracted. When oilseed flakes were being extracted, the flake thickness would be a more important factor instead of size of the flakes. Moisture content in oilseed can affect the extraction results.[93-95] Optimum moisture content of cottonseed meats for extraction was first reported by Reuther et at.[94] to be from 9 to 10%. Work by Arnold and Patel[95] indicated 7 to 10% to be the optimum moisture for cottonseed flakes and very little variation in extraction rate for soybean with moisture content between 8 and 12%.

Wingard and Phillips[96] developed a mathematical model to describe the effect of temperature on extraction rate using a percolation extractor as follows:

$$\log(time, \text{min}) = n\log\left(Temp., {}^{\circ}F\right) + \log k \quad or \quad time = k\left(Temp.\right)^{n} \qquad [14.10.1]$$

where time is defined as the number of minutes required to reach 1% residual oil in the oilseed flakes. For all practical purpose, they concluded that the time in minutes required to reduce the oilseed to 1% residual oil content on a dry basis varied inversely with the square of the extraction temperature in degrees Fahrenheit.

*Evaluation methods*: Except for the pilot plant batch or counter-current extraction described by various labs,[54,74,78] most of the solvent extraction evaluation work found in the literature was done in one or several of the lab scale devices. The percolation batch-extraction apparatus of the Soxhlet type has often been used to evaluate the rate of extraction of hydrocarbon solvents such as the one described by Bull and Hopper.[89] Wingard and Shand[97] described a percolation type of extractor and a co-current batch extractor and claimed to be useful to study the factors influencing equipment design and plant operation as well as fundamental studies contributing to a general understanding of extraction. Wan, et al. modified the design of percolation type extractor to closely simulate a single stage counter current miscella extraction conditions as practiced in the factory.[85] Co-current batch extractor with numerous variations was also frequently applied for the extraction properties of selected solvents which were often operated at room temperature.[75,97] Soxhlet extraction[84,85] and Soxtec System HT6 (Perstorp Analytical, Herndon, VA) were also frequently used to evaluate solvents.[98] Soxhlet extractor allows vaporized and condensed pure solvent to percolate through oilseed sample. The temperature of the condensed solvent is normally lower than its boiling point. Depending upon the cooling efficiency of the condenser and the room temperature, the temperature of the condensed solvent and the temperature of the extracting solvent in the extractor largely varied from lab to lab. This extraction temperature variability was minimized with the Soxtec method by refluxing the oilseed sample in the boiling solvent for 15 minutes followed by Soxhlet type of rinsing for 35 minutes. In theory the Soxtec method is more efficient and better reproduced. However, the Soxtec method only utilized a 3 g oilseed sample. The heterogeneity of an oilseed sample could be a significant source of variation.

Flakes of oilseeds were most frequently used for the solvent extraction studies. Sometimes, ground oilseed kernels through a specified sieve size was used.[98] Residual oil content in the extracted flakes after a certain specified extraction condition or oil content in miscella (mixture of oil and solvent) was examined and the percentage of total oil extracted was calculated.[89-97] The total extractable oil of flakes was determined by four hours Soxhlet extraction. Wan et al.[85] used a precision densitometer to determine the miscella concentration (percent of oil in miscella by weight) after a given time of extraction from which the percentage of oil extracted from cottonseed flakes was calculated. From these data, Wan et al.[85] was also able to estimate the initial rate of extraction and final extraction capacity for each solvent as fresh and at selected initial miscella concentrations up to 30%.

Bull and Hopper[89] conducted extraction of soybean flakes in a stainless steel batch-extraction apparatus of the Soxhlet type with petroleum solvents, Skellysolve F (boiling range, 35 to 58°C) at 28°C and Skellysolve B (boiling range, 63 to 70°C) at 40°C. The extraction was carried out to permit the miscella obtained by each flooding of the flakes with solvent to be recovered separately. Their results showed that iodine number decreased and refractive index increased slightly with the extraction time which implied that more saturated fat was extracted during later stage of the extraction. Oils extracted during the later stages of the extraction were found to contain greater amounts of unsaponifiable matter and were rich in phosphatides, as high as 18% of the last fraction. Skellysolve B which is a hexane rich solvent demonstrated a much faster initial rate of extraction than that of Skellysolve F which is a pentane rich solvent and therefore, it took longer to complete the extraction for Skellysolve F. The fatty acid profile of each fraction showed a slight increase of saturated and a slight decrease of unsaturated fatty acid in the later fractions.

Arnold and Choudhury[82] reported results derived from a lab scale extraction of soybean and cottonseed flakes in a tubular percolation extractor at 135-140°F with pure, high purity and commercial hexane, and reagent grade benzene. They claimed that pure hexane extracted soybean slower than high purity and commercial hexane. During the first 60 minutes of extraction, benzene extracted more oil than the hexanes. However, at the end of 80 minutes, benzene extracted only slightly more than pure hexane but definitely less than the commercial hexanes. Similar results were obtained for the four solvents when cottonseed flakes were extracted.

A laboratory extraction study of cottonseed flakes using various hydrocarbon solvents was reported by Ayers and Dooley.[84] Soxhlet extractor and Waring blender were used for these experiments. Among the petroleum hydrocarbon solvents tested were branched, normal and cyclo-paraffins as well as aromatic hydrocarbons with various degrees of purity. They were: pure grade (99 mole percent purity) n-pentane, isopentane, cyclohexane, benzene, and n-heptane; technical grade (95 mole percent purity) neohexane, diisopropyl, 2-methylpentane, 3-methylpentane, n-hexane, and methylcyclopentane; technical grade (90 mole percent purity) cyclopentane; and commercial grade n-heptane, isohexanes, n-hexane, isoheptane and n-heptane. To assess the performance of these solvents, they used the following empirical formula:

$$\textit{Quality-Efficiency Rating} = 0.4\,(\textit{Oil Yield Factor}) + 0.4\,(\textit{Refining Loss Factor}) + 0.2\,(\textit{Refined and Bleached Oil Color Factor}) \qquad [14.10.2]$$

When comparing the oil yield factor alone, 3-methylpentane was rated the best. When comparing the solvents based on the empirical Quality-Efficiency Rating formula, they concluded that methylpentanes (3- and 2-methylpentane) were superior extraction solvents for cottonseed oil. The normal paraffins, highly-branched isohexanes, cycloparaffins, and aromatics were progressively rated as less efficient than methylpentanes. Therefore, they recommended a tailor-made solvent for the extraction of cottonseed should exclude aromatic hydrocarbons, have low limits on cycloparaffin content, and consist largely of normal and isoparaffin hydrocarbons.

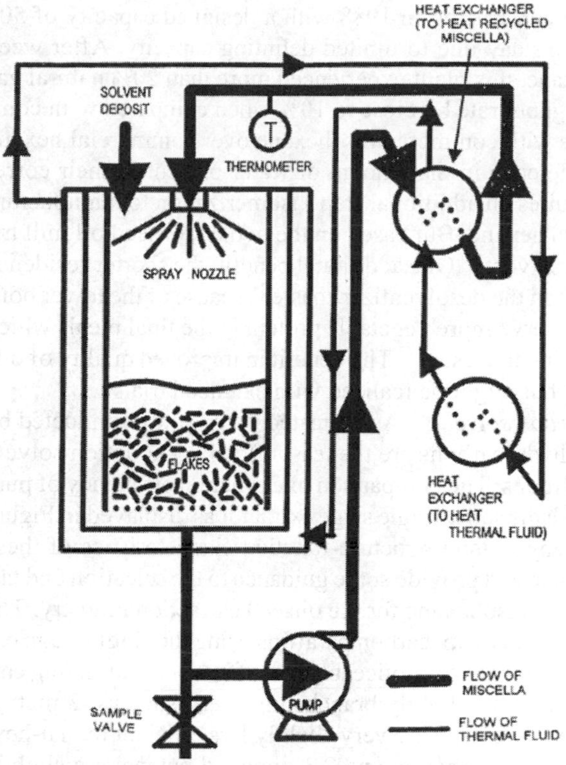

Figure 14.10.3. Schematic of bench-scale dynamic percolation extractor.

A more recent study by Wan et al.[85] using a laboratory scale dynamic percolation type of extractor (Figure 14.10.3) operated at the following conditions such as, temperature (5°C below the boiling point of each solvent) and miscella flow rate (9 gal/min/ft$^2$), similar to those applied in the oil mill practice. Commercial grade hexane, heptane, isohexane, neohexane, cyclohexane, and cyclopentane were used to extract cottonseed flakes which had 5.8% moisture and 31.4% oil. When these solvents were tested near their boiling points, hexane apparently extracted cottonseed oil at a higher initial rate, > 94% oil extracted after 2 minutes, than all other solvents. Both heptane and hexane were able to extract more oil at the end of 10 minutes of extraction. Isohexane demonstrated to have adequate initial extraction rate (80% oil extracted after 2 minutes) and extraction capacity (93% oil extracted after 10 minutes of extraction) but is noticeably less effective than hexane. Similar to findings by Ayers and Dooley,[84] results from the study by Wan et al.[85] also demonstrates that neohexane, cyclohexane and cyclopentane performed distinctly less efficiently than hexane, heptane and isohexane. Conkerton et.al.[98] tested commercial heptane versus hexane in a Soxtec extractor. Under this extraction condition, heptane actually extracted more oil than hexane from ground cottonseed kernel passed through a 20 mesh screen. The oil and meal quality were not appreciably affected by the higher temperature extraction of heptane.

*Plant scale results*: Although hydrocarbon solvents have been used for oilseed extraction since the 1930's, very little in plant operating data are available. During the spring of 1994, Wan et al.[7] conducted plant trials with commercial heptane and isohexane at a 300 tons/day cottonseed crushing plant which routinely used hexane as the extraction solvent. Test results indicated that heptane performed well as an extraction solvent. However, it required extra energy and time to recover and consequently reduced the throughput rate of cottonseed being processed. Isohexane on the other hand was termed as an "easier" solvent by the plant engineers than hexane to operate. The plant also experienced a 40% steam savings and better than 20% throughput increase when it was operating with isohexane.[7] This encouraging result prompted a second plant trial with commercial isohexane.[45] The second plant trial was carried out at a cottonseed oil mill with a relatively new extraction and miscella refining facility which was constructed in 1988 with a designed capacity of 500 tons/day but operated at only 270 tons/day due to limited delinting capacity. After week long testing with commercial isohexane, this plant experienced more than 20% natural gas usage and easily increased the throughput rate by close to 10% when compared with commercial hexane. This energy savings with commercial isohexane over commercial hexane may be largely attributed to the difference in the amount of water present in their corresponding azeotropes. Isohexane requires an additional step - isomerization to manufacture and will always be priced higher than hexane. But based on the two cottonseed oil mill trials, isohexane can be a cost efficient solvent.[99] One additional benefit, the shorter residence time of the extracted cottonseed marc in the desolventizer/toaster because of the lower boiling range of isohexane will likely preserve more vegetable protein in the final meals which has been observed by both plants during the tests.[7,45] The benefit in improved quality of oils were not obvious in both plant trials but might be realized with extended trials.

*Further evaluation of hydrocarbon solvents*: As indicated in the study conducted by Wan et al.,[85] the commercial cyclic hydrocarbons are the least effective extraction solvent than the branched and linear hydrocarbons. The comparison of extraction efficiency of pure isomers of hexane was conducted with the same single stage extractor as displayed in Figure 14.10.3. This was done to identify any unique structure-function characteristics of these pure components of commercial hexane and provide some guidance to the selection and tailored formulation for future commercial isohexane for the oilseed extraction industry. The extraction results for various pure isomers of six carbon paraffins using the single stage extractor indicated the following: (a) cyclohexane is noticeably less efficient in extracting cottonseed flake than all the other isomers; (b) slightly branched isomers, such as, 2-methyl and 3-methyl pentane, and methyl-cyclopentane are very slightly less efficient than n-hexane; and (c) highly branched isomers, 2,2-dimethyl and 2,3-dimethyl butane, are slightly less efficient than slightly branched isomers in extraction (Unpublished data).

## 14.10.5 FUTURE TRENDS

In the future there most likely will be new demands for highly specialized extraction solvents as newly domesticated species that make useful novel oils[30] and other products and new or altered biological products with enhanced nutritional and industrial properties will be developed through conventional breeding and genetic engineering for use as "functional foods"[31] (e.g., phytosterols to achieve cholesterol lowering); as oils with altered lipid profiles[32] (e.g., for lower saturated fat) or with more vitamin E; new drugs/nutraceuticals, industrial chemicals (e.g., fatty acids for lubricants, as cosmetics, coatings, detergents, surfactants, flavors, polymers, etc.); as sources for specialty chemicals; as value added products; etc.[31-38]

Genetically enhanced (GE) /biotech crops make up a growing share of the agricultural output.[39] Biotechnology is the most powerful tool ever put in the hands of agricultural scientists. The ability to breed desirable traits or eliminate problematic ones can yield potentially spectacular benefits, such as various chemicals of importance including improved fats and oils, and vaccines and medicine, improved nutrition (e.g., in casaba, oilseeds, rice, sweet potatoes), and improved yields with the use of less agricultural chemicals. GE/biotech crops could be increasingly developed as biofactories for a wide range of products, including nutrients pharmaceuticals, and plastics. There is much promise for being able to produce products that would protect millions from disease, starvation, and death. However biotechnology and GM crops have become very controversial, and have run into serious problems in Europe, particularly in the UK.[40] Europe,[41] Japan, Korea, and Australia/New Zealand already have some restrictions and require some labeling. The U.S. is reviewing the issue.[42] Thus even though this technology has great promise for increased use of new and existing solvents for extraction of products from diverse biological materials, there are also many potential problems because of misperceptions and misinformation.

In the U.S. EPA is developing Maximum Available Control Technology (MACT) standards for vegetable oil processing that are likely to be finalized in 2001 (see Section 14.10.2.2.1, Hazardous Air Pollutants), with enforcement three years after promulgation.[16,17] Commercial hexane, which is a HAP and a VOC, is presently the solvent used.[16,17] To meet new and existing CAA requirements it is likely that extraction facilities will become much more efficient chemical engineering operations with upgraded equipment,[43] more computerized monitoring and control for better quality management,[44] and better environmental management/stewardship.[19] In addition it is possible that alternate solvents (e.g., isohexane[3,7,45]) or lower n-hexane content commercial hexane (30-50% vs. 64%) will be used to meet these regulations. It is also possible that solvents like acetone, which is not a HAP or VOC and is not on the TRI list, will be more strongly investigated.[16,17,46]

In Europe trans fatty acid labeling of retail foods is required and in the U.S. FDA has proposed to label trans fats as saturated fat on the nutrition labeling panel required on packaged food sold at retail (64 FR 62,764; Nov. 17, 1999). This regulation if, promulgated as proposed, will result in reformulation of many products that could affect the vegetable oil producing and extracting industries.

It is clear that the future has much uncertainty, while at the same time it offers much promise. It appears that there will be many potential changes that will put new demands on extraction solvents. Solvents that are more environmentally friendly, are nontoxic to plants, workers, and consumers, have specialized properties, have high solvent power at low temperatures (are easy to desolventize), etc., may have wider potential use in conventional extraction as well as specialized niche markets.

## REFERENCES


1 **Technology and Solvents for Extracting Oilseeds and Non-petroleum Oils**, P.J. Wan and P.J. Wakelyn, Eds., *AOCS Press*, Champagne, IL, 1997.

2 M.A. Williams and R.J. Hron, **Bailey's Industrial Oils and Fat Products**, 5th edn., Vol. 4: Edible Oil and Fat Products: Processing Technology, Y.H Hui,., Ed., *John Wiley and Sons, Inc.*, 1996, p. 119.

3 P.J. Wan, **Hydrocarbon Solvents**, in Technology and Solvents for Extracting Oilseeds and Non-petroleum Oils, P.J. Wan and P.J. Wakelyn, Eds., *AOCS Press*, Champaign, IL, 1997, p.170-185.

4 R.J. Hron, **Acetone**, in *Ibid*, p.186-191.

5 R.J. Hron, **Ethanol**, in *Ibid*, p.192-197.

6 E.W. Lucas and E. Hernandez, **Isopropyl Alcohol**, in *Ibid*, p. 199-266.

7 P.J. Wan, R.J. Hron, M.K. Dowd, M.S. Kuk, and E.J. Conkerton, *J. Am. Oil Chem. Soc.*, **72**, 661 (1995).

8 Occupational Health and Safety Administration Field Operations Manual, Chapter IV: Violations, C. Health Standards Violations, (OSHA Instruction 2.45B CH-4, Dec.13, 1993), The Bureau of National Affairs, Washington, DC, 1994, pp. 77:2513-18.

9 Occupational Health and Safety Administration Technical Manual, Section I- Sampling, Measurement Methods, and Instruments, Chapter 1 - Personal Sampling for Air Contaminants, Appendix I:1-6. Sampling and Analytical Errors (SAEs) (Issued by OSHA Instruction TED 1.15, September 22, 1995; amended by OSHA Instruction TED 1.15 CH-1, May 24, 1996).

10 1997 TLVs and BEIs, Threshold Limit Values for Chemical Substances and Physical Agents and Biological Exposure Indices, The American Conference of Governmental Industrial Hygienists, Cincinnati, OH, 1997, pp. 12-40.

11 J.B Galvin, C.J. Kirwin, D.W. Kelly, *INFORM*, **6**(8), 951 (1995).

12 H.H. Schaumberg and P.S. Spencer, *Brain*, **99**, 183 (1976).

13 J.B. Galvin, **Toxicity Data for Extraction Solvents Other Than Isohexane/Hexane Isomers, in Technology and Solvents for Extracting Oilseeds and Nonpetroleum oils**, P.J.Wan and P.J.Wakelyn, Eds., *AOCS Press*, Champaign, IL, 1997, p. 75-85.
14 P.J. Wakelyn, *Cotton Gin and Oil Mills Press*, **92**(17), 12 (1991).
15 P.J. Wakelyn and L.A. Forster, Jr., *Oil Mill Gaz.*, **99**(12), 21 (1994).
16 P.J. Wan and P.J. Wakelyn, *INFORM*, **9**(12),1155 (1998).
17 P.J. Wan and P.J. Wakelyn, *Oil Mill Gaz.*, **104**(6), 15 (1998).
18 U.S. EPA, Study of Volatile Organic Compound Emissions from Consumer and Commercial Products, U.S. Environmental Protection Agency, Office of Quality Planning and Standards, Research Triangle Park, NC 27711, EPA - 453/R-94-066-A, March 1995.
19 P.J. Wakelyn and P.K. Adair, Assessment of Risk and Environmental Management, in Emerging Technologies, Current Practices, Quality Control, Technology Transfer, and Environmental Issues, Vol. 1, Proc. of the World Conference on Oilseed and Edible Oil Processing, S.S. Koseoglu, K.C. Rhie, and R.F. Wilson, Eds. AOCS Press, Champaign, IL, 1998, pp. 305-312
20 n-Hexane, U.S. EPA Integrated Risk Information System (IRIS) Substance File, U.S. EPA, 1999 (www.epa.gov/ngispgm3/IRIS/subst/0486.htm).
21 U.S. EPA, Toxic Chemical Release Inventory Reporting Form R and Instructions (Revised 1995 Version), U.S. EPA, Office of Pollution Prevention and Toxics, Washington, DC, EPA 745-K-96-001, March 1996, Table II, p. II-1
22 J.K. Dunnick, Toxicity Studies of n-Hexane in F344/N Rats and B6C3F1 Mice, National Toxicology Programs, U.S. Dept. of Health and Human Services, NTP TOX 2, NIH Publication No. 91-3121, 1991.
23 A. H. Allen, GRAS Self-Determination: Staying Out of the Regulatory Soup, Food Product Design, (April 1996 supplement to Food Product Design, 6 pages) (1996).
24 B.L. Oser and R.A. Ford, *Food Technol.*, **27**(1), 64 (1973).
25 R.L. Hall and B.L. Oser, *Food Technol.*, **19**, 151 (1965).
26 A.M. Galvin, in *Introduction to Fats and Oils Technology*, P.J. Wan, Ed., *AOCS Press*, Champagne, IL, 1991,pp. 137-164.
27 L.A. Jones and C.C. King, in **Bailey's Industrial Oil and Fats Products**, 5th edn., Vol. 2 Edible Oil and Fat Products: Oils and Oil Seeds, Y.H. Hui, Ed., *John Wiley and Sons, Inc.*, 1996, pp. 177-181.
28 H.W. Lawson, **Standards for Fat and Oils**, *The AVI Publishing Co., Inc.*, Westport, CT, 1985, p. 34.
29 S.W. Kuhlmann, M.C. Calhoun, J.E. Huston and B.C. Baldwin, Jr., Total (+)- and (-)- Gossypol in Plasma and Liver of Lambs Fed Cottonseed Meal Processed by Three Methods, *J. Anim. Sci.*, **72** (suppl. 1), 145 (1994).
30 D.J. Murphy, *INFORM*, **11**(1), 112 (2000)
31 M.A. Ryan, *Today's Chemist at Work*, **8**(9), 59 (1999).
32 B.F. Haumann, *INFORM*, **8**(10), 1001 (1997).
33 B. Flickinger and E. Hines, *Food Quality*, **6**(7), 18 (1999).
34 C.T. Hou, Value Added Products from Oils and Fats through Bio-processes, Int. Symp. On New Approaches to Functional Cereals and Oils, Beijing, China, Nov. 9-14, 1997, Chinese Cereals and Oils Association, Beijing, China, 1997, p. 669.
35 D.J. Kyle, New Specialty Oils: Development of a DHA-rich Nutraceutical Product, Ibid, p. 681.
36 J.K. Daum, Modified Fatty Acid Profiles in Canadian Oilseeds, Ibid, pp. 659-668.
37 C.M. Henry, *Chemical Eng. News*, **77**(48), 42 (1999).
38 R. Ohlson, *INFORM*, **10**(7), 722 (1999).
39 G.J. Persley and J.N. Siedow, Applications of Biotechnology to Crops: Benefits and Risks, CAST Issue Paper No. 12, Council for Agricultural Science and Technology, December 1999.
40 M. Heylin, *Chem. Eng. News*, **77**(49), 73 (1999).
41 Anom., *Official J. European Communities*, **43**(L6), 13 (2000).
42 B. Hileman, *Chem. Eng. News*, **77**(50), 31 (1999).
43 P. Delamater, *Oil Mill Gaz.*, **104**(11), 34 (1999).
44 P.J. Wakelyn, P.K. Adair, and S.R. Gregory, *Oil Mill Gaz.*, **103**(6), 23 (1997).
45 M. Horsman, *Oil Mill Gaz.*, **105**(8), 20 (2000).
46 R.J. Hron, P.J. Wan, and P.J. Wakelyn, *INFORM*, In Press, (2000).
47 J.A. Dean, **Dean's Handbook of Chemistry**, Thirteenth Edition, *McGraw-Hill, Inc.*, New York, NY, 1985.
48 R.D. Hagenmaier, **Aqueous Processing**, in Technology and Solvents for Extracting Oilseeds and Non-petroleum Oils, P.J. Wan and P.J. Wakelyn, Eds., *AOCS Press*, Champaign, IL, 1997, p.311-322.
49 L. A. Johnson, **Theoretical, Comparative, and Historical Analyses of Alternative Technologies for Oilseed Extraction**, *Ibid*, p.4-47.

50 D.K. Bredeson, *J. Am. Oil Chem. Soc.*, **60**, 163A (1983).
51 **Cottonseed and Cottonseed Products**, A.E. Bailey, Ed., *Interscience Publishers Inc.*, New York, pp. 5, 615-643,(1948).
52 V.D. Anderson, **U.S. Patent 647,354** (1900).
53 Anonymous. *J. Am. Oil Chem. Soc.*, **54**, 202A (1977).
54 J. Pominski, L.J. Molaison, A.J. Crovetto, R.D. Westbrook, E.L.D'Aquin, and W.F. Guilbeau, *Oil Mill Gaz.*, **51**(12), 33 (1947).
55 O.K. Hildebrandt, *Fette Seifen Anstrichm,* **46**, 350 (1939).
56 M. Bonotto, *Oil & Soap*, **14**, 310 (1937).
57 K.S. Markley, *Oil Mill Gaz.*, **51**(7), 27 (1947).
58 D. Wesson, *Oil & Soap*, **10**, 151 (1933).
59 E. Bernardini, *J. Am. Oil Chem. Soc.*, **53**, 275 (1976).
60 K.W. Becker, *ibid*., **55**, 754 (1978).
61 K.W. Becker, *Oil Mill Gaz.*, **84**, 20 (1980).
62 W.E. Meyerweissflog, *Oil & Soap*, **14**, 10 (1937).
63 W.H. Goss, *ibid*, **23**, 348 (1946).
64 W.H. Goss, *J. Am. Oil Chem. Soc.*, **29**, 253 (1952).
65 R.P. Hutchins, *ibid*, **54**, 202A (1977).
66 G. Karnofsky, *ibid*, **26**, 564 (1949).
67 A.E. MacGee, *Oil and Soap*, **14**, 322 (1937).
68 A.E. MacGee, *ibid*, **14**, 324 (1937).
69 A.E. MacGee, *J. Am. Oil Chem. Soc.*, **26**, 176 (1949).
70 A.E. MacGee, *Oil Mill Gaz.*, **52**, 17,35 (1947).
71 A.E. MacGee, *ibid*, **67**, 22 (1963).
72 H. Rosenthal, and H.P. Trevithick, *Oil and Soap*, **11**, 133 (1934).
73 I.J. Duncan, *J. Am. Oil Chem. Soc.*, **25**, 277 (1948).
74 O.R. Sweeney, and L.K. Arnold, *ibid*, **26**, 697 (1949).
75 A.C. Beckel, P.A., Belter, and A.K. Smith, *ibid*, **25**, 7 (1948).
76 L.L. McKinney, F.B. Weakley, R.E. Campbell, A.C. Eldridge, J.C. Cowan, J.C. Picken, and N.L. Jacobson, *ibid*, **34**, 461 (1957).
77 T.A. Seto, M.O. Shutze, V. Perman, F.W. Bates, and J.M. Saulter, *Agric. Food Chem*., **6**, 49 (1958).
78 F.K. Rao, and L.K. Arnold, *J. Am. Oil Chem. Soc.*, **35**, 277 (1958).
79 E. W. Lusas, L.R. Watkins and K.C. Rhee, in **Edible Fats and Oils Processing: Basic Principles and Modern Practices**, D. R. Erickson, Ed., *AOCS Press*, IL, p. 56, 1990.
80 R.J. Hron, Sr., S. P. Koltun and A. V. Graci, *J. Am. Oil Chem. Soc.*, **59**(9), 674A (1982).
81 R.J. Hron, Sr., M.S. Kuk, G. Abraham and P. J. Wan, *ibid*, **71**(4), 417 (1994).
82 L.K. Arnold and R.B.R. Choudhury, *ibid*, **37**, 458 (1960).
83 K.S. Olson, *Oil Mill Gaz.*, **85**, 20 (1980).
84 A.L. Ayers and J.J. Dooley, *J. Am. Oil Chem. Soc.*, **25**, 372 (1948).
85 P. J. Wan, D. R. Pakarinen, R. J. Hron, Sr., and E. J. Conkerton, *ibid*, **72**(6), 653 (1995).
86 M.P. Doss, **Physical Constants of the Principal Hydrocarbons**, *The Texas Company*, Third Edition, New York. (1942).
87 L.A. Johnson and E. W. Lusas, *J. Am. Oil Chem. Soc.*, **60**(2), 229 (1983).
88 **McGraw-Hill Encyclopedia of Science and Technology**, 10, 71 (1960).
89 W.C. Bull and T.H. Hopper, *Oil and Soap*, **18**, 219 (1941).
90 H.B. Coats and G. Karnofsky, *J. Am. Oil Chem. Soc.*, **27**, 51 (1950).
91 R.D. Good, *Oil Mill Gaz.*, **75**, 14 (1970).
92 H.B. Coats and M.R. Wingard, *J. Am. Oil Chem. Soc.*, **27**, 93 (1950).
93 W.C. Bull, *Oil and Soap*, **20**, 94 (1943).
94 C.G. Reuther, Jr., R.D. Westbrook, W.H. Hoffman, H.L.E. Vix and E.A. Gastrock, *J. Am. Oil Chem. Soc.*, **28**, 146 (1951).
95 L.K. Arnold and D.J. Patel, *ibid*, **30**, 216 (1953).
96 M.R. Wingard and R.C. Phillips, *ibid*, **28**, 149 (1951).
97 M.R. Wingard and W.C. Shand, *ibid*, **26**, 422 (1949).
98 E.J. Conkerton, P.J. Wan and O.A. Richard, *ibid*, **72**, 963 (1995).
99 P.J. Wan, *INFORM*, **7**, 624 (1996).

## 14.11 GROUND TRANSPORTATION

GEORGE WYPYCH
**ChemTec Laboratories, Inc., Toronto, Canada**

The ground transportation industry in the USA is dominated by truck freight (78.6%). Other methods of transportation include: rail (7.9%), water (5.2%), air (4%), pipeline (2.2%), and other (2.1%). Solvents are used and solvent wastes and emissions are generated during refurbishing and maintenance.

Rail car refurbishing involves stripping and painting. Paint is usually removed by mechanical means (steel grit blast system) but solvents are occasionally used. Solid wastes are generated from latex paint wastes but hazardous wastes are also generated from solvent-based paints and thinners. Parts cleaning is mostly done using mineral spirits. Waste solvents are sent off-site for reclamation.

Truck maintenance work usually requires a parts washer which may involve either a heated or ambient temperature solvent, hot tank, or a spray washer. In the solvent tank washer, solvent (usually mineral spirits, petroleum distillates, and naphtha) is recirculated from solvent tank. Spent solvent is usually replaced monthly. Carburetor cleaning compounds contain dichloromethane. Tanker cleaning often involves a solvent spray.

The ground transportation industry employs a large number of people (more than 2 million in the USA). It is one of the less polluting industries. It generates 0.3% of VOC released by all major industries combined (about half of that of the aerospace industry). Most solvents used are of low toxicity. Good system of collection and reclamation of solvent wastes is done effectively and this is the major reason for the relatively good performance of the industry.

### REFERENCES

1 EPA Office of Compliance Sector Notebook Project. Profile of the Ground Transportation Industry. Trucking, Railroad, and Pipeline. US Environmental Protection Agency, 1997.
2 EPA Office of Compliance Sector Notebook Project. Sector Notebook Data Refresh - 1997. US Environmental Protection Agency, 1998.

## 14.12 INORGANIC CHEMICAL INDUSTRY

GEORGE WYPYCH
**ChemTec Laboratories, Inc., Toronto, Canada**

This industry has two major sectors: inorganic chemicals and chlor-alkali. Inorganic chemicals are often of mineral origin processed to basic chemicals such as acids, alkalies, salts, oxidizing agents, halogens, etc. The chlor-alkali sector manufactures chlorine, caustic soda, soda ash, sodium bicarbonate, potassium hydroxide and potassium carbonate. The major processes in this industry do not use solvents but there are many specialized auxiliary processes which use solvents. Tables 14.12.1 and 14.12.2 give information on the reported solvent releases and transfers from inorganic chemical industry.

The tables show that thc industry, which operates almost 1,500 plants and employs over 110,000 people, has minimal impact on global emission of VOCs. Consequently, the industry does not have any major initiative to deal with solvent emissions or wastes. Future safety improvements concentrate on non-solvent issues.

## REFERENCES

1 EPA Office of Compliance Sector Notebook Project. Profile of the Inorganic Chemical Industry. US Environmental Protection Agency, 1995.
2 EPA Office of Compliance Sector Notebook Project. Sector Notebook Data Refresh - 1997. US Environmental Protection Agency, 1998.

**Table 14.12.1. Reported solvent releases from the inorganic chemical plants in 1995 [Data from Ref. 2]**

| Solvent | Amount, kg/year | Solvent | Amount, kg/year |
|---|---|---|---|
| benzene | 33,000 | hexane | 2,000 |
| carbon tetrachloride | 3,000 | methanol | 574,000 |
| chloromethane | 2,600 | methyl ethyl ketone | 460 |
| dichloromethane | 12,500 | N-methyl-2-pyrrolidone | 180 |
| ethyl benzene | 110 | toluene | 12,000 |
| ethylene glycol | 1,800 | xylene | 1,500 |

**Table 14.12.2. Reported solvent transfers from the inorganic chemical plants in 1995 [Data from Ref. 2]**

| Solvent | Amount, kg/year | Solvent | Amount, kg/year |
|---|---|---|---|
| benzene | 780 | methyl ethyl ketone | 9,000 |
| carbon tetrachloride | 6,400 | N-methyl-2-pyrrolidone | 8,700 |
| dichloromethane | 5,000 | toluene | 6,000 |
| ethylene glycol | 12,000 | xylene | 96,000 |

# 14.13 IRON AND STEEL INDUSTRY

GEORGE WYPYCH
**ChemTec Laboratories, Inc., Toronto, Canada**

With almost 1,400 plants, the US iron and steel industry is very diverse industry having total sales of $100 billion and over 400,000 employees. Figure 14.13.1 is a schematic diagram of the iron and steel making process. Only one stage – finishing – employs solvents. The finishing stage includes processes to remove mill scale, rust, oxides, oil, grease and soil prior

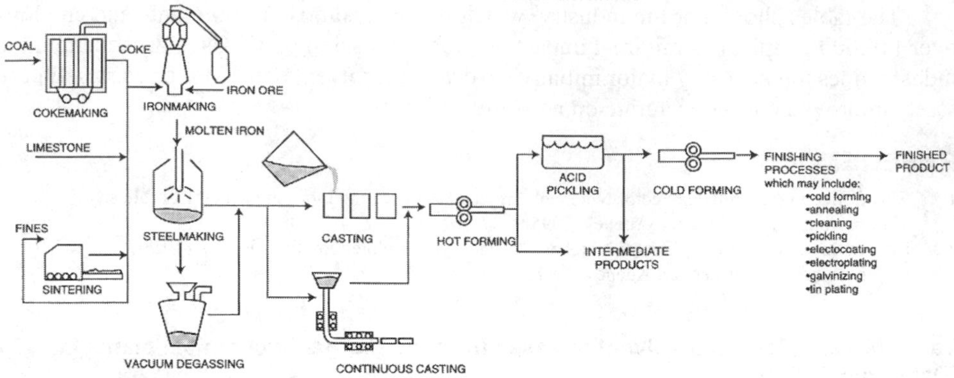


Figure 14.13.1. Schematic diagram of operations in the iron and steel manufacturing process. [Reproduced from EPA Office of Compliance Sector Notebook Project. Reference 1.]

**Table 14.13.1. Reported solvent releases from the iron and steel plants in 1995 [Data from Ref. 2]**

| Solvent | Amount, kg/year | Solvent | Amount, kg/year |
|---|---|---|---|
| benzene | 321,000 | N-methyl-2-pyrrolidone | 3,600 |
| n-butyl alcohol | 26,000 | polycyclic aromatic compounds | 2,400 |
| cresol | 1,800 | tetrachloroethylene | 91,000 |
| dichloromethane | 318,000 | 1,2,4-trimethylbenzene | 17,000 |
| ethylbenzene | 5,000 | trichloroethylene | 620,000 |
| methanol | 241,000 | toluene | 261,000 |
| methyl ethyl ketone | 358,000 | xylene | 168,000 |

to coating. Methods used include solvent cleaning, pressurized water or air blasting, cleaning with abrasives, and alkaline or acid pickling.

Tables 14.13.1 and 14.13.2 give information on the reported solvent releases and transfers from the iron and steel industries. Not all the solvents listed in the tables are used in processing. Some are by-products of coke manufacture from coal. Benzene and polycyclic aromatics compounds are by-products. Strong solvents such as methyl ethyl ketone, toluene, xylene, and trichloroethylene are typical of those used in cleaning processes. There is no program formulated by the industry to reduce amounts of solvents used.

## REFERENCES

1 EPA Office of Compliance Sector Notebook Project. Profile of the Iron and Steel Industry. Trucking, Railroad, and Pipeline. US Environmental Protection Agency, 1995.
2 EPA Office of Compliance Sector Notebook Project. Sector Notebook Data Refresh - 1997. US Environmental Protection Agency, 1998.

**Table 14.13.2. Reported solvent transfers from the iron and steel plants in 1995 [Data from Ref. 2]**

| Solvent | Amount, kg/year | Solvent | Amount, kg/year |
|---|---|---|---|
| benzene | 3,000 | N-methyl-2-pyrrolidone | 11,500 |
| n-butyl alcohol | 1,400 | polycyclic aromatic compounds | 3,820,000 |
| cresol | 12 | tetrachloroethylene | 20,000 |
| dichloromethane | 14,500 | 1,2,4-trimethylbenzene | 3,600 |
| ethylene glycol | 197,000 | trichloroethylene | 165,000 |
| ethylbenzene | 550 | toluene | 11,500 |
| methanol | 25 | xylene | 14,000 |
| methyl ethyl ketone | 66,000 | | |

# 14.14 LUMBER AND WOOD PRODUCTS - WOOD PRESERVATION TREATMENT: SIGNIFICANCE OF SOLVENTS

TILMAN HAHN, KONRAD BOTZENHART, FRITZ SCHWEINSBERG
**Institut für Allgemeine Hygiene und Umwelthygiene**
**Universität Tübingen, Tübingen, Germany**
GERHARD VOLLAND
**Otto-Graf-Institut, Universität Stuttgart, Stuttgart, Germany**

## 14.14.1 GENERAL ASPECTS

Wood preservation is based on various fundamental principles, e.g., construction aspects such as exposure to humidity, selection of different types of wood products according to their durability, and chemistry of wood preservatives. Important groups of chemical wood preservatives are water-soluble and solvent-based substances.[1]

The main requirements of chemical wood preservatives are:[1]

- Stability, especially chemical stability.
- Resistance to environmental conditions, e.g., light or heat.
- Penetration into the wood products.
- Effectiveness against wood attacking agents (e.g., insects, fungi, bacteria).
- Compatibility with other construction components, e.g., paints, adhesives, and fasteners.
- Construction aspects, e.g., corrosion.
- Minimal environmental impact, e.g., minimum emissions or minimum environmental pollution.
- Ability to work with a range of wood products.
- Case applications, e.g. fundamental differences of indoor and outdoor coatings.
- Having favorable visual aspects, e.g., surface properties, color, uniformity, influence on grain pattern, etc.

All requirements cannot be fulfilled completely by the various wood preservatives. Therefore wood preservatives should be selected according to the particular case.

## 14.14.2 ROLE OF SOLVENTS

### 14.14.2.1 Occurrence

Various solvents are added to wood preservatives. Only a limited number of wood preservatives are authorized by governmental agencies, e.g., in Germany “Institut für Bautechnik (DIBt)”.[2] But there is a large grey market for wood preservatives other than the authorized substances. As a result, a large variety of solvents can occur in wood preservatives.

Different systems of classification are used worldwide. In Germany, authorized wood preservative substances are published in an index of wood preservatives (“Holzschutzmittelverzeichnis”)[2] which is elaborated by the DIBt and the central German environmental authority (“Umweltbundesamt”).

A systematic survey of wood preservatives is shown in Table 14.14.1, including wood preservatives containing solvents. Solvents are normally found in wet systems of wood preservatives. Commonly used solvents are substances which are applied in connection with normally used binders (aldehyde resins, acrylates and polyurethanes). Water or the appropriate solvents are added to binders.

**Table 14.14.1. Systems of wood preservatives**

| Purpose and base | Terms | Active components |
|---|---|---|
| Water-soluble agents as preventive treatment against fungi and insects | “CF-salts” | chromium and fluorine compounds |
| | “CFA-salts” | alkali fluorides, alkali arsenate, and bichromate (no longer permitted) |
| | “SF-salts” | silicofluorides |
| | “HF-salts” | hydrogen fluorides |
| | “B-salts” | inorganic boron compounds |
| | “Single CK-salts” | copper salts, bichromate |
| | “CKA-salts” | copper salts, bichromate with arsenic compounds |
| | “CKB-salts” | copper salts, bichromate with boron compounds |
| | “CKF-salts” | copper salts, bichromate with fluorine compounds |
| | “CFB-salts” | chromium, boron and fluorine compounds |
| | collective group | other compounds, e.g., bis(N-cyclohexyl-diazeniumdioxyl)-copper |
| Oily agents as preventive treatment against fungi and insects | tar oil preparations | distillates of bituminous coal tar (carbolineum) |
| | preparations containing solvents | organic fungicides and insecticides |
| | pigment-free preparations containing binders and solvents | organic fungicides and insecticides |
| | preparations with stained pigments containing solvents | organic fungicides and insecticides |
| | special preparations only used in stationary installations | organic fungicides and insecticides |
| | preparations containing coal tar oil | organic agents, special distillates containing coal tar oil, solvents and pigments |
| Preparations used for special applications | pastes | |
| | wood preservatives used in particle board in manufacturing plants | |
| | agents used as preventive treatment against insects contain organic insecticides | |

Solventborne wood preservatives contain mainly nonpolar, organic solvents apart from other substances such as fungicides and insecticides.[1] These solvents are classified as VOCs.

### 14.14.2.2 Technical and environmental aspects

Solvent-based wood preservatives show several advantages, especially in their application and technical effectiveness.[1] They can be applied repeatedly and do not alter the structure of the wood products. Application is faster and the characteristics of the final product are improved, e.g., visual appearance of surface.

Nevertheless, there are some disadvantages, especially environmental ones. Most solvents are released quickly (VOCs) and can cause severe environmental effects. This is especially true if toxic solvents are employed.

Emissions of solvents from wood products are described under various conditions, e.g. indoor air emissions from furniture or emissions in test chambers.[3,4] Solvents can be emitted as primary or reactive products of the wood product or the coating system; solvents can also be investigated as secondary emission products.[3,5] The emission characteristics depend on solvent properties and surrounding conditions, e.g., air velocity and air exchange rate.[6]

In the indoor air, solvents from wood products follow various pathways. Examples of interactions are possible reactions of solvents (e.g., styrene) with air components (e.g., hydroxy radicals),[3] transport into and through indoor materials[7] or sorption processes.[5] The emitted solvents can be reduced by ventilation processes or they may be absorbed by organisms.

For humans, absorption of the wood preservatives or ingredients (e.g., solvents) can cause various toxic effects. It is often difficult to pinpoint the causative agents (see Chapter 20).

### REFERENCES


1 **Ullmann`s Encyclopedia of Industrial Chemistry**, 1998.

2 DIBt (Deutsches Institut für Bautechnik). Holzschutzmittelverzeichnis. Index of wood preservatives (1999).

3 T. Salthammer, A. Schwarz, F. Fuhrmann, *Atmospher. Environ.*, **33**, 75 (1999).

4 T. Salthammer, *Atmospher. Environ.*, **7**, 189 (1997).

5 M. Wensing, H.J. Moriske, T. Salthammer, *Gefahrstoffe Reinhaltung der Luft,* **58**, 463 (1998).

6 E. Uhde, A. Borgschulte, T. Salthammer, *Atmospher. Environ.*, **32**, 773 (1998).

7 R. Meininghaus, T. Salthammer, H. Knoppel, *Atmospher. Environ.*, **33**, 2395 (1999).


## 14.15 MEDICAL APPLICATIONS


George Wypych

**ChemTec Laboratories, Inc., Toronto, Canada**


Industries manufacturing medical devices use a wide variety of technological processes which most likely take advantage of most of the available solvents. The range of solvent use is so wide that a complete description of each solvent and its application is not possible in this book. It is questionable if such analysis is possible given that many processes are guarded by trade secrets where there is no patent disclosure. Some examples are given, more to show that, although solvents do contribute to pollution, they also help to produce materials which are needed for health and well being.

Polyurethanes are materials which have the required properties and biocompatilibity which makes them good candidate for use in medical devices. Common applications include pacemaker leads, peripheral and central catheters, feeding tubes, balloons, condoms, surgical gloves, instrument and appliance covers, wound dressings, and many other.[1-4] Several methods are used to process polyurethanes. These include injection molding, extrusion, and solution processing. In solution processing film casting and dip molding are the most frequent techniques.

Dimethylacetamine, tetrahydrofuran, dichloromethane, methyl ethyl ketone, N,N-dimethylformamide, N-methylpyrrolidone, cyclopentanone, cyclohexanone, dioxane, and chloroform are the most commonly used solvents. Most of these are hazardous but used because they contribute to highly transparent product which is very desirable in medical devices. Transparent materials can only be made from transparent solutions.[1] These solvents can dissolve polymers well and form clear solutions. Ease of solvent removal from the material is very important in formulation design. Obviously, no traces of solvents should remain in the medical devices since even trace amounts may interfere with the treatment and the patient's health. An inappropriate solvent selection may cause the formation of crust as the solvent escapes. This leads to material discontinuity (e.g., pinholes) which renders the product inferior. This brings a discussion of solvent evaporation, the rheological properties of formulation, and formation of multilayer materials.

Good solvents can be used in lower concentration but they result in viscous solutions which, in dip coating, form thick films which have the potential of blistering on evaporation. If the solution is diluted, film continuity suffers which increases the number of pinholes. Rapid evaporation causes a formation of a crust of gelled solidified polymer which makes solvent removal more difficult and damages the integrity of the layer. Also, material does not have time to adjust and leveling suffers. On the other hand slow evaporation may cause dissolution of the layer below the coating in a multilayered products and bubbling between the layers.

The selection of solvents for dip coating is usually a complex process ultimately requiring multicomponent solvent mixtures which include a good solvent, a poor solvent, and a solvent of lower boiling point (sometime called "blush resistor") to balance viscosity and rate of evaporation.[2] In wound dressings, the solvents selected affect the material microstructure which controls the evaporation of exuded body fluids but prevents bacteria and pathogens from entering the wound.[3] In infection-resistant medical devices, the antimicrobial agent must be uniformly distributed over all areas of the medical device which may come into contact with a patient. Otherwise there is a risk of infection.[4] Not all solvents dissolve antimicrobial agents and swell surface of medical device.

Cleaning of penetrable septa, tubing systems, and infusion and dialysis systems is another application in which solvents are used. The solvents which are suitable for elastomer cleaning are dichloromethane, perchloroethylene, halogenated hydrocarbons, and freons.[5] This cleaning method extracts undesirable organic materials from medical devices which might otherwise be extracted by body fluids. Heat treatment of catheters followed by washing with a polar solvent increase its surface lubricity. Catheter with poor surface lubricity often causes frictional pain upon its insertion into the body cavity and damages the mucosal tissue resulting in cross infection.[6] Film dressings contain two types of solvents: solvents to dissolve the polymer and propellant solvents. These must be selected to achieve technological goals related to solubility and compatibility.[7]

These examples show that the many technological considerations place constraints on in solvent selection. Solvent replacement in complex products and technological processes is a long-term, expensive proposition which usually results in a need for complete reformulation of the material with failure to achieve the objective a very possible outcome.

## REFERENCES


1 A J Walder, *Plast. Eng.*, **54**, No.4, 29-31 (1998).
2 M T Shah, **US Patent 5,571,567**, Polygenex International, Inc., 1996.
3 J Delgado, R J Goetz, S F Silver, D H Lucast, **US Patent 5,614,310**, 3M, 1997.
4 S Modak, L Sampath, **US Patent 5,567,495**, Columbia University, 1996.
5 S H Smith, J M Brugger, H W Frey, **US Patent 5,639,810**, COBE Laboratories, Inc., 1997.
6 L Mao, Y Hu, D Piao, **US Patent 5,688,459**, China Rehabilitation Research Center, 1997.
7 A J Tipton, S M Fujita, R L Dunn, **US Patent 5,792,469**, Atrix Laboratories, Inc., 1998.


## 14.16 METAL CASTING


George Wypych
**ChemTec Laboratories, Inc., Toronto, Canada**


The metal casting industry has 3,100 facilities in the USA and employs 250,000 people. Most plants are small and technological processes are very diverse. The processes do share common phases, including pattern making, mold and core preparation, furnace charge and metal melting, mold charging, cooling and finishing. Most steps use solvents. In the mold-making process, a many chemical binding systems are used, some of which contain methanol, benzene, toluene, and cresol. The metal is most often recycled and it typically requires cleaning before it is charged to the furnace. This is accomplished either by precombustion or solvent cleaning. In die casting operations, solvent-based or water-based lubricants are used. Die casters also use die fluxes which contain solvents. Some solvent replacement additives in water-based lubricants contain hazardous solvents. Finishing operations involve casting cleaning to remove scale, rust, oxides, oil, grease, and dirt. Solvents are typically chlorinated solvents, naphtha, toluene, and methanol. Cleaning can also be done by emulsifiers, abrasives, alkaline agents, and acid pickling. The cleaning operation is usually followed by painting which frequently involves solvent-based paints and thinners.

Tables 14.16.1 and 14.16.2 contain information on the reported solvent releases and transfers from metal casting industry. The data show that solvent use is not excessive relative to other industries. The industry plans to further improve its environmental record by developing environmentally improved materials which meet regulations. The solvent cleaning and die lubrication are processes under study.

## REFERENCES


1 EPA Office of Compliance Sector Notebook Project. Profile of the Metal Casting Industry. US Environmental Protection Agency, 1998.
2 EPA Office of Compliance Sector Notebook Project. Sector Notebook Data Refresh - 1997. US Environmental Protection Agency, 1998.

**Table 14.16.1. Reported solvent releases from the metal casting industry in 1995 [Data from Ref. 1]**

| Solvent | Amount, kg/year | Solvent | Amount, kg/year |
|---|---|---|---|
| benzene | 110,000 | methyl ethyl ketone | 22,000 |
| n-butyl alcohol | 15,000 | methyl isobutyl ketone | 22,000 |
| cresol | 20,000 | N-methyl-2-pyrrolidone | 41,000 |
| dichloromethane | 50,000 | tetrachloroethylene | 13,000 |
| ethylbenzene | 10,500 | 1,1,1-trichloroethane | 111,000 |
| ethylene glycol | 64,000 | trichloroethylene | 75,000 |
| hexachloroethane | 16,000 | toluene | 233,000 |
| methanol | 5,860,000 | xylene | 388,000 |

**Table 14.16.2. Reported solvent transfers from the metal casting industry in 1995 [Data from Ref. 1]**

| Solvent | Amount, kg/year | Solvent | Amount, kg/year |
|---|---|---|---|
| benzene | 115 | N-methyl-2-pyrrolidone | 22,000 |
| ethylbenzene | 340 | 1,1,1-trichloroethane | 500 |
| ethylene glycol | 50,000 | trichloroethylene | 1,000 |
| methanol | 10,000 | toluene | 4,000 |
| methyl ethyl ketone | 8,000 | xylene | 82,000 |

## 14.17 MOTOR VEHICLE ASSEMBLY

George Wypych
**ChemTec Laboratories, Inc., Toronto, Canada**

Automotive industry in US alone employs 6.7 million person and it is a large contributor to the gross national product. It uses large quantities of solvents and it is perceived to contribute to pollution by solvents and other materials.[1-6] Solvents are used in a variety of cleaning, preparation, and painting operations. Automotive finishing process may be divided into four main categories: anti-corrosion operations (cleaning, phosphate treatment, and chromic acid treatment), priming operations (electrodeposition of primer, anti-chip coating application, and primer application), joint sealant application, and other finishing operations (color coat, clear coat). These main operations employ many materials which contain solvents.

The cleaning process involves acid/alkaline and solvent cleaning. Typical solvents involved are acetone, xylene, toluene, and 1,1,1-trichloroethylene. The primer bath is water-based but usually some organic solvents are present (5-10%). These solvents are the same as those listed above. After the application of primer, the car body is baked and then undergoes waterproofing with an application of polyvinylchloride sealant which contains a small amount of solvents. Following waterproofing, the automotive body proceeds to the anti-chip booth, where urethane or epoxy solvent-coating systems are applied. This process is followed by application of primer-surfacer coating which is either a polyester or an epoxy ester in a solvent system. The primer-surfacer coating is applied by spraying and provides a durable finish which can be sanded. After the sanding step, the primary color coating is applied also by spraying. These primary color formulations contain about twice as much solvent as the primer-surfacer coating. Solvents are flashed-off (no heating) and a clear coat is applied. Then the entire car body is baked for about 30 min. Solvents used include butanol, isobutanol, methanol, heptane, mineral spirits, butyl acetate ethyl acetate, hexyl acetate, methyl ethyl ketone, acetone, methyl amyl ketone, toluene, and xylene.

Several finishing operations also employ solvents. After baking, a sound-deadener is applied to certain areas of the underbody. It is a solvent based material with a tar-like consistency. A trim is applied with adhesives which contain solvents (see section on adhesives and sealants). After the installation of trim and after the engine is installed, car undergoes an inspection. Some repainting is required in about 2% of the production. If damage is minor then repainting is done by a hand operated spray gun. If the damage is substantial a new body is installed. Equipment cleaning solvents are also used. Spraying equipment is cleaned with a “purge solvent” which may consist of a mixture of dimethylbenzene, 4-methyl-2-pentanone, butyl acetate, naphtha, ethyl benzene, 2-butanone, toluene, and 1-butanol.

Tables 14.17.1 and 14.17.2 contain information on the reported solvent releases and transfers from the motor vehicle assembly industry. The data show that solvent use is very large compared with all industries covered so far in our discussion except for the steel and iron industry. The motor vehicle assembly industry is the sixth largest producer of VOC and also the sixth largest industry in reported emissions and transfers.

The data in Tables 14.17.1 and 14.17.2 are data from 1995 the most recent available. The automotive industry and associated paint companies conduct extensive work on replacement of VOC containing paint systems. These efforts are mainly directed to water-based systems and powder coatings. Until recently, water-based systems were preferred but now attention is shifting to powder coatings which eliminate VOC. There is no status quo. Changes are dynamic and kept protected by trade secrets which makes it difficult to comment on specific progress. Solvent use by the European industry[5] is that the production of one car requires an average of 10 kg of solvents. Solvents use is not the only problem the industry is facing. 16% of the total energy used in car production is required by painting and finishing operations. Both energy conservation and reduction is solvent consumption must be pursued to meet environmental objectives. Not only can these issues be addressed through material reformulation but the design of equipment used in applying and drying the coating can also reduce emission and save energy.

A new trend is apparent as plastics are introduced to automotive production. Plastic parts must also be painted. Paint systems are difficult to select. Chlorinated polyolefins provide good adhesion of paints and reduce VOC but are also under scrutiny because of pres-

**Table 14.17.1. Reported solvent releases from the motor vehicle assembly industry in 1995 [Data from Ref. 2]**

| Solvent | Amount, kg/year | Solvent | Amount, kg/year |
|---|---|---|---|
| benzene | 13,000 | methyl ethyl ketone | 2,320,000 |
| n-butyl alcohol | 2,260,000 | methyl isobutyl ketone | 3,060,000 |
| sec-butyl alcohol | 86,000 | N-methyl-2-pyrrolidone | 193,000 |
| tert-butyl alcohol | 4,200 | methyl tert-butyl ether | 32,000 |
| cyclohexane | 35,000 | tetrachloroethylene | 140,000 |
| dichloromethane | 380,000 | 1,1,1-trichloroethane | 730,000 |
| ethylbenzene | 1,370,000 | trichloroethylene | 1,300,000 |
| ethylene glycol | 180,000 | 1,2,4-trimethylbenzene | 1,120,000 |
| isopropyl alcohol | 9,000 | toluene | 2,610,000 |
| hexane | 95,000 | xylene | 10,800,000 |
| methanol | 1,550,000 | m-xylene | 25,000 |

**Table 14.17.2. Reported solvent transfers from the motor vehicle assembly industry in 1995 [Data from Ref. 2]**

| Solvent | Amount, kg/year | Solvent | Amount, kg/year |
|---|---|---|---|
| benzene | 3,400 | methyl ethyl ketone | 2,100,000 |
| n-butyl alcohol | 1,030,000 | methyl isobutyl ketone | 4,700,000 |
| sec-butyl alcohol | 9,000 | N-methyl-2-pyrrolidone | 330,000 |
| tert-butyl alcohol | 1,000 | methyl tert-butyl ether | 2,300 |
| cyclohexane | 670 | tetrachloroethylene | 49,000 |
| dichloromethane | 450,000 | 1,1,1-trichloroethane | 140,000 |
| ethylbenzene | 1,740,000 | trichloroethylene | 480,000 |
| ethylene glycol | 605,000 | 1,2,4-trimethylbenzene | 330,000 |
| isopropyl alcohol | 2,000 | toluene | 2,020,000 |
| hexane | 25,000 | xylene | 9,200,000 |
| methanol | 760,000 | m-xylene | 2,100 |

ence of chlorine. Powder coatings are available[7] but they require a high energy input. These problems are apparent but the solution to them will take several years to implement due, in large part, to the long term testing needed to confirm coating performance (up to 5 years in Florida).

**Table 14.18.1. Reported solvent releases from the organic chemical industry in 1995 [Data from Ref. 2]**

| Solvent | Amount, kg/year | Solvent | Amount, kg/year |
|---|---|---|---|
| allyl alcohol | 31,000 | ethylbenzene | 370,000 |
| benzene | 690,000 | ethylene glycol | 6,050,000 |
| n-butyl alcohol | 850,000 | hexane | 600,000 |
| sec-butyl alcohol | 63,000 | isopropyl alcohol | 150 |
| tert-butyl alcohol | 430,000 | methanol | 8,750,000 |
| carbon disulfide | 85,000 | methyl ethyl ketone | 260,000 |
| carbon tetrachloride | 10,000 | methyl isobutyl ketone | 520,000 |
| chlorobenzene | 27,000 | N-methyl-2-pyrrolidone | 350,000 |
| chloroform | 7,000 | methyl tert-butyl ether | 64,000 |
| cresol | 280,000 | pyridine | 120,000 |
| m-cresol | 320,000 | tetrachloroethylene | 20,000 |
| o-cresol | 270,000 | toluene | 1,040,000 |
| p-cresol | 162,000 | 1,2,4-trichlorobenzene | 41,000 |
| cyclohexane | 450,000 | 1,1,1-trichloroethane | 130,000 |
| cyclohexanol | 1,100,000 | trichloroethylene | 18,000 |
| dichloroethane | 120,000 | xylene | 350,000 |
| 1,2-dichloroethylene | 70 | m-xylene | 59,000 |
| dichloromethane | 310,000 | o-xylene | 34,000 |
| N,N-dimethylformamide | 25,000 | p-xylene | 660,000 |
| 1,4-dioxane | 12,000 | | |

## REFERENCES


1 EPA Office of Compliance Sector Notebook Project. Profile of the Motor Vehicle Assembly Industry. US Environmental Protection Agency, 1995.
2 EPA Office of Compliance Sector Notebook Project. Sector Notebook Data Refresh - 1997. US Environmental Protection Agency, 1998.
3 G. Wypych, Ed., **Weathering of Plastics. Testing to Mirror Real Life Performance**, *Plastics Design Library, Society of Plastics Engineers*, New York, 1999.
4 M Harsch, M Finkbeiner, D Piwowarczyk, K Saur, P Eyerer, *Automotive Eng.*, **107**, No.2, 211-4 (1999).
5 C A Kondos, C F Kahle, *Automotive Eng.*, **107**, No.1, 99-101 (1999).
6 D C Shepard, *J. Coat. Technol.*, **68**, No.857, 99-102 (1996).
7 T Hosomi, T Umemura, T Takata, Y Mori, US Patent 5,717,055, Mitsubishi Gas Chemical Company, Ltd., 1998.

**Table 14.18.2. Reported solvent transfers from the organic chemical industry in 1995 [Data from Ref. 2]**

| Solvent | Amount, kg/year | Solvent | Amount, kg/year |
|---|---|---|---|
| allyl alcohol | 210,000 | ethylbenzene | 980,000 |
| benzene | 420,000 | ethylene glycol | 6,800,000 |
| n-butyl alcohol | 1,500,000 | hexane | 770,000 |
| sec-butyl alcohol | 1,700,000 | isopropyl alcohol | 85,000 |
| tert-butyl alcohol | 12,500,000 | methanol | 23,000,000 |
| carbon disulfide | 96,000 | methyl ethyl ketone | 800,000 |
| carbon tetrachloride | 12,000 | methyl isobutyl ketone | 390,000 |
| chlorobenzene | 130,000 | N-methyl-2-pyrrolidone | 110,000 |
| chloroform | 92,000 | methyl tert-butyl ether | 210,000 |
| cresol | 430,000 | pyridine | 33,000 |
| m-cresol | 720,000 | tetrachloroethylene | 138,000 |
| o-cresol | 57,000 | toluene | 4,400,000 |
| p-cresol | 870,000 | 1,2,4-trichlorobenzene | 8,000 |
| cyclohexane | 900,000 | 1,1,1-trichloroethane | 290,000 |
| cyclohexanol | 3,700 | trichloroethylene | 42,000 |
| dichloroethane | 230,000 | xylene | 4,000,000 |
| 1,2-dichloroethylene | 1,000 | m-xylene | 51,000 |
| dichloromethane | 870,000 | o-xylene | 460,000 |
| N,N-dimethylformamide | 370,000 | p-xylene | 1,700 |

## 14.18 ORGANIC CHEMICAL INDUSTRY

George Wypych
ChemTec Laboratories, Inc., Toronto, Canada

The chemical industry operates about 1000 plants in the USA with 53 companies producing 50% of the total output of $65 billion in sales and employing 125,000 people. There are point source solvent emissions (e.g., laboratory hoods, distillation units, reactors, storage tanks, vents, etc.), fugitive emissions (e.g., pump valves, flanges, sample collectors, seals, relief devices, tanks), and secondary emissions (waste water treatment units, cooling towers, spills). Organic liquid wastes containing solvent are generated from processes such as equipment washing, surplus chemicals, product purification, product reaction, housekeeping, etc.

Tables 14.18.1 and 14.18.2 give the reported solvent releases and transfers from the organic chemical industry. Large quantities of solvents are involved. The organic chemical industry produced the second largest quantity of VOC and the second largest releases and transfers. The industry is actively working to reduce solvent use because of the high costs (waste treatment, fines, liabilities, etc). There are many efforts under way to reduce environmental emissions and improve safe practices. The initiatives include process modifications such as a reduction in non-reactive materials (e.g., solvents) to improve process efficiency, a reduction in the concentration of chemicals in aqueous solution, and improved R&D and process engineering. Equipment modifications are planned to reduce leaks, prevent equipment breakdown, and improve the efficiency of emission control devices.

## REFERENCES

1 EPA Office of Compliance Sector Notebook Project. Profile of the Organic Chemical Industry. US Environmental Protection Agency, 1995.
2 EPA Office of Compliance Sector Notebook Project. Sector Notebook Data Refresh - 1997. US Environmental Protection Agency, 1998.

# 14.19 PAINTS AND COATINGS

## 14.19.1 ARCHITECTURAL SURFACE COATINGS AND SOLVENTS

Tilman Hahn, Konrad Botzenhart, Fritz Schweinsberg
**Institut für Allgemeine Hygiene und Umwelthygiene**
**Universität Tübingen, Tübingen, Germany**
Gerhard Volland
**Otto-Graf-Institut, Universität Stuttgart, Stuttgart, Germany**

### 14.19.1.1 General aspects

Coating materials and coating techniques can be distinguished and systematized in various ways.

The fundamental principles of common coating systems are:[1]

- *Physical drying*. A solid surface film is formed after the evaporation of water or organic solvents.
- *Physico-chemical drying/curing*. Polycondensation or polyaddition are combined with evaporation of organic solvents.
- *Chemical curing*. Solvents, e.g., styrene or acrylic monomers, react with the curing system.

The actual effects depend on the surrounding conditions and the ingredients of the coating system, e.g., solvents.

Solvents contribute many essential properties to coating systems. Solvents can improve technical factors such as application or surface properties. Solvents also bring negative qualities to coating materials, especially with respect to environmental conditions (e.g., toxic effects of emitted organic solvents).

### 14.19.1.2 Technical aspects and properties of coating materials

Application techniques for coatings can be considered in various ways. The stability and durability of coating is essential. Coatings that have normal wear and tear requirements are mainly based on oils and aldehyde resins. Higher durability or stability can be achieved by the use of one of the following one- or two-component systems.

One-component:

- Bituminous materials
- Chlorinated rubber
- Polyvinyl chloride
- Polyacrylic resin
- Polyethene
- Saturated polyester
- Polyamide

Two-component:

- Epoxy resin
- Polyurethane
- Mixtures of reactive resins and tar

A survey of the performance of different coating materials together with an assessment of various environmental factors is given in Table 14.19.1.1.

**Table 14.19.1.1. Environmental performance of some coating materials**

| Material | Abbreviation | Weathering response | Acid atmosphere | Humidity | Under water | Chemical stress | | Solvent | Temperature <60°C | Mechanical | |
|---|---|---|---|---|---|---|---|---|---|---|---|
| | | | | | | acid | alkali | | | abrasion | scratch |
| plant oil (linseed oil) | OEL | +y,s,f | +/- | +/- | - | - | -s | - | + | - | - |
| alkyd resin | AK | + | +/- | +/- | - | - | -s | - | + | - | - |
| bitumen | B | +/-e,f | + | + | + | +/- | + | - | +/- | - | - |
| chlorinated rubber | RUC | +/-b,c,f | + | + | + | + | + | - | + | +/- | +/- |
| PVC-soft | PVC | +/-e,f | + | +/- | +/- | + | +/- | - | + | +/- | +/- |
| Polyacrylic resin | AY | + | + | + | + | + | + | - | + | + | +/- |
| polyethene | PE | +/-e | + | + | + | + | + | + | + | + | + |
| polyester (saturated) | SP | + | + | +/- | +/- | + | +/-(s) | + | + | + | + |
| epoxy resin | EP | +/-c,g | + | + | + | +/- | + | + | + | + | + |
| polyurethane | PU | + | + | +/- | +/- | + | +/- | +/- | + | + | + |
| epoxy resin tar | EP-T | +/-c,e | + | + | + | +/- | + | +/- | + | + | + |
| polyisobutylene | PLB | +/- | + | + | + | + | + | - | + | + | + |

+ = suitable, +/- = limited suitability, - = unsuitable, () = less distinct; b = bleaching, c = chalking, e = embrittlement, f = acid fragments, g = loss of gloss, s = saponification, y = yellowing

Thc following are examples of architectural surface coatings.

### *14.19.1.2.1 Concrete coating materials*

According to DIN 1045 concrete must be protected against aggressive substances if the chemical attack is severe and long-term. These are the requirements:

- Good adhesion
- Waterproof and resistant to aggressive substances and resistant to the alkalinity of the concrete
- Deformable

To realize these requirements, the following technical solutions can be applied:

- The use of paint coatings based on duroplast or thermoplastic substances
- Surface treatment - impregnation

Protective coatings based on polymers are used in construction with and without fillers and with or without reinforcing materials (fibers).

The following techniques are generally used:

- Film forming paint coatings (brushing, rolling, spraying).
- Coating (filling, pouring).

During application, the coating materials are normally liquid and subsequcntly harden by evaporation of solvents or as a result of chemical reactions.

The common coating systems for concrete are listed in Table 14.19.1.2. Synthetic rcsins (e.g., chlorinated rubber, styrene resins, acrylic resins) contain normally 40-60% solvents and form a thin film. Several coats must be applied. Reactivc rcsins may require little or no solvent.

**Table 14.19.1.2. Survey of often used coating systems for concrete**

| Film-forming agents | Hotmelts | Film-forming agents dissolved | Film-forming agents emulsified | Reaction resin liquid |
|---|---|---|---|---|
| Bituminous substances | + | + | + | |
| Synthetic resin | | + | + | |
| Reactive resins | | (+) | (+) | + |

### *14.19.1.2.2 Coating materials for metals*

Wet coating processes are detailed in Table 14.19.1.3.

**Table 14.19.1.3. Wet coatings - coating materials for metals**

| | | Abbreviation | Application | State | Curing method |
|---|---|---|---|---|---|
| Natural substances | oil | OEL | binders for wet coating | liquid | oxidizing |
| | bitumen | B | | | physical |
| | tar | T | | | physical |
| Plastic materials | alkyd resin | AK | binders for wet coating | | oxidizing |
| | polyurethane | PU | | | chemical |
| | epoxy resin | EP | | | chemical |
| | acrylic resin | AY | | | physical |
| | vinyl resin | PVC | | | physical |
| | chlorinated | RUC | | | physical |
| | rubber | | | | |

The chemical properties of binders are not affected by the drying process if coatings are of physically drying type. After solvent has evaporated, the polymer molecules become intermeshed, thus producing the desired coating properties.

In contrast to physical drying, binders based on reactive resins such as epoxy resins, polyurethanes or polyesters consist of two components (liquid resin and curing agent) which are either mixed shortly before the coat is applied, or, in the case of a one component system, one which is applied as slow reacting mixture. The setting reaction occurs at the surface of the coated material. The final products are normally more resistant and more compact than products based on physically drying binders. Pretreatment of substrate is more critical for applications where chemically hardening products are applied. Coatings can be more or less permeable to water vapor and oxygen. Damage to the metal substrate can occur if water and oxygen reach the reactive surface simultaneously. This is normally impossible if the coatings adhere well and the coated surface is continuous. The adhesion of the coating also prevents the penetration by harmful substances via diffusion processes. Adhesion is enhanced by adsorption and chemical bonds.

### *14.19.1.2.3 Other coating materials*

Other aspects of solvents contained in paint coatings and varnishes are given in Chapter 18.3.

The market offers a wide spectrum of coating systems. Examples of common industrial coating materials are listed in Table 14.19.1.4 and relevant aspects concerning application and environmental or health risks are also included.

**Table 14.19.1.4. Examples of coating materials containing hazardous ingredients, especially solvents[3]**

| Product group/ compounds | Application | Hazardous ingredients, especially solvents | Relevant health and environmental risks |
|---|---|---|---|
| zinc dust coating based on epoxy resin | corrosion protection/primer, application by brush, spray, airless-spraying | xylene 2,5-25%, ethylbenzene 2,5-10%, solvent naphtha 2,5-10%,1-methoxy-2-propanol 1-2,5% | irritations (respiratory, skin, eyes), neurological (narcotic) effects, absorption (skin), flammable, explosive mixtures |
| reactive PUR-polymers containing solvents | corrosion protection/coating | solvent naphtha 10-20%, diphenylmethane isocyanate 2,5-10% | sensitization (respiratory) irritation (skin, eyes, gastrointestinal tract), neurological effects (narcotic effects, coordination), absorption (skin), flammable, water polluting |
| modified polyamine containing solvents | corrosion protection/coatings/rigid system | 4-tert-butylphenol 10-25%, M-phenylenebis 2,5-10%, trimethylhexane, 1,6-diamine 10-25%, nonylphenol, 10-25% xylene 10-25%, ethylbenzene 2,5-10% | sensitization (skin, respiratory), irritations (skin, eye, respiratory, gastrointestinal tract), neurological (narcotic) effects (coordination), flammable, water polluting |
| modified epoxy resin containing solvents | corrosion protection/coatings/ rigid system | bisphenol A (epichlorohydrin) 25-50% oxirane, mono($C_{12}$ -$C_{14}$-alkyloxy)methyl derivatives 2,5-10%, cyclohexanone 1-2,5%, 2-methylpropane-1-ol 2,5-10%, xylene 10-25%, ethylbenzene 2,5-10%, benzyl alcohol 1-2,5%, 4-methyl-pentane-2-one 1-2,5% | irritations (skin), sensitization (skin, contact), neurological (narcotic) effects, absorption (skin), flammable, water polluting |
| modified epoxy resin containing solvent | corrosion protection/top layer | bisphenol A (epichlorohydrin) 50-100%, 3-amino-3,5,5-trimethyl ethylbenzene 10-25%, xylene 10-25% | irritation (skin, eyes, respiratory), sensitization (skin), flammable, explosive gas/air mixtures, hazardous reactions (with acids, oxidizers), water polluting |
| filled and modified epoxy resin | coatings and corrosion prevention/ pore sealer | bisphenol A (epichlorohydrin) 50-100%, P-tert-butylphenyl-1-(2,3-epoxy)propyl-ether 1-2,5% | irritations, sensitization, water polluting |
| filled and modified epoxy resin | coatings and corrosion prevention/ pore sealer | trimethylhexamethylendiazine-1,6-diamine 10-25%, trimethylhexamethylendiamine -1,6-cyanoethylene 50-100% | irritations, sensitization, water polluting |

| Product group/ compounds | Application | Hazardous ingredients, especially solvents | Relevant health and environmental risks |
|---|---|---|---|
| filled and modified epoxy resin | flooring/mortar screen | bisphenol A (epichlorohydrin) 25-50%, benzyl butyl phthalate 2,5-10%, xylene 2,5-10% | irritation (eye, skin, respiratory), sensitization (contact, skin), explosive gas/ air mixtures (with amines, phenols), water polluting |
| modified polyamine | flooring/mortar screen | benzyl alcohol 10-25%, nonylphenol 25-50%, 4,4′-methylenebis(cyclohexylamine) 10-25%, 3,6,9-triazaundecamethylendiamine 10-25% | irritation (eye, skin, respiratory), sensitization (contact, skin), hazardous reactions (with acids, oxidizers), flammable, water polluting |
| filled and modified epoxy resin | corrosion protection/top layer | naphtha 2,5-10%, 2-methoxy-1-methylethylacetate 2,5-10%, 2-methylpropane-1-ol 1-2,5%, xylene 10-25%, ethylbenzene 2,5-10% | irritation (skin, eyes, respiratory), neurological (narcotic) effects, flammable, explosive, hazardous reactions (with oxidizers), water polluting |
| copolymer dispersion | walls, especially fungicidal properties, good adhesion in damp environment, resistant against condensed water | isothiazolone 1-2,5%, 3-methoxybutylacetate 1-2,5% | irritations (skin, eyes, gastrointestinal tract), water polluting |
| coating of synthetic resin containing solvents | corrosion protection/coating (steel) | solvent naphtha (petroleum) 25-50 %, xylene 2,5-10%, ethylbenzene 1-2,5% | irritations (eyes, skin, respiratory, gastrointestinal tract), neurological (narcotic) effects, flammable, explosive gas/air mixtures, hazardous reactions (with oxidizers), water polluting |
| bituminous emulsion (phenol-free, anionic) | corrosion protection/coating (steel, drinking water tanks) | naphtha (petroleum) 25-50% | irritations (skin, eyes, respiratory), neurological effects (coordination), flammable, water polluting |
| modified, filled anthracene oil and polyamine | corrosion protection/top layer | solvent naphtha 2,5-10%, biphenyl-2-ol 2,5-10%, 3-amino-3,5,5-trimethylcyclohexamine 1-2,5%, xylene 1-2,5% | irritations (skin, eyes, respiratory), flammable, water polluting |
| partially neutralized composition of aminoalcohols | sealants and adhesives/elastic products (floor joints, joints) | 2-aminoethanol 2,5-10% | sensitization, irritations, slight water pollution |

| Product group/ compounds | Application | Hazardous ingredients, especially solvents | Relevant health and environmental risks |
|---|---|---|---|
| filled, reactive PUR-polymers | sealants and adhesives/elastic products (floor joints, joints) | 3-isocyanatemethyl-3,5,5-trimethylcyclohexylisocyanate 0,1-1%, N,N-dibenzylidenepolyoxypropylene diamine 1-2,5%, xylene 2,5-10%, ethylbenzene 1-2,5% | sensitization, irritations, hazardous reactions (with amines, alcohols) |
| solvent based composition, based on PVC | coatings and corrosion prevention/impregnation, sealer | solvent naphtha 50-100%, 1-methoxy-2-propanol 2,5-10%, | irritations, flammable, neurological effects |
| acrylate dispersion, water dispersed adhesion, promotor | coatings and corrosion protection/ rigid coat (for concrete and dense mineral substrates) | water-borne so-called solvent-free systems | irritations (long-term contact), slightly water polluting |

## References


1 **Ullmann´s Encyclopedia of Industrial Chemistry**,1998.
2 DIN 1045 (Deutsches Institut für Normung), Concrete and Reinforced Concrete. Beton und Stahlbeton, Beuth Verlag, Berlin (1988), part 1 (1997), part 2 (1999).
3 Sika TechnoBauCD, Sika Chemie GmbH, Stuttgart (2000).


## 14.19.2 RECENT ADVANCES IN COALESCING SOLVENTS FOR WATERBORNE COATINGS


DAVID RANDALL
**Chemoxy International pcl, Cleveland, United Kingdom**


### 14.19.2.1 Introduction

The role of coalescing solvents in coating formulations, the factors which affect the choice of coalescing solvents, and the recent developments in esters of low volatility, low odor and rapid biodegradability for use in coating formulations are discussed.

The term "paint" is widely used to describe a coating applied to a variety of substrates for protective or decorative purposes. "Paint" also implies a pigmented species, whereas "surface coatings" is a broader term for coating systems with or without pigments used for any coating purpose.

All paint systems may be considered as a combination of a small number of constituents. These are: continuous phase or vehicle (polymer and diluent), discontinuous phase or pigment and extender, and additives.

This paper is not concerned with the detail of paint formulation, but the polymer types most widely encountered are alkyds, polyurethanes and nitrocellulose in solvent based systems, and acrylics, styrene-acrylics, and copolymers of vinyl acetate in water-based coatings.

### 14.19.2.2 Water based coatings

Until recent times, most paints were solvent based. Since many polymers were produced in solution, it was natural that the coating system, derived from these polymers, would use the solvent of reaction as the diluent for the polymer. These were almost invariably organic species. The objections to the use of organic solvents in paint formulations were at first confined to those with strong views regarding the loss of significant quantities of organic species to the environment. Latterly, these opinions have been reinforced by the role of some organic species in damage to the ozone layer, or to the production of smog in the lower atmosphere. This is as a result of the reaction between organic species present catalyzed by sunlight and exacerbated by the presence of other pollutants associated with motor vehicles, etc. Flammability of paints also poses a significant problem in storage and use.

All this tended to reinforce the need to develop aqueous based systems, for which these problems would be eliminated.

### 14.19.2.3 Emulsion polymers

Of course, the manufacture of polymers need not be carried out in solution; emulsion polymerization has a long and honorable tradition in the field of macromolecular chemistry. This polymerization technique is performed using water-insoluble monomers, which are caused to form an emulsion with the aqueous phase by the addition of a surfactant. Polymerization may be effected by the use of a variety of radical generating initiating species, and the molecular weight of the polymer is controlled by the use of chain transfer agents and the concentration of initiator employed. The final polymer dispersion is often described as a latex, named after natural rubber, which is also an emulsion polymer!

This polymerization technique allows for the formation of copolymers in which the addition of relatively small quantities of comonomer may have a significant effect on the final properties of the polymer. This is particularly the case with the glass transition temperature, $T_g$, of the polymer. This is a physical transition, which occurs in the polymer when the amorphous structure of that polymer begins to change from a glassy to a rubbery state. At temperatures below a polymer's $T_g$, it will be relatively brittle, and will be unlikely to form a coherent film.

This effect has a major impact on the use of polymers in aqueous systems. In the case of a polymer in solution, the presence of the solvent plasticizes the polymer during film formation. A polymer with a high $T_g$, i.e., one greater than ambient temperature, can, when in solution be applied at temperatures below its $T_g$. In the case of an emulsion polymer, water is a non-solvent in the system and film formation below the polymer $T_g$ is unlikely. The temperature at which a coherent film may be formed from a solution or emulsion based system is known as the Minimum Film Forming Temperature (MFFT or MFT).

Figures 14.19.2.1-14.19.2.4 describe the formation of a discrete film from an emulsion system.

When the film is applied onto the substrate, discrete polymer particles are dispersed in the aqueous phase. The system is stabilized by the presence of the surfactant at the water-particle interface. The particles are spherical, with an average diameter of 0.1-0.2 μm.

As water is lost from the system, the mutual repulsive forces associated with the surfactants present inhibit the close packing of the particles and a cubic arrangement of the particles is formed.

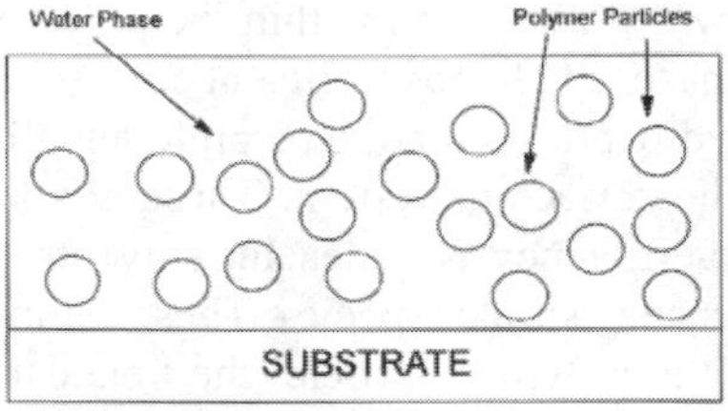


Figure 14.19.2.1. Latex in contact with substrate.

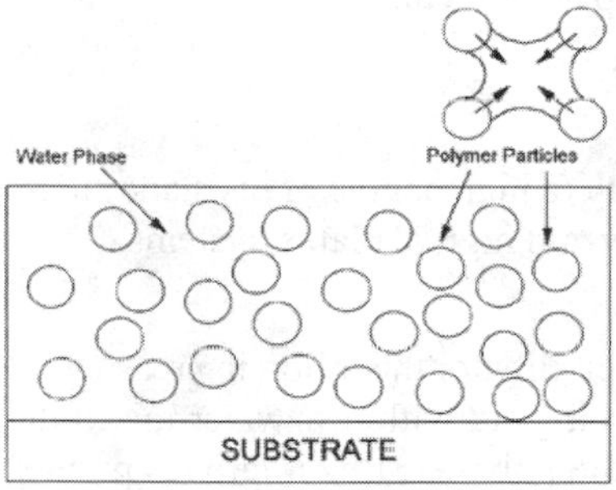


Figure 14.19.2.2. Initial water evaporation and film formation.

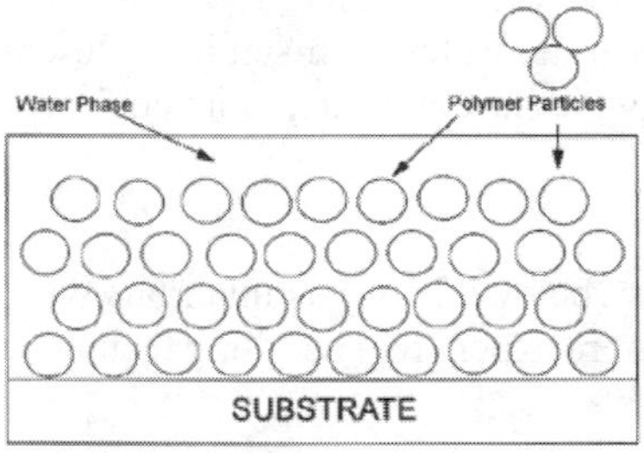


Figure 14.19.2.3. Close packing of latex particles.

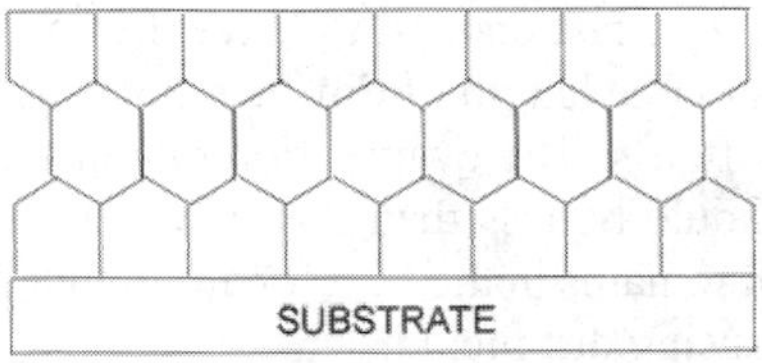


Figure 14.19.2.4. Coalescence to form a polymeric film.

As the water continues to evaporate, the particles become close packed with a solids volume of around 70%. Capillary forces continue to force the particles together.

The final stage is achieved when most of the water is lost from the system. Here, the interparticular repulsive forces are overcome by increasing surface tension and the particles coalesce into a discrete film. This will only occur at temperatures in excess of the MFFT.

### 14.19.2.4 Role of a coalescing solvent

Most coatings based on emulsion polymers are used in environments where they will be expected to form a coherent film at temperatures as low as 0°C. However, other physical properties besides film forming capability are required from the polymer. These include; abrasion resistance, hardness, chemical resistance, impact performance, etc. These can often be impossible to achieve with a polymer of low $T_g$. The polymers, which most clearly meet these criteria are acrylics or copolymers of vinyl acetate or styrene. These would, without additions of coalescing species, be brittle, forming incoherent films with little adhesion to the substrate at normal application temperatures.

Coalescing solvents allow these polymeric systems to form films at ambient or sub-ambient temperatures. The presence of the coalescing solvent has the following effects.

- It reduces the total surface energy of the system by reducing polymer surface area.
- It increases the capillary forces by the controlled evaporation of the water.
- It reduces the repulsive forces between the polymeric particles.
- It allows deformation of the particles in contact with each other by effectively lowering the $T_g$ of the polymer.

An emulsion polymer consists of a dispersion of polymeric particles varying in size from 0.05-1 μm, dispersed in an aqueous environment. The coalescing solvent may be found in several different locations depending on their nature. They are classified according to their preferred location in the aqueous system.

Hydrophobic substances such as hydrocarbons will prefer to be within the polymer particle; these are described as A Group coalescents. These tend to be inefficient coalescing agents. Molecules, which are more hydrophilic than hydrocarbons tend to be preferentially sited at the particle - aqueous interface, along with the surfactant system. These are described as AB Group coalescents. These exhibit the best efficiency as coalescing solvents.

In the third group are the hydrophobic glycol ethers and similar species. These exhibit good coalescing power, but are partitioned between the polymer particle, the boundary layer, and the aqueous phase. More of these are required than AB Group coalescents, they are called ABC Group coalescents. Finally, hydrophilic species such as glycols and the more polar glycol ethers are inefficient coalescing agents, and are more commonly used as freeze thaw stabilizers. These are described as C Group coalescents.

### 14.19.2.5 Properties of coalescing agents

#### *14.19.2.5.1 Hydrolytic stability*

A coalescing agent should have good hydrolytic stability (a high degree of resistance to hydrolysis) so that it can be used successfully in both low and high pH latex systems.

#### *14.19.2.5.2 Water solubility*

It is desirable for a coalescing aid to have low water miscibility for the following reasons:

- When added to a coating formulation, a coalescing aid with low water miscibility partitions into the polymer phase and softens the polymer; this improves pigment binding and polymer fusion.
- There is less tendency for the evaporation of the coalescing aid to be accelerated by evaporation of the water during the early stages of drying.
- The early water resistance of the coating is not adversely affected.
- When a latex coating is applied to a porous substrate, water immiscible coalescing aids are not lost into the porous substrate along with the water; i.e., more coalescing aid will be available to coalesce the coating.

#### *14.19.2.5.3 Freezing point*

The freezing point of a coalescing agent should be low (below -20°C) as materials with a high freezing point may require specialized (therefore more expensive) handling techniques in transport and storage.

#### *14.19.2.5.4 Evaporation rate*

The evaporation rate of a coalescing aid should be slow enough to ensure good film formation of the emulsion coating under a wide range of humidity and temperature conditions; however, it should be fast enough to leave the coating film in a reasonable length of time and not cause excessive film softness. The evaporation rate of the coalescing aid should be less than that of water but not so slow that it remains in the film for an extended period of time causing dirt pickup.

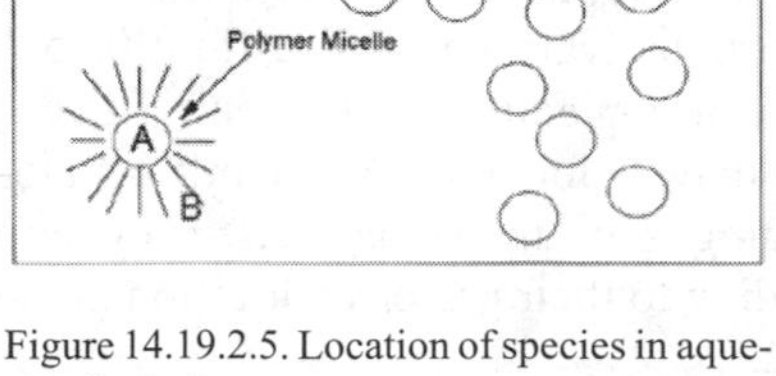

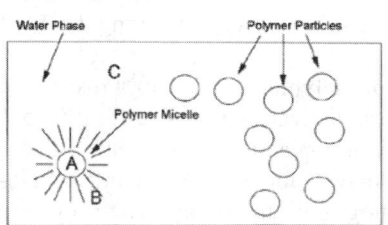


Figure 14.19.2.5. Location of species in aqueous dispersions.

#### *14.19.2.5.5 Odor*

The odor of a coalescing agent should be minimal. This is especially important for interior coatings applications.

#### *14.19.2.5.6 Color*

A coalescing agent should be colorless, to prevent discoloration of the product.

#### *14.19.2.5.7 Coalescing efficiency*

The effectiveness of a coalescing agent is based upon the MFFT test.

#### *14.19.2.5.8 Incorporation*

A good coalescing agent should be capable of addition during any stage of paint manufacture. Any particular coalescing agent may need premixing prior to addition with varying amounts of the water and surfactant used in the paint stage.

#### *14.19.2.5.9 Improvement of physical properties*

A good coalescing agent can provide a significant improvement in scrub resistance, stain resistance, flexibility and weatherability.

#### *14.19.2.5.10 Biodegradability*

A good coalescing agent should exhibit acceptably rapid biodegradation quickly.

#### *14.19.2.5.11 Safety*

The use of the coating system requires that the nature of the coalescing agent should permit use in the domestic environment. The suitability of such a material is therefore obviously dependent on its toxicity, flash point, vapor pressure and VOC classification.

### 14.19.2.6 Comparison of coalescing solvents

From the above, it may readily be seen that AB type coalescing solvents give the best performance in use. These are the high molecular weight esters and ester alcohols. Other species have some of the benefits of these materials but have side effects, which render them less suitable. Minimization of the quantity used tends to be a key financial issue, as well as offering the best environmental option. In virtually all cases, the efficiency of the AB group in application is so marked, that they have become the industry standards. Several other properties, perhaps less important, than efficiency must then be considered when selecting between these.

These products are usually used in enclosed spaces, where product odor will be vital for acceptance to the consumer. The dibasic esters are virtually odor free. For legislative reasons, a solvent not classified as a VOC will be of definite benefit, given the current attitude to organics in consumer products. Again, the dibasic esters have initial boiling points considerably above the VOC threshold of 250°C.

The stability of the product towards hydrolysis is also a significant factor in the selection of the optimum coalescing agent. Effectively, this parameter determines the storage stability of the product. Again, dibasic esters, particularly those produced from higher molecular weight alcohols appear to have advantage.

The efficiency of the AB group of coalescents towards almost all polymer types has already been mentioned. For styrene-acrylics and vinyl acetate - Veova copolymers, the diesters have a significant edge in performance compared with other members of that group. This advantage is shown in the reduction in amount required to attain a particular MFFT or the actual MFFT for a given addition level.

Coasol manufactured by Chemoxy is the blend of di-isobutyl esters of the dibasic acids; adipic, glutaric and succinic. These acids are produced as a by-product in the production of adipic acid in the manufacture of Nylon 6,6. The ratio of the acids is 15-25% adipic,

50-60% glutaric, 20-30% succinic. Esterification using isobutanol is then performed and the product is isolated following distillation. Coasol was introduced in the early part of last decade as a coalescing agent to the aqueous based coatings industry.

Coasol has excellent coalescing properties and is compatible with virtually all paint systems. It has very low vapor pressure at ambient temperature which ensures excellent plasticization throughout the film forming and drying process. Its extremely low odor also significantly improves the general view that aqueous systems have unpleasant odors. This is almost always as a result of the coalescing solvent. Its resistance to hydrolysis allows its use in a wide variety of formulations, including those, which require adjustment of pH to basic conditions.

It rapidly biodegrades, both aerobically and anaerobically, since, as an ester, it is readily attacked by ubiquitous bio-species. Finally, its toxicological profile is virtually benign, so in use it can be handled with confidence in most working environments. Its properties are summarized in Table 14.19.2.1.

**Table 14.19.2.1. Properties of Coasol**

| Physical Characteristic | Coasol | Comments |
|---|---|---|
| Boiling point | > 275°C | Not a VOC |
| Evaporation rate | < 0.01 (Butyl acetate = 1.0) | Slow, allowing excellent coalescence |
| Freezing point | < -55°C | No freeze-thaw issue |
| Water solubility | 600 ppm | |
| Hydrolytic stability | Very Good | No hydrolysis in normal use |
| Color | 5 Hazen units | Imparts no color to coatings |
| Coalescing efficiency | Excellent for most polymer systems | |
| Biodegradability | 80% in 28 days | Biodegradable |
| Odor | None discernible | |
| Toxicology | Oral LD50 (Rat) >16,000 mg/kg | Essentially non toxic |

## 14.19.2.7 RECENT ADVANCES IN DIESTER COALESCING SOLVENTS

We have also prepared the di-isopropyl esters of the higher adipic content stream. This has a vapor pressure similar to that of Coasol, but is slightly more water soluble. Finally, we have manufactured di-isopropyl adipate, which has the highest boiling point, the lowest vapor pressure and the lowest water solubility of all of this range of products. These preparations were undertaken to add to repertoire of products to suit the diverse requirements of the formulators of aqueous based systems.

In virtually all cases, the dibasic esters gave a significant improvement in efficiency in reducing the MFFT for a given quantity of additive. The attempts are made to offer a tailor made solution to each individual polymer system employed in the development of aqueous based systems. The dibasic esters of the AGS acids group offer the opportunity for fine tuning, with the added advantage of low odor, low toxicity and "excellent" VOC status.

### 14.19.2.8 Appendix - Classification of coalescing solvents

| Coalescent Type | Type of Species | Examples | Comments |
|---|---|---|---|
| Type A | Hydrocarbons | White Spirit | |
| Type AB | Diesters | DBE Dimethyl esters<br>DBE Diisobutyl esters<br>Di-isobutyl adipate<br>Di-isopropyl adipate<br>Dibutyl phthalate | Estasol, Du Pont<br>DBE's<br>Coasol, Lusolvan<br>Chemoxy new products |
| Type AB | Ester alcohols | Diol Monoesters | Texanol |
| Type ABC | Glycol esters &<br>Glycol ester ethers | PGDA<br>Butyldiglycol acetate | |
| Type ABC | Ether alcohols<br>& diethers | PnBS<br>2-Butoxyethanol<br>MPG Diethers | Dow Products<br>BASF and others<br>Proglides and glymes |
| Type C | Glycols | DEG<br>DPG<br>TEG | |

## 14.20 PETROLEUM REFINING INDUSTRY

George Wypych
**ChemTec Laboratories, Inc., Toronto, Canada**

The US petroleum refining industry generates sales of over $140 billion with only about 200 plants. It employs 75,000 people. About 90% of the products used in US are fuels of which 43% is gasoline. Figure 14.20.1 illustrates how the products breakdown. The process is described in detail in Chapter 3. Emissions of hydrocarbons to the atmosphere occur at almost every stage of the production process. Solvents are produced in various processes and they are also used to extract aromatics from lube oil feedstock, deasphalting of lubricating base stocks, sulfur recovery from gas stream, production of solvent additives for motor fuels such as methyl tert-butyl ether and tert-amyl methyl ether, and various cleaning operations. Emissions to atmosphere include fugitive emissions of the volatile components of crude oil and its fractions, emissions from incomplete combustion of fuel in heating system, and various refinery processes. Fugitive emissions arise from thousands of valves, pumps, tanks, pressure relief valves, flanges, etc. Individual leaks may be small but their combined quantity results in the petrochemical industry contributing the largest quantity of emissions and transfers.

Tables 14.20.1 and 14.20.2 give solvent releases and transfers data for the petroleum refining industry. Transfers are small fraction of releases which means that most wastes are processed on-site.

In addition to emissions to atmosphere, some plants have caused contamination of ground water by releasing cooling and process water.

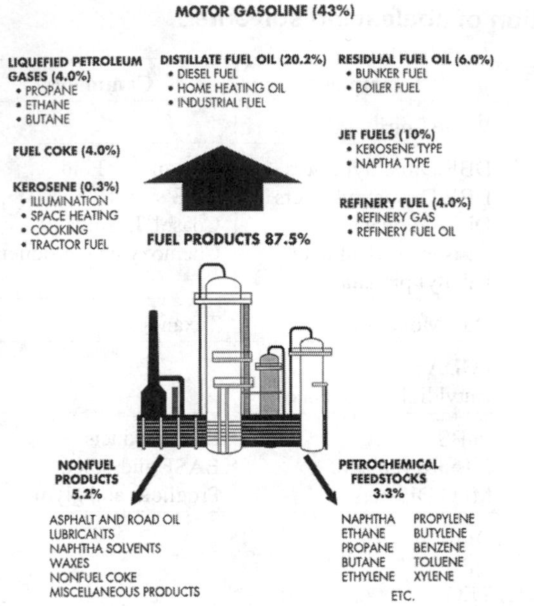


Figure 14.20.1. Diagram of production outputs from refineries. [Reproduced from EPA Office of Compliance Sector Notebook Project. Profile of the Petroleum Refining Industry. US Environmental Protection Agency, 1995.

Toluene, xylenes, and benzene constitute the majority of solvent emissions since they are native components of crude oil. Methyl ethyl ketone is also emitted in large quantities because of its use in lube oil dewaxing.

Pollution prevention will become increasingly important to the petroleum industry as federal, state and municipal regulations become more stringent and waste disposal cost rises. The industry estimates that to comply with 1990 Clean Air Act Amendments it will require investment of $35-40 billion. Actions required to decrease pollution include process equipment modification, waste segregation and separation, recycling, and better training and supervision.

**Table 14.20.1. Reported solvent releases from the petroleum refining industry in 1995 [Data from Ref. 2]**

| Solvent | Amount, kg/year | Solvent | Amount, kg/year |
|---|---|---|---|
| benzene | 1,750,000 | methyl isobutyl ketone | 110,000 |
| n-butyl alcohol | 23,000 | N-methyl-2-pyrrolidone | 280,000 |
| tert-butyl alcohol | 28,000 | methyl tert-butyl ketone | 1,380,000 |
| carbon tetrachloride | 17,000 | tetrachloroethylene | 21,000 |
| cresol | 75,000 | 1,1,1-trichloroethane | 50,000 |
| cyclohexane | 960,000 | trichloroethylene | 730 |
| dichloromethane | 8,000 | 1,2,4-trimethylbenzene | 420,000 |
| ethylbenzene | 600,000 | toluene | 4,360,000 |
| ethylene glycol | 46,000 | xylene | 2,330,000 |
| hexane | 3,000,000 | m-xylene | 170,000 |
| methanol | 540,000 | o-xylene | 150,000 |
| methyl ethyl ketone | 2,100,000 | p-xylene | 1,000,000 |

**Table 14.20.2. Reported solvent transfers from the petroleum refining industry in 1995 [Data from Ref. 2]**

| Solvent | Amount, kg/year | Solvent | Amount, kg/year |
|---|---|---|---|
| benzene | 160,000 | N-methyl-2-pyrrolidone | 4,000 |
| tert-butyl alcohol | 900 | methyl tert-butyl ketone | 34,000 |
| carbon tetrachloride | 1,000 | tetrachloroethylene | 900 |
| cresol | 130,000 | 1,1,1-trichloroethane | 6,500 |
| cyclohexane | 10,000 | 1,2,4-trimethylbenzene | 31,000 |
| ethylbenzene | 61,000 | toluene | 270,000 |
| ethylene glycol | 58,000 | xylene | 340,000 |
| hexane | 13,000 | m-xylene | 11,000 |
| methanol | 180,000 | o-xylene | 30,000 |
| methyl ethyl ketone | 30,000 | p-xylene | 7,000 |
| methyl isobutyl ketone | 3,500 | | |

## REFERENCES

1 EPA Office of Compliance Sector Notebook Project. Profile of the Petroleum Refining Industry. US Environmental Protection Agency, 1995.
2 EPA Office of Compliance Sector Notebook Project. Sector Notebook Data Refresh - 1997. US Environmental Protection Agency, 1998.

# 14.21 PHARMACEUTICAL INDUSTRY

## 14.21.1 USE OF SOLVENTS IN THE MANUFACTURE OF DRUG SUBSTANCES (DS) AND DRUG PRODUCTS (DP)

MICHEL BAUER
**International Analytical Department, Sanofi-Synthélabo, Toulouse, France**
CHRISTINE BARTHÉLÉMY
**Laboratoire de Pharmacie Galénique et Biopharmacie, Faculté des Sciences Pharmaceutiques et Biologiques, Université de Lille 2, Lille, France**

### 14.21.1.1 Introduction

Today the manufacturing of a pharmaceutical drug is almost totally the responsibility of:

- the chemical industry for the preparation of the drug substance (active principle) and the excipients used for preparing the DP (finished product)
- the pharmaceutical industry for the preparation of the DP itself.

For the reader who is not familiar with the pharmaceutical industry, a reminder of the key points is given below.

One or several active compounds (DS) are prepared by organic synthesis, extracted from vegetable, animals, micro-organisms or obtained by biotechnology. The DS is generally associated in the product with several excipients of chemical, mineral or biological in nature either as monomers or as polymers.

The goal is to formulate a processable DP stable over the time and allowing the active substance to be released in vitro and in vivo. Obviously the formulation is designed in relation to the route of administration:

- oral route solid dosage forms (e.g., tablets, capsules, etc.) drinkable solutions etc.
- ORL route (nasal solutions, spray)
- local route (suppositories, transdermal systems, eye-drop formulation, spray)
- intravenous and intramuscular route (injectable solution, lyophilizate, etc.)

At practically every step of the manufacture of the drug substance and the excipients, solvents including water are utilized. This is equally true for the preparation of the pharmaceutical formulations.

Ideally we would like to have available a universal stable solvent, ultrapure, non-toxic and which does not affect the solutes. This is an old dream of the alchemists who searched for a long time for the "Alkahest" or "Menstruum universal" as it was named by Paracelsus.[1]

The fact that a solvent is not a totally inert species allows it to play an important role in chemical equilibria, rates of chemical reactions, appearance of new crystalline forms, etc. and consequently contributes to the great wealth of compounds which the chemists are able to produce.

But of course, there are drawbacks in using solvents. Because there are not totally inert they may favor the formation of undesirable impurities in the intermediates of synthesis and in the DS. Regarding the manufacture of the DP, the solvents, including water, may induce either polymorphic transformations or formation of solvates (hydrates) which, after drying, could lead to a desolvated solvate with quite different physical properties impacting potentially either positively or negatively on the DP performance.[2]

Another crucial aspect which deserves to be discussed, is the notion of purity. Impurities present in solvents could have an impact on the stability of drugs, for example, or on the crystallization process.

Last, but not least, the toxicological aspects should be taken into account. Numerous solvents show different kinds of toxicity and this should be a matter of concern in relation to the health of workers exposed to them.[3] But ultimately residual solvents still present in the DS and DP have to be assessed and systematically limited.

We are now going to consider several aspects of the use of solvents in the manufacture of drug substances (DS) and drug products (DP) including their quality (purity) and influence on the quality, stability and physico-chemical characteristics of pharmaceutical products.

The issue of residual solvents in pharmaceutical products will be considered in Chapter 16.2 and will focus amongst other things on the corresponding ICH Guideline.[4]

## 14.21.1.2 Where are solvents used in the manufacture of pharmaceutical drugs?

### *14.21.1.2.1 Intermediates of synthesis, DS and excipients*

14.21.1.2.1.1 General points

Raw materials are now produced by the chemical industry and involve the use of solvents at different steps in their production. These materials are usually produced by:

- chemical synthesis
- an extraction process, a fermentation process
- or a biotechnology process

The goal of this chapter is, of course, not to deal with the criteria for selection in relationship to their use in particular chemical reactions or extraction processes but rather to stress that impurities present in solvents could have an impact on the purity of the substances obtained, on their stability and potentially on their safety. These three concepts are of paramount importance in the pharmaceutical field.

A list of solvents which are commonly used in the chemical industry[5] is presented in Table 14.21.1.1.

**Table 14.21.1.1. Solvents commonly used in the chemical industry**

| | | |
|---|---|---|
| **Alcohols** | **Ketones** | **Halogenated solvents** |
| Ethanol | Acetone | Ethylene bromide |
| Butanol | Methyl ethyl ketone | Chloroform |
| 2-Ethylhexanol | Methyl isobutyl ketone | Ethylene chloride |
| Isobutanol | Methyl isopropyl ketone | Dichloromethane |
| Isopropanol | Mesityl oxide | Tetrachloroethylene |
| Methanol | | Carbon tetrachloride |
| Propanol | **Ethers** | Trichloroethylene |
| Propyleneglycol | 1.4-Dioxane | |
| | Butyl ether | **Sulphur-containing** |
| **Amide** | Ethyl ether | Dimethylsulfoxide |
| Dimethylformamide | Diisopropyl ether | |
| | Tetrahydrofuran | **Aromatics hydrocarbons** |
| **Amine** | Tert-butyl methyl ether (MTBE) | Toluene |
| Pyridine | | Xylene |
| | **Nitriles** | |
| **Aliphatic hydrocarbons** | Acetonitrile | **Esters** |
| Cyclohexane | | Ethyl acetate |
| Hexane | **Water** | |

It is generally relatively easy to know for pharmaceutical industry the nature of solvents to be looked for in a DS because it produces itself the active component or because it can have by contract an access to the DS Mater File or because there is a compendial monograph giving occasionally some indications (e.g., search for benzene in carbomers).

As a consequence of the ICH Guideline Q3C[4] dealing with the residual solvents in pharmaceutical products (see Chapter 16.2), the use of solvents of class I (solvents to be avoided) like benzene is no more possible. It is known that carbopol resins (carbomer), used to modify the rheology of polar systems and as a binder in sustained release tablets, were up to now polymerized in benzenic medium. In the current quality of poloxamers it was possible to retrieve up to 1000 ppm of benzene. The ICH limit being 2 ppm, it was impossible to achieve this goal. The manufacturers have consequently developed new polymerization media[6] containing either ethyl acetate alone or a mixture of ethyl acetate and cyclohexane, the first one belonging to class III (no safety concern), the second one belonging to class II (ICH limit 3880 ppm).

14.21.1.2.1.2 Criteria of purity

This is a difficult matter. Purity in chemistry is an ideal concept referring to a situation where a product consists of one type of molecules only. This is a theoretical situation which can only be EXPERIMENTALLY approached more or less closely.[7]

The purity of a product is a relative notion and is dependant on the analytical methods used and their performances. More practically the quality finally chosen for a solvent will depend on the specific use for which this solvent is intended to be utilized.[8]

A solvent is considered sufficiently pure if it does not contain impurities able in nature and in quantity to interfere on the admissible quality of the product in the manufacture of which it participates.[7]

14.21.1.2.1.3 Solvents as reaction medium

In this case the solvents should have a range between the melting point and the boiling point as extended as possible and a good thermal and chemical stability. The purity should be of good degree but could depend on the step considered of the global synthesis.

As an example let us consider the case of the dimethylformamide (DMF).[5] If it is used in reactions evolving in anhydrous media, it will be mandatory to control the level of water at the ppm level. The specification regarding the water content will be of course loosened if the DMF in the chemical step considered is used in conjunction with water as reaction media.

We will see further that solvents contain actually a lot of chemical impurities which could be reactive vis a vis the main molecule undergoing the chemical reaction and leading to additional impurities other than those coming from the mechanism of reaction itself.

14.21.1.2.1.4 Solvents for crystallization

They should be carefully chosen in such a way that they show a high solubility at high temperature and a low solubility at low temperature of the substance to be crystallized or recrystallized.

Of course the solvent should be absolutely inert and of the highest achievable purity for at least two reasons:

- The first one being identical to the one mentioned for solvents as reaction media: possibility to produce other impurities.
- The second one being linked to the crystallization process itself. It is well known that the presence of impurities whatever the origin could have serious effects on the nucleation and growth process. We will tell a little bit more about that further in the text.

14.21.1.2.1.5 Solvents used for extraction and preparative chromatography

As in the precedent cases they have to be absolutely inert (as far as it is possible) and with a high degree of purity for the reasons already evoked. In case of preparative chromatography a special care will be taken concerning the chemical inertia to adsorbate[9] of the solvents constituting the mobile phase and the fact that impurities or additives contained in the solvents in a way not under control could impair significantly the reproducibility of the retention times.

14.21.1.2.1.6 Nature and origin of impurities contained in solvents[10]

It should be reminded here that a solvent used at the industrial level is rarely pure (we mean here no impurity analytically detectable).

Industrial solvents may contain:

- impurities coming from their origin or their manufacturing process
- impurities originating from the container during transportation
- stabilizers
- denaturing agents
- impurities resulting from a transformation of the solvent during the chemical reaction

These impurities or side products should be look for as far as it is possible when assessing the purity of the solvent. In fact they could be less volatile than the main solvent and could finally concentrate in the pharmaceutical product.

We will now review shortly the nature of all these kinds of impurities of the most often used solvents.

14.21.1.2.1.6.1 Impurities coming from the origin or the manufacturing process of the solvent[1,10]

**Table 14.21.1.2. Solvent impurities**

| Class of solvents | Possible impurities (according to the manufacturing process) |
|---|---|
| | *Hydrocarbons* |
| Toluene | Methylthiophene, benzene, paraffinic hydrocarbons |
| Xylene | Mixture of ortho, meta and para isomers, paraffinic hydrocarbons, ethyl benzene, sulfur compounds |
| Cyclohexane | Benzene, paraffinic hydrocarbons, carbonyl compounds |
| | *Halogenated compounds* |
| Dichloromethane | Chloroform, carbon tetrachloride, chloromethane |
| Chloroform | Chlorine, carbonyl chloride (phosgene), dichloromethane, carbon tetrachloride, hydrogen chloride |
| Carbon tetrachloride | Chlorides, chlorine, carbon disulfide |
| | *Alcohols* |
| Methanol | Water, acetone, formaldehyde, ethanol, methyl formate, dimethylether, carbon dioxide, ammonia |

| | |
|---|---|
| Ethanol | Aldehydes, ketones, esters, water, ethyl ether, benzene (if anhydrous ethanol) |
| 2-propanol | Water, peroxides |
| N.B.: Some alcohols obtained by fermentation could contain pesticides. It is necessary to obtain from the purchaser some guaranty in requiring limit contents (expressed in Parathion e.g.). | |
| *Aliphatic ethers/cyclic ethers* | |
| Ethylether/isopropyl ether/monoalkylated ethers/ethylene glycol/diethylene glycol/etc. | Alcohols (from which they are prepared), water, corresponding aldehydes, peroxides |
| Tetrahydrofuran | Water, peroxides |
| Dioxane | Acetaldehyde, water, acetic acid, glycol acetal paraldehyde, crotonaldehyde/peroxides |
| *Ketones* | |
| Acetone | Methanol, acetic acid, water |
| *Esters* | |
| Methyl acetate | Acetic acid, water, methanol |
| Ethyl acetate | Acetic acid, ethanol, water |
| *Amides* | |
| Formamide | Formic acid, ammonium formate, water |
| N,N-Dimethylformamide | N-Methylformamide, formic acid, water |
| *Nitriles* | |
| Acetonitrile | Acetamide, ammonium acetate, ammoniac, water, toluene |
| *Nitro compounds* | |
| Nitrobenzene | Nitrotoluene, dinitrothiophene, dinitrobenzene, aniline |

14.21.1.2.1.6.2 Impurities originating from the container during transportation
It relates to contamination coming from tankers or drums not correctly cleaned. These concerns of course solvents of low quality conveyed in industrial quantity. In case of utilization of such solvents, the user has to bear in mind that some incidents or uncommon behavior may find an explanation based on this considerations.
14.21.1.2.1.6.3 Stabilizers
It is of course very difficult to know every stabilizer used. There is here an important problem of confidentiality. We quote thereafter some of them which are well known.

**Table 14.21.1.3 Stabilizers used in selected solvents**

| Solvents | Stabilizers |
|---|---|
| Dichloromethane | Ethanol, 2-methyl-but-2-ene |
| Chloroform | Ethanol (1% V/V) for avoiding the phosgene formation, 2-methyl-but-2-ene |
| Diethylether | 2,6-di-tert-butyl-4 methylphenol (BHT) |
| Tetrahydrofuran | BHT, p-cresol, hydroquinone, calcium hydride |

14.21.1.2.1.6.4 Denaturing agents

This process is relevant primarily to ethanol. Common denaturing agents are: methanol, isopropanol, ethyl acetate, toluene.

14.21.1.2.1.6.5 Transformation of the solvent during the chemical reaction

Solvents are rarely chemically inert. During the reactions where solvents are involved, they can undergo chemical transformation generating impurities which can be found, for example in the DS.

This is a huge field which cannot be exhaustively covered. We give below a few examples of well-known side reactions.

- Acetone in acidic media is easily transformed into mesityl oxide:

$$2\ CH_3\text{-}CO\text{-}CH_3 \xrightarrow{H^{\oplus}} CH_3\text{-}CO\text{-}CH{=}CH{<}^{CH_3}_{CH_3}$$

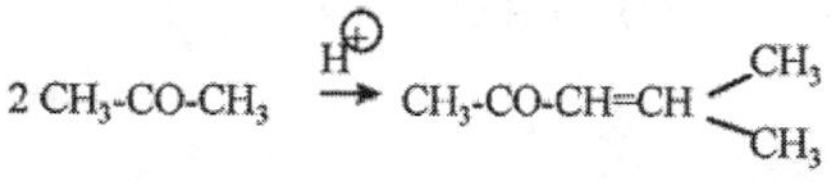

So do not forget to test for it when performing residual solvents analysis on drugs.

- In basic medium the diketone-alcohol is obtained:

$$2\ CH_3\text{-}CO\text{-}CH_3 \xrightarrow{OH^{\ominus}} CH_3\text{-}\underset{O}{\overset{}{C}}\text{-}CH_2\text{-}C{<}^{CH_3}_{OH}\text{-}CH_3$$

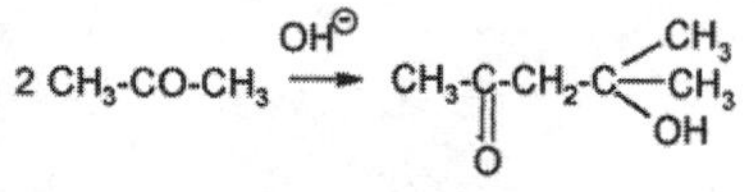

- DMF can be hydrolyzed in presence of hydrochloric acid:

$$HCl + H\text{-}CO\text{-}N{<}^{CH_3}_{CH_3} \longrightarrow HCOOH + CH_3\text{-}\overset{CH_3}{\underset{H}{N^+}}\text{-}H\quad Cl^-$$

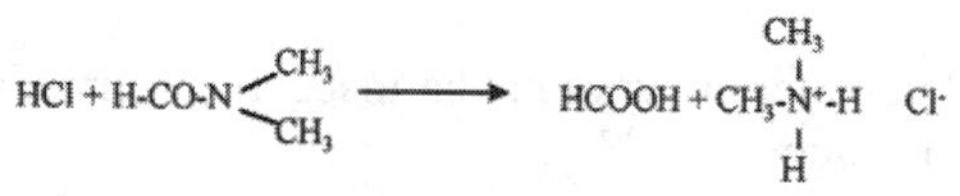

- Acids undergoing reaction in alcoholic media can be partially transformed into esters
- Transesterification reaction. Take care when, for example, recrystallization has to be performed for a molecule containing an ester group:

$$R\text{-}COOR_1 + R_2\text{-}OH \longrightarrow R\text{-}COOR_2 + R_1\,OH$$

- Aldehydes (even ketones) can be transformed in alcoholic solutions into ketals:

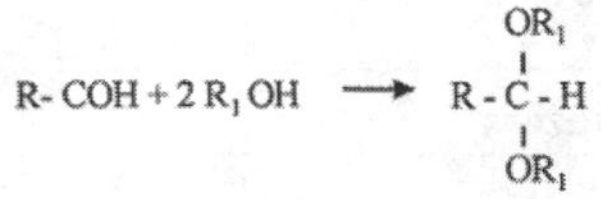

$$R\text{-}COH + 2\,R_1OH \longrightarrow R\text{-}\underset{\underset{OR_1}{|}}{\overset{\overset{OR_1}{|}}{C}}\text{-}H$$

- From time to time the solvent can react in lieu of the reagent.

In a synthesis aimed to prepare 3-chloro-1-methoxy-2-propan-2-ol starting from chloromethyloxirane and methanol, traces of ethanol present in methanol gave the corresponding ethoxylated compound:

Main reaction   $H_2C\text{-}CH\text{-}CH_2\text{-}Cl + CH_3OH \longrightarrow CH_3\text{-}O\text{-}CH_2\text{-}CHOH\text{-}CH_2Cl$

Side reaction   $H_2C\text{-}CH\text{-}CH_2\text{-}Cl + C_2H_5OH \longrightarrow C_2H_5\text{-}O\text{-}CH_2\text{-}CHOH\text{-}CH_2Cl$

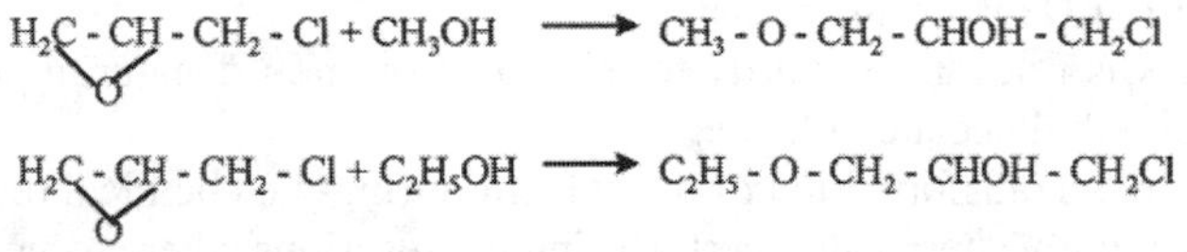

Another and final example concerns the preparation of a urea derivative using the reaction of an amine and isocyanate in presence of isopropanol:

$$R\text{-}NH_2 + O\text{=}C\text{=}N\text{-}CH_3 \xrightarrow[\;CH_3\text{-}CHOH\text{-}CH_3\;]{} R\text{-}NH\text{-}CO\text{-}NH\text{-}CH_3$$

$$+ R\text{-}NH\text{-}CO\text{-}O\text{-}\underset{\underset{CH_3}{|}}{CH}\text{-}CH_3$$

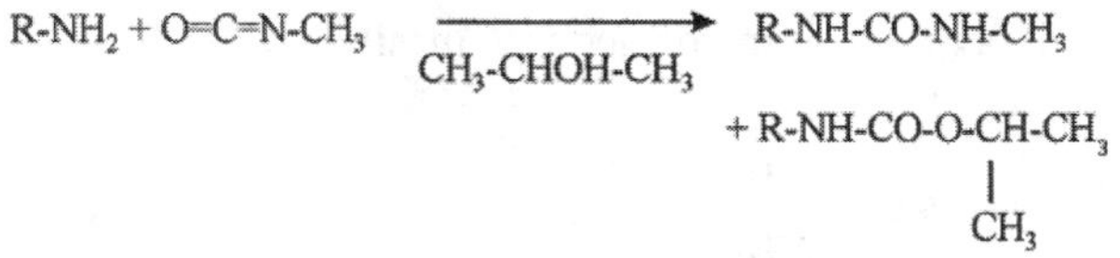

Other examples could be found. These obvious examples stress the need for close collaboration between chemists and analysts when elaborating chemical syntheses and corresponding quality control monographs.

### 14.21.1.2.2 Drug products[11,12]

14.21.1.2.2.1 General points

Because ultimately it is the DP which is administered to the patient, it is necessary to have the quality of the solvents potentially used in the design of pharmaceutical formulations under control.

14.21.1.2.2.2 Areas of utilization

Solvents including water are used in different ways in pharmaceutical formulation:
- either as a part of the final drug product:
  injectables,
  drinkable solutions,
  patches, sprays,
  microemulsions
- or used as an intermediary vehicle which is removed at the end of the process:
  granulation
  coating
  sugar coating
  microencapsulation

We have listed in Table 14.21.1.4. the most commonly used solvents.

**Table 14.21.1.4 Solvents used in formulation**

| | |
|---|---|
| Water | Dichloromethane |
| Ethyl acetate | Chloroform |
| Ethyl alcohol (denaturated with butanol and isopropanol) | Hexane |
| Isopropyl alcohol (denaturated with methyl ethyl ketone) | Cyclohexane |
| Methanol | Polyethylene glycol (low molecular weight) |
| Acetone | |

Manufacturers try progressively to replace the formulations using organic solvents such as chloroform, dichloromethane, cyclohexane belonging to the class 2 (ICH classification see Chapter 16.2), for example by developing aqueous coatings.

14.21.1.2.2.3 What should be the quality?

Taking into account the fact that the solvents used in the DP manufacturing process, either as a component of the formulation or as a residual solvent, will be absorbed by the patient, their quality must be of the highest standard. From a regulatory point of view, in almost every country if not all, it is mandatory to use solvents covered by a pharmacopoeial monograph (e.g., European Pharmacopoeia, USP, JP, local Pharmacopoeias). Some examples are given below.

## 14.21.1.3 Impacts of the nature of solvents and their quality on the physicochemical characteristics of raw materials and DP.

### *14.21.1.3.1 Raw materials (intermediates, DS, excipients)*

The impurities contained in the solvents could have several effects on the raw materials:

- When the solvents are removed, non-volatile or less volatile impurities will be concentrated in raw materials.
- They can induce chemical reactions leading to side products.
- They can affect the stability of the raw material considered.
- They can modify substantially the crystallization process.

14.21.1.3.1.1 Concentration of less volatile impurities

Due to the potential concentration of these impurities, they should be tested for in both DS and excipients and even in intermediates of synthesis if the latter constitute the penultimate step of the synthesis and if solvents belong to class 1 or class 2 solvent (see Chapter 16.2).

14.21.1.3.1.2 Side reactions

This case has already been illustrated (see paragraph 14.21.1.2.1.6.5). The skills of the chemist together with those of the analyst are needed to ensure that the presence of unexpected impurities can be detected. By way of example, the reactions involving the keto-enol tautomerism deserve to be mentioned. The equilibrium is very sensitive to the solvent so that the presence of other solvents as impurities in the main solvent can modify the keto-enol ratio leading to irreproducibility in the chemical process.[13,14]

14.21.1.3.1.3 Consequences for stability

Some solvents, as mentioned in the paragraph 14.21.1.2.1.5, can contain very active entities such as aldehydes and peroxides. For example, if the raw material contains primary or secondary amines and/or is susceptible to oxidation or hydrolysis, it is likely that degradants will be formed over the time, reducing potentially the retest date of the raw material. The same situation could affect the DP and therefore two examples are given in paragraph 14.21.1.3.2.

14.21.1.3.1.4 Solvent purity and crystallization

This issue is may be less known except for chemists working in this specialized area. Because the consequences can be important for the processability, stability and occasionally bioavailability of the drug substance in its formulation, it is relevant to comment on this subject.

14.21.1.3.1.4.1 Role of the nature and the quality of solvent on crystallization

Most of the drugs on the market are obtained as a defined crystalline structure and formulated as solid dosage forms. It is well known that a molecule can crystallize to give different crystalline structures displaying what is called polymorphism. The crystal structures may be anhydrous or may contain a stoichiometric number of solvent molecules leading to the formation of solvates (hydrates in case of water molecules). Pseudopolymorphism is the term used to describe this phenomenon.

Another characteristic which plays a major role in the overall processability of the DS for DP manufacture is the "crystal habit". This term is used[15,16] to describe the overall shape of crystals, in other words, the differing external appearance of solid particles which have the same internal crystalline structure.

Both structures (internal, external) are under the control of different parameters including the nature of the solvent used and its quality. The role of the solvent itself in the overall crystallization process, including the determination of the crystal structure and the crystal habit is well known.[17] But it is equally worth noting that impurities coming from:

- the product to be crystallized
- the solvent used
- the environment

can selectively affect the nucleation process and the growth rates of different crystal faces.[17-21] They can be selectively adsorbed to certain faces of the polymorphs thereby inhibiting their nucleation or retarding their growth to the advantage of others. Crystal shape (habit) can also be modified by a solvent without polymorphic change. Additives or impurities can block, for a defined polymorph, the growth rate of certain faces leading e.g. to needles or plates. It is possible to introduce deliberately additives to "steer" the crystallization process. An interesting example of this crystal engineering strategy have been published for e.g., adipic acid[22] or acetaminophen.[23]

14.21.1.3.1.4.2 Solvent-solid association/overview

After the crystallization of the product, solvents must be removed in order to obtain the minimum amount of residual solvents compatible with safety considerations and/or physicochemical considerations including stability, processability and occasionally microbiological quality (see Chapter 16.2). Different situations can be encountered.

14.21.1.3.1.4.2.1 Solvent outside the crystal

The solvent remains outside the crystals at the time of crystal formation. It is adsorbed on the surface or in the crystal planes. In the first case, the solvent is easily removed. But in the second case, if a cleavage plane exists, the drying process can be very difficult. Two methods can be used to try to remove this type of residual solvent almost completely.

- Displacement by water vapor in an oven, keeping in mind that this method may introduce some degradation leading to processability problems.[11,24]
- Extraction by supercritical $CO_2$[25,26] but the extraction power of $CO_2$ is basically limited to slightly polar solvents.

14.21.1.3.1.4.2.2 Solvent inside the crystal

The solvent remains inside the crystalline structure. Three situations can arise.

14.21.1.3.1.4.2.2.1 Occluded solvents

During rapid crystallization, some degree of disorder (amorphous phases, crystalline defects) can arise, creating pockets where residual solvent can be occluded. Through a process of dissolution/recrystallization this "hole" moves towards the external faces of the crystal releasing the solvent at the end. This phenomenon is more frequent for large crystals (500 μm/600 μm) but rare for smaller crystals (1 - 100 μm). The solvent odor which is detected when opening a drum or a bag containing a substance which was dried in the normal way can be explained by this mechanism.

14.21.1.3.1.4.2.2.2 Solvates

At the end of the crystallization process, the substance can be isolated as a solvate (hydrate), i.e., as a pseudopolymorph. The solvates generally have quite different physicochemical properties from the anhydrous form. Their stability can be questionable and in any case deserves to be investigated. In some cases it is possible to remove the solvent from the crystal without changing the structure of the lattice leading to an isomorphic desolvate which displays a similar X-ray diffraction pattern to that of the parent compound.[2,27] The lattice of the desolvated solvate is in a high energy state relative to the original solvate structure. A better dissolution rate and compressability can be expected,[28] but the drawbacks are hygroscopicity and physico-chemical instability. The lattice could undergo a relaxation process over time which increases the packing efficiency of the substance by reducing the unit cell volume.

When developing a new chemical entity all these aspects have to be considered to avoid unpleasant surprises during development or once the drug is on the market. Due to the need for process scale-up and of making the manufacturing process more industrial, changes are introduced especially in the crystallization and the drying processes, (e.g., change from static drying to dynamic drying). Because the drying is a particularly disturbing process for the integrity of the lattice, defects and/or amorphous phases may be created favoring subsequent polymorphic or pseudopolymorphic transformations of the crystalline form developed so far, if it is not the most stable one.

14.21.1.3.1.4.2.2.3 Clathrates

In contrast to solvates, clathrates do not show any stoichiometric relationship between the number of molecules of the substance and the number of molecules of solvent. Clathrates actually correspond to a physical capture of solvent molecules inside the crystal lattice without any strong bonds including hydrogen bonds. Molecules of one or several solvents can be trapped within the crystalline structure as long as the crystallization has been performed with a pure solvent or a mixture. The case of the sodium salt of warfarin giving "mixed" clathrates with water and isopropyl alcohol is well known and the existence of 8/4/0 or 8/2/2 proportions has been shown.[11]

It is fairly obvious that some powder properties like wettability can be modified by the formation of clathrates. Because their formation is not easy to control, some batch to batch inconsistency may be expected in this situation.

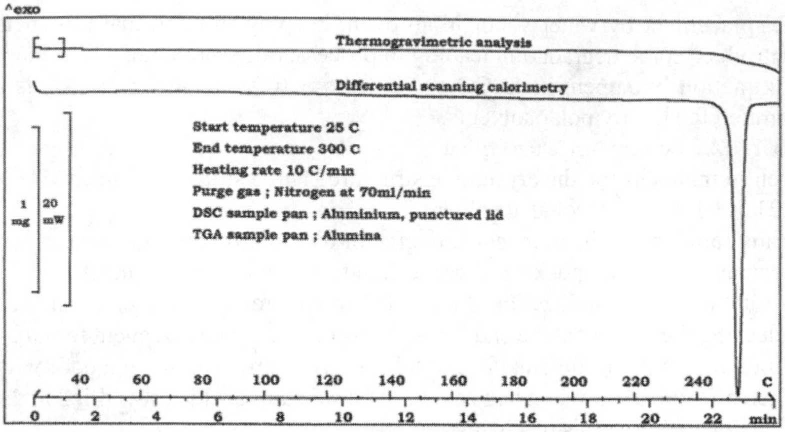


Figure 14.21.1.1. DSC and TG patterns.

In order to demonstrate the internal character of the clathrate relative to the crystal lattice, the example of a molecule developed in the laboratory of one of the authors is given below.

Figure 14.21.1.1 shows the DSC pattern of a molecule with the melting event at 254°C and the corresponding TG pattern obtained at the same temperature scanning rate. At the time the melting occurs, a *loss of weight is observed* corresponding to the loss of 0.2 % of isopropanol. The nature and the amount of the solvent have been confirmed by GC after dissolving the substance.

### *14.21.1.3.2 Drug product*

As for the DS, the solvents used for DP manufacture can produce some negative effects by themselves or through their own impurities.

For liquid or semi liquid formulations, the formulator has to ensure that the solvents themselves do not display chemical interactions with the DS or the excipients. Everything which has been said in paragraphs 14.21.1.2.1.5 and 14.21.1.3.1 remains true here.

14.21.1.3.2.1 Interaction of impurities contained in the solvent

As said in paragraph 14.21.1.3.1 with the DS, impurities contained in the solvent especially if they are strongly reactive, like aldehydes or peroxides, can promote formation of degradants. Regarding aldehydes, the publication of Bindra and all[29] should be mentioned. It relates to the degradation of the o-benzylguanine in an aqueous solution containing polyethylene glycol 400 (PEG 400). This type of solvent very often contains formaldehyde, which can lead to the formation of a precipitate over time:

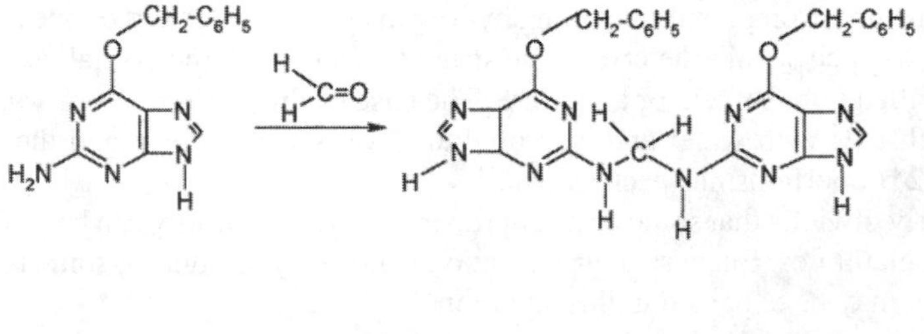

PEG can also contain peroxides which can initiate over time, the formation of degradants via an oxidation process. Several publications have dealt with this phenomenon.[30,31]

14.21.1.3.2.2 Interaction with the container

When the formulation is a solution which is prepared from water and different organic solvents, it is mandatory to investigate possible interactions between the medium and the container especially if the latter is of polymeric nature (PVC-PVDC, polyethylene, etc.) with or without elastomeric stoppers. A thorough investigation is necessary including:

- an examination of the solution for plasticizers, antioxidants, monomers and oligomers, mineral impurities, potentially extracted from the container,
- the evaluation of the absorption by the container of components (DS, excipients) contained in the solution.

In the first case, the migration of impurities into the solution could initiate physicochemical instability and possibly some potential toxicity.

In the second case, a decrease in the content of the DS and/or some excipient (e.g., organic solvents added to promote the solubility) could lead to some loss of therapeutic efficacy and in some case to physical instability (precipitation).

14.21.1.3.2.3 Solvates formation during the solid dosage form manufacture

During the granulation process it is possible that the DS (occasionally the excipient) could transform into a solvated crystalline structure (solvate, hydrate). During the drying process, different situations can occur:

- The solvate is poorly stable and the solvent is easily removed leading to either the original polymorphic form but creating a certain degree of disorder in the crystalline structure or to what is called a "desolvate solvate" form. In this last case, also named "isomorphic desolvate", the desolvated solvate retains the structure of its parent solvated form. The X-ray diffraction patterns look similar between the parent and the daughter forms. In this situation we have the creation of a molecular vacuum which could substantially impact on the stability, hygroscopicity and mechanical characteristics of the DS and finally of the DP.
- The solvate is stable within the formulation: we then have in a sense a new chemical entity. The properties of the solvate could be entirely different (solubility, kinetics of dissolution, stability, processability, etc.) and the consequence could be either positive or negative. The case where the kinetics of dissolution are affected by the formation of solvates should always be investigated. Papers on this subject have been published for molecules such as lorazepam,[32] hydrocortisone,[33] cephalexin,[32] etc.
- Obviously, as in the case of raw materials (14.21.1.3.1.4.2.2.3) clathrate formation should be considered in order to explain possible batch to batch inconsistency.

### *14.21.1.3.3 Conclusions*

We have seen that the solvent, far from being inert, plays a key role by itself and occasionally via its own impurities in different ways which are important for pharmaceutical development. We will now discuss how to set up sound specifications for solvents in relation to their field of use.

### 14.21.1.4 Setting specifications for solvents

#### *14.21.1.4.1 Solvents used for the raw material manufacture*

For the raw materials we should distinguish between solvents:

- used during the synthesis
- and those used for the last step of the manufacture corresponding very often to the crystallization process.

As a rule of thumb, the specifications set for solvents used for the crystallization step will be more stringent than those used during the synthesis.

For the intermediates of synthesis, if the origin of the solvent is under control (e.g., existence of contracts/Quality Assurance audits) a simplified monograph is completely adequate (see Table 14.21.1.5) as long as the supplier provides a detailed certificate of analysis where impurities (including solvents) are properly specified with acceptable limits. If the same solvent is used for the crystallization step additional purity tests are necessary (Table 14.21.1.6).

**Table 14.21.1.5. Examples of monographs**

| |
|---|
| *Solvents used during the synthesis (of well controlled origin)* |
| Character/Appearance |
| Identification (IR, GC or $n_D^{20}$) |
| Purity GC (generally not less than 98%) |

**Table 14.21.1.6. Examples of monographs**

| |
|---|
| *Final crystallization solvents (of well controlled origin)* |
| Character/Appearance |
| Identification (IR, GC or $n_D^{20}$) |
| Tests |
| Water content (0.1 to 0.5 depending on solvent type) |
| Residue on evaporation: not more than 0.01 per cent |
| Purity GC not less than 99 % (examine for denaturing agents and other potential impurities) |

For economic reasons, it may be necessary to recycle solvents. If so, the containers should be fully identified in terms of storage:

- If solvents can be efficiently purified (e.g., by redistillation) they must comply with the same specifications as those of fresh solvents and consequently can be used in any synthesis.
- If they still contain volatile impurities resulting from the reaction they come from, they can be recycled only for this reaction. In this case, the impurities should be identified and their possible impact on the reaction evaluated. In Tables 14.21.1.7 and 14.21.1.8 we have summarized possible specifications for a fresh batch of ethyl acetate used for a defined chemical reaction and those for the recycled solvent.

We recommend working with reliable solvent suppliers who can give every assurance on the quality of solvents provided to avoid any "unpleasant surprises".

Water should be mentioned separately. If it is used during the synthesis of intermediates the quality "drinking water" can be used without any problems. But if water is used during the last step of the process, its quality must be in compliance with the requirements of purified water as they are described in several pharmacopoeias. In Table 14.21.1.9 the requirements for the Ph. Eur and USP are given as examples. Purified water is generally obtained from the drinking water. It undergoes demineralization by either distillation or an ion-exchange process. Particular attention has to be paid to microbiological quality.

**Table 14.21.1.7. Example of monograph applied to new solvent: Ethyl acetate**

**Specifications**

| Controls | Standards |
|---|---|
| Characters | Clear liquid, colorless |
| *Identification* | |
| A - Infrared spectrum | Complies |
| or | |
| B - Refractive index | 1.370 to 1.373 |
| or | |
| C - Gas chromatography | Complies |
| *Assay* | |
| Ethyl acetate (purity) | Not less than 99.5% |

**Table 14.21.1.8. Example of monograph applied to recycled solvent: Ethyl acetate**

**Specifications**

| Controls | Standards |
|---|---|
| Characters | Clear liquid, colorless |
| *Identification* | |
| Gas chromatography | Complies |
| *Tests* | |
| Water content | Not more than 2.0% |
| Related substances | |
| Methanol | Not more than 1.0% |
| Ethanol | Not more than 2.0% |
| Ethyl chloride | Not more than 2.0% |
| Others impurities (sum) | Not more than 2.0% |
| *Assay* | |
| Ethyl acetate (purity) | Not less than 99.5% |

#### *14.21.1.4.2 Solvents used for the DP manufacture*

There is no other choice than to use the quality of solvents defined by a Pharmacopoeia. It is true that there are still discrepancies between the pharmacopoeias of different countries. It is hoped that the ICH process dealing with the harmonization of quality, safety and efficacy amongst three main zones of the world (EU, USA, Japan) will progressively reduce the remaining differences in dossiers submitted to Regulatory Authorities and the way the data are evaluated.

As examples, Tables 14.21.1.10, 14.21.1.11, and 14.21.1.12 summarize specifications for acetone, ethanol and isopropanol given by the Ph. Eur. and USP. As can be seen only the Ph. Eur. monograph makes reference to volatile impurities to be tested for by GC.

### 14.21.1.5 Quality of solvents and analysis

The solvents, including water, are used in almost every area of analytical sciences: spectroscopy, chromatography, potentiometry, electrochemistry. They should be characterized by a set of properties making them suitable for use for their intended purpose.

#### *14.21.1.5.1 Quality of solvents used in spectroscopy*

As a general requirement, the solvents used in spectroscopy should be transparent and stable towards the relevant range of wavelengths. They should be able to dissolve the substance to be examined and not contain impurities affecting the stability of the substance or the validity of the method (selectivity, repeatability, limit of detection, analytical response). Theoretically the solvent chosen should have minimal interaction with the solute. But what could be seen as a disadvantage could also be an important source of structural information. What is called the solvent effect can help in UV, IR and NMR spectroscopies[34] e.g. in struc-

**Table 14.21.1.9. Water quality**

| European Pharmacopoeia- Suppl. 2000 | | USP 24 | |
|---|---|---|---|
| WATER, PURIFIED | | PURIFIED WATER | |
| Controls | Standards | Controls | Standards |
| • Purified water in bulk | | • Purified water produced on site for use in manufacturing | |
| Production | | | |
| . Total viable aerobic count | Not more than 100 micro organisms per ml | | |
| . Total organic carbon | Not more than 0.5 mg/ml | Cf. Tests | |
| . Conductivity | Not more than 4.3 µS-cm$^{-1}$ | Cf. Tests | |
| . Packaging and storage | In conditions designed to prevent growth of micro organisms | | |
| Characters | A clear, colourless and tasteless liquid | | |
| Tests | | | |
| . Nitrates | Not more than 0.2 ppm | | |
| . Heavy metals | Not more than 0.1 ppm | | |
| . Aluminium | Not more than 10 µg/l | | |
| . Bacterial endotoxins | Not more than 0.25 I.U. of endotoxin per ml | | |
| | | . Total organic carbon | Complies |
| | | . Water conductivity | Complies |
| • Purified water in containers | | • Purified water packaged in bulk for industrial use | |
| Characters | A clear, colourless and tasteless liquid | | |
| Tests | | | |
| . Nitrates | Not more than 0.2 ppm | | |
| . Heavy metals | Not more than 0.1 ppm | | |
| . Aluminium | Not more 10 µg/l | | |
| . Bacterial endotoxins | Not more than 0.25 I.U of endotoxin per ml | | |
| . Acidity or alkalinity | Complies | | |
| . Oxidisable substances | Complies | . Oxidisable substances | Complies |
| . Chlorides | Complies | . Chlorides | Complies |
| . Sulphates | Complies | . Sulphates | Complies |
| . Ammonium | Not more than 0.2 ppm | . Ammonia | Complies |
| . Calcium and magnesium | Complies | . Calcium | Complies |
| . Residue on evaporation | Not more than 0.001 % | | |
| . Microbial contamination | Not more than $10^2$ micro organisms per ml | | |
| - | - | . pH | 5.0 to 7.0 |
| - | - | . Carbon dioxide | Complies |
| Packaging and storage | Stored in conditions designed to assure the required microbiological quality | Packaging and storage | Preserve in suitable, tight containers |

ture elucidation. In the book by Reichardt,[1] data regarding the cut-off points of solvents commonly used in UV/visible spectroscopy are provided. The cut-off point is defined as the wavelength in the ultraviolet region at which the absorbance approaches 1.0 using a 1-cm cell path with water as the reference. In the same way the range of transparency for IR-solvents are given. Complete IR-spectra of organic solvents can be found in the "Stadler IR spectra handbook of common organic solvents" . The solvents suppliers usually provide catalogues including a "spectroscopic grade" allowing the user to make a sound choice.

In the case of the NMR spectroscopy problems arise with the residual protonated part of deuterated solvents ($^1$H-NMR) and the $^{13}$C-NMR absorption bands of compounds used as solvents. References can be found[1] where detailed data are given regarding these points.

**Table 14.21.1.10. Specification for acetone**

| European Pharmacopoeia, Suppl. 2000 | | NF 19 | |
|---|---|---|---|
| ACETONE | | | |
| Controls | Standards | Controls | Standards |
| | | | Not less than 99 % calculated on the anhydrous basis |
| Characters | A volatile clear, colourless, miscible with water, alcohol and ether. The vapour is flammable | | |
| Identification | | | |
| . A | Complies | | |
| . B | Complies | | |
| | | IR | Complies |
| Tests | | | |
| . Appearance of solution | Complies | | |
| . Acidity or alkalinity | Complies | | |
| . Relative density | 0.790 to 0.793 | . Specific gravity | Not more than 0.789 |
| . Related substances : | | | |
| - Methanol | Not more than 0.05 % (V/V) | | |
| - 2-propanol | Not more than 0.05 % (V/V) | | |
| - Additional impurity | Not more than 0.05 % (V/V) | | |
| . Matter insoluble in water | Complies | | |
| . Reducing substances | Complies | . Readily oxidizable substances | Complies |
| . Residue on evaporation | Not more than 1 mg (50 ppm) | . Non volatile residue | Not more than 2 mg (0.004 %) |
| . Water (semi-micro determination) | Not more than 3 g/l | . Water (GC) | Not more than 0.5 % |
| | | Assay (GC) | Not less than 99.0 % calculated on the anhydrous basis |
| Storage | Store protected from light | Packaging and storage | Preserve in tight containers remote from fire |
| Impurities | | | |
| . A | Methanol | | |
| . B | 2-propanol | | |

A common problem which can be met regularly in chemistry is the identification of signals from common contaminants in solvents of medium quality (see paragraph 14.21.1.2.1.6.1). Gottlieb and Col[35] have published data on NMR chemical shifts of trace impurities contained in common laboratory solvents, making NMR spectroscopy the instrument of choice as a tool for routine quality control.

### *14.21.1.5.2 Quality of solvents used in chromatography*

The aim of this paragraph is not to focus on strategies for solvent selection in order to achieve extraction or liquid chromatography. A detailed literature review has been published by Barwick[36] on this matter allowing the user to design relevant methodology in any

## Table 14.21.1.11. Specification for ethanol

| European Pharmacopoeia, Suppl. 2000 | | USP 24 | |
|---|---|---|---|
| ETHANOL (96 %) | | ALCOHOL | |
| Controls | Standards | Controls | Standards |
| Definition | Not less than 95.1 % V/V (92.6 % m/m) and not more than 96.9 % V/V (95.2 % m/m) of ethanol at 20°C and water | Definition | Not less than 92.3 % and not more than 93.8 % by weight, corresponding to not less than 94.9 % and not more than 96.0 %, by volume at 15.56° |
| Characters | A colourless, clear, volatile, flammable liquid, hydroscopic, miscible with water and methylene chloride. It burns with a blue, smokeless flame. It boils at about 78°C. | | |
| Identification | | | |
| A - Relative density | 0.8051 to 0.8124 | | |
| B - IR | Complies | | |
| C | Complies | A | Complies |
| D | Complies | B | Complies |
| Tests | | Tests | |
| . Appearance | Complies | | |
| . Acidity or alkalinity | Complies | . Acidity | Complies |
| . Relative density | 0.8051 t o 0.8124 | . Specific gravity | 0.812 to 0.816 indicating between 92.3% and 93.8% by weight or between 94.9% and 96.0% by volume |
| . Absorbance | Complies | | |
| . Volatile impurities (GC) | | | |
| - methanol | Not more than 200 ppm (v/v) | - methanol | Complies |
| - sum of acetaldehyde and acetal | Not more than 10 ppm v/v expressed as acetaldehyde | - aldehydes and other foreigh organic substances | Complies |
| - benzene | Not more than 2 ppm v/v | | |
| - 4-methyl pentan-2-ol | Not more than 300 ppm | | |
| | | . Amyl alcohol and non volatile, carbonizable substance | Complies |
| | | . Limit of acetone and isopropyl alcohol | Complies |
| | | . Limit of non volatile residue | Not more than 1 mg |
| | | . Water insoluble substances | Complies |
| Storage | Store in a well-closed container, protected from light | Packaging and storage | Preserve in tight containers, remote from fire |

Impurities as per European Pharmacopoeia, Supplement 2000:
A - 1,1-diethoxyrethane (acetal); B - Acetaldehyde; C - Acetone; D - Benzene; E - Cyclohexane; F - Methanol; G - Butan-2-one (methyl ethyl ketone); H - 4-methylpentan-2-one (methyl isobutyl ketone); I - Propanol; J - Propan-2-ol; K - Butanol; L - Butan-2-ol; M - 2-methylpropanol (isobutanol); N - Furan-2-carbaldehyde (furfural); O - 2-methylpropan-2-ol (1,1-dimethyl alcohol); P - 2-methylbutan-2-ol; Q - Pentan-2-ol; R - Pentanol; S - Hexanol; T - Heptan-2-ol; U - Hexan-2-ol; V - Hexan-3-ol

**Table 14.21.1.12. Specification for isopropanol**

| European Pharmacopoeia, Suppl. 2000 | | USP 24 | |
|---|---|---|---|
| ISOPROPYL ALCOHOL | | | |
| Controls | Standards | Controls | Standards |
| | | | Not less than 99.0 % |
| Characters | A clear, colourless liquid, miscible with water, alcohol and ether | | |
| Identification | | Identification | |
| A - Relative density | 0.785 to 0.789 | A - Specific gravity | 0.783 to 0.787 |
| B - Refractive index | 1.376 to 1.379 | B - Refractive index | 1.376 to 1.378 |
| C | Complies | | |
| Tests | | | |
| . Appearance | Complies | | |
| . Acidity or alkalinity | Complies | Acidity | Complies |
| . Benzene and related substances (GC) | | | |
| - 2-butanol | Not more than 0.1 % | | |
| - others impurities | Not more than 0.3 % | | |
| - benzene | Not more than 2 ppm | | |
| . Peroxides | Complies | | |
| . Non volatile substances | Not more than 2 mg (20 ppm) | . Limit of non volatile residue | Not more than 2.5 mg (0.005 %) |
| . Water | Not more than 0.5 % | | |
| | | Assay | Not less than 99.0 % |
| Storage | Store protected from light | Packaging and storage | Preserve in tight containers remote from heat |
| Impurities | | | |
| A - | Acetone | | |
| B - | Benzene | | |
| C - | Di isopropyl ether | | |
| D - | Diethyl ether | | |
| E - | Methanol | | |
| F - | Propanol | | |

particular case. It is preferable to focus on the impact of impurities present in the solvent on the chromatographic performances.

Impurities in and additives to solvents can cause several problems and artifacts in liquid and gas chromatography.[37-39] Primarily they can be the origin of irreproducible separations, enhanced UV-background and even of mechanical problems. De Schutter and Col[37] have investigated this problem in purity of solvents used in high-performance thin-layer chromatography.

A way to improve the stability of the chromatographic system by minimizing the role of solvent impurities is to add deliberately a controlled amount of organic modifier. For example, Lauren and Col[40] have applied this technique, using decanol for improving the stability of the LC used for analyzing carotenoids. Middleditch and Zlatkis[41] have listed an impressive range of stabilizers and additives which can be found in solvents which may help the chromatographer in explaining the occurrence of artifacts in chromatography. Zelvensky and Col[38] have determined by gas chromatography the most common impurities contained in solvents for liquid chromatography. The series of solvents investigated include acetonitrile, methanol, ethanol, dichloromethane, formic acid, dimethylformamide, pyridine, tetrahydrofuran and dimethyl sulphoxide. Parsons and Col[39] have performed a search for trace impurities in solvents commonly used for gas chromatographic analysis of environmental samples.

#### *14.21.1.5.3 Quality of solvents used in titrimetry*

For titrimetric determinations performed in an aqueous medium it is highly recommended to use distilled water and to perform a blank determination if necessary.

For determinations performed in non aqueous media, the solvents should be as anhydrous as possible and, of course, inert to the titrant and the substance. Their purity should be such that they do not contain impurities which could react with the substance to be analyzed. A blank titration should be performed if necessary.

### 14.21.1.6 Conclusions

Far from being inert and not affecting the molecules dissolved in it, the solvent can affect the behavior of the solute in different ways. This chapter has aimed to support the idea that it is important for the chemist and the pharmacist to control the quality of the solvents used in the different areas of pharmaceutical activity. As we have tried to show, many pitfalls can be avoided during the development of a drug if a thorough investigation of the quality of the solvents used is carried out.

### References

1 C. Reichardt, **Solvents and solvents effect in organic chemistry**, 2nd edition, 1998, *VCH Verlagsgesellschaft mbH.*
2 G.A. Stephenson, E.G. Groleau, R.L. Kleemann, W. Xu and D.R. Rigsbee, *J. Pharm. Sci.*, **87**, 536 (1998).
3 A. Picot, **Le bon usage des solvants, Information Toxicologique n° 3, Unité de prévention du risque chimique** - *CNRS* (France) (1995).
4 **ICH Harmonised Tripartite Guideline, Impurities: Guideline for Residual Solvents**.
5 M. Gachon, *STP Pharma Pratiques 1*, 531 (1991).
6 **Carbomer Monograph - Eur. Ph. Addendum 2000**, p 494.
7 J.A. Riddik and W.B. Bungler, **Organic Solvents**, *Wiley Intersciences*, New-York, 1970, pp 552-571.
8 G.P.J. Ravissot, *STP Pharma Pratiques*, **9**, 3 (1999).
9 See Reference 1 p 427.
10 M. Debaert, *STP Pharma Pratiques*, ***1***, 253 (1991).
11 A.M. Guyot-Hermann, *STP Pharma Pratiques, 1*, 258 (1991).
12 D. Chulia, M. Deleuil, Y. Pourcelot, **Powder Technology and Pharmaceutical Processes**, *Elsevier*, Amsterdam, 1994.
13 J.L. Burdett and M.T. Rodgers, *J. Am. Chem. Soc.*, **86**, 2105 (1964).
14 M.T. Rodgers and J.L. Burdett, *Can. J. Chem.*, **43**, 1516 (1965).
15 J. Haleblian and W. McCrone, *J. Pharm. Sci.*, **58**, 911 (1969).
16 P. York, *Int. J. Pharm.*, **14**, 1 (1983).
17 S. Khoshkhoo and J. Anwar, *J. Phys. D.; Appl. Phys.*, **26**, B90 (1993).
18 J. Shyh-Ming, *Diss. Abst. Int.*, **57** (10), 6402-B (1997).
19 W. Beckmann and W.H. OTTO, *Chem. E. Res. Des.*, **74**, 750 (1996).
20 N. Rodriguez-Hornedo and D. Murphy, *J. Pharm. Sci.*, **88**, 651 (1999).
21 R. David and D. Giron in **Powder Technology and Pharmaceutical Processes, Handbook of Powder Technology**, *Elsevier Sciences*, Ed., Amsterdam 1994, pp 193-241.
22 A.S. Myerson, S.M. Jang, *J. Cryst. Growth*, **156**, 459 (1995).
23 A.H.L. Chow, D.J.W. Grant, *Int. J. Pharm.*, **42**, 123 (1988).
24 C. Lefebvre-Ringard, A.M. Guyot-Hermann, R. Bouché et J. Ringard, *STP Pharma Pratiques*, **6**, 228 (1990).
25 D.C. Messer, L.T. Taylor, W.N. Moore and W.E. Weiser, *Ther. Drug Monit.*, **15**, 581 (1993).
26 M. Perrut, *Information Chimie*, **321**, 166 (1990).
27 R.R. Pfeiffer, K.S. Yang and M.A. Tucker, *J. Pharm. Sci.*, **59**, 1809 (1970).
28 R. Hüttenrauch, *Pharm. Ind.*, **45**, 435 (1983).
29 D.S. Bindra, T.D. William and V.J. Stella, *Pharm. Res.*, **11**, 1060 (1994).

30 J.W. McGinity and J.A. Hill, *J. Pharm. Sci.*, **64**, 356 (1975).
31 D.M. Johnson and W.F. Taylor, *J. Pharm. Sci.*, **73**, 1414 (1984).
32 J. Joachim, D.D. Opota, G. Joachim, J.P. Reynier, P. Monges and L. Maury, *STP Pharma Sci.*, **5**, 486 (1995).
33 S.R. Byrn, C.T. Lin, P. Perrier, G.C. Clay and P.A. Sutton, *J. Org. Chem.*, **47**, 2978 (1982).
34 J. Wiemann, Y. Pascal, J. Chuche, **Relations entre la structure et les propriétés physiques**, *Masson edit*, Paris, 1965.
35 H.E. Gottlieb, V. Kotlyar and A. Nudelman, *J. Org. Chem.*, **62**, 7512 (1997).
36 V.J. Barwick, *Trends in An. Chem.*, **16**, 293 (1997).
37 J.A. de Schutter, G. Van der Weken, W. Van den Bossche and P. de Moerloose, *Chromatographia*, **20**, 739 (1985).
38 V.Y. Zelvensky, A.S. Lavrenova, S.I. Samolyuk, L.V. Borodai and G.A. Egorenko, *J. Chromatogr.*, **364**, 305 (1986).
39 W.D. Bowers, M.L. Parsons, R.E. Clement, G.A. Eiceman and F.W. Rarasek, *J. Chromatogr.*, **206**, 279 (1981).
40 D.R. Lauren, M.P. Agnew, D.E. McNaughton, *J. Liq. Chromatogr.*, **9**, 1997 (1986).
41 B.S. Middleditch, A. Zlatkis, *J. Chromatogr. Sci.*, **25**, 547 (1987).

## 14.21.2 PREDICTING COSOLVENCY FOR PHARMACEUTICAL AND ENVIRONMENTAL APPLICATIONS

An Li
**School of Public Health, University of Illinois at Chicago, Chicago, IL, USA**

### 14.21.2.1 Introduction

Cosolvency refers to the effects of adding one or more solvents (cosolvents), which are different from the existing solvent in a solution, on the properties of the solution or behavior of the solute. Cosolvency has found its applications in numerous engineering and scientific disciplines. The discussion in this section will be limited to aqueous phase cosolvency (the primary solvent is water), and cosolvents will include only pure organic solvents which are miscible with water either completely (in any proportion) or partially (in only certain proportions). The extent of cosolvency will be quantitatively described by the difference in solute solubilities in pure water and in a mixture of water and cosolvent(s).

Cosolvency has been studied for decades. However, it remains a poorly understood phenomenon due in large measure to our limited awareness of the liquid structure and the intermolecular forces. At present, practical approaches to predicting cosolvency are to develop models based on established theories and to make use of correlation between experimental observations and properties of the substances involved. As with all modeling efforts, it is essential to make judicious simplifications at various levels. The efforts to date have given rise to several models, including the extended regular solution theory[1-3] and its modification,[4] excess free energy model,[5-7] the phenomenological model,[8,9] modified Wilson model,[10,11] the combined nearly ideal binary solvent (NIBS) model,[12] the mixture response surface model,[13] and others. Many of these models, however, are considered to be more descriptive than predictive, because they inevitably involve one or more model parameters which are usually specific to a particular solute/solvent/cosolvent(s) system, and must be estimated from experimental data of solubility obtained for that system. On the other hand, purely empirical models, e.g., the double-log exponential equation,[14] aim at satisfying mathematical descriptions of measured data, and often offer little insight to the process. Comparisons among cosolvency models have been made in several published papers.[15-19]

This discussion is intended to provide an easy approach for predicting the effect of cosolvents, which are frequently involved in pharmaceutical and environmental applications, on the solubility of organic chemicals. The starting point is the widely used log-linear model.

### 14.21.2.2 Applications of cosolvency in pharmaceutical sciences and industry

Although many drugs are formulated and administrated in solid, vapor, powder, or other forms, using solutions as drug delivery vehicle has significant advantages. Most parenterally administered medicines are in the form of liquid solution, and for intravenous injection, liquid form dosage is the only possible form. However, drugs are usually designed with little concern about their level of solubility in solution. It is the task of pharmaceutical formulators to find appropriate forms for the drug to be effectively delivered into biological systems.

The phenomena of cosolvency have been studied for more than a century by pharmaceutical scientists. Numerous experimental data are published in the literature, and most of the models mentioned in the introduction section have been developed from pharmaceutical research. In addition to the need of solubilizing drugs which are poorly water-soluble, controlling the dissolution of drugs administered as solids to optimize therapeutic activity also demands an improved understanding of drug solubilization. Approaches that may be pursued to enhance drug solubility in a liquid dosage formulation include adjusting pH, adding surfactants, cosolvents, or complexation agents. Choice of these techniques depends primarily on the drug's chemical structure and physicochemical properties. For example, control of pH is applicable only when the drug is an electrolyte. To solubilize nonelectrolyte drugs, the use of cosolvents outweighs surfactants and complexing agents.[20]

Cosolvents that are routinely used in drug formulation include ethanol, propylene glycol, polyethylene glycol, and glycerin. Examples of pharmaceutical products containing these cosolvents are summarized in Table 14.21.2.1.

**Table 14.21.2.1. Selected pharmaceutical products containing cosolvents**

| Trade Name | Cosolvent | vol% | Manufacturer | Type |
|---|---|---|---|---|
| Aclovate cream | Propylene glycol | | Schering | Topical |
| Alurate elixir | Ethanol | 20 | Roche | Oral liquid |
| Amidate | Propylene glycol | 35 | Abbott | Parenteral |
| Amphojel | Glycerin | | Wyeth-Ayerst | Oral |
| Apresoline | Propylene glycol | 10 | Ciba | Parenteral |
| Aristocort cream | Propylene glycol | | Fujisawa | Topical |
| Ativan | Polyethylene glycol<br>Propylene glycol | 20<br>80 | Wyeth-Ayerst | Parenteral |
| Bentyl syrup | Propylene glycol | | Lakeside | Oral |
| Brevibioc | Ethanol | 25 | DuPont | Parenteral |
| Cleocin T lotion | Glycerin | | Upjohn | Topical |

| Trade Name | Cosolvent | vol% | Manufacturer | Type |
|---|---|---|---|---|
| Comtrex cough | Ethanol | 20 | Bristol | Oral liquid |
| Cyclocort lotion | Polyethylene glycol | | Lederle | Topical |
| Delsym | Propylene glycol | | McNeil Consumer | Oral |
| Depo-Medrol | Polyethylene glycol | | Upjohn | Parenteral |
| Dilantin | Ethanol<br>Propylene glycol | 10<br>40 | Parke-Davis | Parenteral |
| Dramamine | Propylene glycol | 50 | Searle | Parenteral |
| Elocon lotion | Propylene glycol | | Schering | Topical |
| Entex liquid | Glycerin | | Norwich-Eaton | Oral |
| Fluonid solution | Propylene glycol | | Herbert | Topical |
| Halog cream | Propylene glycol | | Westwood-Squibb | Topical |
| Halog ointment | Polyethylene glycol | | Westwood-Squibb | Topical |
| Kwell cream | Glycerin | | Reed & Carnrick | Topical |
| Lanoxin | Ethanol<br>Propylene glycol | 10<br>40 | Burroughs Wellcome | Parenteral |
| Librium | Propylene glycol | 20 | Roche | Parenteral |
| Lidex | Propylene glycol | | Syntex | Topical |
| Luminal Sod | Propylene glycol | 67.8 | Winthrop | Parenteral |
| MVI-12 | Propylene glycol | 30 | Armour | Parenteral |
| Nembutal | Ethanol<br>Propylene glycol | 10<br>40 | Abbott | Parenteral |
| Neoloid | Propylene glycol | | Lederle | Oral |
| Nitro-BID IV | Ethanol | 70 | Marion | Parenteral |
| Novahistine DH | Glycerin | | Lakeside | Oral |
| Paradione | Ethanol | 65 | Abbott | Oral liquid |
| Pentuss | Propylene glycol | | Fisons | Oral |
| Psorcon ointment | Propylene glycol | | Dermik | Topical |
| Rondec DM | Glycerin | | Ross | Oral |
| S-T Forte syrup | Ethanol | 5 | Scot-Tussin | Oral liquid |
| Sulfoxyl lotion | Propylene glycol | | Stiefel | Topical |
| Tinactin | Polyethylene glycol | | Schering | Topical |
| Trideslon cream | Glycerin | | Miles | Topical |
| Tussar | Propylene glycol | | Rorer | Oral |
| Tussionex | Propylene glycol | | Fisons | Oral |

| Trade Name | Cosolvent | vol% | Manufacturer | Type |
|---|---|---|---|---|
| Tylenol | Propylene glycol | | McNeil Consumer | Oral |
| Valium | Ethanol<br>Propylene glycol | 10<br>40 | Roche | Parenteral |
| Vepesid | Ethanol | 30.5 | Bristol-Myers | Parenteral |

(Data are from reference 21)

### 14.21.2.3 Applications of cosolvency in environmental sciences and engineering

The significance of cosolvency research in environmental sciences stems from the need for accurately modeling the distribution and movement of organic pollutants, and cleaning up polluted soils and sediments. Since the late 1970s, environmental research on the effect of cosolvents has grown steadily.

Most published research papers have focused on the effects of adding cosolvents on the aqueous solubility[18,22-31] and soil sorption[28,32-44] of pollutants of interest. A few researchers have also examined cosolvent effects on liquid phase partitioning.[45,46] In the cases of industrial waste discharges, liquid fuel and paint spills, storage tank leakage, landfill leaching, and illegal dumping, various organic solvents may find their way into the natural environment. These solvents may not only act as pollutants themselves, but also bring substantial changes on the distribution, movement, and fate of other environmental pollutants with high concern. In environmental cosolvency studies, the majority of the solutes are hydrophobic organic compounds (HOCs), including benzene and its derivatives, polycyclic aromatic hydrocarbons (PAHs), polychlorinated biphenyls (PCBs), polychlorinated dibenzo-p-dioxins and furans (PCDDs and PCDFs), and various pesticides. These chemicals are toxic, and many of them are mutagenic and carcinogenic. The 1990 Clean Air Act Amendment has stimulated research on gasoline additives such as methyl t-butyl ether (MTBE) and formulated fuels like gasohol. Their cosolvent effects on the solubility and sorption of a few pollutant groups have also been examined.[25,47]

Meanwhile, environmental engineers have put cosolvents to work in cleaning up contaminated sites. As a consequence of the failure of using traditional pump-and-treat remediation for soils contaminated with organic pollutants, a few new approaches have been experimented since the late 1980s. Among those involving cosolvents, *Ex situ* solvent extraction was developed to treat excavated soils, sediment, or sludge. A typical one is the basic extractive sludge treatment (B.E.S.T.) process certified by USEPA.[48,49] Triethylamine was selected as the extracting solvent due mainly to its inverse miscibility property - it is completely miscible with water below 60°F but separates from water above 90°F. This property makes it easier to recycle the solvents after separating the treated solids from liquids containing the solvent, pollutants, and water. For PCBs in various soils, it is typical to achieve an extraction efficiency higher than 99% using the B.E.S.T technique.

More attractive are *in situ* remediation approaches, which often cost less. Cosolvents promote the mobilization of organic chemicals in soils, thus accelerating the cleanup of contaminated site. Cosolvent flushing has been developed using the same principles as those used in solvent flooding, a technique to enhance petroleum recovery in oil fields. It involves injecting a solvent mixture, mostly water plus a miscible cosolvent, into the vadose or saturated zone upgradient of the contaminated area. The solvent with the removed contaminants is then extracted downgradient and treated above ground. Precise formulations for the water/cosolvent mixture need to be determined by laboratory and pilot studies in order to achieve the desired removal.[50-53]

A few field-scale evaluations of this technique were carried out at Hill Air Force Base, Utah, where the aquifer had been severely contaminated by jet fuel, chlorinated solvents, and pesticides during 1940s and 1950s. These contaminants had formed a complex non-aqueous phase liquid (NAPL) containing more than 200 constituents, which covered the surfaces of soil particles and was trapped in pores and capillaries over the years. One of the evaluations consisted of pumping ternary cosolvent mixture (70% ethanol, 12% n-pentanol, and 18% water) through a hydraulically isolated test cell over a period of 10 days, followed by flushing with water for another 20 days.[54,55] The removal efficiency varied from 90-99% at the top zone to 70-80% at the bottom near a confining clay layer. Similar removal efficiencies were obtained from another test cell using a combination of cosolvent n-pentanol and a surfactant at a total of 5.5 wt % of the flushing solution.[56] In order to remove gasoline residuals at a US Coast Guard base in Traverse City, Michigan, it was demonstrated that the contaminants were mobilized when cosolvent 2-propanol was used at 50% concentration, while methanol at either 20% or 50% showed little effect.[57] Cosolvent flushing was also proven to be effective in treating NAPLs which were denser than water. Methanol, isopropanol, and t-butanol were used in treating soils contaminated with tri- and tetra-chlorinated ethylenes.[58]

The applicability of solvent flushing, however, is often limited by the characteristics of the soil, especially the particle size distribution. While sandy soils may result in uncontrolled fluid migration, clayey soils with particles size less than 60 μm are often considered unsuitable for *in situ* solvent flushing due to low soil permeability. In an attempt to remove PAHs from poorly permeable soils, Li, et al.[59] investigated the possibility of combining cosolvent flushing with the electrokinetic technique. Electrokinetic remediation involves application of a low direct electrical current to electrodes that are inserted into the ground. As water is continuously replenished at anodes, dissolved contaminants are flushed toward the cathode due to electroosmosis, where they can be extracted and further treated by various conventional wastewater treatment methods. Their column experiment of removing phenanthrene from soil was moderately successful with the assistance of cosolvent n-butylamine at 20%(v). Retardation factor (ratio of the water linear velocity to that of the chemical) of phenanthrene was reduced from 753 in pure water to 11 by the presence of n-butylamine, and 43% of the phenanthrene was removed after 127 days or 9 pore volumes. However, significant removal of phenanthrene was not attained in their experiments with acetone and hydrofuran as cosolvents.

### 14.21.2.4 Experimental observations

Numerous experimental data exist in the literature on the solubility of organic solutes, including both drugs and environmental pollutants, in various mixtures of water and cosolvents. Experimental observations are often illustrated by plotting the logarithm of solubility of the solute versus the volume fraction of cosolvent in the solvent mixture. A few examples of solubilization curves are shown in Figure 14.21.2.1, which shows three typical situations for solutes of different hydrophobicity in the mixture of water and ethanol.

The classification of solute/cosolvent/water systems based on their relative polarity was suggested by Yalkowsky and Roseman.[61] Solutes which are less polar than both water

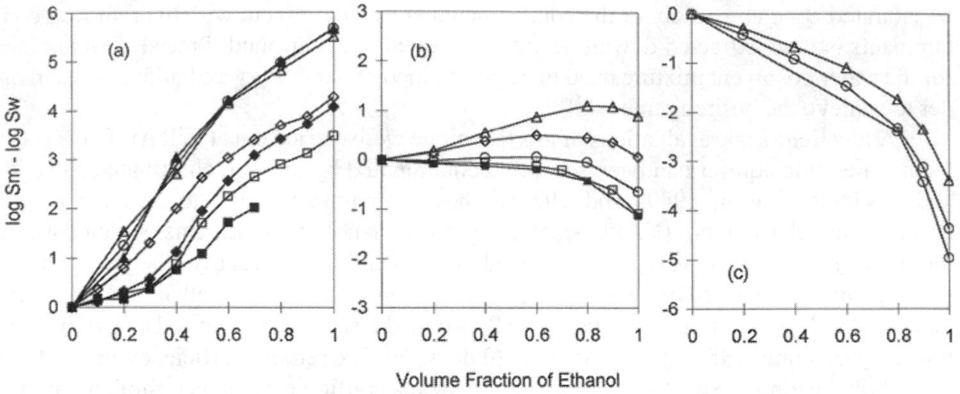


Figure 14.21.2.1. Effects of ethanol on the solubilities of selected organic compounds. (a): ■ benzene, □ naphthalene, ◆ biphenyl, ◇ anthracene, ▲ benzo(a)pyrene, △ perylene, O chrysene; (b) ■ hydantoic acid, □ hydantoin, O methyl hydantoic acic, ◇ 5-ethyl hydantoin, △ 5-isobutyl hydantoin; (c) □ triglycine, O diglycine, △ glycine [Adapted, by permission, from Li and Yalkovsky, *J. Pharm. Sci.*, **83**, 1735 (1994).]

and the cosolvent are considered as "nonpolar", those which have a polarity between those of water and the cosolvent as "semipolar", and those which are more polar than both water and cosolvent as "polar". Figure 14.21.2.1-a illustrates the behavior of relatively hydrophobic compounds, which tend to have monatonically increasing solubilization curves. The solubility enhancement is greater for the more hydrophobic solutes. Curves with opposite trends were mostly observed for polar solutes. The monatonical desolubilization is greater for more hydrophilic solutes, as evidenced by the curves in Figure 14.21.2.1-c. In-between are semipolar solutes with slightly parabolic curves shown in 14.21.2.1-b. The impact of adding cosolvents is much less profound for the semipolars than for the other two groups. On a linear solubility scale, the parabola tends to be more obvious than on the log scale. The same general trends were seen for the cosolvents glycerine[61] and propylene glycol[61,62] and presumably many other water-miscible cosolvents.

It is more difficult to evaluate the effects of cosolvents which have limited miscibility with water. In the literature, such organic solvents have been termed as both cosolvents and cosolutes, and there is no clear criteria for the distinction. Cosolvent is usually miscible with water, or to be used in an attempt to increase the aqueous solubility of the solute. Cosolute, on the other hand, may be organic chemicals which have a similar chemical structure or behave similarly with the solute when they exist in water alone. The effects of cosolutes have been examined in a limited number of published papers.[63-73]

Partially water-miscible organic solvents (PMOSs) may act as either cosolvents or cosolutes, and the research in the past has shown the complexity of their effects.[23,27-30,73-75] It was demonstrated that in order to exert effects on solubility or sorption of HOCs, PMOSs must exist as a component of the solvent mixture in an appreciable amount: Munz and Roberts[23] suggested a mole fraction of greater than 0.005 and Rao and coworkers[27,28] proposed a volume percent of 1% or a concentration above $10^4$ mg/L. Cosolvents with relatively high water solubility are likely to demonstrate observable effects on the solubilities of solutes, up to their solubility limits, in a similar manner to cosolvents of complete miscibility with water. A few experimental examples of the effects of PMOSs include 1-butanol and

1-pentanol acting on PCB congeners[30] and naphthalene.[26] and butanone on anthracene and fluoranthene.[75]

Even more hydrophobic organic solvents produce little or even negative influence on the solubility of HOCs. For instance, the presence of benzene does not increase the aqueous solubility of PCBs up to their saturation concentration.[29] Solubility of a few PCB congeners in water were found to be depressed by dissolved dichloromethane and chloroform.[73] On the other hand, PCB solubility showed little change when cosolvent benzyl alcohol, 1-hexanol, 1-heptanol, or 1-octanol was present.[29,30] Similar "no change" observations were made for naphthalene with cosolvents dichloromethane and chloroform,[73] and for solutes benzene and hexane with cosolvent MTBE.[25] Much of the complexity with hydrophobic cosolvents, or rather, cosolutes, can be explained by the fact that these cosolvents may partition into the solute phase, thus the physical state of the solute is no longer the same as is in pure water. Instead of a basically pure crystalline or liquid phase of solute, the solute and the cosolvent form an organic mixture, and the composition and ideality of this mixture will very much determine the concentrations of its components in the aqueous phase. Such a situation may be better investigated along the line of phase partitioning, where Raoult's law defines an ideal system.

#### 14.21.2.5 Predicting cosolvency in homogeneous liquid systems

The log-linear model

Yalkowsky and Roseman introduced the log-linear model in 1984 to describe the phenomenon of the exponential increase in aqueous solubility for nonpolar organic compounds as the cosolvent concentration is increased.[61] They showed that

$$\log S_m^i = f \log S_c + (1-f)\log S_w \qquad [14.21.2.1]$$

Rearranging equation [14.21.2.1] results in

$$\log(S_m^i / S_w) = f \log(S_c / S_w) = \sigma f \qquad [14.21.2.2]$$

The left side of equation [14.21.2.2] reflects the extent of solubilization; f defines how much cosolvent is required to reach the desired solubilization. The constant σ is the end-to-end slope of the solubilization curve and defined by:

$$\sigma = \log S_c - \log S_w = \log(S_c / S_w) \qquad [14.21.2.3]$$

The model can be extended to systems containing a number of cosolvents:

$$\log(S_m^i / S_w) = \sum \sigma_i f_i \qquad [14.21.2.4]$$

where the subscribe i denotes the i$^{th}$ component of the solvent mixture.

Two measured solubilities will define the value of σ that is specific to a solute/cosolvent pair. The value of σ is also dependent of the solubility unit selected and on whether 10-based or *e*-based logarithm is used. The magnitude of σ reflects the difference in molecular interactions between solute/cosolvent and solute/water. When applied to describe cosolvency, σ is like a microscopic partition coefficient if water and cosolvent are thought of as two independent entities. There had been other definitions of σ, such as the partial derivative $\partial(\log S_m)/\partial f$,[76] or the regressional (not end-to-end) slope of the solubilization curve.[77] The σ defined in these ways will depend on the range of f and on the accuracy of all data points over the entire range of f. These definitions are not desirable because they make σ difficult to predict and interpret in light of the concept of ideal solvent mixture on which the log-linear model is based. Note also that σ is not related to the crystalline structure of the solute, since the contributions from the free energy of melting to the two solubilities cancel out. However, it may change if the solute exists in pure cosolvent with a chemical identity different from that in water, as in the cases where solute degradation, solvation, or solvent-mediated polymorphic transitions occur in either solvent.

Estimation of σ

Laboratory measurements of $S_w$ and $S_c$ can be costly and difficult. Various methods, including group contribution technique and quantitative structure (or property) property relationships (QSPRs or QPPRs),[78] are available to estimate $S_w$ and $S_c$, from which σ values can be derived. A direct approach of predicting σ has also been established based on the dependence of cosolvency on solute hydrophobicity. Among a number of polarity indices, octanol/water partition coefficient, $K_{ow}$, was initially chosen by Yalkowsky and Roseman[61] for correlation with σ, due mainly to the abundance of available experimental $K_{ow}$ data and the wide acceptance of the Hansch-Leo fragment method[79] for its estimation. $K_{ow}$ is a macroscopic property which does not necessarily correlate with micro-scale polarity indices such as dipole moment, and only in a rank order correlates with other macroscopic polarity indicators such as surface tension, dielectric constant, and solubility parameter.

Correlation between σ and solute $K_{ow}$ takes the form:

$$\sigma = a + b\log K_{ow} \qquad [14.21.2.5]$$

where a and b are constants that are specific for the cosolvent but independent of solutes. Their values have been reported for various cosolvents and are summarized in Table 14.21.2.2.

From Table 14.21.2.2, the slopes of equation [14.21.2.5], b, are generally close to unity, with few below 0.6 or above 1.2. Most of the intercepts a are less than one, with a few negative values. In searching for the physical implications of the regression constants a and b, Li and Yalkowsky[80] derived equation [14.21.2.6]:

$$\sigma = \log K_{ow} + \log\left(\gamma_0^{\infty *} / \gamma_c\right) + \log\left(\gamma_w / \gamma_w^{\infty *}\right) + \log\left(V_0^* / V_c\right) \qquad [14.21.2.6]$$

According to this equation, σ~log $K_{ow}$ correlation will indeed have a slope of one and a predictable intercept of log $(V_o^*/V_c)$ if both $\gamma_0^{\infty *}/\gamma_c$ and $\gamma_w/\gamma_w^{\infty *}$ terms equal unity. $V_o^*$ = 0.119 L $mol^{-1}$ based on a solubility of water in octanol of 2.3 mol $L^{-1}$, and $V_c$ ranges from 0.04 to 0.10 $L^3$ $mol^{-1}$, thus log $(V_o^*/V_c)$ is in the range of 0.08 to 0.47, for the solvents included in Table 14.21.2.2 with the exclusion of PEG400. However, both $\gamma_0^{\infty *}/\gamma_c$ and $\gamma_w/\gamma_w^{\infty *}$ are not likely to be unity, and their accurate values are difficult to estimate for many solutes. The ratio $\gamma_0^{\infty *}/\gamma_c$ compares the solute behavior in water-saturated octanol under dilute conditions and in pure cosolvent at saturation, while the $\gamma_w/\gamma_w^{\infty *}$ term reflects both the effect of dissolved octanol on the aqueous activity coefficient and the variation of the activity coefficient with concentration. Furthermore, the magnitudes of both terms will vary from one solute to another, making it unlikely that a unique regression intercept will be observed over a wide range of solutes. Indeed, both a and b were found to be dependent on the range of the solute log $K_{ow}$ used in the regression. For instance, for solutes with log $K_{ow} \leq 0$, 0.01 to 2.99, and $\geq 3$, the correlation of $\sigma$ versus log $K_{ow}$ for cosolvent ethanol have slopes of 0.84, 0.79, and 0.69, respectively, and the corresponding intercepts increase accordingly; the slope of the overall correlation, however, is 0.95. Much of the scattering on the $\sigma$~log $K_{ow}$ regression resides on the region of relatively hydrophilic solutes. Most polar solutes dissociate to some extent in aqueous solutions, and their experimental log $K_{ow}$ values are less reliable. Even less certain is the extent of specific interactions between these polar solutes and the solvent components.

According to equation [14.21.2.6], $\sigma$ may not be a linear function of solute log $K_{ow}$ on a theoretical basis. However, despite the complexities caused by the activity coefficients, quality of the regression of $\sigma$ against log $K_{ow}$ is generally high, as evidenced by the satisfactory $R^2$ values in Table 14.21.2.2. This can be explained by the fact that changes in both $\gamma$ ratios are much less significant compared with the variations of $K_{ow}$ for different solutes. In addition, the two $\gamma$ terms may cancel each other to some degree for many solutes, further reducing their effects on the correlation between $\sigma$ and log $K_{ow}$. It is convenient and reliable to estimate $\sigma$ from known log $K_{ow}$ of the solute of interest, especially when the log $K_{ow}$ of the solute of interest falls within the range used in obtaining the values of a and b.

Dependence of $\sigma$ on the properties of cosolvents has been less investigated than those of solutes. While hundreds of solutes are involved, only about a dozen organic solvents have been investigated for their cosolvency potentials. A few researchers examined the correlations between $\sigma$ and physicochemical properties of cosolvents for specific solutes, for instance, Li et al. for naphthelene,[31] and Rubino and Yalkowsky for drugs benzocaine, diazepam, and phenytoin.[81] In both studies, hydrogen bond donor density (HBD), which is the volume normalized number of proton donor groups of a pure cosolvent, is best for comparing cosolvents and predicting $\sigma$. Second to HBD are the solubility parameter and interfacial tension (as well as log viscosity and $E_T$-30 for naphthalene systems), while log $K_{ow}$, dielectric constant, and surface tension, correlate poorly with $\sigma$. The HBD of a solvent can be readily calculated from the density and molecular mass with the knowledge of the chemical structure using equation [14.21.2.7]. The disadvantage of using HBD is that it cannot distinguish among aprotic solvents which have the same HBD value of zero.

$$HBD = (\text{number of proton donor groups})(\text{density})/(\text{molecular mass}) \qquad [14.21.2.7]$$

In an attempt to generalize over solutes, Li and Yalkowsky[82] investigated the possible correlations between cosolvent properties and slope of the $\sigma$~log $K_{ow}$ regressions (b). Among the properties tested as a single regression variable, octanol-water partition coefficient, interfacial tension, and solubility parameter, are superior to others in correlating with b. Results of multiple linear regression show that the combination of log $K_{ow}$ and HDB of the cosolvent is best (equation [14.21.2.8]). Adding another variable such as solubility parameter does not improve the quality of regression.

$$b = 0.2513 \log K_{ow} - 0.0054\, HBD + 1.1645 \qquad [14.21.2.8]$$

$$(N = 13,\ R^2 = 0.942,\ SE = 0.060,\ F = 81.65)$$

where log $K_{ow}$ (range: -7.6 ~ 0.29) and HBD (range: 0 ~ 41) are those of the cosolvent.

Equation [14.21.2.8] can be helpful to obtain the b values for cosolvents not listed in Table 14.21.2.2. In order to estimate cosolvency for such cosolvents, values of the intercept a are also needed. However, values of a can be found in Table 14.21.2.2 for only about a dozen cosolvents, and there is no reliable method for its estimation. To obtain a for a cosolvent, a reasonable starting value can be log ($V_o^*/V_c$), or -(0.92 + log $V_c$). The average absolute difference between a values listed in Table 14.21.2.2 and log ($V_o^*/V_c$) is 0.18 (N=7) for alcohols and glycols, 0.59 (N = 6) for aprotic cosolvents, and > 1 for n-butylamine and PEG400.[82]

**Table 14.21.2.2. Summary of regression results for relationship between σ and solute log $K_{ow}$**

| Cosolvent | N | a | b | $R^2$ | log $K_{ow}$ range | Ref. |
|---|---|---|---|---|---|---|
| Methanol | 79 | 0.36±0.07 | 0.89±0.02 | 0.96 | -4.53 ~ 7.31 | 80 |
| Methanol | 16 | 1.09 | 0.57 | 0.83 | 2.0 ~ 7.2 | 29 |
| Methanol | 16 | 1.07 | 0.68 | 0.84 | n.a. | 24 |
| Ethanol | 197 | 0.30±0.04 | 0.95±0.02 | 0.95 | -4.90 ~ 8.23 | 80 |
| Ethanol | 107 | 0.40±0.06 | 0.90±0.02 | 0.96 | -4.9 ~ 6.1 | 60 |
| Ethanol | 11 | 0.81 | 0.85 | 0.94 | n.a. | 24 |
| 1-Propanol | 17 | 0.01±0.13 | 1.09±0.05 | 0.97 | -3.73 ~ 7.31 | 80 |
| 2-Propanol | 20 | -0.50±0.18 | 1.11±0.07 | 0.94 | -3.73 ~ 4.49 | 80 |
| 2-Propanol | 9 | 0.63 | 0.89 | 0.85 | n.a. | 24 |
| Acetone | 22 | -0.10±0.24 | 1.14±0.07 | 0.92 | -1.38 ~ 5.66 | 80 |
| Acetone | 14 | 0.48 | 1.00 | 0.93 | 0.6 ~ 5.6* | 24 |
| Acetonitrile | 10 | -0.49±0.42 | 1.16±0.16 | 0.86 | -0.06 ~ 4.49 | 80 |
| Acetonitrile | 8 | 0.35 | 1.03 | 0.90 | n.a. | 24 |
| Dioxane | 23 | 0.40±0.16 | 1.08±0.07 | 0.91 | -4.90 ~ 4.49 | 80 |
| Dimethylacetamide | 11 | 0.75±0.30 | 0.96±0.12 | 0.87 | 0.66 ~ 4.49 | 80 |
| Dimethylacetamide | 7 | 0.89 | 0.86 | 0.95 | n.a. | 24 |
| Dimethylformamide | 11 | 0.92±0.41 | 0.83±0.17 | 0.73 | 0.66 ~ 3.32 | 80 |
| Dimethylformamide | 7 | 0.87 | 0.87 | 0.94 | n.a. | 24 |
| Dimethylsulfoxide | 12 | 0.95±0.43 | 0.79±0.17 | 0.68 | 0.66 ~ 4.49 | 80 |
| Dimethylsulfoxide | 7 | 0.89 | 0.87 | 0.95 | n.a. | 24 |
| Glycerol | 21 | 0.28±0.15 | 0.35±0.05 | 0.72 | -3.28 ~ 4.75 | 80 |
| Ethylene glycol | 13 | 0.37±0.13 | 0.68±0.05 | 0.95 | -3.73 ~ 4.04 | 80 |
| Ethylene glycol | 7 | 1.04 | 0.36 | 0.75 | n.a. | 24 |
| Propylene glycol | 62 | 0.37±0.11 | 0.78±0.04 | 0.89 | -7.91 ~ 7.21 | 80 |

| Cosolvent | N | a | b | $R^2$ | log $K_{ow}$ range | Ref. |
|---|---|---|---|---|---|---|
| Propylene glycol | 47 | 0.03 | 0.89 | 0.99 | -5 ~ 7 | 61 |
| Propylene glycol | 8 | 0.77 | 0.62 | 0.96 | n.a. | 24 |
| PEG400 | 10 | 0.68±0.43 | 0.88±0.16 | 0.79 | -0.10 ~ 4.18 | 80 |
| Butylamine | 4 | 1.86±0.30 | 0.64±0.10 | 0.96 | -1.69 ~ 4.49 | 80 |

*estimated from Figure 3 in Reference 24.
n.a. = not available.

This empirical approach using equations [14.21.2.8], [14.21.2.5], and [14.21.2.2] can produce acceptable estimates of log ($S_m/S_w$) only if the solubilization exhibits a roughly log-linear pattern, such as in some HOC/water/methanol systems. In addition, it is important to limit the use of equations [14.21.2.5] and [14.21.2.8] within the ranges of log $K_{ow}$ used in obtaining the corresponding parameters.

### 14.21.2.6 Predicting cosolvency in non-ideal liquid mixtures

Deviations from the log-linear model

Most solubilization curves, as shown in Figure 14.21.2.1, exhibit significant curvatures which are not accounted for by the log-linear model. A closer look at the solubilization curves in Figure 14.21.2.1 reveals that the deviation can be concave, sigmoidal, or convex. In many cases, especially with amphiprotic cosolvents, a negative deviation from the end-to-end log-linear line is often observed at low cosolvent concentrations, followed by a more significant positive deviation as cosolvent fraction increases.

The extent of the deviation from the log-linear pattern, or the excess solubility, is measured by the difference between the measured and the log-linearly predicted log $S_m$ values:

$$\log\left(S_m / S_m^i\right) = \log S_m - \left(\log S_w + \sum \sigma_i f_i\right) \quad [14.21.2.9]$$

The values of log ($S_m/S^i_m$) for naphthalene, benzocaine, and benzoic acid in selected binary solvent mixtures are presented in Figures 14.21.2.2-a, -b, and -c, respectively.

The log-linear model is based on the presumed ideality of the mixtures of water and cosolvent. The log-linear relationship between log ($S_m/S_w$) and f is exact only if the cosolvent is identical to water, which cannot be the case in reality. Deviation is fortified as any degradation, solvation, dissociation, or solvent mediated polymorphic transitions of the solute occur. The problem is further compounded if the solute dissolves in an amount large enough to exert significant influence on the activity of solvent components. Due to the complexity of the problem, efforts to quantitatively describe the deviations have achieved only limited success.

A generally accepted viewpoint is that the deviation from the log-linear solubilization is mainly caused by the non-ideality of the solvent mixture. This is supported by the similarities in the patterns of observed log $S_m$ and activities of the cosolvent in solvent mixture, when they are graphically presented as functions of f. Based on the supposition that solvent non-ideality is the primary cause for the deviation, Rubino and Yalkowsky[87] examined the correlations between the extent of deviation and various physical properties of solvent mix-

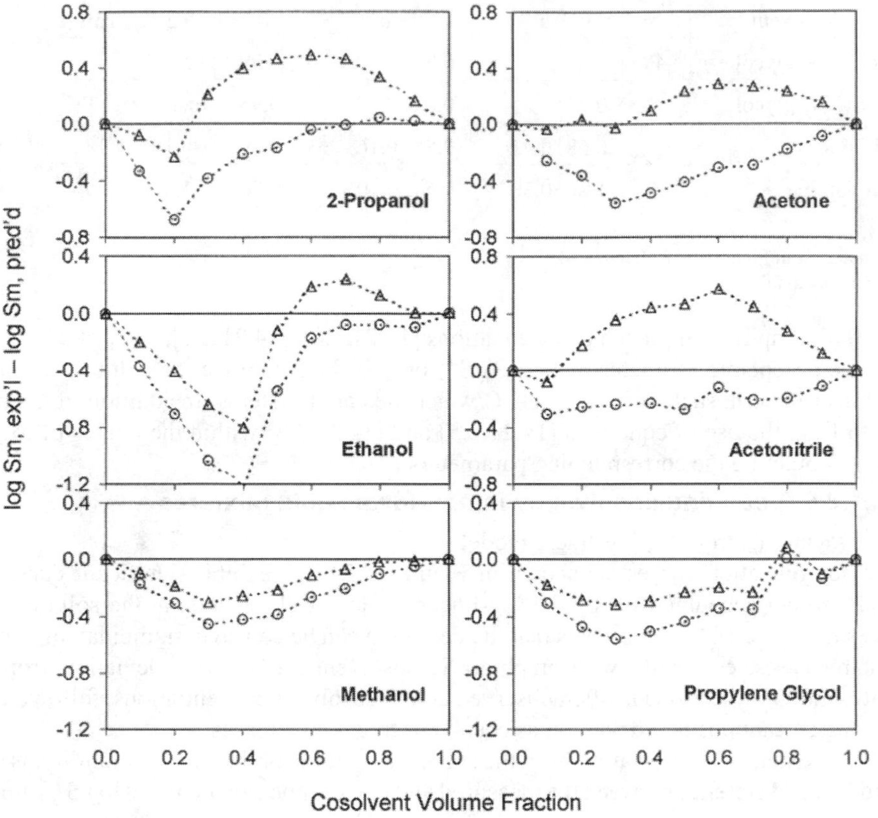


Figure 14.21.2.2a. Deviations from log-linear model (equation [14.21.2.2], triangle) and the extended log-linear model (equation [14.21.2.10], circle) for solute naphthalene in various water cosolvent systems. Experimental data are from Ref. 83.

tures. However, none of the properties consistently predicted the extrema of the deviation, although density corresponded in several cases.

Non-ideality of a mixture is quantitatively measured by the excess free energy of mixing. From this standpoint, Pinal et al.[75] proposed that a term $\Sigma(f_i \ln \gamma_i)$ be added to equation [14.21.2.4] to account for the effect of the non-ideality of solvent mixture:

$$\log\left(S_m^{ii} / S_w\right) = \sum \sigma_i f_i + 2.303 \sum f_i \log \gamma_i \quad [14.21.2.10]$$

where $\gamma_i$ is the activity coefficient of solvent component i in solute-free solvent mixture. Values of $\gamma$'s can be calculated by UNIFAC, a group contribution method for the prediction of activity coefficients in nonelectrolyte, nonpolymeric liquid mixtures.[88] UNIFAC derived activity coefficients are listed in Table 14.21.2.3 for selected cosolvent-water mixtures. They are calculated with UNIFAC group interaction parameters derived from vapor-liquid equilibrium data.[89,90]

The difference between the experimental log $S_m$ and that predicted by the extended log-linear model, i.e., equation [14.21.2.10], is

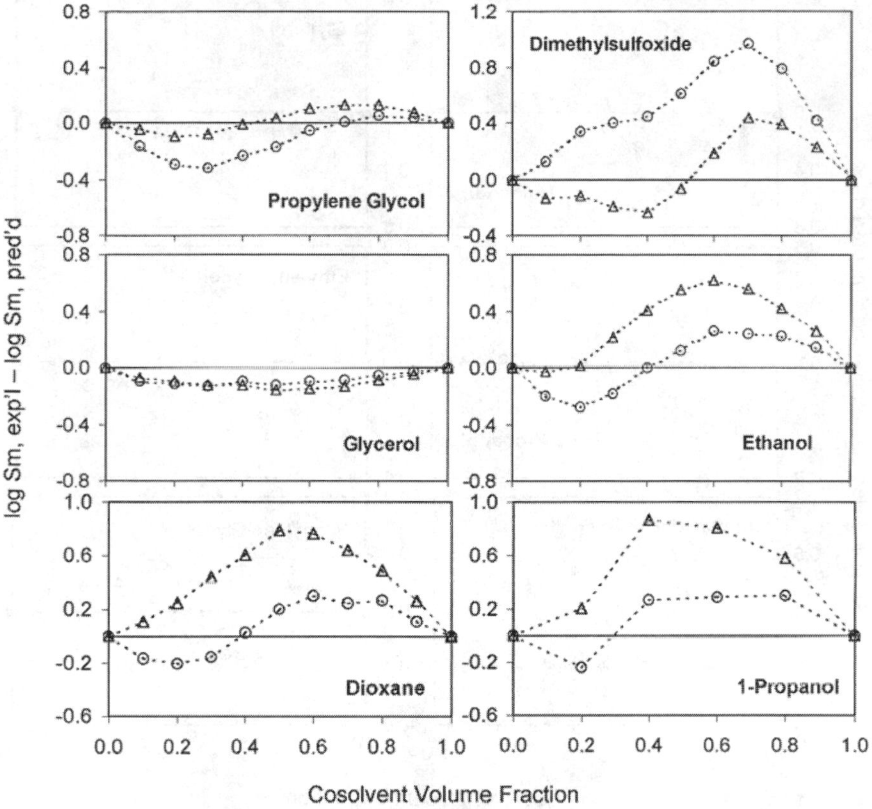


Figure 14.21.2.2b. Deviations from log-linear model (equation [14.21.2.2], triangle) and the extended log-linear model (equation [14.21.2.10], circle) for solute benzocaine in various water cosolvent systems. Experimental data are from Refs. 84 and 85.

$$\log\left(S_m / S_m^{ii}\right) = \log S_m - \left(\log S_w + \sum \sigma_i f_i + 2.303 \sum f_i \log \gamma_i\right) \quad [14.21.2.11]$$

Results of equation [14.21.2.11] for naphthalene, benzocaine, and benzoic acid in selected binary solvent mixtures are also included in Figure 14.21.2.2. A few other examples can be found in Pinal et al.[75]

The extended log-linear model outperforms the log-linear model in more than half of the cases tested for the three solutes in Figure 14.21.2.2. The improvement occurs mostly in regions with relatively high f values. In the low f regions, negative deviations of solubilities from the log-linear pattern are often observed as discussed above, but are not accounted for by the extended log-linear model as presented by equation [14.21.2.10]. In some cases, such as naphthalene in methanol and propylene glycol, and benzoic acid in ethylene glycol, the negative deviations occur over the entire f range of 0~1. In these cases, the extended log-linear model does not offer better estimates than the original log-linear model. With the activity coefficients listed in Table 14.21.2.3, the extended log-linear model generates worse estimates of log $(S_m/S_w)$ than the log-linear model for systems containing dimethylacetamide, dimethylsulfoxide, or dimethylformamide. There is a possibility that

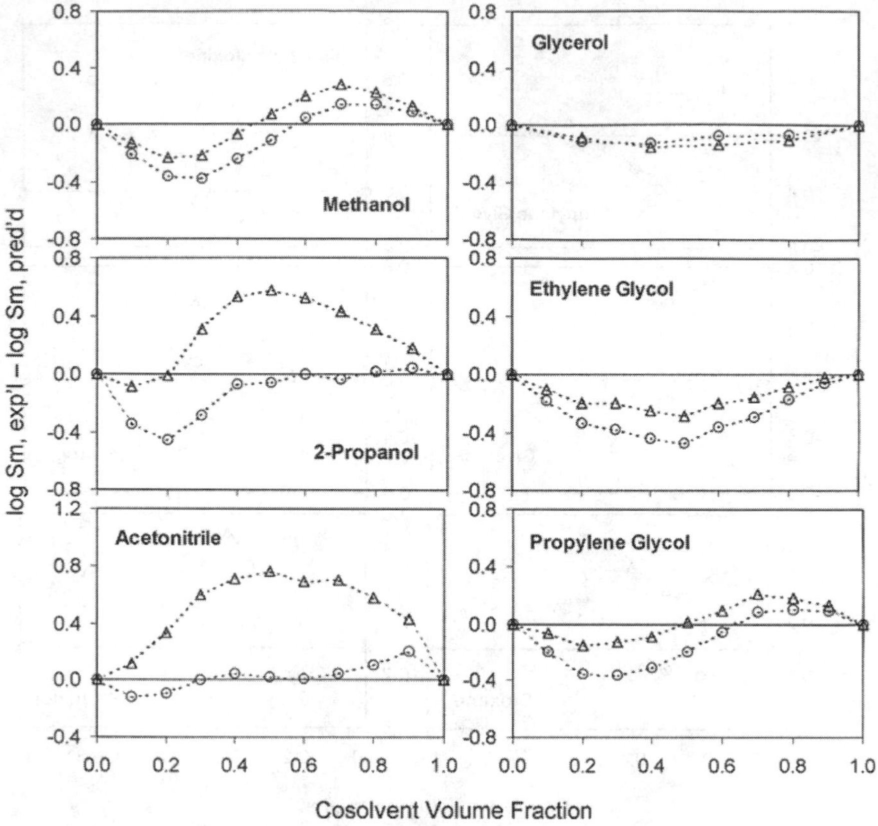


Figure 14.21.2.2c. Deviations from log-linear model (equation [14.21.2.2], triangle) and the extended log-linear model (equation [14.21.2.10], circle) for solute benzoic acid in various water cosolvent systems. Experimental data are from Refs. 61 and 86.

the UNIFAC group interaction parameters involved in these systems are incorrect. With all the systems tested in this study with solute naphthalene, benzocaine, or benzoic acid, it is also found that replacing $f_i$ in the last term of equation [14.21.2.11] with mole fraction $x_i$ offers slight improvement in only a few cases. Dropping the logarithm conversion constant 2.303 results in larger estimation errors for most systems.

An apparent limitation of this modification is the exclusion of any active role the solute may play on the observed deviation. Little understanding of the influence of solute structure and properties on deviations from the log-linear equation has been obtained. Although the patterns of deviations tend to be similar among solutes, as mentioned above, the extent of deviation is solute-dependent. For instance, $C_1$~$C_4$ alkyl esters of p-hydroxybenzoates and p-aminobenzoates demonstrated similar characteristics of solubilization by propylene glycol, with a negative deviation from the log-linear pattern occurring when f is low, followed by a positive one when f increases.[91] The magnitude of the negative deviation, however, was found to be related to the length of the solute alkyl chain in each group, while that of the positive deviation to the type of the polar groups attached.[91]

Both the hydrophobicity and hydrogen bonding property of the solutes seem to be important in influencing the extent of the deviation from the ideal log-linear pattern.

Additional deviations related to the solute's behavior may occur. For organic electrolytes, the acid dissociation constant $K_a$ may decrease as cosolvent fraction f increases.[40,75,92] This, in turn, will affect the patterns of solubilization by cosolvents. Furthermore, a high concentration of solutes may invalidate the log-linear model, which presumes negligible volume fraction of solute and no solute-solute interactions. For solid solutes, solvent induced polymorphism may also bring additional changes in their solubilization profile.

Another approach to quantitatively address the deviations of solubilization from the log-linear model makes use of an empirical parameter β:

$$\beta = \log(S_m / S_w) / \log(S_m^i / S_w) \quad [14.21.2.12]$$

The modified log-linear equation then takes the form:

$$\log(S_m / S_w) = \beta \sum \sigma_i f_i \quad [14.21.2.13]$$

**Table 14.21.2.3. UNIFAC derived activity coefficients for selected binary water-cosolvent systems**

| f | 0.1 | 0.2 | 0.4 | 0.6 | 0.8 | 0.9 |
|---|---|---|---|---|---|---|
| **Methanol** | | MW = 32.04 | | Density = 0.7914 | | |
| mol/L | 2.4700 | 4.9401 | 9.8801 | 14.8202 | 19.7603 | 22.2303 |
| x | 0.0471 | 0.1000 | 0.2286 | 0.4001 | 0.6401 | 0.8001 |
| γ, cosolvent | 1.972 | 1.748 | 1.413 | 1.189 | 1.052 | 1.014 |
| γ, water | 1.003 | 1.013 | 1.055 | 1.14 | 1.298 | 1.424 |
| **Ethanol** | | MW = 46.07 | | Density = 0.7893 | | |
| mol/L | 1.7133 | 3.4265 | 6.8530 | 10.2796 | 13.7061 | 15.4194 |
| x | 0.0331 | 0.0716 | 0.1705 | 0.3163 | 0.5523 | 0.7351 |
| γ, cosolvent | 5.550 | 4.119 | 2.416 | 1.564 | 1.152 | 1.050 |
| γ, water | 1.005 | 1.022 | 1.097 | 1.256 | 1.57 | 1.854 |
| **1-Propanol** | | MW = 60.1 | | Density = 0.8053 | | |
| mol/L | 1.3399 | 2.6799 | 5.3597 | 8.0396 | 10.7195 | 12.0594 |
| x | 0.0261 | 0.0569 | 0.1385 | 0.2657 | 0.4910 | 0.6846 |
| γ, cosolvent | 12.77 | 8.323 | 3.827 | 2.001 | 1.248 | 1.077 |
| γ, water | 1.006 | 1.024 | 1.111 | 1.301 | 1.706 | 2.093 |
| **2-Propanol** | | MW = 60.1 | | Density = 0.7848 | | |
| mol/L | 1.3058 | 2.6116 | 5.2233 | 7.8349 | 10.4466 | 11.7524 |

| f | 0.1 | 0.2 | 0.4 | 0.6 | 0.8 | 0.9 |
|---|---|---|---|---|---|---|
| x | 0.0255 | 0.0555 | 0.1355 | 0.2607 | 0.4846 | 0.6790 |
| γ, cosolvent | 12.93 | 8.488 | 3.921 | 2.040 | 1.258 | 1.080 |
| γ, water | 1.006 | 1.023 | 1.107 | 1.294 | 1.695 | 2.084 |
| **Acetone** | MW = 58.08 | | Density = 0.7899 | | | |
| mol/L | 1.3600 | 2.7200 | 5.4401 | 8.1601 | 10.8802 | 12.2402 |
| x | 0.0265 | 0.0577 | 0.1403 | 0.2686 | 0.4947 | 0.6878 |
| γ, cosolvent | 8.786 | 6.724 | 3.952 | 2.370 | 1.484 | 1.196 |
| γ, water | 1.004 | 1.015 | 1.075 | 1.222 | 1.616 | 2.211 |
| **Acetonitrile** | MW = 41.05 | | Density = 0.7857 | | | |
| mol/L | 1.9140 | 3.8280 | 7.6560 | 11.4840 | 15.3121 | 17.2261 |
| x | 0.0369 | 0.0793 | 0.1868 | 0.3407 | 0.5795 | 0.7561 |
| γ, cosolvent | 10.26 | 7.906 | 4.550 | 2.536 | 1.501 | 1.126 |
| γ, water | 1.005 | 1.021 | 1.11 | 1.366 | 2.076 | 3.571 |
| **Dioxane** | MW = 88.11 | | Density = 1.0329 | | | |
| mol/L | 1.1723 | 2.3446 | 4.6891 | 7.0337 | 9.3783 | 10.5506 |
| x | 0.0229 | 0.0501 | 0.1233 | 0.2404 | 0.4577 | 0.6551 |
| γ, cosolvent | 14.71 | 8.743 | 3.616 | 1.833 | 1.171 | 1.124 |
| γ, water | 1.006 | 1.026 | 1.112 | 1.284 | 1.604 | 1.666 |
| **DMA** | MW = 87.12 | | Density = 0.9429 | | | |
| mol/L | 1.0823 | 2.1646 | 4.3292 | 6.4938 | 8.6584 | 9.7407 |
| x | 0.0212 | 0.0464 | 0.1149 | 0.2261 | 0.4380 | 0.6368 |
| γ, cosolvent | 0.121 | 0.141 | 0.204 | 0.330 | 0.602 | 0.826 |
| γ, water | 0.999 | 0.994 | 0.962 | 0.872 | 0.651 | 0.453 |
| **DMF** | MW = 73.1 | | Density = 0.9445 | | | |
| mol/L | 1.2921 | 2.5841 | 5.1683 | 7.7524 | 10.3365 | 11.6286 |
| x | 0.0252 | 0.0549 | 0.1342 | 0.2586 | 0.4819 | 0.6767 |
| γ, cosolvent | 0.833 | 0.873 | 0.930 | 0.962 | 0.983 | 0.985 |
| γ, water | 0.999 | 0.997 | 0.991 | 0.984 | 0.972 | 0.969 |
| **DMSO** | MW = 78.13 | | Density = 1.10 | | | |
| mol/L | 1.4079 | 2.8158 | 5.6316 | 8.4475 | 11.2633 | 12.6712 |
| x | 0.0274 | 0.0596 | 0.1445 | 0.2754 | 0.5034 | 0.6952 |
| γ, cosolvent | 0.07956 | 0.110 | 0.211 | 0.399 | 0.715 | 0.899 |
| γ, water | 0.996 | 0.981 | 0.913 | 0.774 | 0.540 | 0.386 |

| f | 0.1 | 0.2 | 0.4 | 0.6 | 0.8 | 0.9 |
|---|---|---|---|---|---|---|
| **Glycerol** | | MW = 92.1 | | Density = 1.2611 | | |
| mol/L | 1.3693 | 2.7385 | 5.4771 | 8.2156 | 10.9542 | 12.3235 |
| x | 0.0267 | 0.0580 | 0.1411 | 0.2699 | 0.4964 | 0.6893 |
| γ, cosolvent | 1.257 | 1.066 | 0.903 | 0.899 | 0.969 | 0.996 |
| γ, water | 1.003 | 1.010 | 1.027 | 1.025 | 0.979 | 0.942 |
| **Ethylene glycol** | | MW = 62.07 | | Density = 1.1088 | | |
| mol/L | 1.7864 | 3.5727 | 7.1455 | 10.7182 | 14.2910 | 16.0773 |
| x | 0.0345 | 0.0744 | 0.1765 | 0.3254 | 0.5626 | 0.7432 |
| γ, cosolvent | 2.208 | 1.923 | 1.494 | 1.214 | 1.053 | 1.013 |
| γ, water | 1.002 | 1.01 | 1.047 | 1.12 | 1.247 | 1.338 |
| **Propylene glycol** | | MW = 76.09 | | Density = 1.0361 | | |
| mol/L | 1.3617 | 2.7234 | 5.4467 | 8.1701 | 10.8934 | 12.2551 |
| x | 0.0265 | 0.0577 | 0.1405 | 0.2688 | 0.4951 | 0.6881 |
| γ, cosolvent | 3.392 | 2.498 | 1.567 | 1.177 | 1.044 | 1.019 |
| γ, water | 1.005 | 1.018 | 1.069 | 1.145 | 1.224 | 1.267 |
| **Butylamine** | | MW = 73.14 | | Density = 0.7414 | | |
| mol/L | 1.0137 | 2.0273 | 4.0547 | 6.0820 | 8.1094 | 9.1231 |
| x | 0.0199 | 0.0436 | 0.1084 | 0.2149 | 0.4219 | 0.6215 |
| γ, cosolvent | 6.532 | 4.498 | 2.318 | 1.391 | 1.042 | 0.998 |
| γ, water | 1.004 | 1.016 | 1.071 | 1.175 | 1.326 | 1.384 |

Under the assumptions that the solute is chemically stable and has little influence on the activity of solvent component, β reflects the extent of deviation caused by the nonideality of the solvent mixture, as suggested by Rao et al.[28] However, since β itself is a complicated function of f, equation [14.21.2.13] does not provide additional aid for predicting cosolvency.

### 14.21.2.7 Summary

Applications of cosolvency in pharmaceutical and environmental research and industries are briefly summarized. Using ethanol as an example, the effects of adding a cosolvent on the solubilities of various organic solutes are presented in Figure 14.21.2.1. The log-linear solubilization model, equation [14.21.2.2] or [14.21.2.4], is the simplest theory of cosolvency developed so far. It discovers general trends and major determinant factors of cosolvency, thus providing guidelines for predicting solubility of organic chemicals in mixed solvents. The cosolvency power of a specific cosolvent towards a solute of interest, σ, can be estimated with equation [14.21.2.5] with the knowledge of the solute octanol-water partition coefficient $K_{ow}$. Sources of error associated with this estimation method are discussed based on equation [14.21.2.6]. The slope of the σ~log $K_{ow}$ regression, b, can be estimated from the log $K_{ow}$ and hydrogen bond donor density of the cosolvent, as presented by equation [14.21.2.8]. One of the previously published modifications to the log-linear model, equation [14.21.2.10], is evaluated. The difference between the measured log $S_m$ and those predicted by the log-linear and the extended log-linear model are presented in Figure 14.21.2.2 for solutes naphthalene, benzocaine, and benzoic acids in selected water and cosolvent mixtures.

## Notations

| | |
|---|---|
| a | intercept of $\sigma \sim \log K_{ow}$ regression |
| b | slope of $\sigma \sim \log K_{ow}$ regression |
| f | volume fraction of cosolvent in mixed solvent with water. |
| $K_{ow}$ | n-octanol water partition coefficient |
| $S_c$ | solubility in pure cosolvent |
| $S_m$ | solubility in the mixture of water and cosolvent |
| $S^i_m$ | solubility in the mixture of water and cosolvent, predicted by the log-linear model (Eq. [14.21.2.2]) |
| $S^{ii}_m$ | solubility in the mixture of water and cosolvent, predicted by the extended log-linear model (Eq. [14.21.2.10]) |
| $S_w$ | solubility in pure water |
| $V_o$* | molar volume of 1-octanol saturated with water, 0.119 L $mol^{-1}$ (based on a solubility of water in octanol of 2.3 mol $L^{-1}$) |
| $V_c$ | molar volume of cosolvent |
| $V_w$* | molar volume of water saturated with 1-octanol, ≈ 0.018 L $mol^{-1}$ |
| $V_w$ | molar volume of water, 0.018 L $mol^{-1}$ |
| β | empirically obtained water-cosolvent interaction parameter |
| σ | cosolvency power, $\sigma = \log(S_c/S_w)$ |
| $\gamma_0^{\infty *}$ | infinite dilution activity coefficient of solute in 1-octanol saturated with water |
| $\gamma_c$ | activity coefficient of solute in cosolvent |
| $\gamma_w$ | activity coefficient of solute in water |
| $\gamma_w^{\infty *}$ | infinite dilution activity coefficient of solute in water saturated with 1-octanol |

## References


1 A. Martin, J. Newburger, and A. Adjel, *J. Pharm. Sci.*, **68**, 4 (1979).
2 A. Martin, J. Newburger, and A. Adjel, *J. Pharm. Sci.*, **69**, 487 (1980).
3 A. Martin, A. N. Paruta, and A. Adjel, *J. Pharm. Sci.*, **70**, 1115 (1981).
4 P. Bustamante, B. Escalera, A. Martin, and E. Selles, *Pharm. Pharmacol.*, **45**, 253 (1993).
5 N. A. Williams, and G. L. Amidon, *J. Pharm. Sci.*, **73**, 9 (1984).
6 N. A. Williams, and G. L. Amidon, *J. Pharm. Sci.*, **73**, 14 (1984).
7 N. A. Williams, and G. L. Amidon, *J. Pharm. Sci.*, **73**, 18 (1984).
8 D. Khossrani, and K. A. Connors, *J. Pharm. Sci.*, **82**, 817 (1993).
9 D. Khossravi, and K. A. Connors, *J. Pharm. Sci.*, **81**, 371 (1992).
10 A. Jouyban-Gharamaleki, *Chem. Pharm. Bull.*, **46**, 1058 (1998).
11 W. E. Acree, Jr., J. W. McCargar, A. I. Zvaigzne, and I. L. Teng, *Phys. Chem. Liq.*, **23**, 27 (1991)
12 W. E. Acree, Jr. and A. I. Zvaigzne, *Thermochimica. Acta*, **178**, 151 (1991).
13 A. B. Ochsner, R. J. Belloto Jr., and T. D. Sololoski, *J. Pharm. Sci.*, **74**, 132 (1985).
14 M. Barzegar-Jalali, and J. Hanaee, *Int. J. Pharm.*, **109**, 291 (1994).
15 A. Li, and A. W. Andren, *Environ. Sci. Technol.*, **29**, 3001 (1995).
16 J. K. Fu, and R. G. Luthy, *J. Environ. Eng.*, **112**, 328 (1986).
17 M. Barzegar-Jalali, and A. Jouyban-Gharamaleki, *Int. J. Pharm.*, **140**, 237 (1996).
18 R. M. Dickhut, D. E. Armstrong, and A. W. Andren, *Environ. Toxicol. Chem.*, **10**, 881 (1991).
19 A. Jouyban-Gharamaleki, L. Valaee, M. Barzegar-Jalali, B. J. Clark, and W. E. Acree, Jr., *Intern. J. Pharm.*, **177**, 93 (1999).
20 S. H. Yalkowsky, in **Techniques of solubilization of drugs**; S. H. Yalkowsky, Ed.; *Dekker*, New York, 1984, Chapter 1.

21 S. C. Smolinske, **Handbook of Food, Drug, and Cosmetic Excipients**, *CRC Press*, Ann Arbor, MI, 1992.
22 F. Herzel, and A. S. Murty, *Bull. Environ. Toxicol.*, **32**, 53 (1984).
23 C. Munz, and P. Roberts, *Environ. Sci. Technol.*, **20**, 830 (1986).
24 K. R. Morris, R. Abramowitz, R. Pinal, P. Davis, and S. H. Yalkowsky, *Chemosphere*, **17**, 285 (1988).
25 F. R. Groves, Jr, *Environ. Sci. Technol.*, **22**, 282 (1988).
26 R. M. Dickhut, A. W. Andren, and D. E. Armstrong, *J. Chem. Eng. Data*, **34**, 438 (1989).
27 R. Pinal, P. S. C. Rao, L. S. Lee, P. V. Cline, and S. H. Yalkowsky, *Environ. Sci. Technol.*, **24**, 639 (1990).
28 P. S. C. Rao, L. S. Lee, and A. L. Wood, EPA/600/M-91/009 (1991).
29 A. Li, W. J. Doucette, and A. W. Andren, *Chemosphere*, **24**, 1347 (1992).
30 A. Li, and A. W. Andren, *Environ. Sci. Technol.*, **28**, 47 (1994).
31 A. Li, A. W. Andren, and S. H. Yalkowsky, *Environ. Toxicol. Chem.*, **15**, 2233 (1996).
32 P. Nkedi-Kizza, P. S. C. Rao, and A. G. Hornsby, *Environ. Sci. Technol.*, **19**, 975 (1985).
33 J. K. Fu, and R. G. Luthy, *J. Environ. Eng.*, **112**, 346 (1986).
34 R. W. Walters, and A. Guiseppl-Elie, *Environ. Sci. Technol.*, **22**, 819 (1988).
35 P. Nkedi-Kizza, M. L. Brusseau, P. S. C. Rao, and A. G. Hornsby, *Environ. Sci. Technol.*, **23**, 814 (1989).
36 A. L. Wood, D. C. Bouchard, M. L. Brusseau, and P. S. C. Rao, *Chemosphere*, **21**, 575 (1990).
37 P. S. C. Rao, L. S. Lee, and R. Pinal, *Environ. Sci. Technol.*, **24**, 647 (1990).
38 F. C. Spurlock, and J. W. Biggar, *Environ. Sci. Technol.*, **28**, 1003 (1994).
39 W. J. M. Hegeman, C. H. Van der Weijden, and J. P. G. Loch, *Environ. Sci. Technol.*, **29**, 363 (1995).
40 L. S. Lee, and P. S. C. Rao, *Environ. Sci. Technol.*, **30**, 1533 (1996).
41 V. A. Nzengung, E. A. Voudrias, P. Nkedi-Kizza, J. M. Wampler, and C. E. Weaver, *Environ. Sci. Technol.*, **30**, 89 (1996).
42 R. P. Singh, *Colloids and Surfaces A.*, **122**, 63 (1997).
43 V. A. Nzengung, P. Nkedi-Kizza, R. E. Jessup, and E. A. Voudrias, *Environ. Sci. Technol.*, **31**, 1470 (1997).
44 T. C. Harmon, T. J. Kim, B. K. D. Barre, and C. V. Chrysikopoulos, *J. Environ. Eng.*, January, 87 (1999).
45 M. A. Ei-Zoobi, G. E. Ruch, and F. R. Groves Jr., *Environ. Sci. Technol.*, **24**, 1332 (1990).
46 W. F. Lane, and R. C. Loehr, *Environ. Sci. Technol.*, **26**, 983 (1992).
47 C. S. Chen, and J. J. Delfino, *J. Environ. Eng.*, April 354 (1997).
48 G. R. Jones, *Environ. Prog.*, **11**, 223 (1992).
49 USEPA, Assessment and Remediation of contaminated Sediments (ARCS) Program: Remediation Guidance Document. EPA 905-R94-003, (1994) p.180.
50 USEPA, EPA 542-K-94-006 (1995).
51 C. T. Jafvert, Ground-Water Remediation Technologies Analysis Center, Echnology Evaluation Report, TE-96-02 (1996).
52 R. W. Falta, GWMR, Summer, 94 (1998).
53 D. C. M. Augustijin, R. E. Jessup, P. S. Rao, and A. L. Wood, *J. Environ. Eng.*, **120**, 42 (1994).
54 R. K. Sillan, M. D. Annable, P. S. C. Rao, D. Dai, K. Hatfield, W. D. Graham, A. L. Wood, and C. G. Enfield, *Water Resources Res.*, **34**, 2191 (1998).
55 P. S. C. Rao, M. D. Annable, R. K. Sillan, D. Dai, K. Hatfield, and W. D. Graham, *Water Resources Res.*, **33**, 2673 (1997).
56 J. W. Jawitz, M. D. Annable, P. S. C. Rao, and R. D. Rhue, *Environ. Sci. Technol.*, **32**, 523 (1998).
57 A. T. Kan, M. B. Tomson, and T. A. McRae, Proceedings of the 203rd American Chemical Society National Meeting, San Francisco, CA (1992).
58 D. Brandes, and K. J. Farley, *J. Water Environ. Res.*, **65**, 869 (1993).
59 A. Li, K. A. Cheung, and K. Reddy, *J. Environ. Eng.*, **126**, 527 (2000).
60 A. Li, and S. H. Yalksowsky, *J. Pharm. Sci.*, **83**, 1735 (1994).
61 S. H. Yalkowsky, and T. J. Roseman, in **Techniques of solubilization of drugs**; S. H. Yalkowsky, Ed.; *Dekker*, New York, 1984, Chapter 3.
62 J. T. Rubino, and S. H. Yalkowsky, *J. Pharm. Sci.*, **74**, 416 (1985).
63 P. J. Leinonen, and D. Mackay, *Can. J. Chem. Eng.*, **51**, 230 (1973).
64 R. P. Eganhouse, and J. A. Calder, *Geochimca et Cosmochimica Acta*, **40**, 555 (1976).
65 Y. B. Tewari, D. E. Martire, S. P.; Wasik, and M. M. Miller, *J. Solution Chem.*, **11**, 435 (1982).
66 S. Banerjee, *Environ. Sci. Technol.*, **18**, 587 (1984).
67 D. R. Burris, and W. G. MacIntyre, *Environ. Toxicol. Chem.*, **4**, 371 (1985).
68 H. H. Hooper, S. Michel, and J. M. Prausnitz, *J. Chem. Eng. Data*, **33**, 502 (1988).
69 D. Mackay, *J. Contam. Hydrol.* **8**, 23 (1991).
70 A. Li, and W. J. Doucette, *Environ. Toxicol. Chem.*, **12**, 2031 (1993).
71 S. Lesage, and S. Brown, *J. Contam. Hydrol.*, **15**, 57 (1994).

72 K. Brololm, and S. Feenstra, *Environ. Toxicol. Chem.*, **14**, 9 (1995).
73 G. T. Coyle, T. C. Harmon, and I. H. Suffet, *Environ. Sci. Technol.*, **31**, 384 (1997).
74 P. S. C. Rao, L. S. Lee, P. Nkedi-Kizza, and S. H. Yalkowsky, in **Toxic Organic Chemicals in Porous Media**, Z. Gerstl, Eds. *Springer-Verlag*, New York, 1989, Chapter 8.
75 R. Pinal, L. S. Lee, and P. S. C. Rao, *Chemosphere*, **22**, 939 (1991).
76 S. H. Yalkowsky, and J. T. Rubino, *J. Pharm. Sci.*, **74**, 416 (1985).
77 J. T. Rubino, and S. H. Yalkowsky, *J. Parent. Sci. Technol.*, **41**, 172 (1984).
78 W. J. Lyman, W. f. Reehl, and D. H. Rosenblatt, **Handbook of Chemical Property Estimation Methods: Environmental Behavior of Organic Compounds**. *ACS Publications*, Washington, DC, 1990.
79 C. Hansch, and A. J. Leo, **Substituent constants for Correlation Analysis in Chemistry and Biology**. *John Wiley*, New York, 1979.
80 A. Li, and S. H. Yalkowsky, *Ind. Eng. Chem. Res.*, **37**, 4470 (1998).
81 J. T. Rubino, and S. H. Yalkowsky, *Pharm. Res.*, **4**, 220 (1987).
82 A. Li, and S. H. Yalkowsky, *Ind. Eng. Chem. Res.*, **37**, 4476 (1998).
83 R. Abramowitz, Ph.D. Dissertation, University of Arizona, 1986.
84 A. Li, and S. H. Yalkowsky, unpublished data.
85 J. T. Rubino, Ph.D. Dissertation, University of Arizona, 1984.
86 S. H. Yalkowsky, unpublished data.
87 J. T. Rubino, and S. H. Yalkowsky, *Pharm. Res.*, **4**, 231 (1987).
88 A. Fredenslund, R. L. Jones, and J. M. Prausnitz, *A.I.Ch.E. J.*, **21**, 1086 (1975).
89 J. Gmehling, P. Rasmussen, and A. Fredenslund, *Ind. Eng. Chem. Process Des. Dev.*, **21**, 118 (1982).
90 E. A. Macedo, U. Weidlich, J. Gmehling, and P. Rasmussen, *Ind. Eng. Chem. Process Des. Dev.*, **22**, 678 (1983).
91 J. T. Rubino, and E. K. Obeng, *J. Pharm. Sci.*, **80**, 479 (1991).
92 L. S. Lee, C. A. Bellin, R. Pinal, and P. S. C. Rao, *Environ. Sci. Technol.*, **27**, 165 (1997).

## 14.22 POLYMERS AND MAN-MADE FIBERS

GEORGE WYPYCH
**ChemTec Laboratories, Inc., Toronto, Canada**

The resin production industry has over 450 plants in the USA with total sales of $33 billion/year and about 60,000 employees. The man-made fiber industry has over 90 plants. It employs about 45,000 people and it has sales of $13 billion/year.

In the polymer manufacture industry, production processes are diverse both in technology and equipment design. They have common steps which include preparation of reactants, polymerization, polymer recovery, polymer extrusion (if in pelletized form), and supporting operations. In some preparation operations, solvents are used to dissolve or dilute monomer and reactants. Solvent are also used to facilitate the transportation of the reaction mixture throughout the plant, to improve heat dissipation during the reaction, and to promote uniform mixing. Solvent selection is optimized to increase monomer ratio and to reduce polymerization costs and emissions. The final polymer may or may not be soluble in the solvent. These combinations of polymers and solvents are commonly used: HDPE - isobutane and hexane, LDPE - hydrocarbons, LLDPE - octene, butene, or hexene, polypropylene - hexane, heptane or liquid propylene, polystyrene - styrene or ethylbenzene, acrylic - dimethylacetamide or aqueous inorganic salt solutions. These examples show that there are options available. Excess monomer may replace solvent or water can be used as the solvent. During polymer recovery unreacted monomer and solvents are separated from polymer (monomers and solvents are flashed off by lowering the pressure and sometimes degassing under vacuum), liquids and solids are separated (the polymer may be washed to remove sol-

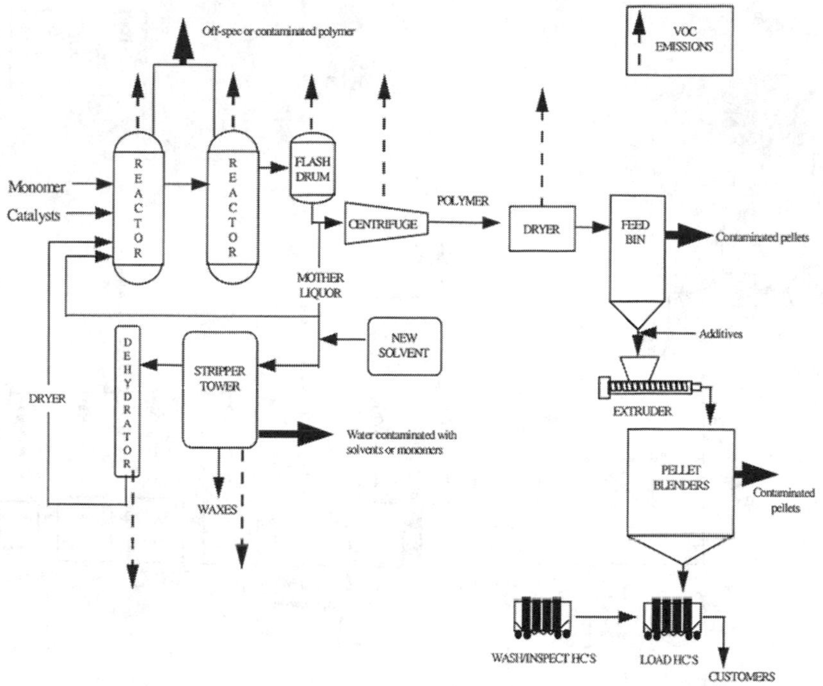


Figure 14.22.1. Schematic diagram of emissions from the polymer manufacturing industry. [Reproduced from EPA Office of Compliance Sector Notebook Project. Profile of the Petroleum Refining Industry. US Environmental Protection Agency, 1995.]

vent), and residual water and solvent are purged during polymer drying. Residual solvents are removed by further drying and extrusion. Solvents are also used in equipment cleaning. Solvents are often stored under a nitrogen blanket to minimize oxidation and contamination. When these systems are vented solvent losses occur. Figure 14.22.1 shows a schematic diagram of potential emissions during polymer manufacture.

Manufacture of man-made fibers involves polymerization (usually the core part of the process), preparation of the solution, spinning, washing and coagulation, drying and other operations. Fibers are formed by forcing the viscous liquid through small-bore orifices. A suitable viscosity can be achieved either by heating or dissolution. The rheological properties of the solution are governed to a large degree by the solvents selected. Wastes generated during the spinning operation include evaporated solvent and wastewater contaminated by solvent. The typical solvents used in the production of fibers are dimethylacetamide (acrylic), acetone or chlorinated hydrocarbon (cellulose acetate), and carbon disulfide (rayon). In the dry spinning process a solution of polymer is first prepared. The solution is then heated above the boiling temperature of the solvent and the solution is extruded through spinneret. The solvent evaporates into the gas stream. With wet spinning the fiber is directly extruded into a coagulation bath where solvent diffuses into the bath liquid and the coagulant diffuses into the fiber. The fiber is washed free of solvent by passing it through an additional bath. Each process step generates emissions or wastewater. Solvents used in production are normally recovered by distillation. Figure 14.22.2 is a schematic diagram of fiber production showing that almost all stages of production generate emissions.

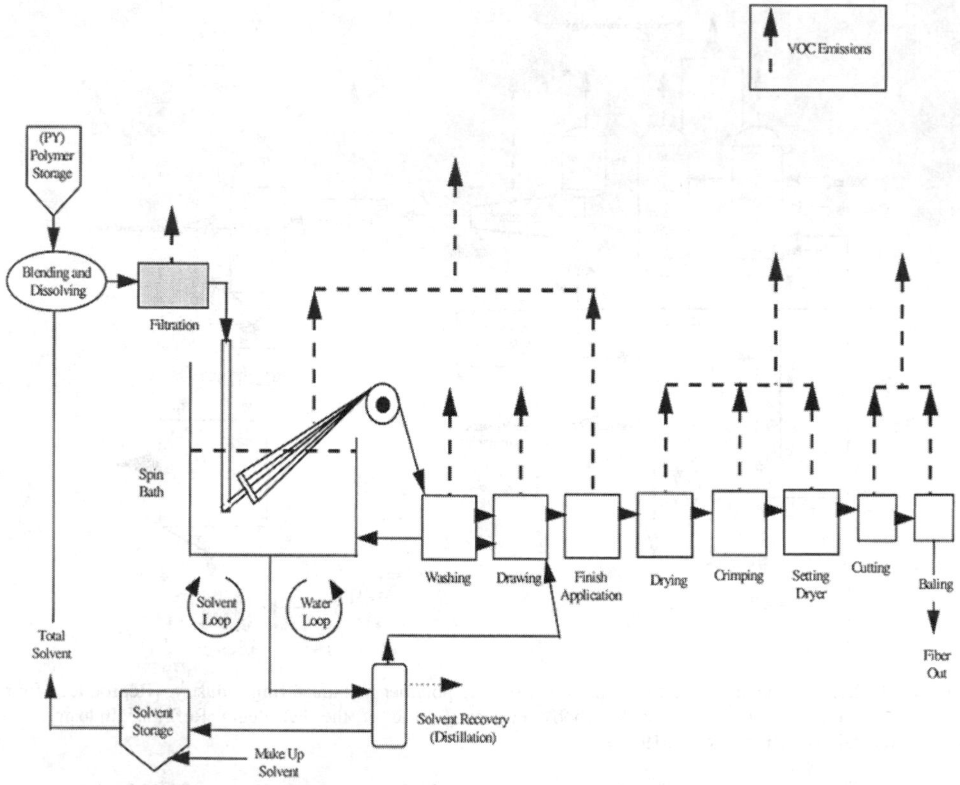


Figure 14.22.2. Schematic diagram of emissions from the man-made fiber manufacturing industry. [Reproduced from EPA Office of Compliance Sector Notebook Project. Profile of the Petroleum Refining Industry. US Environmental Protection Agency, 1995.]

Tables 14.22.1 and 14.22.2 provide data on releases and transfers from both polymer manufacturing and man-made fiber production in the USA. Carbon disulfide, methanol, xylene, and ethylene glycol are used in the largest quantities. Carbon disulfide is used in manufacture of regenerated cellulose and rayon. Ethylene glycol is used in the manufacture of polyethylene terephthalate, the manufacture of alkyd resins, and as cosolvent for cellulose ethers and esters. Methanol is used in several processes, the largest being in the production of polyester. This industry is the 10th largest contributor of VOC and 7th largest in releases and transfers.

There have been many initiatives to reduce emissions and usage of solvents. Man-made fiber manufacturing no longer uses benzene. DuPont eliminated o-xylene and reduced methanol and ethylene glycol use in its Wilmington operation. This change resulted in annual savings of $1 million. Process modification in a polymer processing plant resulted in a decrease in total emissions of 74% and a reduction in the release of cyclohexane by 96%. Monitoring of thousands of valves in Eastman Texas plant resulted in a program of valve replacement which eliminated 99% of the emissions. Plant in Florida eliminated solvents from cleaning and degreasing. These examples show that in many cases pollution can be reduced by better equipment, organization, and care.

**Table 14.22.1 Reported solvent releases from the polymer and man-made fiber industry in 1995 [Data from Ref. 1]**

| Solvent | Amount, kg/year | Solvent | Amount, kg/year |
|---|---|---|---|
| allyl alcohol | 29,000 | 1,4-dioxane | 10,000 |
| benzene | 60,000 | ethylbenzene | 130,000 |
| n-butyl alcohol | 480,000 | ethylene glycol | 1,400,000 |
| sec-butyl alcohol | 25,000 | hexane | 880,000 |
| tert-butyl alcohol | 16,000 | methanol | 3,600,000 |
| carbon disulfide | 27,500,000 | methyl ethyl ketone | 260,000 |
| carbon tetrachloride | 100 | methyl isobutyl ketone | 98,000 |
| chlorobenzene | 19,000 | pyridine | 67,000 |
| chloroform | 14,000 | tetrachloroethylene | 4,000 |
| cresol | 4,000 | 1,1,1-trichloroethane | 120,000 |
| cyclohexane | 98,000 | trichloroethylene | 39,000 |
| 1,2-dichloroethane | 98,000 | 1,2,4-trimethylbenzene | 12,000 |
| dichloromethane | 1,300,000 | toluene | 900,000 |
| N,N-dimethylformamide | 19,000 | xylene | 460,000 |

**Table 14.22.2. Reported solvent transfers from the polymer and man-made fiber industry in 1995 [Data from Ref. 1]**

| Solvent | Amount, kg/year | Solvent | Amount, kg/year |
|---|---|---|---|
| allyl alcohol | 120,000 | ethylbenzene | 880,000 |
| benzene | 160,000 | ethylene glycol | 49,000,000 |
| n-butyl alcohol | 330,000 | hexane | 8,000,000 |
| sec-butyl alcohol | 12,000 | methanol | 5,600,000 |
| tert-butyl alcohol | 160,000 | methyl ethyl ketone | 460,000 |
| carbon disulfide | 14,000 | methyl isobutyl ketone | 43,000 |
| carbon tetrachloride | 200,000 | N-methyl-2-pyrrolidone | 780,000 |
| chlorobenzene | 570,000 | pyridine | 70,000 |
| chloroform | 59,000 | tetrachloroethylene | 330,000 |
| cresol | 20,000 | 1,1,1-trichloroethane | 21,000 |
| cyclohexane | 420,000 | trichloroethylene | 76,000 |
| dichloromethane | 250,000 | 1,2,4-trimethylbenzene | 98,000 |
| N,N-dimethylformamide | 300,000 | toluene | 2,800,000 |
| 1,4-dioxane | 11,000 | xylene | 7,800,000 |

New technology is emerging to reduce solvent use. Recent inventions disclose that, in addition to reducing solvents, the stability of ethylene polymers can be improved with the new developed process.[3] A proper selection of solvent improved a stripping operation and contributed to the better quality of cyclic esters used as monomers.[4] Solvent was used for the recovery of fine particles of polymer which were contaminating water.[5] A new process for producing fiber for cigarette filters uses reduced amounts of solvent.[6] Optical fibers are manufactured by radiation curing which eliminates solvents.[7] A new electrospinning process has been developed which produces unique fibers by the dry spinning method, providing a simpler separation and regeneration of the solvent.[8]

## REFERENCES


1 EPA Office of Compliance Sector Notebook Project. Profile of the Petroleum Refining Industry. US Environmental Protection Agency, 1995.
2 EPA Office of Compliance Sector Notebook Project. Sector Notebook Data Refresh - 1997. US Environmental Protection Agency, 1998.
3 M M Hughes, M E Rowland. C A Strait, **US Patent 5,756,659**, The Dow Chemical Company, 1998.
4 D W Verser, A Cheung, T J Eggeman, W A Evanko, K H Schilling, M Meiser, A E Allen, M E Hillman, G E Cremeans, E S Lipinsky, **US Patent 5,750,732**, Chronopol, Inc., 1998.
5 H Dallmeyer, **US Patent 5,407,974**, Polysar Rubber Corporation, 1995.
6 J N Cannon, **US Patent 5,512,230**, Eastman Chemical Company, 1996.
7 P J Shustack, **US Patent 5,527,835**, Borden, Inc., 1996.
8 A E Zachariades, R S Porter, J Doshi, G Srinivasan, D H Reneker, *Polym. News*, **20**, No.7, 206-7 (1995).


## 14.23 PRINTING INDUSTRY

GEORGE WYPYCH
**ChemTec Laboratories, Inc., Toronto, Canada**

The number of printing and publishing operations in the US is estimated at over 100,000. 1.5 million people are employed. The value of shipments is over $135 billion. 97% of printing is done by lithography, gravure, flexography, letterpress, and screen printing on substrates such as paper, plastic metal, and ceramic. Although, these processes differ, the common feature is the use of cleaning solvents in imaging, platemaking, printing, and finishing operation. Most inks contain solvents and many of the adhesives used in finishing operations also contain solvents. Many processes use the so-called fountain solutions which are applied to enable the non-image area of the printing plate to repel ink. These solutions contain primarily isopropyl alcohol. But the printing operation is, by itself, the largest contributor of VOCs. Each printing process requires inks which differ drastically in rheology. For example, gravure printing requires low viscosity inks which contain a higher solvent concentrations.

Tables 14.23.1 and 14.23.2 provide data on the reported releases and transfers of solvents by the US printing industry. These data show that there are fewer solvents and relatively low releases and transfers compared with other industries. In terms of VOC contribution, the printing industry is 5th and 10th in the total emissions and transfers.

Current literature shows that there is extensive activity within and outside industry to limit VOCs and reduce emissions. Cleaning operations are the major influence on emissions. Shell has developed a new cleaning formulations containing no aromatic or chlorinated hydrocarbons.[3] An additional requirement was to optimize the solvent mixture to prevent swelling of the rubber in blanket cylinders and rollers. It is predicted that the European industry will increase rate of the introduction of radiation-cured inks and eliminate isopropanol from fountain solutions.[4] It is expected that radiation-cured flexographic inks will grow by 30%/year in the next five years.[5] In Germany, 70-80% of emissions or 47,000 ton/year will have to be eliminated by the year 2007.[6] Beginning in 1999, the UK industry must keep VOC concentration below 5 tonnes/year per plant.[7] VOC concentration in outside atmosphere must not exceed 150 $mg/m^3$ (50 $mg/m^3$ if there is more than 5% aromatic solvents). Reactive hot melts are being used in book binding.[8] This will eliminate emissions from currently used solvent adhesives.

Solvent replacements are not the only solution at hand. Solvent-containing systems often give better quality than replacement systems, therefore methods have been developed to make the solvent based materials more acceptable. A soil bed biofiltration system was tested in California with excellent results.[9] This biofilter is a bed of soil impregnated with microorganisms which use VOC as their food. Present California regulations require that such a treatment system has a 67% capture and VOC destruction efficiency. The new method was proven to have 95.8% efficiency. In addition to environmental issues with solvents, the printing industry has addressed the source of their raw materials. Present systems are based on petroleum products which are not considered renewable resources. Terpenes are natural products which are now finding applications in the print industry.[10] In Denmark, of 70% cleaning solvents are vegetable oil based. These and other such innovations will continue to be applied to reduce solvent use and emissions. It is also reported[10] that water-based system replaces fountain solutions.

Other factors are driving changes. Odors in packaging materials and the migration of solvent to foods are unacceptable. Most odors in packaging materials are associated with process and coalescing solvents.[11] Foods which do not contain fat are more susceptible to the retaining the taste of solvents. Printing inks which may be acceptable for foods containing fat may not be suitable for fat-free applications (see more on this subject in Chapter 16.1).[12]

Many recent inventions have also been directed at solving the current environmental problems of printing industry.[13-21] The solvent in gravure printing inks not only contribute to pollution but also to the cost of solvent recovery and/or degradation. A new technology is proposed in which a solvent free ink with a low melting point can be processed in liquid state and then be solidified on cooling.[13] A non-volatile solvent for printing inks was developed based on a cyclic keto-enol tautomer and a drying oil.[14] An alcohol soluble polyamide for rotary letterpress printing inks was developed[15] and subsequently adapted to flexographic/gravure inks.[18] A polyamide was also used in a rotary letterpress ink which enabled low alcohols to be used as the solvent with some addition of an ester.[16] This new ink is compatible with water-based primers and adhesives which could not be used with solvent-based inks. Inks for jet printers are water sensitive. One solvent-based technology was developed using esters and glycols[17] and the other using low alcohols.[20] Another recent invention describes aqueous ink containing some low alcohols.[19] UV and electron beam cured ink concentrates were also developed.[21]

This information from open and patent literature clearly indicates that industry is actively working on the development of new technological processes to reduce emissions of solvents.

**Table 14.23.1. Reported solvent releases from the printing and publishing industry in 1995 [Data from Ref. 2]**

| Solvent | Amount, kg/year | Solvent | Amount, kg/year |
|---|---|---|---|
| n-butyl alcohol | 43,000 | methyl isobutyl ketone | 170,000 |
| dichloromethane | 59,000 | N-methyl-2-pyrrolidone | 31,000 |
| 1,4-dioxane | 8,000 | tetrachloroethylene | 34,000 |
| ethylene glycol | 46,000 | 1,1,1-trichloroethane | 180,000 |
| ethylbenzene | 23,000 | trichloroethylene | 13,000 |
| hexane | 50,000 | 1,2,4-trimethylbenzene | 36,000 |
| isopropyl alcohol | 27,000 | toluene | 12,200,000 |
| methanol | 170,000 | xylene | 700,000 |
| methyl ethyl ketone | 960,000 | | |

**Table 14.23.2. Reported solvent transfers from the printing and publishing industry in 1995 [Data from Ref. 2]**

| Solvent | Amount, kg/year | Solvent | Amount, kg/year |
|---|---|---|---|
| n-butyl alcohol | 7,000 | methyl isobutyl ketone | 63,000 |
| dichloromethane | 43,000 | N-methyl-2-pyrrolidone | 28,000 |
| 1,4-dioxane | 340 | tetrachloroethylene | 27,000 |
| ethylene glycol | 16,000 | 1,1,1-trichloroethane | 39,000 |
| ethylbenzene | 9,200 | trichloroethylene | 4,000 |
| hexane | 12,000 | 1,2,4-trimethylbenzene | 33,000 |
| isopropyl alcohol | 12,000 | toluene | 2,800,000 |
| methanol | 17,000 | xylene | 240,000 |
| methyl ethyl ketone | 700,000 | | |

## REFERENCES

1 EPA Office of Compliance Sector Notebook Project. Profile of the Printing and Publishing Industry. US Environmental Protection Agency, 1995.
2 EPA Office of Compliance Sector Notebook Project. Sector Notebook Data Refresh - 1997. US Environmental Protection Agency, 1998.
3 N C M Beers, M J C M Koppes, L A M Rupert, *Pigment & Resin Technol.*, **27**, No.5, 289-97 (1998).
4 D Blanchard, *Surface Coatings International*, **80**, No.10, 476-8 (1997).
5 B Gain, *Chem. Week*, **160**, No.14, 28-30 (1998).
6 W Fleck, *Coating*, **31**(1), 23-25 (1998).
7 C H Williams, *Converter*, **34**, No.9, 11-2 (1997).

8 Hughes F, TAPPI 1997 Hot Melt Symposium. Conference Proceedings. *TAPPI.* Hilton Head, SC, 15th-18th June 1997, p.15-21.
9 A Mykytiuk, *Paper, Film & Foil Converter*, **72**, No.8, 120-3 (1998).
10 A Harris, *Paper, Film & Foil Converter*, **72**, No.5, 198-9 (1998).
11 R M Podhajny, *Paper, Film & Foil Converter*, **72**, No.12, 24 (1998).
12 T Clark, *Paper, Film & Foil Converter*, **70**, No.11, 48-50 (1996).
13 R Griebel, K A Kocherscheid, K Stammen, **US Patent 5,496,879**, Siegwerk Druckfarben GmbH, 1996.
14 D Westerhoff, **US Patent 5,506,294**, 1996.
15 P D Whyzmuzis, K Breindel, R A Lovald, **US Patent 5,523,335**, Henkel Corporation, 1996.
16 R J Catena, M C Mathew, S E Barreto, N Marinelli, **US Patent 5,658,968**, Sun Chemical Corporation, 1997.
17 J M Kruse, **US Patent 5,663,217**, XAAR Ltd., 1997
18 P D Whymusis, **US Patent 5,714,526**, Henkel Corporation, 1998.
19 H Yanagi, S Wakabayashi, K Kaida, **US Patent 5,736,606**, Kao Corporation, 1998.
20 M Shinozuka, Y Miyazawa, M Fujino, T Ito, O Ishibashi, **US Patent 5,750,592**, Seiko Epson Corporation, 1998.
21 W R Likavec, C R Bradley, **US Patent 5,866,628**, Day-Glo Color Corporation, 1999.

## 14.24 PULP AND PAPER

GEORGE WYPYCH
**ChemTec Laboratories, Inc., Toronto, Canada**

The US pulp and paper industry operates over 550 facilities which employ over 200,000 people. Total shipments are $60 billion with an additional $80 billion in converted products. Several processes contribute to the emission of solvents. These include chemical pulping kraft process (terpenes, alcohols, methanol, acetone, chloroform), bleaching (acetone, dichloromethane, chloroform, methyl ethyl ketone, carbon disulfide, chloromethane, and trichloroethane), wastewater treatment (terpenes, alcohols, methanol, acetone, chloroform and methyl ethyl ketone), and evaporators in chemical recovery systems (alcohols and terpenes).

**Table 14.24.1. Reported solvent releases from the pulp and paper industry in 1995 [Data from Ref. 2]**

| Solvent | Amount, kg/year | Solvent | Amount, kg/year |
|---|---|---|---|
| benzene | 320,000 | methanol | 63,000,000 |
| n-butyl alcohol | 46,000 | methyl ethyl ketone | 700,000 |
| chloroform | 4,500,000 | methyl isobutyl ketone | 10,000 |
| chloromethane | 260,000 | 1,2,4-trimethylbenzene | 17,000 |
| cresol | 410,000 | toluene | 580,000 |
| ethylbenzene | 22,000 | xylene | 49,000 |
| ethylene glycol | 37,000 | o-xylene | 260 |
| hexane | 150,000 | | |

**Table 14.24.2. Reported solvent transfers from the pulp and paper industry in 1995 [Data from Ref. 2]**

| Solvent | Amount, kg/year | Solvent | Amount, kg/year |
|---|---|---|---|
| benzene | 24,000 | hexane | 8,600 |
| n-butyl alcohol | 16,000 | methanol | 23,000,000 |
| chloroform | 150,000 | methyl ethyl ketone | 36,000 |
| chloromethane | 120 | 1,2,4-trimethylbenzene | 1,400 |
| cresol | 3,600 | toluene | 23,000 |
| ethylene glycol | 190,000 | xylene | 4,000 |

**Table 14.25.1. Reported solvent releases from the rubber and plastics industry in 1995 [Data from Ref. 2]**

| Solvent | Amount, kg/year | Solvent | Amount, kg/year |
|---|---|---|---|
| benzene | 5,800 | ethylene glycol | 120,000 |
| n-butyl alcohol | 380,000 | hexane | 1,700,000 |
| sec-butyl alcohol | 17,000 | isopropyl alcohol | 28,000 |
| tert-butyl alcohol | 240 | methanol | 4,000,000 |
| carbon disulfide | 5,500,000 | methyl ethyl ketone | 5,500,000 |
| chlorobenzene | 5,000 | methyl isobutyl ketone | 1,100,000 |
| chloroform | 46,000 | N-methyl-2-pyrrolidone | 32,000 |
| chloromethane | 47,000 | tetrachloroethylene | 160,000 |
| cresol | 9,000 | 1,1,1-trichloroethane | 3,000,000 |
| cyclohexane | 480,000 | trichloroethylene | 660,000 |
| dichloromethane | 11,700,000 | 1,2,4-trimethylbenzene | 260,000 |
| N,N-dimethylformamide | 350,000 | toluene | 7,600,000 |
| 1,4-dioxane | 2,600 | xylene | 2,200,000 |
| ethylbenzene | 210,000 | m-xylene | 6,000 |

Tables 14.24.1 and 14.24.2 give the reported releases and transfers of solvent data for the US pulp and paper industry. If not for the emissions of methanol and chloroform the industry would be a much less serious polluter. It is 7th in VOC contributions and 8th in total releases and transfers.

## REFERENCES


1 EPA Office of Compliance Sector Notebook Project. Profile of the Pulp and Paper Industry. US Environmental Protection Agency, 1995.

2 EPA Office of Compliance Sector Notebook Project. Sector Notebook Data Refresh - 1997. US Environmental Protection Agency, 1998.

**Table 14.25.2. Reported solvent transfers from the rubber and plastics industry in 1995 [Data from Ref. 2]**

| Solvent | Amount, kg/year | Solvent | Amount, kg/year |
|---|---|---|---|
| benzene | 15,000 | hexane | 50,000 |
| n-butyl alcohol | 370,000 | isopropyl alcohol | 14,000 |
| sec-butyl alcohol | 1,100 | methanol | 1,400,000 |
| tert-butyl alcohol | 85,000 | methyl ethyl ketone | 3,500,000 |
| carbon disulfide | 150,000 | methyl isobutyl ketone | 450,000 |
| chloroform | 1,200 | N-methyl-2-pyrrolidone | 120,000 |
| chloromethane | 330 | tetrachloroethylene | 47,000 |
| cresol | 3,000 | 1,1,1-trichloroethane | 160,000 |
| cyclohexane | 350,000 | trichloroethylene | 170,000 |
| dichloromethane | 900,000 | 1,2,4-trimethylbenzene | 14,000 |
| N,N-dimethylformamide | 570,000 | toluene | 2,200,000 |
| 1,4-dioxane | 49,000 | xylene | 940,000 |
| ethylbenzene | 350,000 | m-xylene | 5,700 |
| ethylene glycol | 15,300,000 | | |

## 14.25 RUBBER AND PLASTICS

George Wypych
**ChemTec Laboratories, Inc., Toronto, Canada**

The US rubber and plastics industry employs over 800,000 people and operates over 12,000 plants. Its total production output is estimated at over $90 billion. The industry produces a wide diversity of products some of which do not contain solvents but many of which require the use of process solvents. Solvents are contained in adhesives used in finishing operations. Large quantities of solvents are used for surface cleaning and cleaning of equipment.

Tables 14.25.1 and 14.25.2 provide data on the reported releases and transfers of solvents by the US rubber and plastics industry. These industries contribute small amounts of VOC which are in the range of 0.00001-0.00005 kg VOC/kg of processed rubber. It was the ninth largest contributor to releases and transfers of all US industries. Dichloromethane, toluene, carbon disulfide, methyl ethyl ketone, methanol, 1,1,1-trichloroethane, hexane, methyl isobutyl ketone, and xylene are emitted in very large quantities.

## REFERENCES

1 EPA Office of Compliance Sector Notebook Project. Profile of the Rubber and Plastics Industry. US Environmental Protection Agency, 1995.
2 EPA Office of Compliance Sector Notebook Project. Sector Notebook Data Refresh - 1997. US Environmental Protection Agency, 1998.

# 14.26 USE OF SOLVENTS IN THE SHIPBUILDING AND SHIP REPAIR INDUSTRY

MOHAMED SERAGELDIN
**U.S. Environmental Protection Agency, Research Triangle Park, NC, USA**
DAVE REEVES
**Midwest Research Institute, Cary, NC, USA**

## 14.26.1 INTRODUCTION

The focus of this chapter will be on the use of solvents in the shipbuilding and ship repair industry. This industrial sector is involved in building, repairing, repainting, converting, or alteration of marine and fresh water vessels. These vessels include self-propelled vessels, those propelled by other vessels (barges), military and Coast Guard vessels, commercial cargo and passenger vessels, patrol and pilot boats, and dredges. The industry sector is also involved in repairing and coating navigational aids such as buoys. This chapter begins with an overview of operations in a typical shipbuilding and/or ship repair facility (shipyard), to identify those operations that generate significant volatile organic compound (VOC) emissions and/or hazardous air pollutant (HAP) emissions from the use of organic solvents. Organic solvents that are VOCs contribute to formation of ozone in the troposphere. Other organic solvents such as chlorinated fluorocarbons (CFCs) cause depletion of the ozone layer in the stratosphere. Therefore, VOCs and other air toxics, such as those compounds listed as HAPs, are both indirectly and directly detrimental to the general public's health. Because many solvents are VOCs and often contain large amounts of HAPs, many state agencies[1,2] and the United States Environmental Protection Agency (U.S. EPA) have issued regulations to limit their content in materials used for surface coating and cleaning operations at shipyards.[3-7]

## 14.26.2 SHIPBUILDING AND SHIP REPAIR OPERATIONS

Most facilities engaged in shipbuilding or ship repair activities (shipyards) have several manufacturing areas in common, each including one or more "unit operations". These areas include: (a) surface preparation of primarily steel surfaces, which may include cleaning with multiple organic solvents; (b) assembly operations, which involve assembly of blocks that were constructed from sub-assembled parts (this step involves steel cutting and material movement using heavy equipment such as cranes); (c) cleaning operations (other than surface preparation) such as equipment and parts cleaning; and (d) coating operations.[8,9] There are secondary operations such as chrome plating, asbestos removal, fuel combustion, carpentry, and, to various degrees, polyester lay-up operations (composite materials construction activities). We will next discuss those operations that involve the use of organic cleaning solvents.

## 14.26.3 COATING OPERATIONS

Marine coatings can be applied by the use of spraying equipment, brushes, or rollers. Coating operations at shipyards are typically conducted at two primary locations: (1) outdoor work areas or (2) indoor spray booths. The outdoor work areas can include ship exteriors and interiors. Most shipyards report that typically only a small percentage (10%) of the coating operations are done indoors. However, in large construction yards a larger proportion (up to 30 %) of the coatings are applied indoors.[10] Coating and cleaning operations constitute the major source of VOC and HAP emissions from shipyards. If the metal surface is not well prepared before a coating is applied or if the coating is applied at the wrong ambient conditions, the coating system may fail and the work may have to be redone. The amount of cleaning necessary will depend on the type and extent of the problem and the coating system that is being used.

## 14.26.4 CLEANING OPERATIONS USING ORGANIC SOLVENTS

In most industrial applications involving metal substrates, organic cleaning solvents are used to remove contaminants or undesirable materials from surfaces before a coating is applied to clean equipment and parts utilized to apply the coating or soiled during that operation. Solvents are used for general maintenance of equipment parts. These surfaces are typically made of steel. However, vessels may also be made from natural materials such as wood and synthetic materials such as fiberglass. Therefore, a solvent must be selected that will not attack the substrate being cleaned.

For material accounting purposes, we can classify cleaning (unit) operations as follows:[11,12]

1. Surface preparation of large manufactured components (stage before a coating is applied).
2. Surface preparation of small manufactured components (stage before a coating is applied).
3. Line cleaning (includes piping network and any associated tanks).
4. Gun cleaning (manually or in a machine).
5. Spray booth cleaning (walls and floor).
6. Tank cleaning (mostly inner tank surfaces and any associated pipes).
7. Parts (machine) cleaning (simple dip tanks and large machines).
8. Cleaning of equipment and other items (e.g., bearings, buckets, brushes, contact switches).
9. Floor cleaning (organic solvents are no longer used).

These categories are similar to those found in other industries involved in the application of surface coating. However, the number of cleaning categories varies from one industry to another. For example, the automotive manufacturing industry (SIC code 3711) and the furniture industry are involved to various degrees in all nine types of cleaning operations. On the other hand, the photographic supplies (chemicals) industry will not include the first three listed cleaning operations.[11]

### 14.26.4.1 Surface preparation and initial corrosion protection

Large manufactured ship components are often cleaned with an organic solvent as the first of a number of cleaning steps that are required before a coating is applied. The method of surface preparation is selected to work with a chosen coating system. Surface preparation may include application of chemicals such as etching agents, organic solvents cleaners, and alkaline cleaners. Organic solvents such as mineral spirits, chlorinated solvents, and coal tar solvents are used to remove unwanted materials such as oil and grease.[13] If a ship is being repaired, existing coatings usually need to be removed. Solvents such as dichloromethane are commonly used for removing (stripping off) old or damaged coatings. However, aqueous systems involving caustic compounds are now being used more frequently for such purposes.[14] Pressure washing and hydro blasting are other cleaning techniques used. But, the predominant method is still particulate blasting (using abrasive media), which is used to remove mil scale, extra weld material, rust, and old coatings.

The angle at which the surface is blasted is chosen to generate the desired peaks and valleys on the substrate, that will accommodate the viscosity, chemistry (polar groups) of the primer coating. The surface profiling will also help the primer coating adhere mechanically to the substrate, contributing to the longevity of the coating system.[15] Pre-construction primers are sometimes used immediately following surface preparation (blasting) to prevent steel from oxidizing (rusting). This primer is removed by particulate blasting, before the protective coating system (one or more coatings) is applied to the assembled parts or blocks. Removal of such primers (when they cannot be welded-through) can result in emissions of VOCs and HAPs.

### 14.26.4.2 Cleaning operations after coatings are applied

Surface coating operations at shipyards use predominantly solvent-based coatings. Hence, relatively large amounts of organic solvents are used for cleaning and thinning activities. Table 14.26.1 shows the most common organic solvents used for thinning and cleaning, based on 1992 data.[16] Table 14.26.2 gives examples of solvent products that can be used for both thinning coatings and for cleaning surfaces after coatings are applied and for maintenance cleaning. The solvent products are listed in decreasing order of evaporative rate. Acetone, a ketone solvent is commonly used for cleaning and thinning polyester resins and gel coats. However, it is also used in formulating low-VOC and low-HAP products. Methyl ethyl ketone (MEK) and methyl isobutyl ketone (MIBK) are fast evaporative solvents that are used for thinning and cleaning vinyl coatings, epoxy coatings, and many other high performance coatings. Fast evaporative coatings that can improve application properties for a good finish may also be formulated by blending different solvents. Examples are shown in Table 14.26.2. The fast evaporative mix includes solvents varying in polarity and solubility parameters. They include an oxygenated solvent (MIBK), aromatic hydrocarbon solvents that contain less than 10 percent (by mass) HAPs, and aromatic hydrocarbons like xylene that are 100 percent HAPs as will be shown later. Together they produce the correct solvency for the polymer (resin).

**Table 14.26.1. Predominant solvents used in marine coatings [from ref. 16 ] and EPA regulatory classifications**

| Organic solvent | VOC | HAP, Sec. 112 (d) | Toxic chemicals, Sec. 313 |
|---|---|---|---|
| ALCOHOLS | | | |
| Butyl alcohol | Y | Y | Y |
| Ethyl alcohol | Y | N | N |
| Isopropyl alcohol | Y | N | Y[a] |
| AROMATICS | | | |
| Xylene | Y | Y | Y |
| Toluene | Y | Y | Y |
| Ethyl benzene | Y | Y | Y |

| Organic solvent | VOC | HAP, Sec. 112 (d) | Toxic chemicals, Sec. 313 |
|---|---|---|---|
| ETHERS | | | |
| Ethylene glycol ethers | Y | Y | Y |
| Propylene glycol ethers | Y | N | N |
| KETONES | | | |
| Acetone | N | N | Y |
| Methyl ethyl ketone | Y | Y | Y |
| Methyl isobutyl ketone | Y | Y | Y |
| Methyl amyl ketone | Y | N | N |
| PARAFFINIC | | | |
| Mineral spirits | Y | $Y^b$ | N |
| High-flash naphtha | Y | $Y^b$ | Y |
| n-Hexane | Y | Y | N |

VOC = volatile organic compound; HAP = Hazardous air pollutant; Sec 313 of the Emergency Right-to-know Act (EPCRA), also known as Title III of the Superfund Amendments and Reauthorization Act of 1986 (40 CFR Part 372). [a]Use of strong acid process, no supplier notification. [b]Ligroine (light naphtha), VM&P naphtha, Stoddard solvent, and certain paint thinners are also commonly referred to as mineral spirits. These distillation fractions contain less than 10 % by mass HAPs (see Table 14.26.4).

**Table 14.26.2. Selected products that are used as both solvent thinners and solvent cleaners**

| Thinner & cleaning solvents | Typical coating | Compound, wt% | Solubility parameter, $(cal/cm^3)^{1/2}$ | Relative rate, nBUOAc =1.0 | Vapor press., mmHg @ 20°C | B.P., °C @ 760 mmHg | Surface tension, dynes/cm 20°C | Av. sp. gr, @ 25°C | Viscosity, cP @ 25°C |
|---|---|---|---|---|---|---|---|---|---|
| Acetone | Polyester | Acetone, 100% (approx.) | 9.8 | 6.1 | 186 | 56 | 27.1 | 0.787 | 0.31 |
| MEK* | Vinyl | MEK, 100 % (approx.) | 9.3 | 4.0 | 70 | 80 | 24.2 | 0.806 | 0.43 |
| Spraying thinner & solvent cleaner (fast)[b] | Epoxy | MIBK**, 24 % | 8.58 | 1.7 | 28 | 116 | 23.3 | 0.796 | 0.54 |
| | | N-butyl alc., 24% | 11.6 | 0.44 | 5.5 | 118 | 23.4 | 0.806 | 2.62 |
| | | Toluene, 52% | 8.93 | 2.0 | 22 | 111 | 28.2 | 0.863 | 0.57 |
| Brushing thinner & solvent cleaner (medium)[b] | Epoxy | MIBK, 23 % | 8.58 | 1.7 | 28 | 116 | 23.3 | 0.796 | 0.54 |
| | | EGBE***, 26% | 10.2 | 0.072 | 0.6 | >169 | 26.9 | 0.899 | 3.0 |
| | | AHC****, 30% | 7.7 | 0.16 | 2.0 | >160 | 23.4 | 0.775 | 0.88 |
| | | 1,2,4-Trimethylbenzene, 16% | 8.9 | 19 | 2.1 | 168 | 30.2 | 0.871 | 0.94 |
| | | Xylene (mixed), 2% | 9.9 | 0.77 | 6.0 | >135 | 27.6 | 0.856 | 0.63 |

| Thinner & cleaning solvents | Typical coating | Compound, wt% | Solubility parameter, $(cal/cm^3)^{1/2}$ | Relative rate, nBUOAc =1.0 | Vapor press., mmHg @ 20°C | B.P., °C @ 760 mmHg | Surface tension, dynes/cm 20°C | Av. sp. gr, @ 25°C | Viscosity, cP @ 25°C |
|---|---|---|---|---|---|---|---|---|---|
| Lacquer retarder (thinner) & cleaner (slow)[b] | Lacquer | EGBE, 51% | 10.2 | 0.072 | 0.6 | >169 | 26.9 | 0.899 | 3.0 |
| | | AHC, 30% | 7.7 | 0.16 | 2.0 | >160 | 23.4 | 0.775 | 0.88 |
| | | 1,2,4-Trimethylbenzene, 16% | 8.9 | 19 | 2.1 | 168 | 30.2 | 0.871 | 0.94 |
| | | Xylene (mixed), 1% | 9.9 | 0.77 | 6.0 | >135 | 27.6 | 0.856 | 0.63 |

*MEK=methyl ethyl ketone, **MIBK=methyl isobutyl ketone, ***EGBE=ethylene glycol monobutyl ether, ****AHC=aromatic hydrocarbon solvent; [a]Physical properties mainly from Industrial Solvents Handbook, 110 -114. [b]Material Safety Data Sheet (Mobile Paint Co. Alabama)

The lower specific gravity of ketones (see Table 14.26.2) than other materials such as glycol ethers helps reduce total mass of VOCs (or HAPs) per volume of nonvolatiles (solids) in a container of coating. Glycol ethers are good solvents for epoxies and acrylics. They also have good coupling abilities in blends of poorly miscible solvents[17] and have low evaporative rates. The properties of a solvent product are dependent on the chemical structure and distillation range of the solvent mix in the product. The latter will affect the evaporative rate from a coating or cleaner, affecting the solubility of the resin in the coating and viscosity of the coating and solvent cleaner. Therefore, the viscosity of the solvent product must be close to that of the resin in a coating.[18] The surface tension of a solvent provides a measure of the penetrability of a cleaning solvent. A low surface tension also means the solvent spreads more readily, which is an important property for a cleaning product. However, several properties in Table 14.26.2 come into play in determining the effectiveness of a cleaning solvent.

Most coating operations, due to the size and accessibility of ships, occur in open air in drydocks, graving docks, railway, or other locations throughout a facility. Because of the size of ships, the predominant application method is airless spray guns. The thickness of the coating will determine if the application equipment needs to be cleaned during application of the coating or after the job is completed. The lines from the supply tanks to the spray gun may in some instances exceed 46 m (150 ft) in length. The ensemble of equipment and items that have to do with the application of the coating or "unit operation system (UOS)" is shown schematically in Figure 14.26.1. The representation depicts a layout for outdoor application of coatings. It includes the container used to hold the coating, attached feed pump, line transferring the coating to the spray gun, the spray gun itself, and any other item soiled with a coating that will need to be cleaned with organic solvent before it can be reused.

The need and frequency for cleaning will depend on the individual facility or company cleanliness standard (i.e., requirements) and the number of coating formulation or color changes. Cleaning of spray guns, internal transfer lines, and associated tanks account for a large part of organic solvent usage. At most shipyards, a small percentage of the coatings are applied indoors, in spray-booths. The walls and floors of these booths are cleaned by wiping with a solvent laden cloth. The coating application equipment UOS for most facilities will look very similar to that shown in Figure 14.26.1, except that the coating transfer lines will be shorter if the coating storage tanks are positioned close to the spray booths. The transfer lines, that will need to be cleaned with solvent, will be longer if the coating tanks are located away from the application area. When this is the case, the transfer lines typically run underground at the facility and another representation than the one shown in Figure 14.26.1 will need to be used, to clearly identify the emission points and waste streams for properly quantifying solvent losses. The latter may include a unit for recycling or reclaiming solvents.

Spray gun cleaning procedures may be a once-through type with collection of spent solvent in a container for disposal or reuse. Some facilities use commercial gun washers. Because gun washers are enclosed and recirculate solvent, they can reduce the amount of solvent lost by evaporation. In either case, the emissions are calculated as the difference between the amount used and the amount recovered.

To calculate the emissions associated with cleaning a spray gun it is recommended that a material balance around a "unit operation system" be considered. Several examples are provided in the Alternative Control Techniques (ACT) document on industrial cleaning solvents.[19]

Several types of part cleaners are used at shipbuilding and ship repair facilities. The types used in such facilities vary from the more simple sink and spray systems[20] to more elaborate parts (machine) cleaners of the cold or vapor types.[4] Most of the parts cleaners in shipyards are small — around 1.5 m x 1 m and 1 m deep - usually located in the machine shops, not the paint rooms. Most of the parts are small components being cleaned prior to being joined to other small parts into assemblies and sub-assemblies or being cleaned as part of some type of repair operation. Most of parts cleaners used were basket-type design with the parts loaded into a basket and dropped through the vapor zone several times to clean off the oils and dirt. Some shipyards use contractors to come in and change out the solvent on a routine schedule.

### 14.26.4.3 Maintenance cleaning of equipment items and components

Shipyards also undertake scheduled maintenance cleaning of many ship components such as contacts and switches and equipment items such as bearings and packaging machines. This is mostly done by hand-wiping the parts with organic solvents. These operations will generally consume a relatively small amount of the overall volume of organic solvents used for cleaning in shipyards. Solvents are also used in machine shop areas and thus contribute to the waste stream.

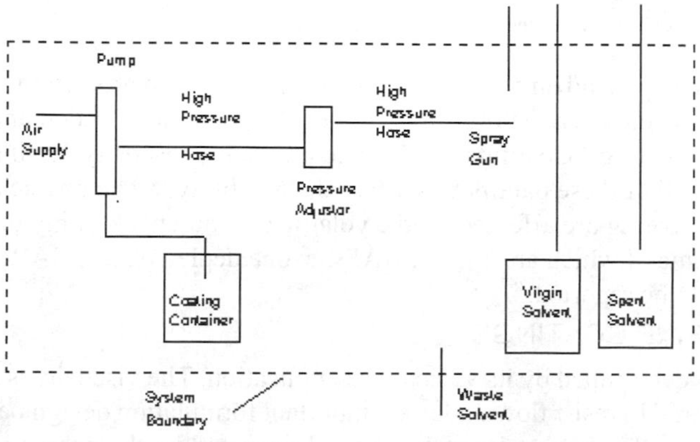

Figure 14.26.1 Schematic diagram of marine coating application equipment.

## 14.26.5 MARINE COATINGS

There are several categories of marine coatings that are used to protect the surface of a ship from the aggressive marine environment and for other performance requirements such as preventing corrosion and fouling; protecting cargoes from contamination; providing safety warnings and informational markings; providing cosmetic and camouflage colors; preventing slipping and sliding on walking surfaces; reducing fire hazards; and providing cathodic protection.[21] The coating systems of marine coatings are selected to meet:

- the type of marine environment to which a vessel will be exposed
- the time a vessel is to remain operational before it needs to be reworked.

General areas of a ship include: (1) underwater hull, (2)superstructures and freeboard, (3) interior habitability areas, (4) exterior deck areas, and (5) fuel, water ballast and cargo tank.[22] The freeboard is the area above water hull. These areas have different characteristics and operational requirements. Table 14.26.3[23] shows the predominant resin and solvent types used on ships based on a 1991/1992 survey of the industry obtained as part of the shipbuilding and ship repair regulation was being developed. The summary table also gives average VOC and HAP content for the various coating category types. Epoxy coatings constitute a large percentage of the coatings used. The epoxy films are strongly resistant to most chemicals and are very good anti-corrosion coatings, and require little surface preparation.

**Table 14.26.3. Summary of marine coating usage (by coating type)[23]**

| Coating types | Average usage in U.S. shipyards, % | Average VOC content, g/L (lb/gal) | | Average HAP content, g/L (lb/gal) | |
|---|---|---|---|---|---|
| *General use types* | | | | | |
| alkyd based | 10 | 474 | (3.95) | 355 | (2.98) |
| epoxy based | 59 | 350 | (2.92) | 56 | (0.47) |
| *Speciality types* | | | | | |
| antifouling (multiple resins)[1] | 11 | 388 | (3.23) | 268 | (2.25) |
| inorganic zinc based | 10 | 545 | (4.54) | 274 | (2.30) |
| other speciality categories | 10 | 400 | (3.33) | 144 | (1.20) |
| TOTAL | 100 | | | | |

[1]resins: epoxy, polyurethane, vinyl, and chlorinated rubber

The coating system used will depend on service requirements. Maximum protection at an economical price can be achieved when the user understands the protection needed and the functions performed by the coatings. Coatings are designed for spray viscosity, drying time, pot-life, and cure profile; all of these parameters affect shelf stability.[24] The physical parameters and properties of a coating are affected by the volatile constituents (mainly organic solvents) in a coating, some of which are VOCs, HAPs, ozone depleters, and SARA 313 toxic chemicals that need to be reported.[25]

## 14.26.6 THINNING OF MARINE COATINGS

The sprayability of a coating is determined by its viscosity at application. The viscosity is a measure of the ability of a material to resist flow and is an important formulation design parameter. Application viscosity is affected by the ambient conditions and by the degree of mixing that occurs within the applicator. The thinner is often the same material as the cleaner (as indicated in Table 14.26.2). The solvent material is often a blend of miscible materials. Together they will dissolve a dry resin that needs to be removed or give the coating certain needed properties such as reduced/increased viscosity or shorter drying times.

Standard spraying equipment will apply coatings up to some maximum viscosity. Above that maximum value, thinning solvents are required. Thinning solvent is sometimes added to enhance brushability or sprayability of a coating. The appropriate viscosity is provided by the coating manufacturer or supplier; it will depend on the solvent content of the coating and temperature at the point of application.[26] Since most coatings are applied outdoors, extreme weather conditions may require adding thinning solvents to the coating. Organic thinning solvents are added to coatings to alter their flowing properties. However, the flow properties of a coating may be altered by using special heaters or a combination of solvent and heat. The effect of a heater on the viscosity of a coating depends on the physical properties of the coating and on the flow rate in the in-line heater. Under cold weather conditions, in-line heaters may provide good viscosity control, but may not be able to solve all application problems that are encountered in the field. Under extremely low temperatures, the substrate surface can act as a heat sink, which may inhibit the setting or curing of the coating. In-line heaters which are used for low volume coatings are not suitable for large volume coatings. As a result, thinning solvents are still needed to transfer the fluid from storage to pumps and hoses.

Under hot and humid weather conditions, certain coatings (e.g., lacquers) can rapidly lose organic solvent prior to and during application. Often under these situations a facility will add solvent blends to make up for the reduction in viscosity and to overcome condensation on the surface (blushing).[27] Evaporative losses can be minimized by adopting good work practices and by using formulations that contain organic solvents with low vapor pressures.

## 14.26.7 SOLVENT EMISSIONS

Several states with their own rules regulating marine coatings have separate rules addressing solvent cleaning operations. While marine coating rules typically address VOC contents and types of application equipment, the cleaning solvent rules are more generic and address cleaning solvents used at any and all metal-related manufacturing operations.

Many types of solvents are used in marine coatings and in their associated cleaning materials as shown in Table 14.26.1. Almost all solvents used at shipyards are VOC and approximately one in three solvents contain HAPs. Of the HAPs reported, several are included on the list of 17 high priority chemicals targeted by U.S. EPA for the 33/50 program.[28] These included xylene (commercial), toluene, and the ketones. Commercial grade xylene represents the major portion of the volatile HAPs reported.

Many of the commonly known solvents are actually petroleum distillation fractions and are composed of a number of compounds (e.g., mineral spirits and naphthas). There are two general types of solvents derived from petroleum, aliphatics or aromatics. Aromatics are stronger solvents than aliphatics since they dissolve a wider variety of resins. Most major solvent suppliers (chemical manufacturers) produce several types and variations of these solvents and the associated HAP contents can vary significantly from manufacturer to manufacturer and from batch to batch. These types of solvents are used extensively and are present in the majority of marine coatings. Table 14.26.4 provides a summary of common petroleum distillate solvents and solvent blends and their associated HAP content. For any solvent or solvent blend that is not listed as specified in Table 14.26.4, another table (Table 14.26.5) was developed to provide solvent groupings and associated HAP component/content values. The HAP values for Tables 14.26.4 and 14.26.5 were adapted from estimates provided in 1998 by the Chemical Manufacturer Association's Solvent's Council.

**Table 14.26.4. HAP content of single solvents and solvent blends. [Adapted, by permission from Chemical Manufacturer Association's Solvent's Council]**

| Solvent/solvent blend | CAS No. | HAP content range, wt% | Average HAP content, wt% | Typical HAP, wt% |
|---|---|---|---|---|
| Toluene | 108-88-3 | 100 | 100 | toluene |
| Xylene(s) | 1330-20-7 | 100 | 100 | xylenes, ethylbenzene[a] |
| Hexane | 110-54-3 | 49-55 | 50 | n-hexane[b] |
| Ethylbenzene | 100-41-4 | 100 | 100 | ethylbenzene |
| Aliphatic 140 | | 0 | 0 | none |
| Aromatic 100 | | <5 | 2 | 1% xylene, 1% cumene |
| Aromatic 150 | | <10 | 9 | naphthalene |
| Aromatic naphtha | 64742-95-6 | <5 | 2 | 1% xylene, 1% cumene |
| Aromatic solvent | 64742-94-5 | <10 | 10 | naphthalene |
| Exempt mineral spirits | 8032-32-4 | 0 | 0 | none |
| Ligroines (VM & P) | 8032-32-4 | 0 | 0 | none |
| VM & P | 8032-32-4 | 0 | 0 | none |
| Lactol spirits | 64742-89-6 | 12-20 | 15 | toluene |
| Low aromatic white spirit | 64742-82-1 | 0 | 0 | none |
| Mineral spirits | 64742-88-7 | <2 | 1 | xylenes |
| Hydrotreated naphtha | 64742-48-9 | <0.01 | 0 | none |
| Hydrotreated light distillate | 64742-47-8 | <0.1 | 0.1 | toluene |
| Stoddard solvent | 8052-41-3 | 1-2 | 1 | xylenes |
| Super high-flash naphtha | 64742-95-6 | 0-6 | 5 | xylenes |
| Varsol® solvent | 8052-49-3 | <1.1 | 1 | 0.5% xylenes, 0.5% ethylbenzene |
| VM & P naphtha | 64742-89-8 | 5-10 | 6 | 3% toluene, 3% xylene |
| Petroleum distillate mixture | 68477-31-6 | 4-9 | 8 | 4% naphthalene, 4% biphenyl |

**Table 14.26.5. Petroleum solvent groups and associated HAP contents. [Adapted, by permission from Chemical Manufacturer Association's Solvent's Council]**

| Solvent Type | Solvent Name | HAP Component | HAP Concentration, wt% |
|---|---|---|---|
| Aliphatic | Mineral spirits 135 | Xylene | 1 |
| | Mineral spirits 150 EC | Toluene | 1 |
| | Naphtha | Ethylbenzene | 1 |
| | Mixed hydrocarbon | | |
| | Aliphatic hydrocarbon | | |
| | Naphthol spirits | | |
| | Petroleum spirits | | |
| | Petroleum oil | | |
| | Petroleum naphtha | | |
| | Solvent naphtha | | |
| | Solvent blend | | |
| Estimated total HAP for each solvent type | | | 3 |
| Aromatic | Medium-flash naphtha | Xylene | 4 |
| | High-flash naphtha | Tolucnc | 1 |
| | Aromatic naphtha | Ethylbenzene | 1 |
| | Light aromatic naphtha | | |
| | Light aromatic hydrocarbons | | |
| | Aromatic hydrocarbons | | |
| | Light aromatic solvent | | |
| Estimated total HAP for each solvent type | | | 6 |

The percentage of solvent used that evaporates during cleaning operations will depend on the volatility of the solvent and on the type cleaning operation. Any materials such as rags or sponges that come into contact with the solvent must also be considered part of the solvent cleaning system and disposed of properly.

Since marine coatings are mostly air dried, it is generally impractical to capture air emissions from coating operations and until recently, to duct air emissions into abatement devices such as incinerators or adsorbers. The use of low solvent (VOC or HAP) containing coatings and high performance coatings provide a suitable means of achieving lower volatile emissions from coating of ships. Such coatings will lead to lower life cycle costs and emissions, since the ship will require less rework. EPA has recently approved the use of technology for capturing and destroying air pollutant emissions from the application of coatings to the hull of a ship.[29]

## 14.26.8 SOLVENT WASTE

The waste cleaning solvent is handled differently by different shipyards. Facilities that use different type solvents for cleaning and thinning, or utilize multiple cleaners may select disposal rather than recycling for re-use. This decision is typically based on cost. Other facilities will first mix their waste solvents with unused coating. Incineration is generally selected as the disposal method when it is determined to be less expensive than recycling.

The type of solvent waste generated from shipbuilding and ship repair facilities does not vary much. Surface preparation and cleaning operations following application of coatings does not represent the most significant part of the organic solvent waste. Certain coatings, such as antifouling coatings, contain toxic pigments (metals) such as chromium, lead, and tributyltin. The waste will likely be hazardous waste if the waste material contained coating residues or other materials that are toxic. Solvents used in cleaning are often mixed with other waste fuels such as hydraulic fluids, waste fuels, and other maintenance by-products. Some of the larger shipyards segregate the solvent waste or recycle it for re-use.

## 14.26.9 REDUCING SOLVENT USAGE, EMISSIONS, AND WASTE

The amount of organic solvent usage for cleaning operations makes up about 20 percent of the total solvent usage in cleaning operations and coating operations at shipyards. The waste from solvent cleaning represents a major portion of the waste in a plant that has surface coating operations. Line cleaning, gun cleaning, and tank cleaning are three operations that utilize most of the cleaning solvents. There are many steps that a facility can undertake to reduce the use of solvent for cleaning. It can reuse some of the solvent, scrape off the paint left on the inside of a mix tank, or modify existing cleaning practices and procedures.[5,30] Some shipyards have modified their coating operations to use less solvent. For planning purposes, a facility needs to know where the major solvent usage, emissions, and waste occur in the facility. This may be achieved by:

- Developing a solvent accounting system for tracking the usage, fate, and costs of organic solvents in coating operations and cleaning operations;
- Identifying actions management will take to reduce solvent content in coatings and reduce usage of cleaning materials containing organic solvents.[5,31]

The steps are not mutually independent and require full management support and financial backing. Obtaining reliable data is one of the major problems that a manager or supervisor involved in pollution reduction faces. Any program adopted should be one that can identify accurate and specific material usage, emission data, and waste data that will be useful for evaluating pollution cost and risk.[32] The "unit operation system" (UOS) approach was used to categorize the universe of cleaning into nine main categories equivalent to those identified here in the section discussing cleaning operations. The concept of a UOS provides the foundation for a standardized material accounting method that provides transportable data, that a secondary user can understand, for setting up a material balance, for comparing alternatives, or evaluating risk. The visual representation of a set up, as was done in Figure 14.26.1, is an essential component of this approach.

When facility managers and operators are able to identify with some confidence, at the source, final solvent disposition and cost of cleaning, they are in a better position to make informed decisions and take the necessary actions for reducing solvent usage. Facilities have the added incentive to reduce worker exposure to toxics contained in solvents because of OSHA requirements.[33] The facility owners and operators also have to abide by state, local, and federal regulations that limit the amount of VOCs and HAPs emitted to the air and water streams. Facilities that generate large amounts of solvent waste streams will have to deal with RCRA hazardous waste regulations.[34]

## 14.26.10 REGULATIONS AND GUIDELINES FOR CLEANING SOLVENTS

The U.S. EPA developed a guideline document to help state and local agencies reduce VOC emissions for the use of cleaning solvents. The document does not provide emission limits, but instead recommends the use of a management system for tracking usage, emissions, and waste from the use of organic solvents and provides a list of definitions that should be used for that purpose.[35] Less than 20 states had regulated VOC emissions from the use of organic cleaning solvents by 1997.[36] Many states only had minimal record-keeping requirements. Some regulations are driving facilities to use solvents with lower vapor pressures. These low vapor pressure solvents evaporate more slowly at ambient temperatures. However, the emission benefits cannot be guaranteed to last forever. An organic solvent should eventually all evaporate unless it is reused, recycled, or sealed in a container.

Many of the cleaning solvents used, such as 1,1,1-trichloroethane, are not considered VOC in the United States. Those solvents are regulated as HAPs and are being phased out because of the Montreal Protocol which aims to reduce ozone depleters. Alternative cleaners are now being used and further developed. To meet regulatory requirements, a facility will often replace or substitute a cleaning product containing HAP material(s) with one or more products containing no HAP solvents (but may contain VOC solvents or vice versa). Acetone, once the preferred solvent for fiberglass boat manufacturers, was replaced starting in 1992 with other solvents such as diacetone alcohol.[37] The latter solvent is as effective as acetone, but has a much lower vapor pressure (0.80 mm Hg at 20°C) and flash point (52°C) to reduce the amount of VOC emitted and hazardous reportable releases. Acetone was recently determined not to have significant photochemical reactivity and is no longer considered a VOC in the United States. As a result, acetone is now making a comeback in cleaners and in some coatings. HCFCs are now being used instead of CFCs, but will soon need to be replaced because of the Montreal Protocol. There are many articles in the literature that discuss alternatives that will reduce VOC and HAP emissions.[38-42] The use of waterborne materials or other material(s) containing no HAPs or VOCs is the ideal solution, when such coatings or cleaners will do the job. Software and internet sites also exist to help in the selection of the appropriate solvent cleaner or method of cleaning.

### Disclaimer

The views expressed in this chapter are those of the authors and do not necessarily reflect those of their affiliation. Mentioning the names of organizations does not constitute an endorsement.

## REFERENCES

1 South Coast Air Quality Management District (SCAQMD), Rule 1106: Marine Coating Operations, 1995.
2 SCAQMD, Rule 1171: Solvent Cleaning Operations, (1996) and Rule 1122: Solvent Degreasers, (1996).
3 40 CFR Part 63, Subpart II, (1995), U.S. Government Printing Office, Washington, DC.
4 40 CFR, Part 63, Subpart T, (1996), U.S. Government Printing Office, Washington, DC.
5 U.S. Environmental Protection Agency (EPA), Control Techniques Guideline Shipbuilding and Ship Repair Operations (Surface Coating), Office of Air Quality Planning and Standards, North Carolina, EPA-453/R-94-015, (NTIS no. PB94-156791).
6 U.S. EPA, Federal Register, August 27, 1996 (vol. 61, no. 167), pp. 44050-44057.
7 M. A. Serageldin in **Electrochemical Society Proceedings**, Volume 95-16, *Electrochemical Society Inc.*, Chicago, Illinois 1997, pp. 1-16.
8 T. J. Snider, An Analysis of Air Pollution Control Technologies for Shipyard Emittted Volatile Organic Compounds (VOCs), National Shipbuilding Research Program, Report No. 0376. March 1993, pp. 3-5.
9 U.S. EPA, EPA Office of Compliance Sector Notebook: Profile of The Shipbuilding and Repairing Industry, EPA Office of Compliance, Washington, D.C, EPA/310-R-97-008, 1997, pp. 75-78.

10 U.S. EPA, Surface Coating Operations at Shipbuilding and Ship Repair Facilities—Background Information for Proposed Standards, OAQPS, North Carolina, EPA-453/D-94-011a, (NTIS no. PB95-197471), 1994, pp. 6-7 to 6-9.
11 U.S. EPA, Alternative Control Techniques Document—Industrial Cleaning Solvents, Office of Air Quality Planning and Standards, North Carolina, EPA-453/R-94-015, 1994, pp. 3-17 to 3-28.
12 M. A. Serageldin in Emission.Inventory Living in a Global Environment, Proc..Special. Conf., December 8-10, 1998, New Orleans, LA, Air and Waste Management Association, Pittsburgh, PA, VIP-88, vol.1, 1999, pp. 341-48.
13 **Metals Handbook**, 9th edition, v.5 Surface Cleaning, Finishing, and Coating, Table 21, *American Society for Metals*, Metal Park, Ohio, W.G. Wood, Ed., 1990, p. 506.
14 U.S. EPA, Guide to Pollution Prevention: The Marine Maintenance and Repair Industry, Office of Research and Development, Washington D.C., EPA/625/7-91/015, 1991, pp. 5-10.
15 **Metals Handbook**, 9th edition, v.5 Surface Cleaning, Finishing, and Coating, *American Society for Metals*, Metal Park, Ohio, W.G. Wood, Ed., 1990, pp. 504 - 505.
16 U.S. EPA, Surface Coating Operations at Shipbuilding and Ship Repair Facilities—Background Information for Proposed Standards, Office of Air Quality Planning and Standards, North Carolina, EPA-453/D-94-011a, (NTIS no. PB95-197471), 1994, p. 318.
17 C. H. Hare, **Protective Coatings Fundamentals of Chemistry and Composition, 26. Solvent Families**, *Technology Publishing Company*, Pittsburgh, PA, 1994, pp. 371-383.
18 H. Burrell, 1995 Official Digest of the Federation of Society for Paint Technology, vol. 27, no. 369, 1955, p. 726.
19 U.S. EPA, Alternative Control Techniques Document—Industrial Cleaning Solvents, Office of Air Quality Planning and Standards, North Carolina, EPA-453/R-94-015, 1994, pp. H-1 to H-6.
20 U.S. EPA, Alternative Control Techniques Document—Industrial Cleaning Solvents, Office of Air Quality Planning and Standards, North Carolina, EPA-453/R-94-015, 1994, pp. C-1 to C-18.
21 H. Bleile and S. Rodgers, *Marine Coatings. Federation Series on Coatings Technology*. 1989. pp. 8.
22 U.S. EPA, Surface Coating Operations at Shipbuilding and Ship Repair Facilities—Background Information for Proposed Standards, OAQPS, North Carolina, EPA-453/D-94-011a, (NTIS no. PB95-197471), 1994, pp. 3-6 to 3-11.
23 U.S. EPA, Surface Coating Operations at Shipbuilding and Ship Repair Facilities—Background Information for Proposed Standards, Office of Air Quality Planning and Standards, North Carolina, EPA-453/D-94-011a, (NTIS no. PB95-197471), 1994, pp 3-18.
24 C. H. Hare in **Protective Coatings Fundamentals of Chemistry and Composition, 1. The Basic Composition of Paint**, *Technology Publishing Company*, Pittsburgh, PA, 1994, pp.1-13.
25 40 CFR 372.65 (1986), U.S. Government Printing Office, Washington, D.C.
26 A. T. Chen and R. T. Wojcik in Metal Finishing , 97' Organic Finishing Guidebook and Directory Issue, *Metal Finishing*, Pittsfield, MA, 1997, pp. 167-179.
27 G. L. Muir in Metal Finishing , 97' Organic Finishing Guidebook and Directory Issue, *Metal Finishing*, Pittsfield, MA, 1997, pp.391-398.
28 U.S. EPA 33/50 Program: The Final Record, Office of Pollution Prevention and Toxics, EPA-745-R-99-004, Washington, D.C., March 1999.
29 U. S. EPA, Evaluation of Application for Approval of an Alternative Methodology for Compliance with the NESHAP for Shipbuilding and Recommended Requirements for Compliance, Research Triangle Park, N.C., EPA-453/R-99-005, 1999.
30 P.M. Randall in Proc: Poll. Prevent. Conf. Low- and No VOC Coating Technol., Air and Engineering Research Lab, RTP, NC, EPA-600/R-94-022, 1994, pp.489-499.
31 M. A. Serageldin, in **Waste Minimization Through Process Design**, A. P. Rossiter, Ed., *McGraw-Hill*, New York, April 1995, pp. 289-303 .
32 M.A. Serageldin, The Unit Operation System for Measuring Plant Environmental Performance (with selected examples), NTIS PB98-124399INZ, November 1996.
33 Occupational Safety and Health Administration, Shipyard Industry, U.S. Department of labor. Document No. 2268, [Online]. URL: Available from http://www.osha-slc.gov/OshDoc/Additional.html#OSHG-HAZ, (revised 1998).
34 42 U.S.C. s/s 321 et seq. (1976)
35 U.S. EPA, Alternative Control Techniques Document—Industrial Cleaning Solvents, OAQPS, North Carolina, EPA-453/R-94-015, 1994, pp. A1- A7.
36 Emission Inventory Improvement Program: Volume III - Area Sources: Chapter 6— Solvent Cleaning, 1997 [Online]. Available from URL: http://www.epa.gov/ttn/chief/eiip/iii06fin.pdf.

37 M. Melody, Boatmakers find solvent substitute cuts emission, costs, *Hazmat World*, February 1992, pp36-39.
38 **1991 UNEP Solvents, Coatings, and Adhesives Technical Options Report**, *UNEP*, Paris, December1991, pp.58-69 and pp. 153-178.
39 P. Milner in **The Coating Agenda America 1955/1996**, *Campden Publishing Limited* (in association with National Paint and Coating Association, U.S.A), London, U.K. pp. 125 and 127.
40 F. Hussey in Advanced Coating Technology, Proc. 3rd, Ann. ESD Adv. Coating. Conf., Dearborn, MI, 9-11 November 1993, ESD, The Engineering Society, Ann Arbor, Michigan, pp. 250 -261.
41 L. Clark, S. Kaldon, E. Reid, and D. Ellicks, Putting Solvent Back in Paint Legally, *Metal Finishing*, 1998, pp 24-28.
42 C. Mouche, Army Tests Enviro-Friendly Solvents Alternative, *Pollution Engineering*, September, 1999, p. 35.

## 14.27 STONE, CLAY, GLASS, AND CONCRETE

George Wypych
**ChemTec Laboratories, Inc., Toronto, Canada**

This is a very diverse industry segment which manufactures glass, glassware, cement, concrete, clay products pottery, gypsum, and plaster products. In the USA the industry employs about 500,000 people and operates over 16,000 facilities. There is a broad range of

**Table 14.27.1. Reported solvent releases from the stone, clay, glass, and concrete industry in 1995 [Data from Ref. 2]**

| Solvent | Amount. kg/year | Solvent | Amount. kg/year |
|---|---|---|---|
| benzene | 4,000 | ethylene glycol | 42,000 |
| n-butyl alcohol | 74,000 | hexane | 12,000 |
| sec-butyl alcohol | 1,700 | isopropyl alcohol | 300 |
| tert-butyl alcohol | 4,000 | methanol | 560,000 |
| carbon disulfide | 5 | methyl ethyl ketone | 700,000 |
| carbon tetrachloride | 14 | methyl isobutyl ketone | 25,000 |
| chlorobenzene | 14 | tetrachloroethylene | 44,000 |
| chloroform | 14,000 | 1,2,4-trichlorobenzene | 46,000 |
| cresol | 610 | 1,1,1-trichloroethane | 330,000 |
| cyclohexane | 77,000 | trichloroethylene | 110,000 |
| 1,4-dichlorobenzene | 14,000 | toluene | 890,000 |
| 1,2-dichloroethane | 17 | xylene | 500,000 |
| dichloromethane | 120,000 | m-xylene | 450 |
| 1,4-dioxane | 230 | o-xylene | 540 |
| ethylbenzene | 60,000 | p-xylene | 110 |

**Table 14.27.2. Reported solvent transfers from the stone, clay, glass, and concrete industry in 1995 [Data from Ref. 2]**

| Solvent | Amount. kg/year | Solvent | Amount. kg/year |
|---|---|---|---|
| benzene | 6,000 | ethylene glycol | 29,000 |
| n-butyl alcohol | 44,000 | hexane | 15,000 |
| sec-butyl alcohol | 600 | isopropyl alcohol | 2,500 |
| tert-butyl alcohol | 2,000 | methanol | 300,000 |
| carbon disulfide | 5 | methyl ethyl ketone | 760,000 |
| carbon tetrachloride | 5,000 | methyl isobutyl ketone | 92,000 |
| chlorobenzene | 4,000 | tetrachloroethylene | 26,000 |
| chloroform | 20,000 | 1,2,4-trichlorobenzene | 10,000 |
| cresol | 7,000 | 1,1,1-trichloroethane | 60,000 |
| cyclohexane | 12,000 | trichloroethylene | 54,000 |
| 1,4-dichlorobenzene | 3,000 | toluene | 800,000 |
| 1,2-dichloroethane | 3,000 | xylene | 470,000 |
| dichloromethane | 20,000 | m-xylene | 600 |
| N,N-dimethylformamide | 5,000 | o-xylene | 1,200 |
| ethylbenzene | 67,000 | p-xylene | 340 |

processes involved which use a wide variety of chemicals. Solvents are used for cleaning and are components of product decoration materials.

Tables 14.27.1 and 14.27.2 give data on the reported releases and transfers of solvents by the US stone, clay, glass, and concrete industry. Numerous solvents are used but most are used in small quantities the only high volume ones being toluene, xylene, methanol, 1,1,1-trichloroethane, and methyl ethyl ketone. Although the industry does not release large quantities of solvents, their total VOC contribution puts them second among the US industries. Total releases and transfers are one of the smallest of all industries.

## REFERENCES

1 EPA Office of Compliance Sector Notebook Project. Profile of the Stone, Clay, Glass, and Concrete Products Industry. US Environmental Protection Agency, 1995.
2 EPA Office of Compliance Sector Notebook Project. Sector Notebook Data Refresh - 1997. US Environmental Protection Agency, 1998.

# 14.28 TEXTILE INDUSTRY

GEORGE WYPYCH
**ChemTec Laboratories, Inc., Toronto, Canada**

The US textile industry operates over 5,500 plants, employs over 600,000 people and has shipments exceeding $70 billion. In addition to conventional methods of processing and finishing, this industry also produces coated fabrics, tire cord and fabrics, and cordage and twine. Numerous operations are employed by the textile manufacturing and finishing but only a few of these operations involve the use of solvents. These include printing (volatile solvents and xylenes), finishing (e.g., methanol), scouring (glycol ethers and mineral spirits), cleaning knit goods, desizing (glycol ethers and methanol in PVA desizing), and coating (methyl ethyl ketone and toluene).

Tables 14.28.1 and 14.28.2 provide data on the reported releases and transfers of solvents by the US textile industry. Methyl ethyl ketone, toluene, and methanol are emitted in the greatest quantities. The total number of solvents used is low. The textile industry is among the lowest contributors to VOC and is the second smallest (after shipbuilding industry) in the amount of emitted and transferred solvents.

## REFERENCES

1 EPA Office of Compliance Sector Notebook Project. Profile of the Textile Industry. US Environmental Protection Agency, 1997.
2 EPA Office of Compliance Sector Notebook Project. Sector Notebook Data Refresh - 1997. US Environmental Protection Agency, 1998.

**Table 14.28.1. Reported solvent releases from the textile industry in 1995 [Data from Ref. 1]**

| Solvent | Amount, kg/year | Solvent | Amount, kg/year |
|---|---|---|---|
| n-butyl alcohol | 900 | methyl isobutyl ketone | 190,000 |
| 1,4-dichlorobenzene | 700 | N-methyl-2-pyrrolidone | 180,000 |
| 1,2-dichloroethane | 4,000 | tetrachloroethylene | 29,000 |
| dichloromethane | 230,000 | 1,2,4-trichlorobenzene | 21,000 |
| N,N-dimethylformamide | 53,000 | 1,1,1-trichloroethane | 150,000 |
| ethylene glycol | 66,000 | trichloroethylene | 130,000 |
| hexane | 59,000 | 1,2,4-trimethylbenzene | 24,000 |
| isopropyl alcohol | 11,000 | toluene | 1,600,000 |
| methanol | 1,300,000 | xylene | 380,000 |
| methyl ethyl ketone | 2,200,000 | | |

**Table 14.28.2. Reported solvent transfers from the textile industry in 1995 [Data from Ref. 1]**

| Solvent | Amount, kg/year | Solvent | Amount, kg/year |
|---|---|---|---|
| 1,2-dichloroethane | 3,500 | tetrachloroethylene | 26,000 |
| dichloromethane | 9,000 | 1,2,4-trichlorobenzene | 51,000 |
| N,N-dimethylformamide | 52,000 | 1,1,1-trichloroethane | 2,000 |
| ethylene glycol | 210,00 | trichloroethylene | 170,000 |
| methanol | 120,000 | 1,2,4-trimethylbenzene | 20,000 |
| methyl ethyl ketone | 640,000 | toluene | 330,000 |
| methyl isobutyl ketone | 61,000 | xylene | 67,000 |
| N-methyl-2-pyrrolidone | 83,000 | | |

## 14.29 TRANSPORTATION EQUIPMENT CLEANING

George Wypych
**ChemTec Laboratories, Inc., Toronto, Canada**

Transportation equipment cleaning includes cleaning of interior of trucks, rail cars, barges, intermodal tank containers, intermediate tank containers, tank interiors, and the exterior of aircraft. These operations rarely involve solvents. Transportation equipment is also cleaned before refurbishing which includes old paint removal, surface cleaning, and painting. These processes usually do require solvents. Aircraft deicing is done using a mixture containing ethylene glycol. There is no data available on the amount of solvents used and emitted from these processes.

## REFERENCES

1 EPA Office of Compliance Sector Notebook Project. Profile of the Transportation Cleaning Industry. US Environmental Protection Agency, 1995.

## 14.30 WATER TRANSPORTATION

George Wypych
**ChemTec Laboratories, Inc., Toronto, Canada**

Water transportation equipment is divided into self-propelled vessels and barges. The US industry is composed of over 13,000 establishments employing over 150,000 people operating about 40,000 vessels. Solvents are used in maintenance for paint removal, painting, degreasing engines, and cleaning carburetors. There is no information available regarding solvent use by the industry.

## REFERENCES

1 EPA Office of Compliance Sector Notebook Project. Profile of the Water Transportation Industry. US Environmental Protection Agency, 1997.

## 14.31 WOOD FURNITURE

George Wypych
**ChemTec Laboratories, Inc., Toronto, Canada**

The US wood furniture industry operates over 11,000 manufacturing facilities which employ over 250,000 people. Total sales are over $22 billion. Several operations use solvents, including veneer application (adhesives), derosination (certain types of wood contain rosin which may interfere with the effectiveness of certain finishes; rosin is removed by application of a mixture of acetone and ammonia), pretreatment (coating contains coalescing solvent (slow evaporating) and diluent (fast evaporating), a new UNICARB® technology uses carbon dioxide in the mixture with a coalescing solvent which eliminates the need for a diluent), several finishing operations (stains, paints, fillers, inks used to print simulated wood grain on plywood, topcoat, and cleaning liquids). Nitrocellulose lacquers are commonly used in the USA; polyurethane and unsaturated polyester finishes are still rarity. Solvents and thinners used in paints include toluene and xylene. Adhesives usually contain solvents such as methyl isobutyl ketone, methyl ethyl ketone, xylene, toluene, and 1,1,1-trichloroethane. Table 14.31.1 shows the relative amounts of VOC emissions from different finishing components depending on furniture type and type of process used. The long process consists of applications of 3 or more coats of stain, a washcoat, a filler, a sealer, a highlight, and two to three topcoat applications. In the short-sequence process two stain coats are applied, followed by a washcoat, a sealer, and two top coat applications. Office furniture uses only one application of a stain, a sealer, and a topcoat.

Tables 14.31.2 and 14.31.3 provide data on the reported releases and transfers of solvents by the US furniture and fixtures industry. Toluene, xylene, methanol, and methyl ethyl ketone are released in the largest quantities. The furniture and fixtures industry is one of lesser contributors of VOCs releases and transfers among the US manufacturing industries.

Several alternative technologies are under development and some are at the field trial stage.[3-12] Waterborne clear varnishes are being used increasingly but the market is still dominated by solvent-based varnishes. One reason may be that waterborne varnishes are 2-3 times more expensive than solvent-based products.[3] It is also not clear if waterborne varnishes have lower odor and can be applied at the same rate as solvent-based products. Waterborne varnishes have one important advantage in clear coatings for they have no color therefore the natural wood color is more vivid. Two-component water-based polyurethane coatings have been introduced recently.[4] They have a very low VOC content and have very good drying speed, gloss, and potlife. The drying rate and potlife are controlled by the ratio of isocyanate and hydroxyl components. Increasing this ratio increases drying speed but decreases potlife. However, at 4 hours, potlife is acceptable. Polyurethane dispersions are used in furniture adhesive applications, and are especially useful in the lamination of kitchen tables.[5]

**Table 14.31.1. Relative VOC emissions. [Reproduced from EPA draft Guidelines for the Control of Volatile Compound Emissions from Wood Furniture Coating Operations]**

| Operation | Percentage of total emissions | | |
|---|---|---|---|
| | Long process | Short process | Office furniture |
| Stain | 26 | 28 | 32 |
| Washcoat | 4 | 4 | - |
| Filler | 3 | - | - |
| Wiping stain/glaze | 8 | - | - |
| Sealer | 18 | 32 | 32 |
| Highlight | 1 | - | - |
| Topcoat | 40 | 36 | 36 |

**Table 14.31.2. Reported solvent releases from the furniture and fixtures industry in 1995 [Data from Ref. 2]**

| Solvent | Amount, kg/year | Solvent | Amount, kg/year |
|---|---|---|---|
| n-butyl alcohol | 940,000 | methyl isobutyl ketone | 590,000 |
| cyclohexane | 10 | tetrachloroethylene | 7,300 |
| dichloromethane | 180,000 | 1,2,4-trichlorobenzene | 90 |
| ethylbenzene | 360,000 | 1,1,1-trichloroethane | 160,000 |
| ethylene glycol | 30 | trichloroethylene | 29,000 |
| hexane | 25,000 | 1,2,4-trimethylbenzene | 140,000 |
| isopropanol | 105,000 | toluene | 5,300,000 |
| methanol | 3,600,000 | xylene | 3,300,000 |
| methyl ethyl ketone | 1,900,000 | | |

One problem related to all water-base coatings is the degree to which they penetrate into wood grain. Waterborne coatings are composed of relatively large emulsified particles which are frequently the size of the pit openings. After the emulsion droplets dry out they become very viscous and penetration becomes even more difficult. The penetration of alkyd waterborne coatings was tested by microautoradiography.[6] The results indicate that the penetration rate is controlled by the viscosity and form of the material (solution vs. emulsion) is less important.

UV curable coatings have also problems in application to wood finishing. These finishes need to have low viscosity but it is difficult to formulate 100% solids curable coatings which have a sufficiently low application viscosity.[7] It possible to formulate waterborne UV curable coatings but their performance is still an issue.

**Table 14.31.3. Reported solvent transfers from the furniture and fixtures industry in 1995 [Data from Ref. 2]**

| Solvent | Amount, kg/year | Solvent | Amount, kg/year |
|---|---|---|---|
| n-butyl alcohol | 53,000 | methyl isobutyl ketone | 145,000 |
| cyclohexane | 110 | 1,2,4-trichlorobenzene | 110 |
| dichloromethane | 28,000 | 1,1,1-trichloroethane | 5,700 |
| ethylbenzene | 340,000 | trichloroethylene | 300 |
| hexane | 16,000 | 1,2,4-trimethylbenzene | 160,000 |
| isopropanol | 11,000 | toluene | 1,030,000 |
| methanol | 570,000 | xylene | 1,700,000 |
| methyl ethyl ketone | 380,000 | | |

Dunlop developed a two-part water-based adhesive which performed well in lamination of wood and foam.[8] Also, a new reactive hot-melt system for attaching moldings in furniture assembly is in use.[9] A water reducible coating for wood finishing has been developed based on copolymer which can be cured by crosslinking.[11] Another solvent-free material was developed to cover scratches and other damage of finished wood articles.[12] Replacement of solvent based adhesives is essential if this industry, which uses 12% of all adhesives, is to significantly reduce its solvent usage.[10]

## REFERENCES


1 EPA Office of Compliance Sector Notebook Project. Profile of the Wood Furniture and Fixtures Industry. US Environmental Protection Agency, 1995.
2 EPA Office of Compliance Sector Notebook Project. Sector Notebook Data Refresh - 1997. US Environmental Protection Agency, 1998.
3 P B Bell, J J Bilancieri, *Paint & Ink International*, **9**, No.5, , 6-10 (1996).
4 M J Dvorchak, *J. Coat. Technol.*, **69**, No.866, 47-52 (1997).
5 M Moss, *Pigment Resin Technol.*, **26**, No.5, 296-9 (1997).
6 R M Nussbaum, E J Sutcliffe, A C Hellgrn, *J. Coat. Technol.*, **70**, No.5, 49-57 (1998)
7 S Peeters, J P Bleus, Z J Wang, J A Arceneaux, J Hall, *Paint & Ink International*, **11**, No.1, 6-9 (1998).
8 *Plast. Rubber Weekly*, No.1691, 1997.
9 Hughes F, TAPPI 1997 Hot Melt Symposium. Conference Proceedings. *TAPPI.* Hilton Head, SC, 15th-18th June 1997, p.15-21.
10 L White, *Eur. Rubber J.*, **179**, No.4, 24-5 (1997).
11 S-H Guo, **US Patent 5.646,225**, ARCO Chemical Technology, 1997.
12 I J Barlow, **US Patent 5,849,838**, SC Johnson & Sons, Inc., 1998.


## 14.32 SUMMARY

The data from the US industry given in the various sections of this chapter allow analysis of releases and transfers of the various solvents and helps us to understand the patterns of solvent use. Also industries can be ranked based on the amounts of solvents released. This information is provided in Tables 14.32.1, 14.32.2, 14.32.3, 14.32.4, 14.32.5, and 14.32.6 which contain total release of solvents from all reporting industries, the total transfers of solvents from all reporting industries, the sum of releases and transfers from these industries and the total release of solvents by each industry, the total transfers of solvents by each in-

**Table 14.32.1. Reported solvent releases from all industries in 1995 [Data from EPA Office of Compliance Sector Notebooks]**

| Solvent | Amount, ton/year | Solvent | Amount, ton/year |
|---|---|---|---|
| acetone | 870 | 2-ethoxyethanol | 19 |
| allyl alcohol | 60 | ethylbenzene | 3,810 |
| benzene | 3,378 | ethylene glycol | 8,122 |
| n-butyl alcohol | 5,640 | hexachloroethane | 16 |
| sec-butyl alcohol | 216 | hexane | 6,571 |
| tert-butyl alcohol | 482 | isopropyl alcohol | 205 |
| carbon disulfide | <u>33,085</u> | methanol | **93,965** |
| carbon tetrachloride | 30 | 2-methoxyethanol | 23 |
| chlorobenzene | 181 | methyl ethyl ketone | 23,315 |
| chloroform | 458 | methyl isobutyl ketone | 6,876 |
| chloromethane | 310 | N-methyl-2-pyrrolidone | 1,111 |
| cresol | 800 | methyl tert-butyl ether | 1,477 |
| m-cresol | 320 | pyridine | 187 |
| o-cresol | 270 | tetrachloroethylene | 1,692 |
| p-cresol | 162 | toluene | *41,015* |
| cyclohexane | 3,001 | 1,2,4-trichlorobenzene | 1,245 |
| cyclohexanone | 303 | 1,1,1-trichloroethane | 8,631 |
| cyclohexanol | 1,100 | trichloroethylene | 6,611 |
| 1,2-dichlorobenzene | 15 | trichlorofluoromethane | 84 |
| 1,4-dichlorobenzene | 15 | 1,2,4-trimethylbenzene | 2,308 |
| 1,2-dichloroethane | 222 | xylene | 26,544 |
| dichloromethane | 16,332 | m-xylene | 269 |
| N,N-dimethylformamide | 447 | o-xylene | 82 |
| 1,4-dioxane | 33 | p-xylene | 1,660 |

dustry, and the total release and transfer of solvents by each industry, respectively. The three highest entries in the tables (either amount, release, or transfer) are marked by **bold** (first), *italic* (second), or <u>underlining</u> (third).

Methanol, ethylene glycol, toluene, xylene, and carbon disulfide occupy the highest places in Tables 14.32.1-14.32.3, followed by methyl ethyl ketone, dichloromethane, 1,1,1-trichloroethane, tert-butyl alcohol, hexane, and methyl isobutyl ketone. These eleven solvents constitute 86% of all 550,738 tons of solvents released and transferred per year. Methanol alone constitutes over a quarter (27.1%) of the released and transferred solvents.

**Table 14.32.2. Reported solvent transfers from all industries in 1995 [Data from EPA Office of Compliance Sector Notebooks]**

| Solvent | Amount, ton/year | Solvent | Amount, ton/year |
|---|---|---|---|
| acetone | 767 | 2-ethoxyethanol | 2 |
| allyl alcohol | 330 | ethylbenzene | 4,691 |
| benzene | 792 | ethylene glycol | **72,549** |
| n-butyl alcohol | 3,636 | hexane | 8,910 |
| sec-butyl alcohol | 1,723 | isopropyl alcohol | 168 |
| tert-butyl alcohol | 12,749 | methanol | *55,287* |
| carbon disulfide | 260 | 2-methoxyethanol | 8 |
| carbon tetrachloride | 224 | methyl ethyl ketone | 15,119 |
| chlorobenzene | 704 | methyl isobutyl ketone | 6,811 |
| chloroform | 322 | N-methyl-2-pyrrolidone | 1,497 |
| chloromethane | 0.5 | methyl tert-butyl ether | 246 |
| cresol | 594 | pyridine | 103 |
| m-cresol | 720 | tetrachloroethylene | 1,189 |
| o-cresol | 57 | toluene | 18,717 |
| p-cresol | 870 | 1,2,4-trichlorobenzene | 69 |
| cyclohexane | 1,711 | 1,1,1-trichloroethane | 1,778 |
| cyclohexanone | 1 | trichloroethylene | 1,967 |
| cyclohexanol | 4 | trichlorofluoromethane | 4 |
| 1,2-dichlorobenzene | 4 | 1,2,4-trimethylbenzene | 729 |
| 1,4-dichlorobenzene | 3 | xylene | 27,000 |
| 1,2-dichloroethane | 3 | m-xylene | 98 |
| dichloromethane | 2,895 | o-xylene | 491 |
| N,N-dimethylformamide | 1,298 | p-xylene | 9 |
| 1,4-dioxane | 60 | | |

Methanol + xylene + toluene constitute over half (52.6%) of all released and transferred solvents by the analyzed industries. This is in fact good news because neither of the five most frequently released and transferred solvents is considered to have carcinogenic effect. But four (methanol, xylene, ethylene glycol, and toluene) contribute to the pollution of lower atmosphere and xylene and toluene cause formation of ozone in lower atmosphere by which it may affect the respiratory system. Carbon disulfide causes formation of hydroxyl radicals which have relatively long half-life (a few days) and thus participate in a variety of photochemical processes. All five solvents are relatively easy to biodegrade.

**Table 14.32.3. Reported solvent releases and transfers from all industries in 1995 [Data from EPA Office of Compliance Sector Notebooks]**

| Solvent | Amount, ton/year | Solvent | Amount, ton/year |
|---|---|---|---|
| acetone | 1,637 | 2-ethoxyethanol | 21 |
| allyl alcohol | 390 | ethylbenzene | 8,501 |
| benzene | 4,170 | ethylene glycol | *80,671* |
| n-butyl alcohol | 9,276 | hexachloroethane | 16 |
| sec-butyl alcohol | 1,939 | hexane | 15,481 |
| tert-butyl alcohol | 13,231 | isopropyl alcohol | 373 |
| carbon disulfide | 33,345 | methanol | **149,252** |
| carbon tetrachloride | 254 | 2-methoxyethanol | 31 |
| chlorobenzene | 885 | methyl ethyl ketone | 38,434 |
| chloroform | 780 | methyl isobutyl ketone | 13,687 |
| chloromethane | 311 | N-methyl-2-pyrrolidone | 2,608 |
| cresol | 1,394 | methyl tert-butyl ether | 1,723 |
| m-cresol | 1,040 | pyridine | 290 |
| o-cresol | 327 | tetrachloroethylene | 2,881 |
| p-cresol | 1,032 | toluene | 59,732 |
| cyclohexane | 4,712 | 1,2,4-trichlorobenzene | 1,314 |
| cyclohexanone | 304 | 1,1,1-trichloroethane | 10,409 |
| cyclohexanol | 1,104 | trichloroethylene | 8,578 |
| 1,2-dichlorobenzene | 19 | trichlorofluoromethane | 88 |
| 1,4-dichlorobenzene | 18 | 1,2,4-trimethylbenzene | 3,037 |
| 1,2-dichloroethane | 225 | xylene | 53,544 |
| dichloromethane | 19,227 | m-xylene | 367 |
| N,N-dimethylformamide | 1,745 | o-xylene | 573 |
| 1,4-dioxane | 93 | p-xylene | 1,669 |

The largest releases and transfers of methanol are from the pulp and paper (57.7%), organic chemical (21.3%), and the polymer and man-made fibers industry (6.2%) which contribute 85.2% of the total methanol releases and transfers. Ethylene glycol releases and transfers are mostly from the polymer and fiber industry (62.5%), the rubber and plastics (19.1%), and the organic chemical industry (15.9%). These three industries contribute 97.5% of all releases and transfers of ethylene glycol from the reporting industries. Toluene and xylene are typical paint, ink, cleaning and process solvents. The largest source of these

**Table 14.32.4. Total releases from different industries in the USA in 1995. [Data from EPA Office of Compliance Sector Notebooks]**

| Industry | Release ton/year | Percent total % | Release/employee kg/year | Release/$1000 of shipment, g |
|---|---|---|---|---|
| Aerospace | 3,483 | 1.1 | 4.3 | 48 |
| Metal fabrication | 24,310 | 7.9 | 221 | **2455** |
| Inorganic chemicals | 643 | 0.2 | 5.8 | 21 |
| Iron and steel | 2,434 | 0.8 | 6.1 | 24 |
| Metal casting | 7,051 | 2.3 | 28 | 237 |
| Motor vehicle | 30,675 | 10.0 | 77 | 153 |
| Organic chemicals | 24,678 | 8.0 | 197 | 380 |
| Petroleum | 19,419 | 6.3 | 259 | 139 |
| Polymer & fiber | 37,646 | 12.3 | **358** | 818 |
| Printing | 14,750 | 4.8 | 9.8 | 109 |
| Pulp & paper | 70,101 | **22.8** | *351* | *876* |
| Rubber & plastics | 45,119 | *14.7* | 56 | 501 |
| Stone, clay, ... | 3,630 | 1.2 | 7.3 | - |
| Textile | 6,629 | 2.2 | 11 | 95 |
| Wood furniture | 16,636 | 5.4 | 67 | 756 |
| **Total** | **307,204** | **100** | **Average = 111** | **Average = 472** |

is from the motor vehicle assembly industry (20%), the printing industry (13.7%), the rubber and plastics industry (11.4%), and the metal fabrication industry (10.7%). In total, these industries contribute over half of all releases and transfers of toluene and xylene (55.8%). The majority of carbon disulfide (82.5%) is contributed by the man-made fiber industry from rayon production.

The solvents ranked below the top five in the releases and transfers are mostly used in the paint and surface cleaning. The majority of emitted methyl ethyl ketone comes from the metal fabricating industry (25.2%), the rubber and plastics industry (23.4%), and the vehicle assembly industry (11.5%) which jointly contribute 60.1% of the total releases and transfers. 1,1,1-trichloroethane is released and transferred from the metal fabricating industry (36.6%), the rubber and plastics industry (30.4%), and the motor vehicle assembly industry (8.4%), totals 75.4% of all releases and transfers of this solvent. Tert-butyl alcohol is almost solely contributed by the organic chemical industry (94.6%). Hexane comes from polymer and fiber (57.4%), petroleum (19.5%), and rubber and plastics industry (11.3%) which together give 88.2% of all releases and transfers. Methyl isobutyl ketone is mostly contributed by the motor vehicle industry (56.7%) but metal fabricating (13%) and rubber and plastics industry (11.3%) also release and transfer large amounts of this solvent. Together all three industries account for 81% of its total release and transfer. Dichloromethane

**Table 14.32.5. Total transfers from different industries in the USA in 1995. [Data from EPA Office of Compliance Sector Notebooks]**

| Industry | Transfer ton/year | Percent total % | Transfer/employee kg/year | Transfer/$1000 of shipment, g |
|---|---|---|---|---|
| Aerospace | 1,080 | 0.4 | 1.3 | 15 |
| Metal fabrication | 13,164 | 5.3 | *120* | *1330* |
| Inorganic chemicals | 324 | 0.1 | 2.9 | 11 |
| Iron and steel | 4,328 | 1.7 | 10.8 | 43 |
| Metal casting | 178 | 0.1 | 0.7 | 6 |
| Motor vehicle | 23,980 | 9.6 | 60 | 60 |
| Organic chemicals | 63,680 | *25.3* | *509* | 980 |
| Petroleum | 1,382 | 0.6 | 18 | 10 |
| Polymer & fiber | 78,584 | **31.4** | **748** | **1708** |
| Printing | 4,051 | 1.6 | 2.7 | 30 |
| Pulp & paper | 23,457 | 9.4 | 117 | 293 |
| Rubber & plastics | 27,215 | 10.9 | 34 | 302 |
| Stone, clay, ... | 2,736 | 1.1 | 5.5 | - |
| Textile | 1,845 | 0.7 | 3.1 | 26 |
| Wood furniture | 4,439 | 1.8 | 18 | 202 |
| **Total** | **250,443** | **100** | **Average = 110** | **Average = 358** |

comes from rubber and plastics (65.5%), metal fabricating (24.8%), and polymer and fibers industry (8.1%) totals 98.4% of dichloromethane.

The second tier solvents are more harmful to health and environment than the first five most frequently released and transferred solvents. Ketones are suspected carcinogens by some sources. Dichloromethane is a probable human carcinogen. 1,1,1-trichloroethane is an ozone depleter. Because of their volatility, these solvents reside mostly in the lower atmosphere causing pollution and participating in radical processes.

Five industries are the largest contributors to the second tier solvents: rubber and plastics (23.8%), metal fabricating (18.1%), motor vehicle assembly (11.8%), organic chemicals (11.7%), and polymer and fiber (9.4%). Their joint contribution accounts for 74.8% of the releases and transfers of these solvents.

Two pertinent observations can be concluded from the above data

- tight controls and applications of new technology in the top three industries using a particular solvent can make big difference
- most releases and transfers stem from the use of paints and coatings.

Tables 14.32.4-14.32.6 analyze releases from different industries. The polymer & fiber, pulp & paper, organic chemicals, rubber & plastics industries pollute the most from the point of view of releases and transfers of solvents.

**Table 14.32.6. Total releases and transfers from different industries in the USA in 1995. [Data from EPA Office of Compliance Sector Notebooks]**

| Industry | Total ton/year | Percent % | Total/employee kg/year | Total/$1000 of shipment, g |
|---|---|---|---|---|
| Aerospace | 4,563 | 0.8 | 5.7 | 63 |
| Metal fabrication | 37,474 | 6.7 | 341 | **3785** |
| Inorganic chemicals | 967 | 0.2 | 8.8 | 32 |
| Iron and steel | 6,762 | 1.2 | 16.9 | 68 |
| Metal casting | 7,229 | 1.3 | 28.9 | 243 |
| Motor vehicle | 54,355 | 9.8 | 136 | 271 |
| Organic chemicals | 88,358 | 15.8 | *707* | 1359 |
| Petroleum | 20,801 | 3.7 | 277 | 158 |
| Polymer & fiber | 116,230 | **20.9** | **1107** | *2527* |
| Printing | 18,801 | 3.4 | 12.5 | 139 |
| Pulp & paper | 93,558 | *16.8* | 467 | 1169 |
| Rubber & plastics | 72,334 | 13.0 | 90 | 803 |
| Stone, clay, ... | 6,366 | 1.1 | 12.7 | - |
| Textile | 8,474 | 1.5 | 14.1 | 121 |
| Wood furniture | 21,075 | 3.8 | 84 | 958 |
| **Total** | **557,347** | **100** | **Average = 230** | **Average = 835** |

It can be speculated from the release data per employee that workers in pulp & paper, rubber & plastics, and polymer & fiber are the most exposed to solvents. Some technologies cost taxpayers more than others. Metal fabrication is one example. In order to produce goods valued at $1,000, the industry releases and transfers almost 3.8 kg of solvents. The cost in health and cleanup may exceed the value of goods manufactured.

It is characteristic that traditional industries which maintain older plants (metal fabrications, polymer & fiber, rubber & plastics, wood furniture, organic chemicals) contribute more to releases and transfers of solvents (as measured by releases plus transfers per $1000 sales), than industries which invest capital in the improvement of equipment, safety, and research and development (e.g., printing, motor vehicle assembly). This suggests that the avoidance of cost of the required investment is one reason for pollution. This reason was illustrated in the discussion of the petroleum industry where one manufacturer almost eliminated pollution by fixing leaking valves.

Industries consisting of smaller individual companies have fewer resources to develop new, non-polluting technologies (or even enforce safe practices). For example, the metal fabrication industry is composed of smaller plants using older technology which do not have the resources to make environmental improvements and remain competitive.

The traditional chemical industries (pulp & paper, rubber & plastics, and polymer & fiber) will hopefully apply their knowledge and chemical know-how to eliminate most of their contribution to pollution. In the future this will become an increasingly important requirement in competing by decreasing the social cost of applied technology. Perhaps more plastics could replace metals if the polymer, rubber, and plastics industries could demonstrate that their cost of pollution per dollar of output is much smaller than that of the metal industry.

# 15

# METHODS OF SOLVENT DETECTION AND TESTING

## 15.1 STANDARD METHODS OF SOLVENT ANALYSIS

GEORGE WYPYCH
**ChemTec Laboratories, Inc., Toronto, Canada**

This section includes information on solvent analysis based on methods included in national and international standards. ASTM standard methods are emphasized because they contain more methods on solvent testing than any other set of methods available. ISO standards are also covered in full detail because they are used in practical applications and are a basis for preparation of national standards. References to other national standards are also given including Australian (AS), Australian/New Zealand (AS/NZS), (British (BS), Canadian (CAN/CGSB), German (DIN), Japanese (JIS), and Finish (SFS). The same number is given in references to all national and international standards for the same solvent property to facilitate searching and referencing. Methods of analysis are grouped by subject in alphabetical order.

### 15.1.1 ALKALINITY AND ACIDITY

Amine acid acceptance by halogenated solvents is the degree to which an amine is capable of absorbing or neutralizing acid present from an external source or generated by the solvent.[1] This method is useful for comparing the effect of an amine with other acid-accepting compounds. The test is performed by the titration of an alkaline water extract from the solvent with 0.2 N hydrochloric acid to pH = 3.9 as detected by pH electrode.

The alkalinity of acetone is determined by a titration with 0.05 N $H_2SO_4$ in the presence of methyl red indicator.[2] The test method provides a measure of acetone alkalinity calculated as ammonia.

The acidity of halogenated solvents can be determined by titration with 0.01 N sodium hydroxide in the presence of a 0.1% solution of bromophenyl blue indicator.[3] Similar to the determination of alkalinity,[1] acidity is determined in water extract. The determination can also be done directly in solvent but the solution of sodium hydroxide should be prepared in methanol. A third option is to determine acidity by a pH-metric titration of a water extract. Prior to determination, the pH of solvent is measured. If the pH is above 7, then an alkalinity measurement is done by the above method.[1] If the pH is below 7, acidity is determined by this method.[3] The method is used to establish purchasing and manufacturing specifications and control the quality of solvents.

The acidity of benzene, toluene, xylenes, naphthas, and other aromatic hydrocarbons is determined by the titration of a water extract with 0.01 N sodium hydroxide in the presence of 0.5% phenolphthalein indicator solution.[4] The method is suitable for setting specifications, internal quality control, and development of solvents. The result indicates the potential corrosivity of solvent.

The acidity of solvents used in paint, varnish, and lacquer is determined by the titration of solvent diluted with water (for water soluble solvents) or isopropyl alcohol or ethanol (for water insoluble solvents) in proportion of 1:1. A water solution of 0.05 N sodium hydroxide in the presence of 0.5% phenolphthalein indicator dissolved in ethanol or isopropanol is used for titration.[5] The method is useful for determination of acidity below 0.05%. Acidity is a result of contamination or decomposition during storage, transportation or manufacture. The method is used to assess compliance with specification.

Solvents which are depleted of stabilizers (amine or alpha epoxide) may become acidic. The following method determines the combined effect of both alkaline (amine) and neutral (usually epoxy) stabilizers.[6] The determination is done in two steps. First solvent is mixed with hydrochlorination reagent (0.1 N HCl), then the excess is titrated with 0.1 N sodium hydroxide in the presence of 0.1% bromophenyl blue as an indicator.

The total acidity of trichlorotrifluoroethane and other halocarbons is determined by titration of a sample diluted with isopropanol with 0.01 N sodium hydroxide in isopropanol in the presence of a 0.05% isopropanol solution of phenolphthalein as an indicator.[7] The method is used for setting specifications and quality control.

### 15.1.2 AUTOIGNITION TEMPERATURE

The autoignition temperature can be determined by the hot and cold flame method.[8] Cool flames occur in vapor-rich mixtures of hydrocarbons and oxygenated hydrocarbons in air. The autoignition temperature is the spontaneous (self-ignition) temperature at which a substance will produce a hot flame without an external ignition source. Autoignition occurs when a hot flame inside a test flask suddenly appears accompanied by a sharp rise in temperature. With cold flame ignition the temperature rise is gradual.

The test equipment shown in a schematic drawing in the method[8] consists of a test flask, a furnace, a temperature controller, a syringe, a thermocouple and other auxiliary parts. The measurement is performed in a dark room for optimum visual detection of cool flames. The results are reported as ignition temperature, time lags (delay between sample insertion and material ignition), and reaction threshold temperature (the lowest flask temperature at which nonluminous pre-flame reactions (e.g., temperature rise) occur).

The results depend on the apparatus employed. The volume of vessel is especially critical. A larger flask will tend to produce lower temperature results. The method is not designed for materials which are solid at the measurement temperature or which undergo exothermic decomposition.

### 15.1.3 BIODEGRADATION POTENTIAL

The method covers a screening procedure which assesses the anaerobic biodegradation of organic materials.[9] The procedure converts organic substances into methane and carbon dioxide which are measured by a gas volumetric pipette – part of the standard apparatus. Other parts include, a biodegradation flask, a magnetic stirrer, a pressure transducer, a syringe, and a water seal. The apparatus may be interfaced with a gas chromatograph to determine quantities of the two gases.

The biodegradation process is conducted in a specially prepared medium inoculated with sludge inoculum. The process occurs under the flow of a mixture of 70% nitrogen and 30% carbon dioxide to provide anaerobic conditions.

The method was developed to screen organic substances for their potential to biodegrade. If a high degree of biodegradability is determined it provides a strong evidence that the test substance will be biodegradable in the anaerobic digestors of a waste treatment plant and in many natural environments. Other references[9] give methods of determining of biological and chemical oxygen demand.

### 15.1.4 BOILING POINT

The boiling point of solvent, its specific heat capacity, and its enthalpy of vaporization determine the energy required for solvent separation in a distillation column. They also determine numerous other properties of solvents (see Chapter 2).

Several methods of determination can be used, but two, distillation and gas chromatography are the most popular. Industrial aromatic hydrocarbons are determined by distillation. The temperature is recorded for the initial boiling point, for the sample which has been distilled at 5%, 10%, then at 10% incrcments up to 90%, then finally at 95%. The temperature should be recorded with precision of 0.1°C.[10] A general test method to determine the distillation range of volatile liquids[11] outlines a similar method of measurement. In addition to the measurements at the intervals given above, the temperature of the dry point (distillation temperature of residual quantities) is also recorded. Results must be reported as specified in the method.

A vacuum distillation procedure is used to determine the amount of solvents in solvent-based paints.[12] The paint sample is diluted with tricresyl phosphate, distilled for a while under normal pressure to evaporate the more volatile solvents, followed by vacuum distillation at 2 mm Hg.

Capillary gas chromatography is used to determine the boiling point of hydrocarbon solvents.[13] The initial boiling point is defined as the point at which the cumulative area of chromatogram equals 0.5% of its final total surface area. The final boiling point is at cumulative area of 95% of the total surface area of chromatogram. The method reports boiling point distribution in 1% intervals over the 1-99% range of the total cumulative surface area of chromatogram as well as the initial and final boiling points. A flame ionization detector is used in the determination and a standard solvent containing 16 known components is used for calibration.

### 15.1.5 BROMINE INDEX

Two methods are used to determine bromine index of aromatic hydrocarbons which contain trace amounts of olefins and are substantially free of materials lighter than isobutane and have distillation end-point lower than 288°C. The methods measure trace amounts of unsaturations in materials which have a bromine index below 500.

The bromine index can be measured by electrometric[14] and coulometric[15] titration. In the electrometric titration method, a sample is titrated with bromide-bromate solution (0.1 N solution of mixture of potassium bromide and potassium bromate) until the end-point increase in potential remains steady for 30 s.[14] In coulometric titration, a potassium bromide solution is used to titrate the solvent until the bromine concentration increases because it is no longer being consumed by the unsaturation of the solvent.[15]

Both methods can be used for setting specification, quality control, and testing of development solvents to find olefinic content. The methods do not differentiate between the types of unsaturations.

### 15.1.6 CALORIFIC VALUE

The heat of combustion of liquid hydrocarbon fuels can be determined with bomb calorimeter.[16] Two definitions are used in result reporting: gross heat of combustion (the quantity of energy released from fuel burned in constant volume with all products gaseous except water which is in liquid state) and net heat of combustion (the same but water is also in a gaseous state). These determinations are useful in assessing the thermal efficiency of equipment used for generation of power or heat. The results are used to estimate the range of an aircraft between refueling stops which is a direct function of heat of combustion. The calorimeter bomb is standardized against benzoic acid standard. Net and gross heats of combustion are reported. A specific method is used for aviation fuels.[17] This method reports results in SI units and the measurements are made under constant pressure. The method is applicable for aviation gasolines or aircraft turbine and jet engine fuels. The method is used when heat of combustion data are not available. An empirical equation was developed which gives net heat of combustion based on the determined values of aniline point (ASTM D 611) and API gravity (ASTM D 287). If the fuel contains sulfur, a correction is applied for sulfur determined according to ASTM D129, D 1266, D 2622, or D 3120 (the method selected depends on the volatility of the sample).

Gross calorific value and ash content of waste materials can be determined by a calorimetric method.[19] After a calorimetric analysis, the bomb washing can be used to determine of mineral content by elemental analysis. The sample is burned under controlled conditions in oxygen. The calorimeter is standardized by burning known amount of benzoic acid. The formation of acids can additionally be determined by titration.

### 15.1.7 CLEANING SOLVENTS

Several standard procedures are available for evaluation of cleaning solvents. The stability of aircraft cleaning compounds is determined after 12 months storage at controlled conditions which may include moderate temperatures, cold storage, and hot storage.[20] Solvent vapor degreasing operations which use halogenated solvents follow standardized procedure.[21] The standard contains information on the location and design of a degreasing installation and operation during startup, degreasing, shutdown, and solvent reclamation. The purpose of the standard is to reduce the probability of accidents and exposure to personnel. A separate standard practice[22] gives reasons for and methods of preventing acid formation in degreasing solvents. The formation of acid is generally related to excessive heat, contaminations, the presence of chlorinated and sulfonated oils, admixture of acids, and solvent mixtures.

### 15.1.8 COLOR

Impurities in benzene, toluene, xylene, naphthas, and industrial aromatic hydrocarbons are determined by a simple colorimetric analysis of an acid wash.[23] A solvent is washed with sulfuric acid and the color of the acid layer is determined by a visual comparison with color standards prepared from solutions of cobalt chloride and ferric chloride.

Aromatic hydrocarbons which melt below 150°C can be subjected to color analysis in the liquid (molten) state using a visual comparison with platinum-cobalt standards.[24] Standards are prepared from a stock solution of $K_2PtCl_2$ to form a scale varying in color intensity. Similar procedure was developed to evaluate color of halogenated organic solvents and their admixtures.[25]

Objective color measurement is based on tristimulus colorimetry.[26] This instrumental method measures tristimulus values of light transmitted by a sample and compares the results to the values transmitted by distilled water. The results can be recalculated to the platinum-cobalt scale referred to in the previous standards.[24,25] The results can also be interpreted by normal methods of color measurement to yellowness index, color, color depth, etc.

## 15.1.9 CORROSION (EFFECT OF SOLVENTS)

The corrosiveness of perchloroethylene to copper is determined using Soxhlet apparatus.[27] Three pre-weighed strips of copper are used, one placed in the bottom flask, the second in the bottom of the Soxhlet attachment, and the third below the condenser. The specimens are exposed to refluxing solvent for 72 h after which the entire apparatus is flushed with distilled water to wash all acidic substances back to the flask. The water layer is titrated with 0.01 N NaOH to determine its acidity and the strips are weighed to determine weight loss. The results indicate quality of solvent. A different method is used to test copper corrosion by aromatic hydrocarbons.[28] Here, a copper strip is immersed in a flask containing solvent and the flask is placed in boiling water for 30 min. Next, the copper strip is compared with ASTM standard corroded copper strips.

If 1,1,1-trichloroethane is not properly stabilized it forms hydrochloric acid in the presence of aluminum. HCl corrodes aluminum. The presence of free water invalidates the result of this test.[29] An aluminum coupon is scratched beneath the surface of a solvent. The coupon is observed for 10 min and 1 h and the degree of corrosion is recorded in form of pass (no reaction) or fail (gas bubbles, color formation, or metal corrosion). The test is important to cleaning operations because aluminum should not be used for parts of machines (pumps, tanks, valves, spray equipment) in contact with corrosive solvent.

## 15.1.10 DENSITY

Density and specific gravity of solvents are discussed together. The difference in their definitions is that specific gravity is the density of material relative to the density of water whereas the density is the weight in vacuo of a unit volume. The density of liquids (including solvents) can be measured by a Bingham pycnometer.[30] The determination includes introduction of the liquid to the tared pycnometer, equilibration of temperature, then weighing. Other standardized method[31] determines the specific density of liquid industrial chemicals by two methods: hydrometer and pycnometer. The pycnometric method is essentially similar to the previously described. It differs in that the water and then solvent are weighed. Thus the density determination error may only be due to an imprecise weighing (the pycnometer calibrated volume does not enter calculations). In the hydrometer method, the calibrated hydrometer is immersed in controlled temperature liquid and direct readings are obtained.

The standard method for determining the specific gravity of halogenated organic solvents[32] involves the use of both a pycnometer and a hydrometer as described above but, in addition, an electronic densitometer is also used. Here, a liquid is placed in U-shaped tube and subjected to electronic excitations. The density changes the mass of tube and frequency of oscillations which is the basis for measurement and display of specific gravity readings.

Two standard tables[33,34] (American and metric system of units) are used to calculate weight and volume of benzene, toluene, xylenes mixture and isomers, styrene, cumene, and ethylbenzene as well as aromatic hydrocarbons and cyclohexane. Tables provide volume corrections for these solvents in a temperature range from -5 to 109°F (-20.5 to 43°C).

## 15.1.11 DILUTION RATIO

The dilution ratio is the maximum number of units of diluent that can be added to unit volume of solvent before precipitation occurs. Cellulose nitrate dissolved in an oxygenated solvent (8 wt% resin) is the most classical method to determine dilution ratio used to evaluate toluene as a standard diluent and to compare different diluents and solvents with a standard solvent (n-butyl acetate). The standard dilution ratio of toluene by n-butyl acetate solution is 2.73-2.83. Two end points are determined. The first occurs when a known amount of diluent forms precipitate after 2 min of vigorous swirling. The second is determined by re-dissolving the precipitate, adding a known volume of solvent (dependent on dilution ratio) and precipitating it again with diluent.[35]

Similar method can be used for any resin, solvent, or diluent.[36] A solution of resin is prepared by the method described in ASTM D 1725 or by dispersing in blender (results may differ). Precipitation with diluent is determined at 25°C. The diluent is added dropwise from burette or weight is controlled throughout the experiment. The resin dilutability is recorded. If more than 100 g of diluent is required for 10 g of solution then diluent is regarded as being infinitely soluble.

Heptane miscibility in lacquer solvents is determined by mixing equal amounts of the specimen (lacquer) and heptane.[37] If a clear solution results after mixing, it indicates good miscibility. If a turbid solution results, either the heptane is immiscible with the tested specimen or water is present in either component.

Water-insoluble admixtures in solvents may affect many uses of solvent. The specimen solvent (primarily acetone, isopropanol, methanol, and many other) is diluted to 10 volumes with water and the resulting turbidity or cloudiness is recorded.[38]

### 15.1.12 DISSOLVING AND EXTRACTION

A standard practice for preparing polymer solution contains information on solvents, their concentration, temperature, pressure, mixing time, and heating.[39] The annex contains information on the best solvents for 75 typical polymers with different degrees of substitution or modification. Frequently, temperature and concentration of solution is also given.

Solvent extraction is used on textile materials to determine naturally occurring oily and waxy materials that have not been completely removed from the fibers.[40] The percentage of extracted material is given in relationship to the dry mass of fiber. Solvents used for extraction including 1,1,2-trichloro-1,2,3-trifluoroethane and dichloromethane but these may be replaced by other solvents by mutual agreement. The Soxhlet extraction and gravimetric determination are used.

### 15.1.13 ELECTRIC PROPERTIES

Specific resistivity is numerically equivalent to the volume resistance between opposite faces of one centimeter cube. The specific resistivity of electronic grade solvents is measured by a low-voltage a-c bridge operating at 1000 Hz providing the specific resistance does not exceed $10^9$ Ω-cm.[41] If the specific resistance is higher than $10^9$ Ω-cm, d-c equipment is used. D-c equipment is capable of measuring up to at least $10^{12}$ Ω-cm.

The electrical conductivity of liquid hydrocarbons can be measured by a precision meter.[42] The method is used to evaluate aviation fuels and other low-conductivity hydrocarbons. The generation and dissipation of electrostatic charges in liquids during handling depends on ionic species which may be characterized by electrical conductivity at rest. The dissipation time of the charges is inversely proportional to the conductivity. The measurement is done by a conductivity meter.

Two standard methods[43,44] were designed to determine the electric breakdown voltage of insulating oils of petroleum origin. These are VDE electrodes method[43] and the method under impulse conditions.[44] The dielectric breakdown voltage measurement allows to estimate the ability of an insulating liquid to withstand electric stress without failure. In the presence of contaminations such as water, dirt, or cellulosic fibers low breakdown voltages are obtained. VDE stands for Verband Deutscher Electrotechniker, an organization which designed brass electrodes used for measurement. The equipment uses a transformer, voltage control equipment, and a voltmeter. The impulse method[44] uses highly divergent field under impulse conditions. The breakdown voltage of fresh oil decreases as the concentration of aromatic hydrocarbons increases. The method can be used for quality control of fresh oil, and determining the effect of service aging, and effect of impurities. The material is placed in test cell containing electrodes which are supplied from an impulse generator controlled by voltage control equipment.

### 15.1.14 ENVIRONMENTAL STRESS CRAZING

Crazing is a group of surface fissures which appear as small cracks after the material has been exposed to solvent and stress. Crazes are usually oriented perpendicular to stress and their appearance depends on the index of refraction and on the angle of viewing. A suitable light source must be used. Transparent plastic materials can be directly tested for crazing.[45] Two variations of method are used to determine stress crazing of transparent plastics: determination of stress required to cause cracking or determination of craze development along the time of stress application. In the first method specimens are exposed to solvent by the direct contact of specimen surface with filter paper wetted with solvent. By selecting different values of stress (using each time new specimen) the range of two stress forces is searched for the largest stress under which specimen does not craze and the smallest stress under which it does craze. In the second variation of the method, sample is tested first without solvent to assure that it does not craze under the selected load. Solvent is then applied and specimen inspected in 15 min time intervals taking note of location of craze front as crazing progresses. If the sample does not craze higher load is selected and *vice versa*. Similar to the first variation the range is determined within which specimen crazes and does not craze.

The crazing effect can be indirectly determined by testing chip impact resistance of specimens which crack either because of weathering or environmental stress cracking.[46]

### 15.1.15 EVAPORATION RATE

The evaporation rate of a solvent is determined to obtain relative value to some standard, selected solvent. The solvent selection depends on reasons for solvent use and the type of solvent and it is usually agreed upon between interested parties. In Europe, diethyl ether is the most frequently used reference solvent and in the US butyl acetate. The evaporation rate of other solvents is determined under identical conditions and the resultant values are used to rank solvents. The most obvious requirement is that the determination is done without excessive drafts and air currents. The evaporation rate is the ratio of the time required to evaporate a test solvent to the time required to evaporate the reference solvent under identical conditions. The results can be expressed either as the percentage evaporated within certain time frame, the time to evaporate a specified amount, or a relative rate. Relative rate is the most common.

For halogenated solvents used in cleaning applications, the relative evaporation rate is compared either to xylene or perchloroethylene.[47] The determination is done on a test panel using 10 ml of solvent. The relative evaporation rate is calculated.

### 15.1.16 FLAMMABILITY LIMITS

These tests cover the methods of determination of the minimum temperature at which vapors in equilibrium with liquid solvent are sufficiently concentrated to form flammable mixtures with air at atmospheric pressure and concentration limits of chemicals. Flammable (explosive) limits are the percent levels, volume by volume, of a flammable vapor or gas mixed in air between which the propagation of a flame or an explosion will occur upon the presence of ignition. The leanest mixture at which this will occur is called the lower flammable limit. The richest mixture at which it will occur is the upper flammable limit. The percent of vapor mixture between the lower and upper limits is known as the flammable range.

The temperature limits of flammability can be determined in an air/vapor mixture above a flammable liquid in a closed vessel.[48] The temperature in vessel is varied until a minimum temperature is reached at which the flame will propagate away from the ignition source. A glass vessel is equipped with an ignition device, a magnetic stirrer, a clamping devices, and a safety glass window. The initial temperature of determination is estimated from closed-cup flash point measurement. If the flash point is below 38°C, the initial temperature should be 8°C below flash point temperature. If the flash point is between 38 and 96°C, the starting temperature should be at least 14°C below the flash point. If the flash point temperature is above 96°C then the initial temperature should be 22 to 44°C below the flash point temperature. Selecting higher initial temperature may result in explosion. The lower temperature limit of flammability is obtained from the test which can be used to determine guidelines for the safe handling of solvents in closed process and storage vessels.

The concentration limits of flammability are determined using another method.[49] The method is limited to atmospheric pressure and temperature of 150°C. Equipment is similar to that used in the previous method. A uniform mixture of vapor and air is ignited and flame propagation from ignition source is noted. The concentration of flammable components is varied until a composition is found which is capable to propagate flame.

## 15.1.17 FLASH POINT

A variety of apparatus such as the small scale closed tester,[50] the tag open cup,[51] the tag closed tester,[52] the Cleveland open cup,[53] the Pensky-Martens closed cup,[54] and the equilibrium method[55] are used to determine the flash point. The selection of method is based suggestions included in separate standard.[56]

The small scale closed tester[50] is a metal cup with a thermometer fitted below the bottom of the internal chamber with a hinge mounted cover having filling orifice. The sample is introduced to the cup and the cup is maintained at a constant temperature by means of temperature controller. After a specific time, a test flame is applied for 2.5 s and an observation is made whether or not flash has occurred. If flash did not occur the cup is cleaned, a new sample is introduced and the temperature is increased by 5°C. The measurements are repeated until the flash point is determined with accuracy of 1°C.

The tag open cup[51] is a larger unit equipped with water-glycol bath for temperature control or a solid carbon dioxide-acetone bath for lower flash points. A much larger sample is used with this equipment and the temperature is gradually increased at a rate of 1°C/min. A taper flame is passed for 1 s in 1 min intervals until the flash point is detected. The tag closed tester[52] can use either a manual or an automated procedure. A sample volume of 50 ml is used. Either a gas flame or an electric ignitor is used. In the automated mode, the equipment is programmed to perform the standard procedure.

The Cleveland open cup[53] is placed on a heated plate which increases temperature at a rate of 5-6°C/min. This method can also be automated. The method is designed for testing petroleum products which are viscous and have flash point above 79°C. The Pensky-Martens[54] closed cup tester was also designed for petroleum products but for those with flash points from 40 to 360°C. This apparatus has its own heating source, stirrer, and cover by which it differs from Cleveland cup. It can be either manual or automated. The equilibrium method[55] uses either a modified tag close cup or the Pensky-Martens apparatus. The modification intended to keep the vapor/air temperature in equilibrium with the liquid temperature. The method is limited to the temperature range from 0 to 110°C.

Depending on the viscosity of liquid and its expected flash point range, one of the above methods is chosen as described in detail elsewhere.[56] It should be additionally noted that if the flash point method uses continuous heating, it is not suitable for testing mixtures of flammable substances because their vapor concentrations are not representative of equilibrium conditions. One of the weaknesses of flash point analysis is that the flame is well above the liquid surface therefore full vapor concentration is not attained. Many cases exist where a flash point cannot be detected but the material does form flammable mixtures. Before a method is chosen and a data interpretation made full information on the test procedure should be studied in detail and the proper authorities should be consulted to define safe practices for a particular material.

### 15.1.18 FREEZING POINT

Freezing point apparatus consists of freezing tube, Dewar flasks to act as cooling and warming baths, stirring mechanisms, absorption tubes, clamps and other auxiliary parts.[57] Freezing point can be obtained precisely from interpretation of time-temperature freezing and melting curves. The determination is made by measuring the electrical resistance of liquid which decreases on cooling and becomes constant when it freezes. This method in conjunction with the testing details described in a separate standard[58] can be used to determine the purity of many hydrocarbon solvents. The data given in the last standard[58] allow for a precise determination of the purity of solvent in percent of pure compound. A simple method was designed to determine solidification point of benzene based on visual observation of formation of solid phase.[59]

### 15.1.19 FREE HALOGENS IN HALOGENATED SOLVENTS

This simple qualitative test involves the extraction by water of free halogens, followed by the reaction of the halogens with potassium iodide in the presence of a starch indicator.[60] The solution color changes to blue in the presence of free halogens.

### 15.1.20 GAS CHROMATOGRAPHY

Gas chromatography provides many tools for the analysis of solvents. In section 15.1.4, a method was discussed which determines the distribution of boiling points of hydrocarbon mixtures. Many uses of gas chromatography for the determination of purity of different solvents are discussed in Section 15.1.25. In this section, some examples of of gas chromatography are included to show its usefulness in the qualitative determination of solvents mixtures[61] the analysis of solvent impurities,[62-64] the determination of solvents in a product by direct injection,[65] and the generation of data to evaluate waste materials to determine their hazardous content.[66]

The relative distribution of aromatic hydrocarbons in xylene products can be quantitatively determined by gas chromatography.[61] A flame ionization or thermal conductivity detector is used with a capillary or packed column containing crosslinked polyethylene glycol as the stationary phase. The peak area of each component is measured and the weight percentage concentration is calculated by dividing the peak area of the component by the sum of the areas of all peaks.

Ortho-xylene concentration and the concentrations of its admixtures and impurities are measured using a flame ionization detector and a polar fused-silica capillary column.[62] An internal standard (iso-octane) is used to increase precision and a standard mixture is used for calibration. A similar method is used with p-xylene but either n-undecane or n-octane are used as the internal standards.[63] In both methods, peak areas are interpreted relative to the peak area of internal standard. The main impurities in benzene are non-aromatics with less than 10 carbons, toluene, 1,4-dioxane and aromatics containing 8 carbon atoms. The method of determination of benzene impurities[64] is similar to two methods described above.[62,63] Normal-nonane is used as an internal standard. In all three methods the internal standards must be at least 99% pure component.

A gas chromatograph equipped with a thermal conductivity or a flame ionization detector and capillary or packed columns is used for direct determination of solvents in paints.[65] Columns are usually packed with either polyethylene glycol (molecular weight 20,000) or a diisodecyl phthalate as liquid phase on diatomaceous earth (60-80 mesh) used as a solid support. Low viscosity paints are drawn into a syringe and injected through the injection port. High viscosity paints are diluted with solvent that does not interfere with the analysis, usually ethyl ether or dichloromethane. A standardized gas chromatographic technique[66] is capable of determining the 67 solvents most frequently found in hazardous wastes. Several detectors are suggested for analysis such as flame ionization, electron capture, thermal conductivity, photoionization, or mass selective. Each waste mixture may contain a large number of solvents. Their detection is facilitated by the use of gas chromatograph interfaced with mass spectrometer. The method is designed to facilitate site assessment, recycling operations, plant control, and pollution programs.

### 15.1.21 LABELING

Warnings, first aid measures and operating instructions are standardized for vapor degreasers.[67] Placards containing this information should be placed close to the degreaser in an area accessible to employees. The placards should contain the information required by applicable federal and local laws and regulations. The placard should include name of the solvent used, and warnings indicating that the vapor is harmful, that breathing the vapor should be avoided, that the machine should only be used with proper ventilation, that swallowing and contact with the skin should be avoided, that cutting or welding should not be performed close to the machine, and that the tank should not be entered unless a proper procedure is followed. In addition, start-up and operation procedures should be available.

Minimum labeling requirements for several halogenated solvents should also follow a standard practice.[68] The label should state the company name, its logo and address, emergency telephone numbers, lot number, the net weight, solvent name, its CAS number, OSHA PEL and ACGIH TLV values, and quantity. In addition, health and safety information, precautions, first aid, and handling and storage information should be provided.

### 15.1.22 ODOR

Strips of rapid qualitative paper are dipped in a standard liquid and in the liquid under the test. Their odor is compared to establish if the odor of sample is more or less acceptable than the standard to the purchaser and the manufacturer.[69] In a similar method, papers dipped in a standard and specimens are allowed to dry at room temperature and tested for residual odor at specified time intervals.

Odor testing may be performed by a selected group of panelist to either determine the effect of various additives on the odor or taste of a medium or to determine the odor or taste sensitivity of a particular group of people.[70] For this purpose, a series of samples is prepared in concentration scale which increases in geometric increments. At each concentration step two samples containing the medium alone are given to a panelist. The panelist should determine which sample is different from the other two samples. The panelist should begin with the lowest concentration selected to be two or three concentration steps below estimated threshold. The method description contains information on sample selection and preparation, result, and precision determination.

A method of evaluation of denatured and undenatured alcohols to assess their acceptability is used.[71] This method is developed specifically to compare methanol, ethanol, isopropanol, and n-propanol. The group of panelist is asked to compare the characteristic and residual odors of evaporation, its intensity by dilution, and its concentration. A similar method was developed for a series of glycols.[72] Here, odor character and intensity are evaluated.

The residual odor of a drycleaning grade of perchloroethylene is determined by comparing treated and untreated samples of bleached cotton fabric. The treated fabric is soaked for 5 min in perchloroethylene and dried for 4 hours at room temperature.[73] Good quality perchloroethylene should leave no odor.

### 15.1.23 PAINTS – STANDARDS RELATED TO SOLVENTS

The paint industry, a major user of solvents, has developed numerous standards. Some are included in Sections 15.1.25, 15.1.31, and 15.1.35. The paint industry also uses many general standards and some specific standards, which have not been included in any other section of this chapter. Details of these are given below.

Sampling and testing requirements for solvents used in the paint industry are summarized in a special standard.[74] This comprehensive list of standards used by paint industry also includes a brief discussion of each method of testing, including sampling, specific gravity, color, distillation range, nonvolatile matter, odor, water, acidity, alkalinity, ester value, copper corrosion test, sulfur, permanganate time test for acetone and methanol, flash point, purity of ketones, solvent power evaluation, water miscibility, analysis of methanol, analysis of ethylene and propylene glycols, acid wash color of aromatic hydrocarbons, paraffins and other nonaromatic hydrocarbons in aromatics, and aromatics in mineral spirits.

The nonvolatile matter in paints is determined by a gravimetric method after drying a 100 ml sample in oven at 105°C.[75] The transfer efficiency of paints is a volume or weight ratio of paint solids deposited to the volume/weight of the paint solids sprayed, expressed in percent. This method[76] can be used to optimize the paint application process. The measurement is done by weighing or measuring the volume of paint used on a certain sprayed surface area and comparing this value with known or predetermined by the above method weight of solids in the paint used for spraying.

In order to determine an ester value for solvents or thinners, the specimen is reacted with aqueous potassium hydroxide, using isopropanol as the mutual solvent.[77] The hydrolysis is conducted at 98°C and the excess potassium hydroxide is determined by titration. From the amount of potassium hydroxide consumed, the ester value is calculated.

### 15.1.24 pH

A method developed for halogenated solvents is applicable for determining the pH of water extracts of solvents.[78] The solvent sample is shaken with distilled or deionized water and the pH is determined either by comparing color upon the addition of Gramercy universal indicator or by using a glass electrode pH meter.

### 15.1.25 PURITY

Several techniques are used to determine purity of solvents. Gas chromatography is the most common and this and other methods are discussed first followed by other analytical methods which include instrumental and simple methods. The aim of these tests is to determine the concentration of the main component but more frequently qualitative and quantitative determination of admixtures. Some methods have already been discussed in Sections 15.1.1, 15.1.5, 15.1.8, 15.1.9, 15.1.18, 15.1.20 and 15.1.23.

Alcohol content and purity of an acetate ester is determined by gas chromatography.[79] The method was applied to ethyl, n-propyl, isopropyl, n-butyl, isobutyl, and 2-ethylhexyl acetates. Water and acetic acid cannot be measured by this method and other methods are used. A thermal conductivity or flame ionization detector is used. A stainless steel column with 80-100 mesh Chromosorb G-HP is used with 9.05% Dow Corning QF-1 and 0.45% Igepal CO-990. The concentration of the main component and the amount of free alcohol are measured by the method.

Traces of benzene in hydrocarbon solvents are measured by capillary gas chromatography.[80] Because of the hazardous nature of benzene, the method was introduced to ensure compliance with the stringent regulations. A flame ionization detector is used with 0.53 mm fused silica capillary columns with bound methyl silicone or polyethylene glycol. Similar method is used to determine benzene content in cyclic products (cyclohexane, toluene, cumene, styrene, etc.).[81] This method does not specify any particular column but the column used should be able to resolve benzene from other components. The method can determine benzene in concentrations of 5 to 300 mg/kg. Traces of thiophene in refined benzene are determined by a flame photometric detector.[82] Several column types given in standard are found satisfactory to overcome potential problem of quenching effects of hydrocarbons on the light emissions from thiophene. High purity benzene for cyclohexane feedstock is tested for several known impurities by capillary gas chromatography.[83] The gas chromatograph is equipped with a flame ionization detector and a splitter injector suitable for fused silica capillary column internally coated with crosslinked methyl silicone. The concentration of benzene and the concentrations of impurities can be adequately determined.

The purity of halogenated solvents is determined using a thermal conductivity or flame ionization detector, a column made from 3.2 mm stainless tubing packed with 30 wt% silicone fluid on 80-100 mesh diatomaceous earth or using capillary column.[84] Admixtures in 1,1,1-trichloroethane are determined using a thermal conductivity or hydrogen flame detector. Column from copper or stainless steel is packed with Chromosorb W HP with 20% polydimethylsiloxane.[85]

Various impurities, such as hydrocarbons, acetone, alcohols and other can be determined using a thermal conductivity or a flame detector.[86] Several columns are specified in the standard mostly using polyethylene glycol on diatomaceous earth. For determination of purity of methyl isobutyl ketone different method is used.[87] In both cases the amounts of determined impurities are subtracted from total mass to give purity of specimen.

The determination of aromatics in mineral spirits is another method that has been developed to ensure compliance with regulations restricting aromatic content.[88] Three methods are given, each capable to determine ethylbenzene and total aromatic content. The methods differ in column type and packing.

A spectrophotometric method[89] for determining thiophene in benzene is available as alternative to gas chromatography.[82] The spectrometer used is capable of detecting absorbance in the range from 400 to 700 nm with a repeatability of 0.005 absorbance units. Thiophene is reacted with isatin to form a colored compound. The quantitative determination is based on reading concentrations from master curve.

The presence of reducing substances in pyridine can be detected by a simple visual observation.[90] The sample of pyridine is mixed with 0.32% potassium permanganate solution and color is observed after 30 min. If color of the potassium permanganate is retained, the sample is free of reducing agents.

Oxidative microcoulometry is used to determine trace quantities of sulphur in aromatic hydrocarbons.[91] An oxidative pyrolysis converts sulfur to sulfur dioxide which is titrated in titration cell with the triiodide ion present in the electrolyte.

The total chloride (organic and inorganic) in aromatic hydrocarbons and cyclohexane can be determined by titration.[92] Bromides and iodides present are recorded as chlorides. The sample is mixed with toluene in a proportion which depends on the expected concentration of chloride. The reagent sodium biphenyl is added to convert organic halogens into inorganic halides. After decomposing the excess reagent with water, the separated aqueous phase is titrated in presence of acetone with a silver nitrate solution. Organic chlorides present in aromatic hydrocarbons can also be determined by microcoulometry.[93] The presence of chlorine compounds may adversely affect equipment and/or reaction therefore their concentration is frequently controlled. A liquid specimen is injected into a combustion tube maintained at 900°C, converted to hydrogen halides, and carried by a carrier gas (50% oxygen, 50% argon) to a titration cell where it reacts with silver ions in the electrolyte.

The determination of peroxides has two goals: one is to monitor peroxide concentration used as initiator and catalysts and the other is to detect formation of hazardous peroxides formed as autoxidation products in ethers, acetals, dienes, and alkylaromatic hydrocarbons. A sample is dissolved in a mixture of acetic acid and chloroform. The solution is deaerated and potassium iodide reagent is added and let to react for 1 h in darkness.[94] The iodine formed in reaction is measured by absorbance at 470 nm and result calculated to active oxygen in the sample. The method can determine hydroperoxides, peroxides, peresters, and ketone peroxides. Oxidizing and reducing agents interfere with the determination.

Mercaptans in motor fuels, kerosene and other petroleum products can be detected by shaking the liquid sample with sodium plumbite solution, adding powdered sulfur and shaking again.[95] If a mercaptan or hydrogen sulfide is present, discoloration of the floating sulfur or liquid phase occurs.

The nonvolatile content of a halogenated solvent is determined by drying the sample in a platinum evaporating dish at 105°C.[96] Depending on boiling point and the concentration of nonvolatile matter three alternate procedures are proposed.

The Karl Fisher method is recommended for general use in solvents to determine the water content.[97] It is not suitable if mercaptans, peroxides, or appreciable quantities of aldehydes and amines are present. Water in halogenated solvents may cause corrosion, spotting, reduce shelf-life of aerosols, or inhibit chemical reactions, thus special method, also based on the Karl Fischer titration, was developed for halogenated solvents.[98]

The titration of ionizable chlorides with mercuric acetate solution in the presence of s-diphenylcarbazone as an indicator is used to determine chloride in trichlorotrifluoroethane.[99] A visual appearance test to detect admixtures in halogenated solvents is based on the observation of suspended particles, sediment, turbidity and free floating water.[100]

The aromatic content of hydrocarbon mixture is estimated from the determination of aniline point.[101] Aromatic hydrocarbons have the lowest and paraffins the highest aniline points. Cycloparaffins and olefins are between the two. Aniline point increases as the molecular weight increases. A mixture of specific aniline and solvent is heated at a controlled rate until it forms one phase. The mixture is then cooled and the temperature at which the miscible liquid separates into two phases is determined. Four methods are discussed in the standard[101] suitable for transparent, non-transparent, easily vaporizing, and measured in small quantities.

The presence of oxidizable materials in acetone and methanol that are associated with contaminations during manufacture and distribution can be evaluated by permanganate time.[102] Oxidizable contaminants may adversely affect catalysts or ligand complexes which are sensitive to oxidation. Oxidizable substances reduce potassium permanganate to manganese oxide which is yellow. The method is designed to measure the time required to change color to the color of a standard.

Small admixtures of acetone in methanol (more than 0.003 wt%) can be detected after a reaction with Nessler's reagent.[103] The reacted sample is compared with a standard, which contains 0.003 wt% acetone and the difference in turbidity is reported.

### 15.1.26 REFRACTIVE INDEX

Refractive index is measured by a standard method.[104] It covers transparent and light colored liquids having a refractive index in the range from 1.33 to 1.50. The refractive index is the ratio of light velocity of a specified wavelength in air to its velocity in the substance under evaluation. The refractive dispersion is the difference between refractive indices for light of two different wavelengths. This value is usually multiplied by 10,000. The method uses a Bausch & Lomb refractometer equipped with a thermostat and with a circulating bath to control the sample temperature with a precision of 0.2°C. Several light sources can be used, including the sodium arc lamp, mercury light lamp, and hydrogen or helium discharge lamp. Light sources are equipped with filters which transmit a specific spectral line. Standardization of equipment is done using a solid reference standard or using liquids standards such as 2,2,4-trimethylpentane (1.39), methylcyclohexane (1.42), or toluene (1.49).

### 15.1.27 RESIDUAL SOLVENTS

Residual solvents may cause odor, off-taste, blocking, and an increased degradation rate in outdoor exposures. A single standard test method has been developed to determine residual solvent levels and it is primarily used for the evaluation of flexible barrier materials.[105] The method is based on gas chromatography. The specimen of the barrier material is enclosed in a container and heated to vaporize the retained solvents into the head space. The vapor from the head space is taken by a gas syringe and injected into a gas chromatograph. The recovery of solvents is compared by the means of response factor which is a peak intensity of the detector in response to a given volume of injected sample. Response factors of different solutions vaporized in the test containers are compared. Round robin tests have demonstrated that this method has a coefficient of variation between laboratories of ±15%. The method does not specify the detector or the columns. Before a flexible barrier material is analyzed, the optimum heating time to recover volatilized solvents is determined.

### 15.1.28 SOLUBILITY

Solvent power of a hydrocarbon solvents is determined by kauri-butanol value.[106] The method applies to solvents having a boiling point above 40°C and a dry point below 300°C. The method is most frequently used to evaluate solvents for applications in paints and lacquers. The kauri-butanol value is the volume of solvent required to produce a defined degree of turbidity when added to 20 g of a standard solution of kauri resin in n-butanol (400 g of resin in 2000 g of solvent). High kauri value indicates a relatively strong dissolving power. The method is standardized using 105±5 ml of toluene.

The solubility of common gases in hydrocarbon liquids is determined to meet requirements of aerospace industry.[107] This test method is based on the Clausius-Clapeyron equation, Henry's law, and the perfect gas law. The results are important in the lubrication of gas compressors where dissolved gas may cause erosion due to cavitation. In fuels, dissolved gases may cause interruption of fuel supply and foaming in tank. The liquid density is determined experimentally. Using this density, the Ostwald coefficient is taken from a chart and used for the calculation of the Bunsen coefficient (solubility of gas). The solubility of the gas or mixture of gases and Henry's law constant are also calculated.

### 15.1.29 SOLVENT PARTITIONING IN SOILS

A procedure is available to determine partitioning organic chemicals between water and soil or sediment.[108] By measuring sorption coefficients for specific solids, a single value is obtained which can be used to predict partitioning under a variety of conditions. The underlining principle of the method is that organic chemicals bind with the surfaces of solids through chemical and physical interactions. However, the sorption coefficient is based on organic carbon content which does not apply to all solvents or all soils.

The sorption coefficient of a particular solvent is measured by equilibrating its aqueous solution, containing a realistic concentration similar to that found in the environment, with a known quantity of soil or sediment. After equilibrium is reached, the concentration of solvent in the water and the soil is measured by a suitable analytical technique. The sorption constants for all solids tested are averaged and reported as a single value. The standard does not define the actual method of determining the concentrations but strategy that should be followed. The data are useful in predicting the migration of chemicals in soil, in estimating their volatility from water and soil, determining their concentration in water, and their propensity to leach through the soil profile.

### 15.1.30 SOLVENT EXTRACTION

Often materials must be extracted from a compounded product to perform testing. Extracted material must then be recovered from solution without degradation to be subjected to testing. The method discussed here was developed as means of recovering asphalt from pavement samples.[109] The solution of extracted asphalt in solvent is distilled by rotating the distillation flask of a rotary evaporator in a heated oil bath. The distillation is carried out under partial pressure in the presence of nitrogen to prevent degradation. The asphalt recovered by this method can be tested in the same manner as were the original asphalt samples.

### 15.1.31 SPECIFICATIONS

Standard specifications are designed to set criteria for commercial solvents which can be used to determine the compliance of a solvent sample. Because applications of solvents differ very widely, the selected criteria are also different for different groups of solvents. Table 15.1.1 is a compiled list of parameters all of which can be found in solvent specifications. The most common parameters used to characterize solvent include acidity, appearance, color, concentration of main component, distillation range, dry point, initial boiling point, and specific gravity. The methods of determining these parameters are found in this chapter.

The list of references includes information on specifications for various solvents. Solvents in this list are arranged into groups: alcohols,[110-120] aromatic hydrocarbons,[121-128] other hydrocarbons,[129-132] ketones,[133-138] esters,[139-145] glycol,[146] and chlorine-containing solvents.[147-152]

### 15.1.32 SUSTAINED BURNING

The sustained burning test was originally developed for British Standard BS-3900 and adapted by ASTM.[153] The purpose of the test is to determine the sustained burning characteristics of solvents by direct experiment rather than by deducing characteristics from flash point data. Mixtures of some flammable liquids (e.g., alcohol and water) are classified as flammable based on the closed-cup flash point method. Some mixtures may be classed as flammable even though they do not sustain burning. The test is performed in a block of aluminum with a concave depression called a well. The liquid under test is heated to a temperature of 49°C and a flame is passed over the well and held in position for 15 s. The specimen is observed to determine if it can sustain burning.

**Table 15.1.1. Typical chemical and physical properties of solvents included in their specifications**

| | |
|---|---|
| Acid acceptance | Distillation range |
| Acidity | Doctor test |
| Acid wash color | Dry point |
| Alkalinity | Flash point |
| Aluminum scratch | Free halogens |
| Aniline point | Initial boiling point |
| Appearance | Iron concentration |
| Bromine index | Kauri-butanol value |
| Color | Non-volatile matter |
| Concentration of admixtures | Permanganate time |
| Concentration of isomers | Residual odor |
| Concentration of main component (or purity) | Specific gravity |
| Concentration of sulfur | Water concentration |
| Copper corrosion | Water miscibility |

## 15.1.33 VAPOR PRESSURE

The vapor pressure is the pressure of the vapor of a substance in equilibrium with the pure liquid at a given temperature. Two procedures are used: an isoteniscope procedure (standard) for measuring vapor pressures from $1 \times 10^{-1}$ to 100 kPa and a gas-saturation pressure for measuring vapor pressures from $1 \times 10^{-11}$ to 1 kPa.[154-155] In the isoteniscope method, a sample is deaerated by heating under reduced pressure. The vapor pressure is then determined by balancing the pressure of the vapor against a known pressure of an inert gas. The vapor pressure is determined at minimum three different temperatures.

In the gas-saturation method, an inert gas is passed through a sufficient amount of compound to maintain saturation. The vapor is then removed from the gas by a sorbent or a cold trap and quantitatively determined by gas chromatography or other suitable technique.

## 15.1.34 VISCOSITY

The viscosity of solvents can be determined by one of three methods: glass viscometer,[156] Saybolt viscometer,[157] and bubble time method.[158] Glass viscometry is applicable to Newtonian, transparent liquids which because of volatility cannot be measured in conventional capillary viscometers. The viscometer uses a purge gas which helps to transfer the measuring liquid from a lower reservoir to the sample bulb. The time of flow is measured for a fixed volume of liquid at a temperature controlled with a precision of 0.01°C. A set of liquids is available as viscosity standards in order to select the standard having closest viscosity to the measured sample.

The Saybolt viscometer was developed for petroleum products. A sample of 60 ml flows through a calibrated orifice and the time of flow is measured at a controlled temperature. Measurements are made at a temperature selected from the range 21-99°C. The instrument is standardized by measuring the flow of a standard oil at both 37.8 and 50°C.

In the bubble method, a standard viscosity tube is filled with a specimen liquid, the temperature of liquid is equilibrated to 25°C in a bath, the level of the meniscus is adjusted to 100 mm line, cork is inserted to end on 108 mm line, sample is hold in thermostating bath for another 20 min. Then the tube is inverted and the time for the bubble to flow from a mark at 27 mm to 100 mm is measured.

### 15.1.35 VOLATILE ORGANIC COMPOUND CONTENT, VOC

Several terms are used in the paint industry to provide data on the VOC content of paints.[159] The percent of solids in paints is calculated either per unit weight or per unit volume. The weight solids content is the weight of non-volatile materials divided by the total weight of the coating. In practice, it is calculated by subtracting the total weight of volatile solvents from the total weight of the coating and dividing the result by the total weight of coating. The final result is multiplied by 100 to express it in percent. Percent of solids by weight is calculated in similar manner.

A solvent, as defined by paint standards, is a volatile liquid that is incorporated primarily for vehicle solvency and control of the application characteristics. The solvent content is the calculated weight of solvents in a specific volume of paint. This definition is not equivalent to the definition of VOC compound because it does include compounds which are excluded by EPA (see more on this subject in Section 18.1).

Volatile organic compound by EPA definition means any compound of carbon, excluding carbon monoxide, carbon dioxide, carbonic acid, metallic carbides or carbonates, and ammonium carbonate, which participates in atmospheric photochemical reactions. This includes any such organic compound other than the following, which have been determined to have negligible photochemical reactivity: methane; ethane; methylene chloride; 1,1,1-trichloroethane; 1,1,2-trichloro-1,2,2-trifluoroethane; trichlorofluoromethane; dichlorodifluoromethane; chlorodifluoromethane; trifluoromethane; dichlorotetrafluoroethane; chloropentafluoroethane; dichlorotrifluoroethane; tetrafluoroethane; dichlorofluoroethane; chlorodifluoroethane; 2-chloro-1,1,1,2-tetrafluoroethane; pentafluoroethane; 1,1,2,2-tetrafluoroethane; 1,1,1-trifluoroethane; 1,1-difluoroethane; perchloroethylene; acetone; parachlorobenzotrifluoride; cyclic, branched, or linear completely methylated siloxanes; and perfluorocarbon compounds which fall into these classes: cyclic, branched, or linear, completely fluorinated alkanes; cyclic, branched, or linear, completely fluorinated ethers with no unsaturations; cyclic, branched, or linear, completely fluorinated tertiary amines with no unsaturations; and sulfur containing perfluorocarbons with no unsaturations and with sulfur bonds only to carbon and fluorine. For purposes of determining compliance with emissions limits, VOC is measured by the test methods in the approved State implementation plan (SIP) or 40 CFR Part 60, Appendix A, as published in (7/1/91) edition, as applicable.

Based on the above definition of VOC, the VOC content is calculated from a formula by excluding from the total solvent content of the paint, the content of water and solvents excluded by the above regulation. The following formula can be used for VOC calculation:

$$VOC = \frac{W_{total\ solvents} - W_w - W_{excluded\ solvents}}{V_{paint} - V_w - V_{excluded\ solvents}} \qquad [15.1.1]$$

where:

| | |
|---|---|
| $W_{total\ solvents}$ | total weight of solvents |
| $W_w$ | weight of water present in formulation |

| | |
|---|---|
| $W_{excluded\ solvents}$ | total weight of solvents excluded by EPA regulation |
| $V_{paint}$ | total volume of paint |
| $V_w$ | volume of water present in formulation |
| $V_{excluded\ solvents}$ | total volume of solvents excluded by EPA regulation |

The above definition may vary in various countries thus the proper definition should be obtained from the appropriate authorities. Examples of other definitions are included in Chapter 18.1.

The weight percent of the volatile content in waterborne aerosol paints is determined by releasing the propellant from can, testing the remaining paint for water content using the Karl Fischer method, and determining non-volatiles.[160] The VOC content in a solventborne automotive paints is determined by the simulation of VOC loss during baking using laboratory panels.[161]

## REFERENCES


1 ASTM D 2106-95. Standard test method for determination of amine acid acceptance (alkalinity) of halogenated organic solvent.

2 ASTM D 1614-95. Standard test method for alkalinity in acetone.
ISO 755-2-81. Butan-1-ol for industrial use - Methods of test - Part 2: Determination of acidity - Titrimetric method.
ISO 756-2-81. Propan-2-ol for industrial use - Methods of test - Part 2: Determination of acidity - Titrimetric method.
ISO 757-2-82. Acetone for industrial use - Methods of test - Part 2: Determination of acidity to phenolphthalein - Titrimetric method.

3 ASTM D 2989-97. Standard test method for acidity-alkalinity of halogenated solvents and their admixtures.
ISO 1393-77. Liquid halogenated hydrocarbons for industrial use - Determination of acidity - Titrimetric method.
ISO 3363-76. Fluorochlorinated hydrocarbons for industrial use - Determination of acidity - Titrimetric method.

4 ASTM D 847-96. Standard test method for acidity of benzene, toluene, xylenes, solvent naphthas, and similar industrial aromatic hydrocarbons.

5 ASTM D 1613-96. Standard test method for acidity in volatile solvents and chemical intermediates used in paint, varnish, lacquer, and related products.

6 ASTM D 2942-96. Standard test method for total acid acceptance of halogenated organic solvents (nonreflux method).

7 ASTM D 3444-95. Standard test method for total acid number of trichlorotrifluoroethane.

8 ASTM E 659-94. Standard test method for autoignition temperature of liquid chemicals.
AS 1896-1976. Method of test for ignition temperature of gases and vapors.

9 ASTM E 1196-92. Standard test method for determining the anaerobic biodegradation potential of organic chemicals.
BS 6068-2.34-88. Water quality. Physical, chemical and biochemical methods. Method for the determination of the chemical oxygen demand.
DIN 38409-41-80. German standard methods for examination of water, waste water and sludge. Summary action and material characteristic parameters (group H). Determination of the chemical oxygen demand (COD) in the range over 15 mg/l.
DIN 38409-43-81. German standard methods for the analysis of water, waste water and sludge. Summary action and material characteristic parameters (group H). Determination of the chemical oxygen demand (COD). Short duration method.
ISO 5815-89. Water quality - Determination of biochemical oxygen demand after 5 days - Dilution and seeding method.
ISO 6060-89. Water quality - Determination of the chemical oxygen demand.
JIS K 0400-20-10-99. Water quality - Determination of the chemical oxygen demand.
JIS K 0400-21-10-99. Water quality - Determination of biochemical oxygen demand after 5 days. Dilution seeding method.
JIS K 3602-90. Apparatus for the estimation of biochemical oxygen demand with microbial sensor.

10 ASTM D 850-93. Standard test method for distillation of industrial aromatic hydrocarbons and related materials.
BS 7392-90. Method for determination of distillation characteristics of petroleum products (ISO 3405).
ISO 918-83. Volatile organic liquids for industrial use - Determination of distillation characteristics.
ISO 3405-88. Petroleum products - Determination of distillation characteristics.
ISO 4626-80. Volatile organic liquids - determination of boiling range of organic solvents used as raw materials.
JIS K 2254-98. Petroleum products - Determination of distillation characteristics.

11 ASTM D 1078-97. Standard test method for distillation range of volatile organic liquids.
BS 4591-90. Method of determination of distillation characteristics of organic liquids (other than petroleum products) (ISO 918).
JIS K 0066-92. Test methods for distillation of chemical products.

12 ASTM D 3272-98. Standard practice for vacuum distillation of solvents from solvent-reducible paints for analysis.

13 ASTM D 5399-95. Standard test method for boiling point distribution of hydrocarbon solvents by gas chromatography.
ISO 3924-99. Petroleum products - Determination of boiling range distribution - Gas chromatography method.

14 ASTM D 5776-98. Standard method for bromine index of aromatic hydrocarbons by electrometric titration.
ISO 3839-96. Petroleum products - Determination of bromine number of distillates and aliphatic olefins - Electrometric method.
JIS K 2605-96. Petroleum distillates and commercial aliphatic olefins - Determination of bromine number - Electrometric method.

15 ASTM D 1492-96. Standard test method for bromine index of aromatic hydrocarbons by coulometric titration.

16 ASTM D 240-97. Standard test method for heat of combustion of liquid hydrocarbon fuels by bomb calorimeter.
BS 7420-91. Guide for determination of calorific values of solid, liquid and gaseous fuels (including definitions).
DIN 5499-72. Gross and net Calorific values. Terms.
DIN 51900-1-89. Determination of gross calorific value of solid and liquid fuels by the bomb calorimeter and calculation of net calorific value. Part 1: Principles, apparatus, methods.
DIN 51900-2-77. Testing of solid and liquid fuels. Determination of gross calorific value by bomb calorimeter and calculation of net calorific value. Method of using isothermal water jacket.
DIN 51900-3-77. Testing of solid and liquid fuels. Determination of gross calorific value by bomb calorimeter and calculation of net calorific value. Method of using adiabatic jacket.
JIS K 2279-93. Crude petroleum and petroleum products - Determination and estimation of heat of combustion.

17 ASTM D 1405-95a. Standard test method for estimation of net heat of combustion of aviation fuels.

18 ASTM D 5468-95. Standard test method for gross calorific value of waste materials.

19 ASTM D 5468-95. Standard test method for gross calorific and ash value of waste materials.

20 ASTM F 1105-95. Standard test method for preparing aircraft cleaning compounds, liquid-type, temperature sensitive, or solvent-based, for storage stability testing.

21 ASTM D 3698-92. Standard practices for solvent vapor degreasing operations.

22 ASTM D 4579-96. Standard practice for handling an acid degreaser or still.

23 ASTM D 848-97. Standard test method for acid wash color of industrial aromatic hydrocarbons.
ISO 5274-79. Aromatic hydrocarbons - Acid wash test.
ISO 755-3-81. Butan-1-ol for industrial use - Methods of test - Part 3: Sulfuric acid color test.
JIS K 0072-98. Testing method of color after treatment with sulfuric acid.

24 ASTM D 1686-96. Standard test method for color of solid aromatic hydrocarbons and related materials in the molten state (platinum-cobalt scale).
BS 5339-76. Method of measurement of color in Hazen units (platinum-cobalt scale) of liquid chemical products (ISO 2211).
ISO 2211-73. Liquid chemical products - Measurement of color in Hazen units (platinum-cobalt scale).
ISO 6271-97. Clear liquids - Estimation of color by the platinum-cobalt scale.
JIS K 0071-98. Testing methods for color of chemical products - Part 1. Estimation of color in Hazen units (platinum-cobalt scale).

25 ASTM D 2108-97. Standard test method for color of halogenated organic solvents and their admixtures (platinum-cobalt scale).
26 ASTM D 5386-93b. Standard test method for color of liquids using tristimulus colorimetry.
27 ASTM D 3316-96. Standard test method for stability of perchloroethylene with copper.
28 ASTM D 849-97. Standard test method for copper strip corrosion by industrial aromatic hydrocarbons.
ISO 2160-98. Petroleum products - Corrosiveness of copper - Copper strip test.
JIS K 2513-91. Petroleum products - Corrosiveness to copper - Copper strip test.
SFS-EN ISO 2160. Petroleum products. Corrosiveness to copper. Copper strip test.
29 ASTM D 2943-96. Standard test method for aluminum scratch of 1,1,1-trichloroethane to determine stability.
30 ASTM D 1217-98. Standard test method for density and relative density (specific gravity) of liquids by Bingham pycnometer.
31 ASTM D 891-95. Standard test method for specific gravity, apparent, of liquid industrial chemicals.
BS 4522-88. Method for determination of absolute density at 20°C of liquid chemical products for industrial use (ISO 758).
JIS K 0061-92. Test methods for density and relative density of chemical products.
32 ASTM D 2111-95. Standard test methods for specific gravity of halogenated organic solvents and their admixtures.
33 ASTM D 1555-95. Standard test method for calculation of volume and weight of industrial aromatic hydrocarbons.
34 ASTM D 1555M-95. Standard test method for calculation of volume and weight of industrial aromatic hydrocarbons (metric).
ISO 5281-80. Aromatic hydrocarbons - Benzene, xylene and toluene - Determination of density at 20°C.
SFS 3773-76. Determination of density of liquids at 20°C.
35 ASTM D 1720-96. Standard test method for dilution ratio of active solvents in cellulose nitrate solutions.
36 ASTM D 5062-96. Standard test method for resin solution dilutability by volumetric/gravimetric determination.
37 ASTM D 1476-96. Standard test method for heptane miscibility of lacquer solvents
38 ASTM D 1722-94. Standard test method for water miscibility of water-soluble solvents.
39 ASTM D 5226-98. Standard practice for dissolving polymer materials.
40 ASTM D 2257-96. Standard test method for extractable mater in textiles.
41 ASTM F 58-96. Standard test method for measuring specific resistivity of electronic grade solvents.
JIS C 0052-95. Environmental testing procedure of electronic and electrical resistance to solvents.
42 ASTM D 4308-95. Standard test method for electrical conductivity of liquid hydrocarbons by precision meter.
43 ASTM D 1816-97. Standard test method for dielectric breakdown voltage of insulating oils of petroleum origin using VDE electrodes.
44 ASTM D 3300-94. Standard test method for dielectric breakdown voltage of insulating oils of petroleum origin under impulse conditions.
45 ASTM F 791-96. Standard test method for stress crazing of transparent plastics.
46 ASTM D 4508-98. Standard test method for chip impact strength of plastics.
47 ASTM D 1901-95. Standard test method for relative evaporation time of halogenated organic solvents and their admixtures.
DIN 53170-91. Solvents for paints and similar coating materials. Determination of the evaporation rate.
48 ASTM E 1232-96. Standard test method for temperature limit of flammability of chemicals.
49 ASTM E 681-98. Test method for concentration limits of flammability of chemicals.
BS 6713-3-86. Explosion protection systems. Method for determination of explosion indices of fuel/air mixtures other than dust/air and gas/air mixtures (ISO 6184-3).
50 ASTM D 3828-97. Standard test method for flash point by small scale closed tester.
AS/NZS 2106.0-99. Methods for determination of flash point of flammable liquids (closed cup) - General.
AS/NZS 2106.1-99. Methods for the determination of the flash point of flammable liquids (closed cup) - Abel closed cup method.
51 ASTM D 1310-97. Standard test method for flash point and fire point of liquids by tag open-cup apparatus.
52 ASTM D 56-97a. Standard test method for flash point by tag closed tester.
53 ASTM D 92-97. Standard test method for flash and fire points by Cleveland open cup.
ISO 2592-73. Petroleum product - Determination of flash and fire points - Cleveland open cup method.

SFS-EN 22592. Petroleum products. Determination of flash and fire points. Cleveland open cup method (ISO 2592).
54 ASTM D 93. Standard test methods for flash-point by Pensky-Martens closed cup tester.
AS/NZS 2106.2-99. Methods for determination of the flash point of flammable liquids (closed cup) - Pensky-Martens closed cup method.
DIN EN 22719-93. Petroleum products and lubricants. Determination of flash point. Pensky-Martens closed cup method (ISO 2719).
ISO 2719-88. Petroleum products and lubricants - Determination of flash point - Pensky-Martens closed cup method.
SFS-EN 22719. Petroleum products and lubricants. Determination of flash point. Pensky-Martens closed cup method (ISO 2719)
55 ASTM D 3941-96. Standard test method for flash point by the equilibrium method with a closed-cup apparatus.
AS/NZS 2106.3-99. Methods for the determination of the flash point of flammable liquids (closed cup) - Flash/no flash test - Rapid equilibrium method.
AS/NZS 2106.4-99. Methods for the determination of the flash point of flammable liquids (closed cup) - Determination of flash point - Rapid equilibrium method.
AS/NZS 2106.5-99. Methods for the determination of the flash point of flammable liquids (closed cup) - Flash/no flash test - Closed cup equilibrium method.
AS/NZS 2106.6-99. Methods for the determination of the flash point of flammable liquids (closed cup) - Determination of flash point - Closed cup equilibrium method.
DIN EN 456-91. Paints, varnishes and related products. Determination of flash point. Rapid equilibrium method (ISO 3679).
ISO 1516-81. Paints, varnishes, petroleum and related products - Flash/no flash test - Closed cup equilibrium method.
ISO 1523-83. Paints, varnishes, petroleum and related products - Determination of flash point - Closed cup equilibrium method.
ISO 3679-83. Paints, varnishes, petroleum and related products - Determination of flash point - Rapid equilibrium method.
ISO 3680-83. Paints, varnishes, petroleum and related products - Flash/no flash test - Rapid equilibrium method.
SFS-EN 456-93. Paints, varnishes, and related products. Determination of flash point. Rapid equilibrium method (ISO 3679).
56 ASTM E 502-94. Standard test method for selection and use of ASTM standards for the determination of flash point of chemicals by closed cup methods.
57 ASTM D 1015-94. Standard test method for freezing points of high-purity hydrocarbons.
JIS K 0065-92. Test method for freezing point of chemical products.
58 ASTM D 1016-94. Standard test method for purity of hydrocarbons from freezing points.
59 ASTM D 852-97. Standard test method for solidification point of benzene.
ISO 5278-80. Benzene - Determination of crystallizing point.
60 ASTM D 4755-95. Standard test method for free halogenated organic solvents and their admixtures.
61 ASTM D 2306-96. Standard test method for $C_8$ hydrocarbon analysis by gas chromatography.
DIN 51437-89. Testing of benzene and benzene homologues. Determination of the content of non-aromatics, toluene and C8-aromatics in benzene. Gas chromatography.
DIN 51448-1-97. Testing of liquid petroleum hydrocarbons. Determination of hydrocarbon types. Part 1: Gas chromatographic analysis by column switching procedure.
62 ASTM D 3797-96. Standard test method for analysis of o-xylene by gas chromatography.
63 ASTM D 3798-96b. Standard test method for analysis of p-xylene by gas chromatography.
64 ASTM D 4492-98. Standard test method for analysis of benzene by gas chromatography.
65 ASTM D 3271-87-93. Standard test method for direct injection of solvent reducible paints into a gas chromatograph for solvent analysis.
DIN 55682-94. Solvents for paints and varnishes. Determination of solvents in water-thinnable coating materials. Gas chromatographic method. (Amendment A1 - DIN 55682/A1-98).
DIN 55683-94. Solvents for paints and varnishes. Determination of solvents in coating materials containing organic solvents only. Gas chromatographic method.
66 ASTM D 5830-95. Standard test method for solvent analysis in hazardous waste using gas chromatography.
67 ASTM D 4757-97. Standard practice for placarding solvent vapor degreasers.

68 ASTM D 3844-96. Standard guide for labeling halogenated hydrocarbon solvent containers.
69 ASTM D 1296-96. Standard test method for odor of volatile solvents and diluents.
ISO 2498-74. Methyl ethyl ketone for industrial use - Examination for residual odor.
70 ASTM E 679-97. Standard practice for determination of odor and taste thresholds by forced-choice ascending concentration series method.
71 ASTM E-769-97. Standard test methods for odor of methanol, ethanol, n-propanol, and isopropanol.
72 ASTM E 1075-97. Standard test methods for odor of ethylene glycol, diethylene glycol, triethylene glycol, propylene, glycol, and dipropylene glycol and taste of propylene glycol.
73 ASTM D 4494-95. Standard test method for detecting residual odor of drycleaning grade perchloroethylene.
74 ASTM D 268-96. Standard guide for sampling and testing volatile solvents and chemical intermediates for use in paint and related coatings and material.
75 ASTM D 1353-96. Standard test method for nonvolatile matter in volatile solvents for use in paint, varnish, lacquer, and related products.
76 ASTM D 5286-95. Standard test methods for determination of transfer efficiency under general conditions for spray application of paints.
77 ASTM D 1617-96. Standard test method for ester value of solvents and thinners.
78 ASTM D 2110-96. Standard test method for pH of water extractions of halogenated organic solvents and their admixtures.
79 ASTM D 3545-95. Standard test method for alcohol content and purity of acetate esters by gas chromatography.
DIN 55686-92. Solvents for paints and varnishes. Acetic esters. Gas chromatographic determination of the degree of purity.
80 ASTM D 6229-98. Standard test method for trace benzene in hydrocarbon solvents by capillary gas chromatography.
81 ASTM D 4534-93. Standard test method for benzene content of cyclic products by gas chromatography.
82 ASTM D 435-96. Standard test method for determination of trace thiophene in refined benzene by gas chromatography.
DIN 51438-91. Testing of benzene and benzene homologues. Determination of the content of thiophene in benzene. Gas chromatography.
83 ASTM D 5713-96. Standard test method for analysis of high purity benzene for cyclohexane feedstock by capillary gas chromatography.
84 ASTM D 3447-96. Standard test method for purity of halogenated organic solvents.
CAN/CGSB 1-GP-71 No. 73-96. Methods of testing paints and pigments. Beilstein test for chlorinated solvents.
85 ASTM D 3742-94. Standard test method for 1,1,1-trichloroethane content.
86 ASTM D 2804-98. Standard test method for purity of methyl ethyl ketone by gas chromatography.
87 ASTM D 3329-94a. Standard test method for purity of methyl isobutyl ketone by gas chromatography.
88 ASTM D 3257-97. Standard test methods for aromatics in mineral spirits by gas chromatography.
89 ASTM D 1685-95. Standard test method for traces of thiophene in benzene by spectrophotometry.
90 ASTM D 2031-97. Standard test method for reducing substances in refined pyridine.
91 ASTM D 3961-98. Standard test method for trace quantities of sulfur in liquid aromatic hydrocarbons by oxidative microcoulometry.
ISO 5282-82. Determination of sulfur content - Pitt-Ruprecht reduction and spectrophotometric determination method.
ISO 8754-92. Petroleum products - Determination of sulfur content - Energy-dispersive X-ray fluorescence method.
SFS-EN ISO 8754. Petroleum products. Determination of sulfur content. Energy-dispersive X-ray fluorescence method.
ISO 14596-98. Petroleum products - Determination of sulfur content - Wavelength-dispersive X-ray fluorescence method.
92 ASTM D5194-96. Standard test method for trace chloride in liquid aromatic hydrocarbons.
93 ASTM D 5808-95. Standard test method for determining organic chloride in aromatic hydrocarbons and related chemicals by microcoulometry.
94 ASTM E 299-97. Standard test method for trace amounts of peroxides in organic solvents.
95 ASTM D 4952-97. Standard test method for qualitative analysis for active sulfur species in fuels and solvents (Doctor Test).
ISO 5275-79. Aromatic hydrocarbons - Test for presence of mercaptans (thiols) - Doctor test.

96 ASTM D 2109-96. Standard test methods for nonvolatile matter in halogenated organic solvents and their admixtures.
BS 5598-3-79. Methods of sampling and test for halogenated hydrocarbons. Determination of residue on evaporation (ISO 2210).
ISO 759-81. Volatile organic liquids for industrial use - Determination of dry residue after evaporation on water bath - General method.
ISO 2210-72. Liquid halogenated hydrocarbons for industrial use - Determination of residue on evaporation.
ISO 5277-81. Aromatic hydrocarbons - Determination of residue on evaporation of products having boiling points up to 150°C.
JIS K 0067-92. Test methods for loss and residue of chemical products.
SFS 3773-77. Determination of residue on evaporation.
97 ASTM D 1364-95. Standard test method for water in volatile solvents (Karl Fischer reagent titration method).
BS 2511-70. Methods for determination of water (Karl Fischer method).
ISO 6191-81. Light olefins for industrial use - Determination of traces of water - Karl Fischer method.
ISO 8917-88. Light olefins for industrial use - Determination of water - Guidelines for use of in-line analyzers.
SFS 3774-76. Determination of water by the Karl Fischer method.
98 ASTM D 3401-97. Standard test methods for water in halogenated organic solvents and their admixtures.
99 ASTM D 3443-95. Standard test method for chloride in trichlorotrifluoroethane.
100 ASTM D 3741-95. Standard test methods for appearance of admixtures containing halogenated organic solvents.
101 ASTM D 611-98. Standard test methods for aniline point and mixed aniline point of petroleum products and hydrocarbon solvents.
ISO 2977-97. Petroleum products and hydrocarbon solvents - Determination of aniline point and mixed aniline point.
102 ASTM D 1363-97. Standard test method for permanganate time of acetone and methanol.
ISO 757-7-83. Acetone for industrial use - Methods of test - Part 4: Determination of permanganate time.
103 ASTM D 1612-95. Standard test method for acetone in methanol.
104 ASTM D 1218-96. Standard test method for refractive index and refractive dispersion of hydrocarbon liquids.
ISO 5661-83. Petroleum products - Hydrocarbon liquids - Determination of refractive index.
JIS K 0062-92. Test methods for refractive index of chemical products.
105 ASTM F 151-97. Standard test method for residual solvents in flexible barrier materials.
106 ASTM D 1133-97. Standard test method for Kauri-butanol value of hydrocarbon solvents.
107 ASTM D 2779-97. Standard test method for estimation of solubility of gases in petroleum liquids.
108 ASTM E 1195-93. Standard test method for determining a sorption constant, $K_{oc}$, for an organic chemical in soil and sediments.
109 ASTM D 5404-97. Standard practice for recovery of asphalt from solution using the rotary evaporator.
110 ASTM D 304-95. Standard specification for n-butyl alcohol (butanol).
BS 508-1-86. Butan-1-ol for industrial use. Specification for butan-1-ol.
BS 508-2-84. Butan-1-ol for industrial use. Methods of test.
JIS K 1504-93. Butanol.
JIS K 8810-96. 1-butanol.
111 ASTM D 319-95. Standard specification for amyl alcohol (synthetic).
112 ASTM D 331-95. Standard specification for 2-ethoxyethanol.
113 ASTM D 770-95. Standard specification for isopropyl alcohol.
BS 1595-1-86. Propan-2-ol (isopropyl alcohol) for industrial use. Specification for propan-2-ol (isopropyl alcohol).
BS 1595-2-84. Propan-2-ol (isopropyl alcohol) for industrial use. Methods of test.
ISO 756-1-81. Propan-2-ol for industrial use - Methods of test - Part 1: General.
JIS K 1522-78. Isopropyl alcohol.
114 ASTM D 1007-95. Standard specification for sec-butyl alcohol.
ISO 2496-73. sec-butyl alcohol for industrial use - List of methods of test.
JIS K 1523-78. 2-butanol.
115 ASTM D 1152-97. Standard specification for methanol (methyl alcohol).
BS 506-1-87. Methanol for industrial use. Specification for methanol.

BS 506-2.84. Methanol for industrial use Methods of test.
CAN/CGSB-3-GP-531M. Methanol, technical.
ISO 1387-82. Methanol for industrial use - Methods of test.
JIS K 1501-93. Methanol.
116 ASTM D 1719-95. Standard specification for isobutyl alcohol.
117 ASTM D 1969-96. Standard specification for 2-ethylhexanol (synthetic).
BS 1835-91. Specification for 2-ethylhexan-1-ol for industrial use.
118 ASTM D 2627-97. Standard specification for diacetone alcohol.
BS 549-70. Specification for diacetone alcohol.
ISO 2517-74. Diacetone alcohol for industrial use - List of methods of test.
119 ASTM D 3128-97. Standard specification for 2-methoxyethanol.
120 ASTM D 3622-95. Standard specification for 1-propanol (n-propyl alcohol).
JIS K 8838-95. 1-propanol.
121 ASTM D 841-97. Standard specification for nitration grade toluene.
AS 3529-1988. Solvent - Toluene.
ISO 5272-79. Toluene for industrial use - Specification.
JIS K 8680-96. Toluene.
122 ASTM D 843-97. Standard specification for nitration grade xylene.
ISO 5280-79. Xylene for industrial use - specification.
JIS K 8271-96. Xylene.
123 ASTM D 4734-98. Standard specification for refined benzene-545.
124 ASTM D 5136-96. Standard specification for high purity p-xylene.
125 ASTM 5211-97. Standard specification for xylenes for p-xylene feedstock.
126 ASTM D 5471-97. Standard specification for o-xylene 980.
127 ASTM D 5871-98. Standard specification for benzene for cyclohexane feedstock.
ISO 5721-79. Benzene for industrial use - Specification.
128 ASTM D 3734-96. Standard specification for high-flash aromatic naphthas.
129 ASTM D 5309-97. Standard specification for cyclohexane 999.
130 ASTM D 235-95. Standard specification for mineral spirits (petroleum spirits) (hydrocarbon dry cleaning solvent).
AS 35-1988. Solvents - Mineral turpentine and white spirit.
BS 245-76. Specification for mineral solvents (white spirit and related hydrocarbon solvents) for paints and other purposes.
CAN/CGSB-53.7-92. Duplicating liquid, Direct process, Spirit type.
131 ASTM D 1836-94. Standard specification for commercial hexanes.
132 ASTM D 3735-96. Standard specification for VM&P naphthas.
133 ASTM D 740-97. Standard specification for methyl ethyl ketone.
134 ASTM D 4360-90. Standard specification for n-amyl ketone.
135 ASTM D 329-95. Standard specification for acetone.
BS 509-1-87. Acetone for industrial use. Specification for acetone.
BS 509-2-84. Acetone for industrial use. Methods of test.
CAN/CGSB-15.50-2. Technical grade acetone.
ISO 757-1-82. Acetone for industrial use - Methods of test - Part 1: General.
JIS K 1503-59. Acetone.
JIS K 8035-95. Acetone.
SFS 3775 ISO/R 757. Acetone for industrial use. Methods of test.
136 ASTM D 740-94. Standard specification for methyl ethyl ketone.
CAN/CGSB-15.52-92. Technical grade methyl ethyl ketone.
ISO 2597-73. Methyl ethyl ketone for industrial use - List of methods of test.
JIS K 1524-78. Methyl ethyl ketone.
137 ASTM D 1153-94. Standard specification for methyl isobutyl ketone.
ISO 2499-74. Methyl isobutyl ketone for industrial use - List of methods of test.
138 ASTM D 2917-94. Standard specification for methyl isoamyl ketone.
ISO 2500-74. Ethyl isoamyl ketone for industrial use - List of methods of test.
139 ASTM D 5137-95. Standard specification for hexyl acetate.
140 ASTM D 1718-94. Standard specification for isobutyl acetate (95% grade).
BS 551-90. Specification for butyl acetate for industrial use.
141 ASTM D 2634-98. Standard specification for methyl amyl acetate (95% grade).

142 ASTM D 3130-95. Standard specification for n-propyl acetate (96% grade).
143 ASTM D 3131-97. Standard specification for isopropyl acetate (99% grade).
BS1834-68. Specification for isopropyl acetate.
144 ASTM D 3540-98. Standard specification for primary amyl acetate, synthetic (98% grade).
145 ASTM D 3728-97. Standard specification for 2-ethoxyethyl acetate (99% grade).
146 ASTM D 5164-96. Standard specification for propylene glycol and dipropylene glycol.
JIS K 1530-78. Propylene glycol.
147 ASTM D 4701-95. Standard specification for technical grade methylene chloride.
ISO 1869-77. Methylene chloride for industrial use - List of methods of test.
JIS K 1516-84. Methyl chloride.
148 ASTM D 4079-95. Standard specification for vapor-degreasing grade methylene chloride.
149 ASTM D 4080-96. Standard specification for trichloroethylene, technical and vapor-degreasing grade.
ISO 2212-72. Trichloroethylene for industrial use - Methods of test.
JIS K 1508-82. Trichloroethylene.
150 ASTM D 4081-95. Standard specification for drycleaning-grade perchloroethylene.
ISO 2213-72. Perchloroethylene for industrial use - Methods of test.
JIS K 1521-82. Perchloroethylene.
151 ASTM D 4126-95. Standard specification for vapor-degreasing grade and general solvent grade 1,1,1-trichloroethane.
AS 2871-1988. Solvents. 1,1,1-trichloroethane (inhibited).
ISO 2755-73. 1,1,1-trichloroethane for industrial use - List of methods of test.
JIS K 1600-81. 1,1,1-trichloroethane.
152 ASTM D 4376-94. Standard specification for vapor-degreasing grade perchloroethylene.
153 ASTM D 4206-96. Standard test method for sustained burning of liquid mixtures using small scale open-cup apparatus.
ISO/TR 9038-91. Paints and varnishes - Determination of the ability of liquid paints to sustain combustion.
154 ASTM E 1194-93. Standard test method for vapor pressure.
DIN EN 12-93. Petroleum products. Determination of Reid vapor pressure. Wet method.
ISO 3007-99. Petroleum products and crude petroleum - Determination of vapor pressure - Reid method.
155 ASTM D 2879-96. Standard test method for vapor pressure-temperature relationship and initial decomposition temperature of liquids by isoteniscope.
156 ASTM D 4486-96. Standard test method for kinematic viscosity of volatile and reactive liquids.
DIN EN ISO 3104. Petroleum products. Transparent and opaque liquids. Determination of kinematic viscosity and calculation of dynamic viscosity.
ISO 3104-94. Petroleum products - Transparent and opaque liquids - Determination of kinematic viscosity and calculation of dynamic viscosity.
SFS-EN ISO 3104. Petroleum products. Transparent and opaque liquids. Determination of kinematic viscosity and calculation of dynamic viscosity.
157 ASTM D 88-94. Standard test method for Saybolt viscosity.
158 ASTM D 1545-93. Standard test method for viscosity of transparent liquids by bubble time method.
159 ASTM D 5201-97. Standard practice for calculating formulation physical constants of paints and coatings.
ISO/FDIS 11890-1-99. Paints and varnishes - Determination of volatile organic compound content (VOC) - Part 1: Difference method.
ISO/FDIS 11890-2-99. Paints and varnishes - Determination of volatile organic compound (VOC) content - Part 2: Gas-chromatographic method.
160 ASTM D 5325-97. Standard test method for determination of weight percent volatile content of water-borne aerosol paints.
161 ASTM D 5087-94. Standard test method for determining amount of volatile organic compound (VOC) released from solventborne automotive coatings and available for removal in a VOC control device (abatement).

# 15.2 SPECIAL METHODS OF SOLVENT ANALYSIS

## 15.2.1 USE OF BREATH MONITORING TO ASSESS EXPOSURES TO VOLATILE ORGANIC SOLVENTS

MYRTO PETREAS
**Hazardous Materials Laboratory, Department of Toxic Substances Control, California Environmental Protection Agency, Berkeley, CA, USA**

### 15.2.1.1 Principles of breath monitoring

Exposure to organic solvents may occur as a result of occupation, diet, lifestyles, hobbies, etc., in a variety of environments (occupational, residential, ambient). Solvents reach the organism through inhalation, ingestion and dermal exposure. The magnitude of exposures, however, may be modulated by factors such as the use of protective equipment (decrease) or physical exertion (increase). All sources and routes of exposure are integrated in the resulting internal dose. The dose at the target organ may trigger mechanisms that eventually may result in irreversible biological changes and disease. Cellular repair mechanisms and individual susceptibility have a significant effect on the onset of disease, resulting in a broad distribution of outcomes. This continuum between exposure to organic solvents and disease can be depicted, in a simplified way, in Figure 15.2.1.1. Biological monitoring focuses on the elucidation of the first step of this process; i.e., the relationship between environmental exposure and internal dose. Whereas the dose at the target organ is the biologically important, concentrations of the solvent (or its metabolites) may be more easily measured in other fluid or tissue samples, such as blood, breath, urine, etc.

Traditional workplace monitoring relied on measurements of airborne contaminant concentrations at the breathing zone, and comparisons to reference values or regulatory benchmarks, such as the Threshold Limit Values (TLVs) and the Permissible Exposure Limits (PELs) in the USA, and similar values promulgated by the European Union and other organizations. Air measurements, however, may not always represent actual personal exposures since workers may be exposed dermally and orally, or they may be using protective equipment. Over the last decade, work-place monitoring programs have expanded to include biological monitoring. Biological monitoring gives an estimate of the dose inside the body (or specific target organ) rather than the concentration of the solvent in the external environment. Internal dose is defined as the amount of the solvent taken up by the whole organism (or a body compartment) over a specified period of time.[1] Depending on the chemical, the time of sampling and the tissue analyzed, the dose may reflect the amount of the chemical recently absorbed, the amount of the chemical stored (body burden), or the amount of the chemical bound to the active sites of a receptor. It is the dose at the receptor or sensitive tissue (the biologically effec-

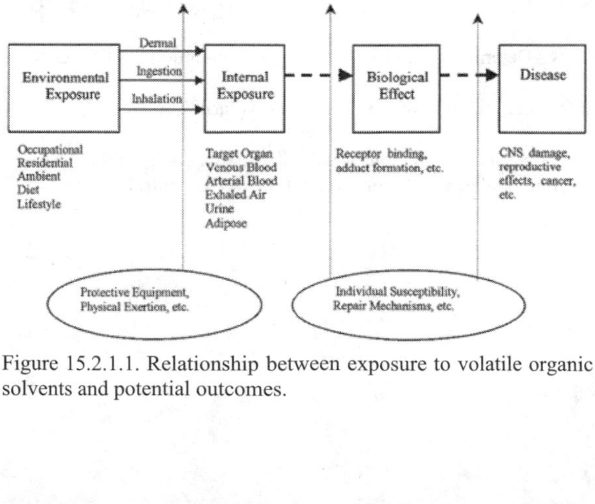


Figure 15.2.1.1. Relationship between exposure to volatile organic solvents and potential outcomes.

tive dose) that is related to toxicity and/or damage. In general, the dose at the receptor can not be measured because, at this time, available methodology is unacceptably invasive to be employed. On the other hand, biological monitoring allows measurements to be made on body fluids or tissues that would serve as surrogate measurements of the dose at the receptor. Certain conditions, however, are required for meaningful interpretations of the analytical results.

1. There should be adequate information on the absorption, distribution, biotransformation, metabolism and elimination of the chemical in the body. This type of information will indicate what tissues to sample, for what compound (parent chemical or a metabolite) and at what time. The latter issue is very important with chemicals that are eliminated rapidly (i.e., those with short half-lives).

2. The concentration of the chemical (or its metabolite) to be measured in the selected medium (breath, blood, urine) must be in equilibrium with the concentration of the chemical at the target organ.

3. The assay has to be sensitive, in order to detect low levels of the compound before any adverse effects take place; specific, in order to link exposure to dose; and accurate and precise, in order to be reliable.

Very few chemicals have been adequately studied to allow biological monitoring according to the above criteria. This is, however, an expanding area, as indicated by the number of substances for which Biological Exposure Indices (BEIs) have been proposed by the American Conference of Governmental Industrial Hygienists (ACGIH). In the 1984-85 Threshold Limit Value (TLV) booklet there were BEIs for 10 chemicals; in the 1998 booklet there were 37 compounds with adopted BEIs.[2] BEIs are considered reference values intended as guidelines and do not determine whether the worker is at risk of disease. They are supplementary to TLVs in evaluating workers' exposures to workplace hazardous agents. The recommended values of BEIs are based on data correlating exposure intensity and biological effects from field studies and/or on pharmacokinetic treatment of data obtained from controlled human exposures. The BEIs developed by the ACGIH, assume 8-hour exposures for 5 days, followed by 2 days of non-exposure, and they strictly specify the time of sample collection.

A clear advantage of biological monitoring over environmental or personal monitoring is that the estimate of the internal dose takes into account all possible routes of entry (inhalation, ingestion and dermal exposure). In addition, exposures other than occupational, such as through diet, hobbies or residential air, all contribute to the internal dose, which is related to adverse effects. When the effect of occupational exposure alone needs to be evaluated, non-occupational exposures need to be assessed and controlled to allow interpretation of biological monitoring results. A special case is the evaluation of protective equipment in reducing workplace exposures.

Biological monitoring is limited by parameters that can affect the exposure and dose relationship. Such parameters include the following: biological conditions (age, sex, obesity, pregnancy, disease), physical workload, or exposure to other agents (other chemicals, tobacco smoking, dietary components) that in some way interfere with the uptake, distribution, metabolism or elimination of the chemical in the body.[3-6] Such parameters may introduce large inter-individual variation in the resulting dose. When simulation techniques were used to evaluate the influence of these parameters, it was shown[6] that up to 100% bias may result from inter-individual variation in the metabolism of aromatic compounds. If these differences occur randomly, they will widen the confidence limits of the biological monitoring results, decreasing the statistical power of a study. This is just one example illustrating the care that must be taken when interpreting biological monitoring results.

In short, biological monitoring should be used with specific objectives, such as to evaluate exceedence of a reference or regulatory value; dermal or oral exposure not assessed by personal monitoring; efficacy of protective equipment; etc. Accordingly, current, recent or cumulative exposure may need to be assessed and, therefore, the appropriate chemical (parent or metabolite) should be monitored in the appropriate specimen, collected at the appropriate time, from the appropriate worker(s).

### 15.2.1.2 Types of samples used for biological monitoring

Of the various tissues/fluids that can be used for biological monitoring, the most common are blood, urine and exhaled air because they are relatively easy to obtain.

Blood collection is the most invasive of the three and the chemical analysis may be subject to interferences because of the complex matrix. Additionally, concern over hepatitis and HIV infection, and confidentiality issues, make blood sampling less attractive for routine collection. Nevertheless, concentrations of a compound of interest in blood are easier to relate to concentrations at the target organ.

Analysis of urine samples allows elimination rates of the chemical or its metabolite(s) to be ascertained. However, measurements in urine require specimen collection under a schedule (first morning urine, or 24-hour sample) that can create logistical difficulties. The concentration of the contaminant, or its metabolite(s), is usually corrected for the dilution of the urine (specific gravity or, more commonly, creatinine correction) to be used as a measure of dose.

Collection of breath samples is the easiest of the three methods. It is non-invasive; it can take place almost anywhere; and the analytical matrix is usually very simple. The basic assumption underlying breath monitoring is the existence of gaseous equilibrium between the concentration of the vapor of interest in the alveolar air and in the arterial blood. This relationship must be established for every solvent for which biological monitoring by exhaled air analysis is considered. The apparent simplicity of breath sampling should not be overstated, however, and the limitations and pitfalls should be identified before any large scale, routine applications of breath monitoring can be envisioned. Since breath is a non-homogeneous mixture of air coming from different regions of the lung with varying ventilation, perfusion, and diffusion characteristics, some portion of the breath should be consistently identified and tested for its relationship to the blood concentration.

The ACGIH only recommends measurements in exhaled air when:[7]

- The chemical is poorly metabolized and, therefore, exhalation is the primary route of elimination;
- Sampling can occur when the effect of time and other circumstantial factors can be controlled;
- The sampling method is well defined.

### 15.2.1.3 Fundamentals of respiratory physiology

In order for any inhaled solvent vapors to enter the blood circulation, they need to reach the alveoli, cross the gas-blood interface and dissolve in the blood. Gas is carried to the gas-blood interface by airways, while blood is carried by blood vessels. Since these functions take place in the lung, the essentials of respiratory physiology will be briefly reviewed.

The primary function of the lungs is to allow oxygen to move from the inhaled air into the arterial blood and to allow carbon dioxide to move from the venous blood to the exhaled air. Both oxygen and carbon dioxide move between air and blood by simple diffusion from an area of high partial pressure to an area of low partial pressure. The barrier between air and blood is less than 0.5 μm in thickness and has an area of between 50 to 100 $m^2$.[8] This barrier is very efficient for gas exchange by molecular diffusion, according to Fick's 2nd law.

The airways consist of a series of tubes that become narrower and shorter as they extend deeper into the lung. The first segment, the trachea, divides into two main bronchi, which in turn divide into lobar and segmental bronchi, which turn into terminal bronchioles. All these segments collectively constitute the anatomical dead space with a volume of about 150 mL for an adult male. The function of this conducting portion of the lung is to transport the inspired air into the gas exchange regions. The terminal bronchioles divide within a distance of 5 mm into respiratory bronchioles and alveolar ducts, which are completely lined with alveoli. This region of the lung is the respiratory zone where the gas exchange takes place. The volume of this zone (about 2,500 mL) makes up most of the lung volume.

#### *15.2.1.3.1 Ventilation*

Air is drawn into the lungs by contractions of the diaphragm and the intercostal muscles, which raise the rib cage, and flows to the terminal bronchioles by bulk flow. After that point, the velocity of the inspired air diminishes as the cross sectional area of the airways increases dramatically. Thereafter, ventilation is carried out by molecular diffusion, which results in rapid exchange of gases.

Similarly to the airways, the pulmonary blood vessels form a series of branching tubes from the pulmonary artery to the capillaries and back to the pulmonary veins. The diameter of a capillary segment is about 10 μm, just large enough for a red blood cell. The capillaries form a dense network in the walls of the alveoli, with the individual capillary segments so short that the blood forms an almost continuous sheet around the alveoli, providing ideal conditions for gas exchange. In about one second, each red blood cell transverses two or three alveoli in the capillary network, achieving complete equilibration of oxygen and carbon dioxide between alveolar gas and capillary blood.

For an adult male at rest, a typical breathing frequency is 15 breaths per minute and a typical exhaled air volume (tidal volume), at rest, is about 500 mL. The total volume of air exhaled in 1 min is, therefore, about 7,500 mL. This is the minute volume total ventilation. The inhaled volume during the same time is slightly greater, since more oxygen is taken up than carbon dioxide released. Not all the inhaled air reaches the alveoli; the anatomical dead space (volume of conducting airways where no gas exchange takes place) contains approximately 150 mL of each breath. Therefore, the volume of fresh inhaled air that reaches the alveoli in 1 min is 5,250 mL [(500-150) mL/breath * 15 breaths/min = 5,250 mL/min].[8] This is the alveolar ventilation and it represents the volume of fresh air available for gas exchange in 1 min. The alveolar ventilation cannot be measured directly, but it can be calculated by measuring the minute volume and subtracting the anatomical dead space. The former is easily measured by collecting all the expired air in a bag. The volume of the anatomical dead space can be assumed to be 150 mL, with minimal variation for all adults, and the dead space ventilation can be calculated given a respiratory frequency. Subtraction of the dead space ventilation from the total ventilation (or minute volume) results in the alveolar ventilation. Pulmonary ventilation may be affected by chronic diseases such as asthma or pulmonary fibrosis, or even by transient hyperventilation or hypoventilation.

#### *15.2.1.3.2 Partition coefficients*

Whereas the rate of transfer of a solvent vapor between alveolar air and capillary blood is determined by its diffusivity, the equilibrium between these matrices is determined by the blood/air partition coefficient ($\lambda$). This is the ratio of the concentration of the vapor in blood and air at 37°C, at equilibrium. Partition coefficients are commonly determined in vitro, and occasionally in vivo.[2,9,10] Although considered constant at a particular temperature, partition coefficients may be affected by the composition of the blood. The blood/air partition coefficients of some industrial solvents were shown to increase by up to 60% after meals,[4] presumably due to an increase in lipid content of the blood.

Depending on the affinity of a gas or vapor for blood, the transfer from the alveoli to the capillaries may be diffusion-limited or perfusion-limited. The difference can be illustrated by examining two solvents, styrene and methyl chloroform. Because of the high solubility of styrene ($\lambda$=52)[11] large amounts of it can be taken up by the blood, and the transfer is only diffusion-limited. On the other hand, methyl chloroform is not very soluble ($\lambda$=1.4)[12] and its partial pressure rises rapidly to that of the alveolar air, at which point no net transfer takes place. The amount of methyl chloroform taken up by the blood will depend exclusively on the amount of available blood flow and not on the diffusion properties of the gas-blood interface. This kind of transfer is perfusion-limited.

It is the concentration in the mixed venous blood, however, that better reflects the concentration at the target organ. It has been shown that:[13]

$$P / Pv = \frac{\lambda}{\lambda + V / Q}$$

where:

| | |
|---|---|
| P | partial pressure in the alveoli and in the arterial blood leaving that region of the lung |
| Pv | partial pressure in the venous blood coming to the lung |
| $\lambda$ | blood/air partition coefficient |
| V | alveolar ventilation |
| Q | blood flow of the lung region under consideration |

The greater the blood/air partition coefficient, the closer the arterial concentration will be to the venous concentration.

The concentration in the alveolar air will reflect both arterial and venous blood if the blood/air partition coefficient of the vapor is greater than 5,[14] a criterion that applies to most industrial solvents. In a normal subject at rest, the ventilation-to-perfusion ratio (V/Q) ranges from 0.7 to 1.0, with an average of about 0.9.[15] Variations in the V/Q ratio exist in various parts of the lung because not all alveoli are ventilated and perfused in ideal proportions. The V/Q ratio may vary from 0.5 at the lung base to 3.0 at the apex,[16] and it becomes more homogeneous with physical exertion.[17] When the partition coefficient exceeds 10, however, even large degrees of ventilation/perfusion imbalance have very little effect on the relationship between alveolar and mixed venous solvent partial pressures.[14]

#### *15.2.1.3.3 Gas exchange*

When a person is exposed to a volatile organic solvent through inhalation, the solvent vapor diffuses very rapidly through the alveolar membranes, the connective tissues and the capillary endothelium and into the red blood cells or plasma. With respiratory gases the whole process takes less than 0.3 seconds.[15] This results in almost instantaneous equilibration between the concentration in alveolar air and in blood and, therefore, the ratio of the solvent concentration in pulmonary blood to that in alveolar air should be approximately equal to the partition coefficient. As the exposure continues, the solvent concentration in the arterial blood exceeds that in the mixed venous blood.[18] The partial pressures in alveolar air, arterial blood, venous blood and body tissues reach equilibrium at steady state. When the exposure stops, any unmetabolized solvent vapors are removed from the systemic circulation through pulmonary clearance. During that period the concentration in the arterial blood is lower than in the mixed venous blood[18] and the solvent concentration in alveolar air will depend on the pulmonary ventilation, the blood flow, the solubility in blood and the concentration in the mixed venous blood. These are the prevailing conditions when breath monitoring takes place after the end of the exposure[17] and therefore, samples of alveolar air after exposure should reflect concentrations in mixed venous blood. During exposure, however, alveolar air should reflect concentrations in arterial blood. Solvent concentrations in arterial and venous blood diverge even more with physical exertion.

To summarize, the uptake of solvent vapors through inhalation will depend on the following factors:[3]

1. Pulmonary ventilation, i.e., the rate at which fresh air (and solvent vapor) enters the lungs. This is determined by the metabolic rate and therefore depends on physical exertion. The concentration in alveolar air approaches that in inhaled air when the physical exertion is great. When exertion is low, the alveolar air concentration approaches the concentration in mixed venous blood.

2. Diffusion of the solvent vapor through the gas-blood interface. According to Fick's 2nd law, the rate of diffusion through the interface depends on the concentration gradient across the tissue membrane. Therefore, the concentration of the vapor leaving the lungs depends on the amount of vapor entering the alveoli, on the solubility in blood (blood/gas partition coefficient) and on the flow rate of the blood through the lungs. For lipophilic chemicals, such as organic solvents that readily cross cellular membranes, diffusion is not the rate-limiting factor.[18]

3. The solubility of the vapor in the blood. The higher the blood/gas partition coefficient, the more rapidly the vapor will diffuse into the blood, until equilibrium is achieved. At equilibrium the net diffusion between blood and air ceases, but the concentrations in air and blood may still be different. A highly soluble vapor will therefore demonstrate a lower alveolar air concentration relative to the inhaled air concentration during exposure, and a higher alveolar concentration relative to the inhaled air concentration after exposure.

4. The circulation of the blood through the lungs and tissues. This depends on the cardiac output and therefore, on physical exertion. For a soluble vapor, high exertion moves blood faster, increasing uptake.

5. The diffusion of the vapor through the tissue membranes, which depends on membrane permeability and is governed by Fick's law, as discussed in #2 above.

6. The solubility of the solvent vapor in the tissue. This reflects the tissue/blood partition coefficient of the solvent vapor and depends largely on the lipid composition of the particular tissue. Lipophilic vapors will exhibit high solubility in tissues with high fat content, freeing blood of vapors and increasing uptake.

Substances that are not well metabolized are eliminated primarily through the lungs[3] and to a lesser extent through the kidneys. During elimination through the lungs the same physiological principles apply in a reverse sequence.

#### *15.2.1.4 Types of exhaled air samples*

Samples obtained from even a single breath may contain different concentrations of the solvent vapor of interest depending on the way the sample was collected. This is a result of the non-homogeneity of the exhaled air discussed previously. Figure 15.2.1.2 depicts a concentration curve of a solvent in exhaled air. When measurement takes place in an environment free of that solvent, the first part of the exhalation, corresponding to air in the anatomical dead space, will be solvent-free (Phase 1). Then the solvent concentration curve rises sharply reflecting exhalation of air from the respiratory zone of the lungs (Phase 2). This phase is followed by a slowly increasing plateau, indicating an alveolar steady state (Phase 3). The level of this plateau depends on the preceding ventilatory state of the subject, being higher for a hypoventilating subject and lower for a hyperventilating subject.[6] Sample col-

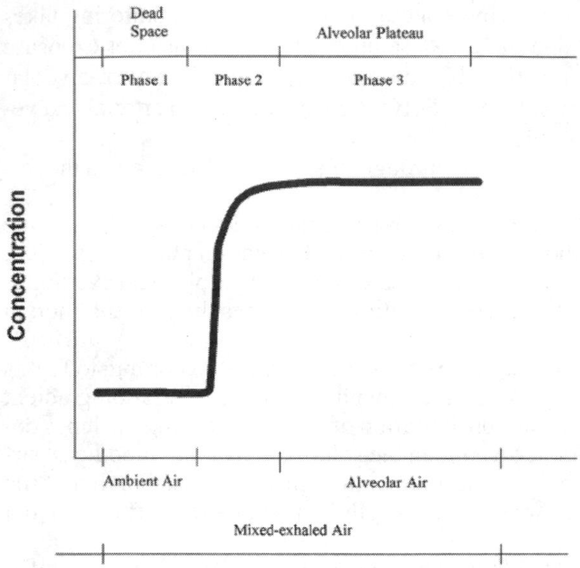


Figure 15.2.2. Solvent concentration in breath during exhalation.

lection at the alveolar plateau constitutes end-exhaled air. When the total expiration is collected (combination of all three phases) the sample is considered mixed-exhaled air. Simultaneous monitoring of $CO_2$,[6,19-21] or oxygen and nitrogen[22] has been suggested to normalize the measured solvent vapor concentrations in breath samples in order to facilitate data interpretation and comparisons.

The various sampling techniques must be evaluated and the one most applicable to the purpose of sampling chosen. The most commonly used sampling techniques based on physiological principles include mixed-exhaled air and end-exhaled air:

Mixed-exhaled air. This technique involves the collection of the entire volume of exhaled air. It corresponds to a mixture of the alveolar air with air from the dead space. The collection apparatus may also contribute to the dead space. Total dead space should be considered and the concentration adjusted, either by subtraction, or by regression against some other technique unaffected by the dead space. Timing of the breath collection is important here since the concentration of the air in the dead space may equal that of the air in the workroom if the sample is taken during exposure, or it may equal zero if taken after the end of exposure.

End-exhaled air. This technique excludes air from the dead space and collects only the last part of the breath, in order to estimate the concentration in the alveolar air which is in equilibrium with the arterial blood. Use of a Haldane-Priestley tube,[23] or simultaneous monitoring of the peaking of the temperature of the breath[19] or its $CO_2$ concentration,[6,19,24] will assure a valid alveolar air sample.[6,25,26]

Two less frequently used techniques include breath holding or rebreathing to homogenize the breath sample:

Breath holding. For compounds that achieve equilibrium slowly, some researchers have used the breath holding technique, which involves holding the breath for 5-30 seconds prior to exhaling into the collection device. This results in a more constant concentration in the exhaled breath. The extent to which the lungs are filled with air (as with a deep inhalation) will affect the results. Kelman[14] showed, however, that breath holding is not necessary for solvents with blood/gas partition coefficients greater than 10.

Rebreathing. With this technique the subject rebreathes his/her exhaled air to provide a more homogenized breath sample. This technique is applicable to solvents that reach steady state very slowly. For prolonged rebreathing, removal of $CO_2$ to avoid pH changes and/or supply of oxygen may be necessary.[27]

### *15.2.1.5 Breath sampling methodology*

The theoretical principles discussed above lead to two choices for collecting breath samples: Should a mixed-exhaled air or an end-exhaled air sample be collected?

End-exhaled air is becoming the standard for breath monitoring. Of the 36 chemicals with established BEIs,[7] end-exhaled air is recommended for 6, while earlier BEIs recommending mixed-exhaled air have been withdrawn. Beyond the regulatory arena, however, mixed-exhaled air continues to be used with satisfactory results.

Methodologically, the difference between these two types of samples is the control over which portion of the breath is collected. Many techniques[28-32] make use of the Haldane-Prestley tube to collect the last portion of the expiration. Earlier work[6,19] used the simultaneous monitoring of $CO_2$ or temperature in the exhaled breath to identify the moment ($CO_2$ or temperature reaching a maximum) when air from the anatomical dead space has been purged and alveolar air can be sampled.

A variety of sampling techniques has been tried for measuring organic vapors in exhaled air. Table 15.2.1.1 shows a classification of these techniques. The main distinction is between direct reading instruments and transportable samples that require analysis in the laboratory. The former group covers the widest range, from simple detector tubes to High Resolution Mass Spectrometry.

**Table 15.2.1.1. Techniques for sampling organic vapors in exhaled air**

| Technique | References |
|---|---|
| A. SAMPLE COLLECTION | |
| A.1 END-EXHALED AIR | |
| Use of Haldane-Prestley tube | 23, 25, 26, 28, 29, 31 |
| Simultaneous $CO_2$ monitoring | 6, 20, 21, 24 |
| Simultaneous temperature monitoring | 19 |
| B. SAMPLE ANALYSIS | |
| B.1 DIRECT READING INSTRUMENTS | |
| B.1.1 DETECTOR TUBES | |
| Colorimetric | 33 |
| B.1.2 GAS CHROMATOGRAPHS with various detectors and modified injector/inlet with mouthpiece | |
| GC/FID | 52 |
| GC/PID | 25, 26 |
| GC/MS | 24, 41, 42, 43 |
| B.2 TRANSPORTABLE SAMPLE COLLECTORS | |
| B.2.1 CONTAINERS | |
| B.2.1.1 Rigid containers | |
| Stainless steel tubes | 24 |
| Glass tubes | 9, 53, 54, 55, 61, 64 |
| Aluminum tubes | 28, 29, 47 |
| B.2.1.2 Flexible containers | |
| Tedlar bags | 6, 49 |
| Saran bags | 54, 55, 60 |
| Mylar bags | |
| B.2.2 ADSORBENTS | |
| B.2.2.1 Tubes/cartridges containing: | 25, 26, 28, 29, 40 |
| Charcoal | 31, 49 |
| Tenax | 6 |
| Silica gel | 48 |
| B.2.2.2 Felt/cloth | |

Specifically, detector tubes operating on colorimetric reaction principles have been used to determine the concentration of alcohol in breath for traffic safety.[33] Alcohol measurements in breath have been thoroughly studied and scrutinized because of the forensic and legal repercussions associated with the results. Portable devices have been developed for field use with on-site analysis of alcohol.[34-36] These portable devices could be adapted for use with other organic solvents of interest in industrial hygiene and occupational health.

Gas chromatographs (GCs) equipped with suitable detectors allow analysis of more than one solvent vapor at a time. The subject exhales through a valve system and the exhaled volume of air, or a fraction of it, is drawn into the GC.[21,26,37-39] Depending on the detector used, one or more cylinders containing appropriate gases under pressure are required to operate the instrument, limiting its field use. An Electron Capture Detector (ECD), which is the most sensitive for polyhalogenated solvents, requires helium and/or nitrogen. A Flame Ionization Detector (FID), which is a universal detector for aliphatic and aromatic hydrocarbons, requires nitrogen, air and hydrogen. The advent of the Photoionization Detector (PID) offered a compromise, in that it can detect all the above classes of organic solvents (with varying sensitivity) requiring only one gas (nitrogen, air or helium) for its operation. A simple system employed a portable GC/PID modified to sample 1 mL of air from a Haldane-Priestley tube into which the subject would exhale.[25,40] The latter assured collection of end-exhaled air and this device was successfully used to assess exposures to styrene[26] and PERC.[40]

Mass spectrometry (MS) offers the advantages of high sensitivity and specificity at the cost of portability and the need of highly trained operators. Direct exhalation into the MS is not possible because the ion source of the detector operates under high vacuum. Two approaches have been taken to overcome this limitation. In the first, an inlet system was designed where the high pressure breath sample was directed over a methyl silicone membrane, permeable to organic molecules but not to air, that served as the entry port to the ion source operating under high vacuum.[41] With the second approach, the inlet system was attached to the source of an Atmospheric Pressure Ionization mass spectrometer (API/MS), that operates under atmospheric pressure.[42,43]

In summary, the basic advantage of the direct reading/analyzing instruments is that there is no sample collection, transportation, storage, handling and analysis where the sample integrity may be compromised. The disadvantages are the need for expensive, specialized equipment, usually requiring highly trained operators and the difficulty, or some times impossibility, of field use.

The other category of breath sampling methodology involves sample collection in the field with subsequent analysis in the laboratory. There are two groups of devices in this category:

1. Sample containers that hold the total volume or part of the exhalation in a gaseous form.
2. Sample adsorbents that trap the chemicals of interest from the exhaled breath.

In the first group belong rigid containers such as stainless steel, aluminum, or glass vials of various sizes, open at both ends. The subject exhales into the vial and then the two openings are closed, either with caps or with stopcocks. Vials of this kind have been used extensively in experimental studies involving exposures to industrial chemicals.[24,28,29,44-47,53-55,61,64]

The advantages of these containers are inertness of the walls, low cost, reusability and simplicity of the technique. The disadvantages are the potential for contamination, photodegradation, or dilution of the collected breath sample with atmospheric air, sorption and/or reaction of the vapor with surfaces and breakage during transportation or storage. The problem of condensation of water vapor has been addressed by heating the tubes,[28,29,47] but this approach increases the complexity and counters the main advantages. Another limitation is that only a small volume of sample can be withdrawn for analysis from a rigid container, with ensuing restrictions on the analytical sensitivity.

The last limitation is overcome when collapsible bags are used for sample collection. The bags are commonly made of Tedlar (clear) or Mylar (aluminized) material. The subject exhales directly into the bag through a valve and when an appropriate volume has been collected, the valve is closed. The bag is transported to the laboratory where an aliquot is withdrawn for analysis through a septum. The collapsibility of the bag allows any volume to be withdrawn without air infiltration. Disadvantages, however, include losses because of permeation through the walls, sorption and/or reaction of the vapor with the wall surfaces and photodegradation. The latter concern is eliminated with the aluminized Mylar bags, which also minimize sorption and permeation through the walls.

The second group of field sampling devices with subsequent analysis involves the use of adsorbents. Extensively used in personal monitoring in industrial hygiene surveys, these devices were naturally considered for exhaled breath sampling. Commercially available glass tubes containing granular adsorbent material such as activated charcoal, Tenax, molecular sieve, etc. could be used with selected organic vapors. However, because of the high pressure drop during exhalation through the adsorbent bed, and the condensation of water vapor, these materials have not been used much for direct sampling of exhaled breath. One device[48] is, basically, a respirator mask with two commercial charcoal inhalation canisters and an exhalation port fitted with a wafer of charcoal cloth. A Wright respirometer connected to the exhalation port records the volume of exhaled air. Another hand-held device employed a custom-made glass tube containing charcoal to collect organic vapors in exhaled air, while the volume collected was measured with a Wright respirometer and adjusted for back-pressure measured with an attached pressure gauge.[25,40]

Commercial tubes or specially made cartridges containing the appropriate adsorbent have been used in conjunction with bag sampling for exhaled breath monitoring.[49] The subject exhales into the bag, the contents of which are subsequently pumped through the tube or canister containing the adsorbent. This combination offers the advantages of bag sampling (low cost, low resistance), while eliminating the disadvantages (losses, bulkiness). The organic vapors can be thermally desorbed from the adsorbent material and the concentrations measured with polymer-coated surface-acoustic-wave detector,[50] GC or GC/MS analysis.[31,49] Alternatively, solvent vapors can be extracted from the adsorbent with an appropriate solvent prior to conventional analysis.

#### *15.2.1.6 When is breath monitoring appropriate?*

Breath monitoring may be used to assess exposures to solvents either qualitatively or quantitatively. Qualitative information may be useful when exposure to a number of solvents is suspected. For qualitative purposes, either mixed-exhaled or end-exhaled air samples may be used. Provided that the sample collection and the analytical techniques are appropriate, a list of solvents may be thus identified and quantitative measurements planned.

Quantitative information is needed for regulatory enforcement situations, where appropriately collected breath samples are compared to BEIs to assess whether exposure limits have been exceeded. BEIs have been developed after thorough studies of pharmacokinetics for a limited number of occupational contaminants. Exhaled air is used as a surrogate for blood when the relationships between exhaled air, alveolar air and (arterial or mixed venous) blood have been established.

Besides regulatory purposes, breath monitoring can provide valid information on solvent exposures when:

- The pharmacokinetics (absorption, distribution, metabolism, elimination) of the solvent are well established.
- The solvent is not soluble in the conducting portion of the lung (anatomical dead space).

- The solvent is not metabolized to a great extent, and it is mainly eliminated via exhalation.
- Sample collection is conducted in a location unaffected by ambient levels of the solvent(s) of concern that may interfere and bias the measurements.
- The concentration in exhaled air is compatible with the analytical method allowing adequate sensitivity.
- Appropriate timing is strictly followed.

### 15.2.1.7 Examples of breath monitoring

Two solvents are used to illustrate the use of breath monitoring in assessing solvent exposure: A solvent with minimal metabolism, tetrachloroethylene (PERC) and a solvent with significant metabolism, styrene. Both solvents are used extensively in industry with a great number of workers potentially exposed. As shown in Table 15.2.1.2, styrene is only minimally excreted through the lungs, whereas exhaled air is the major elimination route for PERC. Accordingly, end-exhaled air is recommended as a BEI for PERC, but not for styrene. Nevertheless, end-exhaled and mixed-exhaled air techniques have been studied in experimental settings and in real-life occupational environments for both solvents. A compilation of such studies is presented below.

**Table 15.2.1.2. Characteristics of styrene and tetrachloroethylene (PERC). [Information compiled from ACGIH, Documentation of Threshold Limit Values and Biological Limit Values, 1991[7]]**

| | STYRENE | PERC |
|---|---|---|
| Eliminated via lungs | ~5% | ~80-100% |
| Eliminated in the urine | ~90% | ~2% |
| Urinary metabolite(s) | Mandelic acid (MA),<br>Phenylglyoxylic acid (PA) | Trichloroacetic acid (TCA) |
| Blood/air partition coefficient (λ) | 32-55 | 9-15 |
| Pulmonary elimination half-lives ($T_{1/2}$) | 13-52 min<br>4-20 hr<br>3 days | 15 min<br>4 hrs<br>4 days |
| TLV (1998) | 50 ppm | 25 ppm |
| STEL (1998) | 100 ppm | 100 ppm |
| BEI (1998)<br>End-exhaled air | Not recommended | 10 ppm end-of-shift, after at least 2 shifts |
| Blood | 0.55 mg/L end-of-shift<br>0.02 mg/L prior-to-next-shift | 1 mg/L end-of-shift,<br>after at least 2 shifts |
| Urine | MA=800 mg/g creatinine, end-of-shift<br>MA=300 mg/g creatinine, prior-to-next-shift<br>PA=240 mg/g creatinine, end-of-shift<br>PA=100 mg/g creatinine, prior-to-next-shift | TCA=7 mg/L |
| Major Industry | Reinforced Plastics | Dry Cleaning |

15.2.1.7.1 PERC

Tetrachloroethylene (PERC) is used extensively in the textile industry as a dry cleaning aid and in metal processing as a metal degreaser. The major concern is on occupational exposures, but elevated indoor air concentrations of PERC have also been documented.[49] As a result, the potentially exposed population is very large, with a wide range of exposures. The uptake of PERC by the lung is high[51,52] because of its high solubility in blood and adipose. Most of it is eliminated via the lung, with very small amounts metabolized to trichloroacetic acid (TCA) and eliminated in the urine.[51,52] Even smaller amounts of trichloroethanol in the urine have been reported.[52]

A number of field studies have been published with similar designs, in which breathing zone air, blood, breath and urine were measured in groups of workers. In addition, chamber studies allowed measurements under predetermined consistent conditions. The emerging patterns formed the basis for recommendations for biologic monitoring.

In a study of thirty two workers,[9] end-exhaled air was measured at the beginning of the first shift of the week, after 15-30 min after the end of the shift on the $3^{rd}$ and $5^{th}$ day of the same week, and at the beginning of the first shift of the subsequent week. The authors required the subjects to hold their breaths for 5 seconds and collected only the last part of the breath to obtain alveolar air. These measurements correlated with personal breathing zone samples collected each day. The highest correlation was found between PERC and TCA in blood at the end of the workweek and the average exposure for the entire week ($R^2$=0.953). Among the non-invasive techniques, PERC in exhaled air collected 15-30 min after the end of the shift at the end of the week had an $R^2$ of 0.931, and was followed by TCA in urine ($R^2$=0.909).

In a field study of workers employed in dry cleaning,[53] twenty-four female and two male subjects were monitored. End-exhaled air and blood samples were collected at the beginning and at 30 minutes after the end of the $3^{rd}$ workday of the week. On that same day, personal, shift-long, breathing zone samples were also collected from each worker. Urine samples were collected before and after the shift on a daily basis and analyzed for TCA. With a mean exposure of 20.8 ppm, no TCA was detected in any of the urine samples (no detection limit was reported). Among the three exposure parameters (breathing zone air, blood and end-exhaled air) only the correlation between the PERC concentration in the blood and in the breath was reported as statistically significant by the authors ($R^2$=0.77). The authors concluded that if the concentration of PERC in blood, 16 hours after the end of exposure, was below 1000 μg/L, the average exposure was likely to have been below 50 ppm.

A group of thirteen male dry cleaners[40] was followed for a week with daily shift-long personal (badge) monitoring; mixed-exhaled air collected daily before- and after-shift; and mixed-exhaled air, blood and urine measurements on the morning after the last shift. Results indicated strong correlations between PERC concentrations in mixed-exhaled air and blood ($R^2$=0.944), as well as mixed-exhaled air and personal air samples collected either after the shift ($R^2$=0.885), or on the following morning ($R^2$=0.770). Mixed-exhaled air measurements collected either at the end of the last shift of the week ($R^2$=0.814) or in the morning following the last shift ($R^2$=0.764) were better indicators of exposure over the entire week than either blood ($R^2$=0.697) or urine ($R^2$=0.678). The authors recommended consideration of mixed-exhaled air as a potential BEI.

In addition to field studies, controlled human exposures to PERC have been reported. In this type of studies, subjects were exposed to various constant levels of PERC for different lengths of time with or without exercise. Samples of blood, exhaled air and urine were collected according to a variety of schedules both during exposure and for varying lengths of time after exposure. One study involved single exposures, whereas in others at least some of the subjects participated more than once.

In one of the first chamber studies, nine male volunteers were exposed to 100-200 ppm of PERC for 80 to 180 minutes. Concentrations of PERC in exhaled air were measured over time.[54] The authors demonstrated that exposures of similar duration resulted in similar decay curves of PERC concentrations in breath after exposure. In addition, the data showed that the length of time that PERC was measurable in exhaled air was proportional to both the concentration and the duration of the exposure.

The same group[55] exposed sixteen males to single and repeated concentrations of 100 ppm PERC for periods of 7 hrs/day over a 5-day workweek. Alveolar air samples were collected over a period of up to 300 hours post-exposure. The authors observed higher alveolar air concentration in subjects with greater body mass. They recommended measurement of PERC in alveolar air 2-16 hours post-exposure to assess the average exposure.

Six healthy males were exposed to 72 and 144 ppm, as well as 142 ppm with 2 half hours of 100 W of exercise.[51] Uptake of PERC was calculated from the minute volume and the concentration of PERC in breath. PERC was measured in breath and blood and TCA was measured in blood and urine. During the first hour of exposure the uptake was higher by 25% than during the last hour and retention decreased with time and with exercise. The mean ratio of the concentration of PERC in venous blood and the concentration of PERC in mixed exhaled air was 23. Assuming a mix-exhaled air to end-exhaled air ratio of 0.71, they estimated a blood/air partition coefficient of 16. After 20 hours from the end of exposure the half-life was estimated to be 12-16 hours, after 50 hours the half-life was estimated to be 30-40 hours and after 100 hours it was estimated at approximately 55 hours. The authors estimated that 80-100% of PERC was excreted unchanged through the lung, and about 2% was excreted as TCA in the urine.

Simple and multiple linear regression analyses were applied to the data collected above to predict the best estimator of exposure to PERC.[56,57] PERC in blood was the best predictor followed by PERC in exhaled air. TCA per gram of creatinine in urine was the least reliable predictor.

Twenty four subjects were exposed to 100, 150 and 200 ppm of PERC for 1 to 8 hours.[52] The concentration of PERC in alveolar air decreased rapidly during the first hours of the post-exposure period, but 2 weeks were required for complete elimination following an exposure of 100 ppm for 8 hours. The authors were the first to demonstrate that the rate of rise and of decay of the alveolar concentration with respect to time was independent of the level of the inspired concentration during exposure.

The experimental data from the above study were used in a pharmacokinetic model[58] that divided the body into four tissue compartments: the Vessel Rich Group (VRG), the Muscle Group (MG), the Fat Group (FG) and the Vessel Poor Group (VPG). No metabolic pathway was considered. After 20 hours from the end of the exposure the concentration of PERC in alveolar air was related to release from the FG with an approximate half-life of 71.5 hours. The model predicted small differences in alveolar concentrations due to differences in body weight, height and fat tissue. The authors provided a nomogram to predict mean exposure given the exposure duration (0-8 hrs), time since end of exposure (0-8 hrs) and post-exposure measurement of the alveolar air concentration.

A perfusion-limited, four compartment model was used to simulate the effect of several confounding factors on the levels of several contaminants in the breath, including PERC.[6] The model, which considered metabolic clearance, consisted of a pulmonary compartment, a vessel-rich group, a group of low-perfused tissues containing muscles and skin, and a group of poorly perfused fatty tissues. The authors predicted that intra-day fluctuations in exposure would have a large effect on breath sampled just after the shift, but that the effect would be negligible the next morning. Breath samples collected the next morning would reflect exposure on preceding days and the measurements would be independent of level of physical exertion. Finally, individual differences in body build and metabolism would have the greatest effect on breath levels, with the largest variations expected for metabolized solvents.

Experimental data were used in a model,[59] which divided the body into four tissue compartments: vessel rich group, muscle group, fat group and liver. Metabolism was assumed to take place in the liver as a combination of a linear metabolic component and a Michaelis-Menten component. Metabolic parameters and partition coefficients determined for rats were scaled for body weight and were used in fitting results for humans. The model fit very well the data reported from other controlled human exposure studies.[51,52,54]

The BEI committee of the ACGIH recommends[2] three indices of PERC exposure:

1) PERC in end-exhaled air collected prior to the shift after at least 2 consecutive workdays.

2) PERC in blood collected prior to the shift after at least 2 consecutive workdays.

3) Trichloroacetic acid (TCA) in urine voided at the end of the workweek. This index is recommended only as a screening test because of the variability associated with urinary excretion of TCA and, as such, no creatinine correction is necessary.

15.2.1.7.2 Styrene

The highest exposures to styrene take place in plants manufacturing glass-reinforced plastics, particularly during lamination, where there is considerable evaporation of styrene. Most of the inhaled styrene is absorbed and retained in the body. The first step in the major metabolic pathway is the formation of styrene-7,8-oxide (phenyloxirane), a reaction catalyzed by Cytochrome P-450 in the liver. The oxide is then hydrated to styrene glycol (1-phenyl-1,2-ethanediol) by microsomal epoxide hydratase. The styrene glycol is conjugated with beta-glucuronic acid, or is oxidized to mandelic acid and further to phenylglyoxylic acid. In humans, less than 5% of absorbed styrene is eliminated via the lung, whereas 90% is eliminated in the urine as mandelic and phenylglyoxylic acids.

A number of chamber and field studies have been reported, where blood, exhaled air and urine have been studied as possible biological indicators. Although conditions and study design differed among studies, certain consistent patterns emerged, which formed the basis for subsequent recommendations.

In an early series of chamber exposures,[60] nine healthy males were exposed to styrene concentrations of 50 to 370 ppm for periods ranging from 1 to 7 hours. Alveolar air samples were collected during and for 8 hours following exposure. The authors showed that the amounts of styrene present in the breath were directly related to the level and the duration of exposure.

In another study,[61] fourteen healthy males were exposed to 50 and 150 ppm of styrene in inspired air during rest and light (50 W) physical exercise. The duration of exposure for each styrene concentration was 30 minutes, with total pulmonary ventilation, cardiac output, and styrene concentration in alveolar air, arterial blood and venous blood measured during and after exposure. During exposure at rest to either 50 or 150 ppm, the alveolar air concentrations were about 20% of the inspired air concentrations. When alveolar ventilation almost tripled with exercise at 50 W, the alveolar air concentrations increased slightly, whereas arterial blood concentrations almost tripled. Alveolar air concentrations reached a plateau within a minute or so, that persisted throughout the 30-minute period. Arterial blood concentrations, however, rose rapidly during each such period and even more with increasing exercise intensity (50, 100 and 150 W). Venous blood concentrations also rose sharply during the 30-minute exposure periods without ever reaching a plateau. No linear relationship was found between exposure duration and arterial and venous blood concentrations, even after log-log transformations. The decay of styrene concentrations in alveolar air and in venous and arterial blood was slower in the first 30 to 50 minutes than later on. The authors concluded that post-exposure alveolar air measurements were poor predictors of exposures to styrene, whereas arterial blood would be the best index of exposure. For practical reasons, they recommended capillary blood samples from fingertips as a surrogate for arterial blood.

In a follow-up study[62] using the same experimental set up as above, seven male subjects were exposed to 50 ppm (210 mg/m$^3$) of styrene in inspired air for 30 minutes at rest, followed by 30 minute exposure periods under exertion at intensities of 50, 100 and 150 W. The mean alveolar air concentration at the end of the first 30-minute period was 16% of the concentration in the inspired air and 23% at the end of the final period. Both the arterial and venous blood concentrations of styrene rose with increasing workload during exposure with no equilibrium achieved between alveolar air and arterial blood during 2 hours of total exposure. Weak correlations were found for the amount of styrene taken up per kg of body weight and the concentrations of styrene in alveolar air 0.5 and 2 hours after exposure. The elimination was faster during periods of exercise than during rest periods, leading to the authors' recommendation for leisure time physical activity for people exposed to solvents as a means to enhance elimination.

In another study,[63] four male volunteers were exposed to 80 ppm of styrene in a chamber for 6 hours. Venous blood samples were collected during exposure and a nearly simultaneous set of 10 blood and mixed exhaled air samples were collected following exposure for up to 40 hours. Blood levels during exposure rose rapidly and reached an almost constant level by the end of the 6 hours. Following exposure, the results showed that styrene was cleared from the blood according to a linear two-compartment pharmacokinetic model, with half-life values of 0.58 and 13 hours, for the rapid and slow elimination phases, respectively.

Retention of styrene was studied in a chamber setting where healthy males were exposed to either constant or fluctuating air concentrations of styrene.[26] A computer-controlled system was used to generate time-varying air concentrations over 4-5 hrs with a mean air concentration of 50 ppm. End-exhaled air measurements taken throughout the exposure period showed styrene retention of 93.5% during constant exposures, but higher retention (96-97%) during fluctuating exposures. The authors speculated that the difference in retention was related to non-steady-state behavior of styrene in the richly perfused tissues.

In summary, pulmonary retention during exposure is about 60-70%[62,64] with exhaled concentrations accounting for 25-35%.[60] Higher retention (96-97%) was observed when volunteers were exposed to fluctuating, rather than constant, air concentrations.[26] Following exposure, the desaturation curve shows two exponential decays with half-lives of 13 minutes[60] to 52 minutes[63] for the rapid elimination phase and 4 hours[60] to 20 hours[63] for the slower elimination phase. A third compartment can be defined to represent elimination from the adipose tissue with a half-life of 3 days.[62]

Measurements of styrene in venous blood at the end of the workshift and/or prior to the next shift have been recommended by the ACGIH.[2] Monitoring of urinary metabolites of styrene has also been recommended by the ACGIH[2] at the end of the week, even though they are not specific to styrene exposure. Styrene in exhaled air after the end of exposure was not recommended as a BEI because the levels would be too low to detect, leading to uncertainties.

## Terminology

| | |
|---|---|
| ACGIH | American Conference of Governmental Industrial Hygienists |
| Alveolar air | Gas in the alveoli |

| | |
|---|---|
| Alveolar ventilation | Inhaled air available for gas exchange in 1 min |
| Anatomical dead space | Volume (~150 mL) of conducting airways where no gas exchange occurs |
| BEI | Biological Exposure Index. Reference value developed by ACGIH |
| Blood/air partition coefficient | Ratio of the vapor concentration in blood and in air at 37°C |
| Breathing frequency | 15 breaths per minute at rest, increases with physical exertion |
| Conductive zone of lungs | First segment of the airways, which includes the trachea, bronchi, terminal bronchioles |
| End-exhaled air | Operationally defined as the last portion of an exhalation to simulate alveolar air |
| Exhaled air volume | Approx. 500 mL at rest, increases with workload |
| GC-MS | Gas Chromatography Mass Spectrometry |
| Minute Volume Ventilation | The volume of air exhaled in 1 min, approx. 7,500 mL |
| Mixed-exhaled air | Entire volume of exhaled air collected from alveolar space and anatomically dead space |
| Respiratory zone | Consists of respiratory bronchioles and alveolar ducts where gas exchange takes place. Approx. 2,500 mL |
| STEL | Short-Term Exposure Limit. The concentration to which most workers may be exposed continuously for a short period of time without adverse effects. Issued by ACGIH |
| Tidal Volume | Volume of exhaled air; ~500 mL at rest |
| TLV | Threshold Limit Value. Airborne concentration of a substance to which most workers may be exposed without adverse health effects. Established by ACGIH and intended as guideline. |

## References


1 R Lauwerys, **Industrial chemical exposure: Guidelines for biological monitoring**. *Biomedical Publications.* Davis, California, 1983.
2 ACGIH, American Conference of Governmental Industrial Hygienists: TLV, Threshold Limit Values and Biological Limit Values. Cincinnati, Ohio, 1998.
3 I Astrand, *Scand J Work Environ Health.* **1**, 199-218, 1975.
4 V Fiserova-Bergerova, J Vlach, JC Cassady, *Br J Ind Med.,* **37**, 42-49, 1980.
5 PJA Borm and B de Barbanson, *J Occup Med.* **30**, 214-223, 1988.
6 PO Droz and MP Guillemin, *J Occup Med.,* **28**, 593-602, 1986.
7 ACGIH, American Conference of Governmental Industrial Hygienists, Documentation of Threshold Limit Values and Biological Limit Values. Cincinnati, Ohio, 1991.
8 JB West, Respiratory Physiology - **The essentials**, *The Williams & Wilkins Co.*, Baltimore, 1981.
9 AC Monster, W Regouin-Peeters, A van Schijndel, J van der Tuin, *Scand J Work Environ Health*, **9**, 273-281, 1983.
10 V Fiserova-Bergerova, **Modeling of inhalation exposure to vapors, Uptake, distribution and elimination**, *CRC Press Inc.*, Boca Raton, Florida, 1983.
11 A Sato and T Nakajima, *Brit. J. Ind. Med.*, **36**, 231-234, 1979.
12 A Morgan, A Black, DR, Belcher, *Ann Occup Hyg.*, **13**, 219-233, 1970.
13 LE Farhi, *Respir Physiol.*, **3**, 1-11, 1967.
14 GR Kelman, *Br J. Ind Med.*, **39**, 259-264, 1982.
15 JE Cotes, **Lung function, Assessment and application in medicine**, *Blackwell Scientific Publications*, Oxford, 1979.
16 J Widdicombe and A Davies, **Respiratory physiology**, *Edward Arnold*, London, 1983.
17 HK Wilson, Breath analysis, *Scand J Work Environ Health*, **12**, 174-192, 1986.
18 V Fiserova-Bergerova, M Tichy, and FJ DiCarlo, *Drug metabolism reviews*, **15**, 1033-1070, 1984.
19 KM Dubowski, *Clin. Chem.*, **20**, 966-972, 1974.
20 HC Niu, DA Schoeller, PD Klein, *J Lab Clin Med.*, **91**, 755-763, 1979.
21 M Guillemin and E Guberan, *Br J Ind Med.*, **39**, 161-168, 1982.
22 TA Robb TA and GP Davidson, *Clinica Chimica Acta*, **111**, 281-285, 1981.
23 JS Haldane, and JG Priestley, *J Physiol.*, **32**, 225-266. 1905.
24 JH Raymer, KW Thomas, SD Cooper, DA Whitaker, ED Pellizzari, *J Anal. Toxicol.*, **14**, 337-344, 1990.
25 SM Rappaport, E Kure, MX Petreas, D Ting, J Woodlee, *Scand. J. Work Envir. Health*, **17**, 195-204, 1991.

26 M Petreas, J Woodlee, CE Becker, SM Rappaport, *Int. Arch. Occup. Environ. Health,* **67**, 27-34, 1995.
27 RH Hill, M Guillemin, PO Droz, in **Methods for Biological Monitoring**, TJ Kneip and JV Crable Eds. *American Public Health Association*, Washington DC, 1988.
28 JF Periago, A Luna, A Morente, A Zambudio, *J. Appl. Toxicol.*, **12**, 91-96, 1992.
29 JF Periago, A Cardona, D Marhuenda, J Roel, et al., *Int. Arch. Occup. Environ. Health*, **65**, 275-278, 1993.
30 L Campbell, AH Jones, HK Wilson, *Am. J. Ind. Med.*, **8**, 143-153, 1985.
31 D Dyne, J Cocker, HK Wilson, *The Sci. Total Envir.*, **199**, 83-89, 1997.
32 L Drummond, R Luck, AS Afacan, HK Wilson, *Br. J. Ind. Med.*, **45**, 256-261, 1988.
33 Dragerwerk AG, *Detector tube handbook*, Lubeck, Germany, 1976.
34 NC Jaim, *J. Chromatogr. Sci.*, **12**, 214-218, 1974.
35 MF Mason and KM Dubowski, *J. Forensic Sci.*, **4**, 9-41, 1975.
36 D Clasing, U Brackmeyer and G Bohn, *Blutalcohol*, **18**, 98-102, 1981.
37 CG Hunter and D Blair, *Am. Occ. Hyg.*, **15**, 193-199, 1972.
38 R Teraniski, TR Mon, AB Robinson, P Cary, L Pauling, *Anal. Chem.*, **44**, 18-20, 1972.
39 PO Droz and JG Fernandez, *Br. J. Ind. Med.*, **35**, 35-42, 1978.
40 MX Petreas, SM Rappaport, B Materna, D Rempel, *J. Exp. Analysis & Env. Epidem., Supplm. 1*, 25-39, 1992
41 HK Wilson and TW Ottley, *Biomed. Mass Spectrom.*, **8**, 606-610, 1981.
42 FM Benoit, WR Davidson, AM Lovett, S Nacson, and A Ngo, *Anal. Chem.*, **55**, 805-807, 1983.
43 FM Benoit, WR Davidson, AM Lovett, S Nacson, and A Ngo, *Int. Arch. Occup. Environ. Health,* **55**, 113-120, 1985.
44 RJ Sherwood, and FWG Carter, The measurement of occupational exposure to benzene vapour. *Ann. Occup. Hyg.*, **13**, 125-146, 1970.
45 DA Pasquini, *Am. Ind. Hyg. Assoc. J.*, **39**, 55-62, 1978.
46 F Brugnone, L Perbellini, GL Faccini, F Pasini, L Romeo, M Gobbi, A Zedde, *Int. Arch. Occup. Environ. Health,* **61**, 303-311, 1989.
47 M Imbriani, S Ghittori, G Pezzagno, E Capodaglio, *G. Ital. Med. Lav.*, **4**, 271-278, 1982.
48 RA Glaser, and JE Arnold, *Am. Ind. Hyg. Assoc. J.*, **50**, 112-121, 1989.
49 L Wallace, R Zweidinger, M Erickson, S Cooper, D Whitaker, E Pellizzari, *Environment International,* **8**, 269-282, 1982.
50 WA Groves and ET Zellers, *Am. Ind. Hyg. Assoc. J.*, **57**, 1103-1108, 1996.
51 AC Monster, G Boersma, H Steenweg, *Int. Arch. Occup. Environ. Health*, **42**, 303-309, 1979.
52 J Fernandez, E Guberan, J Caperos, *Am. Ind. Hyg. Assoc. J.*, **37**, 143-150, 1976.
53 R Lauwerys, J Ferbrand, JP Buchet, A Bernard, J Gaussin, *Int. Arch. Occup. Environ. Health,* **52**, 69-77, 1983.
54 RD Stewart, HH Gay, DS Erley, CL Hake, AW Scaffer, *Arch. Environ. Health*, **2**, 40-46, 1961.
55 RD Stewart, ED Baretta, HC Dodd, TR Torkelson, *Arch. Environ. Health*, **20**, 224-229, 1970.
56 AC Monster, *Int. Arch. Occup. Environ. Health*, **42**, 311-317, 1979.
57 AC Monster, JM Houtkooper, *Int. Arch. Occup. Environ. Health,* **42**, 319-323, 1979.
58 E Guberan and J Fernandez, *Br. J. Ind. Med.*, **31**, 159-167, 1974.
59 RC Ward, CC Travis, DM Hetrick, ME Andersen, ML Gargas, *Tox. Appl. Pharmacol.*, **93**, 108-117, 1988.
60 RD Stewart, HC Dodd, ED Baretta, et al., *Arch. Environ. Health*, **16**, 656-662, 1968.
61 I Astrand, A Kilbom, P Ovrum, I Wahlberg, O Vesterberg, *Work-Environm.& Health*, **11**, 69-85, 1974.
62 J Engstrom, R Bjurstrom, I Astrand P Ovrum, *Scand. J. Work Environ. Health*, **4**, 315-323, 1978.
63 JC Ramsey, JD Young, RJ Karbowski, MB Chenoweth, LP McCarty, WH Braun, *Toxicol. Appl. Pharmacol.*, **53**, 54-63, 1980.
64 E Wigaeus, A Lof, R Bjustrom, MB Nordqvist, *Scand. J. Work. Environ. Health,* **9**, 479-488. 1983.

## 15.2.2 A SIMPLE TEST TO DETERMINE TOXICITY USING BACTERIA

JAMES L. BOTSFORD
**Department of Biology, New Mexico State University, Las Cruces, NM, USA**

### 15.2.2.1 Introduction

The author has developed a simple, inexpensive and rapid method to measure toxicity using a bacterial indicator. This paper describes this test in detail. It also reviews the field of alternative tests for toxicity, tests not involving viable animals.

### 15.2.2.2 Toxicity defined

In toxicology, a compound is defined to be toxic if it damages living organisms. There is no chemical definition of toxicity. Classically, toxicity is determined from the LD50 for animals, the amount of the compound that kills 50% of the test animals (Rodericks, 1992). Typically varying concentrations of the toxin are given to groups of 10 animals and a single test can take 80 to 100 animals. Tests can take several weeks. Animal tests are expensive. The animals must be cared for and trained personnel are required. The animals are force fed the toxin so it passes through the acidic stomach or the toxins are introduced by intraperitoneal injection. This can alter the activity of the toxin. The results from animal tests are difficult to interpret since animals can die of causes other than the toxic chemical and analysis of the data usually requires sophisticated statistical analysis. In the United States it is estimated that 30 million animals die each year testing for toxic chemicals. Many question the value of animal tests (Ruelius, 1987). Results vary dramatically between strains of animals. Results with closely related species, rats and mice, often don't agree well. The drug Thalidomide many years ago was tested in animals and was found to be harmless. The drug had tragic effects on humans. And, of course, animal tests do not note if the animal becomes ill from the toxin. They only note if the toxin kills the animal.

Often it is not known why the chemical is toxic. It is simply observed that animals coming in contact with the toxin die at abnormally high rates. Obviously the first step in a study of a toxic chemical, is simply to determine if it is toxic.

Most animal tests are run with rats and mice, the usual laboratory animals. Rats and mice are rodents and they have a caecum, a chamber that opens off the small intestine. The caecum contains many microorganisms. It is not known what the role of the aecum could be in handling toxic chemicals. This may influence results with rats and mice.

Alternatives to animal tests are sought. The sand flea *Daphnia magna* is used extensively. Tests are inexpensive, most the animal rights advocates are not offended. *Daphnia* tests are difficult to perform and require highly skilled personnel. The tests typically take two days (Stephenson, 1990). Once it has been determined that the toxin kills the juvenile cells, it can be determined if the toxin interferes with maturation of the juveniles, with reproduction in the adults, or if the toxin simply inhibits growth. However, nearly all reports simply deal with death of the animals. *Ceriodaphnia dubia* offers another approach to this sort of test (Jung and Bitton,1997). There is a version of this test called Ceriofast$^{TM}$ that takes only one hour. *Daphnia* testing requires laboratory skills that typically lab workers do not develop. The juvenile insects must be isolated using a dissecting microscope. The animals must be monitored to determine if they are living. It is necessary to watch the culture to know when to isolate the juveniles.

In Europe many tests with animal and human cells have been developed (Clemendson et al., 1996). These tests are effective, provide values comparable to those obtained with animals. But they require that the animal cells be grown in culture. This requires a sophisticated laboratory and well trained personnel. Again the tests take several days. Some tests take up to a week. Damage to the cells is determined in a variety of ways. The ability of cells to reduce the thiazole tetrazolium dye, MTT, is noted; the activity of an enzyme, often lactic dehydrogenase, is determined; the ability of the cell to take up some dyes or to retain other dyes is observed. Often morphological changes are noted to determine if the cells are damaged. These techniques can be quite complicated and require well trained personnel and a sophisticated, well equipped, laboratory.

There are several reports using animal cells to test the MEIC chemicals (Shrivastava et al., 1992; Rommert et al., 1994; Rouget et al., 1993). Shrivastava et al., used freshly isolated rat hepatocyte cells and two transformed cell lines. Comparable results were obtained with the three cell types indicating that rat hepatocytes are no more resistant to toxic chemicals than are other cells. They determined viability of the cells with 1) morphological studies; 2) lactic dehydrogenase activity in the hepatocytes; 3) the ability of the cells to take up the dye tryptan blue. They got much the same results with the three techniques. Rommert et al., simply determined the effect of the toxins on the ability of the cells to replicate. They determined cell numbers with a Coulter Counter after 48 hours. Rouget et al., followed the effect of the toxins by determining the ability of cells to reduce the tetrazolium dye MTT and the ability of cells to take up the dye neutral red. The authors maintained that the two methods provided comparable results.

In Europe tests with rotifers, cysts from aquatic invertebrates (Calleja et al., 1994), development of onion roots have been tried but none have been accepted (Persoone et al., 1994; Snell and Persoone, 1989; Fresjog, 1985; Gaggi et al. 1995). Jaffe (1995) has proposed a method using the survival of a protozoan. Toxic chemicals kill the protozoa. The test requires a Coulter particle counter to determine the numbers of viable protozoa.

Tests using bacteria as the indicator organism are on the market. Microtox™ uses a bioluminescent marine bacterium (Bullich et al., 1990). The bacteria, when growing in high numbers, emit light. Toxic chemicals inhibit the production of light. The test requires a luminometer to measure light production and a refrigerated water bath to grow the cells at 15°C. The results are difficult to interpret. The ability of the cells to emit light gradually decreases and a control to follow production of light in the absence of a toxin must be run. And the value for this enters into the calculations. Typically the calculations require at least a half hour to carry out (Ribo and Kaiser, 1987). A sophisticated computer program is used to analyze the results. The initial cost of the kit is significant, but it is inexpensive to use once this cost is absorbed (Elanabarwy et al., 1988). The test appears to be challenge for an MS level graduate student at our university. It is not certain why toxic chemicals inhibit light production but it is assumed that the chemicals damage the cytoplasmic membrane and reduce electron transport to the pigments responsible for light production. This method is very popular as a "first test", to see if a chemical need be investigated further. The test has been used to determine the toxicity of more than a thousand chemicals (Kaiser and Palabrica, 1991). There is a European version of the same text, Biotox™ (Kahru and Borchardt, 1994). Thomulka et al. (1995) have proposed a variation on this test.

Polytox™ uses a consortium of bacteria from sewage sludge. Toxic chemicals inhibit the oxidation of a carbon source (presumably glucose). Presumably the toxic chemicals damage the cellular membrane and the associated cytochromes involved in electron transport. Oxidation of the substrate is followed with a respirometer or with oxygen electrodes. Both these methods are complicated and expensive to set up and to maintain. Again, at our university, this technique appears to be appropriate for MS level graduate students. This test is used to determine if chemicals will harm sewage treatment plants. It is rarely used to determine the toxicity of compounds. The test provides values comparable to those observed using sewage sludge as the source of bacteria (Sun et al., 1993; Toussaint et al., 1995; Elnabarawy et al., 1988).

Toxitrak™ has been used to determine toxicity in water samples. There is a report (Gupta and Karuppiah, 1995) in which they tested water samples from a river in Maryland using this method, Microtox™, *Ceriodaphnia dubia* (a variation of the standard *Daphnia* test), and this test. With four samples, they got comparable results. The Toxitrak test is not used extensively. Apparently the test follows the reduction of dye resazurin by cells of *Bacillus cereus* (Liu, 1989).

Dozens of tests using bacteria as indicator organisms in toxicity studies have been proposed (Bitton and Dutka, 1986). Very few of these have been characterized extensively. Bacterial tests are inexpensive, inoffensive to animal rights groups, the cells are easy to grow and to prepare. And it is thought there is enough universality at the biochemical level, that a compound that is toxic for a bacterium will be toxic for a higher organism.

Ideally an alternative test for toxic chemicals would be inexpensive. It would not require any specialized equipment, only equipment normally found in a laboratory should be required. The test should not require that the personnel be specially trained in the techniques. It should involve only the procedures that any laboratory worker would know. It should be the sort of test that any laboratory with a tangential interest in toxicology could carry out. Water quality laboratories, industrial safety laboratorics, agricultural research laboratories, should be able to carry out the procedure. Small samples should be tested. Tests with fish often involve quite large volumes of toxic chemicals and this can limit their utility. This has been discussed in detail by Blaise (1991).

The sensitivity of a test can be a problem. Every test for toxic chemicals has some chemicals that the test is very sensitive to. And every test has some toxic chemicals that it is not sensitive to. Often several tests will be run and the median value for toxicity is judged to be representative for the toxicity of that chemical. It can be argued that if a test system is extremely sensitive to a chemical, than it must be concluded that this chemical could be toxic for humans at this very low level. If it damages one life form, it could very well damage another. Tests with live animals, LD50 tests, are usually not as sensitive as tests with *Daphnia*, with animal cells, with bacterial tests. Tests with live animals usually involve injecting the toxin into the stomach and it is not always certain what effect the acidic conditions of the stomach may have on the toxin. And humans rarely come in contact with toxins in this fashion. It can be argued that this sort of animal test is not indicative of the toxicity of the chemical (Ruelius, 1987). On the other hand, chemical analysis is often much more sensitive than any toxicity test. There are reports of chemicals in the range of parts per trillion yet few tests for toxic chemicals can detect less than parts per million or occasionally, parts per billion. Yet we see public policy formulated on the basis of these chemical tests.

#### 15.2.2.3 An alternative

A simple, inexpensive and rapid test to determine the toxicity of chemicals has been developed. (Botsford, 1997, 1998, 1999). The assay uses the bacterium *Rhizobium meliloti* as the indicator organism. This bacterium reduces the thiazole tetrazolium dye, MTT (3-[4,5-dimethylthiazole-2-y]-2,5-biphenyl tetrazolium bromide) readily. The dye is almost completely reduced after a 20 minute incubation at 30°C. The dye turns dark blue when reduced and this can be followed readily with a simple spectrophotometer. Reduction of the dye is inhibited by toxic compounds. More than 200 compounds have been tested and the results obtained with this test are comparable to those obtained with other tests (Botsford, 2000a).

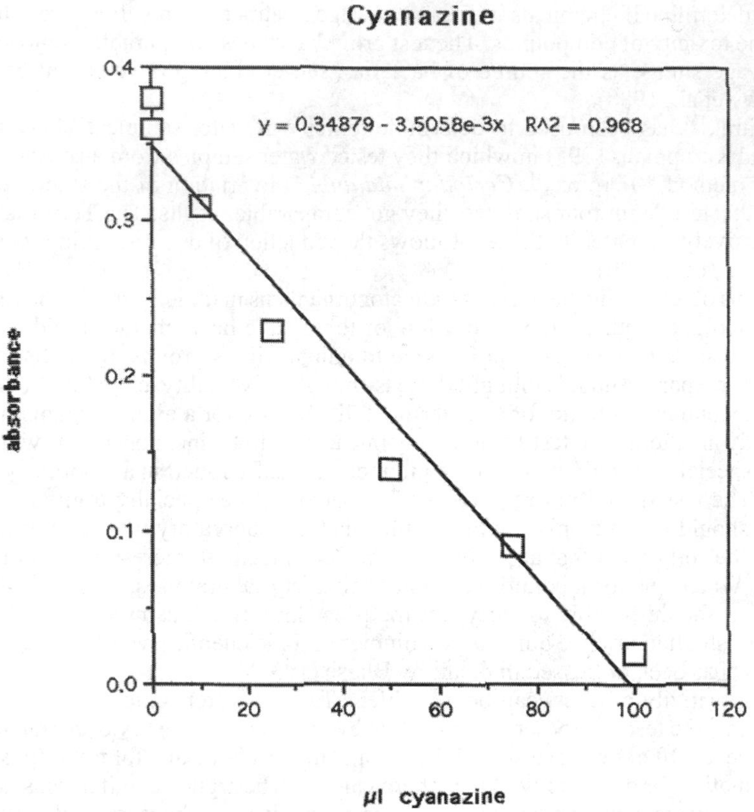


Figure 15.2.2.1. Cyanazine is in units of µl. Cyanazine is a herbicide. The toxicity of cyanazine is: IC50 = [Y/2 -B]/m; Y = 0.35, 0.37 = 0.36, 0.36/2 = 0.18; B = 0.3487 m = 3.5058x$10^{-3}$; IC50 = 48.1; 48.1 µl = ([48.1 x $10^{-6}$ l]/test tube) x([3300 mg]/l) x [1 test tube]/3.3 ml)= 48.1 µg/ml = 48.1 mg/l = 48.1 ppm.

Since alternative tests don't involve killing test animals, results are reported as IC50 values (inhibitory concentration, 50%), the concentration that results in death of 50% of the cells, in 50% inhibition of oxygen consumption, 50% inhibition of reduction of the dye in the *Rhizobium* assay.

Cells are grown overnight in a simple defined medium with 0.1% casamino acids. Cells are collected by centrifugation in a refrigerated preparative centrifuge. Cells are washed once with 0.01 M (pH 7.5) potassium phosphate buffer and are then diluted in this same buffer. Volumes of the toxin (usually 5 to 1000 µl) are put into test tubes. The volume of the toxin is made up to 2.1 ml with water. 0.1 ml of 0.1 M Tris or Bicene buffer, pH 7.5, is added. The absorbancy at 550 nm is determined. 0.1 ml MTT (the tetrazolium dye) is added (3.2 mM). The tubes are incubated 20 minutes at 30°C. The absorbance is read a second time. The difference in the absorbance at time = 0 and at time = 20 min is determined. The results are plotted, the difference in absorbance of each tube with a different volume of the toxic chemical on the X axis versus the absorbance on the Y axis (Figure 15.2.2.1). A regression line is fit to the plot and if the regression coefficient is less than 0.8, the experiment is rejected. From the average value of the controls (tubes with no toxin), the Y intercept forms the regression equation and the slope, the volume of toxin inhibits reduction of the dye 50% can be readily calculated. For some toxins, better results are obtained if the log of the volume of the toxin is plotted. From this the log of the volume corresponding to 50% inhibition of reduction can be determined and the antilog of this value provides the volume. In this laboratory the data is plotted both linearly and logarithmically and the best method providing the regression coefficient closest to 1.00 is chosen. From the concentration of the toxic compound, the volume found to inhibit reduction of the dye could be converted to mg/ml present in the test tube. From this the inhibitory concentration causing 50% inhibition, the IC50, can be calculated. The test has been described in detail (Botsford et al., 1997; Botsford 1998; Botsford, 1999).

A method to lyophilize cells has been developed and cells are active for at least a month after being lyophilized (Robertson, 1996).

This test for toxic chemicals is very inexpensive. It can be carried out by unskilled personnel. It is rapid, taking less than an hour to determine the toxicity of a compound. It provides values comparable to more conventional tests (Botsford, 2000a). It provides a sort of toxicological triage, a method to rapidly determine if something is toxic and should be investigated further. Since it uses bacteria rather than animals as an indicator, it is acceptable with animal rights advocates. The test has been patented and the author is seeking a firm to market the test. In the meantime, it can be carried out by any laboratory equipped to grow bacteria.

Since this test uses a bacterium as the indicator organism. It can be argued that the results with a bacterium are not applicable to humans. But tests with rats and mice may not be applicable to humans (Ruelius, 1987). For example, in a test of 10 chemicals comparing results with mice, rats and five cellular assays, the toxicity of three of compounds differed by at least an order magnitude when rats and mice were compared (Ekwall et al., 1989). Such anomalies been observed with many drugs, that drugs were up to 12× more toxic for humans than for rats and mice (Brodie and Reid, 1967). At a biochemical level, a compound that damages the cytoplasmic membrane of a bacterium will also damage the cytoplasmic membrane of a human or any other organism. This test has been compared with 20 other tests for toxic chemicals and comparable results (similar IC50 values) are obtained (Botsford, 2000a). No one test finds all chemicals to be toxic. Every assay for toxic chemicals is blind to some compounds and conversely every assay for toxic chemicals is much more sensitive to some compounds than are other tests. Thus toxicologists desire "batteries of tests", a compound is not considered to be toxic until it has proven to be toxic in 3 or 4 systems.

#### 15.2.2.4 Chemicals tested

In Table 15.2.2.1, the values for the more than 200 chemicals tested in this laboratory are reported. This includes a series of organic chemicals, chemicals that are reported in the literature as being toxic. It was noted that aromatics, phenol, benzene and toluene are not very toxic. But when they have substituents, a phenoxy group, chlorine, methyl or nitro they become quite toxic. Solvents were tested for toxicity and most solvents are not very toxic. We have found we can dissolve toxic compounds in solvents (methanol and DMSO) and find no change in the toxicity. Apparently the toxin is so much more toxic than the solvent that only the toxin affects the mechanism affording the assay. Many drugs were also tested and it is significant that many of these drugs are toxic for the bacterium and presumably would be toxic for patients receiving them. Often it is noted that when a drug is taken, an individual simply does not feel well. This could be due to toxicity of the drug. It was found that many alkaloids are toxic and then it was shown that many herbs are also toxic. Presumably the herbs contain toxic levels of some alkaloids. Some of these herbs are toxic enough that if mega herb therapy ever becomes popular, there could be problems from the toxicity of the herbs. It was found that divalent cations are toxic for the system. It was found that small amounts of EDTA could relieve this toxicity in soil and water samples. This will be discussed. Neither lead nor arsenic acid were found to be toxic. Finally 30 herbicides were tested for toxicity (Hillaker, 1996). The toxicity determined by this method was compared with the toxicity reported by the manufacturers using rats, ducks, and quail in classical LD50 tests; and with trout fingerlings and with *Daphnia*. The tests with trout fingerlings and with *Daphnia* were much more sensitive than the tests with animals. The tests with *Rhizobium* were less sensitive than with *Daphnia* and the trout fingerlings for most of the herbicides but were always more sensitive than tests with animals.

**Table 15.2.2.1. Toxicity of chemicals tested**

| | n | Ave. | Var. | | n | Ave. | Var. |
|---|---|---|---|---|---|---|---|
| 1,1,1-trichloroethane | 20 | 229 | 10 | pentachlorophenol | 13 | 0.143 | 50 |
| 1,1,1-tricholorethylene | 13 | 66.3 | 27 | 2-methyl resourcinol | 4 | 404 | 19 |
| 1,2-dichloroethane | 2 | 146.5 | | m-cresol (mono methyl) | 6 | 98.9 | 14 |
| 1,10-phenanthroline | 5 | 42.7 | | o-cresol | 22 | 190 | 29 |
| acetic acid | 8 | 429 | 10 | p-cresol | 4 | 386 | 15 |
| acetonitrile | 3 | >3300 | | 2,4-dinitro-o-cresol | 7 | 12.8 | 39 |
| butyl amine | 2 | 271 | | 2,6-dinitro-p-cresol | 7 | 33.5 | 35 |
| coumarin | 7 | 222 | 21 | 4,6-dinitrocresol | 5 | 6.83 | 31 |
| CTAB (detergent) | 7 | 0.90 | 36 | toluene | 4 | 217 | 8 |
| cyclohexyl amine | 9 | 234 | 5 | toluidine | 6 | 1493 | 26 |
| diethyl amine | 2 | 194 | | trihydroxytoluene | 8 | 52.4 | 34 |
| dimethyl amino ethanol | 2 | >1200 | | dinitrotoluene | 4 | 66.7 | 6 |
| dibromethane | 8 | 217 | 7 | trinitrotoluene | 8 | 52.4 | 34 |
| EDTA | 6 | 3788 | 14 | benzene | 4 | 799 | 18 |
| ethyl acetate | 11 | 1107 | 17 | benzyl chloride | 8 | 60.6 | 8 |
| ethyl benzene | 8 | 71 | 6 | chlorobenzene | 12 | 23.9 | 26 |
| ethanol amine | 8 | 251 | 16 | 1,2,4-trimethyl benzene | 14 | 37.6 | 40 |
| hexachlorophene | 7 | 0.077 | 9 | dimethyl amino benzaldehyde | 1 | 236 | |
| hydrogen peroxide | 10 | 164 | 28 | trichlorobenzene | 5 | 14.7 | 18 |
| hydroxyl amine | 13 | 566 | 20 | p-amino benzaldedhyde | 3 | 156 | 4 |
| ibenofuran | 4 | 9.61 | 20 | 1,2-dichlorobenzene | 11 | 37.7 | 5 |
| indole | 3 | 501 | 5 | 1,3-dichlorobenzene | 7 | 41.4 | 11 |
| isonicotinic acid | 5 | 287 | 18 | 1,4-dinitrobenzene | 3 | 313 | 8 |
| octane | 5 | 161 | 20 | pseudocumine | 14 | 37.6 | 40 |
| α-naphthol | 9 | 56.9 | 34 | p-dimethyl amino benzaldehyde | 7 | 177 | 5 |
| potassium ferricyanide | 5 | 3600 | 15 | p-amino benzoic acid | 3 | 291 | 10 |
| pentanol | 4 | 1627 | 22 | p-OH benzoic acid | 5 | 252 | 27 |

| | n | Ave. | Var. | | n | Ave. | Var. |
|---|---|---|---|---|---|---|---|
| phenanthroline | 5 | 43.3 | 20 | 2-OH benzoic acid (salicylic acid) | 3 | 217 | 13 |
| Tris (Trizma, pH 7.5) | 3 | 5447 | 16 | p-amino benzoic acid | 4 | 307 | 2 |
| MOPS 3-[N-morpholino] propane sulfonate | 5 | 13090 | 27 | chlorobenzoic acid | 6 | 233 | 19 |
| tetrachloroethylene | 16 | 91.2 | 21 | 4-chlorobenzoic acid | 5 | 234 | 14 |
| tetrachloromethane | 10 | 1791 | 12 | 3-phenoxybenzoic acid | 4 | 62.7 | 25 |
| salicylic acid | 4 | 198 | 20 | 3(3,5-dichlorophenoxy) benzoic acid | 5 | 41.9 | 20 |
| sulfosalicylic acid | 11 | 1107 | 17 | 3(4-chlorophenoxy) benzoic acid | 7 | 32.9 | 43 |
| xylene | 4 | 132 | 27 | 3(4-methyl phenoxy) benzoic acid | 6 | 92.1 | 9 |
| phenol | 16 | 1223 | 28 | 3(3,4-dichlorophenoxy) benzoic acid | 4 | 45.9 | 24 |
| 4-chlorophenol | 6 | 186 | 23 | **Solvents** | | | |
| 3-chlorophenol | 6 | 3.21 | 45 | acetone | 18 | 123000 | 24 |
| 3-methyl-4-nitro phenol | 4 | 54.9 | 27 | acetic acid | 8 | 429 | 10 |
| 2,4-dichlorophenol | 4 | 68.7 | 5 | coumarin | 7 | 222 | 21 |
| 3,4-dichlorophenol | 5 | 82.3 | 74 | chloroform | 10 | 1550 | 22 |
| 3,5-dichlorophenol | 5 | 174 | 37 | ethanol | 20 | 75581 | 28 |
| 2,3-dimethyl phenol | 10 | 120 | 5 | DMSO | 13 | 87629 | 31 |
| 2,4-dimethyl phenol | 11 | 71.4 | 28 | dichloroethane | 28 | 237 | 38 |
| 2,4,6-trimethyl phenol | 10 | 123 | 9 | dibromomethane | 8 | 217 | 15 |
| 3,4,5-trimethyl phenol | 6 | 49.5 | 4 | isopropyl alcohol | 3 | 57000 | 19 |
| 2,3,6-trimethyl phenol | 6 | 151 | 9 | methanol | 4 | 68172 | 13 |
| dimethyl amino methyl phenol | 10 | 413 | 24 | pentanol | 4 | 1623 | 22 |
| trichlorophenol | 45 | 8.24 | 38 | ethylene glycol | 5 | 212700 | 12 |
| o-nitrophenol | 13 | 37.6 | 33 | | | | |
| **Herbs** | | | | **Detergents** | | | |
| Aloe Vera | 4 | 219 | 4 | CTAB | 7 | 0.90 | 36 |
| Cayenne | 4 | 700 | 1 | SDS | 8 | 31.4 | 14 |
| Echinacea | 7 | 451 | 1 | Tween 20 | 5 | 2896 | 22 |
| Echinacea root | 5 | 939 | 4 | **Insecticides** | | | |
| Elderberry | 4 | 291 | 1 | Lindane | 8 | 41.1 | 28 |
| Evening Primrose | 11 | 35 | 17 | Malathion | 14 | 37.1 | 40 |
| Myrrh | 6 | 71.2 | 3 | Diazanon | 18 | 35.6 | 24 |

| | n | Ave. | Var. |
|---|---|---|---|
| **Herbs continuation** | | | |
| Gensing Drink | 9 | 285 | 9 |
| Royal Jelly | 6 | 1641 | 22 |
| Saw Palmetto | 21 | 149 | 24 |
| Chromium Picolinate | 6 | 206 | 7 |
| St. John's Wort | 20 | 192 | 12 |
| St. John's Wort | 6 | 218 | 4 |
| St. John's Wort | 6 | 639 | 17 |
| Thyme | 4 | 338 | 5 |
| Valarian Root | 12 | 208 | 13 |
| Valarian Root | 5 | 200 | 4 |
| Car's Claw, Garlic Oil, Ginko Bioloba, Ginseng, Golden Seal, and Herb Garlic were found not to be toxic (less than 2000 ppm). Herbs listed more than once were obtained from different suppliers. | | | |
| **Herbicides** | | | |
| 2,4-D | 9 | 347 | 39 |
| Alachlor | 6 | 111 | 8 |
| Bensulide | 7 | 53.7 | 18 |
| Bromoxynil | 8 | 26.0 | 19 |
| Clomazone | 15 | 23.8 | 16 |
| Cyanizine | 13 | 781 | 27 |
| DCPA | 6 | 42719 | |
| Dicamba | 6 | >1200 | |
| Diuron | 9 | 87.0 | 13 |
| EPTC | 9 | 51.4 | 6 |
| Ethalfluralin | 6 | >1200 | |
| Fuazifop-P | 6 | >1200 | |
| Glyphosate | 6 | 18.1 | 12 |
| Imazapyr | 6 | 229 | 1 |
| Imazethapyr | 10 | 35314 | |
| Isoxaben | 16 | 365 | 15 |
| Mepiquat-Cl | 6 | >1200 | |
| Metasulfuron | 6 | >1200 | |

| | n | Ave. | Var. |
|---|---|---|---|
| **Drugs** | | | |
| acetomenaphen | 9 | 249 | 20 |
| acetylsalicylic acid | 8 | 447 | 11 |
| aspirin (Bayer) | 10 | 170 | 21 |
| Amoxicilin | | | |
| Cyprus | 3 | 4412 | 37 |
| Jordan | 5 | 4501 | 51 |
| Mexico | 9 | 4257 | 9 |
| America | 4 | 7393 | 52 |
| chloramphenicol | 5 | 1400 | 24 |
| gentamycin | 5 | 0.68 | 7 |
| GEA-1 (nitroso drug) | 5 | 50.5 | 39 |
| Sin-1 (nitroso drug) | 2 | >1200 | |
| ibuprofen | 6 | 120 | 8 |
| isoproterenol | 1 | 1770 | |
| kanamycin | 5 | 0.85 | 14 |
| naproxin | 3 | 296 | 41 |
| neomycin | 8 | 0.438 | 31 |
| novobiocin | 2 | 2063 | |
| Orudis | 6 | 114 | 14 |
| orphenadrine | 6 | 472 | 32 |
| paparavine | 7 | 95.1 | 15 |
| streptomycin | 5 | 60.4 | 28 |
| Tetracycline | | | |
| America | 4 | 672 | 7 |
| Mexico | 4 | 653 | 2 |
| The national names after the antibiotics indicate the country the antibiotics originated in. | | | |
| **Inhibitors of electron transport** | | | |
| dinitrophenol | 4 | 34.3 | 17 |
| FCCP | 19 | 0.20 | 36 |
| potassium cyanide | 17 | 12.2 | 48 |
| sodium azide | 17 | 630 | 19 |

| | n | Ave. | Var. | | n | Ave. | Var. |
|---|---|---|---|---|---|---|---|
| **Herbicides continuation** | | | | **Alkaloids** | | | |
| Metribuzin | 6 | >1200 | | atropine | 9 | 191 | 11 |
| Naproamide | 10 | 289 | 43 | caffeine | 3 | 3700 | 18 |
| Nicosulfuron | 10 | 267 | 11 | immitine | 11 | 122 | 8 |
| Norflurazon | 8 | 182 | 14 | nicotine | 13 | 990 | 13 |
| Oxadiazon | 6 | 269 | 16 | quinine | 4 | 131 | 14 |
| Quniclorac | 6 | >1200 | | qunidine | 4 | 137 | 15 |
| Sethoxydim | 6 | 2.70 | 17 | scopolomine | 5 | 129 | 5 |
| Thiazopyr | 6 | 43.2 | 4 | | | | |
| Thifensulfuron | 7 | 928 | 4 | | | | |
| Trifluralin | 7 | 10.3 | 18 | | | | |

n = the number of samples tests; ave. = the average of the values obtained, reported as part per million (ppm), mg per liter; var. = the variance in the results, the standard deviation divided by the mean.

### 15.2.2.5 Comparisons with other tests

The *Rhizobium* test has been compared with other tests. This is done by finding values in the literature for various toxins using different tests and plotting the log of these values vs. the log of values from the second assay. A regression line was plotted and the correlation coefficient calculated from the regression coefficient (the correlation coefficient is the square root of the regression coefficient). In Figure 15.2.2.2, values for chemicals from two laboratories

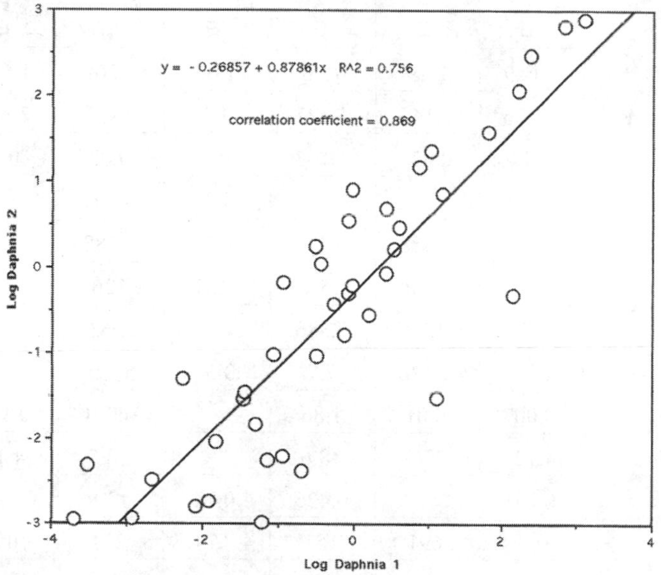


Figure 15.2.2.2. Representative plot of the comparison of two assays. These are data from *Daphnia* 1 (Calleja et al., 1993) and *Daphnia* 2 (Lilius et al.,1994).

(Calleja et al., 1993; Lilius et al., 1994) using the assay involving the sand flea *Daphnia magna* are presented. If two systems give the same results, the correlation coefficient will be 1.00. In the literature there are reports of the toxicity of 50 MEIC chemicals (MEIC, multicenter evaluation of cytotoxicity) tests used in Europe to evaluate different tests for toxicity. When these two tests with *Daphnia* were compared, a correlation coefficient of 0.895 was obtained (Figure 15.2.2.2). This indicates that when a standardized test is run in two laboratories with different personnel, the results are not identical. This discrepancy is noted in tests involving animals. Apparently variations in the strains of animals, differences in care and other factors influence the results of toxicity tests with animals. Often variations of 10x are observed.

This sort of analysis was carried using the *Rhizobium* assay. Published values for the chemicals using Microtox, LD50 reports for rats, IC50 (testing with animal cells), *Daphnia magna*, and HLD (Human Lethal Dose determined from autopsy reports) are included (Table 15.2.2.2). These values for the assay with *Rhizobium* will permit comparisons to be made (Table 15.2.2.3). It should be noted that the values for rats and HLD are much higher than with the other methods. With animal tests, the animals are force fed the chemical, it is injected through a tube into their stomach. Occasionally the toxin is injected intraperitoneally. It is uncertain what happens to the toxic chemical in the acidic stomach. This methodology has been criticized because humans are rarely exposed to toxic chemicals in this fashion. HLD data comes from autopsy reports and it can never be certain how much of the chemical the victim actually came in contact with, it can only be estimated. And this method only provides data for chemicals already in the environment, it is not a good method to predict toxicity.

**Table 15.2.2.2. MEIC chemicals tested. Comparison**

| Chemical tested | Rhizobium, m moles | Microtox, m moles | Rat, m moles | IC50, m moles | Daphnia, m moles | HLD, m moles | rat hepat, m moles |
|---|---|---|---|---|---|---|---|
| acetomenaphen | 1.649 | 2.19 | 15.8 | 1.45 | 0.269 | 1.698 | 10.75 |
| acetylsalicylic acid | 1.106 | 0.145 | 1.122 | 1.7 | 0.932 | 2.818 | 2.662 |
| amitriptyline | 0.0146 | 0.078 | 1.148 | 0.1 | 0.02 | 0.2 | 0.07 |
| barium chloride | 0.109 | | 1.349 | | 0.794 | | 0.47 |
| caffeine | 1.02 | 3.388 | 1 | 1.58 | 3.388 | 1 | 1.596 |
| carbon tetrachloride | 1.91 | 4.786 | 15.1 | 22.1 | 126 | 0.447 | 3.984 |
| chloroform | 5.29 | 12.9 | 7.586 | | 2.63 | 1.738 | 6.198 |
| chlororamphenicol | 4.332 | 1.122 | 7.7 | 0.54 | 5.248 | | 0.402 |
| copper II sulfate | 0.007 | 0.01 | 1.86 | | 0.001 | 0.316 | 0.048 |
| dichloromethane | 4.122 | 37.1 | 18.6 | | 10.5 | 4.17 | 109.1 |
| digoxin | 0.231 | | 0.426 | 0.12 | 12.8 | 0.0014 | 0.271 |
| ethanol | 1643 | 691 | 151 | 158 | 234 | 109.6 | 451 |
| ethylene glycol* | 3427 | 1778 | 75.8 | 322 | 1202 | 25.7 | 358 |

| Chemical tested | Rhizobium, m moles | Microtox, m moles | Rat, m moles | IC50, m moles | Daphnia, m moles | HLD, m moles | rat hepat, m moles |
|---|---|---|---|---|---|---|---|
| hexachlorophene | 0.0002 | 0.019 | 0.138 | 0.0068 | 0.00027 | | 0.002 |
| isopropyl alcohol | 950 | 380 | 83.2 | 90.5 | 155 | 41.7 | 304 |
| iron II sulfate | 0.32 | 0.782 | 2.089 | | 0.302 | 3.020 | 1.621 |
| lindane | 0.141 | 21.9 | 0.263 | 0.36 | 0.0056 | | 0.144 |
| malithion | 0.112 | 0.479 | 0.871 | | | 0.741 | |
| mercury II chloride | 0.00006 | 0.0002 | 3.715 | | 0.00013 | 0.107 | 0.003 |
| methanol | 2130 | 912 | 178 | 673 | 661 | 52.5 | 906 |
| nicotine | 6.1 | 0.224 | 0.309 | 4.52 | 0.0.23 | 0.11 | 3.581 |
| orphenadrine HCI* | 1.168 | 0.005 | 0.832 | | 0.033 | 0.098 | 0.114 |
| paraquat | 0.262 | 2.344 | 0.224 | | 0.1 | 0.166 | 1.176 |
| pentachlorophenol | 0.00054 | 0.02 | 0.1 | 0.025 | 0.0022 | 0.098 | 0.05 |
| phenol | 13 | 0.186 | 3.39 | 4.25 | 0.077 | 1.48 | 0.797 |
| potassium chloride | 290 | 493 | 34.6 | | 490 | 0.288 | 92.22 |
| potassium cyanide | 0.225 | 0.275 | 0.0776 | 1 | 0.0141 | 0.437 | 0.783 |
| quinidine sulfate | 0.422 | 1.202 | 0.617 | 0.036 | 0.0797 | | 0.129 |
| sodium chloride | 288 | 562 | 51.3 | | 60.3 | 17 | 102 |
| sodium oxalate | >13 | 5.428 | | | | | 0.582 |
| theophylline | 1.926 | 13.8 | 1.349 | | 2.63 | 0.724 | 2.175 |
| 1,1,1-trichloroethane | 0.583 | 0.342 | 77.6 | | 6.71 | 1.48 | |
| warfarin | 1.781 | 0.209 | 0.005 | 0.67 | 1.549 | 0.0219 | 0.139 |
| xylene | 1.24 | 0.079 | 34.6 | | 0.851 | 0.468 | 17.47 |

$M_w$ = molecular weight of the compound; n = number of times compound was tested; var = variation in the results, the standard deviation /mean\ toxicity reported as mmoles toxin for the IC50 for the test. The rat test is an LD50. Rhizobium data are from this work. Microtox$^{TM}$, HLD, and Daphnia data are from Calleja, (1993). The IC50 data (pooled data from animal cell tests) is from Halle et al., 1993. Rat hepatocyte data is from Shrivastava (1992). The IC50 samples include only 20 chemicals. The other methods involve about 34 chemicals.

In Table 15.2.2.3 the results of comparisons among these tests are summarized. The data from these 6 determinations were plotted, one assay versus another assay, the regression coefficient noted and the correlation coefficient calculated. The *Rhizobium* assay was the most sensitive for 12 of 33 chemicals. The Microtox assay was most sensitive for 6 of the compounds. The IC50 assay was most sensitive for 6 of the compounds. The *Daphnia* test was most sensitive to 10 of the compounds. The average values for the toxicity of the chemicals was lowest for the IC50, but then it was also the test with the fewest values included.

**Table 15.2.2.3. Comparisons for involving *Rhizobium*, Microtox™, HLD, IC50, *Daphnia magna* and rats**

| | n | $r^2$ | | n | $r^2$ |
|---|---|---|---|---|---|
| Rhizobium x Microtox™ | 34 | 0.875 | Rats x IC50 | 19 | 0.724 |
| Rhizobium x rats | 34 | 0.603 | Rats x HLD | 27 | 0.817 |
| Rhizobium x IC50 | 20 | 0.906 | Rats x Daphnia | 32 | 0.685 |
| Rhizobium x HLD | 34 | 0.628 | Rats x rat hepatocytes | 34 | 0.748 |
| Rhizobium x Daphnia | 34 | 0.888 | IC50 x HLD | 16 | 0.732 |
| Rhizobium x rat hepatocytes | 34 | 0.764 | IC50 x Daphnia | 29 | 0.879 |
| Microtox x rats | 34 | 0.571 | IC50 x rat hepatocytes | 20 | 0.700 |
| Microtox x IC50 | 24 | 0.840 | HLD x rat hepatocytes | 26 | 0.648 |
| Microtox x HLD | 31 | 0.728 | HLD x Daphnia | 29 | 0.702 |
| Microtox x rat hepatocytes | 33 | 0.700 | Daphnia x rat hepatocytes | 33 | 0.814 |
| Microtox x Daphnia | 34 | 0.846 | | | |

n = number of samples in the test. One test was compared with another, the log of the values plotted, a regression line fit and the correlation coefficient determined from the regression coefficient, $r^2$.
The data for rats, Daphnia, HLD is from Calleja et al., 1994. The data for IC50 is from Halle et al., 1992. The data for rat hepatocytes is from Shrivastava et al., 1992.

In Table 15.2.2.4, the results of these comparisons with the results of the *Rhizobium* test are presented. The *Rhizobium* test correlates well with Microtox™, *Daphnia*, and particularly well with the 20 samples in the IC50 test. The comparison of the *Rhizobium* assays with other published results has been examined in detail (Botsford, 2000a). The *Rhizobium* test has been compared with values from several laboratories for the Microtox™ assay and correlation coefficients have varied from 0.750 to 0.893 indicating the two methods provide comparable results. Four comparisons with values for *Daphnia* have provided correlation coefficients from 0.776 to 0.953 indicating that the *Rhizobium* assay agrees well with *Daphnia*.

QSAR provides a method to predict the toxicity of a compound from the structure of the compound and the water/octanol partition coefficient (Nirmilkhandan and Speece, 1988). This method has been compared with assays using sewage sludge and Polytox™ (Sun et al., 1993). The method compares well with both these other techniques. An examination of the data provided by tests for toxic chemicals using these techniques suggested that they are not as sensitive as more direct methods for determining toxicity. It was found that Microtox™, rat hepatocytes, the *Rhizobium* assay, and *Daphnia* all provided more sensitivity to toxic chemicals than did the QSAR estimates (Botsford, 2000a). The IC50 values for these other techniques were lower than those computed using the QSAR method. QSAR data are not often found in the literature.

**Table 15.2.2.4. The *Rhizobium* assay compared with other assay methods**

| | n | cc | Reference |
|---|---|---|---|
| 3T3 cells x Rhizobium | 9 | 0.908 | Ekwall et al., (1989) |
| Asictes cells x Rhizobium | 34 | 0.870 | Romert et al., (1994) |
| Biotox™ x Rhizobium | 24 | 0.928 | Kahru and Bordchardt (1994) |
| B. subtilis x Rhizobium | 20 | 0.619 | Kherzmann (1993) |
| Daphnia 1 x Rhizobium | 35 | 0.891 | Calleja et al., (1993) |
| Daphnia 2 x Rhizobium | 35 | 0.897 | Lilius et al., (1994) |
| pooled Daphnia x Rhizobium | 14 | 0.775 | Calleja, Liilius, Munkitrick |
| E. coli x Rhizobium | 20 | 0.641 | Kherzman (1993) |
| fathead minnow x Rhizobium | 15 | 0.942 | Munkitrick et al., (1991) |
| guppies x Rhizobium | 9 | 0.950 | Konemann (1981) |
| HepG2 cells x Rhizobium | 9 | 0.892 | Ekwali et al., (1989) |
| Microtox™ x Rhizobum | 35 | 0.803 | Kaiser and Palabrica (1991) |
| Microtox™ x Rhizobium | 15 | 0.758 | Munkittrick et al., (1991) |
| mice x Rhizobium | 35 | 0.651 | Calleja et al., (1991) |
| Polytox™ 1 x Rhizobium | 16 | 0.903 | Sun et al., (1993) |
| Polytox™ 2 x Rhizobium | 15 | 0.796 | Elanabarwy et al., (1988) |
| sludge x Rhizobium | 16 | 0.853 | Sun et al., (1993) |
| trout hepatocytes x Rhizobium | 35 | 0.760 | Lilius et al. (1994) |
| trout figerlingx x Rhizobium | 15 | 0.910 | Munkittrick et al., (1991) |
| QSAR x Rhizobium | 19 | 0.779 | Sun (1994) |
| pooled minnow x Rhizobium | 32 | 0.808 | Geiger et al., (1991), Munkitrick et al., (1991) |

n = number of samples in the comparison; cc = the correlation coefficient, the square root of the regression coefficient.

### 15.2.2.6 Toxic herbicides

When the toxic herbicides were studied, it was found that the animal tests supplied by the manufacturers indicated that most of the herbicides were not toxic. The tests run by the manufacturers with trout fingerlings and with *Daphnia* indicated that most of the herbicides were quite toxic. The *Rhizobium* work also showed that most of the herbicides were toxic at levels lower than 1000 ppm. The trout fingerlings showed all but one of the herbicides was toxic at this level and *Daphnia* indicated that all but 2 were toxic (Hillaker, 1998). This shows the necessity of running "batteries of tests," multiple tests with a compound. Every method of determining toxicity has some chemicals it cannot detect as toxic and some chemicals that are detected at very low levels. For example, the *Rhizobium* assay does not detect phenol as being very toxic but detects pentachlorophenol as being extremely toxic, the test is at least an order of magnitude more sensitive to pentachlorophenol than any other test found.

### 15.2.2.7 Toxicity of divalent cations

Reduction of the dye is inhibited by divalent cations in the *Rhizobium* system. Common ions, calcium and magnesium, inhibit the reduction of the dye. The toxicity of the ions is shown in Table 15.2.2.5. Mercury and cadmium, generally thought to be the most toxic minerals were the most toxic with this assay. Calcium and magnesium are also toxic. Water and soil samples typically contain calcium and magnesium so in order to analyze water and soil samples for toxic organic chemicals, a method to eliminate this inhibition by metal ions was sought.

**Table 15.2.2.5. Toxicity of divalent cations**

| Minerals | n | var. | m moles | Minerals | n | var. | m moles |
|---|---|---|---|---|---|---|---|
| $Ba^{+2}$ | 6 | 33 | 0.109 | $Mg^{+2}$ | 9 | 20 | 0.404 |
| $Cd^{+2}$ | 12 | 41 | 0.004 | $Mn^{+2}$ | 11 | 35 | 0.045 |
| $Ca^{+2}$ | 11 | 33 | 0.05 | $Hg^{+2}$ | 10 | 23 | 0.0006 |
| $Co^{+2}$ | 11 | 46 | 0.009 | $Ni^{+2}$ | 11 | 11 | 0.452 |
| $Cu^{+2}$ | 13 | 19 | 0.007 | $Se^{+2}$ | 9 | 39 | 1.849 |
| $Fe^{+2}$ | 7 | 18 | 0.587 | $Zn^{+2}$ | 10 | 7 | 0.062 |
| $Fe^{+3}$ | 14 | 27 | 0.098 | | | | |

n = number of samples tested; var. = variation, standard deviation divided by the mean. Values reported as m moles $l^{-1}$. All minerals except ferrous ion as chloride salts. Several were tested as both chlorides and sulfates with little difference. When ferrous sulfate was tested, it was made up fresh each day before the assay.

EDTA is used routinely in biochemistry to chelate divalent metal ions. It was thought this might chelate the calcium and remove it from the system. EDDA and EGTA are also used and are thought to chelate calcium more effectively than EDTA. Neither of these chelators affected the reaction, both were simply slightly toxic (about 3000 ppm). A series of experiments were run and it was found that the inhibition of reduction caused by 1.4 to 1.6 μmoles of calcium was relieved by 1 μmoles EDTA. There is not a stoichiometric relationship between EDTA and the metal ion. It is not simply chelating the metal ion.

It was found that 2.5 μmoles EDTA would eliminate the toxicity of all the ions at their IC50, the concentration of cation that inhibited reduction of the dye 50%. Thus 2.5 μmoles EDTA would eliminate toxicity from 5.5 μmoles calcium but only 0.006 μmoles mercury. It was observed initially that the toxicity of most organic chemicals could be determined in the presence of 2.5 μmoles EDTA. This suggested there could be two mechanisms involved in the reduction of MTT. One is inhibited by toxic organic chemicals and the second is inhibited by divalent cations.

### 15.2.2.8 Toxicity of organics in the presence of EDTA

A series of experiments were run looking at the toxicity of organic chemicals in the presence of 0.74 μmoles calcium (25 ppm, this concentration inhibits reduction of the dye completely) and 2.5 μmoles EDTA. Four of the 35 chemicals tested had greater toxicity with EDTA and calcium than in the controls. Two chemicals were no longer toxic. The toxicity of 16 of the chemicals was not affected by the calcium and EDTA. The toxicity of 13 chemicals was decreased by at least 10% but was not eliminated by the addition of EDTA and calcium. The toxicity of 4 chemicals was enhanced, was greater, when the calcium and EDTA were present.

The addition of calcium and EDTA at these concentrations had no effect on the apparent toxicity of: 1,4-dinitrobenzene, the herbicide 2,4-D, 2,4-dinitrotoluene, 4-chloro-benzoate, carbon tetrachloride, chloroform, cynazine, hexachlorophene, isonicotinic acid the insecticide Lindane, o-nitrophenol, p-toluidine, the antibiotic Streptomycin, tetrachloroethylene and trichlorophenol.

The addition of calcium and EDTA reduced the toxicity at least 10% of 2,6-dinitrocresol, 2,4-dinitrophenol, 2,6-dinitrophenol, 2,6-dinitrotoluene, 2-methyl resourcinol, 3-phenoxybenzoate, 2,4-dinitrocresol, the detergent CTAB, the antibiotic Neomycin, p-hydroxybenzoate, pentachlorophenol, salicylic acid, and trichloroethylene.

p-amino benzoic acid and p-hydroxy benzoate were no longer toxic with these levels of calcium and EDTA.

3-chlorobenzoate, 3-methyl-4-nitro-phenol, and the detergents sodium lauryl sulfate and Tween 80 had increased toxicity with the addition of calcium and EDTA. No correlation between the structure of the chemical and the effect of EDTA and calcium could be ascertained. For example, streptomycin and neomycin are both amino glycoside antibiotics. Calcium and EDTA did not affect the toxicity of streptomycin but nearly eliminated the toxicity of neomycin.

Several chemicals were tested with 25 μmoles EDTA and 7.4 μmoles of calcium, 10 times the amount used in the experiments reported. The results of this experiment are shown in Table 15.2.2.6. In the presence of high levels of EDTA and calcium, isonicotinic acid is no longer toxic. Dinitrophenol and 3-phenoxy benzoate, in the presence of low levels of EDTA and calcium were not as toxic, in the presence of high levels, it had the same toxicity. In presence of low levels of EDTA and calcium, Streptomycin was comparably toxic, the calcium and EDTA did not affect the toxicity. In the presence of high levels, Streptomycin was extremely toxic. Pentachlorophenol, a common soil contaminant, had reduced toxicity with low levels of EDTA and calcium, had elevated levels of toxicity with high levels of EDTA and calcium. This work shows that the effect of EDTA and specific toxins must be worked out before any conclusions as to the toxicity of the compound in the presence of EDTA and calcium can be established.

Calcium and magnesium are commonly found in water. Obviously if this assay is to be used with water samples, EDTA must be added. It must be determined using water uncontaminated with organic toxins how much EDTA must be used to compensate for the divalent cations. Often the concentration of divalent cations is determined by atomic absorption spectroscopy. However, these values do not agree with the toxicity relieved by EDTA. Soil samples with as much as 5 gm calcium (45 mM) per kg soil have been assayed using 2.5 μmoles EDTA in each sample (Hillaker, 1996). The calcium is complexed with sulfate and phosphate ions and the calcium is not available to the cell, is not seen by the mechanism that reduces the dye. Levels of soluble calcium and magnesium in water are very low. We have found that 2.5 μmoles of EDTA relieves the inhibition caused by divalent cations in all water and soil samples tested thus far (Botsford, 2000b).

**Table 15.2.2.6. Effect of high concentrations of EDTR and calcium on toxicity (Inhibition in %)**

| Compound | 2.5 μmole EDTA, 0.74 μmole Calcium | | 25 μmole EDTA, 7.4 μmole Calcium | |
|---|---|---|---|---|
| | Control | + EDTA, + Ca | Control | + EDTA, + Ca |
| isonicotinic acid | 27.0 | 25.4 | 30.9 | 105 |
| 2,4-dinitrophenol | 44.7 | 63.0 | 58.2 | 50.0 |
| Streptomycin | 55.0 | 48.9 | 47.2 | 0.00 |
| Neomycin | 0.037 | 62.4 | 77.2 | 77.3 |
| pentachlorophenol | 48.0 | 63.7 | 26.0 | 13.3 |
| 3-phenoxybenzoate | 54.3 | 74.0 | 34.5 | 37.8 |

### 15.2.2.9 Mechanism for reduction of the dye

It is not known how the dye is reduced. It is not known why toxic chemicals inhibit the reduction. It is thought that tetrazolium dyes are reduced by cytochromes (Altman, 1975). But this has been questioned (Marshall et al., 1993). In eucaryotic cells, all the cytochromes are in the mitochondria. Marshall's group has found that the dyes are reduced in preparations from cells with the mitochondria removed. It has been found that a mutant of *Escherichia coli* lacking one of two major cytochromes found in this bacterium is unable to reduce MTT (Botsford, unpublished). But in *E. coli*, reduction of the dye is not inhibited by toxic chemicals. This may not be an analogous situation. The dye could be reduced by a different mechanism.

*R. meliloti*, like most other bacteria, has many reductases. Some of these are membrane associated and damage to the membrane could affect the reductase. One of these reductases could be responsible for reduction of the dye. It has been found that the MTT is transported into the cell before it is reduced. The reduced dye is inside the cells. Cells with the dye can be concentrated by centrifugation, the dye appears in the cell pellet. None of the dye is in the supernatant. Toxic chemicals could interfere with the transport of dye into the cell prior to reduction.

Transposon insertion mutants unable to reduce the dye have been obtained and five mutants have been isolated. All grow very slowly in minimal media supplement with 0.1 % casamino acids and obviously all have lost a critical function. With these mutants it should be possible to clone and then to sequence the function responsible for reduction of the dye. From the sequence, the nature of the function can be determined.

In our studies comparing the *Rhizobium* assay with other assays, it was observed that tests using viable animals were almost always less sensitive to toxins. Tests using *Daphnia*, the various animal cell tests and Microtox™ and the *Rhizobium* test seemed to be most sensitive. Tests using fish (fathead minnow, trout fingerlings) give results comparable to tests with the bacterial indicators. Polytox™ and QSAR were less sensitive, had higher IC50 values, but were more sensitive than the tests using viable animals. Were the author asked to recommend a test procedure to indicate if a chemical were toxic, the author would recommend an animal cell test, probably using freshly isolated rat liver hepatocytes, the *Rhizobium* test and the Microtox™ test. Tests with freshly isolated rat hepatocytes would not require that the cells be grown in a laboratory situation and this would be much simpler. These three procedures are much more sensitive than tests involved live animals. These three tests would be simpler than tests with *Daphnia*. This would provide a "battery of tests", would not offend animal rights advocates and should indicate if the chemical is dangerous. All could be performed by personnel with chemical laboratory skills.

### 15.2.2.10 Summary

This work shows that the *Rhizobium* test provides results comparable to other tests. Tests seem particularly comparable to work with *Daphnia magna* and with results from *in vitro* tests with animal cells. The test is simple, unskilled laboratory workers can master it quickly. The test is inexpensive, no specialized equipment is required, given cells, any laboratory able to carry out simple chemical analysis should be able to perform the assay. The test is rapid, a sample can be tested and analyzed in an hour, the test does not take several days. It offers an ideal first test for toxic chemicals (Blaise, 1991)

### References


Altman, F. P. 1976. Tetrazolium salts and formazans. *Progress is Histochemistry and Cytochemistry* **9**:6-52.

Bitton, G., Dutka, B. J. (1986) **Toxicity testing using microorganisms**. *CRC Press Inc*. Boca Raton, Florida 163 pp.

Blaise, C. 1991. Microbiotests in aquatic toxicology. *Environmental Toxicology and Water Quality*. **6**:145-151.

Botsford, J. L., Rivera, J., Navarez, J., Riley, R., Wright T., Baker, R. 1997. Assay for toxic chemicals using bacteria. *Bulletin of Environmental Contamination and Toxicology* **59**:1000-1008.

Botsford, J. L. 1998. A simple assay for toxic chemicals using a bacterial indicator. *World Journal of Microbiology and Biotechnology*. **14**:369-376.

Botsford, J. L. 1999. A simple method for determining toxicity of chemicals using a bacterial indicator organism. *Environmental Toxicology* **99**:285-290.

Botsford, J. L. 2000a. A comparison of alternative tests for toxic chemicals. To be submitted ATLA journal.

Botsford, J. L. 2000b. Role of EDTA in a simple method for determining toxicity using a bacterial indicator organism. World Journal Microbiology and Biotechnology, in press.

Brodie, B., Reid, W.D., 1967. Some pharmacological consequences of species variation in rates of metabolism. *Federation Proceedings* **26**:1062-1070.

Bullich, A. A., Tung, K-K, Scheiber, G. 1990. The luminescent bacteria toxicity test: Its potential as an in vitro alternative. *Journal of Bioluminesence and Chemiluminescenmce* **5**:71-77.

Calleja, M. C., Persoone, G., Geiadi, P. (1994) Comparative acute toxicity of the first 50 multicenter evaluation of in vitro cytotoxicity chemicals to aquatic nonvertebrates. *Archives Enviornmental Contamination and Toxicology* **26**:69-78.

Calleja, M. C., Persoone, G., Gelandi P. 1993. The predictive potential of a batter of exotoxilogical tests for human acute toxicity, as evaluated with the first 50 MEIC chemicals. *ATLA* **21**:330-349.

Clemendson, C., McFarlane-Abdulla, E., Andersson, M., Barile, FA., Calleja, M.G., Chesne, C., Clotheir, R., Cottin, M. Curren, R., Dierickx, P., Ferro, M., Fiskejo G, Garza-Ocanas, L., Gomez-Lecon, M.J., Golden, M., Isomaa, B, Janus, J., Judge, P., Kahru, A., Kemp, R.B., Kerszman, G., Kristen, U. Kunimoto, M., Kaarenlapi, S., Lavrijsen, K., Lewan, L., Lilius, H., Malmsten, A., Ohno, T., Persoone, G., Pettersson, R., Roguet, R., Romert, L., Sandberg, M., Sawyer, T.W., Seibert, H., Shrivastava, R., Sjostrom, Stammati, A., Tanaka, N., Torres-Alanis, O., Voss, J-U. Wakuri,S., Walum, E., Wang, X., Zucco, F., Ekwall, B. (1996). MEIC evaluation of acute systemic toxicity. *ATLA* **24**:273-311.

Ekwall, B, Bondesson, 1, Catell, J.V., Gomez-Lechon, M. J., Heiberg, S., Hogberg, J. Jover, R., Ponsoda, X., Rommert, L., Stenberg, KL., Walum, E. (1989) Cytoxocity evaluation of the first ten MEIC chemicals: Acute lethal toxicity in man predicted by cytotoxicity in five cellular assays and by oral LD50 tests in rodents. *ATLA* **17**:83-100.

Ekwall, B., Johansson, A. 1980. Preliminary studies on the validity of in vitro measurements of drug toxicity using HeLa cells I. Comparative in vitro cytotoxicity of 27 drugs. *Toxicology Letters* **5**:299-307.

Einabarawy, M. T., Robideau, R. R., Beach, S. A. (1988) Comparison of three rapid toxicity test procedures: Microtox$^{TM}$, Polytox$^{TM}$ and activate sludge respiration inhibition. *Toxicity Assessment* **3**:361-370.

Fresjog, G. 1985. The allium test as a standard in environmental monitoring. *Hereditas*, **102**:99-112.

Gaggi, C., Sbrilli G., A.M. Hasab El Naby, Bucci, M., Duccini, M., and Bacci, E. 1994. Toxicity and hazard raking of S-triazine herbicides using Microtox, 2 green algal systems and a marine crustacean. *Environmental Toxicology and Chemistry* **14**:1065-1069.

Geiger, D. L., Brooke, L. T., Call, D. J. editors (1990) Acute toxicities of organic chemicals to fathead minnows (Pimphates promeias). Center for Lake Superior Environmental Studies. University of Wisconsin, Superior, Wisconsin. 900 pp.
Gupta, G., Karuppiah, M. 1996. Toxicity identification of Pocomoke River porewater. *Chemosphere* **33**:939-960.
Halle, W., Baeger, I., Ekwall, B., Spielmann, H. 1991. Correlation between in vitro cytotoxicity and octanol/water partition coefficient of 29 substances from the MEIC program. *ATLA* **19**:338-343.
Hillaker, T. L. 1996. An assay for toxic chemicals using Rhizobium method as the indicator: Use of this test with agricultural herbicides. MS Thesis, Biology. New Mexico State University.
Jaffe, R. L., 1995. Rapid assay of cytotoxicity, using Teratamitus flagellates. *Toxicology and Industrial Health* **11**:543-553.
Jung, K., Bitton, G. 1997. Use of Ceriofast™ for monitoring the toxicity of industrial effluents: Comparison with the 48-H acute Ceriodaphnia toxicity test and Microtox™. *Experimental Toxicology and Chemistry* **16**:2264-2267.
Kahru, A., Borchardt, B. 1994. Toxicity of 39 MEIC chemicals to Bioluminescent photobacteria (The Biotox™ test): Correlation with other test systems. *ATLA* **22**:147160.
Kaiser, K.L.E., Palabrica, V. S., 1991. Photobacterium phosphoreum toxicity data index. *Water Pollution Research Journal of Canada*. **26**:361-431.
Kerszman, G. 1993. Toxicity of the first ten MEIC chemicals to bacteria. *ATLA* **21**:151155.
Kerzman, G. 1993. Of bacteria and men: Toxicity of 30 MEIC chemicals to bacteria and humans. *ATLA* **21**:233-238.
Konemann, H, 1981, Quantitative structure-activity relationship in fish toxicity studies. *Toxiocology* **19**:209-221.
Lilius, H., Isomaa, B., Holstrom, T. A comparison of the toxicity of 50 reference chemicals to freshly isolated rainbow trout hepatocytes and Daphnia magna. *Aquatic Toxicology* **30**:47-60.
Liu, D. 1989. A rapid and simple biochemical test for direct determination of chemical toxicity. *Toxicity Assessment* **4**:389-404.
Marshall, N. J., Goodwin, C. J., Holt, S. J. 1995. A critical assessment of the use of microculture tetrazolium assays to measure cell growth and function. *Growth Regulation* **5**:69-84.
Mossman, T . 1983. Rapid colorimetric assay for cellular growth and survival: Application to proliferation and cytotoxicity assays. *Journal of Immunological Methods* **65**:55-63.
Munkittrick K. R., Power, E. A., Sergy, G. A. 1991. The relative sensitivity of Microtox™, Daphnia, rainbow trout, fathead minnow acute lethality tests. *Environmental Toxicology and Water Qualtity* **6**:35-62.
Nirmalakhandan, N. N., Speece, R. E. 1988. Prediction of aqueous solubility of organic chemicals based on molecular structure. *Environmnental Science and Technology*. **22**:328-338.
Robertson, B. 1996. Developing a technique to lyophilize Rhizobium mefloti MS thesis, Biology, New Mexico State University.
Rodericks, J. V.1992. **Calculated Risks**. Cambridge, *Cabridge University Press* 256 pp
Romert, L., Jansson, T. Jenssen, D. 1994. The cytotoxicity of 50 chemicals from the NEIC study determined by growth inhibition of Ascites Sarcoma BP8 cells: A comparison with acute toxicity data in man and rodents. *Toxicology Letters* **71**:39-46.
Rouguet, R., Cotovia, J. Gaetani, Q., Dossou K. G. Rougier, A. 1993. Cytotoxicity of 28 MEIC chemicals to rat hepatocytes using two viability endpoints: correlation with acute toxicity data in rat and man. *ATLA* **1**:216-224.
Ruelius, H. W. 1987. Extrapolation from animals to man: predictions, pitfalls and perspectives. *Xenobiotica* **17**:255-265.
Shrivastava, R., Deiominie, C., Chevalier, A., John, G., Ekwall, B., Walum, E. Massingham, R. Comparison of in vitro acute lethal potency and in vitro cytotoxicity of 48 chemicals. *Cell Biology and Toxicology* **8**:157-167.
Snell, T. W., Personne, G. 1989. Acute bioassays using rotifers. II. A freshwater test with Brachionus rubens. *Aquatic Toxicology* **14**:81-92.
Stephenson, G. L., Kausik, N. K., Solomon, K. R. 1991. Chronic toxicity of a pure and technical grade pentachlorophenol to Daphnia magna. *Archives Environmental Contamination and Toxicology*. **21**:388-394
Sun, B., Nimalakhandan, N., Hall, E., Wang, X. H., Prakash, J., Maynes, R. (1994) Estimating toxicity of organic chemicals to activated-sludge microorganisms. *Journal of Environmental Engineering* **120**:1459-1469.
Sun, B. (1993). Comparison of interspecies toxicity of organic chemicals and evaluation of QSAR approaches in toxicity prediction. MS Thesis, Environmental Engineering. New Mexico State University.
Thomulka, K. W., McGee, D. J., Lange, J. H. 1993. Detection of biohazzardous materials in water by measuring bioluminescence with the marine organism Vibrio harveyi. *Journal Environmental Science and Health*. **A28**: 2153-2166.
Toussaint, M. W., Shedd, RT. R., van der Schalie, W. H., Leather, G. R. (1992) A comparison of standard acute toxicity tests with rapid-screening toxicity tests. *Environmental Toxicology and Chemistry* **14**:907-915.

## 15.2.3 DESCRIPTION OF AN INNOVATIVE GC METHOD TO ASSESS THE INFLUENCE OF CRYSTAL TEXTURE AND DRYING CONDITIONS ON RESIDUAL SOLVENT CONTENT IN PHARMACEUTICAL PRODUCTS

CHRISTINE BARTHÉLÉMY
**Laboratoire de Pharmacie Galénique et Biopharmacie**
**Faculté des Sciences Pharmaceutiques et Biologiques**
**Université de Lille II, Lille, France**
MICHEL BAUER
**International Analytical Sciences Department**
**Sanofi-Synthélabo Recherche, Toulouse, France**

The presence of residual solvents (RS) in pharmaceutical substances occurs for various reasons. Solvents are involved in all steps of raw material synthesis and pharmaceutical productions. The search for the presence of RS in a pharmaceutical product and their concentrations are now mandatory in any new monographs (as detailed in Chapter 16.2).

The RS remaining in the crystals of pharmaceutical products may be the cause of health disorders because, when a drug is taken every day, chronic toxicity may occur.

The presence of RS may have other consequences, such as modifying stability, organoleptic characters, pharmacotechnical parameters (flow properties, crystalline form, compression ability) and biopharmaceutical characteristics, that may fluctuate according to RS content (as detailed in chapters 14.21.1 and 16.2). It is therefore necessary to reduce the residual solvents contained in crystalline particles as much as possible.

It is well known that solvents can exist in three different states within the crystals:

- Solvents adsorbed on the crystal faces: these are generally easily desorbed during conventional drying because the binding forces between solvents and crystals are very weak.
- Occluded solvents such as microdroplets in the crystal: these are often difficult to extract. Generally, they can escape when the crystal is being dissociated:
  - during grinding, potentially leading to clodding,
  - during storage leading to very compact aggregates.
- Solvents bound to drug molecules in the crystal and known as "solvates". These bound solvents escape at a characteristic temperature, producing desolvated forms; the solvate and the desolvated forms are two different crystalline entities that can exhibit very different mechanical behaviors.

The main objective of any chemist crystallizing pharmaceutical raw materials should be either the total elimination of the organic solvents or the significant reduction of RS level in order to be below the regulatory limits.[1]

### 15.2.3.1 Description of the RS determination method

There are several analytical methods to assess the RS content of drugs. Among them, the gas chromatography (GC) is largely preferred. We refer to the chapter 16.2 for more details. In direct injection methods, the products in which RS are included are usually dissolved in an appropriate solvent and then directly injected into a gas chromatograph (GC). The main problem with these techniques is that non-volatile substances are gradually retained in the column, causing a rapid decrease in its sensitivity and efficiency; this is one of the reasons why headspace techniques are increasingly used instead.

Figure 15.2.3.1. Photograph of the micro-distilling device.

To avoid these problems, we have developed a method consisting initially of the complete dissolution of the substance in an appropriate solvent followed by micro-distillation and then finally of the injection of the distillate into the GC. This technique allows for the complete recovery of RS without the drawbacks mentioned above.[2,3]

The operating conditions are as follows: the powders to be analyzed are poured into a micro-distilling flask. An appropriate solvent is chosen to allow the product as well as the RS to be completely dissolved. The solvent should not interfere with the RS extracted; for example, in the following studied cases: methanol for dioxane, chloroform, tetrahydrofuran, hexane and dichloromethane determinations and 1-butanol for ethanol determination. Assay solutions were obtained by completely dissolving 500 mg of the product to be tested in 5 ml of the appropriate solvent. The micro-distillation is carried out until complete dryness and is followed by condensation of the solvent vapor. The distillate is collected in a gauge glass set on ice. The assembly of the system is photographed on Figure 15.2.3.1.

In this way the non-volatile substances stay entrapped in the micro-distilling flask and the totality of the RS are recovered. The distillate is then directly injected into the GC. The method was validated (specificity, linearity, repeatability, reproducibility...) by distilling, in the same operating conditions, defined standard solutions of solvents. The recovery of RS is very good as can be seen in the example reported on Table 15.2.3.1.

### 15.2.3.2 Application: Influence of crystal texture and drying conditions on RS content

After washing the crystals, drying is the most effective way of lowering the content of organic volatile impurities. The required drying conditions differ greatly according to the solvent state in the crystalline particles, to the thermodynamic events that can occur when the substance is heated, and also to the texture of these particles.

**Table 15.2.3.1. Example of a calibration curve and recovery calculations of dioxane distilled in methanol**

| Dioxane standards | | Dioxane recovered after distillation | |
|---|---|---|---|
| | area under curve and variation coefficient (%) | area under curve and variation coefficient (%) | % and ppm recovered |
| 44.8 ppm | 17161.0±705.5, v.c: 4.11 % | 16819.3±631.5, v.c: 3.75 % | 98.01 %, 43.9 ppm |
| 56 ppm | 25904.7±1493.1, v.c: 5.76 % | 25869.0±748.4, v.c: 2.89 % | 99.86 %, 55.9 ppm |
| 112 ppm | 66147.3±1010.4, v.c: 1.53 % | 66831.7±472.6, v.c: 0.71 % | 101.03 %, 113.2 ppm |
| 168 ppm | 108157.7±2026.6, v.c: 1.53 % | 107361.0±1546.4, v.c: 1.44 % | 99.26 %, 166.8 ppm |
| 224 ppm | 146674.0±1490.4, v.c: 1.02 % | 146722.7±2469.7, v.c: 1.68 % | 100.03 %, 224.1 ppm |
| linear regression | Y = 724.74 X - 14855.6 | Y = 724.29 X - 14889.7 | mean : 99.64±1.11 % v.c : 1.1 % |
| correlation coeff. | 0.9999 | 0.9999 | |

Among all parameters influencing RS content, crystal texture is of utmost importance. It is evident that a crystal exhibiting a porous texture will enable the easy escape of a solvent while a compact and dense crystal will retain the solvent inside its structure whatever the type of particles: monocrystalline (i.e., monoparticular) or "polycrystalline". The term "polycrystalline" particles will be employed to designate elementary particles that can be composed of agglomerates, spherolites or "spherical crystals" according to Kawashima.[4]

To illustrate the importance of the texture of particles on RS content, we can consider some examples taken from our laboratory experiments. Several crystallization processes were investigated on pharmaceutical products with different solvents leading either to compact monocrystalline particles or "sintered-like" (i.e., microcrystallites fitted into each other and partially welded involving a porous texture) or polycrystalline particles which were more or less dense.[2,3] The physical study of these particles such as, optical microscopy, particle size analysis, electron scanning microscopy and thermal analysis, have been used to link the RS contents with the drying conditions of crystals and to demonstrate that the optimal drying conditions differ greatly according to the texture of the particles.

As we shall see in a first example of monocrystalline particles of paracetamol recrystallized in dioxane, a wide open texture is generally favorable to a low RS content after a progressive drying at a moderate temperature to avoid the formation of a superficial crust. In fact, in this case, the drying conditions of crystals highly influence their residual solvent content. Optimal drying conditions seem to be the progressive and moderate ones. In contrast, too drastic drying conditions may hinder the solvent escape by a "crust" effect. This crust is due to a drying temperature which is too high, leading to a melt and a dissolution of the surface of the crystals. Finally, when a desolvation occurs during the drying, it can modify the texture of crystals and form a crust. The intensity of the phenomenon depends on the solvent.

To remove the solvents efficiently, it is then necessary to exceed the desolvation temperature of the solvates that may be produced during crystallization.

With this example we can clearly point out that the crystal texture is a determining factor in the complete escape of the solvent: when the crystalline texture is sintered-like, after the desolvation of solvate crystals, progressive drying is necessary to prevent the "crusting" phenomenon.

The drying conditions should also be adapted to the area offered to the evaporation of solvent, particularly in the case of polycrystalline particles presenting a high porosity and a large surface to be dried. This important surface is due to the disordered rearrangement of very small crystals inside the particles. To illustrate this, we shall see in a second example the cases of spherical crystals of meprobamate and ibuprofen agglomerates. In both cases the RS can escape easily from the large surface of polycrystalline particles.

Lastly, we shall see in a third example, that we must not forget that thermodynamic phenomenon can occur under drying. This is a very particular example because the pharmaceutical product used (paracetamol) presents a polymorphic transition that can occur during the drying phase and lead to a new organization inside the crystal, allowing the escape of the RS. This can be very interesting in the case of products with a low transition temperature. The usual polymorphic form of commercialized paracetamol is the monoclinic form; but in particular cases small amount of metastable orthorhombic form can be obtained. In the case of paracetamol, at the transition temperature (156°C) we observe a brutal solvent escape that could be due to the solid-solid transition undergone by the orthorhombic form into the monoclinic one. The disorder produced and the increase of the specific volume occurring during this first order transition allow the occluded solvents to escape more easily.

#### *15.2.3.2.1 First example: monocrystalline particles of paracetamol*

Preparation of monocrystalline particles

In this first example, paracetamol was recrystallized separately in three different types of solvents (ethanol, water and dioxane) with different boiling points, molecular weights, dielectric constants and paracetamol solubilising power. After the crystallization process,[5] crystals were separated by filtration under vacuum and washed with the same crystallization solvent. Each batch was divided into four fractions to be dried differently.

Particle drying conditions

The paracetamol crystals were submitted to four different drying conditions[2] (Table 15.2.3.2):

Drastic: under vacuum at 100°C for 3 hours.

Drastic: in a ventilated oven at 100°C for 3 hours.

Progressive: in a ventilated oven at 60°C for 1 hour, at 80°C for 1 hour and finally at 100°C for 1 hour.

Very moderate: under ventilated hood at 20°C for 24 hours and then in a ventilated oven at 35°C for 48 hours.

Morphological aspect of the particles

Dried crystals were observed by optical and scanning electron microscopy (SEM) in order to measure their Ferret mean diameter and to determine their habit and texture. The mean diameters of crystals obtained from different solvents and submitted to different drying conditions are reported in Table 15.2.3.3.

**Table 15.2.3.2. Drying conditions studied on the different particles**

| Type of particles | Drugs | Type of drying | Temperature | Time |
|---|---|---|---|---|
| monocrystalline | paracetamol | drastic under vacuum | 100°C | 3 h |
| | | drastic under ventilation | 100°C | 3 h |
| | | progressive | 60°C<br>+ 80°C<br>+ 100°C | 1 h<br>+ 1 h<br>+ 1 h |
| | | very moderate | 20°C<br>+ 35°C | 24 h<br>+ 48 h |
| polycrystalline | meprobamate | drastic | 90°C | 2 h |
| | | progressive | 60°C<br>+ 75°C<br>+ 90°C | 30 min<br>+ 30min<br>+ 1 h |
| | ibuprofen | moderate | 60°C | 2 h |
| | | very moderate | 40°C | 2 h |
| | paracetamol | drastic flash | 156°C | 10 min |
| | | drastic flash | 156°C | 30 min |
| | | drastic | 100°C | 2 h |
| | | drastic + flash | 100°C<br>+ 156°C | 2 h<br>+ 10 min |
| | | progressive | 60°C<br>+ 80°C<br>+ 100°C | 30 min<br>+ 30 min<br>+ 1 h |
| | | progressive + flash | 60°C<br>+ 80°C<br>+ 100°C<br>+ 156°C | 30 min<br>+ 30 min<br>+ 1 h<br>+ 10 min |

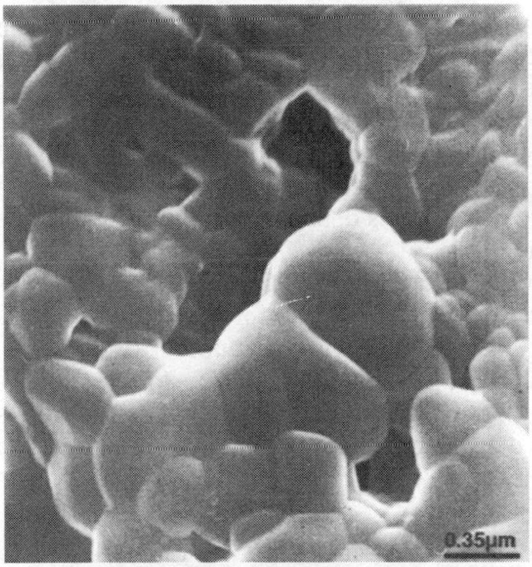


Figure 15.2.3.2. SEM photograph of paracetamol recrystallized in dioxane and submitted to moderate drying (Photograph from reference[2]).

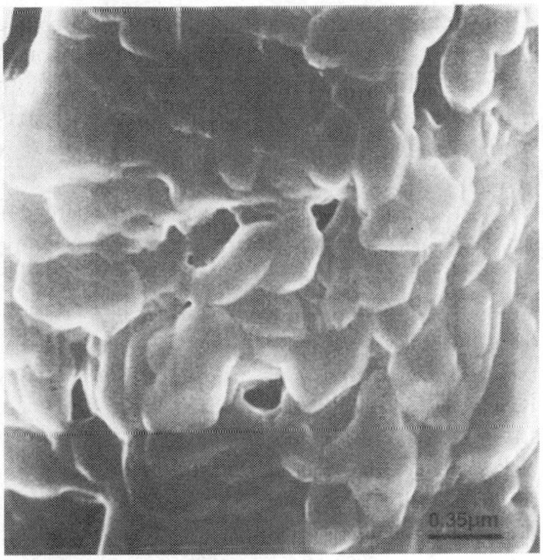


Figure 15.2.3.3. SEM photograph of paracetamol recrystallized in dioxane and submitted to progressive drying (Photograph from reference[2]).

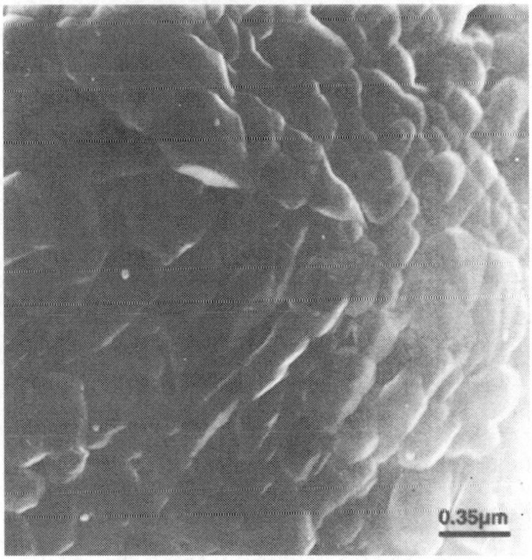


Figure 15.2.3.4. SEM photograph of paracetamol recrystallized in dioxane and submitted to drastic drying (Photograph from reference[2]).

The optical microscopy investigation shows some significant differences in mean diameter and crystal habit due to crystallization conditions.

In the case of water and ethanol, the crystals are transparent and rather regular, their habit is generally parallelepipedal. The only difference to be noted is the mean diameter of these particles; the particles recrystallized in water being 3 times larger. ESM reveals a surface that remains smooth in all drying conditions; their texture seems compact and dense.

In contrast, crystals obtained from dioxane are rather different to those obtained from ethanol and water: they are opaque to transmitted light and their habit is irregular. Their mean diameter is slightly smaller than crystals from ethanol.

It is to be noted that all the crystals obtained from the three solvents present the same polymorphic form: the monoclinic form.

**Table 15.2.3.3: Mean Ferret diameters of particles (Data from references[2,3])**

| Monocrystalline particles | Mean diameter, μm | Polycrystalline particles | Mean diameter, μm |
|---|---|---|---|
| paracetamol / dioxane | 122 | meprobamate spherical crystals | 145 |
| paracetamol / ethanol | 149 | ibuprofen agglomerates | 350 |
| paracetamol / water | 538 | paracetamol spherolites | 113 |

Scanning electron microscopy (SEM) carried out on these crystals reveals different behaviors according to the drying conditions tested. The crystals obtained from dioxane are similar to blocks of sintered particles since they consist of microcrystallites bound to each other, as if partially welded.[5] In fact, the measurement of mercury porosity indicated a very high porosity. This sintered aspect concords with the non-transmission of light through the whole crystal.

In fact, paracetamol forms a solvate with dioxane.[5] Its departure does not modify the crystal habit but the surface can be seen as perforated.

When the drying is very moderate, the crystal surface is perforated by numerous holes through which solvent escapes; these anfractuosities give the impression of a porous sintered-like texture (Figure 15.2.3.2). With progressive drying, the number of holes decreases (Figure 15.2.3.3). On the contrary, when the drying is drastic, the surface is relatively smooth (Figure 15.2.3.4). This can be explained by the too drastic drying conditions involving the formation of a crust at the surface of the crystal preventing the escape of the solvent from the crystal.

Residual solvent determination

Gas Phase Chromatography (for dioxane and ethanol) was performed on a Varian 1440 Chromatograph with a Flame Ionization Detector; packed column Porapack Super Q (Alltech, France), mesh range 80/100, length 1.8 m; internal diameter 2.16 mm; carrier gas was nitrogen (40 ml/min); injector: 210°C; detector: 250°C.

For dioxane: column temperature: isotherm at 170°C. Injection: 10μl. Retention Times (RT): methanol 1.5 min; dioxane 16 min.

For ethanol: column temperature: isotherm at 210°C. Injection: 5μl. RT: ethanol 1.5 min; 1-butanol 4 min.

Determination of residual water on paracetamol crystallized in water: according to the titrimetric direct method of Karl Fischer.

The residual solvent concentrations of the crystals obtained are reported in Table 15.2.3.4.

**Table 15.2.3.4. Residual solvent content of crystals obtained from different solvents and submitted to different drying conditions (Data from reference[2])**

| | Drastic | Drastic under vacuum | Progressive | Moderate |
|---|---|---|---|---|
| Dioxane | 126 ppm | 183 ppm | 25 ppm | 53 ppm |
| Ethanol | 2045 ppm | 2501 ppm | 1495 ppm | 2072 ppm |
| Water | 0.59 % | 0.68 % | 0.51 % | 0.56 % |

Whatever the drying conditions may be, the content of dioxane is always lower than the 380 ppm ICH limit[1,6] (Table 15.2.3.5). Progressive drying always gives lower level residual solvent. In fact, drastic drying, in a ventilated oven and under vacuum, leads to the formation of a superficial "crust" which hinders the solvent escape; this can be clearly visualized on the scanning electron photomicrograph (Figure 15.2.3.4); and can explain the relatively high content of residual solvents.

**Table 15.2.3.5. Solvent class and concentration limits in pharmaceutical products (Data from references[1,6])**

| | Class | Concentration limit, ppm | | Class | Concentration limit, ppm |
|---|---|---|---|---|---|
| Chloroform | 2 | 60 | Hexane | 2 | 290 |
| Dioxane | 2 | 380 | Dichloromethane | 2 | 600 |
| Ethanol | 3 | 5000 | Tetrahydrofuran | 3 | 5000 |

#### *15.2.3.2.2 Second example: polycrystalline particles of meprobamate and ibuprofen*

Preparation of polycrystalline particles

The polycrystalline particles were produced using various crystallization processes and designed to obtain directly compressible particles of pure drug, as tablets cannot be formed by direct compression of the raw materials.

*Spherical crystals of meprobamate*

These spherical particles were prepared following the usual preparation process for spherical crystals described by Guillaume.[7] Spherical crystals of meprobamate appear when stirring a mixture of three liquids in the crystallizer: methanol allowing meprobamate to dissolve; water, as a non-solvent, causing meprobamate precipitation; chloroform as a bridging liquid to gather in its dispersed droplets, meprobamate microcrystallites that finally form "spherical crystals".

*Ibuprofen agglomerates*

Ibuprofen agglomerates were prepared by a phase separation process in a mixture of ethanol and water (50/50 v/v).[8] The saturated solution obtained at 60°C was constantly stirred and cooled down to room temperature.

Particle drying conditions

Depending on the melting point of the drugs, different drying conditions were applied in a ventilated oven. Drastic and progressive temperature conditions were studied for each type of polycrystalline particles (Table 15.2.3.2).

*Spherical crystals of meprobamate*

Taking into account the melting point of meprobamate (105°C), temperatures higher than 90°C must be avoided. The different drying conditions were: drastic drying at 90°C for 2 hours and progressive drying at 60°C for 30 minutes, then at 75°C for 30 minutes and finally at 90°C for one hour.

*Ibuprofen agglomerates*

This is a particular case; a low drying temperature must be applied because of the very low melting point of ibuprofen (76°C). The two drying conditions studied were: 40°C for two hours or 60°C for two hours.

Morphological aspect of the particles

*Spherical crystals of meprobamate*

Meprobamate crystals appear as more or less rounded opaque particles (Table 15.2.3.3). Their consistence is friable. Particle size distribution is very narrow. SEM photograph of meprobamate crystals shows nearly spherical particles; their surface seems apparently smooth (Figure 15.2.3.5). However, at high magnification, the surface appears to be coated with flat crystals (Figure 15.2.3.6).

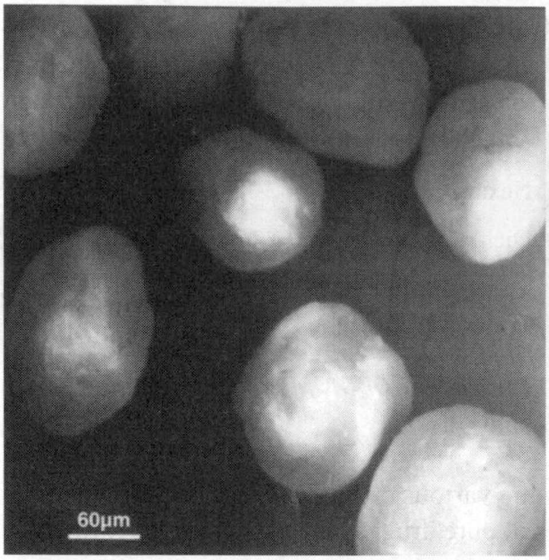


Figure 15.2.3.5. SEM photograph of "spherical crystals" of meprobamate (Photograph from reference[3]).

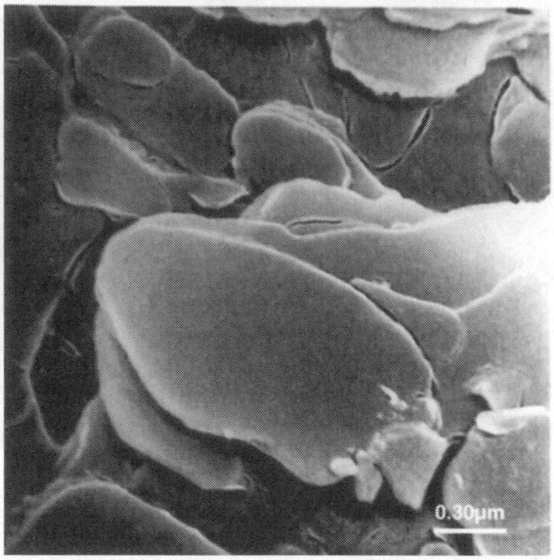


Figure 15.2.3.6. SEM photograph of the surface (high magnification) of "spherical crystals" of meprobamate (Photograph from reference[3]).

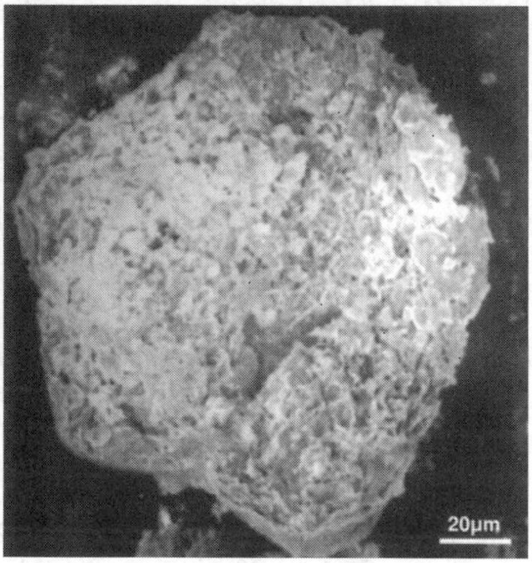


Figure 15.2.3.7. SEM photograph of the cross-section of "spherical crystals" of meprobamate (Photograph from reference[3]).

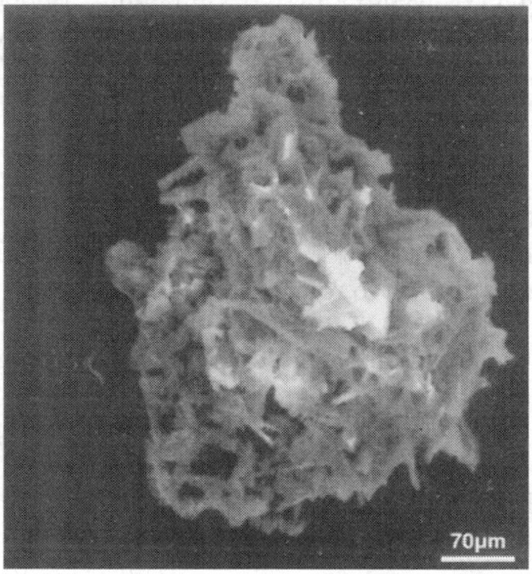


Figure 15.2.3.8. SEM photograph of ibuprofen agglomerates (Photograph from reference[3]).

The cross-section surface shows a dense tangling-up of flat crystals inside the particle (Figure 15.2.3.7). Flat crystals are concentrically disposed at the periphery of the rounded particles, like a shell. The inside of the particle is a disordered arrangement of small flat crystals. Thus, the surface offered to evaporation is very important and this should lead to the easy escape of RS.

*Ibuprofen agglomerates*

Ibuprofen particles are large agglomerates of flat crystals with quite soft consistence and a nearly rounded shape (Table 15.2.3.3).

Ibuprofen crystals observed under SEM are in total disorder (Figure 15.2.3.8). This disordered arrangement of microcrystallites inside the polycrystalline particles gives them a high isotropy of particles arrangement that should improve their compression capacity.

Residual solvent determination

For organic solvents, gas phase chromatography was performed on a Shimadzu GC-14B chromatograph fitted with a Flame Ionization Detector and a CR-6A Shimadzu integrator.

The packed column was Porapack super Q (Alltech, France), mesh range 80/100, length 1.80 m, internal diameter 2.16 mm. Carrier gas: anhydrous nitrogen. Injector temperature: 200°C. Detector temperature: 220°C. The chromatographic conditions were:

- For chloroform in meprobamate crystals: isotherm at 150°C, injection: 5 μl, RT: methanol 0.8 min, chloroform 6.7 min.
- For ethanol in ibuprofen crystals: isotherm at 170°C, injection: 5 μl, RT: ethanol 1.1 min, 1-butanol 5.3 min.

Determination of residual water was carried out using Karl Fischer's titrimetric direct method after calibration with natrium tartrate and dissolution of ibuprofen or meprobamate crystals in methanol.

The residual solvent concentrations of the polycrystalline particles are reported in Tables 15.2.3.6 and 15.2.3.7.

**Table 15.2.3.6. Residual solvent concentration of meprobamate spherical crystals submitted to different drying conditions (Data from reference[3])**

| | Chloroform | Water |
|---|---|---|
| Progressive drying* | 345±22 ppm | 0.71%±0.07% |
| Drastic drying** | 321±8 ppm | 0.21%±0.01% |

*30 min at 60°C + 30 min at 75°C + 1 hour at 90°C; **2 hours at 90°C

**Table 15.2.3.7. Residual solvent concentration of ibuprofen agglomerates submitted to different drying conditions (Data from reference[3])**

| | Ethanol | Water |
|---|---|---|
| 2 hours at 40°C | 42±3 ppm | 0.76%±0.12% |
| 2 hours at 60°C | 21±4 ppm | 0.44%±0.06% |

As far as meprobamate spherical crystals are concerned, no significant differences are to be observed between drastic and progressive drying. No crusting phenomenon appears on meprobamate spherical crystals due to the loose tangling up of crystals, and so the solvent may escape easily between them.

The residual ethanol content of ibuprofen agglomerates is very low because of the open texture of agglomerates. Moreover, the crystallization phenomenon was relatively slow, enabling the solvent to escape from crystals in formation. The higher the temperature, the lower the residual ethanol content.

Both these crystals have a very porous texture. It seems that progressive drying is not essential as far as the polycrystalline particles are concerned.

If we consider the official limits[1,6] reported on Table 15.2.3.5 for residual solvent contents, we can note that the concentration of chloroform in meprobamate spherical crystals is much higher than the limit allowed in any drying conditions. Due to its inherent toxicity, this solvent should be avoided in the recrystallization process of meprobamate. The solvent content obtained for ethanol in ibuprofen agglomerates is very low, in all drying conditions; in all cases it is considerably lower than the tolerated limits.

### *15.2.3.2.3 Third example: polycrystalline particles of paracetamol*

Preparation of polycrystalline particles

Paracetamol agglomerates were prepared by the spontaneous precipitation of paracetamol into a mixture containing hexane, tetrahydrofuran and dichloromethane[9] under stirring. All the crystals obtained were filtered under vacuum.

Particle drying conditions

The melting point of paracetamol being 169°C and the transition temperature being 156°C, different drying conditions were tested (Table 15.2.3.2):

- Drastic flash drying: 156°C for either 10 or 30 minutes,
- Drastic drying: 100°C for 2 hours,
- Progressive drying: 60°C for 30 minutes, then 80°C for 30 minutes, and finally, 100°C for 1 hour.

After drastic drying at 100°C and progressive drying, a test with complementary drying at 156°C was carried out for 10 minutes.

All the recovered crystals were packaged in glass flasks before gas phase chromatography and other analysis.

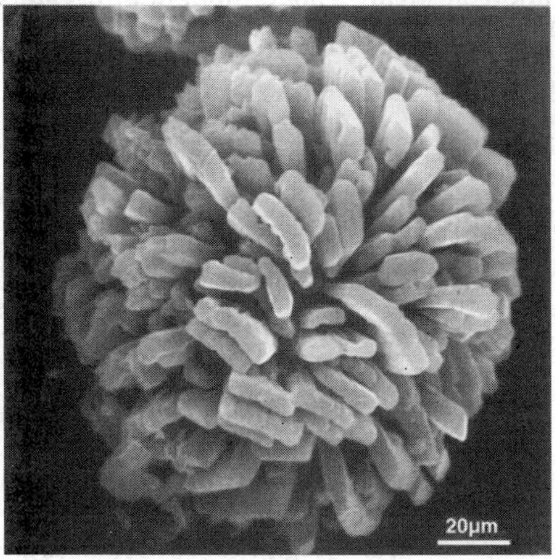

Figure 15.2.3.9. SEM photograph of paracetamol agglomerates (Photograph from reference[3]).

Morphological aspect of particles

Several types of texture and morphology are to be observed in polycrystalline particles, according to solvent proportions. The most interesting and particular example are spherical polycrystalline particles which have a radial texture (spherolites) and appear as urchin-like particles (Table 15.2.3.3). SEM reveals the very typical surface crystallization of the agglomerates (Figure 15.2.3.9). They are made up of parallelepipedal flat crystals arranged perpendicularly to a central nucleus and they are relatively strong.

The implantation of peripherical crystallites is perpendicular to the surface. As it has been clearly demonstrated by Ettabia[10] a nucleus is formed first and then, in a second step, microcrystallites grow on it.

Residual solvent determination

For organic solvents, gas phase chromatography was performed on a Shimadzu GC-14B chromatograph fitted with a Flame Ionization Detector and a CR-6A Shimadzu integrator.

The packed column was Porapack super Q (Alltech, France), mesh range 80/100, length 1.80 m, internal diameter 2.16 mm. Carrier gas: anhydrous nitrogen. Injector temperature: 200°C. Detector temperature: 220°C. The chromatographic conditions were: For dichloromethane, tetrahydrofuran and hexane in paracetamol crystals: isotherm at 150°C, injection: 5 μl, RT: methanol 0.8 min, dichloromethane 3 min, tetrahydrofuran 7.8 min, hexane 11.8 min.

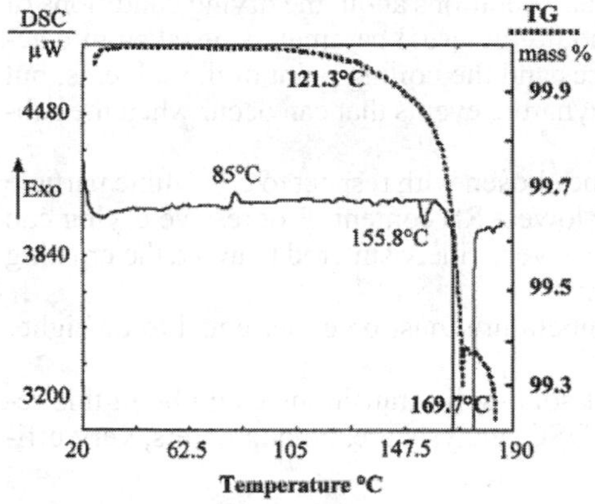


Figure 15.2.3.10. DSC and thermogravimetry curves of paracetamol agglomerates.

The residual solvent concentrations of the polycrystalline particles are reported in Table 15.2.3.8.

The wide-open texture of paracetamol spherolites hinders the crusting phenomenon; it is therefore normal that there should be no difference between the effectiveness of the two types of drying: progressive and drastic (at 100°C for two hours). However, a surprisingly good result is observed when the drying temperature is 156°C whereas a prolonged time at 100°C does not improve the solvent escape in spite of the low boiling points of solvents, all much lower than 100°C (dichloromethane: 39.5°C; tetrahydrofuran: 66°C; hexane: 69°C). Using a DSC method, it can be observed that a slight non constant exotherm at 85°C followed by a small constant endotherm at nearly 156°C occurs before the melting endotherm at 169°C (Figure 15.2.3.10). In fact, the paracetamol obtained by the crystallization process described is mainly the usual monoclinic form. However, as has been demonstrated by Ettabia[10] the formation of a certain amount of the amorphous form, causes the slight exotherm. The amorphous form recrystallized at 85°C into the orthorhombic metastable form, which transits into monoclinic form at 156°C, causing the small endotherm observed at nearly 156°C.

The thermogravimetric analysis shows that no solvation occurs during crystallization. Furthermore, the sudden solvent escape is not specific to one definite solvent, whereas all the contained solvent amounts dramatically decrease at 156°C. The loss of mass is high (about 0.7%) when the substance is about to melt (Figure 15.2.3.10).

**Table 15.2.3.8. Residual solvent concentration of paracetamol agglomerates submitted to different drying conditions (Data from reference[3])**

| | Dichloromethane | Tetrahydrofuran | Hexane |
|---|---|---|---|
| "Flash drying" 156°C | | | |
| for 10 min | < 200 ppm | 1095±5 ppm | 167±1 ppm |
| for 30 min | < 200 ppm | 630±24 ppm | 88±4 ppm |
| PD* | 315±3 ppm | 2066±1 ppm | 305±3 ppm |
| PD*+ "flash" (10 min at 156°C) | < 200 ppm | 883±10 ppm | 145±5 ppm |
| DD ** | 309±3 ppm | 2014±30 ppm | 313±4 ppm |
| DD**+"flash" (10 min at 156°C) | < 200 ppm | 873±2 ppm | 140±1 ppm |

*PD: Progressive drying (30 min at 60°C + 30 min at 80°C + 1 h at 100°C), **DD: Drastic drying (2 h at 100°C)

Paracetamol, containing three different solvents, exhibits low dichloromethane content (200 to 320 ppm). The referencing limits remain higher than the experimental results when these particles are submitted to different drying conditions (Table 15.2.3.5). The same observation is valid for tetrahydrofuran for which the allowed upper limit is 5000 ppm. But the residue in hexane is sometimes above the regulatory threshold. The only way to be definitely below this limit is to heat paracetamol to 156°C; this confirms the advantage of flash drying this substance. As previously described, this temperature is critical for paracetamol recrystallization in the solvents used.

This study enables us to underline some considerations about the drying conditions of crystals. It is important to note that not only should the usual parameters be taken into account, such as the melting point of the substances and the boiling point of the solvents, but also the texture of the particles and the thermodynamic events that can occur when the substance is heated.

The kinetic of temperature increase must be chosen with respect to crystalline particle texture to obtain optimal drying conditions for lowest RS content. Progressive drying can give better solvent elimination when the texture is very finely sintered to avoid the crusting effect, which hinders any ulterior solvent escape.[2]

When a solvate is formed, the drying temperature must be either equal to or higher than the desolvation temperature.

Lastly, the knowledge of the existence of solid-solid transitions could be in this respect of great interest. Thermogravimetry and DSC analysis are, among others, very efficient tools to assess these phenomena.

## References

1 **European Pharmacopoeia**, 3rd edition, addendum 1999, pp. 216-224.
2 C. Barthélémy, P. Di Martino, A-M. Guyot-Hermann, *Die Pharmazie*, **50**, 609 (1995).
3 A. Ettabia, C. Barthélémy, M. Jbilou, A-M. Guyot-Hermann, *Die Pharmazie*, **53**, 565 (1998).
4 Y. Kawashima, M. Okumura, H. Takenaka, *Science*, **216**, 1127 (1982).
5 J-M. Fachaux, A-M. Guyot-Hermann, J-C. Guyot, P. Conflant, M. Drache, S. Veesler, R. Boistelle, *Powder Techn.*, **82**, 2, 123 (1995).
6 Note for Guidance on impurities: Residual solvents, *Drugs made in Germany*, **41**, 98 (1998).
7 F. Guillaume, A-M. Guyot-Hermann, *Il farmaco,* **48**, 473 (1993).
8 M. Jbilou, A. Ettabia, A-M. Guyot-Hermann, J-C. Guyot , *Drug Dev. & Ind. Pharm.*, **25**, 3, 297 (1999).
9 A. Ettabia, E. Joiris, A-M. Guyot-Hermann, J-C. Guyot, *Pharm. Ind.*, **59**, 625 (1997).
10 A. Ettabia, European Thesis, Lille II University, France (8/10/1997).

16

# Residual Solvents in Products

## 16.1 RESIDUAL SOLVENTS IN VARIOUS PRODUCTS

GEORGE WYPYCH
**ChemTec Laboratories, Inc., Toronto, Canada**

There are physical and chemical barriers that control solvent removal from solid-solvent systems. The most basic relation is given by the following equation:

$$W = \frac{P_1}{K_w} \qquad [16.1.1]$$

where:

| | |
|---|---|
| W | equilibrium fraction of residual solvent |
| $P_1$ | partial pressure of solvent in vapor phase |
| $K_w$ | Henry's law constant |

Both the partial pressure and Henry's law constant depend on temperature, pressure, and solvent properties. This relationship does not consider interaction between solute and solvent. In the case of polymers, the Flory-Huggins theory gives a simplified relationship for low concentrations of solvent:

$$\ln \frac{P_1}{P_1^0} = \ln \phi_1 + 1 + \chi \qquad [16.1.2]$$

where:

| | |
|---|---|
| $P_1^0$ | vapor pressure of pure solvent |
| $\phi_1$ | volume fraction of solvent |
| $\chi$ | Flory-Huggins interaction parameter |

Vapor pressures of some solvents can be found in the referenced monograph.[1]

The weight fraction of residual solvent at equilibrium can be calculated from the following equation, which accounts for polymer-solvent interaction:

$$W = \frac{P_1}{P_1^0} \frac{\rho_1}{\rho_2} \exp-(1+\chi) \qquad [16.1.3]$$

where:

| | |
|---|---|
| $\rho_1$ | density of solvent |
| $\rho_2$ | density of polymer |

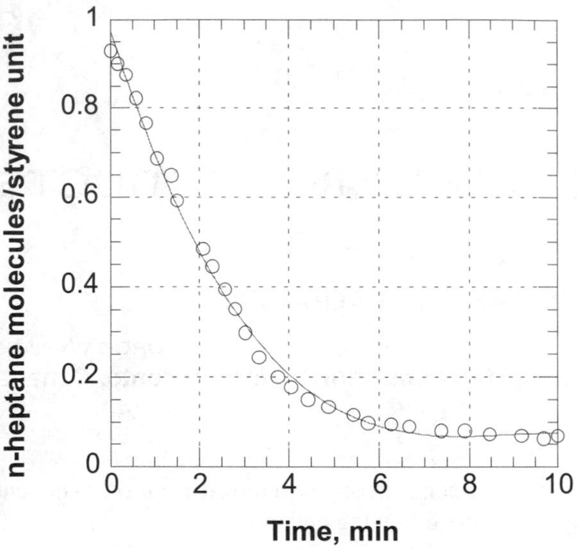


Figure 16.1.1. Number of n-heptane molecules per one mer of polystyrene vs. drying time. [Data from L A Errede, P J Henrich, J N Schrolpfer, *J. Appl. Polym. Sci.*, **54**, 649 (1994).]

The last equation does not give the real values of residual solvents because equilibrium is not attained in real drying processes and the prediction of different interactions by the interaction parameter is too simplistic. The real values are substantially higher and the real barriers of solvent removal more complex. These are discussed below.

In real systems, several phenomena take place. These include chemical interaction between the functional groups of polymer and the solvent. These are mostly related to hydrogen bond formation. The crystalline structure of polymer is responsible for the modification of the diffusion process. Solvent properties determine diffusion. Polymer properties are responsible for the macro-mechanism of solvent removal from a highly viscous polymer. In addition, some real systems make use of stripping solvents which are designed to help in removal of trace quantities of solvents by use of stripping solvent displacing process solvent. These various factors interplay and determine the result. Figure 16.1.1 shows the number of residual solvent molecules per one mer of polystyrene. It is evident that solvent removal has zero-kinetics until its concentration is decreased to about 0.2 molecules of solvent per mer. It is also true that some solvent remains after drying. Even after 24 h drying, 0.06% solvent remains.

These data indicate that there is a different mechanism of removing residual solvent. It is not clear if this is because of interactions, a change in the glass transition temperature, or a change in crystallinity. So far the partial effects of these influences cannot be separated. It is confirmed by experiment that in the last stages of drying, glass transition temperature of polymer changes rapidly. Also, the degree of crystallinity of the polymer increases during drying.[3,4] From studies on polyaniline, it is known that its conductivity depends on the concentration of adsorbed molecules of water.[5] Water interacts by hydrogen bonding with the polymer chain. The activation energy of hydrogen bonding is very low at 3-5 kcal/mol. Drying at 120°C reduced the amount of water molecules from 0.75 to 0.3 molecules per aniline unit. This change in water concentration drastically alters electrical conductivity which decreases by three orders of magnitude. Drying for two hours at 120°C did not result in complete removal of water. Given that the activation energy of hydrogen bonding is very low, the process of interaction is probably not the main barrier to removal of residual moisture. Also the relationship between conductivity and number of molecules of water is linear in the range from 0.15 to 0.75 molecules of water per aniline unit which means that there is no drastic change in the mechanism by which water participates in increasing the conductivity of polyaniline. Its conductivity simply depends on the distance between neighboring adsorbed molecules of water which apparently participate in the charge migration.

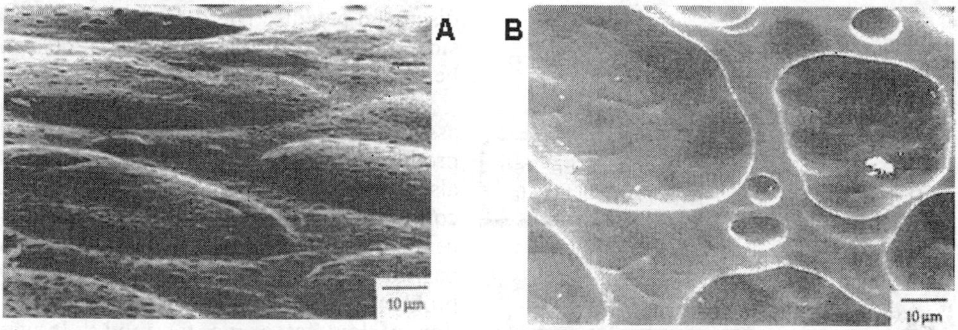

Figure 16.1.2. Blister formation in polyethylene containing originally 4000 ppm hexane. a - lateral surface 2.2 s after extrusion, b - cross-section after 28 s. [Adapted, by permission from R J Albalak, Z Tadmor, Y Talmon, *AIChE J.*, **36**, 1313 (1990).]

SEM studies contribute to an understanding of a major obstacle to residual solvent removal. Figure 16.1.2 shows two photographs of polyethylene strands extruded from a melt which initially contained 4000 ppm hexane. After a short period of time following the extrusion, blisters form which remain in the material and become enlarged until they break and release solvent. This blistering mechanism, determines the rate of residual solvent removal from the material. The rate of removal depends on bubble nucleation, temperature, and polymer rheological properties.[6]

Observing such mechanisms makes it easy to understand the principle involved in stripping solvents which became popular in recent inventions.[7-9] Stripping solvents were used to improve the taste and odor properties and the oxidative thermal stability of thermoplastic ethylene polymers.[7] Volatile components, such as products of degradation, solvent and monomer contribute to taste problems and odor formation. Striping solvents used include highly volatile hydrocarbons (ethylene, propylene, isobutane), inert gases, and supercritical fluids. An addition of at least 0.1% stripping solvent reduces volatiles from the typical levels of between 300-950 ppm to 45 ppm with even as low as 10 ppm possible. A stripping solvent helps in the generation of bubbles and their subsequent breaking by which both the stripping and the residual solvent are removed. In cosmetics and pharmaceutical formulations traces of solvents such as benzene or dichloromethane, used in the synthesis of acrylic acid polymer, disqualify the material. It is not unusual for this polymer to contain up to 1000 ppm of dichloromethane or up to 100 ppm of benzene. The use of mixed ester solvents helps to reduce residual solvent to below 5 ppm.[8] Polycarbonate pellets from normal production may contain up to 500 ppm solvent. This makes processing polycarbonate to optical products very difficult because of bubble formation. Elimination of volatiles renders the product suitable for optical grade articles.[9] These inventions not only demonstrate how to eliminate solvents but also confirm that the mechanism discussed in Figure 16.1.2 operates in industrial processes. The examples also show that large quantities of residual solvents are retained by products in their normal synthesis.

Many standard methods are used to devolatilize materials. Flash devolatilizer or falling strand devolatilizer are synonyms of equipment in which the falling melt is kept below the saturation pressure of volatiles. Styrene-acrylonitrile copolymers devolatilized in flash devolatilizer had a final concentration of ethylbenzene of 0.04-0.06.[1] Devolatilization of LLDPE in a single-screw extruder leaves 100 ppm of hydrocarbon solvent. 500 ppm chlorobenzene remains in similarly extruded polycarbonate.[1] It is estimated that if the polymer contains initially 1-2% solvent, 50-70% of that solvent will be removed through the vacuum port of an extruder.[1] These data seem to corroborate the information included in the above

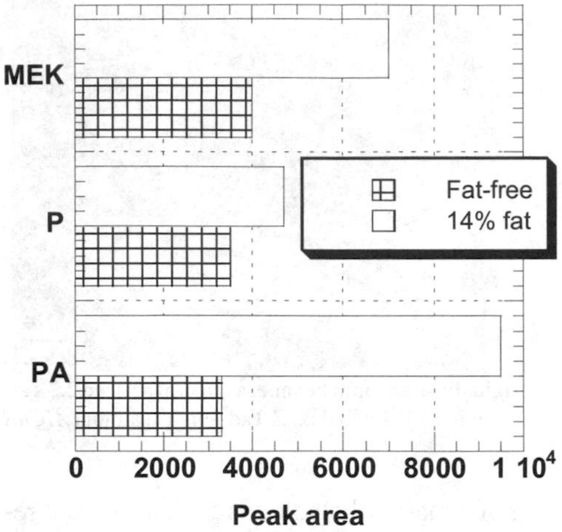


Figure 16.1.3. Volatility of three printing solvents (PA - propyl acetate, P - n-propanol, MEK - methyl ethyl ketone) from fat-free and fat-containing cookies. 20 μg solvent added to 2.5 g cookies. [Data from T Clark, *Paper Film Foil Converter*, **70**, 11, 48 (1996).]

discussed patents. The data show that considerable amounts of residual solvents can be found in polymers and plastic materials.

In the food industry, residual solvents associated with packaging odors enter food products from two sources: packing materials and printing inks.[10] It is estimated that concentrations of residual solvents have recently decreased (from 2000 mg/ream in past to 1000 mg/ream).[10] However, a new problem has become apparent in introduction of low fat or no fat food.[11] It was discovered[11] that more customer complaints about odor were received for these low fat baked goods products. Analysis shows that fat was a good solvent for volatiles (solvents) and consequently solvent odor was not detected because flavor perception is developed relative to the concentration of gaseous flavor compounds. Figure 16.1.3 shows the concentrations of three solvents as detected by gas chromatography. Substantially higher concentrations are detected in solvent-spiked fat-free cookies than in fat-containing cookies.[11] In this experiment solvents were added to the cookies. In another experiment, packaged cookies were exposed to a solvent vapor atmosphere and different trends were recorded for cookies packaged in two different films. If the film had good barrier properties, no difference was noticeable between both types of cookies and the adsorbed quantities of solvents were minimal. If the cookies were packaged in a coextruded film having lower barrier properties, no-fat cookies absorbed 42% more solvent than full fat cookies.

In the construction industry, residual solvent evaporation becomes an increasingly more critical issue, especially in the case of products used for indoor applications. Sealants, adhesives, and paints are now a major focus of this concern since they contribute to indoor pollution. Similar trends are observed in the automotive industry where both solvents and plasticizers are suspected of contributing to a "plastics" odor in car interiors.

In contrast, changes in the solvent evaporation rate may also contribute to product improvement in paints. Small quantities of properly selected solvents can improve physical properties and the appearance of paints. Other applications of residual solvents include time-controlled release of fertilizers and production of materials with controlled morphology.

## REFERENCES

1 R J Albalak, Ed., **Polymer Devolatilization**, *Marcel Dekker, Inc.*, New York, 1996.
2 L A Errede, P J Henrich, J N Schrolpfer, *J. Appl. Polym. Sci.*, **54**, 649 (1994).
3 M O Ngui, S K Mallapragada, *J. Polym. Sci., Polym. Phys.*, **36**, 2771 (1998).
4 H-T Kim, S-K Kim, J-K Park, *Polym. J.*, **31**, 154 (1999).
5 E S Matveeva, *Synthetic Metals*, **79**, 127 (1996).
6 R J Albalak, Z Tadmor, Y Talmon, *AIChE J.*, **36**, 1313 (1990).
7 M M Hughes, M E Rowland, C A Strait, **US Patent 5,756,659**, The Dow Chemical Company, 1998.
8 A Bresciani, **US Patent 5,342,911**, 1994.
9 T Hosomi, T Umemura, T Takata, Y Mori, **US Patent 5,717,055**, Mitsubishi Gas Chemical, Ltd., 1998.
10 R M Podhajny, *Paper Film Foil Converter*, **72**, 12, 24 (1998).
11 T Clark, *Paper Film Foil Converter*, **70**, 11, 48 (1996).

# 16.2 RESIDUAL SOLVENTS IN PHARMACEUTICAL SUBSTANCES

MICHEL BAUER
**International Analytical Department, Sanofi-Synthélabo, Toulouse, France**
CHRISTINE BARTHÉLÉMY
**Laboratoire de Pharmacie Galénique et Biopharmacie, Faculté des Sciences Pharmaceutiques et Biologiques, Université de Lille 2, Lille, France**

## 16.2.1 INTRODUCTION

The need to test for residual solvents (RS) in pharmaceutical substances was recognized in the late 70's when some pharmacopoeias like those of USA (USP XX) or Great-Britain (BP 80 + add 82) introduced specific tests for RS in some monographs. But we had to wait until the early 80's to see a rational approach for establishing specifications from toxicological data. This strategy was developed by a working group of the Italian Pharmacopoeia[1] starting from the threshold limit values for Chemical Substances and Physical Agents in the Work Environment published by the American Congress of Governmental Experts for Industrial hygiene.[2] In the late 80's, RS were definitively classified as impurities *per se*. Methods and specifications appeared in different issues of Pharmacopeial Forum and were submitted for discussion and finally integrated in the USP. At the same time only few such monographs could be found in the European Pharmacopoeia (Eur. Ph.) or British Pharmacopoeia (BP). Interestingly, the notion of a content limit for residual solvents in relation to the daily intake of the drug was introduced (see Eur. Ph. 2nd edition), a concept which was taken up later in the ICH Guideline. Numerous publications have been devoted to this subject.[3-9]

Although the number of papers on RS is immense, they are very often limited to specialized areas such as regulatory aspects or methodology aspects. In this chapter, different topics will be considered, starting with the fundamental question: why look for RS in pharmaceutical products? It is worth noting that Witschi and Doelker[10] published in 1997 a very detailed and up-to-date review stressing the importance of this subject in the pharmaceutical field.

## 16.2.2 WHY SHOULD WE LOOK FOR RS?

As we already have seen in Chapter 14.21.1, RS could have various effects on the drug substances, excipients and drug products.

### 16.2.2.1 Modifying the acceptability of the drug product

The presence of RS could seriously impair customer compliance because of the odor or the taste they can cause in the final pharmaceutical preparation. Rabiant[2] quotes the case of a drug substance having undergone, for technical reasons, a washing with isopropanol not planned in the manufacturing protocol. The oral solution prepared from this batch contained 100 ppm of this solvent and consequently had an odor that the majority of the patients accepted only with reluctance; the batch concerned was finally removed from the market.

### 16.2.2.2 Modifying the physico-chemical properties of drug substances (DS) and drug products (DP)

The role of the quality of the solvents on the stability e.g. of the raw materials, DS and DP (see Section 14.21.1) has already been discussed. It must be remembered that RS (including water) can show different kinds of interactions with solid substances:[11]

- Solvents adsorbed on the crystal surfaces, which generally are easily removed because of the existence of weak physical interactions.
- Occluded solvents and clathrates which are more difficult to extract without impairing the quality of the drug by, for example, excessive drying.
- Solvents bound to drug molecules in the crystal lattice. These solvents present as solvates (hydrates) are lost at a characteristic temperature and may be stable only over a limited range of relative humidity. The solvates and desolvated solvates, whilst being two different chemical entities, can retain the same crystalline structure (similarity of x-ray diffraction pattern) but show different physico-chemical properties.[12-14]

One of the most important effects of organic solvents absorbed on crystal surfaces is the ability to reduce the wettability of the crystals, especially if the solvent concerned is hydrophobic.[15]

Another interesting aspect, particularly in the case of residual water, is its role as an agent of recrystallization of poorly crystalline substances. It is well known that amorphous or partially amorphous products can undergo a recrystallization[16,17] process over time in presence of water. If amorphous phases are an interesting way to promote the dissolution rate of poorly soluble drugs, their main drawback is their physical instability triggering the possible crystallization of the drug and leading to a decrease in dissolution rate and possibly of bioavailability, over a period of time.

Furthermore the residual adsorbed water may have an impact on the flowability of a powder, which is linked to the solubility of the substance and the hydrophilicity of the crystal faces.[15] Other physico-chemical parameters are influenced by RS, like particle size and dissolution properties. For more information we refer to the publications of Doelker[10] and Guyot-Hermann[15] and references quoted therein. Nevertheless, one example deserves to be mentioned here.[10,16] Residual isopropyl alcohol enhanced the water permeability of Eudragit® L films used as tablet coating for protecting water-sensitive drugs as demonstrated by List and Laun.[16] This implies particular conditions for the storage of coated tablets during the film-drying process. The atmosphere should be as dry as possible.

The need to keep the RS level as low as possible can lead to some problems. It has been reported[15,17] that a drug substance displaying a strong odor of residual solvent was submitted to reprocessing, consisting of the displacement of the residual solvent by a stream of water vapor. During this process which slightly modified the surface crystallinity of the particles, a small amount of an impurity was produced. The consequence was an increase of the surface solubility of this drug substance. During the preparation of the DP using an aqueous wet granulation, a liquification of the granulate was observed making the manufacture impossible. After having removed the impurity by purifying the DS a successful manufacture of the DP was achieved.

There are other aspects linked to the manufacturing process and drying conditions which impact on the final RS content. They are discussed in Chapter 15.2.3 of this book (and references cited therein).

Nevertheless before closing this paragraph, another example of the relationship between the manufacturing process and RS is worth mentioning here. It relates to the formation of volatile compounds produced by radiolysis and which could induce odor. Barbarin et al.[18] have investigated this subject in different antibiotics belonging to the cephalosporin group (Cefotaxime, Cefuroxime and Ceftazidine). Using GC-MS and GC-IR they were able to identify carbon monoxide, nitric oxide, carbon disulfide, methanol, acetaldehyde, ethyl formate, methyl acetate and acetaldehyde O-methyloxime. In a subsequent publication[19] on cefotaxime, they demonstrated that some of the radio-induced compounds (such as carbon monoxide sulfide (COS) and carbon disulfide) came from the degradation of the drug itself whereas the formation of others required the RS, present before the irradiation. For instance, acetaldehyde arises from the irradiation of methanol. Incidentally, this is a good way to differentiate between radio-sterilized and non-radio-sterilized products.

It is worth remarking that residual humidity can favor[15] microbiological growth especially in some natural products used as excipients (starch, gelatine, etc.). From a physico-chemical point of view, residual humidity may have an impact on the hardness of the tablets as shown by Chowhan,[20] Down and McMullen.[21]

#### 16.2.2.3 Implications of possible drug/container interactions

It is possible that the RS contained in a powder may migrate up to the interface between the contents and the container facilitating the extraction and migration into the drug of additives used during the container manufacture. On the contrary, solvents may be used during the packaging of a drug. An example of this situation is described by Letavernier et al.[22] Cyclohexanone was used for sealing PVC blisters containing suppositories. After 42 months of storage at ambient temperature up to 0.2 mg to 0.3 mg of cyclohexanone was found per gram of suppository.

#### 16.2.2.4 As a tool for forensic applications

Forensic laboratories are interested in identifying and assessing trace impurities in bulk pharmaceutical products with the idea of using the impurity profile as a "fingerprint" of the manufacturer. Of course, RS can be an important aid in this process.[23,24] It has been demonstrated,[25] for instance, that static headspace GS coupled with mass spectrometry (MS) was able to detect and identify volatile impurities, making possible the characterization of illicit heroin or cocaine samples.

#### 16.2.2.5 As a source of toxicity

##### *16.2.2.5.1 General points*

Toxicity is obviously the main reasons for testing for RS. Besides the toxicity of the drug itself, the related impurities,[26] degradants and the RS obviously can each bring their own contribution. A drug could be prescribed to a patient either for a short period of time or a long one. To minimize either acute toxicity or chronic toxicity resulting from some accumulation process, RS have to be kept at the lowest achievable level. When developing new chemical entities, the presence of RS could bias toxicity studies including mutagenicity and carcinogenicity and cause a risk of wrongly ascribing to the drug substance or the formulation, side effects which are actually due to volatile impurities. Knowing the cost of such studies, it is preferable to have the RS under control.

The long-term exposure to solvents has been recognized for a long time as a possible cause of serious adverse effects in human. Tables of maximal tolerated solvent concentrations in air for defined exposures have been published and used to set limits for residual organic solvents in pharmaceuticals.[1-10] In the late 90's, a set of articles appeared in the Pharmacopeial Forum proposing RS limits from toxicological data,[27] including carcinogenicity, mutagenicity, teratogenicity and neurotoxicity.

**Table 16.2.1. (After from reference 28)**

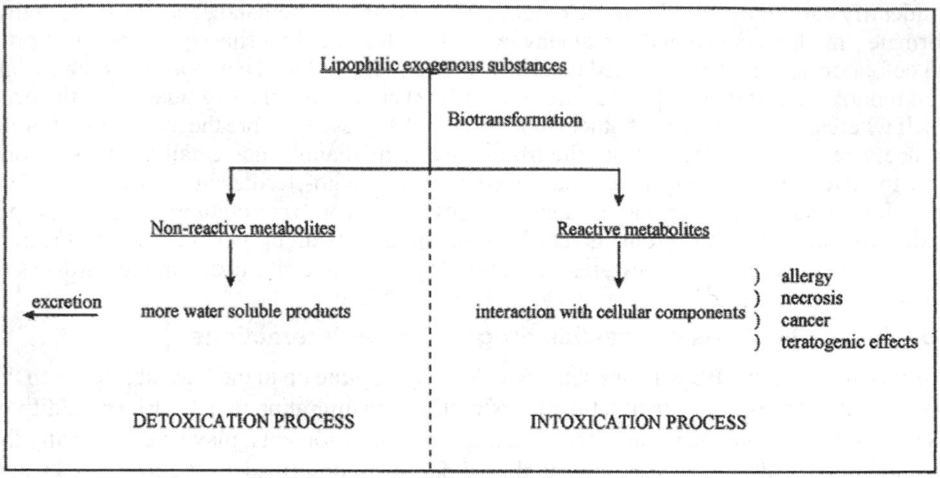


In the early 80's, the "International Conference on Harmonization of Technical Requirements for Registration of Pharmaceuticals for Human use" (ICH) was created. Among the different topics deserving to be harmonized, the need to have an agreement amongst Europe, Japan and the USA on the ways to limit the RS in pharmaceuticals was clearly identified and the topic adopted in June 1994. The final guide was finally adopted in 1998 by the Health Authorities of the three zones and is now in force. This Guideline will be examined in detail in the last paragraph of this part.

### *16.2.2.5.2 Brief overview of the toxicology of solvents*[28,29]

16.2.2.5.2.1 Aspects concerning metabolism

There are four main routes by which solvents can interact with the human body:

- O.R.L. (Otorhinolaryngology)
- transdermal (including the ocular area)
- oral
- injection

Once in the organism, solvents will undergo biotransformation which essentially takes place in the liver. This metabolism very often leads to more water-soluble products than the parent compound[30,31] and, as such, more easily excreted by the kidneys. This detoxication process is beneficial for the individual but varies greatly from one subject to another. Unfortunately, this metabolic detoxication can be complicated by the appearance of reactive intermediates which, if not rapidly inactivated, will destroy the essential constituents of the cells (proteins, nucleic acids, unsaturated lipids) and cause INTOXICATION. The latter ranges from a simple allergic reaction to tissue necrosis or, at worst, to cancer. Table 16.2.1 summarizes the different events which can occur in the organism.

16.2.2.5.2.2 Solvent-related pathology

16.2.2.5.2.2.1 Acute toxicity

It is especially the affinity of the solvents for lipid-rich organs which triggers problems of acute toxicity and these concern primarily the nervous system, the heart, the liver and kidneys. In this acute toxicological process, the molecules act per se without any previous biotransformation. The acute toxicity encompasses:

- nervous toxicity (headache, somnolence, coma, more or less deep, which can extend to death)
- cardiac toxicity[32]
- action on skin and mucous membranes generating irritation (including ocular area)[33]

16.2.2.5.2.2.2 Long-term toxicity

A prolonged exposure, even at low doses, to several liposoluble solvents leads sooner or later to irreversible effects on different organs:

- central nervous system (e.g., toluene could lead to degeneration of the brain)
- peripheral nervous system (e.g., methanol shows a peculiar affinity for the optic nerve leading possibly to blindness)
- liver and kidneys; as solvents are metabolized in the liver and excreted by kidneys, these two organs are, of course, particular targets for these products
- skin and mucous membranes (e.g., dermatitis)
- blood (cyanosis, anaemia, chromosomal abnormality)

With regard to carcinogenicity, benzene has long been recognized as carcinogen in man. It is the reason why its use is strictly limited and not recommended (ICH class I/specifications 2 ppm). Carbon tetrachloride and 1,2 dichloroethane have been demonstrated to be carcinogenic in animals and potentially carcinogenic in man.

The embryotoxicity of solvents must be taken into account. As solvents can cross the placenta, pregnant women should be especially protected.

16.2.2.5.2.2.3 Metabolism of benzene

By way of illustration, Figure 16.2.1 summarizes the metabolic pathway of benzene.[34] Numerous publications dealing with the metabolism of benzene are given in reference 27.2.

## 16.2.3 HOW TO IDENTIFY AND CONTROL RS IN PHARMACEUTICAL SUBSTANCES?

### 16.2.3.1 Loss of weight

Historically this is the first method which appeared in the pharmacopoeias, performed either at normal pressure or under vacuum. This is, of course, an easy method, particularly for routine control but there are several drawbacks:

- lack of specificity
- it is product demanding (1-2 g)
- the limit of detection (LOD) is currently about 0.1 %

This determination can now be done by thermogravimetric analysis (TGA) which makes the method more sensitive (possible LOD 100 ppm) and less product demanding (5 - 20 mg). It can also be used as a hyphenated method linking TGA to a mass spectrometer, allowing the identification of the desorbed solvents (specificity). However, whilst this kind of equipment exists, there are no signs of it replacing gas chromatography (GC) in the near future.

### 16.2.3.2 Miscellaneous methods

Infrared spectroscopy[35] and $^{1}$H-NMR[36] have been used occasionally to identify and to quantify residual solvents, but their sensitivity is rather limited if compared with GC. On the other hand specificity is not always assured. The solvent should display signals well separated from those arising from the product, which is not always the case.

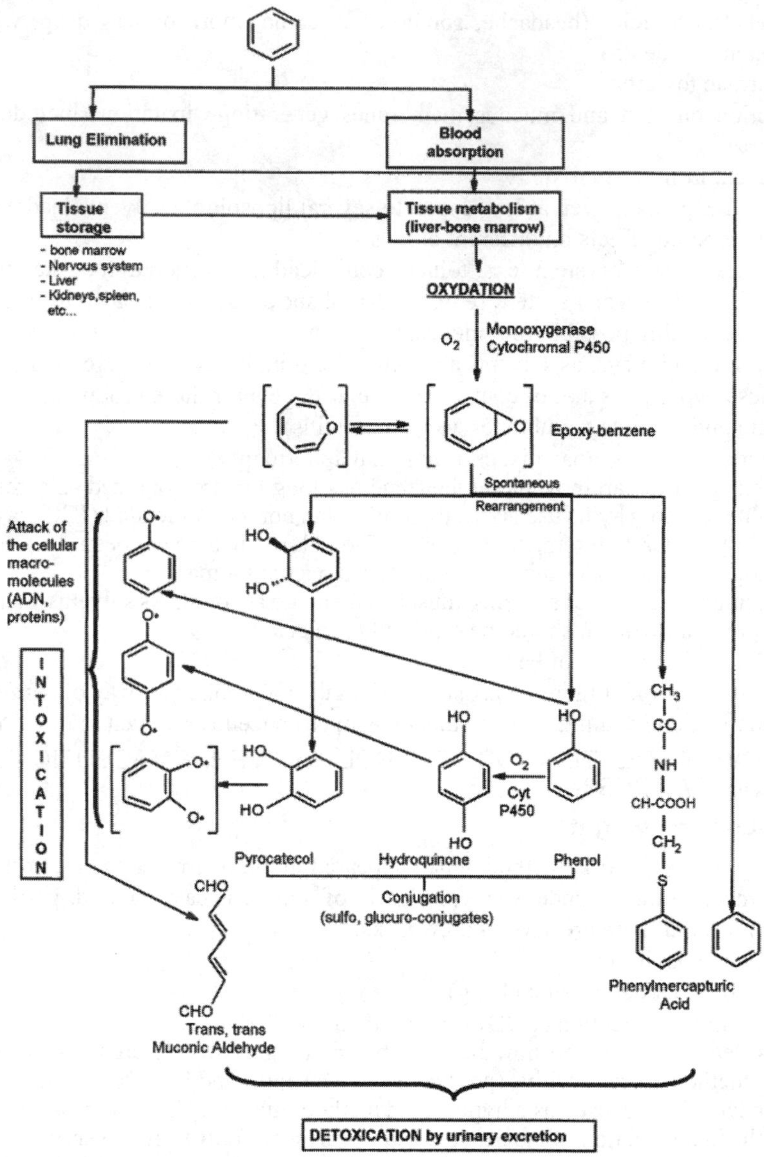


Figure 16.2.1. Overview of benzene metabolism (after reference 34).

## 16.2.3.3 Gas chromatography (GC)

### *16.2.3.3.1 General points*

This is, of course, the method of choice which has long been used to determine RS whatever the area of application (pharmaceuticals, polymers, water analysis, etc.).[37-40] From the late 70's to the beginning of the 80's, there was a large number (or flood) of publications dealing with different possible GC techniques which could be applied for detecting or analyzing residual solvents especially in pharmaceuticals (intermediates of synthesis, drug substances, excipients and drug products). So much has been written that it is difficult to be original. Methodological aspects will be briefly covered in the next paragraph. The paper of Witschi and Doelker[10] is particularly recommended. With 171 references, it represents a worthwhile, up-to-date review of the different GC techniques available. The last part of this chapter concerns pharmacopoeial methods.

#### *16.2.3.3.2 Review of methods*

Regarding GC methodology, four aspects must be examined:

- injection systems
- columns
- detectors
- method validation

16.2.3.3.2.1 Injection systems

16.2.3.3.2.1.1 Direct injection

After having dissolved the substance containing the RS to be looked for in an appropriate solvent, it is possible to directly inject into the system 100% of an aliquot of the solution (if packed columns are used) or partially through a split system (if capillary, narrow-bore and wide-bore columns are used). It is simple, accurate and repeatable (with an internal standard). The main drawback is that samples very often contain non-volatile substances which are retained by the column, leading rapidly to a loss of efficiency and a dramatic decrease in sensitivity.[41] In the current literature dealing with the RS, the direct injection process is less frequently used. Nevertheless, publications have appeared until recently using split/splitless injection.[42,43]

16.2.3.3.2.1.2 Static headspace injector

The solubilized or suspended sample in an appropriate vehicle is heated at a defined temperature in a tightly closed vial until thermodynamic equilibrium is reached between the liquid phase and the gas phase. A known aliquot is then transferred either with the aid of a syringe or by an automatic transfer system onto the column. The main advantage is that only volatile products including solvents are injected into the column greatly improving its lifetime. The sensitivity is good and the system is easily automated.

The main drawback is the existence of matrix effects and the possible non- ideality of the solvents mixture. These imply that ideally one should determine the calibration curve by adding standard solutions of the solvents of interest to the sample matrix, free of solvents. Because it is very difficult to obtain such a sample matrix, the classical standard addition method is recommended. It consists of adding to the sample matrix to be analyzed a known amount of the solvents to be determined. This method requires two analyses for the final calculation but the main advantage is that the matrix effect is overcome. The linearity of the response has, of course, to be demonstrated before the use of the simplified version mentioned above. Nevertheless, if based on a sound validation, external calibration can be used.[41,44]

The nature of the solvent used to prepare the solution (e.g., water, dimethylformamide, dimethylacetamide, 1,3-dimethyl-2-imidazolidinone (DMI)), equilibration temperature, the ratio between the gas phase and the liquid phase and the possible need to promote a salting out effect by adding mineral salts are the parameters, amongst others, which should be investigated and optimized to improve the sensitivity (LOD, LOQ) of the method.[45-47]

Another version of this static headspace chromatography is what has been called by Kolb[48] multiple headspace extraction (MHE) chromatography. This is a multi-step injection technique which was alluded to in the Suzuki publication[39] and more openly developed by MacAuliffe.[49] The principle of this method is the following.[50,51] After the first extraction has been made and the aliquot injected, the gas phase is removed by ventilating the vial and re-establishing the thermodynamic equilibrium. The equilibrium between the analyte in the solid or liquid phase and the gas phase will be displaced each time. After n extractions, the analyte content in the liquid or solid phase becomes negligible. It is then sufficient to sum the peak areas obtained for each extraction (which decrease exponentially) and, from an external calibration curve determine the amount of RS in the substance.

This method is particularly useful for insoluble products or in cases where the partition coefficient of the RS is too favorable relative to the liquid phase. It has been recently successfully applied to the determination of RS in transdermal drug delivery systems.[51]

16.2.3.3.2.1.3 Dynamic headspace injection

In the dynamic headspace method, the sample is put in a thermal desorption unit in order to desorb the RS; a continuous flow of a carrier gas pushes the RS into a trapping system which is refrigerated and where they are accumulated prior to analysis. Then the RS are rapidly desorbed by rapid heating and carried onto the column via the carrier gas. There are different ways to apply this technique.[10,52] The arrangement when purge gas passes through the sample is often called the purge and trap technique (some other equipment uses the acronym DCI (desorption, concentration, injection)). This method is particularly useful for very low concentrations of RS as the total amount of a substance is extracted and can be applied directly to powders without need to dissolve them. The main drawback is that the dynamic headspace methods are not readily automated.[41]

16.2.3.3.2.1.4 Other techniques

Several others techniques dealing with the injection problems have been developed. Among them the solid-phase microextraction method[52-55] (SPME) and the full evaporation technique[56] must be mentioned. According to Camarasu,[53] the SPME technique seems to be very promising for RS determination in pharmaceuticals, with much better sensitivity than the static headspace technique.

16.2.3.3.2.2 Columns

The wealth of publications dealing with RS determination by GC is so impressive that it is difficult to provide an exhaustive review. The interested reader will find plenty of information in the references quoted so far and in others recently published, mentioned below. Packed columns,[57] wide or narrow bore columns, capillary columns, etc. have been used for RS determination. It is true to say that capillary columns and narrow bore columns are the most often mentioned techniques. Today it is almost certain that any user can find in the literature[10] the stationary phase and the relevant conditions to resolve his RS problem, at least in terms of selectivity. By way of example, it has been shown by Brinkmann and Ebel[58,59] how it was possible to screen 65 of the 69 solvents mentioned in the ICH Guideline (discussed in paragraph 16.2.4) using capillary columns filled with two stationary phases (DB 624 and Stabilwax) which basically constitute the strategy proposed by the European Pharmacopoeia.

16.2.3.3.2.3 Detectors

The almost universally used detector is the flame ionization detector (FID) which works with all organic solvents but which is not selective. The mass spectrometer detector (MSD) is now more and more utilized.[24,53,55,60,61] It can be either universal in its electron ionization (EI) mode or selective in its selective ion monitoring mode (SIM). Other detectors, selective and/or universal, can be used.[10,62] Among them the electron capture detector has to be mentioned when looking for chlorinated solvents, even if its use is not straightforward.

16.2.3.3.2.4 Method validation

Whatever the technique used for determining the RS content in pharmaceuticals, a thorough validation of the complete analytical process has to be conducted according to the ICH Guidelines [Text on "validation of analytical procedures" and "validation of analytical procedures: methodology"[63]]. For testing impurities in a quantitative manner the following items have to be completed:

- specificity (or more appropriately selectivity)
- accuracy
- precision
    - repeatability
    - intermediate precision (first part of reproducibility)
- limit of detection (LOD), limit of quantification (LOQ)
- linearity of the response
- the range which is the interval between the upper and lower concentration (amounts) of analyte in the sample (including these concentrations) for which it has been demonstrated that the analytical procedure has a suitable level of precision, accuracy and linearity.

In the publications mentioned above,[24,41-43,51-54,58,59,64] it is possible for reader to find experimental procedures to conduct validation efficiently. Attention is drawn to the fact that, when using the static headspace technique in particular some other parameters have to be investigated, such as:[24,51]

- ratio gas phase/solid or liquid phase
- temperature of equilibrium between the two phases and time to reach it
- temperature of the transfer line
- pressurization time and sampling time when using a fully automated headspace injector

Finally, some additional comments are worth making:

- As regards reproducibility, it is true that the best way to assess it is to set up an inter-laboratory study. It is of course burdensome but this is probably the only way to infer reasonable suitability parameters for the routine quality control monograph (resolution, plate number, tailing factor, repeatability, LOD, LOQ) and specifications. It is worth noting that the latter should take the performance of the analytical method into account. But for those who cannot follow this approach, sound ruggedness testing has to be performed. Maris et al. have designed a method[65] for evaluating the ruggedness of a gas chromatographic method for residual solvents in pharmaceutical substances.
- When using the static headspace injector the possible matrix effect should be studied and can be evaluated by a statistical method.[66,67]
- One of the most important suitability parameters in case of RS determination is, of course, the LOQ (and LOD). In order to avoid an unrealistic value in the QC monograph, it is highly recommended to use a working limit of quantification WLOQ (and WLOD) which consists of determining a reasonable upper limit for LOQ (and LOD). In fact, the LOQ derived from the validation package is obtained in what we can call an ideal or optimized environment.

A way for determining $\alpha$ (> 1) in the relationship:

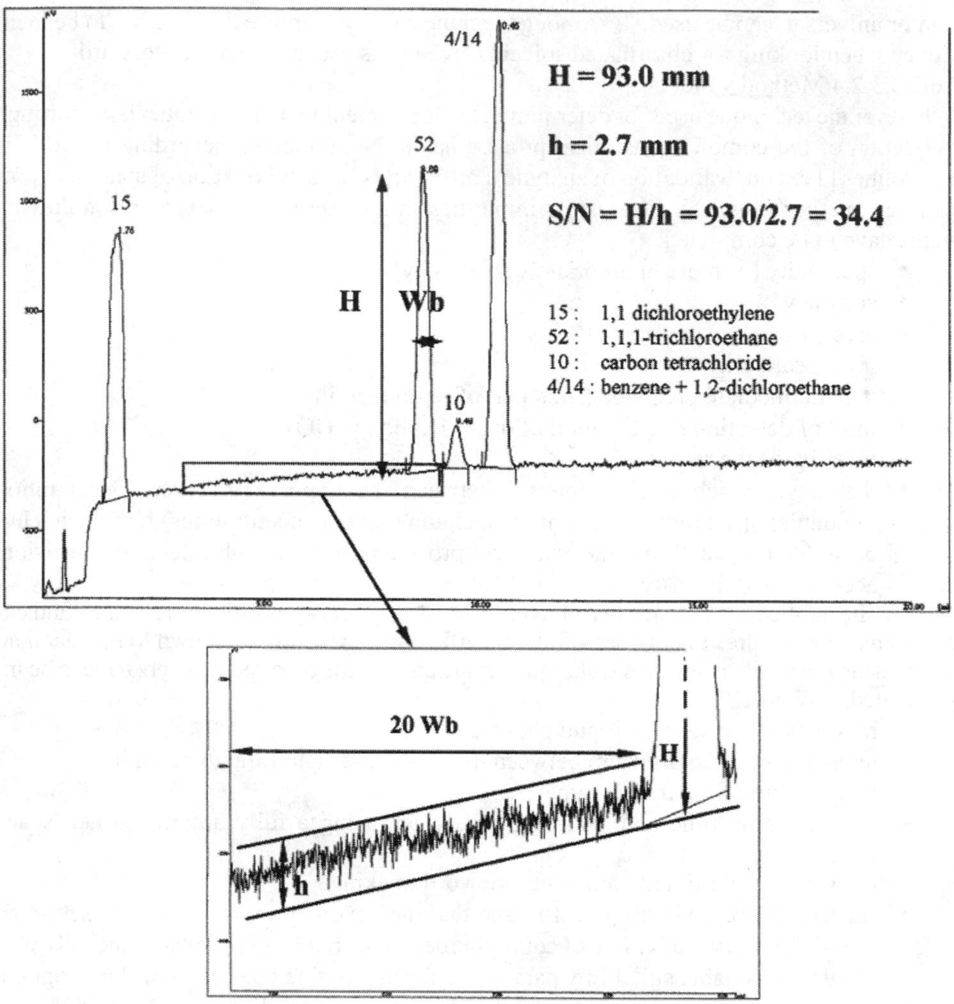


Figure 16.2.2. Head-space injection of the gaseous sample into the chromatographic system. Typical chromatogram of class 1 solvents using the conditions described for System A (European Pharmacopoeia method). Flame-ionization detector: calculation of H/h for 1,1,1-trichloroethane.

$$WLOQ = \alpha\, LOQ$$

is to calculate the LOQ according to the signal to noise ratio S/N method[63,68] (see Figure 16.2.2) and to repeat this determination independently several times (e.g., n = 6) during the intermediate precision determination. If $LOQ_m$ is the mean value and $\sigma$ the standard deviation, a possible definition of WLOQ could be the upper limit of the one side confidence limit at a specified risk $\alpha$ (e.g., 0.05).

$$WLOQ = LOQ_m + t_{n-1,\alpha}\sigma$$

where t is Student's coefficient.

#### *16.2.3.3.3 Official GC methods for RS determination*

Current official GC methods are described in USP XXIII under chapter 467 "Organic volatile impurities". Four methods (I, IV, V, VI) are mentioned. Methods I, V and VI are based on direct injection. They are suitable for water-soluble drugs and V for water insoluble drugs. Method IV describes the static headspace technique and is used for water soluble drugs. Method VI is very general and refers to the individual monograph which describes the chromatographic conditions ( injection, column, conditions) which should be used. The main characteristics of these four methods are summarized in Table 16.2.2.

The European Pharmacopoeia[69] used a two-tiered process based on two different columns:

*System A*. Fused silica capillary or semi capillary column (30 m x 0.32 mm (ID) or 30 m x 0.53 mm (ID)) DB 264 (1.8 µm or 3 µm film thickness of the phase (6 per cent cyanopropylphenyl-94 percent dimethylpolysiloxane)) which is identical to the USP method V.

*System B*. Fused silica capillary or semi-capillary column (30 m x 0.32 mm (ID) or 30 m x 0.53 mm (ID) DB-wax (0.25 µm film thickness of Macrogol 20000R)).

**Table 16.2.2. Gas chromatographic methods described in USP 23 (from ref. 10)**

| Method | Sample | Standardization | Column | Detector |
|---|---|---|---|---|
| USP <467> Method I: Direct GC injection | Dissolved in water or another appropriate solvent | External | 30 m x 0.53 mm ID, fused silica, with 5µm crosslinked G27[a] stationary phase and A 5 m x 0.53 mm ID silica guard column, phenylmethyl siloxane deactivated | FID* |
| USP <467> Method IV: Static HSC | Dissolved in water containing sodium sulphate and heated for 1 h at 80°C before injection of the headspace | External | As USP <467> Method V | FID |
| USP <467> Method V: Direct GC injection | As USP <467> Method I | As in USP <467> Method I | 30 m x 0.53 mm ID, fused silica with 3 µm G43[b] stationary phase and a 5 m x 0.5 mm ID silica guard column, phenylmethyl siloxane deactivated | FID |
| USP <467 Method VI: Direct GC injection ** | As USP <467> Method I | As in USP <467> Method I | One of 9 columns[c], listed under <467>, specified in the monograph | FID |

| Method | Sample | Standardization | Column | Detector |
|---|---|---|---|---|
| USP <467> Method for dichlormethane in coated tablets | The tablet water extract is heated for 20 min at 85°C before headspace injection | Standard addition | As USP <467> Method V | FID |

*To confirm the identity of a peak in the chromatogram, a mass spectrometer can be used or a second validated column, containing a different stationary phase. **Method VI presents a collective of chromatographic systems.
[a]5% phenyl/95% methylpolysiloxane; [b]6% cyanopropylphenyl/94% dimethylpolysiloxane; [c]S2, styrene-divinylbenzene copolymer; S3, copolymer of ethylvinylbenzene and divinylbenzene; S4, styrene-divinylbenzene; G14, polyethyleneglycol ($M_w$ 950-1050); G16, polyethyleneglycol compound (polyethyleneglycol compound 20M or carbowax 20 %; G27, see a; G39, polyethyleneglycol ($M_w$ 1500).

The static headspace injector has been selected and the method developed for water soluble products using water as dissolution medium or using N,N-dimethylformamide for water-insoluble products. If N,N dimethylacetamide and/or N,N-dimethylformamide are suspected in the drug under investigation 1,3-dimethyl 2-imidazolinone (DMI) is used as dissolving medium.

The method has been designed:

- in order to identify most class 1 and class 2 RS potentially present in drug substances, excipients or drug products.
- as a limit test for class 1 and class 2 RS present in drug substances, excipients and drug products.
- to quantify class 2 RS where the content is higher than 1000 ppm or class 3 RS if the need arises.

Figures 16.2.3, 16.2.4, 16.2.5, and 16.2.6 illustrate the separations obtained in the two systems for solvents belonging to class 1 and class 2 solvents. This general method is the outcome of European working party which has been published in Pharmeuropa.[70]

It should be stressed that the use of a general pharmacopeial method is not a reason not to validate the latter when analyzing a particular substance. The matrix effect, in particular, has to be investigated when using the static headspace mode of injection.

## 16.2.4 HOW TO SET SPECIFICATIONS? EXAMINATION OF THE ICH GUIDELINES FOR RESIDUAL SOLVENTS

The introduction briefly summarizes the strategies dealing with the setting of specifications in pharmaceuticals which appeared during the 80's.

From the early 90's onwards, the International Conference on Harmonization was initiated in three important pharmaceutical regions (Europe, Japan, USA) in order to define a common way of preparing a registration file acceptable in the three zones. The topics included in this harmonization process are:

- Quality
- Safety
- Efficacy

Regarding Quality a set of Guidelines have been already adopted. Three of them are particularly relevant with regard to this article:

- Guideline Q3A: Impurities in new active substances
- Guideline Q3B: Impurities in new medicinal products
- Guideline Q3C: Note for guidance on impurities: residual solvents

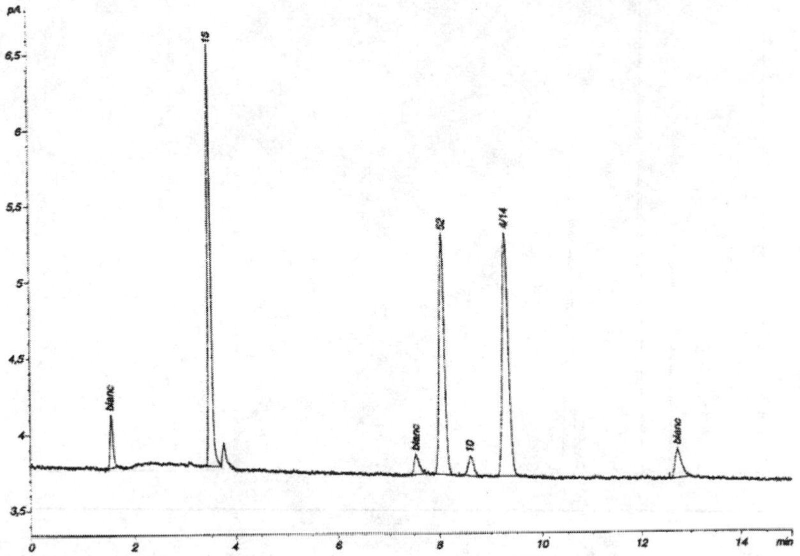

Figure 16.2.3. Typical chromatogram of class 1 solvents using the conditions described for system A and procedure 1. Flame-ionization detector. 4: benzene; 10: carbon tetrachloride; 14: 1,2-dichloroethane; 15: 1,1-dichloroethylene; 52: 1,1,1-trichloroethane. [Adapted, by permission, from **European Pharmacopoeia, Addendum 2000**, pp31-36.] [Please note that information concerning residual solvents are susceptible to be modified in the successive editions of the European Pharmacopeia.]

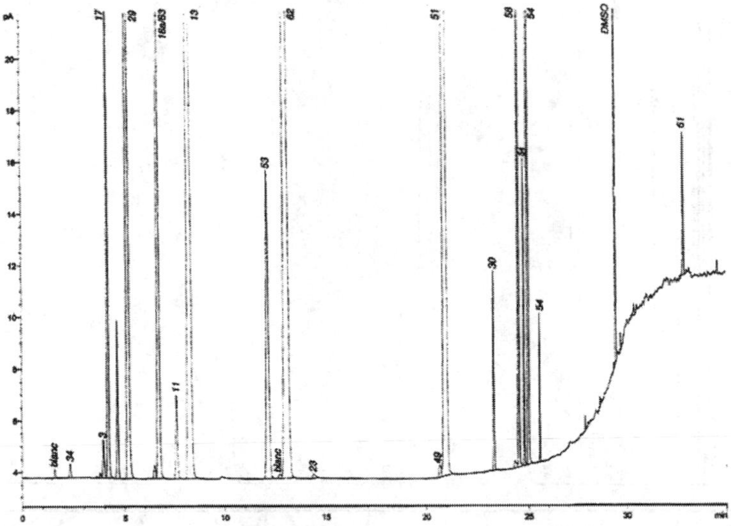

Figure 16.2.4. Chromatogram of class 2 solvents using the conditions described for system A and procedure 1. Flame-ionization detector. 3: acetonitrile; 11: chloroform; 13: cyclohexane; 16a: cis-1,2dichloroethylene; 17: dichloromethane; 29: hexane; 30: 2-hexanone; 34: methanol; 49: pyridine; 51: toluene; 53: 1,1,2-trichloroethylene; 54: xylene ortho, meta, para; 58: chlorobenzene; 61: tetraline; 62: methylcyclohexane: 63; nitromethane; 64: 1,2-dimethoxyethane. [Adapted, by permission, from **European Pharmacopoeia, Addendum 2000**, pp31-36.] [Please note that information concerning residual solvents are susceptible to be modified in the successive editions of the European Pharmacopeia.]

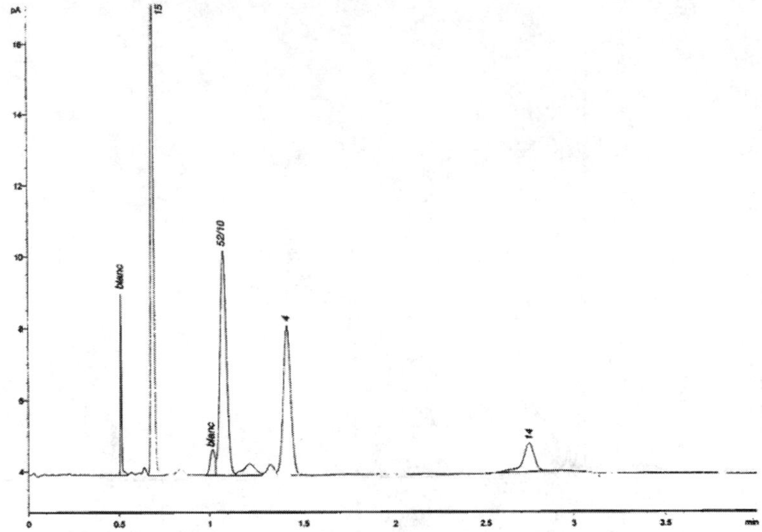


Figure 16.2.5. Chromatogram of class 1 residual solvents using the conditions described for system B and procedure 1. Flame-ionization detector. 4: benzene; 10: carbon tetrachloride; 14: 1,2-dichloroethane; 15: 1,1-dichloroethylene; 52: 1,1,1-trichloroethane. [Adapted, by permission, from **European Pharmacopoeia**, **Addendum 2000**, pp31-36.] [Please note that information concerning residual solvents are susceptible to be modified in the successive editions of the European Pharmacopeia.]

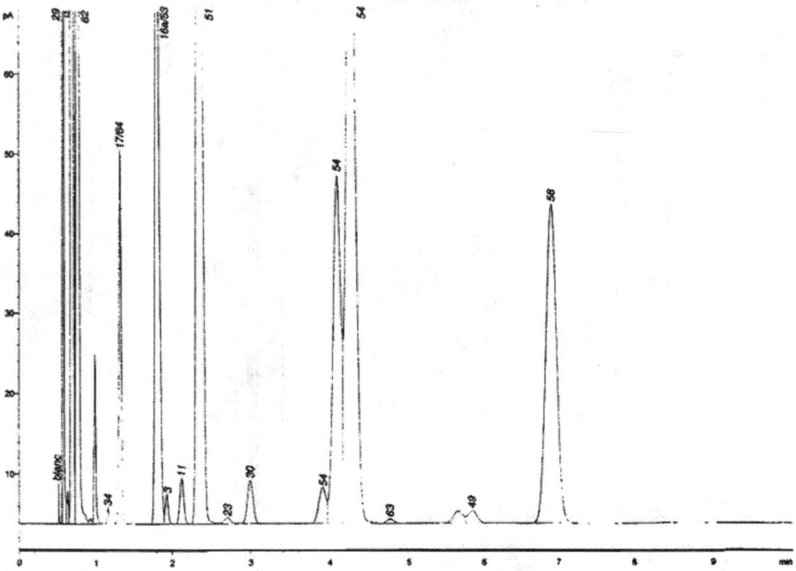


Figure 16.2.6. Typical chromatogram of class 2 residual solvents using the conditions described for system B and procedure 1. Flame ionization detector. 3: acetonitrile; 11: chloroform; 13: cyclohexane; 16a: cis-1,2-dichloroethylene; 17: dichloromethane; 23: 1,4-dioxane; 29: hexane ; 30: 2-hexanone; 34: methanol; 49 : pyridine ; 51 : toluene ; 53 : 1,1,2-trichloroethylene ; 54 : xylene, ortho, meta, para; 58: chlorobenzene; 61: tetralin; 62: methylcyclohexane; 63: nitromethane; 64: 1,2-dimethoxyethane. [Adapted, by permission, from **European Pharmacopoeia**, **Addendum 2000**, pp31-36.] [Please note that information concerning residual solvents are susceptible to be modified in the successive editions of the European Pharmacopeia.]

Comments on the latter are given below. The full text can be found in the US Pharmacopoeia, in the European Pharmacopoeia or in Journals[71] (there is also a website - www.ifpma.org/ich1.htm).

### 16.2.4.1 Introduction

The Guideline recommends acceptable amounts of RS in pharmaceuticals which are safe for the patient. Residual solvents in pharmaceuticals are defined as organic volatile chemicals that are used or produced in the manufacture of active substances or excipients, or in the preparation of medicinal products. It is stated that medicinal products should contain no higher levels of residual solvents than can be supported by safety data. Three classes of solvents have been defined based on risk assessment.

### 16.2.4.2 Classification of residual solvents by risk assessment

*Class 1 solvents*: solvents to be avoided, known as human carcinogens or strongly suspected carcinogens and environmental hazards.

*Class 2 solvents*: solvents to be limited. Nongenotoxic animal carcinogens or possible causative agents of other irreversible toxicities such as neurotoxicity or teratogenicity. Solvents suspected of other significant but reversible toxicities.

*Class 3 solvents*: Solvents with low toxic potential to man: no health-based exposure limit is needed. Class 3 solvents have permitted daily exposures (PDE) of 50 mg or more per day.

### 16.2.4.3 Definition of PDE. Method for establishing exposure limits

The PDE is defined as a pharmaceutically acceptable intakc of RS. The method used to establish PDEs for RS is described in the references cited above.[27,71]

### 16.2.4.4 Limits for residual solvents

**Table 16.2.3. Class 1 solvents in pharmaceutical products (solvents that should be avoided)**

| Solvent | Concentration limit, ppm | Concern |
|---|---|---|
| Benzene | 2 | Carcinogen |
| Carbon tetrachloride | 4 | Toxic and environmental hazard |
| 1.2-Dichloroethane | 5 | Toxic |
| 1.1-Dichloroethene | 8 | Toxic |
| 1.1.1-Trichloroethane | 1500 | Environmental hazard |

*Solvents of class 1* (see Table 16.2.3) should not be employed. However, if their use is unavoidable in order to produce a significant therapeutic advance, then their levels should be restricted as shown in Table 16.2.3, unless otherwise justified.

*Solvents of class 2* (see Table 16.2.4). Two options are available when setting limits for class 2 solvents.

*Option 1*

The concentration limits in ppm stated in Table 16.2.4 can be uscd. They were calculated using the equation [16.2.1] by assuming a product mass of 10 g is administered daily.

$$concentration(ppm) = \frac{1000 \times PDE}{dose}$$

where PDE is given in mg/day and dose is given in g/day.

**Table 16.2.4. Class 2 solvents in pharmaceutical products**

| Solvent | PDE, mg/day | Concentration limit, ppm |
|---|---|---|
| Acetonitrile | 4.1 | 410 |
| Chlorobenzene | 3.6 | 360 |
| Chloroform | 0.6 | 60 |
| Cyclohexane | 38.8 | 3880 |
| 1,2-Dichloroethene | 18.7 | 1870 |
| Dichloromethane | 6.0 | 600 |
| 1,2-Dimethoxyethane | 1.0 | 100 |
| N,N-Dimethylacetamide | 10.9 | 1090 |
| N,N-Dimethylformamide | 8.8 | 880 |
| 1,4-Dioxane | 3.8 | 380 |
| 2-Ethoxyethanol | 1.6 | 160 |
| Ethylene glycol | 6.2 | 620 |
| Formamide | 2.2 | 220 |
| Hexane | 2.9 | 290 |
| Methanol | 30.0 | 3000 |
| 2-Methoxyethanol | 0.5 | 50 |
| Methylbutylketone | 0.5 | 50 |
| Methylcyclohexane | 11.8 | 1180 |
| N-Methylpyrrolidone | 48.4 | 4840 |
| Nitromethane | 0.5 | 50 |
| Pyridine | 2.0 | 200 |
| Sulfolane | 1.6 | 160 |
| Tetralin | 1.0 | 100 |
| Toluene | 8.9 | 890 |
| 1,1,2-Trichloroethene | 0.8 | 80 |
| Xylene | 21.7 | 2170 |

*Option 2*

It is not considered necessary for each component of the medicinal product to comply with the limits given in option 1. The PDE in terms of mg/day as indicated in Table 16.2.4 can be used with the known maximum daily dose and equation [16.2.1] to determine the concentration of residual solvent allowed in the medicinal product. Option 2 may be applied by adding the amounts of RS present in each of the components of the pharmaceutical formulation. The sum of the amounts of solvent per day should be less than that given by the PDE.

**Table 16.2.5. Class 3 solvents which should be limited by GMP or other quality-based requirements**

| | |
|---|---|
| Acetic acid | Heptane |
| Acetone | Isobutyl acetate |
| Anisole | Isopropyl acetate |
| 1-Butanol | Methyl acetate |
| 2-Butanol | 3-Methyl-1-butanol |
| Butyl acetate | Methyl ethyl ketone |
| tert-Butyl methyl cther | Methyl isobutyl ketone |
| Cumene | 2-Methyl-1-propanol |
| Dimethylsulfoxide | Pentane |
| Ethanol | 1-Pentanol |
| Ethyl acetate | 1-Propanol |
| Ethyl ether | 2-Propanol |
| Ethyl formate | Propyl acetate |
| Formic acid | Tetrahydrofuran |

**Table 16.2.6. Solvents for which no adequate toxicological data was found**

| |
|---|
| 1,1-Diethoxypropane |
| 1,1-Dimethoxymethane |
| 2,2-Dimethoxypropane |
| Isooctane |
| Isopropyl ether |
| Methylisopropyl ketone |
| Methyltetrahydrofuran |
| Petroleum ether |
| Trichloroacetic acid |
| Trifluoroacetic acid |

*Solvents with low toxic potential solvents of class 3* (see Table 16.2.5) may be regarded as less toxic and of lower risk to human health. It is considered that amounts of these RS of 50 mg per day or less (corresponding to 5000 ppm or 0.5 % under option 1) would be acceptable without justification. Higher amounts may be acceptable provided they do not have a negative impact on the processability and the stability of the pharmaceutical product.

*Solvents for which no adequate toxicological data was found* (Table 16.2.6). These solvents can be used in the manufacture of drug substances, excipients and drug products, but the manufacturer should supply justification for residual levels of these solvents in pharmaceutical products.

## 16.2.4.5 Analytical procedures

If only class 3 solvents are present, a non-specific method such as loss of drying may be used. In the other cases a selective method (e.g., GC) is required. Especially if solvents of class 2 and class 3 are present at greater than their option 1 limits or 0.5 %, respectively, they should be identified and quantified.

## 16.2.4.6 Conclusions regarding the ICH Guideline

The lists are not exhaustive and other solvents can be used and added later to the lists. Recommended limits of class 1 and class 2 solvents or classification of solvents may change as new safety data become available.

Nevertheless this Guideline is of great interest for those involved in the pharmaceutical development in order to prepare successfully a pharmaceutical dossier acceptable everywhere in the world.

## 16.2.5 CONCLUSIONS

Those who have worked for many years in pharmaceuticals will have observed continuous progress in standards of Quality and Control in manufacturing.

Solvents including water are still used in almost every step of the elaboration of a drug product. Their residues could be detrimental for the processability and stability of the pharmaceutical products and the safety of patients. At the end of this millennium it can be said that the testing and control of RS has been thoroughly assessed and is based on robust and sensitive techniques, for which the limitations are known, resulting in a sound strategy accepted almost everywhere in the world.

## REFERENCES


1 *Societa Italiana di Scienze Farmaceutiche, Cronache Farmaceutiche*, **6**, 227-229 (1980).
2 Treshold limit values for chemical substances and physical agents in the work environment AGGIH 6500 GLENWAY.
3 J. Rabiant, *Ann. Pharm. Fr.*, **42**, 503 (1984).
4 J. Rabiant, *S.T.P. Pharma Pratiques*, **1**, 278 (1991).
5 M. Gachon, *S.T.P. Pharma Pratiques*, **1**, 531 (1991).
6 Procedures for setting limits for organic volatile solvents with methylene chloride as an example of the process, *Pharm For.*, **15**, 5748 (1989).
7 Survey: residual solvents, *Pharmeuropa*, **2**, 142 (1990).
8 H. Stumpf, E. Spiess and M. Habs, *Dtsch. Apoth. Ztg.*, **132**, 508 (1992).
9 A. Eichhorn, T. Gabrio and S. Plagge, *Zent. Bl. Pharm. Pharmakother. Lab. Diagn.*, **128**, 675 (1989).
10 C. Witschi and E. Doelker, *Eur. J. Pharm. Biopharm.*, **43**, 215 (1997).
11 E. Ettabia, C. Barthelemy, M. Jbilou and A.M. Guyot-Hermann, *Pharmazie*, **53**, 8 (1998).
12 G.A. Stephenson, E.G. Groleau, R.L. Kleemann, W. Xu and D.R. Rigsbee, *J. Pharm. Sci.*, **87**, 536 (1998).
13 S.R. Byrn, C.T. Lin, P. Perrier, G.G. Clay and P.A. Sutton, *J. Org. Chem.*, **47**, 2978 (1982).
14 J. Bauer, J. Quick, R. Oheim, *J. Pharm. Sci.*, **74**, 899 (1985).
15 A.M. Guyot-Hermann, *S.T.P. Pharma Pratiques*, **1**, 258 (1991).
16 P.H. List and G. Laun, *Pharm. Ind.*, **42**, 399 (1980).
17 C. Lefebvre Ringard, A.M. Guyot-Hermann, R. Bouche et J. Ringard, *S.T.P. Pharma*, **6**, 228 (1990).
18 N. Barbarin, A.S. Crucq and B. Tilquin, *Radiat. Phys. Chem.*, **48**, 787 (1996).
19 N. Barbarin, B. Rollmann, B. Tilquin, *Int. J. Pharm.*, **178**, 203 (1999).
20 Z.T. Chowhan, *Drug Dev. Ind. Pharm*, **5**, 41 (1979).
21 G.R.B. Down and J.N. McMullen, *Powder Technol.*, **42**, 169 (1985).
22 J.F. Letavernier, M. Aubert, G. Ripoche et F. Pellerin, *Ann. Pharm. Franc.*, **43**, 117, (1985).
23 K.J. Mulligan, T.W. Brueggemeyer, D.F. Crockett, J.B. Shepman, *J. Chrom.*, **686**, 85 (1996).
24 K.J. Mulligan and H. McCaulay, *J. Chrom. Sci.*, **33**, 49 (1995).
25 D.R. Morello and R.P. Meyers, *J. Forensic Sci, JFSCA*, **40**(6), 957 (1995).
26 A.C. Cartwright, *Int. Pharm. J.*, **4**, 146 (1990).
27.1 Procedure for setting limits for organic volatile solvents with methylene chloride as an example of the process, *Pharm. For.*, **15**, 5748 (1989).
27.2 D. Galer, R.H. Ku, C.S. Schwartz, *Pharm. For.*, **17**, 1443 (1991).
27.3 L. Brooks, J.S. Mehring, *Pharm. For.*, **16**, 550 (1990).
27.4 B.D. Naumann, E.V. Sargent, *Pharm. For.*, **16**, 573 (1990).
27.5 G.L. Sprague, S. Beecham-K., D.L. Conine, *Pharm. For.*, **16**, 543 (1990).
28 A. Picot, Information Toxicologique n° 3, Unité de prévention du risque chimique - CNRS (France) (1995).
29 J. Belegaud, Communication personnelle (1997).
30 H.H. Cornish, B.P. Ling and M.L. Barth, *Amer. Ind. Hyg. Ass. J.*, **34**, 487 (1973).
31 K. Morpoth, U. Witting and M. Springorum, *Int. Arch. Arveits Med.*, **33**, 315 (1974).
32 J.W. Hayden and E.G. Comstock, *Clinical Toxicology*, **9**, 164 (1976).
33 M.J. Archieri, H. Janiaut, A. Picot, *L'Actualité Chimique*, Mai-Juin, 241 (1992).
34 Le Goff - Les risques physicochimiques et toxiques des solvants - Communication personnelle
35 M.G. Vachon, J.G. Nain, *J. Microencapsul.*, **12**, 287 (1995).
36 H.W. Ardovich, M.J. Lebelle, C. Savard, W.L. Wilson, *Forensic. Sci. Int.*, **49**, 225 (1991).
37 J. Haslam and A.R. Jeffs, *Analyst*, **83**, 455 (1958).
38 J.A. Hudy, *J. Gas Chromatogr.*, **4**, 350 (1966).

39 M. Suzuki, S. Tsuge and T. Takeuchi, *Anal. Chem.*, **42**, 1705 (1970).
40 K. Grob, *J. Chrom.*, **84**, 255 (1973).
41 T.K. Natishan and Y. Wu, *J. Chromatogr.*, **A800**, 275 (1998).
42 Q. Chanli, K.A. Cohen and G. Zhuang, *J. Chrom. Sci.*, **36**, 119 (1998).
43 T.K. Chen, J.G. Phillips and W. Durr, *J. Chromatogr.*, **A811**, 145 (1998).
44 J.P. Guimbard, M. Person and J.P. Vergnaud, *J. Chromatogr.*, **403**, 109 (1987).
45 Progress report of the working party on residual solvents (technical) of the European Pharmacopoeia Commission, *Pharmeuropa*, **8**, 586 (1996).
46 V.J. Naughton, *Pharm. Forum*, **20**, 7223 (1994).
47 M. de Smet, K. Roels, L. Vanhoof and W. Lauwers, *Pharm. Forum*, **21**, 501 (1995).
48 B. Kolb, *Chromatographia*, **10**, 587 (1982).
49 C. MacAullife, *Chem. Tech.*, **46**, 51 (1971).
50 J.P. Guimbard, J. Besson, S. Beaufort, J. Pittie et M. Gachon, *S.T.P. Pharma Pratiques*, **1**(3), 272 (1991).
51 P. Klaffenbach, C. Brüse, C. Coors, D. Kronenfeld and H.G. Schulz, *LC-GC*, **15**, 1052 (1997).
52 P. Kuran and L. Sojak, *J. Chromatogr.*, **A773**, 119 (1996).
53 C.C. Camarasu, M. Mezei-Szüts, G. Bertok Varga, *J. Pharm. Biomed. Anal.*, **18**, 623 (1998).
54 R.J. Bartelt, *Anal. Chem.*, **69**, 364 (1997).
55 F.J. Santos, M.T. Galceram and D. Fraisse, *J. Chromatogr.*, **742**, 181 (1996).
56 M. Markelov and J.P. Guzowski, *Anal. Chim. Acta*, **276**, 235 (1993).
57 G. Castello, S. Vezzani and T.C. Gerbino, *J. Chromatogr.*, **585**, 273 (1991).
58 K. Brinkmann and S. Ebel, *Pharm. Ind.*, **61**, 263 (1999).
59 K. Brinkmann and S. Ebel, *Pharm. Ind.*, **61**, 372 (1999).
60 W.D. Bowers, M.L. Parsons, R.E. Clement, G.A. Eiceman and F.W. Karaseck, *J. Chromatogr.*, **206**, 279 (1981).
61 C.N. Kunigami, M.S. Sanctos, M. Helena, W. Morelli-Cardoro, *J. High Resol. Chromatogr.*, **22**, 477 (1999).
62 D.G. Westmorland, G.R. Rhodes, *Pure Appl. Chem.*, **61**, 1148 (1989).
63 Validation of analytical methods: Methodology, *Pharmeuropa*, **8**, 114 (1996).
64 R.B. George and P.D. Wright, *Anal. Chem.*, **69**, 2221 (1997).
65 G. Wynia, P. Post, J. Broersten and F.A. Maris, *Chromatographia*, **39**, 355 (1994).
66 R.J. Markovitch, S. Ong and J. Rosen, *J. Chromatogr. Sci.*, **35**, 584 (1997).
67 M. Desmet, K. Roels, L. Vanhoof and W. Lauwers, *Pharm. Forum*, **21**, 501 (1995).
68 C.M. Riley, Statistical parameters and analytical figures of Merit in **Development and Validation of Analytical Methods**, C.M. Riley and T.W. Rosanske, Ed., *Pergamon*, 1996, pp15-72.
69 **European Pharmacopoeia**, **Addendum 2000**, pp31-36. [Please note that information concerning residual solvents are susceptible to be modified in the successive editions of the European Pharmacopeia.]
70 Progress report of the working party on residual solvent (Technical) of the European Pharmacopoeia Commission, *Pharmeuropa*, **8**, 586 (1996).
71 Note for Guidance on impurities: Residual solvents, *Drugs made in Germany*, **41**, 98 (1998).

39 [illegible] Anal. Chem., 42 [illegible]
40 K. Grob, [illegible] 84, 255 (1975).
41 [illegible] Chromatogr. A, [illegible]
42 [illegible] J. Chromatogr. Sci., 36, 119 [illegible]
43 [illegible] Chen, J.G. Phillips and [illegible] Chromatogr. A [illegible]
44 [illegible] Chromatogr. [illegible]
45 [illegible] of the working party on residual solvents, [illegible] of the European Pharmacopoeia Commission, [illegible]
46 [illegible] Pharm. [illegible] (1994).
47 M. [illegible] Pharm. [illegible]
48 [illegible]
49 [illegible]
50 [illegible]
51 [illegible] Schulz [illegible]
52 [illegible]
53 [illegible]
54 [illegible]
55 [illegible] J. Chromatogr. [illegible]
56 [illegible]
57 [illegible] Chromatogr. [illegible]
58 [illegible]
59 [illegible]
60 W.D. [illegible] (1993).
61 [illegible]
62 [illegible]
63 [illegible]
64 [illegible]
65 [illegible] Chromatogr. [illegible]
66 [illegible]
67 [illegible]
68 [illegible] Analytical [illegible]
69 European Pharmacopoeia [illegible] concerning residual [illegible] Pharmacopoeia [illegible]
70 [illegible] Commission [illegible]
71 [illegible] (1998).

17

# Environmental Impact of Solvents

## 17.1 THE ENVIRONMENTAL FATE AND MOVEMENT OF ORGANIC SOLVENTS IN WATER, SOIL, AND AIR[a]

William R. Roy
**Illinois State Geological Survey, Champaign, IL, USA**

### 17.1.1 INTRODUCTION

Organic solvents are released into the environment by air emissions, industrial and waste-treatment effluents, accidental spillages, leaking tanks, and the land disposal of solvent-containing wastes. For example, the polar liquid acetone is used as a solvent and as an intermediate in chemical production. ATSDR[1] estimated that about 82 million kg of acetone was released into the atmosphere from manufacturing and processing facilities in the U.S. in 1990. About 582,000 kg of acetone was discharged to water bodies from the same type of facilities in the U.S. ATSDR[2] estimated that in 1988 about 48,100 kg of tetrachloroethylene was released to land by manufacturing facilities in the U.S.

Once released, there are numerous physical and chemical mechanisms that will control how a solvent will move in the environment. As solvents are released into the environment, they may partition into air, water, and soil phases. While in these phases, solvents may be chemically transformed into other compounds that are less problematic to the environment. Understanding how organic solvents partition and behave in the environment has led to better management approaches to solvents and solvent-containing wastes. There are many published reference books written about the environmental fate of organic chemicals in air, water, and soil.[3-7] The purpose of this section is to summarize the environmental fate of six groups of solvents (Table 17.1.1) in air, water, and soil. A knowledge of the likely pathways for the environmental fate of organic solvents can serve as the technical basis for the management of solvents and solvent-containing wastes.

[a]Publication authorized by the Chief, Illinois State Geological Survey

**Table 17.1.1. The six groups of solvents discussed in this section**

| | |
|---|---|
| *Alcohols*<br>n-Butyl alcohol<br>Isobutyl alcohol<br>Methanol<br><br>*Benzene Derivatives*<br>Benzene<br>Chlorobenzene<br>o-Cresol<br>o-Dichlorobenzene<br>Ethylbenzene<br>Nitrobenzene<br>Toluene<br>o-Xylene<br><br>*Chlorinated Aliphatic Hydrocarbons*<br>Carbon tetrachloride<br>Dichloromethane<br>Tetrachloroethylene<br>1,1,1-Trichloroethane<br>Trichloroethylene | *Chlorinated Fluorocarbons*<br>Trichlorofluoromethane (F-11),<br>1,1,2,2-Tetrachloro-1,2-difluoroethane ( F-112)<br>1,1,2-Trichloro-1,2,2-trifluoroethane (F-113)<br>1,2,-Dichlorotetrafluoroethane (F-114)<br><br>*Ketones*<br>Acetone<br>Cyclohexanone<br>Methyl ethyl ketone<br>Methy isobutyl ketone<br><br>*Others*<br>Carbon disulfide<br>Diethyl ether<br>Ethyl acetate<br>Hexane<br>Decane (a major component of mineral spirits)<br>Pyridine<br>Tetrahydrofuran |

## 17.1.2 WATER

### 17.1.2.1 Solubility

One of the most important properties of an organic solvent is its solubility in water. The greater a compound's solubility, the more likely that a solvent or a solvent-containing waste will dissolve into water and become part of the hydrological cycle. Hence, water solubility can affect the extent of leaching of solvent wastes into groundwater, and the movement of dissolved solvent into rivers and lakes. Aqueous solubility also determines the efficacy of removal from the atmosphere through dissolution into precipitation. The solubility of solvents in water may be affected by temperature, salinity, dissolved organic matter, and the presence of other organic solvents.

### 17.1.2.2 Volatilization

Solvents dissolved in water may volatilize into the atmosphere or soil gases. A Henry's Law constant ($K_H$) can be used to classify the behavior of dissolved solvents. Henry's Law describes the ratio of the partial pressure of the vapor phase of an ideal gas ($P_i$) to its mole fraction ($X_i$) in a dilute solution, viz.,

$$K_{H(i)} = P_i / X_i \qquad [17.1.1]$$

In the absence of measured data, a Henry's Law constant for a given solvent may be estimated by dividing the vapor pressure of the solvent by its solubility in water ($S_i$) at the same temperature;

$$K_{H(i)} = P_i\,(atm) / S_i\,(mol/m^3\ solvent) \qquad [17.1.2]$$

A $K_H$ value of less than $10^{-4}$ atm-mol/m$^3$ suggests that volatilization would probably not be a significant fate mechanism for the dissolved solvent. The rate of volatilization is more complex, and depends on the rate of flow, depth, and turbulence of both the body of water and the atmosphere above it. In the absence of measured values, there are a number of estimation techniques to predict the rate of removal from water.[8]

#### 17.1.2.3 Degradation

The disappearance of a solvent from solution can also be the result of a number of abiotic and biotic processes that transform or degrade the compound into daughter compounds that may have different physicochemical properties from the parent solvent. Hydrolysis, a chemical reaction where an organic solvent reacts with water, is not one reaction, but a family of reactions that can be the most important processes that determine the fate of many organic compounds.[9] Photodegradation is another family of chemical reactions where the solvent in solution may react directly under solar radiation, or with dissolved constituents that have been made reactive by solar radiation. For example, the photolysis of water yields a hydroxyl radical:

$$H_2O + h\nu \rightarrow HO\bullet + H \qquad [17.1.3]$$

Other oxidants such as peroxy radicals ($RO_2\bullet$) and ozone can react with solvents in water. The subject of photodegradation is treated in more detail under atmospheric processes (17.1.4).

Biodegradation is a family of biologically mediated (typically by microorganisms) conversions or transformations of a parent compound. The ultimate end-products of biodegradation are the conversion of organic compounds to inorganic compounds associated with normal metabolic processes.[10] This topic will be addressed under Soil (17.1.3.3).

#### *17.1.2.4 Adsorption*

Adsorption is a physicochemical process whereby a dissolved solvent may be concentrated at solid-liquid interfaces such as water in contact with soil or sediment. In general, the extent of adsorption is inversely proportional to solubility; sparingly soluble solvents have a greater tendency to adsorb or partition to the organic matter in soil or sediment (see Soil, 17.1.3.2).

### 17.1.3 SOIL

#### 17.1.3.1 Volatilization

Volatilization from soil may be an important mechanism for the movement of solvents from spills or from land disposed solvent-containing wastes. The efficacy and rate of volatilization from soil depends on the solvent's vapor pressure, water solubility, and the properties of the soil such as soil-water content, airflow rate, humidity, temperature and the adsorption and diffusion characteristics of the soil.

Organic-solvent vapors move through the unsaturated zone (the interval between the ground surface and the water-saturated zone) in response to two different mechanisms; convection and diffusion. The driving force for convective movement is the gradient of total gas pressure. In the case of diffusion, the driving force is the partial-pressure gradient of each gaseous component in the soil air. The rate of diffusion of a solvent in bulk air can be described by Fick's Law, viz.,

$$Q = -D_f \nabla_a \qquad [17.1.4]$$

where:

| | |
|---|---|
| Q | diffusive flux (mass/area-time) |
| $D_f$ | diffusion coefficient (area/time) |
| $\nabla_a$ | concentration gradient (mass/volume/distance) |

Compared with the relatively unobstructed path for the diffusion of solvents in the atmosphere, diffusion coefficients for solvents in soil air will be less because of the tortuosity of the soil matrix pathways. Several functional relationships have been developed that relate the soil diffusion coefficient ($D_s$) to various soil properties (see Roy and Griffin[11]), such as the Millington Equation[12]

$$D_s = D_f \eta_a^{3.3} / \eta_t^2 \qquad [17.1.5]$$

where:

| | |
|---|---|
| $\eta_a$ | the air-filled porosity, and |
| $\eta_t$ | total soil porosity |

#### 17.1.3.2 Adsorption

As discussed in 17.1.2.4., adsorption by soil components can remove solvents dissolved in water. Furthermore, the rate of movement of dissolved solvents through soil may be retarded by adsorption-desorption reactions between the solvents and the solid phases. The partitioning of solvents between the liquid phase and soil is usually described by an adsorption isotherm. The adsorption of solvents may be described by the Freundlich Equation:

$$x / m = K_f C^{l/n} \qquad [17.1.6]$$

where:

| | |
|---|---|
| x | the mass adsorbed |
| m | mass of sorbent |
| $K_f$ | the Freundlich constant, a soil-specific term |
| C | the equilibrium concentration of the solvent in water, and |
| n | the Freundlich exponent which describes the degree of non-linearity of the isotherm |

When n is equal to one, the Freundlich Equation becomes a relatively simple partition function:

$$x/m = KC \qquad [17.1.7]$$

where K is an adsorption or distribution coefficient which is sometimes written as $K_d$. It has been known since the 1960s that the extent of adsorption of hydrophobic (sparingly soluble in water) solvents often correlates with the amount of organic matter in the soil.[13] When $K_d$ is divided by the amount of organic carbon in the soil, the resulting coefficient is the organic carbon-water partition coefficient ($K_{oc}$):

$$K_d \times 100/organic\ carbon(\%) = K_{oc} \qquad [17.1.8]$$

The organic carbon-water partition coefficient is a compound-specific term that allows the user to estimate the mobility of a solvent in saturated-soil water systems if the amount or organic carbon is known. For hydrophilic solvents, $K_{oc}$ values have been measured for many compounds. Other values were derived from empirical relationships drawn between water solubility or octanol-water partition coefficients.[13]

### 17.1.3.3 Degradation

Solvents may be degraded in soil by the same mechanisms as those in water. In biodegradation, microorganisms utilize the carbon of the solvents for cell growth and maintenance. In general, the more similar a solvent is to one that is naturally occurring, the more likely that it can be biodegraded into other compound(s) because the carbon is more available to the microbes. Moreover, the probability of biodegradation increases with the extent of water solubility of the compound. It is difficult to make generalities about the extent or rate of solvent biodegradation that can be expected in soil. Biodegradation can depend on the concentration of the solvent itself, competing processes that can make the solvent less available to microbes (such as adsorption), the population and diversity of microorganisms, and numerous soil properties such as water content, temperature, and reduction-oxidation potential. The rate and extent of biodegradation reported in studies appears to depend on the conditions under which the measurement was made. Some results, for example, were based on sludge-treatment plant simulations or other biological treatment facilities that had been optimized in terms of nutrient content, microbial acclimation, mechanical mixing of reactants, or temperature. Hence, these results may overestimate the extent of biodegradation in ambient soil in a spill or waste-disposal scenario.

First-order kinetic models are commonly used to describe biodegradation because of their mathematical simplicity. First-order biodegradation is to be expected when the organisms are not increasing in abundance. A first-order model also lends itself to calculating a half-life ($t_{1/2}$) which is a convenient parameter to classify the persistence of a solvent. If a solvent has a soil half-life of 6 months, then about half of the compound will have degraded in six months. After one year, about one fourth the initial amount would still be present, and after 3 half-lives (1.5 years), about 1/8 of the initial amount would be present.

Howard et al.[14] estimated ranges of half-lives for solvents in soil, water, and air. For solvents in soil, the dominant mechanism in the reviewed studies may have been biodegradation, but the overall values are indicative of the general persistence of a solvent without regard to the specific degradation mechanism(s) involved.

## 17.1.4 AIR

### 17.1.4.1 Degradation

As introduced in 17.1.2.3, solvents may be photodegraded in both water and air. Atmospheric chemical reactions have been studied in detail, particularly in the context of smog formation, ozone depletion, and acid rain. The absorption of light by chemical species generates free radicals which are atoms, or groups of atoms that have unpaired electrons. These free radicals are very reactive, and can degrade atmospheric solvents. Atmospheric ozone, which occurs in trace amounts in both the troposphere (sea level to about 11 km) and in the stratosphere (11 km to 50 km elevation), can degrade solvents. Ozone is produced by the photochemical reaction:

$$O_2 + h\nu \rightarrow O + O \quad [17.1.9]$$

$$O + O_2 \rightarrow O_3 + M \quad [17.1.10]$$

where M is another species such as molecular nitrogen that absorbs the excess energy given off by the reaction. Ozone-depleting substances include the chlorofluorocarbons (CFC) and carbon tetrachloride in the stratosphere.

### 17.1.4.2 Atmospheric residence time

Vapor-phase solvents can dissolve into water vapor, and be subject to hydrolysis reactions and ultimately, precipitation (wet deposition), depending on the solubility of the given solvent. The solvents may also be adsorbed by particulate matter, and be subject to dry deposition. Lyman[16] asserted that atmospheric residence time cannot be directly measured; that it must be estimated using simple models of the atmosphere. Howard et al.[14] calculated ranges in half-lives for various organic compounds in the troposphere, and considered reaction rates with hydroxyl radicals, ozone, and by direct photolysis.

## 17.1.5 THE 31 SOLVENTS IN WATER

### 17.1.5.1 Solubility

The solubility of the solvents in Table 17.1.1 ranges from those that are miscible with water to those with solubilities that are less than 0.1 mg/L (Table 17.1.2). Acetone, methanol, pyridine and tetrahydrofuran will readily mix with water in any proportion. The solvents that have an aqueous solubility of greater than 10,000 mg/L are considered relatively hydrophillic as well. Most of the benzene derivatives and chlorinated fluorocarbons are relatively hydrophobic. Hexane and decane are the least soluble of the 31 solvents in Table 17.1.1. Most material safety data sheets for decane indicate that the n-alkane is "insoluble" and that the solubility of hexane is "negligible." How the solubility of each solvent affects its fate in soil, water, and air is illustrated in the following sections.

**Table 17.1.2. The solubility of the solvents in water at 25°C**

| Solubility, mg/L | Solvent (reference) | |
|---|---|---|
| ∞ | Acetone (1)<br>Methanol (1)<br>Pyridine (1)<br>Tetrahydrofuran (1) | Miscible |
| 239,000 | Methyl ethyl ketone (4) | |
| 77,000<br>76,000<br>64,000<br>60,050<br>25,950<br>23,000<br>20,400<br>13,000 | n-Butyl alcohol (4)<br>Isobutyl alcohol (4)<br>Ethyl acetate (4)<br>Diethyl ether (4)<br>o-Cresol (17)<br>Cyclohexanone (4)<br>Methyl isobutyl ketone (4)<br>Dichloromethane (4) | Relatively hydrophillic |
| 2,100<br>1,900<br>1,780<br>1,495<br>1,100<br>1,080 | Carbon disulfide (4)<br>Nitrobenzene (18)<br>Benzene (19)<br>1,1,1-Trichloroethane (4)<br>Trichloroethylene (4)<br>F-11 (4) | |

| Solubility, mg/L | Solvent (reference) | |
|---|---|---|
| 805 | Carbon tetrachloride (4) | Relatively hydrophobic |
| 535 | Toluene (20) | |
| 472 | Chlorobenzene (17) | |
| 175 | o-Xylene (4) | |
| 170 | F-113 (4) | |
| 161 | Ethylbenzene (17) | |
| 156 | o-Dichlorobenzene (17) | |
| 150 | Tetrachloroethylene (4) | |
| 130 | F-114 (4) | |
| 120 | F-112 (4) | |
| 9.5 | Hexane (21) | |
| 0.05 | Decane (22) | "Insoluble" |

### 17.1.5.2 Volatilization from water

Henry's Law constants were compiled for each of the solvents in Table 17.1.1. The numerical values ranged over 7 orders of magnitude (Table 17.1.3). Based on these values, it can be expected that volatilization from water will be a significant fate mechanism for decane, hexane, the chlorinated fluorocarbons, carbon tetrachloride, tetrachloroethylene and trichloroethylene. Many of the solvents in Table 17.1.1 are characterized by $K_H$ values of $10^{-3}$ to $10^{-2}$ atm-$m^3$/mole; volatilization from water can be an important pathway for these solvents, depending on the specific situation. Volatilization may be a relatively slow process for the remaining solvents. The actual rate of volatilization of some solvents from water has been experimentally measured.[4,17] However, experimental data are lacking for some compounds, and the diversity of experimental conditions makes generalizations difficult. Thomas[8] described a two-layer model of the liquid-gas interface that is based on a Henry's Law constant and mass-transfer coefficients. To illustrate the relative volatilities of the solvents in water, the half-lives of each solvent in a shallow stream were compiled (Table 17.1.4). The stream was assumed to be 1 meter deep and flowing at a rate of 1 meter per second. With the exception of hexane, it was also assumed that there was a breeze blowing across the stream at a rate of 3 meters per second. Under these conditions, the predicted half-lives of many of the solvents in Table 16.1.1 are less than 10 hours, indicating that volatilization into the atmosphere can be a relatively rapid pathway for solvents released to surface water. The volatilization of pyridine, isobutyl alcohol, and cyclohexanone may be a slow process, and other fate processes may be more important in water.

### 17.1.5.3 Degradation in water

As mentioned in 17.1.3.3, Howard et al.[14] compiled ranges of half-life values for most of the organic solvents given in Table 17.1.1. If a "rapid" half-life is defined as in the range of 1 to 10 days, then about 12 of the solvents in Table 17.1.1 may degrade rapidly in surface water by primarily biodegradation (Figures 17.1.1 and 17.1.2). Abiotic mechanisms such as photo-oxidation, photolysis, and hydrolysis appear to be either slow or not significant. If "slow degradation" is defined as that taking longer than 100 days, then it appears that F-11 and most of the chlorinated hydrocarbons may be relatively persistent in surface water. The available data suggest that the half-life of nitrobenzene and isobutyl alcohol may be variable. Note that data were not available for all of the solvents listed in Table 17.1.1. In

**Table 17.1.3. Henry's Law constants ($K_H$) for the solvents at 25°C**

| $K_H$, atm-m$^3$/mole | Solvent (reference) |
|---|---|
| 6.98 | Decane (22) |
| 2.8 | F-114 (4) |
| 1.69 | Hexane (21) |
| 0.53 | F-113 (4) |
| $9.74 \times 10^{-2}$ | F-112 (4) |
| $9.70 \times 10^{-2}$ | F-11 (4) |
| $3.04 \times 10^{-2}$ | Carbon tetrachloride (4) |
| $1.49 \times 10^{-2}$ | Tetrachloroethylene (4) |
| $1.03 \times 10^{-2}$ | Trichloroethylene (4) |
| $9.63 \times 10^{-3}$ | Tetrahydrofuran (4) |
| $8.4 \times 10^{-3}$ | Ethylbenzene (17) |
| $8.0 \times 10^{-3}$ | 1,1,1-Trichloroethane (4) |
| $7.0 \times 10^{-3}$ | Pyridine (4) |
| $5.94 \times 10^{-3}$ | Toluene (4) |
| $5.43 \times 10^{-3}$ | Benzene (4) |
| $5.1 \times 10^{-3}$ | o-Xylene (4) |
| $3.58 \times 10^{-3}$ | Chlorobenzene (23) |
| $2.68 \times 10^{-3}$ | Dichloromethane (4) |
| $1.4 \times 10^{-3}$ | Carbon disulfide (4) |
| $1.2 \times 10^{-3}$ | o-Dichlorobenzene (4) |
| $7.48 \times 10^{-4}$ | Diethyl ether (11) |
| $4 \times 10^{-4}$ | Isobutyl alcohol (4) |
| $1.35 \times 10^{-4}$ | Methanol (4) |
| $1.2 \times 10^{-4}$ | Ethyl acetate (4) |
| $9.4 \times 10^{-5}$ | Methyl isobutyl ketone (4) |
| $4.26 \times 10^{-5}$ | Acetone (1) |
| $2.44 \times 10^{-5}$ | Nitrobenzene (2) |
| $1.2 \times 10^{-5}$ | Cyclohexanone (4) |
| $1.05 \times 10^{-5}$ | Methyl ethyl ketone (4) |
| $5.57 \times 10^{-6}$ | n-Butyl alcohol (4) |
| $1.2 \times 10^{-6}$ | o-Cresol (4) |

**Table 17.1.4. Estimated half-lives for the solvents in water at 20°C**

| Half life, h | Solvent |
|---|---|
| 1.6 | Tetrahydrofuran |
| 2.6 | Carbon disulfide |
| 2.7 | Hexane[a] |
| 2.9 | Toluene |
| 3.0 | Dichloromethane |
| 3.1 | Ethylbenzene |
| 3.2 | o-Xylene |
| 3.4 | Trichloroethylene, F-11 |
| 3.7 | Carbon tetrachloride |
| 4.0 | F-112, F-113, F-114 |
| 4.2 | Tetrachloroethylene |
| 4.4 | o-Dichlorobenzene |
| 4.6 | Chlorobenzene |
| 5.3 | Methanol |
| 10 | Ethyl acetate |
| 18 | Acetone |
| 45 | Nitrobenzene |
| 74 | Cyclohexanone |
| 80 | Isobutyl alcohol |
| 90 | Pyridine |

[a]Based on a wind speed of 1 meter per second.[21]

groundwater, the half-life values proposed by Howard et al.[14] appear to be more variable than those for surface water. For example, the half-life of benzene ranges from 10 days in aerobic groundwater to 2 years in anaerobic groundwater.[19] Such ranges in half-lives make meaningful generalizations difficult. However, it appears that methanol, n-butyl alcohol, and other solvents (see Figures 17.1.1 and 17.1.2) may biodegrade in groundwater with a half-life that is less than 60 days. As with surface water, the chlorinated hydrocarbons may be relatively persistent in groundwater. Howard et al.[14] cautioned that some of their proposed half-life generalizations were based on limited data or from screening studies that were extrapolated to surface and groundwater. Scow[10] summarized that it is currently not possible to predict rates of biodegradation because of a lack of standardized experimental methods, and because the variables that control rates are not well understood. Hence, Figures 17.1.1 and 17.1.2 should be viewed as a summary of the potential for each solvent to degrade, pending more site-specific information.

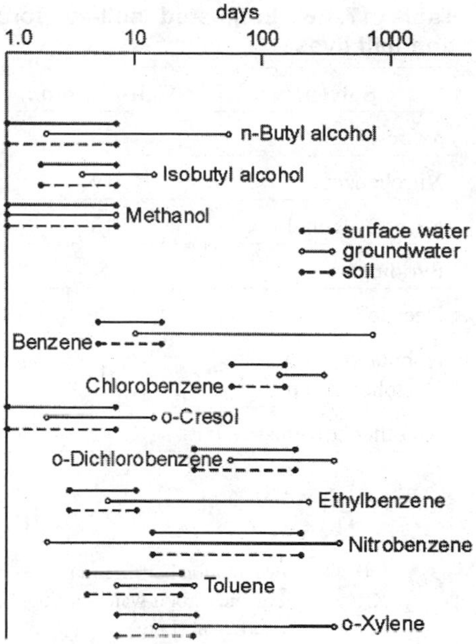


Figure 17.1.1. The ranges in degradation half-lives for the alcohols and benzene derivatives in surface water, groundwater, and soil (data from Howard et al.[14]).

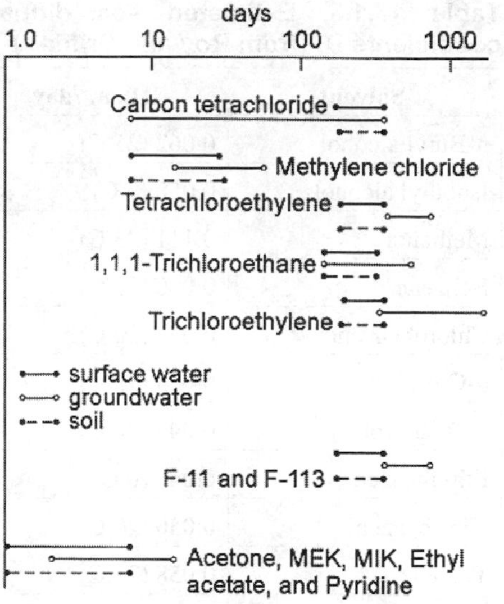


Figure 17.1.2. The ranges in degradation half-lives for the chlorinated aliphatic hydrocarbons, F-11 and F-113, and ketones in surface water, groundwater, and soil (data from Howard et al.[14]).

## 17.1.6 SOIL

### 17.1.6.1 Volatilization

Soil diffusion coefficients were estimated for most of the solvents in Table 17.1.1. Using the Millington Equation, the resulting coefficients (Table 17.1.5) ranged from 0.05 to 0.11 $m^2$/day. Hence, there was little variation in magnitude between the values for these particular solvents. As discussed in Thomas[24] the diffusion of gases and vapors in unsaturated soil is a relatively slow process. The coefficients in Table 17.1.5 do not indicate the rate at which olvents can move in soil; such rates must be either measured experimentally or predicted using models that require input data such as soil porosity, moisture content, and the concentrations of the solvents in the vapor phase to calculate fluxes based solely on advective movement. Variations in water content, for example, will control vapor-phase movement. The presence of water can reduce the air porosity of soil, thereby reducing the soil diffusion coefficient (Eq. 17.1.5). Moreover, relatively water-soluble chemicals may dissolve into water in the vadose zone. Hence, water can act as a barrier to the movement of solvent vapors from the subsurface to the surface.

Solvents spilled onto the surface of soil may volatilize into the atmosphere. The Dow Method[24] was used in this section to estimate half-life values of each solvent if spilled on the surface of a dry soil. The Dow Method is a simple relationship that was derived for the evaporation of pesticides from bare soil;

$$t_{1/2}\ (days) = 1.58 \times 10^{-8}\ (K_{oc}S/P_v) \qquad [17.1.11]$$

**Table 17.1.5. Estimated soil diffusion coefficients $D_s$ (from Roy and Griffin[11])**

| Solvent | $D_s$, m$^2$/day |
|---|---|
| n-Butyl alcohol | 0.062 (25°C) |
| Isobutyl alcohol | 0.050 (0°C) |
| Methanol | 0.111 (25°C) |
| Benzene | 0.060 (15°C) |
| Chlorobenzene | 0.052 (30°C) |
| o-Cresol | 0.053 (15°C) |
| o-Dichlorobenzene | 0.049 (20°C) |
| Ethylbenzene | 0.046 (0°C) |
| Nitrobenzene | 0.050 (20°C) |
| Toluene | 0.058 (25°C) |
| o-Xylene | 0.049 (15°C) |
| Carbon tetrachloride | 0.051 (25°C) |
| Dichloromethane | 0.070 (15°C) |
| Tetrachloroethylene | 0.051 (20°C) |
| 1,1,1-Trichloroethane | 0.075 (20°C) |
| Trichloroethylene | 0.058 (15°C) |
| F-11 | 0.060 (15°C) |
| F-112 | - |
| F-113 | 0.053 (15°C) |
| F-114 | 0.056 (15°C) |
| Acetone | 0.076 (0°C) |
| Cyclohexanone | - |
| Methyl ethyl ketone | - |
| Methyl isobutyl ketone | - |
| Carbon disulfide | 0.074 (25°C) |
| Diethyl ether | 0.054 (0°C) |
| Ethyl acetate | 0.059 (25°C) |
| Hexane | - |
| Mineral spirits | - |
| Pyridine | - |
| Tetrahydrofuran | - |

**Table 17.1.6. Estimated soil-evaporation half lives**

| Solvent | Half-life, min. |
|---|---|
| o-Cresol | 38 |
| Nitrobenzene | 19 |
| n-Butyl alcohol | 18 |
| Pyridine | 8 |
| Decane | 4 |
| Isobutanol<br>Cyclohexanone | 1 |
| All other solvents | <1 |

where:

$t_{1/2}$ evaporation half-life (days)
$K_{oc}$ organic carbon-water partition coefficient (L/kg)
S solubility in water (mg/L), and
$P_v$ vapor pressure (mm Hg at 20°C)

The resulting estimated half-life is inversely proportional to vapor pressure; the greater the vapor pressure, the greater the extent of volatilization. Conversely, the rate of volatilization will be reduced if the solvent readily dissolves into water or is adsorbed by the soil. Organic carbon-water partition coefficients were compiled for each solvent (see 17.1.6.2.), and vapor pressure data (not shown) were collected from Howard.[4] The resulting half-life estimates (Table 17.1.6) indicated that volatilization would be a major pathway if the liquid solvents were spilled on soil; all of the half-life estimates were less than one hour. Thomas[24] cautioned, however, that soil moisture, soil type, temperature, and wind conditions were not incorporated in the simple Dow Model.

**Table 17.1.7. The organic carbon-water partition coefficients ($K_{oc}$) of the solvents at 25°C**

| $K_{oc}$, L/kg | Solvent (reference) | |
|---|---|---|
| <1 | Methanol (13), Tetrahydrofuran[a] | |
| 1 | Acetone (13) | |
| 4 | Methyl ethyl ketone (13) | |
| 7 | Pyridine (13) | Mobile |
| 8 | Ethyl acetate, isobutyl alcohol (13) | |
| 9 | Diethyl ether (13) | |
| 10 | Cyclohexanone (13) | |
| 20 | o-Cresol (17) | |
| 24 | Methyl isobutyl ketone (13) | |
| 25 | Dichloromethane (13) | |
| 63 | Carbon disulfide (13) | |
| 67 | Nitrobenzene (13) | |
| 72 | n-Butyl alcohol (4) | |
| 97 | Benzene (13) | |
| 110 | Carbon tetrachloride (4) | |
| 152 | Trichloroethylene (13) | |
| 155 | 1,1,1-Trichloroethane (13) | |
| 164 | Ethylbenzene (17) | |
| 242 | Toluene (26) | |
| 303 | Tetrachloroethylene (13) | |
| 318 | Chlorobenzene (13) | Relatively mobile |
| 343 | o-Dichlorobenzene (25) | |
| 363 | o-Xylene (13) | |
| 372 | F-113 (13) | |
| 437 | F-114 (13) | |
| 457 | F-112 (13) | |
| 479 | F-11 (13) | |
| 1,950 | Hexane (21) | Relatively Immobile |
| 57,100[a] | Decane | |

[a]Calculated using the relationship $\log K_{oc} = 3.95 - 0.62 \log S$ where S = water solubility in mg/L (see Hassett et al.[25])

### 17.1.6.2 Adsorption

Organic carbon-water partition coefficients were compiled (Table 17.1.7) for each of the solvents in Table 17.1.1. A $K_{oc}$ value is a measure of the affinity of a solvent to partition to organic matter which in turn will control the mobility of the solute in soil and groundwater under convective flow. Although the actual amount of organic matter will determine the extent of adsorption, a solvent with a $K_{oc}$ value of less than 100 L/kg is generally regarded as relatively mobile in saturated materials. Hence, adsorption may not be a significant fate mechanism for 16 of the solvents in Table 17.1.1. In contrast, adsorption by organic matter may be a major fate mechanism controlling the fate of three of the benzene derivatives, and most of the chlorinated compounds. Hexane and particularly decane would likely be relatively immobile. However, when the organic C content of an adsorbent is less than about 1

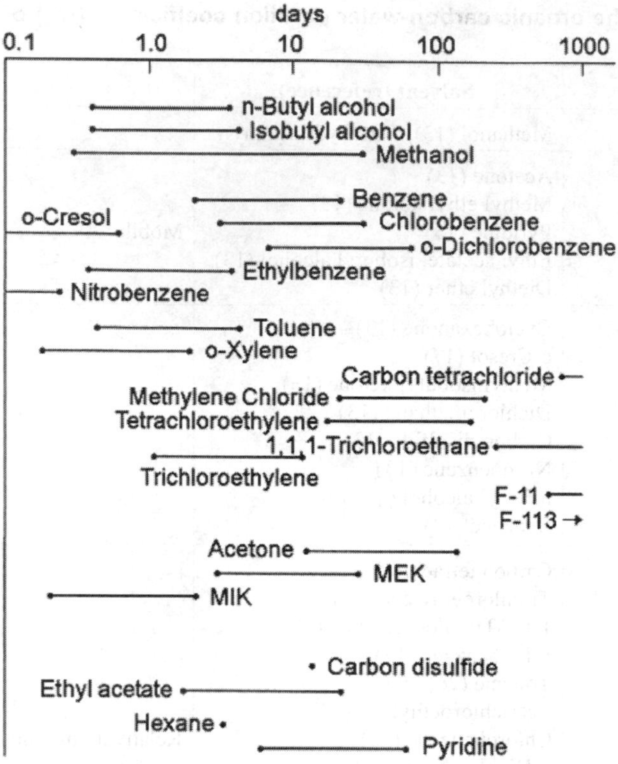


Figure 17.1.3. The ranges in atmospheric half-life of the solvents in Table 17.1.1 (data from Howard et al.[14] and ATSDR[21]).

g/kg, the organic C fraction is not a valid predictor of the partitioning of nonpolar organic compounds,[27] and other properties such as pH, surface area, or surface chemistry contribute to or dominate the extent of adsorption. Moreover, pyridine occurs at a cation (pKa = 5.25) over a wide pH range, and thus it is adsorbed by electrostatic interactions rather than by the hydrophobic mechanisms that are endemic to using $K_{oc}$ values to predict mobility.

The desorption of solvents from soil has not been extensively measured. In the application of advection-dispersion models to predict solute movement, it is generally assumed that adsorption is reversible. However, the adsorption of the solutes in Table 17.1.1 may not be reversible. For example, hysteresis is often observed in pesticide adsorption-desorption studies with soils.[28] The measurement and interpretation of desorption data for solid-liquid systems is not well understood.[29,30] Once adsorbed, some adsorbates may react further to become covalently and irreversibly bound, while others may become physically trapped in the soil matrix.[28] The non-singularity of adsorption-desorption may sometimes result from experimental artifacts.[28,31]

### 17.1.6.3 Degradation

As discussed in 17.1.3.3., Howard et al.[14] also estimated soil half-life values (Figures 17.1.1 and 17.1.2) for the degradation of most of the solvents in Table 17.1.1. Biodegradation was cited as the most rapid process available to degrade solvents in a biologically active soil. The numerical values obtained were often the same as those estimated for surface water.

Consequently, it appears likely that the alcohols, ketones, o-cresol, ethyl acetate, and pyridine will degrade rapidly in soil if rapidly is defined as having a half-life of 10 days or less. Most of the benzene derivatives, F-11, and the chlorinated aliphatic hydrocarbons may be relatively persistent in soil. Analogous information was not located for diethyl ether, hexane, decane, or tetrahydrofuran. ATSDR[21] for example, found that there was little information available for the degradation of n-hexane in soil. It was suggested that n-hexane can degrade to alcohols, aldehydes, and fatty acids under aerobic conditions.

### 17.1.7 AIR

Once released into the atmosphere, the most rapid mechanism to attenuate most of the solvents in Table 17.1.1 appears to be by photo-oxidation by hydroxyl radicals in the troposphere. Based on the estimates by Howard et al.,[14] it appeared that nine of the solvents can be characterized by an atmospheric residence half-life of 10 days or less (Figure 17.1.3). The photo-oxidation of solvents yields products. For example, the reaction of OH radicals with n-hexane can yield aldehydes, ketones, and nitrates.[21]

The reaction of some of the solvents with ozone may be much slower. For example, the half-life for the reaction of benzene with ozone may be longer than 100 years.[19] Solvents such as carbon tetrachloride, 1,1,1-trichloroethane, and the chlorinated fluorocarbons may be relatively resistant to photo-oxidation. The major fate mechanism of atmospheric 1,1,1-trichloroethane, for example, may be wet deposition.[32]

### REFERENCES


1 Agency for Toxic Substances and Disease Registry. Toxicological Profile for Acetone. ATSDR, Atlanta, Georgia, 1994.

2 Agency for Toxic Substances and Disease Registry. Toxicological Profile for Tetrachloroethylene. ATSDR, Atlanta, Georgia, 1991.

3 D. Calamari (ed.) **Chemical Exposure Predictions**, *Lewis Publishers*, 1993.

4 P. H. Howard, **Handbook of Environmental Fate and Exposure Data for Organic Chemicals**. Vol. II Solvents. *Lewis Publishers*, Chelsea, Michigan, 1990.

5 W. J. Lyman, W. F. Reehl, and D. H. Rosenblatt (eds). **Handbook of Chemical Property Estimation Methods**, *American Chemical Society*, Washington, D.C., 1990.

6 R. E. Ney. Fate and Transport of Organic Chemicals in the Environment. 2nd ed. Government Institutes, Inc. Rockville, MD, 1995.

7 B. L. Sawhney and K. Brown (eds.). **Reactions and Movement of Organic Chemicals in Soils**, *Soil Science Society of America*, Special Publication Number 22, 1989.

8 R. G. Thomas, Volatilization From Water, W. J. Lyman, W. F. Reehl, and D. H. Rosenblatt (eds). in **Handbook of Chemical Property Estimation Methods**, *American Chemical Society*, Washington, D.C, Chap. 15, 1990.

9 J. C. Harris. Rate of Hydrolysis, Lyman, W. J., W. F. Reehl, and D. H. Rosenblatt (eds). in **Handbook of Chemical Property Estimation Methods**, *American Chemical Society*, Washington, D.C, Chap. 7, 1990.

10 K. M. Scow, 1990, Rate of Biodegradation, W. J. Lyman, W. F. Reehl, and D. H. Rosenblatt (eds). in **Handbook of Chemical Property Estimation Methods**, *American Chemical Society*, Washington, D.C, Chap. 9, 1990.

11 W. R. Roy and R. A. Griffin, *Environ. Geol. Water Sci.*, **15**, 101 (1990).

12 R. J. Millington, *Science*, **130**, 100 (1959).

13 W. R. Roy and R. A. Griffin, *Environ. Geol. Water Sci.*, 7, 241 (1985).

14 P. H. Howard, R. S. Boethling, W. F. Jarvis, W. M. Meylan, and Edward M. Michalenko. **Handbook of Environmental Degradation Rates**, *Lewis Publishers*, Chelsea, Michigan, 1991.

15 M. Alexander and K. M. Scow, Kinetics of Biodegradation, B. L. Sawhney, and K. Brown (eds.). in **Reactions and Movement of Organic Chemicals in Soils**. *Soil Science Society of America Special Publication*, Number 22, Chap. 10, 1989.

16 W. J. Lyman, Atmospheric Residence Time, W. J. Lyman, W. F. Reehl, and D. H. Rosenblatt (eds). in **Handbook of Chemical Property Estimation Methods**. *American Chemical Society,* Washington, D.C, Chap. 10, 1990.

17 P. H. Howard, **Handbook of Environmental Fate and Exposure Data for Organic Chemicals**. Vol. I. Large Production and Priority Pollutants. *Lewis Publishers*, Chelsea, Michigan, 1989.
18 Agency for Toxic Substances and Disease Registry. Toxicological Profile for Nitrobenzene. ATSDR, Atlanta, Georgia, 1989.
19 Agency for Toxic Substances and Disease Registry. Toxicological Profile for Benzene. ATSDR, Atlanta, Georgia, 1991.
20 Agency for Toxic Substances and Disease Registry. Toxicological Profile for Toluene. ATSDR, Atlanta, Georgia, 1998.
21 Agency for Toxic Substances and Disease Registry. Toxicological Profile for Hexane. ATSDR, Atlanta, Georgia, 1997.
22 D. MacKay and W. Y. Shiu, *J. Phys. Chem. Ref. Data*, **4**, 1175 (1981).
23 Agency for Toxic Substances and Disease Registry. Toxicological Profile for Chlorobenzene. ATSDR, Atlanta, Georgia, 1989.
24 R. G. Thomas. Volatilization from Soil in W. J. Lyman, W. F. Reehl, and D. H. Rosenblatt (eds). **Handbook of Chemical Property Estimation Methods**. *American Chemical Society*, Washington, D.C, Chap. 16, 1990.
25 J. J. Hassett, W. L. Banwart, and R. A. Griffin. Correlation of compound properties with soil sorption characteristics of nonpolar compounds by soils and sediments; concepts and limitations In C. W. Francis and S. I. Auerback (eds), **Environmental and Solid Wastes, Characterization, Treatment, and Disposal**, Chap. 15, p. 161-178, *Butterworth Publishers*, London, 1983.
26 J. M. Gosset, *Environ. Sci. Tech.*, **21**, 202 (1987).
27 T. Stauffer W. G. MacIntyre. 1986, *Tox. Chem.*, **5**, 949 (1986).
28 W. C. Koskinen and S. S. Harper. The retention process, mechanisms. p. 51-77. In **Pesticides in the Soil Environment**. *Soil Science Society of America Book Series*, no. 2, 1990.
29 R. E. Green, J. M. Davidson, and J. W. Biggar. An assessment of methods for determining adsorption-desorption of organic chemicals. p. 73-82. In A. Bainn and U. Kafkafi (eds.), **Agrochemicals in Soils**, *Pergamon Press*, New York, 1980.
30 R. Calvet, *Environ. Health Perspectives*, **83**,145 (1989).
31 B. T. Bowman and W. W. Sans, *J. Environ. Qual.*, **14**, 270 (1985).
32 Agency for Toxic Substances and Disease Registry. Toxicological Profile for 1,1,1-Trichloroethane. ATSDR, Atlanta, Georgia, 1989.

## 17.2 FATE-BASED MANAGEMENT OF ORGANIC SOLVENT-CONTAINING WASTES[a]

WILLIAM R. ROY
**Illinois State Geological Survey, Champaign, IL, USA**

### 17.2.1 INTRODUCTION

The wide spread detection of dissolved organic compounds in groundwater is a major environmental concern, and has led to greater emphasis on incineration and waste minimization when compared with the land disposal of solvent-containing wastes. The movement and environmental fate of dissolved organic solvents from point sources can be approximated by the use of computer-assisted, solute-transport models. These models require information about the composition of leachate plumes, and site-specific hydrogeological and chemical

[a]Publication authorized by the Chief, Illinois State Geological Survey

data for the leachate-site system. A given land-disposal site has a finite capacity to attenuate organic solvents in solution to environmentally acceptable levels. If the attenuation capacity of a site can be estimated, then the resulting information can be used as criteria to make decisions as to what wastes should be landfilled, and what quantities of solvent in a given waste can be safely accepted. The purpose of this section is to summarize studies[1-3] that were conducted that illustrate how knowledge of the environmental fate and movement of the solvents in Section 17.1 can be used in managing solvent-containing wastes. These studies were conducted by using computer simulations to assess the fate of organic compounds in leachate at a waste-disposal site.

#### 17.2.1.1 The waste disposal site

There are three major factors that will ultimately determine the success of a land-disposal site in being protective of the environment with respect to groundwater contamination by organic solvents: (1) the environmental fate and toxicity of the solvent; (2) the mass loading rate, i.e., the amount of solvent entering the subsurface during a given time, and (3) the total amount of solvent available to leach into the groundwater. The environmental fate of the solvents was discussed in 17.1.

The hypothetical waste-disposal site used in this evaluation (Figure 17.2.1) had a single waste trench having an area of 0.4 hectare. Although site-specific dimensions may be assigned with actual sites, this hypothetical site was considered representative of many situations found in the field. The trench was 12.2 meters (40 ft) deep and was constructed with a synthetic/compacted-soil double-liner system. The bottom of the trench was in direct contact with a sandy aquifer that was 6.1 meters (20 ft) thick. The top of the water table was defined as being at the top of the sandy aquifer. Thus, this site was designed as a worst-case scenario. The sandy aquifer directly beneath the hazardous-waste trench would offer little resistance to the movement of contaminants. To further compound a worst-case situation, it was also assumed that the entire trench was saturated with leachate, generating a 12.2 meter (40 ft) hydraulic head through the liner. This could correspond to a situation where the trench had completely filled with leachate because the leachate collection system had either failed or the site had been abandoned.

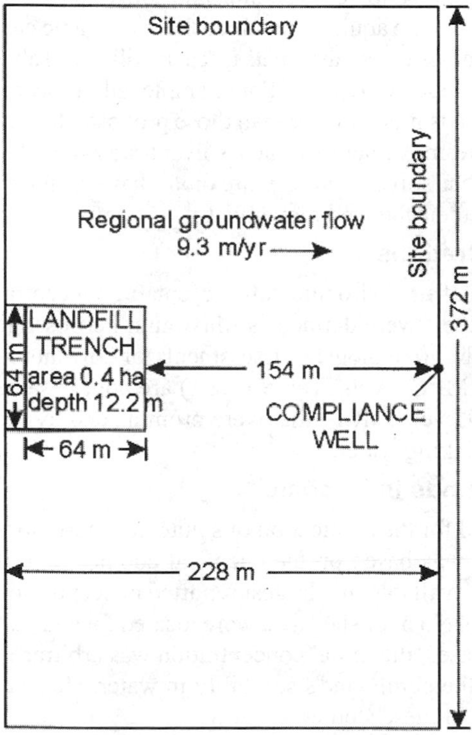


Figure 17.2.1. Design of the waste-disposal site model used in the simulations (Roy et al.[1]).

The following aquifer properties, typical of sandy materials,[1] were used in the study:

saturated hydraulic conductivity = $10^{-3}$ cm/sec
saturated volumetric water content = 0.36 $cm^3/cm^3$
dry bulk density = 1.7 $g/cm^3$
hydraulic gradient = 0.01 cm/cm
mean organic carbon content = 0.18%

These aquifer properties yield a groundwater flow rate of 9.3 meters (30 ft) per year. The direction of groundwater flow is shown in Figure 17.2.1 to be from left to right. The edge of the disposal trench was 154 meters (500 ft) from a monitoring well that was open to the entire thickness of the aquifer. This monitoring well served as a worst-case receptor because it was placed in the center of the flow path at the site boundary and it served as the compliance point for the site. The downgradient concentrations of organic solvents at the compliance well, as predicted by a solute-transport model, were used to evaluate whether the attenuation capacity of the site was adequate to reduce the contaminants to acceptable concentrations before they migrated beyond the compliance point.

#### 17.2.1.2 The advection-dispersion model and the required input

The 2-dimensional, solute-transport computer program PLUME was used to conduct contaminant migration studies. Detailed information about PLUME, including boundary conditions and quantitative estimates of dispersion and groundwater dilution, were summarized by Griffin and Roy.[3] In this relatively simple and conservative approach, PLUME did not take into account volatilization from water. Volatilization is a major process for many of the solvents (see Section 17.1). Adsorption was assumed to be reversible, and soil-water partition coefficients were calculated by assuming that the aquifer contained 0.18% organic carbon (see Roy and Griffin[4]). A degradation half-life was assigned to each solvent (Table 17.2.1). In many cases, conservative half-life values were used. For example, all of the ketones were assigned a half-life of 5 years, which is much longer than those proposed for ketones in groundwater (see Section 17.1). The movement of each solvent was modeled separately whereas it should be recognized that solvents in mixtures may have different chemical properties that can ultimately affect their fate and movement.

#### 17.2.1.3 Maximum permissible concentrations

Central to the type of assessment is a definition of an environmentally acceptable concentration of each contaminant. These acceptable levels were defined as Maximum Permissible Concentrations (MPC), and were based on the toxicological assessments of solvents in drinking water by George and Siegel.[5] These MPC levels (Table 17.2.1) are not the same levels as the current Maximum Contaminant Levels (MCL) that were promulgated by the U.S. Environmental Protection Agency for drinking water.

#### 17.2.1.4 Distribution of organic compounds in leachate

An initial solute concentration must be selected for the application of solute transport models. An initial concentration for each solvent was based on the chemical composition of leachates from hazardous-waste sites.[1] Where available, the largest reported concentration was used in the modeling efforts (Table 17.2.1). No published data were located for some of the solvents such as cyclohexanone. In such cases, the initial concentration was arbitrarily assigned as 1,000 mg/L or it was equated to the compound's solubility in water. Hexane, decane, and tetrahydofuran were not included in these studies.

The amount of mass of each organic compound entering the aquifer via the double-liner system was calculated using these initial leachate concentrations. There was a continuous 12.2-meter head driving the leachate through the liner. Leachate was predicted to break through the liner in 30 years. Under these conditions, approximately 131,720 L/year/acre of leachate would seep through the liner. The assumptions used in deriving this flow estimate were summarized in Roy et al.[1]

**Table 17.2.1. The six groups of solvents discussed in this section, their corresponding Maximum Permissible Concentrations (MPC), the largest reported concentrations in leachate (LC), and the assigned half-lives from Roy et al.[1]**

| | MPC, μg/L | LC, mg/L | Half-life, years |
|---|---|---|---|
| *Alcohols* | | | |
| n-Butyl alcohol | 2,070 | 1,000 | 5 |
| Isobutyl alcohol | 2,070 | 1,000 | 5 |
| Methanol | 3,600 | 42.4 | 5 |
| *Benzene Derivatives* | | | |
| Benzene | 1.6 | 7.37 | 20 |
| Chlorobenzene | 488 | 4.62 | 20 |
| o-Cresol | 304 | 0.21 | 20 |
| o-Dichlorobenzene | 400 | 0.67 | 50 |
| Ethylbenzene | 1,400 | 10.1 | 10 |
| Nitrobenzene | 19,800 | 0.74 | 20 |
| Toluene | 14,300 | 100 | 10 |
| o-Xylene | 14,300 | 19.7 | 10 |
| *Chlorinated Aliphatic Hydrocarbons* | | | |
| Carbon tetrachloride | 0.4 | 25.0 | 50 |
| Dichloromethane | 0.19 | 430 | 20 |
| Tetrachloroethylene | 0.80 | 8.20 | 20 |
| 1,1,1-Trichloroethane | 6.00 | 590 | 50 |
| Trichloroethylene | 2.70 | 260 | 20 |
| *Chlorinated Fluorocarbons* | | | |
| Trichlorofluoromethane (F-11) | 0.19 | 0.14 | 50 |
| 1,1,2,2-Tetrachloro-1,2-difluoroethane ( F-112) | 0.19 | 120 | 50 |
| 1,1,2-Trichloro-1,2,2-trifluoroethane (F-113) | 0.19 | 170 | stable |
| 1,2,-Dichlorotetrafluoroethane (F-114) | 0.19 | 130 | stable |
| *Ketones* | | | |
| Acetone | 35,000 | 62 | 5 |
| Cyclohexanone | 3,500 | 1,000 | 5 |
| Methyl ethyl ketone | 30,000 | 53.0 | 5 |

| | MPC, μg/L | LC, mg/L | Half-life, years |
|---|---|---|---|
| Methyl isobutyl ketone | 143 | 10.0 | 5 |
| Others | | | |
| Carbon disulfide | 830 | 1,000 | 10 |
| Diethyl ether | 55,000 | 1,000 | 5 |
| Ethyl acetate | 55,000 | 1,000 | 5 |
| Pyridine | 207 | 1,000 | 20 |

A mass-loading rate was conservatively calculated for each solvent as,

$$M_{lr} = Q \times C_l \quad [17.2.1]$$

where:

$M_{lr}$ the mass loading rate (mass/time/area),
Q calculated leachate flux (131.7 kL/year/hectare), and
$C_l$ largest concentration of the solvent in leachate (mg/L)

## 17.2.2 MOVEMENT OF SOLVENTS IN GROUNDWATER

Ketones and alcohols have little tendency to be adsorbed by soil materials (see Section 17.1), and would appear at the compliance point only a few years after liner breakthrough (Figure 17.2.2). Because the mass loading rates were held constant, the ketones and alcohols assumed maximum steady-state concentrations after approximately 40 to 50 years (Figure 17.2.3). These two classes of organic solvents degrade readily, reducing their downgradient concentrations. The distribution of the benzene derivatives at the compliance well depended substantially on their soil-water partition coefficients, their tendencies to de-

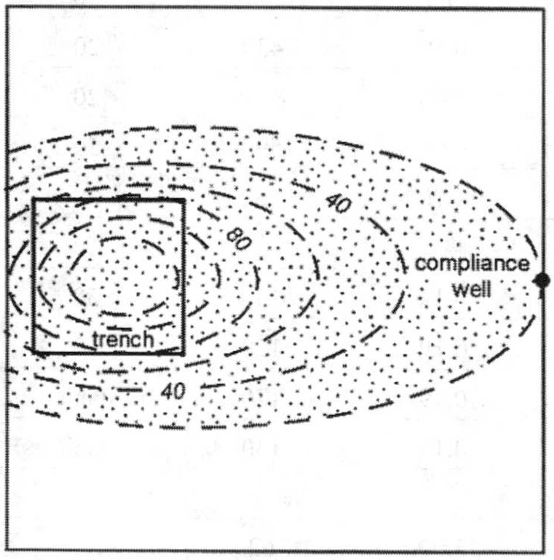


Figure 17.2.2. The predicted distribution of methyl ethyl ketone (mg/L) in the aquifer 100 years after the leachate has broken through the liner (Griffin and Roy[3]).

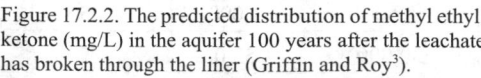

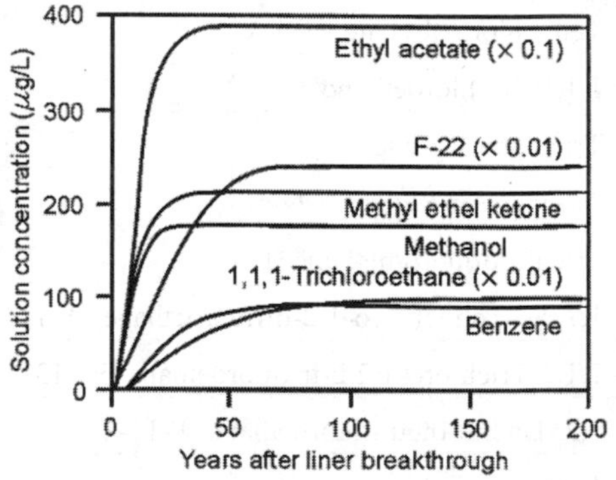


Figure 17.2.3. The predicted concentrations of methyl ethyl ketone, methanol, benzene, 1,1,1-trichloroethane, ethyl acetate, and F-22 at the compliance point as a function of time (Roy et al.[1]).

grade, and the initial concentrations. Under the conditions described, the relative steady-state concentrations of the benzene derivatives were: toluene > benzene > chlorobenzene > p-xylene > nitrobenzene > o-dichlorobenzene > o-cresol > ethyl benzene. Methylene chloride and 1,1,1-trichlorethane would dominate the chlorinated hydrocarbons. Among the group of unrelated organic solvents, the concentration of pyridine at the well was predicted to increase rapidly. Pyridine would eventually dominate this group in the relative order: pyridine > carbon disulfide > ethyl acetate > diethyl ether. The relative order of fluorocarbons at the compliance well in terms of concentration was: F-21, F-22 >> F-12 > F-113 > F-114 > F-112 > R-112a > FC-115 >> F-11.

In brief, the computer simulations predicted that all 28 organic compounds would eventually migrate from the waste trench, and be detected at the compliance well. The predicted concentrations varied by four orders of magnitude, and were largely influenced by the initial concentrations used in calculating the mass loading rate to the aquifer.

### 17.2.3 MASS LIMITATIONS

The next step in this analysis was to determine whether these predicted concentrations would pose an environmental hazard by evaluating whether the site was capable of attenuating the concentrations of the organic compounds to levels that are protective of human health. In Figure 17.2.4 the predicted steady-state concentrations of the organic compounds

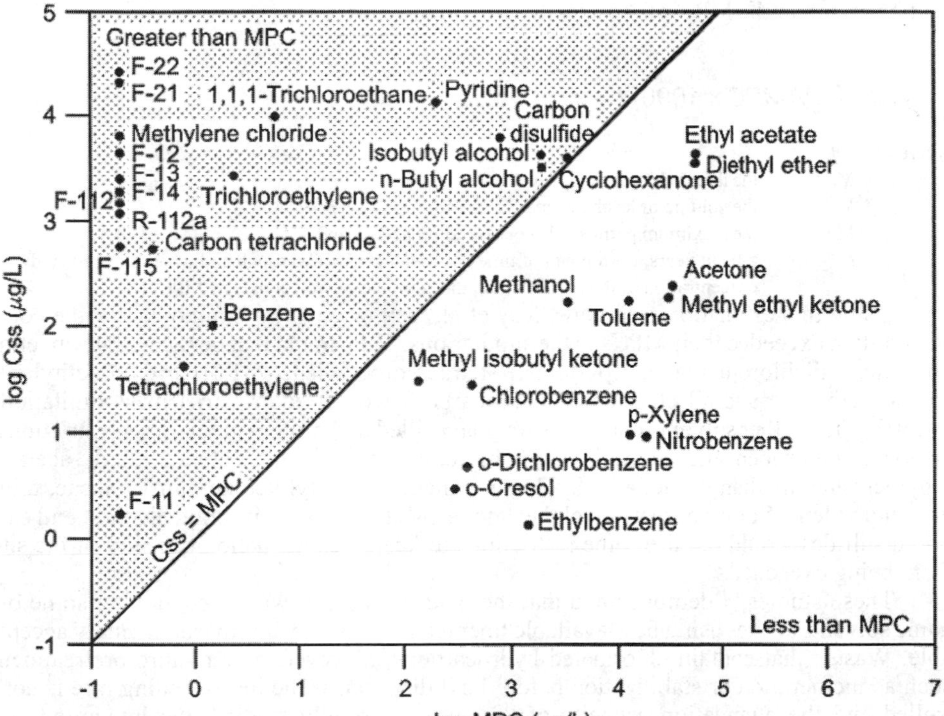


Figure 17.2.4. The predicted steady-state concentrations (Css) of each solvent in groundwater at the compliance point as a function of its Maximum Permissible (MPC) Concentration (Roy et al.[1]).

in groundwater at the compliance well were plotted against their MPCs. The boundary shown in Figure 17.2.4 represents the situation where the steady-state concentration (Css) equals the MPC. Consequently, the predicted Css is less than its corresponding MPC when the Css of a given compound plots in the lower-right side. In this situation, these organic compounds could enter the aquifer at a constant mass loading rate without exceeding the attenuation capacity of the site. The steady-state concentrations of twenty solvents exceeded their corresponding MPCs. The continuous addition of these organic compounds (i.e., a constant mass loading rate) would exceed the site's ability to attenuate them to environmentally acceptable levels in this worst-case scenario. There are two avenues for reducing the steady-state concentrations downgradient from the trench: (1) reduce the mass loading rate, and/or (2) reduce the mass of organic compound available to leach into the aquifer. Because, the RCRA-required double liner was regarded as the state-of-the-art with respect to liner systems, it was not technically feasible to reduce the volume of leachate seeping into the aquifer under the conditions imposed. The worst-case conditions could be relaxed by assuming a lower leachate head in the landfill or by providing a functional leachate-collection system. Either condition would be reasonable and would reduce the mass loading rate. Another alternative is to reduce the mass available for leaching. In the previous simulations, the mass available to enter the aquifer was assumed to be infinite. Solute transport models can be used to estimate threshold values for the amounts of wastes initially landfilled.[2] A threshold mass ($M_t$) can be derived so that the down-gradient, steady-state concentrations will be less than the MPC of the specific compound, viz.,

$$M_t = V(MPC \times 1000)\, t \qquad [17.2.2]$$

where:

| | |
|---|---|
| $M_t$ | the threshold mass in g/hectare |
| V | the volume of leachate entering the aquifer in L/yr/hectare |
| MPC | the maximum permissible concentration as g/L, and |
| t | time in years; the amount of time between liner breakthrough and when the predicted concentration of the compound in the compliance well equals its MPC. |

Using this estimation technique, Roy et al.[1] estimated mass limitations for the compounds that exceeded their MPCs in the simulations. They found that benzene, carbon tetrachloride, dichloromethane, pyridine, tetrachloroethylene, 1,1,1-trichloroethylene, trichloroethylene and all chlorinated fluorocarbons would require strict mass limitations (<250 kg/ha). Other solvents could be safely landfilled at the site without mass restrictions: acetone, chlorobenzene, cresols, o-dichlorobenzene, diethyl ether, ethyl acetate, ethylbenzene, methanol, methyl ethyl ketone, methyl isobutyl ketone, nitrobenzene, toluene, and xylene. Some solvents (cyclohexanone, n-butyl alcohol, isobutyl alcohol, and carbon disulfide) would require some restrictions to keep the attenuation capacity of the site from being exceeded.

These studies,[1-3] demonstrated that the land disposal of wastes containing some organic solvents at sites using best-available liner technology may be environmentally acceptable. Wastes that contain chlorinated hydrocarbons, however, may require pretreatment such as incineration or stabilization before land disposal. If the mass-loading rate is controlled and the attenuation capacity of the site is carefully studied, the integrated and multidisciplinary approach outlined in this section can be applied to the management of solvent-containing wastes.

## REFERENCES

1 W. R. Roy, R. A. Griffin, J. K. Mitchell, and R. A. Mitchell, *Environ. Geol. Water Sci.*, **13**, 225 (1989).
2 W. R. Roy and R. A. Griffin, *J. Haz. Mat.*, **15**, 365 (1987).
3 R. A. Griffin and W. R. Roy. Feasibility of land disposal of organic solvents: Preliminary Assessment. Environmental Institute for Waste Management Studies, Report No. 10, University of Alabama, 1986.
4 W. R. Roy and R. A. Griffin, *Environ. Geol. Water Sci.*, **7**, 241 (1985).
5 W. J. George and P. D. Siegel. Assessment of recommended concentrations of selected organic solvents in drinking water. Environmental Institute for Waste Management Studies, Report No. 15, University of Alabama, 1988.

# 17.3 ENVIRONMENTAL FATE AND ECOTOXICOLOGICAL EFFECTS OF GLYCOL ETHERS

JAMES DEVILLERS
**CTIS, Rillieux La Pape, France**
AURÉLIE CHEZEAU, ANDRÉ CICOLELLA, AND ERIC THYBAUD
**INERIS, Verneuil-en-Halatte, France**

## 17.3.1 INTRODUCTION

Glycol ethers and their acetates are widely used as solvents in the chemical, painting, printing, mining and furniture industries. They are employed in the production of paints, coatings, resins, inks, dyes, varnishes, lacquers, cleaning products, pesticides, deicing additives for gasoline and jet fuel, and so on.[1] In 1997, the world production of glycol ethers was about 900,000 metric tons.[2]

There are two distinct series of glycol ethers namely the ethylene glycol ethers which are produced from ethylene oxide and the propylene glycol ethers derived from propylene oxide. The former series is more produced and used than the latter. Thus, inspection of the 42,000 chemical substances recorded by INRS (France) in the SEPIA data bank, between 1983 and 1998, reveals that 10% of them include ethylene glycol ethers and about 4% propylene glycol ethers.[2] However, due to the reproductive toxicity of some ethylene glycol monoalkyl ethers,[3-5] it is important to note that the worldwide tendency is to replace these chemicals by glycol ethers belonging to the propylenic series.[2]

Given the widespread use of glycol ethers, it is obvious that these chemicals enter the environment in substantial quantities. Thus, for example, the total releases to all environmental media in the United States for ethylene glycol monomethyl ether and ethylene glycol monoethyl ether in 1992 were 1688 and 496 metric tons, respectively.[6] However, despite the potential hazard of these chemicals, the problems of the environmental contaminations with glycol ethers have not received much attention. There are two main reasons for this. First, these chemicals are not classified as priority pollutants, and hence, their occurrence in the different compartments of the environment is not systematically investigated. Thus, for example, there are no glycol ethers on the target list for the Superfund hazardous waste site cleanup program.[6] Second, glycol ethers are moderately volatile colorless liquids with a high water solubility and a high solubility with numerous solvents. Consequently, the classical analytical methods routinely used for detecting the environmental pollutants do not provide reliable results with the glycol ethers, especially in the aquatic environments.

Under these conditions, the aim of this chapter is to review the available literature on the occurrence, environmental fate, and ecotoxicity of glycol ethers.

## 17.3.2 OCCURRENCE

Despite the poor applicability of the most widely used USEPA analytical methods, some ethylene, diethylene, and triethylene glycol ethers have been reported as present in Superfund hazardous waste sites in the US more often than some of the so-called priority pollutants.[6] More specifically, Eckel and co-workers[6] indicated that in Jacksonville (Florida), a landfill received a mixture of household waste and wastes from aircraft maintenance and paint stripping from 1968 to 1970. In 1984, sampling of residential wells in the vicinity revealed concentrations of 0.200, 0.050, and 0.010 mg/l of diethylene glycol diethyl ether, ethylene glycol monobutyl ether, and diethylene glycol monobutyl ether, respectively. One year later, concentrations of 0.050 to 0.100 mg/l of diethylene glycol diethyl ether were found in the most contaminated portion of the site. In 1989, some samples still indicated the presence of diethylene glycol diethyl ether and triethylene glycol dimethyl ether. This case study clearly illustrates that glycol ethers may persist in the environment for many years after a contamination. Concentrations of 0.012 to 0.500 mg/l of ethylene glycol monobutyl ether were also estimated in residential wells on properties near a factory (Union Chemical, Maine, USA) manufacturing furniture stripper containing N,N-dimethylformamide. In addition, in one soil sample located in that site, a concentration of 0.200 mg/kg of ethylene glycol monobutyl ether was also found. In another case study, Eckel and co-workers[6] showed that ethylene glycol diethyl ether was detected with estimated concentrations in the range from 0.002 to 0.031 mg/l in eight residential wells adjacent to a landfill (Ohio) receiving a mixture of municipal waste and various industrial wastes, many of them from the rubber industry. Last, ethylene glycol monomethyl ether was detected in ground-water samples at concentrations of 30 to 42 mg/l (Winthrop landfill, Maine).[6]

In 1991, the high resolution capillary GC-MS analysis of a municipal wastewater collected from the influent of the Asnières-sur-Oise treatment plant located in northern suburban Paris (France) revealed the presence of ethylene glycol monobutyl ether (0.035 mg/l), diethylene glycol monobutyl ether (0.015 mg/l), propylene glycol monomethyl ether (0.070 mg/l), dipropylene glycol monomethyl ether (0.050 mg/l), and tripropylene glycol monoethyl ether (<0.001 mg/l).[7] In the Hayashida River (Japan) mainly polluted by effluents from leather factories, among the pollutants separated by vacuum distillation and identified by GS-MS, ethylene glycol monobutyl ether, ethylene glycol monoethyl ether, and diethylene glycol monobutyl ether were found at concentrations of 5.68, 1.20, and 0.24 mg/l, respectively.[8]

In air samples collected in the pine forest area of Storkow (30 km south east of Berlin, Germany), the "Mediterranean Macchia" of Castel Porziano (Italy), and the Italian station located at the foot of Everest (Nepal), GC-MS analysis showed concentrations of ethylene glycol monobutyl ether of 1.25, 0.40, and 0.10 to 1.59 $\mu g/m^3$, respectively.[9]

### 17.3.3 ENVIRONMENTAL BEHAVIOR

Due to their high water solubilities and low 1-octanol/water partition coefficients (log P), the glycol ethers, after release in the environment, will be preferentially found in the aquatic media and their accumulation in soils, sediments, and biota will be negligible.

The available literature data on the biodegradation of glycol ethers reveal that most of these chemicals are biodegradable under aerobic conditions (Table 17.3.1), suggesting the compounds would not likely persist.

**Table 17.3.1. Aerobic biodegradation of glycol ethers and their acetates in aquatic environments and soils**

| Name [CAS RN], Values | Comments | Ref. |
|---|---|---|
| *Ethylene glycol monomethyl ether [109-86-4]* | | |
| 5 d = 30% bio-oxidation, 10 d = 62%, 15 d = 74%, 20 d = 88% | filtered domestic wastewater, non-acclimated seed, 3, 7, 10 mg/l (at least two), fresh water | 10 |
| 5 d = 6% bio-oxidation, 10 d = 18%, 15 d = 23%, 20 d = 39% | filtered domestic wastewater, non-acclimated seed, 3, 7, 10 mg/l (at least two), salt water | 10 |
| ThOD = 1.68 g/g, $BOD_5$ = 0.12 g/g, %ThOD = 7, COD = 1.69 g/g, %ThOD = 101 | effluent from a biological sanitary waste treatment plant, 20 ± 1°C, unadapted seed, BOD = APHA SM 219, COD = ASTM D 1252-67 | 11 |
| $BOD_5$ = 0.50 g/g, %ThOD = 30 | same conditions, adapted seed | 11 |
| *Ethylene glycol monomethyl ether acetate [110-49-6]* | | |
| ThOD = 1.63 g/g, $BOD_5$ = 0.49 g/g, %ThOD = 30, COD = 1.60 g/g, %ThOD = 98 | effluent from a biological sanitary waste treatment plant, 20 ± 1°C, unadapted seed, BOD = APHA SM 219, COD = ASTM D 1252-67 | 11 |
| *Ethylene glycol monoethyl ether [110-80-5]* | | |
| 5 d = 36% bio-oxidation, 10 d = 88%, 15 d = 92%, 20 d = 100% | filtered domestic wastewater, non-acclimated seed, 3, 7, 10 mg/l (at least two), fresh water | 10 |
| 5 d = 5% bio-oxidation, 10 d = 42%, 15 d = 50%, 20 d = 62% | filtered domestic wastewater, non-acclimated seed, 3, 7, 10 mg/l (at least two), salt water | 10 |
| ThOD = 1.96 g/g, $BOD_5$ = 1.03 g/g, %ThOD = 53, COD = 1.92 g/g, %ThOD = 98 | effluent from a biological sanitary waste treatment plant, 20 ± 1°C, unadapted seed, BOD = APHA SM 219, COD = ASTM D 1252-67 | 11 |
| $BOD_5$ = 1.27 g/g, %ThOD = 65 | same conditions, adapted seed | 11 |
| COD removed = 91.7% | mixed culture, acclimation in a semi-continuous system | 12 |
| $BOD_5$ = 0.353 g/g, BOD/ThOD = 18.1% | fresh water, standard dilution method, 10 mg/l | 13 |
| $BOD_5$ = 0.021 g/g, BOD/ThOD = 1.1% | sea water, 10 mg/l | 13 |
| *Ethylene glycol monoethyl ether acetate [111-15-9]* | | |
| 5 d = 36% bio-oxidation, 10 d = 79%, 15 d = 82%, 20 d = 80% | filtered domestic wastewater, non-acclimated seed, 3, 7, 10 mg/l (at least two), fresh water | 10 |
| 5 d = 10% bio-oxidation, 10 d = 44%, 15 d = 59%, 20 d = 69% | filtered domestic wastewater, non-acclimated seed, 3, 7, 10 mg/l (at least two), salt water | 10 |

| Name [CAS RN], Values | Comments | Ref. |
|---|---|---|
| ThOD = 1.82 g/g, $BOD_5$ = 0.74 g/g, %ThOD = 41, COD = 1.76 g/g, %ThOD = 96 | effluent from a biological sanitary waste treatment plant, 20 ± 1°C, unadapted seed, BOD = APHA SM 219, COD = ASTM D 1252-67 | 11 |
| $BOD_5$ = 0.442 g/g, BOD/ThOD = 24.3% | fresh water, standard dilution method, 10 mg/l | 13 |
| $BOD_5$ = 0.448 g/g, BOD/ThOD = 24.7% | sea water, 10 mg/l | 13 |
| *Ethylene glycol diethyl ether [629-14-1]* | | |
| ThOD = 2.31 g/g, $BOD_{10}$ = 0.10 g/g | | 14 |
| *Ethylene glycol monoisopropyl ether [109-59-1]* | | |
| ThOD = 2.15 g/g, $BOD_5$ = 0.18 g/g, %ThOD = 8, COD = 2.08 g/g, %ThOD = 97 | effluent from a biological sanitary waste treatment plant, 20 ± 1°C, unadapted seed, BOD = APHA SM 219, COD = ASTM D 1252-67 | 11 |
| *Ethylene glycol monobutyl ether [111-76-2]* | | |
| 5 d = 26% bio-oxidation, 10 d = 74%, 15 d = 82%, 20 d = 88% | filtered domestic wastewater, non-acclimated seed, 3, 7, 10 mg/l (at least two), fresh water | 10 |
| 5 d = 29% bio-oxidation, 10 d = 64%, 15 d = 70%, 20 d = 75% | filtered domestic wastewater, non-acclimated seed, 3, 7, 10 mg/l (at least two), salt water | 10 |
| ThOD = 2.31 g/g, $BOD_5$ = 0.71 g/g, %ThOD = 31, COD = 2.20 g/g, %ThOD = 95 | effluent from a biological sanitary waste treatment plant, 20 ± 1°C, unadapted seed, BOD = APHA SM 219, COD = ASTM D 1252-67 | 11 |
| $BOD_5$ = 1.68 g/g, %ThOD = 73 | same conditions, adapted seed | 11 |
| COD removed = 95.2% | mixed culture, acclimation in a semi-continuous system | 12 |
| $BOD_5$ = 0.240 g/g, BOD/ThOD = 10.4% | fresh water, standard dilution method, 10 mg/l | 13 |
| $BOD_5$ = 0.044 g/g, BOD/ThOD = 1.9% | sea water, 10 mg/l | 13 |
| 5 d = 47% bio-oxidation, 15 d = 70%, 28 d = 75% | OECD Closed-Bottle test 301D, no prior acclimation/adaptation of seed, 20°C | 15 |
| 28 d = 95% removal of DOC | OECD 301E, 10 mg/l, non-adapted domestic activated sludge | 16 |
| 28 d = 100% removal | OECD 302B, 500 mg/l, non-adapted domestic activated sludge | 16 |
| 1 d = 22% removal, 3 d = 63%, 5 d = 100% | OECD 302B, 450 mg/l, non-adapted domestic activated sludge | 16 |
| 14 d = 96% BOD of ThOD | 100 mg/l, activated sludge | 16 |
| *Ethylene glycol monobutyl ether acetate [112-07-2]* | | |
| >90% DOC removal in 28 d | OECD 302B, Zahn-Wellens test | 16 |
| *Diethylene glycol monomethyl ether [111-77-3]* | | |
| ThOD = 1.73 g/g, $BOD_5$ = 0.12 g/g, %ThOD = 7, COD = 1.71 g/g, %ThOD = 99 | effluent from a biological sanitary waste treatment plant, 20 ± 1°C, unadapted seed, BOD = APHA SM 219, COD = ASTM D 1252-67 | 11 |

| Name [CAS RN], Values | Comments | Ref. |
|---|---|---|
| $BOD_5$ = 0.95 mmol/mmol | acclimated mixed culture, estimated by linear regression technique from a 20-day test | 17 |
| *Diethylene glycol monomethyl ether acetate [629-38-9]* | | |
| ThOD = 1.68 g/g, $BOD_{10}$ = 1.10 g/g | | 14 |
| *Diethylene glycol monoethyl ether [111-90-0]* | | |
| 5 d = 17% bio-oxidation, 10 d = 71%, 15 d = 75%, 20 d = 87% | filtered domestic wastewater, non-acclimated seed, 3, 7, 10 mg/l (at least two), fresh water | 10 |
| 5 d = 11% bio-oxidation, 10 d = 44%, 15 d = 57%, 20 d = 70% | filtered domestic wastewater, non-acclimated seed, 3, 7, 10 mg/l (at least two), salt water | 10 |
| ThOD = 1.91 g/g, $BOD_5$ = 0.20 g/g, %ThOD = 11, COD = 1.85 g/g, %ThOD = 97 | effluent from a biological sanitary waste treatment plant, 20 ± 1°C, unadapted seed, BOD = APHA SM 219, COD = ASTM D 1252-67 | 11 |
| $BOD_5$ = 0.58 g/g, %ThOD = 30 | same conditions, adapted seed | 11 |
| COD removed = 95.4% | mixed culture, acclimation in a semi-continuous system | 12 |
| $BOD_5$ = 5.50 mmol/mmol | acclimated mixed culture | 17 |
| *Diethylene glycol diethyl ether [112-36-7]* | | |
| ThOD = 2.17 g/g, $BOD_{10}$ = 0.10 g/g | | 14 |
| *Diethylene glycol monobutyl ether [112-34-5]* | | |
| ThOD = 2.17 g/g, $BOD_5$ = 0.25 g/g, %ThOD = 11, COD = 2.08 g/g, %ThOD = 96 | effluent from a biological sanitary waste treatment plant, 20 ± 1°C, unadapted seed, BOD = APHA SM 219, COD = ASTM D 1252-67 | 11 |
| COD removed = 95.3% | mixed culture, acclimation in a semi-continuous system | 12 |
| $BOD_5$ = 5.95 mmol/mmol | acclimated mixed culture | 17 |
| 5 d = 27% bio-oxidation, 10 d = 60%, 15 d = 78%, 20 d = 81% | BOD APHA SM, no prior acclimation/adaptation of seed, 20°C | 15 |
| 5 d = 3% bio-oxidation, 15 d = 70%, 28 d = 88%, | OECD Closed-Bottle test 301D, no prior acclimation/adaptation of seed, 20°C | 15 |
| 28 d >60% removal | OECD 301C, modified MITI test, adapted activated sludge | 16 |
| 28 d = 58% removal | OECD 301C, modified MITI test, adapted activated sludge | 16 |
| 1 d = 14% removal, 3 d = 19%, 5 d = 60%, 6 d = 100% | OECD 302B, modified Zahn-Wellens test, industrial non-adapted activated sludge | 16 |
| 9 d = 100% removal | OECD 302B, modified Zahn-Wellens test, non-adapted activated sludge | 16 |
| 14 d = 94% removal | OECD 301A, domestic secondary effluent sewage, DOC measured | 16 |

| Name [CAS RN], Values | Comments | Ref. |
|---|---|---|
| $BOD_5$ = 0.05 g/g (5.2% ThOD), $BOD_{10}$ = 0.39 g/g (57% ThOD), $BOD_{20}$ = 1.08 g/g (72% ThOD) | | 16 |
| BOD = 0.25 g/g, COD = 2.08 g/g | Dutch standard method, adapted sewage | 16 |
| *Diethylene glycol monobutyl ether acetate [124-17-4]* | | |
| $BOD_5$ = 13.3% ThOD, $BOD_{10}$ = 18.4%, $BOD_{15}$ = 24.6%, $BOD_{20}$ = 67% | | 16 |
| *Triethylene glycol monoethyl ether [112-50-5]* | | |
| 5 d = 8% bio-oxidation, 10 d = 47%, 15 d = 63%, 20 d = 71% | filtered domestic wastewater, non-acclimated seed, 3, 7, 10 mg/l (at least two), fresh water | 10 |
| 5 d = 1% bio-oxidation, 10 d = 10%, 15 d = 12%, 20 d = 22% | filtered domestic wastewater, non-acclimated seed, 3, 7, 10 mg/l (at least two), salt water | 10 |
| ThOD = 1.89 g/g, $BOD_5$ = 0.05 g/g, %ThOD = 3, COD = 1.84 g/g, %ThOD = 97 | effluent from a biological sanitary waste treatment plant, 20 ± 1°C, unadapted seed, BOD = APHA SM 219, COD = ASTM D 1252-67 | 11 |
| COD removed = 96.6% | mixed culture, acclimation in a semi-continuous system | 12 |
| $BOD_5$ = 1.15 mmol/mmol | acclimated mixed culture, estimated by linear regression technique from a 20-day test | 17 |
| *Triethylene glycol monobutyl ether [143-22-6]* | | |
| COD removed = 96.3% | mixed culture, acclimation in a semi-continuous system | 12 |
| *Propylene glycol monomethyl ether [107-98-2]* | | |
| 50% removal: <1 d (0.2 ppm), <2 d (9.9 ppm), <5 d (100 ppm) | sand = 72%, silt = 16%, clay = 12%, OC = 2.5%, 9.9 x $10^6$ bacteria/g of soil | 18 |
| 50% removal: <1 d (0.4 ppm), <7 d (100 ppm) | sand = 74%, silt = 12%, clay = 14%, OC = 2.0%, 5.1 x $10^6$ bacteria/g of soil | 18 |
| 50% removal: <4 d (0.4 ppm), >56 d (100 ppm), <23 d (100 ppm with nutrients) | sand = 94%, silt = 4%, clay = 2%, OC = 0.4%, 9.3 x $10^5$ bacteria/g of soil | 18 |
| $BOD_{20}$ = 58% of ThOD | | 18 |
| *Propylene glycol monomethyl ether acetate [108-65-6]* | | |
| 50% removal: <1 d (2.5 ppm), <1 d (20 ppm), | sand = 72%, silt = 16%, clay = 12%, OC = 2.5%, 9.9 x $10^6$ bacteria/g of soil | 18 |
| 50% removal: <1 d (20 ppm) | sand = 74%, silt = 12%, clay = 14%, OC = 2.0%, 5.1 x $10^6$ bacteria/g of soil | 18 |
| 50% removal: <1 d (20 ppm) | sand = 94%, silt = 4%, clay = 2%, OC = 0.4%, 9.3 x $10^5$ bacteria/g of soil | 18 |
| $BOD_{20}$ = 62% of ThOD | | 18 |

| Name [CAS RN], Values | Comments | Ref. |
|---|---|---|
| *Propylene glycol monophenyl ether [770-35-4]* | | |
| 50% removal: <1 d (1.4 ppm), <2 d (10 ppm), <5 d (107 ppm) | sand = 72%, silt = 16%, clay = 12%, OC = 2.5%, 9.9 x $10^6$ bacteria/g of soil | 18 |
| 50% removal: <1 d (1.5 ppm), <7 d (104 ppm) | sand = 74%, silt = 12%, clay = 14%, OC = 2.0%, 5.1 x $10^6$ bacteria/g of soil | 18 |
| 50% removal: <5 d (1.5 ppm), <23 d (108 ppm) | sand = 94%, silt = 4%, clay = 2%, OC = 0.4%, 9.3 x $10^5$ bacteria/g of soil | 18 |
| $BOD_{20}$ = 52% of ThOD | | 18 |

Note: ThOD = theoretical oxygen demand or the weight ratio of oxygen required per mg of compound for complete conversion of the compound to dioxide and water; BOD = biochemical oxygen demand; COD = chemical oxygen demand; DOC = dissolved organic carbon.

Metabolism pathways involving oxidation of the alcohol functionality and cleavage of the ether bond have been proposed for a reduced number of glycol ethers.[18-20]

While glycol ethers can be considered as biodegradable under aerobic conditions, Eckel et al.[6] have stressed that under anaerobic conditions, such as in the groundwater plume emanating from a landfill, these chemicals may persist for many years. In the same way, due to their physico-chemical properties, glycol ethers can act as cosolvents in mixtures with highly hydrophobic contaminants enhancing the solubility, mobility, and hence, the ecotoxicity of these chemicals.

Abiotic degradation processes for organic chemicals include aqueous photolysis, hydrolysis, and atmospheric photooxidation. The primary abiotic degradation process affecting glycol ethers is atmospheric photooxidation mediated by hydroxyl (OH) radicals formed in the atmosphere.[16] Photooxidation of glycol ethers is generally estimated from quantitative structure-property (QSPR) models due to the scarcity of experimental data.[21-23] Thus, for example, Grosjean[23] estimated atmospheric half-lives of 2 to 20 hours for ethylene glycol ethers (taking OH = $10^6$ molecules $cm^{-3}$).

## 17.3.4 ECOTOXICITY

### 17.3.4.1 Survival and growth

The ecotoxicological effects of glycol ethers and their acetates have been measured on various organisms occupying different trophic levels in the environment. Most of the available data deal with lethality, immobilization of the organisms, inhibition of cell multiplication, or growth (Table 17.3.2).

**Table 17.3.2. Effects of glycol ethers and their acetates on survival and growth of organisms**

| Species | | Results | Comments | Ref. |
|---|---|---|---|---|
| *Ethylene glycol monomethyl ether [109-86-4]* | | | | |
| Bacteria | *Pseudomonas putida* | 16-h TGK >10000 mg/l | toxicity threshold, inhibition of cell multiplication | 24 |
| | *Pseudomonas aeruginosa* | 4-m biocidal = 5-10% | tested in jet fuel and water mixtures | 25 |
| | Sulfate-reducing bacteria | 3-m biocidal = 5-10% | tested in jet fuel and water mixtures | 25 |
| Blue-green algae | *Microcystis aeruginosa* | 8-d TGK = 100 mg/l | toxicity threshold, inhibition of cell multiplication | 26 |
| Algae | *Scenedesmus quadricauda* | 8-d TGK >10000 mg/l | toxicity threshold, inhibition of cell multiplication | 26 |
| Yeasts | *Candida sp.* | 4-m biocidal = 5-10% | tested in jet fuel and water mixtures | 25 |
| Fungi | *Cladosporium resinae* | 4-m biocidal = 10-17%<br>42-d NG = 20% | tested in jet fuel and water mixtures<br>NG = no visible mycelial growth and spore germination, 1% glucose-mineral salts medium, 30°C | 25<br>27 |
| | *Gliomastix sp.* | 4-m biocidal = 17-25% | tested in jet fuel and water mixtures | 25 |
| Protozoa | *Chilomonas paramaecium* | 48-h TGK = 2.2 mg/l | toxicity threshold, inhibition of cell multiplication | 28 |
| | *Uronema parduczi* | 20-h TGK > 10000 mg/l | toxicity threshold, inhibition of cell multiplication | 29 |
| | *Entosiphon sulcatum* | 72-h TGK = 1715 mg/l | toxicity threshold, inhibition of cell multiplication | 30 |
| Coelenterates | *Hydra vulgaris* (syn. *H. attenuata*) | 72-h LC50 = 29000 mg/l | semi-static, adult polyps | 31 |
| Crustacea | *Daphnia magna* | 24-h LC50 >10000 mg/l<br>24-h EC50 >10000 mg/l | static, nominal concentrations<br>static, nominal concentrations, immobilization | 32<br>33 |
| | *Artemia salina* | 24-h TLm >10000 mg/l | static, 24.5°C | 10 |
| Fish | *Carassius auratus* | 24-h TLm >5000 mg/l | static, 20 ± 1°C, measured concentrations | 34 |
| | *Leuciscus idus melanotus* | 48-h LC50 >10000 mg/l<br>48-h LC0 >10000 mg/l<br>48-h LC100 >10000 mg/l | static (Juhnke) | 35 |
| | *Oryzias latipes* | 24-h LC50 >1000 mg/l<br>48-h LC50 >1000 mg/l | static, same results at 10, 20, and 30°C | 36 |

| | Species | Results | Comments | Ref. |
|---|---|---|---|---|
| Fish | *Pimephales promelas* | 20% mortality = 50 mg/l, 0% mortality = 100 mg/l, 30% mortality = 200 mg/l, 70% mortality = 500 mg/l, 70% mortality = 700 mg/l, 100% mortality = 1000 mg/l | static with daily renewal, adults, 7-d of exposure | 37 |
| | *Poecilia reticulata* | 7-d LC50 = 17434 mg/l | semi-static, 22 ± 1°C, rounded value | 38 |
| | *Lepomis macrochirus* | 96-h LC50 >10000 mg/l | static, nominal concentrations, L = 33-75 mm | 39 |
| | *Menidia beryllina* | 96-h LC50 >10000 mg/l | static, sea water, nominal concentrations, L = 40-100 mm | 39 |
| Amphibia | *Xenopus laevis* | 0% mortality=2500 mg/l, 0% mortality – 7500 mg/l, 0% mortality = 15000 mg/l | static with daily renewal, adults, 4-d of exposure | 37 |
| *Ethylene glycol monomethyl ether acetate [110-49-6]* | | | | |
| Fish | *Carassius auratus* | 24-h TLm = 190 mg/l | static, 20 ± 1°C, measured concentrations | 34 |
| | *Lepomis macrochirus* | 96-h LC50 = 45 mg/l | static, nominal concentrations, L = 33-75 mm | 39 |
| | *Menidia beryllina* | 96-h LC50 = 40 mg/l | static, sea water, nominal concentrations, L = 40-100 mm | 39 |
| *Ethylene glycol monoethyl ether [110-80-5]* | | | | |
| Bacteria | *Pseudomonas aeruginosa* | 4-m biocidal = 5-10% | tested in jet fuel and water mixtures | 25 |
| | Sulfate-reducing bacteria | 3-m biocidal = 5-10% | tested in jet fuel and water mixtures | 25 |
| | *Vibrio fischeri* | 5-min EC50 = 376 mg/l, 15-min EC50 = 403 mg/l, 30-min EC50 = 431 mg/l | reduction in light output, 15°C | 40 |
| Yeasts | *Candida sp.* | 4-m biocidal = 2-5% | tested in jet fuel and water mixtures | 25 |
| Fungi | *Cladosporium resinae* | 4-m biocidal = 5-10%<br>42-d NG = 10% | tested in jet fuel and water mixtures<br>NG = no visible mycelial growth and spore germination, 1% glucose-mineral salts medium, 30°C | 25<br>27 |
| | *Gliomastix sp.* | 4-m biocidal = 10-17% | tested in jet fuel and water mixtures | 25 |

| Species | | Results | Comments | Ref. |
|---|---|---|---|---|
| Coelenterates | *Hydra vulgaris* (syn. *H. attenuata*) | 72-h LC50 = 2300 mg/l | semi-static, adult polyps | 31 |
| Crustacea | *Daphnia magna* | 24-h EC50 = 54000 mg/l | static, nominal concentrations, immobilization | 41 |
| | | 48-h EC50 = 7671 mg/l | static, 22 ± 1°C, nominal concentrations, immobilization | 42 |
| | *Artemia salina* | 24-h TLm >10000 mg/l | static, 24.5°C | 10 |
| Fish | *Carassius auratus* | 24-h TLm >5000 mg/l | static, 20 ± 1°C, measured concentrations | 34 |
| | *Poecilia reticulata* | 7-d LC50 = 16399 mg/l | semi-static, 22 ± 1°C, rounded value | 38 |
| | *Lepomis macrochirus* | 96-h LC50 >10000 mg/l | static, nominal concentrations, L = 33-75 mm | 39 |
| | *Menidia beryllina* | 96-h LC50 >10000 mg/l | static, sea water, nominal concentrations, L = 40-100 mm | 39 |
| *Ethylene glycol monoethyl ether acetate [111-15-9]* | | | | |
| Crustacea | *Artemia salina* | 24-h TLm = 4000 mg/l | static, 24.5°C | 10 |
| Fish | *Carassius auratus* | 24-h TLm = 160 mg/l | static, 20 ± 1°C, measured concentrations | 34 |
| *Ethylene glycol monopropyl ether [2807-30-9]* | | | | |
| Bacteria | *Pseudomonas aeruginosa* | 4-m biocidal = 0-2% | tested in jet fuel and water mixtures | 25 |
| | Sulfate-reducing bacteria | 3-m biocidal = 0-2% | tested in jet fuel and water mixtures | 25 |
| Yeasts | *Candida sp.* | 4-m biocidal = 2-5% | tested in jet fuel and water mixtures | 25 |
| Fungi | *Cladosporium resinae* | 4-m biocidal = 2-5% | tested in jet fuel and water mixtures | 25 |
| | *Gliomastix sp.* | 4-m biocidal = 2-5% | tested in jet fuel and water mixtures | 25 |
| *Ethylene glycol monoisopropyl ether [109-59-1]* | | | | |
| Fungi | *Cladosporium resinae* | 42-d NG = 10% | NG = no visible mycelial growth and spore germination, 1% glucose-mineral salts medium, 30°C | 27 |
| Fish | *Carassius auratus* | 24-h TLm >5000 mg/l | static, 20 ± 1°C, measured concentrations | 34 |
| | *Poecilia reticulata* | 7-d LC50 = 5466 mg/l | semi-static, 22 ± 1°C, rounded value | 38 |
| *Ethylene glycol monobutyl ether [111-76-2]* | | | | |
| Bacteria | *Pseudomonas putida* | 16-h TGK = 700 mg/l | toxicity threshold, inhibition of cell multiplication | 24 |
| | *Pseudomonas aeruginosa* | 4-m biocidal = 1-2% | tested in jet fuel and water mixtures | 25 |
| | Sulfate-reducing bacteria | 3-m biocidal = 1-2% | tested in jet fuel and water mixtures | 25 |

| Species | | Results | Comments | Ref. |
|---|---|---|---|---|
| Blue-green algae | *Microcystis aeruginosa* | 8-d TGK = 35 mg/l | toxicity threshold, inhibition of cell multiplication | 26 |
| Algae | *Scenedesmus quadricauda* | 8-d TGK = 900 mg/l | toxicity threshold, inhibition of cell multiplication | 26 |
| | *Selenastrum capricornutum* | 7-d EC50 >1000 mg/l, NOEC = 125 mg/l, LOEC = 250 mg/l | growth rate inhibition | 16 |
| Yeasts | *Candida sp.* | 4-m biocidal = 2-3% | tested in jet fuel and water mixtures | 25 |
| Fungi | *Cladosporium resinae* | 4-m biocidal = 2-3%<br>42-d NG = 5% | tested in jet fuel and water mixtures<br>NG = no visible mycelial growth and spore germination, 1% glucose-mineral salts medium, 30°C | 25<br>27 |
| | *Gliomastix sp.* | 4-m biocidal = 2-3% | tested in jet fuel and water mixtures | 25 |
| Protozoa | *Chilomonas paramaecium* | 48-h TGK = 911 mg/l | toxicity threshold, inhibition of cell multiplication | 28 |
| | *Uronema parduczi* | 20-h TGK = 463 mg/l | toxicity threshold, inhibition of cell multiplication | 29 |
| | *Entosiphon sulcatum* | 72-h TGK = 91 mg/l | toxicity threshold, inhibition of cell multiplication | 30 |
| Coelenterates | *Hydra vulgaris (syn. H. attenuata)* | 72-h LC50 = 690 mg/l | semi-static, adult polyps | 31 |
| Mollusca | *Crassostrea virginica* | 24-h LC50 = 181 mg/l (143 - 228), 48-h LC50 = 160 mg/l (125 - 204), 72-h LC50 = 114 mg/l (93.9 - 138), 96-h LC50 = 89.4 mg/l (72 - 110) | static, 22 ± 1°C, 10 organisms/tank | 43 |
| Crustacea | *Daphnia magna* | 24-h LC50 = 1720 mg/l, 24-h LC0 = 1140 mg/l, 24-h LC100 = 2500 mg/l<br>24-h EC50 = 1815 mg/l (1698-1940), 24-h EC0 = 1283 mg/l, 24-h EC100 = 2500 mg/l<br>48-h EC50 = 835 mg/l | static, nominal concentrations<br>static, nominal concentrations, immobilization<br>static, immobilization | 32<br>33<br>16 |
| | *Artemia salina* | 24-h TLm = 1000 mg/l | static, 24.5°C | 10 |

| Species | | Results | Comments | Ref. |
|---|---|---|---|---|
| Crustacea | *Panaeus setiferus* | 24-h LC50 > 430 mg/l, 48-h LC50 = 173 mg/l (123 - 242), 72-h LC50 = 147 mg/l (116 - 186), 96-h LC50 = 130 mg/l (104 - 162) | static, average size = 1.5 cm | 43 |
| Fish | *Carassius auratus* | 24-h TLm = 1700 mg/l | static, 20 ± 1°C, measured concentrations | 34 |
| | *Leuciscus idus melanotus* | 48-h LC50 = 1575 mg/l, 48-h LC0 = 1350 mg/l 48-h LC100 = 1620 mg/l | static (Juhnke) | 35 |
| | | 48-h LC50 = 1395 mg/l, 48-h LC0 = 1170 mg/l, 48-h LC100 = 1490 mg/l | static (Lüdemann) | |
| | *Pimephales promelas* | 96-h LC50 = 2137 mg/l | static | 16 |
| | *Poecilia reticulata* | 7-d LC50 = 983 mg/l | semi-static, 22 ± 1°C, rounded value | 38 |
| | *Notropis atherinoides* | 72-h LC50 >500 mg/l | static | 16 |
| | *Lepomis macrochirus* | 96-h LC50 = 1490 mg/l | static, nominal concentrations, L = 33-75 mm | 39 |
| | | 24-h LC50 = 2950 mg/l, 96-h LC50 = 2950 mg/l | | 44 |
| | *Cyprinodon variegatus* | 24-h LC50 = 149 mg/l (125 - 176), 48-h LC50 = 126 mg/l (107 - 147), 72-h LC50 = 121 mg/l (105 - 138), 96-h LC50 = 116 mg/l (100 - 133) | static, 22 ± 1°C | 43 |
| | *Menidia beryllina* | 96-h LC50 = 1250 mg/l | static, sea water, nominal concentrations, L = 40-100 mm | 39 |
| *Ethylene glycol dibutyl ether [112-48-1]* | | | | |
| Fungi | *Cladosporium resinae* | 42-d NG = 5% | NG = no visible mycelial growth and spore germination, 1% glucose-mineral salts medium, 30°C | 27 |
| *Diethylene glycol monomethyl ether [111-77-3]* | | | | |
| Bacteria | *Pseudomonas aeruginosa* | 4-m biocidal = 10-17% | tested in jet fuel and water mixtures | 25 |
| | Sulfate-reducing bacteria | 3-m biocidal = 10-17% | tested in jet fuel and water mixtures | 25 |
| Yeasts | *Candida sp.* | 4-m biocidal = 5-10% | tested in jet fuel and water mixtures | 25 |

| | Species | Results | Comments | Ref. |
|---|---|---|---|---|
| Algae | *Scenedesmus subspicatus* | 72-h EC50 > 500 mg/l, 72-h EC20 > 500 mg/l, 72-h EC90 > 500 mg/l | biomass | 45 |
| | *Selenastrum capricornutum* | 96-h EC50 > 1000 mg/l | biomass | 45 |
| Fungi | *Cladosporium resinae* | 4-m biocidal = 10-17%<br>42-d NG = 20% | tested in jet fuel and water mixtures<br>NG = no visible mycelial growth and spore germination, 1% glucose-mineral salts medium, 30°C | 25<br>27 |
| | *Gliomastix sp.* | 4-m biocidal > 25% | tested in jet fuel and water mixtures | 25 |
| Crustacea | *Daphnia magna* | 24-h LC50 = 1495 mg/l (1300 - 1600), 48-h LC50 = 1192 mg/l (1100 - 6500) | static, nominal concentrations | 45 |
| Fish | *Oncorhynchus mykiss* | 96-h LC50 > 1000 mg/l | semi-static, nominal concentrations | 45 |
| | *Carassius auratus* | 24-h TLm >5000 mg/l | static, 20 ± 1°C, measured concentrations | 34 |
| | *Pimephales promelas* | 24-h LC50 = 6400 mg/l (6200 - 6600), 48-h LC50 = 6000 mg/l (6000 -6100), 72-h LC50 = 6000 mg/l (6000 -6100), 96-h LC50 = 5700 mg/l (5600 - 5900) | static | 45 |
| | *Lepomis macrochirus* | 96-h LC50 = 7500 mg/l | static, nominal concentrations, L = 33-75 mm | 39 |
| *Diethylene glycol dimethyl ether [111-96-6]* | | | | |
| Fungi | *Cladosporium resinae* | 42-d NG = 20% | NG = no visible mycelial growth and spore germination, 1% glucose-mineral salts medium, 30°C | 27 |
| *Diethylene glycol monoethyl ether [111-90-0]* | | | | |
| Bacteria | *Vibrio fischeri* | 5-min EC50 = 1000 mg/l,<br>5-min EC50 = 1290 mg/l<br>15-min EC50 = 10954 mg/l (10592.8 - 11327.5) | reduction in light output, nominal concentrations | 46<br><br>47 |
| Fungi | *Cladosporium resinae* | 42-d NG = 20% | NG = no visible mycelial growth and spore germination, 1% glucose-mineral salts medium, 30°C | 27 |
| Coelenterates | *Hydra vulgaris* (syn. *H. attenuata*) | 72-h LC50 = 17000 mg/l | semi-static, adult polyps | 31 |

| Species | | Results | Comments | Ref. |
|---|---|---|---|---|
| Crustacea | *Daphnia magna* | 48-h LC50 = 4670 mg/l (3620 - 6010),<br>48-h LC50 = 3340 mg/l (2120 - 5280) | static, 21.1°C, measured concentrations<br>static, 23.3°C, measured concentrations | 48 |
| | *Orconectes immunis* | 96-h LC50 = 34700 mg/l (29100 - 41400) | flow-through, 15.5°C, 0.47 g, measured concentrations | 48 |
| | *Artemia salina* | 24-h TLm >10000 mg/l | static, 24.5°C | 10 |
| Insecta | *Tanytarsus dissimilis* | 48-h LC50 = 18800 mg/l | static, 22.3°C, measured concentrations | 48 |
| Echinodermata | *Arbacia punctulata* | 4-h EC50 = 10661 mg/l (5576.5 - 20895.9)<br>1-h EC50 = 3370 mg/l (3145.4 - 3610.8)<br>5-h EC50 = 4116 mg/l (3408 - 4907) | static, marine, early embryo growth, nominal concentrations<br>static, marine, sperm cell, nominal concentrations<br>static, marine, increase in DNA, nominal concentrations | 47<br>47<br>49 |
| Fish | *Oncorhynchus mykiss* | 96-h LC50 = 13400 mg/l (11400 - 15700) | flow-through, 14.9°C, 0.68 g, measured concentrations | 48 |
| | *Carassius auratus* | 24-h TLm >5000 mg/l<br>96-h LC50 = 20800 mg/l (15700 - 27500) | static, 20 ± 1°C, measured concentrations<br>flow-through, 20.1°C, 0.92 g, measured concentrations | 34<br>48 |
| | *Gambusia affinis* | 96-h LC50 = 15200 mg/l (12400 - 18700)<br>96-h LC50 = 12900 mg/l (11100 - 15000) | flow-through, 18.3°C, 0.23 g, measured concentrations<br>flow-through, 19.9°C, 0.25 g, measured concentrations | 48 |
| | *Ictalurus punctatus* | 96-h LC50 = 6010 mg/l | flow-through, 17.3°C, 0.72 g, measured concentrations | 48 |
| | *Lepomis macrochirus* | 96-h LC50 >10000 mg/l<br>96-h LC50 = 21400 mg/l (19100 - 23900) | static, nominal concentrations, L = 33-75 mm<br>flow-through, 18.3°C, 0.44 g, measured concentrations | 39<br>48 |
| | *Pimephales promelas* | 96-h LC50 = 13900 mg/l (11600 - 16700)<br>96-h LC50 = 9650 mg/l (7910 - 11800) | flow-through, 24.8°C, 1.44 g, measured concentrations<br>flow-through, 18.1°C, 0.35 g, measured concentrations | 48 |
| | *Menidia beryllina* | 96-h LC50 >10000 mg/l | static, sea water, nominal concentrations, L = 40-100 mm | 39 |
| Amphibia | *Rana catesbeiana* | 96-h LC50 = 20900 mg/l (19400 -22600) | flow-through, 17.2°C, 3.54 g, measured concentrations | 48 |
| *Diethylene glycol diethyl ether [112-36-7]* | | | | |
| Fungi | *Cladosporium resinae* | 42-d NG = 20% | NG = no visible mycelial growth and spore germination, 1% glucose-mineral salts medium, 30°C | 27 |

| Species | | Results | Comments | Ref. |
|---|---|---|---|---|
| *Diethylene glycol monobutyl ether [112-34-5]* | | | | |
| Bacteria | *Pseudomonas putida* | 16-h TGK = 255 mg/l<br>16-h EC10 = 1170 mg/l | toxicity threshold, inhibition of cell multiplication<br>growth inhibition, 25 ± 2°C | 24<br>50 |
| Blue-green algae | *Microcystis aeruginosa* | 8-d TGK = 53 mg/l | toxicity threshold, inhibition of cell multiplication | 26 |
| Algae | *Scenedesmus quadricauda* | 8-d TGK = 1000 mg/l | toxicity threshold, inhibition of cell multiplication | 26 |
| | *Scenedesmus subspicatus* | 96-h EC50 > 100 mg/l | limit test | 50 |
| Fungi | *Cladosporium resinae* | 42-d NG = 5% | NG = no visible mycelial growth and spore germination, 1% glucose-mineral salts medium, 30°C | 27 |
| Protozoa | *Chilomonas paramaecium* | 48-h TGK = 2774 mg/l | toxicity threshold, inhibition of cell multiplication | 28 |
| | *Uronema parduczi* | 20-h TGK = 420 mg/l | toxicity threshold, inhibition of cell multiplication | 29 |
| | *Entosiphon sulcatum* | 72-h TGK = 73 mg/l | toxicity threshold, inhibition of cell multiplication | 30 |
| Crustacea | *Daphnia magna* | 24-h LC50 = 2850 mg/l, 24-h LC0 = 1750 mg/l, 24-h LC100 = 3850 mg/l<br>24-h EC50 = 3200 mg/l (2990-3424), 24-h EC0 = 2333 mg/l, 24-h EC100 = 5000 mg/l<br>48-h EC50 >100 mg/l 48-h NOEC >100 mg/l | static, nominal concentrations<br>static, nominal concentrations, immobilization<br>directive 84/449/ EEC, C2 | 32<br>33<br>16 |
| Fish | *Carassius auratus* | 24-h TLm = 2700 mg/l | static, 20 ± 1°C, measured concentrations | 34 |
| | *Leuciscus idus melanotus* | 48-h LC50 = 1805 mg/l, 48-h LC0 = 1140 mg/l, 48-h LC100 = 2185 mg/l<br>48-h LC50 = 2304 mg/l, 48-h LC0 = 1820 mg/l, 48-h LC100 = 2400 mg/l<br>48-h LC50 = 2750 mg/l | static (Juhnke)<br>static (Lüdemann)<br>static | 35<br>35<br>16 |
| | *Poecilia reticulata* | 7-d LC50 = 1149 mg/l | semi-static, 22±1°C, rounded value | 38 |
| | *Lepomis macrochirus* | 96-h LC50 = 1300 mg/l | static, nominal concentrations, L = 33-75 mm | 39 |
| | *Menidia beryllina* | 96-h LC50 = 2000 mg/l | static, sea water, nominal concentrations, L = 40-100 mm | 39 |

| Species | | Results | Comments | Ref. |
|---|---|---|---|---|
| *Triethylene glycol monomethyl ether [112-35-6]* | | | | |
| Bacteria | *Pseudomonas aeruginosa* | 4-m biocidal = 10-17% | tested in jet fuel and water mixtures | 25 |
| | Sulfate-reducing bacteria | 3-m biocidal = 10-17% | tested in jet fuel and water mixtures | 25 |
| Yeasts | *Candida sp.* | 4-m biocidal = 10-17% | tested in jet fuel and water mixtures | 25 |
| Fungi | *Cladosporium resinae* | 4-m biocidal = 10-17% | tested in jet fuel and water mixtures | 25 |
| | *Gliomastix sp.* | 4-m biocidal = 17-25% | tested in jet fuel and water mixtures | 25 |
| *Triethylene glycol monoethyl ether [112-50-5]* | | | | |
| Bacteria | *Pseudomonas aeruginosa* | 4-m biocidal = 5-10% | tested in jet fuel and water mixtures | 25 |
| | Sulfate-reducing bacteria | 3-m biocidal = 2-5% | tested in jet fuel and water mixtures | 25 |
| Yeasts | *Candida sp.* | 4-m biocidal = 5-10% | tested in jet fuel and water mixtures | 25 |
| Fungi | *Cladosporium resinae* | 4-m biocidal = 10-17% | tested in jet fuel and water mixtures | 25 |
| | *Gliomastix sp.* | 4-m biocidal = 17-25% | tested in jet fuel and water mixtures | 25 |
| Crustacea | *Artemia salina* | 24-h TLm >10000 mg/l | static, 24.5°C | 10 |
| Fish | *Carassius auratus* | 24-h TLm >5000 mg/l | static, 20 ± 1°C, measured concentrations | 34 |
| *Propylene glycol monomethyl ether [107-98-2]* | | | | |
| Fungi | *Cladosporium resinae* | 42-d NG = 20% | NG = no visible mycelial growth and spore germination, 1% glucose-mineral salts medium, 30°C | 27 |
| *Propylene glycol monomethyl ether acetate [108-65-6]* | | | | |
| Fish | *Lepomis macrochirus* | 24-h LC50 = 206 mg/l<br>96-h LC50 = 164 mg/l | | 44 |

It is difficult to draw definitive conclusions from the data listed in Table 17.3.2. Indeed, most of the data have been retrieved from rather old studies performed without GLP protocols. In addition, the toxicity values are generally based on nominal concentrations and the endpoints are different. However, from the data listed in Table 17.3.2, it appears that despite a difference of sensibility among species, glycol ethers do not present acute and subacute ecotoxicological effects to the majority of the tested organisms. However, it is interesting to note that the acetates seem to be more toxic that the corresponding parent compounds.[10,34,39,51] In mammals, the acute toxicity of glycol ethers is also relatively low. The main target organs are the central nervous and haematopoitic systems. However, on the basis of the available data no significant difference exists between the acute toxicity of glycol ethers and their corresponding acetates.[52]

### 17.3.4.2 Reproduction and development

The reproductive and developmental toxicity of the ethylene glycol monomethyl and monoethyl ethers is well documented. Several longer-chain glycol ethers also have been investigated for their reproductive and developmental effects against rodents and rabbits.[53-56] Conversely, there is a lack of information on the reproductive and developmental ecotoxicity of glycol ethers and their acetates. Bowden et al.[31] have tested the teratogenic effects of four glycol ethers through their ability to inhibit the regeneration of isolated digestion regions of *Hydra vulgaris* (syn. *H. attenuata*). They have shown that the concentrations of ethylene glycol monomethyl, monoethyl, monobutyl, and diethylene glycol monoethyl ethers that were 50% inhibitory to regenerating digestive regions (IC50) after 72-h of exposure were 19,000, 1400, 540, and 19,000 mg/l, respectively. More specifically, at 10,000 mg/l of ethylene glycol monomethyl ether, the digestive regions regenerated the mouth and some tentacles. At 19,000 mg/l only tentacle buds were seen, while 38,000 mg/l produced disintegration of the coelenterates. Ethylene glycol monoethyl ether at 900 mg/l allowed the regeneration of the mouth, some tentacles and the basal disc. At 1900 mg/l four digestive regions showed wound healing while the remainder were dead. A concentration of 3700 mg/l was lethal to both polyps and digestive regions. At concentrations up to 370 mg/l of ethylene glycol monobutyl ether, digestive regions regenerated some tentacles and in some cases the basal disc. Normal wound healing only was observed at 740 mg/l while at 920 mg/l the wounds were healed but the region expanded. Last, the digestive regions at 10,000 mg/l of diethylene glycol monoethyl ether regenerated the mouth and some tentacles. At 20,000 mg/l only tentacle buds were seen while a concentration of 40,000 mg/l was lethal to both polyps and digestive regions. Using the LC50 (Table 17.3.2)/IC50 ratio as developmental hazard index, Bowden et al.[31] ranked the four studied glycol ethers as follows: Ethylene glycol ethyl ether (1.7) > ethylene glycol monomethyl ether (1.5) > ethylene glycol monobutyl ether (1.3) > diethylene glycol monoethyl ether (0.9). Johnson et al.[57,58] have also ranked glycol ethers according to the difference between their lowest concentrations overtly toxic to adults (A) and their lowest concentrations interfering with development (D) of the artificial embryos of reaggregated adult *Hydra attenuata* cells. The A/D ratios found by these authors were the following: Ethylene glycol monoethyl ether (5.0) > ethylene glycol monobutyl ether (4.4) > ethylene glycol monophenyl ether (3.3) > diethylene glycol dibutyl ether (2.3) > diethylene glycol monoethyl ether (2.2) > ethylene glycol monomethyl ether (1.3) > ethylene glycol monomethyl ether acetate (1.0) = ethylene glycol monoethyl ether acetate (1.0). Daston et al.[37] have shown that A/D ratios were not constant across species and hence, there was no basis for using this parameter for developmental hazard assessment. Thus, for example, the A/D ratios calculated from the lowest observed effect levels (LOELs) of the ethylene glycol monomethyl ether were 8, >3, 0.5, and <0.3 for the mouse, *Xenopus laevis, Pimephales promelas,* and *Drosophila melanogaster*, respectively. If the A/D ratios were calculated from the NOELs (no observed effect levels), the values became >4, ≥6, 0.4, and ≤0.3 for the mammal, amphibian, fish, and insect, respectively.

Teratogenicity of glycol ethers has been deeply investigated on the fruit fly, *Drosophila melanogaster*. Statistically significant increases in the incidence of wing notches and bent humeral bristles have been observed in *Drosophila melanogaster* exposed during development to ethylene glycol monomethyl ether (12.5, 15, 18, 22, and 25 mg/vial) and ethylene glycol monoethyl ether (54, 59, 65, 71, and 78 mg/vial).[59] Wing notches, rare in control flies, were found in 13.8% of flies treated with ethylene glycol monomethyl ether

(7.5 μl/g).[60] In general, male pupae are much more affected by ethylene glycol monomethyl ether than female pupae. However, teratogenicity appears strain dependent. Higher detoxification occurs with increased alcohol dehydrogenase (ADH) activity. Ethylene glycol monomethyl ether is much more toxic than its oxidation product, methoxyacetic acid, at the level of adult eclosion. Teratogenic effects were observed in an ADH-negative strain in spite of lacking ADH activity suggesting that apparently, ethylene glycol monomethyl ether is a teratogenic compound by itself against *Drosophila melanogaster*.[61] Last, it is interesting to note that recently, Eisses has shown[62] that administration of ethylene glycol monomethyl ether to larvae of fruit fly, containing the highly active alcohol dehydrogenase variant ADH-71k, exposed the mitotic germ cells and the mitotic somatic cells of the imaginal discs simultaneously to the mutagen methoxyacetaldehyde and the teratogen methoxyacetic acid, respectively. Consequently, the chances for specific gene mutations, though non-adaptive, were likely increased by a feedback mechanism.

### 17.3.5 CONCLUSION

Despite their widespread use, glycol ethers and their acetates have received little attention as potential environmental contaminants. Based on their physico-chemical properties, they would tend to remain in the aquatic ecosystems where their bioconcentration, biomagnification and sorption onto sediments will appear negligible. Volatilization from water and hydrolysis or photolysis in the aquatic ecosystems are generally of minimal importance. Glycol ethers are also poorly sorbed to soil and their rapid removal in the atmosphere is expected. While glycol ethers are biodegradable under aerobic conditions, these chemicals may persist for many years under anaerobic conditions.

Based on the available acute ecotoxicity data, glycol ethers and their acetates can be considered as practically non-toxic. However, there is a lack of information on their long-term effects on the biota. This is particularly annoying because the developmental toxicity of some of them has been clearly identified against mammals. Consequently, there is a need for studies dealing with the potential long-term effects of these chemicals against organisms occupying different trophic levels in the environment in order to see whether or not the classical methodological frameworks used for assessing the environmental risk of xenobiotics remain acceptable for this class of chemicals.

### 17.3.6 ACKNOWLEDGMENT

This study was supported by the French Ministry of the Environment as part of the PNETOX program (1998).

### REFERENCES


1 R.J. Smialowicz, *Occup. Hyg.*, **2**, 269 (1996).
2 Anonymous in Ethers de Glycols. Quels Risques pour la Santé?, *INSERM*, Paris, 1999, pp. 1-19.
3 K. Nagano, E. Nakayama, M. Koyano, H. Oobayashi, H. Adachi, and T. Yamada, *Jap. J. Ind. Health*, **21**, 29 (1979).
4 Anonymous in Ethers de Glycols. Quels Risques pour la Santé?, *INSERM*, Paris, 1999, pp. 111-137.
5 A. Cicolella, *Cahiers de Notes Documentaires*, **148**, 359 (1992).
6 W. Eckel, G. Foster, and B. Ross, *Occup. Hyg.*, **2**, 97 (1996).
7 D.K. Nguyen, A. Bruchet, and P. Arpino, *J. High Resol. Chrom.*, **17**, 153 (1994).
8 A. Yasuhara, H. Shiraishi, M. Tsuji, and T. Okuno, *Environ. Sci. Technol.*, **15**, 570 (1981).
9 P. Ciccioli, E. Brancaleoni, A. Cecinato, R. Sparapani, and M. Frattoni, *J. Chromatogr.*, **643**, 55 (1993).
10 K.S. Price, G.T. Waggy, and R.A. Conway, *J.Water Pollut. Control Fed.*, **46**, 63 (1974).
11 A.L. Bridié, C.J.M. Wolff, and M. Winter, *Water Res.*, **13**, 627 (1979).
12 T. Fuka, V. Sykora, and P. Pitter, *Sci. Pap. Inst. Chem. Technol. Praze Technol. Water*, **F25**, 203 (1983) (in Czech).
13 S. Takemoto, Y. Kuge, and M. Nakamoto, *Suishitsu Odaku Kenkyu*, **4**, 22 (1981) (in Japanese).

14 P. Pitter and J. Chudoba, *Biodegradability of Organic Substances in the Aquatic Environment, CRC Press,* Boca Raton, 1990.
15 G.T. Waggy, R.A. Conway, J.L. Hansen, and R.L. Blessing, *Environ. Toxicol. Chem.*, **13**, 1277 (1994).
16 C.A. Staples, R.J. Boatman, and M.L. Cano, *Chemosphere*, **36**, 1585 (1998).
17 L. Babeu and D.D. Vaishnav, *J. Ind. Microbiol.*, **2**, 107 (1987).
18 S.J. Gonsior and R.J. West, *Environ. Toxicol. Chem.*, **14**, 1273 (1995).
19 T. Harada and Y. Nagashima, *J. Ferment. Technol.*, **53**, 218 (1975).
20 F. Kawai, *Appl. Microbiol. Biotechnol.*, **44**, 532 (1995).
21 R. Atkinson, *Int. J. Chem. Kinetics*, **19**, 799 (1987).
22 P.H. Howard, R.S. Boethling, W.F. Jarvis, W.M. Meylan, and E.M. Michalenko, Handbook of Environmental Degradation Rates, *CRC Press*, Boca Raton, 1991.
23 D. Grosjean, *J. Air Waste Manage. Assoc.*, **40**, 1397 (1990).
24 G. Bringmann and R. Kühn, *Z. Wasser Abwasser Forsch.*, **10**, 87 (1977).
25 R.A. Neihof and C.A. Bailey, *Appl. Environ. Microbiol.*, **35**, 698 (1978).
26 G. Bringmann and R. Kühn, *Mitt. Internat. Verein. Limnol.*, **21**, 275 (1978).
27 K.H. Lee and H.A. Wong, *Appl. Environ. Microbiol.*, **38**, 24 (1979).
28 G. Bringmann, R. Kühn, and A. Winter, *Z. Wasser Abwasser Forsch.*, **13**, 170 (1980).
29 G. Bringmann and R. Kühn, *Z. Wasser Abwasser Forsch.*, **13**, 26 (1980).
30 G. Bringmann, *Z. Wasser Abwasser Forsch.*, **11**, 210 (1978).
31 H.C. Bowden, O.K. Wilby, C.A. Botham, P.J. Adam, and F.W. Ross, *Toxic. in Vitro,* **9**, 773 (1995).
32 G. Bringmann and R. Kühn, *Z. Wasser Abwasser Forsch.*, **10**, 161 (1977).
33 G. Bringmann and R. Kühn, *Z. Wasser Abwasser Forsch.*, **15**, 1 (1982).
34 A.L. Bridié, C.J.M. Wolff, and M. Winter, *Water Res.*, **13**, 623 (1979).
35 I. Juhnke and D. Lüdemann, *Z. Wasser Abwasser Forsch.*, **11**, 161 (1978).
36 S. Tsuji, Y. Tonogai, Y. Ito, and S. Kanoh, *Eisei Kagaku*, **32**, 46 (1986) (in Japanese).
37 G.P. Daston, J.M. Rogers, D.J. Versteeg, T.D. Sabourin, D. Baines, and S.S. Marsh, *Fund. Appl. Toxicol.*, **17**, 696 (1991).
38 H. Könemann, *Toxicology*, **19**, 209 (1981).
39 G.W. Dawson, A.L. Jennings, D. Drozdowski, and E. Rider, *J. Hazard. Materials*, **1**, 303 (1975-1977).
40 K.L.E. Kaiser and V.S. Palabrica, *Water Poll. Res. J. Canada*, **26**, 361 (1991).
41 IRCHA, Les Produits Chimiques dans l'Environnement. Registre des Données Normalisées de leurs Effets dans l'Environnement. Classeurs I et II et mises à jour, *IRCHA*, Vert-Le-Petit, 1981-1985.
42 J. Hermens, H. Canton, P. Janssen, and R. de Jong, *Aquat. Toxicol.*, **5**, 143 (1984).
43 Results from the MBA Laboratories, Houston, Texas (1984).
44 W.B. Neely, *Chemosphere*, **13**, 813 (1984).
45 IUCLID Data Set, RIVM/ACT, Substance ID: 111-77-3, 19-Nov-98.
46 C. Curtis, A. Lima, S.J. Lozano, and G.D. Veith, in Aquatic Toxicology and Hazard Assessment. Fifth Conference. ASTM STP 766, J.G. Pearson, R.B. Foster, and W.E. Bishop, Eds., *American Society for Testing and Materials*, Philadelphia, pp. 170-178.
47 D. Nacci, E. Jackim, and R. Walsh, *Environ. Toxicol. Chem.*, **5**, 521 (1986).
48 R.V. Thurston, T.A. Gilfoil, E.L. Meyn, R.K. Zajdel, T.I. Aoki, and G.D. Veith, *Water Res.*, **19**, 1145 (1985).
49 E. Jackim and D. Nacci, *Environ. Toxicol. Chem.* **5**, 561 (1986).
50 IUCLID Data Set, RIVM/ACT, Substance ID: 112-34-5, 19-Nov-98.
51 R.B. Sleet, *Toxicologist*, **11**, 296 (1991).
52 Anonymous in Ethers de Glycols. Quels Risques pour la Santé, *INSERM,* Paris, 1999, pp. 51-67.
53 C.A. Kimmel, *Occup. Hyg.*, **2**, 131 (1996).
54 A. Cicolella, *Santé Publique*, **2**, 157 (1997).
55 Anonymous in Ethers de Glycols. Quels Risques pour la Santé?, *INSERM*, Paris, 1999, pp. 111-137.
56 Anonymous in Ethers de Glycols. Quels Risques pour la Santé?, *INSERM,* Paris, 1999, pp. 139-162.
57 E.M. Johnson, B.E.G. Gabel, and J. Larson, *Environ. Health Perspect.*, **57**, 135 (1984).
58 E.M. Johnson, L.M. Newman, B.E.G. Gabel, T.F. Boerner, and L.A. Dansky, *J. Am. Coll. Toxicol.*, **7**, 111 (1988).
59 D. Lynch and M. Toraason, *Occup. Hyg.*, **2**, 171 (1996).
60 R.L. Schuler, B.D. Hardin, and R.W. Niemeier, *Teratogenesis Carcinog. Mutagen.*, **2**, 293 (1982).
61 K.T. Eisses, *Teratogenesis Carcinog. Mutagen.*, **9**, 315 (1989).
62 K.T. Eisses, *Teratogenesis Carcinog. Mutagen.*, **19**, 183 (1999).

# 17.4 ORGANIC SOLVENT IMPACTS ON TROPOSPHERIC AIR POLLUTION

Michelle Bergin and Armistead Russell
**Georgia Institute of Technology, Atlanta, Georgia, USA**

## 17.4.1 SOURCES AND IMPACTS OF VOLATILE SOLVENTS

Solvents, either by design or default, are often emitted in to the air, and the total mass of emissions of solvents is not small. In a typical city in the United States, solvents can rival automobile exhaust as the largest source category of volatilized organic compound (VOC) emissions into the atmosphere.[1] In the United Kingdom, solvent usage accounted for 36% of the estimated total VOC mass emissions in 1995.[2] Such widespread emissions leads to increased concentrations of many different compounds in the ambient environment, and their release has diverse impacts on air quality.

A large variety of solvent-associated compounds are emitted, many of which are hydrocarbons, oxygenates. Those solvents may have multiple atmospheric impacts. For example, toluene is potentially toxic and can reach relatively high concentrations at small spatial scales, such as in a workplace. Toluene also contributes to the formation of tropospheric ozone at urban scales, while at regional scales toluene can lower the rate of tropospheric ozone formation. Other solvents likewise can have a range of impacts, ranging from local contamination to modification of the global climate system.

This diversity of potential impacts is due, in part, to differences in the chemical properties and reactions that a compound may undergo in the atmosphere, differences in emissions patterns, and differences in the spatial and temporal scales of atmospheric phenomena. Transport and fate of chemical species is closely tied to the speed at which the compound degrades (from seconds to centuries, depending on the compound) as well as to the environmental conditions in which the compound is emitted. If a compound degrades very quickly, it may still have toxic effects near a source where concentrations can be high. In contrast, extremely stable compounds (such as chlorofluorocarbons; CFCs) are able to circumvent the globe, gradually accumulating to non-negligible concentrations.[3]

Of the myriad of solvents emitted into the air, the ones of primary concern are those with the greatest emissions rates, and/or those to which the environment has a high sensitivity. Compounds with very large emissions rates include tri- and tetrachloroethylene (e.g., from dry-cleaning), aromatics (benzene, toluene and xylenes, e.g., from coatings), alcohols, acetone and, historically, CFCs. While those compounds are often emitted from solvent use, other applications lead to their emission as well. For example, gasoline is rich in aromatics and alkanes, and in many cases fuel use dominates emissions of those compounds. CFCs have been used as refrigerants and as blowing agents. This diversity of originating sources makes identifying the relative contribution of solvents to air quality somewhat difficult since there are large uncertainties in our ability to quantify emissions rates from various source categories.

Solvents with a high environmental sensitivity include benzene (a potent carcinogen), xylenes (which are very effective at producing ozone), formaldehyde (both toxic and a strong ozone precursor), and CFCs (ozone depleters and potential greenhouse gases). Most of the solvents of concern in terms of impacting ambient air are organic, either hydrocarbons, oxygenated organics (e.g., ethers, alcohols and ketones) or halogenated organics (e.g., dichlorobenzene). Some roles of these compounds in the atmosphere are discussed below.

While the toxicity of some solvents is uncertain, the role of emissions on direct exposure is not in question. Indoors, vaporized solvents can accumulate to levels of concern for acute and/or chronic exposure. However, the toxicity of solvents outdoors is not typically of as great of concern as indoors except very near sources. Outdoors, solvents have adverse effects other than toxicity. The importance of CFC emissions on stratospheric ozone, for example, is significant, but the problem is well understood and measures are in place to alleviate the problem. Reactive compounds can also aid in the formation of other pollutants, referred to as secondary pollutants because they are not emitted, but formed from directly emitted primary precursors. Of particular concern is tropospheric ozone, a primary constituent of photochemical smog. In the remainder of this chapter, the impacts of solvents on air quality are discussed, with particular attention given to the formation of tropospheric ozone. This emphasis is motivated by current regulatory importance as well as by lingering scientific issues regarding the role of volatile organics in secondary pollution formation.

## 17.4.2 MODES AND SCALES OF IMPACT

Many organic solvents are toxic, and direct exposure to the compound through the atmosphere (e.g., via inhalation) can be harmful. While toxic effects of solvents rely on direct exposure, many solvents also contribute to the formation of secondary pollutants such as tropospheric ozone or particulate matter (PM), which cause health problems and damage the environment on larger spatial scales such as over urban areas and multi-state/country regions. Very slowly reacting solvent compounds also impact the atmosphere on the global scale, which may cause imbalances in living systems and in the environment. While some mechanisms of environmental imbalance are understood, the risks associated with global atmospheric impacts are highly uncertain.

Transport of solvents in the atmosphere is similar to most other gaseous pollutants, and is dominated by the wind and turbulent diffusion. There is little difference between the transport of different solvent compounds, and the fact that most solvents have much higher molecular weights than air does not lead to enhanced levels at the ground. Heavy solvents are, for the most part, as readily diffused as lighter solvents, although they may not vaporize as fast. The higher levels of many solvents measured near the ground are due to proximity to emissions sources, which are near the surface, and the fact that most solvents degrade chemically as they mix upwards. A major difference in the evolution of various solvents is how fast they react chemically. Some, such as formaldehyde, have very short lifetimes while others, such as CFCs, last decades.

### 17.4.2.1 Direct exposure

Volatilization of solvents allows air to serve as a mode of direct exposure to many compounds known to be toxic. Generally, direct exposure is a risk near strong or contained sources, and can cause both acute and chronic responses. Most of the non-workplace exposure to solvents occurs indoors. This is not surprising since, on average, people spend a vast majority of their time indoors, and solvents are often used indoors. Outdoors, solvents rapidly disperse and can oxidize, leading to markedly lower levels than what is found indoors near a source. For example, indoor formaldehyde levels are often orders of magnitude greater than outdoors. There still are cases when outdoor exposure may be non-negligible, such as if one spends a significant amount of time near a major source. Toxic effects of solvents are fairly well understood, and many countries have developed regulatory structures to protect people from direct exposure. The toxic effects of solvent emissions on ecosystems are less well understood, but are of growing concern.

### 17.4.2.2 Formation of secondary compounds

In addition to transport, organic compounds emitted into the air may also participate in complex sets of chemical reactions. While many of these reactions "cleanse" the atmosphere (most organic compounds ultimately react to form carbon dioxide), a number of undesirable side effects may also occur. Such adverse impacts include the formation of respiratory irritants and the destruction of protective components of the atmosphere. Ozone is a classic example of the complexity of secondary atmospheric impacts. Ozone is a highly reactive molecule consisting of three oxygen atoms ($O_3$). In one part of the atmosphere ozone is beneficial, in another, it is a pollutant of major concern. Solvents and other organic emissions may either increase or decrease ozone concentrations, depending on the compound, location of reaction, and background chemistry. The mechanisms of some adverse secondary responses are discussed below.

### 17.4.2.3 Spatial scales of secondary effects

Two layers of the Earth's atmosphere are known to be adversely impacted by solvents - the troposphere and the stratosphere. These two layers are closest to Earth, and have distinct chemical and physical properties. The troposphere (our breathable atmosphere) is the closest layer, extending from the Earth to a height of between 10 to 15 km. The rate of chemical reaction generally determines the spatial scale over which emissions have an impact in the troposphere. Most non-halogenated solvents have lifetimes of a week or less, and elevated concentrations will only be found near the sources.[4] Compounds that do not react rapidly in the troposphere (e.g., CFCs) are relatively uniformly distributed, and may eventually reach the stratosphere. The stratosphere is the next vertical layer of the atmosphere, extending from the tropopause (the top of the troposphere) to about 50 km in altitude. Little vertical mixing occurs in the stratosphere, and mixing between the troposphere and the stratosphere is slow.

Impacts on the stratosphere can be considered global in scale, while impacts on the troposphere are generally urban or regional in scale. Distinct chemical systems of interest concerning solvents in the atmosphere are stratospheric ozone depletion, global climate change, and tropospheric photochemistry leading to enhanced production of ozone, particulate matter, and other secondary pollutants such as organonitrates.

#### *17.4.2.3.1 Global impacts*

Because some solvent compounds are nearly inert, they can eventually reach the stratosphere where they participate in global scale atmospheric dynamics such as the destruction of stratospheric ozone and unnatural forcing of the climate system. Stratospheric ozone depletion by chlorofluorocarbons (CFCs) is a well-known example of global scale impacts. CFCs were initially viewed as environmentally superior to organic solvents. They are generally less toxic than other similarly acting compounds, less flammable and are virtually inert in the troposphere. Replacing solvents using volatile organic compounds (VOCs) with CFCs was hoped to reduce the formation of tropospheric ozone and other secondary pollutants. Because of their inert properties, there are no effective routes for the troposphere to remove CFCs, and, over the decades, emissions of CFCs have caused their accumulation, enabling them to slowly leak into the stratosphere. In the stratosphere, the strong ultraviolet (UV) light photodissociates CFCs, releasing chlorine, which then catalytically attacks ozone. CFC use has been largely eliminated for that reason. Partially halogenated organic solvents do not contribute as seriously to this problem since they react faster in the troposphere than CFCs, so the associated chlorine does not reach the stratosphere as efficiently. CFCs and other solvent compounds also have a potential impact on global climate change.

#### *17.4.2.3.2 Stratospheric ozone depletion*

Natural concentrations of stratospheric $O_3$ are balanced by the production of ozone via photolysis of oxygen by strong UV light, and destruction by a number of pathways, including reactions with nitrogen oxides and oxidized hydrogen products that are present. Photolysis of an oxygen molecule leads to the production of two free oxygen atoms:

$$O_2 + h\nu \longrightarrow O + O \quad [17.4.1]$$

Each oxygen atom can then combine with an oxygen molecule to form ozone:

$$O + O_2 \longrightarrow O_3 \quad [17.4.2]$$

Ozone is then destroyed when it reacts with some other compound, e.g., with NO:

$$O_3 + NO \longrightarrow NO_2 + O_2 \quad [17.4.3]$$

Addition of either chlorine or bromine atoms leads to extra, and very efficient, pathways for ozone destruction. The free chlorine (or bromine) atom reacts with ozone, and the product of that reaction removes a free oxygen atom:

$$Cl + O_3 \longrightarrow ClO + O_2 \quad [17.4.4]$$

$$\underline{ClO + O \longrightarrow Cl + O_2} \quad [17.4.5]$$

*Net (reactions* 4+5 *together)*: $$O_3 + O \longrightarrow 2\,O_2 \quad [17.4.6]$$

Removing a free oxygen atoms also reduce ozone since one less ozone molecule will be formed via reaction 17.4.2. Thus, the chlorine atom reactions effectively remove two ozone molecules by destroying one and preventing the formation of another. Additionally, the original chlorine atom is regenerated to catalytically destroy more ozone. This reaction cycle can proceed thousands of times, destroying up to 100,000 molecules of $O_3$ before the chlorine is removed from the system (e.g., by the formation of HCl).

Reduction of ozone is greatly enhanced over the poles by a combination of extremely low temperatures, decreased transport and mixing, and the presence of polar stratospheric clouds that provide heterogeneous chemical pathways for the regeneration of atomic chlorine. The resulting rate of $O_3$ destruction is much greater than the rate at which it can be naturally replenished.

Current elevated levels of CFCs in the troposphere will provide a source of chlorine to the stratosphere for decades, such that the recent actions taken to reduce CFC emissions (through the Montreal Protocol) will have a delayed impact.

#### *17.4.2.3 Global climate forcing*

Over the past decade, the potential for non-negligible changes in climate caused by human activity has been an issue of great concern. Very large uncertainties are associated with both estimations of possible effects on climate as well as estimations of the potential impacts of changes in climate. However, current consensus in the international scientific community is that observations suggest "a discernible human influence on global climate".[3]

Solvent compounds, especially CFCs and their replacements, participate in climate change as "greenhouse gases". Greenhouse gases allow short-wave solar radiation to pass through, much of which the earth absorbs and re-radiates as long-wave radiation. Greenhouse gases absorb the long-wave radiation, causing the atmosphere to heat up, thereby acting as a blanket to trap radiation that would normally vent back to space. Climate change is a controversial and complex issue, but it is likely that restrictions such as those from the Kyoto Protocol will be adopted for emissions of compounds strongly suspected of exacerbating climate change. Many countries have already adopted stringent policies to reduce greenhouse gas emissions.

#### *17.4.2.4 Urban and regional scales*

Another area of concern regarding outdoor air is exposure to secondary pollutants that are due, in part, to chemical reactions involving solvent compounds. Examples include the formation of elevated levels of ozone, formaldehyde, organonitrates, and particulate matter. Formaldehyde, a suspected carcinogen, is an oxidation product of organic compounds. Tropospheric ozone and organonitrates, as discussed below, are formed from a series of reactions of organic gases and nitrogen oxides in the presence of sunlight. Particulate matter formation is linked to ozone, and some solvents may react to form particulate matter. The particulate matter of concern is small (generally less than 2.5 μm in diameter) usually formed by gas-to-aerosol condensation of compounds via atmospheric chemical reactions. Ozone and particulate matter are both regulated as "criteria" pollutants in the United States because they have been identified as risks to human health. Ozone is believed to cause respiratory problems and trigger asthma attacks, and PM has a variety of suspected adverse health outcomes (e.g., respiratory and coronary stress and failure). Many organonitrates, such as peroxyacetyl nitrate, are eye irritants and phytotoxins. Currently, the formation and effects of ozone are better understood than those of fine particulate matter and organonitrates. The following section of this chapter discusses the effects, formation, and control of tropospheric ozone. The role of solvents in forming particulate matter is currently viewed as less urgent.

## 17.4.3 TROPOSPHERIC OZONE

Tropospheric ozone, a primary constituent of photochemical smog, is naturally present at concentrations on the order of 20-40 parts per billion (ppb).[4] However, elevated levels of ground-level ozone are now found virtually worldwide, reaching in some cities concentrations of up to 10 times the natural background.

### 17.4.3.1 Effects

Ozone is believed to be responsible for both acute (short-term) and chronic (long-term) impacts on human health, especially on lung functions. Major acute effects of ozone are decreased lung function and increased susceptibility to respiratory problems such as asthma attacks and pulmonary infection. Short-term exposure can also cause eye irritation, coughing, and breathing discomfort.[5-7] Evidence of acute effects of ozone is believed to be "clear and compelling".[8] Chronic health effects may present a potentially far more serious problem; however, definitive evidence is difficult to obtain. Recent studies do suggest that ambient levels of ozone induce inflammation in human lungs, which is generally accepted as a precursor to irreversible lung damage,[6] and chronic animal exposure studies at concentrations within current ambient peak levels indicate progressive and persistent lung function and structural abnormalities.[5,8]

Crop damage caused by air pollution has also received much attention. It is estimated that 10% to 35% of the world's grain production occurs in regions where ozone pollution likely reduces crop yields.[9] Air pollution accounts for an estimated several billion dollar crop loss every year in the United States alone, and research and analysis suggests that about 90% of this crop loss can be directly or indirectly attributed to ozone.[10] Evidence also indicates that ozone may cause short- and long-term damage to the growth of forest trees,[11] as well as altering the biogenic hydrocarbon emissions of vegetation.[12]

#### 17.4.3.2 Tropospheric photochemistry and ozone formation

In the lowest part of the atmosphere, chemical interactions are very complex. A large number of chemical compounds are present, the levels of many of these compounds are greatly elevated, and emissions vary rapidly due to both natural and anthropogenic sources. Ozone formation in the troposphere results from non-linear interactions between $NO_x$, VOCs, and sunlight.[4,13] In remote regions, ozone formation is driven essentially by methane,[14] however elsewhere most VOCs participate in ozone generation. For example, measurements of non-methane organic compounds in the South Coast Air Basin of California during the 1987 Southern California Air Quality Study, identified more than 280 ambient hydrocarbon and oxygenated organic species,[15] many of which originated from solvents and contribute in differing degrees to ozone generation.

The only significant process forming $O_3$ in the lower atmosphere is the photolysis of $NO_2$ (reaction with sunlight), followed by the rapid reactions of the oxygen atoms formed with $O_2$.

The only significant process forming $O_3$ in the lower atmosphere is the photolysis of $NO_2$ (reaction with sunlight), followed by the rapid reactions of the oxygen atoms formed with $O_2$.

$$NO_2 + h\nu \rightarrow O(^3P) + NO \qquad [17.4.7]$$

$$O(^3P) + O_2 + M \rightarrow O_3 + M$$

This is reversed by the rapid reaction of $O_3$ with NO,

$$O_3 + NO \rightarrow NO + O_2 \qquad [17.4.8]$$

This reaction cycle results in a photostationary state for $O_3$, where concentrations only depend on the amount of sunlight available, dictated by the $NO_2$ photolysis rate ($k_1$) and the $[NO_2]/[NO]$ concentration ratio.

$$[O_3]_{steady-state} = \frac{k_1[NO_2]}{k_2[NO]} \qquad [17.4.9]$$

Because of this photostationary state, ozone levels generally rise and fall with the sun, behavior that is referred to as "diurnal."

If the above $NO_x$ cycle were the only chemical process at work, the steady-state concentrations of ozone would be relatively low. However, when VOCs such as organic solvent compounds are present, they react to form radicals that may either (1) consume NO or (2) convert NO to $NO_2$. This additional reaction cycle combined with the above photostationary state relationship causes $O_3$ to increase.

Although many types of reactions are involved,[4,13,16,17] the major processes for most VOCs can be summarized as follows:

$$VOC + OH \rightarrow RO_2 + products \quad [17.4.10a]$$

$$RO_2 + NO \rightarrow NO_2 + radicals \quad [17.4.10b]$$

$$radicals \rightarrow ... \rightarrow OH + products \quad [17.4.10c]$$

$$products \rightarrow ... \rightarrow ... + CO_2 \quad [17.4.10d]$$

The last two pseudo-reactions given comprise many steps, and the products often include formaldehyde, carbon monoxide and organonitrates. The rate of ozone increase caused by these processes depends on the amount of VOCs present, the type of VOCs present, and the level of OH radicals and other species with which the VOCs can react. One of the major determinants of a compound's impact on ozone is the rate of the reaction of the particular VOC with the hydroxyl radical via reaction [17.4.10a], above. The total amount of ozone formed is largely determined by the amount of VOC and $NO_x$ available.

The dependence of $O_3$ production on the initial amounts of VOC and $NO_x$ is frequently represented by means of an *ozone isopleth diagram*. An example of such a diagram is shown in Figure 17.4.1. The diagram is a contour plot of ozone maxima obtained from a large number of air quality model simulations using an atmospheric chemical mechanism. Initial concentrations of VOC and $NO_x$ are varied; all other variables are held constant. Notice that there is a "ridge" along a certain VOC-to-$NO_x$ ratio where the highest ozone concentrations occur at given VOC levels. This is referred to as the "optimum" VOC-to-$NO_x$ ratio. While the atmosphere is more complicated than this idealized system, important features are very similar.

VOC-to-$NO_x$ ratios sufficiently low to retard ozone formation from an optimum ratio (represented in the upper left quadrant of Figure 17.4.1) can occur in central cities and in plumes immediately downwind of strong $NO_x$ sources. Rural environments tend to be characterized by fairly high VOC-to-$NO_x$ ratios because of the relatively rapid removal of $NO_x$

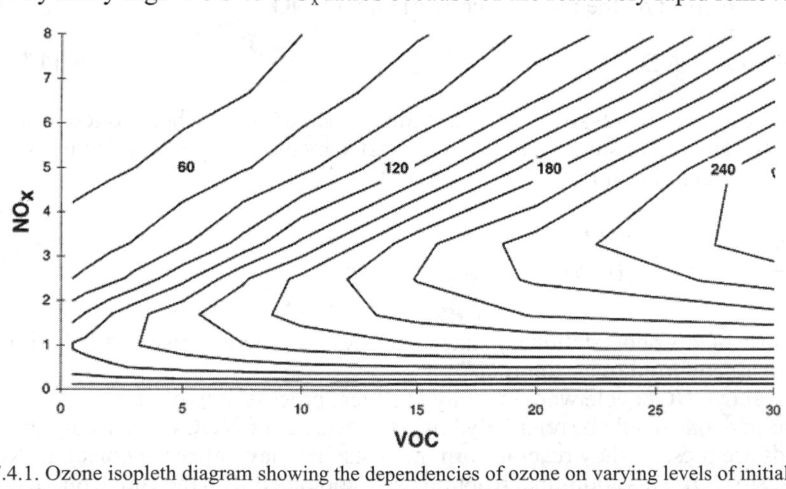


Figure 17.4.1. Ozone isopleth diagram showing the dependencies of ozone on varying levels of initial VOCs and $NO_x$. Concentrations are given in ppb. [Adapted from M.S. Bergin et al., *Enc. of Env. Analysis and Remediation*, **29**, 3029, (1998)]

from non-local sources as compared to that of VOCs, coupled with the usual absence of strong local $NO_x$ sources and the presence of natural VOC sources. In such rural environments, the formation of ozone is limited more by the absence emissions of $NO_x$, and most ozone present was directly transported from upwind. Indeed, in most of the troposphere, except in areas of strong $NO_x$ sources, the availability of $NO_x$ governs ozone production.

#### 17.4.3.3 Assessing solvent impacts on ozone and VOC reactivity

As mentioned previously, the contribution of solvents to the VOC levels, and hence ozone formation, is significant. For example, in Los Angeles, about 25% of the VOC mass is from solvent use.[1] This fraction is down from earlier years due to various controls such as using water-based paints and enclosing/controlling paint spraying operations. On the other hand, reduction in the use of CFCs as propellants has led to an increase in organic emissions from substituted compounds.[2] However, the impact on ozone formation by a specific source is not directly proportional to the amount of VOC emitted by that source. A major determinant of the ozone forming potential is the reactivity of the compound or compound mixture emitted. Reactivity can be viewed as the propensity for a compound to form ozone, and this propensity varies dramatically between compounds and between environments.[18,19]

As seen in Table 17.4.1, 'box' model (single cell) simulations designed to represent summertime conditions in Los Angeles, California indicate that the amount of carbon associated with each class of compound only roughly corresponds to the amount of ozone formed from those compounds. Methane, which reacts very slowly but comprises most of the carbon, contributes little to ozone formation. Alkenes and aromatics are only a small part of the total carbon, but lead to much of the ozone formation.

**Table 17.4.1. Percentage of ozone production attributable to each organic. The percentages shown should be viewed as only approximate, and will depend upon local emissions characteristics. (*While not considered organic carbon, carbon monoxide acts to facilitate ozone formation similar to organic compounds.) [Adapted from F.M. Bowman and J.H. Seinfeld, *J. Geophys. Res.*, 99, 5309, (1994) and M.S. Bergin et al., *Env. Sci. Technol.*, 29, 3029 (1998)]**

| Compound Class | Percent of carbon in each specified class | Percent of ozone due to specified organic class |
|---|---|---|
| carbon monoxide* | 35 | 6 |
| methane | 40 | 1 |
| aldehydes and ketones | 1 | 3 |
| non-methane alkanes, ~4C | 8 | 17 |
| non-methane alkanes, ~8C | 5 | 16 |
| aromatics, including toluene | 3 | 5 |
| aromatics, including xylenes and others | 3 | 13 |
| ethene | 2 | 12 |
| biogenic alkenes ans isoprene | 1 | 10 |
| other alkenes | 2 | 17 |

### *17.4.3.3.1 Quantification of solvent emissions on ozone formation*

Two methods are generally employed to quantify the role pollutants play in forming ozone: experimental and computational. Both types of estimation approaches have their limitations. In the case of physical experiments, it is difficult to fully simulate ambient conditions, so the results do not have general applicability. In the case of computational approaches, uncertainties and approximations in the model for airshed conditions, in its formulation, and in the chemical mechanism cause uncertainties in the predicted ozone impacts. For these reasons, modeling predictions and experimental measurements are used together.

17.4.3.3.1.1 Experimental analysis

Experimental analysis is performed using environmental 'smog' chambers, either with a series of single hydrocarbons irradiated in the presence of $NO_x$ or using complex mixtures to simulate, for example, automobile exhaust emitted into characteristic urban ambient conditions. Such chambers are large reaction vessels (some with internal volumes of cubic meters), in which air and small amounts of hydrocarbons and $NO_x$ are injected, and then irradiated with real or artificial light. Both indoor and outdoor chambers are used so behaviors can be evaluated under natural radiative conditions and under controlled conditions. While these experiments[18-23] clearly indicate differences in ozone formation from individual hydrocarbons, they do not represent some important physical systems of urban pollution such as the mixing processes and continuing emissions cycles. Such experiments have focused both on groups of compounds as well as specific VOCs, including solvents. A particular limitation has been studying very low vapor pressure solvents because it is difficult to get enough of the compound into the vapor phase in the chamber to appreciably change the ozone levels. Another limitation is the expense of using smog chambers to simulate a large range of conditions that might occur in the atmosphere. On the other hand, smog chambers are very powerful, if not fundamental, for developing chemical mechanisms that describe the reaction pathways that can be used in computational approaches.

17.4.3.3.1.2 Computational analysis (air quality models)

Given the limitations of physical experiments to simulate atmospheric conditions, computer models have been developed to assess the impact of emissions on ozone. These models, called airshed models, are computerized representations of the atmospheric processes responsible for air pollution, and are core to air quality management.[23] They have been applied in two fashions to assess how solvents affect ozone. One approach is to conduct a number of simulations with varying levels of solvent emissions.[2] The second approach is to evaluate individual compounds and then calculate the incremental reactivity of solvent mixtures.[19,21,24-28]

Derwent and Pearson[2] examined the impact of solvent emissions on ozone by simulating air parcel trajectories ending in the United Kingdom and perturbing the emissions to account for an anticipated 30% mass reduction in VOCs from solvents between 1995 and 2007. They found a small decrease in ozone-from 78 to 77 ppb in the mean peak ozone in the UK, and a 9 ppb reduction from 129 ppb outside of London. A more substantial decrease of 33 ppb from the 129 ppb peak outside of London was found from reducing non-solvent mass VOC emissions by 30% outside of the UK and 40% within the UK. This suggests that the VOC emissions from sources other than solvents have a higher average reactivity, as is discussed by McBride et al.[29]

While the types of simulations conducted by Derwent and Pearson[2] are important to understanding the net effect of solvent emissions on ozone, there is an unanswered associated and important question, that being which specific solvents have the greatest impacts.

This question is critical to assessing if one solvent leads to significantly more ozone formation than a viable substitute (or *vice versa*).

To evaluate the contribution of individual organic compounds to ozone formation, the use of *incremental reactivities* (IR) was proposed,[18-21] defined as the change in ozone caused by a change in the emissions of a VOC in an air pollution episode. To remove the dependence on the amount of VOC added, incremental reactivity is defined by equation [17.4.11] as the limit as the amount of VOC added approaches zero, i.e., as the derivative of ozone with respect to VOC:

$$IR_i = \frac{\partial[O_3]}{\partial[VOC_i]} \qquad [17.4.11]$$

Here, $IR_i$ is the incremental reactivity and the subscript i denotes the VOC being examined. This definition takes into account the effects of all aspects of the organic's reaction mechanism and the effects of the environment where the VOC is emitted. A similar quantity is the relative reactivity,[23] $RR_i$:

$$RR_i = \frac{R_i}{\sum_{i=1}^{N} F_{B_i} R_i} \qquad [17.4.12]$$

where:

$F_{B_i}$ mass fraction of compound i in the reference mixture
$IR_i$ incremental reactivity of species i (grams ozone formed per gram compound i emitted)

In this case, the incremental reactivity is normalized by the reactivities of a suite of organics, thus removing much of the environmental dependencies found when using IRs defined by [17.4.11]. This metric provides a means for directly comparing individual compounds to each other in terms of their likely impact on ozone.

A number of investigators have performed calculations to quantify incremental and/or relative reactivities for various solvents and other organics[23-28,30] and references therein. Those studies found very similar results for the relative reactivities of most compounds found in solvents. Figure 17.4.2 (based on references 23, 24 and 26) shows the relative reactivities for some of the more common compounds, as well as possible solvent substitutes and isoprene, a naturally emitted organic. (For a more extensive list of relative reactivities, see 19, 27 and 30.) As can be seen, even normalized compound reactivities can vary by orders of magnitude. Some compounds even exhibit "negative" reactivities, that is that their emission can lead to ozone decreases under specific conditions. In particular, negative reactivities are found most commonly when the levels of $NO_x$ are low, e.g., in non-urban locations. For example, Kahn et al.,[26] found that a solvent can promote ozone formation in one area (e.g., near downtown Los Angeles), but retard ozone formation further downwind. Kahn also found that the relative reactivities of the eight different solvents studied were similar in very different locations, e.g., Los Angeles, Switzerland and Mexico City.

Looking at Figure 17.4.2, it is apparent that alkenes and aromatic hydrocarbons with multiple alkyl substitutions (e.g., xylenes and tri-methyl benzene) have relatively high reactivities. Alcohols, ethers and alkanes have lower reactivities. Halogenated organics have some of the lowest reactivities, so low that they are often considered unreactive. This suggests that there are two ways to mitigate how solvents contribute to air quality problems. The more traditional method is to reduce the mass of organic solvent emissions (e.g., by using water-based paints). A second approach is to reduce the overall reactivity of the solvent

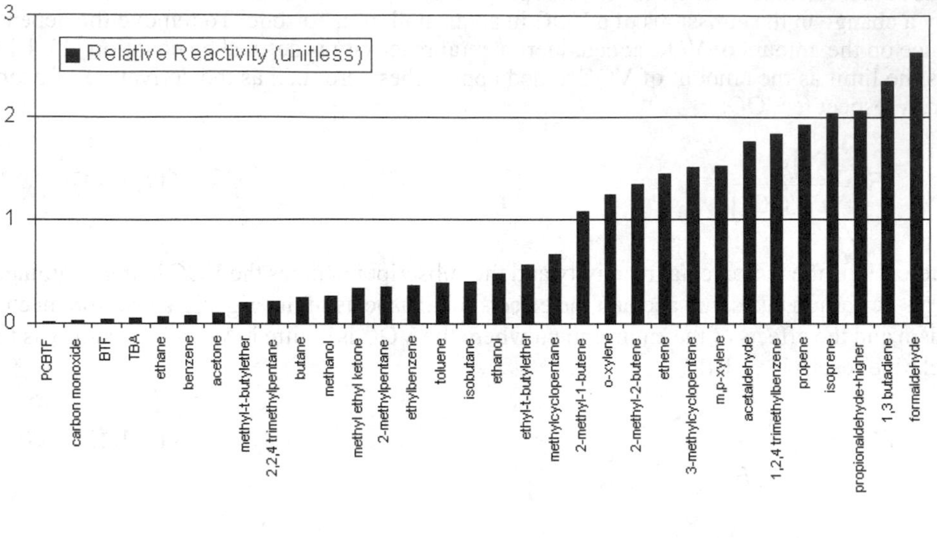


Figure 17.4.2. Solvent relative reactivities based on mass of ozone formed per gram of solvent emitted into the gas phase. PCBTF is para-chlorobenzo-trifluoride, BTF is benzo-trifluoride and TBA is tertiary butyl acetate. [Adapted from M.S. Bergin et al., *Env. Sci. Technol.*, **29**, 3029 (1998) and M. Khan et al., *Atmos. Env.*, **33**, 1085 (1999)].

used, e.g., by switching to ethers, alcohols, alkanes or halogenated compounds. Solvent substitution, however, is complicated by the need to maintain product quality.

## 17.4.4 REGULATORY APPROACHES TO OZONE CONTROL AND SOLVENTS

Historically, regulatory approaches to reducing ozone concentrations have relied reducing the mass emissions of VOCs,[2,4] and this has led to stringent controls on solvents. Two factors are important in determining if an organic solvent is considered a VOC: its reactivity (discussed above) and its vapor pressure. In the U.S., traditionally, if a compound was less reactive than ethane, it was considered unreactive. Such compounds include many halogenated species and some acetates and ethers. Recently, acetone was also added as an unreactive compound. A vapor pressure threshold is also used in many areas (e.g., Europe) since it is viewed that compounds with very low vapor pressures will not be emitted rapidly into the atmosphere. It has been argued that a vapor pressure limit may not be appropriate since, given time, even lower vapor pressure compounds will have ample time to evaporate. Just recently, California is considering regulations that more fully account for the full range of reactivities that solvents possess. This is due, in part, to make it easier for manufacturers to meet stringent regulations being adopted in that state to help them meet their air quality goals. It is likely that other areas will also have to employ increasingly more stringent regulations, to both lower ozone and alleviate other environmental damage.

In many countries, greater focus is now being placed on reducing $NO_x$ emissions to mitigate ozone formation. This has important ramifications for solvent use, indicating the regulatory focus is now turning from VOCs towards $NO_x$, the other main precursor to ozone. Another imminent regulatory issue is the control of ambient fine particulate matter. While the role of solvent emissions in forming particulate matter is not well understood, studies to date do not suggest they are a major contributor.

### 17.4.5 SUMMARY

Solvents are, and will continue to be, one of the major classes of organic compounds emitted into the atmosphere. These compounds have a wide range of air quality impacts. Accumulation of toxic compounds indoors is of concern, although outdoors the concern of toxicity is significantly less substantial due to rapid dilution. In the stratosphere, some of the halogenated solvents lead to depletion of the protective layer of ozone, while in the troposphere solvents generally lead to increased ozone levels, where it adversely affects health and the environment. The former has led to regulations of CFCs, and the latter to regulations of organic solvents. Some solvents are also considered to be precursors to the formation of secondary tropospheric pollutants other than ozone, such as particulate matter, however these relationships are currently less certain.

In the aggregate, total VOC emissions from solvents in the U.S. are the second largest single source category in polluted urban areas, falling just behind motor vehicle VOC emissions both in terms of mass and urban ozone production. For now, regulations are designed to reduce the loss of ozone in the stratosphere and the formation of excess ozone in the troposphere. However, while some solvents are very reactive, others are substantially less reactive, suggesting that there is considerable opportunity to reduce urban ozone formation from solvents by utilizing substitutes with low ozone forming potentials. Currently, most regulations are targeted at reducing the mass of VOC emissions, not their relative impacts on ozone.

### REFERENCES


1 SCAQMD (South Coast Air Quality Management District). (1996). 1997 Air Quality Management Plan. November.

2 R.G. Derwent and J.K. Pearson, *Environ. Technol.*, **18**, 1029 (1997).

3 **Climate Change 1995: The Science of Climate Change**. Contribution of Working Group I to the Second Assessment Report of the Intergovernmental Panel on Climate Change (IPCC). Ed. J.T. Houghton, et al. *Cambridge University Press*, Cambridge, Great Britain. 1996.

4 NRC (National Research Council). **Rethinking the Ozone Problem in Urban and Regional Air Pollution**. *National Academy Press*, Washington, DC. (1991).

5 OTA (U.S. Congress, Office of Technology Assessment). **Catching Our Breath: Next Steps for Reducing Urban Ozone**, OTA-O-412, *U.S. Government Printing Office*, Washington, DC. July,1989.

6 S.M. Horvath and D.J.McKee (1994). In D.J. McKee, ed., **Tropospheric Ozone: Human Health and Agricultural Impacts**, *CRC Press/Lewis Publishers*, Boca Raton, FL, Chapter 3.

7 M. Lippmann, *Environ. Sci. & Technol.*, **25**(12), 1954, (1991).

8 M. Lippmann, *J. of Exposure Analysis and Environmental Epidemiology*, **3**(1), 103, (1993).

9 W.L.Chameides, P.S. Kasibhatla, J Yienger, and H. Levy II., *Science*, **264**(5155), 74, (1994).

10 D.T. Tingey, D.M. Olszyk, A.A.Herstrom, and E.H. Lee, (1994). "Effects of Ozone on Crops" in D.J. McKee, ed., **Tropospheric Ozone: Human Health and Agricultural Impacts**, *CRC Press/Lewis Publishers*, Boca Raton, FL, Chapter 6.

11 S.B.McLaughlin and L.J. Downing, *Nature*, **374**(6519), 252, (1995).

12 W. Mehlhorn, *Nature*, **327**, 417, (1989).

13 J.H. Seinfeld and S.N. Pandis, **Atmospheric Chemistry and Physic: From Air Pollution to Climate Change**, New York: *J. Wiley & Sons*, (1998).

14 J.A. Logan, M.J. Pather, S.C. Wofsy, and M.B. McElroy, *J. Geophys. Res.*, **86**, 7210, (1981).

15 F.W.Lurmann and H.H. Main, Analysis of the Ambient VOC Data Collected in the Southern California Air Quality Study, final report, Contract A832-130, California Air Resources Board, Sacramento, CA., 1992.

16 R. Atkinson, *Atmos. Environ.*, **24A**, 1, (1990).

17 R. Atkinson, *J. Phys. Chem. Ref. Data*, Monograph No. 2, (1994).
18 W.P.L. Carter, *Atmos. Environ.*, **24A**, 481, (1990).
19 W.P.L.Carter, (1991). Development of Ozone Reactivity Scales for Volatile Organic Compounds, EPA 600/3-91-050. U.S. Environmental Protection Agency, Research Triangle Park, NC (August).
20 W.P.L. Carter and R. Atkinson, *Environ. Sci. and Technol.*, **23**, 864, (1989).
21 W.P.L. Carter, *Atmos. Environ.*, **29**, 2513, (1995).
22 N.A. Kelly and P. Wang, (1996) Part I: Indoor Smog Chamber Study of Reactivity in Kelly, N.A.; Wang, P.; Japar, S.M.; Hurley, M.D.; and Wallington, T.J. (1996). Measurement of the Atmosphere Reactivity of Emissions from Gasoline and Alternative-Fueled Vehicles: Assessment of Available Methodologies, Second-Year Final Report, CRC Contract No. AQ-6-1-92 and NREL Contract No. AF-2-112961. Environmental Research Consortium, (September).
23 M.S. Bergin, A.G. Russell, W.P.L. Carter, B.E. Croes, and J.H. Seinfeld, Ozone Control and VOC Reactivity, in the **Encyclopedia of Environmental Analysis and Remediation**, Ed. R.A. Meyers, *J. Wiley & Sons, Inc*, New York, NY. 1998.
24 M.S. Bergin et al., *Env. Sci Technol.*, **29**, 3029 (1998)
25 F.M. Bowman and J.H. Seinfeld, *J. Geophys. Res.*, **99**, 5309, (1994a).
26 M. Khan et al., *Atmos. Env.*, **33**, 1085 (1999).
27 W.P.L. Carter, *J. Air and Waste Mgmt. Assoc.*, **44**, 881, (1994a).
28 R.G. Derwent, and M.E. Jenkin, *Atmos. Environ.*, **25**(A):1661-1673 (1991).
29 S. B. McBride et al., *Env. Sci. Technol.*, **31**, 238a, (1997)
30 http://www.cert.ucr.edu/~carter/bycarter.htm

18

# Concentration of Solvents in Various Industrial Environments

## 18.1 MEASUREMENT AND ESTIMATION OF SOLVENTS EMISSION AND ODOR

Margot Scheithauer
**Institut für Holztechnologie Dresden, Germany**

### 18.1.1 DEFINITION "SOLVENT" AND "VOLATILE ORGANIC COMPOUNDS" (VOC)

Solvents are generally understood to be substances that can physically dissolve other substances, more narrowly they are inorganic and organic liquids able to dissolve other gaseous, liquid, or solid substances. A qualifier for the suitability as a solvent is that, during the solution, neither the solvent nor the dissolved substance undergoes chemical change, i.e., the components of the solution may be recovered in their original form by physical separation processes, such as distillation, crystallization, sublimation, evaporation, adsorption.

From a chemical point of view, solvents or volatile organic compounds, VOCs, vary widely. They are often classified using their boiling point, and vapor pressure. These properties do not define their suitability as solvents.

In the context of VOCs, only organic solvents are of relevance. Therefore, in the following discussion, only organic solvents will be discussed.

**Definition of solvent according to Council Directive 1999/13/EC of 11 March 1999**[1]

Organic solvent shall mean any VOC which is used alone or in combination with other agents, and without undergoing a chemical change, to dissolve raw materials, products or waste materials, or is used as a cleaning agent to dissolve contaminants, or as a dissolver, or as a dispersion medium, or as a viscosity adjuster, or as a surface tension adjuster, or a plasticizer, or as a preservative.

**Definition of solvent according to ISO/DIS 4618-4: 1999**[2]

Paints and varnishes - terms and definition for coating materials: Solvent: A single liquid or blends of liquids, volatile under specified drying conditions, and in which the binder is completely soluble.

For certain application purposes, solvents are defined more specifically:

**Technische Regeln für Gefahrstoffe TRGS 610 (Technical Regulations for Hazardous Substances)** as applied to high solvent-containing primers and flooring glues the definition is:

Solvents are volatile organic compounds as well as mixtures thereof at a boiling point < 200°C, which under normal conditions (20°C and 1013 hPa) are liquid and are applied for dissolving and diluting other substances without undergoing chemical change.[3]

As can be seen, there are clear differences in the way VOCs are defined.[4] The most general, and hence least disputable, one is:

**Definition of volatile organic compounds according to DIN ISO 11890/1,2,**[5,6]

Solvent is generally any organic liquid and/or any organic solid substance, which evaporates by itself under prevailing conditions (temperature and pressure).

**Definition of VOC according to Council Directive 1999/13/EC**[1]

Volatile organic compound, VOC, shall mean any organic compound having at 293.15 K a vapor pressure of 0.01 kPa or more, or having a corresponding volatility under the particular conditions of use. For the purpose of this Directive, the fraction of creosote, which exceeds this value of vapor pressure at 293.15 K shall be considered as a VOC.

In the **Lösemittelverordnung Österreichs (Austrian Regulations on Solvents)** of 1995,[7] which on this issue corresponds to the **österreichische Lackieranlagen-Verordnung (Austrian Regulation on Varnishing Plants)** of 1995,[8] VOCs have a maximum boiling point of 200°C. Hence, the volatile compounds at a boiling point > 200°C are not included. These comprise, e.g., some film forming media, such as butyldiglycol, butyldiglycolacetate, and texanol. Similarly, reactive solvents are excluded from this regulation.

**Definitions of VOCs in the USA according to ASTM D 3960-1**[9]

Volatile Organic Compound (VOC), means any compound of carbon, excluding carbon monoxide, carbon dioxide, carbonic acid, metallic carbides or carbonates, and ammonium carbonate, which participates in atmospheric photochemical reactions. This includes any such organic compound other than the following, which have been determined to have negligible photochemical reactivity: Methane; ethane; methylene chloride (dichloromethane); 1.1.1-trichloroethane (methyl chloroform); 1.1.1-trichloro-2.2.2-trifluorethane (CFC-113), furthermore cyclic, branched or linear completely methylated siloxanes: acetone: and perfluorocarbon compounds.

The regulatory definition under the control of the U.S. EPA may change.

Certain organic compounds that may be released under the specified bake conditions are not classified as VOC as they do not participate in atmospheric photochemical reactions. Such non-photochemically active compounds are referred to as exempt volatile compounds in the practice.

An example of an exempt compound in the USA according to the EPA is tertiary-butyl-acetate (TBA, B.p. of 98 °C). It is described as a substance not harmful in air.[10]

Substances contained in air are differentiated in their mixtures according to their volatility:[11]

| Abbreviation | Compound class | Range of boiling point in °C |
|---|---|---|
| VVOC | very volatile organic compounds | < 0 up to 50 ÷ 100 |
| VOC | volatile organic compounds | 50 ÷ 100 up to 250 ÷ 260 |
| SVOC | semi-volatile organic compounds | 250 ÷ 260 up to 380 ÷ 500 |

TVOC means "total volatile organic compounds".

## 18.1.2 REVIEW OF SOURCES OF SOLVENT EMISSIONS

### 18.1.2.1 Causes for emissions

The basic human needs include: eating, drinking and breathing. In a 60-year lifetime, human being takes up about 30 t of food, 60 t of drink and 300 t of air.[12] Thus air pollution may have a large influence on human health. In comparison with outside air and the ambient air within public transportation vehicles and terminals, room air quality is of the greatest interest, since the population in North American cities stays indoors more than 93% time according to findings by Szalai[13] and more than 91% according to Chapin,[14] of this time > 67 % is spent in living space and about 4-5 % in the workplace.

Sources for volatile substances in indoor air are diverse and originate, apart from that brought in with outside air and produced by the living occupants (people, pets) and their activities (e.g., smoking), from the materials themselves which have been applied for building and furnishing the room. These include:

- building materials
- glues
- floor covering
- wallpaper
- internal decorative textiles
- paints
- furniture, upholstered furniture, etc.

These materials may contain volatile organic, but also inorganic, compounds, which are evaporated during use. The emission of these volatile compounds is influenced by a large number of factors, such as:

- chemical/physical structure of the material (thickness, surface structure)
- volatility, polarity of the volatile compounds
- room temperature and humidity
- room load, i.e., the ratio of the area of emitting surface and the volume of room air
- ventilation, rate of air flow
- load of the external air entering (dust, substances)

The residual monomers from plastics, reactive products (e.g., formaldehyde), degradation products, flame protection media, softeners etc. may be emitted from building materials.

### 18.1.2.2 Emissions of VOCs from varnishes and paints

A substantial part of emitted materials are solvents. They preferably originate from varnishes, paints and glues. Table 18.1.1 shows a survey of the main types of varnishes referring to their average solvent content.

**Table 18.1.1. Solvent share in various types of varnishes**

| Material | Solvent content, wt% |
|---|---|
| Stain | ~95 |
| Cellulose nitrate varnish | ~75 |
| PUR varnish | 35-70 |
| Polyester varnish | 35 |
| UV roller varnish based on acrylate | 2-10 |

Despite the introduction of water based and powder varnishes, cellulose nitrate varnishes with their high solvent content, are still widely used due to their easy application and their low prices. The current varnish application still requires the use of solvents. The chosen coating procedure determines the necessary processing viscosity, which may be adjusted in different ways:

- Thinning the binding agent/varnish with a solvent consisting of one or more organic compounds without altering the binding agent chemically. Nowadays, the solvent can partly or almost totally be replaced by water.
- Thinning the binding agent with a reactive thinner, i.e., one or more monomers/oligomers included in the system, which react with the binding agent during the hardening process and is built into a molecule. According to definitions, they do not have characteristics of solvents. They partly take over the solvent function while they remain in the liquid state.
- Temperature increase of the coating system, which limits the required amount of solvents.
- Liquid application of a melted solid system

Profiled construction components form parts of complicated shapes, which must rely on the spraying of coatings for decorative and protective purposes. The spraying requires low-viscosity material. Viscosity reduction is usually achieved by adding solvents but sometimes by increasing the material temperature. Parts with large flat surfaces however, may be coated by roller-coating, which tolerates higher viscosity materials. After fulfilling their function as solvents, they are expected to completely evaporate from the varnish system.

Emission of solvents from a varnish system occurs in the course of the "life-cycle" of a varnish in several different locations as the following survey shows (Figure 18.1.1). As early as in the production stage of the varnish system, solvent emissions contaminate the ambient air of the factory. About 90 % of the solvents contained in the varnish system evaporates during its application which affects the air quality in the workplaces. These emissions are in the milligram per cubic meter of air range. The thinning effect decreases the solvent concentration in the ambient air at the varnisher's workplace into the microgram to nanogram per cubic meter.

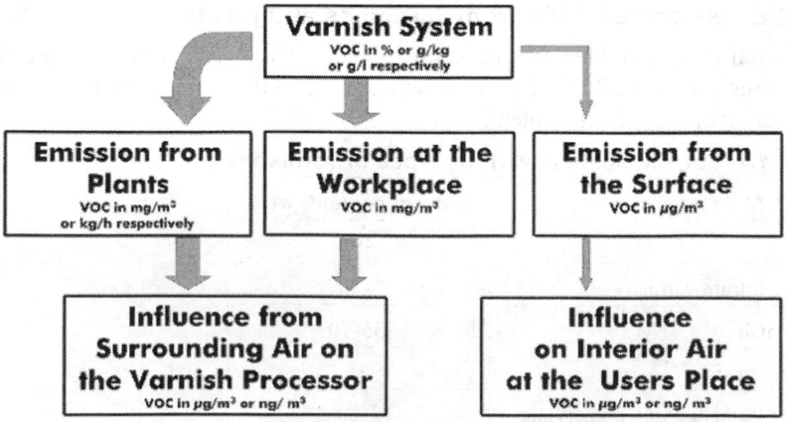


Figure 18.1.1. Emission of VOCs in the life-cycle of a varnish.

About 10 % of the solvent remains in the varnish and it is emitted, depending on the type of solvent and varnish, with variable rate on their way to the consumers and in their living spaces.

According to the European Commission, the furniture industry alone used more than 185,000 tons of solvents/year in 1994.[15]

Not all the VOCs enter the air, however, since remedial efforts, such as absorption and solvent reclaim, after-burning of solvents, the use of bio-filters, result in emission reduction.

Nevertheless, it is still evident that limitation of solvents, VOCs, reduction in varnishes and coatings may help in reducing emissions.

#### 18.1.2.3 VOC emissions from emulsion paints

Although emulsion paints and plasters contain less than 2% VOCs these are produced in such large quantities (3.5 Million t in 1995)[16] and applied over such an extensive surface that 100,000 t of VOCs are emitted from these materials per year. These enter the environment in an uncontrolled way.

Prior to an interior application of the coating, there is no VOC in room atmosphere, unless sources already exist in the room. Then the load - from the beginning of the application - increases very rapidly. After the application, the VOC concentration in the ambient air decreases exponentially. If these emissions are to be monitored it is desirable to have methods available to record VOC concentrations typical of emitting stages.[17]

### 18.1.3 MEASURING OF VOC-CONTENT IN PAINTS AND VARNISHES

#### 18.1.3.1 Definition of low-emissive coating materials

In order to limit the sources of solvent emission in ambient air in the most effective manner, primary goal must be to reduce the VOC content in coating materials. In Germany, there is a draft for the Varnishing Plant Regulation[18] - regulating the application of low-emissive varnishes, which will require that coating materials must initially not contain more than 420 g/l VOC. Four years after the introduction of the regulation this is reduced to 250 g/l VOC.

The "Jury Umweltzeichen" (Environmental Trade Mark), which, since 1999, rewards certain products with the environmental trade mark goes even further: For flat, plain materials (e.g., interior doors, panels, floors with varnished surfaces, pre-fabricated parquet) the coating systems applied must not exceed the VOC content = max. 250 g/l.[19]

Varnishing plants are exempt if they employ waste gas purification facilities corresponding to the EU VOC Guideline.[20]

The draft of the Varnishing Plant Regulation follows the US method according to ASTM D 3960-121 in determining the VOC content.

#### 18.1.3.2 Determination of the VOC content according to ASTM D 3960-1

ASTM D 3960-1 is the method used in the US to determine the VOC content of coating materials. The VOC value is calculated as follows:

$$VOC\ value\,[g/l] = \frac{mass\ of\ volatile\ components\,[g] - mass\ of\ water\,[g]}{volume\ of\ varnish\,[l] - volume\ of\ water\,[l]} \quad [18.1.1]$$

For varnishes exclusively containing solvents, this formula results in high values. This calculation is, however, of dubious value for water-containing varnishes. The unit g/l does not refer to 1 liter of an actual varnish, but to a hypothetical product, which only expresses the binding agent in liters.

This is demonstrated in Example 1, which will then also be calculated according to other methods.

*Example 1*:

Water spray varnish: 30% solid matter content (binding agent and pigments)
10% solvent content
60% water content

varnish density: 1.0 g/cm$^3$

VOC (g/l) = (700 g - 600 g)/(1 l - 0.6 l) = 250 g/l

#### 18.1.3.3 Determination of the VOC content according to ISO/DIS 11 890/1[5] and 2[6]

##### *18.1.3.3.1 VOC content > 15%*

The calculation according to ISO/DIS 11 890/1 gives a more logical result. VOCs are given here as a mass share in % of the ready-to-use varnish. At a VOC content > 15 % (of the mass share), the determination is made by the differential approach:

$$VOC\ [\%\ of\ the\ mass\ share] = 100 - nfA - m_w \qquad [18.1.2]$$

where:

nfA non-volatile parts determined by ISO 3251[22] (1 h/110°C)

$m_w$ mass as water in % of the mass parts, determined according to ISO 760[23] (after Karl-Fischer)

Many more recently available techniques for water determination are not sufficiently common and thus are not considered.

*Example 2*:

Water solvent stain: 5% solid matter content = non-volatile parts
25% solvent
70% water share

VOC (mass %) = 100 - 5 - 70 = 25 %

For density ≠ 1

$$VOC\ [g/l] = 10(100 - nfA - m_w)\,\rho_s \qquad [18.1.3]$$

where:

$\rho_s$ varnish density in g/ml at 23°C

In this case, the VOC content refers to 1 l of actual varnish.

*Example 3*:

Solvent stain: 5% solid matter content
95% solvent share

VOC (g/l) = (100 - 5 - 0) × 0.9 = 855 g/l

*Example 4*:

Solvent/water stain: 5% solid matter content
70% solvent
25% water share

Density: 0.9 g/cm$^3$

VOC (g/l) = (100 - 5 - 25) × 0.9 × 10 = 630 g/l

DIN ISO 11 890 also allows the calculation to be made using the method prescribed by ASTM D 3960:

$$VOC_{lw}\left[g/l\right]=1000\frac{100-nfAm_w}{100-\rho_s\dfrac{m_w}{\rho_w}}\rho_s \qquad [18.1.4]$$

where:

| | |
|---|---|
| $\rho_s$ | varnish density in g/ml at 23°C |
| nfA | non-volatile parts determined by ISO 3251[22] (1 h/110°C) |
| $m_w$ | mass as water in % mass parts, determined acc. to ISO 760[23] (after Karl-Fischer) |
| $\rho_w$ | density, in g/ml, of water at 23°C ($\rho_w$ = 0.997537 g/ml at 23°C) |

*Example 1*

Water spray varnish: 30% solid matter content
10% solvent
60% water content

varnish density 1.0 g/cm$^3$

water density 1.0 g/cm$^3$

$VOC_{lw}$ = [(100 - 30 - 60)/(100 - 1.0 × 60)] × 1.0 × 1000 = 250 g/l

*Example 1*:

Water spray varnish but with a density of 0.9 g/cm$^3$

$VOC_{lw}$ = [(100 - 3 - 60)/(100 - 0.9 × 60)] × 0.9 × 1000 = 196 g/l

Or another example:

*Example 5*:

Water varnish primer: 15% solid matter content
10% solvent
75% water

varnish density 1.0 g/cm$^3$

$VOC_{lw}$ = [(100 - 15 -75)/(100 - 1.0 × 75)] × 1.0 × 1000 = 400 g/l

The calculation permits the deduction of exempt compounds, which are defined as organic compounds which do not participate in photochemical reactions in the atmosphere.

$$VOC_{lwe}\left[g/l\right]=1000\frac{100-nfA-m_w\sum_{i=1}^{i=n}m_{eci}}{100-\rho_s\dfrac{m_w}{\rho_w}-\rho_s\sum_{i=1}^{i=n}\dfrac{m_{eci}}{\rho_{eci}}}\rho_s \qquad [18.1.5]$$

where:

| | |
|---|---|
| $VOC_{lwe}$ | VOC content, in g/l, contained in the product in its ready-to-use condition minus water and minus exempt compounds |
| nfA | content of non-solvent parts, as a mass share in % |
| $m_w$ | water content, as mass share in % |
| $m_{eci}$ | content, as mass share in %, of the exempt compound i |
| $\rho_s$ | density, in g/ml, of sample at 23°C |

| | |
|---|---|
| $\rho_w$ | density, in g/ml, of water at 23°C ($\rho_w$ = 0.997537 g/ml at 23°C); |
| $\rho_{eci}$ | density, in g/ml, of the exempt compound i |
| 1000 | conversion factor × 1000 in g/l |

*Example 4*:

Exempt compounds are propanol and ethanol

| | |
|---|---|
| Solvent water stain | 5% solid matter content |
| (density 0.9 cm$^3$) | 20% propanol (density 0.8 g/cm$^3$) |
| | 10% ethanol (density 0.79 g/cm$^3$) |
| | 40% other solvent |
| | 25% water |

$VOC_{lwe}$ = [(100-5-25-(10))/{100-0.9(25/1.0)-0.9[(20/0.8)+(10/0.79)]}]×0.9×1000=825 g/l

Calculated according to DIN ISO 11 890-1, for the same water stain applies (example 4)

after approach 1: 70%

after approach 2: 630 g/l

(propanol and ethanol are treated as VOC)

#### *18.1.3.3.2 VOC content > 0.1 and < 15 %*

For VOC content > 0,1 (by mass) and < 15 % the determination is made using gas chromatography (ISO/DIS 11 890-2).[6] A hot or the cold injection is used depending on the sample properties.

After assessing the peak areas, the quantitative assessment and evaluation of the VOC content in the product in its ready-to-use condition, the calculation is made in the simplest case according to:

$$VOC[g/l] = \sum_{i=1}^{i=n} m_i \rho_s 1000 \qquad [18.1.6]$$

where:

| | |
|---|---|
| VOC | VOC content, in g/l, of the ready-to-use product; |
| $m_i$ | mass, in g, of compound i in 1 g of the sample |
| $\rho_s$ | density, in g/ml, of the sample at 23°C |
| 1000 | conversion factor × 1000 in g/l. |

Table 18.1.2 shows how the application of the various formulae affects the result of calculation. The VOC contents, calculated after ISO/DIS 11 890, are credible for Examples 2, 3, 4. Values calculated according to ASTM, using ratio solvent - water in Examples 2 and 4, are almost the same as VOC contents calculated according to ISO/DIS.

Only by increasing the solid matter content, the VOC content can be reduced in examples calculated according to ASTM (cf. Examples 5 and 1), while an increase in the water content, in spite of the same solvent share, leads to an increase of the ASTM value, because a higher water content increases the sum of the denominator and consequently increases the VOC content.

### 18.1.3.4 Determination of VOC-content in water-thinnable emulsion paints (in-can VOC)

DIN 55 649[24] describes an approach, which, via the so-called total evaporation method, assesses the content of volatile organic compounds in water-thinnable emulsion paints (in-can VOC).

**Table 18.1.2. Examples for the VOC content calculated by various methods**

| Ex-ample | Coating material wt% | Non-vola-tile parts wt% | Solvent fraction wt% | Water fraction wt% | Density g/cm$^3$ | VOC content according to | | |
|---|---|---|---|---|---|---|---|---|
| | | | | | | ASTM D 3960 g/l | DIN ISO 11 890/1 % | DIN ISO 11 890/1 g/l |
| 1 | Water spray varnish | 30 | 10 | 60 | 1.000<br>0.900 | 250<br>196 | 10<br>10 | 100<br>90 |
| 2 | Water/solvent stain | 5 | 25 | 70 | 1.000 | 833 | 25 | 250 |
| 3 | Solvent stain | 5 | 95 | 0 | 0.900 | 855 | 95 | 855 |
| 4 | Solvent/water stain | 5 | 70 | 25 | 0.900 | 839 | 70 | 630 |
| 5 | Water varnish primer | 15 | 10 | 75 | 1.000 | 400 | 10 | 100 |

*Measuring principle*:

The VOCs are totally evaporated from a very small amount (a few milliliters) of the thinned original sample by means of a head-space injector and subsequent gas chromatographic analysis.

All components, whose retention time is lower than the retention time of tetradecane (boiling point 252.6°C) are included as VOCs.

VOC content calculation:

$$VOC[mg/kg] = 1000\frac{m_{VOC}}{E_\rho} \quad [18.1.7]$$

where:

$m_{VOC}$ mass in mg VOC, related to originally weighted-in quantity of the original sample $E_\rho$

$E_\rho$ originally weighted-in quantity in g of the original sample

$$VOC[g/l] = \frac{VOC_{[mg/kg]}\rho_{df}}{100} \quad [18.1.8]$$

where:

$\rho_{df}$ density in g/ml of the original sample (emulsion paint)

## 18.1.4 MEASUREMENT OF SOLVENT EMISSIONS IN INDUSTRIAL PLANTS

### 18.1.4.1 Plant requirements

In spite of considerable reductions in the use of organic solvents, the worldwide-adopted ozone values in the troposphere are not being attained. It is crucial to apply consistent measures worldwide to reduce solvent emissions.

In Europe, for example, the Council Directive 1999/13/EG[1] "On limiting emissions of volatile organic compounds, due to the use of organic solvents in certain activities and installations" [EU-VOC-Richtlinie] came into force on March 11, 1999. It has to be transferred into national laws within two years. The application area refers to activities in applying solvents, as enlisted in Appendix IIA of the EU-VOC-Richtlinie (Table 18.1.3).

**Table 18.1.3 Threshold and emission controls**

| [a]Activity | [b]Threshold | [c]Emission limit, mgC/Nm³ | [d]Fugitive emission values, % | | [e]Total emission | | Special provisions |
|---|---|---|---|---|---|---|---|
| | | | New | Exist. | New | Exist. | |
| Web offset printing (>15) | 15-25<br>>25 | 100<br>20 | 30(1)<br>30(1) | | | | (1)Solvent residue in finished product is not to be considered as part of fugitive emissions |
| Publication rotogravure (>25) | | 75 | 10 | 15 | | | |
| Other rotogravure, flexography, rotary screen, printing, laminating or varnishing units (>15) rotary screen printing on textile/cardboard (>30) | 15-25<br>>25<br>>30(1) | 100<br>100<br>100 | 25<br>20<br>20 | | | | (1)Threshold for rotary screen printing on textile and on cardboard |
| Surface cleaning(1) (>1) | 1-5<br>>5 | 20(2)<br>20(2) | 15<br>10 | | | | (1)Using compounds specified in Article 5(6) and (8); (2)Limit refers to mass of compounds in mg/Nm³, and not to total carbon |
| Other surface cleaning (>2) | 2-10<br>>10 | 75(1)<br>75(1) | 20(1)<br>15(1) | | | | (1)Installations which demonstrate to the competent authority that the average organic solvent content of all cleaning materials used does not exceed 30 wt% are exempt from application of these values |
| Vehicle coating (<15) and vehicle refinishing | >0,5 | 50(1) | 25 | | | | (1)Compliance in accordance with Article 9(3) should be demonstrated based on 15 minute average measurements |
| Coil coating (>25) | | 50(1) | 5 | 10 | | | (1)For installations which use techniques which allow reuse of recovered solvents, the emission limit shall be 150 |

| [a]Activity | [b]Threshold | [c]Emission limit, mgC/Nm³ | [d]Fugitive emission values, % New Exist. | [e]Total emission New Exist. | Special provisions |
|---|---|---|---|---|---|
| Other coating, including metal, plastic, textile(5), fabric, film and paper coating (>5) | 5-15<br>>15 | 100(1)(4)<br>50/75(2)(3)(4) | 20(4)<br>20(4) | | (1)Emission limit value applies to coating application and drying processes operated under contained conditions. (2)The first emission limit value applies to drying processes, the second to coating application processes. (3)For textile coating installations which use techniques which allow reuse of recovered solvents, the emission limit applied to coating application and drying processes taken together shall be 150. (4)Coating activities which cannot be applied under contained conditions (such as shipbuilding, aircraft painting) may be exempted from these values, in accordance with Article 5(3)(b). (5)Rotary screen printing on textile is covered by activity No 3 |
| Winding wire coating (>5) | | | | 10 g/kg(1)<br>5 g/kg(2) | (1)Applies to installations where average diameter of wire ≤0.1mm. (2)Applies to all other installations |
| Coating of wooden surfaces (>15) | 15-25<br>>25 | 100(1)<br>50/70(2) | 25<br>20 | | (1)Emission limit applies to coating application and drying processes operated under contained conditions. (2)The first value applies to drying processes, the second to coating application processes |

| [a]Activity | [b]Threshold | [c]Emission limit, $mgC/Nm^3$ | [d]Fugitive emission values, %<br>New Exist. | [e]Total emission<br>New Exist. | Special provisions |
|---|---|---|---|---|---|
| Dry cleaning | | | | 20 g/kg[(1)(2)(3)] | [(1)]Expressed in mass of solvent emitted per kilogram of product cleaned and dried. [(2)]The emission limit in Article 5(8) does not apply to this sector. [(3)]The following exemption refers only to Greece: the total emission limit value does not apply, for a period of 12 years after the date on which this Directive is brought into effect, to existing installations located in remote areas and/or islands, with a population of no more than 2000 permanent inhabitants where the use of advanced technology equipment is not economically feasible |
| Wood impregnations (>25) | | 100[(1)] | 45 | 11 $kg/m^3$ | [(1)]Does not apply to impregnation with creosote |
| Coating of leather (>10) | 10-25<br>>25<br>(>10)[(1)] | | | 85 $g/m^2$<br>75 $g/m^2$<br>150 $g/m^2$ | Emission limits are expressed in grams of solvent emitted per $m^2$ of product produced. [(1)]For leather coating activities in furnishing and particular leather goods used as small consumer goods like bags, belts, wallets, etc. |
| Footwear manufacture (>5) | | | | 25 g per pair | Total emission limit values are expressed in grams of solvent emitted per pair of complete footwear produced |
| Wood and plastic lamination (>5) | | | | 30 $g/m^2$ | |
| Adhesive coating (>5) | 5-15<br>>15 | 50[(1)]<br>50[(1)] | 25<br>20 | | [(1)]If techniques are used which allow reuse of recovered solvent, the emission limit value in waste gases shall be 150 |

| [a]Activity | [b]Threshold | [c]Emission limit, mgC/Nm³ | [d]Fugitive emission values, % New | Exist. | [e]Total emission New | Exist. | Special provisions |
|---|---|---|---|---|---|---|---|
| Manufacture of coating preparations, varnishes, inks and adhesives (>100) | 100-1000<br>>1000 | 150<br>150 | 5<br>3 | | 5% of solvent input | | The fugitive emission value does not include solvent sold as part of a coatings preparation in a sealed container |
| Rubber conversion (>15) | | $20^{(1)}$ | $25^{(2)}$ | | 25% of solvent input | | (1)If techniques are used which allow reuse of recovered solvent, the emission limit value in waste gases shall be 150. (2)The fugitive emission value does not include solvent sold as part of products or preparations in a sealed container |
| Vegetable oil and animal fat extraction and vegetable oil refining activities (>10) | | | | | [f]see below the table | | (1)Total emission limit values for installations processing individual batches of seeds and other vegetable matter should be set by the competent authority on a case-by-case basis, applying the best available techniques. (2)Applies to all fractionation processes excluding de-gumming (the removal of gums from the oil). (3)Applics to de-gumming |
| Manufacturing of pharmaceutical products (>50) | | $20^{(1)}$ | $5^{(2)}$ | $15^{(2)}$ | 5% of solvent input | 15% | (1)If techniques are used which allow reuse of recovered solvent, the emission limit value in waste gases shall be 150. (2)The fugitive emission limit value does not include solvent sold as part of products or preparations in a sealed container |

[a](solvent consumption threshold in tonnes/year); [b](solvent consumption threshold in tonnes/year); [c]values in waste gases; [d](percentage of solvent input); [e]limit values; [f]Animal fat: 1.5 kg/tonne, Castor: 3 kg/tonne, Rape seed: 1 kg/tonne, Sunflower seed: 1 kg/tonne, Soya beans (normal crush): 0.8 kg/tonne, Soya beans (white flakes): 1.2 kg/tonne, Other seeds and other vegetable matter: 3 kg/tonne$^{(1)}$ 1.5 kg/tonne$^{(2)}$ 4 kg/tonne$^{(3)}$

The quoted threshold levels for the solvent consumption in tonnes/year is of decisive importance as to whether a plant falls under this directive. This value changes depending on the technical feasibility within the industry.

For example, the threshold level for solvent consumption is for illustration-gravure-printing > 25 tonnes/year, for wood coating > 15 tonnes/year.

An analysis of the activity requires data on solvent consumption as calculated from the amount of solvent purchased both as pure solvent and included in solvent containing materials less the amount of solvent retained and/or contained in waste. The calculated difference includes all emissions including diffuse emissions such as the solvent loss from drying racks or solvent initially retained by pained, varnished, printed or dry cleaned articles. These measurements are essential if the statutes are to be enforced and emissions are to be effectively reduced.

Emissive limit values are stipulated for plants covered under this directive for exhaust-gases in mgC/Nm³ as well as limits for diffuse emissions in % solvents input. If these data are not available, total emission limit values are used.

Definition of "Emission Limit Value": The "Emission Limit Value" is understood as the mass of volatile organic compounds, the concentration, the percentage and/or the amount of emission - ascertained under normal conditions - expressed in certain specific parameters, which in one or several time periods must not be exceeded [EU-VOC-Richtlinie].[1]

Plant measurement must be made to provide data for analyzing the actual situation as well as for being able to prove the emission reduction which business must attain.

### 18.1.4.2 The determination of the total carbon content in mg C/Nm³

#### *18.1.4.2.1 Flame ionization detector (FID)*

The total C/Nm³ is assessed according to the Guideline VDI 3481/page 3[25] by means of a flame ionization detector (FID). This device is the component of a mobile device for random sample tests or a continuously measuring device for total carbon concentration measurement in an exhaust-gas flow. This approach measures the total organic substance in an exhaust gas. Should the composition of the solvents contained in the exhaust air not be known, their concentration may be quoted in carbon equivalents as mgC/m³.

Measuring principle:

An FID detects ionized organically compounded C atoms in a hydrogen flame. The ion flow developed in the induced electrical field is electrically amplified and measured. The ion flow arising when burning carbon compounds is proportional to the mass of carbon atoms exposed to the flame per time unit.

The detection of organic compounds with heteroatoms, e.g., N, O, S, Cl, is generally less sensitive. The calibration of the device is done in most cases with propane as the test gas. Procedural data (measuring ranges, proof limits, etc.) by FIDs of various types are compiled in a table in the Guidelines VDI. Figure 18.1.2 shows a measuring arrangement for assessing emissions from a drying plant by means of an FID.

Figure 18.1.2. Emission measurement by FID.

#### *18.1.4.2.2 Silica gel approach*

The determination of the total carbon concentration in an exhaust gas by means of the silica gel approach is effected according to the Guideline VDI 3481, p. 2.[26] A partial flow of the exhaust air to be tested is guided through a sorption pipe filled with silica gel. The organic compounds are absorbed by the silica gel. Subsequently, the organic compounds are desorbed in an oxygen flow at an increased temperature and burnt to carbon dioxide ($CO_2$), which is determined quantitatively. This is a discontinuous method. The data reflect concentration in the measured time intervals and the causes of deviations are difficult to ascertain.

### 18.1.4.3 Qualitative and quantitative assessment of individual components in the exhaust-gas

While the methods mentioned under 18.1.4.2 reflect the total carbon concentration as a summary parameter, it is still necessary to know concentrations of the individual solvents to assess the plant emission.

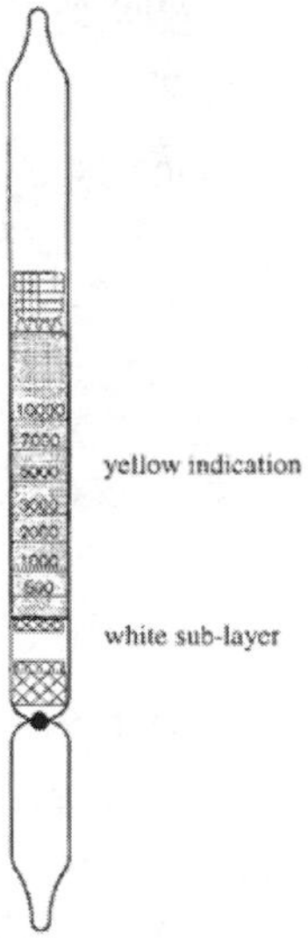

Figure 18.1.3. Indication tube for ethyl acetate 500/a-D.

#### *18.1.4.3.1 Indicator tubes*

For many solvents and other substances directly indicating detector tubes are available.[27] They are used for rapid assessment of emission. Since they often do not react specifically with a certain compound alone, errors have to be considered due to cross-sensitivity.

Detector tubes contain sorption-active agents. A gas flow of a defined rate is fed through and the substance in the gas flow produces a color reaction with the agent in the tube. The color intensity is checked against the scale on the test tube and the ppm value can be registered.

Figure 18.1.3 shows such a tube for determining ethyl acetate.[27] The color change in this case is from yellow to light green.

Reaction principle:

$CH_3\text{-}COOC_2H_5 + Cr^{VI} \rightarrow Cr^{III} +$ various oxidation products

This reaction is not substance-specific. Other alcohols, at varying cross-sensitivity as well as methyl ethyl ketone, are reactive as well.

#### *18.1.4.3.2 Quantitative solvent determination in exhaust gas of plants by means of gas-chromatography*

A defined amount of exhaust gas is fed into an adsorption medium (activated carbon, thenax, etc.) and its concentration is enriched. There is a subsequent extraction by means of a solvent. The mixture is then injected into a gas chromatograph. The individual components are determined according to Guideline VDI 34 82, page 1.[28]

Measuring principle:

A heatable capillary coated on the inside with various materials (polyester or silicon resins, silica gel) is used as a separating unit. The gaseous substance mixture is transported by means of an inert gas (nitrogen, argon, etc.). The individual solvents remain in the column for different time periods depending on their polarity. After leaving the column solvent is detected by sensitive physical methods. By combining the highly resolving capillary gas chromatography with a mass spectrometer even unknown substances may be identified in the mixtures. The quantitative evaluation of the gas chromatograms is automatically done with appropriate software.

*1st Example: Emission measurement in an industrial varnishing plant*

The measurement was done in a furniture company, which applies different varnish systems on several varnishing lines.[29]

For preliminary investigations, four chimneys were selected for the measurement as typical sources of emission.

Based on the analysis of the varnish materials and measurements performed over extended period of time, it is certain that the materials and concentrations processed during the measuring gave a representative cross-section of the production activity. Table 18.1.4 shows a summary of the varnish and solvent amounts processed per day, which were emitted from chimneys 1 to 4.

**Table 18.1.4. Summary of the varnish and solvent amounts processed in the 1st example**

| Measuring point chimney | Varnish | Solvent share, % | Varnish processed, l/day | Total solvent, l/day |
|---|---|---|---|---|
| 1 | 1<br>2<br>3<br>3<br>4 | 34<br>76<br>80<br>80<br>80 | 10 - 15<br><br>30 - 35<br>5<br>5 | ~40 - 45 |
| 2 | 5<br>6 | 56<br>34 | 50 - 60<br>10 | 37 |
| 3 | 7<br>6 | 38<br>34 | 30<br>3 | 12.5 |
| 4 | 1<br>8<br>6 | 34<br>52.88<br>34 | ~240<br>120-130<br>50 | ~170 |

At the chimneys, the following exhaust air parameters were established:

| | |
|---|---|
| Chimney diameter | mm |
| Exhaust air flow rate | m/s |
| Exhaust air temperature | °C |
| Humidity | % |
| Volume flow (operation) | m³/h |
| Volume flow | Nm³/h |
| Heat flow | MW |

Table 18.1.5 contains a summary of the established emission values expressed in mg C total/Nm³ as well as the mass flow g C/h.

**Table 18.1.5. Measuring results for total carbon concentration at different measuring points**

| Measuring point chimney | Time | Volume flow, Nm³/h | Carbon concentration, mgC/Nm³ | Mass flow, gC/h | Odorant concentration[(1)], GE/m³ | Mass flow, TGE/m³ |
|---|---|---|---|---|---|---|
| 1 | 10.35 - 11.20<br>11.45 - 12.00<br>13.45 - 14.35 | 35 200 | 112.6<br>96.5<br>128.6 | 3 964<br>3 397<br>4 527 | 93<br>87 | 3 519<br>3 292 |

| Measuring point chimney | Time | Volume flow, $Nm^3/h$ | Carbon concentration, $mgC/Nm^3$ | Mass flow, gC/h | Odorant concentration[(1)], $GE/m^3$ | Mass flow, $TGE/m^3$ |
|---|---|---|---|---|---|---|
| | 11.25 - 12.00 | | 72.4 | 2 841 | 91 | 4 151 |
| 2 | 12.45 - 13.10 | 39 240 | 56.3 | 2 209 | 87 | 3 969 |
| | 13.10 - 14.30 | | 56.3 | 2 209 | 93 | 4 242 |
| | 13.50 - 14.40 | | 48.2 | 1 634 | 100 | 3 747 |
| 3 | 09.25 - 09.55 | 33 900 | 56.3 | 1 908 | 115 | 4 309 |
| | 09.55 - 10.25 | | 64.3 | 2 180 | | |
| | 09.15 - 09.45 | | 112.5 | 12 223 | 98 | 12 837 |
| 4 | 09.45 - 10.15 | 108 650 | 152.8 | 16 602 | 213 | 27 900 |
| | 10.15 - 10.45 | | 160.8 | 17 471 | | |

[(1)]The odorant concentration does not refer to the times stated.

The EU VOC Directive for plants for wood coating with a solvent consumption of 15 - 25 t/year requires an emission limit of 100 mg $C/Nm^3$.

In the example, this is only met at chimneys 2 and 3.

As a parallel to these measurements, using activated carbon tubes and subsequent gas chromatographic evaluation it was determined that the main components of VOCs were acetone, ethyl acetate, toluene, butyl acetate, xylene, ethylbenzene (Table 18.1.6). Butyl acetate and ethyl acetate were the main components in the respective solvent mixtures.

**Table 18.1.6. Measured results for individual components at different measuring points**

| MPC* | Acetone | | Ethyl acetate | | Toluene | | Butyl acetate | | Ethyl benzene | | Xylene | |
|---|---|---|---|---|---|---|---|---|---|---|---|---|
| | Conc[a] | MF[b] | Conc[a] | MF[b] | Conc[a] | MF[b] | Conc[a] | MF[b] | Conc[a] | MF[b] | Conc[a] | MF[b] |
| 1 | 6.5 | 229 | 15.5 | 546 | 9.0 | 317 | 15.5 | 545 | 1.0 | 352 | 4.2 | 158 |
| | 4.5 | 158 | 11.0 | 387 | 5.5 | 194 | 11.0 | 387 | 1.0 | 352 | 2.5 | 88 |
| 2 | 11.0 | 432 | 24.5 | 961 | 14.5 | 569 | 24.5 | 961 | 1.0 | 39 | 6.5 | 255 |
| | 4.5 | 176 | 9.0 | 353 | 6.5 | 255 | 14.5 | 569 | - | - | - | - |
| 3 | - | - | 4.5 | 152 | 2.0 | 68 | 3.5 | 118 | 0.5 | 17 | 2.0 | 68 |
| | - | - | 15.5 | 525 | 5.5 | 186 | 8.0 | 271 | 1.0 | 34 | 5.5 | 186 |
| 4 | 9.0 | 978 | 20.0 | 2 173 | 9.0 | 978 | 17.0 | 1 847 | 2.0 | 217 | 8.0 | 869 |
| | 5.5 | 597 | 27.0 | 2 933 | 18.0 | 1 956 | 37.0 | 4 020 | 3.5 | 380 | 17.0 | 1 847 |

*MPC = Measuring point chimney; [a]concentration in $mg/Nm^3$; [b]MF = mass flow in g/h

*2nd Example: Emissions measured in a spray-room*

The usual technique of measuring VOC emissions in plants by means of the FID according to VDI 3481/page 3 has the shortcoming that individual components in the gas mixture cannot be measured separately, but only determined as total carbon. Also the device needs to be placed in the immediate vicinity of the measuring point and requires frequent calibration.

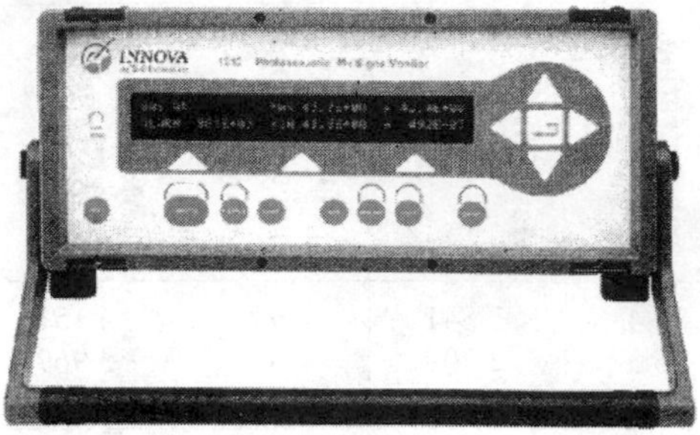

Figure 18.1.4. Photoacoustic Multigas Monitor 1312.

A portable multi-gas monitor 1312 [Bruel & Kjaer][30] weighing 9 kg (Figure 18.1.4) enables sampling in six places up to 50 m away from the monitor. It can determine at each measuring point the concentration of up to a maximum 5 components as well as the humidity. This is done by a multi-point sampler with a dosing apparatus controlled by system software. The device needs calibration only four times a year. Beyond that, the device may also be used for determining the total C concentration.

Measuring principle (Figure 18.1.5)

1. The "new" air sample is hermetically sealed in the analysis cell by closing the inlet and outlet valves.

2. Light from an infrared light source is reflected by a mirror, passed through a mechanical chopper, which pulsates it, and then passes through one of the optical filters in the filter carousel.

3. The light transmitted by the optical filter is selectively absorbed by the gas monitored, causing the temperature of the gas to increase. Because the light is pulsating, the gas temperature increases and decreases, causing an equivalent increase and decrease in the pressure of the gas (an acoustic signal) in the closed cell.

4. Two microphones mounted in the cell wall measure this acoustic signal, which is directly proportional to the concentration of the monitored gas present in the cell.

5. The filter carousel turns so that light is transmitted through the next optical filter, and the new signal is measured. The number of times this step is repeated depends on the number of gases being measured.

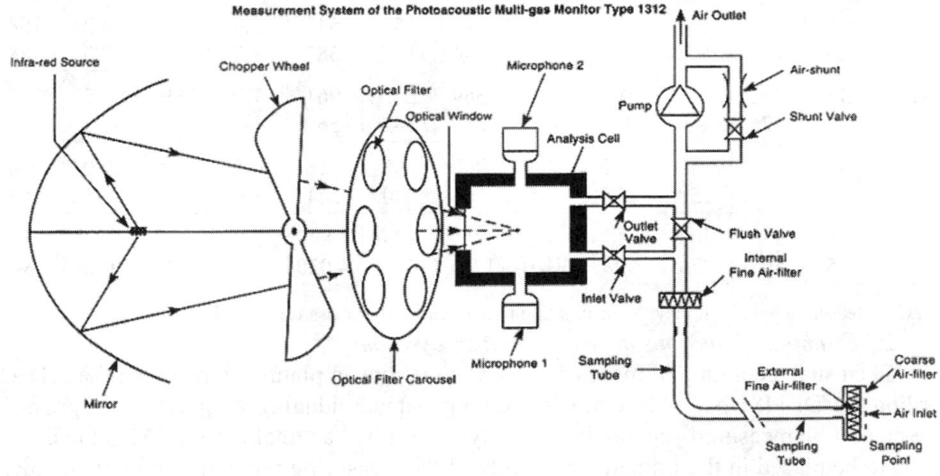


Figure 18.1.5. Measurement principle of Photoacoustic Multigas Monitor Type 1312.

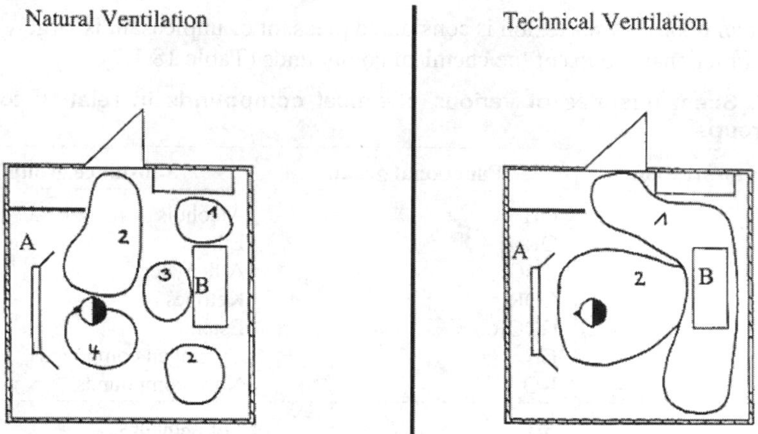


Natural Ventilation

| area | mg/m³ |
|---|---|
| 1 | 3650 |
| 2 | 4430 |
| | 4910 |
| | 4530 |
| 3 | 5580 |
| 4 | 6690 |

Technical Ventilation

| area | mg/m³ |
|---|---|
| 1 | 489 |
| | 483 |
| | 499 |
| 2 | 662 |
| | 614 |
| | 564 |

Figure 18.1.6. Maximum concentration of n-butyl acetate in the spray-room of a joiner's with windows and door closed. A - Spray wall, B - Rack track for shelving the varnished parts.

The response time is only 13 s for a single gas or water-vapor, or approx. 40 s if 5 gases and water-vapor are measured.

The measurements are quasi-continuous. When measuring a single gas, the measuring value is updated every 30 s, with five gases and humidity the update time is 105 s. The measured values are stored, statistically evaluated and numerical or graphical output can be obtained either immediately or on demand. Such devices make the emission measurement effort much easier than with the older, less automated equipment.

Figure 18.1.6 shows the results of emission investigations in the spray-room of a joiner's with and without ventilation.[31] Windows and doors were closed. The varnisher was standing in front of a spray-wall, opposite to him was placed a platform rack truck for shelving the varnished parts. Those areas in the joiner's, which showed roughly the same emission values of n-butyl acetate, are marked on each drawing.

The emission load in the room was decreased by the technical ventilation to about a tenth of the values.

## 18.1.5 "ODOR" DEFINITION

One of the senses of the living organism is the sense of smell. Smell is caused by a substance whose effect is largely dependent on its chemical structure.

The odorous substances perceived by human are suspended in the air as volatile substances. The degree of response to a substance depends on its vapor pressure, thus, in turn on its concentration in the air and its molecular weight. An odorant also has to be sufficiently water and fat-soluble, which enables it to interact with the olfactory receptors.

Whether an odorous impression is considered pleasant or unpleasant is largely determined by the functional groups of the chemical compounds (Table 18.1.7).

**Table 18.1.7. Scent qualities of various chemical compounds in relation to their functional groups[33,34]**

| Odorous impression | Functional group | Substance group |
|---|---|---|
| Pleasant | -OH<br>-OR<br>-CHO<br>-COR<br>-COOR<br>-CN<br>$-NO_2$ | Alcohols<br>Ethers<br>Aldehydes<br>Ketones<br>Esters<br>Cyanogen compounds<br>Nitro compounds |
| Unpleasant | -SH<br>-SR<br>-CSR<br>-NC<br>$-NH_2$ | Merkaptanes<br>Thioethers<br>Thioketones<br>Nitriles<br>Amines |

This division should be, however, only regarded as a rough guideline, since the so-called "pleasant" odor, at higher concentrations, can easily be perceived as very "unpleasant".

Almost all solvents, such as, ketones, esters, glycols, alcohols, aromatic and aliphatic hydrocarbons, contribute to a more or less intensive smell.

"Odor" is not a parameter of substance, but a summarized parameter of effects. Its determination is based on the fact that the sense of a smell can be used subjectively to evaluate certain substances (odorants). The concentration of those substances suspended in the air can be determined, which is called odorant concentration.

<u>Odor threshold or perception threshold</u>[35-38]

The odor threshold is a concentration of the odorant in the air, given in $mg/m^3$. The odor threshold corresponds to an odor unit (GE). It is the amount of odorous substance, which - distributed in 1 $m^3$ of scent-neutral synthetic air - initiates in just 50% of the evaluators a perception of smell and in the other 50%, no response.

The odor threshold is very specific to substance. It is determined in several measuring series and the results form a Gaussian distribution curve. Since this is ultimately a subjective evaluation, one should not be surprised to find more as well as, less, reliable data in literature. Examples are shown in Table 18.1.8.

<u>Odorant concentration</u>

The odorant concentration of a sample is the multiple of the odor threshold and it is determined in odor units (GE) per 1 $m^3$ neutral air.

<u>Odor intensity</u>[39]

Since the odor threshold alone is an insufficient evaluation criterion for an odorant, the increase in response with increasing odorant concentration may additionally be taken into account as a scale of reference.

The increase in response is mainly material and/or mixture dependent for a given odorant concentration and is called odor intensity. [Schön, p. 68][32]

**Table 18.1.8. Odor thresholds of selected compounds from literature [Geruchs-Immissions-Richtlinie][34]**

| Compound | Odor threshold, mg/m³ | Compound | Odor threshold, mg/m³ |
|---|---|---|---|
| Butyl acetate | 0.03 | Ethanol | 19.1 |
| Ethyl acetate | 22 | 1-butanol | 0.4 |
| Benzene | 16.2 | Acetone | 48 |
| Toluene | 7.6 | Ethyl acrylate | 0.002 |
| Xylene | 0.35 | Dichloromethane | 706 |

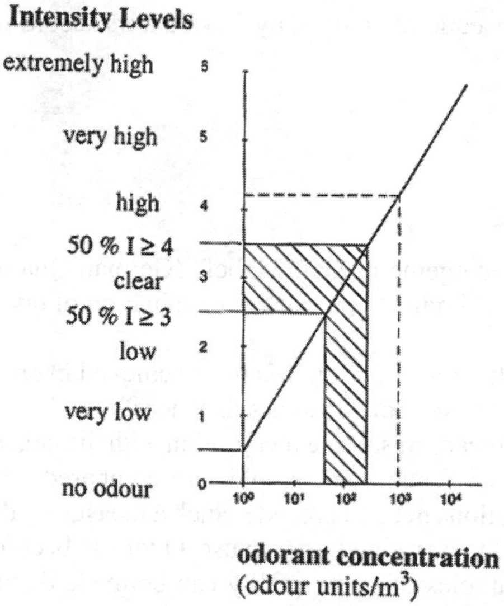


Figure 18.1.7. Interdependence of odorant concentration and odor intensity in an artificial example.

Figure 18.1.7 shows the interaction between odorant concentration and odor intensity in an artificial example.[41]

Odor intensity is assessed by means of a scale from 1 to 6 [VDI 3882/1][39] (Table 18.1.9).

**Table 18.1.9. Correlation of odor with an intensity level**

| Odor | Intensity level |
|---|---|
| Extremely high | 6 |
| Very high | 5 |
| High | 4 |
| Moderate | 3 |
| Low | 2 |
| Very low | 1 |
| No scent | 0 |

Hedonic odor tone

An odor may be quantified by determining the odor concentration, the character of the odor (pleasant or unpleasant) is not considered. Unpleasant odors may result in deteriorating health and should therefore be avoided.

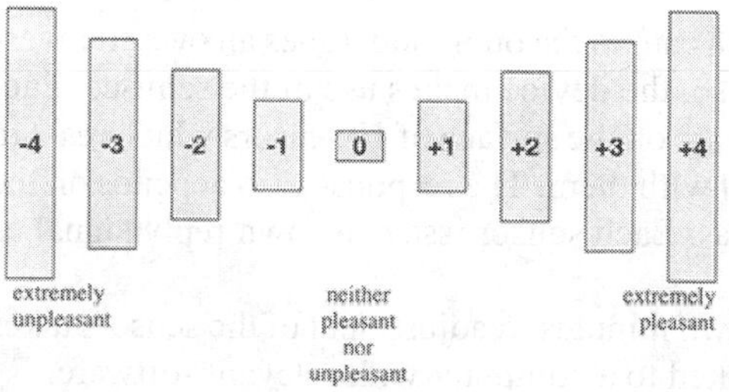


Figure 18.1.8. Evaluation of the hedonic effect of an odor sample [VDI 3882/2].[40]

By determining the hedonic odor effect, the emotional reaction initiated by an irritation to the sense of smell may be included. It should not be confused with the kind of smell (it smells like ...) or with the odor intensity (it smells "strong" to "weak"). It may be determined not only for a single odorant, but also for an odorant mixture. Odor samples of an odorant concentration above the odor threshold are ranked by evaluators according to the odorants' position in the following response range between "extremely unpleasant" to "extremely pleasant".

In many cases of odor evaluation, there are big differences between the results among the individual evaluators, since they have a widely varied background of experience.

## 18.1.6 MEASUREMENT OF ODOR IN MATERIALS AND INDUSTRIAL PLANTS

### 18.1.6.1 Introduction

Various methods are known for determining odors. One of particular interest is pupil dilatation. This is based on the fact, that pupils dilatate when a strong odor is sensed, as may be caused by concentrated ammonia.

A simple sampling of odors is based on the Öko-Tex Standard 200.[42] The samples are conditioned in a desiccator and their odor is subsequently judged by 6 evaluators according to a five-point scale:

1 odorless
2 low odor
3 bearable odor
4 annoying odor
5 unbearable odor

Such a scale also serves the "Deutsche Gütegemeinschaft Möbel" (German Quality Association of Furniture Manufacturers) as a methodical approach for evaluation of odors of furniture, cover fabric, leather, etc.

Level 3, however, is defined as "commodity typical". It needs to be achieved in order for the respective product to have the attribute "tested for noxious substances".

In principle, odors may be determined by means of sample recognition with the help of arrays of gas sensors, so-called electronic noses. Unknown samples are compared with known samples. Hence, olfactometric investigations need to precede. Such a measuring device is applicable only in a specific case and has to be trained prior to use. Odor can become a controllable quality feature of a product. Samples of good quality can be made distinguishable from samples of bad quality.

In practice, odor determination by means of the olfactometer has been widely applied and generally adopted.

#### *18.1.6.2 Odor determination by means of the "electronic nose"*

The principle of odor determination in different fields of application has been discussed in detail by Moy and Collins[43] and Schulz.[44]

<u>Measuring principle</u>:

The substance mixture in question, which causes the odor, undergoes an overall investigation by means of a sensor array. In doing so, the device makes use of the semi-conductive properties of various metal oxides, which are on the surface of 12 sensors which react to the gaseous substances which come in contact with them. The response also depends on the temperature, humidity and flow rate of the gas. Each sensor issues its own reply signal as soon as the sample of air touches its surface.

The measurement is done within about two minutes, readjustment of the sensors takes about 4 minutes. The measuring device is linked to a computer with relevant software.

Each odor is, according to the twelve sensors, represented by 12 graphs showing a characteristic "profile" ("fingerprint"). These sensor-specific, time-dependent, series of electric conductivity data, which are produced by the presence of the odorous substances provide the data for the evaluation. They are processed by means of neuronal networks.

The available odor samples can be shown as two or three-dimensional. Comparative or referential samples are treated in the same way. By means of sample series, they may serve the purposes of the identification of new samples.

It remains an issue, that odor-relevant compounds exist in much lower concentration in most cases than the less odorous compounds, thus they are also less influential on the sensor signal. Much research effort has yet to be undertaken to apply the electronic nose in the future.

### 18.1.6.3 Odor determination by means of the olfactometer

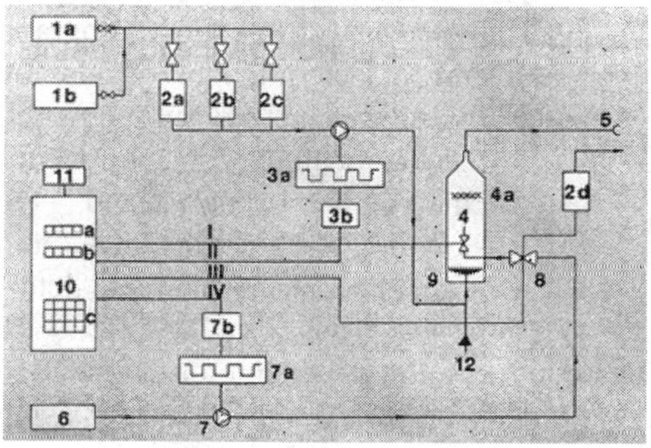


Figure 18.1.9. Principle of an olfactometer. 1 diluent air a) outer air, b) synthetic air (bottle); 2 acitvated charcoal filter air from 2d to rinse by-pass, 3 glass bulb pump with electronically controlled drive (a,b); 4 mix chamber (with installation a); 5 evaluator's mask; 6 sample air (if necessary pre-mixed); 7 glass bulb pump with electronically controlled drive (a,b); 8 by-pass valve; 9 mix chamber valve; 10 control terminal, a) display (programme step), b) display, c) terminal keyboard; 11 printer: I - IV control line; 12 additional pump (24 l x min-1).

The method of odor determination by means of the olfactometer is based on the guidelines Richtlinien VDI 3881, pages 1 to 4[35-38] and VDI 3882, pages 1 and 2.[39,40]

Measuring principle:

An odorant sample is diluted in a defined way with neutral air in an olfactometer and offered to test persons as an odor sample. The test persons are exposed to several dilution levels. Should an odor be perceived, it shall be confirmed by pressing a button. An olfactometric determination requires at least 5, and preferably 8 evaluators. These evaluators need to pass a suitability test in accordance with Guideline VDI 3882, p. 2.[40] The principle of an olfactometer[45] is shown in Figure 18.1.9.

The results are presented in odor units/m$^3$ of neutral air and they are automatically displayed as the averaged data of the panel's evaluation.

### 18.1.6.4 Example for odor determination for selected materials: Determination of odorant concentration in varnished furniture surfaces

The odor potential of furniture is determined primarily by the applied varnishes although adhesives also play a minor role. Furniture varnishes may contain up to 80% solvents. The residual solvents remaining in the varnish thus determine the VOC and odorant concentration of furniture surfaces.

A specific test method has been developed for determining the VOC and odorant concentration of coated surfaces,[46-48] as shown schematically in Figure 18.1.10. Similar to formaldehyde determination,[49] sample testing is done using a test chamber approach. After some atmospheric conditioning, of the varnished furniture, the samples are stored in a test chamber (typically 1 m$^3$) under the following conditions: 23°C±0.5; 45±3% relative humidity; charged with 1 = 1 m$^2$ of emitting sample surface per 1 m$^3$ test chamber volume air flow rate at sample surface: between 0.1 and 0.3 m/s. The samples remain in the test chamber for

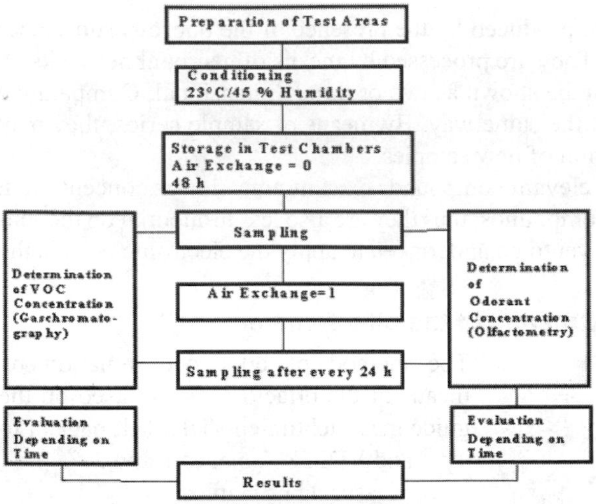


Figure 18.1.10. Test approach for the determination of VOC and odorant concentration in coated surfaces.

an initial 48 hours without an air exchange in the test chamber, in order to simulate a "worst case scenario", i.e., that, for instance, the new furniture is sealed off air-tight during transport, so that residual solvents accumulate.

After 48 h of storage time without an air exchange, air samples are taken in parallel for the determination of VOC emission by means of gas chromatograph and for determination of odor by the olfactometer. Subsequently, there is an air exchange of one air change per hour, and the slow-down curve of VOC and odorant concentration is determined.

Figure 18.1.11 shows the decreasing VOC concentration depending on the storage time in the test chamber with an air exchange 1,[50,46] and Figure 18.1.12 shows the decreasing odorant under the same test conditions.[50,46]

The curves for VOC and the odorant concentration follow a similar pattern, they are, however, not identical. While the VOC concentration in water varnish is the lowest after 11 days, the odorant concentration after this time is equally high with both DD-varnish and water varnish. This, however, does not generally apply to these types of varnish. Also DD-varnishes of different origins may differ greatly in emitting residual solvents.

In single-solvent systems, it is easy to see the interdependence between VOC and odorant concentration. This does not necessarily apply to solvent mixtures. This is due to the large variations in odor thresholds and the different evaporating behavior of the various solvents.

Acrylate varnishes, for example, need contain only a few μg of ethyl acrylate in order to produce odor, since odor threshold level is at 0.002 mg/m$^3$. Thus the determination of the

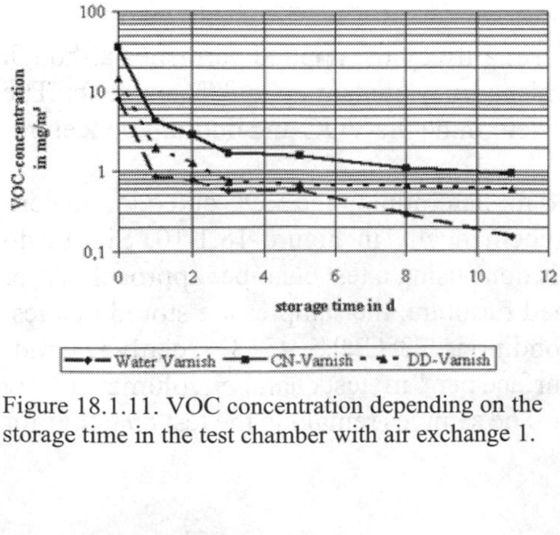


Figure 18.1.11. VOC concentration depending on the storage time in the test chamber with air exchange 1.

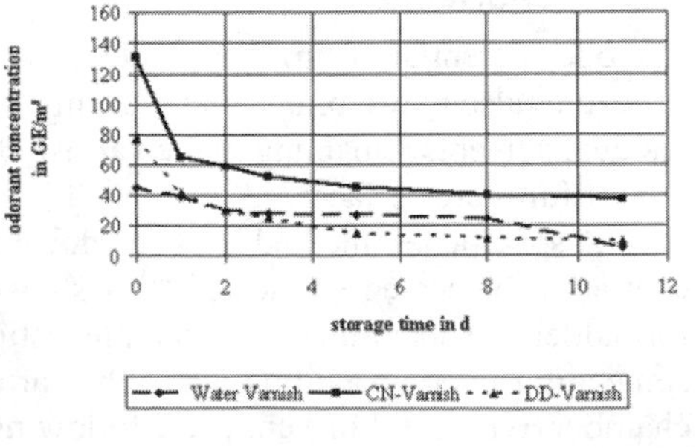


Figure 18.1.12. Decreasing odorant concentration depending on the storage time in the test chamber with air exchange 1.

VOC concentration alone is not sufficient for determining the quality of coated surfaces must also be evaluated.

#### 18.1.6.5 Example of odor determination in industrial plants: Odor measurement in an industrial varnishing plant

In a study parallel to the measurements of the VOC emission at the varnishing lines of a furniture company, as described in Section 18.1.4.3.2, 1st Example, odorant concentrations were sampled in the exhaust air from the individual chimneys by means of olfactometry.[29] All conditions of measurement described in Section 18.1.4.3.2 also apply to the determination of the odorant concentration.

The results are contained in Table 18.1.5 as odorant concentration in GE (odorant units)/m$^3$ and in the mass flow in TGE/h (one thousand odorant units per hour). The evaluation of the analytical determination of individual components resulted in the main components being butyl acetate and ethyl acetate. Due to the very low odorant threshold of butyl acetate (0.03 mg/m$^3$), as compared to other available substances, (ethyl acetate 22 mg/m$^3$), butyl acetate may be assumed to be responsible for for the bulk of detectable odor.

An immediate comparison of samples taken at the same point in time resulted in the following data:

concentration of butyl acetate: 8 mg/m$^3$
odorant concentration: 115 GE(odorant units)/m$^3$

Adopting the value quoted in the literature for the odorous threshold of butyl acetate at 0.03 mg/m$^3$, which equals one odorant unit, the odorant concentration estimates a butyl acetate concentration of about 4 mg/m$^3$. This value is in the same range as the value established for butyl acetate by gas chromatography. In these comparisons one has to consider that odorant thresholds may deviate by one to two exponents to even ten, according to literature sources.[51]

Measurements taken in the housing area about 300 m away from the chimneys of the varnishing plants could not establish any solvent-typical components in the gas analysis. The maximum concentration value for TVOC was at 0.004 mg/m$^3$.

Also the spreading calculation, which was carried out on the basis of the determined solvent concentration, showed that there should be no significant odor annoyance in the vicinity of the emission source. The results of the spreading calculation on the basis of measured solvent concentrations are considerably more precise, since solvent concentrations may be determined more exactly than odorant concentrations.

The spreading calculation was done using the software package "IMMPROG-P" by AIRINFO AG, Switzerland, and carried out according to a method of the Odor Emission Guideline of the German State of Northrhine-Westfalia/Geruchsimmissions-Richtlinie des deutschen Bundeslandes Nordrhein-Westfalen.[34]

## REFERENCES

1 Council Directive 1999/13/EC of March 11, 1999 of the limitation of emissions of volatile organic compounds due to the use of organic solvents in certain activities and installations.
2 ISO/DIS 4618-4: 1999-12-03, Paints and varnishes - Terms and definitions for coating materials - Part 4: Terms relating to raw materials.
3 Technische Regeln für Gefahrstoffe TRGS 610 "Ersatzstoffe, Ersatzverfahren und Verwendungsbeschränkungen für stark lösemittelhaltige Vorstriche und Bodenbelagsklebstoffe"; Absatz 2.7, 04.06.1992.
4 M. Scheithauer, I-Lack 66, 325 - 331 (1998).

5 ISO/DIS 11890-1, Draft 03/1998, Paints and varnishes - Determination of volatile organic compound (VOC) content. Part 1: Difference method (DIN ISO 11890-1).
6 ISO/DIS 11890-2, draft 03/1998, Paints and varnishes - Determination of volatile organic compound (VOC) content - Part 2: Gas chromatographic method (DIN ISO 11890-2).
7 Lösemittelverordnung 1995 in Österreich 872.VO im Bundesgesetzblatt für die Republik Österreich.
8 Lackieranlagen-Verordnung 1995 in Österreich 873.VO im Bundesgesetzblatt für die Republik Österreich.
9 ASTM D 3960-98 (1998-11-10). Standard Practice for Determining Volatile Organic Compound (VOC) Content of Paints and Related Coatings.
10 *Farbe und Lack*, **105**, 12/99, 111 (1999).
11 B. Seifert, **Richtwerte für die Innenraumluft; Bundesgesundheitsblatt - Gesundheitsforschung - Gesundheitsschutz 3**; *Springer-Verlag*, Berlin, 1999, pp. 270 - 278.
12 J. Witthauer, 4. Freiberger Polymertag, Freiberg, May 27-28, 1999, conference paper A/19.
13 A. Szalai, **The use of Time: Daily activities of urban and suburban populations in twelve countries**, Den Haag, Paris: *Mouton*, 1972.
14 F. S. Chapin, **Human activity patterns in the city**, *Wiley - Interscience*, New York, 1974.
15 EUWID Möbel 24, 5 (1994).
16 F. Busato, *Mod. Paint Coat.*, March 97, 30 - 33 (1997).
17 H. Zeh, 4. Freiberger Polymertag, Freiberg, May 27-28, 1999, Conference paper P/2.
18 Lackieranlagen-Verordnung in Deutschland (Varnishing Plant Regulation): Umweltbundesamt Deutschland III 3.2 - 52337/7); draft April 1, 1996.
19 Deutsches Institut für Gütesicherung und Kennzeichnung e.V.: "RAL-UZ 38: Emissionsarme Produkte aus Holz und Holzwerkstoffen" Eigenverlag, St. Augustin, 1999, p.4.
20 M. Große Ophoff, *Holz- und Kunststoffverarbeitung*, **12/96**, 52-55 (1996).
21 ASTM D 3960 - 98: Standard Practice for Determining Volatile Organic Compound (VOC) Content of Paints and Related Coatings, 1998.
22 ISO 3251: 1993. Paints and varnishes - Determination of non-volatile matter of paints, varnishes and binders for paints and varnishes.
23 ISO 760: 1978. Determination of water - Karl-Fischer method (general method).
24 E DIN 55 649: 1998 - 10. Paints and varnishes - Determination of volatile organic compounds content in waterthinnable emulsion paints (In-can VOC); draft: October 1998.
25 Richtlinie VDI 3481 / Blatt 3: Gaseous emission measurement - Determination of volatile organic compounds, especially solvent, flame ionisation detector (FID), 10.95.
26 Richtlinie VDI 3481 / Blatt 2 E: Messen gasförmiger Emissionen (Gaseous emission measurement) Bestimmung des durch Adsorption an Kieselgel erfassbaren organisch gebundenen Kohlenstoffs in Abgasen; 11.96.
27 Dräger, **Dräger-Röhrchen Handbuch**, Lübeck, 1991, 249.
28 Richtlinie VDI 3482/Blatt 1. Gaseous air pollution measurement; gas-chromatographic determination of organic compounds, fundamentals; 02.86.
29 M. Broege, Gutachten zu von Lackieranlagen verursachten Geruchsimmissionen, Institut für Holztechnologie Dresden, 1993.
30 1312 Photoacoustic Multi-gas Monitor, Product Data from INNOVA, Air Tech Instruments A/S, Denmark, 04/97.
31 R. Kusian, M. Henkel, Forschungsbericht des Institutes für Holztechnologie Dresden, Germany (1997): Untersuchung der Emissionsverhältnisse bei der Oberflächenbehandlung von Holz und Holzwerkstoffen.
32 M. Schön, R. Hübner, **Geruch-Messung und Beseitigung**, *Vogel-Buchverlag*, Würzburg, 1996.
33 A. L. Lehninger, **Biochemie**,Weinheim-New York, *Verlag Chemie*, 1979, p. 19.
34 Feststellung und Beurteilung von Geruchsimmissionen (Geruchsimmissionsrichtlinie, GJR); Der Minister für Umwelt, Raumordnung und Landwirtschaft des Landes Nordrhein-Westfalen, Stand 15.02.1993, Anhang B.Richtlinie VDI 3881.
35 Blatt 1: Olfactometry; odor threshold determination; fundamentals; 05 / 1986.
36 Blatt 2: Olfactometry, odor threshold determination; sampling; 01/1987.
37 Blatt 3: Olfactometry; odor threshold determination; olfactometers with gas jet dilution; 11/1986.
38 Blatt 4: Olfaktometrie; Geruchsschwellenbestimmung; Anwendungsvorschriften und Verfahrenskenngrößen; 12/86 Richtlinie VDI 3882.
39 Blatt 1: Olfactometry; determination of odor intensity; October 1992.
40 Blatt 2: Olfactometry - Determination of hedonic odor tone, September 1994.
41 M. Paduch, VDI-Berichte 1059, "Aktuelle Aufgaben der Messtechnik in der Luftreinhaltung" zum Kolloquium Heidelberg, Düsseldorf, Juni 2-4, 1993, pp. 593-607.

42 Öko-Tex Standard 200: Prüfverfahren für die Vergabe der Berechtigung zur Kennzeichnung von Teppichböden mit "Schadstoff geprüft nach Öko-Tex Standard 100" (Österreichisches Forschungsinstitut) 1992.
43 L. Moy, M. Collins, *LaborPraxis - Journal für Labor und Analytik*, **20**/5, 14-18 (1996).
44 H. Schulz, 4. Freiberger Polymertag, Freiberg, May 27-28, 1999, conference paper C 4/5.
45 Ströhlein, Labor-, Mess- und Umwelttechnik, Kaarst, Germany, Product Data p. 25, principle of an olfactometer.
46 M. Scheithauer, K. Aehlig, M. Broege, *Holz- und Kunststoffverarbeitung*, **1/96**, 58-61 (1996).
47 M. Scheithauer, K. Aehlig, Konferenz im ITD, Poznan (Poland), 1995.
48 K. Aehlig, M. Scheithauer, M. Broege, *Holz*, **5**, 26-32; (1996).
49 prENV 717-1: 1998: Holzwerkstoffe Bestimmung der Formaldehydabgabe Teil 1: Formaldehydabgabe nach der Prüfkammer-Methode (1998).
50 M. Broege, K. Aehlig, 4. Freiberger Polymertag, Freiberg, Mai 27-28, 1999, conference paper R 4/5.
51 G. Scharfenberger, *Chemie in Labor und Biotechnik*, **42**, 498-502 (1991).

## 18.2 PREDICTION OF ORGANIC SOLVENTS EMISSION DURING TECHNOLOGICAL PROCESSES

KRZYSZTOF M. BENCZEK, JOANNA KURPIEWSKA
**Central Institute for Labor Protection, Warsaw, Poland**

### 18.2.1 INTRODUCTION

The concentration of toxic substances in air during technological process is very important factor for occupational safety. Typical examples of processes, which have the potential to harm workers, are metal degreasing, painting, and wood impregnation.

If metal processing involves several steps some of which may be done in more than one manufacturing facility, the semi-processed metal parts must be protected during transportation and storage. Such protective coatings of grease and rust preventatives must be removed in degreasing operation. For many processes (e.g., painting, galvanic metal deposition), clean surface is an important requirement. The cleaning process may be done in automated and enclosed equipment or it may be done manually in the open. The degreasing agent may be an organic solvent, a solvent blend, or a water solution, usually alkaline in nature.

Depending on the process used the operation may pose no risk to the worker or be a serious occupational hazard. Similarly environmental emissions may be negligible or of serious concern.

We present a method of evaluating the quantities of emissions from such processes which involve solvents. The method may be applied to such diverse operations as painting, wood preservation, impregnation of porous materials, gluing, cleaning, filing open tanks, general solvent handling operations, and many others.

We have selected metal degreasing as a representative example to demonstrate how the method may be applied.

### 18.2.2 METHODS OF DEGREASING

Six methods can be identified which differ in the degreasing agent used:

- degreasing in liquid organic solvents, such as naphtha, petroleum, chlorinated aliphatic hydrocarbons, etc.
- degreasing in hot vapors of halogenated solvents such as, trichloroethylene, perchloroethylene, fluorochloroethane and so on
- degreasing in alkaline water solutions of hydroxides, phosphates, surfactants, emulsifiers, common inhibitors, etc.
- degreasing in an emulsion of organic solvents in water
- degreasing in water steam
- supersonic degreasing in stabilized chlorinated hydrocarbons.

Degreasing process can be conducted automatically in different ways by:

- dipping
- spraying
- using high pressure
- supersonic
- pulsating washing

Process can be conducted in open or enclosed equipment. Manual degreasing is still very popular but it is very time and labor consuming, expensive, and large amounts of solvents are lost.

### 18.2.3 SOLVENTS

In metal degreasing these solvents are most frequently used:

- naphtha solvent,
- naphtha anti-corrosive
- 1-butanol,
- 1,1,2-trichloroethylene,
- 1,1,1- trichloroethane,
- extraction naphtha,
- petroleum(mineral) spirits.

### 18.2.4 IDENTIFICATION OF THE EMITTED COMPOUNDS

In many cases, solvent mixtures are used. Their composition must be identified. Rodofos is one example of such solvent used in Poland. Its composition was determined by gas chromatography. Analyses were performed using a Hewlett-Packard gas chromatograph model 5890 coupled with computerized mass spectrometer instrument, model 5970.

Capillary column 50 m x 0.32 mm i.d., d.f.= 0.52 μm FFAP and helium as a carrier gas was used at temperature of 40°C. Ions from 20 to 400 amu were counted, delay time was 3.5 min.

Samples were collected by drawing a known volume of air through a bubbler containing 1 ml of carbon disulfide. Volume of the injected sample was 1-5 μl.

Chromatograms are presented in Figure 18.2.1.

### 18.2.5 EMISSION OF ORGANIC SOLVENTS DURING TECHNOLOGICAL PROCESSES

The concentration of substances emitted to the air during the degreasing processes reached the steady-state constant value:

$$C = E/q \qquad [18.2.1]$$

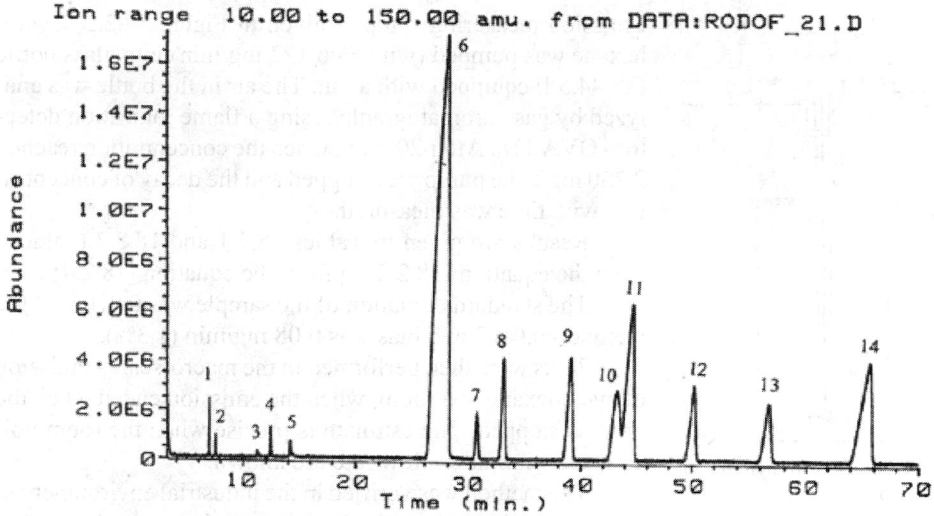


Figure 18.2.1. Chromatogram of substances emitted during degreasing. Oven temperature 40°C. 1 - benzene, 2 - 2,2-dichloromethylpropane, 3 - butyl ester of formic acid, 4 - methylbenzene, 5 - water, 6 - 1-butanol, 7, 8, 10, 11, 12, 13, 14 - derivatives of methyl ethyl benzene, 9 - propyl benzene.

where:

| | |
|---|---|
| C | concentration, $mg/m^3$ |
| q | sum of ventilation flow rates, $m^3/min$ |
| E | emission, mg/min. |

Thus,

$$E = C \times q \quad [18.2.2]$$

The value C can be measured, but a value for q is difficult to estimate, because it includes mechanical and gravitational ventilation (central air conditioning, influence of open doors, windows, fans and natural ventilation).

When the process of degreasing ends, the decay of concentration can be described by:

$$C_t = C_e \times \exp(-t \times q/V) \quad [18.2.3]$$

where:

| | |
|---|---|
| $C_e$ | concentration of the emission at the termination of the process, $mg/m^3$, |
| t | duration of process, min, |
| V | room volume, $m^3$. |

After a transformation:

$$q = (\ln C_e - \ln C_t) \times V/t \quad [18.2.4]$$

The quantity emitted to a room of known volume depends only on the changes of concentration at time (t).

## 18.2.6 VERIFICATION OF THE METHOD

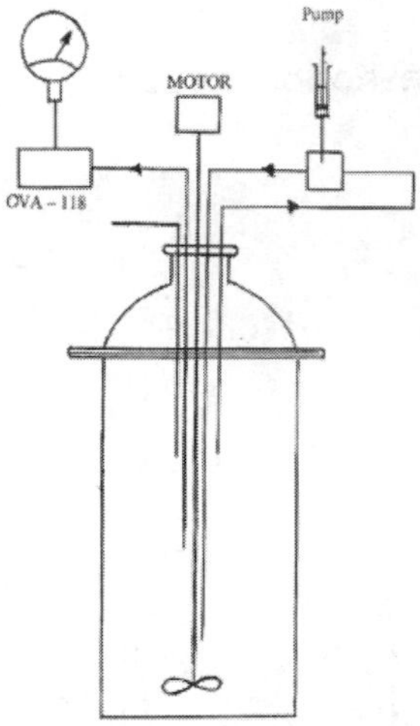


Figure 18.2.2. Emission measuring setup.

Emission measuring setup is given in Figure 18.2.2. Cyclohexane was pumped (with E=6.123 mg/min) into glass bottle (V=44.5 l) equipped with a fan. The air in the bottle was analyzed by gas chromatography using a flame ionization detector - OVA 118. After 20 min., when the concentration reached 2.250 mg/l, the pump was stopped and the decay of concentration with time was measured.

Results are given in Tables 18.2.1 and 18.2.2 (value E from the equation [18.2.2], q from the equation [18.2.4]).

The standard deviation of the sample was 0.0106, of the method - 0.0102 and bias was 0.08 mg/min (1.3%).

Tests were then performed in the macro-scale - emission of cyclohexane in a room, when the emission ended when the process stopped. The estimate is precise when the room volume and ventilation flow rate are known.

The method was verified in the industrial environment in automotive plant. Here, the hazardous substances continue to be emitted to the air after the process had stopped. The concentration measured near the outlet of exhaust was 620 mg/m$^3$, and because the exhaust flow rate was known to be 50 m$^3$/min, the emission was precisely estimated according equation [18.2.2] as 31 g/min or 312.48 kg/week. This value was comparable with the average solvent loss - 326.22 kg/week.

**Table 18.2.1. Emission of cyclohexane**

| t, min | $C_t$, mg/l | E, mg/min |
|---|---|---|
| 1 | 0.138 | 6.210 |
| 2 | 0.273 | 6.212 |
| 3 | 0.405 | 6.213 |
| 4 | 0.533 | 6.201 |
| 5 | 0.659 | 6.201 |
| 6 | 0.780 | 6.184 |
| 8 | 1.020 | 6.200 |
| 10 | 1.250 | 6.212 |
| 12 | 1.465 | 6.200 |
| 15 | 1.770 | 6.188 |
| 20 | 2.250 | 6.218 |
| E ave. - 6.203, $\sigma_n$= 0.0102, $\sigma_{n-1}$= 0.0106 | | |

**Table 18.2.2. Ventilation flow rate**

| t, min | $C_t$, mg/l | Q, l/min |
|---|---|---|
| 1 | 2.200 | 1.000 |
| 2 | 2.215 | 1.011 |
| 3 | 2.100 | 1.024 |
| 4 | 2.055 | 1.008 |
| 5 | 2.010 | 1.004 |
| 6 | 1.965 | 0.993 |
| 8 | 1.880 | 0.999 |
| 10 | 1.800 | 0.992 |
| 12 | 1.720 | 0.996 |
| 15 | 1.605 | 1.002 |
| 20 | 1.435 | 1.000 |
| q ave. = 1.0026, $\sigma_n = 8.7103\times10^{-3}$, $\sigma_{n-1} = 9.1354\times10^{-3}$ | | |

## 18.2.7 RELATIONSHIPS BETWEEN EMISSION AND TECHNOLOGICAL PARAMETERS

Emission of solvents depends on the evaporation rate of the solvent in the process. The evaporation rate from the surface depends on the concentration in the layer on the surface and the coefficient of mass transfer on the air-side. This relation is approximately true for degreasing operations using both liquid and vapors of organic solvents.

The concentration of solvent in a saturated vapor layer depends on temperature and vapor pressure. The coefficient of mass transfer on the air-side depends on the air velocity in the layer on the surface and Schmidt's number (includes dynamic vapor viscosity, vapor density, and diffusion coefficient). Emissions are measured in mass unit per unit of time and the amount depends on surface area and the rate of evaporation, which, in turn, depends on temperature, air velocity over the surface of solvent and the mass of solvent carried out on the wetted parts which have been degreased.

### 18.2.7.1 Laboratory test stand

A thin-metal, flat dish 6 cm in diameter was filled with solvent up to 2-3 mm from the upper edge and placed on a laboratory balance (Figure 18.2.3). The amount of evaporated solvent was measured with (precision 0.1 mg) as the difference between the mass of the dish and solvent at the start of the test and the mass of the dish with solvent after pre-determined period of time. The test was repeated under different conditions of temperature and air velocity near the surface of the solvent. The results were reported as the evaporated mass per 1 minute.

### 18.2.7.2 The influence of temperature on emission

Temperature was measured with a mercury thermometer with a range from 0 to 30°C. Air velocity was 0.3 to 0.4 m/s in this temperature region. The relationship is linear and the equations expressing emission [g/m$^3$/h] relative to temperature [°C] for different solvents are:

| | |
|---|---|
| trichloroethylene | $E = 63.6 \times t + 699.6$ |
| naphtha solvent | $E = 9.3 \times t + 4.9$ |
| naphtha anti-corrosive | $E = 0.55 \times t + 5.65$ |
| 1-butanol | $E = 6.29 \times t + 16.9$ |
| 1,1,1- trichloroethane | $E = 94.7 \times t + 805.6$ |
| extraction naphtha | $E = 49.5 \times t + 1147$ |
| petroleum(mineral) spirits | $E = 11.9 \times t + 76.7$ |
| wood preservatives | $E = 2.0 \times t + 42$ |

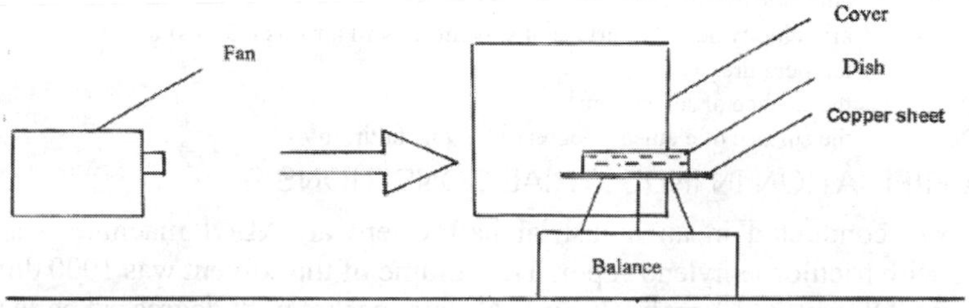


Figure 18.2.3. Test laboratory stand.

### 18.2.7.3 The influence of air velocity on emission

Air velocity was measured with a TSI air velocity meter in range 0.2- 1.5 m/s. Temperature was 20°C. The relationship has linear character and the equations expressing emissions [g/m$^3$/h] relative to air velocity [m/s] for different solvents are:

| | |
|---|---|
| trichloroethylene | $E= 657 \times v + 1766$ |
| naphtha solvent | $E= 109 \times v + 159$ |
| naphtha anti-corrosive | $E= 9.1 \times v + 14.53$ |
| 1-butanol | $E= 72 \times v + 121$ |
| 1,1,1- trichloroethane | $E= 1403 \times v + 2120$ |
| extraction naphtha | $E= 632 \times v + 830$ |
| petroleum(mineral) spirits | $E= 160 \times v + 242$ |
| wood preservatives | $E = 516 \times v$ |

### 18.2.7.4 The relationship between the mass of solvent on wet parts and emissions

The amount of solvent carried out on a degreased part depends on its surface, shape and roughness and on the viscosity and density of the solvent. The amount of solvent carried out on a degreased surface was measured by gravimetric method. Tests show that the amount of solvent retained on 1 m$^2$ of flat surface is:

| | |
|---|---|
| trichloroethylene | 34.7 g |
| naphtha solvent | 26.0 g |
| naphtha anti-corrosive | 34.5 g |
| 1-butanol | 31.3 g |
| 1,1,1- trichloroethane | 38.5 g |
| extraction naphtha | 19.0 g |
| petroleum(mineral) spirits | 19.2 g |

## 18.2.8 EMISSION OF SOLVENTS

Combining these factors, the final equations for the evaluated solvents are:

| | | |
|---|---|---|
| trichloroethylene | $E= (657v + 63.6t + 699.6) \times P_r + 34.7P$ | [18.2.5] |
| naphtha solvent | $E= (109v + 9.3t + 4.9) \times P_r + 26P$ | [18.2.6] |
| naphtha anti-corrosive | $E= (9.1v + 0.55t + 5.65) \times P_r + 34.5P$ | [18.2.7] |
| 1-butanol | $E= (72v + 6.3t + 17) \times P_r + 31.3P$ | [18.2.8] |
| 1,1,1- trichloroethane | $E= (1403v + 94.7t + 805.6) \times P_r + 38.5P$ | [18.2.9] |
| extraction naphtha | $E= (632v + 49.5t + 1147) \times P_r + 19P$ | [18.2.10] |
| petroleum(mineral) spirits | $E= (160v + 12t + 76.7) \times P_r + 19.2P$ | [18.2.11] |
| wood preservatives | $E= (516v + 2t + 42) \times P_r$ | [18.2.12] |

where:

| | |
|---|---|
| E | emission of the solvent, g/h |
| v | air velocity near the surface of solvent (in still air it is 0.3 m/s) |
| t | temperature, °C |
| $P_r$ | the surface of solvent, m$^2$ |
| P | the surface of elements degreased in one hour, m$^2$/h. |

## 18.2.9 VERIFICATION IN INDUSTRIAL CONDITIONS

Research was conducted in an industrial hall where an ALDI machine was used for degreasing with trichloroethylene vapor. The volume of the solvent was 1000 dm$^3$, the surface area was 0.825 m$^2$ (110 cm by 75 cm). A metal basket with degreased parts was introduced into the degreasing compartment and vapors of trichloroethylene condensed on surface of the parts. After 30 s, the basket was removed from the machine. The average temperature at the surface was 24°C. There was an exhaust in the hall with a flow rate of 50 $m^3$/min. The solvent loss was 30 l/day and it was independent of the number of degreased elements. The solvent loss of 30 l/day equals 1.25 l/h or 1.825 kg/h with density of trichloroethylene equal 1.46 g/$cm^3$.

The average concentration of trichloroethylene measured near the exhaust by portable IR spectrophotometer-Miran was 680 mg/$m^3$. Calculation (equation [18.2.2]) with q=50 $m^3$/min gives emission of 34 g/min or 2040 g/h.

According to equation [18.2.5] for trichloroethylene:

$E = (657v + 63.6t + 699.6) \times P_r + 34.7P$

v=0.3 m/s (still air near the surface)

t=24°C

$P_r$=0.825 $m^2$

In vapor degreasing, the coefficient of mass carried out on wet details ready to ship P=0. Then

$E = (657 \times 0.3 + 63.6 \times 24 + 699.6) \times 0{,}825 = 1999$ g/h

The measured value of 2040 g/h and the calculated value of 1999 g/h are in good agreement, meaning that the equations can be successfully applied to predict the organic solvents emission during process of automatic degreasing.

Process of manual degreasing was verified for washing motor parts in a metal dish - 0.72 $m^2$ filled with 500 l of extraction naphtha. The surface of parts was 0.227 $m^2$. This process took place in an open garage at temperature 14°C. Air velocity in vicinity of parts was 0.48 m/s. The residual solvent volume after degreasing was 4.58 l. The process lasted 7 minutes. Emission estimated according the loss of the solvent during the process was 420 ml. Taking into account the weight density of naphtha - 0.72 g/$cm^3$, emission was equal 302.4 g/7 min or 2592 g/h.

Applying the equation [18.2.10] for extraction naphtha

$E = (632v + 49.5t + 1147) \times P_r + 19P$

and:

v = 0.48 m/s

t = 14°C

$P_r = 0.72 + 0.227 = 0.947$ $m^2$, surface of the dish and details

P = 0.947 $m^2$/7min [8.12 $m^2$/h], solvent carried out on the details and on the surface of the dish after pouring out extraction naphtha

Thus:

$E = (632 \times 0.48 + 49.5 \times 14 + 1147) \times 0.947 + 19 \times 8.12 = 2184$ g/h

Again measured value of 2592 g/h and the calculated value of 2184 g/h were in good agreement thus the equation can be applied for predicting the organic solvents emission during the processes of manual degreasing.

# 18.3 INDOOR AIR POLLUTION BY SOLVENTS CONTAINED IN PAINTS AND VARNISHES

TILMAN HAHN, KONRAD BOTZENHART, FRITZ SCHWEINSBERG
**Institut für Allgemeine Hygiene und Umwelthygiene**
**Universität Tübingen, Tübingen, Germany**
GERHARD VOLLAND
**Otto-Graf-Institut, Universität Stuttgart, Stuttgart, Germany**

## 18.3.1 COMPOSITION - SOLVENTS IN PAINTS AND VARNISHES. THEORETICAL ASPECTS

Solvents are normally defined as fluids having a maximum b.p. of 250°C, which are able to dissolve other components of paints, especially binders. They evaporate under drying conditions when paint films are formed.[1,2,3,4] Solvents must not react with the painted or varnished product.

The composition of paints and varnishes is determined by application requirements, drying temperature, and drying time.[2,5] Depending on the properties of paints and varnishes, different mixtures of solvents are added.

Functions and properties of solvents in paints and varnishes:[2,5]

- Dissolve several components, especially binders
- Influence and control paint viscosity
- Wet pigments (influence on solubility, hydrogen bonding by solvents, prevent the separation of pigment)
- Influence and control flow properties (e.g., butyl acetate, butanol, glycol ethers)
- Influence skin formation.[6] The aim is to produce a homogeneous cure when the paint or varnish film hardens without the formation of a stable surface film during the drying period. The correct composition of the solvent will avoid trapping solvents under the surface film.
- Influence drying process. Acceleration by low boilers, production of a flawless surface by medium and high boilers (chemical and physical drying processes)
- Influence surface tension (e.g., increase by rapid evaporation of solvents)
- Influence mechanical properties of paints (e.g., adhesion properties)
- Influence blushing or blooming of paints by preventing the absorption of condensed water by various solvents, e.g., ethanol and glycol ether
- Influence gloss (e.g., improvement with high boilers)
- Prevent defects particularly in varnish coating (e.g., background wetting)
- Influence electrostatic properties (spray painting)
- Influence defined surface properties (structural change)
- Influence durability of paints and varnishes
- Influence product suitability, e.g., spraying and dipping lacquers which need to dry at room temperature

In addition to their effect on performance and properties solvents interact with other components in paints and varnishes in significant ways.

Interactions between binders and solvents in paints and varnishes are very important.[2,5] With the aid of solubility parameters solvents or mixtures of solvents which produce the required properties may be selected. The influence extends to dissolving of binders, reduction of paint viscosity, pigment wetting, etc. Optimized dissolving of binders can be achieved by selecting the appropriate solvent mixtures, in which the density approaches that of the binder solubility range. Nevertheless the selection of an optimized solvent mixtures is difficult because there are conflicting requirements and outcomes.[2,5] On one hand, the chemical nature of the solvents should be similar to those of the binder to improve the flow but on the other hand, the solubility and hydrogen-bonding abilities of the solvents should be at the edge of the binder solubility range, because this results in rapid drying with low retention of solvents.

## 18.3.2 OCCURRENCE OF SOLVENTS IN PAINTS AND VARNISHES

### 18.3.2.1 Solvents in products

Classification

Authorized commissions in Germany describe various groups of solvents in paints and varnishes:

The Association of Varnish Industry VdL (Verband der Lackindustrie) classifies solvents generally used in paints and varnishes in the following groups:[4] Aliphatics, alcohols, aromatic hydrocarbons, esters, glycols, glycol ethers, ketones, terpenes, gasoline, water.

The commission of Hygiene, Health and Environmental Protection (Hygiene, Gesundheit und Umweltschutz) which belongs to the German Standards Commission DIN (Deutsches Institut für Normung e.V.) also describes solvents which may be present in paints as follows:[10]

- Aliphatics (white spirit, turpentine substitutes, cyclohexane)
- Aromatic hydrocarbons (toluene, xylenes, naphtha, styrene)
- Alcohols (methyl alcohol, ethyl alcohol, propyl alcohol, butyl alcohol, glycols)
- Ketones (acetone, methyl ethyl ketone, methyl isobutyl ketone, cyclohexanone)
- Esters (methyl acetate, ethyl acetate, butyl acetate)
- Others: methylpyrrolidone, oils of turpentine

Chlorinated hydrocarbons are not contained in the above list. Benzene is not included as it is obsolete, although it is sometimes found in some petroleum solvent-based paints.

Shortcomings

The description of paints and varnishes is usually neither complete nor reliable:

- Instructions and comments from manufacturers and suppliers differ in quantity and quality. Usually manufacturers or suppliers are not committed to indicate the exact details of their product's composition. They often omit information (e.g., information concerning product components, additives or by-products).
- Lack of standards. In Europe there are some mandatory standards concerning products used in construction (e.g., asbestos, formaldehyde in particle board, organic solvents in certain activities and installations are prohibited).
- Solvents or other organic compounds may be present, in low concentrations, in some products (even in water-based paints), e.g., as additives or by-products although there is no reference to them.
- Additives (e.g., low molecular compounds such as esters or glycol ethers with high boiling points) serve as aids for the formation of films, in repelling water, in assisting plasticization or for other functions. In solvent-based paints, additives or diluting agents are often intentionally mixed in.

**Table 18.3.1. Systematology of paint and varnish products (after reference 10)**

| |
|---|
| 1. Paint coatings |
| 1.1 Coloring agents |
| 1.1.1 soluble pigments |
| 1.1.1.1 natural pigments |
| 1.1.1.2 synthetic pigments |
| 1.1.2 insoluble pigments |
| 1.1.2.1 inorganic pigments |
| 1.1.2.2 organic pigments |
| 1.1.2.2.1 animal and vegetable pigments |
| 1.1.2.2.2 synthetic pigments |
| 1.2 Binders |
| 1.2.1 water dilutable binders |
| 1.2.1.1 slaked lime (lime colors) |
| 1.2.1.2 standard cement (cement colors) |
| 1.2.1.3 sodium silicate (colors of 1 or 2 components) |
| 1.2.1.4 vegetable glues (limewash) |
| 1.2.1.5 casein (lime-casein products, alkali-lime products) |
| 1.2.1.6 dispersions |
| 1.2.1.6.1 natural resin emulsion paints |
| 1.2.1.6.2 plastomer emulsion paints (PVAC (homopolymers, copolymers), PVP, polyacrylates (PMMA, styrol-acetate)) |
| 1.2.1.6.3 water emulsifiable varnish systems (aqueous acrylate systems, aqueous polyurethane systems) |
| 1.2.2 solvent dilutable binders |
| 1.2.2.1 oil paints |
| 1.2.2.2 varnishes |
| 1.2.2.2.1 products drying by air oxidation (nitrocellulose varnishes, aldehyde resin varnishes, oil varnishes) |
| 1.2.2.2.2 physical drying products (polymer resin varnishes, polyvinyl chloride varnishes, polyvinyl acetate varnishes, polyacrylate varnishes, chlorinated rubber varnishes) |
| 1.2.2.2.3 chemical curing products (phenolic varnishes, aminoplast varnishes, melamine resin varnishes) |
| 1.2.2.2.4 acid-curing varnishes |
| 1.2.2.2.5 epoxy resin varnishes |
| 1.2.2.2.6 polyurethane varnishes |
| 1.2.2.2.7 unsaturated polyester varnishes |

- Residuals or by-products can result from various processes (e.g., residual monomers in a polymerization processes).
- Substances which can be classified as solvents are not always exactly defined, e.g., substances in paints with maximum boiling point above 250°C. Plasticizers and solvents cannot often be strictly separated (e.g., various SVOCs).
- Analytical problems and difficulties in assessment (e.g., mixtures of substances, very low concentrations).
- Even new products are often not well documented (e.g., concerning composition, see above).

### 18.3.2.2 Paints and varnishes

Definitions and systematology of paints and varnishes are given by various authorized commissions, e.g., European Committee for Standardization (CEN)[1] and DIN-Commissions[9] (see also Table 18.3.1).

Various databases list the paint and varnish products and their composition generally applied in Europe, see Table 18.3.2).

**Table 18.3.2. Types of paints and varnishes mainly used in Europe - product groups, important compounds, applications (after references 2,5,11,23,25)**

| Product groups | Compounds | Application |
|---|---|---|
| Silicate paints/products, emulsion paints DIN 18363 (M-SK01) | potash-waterglass (binder), inorganic/organic pigments, mineral fillers, synthetic resins (dispersions, stabilizers), water (dispersing agent) | wall paints, outside house paints |
| Silicate paints/products, (M-SK02) | potassium silicate (binder, fixing solution), inorganic pigments, adjuvants | water- and weatherproof painting, interior coating (resistant to chemicals) |
| Emulsion paints (M-DF02) | synthetic resins (dispersions), mineral fillers, inorganic/organic pigments, water (dispersing agent), additives (film-forming agents < 3 %: e.g., glycol-ethers, esters, glycols, hydrocarbons), formaldehyde < 0,1 % | outside house paints, interior coating (wall paints) |
| Emulsion paints, solvent-free (M-DF01) | synthetic resins (dispersions), mineral fillers, inorganic/organic pigments, water (dispersing agent), additives (film-forming agents: no solvents), formaldehyde < 0,1 % | interior coating (wall paint) |
| Emulsion paints, outside house paint, water dilutable | emulsions of plastic materials (acrylate, vinyl), pigments, water (dispersing agent), organic solvents: < 0,1-4 %, (glycols, glycol-ethers, mineral varnish) | exterior coating |
| Emulsion paints, varnishes (M-LW01) | synthetic resins, alkyd resins, copolymerizates, polyurethane resins, inorganic/organic pigments, mineral fillers, additives, water (dispersing agents), organic solvents: 5-10 %, glycols, glycol-ethers, esters, mineral varnish | various applications |
| Aldehyde resin varnishes, aromatic compounds (M-LL03) | aldehyde resins (binders), inorganic/organic pigments, fillers, 30-55 % solvents: mineral varnish (mixture of hydrocarbons), other solvents (< 10 % esters, ethers, alcohols) | covering varnishes, primers |
| Aldehyde resin varnishes, low levels of aromatic compounds (M-LL02) | aldehyde resins (binders), inorganic/organic pigments, fillers, 30-55 % solvents: mineral varnish (mixture of hydrocarbons), other solvents (< 10 % esters, ethers, alcohols) | covering varnishes, primers |
| Aldehyde resin no aromatic substances (M-LL01) | aldehyde resins (binders), inorganic/organic pigments, fillers, 30-55 % solvents: mineral varnish (mixture of hydrocarbons), other solvents (< 10 % esters, ethers, alcohols) | covering varnishes, primers |

| Product groups | Compounds | Application |
|---|---|---|
| Polymer resin paints, high levels of aromatic compounds (M-PL03) | copolymers, inorganic/organic pigments, fillers, 35-50 % solvents: mineral varnish (mixtures of hydrocarbons), other solvents (< 10 %, esters, ethers, alcohols) | outside paints (mineral background) |
| Polymer resin paints, low levels of aromatic compounds (M-PL02) | copolymers, inorganic/organic pigments, fillers, 35-50 % solvents: mineral varnish (mixtures of hydrocarbons), other solvents (< 10 %, esters, ethers, alcohols) | outside paints (mineral background) |
| Polymer resin paints, no aromatic compounds (M-PL01) | copolymers, inorganic/organic pigments, fillers, 35-50 % solvents: mineral varnish (mixtures of hydrocarbons), other solvents (< 10 %, esters, ethers, alcohols) | outside paints (mineral background) |
| Polymer resin paints, dilutable with solvents (M-PL04) | copolymers, inorganic/organic pigments, fillers, 35-50 % solvents: mineral varnish (mixtures of hydrocarbons), other solvents (< 10 %, esters, ethers, alcohols) | outside paints (mineral background) |
| Natural resin paints | natural resins (e.g., shellac) or chemical modified natural resins (e.g., colophonium derivates), additions (e.g., methyl cellulose, natural latex, casein), inorganic, organic pigments (mainly natural origin), mineral fillers, additives (organic solvents: alcohols, terpenes, oil of turpentine, limonenes), essential oils (eucalyptus oil, oil of rosemary, oil of bergamot) | various applications |
| Natural resin paints, solvent-free | natural resins (e.g., shellac) or chemical modified natural resins (e.g., colophonium derivates), additions (e.g., methyl cellulose, natural latex, casein), inorganic, organic pigments (mainly natural origin), mineral fillers, additives (see above): < 1 % | various applications |
| Oil paints, terpene products (M-LL04) | oils (linseed oil, wood oil, soya oil), natural resins and modified natural resins, mineral pigments, wetting agent, flow improver, solvents: oils of turpentine, isoaliphatics, terpenes (citrus, orange) | covering varnishes, primers |
| Oil paints, terpene-free (M-LL05) | oils (linseed oil, wood oil, soya oil), natural resins and modified natural resins, mineral pigments, wetting agents, flow improvers, solvents: isoaliphatics (dearomatized) | covering varnishes, primers |
| Oil paints solvent-free No. 665 | standard oils, calcium carbonate, pigments, siccatives, lemon oil water-soluble | exterior coating (paneling) |
| Clear lacquers/glazing composition (wood), low levels of aromatic compounds (M-KH03) | alkyd resins, nitro resins, polymer resins, pigments, fillers, 40-50 % solvents: mineral varnish (mixtures of hydrocarbons), other solvents (< 10 %, esters, ethers, alcohols) | interior coating (clear lacquers), exterior and interior coatings (glazing compositions, wood) |
| Lead chromate enamels, aromatic compounds | synthetic resins (e.g., aldehyde resins, PVC-polymerizates) inorganic/ organic pigments (lead chromate up to 20 %), fillers, 30-55 % solvents (mixtures of mineral varnish, glycol-ethers, aromatic compounds) | priming coat (steel, aluminum, zinc) |

| Product groups | Compounds | Application |
|---|---|---|
| Silicone resin products, water dilutable (M-SF01) | emulsions of silicone resins mineral fillers, inorganic/organic pigments, water (dispersing agent), additives (film forming agents < 3%) | |

Solvent composition is an important aspect in classifying paints and varnishes.[5] Main groups are:

- Solvent products. These products contain solvents of various mixtures, types and concentrations depending on the properties desired (e.g., application method, surface film or skin formation, see above 18.3.1, 18.3.2.1, Table 18.3.1 and 18.3.2). Solvents are normally the main components of these products (for example up to 80 % in nitrocellulose varnishes, low solids).
- Solvent reduced products. These contain solvents in lower concentrations compared to conventional products and hence have a higher content of solids.[1,2] The main groups of solvent reduced paints are medium solid contents (solids 55-65 %) and high solid contents (solids 60-80 %).
- Waterborne coatings. This group contains deionized water as a dispersing agent. Normally cosolvents are added (up to 25 %). The term "waterborne coatings" is mainly applied to industrial coating materials, which differ from silicate colors, wood preservative varnishes and emulsion paints.
- Solvent-free products. The products of this group are produced and applied without organic solvents: powder coatings, radiation curing systems, solvent-free water coating (without cosolvents).

The use of paints and varnishes containing high solvent concentrations is becoming less common, solvent-reduced products, waterborne coatings and solvent-free products are applied more often.

Whereas environmental and health-related concerns call for the reduction of solvents in paints and varnish products (see 18.3.3.2 and 18.3.4), qualitative aspects still demand the use of solvents in some fields of application.[12] Solvent-reduced products should achieve the same qualitative properties as solvent-containing products (e.g., application properties, periods of guarantee, limited costs, loading capacities, surface properties). The development of solvent-reduced or solvent-free varnish products with high quality (e.g., durable, good application properties) and limited costs must be encouraged if not mandated. In some fields of application (e.g., waterborne coatings, high solids in varnish coatings of vehicles) a lot of developmental work has already been done. Additionally, other components in varnishes apart from solvents or modifications of application techniques can improve the properties of solvent-reduced paint and varnish products. Nevertheless, a wide rage of quality exists in paints and varnish products which are offered commercially and, in some cases (e.g., concerning film forming processes, processibility, corrosion protection, purification, special applications or wood preservation), solvent-based products are still preferred. In the field of wood preservation especially, solvent-based products (alkyd resins) are used because of some technical advantages (e.g., more solid surfaces) but alternative high solid systems are available.[12]

## 18.3.3 EMISSION OF SOLVENTS

### 18.3.3.1 Emission

Solvents are usually the most significant emission products coming from building materials and interior furnishings.[2,5,6,13] All painted products are potential sources of emission. Even the so-called "bio" paints or natural paints emit various substances,[5] e.g., mineral varnishes, natural oils and even synthetic terpene-like compounds.

Depending on the products and the components which make them up, the various parameters listed below can determine the emission and behavior of solvents in ambient air from paints and varnishes:[2,5,6]

- Film formation. During the film formation stage solvents are emitted. The emission rate is directly proportional to the VOC concentrations in the product and inversely proportional to the film thickness (first order of kinetics). When the film has completely formed, the emission is controlled by diffusion processes, the emission rate is now inversely proportional to the square of the film thickness.
- Application of the paints and varnishes, methods of application of the paint or varnish, e.g. speed of application of the paint.
- Characteristics of solvents in paints and varnishes, e.g., volatility (boiling point), dynamic characteristics of evaporation and concentrations. Substances having a low boiling point evaporate fast, mostly during application and cause a rapid skin formation. Thus the risk of exposure is mainly with the painters. Medium boilers allow the surface to remain open for a while (evaporation of volatile products). The evaporation of substances with a high boiling point is slow, taking several weeks or months after application, resulting in exposure to the building occupants.
- Characteristics of other compounds in paints and varnishes (e.g., relationship of binders to solvents, possible reactions between solvents and other compounds).
- Characteristics of surfaces which have been painted (e.g., area, structure of surface).
- Characteristics of emission processes, e.g., type of emissions (e.g., diffusion), dynamics of emissions (constant of evaporation), interrelations (e.g., diffusion and back diffusion).

The quantitative assessment of emission processes can be described with various models. The usefulness of these models differs. Some models describe these processes very well, as proven by various experiments or measurements (e.g., test chambers). Basic equations which describe emission processes are shown in Table 18.3.3.

The emission processes of solvents from paints and varnishes can be divided into two phases:[2,5,6]

1. Emissions during application of paints. This deals with complex interrelations dependent on various parameters.

2. Emissions after application process. Here the course is governed by complex emission processes dependent on various parameters (e.g., film formation, surface area).

Most solvent products, especially organic solvents and some additives, emitted from paints and varnishes are VOCs. The largest components of VOCs are solvents, e.g., aliphatic and aromatic hydrocarbons, alcohols, amines, acids, aldehydes, esters, ketones, terpenes. The definition of the term VOC varies, a standard definition is published by CEN (European Committee for Standardization):[1] VOCs are any organic liquids and/or solids that evaporate spontaneously at the prevailing temperature and pressure of the atmosphere.

VOCC (volatile organic compound content) is defined as follows:[1] Mass of the volatile organic compounds in a coating material, as determined under specified conditions.

**Table 18.3.3. Example for basic calculations of VOC-emissions during application of emulsion paints (after references 2,5,6,13)**

1. At the beginning of the application process (t=0) the mass of VOC changes positively ($v_{ST}d_sc_w$) , on the other hand VOC evaporates (first order of kinetics).
   $$dm_w/dt = v_{ST}d_sc_w - k_1m_w/d_s \qquad [18.3.1]$$
2. In the ambient air the mass of VOC increases because of the evaporation out of the wall and decreases according to the ventilation rates.
   $$dm_L/dt = k_1m_w/d_s - k_2m_L \qquad [18.3.2]$$
3. If the connected differential equations are solved and integrated (from t=0 until the end of application $t=A/v_{ST}$), the following equations are received:
   $$m_w(t) = v_{ST}d_s^2c_w/k_1(1-\exp(-k_1t/d_s)) \qquad [18.3.3]$$
   and
   $$m_L(t) = ((1-\exp(Bt)/B - (1-\exp(k_2t)/k_2)v_{ST}d_sc_w\exp(-k_2t)) \qquad [18.3.4]$$
   with
   $$B = k_2 - k_1/d_2 \qquad [18.3.5]$$
4. After finishing application only evaporation is relevant (equation [18.3.1] is simplified):
   $$dm_w/dt = - k_1m_W/d_s \qquad [18.3.6]$$
5. The course of VOC in the ambient air does not change (equation [18.3.2] corresponds to equation [18.3.7]):
   $$dm_L/dt = k_1m_w/d_s - k_2m_L \qquad [18.3.7]$$
6. The solution of these differential equations describes the quantities of VOC in the wall (equation [18.3.8] and the course of VOC in the ambient air [18.3.9]:
   $$m_w(t) = m_{w,AE}\exp(-k_1(t-t_{AE})/d_s) \qquad [18.3.8]$$
   $$m_L(t) = ((k_1m_{w,AE}/d_s)(\exp(B(t- t_{AE}))-1)/B + m_{L,AE})\ \exp(-k_2(t-t_{AE})) \qquad [18.3.9]$$
   with
   $$B = k_2 - k_1/d_s \qquad [18.3.5]$$

where:

| | |
|---|---|
| A | area of the wall |
| a | coating thickness |
| B | fraction of binder |
| $c_W$ | VOC-concentration in the wall |
| $c_L$ | VOC-concentration in the ambient air |
| D | density |
| $d_s$ | thickness of the layer of the paint application (=a/D) |
| $k_1$ | constant of evaporation |
| $k_2$ | ventilation rate of the indoor air |
| $m_L$ | mass of VOC in the indoor air |
| $m_{L,AE}$ | $m_L$ at the end of the application |
| $m_w$ | mass of VOC in the wall |
| $m_{W,AE}$ | $m_W$ at the end of the application |
| $R_M$ | VOC-content in the dispersion |
| t | time |
| V | volume of the indoor air |
| $v_{ST}$ | spreading velocity |

In VOCs, especially these emitted from coating materials, the evaporation temperature is specified by European regulations: b.p. max. 250°C (according to DIN ISO 11890-1,-2 or 96/13/EC)[14] or b.p. max. 260°C (according to VDI Guidelines 4300-6).[15] In some US governmental legislation VOCs are defined solely as those compounds that are photochemically active in the atmosphere (ASTM D 3960).[1]

Regulations and assessment of VOCs are under discussion in Germany, e.g., discussion of various threshold limit values of TVOC (total volatile organic compounds): < 200 $\mu g/m^3$ according studies of Molhave[16] and 300 $\mu g/m^3$ according to BGA.[17]

The percentage of VOC emissions caused by solvents has increased recently mainly because of the reduction of VOC coming from vehicular traffic, e.g., in Germany the estimated anthropogenic VOC emissions (without methane) caused by solvents in 1994 was about 51% compared to 37% in 1988.[12] The solvent VOC emissions are mainly connected with coating materials which are responsible for more than 50% of solvent-based VOC emissions in Germany in 1995 (about 38% caused by varnishes, 4 % by building materials, 19% by other processes such as metal cutting).[12] The main source of emissions of solvents during varnish processing are from equipment without licence requirements (about 60%) and from trade and paint work (about 15%); equipment which has licence requirements (manufacturing plants) contributes only 10% of VOC emissions.[12]

Therefore, solvents released during the application processes of paints and varnishes must be reduced. New EC regulations concerning reduction of solvents in special industrial plants or working processes (e.g., varnish coating of vehicles) have already been formulated.[8] Similar regulations concerning other working fields (e.g., the use of solvent-based paints in home workshops) have been prepared throughout Europe and are already in force in some countries.

### 18.3.3.2 Immission

In Europe, various regulations and schemes for the assessment of VOCs are in use according to special conditions, e.g.:

- Indoor air: Schemes and values proposed by central environmental institutions (Sachverständigenrat für Umweltfragen),[18,19] commissions of UBA (Umweltbundesamt).[20]
- Place of work: Special values (e.g., TLV-, MAK-, TRK-values) proposed by special institutions, e.g., commissions of ACIGH (American Conference of Governmental Industrial Hygienists)[21] or DFG (Deutsche Forschungsgemeinschaft).[22] TLV (Threshold Limit Values) refer to airborne concentrations of substances. They represent conditions to which workers may be repeatedly exposed during an 8-hour workday in a 40-hour week. MAK-values (Maximale Arbeitsplatzkonzentration) describe maximum concentrations of individual substances allowed in the work place (8 hours per day or 40 hours per week with some exceptions allowed). MAK-values are obligatory limits in Germany but in cases where MAK values cannot be evaluated (e.g., carcinogenic compounds) TRK values (Technische Richtkonzentrationen) are used. These recommendations are based on current technical knowledge.[22]
- Atmosphere, especially the significance of photooxidation. Various measures and regulations (e.g., reduction of VOCs)[8] intend to limit substances which play a part in photooxidizing processes (e.g., ozone formation).

When defining threshold limit values, it is important that prevailing conditions and methods are specified, e.g., ventilation rates, methods of sampling, determination and calculation.[1,17,23]

Immission processes of solvents from paints and varnishes are determined by surrounding conditions: e.g., parameters of the ambient air, indoor air parameters, e.g., ventilation (ventilation rates), air distribution, air movements, composition of ambient air, concentrations of air compounds, humidity, temperature, volume of the indoor air, extreme conditions (e.g., worst case).

## 18.3.4 EFFECTS ON HEALTH OF SOLVENTS FROM PAINTS AND VARNISHES

Only the effects resulting directly from the exposure to paints and varnish solvent are discussed in this chapter. For the effects of other solvents, see Chapter 20.1.

### 18.3.4.1 Exposure

Usually the effects of solvents in paints and varnishes on health are dependent on concentration and exposure time. Adverse health effects may follow exposure to paints, varnishes and their solvents at the workplace. The conditions at paint and lacquer manufacturing sites (e.g., manufacturing methods, use of exhaust hoods, etc.) are responsible for the levels of evaporated solvents measured in the air.[24] Adverse health effects depend on how the paint is applied, paint properties and working conditions (e.g., increased risk with spray painting). There is sufficient evidence to substantiate the fact that solvents to which painters have been exposed, are responsible for incidences of cancer. However, within paint manufacturing plants, this evidence is inadequate.[25]

Nevertheless, person working at home, occupants of painted rooms and children of parents which have been exposed are also at risk. Those involved with the abuse of solvents as a psychoactive substance (e.g., aromatic hydrocarbons in spray paints, mainly semi-volatile or nonvolatile components) are at a much higher risk.

The main path of entry of solvents from paints and varnishes to the body is by inhalation. Volatile paint compounds present a particularly high risk as do some forms of paint application (e.g., spray painting with the risk of inhalation of even less volatile and nonvolatile paint components). Other pathways should, however, also be considered as dermal contact.

In addition to solvents, other compounds from paints and varnishes can cause various diseases, often similar to the effects caused by solvents (e.g., asbestos as paint filler or in the construction and shipyard industry, silica, dusts, thermal decomposition products, contaminations of solvents, chromium, iron and lead compounds in paint pigments). It is often difficult to associate a particular components of paints and varnishes with adverse health effects. In most cases, the paint and varnish products were found to be a relevant cause of illness, but their individual compounds were not.

### 18.3.4.2 Health effects

#### *18.3.4.2.1 Toxic responses of skin and mucose membranes*

The following symptoms involving the skin and mucose membranes may occur as a result of using paints and varnishes:

- Irritations of skin and mucose membranes
- Allergic diseases of skin and mucose membranes
- Removal of grease from skin (removal of sebaceous matter, with subsequent adverse skin conditions such as infection)
- Changes in the lens of the eye and corneal changes
- Absorption of solvents through the skin (e.g, benzene, toluene, xylene, methyl alcohol, methyl ethyl ketone, glycol ethers)

Workers in the paint manufacturing industry and painters have experienced occupational diseases, especially dermatosis affecting the hands and arms.[26-28] Whereas several paint components have been shown to cause non-allergic and allergic contact eczema, organic solvents were shown to provoke mainly non-allergic contact eczema and some solvents cause only irritation (e.g., some ketones and esters).

In another study by Mancuso et al.[28] it was shown that the occupational contact dermatitis of workers in shoe factories was probably a result of solvents in varnishes and adhesives. This study was based on interviews, medical examinations and patch test series.

Eye diseases stemmed from both non-allergic and allergic reactions, and in some cases corneal and lens changes were noted.[29]

In a further study,[30] with water based paints there was a significant reduction in eye and skin disease and worker discomfort on the job.

#### *18.3.4.2.2 Neurological disorders*

Indoor air immissions of organic solvents from paints and varnishes can cause neurological disorders:

- Neuro-psychological and neuro-behavioral symptoms (e.g., subjective symptoms, multiple chemical sensitivity - MCS)
- Neurophysiological symptoms
- Neurological diseases (e.g., polyneuropathy)
- Neuropsychiatric diseases

Other compounds in paints and varnishes apart from solvents can cause neurological disorders (e.g., lead). However, it is difficult to prove that solvents specifically cause neurological changes.

A study of production plants producing dyes and varnishes,[31] showed that mixtures of organic solvents are responsible for several neurological and neurophysiological symptoms: headache, dizziness, increased emotional excitability, memory and concentration disorders, mood instability, fatigue. Neurological examinations, however, showed no significant changes in the central and peripheral nervous system but EEG and VEP anomalies were seen.

In contrast to the studies mentioned above, a study of shipyard spray painters[32] exposed to xylene and mixed organic solvents described neurophysiological changes, e.g., decreased nerve function and, in addition, neuropsychological symptoms, e.g., mood changes and fatigue. Similar results including reduced nerve conduction were shown by workers exposed to styrene. Other studies found several dose-response relationships between solvent mixtures and neuro-behavioral effects among paint manufacturing employees and painters.[33,34] Significant relationships concerning the total amounts of hydrocarbons, lifetime exposure and lifetime-weighted average were described.

In earlier reviews and cross-sectional studies, various symptoms and neuro-behavioral effects were described for workers in the paint manufacturing industry, house painters, car and industry painters, and shipyard painters.[35] Subjective symptoms (fatigue, loss of concentration, emotional instability, short-term memory disorders, headache) or effects on psychomotoric performance are examples of these symptoms. However, similar former studies did not find symptoms in house painters using mainly water-based paints.[36]

Different results were found concerning neuro-physiological changes and neurological diseases. Electroencephalographic changes and a slight decrease in cerebral blood flow of paint industry workers was noted by Oerbaek et al.[37] and there were occasional cases of clinical polyneuropathy in spray painters exposed mainly to methyl n-butyl ketone as described by Mallov.[38] Furthermore, slight neurological impairments of car and industry painters was noted by Elofsson et al.[29] and Maizlish et al.[39] and in house painters by Askergren et al.[36] Nevertheless, no similar effects were described in other studies.[42,43] In particular, no effects on the peripheral nervous system were reported in workers using water-based paints.[36]

Inconsistent results were reported for severe diseases of the central nervous system (neuropsychiatric diseases, encephalopathy) for painters and other persons exposed to solvents.[42-44,47]

Environmental exposure to organic solvents is supposed to be one cause of multiple chemical sensitivity which characterizes neuropsychological disorders. Organic solvents from paints seem to play a role.[45,46] The relevant exposure characteristics (dose, time, possible dose-response relationship) are also discussed.[45]

#### *18.3.4.2.3 Carcinogenic effects*

Various solvents which occur in paints and varnishes exhibit a carcinogenic potential. Benzene and all isomers of dinitrobenzenes are particular examples (see also Chapter 20.1). This was confirmed by several extensive studies of occupational exposures in paint manufacturing plants and in painters.[25] These data show the risk of contracting cancer to be about 20% above the national average.

In many studies, increased risks were described mainly for lung cancer (about 40% above the national average), leukemia, bladder cancer, liver cancer, and childhood cancers where there has been parental exposure.[25]

In a meta-analysis using standardized mortality ratios the relationship between painting exposure[48] and cancer mortality showed the highest risks for leukemia and liver cancer. Elevated risks were also predicted for lung cancer, oesophageal cancer, stomach cancer and bladder cancer. The development of leukemia, especially in the case of benzene mixed with other organic solvents, being most prominent and the development of lung cancer (main risks being lead chromate and asbestos) to a lesser degree.

It should be noted, however, that more critical risk factors such as smoking and alcohol can obscure such correlations.

Although high risks for cancer resulting from occupational exposures were mentioned, no significant information concerning the occupation of painted rooms could be noted. This area has been inadequately studied.

As mentioned above, lung cancer is a major concern. Painters, as opposed to those involved in paint manufacture were shown to be at greatest risk for contracting lung cancer.[25,48] In fact, there was no evidence of increased risk in persons involved in the manufacture of paints.[25,49]

Other cancers of the respiratory tract are documented also, e.g., cancer of the nasal cavity,[50] pleural mesothelioma with high incidences in painters and paper-hangers[51] and cancers of the larynx.[52]

The evidence of several types of leukemia is convincing with regard to occupational exposures to paints and solvents.[25,48] Increased risks for contracting other haematopoietic neoplasmas[25] such as Hodgkin´s disease,[53] non-Hodgkin´s lymphoma,[54] multiple myeloma,[55] reticulum-cell sarcoma and lymphosarcoma[56] were also reported. Probably all neoplasmas were caused mainly by organic solvents. Some former studies, however, included solvents which are rarely used now, especially benzene.[25]

Davico et al.[57] and Crane[58] showed clonal aberrations in chromosome 8 of workers exposed to paints which is correlated with higher risks of acute myelocytic leukemia.

Occupational exposure to dyestuff manufacture, paints, solvents and inks are an important risk factor for cancer development of the urinary tract (especially aromatic amines in bladder cancer).[25,59] A major risk factor responsible for about 50% of bladder cancer cases in western countries is smoking. A second risk factor is occupational exposure (including exposure to paints and solvents).[59] Some studies showed greater risks according to certain exposures: e.g., spray painting,[60,61] lacquering and painting of furniture and cars or sign-post painting.[62]

An increased risk of cancer of the prostate was found for workers in paint manufacturing plants, but no increase in risks could be evaluated for exposure to paints.[63] In one study an increased risk for testicular cancer was shown.[25]

Inconsistent results were noted for cancers of the gastrointestinal tract.[25] Some studies showed higher rates of stomach and intestinal cancers, but other risk factors apart from paints cannot be excluded. Higher risks for liver cancer was found by Chen and Seaton.[48] Norell et al.[64] noted an excess of pancreatic cancers, especially as a result of exposure to paint thinners. No significant risks were described concerning cancer of the biliary tract.[65]

A distinct relation between parental occupational exposure and childhood cancer was shown for solvents and paints. High parental exposure resulted in higher incidences of childhood cancers.[66] In the same study, however, generally more cancers were found as a result of parental use of alcohol and tobacco smoke. Childhood leukemia and nervous system cancers, in particular, are the types suspected to be caused by parental exposure to paints and solvents.[67] Kishi et al.[68] described an elevated risk for acute lymphatic leukemia in children of mothers with prenatal exposures to benzene and to paints. In former studies with small numbers of children these tendencies could also be shown, mainly in male painters whose children showed a higher incidence of childhood leukemia and brain tumors.[25]

#### *18.3.4.2.4 Respiratory effects*

The following respiratory symptoms are provoked by solvents:

- Irritations
- Allergic reactions
- Changes in lung function parameters (mainly obstructive ventilation)
- Pathohistological changes

Irritations of the air passages were described in people exposed to fumes in paint factories.[27,69]

VOCs in paints can provoke respiratory symptoms (wheezing, breathlessness) in asthmatics. Conventional water-based paints with only small amounts of VOCs have also been shown to cause such symptoms, but there were no effects using VOC-free paints.[70] No differences were found in the same study looking at lung function and airway responsiveness. Toxic pyrolysis products in paints and polymer films probably evoke asthma-like symptoms similar to PVC pyrolysis.[71]

A decrease in forced vital capacity, expiratory volume in one second and of peak exspiratory flow was observed after exposure to water-based paints.[72] Bronchial obstruction in painters was confirmed by White and Baker,[69] but other studies could not detect changes in lung function parameters in house painters who were exposed to solvent-based and water-based paints.[36] Beving et al.[73] did not find obstructive effects in car painters. An increase in mortality from chronic obstructive bronchitis of painters was recorded by Engholm and Englund,[74] but not confirmed in other studies.[25]

Histological changes in the nasal mucosa of spray painters were also noted in a study of Hellquist et al.[75]

### *18.3.4.2.5 Toxic responses of the blood*

A case-control study of Guiguet et al.[76] observed an increased risk for aplastic anaemia after exposure to paints (effective compounds in the paints are unknown), but no increased risk after exposure to solvents alone. Reduction of blood hemoglobin levels was shown after exposure to gasoline, car spray paint (xylene) and various solvents (house painters).[25,77,78] An increased Hb level was noted, however, in a study of car and industrial spray painters.[29]

Alterations of several blood components (production of auto-antibodies) and of vascular endothelial cells were described for workers exposed to hydrocarbon-based paints or to mixed solvents.[79]

Even a group experiencing low exposure (house painters exposed to alkyd paints) showed hematological changes. Higher platelet counts and higher resistivity calculated from the the impedance of the whole blood were observed.[73]

Some studies showed decreases of thrombocyte counts in car painters and paint industry workers, other studies, however, recorded no changes.[25] Studies concerning effects of paints on the white cell and thrombocyte counts were inconsistent. A decrease of white cells was described in several studies[11] and a lymphocytosis was noted by Angerer and Wulf.[77] Elofsson et al.,[29] however, recorded no changes in white cell counts. Additionally, myelotoxic effects of solvents were shown, especially for benzene.

### *18.3.4.2.6 Toxic responses of the reproductive system*

In interviews with pregnant women, the effects of paints, varnishes and solvents were evaluated. No effects were found concerning congenital heart disease of infants with Down Syndrome (trisomie 21).[80] The maternal risk was associated with smoking.

Maternal toluene sniffing used as an organic solvent in acrylic paints, varnishes and other sources is associated with premature births and, in one case, renal tubular acidosis.[81]

Other studies of female painters showed tendencies towards an increase of spontaneous abortions[82] or infant mortality.[83] Other studies[84] did not confirm these tendencies.

McDowall reported an increase in malformations in children (polydactyly, syndactyly, spina bifida and anencephalus) whose parents were painters, assemblers or had related occupations.[83] A study by Olson[85] showed congenital malformations of the central nervous system for paternal exposure to paints.

### *18.3.4.2.7 Toxic responses on other organ systems*

Contradictory effects of solvents in paints on the kidneys have been described.[25] Lauwerys et al., for example, have described some solvents used in paints as nephrotoxic (e.g., toluene).[88] Minor effects such as slight hematuria and albuminuria or small effects on the glomeruli or even no effects were detected in other studies.[25]

In some studies of spray painters in the automobile and airplane industries and of house painters mortality from liver cirrhosis was increased.[25,87] Another study investigating house painters, however, did not show an increase.[42]

There seem to be no increased risk for cardiovascular diseases according to some studies of paint industry workers.[25,89] Nevertheless, effects of solvents on muscles and vessels could be shown (e.g., increased serum creatinine levels).[25]

## 18.3.5 METHODS FOR THE EXAMINATION OF SOLVENTS IN PAINTS AND VARNISHES

### 18.3.5.1 Environmental monitoring

There are several test procedures generally used in Europe. Procedures are still being developed to determine the presence of solvents in paints and varnishes.

#### *18.3.5.1.1 Solvents in products*

Officially approved test procedures are:

- Gravimetry: Determination of volatile organic compounds (DIN ISO 11 890-1[7] - VOCs > 15 %).
- Gas chromatography: Determination of volatile organic compounds (DIN ISO 11 890-2[7] - VOCs 0,1 % -15 %).

#### *18.3.5.1.2 Emission of solvents*

Test chamber[7,92,93] (volume 1 $m^3$, defined conditions: temperature, humidity, air changing rates, air velocity). This is usually carried out 3 days after application of paints or varnishes by placing the products into a test chamber. The test series starts with sampling emissions of the products on defined absorption materials (Tenax, charcoal filters, activated charcoal), reconditioning and finally identification and classification of VOCs with gas chromatography. The test methods are repeated after 14 and 28 days. These test criteria permit determination of the behavior of emissions from finished products under defined conditions.

Emission test cells.[7,92,93] The product under investigation is hermetically sealed into the emission test cells. The emitted compounds are sampled by absorption materials and analyzed. This method can be used under laboratory conditions or *in situ*.

### 18.3.5.2 Biological monitoring of solvents in human body fluids

#### *18.3.5.2.1 Solvents and metabolites in human body fluids and tissues*

The biological monitoring of solvents emitted from paints or varnishes on humans is not well developed. In two studies,[94,95] solvents from paints and varnishes were determined in blood, urine and internal breath. Blood and urine analysis is less sensitive than internal breath measurements. This was carried out in a study on exposure to paints in aircraft maintenance.[94]

Kramer et al.[95] found xylenes during paint production and paint-spraying in ambient air and in the blood and urine of workers. Threshold limit values (TLV) and biological exposure indices (BEI) were not exceeded.[22,23]

#### *18.3.5.2.2 Biomarkers*

Even low air concentrations of solvents emitted from paints have an influence on the human organism, e.g. the induction of hepatic enzymes. This effect was shown in workers exposed to butyl glycol from paints in an electrophoresis painting plant where the exposure value was shown to be less than 0.3 times of the average limit. In these cases D-glucaric acid in urine, which reflects the D-glucarid acid enzyme pathway, was increased.[96]

In some studies the genetic effects on paint industry workers could not be detected when chromosomal aberrations and sister chromatid exchanges were studied.[86,87] In other studies[94,95] it was shown that the frequency of micronuclei and sister chromatid exchanges increased.

## REFERENCES


1 CEN (European Committee for Standardization), Paints and varnishes. Terms and definitions for coating materials. Part 1: General terms. EN 971, 1996.
2 **Ullmann`s Encyclopedia of Industrial Chemistry**, 1998.
3 DIN 55945 (Deutsches Institut für Normung), Coating materials (Varnishes, Painting Materials and other Materials), Terms. Beschichtungsstoffe (Lacke, Anstrichstoffe und ähnliche Stoffe); Begriffe, Beuth Verlag, Berlin, 1988.
4 VdL (Verband der Lackindustrie/ Association of the Varnish Industry in Europe)- VdL-Guidelines Building Coating Materials. VdL-Richtlinie Bautenanstrichstoffe, RL 01, 1997.
5 **Römpp Lexikon, Varnishes and Paints. Lacke und Druckfarben**, *Thieme*, 1998.
6 H. Zeh, M. Kroll, and K. Kohlhammer, VDI Berichte, 1122, pp. 455-475 (1994).
7 DIN (Deutsches Institut für Normung), ISO/ DIN 11890. Paints and varnishes - Determination of volatile organic compound (VOC) content, Beuth Verlag, Berlin, 1998.
8 European Commission, ECA-IAQ, Total Volatile Organic Compounds (TVOC) in Indoor Air Quality Investigations, EUR 17675 EN, Report No. 19, 1997.
9 1999/13/EC, Council Directive of 11 March 1999 on the limitations of emission of volatile organic compounds due to the use of organic solvents in certain activities and installations. ABl. EG Nr. L 085, 29.3.99, 1999.
10 DIN (Deutsches Institut für Normung)(Normenausschuss Bauwesen NABau), Koordinierungsausschuss KOA 03 "Hygiene, Gesundheit und Umweltschutz": Guidelines for Health Assessment of Building Products. Leitfaden zur Beurteilung von Bauprodukten unter Gesundheitsaspekten, KRdL-CEN/264/7 N 1999-07, 1999.
11 GISBAU, WINGIS, Bau-Berufsgenossenschaften (Professional Associations of the Building Industry in Germany), 1999.
12 Deutsches Lackinstitut Frankfurt a.M. (German Institute of Varnishes Frankfurt a.M.), Solvent reduction in the trade of painters and lacquerers. Documents concenring varnishes and paints. Magazine No. 7. Lösemittel-Reduzierung im Maler- und Lackiererhandwerk. Dokumente zu lacken und Farben. Heft Nr. 7, DeutschesLackinstitut, Frankfurt a.M., 1999.
13 T. Salthammer, *Chemie in unserer Zeit*, **28**, 280 (1994).
14 DIN (Deutsches Institut für Normung), DIN ISO 11890-1,-2, Beuth Verlag, Berlin, 1998.
15 VDI (Verein Deutscher Ingenieure), VDI 4300, VDI, Düsseldorf, 1999.
16 L. Molhave, *Indoor Air*, **1**, 357 (1995).
17 Bundesgesundheitsamt (Public Health Department of the German Government), *Bundesgesundheitsblatt*, **36** (3), 117-118 (1993).
18 SRU (Environmental Council of the German Government)(Sachverständigenrat für Umweltfragen), Air pollutions in rooms, Special Report, Metzler-Poeschel, Stuttgart, 1987.
19 SRU (Environmental Council of the German Government)(Sachverständigenrat für Umweltfragen), Environmental Report 1996, Metzler-Poeschel, Stuttgart, 1996.
20 Anonymous, *Bundesgesundheitsblatt*, **11**, 422 (1996).
21 ACGIH (American Conference of Governmental Industrial Hygienists), TLVs and BEIs. Threshold Limit Values for Chemical Substances and Physical Agents. ACGIH, 1998.
22 DFG (Deutsche Forschungsgemeinschaft), MAK- und BAT-Werte-Liste 1999, Wiley-VCH, 1999.
23 M. Fischer, and E. Böhm, Evaluation and assessment of emission of noxes caused by furniture varnishes. Erkennung und Bewertung von Schadstoffemissionen aus Möbellacken. Schadstoffe und Umwelt, Erich Schmidt Verlag, Berlin, 1994.
24 W. Wesolowski, and J.P. Gromiec, *Int. J. Occup. Med. Environ. Health*, **10** (1), 79 (1997).
25 IARC (International Agency for Research on Cancer), IARC Monographs on the evaluation of carcinogenic risks to humans. Some organic solvents, resin monomers, resin monomers and related compounds, pigments and occupational exposures in paint manufacture and painting. WHO, 47, IARC, Lyon, 1989.
26 U. Ulfvarson, Int. Symp. on the Control of Air Pollution in the Working Environment, Stockholm, September 6-8, Swedish Work Environment Fund/ International Labour Office, Part II, 1977, 63-75.
27 R.V. Winchester and V.M. Madjar, *Ann. Occup. Med.*, **4**, 221 (1986).
28 G. Mancuso, R.M. Reggiani, and R.M. Berdandini, *Contact Dermatitis*, **34** (1), 17 (1996).
29 S.A. Elofsson, F. Gamberale, T. Hindmarsh, A. Iregren, A. Isaksson, I. Johnsson, B. Knave, E. Lydahl, P. Mindus, H.E. Persson, B. Philipson, M. Steby, G. Struwe, E. Söderman, A. Wennberg, and L. Widen, *Scand. J. Work Environ. Health*, **6**, 239 (1980).
30 G. Wieslander, D. Norback, and C. Edling, *Occup. Environ. Med.*, **51**(3), 181 (1994).

31 J.A. Indulski, H. Sinczuk-Walczak, M. Szymczak, and W. Wesolowski, *Int. J. Occup. Med. Environ. Health*, **9**(3), 235 (1996).
32 M.W. Ruijten, J.T. Hooisma, J.T. Brons, C.E. Habets, H.H. Emmen, and H. Muijser, *Neurotoxicology*, **15** (3), 613 (1994).
33 F. Fung, and R.F. Clark, *J. Toxicol. Clin. Toxicol.*, **37**(1), 91 (1999).
34 A. Seeber, B. Sietmann, and M. Zupanic, *Int. J. Occup. Med. Environ. Health*, **9** (3), 235 (1996).
35 NIOSH (National Institute for Occupational Safety and Health), Organic solvent neurotoxicity, Current Intelligence Bulletin No. 48, Cincinnati, 1987.
36 A. Askergren, H. Beving, M. Hagman, J. Kristensson, K. Linroth, O. Vesterberg, and A. Wennberg, *Arb. Hälsa*, **4**, 1 (1988).
37 P. Oerbak, J. Risberg, I. Rosen, B. Haeger-Aronsen, S. Hagstadius, U. Hjortsberg, G. Regnell, S. Rehnström, K. Svensson, and H. Welinder, *Scand. J. Work Environ. Health*, **11** (Suppl. 2), 1 (1985).
38 J.S. Mallov, MBK neuropathy among spray painters, *J. Am. Med. Ass.*, **235**, 1455 (1976).
39 N.A. Maizlish, G.D. Langolf, L.W. Whitehead, L.J. Fine, J.W. Albers, J. Goldberg, and P. Smith, *Br. J. Ind. Med.*, **42**, 579 (1985).
40 N. Cherry, H. Hutchins, T. Pace, H.A. Waldron, *Br. J. Ind. Med.*, **42**, 291 (1985).
41 D. Triebig, D. Claus, I. Csuzda, K. Druschky, P. Holler, W. Kinzel, S. Lehrl, P. Reichwein, W. Weidenhammer, W.U. Weitbrechst, D. Weltle, K.H. Schaller, and H. Valentin, *Int. Arch. Occup. Environ. Health*, **60**, 223 (1988).
42 S. Mikkelsen, *Scand. J. Soc. Med. Suppl.*, **16**, 34 (1980).
43 H. Rasmussen, J. Olsen, and L. Lawitsen, *J. Occup. Med.*, **27**, 561 (1985).
44 C. Van Vliet, G. Swaen, J. Slangern, T. Border, F. Stirman, *Int Arch. Occup. Environ. Health*, **59**, 493 (1987).
45 E. Kiesswetter, *Zbl. Hyg. Umweltmed.*, **202**, 191 (1998).
46 N. Fiedler and H. Kiepen, *Environ. Health Perspect.*, **105** Suppl. 2, 409 (1997).
47 S. Mikkelsen, M. Jorgensen, E. Browne, and C. Gyldensted, *Acta Neurol. Scand. Suppl.*, **118**, 1(1988).
48 R.L. Chen, and A. Seaton, *Cancer detection and prevention*, **22** (6), 533 (1998).
49 K.L. Milne, D.P. Sandler, R.B. Everson, S.M. Brown, *Am. J. Ind. Med.*, **4**, 565 (1983).
50 S. Hernberg, P. Westerholm, K. Schultz-Larsen, R. Degerth, E. Kuosma, A. Englund, U. Engzell, H. Sand Hansen, and P. Mutanen, *Scand. J. Work Environ. Health*, **9**, 315 (1983).
51 H.S.R. Malker, J.K. McLaughlin, B.K. Malker, B.J. Stone, J.A. Weiner, J.L.E. Erickson, and W.J. Blot, *J. Natl. Cancer Inst.*, **74**, 61 (1985).
52 L.M. Brown, T.J. Mason, L.W. Pickle, P.A. Stewart, BP. Auffler, K. Burau, R.G. Ziegler, and F. J. Fraumeni, *Cancer Res.*, **48**, 1960 (1988).
53 H. Olssen, and L. Brandt, *Scand. J. Work Environ. Health*, **14**, 246 (1980).
54 H. Olssen, L. Brandt, *Scand. J. Work Environ. Health*, **6**, 302 (1988).
55 J. Cuzik, and B. De Stavola, *Br. J. Cancer*, **57**, 516 (1988).
56 N.J. Vianna, and A. Polan, *Lancet*, 1394 (1979).
57 L. Davico, C. Sacerdote, G. Ciccone, L. Pegoraro, S. Kerim, G. Ponzio, and P. Vineis, *Cancer Epidemiol. Biomarkers Prev.*, **7** (12), 1123 (1998).
58 M.M. Crane, *Cancer Epidemiol. Biomark. Prev.*, **5**, 639 (1996).
59 R. Piratsu, I. Iavarone, and P. Comba, *Ann. Ist. Super Sanita*, **32** (1), 3 (1996).
60 J. Claude, E. Kunze, R. Frentzel-Beyme, K. Paczkowski, J. Schneider, H. Schubert, *Am. J. Epidemiol.*, **124**, 578 (1986).
61 J. Claude, R. Frentzel-Beyme, E. Kunze, *Int. J. Cancer*, **41**, 371-379 (1988).
62 O.M. Jensen, J. Wahrendorf, J.B. Knudsen, B.L. Sorensen, *Scand. J. Work Environ. Health*, **13**, 129 (1987).
63 R.C. Brownson, J.C. Chang, J.R. Davis, and J.R. Bagby, *J. Occup. Med.*, **30**, 523 (1988).
64 S. Norell, A. Ahlbom, R. Olin, R. Ewald, G. Jacobsen, I. Lindberg,-Navier, and K. Wiechel, *Br. J. Ind. Med.*, **43**, 775 (1986).
65 H.S.R. Malker, J.K. McLaughlin, B.K. Malker, B.J. Stone, J.A. Weiner, J.L.E. Erickson, and W.J. Blot, *Br. J. Ind. Med.*, **43**, 257 (1986).
66 M.L. McBride, *Can. J. Public Health*, **89** (1), 53 (1998).
67 J.S. Colt, and A. Blair, *Environ. Health Perspect.*, **106** (3), 909 (1998).
68 R. Kishi, Y. Katakura, J. Yuasa, and H. Miyake, *Sangyo Igaku*, **35** (6), 5115 (1993).
69 M.C. White and E.L. Baker, *Br. J. Ind. Med.*, **45**, 523 (1988).
70 J.R. Beach, J. Raven, C. Ingram, M. Bailey, D. Johns, E.H. Walters, and M. Abramson, *Eur. Respir. J.*, **10** (3), 563 (1997).
71 J.E. Peterson, *Occup. Med.*, **8**(3), 533 (1993).

72 U. Ulfvarson, R. Alexandersson, M. Dahlqvist, U. Ekholm, B. Bergstrom, and J. Scullman, *Scan. J. Work. Environ. Health*, **18** (6), 376 (1992).
73 H. Beving, B. Tedner, and L.E. Eriksson, *Int. Arch. Occup. Environ. Health*, **63** (6), 383 (1992).
74 G. Engholm and A. Englund in **Advances in Modern Environmental Toxicology, Vol. II, Occupational Health Hazards of Solvents**, A. Englund, K. Ringen, and M.A. Mehlman, eds., *Princeton Scientific Publishers*, Princeton, N.J., pp. 173-185 (1982).
75 H. Hellquist, K. Irander, C. Edling, and L.M. Ödkvist, *Acta Otolaryngol.*, **96**, 495 (1983).
76 M. Guiguet, E. Baumelou, and J.Y. Mary, *Int. J. Epidemiol.*, **24** (5), 993 (1995).
77 J. Angerer, and H. Wulf, *Int. Arch. Occup. Environ. Health*, **56**, 307 (1985).
78 M. Hane, O. Axelson, J. Blume, C. Hogstedt, L. Sundell, B. Ydreborg, *Scand. J. Work Environ. Health*, **3**, 91 (1977).
79 A. Stevenson, M. Yaqoob, H. Mason, P. Pai, and G.M. Bell, *QJM*, **88** (1), 23 (1995).
80 D.E. Fixler, N. Threlkeld, *Teratology*, **58** (1), 6 (1998).
81 J. Erramouspe, R. Galvez, and D.R. Fischler, *J. Psychoactive Drugs*, **28** (2), 201 (1996).
82 L.Z. Heidam, *J. Epidemiol. Commun. Health*, **38**, 149 (1984).
83 M.E. McDowall, Occupational Reproductive Epidemiology: The Use of Routinely Collected Statistics in England and Wales, 1980-82, Studies on Medical and Populations Subjects No. 50, Her Majesty's Stationary Office, Office of Population Censuses and Surveys, London, 1980.
84 W.E. Daniell, and T.L. Vaughan, *Br. J. Ind. Med.*, **45**, 193 (1988).
85 J.H. Olsen, *Dan. Med. Bull.*, **30**, 24 (1983).
86 U. Haglund, I. Lundberg, and L. Zech, *Scand. J. Work Environ. Health*, **6**, 291 (1980).
87 K.T. Kelsey, J.K. Wiencke, F.F. Little, E.L. Baker, J.B. Little, *Environ. Mol. Mutagenesis*, **11**, 389 (1988).
88 R. Lauwerys, A. Bernard, C. Viau, and J.P. Buchet, *Scand. J. Work Environ. Health*, **11** (1), 83 (1985).
89 G. Engholm, A. Englund, H. Löwing, *Scand. J. Work Environ. Health*,. **13**,. 181 (1987).
90 R.W. Morgan, K.W. Claxton, S.D. Kaplan, J.M. Parsons, and O. Wong, *J. Occup. Med.*, **27**, 377 (1985).
91 I. Lundberg, *Scand. J. Work Environ. Health.*, **12**, 108 (1986).
92 European Collaborative Actions (ECA), Indoor Air Quality & Ist Impact On Man. Evaluation of VOC Emissions from Building Products - Solid Flooring Materials, ECA Report No. 18, 1997.
93 Anonymous, Building products - Determination of volatile organic compounds Part 1: Emission test chamber method. prENV 13419-1, draft.
94 G.K. Lemasters, J.E.Lockey, D.M. Olsen, S.G. Selevan, M.W. Tabor, and G.K. Livingston, *Drug and Chemical Toxicology*, **22** (1), 181 (1999).
95 A. Kramer, M. Linnert, R. Wrbitzky, and J. Angerer, *Int. Arch. Occup. Environ. Health*, **72**(1), 52 (1999).
96 J.P. Collinot, J.C. Collinot, F. Deschamps, D. Decolin, G. Siest, and M.M. Galteau, *J. Toxicol. Environ. Health*, **48** (4), 349 (1996).


## 18.4 SOLVENT USES WITH EXPOSURE RISKS


Pentti Kalliokoski

**University of Kuopio, Kuopio, Finland**

Kai Savolainen

**Finnish Institute of Occupational Health, Helsinki, Finland**


### 18.4.1 INTRODUCTION

This chapter deals merely with exposure to common organic solvents which are used in large quantities to dissolve fats, resins, and other materials. Very dangerous chemicals, such as benzene, which are no more used as solvents due to their toxic properties, will not be discussed even though they may have had even extensive uses as solvents earlier and even though those may be still important chemicals as petroleum components or as intermediates for other chemicals. Also, solvents with very specialized uses, such as carbon disulfide the use of which is practically limited to viscose rayon industry and laboratories, are only shortly considered. Exposure data presented originate mainly from literature; however, some unpublished data obtained from the Finnish Institute of Occupational Health (FIOH) has been added to indicate the order of magnitude of current exposure levels in industrialized countries (data cover the years 1994-1996).

Exposure to organic solvents should be avoided mainly due to the risk of neurotoxic effects. Acute effects are narcotic resembling those caused by the use of alcohol. Those appear, for example, as a decrease in reaction time and impairment in psychological performance. Chronic neurotoxic effects are often called as the organic psychosyndrome (OPS) including memory disturbances, excessive tiredness, personality changes, irritability and affect lability. Intellectual reduction may occur but it seems that development of real dementia would require simultaneously heavy consumption of alcohol.[1] Increased risk of sleep apnoea has been observed among men exposed occupationally to organic solvents.[2] Exposure to toluene, xylene, and styrene has been found to contribute to the development of noise-induced hearing loss. This interaction has been suggested to be due to neurotoxic injuries caused by solvents in the cochlea.[3]

Long-term exposure to carbon disulfide and n-hexane may result in peripheral neuropathy. It should be noted that n-heptane and n-octane have not been shown to cause the type of peripheral neuropathy (numbness, weakness, and pain in extremities) associated with n-hexane.[4] Otherwise than the other alkanes, n-hexane is metabolized to a reactive, toxic compound (2,5-hexanedione).[5]

Organic solvents are generally skin irritants. Repeated or prolonged contact may cause erythema and dryness of the skin. Defatting may lead to cracked skin. Many solvents also readily absorb through the skin (see 18.4.2). Reactive solvents, such as styrene and vinyltoluene, may cause contact dermatitis.[6] The annual incidence of occupational diseases caused by organic solvents was 20.6 cases/10,000 exposed workers in Finland in the 1980's. More than half of the cases (64%) were dermatoses.[7]

Benzene is a well-known human carcinogen. An association with exposure to benzene and leukemia was detected already in 1920's.[8] There is some evidence that exposure to other organic solvents may constitute a carcinogenic risk. The International Agency for Research on Cancer (IARC) has concluded that there is sufficient evidence for the carcinogenicity of painters' occupational exposure.[9] Epidemiological studies conducted among dry cleaning and metal degreasing workers suggest an increased cancer risk due to exposure to chlorinated solvents.[10] The carcinogenicity of aromatic solvents (styrene, toluene, and xylene) has been investigated in many large epidemiological studies. Although the results have been inconclusive, certain site-specific associations (e.g., for lymphohematopoietic tissues) have appeared in some of the studies.[11] Only styrene has shown to have some genotoxic activity in animal studies.[9]

## 18.4.2 EXPOSURE ASSESSMENT

Inhalation is usually clearly the most important route for occupational exposure to solvents. Organic solvents also enter the body through the skin.[12] For some solvents, especially for alkoxyalcohols (glycol ethers) and their acetates, this is even the main route of absorption. Even the skin uptake of vapor can be significant for these compounds.[13] Dermal absorption of vapors is, however, usually negligible and contact with liquid is required. Increased workload and heat enhance both inhalation and dermal absorption.[14] Skin contamination may also lead to oral uptake due to eating and smoking but this is of minor importance for volatile solvents.

Most industrial countries have occupational exposure limits (OELs) for airborne concentrations to prevent excessive exposures. The threshold limit values (TLVs) published by the American Conference of Governmental Industrial Hygienists (ACGIH[15]) are unofficial but have had a great impact on OELs in the Western countries. These are updated annually based on new epidemiological and toxicological data. Organic solvents are often used as mixtures. Because the OELs for most solvents base on their neurotoxic properties, their effect is considered to be additive and the combined exposure levels (or hygienic effects) are calculated as the sum of the fraction of the OEL that each solvent represents. If the sum is larger than one, a noncompliance situation exists. Compounds which may be dangerous by uptake through the skin carry a skin notation. However, no quantitative dermal exposure limits are available yet. Methanol, turpentine, glycol ethers, and many chlorinated and aromatic hydrocarbons have the skin notation.

Exposures can also be assessed by analyzing biological specimens, such as blood or urine. Metabolites of compounds are usually analyzed from urine samples. The amount of biological exposure limits or indices (BEIs) is, however, much smaller than that for airborne concentrations. However, biological monitoring may be useful although no BEI has been established to ascertain effectiveness of personal protection or to follow exposure trend. Biological monitoring reflects exposure via all routes. This is beneficial for individual exposure assessment. Biological monitoring also reveals possible accumulation of a compound. This is, however, of minor importance for most solvents because of their short half-lives. The largest disadvantage of biological monitoring is that it does not provide any information on the reasons of exposure. About twenty solvents have an ACGIH BEI (e.g., acetone, carbon disulfide, 2-ethoxyethanol, n-hexane, methanol, methyl ethyl ketone, perchloroethylene, styrene, toluene, trichloroethylene, and xylene).[15]

Airborne concentrations of solvents are most commonly determined by taking samples on adsorbents, such as activated carbon. Sampling can be done with a pump or passively based on diffusion. In industrial working places where airborne concentrations of solvents are on ppm-level, samples are generally extracted with a solvent (carbon disulfide is the most common one). In offices and other nonindustrial environments where concentrations are at ppb-level, samples are taken onto Tenax adsorption tubes which are desorbed thermally. In both cases, gas chromatography is the most common analytical method. Because the sampling time generally varies, it is important to calculate the time-weighted avcragc (TWA) concentration. TWA concentration is obtained by using the sampling times as weights. Sampling and analysis has been reviewed e.g. by Soule.[16] Direct-reading instruments, such as infrared or photoionic analyzers are good if only one solvent is present but interferences may be a problem when solvent mixtures are analyzed.

Monitoring methods for dermal exposure have not been standardized as well as the methods to assess inhalation exposure. Adsorptive pads of activated carbon cloth can be used for monitoring of dermal exposure to organic solvents. Patches are attached on various parts of the body under the clothing and the amounts analyzed are then multiplied with the areas of each body region.[17] Exposure assessment is complicated by the contribution of solvent vapor on the samples. Because the adsorptive surface of activated carbon is much larger than that of the skin, pads will adsorb much more solvents than the skin. However, only exposure to liquid phase is usually significant; therefore, the adsorbed vapor should be subtracted from the total mass analyzed.[18] However, this makes the exposure assessment quite complicated, and if biological monitoring methods are available those usually provide a more practical alternative.

Urine and blood are the most commonly used biological samples. Exhaled air samples can also be used. Sampling is usually carried out at the end of the shift. Because the amount of water consumed affects the concentration of the solvent or its metabolite in urine, creatinine (a normal constituent of urine) correction is often applied (mass analyzed is presented per gram of creatinine). Alternatively, the correction can be made for the relative density of urine. A density of urine of 1.024 is usually applied for this purpose. In principle, collection of 24-hour urine would be an ideal approach for biological monitoring because then the actually excreted amount of the metabolite would be measured. Because this is very difficult to carry out creatinine or density correction is generally chosen as a more practical alternative. Unfortunately, not only several ways of correction for the density of urine are being used but also several different units have been adopted. For example, the ACGIH BEIs are usually given as mg/g creatinine but the values of the FIOH and many other European organizations are presented as μmol/l of urine. In addition, other units, such as μmol/mol of creatinine, are used. The mass units given in mg can be converted to mmol by dividing by the molecular weight of the compound (molecular weight of creatinine is 113). On the other hand, because the concentration of creatinine in urine varies and it is determined separately for each individual sample, values given with and without creatinine correction (or density correction) cannot be compared directly. That is the reason why the BEIs given by different institutions or agencies have not been converted to the same units in the text. As a rule of thumb, the concentration per gram of creatinine can be obtained by multiplying the concentration per liter of urine with a factor of 0.5 - 1.0.

Airborne solvent concentrations usually vary much with time. Although repeated random personal sampling is theoretically the optimal method for inhalation exposure assessment, it is very time-consuming and does not necessarily reveal the reasons for exposure. Often, occupational hygiene surveys are conducted only to make certain that the concentrations of air impurities are in compliance with the OELs. Also, the European standard (EN 689/95) for occupational exposure assessment is primarily issued for this purpose. It would, however, be practical if the measurements conducted would also provide useful information for planning of remedial measures if the measurements reveal those to be necessary. For this purpose, it is important to recognize the difference between manual tasks and process industries.

In manual tasks, emissions are released very close to the worker. Most problematic solvent exposures occur while performing manual tasks, such as painting, gluing, degreasing, and cleaning. Batch processes in paint and printing ink manufacture also contain many manual tasks. Manual lay-up methods are common in the reinforced plastics industry. The workers often perform different tasks with different exposure levels. All major tasks should be investigated under various conditions. Smoke tube tests provide useful information on the spreading of solvent vapor and the efficiency of local exhausts. Good enclosure for the emission source is important for successful exposure control. Detailed instructions for ventilation arrangements in various industries are available (references will be given later).

Rotogravure printing is an example of a process industry with solvent exposures. The process is, in principle, closed but emissions take place from the openings for the paper web. The workers do not need to stay in the immediate vicinity of the emission sources but can often spend most of their time in the clean control room. Thus, the workers' exposure levels depend mainly on their moving pattern during the work shift. In addition, the emission rate depends on the production rate, i.e. on the consumption rate of the solvent (toluene). Stationary sampling can be applied for exposure assessment when it is combined with questionnaire on use of time in various areas. This can be done with a direct-reading instrument because the airborne concentration level remains quite stable.[20]

The possible skin exposure should also be taken into consideration. If skin contamination seems possible adequate protective clothing should be used. Glove selection is an important but difficult issue. Glove materials often tolerate organic solvents poorly. Glove manufacturers have useful information for selecting gloves for individual solvents. Especially difficult is, however, to find gloves protecting efficiently penetration of solvent mixtures. Nitrile (butadiene-acrylonitrile copolymer) gloves are often chosen in such a situation. There is, however, considerable intermanufacturer and even batch lot variability in penetration of solvents through nitrile gloves.[19] The workers should also be instructed not to use thinners for hand and skin cleaning.

Total quality management (TQM) is an effective way to ensure also a high quality of the working environment. The guidance for right and safe working practices should be subjoined to all working instructions. Employee participation is an essential feature of a well working TQM, and also greatly assists the achievement of the hygienic goals set.

### 18.4.3 PRODUCTION OF PAINTS AND PRINTING INKS

Painting technology has changed over the years. The exposure levels were generally highest between the mid-1950's and mid-1960's when solvent-based paints were used extensively both as construction and industrial paints, and exposure control technology was still undeveloped.[21] Rotogravure and silkscreen printing inks contain organic solvents. Until 1950's benzene was used as the solvent in rotogravure inks and it remained as an impurity in toluene until 1960's.[22] Today, toluene used in rotogravure inks does not contain benzene. Alkoxyalcohols and their acetates are used as silkscreen ink solvents. The development of safer solvents has been started quite recently. First, alkoxyethanols were replaced by their acetates which have lower vapor pressures. This did not, however, improve the safety much because the skin is the main route of absorption for both alkoxyethanols and alkoxyethylacetates, and they are considered to be equally toxic. Their substitution by alkoxypropanols and their acetates is, however, a significant improvement because those do not metabolize to toxic alkoxyacetic acids.[23]

High exposures remained common in paint and printing ink industry still in 1970's; for example, most solvent measurements conducted in these industries in Finland exceeded the present OELs.[24] Water-based paints are today clearly most common in construction painting. Alkyd-based construction paints with white spirit (Stoddard solvent, mineral spirit or solvent naphtha) as a solvent are, however, still produced. Solvent-based paints have remained most common in industrial painting, even though the solventless powder paints are also produced in large quantities.

The main products are nowadays manufactured in automated processes. On the other hand, special products are also usually made in batches and include several manual tasks. In addition to paints and inks, thinners are often canned. When the processes are provided with proper enclosures and local exhausts, the airborne solvent concentrations can be kept well below the OELs. An easy but important control measure is to keep all solvent containing pots covered. Xylene (TLV 100 ppm, the Finnish OEL 100 ppm with skin notation), toluene

(TLV and the Finnish OEL 50 ppm with skin notation), butanol (TLV 50 ppm as ceiling value and with skin notation, the Finnish OEL 50 ppm with skin notation), and white spirit (TLV 100 ppm, the Finnish OEL 770 mg/m$^3$ or 135 ppm) are usually the main concerns in paint manufacturing plants.[24,25] In the 1980's, the combined solvent exposure levels were generally below the OEL but high exposure levels were detected in pot cleaning.[26] The situation has further improved to some extent in Finland in 1990's. The measurements conducted in a large Finnish paint factory in 1996-7 yielded the following mean combined personal solvent exposure levels: batch production 0.34 (range 0.14 -0.72), automated production lines 0.17, filling 0.41 (range 0.14-1.0), and pot cleaning 1.51 (range 0.28-3.06). Thus, pot cleaning remains as a problematic task. Exposures to xylene and 1-methoxy-2-propanol caused the largest contributions in the hygienic effects. Xylene was the solvent with the highest airborne concentration (mean 26 ppm; range 0.1-359 ppm) compared to the OELs in the Finnish paint manufacturing plants in general in the mid 1990's. It has been reported that no profound changes in exposures have occurred in the Swedish paint industry since the mid-1980's.[21]

In the Finnish printing ink plants, the combined solvent levels were still in the 1980's often out of compliance when compared to the OELs.[26] Significant airborne concentrations were observed for toluene, ethyl acetate (TLV 400 ppm, the Finnish OEL 300 ppm), aromatic solvent naphtha ( the Finnish OEL 240 mg/m$^3$), and acetone (TLV and the Finnish OEL 500 ppm). The cleaning of vessels of barrels was again an especially problematic task. If cleaning is done manually, it is difficult to control the exposure sufficiently well with local ventilation but respiratory protection is also needed.

### 18.4.4 PAINTING

Painters are probably the largest worker group exposed to solvents. This may also be the reason why much data on occupational health risks due to solvent exposure originate from painting work although the exposure levels have been generally lower than e.g. in paint manufacture. On the other hand, chronic neurotoxic effects and especially cancer require long exposure time (more than 10 years) and thus the patients with solvent-related disorders have been exposed long ago when the exposure levels, especially in construction painting, were considerably higher than nowadays. In addition, a solvent induced mild toxic encephalopathy is often a progressive disease after the cessation of exposure. There is, however, a great individual variability both in susceptibility and prognosis of the disease.[27] Several studies have indicated an interaction between solvent exposure and high alcohol consumption. It has even been suggested that solvent exposure and use of alcohol are acting synergistically.[28]

The changes in paint technology have had a great impact on exposure to solvents in construction painting where the influence has been much more significant than in paint production because of poor ventilation. Period of high solvent exposure lasted from mid-1950's to mid-1970's.[28] At that time, even acute intoxications occurred. Painting and lacquering using epoxyester formulations often caused very high exposure levels.[24] The average exposure level to white spirit was estimated to be 130 ppm for painters when alkyd paints were extensively used. During actual painting situations, the exposure levels could rise to 300 ppm. Levels of about 200 ppm are, however, more typical in conditions of poor ventilation. If doors and windows can be kept open, the airborne white spirit concentration is significantly reduced, about to 40 ppm. In early 1980's, the average exposure levels to white spirit were reduced to about 40 ppm in Finland.[29] Nowadays, water-based paints are also commonly used in repair and maintenance painting and the average exposure levels have become generally low.[28] Occasionally, high peak exposures continue to take place when solvent-based paints are used in poorly ventilated spaces.

When small items are painted industrially, the worker can stay outside the spray booth and exposure to solvents can be effectively controlled. Measurements conducted in the 1970's already indicated that no serious problems existed in that kind of work.[29] Video terminal painting is a modern example of such a situation.[30] On the other hand, when large metal products, such as cars and trailers, are painted, the worker must enter the booth and control is much more difficult. Solvents are not, however, the biggest hygienic problem in car painting but the paint mist (overspray). The measurements carried out in Finland in the 1980's indicated that the combined solvent concentration was usually less than 50% of the OEL.[26] On the other hand, the paint mist concentrations exceeded clearly the standards.[29] If painting is done in a poorly ventilated booth, high solvent concentrations, however, appear.[30] Isocyanate-based urethane paints are nowadays common in car painting. They contain 1,6-hexamethylene diisocyanate (HDI) based polyisocyanates, which have become a major cause of occupational diseases, especially asthma.[31,32] Downdraft spray-painting booths provide the best overspray control. Air velocity, flow direction, and flow homogeneity are the ventilation parameters having the largest effect on booth performance. Spray-painting booths do not, however, completely control exposure to paint mist and isocyanates.[33,34]

In addition to actual painting, the workers may become exposed to solvents during other tasks, such as solvent cleaning. For example, isopropanol (TLV 400 ppm, the Finnish OEL 200 ppm) is used for wiping cars before the application of the primer. If this is performed in a well-ventilated room, the concentrations remain below the OEL. As high concentrations as 130 ppm were, however, measured in a room provided with a ventilation of 50 air changes per hour.[35]

Solvent containing formaldehyde resin paints and lacquers have been used extensively in Nordic furniture and wood product industry. In the early 1980's, combined solvent concentration and especially formaldehyde levels often exceeded the OELs. The OEL violations became rare in the late 1980's.[36] The recent concentrations of other solvents than ethanol (even its mean concentration was only 17 ppm) have been below 10 ppm in Finland. Nowadays, solventless acrylics are mainly used for industrial wood coatings. This substitution has, however, created a new occupational health problem. The new products have caused many cases of dermal sensitization among exposed workers.

### 18.4.5 PRINTING

The exposure to toluene has often been extensive in the past. As high exposure estimates as 450 ppm have been given for Swedish rotogravure printing workers in the 1950's.[37] A linear correlation has been observed between airborne toluene concentration in the pressroom air and the consumption rate of toluene.[20] Large day-to-day variation in the airborne concentration is, therefore, common. The long-term mean concentrations were 63-186 ppm in two Nordic studies.[24,38] At least in Finland, the exposure to toluene remained rather stable between 1960-1980 because the effect of ventilation improvements was outweighed by increased production. In the early 1980's, the Finnish rotogravure printing plants were modernized and effective ventilation systems were installed. The new presses had better enclosures and the contaminated zones around the presses were separated from the other parts of the pressroom.[39] The mean exposure level of toluene can be less than 20 ppm in a modern rotogravure printing plant.[37]

As stated earlier, toxic alkoxyethanols and their acetates have been used commonly in silkscreen printing inks. Those have no warning odor and are, therefore, easily considered to be safe. Airborne concentrations of 2-ethoxyethanol and 2-ethoxyethyl acetate from 3 to 14 ppm (both have a TLV of 5 ppm, the Finnish OEL 2 ppm) have been detected during printing.[40,41] Because the inhalation exposure consists only a part of the total exposure and absorption through the skin is of equal or greater importance, biological monitoring is the preferred exposure assessment method. The concentration of the toxic metabolite, 2-ethoxyacetic acid, in urine is the most commonly used method. Biomonitoring is also important because, otherwise than most other solvents, 2-ethoxyacetic acid has a long half-life and, therefore, it accumulates in the body during the workweek. Its ACGIH BEI is 100 mg/g creatinine (109 mmol/mol creatinine). A lower value of 30 mmol/mol creatinine has been proposed in Finland.[42] The average urinary 2-ethoxyacetic acid concentrations were found to increase from 40 to near 90 mmol/mol of creatinine during a workweek.[42]

Other alkoxyethanols, especially 2-methoxyethanol and 2-butoxyethanol and their acetates have also been used in silkscreen printing inks. The TLV for 2-methoxyethanol and its acetate is 5 ppm (their Finnish OEL is 0.5 ppm) and 25 ppm for 2-butoxyethanol (the Finnish OEL is 20 ppm for 2-butoxyethanol and its acetate but a value of 5 ppm has been proposed.[42] ACGIH has not set BEIs for these compounds. Limits of 3 mmol methoxyacetic acid/mol creatinine and 60 mmol butoxyacetic acid/mol creatinine has been recommended in Finland for the concentrations of the urinary metabolites.[41] Mean inhalation exposure levels to 2-methoxyethyl acetate from 0.2 to 1.9 ppm have ben reported.[40,42] The urinary methoxyacetic acid concentrations have ranged from 0.1 to 10.7 mmol/mol creatinine.[42] Airborne mean 2-butoxyethanol and 2-butoxyethylacetate concentrations from 0.1 to 3.2 ppm have been observed.[40,42]

The use of safer alkoxypropanols and their acetates is rather new in silkscreen printing inks. Among them, only 1-methoxy-2-propanol has a TLV (100 ppm). Its German MAK value is 50 ppm (also given for its acetate). A German MAK value of 20 ppm have been set for 2-methoxy-1-propanol and its acetate (this is also the Norwegian OEL). These have no BEIs or official biological action limits in any country. A value of 3 mmol 2-methoxypropionic acid/mol creatinine has been found to correspond to inhalation exposure to 1-methoxy-2-propyl acetate at the German MAK level.[42] The corresponding value for 1-ethoxypropionic acid is 40 mmol/mol creatinine.[42] Even though alkoxypropanols are less toxic than alkoxyethanols and their dermal uptakes are lower, they have the disadvantage of having higher vapor pressures. However, if enclosed and automatic machines have been used airborne levels of alkoxypropanols have remained well below the German MAK levels.[42] Screen printing inks also contain other solvents, e.g., toluene, xylene, and cyclohexanone (TLV 25 ppm with skin notation, the Finnish OEL 25 ppm as a ceiling value and with skin notation).

The printers are also exposed to solvents while cleaning the press. The rollers are cleaned regularly. Occasionally, ink stains are also removed from other parts of the presses and floor (see 18.4.6).

### 18.4.6 DEGREASING, PRESS CLEANING AND PAINT REMOVAL

Chlorinated solvents have traditionally been used as metal degreasing agents. Trichloroethylene (TLV 50 ppm; the Finnish OEL 30 ppm) and 1,1,1-trichloroethane (methyl chloroform; TLV 350 ppm; the Finnish OEL 100 ppm) have been the most common solvents for this purpose. The use of 1,1,1-trichloroethane has, however, recently been restricted in many countries due to environmental reasons. Urinary trichloroacetic acid and trichloroethanol concentrations can be used for biological monitoring for both solvents. The ACGIH BEI for trichloroacetic acid is 100 mg/g creatinine for trichloroethylene exposure and 10 mg/l for 1,1,1-trichloroethane.[15] The BEI given by the Finnish Institute of Occupational Health (FIOH) is 360 μmol/l for trichloroethylene. In Finland, the concentration of 1,1,1-trichloroethane in blood (FIOH BEI 2 μmol/l) is recommended for biological monitoring of 1,1,1-trichloroethane because less than 10% of this solvent is metabolized. The ACGIH BEI for urinary trichloroethanol is 30 mg/l for 1,1,1-trichloroethane. For trichloroethylene, the sum of urinary trichloroethanol and trichloroethanol is used (ACGIH BEI 300 mg/g creatinine; the FIOH BEI 1000 μmol/l). Chlorofluorocarbons (CFCs) are also used but their use is decreasing due to their contribution to the ozone depletion in the stratosphere.[43] 1,1,2-trichloro-1,2,2-trifluoroethane (Freon 113; TLV and the Finnish OEL 1000 ppm) is used especially for ultrasonic cleaning of small metal parts. Alternative degreasing methods using either alkaline aqueous solutions or citrus oil (D-limonene) have been developed.[43] Exposure to trichloroethylene and 1,1,1-trichloroethane should be avoided especially due to their neurotoxic properties.[44,45] Some evidence is on the carcinogenicity of trichloroethylene.[9] 1,1,1-Trichloroethane was earlier stabilized with carcinogenic 1,4-dioxane. Nowadays, it has usually been replaced with other stabilizers.[46] 1,1,1-Trichloroethane has the skin notation in the Finnish OEL list.

Metal degreasing is usually performed as vapor phase operation, where solvent is heated and the vapor condenses on the metal part. An extensive occupational hygiene survey on the US vapor degreasing operations has been performed in the 1970's. Typical sample concentrations were 100-400 ppm.[50] Much lower airborne trichloroethylene concentrations were observed in Finland in the 1980's. The mean 8 h concentration was 7 ppm (range <1-30 ppm).[46] The biological monitoring, however, revealed higher exposure levels. The mean concentration of trichloroacetic acid in urine was 153 μmol/l. The highest level was 860 μmol/l.[46] Airborne 1,1,1-trichloroethane levels from 34 to 85 ppm has been detected in Finland during metal degreasing in the 1980's.[46] The mean airborne concentration of trichloroethylene was 22 ppm (range 9-45 ppm) in Finland in the mid 1990's. The level of 1,1,1-trichloroethane has varied from less than 0.1 to 1.8 μmol/l (mean 0.3 μmol/l) in blood.[46] The concentrations of 1,1,2-trichloro-1,2,2-trifluoroethane have been low in Finland. The mean airborne concentration was 25 ppm (range 5-85 ppm) in the 1980's.[46]

Methylene chloride (dichloromethane) is most commonly used for paint removal. The stripping solution contains 55-85% methylene chloride.[47] Ethanol (TLV and the Finnish OEL 1000 ppm), methanol (TLV and the Finnish OEL 200 ppm with skin notation), toluene and formic acid are other common constituents.[46,47] US National Institute of Occupational Safety and Health has recommended that methylene chloride be regarded as a potential carcinogen and that it should be controlled to the lowest feasible limit.[48] Its TLV is 50 ppm (the Finnish OEL 100 ppm).

Very high airborne methylene chloride concentrations have been detected in the USA during furniture stripping. The TWA concentration was above 50 ppm in all five plants investigated. The highest TWA level was as high as 854 ppm. In individual samples, concentrations up to 2160 ppm were detected.[51] In Finland, the 8 h TWA concentration has varied from 26 to 380 ppm. The highest sample concentration was 850 ppm.[46] Local exhaust system based on the ACGIH manual criteria for open surface tanks[51] has been found to be effective for stripping tanks.[47] When different versions of such a hood was installed in an existing stripping tank, considerable improvement in airborne concentrations was achieved. Methylene concentrations were reduced from 600-1150 ppm to 28-34 ppm. By the use of side baffles and larger room area which would allow more even supply of make-up air even lower airborne levels would be possible to achieve.[47]

Toluene is used to clean the rotogravure presses. High exposure levels, up to 700 ppm, may occur during this task.[49] Offset presses are usually cleaned with chlorinated solvents. 1,1,1-Trichloroethane and Freons were earlier commonly used as cleaning solvents.[46] The mean concentration of 1,1,1-trichloroethane has been 23 ppm. The highest concentration observed in Finland has been 117 ppm.[46] Solvent cleaning is a task where exposure is difficult to control with ventilation arrangements but the use of a respirator is needed. Because it usually lasts only for a short time (typically about 1 hour/week) its use is neither too heavy physically.

## 18.4.7 DRY CLEANING

Carbon tetrachloride was the first solvent used in dry cleaning. It was introduced in the 1920's and used until the 1950's. The use of trichloroethylene started in the 1930's and it is still used to a limited extent. Perchloroethylene (tetrachloroethylene) was introduced in the 1950's and is today by far the most common dry cleaning solvent.[10] Perchloroethylene is neurotoxic and NIOSH considers it as potential human carcinogen.[53] Its TLV is 25 ppm and the Finnish OEL 50 ppm. The concentration of perchloroethylene in blood is used for biological monitoring. The ACGIH BEI is 0.5 mg/l and FIOH BEI 6 mol/l (1 mg/l). Petroleum solvents are a safer alternative for dry cleaning.[53] Emulsion cleaning, where an alkaline mixture of surfactants is used together with a small amount of nonchlorinated organic solvent in water, can be used for textiles that tolerate water.[43]

Dry cleaning machines have developed safer over the time. Five generations of machines have appeared. The first generation machines, where solvent-laden clothing was manually transferred to a separate dryer, were used until the late 1960's. This transfer was eliminated in the machines of the second generation. Residual solvents were still vented to the atmosphere. Condensers were introduced to the non-vented, third generation machines in the late 1970's. The fourth generation machines have in addition to the refrigerated condenser an activated carbon absorber to control solvent vapor at the cylinder outlet at the end of the drying cycle. The fifth generation machines are also provided with a monitor and an interlocking system to ensure that the concentration of perchloroethylene is below 300 ppm when the door is opened.[54] The TWA exposure level was 40-60 ppm when the first generation machines were used. With the second and third generation machines the TWA exposure level was 15-20 ppm. All these machines, however, created short-term peak concentrations up to 1000-4000 ppm. The fourth generation machines reduced the TWA exposure level below 3 ppm, and when the fifth generation machines are used TWA concentration remains below 2 ppm. With the fourth and fifth generation machines, the peak concentrations are 10-300 ppm.[54] Local ventilation hoods can be used above the door as an additional control measure.[55] In addition to the cleaners, pressers, sewers, counters, and maintenance workers may be exposed to some extent.[56]

The mean 8 h TWA concentration of perchloroethylene was 13 ppm (range 3-29 ppm) in Finland. The mean concentration of perchloroethylene in the blood of Finnish dry cleaning workers was 1.3 μmol/l (range <0.1 - 14.0 μmol/l).[46] The average value for maintenance men was 0.3 μmol/l. Their highest value was 1.3 μmol/l.[46]

### 18.4.8 REINFORCED PLASTICS INDUSTRY

Exposure to styrene is the main occupational hygiene problem in reinforced plastics industry, where it is used as a crosslinking agent and solvent in unsaturated polyester resins. In addition, workers are exposed to acetone which is used as a clean-up solvent. Other solvents, such as methylene chloride, toluene, xylene, heptane (TLV 400 ppm, the Finnish OEL 300 ppm), methylcyclohexane (TLV and the Finnish OEL 400 ppm), and butyl acetate (TLV and the Finnish OEL 150 ppm) may also be used.[57] Styrene is neurotoxic.[58] Styrene is also a suspected carcinogen because it is metabolized via styrene-7,8-oxide.[59] The TLV and the Finnish OEL of styrene is 20 ppm. Urinary mandelic acid concentration is the most common biological monitoring method for styrene. The ACGIH BEI is 800 mg/g creatinine and the FIOH BEI 3.2 mmol/l.

Open mold methods (hand lay-up and spray-up methods) are most commonly used. In addition, the products often have large surface areas and large styrene vapor emissions take place. The emission depends on the type of resin used. While the mean evaporation loss of styrene was 11.6% for conventional resins, it was only 4.4% for the low styrene emission (LSE) resins containing volatilization inhibitors.[60]

The reinforced plastics industry began in 1950's. The airborne concentrations of styrene were typically 100-200 ppm until the 1980's.[61] Styrene concentrations of about 50 ppm were found possible to achieve during the production of small and medium-sized products in the 1980's by using a mobile fan or a set of successive air jets to blow fresh air from behind the worker towards the mold. The prevention of eddy formation over the mold was found to be of crucial importance.[62] When a large occupational hygiene survey was conducted in Finland in the early 1990's, it appeared that styrene concentrations had remained high. As many as 77% of the 8-h TWA concentrations exceeded the Finnish OEL of 20 ppm. The mean TWA concentration was 43 ppm and the highest TWA value 123 ppm.[63] The average concentrations of styrene oxide were low, on an average one pre mil of the styrene concentration (mean concentration was 0.1 ppm).[63] The mean concentration of acetone was 78 ppm (range 3-565 ppm).[63] Airborne levels of styrene were still high in the Finnish boat manufacturing plants in the mid 1990's (mean concentration 51 ppm, range 15-89 ppm). The mean acetone concentration was 25 ppm (range 9-53 ppm). Lower concentrations of about 10 ppm were reported from Sweden.[64] However, no detailed process and ventilation description was given.

The mean urinary mandelic acid concentration was about 5 mmol/l in the late 1970's in Finland. The level was reduced to about 4 mmol/l in mid-1980's and to 35 mmol/l in late 1980's. However, the maximum levels remained above 30 mmol/l throughout the whole period.[65]

Recent studies have indicated that sophisticated ventilation arrangements are needed to reduce the concentration of styrene below 20 ppm. In hand lay-up work, this is possible by the use of a mobile supply ventilation and exhaust ventilation.[66] In the spray-up work, concentrations below 20 ppm can be achieved by using a booth provided with mobile curtains to reduce the open face area and by maintaining a control velocity of 0.35 m/s.[66]

### 18.4.9 GLUING

Shoe manufacturing is an industry where toxic solvents have been used. Benzene containing glues were used until mid-1960's, at least in Italy.[67] A significant part of the information on the toxic and carcinogenic effects of benzene on the hematopoietic system originates from this industry.[68] After benzene, n-hexane became the most important solvent.[67] Epidemiological evidence on neuropathy caused by exposure to n-hexane comes again largely from shoe manufacturing.[5] The TLV and the Finnish OEL of hexane is 50 ppm. The concentration of the toxic metabolite, 2,5-hexanedione is used for biological monitoring of exposure to n-hexane. The BEI given by the FIOH is 12 μmol/l. The ACGIH BEI is 5 mg/g creatinine. Shoe glues may also contain toluene, methyl ethyl ketone, and acetone.[69]

The mean airborne concentration of n-hexane was 38 ppm (range 4-90 ppm) in the Finnish shoe industry in the 1980's.[26] The mean concentrations of acetone and toluene were 131 ppm and 14 ppm.[26] The mean exposure levels in two US shoe manufacturing plants in the late 1970's were: toluene 22-50 ppm, methyl ethyl ketone 133-153 ppm, acetone 46-223 ppm, and n-hexane 22-55 ppm.[69] The exposure is, however, easy to control by doing the gluing in enclosure hoods.

Contact glues contain usually white spirit with low boiling point range (60-90°C). Its Finnish OEL is 350 mg/m$^3$. The solvent contains n-hexane, earlier usually 5% or more. In the 1990's, white spirits containing only ca. 2% n-hexane have become available. Contact glues are used to fasten seals and strips e.g., in construction and transportation vehicle industries. Levels have usually been below the OEL in Finland. If contact glues are used for floor coverings, very high exposures may occur. These glues have, however, largely been replaced by ethanol- and water-based glues.[24]

### 18.4.10 OTHER

In the production of viscose, the workers are exposed to carbon disulfide. In addition to inhalation exposure, absorption through the skin may be important.[70] It is strongly neurotoxic.[71] Its TLV is 10 ppm and the Finnish OEL 5 ppm. Both carry the skin notation. Urinary 2-thiothiazolidine-4-carboxylic acid concentration can be used for biological monitoring of exposure to carbon disulfide. Its ACGIH BEI is 5 mg/g creatinine (4 mmol/mol creatinine) and the BEI given by the FIOH 2.0 mmol/mol creatinine.

Airborne concentrations from less than 0.2 to 65.7 ppm were detected in measurements conducted in a German plant in the 1990's. The urinary 2-thiothiazolidine-4-carboxylic acid levels ranged from less than 0.16 to 11.6 mg/g creatinine.[70] A long time average of about 8 ppm has been reported in a Dutch factory.[72]

In acetate fiber plants, the workers are exposed to acetone. The mean concentration in the air of three Japanese factories was 372 ppm. Some workers' exposure levels exceeded 1000 ppm.[73]

Exposure to various solvents takes place in pharmaceutical industry. The airborne 8-h TWA concentrations of methylene chloride were 16-167 ppm in a Finnish factory in the 1980's.[46]

Even though the active ingredients cause the main health risk while pesticides are used, but the exposure to solvents may also be important. As many as 71 different solvents were used in the 8000 pesticide formulations sold in Italy in the 1990's. Among the solvents, carcinogens such as benzene, carbon tetrachloride, and chloroform were found.[74]

Cyclohexanone is used as a coating solvent in audio and video tape production. Airborne TWA concentrations from 2 to 20 ppm were detected in factory in Singapore.[75] Urinary cyclohexanol level of 54.5 mg/l (23.3 mg/g creatinine) was found to correspond inhalation exposure to 25 ppm of cyclohexanone.[75]

### 18.4.11 SUMMARY

Even though organic solvents have been increasingly replaced by water-based and solid formulations, they are still widely used. About 3% of the Finnish working population has estimated to have significant solvent exposure.[26] It should also be noted that solventless products may cause new health problems and the knowledge of the health risks of the substitutions is often poor.[43]

An extensive survey of chemical exposure conducted by FIOH in Finland in the 1980's. The highest exposure levels were detected in shoe gluing, painting of furniture and other wood products, silkscreen and rotogravure printing, floor lacquering, reinforced plastics production, paint removing, metal degreasing, and various solvent cleaning operations.[25,45,64]

Mostly, inhalation exposure can be effectively controlled by enclosures and ventilation. Open mold methods in the reinforced plastics industry and paint stripping, however, require sophisticated ventilation arrangements. High exposures often occur during solvent cleaning operations. Exposure levels may be very high while cleaning pots in paint and printing ink industry. The use of a proper respirator is necessary during such a work. The avoidance of high peak exposures is important also to prevent development of addiction. The workers may adopt working habits that cause unnecessary exposure. In extreme causes, even this may lead to solvent sniffing.[24]

Recent studies have revealed that percutaneous absorption is often an important route of solvent exposure. As the control of respiratory exposures develops, the relative importance of dermal exposures increases.

### REFERENCES


1 C. Hogstedt, *Scand. J. Work Environ. Health*, **20**, 59 (1994).
2 C. Edling, A. Lindberg, and J. Ulfberg, *Br. J. Ind. Med.*, **50**, 276 (1993).
3 L. Barregård and A. Axelsson, *Scand. Audiol.*, **13**, 151 (1984).
4 L. Low, J. Meeks, and C. Mackerer in Ethel Browning's **Toxicity and Metabolism of Industrial Solvents**, R. Snyder, Ed., *Elsevier*, Amsterdam, 1987, pp. 297-311.
5. D. Couri and M.Milks, *Annual Review of Pharmacology and Toxicology*, **22**, 145 (1982).
6 S. Sjöberg., J .Dahlqvist, S. Fregert, *Contact Dermatitis*, **8**, 207 (1982).
7 A. Tossavainen and J. Jaakkola, *Appl. Occup. Environ. Hyg.*, **9**, 28 (1994).
8 P. Delore and C. Borgomano, *J. Med. Lyon*, **9**, 227 (1928).
9 IARC Monographs on the Evaluation of Carcinogenic Risks to Humans, Vol. 47, Lyon, 1989.
10 D. McGregor, E. Heseltine, and H. Moller, *Scand. J. Work Environ. Health*, **21**, 310 (1995).
11 A. Anttila, E. Pukkala, R. Riala et al., *Int. Arch. Occup. Environ. Health*, **71**, 187 (1998).
12 H-W. Leung and D. Paustenbach, *Appl. Occup. Environ. Hyg.*, **9**, 187 (1994).
13 G. Johanson and A. Boman, *Br. J. Ind. Med.*, **48**, 788 (1991).
14 V. Fiserova-Bergerova, *Ann. Occup. Hyg.*, **34**, 639 (1990).
15 ACGIH, TLVs and BEIs, 1999
16 R. Soule in **Patty's Industrial Hygiene and Toxicology**, G. Clayton and F. Clayton, Eds, Vol.1, Part A, *John Wiley and Sons*, New York, 1991, pp. 137-194.
17 E. Andersson, N. Browne, S. Duletsky et al. Development of Statistical Distributions of Ranges of Standard Factors Used in Exposure Assessments, National Technical Information Service, U.S. Department of Commerce, 1985.
18 B-S. Cohen and W. Popendorf, *Am. Ind. Hyg. Assoc. J.*, **50**, 216 (1989).
19 J. Perkins and B. Pool, *Am. Ind. Hyg. Assoc. J.*, **58**, 474 (1997).
20 P. Kalliokoski, *Am. Ind. Hyg. Assoc. J.*, **51**, 310 (1990).
21 I. Lundberg and R. Milatou-Smith, *Scand. J. Work Environ.& Health*, **24**, 270 (1998).

22 B. Svensson, G. Nise, V. Englander et al., *Br. J. Ind. Med.*, **47**, 372 (1990).
23 J. Laitinen, *Int. Arch. Occup. Health*, **71**, 117 (1998).
24 P. Kalliokoski in **Safety and Health Aspects of Organic Solvents**, V. Riihimäki and U. Ulfvarson, Eds., *Alan R. Liss*, New York, 1986, pp. 21-30.
25 D. Rees, N. Soderlund, R. Cronje et al., *Scand. J. Work Environ. Health*, **19**, 236 (1993).
26 H. Riipinen, K. Rantala, A. Anttila, Organic Solvents, Finnish Institute of Occupational Health, Helsinki, 1991 (in Finnish).
27 C. Edling, K. Ekberg, G. Ahlborg et al., *Br. J. Ind. Med.*, **47**, 75 (1990).
28 I. Lundberg, H. Michelsen, G. Nise et al., *Scand. J. Work Environ. Health*, **21**, suppl.1 (1995).
29 R. Riala, P. Kalliokoski, G. Wickström et al., *Scand. J. Work Environ. Health*, **10**, 263 (1984).
30 J-D. Chen, J-D. Wang, J-P. Jang et al., *Br. J. Ind. Med.*, **48**, 696 (1991).
31 A. Musk, J. Peters, D. Wegman, *Am. J. Ind. Med.*, **13**, 331 (1988).
32 M. Janko, G. McCarthy, M. Fajer et al., *Am. Ind. Hyg. Assoc. J.*, **53**, 331 (1992).
33 W. Heitbrink, M. Wallace, C. Bryant, *Am. Ind. Hyg. Assoc. J.*, **56**, 1023 (1995).
34 N. Goyer, *Am. Ind. Hyg. Assoc. J.*, **56**, 258 (1995).
35 D. George, M. Flynn, R. Harris, *Am. Ind. Hyg. Assoc. J.*, **56**, 1187 (1995).
36 E. Priha, Exposure to Gaseous and Material-bound Formaldehyde in the Wood and Textile Industries, Ph.D Thesis, University of Kuopio, 1996.
37 B-G. Svensson, G. Nise, E. Erfurth et al., *Br. J. Ind. Med.*, **49**, 402 (1992).
38 P. Övrum, M. Hultengren, T. Lindqvist, *Scand. J. Work Environ. Health*, **4**, 237 (1978).
39 P. Kalliokoski in **Ventilation '88**, J. Vincent, Ed., *Pergamon Press*, London, 1988, pp. 197-200.
40 H. Veulemans, D. Groeseneken, R. Masschelein et al., *Scand. J. Work Environ. Health,* **13**, 239 (1987).
41 R. Vincent, B. Rieger, I. Subra et al., *Occup. Hyg.*, **2**, 79 (1996).
42 J. Laitinen, Biomonitoring and Renal Effects of Alkoxyalcohols and Their Acetates Among Silkscreen Printers, PhD Thesis, University of Kuopio, 1998.
43 A-B. Antonsson, *Am. Ind. Hyg. Assoc. J.*, **56**, 394 (1995).
44 A. Seppäläinen and M. Antti-Poika, *Scand. J. Work Environ. Health*, **9**, 15 (1983).
45 M Ruijten, M. Verberk, H. Salle, *Br. J. Ind. Med.*, **48**, 87 (1990).
46 K. Rantala, H. Riipinen, A. Anttila, Halogenated Hydrocarbons, Finnish Institute of Occupational Health, Helsinki, 1992 (in Finnish).
47 C. Estill and A. Spencer, *Am. Ind. Hyg. Assoc. J.*, **57**, 43 (1996).
48 NIOSH, Current Intelligence Bulletin 46: Methylene Chloride, DHHS (NIOSH) Publication No. 86-114, Cinicinnati, 1986.
49 P. Kalliokoski, Toluene Exposure in Finnish Publication Rotogravure Plants, PhD Thesis, University of Minnesota, 1979.
50 W. Burgess, **Recognition of Health Hazards in Industry**, *John Wiley & Sons*, New York, 1981.
51 C. McCammon, R. Glaser, V. Wells et al., *Appl. Occup. Environ. Hyg.*, **6**, 371 (1991).
52 ACGIH, Industrial Ventilation, A Manual of Recommended Practice, Cincinnati, 1988.
53 NIOSH, Hazard Controls, HC 17, DHHS (NIOSH) Publication No. 97-155, Cincinnati,1997.
54 NIOSH, Hazard Controls, HC 18, DHHS (NIOSH) Publication No. 97-156, Cincinnati, 1997.
55 NIOSH, Hazard Controls, HC 19, DHHS (NIOSH) Publication No. 97-157, Cincinnati, 1997.
56 A. Blair, P. Stewart, P. Tolbert et al., *Br. J. Ind. Med.*, **47**, 162 (1990).
57 G. Triebig, S. Lehrl, D. Weltle et al., *Br. J. Ind. Med.*, **46**, 799 (1989).
58 N. Cherry and D. Gautrin, *Br. J. Ind. Med.*, **47**, 29 (1990).
59 M. Severi, W. Pauwels, P. Van Hummelen et al., *Scand. J. Work Environ. Health*, **20**, 451, 1994.
60 P. Kalliokoski, R. Niemelä, A. Säämänen et al. In Ventilation '94, A. Janson and L. Olander, Eds., Stockholm, 1994, pp. 207-210.
61 WHO, Styrene, Environmental Health Criteria 26, Geneva, 1983.
62 P. Kalliokoski, A. Säämänen, L. Ivalo et al., *Am. Ind. Hyg. Assoc. J.*, **49**, 6, 1988.
63 P. Pfäffli, L. Nylander, A. Säämänen et al., *Työ ja Ihminen*, **2**, 213, 1992 (in Finnish, English summary).
64 U. Flodin, K. Ekberg, L. Andersson, *Br. J. Ind. Med.*, **46**, 805, 1989.
65 A. Säämänen, A. Anttila, P. Pfäffli, Styrene, Finnish Institute of Occupational Health, Helsinki, 1991 (in Finnish).
66 A. Säämänen, Methods to Control Styrene Exposure in the Reinforced Plastics Industry, Publications of the Technical Research Centre in Finland No. 354, Espoo, 1998.
67 E. Paci, E. Buiatti, A. Constantini et al., *Scan. J. Work Environ. Health*, **15**, 313, 1989.
68 E. Vigliani and G. Saita, *N. Engl. J. Med.*, **271**, 872, 1964.
69 J. Walker, T. Bloom, F. Stern et al., *Scand. J. Work Environ. Health*, **19**, 89, 1993.

70 H. Drexler, Th. Göen, J. Angerer, *Int. Arch. Occup. Environ. Health*, **67**, 5, 1995.
71 WHO, Carbon disulfide, Environmental Health Criteria 10, Geneva, 1979.
72 M. Ruijten, H. Salle, M. Verberk, *Br. J. Ind. Med.*, **50**, 301, 1993.
73 A. Fujino, T. Satoh, T. Takebayashi et al., *Br. J. Ind. Med.*, **49**, 654, 1992.
74 G. Petrelli, G. Siepi, L. Miligi et al., *Scand. J. Work Environ. Health*, **19**, 63, 1993.
75 C-N. Ong, S-E Chia, W-H. Phoon et al., *Scand. J. Work Environ. Health*, **17**, 430, 1991.

# 19

# REGULATIONS

CARLOS M. NUÑEZ
**U.S. Environmental Protection Agency**
**National Risk Management Research Laboratory**
**Research Triangle Park, NC, USA**

## 19.1 INTRODUCTION

Since the Clean Air Act Amendments of 1970, when a war against pollution was declared, solvents have been on the U.S. regulatory radar screen.[1] This represented the beginning of new environmental policies which led to today's more stringent regulations with a greater focus on toxic substances.[2] In the U.S., historically, these policies and regulations have been divided by medium (air, water, and land), which has represented a challenge to the regulated industries and the Federal, State, and local regulatory communities. In recent years, however, the U.S. Congress and the Environmental Protection Agency (EPA) have recognized the need to address environmental problems from a more holistic approach considering multimedia and innovative environmental management strategies.[3] Hence, various programs and initiatives have emerged, which have proven to be extremely successful, encouraging voluntary industry participation rather than the old command and control approach.[4]

Almost all solvents are volatile organic compounds (VOCs) and hazardous air pollutants (HAPs), and their evaporation creates environmental problems that have become the focus of many domestic and international regulations and initiatives. A VOC solvent is defined by EPA as any compound of carbon, excluding carbon monoxide, carbon dioxide, carbonic acid, metallic carbides or carbonates, and ammonium carbonate, which is emitted or evaporated into the atmosphere.[5] In 1996, an estimated 19 million short tons of VOCs were emitted in the U.S. of which 33 percent came from solvent utilization (see Table 19.1).[6] In the lower atmosphere VOC solvents participate in photochemical reactions to form, to varying degrees, ground level ozone and other oxidants which affect health, as well as cause damage to materials, crops, and forests. Ozone impairs normal functioning of the lungs and reduces the ability to perform physical exercise. Such effects are more severe in individuals with sensitive respiratory systems. Even healthy adults can experience symptoms and reduction in lung function during moderate exercise at ozone levels below the current ozone standard.

Some solvents are also toxic and/or carcinogenic which contributes to direct health problems. They are pollutants that have been associated with serious health effects such as cancer, liver or kidney damage, reproductive disorders, and developmental or neurological problems.[7,8] They also have detrimental environmental effects on wildlife and degrade water or habitat quality. The Clean Air Act of 1990 in section 112(b) lists 188 HAPs,[9] some of which are solvents (see Table 19.2),[10-14] for which sources are identified and regulated. The contents of Table 19.2 will be discussed in more detail later.

**Table 19.1. Major categories in the solvent utilization sector and their estimated VOC emissions[6]**

| Source category | Estimated Emissions, 1996 | |
|---|---|---|
| | Quantity, tons | distribution in solvent utilization sector, % |
| Degreasing | 661,000 | 11 |
| Graphic arts | 389,000 | 6 |
| Dry cleaning | 190,000 | 3 |
| Surface coating | 2,881,000 | 46 |
| Other industrial and non-industrial processes | 2,153,000 | 34 |
| Total emissions | 6,274,000 | 100 |

**Table 19.2. Classification of solvents by their codes, exposure limits, and environmental effects**

| CAS No. | Chemical name | EPA Code[10] | †Permissible Exposure Limits (PEL)[13] | | ‡TLV[14] ppmv | Environmental Effect | | | |
|---|---|---|---|---|---|---|---|---|---|
| | | | ppmv | mg/m³ | | O | T | D | G |
| 50-00-0 | Formaldehyde | K009; K010; K040; K156; K157 | see 29 CFR 1910.1048 | | | ✓ | ✓ | | |
| 50-21-5 | Lactic acid | | | | | ✓ | | | |
| 50-70-4 | Sorbitol | | | | | ✓ | | | |
| 56-23-5 | Carbon tetrachloride | F001; F024; F025; K016; K019; K020; K021; K073; K116; K150; K151; K157; U211; D019 | | (2) | 5 | ✓ | ✓ | ✓ | ✓ |
| 56-38-2 | Parathion | P089 | | (S) 0.1 | 0.1 mg/m³ | ✓ | ✓ | | |
| 56-81-5 | Glycerine | | | | 10 mg/m³ | ✓ | | | |
| 56-93-9 | Benzyltrimethylammonium chloride | | | | | ✓ | | | |
| 57-14-7 | 1,1-Dimethyl hydrazine | K017; K108; K109; K110; U098 | (S) 0.5 | (S) 1 | | ✓ | ✓ | | |
| 57-55-6 | Propylene glycol | | | | | ✓ | | | |
| 57-57-8 | β-Propiolactone | | see 29 CFR 1910.1013 | | 0.5 | ✓ | ✓ | | |
| 60-29-7 | Ethyl ether | U117 | 400 | 1200 | | ✓ | | | |

| CAS No. | Chemical name | EPA Code[10] | †Permissible Exposure Limits (PEL)[13] | | ‡TLV[14] ppmv | Environmental Effect | | | |
|---|---|---|---|---|---|---|---|---|---|
| | | | ppmv | mg/m³ | | O | T | D | G |
| 60-34-4 | Methyl hydrazine | P068 | (C) (S) 0.2 | (C)(S)0.35 | 0.01 | ✓ | ✓ | | |
| 62-53-3 | Aniline | K083; K103; K104; U012 | (S) 5 | (S) 19 | 2 | ✓ | ✓ | | |
| 62-75-9 | N-Nitrosodimethylamine | P082 | see 29 CFR 1910.1016 | | (S) | ✓ | ✓ | | |
| 64-17-5 | Ethyl alcohol | | 1000 | 1900 | | ✓ | | | |
| 64-18-6 | Formic acid | K009; K010; U123 | 5 | 9 | 5 | ✓ | | | |
| 64-19-7 | Acetic acid | | 10 | 25 | 10 | ✓ | | | |
| 64-67-5 | Diethyl sulfate | | | | | ✓ | ✓ | | |
| 67-56-1 | Methanol | U154 | 200 | 260 | 200 | ✓ | ✓ | | |
| 67-63-0 | Isopropyl alcohol | | 400 | 980 | 400 | ✓ | | | |
| 67-64-1 | Acetone | U002 | 1000 | 2400 | 500 | ε | | | |
| 67-66-3 | Chloroform | K009; K010; K019; K020; K021; K029; K073; K116; K149; K150; K151; U044; D022 | (C) 50 | (C) 240 | 10 | ✓ | ✓ | | |
| 67-68-5 | Dimethyl sulfoxide | | | | | ✓ | | | |
| 68-12-2 | Dimethyl formamide | | (S) 10 | (S) 30 | 10 | ✓ | ✓ | | |
| 71-23-8 | n-Propyl alcohol | | 200 | 500 | | ✓ | | | |
| 71-36-3 | n-Butyl alcohol | U031 | 100 | 300 | | ✓ | | | |
| 71-43-2 | Benzene | F005; F024; F025; F037; F038; K085; K104; K105; K141; K142; K143; K144; K145; K147; K151; K159; K169; U019; D018 | see 29 CFR 1910.1028 Table Z-2 | | 0.5 | ✓ | ✓ | | |
| 71-55-6 | 1,1,1-Trichloroethane | F001, F002, F024, F025, K019, K020, K028, K029, U226 | 350 | 1900 | 350 | ε | ✓ | ✓ | ✓ |
| 74-88-4 | Methyl iodide | U138 | (S) 5 | (S) 28 | 2 | ✓ | ✓ | | |
| 74-96-4 | Ethyl bromide | | 200 | 890 | 5 | ✓ | | | |
| 75-00-3 | Ethyl chloride | K018 | 1000 | 2600 | 100 | ✓ | ✓ | | |
| 75-04-7 | Ethylamine | | 10 | 18 | 5 | ✓ | | | |

| CAS No. | Chemical name | EPA Code[10] | †Permissible Exposure Limits (PEL)[13] | | ‡TLV[14] ppmv | Environmental Effect | | | |
|---|---|---|---|---|---|---|---|---|---|
| | | | ppmv | mg/m³ | | O | T | D | G |
| 75-05-8 | Acetonitrile | K011; K013; K014; U003 | 40 | 70 | 40 | ✓ | ✓ | | |
| 75-07-0 | Acetaldehyde | U001 | 200 | 360 | | ✓ | ✓ | | |
| 75-08-1 | Ethyl mercaptan | | (C) 10 | (C) 25 | | ✓ | | | |
| 75-09-2 | Dichloromethane | F001; F002; F024; F025; K009; K010; K156; K158; U080 | | (2) | 50 | ε | ✓ | | |
| 75-15-0 | Carbon disulfide | F005; P022 | | (2) | 10 | ✓ | ✓ | | |
| 75-18-3 | Dimethyl sulfide | | | | | ✓ | | | |
| 75-25-2 | Bromoform | U225 | (S) 0.5 | (S) 5 | 0.5 | ✓ | ✓ | | |
| 75-31-0 | Isopropylamine | | 5 | 12 | 5 | ✓ | | | |
| 75-33-2 | Isopropyl mercaptan | | | | | ✓ | | | |
| 75-34-3 | 1,1-Dichloroethane | F024, F025; U076 | 100 | 400 | 100 | ✓ | ✓ | | |
| 75-35-4 | Vinylidene chloride | F024; F025; K019; K020; K029; U078; D029 | | | 5 | ✓ | ✓ | | |
| 75-52-5 | Nitromethane | | 100 | 250 | 20 | ✓ | | | |
| 75-55-8 | Propyleneimine | P067 | 2 | 5 | 2 | ✓ | ✓ | | |
| 75-56-9 | Propylene oxide | | 100 | 240 | 20 | ✓ | ✓ | | |
| 75-64-9 | tert-Butylamine | | | | | ✓ | | | |
| 75-65-0 | tert-Butyl alcohol | | 100 | 300 | 100 | ✓ | | | |
| 75-68-3 | 1-chloro-1,1-difluoroethane (HCFC-142b) | F001 | | | | ε | | ✓ | ✓ |
| 75-69-4 | Trichlorofluoromethane (CFC-11) | F001; F002; F024; F025; U121 | | | | ε | | ✓ | ✓ |
| 75-83-2 | Neohexane | | | | | ✓ | | | |
| 76-13-1 | 1,1,2-Trichloro-1,2,2-trifluoroethane (CFC-113) | F001; F002 | 1000 | 7600 | 1000 | ε | | ✓ | ✓ |
| 76-14-2 | 1,2-Dichloro-1,1,2,2-tetrafluoroethane(CFC-114) | F001 | 1000 | 7000 | 1000 | ε | | ✓ | ✓ |
| 77-47-4 | Hexachlorocyclopentadiene | F024; F025; K032; K033; K034; U130 | | | 0.01 | ✓ | ✓ | | |
| 77-78-1 | Dimethyl sulfate | K131; U103 | (S) 1 | (S) 5 | 0.1 | ✓ | ✓ | | |
| 78-59-1 | Isophorone | | 25 | 140 | | ✓ | ✓ | | |

| CAS No. | Chemical name | EPA Code[10] | †Permissible Exposure Limits (PEL)[13] | | ‡TLV[14] ppmv | Environmental Effect | | | |
|---|---|---|---|---|---|---|---|---|---|
| | | | ppmv | mg/m³ | | O | T | D | G |
| 78-78-4 | Isopentane | | | | | ✓ | | | |
| 78-81-9 | Isobutylamine | | | | | ✓ | | | |
| 78-83-1 | Isobutyl alcohol | F005; U140 | 100 | 300 | 50 | ✓ | | | |
| 78-84-2 | Isobutyraldehyde | | | | | ✓ | | | |
| 78-87-5 | 1,2-Dichloropropane | U083 | 75 | 350 | 75 | ✓ | ✓ | | |
| 78-88-6 | 2,3-Dichloropropene | F024; F025 | | | | ✓ | | | |
| 78-92-2 | sec-Butyl alcohol | | 150 | 450 | 100 | ✓ | | | |
| 78-93-3 | Methyl ethyl ketone | F005; U159; D035 | 200 | 590 | 200 | ✓ | ✓ | | |
| 78-96-6 | Monoisopropanolamine | | | | | ✓ | | | |
| 79-01-6 | Trichloroethylene | F001; F002; F024; F025; K018; K019; K020; U228; D040 | | (2) | | ✓ | ✓ | | |
| 79-04-9 | Chloroacetyl chloride | | | | 0.05 (S) | ✓ | | | |
| 79-09-4 | Propionic acid | | | | 10 | ✓ | | | |
| 79-10-7 | Acrylic acid | U008 | | | 2 | ✓ | ✓ | | |
| 79-20-9 | Methyl acetate | | 200 | 610 | 250 | ε | | | |
| 79-21-0 | Peracetic acid | | | | | ✓ | | | |
| 79-24-3 | Nitroethane | | 100 | 310 | 100 | ✓ | | | |
| 79-34-5 | 1,1,2,2-Tetrachloroethane | F001; F024; F025; K019; K020; K030; K073; K095; K150; U209 | 5 | 35 | 1 | ε | ✓ | | |
| 79-44-7 | Dimethyl carbamoyl chloride | U097 | | | | ✓ | ✓ | | |
| 79-46-9 | 2-Nitropropane | F005; U171 | 25 | 90 | 10 | ✓ | ✓ | | |
| 80-62-6 | Methyl methacrylate | U162 | 100 | 410 | 100 | ✓ | ✓ | | |
| 84-66-2 | Diethyl phthalate | U088 | | | 5 | ✓ | | | |
| 84-74-2 | Dibutyl phthalate | U069 | | 5 | 5 mg/m³ | ✓ | ✓ | | |
| 85-68-7 | Butyl benzyl phthalate | | | | | ✓ | | | |
| 87-68-3 | Hexachlorobutadiene | F024; F025; K016; K018; K030; D033 | | | 0.02 | ✓ | ✓ | | |
| 90-02-8 | Salicyl aldehyde | | | | | ✓ | | | |
| 91-17-8 | Decahydronapthalene | | | | | ✓ | | | |
| 91-22-5 | Quinoline | | | | | ✓ | ✓ | | |

| CAS No. | Chemical name | EPA Code[10] | †Permissible Exposure Limits (PEL)[13] | | ‡TLV[14] ppmv | Environmental Effect | | | |
|---|---|---|---|---|---|---|---|---|---|
| | | | ppmv | mg/m$^3$ | | O | T | D | G |
| 95-47-6 | o-Xylene | U239 | 100 | 435 | 100 | ✓ | ✓ | | |
| 95-50-1 | 1,2-Dichlorobenzene | F024; F025; K042; K085; K105; U070 | (C) 50 | (C) 300 | 25 | ✓ | | | |
| 96-09-3 | Styrene oxide | | | | | ✓ | ✓ | | |
| 96-12-8 | 1,2-Dibromo-3-chloropropane | U066 | see 29 CFR 1910.1044 | | | ✓ | ✓ | | |
| 96-33-3 | Methyl acrylate | | 10 | 35 | | ✓ | | | |
| 96-37-7 | Methyl cyclopentane | | | | | ✓ | | | |
| 97-63-2 | Ethyl methacrylate | | | | | ✓ | | | |
| 97-64-3 | Ethyl lactate | | | | | ✓ | | | |
| 97-88-1 | n-Butyl methacrylate | | | | | ✓ | | | |
| 98-00-0 | Furfuryl alcohol | | 50 | 200 | 10 | ✓ | | | |
| 98-01-1 | Furfural | U125 | (S) 5 | (S) 20 | 2 | ✓ | | | |
| 98-07-7 | Benzotrichloride | K015; K149; U023 | | | | ✓ | ✓ | | |
| 98-56-6 | p-Chlorobenzotrifluoride | | | | | ε | | | |
| 98-82-8 | Cumene | U055 | (S) 50 | (S) 245 | 50 | ✓ | ✓ | | |
| 98-95-3 | Nitrobenzene | F004; K083; K103; K104; U169; D036 | (S) 1 | (S) 5 | 1 | ✓ | ✓ | | |
| 99-87-6 | p-Cymene | | | | | ✓ | | | |
| 100-37-8 | Diethylethanolamine | | | | | ✓ | | | |
| 100-39-0 | Benzyl bromide | | | | | ✓ | | | |
| 100-41-4 | Ethyl benzene | | 100 | 435 | 100 | ✓ | ✓ | | |
| 100-42-5 | Styrene | | | (2) | 20 | ✓ | ✓ | | |
| 100-44-7 | Benzyl chloride | K015; K149; K085; P028 | 1 | 5 | | ✓ | ✓ | | |
| 100-46-9 | Benzylamine | | | | | ✓ | | | |
| 100-47-0 | Benzonitrile | | | | | ✓ | | | |
| 100-51-6 | Benzyl alcohol | | | | | ✓ | | | |
| 100-52-7 | Benzaldehyde | | | | | ✓ | | | |
| 100-61-8 | N-Methylaniline | | 2 | 9 | 0.5 | ✓ | | | |
| 101-84-8 | Diphenyl ether | | | | | ✓ | | | |
| 103-11-7 | 2-Ethylhexyl acrylate | | | | | ✓ | | | |

| CAS No. | Chemical name | EPA Code[10] | †Permissible Exposure Limits (PEL)[13] | | ‡TLV[14] ppmv | Environmental Effect | | | |
|---|---|---|---|---|---|---|---|---|---|
| | | | ppmv | mg/m³ | | O | T | D | G |
| 103-23-1 | Dioctyl adipate | | | | | ✓ | | | |
| 104-72-3 | Decylbenzene | | | | | ✓ | | | |
| 104-76-7 | 2-Ethyl hexanol | | | | | ✓ | | | |
| 105-39-5 | Ethyl chloroacetate | | | | | ✓ | | | |
| 105-46-4 | sec-Butyl acetate | | 200 | 950 | 200 | ✓ | | | |
| 105-54-4 | Ethyl butyrate | | | | | ✓ | | | |
| 105-58-8 | Diethyl carbonate | | | | | ✓ | | | |
| 106-35-4 | Ethyl butyl ketone | | 50 | 230 | 50 | ✓ | | | |
| 106-42-3 | p-Xylene | U239 | 100 | 435 | 100 | ✓ | ✓ | | |
| 106-43-4 | p-Chlorotoluene | | | | | ✓ | | | |
| 106-63-8 | Isobutyl acrylate | | | | | ✓ | | | |
| 106-88-7 | 1,2-Epoxybutane | | | | 0.3 | ✓ | ✓ | | |
| 106-89-8 | Epichlorohydrin | K017; U041 | (S) 5 | (S) 19 | | ✓ | ✓ | | |
| 106-91-2 | Glycidyl methacrylate | | | | | ✓ | | | |
| 106-93-4 | Dibromoethane | U067 | | (2) | | ✓ | ✓ | | |
| 106-99-0 | Butadiene | | 1000 | 2200 | | ✓ | ✓ | | |
| 107-02-8 | Acrolein | P003 | 0.1 | 0.25 | | ✓ | ✓ | | |
| 107-03-9 | n-Propyl mercaptan | | | | | ✓ | | | |
| 107-05-1 | Allyl chloride | F024, F025 | 1 | 3 | 1 | ✓ | ✓ | | |
| 107-06-2 | 1,2-Dichloroethane | F024; F025; K019; K020; K029; K030; U077; D028 | | (2) | 10 | ✓ | ✓ | | |
| 107-07-3 | Ethylene chlorohydrin | | 5 | 16 | (S) | ✓ | | | |
| 107-10-8 | n-Propylamine | | | | | ✓ | | | |
| 107-13-1 | Acrylonitrile | K011; K013; U009 | see 29 CFR 1910.1045 | | 2 | ✓ | ✓ | | |
| 107-15-3 | Ethylenediamine | | | | | ✓ | | | |
| 107-18-6 | Allyl alcohol | P005 | (S) 2 | (S) 5 | (2) | ✓ | | | |
| 107-21-1 | Ethylene glycol | | | | | ✓ | ✓ | | |
| 107-30-2 | Chloromethyl methyl ether | U046 | see 29 CFR 1910.1006 | | | ✓ | ✓ | | |
| 107-31-3 | Methyl formate | | 100 | 250 | 100 | ✓ | | | |

| CAS No. | Chemical name | EPA Code[10] | †Permissible Exposure Limits (PEL)[13] | | ‡TLV[14] ppmv | Environmental Effect | | | |
|---|---|---|---|---|---|---|---|---|---|
| | | | ppmv | mg/m$^3$ | | O | T | D | G |
| 107-39-1 | Diisobutylene | | | | | ✓ | | | |
| 107-41-5 | Hexylene glycol | | | | | ✓ | | | |
| 107-66-4 | Dibutylphosphate | | 1 | 5 | 1 | ✓ | | | |
| 107-83-5 | Isohexane | | | | | ✓ | | | |
| 107-87-9 | Methyl propyl ketone | | 200 | 700 | 200 | ✓ | | | |
| 107-92-6 | n-Butyric acid | | | | | ✓ | | | |
| 107-98-2 | Propylene glycol methyl ether | | | | 100 | ✓ | | | |
| 108-03-2 | 1-Nitropropane | | 25 | 90 | 25 | ✓ | | | |
| 108-10-1 | Methyl isobutyl ketone | U161 | 100 | 410 | | ✓ | ✓ | | |
| 108-11-2 | Methylamyl alcohol | | (S) 25 | (S) 100 | | ✓ | | | |
| 108-18-9 | Diisopropylamine | | (S) 5 | (S) 20 | 5 | ✓ | | | |
| 108-20-3 | Isopropyl ether | | 500 | 2100 | 250 | ✓ | | | |
| 108-21-4 | Isopropyl acetate | | 250 | 950 | 250 | ✓ | | | |
| 108-38-3 | m-Xylene | U239 | 100 | 435 | 100 | ✓ | ✓ | | |
| 108-39-4 | m-Cresol | F004; U052; D024 | (S) 5 | (S) 22 | 5 | ✓ | ✓ | | |
| 108-82-7 | Diisobutylcarbinol | | | | | ✓ | | | |
| 108-83-8 | Diisobutyl ketone | | 50 | 290 | 25 | ✓ | | | |
| 108-84-9 | Methyl amyl acetate | | | | | ✓ | | | |
| 108-86-1 | Bromobenzene | | | | | ✓ | | | |
| 108-88-3 | Toluene | F005; F024; F025; K015; K027; K036; K037; K149; U220 | | (2) | 50 | ✓ | ✓ | | |
| 108-90-7 | Chlorobenzene | F002; F024; F025; K015; K085; K105; U037; D021 | 75 | 350 | 10 | ✓ | ✓ | | |
| 108-91-8 | Cyclohexylamine | | | | 10 | ✓ | | | |
| 108-93-0 | Cyclohexanol | | 50 | 200 | 50 | ✓ | | | |
| 108-94-1 | Cyclohexanone | U057 | 50 | 200 | 25 | ✓ | | | |
| 108-95-2 | Phenol | K001; K022; K050; K060; K087; U188 | (S) 5 | (S) 19 | 5 | ✓ | ✓ | | |
| 109-06-8 | 2-Methylpyridine | U191 | | | | ✓ | | | |
| 109-60-4 | n-Propyl acetate | | 200 | 840 | 200 | ✓ | | | |
| 109-66-0 | Pentane | | 1000 | 2950 | 600 | ✓ | | | |

| CAS No. | Chemical name | EPA Code[10] | †Permissible Exposure Limits (PEL)[13] | | ‡TLV[14] ppmv | Environmental Effect | | | |
|---|---|---|---|---|---|---|---|---|---|
| | | | ppmv | mg/m³ | | O | T | D | G |
| 109-67-1 | 1-Pentene | | | | | ✓ | | | |
| 109-73-9 | Butylamine | | (C) (S) 5 | (C) (S) 15 | | ✓ | | | |
| 109-78-4 | Ethylene cyanohydrin | | | | | ✓ | | | |
| 109-79-5 | n-Butyl mercaptan | | 10 | 35 | 0.5 | ✓ | | | |
| 109-86-4 | Ethylene glycol monomethyl ether | | | | | ✓ | | | |
| 109-86-4 | Methyl cellosolve | | (S) 25 | (S) 80 | 5 | ✓ | | | |
| 109-87-5 | Dimethoxymethane | | 1000 | 3100 | | ✓ | | | |
| 109-87-5 | Methyl formal | | | | | ✓ | | | |
| 109-89-7 | Diethylamine | | 25 | 75 | 5 (S) | ✓ | | | |
| 109-94-4 | Ethyl formate | | 100 | 300 | | ✓ | | | |
| 110-12-3 | Methyl isoamyl ketone | | 100 | 475 | 50 | ✓ | | | |
| 110-19-0 | Isobutyl acetate | | 150 | 700 | 250 | ✓ | | | |
| 110-43-0 | Methyl n-amyl ketone | | 100 | 465 | 50 | ✓ | | | |
| 110-49-6 | Methyl cellosolve acetate | | (S) 25 | (S) 120 | 5 | ✓ | | | |
| 110-54-3 | Hexane | | 500 | 1800 | 50 | ✓ | ✓ | | |
| 110-63-4 | 1,4 - Butanediol | | | | | ✓ | | | |
| 110-66-7 | n-Amyl mercaptan | | | | | ✓ | | | |
| 110-71-4 | Ethylene glycol dimethyl ether | | | | | ✓ | | | |
| 110-80-5 | Cellosolve | F005; U359 | (S) 200 | (S) 740 | | ✓ | | | |
| 110-82-7 | Cyclohexane | U056 | 300 | 1050 | (300) | ✓ | | | |
| 110-86-1 | Pyridine | F005; K026; K157; U196; D038 | 5 | 15 | 5 | ✓ | | | |
| 110-87-2 | Dihydropyran | | | | | ✓ | | | |
| 110-91-8 | Morpholine | | (S) 20 | (S) 70 | 20 | ✓ | | | |
| 111-15-9 | Cellosolve acetate | | (S) 100 | (S) 540 | 5 | ✓ | | | |
| 111-15-9 | Ethylene glycol monoethyl ether acetate | | | | | ✓ | | | |
| 111-27-3 | n-Hexanol | | | | | ✓ | | | |
| 111-40-0 | Diethylenetriamine | | | | 1 (S) | ✓ | | | |
| 111-42-2 | Diethanolamine | | | | 2 mg/m³ | ✓ | ✓ | | |

| CAS No. | Chemical name | EPA Code[10] | †Permissible Exposure Limits (PEL)[13] | | ‡TLV[14] ppmv | Environmental Effect | | | |
|---|---|---|---|---|---|---|---|---|---|
| | | | ppmv | mg/m³ | | O | T | D | G |
| 111-46-6 | Diethylene glycol | | | | | ✓ | | | |
| 111-49-9 | Hexamethyleneimine | | | | | ✓ | | | |
| 111-55-7 | Ethylene glycol diacetate | | | | | ✓ | | | |
| 111-65-9 | Octane | | 500 | 2350 | 300 | ✓ | | | |
| 111-66-0 | 1-Octene | | | | | ✓ | | | |
| 111-76-2 | 2-Butoxyethanol | | (S) 50 | (S) 240 | (25) | ✓ | | | |
| 111-76-2 | Ethylene glycol monobutyl ether | | | | | ✓ | | | |
| 111-77-3 | Diethylene glycol monomethyl ether | | | | | ✓ | | | |
| 111-84-2 | Nonane | | | | 200 | ✓ | | | |
| 111-90-0 | Diethylene glycol monoethyl ether | | | | | ✓ | | | |
| 111-92-2 | Di-n-butylamine | | | | | ✓ | | | |
| 111-96-6 | Diethylene glycol dimethyl ether | | | | | ✓ | | | |
| 112-07-2 | Ethylene glycol monobutyl ether acetate | | | | | ✓ | | | |
| 112-30-1 | n-Decyl alcohol | | | | | ✓ | | | |
| 112-34-5 | Diethylene glycol monobutyl ether | | | | | ✓ | | | |
| 112-41-4 | 1-Dodecene | | | | | ✓ | | | |
| 112-50-5 | Ethoxy triglycol | | | | | ✓ | | | |
| 112-55-0 | Lauryl mercaptan | | | | | ✓ | | | |
| 112-80-1 | Oleic acid | | | | | ✓ | | | |
| 115-10-6 | Dimethyl ether | | | | | ✓ | | | |
| 17-81-7 | Bis(2-ethylhexyl)phthalate | U028 | | 5 | | ✓ | ✓ | | |
| 121-44-8 | Triethylamine | K156; K157; U404 | 25 | 100 | 1 | ✓ | ✓ | | |
| 121-69-7 | N,N-dimethylaniline | | (S) 5 | (S) 25 | 5 | ✓ | ✓ | | |
| 123-38-6 | Propionaldehyde | | see 29 CFR 1910.1013 | | | ✓ | ✓ | | |
| 123-42-2 | Diacetone alcohol | | 50 | 240 | 50 | ✓ | | | |
| 123-51-3 | Isoamyl alcohol | | 100 | 360 | 100 | ✓ | | | |

| CAS No. | Chemical name | EPA Code[10] | †Permissible Exposure Limits (PEL)[13] | | ‡TLV[14] ppmv | Environmental Effect | | | |
|---|---|---|---|---|---|---|---|---|---|
| | | | ppmv | mg/m³ | | O | T | D | G |
| 123-62-6 | Propionic anhydride | | | | | ✓ | | | |
| 123-63-7 | Paraldehyde | K009; K010; K026; U182 | | | | ✓ | | | |
| 123-72-8 | n-Butyraldehyde | | | | | ✓ | | | |
| 123-86-4 | n-Butyl acetate | | 150 | 710 | 150 | ✓ | | | |
| 123-91-1 | 1,4-Dioxane | U108 | (S) 100 | (S) 360 | | ✓ | ✓ | | |
| 123-92-2 | Isoamyl acetate | | 100 | 525 | | ✓ | | | |
| 124-11-8 | 1-Nonene | | | | | ✓ | | | |
| 124-17-4 | Diethylene glycol monobutyl ether acetate | | | | | ✓ | | | |
| 124-40-3 | Dimethylamine | U092 | 10 | 18 | 5 | ✓ | | | |
| 126-33-0 | Sulfolane | | | | | ✓ | | | |
| 126-99-8 | Chloroprene | | (S) 25 | (S) 90 | 10 | ✓ | ✓ | | |
| 127-18-4 | Tetrachloroethylene | F001; F002; F024; F025; K116; K019; K020; K073; K150; K151; U210; D039 | | (2) | 25 | ε | ✓ | | |
| 127-19-5 | N,N-Dimethylacetamide | | (S) 10 | (S) 35 | 10 | ✓ | | | |
| 131-11-3 | Dimethyl phthalate | U102 | | 5 | 5 mg/m³ | ✓ | ✓ | | |
| 131-18-0 | Di-n-amyl phthalate | | | | | ✓ | | | |
| 140-88-5 | Ethyl acrylate | U113 | (S) 25 | (S) 100 | 5 | ✓ | ✓ | | |
| 141-32-2 | n-Butyl acrylate | | | | 10 | ✓ | | | |
| 141-43-5 | Ethanolamine | | 3 | 6 | | ✓ | | | |
| 141-43-5 | Monoethanolamine | | | | | ✓ | | | |
| 141-78-6 | Ethyl acetate | U112 | 400 | 1400 | 400 | ✓ | | | |
| 141-79-7 | Mesityl oxide | | 25 | 100 | 15 | ✓ | | | |
| 141-97-9 | Ethyl acetoacetate | | | | | ✓ | | | |
| 142-82-5 | Heptane | | 500 | 2000 | 400 | ✓ | | | |
| 142-84-7 | Di-n-propylamine | U110 | | | | ✓ | | | |
| 142-96-1 | Di-n-butyl ether | | | | | ✓ | | | |
| 151-56-4 | Aziridine | P054 | see 29 CFR 1910.1012 | | 0.5 | ✓ | ✓ | | |
| 287-92-3 | Cyclopentane | | | | 600 | ✓ | | | |

| CAS No. | Chemical name | EPA Code[10] | †Permissible Exposure Limits (PEL)[13] | | ‡TLV[14] ppmv | Environmental Effect | | | |
|---|---|---|---|---|---|---|---|---|---|
| | | | ppmv | mg/m$^3$ | | O | T | D | G |
| 298-07-7 | Di-(2-ethylhexyl)phosphoric acid | | | | | ✓ | | | |
| 302-01-2 | Hydrazine | U133 | (S) 1 | (S) 1.3 | 0.01 | ✓ | ✓ | | |
| 306-83-2 | 2,2-Dichloro-1,1,1-trifluoro ethane (HCFC-123) | F001 | | | | ε | | ✓ | ✓ |
| 502-56-7 | Di-n-butyl ketone | | | | | ✓ | | | |
| 510-15-6 | Chlorobenzilate | U038 | | | | ✓ | ✓ | | |
| 513-37-1 | Methallyl chloride | | | | | ✓ | | | |
| 540-59-0 | 1,2-Dichloroethylene | F024; F025; K073 | 200 | 790 | 200 | ✓ | | | |
| 540-84-1 | 2,2,4-Trimethylpentane | | | | | ✓ | ✓ | | |
| 540-88-5 | tert-Butyl acetate | | 200 | 950 | 200 | ✓ | | | |
| 541-41-3 | Ethyl chloroformate | | | | | ✓ | | | |
| 541-85-5 | Ethyl amyl ketone | | 25 | 130 | 25 | ✓ | | | |
| 542-75-6 | 1,3-Dichloropropene | F024; F025; U084 | | | 1 | ✓ | ✓ | | |
| 542-88-1 | Bis(chloromethyl)ether | K017; P016 | | | | ✓ | ✓ | | |
| 543-59-9 | n-Amyl chloride | | | | | ✓ | | | |
| 584-84-9 | 2,4-Toluene diisocyanate | K027 | (C) 0.02 | (C) 0.14 | 0.005 | ✓ | ✓ | | |
| 591-78-6 | Methyl n-butyl ketone | | 100 | 410 | 10 | ✓ | | | |
| 592-41-6 | 1-Hexene | | | | 30 | ✓ | | | |
| 592-76-7 | 1-Heptene | | | | | ✓ | | | |
| 594-42-3 | Perchloromethyl mercaptan | | 0.1 | 0.8 | 0.1 | ✓ | | | |
| 624-83-9 | Methyl isocyanate | | (S) 0.02 | (S) 0.05 | 0.02 | ✓ | ✓ | | |
| 628-63-7 | n-Amyl acetate | | 100 | 525 | 100 | ✓ | | | |
| 626-38-0 | sec-Amyl acetate | | 125 | 650 | 125 | ✓ | | | |
| 629-14-1 | Ethylene glycol diethyl ether | | | | | ✓ | | | |
| 629-76-5 | Pentadecanol | | | | | ✓ | | | |
| 680-31-9 | Hexamethylphosphoramide | | | | | ✓ | ✓ | | |
| 696-28-6 | Phenyldichloroarsine | | | | | ✓ | | | |
| 763-29-1 | 2-Methyl-1-pentene | | | | | ✓ | | | |
| 822-06-0 | Hexamethylene-1,6-diisocyanate | | | | 0.005 | ✓ | ✓ | | |
| 872-05-9 | 1-Decene | | | | | ✓ | | | |

| CAS No. | Chemical name | EPA Code[10] | †Permissible Exposure Limits (PEL)[13] | | ‡TLV[14] ppmv | Environmental Effect | | | |
|---|---|---|---|---|---|---|---|---|---|
| | | | ppmv | mg/m³ | | O | T | D | G |
| 872-50-4 | N-Methylpyrrolidinone | | | | | ✓ | | | |
| 1191-17-9 | 2,2-Dichloroethyl ether | | | | | ✓ | | | |
| 1300-71-6 | Xylenol | | | | | ✓ | | | |
| 1330-20-7 | Xylene (isomer mixtures) | U239 | 100 | 435 | 100 | ✓ | ✓ | | |
| 1336-36-3 | Polychlorinated biphenyls | | | | | ✓ | ✓ | | |
| 1338-24-5 | Naphthenic acids mixtures | | | | | ✓ | | | |
| 1634-04-4 | Methyl tert-butyl ether | | | | | ✓ | ✓ | | |
| 1717-00-6 | 1,1-dichloro-1-fluoroethane (HCFC-141b) | F001 | | | | ε | | ✓ | ✓ |
| 2837-89-0 | 2-chloro-1,1,1,2-tetrafluoro ethane (HCFC-124) | F001 | | | | ε | | ✓ | ✓ |
| 4170-30-3 | Crotonaldehyde | U053 | | | | ✓ | | | |
| 7664-39-3 | Hydrofluoric acid | U134 | | | | ✓ | ✓ | | |
| 7785-26-4 | Dipentene | | | | | ✓ | | | |
| 8030-30-6 | Naphtha: coal tar | K022 | 100 | 400 | | ✓ | | | |
| 8032-32-4 | Petroleum naphtha or naphtha: VM & P | | | | 300 | ✓ | | | |
| 8052-41-3 | Naphtha: Stoddard solvent | | 500 | 2900 | 100 | ✓ | | | |
| 13360-63-9 | N-Ethyl-n-butylaminc | | | | | ✓ | | | |
| 13952-84-6 | sec-Butylamine | | | | | ✓ | | | |
| 25154-52-3 | Nonylphenol | | | | | ✓ | | | |
| 25265-71-8 | Dipropylene glycol | | | | | ✓ | | | |
| 25322-69-4 | Polypropylene glycol | | | | | ✓ | | | |
| 25340-17-4 | Diethylbenzene | | | | | ✓ | | | |
| 25378-22-7 | Dodecene | | | | | ✓ | | | |
| 26761-40-0 | Diisodecyl phthalate | | | | | ✓ | | | |
| 26952-21-6 | Isooctyl alcohol | | | | 50 | ✓ | | | |
| 27215-95-8 | Nonene | | | | | ✓ | | | |
| 28473-21-3 | Nonanol | | | | | ✓ | | | |
| 29063-28-3 | Octanol | | | | | ✓ | | | |
| 34590-94-8 | Dipropylene glycol methyl ether | | (S) 100 | (S) 600 | 100 | ✓ | | | |

| CAS No. | Chemical name | EPA Code[10] | †Permissible Exposure Limits (PEL)[13] | | ‡TLV[14] ppmv | Environmental Effect | | | |
|---|---|---|---|---|---|---|---|---|---|
| | | | ppmv | mg/m$^3$ | | O | T | D | G |
| 53535-33-4 | Heptanol | | | | | ✓ | | | |
| 64742-81-0 | Kerosene | | | | | ✓ | | | |
| 74806-04-5 | Carene | | | | | ✓ | | | |

(ε) Exempt VOC.
†Permissible Exposure Limits (PELs) are 8-hour time weighted averages (TWAs) unless a number is preceded with a (C) which denotes a ceiling limit. When entry is in the mg/m$^3$ column only, the value is exact; otherwise, it is approximate.
‡Threshold Limit Values (TLVs) are the time weighted average concentrations for a conventional 8-hour workday and a 40-hour workweek.
(C) Indicates ceiling limit, or a concentration that should not be exceeded during any part of the working exposure.
(S) Denotes skin designation as the basis for the value.
O Classified as a volatile organic compound (VOC). VOC can react in the lower atmosphere to form ozone and other oxidants. VOC means any compound of carbon, excluding carbon monoxide, carbon dioxide, carbonic acid, metallic carbides or carbonates, and ammonium carbonate, which participates in atmospheric photochemical reactions. Some compounds are specifically exempted from this definition which is found in 40 C.F.R. § 51.100(s).
T Considered a hazardous air pollutant (HAP) and listed in Title III of the Clean Air Act Amendments of 1990.
D A regulated stratospheric ozone layer depleter.
G May impact climate change.

Other solvents, known as chlorofluorocarbons (CFCs), cause stratospheric ozone depletion. They gradually release chlorine and other halogens into the atmosphere, which are effective at destroying the ozone layer,[15] our natural protection from damaging ultraviolet light (UV). They also have high global warming potentials (GWPs).[16] These solvents are generally not considered significant precursors of ground level ozone formation. The production and use of many CFCs have been banned[17,18] and new chemicals have been synthesized and are becoming available for various applications (see Table 19.3).[19,20]

**Table 19.3. List of acceptable replacement chemicals for ozone-depleting solvents (ODS)[20]**

| Substitutes | CFC-113[a] | | | | Methyl chloroform[a] | | | | HCFC-141b[b] | | | |
|---|---|---|---|---|---|---|---|---|---|---|---|---|
| | C | E | M | P | C | E | M | P | C | E | M | P |
| Alternative Technologies (e.g., powder, hot melt, thermoplastic plasma spray, radiation-cured, moisture-cured, chemical-cured, and reactive liquid) | c | | | | | | | | | | | |
| Aqueous cleaners | | c | c | c | | c | c | c | | | | |
| Benzotrifluorides | d | d | d | d | d | d | d | d | | | | |
| HFC-123 | | | | c | | | | c | | | | |
| HFC-225ca/cb | | d | | d | | d | | d | | | | |
| HFC-4310mee | | d | c | c | | d | c | c | | c | c | c |
| High-solids formulations | c | | | | | | | | | | | |

| Substitutes | CFC-113[a] | | | | Methyl chloroform[a] | | | | HCFC-141b[b] | | | |
|---|---|---|---|---|---|---|---|---|---|---|---|---|
| | C | E | M | P | C | E | M | P | C | E | M | P |
| Hydrofluorether (HFE) | | c | c | c | | c | c | c | | c | c | c |
| Methoxynanofluorobutane, iso and normal | | c | c | c | | c | c | c | | c | c | c |
| Methylene chloride | c | c | c | c | | c | c | c | | | | |
| Monochlorotoluenes | d | d | d | d | d | d | d | d | | | | |
| No clean alternatives | | c | | | | c | | | | | | |
| Oxygenated solvents (alcohols, ketones, esters, ethers) | c | | | | | | | | | | | |
| Perchloroethylene | c | c | c | c | | c | c | c | | | | |
| Perfluorocarbons (C5F12, C6F12, C6F14, C7F16, C8F18, C5F11NO, C6F13NO, C7F15NO, and C8F16) | | e | | e | | e | | e | | | | |
| Perfluoropolyethers | | e | | e | | e | | e | | | | |
| Petroleum hydrocarbons | c | | | | | | | | | | | |
| Plasma cleaning | | c | | c | | c | | c | | | | |
| Semi aqueous cleaners | | c | c | c | | c | c | c | | | | |
| Straight organic solvent cleaning | | c | c | c | | c | c | c | | | | |
| Supercritical fluids | | c | c | c | | c | c | c | | | | |
| Terpenes | c | | | | | | | | | | | |
| Trans-1,2-dichloroethylene | d | c | c | c | d | c | c | c | | | | |
| Trichloroethylene | c | c | c | c | | c | c | c | | | | |
| UV/ozone cleaning | | c | | c | | c | | c | | | | |
| Vanishing oils | | | c | | | | c | | | | | |
| Volatile methyl siloxanes | | c | c | c | | c | c | c | | | | |
| Water-based formulations | c | | | | | | | | | | | |

where:

| | |
|---|---|
| a | Class I ODS - substance with an ozone depletion potential of 0.2 or higher |
| b | Class II ODS - substance with an ozone depletion potential of less than 0.2 |
| c | Acceptable |
| d | Acceptable, subject to use conditions |
| e | Acceptable, subject to narrowed use limit |
| C | Adhesive, coating, and ink applications |
| E | Electronics cleaning |
| M | Metals cleaning |
| P | Precision cleaning |

Solvents are used commercially and industrially for numerous applications. Some of those applications include polymer synthesis, coating formulations, cleaning (e.g., electronic circuit board, fabrics, ink supply lines, application equipment, and degreasing), and chemical processes. More than 300 solvents are currently used in the U.S.[21]

Organic chemical solvents are also widely used as ingredients in household products. Paints, varnishes, and wax all contain organic solvents, as well as many cleaning, disinfecting, cosmetic, degreasing, and hobby products.[21] These consumer products can release organic compounds while in use and even when stored. Their release indoors can cause health problems, especially considering that we spend around 90 percent of our time indoors.[22]

All of the environmental and health problems aforementioned have prompted a number of regulations specifically designed to promote and maintain a cleaner and sustainable environment as well as a healthier and safer workplace. This chapter discusses the environmental and safety laws and regulations and the regulatory process in the U.S. Some international regulatory perspectives will also be provided. The purpose of the chapter is to present some key elements of relevant laws and regulations affecting the use of solvents. Solvent users can use this information to understand the range and basic structure of laws and regulations that may affect solvent use. State and federal regulators can use this information to gain a better understanding of the full scope of requirements across all media. In sections 19.2 to 19.4, the discussion follows the single-medium-based regulatory approach used in the U.S. Recognizing, however, that environmental problems may cross medium lines, section 19.5 of this chapter provides a brief discussion of multimedia laws and regulations. Section 19.6 discusses laws and regulations to protect workers from exposure to hazardous chemicals. An overview of international environmental laws and regulations is given in section 19.7.

## 19.2 AIR LAWS AND REGULATIONS

### 19.2.1 CLEAN AIR ACT AMENDMENTS OF 1990[9]

#### 19.2.1.1 Background

In the seventies, the Air Quality Act of 1967 was amended twice: first in 1970 and then in 1977. These amendments brought, among other things, the establishment of national ambient air quality standards (NAAQS), new source performance standards (NSPS) for major new or modified stationary sources, and the national emission standards for hazardous air pollutants (NESHAPS). Also in 1970, the decentralized environmental quality programs and oversight were integrated with the founding of EPA.[23] EPA became the federal agency with primary environmental regulatory responsibility in the U.S with 10 Regional Offices located in major cities (see Figure 19.1 and Table 19.4). The Clean Air Act was amended once again in 1990, building on earlier editions while incorporating some significant changes. The old command and control policy changed to a more cooperative effort on the part of the federal government, state agencies, local municipalities, and industry. Similar approaches were implemented as early as 1979. Then, EPA tried to develop market-based concepts in an attempt to limit and control pollution within a given area, also known as a bubble, while permitting some level of flexibility for the affected facilities. Facilities achieving greater control than required, were allowed to sell their excess pollution reduction credit to other facilities within that area provided the overall pollution level was not exceeded.[24] These attempts did not achieve the desired pollution reduction, which prompted revisions and amendments to the CAA.

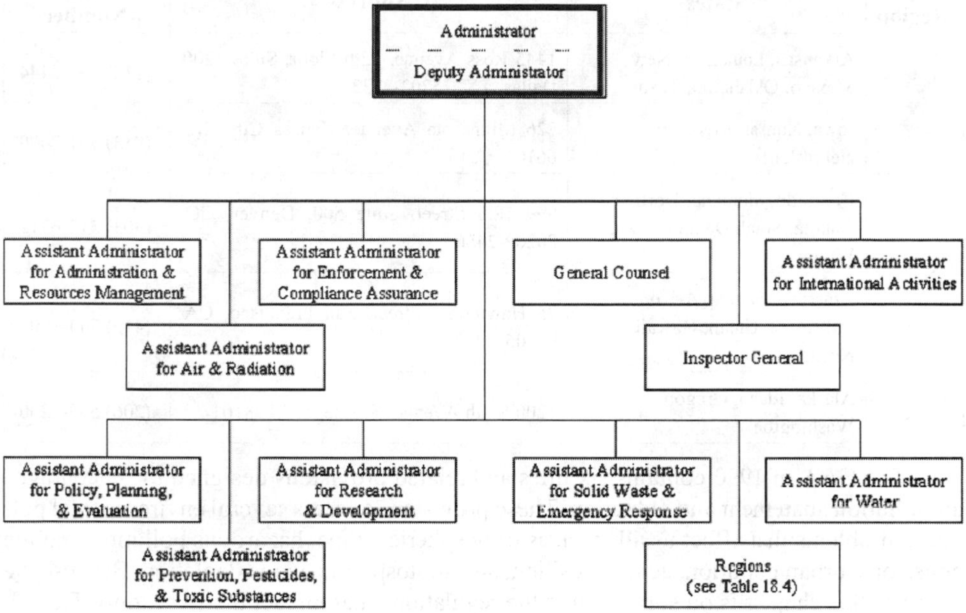


Figure 19.1. EPA organizational structure.

Despite substantial revisions and amendments to the CAA, prior EPA, state, or local laws or regulations were not repealed. In fact, even today, many states still follow prior CAA amendments.[25]

**Table 19.4. List of EPA's regional offices**

| EPA Region | States | Address | Phone Number |
|---|---|---|---|
| 1 | Connecticut, Maine, Massachusetts, New Hampshire, Rhode Island, Vermont | John F. Kennedy Federal Bldg., One Congress Street, Suite 1100, Boston, MA 02114-2023 | (617) 918-1111 |
| 2 | New Jersey, New York, Puerto Rico, Virgin Islands | 290 Broadway, New York, NY 10007-1866 | (212) 637-3000 |
| 3 | Delaware, District of Columbia, Maryland, Pennsylvania, Virginia, West Virginia | 1650 Arch Street, Philadelphia, PA 19103-2029 | (215) 814-5000 |
| 4 | Alabama, Florida, Georgia, Kentucky, Mississippi, North Carolina, South Carolina, Tennessee | 61 Forsyth Street, SW, Atlanta, GA 30303-3104 | (404) 562-9900 |
| 5 | Illinois, Indiana, Michigan, Minnesota, Ohio, Wisconsin | 77 West Jackson Boulevard, Chicago, IL 60604-3507 | (312) 353-2000 |

| EPA Region | States | Address | Phone Number |
|---|---|---|---|
| 6 | Arkansas, Louisiana, New Mexico, Oklahoma, Texas | 1445 Ross Avenue, 12th Floor, Suite 1200, Dallas, TX 75202-2733 | (214) 665-6444 |
| 7 | Iowa, Kansas, Missouri, Nebraska | 726 Minnesota Avenue, Kansas City, KS 66101 | (913) 551-7000 |
| 8 | Colorado, Montana, North Dakota, South Dakota, Utah, Wyoming | 999 18th Street, Suite 500, Denver, CO 80202-2466 | (303) 312-6312 |
| 9 | American Samoa, Arizona, California, Guam, Hawaii, Nevada | 75 Hawthorne Street, San Francisco, CA 94105 | (415) 744-1702 |
| 10 | Alaska, Idaho, Oregon, Washington | 1200 Sixth Avenue, Seattle, WA 98101 | (206) 553-1200 |

The CAA of 1990 contains six titles and related provisions designed to "encourage" air pollution abatement and reduction. These provisions address several environmental pollution problems that affect us all, such as tropospheric ozone, hazardous pollution, mobile emissions, urban pollution, acid deposition, and stratospheric ozone depletion. Because the scope of this chapter is on solvents and the regulations that impact their use, only Titles I, III, V, and VI and their relevance to solvents will be discussed.

### 19.2.1.2 Title I - Provisions for Attainment and Maintenance of National Ambient Air Quality Standards

Under this Title (Section 108), EPA has issued National Ambient Air Quality Standards (NAAQS) for six criteria pollutants - referred to as traditional pollutants in Canada and the European Union (EU):[26] ground level ozone, nitrogen oxides, carbon monoxide, sulfur dioxide, particulate matter less than 10 μm in aerodynamic diameter ($PM_{10}$), and lead. They are called criteria pollutants because EPA has developed health-based criteria or science-based guidelines as the basis for establishing permissible levels. The air quality criteria established by EPA must "accurately reflect the latest scientific knowledge useful in indicating the kind and extent of all identifiable effects on public health and welfare which may be expected from the presence of such pollutants in the ambient air."[9] These ambient air pollutants or NAAQS pollutants are generally present nationally and may be detected through the use of ambient monitoring detectors. Other air pollutants may be regulated as HAPs under Section 112 of Title III if they pose danger to human health but their effect is confined to a localized area.[25] NAAQS include standards to allow for an adequate protection of health (primary) and standards to prevent environmental and property damage from anticipated adverse effects (secondary). In July 1997, EPA published its revision to the ozone NAAQS. The new ozone NAAQS limit is 0.08 ppm and is based on an 8-hour air measurement average.

Areas that do not meet the established NAAQS or that affect the ambient air quality in a nearby area that does not meet NAAQS for the pollutant are known as non-attainment areas.[9] In the U.S., an estimated 90 million people live in these areas. These areas may be further classified based on the severity of non-attainment and the availability and flexibility of the pollution control measures believed necessary to achieve attainment.[9] A State Imple-

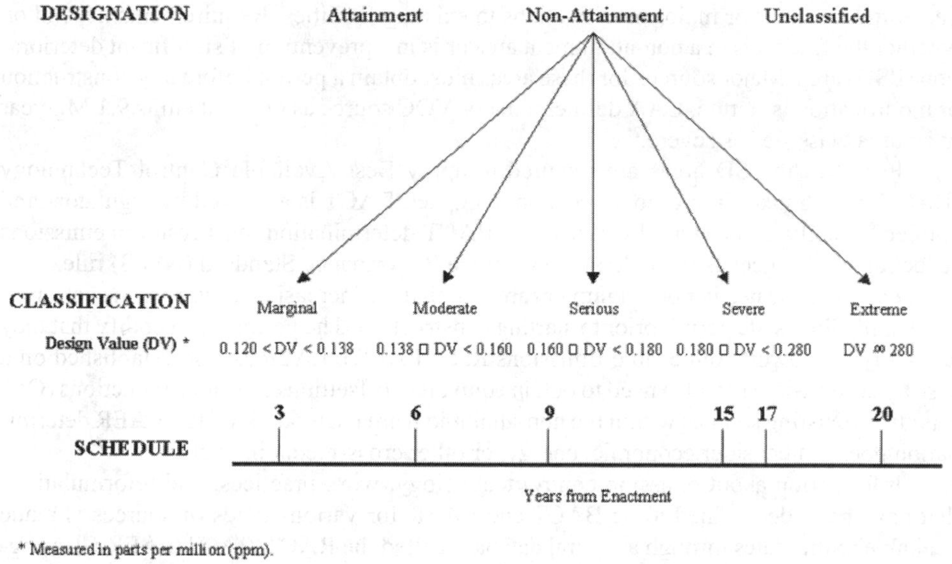


Figure 19.2. Designation and classification of ozone areas.

mentation Plan (SIP) must be prepared and submitted by each state with a non-attainment area. The SIP must contain strategies to achieve compliance within the state's borders and "Air Quality Control Regions." Once a non-attainment area meets the NAAQS for a given pollutant, it is then classified as a maintenance area.

Ozone non-attainment areas are classified according to the severity of the pollution problem (see Figure 19.2). These areas also have to achieve VOC reduction as determined by the non-attainment classification. A 15 percent VOC reduction in non-attainment areas, classified as moderate and above, was required by 1996. Also, Section 182(c)(2)(B) requires that non-attainment areas classified as serious and above follow with a 3 percent VOC reduction per year.[9]

Each state with a non-attainment area selects reasonable control measures, known as Reasonably Available Control Technologies (RACTs), for major sources to achieve the required reduction. Specific information about RACTs can be found in the EPA published Control Technology Guideline (CTG) documents, which are designed to assist state agencies achieve VOC reduction and ozone NAAQS compliance. States are primarily responsible for meeting NAAQS following measures laid out in their SIPs. States work with stationary and mobile sources to ensure that criteria pollutant levels (ozone in the case of solvents) allow the states to meet NAAQS requirements. If a state is not doing an adequate job in improving and maintaining air quality through the activities identified in the SIP, EPA may step in and insist on more stringent measures. EPA promulgates Federal Implementation Plans (FIPs) to correct inadequacies in SIPs. A FIP includes enforceable emission limitations or other control measures and techniques or economic incentives so compliance with the NAAQS is achieved.

In the case of new major stationary sources, the states try to stay or come into attainment through the New Source Review (NSR) program. The NSR program, sometimes called preconstruction permitting, addresses emissions from new major stationary sources;

i.e., entire facilities or major modifications to existing facilities. Requirements depend on whether the facility is in a non-attainment area or is in a prevention of significant deterioration (PSD) area. Major sources for these areas must obtain a permit before any construction or modification is started. CAA defines a major VOC source as one that emits 9.1 Mg/year in an area classified as severe.[26]

Facilities in PSD areas are required to apply Best Available Control Technology (BACT) which takes into account economic impact. BACT is analyzed by regulators and applied to facilities on a case by case basis. BACT determination must result in emissions reduction as stringent as the federal New Source Performance Standard (NSPS) rule.

Facilities located in non-attainment areas or that are increasing emissions in such areas must apply for a state permit prior to starting construction. The permit must codify that they are applying Lowest Achievable Emissions Rate (LAER). LAER is also established on a case by case basis. They also need to obtain equivalent offsetting emissions reductions (Offsets) from existing sources within the non-attainment area. Unlike BACT, a LAER determination does not consider economic, energy, or other cross-media impacts.

Information about emission control technology, work practices, and reformulations that have been determined to be BACT and LAER for various types of sources is made available to the states through a central database called the RACT/BACT/LAER Clearinghouse.

Most organic solvents in this book are classified as VOCs and are found in commercial and consumer products such as paints, coatings, containers, and packaging (see Table 19.2 for a list of these VOC solvents and exempt solvents). They contribute to the formation of ozone and are regulated under this Title, Section 183(e).[9] Fuels and fuel additives are not the focus of Section 183(e). VOCs, nitrogen oxides, and oxygen undergo a photochemical reaction in the presence of sunlight resulting in the formation of tropospheric or ground level ozone, the primary component of smog. Heat also increases the reaction rates of these reactions. The contribution of some solvents to photochemical reaction may be considered negligible. As a result, some VOC solvents may be reclassified as exempt solvents; e.g., acetone. This reclassification begins with industry submitting a petition to EPA for a solvent or solvents to be reclassified as exempt. As of December 3, 1999, EPA had a list of 18 compounds for which petitions requesting VOC exempt status (see Table 19.5) had been received.[27]

**Table 19.5. List of VOCs with requested VOC exempt status**

| Compound | Submitting Organization | Submitting Date | Proposed date |
|---|---|---|---|
| Chlorobromomethane | ICF Kaiser | 11/10/95 | |
| Bromopropane | Enviro Tech International | 05/10/96 | |
| | Albemarle Corp. | 11/18/97 | |
| Methyl bromide | Chemical Manufacturers Association | 07/19/96 | |
| n-Alkanes ($C_{12}$ - $C_{18}$) | The Aluminum Association | 11/27/96 | |
| Technical white oils | The Printing Industries of America | 12/20/96 | |
| | Penzoil Products Company | 12/20/96 | |

| Compound | Submitting | | Proposed date |
|---|---|---|---|
| | Organization | Date | |
| t-Butyl acetate | ARCO Chemical Company (now Lyondell) | 01/17/97 | 9/30/99 64 FR 52731 |
| Benzotrifluoride | Occidental Chemical Company | 03/11/97 | |
| Carbonyl sulfide | E.I. du Pont de Nemours and Company | 08/11/97 | |
| | Texas Mid-Continent Oil & Gas Association | 12/05/97 | |
| trans-1,2-Dichloroethylene | 3M Corporation | 10/08/97 | |
| Dimethyl succinate and dimethyl glutarate | Dibasic Esters Group, affiliated with the Synthetic Organic Chemical Manufacturers Association, Inc. | 10/14/97 | |
| Carbon disulfide | Texas Mid-Continent Oil & Gas Association | 12/05/97 | |
| Acetonitrile | BP Chemicals | 01/21/98 | |
| | GNI Chemicals Corporation | 01/21/98 | |
| Toluene diisocyanate (TDI) | Chemical Manufacturers Association | 01/22/98 | |
| 1,1,1,2,3,3,3-heptafluoropropane (HFC-227) | Great Lakes Chemical Corporation | 02/18/98 | |
| Methylene diphenyl diisocyanate (MDI) | Chemical Manufacturers Association | 08/19/98 | |
| 1,1,1,2,2,3,3-heptafluoro-3-methoxy-propane | 3M Performance and Fluids Division | 02/05/99 | |
| Propylene carbonate | Huntsman Corporation | 07/27/99 | |
| Methyl pivalate | Exxon Chemical Company | 11/22/99 | |

Note: As of December 3, 1999, EPA had not published final actions on these chemicals.

In 1995, under Section 183(e), EPA generated a list of consumer and commercial products categories to be regulated which accounted for at least 80 percent of the VOC emissions from such products in ozone non-attainment areas.[28] The list was divided into four groups, and each group is to be regulated every 2 years (starting in 1997) based on the best available controls (BACs). These four groups are listed in Table 19.6.[29] The term control is broad, including not only end-of-pipe control technology, but also source reduction alternatives. The product categories were selected for regulation based on eight criteria developed to address five factors outlined in Section 183(e)(2):

1. utility
2. commercial demand
3. health or safety functions
4. content of highly reactive compounds
5. availability of alternatives
6. cost effectiveness of controls
7. magnitude of annual VOC emissions
8. regulatory efficiency and program considerations

**Table 19.6. Consumer and commercial products categories**

| |
|---|
| **Group I (1997)**<br>Aerospace Coatings, Architectural Coatings, Autobody Refinishing Coatings, Consumer Products (24 categories), Shipbuilding and Repair Coatings, Wood Furniture Coatings |
| **Group II (1999)**<br>Flexible Packaging Printing Materials |
| **Group III (2001)**<br>Aerosol Spray Paints, Industrial Cleaning Solvents, Flat Wood Paneling Coating, Lithographic Printing Materials |
| **Group IV (2003)**<br>Auto and Light Duty Truck Assembly Coatings, Fiberglass Boat Manufacturing Materials, Large Appliance Coatings, Letterpress Printing Materials, Metal Furniture Coatings, Miscellaneous Industrial Adhesives, Miscellaneous Metal Products Coatings, Paper, Film, and Foil Coatings, Petroleum Drycleaning Solvents, Plastic Parts Coatings |

This table reflects the regulatory schedule revised on March 18, 1999 Federal Register (64 FR 13422).

### 19.2.1.3 Title III - Hazardous Air Pollutants

Section 112 of Title III of the CAA of 1990 lists 188 HAP compounds or groups of compounds (the original list had 189, but caprolactam was removed from the list) which are believed to be carcinogens or can otherwise pose serious health effects. HAP solvents can cause a number of health problems such as cancer, kidney and liver damage, developmental problems in children, nervous system problems, respiratory irritation, birth defects, and miscarriages.[7,8] It may take minutes or years for these health effects to manifest themselves. Approximately half (90 compounds) of the HAP compounds are solvents and are found in many of the source categories regulated under Title III. EPA has the authority to delete chemicals from or add new chemicals to this original list, but no criteria pollutant can be added, only its precursors. If a petition to add or delete a chemical is submitted to EPA, EPA has 18 months to respond with a written explanation of its denial or concurrence. Before the 1990 CAA, EPA listed only eight substances as HAPs (beryllium, mercury, vinyl chloride, asbestos, benzene, radionuclides, arsenic, and coke oven emissions) and issued standards for seven (coke oven emissions were not regulated before the 1990 CAA). After the enactment of the CAA of 1990, a substantially greater number (174) of source categories were identified to be regulated within a period of 10 years. In other words, EPA would promulgate regulations for 174 source categories in half the time that it took to promulgate regulations for only seven. This number of categories to be regulated has been reduced because of subcategorization or deletion of categories.

Many of these categories are major stationary sources which emit HAP or toxic solvents to the atmosphere (see Table 19.2) and are scheduled to be regulated by the year 2000. These regulations are technology-based standards, known as Maximum Achievable Control Technology (MACT) standards. MACT standards are EPA's regulatory means to establish emission limits for air toxics. MACT standards have already been promulgated for 39 source categories (see Table 19.7). EPA will not be developing MACT standards for seven of the originally listed major source categories. There are 66 source categories that remain to be regulated.

**Table 19.7. Categories of sources of HAPs for which Maximum Achievable Control Technology (MACT) standards have already been developed[30]**

| Schedule | Source Categories | Compliance Date |
|---|---|---|
| 2-Year | Dry Cleaning | 09/23/96 |
| | Hazardous Organic NESHAP | F/G - 05/14/01; H - 05/12/99; New Sources - 05/12/98 |
| 4-Year | Aerospace Industry | 09/01/98 |
| | Asbestos | Delisted |
| | Chromium Electroplating | Deco - 0 1/25/96; Others -01/25/97 |
| | Coke Ovens | Not available |
| | Commercial Sterilizers | 12/06/98 |
| | Degreasing Organic Cleaners (Halogenated Solvent Cleaning) | 12/02/97 |
| | Gasoline Distribution (Stage 1) | 12/15/97 |
| | Hazardous Waste Combustion | 06/19/01 |
| | Industrial Cooling Towers | 03/08/95 |
| | Magnetic Tape | w/o new control devices -12/15/96; w/ new control devices - 12/15/97 |
| | Marine Vessel Loading Operations | MACT - 09/19/99; RACT -09/19/98 |
| | Off-Site Waste Recovery Operations | 07/01/00 |
| | Petroleum Refineries | 08/18/98 |
| | Polymers & Resins I: Butyl Rubber, Epichlorohydrin Elastomers, Ethylene Propylene Rubber, Hypalon (TM) Production, Neoprene Production, Nitrile Butadiene Rubber, Polybutadiene Rubber, Polysulfide Rubber, Styrene-Butadiene Rubber & Latex | 07/31/97 |
| | Polymers & Resins II: Epoxy Resins Production, Non-Nylon Polyamides Production | 03/03/98 |
| | Polymers & Resins IV: Acrylontrile-Butadiene-Styrene, Methyl Methacrylate-Acrylonitrile, Methyl Methacrylate-Butadiene, Polystyrene, Styrene Acrylonitrile, Polyethylene Terephthalate | 07/31/97 |
| | Printing & Publishing | 05/30/99 |
| | Secondary Lead Smelters | 06/23/97 |
| | Shipbuilding & Ship Repair | 12/16/96 |
| | Wood Furniture | 11/21/97 |
| 7-Year | Chromium Chemical Manufacturing | Delisted |
| | Electro Arc Furnace: Stainless & Non-Stainless Steel | Delisted |

| Schedule | Source Categories | Compliance Date |
|---|---|---|
| 7-Year | Ferroalloys Production | 05/20/01 |
| | Flexible Polyurethane Foam Production | 10/08/01 |
| | Generic MACT: Acetal Resins, Hydrogen Fluoride, Polycarbonates Production, Acrylic/Modacrylic Fibers | 06/29/02 |
| | Mineral Wool Production | 06/01/02 |
| | Nylon 66 Production | Delisted |
| | Oil & Natural Gas Production | 06/17/02 |
| | Pesticide Active Ingredient Production: 4-Chlor-2-Methyl Acid Production; 2,4 Salts & Esters Production; 4,6-dinitro-o-cresol Production; Butadiene Furfural Cotrimer; Captafol Production; Captan Production; Chloroneb Production; Dacthal$^{TM}$ Production; Sodium Pentachlorophenate Production; TordonTM Acid Production | 06/30/02 |
| | Pharmaceutical Production | 09/21/01 |
| | Phosphoric Acid / Phosphate Fertilizers | 06/10/02 |
| | Polyether Polyols Production | 06/01/02 |
| | Portland Cement Manufacturing | Not available |
| | Primary Aluminum Production | 10/07/99 |
| | Primary Lead Smelting | 06/04/02 |
| | Pulp & Paper (non-combust) - MACT I | 04/15/01 |
| | Pulp & Paper (non-combust) - MACT III | 04/16/01 |
| | Steel Pickling - HCl Process | 06/22/01 |
| | Tetrahydrobenzaldehyde Manufacture (formerly known as Butadiene Dimers Production) | 05/12/01 |
| | Wood Treatment MACT | Delisted |
| | Wool Fiberglass Manufacturing | 06/14/01 |
| 10-Year | Cyanuric Chloride Production | Delisted |
| | Lead Acid Battery Manufacturing | Delisted |
| | Natural Gas Transmission & Storage | 06/17/02 |

Under Section 112 of Title III of the CAA of 1990, EPA will regulate all major sources and area sources deemed appropriate. A major source is a stationary source located within a contiguous area under common control that emits or has the potential to emit ≥ 9.1 Mg/yr (10 tons/yr) of a single HAP or ≥23.7 Mg/yr (25 tons/yr) of a combination of HAPs. An area source is a non-major stationary source. Under Section 112, EPA is required to promulgate 40 categories by 1992, 25 percent of the listed categories by 1994, an additional 25 percent by 1997, and all categories by the year 2000. A list of sources was developed by EPA in 1992[31] and modified 4 years later.[32] Each MACT standard allows existing sources 3 years to comply. Additional years may be granted to complete the installation of controls, to dry and cover mining waste, and for coke ovens. A new source that began construction after the date a MACT standard was proposed will begin compliance at startup or by the date the final standard is issued, whichever is later. Under the early reduction program, additional time may be given to a facility to meet HAP reduction requirements if about 90 percent of the HAP emissions are reduced before a MACT standard is in effect. Once MACT standards are promulgated, the CAA requires EPA to conduct a risk assessment, if considered necessary for public health, to:

- assess the level of risk or "residual risk" remaining to public health,
- estimate the public health significance of the estimated residual risk,
- identify the technological and commercially available methods and costs to reduce the residual risk,
- estimate the actual health effects in the vicinity of sources,
- identify any health studies and uncertainties in the available risk assessment methodologies,
- determine any negative health and environmental impacts resulting from the reduction of the residual risk, and
- produce recommendations leading to legislation for the residual risk.

This may result in even more stringent regulations for solvent users. Figure 19.3 outlines basic steps involved in the development of a MACT regulatory standard.[28]

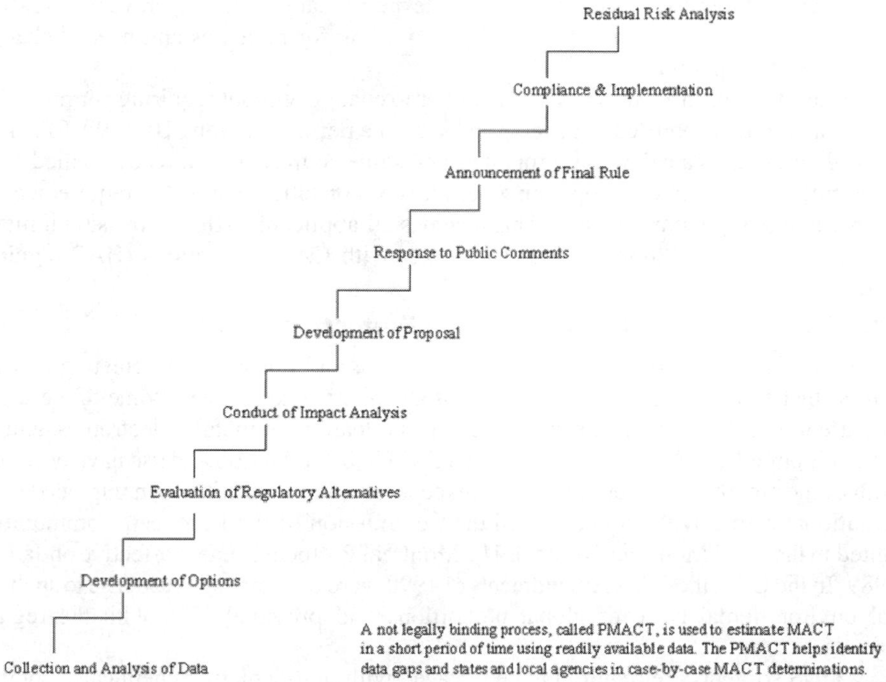
Residual Risk Analysis
Compliance & Implementation
Announcement of Final Rule
Response to Public Comments
Development of Proposal
Conduct of Impact Analysis
Evaluation of Regulatory Alternatives
Development of Options
Collection and Analysis of Data
A not legally binding process, called PMACT, is used to estimate MACT in a short period of time using readily available data. The PMACT helps identify data gaps and states and local agencies in case-by-case MACT determinations.

Figure 19.3. Steps for the development of MACT regulations.

The CAA provides for potential delays in the promulgation of MACTs for major HAP sources; i.e., Section 112(j). This is known also as the MACT Hammer.[25] This provision requires major HAP sources to apply for a permit 18 months after the MACT standard scheduled promulgation date. The State must then make a case-by-case determination of MACT for the source. If the delayed MACT standard is later promulgated, the permit will be revised to reflect any new emission limitation requirements and the affected source may be given up to 8 years to comply.[9]

Air toxics have also been regulated at the state level under various air toxic state programs. In the U.S., the State of California has one of the more stringent and complex air toxics programs. California leads with air toxic programs such as Assembly Bill 1807 (AB 1807) that targets emissions that pose greatest risk to public health and Senate Bill 1731 (SB 1731) that regulates air toxics through facility-based controls rather than risk-based controls.[33]

#### 19.2.1.4 Title V - Permits

Title V was added to the CAA, in the 1990 CAA Amendments, to establish a permit program, federally mandated, to be implemented by the states. Under this Title, EPA has established 10 minimum "elements" for an operating permit program to be administered by the states. These elements serve as basic guidelines to be used when establishing an operating permit program. They include requirements for applications, monitoring and reporting, and annual fees to the owner or operator of sources subject to a permit. They also include the following elements to ensure the effectiveness of an operating permit program:

- adequate personnel, funding, and permitting authority to administer the program;
- procedures to expeditiously evaluate permit applications, prevent unreasonable delay by the permitting authority, and allow for revisions and needed changes to existing permits; and
- provisions to allow permitted facilities to change without requiring permit revision.[9]

Each state is required to develop and submit a permit program. By 1997, EPA had approved nearly all states' permit programs, and some sources have already applied for Title V permits.[25] The idea of an operating permit is to consolidate all CAA requirements for a source in one regulatory vehicle. This includes all applicable SIP air emission limitations, monitoring, and reporting requirements along with CAA regulations (HAP regulations, NSPS, etc.).[25]

#### 19.2.1.5 Title VI - Stratospheric Ozone Protection

Title VI of the CAA deals with chemicals, CFCs, and other ozone depleting substances (ODS) that can cause deterioration of the stratospheric ozone layer. Some of these chemicals are used as solvents in cleaning operations (cleaning of metals, electronics, and precision equipment), coatings, adhesives, and inks. Their production and use have been banned both domestically and internationally[34] (except for a few countries). An unprecedented international effort by 20 countries and the Commission of the European Communities resulted in the 1987 Montreal Protocol. The Montreal Protocol became effective on January 1, 1989. In the U.S., the CAA Amendments of 1990 were the domestic response to such a critical environmental issue of global proportions and provided EPA with the regulatory agenda.

ODS solvents are required to be replaced with non-depleting chemicals. Controlling these substances is not an alternative. ODS solvents are listed in Table 19.2. Substitute solvents deemed acceptable have been identified for some of these ODS solvents under EPA's Significant New Alternatives Policy (SNAP), but other health concerns have limited their use. Some exemptions to the use of phase out ODS exist for essential uses as long as they are consistent with the Montreal Protocol (to date, none has been authorized by EPA).[25]

The SNAP program was established to control the commercialization of ODS substitutes and to ensure that such substitutes do not pose greater harm to the environment than the original chemicals. Through the SNAP program, EPA identifies, classifies, restricts, or prohibits the use of ODS substitutes.

## 19.3 WATER LAWS AND REGULATIONS

### 19.3.1 CLEAN WATER ACT

#### 19.3.1.1 Background

In 1972, Congress enacted the Federal Water Pollution and Control Act (FWPCA), also known as the Clean Water Act (CWA).[35] The CWA is the primary federal legislation that regulates the quality of streams, lakes, estuaries, and coastal waters in the U.S. The CWA has been amended three times: 1977, 1982, and 1987.[36,37,38,39] The CWA is divided into six titles. Its purpose is to restore and maintain the chemical, physical, and biological integrity of the Nation's waters by eliminating the discharge of pollutants into navigable waters or municipal sewer systems.[39] Waters can become contaminated by pollutants discharged from point sources, non-point sources, and storm water and by pollutants from the degradation of wetlands. Point source discharges originate from industrial and sewage treatment plants while non-point discharges are generated from diffuse sources such as sediment runoff from construction sites, leaks from faulty septic systems, and fertilizer and herbicide runoff from agricultural and residential activities. Point sources are "discernable, confined, and discrete conveyances."[38] Examples of a point source include pipes, ditches, channels, tunnels, conduits, discrete fissures, or containers. These sources are regulated and control through the National Pollutant Discharge Elimination System (NPDES) permit program.

In 1976, EPA's focus changed from control of conventional pollutants (biological oxygen demand and suspended solids) to control of toxic pollutants. This shift in EPA's focus was the result of a lawsuit in 1976 which resulted in the Flannery Decree and in the 1977 amendments.[36,40] These amendments established a permit program for point source discharges to focus on 21 major industrial categories and 65 priority pollutants (see Table 19.2 for list of priority pollutant solvents). Today's list of priority pollutants has been expanded to include 129 toxic pollutants from 34 industrial categories.

In the 1987 amendments of the CWA, Congress incorporated Section 319 to establish a national program to control non-point source discharges which provided a regulatory schedule for storm water. These amendments also established a revolving loan fund for construction of sewage treatment plants and provided EPA with enhanced enforcement tools.[38]

#### 19.3.1.2 Effluent Limitations

Under Title III of the CWA, effluent limitations provide the control conditions for a facility's wastewater discharge under an NPDES permit which regulates the quantity and rate of discharges to navigable waters. Guidelines, called the effluent limitations guidelines, have been developed to establish technology-based limits for all types of industrial discharges. Over 50 industry-specific technology-based effluent guidelines have already been promulgated.[40] There are five technology-based treatment standards currently established by the

CWA to establish the minimum level of control to be required by a permit. They are the Best Practicable Control Technology (BPT), Best Available Technology Economically Achievable (BAT), Best Conventional Pollutant Control Technology (BCT), Best Available Demonstrated Technology (BDT), and Best Management Practices (BMPs). Industries that do not discharge directly into surface waters but into publicly operated treatment works (POTWs) are subject to pretreatment standards under Section 307(b) of the CWA.[38]

#### 19.3.1.3 Permit Program

The CWA, under Section 301 (effluent limitations), prohibits the discharge of any pollutant unless the source obtains a permit and ensures compliance. The NPDES permit program, pursuant to Section 402 of the CWA, requires that commercial and industrial facilities and POTWs apply for permits issued by EPA or a designated state. A permit regulates the amount, concentration, and rate of discharge of each regulated pollutant. According to the CWA, a pollutant includes dredged soil, solid waste, incinerator residue, sewage, garbage, sewage sludge, munitions, chemical wastes, biological and radioactive materials, wrecked or discarded equipment, rock, sand, cellar dirt, and industrial, municipal, and agricultural waste.[39] These pollutants are further classified in three categories: conventional, toxic, and non-conventional. Virtually, any material and characteristics such as toxicity or acidity is considered a pollutant.[40] Solvents fall under the category of toxic pollutants.

An NPDES permit can be issued by EPA or by a state that has received EPA permitting approval. As of 1996, 40 states and territories had received EPA's approval.[40] States and territories that are not authorized to administer the NPDES program are Alaska, Arizona, District of Columbia, Idaho, Louisiana, Maine, Massachusetts, New Hampshire, New Mexico, the Pacific Territories, Puerto Rico, and Texas. For these states and territories, the 10 EPA Regional Offices will issue and administer the NPDES permits. The permit process is elaborate and requires extensive information and numerous steps. First a number of forms need to be completed to provide information about the facility, its operation, and the nature of the discharges. This is followed by a period of discussion between EPA and the discharging facility, and the draft permit is announced for comments from the public for 30 days. Once the comments are reviewed, responded to, and implemented, then the final NPDES permit is issued.

### 19.3.2 SAFE DRINKING WATER ACT

#### 19.3.2.1 Background

Enacted in 1974, the Safe Drinking Water Act (SDWA) was established to ensure safe drinking water in public water systems. It required that EPA identify substances in drinking water which could adversely affect public health.[41] SDWA safeguards drinking water with two standards. Primary drinking water standards with a maximum contaminant level (MCL) designed to protect human health and secondary drinking water standards to protect public welfare. The secondary standards involve physical characteristics of drinking water such as color, taste, and smell. A total of 83 contaminants are regulated under the SDWA: 49 volatile and synthetic organic chemicals, 23 inorganic chemicals, 6 microbiological contaminants, and 5 radiological contaminants.

In 1986, the SDWA was amended to greatly increase the responsibilities of EPA and state agencies. Under the 1986 amendments, EPA is directed to schedule the promulgation of primary public drinking water regulations, impose civil and criminal penalties for tampering with public water systems, and enforce more stringent standards.[2] Also, under the

1986 amendments, the wellhead protection program was established to focus on potential contamination of surface and subsurface surrounding a well.[42] In 1988, Congress passed the Lead Contamination Control Act (LCCA) as an additional amendment to the SDWA to protect the public against the contamination of drinking water by lead.

In 1996, the SDWA was amended again to place emphasis on sound science and a risk-based standard approach, small water supply systems, and water infrastructure assistance through a revolving loan fund.[43]

#### 19.3.2.2 National Primary Drinking Water Regulations

National Primary Drinking Water Regulations (NPDWRs) are enforceable, unlike secondary regulations that are only considered advisable, and established by EPA for contaminants that may cause adverse health effects. The promulgation of the NPDWRs begins with the publication of the advanced notice of the proposed rule. This notice presents the current scientific and technical knowledge about the contaminant and the approach EPA is taking toward its regulation. During this period, EPA seeks comments and any additional information through public meetings, workshops, etc. Comments are then reviewed, and a proposed rule is announced in the Federal Register which includes MCLs and non-enforceable health goals called maximum contaminant level goals (MCLGs). After additional public comments, the final regulation is announced and promulgated. The schedule for this regulatory development is contained in the SDWA amendments of 1986. The first 9 contaminants were to be regulated by EPA within 1 year of enactment, followed by another 40 contaminants within 2 years of enactment, and 34 contaminants within 3 years of enactment for a total of 83 by 1989. The SDWA of 1986 also required EPA to regulate 25 additional contaminants every 3 years. EPA regulated these 83 contaminants except for arsenic, sulfate, and radionuclides. These three are addressed under the 1996 amendments. With the 1996 amendments of the SDWA, beginning in 2001, EPA is required to make regulatory decisions for a minimum of five listed contaminants every 5 years.

Along with the regulatory development, EPA prepares criteria documents which provide technical support for the final rules. These criteria documents identify fundamental information regarding health effects of contaminants in drinking water. Health-effects-related data such as physical and chemical properties, toxicokinetics, human exposure, health effects in animals and humans, mechanisms of toxicity, and quantification of toxicological effects are evaluated and serve as supporting documentation for the MCLGs.[44]

## 19.4 LAND LAWS & REGULATIONS

### 19.4.1 RESOURCE CONSERVATION AND RECOVERY ACT (RCRA)[10]

#### 19.4.1.1 Background

The law that regulates solid and hazardous waste management is called RCRA. It was passed as a law in 1976 as an amendment to the Solid Waste Disposal Act (SWDA), and amended in 1980 and 1984, to regulate solid hazardous waste from its creation all the way to its disposal. The 1984 amendments are called the Hazardous and Solid Waste Amendments (HSWAs). These amendments made RCRA a more stringent law giving EPA more increased authority to enforce tighter hazardous waste management standards. They also provided for EPA to revisit problems associated with underground storage tanks. More specifically, Subtitle C regulates hazardous waste while Section D regulates non-hazardous waste such as municipal solid waste and sewage sludge.[45] RCRA requirements were also amended to include small businesses that manage specified amounts of hazardous waste and show an increase in the number of hazardous wastes.[46]

#### 19.4.1.2 RCRA, Subtitle C - Hazardous Waste

Subtitle C is RCRA's regulatory arm to deal with management of hazardous solid waste. A waste is considered hazardous if it appears on EPA's list of hazardous wastes or if hazardous characteristics can be identified. However, the person generating the solid waste is responsible for determining if the generated solid waste is hazardous. Over 500 solid wastes have been identified and listed by EPA as hazardous. Once a waste appears on the EPA hazardous list (wastes classified as F, K, P, U, or so-called "listed" wastes), such a waste is always considered hazardous regardless of its chemical composition. However, EPA is currently working on developing "exit criteria" for certain chemicals in listed wastes that would allow these wastes to be managed as non-hazardous solid wastes if the chemical concentrations are below specified levels. In Part 26, EPA classifies hazardous wastes by the following codes[10] (see Table 19.2 for EPA's hazardous codes for solvents):

F wastes from non-specific sources
K wastes from specific sources
P discarded acutely hazardous commercial chemical products
U discarded commercial chemical products
D wastes for which the above codes do not apply – considered hazardous only if they exhibit ignitability (I), corrosivity (C), reactivity (R), and/or toxicity (E/H/T).

A solid waste considered hazardous will always maintain its listed classification even if it is found mixed with any other waste or a residue derived from a listed hazardous waste.[10] In addition, land disposal of a hazardous waste is prohibited by RCRA unless it is demonstrated that hazardous constituents will not migrate from the disposal unit, causing harm to the environment and public health (40 CFR 268).[10] Hazardous solid waste generators are classified as small quantity generators (SQGs) (i.e., 100 to 1000 kg/month) or large quantity generators (LQGs) (i.e., ≥ 1000 kg/month or > 1 kg of an acutely toxic hazardous waste). These waste quantities apply to each month, not the average over the year. For example, if a facility generates 5 kg in each of the first 11 months and 1200 kg in the last month, such a facility would be considered a large quantity generator (LQG) even though its average is less than 1000 kg/month. Generators can be further classified as conditionally exempt small quantity generators (CESQGs) if they produce < 100 kg/month of hazardous waste and do not accumulate more than 1000 kg of hazardous waste. CESQGs are exempt from all RCRA notification, reporting, and manifesting requirements. Waste from CESQGs must be sent to properly permitted, licensed, or registered treatment, storage, and disposal (TSD) facilities, recycling/reclamation facilities, or facilities authorized to manage industrial or municipal solid waste.[2]

RCRA regulations also provide for states to have primary responsibility for managing and implementing the RCRA hazardous waste program (40 CFR 271). Currently, most states and territories have been granted this authority. States can receive an interim authorization (40 CFR 271, Subpart B) as a temporary vehicle while developing a program for final authorization, which is fully equivalent with the federal program. Hence, for a state to obtain final authorization, it must also be consistent with and no less stringent than the Federal program.

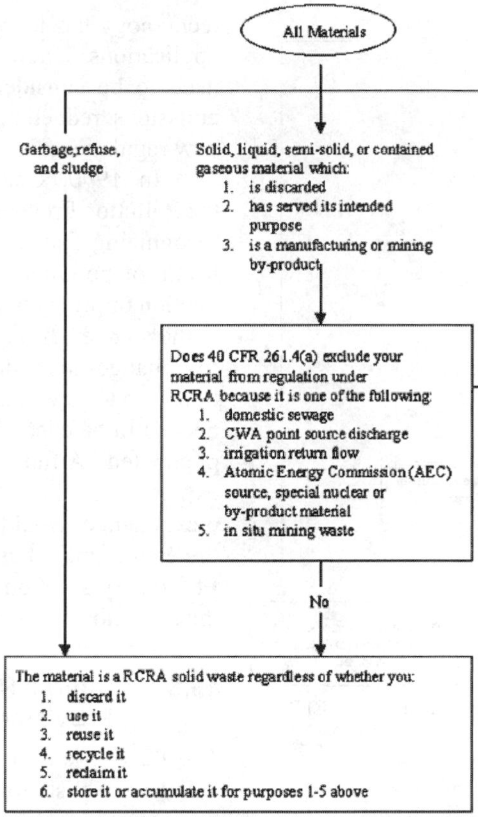

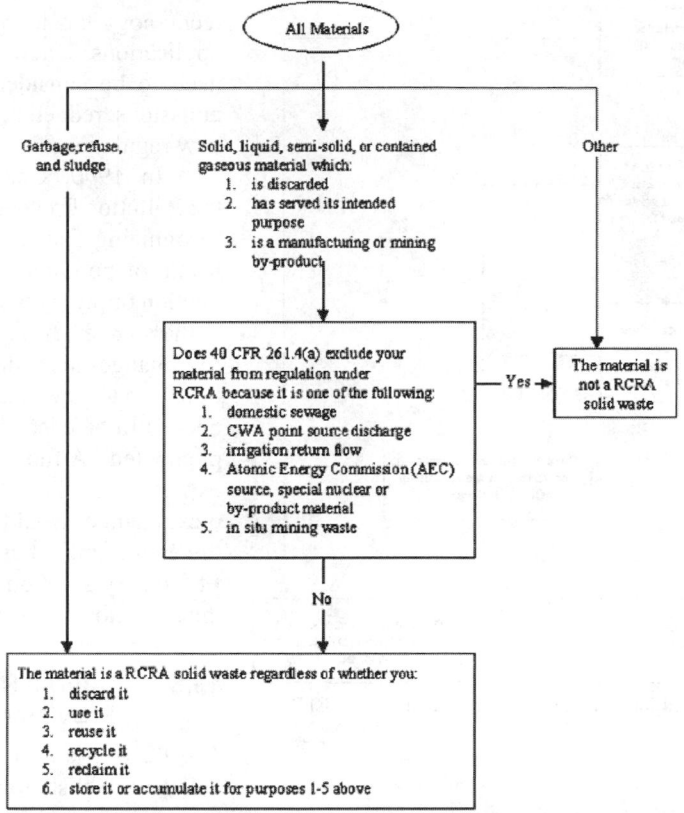

Figure 19.4. Definition of a solid waste (Figure 1 - 40 CFR 260, Appendix I).

RCRA also addresses the transportation of hazardous waste (40 CFR 263). Any transporter of hazardous waste must also comply with the Department of Transportation (DOT) rules under the Hazardous Materials Transportation Act (HMTA) regarding proper labeling, packaging, handling, and placarding.[47]

Finally, a facility generating waste can determine if its waste is regulated under Subtitles C or D by following the steps outlined in the Code of Federal Regulations, Title 40 Part 260, Appendix I (see Figures 19.4, 19.5, and 19.6).

Other provisions exist if the solid waste will be incinerated.

## 19.5 MULTIMEDIA LAWS AND REGULATIONS

### 19.5.1 POLLUTION PREVENTION ACT OF 1990[3]

#### 19.5.1.1 Background

Before 1990, most regulations in the U.S. were issued based on the use of conventional pollution management practices usually considering waste management and end-of-pipe control technologies. Only a handful of government activities were considering reduction of pollution at the source. In the area of waste minimization, recycling approaches were employed in addition to add-on control technology to achieve pollution reduction. However, reducing pollution at the source refers to the top of the hierarchy. Although add-on control

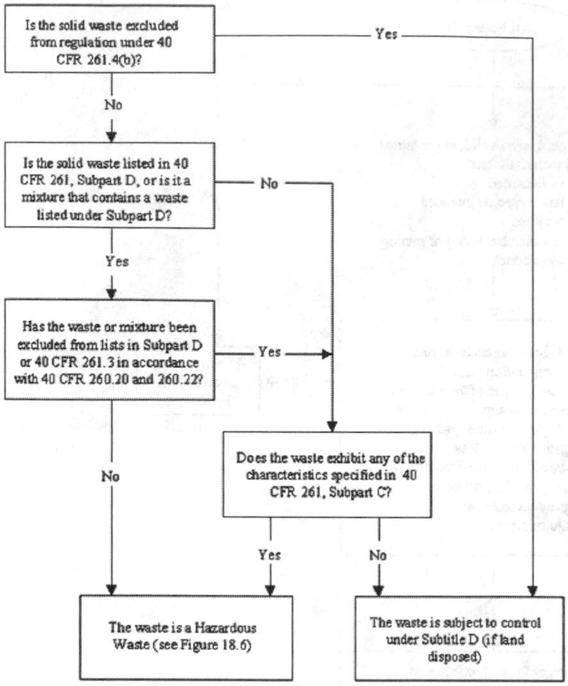

Figure 19.5. Definition of a hazardous waste (Figure 2 - 40 CFR 260, Appendix I).

technology has its place in many applications, new approaches need to be considered to obtain emissions reduction required by new regulations.

In 1990, Congress passed the Pollution Prevention (P2) Act, recognizing that to attain greater levels of pollution reduction, reduction or prevention of pollution at the source "through cost-effective changes in production, operation, and raw materials use" needed to be encouraged and implemented.[3] A fundamental belief expressed in the P2 Act, is that these changes would not only help the environment but would benefit industry's bottom line through the reduction of raw material, pollution control, and liability costs.[3]

### 19.5.1.2 Source Reduction Provisions

The P2 Act is an attempt by the U.S. to establish a framework for integrated multimedia environmental initiatives. The P2 Act provides for a multimedia, integrated, and cost-effective approach to solving environmental problems while encouraging sustainable development. P2 or source reduction includes any practice that reduces or eliminates the creation of pollutants through increased efficiency in the use of raw materials, energy, water, or other resources, or protection of natural resources by conservation. Pollution may be reduced by modifying equipment or technology, modifying processes, reformulating products, substituting raw materials, and improving housekeeping, maintenance, training, or inventory control.[3] The P2 Act encourages the reduction or elimination of wastes of all types and requires that facilities reporting to the Toxic Release Inventory (TRI) provide documentation of their P2/waste minimization or reuse efforts for TRI-reportable chemicals.

The P2 Act required EPA to establish an Office of Pollution Prevention with the authority of promoting source reduction through a multimedia perspective. The Office of Pollution Prevention already existed under The Office of Policy, Planning, and Evaluation (OPPE). It was established by EPA 2 years prior the enactment of the P2 Act. In response to the P2 Act, in 1991, this office developed a source reduction strategy.

The P2 Act also includes provisions aimed at improving the collection and public access of data. Under the Emergency Planning and Community Right-to-Know Act of 1986, industrial facilities are required to report on their annual releases of toxic chemicals to the

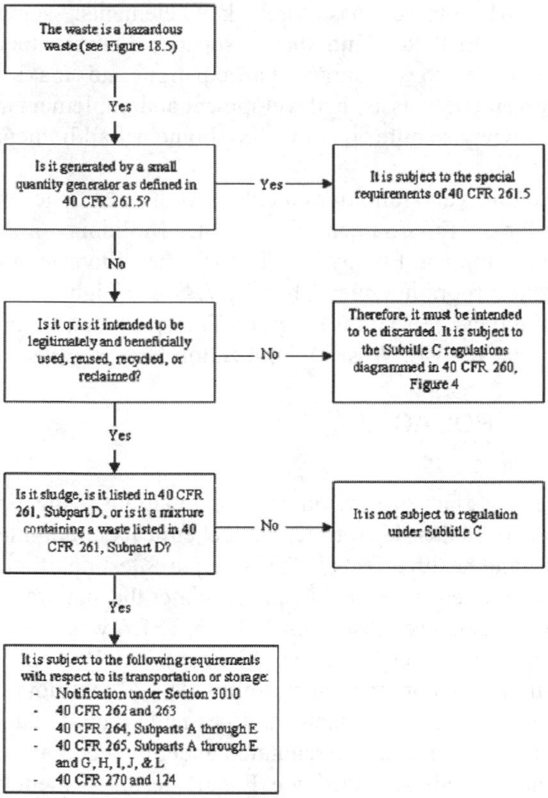


Figure 19.6. Special provisions for certain hazardous wastes (Figure 3 - 40 CFR 260, Appendix I).

environment. In response to the P2 Act, the report must include information on their efforts on source reduction and recycling.

In response to the P2 Act and to help promote source reduction, EPA established various prevention programs and initiatives:

- 33/50 Program - A very successful EPA experimental program based on voluntary industry participation. Companies that reported to TRI any of the 17 target chemicals were encouraged to participate voluntarily to meet a 33 and 50 percent reduction by 1992 and 1995, respectively. Over 1000 companies participated in this program achieving a 55 percent reduction, which translates to an overall reduction of 340 x 106 kg (750 x 106 lb) of toxic chemical releases and transfers.[4] More information about this program can be obtained by contacting the EPA's Office of Pollution Prevention and Toxics website at http://www.epa.gov/opptintr/3350/.
- Common Sense Initiative (CSI) – This initiative consisted of six pilot sectors: automobile manufacturing, computer and electronics, iron and steel, metal finishing, printing, and petroleum refining. The CSI idea was to focus on environmental management by industrial sector rather than by environmental medium.[48] More information can be found at http://www.epa.gov/commonsense.
- Source Reduction Review Project (SRRP)[49] – This was a pilot program designed to ensure that source reduction measures and multimedia issues were considered during the development of air, water, and hazardous waste regulations affecting 17 industrial categories. The SRRP was established to provide a model for the regulatory development process throughout EPA. Some progress has been made, but most regulations are still developed primarily based on a single medium and, while source reduction is encouraged, add-on control technologies are the basis of the regulations.
- Project XL (eXcellence and Leadership) – This is another pilot program recently established by EPA to promote the use of innovative approaches rather than conventional regulatory and policy strategies to achieve more cost-effective

environmental results. The XL project has three key elements: superior environmental performance (SEP) to anticipate superior environmental performance, regulatory flexibility to encourage participation and maximize success, and stakeholder involvement to assist in development and implementation of the project.[50] Visit http://www.yosemite.epa.gov/xl/xl_home.nsf/all/homepage for more information.

Other examples of voluntary programs and initiatives include: AgStar, Climate Wise, Coalbed Methane Outreach, Design for the Environment, Energy Star Builidings, Energy Star Residential, Energy Star Office Equipment, Energy Star Transformer, Environmental Accounting, Environmental Leadership Program, Green Chemistry, Green Lights, Indoor Environments, P2 Information Clearinghouse, P2 Grant Program, Transportation Partners, U.S. Initiative on Joint Implementation, WAVE, Waste Minimization National Plan, and WasteWi$e.

## 19.5.2 TOXIC SUBSTANCES CONTROL ACT

### 19.5.2.1 Background

Also referred to by some as the sleeping giant of environmental regulations,[51] the Toxic Substances Control Act (TSCA) was enacted in 1976 to regulate chemicals in commerce that may cause adverse environmental and health effects.[52] TSCA requires testing of manufactured substances to determine the character of their effect and regulates the manufacture, distribution, use, and disposal of new and existing substances. In 1986, TSCA was amended to include the Asbestos Hazardous Emergency Response Act (AHERA). The 1990 amendments provided for coverage of all public and commercial buildings. TSCA does not regulate food and food products, tobacco or tobacco products, and pesticide manufacturing, processing, or distribution in commerce. These areas are regulated under the Federal Food, Drug, and Cosmetic Act (FFDCA) and the Federal Insecticide, Fungicide, and Rodenticide Act (FIFRA). TSCA's regulatory responsibility is not delegated to states, as may be the case with the other regulations previously mentioned. Since the focus of this book is on solvents, only some key provisions of Title I will be briefly discussed.

### 19.5.2.2 Controlling Toxic Substances

Under Title I of TSCA (Section 4), manufacturers, importers, and processors of chemical substances and mixtures may be required by EPA to obtain health and environmental data on their health and environmental effects if:

their manufacture, processing, use, distribution in commerce, and disposal, or any combination of such activities, poses an unreasonable risk to human health or the environment;

- they are or will be produced and enter the environment in substantial quantities;
- human exposure is significant; and
- data and experience to determine the chemicals' potential impact on human health and the environment are insufficient.

TSCA (Section 5) also provides EPA with the authority to regulate and control the introduction of new chemicals either through manufacture or import, or the processing of an existing chemical for a significant new use. Such manufacturers and importers must file a Premanufacture Notification (PMN) 90 days before producing or importing the chemical. Once a PMN is filed, EPA assesses the information and determines if the chemical poses an unreasonable risk of injury to health or the environment. An additional 90 days may be required by EPA to complete the revision of the PMN. If during the first 90 days, EPA does not respond to the PMN, the manufacture or import of the chemical can begin. EPA has the authority to limit, prohibit, or ban the production of chemicals based on insufficient evidence to conduct a reasonable evaluation of the risk or because the chemical poses an unreasonable risk to human health and the environment (Section 6). Examples of prohibited substances by TSCA include asbestos, halogenated chlorofluoroalkanes, and polychlorinated biphenyls (PCBs).

TSCA provides specific guidance and requirements for data gathering and reporting. This includes information such as production volume, plant size, amount lost to the environment during production or import, quantity of releases, and worker exposure information pertaining to controlled or uncontrolled releases. Companies are also required to maintain records of allegations of significant health or environmental effects (Section 8).

Since 1979, EPA has maintained and published a list of chemical substances manufactured, imported, or processed for commercial purposes. This list, known as the TSCA Inventory, currently has almost 80,000 chemicals.

## 19.6 OCCUPATIONAL LAWS AND REGULATIONS

### 19.6.1 OCCUPATIONAL SAFETY AND HEALTH ACT[53]

#### 19.6.1.1 Background

In April 1971, the Occupational Safety and Health (OSH) Act of 1970 was enacted and resulted in the creation of the Occupational Safety and Health Administration (OSHA) and the National Institute for Occupational Safety and Health (NIOSH). The purpose of this Act is to ensure that workers are provided with workplaces free from recognized safety, health, and environmental hazards. OSHA is part of the Department of Labor (DOL) and responsible for developing and enforcing rules to ensure workplace safety and health. NIOSH is a research agency, part of the Centers for Disease Control and Prevention (CDC), under the Department of Health and Human Services. NIOSH is responsible for developing and establishing new and improved OSH standards and identify potential hazards of new work technologies and practices.

Since 1970, OSHA has issued more than 500 rules dealing with all aspects of worker safety and protection in the workplace. Recent amendments to the 1970 Act require employers to notify employees of potential workplace health hazards, including chemicals. OSHA regulates exposure to about 400 hazardous and toxic substances in the workplace that can cause harm.[54] Under Section 18 of the OSH Act, states are allowed to assume the responsibility for developing and enforcing their own safety and health programs.[53] To date, 25 states have OSHA-approved state plans.[55]

Under the OSH Act, employers are required to provide employees with a workplace free from "recognized hazards." Employers are also required to maintain accurate records of employees' exposure to potentially toxic materials required to be monitored or measured, conduct periodic inspections, and notify employees if they are exposed to toxic materials at higher levels than those prescribed by existing health and safety regulations.

#### 19.6.1.2 Air Contaminants Exposure Limits[12-14]

Information on exposure limits to solvent emissions in the workplace is provided in Table 19.2. These permissible exposure limit (PEL) values were obtained from OSHA's Tables Z-1, Z-2, and Z-3.[13] They are the maximum allowed PELs legally enforceable by OSHA.

For comparison purposes, Table 19.2 provides exposure limit values generated by the American Conference of Governmental Industrial Hygienists (ACGIH) which for many of the listed chemicals are the only values currently available.[14] Usually, the ACGIH values are more stringent than the OSHA values. Both ACGIH and NIOSH have proposed more stringent PELs for over 35 and 25 solvents, respectively.[14,56] Although Table Z-1 has not been formally updated since 1987, new toxicological data on specific chemicals could be used by OSHA to produce directives intended to amend old exposure limit requirements or to add new chemicals to the regulated list. An OSHA PEL is based on an 8-hour, time weighted average (TWA) concentration. These were initially based on ACGHI TLV/TWA values in place since 1971, and many have not been revised since.[57] The lower the value, the more toxic the solvent. PEL values in Table 19.2 are given in parts per million by volume (ppmv) of contaminated air volume and the equivalent milligrams of solvent in a cubic meter of air (mg/m$^3$). These values are focused only on inhalation exposure, which has been strongly criticized.[54] Although this is the most common exposure route to solvent emissions in the workplace, other exposure risks exist such as skin absorption and eye contamination. It is then very important to observe toxicological data compiled in the Material Safety and Data Sheets (MSDSs). Chemical manufacturers and importers are responsible to produce or obtain an MSDS for each hazardous chemical they produce or import, distributors must provide their customers with a copy of MSDSs, and employers are responsible to maintain a copy of MSDSs for each hazardous chemical used in their facility and make them available to their employees.

There are various sources of additional helpful information on the safe use of solvents in the workplace. The Hazard Evaluation System and Information Service (HESIS) of California's Department of Health Services has a 50-page document (Using Solvent Safely) that describes methods of using solvents in the workplace. This document includes information on solvents that pose reproductive problems to both women and men. For more information, visit their website on http://www.ohb.org/solvmenu.htm.[58]

#### 19.6.1.3 Hazard Communication Standard

The Hazard Communication Standard (HCS) was established to provide workers the "right-to-know" of potential hazards associated with their jobs. HCS is a proactive measure to disseminate information to workers and employers about the health hazards pertaining to the chemicals they handle and the protection required. Employers are required to prepare and implement a hazard communication (HAZCOM) program, ensure that all containers have appropriate labels, provide employees easy access to MSDSs, and conduct training programs. The HAZCOM program is the written plan of action that describes the standard implementation strategy for a given facility.

## 19.7 INTERNATIONAL PERSPECTIVE

Similar environmental efforts have occurred in other countries as they have made the environment and human health a priority. Many countries around the world have developed policies, laws, and regulations in an effort to improve and maintain a cleaner environment and human health while providing for a sustainable future. This section provides a view of such efforts.

### 19.7.1 CANADA[59]

In Canada, Environment Canada, which was created in 1971, is the organization responsible for environmental regulation and protection. Environment Canada is divided into five Regional Offices: Ontario, Atlantic, Prairie and Northern Region, Quebec, and Pacific and Yukon. Its environmental services are administered through the Canadian Environment Protection Act (CEPA) which includes the Air Quality Act, the Canadian Water Act, the Ocean Dumping Act, the Environment Contaminant Act, and the Department of the Environment Act. CEPA was recently amended (September 1999), placing more emphasis on pollution prevention and toxic substances. It also provides an increased enforcement authority and resources to carry out necessary research and development activities.

In 1978, Canada signed the Great Lakes Water Quality Agreement with the U.S. which identified specific objectives for minimum levels of water quality for more than 35 substances and compounds.[60] Although this is an international agreement, it has certainly impacted and, to some extent, shaped Canada's domestic environmental laws.

As a result of these statutes, the federal government establishes objectives, guidelines, and emission standards for five national air quality pollutants (carbon monoxide, nitrogen oxides, ozone, sulfur dioxide, and suspended particulate matter). In the case of stationary sources, the regulatory responsibility usually falls under the jurisdiction of 10 provinces. The jurisdiction of water quality management is divided between the provinces and the federal government. The provinces' primary responsibility is to regulate fresh water resources, while the federal government's jurisdiction is over seacoast and inland fisheries, navigation and fisheries, and trade and commerce. The 10 provinces are also primarily responsible for regulating the management of household, non-hazardous, and hazardous solid waste.

### 19.7.2 EUROPEAN UNION

In Europe, the European Community (EC) was formed by the union of three organizations: the European Coal and Steel Community (ECSC), established in 1951; the European Economic Community (EEC), established by the Treaty of Rome in 1957; and the European Atomic Energy Community (EURATOM), established in 1957.[60,61] In 1967, with the Merger Treaty, these organizations merged to form the EC. After the Maastricht Treaty in 1992, the EC became the European Union (EU) which consists of the European Commission, the European Council, the European Parliament, and the European Court of Justice.[61]

Environmental policy was structured as part of the formation of the integrated European states. This effort was extended and refined later with the Maastricht Treaty of 1992. Over the years, however, EC has taken steps to provide a better environment and health for its member states and citizens. Some of these steps have been through the implementation of comprehensive Environmental Action Programmes (EAPs) and specific legislative measures or directives.[62,63] The fifth EAP started in 1993 with an emphasis on sustainable environmental development and a variety of environmental issues such as climate change, acidification and air pollution, depletion of natural resources and biodiversity, depletion and pollution of water resources, deterioration of the urban environment and coastal zones, and waste.[60] Also in 1993, the European Environment Agency (EEA) was established to collect, organize, and disseminate technical, scientific, and economic information pertaining to the quality of the environment in Europe.

Air legislation is targeting emissions from industrial operations, greenhouse gases, lead, motor vehicle emissions, nitrogen oxides, ODS, sulfur dioxide, and suspended particulate matter. Water legislation in the EU is divided into effect- and source-oriented directives. Four effect-oriented directives provide objectives for water with specific end uses: Bathing Water, Drinking Water, Fish Water and Shellfish Water, and Fresh Water Quality Information Exchange. Seven source-oriented directives focus on the elimination or reduction of pollution at the source: Asbestos, Dangerous Substances, Groundwater, Nitrate, Titanium Dioxide, Urban Wastewater, and Water Pollution Information Exchange.[64]

Solid and hazardous waste has received great attention in recent years, and directives have been developed to address their definition, classification, generation, management, and transport across frontiers. In 1993, EU implemented the European Waste Catalogue which defines 15 categories of waste and a residual category intended to capture any materials, substances, or products not included in those categories. There are 27 substances defined as toxic and hazardous waste for which specific information had to be provided during transport: nature, composition, quantity of waste, and sender's and receiver's name.[65]

## 19.8 TOOLS AND RESOURCES FOR SOLVENTS

Various software tools have been developed to identify environmentally benign alternative solvents or equipment modifications to reduce the amount of toxic and volatile solvents. Some tools have attempted to consider life cycle impact in their selection methodologies, but these are the exceptions rather than the rule. This section provides a brief description of some of the tools and resources available to assist in solvent and equipment replacement. In addition, a hotline listing is provided (Table 19.8). Some of these tools are the result of years of research and development by the National Risk Management Research Laboratory of EPA's Office of Research and Development. This laboratory has been researching solvent abatement,[66] replacement, reduction,[67] complete elimination for cleaning and coating operations, and developing tools[68] to assess the overall environmental impact of alternative approaches.

**Table 19.8. List of hotlines in the U.S.[69]**

| Name | Description | E-mail | Phone No. |
|---|---|---|---|
| Air RISC Hotline | Information on health effects, urban toxics, risk assessment, human health, and exposure. | air.risc@epa.gov | 919-541-0888 |
| Hazardous Waste Ombudsman | Assists public and regulatory community in resolving problems associated with the Hazardous Waste Program. | N/A | 800-262-7937<br>202-260-9361 |
| Pollution Prevention Information Clearinghouse (PPIC) | Provides answers and referrals in response to questions about pollution prevention. | ppic@epamail.epa.gov | 800-424-9346 |
| Office of Pollution Prevention and Toxics (OPPT) | OPPT has the Non-confidential Information Center with public dockets for TSCA and Toxic Release Inventory (TRI) rulemaking actions, TSCA administrative record, and non-confidential case files for documents submitted under TSCA. | N/A | 202-260-7099 |

| Name | Description | E-mail | Phone No. |
|---|---|---|---|
| RACT/BACT/LAER Clearinghouse | Provides information on air pollution prevention and control technologies, permit requirements at stationary air pollution sources, and related Federal air pollution emission standards. | N/A | 919-541-0800 |
| RCRA Information Center | Provides access to all regulatory materials supporting EPA's actions under RCRA. | N/A | 703-603-9230<br>800-424-9346 |
| RCRA Information Hotline | Responds to requests about hazardous waste concerning identification, generators, transporters, treatment, storage and disposal facilities, recycling sites, export, and import. | N/A | 703-603-9230 |
| Safe Drinking Water Hotline | Provides information and assistance regarding drinking water regulations, the wellhead protection program, source water protection and guidance, and education materials. | hotline-sdwa@epamail.epa.gov | 800-426-4791 |
| Subsurface Remediation Information Center | Provides technical and scientific information on groundwater protection and remediation. | N/A | 580-436-8651 |
| TSCA Assistance Information Service | Provides TSCA regulation information. | tsca-hotline@epamail.epa.gov | 202-554-1404 |
| WasteWi$e Helpline | WasteWi$e is a voluntary program to encourage businesses to reduce solid waste. | ww@cais.net | 800-372-9473 |

- Solvent Alternatives Guide (SAGE): SAGE is an Internet-based tool developed by the Surface Cleaning Program at Research Triangle Institute (RTI) in cooperation with EPA's Air Pollution Prevention and Control Division to identify solvent and process alternatives for parts cleaning and degreasing. SAGE works as an expert system asking the user a series of questions concerning the part's size and volume, nature of the soil to be removed, production rate, etc. SAGE then produces a ranked list of candidate processes and chemistries most likely to work for a given situation. Since SAGE is based on readily available and proven processes and chemistries, it does not assist in the design of new solvents. SAGE is available at http://clean.rti.org/.
- Coating Alternatives Guide (CAGE): CAGE is based on similar principles found in SAGE but focuses on identifying alternatives for paint and coating formulations. CAGE was also developed by RTI in cooperation with EPA. It is an expert system designed to provide recommendations on low VOC/HAP coating alternatives for various substrates. It also provides the user with links to other useful websites. CAGE is available at http://cage.rti.org/.
- Enviro$en$e, first opened to the public in 1994, is a central reservoir of pollution prevention and cleaner production information and databases. It provides the user with a solvent substitution data system, compliance and enforcement assistance information, and a site for communication and exchange of information (Enviro$en$e Cooperatives). Enviro$en$e can be reached at http://es.epa.gov/.

- Program for Assisting the Replacement of Industrial Solvents (PARIS): PARIS is a solvent design software system developed by EPA, the National Research Council, and RTI to substitute offending solvents with a single chemical or a mixture of solvents based on physical and chemical properties and activity coefficients. PARIS also evaluates and considers the environmental properties of the substitute chemicals. The program's recommended alternatives may require testing to validate their performance.[70]
- Computer-Aided Molecular Design (CAMD): CAMD was developed by the Department of Chemical Engineering's Computer-Aided Process Engineering Centre at the Technical University of Denmark. CAMD can be used to select and design new solvents based on thermodynamic properties. It contains a database with thousands of chemicals which can be accessed to select the desired chemical. If the chemical does not exist, CAMD uses computational chemistry to build the chemical configuration of the new chemical.[70]
- EPA's Environmental Technology Verification (ETV) program was established to verify the performance of innovative technical solutions to environmental and human health problems. Companies with new commercial-ready environmental technologies can participate in this program. For more information visit ETV's website at http://www.epa.gov/etvprgrm/index.htm. Canada also instituted an ETV program to foster Canada's environmental technologies (http://www2.ec.gc.ca/etad/etv_e.html).
- South Coast Air Quality Management District (SCAQMD) has established a program to certify clean air solvents for industrial cleaning. Large and small industries can be exempted from record-keeping requirements and emission fees if they use clean air solvents. For more information about SCAQMD's Clean Air Solvent (CAS) Certification program visit http://www.aqmd.gov/tao/cas/cas.html.

## 19.9 SUMMARY

Solvent releases can affect the quality of air, water, and soil which can then have adverse effects on human health and the environment. More stringent environmental laws and regulations have been established to control their utilization and ensure a safer and healthier environment and a sustainable future. They are placing greater emphasis on the elimination or reduction of such releases at the source and the preservation of limited natural resources. However, replacing offensive solvents requires a comprehensive assessment of their overall environmental impact. This will ensure that substitute chemicals will not impose more stress on our environment and human health by transferring the problem to other media.

This chapter was intended to provide a "bird's eye view" of key environmental laws and regulations for solvents. This chapter will not serve as a replacement of the laws discussed herein.

## REFERENCES

1 Clean Air Act Amendments of 1970, Public Law 91-604, December 31, 1970.
2 N.P. Cheremisinoff and M.L. Graffia, **Environmental and Health & Safety Management: A Guide to Compliance**, *Noyes Publications*, New York, NY, 1995.
3 Pollution Prevention Act of 1990,42 U.S.C. §13101, et seq.

4 U.S. Environmental Protection Agency, 33/50 Program: The Final Record, EPA-745/R-99-004, Office of Pollution Prevention and Toxics, Washington, DC, 1999.
5 Requirements for Preparation, Adoption, and Submittal for Implementation Plans, 40 Code of Federal Regulations § 51.100(s).
6 S.V. Nizich, T. Pierce, A.A. Pope, P. Carlson, and B. Barnard, National Air Pollutant Emission Trends, 1900-1996, EPA-454/R-97-011, NTIS PB98-153158, Office of Air Quality Planning and Standards, Research Triangle Park, NC, December 1997.
7 National Institute for Occupational Safety and Health Pocket Guide to Chemical Hazards (Publication No. 94-116), Cincinnati, OH (1994).
8 I. Lundberg, G. Nise, G. Hedenberg, M. Högberg, and O. Vesterberg, Liver function tests and urinary albumin in house painters with previous heavy exposure to organic solvents, *Occup. and Environ. Med.*, **51**, 347 (1994).
9 Clean Air Act Amendments of 1990, Public Law 101-549, November 15, 1990.
10 Identification and Listing of Hazardous Waste, 40 Code of Federal Regulations 261.
11 1996 North American Emergency Response Guidebook, J.J. Keller & Associates, Inc., Washington, DC, 1996.
12 U.S. Environmental Protection Agency, Title III List of Lists, Consolidated List of Chemicals Subject to the Emergency Planning and Community Right-to-Know Act (EPCRA) and Section 112(r) of the Clean Air Act, as Amended, EPA-550-B-98-017, Office of Solid Waste and Emergency Response, Washington, DC, November 1998.
13 Occupational Safety and Health Standards, 29 Code of Federal Regulations, Chapter XVII, Part 1910.
14 1999 TLVs and BEIs, Threshold Limit Values for Chemical Substances and Physical Agents Biological Exposure Indices, American Conference of Governmental Industrial Hygienists, Cincinnati, OH (1999).
15 S. Solomon and D. Albritton, Time-Dependent Ozone Depletion Potentials for Long- and Short-Term Forecasts, *Nature*, **357**, 33 (1992).
16 J.T. Houghton, L.G. Meira Filho, B.A. Callander, N. Harris, A. Kattenberg, and K. Maskell, Eds, **Climate Change 1995, The Science of Climate Change, Contribution of Working Group 1 to the Second Assessment Report of the Intergovernmental Panel on Climate Change**, *Cambridge University Press*, Cambridge, UK (1996).
17 Montreal Protocol, *The International Lawyer*, **26**, 148 (1992).
18 The Montreal Protocol, *Harvard Environmental Law Review*, **15**, 275 (1991).
19 J.S. Nimitz, Development of Nonflammable, Environmentally Compliant Fluoroiodocarbon Solvents: Phase I Final Report, Prepared for U.S.A.F. Wright Laboratories under contract F33615-94-C-5003, ETEC 95-2 (1995).
20 U.S. Environmental Protection Agency, List of Substitutes for Ozone-Depleting Substances, last update May 1999, http://www.epa.gov/spdpublc/title6/snap/lists/index.html. Stratospheric Protection Division, Washington, DC, June 1999.
21 Technical Guide No. 6, **Handbook of Organic Industrial Solvents**, *American Mutual Insurance Alliance*, Chicago, IL, 1980.
22 J.P. Robinson and W.C. Nelson, National Human Activity Pattern Survey Data Base, U.S. EPA, Atmospheric Research and Exposure Assessment Laboratory, Research Triangle Park, NC, 1995.
23 Reorganization Plan No. 3 of 1970, 3 CFR 1072.
24 M. Seralgeldin, Symp. Environmentally Acceptable Inhibitors and Coatings, The Electrochemical Society, Impact of the Regulatory Environment on Green Technologies, Pennington, NJ, 1997, pp. 1-16.
25 R.J. Martineau, Jr. and D.P. Novello, **The Clean Air Act Handbook**, *American Bar Association*, Chicago, IL, 1998.
26 P.E.T. Douben and M.A. Serageldin, **Pollution Risk Assessment and Management**, P.E.T. Douben, Ed., *John Wiley & Sons Ltd.,* New York, NY, 1998, pp. 49-91.
27 Communication with William Johnson on December 7, 1999, EPA's Office of Air Quality Planning and Standards, Research Triangle Park, NC (1999).
28 Coating Regulations Workshop, April 8-9, 1997, EPA's Office of Air Quality Planning and Standards, Durham, NC.
29 Federal Register 64: 13422, Consumer and Commercial Products: Schedule for Regulation, Notice of revisions to schedule for regulation (1999).
30 U.S. Environemental Protection Agency's Technology Transfer Network, Unified Air Toxics Website: Rules and Implementation, http://www.epa.gov/ttn/uatw/eparules.html, Office of Air Quality Planning and Standards, Research Triangle Park, NC, December 21, 1999.

31 Federal Register 57: 31576, Initial list of Categories of Sources Under Section 112(c)(1) of the Clean Air Act Amendments of 1990, Notice of initial list of categories of major and area sources. U.S. Environmental Protection Agency (1992).
32 Federal Register 61: 28197 and 37542, Revision of list of Categories of Sources and Schedules for Standards Under 112(c) and (e) of the Clean Air Act Amendments of 1990, Notice of revisions of initial list.
33 G.S. Koch and P.R. Ammann, Current Trends in Federal Regulation of Hazardous Air Pollutants, *J. Environ. Reg.*, **4**, 25 (1994).
34 P. Zurer, Ozone Treaty Tightened, CFC Substitutes Controlled, *Chemical & Engineering News,* **70**, 5 (1992).
35 Federal Water Pollution Control Act, Public Law 92-500 (1972).
36 Clean Water Act Amendments of 1977, Public Law 95-217 (1977).
37 Clean Water Act Amenments of 1982, Public Law 97-117 (1982).
38 Clean Water Act of 1987 (Clean Water Act), Public Law 100-4 (1987).
39 Navigation and Navigable Waters, 33 U.S.C. 1251-1387.
40 L.M. Gallagher and L.A. Miller, **Clean Water Handbook**, *Government Institutes, Inc*., Rockville, MD, 1996.
41 Safe Drinking Water Act, 42 U.S.C 300f et seq. (1974).
42 Safe Drinking Water Act Amendments of 1986, Public Law 99-339 (1986).
43 Safe Drinking Water Act Amendments of 1996, Public Law 104-182 (1996).
44 E.J. Calabrese, C.E. Gilbert, and H. Pastides, **Safe Drinking Water Act: Amendments, Regulations and Standards**, *Lewis Publishers*, Chelsea, MI, 1989.
45 RCRA: Reducing Risk from Waste, EPA-530/K-97-004, Office of Solid Waste and Emergency Response, Washington, DC, September 1997.
46 Understanding the Hazardous Waste Rules: A Handbook for Small Businesses - 1996 Update, EPA-530/K-95-001, Office of Solid Waste and Emergency Response, Washington, DC, June 1996.
47 Regulations Relating to Transportation, 49 Code of Federal Regulations 171-195.
48 U.S. Environmental Protection Agency, The Common Sense Initiative, Lessons Learned About Protecting the Environment in Common Sense, Cost Effective Ways, EPA-100/R-98-011, Office of Reinvention, Washington, DC, December 1998.
49 U.S. Environmental Protection Agency, Source Reduction Review Project, EPA-100/R-92/002, Office of Pollution Prevention and Toxics, Washington, DC, August 1992.
50 U.S. Environmental Protection Agency, Project XL 1999 Comprehensive Report, EPA-100/R-99-008, Office of the Administrator, Washington, DC, 1999.
51 J.W. Vincoli, **Basic Guide to Environmental Compliance**, *Van Nostrand Reinhold*, New York, NY, 1993.
52 Commerce and Trade, 15 U.S.C. 2601, et seq.
53 Occupational Safety and Health Act of 1970, 29 U.S.C. 651 et seq.
54 B. Plog, Fundamentals of Industrial Hygiene, National Safety Council, Itasca, IL, 1996.
55 OSHA. State Occupational Safety and Health Plans. http://www.osha-slc.gov/fso/osp/. January 18, 2000.
56 National Institute for Occupational Safety and Health recommendations for occupational safety and health: Compendium of policy documents and statements, Publication No. 92-100, Cincinnati, OH (1992).
57 G. Noll, M. Hildebrand, and J. Yvirra, Hazardous Materials: Managing the Incident, Oklahoma State University, Stillwater, OK, 1995.
58 California's Department of Health Services. http://www.ohb.org/solvmenu.htm, Oakland, CA, June 1, 1997.
59 Commission for Environmental Cooperation. Summary of Environmental Law in North America, Last update October 1, 1995, htttp://www.cec.org/infobases/law, Montreal, Canada.
60 L.K. Caldwell, **International Environmental Policy**, *Duke University Press*, Durham, NC, 1996.
61 B.J.S. Hoetjes, **Environmental Law and Policy in the European Union and the United States**, R. Baker, Ed., *Praeger Publishers*,Westport, CT, 1997, pp. 31-44.
62 K.Hanf, **Environmental Law and Policy in the European Union and the United States**, R. Baker, Ed., *Praeger Publishers*, Westport, CT, 1997, pp. 125-146.
63 J. Ebbesson, **Compatibility of International and National Environmental Law**, *Kluwer Law International*, 1996.
64 C.H.V. de Villeneuve, **Environmental Law and Policy in the European Union and the United States**, R. Baker, Ed., *Praeger Publishers*, Westport, CT, 1997, pp. 159-168.
65 I.J. Koppen, **Environmental Law and Policy in the European Union and the United States**, R. Baker, Ed., *Praeger Publishers*, Westport, CT, 1997, pp. 234-247.
66 C.M. Nunez, G.H. Ramsey, W.H. Ponder, J.H. Abbott, L.E. Hamel, and P.H. Kariher, Corona Destruction: An Innovative Control Technology for VOCs and Air Toxics, *J. Air & Waste Management Assoc*., **43**, 242-247 (1993).

67 C.M. Nunez, A.L. Andrady, R.K. Guo, J.N. Baskir, and D.R. Morgan, Mechanical Properties of Blends of PAMAM Dendrimers with Poly(vinyl chloride) and Poly(vinyl acetate), *J. Polymer Sci., Part A - Polymer Chemistry*, **36**, 2111-2117 (1998).
68 C.M. Nunez, G.H. Ramsey, M.A. Bahner, and C.A. Clayton, An Empirical Model to Predict Styrene Emissions from Fiber-Reinforced Plastics Fabrication Processes, *J. Air & Waste Management Assoc.*, **49**, 1168-1178 (1999).
69 U.S. Environmental Protection Agency, Headquarters Telephone Directory, EPA/208-B-99-001, Office of Information Resources Management, Washington, DC, April 1999.
70 U.S. Environmental Protection Agency, Headquarters Telephone Directory, EPA/625/R-99/005, Office of Research and Development, Washington, DC, July 1999.

## ACRONYMS

| | |
|---|---|
| ACGIH | American Conference of Governmental Industrial Hygienists |
| AEC | Atomic Energy Commission |
| AHERA | Asbestos Hazardous Emergency Response Act |
| BAC | Best Available Control |
| BACT | Best Available Control Technology |
| BAT | Best Available Technology Economically Achievable |
| BCT | Best Conventional Pollutant Control Technology |
| BDT | Best Available Demonstrated Technology |
| BMP | Best Management Practices |
| BPT | Best Practicable Control Technology |
| CAA | Clean Air Act |
| CAGE | Coating Alternatives Guide |
| CAMD | Computer-Aided Molecular Design |
| CAS | Clean Air Solvent |
| CDC | Centers for Disease Control and Prevention |
| CEPA | Canadian Environment Protection Act |
| CESQG | conditionally exempt small quantity generator |
| CFC | chlorofluorocarbon |
| CSI | Common Sense Initiative |
| CTG | Control Technology Guideline |
| CWA | Clean Water Act |
| DEPH | bis (2-ethylhexyl)phthalate |
| DOL | Department of Labor |
| DOT | Department of Transportation |
| DV | Design Value |
| EAP | Environmental Action Programme |
| EC | European Community |
| ECSC | European Coal and Steel Community |
| EEA | European Environment Agency |
| EEC | European Economic Community |
| ENVIRO$EN$E | public environmental information system |
| EPA | U.S. Environmental Protection Agency |
| ETV | Environmental Technology Verification |
| EU | European Union |
| EURATOM | European Atomic Energy Community |
| FFDCA | Federal Food, Drug, and Cosmetic Act |
| FIFRA | Federal Insecticide, Fungicide, and Rodenticide Act |
| FIP | Federal Implementation Plan |
| FWPCA | Federal Water Pollution and Control Act |
| GWP | global warming potential |
| HAP | hazardous air pollutant |
| HAZCOM | hazard communication |
| HCFC | hydrochlorofluorocarbon |
| HCS | Hazard Communication Standard |
| HESIS | Hazard Evaluation System and Information Service |
| HFE | hydrofluoroether |
| HMTA | Hazardous Materials Transportation Act |

| | |
|---|---|
| HSWA | Hazardous and Solid Waste Amendments |
| LAER | Lowest Achievable Emissions Rate |
| LCCA | Lead Contamination Control Act |
| LQG | large quantity generator |
| MACT | Maximum Achievable Control Technology |
| MCL | maximum contaminant level |
| MCLG | maximum contaminant level goal |
| MSDS | Material Safety and Data Sheet |
| NAAQS | national ambient air quality standards |
| NESHAP | national emission standards for hazardous air pollutants |
| NIOSH | National Institute for Occupational Safety and Health |
| NPDES | National Pollutant Discharge Elimination System |
| NPDWR | National Primary Drinking Water Regulations |
| NSPS | new source performance standards |
| NSR | New Source Review |
| ODS | ozone depleting substance |
| OPPE | Office of Policy, Planning, and Evaluation |
| OPPT | Office of Pollution Prevention and Toxics |
| OSH | Occupational Safety and Health |
| OSHA | Occupational Safety and Health Administration |
| P2 | Pollution Prevention |
| PARIS | Program for Assisting the Replacement of Industrial Solvents |
| PCB | polychlorinated biphenyl |
| PCBTF | p-chlorobenzotrifluoride |
| PEL | permissible exposure limit |
| $PM_{10}$ | particulate matter of 10 micrometers in aerodynamic diameter or smaller |
| PMACT | Preliminary Maximum Achievable Control Technology |
| PMN | Premanufacture Notification |
| POTW | publicly operated treatment works |
| PPIC | Pollution Prevention Information Clearinghouse |
| PSD | prevention of significant deterioration |
| RACT | Reasonable Available Control Technology |
| RCRA | Resource Conservation and Recovery Act |
| RISC | Risk Information Support Center |
| SAGE | Solvent Alternatives Guide |
| SCAQMD | South Coast Air Quality Management District |
| SDWA | Safe Drinking Water Act |
| SEP | superior environmental performance |
| SIP | State Implementation Plan |
| SNAP | Significant New Alternatives Policy |
| SQG | small quantity generator |
| SRRP | Source Reduction Review Project |
| SWDA | Solid Waste Disposal Act |
| TLV® | Threshold Limit Value |
| TRI | Toxic Release Inventory |
| TSCA | Toxic Substances Control Act |
| TSD | treatment, storage, and disposal |
| TWA | time weighted average |
| UN | United Nations |
| UV | ultraviolet light |
| VOC | volatile organic compound |
| XL | eXcellence and Leadership |

# 19.10 REGULATIONS IN EUROPE

TILMAN HAHN, KONRAD BOTZENHART, FRITZ SCHWEINSBERG
**Institut für Allgemeine Hygiene und Umwelthygiene**
**Universität Tübingen, Tübingen, Germany**

The legislative and executive regulations in Europe are generally based on EEC regulations which are converted to and coordinated by national regulations as it is a rule in Europe. EEC regulations do not exist concerning all aspects, or they are not always converted effectively in the EEC Member States.

In Europe environmental law, especially concerning chemical and hazardous substances, was first realized about 30-40 years ago.[1] Main aims of EEC regulations are the registration and the classification of a wide variety of chemical substances, the environmental protection, the health protection, harmonization of national laws, and the liberalization of the market. For these reasons several EEC regulations were created and completed (see 19.10.1). Changes of regulations intended to optimize some aspects, e.g., environmental and health aspects by the obligation of registration and test procedures.[7] The list 19.10.1 shows all relevant regulations in Europe concerning solvents.

Relevant German regulations concerning solvents are listed in 19.10.2 as an example of the conversion of EEC regulations in Member States. In Germany the application of chemical and hazardous substances are based on a few regulations.[28,30] Special and practical applications of these basic regulations are put in concrete terms in different standards, e.g., TRGS[29] or DIN.[31-56]

## 19.10.1 EEC REGULATIONS

1 67/548/EEC, Council Directive of 27 June 1967 on the approximation of the laws, regulations and administrative provisions of the Member States relating to the classification, packaging and labeling of dangerous substances, ABl. EG Nr. L 196/1, 1967.

2 73/173/EEC, Council Directive of 4 June 1973 on the approximation of the laws, regulations and administrative provisions of the Member States relating to the classification, packaging and labeling of dangerous preparations (solvents), ABl. EEC Nr. L 189, 1973.

3 Council Declaration (EC) and of the Member States of 22 November 1973 concerning an environmental program (EC), ABl. EG Nr. C 112/1, 1973.

4 75/442/EEC, Council Directive of 15 July 1975 on waste, ABl. EEC Nr. L 194/47, 1975.

5 76/769/EEC, Council Directive of 27 July 1976 on the approximation of the laws, regulations and administrative provisions of the Member States relating to limitations of the use of certain dangerous substances and preparations, ABl. EEC Nr. L 262/201, 1976.

6 77/728/EWG, Council Directive of 7 November 1977 on the approximation of the laws, regulations and administrative provisions of the Member States relating to the classification, packaging and labeling of paints, varnishes, printing paints, adhesives etc., ABl. EEC Nr. L 303/23, 1977.

7 79/831/EEC, Council Directive of 18 September 1979 on the approximation of the laws, regulations and administrative provisions of the Member States relating to the classification, packaging and labeling of dangerous preparations, ABl. EEC Nr. L 154/1, 1979.

8 88/379/EEC, Council Directive of 7 June 1988 on the approximation of the laws, regulations and administrative provisions of the Member States relating to the classification, packaging and labeling of dangerous preparations, ABl. EEC Nr. L 187/14, 1988.

9 88/344/EEC, Council Directive of 13 June 1988 on the approximation of the law of the Member States on extraction solvents used in the production of foodstuffs and food ingredients, ABl. EEC Nr. L 57/28, 1988.

10 89/349/EEC, Commission Recommendation of 13 April 1989 on the reduction of chlorofluoro-carbons by the aerosol industry, Abl. EG Nr. L 144/56, 1989.

11 91/155/EEC, Commission Directive of 5 March 1991 defining and laying down the detailed arrangements of the system of specific information relating to dangerous preparations in implementation of Article 10 of Directive 88/379/EEC, ABl. EEC Nr. L 76/35, 1991.
12 91/689/EEG, Council Directive of 12 December 1991 on hazardous waste, ABl. EG Nr. L 23/29, 1993.
13 92/2455/EEC, Council Regulation (EEC) of 23 July 1992 concerning the export and import of certain dangerous chemicals, ABl. EEC Nr. L 251/13, 1992.
14 93/793/EEC, Corrigendum to Council Regulation (EEC) No. 793/93 of 23 March 1993 on the evaluation and control of the risks of existing substances, ABl. EEC Nr. L 224/34, 1993.
15 93/67/EEC, Commission Directive of 20 July 1993 laying down the principles for assessment of risks to man and the environment of substances notified in accordance with Council Directive 67/548/EEC, ABl. EEC Nr. L 227/9, 1994.
16 Council Regulation (EEC) No. 259/93 of 1 February 1993 on the supervision and control of waste within, into and out of the European Community, ABl. EG Nr. L 30/1, 1993,
17 Commission decision of 20 December 1993 concerning a list of hazardous waste pursuant to Article 1(a) of Council Directive 91/689/EEC on hazardous waste, ABl. EG Nr. L5/15, 1994.
18 94/1179/EEC, Commission Regulation of 25 May 1994 concerning the first list of priority substances as foreseen under Council Regulation (EEC) No. 793/93, ABl. EEC Nr. L 161/3, 1994.
19 94/1488/EEC, Commission Regulation of 28 June 1994 laying down the principles of the assessment of risks to mass and the environment of existing substances in accordance with Council Regulation No. 793/93, ABl. EEC Nr. L 161/3, 1994.
20 Council Regulation (EEC) No. 3093/94 of 15 December 1994 on substances that deplete the ozone layer, ABl. EG Nr. L 333/1, 1994.
21 94/904/EEC, Council Decision of 22 December 1994 establishing a list of hazardous waste pursuant to Article 1(4) of Council Directive 91/689/EEC on hazardous waste, ABl. EG Nr. L 356/4, 1994.
22 95/365/EEC, Commission Decision of 25 July 1995 establishing the ecological criteria for the award of the Community eco-label to laundry detergents, ABl. EG Nr. L 217/14, 1995.
23 95/2268/EEC, Commission Regulation of 27 September 1995 concerning the second list of priority substances as foreseen under Council Regulation No. 793/93, ABl. EEC Nr. L 231/18, 1995.
24 96/61/EEC, Council Directive of 24 September 1996 concerning integrated pollution prevention and control, ABl. EG Nr. L 257/26, 1996.
25 97/142/EEC, Commission Regulation of 27 January 1997 concerning the delivery of information about certain existing substances, ABl. EEC Nr. L 25/11, 1997.

## 19.10.2 GERMAN REGULATIONS

26 Verordnung über Höchstmengen an bestimmten Lösemitteln in Lebensmitteln (Lösemittel-HöchstmengenVerordnung - LHmV) vom 27. Juli 1989. Regulation concerning maximum quantities of certain solvents in foods (LHmV) of 27 July 1989, BGBl., 1568 (1989).
27 Verordnung über die Entsorgung gebrauchter halogenierter Lösemittel (HKWAbfV) vom 23. Oktober 1989. Regulation concerning the waste management of halogenated solvents (HKWAbfV) of 23 October 1989, BGBl., 1918 (1989).
28 Verordnung zum Schutz vor gefährlichen Stoffen (Gefahrstoffverordnung - GefStoffV) vom 26.10.1993. Regulation concerning the protection against hazardous substances of 26 October 1993.
29 Technische Regeln für Gefahrstoffe (TRGS). Technical Regulations concerning hazardous substances (TRGS), e.g. TRGS 002, TRGS 003, TRGS 101, TRGS 102, TRGS 150, TRGS 220, TRGS 222, TRGS 400, TRGS 402, TRGS 403, TRGS 404, TRGS 415, TRGS 420, TRGS 512, TRGS 519, TRGS 531, TRGS 900, TRGS 903, TRGS 905, TRGS 906, TRGS 910.
30 Gesetz zum Schutz vor gefährlichen Stoffen (Chemikaliengesetz - ChemG) vom 25. Juli 1994. Law concerning the protection against hazardous substances of 25 July 1994, BGBl. I, 1703 (1994).
31 DIN 53169, Solvents for paints and varnishes; determination of density, refractive index, flash point, acid value, saponification value, olefinic and aromatic content in hydrocarbons, 1991.
32 DIN 53170, Solvents for paints and similar coating materials; determination of the evaporation rate, 1991.
33 DIN 53171, Solvents for paints and varnishes; determination of distillation characteristics (boiling range and boiling temperature as a function of distilled volume), 1991.
34 DIN 53172, Solvents for paints and varnishes; determination of evaporation residue, 1993.
35 DIN 53173, Solvents for paints and varnishes; determination of carbonyl value, 1991.
36 DIN 53174-1, Solvents for paints and varnishes - Methods of test for solvent mixtures - Part 1: General references and survey, 1995.

37 DIN 53174-2, Solvents for paints and varnishes and similar coating materials; methods of analysis for solvent mixtures; gas chromatographic method, 1992.

38 DIN 53175, Binders for paints, varnishes and similar coating materials; determination of the solidification point (titer) of fatty acids, 1991.

39 DIN 53245, Solvents for paints and varnishes; alcohols; supply specification, further properties and methods of test, 1994.

40 DIN 53246, Solvents for paints and varnishes - Acetic esters - Delivery specification, further requirements and methods of test, 1997.

41 DIN 53247, Solvents for paints and varnishes - Ketones - Supply specifications, further requirements and methods of test, 1997.

42 DIN 53248, Solvents for paints, varnishes and similar coating materials - Gum spirit of turpentine and wood turpentines - Requirements and methods of test, 1995.

43 DIN 53249, Solvents for paints, varnishes and similar coating materials - Dipentene - Requirements and methods of test, 1995.

44 DIN 55651, Solvents for paints and varnishes - Symbols, 1997.

45 DIN 55681, Solvents; stability testing of trichloroethylene, 1985.

46 DIN 55682, Solvents for paints and varnishes - Determination of solvents in water-thinnable coating materials - Gas chromatographic method, 1994.

47 DIN 55682/A1, Solvents for paints and varnishes - Determination of solvents in water-thinnable coating materials - Gas chromatographic method; Amendment A1, 1998.

48 DIN 55683, Solvents for paints and varnishes - Determination of solvents in coating materials containing organic solvents only - Gas chromatographic method, 1994.

49 DIN 55685, Solvents for paints and varnishes; alcohols; gas chromatographic determination of the degree of purity, 1992.

50 DIN 55686, Solvents for paints and varnishes; acetic esters; gas chromatographic determination of the degree of purity, 1992.

51 DIN 55687, Solvents for paints and varnishes; ketones; gas chromatographic determination of the degree of purity, 1992.

52 DIN 55688, Solvents for paints and varnishes - Ethylene glycol ethers - Gas chromatographic determination of the degree of purity, 1995.

53 DIN 55689, Solvents for paints and varnishes - Propylene glycol ethers - Gas chromatographic determination of the degree of purity, 1995.

54 DIN 55997, Solvents for paints and varnishes - Deionized water - Requirements and methods of test, 1998.

55 DIN 55998, Solvents for paints and varnishes - Propylene glycol ethers - Supply specification, further requirements and methods of test, 1998.

56 DIN 55999, Solvents for paints and varnishes, ethylene glycol ethers; supply specification, further properties and methods of test, 1994.

# 20

# Toxic Effects of Solvent Exposure

## 20.1 TOXICOKINETICS, TOXICODYNAMICS, AND TOXICOLOGY

Tilman Hahn, Konrad Botzenhart, Fritz Schweinsberg
**Institut für Allgemeine Hygiene und Umwelthygiene**
**Universität Tübingen, Tübingen, Germany**

### 20.1.1 TOXICOKINETICS AND TOXICODYNAMICS

#### 20.1.1.1 Exposure

Highest exposures can be found in workplace (e.g., evaporation of solvents) or during special processes (e.g., leaks of normally closed systems). Acute and severe solvent accidents often happen in workplaces (high solvent concentrations, intermittent high-level exposures, high duration of exposure). Apart from working sites, various other emission sources of solvents should be considered, e.g., consumer products.

The description of exposure parameters (type of solvents, concentrations, duration, routes of exposure) are important for the evaluation of toxicokinetics. Solvents and other chemicals are usually emitted as a mixture of various substances. Therefore, the risk assessment of emitted solvents is difficult to ascertain.[1,2] Solvent concentrations and duration of exposure vary in most cases (intermittent high-value peaks, periods of low exposure). The exposure is influenced essentially by surrounding occupational and environmental conditions, such as working climate, protective equipment and by individual parameters such as eating habits.

The exposure to solvents is regulated by relevant threshold limit values.[1,2] Exposure and exposure values can be controlled by defined methods (e.g., ambient and biological monitoring).

#### 20.1.1.2 Uptake

Relevant uptake routes of solvents are absorption from the lung and percutaneous absorption. The intestinal uptake is usually caused by accidents or by intent. The absorption rate is influenced by various factors.

### *20.1.1.2.1 Inhalation*

Inhalation is the most common pathway of solvent absorption, especially at working sites. The pulmonal absorption of solvents depends on the following parameters:[3-6]

- Exposure (concentrations and concentration fluctuations in the ambient air, exposure time, physical exertion). The alveolar concentration of solvents or the difference between air and blood concentration levels determine the diffusion process into alveolar blood vessels. Physical exertion influences lung parameters, especially ventilation, and consequently alveolar and blood concentrations.
- Lung parameters (pulmonary and alveolar ventilation, pulmonary perfusion, air-blood coefficient, blood-tissue coefficient). These coefficients describe the amount of solvents which can diffuse. The blood-tissue partition coefficient influences the tissue equilibrium concentrations. Solvents with stronger hydrophobic properties (e.g., toluene) reach equilibrium more rapidly because of a low tissue-blood coefficient. Intraindividual differences such as child/adult are also of significance.
- Physicochemical characteristics of solvents (solubility such as hydrophobic and hydrophilic properties, state such as liquid or gaseous and degree of volatility).

### *20.1.1.2.2 Dermal uptake*

Dermal uptake of solvents requires skin contact and depends on the area of contact, skin thickness, dermal state (e.g., eczema and defects in the stratum corneum), exposure parameters (contact time, etc.) and solvent properties.[7,8]

The main barrier against percutaneous uptake of solvents are structures of the stratum corneum, especially intercellular lipids and fibrous keratin. Removal of lipids by polar solvents such as ethanol or hydration in the stratum corneum is associated with an increase of skin permeability. Defects or lack of stratum corneum that may occur in skin diseases, at particular skin locations such as hair follicles or glandula regions enhance the percutaneous movement of solvents. The absorption through mucosa membranes is facilitated because of the lack of the stratum corneum.

Skin defects or diseases can be provoked by solvents which cause irritation, cellular hyperplasia and swelling, or removal of lipids. Skin defects are provoked mainly by frequent use of solvents thus enhancing their absorption.

Other characteristics, which influence percutaneous absorption, are solvent concentration gradients, solvent partitioning (water/lipid partition coefficient) and permeability constants.

Lipophilic chemicals are absorbed most easily (for example, benzene). These can include liquid solvents or solvents having low vapor pressure.[9-11] Vapors absorbed by dermal uptake can significantly contribute to the body burden as a result of the whole body exposure: e.g. 1-2 % of xylene or toluene, up to 5-10 % 1-methoxypropane-2-ol.[10] For other substances, much higher skin absorption rates were measured after the whole body exposure: 2-methoxyethanol up to 55 %, 2-ethoxy-ethanol up to 42 %.[12]

It is important to consider that the dermal uptake of vapors is especially significant when using a gas-mask.[10] In addition to inhalation measurements, measurement of percutaneous absorption is an important method for assessing health or environmental risks.

Dermal absorption of solvents is shown in Table 20.1.1.

**Table 20.1.1. Dermal uptake of solvents according to the German MAK-list.[2,64]**

| | | | |
|---|---|---|---|
| Benzene | Cyclohexanone | 2-Hexanone | 1,1,2,2-Tetrachloroethane |
| Bromomethane | Dimethylformamide | Methanol | Tetrachloroethene |
| 2-Butanone | Dimethylsulfoxide | 2-Methoxyethanol | Tetrachloromethane |
| 2-Butoxyethanol | 1,4-Dioxane | Methyl formate | Toluene |
| Carbon disulfide | 2-Ethoxyethanol | Nitrobenzene | Toluidine(s) |
| 2-Chloroethanol | Ethylbenzene | Nitrotoluene(s) | 1,1,2-Trichloroethane |
| Chloromethane | Ethyl formate | Phenol | Trichloromethane |
| Cresol(s) | Ethylene glycol | iso-Propyl benzene | Xylene(s) |
| Cycolhexanol | n-Hexane | n-Propanol (from ACGIH[1]) | |

### 20.1.1.2 Metabolism, distribution, excretion

Specific toxicity of solvents is directly related to their metabolism which is predominantly catalyzed by cytochrome P-450 mixed-function oxidases in the liver or other tissues.

Relevant examples of specific metabolism are toxic epoxides of benzene (hemopoietic toxicity), n-hexane 2,5-hexanedione (peripheral neurotoxic effects), metabolites of ethylene glycol ethers (reproductive toxicity), and unidentified metabolites from trichloroethylene (renal-toxic effects).[13] It should be emphasized that only the metabolites of these solvents are associated with toxic effects.

Other relevant metabolic pathways result in detoxified substances, such as biotransformation processes in the liver – conjugation with glycine, glucuronic acid and sulphuric acid (e.g., via hydroxylation of toluene) or biotransformation by hydrolysis, oxidation and conjugation (e.g., glycol ethers).

It should be noted that metabolism processes vary according to the following conditions:[14]

- Species, sex, age, genetics, e.g., variability in enzymatic factors such as polymorphisms (cytochrome systems) or tissue repair mechanisms[15]
- Life style – diet, smoking, drug consumption, physical activity
- Saturation. Massive concentrations of solvents result in saturation of metabolic pathways. This is important with regard to detoxification
- Induction of enzymes. Specific induction of enzyme systems by chemicals (solvents as well as other chemicals such as drugs) may consequently provoke an increase or decrease of solvent toxicity
- Interactions may be involved in enhancing or reducing toxicity of solvents. For example Bloch et al.[16] showed that in cases of alcohol abuse an increase in the toxic effects of benzene and other lipophilic petroleum derivatives occurs. Also, it has been shown that benzene inhibits the metabolism of toluene.[17]

Solvents can be excreted via various pathways:

- Exhalation (unchanged)
- Urine tract and biliary tract (unchanged or metabolites, e.g. water-soluble conjugates)

### 20.1.1.3 Modeling of toxicokinetics and modifying factors

The complexity of toxicokinetic processes of solvents can be described in models, e.g., predicting exposure situations and distribution phenomena in the human body and quantifying these processes (e.g. dose-effect response relationships). This applies especially to simulation of physiological and physicochemical parameters[18] or to assessing low exposures to complex chemical mixtures.[19]

## 20.1.2 TOXICOLOGY

### 20.1.2.1 General effects

General effects of solvents concern primarily acute exposures to high solvent concentrations. Despite some variations of symptoms, the resulting effects on the central nervous system (CNS) are rather stereotypical.[20]

Several solvents have depressant or narcotic effects, and hence, some solvents are used as anesthetics.[21] The main acute health hazards result from the narcotic effects. Their intensity is proportional to the solvent concentrations in brain tissue and is caused by the solvents themselves (physical and chemical interactions with neural membranes, nerve cells or neurotransmitters of the CNS).

General CNS dysfunctions after solvent exposure, are initially euphoria and disinhibition, higher exposures result in pre-narcotic symptoms such as dizziness, euphoria, disorientation and confusion, nausea, headache, vomiting, ataxia, paresthesia, increased salivation and tachycardia.[22,23] The symptoms are rapidly reversible when the solvents are removed.

In addition to the non-specific acute narcotic effects of solvents mentioned above, alterations of behavioral, cognitive and psychomotoric functions are typically found after short-term exposure to solvent levels close to the TLV. Overexposure leads to convulsions, coma and death. Typical changes are paresthesias, visual and auditory deficits, cognitive deficits (short-term and long-term memory loss), confusion, disorientation, affective deficits (nervousness, irritability, depression, apathy, compulsive behavior) and motor deficits (weakness in extremities, incoordination, fatigue, tremor).[24,25]

It is difficult to develop useful methods and models for testing these behavioral effects of solvents but for this purpose tests of attention and reaction, cognitive tests and other test systems are used.[26,27]

Acute CNS dysfunction diseases can show mild (organic affective syndrome), moderate or severe (acute toxic encephalopathy) symptoms.[28,29]

Unspecific irritations of skin and mucosa membrane structures can be caused by solvents. Various solvents are significant occupational irritants, e.g., solvents which cause irritant contact dermatitis.[30] Intact skin structures can be destroyed by solvents which dissolve grease and fat. Typically, the dermatitis is characterized by dryness, scaling and fissuring and is usually located on the hands. It is often caused by handling solvent-contaminated products or by cleaning procedures.[31,32]

Unspecific irritation of mucous membranes is often caused by solvent vapors, e.g., irritation of the eyes and various sections of the airways.

### 20.1.2.2 Specific non-immunological effects

Table 20.1.2 summarizes the main specific effects of solvents:[33-47]

- Hepatotoxicity
- Nephrotoxicity
- Reproductive toxicity
- Hemopoietic toxicity
- Neurotoxicity
- Ocular toxicity

**Table 20.1.2. Examples for specific effects of selected solvents**

| Organ-system | Solvents | Symptoms |
|---|---|---|
| Liver | halogenated hydrocarbons (e.g., carbon tetrachloride, tetrachloroethane, chloroform), ethanol, 1,1,1-trichloroethane, trichloroethylene, bromobenzene, dimethylformamide | acute (necrosis, steatosis) and chronic (cirrhosis) hepatotoxic symptoms |
| Kidney | halogenated hydrocarbons (e.g., carbon tetrachloride), toluene, dioxane, diethylene glycol, ethylene glycol, glycol ethers, conjugates of trichloroethylene | acute tubular necrosis, glomerular and tubular dysfunctions (e.g., albuminuria, proteinuria), glomerulonephritis, note: modification of solvent effects caused by renal dysfunctions possible |
| Reproductive system | carbon disulfide, benzene, glycol ethers, nitrobenzene | disturbance of menstrual cycle; reduced sperm counts, embryotoxic effects |
| Hemopoietic system | benzene metabolites (e.g., benzoquinone, hydroquinone) | marrow depression, myelotoxic effects |
| Nerval system | n-hexane, ethanol, styrene, tetrachloroethylene | peripheral neuropathy (especially distal axons, axon swelling and degeneration, loss of sensibility, muscular atrophy, loss of tendon reflexes) |
| Eye | methanol | impaired vision |

Note: the data shown come predominantly from data of occupational exposure.

### 20.1.2.3 Immunological effects

Various solvents have well-known allergic potentials. Allergic symptoms of the respiratory tract (rhinitis, tracheitis, bronchitis, asthma), allergic contact dermatitis and conjunctivitis can be provoked by solvents. The allergic effects of solvents can also contribute to other diseases such as MCS, autoimmune diseases.

Nowadays, solvents or by-products with allergic potential occur mainly at workplaces and, to a lesser degree, in consumer products. According to EG regulations, solvent ingredients of some consumer products, e.g., cosmetic products, must be labeled. It is often difficult to detect the causative solvent allergen (allergens which cause cross allergies, secondary products of solvents such as oxygenated terpenes, unknown allergens). Various specific test systems are available for carrying out individual test diagnoses: e.g., chamber tests,[48] skin tests such as patch-tests[49] and special applications of biological monitoring.

Solvent-induced allergies can occur at a variety of working sites, e.g., in shoe factories,[50] in electronic industries,[51] in synthetic chemical industries,[52] in metal industries[53] or in perfume and potter industries (oil of turpentine and other solvents).[54] Similar occurrence of solvents can be found in consumer products, e.g., in nail polishes (e.g., toluene).[55] Allergic solvent substances are listed in various catalogues and databases.[1,2,49]

Examples of allergic solvents are terpene products with high sensitivity potential, which can cause positive test reactions (patch-test) or even allergic diseases (contact sensitization and dermatitis). Allergic dermatitis can even be provoked by d-limonene in the air.[56] Terpenes and terpenoid substances are found especially in "natural products", e.g., cosmetic products, foods, and plants (oilseed rape).[57,58]

Allergic potential of solvent products depends on the typical solvent structure. For example, in glycol ethers their allergic potential is proportional to the charge of interacting molecules.[59]

Allergic effects can also be associated with other skin conditions caused by solvents such as irritations. Multiple areas of skin damage, including solvent allergies, can change the skin structure and provoke severe skin disease.[60]

In addition to other substances (pesticides, food additives, dust, smoke, etc.), allergic effects of solvents are discussed as an initial cause of MCS.[61]

Organic solvents are associated with human autoimmune diseases, but defined pathomechanisms of these solvents have not yet been detected (role of solvents in the initiation or progression of autoimmune diseases).[62]

#### 20.1.2.4 Toxic effects of solvents on other organisms

In addition to humans, microorganisms animals and plants are also exposed to solvents. The interaction between organisms and solvents are often specific. For example, the reactions elicited by certain solvents depend on the species and abilities of the particular organism affected.

Hydrophobic organic solvents, in particular, are toxic to living organisms, primarily because they disrupt cell membrane structure and mechanisms. Some living organisms especially certain bacterial species, are able to adapt to these solvents by invoking mechanisms such as accelerating repair processes (through changes in the rate of phospholipid biosynthesis), reduction of the diffusion rate of the solvent and active reduction of the intracellular concentration of the solvent. More information and examples are shown in Chapter 14.4.2.

#### 20.1.2.5 Carcinogenicity

The term carcinogenicity is used for toxicants that are able to induce malignant neoplasms. Carcinogens can be effective at different stages of the carcinogenic process, e. g., initiation, promotion and progression. They may interact with other noxes and thereby enhance tumor development. Interactive carcinogenesis can be described as co- and syn-carcinogenesis. A co-carcinogen is defined as a non-carcinogenic compound that is able to enhance tumor development induced by a given carcinogen. In syn-carcinogenesis two or more carcinogens, each occurring in small amounts that are usually not sufficient to induce a tumor in a specific target organ, may interact to lead to tumor formation in that organ.

As with all carcinogens the carcinogenic potency of solvents has been assessed by short-term in vitro tests, e. g., Ames assay, by long-term tumor induction experiments in animals and - especially important for the evaluation of the carcinogenic action in humans - prospective and retrospective epidemiological studies, for solvent exposure mainly in work places.

From this data it is generally not possible to evaluate the carcinogenic action of solvent mixtures, which occur in the majority of exposure situations. It is also important to note, that for a number of reasons, e. g., very long latency period of tumor generation, accumulation of single hits in the target cells, significance of repair mechanisms it is not possible to define TLVs for carcinogens.

In accordance with the evidence available, different classes for chemical carcinogens have been developed by health authority organizations.[1,2,34-36] Examples of the classification of carcinogenic solvents are presented in Table 20.1.3.

**Table 20.1.3. Carcinogenicity - Survey of selected solvents**

| Solvent | Organ-System | Category* | | | | |
|---|---|---|---|---|---|---|
| | | MAK | EG | ACGIH | IARC | NTP |
| Benzene | *hemopoietic system* | 1 | K1 | A1 | 1 | K |
| Bromomethane | upper gastrointestinal, tract and respiratory, tract (animals) | 3 | K3 | n.c.** | 3 | n.l.** |
| Carbon tetrachloride | lymphatic system, liver (mice, rats), mamma (rats), suprarenal gland (mice) | 3 | K3 | A2 | 2B | R |
| Epichlorohydrin | *lung, CNS*, forestomach (rats), nasal cavity, skin (mice) | 2 | K2 | A3 | 2A | R |
| Chloroethane | uterus (mice) | 3 | K3 | n.l. | 3 | n.l. |
| Cyclohexanone | suprarenal gland (rats) | 3 | n.c. | A4 | 3 | n.l. |
| 1,2-Dibromoethane | forestomach (mice), lung (mice, rats), nasal cavity, peritoneum, mamma, connective tissue (rats) | 2 | K2 | A3 | 2A | R |
| 1,2-Dichloroethane | *brain, lymphatic and hemopoietic system, stomach, pancreas*; lung, mamma, stomach (mice, rats), lymphatic system (mice) | 2 | K2 | A4 | 2B | R |
| Dichloromethane | liver, lung (mice, rats), mamma (rats), lymphosarcomas (mice) | 3 | K3 | A3 | 2B | R |
| 1,2-Dichloropropane | liver (mice), mamma (rats) | 3 | K3 | A4 | 3 | n.l. |
| Dimethylformamide | *testes* | n.c. | n.c. | A4 | 3 | n.l. |
| 1,4-Dioxane | liver (rats, guinea pigs), biliary tract (guinea pigs), mamma , peritoneum (rats), nasal cavity (mice) | 4 | n.i.** | A3 | 2B | R |
| 1,2-Epoxypropane | mamma, upper respiratory tract, thyroid gland (mice, rats) | 2 | K2 | A3 | n.l. | n.l. |
| Hexamethyl phosphoramide | nasal cavity, lung (rats) | n.l. | n.i. | A3 | 2B | R |
| 2-Nitropropane | liver (rats) | 2 | n.l. | A3 | 2B | R |
| Nitrobenzene | lung, thyroid gland, mamma (mice), liver, kidney, uterus (rats) | 3 | K3 | A3 | 2B | n.l. |
| 2- Nitrotoluene | epididymis (rats) | 2 | K2 | n.c.,BEI** | 3 | n.l. |
| Phenol | lymphatic system, hemopoietic system suprarenal gland, thyroid gland, skin (mice, rats) | 3 | n.c. | n.c.,BEI | 3 | n.l. |
| Tetrachloroethane | liver (mice) | 3 | K3 | A3 | 3 | n.l. |
| Tetrachloroethylene | *oesophagus, kidney, hemopoietic system, lymphatic system*; liver (mice), hemopoietic system (rats) | 3 | K3 | A3 | 2A | R |

| Solvent | Organ-System | Category* | | | | |
|---|---|---|---|---|---|---|
| | | MAK | EG | ACGIH | IARC | NTP |
| Tetrachloromethane | stomach, liver, kidney, thyroid gland (rats, mice) | 3 | K3 | A2 | n.l. | n.l. |
| o-Toluidine | mamma, skin, bladder, liver, spleen, peritoneum, connective tissue (rats), vessels (mice) | 2 | n.i. | A3 | n.l. | R |
| 1,1,2-Trichlorethane | liver, suprarenal gland (mice) | 3 | K3 | A4 | 3 | n.l. |
| Trichloroethylene | *kidney*; liver, biliary tract, kidney, lung, cervix, testes, lymphatic system (rats, mice) | 1 | K3 | A5 | 2A | n.l. |
| Chloroform | stomach, liver, kidney, thyroid gland (mice, rats) | 4 | K3 | A3 | n.l. | R |
| 1,2,3-Trichloropropane | oral mucosa (mice, rats), uterus (mice), liver, pancreas, forestomach, kidney, mamma (rats) | 2 | n.i. | A3 | 2A | R |

*Categories

MAK (German regulations)[2]

**1**: substances that cause cancer in humans and can be assumed to make a significant contribution to cancer risk. Epidemiological studies provide adequate evidence of a positive correlation between the exposure of humans and the occurrence of cancer. Limited epidemiological data can be substantiated by evidence that the substance causes cancer by a mode of action that is relevant to humans.

**2**: substances that are considered to be carcinogenic for humans because sufficient data from long-term animal studies or limited evidence from animal studies substantiated by evidence from epidemiological studies indicate that they can make a significant contribution to cancer risk. Limited data from animal studies can be supported by evidence that the substance causes cancer by a mode of action that is relevant to humans and by results of in vitro tests and short-term animal studies.

**3**: substances that cause concern that they could be carcinogenic for humans but cannot be assessed conclusively because of lack of data. In vitro tests or animal studies have yielded evidence in one of the other categories. The classification in Category 3 is provisional. Further studies are required before a final decision can be made. A MAK value can be established provided no genotoxic effects have been detected.

**4**: substances with carcinogenic potential for which genotoxicity plays no or at most a minor role. No significant contribution to human cancer risk is expected provided the MAK value is observed. The classification is supported especially by evidence that increases in cellular proliferation or changes in cellular differentiation are important in the mode of action. To characterize the cancer risk, the manifold mechanisms contributing to carcinogenesis and their characteristic dose-time-response relationships are taken into consideration.

**5**: substances with carcinogenic and genotoxic potential, the potency of which is considered to be so low that, provided the MAK value is observed, no significant contribution to human cancer risk is to be expected. The classification is supported by information on the mode of action, dose-dependence and toxicokinetic data pertinent to species comparison.

EG[65]

- **K1**: confirmed human carcinogen
- **K2**: compounds which should be considered as carcinogen
- **K3**: compounds with possible carcinogenic evidence

ACGIH[1]

- **A1**: confirmed human carcinogen
- **A2**: suspected human carcinogen
- **A3**: confirmed animal carcinogen with unknown relevance to humans
- **A4**: not classifiable as a human carcinogen
- **A5**: not suspected as a human carcinogen

IARC[34-36]

**1**: carcinogenic to humans
**2A**: probably carcinogenic to humans
**2B**: possibly carcinogenic to humans
**3**: not classifiable as to its carcinogenicity to humans

NTP[66]

**K**: Known to be a Human Carcinogen
**R**: Reasonably Anticipated to be a Human Carcinogen (RAHC)

**Notes:

italic: cancer in humans
n.c.: not classified as carcinogenic
n.i.: no information available
n.l.: not listed
BEI: not classified as carcinogenic but biological monitoring is recommended

### 20.1.2.6 Risk assessment

For risk assessment of solvent exposure, and in addition to factors for general risk assessment (age, gender, race, diet, physical activity, stress, physical noxes, etc.) it is important to consider:

- Occupational exposure (high doses) and environmental exposure (low doses) to solvents separately.
- The effect of exposure time, e. g., life long environmental low exposure or occupational intermittent high exposure.
- Exposure assessment (generally the most neglected aspect in risk assessment). This involves extensive ambient monitoring over a long period of time. Only a small amount of data on biological monitoring of solvents and/or metabolites (representing the "effective" dose) is available.
- The high volatility of solvents, e. g., VOCs and the fast biotransformation rate (in the environment and within the human body) for most of the solvents.
- Complex mixtures and numerous sources of environmental exposure.
- Especially for environmental solvent exposure: High-to-low-dose extrapolation for evaluation of adverse health effects may be misleading.
- Confounding factors, e.g., smoking and alcohol consumption, as adverse health effects which may dominate in cases of low solvent exposure.
- Risk in this context is defined in terms of the probability as occurrence of a particular adverse health effect, e. g. 1 in $10^6$.
- Finally, as in general risk assessment, definition of a risk level that is acceptable.

### 20.1.3 CONCLUSIONS

- For solvent exposure at workplaces considerable amount of evidence for adverse health effects has been gathered.
- In this regard, specific and carcinogenic effects in particular have been discussed (see Table 20.1.2 and 20.1.3).
- For environmental solvent exposure only a few examples of adverse health effects have been documented.
- It is rather unlikely that potentially toxic environmental solvent exposures, e. g., benzene or halogenated hydrocarbons, can be prevented in the near future.

- Many suspicions, but only a small amount of scientific data demonstrate a correlation between "environmental diseases", e. g., sick building syndrome and solvent exposure.
- It has been hypothesized that - as a rule - exposure to mixtures of solvents at low non-toxic doses of the individual constituent represents no danger to health.[63]
- There exists overwhelming evidence of adverse health effects caused by accepted environmental noxes such as tobacco smoke and the consumption of alcoholic beverages.

## REFERENCES

1 ACGIH, TLV's and BEI's, ACGIH, Cincinnati, 1998.
2 DFG, MAK- und BAT-Werte-Liste, VCH, Weinheim, 1999.
3 I. Astrand, *Scand. J. Work Environ. Health*, **1**, 199, 1975.
4 I. Astrand in **Occupational Health Hazards of Solvents**, A. England, K. Ringen, M.A. Mehlman, Eds., *Princeton*, NJ, 1986, pp. 141-142.
5 K.H. Cohr, in **Safety and Health Aspects of Organic Solvents**, V. Riihimäki, U. Ulfvarson, Eds., *Alan R. Liss*, N.Y., 1986, pp. 45-60.
6 J.J.G. Opdam, *Br. J. Ind. Med.*, **46**, 831 (1989).
7 M.K. Bahl, *J. Soc. Cosmet. Chem.*, **36**, 287 (1985).
8 M. Bird, *Ann. Occup. Hyg.*, **24**, 235 (1981).
9 J. Angerer, E. Lichterbeck, J. Bergerow, S. Jekel, G. Lehnert, *Int. Arch. Occup. Environ. Health*, **62**, 123 (1990).
10 I. Brooke, J. Cocker, I. Delic, M. Payne, K. Jones, N.C. Gregg, D. Dyne, *Ann Occup. Hyg.*, **42**, 531 (1998).
11 G. Johanson, *Toxicol. Lett.*, **43**, 5 (1988).
12 S. Kezic, K. Mahieu, A.C. Monster, F.A. de Wolff, *Occup. Environ. Med.*, **54**, 38 (1997).
13 E.A. Lock, *Crit. Rev. Toxicol.*, **19**, 23 (1988).
14 A. Lof, G. Johanson, *Crit. Rev. Toxicol.*, **28**, 571 (1998).
15 H.M. Mehendale, *Toxicology*, **105**, 251 (1995).
16 P. Bloch, A. Kulig, M. Paradowski, T. Wybrzak-Wrobel, *Pol. J. Occup. Med.*, **3**, 69 (1990).
17 O. Inoue, K. Seiji, T. Watanabe, M. Kasahara, H. Nakatsuka, S.N. Yin, G.L. Li, S.X. Cai, C. Jin, M. Ikeda, *Int. Arch. Occup. Environ. Health*, **60**, 15 (1988).
18 V. Fiserova-Bergerova, *Scand. J. Work Environ. Health*, **11**, 7 (1985).
19 S. Haddad, K. Krishnan, *Environ. Health Perspect.*, **106**, 1377 (1998).
20 W.K. Anger in **Neurobehavioral Toxicology**, Z. Annau, Ed., *John Hopkins University Press*, Baltimore, MD, 1986, pp. 331-347.
21 A. Laine, V. Riihimäki in **Safety and Health Aspects of Organic Solvents**, V. Riihimäki, U. Ulfvarson, Eds., *Alan R. Liss*, N.Y., 1986, pp. 123-126.
22 E. Browning, **Toxicity and Metabolism of Industrial Solvents**, *Elsevier Publishing Co.*, N. Y., 1965.
23 R.E. Gosselin, R.P. Smith, H.E. Hodge, **Clinical Toxicology of Commercial Products**, *Williams and Wilkins*, Baltimore, 1984.
24 E.L. Baker, *Ann. Rev. Public Health*, **9**, 233 (1988).
25 P. Grasso, M. Sharratt, D. M., Davies, D. Irvine, *Food Chem. Toxicol.*, **22**, 819 (1984).
26 R.B. Dick, *Neurotoxicol. Teratol.*, **10**, 39 (1988).
27 W.K. Anger, *Neurotoxicology*, **11**, 627 (1990).
28 P. Arlien-Soborg, L. Hansen, O. Ladefoged, L. Simonsen, *Neurotoxicol. Teratol.*, **14**, 81 (1992).
29 WHO, Organic solvents and the central nervous system, WHO European Office Copenhagen, (1985).
30 J.F.Fowler, *Dermatology*, **10**, 216 (1998).
31 K.E. Andersen in **Safety and Health Aspects of Organic Solvents**, V. Riihimäki, U. Ulfvarson, Eds., *Alan R. Liss*, N. Y., 1986, pp. 133-138, 1986.
32 C.G.T. Mathias, *Occup. Med. State of the Art Rev.*, **1**, 205 (1986).
33 M. Hodgson, A.E. Heyl, D.H. Van Thiel, *Arch. Intern. Med.*, **149**, 1793 (1989):
34 IARC, IARC Monographs on the evaluation of carcinogenic risks to humans. Some organic solvents, resin monomers and related compounds, pigments and occupational exposures in paint manufacture and painting, WHO, 47, IARC, Lyon, 1989.
35 IARC, IARC Monographs on the evaluation of carcinogenic risks to humans. Dry cleaning, some chlorinated solvents and other industrial chemicals, WHO, 63, IARC, Lyon, 1995.

36 IARC, IARC Monographs on the evaluation of carcinogenic risks to humans. Re-evaluation of some organic chemicals, hydrazine and hydrogen peroxide. WHO, 71, IARC, Lyon, 1999.
37 DFG, Gesundheitsschädliche Arbeitsstoffe. Toxikologisch-arbeitsmedizinische Begründungen von MAK-Werten, VCH, Weinheim, 1999.
38 R.R. Lauwerys, A. Bernard, C. Viau, J.P. Buchet, *Scand. J. Work Environ. Health*, **11**, 83 (1985).
39 E.A. Lock, *Crit. Rev. Toxicol.*, **19**, 23 (1988).
40 N.A. Nelson, T.G. Robins, F.K. Port, *Am. J. Nephrol.*, **10**, 10 (1990).
41 H.J. Mason, A.J. Stevenson, G.M. Bell, *Ren. Fail.*, **21**, 413 (1999).
42 O. Ladefoged, H.R. Lam, G. Ostergaard, E.V. Hansen, U. Hass, S.P. Lund, L. Simonsen, *Neurotoxicology*, **19**, 721 (1998).
43 A.M. Seppalainen, *Crit. Rev. Toxicol.*, **18**, 245 (1988).
44 P.S. Spencer, H.H. Schaumburg, *Scand. J. Work Environ. Health*, **11**, 53 (1985).
45 L.H. Welch, S.M. Schrader, T.W. Turner, M.R. Cullen, *Am. J. Ind. Med.*, **14**, 509 (1988).
46 I.J Yu, J.Y. Lee, Y.H. Chung, K.J. Kim, J.H. Han, G.Y. Cha, W.G. Chung, Y.M. Cha, J.D. Park, Y.M. Lee, Y.H. Moon, *Toxicol. Letters*, **109**, 11 (1999).
47 W.G. Chung, I.J. Yu, C.S. Park, K.H. Lee, H.K. Roh, Y.N. Cha, *Toxicol. Letters*, **104**, 143 (1999).
48 J.C.Selner, *Regul. Toxicol. Pharmacol.*, **24**, 87 (1996).
49 K.E. Andersen, S.C. Rastogi, L. Carlsen, *Acta Derm. Venereol.*, **76**, 136 (1996).
50 G. Mancuso, M. Reggiani, R.M. Berdodini, *Contact Dermatitis*, **34**, 17 (1996).
51 H.H. Tau, M. Tsu Li-Chan, C.L. Goh, *Am. J. Contact. Dermat.*, **8**, 210 (1997).
52 T. Chida, T. Uehata, *Sangyo Igaku,* **29**, 358 (1987).
53 P.J. Coenraads, S.C. Foo, W.O. Phoon, K.C. Lun, *Contact Dermatitis*, **12**, 155 (1985).
54 J.T. Lear, A.H. Heagerty, B.B. Tan, A.G. Smith, J.S. English, *Contact Dermatitis*, **35**, 169 (1996).
55 E.L. Sainio, K. Engstrom, M.L. HenriksEckerman, L. Kanerva, *Contact Dermatitis*, **37**, 155 (1997).
56 A.T. Karlberg, A. DoomsGoossens, A., *Contact Dermatitis*, **36**, 201 (1997).
57 D.M. Rubel, S. Freeman, I.A. Southwell, *Australas J. Dermatol.*, **39**, 244 (1998).
58 M. McEwan, W.H. McFarlane-Smith, *Clin. Exp. Allergy*, **28**, 332 (1998).
59 G. Angelini, L. Rigano, C. Foti, M. Grandolfo, G.A. Vena, D. Bonamonte, L. Soleo, A.A. Scorpiniti, A.A., *Contact Dermatitis*, **35**, 11 (1996).
60 J. van de Walter, S.A. Jimenez, M.E. Gershwin, *Int. Rev. Immunol.*, **12**, 201 (1995).
61 D. Eis, *Allergologie*, **22**, 538 (1999).
62 J.J. Powell, J. Van-de-Water, M.E. Gershwin, *Environ. Health Perspect.*, **197**, 667 (1999).
63 F.R. Cassee, *Crit. Rev. Toxicol.*, **28**, 73 (1998).
64 **Römpp, Lexikon Chemie**, J. Falbe, M. Regitz, Eds., *Thieme*, Stuttgart (1999).
65 GISBAU, WINGIS, Bau Berufsgenossenschaften (Professional Associations of the Building Industry in Germany), 1999.
66 U. S. Department of Health and Human Services, National Toxicology Program, The 8th Report on Carcinogens, 1998.

# 20.2 COGNITIVE AND PSYCHOSOCIAL OUTCOME OF CHRONIC OCCUPATIONAL SOLVENT NEUROTOXICITY

JENNI A OGDEN
**Department of Psychology, University of Auckland, Auckland, New Zealand**

## 20.2.1 INTRODUCTION

Many organic solvents used in industry are neurotoxic, and may lead to a range of largely irreversible cognitive and psychological or psychiatric impairments in workers who are exposed over long periods of time, or who have had a peak exposure (an episode in which they were briefly exposed to a larger than normal level of solvent). The most vulnerable workers are those who work in the spray painting, boat building, printing, textile, plastic, agricultural and pharmaceutical industries. Often self-employed workers or those in small businesses are more at risk because the safety measures they take are not as closely monitored, and peer pressure to use safety equipment even when it is unwieldy, restrictive or expensive, is unlikely to be as strong as in large workshops. In addition they may be less well educated regarding the neurotoxic effects of the solvents they work with. The great majority of workers diagnosed with OSN are men, presumably because men make up the bulk of the workforce in trades and industries that use neurotoxic solvents.

The chronic, and often slow and insidious effects of occupational solvent neurotoxicity (OSN) include psychological and psychiatric symptoms, impairments in cognitive functioning, and negative psychosocial consequences. The Scandinavian countries are the research leaders in this field, and in recent years health professionals and industries in the United States and other major industrialized countries have become increasingly aware of the debilitating symptoms that can affect workers exposed to neurotoxins over a long time.[1] There have been allegations that OSN is often over-diagnosed by health professionals who are zealous believers, and that a significant number of workers who complain of OSN symptoms are malingering in the hope of obtaining financial compensation.[2] While these allegations almost certainly have some credibility, especially in countries such as the USA, where civil litigation has resulted in large settlements and the existence of OSN is now enshrined in legal precedent,[2] there is ample evidence that the OSN syndrome does exist and is a major health problem for workers in industries that utilize neurotoxic solvents. A number of research studies establishing the existence of OSN have been conducted in countries where there is only limited, if any, financial gain to be made from diagnosing OSN, including Hong Kong[3] and New Zealand.[4]

One of the primary difficulties researchers and health professionals face when trying to ensure that the symptoms they are observing are indeed the result of OSN, lies in the fact that the neurological damage resulting from chronic neurotoxin exposure tends to be diffuse, or may, for example, involve a neurotransmitter imbalance. It is therefore unlikely to be evident on a Computerized Tomograph (CT) or Magnetic Resonance Image (MRI) of the brain. A neurological examination is rarely helpful,[5] and in many cases the psychological and cognitive impairments are the only clear indicators of neurotoxicity. A neuropsychological assessment which utilizes a range of tests to assess cognitive abilities including attention, concentration, psychomotor speed, memory and visuospatial skills, along with a psychological interview or questionnaire assessing depression, irritability, motivation and fatigue, thus plays a major role in diagnosing chronic OSN.[6] The World Health Organization (WHO) and the Nordic and New Zealand Governments all require that a neuropsychological assessment be used in the diagnosis of solvent neurotoxicity.[7-9]

Many victims of OSN do not realize that their chronic fatigue, irritability, poor memory and other problems may be associated with the solvents in their workplace, and by the time they seek help from their doctor, psychologist or marriage guidance counsellor, the OSN symptoms are likely to be compounded and masked by other work and relationship problems (themselves possibly a consequence of the OSN symptoms).[6] Identification of OSN as the primary cause of the problems is therefore even more difficult, and proving cause and effect usually impossible. That OSN is a significant cause of the person's problems, can, however, often be established beyond reasonable doubt, provided that some guidelines are followed. The individual must clearly have been exposed to neurotoxins over a long period (usually set, rather arbitrarily, at 10 years or more of occupational exposure), or have suffered a peak exposure. Other major contributors to neurological impairment should be excluded (e.g., significant traumatic brain injury, or alcohol addiction), there should be no evidence of malingering, and the pattern of cognitive impairments and psychological symptoms should be typical of OSN.

### 20.2.2 ACUTE SYMPTOMS OF SOLVENT NEUROTOXICITY

Neurotoxic solvent exposure can result in some workers experiencing nausea, vomiting, loss of appetite, severe headaches, confusion, light-headedness and dermatitis. The solvent may be detectable on their breath and skin for hours and even days after they have left the solvent environment. Most of these symptoms resolve when they stop working with solvents but return when they come into contact with solvents again. Workers who suffer these acute symptoms do not necessarily go on to develop the chronic syndrome of OSN, perhaps in many cases because they are so disabled by the acute symptoms they stop working before irreversible damage occurs. Some workers who suffer acute symptoms do remain in the work environment, sometimes because of financial necessity, or because they do not realize the solvents are the cause of their problems.[10] Some workers who develop a chronic OSN syndrome have suffered from acute symptoms, but others have not. The reason for these individual differences is not clear.

### 20.2.3 CATEGORIZATION OF OSN

The 1985 International Solvent Workshop[11] proposed three types of OSN, as follows:

- Type 1 OSN: Characterized by subjective complaints of fatigue, irritability, depression and episodes of anxiety. No cognitive impairments are demonstrable on neuropsychological testing, and the psychological symptoms resolve on removal from the solvents. This is also known as the organic affective syndrome, or the neurasthenic syndrome.
- Type 2 OSN: A more severe and chronic form than Type 1 in which many of the symptoms and cognitive impairments are thought to be irreversible when the worker is removed from the solvent environment. It is also known as mild toxic encephalopathy. Type 2 has been divided further into two sub-types based on psychological symptoms (Type 2A) and cognitive impairments (Type 2B). Type 2A sufferers have a range of symptoms which may include sustained personality and mood disturbances, fatigue, poor impulse control and poor motivation. Type 2B symptoms include poor concentration, impairments of new verbal and visual

learning and memory, psychomotor slowing, and in more severe cases, executive (or frontal-lobe) impairments. These can include impoverished verbal fluency, difficulties with abstract thinking, and impairments in the ability to make plans and organize tasks logically. These cognitive symptoms must be demonstrable on neuropsychological tests following a solvent-free period. There is some research which indicates that this separation of Type 2 OSN into psychological and cognitive impairment profiles is largely unrealistic, as most workers with Type 2 OSN have symptoms of both types.[10,12] Type 2 OSN is the primary focus of this section given its largely irreversible nature and its frequency in the workplace.

- Type 3 OSN: This is the most severe form of OSN and signals an irreversible dementia with severe impairment across most cognitive and emotional domains. It is also known as severe toxic encephalopathy, and is fortunately rare in occupational situations. It is more likely to occur in long-term recreational solvent abusers.

### 20.2.4 ASSESSMENT OF OSN

There have been a few studies reporting specific symptoms caused by a specific solvent. The widely used industrial solvent trichloroethylene (TCE), has, for example, been reported to result in severe agitated depression, sometimes accompanied by violent behaviors towards self and others.[13] Toluene and TCE can cause peripheral neuropathy, and TCE can damage the trigeminal or fifth cranial nerve, resulting in a loss of sensation to the face, mouth and teeth.[1] It is, however, rare to be able to pinpoint a specific solvent as the cause of specific cognitive or psychological symptoms, and most research on occupational solvent neurotoxicity has been carried out on workers exposed to a mixture of solvents. A core neuropsychological battery has been developed by the WHO/Nordic Council,[8] and most other formal and informal batteries developed for the assessment of OSN include a similar range of tests, as these are the tests most sensitive to the common neuropsychological impairments of OSN.[9,14,15,16] Specific tests used in these batteries will not be listed here, as neuropsychologists qualified to administer, score, and interpret these tests can find specialist information in texts written on OSN assessment.[1]

The assessment of OSN may be initiated if a worker receives a poor score on a screening workplace questionnaire designed to assess the frequency of self-reported problems such as irritability and poor memory.[12] In other cases the worker comes to the attention of a health professional because of interpersonal or memory problems which concern the worker, family, or work colleagues. In New Zealand, in 1993 the Occupational Safety and Health Service (OSH) of the Government Department of Labour, established a panel of experts to develop national guidelines for the diagnosis of OSN.[4,9] Workers who are diagnosed as suffering from OSN are registered as part of the Notifiable Occupational Disease System. Other panels provide a similar function for other occupational diseases such as asthma and asbestos-related disorders. Following is a description of the procedures for diagnosing OSN that the New Zealand panel has developed and tested since 1993.[4,9]

Individuals, industries, industrial health workers, or general practitioners can notify a possible case of OSN to the panel. Occupational hygienists then attempt to measure the types and levels of solvents the worker has been potentially exposed to throughout his or her working life. This is easier if the worker is currently in the solvent environment, but estimates only can be made of solvent levels in previous workplaces, and of the workplace and worker's appropriate use of protective equipment over the years. If there is reason to suspect that the worker has been exposed to neurotoxic solvents for 10 years or more, or has suffered peak exposures, the occupational physician will examine and interview the worker (and where possible a close family member) and make an initial assessment regarding the worker's symptoms. It is not uncommon at this interview stage for the worker, often a middle-aged tradesman not accustomed to talking about his cognitive or emotional problems, to break down in tears. Most health professionals experienced in assessing OSN are in no doubt that it is a real syndrome with devastating consequences for the worker and family.[6]

If the symptom complex generally fits with that typical of OSN, the symptoms are significant enough to be causing the worker or his family concern, and other possible causes have been explored and considered to be unlikely as the primary cause of the problems, the worker will proceed to a neuropsychological assessment. Whenever possible, this should take place following two or more weeks away from solvents. This is again a somewhat arbitrary time period, arrived at in an attempt to find a balance between the real time it takes for any acute effects of a mixture and range of solvents to resolve, and the amount of time (usually unpaid) an undiagnosed worker is willing or able to take away from his workplace. The assessment usually commences with a psychological assessment, which may include both an interview and standard questionnaires on mood, fatigue levels, motivation, memory problems in daily life and so on. Often, with the worker's permission, information is also obtained from family members and work colleagues. Not only does this allow an assessment of the problems the worker is experiencing at work and at home, but also gives the neuropsychologist some idea of the time course of these problems. Other possible confounding psychosocial factors are checked out at this point. Whilst factors such as a high use of alcohol, or a series of minor head injuries whilst playing sport 10 years previously, or a recent marriage breakup, may not negate the possibility of the worker being diagnosed as suffering from OSN, clearly these factors must be taken into account in making the diagnosis and the confidence that can be placed in that diagnosis, as well as when designing an intervention or rehabilitation program for the worker.

Having ascertained that the worker's exposure levels and psychological and subjective cognitive symptoms (e.g., complaints of memory problems) meet the criteria for possible OSN, a battery of carefully chosen neuropsychological tests is then given. This is often scheduled for a later session, given the distress that the worker may have expressed during the interview, and the high fatigue levels that are a common consequence of OSN. This battery should include one or more tests which can, along with education and occupational history, provide an estimate of the worker's cognitive ability level prior to working with solvents. Also included should be some tests which one would not expect to be impaired by solvents, such as well-established vocabulary (meanings of words). Tests which are included because of their sensitivity to OSN symptoms include tests of concentration and attention, new verbal and visuospatial learning and memorizing (old, well-established memories are rarely impaired), reaction time, psychomotor speed, and planning, organizational and abstraction abilities.

If the pattern of spared and impaired psychological and neuropsychological test results is typical of OSN, and other factors can be ruled out as the primary cause of this profile, the worker will be diagnosed as having OSN.[6,10] This pattern analysis provides one way of guarding against malingering, as the worker does not know which tests he or she should remain unimpaired on and which are commonly impaired following OSN. In addition, on many tests, it is very difficult or impossible for the malingerer to perform in a way that is consistent with true organic impairment, even if he or she has been coached on how to perform poorly on the tests. For example, if an individual was unable to remember any new visual stimuli (an extremely rare condition), when given a memory test where the worker is shown 50 photographs of unknown faces, and is then shown fifty pairs of faces and must choose from each pair the face which he or she has previously seen, he or she should obtain a score of approximately 50% (chance level) correct. If the score was considerably worse than that, malingering or exaggerating might reasonably be suspected. Tests which measure reaction or response times for increasingly complex tasks are also difficult to malinger successfully on as humans are not good at estimating response times in milliseconds, or even seconds.

A diagnosis of Type 2 OSN is based on score deficits (measured by the number of Standard Deviations (SD) below the worker's estimated premorbid ability level) on those tests commonly impaired by OSN. At least three neuropsychological test scores must fall more than 1 SD below the scores expected for that worker to be categorized as mild Type 2 OSN, three test scores below 2 SDs for moderate Type 2 OSN, and three or more test scores more than 3 SDs below the expected levels for moderate-severe Type 2 OSN.[4] The presence and severity of typical psychological symptoms are also taken into account, and in clear cases in which either psychological or cognitive symptoms are very dominant, this information informs a decision regarding Type 2A or Type 2B OSN. Whilst psychological symptoms are the reason most workers come to the attention of health professionals, because of the difficulty of measuring the severity of these symptoms and of attributing them to a neurological syndrome, only workers who demonstrate neuropsychological impairments on testing are positively diagnosed with OSN. The New Zealand experience has, however, demonstrated that the vast majority of workers with significant solvent exposure histories and severe psychological problems do demonstrate neuropsychological impairments, and *vice versa*.[10]

### 20.2.5 DO THE SYMPTOMS OF TYPE 2 OSN RESOLVE?

Occasionally after an extended period away from solvents (perhaps 6 months to a year), the worker's psychological symptoms resolve, and on re-testing it is found that his or her neuropsychological impairments have also resolved. In these cases the classification is changed to Type 1 OSN (resolved). A recent New Zealand study re-assessed 21 men with confirmed cases of OSN 6 to 41 (mean 27) months after ceasing exposure.[17] An exposure score was calculated for each worker by using the formula AxBxC, where A = years of solvent exposure, B = a weighting for the occupational group (where boat builders, spray painters and floorlayers had the highest weighting of 3), and C = a weighting reflecting the lack of safety precautions taken by the worker relative to other workers in the same job. Neuropsychological and psychological symptoms at the initial and follow-up assessments were categorized as mild, moderate or moderate-severe (using the system described above) by a neuropsychologist blind to the men's initial diagnosis or exposure history. Twelve men (57%) showed no improvement (or in one case a slight worsening) on cognitive and psychological assessment. Seven men showed some improvement on cognitive tests (but not to "normal" levels), only three of whom also improved on psychological assessment. A further two men showed an improvement in psychological functioning only. Men given a more severe OSN diagnosis at their initial assessment were more likely to improve than men with milder symptoms at the time of their first assessment. Possible explanations for this include the likelihood that some of the more severe symptoms on initial assessment were exacerbated by the lingering effects of acute solvent exposure, or that those with mild OSN were misdiagnosed and their "symptoms" were due to some other cause or were "normal" for them, or that there were psychosocial difficulties present at the first assessment which exacerbated the organic symptoms and resolved with rehabilitation. The disturbing message is, however, that the symptoms of Type 2 OSN are often persistent, and in these cases probably permanent. Even in those individuals where improvement occurs, their symptoms rarely resolve completely.

There was no association between improvement on neuropsychological tests and either time between the two assessments or total time away from solvents. There was no correlation between the exposure score and severity at diagnosis or extent of recovery, and there was no association between a past history of peak exposures and either severity at initial diagnosis or change on neuropsychological assessment. A recent review[18] of studies looking at whether the degree of impairment is related to the dose severity concludes that although several studies have demonstrated significant dose response relationships, there are disturbing inconsistencies, with some studies showing no relationship,[19,20] and one study showing a dose response relationship in painters with levels of exposure considerably lower than the negative studies.[21] Methodological problems and differences and different research populations probably account for these inconsistent findings, and more research is clearly required.

### 20.2.6 INDIVIDUAL DIFFERENCES IN SUSCEPTIBILITY TO OSN

One possible reason for the inconsistent findings both across and within studies examining the relationship between exposure levels and OSN symptoms, may be that individuals have different susceptibilities to solvents. It is not uncommon to diagnose one worker with moderate Type 2 OSN, yet find no symptoms or serious complaints whatsoever in his workmate who has worked by his side in the same spray painting workshop for twenty years. On closer assessment it may be discovered that the affected worker sustained a number of minor head injuries in his younger football-playing days, or has smoked a marijuana joint every weekend for the past 20 years. Subclinical neuronal damage caused by previous insults, or even by normal aging, may make an individual more susceptible to OSN. Another possibility is that some people are biologically, and even genetically more susceptible to solvents. In this sense, OSN can be likened to the post-concussional syndrome following a mild to moderate traumatic brain injury.[6] Not only are the psychological and neuropsychological symptoms very similar, but for reasons which cannot be explained simply by lifestyle differences or malingering, individuals appear to differ widely regarding their susceptibility to developing a post-concussional syndrome. In illustration of this, a recent study reports varying outcomes from apparently equivalent head injuries in a group of athletes.[22]

### 20.2.7 PSYCHOSOCIAL CONSEQUENCES OF OSN, AND REHABILITATION

The common psychological and physical symptoms of OSN of fatigue, irritability, depression, sometimes aggression and violence, headaches, and hypersensitivity to noise and alcohol, along with memory difficulties, poor concentration, poor motivation, and slowed thinking, are a recipe for disaster in interpersonal relationships. Thus it is not uncommon for workers to be diagnosed and treated first for a psychiatric disorder (especially clinical depression) and for their marriages to break up, before they are even suspected of having OSN.[6,10] Once OSN is diagnosed, the prospect of losing their job is a grim one for most victims, most of whom are tradesmen in middle age or older who may have difficulty obtaining or even training for another occupation, especially given their memory, motivation, and concentration problems.

Rehabilitation programmes[6,10] begin with psychoeducation for the worker and his family about the effects of solvents and the importance of protecting himself from exposure in the future. Family members can be taught strategies to reduce the stress on the victim, such as encouraging him to have a rest in the afternoon, and limit his alcohol intake, and by helping him avoid noisy environments such as parties and the family room in the early evening when children are irritable and hungry. Counseling and therapy for the victim and family can be helpful in assisting them to vent their anger at the unfairness of their situation, grieve for their lifestyle and cognitive abilities lost, and come to terms with a "different" person (whose memory may be permanently impaired, and concentration span and motivation lowered). Financial and practical assistance is more often than not of extreme importance, as it is difficult to find the motivation to work on one's psychological and family problems when one is worried about feeding and clothing the children. Antidepressant or anti-anxiety medications may be of assistance in severe cases of mood disorder. In some cases both the neurological damage and the psychological overlay can result in aggressive and violent behaviors not typical of the worker in his younger days. In these cases it is important to first attend to the safety of family members, and then to try and involve the worker in anger management programs, or other therapy with the goal of helping him understand how to control his aggressive or violent behaviors. Similarly, alcohol may be a problem given that it seems likely that neurotoxic solvents damage the pre-frontal lobes, thus resulting in a heightened susceptibility to intoxication. A rehabilitation program aimed at reducing alcohol intake will be important in this case.

Vocational counseling and training are important not only to guide the worker towards a new occupation where solvents are preferably absent, or where protection from solvent exposure is good, but it is also important for the victim's self-esteem and mood. Unfortunately, in many countries where unemployment is high, the prospects of finding a satisfying new career in middle-age are bleak. The task for the rehabilitation therapist in these sad cases is to encourage the worker to take up new hobbies and recreational activities, to spend more quality time with family and friends, and to try and live on a sickness benefit or unemployment benefit without losing self-respect.

## REFERENCES


1 D.E.Hartman, **Neuropsychological Toxicology: Identification And Assessment of Human Neurotoxic Syndromes**. 2nd Ed. *Plenum Press*, New York, 1995.

2 P.R.Lees-Haley, and C.W.Williams, *J.Clin.Psychol.*, **53**, 699-712 (1997).

3 T.P.Ng, S.G.Ong, W.K.Lam, and G.M.Jones, *Arch.Environ, Health*, **12**, 661-664 (1990).

4. E.W.Dryson, and J.A.Ogden, *N.Z. Med.J.*, **111**, 425-427 (1998).

5 J.Juntunen in *Neurobehavioral Methods In Occupational Health*, R.Gilioli, M.G.Cassitto, and V.Foa, Eds., *Pergamon Press*, Oxford, 1983, pp. 3-10

6 J.A.Ogden, **Fractured Minds. A Case-Study Approach To Clinical Neuropsychology**. *Oxford University Press*, New York, 1996, pp. 174-184; 199-213.

7 World Health Organization and Commission of the European Communities, Environmental Health Document 6: Neurobehavioral Methods In Occupational and Environmental Health: Symposium Report. WHO Regional Office for Europe and Commission of the European Communities, Copenhagen, 1985.

8 World Health Organization, Nordic Council of Ministers, Organic Solvents And The Central Nervous System, EH5, WHO, Copenhagen, 1985.

9 E.W.Dryson, and J.A.Ogden, Chronic Organic Solvent Neurotoxicty: Diagnostic Criteria. Department of Labour, Wellington, 1992.

10 J.A.Ogden, *N.Z. J. Psychol.,* **22**, 82-93 (1993).

11 E.L.Baker, and A.M.Seppalainen, *Neurotoxicology*, **7**, 43-56 (1986).
12 T.L.Pauling, and J.A.Ogden, Int. *J. Occup. Environ. Health*, **2**, 286-293 (1996).
13 R.F.White, R.G.Feldman, and P.H.Travers, *Clin. Neuropharm.*, **13**, 392-412 (1990).
14 H.Hanninen in **Neurobehavioral Methods In Occupational Health**, R.Gilioli, M.G.Cassitto, and V.Foa, Eds., *Pergamon Press*, Oxford, 1983, pp. 123-129
15 L.A.Morrow, C.M.Ryan, M.J.Hodgson, and N.Robin, *J Nerv. Ment. Dis.*, **179**, 540-545, (1991).
16 C.M.Ryan, L.A.Morrow, E.J. Bromet, *J. Clin. Exp. Neuropsychol.*, **9**, 665-679, (1987).
17 E.W.Dryson, and J.A.Ogden, Organic solvent induced chronic toxic encephalopathy: Extent of recovery and associated factors following cessation of exposure. Submitted.
18 S.Mikkelsen, *Environ. Res.*, **73**, 101-112, (1997).
19 J.Hooisma, H.Hanninen, H.H.Emmen, and B.M.Kulig, *Neurotoxicol. Teratol.*, **15**, 397-406, (1993).
20 A.Spurgeon, D.C.Glass, I.A.Calvert, M.Cunningham-Hill, and J.M.Harrington, *J. Occup. Environ. Med.*, **51**, 626-630, (1994).
21 M.L.Bleecker, K.I.Bolla, J.Agnew, B.S.Schwartz, and D.P.Ford, *Am. J. Ind. Med.*, **19**, 715-728, (1991).
22 S.N.Macciocchi, J.T.Barth, and L.M.Littlefield, *Clin. Sports Med.*, **17**, 27-36, (1998).


## 20.3 PREGNANCY OUTCOME FOLLOWING MATERNAL ORGANIC SOLVENT EXPOSURE


Kristen I. McMartin and Gideon Koren

**The Motherisk Program, Division of Clinical Pharmacology and Toxicology, Hospital for Sick Children, Toronto, Canada**


### 20.3.1 INTRODUCTION

Organic solvents are a structurally diverse group of low molecular weight liquids that are able to dissolve other organic substances.[1] Chemicals in the solvent class include aliphatic hydrocarbons, aromatic hydrocarbons, halogenated hydrocarbons, aliphatic alcohols, glycols, and glycol ethers. Fuels are a mixture of various hydrocarbons. They are generally ubiquitous in industrialized society, both at work and at the home. They may be encountered as individual agents or in complex mixtures such as gasoline. Incidental exposures may include vapors from gasoline, lighter fluid, spot removers, aerosol sprays and paints. These short duration and low level exposures may often go undetected. More serious exposures occur mainly in the industrial or laboratory settings during manufacturing and processing operations such as dry cleaning, regular working with paint removers, thinners, floor and tile cleaners, glue and as laboratory reagents. Gasoline sniffing or glue sniffing, albeit not occurring in the occupational setting, is another source of exposure to organic solvents during pregnancy.

Counseling pregnant women who are occupationally exposed to numerous chemicals (mostly organic solvents) is difficult because it is hard to estimate the predominant chemicals and their by-products. Even after identifying the more toxic agents, it is still difficult to assess the circumstances of exposure as for many chemicals one can measure neither airborne nor blood levels. Smelling the odor of organic solvents is not indicative of a significant exposure as the olfactory nerve can detect levels as low as several parts per million which are not necessarily associated with toxicity. As an example, the odor threshold of toluene is 0.8 parts per million whereas the TLV-TWA (threshold limit value-time weighted average) is 50 parts per million. In addition, reproductive information on many individual solvents is at best sparse, either limited to animal studies or nonexistent.

Many organic solvents are teratogenic and embryotoxic in laboratory animals depending on the specific solvent, dose, route of administration and particular animal species.[1] The various malformations described include hydrocephaly, exencephaly, skeletal defects, cardiovascular abnormalities and blood changes. Also, some studies suggest poor fetal development and neurodevelopmental deficits. In a portion of these studies exposure levels were high enough to induce maternal toxicity.

Organic solvents are a diverse, complex group and because exposure usually involves more than one agent and different circumstances, adequate human epidemiological studies are difficult to interpret. Many studies are subject to recall and response bias and are not always controlled for other risk factors such as age, smoking, ethanol, and concurrent drug ingestion. It is hard to prove or quantify the suspicion that organic solvents are a reproductive hazard. One may even expect that a ubiquitous exposure to solvents would by chance alone be associated with an increase in birth defects or spontaneous abortions, which may differ from one study to another. While fetal toxicity is biologically sensible in cases of intoxicated mothers, evidence of fetal damage from levels that are not toxic to the mother is scanty, inconsistent or missing.

This chapter will review the reproductive toxicology of organic solvents with particular focus on exposure during pregnancy. Firstly, examples of animal studies with regard to three organic solvents will be discussed. This will be followed by information obtained from human studies including: a meta-analysis of pregnancy outcome following maternal organic solvent exposure; results from the first prospective study by the Motherisk Program at the Hospital for Sick Children on gestational exposure during pregnancy; and finally, a proactive approach for the evaluation of fetal safety in chemical industries.

## 20.3.2 ANIMAL STUDIES

There are numerous experimental studies that examine the reproductive effects of organic solvents in animals. The reproductive effects of maternal organic solvent exposure will be summarized using three organic solvents as examples. The solvents discussed will be benzene, toluene and tetrachloroethylene.

### Benzene

Watanabe and Yoshida[2] were the first to claim teratogenic effects of benzene after administration during organogenesis only. Groups of 15 mice were given single subcutaneous injections of 3 ml benzene/kg on one of days 11-15 of pregnancy. This dose caused leukopenia lasting 24-48 hours but had no effect on body weight in the dams. Litter size ranged form a average of 6.5-8.5 in the 4 treatment groups. Malformations were seen in most treated groups; cleft palate occurred in 5.5% of fetuses exposed on day 13 and in 1.0% of fetuses exposed on day 14 and agnathia or micrognathia was seen on 0.9%, 2.4% and 1.0% of fetuses exposed on days 11, 13 and 14 respectively. Extra 14th ribs were seen in 10-16% of fetuses in all treated groups. Fetuses from 5 dams treated on day 15 had no malformations but 24% had extra 14th ribs. In the absence of any control data it is not known if these represent significant increases in malformations and anomaly rates. Extra 14th ribs for example, can be a common skeletal variant in some strains of mice and rats.[9]

Matsumoto et al.[3] have given groups of 8-11 mice subcutaneous injections of 0, 2, or 4 ml of benzene/kg on days 8 and 9 or 12 and 13 of pregnancy. Fetuses were examined externally and for skeletal defects only; internal soft tissues were not examined. They claim that fetal weight was significantly decreased in both groups given 4 ml/kg and placental weight significantly reduced in those given 4 ml/kg on days 12 and 13 of pregnancy. However, reworking of the data shows p values of >0.4 in all cases.[9] Sporadic malformations (cleft palate and open eye) did not differ significantly between treated and control groups, neither did the incidence of dead or resorbed embryos and fetuses. A small degree of retarded ossification was seen in fetuses from dams given 4 ml/kg.

Nawrot and Staples[4] investigated the effects of oral administration by gavage of 0.3, 0.5 or 1.0 ml/kg on days 6-15 of pregnancy or 1.0 ml/kg on days 12-15 of pregnancy in the mouse. After dosing on days 6-15, 0.5 and 1.0 ml/kg caused some maternal mortality and embryolethality. Fetal weight was significantly reduced at all 3 dose levels but no increase in malformations was seen. There were similar findings after dosing on days 12-15 except that resorptions occurred later in gestation. The study is reported in abstract only and no further details are given.

Murray et al.[5] exposed groups of 35-37 mice to 0 or 500 ppm benzene for 7 hr/day on days 6-15 of pregnancy. Acceptable teratological methods were used.[9] There was no evidence of maternal toxicity. There were no effects on implants/dam, live fetuses/dam, resorptions/dam or malformation rates. Fetal body weight was significantly reduced and delayed ossification significantly increased in fetuses from the benzene group.

Iwanaga et al.[6] demonstrated an increased postnatal susceptibility to benzene toxicity in mice exposed prenatally to benzene by injection of the dams with 4 ml benzene/kg on day 9 or 12 of gestation. At 10 weeks of age the offspring were injected with 5 daily doses of 0.1 ml benzene/kg and the effects on erythrocytes, leukocytes, body weight, thymus and spleen were more marked than in non-prenatally exposed controls.

There have been several inhalational studies on benzene in the rat. In an unpublished study summarized by Murray et al.,[5] teratogenic effects were observed at 500 ppm when rats were exposed to 0, 10, 50 or 500 ppm benzene for 7 hr/day on days 6-15 of pregnancy and a low incidence of exencephaly, kinked ribs and abnormal ossification of the forepaws was noted at 500 ppm. In another unpublished study quoted by Murray et al.[5] no teratogenicity but increased embryoloethality was seen after exposure to 10 or 40 ppm for 6 hours/day on days 6-15 of pregnancy in the rat.

Hudak and Ungvary[7] exposed groups of 19-26 rats to 0 or 313 ppm benzene for 24 hours/day on days 9-14 of pregnancy. Acceptable teratological methods were used.[9] There was no maternal mortality but maternal weight gain was significantly reduced. There were no significant effects on live fetuses/dam, resorbed or dead fetuses/dam or malformation rate. Mean fetal weight was significantly reduced and retarded ossification, abnormal fusion of sternebrae and extra ribs were all significantly increased in the benzene-exposed group.

Green et al.[8] exposed groups of 14-18 rats to 100, 300 or 2200 ppm benzene for 6 hours/day on days 6-15 of pregnancy, each benzene-exposed group having a concurrent 0 ppm control group. Maternal weight gain was significantly reduced in the 2200 ppm group, but not at lower exposure levels. There were no significant effects on implants/dam, live fetuses/dam, resorptions/dam or malformation rates. There was a significant 10% reduction in fetal weight in the 2200 ppm benzene group and skeletal anomalies were sporadically increased in benzene-exposed groups (missing sternebrae at 100 ppm, delayed ossification of sternebrae in female offspring only at 300 ppm and 2200 ppm and missing sternebrae at 2200 ppm). The authors suggest the higher number of affected female fetuses is in accordance with other observations on the increased susceptibility of females to benzene toxicity.[9] In addition, they observed a non-significant low incidence of hemorrhages in all 3 benzene-exposed groups which were not seen in control fetuses.

In conclusion, embryolethal and teratogenic effects are not seen even at maternally toxic doses but significant fetotoxicity in terms of reduced body weight sometimes accompanied by increases in skeletal variants and delayed ossification is seen at doses which are not necessarily toxic to the dam. The absence of any such effects in a large number of adequately conducted studies reported in full suggests these observations may be of no biological significance. The role that benzene-induced maternal anemia may play in any adverse effects on the offspring is not known.[9]

Toluene

Euler[10] exposed mice to a mixture of toluene and trichloroethylene similar to that which has been used in the soling of shoes. The mixture was composed of 32 ppm (120 mg/m$^3$) toluene and 64 ppm (340 mg/m$^3$) trichloroethylene, equivalent to inhaling 157 mg/kg toluene and 406 mg/kg trichloroethylene in the mice. They inhaled the mixture for 10 days before mating or during part or the whole of pregnancy. Differences were noted between treated and control groups in pregnancy rates, length of pregnancy, damaged embryos, birth weights and neonatal mortality but the direction and magnitude of these differences is not stated. No groups were exposed to toluene alone.

Nawrot and Staples[4] gave mice 0.3, 0.5, or 1.0 ml toluene/kg orally by gavage on days 6-15 of pregnancy or 1.0 ml/kg on days 12-15 of pregnancy. There was no maternal toxicity except a decrease in maternal weight gain in those dosed on days 12-15. There was a significant increase in embrylolethality at all 3 dose levels and a significant reduction in fetal weight in the 0.5 and 1.0 ml/kg groups after dosing on days 6-15. Those dosed with 1.0 ml/kg on days 6-15 had a significant increase in numbers of fetuses with cleft palate which was not simply due to general growth retardation. Treatment on days 12-15 only had no adverse effects on the offspring. The study is reported in abstract only and no further details are given.

Teratological investigations on inhaled toluene in mice and rats have been carried out by Hudak et al.[7] Mice were exposed to 0, 133 or 399 ppm (500 or 1500 mg/m$^3$) toluene for 24 hr/day on days 6-13 of pregnancy. In the high dose group all 15 exposed dams died within the first 24 hr of exposure. No maternal deaths occurred in the 11 mice exposed to 133 ppm and there were no effects on implants/dam, live fetuses/dam, dead and resorbed fetuses/dam, malformations or anomaly rates, but fetal weight was significantly reduced by 10% in comparison with controls. It is not stated whether 133 ppm had any effect on maternal weight gain.[7]

In conclusion, similar to benzene, toluene does not appear to be teratogenic. It is fetotoxic, causing a reduction in fetal weight in mice and rats and retarded ossification and some increase in skeletal anomalies in rats at doses that are below those toxic to the dam as well as at toxic doses.[9] Embryolethality has also been seen with inhalation of very high concentrations lethal to some of the dams or following oral administration of non-toxic doses.[9]

Tetrachloroethylene

Schwetz et al.[11] exposed rats and mice to 300 ppm tetrachloroethylene for 7 h/day on days 6-15 of pregnancy. The dams were killed just before term and the fetuses examined by acceptable teratological methods but results are given on a per litter basis only. The number of treated animals in each case was 17 and the number of controls (air exposed) 30 for both rat and mouse studies.

Effects of tetrachloroethylene on the dams varied between species.[11] In the mouse relative liver weight was significantly increased and the absolute liver weight increased but not significantly and with no effect on maternal body weight. In the rat there was a non-significant decrease in absolute and relative liver weights and a significant 4-5% decrease in mean body weight. Food consumption was unaffected.

Effects on the embryo and fetus also differed.[11] In the mouse there was no effect on implantation sites, live fetuses or resorption rates but mean fetal weight was significantly reduced, 59% of litters containing runts (weight less than 3 standard deviations below the mean) compared with 38% of control litters. Whereas in the rat, resorption rate was significantly increased from 4% in controls to 9% in the exposed group, while fetal body was unaffected (mean slightly higher than controls).

In the mouse, examination for anomalies revealed an increase in delayed ossification of the skull bones (significant) and of the sternebrae (nonsignificant) as might be expected from the fetal weight data. There were also significant increases in the incidence of split sterenbrae and subcutaneous edema. No gross malformations were found. In the rat, gross malformations (short tail) were reported but the incidence did not differ significantly from that in controls. There were no other significant differences in soft tissue or skeletal abnormalities.[11]

The results of this study are difficult to assess, partly because no indication of the numbers of fetuses affected within affected litters is given and partly because of the uncertain nature of the "subcutaneous edema" reported.[9,11] Exposure to tetrachloroethylene and the concurrent controls were part of a large study on four different solvents. The incidence of subcutaneous edema in the mouse ranged from 8-59% of litters affected which seems very high and while the incidence in the tetrachloroethylene group was highest at 59%, it was as high as 45% in nonconcurrent controls (27% in concurrent controls).[11] In the rat, the incidence of this particular anomaly also varied enormously between groups from 0% (tetrachloroethylene group) to 28% (trichloroethylene group).[11] It is therefore important to know how strict were the criteria for designation of "subcutaneous edema" and in particular whether the designation was made before or after fixing, subcutaneous edema being a common fixative artifact.[9] However, the retardation of growth and ossification and the increased incidence of split sternebrae in fetal mice exposed to tetrachloroethylene were clear effects and in the absence of any effect on maternal body weight, suggest that tetrachloroethylene has some maternal hepatotoxicity but has no effect in the rat where there is no hepatotoxicity at 300 ppm.[11]

The results of a behavioral teratology study in the rat by Nelson et al. have been reported.[12] Rats were exposed to 0 or 900 ppm tetrachloroethylene for 7 hours/day on days 7-13 or 14-20 of pregnancy (9-16 rats per group). The dams were affected by this level, showing reduced food consumption and lower weight gain during exposure but histopathological examination of the maternal liver and kidney in dams sacrificed on day 21 of pregnancy revealed no abnormalities.[12]

Postnatally, offspring were tested for olfaction, neuromuscular ability, exploratory and circadian activity, aversive and appetitive learning.[12] There was evidence of impaired neuromuscular ability.[12] Offspring from dams exposed on days 7-13 were poorer than controls in ascent of a wire mesh screen during the second week of life and were poorer than controls on a rotorod test on one of the 3 days tested in the fourth week of life.[12] Offspring from dams exposed on days 14-20 performed less well in ascent of a wire mesh screen.

However, the latter group were consistently superior to controls on the rotorod later in development.[12] Both exposed groups were generally more active in open field tests than controls but only those exposed on days 14-20 of gestation differed significantly from controls.[12] Biochemical analyses of whole brain neurotransmitter levels showed no effects in newborns but significant reductions in acetylcholine levels at 21 days of age in both exposed groups of offspring and reduced dopamine levels at 21 days of age in those from dams exposed on days 7-13.[12] There were no significant differences between exposed and control groups on any other of the tests.[12] Exposure of offspring to 100 ppm on days 14-20 of gestation showed no significant differences from controls on any of the above behavioral tests.[12] It was not stated whether neurotransmitter levels were measured in this low-dose group.[9,12]

In view of these results, suggesting some fetotoxicity in the mouse but not the rat at 300 ppm and postnatal effects in the rat at 900 ppm but not 100 ppm, there is a need for further studies at low levels between 900 and 100 ppm to establish a more accurate no-effect-level.[9]

### 20.3.3 PREGNANCY OUTCOME FOLLOWING MATERNAL ORGANIC SOLVENT EXPOSURE: A META-ANALYSIS OF EPIDEMIOLOGIC STUDIES

[Adapted, by permission, from K.I. McMartin, M. Chu, E. Kopecky, T.R. Einarson and G. Koren, *Am. J. Ind. Med.*, 34, 288 (1998) Copyright 1998 *John Wiley & Sons, Inc.* Reprinted by permission of Wiley-Liss, Inc. a division of John Wiley & Sons, Inc.]

Introduction

Evidence of fetal damage or demise from organic solvent levels that are not toxic to the pregnant woman is inconsistent in the medical literature. A mathematical method has been previously developed and utilized to help overcome bias and arrive at a single overall value that describes the exposure-outcome relationship; namely, meta-analysis.[15]

The risk for major malformations and spontaneous abortion from maternal inhalational organic solvent exposure during pregnancy is summarized using meta-analysis.[31] Besides being more objective than the traditional methods of literature review, it has the ability to pool research results from various studies thereby increasing the statistical strength/power of the analysis. This is especially useful in epidemiologic studies, such as cohort studies or case control studies since very often large numbers of subjects are required in order for any problem to be significantly addressed. This is particularly true for teratogenic studies where the frequencies of malformation are often very low.

Methods

A literature search was conducted to collect studies for the meta-analysis. Using Medline, Toxline and Dissertation Abstracts databases spanning 1966-1994, literature was identified concerning the problem in question. In addition, external colleagues were consulted (regarding unpublished studies) whose area of interest is in occupational exposure and reproductive toxicology. All references from the extracted papers and case reports were investigated. Standard textbooks containing summaries of teratogenicity data were consulted for further undetected references.

Inclusion criteria consisted of human studies of any language which were 1) case control or cohort study in design; 2) included maternal inhalational, occupational, organic solvent exposure; 3) had an outcome of major malformation and/or spontaneous abortion; and 4) included first trimester pregnancy exposure. Exclusion criteria consisted of animal studies, non-inhalational exposure, case reports, letters, editorials, review articles and studies that did not permit extraction of data. For subgroup analysis, we also identified and analyzed cohort and case-control studies specifically involving solvent exposure. Major malformations were defined as malformations which were either potentially life threatening or a major cosmetic defect.[13] Spontaneous abortion was defined as the spontaneous termination of pregnancy before 20 weeks gestation based upon the date of the first day of the last normal menses.[14]

To obtain an estimate of the risk ratio for major malformations and spontaneous abortion in exposed versus unexposed infants, an overall summary odds ratio (ORs) was calculated according to the protocol established by Einarson et al.[15] Additionally, homogeneity of the included studies, power analysis and the extent of publication bias were also examined as described by Einarson et al.[15]

Results and discussion

The literature search yielded 559 articles. Of these, 549 in total were rejected for various reasons. The types of papers rejected were: animal studies (298), case reports/series (28), review articles (58), editorials (13), duplicate articles (10), not relevant (62), malformation not specified (29), spontaneous abortion not defined (31), unable to extract data (4), no indication of timing of exposure (16). Five papers were included into the major malformation analysis (Table 20.3.1) and 5 papers were included into the spontaneous abortion analysis (Table 20.3.2).

**Table 20.3.1. Studies of teratogenicity of organic solvents meeting criteria for meta-analysis [Adapted, by permission, from K.I. McMartin, M. Chu, E. Kopecky, T.R. Einarson and G. Koren, *Am. J. Ind. Med.*, 34, 288 (1998) Copyright 1998 *John Wiley & Sons, Inc.* Reprinted by permission of Wiley-Liss, Inc. a division of John Wiley & Sons, Inc.]**

| Authors | Study type | Data collection | Malformation described |
|---|---|---|---|
| Axelsson et al.[16] | C | R | "serious malformations" |
| Tikkanen et al.[17] | CC | R | cardiac malformations |
| Holmberg et al.[18] | CC | R | CNS, oral clefts, musculoskeletal, cardiac defects |
| Cordier et al.[19] | CC | R | "major malformations" |
| Lemasters[20] | C | R | "major malformations" |

CC=Case control; C=Cohort; R=Retrospective

A. Malformations

In total 5 studies describing results from organic solvent exposure were identified (Table 20.3.3). The summary odds ratio obtained was 1.64 (95% CI: 1.16 - 2.30) which indicates that maternal inhalational occupational exposure to organic solvents is associated with an increased risk for major malformations. The test for homogeneity yielded a chi square of 2.98 (df=4, p=0.56). When studies were analyzed separately according to study type, the chi square value from the test for homogeneity of effect for cohort studies was 0.52 (df=1, p=0.47) and for case control studies it was 0.01 (df=2, p=0.99). Their combinability remains justified on the basis of the lack of finding heterogeneity among the results.

Meta-analysis of both the cohort studies and case-control studies produced similar results, i.e., they demonstrate a statistically significant relationship between organic solvent exposure in the first trimester of pregnancy and fetal malformation. The summary odds ratio for cohort studies was 1.73 (95% CI: 0.74 - 4.08) and 1.62 (95% CI: 1.12 - 2.35) for case-control studies.

**Table 20.3.2. Studies of spontaneous abortion of organic solvents meeting criteria for meta-analysis. [Adapted, by permission, from K.I. McMartin, M. Chu, E. Kopecky, T.R. Einarson and G. Koren, *Am. J. Ind. Med.*, 34, 288 (1998) Copyright 1998 *John Wiley & Sons, Inc.* Reprinted by permission of Wiley-Liss, Inc. a division of John Wiley & Sons, Inc.]**

| Authors | Study type | Data collection |
|---|---|---|
| Windham et al.[21] | CC | R |
| Lipscomb et al.[22] | C | R |
| Shenker et al.[23] | C | P |
| Pinney[24] | C | R |
| Eskenazi et al.[25] | C | P |

CC=Case control, C=Cohort, R=Retrospective, P=Prospective

**Table 20.3.3. Results of studies comparing outcomes of fetuses exposed or not exposed to organic solvents. [Adapted, by permission, from K.I. McMartin, M. Chu, E. Kopecky, T.R. Einarson and G. Koren, *Am. J. Ind. Med.*, 34, 288 (1998) Copyright 1998 *John Wiley & Sons, Inc.* Reprinted by permission of Wiley-Liss, Inc. a division of John Wiley & Sons, Inc.]**

| Reference | Exposure | | Congenital Defect | | |
|---|---|---|---|---|---|
| | | | Yes | No | Total |
| Axelsson et al.[16] | organic solvents | yes | 3 | 489 | 492 |
| | | no | 4 | 492 | 496 |
| | | total | 7 | 981 | 988 |
| Tikkanen et al.[17] | organic solvents | yes | 23 | 26 | 49 |
| | | no | 546 | 1026 | 1572 |
| | | total | 569 | 1052 | 1621 |
| Holmberg et al.[18] | organic solvents | yes | 11 | 7 | 18 |
| | | no | 1464 | 1438 | 2902 |
| | | total | 1475 | 1475 | 2950 |
| Cordier et al.[19] | organic solvents | yes | 29 | 22 | 51 |
| | | no | 234 | 285 | 519 |
| | | total | 263 | 307 | 570 |
| Lemasters[20] | styrene | yes | 4 | 68 | 72 |
| | | no | 13 | 822 | 835 |
| | | total | 17 | 890 | 907 |
| TOTAL | | yes | 70 | 612 | 682 |
| | | no | 2261 | 4100 | 6354 |
| | | total | 2331 | 4712 | 7036 |

In this meta-analysis, major malformations were defined as "potentially life threatening or a major cosmetic defect".[13] In the general population there is a 1-3% baseline risk for major malformations. Estimate incidence via cohort studies indicated 2 studies with a total of 7 malformations in 564 exposures or 1.2% rate of malformations which falls within the baseline risk for major malformations.

Publication bias is the tendency for statistically significant studies to be submitted and accepted for publication in preference to studies that do not produce statistical significance.[15] This may be the case for solvent exposure and major malformations. Determining the extent of possible publication bias (file drawer analysis) is not unlike power analysis for nonsignificant results. Each provides some quantitative measure of the magnitude of the findings with respect to disproving them and requires judgment for interpretation. In order to perform a file drawer analysis effect sizes must be calculated from the summary statistic. Effect sizes represent the magnitude of the relationship between two variables. Unlike statistical significance, which is directly related to sample size, an effect size may be thought of as significance without the influence of sample size. In other words, effect size represents the "true" impact of an intervention. Cohen has determined that an effect size d=0.2 is considered small, 0.5 is medium and 0.8 is large.[15]

The result from this file drawer analysis indicates that one would have to obtain 2 articles with a small effect size (d=0.001) to bring the study's overall effect size (d=0.071) to a smaller effect size of 0.05. One of the acceptable studies achieved such a small effect size. The smallest effect size was d=0.000682.[16] It would therefore seem probable to have some studies stored away in file drawers with very small effect sizes (lack of statistical significance). Unfortunately, no statistical test yet exists to precisely determine such a probability and one must therefore exercise judgment.

There are some considerations to bear in mind when interpreting results of this meta-analysis:

1. Environmental exposure in pregnancy is seldom an isolated phenomenon, therefore, analysis of human teratogenicity data may require stratification for a number of factors depending on the intended focus of the analysis.

2. Organic solvents belong to many classes of chemicals. Not all of the studies have examined the exact same groups of solvents in terms of both extent and range of solvents as well as frequency and duration of exposure.

3. The malformations listed in each of the papers seems to reflect a diverse range of anomalies. One might expect to notice a particular trend in malformations between studies, however, this does not appear to be the case.

Certain factors should be kept in mind when evaluating the results such that a number of studies were case control in design. Certain factors inherent in this study design may affect the interpretation of their results, including recall of events during pregnancy, selection of samples based on volunteer reporting and a change in the knowledge over time regarding factors considered to significantly affect the fetus. Mothers of malformed children may understandably report exposure more often than mothers of healthy children. The recall of the exact name of the chemical, amount of exposure, starting and stopping date of exposure are also difficult to establish retrospectively. Recall may be affected by the method of questioning; when asked open ended questions, women may not recall details as well as when questioned with respect to specific chemical exposure. As a result, there could be systematic bias toward reporting exposure.

It is important to consider the criteria or "proof" for human teratogenicity as established by Shepard:[26]

1. Proven exposure to agent at critical time(s) in prenatal development. One of the inclusion criteria for this meta-analysis, with malformations as the outcome of exposure, was first trimester exposure to organic solvents.

2. Consistent findings by two or more epidemiologic studies of high quality including: control of confounding factors, sufficient numbers, exclusion of positive and negative bias factors, prospective studies if possible, and studies with a relative risk of six or more.

When this happens it is unlikely that methodological problems or systematic biases can influence the results of the studies conducted in different contexts and different study designs. The studies included in this meta-analysis usually controlled for such items as geographical location and date of birth, however, other potential confounding factors such as maternal age, alcohol, and smoking that could lead to subsequent problems in outcome presentation were not consistently reported.

In addition, this meta-analysis included studies that were contained within large databases spanning many years. The majority of information about occupational exposure in general during pregnancy originates from Scandinavia, namely, the Institute of Occupational Health in Helsinki. For example, Finland monitors spontaneous abortions through the spontaneous abortion registry. The registry contains all information about women who were hospitalized with spontaneous abortions covering approximately 90% of all spontaneous abortions in Finland. Finland also monitors births via the Finnish Register of Congenital Malformations. All new mothers in Finland are interviewed during their first prenatal visit, at 3 months post-delivery, at Maternity Care Centers located in every province throughout Finland.

When scanning the literature, there are no studies that prospectively examine occupational exposure to organic solvents during pregnancy and pregnancy outcome with regard to malformations. The studies are retrospective, either case-control or cohort in design. In contrast, however, there are a number of studies that prospectively examine occupational exposure during pregnancy and pregnancy outcome with regard to spontaneous abortion.

In all the studies there was an attempt to ascertain the occupational exposure by an industrial hygienist who blindly assessed the group exposure information. In addition, the individual studies included in the meta-analysis did not obtain an odds ratio or relative risk of 6.0 or more with a significant 95% confidence interval. The larger the value of the relative risk, the less likely the association is to be spurious. If the association between a teratogen is weak and the relative risk small (i.e., range 1.1-2.0), it is possible to think that the association is indeed due to unknown confounding factors and not to the teratogen under study. However, weak associations may be due to misclassification of exposure or disease. They may also indicate an overall low risk but the presence of a special subgroup at risk of teratogenesis within the exposed group.

3. Careful delineation of the clinical cases. A specific defect or syndrome, if present, is very helpful. If the teratogen is associated only to one or a few specific birth defects, the possibility of a spurious association becomes smaller. In this meta-analysis, the malformations were variable with no specific trend apparent.

4. Rare environmental exposure associated with rare defect.

5. Teratogenicity in experimental animals important but not essential.

6. The association should make biologic sense.

When a chemical or any other environmental factor caused a malformation in the experimental animals and/or the biological mechanism is understood, the observation of an association in humans becomes more plausible. Although the statistical association must be present before any relationship can be said to exist, only biological plausible associations can result in "biological significance".

The mechanisms by which many solvents exert their toxicity are unclear and may vary from one solvent to another. Halogenated hydrocarbons such as carbon tetrachloride may generate free radicals.[27] Simple aromatic compounds such as benzene may disrupt polyribosomes, whereas some solvents are thought to affect lipid membranes and to penetrate tissues such as the brain.[27]

In 1979 a syndrome of anomalies (hypertonia, scaphocephaly, mental retardation and other CNS effects) was suggested in two children in a small American Indian community where gasoline sniffing and alcohol abuse are common.[28] Four other children had similar abnormalities, however, in these cases it was impossible to verify gasoline sniffing. Also, it is unclear what was the contribution of the lead in the gasoline or the alcohol abuse in producing these abnormalities. It is important to remember that the mothers in many of these cases showed signs of solvent toxicity indicating heavy exposure. This is not the case in most occupational exposures during pregnancy. While fetal toxicity is biologically sensible in cases of intoxicated mothers, the evidence of fetal damage from levels that are not toxic to the mother is scanty and inconsistent.

7. Proof in an experimental system that the agent acts in an unaltered state.
8. Important information for prevention.

Several lists of criteria for human teratogenicity have included the dose (or concentration) response relationship.[1] Although a dose response may be considered essential in establishing teratogenicity in animals it is extremely uncommon to have sufficient data in human studies. Another criterion which is comforting to have but not very often fulfilled is biologic plausibility for the cause. Shepard states that at present there is no biologically plausible explanation for thalidomide embryopathy and that at least one half of all human teratogens do not fit this criterion.[26]

B. Spontaneous abortion

Estimates for clinically recognized spontaneous abortions as a proportion of all pregnancies vary markedly. In ten descriptive studies reviewed by Axelsson,[29] the proportion of spontaneous abortions varied from 9% to 15% in different populations. The variation depended not only on the characteristics of the population but on the methods used in the study, i.e., the selection of the study population, the source of pregnancy data, the definition of spontaneous abortion, the occurrence of induced abortions and their inclusion or otherwise in the data. The weaknesses of the studies using interviews or questionnaires pertain to the possibility of differential recognition and recall (or reporting) of spontaneous abortions and of differential response. Both exposure and the outcome of pregnancy may influence the willingness of subjects to respond to a study. One advantage of interview data is that it is more likely to provide information on early spontaneous abortion than medical records. However, the validity of information on early abortion which may be difficult to distinguish from a skipped or delayed menstruation has been suspect. Spontaneous abortions which have come to medical attention are probably better defined than self-reported abortions.

The feasibility of using medical records as a source of data depends on the pattern of use of medical facilities in the community and the coverage and correctness of the records.

Of concern is the potential selection bias due to differing patterns of use of medical services. The primary determinant for seeking medical care is probably gestational age so that earlier abortions are less likely to be medically recorded than later abortions.[29] The advantage of data on medically diagnosed spontaneous abortions, compared to interview data is that the former are independent of an individuals own definition, recognition and reporting.

In total, 5 papers describing results from organic solvent exposure were identified (Table 20.3.4). The summary odds ratio obtained was 1.25 (95% CI: 0.99 - 1.58). The test for homogeneity yielded a chi square=4.88 (df=4, p=0.300). When studies were analyzed separately according to study type, the chi-square value for homogeneity of effect for cohort studies was 4.20 (df=3, p=0.241). Meta-analysis of both cohort and case-control studies produced similar results, i.e., they do not demonstrate a statistically significant relationship between organic solvent exposure in pregnancy and spontaneous abortion. The summary odds ratio for cohort studies was 1.39 (95% CI: 0.95 - 2.04) and 1.17 (95% CI: 0.87 - 1.58) for case control studies. Their combinability seems justified on the basis of the lack of finding heterogeneity among the results.

**Table 20.3.4. Results of studies comparing outcomes of fetuses exposed or not exposed to organic solvents. [Adapted, by permission, from K.I. McMartin, M. Chu, E. Kopecky, T.R. Einarson and G. Koren, *Am. J. Ind. Med.*, 34, 288 (1998) Copyright 1998 *John Wiley & Sons, Inc.* Reprinted by permission of Wiley-Liss, Inc. a division of John Wiley & Sons, Inc.]**

| Reference | Exposure | | Spontaneous Abortion | | |
|---|---|---|---|---|---|
| | | | yes | no | total |
| Windham et al.[21] | any solvent product | yes | 89 | 160 | 249 |
| | | no | 272 | 575 | 847 |
| | | total | 361 | 735 | 1096 |
| Lipscomb et al.[22] | organic solvent | yes | 10 | 39 | 49 |
| | | no | 87 | 854 | 941 |
| | | total | 97 | 893 | 990 |
| Schenker et al.[23] | organic solvents | yes | 12 | 8 | 20 |
| | | no | 16 | 21 | 37 |
| | | total | 28 | 29 | 57 |
| Pinney[24] | organic solvents | yes | 35 | 228 | 263 |
| | | no | 25 | 166 | 191 |
| | | total | 60 | 394 | 454 |
| Eskenazi et al.[25] | organic solvents | yes | 4 | 97 | 101 |
| | | no | 7 | 194 | 201 |
| | | total | 11 | 291 | 302 |
| TOTAL | | yes | 150 | 532 | 682 |
| | | no | 407 | 1810 | 2217 |
| | | total | 557 | 2342 | 2899 |

The overall ORs of 1.25 indicates that maternal inhalational occupational exposure to organic solvents is associated with a tendency towards a small increased risk for spontaneous abortion. The addition of one study of similar effect size would have rendered this trend statistically significant.

Traditionally, a power analysis would be conducted to determine the number of subjects or in this situation the number of "studies" that need to be added to produce a significant result. In order to perform a power analysis effect sizes must be calculated from the summary statistic. The result from this power analysis indicates that one would have to obtain 2 studies with a medium effect size (0.5) to bring this study's overall effect size (d=0.095) to a small effect size of 0.2. Similarly, 5 articles with an effect size of d=0.3 are needed to bring the study's overall effect size to 0.2. The largest effect size in the spontaneous abortion analysis was d=0.2. None of the acceptable studies achieved such a large effect size as 0.5. It may be improbable because one would expect that such results would undoubtedly have been published. Unfortunately, no statistical test yet exists to precisely determine such a probability and one must therefore exercise judgment.

This meta-analysis addresses the use of organic solvents in pregnancy. Organic solvent is a very broad term that includes many classes of chemicals. There may still exist rates of abortion higher than the value reported with certain groups of solvents. However, a detailed analysis of classes of solvents is in order to incriminate a particular solvent. Not all of the studies have examined the same groups of solvents in terms of both extent and range of solvents as well as frequency and duration of exposure. Hence it would be very difficult to obtain any clear estimate of risk for a given solvent given the limited number of studies available.

Conclusion

The meta-analysis examining organic solvent use in pregnancy did not appear to find a positive association between organic solvent exposure and spontaneous abortions (ORs = 1.25, confidence interval 0.99 - 1.58). The results from the meta-analysis examining organic solvent use in the first trimester of pregnancy and major malformations indicate that solvents are associated with an increased risk for major malformations (ORs = 1.64, confidence interval 1.16 - 2.30). Because of the potential implications of this review to a large number of women of reproductive age occupationally exposed to organic solvents, it is important to verify this cumulative risk estimate by a prospective study. Similarly, it is prudent to minimize women's exposure to organic solvents by ensuring appropriate ventilation systems and protective equipment.

Meta-analysis can be a key element for improving individual research efforts and their reporting in the literature. This is particularly important with regard to an estimate of dose in occupational studies as better reporting of the quantification of solvent exposure is needed in the reproductive toxicology literature.

## 20.3.4 PREGNANCY OUTCOME FOLLOWING GESTATIONAL EXPOSURE TO ORGANIC SOLVENTS: A PROSPECTIVE CONTROLLED STUDY

[Adapted, by permission, from S. Khattak, G. K-Moghtader, K. McMartin, M. Barrera, D. Kennedy and G. Koren, *JAMA.*, **281**, 1106 (1999) Copyright 1999, *American Medical Association*]

The Motherisk Program at the Hospital for Sick Children was the first to prospectively evaluate pregnancy and fetal outcome following maternal occupational exposure to organic solvents with malformations being the primary outcome of interest.[30]

Methods

The study group consisted of all pregnant women occupationally exposed to organic solvents and counseled between 1987-1996 by the Motherisk Program at the Hospital for Sick Children. Details concerning the time of exposure to organic solvents were recorded for determination of temporal relationship between exposure and conception. The details on chemical exposure were recorded, including occupation, type of protective equipment used, and other safety features, including ventilation fans. Adverse effects were defined as those known to be caused by organic solvents (e.g., irritation of the eyes or respiratory system, breathing difficulty, headache). Temporal relationship to exposure was investigated to separate these symptoms from those associated with pregnancy. One hundred twenty-five pregnant women who were exposed occupationally to organic solvents and seen during the first trimester between 1987and 1996. Each pregnant woman who was exposed to organic solvents was matched to a pregnant woman who was exposed to a nonteratogenic agent on age (+/- 4 years), gravidity (+/- 1) and smoking and drinking status.

The primary outcome of interest was major malformations. A major malformation was defined as any anomaly that has an adverse effect on either the function or the social acceptability of the child. The expected rate of major malformations is between 1% to 3%.

Results and discussion

Significantly more major malformations occurred among fetuses of women exposed to organic solvents than controls (13 vs 1; relative risk, 13.0; 95% confidence interval, 1.8-99.5). Twelve malformations occurred among the 75 women who had symptoms temporally associated with their exposure, while none occurred among 43 asymptomatic exposed women ($p<0.001$). (One malformation occurred in a woman for whom such information was missing.) More of these exposed women had previous miscarriage while working with organic solvents than controls (54/117 [46.2%] vs 24/125 [19.2%]; $p<0.001$). However, exposed women who had a previous miscarriage had rates of major malformation that were similar to exposed women who had no previous miscarriage.

The Motherisk protocol allowed us to record in a systematic manner all exposure data and other maternal and paternal medical details at the time of exposure during the first trimester of pregnancy and to follow up pregnancy outcomes prospectively in this cohort. The control group was assessed in an identical manner.

This prospective study confirms the results of our recent meta-analysis.[31] Women occupationally exposed to organic solvents had a 13 fold risk of major malformations as well as an increased risk for miscarriages in previous pregnancies while working with organic solvents. Moreover, women reporting symptoms associated with organic solvents during early pregnancy had a significantly higher risk of major malformations than those who were asymptomatic suggesting a dose-response relationship. Other factors, for example, type of solvent, might have accounted for the presence of symptoms in some women.

Although some human teratogens have been shown to cause a homogeneous pattern of malformation(s), in other cases no specific syndrome has been described.[32] No homogenous pattern of malformations is obvious from the prospective study. However, organic solvents although traditionally clustered together, are a diverse group of compounds that should not be expected to cause similar patterns of reproductive toxic effects. Although more prospective studies will be needed to confirm our results, it is prudent to minimize women's exposure to organic solvents during pregnancy. This is most important during the first trimester of pregnancy.

## 20.3.5 A PROACTIVE APPROACH FOR THE EVALUATION OF FETAL SAFETY IN CHEMICAL INDUSTRIES

[Adapted, by permission from K.I. McMartin and G. Koren, *Teratology*, **60**, 130 (1999) Copyright 1999 *John Wiley & Sons, Inc.* Reprinted by permission of Wiley-Liss, Inc. a division of John Wiley & Sons, Inc.]

Introduction

Women, their families and employers are concerned about potential fetal risks that may be associated with occupational exposure to chemicals. To be able to assess such risks in a particular plant, one has to quantify local exposure and contrast it with evidence-based literature data. There are, however, numerous obstacles that prevent such risk assessment from being routinely performed. In the reproductive literature there are few studies that actually quantify exposure levels. In the instance where authors attempt to quantify or stratify exposure, the exposure frequencies and the exposure doses are inconsistent between studies.

For many chemicals one can measure neither airborne nor blood levels. Smelling the odor of organic solvents is not indicative of a significant exposure as the olfactory nerve can detect levels lower than several parts per billion, which are not necessarily associated with toxicity. Odor thresholds for some solvents are far below several parts per million (ppm). Examples of some odor thresholds[33] include carbon disulfide (0.001 ppm vs. TLV-TWA (skin) [Threshold Limit Value-Time Weighted Average] 10 ppm), acetaldehyde (0.03 ppm vs. TLV-TWA 25 ppm), and ethyl mercaptan ($2x10^{-5}$ ppm vs. TLV-TWA 0.5 ppm).[34] In the workplace, exposure is usually to several chemicals that may change between working days or even within a single day. The amounts of chemicals absorbed are often unknown, and the circumstances of exposure may vary from workplace to workplace or even within the same operation.

Typically, investigations into fetal safety are induced by single or clusters of specific malformations, or by symptoms in exposed women. We recently reported a proactive consultation process where, for a selected chemical compound to which women working in the Products and Chemicals Divisions at Imperial Oil Limited (IOL) may be exposed, actual exposure data were contrasted with literature values and a risk assessment was constructed.[35]

Methods

An agent inventory list was used to analyze the component (the name of material or agent), exposure group, the number of employees within an exposure group, and the routine rating factor for routine work. Exposure group is defined as a group of employees who have similar exposures to chemical, physical, and/or biological agents when: 1) holding different jobs but working continuously in the same area (e.g., process workers), or 2) holding unique jobs in an area or moving frequently between areas (e.g., maintenance workers). The routine rating factor for routine work (work which is part of the normal repetitive duty for an exposure group) is defined as follows:

| *Rating Factor (RF)* | *Definition* |
|---|---|
| 0 | No reasonable chance for exposure |
| 1-5 | Minimal, exposure not expected to exceed 10% of the occupational exposure limit (OEL) |
| 6-9 | Some daily routine exposures may be expected between 10% and 50% of the OEL |
| 10-15 | Some daily routine exposures may exceed 50% of the OEL |

The rating factor (RF) can be assessed using industrial hygiene professional judgment or monitored data. NRRF is the non-routine rating factor for non-routine work defined as job task or activities which are done seasonally, occasionally or cyclical. The definitions listed for RRF apply.

For each component a listing was created with respect to individual chemicals, including rating factors, for female exposure in the Products and Chemicals Divisions. In addition, a literature search was performed for each chemical that incorporated female occupational exposure during pregnancy with human teratogenicity and spontaneous abortion as pregnancy outcomes. Teratogenicity and spontaneous abortion were chosen as the outcomes of interest as they represent the majority of endpoints examined in studies focusing on female occupational exposure during pregnancy.

Most of the selected female reproductive toxicology studies examined explicitly stated chemical exposure levels: either as parts per million, stratifying as to number of days of exposure, or as estimates of the percentage of the threshold limit values. Medline, Toxline, and Dissertation Abstracts databases were utilized to search for all research papers published in any language from 1966 to 1996. In total, 559 studies were obtained from the literature search. Of these, only 21 studies explicitly stated some sort of exposure level for the various chemicals. These chemical exposure levels in the literature and subsequent pregnancy outcomes were compared to IOL chemical exposure indices. The following is an example of one of the many chemical exposures encountered, namely exposure to toluene. For other compounds, Table 20.3.5 contrasts values in the literature with IOL indices of chemical exposure.

**Table 20.3.5. Examples of IOL compound exposure indices contrasted to literature values. [Adapted, by permission from K.I. McMartin and G. Koren, *Teratology*, 60, 130 (1999) Copyright 1999 *John Wiley & Sons, Inc.* Reprinted by permission of Wiley-Liss, Inc. a division of John Wiley & Sons, Inc.]**

| Chemical | Reference | Literature Exposure Levels | IOL Exposure Levels |
|---|---|---|---|
| Aniline | Posluzhnyi[42] | "low exposure area" | "no reasonable chance for exposure to minimal exposure not expected to exceed 10% OEL" TLV-TWA: 2 ppm |
| Benzene | Mukhametova and Vozovaya[43] | "within or lower than the maximum permissible levels" | "no reasonable chance for exposure to some daily exposures may be expected between 10-50% OEL" TLV-TWA:10 ppm |
| Chloroform | Taskinen et al.[44] | <once a week<br>>once a week | "no reasonable chance for exposure to minimal exposure not expected to exceed 10% OEL" TLV-TWA: 10 ppm |
| Dichloromethane | Taskinen et al.[44]<br>Windham et al.[21] | >once a week<br><once a week<br>>10 hrs a week<br><10 hrs a week | "no reasonable chance for exposure to minimal exposure not expected to exceed 10% OEL" TLV-TWA: 50 ppm |

| Chemical | Reference | Literature Exposure Levels | IOL Exposure Levels |
|---|---|---|---|
| Styrene | Saamanen[45]<br>Harkonen[46] | 70-100 ppm<br>20-300 ppm | "no reasonable chance for exposure to minimal exposure not expected to exceed 10% OEL" TLV-TWA: 50 ppm, TLV-STEL: 100 ppm |
| Toluene | Euler[10]<br>Syrovadko[36]<br>Ng et al.[39] | 298 ppm<br>13-120 ppm<br>50-150 ppm | "no chance for exposure to some daily exposure exceeding 50% of the OEL" TLV-TWA: 50 ppm |

IOL: Imperial Oil Limited, OEL: Occupational Exposure Limit

Results and discussion

Six studies were found that quantified toluene concentrations. The countries that reported these observations included Germany, Russia, Finland and Singapore. In general, IOL toluene levels are lower than those reported in the literature.

A few case reports of malformations in association with toluene exposure have appeared. Euler[10] reported 2 cases of multiple malformations where the anomalies were similar in children born to women who worked in shoemaking and were exposed to a soling solution containing toluene and trichloroethylenc. The average concentration of toluene in the air was 298 ppm (1.12 mg/l) and of trichloroethylene 230 ppm (1.22 mg/l). No further details of these cases were given.

Toutant and Lippmann[28] reported a single case of adverse pregnancy outcome in a woman addicted to solvents (primarily toluene). The woman, aged 20 years, had a 14-year history of daily heavy solvent abuse. On admission to the hospital, she had ataxia, tremors, mild diffuse sensory deficits, short-term memory loss, blunted affect, and poor intellectual functioning compatible with severe solvent and/or alcohol abuse. The male child born at term was microcephalic with a flat nasal bridge, hypoplastic mandible, short palpebral fissures, mildly low-set ears, pronounced sacral dimple, sloping forehead and incoordination of arm movements with unusual angulation of the left shoulder and elbow. There was a poor sucking reflex and movements were jerky at 2-4 days of age, although this improved spontaneously. The authors of this report point out the similarities between this case and fetal alcohol syndrome and suggest that there may be an analogous "fetal solvent syndrome" or that excessive solvent intake may cnhance the toxicity of alcohol.

Syrovadko[36] studied the outcome of pregnancy in a substantial number of women exposed to toluene. Toluene exposure averaged 55 ppm (range 13-120 ppm). The factory had its own maternity section where the women had their deliveries. Records of labor and newborns wcre examined for 133 women in contact with toluene and for 201 controls from the factory offices. There was no detectable effect on fertility. In the exposed group, records showed a mean pregnancy rate of 3.2/worker compared with 2.6/worker in the control group. There were no significant differences between exposed and control groups in the mortality or adverse effects on the newborn.

In the Finnish study of Holmberg[37] on central nervous system defects in children born to mothers exposed to organic solvents during pregnancy, 3 of the cases were exposed to toluene, or toluene in combination with other solvents. In one case with hydranencephaly and death 24 days after birth, there was exposure to toluene, xylene, white spirit and methyl ethyl ketone from rubber products manufacture. The second case had multiple abnormalities of hydrocephalus, agenesis of the corpus callosum, pulmonary hypoplasia and diaphragmatic hernia, and died 2 hours after birth. In the second case, the mother was exposed to toluene while manufacturing metal products. The third case had lumbar meningomyelocele and survived. The mother was exposed to toluene and white spirit. Toluene air concentrations were not stated.

A case-referent study concerning selected exposures during pregnancy among mothers of children born with oral clefts was conducted in Finland.[38] The study covered the initial 3.5 years' material and was a more detailed extension of earlier retrospective studies concerning environmental factors in the causation of oral clefts, using cases accumulated from the Finnish Register of Congenital Malformations. More case mothers (14) than referent mothers (3) had been exposed to organic solvents during the first trimester of pregnancy. The mothers were considered "substantially" exposed if their estimated continuous exposure had been at least one-third of the current TLV concentration or if the estimated peak exposure had reached the TLV concentration, e.g., during home painting in confined spaces. Various solvents included: lacquer petrol, xylene, toluene, acetates, alcohols, denatured alcohol, methyl ethyl ketone, dichloromethane, turpentine, styrene, and aromatic solvent naphtha ($C_4$-$C_{14}$ aromatics).

Ng et al.[39] examined the risk of spontaneous abortion in workers exposed to toluene. Rates of spontaneous abortions were determined using a questionnaire administered by personal interview to 55 married women with 105 pregnancies. The women were employed in an audio speaker factory and were exposed to high concentrations of toluene (mean 88 ppm, range 50-150 ppm). These rates of spontaneous abortion were compared with those among 31 women (68 pregnancies) who worked in other departments in the same factory and had little or no exposure to toluene (0-25 ppm) as well as with a community control group of women who underwent routine antenatal and postnatal care at public maternal health clinics. Significantly higher rates for spontaneous abortions were noted in the group with higher exposure to toluene (12.4 per 100 pregnancies) compared with those in the internal control group (2.9 per 100 pregnancies) and in the external control group (4.5 per 100 pregnancies). Among the exposed women, significant differences were also noted in the rates of spontaneous abortion before employment (2.9 per 100 pregnancies) and after employment in the factory (12.6 per 100 pregnancies).

Tikkanen et al.[17] performed a study to explore for possible associations between occupational factors and cardiovascular malformations. Information on the parents of 160 infants with cardiovascular malformations and 160 control parents were studied. The mother was considered "substantially" exposed to "organic solvents" if the estimated continuous exposure was at least one third of the ACGIH threshold limit value concentration or the estimated short term exposure reached the TLV concentration (while painting kitchen walls). Organic solvents were categorized as 1) "hydrocarbons", 2) "alcohols" and 3) "miscellaneous". Hygiene assessments of exposures were classified as i) "any exposure intensity" (at any period in pregnancy and in the first trimester only) and ii) "substantial exposure intensity" (at any period in pregnancy and in the first trimester only).

Of the 320 mothers, 41 case and 40 control mothers reported an exposure to organic solvents.[16] The hygiene assessment indicated some solvent exposure in 27 case and 25 control mothers. Twenty-one case and 16 control mothers had been exposed in the first trimester. Of these, substantial exposure to hydrocarbons occurred for 6 case and 2 control mothers; one case and one control mother to toluene at work and five cases and one control mother to lacquer petrol while painting indoors at home for 1 to 2 days.

Lindbohm et al.[40] investigated the association between medically diagnosed spontaneous abortions and occupational exposure to organic solvents (case-control design). The study population was composed of women who were biologically monitored for solvents. The workers were classified into exposure categories on the basis of work description and the use of solvents as reported in the questionnaires and on measurements of biological exposure. Three exposure levels were distinguished: high, low, and none. The level of exposure was assessed on the basis of the reported frequency of solvent use and the available information on typical levels of exposure in that particular job, as based on industrial hygiene knowledge.

The feasibility of biological monitoring data for classification of exposure was limited because the solvent measurements describe only short-term exposure (from 2 hours to a few days) and only 5% of the workers had been measured during the first trimester of pregnancy. Therefore, the exposure classification was based mainly on the work task description and reported solvent usage. Exposure was defined as "high" if the worker handled the solvents daily or 1-4 days a week and the level of exposure was high according to biological exposure measurements or industrial hygiene measurements available at the Institute of Occupational Health. Exposure was defined as "low" if the worker handled solvents 1-4 days a week and the level of exposure according to the measurements of the Institute was low or if the worker handled solvents less than once a week. Otherwise, the level of exposure was defined as "none". After classification, the work tasks and the related exposures were listed by the level of exposure which was checked by an independent, experienced industrial hygienist. The final population for the analysis was restricted to the matched case-control sets who confirmed their pregnancy and reported in detail their occupational exposures during early pregnancy (73 cases and 167 controls).

The odds ratios for tetrachloroethylene and aliphatic hydrocarbons, adjusted for potentially confounding factors, increased with the level of exposure.[40] For toluene the reverse was the case. Aliphatic hydrocarbons had not been biologically monitored, but industrial hygiene measurements had been performed by the Institute of Occupational Health in two printing houses which contributed subjects to this study. In two of four measurements, the concentrations of white spirit in air exceeded, during the cleaning of the printing machine, the Finnish Threshold Limit Value (150 ppm). All the printers included in this study reported that their work included cleaning of the machine.

The association of tetrachloroethylene, toluene and aliphatic hydrocarbons with spontaneous abortions was also examined by detailed records of occupational task.[40] The odds ratio of spontaneous abortion for aliphatic hydrocarbons was increased among graphic workers [5.2 (1.3-20.8)] and painters [2.4 (0.5-13.0)] but not among other workers. However, in the latter group the proportion of highly exposed workers was only 30%, whereas it was 69% in the two former groups. The odds ratio was increased also among toluene-exposed shoe workers [odds ratio 9.3 (1.0-84.7)] and dry cleaners exposed to tetrachloroethylene [odds ratio 2.7 (0.7-11.2)].

The results of the study by Lindbohm et al.[40] support the hypothesis of a positive association between spontaneous abortion and exposure to organic solvents during pregnancy and suggest that exposure, especially to aliphatic hydrocarbons, increases the risk of abortion. The highest risk for aliphatic hydrocarbons was found among graphic workers who were employed as offset printing workers or printing trade workers. They used the solvents for cleaning the printing machines and as diluent for printing ink. In cleaning the machines, exposure to mixtures of nonaromatic mineral oil distillates with 0-15% aromatic compounds may reach a high level for a short period.[40] The workers were also exposed among other things to toluene, 1,1,1-trichloroethane, thinner, and xylene. Although the data suggest that the findings are due to aliphatic hydrocarbons, combined solvent effects cannot be excluded because of the multiple exposures to different solvents.[40]

The mean measured level of blood toluene among the shoe workers was slightly higher (0.51 mmol/L, 13 morning samples) than the mean among the other toluene-exposed workers (0.38 mmol/L, 10 morning samples).[40] The shoe workers also reported use of toluene more frequently than the other toluene-exposed workers. Industrial hygiene measurements had been performed in three of the five work places of the shoe workers. The concentration of toluene in air varied from 1 ppm to 33 ppm. Other solvents detected were acetone and hexane. In two of the three shoe factories from which industrial hygiene measurements were available, relatively high levels of hexane (33-56 ppm) were measured. Hexane, being an aliphatic compound, may have contributed to the excess of spontaneous abortions.[40]

Comparison with IOL levels

The routine rating factor and non-routine rating factor from the Products and Chemical Divisions range from 00 to 11 indicating no reasonable chance for exposure to some daily exposures exceeding 50% of the OEL. The TLV-TWA for toluene is 50 ppm.[34] The Euler case reports documented air concentrations of 298 ppm for toluene and 230 ppm for trichloroethylene.[10] Both of these air concentrations exceed current standards, but no further details of these cases were given. Syrovadko reported a toluene exposure of 55 ppm (range 13-120 ppm), again, exceeding current standards.[36]

Holmberg et al.[38] and Tikkanen[17] considered workers "substantially" exposed if their estimated continuous exposure had been at least one-third of the current TLV concentration or if the estimated peak exposure had reached the TLV concentration. Similarly, Ng[38] described high concentrations of toluene (mean 88, range 50-150 ppm) exceeding current standards. All these exposure levels for toluene exceed the current threshold limit value. IOL toluene exposure levels are considerably lower than any value reported in the literature.

Lindbohm et al.,[40] for two of four air measurements, reported concentrations of white spirit exceeded the Finnish Threshold Limit Value (150 ppm) during the cleaning of the printing machine. Industrial hygiene measurements were performed in three of the five work places of the shoe workers. The concentration of toluene in air varied from 1 ppm to 33 ppm. Other solvents detected were acetone and hexane. In two of the three shoe factories from which industrial hygiene measurements were available, relatively high levels of hexane (33-56 ppm) were noted.

The routine rating factor and non-routine rating factor from the Products and Chemicals Divisions for hexane isomers range from 00 to 05 indicating no reasonable chance for exposure or minimal exposure not expected to exceed 10% of the occupational exposure limit (OEL). The routine rating factor and non-routine rating factor from the Products and Chemicals Divisions for n-hexane range from 00 to 07 indicating no reasonable chance for exposure or some daily exposures between 10% and 50% of the OEL. The TLV-TWA of n-hexane is 50 ppm or 176 $mg/m^3$.[34] The TLV-TWA of other hexane isomers is 500 ppm or 1760 $mg/m^3$ and the TLV-STEL is 1000 ppm or 3500 $mg/m^3$.[34] In comparison with the previous hexane levels reported in the literature, IOL hexane exposure levels are substantially lower.

In mice levels of inhalation exposure to toluene have included 100 to 2,000 ppm at various times during gestation as well as at various durations of exposure (6-24 hours/day).[41] Growth and skeletal retardation were noted at lower levels (133 ppm and 266 ppm, respectively) when such exposures were of a 12-24 hour duration for at least half of the gestation period.[41] Human levels of inhaled toluene exposure that would be comparable would be those obtained by chronic abusers (5,000-12,000 ppm). The only noted malformations were an increase in the frequency of 14th ribs, which was noted at 1,000 ppm on days 1-17 of gestation for 6 hours/day. As Wilkins-Haug[41] notes this has been the highest exposure studied in the mouse model and is comparable to the inhaled toluene exposure which produces euphoria in humans (500 ppm).

In 1991 we were approached by the medical department of Imperial Oil Limited to develop a proactive approach of risk evaluation of their female workers. The paradigm developed and used by us could be extrapolated to any other chemical operation. Its advantage is in its proactive nature, which aims at informing workers and preventing potential fetal risks, while also preventing unjustified fears which may lead women to quit their jobs or, in extreme cases, even consider termination of otherwise wanted pregnancies.

Upon comparing the occupational literature that presented any quantifiable chemical exposure dose or estimate of dose for any chemical with the IOL routine rating factors in the Products and Chemicals Divisions, we could conclude that IOL chemical exposure levels overall were lower than those reported in the literature. Of utmost importance is the need in published occupational reports for at least some industrial hygiene documentation, namely improved reporting of a quantifiable chemical exposure dose (for example, as implemented and currently utilized by IOL) and ideally a standard and consistent way of reporting this in the occupational literature.

### 20.3.6 OVERALL CONCLUSION

The Motherisk program is an information and consultation service for women, their families and health professionals on the safety/risk of exposure to drugs, chemicals, radiation and infection during pregnancy and lactation. Chemical exposure in the workplace is a common source of concern among our patients and health professionals.

Occupational exposure to organic solvents during pregnancy is associated with an increased risk of major fetal malformations. This risk appears to be increased among women who report symptoms associated with organic solvent exposure. Although more prospective studies will be needed to confirm our results, it is prudent to minimize women's exposure to organic solvents during pregnancy. This is most important during the first trimester of pregnancy. Moreover, symptomatic exposure appears to confer an unacceptable level of fetal exposure and should be avoided by appropriate protection and ventilation. Health care professionals who counsel families of reproductive age should inform their patients that some types of employment may influence reproductive outcomes.

Of utmost importance is the need in published occupational reports for some industrial hygiene documentation. Specifically, improved reporting of a quantifiable chemical exposure dose (for example, as implemented and currently utilized by IOL) and ideally a standard and consistent way of reporting this in the occupational literature pertaining to human reproductive toxicology.

## REFERENCES


1 J. Schardein, in **Chemically Induced Birth Defects**, *Marcel Dekker,* New York, 1985, pp. 645-658.
2 G. Watanabe and S. Yoshida, *Acta. Medica. Biol. Niigata.*, **12**, 285 (1970).
3 W. Masumoto, S. Ijima and H. Katsunuma, *Congenital Anomalies*, **15**, 47 (1975).
4 P.S. Nawrot and R.E. Staples, *Teratology*, **19**, 41A (1979).
5 F.J. Murray, J.A. John, L.W. Rampy, R.A. Kuna and B.A. Schwetz, *Ind. Hyg. Ass. J.*, **40**, 993 (1979).
6 R. Iwanaga, T. Suzuki and A. Koizumi, *Jap. J. Hyg.*, **25**, 438 (1970).
7 A. Hudak and G. Ungvary, *Toxicology*, **11**, 55 (1978).
8 J.D. Green, B.K.J. Leong and S. Laskin, *Toxicol. Appl. Pharmacol.*, **46**, 9 (1978).
9 S.M. Barlow and F.M. Sullivan, in **Reproductive Hazards of Industrial Chemicals**, *Academic Press*, London, 1982.
10 H.H. Euler, *Arch. Gynakol.*, **204**, 258 (1967).
11 B.A. Schwetz, B.M.J. Leong and B.J. Gehring, *Toxicol. Appl. Pharmacol.*, **32**, 84 (1975).
12 B.K. Nelson, B.J. Taylor, J.V. Setzer and R.W. Hornung, *J. Environ. Pathol. Toxicol.*, **3**, 233 (1980).
13 O. Heinonen, in **Birth Defects and Drugs in Pregnancy**, *PSG Publishing*, Littleton, MA, 1977, pp. 65-81.
14 F.G. Cunningham, P.C. McDonald and N. Gant in Williams Obstetrics, Appleton and Lange, Norwalk, Connecticut, 1989, pp. 489-509.
15 T.R. Einarson, J.S. Leeder and G. Koren, *Drug. Intell. Clin. Pharm.*, **22**, 813 (1988).
16 G. Axelsson, C. Liutz and R. Rylander. *Br. J. Ind. Med.*, **41**, 305 (1984).
17 J. Tikkanen and O. Heinonen, *Am. J. Ind. Med.*, **14**, 1 (1988).
18 P.C. Holmberg, K. Kurppa, R. Riala, K. Rantala and E. Kuosma, *Prog. Clin. Biol. Res.*, **220**, 179 (1986).
19 S. Cordier, M.C. Ha, S. Ayme and J. Goujard, *Scand. J. Work Environ. Health.*, **18**, 11 (1992).
20 G.K. Lemasters, An epidemiological study of pregnant workers in the reinforced plastics industry assessing outcomes associated with live births, University of Cincinnati, Cincinnati, 1983.
21 G.C. Windham, D. Shusterman, S.H. Swan, L. Fenster and B. Eskenazi, *Am. J. Ind. Med.*, **20**, 241 (1991).
22 J.A. Lipscomb, L. Fenster, M. Wrensch, D. Shusterman and S. Swan, *J. Occup. Med.*, **33**, 597, (1991).
23 M.B. Schenker, E.B. Gold, J.J. Beaumont, B. Eskenazi, S.K. Hammond, B.L. Lasley, S.A. McCurdy, S.J. Samuels, C.L. Saiki and S.H. Swan, Final report to the Semiconductor Industry Association. Epidemiologic study of reproductive and other health effects among workers employed in the manufacture of semiconductors, University of California at Davis, 1992.
24 S.M. Pinney, An epidemiological study of spontaneous abortions and stillbirths on semiconductor employees, University of Cincinnati, Cincinnati, 1990.
25 B. Eskenazi, M.B. Bracken, T.R. Holford and J. Crady, *Am. J. Ind. Med.*, **14**, 177, (1988).
26 T.H. Shepard, *Teratology*, **50**, 97, (1994).
27 Y. Bentur in **Maternal Fetal Toxicology**, G. Koren, Ed., *Marcel Dekker*, New York, 1994, pp. 425-445.
28 C. Toutant, and S. Lippmann, *Lancet*, **1**, 1356, (1979).
29 G. Axelson, R. Rylander, *Int. J. Epidemiol.*, **13**, 94, (1984).
30 S. Khattak, G. K-Moghtader, K. McMartin, M. Barrera, D. Kennedy and G. Koren, *JAMA.*, **281**, 1106 (1999).
31 K.I. McMartin, M. Liau, E. Kopecky, T.R. Einarson and G. Koren, *Am. J. Ind. Med.*, **34**, 288 (1998).
32 G. Koren, A. Pastuszak and S. Ito, *N. Engl. J. Med.*, **338**, 1128 (1998).
33 M.J. Ellenhorn and D.G. Barceloux, **Medical Toxicology: Diagnosis and Treatment of Human Poisoning**. *Elsevier Science Publishing Company Inc.*, New York, 1988, pp. 1412-1413.
34 Anonymous, Threshold limit values for chemical substances and physical agents and biological exposures, American Conference of Governmental Industrial Hygienists, Cincinnati, 1993, pp. 12-35.
35 K.I. McMartin and G. Koren, *Teratology*, **60**, 130 (1999).
36 O.N. Syrovadko, *Gig Tr Prof Zabol.*, **21**, 15 (1977).
37 P.C. Holmberg, *Lancet*, **2**, 177, (1979).
38 P.C. Holmberg, S. Hernberg, K. Kurppa, K. Rantala, and R. Riala, *Int. Arch. Occup. Environ. Health*, **50**, 371 (1982).
39 T.P. Ng, S.C. Foo, and T. Yoong, *Br. J. Ind. Med.*, **49**, 804, (1992).
40 M.L. Lindbohm, H. Taskinen, M. Sallmen, and K. Hemminki, *Am. J. Ind. Med.*, **17**, 447, (1990).
41 L. Wilkins-Haug, *Teratology*, **55**, 145, (1997).
42 P.A. Podluzhnyi, *Gig. Sanit.*, **1**, 44, (1979).
43 G.M. Mukhametova, and M.A. Vozovaya, *Gig. Tr. Prof. Zabol.*, **16**, 6 (1972).
44 H. Taskinen, M.L. Lindbohm, and K. Hemminki, *Br. J. Ind. Med.*, **43**, 199 (1986).
45 A. Saamanen, Styreeni, Institute of Occupational Health, Helsinki, 1991.
46 H. Harkonen, S. Tola, M.L. Korkala, and S. Hernberg, *Ann. Acad. Med. Singapore*, **13**, 404 (1984).

# 20.4 INDUSTRIAL SOLVENTS AND KIDNEY DISEASE

NACHMAN BRAUTBAR
**University of Southern California, School of Medicine, Department of Medicine, Los Angeles, CA, USA**

## 20.4.1 INTRODUCTION

Industrial solvents are used extensively in the industry, as well as modern living. The principle class of components are the chlorinated and non-chlorinated hydrocarbons. The various types of commonly used hydrocarbons are presented in Figure 20.4.1.

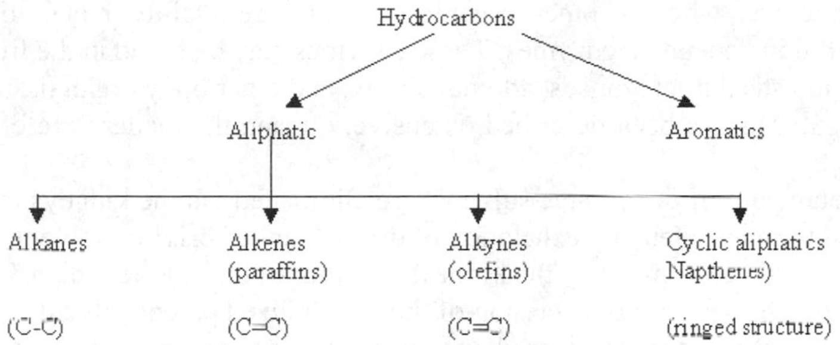


Figure 20.4.1. General classification of hydrocarbons based on general structure.[Adapted, by permission, from A.T. Roy, *Nephron*, **58**, 385, 1991.]

Solvents are absorbed into the human body through several routes including 1) inhalation through the lungs, 2) absorption through the skin, 3) ingestion (in rare cases). The main route of absorption is commonly pulmonary, the lung, and this depends on several factors including the frequency of breathing, diffusion of solvent vapors across the alveolar membrane, partial pressure of solvent vapor in inspired air and blood, and solubility of the solvent in blood as the result of to air partition coefficient, and blood flow through the lungs.[1,2,3] Once in the circulation, 25% of the cardiac output which is about 1200 cc of blood per minute passes through the kidneys. Therefore it is no surprise that with this amount of blood passing through the kidney and carrying solvents (from either industrial inhalation, skin absorption, and on rare occasions ingestion) the effects of solvents on the kidney has become a practical clinical one.

Since inhaled hydrocarbons are readily absorbed into the blood stream and become lipophilic and readily pass across the lipid membranes. In addition to reaching the kidney, the solvents reach the brain (as does the most ancient solvent, alcohol) and enter the blood brain barrier in high concentration.

Skin absorption is the second most important route for solvent entry into the body and at times is much more significant than inhalation. The reason is that absorption of organic solvent vapors by inhalation at the threshold limit value is insignificant and is less than 2% of the amount absorbed via inhalation under the same exposure conditions.[3] In contrast, solvents may be absorbed through the skin in significant amounts even at below the threshold limit value.[3] Factors that effect the skin absorption of solvents include the composition of the skin, whether the skin is healthy or not (there is increased absorption if the skin has reduced cellular membrane), and the lipid solubility of the solvent.

As far as the gastrointestinal tract, commonly this is not a significant route of absorption. Solvents absorbed via the gastrointestinal tract are removed immediately by the liver through the first-pass metabolism. If the amount of solvents and quantity of solvents ingested is increased significantly and exceeds the capacity of the liver to metabolize the solvents, then the gastrointestinal tract route will become significant.[4,5,6]

The distribution of an organic solvent in the human body depends upon its partial pressure in the arterial blood and the solubility of the solvent in the tissue, as well as the blood flow rate through the tissue.[7] Data on tissue distribution of various solvents are limited at best.

The metabolism of solvents depend on the solvent. Alcohols are metabolized via alcohol dehydrogenase, whereas other organic solvents are mainly metabolized by the cytochrome P-450-dependent enzymes. These enzymes may be found in the liver, kidneys, lungs, gastrointestinal tract, gonads, adrenal cortex, and other body organ tissues. The metabolism of solvents has been described extensively,[8,9] and the reader is referred to those writings.

The metabolites of the organic solvents are eliminated via the kidneys through urine excretion and to some extent, by exhalation of the unchanged original solvent. Commonly the parent solvent is eliminated by the kidneys and this amounts to less than 1%. The metabolites are the main source of excretion of the metabolized parent solvent.

In the last several decades, there have been several studies in experimental animals, case reports in humans, case studies in humans, and epidemiological studies in humans on the effects of solvents on the kidney, both acutely and chronically. The scope of this chapter is the clinical chronic effects of solvents on the kidney (chronic nephrotoxicology).

### 20.4.2 EXPERIMENTAL ANIMAL STUDIES

The toxic effects of organic solvents on the kidneys has been studied in several experimental species, especially mice and rats. Damage to the kidney has been shown in these experimental animals in the form of acute damage to various parts of the nephron, especially the tubules. This has usually been described as tubular degeneration with regenerative epithelium, deposits of mineral crystals and of intralobular proteins, and interstitial inflammation.[8,10-15] Several studies have shown glomerular damage in experimental animal[16,17] and have suggested that long-term solvent exposure alters the immune system and leads to the glomerulopathy with mesangial IgA deposits.

While the exact mechanism is not known and various mechanisms have been postulated, it is reasonable to accept a mechanistic approach which takes into account genetic, environmental, susceptibility (such as pre-existing diseases including hypertensive kidney disease and diabetes), direct tubular toxicity, permeability changes and immunosuppression.

### 20.4.3 CASE REPORTS

The earlier documentation of chronic renal disease and hydrocarbon exposure consists of case reports, and this data was summarized by Churchill et al.[18] describing Goodpasture's syndrome in 15 adults, epimembranous glomerulonephritis in 5 adults, and subacute proliferative glomerulonephritis in one adult. The hydrocarbon exposures were for solvents in 12 patients, gasoline in 4, gasoline-based paint in 3, jet fuel, mineral turpentine and unspecified in 1 case report. These case reports were previously summarized by us in a previous publication[19] and are represented in the following table.

**Table 20.4.1. Case series report of glomerulonephritis and hydrocarbon exposure. [Data from reference number 19]**

| Investigator | n | Diagnosis | Agent |
|---|---|---|---|
| Sperace[20] | 2 | Goodpasture's | Gasoline |
| Heale, et al.[21] | 1 | Goodpasture's | Gasoline |
| Klavis and Drommer[22] | 1 | Goodpasture's | Gasoline-based paint spray |
| Beirne and Brennan[23] | 5<br>1 | Goodpasture's<br>RPGN | Degreasing and paint<br>Solvents and jet fuel |
| D'Apice, et al.[24] | 2 | Goodpasture's | Gasoline mineral turpentine |
| Kleinknecht, et al.[25] | 2 | anti-GBM nephritis | Organic solvent vapors |
| Daniell, et al.[26] | 1 | anti-GBM nephritis | Stoddard solvent |
| Von Scheele, et al.[27] | 1 | subacute GN | Paint solvent |
| Ehrenreicht, et al.[28] | 4 | epimembranous GN | Solvents |
| Cagnoli, et al.[29] | 1 | epimembranous GN | ? |

GBM = Glomerular basement membrane; RPGN = rapidly progressive glomerulonephritis; GN= glomerulonephritis

While these studies represent case reports, they suggest an association between exposure and the development of chronic glomerular disease.

## 20.4.4 CASE CONTROL STUDIES

Several case-control studies have examined the role of organic solvent exposure in a population of patients with glomerulonephritis. A total of 14 case control studies examining human exposure to solvents and glomerulonephritis have been conducted and are documented here in Table 20.4.2.[30]

**Table 20.4.2. Glomerulonephritis and organic solvents: Case-control studies summarized. [Data from reference number 30]**

| Investigator | Increased risk factor | Investigator | Increased risk factor |
|---|---|---|---|
| Lagrue, et al.[31,32] | 4.9*, 5.2* | Nuyts, et al.[39] | 1.1* |
| Bell, et al.[33] | increased | Zimmerman, et al.[40] | increased |
| Ravnskov, et al.[34] | 3.9* | Ravnskov[41] | increased |
| Ravnskov, et al.[35] | 2.8* | Finn, et al.[42] | 3.6*, 3.2* |
| Porro, et al.[36] | 3.9* | Van der Laan[43] | 1.1 |
| Yaqoob, et al.[37] | aliphatic 15.5,<br>halogenated 5.3,<br>aromatic-oxygenated 2.0 | Harrison, et al.[44] | 8.9* |
| Yaqoob, et al.[38] | increased | *P<0.05, statistically significant | |

The study by Lagure, et al,[31,32] showed significantly increased risk of solvent related glomerulonephritis of 4.9. That this increased risk of glomerulonephritis follows a dose-response relationship was shown in the study of the populations examined by Ravnskov, et al.,[34,35] Bell, et al.,[33] Porro, et al.,[36] Yaqoob, et al.,[37,38] Nuyts, et al.,[39] and demonstrates: 1) temporal relationship between exposure to solvents and the development of kidney disease, 2) a dose-response relationship, strongly showing the causal link between solvent exposure and glomerulonephritis. The study by Nuyts, et al.,[39] examined a large population of 272 patients with chronic renal failure and assessed several occupational exposures, among those were hydrocarbons. The increased risk of chronic kidney disease in the form of renal failure in patients exposed to solvents was 5.45. The study of Askergren et al.[45] looked into kidney functions in patients exposed to various organic solvents, specifically excretion of red blood cells in the urine in 101 patients exposed to solvents as compared to 39 non-exposed controls. Those who were exposed to organic solvents significantly excreted more cells than the ones who were not exposed. These studies showed the role for organic solvent exposure in the development of damage to the glomerules since excretion of red blood cells represents damage to the glomerules rather than tubules. That exposure to solvents is associated with glomerular damage rather than tubular damage fits with the various case reports and case-control studies and further suggest a plausible causal connection between exposure to industrial solvents and glomerular damage leading later on to chronic glomerulonephritis. The study by Bell et al.[33] studied 50 patients who had organic solvent exposure and biopsy-proven proliferative glomerulonephritis. They have shown that none of these patients had evidence of any other systemic disease or preexisting infection, and compared those with 100 control subjects matched for age, sex and social class. This study is important since exposure assessment was done and showed significantly greater exposure scores in patients with glomerulonephritis compared to the control subjects. Furthermore, the degree of exposure was significantly higher in those patients who have more severe glomerulonephritis than those who have less severe glomerulonephritis, further indicative of a dose response relationship. This is a study which demonstrates significant statistical association, as well as dose response relationship between solvent exposure and kidney damage in the form of glomerular lesion and end-stage glomerulonephritis, ranging from mild to chronic severe glomerulonephritis. The study by Daniell et al.[26] evaluated the risk of developing glomerular lesion associated with hydrocarbon exposure and showed a does-response relationship and variations in disease severity in relation to the exposure intensity. They showed an increase risk of developing glomerular nephritis, ranging from 2.8 to 8.9 fold increase as compared to the non-exposed population. There was clear temporal relationship between the exposure, absence of any other causes, a dose-response relationship which further validated the observations of Bell et al.[33] and conclude that intense or long-term exposure (low-level but long-term or short-term and high levels) to commonly used industrial solvents played a causal role in the development of glomerular damage and chronic glomerulonephritis.

In a comprehensive study, Yaqoob et al.[46] performed a population study which looked into 3 groups of healthy men working in 3 different areas of a major car manufacturing plant. They have studied 3 groups, Group 1 included 112 paint sprayers exposed to a paint-based mixture of hydrocarbons, Group 2 which was composed of 101 transmission shop workers with exposure to petroleum-based mineral oils, and Group 3 which was comprised of 92 automated press operators with minimal background exposure to lubricating oils and who acted as internal controls. The 3 groups studied were comparable in age, duration of employment, duration of hydrocarbon exposure, and other factors. The cumulative exposure to hydrocarbons was evaluated. The hydrocarbon exposure scores were significantly higher in Groups 1 and 2, as compared to Group 3 (which served as an internal control and epidemiologically is a good working population control group, since this method takes into account the healthy worker). The principal hydrocarbons used throughout the period of time of the study were toluene, xylene, and n-butyl alcohol in paints and various petroleum fractions in the mineral oils. The study evaluated markers of kidney dysfunction in the subjects chronically exposed to hydrocarbons at the described work site. The authors concluded that paint exposure in the long-term is associated with renal impairment and micro-proteinuria without elevation in serum creatinine (which indicates that the kidney functions from a creatinine clearance point of view are still intact, and are less sensitive as a biological marker of glomerular damage) is a feature of workers chronically exposed to petroleum based mineral oils. The investigators also reported significant urinary excretion of protein which also indicated early glomerular damage in susceptible individuals. The authors concluded from these studies that chronic hydrocarbon exposure can be associated with renal impairment. They further concluded that the significance of the early markers of renal damage can predict progressive deterioration in renal functions. These data indicate that chronic hydrocarbon exposure may be associated with early and sub-clinical renal dysfunction leading to a chronic glomerulonephritis.

Porro et al.[36] performed a case referent study and they looked into a group of 60 patients with chronic glomerulonephritis established by biopsy, with no evidence of any other systemic diseases, and was compared to 120 control subjects who were not exposed to solvent vapors. Exposure assessment was based on scores from questionnaires. Exposure was significantly higher in the case group studies than in the reference control group for both total and occupational solvent exposure. They further found that the odds ratio of chronic glomerulonephritis for patient's occupationally exposed to solvents was 3.9 and using a logistic regression model and they showed a dose-response effect of occupational exposure to solvents and glomerulonephritis. Histological studies of the 60 patients with chronic glomerulonephritis ruled out other systemic disease and demonstrated the whole-spectrum of glomerular diseases, the most common one is IgA nephropathy. When the sub-group of patients with IgA nephropathy and their matched controls were separately examined, the cases appeared to be significantly more exposed than the patients with other non-glomerular diseases such as kidney stones. Based on their findings, the investigators concluded that their results are in agreement with the hypothesis that the onset of glomerulonephritis could be related to a non-acute exposure to solvents even of light intensity.

The work of De Broe et al.[47] looked into occupational renal diseases and solvent exposure. They have concluded that the relation between hydrocarbon exposure and glomerulonephritis seems to be well-defined from an epidemiological point of view. They further show, in a case-control study of a group of patients with diabetic nephropathy, that hydrocarbon exposure was found in 39% of the patients with that particular form of kidney disease. They find that this was in agreement with the findings of Yaqoob et al.[37] who found higher levels of hydrocarbon exposure in patients with incipient and overt diabetic nephropathy than in diabetic patients without clinical evidence of nephropathy. These data indicate a particular sensitivity of patients with diabetic kidney toward the damaging effects of the hydrocarbons. The findings of these investigators are agreement with the study of

Goyer,[48] who showed that existing renal diseases, particularly hypertensive and diabetic nephropathies, are clear risk factors predisposing to abnormal accumulation and excess blood levels of any nephrotoxic drugs and chemicals, as well as solvents. Indeed this observation makes a lot of scientific and clinical sense, since it is known that the ability of the kidney to excrete the breakdown metabolites of various materials including industrial solvents is reduced with any incremental reduction of kidney function, and there would certainly be more accumulation of these breakdown products, as well as the parental solvents in the kidney tissue, and as such, it makes sense that these individuals with underlying kidney disease such as hypertensive kidney disease, diabetic kidney disease, or interstitial kidney disease which may not yet be clinically overt, are at a significantly increased risk of developing chronic kidney disease as a result of the documented damaging effects of solvents on the kidney.

### 20.4.5. EPIDEMIOLOGICAL ASSESSMENT

The epidemiological diagnostic criteria for most cases of end-stage kidney disease is deficient since no etiologic information is available in the majority of the cases. Fewer than 10% of the end-stage renal disease cases are characterized etiologically.[49] Clinically, many patients are classified histologically such as glomerulonephritis, but little effort is made to look for toxic factors. Indeed, the majority of the clinicians seeing patients with end-stage kidney disease are not trained to look into occupational, environmental, or toxicological issues and end-stage renal disease. Many patients are listed as having hypertensive end-stage kidney disease and are presumed therefore to be "idiopathic" in origin, however, these cases may very well be the result of other industrial and/or environmental factors, among them, solvent exposure. Many of the problems in the epidemiological analysis is the result of a great reserve capacity of the kidney that can function relatively adequately despite slowly progressive damage. End-stage kidney disease is typically not diagnosed until considerable kidney damage has already occurred at the time when the patient seeks clinical attention. Furthermore, kidney biopsy and post-mortem examination, almost always find small kidneys, inadequate to help in the histopathological assessment, and therefore the etiology is either missed or is misclassified as "idiopathic" or "unknown".

Indeed the study by Stengel. et al.[50] looked at organic solvent exposure and the risk of IgA nephropathy. These investigators have shown that the risk of IgA nephropathy is highest among the most exposed group to oxygenated solvents. The study by Yaqoob et al.[37] showed an increased risk factor of 15.5 for development of glomerulonephritis in patients exposed to aliphatic hydrocarbons and a risk factor of 5.3 in patients exposed to halogenated hydrocarbons. These epidemiological data further supports observations made in the case reports, case studies and experimental animal studies. The epidemiological studies by Steenland et al.[51] had evaluated the risks and causes of end-stage kidney disease and concluded that regular exposure to industrial solvents played a significant role in the development of chronic end-stage kidney disease.

Based on the current literature from experimental animal studies, case reports, case-control studies, and epidemiological studies, one can conclude that the studies show:

1. Biological plausibility.
2. A temporal relationship between exposure to industrial solvents and the development of chronic kidney disease (glomerulonephritis).
3. A dose-response relationship.
4. Consistency of association.

5. Statistical association in the majority of the studies.

These criteria fulfill the Bradford-Hill criteria,[52] and establish the basic criteria required for causation.

## 20.4.6. MECHANISM

Immune-mediated mechanisms play a major role in the pathogenesis of glomerular disease, in general. In the vast majority of the cases, antigen-antibody reaction and immune complexes form in the kidney, mainly around the glomerular capillary wall and mesangium. Cellular antigens, both endogenous such as DNA and tumor antigens, as well as exogenous such as viral antigen hepatitis B and C, drugs, and bacteria have been shown to be causative factors in human glomerular immune-mediated diseases. The most common pathological process described in association with solvent exposure and chronic glomerular nephritis has been that of IgA nephropathy, Good Pasture's syndrome, and proliferative glomerulonephritis.

Unlike acute renal failure caused by hydrocarbons, where the renal damage is secondary to the nephrotoxins and mainly cause damage of the proximal tubule acute renal failure, the glomerular chronic renal failure, appears to be immunologically mediated. Among others, genetic factors may be involved in the pathogenesis of hydrocarbon induced nephropathy. It has been suggested that the propensity to develop this autoimmune disease depends on a combination of a genetic component and predilection, and environmental component.[24] Individuals susceptible to glomerular or tubular injury by hydrocarbons may develop chronic kidney disease through three possible mechanisms. The first mechanisms is direct tubular toxicity which is commonly the cause of acute renal failure. While it is true that the initial injury of acute renal failure is directed toward the tubule of the nephron, glomerulonephritis may be the result of an autoimmune reaction to the tubulotoxins.[53,54] The second mechanism mainly involves immunosuppression. Ravnskov,[53] in a review of the pathogenesis of hydrocarbon associated glomerulonephritis, suggested that hydrocarbons are immunosuppressives and this effect is noted in several locations in the immunological cascade. This includes leukocyte mobility and phagocytosis suppression such as shown in the benzene effects in mice.[55] This suppression of the normal immune response by hydrocarbons may play a role in the pathogenesis of immune-mediated glomerular lesions. The third mechanism involves alteration in membrane permeability. Good Pasture's syndrome is mediated by antibodies reactive with the glomerular basement membrane and alveolar basement membranes. Antibodies in experimental models can usually bind to alveolar basement membranes in vitro by indirect immunofluorescence. Experimental studies suggested that hydrocarbons alter the permeability of pulmonary capillaries, thereby allowing anti-glomerular basement membranes to bind to the alveolar basement membranes.[56] This etiology is further supported by the observation that differential sensitivity to exposures due to genetic factors since DR3 and DR4 antigens are more frequent in patients with toxic nephritis than in the general population.[57,58] The study by Zimmerman et al.[40] have shown that in 6 of 8 patients with Good Pasture's syndrome had extensive occupational exposure to solvents ranging from 4 months to 10 years. The results of this study suggested that interaction between the inhaled hydrocarbons and the lung and kidney basement membranes could induce autoantibodies to these membranes. Goyer[48] suggested an autoimmune mechanisms responsible for glomerular lesions following chronic exposure to solvents. Based on case studies and case reports, it is proposed that chronic exposure to low levels of solvents in susceptible individuals induces an initial cell injury sufficient to damage cell membranes and to provide the antigen triggering the immune response, accelerating a cascade of a reaction ending with glomerulonephritis.

## REFERENCES


1 Morrison, RR, and Boyd RN, **Organic Chemistry**, 5th Edition, Morrison RT, Boyd RN, Eds, *Allyn & Bacon*, Boston (1987).

2 Domask, WG, **Renal Effects Of Petroleum Hydrocarbons**, Mehlman, MA, Hemstreet GP III, Thorpe JJ, Weaver NK, Eds, *Princeton Scientific Publishers*, Princeton, 1-25.

3 Pederen, LM., *Pharmacol Toxicol*, **3**, 1-38 (1987).

4 Ervin ME, **Clinical Management of Poisoning and Drug Overdose**, Addad LM, Winchester JF, Eds, *Saunders*, Philadelphia, 771 (1983).

5 Wolfsdorf J, *J Pediatr*, **88**, 1037 (1976).

6 Janssen S, Van der Geest S, Meijer S, and Uges DRA, *Intensive Care Med*, **14**, 238-240 (1988).

7 Smith TC, and Wollman H, **The Pharmacological Basis of Therapeutics**, Goodman A, Gilman L, Eds, *MacMillan*, New York, 260-275 (1985).

8 Clayton GD, and Clayton EE (Eds), **Patty's Industrial Hygiene and Toxicology**, *Wiley*, New York, 26 (1981) and 2C (1982).

9 World Health Organization and Nordic Council of Ministers, Organic Solvents and the Central Nervous System, Environmental Health 5, Copenhagen and Oslo, 1-135 (1985).

10 Browning E, **Toxicity & Metabolism of Industrial Solvents**, *Elsevier Publishing Company*, Amsterdam-London-New York (1965).

11 Mehlman MA, Hemstreet GP III, Thorpe JJ, and Weaver NK, Series: Advances in Modern Environmental Toxicology: Volume VII- **Renal Effects of Petroleum Hydrocarbons**, *Princeton Scientific Publishers*, Princeton (1984).

12 Carpenter CP, Geary DL Jr Myers, et al, *Toxicol Appl Pharmacol*, **41**, 251-260 (1977).

13 Gibson JE, and Bus JS, *Ann NY Acad Sci,* **534**, 481-485 (1988).

14 Thomas FB, and Halder CA, Holdsworth CE, Cockrell By, **Renal Heterogeneity and Target Cell Toxicity**, Back PH, Lock EA, Eds, Chichester, 477-480 (1985).

15 Halder CA, Van Gorp GS, Hatoum NS, and Warne TM, *Am Ind Hyg Assoc J*, **47**, 164-172 (1986).

16 Coppor, et al, *American Journal of Kidney Disease*, **12**, 420-424 (1988).

17 Emancipator SN, *Kidney International,* **38**, 1216-1229 (1990).

18 Churchill DN, Fina A, and Gault MH, *Nephron*, **33**, 169-172 (1983).

19 Roy AT, Brautbar N, and Lee DBN, *Nephron*, **58**, 385-392 (1991)

20 Sperace GA, *Am Rev Resp In Dis*, **88**, 330-337 (1963).

21 Hele WF, Matthisson Am, and Niall JF, *Med J Aust*, 355-357 (1969).

22 Klavis G, and Drommer W, *Arch Toxicol*, **26**, 40-50 (1970).

23 Beirne GJ, and Brennan JT, *Arch Environ Health*, **25**, 365-369 (1972).

24 D'Apice AJF, et al, *Ann Intern Med*, **88**, 61-62 (1980).

25 Kleinknecht D, et al, *Arch Intern Med*, **140**, 230-232 (1980).

26 Daniell WE, Couser WG, and Rosenstock L, *JAMA*, **259**, 2280-2283 (1988).

27 Von Scheele C, Althoff P, Kempni V, and Schelin H, *Acta Med Scan*, **200**, 427-429 (1976).

28 Ehrenreicht T, Yunis SL, and Churg J, *Environ Res*, **24**, 35-45 (1977).

29 Cagnoli L, et al, *Lancet,* 1068-1069 (1980).

30 Brautbar N and Barnett A, *Environmental Epidemiology and Toxicology*, **2**, 163-166 (1999).

31 Lagrue G, et al, *J Urol Nephrol*, **4-5**, 323-329 (1977).

32 Lagrue G, et al, *Nouv Press med*, **6**, 3609-3613 (1977).

33 Bell GM, et al, *Nephron*, **40**, 161-165 (1985).

34 Ravnskov U, Forsbergg B, and Skerfving S, *Acta Med Scand,* **205**, 575-579 (1979).

35 Ravnskov U, Lundstrom S, and Norden A, *Lancet,* 1214-1216 (1983).

36 Porro A, et al, *Br J Ind Med*, **49**, 738-742 (1992).

37 Yaqoob M, Bell GM, Percy D, and Finn R, *Q J Med,* **301**, 409-418 (1992).

38 Yaqoob M, et al, *Diabet Med*, **11**, 789-793 (1994).

39 Nuyts GD, et al, *Lancet*, **346**, 7-11 (1994).

40 Zimmerman SW, Groehler K, and Beirne GJ, *Lancet*, 199-201 (1975).

41 Ravnskov U, *Acta Med Scand*, **203**, 351-356 (1978).

42 Finn R, Fennerty RG, and Ahmad R, *Clin Nephrol*, **14**, 173-175 (1980).

43 Van der Laan G, *Int Arch Occup Envion Health*, **47**, 1-8 (1980).

44 Harrison DJ, Thompson D, and MacDonald MK, *J Clin Pathol*, **39**, 167-171 (1986).
45 Askergren A, *Acta Med Scand*, **210**, 103-106 (1981).
46 Yaqoob M, et al, *Q J Med*, **86**, 165-174 (1993).
47 De Broe ME, D'Haese PC, Nuyts GD, and Elseviers MM, *Curr Opin Nephrol Hypertens*, **5**, 114-121 (1996).
48 Goyer RA, *Med Clin of North America*, **74**, 2, 377-389 (March 1990).
49 Landrigan P, et al, *Arch Environ Health*, **39**, 3, 225-230 (1984).
50 Stengels B, et al, *Int J Epidemiol*, **24**, 427-434 (1995).
51 Steenland NK, Thun MJ, Ferguson BA, and Port FK, *AJPH*, **80**, 2, 153-156 (1990).
52 Hill AB, The Environment and Disease: Association or Causation? President's Address. Proc Royal Soc Med, 9, 295-300 (1965).
53 Ravnskov U, *Clin Nephrol*, **23**, 294-298 (1985).
54 Tubbs RR, et al, *Am J Clin Pathol*, **77**, 409 (1982).
55 Horiguchi S, Okada H, and Horiuchi K, *Osaka City Med J*, **18**, 1 (1972).
56 Yamamoto T, and Wilson CB, *Am J Pathol*, **126**, 497-505 (1987).
57 Emery P, et al, *J Rheumatol*, **11**, 626-632 (1984).
58 Batchelor JR, et al, *Lancet*, *i*, 1107-1109 (1984).

## 20.5 LYMPHOHEMATOPOIETIC STUDY OF WORKERS EXPOSED TO BENZENE INCLUDING MULTIPLE MYELOMA, LYMPHOMA AND CHRONIC LYMPHATIC LEUKEMIA

NACHMAN BRAUTBAR
**University of Southern California, School of Medicine, Department of Medicine, Los Angeles, CA, USA**

### 20.5.1 INTRODUCTION

Benzene, is an aromatic hydrocarbon and historically has been produced during the process of coal tar distillation and coke production, while today benzene is produced mainly by the petrochemical industry. Based on the National Institute of Occupational Safety and Health (NIOSH), in the United States, it has been estimated that in 1976 two million Americans were exposed occupationally to benzene.[1] Worldwide production of benzene is approximately 15 million tons[2] and the production in the United States is estimated to be increasing at least 3% annually,[3] approaching 6 million tons of benzene produced in the United States in 1990 and 6.36 million tons produced in 1993.[2] Benzene has been described as a clear, colorless, non-corrosive, and flammable liquid with a strong odor.

Benzene is used as an excellent solvent and degreasing agent, and as a basic aromatic unit in the synthetic process of other chemicals.[4] Exposure to benzene in the occupational setting most commonly occurs in the chemical, printing, rubber, paint, and petroleum industry. Among other sources of exposure to benzene, non-occupational exposures in the form of cigarette smoking and exposure to gasoline and its vapors during fueling motor vehicles.[5]

### 20.5.2 ROUTES OF EXPOSURE

The major route of exposure to benzene is inhalation through the lungs of benzene vapors, however skin absorption of benzene has been shown to be significant depending on the circumstances of the exposure such as time of contact between the benzene and the skin.[1,6] Benzene absorption, as other solvents, through the skin is enhanced if the skin condition is altered by either disease or loss of skin or cracking of the skin.[4] The dermal absorption of benzene deserves much more attention than previously described in various texts of toxicology and occupational medicine, as well as environmental and industrial health. Dermal absorption of benzene in workers who use either toluene containing benzene or other solvents containing benzene, is a significant factor in calculating dosimetry and absorption of benzene and can be calculated utilizing standard accepted methodology. A recent study by Dr. Brenner et al.[6] described chronic myelogenous leukemia due to skin absorption of benzene as a contaminant of other solvents. The investigators in that study[6] concluded that the total benzene absorbed dose via skin and inhalation was equivalent to an accumulated vapor exposure of 196.4 + 42 ppm-years. Dermal exposure accounted for 97% of the total absorbed dose of benzene. Inhalation of benzene from occupational, smoking and ambient non-occupational sources accounted for only 3% of the benzene dose. The authors presented the reports of dermal absorption of benzene in the following table.

**Table 20.5.1. Summary of benzene dose expressed as equivalent ppm-years. [Adapted, by permission, from D. Brenner, *Eur. J. Oncol.*, 3(4), 399-405, 1998.]**

| Case | Solvent | Dermal | Occupational inhalation | Cigarette smoke | Ambient inhalation | Total |
|---|---|---|---|---|---|---|
| 1997 | Toluene, MEK, Acetone | 170.4 | 19.2 | 0.05* | 0.29 | 189.9 |
| 1998 Case 1 | Mineral spirits | 41.1 | 17.8 | 0.1* | 0.23 | 59.3 |
| 1998 Case 2 | Refinery process streams | 19.0 | 4.6 | 1.4 | 0.4 | 196.4 |

*Second hand cigarette smoke

Therefore, workers who are exposed to solvents containing benzene should be evaluated for skin absorption dosimetry, in addition to other sources such as inhalation, to address the range of levels of exposure.

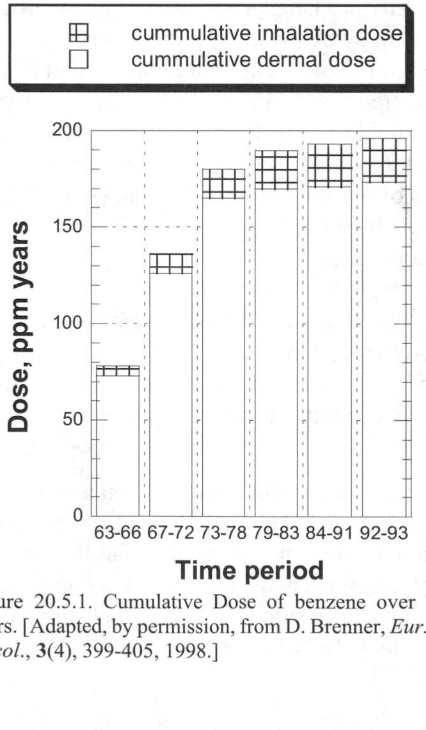

Figure 20.5.1. Cumulative Dose of benzene over 30 years. [Adapted, by permission, from D. Brenner, *Eur. J. Oncol.*, **3**(4), 399-405, 1998.]

Therefore, dermal absorption of benzene, especially in connection with benzene as a byproduct in other solvents, is of extreme importance in dosimetry analysis.

Once benzene reaches the blood, it is metabolized mainly in the liver. The metabolic products are excreted in the urine within 48 hours from absorption. Several metabolites have been found in the urine after benzene exposure, among those are phenol, quinone, hydroquinone, and muconic acid.[7] The liver utilizes the cytochrome P450 and oxidization system for the metabolism of benzene.[8,9]

Specific cellular toxic effects from benzene have been described and those include, among others, the central nervous system (doses of over 100 ppm), liver, kidney, skin, immunological, and carcinogenic. The various toxicological effects of benzene will not be discussed in this chapter since the scope of this chapter is the hematopoietic effects, the reader is referred to other sources.[3]

### 20.5.3 HEMATOPOIETIC EFFECTS OF BENZENE

Benzene is a proven human carcinogen. The toxicity of benzene has been known since the 19th Century when aplastic anemia was first reported.[4,10] Indeed the causal link between benzene and bone marrow toxicity in the form of hematoxicity and bone marrow suppression was described already in 1897.[11] In 1928, Delore et al. described leukemia as a result of benzene exposure.[12] In 1932 Lignac[13] reported lymphoblastoma in association with benzene exposure. Several studies have reported the association between exposure to benzene and hematopoietic toxicity and leukemia.[14,15] Acute myeloid leukemia has been the most frequent form of leukemia found to be related to benzene exposure. Other forms of leukemia have been described in association with benzene exposure, such as erythroleukemia, thrombocytopenia, acute myeloid leukemia, myelodysplastic syndrome, acute lymphoblastic leukemia, chronic lymphocytic leukemia, and Hodgkin's and non-Hodgkin's lymphoma. As a result of the high toxicity of benzene the American Petroleum Institute in their paper on benzene exposure in 1948 have concluded that the only safe level of exposure to benzene is no exposure at all.[16] The language utilized was as follows, "In as much as the body develops no tolerance to benzene, and as is there is a wide variation in individual susceptibility, it is generally considered that the only absolutely safe concentration for benzene is zero."

The following hematological conditions have been described in association with benzene.[17-40]

1. Acute myelogenous leukemia.
2. Erythroleukemia.
3. Aplastic anemia.
4. Acute monocytic leukemia.
5. Chronic myelogenous leukemia.
6. Myelofibrosis and myeloid metaplasia.
7. Thrombocythemia.
8. Acute lymphoblastic leukemia.
9. Chronic lymphocytic leukemia.
10. Lymphomas and related disorders.
11. Multiple myeloma
12. Myelodysplastic syndrome.

### 20.5.4 CARCINOGENIC EFFECTS OF BENZENE

Several well conducted epidemiological scientific studies and data have provided the epidemiological basis for benzene as a hematopoietic and lymphopoietic cancer. In his paper entitled "Benzene Health Effects", Mehlman described a wide range of the hematotoxicity of benzene.[41] Nilsson et al.[42] described leukemia, lymphoma and multiple myeloma in seamen exposed to benzene in tankers. In this study, an increased incidence of lymphatic and hematopoietic malignancies was described and, while it is true that the cargo vapors from gasoline and other light petroleum products and chemicals have been studied, benzene exposure during loading, unloading and tank cleaning operations was concluded to be the likely source of the carcinogenic exposure. Rinsky et al.[43] described various hematological malignancies in their study of benzene exposure and showed that the overall standardized mortality ratio for leukemia and multiple myeloma were increased significantly. The investigators of this study concluded that there is a quantitative association between benzene exposure and development of leukemia. Wong[44] evaluated a mortality study of chemical workers occupationally exposed to benzene and found a significantly increased risk for lymphohematopoietic malignancies. Linet et al.[45] studied hematopoietic malignancies and related disorders among benzene exposed workers in China and showed a wide spectrum of hematopoietic malignancies.

Song-Nian Yin et al.[46] in a cohort study of cancer among benzene exposed workers in China, studied workers employed in a variety of occupations and showed a statistically significant increased deaths among benzene exposed subjects for leukemia, malignant lymphoma, neoplastic diseases of the blood, and other malignancies. The rates were significantly elevated for the incidence of lymphohematopoietic malignancies risk ratio of 2.6, malignant lymphoma risk ratio of 3.5, and acute leukemia risk ratio 2.6. A significant excess risk was also found for aplastic anemia and myelodysplastic syndrome. These investigators concluded that employment in benzene exposure occupations is associated with a wide spectrum of myelogenous and lymphatic malignant diseases and related disorders of the hematopoietic lymphatic system.

Hayes et al.[47] in one of the largest epidemiological studies on benzene exposure, showed a wide spectrum of hematological neoplasms and their related disorders in humans. The risk for these conditions is elevated at average benzene exposure levels of less than 10 ppm. These investigators further concluded that the pattern of benzene exposure appears to be important in determining the risk of developing specific diseases. Wong[48] studied a cohort of 7,676 male chemical workers from seven plants who were occupationally exposed continuously or intermittently to benzene for at least 6 months, and compared them to a group of male chemical workers from the same plant who had been employed for at least 6 months during the same period but were never occupationally exposed to benzene and showed a significantly increased risk of lymphohematopoietic malignancies.

In experimental animals benzene has been shown to be associated, in rats, with cancers of zymbal gland, oral cavity, nasal cavities, skin, forestomach, mammary gland, Harderian gland, preputial glands, ovary, uterus, angiosarcoma of liver, hemolymphoreticular neoplasia, lung cancers and leukemia.[49] The ability of benzene metabolites to effect lymphocytic growth and function in vitro, is shown to correlate with the oxidation capacity and concentration of the metabolites at the target site. Benzene also effects macrophages, as well as lymphocytes.[36,50,51] Kalf and Smith[52,53] have shown that benzene exposure reduces the ability of marrow stromal cells to support normal stem cell differentiation. From these experimental animal studies, and in vitro studies a wide range of bone marrow effects of benzene metabolites is shown. It has been concluded that the hematotoxicity of benzene depends on the breakdown metabolites and can effect the stem cell at any point in time, for instance myeloid, erythroid, macrophages, lymphocytic stem cells, and therefore benzene has been named as a pleural potential stem cell toxicant.[54]

In addition to carcinogenic effects, animal studies have shown the effects of benzene exposure on the immune system. Reid et al.[55] showed a significant decrease in splenic cell proliferation in mice exposed to benzene for 14 days. Experimental animal studies also reported reduced circulating white blood cells, as well as changes in spleen morphology and weight in various experimental animal studies.[54] These experimental animal studies further support the observation from 1913 by Winternits and Hirschfelder[56] that rabbits exposed to benzene showed an increased susceptibility to pneumonia and tuberculosis. The experimental animal data and the epidemiological studies clearly show that 1) benzene is a carcinogen for the lymphohematopoietic system, 2) benzene has a direct effect on the immune system, 3) benzene has a direct effect on the early development of the blood cells, and 4) benzene is a pluripotent hematological carcinogen.

### 20.5.5 RISK ASSESSMENT ESTIMATES

The United States EPA has used several databases in their estimates for benzene exposure and risk. (Environmental Protection Agency, 5.0 Benzene, 5.1. Chemical and Physical Properties, EPA, 1988) The data utilized by the EPA to assess the risk included the study by Rinsky et al. in 1981[57] where the duration of exposure was at least 24 years and exposure levels are between 10 to 100 ppm (8 hour TWA) with a statistically significant increase incidence of leukemia. The study of Ott et al.[57] in 1978 showed levels of anywhere from 2 to 25 ppm (8 hour TWA) with increased incidence of leukemia; and Wong et al.[57] 1983, where the exposure was at least 6 months, levels were from 1 ppm to 50 ppm, and there was a statistically significant increase in the incidence of leukemia, lymphatic and hematopoietic cancers.

The International Agency for Research on Cancer (IARC) has classified benzene as a Group 1 carcinogen. [58] A Group 1 carcinogen is defined as an agent that is carcinogenic to humans. This classification is based on sufficient evidence for carcinogenicity in humans. IARC based this conclusion on the fact that numerous case reports and follow-up studies have suggested a relationship between exposure to benzene and the occurrence of various types of leukemia. In addition, IARC considers the evidence for carcinogenicity to animals to be sufficient. No unit risk was determined by IARC for benzene.

The regulatory agencies in the estimated risk of increased benzene related cancers rely mainly on the study by Rinsky, 1987, which concluded that the mean annual cumulative exposure level of less than 1 ppm accumulated over 40 years working life-time would not be associated with increased death from leukemia. This epidemiological study showed an exponential decrease in the risk of death from leukemia which could be achieved by lowering occupational exposure to benzene. According to the model derived in this study, a worker occupationally exposed to benzene at an average exposure level of 10 ppm for 40 years would have an increased risk of death form leukemia 154.5. If the average exposure was lowered to 1 ppm, that excess risk would decrease to 1.7. At 0.1 ppm times 40 years cumulative exposure the risk be virtually equivalent to background risk,[26] Infante et al.[59] have shown a relative risk of 5.6 with an estimated cohort exposure of 10 to 100 ppm over 8.5 years average, and Vigliani[60] showed a relative risk of 20 estimated cohort exposure to 200 to 500 ppm over 9 years average, and Aksoy showed a relative risk of 25 with an estimated cohort exposure of 150 to 210 ppm over 8.7 years average.[20,40] These studies clearly show that the risk of developing lymphohematopoietic cancers is significant, and that benzene is carcinogenic from an epidemiological point of view at very low levels of 0.1 ppm.

### 20.5.6 LEVELS OF EXPOSURE

Regulatory levels of exposure to various chemicals, among them solvents and benzene, have been a subject for constant pressure from industry manufacturers on one hand, regulatory agencies, health care, and patients on the other hand. The most common question asked is "Is there a safe level of exposure to benzene?" The answer to that question has been given by the American Petroleum Institute in their paper on benzene, 1948,[16] and their statement

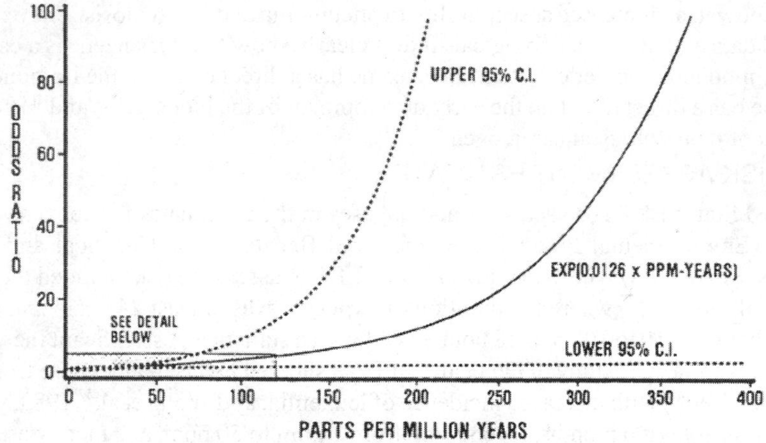


Figure 20.5.2. Extrapolations of levels of exposure to benzene. [Adapted, by permission, from R.A. Rinsky, *New England J. Med.*, 1987.]

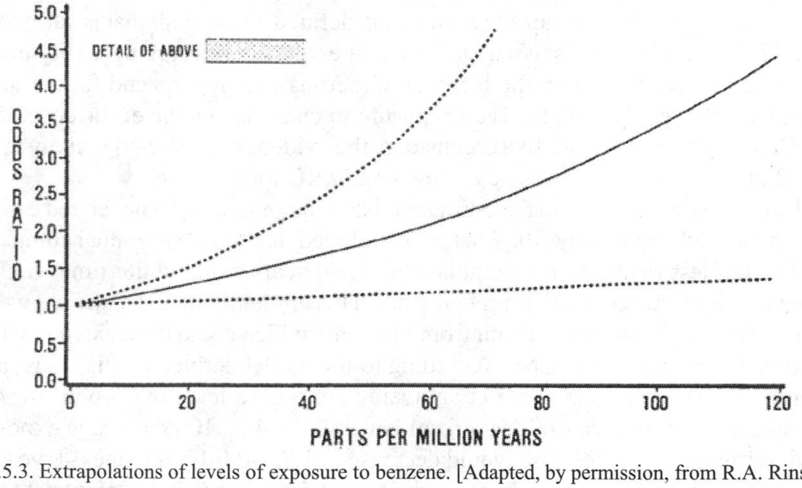


Figure 20.5.3. Extrapolations of levels of exposure to benzene. [Adapted, by permission, from R.A. Rinsky, *New England J. Med.*, 1987.]

that the only safe level of exposure to benzene is no exposure at all. Rightly so, the paper based that opinion on the fact that the body develops no tolerance to benzene and there is wide variation in individual susceptibility and therefore the only absolutely safe concentration for benzene is zero. This approach has been confirmed from a point of view of epidemiological studies and experimental animal studies showing benzene to cause cancer in experimental animals, and case reports and epidemiological studies in humans. Based on epidemiological studies and extrapolation from experimental animal studies, IARC's position is that a linear regression line should be applied for the dose response crossing the zero point for low level exposure of benzene. The EPA concurred that at low levels of exposure (since no epidemiological studies are available at low levels of exposure) linear dose response is indicated. The modalities of exponential dose response relationship for low levels is not applicable here based on the most recent EPA and IARC positions.[57,58] This linear, non-threshold model assumes that every increment of dose is accompanied by a commensurate increment in the excess cancer risk. The use of this toxicological model allows extrapolation of risks from relatively high dose levels, where cancer responses can be measured, to relatively low dose levels, where such risks are too small to be measure directly through epidemiological studies.[61] Figures 20.5.2 and 20.5.3 demonstrate the extrapolation from high levels of exposure to low levels of exposure utilizing the linear modality.

Indeed, in its most recent publication the U.S. EPA, 10/14/98,[62] further supports that approach and the panel members who evaluated the data felt that for the leukemogenic effects of benzene the linear model is consistent with the spirit of the proposed cancer risk assessment guidelines.

## 20.5.7 CELL TYPES: HEMATOLYMPHOPROLIFERATIVE EFFECTS OF BENZENE

The hematopoietic cell type toxicity from benzene have been described in animal data, case reports, case studies, animal data, and epidemiological data. Essentially, Wong in his OSHA testimony[63] concludes that for the continuously exposed group the lymphohematopoietic cancer risk ratio was 3.2 with a statistical significance of $p<0.05$, the risk ratio for non-Hodgkin's and other lymphopoietic cancer, i.e., all lymphopoietic cancers minus Hodgkin's disease, for the continuously exposed group was 3.77. The data demonstrated a statistical significant dose response relationship between cumulative exposure to benzene and mortality from all lymphopoietic cancer combined with leukemia. Wong further stated that it would be appropriate to combine lymphoma and leukemia in some of the analyses. This approach has also been recommended recently by other investigators[64] and therefore the agency conducting the hearing felt that the analysis based on the revised grouping of lymphopoietic cancers with Hodgkin's disease separated out was appropriate.[63] In the documents submitted to the OSHA hearing, Wong concluded that a dose response relationship for all lymphatic and hematopoietic cancers has been demonstrated.

Based on the case reports, case studies, and epidemiological studies, the sub-classification to cell types is indicated from a medical point of view to treat various hematopoietic diseases and cancers with the various appropriate treatments per each type of cell injured, however the data from the experimental data and clinical analysis of benzene cases clearly show that benzene causes damage to the stem cell and therefore it is a pluripotent toxin, causing a wide range of lymphohematopoietic malignancies.

## 20.5.8 EPIDEMIOLOGICAL STUDIES

Wong,[48] studied a cohort of 7,676 male chemical workers from seven plants who were occupationally exposed continuously or intermittently to benzene for at least 6 months, and compared them to a group of male chemical workers from the same plant who had been employed for at least 6 months during the same period but were never occupationally exposed to benzene. When the group with no occupational exposure was used for direct comparison, the continuously exposed group experienced a relative risk from lymphohematopoietic cancer of 3.2 with a statistical significance of $p<0.05$. That paper further concluded that the medical problems are replete with reports documenting the transition from certain lymphomas and multiple myelomas to leukemias. It was concluded that the transitions or progressions from lymphoma to leukemia are further complicated by the historical changes in nomenclature and in diagnostic overlap between the 2 disorders. It was felt, based on the work of others, that the major clone in chronic myelocytic leukemia affected cells capable of becoming lymphocyte, granulocyte, and erythrocyte differentiations leading to the conclusion that transformation events occur at an early multipotent stem cell level.

Nilsson et al.[42] investigated Swedish seamen, 20-64 years of age, who had been exposed to cargo vapors for at least 1 month on chemical or product tankers, had an increased risk of lymphatic and hematopoietic malignancies odds ratio of 2.6 with 95% confidence interval, with a significant exposure response relation. The odds ratio was increased for non-Hodgkin's lymphoma at 3.3 with 95% confidence interval and was statistically significant. Rinsky et al.[43] studied a cohort of 11,065 white men with at least 1 ppm per day of cumulative exposure to benzene. They have demonstrated that there was a statistical significant increase in death from all lymphatic and hematopoietic neoplasms, 15 observed versus 6.6 expected standard mortality ratio which is 227, 95% confidence interval, further demonstrating that benzene is toxic to all cell types.

Hayes et al.[47] studied a cohort of 74,821 benzene exposed and 35,805 unexposed workers from 1972 until 1987 in 12 cities in China. By and large this is the largest and most significant cohort of benzene workers studied and published. The investigators found that 1) benzene exposure is associated with a spectrum of hematological neoplasms, 2) workers with 10 or more years of benzene exposure had a risk ratio of developing non-Hodgkin's lymphoma of 4.2 with 95% confidence interval, and the development of this neoplasm was linked most strongly to exposure that had occurred at least 10 years before the diagnosis, and 3) the risk for the combination of acute non-lymphocytic leukemia and related myelodysplastic syndromes was significantly increased among those with more recent benzene exposure. These studies confirm the previous studies proving that the damage from benzene is to all cell type.

Linet et al.[45] studied hematopoietic malignancies and related disorders among benzene exposed workers in China and showed a wide range of hematopoietic malignancies. Yin et al.[46] examined a large cohort of benzene workers and concluded that benzene exposed workers have a statistically significant excess death due to leukemia, risk ratio of 2.3 with 95% confidence interval; malignant lymphoma, risk ratio of 4.5 with 95% confidence interval; and non-neoplastic diseases of the blood.

In summary, these epidemiological studies published in the peer-reviewed scientific literature and relied on by scientific and governmental agencies clearly show 1) significant statistical association between benzene exposure and lymphohematopoietic cancers of all cell types, 2) an increased risk and/or increased standard mortality rate over a factor of 2 in patients exposed to benzene with the development of non-Hodgkin's lymphoma, leukemias, and other lymphohematopoietic malignancies, and 3) benzene is carcinogenic with a linear dose response demonstrating no threshold.

## 20.5.9 SOLVENTS AND BENZENE

Solvents commonly used in the industry have been shown to contain benzene. Elkins, et al.[65] found that from time to time analyzed solvents for benzene content showed anywhere from 1% to 2% benzene. In that paper, which was published in 1956, the authors state that the TLV value from a regulatory point of view, at that time, was 35 ppm compared with 100 ppm previously. According to their calculations, they found that a benzene content below 3.5% will be necessary, for instance in solvents containing naphtha, hexane, and toluene, otherwise the permissible level for benzene vapor will be exceeded over 35 ppm, which we now know is extremely and significantly higher than the standard allowed today from a regulatory point of view. At the request of the petrochemical companies, the authors decided to reevaluate the content of benzene in solvents and for this purpose a total of 8 samples of low boiling petroleum naphtha were obtained. After utilizing methodology which included, among others, mass spectrometry for benzene content, the authors concluded that, in general, the benzene content of solvents ranged from 1% to 4% in volume. They have further shown that in the air of one plant where hexane with a relatively low benzene content (1.5%) was used as a solvent in a fabric-spreading operation, a benzene vapor concentration of 1 ppm was found. Since exposure to benzene is cumulative, if a worker is exposed to hexane containing 1.5% benzene, with both inhalation and skin contact, the cumulative exposure over a certain period of time increases the risk of developing benzene related cancers as described by the Rinsky model. The study by Pagnotto,[66] looked at and analyzed 32 naphtha solvents. The benzene concentration ranged from 1.5% to 9.3% by weight. Excessive benzene exposure was found at 3 out of 4 plants during their operations on a daily basis. On one occasion the concentration of benzene vapor was as high as 125 ppm (extremely high), and the urinary phenol excretion of the workers in these 3 plants were the highest that these investigators report ranging from 370 to 917 mg per liter of urine. These study indicate that solvents which contain benzene, even at levels of 1.5% per volume, can be associated with significant atmospheric exposure to benzene, shown as causing human exposure with significant excretion of phenol in the urine indicative of heavy benzene exposure. The investigators recommended additional ventilation, and on a follow-up visit the benzene exposure was found to have been reduced to about 70 ppm with urinary phenols of less than 70 mg per liter, still significantly elevated and considered a significant risk. These investigators also looked at blood examinations of 47 men at these plants. Five employees showed lower hemoglobin. One man showed a low hemoglobin at the age of 28, having been employed for 3 years in the environment preparing mixes for the saturating machine. While leukemia was suspected due to bone marrow disease, the patient was treated with iron and recovered. The authors conclude that excessive benzene exposures is consistently found on saturators using naphtha containing more than 3% benzene. This study further shows the importance of assessing benzene concentration is other solvents, from a dosimetry point of view.

The manuscript entitled A Recommended Standard for Occupational Exposure to Refined Petroleum Solvents from the U.S. Department of Health, Education and Welfare, Public Health Service, Center for Disease Control, National Institute for Occupational Safety and Health, July 1977,[67] recommended standards to be applied to occupational exposure of workers to the following refinery petroleum solvents: petroleum ether, rubber solvent, varnish maker's and painter's naphtha, mineral spirits, Stoddard solvents, and kerosene are all included in the term refined petroleum solvents. According to these standards petroleum ether and rubber solvents which contain 1.5% benzene, varnish maker's and painter's naphtha which contain 1.5% benzene, mineral spirits which contain 13-19% aromatics, Stoddard solvent which contains 0.1% benzene, 140 Flash Aliphatic Solvent which contains 0.7% benzene, kerosene. NIOSH indicated that some of the refined petroleum solvents contain aromatic hydrocarbons including, in some cases, benzene. Standards were applied, among others, to reduce the benzene exposure. Among others, the use of respirators and skin protective devices were required to protect from the effects of the solvents, as well as the benzene component.[67] In his testimony in front of the Occupational Safety and Health

Administration,[68] Proctor testified that, among others, refining operations are continuously changing. Many refineries obtain crude of differing characteristics from various producing areas which sometimes must be processed individually due to crude incompatibility and produce requirements. This means the operation of a crude fractionation unit is altered frequently; a single crude run may be as short as 2 days. Consequently, the crude tankage, crude fractionation units, and all downstream processing units frequently contain benzene. Proctor further stated that it should be clear by now that benzene is a naturally occurring compound in crude oil and is also found in the catalytic and cracking process, and therefore will always be a contaminant of these solvents. Benzene levels in gasoline today are running about 1.1% on the average across the nation but occasionally may reach 4% on individual samples. Reduction of benzene levels in gasoline is technically possible through employment of a number of physical processing schemes to the various gasoline component streams. He further testified that any attempt of physical separation of hydrocarbons, such as distillation, solvent extraction, or adsorption, the separation is not 100% complete. Therefore, some residuals of benzene will always be present in the remaining fraction. Therefore he recommended that benzene should be converted to cyclohexane by hydrogenation which would require an expensive catalyst, expensive high pressure reaction vessels, and consumption of valuable hydrogen. The testimony further indicates that it is believed that it is almost physically impossible to reduce these streams below 0.1% benzene. The 1978 OSHA[69] indicate that "The record establishes that there is a wide variation in the benzene content of petroleum solvents used in the rubber, paints, coatings, adhesives, sealants, and other downstream industries. As reported by Smith, reporting on behalf of MCA,[69] the benzene content of petroleum solvents of all types generally range from under 0.1% to 4%. Data submitted by downstream industries confirms that benzene is present in virtually all petroleum solvents, at levels which approach and even exceed 3.5% in some cases." It was stated that in the rubber industry, solvent benzene content appears to range from 0.1% to 0.7% or slightly higher. Similarly, solvents used by adhesive manufacturers show broad variations from less than 0.1% to 3.5%.[69] Representatives of the paint industry report variations from under 0.1% to as high as 3.7%.[69] Smith in his testimony emphasized that solvent benzene content is likely to vary substantially among supplies, among different plants of the same supplier, and among deliveries from the same plant. Because refinery processes are not designed to precisely control benzene content, variations will inevitably occur.

These data clearly show that benzene contents in solvents are difficult to control and vary depending on the sources, processes and therefore solvent exposure must take into account the level of benzene concentration in these solvents. These data, taken together with the most recent study of Brenner et al.[6] show that industrial toluene solvent does contain benzene and contributed significantly to the exposure via the skin to benzene. One must remember the importance of benzene exposure through the use of solvents produced through the petrochemical refining processes.

## 20.5.10 GENETIC FINGERPRINT THEORY

Benzene and its metabolites have long been known to cause chromosomal aberrations of various types of cell cultures of exposed humans. (To be discussed in Chapter 20.6 in this book entitled as Benzene Exposure and Sister Chromatoid Changes.) While it is true that genetic changes have been described and frequently effect chromosomes 5 and 7, and others, there is no scientific evidence that these are required for the diagnosis of benzene exposure related cancers. Specifically, many cases of patients who have been exposed to benzene and have developed hematopoietic malignancies do not have changes in chromosomes, therefore the chromosomal changes cannot be used as a "genetic fingerprint". Indeed, Irons' publication in the Journal of Toxicology and Environmental Health[70] concluded that the significance of these chromosomal alterations with respect to bone marrow damage or leukemogenesis of benzene is unclear. It is not possible today to determine whether leukemia is caused by benzene based on changes in chromosomes, specifically chromosomes 5 and 7. Smith from the University of California at Berkley, who is a leading authority in the biological markers of benzene exposure, opined that the data which supposedly suggest that one must have changes in chromosomes 5 and 7 to assume benzene causation is unreliable and obsolete.[71] In summary, while it is true that benzene and related products have been described with changes in chromosomes, DNA adducts, and cell cycle, by no means can they not be used as a diagnostic tool to address benzene causation or not.

## REFERENCES


1 NIOSH, Revised Recommendation for Occupational Exposure Standards for Benzene, Cincinnati, OH, DHEW Publications (NIOSH), 76-76-137-A (1976).
2 Fishbein L, *Scan J Work Environ Health*, **8**, Supplement 1, 5-16 (1992)
3 ATSDR Toxicological Profile for Benzene- Update, U.S. Department of Health & Human Services, Atlanta, Georgia (1996)
4 Browning E. **Toxicity and Metabolism of Industrial Solvents**, Amsterdam, *Elsevier Publishing*, Chapter 1 (1965)
5 Wallace, LA, *Environ Health Perspect,* **82**, 165-169 (1999)
6 Brenner D, et al., *Eur J Oncol,* **3**(4), 399-405 (1998)
7 Henderson RF, et al., *Environ Health Perspect,* 82, 9-17 (1989).
8 Schneider CA, et al., *Toxiocology & Applied Pharmacology*, **54**, 323-331 (1980)
9 Schneider R, Damitriadis E, and Guy R, *Environ Health Perspect,* **8**, 31-35 (1989)
10 Goldstein BD, *Adv Mod Environ Toxicol*, **4**, 51-61 (1983)
11 Santesson CG, *Arch Hyg,*, **31**, 336-349 (1897)
12 Delore P and Borgomano C, *Journal Medicine Lyon*, **9**, 227-233 (1928)
13 Lignac GOE, *Krankeitsforsch*, **9**, 426-453 (1932)
14 Vigliani EC, and Saita G, *New England Journal of Medicine*, **271**, 872-876 (1974)
15 Aksoy M, Erdem S, and Dincol G, *Blood*, **44**, 837-841 (1974)
16 API Toxicological Review, Benzene, September, (1948)
17 Maltoni C, **Myths and Facts in the History of Benzene Carcinogenicity : in Advances in Modern Environmental Toxicology**, Volume IV, M. Mehlman (Ed). *Princeton Scientific Publishing Company,* Princeton, 1-15 (1983)
18 NIOSH, Criteria for a Recommended Standard. Occupational Exposure to Benzene. U.S. Department of Health, Education and Welfare, Public Health Service, Center for Disease Control, Cincinnati, OH, Pub. No.74-137 (1974)
19 NIOSH, Occupational Safety and Health Guidelines for Benzene: Potential Human Carcinogen. U.S. Department of Health and Human Services, Public Health Services, Cincinnati, OH, Pub No. 89-104, Supp II (1988)
20 Aksoy M, *Environ Res*, **23**, 181-190, (1980)
21 Aksoy M, *Environ Health Perspect*, **82**, 1931-198 (1989)
22 USEPA, Ambient Water Quality Criteria for Benzene, Environmental Criteria and Assessment Office, Cincinnati, OH, EPA 440/5-80-018, NTIS PB81-117293 (1980)
23 Goldstein BD and Snyder CA, *Environ Sci Res*, **25**, 277-289 (1982)
24 IARC and IACR, In: IARC Monographs on the Evaluation of the Carcinogenic Risk of Chemicals to Humans: Some Industrial Chemicals and Dyestuffs. IARC, Lyon, France, 28, 183-225 (1982)
25 Aksoy M, et al., *Br J Ind Med*, **44**, 785-787 (1987)
26 Rinksy RA, et al., *N Engl J Med*, **316**, 1044-1050 (1987)
27 Goldstein BD, **Benzene Toxicity, In: Occupational Medicine**: State of the Art Reviews, 3(3), 541-554, NK Weaver (ed), *Hanley and Belfus, Inc*, Philadelphia PA (1988)
28 Paci E, et al., *Scand J Work Environ Health*, **15**, 313-318 (1989)
29 Rinksy RA, *Environ Health Perspect,* **82**, 189-192 (1989)

30 Ciranni R, Barale R, and Adler ID, *Mutagenesis*, **6**, 417-421 (1991)
31 Cox LA, *Risk Anal,* **11**, 453-464 (1991)
32 Ruis MA, Vassallo J, and Desouza CA, *J Occup Med*, **33**, 83 (1991)
33 Midzenski MA, et al., *Am J Ind Med*, **22**, 553-565 (1992)
34 ATSDR Toxicological Profile for Benzene- Update, U.S. Department of Health & Human Services, Atlanta, Georgia (1993)
34 Medinsky MA, Schlosser PM and Bond JA, *Environ Health Perspect*, **102** Suppl 9:119-124 (1994)
35 Travis LB, et al., *Leuk Lymphoma*, **14**(1-2), 92-102 (1994)
36 Niculescu R and Kalf GF, *Arch Toxicol*, **69**(3),141-148 (1995)
37 Wintrobe MM, Lee GR and Boggs DR, **Wintrobe's Clinical Hematology**, 554, *Lea and Febiger*, Philadelphia, PA (1981)
38 Schottenfeld D and Fraaumeni JF Jr, **Cancer: Epidemiology and Prevention,** *Saunders*, Philadelphia, PA (1982)
39 Zoloth SR, et al., *J Natl Cancer Inst.*, **76**(6),1047-1051 (1986)
40 Mehlman MA (ed.) **Advanced in Modern Environmental Toxicology**, Volume IV, *Princeton Scientific Publishing Company*, Princeton, NJ (1983)
41 Mehlman MA, *American Journal of Industrial Medicine*, **20**, 707-711 (1991)
42 Nilsson RI, et al., *Occupational & Environmental Medicine*, **55**, 517-521 (1998)
43 Rinsky A, et al., *New England Journal of Medicine*, **316**, 17, 1044-1050 (April 23, 1987)
44. Wong O, *British Journal of Industrial Medicine*, **44**, 382-395 (1987)
45 Linet MS, et al., *Environmental Health Perspectives*, **104**, Supplement 6, 1353-1364 (December 1996)
46 Yin SN, et al., *American Journal of Industrial Medicine*, **29**, 227-235 (1996)
47 Hayes RB, et al., *Journal of the National Cancer Institute*, **89**, 1065-1071 (1997
48 Wong O, *British Journal of Industrial Medicine*, **44**, 365-381 (1987)
49 Huff JE, et al., *Environ Health Perspec.*, **82,** 125-163 (1989)
50 Lewis JG, Odom B and Adams DO, *Toxicol Appl Pharmacol*, **92**, 246-254 (1988)
51 Thomas DJ, Reasor MJ and Wierda D, *Toxicol Appl Pharmacol*, **97**, 440-453 (1989)
52 Kalf GF, *Crit Rev Toxicol*, **18**, 141-159 (1987)
53 Smith MT, et al., *Environ Health Perspect*, **82**, 23-29 (1989)
54 Gist GL and Burg JR, *Toxicol and Industrial Health*, **13**(6), 661-714 (1997)
55 Reid LL, et al., *Drug Chem Toxicol,* **17**(1):1-14 (1994)
56 Winternitz MC and Hirschfelder AD, *J Exp Med*, **17**, 657-664 (1913)
57 USEPA, 5.0 Benzene, 5.1 Chemical and Physical Properties, EPA (1988)
58 International Agency for Research on Cancer (IARC), On the evaluation of carcinogenic risk of chemicals to humans: overall evaluations of carcinogenicity: An updating of IARC monographs 1 to 42, IARC, Lyon, France, Suppl 7 (1987)
59 Infante PF, et al., *Lancet*, **2**:76:1977
60 Vigliani EC and Forni A, *J Occup Med*, **11**, 148 (1969)
61 McMichael AJ, Carcinogenicity of benzene, toluene, and xylene: Epidemiological and Experimental Evidence, IARC, Environmental carcinogens methods of analysis and exposure measurement, Volume 10-**Benzene and Alkylated Benzens**, Fishbein L and O'Neill IK (eds) *IARC Scientific Publications*, No. 85 (1988)
62 USEPA, Attachment to IRIS file for benzene, 10/14/98, Response to the Peer-Review Draft Carcinogenic Effects of Benzene: An Update, EPA/600/P-97/001A, June 1997
63 Statement Submitted to the OSHA Benzene Hearing by Otto Wong, ScD., FACE., Environmental Health Associates, Inc, 03/04/86.
64 Decoufle P, Blattner WA and Blair A, *Environ Res*, **30**, 16-25 (1983)
65 Elkins HB, and Pagnotto LD, *Archives of Industrial Health*, **13**, 51-54 (1956)
66 Pagnotto LD, et al., *Industrial Hygiene Journal*, 417-421 (December 1961)
67 NIOSH. A Recommended Standard for Occupational Exposure to Refined Petroleum Solvents from the U.S. Department of Health, Education and Welfare, Public Health Service, Center for Disease Control, National Institute for Occupational Safety and Health (July 1977)
68 Testimony of R.S. Proctor in regards to Proposed Revised Permanent Standard for Occupational Exposure to Benzene, OSHA Docket No. H-059A, (July 11, 1977)
69 Post-Hearing Comments and Records Designations in OSHA Docket H-059 In re: Proposed Amendment to the Permanent Standard for Occupational Exposure to Benzene (June 12, 1978)
70 Irons R, *Toxicology and Environmental Health*, **16**, 673-678 (1985)
71 Kathleen Lavender, et al., v Bayer Corporation, et al., Civil Action No. 93-C-226K, in the Circuit Court of Marshall County, West Virginia

# 20.6 CHROMOSOMAL ABERRATIONS AND SISTER CHROMATOID EXCHANGES

Nachman Brautbar
**University of Southern California, School of Medicine, Department of Medicine, Los Angeles, CA, USA**

Several technologies have developed in the last 10 years to look at chromosomal changes and DNA changes caused by environmental exposures, as well as a marker of environmental exposures. The use of chromosomal translocation as a biological marker of exposure in humans have become an important tool in the research, as well as in some instance a marker of exposure. Several methodologies have utilized and include structural chromosomal aberrations, sister chromatoid exchanges (SCEs) and micronuclear changes. These are markers of changes in the cellular genetic materials, and represent damage induced by chemicals. These methodologies are viewed as cytogenetic assays, and by themselves cannot provide a diagnosis, but they complement other methodologies which include gene mutation analysis, and DNA changes. Among the important uses of cytogenetics as a biomarker is the relationship between chromosomal aberrations secondary to chemicals and carcinogeneses. Since the scope of this chapter is not addressing mechanisms of carcinogeneses, the reader is referred to other sources.[1]

Studies in patients with acute non-lymphocytic leukemia whose bone marrow analyzed for chromosomal changes have shown that 50% of them had changes in chromosomes 5, 7, 8 and 21.[2] Mitelman et al.[3] have looked at patients with acute non-lymphocytic leukemia and looked at the occupational history. They have studied a group of 56 patients. Twenty-three out of the 56 patients had a history of exposure to chemical solvents, insecticides and petroleum products. They have further shown that in males with acute non-lymphocytic leukemia the frequency of exposure to petroleum products was as high as 36%. In their study, Mitelman et al.[3] found a striking differences between the chromosomal findings in non-exposed versus exposed groups. In the non-exposed group only 24.2% of the patients had chromosomal aberrations in their bone marrow cells, while 82.6% of the exposed patients had chromosomal aberrations. The authors concluded, based on these studies that the difference between the exposed and non-exposed group strongly indicates that the karyotypic pattern of the leukemic cells were, in fact, influenced by the exposure. The authors further suggested that the prognosis in those patients with normal chromosomes was significantly better than those with abnormal chromosomes, something which has been suggested by previous investigators.[4,5,6] Based on their observations and others,[7-11] the authors suggested that certain chromosomal regions possible being specifically vulnerable to the chromosome damaging actions of different chemicals which are carcinogens. Indeed, the changes of chromosomes 5, 7, 8 and 21 in workers exposed to different chemical solvents, among them also petrol and pesticides, supports this hypophysis. Studies in cultured lymphocytes from 73 workers in chemical laboratories and the printing industry were found to have a significantly increased frequency of chromatoid and sister-chromatoid breaks in comparison to 49 control subjects.[12] The authors suggested that the observed cytogenetic changes is reasonably assumed to be the result of strong factors in the working environment which induced chromosome breaks and sister-chromatoid exchange. The considerable variation in cytogenic changes between subjects suggested varying degrees of exposure to mutagenic agents. Brandt et al.[13] studied the effects of exposure to organic solvents and chromosomal aberrations. Ten patients had a history of daily handling of organic solvents for at least one year preceding the diagnoses of non-Hodgkin's lymphoma. All have been exposed to a variety of solvents to include aromatic and aliphatic compounds. Forty-four patients for only a shorter period of time worked with organic solvents, and therefore served as the control group. There was a statistically significant increase in chromosomal changes in the exposed group compared to the non-exposed group. The authors suggested based on their results, that at diagnosis of non-Hodgkin's lymphoma, patients with a history of significant occupational exposure to organic solvents tends to have a larger number of chromosomal aberrations in the lymphoma cells. Furthermore, certain aberrations may be characteristic for the exposed patients. They have concluded that the over representation of certain chromosome aberrations in non-Hodgkin's lymphoma patients occupationally exposed to organic solvents supports the concept that these exposure may be relevant for the subsequent development of non-Hodgkin's lymphoma. Their data are in agreement with the studies published previously, indicating that workers handling organic solvents and other petroleum products have an increased frequency of chromosomal aberrations in lymphocytes.[12,14,15,16] Indeed, association between exposure to solvents and other chromosomal changes in the cells have been studied by other investigators, describing chromosomal changes in acute non-lymphocytic leukemia.[3,17,18,19] In cultured cells, Koizumi[20] examined the effects of benzene on DNA syntheses in chromosomes of cultured human lymphocytes. They have shown an increased incidents of chromatoid gaps and breaks in cultures treated with benzene, compared to those who were untreated with benzene. They also showed changes in DNA metabolism in the form of inhibition of syntheses in those samples which were treated with benzene. This observation was confirmed by other investigators[21] who have shown an increased chromosomal aberrations in cultured human leukocytes exposed to benzene. From the in vitro to the in vivo experimental animals, it was shown that treated rats with benzene, were found to have increased chromosomal changes taken from the bone marrow as compared to the non-treated animals.[22] It was of interest that the degree of chromosomal changes that were induced by benzene, were similar to that induced by toluene[23] (probably secondary to the benzene in toluene). Experimental rabbits treated with benzene also showed a significant amount of bone marrow chromosomal changes persisting up to 60 days after the end dosing with benzene.[24]

A patient who has developed aplastic anemia after exposure to benzene, was shown to have significant chromatoid fragments.[25] A cytogenic study which was carried out later,[26] on a patient who developed leukemia after 22 years of continuous exposure to a high concentration of benzene, showed that later in the process there were changes in 47 chromosomes in the bone marrow. Sellyei et al.[27] studied patients who developed pancytopenia after having been exposed for 18 months to benzene. Significant chromosomal changes were detected even 7 years after remission from the anemia and the presentation of leukemia. In line with these changes, Forni et al.[28] have studied 25 subjects with a history of hematopoietic abnormalities and benzene exposure, and compared these to 25 matched controls. They have shown that 18 years after clinical and hematological symptoms chromosomal aberrations were increased as compared to the control group. In 1965, Tough et al.[29] have studied chromosomes of workers exposed to benzene for periods varying from 1 to 18 years. They have also shown a small but significant increase in chromosomal changes compared to a control group. These same investigators looked at workers exposed to benzene levels from 25 to 120 ppm, and found that they had significant chromosomal aberrations as compared to the normal population (which has a general background exposure to benzene levels). The study of Forni et al.[14] showed significant chromosomal aberrations in those patients who were exposed mainly to benzene, and not to those who were exposed to toluene only. Hartwich et al.[30] looked at 9 healthy refinery workers who were exposed to low levels of benzene, and also found significantly increased chromosomal changes compared to the control group. The National Research Council Advisory Center and Toxicology Study[31] concluded that close correlation between occupational exposure to benzene and persistence of chromosomal aberrations can be discussed only when there is an association between benzene induced hematopoietic disease and chromosomal aberrations, however, the absence of chromosomal changes, cannot be a determinant in the temporal relationship between exposure to benzene and hematopoietic diseases.[31] While some studies suggested that the chromosomal changes require heavy exposure to benzene, the study by Picciano, 1979,[32] looked at chromosomal changes in 52 workers exposed to a mean benzene concentration of less than 10 ppm compared to non-exposed controls. There was a statistical significant increase in chromosomal aberrations in exposure as low as less than 10 ppm. Furthermore, these same investigators[33] reported a dose response increase in the aberrations when the exposed workers were divided into smaller groups by the exposure levels (less than 1 ppm, 1-2.5 ppm, and 2.5-10 ppm). Drivers of petrol tankers and crew members of gasoline tankers, ships and petrol station attendants were studied for chromosomal changes.[15] The degree of exposure to benzene of the three groups was estimated to be at a mean of 0.4 ppm, and the crew members were estimated to be at 6.56 ppm while engaging in handling of gasoline. The frequency of chromosomal and chromatoid aberrations in the petrol tanker drivers was significantly greater than in those of petrol attendants and the crew members. The effects of long-term benzene exposure from the incidents of chromosomal changes were studied in 16 female workers who were exposed to a maximum of 40 ppm benzene between 1-20 years.[34] The cytogenetic study was conducted 6 months after benzene was eliminated from the work environment, and they have found no significant increase in chromosomal changes. Clare et al.[35] looked at chromosomal changes in the peripheral lymphocytes of workers after a single, one exposure to benzene. Exposure levels were described as high after a spillage of a large amount of benzene during the loading of a ship. Three months after the incident, chromosomal analysis showed no significant abnormalities. The authors concluded that there was no evidence of lasting chromosomal damage in the peripheral blood lymphocytes of these exposed workers. Golomb et al.[18] reviewed the literature and reported the results in regards to exposure to benzene and chromosomal changes. They have stated that they studied exposure data on 74 patients with acute leukemia. They describe that 75% of the exposed patients had an abnormal karyotype, whereas only 43% of the patients characterized as non-exposed had an abnormal karyotype. While it is true that these findings are in agreement with previous studies[18] they still could not explain the 43% of the patients who were not exposed, and still had abnormal chromosomal changes. This is a very important observation, since some investigators in the field claimed that the "absence of chromosomal changes" in benzene exposed individuals negates the clinical causative diagnosis of benzene induced hematopoietic disease. Essentially, all of the studies show that benzene can cause chromosomal changes, but does not cause it in all the patients, and the absence of chromosomal changes cannot and does not rule out the exposure to benzene as a causative factor. In this same paper, Golomb et al.[18] looked at the chromosomal changes of patients treated with chemotherapeutic agents for other malignancies. Essentially, they looked at a secondary leukemia developing as a result of alkylating agents. For some reason, they have proposed that losses of part or all of chromosomes numbers 5 and 7 are the specific change resulting from mutageneses, leukemogeneses associated with various chemicals including insecticides, petroleum products and alkylating agents.

While this interpretation is compatible with the various animal studies, as well as observations in patients, there is certainly a lack of scientific connection between the benzene exposed chromosomal changes, and the chromosomal changes reported in patients treating with alkylating agents.

Smith, in a recent paper[36] suggested that oxidation of benzene to multiple metabolites plays a role in producing benzene induced toxicity of DNA damage in bone marrow, and adds further weight to the hypophysis that multiple metabolites are involved in benzene toxicity. They also described DNA changes which have been shown to be cause-point mutation. The investigators measured mutation frequency in 24 workers heavily exposed to benzene, and 23 matched controls. They found that benzene caused a highly significant increase in one variant of mutation, suggesting that benzene produces gene duplicating mutations, but no gene inactivating ones. They suggested that the most-likely consequence of aberrant recombination caused by benzene metabolites is the production of stable chromosomal translocation. Indeed, there are several chromosomal abnormalities shown in leukemic cells. This includes Philadelphia chromosome which results from reciprocal translocation between chromosome 9 and 22, and has been associated with chronic myeloid leukemia, and reciprocal translocation between chromosomes 8 and 21. From these studies it is concluded that benzene is a genotoxic carcinogen, but that other genetic phenomena may mediate benzene induced hematopoietic toxicity. Based on the available data up to date, it is proposed[36] that benzene is a carcinogen that does not produce cancer through simple gene mutations, but rather through a separate class of carcinogens (metabolites of benzene) that act by a similar mechanism.

In summary, the studies in experimental animals, in vitro, and patients show that benzene and a wide range of organic solvents are associated with changes in chromosomes and DNA adducts. While these changes may be helpful in epidemiological studies, the absence or presence of genetic changes or DNA adducts, cannot be used in a specific case to rule out or establish causation. The biomarkers described in this chapter in the form of genetic biomarkers, can be helpful in identifying individual susceptibility, and in some cases understanding of the mechanism of the disease process. They have a significant number of limitations, and these include measurements, errors and confounding factors.

## REFERENCES


1 **Biomarkers: Medical and Workplace Applications**, Medelson ML, Moor LC and Peeters JP (eds), *John Henry Press,* Washington, D.C., (1998)

2 Rowley JD and Potter D, *Blood*, **47**:705, (1976)

3 Mitelman F, Brandt L and Nilsson PG, *Blood*, **52**(6):1229-1237, (December 1978)

4 First International Workshop on Chromosomes in Leukaemia: Chromosomes in acute non-lymphocytic leukaemia. *Br J Haematol*, **39**:311, (1978)

5 Golomb HM, Vardiman J and Rowley JD, *Blood*, **48**:9, (1976)

6 Nilsson PG, Brandt L and Mitelman R, *Leukemia Res*, **1**:31, (1977)

7 Mitelman F, et al., *Science*, **176**:1340, (1972)

8 Mitelman F, The Rous sarcoma virus story: Cytogenetics of Tumors induced by RSV in German J (ed): **Chromosomes and Cancer**, New York, *Wiley*, page 675, (1974)
9 Levan G and Levan A, *Hereditas*, **79**:161, (1975)
10 Levan G and Mitelman F, *Hereditas*, **84**:1, (1976)
11 Sugiyama T, et al., *J Natl Cancer Inst*, **60**:153 (1978)
12 Funes-Cravioto F, et al., *The Lancet,* 322-325, (August 13, 1977)
13 Brandt L, et al., *European Journal of Haematol*, **42**:298-302, (1989)
14 Forni A, Pacifico E, and Limonta A, *Arch Environ Health*, **22**:373-378, (1971)
15 Fredga K, Reitalu J, and Berlin M, Chromosome studies in workers exposed to benzene. In: Berg K (ed), **Genetic Damage in Man Caused by Environmental Agents**, New York, *Academic Press*, 187-203, (1979)
16 Hogstedt B, et al., *Hereditas*, **94**:179-184, (1981)
17 Mitelman F, et al., *Cancer Genet Cytogenet*, **4**:194-214, (1981)
18 Golomb HM, et al., *Blood*, **60**(2):404-411, (1982)
19 Fourth International Workshop on Chromosomes in Leukemia, 1982: The correlation of karyotype and occupational exposure to potential mutagenic/carcinogenic agents in acute nonlymphocytic leukemia. *Cancer Genet Cytogenet*, **11**:326-331, (1984)
20 Koizumi A, et al., *Jap J Ind Health*, **12**:23-29, (1974)
21 Morimoto K, *Jap J Ind Health,* **17**:106-107, (1975)
22 Philip P and Jensen MK, *Acta Pathol Microbiol Scand, Section A*, **78**:489-490, (1970)
23 Dobrokhotov VB, *Gig Sanit,* **37**:36-39, (1972)
24 Kissling M and Speck B, *Helv Med Acta*, **36**:59-66, (1971)
25 Pollini G and Colombi R, *Med Lav*, **55**:244-255, (1964)
26 Forni A and Moreo L, *Eur J Cancer,* **3**:251-255, (1967)
27 Sellyei M and Kelemen E, *Eur J Cancer*, **7**:83-85, (1971)
28 Forni AM, et al., *Arch Environ Health*, **23**:385-391, (1971)
29 Tough IM and Court Brown WM, *Lancet*, **I**, 684, (1965)
30 Hartwich G and Schwanitc G, *Dtsch. Med. Wschr*, **87**:45-49, (1972)
31 National Research Council Advisory Centre on Toxicology, Washington D.C., Health Effects of Benzene: A Review. Prepared for the Environmental Protection Agency. Report No. NAS/ACT/P-829, (June, 1976)
32 Picciano DJ, *Environ Res*, **19**:33-38, (1979)
33 Picciano DJ, Monitoring Industrial Populations by Cytogenetic Procedures. In: Infante PF and Legator MS (eds). Proceedings of the Workshop on Methodology for Assessing Reproductive Hazards in the Workplace, U.S. Government Printing Office, Washington DC, pages 293-306, (1980)
34 Watanabe T, et al., *Environ Health*, **46**:31-41, (1980)
35 Clare MG, et al., *Br J Ind Med*, **41**:249-253, (1984)
36 Smith MT, *Environ Health Persp*, **104**(6):1219-1225, (December 1996)

## 20.7 HEPATOTOXICITY

NACHMAN BRAUTBAR
**University of Southern California, School of Medicine**
**Department of Medicine, Los Angeles, CA, USA**

### 20.7.1 INTRODUCTION

Solvents which are inhaled or gain access to the blood circulation via skin absorption or at times ingestion largely are metabolized by the liver. The liver has a complex mechanism composed of the cytochrome P450 enzyme, and other enzymes related to conjugation pathways such as glutathione conjugation. This is represented schematically in Figure 20.7.1

It is therefore no surprise that in the occupational setting as well as non-occupational setting, liver damage anywhere from transient to subacute to chronic, and at times terminal liver damage has been described.

The circumstances of exposure to hepatotoxic agents are divided to:

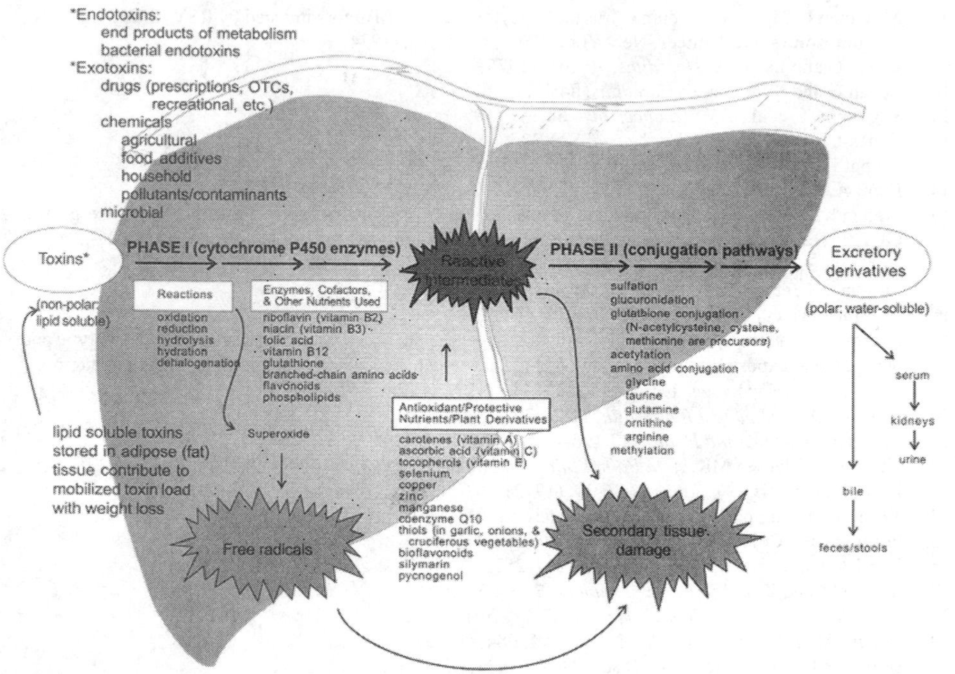


Figure 20.7.1. The phases of detoxification. [Adapted, by permission, from DS Jones, Institute for Functional Medicine, Inc., 1999]

1. Occupational: Either through a routine daily inhalation or skin absorption of solvents which have been shown to be toxic to the liver (accidental exposure).

2. Domestic during either accidental or intentional suicidal exposure, ingestion in foods or as a toxic contaminant of food, exposure to toxic agents such as in the form of glue sniffing.

3. Environmental, most commonly exposure through contaminated water with solvents (drinking water contamination) or through atmospheric pollution such as release to the environment from plants utilizing solvents.

Historically the first cases attributing chloroform to liver toxicity were described in 1887, 1889 and 1904.[2,3,4] The role of carbon tetrachloride and liver injury has been originally described in 1967 and 1973.[5,6] In general, the understanding of hepatotoxicity is extremely complex, and the reader is referred to the outstanding text by Hyman J. Zimmerman.[7] A typical example of how metabolism and toxicity of a water takes place is the aromatic chemical such as benzene attached to bromine. The effect on the liver has been originally studied by Mitchell in 1975[8] who have shown that a change in the rate of the metabolism of this compound is required to create its toxic products. While bromobenzene and carbon tetrachloride share a similar place of metabolism in the liver, the toxicity of bromobenzene and carbon tetrachloride are different, since the bromobenzene toxicity is related to the metabolic capacity of the liver, while that of carbon tetrachloride is not.

Several factors contribute to the handling of the solvents by the liver and affect the final toxicity, including species differences. For instance, rats are vulnerable to a wide variety of toxic agents such as carbon tetrachloride and bromobenzene due to the ability of the liver to convert these agents to their respective toxic metabolites.[6,9] Among other mechanisms responsible for the species differences are liver blood flow, protein binding, and the points of binding intracellularly. Genetic factors in humans are of extreme importance. Genetic factors are most-likely responsible for the various levels of adverse effects of alcohol in different individuals due to induced activity of detoxification enzymes in the liver in some and lack of those or reduced activity in others. Another important factor is age. The effect of age on the susceptibility has been shown in experimental animals. For instance, the neonatal rats are less susceptible to carbon tetrachloride and bromobenzene toxicity as compared to adult animals.[9] In humans, liver necrosis after the administration of Halothane was rare in children, but more common in more elderly patients. Factors such as sex and endocrine status have also influence and different toxic effects of solvents in this case. Nutritional status is a major factor in the effects of solvents on the liver. For instance, protein malnutrition leads to reduced activity of cytochrome P450. Increasing the percentage of fat in the liver has been shown to increase the susceptibility of toxicity to such agents as carbon tetrachloride. Several studies have looked into the histopathological injury of some solvents and solvent-like agents and the liver, and are shown in Table 20.7.1.

**Table 20.7.1. Partial list of agents that produce hepatic necrosis in experimental animals. [Adapted, by permission, from HJ Zimmerman, Hepatotoxicity, 1978]**

| | Site of necrosis | | | | |
|---|---|---|---|---|---|
| | Centrizonal | Midzonal | Peripheral Zone | Massive | Steatosis |
| Bromobenzene | + | | | | (+) |
| Bromotrichloromethane | + | | | | + |
| $CCl_4$ | + | | | | + |
| Chlorobenzenes | + | | | (+)? | |
| Chloroform | + | | | | + |
| Dichloropropane | + | | | | + |
| Dinitrobenzene | + | | | (+) | + |
| Dinitrotoluene | + | | | (+) | + |
| Ethylene dichloride | + | | | | + |
| Methylene chloride | (+) | | | (+) | (+) |
| Naphthalene | + | | | | + |

The specific mechanism of hepatotoxicity of many solvents are unknown, however, the knowledge have been gathered from experimental studies now available for the reader for review.[10,11,12,13]

Due to the lack of specific information for many solvents, I have decided to discuss in this chapter some of the most typical ones which have been used in the past heavily, or are used currently.

**Table 20.7.2. Some occupational solvents that produce acute and chronic liver disease**

| | | |
|---|---|---|
| Carbon tetrachloride | Halothane | Ethyl alcohol |
| Chloroform | Trichlorodiphenyl | A mixture of solvents such as toluene and xylene |
| Trichloroethylene | Trichloroethene | |
| Tetrachloroethylene | Trinitrotoluene | Dichloromethane |
| Dinitrobenzene | | |

Tables 20.7.3 and 20.7.4 describe two major phases in the liver metabolism and detoxification of drugs, foreign agents and solvents.

Carbon tetrachloride was found to cause liver injury in man and in experimental animals.[14] Carbon tetrachloride is a known and potent liver toxicant, and therefore has been studied extensively in experimental animals. Recknagel[6] and Reynolds[15] have shown that single doses will lead to areas of necrosis in the liver within a few minutes. This has been shown to be associated with changes in liver enzymes which are known to indicate liver damages.[10,11] Prolonged exposure to carbon tetrachloride has been shown to lead to liver cirrhosis and to liver cancers. In order to become toxic the carbon tetrachloride has to undergo metabolic changes in the liver.[6,16] The lesions described initially in animals have been shown in humans poisoned with carbon tetrachloride.[14,17] It has also been shown that alcohol enhances the susceptibility to carbon chloride toxicity.[18] Several factors play a role in the susceptibility to toxicity by carbon tetrachloride, among them are sex, age, diet, underlying preexisting liver dysfunction and alcoholism. Over the years in both clinical and experimental studies and observations, it has been shown that carbon tetrachloride induced liver damage is divided to fatty metamorphosis, and independently liver necrosis. Fat starts to accumulate in experimental animals as early as one hour after administration of a high dose of carbon tetrachloride. Liver necrosis occurs as early as 6 to 12 hours, and a maximum of 24 to 36 hours.

The concept of steatosis (fat accumulation in the liver) is a common one for looking at the effect of solvents (those which are known to be toxic to the liver). The fat accumulation is the result of abnormal transport of lipids and as a result, accumulation of lipids in the liver. Therefore clinically industrial exposure to hepatotoxic solvents is associated with liver steatosis, among others.

Necrosis, which is the second most common effect of liver damage of solvents toxic to the liver, is the result of destruction of the cell architecture as well as biochemical pathways. It has been shown in many experimental studies that the toxicity of carbon tetrachloride (and some of the other solvents which are toxic to the liver and other organs, such as benzene and the hematopoietic system) requires several reactions, in order to produce the toxic metabolites which are causing the damage to the liver. Most studies point to the responsibility of cytochrome P450 system.[6] The metabolite responsible for the liver damaging effect of carbon tetrachloride is a C Chloride III which is formed from carbon tetrachloride.[6,19] There is however also information that non-metabolized carbon tetrachloride contributes to the injury, especially that of the cell membrane,[20,21] something which is logical since solvents are a mechanistic injury to various tissues is through the effects on the cell membrane which is either dissolved or damaged by the solvent. A consolidation of the data available and views on the pathogenesis of carbon tetrachloride liver damage has been eloquently described by Zimmerman.[7]

**Table 20.7.3. Phase I reactions. [Adapted, by permission, from HJ Zimmerman, Hepatotoxicity, 1978]**

| | |
|---|---|
| ***I. OXIDATIONS*** | |
| A. MFO-dependent | |
| Aromatic hydroxylations | 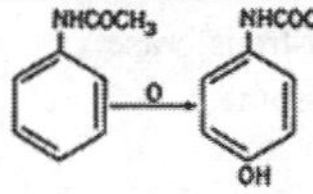 |
| Aliphatic hydroxylations | |
| Primary alcohol | $RCH_2CH_2CH_3 \xrightarrow{O} RCH_2CH_2CH_2OH$ |
| Secondary alcohol | $RCH_2CH_2CH_3 \xrightarrow{O} RCH_2CH(OH)—CH_3$ |
| Tertiary alcohol | $RCH_2CH(R')—CH_3 \xrightarrow{O} R—CH_2—C(R')(OH)—CH_3$ |
| Epoxidation | 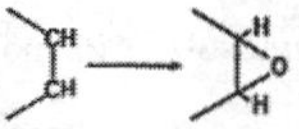 |
| Arene oxide formation | 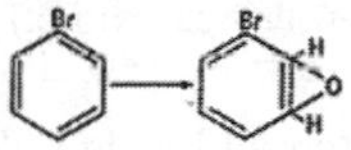 |
| Oxidative dealkylation | |
| N—dealkylation | $R—NH—CH_3 \rightarrow [R—NH—CH_2OH] \rightarrow RNH_2 + HCHO$ |
| O—dealkylation | $R—O—CH_3 \rightarrow [R—O—CH_2OH] \rightarrow ROH + HCHO$ |
| S—dealkylation | $R—S—CH_3 \rightarrow [R—S—CH_2OH] \rightarrow RSH + HCHO$ |
| N—Oxidation | |
| Primary amine | $RNH_2 \rightarrow RNHOH$ |
| Secondary amine | $(R)(R^1)NH \rightarrow (R)(R^1)NOH$ |
| Tertiary amine | $R—N(R^1)(R'') \rightarrow R—N(R^1)(R'')=O$ |
| S—oxidation | 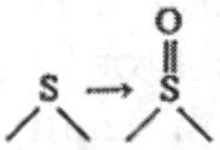 |
| Deamination | $RCH(NH_2)—CH_3 \rightarrow [R—C(OH)(NH_2)—CH_3] \rightarrow R—C(=O)—CH_3 + NH_3$ |
| Desulfuration | $RSH \rightarrow ROH$ |
| Dechlorination | $CCl_4 \rightarrow [CCl_3^{\cdot}] \rightarrow CHCl_3$ |
| B. Amine oxidase | $RCH_2—NH_2 \rightarrow RCHO + NH_3$ |
| C. Dehydrogenation | |
| Alcohols and aldehydes | $CH_3CH_2OH \rightarrow CH_3CHO \rightarrow CH_3COOH \rightarrow CO_2 + H_2O$ |
| ***II. REDUCTION REACTIONS*** | |
| Azo reductions | $RN=NR' \rightarrow RNH—NHR^1 \rightarrow RNH_2 + R'NH_2$ |
| Nitro reductions | $RNO_2 \rightarrow RNO \rightarrow RNHOH \rightarrow RNH_2$ |
| Carbonyl reductions | $RC(=O)—R' \rightarrow RCH(OH)—R'$ |
| ***III. HYDROLYSIS REACTIONS*** | |
| Esters | $RCOOR^1 \rightarrow RCOOH + R^1OH$ |
| Amides | $RCONH_2 \rightarrow RCOOH + NH_3$ |

**Table 20.7.4. Phase II (conjugation) reactions. [Adapted by permission, from HJ Zimmerman, Hepatotoxicity, 1978]**

| Type of conjugation | Endogenous substance | Transferring enzyme and location | Type of xenobiotics & metabolites conjugated |
|---|---|---|---|
| Glucuronidation | UDP-glucuronic acid | UDPG-transferase (microsomes) | Phenols; alcohols; carboxylic acids; primary amines; hydroxylamines; sulfonamides, etc. |
| Dihydrodiol formation | Water | (Epoxide) hydrase (cytosol) | Epoxides and arene oxides |
| GSH conjugation | GSH | GSH-S-transferase (cytosol) | Epoxides; arene oxides; halides; nitro groups; hydroxylamines, etc. |
| Glycine conjugation | Glycine | Acyl CoA-glycine transferase (mitochondria) | CoA derivatives of carboxylic acids |
| Sulfate conjugation | PAP-sulfate | Sulfotransferase (cytosol) | Phenols; alcohols; aromatic amines |
| Methylation | S-adenosyl-methionine | Transmethylase (cytosol) | Catechols; phenols; amines; histamine |

The carcinogenic effect of carbon tetrachloride will not be discussed in this chapter, and the reader is referred to other texts.[22]

## 20.7.2 INDIVIDUAL VARIABILITY AND HEPATOTOXICITY OF SOLVENTS

The issue of individual variability based on various factors as described above is important, especially in medical monitoring and risk assessment in occupationally exposed patients. The fact that workers in industrial environments are not the same and are subject for differences such as body build, underlying kidney function differences (genetic or acquired), exposure to other solvents or other liver toxicants may effect the results. A recent study[23] evaluated a population response to solvent exposure. These investigators have shown that body fat is the most important body compartment for fat soluble solvents. Body fluids and physical work load effect the blood flow, alveoli ventilation and therefore will effect the amount of solvent inhaled as well as absorbed through the blood and delivered to the liver. They have developed the physiological model which takes into account variability in the form of exposure, physical overload, body build, liver function and renal function. Other factors which have been taken into account are solubility in blood and tissue. Investigators suggested that such a model should be useful in improving our understanding of the complex and multifactorial system and to generate a hypothesis, and to improve our assessment of occupational exposure. This has significance from a clinical toxicology point of view. A patient who has an increased body fat will be at a higher risk of solvent toxicity. If this same patient also has a habit of heavy alcohol consumption the risk for solvent liver toxicity is significantly increased. Epidemiological studies are commonly not designed to evaluate the individual hepatotoxicity of solvents and therefore the issue of cause and effect must be viewed taking into account individual variability, other risk factors, and medical common sense, following the well established criteria by practitioners of medicine.

Chloroform is another haloalkane which has been typically used as an example to understand and study the toxic effects on the liver. Studies in experimental animals in 1866[24] have shown that the chloroform causes liver toxicity. In 1923 Meyer et al.[14] have shown that toxicity of chloroform to the liver in humans. Chloroform was used years ago as an anesthetic, and has been used successfully, however, due to its' toxicity the use of it has been abandoned. Acute exposure and toxicity has been associated with liver necrosis, liver steatosis, and chronic exposure has been associated with liver cirrhosis. The mechanism of injury most-likely is the result of metabolic changes of chloroform by the liver. Different effects and level of toxicity between carbon tetrachloride and chloroform is most-likely the result from the solubility in lipid and water, and the mechanisms by which these two agents are metabolized and then cause liver toxicity. Table 20.7.5 summarizes liver damage described in the literature as a result of halogenated aliphatic hydrocarbons.

**Table 20.7.5. Lesions produced by halogenated aliphatic hydrocarbons. [Adapted, by permission, from HJ Zimmerman, Hepatotoxicity, 1978]**

| Steatosis and centrizonal necrosis | Steatosis only | Slight steatosis or no injury |
|---|---|---|
| $CCl_4$ | $CH_2ClBr$ | $CH_3Cl$ |
| $CI_4$ | $CH_2Cl_2$ | $CH_3Br$ |
| $CCl_3Br$ | $CHCl{=}CHCl$ (cis) | $CH_3I$ |
| $CHCl_3$ | $CCl_2{=}CCl_2$ | $CCl_2F_2$ |
| $CHI_3$ | $CH_3CH_2CHClCH_3$ | $CHCl{=}CHCl$ (trans) |
| $CHBr_3$ | | $CH_3CH_2Cl$ |
| $CHCl_2CHCl_2$ | | $CH_3CH_2I$ |
| $CH_2ClCH_2Cl$ | | $CH_3CH_2Br$ |
| $CH_2BrCH_2Br$ | | $CH_3CH_2CH_2CH_2Cl$ |
| $CH_3CCl_3$ | | |
| $CHCl_2CCl_3$ | | |
| $CHCl{=}CCl_2$ | | |
| $CH_3CHClCH_3$ | | |
| $CH_3CHClCH_2Cl$ | | |

## 20.7.3 NON-HALOGENATED SOLVENTS

While the halogenated hydrocarbons discussed here include carbon tetrachloride, chloroform, 1,1,1-trichloroethene, trichloroethylene are significantly hepatotoxic, the literature on the toxicity of the non-halogenated hydrocarbons is a combination of positive and negative studies. Several studies looking into the hepatotoxicity of both aliphatic solvents such as kerosene, hexane and aromatics such as xylene, toluene and styrene have reported mixed results. Xylene is an aromatic hydrocarbon which is used heavily in the industry, as well as medical technology as a solvent.

Xylene commonly has been reported with impurities in varying amounts which include ethyl benzene, prime ethyl benzene, phenol, benzene and toluene.[25] To evaluate the effects of xylene in experimental animals, Toftgard et al.[26] studied rats who were exposed for three days by inhalation to xylene and to a mix of xylene isomers. Hepatic cytochrome P450 concentration increased as well as C reductase activity, and NADPH cytochrome C reductase activity. Furthermore, xylene and its isomers were able to modify the metabolism of other potentially toxic substances. In addition to these biochemical changes, the investigators found that xylene increased liver body weight, most-likely secondary to proliferation of the endoplastic reticulum. These studies show that at these levels xylene induced cytochrome P450 activity and NADPH cytochrome C reductase activity, but was not associated with significant pathological abnormalities. On the other hand, these same investigators, concluded that the capacity of xylene and xylene isomers to induce hepatic cytochrome P450 suggests the possibility of synergistic toxic effects from simultaneous exposure to xylene and substances metabolically activated by these cytochromes P450. Therefore, exposure to a mixture of solvents which also include xylene will increase the toxicity and xylene is a synergistic liver toxicant. For instance, the formation of 2-hexanoyl a metabolic precursor of 2,5-hexanedione, which is the main metabolite found in urine of workers exposed to n-hexane, is increased following xylene treatment.[27]

## 20.7.4 SOLVENT MIXTURES

Various solvent mixtures have been reported as hepatotoxic, Fishbien et al.[28,29] reported abnormal liver functions in chemical workers exposed to a mixture of solvents. It was suggested that bile acids as indicators of hepatic function will be utilized as markers of injury. Franco et al.[30] examined a group of workers exposed to organic solvents, and used the criteria addressing exposure to solvent mixtures for over two years, daily ethanol consumption less than 50 grams, no history of hepatic disease, no drug intake in the previous three months. Workers were exposed to between 6 and 9 solvents, mostly toluene, xylene, acetone, methyl acetate, and butanol ethyl acetate. The mean levels of liver enzyme activities in the exposed and the control group were similar. The mean serum bile acid contents was statistically and significantly increased in the exposed group compared to the controlled group. The authors concluded that the observation of higher serum bile acid levels in the group of workers currently exposed to organic solvents might be explained by a change in hepatocyte function, and that the commonly followed parameters of liver enzymes may be insensitive to these preliminary initial changes in liver function. Conventional liver function tests seem to be rather insensitive to early detect liver damage from solvents exposure. Early detection is crucial, but is probably missed since the standard liver functions tests are not sensitive enough to detect early liver damage from solvents. What that means is that by the time the patient is seen by the doctor with liver fibrosis or necrosis and solvent toxicity, it is already in advanced stages. Joung-Dar-Chen et al.[31] evaluated the effects of solvents exposure on liver functions, specifically looking at gamma glutamyl transferase activity. They have studied the effects of xylene and toluene. The median air concentration was evaluated in the exposed workers who used mixed solvents in the process of spray booth car painting. These investigators showed that gamma glutamyl transferase activities increased independently with both an increased consumption of alcohol and exposure to a mixture of solvents. They have concluded that an increase in GGT activity may be a form of enzyme induction rather than a marker for cellular damage. Kurpper et al.[32] examined the effect of mixed organic solvents on liver enzymes activity in car painters, and found that at the exposure level of that study (at that time was 1/2 the level recommended by the regulatory agencies) the liver enzyme activities of car painters were not effected by exposure to mixed solvents. The continuation of the previous study by Franco et al.[33] examined serum bile acid concentrations as a marker of liver functions in a group of workers exposed to organic solvents. They have shown a significantly increased concentration of serum bile acids with normal liver enzymes, and concluded that this indicated a very sensitive and early marker of liver function abnormality in patients exposed to mixed solvents, and might be explained as an early sign of liver dysfunction. While it was not possible to state which solvent caused what type of abnormality, the authors concluded that conventional liver function tests seem to be rather insensitive for early liver disease detection, and normal measurement does not rule out the existence of subclinical disease, and therefore, an elevation of serum bile acid indicates early potential hepatotoxicity from mixed organic solvents. From a clinical point of view this study suggests that exposure to mixture of solvents in an industrial setting may cause liver damage which is subclinical and initially undetected, unless liver enzymes and bile acids are measured. It seems likely that this initial stage of liver damage in patients exposed to solvent mixtures is commonly missed.

## 20.7.5 TRICHLOROETHYLENE

The liver is a target organ toxicity for trichloroethylene in experimental animals. Data in humans are limited.[34] Case reports describe trichloroethylene induced hepatitis and liver necrosis.[35] Guzelian et al. described both hepatic necrosis and fatty metamorphasis.[36] As early as 1962, the hepatotoxicity of trichloroethylene has been studied in humans. Trichloroethylene has been found to cause liver damage after both acute and chronic exposures.[37,38,39,40,41] Several studies reported a history of pathological changes including individual or focal necrosis after treatment of experimental animals with trichloroethylene.[42] In addition to these histopathological changes Berman also found a dose response relationship to the histopathological changes. Since trichloroethylene is commonly present as a contaminant in ground water (from degreasing, paint thinning and plastic metal processes) Barton has evaluated risks assessment of trichloroethylene and liver toxicity,[43] and showed that exposure was associated with continuous response in the form of liver toxicity. There was a connection between increased liver weight over body weight and liver toxicity, this effect also appears to be a sensitive indicator of liver toxicity. The authors concluded that trichloroethylene is toxic to the liver, based on their analysis of their findings, and that the use of liver enzymes by themselves may miss the early signs of toxicity to the liver by trichloroethylene. From these studies on trichloroethylene, and the studies on solvent mixtures described above, it seems reasonable to conclude that the early subclinical stages of solvent hepatotoxicity are commonly missed. When patients present to the doctor they will already be in more advanced stages such as liver steatosis and/or chronic hepatitis, and at times liver fibrosis.

The toxicity of trichloroethylene is dependent upon metabolism and induction of cytochrome P450. Trichloroethylene is metabolized through chloral hydrate to compounds including trichloroacetic acid and dichloroacetic acid which alter intercellular communication, induce peroxisome proliferation and may promote tumor production.[44] Significant variability in trichloroethylene metabolism in 23 human haptic microsomal samples was reported by Lipscomb et al.[44] It was also demonstrated that the trichloroethylene metabolism is dependent on enzymatic activities of the cytochrome system, and they conclude that their data indicates that humans are not uniform in their capacity for CPY dependent metabolism of trichloroethylene and increased activity may increase susceptibility to trichloroethylene induced toxicity in humans. These observations are compatible with the variability reaction which is depending on nutritional factors, enzyme induction factors, hormonal factors and interaction with other environmental chemicals, prescription medications and general health conditions, and explains the variable reports as far as trichloroethylene and level of liver toxicity in the various individuals studied.

In a predisposed individual (for example, a patient who is on medications or alcohol) it is highly likely that exposure to trichloroethylene will be a substantial factor in the genesis of a wide variety of liver diseases.

## 20.7.6 TETRACHLOROETHYLENE

Tetrachloroethylene is synthetic chemical used for dry cleaning fabrics, and has also been named as perchloroethylene and tetrachloroethene. The liver is the target organ in humans, mainly in reports of accidental exposure to a high concentration. Meckler et.al.[45] has shown liver damage in a woman exposed occupationally to tetrachloroethylene fumes documented by a liver biopsy. Other investigators also have shown elevation of liver enzymes, jaundice and enlarged liver.[46,47] Experimental animal studies also have shown liver damage by inhalation of tetrachloroethylene.[48,49,50,51] Liver necrosis occurred in experimental mice exposed to 100 and 200 ppm of tetrachloroethylene for 103 weeks.[52] Experimental animals exposed orally to tetrachloroethylene have been shown to develop liver changes similar to those produced by inhalation studies, and mice are more sensitive than rats to tetrachloroethylene induced liver toxicity. Humans exposed by oral routes to tetrachloroethylene except for heavy doses commonly have not shown significant changes other than obstructive jaundice and enlarged liver reported in an infant exposed to tetrachloroethylene via breast milk.[53] Issues of carcinogenicity will not be addressed in this chapter, and the interested reader is referred to the toxicological profile for tetrachloroethylene.[54] It is highly likely that tetrachlorethylene is a hepatotoxic agent in high doses, and probably in low doses in susceptible individuals with either other environmental exposures, prescription medications, alcoholism, nutritional and/or genetic factors, and preexisting disease of the liver.

## 20.7.7 TOLUENE

Industrial use of toluene is wide, commonly in paint, paint thinners, fingernail polish, lacquers, adhesives, rubbers and in the printing letter industry. Toluene is extensively metabolized by the liver; however, the liver does not appear to be a primary target for toluene toxicity. A study of printing factory workers who were exposed to toluene at a concentration of less than 200 ppm, showed minimal changes of liver enzymes.[55] The cohorts included 289 men, of which 8 showed elevated liver enzymes, and 6 of them had enlarged livers. Seven of those patients had liver biopsies which showed some centrally lobular and periportal fat accumulation, and Kupffer cell hyperplasia. The study by Svensson et al.[56] has looked at 47 rotogravure workers occupationally exposed to toluene at a concentration of 80 ppm for 3-39 years, and showed a significant elevation of liver enzymes, finding of chemical hepatitis. Seiji et al.[57] has examined a group of 157 female shoemakers who were exposed to toluene at levels of 7-324 ppm from 2-14 months, and showed no significant elevation of commonly measured liver functions. Another study that looked at 47 Swedish paint industry workers who were exposed for more than a 10 year period of time to organic solvents which included xylene, toluene, isobutanol, n-butanol, mineral spirits, methyl acetates, dichloromethane, methyl ethyl ketone and isopropanol did not show any changes in liver enzymes.[58] However, this study cannot be relied upon as a specific study, since the cohort size was small, and there were multiple exposures to multiple solvents, and therefore, the study had only limited power to detect the effects of toluene on the liver of exposed workers. Experimental animals exposed to toluene at concentrations of 533 to 800 ppm for 7 days showed increased liver weights, but no significant morphological changes by microscopy. Electron microscopical examination revealed ultrastructural changes which were compatible with changes in the cytochrome P-450 concentrations. Others have shown no effect on liver size or liver functions.[59,60,61,62] Overall the data seems to suggest that toluene may cause liver damage in certain industries, and especially in synergism with other solvents. It is highly likely that, in predisposed individuals, toluene can cause liver damage from chemical hepatitis to necrosis and fibrosis.

### 20.7.8 DICHLOROMETHANE

Dichloromethane also called dichloromethane, is a colorless liquid that has a mild sweet odor. It is used widely in the industry as a solvent and a paint stripper. It is commonly found in spray painting operations, automotive degreasing, in cleaners and in household products. Stewart et al.[63] showed no changes in liver enzymes in patients exposed for a period of 6 weeks to levels of dichloromethane via inhalation from 50-500 ppm. On the other hand, Ott et al.[64] has shown an elevation in bilirubin in workers exposed to dichloromethane up to 475 ppm. In experimental animals dichloromethane exposure has been associated with fatty changes of the liver and elevated liver enzymes. Norpoth et al.[65] have shown hepatic microsomal enzyme elevation at 500 ppm of dichloromethane exposure for 10 days, and others have shown significant fatty changes of the liver upon exposure of mice and rats for 100 days to 75-100 ppm of trichloroethylene.[66,67,68] When exposure to dichloromethane continues for 2 years there was increased evidence of pathological changes and fatty liver changes.[69,70,71] The overall weight of the data supports a hepatotoxic effect of dichloromethane on the liver.

### 20.7.9 STODDARD SOLVENT

Stoddard solvent is a widely used organic solvent synthetically made, and comes from the refining of crude oil. It is a petroleum mixture made from distilled alkanes, cycloalkanes (naphthenes), and aromatic compounds. In addition, it goes by other names such as Varsol 1, Texsolve S and others. It is commonly used as a paint thinner, as solvents in some types of photocopier toners, printing ink, adhesives, dry cleaning and as a general cleaner and degreaser. Twelve men exposed to 610 mg per cubic meter of vaporized Stoddard solvents for a period of 6 hours revealed no changes in serum glucose, triglycerides, cholesterol or urate.[72] Dossing et al.[73] described painters who were exposed to non-specified levels of Stoddard solvents and other chemicals for chronic periods, and elevated levels of serum alanine aminotransferase, but other functions were normal and normal liver biopsies. Flodin et al.[74] has studied a group of patients exposed to a variety of solvents, including Stoddard solvents and showed normal liver function tests, but an elevated gamma glutamyl transferase. Hane et al.[75] has shown that a group of painters exposed to Stoddard solvents had no significant abnormality of liver enzymes. Studies in experimental animals showed minimal fatty changes of the liver, Jenkins et al.[76] as did the studies by Carpenter and Phillips.[77,78,79] The data from experimental animals and humans suggests a potential hepatotoxic effect of Stoddard solvents, but additional studies and a case by case evaluation is required.

### 20.7.10 1,1,1-TRICHLOROETHANE

1,1,1,-Trichloroethane is a colorless solvent which is manmade. It is produced by industry and is used in commercial products. It is used as a solvent, and is heavily used in glue and paint, as well as a degreaser and metal parts manufacturing. It is also used in some household products such as spot cleaners, glues and aerosol sprays. It is commonly found in soil and water as a contaminant. Brief single exposures to very high levels of 1,1,1-trichloroethane and a moderate high concentration have been shown to cause elevation of urobilinogen in patients.[80] This type of finding indicates reduced bile excretion and some intrinsic liver damage. Stewart et al.[81] showed in patients accidentally exposed to a high concentration of 1,1,1-trichloroethane increased levels of urinary urobilinogen for approximately 4 days following the exposure. While looking at enzyme levels in the blood after exposure to 1,1,1-trichloroethane, there was no evidence of elevated serum enzyme levels.[80,82,83,84] Case studies of people exposed to a high concentration of 1,1,1-trichloroethane did not show elevated liver enzymes.[81,85] Histopathological examination of the liver of patients who died following inhalation of a high concentration of 1,1,1-trichloroethane showed minimal changes, mainly those of mild fatty changes of the liver.[86,87] Kramer et al.[88] studied humans at low levels of exposure and found minor changes in liver enzymes. Experimental animal studies showed mild histopathological changes and effects on liver enzymes.[89,90] Truffert et al.[91] showed that intermittent duration and intermittent exposure to a low concentration of 1,1,1-trichloroethane produced a 67% increase in the synthesis of DNA of the livers of exposed rats, and concluded that the DNA synthesis measurements may be a more sensitive indicator of liver damage than just measurements of liver enzymes. McNutt et al.[92] showed histological damage following exposure to 1,1,1-trichloroethane with hepatocyte necrosis. The most commonly reported effects of 1,1,1-trichloroethane on the liver in experimental animals is increased fat accumulation.[93,94,95] The function of duration of exposure played an important role in experimental rats, and was seen in those who were exposed for 7 hours, but was not seen in those exposed for 2 hours in high levels. Savolainen et al.[96] showed that exposure to a moderate concentration in experimental animals caused decreased microsomal cytochrome P450 enzyme activity. Overall, the animal studies and human studies suggest an effect of 1,1,1-trichloroethane on the liver, but the severity appears to be related to the dose and duration of the exposure.

## 20.7.11 SUMMARY

In summary, from the available data, it is clear that exposure to solvents and hepatotoxicity must be evaluated in context of the individual variability, exposure to mixture of solvents, and synergistic toxicity.

## REFERENCES


1 Jones DS, ed, **Detoxification: A Clinical Monograph**, *Institute for Functional Medicine, Inc.*, 1999
2 Ungar, *Viertelsjarisch f. gericht. Med.*, **47**:98 (1887)
3 Ostertag R, *Virchows Arch.*, **118**:250 (1889)
4 Stiles HJ and McDonald S, *Scott Med Surg J*, **15**:97 (1904)
5 Recknagal RO, *Pharmacol Rev*, **19**:145 (1967)
6 Recknagel RO and Glinde EA, *CRC Crit Rev Toxicol*, **2**:263 (1973)
7 Zimmerman HJ, **Hepatotoxicity: The Adverse Effects of Drugs and Other Chemicals on the Liver**, *Appleton-Century-Crofts*, New York, 1978
8 Mitchell JR, et al. In **Concepts in Biochemical Pharmacology**, Part 3, JR Gillette, JR Mitchell and PS Randall (eds), *Springer-Verlag*, Berlin, 383-419 (1975)
9 Mitchell JR, et al., *Drug Met Disp*, **1**:418 (1973)
10 Rouiller CH, In **The Liver**, Volume II, CH Rouiller (ed), *Academic Press*, New York, 335-476 (1964)
11 Von Oettingen WF, **The Halogenated Hydrocarbons of Industrial and Toxicological Importance**, *Elsevier*, Amsterdam (1964)
12 Browning E, **Toxicology and Metabolism of Industrial Solvents**, *Elsevier*, Amerstdam (1965)
13 Von Oettingen WF, The Halogenated Aliphatic, Olephinic Cyclic, Aromatic and Aliphatic-Aromatic Hydrocarbons Including the Halogenated Insecticides. Their Toxicity and Potential Dangers. U.S. Dept. HEW, U.S. Govt Printing Office, Washington, D.C. (1955)
14 Meyer J and Pessoa SB, *Am J Trop Med*, **3**:177 (1923)
15 Reynolds ES, *Biochem Pharmacol*, **21**:2255 (1972)
16 Slater TF, *Nature* (Lond), **209**:36 (1966)
17 Jennings RB, *Arch Pathol*, **55**:269 (1955)
18 Klatskin G, **Toxic and Drug Induced Hepatitis. In Diseases of the liver**, 4th Ed, L Schiff (ed), *JB Lippincott*, Philadelphia, 604-710 (1975)

19 Slater TF, **Free Radical Mechanisms in Tissue Injury**, *JW Arrowsmith, Ltd*, Bristol, 118-163 (1972)
20 Le Page RN and Dorling PR, *Aust J Exp Biol*, **49**:345 (1971)
21 Rufeger U and Frimmer M, *Arch Pharmacol*, **293**:187 (1976)
22 Agency for Toxic Substances and Disease Registry (ATSDR), Toxicological Profile for Carbon Tetrachloride, Atlanta, Georgia (1997)
23 Droz PO, Wu MM, Cumberland WG and Berode M, *Br J Ind Med*, **46**(7):447-460 (July 1989)
24 Nothnagel, *Berlin Klin Wchnschr*, **3**:31 (1866)
25 International Agency for Research on Cancer (IARC) Monographs on the Evaluation of Carcinogenic Risks to Humans, Volumed 47:125-156; IARC, Lyon, France Volumes 43-48 (1987-1990)
26 Toftgard R and Nilsen OG, *Toxicology*, **23**(2-3):197-212 (1982)
27 Frommer U, Ullrich V and Orrenius S, *FEBS Lett*, **41**(1):14-16 (Apr 15, 1974)
28 Sotaniemi EA, et al., *Acta Med Scand*, **212**:207-215 (1982)
29 Fisbien A, et al., *Lancet*, **1**:129 (1983)
30 Franco G, et al., *Br J Ind Med*, **46**:141-142 (1989)
31 Chen JD, et al., *Brit J Ind Med*, **48**:676-701 (1991)
32 Kurppa K and Husman K, *Scand J Work Environ Health*, **8**:137-140 (1982)
33 Franco G, et al., *Int Arch Occup Environ Health,* **58**:157-164 (1986)
34 Davidson IW and Beliles RP, *Drug Metabolism Review*, **23**:493-599 (1991)
35 Bond GR, *Clinical Toxicology*, **34**(4):461-466 (1996)
36 Guzelian PS, Disorders of the Liver, In **Principles and Practice of Environmental Medicine**, Aylce Bezman Tacher (eds), *Plenum Medical Book Company*, New York, 319-333, (1992)
37 Baerg RD and Kimberg DV, *Annals of Internal Medicine*, **73**:713-720 (1970)
38 Clearfield HS, *Digestive Digest,* **15**:851-856 (1970)
39 Fielder RJ, Loweing RK and Shillaker RO, *Toxic Rev*, **6**:1-70 (1982)
40 Litt I and Cohen M, *New England Journal of Medicine*, **281**:543-544 (1969)
41 Sax N, **Dangerous Properties of Industrial Chemicals**, 4th Edition, *Van Nostrand Reinhold Co*, Toronto, Canada, 1186 (1975)
42 Berman E, et al., *J Tox Environ Health,* **45**(2):127-143 (June 1995)
43 Barton HA, et al., *Regulatory Toxicology & Pharmacology*, **24**:269-285 (1986)
44 Lipscomb JC, et al., *Toxicology & Applied Pharmacology*, **142**:311-318-1997
45 Meckler LC and Phelps PK, *J Am Med Assoc*, **197**:662-663 (1966)
46 Coler HR and Rossmiller HR, *AMA Arch Ind Hyg Occup Med*, **8**:227-233 (1953)
47 Hake CL and Stewart RD, *Environ Health Perspect*, **21**:231-238 (1977)
48 Kyline B, et al., *Acta Pharmacol Toxicol*, **20**:16-26 (1963)
49 Kyline B, Sumegi I and Yllner S, *Acta Pharmacol Toxicol*, **22**:379-385 (1965)
50 Carpenter CP, *J Ind Hyg Toxicol*, **19**:323-336 (1937)
51 Rowe VK, et al., *AMA Arch Ind Hyg Occup Med*, **5**:566-579 (1952)
52 National Toxicology Program (NTP) - Technical Report Series No. 311, Toxicology and Carcinogenesis Studies of Tetrachloroethylene (Perchloroethylene) (CAS No. 127-18-4) in F344/N rats and B6C3F1 mice (inhalation studies), Research Triangle Park, NC, U.S. Dept of Health & Human Services, Public Health Service, National Institute of Health, NIH Publication No. 86-2567 (1986)
53 Bagnell PC and Ellenberger HA, *Can Med Assoc J*, **117**:1047-1048 (1977)
54 Agency for Toxic Substances and Disease Registry (ATSDR), Toxicological Profile for Tetrachloroethylene, Atlanta, Georgia (1993)
55 Guzelian P, Mills S and Fallon HJ, *J Occup Med*, **30**:791-796 (1988)
56 Svensson BG, et al., *Br J Ind Med*, **49**:402-408 (1992)
57 Seiji K, et al., *Ind Health*, **25**:163-168 (1987)
58 Lundberg I and Hakansson M, *Br J Ind Med*, **42**:596-600 (1985)
59 Bruckner JV and Peterson RG, *Toxicol Appl Pharmacol*, **61**:27-38 (1981)
60 Bruckner JV and Peterson RG, *Toxicol Appl Pharmacol*, **61**:302-312 (1981)
61 Kjellstrand P, et al., *Acta Pharmacol Toxicol,* **57**:242-249 (1985)
62 National Toxicology Program (NTP) - Technical Report Series, Toxicology and Carcinogenesis Studies of Toluene (CAS No. 108-88-3) in F344/N rats and 86C3F mice (inhalation studies), Research Triangle Park, NC, U.S. Environmental Protection Agency, U.S. Dept of Health & Human Services, No. 371. PB90-256371 (1990)
63 Stewart RD, et al., Methylene Chloride: Development of Biologic Standard for the Industrial Worker by Breath Analysis. Report of the National Institute of Occupational Safety and Health, Cincinnatti, OH, by the Medical College of Wisconsin, Milwaukee, Wisconsin, NTIS No. PB83-245860 (1974)

64 Ott MG, et al., *Scand J Work Environ Health*, **9**(Supple 1):17-25 (1983)
65 Norpoth K, Witting U and Springorum M, et al., *Int Arch Arbeitsmed*, **33**:315-321 (1974)
66 Haun CC, et al., Continuous Animal Exposure to Low Levels of Dichloromethane. In: Proceedings of the Third Annual Conference on Environmental Toxicology. Wright Patterson Air Force Base, OH: Aerospace Medical Research Laboratory, 199-208, AMRL-TR-72-130, (1982)
67 Kjellstrand P, et al., *Acta Pharmacol Toxiocl (Copenh)*, **59**:73-79 (1986)
68 Weinstein RS and Diamond SS, Hepatotoxicity of dichloromethane (methylene chloride) with continuous inhalation exposure at a low dose level. In: Proceedings of the Third Annual Conference on Environmental Toxicology. Wright Patterson Air Force Base, OH: Aerospace Medical Research Laboratory, 209-220, AMRL-TR-72-130, (1972)
69 Burek JD, et al., *Fund Appl Toxicol*, **4**:30-47 (1984)
70 Nitschke KD, et al., *Fundam Appl Toxicol*, **11**:48-59 (1988)
71 National Toxicology Program (NTP) - Technical Report Series No. 306, Toxicology and Carcinogenesis Studies of Dichloromethane (Methylene Chloride) (CAS No. 75-09-2) in F344/N rats and B6C3F1 mice (inhalation studies), Research Triangle Park, NC, U.S. Dept of Health & Human Services, Public Health Services, Centers for Disease Control, National Institute of Health, (1986)
72 Pederson LM and Cohr KH, *Acta Pharmacol Toxicol,* **55**:317-324 (1984)
73 Dossing M, Arlien-Soborg P and Petersen LM, *Eur J Clin Invest*, **13**:151-158 (1983)
74 Flodin U, Edling C and Axelson O, *Am J Ind Med*, **5**:287-295 (1984)
75 Hane M, et al., *Scand J Work Environ Health*, **3**:91-99 (1977)
76 Jenkins LJ, et al., *Toxicol Appl Pharmacol*, **18**:53-59 (1971)
77 Carpenter CP, et al., *Toxicol Appl Pharmacol*, **32**:246-262 (1975)
78 Carpenter CP, et al., *Toxicol Appl Pharmacol*, **32**:282-297 (1975)
79 Phillips RD and Egan GF, *Fundam Appl Toxicol*, **4**:808-818 (1984)
80 Stewart RD, et al., *Am Ind Hygn Assoc J*, **22**: 252-262 (1961)
81 Stewart RD, *J Am Med Assoc*, **215**:1789-1792 (1971)
82 Stewart RD, et al., *Arch Environ Health*, **19**:467-472 (1969)
83 Dornette WHL and Jones JP, *Anesthesia and Analgesia*, **39**:249-252 (1960)
84 Torkelson TR, et al., *Am Ind Hygn Assoc J*, **19**:353-362 (1958)
85 Wright MF and Strobl DJ, *J Am Osteopath Assoc,* **84**:285-288 (1984)
86 Caplan YH, Backer RC and Whitaker JQ, *Clin Toxicol*, **9**:69-74 (1976)
87 Hall FB and Hine CH, *J Forensic Sci*, **11**:404-413 (1966)
88 Kramer CG, et al., *Arch Environ Health*, **33**:331-342 (1978)
89 Carlson GP, *Lif Sci (United States)*, **13**:67-73 (1973)
90 Gehring PJ, *Toxicol Appl Pharmacol*, **13**:287-298 (1968)
91 Truffert L, et al., *Arch Mal Prof Med Trav Secur Soc*, **38**:261-263 (1977)
92 McNutt NS, et al., *Lab Invest*, **32**:642-654 (1975)
93 Takahara K, *Okayama Igakkai Zasshi*, **98**:1099-1110 (Japanese) (1986)
94 Adams EM, et al., *Am Med Assoc Arch Ind Hyg Occup Med*, **1**:225-236 (1950)
95 Herd PA, Lipsky M and Martin HF, *Arch Environ Health*, **28**:227-3 (1974)
96 Savolainen H, et al., *Arch Toxicol*, **38**:229-237 (1977)

# 20.8 SOLVENTS AND THE LIVER

DAVID K. BONAUTO
**Occupational Medicine, University of Washington**
**Seattle, Washington, USA**
C. ANDREW BRODKIN
**Department of Internal Medicine and Department of Environmental Health,**
**University of Washington, Seattle, Washington, USA**
WILLIAM O. ROBERTSON
**Washington Poison Center, University of Washington**
**Seattle, Washington, USA**

The toxic effects of organic solvent compounds on the liver are dependent on the intensity and duration of exposure, route of exposure, the intrinsic toxicity of the specific compound, as well as individual susceptibility factors.[1] There are a number of pathologic manifestations of solvent induced hepatotoxicity, including inflammation, fat accumulation in the liver (steatosis), hepatocellular necrosis and carcinogenesis. Functional disturbances in liver physiology have also been associated with solvent exposure.

The purpose of this chapter is to review the known hepatotoxicity of commonly used industrial solvents.[2] A brief review of normal anatomic and physiologic function of the liver will be provided as a background for understanding histopathologic and biochemical changes associated with solvent toxicity. The final segment includes a discussion of solvents known to cause liver injury with a review of the available medical evidence suggestive of human hepatotoxicity of solvents at present day exposure levels. Solvent induced hepatotoxicity is almost exclusively encountered in an occupational setting and thus this review will focus on evidence culled from that setting.

## 20.8.1 NORMAL ANATOMIC AND PHYSIOLOGIC FUNCTION OF THE LIVER

The liver is the largest internal organ and is involved in many physiologic processes including nutrient homeostasis, synthesis and excretion of bile, lipid metabolism and lipoprotein and protein synthesis.[3] Most importantly for purposes of this chapter, the liver is the site of the biotransformation of a wide variety of endogenous and exogenous toxins.[4] The ability of the liver to biotransform various chemicals is due to the multiple different enzyme systems contained within the hepatocytes.[3,5] One such enzyme system is the cytochrome p 450 enzyme system. It consists of a large group of enzymes which biotransform many different substances by either oxidation or reduction to facilitate excretion from the body. Specifically different members of the cytochrome p 450 family catalyze reactions involving aromatic and aliphatic hydroxylation, epoxidation, dehalogenation, dealkylation, N-, S-oxidation as well as O-, N-, S- dealkylation reactions.[3,5]

The diverse metabolic activities of the liver make it susceptible to solvent induced injury, particularly from reactive intermediates which damage cellular macromolecules. The microscopic anatomy of the liver provides an explanation for this susceptibility. The basic unit of the liver is the hepatic lobule which consists of a central vein surrounded radially by sinusoids of liver cells (hepatocytes). Portal triads consisting of a hepatic artery, a hepatic vein and a bile canniliculus are located at the periphery. Liver cells closest to the vascular supply or the portal triad, zone one, are more resistant to oxidative stress, while hepatocytes near the central vein, zone three or centrizonal region, are most susceptible to solvent induced injury.

#### 20.8.1.1 Factors Influencing Solvent Hepatotoxicity

Bioavailability: The physical and chemical properties of a solvent and its toxicokinetics determine its availability to hepatic tissues. The primary route of absorption of most solvents which cause hepatotoxicity into human biological systems is via the lung. Therefore, the greater the volatility of the solvent, the greater its concentration in the air, and subsequently the larger the potential dose.[4] While accidental or intentional ingestion of solvents is reported in the medical literature, it is an uncommon route of exposure in the occupational setting. Dermal absorption should be considered a significant route of exposure for most solvent compounds based on their lipid solubility. The degree of exposure can often be modified by the use of personal protective equipment such as gloves or a respirator and engineered exposure controls such as building ventilation.

The lipid solubility of solvents also favors their deposition of into lipid rich organ systems such as the liver. The toxicity of a particular solvent may be enhanced by its long residence time in the liver.[6]

Genetic and environmental factors: While some solvents are directly hepatotoxic, frequently biotransformation of solvents by hepatic mixed function oxygenases, such as the cytochrome p-450 system, result in toxic intermediates.[7] A variety of genetic and environmental factors inhibit or induce the activity of these hepatic enzyme systems, effecting the biotransformation and resulting toxicity. Genetic factors thought to determine the activity or even the presence of an enzyme within an organism center around human polymorphisms or variations in the genetic code.[7] As the activities of the liver enzymes are changed so will the rate of formation of the metabolite thus increasing or decreasing the toxicity of the foreign substance.[3] Environmental factors which determine the activity of liver enzyme systems include co-exposure to other drugs and toxins or characteristics of the individual particularly disease states which induce or inhibit the activity of liver biotransformation enzymes.[1,3,7-9] Individual characteristics such as age, nutritional status, pregnancy or disease states such as diabetes or obesity may also change the activity of cytochrome p-450 enzymes.[7,9] The assessment of an individual's susceptibility to exposure should attempt to account for these variables in determining risk.

#### 20.8.1.2 Microscopic, Biochemical and Clinical Findings Associated with Liver Injury due to Solvents

Hepatotoxic manifestations associated with acute solvent exposure are dose dependent. Acute cytotoxic injury of a solvent directly or by its metabolites causes an alteration in the normal physiologic function leading to ballooning fatty change and ultimately cellular necrosis. If the dose is minimal and doesn't exceed the 'regenerative capacity' of the liver, inflammatory changes generally resolve within two weeks to several months. Metabolic derangement results in the accumulation of fats in the liver, termed steatosis.[10] A less common form of acute cytotoxic injury is related to cholestatic injury with disruption of normal biliary flow.[11] Severe long term exposures can lead to fibrosis or scarring and cirrhosis which distort the hepatic architecture and lead to altered liver function.

Current research focuses on the effect of low doses of solvents on the liver, with concern that low grade prolonged solvent exposure could lead to chronic injury and eventual impairment.[5,12]

Tests used to evaluate and screen for liver injury can be divided into three general categories: serum biomarkers of disease, tests of hepatic clearance, and anatomic evaluation.[11] The hepatic enzymes most commonly screened for related to hepatocellular necrosis and inflammatory changes are aspartate aminotransferase (AST), and alanine aminotransferase (ALT). Elevation of these enzymes in the setting of significant exposure is indicative of hepatotoxic injury, though alternative causes such as alcohol and viral hepatitis should be excluded. Importantly, serum hepatic transaminase levels indicate hepatocellular necrosis or inflammation, but may not indicate more subtle metabolic alterations in hepatic function. Measures of other hepatic enzymes, gamma glutamyl transpeptidase (GGT), alkaline phosphatase (Alk Phos), total and direct bilirubins may also be suggestive of solvent induced hepatotoxicity. Specifically if hepatic excretion of bile, is diminished, the resulting intrahepatic cholestasis is associated with elevations in GGT, Alk Phos and serum bile acids. Significant elevations of bilirubins leads to the clinical observation of jaundice or yellowing of the skin. However, pathologic obstruction of the biliary tract is not a common finding in solvent induced hepatotoxicity.[5]

Clearance tests of liver function assess a number of physiologic activities including hepatic uptake, hepatic metabolism, and hepatic excretion. Typical clearance tests of liver function include indocyanine green (ICG), antipyrine clearance test and $^{14}C$ aminopyrine breath test. These tests give an estimation of the ability of the liver to extract and detoxify exogenous toxins (xenobiotics). Measuring the excretion of endogenously produced serum bile acids is an additional measure of hepatic clearance and has been used as a sensitive measure of early solvent hepatotoxicity.[13,14]

Anatomic evaluation of solvent hepatotoxicity centers on physical examination of the liver, radiologic study and liver biopsy. Physical exam is nonspecific as to the cause and characterization of the disease. Radiologic studies such as ultrasound can identify hepatobiliary disease and liver parenchymal disease, namely steatosis and fibrosis. Steatosis and fibrosis are noted on ultrasound by a change in the echogenicity of the liver. While liver biopsy is the 'gold standard' for anatomic evaluation of the liver, the invasiveness of the test, the morbidity and discomfort of the procedure, and its cost make it prohibitive for routine screening. It is usually reserved for definitive diagnostic and prognostic purposes. Algorithmic strategies for screening for liver injury and evaluation of abnormal results have been reported in several references.[11,12,15,16]

## 20.8.2 HEPATOTOXICITY ASSOCIATED WITH SPECIFIC SOLVENTS

The following section presents specific classes of organic solvents strongly associated with hepatotoxicity in human populations or animal studies. While there is more limited evidence of hepatotoxicity related to inhalational and dermal exposure to aliphatic hydrocarbons, ketones, alcohols, aldehydes, esters and ethers, potential hepatotoxicity related to these agents must be assessed on an individual basis with regard to concentration, duration, and bioavailability of exposure.[5,11] Variations in individual susceptibility must also be considered with regard to concurrent use of alcohol, mixed solvent exposure, underlying liver diseases (e.g., viral hepatitis, hemochromatosis, hypertriglyceridemia and diabetes) as well as demographic differences in hepatic metabolism.[11] Given these limitations, the organic solvents of primary concern with regard to hepatotoxicity are the haloalkanes, haloalkenes, dimethylformamide, and nitroparaffins.[5,11] Other agents such as styrene have been associated with hepatotoxicity in some studies.[17,18] Potential interactive effects of solvent mixtures should always be considered in the assessment of hepatotoxicity, even if composed of solvents not commonly associated with hepatotoxicity.

### 20.8.2.1 Haloalkanes and haloalkenes

Some of the most extensively studied and most concerning hepatoxins are the haloalkane solvents. Major haloalkanes encountered industrially, with documented animal and human hepatotoxicity, are carbon tetrachloride, chloroform, 1,1,2,2-tetrachlorethane, methyl chloroform and 1,1,2-trichloroethane, tetrachloroethylene, and trichloroethylene.[5,6,19] The relative hepatotoxicity of each is correlated inversely with the carbon chain length, and carbon halogen bond energy and correlated directly with the number of halogens on the molecule and the atomic number of the halogen.[5,20] Some of these solvents have been eliminated from common industrial use due to their deleterious environmental and human effects, though they may still be encountered in specific processes and regions (e.g., developing countries). Carbon tetrachloride is the most extensively studied and serves as a model for hepatotoxicity for other haloalkanes.[21,22]

### 20.8.2.2 Carbon tetrachloride

Carbon tetrachloride hepatotoxicity has been reported since the early twentieth century.[23] The toxicological literature is extensive with regard to carbon tetrachloride hepatotoxicity in animals. [21,23] Human toxicological information derives primarily from accidental or intentional ingestion in humans or by inhalational exposure in groups of workers.[21] The industrial use of carbon tetrachloride has declined precipitously, due to the recognized health effects and regulatory policy.[21] Historically, it was used as a solvent in the manufacture of industrial chemicals, in the dry cleaning industry and even as an antiparasitic medication.[5] Presently the main means of exposure is in research laboratory settings, or as low level environmental contaminant.[21] Because it is so volatile, the main mode of carbon tetrachloride exposure in occupational setting is via inhalation, although exposure by the dermal route also occurs.

Animal and human susceptibility to carbon tetrachloride hepatotoxicity is dependent on many different factors. There is substantial interspecies variation in carbon tetrachloride induced hepatotoxicity in animals due to differences in metabolic pathways among species.[21] Based on animal models, hepatotoxicity in humans is most likely mediated from the trichloromethyl radical formed from the metabolism of carbon tetrachloride by hepatic cytochrome p 450 2E1.[24] Animal studies suggest differential hepatotoxicity based upon the animal's age and gender, with greater toxicity demonstrated in adult rats compared to newborns,[25,26] and males compared with females.[5] Cytochrome p-450 enzyme systems are present in the human fetus suggesting a potential for in utero liver toxicity.[27] Human gender differences in the metabolism of carbon tetrachloride have not been demonstrated despite potential sex steroid influences on the cytochrome p-450 system.[28]

The hepatotoxic effects of carbon tetrachloride are more severe in the setting of alcohol consumption.[21,29,30] Animal studies demonstrate that the temporal relationship between ethanol ingestion and carbon tetrachloride exposure determines the severity of toxicity.[31-33] Maximal hepatotoxicity is derived from ethanol ingested eighteen hours preceding exposure to carbon tetrachloride,[32,33] whereas exposure to ethanol three hours prior to carbon tetrachloride exposure leads to minimal hepatotoxicity.[33] The mechanism for this interaction is the induction of cytochrome p-450 enzymes leading to greater formation of toxic intermediates.[34] In contrast, exposure to ethanol immediately preceding carbon tetrachloride exposure leads to competitive inhibition of carbon tetrachloride metabolism.[34] Several other alcohols (e.g., isopropanol,[32,35] t-butyl alcohol[36]) and ketones[37] potentiate the effect of carbon tetrachloride hepatotoxicity. Exposures to other haloalkanes[38] or haloalkenes[39] potentiate carbon tetrachloride hepatotoxicity while carbon disulfide exposure is 'protective' of carbon tetrachloride hepatotototoxicity.[40] Dietary factors, medications, chronic diseases and persistent halogenated environmental contaminants such as PCB's and DDT have been shown to modulate or exacerbate the hepatotoxicity of carbon tetrachloride.[5,21]

Cellular disruption leading to hepatocellular necrosis results from damage to cellular macromolecules by trichloromethyl radicals.[24] Cellular disruption involves alteration of calcium homeostasis,[41] impaired oxidative phosphorylation,[42] and trichloromethyl radical binding to cellular proteins, nucleic acids, and induction of lipid peroxidation.[6,21] Histologically there is preferential necrosis of zone three hepatocytes in the liver acinus so called centrizonal necrosis as well as zone three steatosis.

As with other halogenated hydrocarbons, carbon tetrachloride is an intrinsic hepatotoxin with adverse effects occurring at predictable exposure levels. The American Conference of Governmental and Industrial Hygienists (ACGIH) Threshold Limit Value (TLV), for carbon tetrachloride, based upon animal and human exposure data where limited adverse health effects are observed, is 5 ppm over an 8 hour time weighted average and a 40 hour work week for carbon tetrachloride.[43] In human population studies, elevations in hepatic transaminase levels occur at carbon tetrachloride concentrations averaging 200 ppm, with small but significant elevations of ALT, AST, Alk Phos and GGT occurring at exposure levels below the TLV.[44-46]

Carbon tetrachloride also affects many other organ systems, specifically the central nervous system, the gastrointestinal tract, the liver and the kidney.[5,6] Hepatic manifestations of carbon tetrachloride include serum AST and ALT elevations as early as three hours following exposure. Clinical evidence of hepatic disease occurs approximately twenty four hours following exposure, and is manifest in half the cases as jaundice accompanied by hepatic enlargement. In severe poisonings, progressive hepatic injury leads to coma and death within a week of exposure. Fortunately, non lethal exposures are often associated with significant clinical recovery in two to three weeks. Treatment is limited to supportive care, in a hospital setting. Chronic exposures to carbon tetrachloride have been associated with hepatic fibrosis and cirrhosis in animals and documented as well in several case reports in humans.[47-50]

### 20.8.2.3 Chloroform

Medical and industrial use of chloroform has also declined significantly.[51] Today, industrial use is limited to the manufacture of refrigerants and fluoropolymers.[51] Chloroform metabolism involves the same cytochrome p-450 2E1 as carbon tetrachloride but with oxidation of chloroform to trichloromethanol with spontaneous formation of phosgene via the elimination of hydrochloric acid.[3] In turn, phosgene reacts with hepatic lipids and microsomal proteins and depletes cellular glutathione, a cellular antioxidant.[7,51] Factors potentiating chloroform hepatotoxicity include ethanol and other alcohols,[52-54] hypoxia,[53] ketones,[55] fasting state,[56,57] concomitant chronic medical disease and chronic medication use, or those with repeated exposures to chloroform.[5] The pattern of human liver injury associated from chloroform poisoning is centrilobular necrosis and steatosis.[23]

The ACGIH has set a TLV of 10 ppm over an 8 hour time weighted average and a 40 hour work week for chloroform. Because chloroform is a potential carcinogen, the lowest possible exposure is recommended. Occupational hepatotoxicity below the ACGIH TLV has been demonstrated, with evidence of adverse effects between 2 and 10 ppm.[51,58]

Clinical manifestations of chloroform toxicity involve multi-organ system effects including damage to the central nervous system, the kidney and lung as well as the liver.[59] Fulminant toxic hepatitis appears within one to three days following exposure, with death at approximately one week in severe poisonings.[57] In nonfatal cases, hepatic inflammatory changes,with hepatomegaly and transaminitis can occur within hours.[57] Ingestion or significant inhalational exposure should be managed in a closely monitored hospital setting.

### 20.8.2.4 Dichloromethane

Dichloromethane is commonly used as a degreaser and a paint stripper. It is metabolized in the liver by the cytochrome p-450 pathway to produce carbon monoxide.[60] An independent pathway of metabolism occurs via conjugation with glutathione.[60] Animal experimentation has demonstrated hepatotoxicity at near lethal concentrations of dichloromethane.[61,62] Dichloromethane potentiates carbon tetrachloride hepatotoxicity in rat livers.[38] Short term exposure to both ethanol and dichloromethane demonstrate an antagonistic relationship, while chronic exposure potentiates hepatotoxicity.[63]

Cases of human hepatotoxicity to dichloromethane have been reported.[62,64] Workers in an acetate fiber production plant, exposed to 140 to 475 ppm of dichloromethane, with concomitant exposures to acetone and methanol, were observed to have elevated bilirubin and ALT levels relative to workers exposed to acetone alone.[64] Bilirubin elevations were dependent on the level of dichloromethane exposure.[64] Other studies have shown no significant effects in the range of 5 to 330 ppm of dichloromethane.[65] Chronic exposure (greater than 10 years) to dichloromethane levels greater than 475 ppm was not associated with significant elevations in liver function tests.[66] There is minimal evidence of human hepatotoxicity of dichloromethane less than the ACGIH TLV of 50 ppm over an 8 hour time weighted average.[43]

### 20.8.2.5 Trichloroethanes

There are two isomers of trichloroethane, namely methyl chloroform and 1,1,2-trichloroethane. Animal hepatotoxicity to 1,1,2-trichloroethane is documented in the literature[67] with potentiation of toxicity in association with acetone,[68] isopropyl alcohol[69] and ethanol.[70] Hepatotoxicity, with steatosis, necrosis, elevated serum enzymes, and increased liver weight have been observed in animal models exposed to 1000 ppm of methyl chloroform.[71] Human studies consist of case reports documenting hepatotoxicity, with elevated serum transaminases and fatty liver disease related to 1,1,1-trichloroethane exposure.[72,73] Epidemiologic evidence suggests little hepatotoxicity related to this agent at exposure levels <350 ppm.[74,75]

### 20.8.2.6 1,1,2,2-Tetrachloroethane

Though rarely used in current practice, this solvent was an important cause of hepatotoxicity in the past. Its hepatotoxic potential was first identified during its use in the first World War.[5] Animal hepatotoxicity with fatty degeneration of the liver has been documented in multiple species.[76] Human inhalational exposures manifest in liver enlargement, jaundice, steatosis with subsequent liver failure in severe poisonings.[77,78] Subacute exposure periods of weeks to months is generally required for hepatic injury.[77] Liver regeneration occurs after nonfatal exposures.[77] The mechanism of hepatotoxicity has not been elucidated in humans but the reactive metabolites 1,1-dichloroacetyl chloride with binding to hepatic macromolecules may play a role.[76] Metabolism of 1,1,2,2-tetrachloroethane is potentiated by fasting and ethanol in rats.[79,80] There is little documentation of any precise inhalational exposure levels necessary to cause hepatotoxicity.

#### 20.8.2.7 Tetrachloroethylene and trichloroethylene

This widely used dry cleaning agent and degreasing agent is associated with hepatotoxic effects.[81,82] Cases of human hepatotoxicity to tetrachloroethylene at exposure levels greater than 100 ppm have been reported in the literature.[83,84] Humans exposed to tetrachloroethylene at dosages up to 150 ppm for durations of one to five 8 hour shifts had no difference in hepatic enzyme levels from baseline levels.[85] Studies of workers chronically exposed to concentrations of tetrachloroethylene less than 50 ppm showed no difference in liver enzyme levels, relative to groups of workers who did not have the exposure.[82] However, dry cleaning workers chronically exposed to low levels of tetrachloroethylene at less than 25 ppm had evidence of an alteration in hepatic echogenicity relative to non-exposed workers.[80] This is suggestive evidence that steatosis may occur at levels below the ACGIH TLV, without associated alterations in serum hepatic enzymes. The long term effects of exposures have not been well characterized.

Wide spread use of trichloroethylene occurs in the dry cleaning industry and industrially as a degreasing agent. Historical use as an anesthetic generally suggests little acute hepatotoxicity.[86,87] Longer term exposures in an occupational setting are associated with elevations in serum transaminases, with variable findings in epidemiologic studies.[88-90] Exposures below the ACGIH TLV of 50 ppm in workers using trichloroethylene as a cleaning agent found elevated levels of serum bile acids.[45,91] Hepatotoxicity is potentiated by alcohol,[92] isopropanol and acetone.[69] The long term effects of subclinical exposures are not known.

#### 20.8.2.8 Other halogenated hydrocarbons

Vinyl chloride, a gas at normal temperature and pressure, has solvent properties at high pressures; its industrial use as a monomer in the manufacture of polyvinylchloride and hepatotoxicity with chronic exposure make it an important public health risk. Vinyl chloride is associated with angiosarcoma,[93,94] a rare highly malignant hepatic tumor, hepatic fibrosis,[6] hepatocellular injury[95] and hepatoportal sclerosis, a form of noncirrhotic portal hypertension.[96,97] Appearance of angiosarcoma and hepatoportal sclerosis occurred in workers after decade long exposures.[94,98] Measures to limit both occupational and environmental exposures have been instituted to decrease potential hepatic outcomes, with effective screening programs using indocyanine green clearance tests.[6]

Haloalkanes other than the chloroalkanes, especially those with structural homology to known hepatotoxic chloroalkanes, should be considered potentially hepatotoxic despite little industrial use as solvents.[5,20] Case reports of bromoethane and hydrochlorofluorocarbon poisonings with hepatotoxicity have been reported in the literature.[99-102]

#### 20.8.2.9 Styrene and aromatic hydrocarbons

Styrene is not only used as a monomer in the production of polystyrene but also as a reactive solvent in the manufacture of unsaturated polyester resins.[103] The hepatic metabolism of styrene involves the formation of the reactive intermediate styrene 7,8-oxide.[104] In rat models, styrene 7,8-oxide binds to hepatic macromolecules and lipids causing hepatocellular injury.[105,106] Epidemiologic investigations of workers exposed to high concentrations (greater than 50 ppm) of styrene have shown elevations in GGT, AST, ALT,[107-109] and serum bilirubin levels.[110] At the ACGIH TLV of 50 ppm or less, evidence of transaminase and GGT elevations[111,112] are lacking but elevated levels of serum bilirubins[113] and bile acids[17,110] have been demonstrated. There is no evidence of alterations in hepatic echogenicity at exposure levels less than 50 ppm.[18]

Toluene, benzene and xylenes are generally considered to have limited hepatotoxicity.[6,114-118] Exposure to xylene is reported to cause mild steatosis.[5] Exposure to a mixture of solvents, inclusive of xylene and toluene have been reported to produce elevated serum bile acids.[13]

### 20.8.2.10 N-substituted amides

Two important N-substituted amides are dimethylformamide and dimethylacetamide. Dimethylformamide is used in the fabrication of synthetic textiles such as rayon. Its hepatotoxicity has been well demonstrated in occupational settings.[119-121] Evidence of dose dependent alcohol intolerance and subjective gastrointestinal symptoms (abdominal pain, anorexia and nausea) have been described.[122] Objective clinical and biochemical signs include elevations of transaminases, AST and ALT, hepatomegaly and abnormal liver biopsy findings demonstrating hepatocellular necrosis and steatosis.[120,123] Workers with acute toxicity related to DMF have more severe symptoms and higher transaminase levels than workers with toxicity related to chronic exposures.[124] Of significance, symptoms may occur under the ACGIH TLV of 10 ppm. Dermal absorption is a main exposure pathway in addition to inhalation.[119-121]

Dimethylacetamide is used as a solvent in the manufacture of plastics and as a paint remover. Occupational poisoning and hepatotoxicity to extreme concentrations of dimethylacetamide (DMA) are reported in the medical literature.[125] Decreases in hepatic clearance measures and alterations in hepatic transaminases with hepatomegaly have been reported at lower doses.[126] Like dimethylformamide, DMA is readily absorbed through the skin. Chronic exposures in workers exposed to low air concentrations of DMA of less than 3 ppm and with biological monitoring assessments to measure dosages by dermal absorption demonstrated little evidence of hepatotoxicity by clinical chemistries.[127]

### 20.8.2.11 Nitroparaffins

The well known hepatotoxicity of nitroaromatic compounds such as trinitrotoluene lends suspicion to the hepatotoxicity of the nitroparaffins.[114] Nitromethane and nitroethane produce steatosis in animal models, but there is limited evidence of hepatotoxicity of these agents in humans.[114] Evidence for the hepatoxicity of 2-nitropropane has been raised by case reports and case series of occupational fatalities in settings of severe exposure.[128,129] In these cases the lack of appropriate industrial hygienic measures such as adequate ventilation, and personal protective equipment contributed to the severity of the exposures.[130] Autopsies of the fatal cases revealed hepatocellular necrosis and fatty infiltration of the liver.[128] No significant evidence of hepatotoxicity has been demonstrated below the ACGIH TLV of 10 ppm.[131] Medical surveillance of workers exposed to less than 25 ppm of 2-nitropropane have not shown alterations in liver chemistries.[132]

### 20.8.2.12 Other solvents and mixed solvents

Suggestive evidence for hepatotoxicity of many compounds exist in the literature.[5] Two solvents with some suspicion for hepatotoxic potential in humans are tetrahydrofuran and 1,4-dioxane, both solvents used in industry.[133-135] Cases of tetrahydrofuran induced hepatotoxicity have been reported in the literature.[133] Tetrohydrofuran's inhibition of the cytochrome p-450 enzyme system lends biologic credibility to it being a hepatotoxin.[134] 1,4-dioxane is reported to be hepatotoxic but epidemiologic evidence in human populations for this is limited.[135]

Rarely do solvents exist in isolation and thus evaluation of hepatotoxicity must consider the effects of mixtures of solvents.[136] Alterations in the hepatotoxic potential of a chemical may exist, especially when the biotransforming enzymes are modulated or effected by various components of the mixture. Usual mechanisms for the potentiation of toxicity by alcohols, ketones may be altered when solvents are mixed. In such settings hepatotoxicity may occur below recommended levels.[137]

Much remains unknown regarding the hepatotoxic effects of compounds. For this reason, vigilance regarding the potential adverse hepatic effects of chemicals is appropriate. Maintaining active surveillance for solvent induced hepatotoxicity is important in protecting workers' health, and will further our knowledge of the hepatotoxic effects of solvents. With emerging knowledge, occupational and environmental standards can be refined to further protect the health of workers and the public.

## REFERENCES


1 H. Zimmerman in **Schiff's Diseases of the Liver**, E. Schiff, M. Sorrell and W. Maddrey, Eds., *Lippincott-Raven Publishers,* Philadelphia, 1999, pp.973 -1064.
2 J. Rosenberg in **Occupational and Environmental Medicine**, 2nd ed., J. LaDou. Ed., *Appleton and Lange*, Stamford, Conn., 1997, pp. 359-386.
3 A. Parkinson in **Casarrett and Doull's Toxicology; The Basic Science of Poisons**, 5th ed., C. Klaasen, Ed., *McGraw-Hill*, New York, 1996, pp. 113-186.
4 M. Ellenhorn, S. Schonwald,G. Ordog, and J. Wasserberger eds., **Ellenhorn's Medical Toxicology: Diagnosis and Treatment of Human Poisoning**. *Williams and Wilkins.* Baltimore, 1997.
5 H. Zimmerman, **Hepatotoxicity: The Adverse Effects of Drugs and Other Chemicals on the Liver**. 2nd Ed., *Lippincott, Williams and Wilkins, Philadelphia*, 1999.
6 N. Gitlin in **Hepatology; A Textbook of Liver Disease**. 3rd ed., D. Zakim and T. Boyer, Eds., *WB Saunders Co.*, Philadelphia, 1996, pp. 1018-1050.
7 J. Raucy, J. Kraner, and J. Lasker, *Crit. Rev. Toxicol.*, **23**, 1 (1993).
8 P. Watkins, *Semin. Liver Dis.*, **10**, 235 (1990).
9 D. Vessey in **Hepatology; A Textbook of Liver Disease**. 3rd ed., D. Zakim and T. Boyer, Eds., *WB Saunders Co.*, Philadelphia, 1996, pp. 257-305.
10 G. Lundqvist, U. Flodin, and O. Axelson, *Am. J. Ind. Med.*, **35**, 132 (1999).
11 C. Redlich and C. Brodkin in **Textbook of Clinical Occupational and Environmental Medicine**, L. Rosenstock and M. Cullen, Eds., *WB Saunders*, Philadelphia, 1994, pp. 423-436.
12 C. Tamburro and G. Liss, *J. Occup. Med.*, **28**, 1034 (1986).
13 G. Franco, R. Fonte, G. Tempini, and F. Candura, *Int. Arch. Occup. Environ. Health*, **58**, 157 (1986).
14 G. Franco, *Br. J. Ind. Med.*, **48**, 557 (1991).
15 D. Herip, *Am. J. Ind. Med.*, **21**, 331 (1992).
16 R. Harrison, *Occupational Medicine: State of the Art Reviews*, **5**, 515 (1990).
17 C. Edling and C. Tagesson, *Br. J. Ind. Med.*, **41**, 257 (1984).
18 C. Brodkin, J. Moon, D. Echeverria, K. Wang, and H. Checkoway, ICOH 2000 Congress, (Abstract In Press).
19 R. Harbison, **Hamilton and Hardy's Industrial Toxicology**. 5th ed., *Mosby*, St. Louis, 1998.
20 G. Plaa and W. Hewitt in **Toxicology of the Liver**, G. Plaa and W. Hewitt, Eds., *Raven Press,* New York, 1982, pp. 103-120.

21 O. Faroon, J. Riddle, Y. Hales and W. Brattin, **Toxicological Profile for Carbon Tetrachloride**. Agency for Toxic Substances and Disease Registry (ATSDR), *US Department of Health and Human Services*, Atlanta, 1994.
22 R. Recknagel, *Pharmacol. Rev.*, **19**, 145 (1967).
23 W. Von Oettingen, **The Halogenated Hydrocarbons of Industrial and Toxicological Importance**, *Elsevier*, Amsterdam, 1964.
24 R. Recknagel and E. Glende, Jr., *Crit. Rev. Toxicol.*, **2**, 236 (1973).
25 M. Dawkins, *J. Pathol. Bacteriol.*, **85**, 189 (1963).
26 S. Cagen and C. Klaasen, *Toxicol. Appl. Pharmacol.*, **50**, 347 (1979).
27 G. Mannering, *Fed. Proc.*, **44**, 2302 (1985).
28 J. Gustaffson, *Annu. Rev. Physiol.*, **45**, 51 (1983).
29 M. Manno, M. Rezzadore, M. Grossi, and C. Sbrana, *Hum. Exp. Toxicol.*, **15**, 294 (1996).
30 P. New, G. Lubash, L. Scherr, and A. Rubin, *JAMA*, **181**, 903 (1962)
31 H. Ikatsu, T. Okino, and T. Nakajima, *Br. J. Ind. Med.*, **48**, 636 (1991).
32 G. Traiger and G. Plaa, *Toxicol. Appl. Pharmacol.*, **20**, 105 (1971).
33 H. Cornish and J. Adefuin, *Am. Ind. Hyg. Assoc. J.*, **27**, 57 (1966).
34 T. Castillo, D. Koop, S. Kamimura, G. Triadafilopoulos, and H. Tsukamoto, *Hepatology*, **16**, 992 (1992).
35 D. Folland, W. Schaffner, H. Ginn, O. Crofford, and D. McMurray, *JAMA*, **236**, 1853 (1976).
36 R. Harris and M. Anders, *Toxicol. Appl. Pharmacol.*, **56**, 191 (1980).
37 G. Plaa, *Fundam. Appl. Toxicol.*, **10**, 563 (1988).
38 Y. Kim, *Fundam. Appl. Toxicol.*, **35**, 138 (1997).
39 D. Pessayre, B. Cobert, V. Descatoire, C. Degott, G. Babany, C. Funck-Brentano, M. Delaforge and D. Larrey, *Gastroenterology*, **83**, 761 (1982).
40 Y. Masuda and N. Nakayama, *Biochem. Pharmacol.*, **31**, 2713 (1982).
41 E. Glende and R. Recknagel, *Res. Commun. Chem. Pathol. Pharmacol.*, **73**, 41 (1991).
42 G. Christie and J. Judah, *Proc. Roy. Soc. Lond. B.*, **142**, 241 (1954).
43 American Conference of Governmental Industrial Hygienists (ACGIH), Threshold Limit Values(TLVs) for chemical substances and physical agents and Biological Exposure Indices (BEIs), American Conference of Governmental Industrial Hygienists, Cincinnati, 1996.
44 R. Barnes and R. Jones, *Am. Ind. Hyg. Assoc. J.*, **28**, 557 (1967).
45 T. Driscoll, H. Hamdan, G. Wang, P. Wright, and N. Stacey, *Br. J. Ind. Med.*, **49**, 700 (1992).
46 J. Tomenson, C. Baron, J. O'Sullivan, J. Edwards, M. Stonard, R. Walker, D. Fearnley, *Occup. Env. Med.*, **52**, 508 (1995).
47 G. Cameron and W. Karunatne, *J. Pathol. Bacteriol.*, **42**, 1 (1936).
48 C. Poindexter and C. Greene, *JAMA*, **102**, 2015 (1934).
49 N. Gitlin, *S.A. Med. J.*, **58**, 872 (1980).
50 T. Paerez, *Hepatology*, **3**, 112 (1983).
51 S. Chou, W. Spoo, **Toxicological Profile for Chloroform**. Agency for Toxic Substances and Disease Registry (ATSDR), *US Department of Health and Human Services*, Atlanta, 1997.
52 S. Kutob and G. Plaa, *J. Pharmacol. Exposure Therap.*, **135**, 245 (1962).
53 K.Hutchens and M. Kung, *Am. J. Med.*, **78**, 715 (1985).
54 S. Ray and H. Mehendale, *Fundam. Appl. Toxicol.*, **15**, 429 (1990).
55 J. Brady, D. Li, H. Ishizaki, M. Lee, S. Ning, F. Xiao, and C. Yang, *Toxicol. Appl. Pharmacol.*, **100**, 342 (1989).
56 D. McMartin, J. O'Connor Jr., and L. Kaminsky, *Res. Commun. Chem. Pathol. Pharmacol.*, **31**, 99 (1981).
57 S. Winslow and H. Gerstner, *Drug Chem. Toxicol.*, **1**, 259 (1978).
58 L. Li, X. Jiang, Y. Liang, Z. Chen, Y. Zhou, and Y. Wang, *Biomed. Environ. Sci.*, **6**, 179 (1993).
59 Occupational Safety and Health Administration(OSHA), Occupational Safety and Health Guidelines for Chloroform, available at http://www.osha.slc.gov/SLTC/ healthguidelines/chloroform/recognition.html, 2000.
60 R. Snyder and L. Andrews in **Casarrett and Doull's Toxicology; The Basic Science of Poisons**, 5th ed., C. Klaasen, Ed., *McGraw-Hill*, New York, 1996, pp. 737-771.
61 K. Mizutani, K. Shinomiya, and T. Shinomiya, *Forensic Sci. Int.*, **38**, 113 (1988).
62 Occupational Safety and Health Administration, OSHA Federal Register: Occupational Exposure to dichloromethane-62: 1494-1619, available at http://www.osha-slc.gov/FedReg_osha_data/FED19970110.html, 2000.
63 M. Balmer, F. Smith, L. Leach, and C. Yuile, *Am. Ind. Hyg. Assoc. J.*, **37**, 345 (1976).
64 M. Ott, L. Skory, B. Holder, J. Bronson, and P. Williams, *Scand. J. Work Environ. Health*, **9** S(1), 1 (1983).

65 H. Anundi, M. Lind, L. Friis, N. Itkes, S. Langworth, and C. Edling, *Int. Arch. Occup. Env. Health*, **65**, 247 (1993).
66 K. Soden, *J. Occup. Med.,* **35**, 282 (1993).
67 Syracuse Research Corporation, Toxicological Profile for 1,1,2-Trichloroethane, Agency for Toxic Substances and Disease Registry (ATSDR), US Department of Health and Human Services, Atlanta, 1989.
68 J. MacDonald, A. Gandolfi, I. Sipes, and J. MacDonald, *Toxicol. Lett.*, **13**, 57 (1982).
69 G. Traiger and G. Plaa, *Arch. Environ. Health*, **28**, 276 (1974).
70 C. Klaasen and G. Plaa, *Toxicol. Appl. Pharmacol.*, **9**, 139 (1967).
71 M. Williams and F. Llados, Toxicological Profile for 1,1,1-Trichloroethane. Agency for Toxic Substances and Disease Registry (ATSDR), US Department of Health and Human Services, Atlanta, 1995.
72 M. Hodgson, A. Heyl, and D. Van Thiel, *Arch. Intern. Med.*, **149**, 1793 (1989).
73 C. Cohen and A. Frank, *Am. J. Ind. Med.*, **26**, 237 (1994).
74 C. Kramer, H. Imbus, M. Ott, J. Fulkerson, and N Hicks, *Arch. Environ. Health*, **38**, 331 (1978).
75 R. Stewart, H. Gay, A. Schaffer, D. Erley, and V. Rowe, *Arch. Environ. Health*, **19**, 467 (1969).
76 L. Smith and J. Mathews, Toxicological Profile for 1,1,1,2-Tetrachloroethane, Agency for Toxic Substances and Disease Registry (ATSDR), US Department of Health and Human Services, Atlanta, 1994.
77 H. Coyer, *Ind. Med.*, **13**, 230 (1944).
78 R. Gurney, *Gastroenterology*, **1**, 1112 (1943).
79 T. Nakajima and A. Sato, *Toxicol. Appl. Pharmacol.*, **50**, 549 (1979).
80 A. Sato, T. Nakajima, and Y. Koyama., *Br. J. Ind. Med.*, **37**, 382 (1980).
81 C. Brodkin, W. Daniell, H. Checkoway, D. Echeverria, J. Johnson, K. Wang, R. Sohaey, D. Green, C. Redlich, D. Gretch, and L. Rosenstock, *Occup. Env. Med.*, **52**, 679 (1995).
82 P. Gennari, M. Naldi, R. Motta, M. Nucci, C. Giacomini, F. Violante, and G. Raffi, *Am. J. Ind. Med.*, **21**, 661 (1992).
83 G. Saland, *N.Y.S. J. Med.*, **67**, 2359 (1966).
84 L. Meckler, and D. Phelps, *JAMA*, **197**, 662 (1966).
85 R. Lauwerys, J. Herbrand, J. Buchet, A. Bernard, and J. Gaussin, *Int. Arch. Occup. Environ. Health*, **52**, 69 (1983).
86 G. Smith, *Br. J. Ind. Med.*, **23**, 249 (1966).
87 R. Defalque, *Clin. Pharmacol. Ther.*, **2**, 665 (1961).
88 R. McCunney, *Br. J. Ind. Med.*, **45**, 122 (1988).
89 G. Bond, *J. Toxicol. Clin. Toxicol.*, **34**, 461 (1996).
90 T. Nagaya, N. Ishikawa, H. Hata, and T. Otobe, *Int. Arch. Occup. Environ. Health*, **64**, 561 (1993).
91 M. Neghab, S. Qu, C. Bai, J. Caples, and N. Stacey, *Int. Arch. Occup. Environ. Health*, **70**, 187 (1997).
92 G. Müller, M. Spassowski, and D. Henschler, *Arch. Toxicol.*, **33**, 173 (1975).
93 J. Creech Jr., and M. Johnson, *J. Occup. Med.*, **16**, 150 (1974).
94 F. Lee, P. Smith, B. Bennett, and B. Williams, *Gut*, **39**, 312 (1996).
95 S. Ho, W. Phoon, S. Gan, and Y. Chan, *J. Soc. Occup. Med.*, **41**, 10 (1991).
96 P. Smith, I. Crossley and D. Williams, *Lancet*, **2** (7986), 602 (1976).
97 P. Bioulac-Sage, B. LeBail, P. Bernard, and C. Balabaud, *Semin. Liver Dis.*, **15**, 329 (1995).
98 W. Lelbach, *Am. J. Ind. Med.*, **29**, 446 (1996).
99 A. Van Haaften, *Am. Ind. Hyg. Assoc. J.*, **30**, 251 (1969).
100 P. Hoet, M. Graf, M. Bourdi, L. Pohl, P. Duray, W. Chen, R. Peter, S. Nelson, N. Verlinden, and D. Lison, *Lancet.*, **350**, 556 (1997).
101 D. Anders and W. Dekant, *Lancet*, **350**, 1249 (1997).
102 G. Rusch, *Lancet*, **350**, 1248 (1997).
103 P. Pfaffli and A. Saamanen in **Butadiene and Styrene: Assessment of Health Hazards**, *IARC Scientific Publications* No 127, M. Sorsa, K. Peltonen, H. Vainio, and K. Hemmiki Eds. International Agency for Research on Cancer, Lyon, 1993, pp. 15 - 33.
104 J. Bond, *Crit. Rev. Toxicol.*, **19**, 227 (1989).
105 J. Marniemi in **Microsomes and Drug Oxidations**, V. Ullrich ed., *Pergamon*, New York, 1977, pp. 698-702.
106 J. Van Anda, B. Smith, J. Fouts, and J. Bend, *J. Pharmacol. Exp. Ther.*, **211**, 207 (1979).
107 O. Axelson O and J. Gustavson, *Scand. J. Work Environ. Health*, **4**, 215 (1979).
108 G. Triebig, S. Lehrl, D. Weltle, K. Schaller, and H. Valentin, *Br. J. Ind. Med.*, **46**, 799 (1989).
109 A. Thiess, and M. Friedheim, *Scand. J. Work Environ. Health*, **4** (S2), 220 (1978).
110 R. Vihko in **Biological Monitoring And Surveillance Of Workers Exposed To Chemicals**, A. Aitio, V. Riihimäki, and H. Vainio Eds., *Hemisphere Publishing Corp*, Washington, D.C., 1984, pp. 309-313.

111 W. Lorimer, R. Lilis, W. Nicholson,, H. Anderson, A. Fischbein, S. Daum, W. Rom, C. Rice, and I. Selikoff, *Environ. Health Perspect.*, **17**, 171 (1976).
112 H. Harkonen, A. Lehtniemi, and A. Aitio, *Scand. J. Work Environ. Health*, **10**, 59 (1984).
113 C. Brodkin, Personal Communication.
114 H. Zimmerman and J. Lewis, *Gastroenterol. Clin. North. Am.*, **24**, 1027 (1995).
115 L. Low, J. Meeks, and C. Mackerer, *Toxicol. Ind. Health*, **4**, 49 (1988).
116 P. Guzelian, S. Mills, and H. Fallon, *J. Occup. Med.*, **30**, 791 (1988).
117 R. Morris, *J. Occup. Med.*, **31**, 1014 (1989).
118 C. Boewer, G. Enderlein, U. Wollgast, S. Nawka, H. Palowski, and R. Bleiber, *Int. Arch. Occup. Environ. Health*, **60**, 181 (1988).
119 V. Scailteur and R. Lauwerys, *Toxicology*, **43**, 231 (1987).
120 C. Redlich, W. Beckett, J. Sparer, K. Barwick, C. Riely, H. Miller, S. Sigal, S. Shalat and M. Cullen, *Ann. Int. Med.*, **108**, 680 (1988).
121 A. Fiorito, F. Larese, S. Molinari, and T. Zanin, *Am. J. Ind. Med.*, **32**, 255 (1997).
122 S. Cai, M. Huang, L. Xi, Y. Li, J. Qu, T. Kawai, T. Yasugi, K. Mizunuma, T. Watanabe, and M. Ikeda, *Int. Arch. Occup. Environ. Health,* **63**, 461 (1992).
123 L. Fleming, S. Shalat, and C. Redlich, *Scand. J. Work Environ. Health*, **16**, 289 (1990).
124 C. Redlich, A. West, L. Fleming, L. True, M. Cullen, and C. Riely, *Gastroenterology*, **99**, 748 (1990).
125 G. Marino, H. Anastopoulos, and A. Woolf, *J. Occup. Med.*, **36**, 637 (1994).
126 G. Corsi, *Med Lav*, **62**, 28 (1971).
127 G. Spies, R. Rhyne Jr., R. Evans, K. Wetzel, D. Ragland, H. Turney, T. Leet, and J. Oglesby, *J. Occup. Med.*, **37**, 1102 (1995).
128 C. Hine, A. Pasi, and B. Stephens, *J. Occup. Med.*, **20**, 333 (1978).
129 R. Harrison, G. Letz, G. Pasternak, and P. Blanc, *Ann. Int. Med.,* **107**, 466 (1987).
130 D. Hryhorczuk, S. Aks, and J. Turk, *Occup Med: State of Art Reviews*, **7**, 567 (1992).
131 T. Lewis, C. Ulrich, and W. Busey, *J. Environ. Pathol. Toxicol.*, **2**, 233 (1979).
132 G. Crawford, R. Garrison, and D. McFee, *Am. Ind. Hyg. Assoc. J.*, **46**, 45 (1985).
133 R. Garnier, N. Rosenberg, J. Puissant, J. Chauvet, and M. Efthymiou, *Br. J. Ind. Med.*, **46**, 677 (1989).
134 D. Moody, *Drug Chem. Toxicol.*, **14**, 319 (1991).
135 C. DeRosa, S. Wilbur, J. Holler, P. Richter, and Y. Stevens, *Toxicol. Ind. Health.*, **12**, 1 (1996).
136 F. Tomei, P. Giuntoli, M. Biagi, T. Baccolo, E. Tomao, and M. Rosati, *Am. J. Ind. Med.*, **36**, 54 (1999).
137 E. Sotaniemi, S. Sutinen, S. Sutinen, A. Arranto, and R. Pelkonen, *Acta Med. Scand.*, **212**, 207 (1982).

## 20.9 TOXICITY OF ENVIRONMENTAL SOLVENT EXPOSURE FOR BRAIN, LUNG AND HEART

Kaye H. Kilburn
**School of Medicine, University of Southern California**
**Los Angeles, CA, USA**

This chapter considers the neurobehavioral effects of environmental exposures to organic solvents. Much information applicable to environmental or community exposures usually at home came from animal experiments, brief human exposures in chambers and prolonged workplace exposures. The mode of entry of solvent chemicals into the body is almost always by inhalation not by contact or ingestion.[1] While inhalational exposures to single chemicals occur in the community mixtures are usual making measurements more complex. Effects from animal experiments, and human exposures in chambers, and workplace exposures are usually consistent and help predict environmental effects. The major categories of environmental exposures to solvents are from petroleum refining to consumer use indoors, Table 20.9.1. Sometimes adverse human effects are from surprisingly small environmental doses, an order of magnitude or two lower than those needed for workplace effects. One possible explanation is greater sensitivity of measurements but many of the methods were adapted from studies of workers.[2]

**Table 20.9.1. Sources of environmental exposure to solvents**

| Processes | Chemicals | Media | Example |
|---|---|---|---|
| Losses during refining and chemical production | MTBE | air | Seymour, IN, Santa Maria, CA |
| Losses from use in industry | TCE + toluene | water surface, ground water, air, water | Phoenix, AZ, Motorola, Printers, Baton Rouge, LA, Abuse-glue sniffers |
| Leaks and spills during transportation (pipeline, truck, rail, ship) | toluene, xylene, PAH | air | Avila Beach, CA, Livington Parish, LA |
| Combustion:<br>a. Fires as incidents,<br>b. Incineration of fuel air pollution<br>c. Incineration of garbage | hydrocarbon particles | air | Wilmington, CA,<br>San Bernardino, CA, Los Angeles, Houston, TX, Mexico City<br>Oak Ridge, TN, Walker, LA |
| Contaminated sites | TCE | water, ground water | 800 national hazardous acid pits |
| Outgassing indoors of forest products-like particle board, carpets, drapes, adhesives | TCA | air | Indoor air incidents, Sick building syndrome |

+ - other chlorinated solvents; TCA - trichloroethane; MTBE - methyl ter butyl ether, additive in gasoline; PAH - polyaromatic hydrocarbons, example benzo(a)pyrene

**Table 20.9.2. Differences in environmental and occupational toxicology**

| | Occupational | Environmental |
|---|---|---|
| Subjects<br>health<br>age, years<br>selection positive | <br>healthy<br>18-60<br>selected for employment<br>attenuated by losses of sick | <br>chronic illness<br>0-100<br>unselected<br>sick collect |
| Duration<br>in a week<br>total | <br>40 hours<br>years | <br>168 hours<br>lifetime |
| Source | raw materials<br>processes | leaks and spills (gasoline)<br>outgassing of consumer goods<br>fuel combustion |
| Chemical exposure | one or few | many |
| Monitoring exposure | area or personal | rarely possible |
| Environmental transformation of agents | unusual | frequent |

Differences between exposure at work and in the community are important, Table 20.9.2. Most worker groups were younger and healthier, met job-entry criteria, have more reserve function so are less likely to manifest damage. Workers have had selective attrition of affected or less fit people to accentuate the difference.[3] In contrast community people are unselected and include more susceptible groups: infants, children, the aged and the unwell.[1] Some differences in exposure are obvious, work exposure is rarely longer than 40 of the weeks 168 hours. This time away allows work acquired body burdens of chemicals (and their effects) to diminish or disappear while workers are at home. In contrast, home exposures may be continuous or nearly so.[4] At work the time that elapsed between exposure and effect is short, making measurement of the dose of a toxic agent easy. It is less obvious what should be measured in community exposures.

Good detective work is needed to specify the chemicals to search for and measure in air, water or soil. Community effects may take years to be recognized as a problem. Opportunities for pertinent environmental measurements were overlooked and have disappeared with time, often simply evaporated. Thus measurement of relevant doses are seldom possible and dose-response curves can rarely be constructed. The logical surrogates for dose such as distance and direction from a chemical source and duration are rarely satisfactory.[1] Thus plausible estimates of dose are needed to focus the association of measured effects and the chemicals that are probably responsible.

The realistic starting place is people's symptoms-complaints that indicate perception of irritation from chemicals.[3] These serve as sentinels to alert one to a problem but cannot be interpreted as impairment or damage without measurements of brain functions. The inability to characterize exposure should not postpone or prevent adequate investigation for adverse human health effects. It is intuitive and ethical to suggest that absent of adverse human effects should be the only reason for stopping inquiry. People's complaints and upset moods (anxiety, depression, anger, confusion and fatigue) frequently reflect or parallel impairment. The question then becomes how to measure effects on the brain to decide whether it is damaged and if so how much, Table 20.9.3.

**Table 20.9.3. Useful tests of evaluation of brain damage from solvents**

| Tests | Part of brain measured |
|---|---|
| Simple reaction time & visual two choice reaction time | retina, optic nerve and cortex integrative radiation to motor cortex |
| Sway-balance | inputs: ascending proprioceptive tracts, vestibular division 8th cranial nerve, cerebellum, vision, visual integrative and motor tracts |
| Blink reflex latency | sensory upper division trigeminal nerves (V), pons, facial nerves (VII) |
| Color confusion index | center macular area of retina, with optic cones, optic nerve, optic occipital cortex |
| Visual fields | retina-optic nerve-optic cortex occipital lobe |
| Hearing | auditory division of 8th cranial nerve |
| Verbal recall memory | limbic system of temporal lobe, smell brain |
| Problem solving culture fair & digit symbol | cerebral cortex: optic-occipital and parietal lobe cortex |

| Tests | Part of brain measured |
|---|---|
| Vocabulary | long-term memory, frontal lobes |
| Information, picture completion & similarities | long-term memory, frontal lobes |
| Pegboard performance | optic cortex to motor cortex |
| Trail making A & B | (eye-hand coordination) |
| Fingertip number writing | parietal lobe, sensory area of pre-Rolandic fissure |
| Profile of mood states | limbic system for emotional memory |

The effects of ethyl alcohol are familiar to most people so I will start with this best studied of mind altering solvents. Measurements of alcoholic patients in the mid-twentieth century at New York's Bellevue Hospital helped David Wechsler formulate his adult intelligence scale, 11 tests that measure attention, problem solving, concept juggling and memory including vocabulary.[5] Many other tests were devised to estimate intelligence, how the mind works as defined by AR Luria and others.[6] Ward Halstead assembled and created function tests to measure the effects of traumatic damage to the brain by wartime missiles and by neurosurgery, prefrontal lobotomy.[7] Application of these tests, by Reitan,[8] helped differentiate the organic brain disorders from schizophrenia and other mental illnesses. Thus the starting place for testing became brain diseases recognized by the neurologist using simple bedside qualitative tests. The tests were not used to detect impairment before it was clinically recognized. The first steps were Benjamin Franklin's recollections of his own brain poisoning by lead while he was a printer and Lewis Carroll's mad hatter, from mercury used in felting beaver hair. The next steps were taken in Nordic countries in the 1960's.[9]

Carbon disulfide was the first solvent studied and had adverse effects observed by Delpech in 1863. Neuropsychiatric abnormalities were described 13 years later by Eulenberg in workers in the rubber and viscose rayon industries.[2] A Finnish psychologist, Helen Hanninen tested 100 carbon disulfide exposed workers in 1970, 50 were poisoned, 50 exposed and compared them to 50 unexposed.[10] She found intelligence, tasks of attention, motor skill vigilance and memory were impaired in clinically poisoned and exposed men compared to unexposed. Digit symbol substitution from the Wechsler's scale[5] showed the most effect of exposure. Additional studies of spray painters in the 1970's and compared to computer augmented tomography (CT) scans of the brain and function tests. Symptomatic painters after 20 years or more of exposure had brain atrophy associated with impairment.[11-14]

The key to progress in this field was sensitive tests to measure brain function, Table 20.9.3. Fortunately, the Finnish, Danish and Swedish occupational-environmental health centers units included cooperative neurologists, neurophysiologists and psychologists who did not defend disciplines to limit activities. The obvious reality that the nervous system regulates and controls many essential functions helped select measurements to assess vision, hearing, vibration, odor perception, balance, reaction time including automatic responses that are measured as blink reflex latency,[15,16] heart rate variation[17] and peripheral nerve conduction and Hoffmann's (H) reflex.[18] Tests must be sensitive and reliable, easily understood and economical of time, taking 3 to 4 hours with rest periods.

Sensitivity's main dimensions are time and mapping. For example, balance is measured using the classic Romberg stance (1850) standing feet together with eyes ahead open and then closed and using a force (displacement) platform or even simpler the position of the head from a sound emitter secured to a headband and recorded by two microphones to inscribe the distance swayed and the speed, in centimeters per second.[19] From three performances for 20 seconds with the eyes closed alternating with the eyes open the minimal speed of sway is selected. The inscribed path, the map, may provide more information but how to interpret this is unclear.

Eye-hand choice reaction time is tested as the speed to cancel by tapping a keypad, a 4 inch letter that appears on the screen of a laptop computer.[20] Twenty trials repeated twice and the median time of last 7 trials in each run is recorded. Simple, same letter, reaction time takes 1/4 of a second, 250 ms while choice between 2 letters takes twice as long, 500 ms. Many tests are faster in women, most deteriorate with aging after 25 years and for people with more years of educational attainment scores are higher.[21]

Vision is measured by mapping for color perception which is a central retinal cone function. This consists of placing 15 pale colors in a spectral array, the Lanthony desaturated hue test. Retinal rod function which is light perception was mapped for the central 30° of each visual field at 80 points using an automated perimeter recording to a laptop computer.[22] This standardized and speeded up the fields that had been done by the tangent screen and a skilled operator for 100 years.

It was logical to consider the 12 nerves of the head, cranial nerves as the scaffold for organizing tests and for reviewing brain functions that are adversely affected by chemicals. Smell (Nerve I, olfactory), is tested by recognition of familiar odors and of threshold concentration for detecting them. Smell disorders include loss and disturbed perception. Nerve II, the optic was described above. Nerves III, IV and VI move the eyes and rarely show effects of chemicals. An exception is the optokinetic effects of styrene. In contrast the faces sensory nerve, the trigeminal, number V and motor nerve, the facial, number VII are needed to blink and are tested by blink reflex latency which is measured electromyographically in milliseconds (10 to 15 ms) after stimulation by a tap, that is mechanical or an electrical impulse. Blink is slowed by exposures to chlorinated solvents like trichloroethylene (TCE), by chlorine and by arsenic. Nerve VIII has hearing and vestibular (balance) divisions which are tested by audiometry and by sway speed for balance. Nerve IX, the glossopharyngeal innervates the throat and is needed for the gag reflex and baroreceptor. Nerve X, the vagus X is evaluated by recording variations of heart rate with breathing. Nerves XI, spinal accessory is tested as strength of neck muscles and XII tongue's hypoglossal nerve by speech.

Using these tests implies comparing scores observed to a standard, an expected value. Ideally that would be to the same subject which is rarely possible, although it works for before and after exposures of workers. The next best comparisons are to suitable unexposed normal subjects who can be called controls.[21] We developed over several years a national sample of unexposed people, tested their performance and calculated expected values using prediction equations with coefficients for age, sex, education and other factors such as height and weight that affected some tests. Thus individual observed values for each subject are compared to predicted values (observed/predicted x100) equal percent predicted. Frequently, we needed to be sure that comparison groups of apparently unexposed control people were normal because adverse effects are widespread. Next from the standard deviations of the mean, each tests confidence intervals were developed that included 95% of values, excluded as abnormal approximately 5% of unexposed subjects on each test that defined abnormal precisely. For these tests they were values outside the mean plus 1.5 x standard deviations (sd) that defined normal. The next concern what was the best summary for each subject. The number of abnormalities was best adjusting balance and vision above other tests and given grip strength, blink reflex and color discrimination 0.5 for right and left sides of the body.

The attributes of plausible association leading to attribution of effect include temporal order, strength of association, exposure intensity and duration, specificity, consistency of findings and coherence and plausibility.[1] As noted earlier the fact of exposure or suspected exposure may be the only certainty about exposure so its plausibility is important based on chemical properties, experiments and studies of workers. Consistency with results of occupational exposures and animal experiments is helpful. Koch's 4 postulates developed to judge causation of infectious agents (1, organism present in every case; 2, grown in pure culture; 3, produces the disease when inoculated and 4, recovery and growth in pure culture) are usually inapplicable. This reality is discomforting to some interpreters of the new observations.

The next section reviews the neurobehavioral affects of solvents found in the environment in the order of importance.[2,4] We begin with trichloroethylene (TCE) and related short chain chlorinated agents.[22] Next are ring compounds toluene including related xylene and styrene with comments on creosols or phenols. The chlorinated ring compounds follow: dichlorophenol and polychlorinated biphenyls and their highly neurotoxic derivatives, the dibenzofurans. Other straight chain solvents leading off with n-hexane move through white solvent (paint thinner) and solvent mixtures.

Before studies of effects of TCE on many brain functions came the measurement of blink reflex latency in 22 people exposed at home to solvents rich in TCE at Woburn, MA. They showed significant delay of blink but no other functions were measured.[16] In France about this time workers exposed to TCE had similar delays of blink.[24] Earlier experimental exposure of 12 subjects to TCE at 1,000 parts per million (ppm) for 2 hours in a chamber had produced rapid flickering eye movements when following figures on a rotating drum (optokinetic nystagmus), a lowered fusion limit.[22] Thus TCE induced dysfunction of several cranial nerves VI (with III, IV) and V and VII. Nystagmus normalized after a washout and recovery time. Blink is the easier and quicker measurement.[16]

A community within Tucson, AZ of over 10,000 people who depended on well-water for drinking and bathing had developed many complaints and had excesses of birth defects and cancers that associated with TCE in their water. The source was metal cleaning that included stripping off protective plastic coatings, from demothballing aircraft stored on the desert with TCE. This had dumped vast quantities of TCE on the porous, desert floor that drained into the shallow Santa Cruz River aquifer. Testing of 544 people from this water exposure zone showed increased blink reflex latency, impaired balance, slowed simple and choice reaction times, reduced recall, poor color discrimination, and impaired problem solving in making designs with blocks, digit symbol substitution and Culture Fair (consisting of 4 subtests: selection of designs for serial order, for difference, for pattern completion and refining defined relationships).[25,26] Also peg placement in a slotted board and trail making A (connecting 25 numbers in ascending sequence and B connecting numbers alternating with letters).[27] TCE concentrations at the well heads and distribution pipes to homes were measured to calculate with duration of exposure, lifetime peak levels, lifetime averages, and cumulative exposure. Possible relationships of neurobehavioral test scores to these surrogates for dose were searched by regression analysis. No relationships were found for these dose surrogates which was disappointing.

TCE dominated the mixture of chlorinated aliphatic solvents in air and water of northeast Phoenix around two Motorola microchip manufacturing plants that began production in 1957. Neighbor's complaints of adverse health effects started amelioration effects in 1983. An underwater TCE solvent plume spread west and south of the plants in the Salt River aquifer. Test wells showed concentrations of TCE from 50 ppm to 1.4%. Also air dispersal was important for direct exposure as TCE escaped into the air from the plant. It drained into dry wells, sewers and into a canal running northwest through the neighborhood. In 1993, 236 exposed adults were compared to 161 unexposed ones from a town 80 km northwest across the mountains at a higher elevation. The exposed group showed delayed blink reflex, faster sway speed, slowed reaction time, impaired color discrimination and reduced cognitive function and perceptual motor speed and reduced recall. Airway obstruction was shown by pulmonary function testing. Adverse mood state scores and frequencies of 32 symptoms were also increased.[4,28]

Remedial efforts directed at dumping and ground water had not reduced the effect suggesting either these were ineffective or impairment was permanent and had developed after 1983. Additional groups of subjects on plume but not in the lawsuit were not different from clients so there was no client bias. Phoenix residents off the plume had only abnormal slowing of blink interpreted as due to TCE and abnormal airway obstruction compared to the unexposed population of Wickenburg, AZ. Airway obstruction was attributed to Phoenix wide air pollution. Proximity within 1.6 km seemed to increase impairment.

In 1998 retesting of 26 people from original groups showed improvement with faster blink reflexes but worse airway obstruction that had persisted (ref. 4 and unpublished). The improvement to normal in blink paralleled that seen in chlorine exposed people 3 years after exposure and first evaluations that were abnormal.[4] The perceptual motor tests, trail making A and B and peg placement were improved, as were cognitive function measured as Culture Fair and verbal recall. We deduced that diminished TCE releases from Motorola after 1993 allowed recovery of cranial nerves V and VII so blink latency decreased, accompanied by some improvement in vigilance and tracking for the better scores. A possible reversal of effect is so important that these observations should be verified in other groups.

Workers welding and grinding on jet engines in a repair shop were an unusual way to focus attention on the Gerber-Wellington aquifer in Oklahoma.[28] Our attention was temporarily on metals in alloys but when testing of 154 workers showed impaired balance, slowed choice reaction times and impaired color discrimination compared to 112 unexposed subjects, the priority became effects on the brain. These worker's cognitive function, perceptual motor and recall were all abnormal using the same tests as in Phoenix. We probed for their exposures after observing these effects and found that these workers had used TCE, trichloroethane, methanol and Freon FC-113 in metal cleaning. The 112 control subjects had not worked with solvents or TCE, but many people, both workers and control groups, lived on and drew well water from the Gerber-Wellington aquifer that is contaminated with TCE. Mapping the blink reflex latencies in the control group showed the people with normal blink lived outside the aquifer. The aquifer was TCE contaminated (national priority list 1990) Thus here we had an example of a probable environmental exposure to TCE with a superimposed occupational one.[4]

TCE leakage had produced these same effects in Joplin, MO neighbors of a company manufacturing ball bearings and cleaning with this solvent. The companies decision to clean and reuse TCE rather than dump it ameliorated the effects.[4] In San Gabriel and San Fernando Valleys in California similar observations confirm adverse effects from TCE contamination of groundwater and air. Clearly the observations were replicated and each time TCE was associated with reasonable timing and proximity. More than half of the federal superfund sites in the US are contaminated with TCE suggesting 800 potential replications.[29] Experience with several patients have shown me that the effects of dichloroethylene and of 1,1,1-trichloroethane are indistinguishable from those of TCE.

Toluene is the most toxic and best studied of the aromatic ring compound solvents. Both acute and chronic effects were observed by 1961 from inhaling "huffing" toluene[30,31] or lacquer thinner,[32] especially in children sniffing airplane glue.[30] Chronic impairment was shown shoemakers[33] and rotogravure printing workers[34] using neurobehavioral testing. Toluene exposed experimental animals, mainly rats and mice showed enhanced motor activity, abnormal movements, altered sleep patterns and electroencephalograph (EEG) changes from an integrative brain loop, the hippocampus.[2] Occupational exposures produced memory disturbances, poorer performance on block design assembly, embedded figures, visual memory and eye-hand coordination. CT scans showed some generalized brain swelling[34] that correlated with impaired psychological functions.[35] Women working in electronic assembly had environmental air levels of toluene of 88 ppm compared to 13 ppm for controls and comparable differences in blood levels. These workers were less apt at placing pegs in a grooved board, at trail making, digit symbol, visual retention and reproduction and verbal memory.[36] They were tested during the day after being away from exposure for at least 16 hours.

Protracted sniffing of solvents alone or in glue has produced intention tremor and titubating gait[30] consistent with cerebellar degeneration which continued after 5 years with ataxia, EEG slowing and cerebral atrophy.[31] Polyneuropathy was observed in 2 glue sniffers in Japan whose exposures were to n-hexane and toluene.[37] Many such descriptions[2] outweigh one epidemiological study that found no differences in performance when comparing 12 glue sniffing boys, ages 11 to 15, mean 13.8 years and 21 controls, ages 11 to 15, mean 12.6 years. Four non-standard tests and the Benton visual retention and design reproduction test were used but the exposed group was 1.2 years older and should have outperformed younger controls whose skills were less developed.[38]

Some published data are difficult to interpret. For example, 26 men were exposed in tanks and holds of two merchant vessels being painted (solvents) and sprayed with malathion 20% and pyrethrin 1.5%, with piperonyl butoxide in toluene. They showed losses of concentration, unawareness of danger and unconsciousness at toluene levels estimated as 10,000 to 12,000 ppm and up to 30,000 ppm below waist level.[39] Additive effects of the neurotoxic insecticides were not discussed.

Effects of toluene in 52 men and paint solvents in 44 men were contrasted with unexposed men. Painters had impaired reading scores, trails B, visual search, block design, grooved pegboard, simple reaction time and verbal memory.[40] Toluene exposed men had only abnormal reading scores reduced significantly, although scores on all tests were lower.

Levels of toluene were less than 200 ppm for 4 years prior to this study, although above 500 ppm earlier.

The axiom that environmental exposures are "never" to pure chemicals is matched by another that the mixtures are frequently so complex as to defy description. The few observations suggest that effects seen from mixtures may be due to one or two specific neurotoxic agents. Judgment must be exercised to curb bias and accept the most plausible attribution as the above studies illustrate.

Studies of a population exposed in Louisiana to toluene rich solvents and other chemicals distilled from a site for 17 years were contrasted to unexposed people living 55 km to the east.[41] The Combustion site accepted 9 million gallons of used motor oil in 1975-1976 and 3 to 4 million gallons from 177 to 1983. Tons of liquid chemical waste from over 100 chemical factories was consigned to this site including toluene, xylene, styrene and benzene, many chlorinated aliphatics solvents like TCE and chlorinated aromatics including PCBs and dibenzofurans. Lead, cadmium, mercury and other metals were present in samples of sludge in ponds after the site closed in 1983 but they were rich in toluene, benzene and other aromatics. Modeling based on toluene and benzene and using standard Environmental Protection Agency assumptions and a windrose showed symmetrical spread eastward. Excesses of leukemia in school children, cancers and neurobehavioral symptoms in the about 5,000 neighbors of the site led to neurobehavioral testing for impairment in 131 adult subjects within 2 km of the site and 66 adult controls from voter registration rolls of a town 50 km east. The exposed group matched controls for age but were 1.4 years less educated.

There were adverse effects from exposures while living within 2 km from the site for 4 to 17 years.[41] that were shown by slowed simple and choice reaction times and abnormal sway speeds. Cognitive function in Culture Fair and block design was decreased and pegboard and trail making A and B were diminished, as was recall of stories. Profile of mood states (POMS) scores were 2.5 fold increased with low vigor and high depression, tension, confusion, anger and fatigue. Thirty of 32 symptoms inquiring about chest complaints, irritation, nausea and appetite associated, balance, mood, sleep, memory and limbic brain were significantly more frequent in exposed people and the other two were rare in both groups. When differences were adjusted for age, color discrimination and similarities became abnormal were added and trail making A became normal.

The second study was designed to answer how large an area-population was affected, was direction important and were abnormalities related to the duration of exposure.[42] I examined 408 subjects selected to fill 3 distances outward to 1.6, 3.2 and 4.8 km in 8 compass octants, thus 24 sectors. The same tests were given by the same staff and results replicated the earlier study. Regression analysis of each test against distance showed no significant coefficient and comparison of inner and other sectors found no differences, thus there was no evidence of a diminished effect from distance. There were no effects of direction. A possible lessening of effect for durations of exposure of less the 3 years was seen only for peg placement and trail making B scores. Distance, direction and duration as surrogates for exposure did not influence impairment as measured.

We concluded that the periphery of effect was beyond 4.8 km meaning a health impact area larger than 75 $km^2$. There was no gradient of effect from the distilling plant outward suggesting airborne spread and mixing had produced even dosing from a large "cloud".[4] Peoples migration inside the exposure zone did not influence effects. Bias of examiners was unlikely. Was the control group suitable? Their average measurements and the distribution were like three other groups in different parts of the country. The possibility of confounding exposures was considered from two sites that were beyond 4.8 km and to the south. Unfortunately, the resources were unavailable to extend testing beyond 4.8 km to find the rim of Combustion's effect on people and detect effects of other nearby sites.

Xylene is a solvent for paints, lacquers and adhesives and is a component of gasoline. In human volunteers in exposure chambers xylene at 70 ppm for 2 hours had no effect on reaction time or recall memory but levels of 100 to 400 ppm for 2 hours impaired body balance, memory span, critical flicker fusion and cause eye irritation.[43-45] Alcohol and 1,1,1-trichloroethane had adverse effects on balance that show synergism with xylene.[46] and increased the latencies for visual and auditory evoked potentials.[47] Occupational studies have focused on psychiatric symptoms in photogravure workers who also showed headache, nausea, vomiting and dizziness.[48] Only one study showed impairment for recall memory attributed to xylene but workers were also exposed to formaldehyde.[49]

Xylene toxicity has received less study than that of toluene, but appears considerable less which supports attributed the neurotoxicity to toluene of mixtures of xylene, benzene and toluene with straight chain hydrocarbons such as gasoline.

Styrene's major use is in reinforced fiberglass plastics in constructing boats and bathtubs and showers and in styrene-butadiene rubber.[2] Small amounts are used in polystyrene foam cups and packing materials. Styrene inhalation increased locomotion activity in rats and grip strength at the highest 700 to 1,400 ppm concentrations.[2] Studies of workers showed hearing loss (increased high frequency hearing thresholds at 16 kHz).[50] Color discrimination is also reduced.[51] Other observers found abnormal hearing and by posturography-larger sway areas and poor rotary visual suppression-inhibition or vestibulatory nystagmus.[52] In 25 studies of workers[2] some showed slowing of reaction time, poor performance on block design, short-term memory, EEG abnormalities and neuropathy.

These relatively mild effects made me predict less than the severe impairment than observed in 4 women from a factory making styrene-fiberglass shower-bathtubs. Two sprayed styrene and the other 2 who had developed skin and airway symptoms on initial exposure did lay-up and assembly. Five weeks after her first exposure one woman became lightheaded and dizzy, felt hot and her vision blacked out. On testing reaction times were slow, sway speed was increased. Problem solving was impaired as was verbal recall and POMS scores were elevated. She left work stopping exposure. Ten days later, on a trip to the mountains 4,000 feet above sea level she collapsed and became unconscious. Retesting showed constricted visual fields and worse performance of the above tests. Testing on the second woman who had developed asthma showed multiple blind spots in her visual fields, diminished problem solving ability, grip strength, excessive fingertip writing errors and failure to recall stories after 30 minutes. A third woman also had asthma and severe airway obstruction showed abnormal balance with eyes open and closed, diminished hearing, bilaterally constricted visual fields and decreased vibration sense.

The fourth woman had a skin rash and red welts that had kept her away from direct contact with epoxy and styrene. She had abnormal color discrimination, decreased vibration sensation, a blind spot in the retina of the left eye and decreased recall of stories. She was the least impaired although her POMS score and symptom frequencies were increased.

Air sampling was not permitted, concentrations of styrene are unknown and contributions of other chemicals to this exposure cannot be excluded. However, exposures to formaldehyde and phenol are unlikely as these workers did not "lay-up" fiberglass resin. Inhalation of sprayed styrene is the most attribution for the neurobehavioral impairments. The impairment exceeded that found in a review of boat building and other studies but tests were more sensitive and the styrene levels may have been higher. We encourage more neurobehavioral evaluations of styrene spraying workers using such sensitive tests.

Polychlorinated biphenyls (PCBs), the ultimate (poly)chlorinated solvents are 2 membered ring compounds that when heated to 270° produce dibenzofurans (DBFs) that are 1,000 or more times as neurotoxic.[53] Initial evaluations were of a few PCB exposed individuals and 14 firemen exposed to DBFs who showed severe impairment measured after a medical schools power plant transformers cooked and exploded. Most of the firemen could not pass the physical, balance and truck driving requirements to return to duty and were retired on disability.[4,54]

A community study explores effects of environmental exposures. PCBs were used as pump lubricants in natural gas pipelines running north from Texas and Louisiana from 1950 to the middle 1970's. One pumping station was at Lobelville, TN and at least 16 other US communities had them.[4] Ninety-eight adult village dwellers were compared to 58 unexposed subjects from 80 km east or 35 km north. The exposed people were the most abnormal group I have studied. They had abnormal simple and choice reaction times, balance, hearing, grip strength and the visual function of color discrimination, contract sensitivity and visual field performance. The cognitive functions of Culture Fair, digit symbol were abnormal as were vocabulary, information, picture completion and similarities. Story recall was diminished and peg placement and trail making A and B and fingertip number writing errors were decreased. Other possible associations were ruled out and there were no other causes of impairment. This exposure had caused the most severe neurobehavioral impairment for these people that I have observed.[4] It exceeded that from distilling chemical waste rich in toluene, from TCE and from other solvents.

n-hexane by inhalational or through the skin causes peripheral nerves to die-back. Glue sniffing exposure frequently combines n-hexane and toluene. Twenty-five percent of workers using glue in shoes and leather goods with n-hexane, 40 to 99.5% had symptomatic polyneuropathy, slowed nerve conduction and neurological signs.[55] Abnormal findings increased with age and durations of exposure and were accompanied by lower limb weakness and pain, abnormal sensations (parenthesis) in the hands and muscles spasm. In another shoe plant exposure group upper extremity nerve conduction was slowed, frequently after 5 years of exposure.[56] Sensormotor distal neuropathy characterized 98 of 654 workers in the Italian shoe industry, 47 had decreased motor conduction velocity with headache, insomnia, nausea and vomiting irritability and epigastric pain.[57] Most workers improved when removed from exposure.[56] In Japan beginning in 1964 several studies found polyneuropathy in polyethylene laminating printers[58] and makers of sandals and slippers.[59] A major metabolite of n-hexane and of methyl butyl ketone is 2,5-hexanedione that is more neurotoxic than these precursors causing swelling of nerve axons and accumulations of neurofilaments in mid-portions of peripheral nerves.[60] Methyl ethyl ketone studied in workers lengthened choice reaction time and motor nerve conduction and decreased vibration sensation signs of neuropathy. These effects were also seen in glue sniffers.[2]

The focus on polyneuropathy in 14 studies[2] has usurped studies of central nervous system (CNS) functions except for one showing increased latencies of visual and auditory evoked potentials.[61] Neurophysiological and psychological assessments[62] show narcotic effects that match those in animals.[2] Until restricted from foods n-hexane was used to extract oil from soybean meal and exposed US workers. They had headache, dysesthesia, insomnia, somnolence and memory loss but testing for appraise brain damage was not done.

Gasoline is a mixture of aliphatic straight and branched chains and aromatic hydrocarbons with toxicity attributable to toluene, xylene and perhaps hexane and additives including methyl ter butyl ether and tri-orthocresyl phosphate.[63]

Effects of the lungs of inhaled solvents simplify to the consideration of agents affecting airway cells that include n-hexane and PCBs. Both cause proliferation and transformation of distal airway lining cells to produce mucus and obstruct airways[64-66] and cause inflammatory cells to pour into the lungs distal alveolar spaces interfering with for gas exchange.

Cardiac effects of solvents are of three types A, alterations in rhythm B, cardiomyopathy and C, hypertension directly and via renal changes as in interstitial nephritis, glomerulonephritis and Goodpasture's syndrome.[67]

Alterations in heart rhythm have been attributed to anesthesia with TCE and were serious, especially when administered in soda lime $CO_2$ absorbing anesthesia machines to stop the use of TCE for anesthesia in the 1960's.[4] Rhythm disturbances have also been observed in some groups of workers exposed to TCE. Knowing this we did electrocardiograms (ECG's) on the Tucson TCE exposed population and found no arrhythmias. A loss of respiratory variation in heart rate has been associated with exposure to organic solvents, including carbon disulfide, acrylamide and alcohol but not toluene and with diabetes mellitus and syndromes of autonomic nervous system dysfunction.[17,68,69] Later freons, volatile chlorofluorocarbons were associated with arrhythmias and withdrawn from use to propel therapeutic aerosols used for asthma.[70] Cardiomyopathy, heart muscle dysfunction and enlargement have been associated with alcohol ingestion. Two epidemics were ascribed to cobalt used to color beer.[9] But in the most common cardiac muscle disorder from alcohol, cobalt is not incriminated.

Hypertension has been associated with solvent exposure in workers, an association that needs further study. Associations with hypertension were absent in the authors studies of TCE, toluene rich waste and PCBs discussed earlier.

Workers who used methylene chloride in making acetate film had sleepiness and fatigue and decreased digit symbol substitution scores and lengthened reaction time.[71] Use of methylene chloride in closed spaces has been fatal with brain edema, elevated blood levels of carboxyhemoglobin and caused temporary right hemispheric paralysis and/or unconsciousness.[2] Chronic exposure has been associated with dementia, headache, dizziness and disturbed gait.[72]

Chloromethane exposures from foam production caused tremor and decreased attention and ability to do arithmetic.[73] Environmental exposures from leaks in refrigerating systems[2] caused deaths, convulsions, myoclonus and personality changes. One fishing boat exposure of 15 men left profound neurological residuals, fatigue, depression and alcohol intolerance.[2] The effects resemble those of methyl bromide poisoning.

Methanol has profound and specific toxic effects on the optic nerve and vision causing central blind spots and ingestion of methanol for the intoxicating effects of ethanol has caused blindness.[74] Its metabolism to formic acid suggests the possibility that of other central nervous system effects.

White spirit is a mixture of straight and branched chain paraffins, naphthalenes and alkyl aromatic hydrocarbons is used widely as a paint solvent. Ten studies of painters, mostly in Nordic countries, have shown increased neurobehavioral symptoms and several showed decreased performance on psychological tests.[11-14] Longitudinal studies showed an almost doubled risk for neuropsychiatric disability pension in painters compared to construction workers. Several such studies support the concept of neuropsychiatric impairment and disability linked to the painting trade in many countries.[75-78] Several women in my consulting practice had profound neurobehavioral impairment after entering their homes during spray painting including unconsciousness which suggest there may be a considerable problem from environmental exposures.

Many industrial painters exposures are to solvent mixtures. Those painting airplanes where dust and hence fume exposure is limited by strict cleanliness, which means good air hygiene for the workers, have little trouble compared to symptoms, impairment, disability and brain atrophy with dementia in car and refrigerator painters. These groups supplied the clear evidence of solvent effects in workers in Nordic countries that established how to assess human subject's neurobehavioral status and detect impairment that were discussed early in this chapter. Many cross sectional studies showed adverse effects, excessive neuropsychiatric symptoms and several longitudinal studies show greatly increased likelihood of receiving a pension for neuropsychiatric disability.[2,78-80]

Chemical companies fight the concept that chemicals damage human subjects. They are more combative and better defended than are bacteria and other infectious agents. In the past 25 years companies learned from asbestos litigation, the bankruptcy of Johns Manville Company and the banning of asbestos to contest observations and their scientific basis and frequently hire scientists to support their position of null effects-not harm and sponsor environmental meeting and advertise their concern and sense of responsibility. They avoid or shift responsibility for damage to the victim or community and the social security system. The necessary banning of PCBs and chlordane enforced their strategy of "controversy" even about incontrovertible facts. Perhaps, they count on having the 50 years that tobacco companies enjoyed before having to accept responsibility for adverse effects of tobacco smoking.

## REFERENCES


1 K Kilburn, and R Warshaw, **Epidemiology of adverse health effect from environmental chemicals**, *Princeton Scientific Publishing Co.*, 1995, pp 33-53.
2 P Arlien-Soberg, **Solvent Neurotoxicity**, *CRC Press*, Boca Raton, 1992.
3 J Angerer, *Scan. J. Work. Environ. Hlth.*, **11** (Suppl 1),45(1985).
4 K Kilburn, **Chemical Brain Injury**, *John Wiley & Sons*, New York, 1998.
5 D Wechsler, **Adult Intelligence Scale Manual**, (revised), *The Psychological Corporation*, New York, 1981.
6 AR Luria, **Higher Cortical Function in Man**, *Travistock*, London, 1966.
7 W Halstead, **Brain and Intelligence**, *The University of Chicago Press*, Chicago, 1947.
8 R Reitan, *Percept. Motor Skills,* **8**, 271(1958).
9 D Hunter, **Diseases of Occupations**, 4th Edition, *Little Brown*, Boston, 1969.
10 H Hanninen, *Brit. J. Indust. Med.*, **28**, 374(1971).
11 B Knave, B Kolmodin-Hedman, H Persson, and J Goldberg, *Work Environ, Health*, **11**, 49(1974).
12 S Elofsson, F Gamberale, T Hindmarsh, et al., *Scand. J. Work Environ. Health*, **6**, 239(1980).
13 O Axelson, M Hane, and C Hogstedt, *Scand. J. Work Environ. Health*, **2**, 14(1976).
14 P Arlien-Soberg, P Bruhm, C Gyldensted, and B Melgaard, *Acta Neurol. Scand.*, **60**, 149(1979).
15 R Feldman, J Chirico-Post, and S Proctor, *Arch. Environ. Health*, **43**, 143(1988).

16 K Kilburn, J Thornton, and B Hanscom, *Electromyograph Clin. Neurophysiol.*, **38**, 25(1998).
17 E Matikainen, and J Juntunen, **Neurobehavioral Methods in Occupational and Environmental Health**. *World Health Organization*, Copenhagen, 1985, pp. 57-60.
18 S Oh, **Clinical Electromyography; nerve conduction studies**, 2nd Edition, *Wilkins and Wilkins*, Baltimore, 1993, pp.89.
19 K Kilburn, and R Warshaw, *Occup. Environ. Med.*, **51**, 381(1994).
20 J Miller, G Cohen, R Warshaw, J Thornton, and K Kilburn, *Am. J. Indust. Med.*, **15**, 687(1989).
21 K Kilburn, J Thornton, and B Hanscom, *Arch. Environ. Health*. **53**, 257(1998).
22 B Kylin, K Axell, H Samuel, and A Lindborg, *Arch. Environ. Health*, **15**, 48(1967).
23 K Kilburn, *Neurotoxicology*, **21**, (2000).
24 L Barrett, S Garrel, V Danel, and J Debru, *Arch. Environ. Health*, **42**, 297(1987).
25 R Cattell, S Feingold, and S Sarason, *J. Educational Psych.*, **32**, 81(1941).
26 R Cattell, *J. Consulting Psych.*, **15**, 154(1951).
27 K Kilburn, and R Warshaw, *J Toxicol. Environ. Health*, **39**, 483(1993).
28 K Kilburn, *Environ. Res.*, **80**, 244(1999).
29 National Research Council, **Public Health and Hazardous Waste, Environ. Epid**., *National Academy Press*, Washington DC, 1991.
30 D Grabski, *Am. J. Psychiatry*, **118**, 461(1961).
31 J Knox, and J Nelson, *New Eng. J. Med.*, **275**, 1494(1966).
32 L Prockop, M Alt, and J Tison, *JAMA*, **229**, 1083(1974).
33 T Matsushita, Y Arimatsu, A Ueda, K Satoh, and S Nomura, *Ind. Health*, **13**, 115(1975).
34 J Juntunen, E Matikaninen, M Antti-Poika, H Suouanta, and M Valle, *Acta Neurol. Scand.*, **72**, 512(1985).
35 H Hanninen, M Antti-Poika, and P Savolainen, *Int. Arch. Occup. Environ. Health*, **59**, 475(1987).
36 S Foo, J Jeyaratnam and D Koh, *Brit. J. Indust. Med.*, **47**, 480(1990).
37 T Shirabe, T Tsuda, A Terao, and S Araki, *J. Neurol. Sci.*, **21**, 101(1974).
38 J Dodds, and S Santostefano, *J. Pediatr.*, **64**, 565(1964).
39 E Longley, A Jones, R Welch, and Lomaev, *Arch. Environ. Hlth.*, **14**, 481(1967).
40 N Cherry, H Hutchins, T Pace, and H Waldron, *Brit. J. Indust. Med.*, **42**, 291(1985).
41 K Kilburn, and R Warshaw, *Neurotoxicol. Terato.*, **17**, 89(1995).
42 K Kilburn, *Environ. Res.*, **81**, 92(1999).
43 C Carpenter, E Kinkead, D Geary Jr, L Sullivan, and J King, *Toxicol. Appl. Pharmacol.*, **33**, 543(1976).
44 K Savolainen, V Riihimaki, and M Linnoila, *Int. Arch. Occup. Environ. Hlth.*, **44**, 201(1979).
45 K Savolainen, V Riihimaki, Omuona, and J Kekoni, A Laine, *Arch. Toxicol., Suppl.*, **5**, 96(1982).
46 K Savolainen, V Riihimaki, O Muona, J Kekoni, R Luukkonen, and A Laine, *Acta Pharmacol. Toxicol.*, **57**, 67(1985).
47 A Seppalainen, K Savolainen, and T Kovala, *Electroencephalo. Clin. Neurophysiol.*, **51**, 148(1981).
48 D Klaucke, M Johansen, and R Vogt, *Am. J. Induct. Med.*, **3**, 173(1982).
49 K Kilburn, R Warshaw, and J Thornton, *Arch. Environ. Health*, **42**, 117(1987).
50 H Muijser, E Hoogendijk, and J Hooisma, *Toxicol.*, **49**, 331(1988).
51 F Gobba, and A Cavalleri, In: **Butadiene and Styrene: Assessment of Health Hazards**, *IARC Scientific Publications* No. 127, Lyon, France, 1993.
52 C Moller, L Odkvist, B Larsby, R Tham, T Ledin, and L Bergholtz, *Scand. J. Work. Environ. Hlth.*, **16**, 189(1990).
53 O Hutzinger, G Choudhry, B Chittim, and L Johnston, *Environ. Health Perspect.*, **60**, 3 (1985).
54 K Kilburn, R Warshaw, and M Shields, *Arch. Environ. Health*, **44**, 345(1989).
55 E Buiatti, S Cecchini, O Ronchi, P Dolara, and G Bulgarelli, *Brit. J. Ind. Med.*, **35**, 168(1978).
56 I Aiello, G Rosati, G Serra, and M Manca, *Acta Neurol.*, **35**, 285(1980).
57 S Passero, N Battistini, R Cioni, F Giannini, C Paradiso, F Battista, F Carboncini, and E Sartorelli, *Ital. J. Neurol. Sci.*, **4**, 463(1983).
58 S Yamada, *Jpn. J. Ind. Health*, **6**, 192(1964).
59 Y Yamamura, *Folia Psychiatr. Neurol. Jpn.*, **23**, 45(1969).
60 P Spencer, and H Schaumburg, *J. Neurol. Neurosurg. Psychiatr.*, **38**, 771(1975).
61 A Seppalainen, C Raitta, and M Huuskonen, *Electroencephalogr. Clin. Neurophysiol.*, **47**, 492(1979).
62 S Sanagi, Y Seki, K Sugimoto, and M Hirata, *Int. Arch. Occup. Environ. Health*, **47**, 69(1980).
63 T Burbacher, *Environ. Hlth. Perspect.*, **101**(Suppl 6),133 (1993).
64 R Warshaw, A Fischbein, J Thornton, A Miller, and I Selikoff, *Ann. NY Acad. Sci.*, **320**, 277(1979).
65 P Houch, D Nebel, and S Milham Jr, *Am. J. Indust. Med.*, **22**, 109(1992).
66 N Shigematsu, S Ishimaru, and R Saito, *Environ. Res.*, **16**, 92(1978).

67 N Benowitz, **Cardiotoxicity in the workplace, Occup. Med.**, Vol 7, No. 3, *Hanley & Belfus, Inc.*, Philadelphia, 1992.
68 Y Kuriowa, Y Shimada, and Y Toyokura, *Neurology*, **33**, 463-467(1983).
69 K Murata, S Araki, K Yokoyama, K Yamashita, F Okajima, and K Nakaaki, *NeuroToxicol.*, **15**, 867-876(1994).
70 W Harris, *Arch. Intern. Med.*, **131**, 1621(1973).
71 N Cherry, H Venables, H Waldron, and G Wells, *Brit. J. Ind. Med.*, **38**, 351(1981).
72 B Friedlander, T Hearne, and S Hall, *J. Occup. Med.*, **20**, 657(1978).
73 J Repko, P Jones, L Garcia, E Schneider, E Roseman, and C Corum, **Behavioral and neurological effects of methyl chloride,** CDC-99-74-20, *National Institute for Occupational Safety and Health*, 1976.
74 W Grant, **Toxicology of the Eye**, *Charles C Thomas Publisher,* Springfield, Illinois, 1974.
75 S Mikkelsen, *Scand. J. Soc. Med.*, (Suppl. **16**), 34,(1980).
76 H Rasmussen, J Olsen, and J Lauritsen, *J. Occup. Med.*, **27**, 561(1985).
77 K Lindstrom, *Scand. J. Work Environ. Health*, **7** (Suppl. 4), 48(1981).
78 J Olsen, and S Sabroe, *Scand. J. Soc. Med.*, **44** (Suppl. 16), 44,(1980).
79 K Lindstrom, H Riihimaki, and K Hanninen, *Scand. J. Work Environ. Health*, **10**, 321(1984).
80 T Riise, and B Moen, *Acta Neurol. Scand.*, **77** (Suppl. 116), 104(1988).
Note: New general source for neurotoxicity for solvents. R Feldman, **Occupational and Environmental Neurotoxicology**, *Leppincott-Raven*, Hagerstown, Maryland, 1999.

# 21

# SUBSTITUTION OF SOLVENTS BY SAFER PRODUCTS AND PROCESSES

## 21.1 SUPERCRITICAL SOLVENTS

AYDIN K. SUNOL AND SERMIN G. SUNOL
**Department of Chemical Engineering**
**University of South Florida, Tampa, FL, USA**

### 21.1.1 INTRODUCTION

Significant and steady inroads towards wider and more effective utilization of supercritical fluids have been made over the last two decades, especially for high value added differenti-

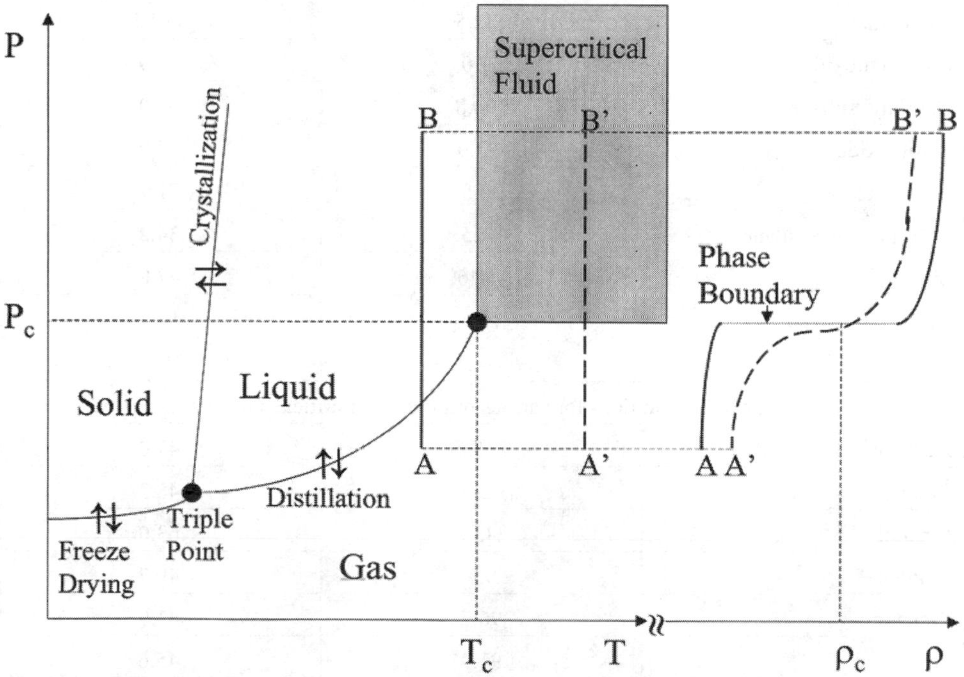


Figure 21.1.1. Pressure-temperature and pressure-density behavior of matter.

ated products. Furthermore, a new wave of second-generation supercritical technologies started to emerge, bringing forth new roles for dense gases. The motivation of this chapter is to assess the current status of the technology in an effort to extricate the challenges both in the current practice of processing with supercritical fluids and potential areas into which we have as yet to venture. The foundations including transport phenomena, reactions and thermodynamics as well as generic processing principles will be highlighted.

"Supercritical" refer to the state of the matter where the temperature and pressure of a single component fluid is above the critical point at which the phase boundaries diminish. A portfolio of chemical and physical operations carried out in the vicinity of this region defines Supercritical Fluid Technology (SFT). The pressure-temperature-volume (PVT) behavior of a substance can be best depicted by pressure-temperature and pressure-density (volume) projections, as shown in Figure 21.1.1. Pressure-temperature diagram identifies the supercritical fluid region, alternative separation techniques that involve phase transition including the associated phase boundaries, and the fact that an isotherm below critical (AB) involves phase transition while one above (A'B') does not. Pressure density projections illustrate tunability of the solvent density at supercritical conditions (A'B') and again the continuity of the isotherm that does not encounter any phase transition. The aforementioned behavior is for a pure component, solvent. The critical properties of various organic and inorganic substances are shown in Table 21.1.1.

**Table 21.1.1. The critical properties of solvents**

| Solvents | Critical temperature, °C | Critical pressure, atm |
|---|---|---|
| Critical conditions for various inorganic supercritical solvents | | |
| Ammonia | 132.5 | 112.5 |
| Carbon dioxide | 31.0 | 72.9 |
| Carbonyl sulfide | 104.8 | 65.0 |
| Nitric oxide | -93.0 | 64.0 |
| Nitrous oxide | 36.5 | 71.7 |
| Chlorotrifluoro silane | 34.5 | 34.2 |
| Silane | -3.46 | 47.8 |
| Xenon | 16.6 | 58.0 |
| Water | 374.1 | 218.3 |
| Critical conditions for various organic supercritical solvents | | |
| Acetone | 235.5 | 47.0 |
| Ethane | 32.3 | 48.2 |
| Ethanol | 243.0 | 63.0 |
| Ethylene | 9.3 | 49.7 |
| Propane | 96.7 | 41.9 |
| Propylene | 91.9 | 45.6 |

| Solvents | Critical temperature, °C | Critical pressure, atm |
|---|---|---|
| Cyclohexane | 280.3 | 40.2 |
| Isopropanol | 235.2 | 47.0 |
| Benzene | 289.0 | 48.3 |
| Toluene | 318.6 | 40.6 |
| p-Xylene | 343.1 | 34.7 |
| Chlorofluoromethane | 28.9 | 38.7 |
| Trichlorofluoromethane | 198.1 | 43.5 |

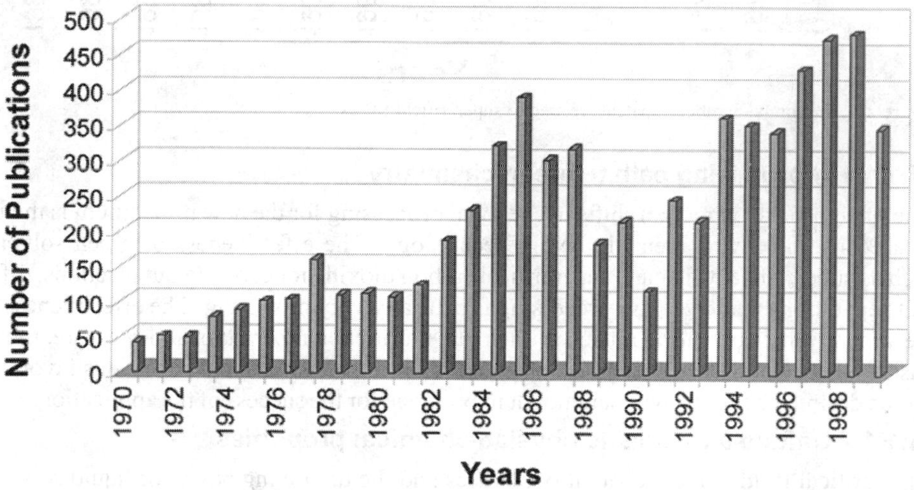


Figure 21.1.2. Annual number of publications related to supercritical fluids.

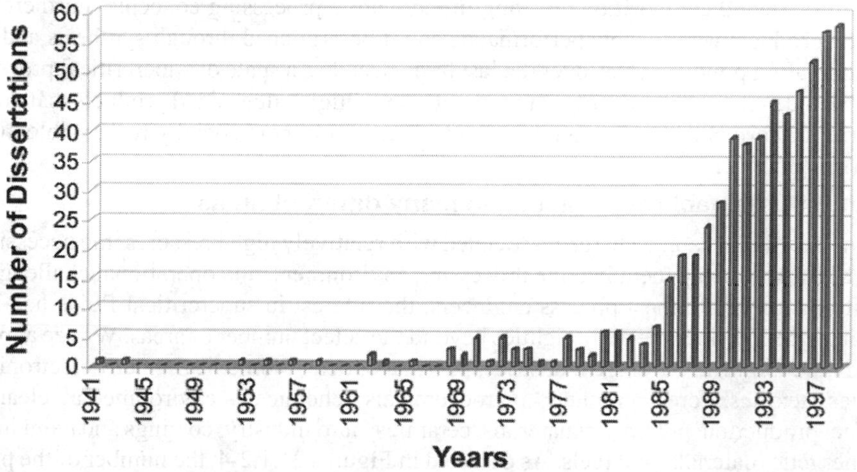


Figure 21.1.3. Annual number of dissertation related to supercritical fluids.

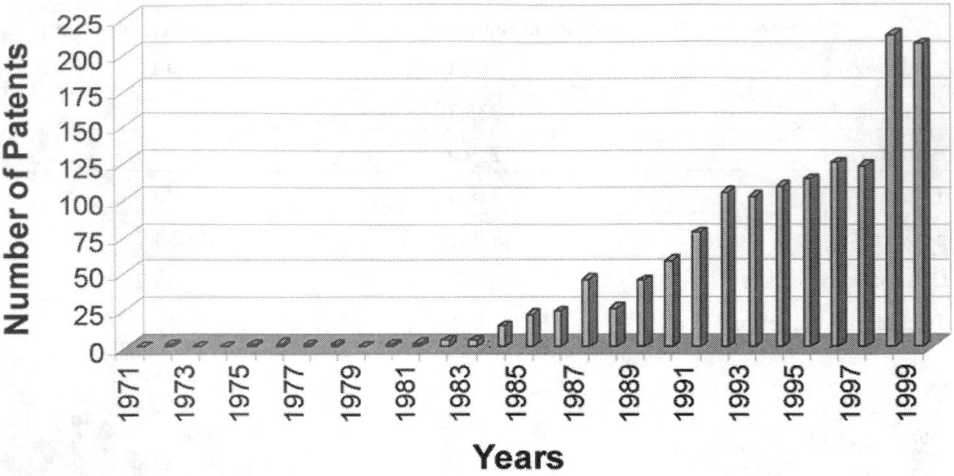

Figure 21.1.4. Annual number of patents related to supercritical fluids.

### 21.1.1.1 A promising path to green chemistry

One of significant paradigm shifts in chemical processing for the new millennium is the increased use of environmentally benign technology. The effectiveness of green solvents such as supercritical water and supercritical carbon dioxide for carrying out reactions, difficult separations, and materials processing is naturally very promising. The effectiveness of the supercritical solvents is related to their state, critical temperature, and pressure. Obviously, synergy between the physical characteristics of solvent and the conditions favorable for the desired chemistry is of paramount importance for the success of the application.

### 21.1.1.2 Unique and tunable physico-chemical properties

Supercritical fluids have the mobility of gases and the dissolving power of liquid solvents resulting in efficient penetration into porous matrices, high mass transfer rates, and high solvency. Furthermore, these properties are extremely sensitive to perturbations in temperature, pressure and composition resulting in innovative processing concepts. Furthermore, tailored products with tunable performance could be produced through synthesis and creative operating policies. Thus, over the last three decades, a spate of supercritical processes have been developed particularly for manufacture of high value added products that are superior in performance and exhibit conscious regard for a more socially responsible manufacturing practice.

### 21.1.1.3 Sustainable applications in many different areas

Despite the higher capital charges associated with relatively high pressures, the necessity to often add a new component into the processing environment, and operational challenges at conditions foreign to most process engineers, the interest in supercritical fluids had been growing steadily since the early eighties beyond the select number of areas. We see applications in the food and beverage industries, pharmaceutical, biomedical, micro-electronic industries, textiles, forest products, petrochemicals, chemicals, environmental clean-up, syn-fuel production, polymeric materials, ceramics, auto industry, coatings and paint industry, energetic materials, and fuels. As depicted in Figures 21.1.2-4, the number of the publications, dissertations, and patents in the area are still growing from one year to next.

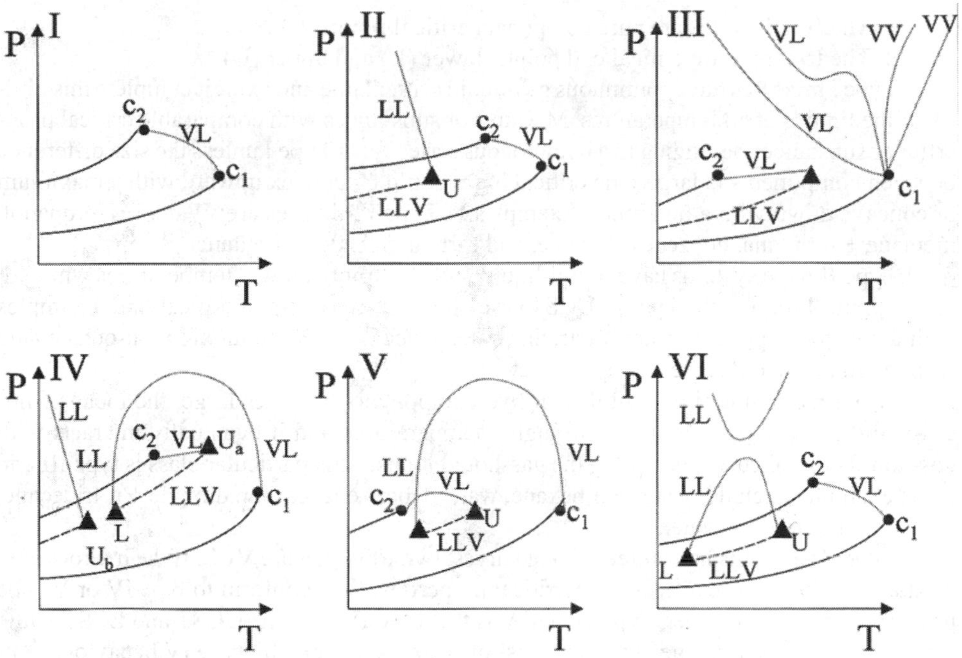


Figure 21.1.5 Classification of vapor-liquid phase behavior of binary systems.

## 21.1.2 FUNDAMENTALS

### 21.1.2.1 Phase behavior with supercritical solvents

The first recorded observation of supercritical fluid's ability to dissolve solids with low volatility goes back to 1879.[1] Since then, better understanding of high-pressure phase equilibrium emerged partially as a result of experiments of the last century, especially those of Timmermans,[2] Schneider,[3,4] Franck[5] on aqueous systems, McHugh[6] on polymeric systems and more recently the Delft group in Holland.[7,8] In classification of phase diagrams, one of the most important analytic contributions is due to Scott and Van Konynenberg.[9] They demonstrated that most of the experimental observations could be described qualitatively by the van der Waals equation of state. Their analysis of critical and three phase boundaries in temperature-pressure projection of the phase behavior led to their categorization of possible fluid phase equilibrium into five classes. Rowlinson and Swinton[10] later added a sixth class that occurs in some aqueous systems but is not predicted by the van der Waals equation of state. The classes described by Scott illustrate only a few of the known types of phase equilibrium and are shown in Figure 21.1.5. This classification scheme illustrates the principal lines (degrees of freedom = 1) and end points (degrees of freedom–0) that form the boundaries of pressure-temperature space of the surfaces that describe the equilibrium between two components. The types of boundaries are:

1. Solid lines (______) are pure component vapor pressure curves.
2. The dashed lined (_ _ _ _ _) are for three phase lines.
3. The dotted lines (.........) are for critical lines.

The types of points are:

1. The filled circles are pure component critical points
2. The triangles are critical end points, lower (L) and upper (U)

Type I mixtures have continuous gas-liquid critical line and exhibit complete miscibility of the liquids at all temperatures. Mixtures of substances with comparable critical properties or substances belonging to a homologous series form Type I unless the size difference between components is large. The critical locus could be convex upward with a maximum or concave down with a minimum. Examples of Type I mixtures are: Water + 1-propanol, methane + n-butane, benzene + toluene, and carbon dioxide + n-butane.

Type II have systems have liquid-liquid immiscibility at lower temperatures while locus of liquid-liquid critical point (UCST) is distinct from gas liquid critical line. Examples include: water + phenol, water + tetralin, water + decalin, carbon dioxide + n-octane, and carbon dioxide + n-decane.

When the mutual immiscibility of two components becomes large, the locus of liquid-liquid critical solution moves to higher temperatures and it eventually interacts with gas-liquid critical curve disrupting the gas-liquid locus. This particular class is type III and some examples include: water + n-hexane, water + benzene, carbon dioxide + n-tridecane, and carbon dioxide + water.

Type IV systems have three critical curves, two of which are VLL. If the hydrocarbon mixtures differ significantly in their critical properties, they conform to type IV or V. The primary difference between Type IV and V is that type IV exhibits UCST and LCST while type V has LCST only. One important class of systems that exhibit type IV behavior is solvent polymer mixtures such as cyclohexane + polystyrene. Other examples of type IV include carbon dioxide + nitrobenzene and methane + n-hexane while ethane with ethanol or 1-propanol or 1-butanol exhibit type V behavior.

Type VI systems are composed of complex molecules with hydrogen bonding or other strong intermolecular forces and result in behavior where LCST and UCST are at temperatures well removed from gas-liquid critical temperature of the more volatile component.

Types IV and I are of particular interest in representing behavior exploited in supercritical extraction. There are many more possible classes and subclasses, especially where azeotropic behavior variations are involved, as discussed in the works of Rowlinson[10] and King.[11] More on the phase behavior applicable to dense gases can be found in reviews[12,13] and specialized texts[14] in this area.

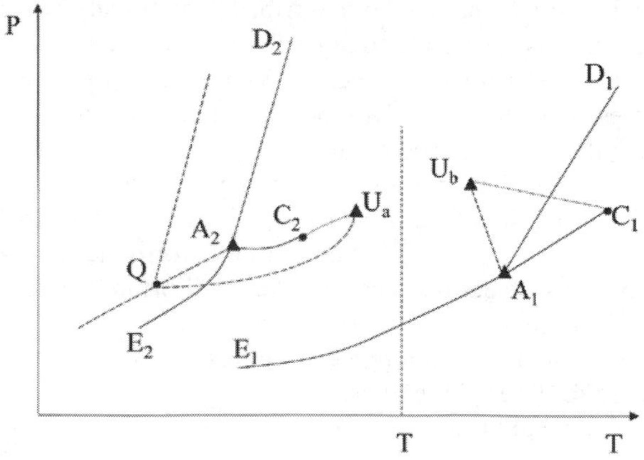


Figure 21.1.6 Phase behavior of dissimilar systems.

Supercritical extraction often involves separation of relatively non-volatile components, often in the solid phase, through selective solubility in the supercritical gasses. Thus, the critical temperatures of the pure components are likely to be significantly different and the critical temperature of the solvent is likely to be lower than the triple point temperature of the solute. The implication is that there is no common temperature range where both

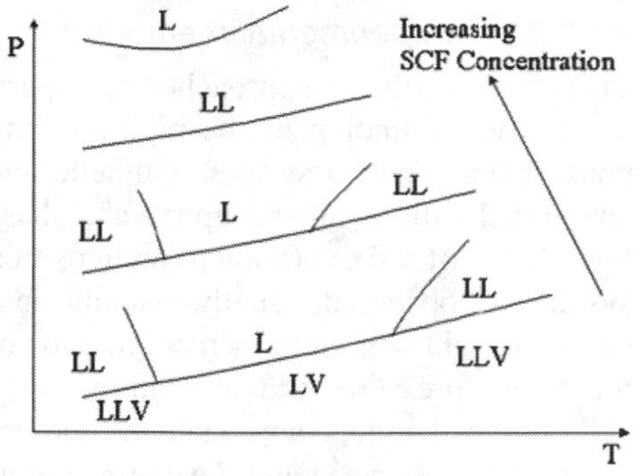


Figure 21.1.7. The effect of supercritical fluid concentration on polymer-solvent phase behavior.

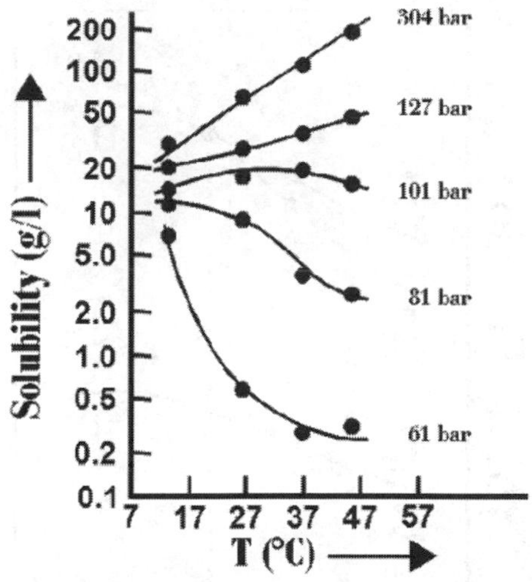


Figure 21.1.8 Solubility of naphthalene in supercritical ethylene.

components will be liquid. The phase diagram of systems with solid and supercritical fluid is shown in Figure 21.1.6.[15] The full lines are the sublimation, melting, and vapor pressure curves of the pure components, $C_1$ and $C_2$. Point Q is the quadruple point, where four phases $S_1$, $S_2$, L, and G are in equilibrium. If the temperature of $C_2$ lies far below of that $A_1$ (triple point of $C_1$), the three-phase region $S_2$, L, G curves upward and intersects the gas liquid line to form critical end points $U_a$ and $U_b$. No liquid phase exits between the temperatures of $U_a$ and $U_b$. The three-sided region $A_1C_1U_b$ is a region of gas-liquid phase separation. At temperatures between this region and $U_a$, the solid phase of component $C_1$ is in equilibrium with $C_2$ rich gas phase.

The polymer solution behavior is of significant importance due to large number of applications in the field. The supercritical fluids primarily follow class four behavior. An example related to the effect of supercritical solvent on phase behavior of such systems is shown in Figure 21.1.7. More comprehensive coverage of the field can be found in texts[6] and review papers.[16]

The unique behaviors of supercritical fluids relate to solubility enhancement and how the solubility varies with operating conditions. As can be seen from Figure 21.1.8, the solubility of a solid solute is enhanced several orders of magnitude over the ideal solubility.[17] Furthermore, the solubility decreases with increasing temperature at supercritical pressures somewhat above the critical, while at supercritical pressures above a "cross-over pressure", solubility of a solute increases with temperature. The crossover pressures are different for different solutes. As depicted in Figure 21.1.9, complete fractionation of solutes solubilized in the supercritical solvent is possible by slight increase in temperature.[18] Another unique behavior, called "barotropic effect", exists when highly compressed dense gas reaches pressures higher than the incompressible liquid. In such instances, dense gas phase is below the liquid or the second gas phase.

There are three essential elements that make up the thermodynamic foundations of supercritical fluids. These elements are experimental and identification techniques for elucidating the phase behavior, models for dense gases, and computation methods.

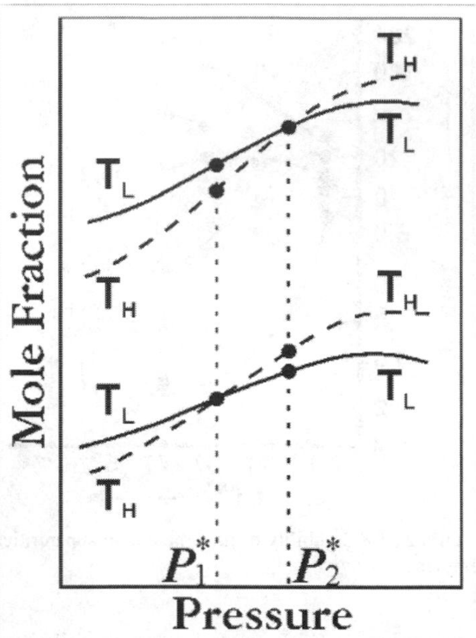


Figure 21.1.9 The crossover pressure and retrograde behavior.

*21.1.2.1.1 Experimental methods*

There are two basic approaches to experimental determination of the high-pressure phase behavior of a system, synthetic and analytic. In the synthetic approach, phase boundaries of a fixed (known) composition system are observed, usually visually, in a cell with sight windows, by manipulation of the system pressure and temperature.

In the analytical approach, the components are equilibrated and the compositions of the co-existing phases are determined either through sampling and off-line analysis or on-line spectroscopic techniques. A typical system for determination of vapor/liquid/liquid phase behavior is shown in Figure 21.1.10. The system is a double re-circulation variable volume system. A dynamic variant of this method is for the determination of the solubility of solids in a supercritical fluid. The stationary solid phase is contacted with a continuous supply of

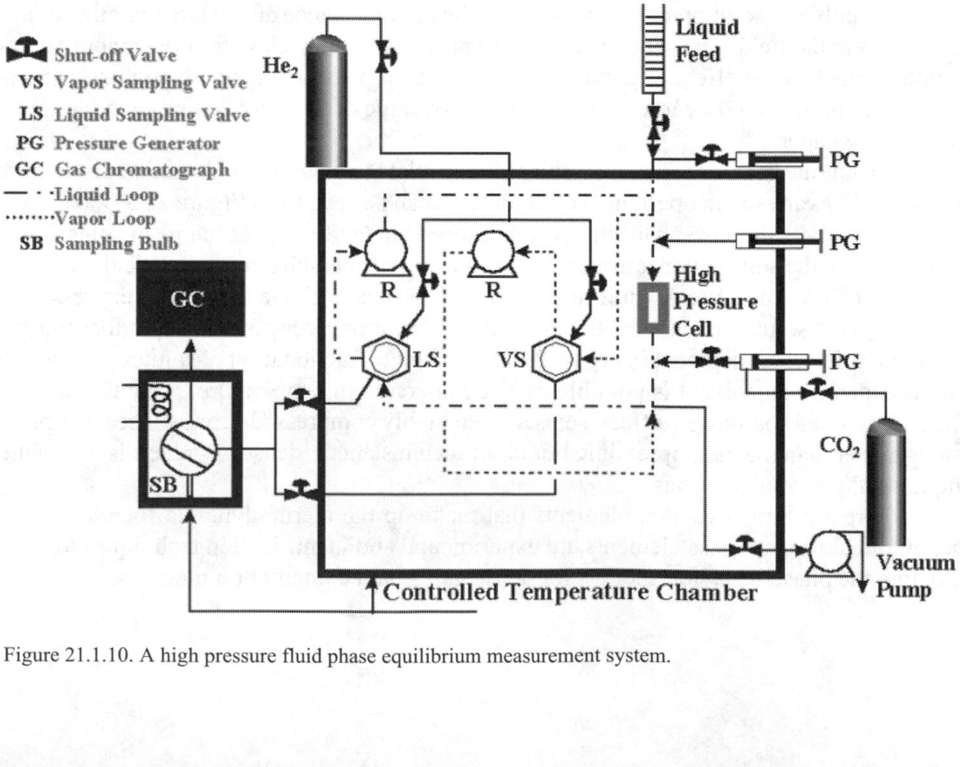


Figure 21.1.10. A high pressure fluid phase equilibrium measurement system.

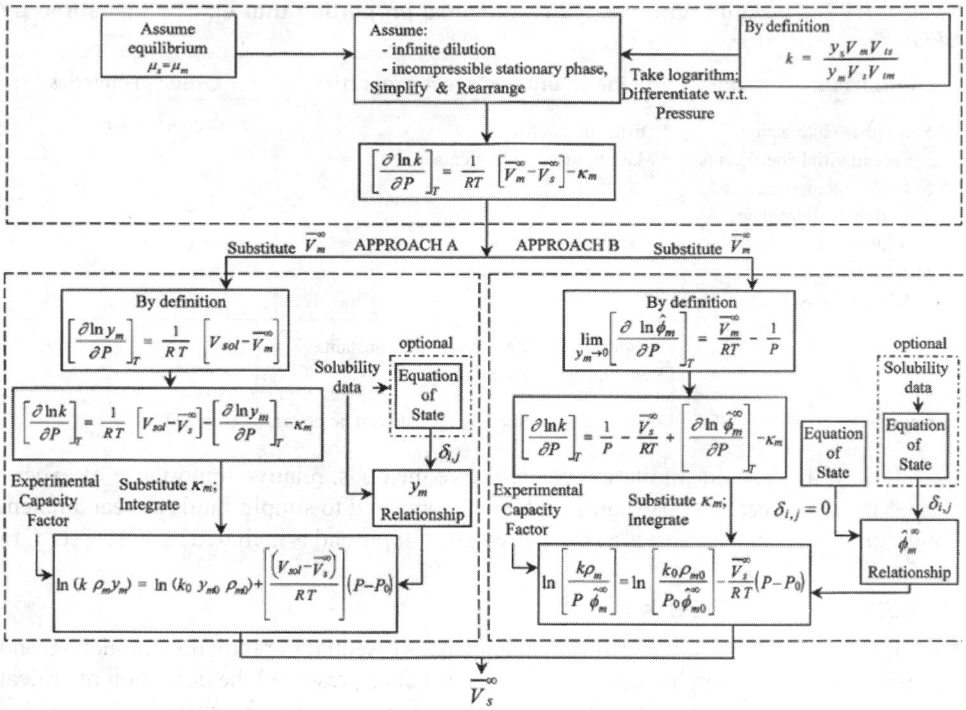


Figure 21.1.11. Estimation of solubility and partial molar volume using supercritical fluid chromatography.

supercritical fluid that is ensured to exit with an equilibrium amount of the solid.

The chromatographic and calorimetric information can also be used to infer phase behavior information. For example, solubility measurements can be performed on a Supercritical Fluid Chromatograph (SFC) where a relatively short fused silica capillary tubing system replaces the usual column. Sample is coated into this tubing using the same techniques that are used to coat a stationary phase on an analytical capillary column. Raising the solvent pressure stepwise from a selected starting point allows isothermal measurements of solubility. At each step, the ion current in the mass spectrometer, which is used as the detector, is monitored to determine the quantity of solute in solution. Also, solubility may be estimated from the last equations in Approach A and Approach B summarized in Figure 21.1.11.[19] This is true provided that the adsorbent is a solid and partial molar volume of the solute (e.g., naphthalene) in the stationary phase is equal to the solid molar volume of the solute.

It is worth noting that Supercritical Fluid Chromatography is an important experimental tool for measurement of a portfolio of physical-chemical properties as depicted in Table 21.1.2.

However, indirect approaches and methods that involve sampling suffer from magnification of propagated error due to the high sensitivity and non-linear nature of governing fundamental relations in the critical region.

**Table 21.1.2. Classification of physicochemical properties that can be measured by SFC**

| Equilibrium properties | Kinetic and transport properties | Other properties |
|---|---|---|
| * Fluid phase interactions<br>- Second virial coefficients<br>* Solution interactions<br>- Partial molar volumes<br>- Solubilities<br>* Surface interactions<br>- Adsorption isotherms | * Diffusion coefficients<br>* Mass transfer coefficients | * Molecular masses |
| | °Adsorption and desorption rate constants<br>°Reaction rate constants | |

*Properties determined by SFC; °Properties determined by GC and can be determined by SFC

Due to the inherent limitation of predictive methods, relative importance of experimental methods become very significant when compared to simple fluids at near ambient condition. Therefore, the area is a popular review[3,20] topic and is included in most texts[6,21] in the field.

### *21.1.2.1.2 Computational aspects*

Computation and definition of critical phenomena is essential in prediction, modeling, and identification of various phases at high pressures. Gibbs presented the definition of critical state in his pioneering paper "On the Equilibrium of Heterogeneous Substances" in 1876.[22] For mixtures, mechanical and thermal stability criteria are not sufficient to describe the critical behavior and diffusional (material) stability criteria are required. Various investigators have transformed Gibbs' criteria to other variable sets and Reid and Beegle[23] have developed Gibbs' criteria in Legendre transforms with suggestions to overcome indeterminacy. Works of Michelsen[24] and Heidemann[25] combine algorithms with more convenient representation of Gibbs criteria.

Generation of complete phase diagrams is not a trivial task and may require utilization of either insight and heuristic guidance or global approaches coupled with Mixed Integer Non-linear Programming (MINLP) or intelligent search algorithms such as genetic algorithms and simulated annealing.

The methods used in generation of phase diagrams may either employ the more popular integral approach that is suitable for design purposes or the more insightful differential approach that is preferred for generation of phase diagrams.[26] The basic elements of integral approach are the determination of fluid phase partition coefficient, K, for each component as well as stability criterion coupled with a computational algorithm.

For determination of fluid phase partition coefficients, K, one may take a symmetric approach that uses Equation of States (EOS) for all fluid phases or an unsymmetrical approach that makes use of liquid activity models for the liquid phase retaining the equation of state models for the gas phase.

For symmetric approach:

$$K_i = \frac{\hat{\varphi}_i^{\beta}(T,P,x_i)}{\hat{\varphi}_i^{\alpha}(T,P,y_i)}$$

For the unsymmetrical approach:

$$K_i \frac{\gamma_i^{\beta}(T,P,x_i)f_i(T,P)}{\hat{\varphi}_i^{\alpha}(T,P,y_i)}$$

where:

| | |
|---|---|
| $K_i$ | partition coefficient of component i in phases α and β |
| $\hat{\varphi}_i^{\beta}$ | fugacity coefficient of component i in mixture for phase β |
| $\hat{\varphi}_i^{\alpha}$ | fugacity coefficient of component i in mixture for phase α |
| $\gamma_i^{\beta}$ | activity coefficient of component i in mixture for phase β |
| $f_i$ | fugacity of pure component i |

If one assumes no solvent in the solid phase, the relation for solubility of the solid in the supercritical solvent reduces to:

$$y_i = \left(\frac{P_i^{sub}(T)}{P}\right)\left(\frac{1}{\hat{\varphi}_i^{DG}}\right)\exp\left(\frac{\overline{V}_i^{s}\left(P - P^{sub}(T)\right)}{RT}\right)$$

A B C

where:

| | |
|---|---|
| $y_i$ | mole fraction of component i in gas phase |
| $P_i^{sub}$ | sublimation pressure of component i |
| P | total pressure |
| $\hat{\varphi}_i^{DG}$ | fugacity coefficient of component i in mixture for dense gas phase |
| $V_i^{s}$ | partial molar volume of component i in solid phase |
| R | universal gas constant |
| T | absolute temperature |

The first term A is for ideal gas solubility, the second term B accounts for the non-ideality, while the third term C (Poynting correction) accounts for the pressure effects. The product of the second and third terms is often referred as the Enhancement Factor (over the ideal). In the absence of data, the sublimation pressure may be approximated through extrapolation of the vapor pressure information.

#### *21.1.2.1.3 Modeling*

The approaches to modeling the high pressure phase behavior follow either use of the equation of state models for fugacity coefficient determination or the treatment of the dense gas as an "expanded liquid" through an activity coefficient model, the former approach being more popular.

The macroscopic dense gas modeling approaches include van der Waals family of equation of states, virial family of equation of states, and non-classical approaches. The molecular/theoretical approaches and considerations contribute not only to more comprehensive models but also provide insight bringing forth sound parameters/terms to the macroscopic models. Computer simulation (molecular) can also be used to directly compute phase behavior with some success.[27,28] The virial family of equation of states finds limited use in supercritical applications due to the necessity for a large number of terms and their inadequacy in representing the critical point. The non-classical scaling approaches[29,30] fail when extended beyond the very narrow critical region and their combinations with equations of state (crossover EOSs) are still far from having practical applicability. Therefore, we will focus on the van der Waals family of Equation of States.[31]

These equations of states have both a repulsive and an attractive term. The most popular variants differ in how the attractive terms are modeled. The success of more recent Peng-Robinson[32] (PR) and Soave-Redlich-Kwong[33] (SRK) equation of states in phase behavior representation is due to the added parameter, accentric factor that incorporates vapor pressure information into the model. The shortcomings around the critical point are mainly due to the enhanced contribution of repulsive forces that can be modified through Perturbed Hard Chain Theory (PHCT).[34] Naturally, one other inherent difficulty is due to the instability of high molecular weight compounds at their critical point. This implies use of other parameters or pseudo critical properties to model the compounds and their mixtures. These equations of states are better in representing molar gas volume then the liquid. The accuracy of molar liquid volume can be improved through volume translation.[35] The added terms and parameters naturally complicate the equation of state since the order of the equation increases above cubic.

Yet another essential dimension of the equation of states is their representation of mixture properties, which is achieved through mixing rules (models) that often include adjustable binary interaction parameters determined from multi-component, ideally binary, data. Most classical approaches fail to model mixing behavior of systems with dissimilar size components and hydrogen bonding, particularly if one expects to apply them throughout a wide pressure and temperature range that extends from very low to high pressures. Some of the empirical density dependent mixing rules such as Panagiotopoulos-Reid[36] can account for the relative size of the solute and the solvent but have theoretical shortcomings violating one fluid model. There are also approaches that combine equation of state with activity models by forcing the mixture EOS to behave like the activity models ($G_E$) at liquid densities.[31] EOS-$G_E$ models can be combined with group contribution models to make them predictive. More recently, Wong-Sandler[37] mixing rules provide an avenue to incorporate more readily available low-pressure information into the equation of states extrapolate throughout the entire pressure range, as well as ensuring fundamentally correct boundary conditions.[38]

Considering the difficulty associated with representing fugacity coefficients that may vary over ten orders of magnitude by mere change in temperature, pressure and composition; a simple equation of state such as Peng-Robinson, which is given below, do very well in correlating the experimental data and representing the phase behavior well.

$$p = \underbrace{\frac{RT}{V_m - b}}_{\text{repulsive term}} - \underbrace{\frac{a}{V_m(V_m + b) + b(V_m - b)}}_{\text{attractive term}}$$

where:

$$b = \sum_i x_i b_i$$

$$a = \sum_i \sum_j x_i x_j (a_i a_j)^{0.5} (1 - k_{ij})$$

$$b_i = 0.07780\ (RT_{ci}/p_{ci})$$

$$a_i = \alpha_i 0.45724 (R^2T^2_{ci}/p_{ci})$$
$$\alpha_i = [1 + m_i(1 - T^{1/2}_{ri})]^2$$
$$m_i = 0.37464 + 1.54226\omega_i - 0.26992\omega_i^2$$

where:

| | |
|---|---|
| $p$ | pressure |
| $R$ | universal gas constant |
| $T$ | temperature |
| $V_m$ | molar volume |
| $b$ | function of $T_{ci}$ and $P_{ci}$ and $x_i$ |
| $a$ | function of $T_{ci}$, $P_{ci}$, $x_i$ and $k_{ij}$ |
| $\alpha_i$ | a function of reduced temperature, $T_{ri}$ and accentric factor, $\omega_i$ |
| $m_i$ | a function of accentric factor of component i |
| $\omega_i$ | accentric factor of component i |
| $k_{ij}$ | the interaction coefficient of the molecules i and j |
| $T_{ci}$, $P_{ci}$ | critical temperature and pressure of component i |
| $x_i$ | mole fraction of component i |
| $T_{ri}$ | reduced temperature, $T/T_{ci}$ |

The polymer solutions[39] warrant use of a special class of lattice models such as Florry-Huggins. For correlation purposes Sanchez-Lacombe[40] method is sufficient but one may also use Statistical Association Fluid Theory[41] (SAFT) models to obtain a better representation.

#### 21.1.2.2 Transport properties of supercritical solvents

The transport properties of the supercritical fluids fall somewhat in between the gas and the liquid and also depend on how removed one is from the critical point. Dense gasses have the solubilizing power of liquids and the mobility of gasses as depicted in Table 21.1.3. There are quite a few empirical correlations and theoretical models, which are primarily extensions of corresponding low-pressure liquid and gas counter parts. Similarly, of the classical experimental methods can be used for measurement of transport properties of supercritical fluids. A rather brief overview of the methods applicable for supercritical fluids will be presented since specialized reviews in the area give a good account of the state of the art.[30,42] For engineering purposes, one can use applicable property estimation methods available in flowsheet simulators[43] such as ASPEN PLUS, PROII, HYSIM, and CHEMCAD. These methods are discussed in a text classical in the field.[44]

**Table 21.1.3. Transport properties of gases, liquids, and supercritical fluids**

| State | Condition | Property | | |
|---|---|---|---|---|
| | | Density, $g/cm^3$ | Diffusivity, $cm^2/s$ | Viscosity, g/cm-s |
| Gas | 1 atm, 25°C | $0.6\text{-}2\times10^{-3}$ | $1\text{-}4\times10^{-1}$ | $1\text{-}3\times10^{-4}$ |
| Liquid | 1 atm, 25°C | 0.6-1.6 | $0.2\text{-}2\times10^{-5}$ | $0.2\text{-}3\times10^{-2}$ |
| SC Fluid | $T_c$, $P_c$ | 0.2-0.5 | $0.5\text{-}4\times10^{-3}$ | $1\text{-}3\times10^{-4}$ |
| SC Fluid | $T_c$, $4P_c$ | 0.4-0.9 | $0.1\text{-}1\times10^{-3}$ | $3\text{-}9\times10^{-4}$ |

##### *21.1.2.2.1 Viscosity*

Both the capillary viscometer (providing about 0.7% accuracy), the theory of which is based on the Hagen-Poiseuille equation and the oscillating disc viscometer (providing about 0.2% accuracy) are applicable to experimental determination of viscosity at high pressures and temperatures.

Most of the theoretically based estimation methods for viscosity of dense gases rely on modified Enskog theory. Corresponding states based methods are also popular but should be used with care due to their empiric nature.

### *21.1.2.2.2 Diffusivity*

The experimental techniques used to measure the diffusion are based upon chromatographic methods,[19] NMR (for self-diffusion), and photon correlation spectroscopy.[45]

The so called chromatographic method for the measurement of diffusion coefficients, strictly speaking, is not a chromatographic method since no adsorption/desorption or retention due to partition in two phases are involved in the method. The experimental system used is however a chromatograph. Diffusion occurs in an empty, inert column on which the fluid phase is not supposed to be adsorbed. The fluid, in which the solute diffuses flows, continuously through the empty column and the solute, which is introduced into the column at one end, is detected at the other end as effluent concentration.

The theory of diffusion in flowing fluids is first given by Taylor[46] and Aris.[47] According to Aris, a sharp band of solute, which is allowed to dissolve in a solvent flowing laminarly in an empty tube, can be described in the limit of a long column as a Gaussian distribution, the variance of which, $\sigma^2$, in length units is:

$$\sigma(x)^2 = 2D_{eff}t$$

where:

t time of migration of the peak

$D_{eff}$ effective diffusion coefficient given by,

$$D_{eff} = D_{12} + \frac{r^2u^2}{48D_{12}}$$

where:

u average solvent velocity

r inner radius of the tube

$D_{12}$ binary diffusion coefficient.

The first term describes the longitudinal diffusion in the axial direction. The second term is called the Taylor diffusion coefficient and describes band broadening due to the parabolic flow profile and therefore radial diffusion. The height equivalent to a theoretical plate, H, is a measure of the relative peak broadening and is defined as

$$H = \sigma(x)^2 / L$$

where:

L length of the tubing.

Combining the above equations yields

$$H = \frac{2D_{12}}{u} + \frac{r^2u}{24D_{12}}$$

Bulk diffusivity can be calculated from the above equation.

The estimation of low pressure diffusivity is based on the corresponding states theory. The dense gas diffusion coefficient estimation is based on the Enskog theory. The binary diffusion coefficient $D^v_{ij}$ at high pressures as modeled by the Dawson-Khoury-Kobayashi correlation, is next given as a representative model. For a binary system, the equations are:

$$D^v_{ij}\rho^v_m = \left[1 + a_1\rho^v_{rm} + a_2\left(\rho^v_{rm}\right)^2 + a_3\left(\rho^v_{rm}\right)^3\right] D^v_{ij}(p=0)\rho_m(p=1\,atm)$$

$$\rho^v_{rm} = V_{cm}/V^v_m, \quad \rho^v_m = 1/V^v_m, \quad V_{cm} = \frac{y_i V^*_{ci} + y_j V^*_{cj}}{y_i + y_j}$$

where:

| | |
|---|---|
| $\rho_m$ | molar density |
| $\rho_{rm}$ | reduced molar density |
| $a_1,a_2,a_3$ | constants |
| $V_{cm}$ | critical molar volume |
| $V_m$ | molar volume |
| $V_{ci}$ | critical volume of component i |
| $y_i$ | mole fraction of component i |

$D^v_{ij}$ (p=0) is the low pressure binary diffusion coefficient obtained from the Chapman-Enskog-Wilke-Lee model. The molar volume and the molar density are calculated with an equation of state model.

### *21.1.2.2.3 Thermal conductivity*

Transient hot-wire and co-axial cylinder methods are typically applied for measurement of the thermal conductivity of supercritical fluids.[48]

The theory of thermal conductivity is based on the Enskog theory and a typical model for both vapor and liquid thermal conductivity is given by Chung-Lee-Starling. This model makes use of the PRWS (Peng-Robinson with the Wong-Sandler mixing rules) to provide a flexible and predictive equation of state. The main equation for the Chung-Lee-Starling thermal conductivity model is:

$$\lambda = \frac{31.2\eta(p=0)}{M} f_1 + f_2$$

where:

$$f_1 = \mathrm{fcn}(\rho_{rm}, \omega, p_r, \kappa)$$
$$f_2 = \mathrm{fcn}(T_c, M, V_{cm}, \rho_{rm}, \omega, p_r, \kappa)$$
$$\psi = \mathrm{fcn}(C_v, \omega, T_r)$$

$\eta(p=0)$ can be calculated by the low pressure Chung-Lee-Starling model. The parameter $p_r$ is the reduced dipole moment given by:

$$p_r = 4.152\frac{p}{\left(V_{cm}T_c^{1/2}\right)}$$

where:

| | |
|---|---|
| $\lambda$ | thermal conductivity |
| $\eta$ | viscosity |
| M | molecular mass |
| $\rho_{rm}$ | reduced molar density |
| $\omega$ | accentric factor |
| p | dipole moment |
| $p_r$ | reduced dipole moment |
| $V_{cm}$ | critical molar volume |
| $T_c$, $T_r$ | critical and reduced temperature |
| $\kappa$ | polar parameter |

For low pressures, $f_1$ is reduced to 1.0 and $f_2$ is reduced to zero. This gives the Chung-Lee-Starling expression for thermal conductivity of low pressure gases. The molar density, $\rho_{rm}$, can be calculated using and equation of state model (for example, the Peng-Robinson-Wong-Sandler equation of state) where the mixing rule for b is obtained as follows. The second virial coefficient must depend quadratically on the mole fraction:

$$B(T) = \sum_i \sum_j x_i x_j B_{ij}$$

with:

$$B_{ij} = \frac{\left(B_{ii} + B_{ij}\right)}{2}\left(1 - k_{ij}\right)$$

The relationships between the equation of state at low pressure and the virial coefficient are:

$$B = b - \frac{a}{RT}; B_{ii} = b_i - \frac{a_i}{RT}$$

Wong and Sandler has shown that the following mixing rule does satisfy the second virial coefficient equation:

$$b = \frac{\sum_i \sum_j x_i x_j B_{ij}}{1 - \frac{A_m^E (p = \infty)}{\Lambda RT} - \sum_i x_i B_{ii}}$$

where:

| | |
|---|---|
| B | the second virial coefficient |
| $k_{ij}$ | the interaction coefficient of the molecules i and j |
| b | function of $T_{ci}$ and $P_{ci}$ and $x_i$ |
| a | function of $T_{ci}$, $P_{ci}$, $x_i$ and $k_{ij}$ |
| $A^E_m$ | Helmholtz free energy |

The pressure correction to the thermal conductivity for a pure component or mixture at low pressure is given by:

$$\lambda^v = fcn\left(\lambda^v (p = 0), \rho_{rm}, y_i, M_i, T_{ci}, V_{ci}, Z_{ci}\right)$$

where:

$$\rho_{rm} = \sum_i y_i \frac{V_{ci}}{V_m^v}$$

where:

| | |
|---|---|
| $\lambda^v$ | thermal conductivity |
| $\rho_{rm}$ | reduced molar density for mixtures |
| $y_i$ | mole fraction of component i |
| $M_i$ | molecular weight of component i |
| $T_{ci}$ | critical temperature of component i |
| $V_{ci}$ | critical volume of component i |
| $Z_{ci}$ | critical compressibility of component i |

The vapor molar volume, $V^v_m$ can be obtained from the equation of state models as described above.

#### *21.1.2.2.4 Surface tension*

Stability of phase boundaries depends on the surface tension. Surface tension in a supercritical fluid system is of major importance for drying, surfactant efficacy, and extraction. The surface tension of a gas increases with pressure and approaches zero at the critical point while the surface tension of liquid decreases with pressure resulting in dissolution of supercritical components in the liquid phase. The methods useful in correlating surface tension include Macleod-Sugden correlation and corresponding states theory.[21]

### 21.1.2.3 Entrainer (co-solvent effects) of supercritical solvents

Entrainers, modifiers, and co-solvents are basically mixed solvent systems and provide another dimension to supercritical fluid extraction. The entrainers enhance the solubility of the low volatile substance in the solvent, provide selective solubility in multi-solute instances, and enhance the sensitivity of the solubility and selectivity to temperature, pressure, and composition. The entrainers may be reactive and are also useful as slurrying media. Table 21.1.4 shows representative data on the effect of entrainers on vapor liquid systems that has been systematically studied by Brunner's group. Kurnik and Reid[50] as well as Johnston's group[51] present data for dense gas-solid systems.

**Table 21.1.4. The effect of entrainers on separation factor (Adapted from Brunner[21])**

| Entrainer | $\alpha$ | Entrainer | $\alpha$ |
|---|---|---|---|
| Hexadecanol(1) - octadecane(2) - entrainer - nitrous oxide (120 bar, 70°C, $y_0$ = 1.5 wt%, $a_0$ = 2, $y_{(1)+(2)}$ = 2 wt%) | | Octadecane(1) - salicylic acid phenal ester(2) - entrainer - nitrous oxide (120 bar, 70 °C, $y_0$ = 1.5 wt%, $a_0$ = 2, $y_{(1)+(2)}$ = 1.7 wt%) | |
| acetone | 3 | methanol | 1.65 |
| methanol | 2.4 | dichloromethane | 1.5 |
| dichloromethane | 2.1 | benzene | 1.45 |
| benzene | 2 | cyclohexane | 1.4 |
| methyl acetate | 2 | acetone | 1.4 |
| hexane | 1.9 | hexane | 1.3 |
| cyclohexane | 1.7 | methyl acetate | 1.2 |

| Entrainer | $\alpha$ | Entrainer | $\alpha$ |
|---|---|---|---|
| Hexadecanol(1) - octadecane(2) - entrainer - ethane (120 bar, 70°C, $y_0$ = 15 wt%, $a_0$ = 2.5, $y_{(1)+(2)}$ = 30 wt%) | | Hexadecanol(1) - octadecane(2) - entrainer - carbon dioxide (120 bar, 70 °C, $y_0$ = 0.4 wt%, $a_0$ = 2.5, $y_{(1)+(2)}$ = 1 wt%) | |
| methanol | 2.2 | methanol | 3.4 |
| acetone | 1.6 | acetone | 1.8 |
| hexane | 1.6 | methyl acetate | 1.6 |
| (120 bar, 90°C, $y_{(1)+(2)}$ = 20 wt%) | | hexane | 1.4 |
| methyl acetate | 1.8 | | |
| benzene | 1.2 | | |
| dichloromethane | 1.1 | | |

Separation factor, $\alpha = (y_2/x_2)/(y_1/x_2)$

Although entrainers provide the aforementioned added advantages, they also bring forth more complex process flowsheets. The separation of extract from entrainer and entrainer from the supercritical component are not as easy and as sharp as in the instances without entrainer. The selection of entrainer is based on thermodynamic, environmental, and economic considerations. The solute-entrainer and entrainer-supercritical separations are the key from the processing perspective.[52]

#### 21.1.2.4 Reaction rate implication in supercritical solvents

Reactions in supercritical media utilize high pressures. Therefore, the effect of pressure on reaction equilibrium as well as reaction rate plays an important role in supercritical phase reactions.[53]

The kinetics of the reaction can be explained in terms of the transition-state theory. According to the theory, the reaction occurs via a transition state species M* and the generic elementary reaction can be written as:

aA + bB + ........... $\leftrightarrow$ M* $\rightarrow$ Products

The effect of pressure on the rate constant is given as:

$$\left(\frac{\partial \ln k_x}{\partial P}\right)_T = \left(\frac{\partial \ln K_x^*}{\partial P}\right)_T + \left(\frac{\partial \ln \kappa}{\partial P}\right)_T = \frac{\Delta \overline{V}^*}{RT} + \left(\frac{\partial \ln \kappa}{\partial P}\right)_T$$

where:

| | |
|---|---|
| $k_x$ | rate constant in mole fraction units |
| P | pressure |
| T | temperature |
| $K^*_x$ | mole fraction based equilibrium constant for reaction involving reactants and transition state |
| $\kappa_T$ | isothermal compressibility |
| $\Delta\overline{V}^*$ | activation volume (difference between partial molar volumes of activated complex and reactants), $\Delta\overline{V}^* = \overline{V}^* - a\overline{V}_A - b\overline{V}_B - ...$ |
| R | universal gas constant |

The rate constant in the above equation is expressed in terms of pressure independent units (mole fraction). If the rate constant is expressed in terms of pressure dependent units (such as concentration), the relevant equation is:

$$\left(\frac{\partial \ln k}{\partial P}\right)_T = -\frac{\Delta \overline{V}^*}{RT} + \left(\frac{\partial \ln \kappa}{\partial P}\right)_T + \kappa_T(1-a-b-\cdots)$$

If the volume of activation is positive, the reaction is hindered by pressure. However, for high negative values of the volume of activation, the pressure enhances the rate of the reaction. Therefore, supercritical fluids that exhibit very high negative activation volumes for certain reactions will improve the rate of the reaction.

The volume of reaction, rather than activation, is crucial in determining the effect of pressure on the equilibrium constant.

$$\left(\frac{\partial \ln K_x}{\partial P}\right)_{T,x} = -\frac{\Delta \overline{V}_f}{RT}$$

where:

$\Delta \overline{V}_f$ reaction volume (difference between partial molar volumes of products and reactants)

If the equilibrium constant is expressed in terms of pressure dependent units (such as concentration), the relevant equation is:

$$\left(\frac{\partial \ln K}{\partial P}\right)_{T,x} = -\frac{\Delta \overline{V}_r}{RT} + \kappa_T \sum v_i$$

where:

$v_i$ stoichiometric coefficient

As the above equation implies, supercritical fluids that exhibit very high negative activation volumes for certain reactions will improve the equilibrium conversion of the reaction.

### 21.1.2.5 Sorption behavior of supercritical solvents

Both adsorption from a supercritical fluid to an adsorbent and desorption from an adsorbent find applications in supercritical fluid processing.[54,55] The extrapolation of classical sorption theory to supercritical conditions has merits. The supercritical conditions are believed to necessitate monolayer coverage and density dependent isotherms. Considerable success has been observed by the authors in working with an equation of state based upon the Toth isoterm.[56] It is also important to note that the retrograde behavior observed for vapor-liquid phase equilibrium is experimentally observed and predicted for sorptive systems.

### 21.1.2.6 Swelling with supercritical solvents

As the pressure of the gas is increased, the solubility of the supercritical gas in the solid polymer increases resulting in swelling, a phenomena that could be advantageous in certain applications while its deleterious impact should be minimized, if not totally eliminated, in other instances.

The sorption of supercritical solvent and the resulting swelling could be very high, for example around 30% and 20% respectively for carbon dioxide in polymethylmethacrylate (PMMA). In such instances, the experimental information could be summarized using polymer equation of states such as Sanchez-Lacombe where a single mixture fitting parameter is used.[6]

Swelling can be advantageous in that it enables permeation and diffusion of the supercritical fluid into the polymer network. Fragrances, dyes, or medicinal substances loaded in the supercritical fluid can readily impregnate into the polymer and load the polymer with the aforementioned additives. Upon release of the pressure, only the supercritical fluid (i.e., carbon dioxide) flashes off. This avenue has led to a plethora of controlled (timed) release products.

Naturally, swelling may be undesirable in many instances as well. For example, swelling of organic polymer based membranes decreased selectivity. Other possible deleterious effects could include malfunctions due to solubilization and swelling of sealants such as gaskets or o-rings.

#### 21.1.2.7 Surfactants and micro-emulsions

Most highly polar and ionic species are not amenable to processing with desirable solvents such as carbon dioxide or any other solvent such as water that has a higher critical temperature well above the decomposition temperature of many solutes. In such instances, the combination of the unique properties of supercritical fluids with those of micro-emulsions can be used to increase the range of applications of supercritical fluids.[57] The resulting thermodynamically stable systems generally contain water, a surfactant and a supercritical fluid (as opposed to a non-polar liquid in liquid micro-emulsions). The possible supercritical fluids that could be used in these systems include carbon dioxide, ethylene, ethane, propane, propylene, n-butane, and n-pentane while many ionic and non-ionic surfactants can be used. The major difference between the liquid based emulsions and the supercritical ones is the effect of pressure. The pressure affects the miscibility gaps as well as the microstructure of the micro-emulsion phase.

The incorporation of the micro-emulsion phase creates interesting potential advantages for reactions as well as separations. Isolation of components from fermentation broths and garment cleaning appear to be two of the more competitive applications of these systems.

### 21.1.3 SEPARATION WITH SUPERCRITICAL SOLVENTS

Supercritical fluids are effective at much lower temperatures than distillation, and their application in separation avoids degradation and decomposition of heat-labile compounds. Attractiveness of supercritical extraction processes are due to the sensitivity of responses to process variables, promise of complete and versatile regeneration of solvents, energy savings, enhanced solute volatilities, solvent selectivities, favorable transport properties for solvents, and state governed effectiveness of solvents which enables the use of low cost, non-toxic, environmentally acceptable solvents. The impact of inherent characteristics of supercritical fluids on separations is summarized in Table 21.1.5.

**Table 21.1.5 The characteristics and challenges with supercritical separations**

| Inherent characteristics of systems @ supercritical conditions | Resulting promise | Challenges to be met |
|---|---|---|
| Enhanced solubility | Effective at lower temperatures | Recovery of substances inadvertently extracted |
| Favorable transport properties | High mass transfer rates especially in porous media | |

| Inherent characteristics of systems @ supercritical conditions | Resulting promise | Challenges to be met |
|---|---|---|
| Sensitivity of responses to process variables | Effective recovery of solvents, and innovative fractionation possibilities | Control and optimization |
| High pressure | | High capital cost and need for recovery of mechanical energy |
| Dominance of physical characteristic on solvency over chemical | Expansion of solvent spate to environmentally benign, inexpensive, and non flammable solvents and their mixtures | |

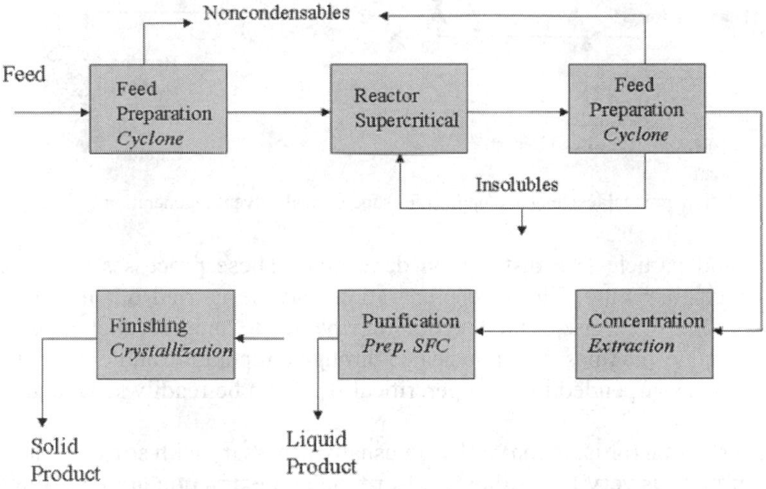


Figure 21.1.12. A generic process flow sheet.

There have been many useful attempts made to classify separation technologies. Supercritical fluids are applicable with both intra-phase and inter-phase separations. Due to the ease and flexibility in which a new phase can be formed for regeneration of the solvent, inter-phase is the more common. Furthermore, material solubility and swelling problems, particularly with organic-component based membranes, limit inter-phase separations. This is due to the enhanced solubility of these components in supercritical solvents.

Generic steps involved in a typical chemical process are shown in Figure 21.1.12, with each separation sub task type identified. Environmental processes are usually dominated by feed preparation tasks while biochemical processes utilize all of the four subtasks. Blending is considered a finishing task while splitting is considered a feed preparation task. The objective criteria used for the separation may vary according to the subtask. For example, in feed preparation, technical feasibility such as removal of the fines is more appropriate while the usual economic objectives subject to technical feasibility constraints become the goal for the purification and finishing stages.

The feed preparation task involves removal of insoluble components and at times non-condensable compounds. The technology selection and the scheduling of tasks are con-

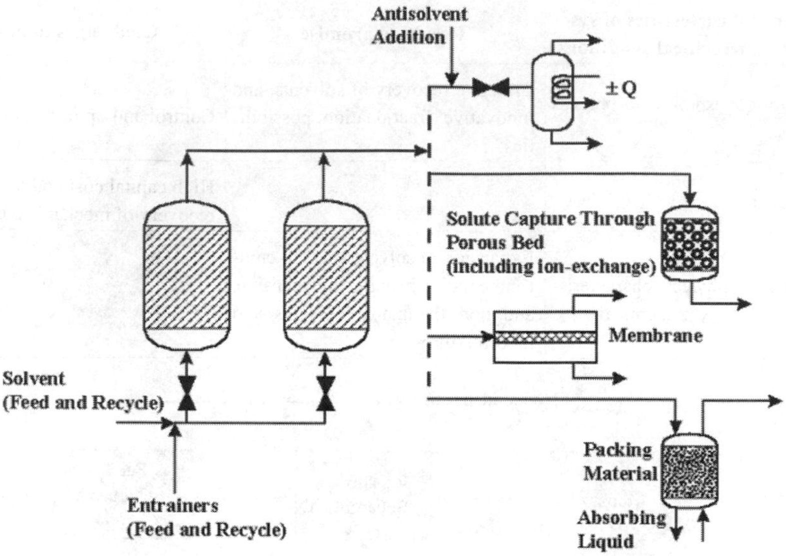


Figure 21.1.13. Supercritical leaching and options for supercritical solvent regeneration.

centration and particle size distribution dependent. These processes are usually physical/mechanical in nature. The feed preparation may be carried out in a unit designed particularly for the purpose or the function may be tied to another function within a single unit. For example, the fines, either nucleated through temperature and/or pressure perturbation, or particles suspended in the supercritical fluid can be readily separated through cyclones.

Concentration (or isolation) tasks are usually necessary with streams where the product concentration is very low (dilute). Adsorption and extraction are typical technologies utilized here. Energy considerations dominate the technology selection subject to product contamination and stability constraints. Thus, the base technology depends on the availability of the Mass Separating Agent (e.g., adsorbent or solvent) and its efficient recycle. Supercritical extraction of organic compounds from aqueous streams is a typical example of isolation step.

The candidate technologies for purification are many. Distillation, the work-horse of the chemical processes, leads the pack. Most of the synthesis effort to date has concentrated on the product purification step. This step is often the last step for liquid products especially in the chemical and petrochemical industries. The biochemical industry utilizes membrane and chromatographic processes more than the other industries due to the thermal stability and purity requirements. In the electronic industry, membrane processes are more prevalent due the ultra-purities necessary. Supercritical fractionation of alcohol water systems with the aid of a dense gas is an example of a purification step.

Crystallization, drying, and blending are typical operations necessary to polish (finish) the product to its final form. Particle nucleation from a supercritical fluid solution is an example of finishing. Each supercritical separation technique is identified on this generic process.

### 21.1.3.1 Leaching - generic application

Leaching is usually incorrectly referred to as extraction of a solid substrate, and is commercially the most significant application of supercritical fluids. A generic flowsheet, shown in Figure 21.1.13, illustrates the basic processing concepts. A heat labile natural substrate is usually contacted with a flowing supercritical fluid in a semi-continuous mode. The supercritical fluid may optionally be doped with an entrainer. Due to the stationary (batch) nature of the leaching stage, two or more parallel vessels loaded with the substrate may be utilized. The material leached is recovered and fractionated through pressure reduction, temperature perturbation, adsorption, membranes or through absorption. During the initial stages of leaching, the solute loading is high and may approach the equilibrium solubility limit, while at the latter stages kinetic and mass transfer limitations are responsible from relatively lower the concentration of the solute. Both coffee decaffeination[59] and hops extraction[60] are popular commercial successes of this application. Further, typical applications are given in Table 21.1.5.

**Table 21.1.6. Leaching applications of supercritical fluids**

| Purpose | Supercritical medium | Substrate |
|---|---|---|
| | **Food processing[61]** | |
| Food purification | $CO_2$ | Coffee decaffeination<br>Cholesterol removal from egg yolk |
| | **Essential oils[62]** | |
| Leaching of natural products | $CO_2$ | Oil from soybean, sunflower, spearmint, Camomile, thyme, rosemary sage, chervil |
| Leaching of natural products | $CO_2$, $N_2O$, $CHF_3$, $SF_6$ | Fungal oil |
| Leaching of natural products | Methanol, ethanol, acetone | Liquid from hazelnut seed coat |
| | **Petrochemical** | |
| Obtaining liquids from solid fuels[63] | Toluene, water, acetone, methanol, THF, tetralin, ethanol | Liquid from coal, lignite |
| Cleaning and upgrading solid fuels[64] | Toluene, methanol, ethanol, water, acetone, tetralin, $CO_2$ | Sulfur, nitrogen, oxygen removal |
| | **Chemical[65]** | |
| Cleaning of parts | $CO_2$ | Oil from contaminated glass grinds, organic additives from extrusion-molded ceramic parts |
| Cleaning of parts | $CO_2$, fluorocarbon solvents | Particles from precision devices |
| | **Biochemical[66]** | |
| Vitamin, pharmaceutical compounds | $CO_2$ | Natural plants |
| | **Environmental[67]** | |
| Extraction of soils | $CO_2$ + modifier | Polycyclic-aromatic hydrocarbons, pesticides |

| Purpose | Supercritical medium | Substrate |
|---|---|---|
| Extraction of soils | Nitrous oxide | Aromatic amines, pesticides |
| Extraction of soils | Water | Polycyclic-aromatic hydrocarbons |
| | **Polymer processing**[6] | |
| Removal of trace compounds | $CO_2$ | Siloxane oligomer, cyclic trimer, organic compounds from plastic, carbon tetrachloride from polyisoprene |

### 21.1.3.2 Extraction - generic applications

Liquid streams containing close-boiling and heat-labile component can be fractionated with the aid of supercritical fluids or valuable components of aqueous streams could be isolated and concentrated using supercritical solvents. Entrainers could also be added to the solvent to enhance the selectivity, particularly in fractionation instances. The main reason for using entrainers for concentration purposes is to enhance solubility. One can also use membranes along with supercritical fractionator.[68] Typical applications of this process technology are given in Table 21.1.7 and Table 21.1.8 while several review papers and texts provide useful design information.[69-72]

**Table 21.1.7. Extraction applications of supercritical fluids**

| Purpose | Supercritical medium | Substrate |
|---|---|---|
| | **Food processing** | |
| Food purification | $CO_2$ | Oil purification<br>Cholesterol removal from milk fat |
| Flavor extraction | $CO_2$ | Beer, wine and fruit flavor concentrates |
| Fractionation of food | $CO_2$ | Oil, milk fat |
| | **Forest products** | |
| Extraction of lignocellulosic materials and biomass | $CO_2$, $SO_2$, $N_2O$, $H_2O$, methylamine, ethanol, ethylene | Cellulose or ligno-cellulosic materials from wood or wood pulp |
| | **Biochemical** | |
| Medicinal compound | $CO_2$ | Fermentation broth |
| | **Petrochemical** | |
| Purifying liquid fuels | $CO_2$ | Deasphaltation, demetallization |
| | **Environmental** | |
| Waste water extraction[73] | $CO_2$+entrainer | Aqueous toxic wastes, phenol, metabolic wastes |
| | **Essential oils** | |
| Purification of essential oils | $CO_2$ | Terpenes from orange, citrus and bergamot peel oil |

**Table 21.1.8. Fractionation applications of supercritical fluids**

| Purpose | Supercritical medium | Substrate |
|---|---|---|
| | | **Essential oils** |
| Fractionation[74] | $CO_2$ | Glycerol, fatty acids, glycerides |
| | | **Chemical** |
| Fractionation | $CO_2$ | Xylenes, glycerides, alkanes, aromatics |
| Fractionation | $CO_2$ + entrainer | Methyl esters, fullerenes |
| | | **Petrochemical** |
| Fractionation | Toluene, pentane, propane, | Petroleum pitch, asphalt |
| | | **Polymer processing** |
| | $CO_2$ | Polydimethylsiloxane, fluoropolymers, polypropylene |

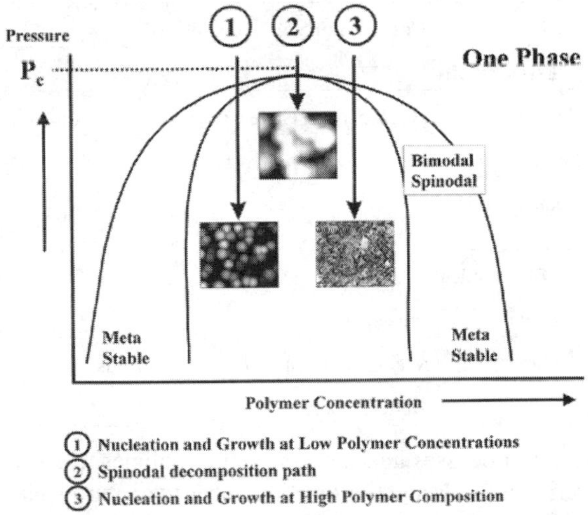


Figure 21.1.14. The pressure reduction path for nucleation of particles from supercritical fluids.

The methods useful in design of extraction systems are the same as those traditional methods applicable to liquid and gaseous systems and include Kremser, McCabe Thiele, Panchon Savarit (including Janecke), and more rigorous rate-based as well as equilibrium based methods. Due to the highly non-linear nature of the physical-chemical properties in the critical region, great care has to be taken in representing both the properties and the solution algorithms.

### 21.1.3.3 Crystallization - generic applications

Particles nucleated from supercritical fluids have unique particle size, shape, and particle size distributions.[75] The means of achieving nucleation and growth are many, and include pressure reduction, temperature perturbation, and addition of anti-solvents. Time profiling and staging have profound effect on the particle characteristics. The effect of pressure reduction on the resulting nucleated material for different solute concentrations[16] is illustrated in Figure 21.1.14.

### 21.1.3.4 Sorption - generic applications

The sorption applications[21] include regeneration of porous beds,[76] preparative scale supercritical chromatography,[21] simulated moving beds,[21] thermal swing schemes,[77] and adsorption/desorptions cycles. Although initial applications of supercritical fluids in this domain were on regeneration of porous beds, more recent emphasis on fractionation best reflects where the true potential of this technique lies. Typical examples are provided in Table 21.1.9.

**Table 21.1.9 Sorption applications of supercritical fluids[21,72]**

| Purpose | Supercritical medium | Substrate |
|---|---|---|
| | **Environmental** | |
| Soil remediation | $CO_2$ + entrainers | Soil |
| Regeneration of adsorbents | $CO_2$ | Polychlorinated biphenyls from adsorbents |
| | **Petrochemical** | |
| Regeneration of adsorbents | $CO_2$ | Benzene, toluene, ethyl acetate from activated carbon |
| Regeneration of catalysts | $CO_2$, THF, hexane, ethylbenzene, benzene | Hydrocarbon, coke removal from catalysts |
| Chromatographic fractionation | $CO_2$ | Coal tar, mineral oil distillate |
| | **Essential oils[78]** | |
| Chromatographic fractionation | $CO_2$ + alcohol | Black pepper, clove extract |
| | **Polymer processing** | |
| Chromatographic fractionation | $CO_2$, n-pentane+methanol | Styrene oligomer |
| | **Biochemical** | |
| Chromatographic fractionation | $CO_2$ + alcohol | Prostaglandis |
| Chromatographic fractionation | $CO_2$ + alcohol | Vitamins, polyunsaturated fatty acids |
| Chromatographic chiral separation | $CO_2$ | Optical isomers |

In synthesis of supercritical sorption processes and their operating policies, the synergy between the characteristics supercritical fluids and the sorption needs to captured. Naturally, the domain requires good distributed process models and solvers in addition to physical property models.

## 21.1.4 REACTIONS IN SUPERCRITICAL SOLVENTS

As stated in the previous sections, supercritical solvents are widely used for separation, especially for extraction of thermally labile compounds. Although the unique properties of supercritical fluids make them attractive as a reaction medium as well, the use of supercritical solvents as reaction medium is becoming more and more popular only in the recent years. The first review on reactions in supercritical fluids was presented by Subramaniam and McHugh,[79] whereas a more recent and through review was given by Savage et al.[80]

The motivations for using supercritical solvents in chemical reactions are many.[81,82] In this section, the motivations will be stated and explained with examples from each type of reaction. Emphasis in this chapter will be given to more recent examples, which were not included in the previous review papers.

**Reasons for carrying out reactions in supercritical fluids**:
There are several reasons for carrying out reactions in the supercritical phase. Naturally, some of the reasons are coupled. Nevertheless, they, in general, relate to control, favorable mass transfer and kinetic considerations.

- Enhanced reaction rate

The effect of pressure on the reaction rate and equilibrium constant at high pressures is described in Section 21.1.2.4. As can be perceived from this section, supercritical fluids that exhibit very high negative activation volumes for certain reactions will improve the rate and equilibrium conversion of the reaction.

- Homogenization

Reactions that otherwise would be carried out in more than one phase (heterogeneous reactions) can be transformed to homogeneous ones with the aid of supercritical fluids, so that inter-phase transport limitations are eliminated. This is realized due to enhanced solubilities of the reaction components in the supercritical fluids. Typical examples are reactions in water (supercritical water can solubilize organic compounds), homogeneous catalytic reactions, and reactions of organometallic compounds. Homogenizing one compound more than the others in a system may also affect relative rates in complex reactions and enhance the selectivity.

- Enhanced mass transfer

In many instances, reaction rates are limited by diffusion in the liquid phase. The rate of these reactions can be increased if the reaction is carried out in the supercritical phase. Typical examples are enzyme catalyzed reactions as well as some very fast reactions such as certain free radical reactions. Selectivity considerations usually dominate in complex reactions. If some steps of the complex reaction are controlled by diffusion, changing the diffusivity changes the relative rates of the reaction steps and affects the selectivity.

- Ease of down-stream separation

Another reason for using supercritical fluids as the reaction medium is to fractionate products, to purify the products or to remove unreacted reactants from the product stream. Supercritical fluids can be used as either a solvent or anti-solvent in these instances.

- Increased catalyst activity

Some heterogeneous catalytic reactions are carried out in the supercritical phase, in order to increase catalyst activity and life through in-situ regeneration of surfaces with tuning of operation conditions. For example, supercritical fluids are capable of dissolving carbon that may otherwise be deposited on the catalyst in the absence of the supercritical solvent.

- Tunable reaction rates through dielectric constant

Some properties of supercritical fluids can be monitored (manipulated) continuously by adjusting the density of the fluid. Dielectric constant is such a property and the solvent's dielectric constant can influence the rate of the reaction.

### 21.1.4.1 Homogenous reactions in supercritical solvents - examples

Homogeneous reactions carried out in supercritical fluids can be either catalytic or non-catalytic. The objective of carrying out both catalytic and non-catalytic reactions in supercritical fluids is to increase the overall rate of the reaction by eliminating the inter-phase transport effects.

#### *21.1.4.1.1 Homogeneous reactions catalyzed by organometallic compounds*

Homogeneous catalysts have advantages over heterogeneous catalysts such as possibility of carrying out the reaction at milder conditions, higher activity, and selectivity, ease of spectroscopic monitoring, and controlled and tunable reaction sites.

Organic reactants and products are not soluble in water while most catalytic materials are soluble in water. Therefore, homogenization of organic systems utilized environmentally undesirable organic solvents. An alternative to environmentally unacceptable organic solvents is a supercritical solvent that has added advantages over organic solvents such as increased reaction rate, higher selectivity and easy separation of reactants and products as well as of the catalyst after the reaction. Since the properties of supercritical solvents can be adjusted by manipulating the operating conditions, reaction rate and selectivity are better tunable in reactions carried out in supercritical solvents.

Carbon dioxide is the supercritical solvent that is most commonly used in homogeneous catalytic reactions. In addition to being environmentally acceptable (nontoxic, nonflammable), inexpensive, and available in large quantities, carbon dioxide does not participate in most reactions. It also has an ambient critical temperature. Although, supercritical carbon dioxide is more effective in dissolution of non-polar, nonionic and low molecular mass compounds, addition of co-solvents enhances the solubility of many otherwise insoluble compounds in supercritical carbon dioxide. A recent review by Noyori et al.[83] discusses homogeneous catalytic reactions under supercritical conditions.

When homogeneous reactions are carried out under supercritical conditions, gas/liquid interfacial transport is eliminated, which is an advantage for reactions such as hydrogenation, where diffusion of gas into the liquid may be limiting the reaction rate. In asymmetric hydrogenation reactions, hydrogen and the supercritical solvent are miscible and this results in better enantioselectivity. In Diels-Alder reactions, the advantage of the supercritical solvent is the higher selectivity obtained rather than increased rate of the reaction due to the solvent. Most of the oxidation reactions are carried out in supercritical water. Some heterogeneously catalyzed reactions are also carried out in supercritical carbon dioxide. Recently, homogeneously catalyzed reactions carried out in supercritical carbon dioxide have been reported. Examples of homogeneous catalytic reactions carried out under supercritical conditions are summarized in Table 21.1.10.

**Table 21.1.10. Homogenous reactions in supercritical carbon dioxide catalyzed by organametallic compounds**

| Reaction | Catalyst |
|---|---|
| **Isomerization** | |
| 1-hexene to 2-hexene | Iron catalyst |
| Hydrogenation | |
| $CO_2$ to formic acid | Ruthenium(II) phosphine complex |
| Asymmetric hydrogenation of tiglic acid | Ruthenium catalyst |
| Asymmetric hydrogenation of enamides | Cationic rhodium complex |
| Cyclopropene | Manganese catalyst |

| Reaction | Catalyst |
|---|---|
| **Hydroformylation** | |
| 1-Octene | Rhodium catalyst |
| Propylene | Cobalt carbonyl |
| **Olefin metathesis** | |
| Ring opening metathesis polymerization | Ruthenium catalyst |
| Ring closing metathesis of dienes to cyclic olefins | Ruthenium catalyst |
| **Diels-Alder reactions** | |
| Synthesis of 2-pyrones | Nickel catalyst |
| Synthesis of cyclopentones | Cobalt catalyst |
| Cyclotrimerization of alkynes to substituted benzene derivatives | Cobalt catalyst |
| **Oxidation** | |
| Alkene epoxidation | Molybdenum catalyst |
| 2,3-Dimethylbutene epoxidation | Molybdenum catalyst |
| Cyclooctene epoxidation | Molybdenum catalyst |
| Cyclohexene epoxidation | Molybdenum catalyst |

#### *21.1.4.1.2 Homogeneous reactions of supercritical water*

Homogeneous reactions carried out in supercritical fluids are reactions in supercritical water, organo-metallic reactions and Diels Alder reactions. Reactions in supercritical water are well studied[84] and will be described in the following section.

Despite the higher temperature and pressures required in supercritical water applications, this solvent possesses unique properties that make it attractive as a reaction medium. Supercritical water has a lower dielectric constant as compared to liquid water, and the dielectric constant of supercritical water changes significantly with the density. Also, the effect of hydrogen bonding is less pronounced at supercritical conditions, one consequence of which is the high solubility of organics in supercritical water. The reactions in the supercritical water medium are carried out in a single phase, which implies high reactant concentration and negligible inter-phase mass transfer resistance. Also, the ion dissociation constant of water is higher in the critical region and is lower as supercritical conditions are accessed. These properties also vary continuously in the supercritical region, so that they can be tuned during the reaction by changing the temperature and/or pressure.

Examples of the homogeneous reactions are given in Table 21.1.11. In some of the reactions discussed in this category, acid or base catalysts were used to enhance the rates.[84]

**Table 21.1.11. Homogeneous reactions of supercritical water**

| Reaction | Catalyst |
|---|---|
| **C-C Bond formation** | |
| Phenol and p-cresol alkylation | None |

| Reaction | Catalyst |
|---|---|
| Diels-Alder cycloaddition | None/NaOH |
| Ring opening of 2,5-dimethylfuran | Acid |
| **Hydration/dehydration** | |
| Conversion of tert-butyl alcohol to isobutylene | None/$H_2SO_4$/NaOH |
| Dehydration of cyclohexanol, 2-methyl cyclohexanol, 2-phenylethanol | Acid |
| **Hydrolysis** | |
| Esters to carboxylic acids and alcohols | Autocatalytic |
| Nitriles to amides and then to acid | Autocatalytic |
| Butyronitrile | Autocatalytic |
| Polyethyleneteraphthalate and polyurethane | None |
| Diaryl ether to hydroxyarene | None |
| Triglycerides into fatty acid | None |
| **Decomposition** | |
| Cellulose and glucose decomposition | None |
| Nitrobenzene | None |
| 4-Nitroaniline | None |
| 4-Nitrotoluene | None |
| **Oxidation** | |
| Phenols | None |
| Ethanol | None |
| 2-Propanol | None |
| 2-Butanol | None |
| Chlorinated hydrocarbons | None |

### *21.1.4.1.3 Homogeneous non-catalytic reactions in supercritical solvents*

The use of supercritical fluids as reaction media for organometallic species has also been investigated.[85] Reactions include photochemical replacement of carbon monoxide with $N_2$ and $H_2$ in metal carbonyls, where the reaction medium is supercritical xenon. Also, photochemical activation of C-H bonds by organometallic complexes in supercritical carbon dioxide has also been investigated. More recent studies on photochemical reactions also include laser flash photolysis of metal carbonyls in supercritical carbon dioxide and ethane[86] and laser flash photolysis of the hydrogen abstraction reaction of triplet benxophenone[87] in supercritical ethane and $CHF_3$.

## 21.1.4.2 Heterogeneous reactions in supercritical solvents - examples

Heterogeneous reactions in supercritical fluids can be catalytic or non-catalytic. Catalytic heterogeneous reactions are reactions carried out on solid catalysts and are of great importance in the chemical process industries. As described in the next section, the advantages of carrying out these reactions in a supercritical medium include enhanced inter-phase and intra-particle mass and heat transfer and in-situ regeneration of catalyst. Catalytic supercritical water oxidation will also be discussed. Other heterogeneous reactions that will be described are fuels processing and treatment of biomass.

#### *21.1.4.2.1 Heterogeneous catalytic reactions in supercritical solvents*

Obviously, a solid catalyzed reaction takes place only on the active sites of the porous catalyst with the implication of some mass and heat transport steps prior to and after the reaction. The first step is the diffusion of the reactants through the film surrounding the catalyst particle to the external surface of the catalyst, followed by diffusion of the reactants into the catalyst pore to the active site in the pores. These steps are limited by the diffusivity and viscosity of the reactants. In the case of a supercritical fluid phase reaction, the diffusivity is higher than the liquid diffusivity, viscosity is less than the liquid viscosity and therefore, the rate of transfer to the active site will be higher. After the adsorption, reaction and desorption steps, the products have to diffuse out of the pore, and again through the film surrounding the particle into the bulk fluid. Rates of these steps can be accelerated utilizing a supercritical medium for the reaction. Heat transfer effects are also important in a solid catalyzed reaction. Higher thermal conductivity of supercritical fluids is an advantage as well.[88]

For two-phase reactions (typically hydrogenation and oxidation reactions), the reaction steps include the diffusion of the gas reactant to and through the gas-liquid interface and then into the bulk liquid. This mass transfer limitation is also eliminated if the reaction is carried out in a supercritical medium where the reaction takes place in a single phase.[88]

Supercritical fluids bring other benefits to the solid catalyzed reaction rate besides eliminating or minimizing mass and heat transfer resistance. Supercritical solvents have the ability to regenerate the catalyst during the course of the reaction, which increases the catalyst life and activity, since undesirable deposits on the catalyst, such as carbon deposits, are soluble in the supercritical fluids. The rate of the intrinsic reaction is increased in supercritical fluids and tuning the properties of the supercritical medium can control the selectivity.[88]

Supercritical fluids may also bring opportunities in downstream separation of the reactants and products. Examples of solid catalyzed reactions in supercritical fluids are given in Table 21.1.12.[88]

**Table 21.1.12. Solid catalyzed reactions in supercritical solvents**

| Reaction | Supercritical medium | Catalyst |
|---|---|---|
| **Hydrogenation** | | |
| Fats and oils | Propane, $CO_2$ | Supported platinum, palladium catalysts |
| Acetophenone, cyclohexene | $CO_2$ | Palladium on polysiloxane |
| Fischer-Tropsch synthesis | n-hexane, n-pentane, propane | Fe, Ru, $Co/Al_2O_3$, $SiO_2$ |
| **Oxidation** | | |
| Toluene | $CO_2$ | $Co/Al_2O_3$ |
| Propene | SC reactant | $CaI_2$, CuI, $Cu_2$/MgO or $Al_2O_3$ |
| Isobutane | SC reactant | $SiO_2$,$TiO_3$, Pd/carbon |

| Reaction | Supercritical medium | Catalyst |
|---|---|---|
| | **Cracking** | |
| Heptane | $CO_2$ | Zeolite |
| | **Isomerization** | |
| 1-Hexene | $CO_2$, and co-solvents | $Pt/Al_2O_3$ |
| Xylene | SC reactant | Solid acid catalyst |
| | **Alkylation** | |
| Benzene, ethylene, isopentane, isobutene, isobutane | $CO_2$ or SC reactant | Zeolite |
| Mesitylene, propene, propan-2-ol | $CO_2$ | |
| | **Disproportionation** | |
| Toluene to p-xylene, benzene | SC reactant | Zeolite |
| Ethylbenzene to benzene and diethylbenzene | Butane, pentane | Zeolite |

Catalytic Supercritical Water Oxidation is an important class of solid catalyzed reactions that utilize advantageous solution properties of supercritical water (dielectric constant, electrolytic conductance, dissociation constant, hydrogen bonding) as well as the superior transport properties of the supercritical medium (viscosity, heat capacity, diffusion coefficient, density). The most commonly encountered oxidation reactions carried out in supercritical water are oxidation of alcohols, acetic acid, ammonia, benzene, benzoic acid, butanol, chlorophenol, dichlorobenzene, phenol, 2-propanol, (catalyzed by metal oxide catalysts such as CuO/ZnO, $TiO_2$, $MnO_2$, $KMnO_4$, $V_2O_5$, $Cr_2O_3$) and 2,4-dichlorophenol, MEK, and pyridine (catalyzed by supported noble metal catalysts such as supported platinum).[89]

### *21.1.4.2.2 Heterogeneous non-catalytic reactions in supercritical solvents*

Use of the supercritical fluids as the reaction medium in synfuel processing is one of the earliest applications in the field. The advantage of the supercritical fluid as the reaction medium are again three-fold. During thermal degradation of fuels (oil-shale, coal), primary pyrolysis products usually undergo secondary reactions yielding to repolymerization (coking) or cracking into gas phase. Both reactions decrease the yield of the desired product (oil). To overcome this problem, a dense (supercritical) hydrogen donor (tetralin), or non-hydrogen donor (toluene), or an inorganic (water) medium is used.[63] Also, a supercritical medium provides ease of transport in the porous coal matrix. Finally, downstream processing (separation of the products) becomes an easier task when a supercritical medium is used. Kershaw[90] reviews the use of supercritical fluids in coal processing, while Sunol discusses the mechanism.[91]

The forest product applications in this category include biomass conversion[92] and delignification for pulping purposes.[93]

### 21.1.4.3 Biochemical reactions - examples

Due to their tunable properties, supercritical solvents provide a useful medium for enzyme catalyzed reactions.[94] The mechanism of enzyme-catalyzed reactions is similar to the mechanism described for solid catalyzed reactions. External as well as internal transport effects may limit the reaction rate. Utilizing supercritical fluids enhances external transport rate due to an increase in the diffusivity and therefore mass transfer coefficient. Internal transport rate depends on the fluid medium as well as the morphology of the enzyme. Supercritical fluids can alter both.

Water is known to be essential for the enzyme activity. Small amounts of water enhance enzyme activity, however excess water hinders the rate of some enzyme catalyzed reactions. The active site concentration on enzymes, hence the enzyme activity is found to be higher in the presence of hydrophobic supercritical fluids (ethane, ethylene) as compared to hydrophilic supercritical carbon dioxide.

The effect of pressure on enzyme catalyzed reactions can be explained in terms of the transition theory. Supercritical fluids that exhibit very high negative activation volumes for certain reactions are expected to improve the rate of the reaction.

Although, supercritical carbon dioxide has the advantage of being non-toxic and abundant, it is practically immiscible with water. Therefore, supercritical fluids used as the reaction medium in enzyme catalyzed reactions include fluoroform, sulfur hexafluoride and ethane, while lipases are the enzymes utilized in such reactions.[95]

### 21.1.4.4 Polymerization reactions - examples

Supercritical carbon dioxide is a promising green alternative to traditional solvents in polymer synthesis due to gas-like transport properties and liquid-like solubility. Supercritical carbon dioxide can be removed easily from the polymer solution by depressurization during drying of the polymer. Supercritical carbon dioxide provides easy separation of the polymer from the unreacted monomers and catalysts. Finally, supercritical carbon dioxide also exhibits Lewis acid-base interactions with electron donating functional groups of polymer chains.[96] Examples of homogeneous and heterogeneous polymerization reactions carried out in supercritical carbon dioxide are given in Table 21.1.13.

**Table 21.1.13. Polymerization reactions in supercritical fluids**

| Polymerization mechanism | Substrate |
|---|---|
| Homogeneous, free radical/cationic polymerization | Amorphous fluoropolymers |
| Precipitation, free radical polymerization | Vinyl polymer, semicrystalline fluoropolymers |
| Dispersion, free radical polymerization | Polyvinyl acetate and ethylene vinyl acetate copolymer |
| Dispersion, cationic polymerization | Isobutylene polymer |
| Homogeneous/precipitation, cationic polymerization | Vinyl ether polymer |
| Transition metal catalyzed, ring opening methathesis polymerization | Norbornene polymer, polycarbonate |

Butane, pentane, and propane are also used as the reaction medium in polymer synthesis.[97] Furthermore, some polymerization reactions (such as polyethylene synthesis) are carried out under supercritical conditions of the monomer.

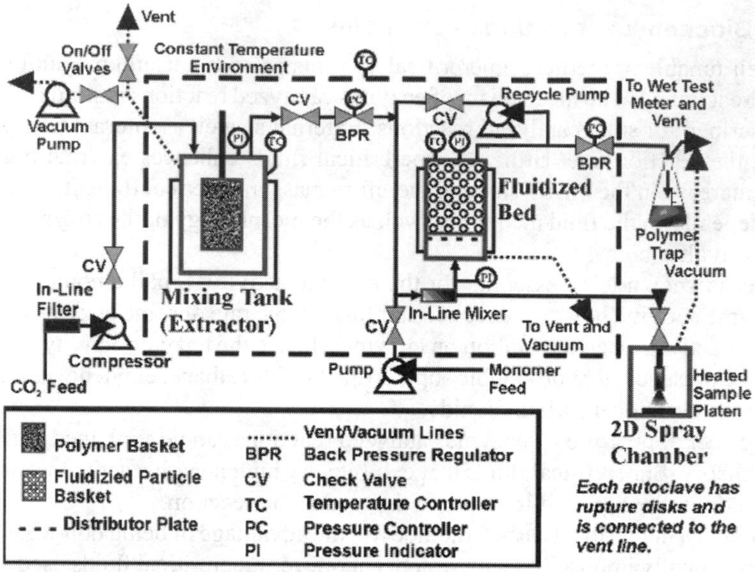


Figure 21.1.15. A flexible pilot plant for supercritical fluid aided materials processing.

### 21.1.4.5 Materials processing with supercritical solvents

The material field related applications of supercritical fluids are rapidly developing with exciting innovative developments continually emerging. The recent increase of activity in the supercritical field at large is partially due to these new horizons. Supercritical fluids are used to make highly porous material, aids in making uniform porous or non-porous films, to deposit solubilized components by diffusion and permeation into porous media resulting in composites with tailored characteristics or functionalized surfaces, is used to make particles with desired shape and particle size distribution, is used to encapsulate particles for time release applications, is used to fractionate macromolecules to desired dispersity and molecular weight distributions, and soon may be part of our household for cleaning/washing.

What makes supercritical fluids so attractive in this domain is their sensitivity to a large number of processing variables in a region where transition from a single or multiphase system into another is rather simple through a variety of paths.

One can start with a homogeneous phase and use pressure, temperature, mass separating agents, other external fields such as electromagnetic or irradiation, to nucleate and grow, or react or fractionate, to form new material with unique performance characteristics. In the homogenization step, supercritical fluids are used to solubilize. If solubilization in the supercritical fluid is not possible, the supercritical fluid can be used to induce phase separation as an anti-solvent in a subsequent step.

The supercritical fluids are effective in heterogeneous environments as well. They penetrate into porous environment loaded with additives or, used as a pure supercritical fluid to clean, dry (extract), coat, impregnate and process (e.g. extrude) a low viscosity solution.

A flexible pilot plant that addresses all the materials processing demands excluding extrusion is shown in Figure 21.1.15 while the modes for encapsulation and aerogel/impregnation are expanded upon Figures 21.1.16 and 21.1.17 respectively.

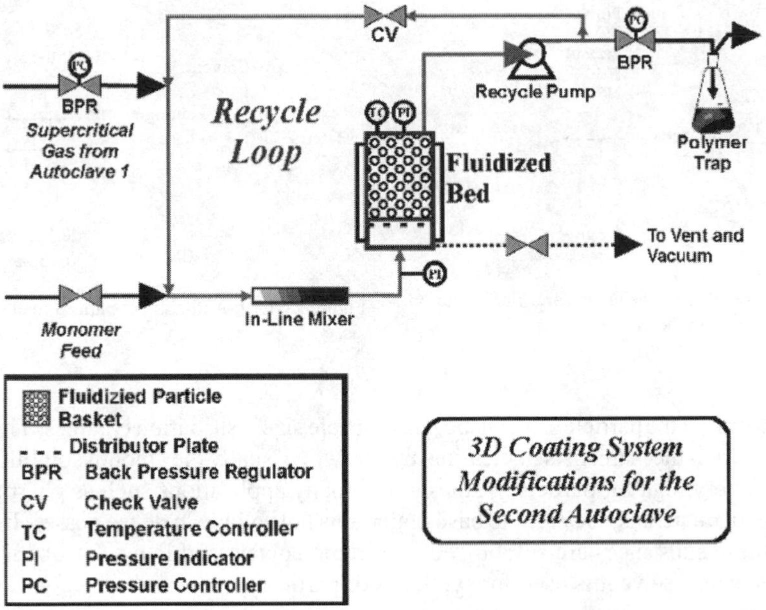


Figure 21.1.16. Particle coating section of the plant.

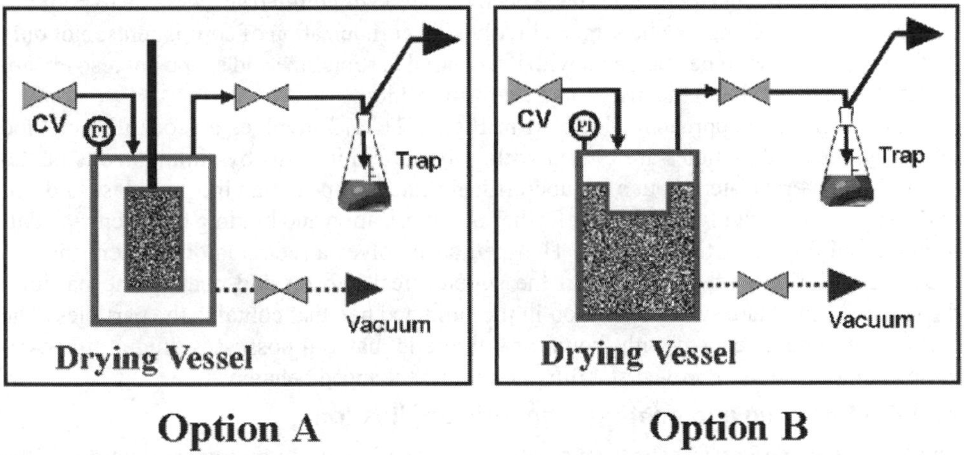


Figure 21.1.17. Supercritical drying section of the plant.

### 21.1.4.6 Particle synthesis - generic application

The routes to particle synthesis via supercritical fluids basically follow two paths, Rapid Expansion of Supercritical Solution (RESS) and Supercritical Anti-Solvent (SAS).[6,98] The basic processing steps are outlined in Figure 21.18 and Figure 21.19, respectively. RESS involves homogenization of the particles raw material in the supercritical fluid followed by rapid expansion of the solution through an expansion device such as a nozzle. Depending on the nozzle design, time-temperature and time-pressure profiles, and whether one uses

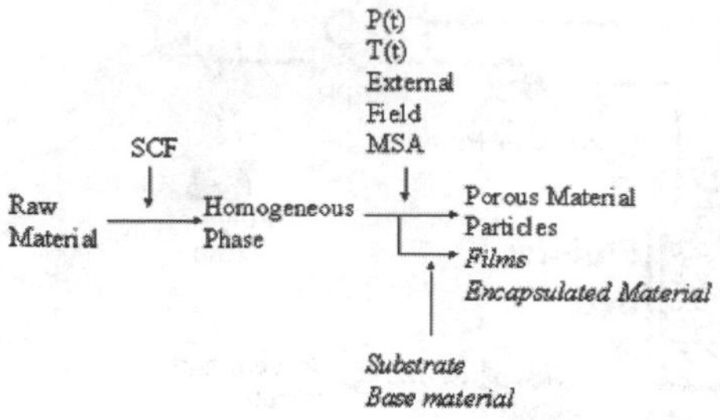


Figure 21.1.18. Homogeneous supercritical solution path to materials synthesis.

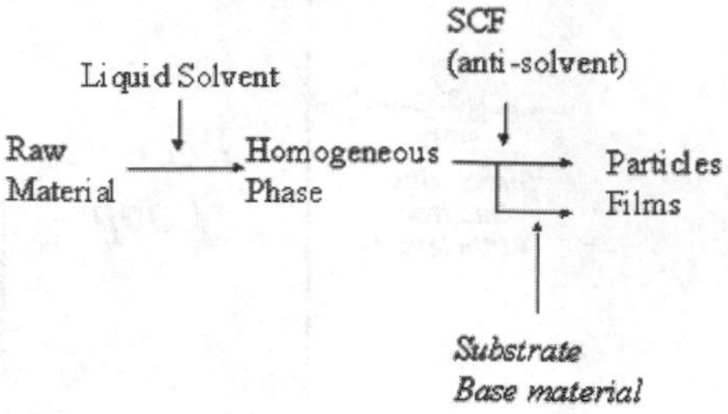


Figure 21.1.19. Anti-solvent path to materials synthesis.

entrainers or not, the particle size, shape, and particle size distribution changes. The primary particle creation mechanism is nucleation and growth. Usually the amount solubilized in the SCF is not very high for particle synthesis. For many applications such as pharmaceutical and energetic materials, the particle base material is not soluble in dense gasses. In those instances, these substances are solubilized in a liquid solvent and in a subsequent stage the supercritical anti-solvent is used for synthesis of particles.

#### 21.1.4.7 Encapsulation - generic application

Micro-encapsulation of drug-polymer systems using the RESS (Rapid Expansion of Supercritical Fluid Solutions) techniques have been initiated with limited success due to poor understanding of the complex phenomena involved in co-nucleation of components. Not only do the particles have to be nucleated with the desired particle size and shape but also encapsulate the material simultaneously in an uniform fashion.

An alternative approach[99] depicted in Figure 21.1.17 involves a sequential method where synthesized particles are coated with polymeric thin films by simultaneous nucleation of polymeric material out of a supercritical fluid, encapsulating the particles fluidized in the supercritical fluid, followed by further polymerization and binding of the encapsulating material on the particle surface. The method involves a recirculatory system that includes dissolution of the polymer in the supercritical solvent and coating the particles through a temperature swing operation in the fluidized bed that contains the particles. The particulate material coated with the tailor polymeric material possesses unique timed-release characteristics, improved stability, and often-enhanced behavior.

#### 21.1.4.8 Spraying and coating - generic application

The RESS and SAS approaches can be applied to coating and spraying applications if the inhomogenization step is exercised on a surface (a base material).

#### 21.1.4.9 Extrusion - generic application

The addition of a supercritical fluid to a Newtonian or non-Newtonian fluid reduces the viscosity of the fluid improving its processibility and end materials morphology. Supercritical fluids can aid in food and polymer extrusion.

#### 21.1.4.10 Perfusion (impregnation) - generic application

Excellent transport characteristics, solubilization power, and sensitivity to process variables all contribute to the success of the methodology. The generic scheme as shown in Fig-

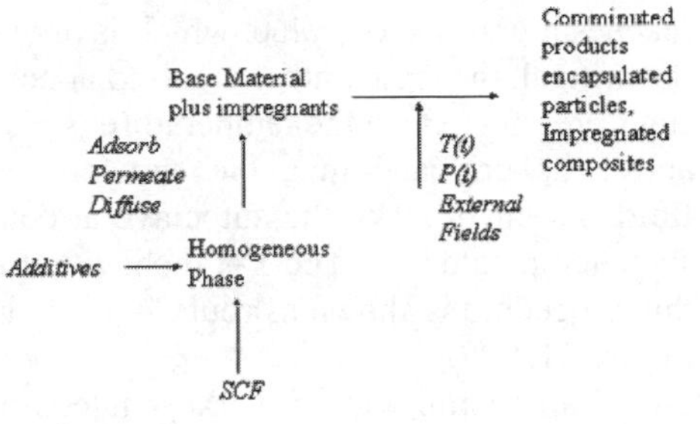


Figure 21.1.20. Supercritical fluid impregnation processes.

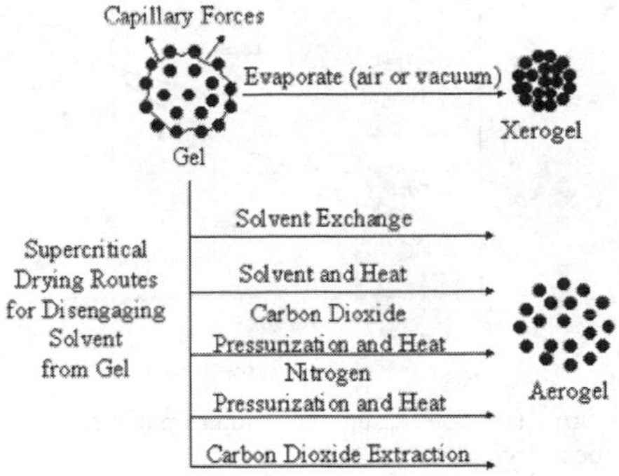


Figure 21.1.21. Sol-gel drying methods.

ure 21.1.20 involves homogenization, impregnation, deposition, and optional post treatment steps such as curing reactions. The mechanism of transport in the porous matrix is permeation and diffusion while the primary mechanisms for deposition are pressure reduction, temperature swing, sorption, and reaction (i.e., polymerization).

This process is a simplified version of impregnation process.[100] The impregnating solvent does not contain any material to be deposited and the pressure release causes disintegration of the impregnated material.[6,72]

#### 21.1.4.11 Parts cleaning - generic application

In essence, the part cleaning process is basically extraction/leaching with or without surfactants. The basic steps are shown in Figure 21.1.13. Drying in the absence of capillary forces and solvent residue free substrate makes the technology attractive. The technology is comprehensive covered in a monograph[65] and there are many existing commercial applications.

#### 21.1.4.12 Drying - generic application

A pilot plant used in drying is shown in Figures 21.1.15 and 21.1.17. The supercritical drying routes are particularly attractive in their ability to eliminate or at least minimize the capillary effect that cause non-uniformities in films as well as shrinkage and collapsing in pore structure. These supercritical avenues permit successful creation of highly porous structures such as foams, aerogels, coatings, and films.

The solvent can be removed from the wet gel using different methods. These methods and resulting gels are shown in Figure 21.1.21. If solvent is evaporated slowly from the gel, a xerogel is obtained. During evaporation, large capillary forces are exerted as the liquid-vapor interface moves through the gel. These forces cause shrinkage of the pores within the gel. Removal of the solvent (alcohol) from the gel under supercritical conditions results in the formation of the aerogel. Since this drying procedure eliminates the liquid-vapor interface, aerogels are formed in the absence of capillary forces. Aerogels retain the morphology of the original alcogel.

There are several methods developed for removing the solvent from the gel under supercritical conditions. The first one is the one suggested in the pioneering work by Kistler,[101] in which the solvent is brought to supercritical conditions in an autoclave and evacuated under these conditions. In order to pressurize the autoclave to a pressure above the critical value for the alcohol, more alcohol is added to the autoclave. Supercritical conditions of the solvent are reached by supplying heat to the autoclave. After the pressure

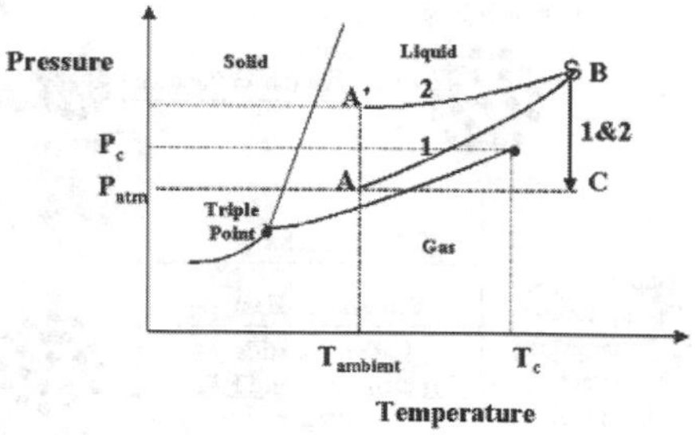


Figure 21.1.22. Pressure-temperature paths for supercritical drying.

reaches a preselected value, which is above the critical, the temperature is raised at constant pressure. Once the temperature is also above its critical value, the supercritical fluid is vented out of the autoclave at constant temperature.[102] The P-T behavior for this procedure is shown as route A-B (1) in Figure 21.1.22.

Van Lierop and co-workers later improved this method. In this improved version of the method, any boiling of the solvent was completely suppressed. This was done by introducing pressure of an inert gas into the autoclave prior to any heat treatment and furthermore, the autoclave was kept closed during heating until the temperature reached the predetermined value above the critical. The P-T behavior for this process is shown as route A'-B (2) in Figure 21.1.22.

In another method, developed by Jacobucci[104] and co-workers, alcohol is removed by supercritical $CO_2$ extraction in a semi-continuous system.

Solvent exchange is another method utilized to dry alcogels. The liquid in the pores (excess water of the hydrolysis) of the wet-gel is first displaced by alcohol and then alcohol is displaced by liquid $CO_2$ at about 16-20°C and 100 bar. After solvent exchange, the temperature of the system is increased to 40°C and supercritical carbon dioxide is vented out of the system.[105]

In yet another method, the system is pressurized with carbon dioxide prior to drying. Through partial solvent exchange during heating, drying of the gel at temperatures much lower than the critical temperature of the solvent (alcohol) can be achieved. Also, gels can be dried at a wide range of temperature and pressure combinations where optimum tunable conditions (resulting in optimum pore and surface characteristics) can be determined for a variety of objectives and cases.[106]

## REFERENCES


1 Hannay, and Hogarth, *Proc. Roy. Soc. A*, **29**, 324 (1879).
2 J. Timmermans, *J. Chim. Phys.*, **20**, 491 (1923).
3 G.M. Schneider, in **Supercritical Fluids, Fundamentals for Application**, E. Kiran, and J.M.H. Levent Sengers, Eds., *Kluver Academic Publishers*, Dordrecht, 1994, pp. 91-115.
4 G.M. Schneider, in **Water-A Comprehensive Treatise**, vol. 2, F. Frank Ed., *Plenum Press*, New York, 1973, Chp. 6.
5 E.U. Franck, in **Physical Chemistry: Advanced Treatise**, H. Eyring, D. Henderson, and W. Jost, Eds., Vol. 1, *Academic Press*, New York, 1971
6 M.A. McHugh, V.J. Krukonis, **Supercritical Fluid Extraction, Principles and Practice**, *Butterworth-Heinemann*, Boston, 1994.
7 C. J. Peters, in **Supercritical Fluids, Fundamentals for Application**, E. Kiran, and J. M.H. Levent Sengers, Eds., *Kluver Academic Publishers*, Dordrecht, 1994, pp. 117-146.
8 J. DeSwaan Arons, and T.W. deLoss, in **Models for Thermodynamic and Phase Equilibria Calculations**, S.I. Sandler, Ed., *Marcel Dekker*, New York, 1994, pp. 363-505.
9 P.H. Van Konynenburg, and R.L. Scott, *Philos. Trans. R. Soc.*, **298**, 495 (1980).
10 J.S. Rowlinson, F.L. Swinton, **Liquids and Liquid Mixtures**, *Butterworths*, London, 1982.
11 M.B. King, *Phase Equilibria in Mixtures*, *Pergamon Press*, Oxford, 1969.
12 J.C. Rainwater, in **Supercritical Fluid Technology**, T.J. Bruno, and J.F. Ely, Eds., *CRC Press*, London, 1991, pp. 57-162.

13 M.P. Ekart, J.F. Brenneke, and C.A. Eckert, in **Supercritical Fluid Technology**, T.J. Bruno, and J.F. Ely, Eds., *CRC Press*, London, 1991, pp. 163-192.

14 R. J. Sadus, **High Pressure Phase Behaviour of Multicomponent Fluid Mixtures**, *Elsevier*, Amsterdam, 1992.

15 W.B. Street, in **Chemical Engineering at Supercritical Fluid Conditions**, M.E. Paulatis, J.M.L. Penninger, R.D. Gray, and P. Davidson, Eds., *Ann Arbor Science*, Ann Arbor, MI, 1983.

16 E. Kiran, and W. Zhuang, in **Supercritical Fluids, Extraction and Pollution Prevention**, M.A. Abraham, and A.K. Sunol, Eds., *American Chemical Society*, Washington, DC, 1997, pp. 2-36.

17 Y. Tsekhankaya, M.B. Jomtev, and E.V. Mashkina, *Z. Fiz. Chim.*, **38**, 2166 (1964).

18 E.H. Chimowitz and K.J. Pennisi, *AICHE J.*, **32**, 1665 (1986).

19 S.G. Sunol, B.Mierau, I. Serifoglu, and A.K. Sunol, in **Supercritical Fluids, Extraction and Pollution Prevention**, M.A. Abraham, and A.K. Sunol Eds., *American Chemical Society*, Washington, DC, 1997, pp.188-206.

20 T.J. Bruno, in **Supercritical Fluid Technology**, T.J. Bruno, and J.F. Ely, Eds., *CRC Press,* London, 1991, pp. 293-324.

21 G. Brunner, **Gas Extraction**, *Springer*, New York, 1994.

22 J.W. Gibbs, *Trans. Conn. Acad. III*, **108** (1876), 343 (1878).

23 R.C. Reid, and B.L. Beegle, *AIChE J.*, **23**, (1977).

24 M.L. Michelsen, and R.A. Heidemann, *AIChE J.*, **27**, 521 (1981).

25 R.A. Heidemann, in **Supercritical Fluids, Fundamentals for Application**, E. Kiran, and J.M.H. Levent Sengers, Eds., *Kluver Academic Publishers*, Dordrecht, 1994, pp. 39-64.

26 J.W. Tester, and M. Modell, **Thermodynamics and Its Applications**, *Prentice Hall PTR*, New Jersey, 1997.

27 A.Z. Panagiotopoulos, in **Supercritical Fluids, Fundamentals for Application**, E. Kiran, and J.M.H. Levent Sengers, Eds., *Kluver Academic Publishers*, Dordrecht, 1994, pp. 411-437.

28 K.S. Shing, in **Supercritical Fluid Technology**, T.J. Bruno, and J.F. Ely, Eds., *CRC Press*, London, 1991, pp. 227-244.

29 J.M.H. Sengers, in **Supercritical Fluids, Fundamentals for Application**, E. Kiran, and J.M.H. Levelt Sengers, Eds., *Kluver Academic Publishers*, Dordrecht, 1994, pp. 3-38.

30 J.V. Sengers, in **Supercritical Fluids, Fundamentals for Application**, E. Kiran, and J.M.H. Levelt Sengers, Eds., *Kluver Academic Publishers*, Dordrecht, 1994, pp. 231-272.

31 S.I. Sandler, H. Orbey, and B.-I. Lee, in **Models for Thermodynamic and Phase Equilibria Calculations**, S.I. Sandler, Ed., *Marcel Dekker,* New York, 1994, pp. 87-186.

32 D.Y. Peng, and D.B. Robinson, *Ind. Eng. Chem. Fundam.*, **15**, 59 (1976).

33 G.S. Soave, *Chem. Eng. Sci.*, **27**, 1197 (1972).

34 J.M. Prausnitz, R.N. Lichtenthaler, E.G. de Azevedo, 3rd ed., *Prentice Hall* PTR, New Jersey, 1999.

35 J.J. Martin, *Ind. Eng. Chem. Fundam.*, **18**, 81 (1979).

36 A.Z. Panagiotopoulos and R.C. Reid, *Fluid Phase Equilibria*, **29**, 525 (1986).

37 D.S.H. Wong and S.L. Sandler, *AICHE J.*, **38**, 671 (1992).

38 H. Orbey, and S.I. Sandler, **Modeling of Vapor-Liquid Equilibria**, *Cambridge University Press*, Cambridge, 1998.

39 I.C. Sanchez, and C. Panayiotou, in **Models for Thermodynamic and Phase Equilibria Calculations**, S.I. Sandler, Ed., *Marcel Dekker,* New York, 1994, pp. 187-285.

40 I.C. Sanchez, and R.H. Lacombe, *J. Phys. Chem.*, **80**, 2352, 2568 (1976).

41 W.G. Chapman, G. Jackson, and K.E. Gubbins, *Molec. Phys.*, **65**, 1057 (1988).

42 V. Vesovic, and W.A. Wakeham, in **Supercritical Fluid Technology**, T.J. Bruno, and J.F. Ely, Eds., *CRC Press*, London, 1991, pp. 245-289.

43 W.D. Seider, J.D. Seader, and D.R. Lewin, **Process Design Principles**, *Wiley*, New York, 1999.

44 R.C. Reid, J.M. Prausnitz, and B Poiling, **The Properties of Gases and Liquids**, 4th ed., *McGraw Hill*, New York (1987).

45 H. Saad and E. Gulari, *Ber. Bunsenges. Phys. Chem.*, **88**, 834 (1984).

46 G. Taylor, *Proc. Roy. Soc. A*, **219**, 186 (1953).

47 R. Aris, *Proc. Roy. Soc. A*, **235**, 67 (1956).

48 C.A.N. deCastro, in **Supercritical Fluid Technology**, T.J. Bruno, and J.F. Ely, Eds., *CRC Press*, London, 1991, pp. 335-363.

49 S. Peter, A. Blaha-Schnabel, H. Schiemann, and E. Weidner, in **Supercritical Fluids, Fundamentals for Application**, E. Kiran, and J.M.H. Levent Sengers, Eds., *Kluver Academic Publishers*, Dordrecht, 1994, pp. 731-738

50 R.T. Kurnik, and R.C. Reid, *J. Fluid Phase Equilib.*, **8**, 93 (1982).

51 J.M. Dobbs, J.M. Wong, and K.P. Johnston, *Ind. Eng. Chem. Res.*, **26**, 56 (1987)

52 A.K. Sunol, B. Hagh, and S. Chen, in **Supercritical Fluid Technology**, J.M.L. Penniger, M. Radosz, M.A. McHugh, and V.J. Krukonis, Eds., *Elsevier*, Amsterdam, 1985, pp. 451-464.

53 C.D. Hubbard and R. van Eldik, in **Chemistry Under Extreme or Non-Classical Conditions**, R. van Eldik, and C.D. Hubbard, Eds., *Wiley*, New York, 1997, pp. 53-102.

54 C.-S. Tan, and D.-C. Liou, *Ind. Eng. Chem. Res.*, **29**, 1412 (1990).

55 C.-S. Tan, and D.-C. Liou, *Ind. Eng. Chem. Res.*, **27**, 988 (1988).

56 U. Akman, A.K. Sunol, *AIChE J.*, **37**, 215 (1991).

57 E.J. Beckman, J.L. Fulton, and R.D. Smith, in **Supercritical Fluid Technology**, T.J. Bruno, and J.F. Ely, Eds., *CRC Press*, London, 1991, pp. 405-449.

58 J.L. Bravo, J.R. Fair, J.L. Humphrey, C.L. Martin, A.F. Seibert, and S. Joshi, **Fluid Mixture Separation Technologies for Cost Reduction and Process Improvement**, *Noyes Pub*., New Jersey, 1986.

59 E. Lack, and H. Seidlitz, in **Extraction of Natural Products Using Near-Critical Solvents**, M.B. King, and T.R. Bott, Eds., *Blackie Academic & Professional*, Glasgow, 1993, pp. 101-139.

60 D.S. Gardner, in **Extraction of Natural Products Using Near-Critical Solvents**, M.B. King, and T.R. Bott, Eds., *Blackie Academic & Professional,* Glasgow, 1993, pp. 84-100.

61 N. Sanders, in **Extraction of Natural Products Using Near-Critical Solvents**, M.B. King, and T.R. Bott, Eds., *Blackie Academic & Professional*, Glasgow, 1993, pp. 34-49.

62 D.A. Moyler, in **Extraction of Natural Products Using Near-Critical Solvents**, M.B. King, and T.R. Bott, Eds., *Blackie Academic & Professional*, Glasgow, 1993, pp. 140-183.

63 R.R. Maddox, J. Gibson, and D.F. Williams, *CEP*, June (1979).

64 A. Wilhelm, and K. Hedden, in **Supercritical Fluid Technology**, J.M.L. Penniger, M. Radosz, M.A. McHugh, and V.J. Krukonis, Eds., *Elsevier*, Amsterdam, 1985, pp. 357-375.

65 J. McHardy and S. P. Sawan Eds. **Supercritical Fluid Cleaning**, *Noyes Pub*., New Jersey, 1998

66 U. Nguyen, D.A. Evans, and G. Frakman, in **Supercritical Fluid Processing of Food and Biomaterials**, H. Rizvi, Ed., *Blackie Academic & Professional,* London, 1994, pp. 103-113.

67 A. Akgerman, G. Madras, in **Supercritical Fluids, Fundamentals for Application**, E. Kiran, and J.M.H. Levent Sengers, Eds., *Kluver Academic Publishers*, Dordrecht, 1994, pp. 669-695.

68 J.-H. Hsu, and C.-S. Tan, in., **Supercritical Fluid Processing of Food and Biomaterials**, H. Rizvi, Ed. *Blackie Academic & Professional*, London, 1994, pp. 114-122.

69 M.B. King, and O. Catchpole, in **Extraction of Natural Products Using Near-Critical Solvents**, M.B. King, and T.R. Bott, Eds., *Blackie Academic & Professional*, Glasgow, 1993, pp. 184-231.

70 R. Eggers, in **Extraction of Natural Products Using Near-Critical Solvents**, M.B. King, and T.R. Bott, Eds., *Blackie Academic & Professional,* Glasgow, 1993, pp. 232-260.

71 G. Vetter, in **Extraction of Natural Products Using Near-Critical Solvents**, M.B. King, and T.R. Bott, Eds., *Blackie Academic & Professional,* Glasgow, 1993, pp. 261-298.

72 E. Stahl, K.W. Quirin, D. Gerard, **Dense Gases for Extraction and Refining**, *Springer-Verlag*, Berlin, 1988.

73 H. Schmieder, N. Dahmen, J. Schon, and G.Wiegand, in **Chemistry Under Extreme or Non-Classical Conditions**, R. van Eldik, and C.D. Hubbard, Eds., Wiley, New York, 1997, pp. 273-316.

74 Y. Ikushima, N. Saito, K. Hatakeda, and S. Ito, in **Supercritical Fluid Processing of Food and Biomaterials**, H. Rizvi, Ed., *Blackie Academic & Professional*, London, 1994, pp. 244-254.

75 M.H.M. Caralp, A.A. Clifford, and S.E. Coleby, in **Extraction of Natural Products Using Near-Critical Solvents**, M.B. King, and T.R. Bott, Eds., *Blackie Academic & Professional,* Glasgow, 1993, pp. 50-83.

76 A. Akgerman, in **Supercritical Fluids, Extraction and Pollution Prevention**, M.A. Abraham, and A.K. Sunol Eds., *American Chemical Society*, Washington, DC, 1997, pp. 208-231.

77 B.Mierau, and A.K. Sunol, in **High Pressure Chemical Engineering**, Ph. R. von Ruhr, and Ch. Trepp Eds., *Elsevier*, 1996, Amsterdam, pp. 321-326.

78 I. Flament, U. Keller, and L. Wunsche, in **Supercritical Fluid Processing of Food and Biomaterials**, H. Rizvi, Ed., *Blackie Academic & Professional,* London, 1994, pp. 62-74.

79 B. Subramanian, and M.A. McHugh, , *Ind. Eng. Chem. Res*., **37**, 4203 (1998).

80 P.E. Savage, S. Gopalan, T.I. Mizan, C.J. Martino, and E.E. Brock, *AIChE J*., **41**, 1723 (1995).

81 A.A. Clifford, in **Supercritical Fluids, Fundamentals for Application**, E. Kiran, and J.M.H. Levent Sengers, Eds., *Kluver Academic Publishers*, Dordrecht, 1994, pp. 449-479.

82 B.C. Wu, S.C. Paspek, M.T. Klein, and C. LaMarka, in **Supercritical Fluid Technology**, T.J. Bruno, and J.F. Ely, Eds., *CRC Press*, London, 1991, pp. 511-524.

83 G.P. Jessop, T. Ikariya, and R. Noyori, *Chem. Rev*., **99**, 474 (1999).

84 P.E. Savage, *Chem. Rev*., **99**, 603 (1999).

85 M. Poliakoff, M.W. George, and S.M. Howdle, in **Chemistry Under Extreme or Non-Classical Conditions**, R. van Eldik, and C.D. Hubbard, Eds., *Wiley*, New York, 1997, pp. 189-218.
86 Q. Ji, E.M. Eyring, R. van Eldik, K.P. Johnston, S.R. Goates, M.L. Lee, *J. Phys. Chem.*, **99**, 13461 (1995).
87 C.B. Roberts, J.F. Brennecke, and J.E. Chateauneuf, *AIChE J.*, **41**, 1306 (1995).
88 A. Baiker, *Chem. Rev.*, **99**, 443 (1999).
89 Z.Y. Ding, M.A. Frisch, L. Li, and E.F. Gloyna, *Ind. Eng. Chem. Res.*, **35**, 3257 (1996).
90 J.R. Kershaw, *J. Supercrit. Fluids*, **2**, 35 (1989).
91 A.K. Sunol, and G.H. Beyer, *Ind. Eng. Chem. Res.*, **29**, 842 (1990).
92 J.R. Vick Roy, and A.O. Converse, in **Supercritical Fluid Technology**, J.M.L. Penniger, M. Radosz, M.A. McHugh, and V.J. Krukonis, Eds., *Elsevier,* Amsterdam, 1985, pp. 397-414.
93 K.A. Sunol, **US Patent 5,041,192**.
94 K. Nakamura, in **Supercritical Fluid Processing of Food and Biomaterials**, S.S,H. Rizvi, Ed., *Blackie Academic & Professional*, London, 1994, pp. 54-61.
95 A.J. Mesiano, E.J. Beckman, and A.J. Russell, *Chem. Rev.*, **99**, 623 (1999).
96 J.L. Kendall, D.A. Canelas, J.L. Young, and J.M. DeSimone, *Chem. Rev.*, **99**, 543 (1999).
97 G. Sirinivasan, J.R. Elliot, *Ind. Eng. Chem. Res.*, **31**, 1414 (1992).
98 P.G. Debedenetti, in **Supercritical Fluids, Fundamentals for Application**, E. Kiran, and J.M.H. Levent Sengers, Eds., *Kluver Academic Publishers*, Dordrecht, 1994, pp. 719-729.
99 A.K. Sunol, J. Kosky, M. Murphy, E. Hansen, J. Jones, B. Mierau, S. G. Sunol, in Proceedings of the 5th International Symposium on Supercritical Fluids, Tome 1, ISAFS , Nice, 1998.
100 K.A. Sunol, **US Patent 4,992,308**.
101 S.S. Kistler, *J. Physical Chemistry*, **36** (1932) 52.
102 W.J. Schmitt, R.A. Grieger-Block, T.W. Chapman, in **Chemical Engineering at Supercritical Fluid Conditions**, M.E. Paulatis, J.M.L. Penninger, R.D. Gray, and P. Davidson, Eds., *Ann Arbor Science*, Ann Arbor, MI, 1983, p.445-460.
103 J.G. Van Lierop, A. Huitzing, W.C.P.M. Meerman and C.A.M. Mulder, *J. Non-Crystalline Solids*, **82**, 265 (1986).
104 R.J. Ayen, P.A. Iaobucci, *Reviews in Chemical Engineering*, **5**, 157 (1988).
105 B. Rangarajan, C.R. Lira, *J. Supercritical Fluids*, **4**, 1 (1991).
106 S.G. Sunol, A.K. Sunol, O.Keskin, O. Guney, **Innovations in Supercritical Fluids Science and Technology**, K. W. Hutchenson and N. R. Foster Eds., *American Chemical Society*, Washington, DC, 1995, pp. 258-268.

## 21.2 IONIC LIQUIDS

D.W. Rooney, K.R. Seddon
**School of Chemistry, The Queen's University of Belfast**
**Belfast, Northern Ireland**

### 21.2.1 INTRODUCTION

The first question one would ask is "What is an ionic liquid?" Ionic liquids can be described, in the crudest terms, as room temperature molten salts. The term "ionic liquid" could therefore be applied to all molten salt systems such as cryolite ($Na_3AlF_6$) used in aluminum production or even molten table salt (NaCl). However the use of the word "molten" conjures up images of high temperature processes which are highly corrosive and difficult to design. In contrast to this, the ionic liquids discussed here are generally benign solvents which can be applied to a significant number of industrial processes leading to enhanced yields, greater recyclability and processes with an overall reduced environmental impact. Therefore an ionic liquid is normally described as a molten salts which is fluid at room temperature, or close to room temperature (salts melting below 100°C are often considered in this category). One only has to look at recent patent publications to discover that these solvents are finding application in commercial sectors as diverse as the nuclear industry,[1,2] pharmaceuticals and fine chemicals,[3,4] as well as in mainstream petrochemical processes.[5-7] This contrasts with the unique perspective of Takahashi et al. who state that the two major applications are batteries and electrolytes, clearly unaware of the modern literature.[8] In a number of indicated cases, these processes have been taken through the development process to a point of industrial commercialization and represent first generation ionic liquid processes, principally based on chloroaluminate(III) ionic liquids which are currently ready for industrial uptake. Following this line, second generation ionic liquid processes based on other, more benign, ionic liquids are currently under investigation and development in a variety of laboratories around the world. Many of these processes utilize the ability of a range of ionic liquids to selectively immobilize transition metal catalysts for liquid-liquid two-phase catalysis while permitting easy, often trivial, extraction of products.[9-14]

So why have these solvents not been used to rapidly replace the current volatile organics currently found in industry? There are a number of answers to this, but the most significant reason is economics. By looking through any chemical catalogue, it is obvious that 1-methylimidizole (a precursor for the manufacture of ionic liquids) is considerably more expensive than standard solvents. By the time this compound is processed to the final ionic liquid product, its cost will have increased to many times that of normal solvents. This is not helped by the fact that at present there is a very limited market for these compounds, keeping retail prices high. It is envisaged that if ionic liquid technology does become widely accepted, then the cost of production will decrease rapidly. Other quaternary ammonium salts and those that are based around pyridinium are cheaper alternatives (principally due to the scale of manufacture), but economics is not the only problem. As yet, there is a significant deficit of raw physical property data for engineers to use when designing new processes or retrofitting old plants.

For the efficient design of any new industrial process incorporating ionic liquid technology, a complete understanding of the behavior of the solvent during operation is necessary. Physical properties such as viscosity, density, heat capacity and surface tension are all important during these early design stages. Others like electrochemical windows and electrical conductivities, will be important for more specific applications. With the advent of computers, chemical engineers have been able to use powerful process simulation software packages to estimate how a particular process will behave under certain operating conditions. These software packages, like the manual calculations which preceded them, predict the physical properties of organic solvents by using a number of empirical and semi-empirical equations which are available. Unfortunately these equations where developed for molecular compounds and tend to require the critical temperature and pressure data, information which does not apply to ionic liquids. In addition, other techniques for predicting physical properties like surface tension involve using group contribution methods, but again these fail to account for organic salts and therefore cannot

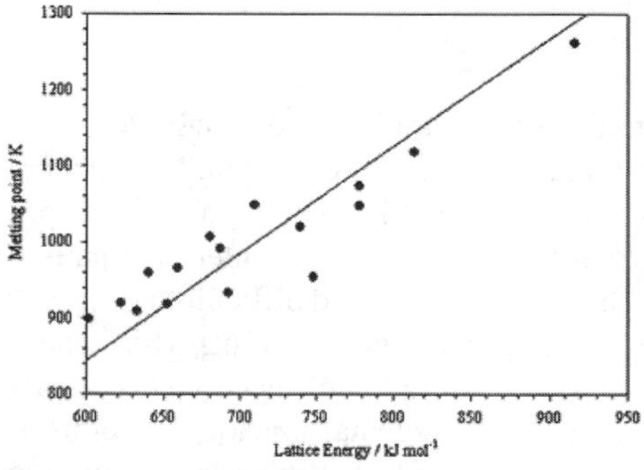

Figure 21.2.1. Showing the relationship between lattice energy[26] and melting point[27] for the Group 1 halide salts.

be used. This lack of predictive power and the overall general lack of physical property data in the literature will inevitably slow the transfer of ionic liquid technology into industrial processes.

## 21.2.2 FUNDAMENTAL PRINCIPLES OF THE FORMATION OF ROOM TEMPERATURE IONIC LIQUIDS

A number of detailed reviews and articles have been published recently on the theory and development behind ionic liquid technology.[15-25] However in order to develop a greater understanding and appreciation of these solvents, we will summarize the main points here. Therefore the purpose of this section is to give an introduction to the nature and properties of room-temperature ionic liquids, with particular emphasis being placed upon their potential as solvents for industrially relevant catalytic reactions, and (more generally) for clean technology.

### 21.2.2.1 Development of ionic liquids

To begin with, the melting point of a salt is related to its lattice energy. In fact, if one was to plot the lattice energy of a series of salts, for instance the Group 1 halides, against the melting points in Kelvin, then one can see that there is reasonable linearity between the melting point and the lattice energy (see Figure 21.2.1).

Although not particularly accurate, this simple approach will give an estimated melting point of the Group 1 salts if the lattice energy is known. The deviations from this treatment are frequent and are usually due to other forms of bonding within the structure, as shall be explained later. The first theoretical treatment of lattice energy began with Born and Landé, and was then further developed by Kapustinskii[28] into the what is know as the "Kapustinskii equation" (eqn. [21.2.1]):

$$U = \frac{287.2 v Z^+ Z^-}{r_0}\left(1 - \frac{0.345}{r_0}\right) \qquad [21.2.1]$$

where:

| | |
|---|---|
| U | lattice energy |
| v | number of ions per molecule |
| $r_0$ | sum of ionic radii |
| $Z^+$, $Z^-$ | charge of the ionic species |

Therefore by increasing the value of $r_0$, i.e., by using larger anionic and cationic components in the salt, it is possible to lower this energy and therefore reduce the melting point. This effect is shown Table 21.2.1.

**Table 21.2.1. Melting points[27] of various inorganic salts, melting points given in °C**

| Anion | Lithium | Sodium | Potassium | Rubidium | Caesium |
|---|---|---|---|---|---|
| Fluoride | 842 | 988 | 846 | 775 | 682 |
| Chloride | 614 | 801 | 776 | 718 | 645 |
| Bromide | 550 | 747 | 734 | 693 | 636 |
| Iodide | 450 | 661 | 686 | 647 | 626 |

From this one can clearly see that as we increase both the size of the anion and the cation, the melting point decreases. From the Kapustinskii equation one must also note that by

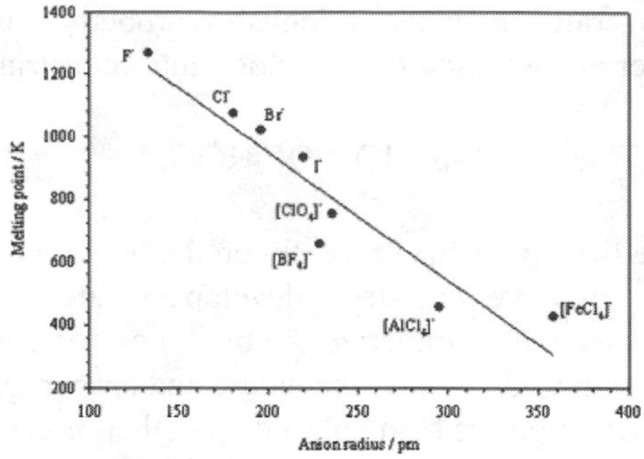


Figure 21.2.2. Showing the relationship between the size of similar anions containing halides and melting point of various sodium salts.

increasing the ionic charge will tend to increase the lattice energy of the crystal. However the effect on melting point is complicated by the fact that according to Fajans' rules an increasing charge also results in increasing covalency particularly for small cations and large anions. The effect of size can even be observed in the melting points of the lithium halide series, where the small size of the lithium ion leads to a greater covalent character in the lithium halide bond, reducing the melting point when compared to the other Group 1 salts.

For a given cation, and singly charged cations and ions, equation [21.2.1] demonstrates that the lattice energy will become only be a function of $r_0$, and since $r_0$ is the sum of $r^+$ and $r^-$, then the lattice energy is only dependant on anion size. The radius for simple anions can be found in a number of different sources.[27,29] However for more complex anions it becomes more difficult to assign unique values to the radii. Yatsimirskii[30] has shown that it is possible to determine the apparent values of the ionic radii of these ions indirectly from the lattice energy of the compounds containing them. These values are generally referred to as the thermochemical radii and a number are shown in Table 21.2.2. In most cases the fact that ions such as $[CNS]^-$ and $[CH_3COO]^-$, are markedly non-spherical makes these radii of limited use. However for the case of tetrahedral ions, the symmetry is sufficiently high enough for comparison purposes. If one were to plot the melting points a number of sodium salts against the opposing anions' thermochemical radius a clear relationship is observed (see Figure 21.2.2)

**Table 21.2.2. Anionic and thermochemical radii**

| Name | Anion | r, pm | Name | Anion | r, pm |
|---|---|---|---|---|---|
| Tetrachloroferrate[31] | $[FeCl_4]^-$ | 358 | Chlorate[33] | $[ClO_3]^-$ | 201 |
| Tetrachloroborate[31] | $[BCl_4]^-$ | 310 | Thiocyanate[33] | $[CNS]^-$ | 195 |
| Tetrachloroaluminate[31] | $[AlCl_4]^-$ | 295 | Borate[33] | $[BiO_4]^-$ | 191 |
| Tetrachlorogallate[31] | $[GaCl_4]^-$ | 289 | Bromate[33] | $[BrO_3]^-$ | 191 |
| Trifluoromethylsulfonate[32] | $[CF_3SO_3]^-$ | 267 | Nitrate[33] | $[NO_3]^-$ | 188 |
| Hexafluoroarsenate[32] | $[AsF_6]^-$ | 259 | Cyanide[33] | $[CN]^-$ | 181 |
| Hexafluorophosphate[32] | $[PF_6]^-$ | 253 | Iodate[33] | $[IO_3]^-$ | 181 |
| Periodate[33] | $[IO_4]^-$ | 249 | Cyanate[33] | $[CNO]^-$ | 160 |
| Perchlorate[33] | $[ClO_4]^-$ | 236 | Formate[33] | $[HCOO]^-$ | 160 |
| Permanganate[33] | $[MnO_4]^-$ | 240 | Acetate[33] | $[CH_3CO_2]^-$ | 156 |
| Tetrafluoroborate[33] | $[BF_4]^-$ | 229 | Nitrite[33] | $[NO_2]^-$ | 156 |
| Trinitrophenoxide[33] | $[C_6H_2(NO_2)_3O]^-$ | 222 | Amide[33] | $[NH_2]^-$ | 129 |

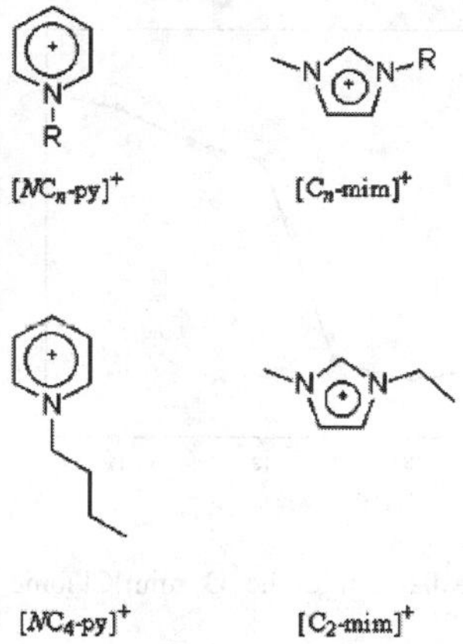


Figure 21.2.3. The aromatic heterocyclic N-butylpyridinium, 1-ethyl methylimidazolium cations, and general structural information for $[NC_n\text{-py}]^+$ and $[C_n\text{-mim}]^+$ based ionic liquids.

**Table 21.2.3. Some melting points for various [$C_2$-mim] salts**

| Salt | Melting point, °C |
|---|---|
| $[C_2\text{-mim}]Cl$[35] | 84 |
| $[C_2\text{-mim}]Br$[36,37] | 81 |
| $[C_2\text{-mim}]I$[36,37] | 79-81 |
| $[C_2\text{-mim}][PF_6]$[38] | 62 |
| $[C_2\text{-mim}][NO_2]$[39] | 55 |
| $[C_2\text{-mim}][GaCl_4]$[40] | 47 |
| $[C_2\text{-mim}][NO_3]$[39] | 38 |
| $[C_2\text{-mim}][AlCl_4]$[35] | 7 |
| $[C_2\text{-mim}][(CF_3SO_2)_2N]$[38] | -15 |

It should be noted that if one plotted the melting points and radii of the planar anions in Figure 21.2.2, they would not fall on the trend line. Since our desired goal is to produce a salt which melts at or around room temperature, we extrapolate that we would require an anion with a radius of at least 400 pm. The number of such large anions is limited, and they also have a relatively high degree of covalence associated with the structure. However, as clear from above, it is also possible to reduce the lattice energy by increasing the size of the cation. By moving to organic salts the effect can be demonstrated more easily. For example the organic salt $[EtNH_3][NO_3]$ was shown in 1914 to have a melting point of 12°C and was hence the first room temperature ionic liquid.[34] This concept can be taken further and applied to other organic systems, such as, the cations "butylpyridinium" and "1-ethyl 3-methylimidazolium" ($[NC_4\text{-py}]^+$ $[C_2\text{-mim}]^+$), (see Figure 21.2.3). For the purpose of clarity we will simply refer to the length of the alkyl chain by the number of carbon atoms in that chain, hence ethyl is represented by $C_2$, butyl by $C_4$, etc.

In the case of the $[C_2\text{-mim}]^+$ cation it can be observed that it has the lowest possible symmetry making it even more difficult for a crystal to form. Given this argument one would expect the melting point of salts based around this cation to be low. Indeed Wilkes et al.[35] have shown that the melting point for $[C_2\text{-mim}]Cl$ is 84°C significantly lower than any of the Group 1 halides. The melting points of a number of $[C_2\text{-mim}]^+$ salts have been reported over recent years, and some of these are summarized in Table 21.2.3.

In order to develop a greater understanding of these salts, the nature of the $[C_2\text{-mim}]^+$ cation and how it interacts with various anions has been explored in both solution[41] and solid[42] phases and via theoretical studies.[43] It is now possible to produce a similar plot to Figure 21.2.2 showing the effect of anion on the melting point of the $[C_2\text{-mim}]^+$ salts using the data in Table 21.2.3.

The $[AlCl_4]^-$ anomaly can be partly explained by the fact that in the ionic liquid an equilibrium existing between the $[AlCl_4]^-$ anion and the larger $[Al_2Cl_7]^-$ anion, is expected to be much higher than for the $[GaCl_4]^-$ ionic liquid. In addition results from our own research

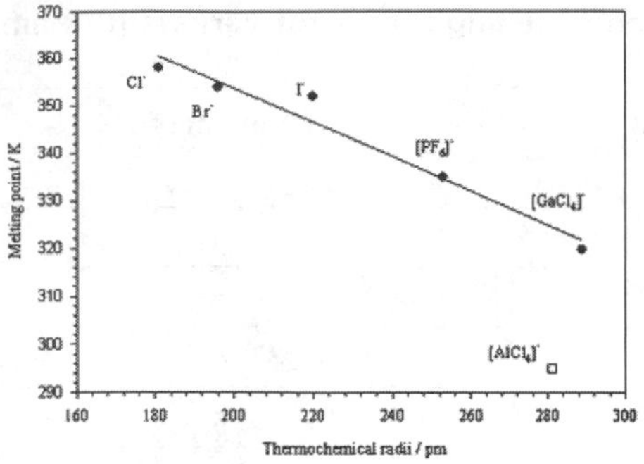


Figure 21.2.4. Relationship between the size of similar anions containing halides and the melting point of their [$C_2$-mim] salts.

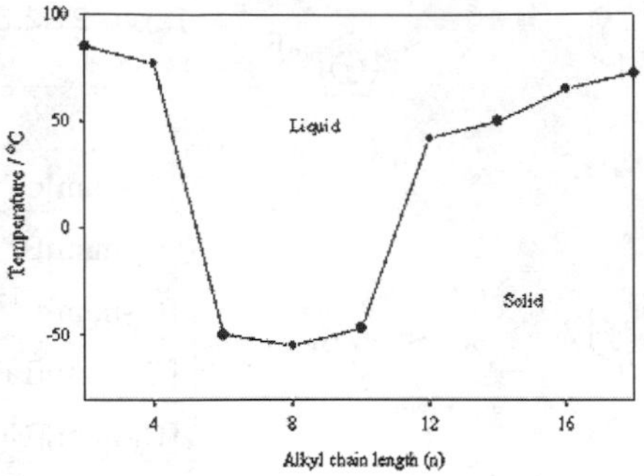


Figure 21.2.5. Phase diagram for the [$C_n$-mim]Cl ionic liquids.[44]

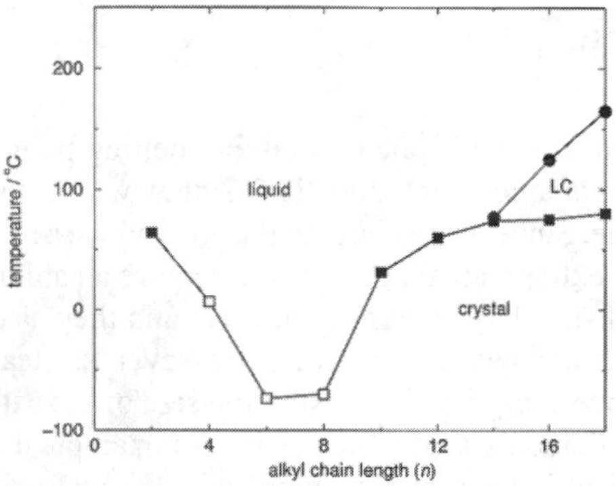


Figure 21.2.6. Melting point phase diagram for [$C_n$-mim][$PF_6$] ionic liquids as a function of alkyl chain length n showing the melting transitions from crystalline (closed square) and glassy (open square) materials and the clearing transition (circle) of the liquid crystalline (LC) terms.

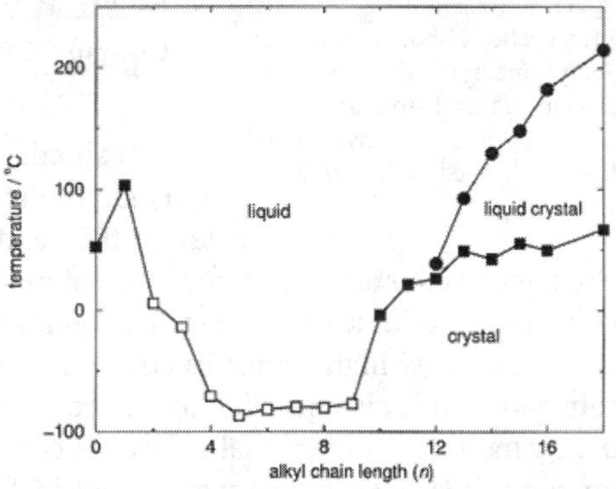


Figure 21.2.7. Melting point phase diagram for [$C_n$-mim][$BF_4$] ionic liquids as a function of alkyl chain length n showing the melting transitions from crystalline (closed square) and glassy (open square) materials and the clearing transition (circle) of the liquid crystalline terms.

show that the melting point for the [$FeCl_4$]$^-$ ionic liquid does appear to follow the given trend.

In order to expand the number of potential ionic liquids the possibility also exists to increase the length of the alkyl group thereby further decreasing the lattice energy. For the [$C_n$-mim] chlorides the result is a range of salts all of which have low melting points (some of which are below room temperature) as shown in Figure 21.2.5, though often at the cost of increased viscosity.

One can observe that there is a maximum chain length allowed before other forms of bonding begin to dominate and the melting point increases. The melting points indicated below the 0°C line are in fact glass transition temperatures rather than true melting points. As stated above, one point to note is that the viscosity of these room temperature "chloride" liquids is very high and they would therefore have to be used at relatively high temperatures if required in a process.

Figures 21.2.6 and 21.2.7 illustrate the significant variation in melting point which can be induced by simply changing the anion from [$C_n$-mim][Cl] to the [$C_n$-mim][$PF_6$] and

**Table 21.2.4 Selected melting points of the Group 1 chloride mixtures[48]**

| System | Mole ratio | m.p., °C |
|---|---|---|
| LiCl-LiF | 70:30 | 496 |
| LiCl-LiI | 35:65 | 368 |
| LiCl-NaCl | 25:75 | 551 |
| LiCl-CsCl | 60:40 | 355 |
| NaCl-KCl | 50:50 | 658 |
| CsCl-KCl | 35:65 | 610 |

**Table 21.2.5. The melting points of selected tetrachloroaluminate(III) salts[48]**

| System | Mole ratio | m.p., °C |
|---|---|---|
| LiCl-$AlCl_3$ | 35:65 | 80 |
| LiCl-$AlCl_3$ | 50:50 | 132 |
| NaCl-$AlCl_3$ | 39:61 | 108 |
| NaCl-$AlCl_3$ | 50:50 | 151 |
| RbCl-$AlCl_3$ | 30:70 | 148 |
| RbCl-$AlCl_3$ | 50:50 | 336 |

[$C_n$-mim][$BF_4$] ionic liquids.[45-47] This procedure has produced a lower melting point salt for example the [$C_4$-mim][$PF_6$] has a melting point of 5°C whereas the [$C_4$-mim][Cl] has a melting point of 80°C. These lower melting point liquids with the shorter alkyl chains lead to a much more fluid and easily managed liquid.

An interesting feature of these phase diagrams is the appearance of liquid crystalline phases with the longer alkyl chains, and this is confirmed when their optical textures are examined.[35] The implication of the existence of these stable phases has still to be explored in terms of stereochemical control of reactions.

From these figures it can be seen that those salts with short alkyl chains (n=2-10) are isotropic ionic liquids at room temperature and exhibit a wide liquid range, whereas the longer chain analogues are low melting mesomorphic crystalline solids which display an enantiotropic smectic A mesophase. The thermal range of the mesophase increases with increasing chain length and in the case of the longest chain salt prepared, [$C_{18}$-mim][$BF_4$], the mesophase range is ca. 150°C.

### 21.2.2.2 Binary ionic liquid systems

In Table 21.2.1 it was shown that the melting points of the Group 1 salts are significantly above room temperature, and far too high to form a generic medium for reactive chemistry. However by both increasing the size of the cation and the anion it has become possible to produce salts that are liquid at room temperature. It is well known that mixing together different salts deforms the crystal structure, leading to a lower lattice energy, and hence a lower melting point. At certain concentrations, referred to as eutectic points, the melting point has reached its minimum. This effect on melting point, obtained by combining the Group 1 chlorides into various mixtures can be seen in Table 21.2.4.

The melting points of the simple tetrachloroaluminate(III) salts of both the sodium and [$C_2$-mim] cations have been shown to be significantly lower than their respective chloride salts, indeed [$C_2$-mim][$AlCl_4$] was shown to be a liquid at room temperature. Such salts are produced by combining equimolar quantities of either NaCl or [$C_2$-mim]Cl with $AlCl_3$. However, as can be seen from Table 21.2.5, the 50:50 mole ratio does not usually correspond to the lowest melting point.

For the inorganic salts we now have melting points which are in the maximum range of some of the high-boiling organic solvents (e.g., 1,4-dichlorobenzene, b.p. 174°C). Hence given that the 50:50 mole ratio does not correspond to the lowest melting point of the inorganic salts, one would expect the same to be true for the organic cations. Figures 21.2.8 and

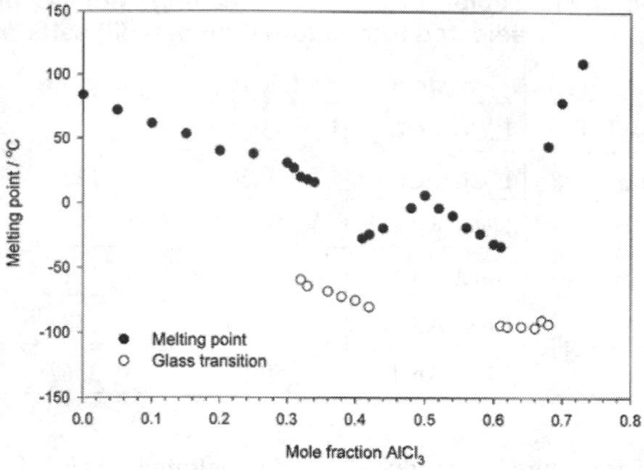


Figure 21.2.8. Phase diagram for the $[C_2\text{-mim}]Cl\text{-}AlCl_3$ system.[35]

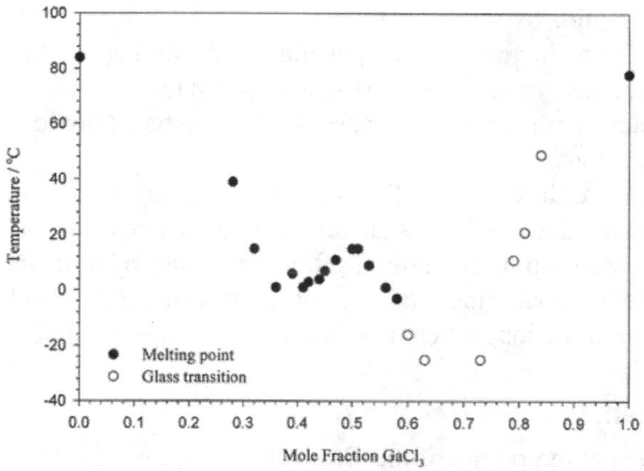


Figure 21.2.9. Phase diagram for the $[C_2\text{-mim}]Cl\text{-}GaCl_3$ system.[40]

21.2.9 show the phase diagrams of $[C_2\text{-mim}]Cl$ when combined with varying quantities of $AlCl_3$ and $GaCl_3$.

Thus a large number of ambient temperature ionic liquids can be formed from a mixture of solid aluminium(III) chloride and solid 1-ethyl-3-methylimidazolium chloride. Mixing these two salts together results in an exothermic reaction and generation of a clear, colorless liquid, and an ambient temperature molten salt with low viscosity. We can see that depending on the proportions of the two components it is possible to obtain eutectics with melting points as low as -96°C.[35] These liquids are also thermally stable in excess of 200°C allowing for a tremendous liquidus range. When the molar proportions are equal, the system is a neutral stoichiometric salt 1-ethyl-3-methylimidazolium tetrachloroaluminate(III) chloride $[C_2\text{-mim}][AlCl_4]$ which melts at its congruent melting point of about 7°C. The lowest melting point is achieved when the molar ratio of the system is 1:2 $[C_2\text{-mim}]Cl:AlCl_3$. Qualitatively these phase diagrams are similar to that reported by Hurley and Weir for the 1-ethylpyridinium bromide-$AlCl_3$ binaries[49] in that they all show a characteristic "W" shape.

It has been shown that if the butyl and ethyl groups are exchanged for a generic linear alkyl function then a series of cations can be generated. These can then be balanced against any one of a number of opposing anions to produce a final salt. Given that the range of available anions and cations has expanded enormously in the past decade it is possible to produce a number of combinations. Indeed, it is our best estimate that, if binary and ternary mixtures are included (and there are very good practical and economic reasons for doing that), there are approximately one trillion ($10^{18}$) accessible room temperature ionic liquids.[50]

## 21.2.3 CATALYSIS IN IONIC LIQUIDS

The wide range of reactions that have been undertaken in low temperature ionic liquid solvents is quite remarkable. It is limited simply by ones imagination. The specific and tuneable solvent properties of ionic liquids are a key feature for their use as solvents and have been utilized, especially in combination with the catalytic properties of the chloroaluminate(III) ionic liquids, to develop a range of synthetically important catalytic reactions. Currently a number of these are being investigated as economically and environmentally viable alternatives to existing industrial processes. A number of reactions are summarized below; for a more detailed discussion in this area readers are recommended to the recent review by Holbrey and Seddon,[25] and to a forthcoming book in the NATO ARW series.

### 21.2.3.1 Reactions involving first generation chloroaluminate(III) ionic liquids

The chloroaluminate ionic liquid mixtures are governed by the following primary equilibrium, equation:[51]

$$2[AlCl_4]^- \leftrightarrow [Al_2Cl_7]^- + Cl^-; \; k \approx 10^{-16} \qquad [21.2.2]$$

This is an acid base equilibrium under the Franklin definition. The $[Al_2Cl_7]^-$ species is the "acid" and the $Cl^-$ is the base. Note that this is an aprotic equilibrium. Therefore if the mole ratio of $[C_2\text{-mim}]Cl{:}AlCl_3$ is greater than, less than, or exactly equal to 50:50, the solvent behavior can be described as Franklin basic, Franklin acidic, or neutral.

Considering that aluminum trichloride is a very important commercial catalyst with over 25,000 tonnes produced annually in the USA alone, such liquids containing aluminum trichloride and allowing for differing levels of acidity have been extensively studied as first generation ionic catalytic solvents in a wide variety of synthetic and catalytic processes. Ionic liquids could therefore be used as substitutes for conventional solid or suspended sources of aluminum(III) chloride. As liquid phase catalysts, they allow for tremendous control of reactor inventories and can be cleaned and recycled with ease. Therefore, ionic liquids, in ideal cases, have no waste associated with them, whereas the supported aluminum(III) chloride catalysts will require large (and annually rising) waste disposal costs.

An industrial example of the use of chloroaluminate ionic liquids in alkene catalysts is the recent development of the IFP Difasol process[52] which is widely used industrially for alkene dimerization (typically propene and butanes). It was observed by Chauvin and coworkers[53-60] that chloroaluminate(III) ionic liquids would be good solvents for the nickel catalyst used in the reaction, and discovered that by using a ternary ionic liquid system ($[C_4\text{-mim}]Cl\text{-}AlCl_3\text{-}EtAlCl_2$) it was possible to form the active catalyst from a $NiCl_2L_2$ precursor and that, the ionic liquid solvent stabilized the active nickel species.

Overall they found that the nickel catalyst remains selectively dissolved in the ionic liquid phase to the extent that over 250 kg of propene could be dimerized per 1 g of nickel catalyst. In addition the product was insoluble in the ionic liquid, which made product recovery facile.

Other studies have shown that a wide range of acidic chloroaluminate(III) and alkylchloroaluminate(III) ionic liquids can be used to catalyze the dimerization and oligomerization of olefins.[61,62] In the reaction, the olefinic feedstock may be mixed with, or simply bubbled through, the ionic liquid catalyst to produce oligomeric products. A significant outcome of this reaction is that the product has a low solubility in the ionic liquid and separates as a less dense organic phase which is readily removed.

In addition to the above reactions it was shown that isobutene can be polymerized in an acidic ionic liquid to polyisobutene with a higher molecular weight than is formed using other polymerization processes. Polyisobutene, traditionally prepared by the Cosden process, is a valuable lubricant, and also a route to higher value-added materials. In general it was observed that the catalytic activity of the ionic liquids increases towards higher degrees of polymerization from short-chain oligomers as the alkylchain length of the 1-alkyl-3-methylimidazolium or N-alkylpyridinium cation is increased.[63]

The ionic liquid process has a number of significant advantages over the industrial Cosden process. This system uses a supported or liquid phase aluminum(III) chloride catalyst.[63] Using the ionic liquid process, the polymer forms a separate layer, which is substantially free of catalyst and ionic liquid solvent. This effect greatly enhances the degree of control available to reduce undesirable secondary reactions (i.e., isomerization) without requiring alkali quenching of the reaction.

In addition Ziegler-Natta polymerization reactions have also shown some success when carried out in ionic liquids.[64] The most common production methods for this form of polymerization involve the use of triethylaluminium catalysts at ca. 100°C and 100 atmospheres pressure. Advances have been developed through the use of organometallic transition metal catalysts, typically nickel or titanium. Given the solvent characteristics of ionic liquids it should be possible to effectively immobilize such catalysts in an ionic liquid solvent. Indeed, Carlin and Wilkes[64] have reported the Ziegler-Natta polymerization of ethene in an ionic liquid solvent. In these reactions an acidic $[C_2\text{-mim}]Cl\text{-}AlCl_3$ ionic liquid solvent was used to support dichlorobis($\eta^5$-cyclopentadienyl)titanium(IV) with an alkylchloroaluminium(III) co-catalyst.

Electrophilic substitution[65] and other reactions of naphthalenes (alkylation, acylation, condensation and migration in acidic ionic liquids[66,67] have been reported. Anthracene undergoes photochemical [4+4] cycloaddition reactions[68,69] in acidic chloroaluminate(III) ionic liquids. One interesting study included a one-pot synthesis of anthraquinone from benzene giving a 94% yield. In general a much wider range of redox products are formed than occur in conventional solvents; the strong Brønsted acidity of the ionic liquid induces protonation of anthracene, by residual traces of HCl, to form an anthracenium species which couples readily via photochemically driven electron transfer mechanisms.

Both the Friedel-Crafts alkylations and acylations are of great importance to the fine chemical and pharmaceutical industries. Typically, these reactions are run in an inert solvent with suspended or dissolved aluminum(III) chloride as a catalyst, and may take six hours and go only to 80% completion giving a mixture of isomeric products. In addition, there are a number of problems, especially with misnamed "catalytic" Friedel-Crafts acylation reactions, which are actually stoichiometric, consuming 1 mole of $AlCl_3$ per mole of reactant. Annually, massive quantities of aluminum(III) chloride are consumed in these reactions causing a number environmental problems through waste disposal. Both alkylation and acylation reactions under Friedel-Crafts conditions have been demonstrated using chloroaluminate(III) ionic liquids as both solvent and catalysts.[66-77] Here it has been shown that reaction rates are much faster with total reagent conversion and often with surprising specificity to a single product.

Boon et al. have reported the alkylation of benzene with a wide number of alkyl halides in acidic chloroaluminate(III) ionic liquids[73] and general organic reactions in low melting chloroaluminate ionic liquids have also been described,[75,78] which include chlorinations and nitrations in acidic ionic liquids.[72,73,76]

One specific example is the alkylation of benzene with chloroethane which gives a mixture of mono- to hexa-substituted products. The ionic liquid solvent/catalyst activates the reaction and the alkylation can be performed even at temperatures as low as -20°C in the ionic liquid solvent. Again it was shown that the products have a low solubility in the ionic liquid leading to facile separation.

Many acylation reactions have been demonstrated in acidic chloroaluminate(III) ionic liquids liquids.[70,74] However as described above these processes are essentially non-catalytic in aluminum(III) chloride which necessitates destroying the ionic liquid catalyst by quenching with water to extract the products. However, regioselectivity and reaction rates observed from acylation reactions in ionic liquids are equal to the best published results. The Friedel-Crafts acylation of benzene has been shown to be promoted by acidic chloroaluminate(III) ionic liquids.[74] It was observed that the acylated products of these reactions have high selectivities to a single isomer for example toluene, chlorobenzene and anisole are acylated in the 4-position with 98% specificity. Naphthalene is acylated in the 1-position which is the thermodynamically unfavored product. In addition to benzene and other simple aromatic rings, a range of organic and organometallic substrates (e.g., ferrocene)[79,80] have been acylated in acidic chloroaluminate(III) ionic liquids.

### 21.2.3.2 Reactions in neutral or second generation ionic liquids

Acidic $[C_2\text{-mim}]Cl\text{-}AlCl_3$ mixtures have been shown above to be very useful catalytic solvents for a number of industrially relevant reactions. However, reactions catalyzed by aluminum trichloride, whether used in an ionic liquid or as a solid phase reactions have one significant disadvantage, they are air and moisture sensitive. Using such liquids within an industrial process will therefore require that reactants are kept as dry as possible, adding to the overall process cost. We have seen how the activity and properties of the liquid can be readily controlled by both changing the anion or the cations present. By switching to neutral ionic liquids containing for example $[BF_4]^-$, $[PF_6]^-$ and $[SbF_6]^-$ anions, the reactive polymerization and oligomerization reactions of olefins catalyzed by acidic anions is not observed and more controlled, specific reactions can be catalyzed.

Since these ionic liquids cannot support the existence of reactive Lewis acid conjugate anions (such as $[Al_2Cl_7]^-$ i.e. there is no analogous mechanism to support $2[BF_4]^- \leftrightarrow [B_2F_7]^- + F^-$), they are much less reactive when used as solvents. Under these conditions, many conventional transition metal catalysts can be utilized. Again modification of the ionic liquid solvents allows the potential to immobilize catalysts, stabilizing the active species and to optimize reactant/product solubilities and to permit facile extraction of products. In general it has been found that charged, especially cationic transition metal complexes are most effectively "immobilized" in the ionic liquid solvents. In addition ionic liquids can immobilize less complex catalysts than two-phase aqueous-organic systems, where expensive, unstable, synthetically challenging ligands are often required.

Hydrogenation of olefins catalyzed by transition metal complexes dissolved in ionic liquid solvents have been reported using rhodium-,[81] and ruthenium- and cobalt-containing catalysts.[82] In these studies it was shown that hydrogenation rates where up to five times higher than the comparable reactions when carried out in propanone. The solubilities of the alkene reagents, TOFs, and product distributions where all strongly influenced by the nature of the anion in the ionic liquid solvent.

Chauvin et al. reported the asymmetric hydrogenation of acetamidocinnamic acid[83] to (S)-phenylalanine with a cationic chiral rhodium catalyst in [$C_4$-mim][$SbF_6$] ionic liquid, more recently the 2-arylacrylic acid has been produced with a reasonable 64% yield[84] using a chiral ruthenium catalysts in [$C_4$-mim][$BF_4$] ionic liquids. Palladium catalysts[85] immobilized in an ionic liquid-polymer gel membrane[86] containing either [$C_2$-mim][$CF_3SO_3$] or [$C_2$-mim][$BF_4$] have also been reported as catalysts for heterogeneous hydrogenation reactions.

Fuller et al have also reported the hydroformylation[86] of pent-1-ene in [$C_4$-mim][$PF_6$] using a rhodium catalysts showing both a high catalytic activity and product separation as a second organic phase. However it was observed that a small quantity of the neutral catalyst leached into the organic phase.

Other commercially important reactions include the hydrodimerization of 1,3-butadiene to octa-2,7-dien-1-ol[87,88] carried out using palladium catalysts in [$C_4$-mim][$BF_4$]. The catalyst precursor [Pd(mim)$_2$Cl$_2$] was prepared *in situ* from an imidazolium tetrachloropalladate(II) salt, [$C_4$-mim]$_2$[$PdCl_4$], dissolved in the ionic liquid solvent. The reaction proceeds in a liquid-liquid two-phase system, where the products separate from the catalytic reaction mixture as a separate layer on cooling.

Ionic liquids have been demonstrated as effective solvents for Diels-Alder reactions[89-91] where they have shown significant rate enhancements as well as high yields and selectivities when compared with the best results obtained in conventional solvents. To date, the biggest developments in Diels-Alder chemistry have come through reactions in Li[$ClO_4$]-$Et_2O$, where the high electrolyte concentrations are cited as beneficial through "salt-effects" and the high internal pressure of the solvent. However the use of such mixtures of perchlorate salts with organic molecules could cause a number of hazards when used on an industrial scale. Hence there is considerable potential for ionic liquids in this area.

The stability of the neutral ionic liquids allows them to be used in environments unsuitable to the tetrachloroaluminate(III) based ionic liquids. As such they offer considerable advantages allowing them to be used with "wet" process streams. The use of the fluorinated anions as shown in Figures 21.2.6 and 21.2.7, i.e., [$PF_6$] and [$BF_4$], introduces the concept of hydrophilic ionic liquids which are partially immiscible with water. In fact all [$PF_6$] liquids are found to be immiscible with water and all [$BF_4$] liquids with chain lengths greater than $C_4$ will form two separate phases with water at sufficient concentrations. This property has prompted investigation into the application of these solvents as extraction solvents for a number of materials. Rogers et al.[92] has studied the partition of benzene with water and has recently studied the relationship between pH and extraction efficiency.[93] Figure 21.2.10 il-

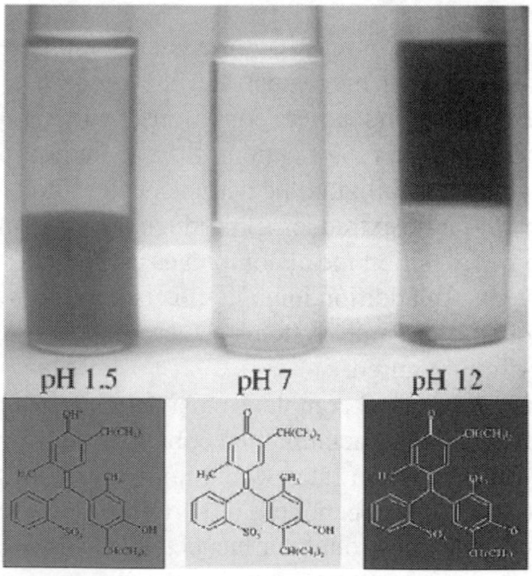


Figure 21.2.10. The phase preference of the three forms of thymol blue in [$C_4$-mim][PF6].[93]

lustrates the qualitative partitioning of thymol blue in its three forms between aqueous (top) and [$C_4$-mim][$PF_6$] (bottom) phases, as aqueous phase pH is changed from very acidic, to very basic. At low pH the thymol blue exists in its red form as a neutral zwitterion which prefers the ionic liquid phase. As the pH is increased via the addition of NaOH, the yellow monoanion forms with some detectable increase in concentration in the aqueous phase. The blue dianion, above pH = 10, partitions quantitatively to the aqueous phase. The same result was also obtained by bubbling $CO_2$ and $NH_3$ through the ionic liquid phase.

Lye[94] has also studied the use of ionic liquids in the extraction of erythromycin-A for the *Rhodococcus* R312 catalysed biotransformation of 1,3-dicyanobenzene (1,3-DCB) in a liquid-liquid two-phase system. Here it was found that the ionic solvent was less harmful to the system than conventional molecular solvents with the specific activity of the biocatalyst in the water-[$C_4$-mim][$PF_6$] system being almost an order of magnitude greater than in the water-toluene system. In addition, recent reports have effectively shown the potential for combining ionic liquid-and supercritical $CO_2$ systems for product extraction and separation[95] of naphthalene. Here it was found that the $CO_2$ was highly soluble in [$C_4$-mim][$PF_6$] reaching a mole fraction of 0.6 at 8 MPa and after separation there was no detectable ionic liquid in the extract. Another recent example is a paper by Armstrong et al.,[96] where ionic liquids ([$C_4$-mim][$PF_6$] and [$C_4$-mim][$BF_4$]) where used as stationary phases for gas chromatography. From these studies it was concluded that the ionic liquids act as nonpolar stationary phases when separating nonpolar analytes, however they are highly interactive and retentive when used to separate molecules with somewhat acidic or basic functional groups.

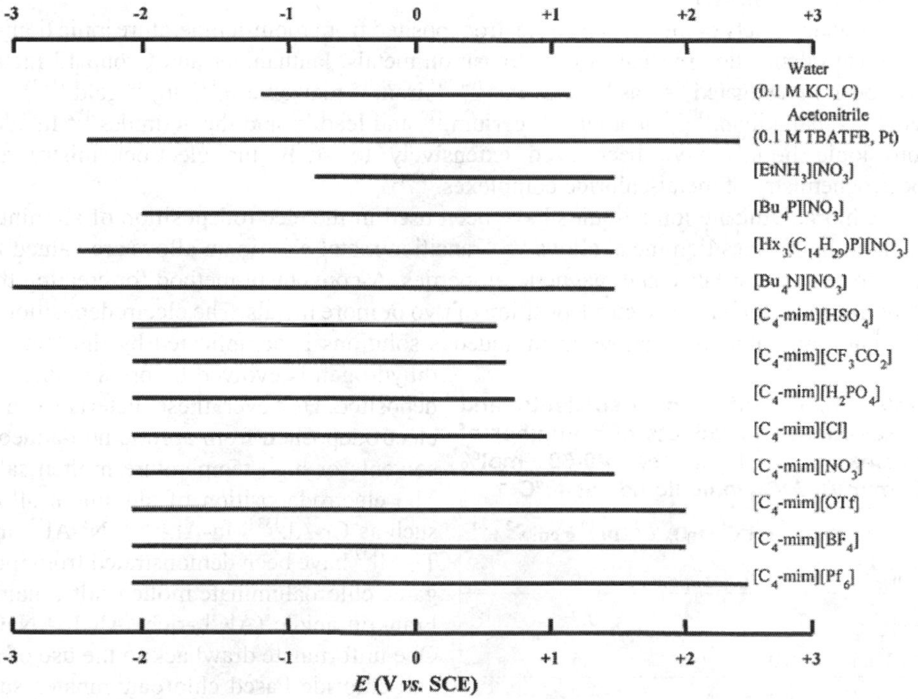


Figure 21.2.11. The electrochemical windows of a number of ionic liquids.[97]

## 21.2.4 ELECTROCHEMICAL APPLICATIONS

Since ionic liquid where first developed as an adjunct to the U.S. Naval program for use as battery electrolytes it seems only fair to begin this discussion by firstly looking at the electrochemical properties. One of the most important electrochemical properties of a solvent is its electrochemical window, the range of potentials over which a solvent system (including the solvent and any supporting electrolyte) is inert. The cathodic limit results from reduction of the solvent system; the anodic limit from its oxidation. Electrochemical windows are somewhat arbitrary as they are dependent upon the amount of background current which can be tolerated. Figure 21.2.11 shows the electrochemical windows of a range of ionic liquids as well as more commonly used solvents such as water and acetonitrile. The electrochemical windows shown in Figure 21.2.11 were determined using cyclic voltammetry at a glassy carbon disc electrode and are referenced against the Ag(I)/Ag couple (versus Ag(I)/Ag) and converted to the equivalent SCE potential. The reference electrode consisted of a silver wire immersed in a 0.1 M solution of $AgNO_3$ dissolved in [$C_6$-mim][$NO_3$]. In chloroaluminate ionic liquids, the most common reference electrode is an aluminum wire immersed in neat acidic ionic liquid (versus Al(III)/Al).

It can be seen from Figure 21.2.11 that the [$C_4$-mim] cation has a cathodic limit of approximately -2 V versus SCE and that this value is essentially the same for all of the [$C_n$-mim] cations. Given that the deposition potentials for many metals will fall positive of this potential, it becomes possible to use ionic liquids as electrolytes for metal plating and other similar processes. The broad electrochemical windows (in some cases, over 4 V) indicate that a variety of organic and inorganic electrochemical oxidations and reduction should be possible in ionic liquids.

A wide variety of metals can be electrodeposited from room-temperature ionic liquids. The electrodeposition mechanisms of transition metals, lanthanides and Group 13 metals have been investigated[98-103] as have cobalt,[104-107] iron,[108] manganese,[109] tin,[110] gold,[111,112] silver,[113-116] palladium,[117,118] mercury,[119] cerium,[120] and lead,[121] and the actinides.[122] In addition, ionic liquids have been used extensively to study the electrochemistry and spectrochemistry of metal-chloride complexes.[123-131]

Chloroaluminate ionic liquids have been used in the electrodeposition of aluminum and aluminum-transition metal alloys.[132] Transition metal-aluminum alloys are valued for their corrosion resistance and magnetic properties. A convenient method for creating thin alloy films is through the electrodeposition of two or more metals. The electrodeposition of aluminum and aluminum alloys from aqueous solutions is complicated by the fact that dihydrogen is evolved before aluminum is deposited. However, these materials can be electrodeposited from aprotic non-aqueous solvents or high temperature molten salts. The electrodeposition of aluminum alloys such as Cr-Al,[133] Mn-Al,[134,135] Ni-Al[136] and Ti-Al[137] have been demonstrated from inorganic chloroaluminate molten salts, mainly Franklin acidic ($AlCl_3$-rich) $AlCl_3$ - NaCl. One unfortunate drawback to the use of alkali-chloride based chloroaluminates such as $AlCl_3$-NaCl for the electroplating of

**Table 21.2.6. Diffusion coefficients and Stokes-Einstein products of a number of species studied in the 40-60 mol% [$C_2$-mim]Cl-$AlCl_3$ ionic liquids at 40°C**

| | $D_o$ $10^{-7}$ $cm^2s^{-1}$ | $\eta D_o T^{-1}$, $10^{-10}$ g cm $s^{-2}$ $K^{-1}$ |
|---|---|---|
| Co(II) | 6.6 | 2.6 |
| Cu(I) | 2 | 0.7 |
| Ni(II) | 10 | 3.5 |
| Zn(II) | 6.7 | 2.3 |

these alloys is the substantial vapor pressure of $Al_2Cl_6$ associated with the acidic region of these melts.

This is of course not the case when working with room temperature ionic liquid systems. Electrochemical and spectroscopic studies of cobalt,[138] copper,[139] and nickel,[140] have been carried out in the $AlCl_3$-butylpyridinium chloride molten salt system. The direct current and pulsed current electrodeposition of Ni-Al alloys has also been shown in acidic $AlCl_3$-butylpyridinium chloride ionic liquids. This particular alloy has also been shown to be successful in $AlCl_3$-[$C_2$-mim]Cl as have Co-Al and Cu-Al.[141] Electrochemical techniques can also be used to calculate the diffusion coefficients of metal ions. Table 21.2.6 shows the calculated diffusion coefficients and stokes-Einstein products of cobalt(II),[104] copper(I),[142] nickel(II)[143] and zinc(II)[144] in the 40-60 mol% [$C_2$-mim]Cl-$AlCl_3$ ionic liquid.

#### 21.2.4.1 Electrosynthesis

Electrosynthesis provides an attractive alternative to conventional methods used for performing synthetic chemistry. It can effect the clean, complete conversion of starting material to product without using hazardous or toxic experimental conditions. Ionic liquids possess many advantages over conventional solvents typically used in electrochemical experiments. Their polar nature allows them to dissolve large concentrations of a wide variety of organic and inorganic compounds. They possess no vapor pressure, are stable on heating and because they are completely ionized the need for a supporting electrolyte is eliminated. Such benefits are not associated with many of the non-aqueous systems employed in electrochemical applications.

Electrochemical studies in ionic liquids are common;[145-148] however the application of this technology to molecular synthesis remains largely unexplored. Some of the examples to date deal with electro-polymerizations. Osteryoung et al. prepared polyaniline on platinum and glassy carbon electrodes by anodic oxidation of the monomer in imidazolium based chloroaluminate ionic liquids.[149] Elsewhere the formation of polyfluorene was achieved by Janiszewska and Osteryoung.[150] Actual transformation of an organic moiety was achieved electrochemically in 1-methyl-3-butylimidazolium chloroaluminate systems. The reactions studied dealt with the reduction of aromatic ketones.[151] Reductions of perylene and phenazine have also been achieved electrochemically in basic mixtures of aluminum trichloride and 1-ethyl-3-methylimidazolium chloride.[152,153] The use of a neat ionic liquid 1-ethyl-3-methylimidazolium hydrogen dichloride [$C_2$-mim][$HCl_2$] was also investigated with regard to the reduction of phenazine.[152] Most of the work in electrosynthesis has been carried out in air- and moisture-sensitive ionic liquids, those containing chloroaluminates and [$HCl_2$]$^-$; little investigation into the electrosynthetic possibilities presented by air and moisture stable ionic liquids has been attempted.[154]

### 21.2.5 PHYSICAL CHARACTERIZATION

#### 21.2.5.1 Viscosity

Viscosity is probably the most important physical property for initially determining the "processability" of a solvent. It is necessary for all calculations involving fluid flow, (pumping, mixing, etc.) as well as in the estimation of heat transfer and diffusion coefficients. Information on the change in viscosity as a function of temperature, solvent or reagent addition is required for the efficient and design of any handling equipment. Ideally one would like the viscosity of a fluid to be as low as possible allowing for the fluid to be pumped easily. In addition it is desired for the fluid to have only small changes in viscosity through the normal operating temperature range.

In general, the published data on the viscosity of ionic liquids is scarce. Most of this published literature on ionic liquids viscosity deals with the first generation ionic liquids. The viscosity of any fluid is highly dependent on both the measuring technique used and the purity of the samples. Given this difficulty the reported values in the literature are often neither comparable or reproducible.

Once again, the first studies of the viscosity where related to the development of ionic liquids as nonaqueous battery electrolytes. Hussey et al.[155] reported the viscosity of several N-alkylpyridinium chloroaluminate salts over a temperature range of 25 to 75°C where they observed an increase in viscosity with alkyl chain length and concluded that the temperature dependence of the viscosity could be fitted to the Arrhenius type equation, i.e.,

$$\eta = \eta_0 \exp\left(E_\eta / RT\right) \qquad [21.2.3]$$

where:

$E_\eta$ is the energy of activation for viscous flow
R is the gas constant
T is the absolute temperature

This conclusion was supported by the linearity of the ln η vs. 1/T plots for the studied temperature range. The above equation can be modified to give;

$$\ln\eta = A_\eta + B_\eta / T \qquad [21.2.4]$$

where:

$A_\eta$ = ln $\eta_0$
$B_\eta$ = $E_\eta/R$

The fitted parameters, along with the values of $E_\eta$, appear in Table 21.2.7.

**Table 21.2.7. Parameters for Arrhenius equations for viscosity**

| Melt | Mol ratio | $-A_\eta$ | $E_\eta$ |
|---|---|---|---|
| $[NC_1\text{-py}]Cl\text{-}AlCl_3$ | 1:2 | 4.801 | 4.639 |
| $[NC_2\text{-py}]Cl\text{-}AlCl_3$ | 1:2 | 4.355 | 4.276 |
| $[NC_3\text{-py}]Cl\text{-}AlCl_3$ | 1:2 | 4.734 | 4.523 |
| $[NC_4\text{-py}]Cl\text{-}AlCl_3$ | 1:2 | 5.314 | 4.950 |
| $[NC_2\text{-py}]Br\text{-}AlCl_3$ | 1:2 | 5.575 | 5.136 |

The viscosities of the 1,3-dialkylimidazoilium aluminium chloride[156] and 1-methyl-3-ethylimidazolium aluminium bromide[157] ionic liquids have also been reported for different compositions and temperatures. For both the chloroaluminate and bromoaluminate ionic liquids the temperature dependence was found not to have an Arrhenious type curve, with non-linear plots of ln η vs. 1/T. In these studies the temperature range used was wider than that of the N-alkylpyridinium. This non-Arrhenius behavior is characteristic of glass forming melts. Here the three parameter Vogel-Tammann-Fulcher (VFT) equation:

$$\ln\eta = K_\eta / (T - T_0) + ½\ln T + \ln A_\eta \qquad [21.2.5]$$

where:

$A_\eta$ is a scaling factor
$T_0$ is the "ideal transition temperature" and
$K_\eta$ is a constant characteristic of the material

can be used for such fluids, with $K_\eta$ being the formal analogue of the Arrhenius activation energy. The parameter $T_0$ has been given a quasi-theoretical significance by the free volume theory of Cohen and Turnbull. According to this theory, $T_0$ represents the temperature at which the free volume of the liquid disappears and liquid transport becomes impossible.[155,157] Alternatively, it has been interpreted by Adams and Gibbs as the temperature at which the configuration entropy of the supercooled liquid becomes zero.[155,157]

The fitted parameters for the VFT equation are displayed in Table 21.2.8 for the 1,3-dialkylimidazolium chloride-aluminium chloride systems.

**Table 21.2.8. Fitted parameters for the VFT equation for some $AlCl_3$ systems[156]**

| Melt | Mol ratio | $T_0$, K | $K_\eta$, K | - ln $A_\eta$ |
|---|---|---|---|---|
| $[C_1\text{-mim}]Cl\text{-}AlCl_3$ | 1:2 | 109.1 | 1024 | 5.444 |
| | 2:1 | 194.8 | 811 | 4.739 |
| $[C_3\text{-mim}]Cl\text{-}AlCl_3$ | 1:1 | 141.5 | 879 | 5.157 |
| | 1:2 | 160.0 | 608 | 4.351 |
| | 2:1 | 185.0 | 958 | 5.208 |
| $[C_4\text{-mim}]Cl\text{-}AlCl_3$ | 1:1 | 184.5 | 484 | 3.819 |
| | 1:2 | 154.8 | 666 | 4.537 |
| | 2:1 | 165.4 | 1261 | 5.793 |
| $[C_4\text{-bim}]Cl\text{-}AlCl_3$ | 1:1 | 156.5 | 792 | 4.811 |
| | 1:2 | 150.7 | 750 | 4.765 |

These systems show a strong dependence of the composition for the basic range, with an increase in viscosity as the aluminum trichloride mole fraction decreases. In the acidic range the viscosity seems almost independent of the composition, nevertheless showing a slight decrease with increasing aluminum trichloride mole fraction. However it was also observed that increased viscosities where associated with longer alkyl chain lengths.

In the study of the 1-methyl-3-ethylimidazolium bromide - aluminum bromide systems, the VFT parameters for the different compositions were fitted as a function of X ($AlBr_3$ mole fraction) to cubic polynomial equations of the form

$$F(x) = c_0 + c_1X + c_2X^2 + c_3X^3 \qquad [21.2.6]$$

in neutral and basic melts and to second order equations of the form

$$F(x) = c_0 + c_1X^{-1} + c_2X^{-2} \qquad [21.2.7]$$

in acidic melts. The values generated for these constants are displayed in Table 21.2.9.

**Table 21.2.9. Parameters for 1-ethyl-3-methylimidazolium bromide/aluminum bromide systems[157]**

| Melt composition | VFT Parameter | $c_0$ | $c_1$ | $c_2$ | $c_3$ |
|---|---|---|---|---|---|
| Neutral and basic melts $0.35 \le X \le 0.50$ | $T_0$ | $2.1405 \times 10^3$ | $-1.5128 \times 10^4$ | $3.9396 \times 10^4$ | $-3.4596 \times 10^4$ |
| | $K_\eta$ | $-1.4147 \times 10^4$ | $1.1709 \times 10^5$ | $-3.0743 \times 10^5$ | $2.6769 \times 10^5$ |
| | $\ln A_\eta$ | $-1.3581 \times 10^1$ | $1.8522 \times 10^2$ | $-5.1772 \times 10^2$ | $4.7652 \times 10^2$ |
| Acidic melts $0.50 < X \le 0.75$ | $T_0$ | $-3.8691 \times 10^2$ | $7.6383 \times 10^2$ | $-2.6176 \times 10^2$ | - |
| | $K_\eta$ | $6.6201 \times 10^3$ | $-7.8145 \times 10^3$ | $-2.5214 \times 10^3$ | - |
| | $\ln A_\eta$ | $2.5515 \times 10^1$ | $-2.2656 \times 10^1$ | 7.3019 | - |

In this case, the viscosity showed a minimum at X =0.50, with the viscosity rising as X is either increased or decreased; displaying an irregular increase in the acidic range. This result is somewhat different for the chloroaluminate analogue discussed above.

It is important to note that in all the studies described above, different types of capillary viscometers were used for measuring the viscosity. This technique generates a kinematic viscosity, ν, that has to be multiplied by the density, ρ, of the melt to obtain the absolute viscosity, η.

Other viscosity studies in ionic liquids were carried out by Grätzel et al.[158] as part of a research program for the application of ionic liquids as solvents for a dye-sensitized nanocrystalline solar cell.

A series of imidazolium based salts with various alkyl substituents and different anions were characterized in terms of viscosity in order to establish a correlation between chemical structure and physical properties. The results are summarized in Table 21.2.10.

**Table 21.2.10. Viscosity at 20°C, cps. $TfO^-$: triflate; $NfO^-$: nonaflate; $Tf_2N^-$: bis(tryfil)amide; $TA^-$: trifluoroacetate; $HB^-$: heptafluorobutanoate; $AcO^-$: acetate[158]**

| Imidazolium cation | | $TfO^-$ | $NfO^-$ | $Tf_2N^-$ | $TA^-$ | $HB^-$ | $AcO^-$ |
|---|---|---|---|---|---|---|---|
| 3-Methyl | 1-Methyl | | | 44 | | | |
| | 1-Ethyl | 45 | | 34 | 35 | 105 | 162 |
| | 1-Butyl | 90 | 373 | 52 | 73 | 182 | |
| | 1-i-Butyl | | | 83 | | | |
| | 1-MeOEt | 74 | | 54 | | | |
| | 1-$CF_3CH_2$ | | | 248 | | | |
| 3-Ethyl | 1-Ethyl | 53 | | 35 | 43 | | |
| | 1-Butyl | | 323 | 48 | 89 | | |
| 1-Ethyl-2-methyl | 3-Methyl | | | 88 | | | |
| 1-Ethyl-5-methyl | 3-Methyl | 51 | | 37 | | | |
| | 3-Ethyl | | | 36 | | | |

According to this study, the viscosity of ionic liquids is mainly controlled by hydrogen bonding, van der Waals forces, molecular weight, and mobility.

In respect to the cation structure, alkyl lengthening or fluorination make the salt more viscous. Also the reduction of freedom of rotation (from butyl to isobutyl) increased the viscosity. Unexpectedly, methylation at the $C_2$(H) of the imidazolium ring increased viscosity even if it suppressed the position most likely to hydrogen bond with the anion.[158]

In respect to the anion, from $TfO^-$ to $NfO^-$ and from $TA^-$ to $HB^-$, the increase of van der Waals attraction dominates over the H-bonding decrease due to the better charge delocalization. However, from $TfO^-$ to $Tf_2N^-$, the almost complete suppression of H-bonding seems to influence more than the increase in Van der Waals attractions smaller anion weight of $TfO^-$. On the other hand, comparing $TA^-$ and $AcO^-$, the H-bonding strength of $AcO^-$ is more important than its smaller size. The salts with lower viscosity are $TA^-$, with minimal anion weight and moderate basicity, and $Tf_2N^-$, with minimal basicity and moderate anion weight.[158]

**Table 21.2.11. Viscosity of some ionic liquids at 25°C. For the binary mixtures the ratio means the proportion of $AlX_3$ to ImX or TMSuX**

| Cation | Anion | η, cps at 25°C |
|---|---|---|
| [$C_2$-py] | Br/$AlCl_3$ (ratio 2:1) | 22.5[155] |
| | Br/$AlBr_3$ (ratio 2:1) | ~50[159] |
| [$C_2$-mim] | Cl/$AlCl_3$ (ratio 2:1) | 13.5[156] |
| | Br/$AlBr_3$ (ratio 2:1) | 31[157] |
| | $BF_4$ | 37.7 (22°C)[160] |
| | Triflate ($CF_3SO_3$) | 43[161] |
| | Mesylate ($CH_3SO_3$) | 160[161] |
| [$C_4$-mim] | $BF_4$ | 233 (30°C)[162] |
| | $PF_6$ | 312 (30°C)[162] |
| Trimethylsulfonium | Cl/$AlCl_3$ (ratio 2) | 39.3[163] |
| | Br/$AlCl_3$ (ratio 2) | 54.9[163] |
| | Br/$AlBr_3$ (ratio 2) | 138[163] |

Clearly, many structural parameters affect the viscosity of ionic liquids and a more exhaustive study is needed to rationalize the different trends and to establish a correlation model for prediction.

Finally, there are some scattered data on viscosity of other ionic liquids which are summarized in Table 21.2.11, together with the most relevant data from the studies cited above.

As it was mentioned above, care must be taken when comparing or using these data as different measuring techniques can yield different viscosity results and furthermore, our recent investigations (to be published shortly) have shown that the presence of traces of water or other impurities like $Cl^-$ can have a dramatic effect on viscosity.

**Table 21.2.12. Densities of the $[C_4\text{-mim}][PF_6]$ and $[C_4\text{-mim}][BF_4]$ ionic liquids when compared against other solvents at room temperature**

| Solvent | ρ, kg $m^{-3}$ |
|---|---|
| trichloromethane | 1483 |
| $[C_4\text{-mim}][PF_6]$[164] | 1372 |
| dichloromethane | 1327 |
| $[C_4\text{-mim}][BF_4]$[164] | 1208 |
| water | 1000 |
| diethyl ether | 714 |
| toluene | 867 |
| 2-propanol | 786 |
| benzene | 877 |

### 21.2.5.2 Density

Information on solvent density is also another important physical property. This is a particularly used in fluid flow calculations and for the design of liquid/liquid two phase mixer-settler units. Given that many of the ionic liquids have a "heavy" anion it would be expected that their density would be relatively high when compared to normal industrial solvents. However it can be seen from the Table 21.2.12 that the density of the $[C_4\text{-mim}][PF_6]$ and $[C_4\text{-mim}][BF_4]$ ionic liquids fall within the region of the chlorinated solvents.

Overall the density of ionic liquids is somewhat easier to model than the viscosity. In general the change of density with temperature has been fitted to linear equations of the form

$$\rho = a + b \times T \qquad [21.2.8]$$

where a and b are constants.[155-157]

The fitted parameters for N-alkylpyridinium are displayed in Table 21.2.13. The bromide containing melts showed a much higher density than the chloride ones which is to be expected. It has also been shown that the lengthening of the alkyl chain smoothly decreases the density.[155,156]

**Table 21.2.13. Fitted parameters for the density of N-alkylpyridinium[155]**

| Melt | Mol ratio | a | $-b \times 10^3$ | Temp range, °C |
|---|---|---|---|---|
| $[NC_1\text{-py}]Cl\text{-}AlCl_3$ | 1:2 | 1.4625 | 0.87103 | 25-75 |
| $[NC_2\text{-py}]Cl\text{-}AlCl_3$ | 1:2 | 1.4307 | 0.91446 | 25-67 |
| $[NC_3\text{-py}]Cl\text{-}AlCl_3$ | 1:2 | 1.3973 | 0.8847 | 26-75 |
| $[NC_4\text{-py}]Cl\text{-}AlCl_3$ | 1:2 | 1.3680 | 0.86042 | 31-76 |
| $[NC_2\text{-py}]Br\text{-}AlCl_3$ | 1:2 | 1.5473 | 0.93059 | 25-32 |

The variation of density with different concentrations of aluminum trichloride in 1,3-dialkylimidazolium chloride- aluminum chloride melts is not however a smooth function.[156] In the study carried out by Wilkes et al. on the density of 1-methyl-3-ethylimidazolium bromide aluminum bromide systems, the change of the parameters a and b with composition, X, were successfully fitted to polynomials of third order. The constants obtained from this study are shown in Table 21.2.14.

**Table 21.2.14. Calculated constants for third order polynomial**

| | | $c_0$ | $c_1$ | $c_2$ | $c_3$ |
|---|---|---|---|---|---|
| Neutral and basic melts $0.35 \le X \le 0.50$ | a | 1.0588 | 3.8594 | -2.8945 | 0.00 |
| | b | $1.2907\times 10^{-3}$ | $-1.0448\times 10^{-2}$ | $1.1519\times 10^{-2}$ | 0.00 |

| | | $c_0$ | $c_1$ | $c_2$ | $c_3$ |
|---|---|---|---|---|---|
| Acidic melts<br>0.5 0< X ≤ 0.75 | a | -5.8410 | 37.009× $10^1$ | -56.829× $10^1$ | 30.474× $10^1$ |
| | b | 1.7946× $10^{-2}$ | -9.0089× $10^{-2}$ | 1.4121× $10^{-1}$ | -7.4099× $10^{-2}$ |

Grätzel et al.[158] has also measured the densities for the different ionic liquids with the results are shown in Table 21.2.15.

**Table 21.2.15. Density at 22°C or indicated temperature, g cm$^{-3}$. TfO$^-$: triflate; NfO$^-$: nonaflate; Tf$_2$N$^-$: bis(tryfil)amide; TA$^-$: trifluoroacetate; HB$^-$: heptafluorobutanoate; AcO$^-$: acetate[158]**

| Imidazolium cation | | TfO$^-$ | NfO$^-$ | Tf$_2$N$^-$ | TA$^-$ | HB$^-$ |
|---|---|---|---|---|---|---|
| 3-Me | 1-Me | | | 1.559 | | |
| | 1-Et | 1.390 | | 1.520 | 1.285 | 1.450 |
| | 1-Bu | 1.290$^{(20)}$ | 1.473$^{(18)}$ | 1.429$^{(19)}$ | 1.209$^{(21)}$ | 1.333 |
| | 1-i-Bu | | | 1.428$^{(20)}$ | | |
| | 1-MeOEt | 1.364 | | 1.496 | | |
| | 1-$CF_3CH_2$ | | | 1.656$^{(20)}$ | | |
| 3-Et | 1-Et | 1.330 | | 1.452$^{(21)}$ | 1.250 | |
| | 1-Bu | | 1.427$^{(18)}$ | 1.404$^{(19)}$ | 1.183$^{(23)}$ | |
| 1-Et-2-Me | 3-Me | | | 1.495$^{(21)}$ | | |
| 1-Et-5-Me | 3-Me | 1.334 | | 1.470 | | |
| | 3-Et | | | 1.432$^{(23)}$ | | |

As with viscosity, there are some scattered data on density of other ionic liquids which is summarized in Table 21.2.16, together with the most relevant data from the studies cited above.

**Table 21.2.16. Density of some ionic liquids at 25°C. For the binary mixtures the ratio means the proportion of AlX$_3$ to ImX or TMSuX**

| Cation | Anion | ρ, g cm$^{-3}$ at 25°C |
|---|---|---|
| 1-ethylpyridinium | Br/$AlCl_3$ (ratio 2) | 1.52[155] |
| | Br/$AlBr_3$ (ratio 2) | 2.20[159] |
| 1-ethyl-3-methylimidazolium | Cl/$AlCl_3$ (ratio 2) | 1.39[156] |
| | Br/$AlBr_3$ (ratio 2) | 2.22[157] |
| | $BF_4$ | 1.24 (22°C)[160] |
| | Triflate ($CF_3SO_3$) | 1.38[161] |
| | Mesylate ($CH_3SO_3$) | 1.24[161] |
| 1-butyl-3-methylimidazolium | $BF_4$ | 1.17 (30°C)[162] |
| | $PF_6$ | 1.37 (30°C)[162] |

| Cation | Anion | ρ, g $cm^{-3}$ at 25°C |
|---|---|---|
| 1-butyl-3-ethylimidazolium | Triflate ($CF_3SO_3$) | 1.27[161] |
| | Mesylate ($CH_3SO_3$) | 1.14[161] |
| 1-dodecyl-3-ethylimidazolium | Triflate ($CF_3SO_3$) | 1.10[161] |
| Trimethylsulfonium | Cl/$AlCl_3$ (ratio 2) | 1.40[163] |
| | Br/$AlCl_3$ (ratio 2) | 1.59[163] |
| | Br/$AlBr_3$(ratio 2) | 2.40[163] |

## 21.2.6 SUMMARY

In summary ionic liquids have a proven ability to enhance current industrial chemistry. Within current industrial processes using conventional solvents, selectivities, TOF and reaction rates are effectively uncontrolled: however by using ionic liquid media for the catalysis, it is possible to have a profound influence on all these factors. Tailoring of the ionic liquid by a combination of subtle (i.e., changing cation substitution patterns) and gross (anion type) modifications can permit very precise tuning of reactions.

Ionic liquids have a proven ability to be used as effective solvents and catalysts for clean chemical reactions; as replacements for volatile organic and dipolar aprotic solvents (i.e. DMF, DMSO) and solid acid catalysts in reactions whether being used at the laboratory or industrial scale. Many of their physical properties are in the same region as organic solvents and as such could easily be incorporated into any future processes. Properties such as negligible vapor pressure and high density/variable interfacial tension offer considerable advantages in the design of unit operations such as evaporators, mixer-settlers and reactors. It is envisaged that as more information is gathered on their physical properties it will be possible to accurately design and optimize entire plants around these novel solvent systems. In a time where considerable emphasis is being placed on clean reactions and processes with minimal waste and efficient product extraction ionic liquids have emerged as a novel and exciting alternative to modern production methods. They are truly designer solvents.

## REFERENCES


1 W. Pitner, D. Rooney, K. Seddon, R. Thied, **World Patent, WO9941752**.
2 W. Pitner, M. Fields, D. Rooney, K. Seddon, R. Thied, **World Patent, WO9914160**.
3 P. Davey, M. Earle, C. Newman, K. Seddon, **World Patent, WO9919288**.
4 C. Lok, M. Earle, J. Hamill, G. Roberts, C. Adams, K. Seddon, **World Patent, WO9807680**.
5 C. Greco, F. Sherif, L. Shyu, **U.S. Patent, US5824832**.
6 W. Keim, P. Wasserscheid, **World Patent, WO9847616**.
7 O. Hodgson, M. Morgan, B. Ellis, A. Abdul-Sada, M. Atkins, K. Seddon, **U.S. Patent, US5994602**.
8 S. Takahashi, N. Koura, S. Kohara, M.L. Saboungi, L.A. Curtis, *Plasmas and ions*, 2, (1999).
9 A.Carmichael, M. Earle, J. Holbrey, P. McCormac, K. Seddon, *Organic letters*, **1**, 7, 997-1000, (1999).
10 C. DeBellefon, E. Pollet, P. Grenouillet, *J. Mol. Cat. A-Chemical*, 145, (1999).
11 W. Chen, L. Xu, C. Chatterton, J. Xiao, *Chem, Com.*, 13, (1999).
12 L. Simon, J. Dupont, R. deSouza, *Applied Catalysis A-General*, 175, (1998).
13 S. Silva, P. Suarez, R. deSouza, J. Dupont. *Polymer Bulletin*, **40**, 4-5, (1998).
14 Y. Chauvin, H. Olivier, C.N. Wyrvalski, L.C. Simon, R.F. De Souza, *J. Catalysis*, **165**, 2, (1997).
15 Y. Chauvin. and H. Olivier-Bourbigou, *Chemtech*, 25, (1995).
16 Y. Chauvin, *Actualite Chimique*, (1996).
17 M. Freemantle, *Chem. Eng. News*, **76**, (30th March 1998).
18 M. Freemantle, *Chem. Eng. News*, **76**, (24th August 1998).
19 M. Freemantle, *Chem. Eng. News*, **77**, (4th January 1999).
20 C.L. Hussey, *Adv. Molten Salt Chem.*, 5, (1983).

21 C.L. Hussey, *Pure Appl. Chem.*, **60**, 1, (1988).
22 K.R. Seddon, *Kinetics and Catalysis*, **37**, 693, (1996).
23 K.R. Seddon, *J. Chemical Tech. Biotech.*, 68, (1997).
24 K.R. Seddon, Molten Salt Forum: Proceedings of the 5th International Conference on Molten Salt Chemistry and Technology, (1998).
25 J. D. Holbury and K.R. Seddon, *Clean products and processes*, 1, (1999)
26 C.L. Huheey, **Inorganic chemistry, Principles of structure and reactivity**, *Harper and Row*, (1972)
27 **CRC handbook**, 53rd ed, *CRC Press*, (1972).
28 A.F. Kapustinskii, *Z. Physik. Chem. (Leipzig)*, **B22**, 257, (1933).
29 J. Emsley, **The Elements**, *Oxford*, (1989).
30 K.B. Yatsimirskii, *Izvest, Akad. Nauk SSSR, Otdel, Khim. Nauk.*, 453, (1947).
31 R.D. Shannon, *Acta Crystallogr., Sect. A: Found. Crystallogr.*, **32**, 751 (1976).
32 A. B. McEwen, H. L. Ngo, K. LeCompte, J. L. Goldman, *J. Elect. Chem. Soc.*, **146**, 5, (1999).
33 T.C. Waddington, *Adv. Inorg. Chem. Radiochem.*, 1, (1959).
34 P. Walden, *Bull. Acad. Imper. Sci. (St. Petersburg)*, 1800, (1914).
35 A. Fannin, D. Floreani, L. King, J. Landers, B. Piersma, D. Stech, R. Vaughn, J. Wilkes, J. Williams, *J. Phys. Chem*, 88, (1984).
36 A.Elaiwi, P. Hitchcock, K. Seddon, N. Srinivasan, Y. Tan, T. Welton and J. Zora, *J. Chem. Soc. Dalton Trans*, 3467, (1995).
37 B. Chan, N. Chang and M. Grimmel, *Aust. J. Chem.*, 30, (1977).
38 A. McEwen, H. Ngo, K. LeCompte, J. Goldman, *J. Electrochem. Soc*, **146**, 5, (1999).
39 J. Wilkes and M. Zaworotko, *J. Chem. Soc., Chem. Com.*, 965, (1992).
40 S. Wicelinski, R. Gale, *Themochemica Acta*, 126, (1988).
41 S. Tait and R. Osteryoung, *Inorg, Chem.*, 23, (1984).
42 A. Abdul-Sala, A. greenway, P. Hitchcock, T. Mohammed, K. Seddon and J. Zora, *J. Chem. Soc., Chem. Commun.*, 1753, (1986) .
43 K. Dieter, C. Dymek, N. Heimer, J. Rovang, Jwilkes, *J. Am. Chem. Soc.*, **10**, 2711, (1988).
44 T. Bradley, PhD thesis, Queens University of Belfast, Unpublished results, (2000).
45 C.J. Bowlas, D.W. Bruce and K.R. Seddon, K.R, *J. Chem. Soc., Chem. Commun.*, (1996).
46 C.M. Gordon, J.D. Holbrey, A. Kennedy and K.R. Seddon, *J. Mater. Chem.*, 8, (1998).
47 J.D. Holbrey and K.R. Seddon, *J. Chem. Soc., Dalton Trans.*, (1999).
48 V. Posypaiko, E. Alekseeva, **Phase Equilibria in binary halides**, *IFI/Plenum*, (1987).
49 F. Hurley, T. Weir, *J. Electrochem. Soc.*, **98**, 203, (1951).
50 K. Seddon, in The International George Papatheodorou Symposium: Proceedings, Eds. S. Boghosian, V. Dracopoulos, C.G. Kontoyannis and G.A. Voyiatzis, (1999).
51 R.A. Osteryoung, in **Molten Salt Chemistry: An Introduction and Selected Applications** (Eds. Mamantov, G. and Marassi, R.), NATO ASI Series C: *Mathematical and Physical Sciences*, 202, 329, (1987).
52 D. Commereuc., Y. Chauvin, G. Leger, and J. Gaillard, *Revue de L'Institut Francais du Petrole*, **37**, 639, (1982).
53 Y. Chauvin, D. Commereuc, A. Hirschauer, F. Hugues, and L. Saussine, **French Patent, FR 2,611,700**.
54 Y. Chauvin, F. DiMarco, H.Olivier, and B. Gilbert, Ninth International Symposium on Molten Salts, C.L. Hussey, D.S. Newman, G. Mamantov and Y. Ito (Eds.), The Electrochemical Society, Inc., San Francisco, California. USA, 617, (1994).
55 Y. Chauvin, S. Einloft and H. Olivier, *Industrial & Engineering Chemistry Research*, **34**, 1149, (1995).
56 Y. Chauvin, B. Gilbert and I. Guibard, *J. Chem. Soc., Chem. Commun.*, 1715, (1990).
57 Y. Chauvin, B. Gilbert and I. Guibard, in Seventh International Symposium on Molten Salts, C.L. Hussey, S.N. Flengas, J.S. Wilkes and Y. Ito (Eds.), The Electrochemical Society, Montreal, Canada, 822, (1990).
58 Y. Chauvin, H. Olivier, C.N. Wyrvalski, L.C. Simon, R. De Souza and J. Dupont, *J. Catal.*, **165**, 275, (1997).
59 S. Einloft, F.K. Dietrich, R.F. de Souza and J. Dupont, *Polyhedron*, **15**, 3257, (1996).
60 H. Olivier, Y. Chauvin, and A. Hirschauer, *Abst. Papers Am. Chem. Soc.*, **203**, 75, (1992).
61 A.K.Abdul-Sada, P.W.Ambler, P.K.G.Hodgson, K.R.Seddon, N.J.Steward, **World Patent, WO 9521871**.
62 P.W. Ambler, P.K.G. Hodgson and N.J. Stewart, Butene Polymers, **European Patent Application, EP/0558187A0558181**, (1996).
63 K. Weissermel, and H.J. Arpe, **Industrial Organic Chemistry**, *VCH*, Weinheim, 3rd edition, (1997).
64 R.T. Carlin and J.S. Wilkes, *J. Mol. Catal.*, **63**, 125, (1990).
65 L.M. Skrzynecki-Cooke and S.W. Lander,. *Abst. Papers Am. Chem. Soc.*, **193**, 72, (1987).

66 E. Ota, in Joint (Sixth) International Symposium on Molten Salts, G. Mamantov, M. Blander, C.L. Hussey, C. Mamantov, M.L. Saboungi and J.S. Wilkes (Eds.), The Electrochemical Society, Inc., 1002, (1987).
67 E. Ota, *J. Electrochem. Soc.*, **134**, C512, (1987).
68 G. Hondrogiannis, C.W. Lee, R.M. Pagni and G. Mamantov, *J. Am. Chem. Soc.*, **115**, 9828, (1993).
69 R..M. Pagni, G. Mamantov, C.W. Lee and G. Hondrogannis, in Ninth International Symposium on Molten Salts, C.L. Hussey, D.S. Newman, G. Mamantov and Y. Ito (Eds.), The Electrochemical Society, Inc., San Francisco, California. USA, 638, (1994).
70 C.J.Adams, M.J.Earle, G.Roberts, K.R.Seddon, *Chem. Commun.*, 207, (1998).
71 J.A. Boon, S.W. Lander, J.A. Levisky, J.L. Pflug, L.M. Skrynecki-Cooke and J.S. Wilkes, in Joint (Sixth) International Symposium on Molten Salts, G. Mamantov, M. Blander, C.L. Hussey, C. Mamantov, M.L. Saboungi and J.S. Wilkes (Eds.), The Electrochemical Society, Inc., 979, (1987).
72 J.A. Boon, S.W. Lander, J.A. Levisky, J.L. Pflug, L.M. Skrynecki-Cooke and J.S. Wilkes: Catalysis and reactivity of electrophilic reactions in room-temperature chloroaluminate molten-salts, *J. Electrochem. Soc.*, **134**, 510, (1987).
73 J.A. Boon, J.A. Levisky, J.L. Pflug and J.S. Wilkes, *J. Org. Chem.*, **51**, 480, (1986).
74 M.B. Jones, D.E. Bartak and D.C. Stanley, *Abst. Papers Am. Chem. Soc.*, **190**, 58, (1985).
75 J.A. Levisky, J.L. Pflug and J.S. Wilkes, in The Fourth International Symposium on Molten Salts, M. Blander, D.S. Newman, M.L. Saboungi, G. Mamantov and K. Johnson (Eds.), The Electrochemical Society, Inc., San Francisco, 174, (1984).
76 B.J. Piersma and M. Merchant, in Seventh International Symposium on Molten Salts, C.L. Hussey, S.N. Flengas, J.S. Wilkes and Y. Ito (Eds.), The Electrochemical Society, Montreal, Canada, 805, (1990).
77 J.S. Wilkes, in **Molten salt chemistry: An introduction and selected applications**, G. Mamantov and R. Marassi (Eds.), *D. Reichel Publishing Company*, Dordrecht, 405-416, (1987).
78 J.A. Levisky, J.L. Pflug and J.S. Wilkes, *J. Electrochem. Soc.*, **130**, C127, (1983)
79 P.J. Dyson, M.C. Grossel, N. Srinivasan, T. Vine, T. Welton, D.J. Williams, A.J.P. White and T Zigras,: *J. Chem. Soc., J. Chem. Soc., Dalton Trans.*, 3465, (1997).
80 J.K.D. Surette, L. Green and R.D. Singer, *J. Chem. Soc., Chem. Commun*, 2753, (1996).
81 P.A.Z. Suarez, J.E.L. Dullius, S. Einloft, R.F. De Souza and J. Dupont, *Polyhedron*, **15**, 1217, (1996).
82 P.A.Z. Suarez, J.E.L. Dullius, S. Einloft, R.F. De Souza and J. Dupont, *Inorg. Chim. Acta*, **255**, 207, (1997).
83 Y. Chauvin, L. Mussmann and H. Olivier, *Angew. Chem. Int. Ed. Engl.*, 2698, (1995).
84 A.L. Monteiro, F.K. Zinn, R.F. De Souza and J. Dupont, *Tetrahedron Asymmetry*, **8**, 177, (1997).
85 R.T. Carlin and J. Fuller, *J. Chem. Soc., Chem. Commun,* 1345, (1997).
86 J. Fuller, A.C. Breda and R.T. Carlin, *J. Electrochem. Soc.*, 144, (1997).
87 J.E.L. Dullius, P.A.Z. Suarez, S. Einloft, R.F. de Souza, J. Dupont, J. Fischer and A. DeCian, *Organometallics*, **17**, 815, (1998).
88 S.M. Silva, P.A.Z. Suarez, R.F. De Souza, *Polymer Bulletin*, **40**, 401, (1998).
89 D.A. Jaeger and C.E. Tucker, *Tetrahedron Lett.*, **30**, 1785, (1989).
90 M.J. Earle, P.B. McCormac and K.R. Seddon, *Green Chem.*, 1, (1999)
91 T. Fischer, T. Sethi, T. Welton and J. Woolf , *Tet. Lett.*, **40**, 793, (1999).
92 J.G. Huddleston, H.D. Willauer, R.P. Swatloski, A.E. Visser, R.D. Rogers. *Chem Commun.*, 1765, (1998).
93 A.E. Visser, R.P. Swatloski and R.D. Rogers, *Green Chemistry*, **2**, 1, (2000).
94 S.G.Cull, J.D. Holbrey, V. Vargas-Mora, K.R. Seddon and G.J. Lye, *Biotechnol. Bioeng.*, (in press)
95 L. A. Blanchard, D. Hancu, E. J. Beckman, J. F. Brennecke, *Nature*, **399**, 6th May, 28, (1999).
96 D. Armstrong, L. He, Y, Liu, *Analytical chemistry*, **71**, 17, (1999)
97 W.R. Pitner, Unpublished Results, Queens University of Belfast, 2000
98 Y. Liu, P. Y. Chen, I. W. Sun, C. L.Hussey. *J. Electrochem. Soc.*, **144**, 7, 1997.
99 Y.F. Lin, I.W. Sun, *Electrochimica Acta*, **44**, 16, (1999).
100 Y.F. Lin, I.W. Sun, *J. Electrochem. Soc.*, **146**, 3, (1999).
101 W. J. Gau, I.W. Sun, *J. Electrochem. Soc.*, **143**, 1, (1996).
102 E.G.S. Jeng, I.W. Sun, *J. Electrochem. Soc.*, **145**, 4, (1998).
103 W.J. Gau, I.W. Sun, *J. Electrochem. Soc.*, **143**, 3, (1996).
104 J.A. Mitchell, W. R. Pitner, C. L. Hussey, and G. R. Stafford, *Journal of the Electrochemical Society*, **143**, 11, (1996).
105 M.R. Ali, A. Nishikata, T. Tsuru, *Electrochimica Acta*, **42**, 12, (1997).
106 R.T. Carlin, P.C. Trulove, H.C. DeLong, *J. Electrochem. Soc.*, **143**, 9, (1996).
107 R.T. Carlin, H.C. DeLong, J. Fuller and P.C. Trulove, *J. Electrochem. Soc.*, **145**, 5, (1998).
108 S. Pye, J. Winnick, P.A. Kohl, *J. Electrochem. Soc*, **144**, 6, (1997).
109 H.C. Delong, J.A. Mitchell, P.C. Trulove, *High Temperature Material Processes*, **2**, 4, (1998).

110 G.P. Ling, N. Koura, *Denki Kagaku*, **65**, 2, (1997).
111 X.H. Xu, C.L. Hussey, *J. Electrochem. Soc*, **139**, 11, (1992).
112 M. Hasan, I.V. Kozhevnikov, M.R.H. Siddiqui, A. Steiner, N. Winterton, *Inorganic Chemistry*, **38**, 25, (1999).
113 X.H. Xu, C.L. Hussey, *J. Electrochem. Soc,* **139**, 5, (1992).
114 F. Endres, W. Freyland, B. Gilbert, *Berichte Der Bunsen-Gesellschaft-Physical Chemistry Chemical Physics*, **101**, 7, (1997).
115 C.A. Zell, F. Endres, W. Freyland, *Physical Chemistry Chemical Physics*, **1**, 4, (1999).
116 F. Endres, W. Freyland, *Journal Of Physical Chemistry B*, **102**, 50, (1998).
117 I.W. Sun, C.L. Hussey, *J. Electroanalytical Chem.*, **274**, 1-2, (1989).
118 H.C. Delong, J.S. Wilkes, R.T. Carlin, *J. Electrochem. Soc*, **141**, 4, (1994).
119 X.H. Xu, C.L. Hussey, *J. Electrochem. Soc*, **140**, 5, (1993).
120 F.M. Lin, C.L. Hussey, *J. Electrochem. Soc*, **140**, 11, (1993).
121 C.L. Hussey, X.H. Xu, *J. Electrochem. Soc*, **138**, 7, (1991).
122 D. Costa, W.H. Smith, *Abstracts Of Papers Of The American Chemical Society*, **216**, 2, (1998)
123 S.K.D. Strubinger, I.W. Sun, W.E. Cleland, C.L. Hussey, *Inorganic Chemistry*, **29**, 5, (1990).
124 P.A. Barnard, C.L. Hussey, *J. Electrochem. Soc.*, **137**, 3, (1990).
125 S.K.D. Strubinger, I.W. Sun, W.E. Cleland, C.L. Hussey, *Inorganic Chemistry*, **29**, 21, (1990).
126 D. Appleby, P.B. Hitchcock, K.R. Seddon, J.E. Turp, J.A. Zora, C.L. Hussey, J.R. Sanders, T.A. Ryan, *Dalton Trans.*, 6, (1990).
127 S.K.D Strubinger, C.L. Hussey, W.E. Cleland, *Inorganic Chemistry*, **30**, 22, (1991).
128 C.L. Hussey, P.A. Barnard, I.W. Sun, D. Appleby, P.B. Hitchcock, K.R. Seddon, T. Welton, J.A. Zora, *J. Electrochem. Soc.*, **138**, 9, (1991).
129 R. Quigley, P.A. Barnard, C.L. Hussey, K.R. Seddon, *Inorganic Chemistry*, **31**, 7, (1992).
130 C.L. Hussey, R. Quigley, K.R. Seddon, *Inorganic Chemistry*, **34**, 1, (1995).
131 R.I. Crisp, C.L. Hussey, K.R. Seddon, *Polyhedron*, **14**, 19, (1995).
132 Q. Liao, W.R. Pitner, G. Stewart, C.L. Hussey, and G.R. Stafford, *J. Electrochem. Soc.*, **144**, 3, (1997).
133 T.P. Moffat, *J. Electrochem. Soc.*, 141, (1994)
134 G. R. Stafford, *ibid.*, 136, (1989)
135 G. R. Stafford, B. Grushko, and R. D. Mc Michale, *J. Alloys Compd.*, 200, (1993)
136 T. P. Moffat, *J. Electrochem. Soc.*, 141, (1994).
137 G. R. Stafford, *ibid.*, 945, (1994)
138 C. L. Hussey and T. M. Laher, *ibid.*, 20, (1981)
139 C. L. Hussey, L. A. King and R. A. Carpio, *J. Electrochem. Soc.*, 126, (1979)
140 R. J. Gale, B. Gilbert and R. A. Osteryoung, *Inorg. Chem.*, 18, (1979)
141 W. R. Pitner , PhD Thesis, University of Mississippi, (1997)
142 B. J. Tierney, W. R. Pitner, C. L. Hussey, and G. R. Stafford. *J. Electrochem. Soc.*, 145, (1998).
143 W. R. Pitner, C. L. Hussey, and G. R. Stafford, *J. Electrochem. Soc.*, **143**, 1, (1996).
144 W. R. Pitner and C. L. Hussey, *J. Electrochem. Soc.*, **144**, 9, (1997).
145 R.T. Carlin, P.C. Truelove, *Electrochimica Acta*, **37**, 14, (1992).
146 G.T. Cheek, R.A. Osteryoung, *J. Electrochem. Soc.*, **129**, 11, (1982).
147 M. Lipsztajn, R.A. Osteryoung, *Inorganic Chemistry*, **24**, 5, (1985).
148 C.L. Hussey, L.A. King, The Second International Symposium on Molten Salts, Pittsburgh, USA, The Electrochemical Society, Inc., (1978).
149 J. S. Tang, R.A. Osteryoung, *Synthetic Metals*, **45**, 1, (1991).
150 L. Janiszewska, R.A. Osteryoung, *J. Electrochem. Soc.*, **135**, 1, (1988).
151 G.T. Cheek, R.B. Herzog, The Fourth International Symposium on Molten Salts, San Francisco, The Electrochemical Society, Inc., (1984).
152 J.E. Coffield, G. Mamantov, *J. Electrochem. Soc.*, **138**, 9, (1991).
153 J.E. Coffield, G. Mamantov, *J. Electrochem. Soc.*, **139**, 2, (1992).
154 F. Fuller, R.T. Carlin, R.A. Osteryoung, *J. Electrochem. Soc.*, **144**, 11, (1997).
155 R.A. Carpio, A.K. Lowell, R.E. Lindstrom, J.C. Nardi and C.L. Hussey, *J.Electrochem.Soc*, 126; (1979).
156 A.A. Fannin, D.A. Floreani, L.A. King, J.S. Landers; B.J. Piersma, D.J. Stech, R.L. Vaughn J.S. Wilkes and J.L. Williams, *J.Phys.Chem*; 88, (1984).
157 J.R. Sanders, E.H. Ward and C.L. Hussey, *J.Electrochem.Soc.*; 133; (1986).
158 P. Bônhote, A. Dias, N. Papageorgiou, K. Kalyaanaasundaram and Grätzel, *Inorg.Chem.*; 35, (1996).
159 J. Robinson, R.C. Bugle, H.L. Chum, D. Koran, and R.A. Osteryoung, *J.Am.Chem.Soc*, 101, (1979).
160 J. Fuller, R.T. Carlin and R.A. Osteryoung, *J. Electrochem.Soc*; **144**, 11, (1997).

161 E.I. Cooper, E.J.M. O'Sullivan, *Proceedings of the 8th international symposium on molten salts*; **92**, 16, (1992).
162 P.A.Z. Suarez, S. Einloft, J.E.L. Dullius, R.F. de Souza, and J. Dupont, *J.Chem.Phys.*, **95**, (1998).
163 M. Ma and K.E. Johnsom, *Can. J. Chem.*, **73**, (1995).
164 M.Torres, Unpublished results, Queens University of Belfast, (2000).

## 21.3 OXIDE SOLUBILITIES IN IONIC MELTS

VICTOR CHERGINETS
**Institute for Single Crystals, Kharkov, Ukraine**

### 21.3.1 METHODS USED FOR SOLUBILITY ESTIMATIONS IN IONIC MELTS

Processes of the dissolution of metal oxides in ionic melts are accompanied by interactions between ions of dissolved substance with ions of the melt (solvation). The superimposition of the mentioned processes results in the formation of metal complexes with the melt anions and cation complexes with oxide-ions. Therewithal, the definite part of the oxide passes into the solution without dissociation as uncharged particles. Thus, in saturated solution of oxide the following equilibria take place:

$$MeO_s = MeO_l \quad [21.3.1]$$

$$MeO_l = Me^{2+} + O^{2-} \quad [21.3.2]$$

$$Me^{2+} + iX^{k-} = MeX_i^{2-ik} \quad [21.3.3]$$

$$O^{2-} + nKt^{m+} = Kt_nO^{nm-2} \quad [21.3.4]$$

where:

| | |
|---|---|
| s and l | subscripts denoting solid and dissolved oxide, respectively, |
| Me | the designation of two-valent metal |
| $Kt^{m+}$ | the melt cation (such as $Cs^+$, $K^+$, $Na^+$, $Ba^{2+}$, $Ca^{2+}$, etc.), |
| $X^{k-}$ | the melt anion ($Cl^-$, $Br^-$, $I^-$, $SO_4^{2-}$, $PO^{3-}$, etc.) |

Since at the dissolution of any oxide in melts a degree of interaction "$Kt^{m+}$ - $O^{2-}$" should be the same and the complexation with melt anions for cations of oxide may be assumed as closed,[1] therefore, it can be believed that oxide solubilities depend mainly on the degree of interactions [21.3.1] and [21.3.2], latter may be considered as an acid-base interactions as proposed by Lux.[2] The constant of [21.3.2]

$$K_{MeO} = \frac{a_{Me^{2+}} a_{O^{2-}}}{a_{MeO}} \approx \frac{m_{Me^{2+}} m_{O^{2-}}}{m_{MeO}} \quad [21.3.5]$$

where:

a and m activities and molarities of the particles noted in the subscripts

may be considered as a measure of acidic properties of the cation if reaction [21.3.2] is homogeneous. However, a majority of oxides possess limited solubilities in molten salts and the excess of the oxide should precipitate. In this case the fixation and removing oxide ions from melt took place not only because of the cation acidity but owing to formation of new phase - the precipitate of the oxide. Therefore, works in which the estimations of cation acidities were based on measurements of changes in oxide concentration before and after cation addition[3] or by interactions between, e.g., carbonates with Lux acids[4,5] contained distorted results. The latter works[4,5] contain the additional error - insoluble metal carbonates $MeCO_3$ after interaction with acid $K_2Cr_2O_7$[4] or $NaPO_3$[5] were transformed into insoluble chromates or phosphates - all reactions were heterogeneous. The above may be also referred to works of Slama[6,7] where cation acidities were estimated on the base of reactions

$$Me^{2+} + NO_3^- = MeO\downarrow + NO_2^+ \qquad [21.3.6]$$

$$NO_2^+ + NO_3^- = 2NO_2\uparrow + \tfrac{1}{2}O_2 \qquad [21.3.7]$$

there was no evidence of homogeneity of reaction [21.3.6].

The mentioned method seems to have no wide usage for studying behavior of oxides in molten salts, solubility studies by isothermal saturation and potentiometric titration methods are more precise and informative.

From the listed above scheme of equilibria [21.3.1-21.3.4] it follows that the fixation of oxide ions by metal cations is in common case a heterogeneous process resulting in the formation of a new phase and cation acidity may have no connection with the completeness of interaction $Me^{2+}$ - $O^{2-}$ in molten salts.

From eq. [21.3.1] it is clear that molecular oxide concentration in the saturated solution is not dependent on the constituent ion concentrations but connected with the precipitate properties, mainly with the molar surface of the deposit, σ.

Since the concentration of the non-dissociated oxide in the saturated solution, as a rule, is hardly determined, for practical purposes the solubility product, the magnitudes of PMeO (used below pP≡-logP),

$$P_{MeO} = a_{Me^{2+}} a_{O^{2-}} \approx m_{Me^{2+}} m_{O^{2-}} \qquad [21.3.8]$$

are usually employed for the description of the saturated solutions. The known values of $P_{MeO}$ and $K_{MeO}$ give possibility to estimate the concentration of the non-dissociated oxide, $s_{MeO}$, in the saturated solution:

$$s_{MeO} = \frac{P_{MeO}}{K_{MeO}} \qquad [21.3.9]$$

It should be noted, however, that there was no reliable method for determining dissociation constants and $s_{MeO}$. Let us consider two generally accepted methods for oxide solubility determinations.

### 21.3.1.1 Isothermal saturation method

This method is simple enough and includes some modifications:

1. placing baked sample of the oxide in the melt-solvent and regular tests of the metal concentration in the melt up to the saturation. Known test routines are either radiochemical[8] or complexonometric[9] analysis of cooled samples of the saturated solutions;

2. the addition of known weights of oxide to the melt up to the saturation detected by a potentiometric technique.[10,11]

The sum of molecular and ionic form concentrations in the saturated solution, $\Sigma s_{MeO}$, is obtained as the main result of such investigations, this magnitude may be expressed by the following equation:

$$\sum s_{MeO} = \sqrt{P_{MeO}} + \frac{P_{MeO}}{K_{MeO}} \qquad [21.3.10]$$

Since $\Sigma s_{MeO}$ is the main result of this method, there exists a possibility to determine the dissociation constant according to routine 2 if an oxygen electrode was preliminary calibrated by known additions of the completely dissociated Lux base. But there is no information about such studies.

Studies connected with the analysis of cooled samples allow to obtain thermal dependencies of oxide solubilities by analysis of the saturated melt heated to the definite temperature. Main disadvantages of isothermal saturation method are:

- the inclusion of suspended particles of oxide in the sample for analysis and following overvaluing of the results;
- the magnitude determined is the concentration but not its logarithm; range of oxides available for this routine is essentially narrowed and this method is unsuitable for studies of practically insoluble substances. The presence of oxygen impurities in the melt studied leads to lowering the calculated solubilities, especially, if oxide solubility is of the same order as initial oxide concentration since oxide admixtures suppress the oxide dissolution.

### 21.3.1.2 Potentiometric titration method

This method is more frequently used than the previously described one. It may be employed for oxide solubility studies in a wide range of concentrations of saturated solutions. This method allows to determine solubility products of oxides and in some cases (the existence of the non-saturated solution region) dissociation constants may be calculated.

As compared with isothermal saturation a potentiometric method possesses the following advantages:

- e.m.f. values are linearly dependent on logarithm of $O^{2-}$ concentration; it is possible to measure the latter even in solutions with extremely low potential-determining ion concentrations;
- simplicity of e.m.f. measurements and good reproducibility of the experimental values;
- possibility to made studies *in situ*, i.e., without any effect on the process studied.

Main limitations of the potentiometric titration are due to the potential of liquid junction, however, the latter is negligible in molten salts,[10] and the absence of reliable and generally accepted methods for calculation of activity coefficients for oxide ions and cations.

As for determination of solubility and dissociation parameters it should be noted that this method allows to determine values $K_{MeO}$ only if the potentiometric curve contains the non-saturated solution region. Studies in the wide temperature range are impeded as compared with the isothermal saturation method.

## 21.3.2 OXYGEN-CONTAINING MELTS

Frederics and Temple studied CuO, MgO, PbO and ZnO solubilities in molten equimolar mixture $KNO_3$-$NaNO_3$ in the range 290-320°C.[12] The solubilities were determined by isothermal saturation method with the potentiometric control of solubilities. Solubility products were $2.24\times10^{-13}$ for CuO, $2.2\times10^{-13}$ for ZnO, $2.16\times10^{-14}$ for MgO and $4.34\times10^{-12}$ for PbO. The increase of the temperature to 320°C led to the increase of the solubilities by 6-10 times.[12]

Individual Lux acids derived from $B_2O_3$ (borax) and $P_2O_5$ (sodium metaphosphate) are interesting since they are often used as acidic components of different fluxes. Delimarsky and Andreeva[10] determined PbO solubility in molten $NaPO_3$ at 720°C by isothermal saturation method with the potentiometric control of saturation. The concentration cell with oxygen electrodes $Pt(O_2)$ was used. PbO solubility was estimated as 31.6 mol%.

Nan and Delimarsky investigated solubilities of acidic metal oxides ($MoO_3$, $WO_3$, $TiO_2$) in molten borax at 900°C by a similar method.[13] The control of saturation was performed by a gravimetric analysis. Solubilities of the oxides deviated from those predicted by Shreder's equation, because of chemical interactions between the substances dissolved and the solvent. Oxide solubilities increased with the reduction of melting point of oxide. Solubilities were 66.1 mol% for $MoO_3$, 63.2 mol% for $WO_3$ and 21.2 mol% for $TiO_3$.

## 21.3.3 HALIDE MELTS

### 21.3.3.1 The eutectic mixture KCl-LiCl (0.41:0.59)

Delarue[14,15] performed qualitative studies on oxide solubilities in the melt at t~500°C by visual method. The experiments consisted of addition of potassium hydroxide, KOH, to the metal chloride solution in KCl-LiCl. If there were no formation of the oxide deposit then the oxide was considered soluble, in other cases, oxide was referred to slightly soluble or insoluble depending on the amount of oxide deposited. CdO, PbO, BaO, CaO and $Ag_2O$ belonged to the group of soluble oxides, CoO, NiO, ZnO were examples of slightly soluble oxides and MgO, BeO and $Al_2O_3$ were insoluble. Copper (II) oxide was unstable in molten chlorides since its dissolution was accompanied by the reduction of $Cu^{II}$ to $Cu^{I}$. This transformation has been demonstrated[16] to be substantial even under the partial pressure of chlorine equal to 1 atm. The similar behavior is typical of other transition cations in the highest degree of oxidation, e.g., $Tl^{III}$, $Fe^{III}$, and $Au^{III}$. The addition of corresponding chlorides to the saturated solutions formed by non-oxidizing cations resulted in dissolution of latter due to reactions similar to:

$$2CuCl_2 + ZnO\downarrow \rightarrow 2CuCl + ZnCl_2 + \tfrac{1}{2}O_2\uparrow \qquad [21.3.11]$$

An important feature of oxide solubilities[14,15] was the essential difference of solubility values of "powdered" (added as powder) and prepared *in situ* oxide solubilities. Solubilities of latter samples have been found to be considerably greater (CaO, CdO), although there was no explanation to this observation. Quantitative characteristics of oxides solubilities in KCl-LiCl were not presented.[14,15]

Laitinen and Bhitia[17] determined solubility products for certain oxides at 450°C in order to evaluate possibility to use corresponding metal-oxide electrodes for oxoacidity measurements in molten salts. Oxide solubilities were as the following (mole/kg): NiO - $3.3\times10^{-4}$, BiOCl - $6.8\times10^{-4}$, PdO - $9.4\times10^{-3}$, PtO - $3.32\times10^{-2}$, $Cu_2O$ - $3.8\times10^{-2}$.

The important fact found was the relatively high PtO solubility, therefore, platinum gas oxygen electrode, sometimes considered as metal-oxide one, cannot be used for measurement of pO in strongly acidic media because of complete dissolution of the oxide film over its surface. Pt|PtO electrode was attempted for pO index measurements in buffer solutions $Ni^{2+}/NiO$.[18]

Shapoval et al.[19-21] studied saturated solutions of different oxides by polarographic technique. A degree of interaction between oxide dissolved and the melt has been estimated taking into account potential and polarogram shifts and their deviations from theoretical ones. Rate of cation reduction has been found to be limited by the acid-base dissociation stage. Stability constants of some oxides have been found to be (in mol%): CoO - $(9.9\pm1.9)\times10$, NiO - $(6.8\pm0.8)\times10^3$, PbO - $(3.4\pm1.3)\times10^3$, $Bi_2O_3$ - $(8.7\pm3.5)\times10^3$. Cadmium, copper (I) and silver oxides have been shown to completely dissociate under the experimental conditions (400-600°C) this was in a good agreement with Delarue's results.[14,15]

Stabilities of iron (II) and (III) oxides in the chloride melt at 470°C have been investigated.[22,23] Oxide precipitation by carbonate $CO_3^{2-} \equiv O^{2-}$ has been shown by potentiometric and diffractometric data to result in formation of FeO from $Fe^{II}$ solutions. $Fe^{III}$ precipitation led to formation of solid solutions $LiFeO_2$-$Li_yFe_{1-y}O$. Solubility products in molarity scale have been determined as FeO-$10^{-5.4}$, $Fe_3O_4$-$10^{-36.3}$, $Fe_2O_3$-$10^{-29.16}$. Cations $Fe^{3+}$ have been found to oxidize chloride melt with chlorine evolution, that is in good agreement with the results of other studies.[14,15] Carbonate ion as oxide donor had its dissociation to $O^{2-}$ essentially incomplete and, therefore, obtained solubility products gave systematic error.[20,21]

Cherginets and Rebrova[24,25] studied oxide solubilities in this melt at 700°C. Solubility products of CoO (pP=4.43±0.11) NiO (pP=5.34±0.2) and MgO (pP=5.87±0.05) were higher than those in molten eutectic KCl-NaCl. The shift of the solubilities in molar fraction scale was ~3.36 log units and allowed to estimate solubilities of other MeO oxides on the base of the known values in KCl-NaCl. It has been shown that acidic properties of $Pb^{2+}$ and $Cd^{2+}$ cations were suppressed by those of $Li^+$, therefore, these cations and $Ba^{2+}$ and $Sr^{2+}$ did not change oxoacidic properties of the molten KCl-LiCl.

### 21.3.3.2 Molten KCl-NaCl (0.50:0.50)

One of the first works devoted to oxide solubilities determination in this melt has been performed by Naumann and Reinhardt.[8] CaO, SrO and BaO solubilities have been determined by isothermal saturation technique with isotopic control of the saturation in temperature range from melting point to 900°C. KCl-NaCl and individual molten chlorides, KCl and NaCl, have been used as solvents. Oxide solubilities have been stated to increase in the sequence CaO<SrO<BaO and solubility product values in KCl-NaCl and KCl were similar, and the values for NaCl were higher. A comparison of experimental and thermodynamic data for MeO and $Me^{2+}+O^{2-}$ solutions have shown that the solubility values are between values calculated for solutions of completely dissociated and completely molecular oxide. It means that these constituents were present simultaneously.

Similar investigations have been done in the temperature range 700-800°C by Volkovich.[9] Saturated solutions were prepared by placing pressed oxide tablets into the melt and holding them for 2-3 h. The analysis of solution samples was performed by complexometric titration.

Potentiometric studies of cations $Ca^{2+}$, $Li^+$ and $Ba^{2+}$ were conducted to determine acidic properties at 700°C; the calculated dissociation constants (in mol%) were $2.8\times10^{-3}$ for

$Li_2O$, and $1.2\times10^{-2}$ for BaO, these values showed that all studied cations had considerable acidic properties.[26,27]

Ovsyannikova and Rybkin[3] developed the cation acidity scale in molten KCl-NaCl at 700°C on the basis of e.m.f. shifts after addition of metal sulfates. Acidic properties of the main subgroup elements decreased with the increase of their atomic numbers and there was no similar relationships for side groups and transition metals. Quantitative data elucidation according to acid-base equilibrium [21.3.2] in this melt is affected by additional reaction:

$$SO_4^{2-} = SO_3\uparrow + O^{2-} \qquad [21.3.12]$$

caused by additions to chloride melts of the corresponding sulfates. The use of sulfates may cause $SO_3$ formation from highly acidic cation solutions. In particular, $Na^+$ is the most acidic cation of KCl-NaCl melt, therefore, additions of more basic oxides than $Na_2O$ should result in oxide ion exchange:

$$Me^{I}_2O + 2Na^+ \rightarrow 2Me^+ + Na_2O \qquad [21.3.13]$$

which is shifted to the right. Similar considerations have been made for $K^+$ - $Na^+$ - $O^{2-}$ system in molten KCl-NaCl.[28] From [21.3.13] it follows that any solution of cations in Na-based melts cannot create oxide ion concentration exceeding that of equimolar addition of corresponding salt of the most acidic cation of the melt ($Na^+$). However, the oxide ion concentration in $BaSO_4$ solution has been shown[3] to be 10 times higher than in $K_2SO_4$ and $Na_2SO_4$ ($\Delta E$=0.11 V). Similar deviations for other cations are smaller, e.g., Cs - 0.49 V, Sr - 0.30 V, Rb - 0.1 V.

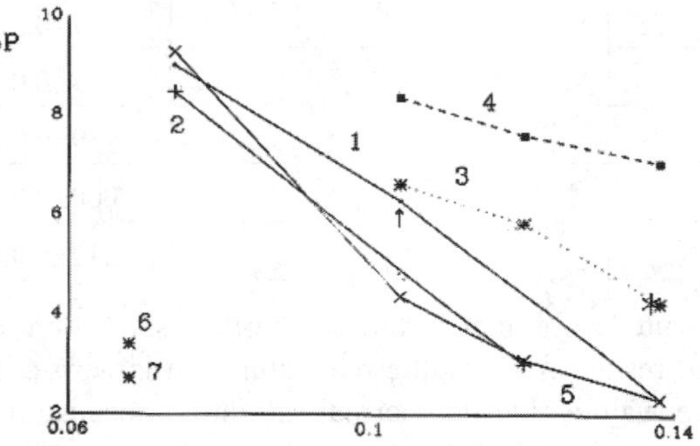

Figure 21.3.1. Alkaline oxide solubilities in KCl-NaCl: 1[11,28-32] at 1000K; 2;[29] 3;[8] 4;[9] 5;[33,38] 6;[40,41] 7;[42] at 700°C.

A potentiometric study of ZnO, MgO, NiO and SrO solubilities at 700°C have been conducted[29] by direct and reverse titrations of cation by KOH. These measurements have resulted in a set of characteristics corresponding (by the calculation formula) to solubility products and dissociation constants. The averaging or another procedure of data treatment were not included. Oxide solubilities have been found to increase in sequence: MgO<NiO<ZnO<SrO.

Oxide solubilities in the chloride melt at 1000 K were studied.[11,28-32] MgO is neutralized in two stages: first stage product was $Mg_2O^{2+}$ and end product was MgO.[11] It should be noted, however, that although MgO solubilities were investigated in similar manner in KCl-NaCl,[29,33] CsCl-KCl-NaCl[34-36] and $BaCl_2$-$CaCl_2$-NaCl[37] no pronounced first stage was detected. Thermal dependence of CaO solubility has been determined by isothermal saturation method[31,32] to be approximated by the following plot $pK_{s,CaO}$=10800/T-5.8. We have studied solubilities of 11 oxides in this melt at 700°C and developed the methods of saturation detection to determine in some cases dissociation constants.[33,38]

Thermal dependence of $ZrO_2$ solubility (molar fraction, N) in molten KCl-NaCl was evaluated in the temperature range of 973-1174 K:[39] N=(-5.7±3.1)×10-6 + (7±3)×$10^{-9}$×T. $ZrO_2$ solubility at t~700°C was negligible.

Barbin et al.[40,41] determined thermal dependence of $Li_2O$ solubility in equimolar mixture KCl-NaCl by isothermal saturation technique in temperature range of 973-1073 K: N=0.107-5.221/T, in this range the solubility varied from 0.52 to 0.86 mol%. In earlier work by Kaneko and Kojima,[42] the solubility at 973 K was lower (0.31 mol%). The solubility of lithium oxide was close to that for BaO.

Solubility products of some oxides in molten KCl-NaCl in molarity scale are presented in Table 21.3.1 and plotted in Figure 21.3.1 (data obtained by same authors are connected by lines).

**Table 21.3.1. Oxide solubilities in molten KCl-NaCl at 700°C (-log P, molarities)**

| Oxide | pP[11,28-32] | pP[29] | pP[8] | pP[9] | pP[33,38] |
|---|---|---|---|---|---|
| MgO | 9.00±0.15 | 8.46 | | | 9.27±0.06 |
| CaO | ~6.29 | | 6.62 | 8.36 | 4.36±0.06 |
| SrO | | 3.00 | 5.84 | 7.60 | 3.08±0.40 |
| BaO | 2.31±0.05 | | 4.22 | 7.05 | 2.30±0.15 |
| NiO | 11.2 | 8.32 | | | 9.03±0.06 |
| ZnO | | 6.18 | | | 6.93±0.20 |
| $Cu_2O$ | 5.4 | | | | 4.17±0.30 |
| MnO | | | | | 6.78±0.05 |
| CoO | | | | | 7.89±0.03 |
| CdO | | | | | 5.00±0.03 |
| PbO | | | | | 5.12±0.05 |

Two groups of values are presented: obtained from isothermal saturation and potentiometric results. The precision of results from isothermal saturation is worse. The potentiometric results should give lower values than isothermal saturation because it includes molecular oxide concentration. Figure 21.3.1 shows that results have opposite trend to [21.3.10].

The following explanation of the above discrepancy is proposed.[43,44] Let us consider the chemical potential, μ, of oxide in its saturated solution:

$$\mu_{MeO,s} = \mu^0_{MeO,l} + RT\ln s_{MeO,l} \qquad [21.3.14]$$

From this equation it follows that the oxide solubility should be constant. But every precipitate possesses a definite surface, hence, the effect of latter should be taken into account, too:

$$\mu_{MeO,s} + \sigma S = \mu^0_{MeO,l} + RT\ln s_{MeO,l} \qquad [21.3.15]$$

where:

$\sigma$ the surface energy

S molar surface square of the precipitate

Since for the same oxide σ is constant, the increase of the precipitate square should result in the increase of precipitate solubility.

Similar considerations were included in well-known Ostvald-Freundlich equation.[45] For substances having 1:1 dissociation, a similar equation can be written in the following form:

$$\frac{RT}{M}\ln\frac{s_1}{s_2} = \frac{RT}{M}\ln\frac{P_1}{P_2} = \frac{\sigma}{d}\left(\frac{1}{r_2} - \frac{1}{r_1}\right) \quad [21.3.16]$$

where:

| | |
|---|---|
| $s_1, s_2$ | the solubility of crystals with radii $r_1$ and $r_2$, respectively |
| $P_1, P_2$ | corresponding solubility products |
| M | molecular weight |
| d | density |

From this equation, it follows that the increase of crystal size reduces its solubility. Oxide particles deposited from more concentrated cation solution, should possess larger sizes due to the so-called "deposit ageing", than those obtained from mere diluted solutions. Therefore, results of Delimarsky et al.[29] obtained from 0.01 mole/kg solutions should be higher that those obtained in other works[11,33,38] using 0.05 mole/kg solutions. In the first case solubility was greater approximately by half-order of the magnitude than in second one. Oxide formed from more concentrated solution should have less surface and, hence, less solubility. Data obtained in works[11,33,38] practically coincide although we[33,38] have used NaOH as titrant similarly to other study.[29]

Sedimentational titration results in formation of fine dispersed oxide which immediately begins to age because of the recrystallization, i.e., the growth of larger crystals and disappearance of smaller ones, this process leads to the surface energy, σS, decrease. Continuous holding at high temperature and high solution concentration favors this process. In particular, samples used for isothermal saturation technique studies exposed to high temperatures for a long time, and tablets obtained are held in the contact with the melt for some hours. High concentration solutions favor transfer of substance from small to large crystals (diffusion). The differences between data in references 8 and 9 may be explained by different conditions of the powder calcination. During the annealing of the oxide sample recrystallization processes occur, which lead to the reduction of surface square of the powder and the decrease of solubility. The differences in data for KCl-NaCl are caused by the differences in crystal sizes of solid oxide being in equilibrium state with the saturated solution.

### 21.3.3.3 Other chloride-based melts

$ZrO_2$ solubilities in molten mixtures KCl-$KPO_3$ at 800°C have been investigated.[46] $Zr^{IV}$ concentration in pure $KPO_3$ was 1.34 wt%, and in the equimolar mixture KCl-$KPO_3$ concentration was 3.25 wt%. A reason for solubility increase is in the depolymerization of $PO^{3-}$ which may be schematically described by the following equation:

$$(PO_3)_n^{n-} = nPO_3^- \quad [21.3.17]$$

and increase in acidic properties of melt. The increase of KCl molar fraction over 0.6 led to sharp reduction of $ZrO_2$ solubility because the decrease of acid ($PO^{3-}$) concentration could not be compensated by the depolymerization process [21.3.17].

Deanhardt and Stern[47,48] studied solubilities of NiO and $Y_2O_3$ in molten NaCl and $Na_2SO_4$ at 1100°C. $Y_2O_3$ solubility product was $(1.4\text{-}2.2)\times10^{-36}$ and $(4.5\text{-}6.4)\times10^{-31}$ in the chloride and the sulphate melt respectively. The formation of $YO^{2-}$ was observed when oxide ion was in excess. The formation of peroxide ions in the presence of $O_2$ over melt was favored by thermodynamic properties. Watson and Perry[49] have studied ZnO solubility in molten KCl and found that it was $2.3\times10^{-8}$ mole/kg for ZnO and $3.2\times10^{-12}$ for $K_2ZnO_2$.

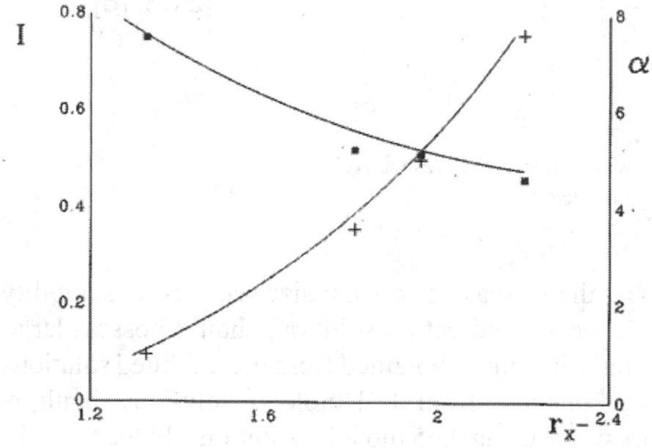


Figure 21.3.2. The formal ionic moments (I, $r_X^{-1}$, Å$^{-1}$) and polarizations (α) of halide-ions.

We have studied oxide solubilities in molten CsCl-KCl-NaCl at 600 and 700°C in order to estimate oxide solubility changes with temperature.[34-36,50] Oxide solubilities were lower than those predicted by Shreder's equation but estimations of their thermal changes usually were in a good agreement with those calculated from Shreder's equation. Oxides solubilities (in molar fractions) in KCl-NaCl and CsCl-KCl-NaCl have been found to be close at the same temperature.

Some works are related to oxo-acidity studies in alkaline chloride melts with cations of high acidity: MgO-$BaCl_2$-$CaCl_2$-NaCl,[37] CaO-$CaCl_2$-KCl-NaCl,[51] $CeCl_3$-KCl-NaCl,[52] ZnO-NaCl-ZnCl.[2.53,54] Acidic cations have been found to affect acid-base interactions by fixing oxide ions. Oxide solubilities[37,51] were higher than those for alkaline chloride mixtures without acidic admixtures. The same conclusions came from studies[55] on MgO solubility in $CaCl_2$-CaO mixture, $K_s=10^{-6.2}$ at 1133 K. The oxide solubility products in Ca-based chlorides are increased by 4 orders of magnitude.

Solubilities (as pP) of MgO (7,61±0,06), NiO (7,44±0,09), CoO (6,25±0,12), MnO (5,20±0,17), ZnO (4,97±0,07) and CdO (3,21±0,07) in molten eutectic $SrCl_2$-KCl-NaCl (0.22:0.42:0.36) at 700°C have been determined.[56] The precipitation of PbO from solutions containing $Pb^{2+}$ was not observed although these cations demonstrated appreciable acidic properties (pK=2,58±0,05). The oxide solubility products in molar fractions have been increased by 1.95 orders of the magnitude for all the oxides.

A similar study was performed in two Ba-based melts: $BaCl_2$-KCl-NaCl (0.43:0.29:0.28) and $BaCl_2$-KCl (0.26:0.74) at 700°C.[57] Solubility products in molar fractions of CdO were 5.43 and 5.62, for ZnO these magnitudes were 6.97 and 7.14 in molten $BaCl_2$-KCl-NaCl and $BaCl_2$-KCl, respectively. The corresponding pP in KCl-NaCl were 2.01 and 1.83. The oxide solubility products in alkaline earth and Li-based chlorides are presented in Table 21.3.2.

**Table 21.3.2. Solubility products of some oxides in melts based on moderately acidic cations at 700°C (molar fractions, at the confidence level 0.95)**

| Oxide/composition | KCl-LiCl[24,25] | $SrCl_2$-KCl-NaCl[56] | $BaCl_2$-KCl-NaCl[57] | $BaCl_2$-KCl[57] |
|---|---|---|---|---|
| | 0.4:0.6 | 0.22:0.42:0.36 | 0.43:0.29:0.28 | 0.26:0.74 |
| MgO | 8.38±0.05 | 9.63±0.06 | 9.76±0.12 | |
| MnO | | 7.32±0.17 | 7.38±0.06 | |
| CoO | 6.94±0.11 | 8.37±0.12 | 8.12±0.12 | |
| NiO | 7.85±0.2 | 9.56±0.09 | 9.26±0.11 | |
| ZnO | | 7.09±0.07 | 6.97±0.1 | 7.11±0.1 |
| CdO | | 5.33±0.07 | 5.43±0.2 | 5.62±0.2 |

## 21.3.3.4 Other alkaline halides

NiO solubility in molten KF-LiF at 550°C determined by potentiometric method is $1.3\times10^{-6}$.[58] This value considerably exceeds values obtained for the above chloride melts. The increase of NiO solubility occurred due to formation of fluoride complexes which are more stable than chloride ones.

Demirskaya, Cherginets and Khailova conducted oxide solubility studies in molten CsBr·2KBr[59-61] and CsI[62] at 700°C. Results are presented in Table 21.3.3. These data are useful to evaluate the effect of halide ion of melt on changes of oxide solubilities. There is a significant reduction of oxide solubility with anion exchange $Cl^-\rightarrow Br^-$, pP changes were of order of 1-2 units for all oxides studied, a degree of dissociation decreases simultaneously. Oxide solubilities in molten CsI possess intermediate values (Table 21.3.3.).

The changes of solubility in halide melts may be explained by changes of complexation abilities of halide ions. In general, the complexation ability increases together with the formal ionic moment, $r_{X^-}^{-1}$, $Å^{-1}$, and the polarization, α, $Å^3$, of anions. In the sequence $F^- > Cl^- > Br^- > I^-$ the formal ionic moments decrease and polarization increases as in Figure 21.3.2. The superimposition of these factors leads to the extremum in the bonding energy in the halide complexes, therefore, its minimum for chloride or bromide complexes should be expected. The studies[33,59-62] show that this leads to minimal solubilities in molten bromides.

**Table 21.3.3. Solubility products, of some oxides in molten alkaline halides (molar fractions).[After references 33,59-62]**

| Oxide | CsCl-KCl-NaCl (0.45:0.25:0.30) | | CsBr-KBr (0.66:0.34) | CsI |
|---|---|---|---|---|
| MgO | 12.68±0.11 | 11.68±0.04 | 12.76±0.06 | |
| CaO | 7.73±0.05 | 6.86±0.13 | 9.93±0.06 | 7.50±0.14 |
| SrO | 5.81±0.40 | 5.54±0.13 | 6.82±0.12 | 6.28±0.30 |
| BaO | 4.75±0.10 | 4.50±0.20 | 5.16±0.30 | 4.98±0.15 |
| MnO | 9.73±0.15 | 9.49±0.14 | 10.70±0.04 | |
| CoO | 10.92±0.06 | 10.50±0.03 | 11.33±0.20 | |
| NiO | 12.60±0.08 | 11.42±0.06 | 12.71±0.30 | |

| Oxide | CsCl-KCl-NaCl (0.45:0.25:0.30) | | CsBr-KBr (0.66:0.34) | CsI |
|---|---|---|---|---|
| $Cu_2O$ | 6.95±0.10 | 6.35±0.05 | | |
| ZnO | 8.80±0.07 | 8.15±0.03 | 11.03±0.16 | |
| CdO | 7.54±0.48 | 7.09±0.05 | 9.35±0.16 | 8.00±0.06 |
| SnO | 11.28±0.05 | | | |
| PbO | 8.21±0.05 | 7.04±0.05 | 8.52±0.07 | 7.80±0.10 |

## 21.3.4 ON THE POSSIBILITY TO PREDICT OXIDE SOLUBILITIES ON THE BASE OF THE EXISTING DATA

Studies[33,59-62] show that the estimation of oxide solubilities in non-studied halide melts are possible, at least, for 1:1 oxides, i.e., MeO.

### 21.3.4.1 The estimation of effect of anion

It has been found that oxide solubilities (pP) are related to cation radius.[33] For Na- and K-based halide melts, a plot for solubility products expressed in molar fractions is described by the following equation:

$$pP_N = pP_{N,0} + a r_{Me^{2+}}^{-2} \qquad [21.3.20]$$

where:

$r_{Me^{2+}}$ the cation of oxide radius, nm.

The value of a is ~0.053 and values of $pP_{N,0}$ are 1.8, 3.2 and 2.6 for chloride, bromide and iodide melts, respectively. The accurate values are presented in Table 21.3.4.

**Table 21.3.4. Coefficients of plots [21.3.20] for molten alkaline halides at the confidence level 0.95. After references [38,59-62]**

| Melt | t, °C | $pP_{N,0}$ | a |
|---|---|---|---|
| KCl-NaCl | 700 | 1.8±0.9 | 0.053±0.01 |
| CsCl-KCl-NaCl | 600 | 1.7±2.0 | 0.057±0.01 |
| CsCl-KCl-NaCl | 700 | 1.7±0.3 | 0.054±0.02 |
| CsBr-KBr | 700 | 3.2±1.8 | 0.053±0.01 |
| CsI | 700 | 2.65±0.58 | 0.052±0.01 |

### 21.3.4.2 The estimation of effect of melt acidity

Papers[25,26,57,58] give us the possibility to estimate the oxide solubilities for melts with cations more acidic than $Na^+$. For acidic melts the following equation applies:

$$pP_N = pP_{N,KCl-NaCl} - pI_l \qquad [21.3.21]$$

where $pI_l$ is ~3.5 for KCl-NaCl, ~2 for $SrCl_2$-KCl-NaCl, ~4 for $CaCl_2$-KCl-NaCl and ~1,8 for $BaCl_2$-KCl. If the composition of acidic cation is different from that of melts described in the previous section the value $pI_l$ in [21.3.21] can be corrected as follows:

$$pI_{I_I} = pI_I + \log \frac{N_{Me^{n+},I_I}}{N_{Me^{n+},I}} \quad [21.3.22]$$

where:

$pN_{Me^{n+},I_I}$, $N_{Me^{n+},I}$ molar fractions of the most acidic cation in non-studied and the studied melts, respectively.

### 21.3.4.3 The estimation of effect of temperature

The pP value obtained according the above equations can be corrected for temperature if it differs from 700°C. It has been shown that at relatively small temperature changes oxide solubility is close to value predicted by the Shreder's equation.[37] This allows one to estimate the solubility products at different temperatures using the following equation:

$$pP_{T_2} = pP_{T^*} - \frac{5.2T_{mp}(T^* - T_2)}{(T^* + T_2)} \quad [21.3.23]$$

where:

$T^*$ temperature at which $pP_{T^*}$ is known
$T_2$ temperature at which pP should be estimated
$T_{mp}$ melting point of the oxide

## 21.3.5 CONCLUSIONS

The existing data allow

- to estimate level of contamination of melts by oxide containing impurities
- to predict oxide solubility in molten salts having different anion composition and acidic properties

It should be emphasized, however, that there are problems which require further work:

- quantitative estimation of surface effect on oxide solubility
- the removal of oxide ion admixtures by the their conversion to oxide deposit.

## REFERENCES

1 S.E.Lumkis, *Izv.AN SSSR, Otd.Tekh.Nauk*, **N12**, 100 (1958).
2 H.Lux, *Z. Elektrochem*., **45**, 303 (1939).
3 N.N.Ovsyannikova and Yu.F.Rybkin, *Ukr.Khim.Zhurn*., **42**, 151 (1976).
4 A.M.Shams El Din and A.A.El Hosary, *J.Electroanal.Chem*., **16**, 551 (1968).
5 A.M.Shams El Din, A.A.El Hosary, H.D. Taki El Din, *Electrochim.Acta*, **13**, 407 (1968).
6 I.Slama, *Coll.Czechoslov.Chem.Commun*., **28**, 985 (1963).
7 I.Slama, *Coll.Czechoslov.Chem.Commun*., **28**, 1069 (1963).
8 D.Naumann and G.Reinhardt, *Z.anorg.allg.Chem*., **343**, 165 (1966).
9 A.V.Volkovich, *Rasplavy* , **N4**, 24 (1991).
10 Yu.K.Delimarsky and V.N.Andreeva, *Zhurn.Neorg.Khim*, **5**, 1123 (1960).
11 R.Combes, F.De Andrade, A.De Barros and H.Ferreira, *Electrochim. Acta*, **25**, 371 (1980).
12 M.Frederics and R.B.Temple, *Inorg.Chem*., **11**, 968 (1972).
13 Shen' Tsin Nan and Y.K.Delimarsky, *Ukr.Khim.Zhurn*., **27**, 454 (1961).
14 G.Delarue, *J.Electroanalyt.Chem*., **1**, 13 (1959).
15 G.Delarue, *Bull.Soc.Chim.France*., **N8**, 1654 (1960).
16 A.N.Shvab, A.P.Timchenko and A.V.Gorodysky, *Ukr.khim.zhurn*., **40**, 90 (1974).
17 H.A.Laitinen and B.B.Bhatia, *J.Electrochem.Soc*., **107**, 705 (1960).
18 V.L.Cherginets, *Rasplavy* , **N1**, 62 (1991).
19 V.I.Shapoval and O.F.Pertchik, *Elektrokhimiya*, **10**, 1241 (1974).
20 V.I.Shapoval, V.F.Makogon and O.F.Pertchik, *Ukr.khim.zhurn*., **45**, 7 (1979).
21 V.I.Shapoval and V.F.Makogon, *Ukr.khim.zhurn*., **45**, 201 (1979).

22 J.Picard, F.Seon and B.Tremillon, *Electrochim.Acta*, **22**, 1291 (1978).
23 F.Seon, J.Picard and B.Tremillon, *J.Electroanalyt.Chem.*, **138**, 315(1982).
24 V.L.Cherginets and T.P.Rebrova, *Zhurn.Fiz.Khim.*, **73**, 687 (1999).
25 V.L.Cherginets and T.P.Rebrova, *Electrochim.Acta,* **45**, 471 (1999).
26 V.I.Shapoval and O.G.Tsiklauri, Conf."Phys. chemistry and electrochemistry of molten salts and solid electrolytes", Sverdlovsk, June 5-7, 1973, Part II, Sverdlovsk,1973,pp.32-33.
27 V.I.Shapoval, O.G.Tsiklauri and N.A.Gasviani, *Soobsch.AN Gruz.SSR*, **89**, 101 (1978).
28 R.Combes, J.Vedel and B.Tremillon, *Electrochim.Acta*, **20**, 191 (1975).
29 Y.K.Delimarsky,V.I.Shaponal and N.N.Ovsyannikova, *Ukr.khim.zhurn.*, **43**, 115 (1977).
30 R.Combes, J.Vedel and B.Tremillon, *Anal.Lett.*, **3**, 523 (1970).
31 F.De Andrade, R.Combes and B.Tremillon, *C.R.Acad.Sci.*, **C280**, 945 (1975).
32 R.Combes, B.Tremillon and F.De Andrade, *J.Electroanalyt.Chem.*, **83**, 297 (1987).
33 V.L.Cherginets and V.V.Banik, *Rasplavy* , **N1**, 66 (1991).
34 T.P.Boyarchuk, E.G.Khailova and V.L.Cherginets, *Electrochim.Acta*, **38**, 1481 (1993).
35 V.L.Cherginets and E.G.Khailova, *Zhurn.neorg.khim.*, **38**, 1281 (1993).
36 V.L.Cherginets and E.G.Khailova, *Electrochim.Acta*, **39**, 823 (1994).
37 P.Bocage, D.Ferry and J.Picard, *Electrochim.Acta*, **36**, 155 (1991).
38 V.L.Cherginets and A.B.Blank, *Ukr.Khim.Zhurn*, **57**, 936 (1991).
39 V.E.Komarov and V.E.Krotov, *Coll.Inst.Electrochem. of Ural Sci.Centre of Acad.Sci.USSR*, **N27**, 61 (1978).
40 N.M.Barbin, V.N.Nekrasov, L.E.Ivanovsky, P.N.Vinogradov and V.E.Petukhov, *Rasplavy*, **N2**, 117 (1990).
41 N.M.Barbin, V.N.Nekrasov, *Electrochim.Acta*, **44**, 4479 (1999)
42 Y.Kaneko and H.Kojima, *Denki Kagaku*, **42**, 304 (1974).
43 V.L.Cherginets, *Uspekhi Khimii*, **66**, 661 (1997).
44 V.L.Cherginets, *Electrochim.Acta*, **42**, 3619 (1997).
45 H.A.Laitinen and W.E.Harris, **Chemical analysis**, 2nd edition, *McGraw-Hill Book Company*, London, 1975.
46 F.F.Grigorenko, A.V.Molodkina, V.M.Solomaha and M.S.Slobodyanik, *Visnyk Kyivskogo Universitetu.Ser.Khim.*, **N14**, 38 (1973).
47 M.L.Deanhardt and K.H.Stern, *J.Electrochem.Soc.*, **128**, 2577 (1981).
48 M.L.Deanhardt and K.H.Stern, *J.Electrochem.Soc.*, **129**, 2228 (1982).
49 R.F.Watson and G.S.Perry, *J.Chem.Soc.Faraday Trans.*, **87**, 2955 (1991).
50 E.G.Khailova,T.A.Lysenko and V.L.Cherginets, *Zhurn.Neorg.khim.*, **38**, 175 (1993).
51 G.T.Kosnyrev, V.N.Desyatnik, N.A.Kern and E.N.Nosonova, *Rasplavy* , **N2**, 121 (1990).
52 R.Combes, M.N.Levelut and B.Tremillon, *Electrochim.Acta*, **23**, 1291 (1978).
53 Y.Ito, H.Kotohda, J.Uchida and S.Yoshizawa, *J.Chem.Soc.Jap. Chem.and Ind.Chem.*, **6**, 1034 (1982).
54 D.Ferry,Y.Castrillejo and G.Picard, *Electrochim.Acta*, **34**, 313 (1989).
55 G.N.Kucera and M.-L.Saboungi, *Met.Trans.*, **B7**, 213 (1976).
56 V.L.Cherginets and T.P.Rebrova, *Zhurn.Fiz.Khim.*, **74**, 244 (2000).
57 V.L.Cherginets and T.P.Rebrova, *Electrochem.Commun.*, **1**, 590 (1999).
58 Y.Ito,H.Hayashi,Y.Itoh and S.Yoshizawa, *Bull.Chem.Soc.Jpn.*, **58** 3172 (1985).
59 O.V.Demirskaya,E.G.Khailova and V.L.Cherginets, *Zhurn.Fiz.khim.*, **69**, 1658 (1995).
60 V.L.Cherginets, E.G.Khailova and O.V.Demirskaya, *Electrochim.Acta*, **41**, 463 (1996).
61 V.L.Cherginets and E.G.Khailova, *Ukr.Khim.Zhurn.*,**62**, 90 (1996).
62 V.L.Cherginets, E.G.Khailova and O.V.Demirskaya, *Zhurn.Fiz.khim.*,1997, 371 (1997).

# 21.4 ALTERNATIVE CLEANING TECHNOLOGIES/DRYCLEANING INSTALLATIONS

KASPAR D. HASENCLEVER
**Kreussler & Co.GmbH, Wiesbaden, Germany**

## 21.4.1 DRYCLEANING WITH LIQUID CARBON DIOXIDE (LCD)

### 21.4.1.1 Basics

The use of compressed $CO_2$ for dissolving oils and fats from different substrates under industrial conditions has been published in German journals since 1982 (Quirin, KW; FSA 84; 460-468).

The activity of surfactants under pressure and the formulation of reverse micelles has been published by Johnston, Lemert and McFann in Am. Chem. Soc. Ser. 406 (1989).

The structure of reverse micelle and microemulsion phases in near critical and supercritical fluid as determined from dynamic light scattering studies has been published by Johnston and Penninger in: Supercritical Fluid Science and Technology, ACS Symp. Ser. Washington DC, 1989.

The use of perfluoropolyether microemulsions in liquid and supercritical $CO_2$ has been published by Chittofrati, Lenti, Sanguinetti, Visca, Gambi, Senatra and Zhou in Progr. Colloid & Polym. Sci., 79, 218-225 (1989).

A process for cleaning or washing of clothing in liquid and supercritical $CO_2$ is the issue of the German Patent DP 39 04 514 A1 by Schollmeyer and Knittel of 23.08.1990.

$CO_2$ is a slightly toxic, colorless gas with a pungent, acid smell. It will not burn or support combustion. The gas is 1.4 times heavier than air and sublimes at atmospheric pressure at minus 78°C. $CO_2$ is not corrosive to steel, as long as it is free of water. With water it reacts to $H_2CO_3$, which can cause rapid corrosion to steel. Chromesteel or aluminum should be used if contact with water is unavoidable. $CO_2$ will react violently with strong bases, ammonia and amines.

The critical data of $CO_2$ are:

| | |
|---|---|
| pressure | 73.81 bar |
| temperature | 31.3°C |
| volume | 0,096 l |
| density | 0.468 g/ml |

Above $T_{crit}$, $CO_2$ cannot be liquefied not even under highest pressure. Above critical data substances are in supercritical condition. (Supercritical $CO_2$ = ScCD).

At 20°C and 55.4 bar $CO_2$ is liquid (LCD). The physical properties are:

| | |
|---|---|
| density | 0.77 g/ml |
| viscosity | 0.1 mPas |
| surface tension | 5 mN/m |
| solvent power | about 20 (K.B.) |

LCD is a solvent for apolar substances. Its activity can be widened by combination with surfactants. Micro-emulsions in LCD can be created with different surfactants (AOT/F-Surf.) and water.

$CO_2$ is a natural resource, non-flammable, non-smog producing, physically stable. LCD can be stored and transported under pressure without harm. Containers must be treated in accordance to the national regulations (in Germany Druckbehälter Verordnung certification). Tanks and/or containers for LCD have three to four times the weight of their net capacity.

#### 21.4.1.2 State of the art

The stability of textiles, textile dyes, buttons, zips, interlinings under influence of high pressure in LCD and ScCD was tested at DTNW, Krefeld, DWI/TH Aachen. Both, LCD and ScCD, if they are pure, do not harm textiles and dyes. After treatment in ScCD buttons and plastic zips are destroyed, when the decompression of $CO_2$ runs fast.

LCD and ScCD are able to penetrate into apolar polymers such as polyester and polyamide, and plasticize the material so that dispersed substances can migrate into them. This behavior is used for dyeing processes with dispersion dyestuffs in ScCD, which allows textile dyeing without waste. The same procedure can cause greying in drycleaning, if resolved pigments are dispersed in LCD cleaning fluids.

In Journal of Supercritical Fluids, (1990), 3, 51-65 Consani and Smith of Battelle Pacific Northwest Laboratories report the Observation on the Solubility of Surfactants and Related Molecules in Carbon Dioxide at 50°C. Nearly all known surfactants are classified.

In US patent 5,467,492 Hughes Aircraft Company claims a dry cleaning in LCD with the distinguishing feature of a non moving basket together with jet agitation to the load by current circulation of the LCD cleaning fluid. In order to reach this aim, the pressure vessel contains a cylindrical perforated basket to take the load. After the pressure vessel is totally filled with LCD, the load is then set into motion and agitated by high velocity fluid jets. A dry cleaning machine presented at Las Vegas in 1997 was equipped with a 135 l pressure cylinder and the capability to clean a load of textiles of about 10 kg.

At the same exhibition Miccel, Technologies, Inc.; North Carolina State University presented the MiCARE Garment Cleaning Fluid System, which consists of LCD together with patented surfactants as cleaning fluid and a dry cleaning machine for LCD, equipped with a moving basket, creating agitation in the conventional manner.

In Science, Feb. 1996, E. Goldbaum published a report on the development of a team of scientists of the University of Texas, the University of Nottingham and the University of Colorado, using water in $CO_2$ microemulsions with fluorinated surfactants in place of conventional solvents, such as chlorinated hydrocarbons or hydrocarbons. This research was funded by a Department of Energy grant of the US.

The Research Institute Hohenstein together with other German Research Institutes and industrial Partners (Kreussler) are working on a basic research project, to define the interactions between LCD, textiles, surfactants, dissolved, emulsified and/or dispersed matter in this system.

The Dutch Research Institute TNO, Delft works together with industrial partners on a project to develop the complete background for textile cleaning in LCD under practical conditions. As part of this project Kreussler is responsible for the research and development of detergents and the cleaning process.

#### 21.4.1.3 Process technology

Outer garments - apparel - in average of European countries contain about 15 g/kg soiling, when they are brought to dry cleaning. This "soiling" contains:

| | |
|---|---|
| 50% pigments | aerosols, carbon black, iron oxides, dust; |
| 30% water soluble | perspiration, salt, body excrements; |

10% polymers albumin (blood, milk), starch (food);
10% solvent soluble oils, fat, wax, grease.

The main problem in dry cleaning is the removal of pigments, water soluble material and polymers. Apolar solvents as perc, hydrocarbons or LCD are not able to achieve this.

Drycleaning detergents (DD) will widen the activity of the drycleaning process in removing pigments and polar substances. In order to gain a maximal benefit from DD, the process technology must meet the special requirements. DD's offer dispersing and emulsifying activity to solvents, they activate water additions into micro-emulsions in order to achieve removal of polar matter in apolar solvent. To optimize this action, a multi bath process is used, which works under the conditions of extraction from high to low contaminated cleaning fluids. Regeneration of cleaning fluids is realized by filtration and distillation. For filtration a drum-pump-filter-drum circuit is used. For distillation a distilling vessel, condensing equipment and clean solvent tank are necessary. The dimensions must be constructed with regard to the requirements of the quantity of removed "soil" during a full working day, which means with a 10 kg machine 750 g pigments, 600 g "salts" and polymers, 100 g oils and fats, 1000 g DD and about 10 l of water. In order to save costs, the cleaning baths are as short as possible, which means, about 3 l of solvent per kg load, so that per process one full bath can be distilled. In order to separate clean, average and high contaminated cleaning fluid, most drycleaning machines are equipped with 3 tanks, a filter and a distilling vessel with the capacity of the biggest tank.

In order to build less expensive machines, sometimes drycleaning machines are offered which are equipped with insufficient size for distilling and less than 3 tanks. In this case the possible cleaning result will not meet the necessary hygienic and esthetic requirements of customers.

The same rules apply to LCD as to conventional solvents. A one tank machine and distilling from tank to tank instead from cage to clean tank, combined with an insufficient filter size, cannot meet the minimum requirements, even if the solvent is LCD.

### 21.4.1.4 Risks

The solubility of water in LCD is low (0.1%); the solubility of $CO_2$ in water is high; at 4 bar and 20°C 1 l of water dissolves 4 l of $CO_2$ gas, of which about 0.1 % reacts to $H_2CO_3$, a corrosive substance to iron and steel.

Up to now the interaction in the system LCD, surfactants, water and textiles together with solved and dispersed contaminants has not yet been studied under practical conditions. The influence of surfactants on the equilibrium of water in the system of high pressure without airspace (free water, water in LCD emulsion, water adsorbed on textiles) is not yet known. In order to achieve the removal of polar substances from textiles, in conventional cleaning systems a desorption of moisture from textiles is necessary. If this method is compatible with LCD is not yet known.

The low risk of shrinkage on natural textiles in dry cleaning is due to the fact, that the adsorption of moisture is reduced, after fibers are soaked with apolar solvent beyond condensation conditions of water. In presence of moisture within condensation conditions natural textiles will shrink in dry cleaning more than in aqueous processes. The LCD cleaning process works completely within condensation conditions of water. This means, that LCD cleaning can bear a high risk of shrinkage, in particular, when water additions are used in order to remove apolar soiling.

The material and manufacturing costs of LCD machines are more expensive than normal drycleaning machines, and substantially more expensive than wet cleaning machines. The operating costs depend on the expenses of the distribution of LCD, which seems to need a total new network. With costs of about DM 2.50 per kg ($US 1.20/kg), LCD is more expensive than perc and similar to HCS. At the same time the consumption of LCD seems to be four times larger than perc and eight times larger than HCS. Not to mention water in wet cleaning.

Present Care Labeling considers perc, HCS and water, no LCD. If any textile damage will occur, it will be the user's responsibility.

The investment climate in drycleaning is very weak, so that there is no tendency to invest into an unknown technology.

### 21.4.1.5 Competition

LCD stands in competition to dry cleaning in TCE and in HCS as well as to wet cleaning.

The following comparison is done under the supposition of a capacity of 20 kg/h for investment costs and a quantity of 100 kg for consumption costs (Table 21.4.1). All costs in DEM.

The result of this comparison is very clear: most competitive is Wet Cleaning, followed by HCS dry cleaning. In order to avoid garment risks on sensitive textiles in wet cleaning, a combined installation with 70% wet cleaning and 30% HCS cleaning would be the optimum.

In order to make LCD competitive, solvent costs must be reduced rapidly and basic research work has to be done in order to increase garment safety, cleaning results and the technical reliability.

**Table 21.4.1. Cost comparison of textile cleaning methods**

| Costs/Properties | TCE | HCS | Wet Clean | LCD |
|---|---|---|---|---|
| Machine | 120,000.00 | 80,000.00 | 40,000.00 | 160,000.00 |
| Space, $m^3$ | 2.20 | 2.50 | 1.80 | 5.00 |
| Solvent | 5.40 | 5.80 | 2.00 | 30.00 |
| Additives | 12.00 | 8.00 | 16.00 | 12.00 |
| Energy | 23.80 | 21.40 | 11.20 | 15.00 |
| Waste | 12.50 | 8.00 | 0.00 | 4.00 |
| Consumables total | 53.70 | 43.20 | 29.20 | 61.00 |
| Reliability | good | sufficient | very good | high risk |
| Maintenance | good | good | very good | not known |
| Cleaning results | good | sufficient | very good | not known |
| Garment risks | low | very low | sufficient | not known |
| Environment | high risk | low risk | no risk | no risk |

Figure 21.4.1. Miele System Kreussler washing machine and dryer.

## 21.4.2 WET CLEANING

The new technique results from intensive research work of Kreussler. The processing methods and in particular their translation into production-ready machine technology was developed in co-operation of Kreussler with Miele of Gütersloh.

What does this new textile cleaning system look like, what prospects does it offer, what are its limitations and how it can be integrated into the practical operations of a textile cleaning company?

### 21.4.2.1 Kreussler textile cleaning system

For achieving drycleaning with water it is essential to use a washing machine with at least 80 cm drum diameter, centrifugal acceleration of 0.9 g-force in the wash cycle and 450 g-force during spinning (at least 1000 rpm), appropriate drum and lifter rib design, and accurate regulation of temperature and liquor flow, as well as precise control of mechanical action and dosing.

Also necessary is a large-capacity tumble dryer, with drum at least 100 cm in diameter, having precisely controlled centrifugal acceleration; horizontally directed air flow, parallel to the drum axle; and precision electronics to regulate temperature and residual moisture level.

These requirements, comprising washer-extractor and tumble dryer together with other important know-how, are provided by the "Miele System Kreussler".

Miele System Kreussler

The washer-extractor (Miele WS 5220 TR), with a drum volume of 220 litres, has a loading capacity of 8-10 kg for delicate outerwear, 12-15 kg of ordinary outerwear and 20-22 kg of normal textiles designated washable by the care label. The special outerwear cleaning is carried out using the LANADOL process, for which approx. 140 litres of water are used per load. Cycle time is about 25 minutes.

The dryer (Miele 6559 TR) has a drum diameter of 110 cm, with a 550 litre volume. The drying air flows through the load horizontally - parallel to the drum axle - from the rear towards the door, thus achieving optimum evaporation level with a short drying time. The necessary temperature progression and precise residual moisture is ensured with precision electronics.

A significant aspect is that not only the LANADOL process but all other wash programs can be carried out in the Miele System Kreussler machines, providing comprehensive textile cleaning facilities.

LANADOL processing technique

Most textiles offered to drycleaners for processing can be handled with three basic processes:

- Extra: Garments without wash symbol
- Normal: Garments with 30/40°C wash symbol
- Proofing: Poplin and sportswear

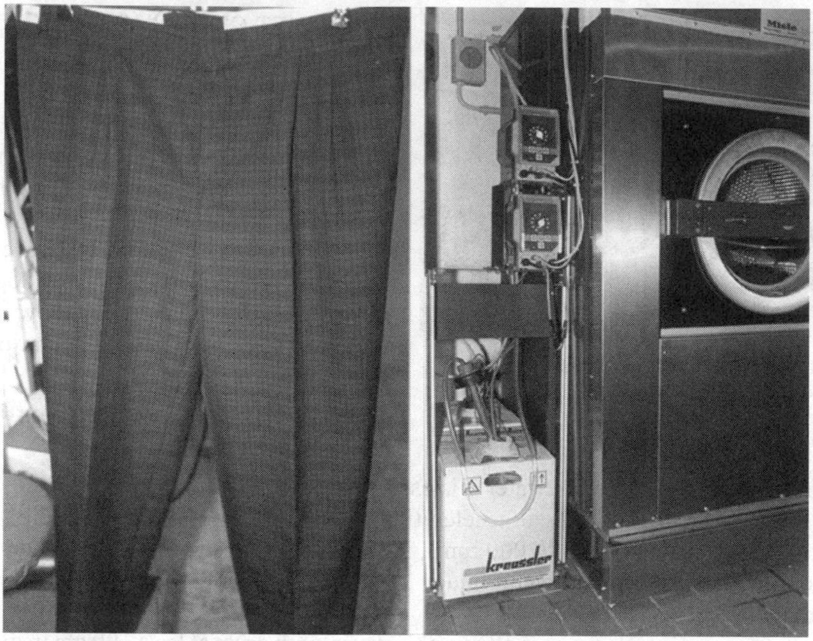


Figure 21.4.2. (left) Polyester/wool trousers after drying.(right) Dosing method for polyester/wool trousers after drying LANADOL AKTIV and LANADOL APRET.

Figure 21.4.3. Knitwear and silk, unfinished after drying.

Within each of these processing groups, the work should be sorted into light/medium/heavy weight, the best load combination consisting of items with the same or similar drying characteristics. It would be preferable, too, to differentiate roughly between light and dark colors.

Heavily soiled or grease stained areas should be pre-treated by simple pre-spotting with LANADOL AVANT, using PRENETT A-B-C for special staining. Where non-color-fastness is suspected, a dye-fastness check (seam test) should be made with a damp white cotton cloth.

After the relevant program is selected, the process runs fully automatically. The exact amount required of the special cleaning and fibre protection agent LANADOL AKTIV is added automatically at precisely the right time.

A fibre protection agent, providing retexturing and antistatic finish for drying, is also added automatically.

Following spin-drying, woollen textiles have a particularly low residual moisture content of about 30%. After selection of the program suitable for the

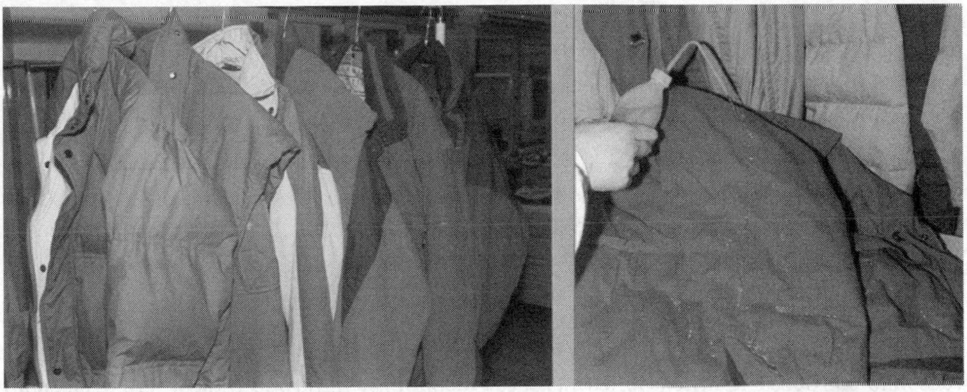

Figure 21.4.4 Down-filled anoraks and ski clothing, washed... ... and impregnated.

load make-up, the tumbler will dry to the pre-determined residual moisture level within 15 - 25 minutes.

With garments structured in several layers - such as men's jackets, ladies' suit jackets and coats - further drying at room temperature or in a drying cabinet is beneficial, before finishing on a hot air - steam garment former. All other garments can be finished normally.

### 21.4.2.2 Possibilities

We will not dwell here on the possibilities for processing textiles which are washable according to their care labels. On removal from the dryer, the items are quite smooth, needing little effort to finish. Treatment of washable outerwear with the LANADOL process is so problem-free, so simple and rapid, the work is so clean, bright and fresh, with such a pleasant handle and freedom from static, that one would not wish to handle this classification in any other way. One immediately realizes that conventional drycleaning would be just second best for such textiles. Customers will also come to appreciate this. Even more interesting, however, is the use of the LANADOL process for textiles which are not washable according to their care label.

According to current processing knowledge, the progression from “ideal” to “difficult” garments can be listed as follows:

Silk, knitwear, cotton, wool, viscose, linen and structured garments (jackets). “Difficult” in this context refers mainly to finishing requirements. It can be quite possible that a man’s jacket, of which the interlining has shrunk, could in addition to treatment on the garment former, need additional hand ironing for up to 10 minutes for perfect results. Usually, however, suit jackets can be processed without problems, although not particularly fast. Without doubt, men’s and ladies’ jackets can be drycleaned faster, simpler and with fewer problems. After the LANADOL process, however, they will be fresher, cleaner and nicer to wear.

Other classifications which are not indicated by the care label to be suitable for domestic washing can generally be treated without problems by the LANADOL process, with which processing risks can in most cases be assessed as less than for drycleaning.

Of special interest is treatment by the LANADOL process of textiles which are problematic for drycleaning. Down-filled anoraks, raincoats, impregnated goods - and also covers for rheumatism sufferers, bulky textiles and glittery items. The LANADOL process is ideal for these - quick and safe with excellent results. Wet proofing is just in a class of its own. Bulky textiles without sweals, without perc residues, without peculiar odors but clean, fresh, fluffy and soft, and all after a short process demonstrate particularly well LANADOL processing's superiority for this group of classifications.

#### 21.4.2.3 Limitations

Where there is light, there is also shadow. To evaluate the new system, coming to terms with the shadowy side is vital. There are darker aspects to the method in the high demands it places on operators and the limits, often not obvious, to its care possibilities for certain textiles.

System limits

The compromise necessary to achieve low textile shrinkage, good cleaning results and reasonable processing time implies a process-technology tightrope walk where even small deviations from the norm can lead to a fall. For this, read damaged textiles. It is essential, therefore, to adhere to all the specified parameters. This involves not only the equipment together with the types and amounts of chemicals used, but also application of the correct process for the classification to be treated. At the moment, we will not risk even a transfer of the process to other machine sizes of the same make.

Skill limits

To sort work into classifications for the appropriate process methods, personnel with knowledge of textiles gained after comprehensive training in textile cleaning are required. Furthermore, experienced finishing personnel are needed, particularly if textiles not washable according to their care labels are to be treated with the LANADOL process, even more so if jackets are to be cleaned.

Limits set by textiles

The main problem item, the jacket, has already been mentioned several times. Problems can be caused by:

- Seam shrinkage
- Interlining shrinkage
- Lining shrinkage

In most cases seam shrinkage can be remedied by expert finishing methods. Remedying shrunk interlining can be simple or very difficult. The amount of finishing work required will vary considerably, according to the adhesive and fabric construction used. Pure viscose lining material can shrink considerably in some cases. If the prescribed residual moisture level was kept to precisely during the drying process, this shrinkage can be remedied. However, if there was over-drying then recovery is frequently impossible.

If all three eventualities have occurred on the same jacket, then it is a goner. On the other hand, this really happens very rarely. The limits outlined here accurately reflect experiences to date. With wider use, the catalogue might possibly grow more extensive - but so also would the experience to deal with such problems.

#### 21.4.2.4 Adapting to working practices

The LANADOL process is ideal to supplement an existing drycleaning process, so that one can readily envisage the proportion growing a step at a time. With increasing confidence in this processing method, soon not only those textiles which are washable according to their care label will be cleaned in water, but also more general outerwear. It is then only a question of time in a textile care company before most work will be cleaned with water and the smaller part, comprising articles requiring more complicated treatment, will be treated with solvent.

Not ideal, but nevertheless possible, is a situation where an attractive plant location uses the LANADOL process exclusively, should local conditions prohibit use of solvent equipment or permit it only with great difficulty. In this case, with a well equipped finishing department and qualified operators, a drycleaner's complete service range is feasible, with possibly just very few exceptions.

### 21.4.3 FUTURE

Dry cleaning with TCR is at the end of its development. HCS dry cleaning at present is in a phase of consolidation. Wet cleaning is in its early beginnings and LCD cleaning isn't in practice yet.

TCE will probably soon be rendered a "knock out property" by its environmental behavior. Also the operating costs will increase rapidly because of more specific regulations with regard to storage and handling as well as to the waste removal of TCE. Today 80% of dry cleaning is done in TCE. Possibly more than 90% of existing TCE machines will be replaced by other systems within the next 10 years.

HCS has no probable "knock out property". Possible new developments may reduce the solvent consumption and increase the cleaning results. Because of low operating costs HCS will be the most important replacement for perc machines in the near future.

Wet Cleaning has no "knock out property". New developments will reduce the garment risks, the finishing work and the operating speed. Minimizing the water consumption is already solved technically though these technologies have not been put to use due to the low cost availability of water resources. New services to customers and the revitalization of professional textile care will move on with Wet Cleaning. Most of the new operations will be installed with wet cleaning equipment only or with a combination of wet- and HCS cleaning.

LCD bears the risk of three "knock out properties": reliability, cleaning result, and garment damage. In order to solve the problems connected with the removal of polar "soiling" and the prevention against greying on synthetic fibers and shrinking of natural fibers, a lot of basic research has to be done. Additionally a lot of developmental work has to be done in order to make LCD economically competitive to existing processes. That means not only a reduction of LCD consumption without increasing the process time and the use of energy, but also build up of a simple distribution of this solvent.

# Solvent Recycling, Removal, and Degradation

## 22.1 ABSORPTIVE SOLVENT RECOVERY

Klaus-Dirk Henning
**CarboTech Aktivkohlen GmbH, Essen, Germany**

### 22.1.1 INTRODUCTION

A variety of waste gas cleaning processes[1-4] are commercially available. They are used to remove organic vapors (solvents) from air emissions from various industries and product lines. The main applications of solvents in various industries are given in Table 22.1.1.

**Table 22.1.1. Application of organic solvents. [After references 2-4]**

| Application | Solvents |
|---|---|
| Coating plants: Adhesive tapes, Magnetic tapes, Adhesive films, Photographic papers, Adhesive labels | Naphtha, toluene, alcohols, esters, ketones (cyclohexanone, methyl ethyl ketone), dimethylformamide, tetrahydrofuran, chlorinated hydrocarbons, xylene, dioxane |
| Chemical, pharmaceutical, food industries | Alcohols, aliphatic, aromatic and halogenated hydrocarbons, esters, ketones, ethers, glycol ether, dichloromethane |
| Fibre production | |
| Acetate fibers | Acetone, ethanol, esters |
| Viscose fibers | carbon disulfide |
| Polyacrylonitrile | dimethylformamide |
| Synthetic leather production | Acetone, alcohols, ethyl ether |
| Film manufacture | Alcohols, acetone, ether, dichloromethane |
| Painting shops: Cellophane, Plastic films, Metal foils, Vehicles, Hard papers, Cardboard, Pencils | Amyl acetate, ethyl acetate, butyl acetate, dichloromethane, ketones (acetone, methyl ethyl ketone), butanol, ethanol, tetrahydrofuran, toluene, benzene, methyl acetate |
| Cosmetics industry | Alcohols, esters |
| Rotogravure printing shops | Toluene, xylene, heptane, hexane |
| Textile printing | Hydrocarbons |
| Drycleaning shops | Hydrocarbons, tetrachloroethane |

| Application | Solvents |
|---|---|
| Sealing board manufacture | Hydrocarbons, toluene, trichloroethane |
| Powder and explosives production | Ether, alcohol, benzene |
| Degreasing and scouring applications<br>Metal components<br>Leather<br>Wool | <br>Hydrocarbons, trichloroethylene, tetrachloroethylene, dichloromethane<br>Tetrachloroethylene<br>Tetrachloroethane |
| Loading and reloading facilities (tank venting gases) | Gasoline, benzene, toluene, xylene |

The basic principles of some generally accepted gas cleaning processes for solvent removal are given in Table 22.1.2. For more than 70 years adsorption processes using activated carbon, in addition to absorption and condensation processes, have been used in adsorptive removal and recovery of solvents. The first solvent recovery plant for acetone was commissioned for economic reason in 1917 by Bayer. In the decades that followed solvent recovery plants were built and operated only if the value of the recovered solvents exceeded the operation costs and depreciation of the plant. Today such plants are used for adsorptive purification of exhaust air streams, even if the return is insufficient, to meet environmental and legal requirements.

**Table 22.1.2. Cleaning processes for the removal and recovering of solvents**

| Process | Basic principle | Remarks |
|---|---|---|
| Incineration | The solvents are completely destroyed by:<br>thermal oxidation (>10 $g/m^3$)<br>catalytic oxidation (3-10 $g/m^3$) | Only waste heat recovery possible |
| Condensation | High concentration of solvents (>50 $g/m^3$) are lowered by direct or indirect condensation with refrigerated condensers | - Effective in reducing heavy emissions,<br>- good quality of the recovered solvents,<br>- difficult to achieve low discharge levels |
| Absorption | Scrubbers using non-volatile organics as scrubbing medium for the solvent removal. Desorption of the spent scrubbing fluid | Advantageous if:<br>- scrubbing fluid can be reused directly without desorption of the absorbed solvent<br>- scrubbing with the subcooled solvent itself possible |
| Membrane permeation | Membrane permeation processes are principle suited for small volumes with high concentrations e.g. to reduce hydrocarbon concentration from several hundred $g/m^3$ down to 5-10 $g/m^3$ | To meet the environmental protection requirements a combination with other cleaning processes (e.g. activated carbon adsorber) is necessary |
| Biological treatment | Biological decomposition by means of bacteria and using of bio-scrubber or bio-filter | Sensitive against temperature and solvent concentration |
| Adsorption | Adsorption of the solvents by passing the waste air through an adsorber filled with activated carbon. Solvent recovery by steam or hot gas desorption | By steam desorption the resulting desorbate has to be separated in a water and an organic phase using a gravity separator (or a rectification step) |

This contribution deals with adsorptive solvent removal[2,12,13] and recovery and some other hybrid processes using activated carbon in combination with other purification steps.

Due to more stringent environmental protection requirements, further increases in the number of adsorption plants for exhaust air purification may be expected.

## 22.1.2 BASIC PRINCIPLES

### 22.1.2.1 Fundamentals of adsorption

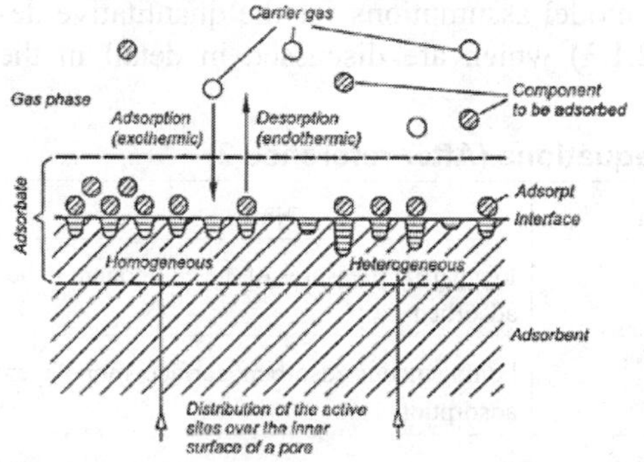


Figure 22.1.1. Adsorption terms (after reference 2).

To understand the adsorptive solvent recovery we have to consider some of the fundamentals of adsorption and desorption[1-9] (Figure 22.1.1). Adsorption is the term for the enrichment of gaseous or dissolved substances on the boundary surface of a solid (the adsorbent). The surfaces of adsorbents have what we call active sites where the binding forces between the individual atoms of the solid structure are not completely saturated by neighboring atoms. These active sites can bind foreign molecules which, when bound, are referred to as adsorpt. The overall interphase boundary is referred to as adsorbate. The term, desorption, designates the liberation of the adsorbed molecules, which is a reversal of the adsorption process. The desorption process generates desorbate which consists of the desorpt and the desorption medium.[2]

A distinction is made between two types of adsorption mechanisms: chemisorption and physisorption. One, physisorption, is reversible and involves exclusively physical interaction forces (van der Waals forces) between the component to be adsorbed and the adsorbent. The other, chemisorption, is characterized by greater interaction energies which result in a chemical modification of the component adsorbed along with its reversible or irreversible adsorption.[2]

Apart from diffusion through the gas-side boundary film, adsorption kinetics are predominantly determined by the diffusion rate through the pore structure. The diffusion coefficient is governed by both the ratio of pore diameter to the radius of the molecule being adsorbed and other characteristics of the component being adsorbed. The rate of adsorption is governed primarily by diffusional resistance. Pollutants must first diffuse from the bulk gas stream across the gas film to the exterior surface of the adsorbent (gas film resistance).

Due to the highly porous nature of the adsorbent, its interior contains by far the majority of the free surface area. For this reason, the molecule to be adsorbed must diffuse into the pore (pore diffusion). After their diffusion into the pores, the molecules must then physically adhere to the surface, thus losing the bulk of their kinetic energy.

Adsorption is an exothermic process. The adsorption enthalpy, decreases as the load of adsorbed molecules increases. In activated carbon adsorption systems for solvent recovery, the liberated adsorption enthalpy normally amounts to 1.5 times the evaporation enthalpy at the standard working capacities which can result in a 20 K or more temperature increase. In the process, exothermic adsorption mechanisms may coincide with endothermic desorption mechanisms.[2]

### 22.1.2.2 Adsorption capacity

The adsorption capacity (adsorptive power, loading) of an adsorbent resulting from the size and the structure of its inner surface area for a defined component is normally represented as a function of the component concentration c in the carrier gas for the equilibrium conditions at constant temperature. This is known as the adsorption isotherm $x = f(c)_T$. There are variety of approaches proceeding from different model assumptions for the quantitative description of adsorption isotherms (Table 22.1.3) which are discussed in detail in the literature.[2,9,11]

**Table 22.1.3. Overview of major isotherm equations (After reference 2)**

| Adsorption isotherm by | Isotherm equation | Notes |
|---|---|---|
| Freundlich | $x = K \times p_i^n$ | low partial pressures of the component to be adsorbed |
| Langmuir | $x = \frac{x_{max} b p_i}{1 + b p_i}$ | homogeneous adsorbent surface, monolayer adsorption |
| Brunauer, Emmet, Teller (BET) | $\frac{p_i}{V(p_s - p_i)} = \frac{1}{V_{MB} C} + \frac{C-1}{V_{MB} C} \frac{p_i}{p_s}$ | homogeneous surface, multilayer adsorption, capillary condensation |
| Dubinin, Raduskevic | $\log V = \log V_s - 2.3\left(\frac{RT}{\varepsilon_o \beta}\right)\left(\log \frac{p_s}{p_i}\right)^n$ | potential theory; n = 1,2,3; e.g., n = 2 for the adsorption of organic solvent vapors on micro-porous activated carbon grades |

where: K, n - specific constant of the component to be adsorbed; x - adsorbed mass, g/100 g; $x_{max}$ - saturation value of the isotherm with monomolecular coverage, g/100 g; b - coefficient for component to be adsorbed, Pa; C - constant; T - temperature, K; $p_i$ - partial pressure of component to be adsorbed, Pa; $p_s$ - saturation vapor pressure of component to be adsorbed, Pa; V - adsorpt volume, ml; $V_{MB}$ - adsorpt volume with monomolecular coverage, ml; $V_s$ - adsorpt volume on saturation, ml; $V_M$ - molar volume of component to be adsorbed, l/mol; R - gas constant, J(kg K); $\varepsilon_o$ - characteristic adsorption energy, J/kg; β - affinity coefficient

The isotherm equation developed by Freundlich on the basis of empirical data can often be helpful. According to the Freundlich isotherm, the logarithm of the adsorbent loading increases linearly with the concentration of the component to be adsorbed in the carrier gas. This is also the result of the Langmuir isotherm which assumes that the bonding energy rises exponentially as loading decreases.

However, commercial adsorbents do not have a smooth surface but rather are highly porous solids with a very irregular and rugged inner surface. This fact is taken into account by the potential theory which forms the basis of the isotherm equation introduced by Dubinin.

At adsorption temperatures below the critical temperature of the component to be adsorbed, the adsorbent pores may fill up with liquid adsorpt. This phenomenon is known as capillary condensation and enhances the adsorption capacity of the adsorbent. Assuming cylindrical pores, capillary condensation can be quantitatively described with the aid of the Kelvin equation, the degree of pore filling being inversely proportional to the pore radius.

If several compounds are contained in a gas stream the substances compete for the adsorption area available. In that case compounds with large interacting forces may displace less readily adsorbed compounds from the internal surface of the activated carbon.

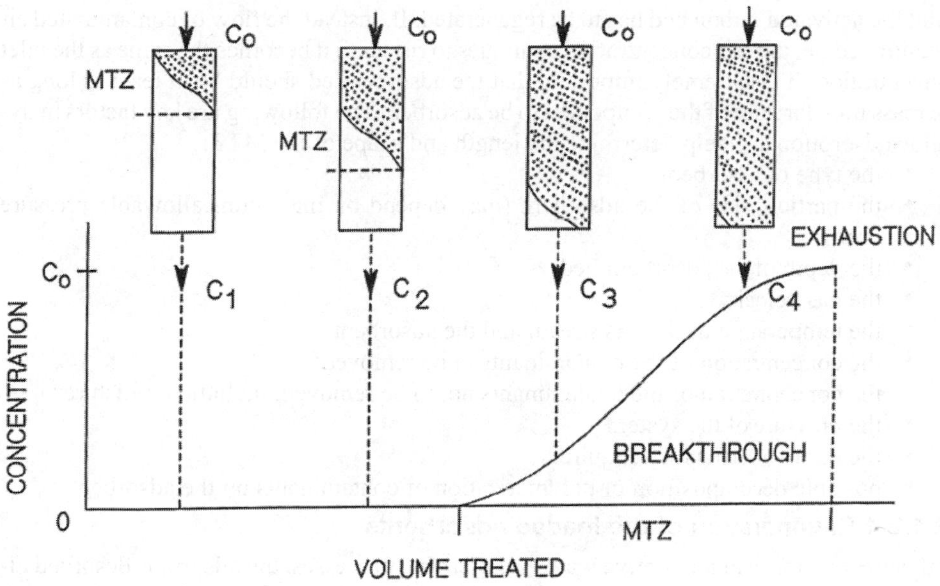

Figure 22.1.2. Idealized breakthrough curve of a fixed bed adsorber.

As previously stated, the theories of adsorption are complex, with many empirically determined constants. For this reason, pilot data should always be obtained on the specific pollutant adsorbent combination prior to full-scale engineering design.

### 22.1.2.3 Dynamic adsorption in adsorber beds

For commercial adsorption processes such as solvent recovery not only the equilibrium load, but also the rate at which it is achieved (adsorption kinetics) is of decisive importance.[9-15] The adsorption kinetics is determined by the rates of the following series of individual steps:

- transfer of molecules to the external surface of the adsorbent (border layer film diffusion)
- diffusion into the particle
- adsorption

The adsorbent is generally in a fixed adsorber bed and the contaminated air is passed through the adsorber bed (Figure 22.1.2). When the contaminated air first enters through the adsorber bed most of the adsorbate is initially adsorbed at the inlet of the bed and the air passes on with little further adsorption taking place. Later, when the inlet of the absorber becomes saturated, adsorption takes place deeper in the bed. After a short working time the activated carbon bed becomes divided into three zones:

- the inlet zone where the equilibrium load corresponds to the inlet concentration, $c_o$
- the adsorption or mass transfer zone, MTZ, where the adsorbate concentration in the air stream is reduced
- the outlet zone where the adsorbent remains incompletely loaded

In time the adsorption zone moves through the activated carbon bed.[18-21] When the MTZ reaches the adsorber outlet zone, breakthrough occurs. The adsorption process must be stopped if a pre-determined solvent concentration in the exhaust gas is exceeded. At this point the activated carbon bed has to be regenerated. If, instead the flow of contaminated air is continued on, the exit concentration continues to rise until it becomes the same as the inlet concentration. It is extremely important that the adsorber bed should be at least as long as the mass transfer zone of the component to be adsorbed. The following are key factors in dynamic adsorption and help determine the length and shape of the MTZ:

- the type of adsorbent
- the particle size of the adsorbent (may depend on maximum allowable pressure drop)
- the depth of the adsorbent bed
- the gas velocity
- the temperature of the gas stream and the adsorbent
- the concentration of the contaminants to be removed
- the concentration of the contaminants not to be removed, including moisture
- the pressure of the system
- the removal efficiency required
- possible decomposition or polymerization of contaminants on the adsorbent

### 22.1.2.4 Regeneration of the loaded adsorbents

In the great majority of adsorptive waste gas cleaning processes, the adsorpt is desorbed after the breakthrough loading has been attained and the adsorbent reused for pollutant removal. Several aspects must be considered when establishing the conditions of regeneration for an adsorber system.[2-4,6,13-16] Very often, the main factor is an economic one, to establish that an in-place regeneration is or is not preferred to the replacement of the entire adsorbent charge. Aside from this factor, it is important to determine if the recovery of the contaminant is worthwhile, or if only regeneration of the adsorbent is required. The process steps required for this purpose normally dictate the overall concept of the adsorption unit.

Regeneration, or desorption, is usually achieved by changing the conditions in the adsorber to bring about a lower equilibrium-loading capacity. This is done by either increasing the temperature or decreasing the partial pressure. For regeneration of spent activated carbon from waste air cleaning processes the following regeneration methods are available (Table 22.1.4).

**Table 22.1.4. Activated carbon regeneration processes (After references 2,16)**

| Regeneration method | Principle | Application examples | Special features |
|---|---|---|---|
| Pressure-swing process | Alternating between elevated pressure during adsorption and pressure reduction during desorption | Gas separation, e.g., $N_2$ from air, $CH_4$ from biogas, $CH_4$ from hydrogen | Raw gas compression; lean gas stream in addition to the product gas stream |
| Temperature-swing process: | | | |
| Desorption | Steam desorption or inert gas desorption at temperatures of < 500°C | Solvent recovery, process waste gas cleanup | Reprocessing of desorbate |
| Reactivation | Partial gasification at 800 to 900°C with steam or other suitable oxidants | All organic compounds adsorbed in gas cleaning applications | Post-combustion, if required scrubbing of flue gas generated |

| Regeneration method | Principle | Application examples | Special features |
|---|---|---|---|
| Extraction with: | | | |
| Organic solvents | Extraction with carbon disulfide or other solvents | Sulphur extraction Sulfosorbon process | Desorbate treatment by distillation, steam desorption of solvent |
| Caustic soda solution | Percolation with caustic soda | Phenol-loads activated carbon | Phenol separation with subsequent purging |
| Supercritical gases | e.g. extraction with super-critical $CO_2$ | Organic compounds | Separation of $CO_2$/or-ganic compounds |

- Pressure-swing process
- Temperature-swing process
- Extraction process

In solvent recovery plants, temperature-swing processes are most frequently used. The loaded adsorbent is direct heated by steam or hot inert gas, which at the same time serves as a transport medium to discharge the desorbed vapor and reduce the partial pressure of the gas-phase desorpt. As complete desorption of the adsorpt cannot be accomplished in a reasonable time in commercial-scale systems, there is always heel remaining which reduces the adsorbent working capacity.

Oxidic adsorbents offer the advantage that their adsorptive power can be restored by controlled oxidation with air or oxygen at high temperatures.

### 22.1.3 COMMERCIALLY AVAILABLE ADSORBENTS

Adsorbers used for air pollution control and solvent recovery predominantly employ activated carbon. Molecular sieve zeolites are also used. Polymeric adsorbents can be used but are seldom seen.

#### 22.1.3.1 Activated carbon

Activated carbon[1-8] is the trade name for a carbonaceous adsorbent which is defined as follows: Activated carbons[5] are non-hazardous, processed, carbonaceous products, having a porous structure and a large internal surface area. These materials can adsorb a wide variety of substances, i.e., they are able to attract molecules to their internal surface, and are therefore called adsorbents. The volume of pores of the activated carbons is generally greater than 0.2 $mlg^{-1}$. The internal surface area is generally greater than 400 $m^2g^{-1}$. The width of the pores ranges from 0.3 to several thousand nm.

All activated carbons are characterized by their ramified pore system within which various mesopores ($r = 1$-$25$ nm), micropores ($r = 0.4$-$1.0$ nm) and submicropores ($r < 0.4$ nm) each of which branch off from macropores ($r > 25$ nm).

Figure 22.1.3. shows schematically the pore system important for adsorption and desorption.[17] The large specific surfaces are created predominantly by the micropores. Activated carbon is commercially available in shaped (cylindrical pellets), granular, or powdered form.

Due to its predominantly hydrophobic surface properties, activated carbon preferentially adsorbs organic substances and other non-polar compounds from gas and liquid phases. Activated alumina, silica gel and molecular sieves will adsorb water preferentially from a gas-phase mixture of water vapor and an organic contaminant. In Europe cylindrically-shaped activated carbon pellets with a diameter of 3 or 4 mm are used for solvent recovery, because they assure a low pressure drop across the adsorber system. Physical and

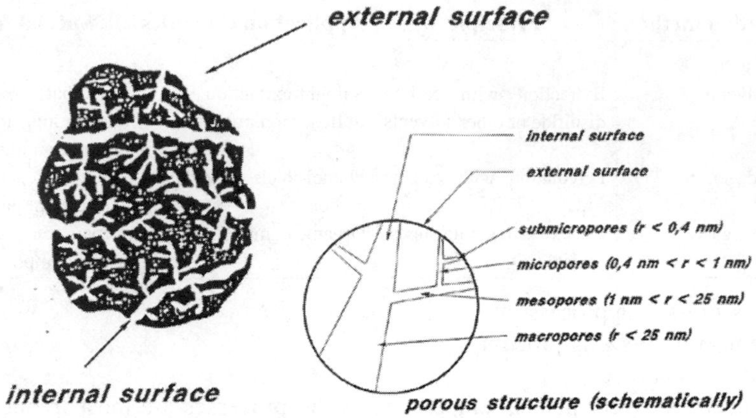


Figure 22.1.3. Schematic activated carbon model (After reference 17).

chemical properties of three typical activated carbon types used for solvent recovery are given in Table 22.1.5.

**Table 22.1.5 Activated carbon types for waste air cleaning (After reference 2)**

| Adsorbent | Applications (typical examples) | Compacted density, kg/m$^3$ | Pore volume for pore size d < 20 nm, ml/g | Pore volume for pore size d > 20 nm, ml/g | Specific surface area, m$^2$/g | Specific heat capacity, J/kgK |
|---|---|---|---|---|---|---|
| Activated carbon, fine-pore | Intake air and exhaust air cleanup, odor control, Adsorption of low-boiling hydrocarbons | 400 - 500 | 0.5 - 0.7 | 0.3 - 0.5 | 1000-1200 | 850 |
| Activated carbon, medium-pore | Solvent recovery, Adsorption of medium-high boiling hydrocarbons | 350 - 450 | 0.4 - 0.6 | 0.5 - 0.7 | 1200-1400 | 850 |
| Activated carbon, wide-pore | Adsorption and recovery of high-boiling hydrocarbons | 300 - 400 | 0.3 - 0.5 | 0.5 - 1.1 | 1000-1500 | 850 |

### 22.1.3.2 Molecular sieve zeolites

Molecular sieve zeolites[2,10,11] are hydrated, crystalline aluminosilicates which give off their crystal water without changing their crystal structure so that the original water sites are free for the adsorption of other compounds. Activation of zeolites is a dehydration process accomplished by the application of heat in a high vacuum. Some zeolite crystals show behavior opposite to that of activated carbon in that they selectively adsorb water in the presence of nonpolar solvents. Zeolites can be made to have specific pore sizes that impose limits on the size and orientation of molecules that can be adsorbed. Molecules above a specific size cannot enter the pores and therefore cannot be adsorbed (steric separation effect).

Substitution of the greater part of the $Al_2O_3$ by $SiO_2$ yields zeolites suitable for the selective adsorption of organic compounds from high-moisture waste gases.

#### 22.1.3.3 Polymeric adsorbents

The spectrum of commercial adsorbents[2] for use in air pollution control also includes beaded, hydrophobic adsorber resins consisting of nonpolar, macroporous polymers produced for the specific application by polymerizing styrene in the presence of a crosslinking agent. Crosslinked styrene divinylbenzene resins are available on the market under different trade names. Their structure-inherent fast kinetics offers the advantage of relatively low desorption temperatures. They are insoluble in water, acids, lye and a large number of organic solvents.

### 22.1.4 ADSORPTIVE SOLVENT RECOVERY SYSTEMS

Stricter regulations regarding air pollution control, water pollution control and waste management have forced companies to remove volatile organics from atmospheric emissions and workplace environments. But apart from compliance with these requirements, economic factors are decisive in solvent recovery. Reuse of solvent in production not only reduces operating cost drastically but may even allow profitable operation of a recovery system.

#### 22.1.4.1 Basic arrangement of adsorptive solvent recovery with steam desorption

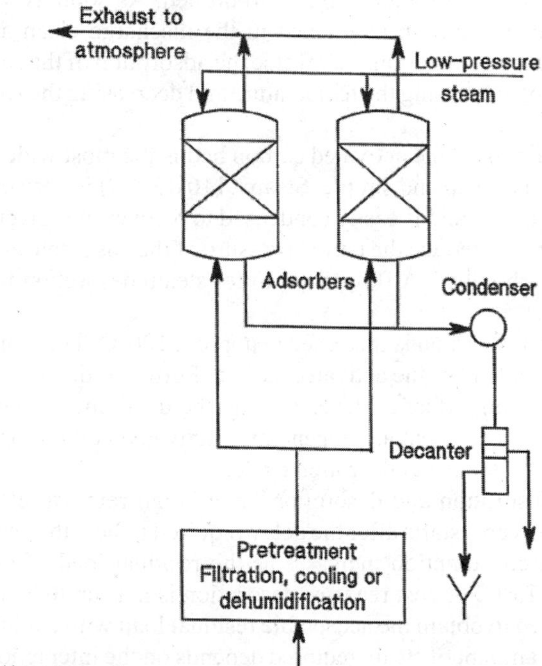


Figure 22.1.4. Flow sheet of a solvent recovery system with steam desorption.

Adsorptive solvent recovery units have at least two, but usually three or four parallel-connected fix-bed adsorbers which pass successively through the four stages of the operation cycle.[1-4,18-23]

- Adsorption
- Desorption
- Drying
- Cooling

Whilst adsorption takes place in one or more of them, desorption, drying and cooling takes place in the others. The most common adsorbent is activated carbon in the shape of 3 or 4 mm pellets or as granular type with a particle size of 2 to 5 mm (4 x 10 mesh). A schematic flow sheet of a two adsorber system for the removal of water-insoluble solvents is shown in Fig 22.1.4.

The adsorber feed is pre-treated if necessary to remove solids (dust), liquids (droplets or

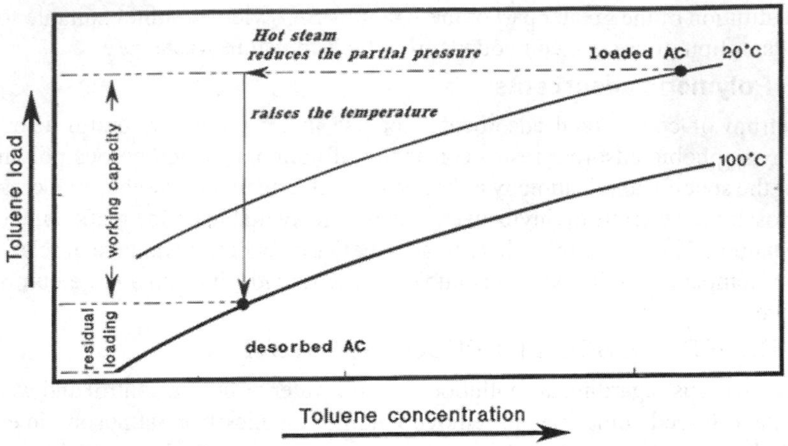


Figure 22.1.5. Principle of steam desorption.

aerosols) or high-boiling components as these can hamper performance. Frequently, the exhaust air requires cooling.

To prevent an excessive temperature increase across the bed due to the heat of adsorption, inlet solvent concentrations are usually limited to about 50 g/m$^3$. In most systems the solvent-laden air stream is directed upwards through a fixed carbon bed. As soon as the maximum permissible breakthrough concentration is attained in the discharge clean air stream, the loaded adsorber is switched to regeneration. To reverse the adsorption of the solvent, the equilibrium must be reversed by increasing the temperature and decreasing the solvent concentration by purging.

For solvent desorption direct steaming of the activated carbon bed is the most widely used regeneration technique because it is cheap and simple. Steam (110-130°C) is very effective in raising the bed temperature quickly and is easily condensed to recover the solvent as a liquid. A certain flow is also required to reduce the partial pressure of the adsorbate and carry the solvent out of the activated carbon bed. A flow diagram for steam desorption for toluene recovery is given in Figure 22.1.5.

First, the temperature of the activated carbon is increased to approx. 100°C. This temperature increase reduces the equilibrium load of the activated carbon. Further reduction of the residual load is obtained by the flushing effect of the steam and the declining toluene partial pressure. The load difference between spent and regenerated activated carbon - the "working capacity" - is then available for the next adsorption cycle.

The counter-current pattern of adsorption and desorption favors high removal efficiencies. Desorption of the adsorbed solvents starts after the delay required to heat the activated carbon bed. The specific steam consumption increases as the residual load of the activated carbon decreases (Figure 22.1.6). For cost reasons, desorption is not run to completion. The desorption time is optimized to obtain the acceptable residual load with a minimum specific steam consumption. The amount of steam required depends on the interaction forces between the solvent and the activated carbon. The mixture of steam and solvent vapor from the adsorber is condensed in a condenser. If the solvent is immiscible with water the condensate is led to a gravity separator (making use of the density differential) where it is separated into a aqueous and solvent fraction.

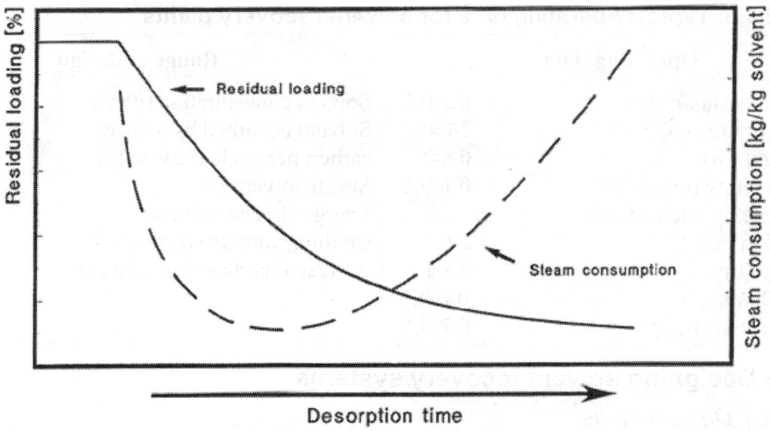


Figure 22.1.6. Desorption efficiency and steam consumption.

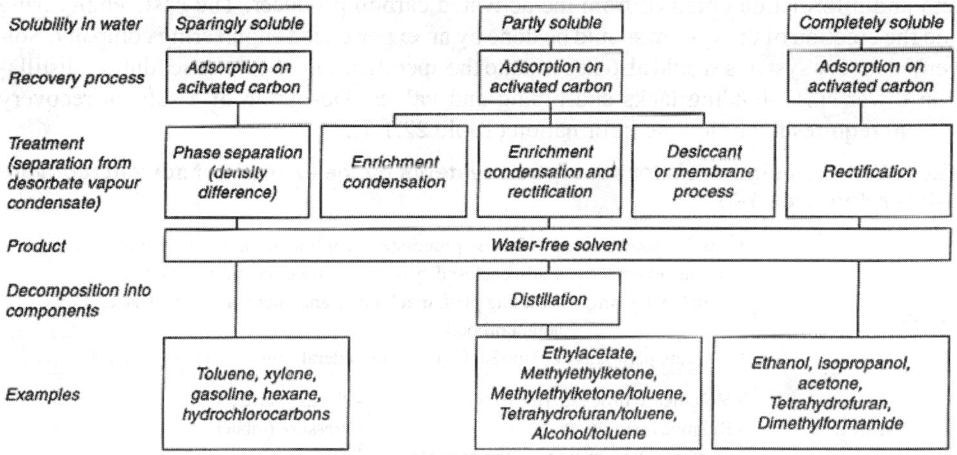


Figure 22.1.7. Options for solvent recovery by steam desorption (After references 2, 3,12).

If the solvents are partly or completely miscible with water, additional special processes are necessary for solvent separation and treatment (Figure 22.1.7). For partly miscible solvents the desorpt steam/solvent mixture is separated by enrichment condensation, rectification or by a membrane processes. The separation of completely water-soluble solvents requires a rectification column.[2]

After steam regeneration, the hot and wet carbon bed will not remove organics from air effectively, because high temperature and humidity do not favor complete adsorption. The activated carbon is therefore dried by a hot air stream. Before starting the next adsorption cycle the activated carbon bed is cooled to ambient temperature with air. Typical operating data for solvent recovery plants and design ranges are given in Table 22.1.6.

**Table 22.1.6. Typical operating data for solvent recovery plants**

| Operating data | | Range of design | |
|---|---|---|---|
| Air velocity (m/s): | 0.2-0.4 | Solvent concentration (g/m$^3$): | 1-20 |
| Air temperature (°C): | 20-40 | Solvent adsorbed by activated carbon per cycle (% weight): | 10-20 |
| Bed height (m): | 0.8-2 | Steam/solvent ratio: | 2-5:1 |
| Steam velocity (m/s): | 0.1-0.2 | Energy (kWh/t solvent): | 50-600 |
| Time cycle for each adsorber | | Cooling water (m$^3$/t solvent): | 30-100 |
| Adsorption (h): | 2-6 | Activated carbon (kg/t solvent): | 0.5-1 |
| Steaming (h): | 0.5-1 | | |
| Drying (hot air, h): | 0.2-0.5 | | |
| Cooling (cold air, h): | 0.2-0.5 | | |

## 22.1.4.2 Designing solvent recovery systems

### *22.1.4.2.1 Design basis*[1-3,18-23]

Designing an activated carbon system should be based on pilot data for the particular system and information obtained from the activated carbon producer. The basic engineering and the erection of the system should be done by an experienced engineering company. Solvent recovery systems would also necessitate the specification of condenser duties, distillation tower sizes, holding tanks and piping and valves. Designing of a solvent recovery system requires the following information (Table 22.1.7).

**Table 22.1.7. Design basis for adsorption systems by the example of solvent recovery (After references 1-3)**

| | |
|---|---|
| Site-specific conditions | Emission sources: Production process(es), continuous/shift operation<br>Suction system: Fully enclosed system, negative pressure control<br>Available drawings: Building plan, machine arrangement, available space, neighborhood<br>Operating permit as per BimSchG (German Federal Immission Control Act) |
| Exhaust air data (max., min., normal) | - Volumetric flow rate (m$^3$/h) - Temperature (°C)<br>- Relative humidity (%) - Pressure (mbar)<br>- Solvent concentration by components (g/m$^3$)<br>- Concentration of other pollutants and condition (fibers, dust) |
| Characterization of solvents | - Molecular weight, density - Boiling point, boiling range<br>- Critical temperature - Explosion limits<br>- Water solubility - Corrosiveness<br>- Stability (chemical, thermal) - Maximum exposure limits (MEL)<br>- Safety instructions as per German Hazardous Substances Regulations (description of hazards, safety instructions, safety data sheets)<br>- Physiological effects: irritating, allergenic, toxic, cancerogenic |
| Utilities (type and availability) | - Electric power - Steam - Cooling water<br>- Cooling brine - Compressed air, instrument air - Inert gas |
| Waste streams (avoidance, recycling, disposal) | - Solvent - Steam condensate - Adsorbent - Filter dust |

| | |
|---|---|
| Guaranteed performance data at rated load | - Emission levels: pollutants, sound pressure level (dB(A))<br>- Utility consumption, materials- Availability |

After this information has been obtained, the cycle time of the system must be determined. The overall size of the unit is determined primarily by economic considerations, balancing low operating costs against the capital costs (smaller system → shorter cycle time; bigger adsorber → long cycle time). The limiting factor for the adsorber diameter is normally the maximum allowable flow velocity under continuous operating conditions. The latter is related to the free vessel cross-section and, taking into account fluctuations in the gas throughput, selected such (0.1 to 0.5 m/s) that it is sufficiently below the disengagement point, i.e., the point at which the adsorbent particles in the upper bed area are set into motion. Once the adsorber diameter has been selected, the adsorber height can be determined from the required adsorbent volume plus the support tray and the free flow cross-sections to be provided above and below the adsorbent bed. The required adsorbent volume is dictated by the pollutant concentration to be removed per hour and the achievable adsorbent working capacity.

From the view point of adsorption theory, it is important that the bed be deeper than the length of the transfer zone which is unsaturated. Any multiplication of the minimum bed depth gives more than proportional increase in capacity. In general, it is advantageous to size the adsorbent bed to the maximum length allowed by pressure drop considerations. Usually fixed-bed adsorber has a bed thickness of 0.8 - 2 m.

#### *22.1.4.2.2 Adsorber types*

Adsorbers for solvent recovery[1-3] units (Figure 22.1.8) are build in several configurations:

- vertical cylindrical fixed-bed adsorber
- horizontal cylindrical fixed-bed adsorber
- vertical annular bed adsorber
- moving-bed adsorber
- rotor adsorber (axial or radial)

For the cleanup of high volume waste gas streams, multi adsorber systems are preferred. Vertical adsorbers are most common. Horizontal adsorbers, which require less overhead space, are used mainly for large units. The activated carbon is supported on a cast-iron grid which is sometimes covered by a mesh. Beneath the bed, a 100 mm layer of 5 - 50 mm size gravel (or $Al_2O_3$ pellets) is placed which acts as a heat accumulator and its store part of the desorption energy input for the subsequent drying step. The depth of the carbon bed, chosen in accordance with the process conditions, is from 0.8 to 2 m. Adsorbers with an annular bed of activated carbon are less common. The carbon in the adsorber is contained between vertical, cylindrical screens. The solvent laden air passes from the outer annulus through the outer screen, the carbon bed and the inner screen, then exhausts to atmosphere through the inner annulus. The steam for regeneration of the carbon flows in the opposite direction to the airstream. Figure 22.1.8 illustrates the operating principle of a counter-current flow vertical moving bed adsorber. The waste gas is routed in counter-current to the moving adsorbent and such adsorption systems have a large number of single moving beds. The discharge system ensures an uniform discharge (mass flow) of the loaded adsorbent. In this moving bed system adsorption and desorption are performed in a separate column or column sections. The process relies on an abrasion resistant adsorbent with good flow proper-

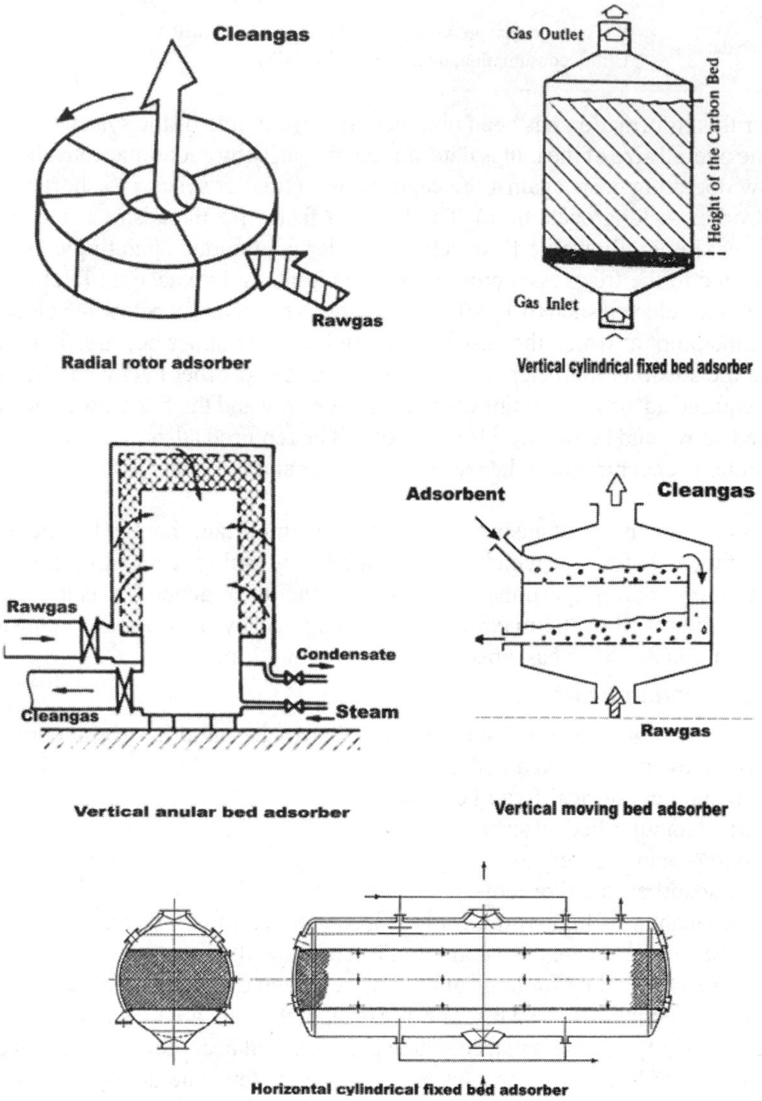


Figure 22.1.8. Adsorber construction forms.

ties (beaded carbon or adsorber resins). After the adsorption stage of a multi-stage fluidized-bed system, the loaded adsorbent is regenerated by means of steam or hot inert gas in a moving-bed regenerator and pneumatically conveyed back to the adsorption section.[2] When using polymeric adsorbents in a comparable system, desorption can be carried out with heated air because the structure-inherent faster kinetics of these adsorbents require desorption temperatures of only 80 to 100°C.

Rotor adsorbers allow simultaneous adsorption and desorption in different zones of the same vessel without the need for adsorbent transport and the associated mechanical stresses. Rotor adsorbers can be arranged vertically or horizontally and can be designed for axial or radial gas flow, so that they offer great flexibility in the design of the adsorption system.

The choice of the rotor design[2,27,33] depends on the type of adsorbent employed. Rotor adsorbers accommodating pellets or granular carbon beds come as wheel type baskets divided into segments to separate the adsorption zone from the regeneration zone. Alternatively, they may be equipped with a structured carrier (e.g., ceramics, cellulose) coated with or incorporating different adsorbents such as

- activated carbon fibers
- beaded activated carbon
- activated carbon particles
- particles of hydrophobic zeolite

The faster changeover from adsorption to desorption compared with fixed-bed adsorbers normally translates into a smaller bed volume and hence, in a reduced pressure loss. Hot air desorption is also the preferred method for rotor adsorbers. Hot air rates are about one tenth of the exhaust air rate. The desorbed solvents can either be condensed by chilling or disposed of by high-temperature or catalytic combustion.

In some cases, removal of the concentrated solvents from the desorption air in a conventional adsorption system and subsequent recovery for reuse may be a viable option. The structural design of the adsorber is normally governed by the operating conditions during the desorption cycle when elevated temperatures may cause pressure loads on the materials. Some solvents when adsorbed on activated carbon, or in the presence of steam, oxidize or hydrolyze to a small extent and produce small amounts of corrosive materials. As a general rule, hydrocarbons and alcohols are unaffected and carbon steel can be used. In the case of ketones and esters, stainless steel or other corrosion resistant materials are required for the solvent-wetted parts of the adsorbers, pipework and condensers, etc. For carbon tetrachloride and 1,1,1-trichloroethane and other chlorinated solvents, titanium has a high degree of resistance to the traces of acidity formed during recovery.

#### *22.1.4.2.3 Regeneration*

The key to the effectiveness of solvent recovery is the residual adsorption capacity (working capacity) after in-place regeneration. The basic principles of regeneration are given in Sections 22.1.2.4 and 22.1.4.1. After the adsorption step has concentrated the solvent, the design of the regeneration system[1-3,13-16] depends on the application and the downstream use of the desorbed solvents whether it be collection and reuse or destruction.Table 22.1.8 may give some indications for choosing a suitable regeneration step.

**Table 22.1.8 Basic principle of regeneration**

| Regeneration step | Basic principle | Typical application |
|---|---|---|
| Off-site reactivation | Transportation of the spent carbon to a reactivation plant. Partial gasification at 800-900°C with steam in a reactivation reactor. Reuse of the regenerated activated carbon | - low inlet concentration<br>- odor control application<br>- solvents with boiling points above 200 °C<br>- polymerization of solvents |

| Regeneration step | Basic principle | Typical application |
|---|---|---|
| Steam desorption | Increasing the bed temperature by direct steaming and desorption of solvent. The steam is condensed and solvent recovered. | Most widely used regeneration technique for solvent recovery plants. Concentration range from 1-50 g/m$^3$ |
| Hot air desorption | Rotor adsorber are often desorbed by hot air | Concentration by adsorption and incineration of the high calorific value desorbate |
| Hot inert gas desorption | Desorption by means of an hot inert gas recirculation. Solvent recovery can be achieved with refrigerated condenser | Recovery of partly or completely water soluble solvents. Recovery of reactive solvents (ketones) or solvents which react with hot steam (chlorinated hydrocarbons) |
| Pressure swing adsorption/vacuum regeneration | Atmospheric pressure adsorption and heatless vacuum regeneration. The revaporized and desorbed vapor is transported from the adsorption bed via a vacuum pump. | Low-boiling and medium-boiling hydrocarbon vapor mixes from highly concentrated tank venting or displacement gases in large tank farms. Recovery of monomers in the polymer industry. |
| Reduced pressure and low temperature steam desorption | Steam desorption at low temperature and reduced pressure. | Recovery of reactive solvents of the ketone-type. Minimizing of side-reactions like oxidizing, decomposing or polymerizing. |

#### 22.1.4.2.4 Safety requirements

Adsorption systems including all the auxiliary and ancillary equipment needed for their operation are subject to the applicable safety legislation and the safety requirements of the trade associations and have to be operated in accordance with the respective regulations.[2,19]

Moreover, the manufacturer's instructions have to be observed to minimize safety risks. Activated carbon or its impurities catalyze the decomposition of some organics, such as ketones, aldehydes and esters. These exothermic decomposition reactions[25,26] can generate localized hot spots and/or bed fires within an adsorber if the heat is allowed to build up. These hazards will crop up if the flow is low and the inlet concentrations are high, or if an adsorber is left dormant without being completely regenerated. To reduce the hazard of bed fires, the following procedures are usually recommended:

- Adsorption of readily oxidizing solvents (e.g., cyclohexanone) require increased safety precautions. Instrumentation, including alarms, should be installed to monitor the temperature change across the adsorber bed and the outlet $CO/CO_2$ concentrations. The instrumentation should signal the first signs of decomposition, so that any acceleration leading to a bed fire can be forestalled. Design parameters should be set so as to avoid high inlet concentrations and low flow. A minimum gas velocity of approximately 0.2 m/s should be maintained in the fixed-bed adsorber at all times to ensure proper heat dissipation.
- A virgin bed should be steamed before the first adsorption cycle. Residual condensate will remove heat.

- Loaded activated carbon beds require constant observation because of the risk of hot spot formation. The bed should never be left dormant unless it has been thoroughly regenerated.
- After each desorption cycle, the activated carbon should be properly cooled before starting a new adsorption cycle.
- Accumulation of carbon fines should be avoided due to the risk of local bed plugging which may lead to heat build up even with weakly exothermic reactions.
- CO and temperature monitors with alarm function should be provided at the clean gas outlet.
- The design should also minimize the possibility of explosion hazard
- Measurement of the internal adsorber pressure during the desorption cycle is useful for monitoring the valve positions.
- Measurement of the pressure loss across the adsorber provides information on particle abrasion and blockages in the support tray.
- Adsorption systems must be electrically grounded.

### 22.1.4.3 Special process conditions

#### *22.1.4.3.1 Selection of the adsorbent*

The adsorption of solvents on activated carbon[1-4] is controlled by the properties of both the carbon and the solvent and the contacting conditions. Generally, the following factors, which characterize the solvent-containing waste air stream, are to be considered when selecting the most well suited activated carbon quality for waste air cleaning:

*for solvent/waste air*:

- solvent type (aliphatic/aromatic/polar solvents)
- concentration, partial pressure
- molecular weight
- density
- boiling point, boiling range
- critical temperature
- desorbability
- explosion limits
- thermal and chemical stability
- water solubility
- adsorption temperature
- adsorption pressure
- solvent mixture composition
- solvent concentration by components
- humidity
- impurities (e.g., dust) in the gas stream

*for the activated carbon type*:

General properties:

- apparent density
- particle size distribution
- hardness
- surface area

- activity for $CCl_4$/benzene
- the pore volume distribution curve

For the selected solvent recovery task:

- the form of the adsorption isotherm
- the working capacity
- and the steam consumption

The surface area and the pore size distribution are factors of primary importance in the adsorption process. In general, the greater the surface area, the higher the adsorption capacity will be. However, that surface area within the activated carbon must be accessible. At low concentration (small molecules), the surface area in the smallest pores, into which the solvent can enter, is the most efficient surface. With higher concentrations (larger molecules) the larger pores become more efficient. At higher concentrations, capillary condensation will take place within the pores and the total micropore volume will become the limiting factor. These molecules are retained at the surface in the liquid state, because of intermolecular or van der Waals forces. Figure 22.1.9 shows the relationship between maximum effective pore size and concentration for the adsorption of toluene according to the Kelvin theory.

It is evident that the most valuable information concerning the adsorption capacity of a given activated carbon is its adsorption isotherm for the solvent being adsorbed and its pore volume distribution curve. Figure 22.1.10 presents idealized toluene adsorption isotherms for three carbon types:

- large pores predominant
- medium pores predominant
- small pores predominant

The adsorption lines intersect at different concentrations, depending on the pore size distribution of the carbon. The following applies to all types of activated carbon: As the toluene concentration in the exhaust air increases, the activated carbon load increases.

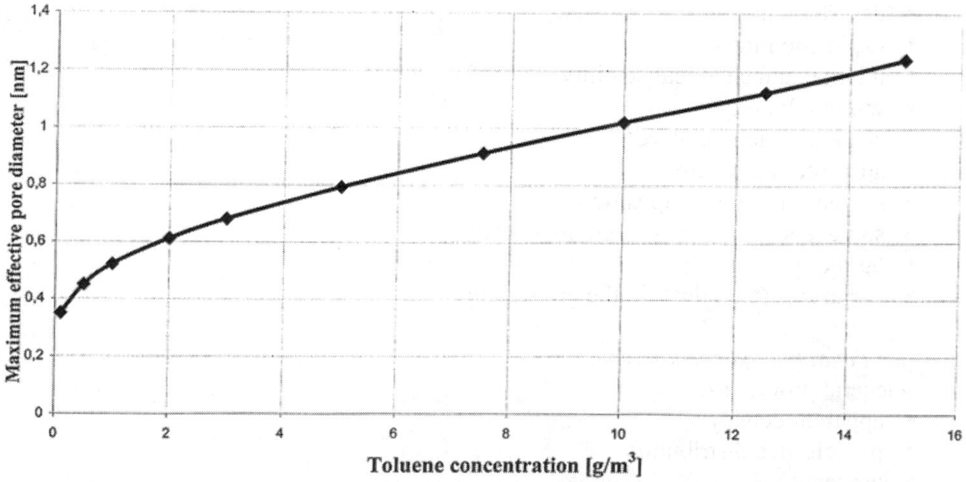


Figure 22.1.9. Relationship between maximum effective pore diameter and toluene concentration.

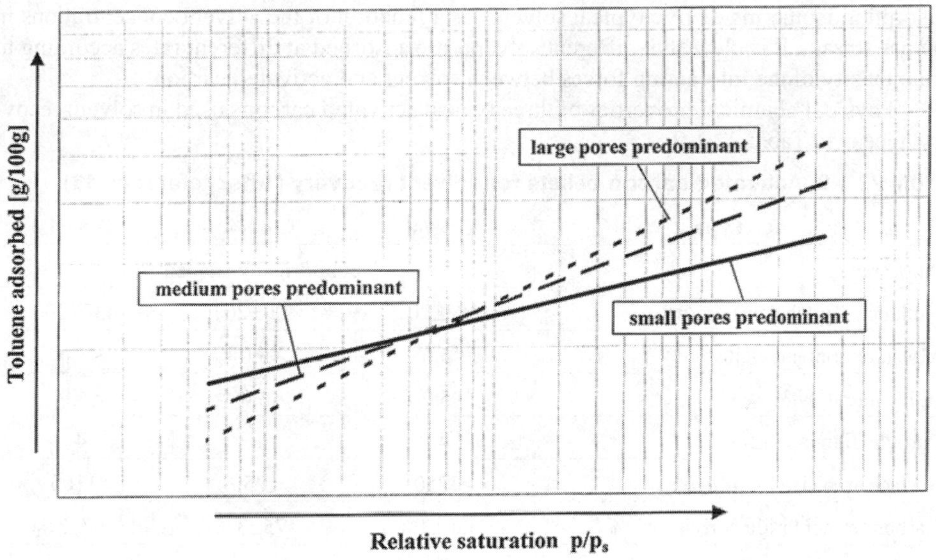


Figure 22.1.10. Idealized toluene adsorption isotherms.

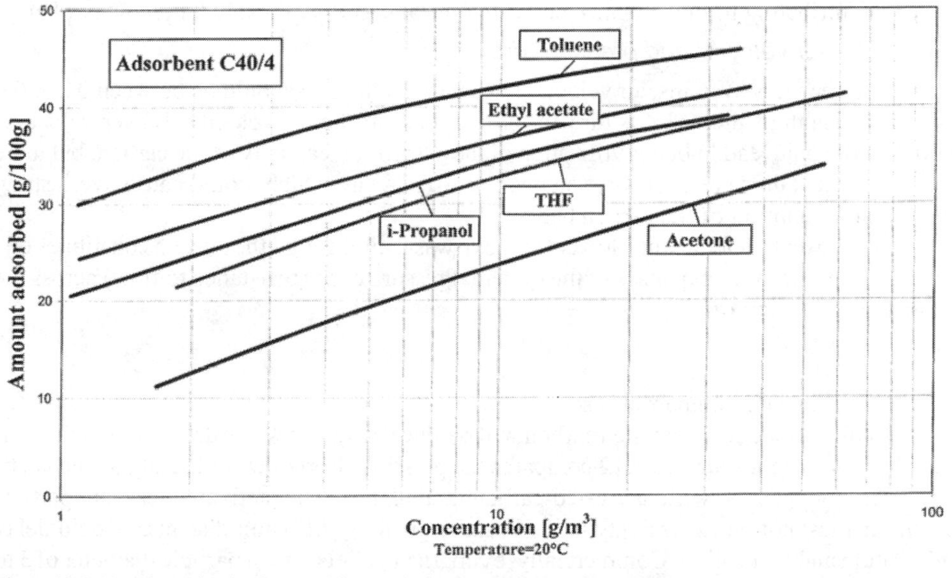


Figure 22.1.11. Adsorption isotherm of five typical solvents (After reference 13).

From the viewpoint of adsorption technology, high toluene concentration in the exhaust air is more economic than a low concentration. But to ensure safe operation, the toluene concentration should not exceed 40% of the lower explosive limit. There is, however, an optimum activated carbon type for each toluene concentration (Figure 21.1.10).

The pore structure of the activated carbon must be matched to the solvent and the solvent concentration for each waste air cleaning problem. Figure 22.1.11 shows Freundlich adsorption isotherms of five typical solvents as a function of the solvent concentrations in the gas stream. It is clear that different solvents are adsorbed at different rates according to the intensity of the interacting forces between solvent and activated carbon.

Physical-chemical properties of three typical activated carbons used in solvent recovery appear in Table 22.1.9.

**Table 22.1.9. Activated carbon pellets for solvent recovery (After reference 13)**

| AC-Type | C 38/4 | C 40/4 | D 43/4 |
|---|---|---|---|
| Appearance | cylindrically shaped | | |
| Bulk density (shaken), $kg/m^3$ | 380±20 | 400±20 | 430±20 |
| Moisture content, wt% | <5.0 | <5.0 | <5.0 |
| Ash content, wt% | <6.0 | <6.0 | <5.0 |
| Particle diameter, mm | 4 | 4 | 4 |
| Surface area (BET), $m^2/g$ | >1250 | >1250 | >1100 |
| Carbon tetrachloride activity, wt% | 80±3 | 75±3 | 67±3 |

Producers[13] of activated carbon are in the best position to provide technical advice on

- selecting the right activated carbon type
- contributing to the technical and economical success of a solvent recovery plant

### *22.1.4.3.2 Air velocity and pressure drop*

In solvent recovery systems, air velocity rates through the bed should be between 0.2 to 0.4 m/s. The length of the MTZ is directly proportional to the air velocity. Lower velocities (<0.2 m/s) would lead to better utilization of the adsorption capacity of the carbon, but there is a danger that the heat of adsorption not be carried away which would cause overheating and possibly ignition of the carbon bed.

The power to operate the blowers to move waste air through the system constitutes one of the major operating expenses of the system. Pressure drop (resistance to flow) across the system is a function of

- air velocity
- bed depth
- activated carbon particle size

Small particle size activated carbon will produce a high pressure drop through the activated-carbon bed. Figure 22.1.12 compares the pressure drop of cylindrically-shaped activated carbon pellets with activated-carbon granulates. The activated carbon particle diameter must not be excessively large, because the long diffusion distances would delay adsorption and desorption. Commercially, cylindrical pellets with a particle diameter of 3 to 4 mm have been most efficient.

### *22.1.4.3.3 Effects of solvent-concentration, adsorption temperature and pressure*

For safety reasons the concentration of combustible solvent vapors should be less than 50 % of the lower explosive limit. The adsorption capacity of adsorbents increases as the concentration of the solvents increases. But the length of the MTZ is proportional to the solvent concentration. Because of the adsorption heat, as the adsorption front moves through the

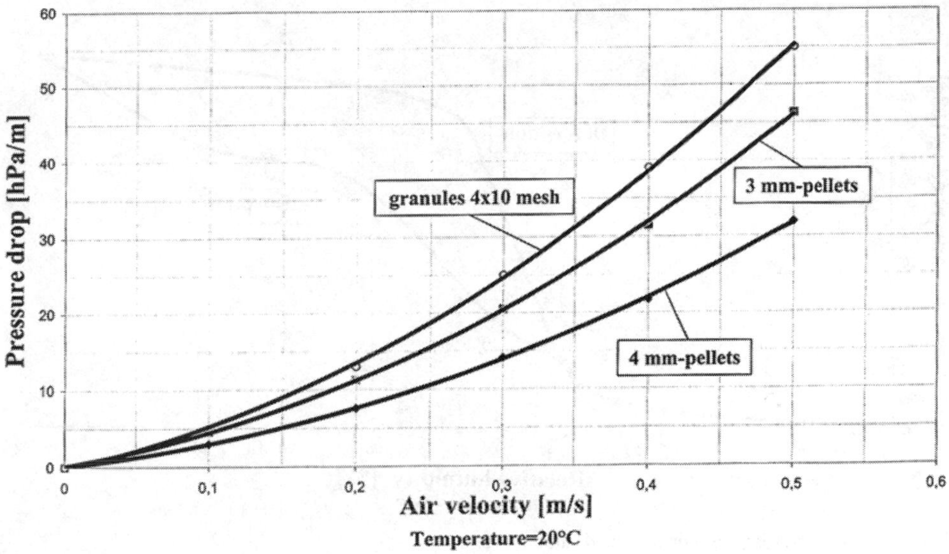


Figure 22.1.12. Pressure drop for various activated carbon types.

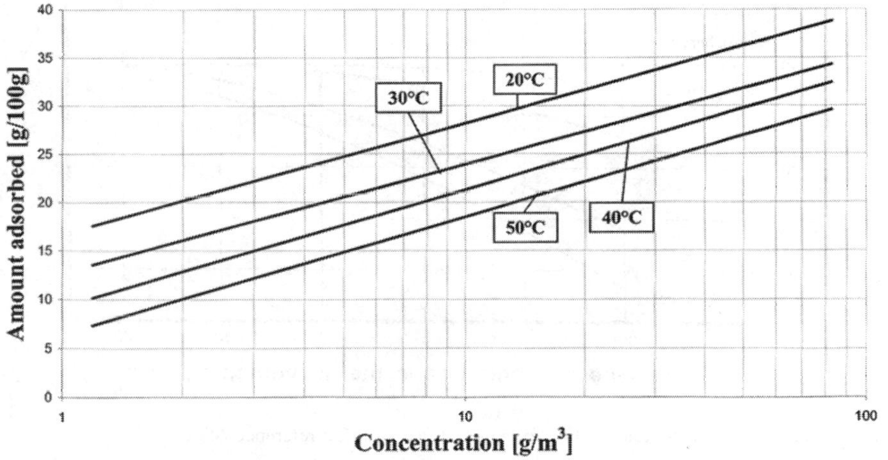


Figure 22.1.13. Adsorption isotherm of tetrahydrofuran for several temperatures.

bed also a temperature front follows in the same direction. To deal with the adsorption heat the inlet solvent concentration is usually limited to about 50 g/m$^3$.

The adsorption capacity of the adsorbent increases with pressure because the partial pressure of the solvent increases. An increase in adsorber temperature causes a reduction in adsorption capacity. Because the equilibrium capacity is lower at higher temperatures, the dynamic capacity (working capacity) of the activated carbon adsorber will also be lower. To enhance adsorption, the inlet temperature of the adsorber should be in the range of 20-40°C. In Figure 22.1.13 the adsorption isotherms of tetrahydrofuran on activated carbon D43/3 for several temperatures are shown.[13]

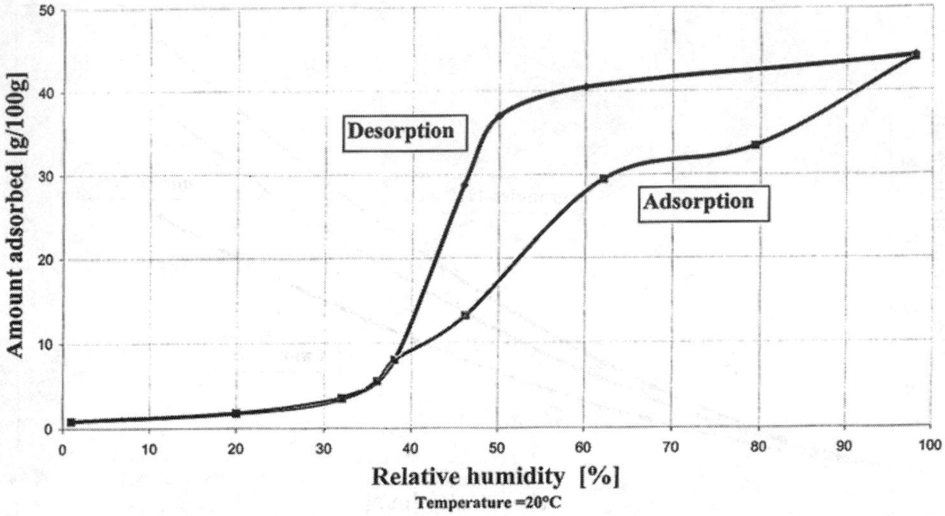


Figure 22.1.14. Water isotherm on activated carbon D43/4.

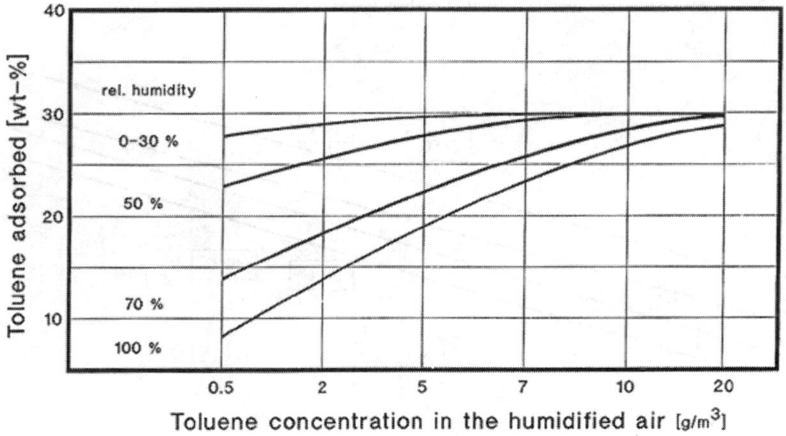


Figure 22.1.15. Dynamic toluene adsorption from humidified air (after reference 24).

### *22.1.4.3.4 Influence of humidity*

Activated carbon is basically hydrophobic; it adsorbs preferably organic solvents. The water adsorption isotherm (Figure 22.1.14) reflects its hydrophobic character. Below a relative gas humidity of about 40% co-adsorption of water can be neglected in most applications. However, higher humidity of the waste gases may affect the adsorption capacity of the activated carbon. Figure 22.1.15 demonstrates, using toluene as an example, how the relative humidity of the exhaust air influences the activated carbon loads.

Experimental work[24] has shown that

- relative humidity rates below 30% will neither reduce the adsorption capacity nor the adsorption time,
- relative humidity rates above 70% substantially reduce the adsorption capacity,
- the humidity of the gas will affect the adsorption capacity much more with low toluene concentrations than with high concentrations,
- with toluene concentrations between 10 and 20 $g/m^3$ the negative influence of the humidity is small.

The water content of the activated carbon after desorption may constitute another problem. The purification efficiency of each activated carbon is the better the less water is present after desorption. Unfortunately, the desorbed activated carbon in the vicinity of the adsorber walls usually contains high proportions of water (approx. 10 to 20%). With such a high water content it is difficult to remove all the water even when drying with hot-air over longer periods. The wet and poorly-regenerated activated carbons in these zones frequently lead to higher solvent concentrations in the purified air, and this is even at the beginning of the adsorption cycle. Some proposals for improvement include:

- good insulation of the adsorber walls,
- sufficient drying after each regeneration cycle,
- cycle the desorption steam and the drying air counter-current to the direction of the adsorptive stream.

#### *22.1.4.3.5 Interactions between solvents and activated carbon*

The majority of solvents are effectively recovered by adsorption on activated carbon and, when this is the case, the operation of the plant is straightforward. But some solvents may decompose, react or polymerize when in contact with activated carbon during the adsorption step and the subsequent steam desorption.[19,25,26]

*Chlorinated hydrocarbon* solvents can undergo hydrolysis to varying degrees on carbon surfaces, resulting in the formation of hydrogen chloride. Each activated carbon particle which is in contact with a metal screen or other constructional component may then act as a potential galvanic cell in the presence of moisture and chloride ions.

*Carbon disulfide* can be catalytically oxidized to sulphur which remains on the internal surface of the activated carbon.

*Esters* such as ethyl acetate are particularly corrosive because their hydrolysis results in the formation of e.g. acetic acid.

*Aldehydes* and phenol or styrene will undergo some degree of polymerization when in contact with hot activated carbon.

During the adsorptive removal of ketones such as methyl ethyl ketone or cyclohexanone a reduced adsorption capacity has been measured. Corrosion problems were also apparent and in some cases even spontaneous ignition of the activated carbon occurred.

Activated carbon acts, in the presence of oxygen, as a catalyst during adsorption and even more frequently during desorption because of higher temperature.[25,26] Figure 22.1.16 shows that the catalytic reactions occurring with cyclohexanone are predominantly of the oxidation type. Adipic acid is first formed from cyclohexanone. Adipic acid has a high boiling point (213°C at a 13 mbar vacuum). Further products identified were: cyclopentanone as a degradation product from adipic acid, phenol, toluene, dibenzofuran, aliphatic hydrocarbons and carbon dioxide. The high-boiling adipic acid cannot be desorbed from the activated carbon in a usual steam desorption. Consequently the life of the activated carbon becomes further reduced after each adsorption/desorption cycle.

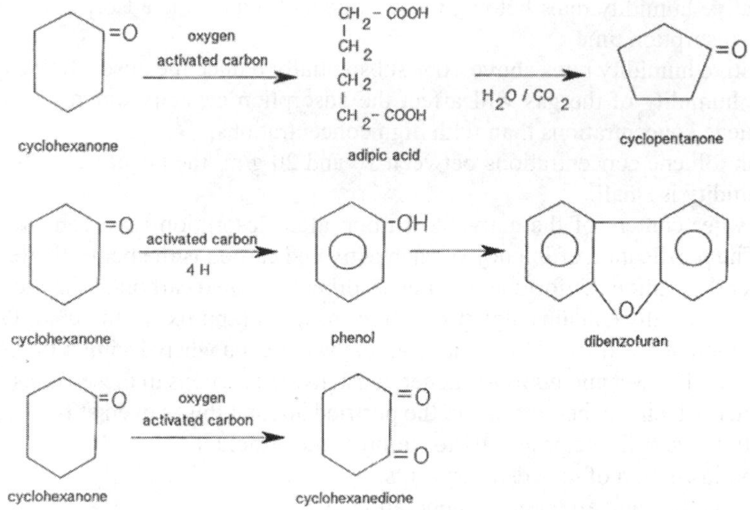


Figure 22.1.16. Surface reaction of cyclohexanone on activated carbon (After reference 25).

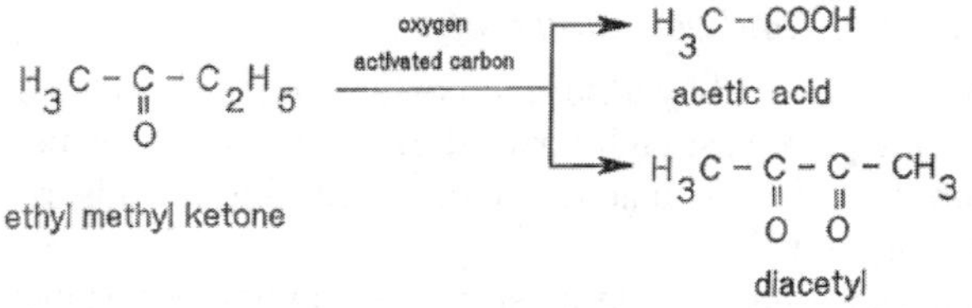


Figure 22.1.17. Surface reaction of methyl ethyl ketone on activated carbon (After reference 25).

During adsorption of methyl ethyl ketone in the presence of oxygen, some catalytic surface reactions occur (Figure 22.1.17). The reaction products are acetic acid, di-acetyl, and presumably also di-acetyl peroxide. Di-acetyl has an intensive green coloration and a distinctive odor. Important for the heat balance of the adsorber is that all of the reactions are strongly exothermic.

Unlike the side products of cyclohexanone, acetic acid the reaction product of the MEK (corrosive!) will be removed from the activated carbon during steam desorption. Adsorption performance is thus maintained even after a number of cycles.

Operators of ketones recovery plants should adhere to certain rules and also take some precautions[25]

- Adsorption temperature should not be higher than 30°C (the surface reaction rates increase exponentially with temperature).
- The above requirement of low-adsorption temperatures includes:
  a) adequate cooling after the desorption step;
  b) a flow velocity of at least 0.2 m/s in order to remove the heat of adsorption.

Since ketones will, even in an adsorbed state, oxidize on the activated carbon due to the presence of oxygen, these further precautions should be taken:

- keep the adsorption/desorption cycles as short as possible,
- desorb the loaded activated carbon immediately at the lowest possible temperatures,
- desorb, cool and blanket the equipment with inert gas before an extended shut down.

#### *22.1.4.3.6 Activated carbon service life*

Activated carbon in solvent recovery service will have a useful service life of 1 to 10 years, depending on the attrition rate and reduction in adsorption capacity.

Attrition rates are usually less than 1-3% per year. The actual rate will depend on carbon hardness. Particle abrasion and the resulting bed compaction leads to an increased pressure loss after several years of service. After 3-5 years of service screening to its original size is necessary.

Adsorption capacity can be reduced by traces of certain high-boiling materials (resins, volatile organosilicone compounds) in the waste air which are not removed during the desorption cycle. Some solvents may decompose, react or polymerize when in contact with activated carbon and steam (Section 22.1.4.3.5).

This gradual loading will decrease the carbon's activity. While carbon can remain in service in a reduced capacity state, it represents a non-optimal operation of the system that results in

- reduced amount of solvent removed per cycle
- increased steam consumption
- higher emissions
- shorter adsorption cycles

Eventually, recovering capacity will diminish; it is more economical to replace carbon than continue its use in a deteriorated state. In replacing spent activated carbon there are two options:

- Replacing with virgin carbon and disposal of the spent carbon.
- Off-site reactivation of the spent carbon to about 95% of its virgin activity for about one-half of the cost of new carbon

Some activated carbon producers offer a complete service including carbon testing, off-site reactivation, transportation, adsorber-filling and carbon make-up.[13]

## 22.1.5 EXAMPLES FROM DIFFERENT INDUSTRIES

Some examples of solvent recovery systems in different industries show that reliable and trouble-free systems are available.

### 22.1.5.1 Rotogravure printing shops

Process principle

Adsorptive solvent recovery with steam desorption and condensation units with gravity separator and a stripper have become a standard practice in modern production plants. The solvents-laden air (toluene, xylene) is collected from emission points, e.g., rotogravure printing machines, drying ducts by means of a blower and passed through the recovery plant.

Design example[30]

In 1995, one of Europe's most modern printing shops in Dresden was equipped with an adsorptive solvent recovery system (Supersorbon® process (Figure 22.1.18.)).

Design data

| | |
|---|---|
| (First stage of completion) | |
| Exhaust air flow rate | 240,000 $m^3/h$ |
| (expandable to | 400,000 $m^3/h$) |
| Solvent capacity | 2,400 kg/h (toluene) |
| Exhaust air temperature | 40°C |
| Toluene concentration in clean air (half-hour mean) | max. 50 $mg/m^3$ |

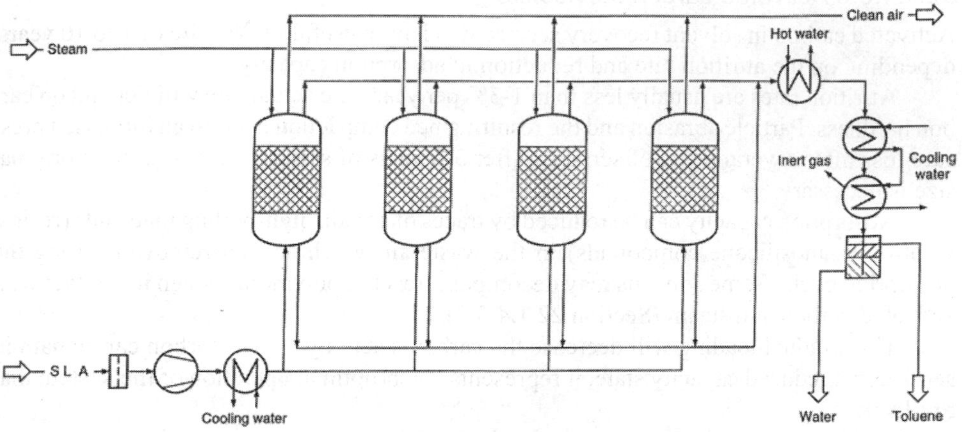


Figure 22.1.18. Supersorbon® process for toluene recovery (After reference 30).

Solvent-laden air is exhausted at the three rotogravure printing presses by several fans operating in parallel and is routed in an upward flow through four adsorbers packed with Supersorbon® activated carbon. The solvent contained in the air is adsorbed on the activated carbon bed. Adsorption continues until breakthrough, when the full retentive capacity of the adsorbent for solvent vapors is used up.

The purity of the clean exhaust air complies with the regulation for emission level of less than 50 mg/m$^3$. Regeneration of the adsorbent takes place by desorbing the solvent with a countercurrent flow of steam. The mixture of water and toluene vapors is condensed and the toluene, being almost insoluble in water, is separated by reason of its different density in a gravity separator. The recovered toluene can be re-used in the printing presses without further treatment. Very small amounts of toluene dissolved in the wastewater are removed by stripping with air and returned to the adsorption system inlet together with the stripping air flow. The purified condensate from the steam is employed as make-up water in the cooling towers.

Operating experience

The adsorption system is distinguished by an extremely economical operation with a good ratio of energy consumption to toluene recovery. Solvent recovery rate is about 99.5 %.

### 22.1.5.2 Packaging printing industry

In the packaging printing industry solvents like ethyl acetate, ethanol, ketones, tetrahydrofuran (THF), hexane and toluene are used in printing. The solvent laden air generally contains between 2 and 15 g/m$^3$ of solvent and, depending on the season, 5-18 g/m$^3$ of water.

Adsorptive recovery with steam desorption was widely used for many years but its disadvantage is that the solvents are recovered in various mixtures containing large amounts of water. The recovery of solvents is complicated because most solvents produce azeotropes with water which are not easily separated and, consequently, the waste water fails to meet the more stringent environmental regulations. In the last 10 years the adsorbent regeneration with hot inert gas has became the more favored process in the packaging printing industry.

### 22.1.5.2.1 Fixed bed adsorption with circulating hot gas desorption[31,32]

Activated carbon preferentially adsorbs non-polar organic solvents, while adsorbing relatively little water. Thus activated carbon allows the passage of 95-98 % of the water while retaining the solvents from the gas stream. During the adsorption step solvent laden air or gas is delivered to the bed through the appropriate valves. As adsorption starts on a fresh bed, the effluent gas contains traces of solvent. Through time, the solvent level increases and when it reaches a predetermined value, the adsorption is stopped by closing the feed gas valves. The adsorption step usually lasts for 4-16 hours. An example of a solvent recovery plant with 5 activated carbon beds and hot gas desorption is shown in Figure 22.1.19.

If necessary, feed air is cooled to 35°C and sent by the blower V-1 to the main heater 6 from which it enters 4 beds simultaneously, while one bed is being regenerated. The cleaned air is collected in header 7 and vented.

After the adsorption step, the air is displaced from the bed by an inert gas such as nitrogen. A circulating inert gas at 120-240°C serves for heating and stripping the solvent from the carbon. The solvent is recovered by cooling and chilling the circulating gas. The optimum chilling temperature depends on the boiling point of the solvent. Generally, chilling temperatures are between +10 and -30°C.

During the heating step, the temperature of the effluent gas gradually increases until a predetermined value (for example, 150°C) is reached, after which the heater is bypassed and the bed is cooled down by cold gas. Typical solvent recovery plants will have from 2 to 8 beds. All beds go through the same adsorption, inertization, heating and cooling steps but each at different time.

Referring to the flow sheet in Figure 22.1.19 the regeneration gas is circulated by blower V-2 to the gas heater, the header 8 to the bed being regenerated. The effluent gas is passed to the header 9, cooler, molecular sieves (or water condenser), to the chiller for sol-

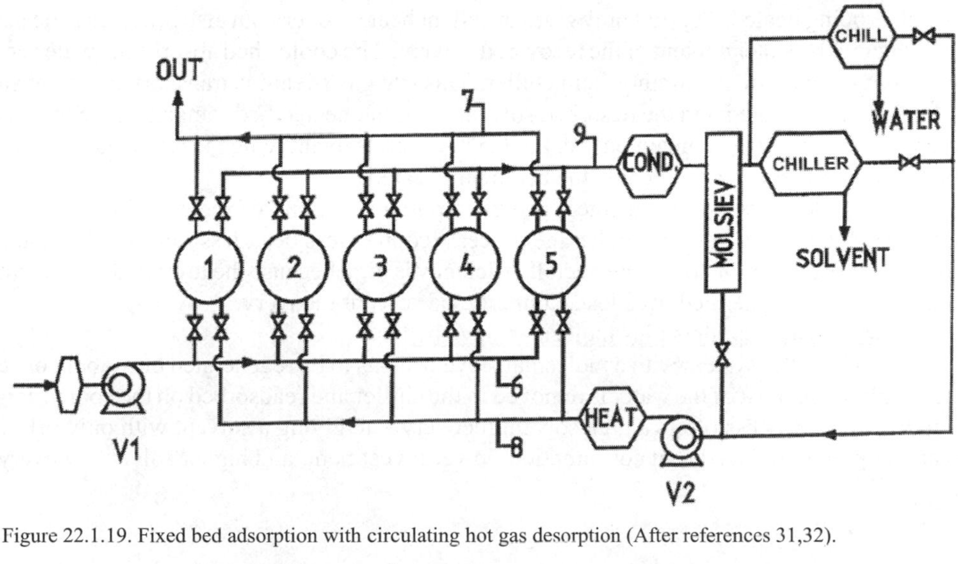


Figure 22.1.19. Fixed bed adsorption with circulating hot gas desorption (After references 31,32).

vent condensation and back to the blower V-2. At the end of the adsorption step the bed contains 10-30% solvent and 1-2% water.

There are various ways to remove the water separately and to recover a solvent containing between 0.1 and 1.5 % water (Table 22.1.10).

**Table 22.1.10. Process conditions and water content in the solvent (After references 32,33)**

| | Water in the raw solvent | Investment cost, % | Heat needed, kWh/kg S | Power, kWh/kg S |
|---|---|---|---|---|
| Basic process with molsieve beds | 0.1 | 100 | 2.2 | 1.0 |
| Chiller instead of molsieve | 1.5 | 91 | 1.5 | 1.0 |
| Chiller and 1 carbon bed | 0.4 - 1.0 | 93 | 1.1 | 1.0 |
| Chiller, and 1 carbon bed molsieve | 0.1 | 96 | 1.1 | 1.0 |

Using molecular sieves

Beds of molecular sieves are used to recover a solvent with 0.1%. This process requires around 2-2.5 kWh heat/kg of recovered solvent. Regeneration loop pressure drops are high, because of the additional molsieve beds and valves.

Separate condensation of water

Initially, when regenerative heating of the bed starts, very little solvent is desorbed, but much of the water (about 87%) is desorbed. The circulating gas is first passed through a cooler and then a separate chiller for the condensation or freezing of the water. Subsequently the gas is passed to the chiller where the water is condensed or frozen. From 1 to 1.8% water is recovered. This process is simpler than the molecular sieves process.

Two separate chillers

Two separate chillers are used in another adsorption process which offers further substantial improvements. A third activated carbon bed is added so that while two of the beds are being regenerated one of these is being cooled and transferring its heat to the other bed which is being heated. This not only gives excellent heat recovery but also provides a means of reducing the water content of the recovered solvent. The cooled bed absorbs the water vapor from the gas stream coming from chiller. This dry gas stream is transported to the bed which is being heated and the desorbed solvent from the heated bed remains dry. Solvents recovered at this stage contain only 0.4 to 0.5% water. Solvent which is removed from the bed being cooled is readsorbed on the bed being heated.

The process, which is patented, brings important benefits of regeneration gas flow rates. Such low flow rates mean that the process requires less heat, less refrigeration and a lower cost regeneration loop. The overall efficiency is high because the low residual solvent loading of the regenerated beds leads to increased solvent recovery.

Two separate chillers plus molecular sieve bed

A two chiller system with a molecular sieve bed has to be regenerated only about once per week, since most of the water is removed in the chiller and readsorbed on the cooled carbon bed. This process offers a high solvent recovery rate giving a solvent with only 0.1 % water at greatly reduced heat consumption, lower investment, and higher solvent recovery rate.

Key plants with all the above mentioned processes are offered by special engineering companies.[31,32] The investment cost for a typical recovery plant with hot gas regeneration can be estimated by the following equation:

$$\text{COST (in 1000 Euro)} = 200 + 0.4 \times A^{0.7} + 15 \times S^{0.7}$$

where:

A feed air rate in $Nm^3/h$

S solvent flow rate in kg/h

### *22.1.5.2.2 Solvent recovery with adsorption wheels*

Adsorption wheels[33] are used for the continuous purification of large volumes of exhaust air containing relatively low solvent concentrations.

The adsorption wheel consists of a number of identical chambers arranged axially around a vertical axis. All chambers contain adsorbent. The wheel rotates and each chamber passes in sequence over an exhaust air duct and solvent molecules are adsorbed. As wheel continues to rotate the chamber in which adsorption had occurred now moves into a desorption position in which hot air is passed through the chamber and over the adsorbent. This removes the adsorbed solvent. The hot air flow in the desorption sector is at relatively low flow rate compared to that of contaminated gas stream. The number of chambers in the desorption zone is much greater than the number in the absorption zone so the desorption stream is many times more concentrated in solvent than was the exhaust stream. At this higher concentration, the desorption stream can now be economically treated in one of the ways:

- The adsorption stream is purified by means of recuperative or regenerative oxidation. This approach is advantageous in painting applications with solvent mixtures that cannot be reused in production.
- For solvent recovery by condensation, the desorption stream is cooled down in a cooling aggregate and the liquid solvent is recovered for reuse. Water contents below 1% can be achieved without further purification.

Example:

A manufacturer who specializes in flexible packaging, i.e., for confectionary, always adds the same solvent mixture (ethanol, ethyl acetate, ethoxypropanol) to his printing ink.

Adsorptive solvent removal of the solvent-mixture by use of an adsorption wheel (Figure 22.1.20) and solvent recovery via condensation proved the technically and cost effective. 10,000 to 65,000 $Nm^3/h$ of exhaust air from printing machines and washing plants at a temperature maximum of 45°C and a maximum solvent loading of 4.6 $g/m^3$ is to be cleaned. Depending on the air volume two adsorption wheels with a capacity of 26,000 $Nm^3/h$ and 39,000 $Nm^3/h$ and separate desorption circuits are used either alternatively or together. The total desorption air of max. 11,000 $Nm^3/h$ is being concentrated to max. 27 $g/m^3$, which corresponds to 50% of the lower explosion limit. The condensation unit for the solvent recovery process is gradually adjusted to small, medium or large exhaust air volumes. The recovered solvents are stored and returned to the production process.

### **22.1.5.3 Viscose industry**

In plants which produce viscose fibre (stable fibre), viscose filament yarn (rayon) and viscose film (cellophane), large volumes of exhaust air contaminated with carbon disulfide

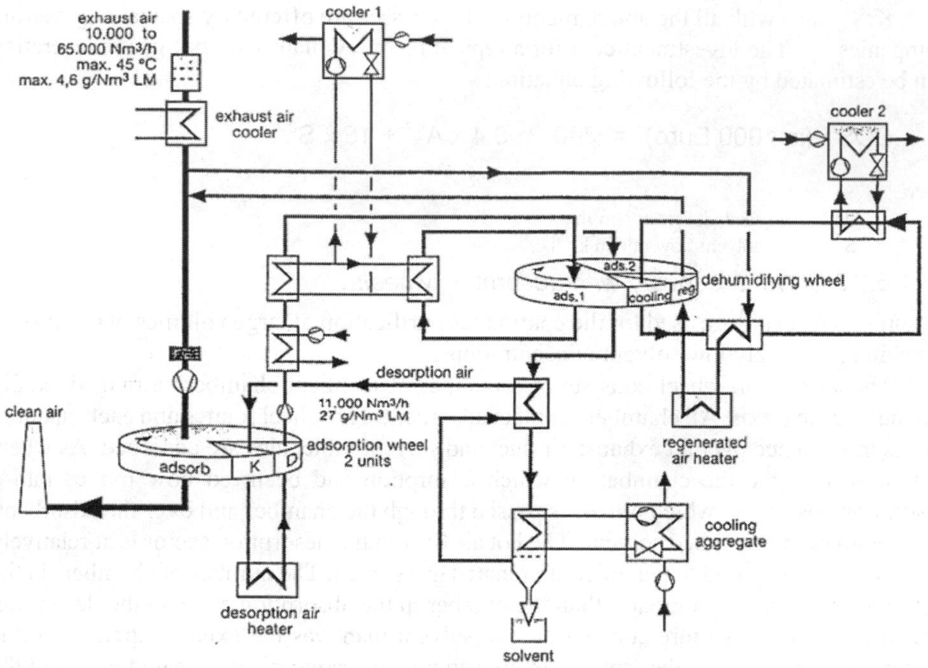


Figure 22.1.20. Adsorption wheel for solvent recovery in packaging printing (After reference 33).

($CS_2$) and hydrogen sulphide ($H_2S$) have to be cleaned. Typical waste gas volumes and concentration of $CS_2$ and $H_2S$ are given in Table 22.1.11.

Different cleaning/recovery processes are available for the removal of each of the two sulfur containing compounds:[2,17,34,35]

<u>Hydrogen sulfide</u>

- Absorption of $H_2S$ in a NaOH-scrubber
- Catalytic oxidation of $H_2S$ to elemental sulphur on iodine impregnated activated carbon (Sulfosorbon-process)

$$2\,H_2S + O_2 \rightarrow 2\,S + 2\,H_2O$$

The sulphur is adsorbed at the internal surface at the carbon-catalyst.

**Table 22.1.11. Waste gas in the viscose industry (After reference 35)**

| Product | Spec. waste gas volume, $m^3/t$ | Concentration ($mg/m^3$) | |
|---|---|---|---|
| | | $H_2S$ | $CS_2$ |
| Rayon | 400,000 - 700,000 | 60 -130 | 300 - 800 |
| Staple fibre | 50,000 - 90,000 | 700 -1800 | 2300 - 4000 |
| Viscose fibre | 100,000 -150,000 | 280 - 400 | 1000 - 3000 |

Carbon disulfide

Adsorptive removal on activated carbon and recovery by steam desorption. For simultaneous $H_2S$ and $CS_2$ removal the Sulfosorbon-process uses adsorbers packed with two different activated carbon types

- Iodine impregnated wide-pore activated carbon for $H_2S$ oxidation in the bottom part (gas inlet) of the adsorber
- Medium-pore activated carbon for the adsorption of $CS_2$ in the upper layer of the fixed bed.

As soon as the $CS_2$ concentration in the treated air approaches the emission limit, the exhaust air stream is directed to a regenerated adsorber. After an inert gas purge, the carbon disulfide is desorbed with steam at 110 to 130°C. The resulting $CS_2$/steam mixture is routed through a condenser and cooler, before entering a gravity separator where phase separation occurs.

At the usual $CS_2$ and $H_2S$ concentrations in viscose production exhaust air, the $CS_2$ adsorption/desorption cycles can be run for a prolonged periods before the first layer loaded with elemental sulphur has to be regenerated.

The regeneration process of the sulphur loaded activated carbon involves the following steps:

- washing out of sulphuric acid with water
- extraction of elemental sulphur with carbon disulfide
- desorption of carbon disulfide with steam
- air drying and cooling of activated carbon

The elemental sulphur present in the carbon disulfide in dissolved form can be separated by distillation and recovered as high-purity sulphur.

Design example[34]

In the production of viscose filament yarn a Supersorbon®-system combined with a NaOH-scrubber system (Figure 22.1.21) has been in use since 1997. A very high standard of safety engineering is implemented in the treatment system owing to the flammability of the $CS_2$ and the toxic nature of the constituents to be removed from the exhaust air.

System concept

First treatment stage:

Absorption of $H_2S$ in two NaOH jet scrubbers and one water-operated centrifugal scrubber.

Second treatment stage:

Fixed-bed adsorption for purification of exhaust air and $CS_2$ recovery.

Treated air stream

Exhaust air from viscose filament yarn plant.

Design data:

| | |
|---|---|
| Exhaust air flow rate | 12,000 $Nm^3/h$ |
| Solvent capacity | 27 kg $CS_2$/h and 7.5 kg $H_2S$/h |
| Exhaust air temperature | 15 - 30°C |
| Clean air solvent concentration | |
| $CS_2$: max. | 100 $mg/Nm^3$ (24-hour mean) |
| $H_2S$: max. | 5 $mg/Nm^3$ (24-hour mean) |

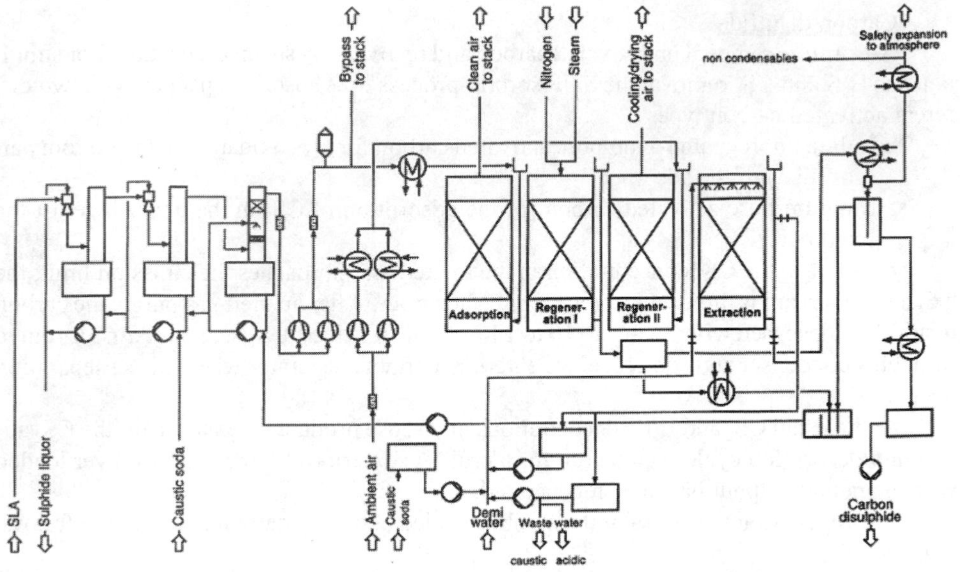


Figure 22.1.21. $H_2S$ and $CS_2$ removal in the filament yarn production (After reference 34).

$H_2S$-removal

The exhaust air, saturated with water vapor, is first treated in an absorption unit for removal of $H_2S$ by routing it through two successive jet scrubbers working with dilute caustic soda solution. Downstream of these, a centrifugal scrubber is installed as an entrainment separator. The sulphide-containing solution rejected from the scrubbers is used to precipitate zinc in the waste water treatment system of the production plant. By products adsorbed on the activated carbon, such as sulphuric acid and elemental sulphur, are removed periodically by water and alkaline extraction.

$CS_2$-removal and recovery

The pre-cleaned airstream containing carbon disulfide vapors passes through two or three parallel adsorbers, in which the $CS_2$ is absorbed on a bed of Supersorbon® activated carbon. As soon as the adsorbent is saturated it is regenerated by desorbing the solvent with a countercurrent flow of steam. The resulting mixture of water and $CS_2$ vapors is condensed and separated in gravity settlers. The recovered $CS_2$ is returned to the viscose production without further treatment. The condensed steam is stripped of residual $CS_2$ in the centrifugal scrubber and used as dilution water for operation of the jet scrubbers.

Operating experience

The specified purity of the treated exhaust air is reliably achieved with the Supersorbon® process[34] with respect to both $CS_2$ and $H_2S$. The $CS_2$ recovery rate is about 95%. The purity of the recovered solvent meets viscose production specifications. A widely varying $CS_2$ concentration in the exhaust air has not adversely affected operation of the adsorption system.

### 22.1.5.4 Refrigerator recycling

Condensation process

Condensation processes are especially suitable for the cleaning of low flow highly concentrated streams of exhaust gas.[36] The entire waste gas stream is cooled below the dew point of the vapors contained therein, so that these can condense on the surface of the heat exchanger (partial condensation). Theoretically, the achievable recovery rates depend only on the initial concentration, the purification temperature and the vapor pressure of the condensables at that temperature. In practice however, flow velocities, temperature profiles, the geometry of the equipment, etc. play decisive roles, as effects such as mist formation (aerosols), uneven flow in the condensers and uncontrolled ice formation interfere with the process of condensation and prevent an equilibrium concentration from being reached at the low temperatures.

The Rekusolv Process[37] uses liquid nitrogen to liquefy or freeze vapors contained in the exhaust gas stream. In order to reduce the residual concentrations in the exhaust to the legally required limits, it is often necessary to resort to temperatures below minus 100°C. The Rekusolv process is quite commonly used in the chemical and pharmaceutical industry and at recycling plants for solvent recovery.

Example of solvent recovery in refrigerator recycling[36,37]

In refrigerator recycling plants R11 or pentane is released from the insulating foam of the refrigerator during shredding. Figure 22.1.22. shows the Rekusolv process as it is used by a refrigerator recycling company. At this plant around 25 refrigerators per hour are recycled thereby generating 8 kg/h of polluting gases. The Rekusolv plant is capable of condensing almost all of this. The unit is designed to operate for 10 to 12 hours before it has to be defrosted. The plant is operated during the day and is automatically defrosted at night. The concentration of pollutants in the exhaust gas is reduced from 20 to 40 g/m$^3$ to 0.1 g/m$^3$ - a recovery rate of more than 99.5%.

### 22.1.5.5 Petrochemical industry and tank farms

Vapor recovery units are installed at petrochemical plants, tank farms and distribution terminals of refineries. Tank venting gases are normally small in volume, discharged intermittently at ambient temperature and pressure and loaded with high concentrations of organic vapors.[2]

Processes used for the cleanup of such waste air streams with organic vapor up to saturation point are often combined processes:[2,38,39]

- absorption and pressure-swing adsorption
- membrane permeation and pressure-swing adsorption
- condensation and adsorption

Adsorption on activated carbon and vacuum regeneration

Figure 22.1.23 represents the basic principle of the systems which are used to recover vapors displaced from tank farms and loading stations and blend them back into the liquids being loaded. The plant consists of two fixed-bed adsorbers packed with activated carbon which operate alternatively. The vapor pressure gradients required as a driving force for desorption is generated by vacuum pumps. As the raw gas entering the adsorber is saturated with hydrocarbon vapors, the heat of adsorption causes local heating of the well insulated adsorbent bed. The higher temperature of the spent activated carbon supports subsequent desorption as the pressure is being lowered by means of a vacuum pump.

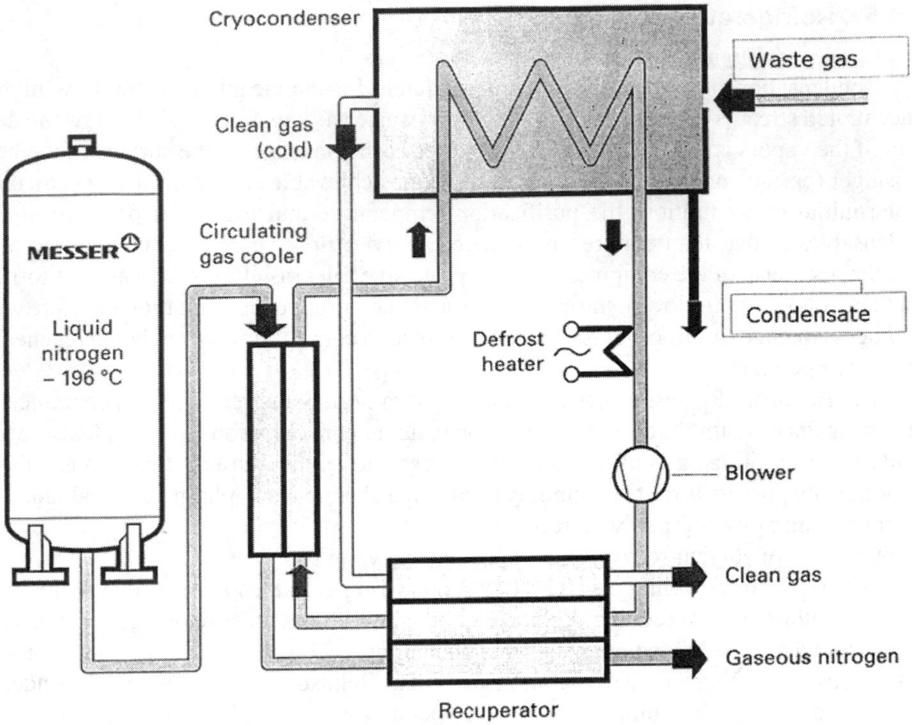


Figure 22.1.22. Block diagram of the Rekusolv process (After references 36,37).

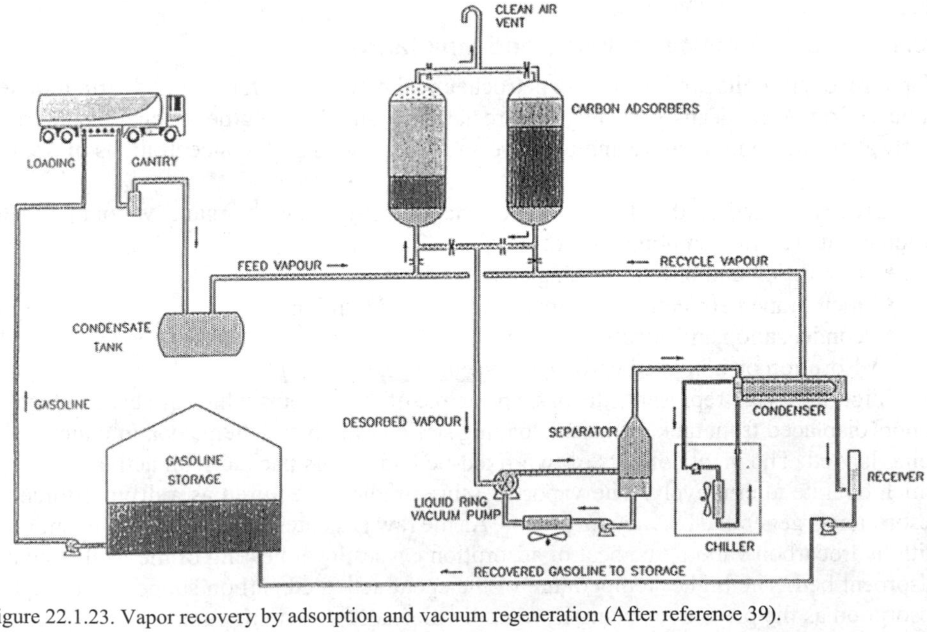


Figure 22.1.23. Vapor recovery by adsorption and vacuum regeneration (After reference 39).

The desorbed vapor is transported from the adsorption bed via a vacuum pump. A liquid ring pump using water for the seal fluid is normally applied for safety purposes in operations where air is allowed to enter the adsorbers.

The desorption stream is processed through a cooler and absorber. The absorber uses gasoline (or the recovered solvent) as a sorbent and serves as a hydrocarbon reduction stage for the entire system. Loading facilities with the inlet vapor concentrations of over 10% by volume can demonstrate recovery efficiencies of over 99%.

### 22.1.5.6 Chemical industry

In the chemical industry a variety of solvents and solvent mixtures are in use. For many gas cleaning operations special adsorption processes are required.

Two-stage adsorption[40]

In many cases, a two-stage process offers significant advantages. This process makes use of two adsorbers that are operating alternately in series. Four-way reverse valves ensure safe change-over. During operation, the air passages of the valves are always open so that any blocking of the flow is impossible. Solvent laden air first passes into one of the adsorbers. Purified air is released into the atmosphere only during the time when the second adsorber is being steamed. During this phase of the process, the on-line adsorber is only slightly charged with solvent which means that it will be completely adsorbed. In general the time required for steaming an adsorber is at most 50% of the time provided for adsorption.

After the adsorber has been steamed, the outlet valve is switched over. The steamed adsorber is then dried with solvent-free air taken from the laden adsorber. A feature of this process is that the steamed adsorber is always dried with solvent-free air. An advantage of this two-stage process is that the laden adsorber can be loaded beyond the breakthrough capacity since the second adsorber is unladen and can therefore accept any excess solvent.

The condensation phase and reprocessing of the desorbate are carried out in the same manner as described for the single-stage plant. For example, a two-stage adsorption plant for the recovery of dichloromethane is able to clean a gas stream of 3000 $m^3/h$ to a residual solvent content of less than 20 $mg/m^3$.

## REFERENCES

1 J.W. Leatherdale, **AirPollution Control by Adsorption, Carbon Adsorption Handbook**, *Ann Arbor Science Publisher Inc.*, Michigan, 1978 P.N. edited by Cheremisinoff, F. Ellerbusch.
2 VDI 3674, Waste gas cleaning by adsorption, Process and waste gas cleaning, Beuth Verlag GmbH, 10772 Berlin, 1998.
3 H. Menig, **Luftreinhaltung durch Adsorption, Absorption und Oxidation**, *Deutscher Fachschriften-Verlag,* Wiesbaden, 1977.
4 H. von Kienle, E. Bäder, **Aktivkohle und ihre industrielle Anwendung**, *Ferdinand Enke Verlag*, Stuttgart, 1980.
5 Test methods for activated carbon. European Council of Chemical Manufacturers` Federations/CEFIC Brussels (1986) 7.
6 A. Ychaskel, **Activated Carbon, Manufacture and Regeneration**, *Noyes DATA Corporation*, Park Ridge, New Jersey, USA, 1978.
7 M. Smisek, S. Cerny, **Activated Carbon**. *Elsevier Publishing Company*, Amsterdam-London- New York, 1970.
8 R.Ch. Bansal, J.-B. Donnet and F. Stoeckli, **Acitvated Carbon**, *Marcel Dekker Inc.*, New York, 1988.
9 H. Jüntgen, Grundlagen der Adsorption, *Staub-Reinhaltung Luft*, **36**, (1976) No. 7, pp. 281-324.
10 H. Seewald, Technische verfügbare Adsorbentien. Vortragsveröffentlichung Haus der Technik, H. 404, pp. 24-34, Vulkan-Verlag, Essen, 1977.
11 W. Kast, **Adsorption aus der Gasphase**. *VCH-Verlag*, Weinheim, 1988.

12 H. Krill, Adsorptive Abgasreinigung - eine Bestandsaufnahme, VDI-Berichte 1034, pp. 339-371, VDI-Verlag, Düsseldorf, 1993.
13 K.-D. Henning, Activated Carbon for Solvent Recovery, Company booklet CarboTech Aktivkohlen GmbH, D-45307 Essen, Germany.
14 A. Mersmann, G.G. Börger, S. Scholl, Abtrennung und Rückgewinnung von gasförmigen Stoffen durch Adsorption, *Chem.-Ing.-Techn.*, **63**, No. 9, pp. 892-903.
15 H. Jüntgen, Physikalisch-chemische und verfahrenstechnische Grundlagen von Adsorptionsverfahren, Vortragsveröffentlichung Haus der Technik, H. 404, pp. 5-23, Vulkan-Verlag, Essen, 1977.
16 K.-D. Henning, Science and Technology of Carbon, Eurocarbon, Strasbourg/France, July 5-9, 1998.
17 K.-D. Henning, S. Schäfer, Impregnated activated carbon for environmental protection, *Gas Separation & Purification*, 1993, Vol. **7**, No. 4, p. 235.
18 P.N. Cheremisinoff, Volatile Organic Compounds, *Pollution Engineering*, March 1985, p.30
19 C.S. Parmele, W.L. O`Connell, H.S. Basdekis, Vapor-phase cuts, *Chemical Engineering*, Dec. 31, 1979, p. 59.
20 M.G. Howard, Organic solvents recovered efficiently by dry processes, *Polymers Paint Colour Journal*, May 18, 1983, p. 320.
21 W.D. Lovett, F.T. Cunniff, Air Pollution Control by Activated Carbon, *Chemical Engineering Progress,* Vol. **70**, No. 5, May 1974, p. 43.
22 H.L. Barnebey, Activated Charcoal in the Petrochemical Industry, *Chemical Engineering Progress*, Vol. **67**, No. 11, Nov. 1971, p. 45.
23 K.-D. Henning, W. Bongartz, J. Degel, Adsorptive Removal and Recovery of organic Solvents, VfT Symposium on Non-Waste Technology, Espo/Finnland, June 20-23, 1988.
24 G.G. Börger, A. Jonas, Bessere Abflugtreinigung in Aktivkohle-Adsorben durch weniger Wasser, *Chem.-Ing.-Techn.*, **58** (1986), p. 610-611, Synopse 15.12.
25 K.-D. Henning, W. Bongartz, J. Degel, Adsorptive Recovery of problematic solvents, Paper presented at Nineteenth Biennial Conference on Carbon, June 25-30, 1989, Pennstate University/Pennsylvaia/USA.
26 A.A. Naujokas, *Loss Preventation*, **12** (1979) p. 128.
27 G. Konrad, G. Eigenberger, Rotoradsorber zur Abluftreinigung und Lösemittelrückgewinnung. *Chem.-Ing.-Techn.*, **66** (1994) No. 3, pp. 321-331.
28 H.W. Bräuer, Abluftreinigung beim Flugzeugbau, Lösemittelrückgewinnung in der Wirbelschicht, *UTA Umwelt Technologie Aktuell* (1993),No. 2, pp. 115-120.
29 W.N. Tuttle, Recent developments in vacuum regenerated activated carbon based hydrocarbon vapour recovery systems, *Port Technology International*, 1, pp. 143-146, 1995.
30 Exhaust air purification and solvent recovery by the Supersorbon(r) Process, Company booklet, Lurgi Aktivkohlen GmbH, D-60388 Frankfurt, Gwinnerstr. 27-33, Germany.
31 G. Caroprese, Company booklet, Dammer s.r.l., I-20017 Passirana die Rhol (Mi), Italy.
32 R. Peinze, Company booklet, Air Engineering, CH-5465 Mellikon, Switzerland.
33 Environmental Technology, Company booklet, Eisenmann, D-71002 Böblingen, Germany.
34 Exhaust air purification and $CS_2$ recovery by the Supersorbon® process with front- and $H_2S$ absorption, Company booklet, Lurgi Aktivkohlen GmbH, D-60388 Frankfurt, Germany.
35 U. Stöcker, Abgasreinigung in der Viskose-Herstellung und -Verarbeitung, *Chem.-Ing.-Techn.*, **48**, 1976, No. 10.
36 Exhaust gas purification, Company booklet, Messer Griesheim GmbH, D-47805 Krefeld, Germany.
37 F. Herzog, J. Busse, S. Terkatz, Refrigerator recycling - how to deal with CFCs and pentane, Focus on gas 13, Company booklet, Messer Griesheim GmbH, D-47805 Krefeld, Germany.
38 C.J. Cantrell, Vapour recovery for refineries and petrochemical plants, *CEP*, October 1982, pp. 56-60.
39 Kaldair Vapor Recovery Process, Company booklet, Kaldair Ltd., Langley Berks SL3 6EY, England.
40 Waste Air Purification, Company booklet, Silica Verfahrenstechnik GmbH, D-13509 Berlin, Germany.

# 22.2 SOLVENT RECOVERY

Isao Kimura
**Kanken Techno Co., Ltd., Osaka, Japan**

## 22.2.1 ACTIVATED CARBON IN FLUIDIZED BED ADSORPTION METHOD

This section discusses a solvent recovery process developed by Kureha Engineering Co. Ltd., Japan.[1] In this continuous process, spherical particles of activated carbon (AC) circulate in the adsorption and desorption columns by fluidization. In the adsorption column, the particles form fluidized beds on multi-trays to adsorb the solvent in counter-current contact with the feed gas. The cleaned gas is released from the top of the adsorption-column to the atmosphere.

The carbon absorber with adsorbed solvent is electrically heated (150-250°C) at the upper part of the desorption column and solvent is desorbed by nitrogen, which is supplied from the bottom of the desorption column. Nitrogen is continuously recycled in the desorption column. The solvent in the carrier gas is fed to a condenser for its recovery.

Features of the process:

- Atmospheric condensate is the only waste product because nitrogen (or air) is used as the carrier gas.
- High flow rate per unit area because the process is conducted by fluidization.
- Low thermal energy loss because both the adsorption and desorption are conducted at constant temperature.
- Low electric energy consumption because the fluidization needs less blower driving power.

Figure 22.2.1 shows the process diagram and Table 22.2.1 shows application data.

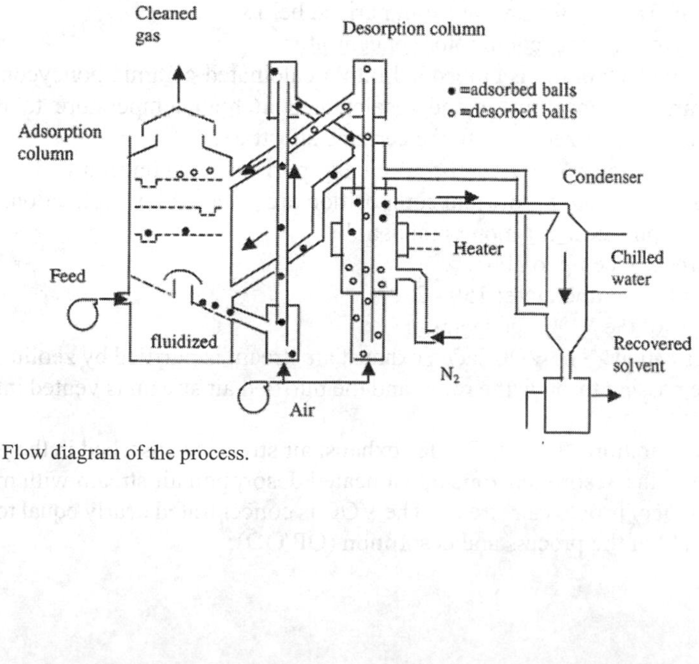


Figure 22.2.1. Flow diagram of the process.

**Table 22.2.1. Application data. Activated carbon fluidized bed adsorption process**

| Solvents | Manufacturing plants | Flow rate, $m^3$/min | Feed conc., ppm |
|---|---|---|---|
| phenol, naphthalene | semiconductors | 50 | 100* |
| curry odor | food processing | 240 | 4,000 |
| IPA, phenol, acetone, ethanol | electronics | 120 | 220 |
| terpineol | IC | 300 | 30 |
| dichloromethane | semiconductors | 80 | 500 |
| IPA, xylene | semiconductors | 150 | 110 |
| IPA | semiconductors | 80 | 200 |
| manure odor | fertilizers | 600 | 3,000* |
| phenol, formaldehyde, ammonia | shell mold casting | 2,000 | 4 |
| butyl acetate, IPA, etc. | semiconductors | 420 | 50 |
| terpineol | IC | 300 | 30 |
| butyl acetate, etc. | solvents | 550 | 50 |

*odor concentration

## 22.2.2 APPLICATION OF MOLECULAR SIEVES

Molecular sieves is a term synonymous in this context to aluminosilicate hydrate also called zeolite. The zeolite is applied to deodorization, gas separation and some other processes. Since this material is, unlike activated carbon, non-combustible, it has increasingly been applied to concentrate volatile organic compounds, VOC. The concentration process discussed here applies to the "VOC concentrator" developed by Seibu Giken Co. Ltd. Japan.[2] The element referred to here as the VOC concentrator is non-combustible and has a honeycomb structure. The key features are summarized below.

<u>Key features of honeycomb rotor concentrator</u>

- Hydrophobic zeolite is embedded into a calcinated ceramic honeycomb substrate with an inorganic binder and recalcinated at high temperature to improve the compatibility of zeolite with the ceramic substrate.
- Various types of VOCs can be efficiently purified and concentrated.
- The most suitable zeolite and composition are selected for application.
- Absorption medium is non-combustible
- Heat resistance up to 500°C
- Desorption temperature: 150 - 220°C

<u>Functions of the VOC concentrator</u>

(1) Purification - The VOC laden exhaust air stream is purified by zeolite or other adsorbent while passing through the rotor, and the purified air stream is vented into the atmosphere.

(2) Concentration - The VOC laden exhaust air stream is adsorbed in the process zone and desorbed in the desorption zone into a heated desorption air stream with much less air volume than that of process air stream. The VOC is concentrated nearly equal to the air volume ratio (5-15) of the process and desorption (QP/QD).

Basic design of VOC concentrator

VOC concentrator consists of a VOC rotor, a rotor driving device, a rotor casing with a set of seals, a pair of chambers (front/rear) with a zone partition wall, and a desorption heater as shown in Figure 22.2.2.

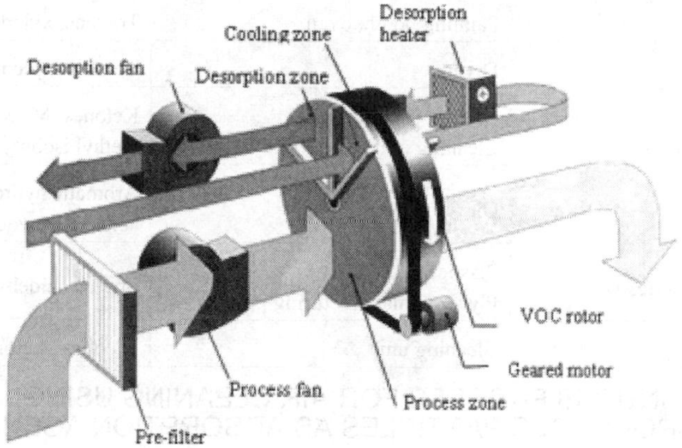

Figure 22.2.2. VOC concentrator.

Advantages

VOC concentrators, outstanding in their purifying and concentrating efficiency, have been used worldwide as a high-quality and safe product owing to material engineering expertise. By being combined with oxidizers, and other recovery equipment, the initial investment and running cost of the total system can be drastically diminished. Even for such VOCs as cyclohexane/MEK/ammonia/methyl alcohol/ethyl alcohol/styrene/formaldehyde/isophorone/phenol, which have been found unsuitable for the existing technology using activated carbon, due to economic or safety concerns, the VOC concentrator is cost-effective and safe.

- *Non-combustible components* - The development of non-flammable rotors has been achieved by engineering the appropriate materials for the honeycomb substrate, the adsorbent, and the binder and by application of special material processing technique.
- *Treatment of VOCs which have high boiling point* - Taking advantages of zeolite rotors, non-combustible component, and high heat resistance, concentrators can use adsorption air at high temperatures. Accordingly, VOCs that could not be treated by carbon material due to its desorption temperature limit have turned out to be easily treated.
- *Inertness* - VOCs that can easily polymerize, such as, styrene, cyclohexanone, etc., can be effectively treated by hydrophobic zeolite.
- *Cleaning and reactivation* - Zeolite rotors can be calcinated under high temperature, due to their all inorganic content (including binder). The rotors can be easily washed. Zeolite rotors can be reactivated by heat treatment.

Table 22.2.2 shows the typical applications of the VOC concentrator.

**Table 22.2.2. Typical applications of the VOC concentrator**

| Industry | Facilities | Treated VOCs |
|---|---|---|
| Automobile, Steel structure manufacturing | Painting booth | Toluene, xylene, esters, alcohols |
| Steel furniture | Painting booth, oven | Toluene, xylene, esters, alcohols |
| Printing | Dryer | Toluene, xylene, esters, alcohols |
| Adhesive tape<br>Hook and loop fastener | Coating process<br>Cleaning unit | Ketones (MEK, cyclohexanone, methyl isobutyl ketone, etc.) |
| Chemicals | Oil refinery, reactors | Aromatic hydrocarbons, organic acids, aldehydes, alcohols |
| Synthetic resin adhesive | Plastics<br>Plywood manufacturing process | Styrene, aldehydes, esters |
| Semiconductor | Cleaning unit | Alcohols, ketones, amines |

## 22.2.3 CONTINUOUS PROCESS FOR AIR CLEANING USING MACROPOROUS PARTICLES AS ADSORPTION AGENTS

In the photoresist segment of the semiconductor process a very low concentration of VOC is generated. These low concentration VOCs can be removed by a concentration step followed by a catalytic oxidation process. The industrial application of this method is discussed below. Dilute VOC is fed to the rotating zeolite concentrator discussed in Section 22.2.2 (honeycomb rotor) where it is concentrated to 10 times of the original concentration and burned on catalysts at relatively low temperature. This saves fuel cost. The VOC is a mixture of IPA, acetone, MEK, toluene, ethyl acetate, n-hexane, propylene glycol monoethyl ether, propylene glycol monoethyl ether acetate, and tetramethyl ammonium hydroxide. The mix-

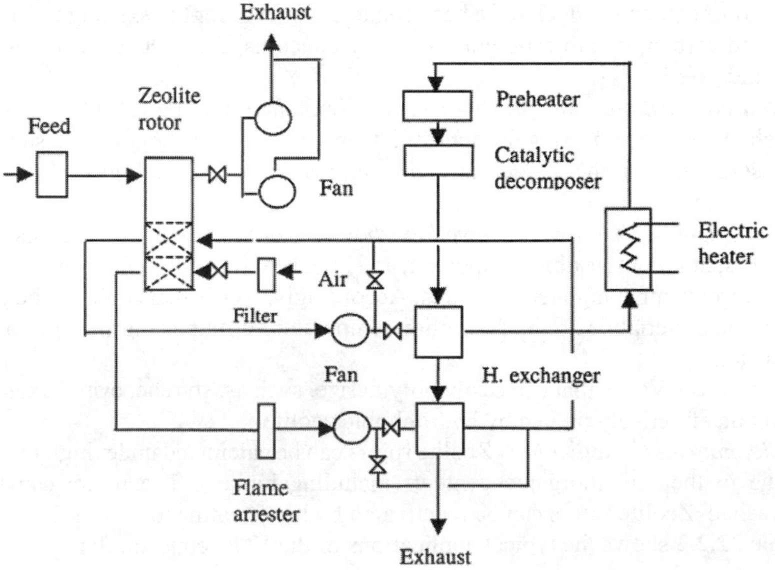


Figure 22.2.3. Flow diagram of continuous air cleaning process by zeolite concentrator.

Figure 22.2.4. Continuous air cleaning unit by zeolite concentrator.

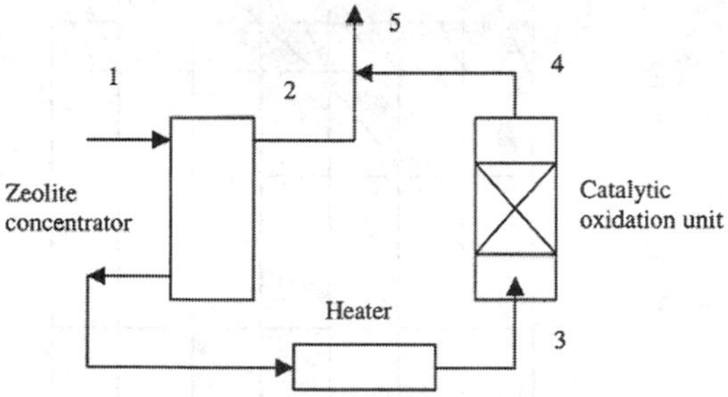


Figure 22.2.5. Block diagram. Continuous air cleaning process by zeolite concentrator.

ture is exhausted from the photoresist at approximately 100 Nm$^3$/min, 20°C and 60% RH. Figure 22.2.3 shows the process flow diagram. The VOC laden feed gas (exhausted from the photoresist) is pre-treated by filters in two steps, first by a bag filter and then by activated carbon. The former removes dusts and the latter removes any readily polymerizable vapors such as, acrylic acid, organic silicon compounds e.g. hexamethylene disilazane (HMDS), phosphorous-compounds and halogen-compounds which are all catalyst poisons. The pretreated gas is fed to the zeolite honeycomb rotor for concentration. (Figure 22.2.4)

Table 22.2.3 shows measurements at the designated points in the block diagram (see Figure 22.2.5). VOC removal reached 91 % at outlet of the rotor and 98 % at outlet of the catalytic oxidation unit.

**Table 22.2.3. Measurements according to block diagram (Figure 22.2.5). Continuous air cleaning process by zeolite concentrator**

| No. | Point of detection | Temp., °C | Conc., ppm | Flow rate, Nm$^3$/min |
|---|---|---|---|---|
| 1 | Concentrator, inlet | 26 | 430 | 102 |
| 2 | Concentrator, outlet | 33 | 40 | 116 |
| 3 | Cat. oxidation, inlet | 450 | 3,700 | |
| 4 | Cat. oxidation, outlet | 490 | 83 | |
| 5 | Exhaust | | 39 | |

## 22.2.4 SOLVENT RECOVERY FROM HAZARDOUS WASTES

**Table 22.2.4. Properties of cycloparaffins as a cleaning solvent**

| Property | Value |
|---|---|
| Appearance | transparent |
| Color (Seibolt) | +30 |
| Specific gravity | 0.839 |
| Viscosity, 40°C Cst | 1.6 |
| Flash point, °C | 76 |
| Aniline point | 58 |
| Cycloparaffins content, % | 80 |
| Alkylbenzene content, % | 10 |
| Aromatics content, % | 1 |

**Table 22.2.5. Cleaning performance of solvents**

| Cleaning solvent | Oils | Residual oil, mg/cm$^2$ |
|---|---|---|
| cycloparaffins | press oil<br>fan press oil<br>bender oil | 0.0017<br>0.0015<br>0.0017 |
| trichloroethane | press oil<br>fan press oil<br>bender oil | 0.0021<br>0.0013<br>0.0013 |

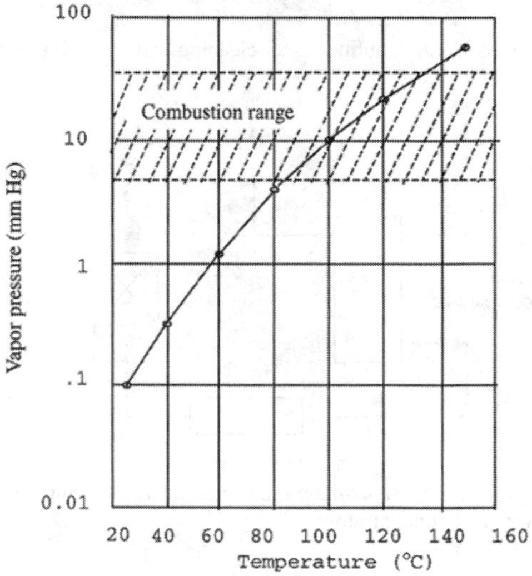


Figure 22.2.6. Vapor pressure and combustion range of cycloparaffins.

The process was previously used for chlorine compounds represented by perchloroethylene and trichloroethylene and for fluorine compounds used for cleaning of press stamped metal parts. However, these compounds are now prohibited for use and production by environmental regulations. The current replacements include emulsions, organic and inorganic solutions, etc. Cycloparaffins originating from petroleum products were quali-

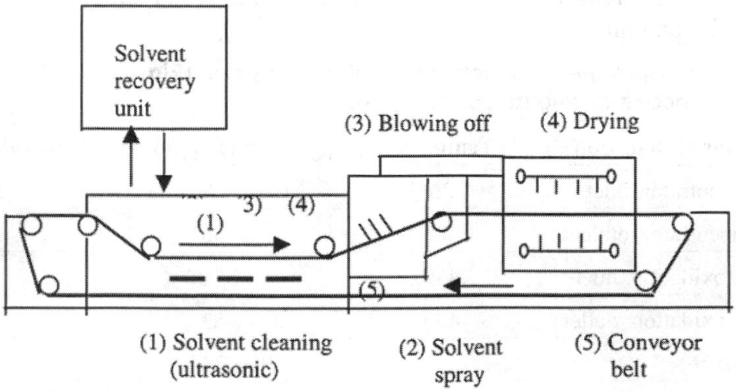


Figure 22.2.7. Metal parts cleaning employing cycloparaffins.

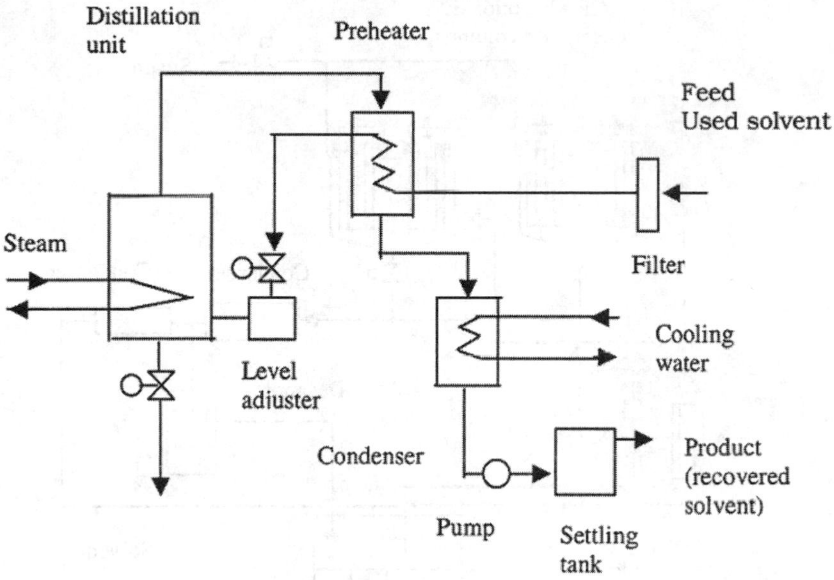


Figure 22.2.8. Flow diagram of solvent recovery process.

**Table 22.2.6. Brief specification of the cycloparaffins recovery unit**

| Item | Specification |
|---|---|
| Solvent | cycloparaffins |
| Capacity | 70 l/min |
| Dimensions | 1,050 x 533 x 1,355 mm |
| Weight | 250 kg (approx.) |
| Steam pressure | 6.0 MPa |
| Steam consumption | 25 kg/h |
| Cooling water | <30°C, >2MPa, >20 l/min |

fied as a good cleaning solvent (see Table 22.2.4).

Table 22.2.5 shows the cleaning performance of the solvent. Figure 22.2.6 shows the vapor pressure and combustion range of the solvent. Figure 22.2.7 shows a process diagram of cycloparaffins solvent use in cleaning stamped metal parts. The solvent is recycled after recovery by distillation. Combining the cleaning and the distillation process gives a continuous process.

Figure 22.2.8 shows a schematic diagram of distillation process. Table 22.2.6 shows specification of the recovery unit.

## 22.2.5 HALOGENATED SOLVENT RECOVERY

### 22.2.5.1 Coating process

A process to recover ethylene dichloride, EDC, and methylene dichloride, MDC, using activated carbon fibers, ACF, is discussed in this section. The gas at approximately 50°C, containing EDC and MDC is received from a painting process. It is pretreated by a filter to remove dusts and passed through a heat exchanger to cool it down to approximately 30-40°C. The pre-treated gas is fed to adsorption columns filled with ACF where the gas is cleaned sufficiently for it to be safely released to atmosphere.

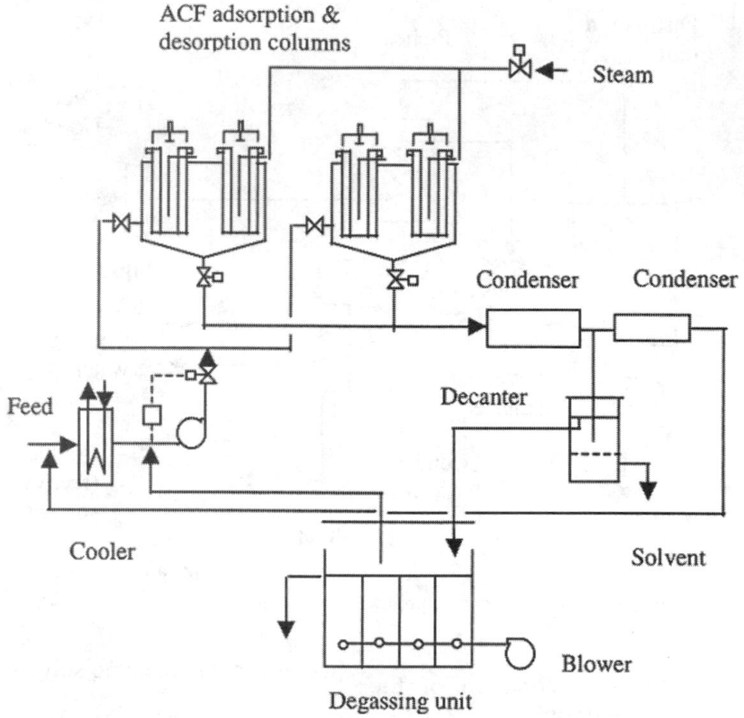


Figure 22.2.9. Flow diagram of ACF solvent recovery process.

Figure 22.2.10. ACF solvent recovery unit.

Figure 22.2.9 shows the process flow diagram. The process consists of two (or more) adsorption columns to ensure a continuous run of adsorption and desorption cycles. Figure 22.2.10 shows an industrial application of the system.

**Table 22.2.7. Performance data for the ACF solvent recovery unit**

| Item | Operational & performance data |
|---|---|
| Flow rate | 100 $m^3$/min, 20°C |
| Solvent | |
| EDC | 46.7 l/h |
| MDC | 23.3 l/h |
| Gas concentration | |
| Inlet | 2,900 ppm |
| Outlet | 58 ppm |
| Adsorption efficiency | 98 % |
| ACF specific surface area | 900 - 2,000 $m^2$/g |
| Adsorption & desorption cycle | 10 min |

The desorption is done by steam. The desorbed stream is fed to a series of two condensers in which both the steam and solvent are condensed. The condensate is fed to a decanter to separate water and solvent by gravity. Water containing approximately 1-2% solvent is fed to a degassing unit and the vaporized solvent is joined to the feed. The decanted solvent is refined, stored in a tank, analyzed, and adjusted to its appropriate EDC/MDC ratio. The refined solvent is recycled for painting.

Water from the degassing unit contains solvent in a concentration relative to the solvent's solubility. For example, if the solubility of perchloroethylenene in water is designated as 1, the relative solubility of EDC in water would be 58 and of MDC in water would be 88. In degassing at 60°C and 180 l/min x 30 min the residual EDC content in water is brought below 0.01 ppm (the environmental standard limit is 0.04 ppm) but in the case of MDC it remains a problem to reach to the environmental standard of 0.2 ppm. A treatment by activated carbon may be one of the solutions. Table 22.2.7 shows performance data in the industrial application.

The performance of the ACF solvent recovery process can be summarized as follows: Since ACF has lager surface area and smaller bulk density than particulate activated carbon, PAC, ACF's filling density becomes 1/10 of PAC's.

Advantages:

- *High recovery rate* - The solvent recovery rate of the ACF process is higher than that of the PAC, especially in the case of lower boiling point solvents, such as, MDC, benzene, trichloroethane, etc.
- *High quality* - ACF has low catalytic activity and a short adsorption and desorption cycle time (approx. 10 min). The solvent is less decomposed in the process. In the recovery of chlorine-containing solvents, it produces less acids as decomposition products. Thus it is less corrosive to the materials of construction. Also, the recovered solvent is of better quality.
- *Broader application* - It is applicable to polymerizing monomers and high boiling point compounds.
- *Light weight, compact and safe* - The process unit can be compact because of the short adsorption and desorption cycles and it is safer because there is less heat accumulation.
- *Energy savings* - The process needs less steam consumption in the adsorption and desorption.

Disadvantages:

- *Less flexibility regarding capacity* - ACF's filling volume cannot be readily changed.

- *Difficult in decreasing outlet temperature* - The short cycle in adsorption and desorption makes the cooling process difficult to synchronize.
- *Expensive* - ACF are more expensive than PAC, thus the unit modification cost is more expensive.

### 22.2.5.2 Tableting process of pharmaceutical products

In a process which produces tablets, a mixture of solvents, containing MDC and methanol and binders is used. Solvents are emitted from the tableting unit. The following process is used to recover the emitted solvent.

Feed gas at 50°C from the tableting process is pre-treated by a filter to remove dusts and by a rotary wet scrubber to absorb mainly methanol. During the scrubbing MDC also dissolves in water, although at low concentration. This water is treated in two steps, by a degassing unit at room temperature and by an activated carbon filter so that the concentration of residual solvents is decreased below that of the environmental standard. The gas from the degassing unit is joined to the feed gas at the outlet of the demister of the MDC recovery unit. The gas, the pretreated feed, and gas from the degassing are cooled down to 30 - 40°C before feeding to the MDC recovery unit which consists of at least two columns for adsorption and desorption cycles. After processing the gas is released to the atmosphere.

Because little methanol is vaporized at room temperature, the water stream is fed to another degassing unit heated to 65°C to vaporize methanol. For safety purposes, the degassing air flow rate is controlled at 10 $m^3$/min. The methanol laden gas from the heated degassing unit is fed to a catalytic oxidation unit where methanol is decomposed before exhausting the gas stream to the atmosphere. Table 22.2.8 shows the process performance.

**Table 22.2.8. Performance of the solvent recovery in the tableting process**

| Item | Performance |
|---|---|
| Flow rate at rotary scrubber | 1,000 l/min |
| Feed concentration | MeOH 10,000 ppm<br>MDC 2,000 ppm |
| Exhaust concentration | MeOH 150 ppm<br>MDC 3 ppm |
| Catalytic oxidation flow rate | 10,000 l/min |
| Feed concentration | MeOH 8,500 ppm<br>MDC 1 ppm |
| Exhaust concentration | MeOH 10 ppm |

This is a unique example of a process in which the recovery and the catalytic oxidation steps are combined. In addition to its use in the pharmaceutical industry, the adsorption is a suitable method for the separation of chlorine-containing solvents. Figure 22.2.11 shows the process flow diagram.

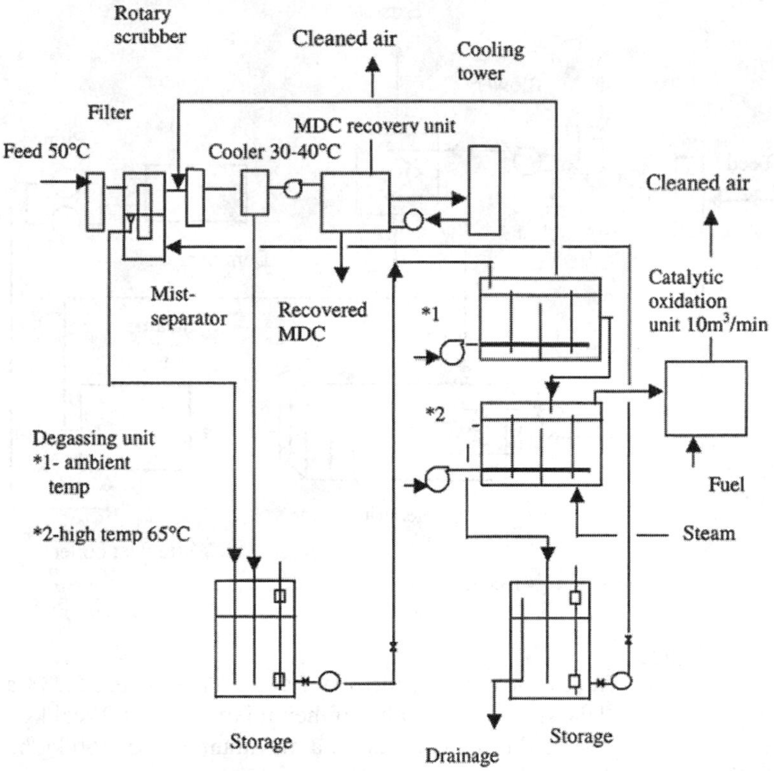


Figure 22.2.11. Flow diagram of tablet binder recovery process.

## 22.2.6 ENERGY RECOVERY FROM WASTE SOLVENT

**Table 22.2.9. Specifications of the VOC oxidation unit**

| | |
|---|---|
| Type of unit | Direct firing |
| Total gas flow rate | 2,000 kg/h |
| Solvent flow rate | 160 kg/h |
| Solvent | Acetone, methanol, IPA |
| Feed temp. | 60 - 100°C |
| Oxidation temp. | 800 °C |
| Performance, THC meter | 97 % |
| Supplemental fuel | Heavy oil, max170 l/min |

**Table 22.2.10. Specifications of the waste heat boiler**

| | |
|---|---|
| Type of unit | Horizontal, multi-tube |
| Use for | Steam as heating medium |
| Waste gas flow rate | 5,000 $Nm^3/h$ |
| Temp., waste gas | Inlet: 800, outlet: 350°C |
| Supply water temp. | 20°C |
| Steam generated | 1,300 kg/h |

It is not always economically justified to recover solvents if they are at very low concentrations and in mixtures. It may be more economical to burn the solvent blend in a boiler to save fuel. Tables 22.2.9 and 22.2.10 contain the specifications of industrial combustion units.

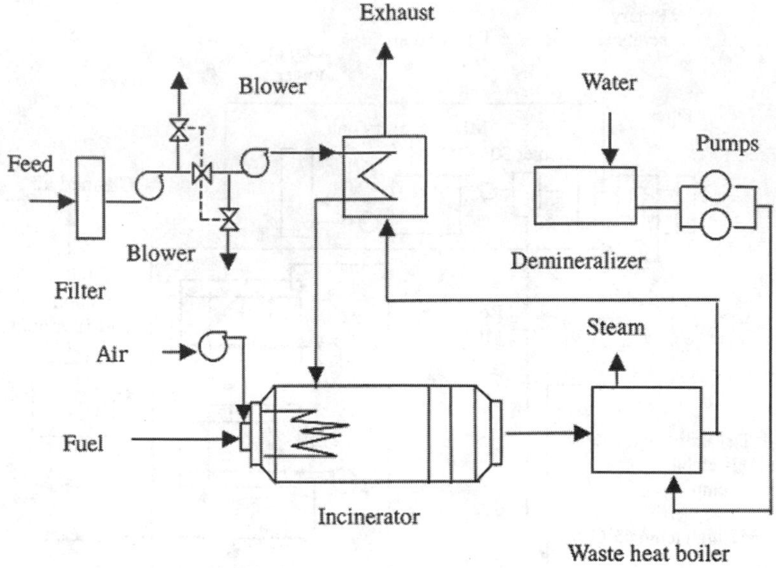


Figure 22.2.12. Flow diagram of energy recovery process.

Savings of energy

The net calorific values for acetone, methanol and IPA is 6,962, 5,238 and 7,513 kcal/kg, respectively, and the simple mean value of their mixture is 6,600 kcal/kg. Let us assume that solvents are generated from a process at a maximum rate of 160 kg/h. The total heat generation of the VOC oxidation unit is as follows:

6,600 x 160 = 1,056 x $10^3$ kcal/h

Heavy oil = 10,200 kcal/kg; specific gravity of heavy oil = 0.86

The saved energy is as follows:

1,056 x $10^3$ / 10,200 = 103.5 kg/h

103.5 / 0.86 = 120 l/h [1]

If no solvent is supplied, the required heavy oil is 170 l/h [2]

If solvent is supplied at max. rate, the required heavy oil is only 50 l/h.

[1] - [2] = 50 l/h

Thus, energy saving performance is 71% ([1]/[2] x 100 = 71 %)

Figure 22.2.12 shows the process flow diagram and Figure 22.2.13 shows an industrial application of the energy recovery unit.

The VOC laden gas from a painting plant is supplied through a filter and is blown to a heat exchanger and then to succeeding units. The fan drive motors are controlled by inverters to adjust the flow rate. In the heat exchanger the feed is preheated with a waste heat boiler's off gas, at a temperature of 350°C, to 60 - 100°C before it is fed to a VOC oxidation unit. In the oxidation unit, heavy oil and VOC are mixed and burnt. At approximately 800°C, VOC is burnt out with yield of more than 97 %. The off gas at 800°C is fed to a waste heat boiler where it generates 1,300 kg/h steam at 8 MPa. The heat transfer gas at 350°C is used for the preheating the feed gas.

Figure 22.2.13. Energy recovery unit.

## REFERENCES


1 T.Takahashi and S.Ito, *Kogyo Toso*, No. **150**, 75-79, 1998.
2 Seibu Giken's Catalog. VOC Concentrator (1997).


# 22.3 SOLVENT TREATMENT IN A PAINTS AND COATING PLANT


DENIS KARGOL
**OFRU Recycling GmbH & Co. KG, Babenhausen, Germany**


The Swiss coating manufacturer Karl Bubenhofer AG was planning to acquire various wash plants for cleaning batch, mixing and multi-trip containers efficiently. The contaminated solvents obtained during the wash process were then to be treated automatically by distillation. The distillation plant had to have a suitable output, operate continuously and be simple to operate. Another requirement was the ability to combine this plant with the wash plants in a new building complex. A distillation plant suitable for this purpose was supplied by OFRU Recycling in Babenhausen/Germany.

Karl Bubenhofer AG, known by the name, "Kabe-Farben", is regarded as one of the largest paint and coating producers in Switzerland and has been producing coating materials since 1926. The product range covers architectural paints and plasters, powder coatings and industrial liquid paints. The company´s wide range of customers includes industrial, commercial and contract paint companies as well as resellers at home at abroad.

Kabe offers a special service in cooperation with regional painters' associations. For a small charge customers and painters can offload their waste paints and contaminated solvents to Kabe in special multi-trip containers. About 160 tonnes per year of spent solvents

Figure 22.3.1. Soiled multi-trip container.

are obtained by way of this disposal system. Even before the planning phase, these contaminated spent solvents were treated with the company's own small distillation plant.

Until now, manual cleaning of these multi-trip containers and large batching and mixing vessels from production was particularly inconvenient and labor-intensive for Kabe. A wash plant for mobile large containers (250-1800 l volume) and for small drums pails (about 80-100 per day) would obviate the unpleasant manual cleaning task in the future. A combination of container wash following by distillation therefore seemed very practicable. The treated clean solvent would feed the wash plants and ensure thorough cleaning. The company was also interested in further processing a part of the distillate for Kabe products.

When the solvent requirement of the container wash plant and the customers's spent solvent quantities were added together, the total was about 200 tonnes of contaminated solvents per year. Solvents such as toluene, ketones, esters, ethyl acetate, xylene, glycol and petroleum spirits were to be separated almost completely from solids such as alkyd, 2-pack, epoxy and polyester resins.

The solvent recovery plant had to meet the following requirements:

1. Combination with wash plants and storage tanks
2. Distillation with a high rate of recovery
3. Continuous operation (24 hour)
4. Distillation output of 200-400 l/h
5. Low maintenance
6. Minimum operating and control times
7. Residue to be contaminated only slightly with residual solvent

Kabe-Farben found the solution Rio Beer, a wash plant producer from Switzerland. Rio Beer supplied not only the wash plant design but also integrated OFRU Recycling from Babenhausen in Germany with its solvent recovery plants. OFRU Recycling provided the necessary expertise from the paints and coatings sector.

Kabe-Farben visited the OFRU pilot plant in Babenhausen and was able to recover 224 kg of clean solvent within a few hours from 300 kg of contaminated solvent. A quantity of 76 kg of solid paint slurry was then discharged from the distillation unit. An analysis afterwards showed that the residue contained only 5.5% solvent. More than 74% of the solvent could therefore be recovered in the first instance.

The solution

Kabe decided in favor of a vacuum evaporator with an integrated stirrer. The ASC-200 model together with three solvent tanks stand together in an explosion-proof room. The cus-

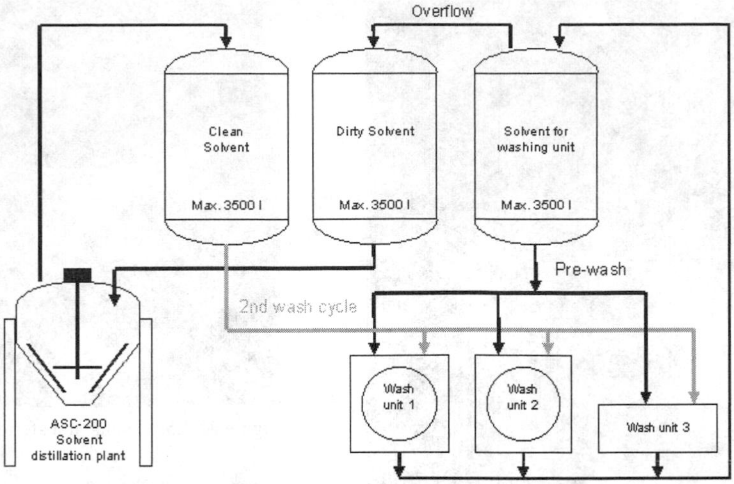


Figure 22.3.2. Solvent circuit and treatment.

Figure 22.3.3. Wash plant for mixing vessels.

tomer's contaminated spent solvents are poured in through a large sieve by means of a drum tilting device. Twin diaphragm pumps then convey the contaminated solvent to the first soiled solvent tank. In view of the large proportion of paint in the solvent (5-20 vol.%), the solvent is kept homogeneously in motion in the tank by means of the stirrer. Another soiled solvent tank is connected up to the wash circuit and supplies the solvent for pre-washing.

The ASC-200 model used by Kabe operates entirely under vacuum. A considerable increase in capacity is achieved by means of the vacuum equipment; moreover, distillation is milder and residue is largely solvent free.

The contaminated product is introduced automatically by suction into the evaporator by means of a magnetic valve and the vacuum until it reaches the height of a capacitive level controller, which keeps the solvent level constant during continuous operation. The filling operation is adjusted in such a way that no excessive cooling or interruption in the distillation process takes place.

The solvent is usually heated by means of a double jacket with its own thermal oil facility. Kabe, on the other hand, operates a catalytic waste air purification plant. The solvent-air mixture obtained in the plant is converted catalytically to heat and stored as an energy store. A small portion of the heat thereby obtained is absorbed in our case by a thermal oil system and released to the distillation plant.

A stirrer is fitted inside the ASC-200 distillation plant. Its scrapers rotate inside the inner wall of the vessel. They keep the heat transfer surface free from encrustation and guarantee uniform heat transfer into the boiling solvent. The stirrer keeps the liquid in motion

Figure 22.3.4. ASC-200 distillation plant in the explosion-proof room.

Figure 22.3.5. Residue transferred to PE pail.

and thereby promotes evaporation of the solvent. Finally, the solvent vapor reaches a water-cooler tubular condenser provided with a cooling water control unit. The condensed, clean distillate is pumped from here to the distillate tank. This tank has a level control and now feeds the clean solvent directly into the wash plant for a second wash. The distillation process takes place fully automatically and lasts about 2 days. Distillation of the bottom product takes place independently afterwards. After the residue has been thickened to a predetermined viscosity, the equipment shuts down automatically. The still is drained by means of a discharge valve at the bottom of the cone. The 200-300 kg of compact greyish residues are transferred to 20 completely empty 20 liter plastic pails and then fed to an inexpensive solid waste incineration process. No further cleaning operations are required at the plant so the stirred distillation unit is ready for use again immediately.

The plant used at Kabe treats about 200 liters of solvent per hour. At about 200 tonnes per year it is therefore being used to 60-70% of its capacity at present. Since its installation, the distillation plant has been operating for 5 days per week with very little maintenance. A few mechanics maintain the plant twice a year without external support.

# 22.4 APPLICATION OF SOLAR PHOTOCATALYTIC OXIDATION TO VOC-CONTAINING AIRSTREAMS

K. A. Magrini, A. S. Watt, L. C. Boyd, E. J. Wolfrum, S. A. Larson, C. Roth, and G. C. Glatzmaier
**National Renewable Energy Laboratory, Golden, CO, USA**

## 22.4.1 SOLVENT DEGRADATION BY PHOTOCATALYTIC OXIDATION

Photocatalytic solvent oxidation had been demonstrated at the pilot scale in two recent field tests located at McClellan Air Force Base (AFB) in Sacramento, California and at the Fort Carson U. S. Army Installation in Colorado Springs, Colorado (Watt et al., 1999; Magrini et al., 1998). The objective of the tests was to determine the effectiveness of solar-powered photocatalytic oxidation (PCO) treatment units for destroying emissions of chlorinated organic compounds (trichloroethylene and dichloroethylenes) from an air stripper at ambient temperature and destroying paint solvent emissions (toluene and MEK) from a painting facility at higher temperatures. Goals for field testing these solar-driven systems were to gather real-world treatability data and establish that the systems maintained performance during the duration of the testing.

The photocatalytic oxidation process can effectively destroy hazardous organic pollutants in air and water streams. Although treatment systems will vary depending on the type of stream being treated, the basic process remains the same. The key ingredient is the photoactive catalyst titanium dioxide ($TiO_2$), which is an inexpensive, non-toxic material commonly used as a paint pigment. When $TiO_2$ is illuminated with lamps or natural sunlight, powerful oxidizing species called hydroxyl radicals form. These radicals then react with the organic pollutant to tear it apart and ultimately form carbon dioxide ($CO_2$) and water (Phillips and Raupp, 1992). When halogenated organics are treated, dilute mineral acids like HCl form. The process works at both ambient and mildly elevated temperatures (>200°C) (Fu et. al., 1995; Falconer and Magrini, 1998).

Researchers throughout the world have been investigating PCO as an advanced oxidation technology for treating air and water streams contaminated with a variety of organic and inorganic compounds (Blake, 1996; Cummings et al., 1996). The susceptibility of an organic species to complete oxidation is typically reported in terms of photoefficiency, defined as the number of molecules of contaminant oxidized to carbon dioxide, water, and simple mineral acids divided by the number of photons incident on the catalyst. These values vary widely, depending on the reactor design, catalyst geometry and the compound of interest.

Much of the work on photocatalytic oxidation focuses on treating the halogenated organics trichloroethylene (TCE) and perchloroethylene, contaminants commonly found in ground water sources. These compounds and other chlorinated ethylenes typically react rapidly with $TiO_2$ and photons at efficiencies greater than 100%. These rates are likely due to chain reactions propagated by chlorine radicals (Luo and Ollis, 1996; Nimlos et. al. 1993; Yamazaki-Nishida, 1996). Paint solvent emissions generally consist of toluene, xylenes, ketones and acetate vapors. Measured photoefficiencies for benzene and other aromatic compounds like toluene are typically less than 5% (Gratson et. al., 1995; d'Hennezel and Ollis, 1997). Besides exhibiting low photoefficiencies, aromatic species tend to form less reactive, nonvolatile intermediates during gas phase PCO. These intermediates build up on the catalyst surface and block or inhibit the active catalytic sites for further reaction (Larson and Falconer, 1997). The addition of heat and small amounts of platinum to the $TiO_2$ catalyst overcome these problems (Falconer and Magrini, 1998; Fu et. al., 1996). Oxygenated organics such as ethanol and acetone have photoefficiencies typically around 1%-10% (Peral and Ollis, 1992).

Several field demonstrations of PCO using sunlight to treat groundwater contaminated with TCE have been reported (Mehos and Turchi, 1993; Goswami et. al., 1993). These field tests found that nontoxic constituents in the water can non-productively react with or "scavenge" the photogenerated hydroxyl radicals and reduce the rate of the desired reaction. Common scavengers such as humic substances and bicarbonate ions increase treatment costs for the technology (Bekbolet and Balcioglu, 1996). Turchi and co-workers, (1994) found that by air stripping the volatile contaminants from the water stream, the regulated compounds at many contaminated sites could be transferred to the air, leaving the radical scavengers behind. The water can then be safely discharged and the air effectively treated with PCO. The improved photoefficiency reduces treatment costs and more than offsets the added cost of air stripping these contaminants from water. Read et. al., (1996) successfully field tested a lamp-driven, PCO system on chloroethylene vapors from a soil vapor extraction well located at DOE's Savannah River Site. Magrini et al., 1996, and Kittrell et al., 1996, both used modified $TiO_2$ catalysts and lamp-driven reactors to treat VOCs representative of semiconductor manufacturing and contact lens production, respectively. 1,2-dichloroethane, stripped from contaminated groundwater, was successfully treated in a pilot scale PCO demonstration at Dover AFB (Rosansky et al., 1998).

The second field test assessed PCO to treat paint solvent vapors. Painting operations for military and civilian vehicles are conducted in ventilated enclosures called paint booths. Filters in the exhaust ducts trap paint droplets from the paint overspray while the VOC-laden air is typically exhausted through roof vents. The vent emissions can contain several hundred parts per million (ppm) of the paint solvents, which continue to evaporate from the vehicle after painting is complete. Most types of paint generally contain significant amounts of VOCs such as toluene, a suspected carcinogen, as well as other hazardous solvents such as methyl ethyl ketone, methyl isobutyl ketone, hexanes, xylenes, n-butyl acetate, and other components in lesser amounts.

Current technologies for treating these emissions include catalytic combustion of the vapors over supported Cu and Cr-oxides at temperatures of 350°C (Estropov et. al, 1989); air-flow reduction and recirculation strategies (Ayer and Darvin, 1995); and regenerative thermal oxidation at near incineration temperatures (Mueller, 1988). The use of platinized $TiO_2$ and temperatures of 180-200°C provide significant energy savings in treating these emissions. Our goals for testing the paint booth emissions was to gather real-world treatability data and establish that the system maintained performance during the duration of the testing.

## 22.4.2 PCO PILOT SCALE SYSTEMS

### 22.4.2.1 Air stripper application

Figure 1 shows a schematic of the pilot-scale system used at McClellan AFB. This system, scaled from an optimized laboratory reactor, was fabricated by Industrial Solar Technology

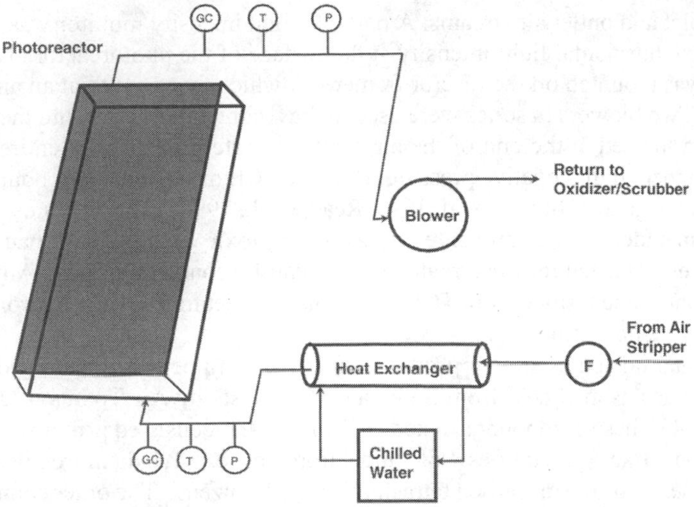


Figure 22.4.1. Schematic of the 10 SCFM solar photoreactor system used at McLellan AFB.

Corporation (Golden, CO). The reactor, 1.22 m wide by 2.44 m, consisted of a fiberglass-reinforced plastic I-beam frame. A transparent fluoropolymer film, treated to accept adhesives, formed the front and back windows of the reactor. The film windows were attached to the reactor frame with foam tape coated on both sides with an acrylic adhesive. The catalyst, titanium dioxide (Degussa P25), was coated onto a structured, perforated polypropylene tubular packing commonly used in oil-water separators. The $TiO_2$ was suspended in water as a slurry and sprayed onto the tubular supports with a new paint sprayer until the supports were opaque. Fluid modeling of airflow through the reactor showed that a 5.1 cm PVC manifold, located at the inlet and outlet of the reactor, would provide an even flow distribution. Small (0.6 cm) holes, drilled into the manifold at one inch intervals, provided the even flow distribution required for efficient contact of the contaminated air with the catalyst.

A two-inch diameter pipe from the outlet of the air stripper supplied the contaminated air stream to the reactor. The air stream was first passed through to a tube-in-shell heat exchanger for partial dehumidification. A chiller circulated 2°C ethylene glycol through the tube side of the heat exchanger. Chilling, which reduced the relative humidity of the air stream from near saturation to less than 20% at 20°C, was required because high humidity reduces the TCE destruction rate, likely because of competitive adsorption between moisture and TCE molecules at the catalyst surface (Fan and Yates, 1996). By lowering the humidity to less than 20% (@ 20°C), the reaction rate was sufficient to ensure complete TCE destruction. This effect was observed by Hung and Marinas (1997) in a lamp-illuminated annular reactor in which TCE conversion was not affected by relative humidity up to 20%; conversion deteriorated as the humidity level reached saturation. Laboratory tests of the system demonstrated that at TCE concentrations of near 100 ppmv, less than 15% of the TCE in the airstream transferred to the condensate formed in the heat exchanger. The condensate (approx. 1-2 liters per day) could be fed back into the air stripper for treatment.

On the reactor, GC sample ports and temperature and pressure sensors provided monitoring of reactor inlet and outlet VOC concentrations and temperature and pressure drop throughout the system. A portable gas chromatograph provided on-line VOC organic analysis of the inlet and outlet air streams. A portable light intensity monitor was used to measure the global horizontal light intensity at the surface of the photoreactor. The calibrated radiometer was mounted on the reactor framework which faced south at an angle of 40° to the horizon. Two blowers in series were used to draw contaminated air into the system. The blowers were located at the end of the air handling system so that the entire system was maintained under slight negative pressure. Because TCE oxidation can produce intermediates such as phosgene, (Nimlos et. al, 1993; Read et. al., 1996; Fan and Yates, 1996), keeping the system under vacuum prevented release of any toxic vapors. The exhaust gases from the blower were returned for final treatment by a caustic contacting tower. Any hydrochloric acid and phosgene formed from TCE oxidation were neutralized and hydrolyzed respectively by the caustic scrubber.

The contaminated airstream provided by the air stripper unit contained volatile organic compounds as analyzed from Summa canister tests by Air Toxics, Ltd. using EPA method TO-14. These compounds, listed in Table 22.4.1, consisted primarily of trichloroethylene and trace quantities of dichloromethane, 1,1-dichloroethene, cis-1,2-dichloroethane, chloroform, carbon tetrachloride, and benzene. The other compounds present in the outlet presumably formed during treatment.

**Table 22.4.1. Results from the Summa canister tests of inlet and outlet reactor air streams during the McClellan tests. The two inlet and two outlet samples were taken sequentially and represent approximately replicate samples**

| Compound | Inlet 1, ppbv | Inlet 2, ppbv | Outlet 1, ppbv | Outlet 2, ppbv | Detection limit, ppbv |
|---|---|---|---|---|---|
| Chloromethane | ND | ND | 4.8 | 4.1 | 1.5-30 |
| Dichloromethane | 36 | 36 | 30 | 32 | 1.5-30 |
| 1,1-Dichloroethane | ND | ND | 17 | 17 | 1.5-30 |
| 1,1-Dichloroethene | 200 | 200 | ND | ND | 1.5-30 |
| cis-1,2-Dichloroethane | 220 | 240 | 6 | 2.2 | 1.5-30 |
| Chloroform | 250 | 260 | 280 | 280 | 1.5-30 |
| Carbon tetrachloride | 700 | 730 | 790 | 790 | 1.5-30 |
| Benzene | 170 | 190 | 120 | 110 | 1.5-30 |
| 1,2-Dichloroethane | ND | ND | 12 | ND | 1.5-30 |
| Trichloroethylene | 13000 | 13000 | 660 | 280 | 1.5-30 |
| Acetone | ND | ND | 84 | 69 | 9.8-120 |
| 1,4-Dioxane | ND | ND | 10 | ND | 9.8-120 |

### 22.4.2.2 Paint booth application

For the paint solvent application, a slipstream of a paint booth vent located at Fort Carson was analyzed to determine the components present with rapid scan gas chromatograph. On any operational day, several paint types are sprayed. The solvent emissions content thus varies during painting operations. In general, the amounts and types of solvent emissions from the Fort Carson paint booth are listed in Table 22.4.2.

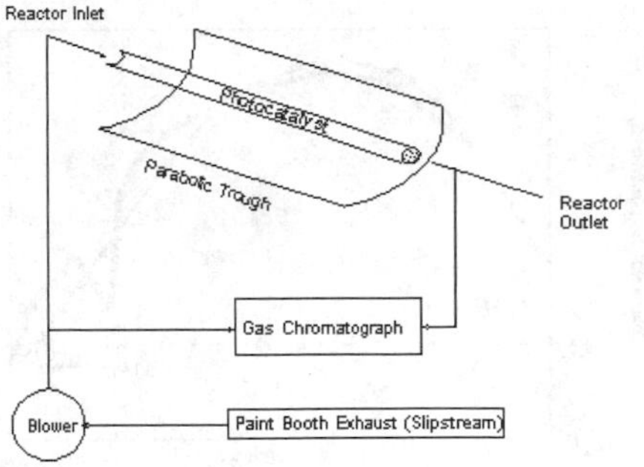


Figure 22.4.2. Schematic of the 10 SCFM parabolic trough, solar photoreactor system used at Fort Carson U. S. Army Installation.

**Table 22.4.2. Breakdown of solvent emissions from painting operations at Fort Carson in 1994. VOC refers to volatile organic compounds, HAP to hazardous air pollutants, MEK to methyl ethyl ketone, and MIBK to methyl isobutyl ketone**

| Paint Type | Total usage, l/yr | VOC, lbs/yr | HAP, lbs/yr | MEK, lbs/yr | Toluene, lbs/yr | MIBK, lbs/yr | Xylene, lbs/yr |
|---|---|---|---|---|---|---|---|
| Solvent base | 19,656 | 110,074 | 5405 | 3931 | 491 | 0 | 983 |
| Thinner | 983 | 7235 | 722 | 552 | 0 | 174 | 0 |

A small field test system, consisting of a photoreactor packed with Degussa $TiO_2$-coated glass beads, was used to assess treatability of the slipstream and provide design data for the pilot scale system. The small reactor operated with a flow rate of 20 liters/minute provided by a pump connected to the paint booth exhaust vent line and variable reactor temperature (ambient to 200°C). Inlet and outlet VOC concentrations were simultaneously measured by two gas chromatographs directly connected to both lines. This arrangement provided real time monitoring of the inlet and outlet VOC concentrations during painting operations.

The resultant treatability data was used to size a 5-SCFM pilot-scale system (Figure 22.4.2) also built by Industrial Solar Technologies. This system consists of a parabolic trough reflector to focus incident sunlight onto a receiver tube containing the catalyst. The trough has a 91 cm aperture width and is 243 cm long. The reflective surface is covered with SA-85 (3M Company), a polymer film that reflects greater than 90% of the incident radiation onto the catalyst tube, providing both heat and light for reaction. The catalyst-containing receiver tube is a 3.81 cm I.D. borosilicate tube packed with 1.9 cm ceramic berl saddles coated with 2.0 wt% platinum on $TiO_2$. A higher weight loading of platinum was used in the field test because of the variable VOC composition of the paint booth stream. Operating temperatures of 200°C were obtained, depending on cloud cover, during most of the day. Note that a thermally heated, lamp-illuminated system will provide similar performance as the solar-based system described here. The reflective trough rotates around the fixed receiver tubes by means of a gear motor. A control and photodiode arrangement activate the gear motor, permitting the trough to track the sun throughout the day. Paint booth emissions were sampled directly from the exhaust stream via 10.1 cm aluminum ducting. On-line analysis of the inlet and outlet concentrations was provided by sample lines connected to the MTI gas samplers. Reactor temperature and inlet flows were measured continuously.

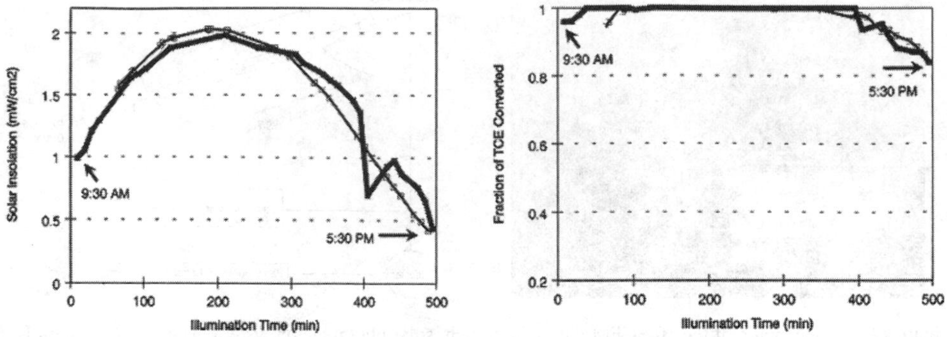


Figure 22.4.3. Solar insolation (left) and fractional TCE conversion (right) versus illumination time. Flow rate and relative humidity of the incoming air stream was 10 SCFM at 20% RH. Data taken on April 8-9, 1997.

## 22.4.3 FIELD TEST RESULTS

### 22.4.3.1 Air stripper application

The field test at McClellan AFB took place during a 3-week period in April, 1997. System operating conditions included two flow rates (10 and 20 SCFM) and variable light intensity. The photoreactor system, designed to perform optimally at 10 SCFM with natural sunlight, was also operated at 20 SCFM to challenge the system. Because of varying solar illumination levels during the day, the system can operate for longer periods at lower flow rates. Residence times in the reactor at 10 and 20 SCFM were 0.50 sec and 0.25 sec, respectively.

Figure 22.4.3 shows TCE conversion and UV intensity versus illumination time at 10 SCFM for two days of operation. UV intensities were measured with a UV radiometer (UHP, San Gabriel, CA, Model UHX). TCE inlet concentrations varied from 10-15 ppmv. Detection limits for the gas chromatograph were measured at 0.5 ppmv. Figure 22.4.3 indicates that conversions greater than 95% at 10 SCFM are achieved with UV intensities at or greater than 1.5 mW/cm$^2$. At 10 SCFM, continuous destruction greater than 95% was achieved for at least 6 hours per day on two clear days in April at this latitude (37°N). The length of operation for a solar-driven system will vary with time of year, latitude, and altitude due to the effect each parameter has on available solar-UV intensity.

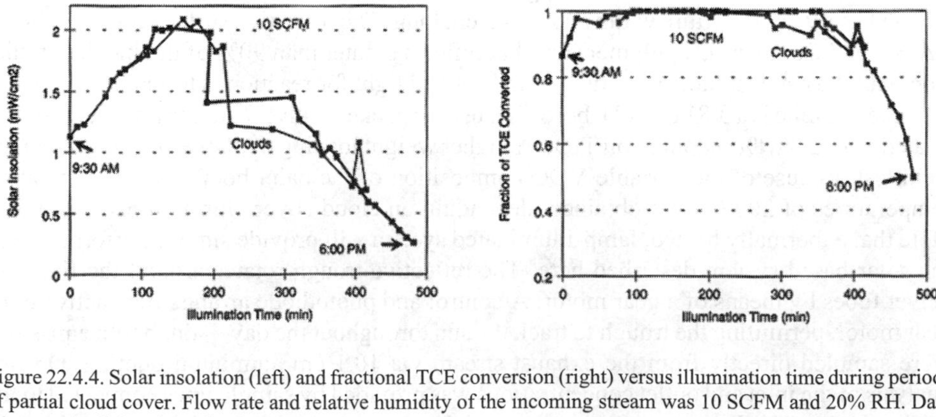


Figure 22.4.4. Solar insolation (left) and fractional TCE conversion (right) versus illumination time during periods of partial cloud cover. Flow rate and relative humidity of the incoming stream was 10 SCFM and 20% RH. Data taken on April 7 and April 16, 1997.

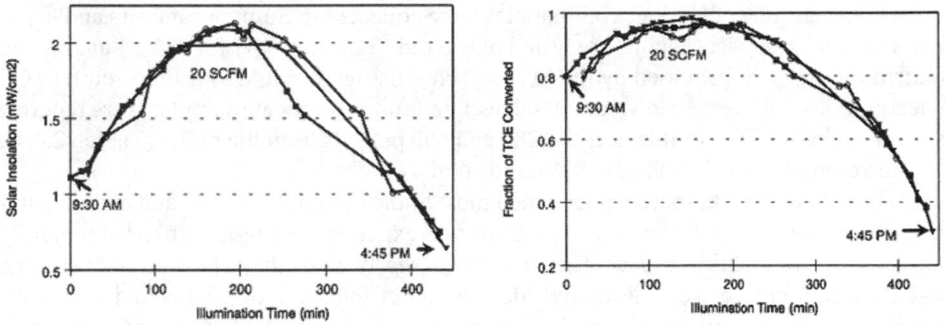


Figure 22.4.5. Solar insolation (left) and fractional TCE conversion (right) versus illumination time. Flow rate and relative humidity of the incoming air stream was 20 SCFM at 20% RH. Data taken on April 10, 14, and 15, 1997.

Figure 22.4.4 shows the effect of clouds on solar insolation (left) and TCE conversion (right) as a function of illumination time at 10 SCFM for two separate days. Again TCE conversions greater than 95% were obtained when the insolation levels were greater than 1.5 $mW/cm^2$. In general, clear skies had UV intensities of 2 $mW/cm^2$ or better. Haze and high clouds reduced the intensity to 1.5 $mW/cm^2$, and thicker clouds reduced available intensity to about 1 $mW/cm^2$. Conversions greater than 90% were still achievable during periods of significant cloud cover due to the photoreactor's ability to use global horizontal (diffuse) UV illumination.

TCE conversion versus solar insolation for three days of operation at 20 SCFM is shown in Figure 22.4.5. Doubling the flow rate halves the residence time of the gas in the photoreactor. Conversions greater than 95% are achieved with UV intensities greater than 2.0 $mW/cm^2$. The reproducibility of TCE conversion indicates that the little or no catalyst deactivation occurred during the three weeks of testing.

Pooling all of the data taken at the two flow rates provides a relation between TCE conversion and UV intensity (Figure 22.4.6). Significant scatter, which exists for both flow rates, is likely due to the rapidly varying UV intensity levels during cloud events. The variable illumination makes it difficult to establish a precise correlation between an isolated UV measurement and the UV exposure received by the gas as it flows through and reacts with the photocatalyst bed. The 10 SCFM data suggests a square root relationship between UV intensity and conversion. The square root dependence was observed in a continuous flow reactor when TCE concentrations were less than 60 ppmv (Nimlos et. al., 1993). Higher concentrations yielded an approximately linear dependence.

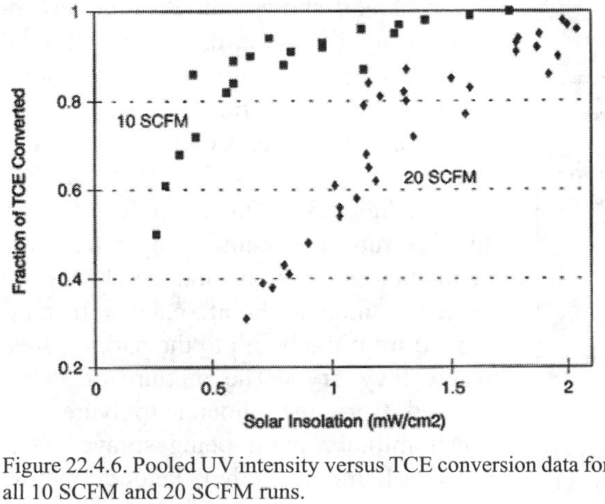


Figure 22.4.6. Pooled UV intensity versus TCE conversion data for all 10 SCFM and 20 SCFM runs.

On the last day of testing, grab samples were collected in Summa canisters and delivered to a local analytical laboratory (Air Toxics, Ltd., Folsom, CA) for TO-14 analysis to confirm the analyses provided by the MTI gas chromatographs and to provide outlet TCE concentrations that were below our GCs detection limits. Replicate samples were reported as 13000 ppbv for TCE in inlet air, and 280 and 660 ppbv TCE in the outlet (Table 22.4.1). This corresponds to 97.8% and 94.9% removal of TCE.

The reactor outlet airstream contained most of the dichloromethane and carbon tetrachloride present in the inlet stream. This result was expected, as single carbon haloorganics do not easily photo-oxidize (Jacoby et. al, 1994). Nearly all of the 1,1-dichloroethene and cis-1,2-dichloroethane were destroyed (destruction efficiencies of 97.3% and 99.1% respectively). This too was expected, as dichlorinated ethanes and ethenes photooxidize almost as rapidly as TCE.

Small amounts (<20 ppbv) of chloromethane, 1,1-dichloroethane, 1,2-dichloroethane and acetone, and 1,4-dioxane appeared in the reactor outlet and are likely reaction intermediates from TCE oxidation.

About 40% of the benzene in the inlet stream did not come out of the reactor. This result is in line with recent results in which benzene was oxidized with a sol-gel $TiO_2$ catalyst (Fu, X. et. al., 1995). Though not all of the benzene reacted, that which did formed only $CO_2$ and $H_2O$. Adding platinum to the $TiO_2$ sol significantly improved benzene oxidation. Transient reaction studies of benzene and Degussa $TiO_2$ thin films indicate that benzene oxidizes rapidly at 300K to form strongly bound surface intermediates that oxidize very slowly (Larson and Falconer, 1997). Other work with thin films of $TiO_2$ and benzene shows that benzene does not form intermediates when water vapor is present. In the absence of water, acetone forms and the catalyst becomes coated with a durable brown material (Sitkiewitz, S. and Heller, A., 1996). Because these tests were run at an average of 20% RH and the catalyst exhibited stable activity, it is likely that benzene was oxidized. Thus, since the catalyst activity did not deteriorate over the course of testing, it may be possible that low concentrations of aromatic species can be tolerated and treated in a solar photocatalytic reactor. It is also possible that the benzene was absorbed by the catalyst support material.

### 22.4.3.2 Paint booth application

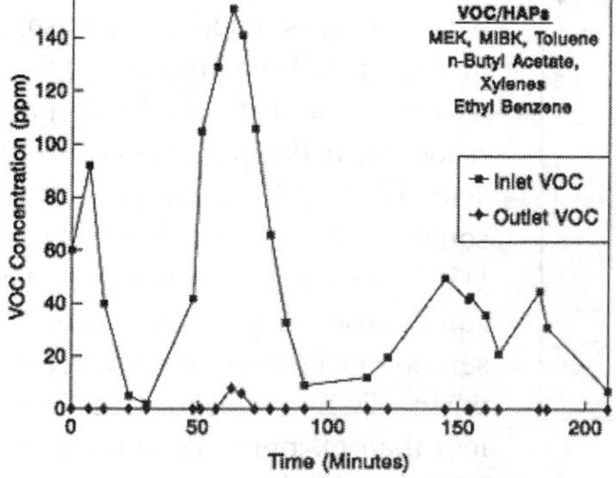


Figure 22.4.7. Inlet and outlet VOC concentrations from the Fort Carson paint booth using the small, lamp-driven, field test reactor.

Figure 22.4.7 shows the real time test results using the lamp-driven, small field scale reactor for treating VOC emissions from painting operations at Fort Carson. The inlet and outlet VOC concentrations are given in ppm. Test data begins at 8:10 AM on June 25, 1996 and ended at 11:55 for this run. The painting operations are conducted in a batch mode. Vehicles are placed in the paint booth, painted, then removed from the booth to the parking area, where they dry. The mixture of VOCs emitted from the aliphatic polyurethane green military paint being sprayed consisted of methyl ethyl ketone, methyl

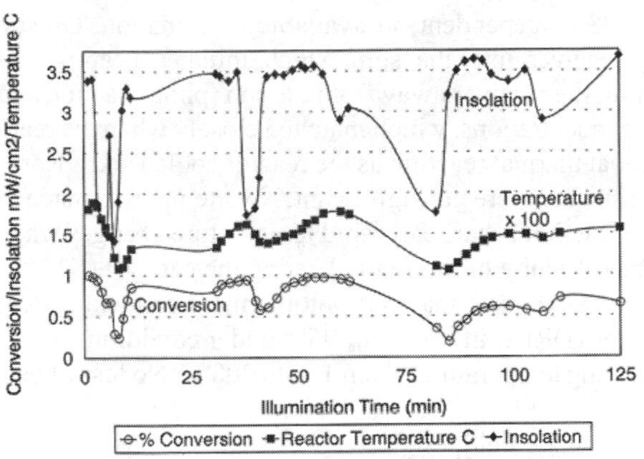


Figure 22.4.8. The effect of sunlight fluctuations on the photothermal destruction of methyl ethyl ketone in the parabolic trough, photothermal reactor. MEK conversion, insolation level, and reactor temperature are plotted as a function of continuous outdoor illumination time.

isobutyl ketone, toluene, n-butyl acetate, xylenes, and ethyl benzene.

When the reactor reached an operating temperature of 170°C, complete VOC destruction was achieved for the entire run except for a period of peak VOC emissions beginning at 9:30 and ending at 9:45. Thus, during this four hour period, the PCO process destroyed virtually all of the VOCs in the slip stream from the paint booth exhaust. The total time averaged conversion is estimated to be greater than 95%. These data were used to design the solar-driven, photothermal reactor system.

Based on the lamp tests, the 5-SCFM thermal photoreactor system was tested with methyl ethyl ketone (MEK), a paint solvent surrogate. Figure 22.4.8 shows the results of on-sun testing with this system. The Y-axis plots three factors: temperature in °C x100 MEK, conversion as a fraction with 1 equaling 100% conversion, and solar insolation as mW/cm$^2$. The X-axis plots illumination time in minutes. This test was performed on a hazy day with numerous clouds to evaluate the system under challenging conditions. Though conversion and reactor temperature parallel the insolation, they do so in a less than 1:1 ratio. This is likely due to the square root dependence of reaction rate on light intensity and to the heat transfer (insulative) characteristics of the ceramic support. 100% conversion was achieved when insolation levels were above 3.5 mW/cm$^2$ and reactor temperatures above 175°C.

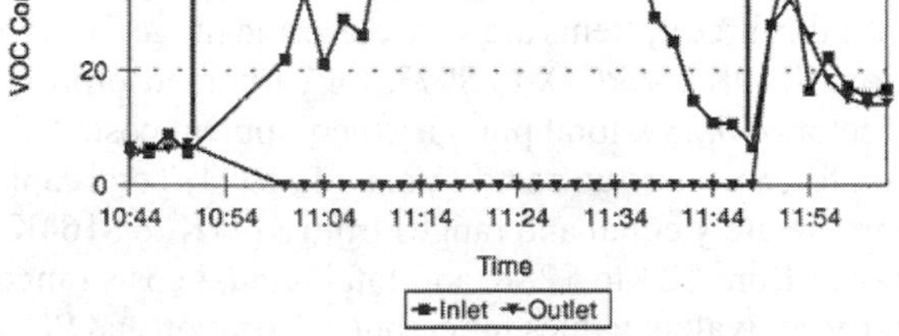


Figure 22.4.9. VOC concentration from paint solvents as a function of illumination time during photocatalytic treatment with a concentrating, parabolic trough, solar photoreactor.

The photothermal reactor was tested in July, 1997 for three weeks on paint booth emissions using natural sunlight. A representative data set for July 22, 1997 is shown in Figure 22.4.9 in which the total VOC concentration for the inlet and the outlet of the parabolic trough, photothermal reactor is plotted against illumination time. The solvent stream, which varied with time, generally contained about 140 ppmv of toluene, methyl ethyl ketone, methyl isobutyl ketone, xylenes, butyl acetate and ethyl benzene. The operating temperature was from 190 to 208°C, dependent on available illumination. On-sun refers to turning the concentrating receiver into the sun, which initiates heating and photoreaction. Off-sun refers to turning the receiver away from the sun (photoreaction and heating stops). Inlet and outlet VOC concentrations, which matched closely when the reactor was off-sun, were taken to ensure that thermal reaction as the reactor cooled and adsorption of contaminants onto the catalyst surface were not significant. No attempt was made to decouple photoreaction from thermal reaction because of rapid temperature changes when the trough was rotated off-sun and the difficulty of thermally heating the catalyst.

The results of 3 weeks of testing demonstrated that the photothermal treatment system has destruction and removal efficiencies (DREs) in excess of 95% under conditions where incident UV irradiation provided operating temperatures from 150 to 200°C. No loss of performance of the catalyst occurred.

Significant results from these pilot-scale field tests:

- A non-tracking, ambient-temperature, solar-powered photocatalytic reactor consistently achieved greater than 95% destruction of 10-15 ppmv TCE for 4 to 6 hours of operation per day during April at a latitude of 37°N. Catalyst performance remained constant during three weeks of testing.
- 95% destruction was achieved at 10 SCFM with UV intensities greater than 1.5 $mW/cm^2$ and at 20 SCFM at UV intensities greater than 2.0 $mW/cm^2$.
- Greater than 95% destruction of a variable paint solvent stream (140 ppm total VOC) was achieved with a concentrating solar photoreactor operated with ambient sunlight at temperatures of 180-200°C. These operating temperatures were achievable for 4-8 hours of operation per day in July at a latitude of 39.5°N. Catalyst performance remained constant during three weeks of testing.

## 22.4.4 COMPARISON WITH OTHER TREATMENT SYSTEMS

The design and fabrication of the photocatalytic reactor system used in the McLellan field test was provided by NREL engineers and the Industrial Solar Technology Corporation (IST), Golden, CO. IST estimates that the cost for a 200-SCFM system is about $14,000.00, depending on the site requirements for hours of operation per day and the site solar potential. Costs to construct the reactors could be significantly reduced through economies of scale and large scale production likely could reduce reactor costs by half.[21] Other purchased equipment costs for the photoreactor include a condenser/chiller and an instrumented blower/scrubber skid system for air handling. The level of instrumentation on the skid directly affects cost. The system described here was instrumented more than would be required for a commercial system. Equipment and installation costs for comparably sized carbon adsorption and thermal oxidation systems were calculated using EPAs HAPPRO v.2.1 software. Thus, total direct costs for the three systems are very close and range from an estimate of $131K for thermal oxidation, $119K for PCO, to $97K for carbon adsorption. Indirect costs were estimated using a factor of 0.28 x total purchased equipment cost. This value includes engineering, construction fee, and start-up costs. (Vatavuk, 1994) Total capital required for all three systems is approximately equal and ranges from $121K to $164K. Operating costs are also similar and range from $23 to $28K and total annual costs range from $37K for thermal oxidation to nearly equivalent values for carbon adsorption and PCO at $32K and $33K, respectively. No fuel is required for PCO and carbon adsorption. Details of the calculations are provided in the footnotes following Table 22.4.3.

**Table 22.4.3. Comparison of estimated costs for similarly sized PCO, carbon adsorption, and thermal oxidation VOC treatment systems**

| Direct Costs | Solar PCO non-concentrating | Carbon Adsorption | Thermal Oxidation |
|---|---|---|---|
| Purchased Equipment (PE) | | | |
| Photoreactor[1] | | | |
| Condenser/Chiller[2] | 14,000 | | |
| Instrumented | 10,625 | | |
| Blower/Scrubber Skid[3] | 60,000 | | |
| Total PE (TPE)[4] | 91,395 | 74,500[5] | 100,500[5] |
| Installation | 27,419 | 22,350 | 30,150 |
| Total Direct Costs | 118,814 | 96,850 | 130,650 |
| Indirect Costs<br>Indirect Installation Costs[7] | 29,923 | 24,391 | 32,904 |
| Total Capital Required[8] | 148,736 | 121,241 | 163,554 |
| Operating Costs[9] | | | |
| Electricity | 2,880 | 1,440 | 1,440 |
| Fuel[10] | 0 | 0 | 5,000 |
| Maintenance, Labor and Material | 20,376 | 19,308 | 20,952 |
| Direct Annual Costs | 23,256 | 25,748 | 27,392 |
| Indirect Annual Costs | 8,664 | 7,063 | 9,527 |
| Total Annual Costs | 31,920 | 32,811 | 36,919 |
| Levelized Cost[12] | 56,135 | 25,549 | 63,456 |
| Levelized Cost/CFM | 281 | 263 | 318 |

[1]IST Corp., Golden, CO. Non-concentrating reactor cost is current and estimated price reductions anticipated from economies of scale at higher production levels.
[2]Peters and Timmerhaus, 1991.
[3]Zentox Corp., Ocala, FL. Concentrating vs. non-concentrating is primarily a function of instrumentation level.
[4]Includes 0.08 x PE for taxes and freight (Vatavuk, 1994).
[5]Calculated using EPAs HAPPRO v.2.1 software.
[6]0.38 x PE (Vatavuk, 1994).
[7]0.28 x TPE, including engineering costs, construction fee, and start-up costs (Vatavuk, 1994)
[8]Includes 6% contingency.
[9]Based on $0.08/kWh, 10 kW blower and chiller costs.
[10]Based on $0.33/ccf natural gas.
[11]0.06 x Total Capital Required for taxes, insurance, and administration.
[12]Based on 10% interest rate, 10 year depreciation.

## REFERENCES


Ayer, J. and Darvin, C. H., 1995, "Cost Effective VOC Emission Control Strategies for Military, Aerospace, and Industrial Paint Spray Booth Operations: Combining Improved Ventilation Systems with Innovative, Low-Cost Emission Control Technologies", Proceedings of the 88th Annual Meeting of the Air and Waste Management Association, Vol. 4B, 95-WA77A.02.

Blake, D.M., 1996, Bibliography of Work on the Heterogenous Photocatalytic Removal of Hazardous Compounds from Water and Air, Update Number 2 to October 1996. NREL/TP-430-22197. Golden, CO: National Renewable Energy Laboratory.

Bekbolet, M. and Balcioglu, I., 1996, *Water Science and Technology*, Vol. **34**, No. 9, pp. 73-80.

Cummings, M. and Booth, S. R., Proc. Int. Conf. Incineration Therm.Treat. Technol. (1996), pp. 701-713.
d'Hennezel, O. and Ollis D. F., 1997, *Journal of Catalysis*, Vol. **167**, No. 1, pp. 118-126.
Evstropov, A. A., Belyankin, I. A., Krainov, N. V., Mikheeva, T. Ya., Kvasov, A. A., 1989, "Catalytic Combustion of Gas emissions from A Spray-Painting Chamber", *Lakokras. Mater. Ikh. Primen*, Issue 1, pp. 115-118.
Falconer, J. L. F. and Magrini-Bair, K. A., 1998, "Photocatalytic and Thermal Catalytic Oxidation of Acetaldehyde on Pt/$TiO_2$", accepted for publication in the Journal of Catalysis.
Fan, J. and Yates, J. T., *Journal of the American Chemical Society*, Vol. **118**, pp. 4686-4692.
Fu, X., Clark, L. A., Zeltner, W. A., Anderson, M. A., 1996, "Effects of Reaction Temperature and Water Vapor Content on the Heterogeneous Photocatalytic Oxidation of Ethylene", *Journal of Photochemistry and Photobiology A*, Vol. **97**, No. 3, pp. 181-186.
Fu, X. , Zeltner, W. A., and Anderson, M. A., 1995, *Applied Catalysis B: Environmental*, Vol. **6**, pp. 209-224.
Gratson, D.A.; Nimlos, M.R.; Wolfrum, E.J., 1995, Proceedings of the Air and Waste Management Association.
Hung, C.-H. and Marinas, B. J. , 1997, *Environmental Science and Technology*, Vol. **31**, pp. 1440-1445.
Jacoby, W.A.; Nimlos, M.R.; Blake, D.M.; Noble, R.D.; Koval, C.A., 1994, *Environmental Science and Technology*, Vol. **28**, No. (9), pp. 1661-1668.
Kittrell, J. R., Quinlan, C. W., Shepanzyk, J. W., "Photocatalytic Destruction of Hexane Eliminates Emissions in Contact Lens Manufacture", Emerging Solutions in VOC Air Toxics Control, Proc. Spec. Conf. (1996), pp. 389-398.
Larson, S. A. and Falconer J. L., 1997, *Catalysis Letters*, Vol. **44**, No. 9, pp. 57-65.
Luo, Y. and Ollis, D. F., 1996, *Journal of Catalysis*, Vol. **163**, No. 1, pp. 1-11.
Magrini, K. A., Watt, A. S., Boyd, L. C., Wolfrum, E. J., Larson, S. A., Roth, C., and Glatzmaier, G. C., "Application of Solar Photocatalytic Oxidation to VOC-Containing Airstreams", ASME Proceedings of the Renewable and Advanced Energy Systems for the 21st Century, pp. 81-90, Lahaina, Maui, April 1999.
Magrini, K. A., Rabago, R., Larson, S. A., and Turchi, C. S., "Control of Gaseous Solvent Emissions using Photocatalytic Oxidation", Proc. Annu. Meet. Air Waste Manage. Assoc., 89th, (1996) pp. 1-8.
Mueller, J. H., (1988), "Spray Paint Booth VOC Control with a Regenerative Thermal Oxidizer", Proceedings-APCA Annual Meeting 81st(5), Paper 88/84.3, 16 pp.
Nimlos, M. R., Jacoby, W. A., Blake, D. M. and Milne,T. A., 1993, *Environmental Science and Technology*, vol. **27**, p. 732.
Peral, J. and Ollis, D.F., 1992, *Journal of Catalysis*, Vol. **136**, pp. 554-565.
Phillips, L.A.; Raupp, G.B., 1992, *Journal of Molecular Catalysis*, Vol. **7**, pp. 297-311.
Read, H. W., Fu, X., Clark, L. A., Anderson, M. A., Jarosch, T., 1996, "Field Trials of a $TiO_2$ Pellet-Based Photocatalytic Reactor for Off-Gas Treatment at a Soil Vapor Extraction Well", *Journal of Soil Contamination*, Vol. **5**, No. 2, pp. 187-202.
Rosansky, S. H., Gavaskar, A. R., Kim, B. C., Drescher, E., Cummings, C. A., and Ong, S. K., "Innovative Air Stripping and Catalytic Oxidation Techniques for Treatment of Chlorinated Solvent Plumes", Therm. Technol., Int. Conf., Rem. Chlorinated Recalcitrant Compounds., 1st (1998) pp. 415-424.
Sitkiewitz, S. and Heller, A., 1996, *New Journal of Chemistry*, Vol. **20**, No. 2, pp. 233-241.
Turchi, C.S., Wolfrum, E., Miller, R.A., 1994, Gas-Phase Photocatalytic Oxidation: Cost Comparison with Other Air Pollution Control Technologies, NREL/TP-471-7014. Golden, CO: National Renewable Energy Laboratory.
Watt, A. S., Magrini, K. A., Carlson-Boyd, L. C., Wolfrum, E. J., Larson, S. A., Roth, C., and Glatzmaier, G. C., "A Pilot-Scale Demonstration of an Innovative Treatment for Vapor Emissions", *J. of Air and Waste Management*, November, 1999.
Yamazaki-Nishida, S., Fu, X., Anderson, M. A. and Hori, K., 1996, *Journal of Photochemistry and Photobiology. A: Chemistry*, Vol. **97**, pp. 175-179.

# Contamination Cleanup: Natural Attenuation and Advanced Remediation Technologies

## 23.1 NATURAL ATTENUATION OF CHLORINATED SOLVENTS IN GROUND WATER

Hanadi S. Rifai
**Civil and Environmental Engineering, University of Houston, Houston, Texas, USA**
Charles J. Newell
**Groundwater Services, Inc., Houston, Texas, USA**
Todd H. Wiedemeier
**Parson Engineering Science, Denver, CO, USA**

### 23.1.1 INTRODUCTION

Chlorinated solvents were first produced some 100 years ago and came into common usage in the 1940's. Chlorinated solvents are excellent degreasing agents and they are nearly non-flammable and non-corrosive. These properties have resulted in their widespread use in many industrial processes such as cleaning and degreasing rockets, electronics and clothing (used as dry-cleaning agents). Chlorinated solvent compounds and their natural degradation or progeny products have become some of the most prevalent organic contaminants found in the shallow groundwater of the United States. The most commonly used chlorinated solvents are perchloroethene (PCE), trichloroethene (TCE), 1,1,1-trichloroethane (TCA), and carbon tetrachloride (CT).[1]

Chlorinated solvents (CS) undergo the same natural attenuation processes as many other ground water contaminants such as advection, dispersion, sorption, volatilization and biodegradation. In addition, CS are subject to abiotic reactions such as hydrolysis and dehydrohalogenation and abiotic reduction reactions. While many of the physical and chemical reactions affecting chlorinated solvents have been extensively studied, their biodegradation is not as well understood as perhaps it is for petroleum hydrocarbons. Researchers are just beginning to understand the microbial degradation of chlorinated solvents with many degradation pathways remaining to be discovered. Unlike petroleum hydrocarbons, which can be oxidized by microorganisms under either aerobic or anaerobic conditions, most chlorinated solvents are degraded only under specific ranges of oxidation-reduction potential. For example, it is currently believed that PCE is biologically degraded through use as a primary growth substrate only under strongly reducing anaerobic conditions.

This chapter is focused on the natural attenuation behavior of CS at the field scale. The first part of the chapter reviews many of the physical, chemical and abiotic natural attenuation processes that attenuate CS concentrations in ground water. Some of these processes have been described in more detail in previous chapters in the handbook and are therefore only reviewed in brief. In the second part of this chapter, we will review the biological processes that bring about the degradation of the most common chlorinated solvents, present conceptual models of chlorinated solvent plumes, and summarize data from field studies with chlorinated solvent contamination.

## 23.1.2 NATURAL ATTENUATION PROCESSES AFFECTING CHLORINATED SOLVENT PLUMES

Many abiotic mechanisms affect the fate and transport of organic compounds dissolved in ground water. Physical processes include advection and dispersion while chemical processes include sorption, volatilization and hydrolysis. Advection transports chemicals along ground water flow paths and in general does not cause a reduction in contaminant mass or concentration. Dispersion or mixing effects, on the other hand, will reduce contaminant concentrations but will not cause a reduction in the total mass of chemicals in the aquifer. Sorption or partitioning between the aquifer matrix and the ground water, much like dispersion, will not cause a reduction in contaminant mass. Volatilization and hydrolysis both will result in lower concentrations of the contaminant in ground water. The majority of these processes, with the exception of hydrolysis and dehydrohalogenation chemical reactions, do not break down or destroy the contaminants in the subsurface.

Chlorinated solvents are advected, dispersed, and sorbed in ground water systems. They also volatilize although their different components have varying degrees of volatility. Chlorinated solvents additionally hydrolyze and undergo other chemical reactions such as dehydrohalogenation or elimination and oxidation and reduction. These abiotic reactions, as will be seen later in the chapter, are typically not complete and often result in the formation of an intermediate that may be at least as toxic as the original contaminant.

### 23.1.2.1 Advection

Advective transport is the transport of solutes by the bulk movement of ground water. Advection is the most important process driving dissolved contaminant migration in the subsurface and is given by:

$$v_x = -\frac{K}{n_e}\frac{dH}{dL} \qquad [23.1.1]$$

where:

| | |
|---|---|
| $V_x$ | seepage velocity [L/T] |
| K | hydraulic conductivity [L/T] |
| $n_e$ | effective porosity [$L^3/L^3$] |
| dH/dL | hydraulic gradient [L/L] |

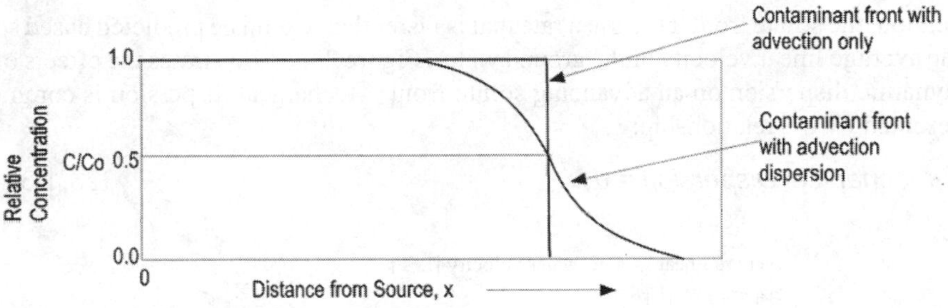


Figure 23.1.1. Breakthrough curve in one dimension showing plug flow with continuous source resulting from advection only and from the combined processes of advection and hydrodynamic dispersion. [From T.H. Wiedemeier, H. S. Rifai, C. J. Newell and J.T. Wilson, **Natural Attenuation of Fuels and Chlorinated Solvents in the Subsurface**. Copyright © 1999 *John Wiley & Sons*. Reprinted by permission of John Wiley & Sons.]

Typical velocities range between $10^{-7}$ and $10^{3}$ ft/day[2] with a median national average of 0.24 ft/day. The seepage velocity is a key parameter in natural attenuation studies since it can be used to estimate the time of travel of the contaminant front:

$$t = \frac{x}{v_x} \quad [23.1.2]$$

where:

x travel distance (ft or m)
t time

Solute transport by advection alone yields a sharp solute concentration front as shown in Figure 23.1.1. In reality, the advancing front spreads out due to the processes of dispersion and diffusion as shown in Figure 23.1.1, and is retarded by sorption (Figure 23.1.2) and biodegradation.

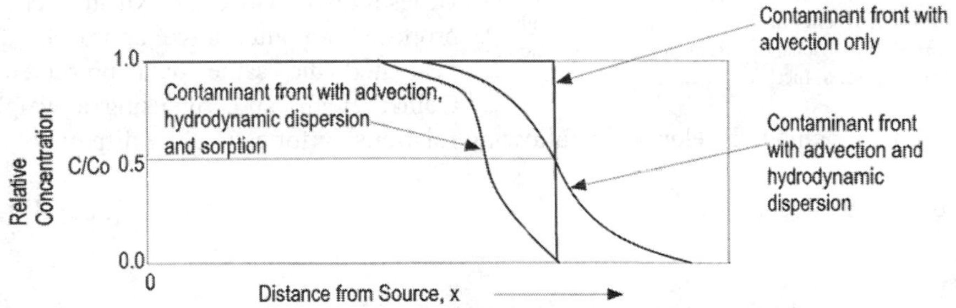


Figure 23.1.2. Breakthrough curve in one dimension showing plug flow with continuous source resulting from advection only; from the combined processes of advection and hydrodynamic dispersion; and from the combined processes of advection, hydrodynamic dispersion, and sorption. [From T.H. Wiedemeier, H. S. Rifai, C. J. Newell and J.T. Wilson, **Natural Attenuation of Fuels and Chlorinated Solvents in the Subsurface**. Copyright © 1999 *John Wiley & Sons, Inc*. Reprinted by permission of John Wiley & Sons, Inc.]

### 23.1.2.2 Dispersion

Hydrodynamic dispersion causes a contaminant plume to spread out from the main direction of ground water flow. Dispersion dilutes the concentrations of the contaminant, and introduces the contaminant into relatively pristine portions of the aquifer where it mixes with more electron acceptors crossgradient to the direction of ground-water flow. As a result of dispersion, the solute front travels at a rate that is faster than would be predicted based solely on the average linear velocity of the ground water. Figure 23.1.1 illustrates the effects of hydrodynamic dispersion on an advancing solute front. Mechanical dispersion is commonly represented by the relationship:

$$\textit{Mechanical Dispersion} = \alpha_x v_x \qquad [23.1.3]$$

where:

- $v_x$ average linear ground-water velocity [L/T]
- $\alpha_x$ dispersivity [L]

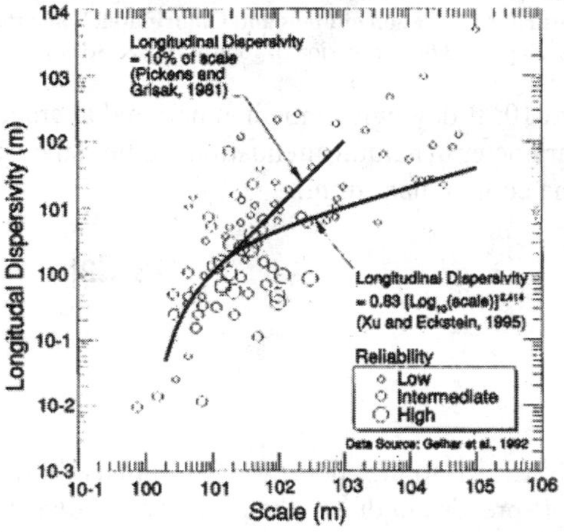


Figure 23.1.3. Relationship between dispersivity and scale. [From T.H. Wiedemeier, H. S. Rifai, C. J. Newell and J.T. Wilson, **Natural Attenuation of Fuels and Chlorinated Solvents in the Subsurface**. Copyright © 1999 *John Wiley & Sons, Inc.* Reprinted by permission of John Wiley & Sons, Inc.]

Dispersivity represents the spreading of a contaminant over a given length of flow and is characteristic of the porous medium through which the contaminant migrates. It is commonly accepted that dispersivity is scale-dependent, and that at a given scale, dispersivity may vary over three orders of magnitude[3,4] as shown in Figure 23.1.3.

Several approaches can be used to estimate longitudinal dispersivity, $\alpha_x$, at the field scale. One technique involves conducting a tracer test but this method is time consuming and costly. Another method commonly used in solute transport modeling is to start with a longitudinal dispersivity of 0.1 times the plume lengths.[5-7] This assumes that dispersivity varies linearly with scale. Xu and Eckstein[8] proposed an alternative approach. They evaluated the same data presented by Gelhar et al.[4] and, by using a weighted least-squares method, developed the following relationship for estimating dispersivity:

$$\alpha_x = 0.83\left(\log_{10} L_p\right)^{2.414} \qquad [23.1.4]$$

where:

- $\alpha_x$ longitudinal dispersivity [L]
- $L_p$ plume length [L]

Both relationships are shown on Figure 23.1.3.

In addition to estimating longitudinal dispersivity, it may be necessary to estimate the transverse and vertical dispersivities ($\alpha_T$ and $\alpha_Z$, respectively) for a given site. Commonly, $\alpha_T$ is estimated as $0.1\alpha_x$ (based on data[4]), or as $0.33\alpha_x$.[9,10] Vertical dispersivity ($\alpha_Z$) may be estimated as $0.05\alpha_x$,[9] or as 0.025 to $0.1\alpha_x$.[10]

### 23.1.2.3 Sorption

Many organic contaminants, including chlorinated solvents, are removed from solution by sorption onto the aquifer matrix. Sorption of dissolved contamination onto the aquifer matrix results in slowing (retardation) of the contaminant relative to the average advective ground-water flow velocity and a reduction in dissolved organic concentrations in ground water. Sorption can also influence the relative importance of volatilization and biodegradation.[11] Figure 23.1.2 illustrates the effects of sorption on an advancing solute front. Sorption is a reversible reaction; at given solute concentrations, some portion of the solute is partitioning to the aquifer matrix and some portion is also desorbing, and reentering solution.

Sorption can be determined from bench-scale experiments. These are typically performed by mixing water-contaminant solutions of various concentrations with aquifer materials containing various amounts of organic carbon and clay minerals. The solutions are then sealed with no headspace and left until equilibrium between the various phases is reached. The amount of contaminant left in solution is then measured.

The results are commonly expressed in the form of a sorption isotherm or a plot of the concentration of chemical sorbed (μg/g) versus the concentration remaining in solution (μg/L). Sorption isotherms generally exhibit one of three characteristic shapes depending on the sorption mechanism. These isotherms are referred to as the Langmuir isotherm, the Freundlich isotherm, and the linear isotherm (a special case of the Freundlich isotherm). The reader is referred to ref. 1 for more details on sorption isotherms.

Since sorption tends to slow the transport velocity of contaminants dissolved in ground water, the contaminant is said to be "retarded." The coefficient of retardation, R, is defined as:

$$R = \frac{v_x}{v_c} \quad [23.1.5]$$

where:

| | |
|---|---|
| R | coefficient of retardation |
| $v_x$ | average linear ground-water velocity parallel to ground-water flow |
| $v_c$ | average velocity of contaminant parallel to ground-water flow |

The ratio $v_x/v_c$ describes the relative velocity between the ground water and the dissolved contaminant. The coefficient of retardation for a dissolved contaminant (for saturated flow) assuming linear sorption is determined from the distribution coefficient using the relationship:

$$R = 1 + \frac{\rho_b K_d}{N} \quad [23.1.6]$$

where:

| | |
|---|---|
| R | coefficient of retardation [dimensionless] |
| $\rho_b$ | bulk density of aquifer [$M/L^3$] |
| $K_d$ | distribution coefficient [$L^3/M$] |
| N | porosity [$L^3/L^3$] |

The bulk density, $\rho_b$, of a soil is the ratio of the soil mass to its field volume. In sandy soils, $\rho_b$ can be as high as 1.81 g/cm$^3$, whereas, in aggregated loams and clayey soils, $\rho_b$ can be as low as 1.1 g/cm$^3$.

The distribution coefficient is a measure of the sorption/desorption potential and characterizes the tendency of an organic compound to be sorbed to the aquifer matrix. The higher the distribution coefficient, the greater the potential for sorption to the aquifer matrix. The distribution coefficient, $K_d$, is given by:

$$K_d = \frac{C_a}{C_l} \qquad [23.1.7]$$

where:

| | |
|---|---|
| $K_d$ | distribution coefficient (slope of the sorption isotherm, mL/g). |
| $C_a$ | a sorbed concentration (mass contaminant/mass soil or μg/g) |
| $C_l$ | dissolved concentration (mass contaminant/volume solution or μg/mL) |

Several researchers have found that if the distribution coefficient is normalized relative to the aquifer matrix total organic carbon content, much of the variation in observed $K_d$ values between different soils is eliminated.[12-19] Distribution coefficients normalized to total organic carbon content are expressed as $K_{oc}$. The following equation gives the expression relating $K_d$ to $K_{oc}$:

$$K_{oc} = \frac{K_d}{f_{oc}} \qquad [23.1.8]$$

where:

| | |
|---|---|
| $K_{oc}$ | soil sorption coefficient normalized for total organic carbon content |
| $K_d$ | distribution coefficient |
| $f_{oc}$ | fraction total organic carbon (mg organic carbon/mg soil) |

Table 23.1.1 presents calculated retardation factors for several LNAPL and DNAPL related chemicals as a function of the fraction of organic carbon content of the soil. It can be seen from Table 23.1.1 that R can vary over two orders of magnitude at a site depending on the chemical in question and the estimated value of porosity and soil bulk density.

**Table 23.1.1. Calculated retardation factors for several chlorinated solvent-related chemicals**

| Compound | log($K_{oc}$) | Fraction of organic compound in aquifer ($f_{oc}$) | | | |
|---|---|---|---|---|---|
| | | 0.0001 | 0.001 | 0.01 | 0.1 |
| Carbon tetrachloride | 2.67 | 1.2 | 3.3 | 24.0 | 231.2 |
| 1,1,1-TCA | 2.45 | 1.1 | 2.4 | 14.9 | 139.7 |
| PCE | 2.42 | 1.1 | 2.3 | 13.9 | 130.4 |
| 1,1- or 1,2-DCA | 1.76 | 1.0 | 1.3 | 3.8 | 29.3 |
| trans 1,2-DCE | 1.42 | 1.0 | 1.1 | 2.3 | 13.9 |
| cis 1,2-DCE | 1.38 | 1.0 | 1.1 | 2.2 | 12.8 |
| TCE | 1.26 | 1.0 | 1.1 | 1.9 | 10.0 |
| Chloroethane | 1.25 | 1.0 | 1.1 | 1.9 | 9.8 |
| Dichloromethane | 1.23 | 1.0 | 1.1 | 1.8 | 9.4 |
| Vinyl chloride | 0.06 | 1.0 | 1.0 | 1.1 | 1.6 |

Notes: Units of $f_{oc}$: g naturally-occurring organic carbon per g dry soil. Assumed porosity and bulk density: 0.35 and 1.72, respectively. [From T.H. Wiedemeier, H. S. Rifai, C. J. Newell and J.T. Wilson, **Natural Attenuation of Fuels and Chlorinated Solvents in the Subsurface**. Copyright © 1999 *John Wiley & Sons, Inc.* Reprinted by permission of John Wiley & Sons, Inc.]

### 23.1.2.4 One-dimensional advection-dispersion equation with retardation

In one dimension, the advection-dispersion equation is given by:

$$R\frac{\partial C}{\partial t} = D_x\frac{\partial^2 C}{\partial x^2} - v_x\frac{\partial C}{\partial x} \quad [23.1.9]$$

where:

| | |
|---|---|
| $v_x$ | average linear ground-water velocity [L/T] |
| R | coefficient of retardation [dimensionless] |
| C | contaminant concentration [$M/L^3$] |
| $D_x$ | hydrodynamic dispersion [$L^2/T$] |
| T | time [T] |
| x | distance along flow path [L] |

### 23.1.2.5 Dilution (recharge)

Ground water recharge can be defined as the entry into the saturated zone of water made available at the water-table surface.[20] Recharge may therefore include precipitation that infiltrates through the vadose zone and water entering the ground-water system due to discharge from surface water bodies (i.e., streams and lakes). Recharge of a water table aquifer has two effects on the natural attenuation of a dissolved contaminant plume. Additional water entering the system due to infiltration of precipitation or from surface water will contribute to dilution of the plume, and the influx of relatively fresh, electron acceptor-charged water will alter geochemical processes and in some cases facilitate additional biodegradation.

Wiedemeier et al.[1] present the following relationship for estimating the amount of dilution caused by recharge:

$$C_L = C_o \exp-\left[\frac{RW\left(\frac{L}{V_D}\right)}{WThV_D}\right] \quad [23.1.10]$$

eliminating the width and rearranging, gives:

$$C_L = C_o \exp\left(-\frac{RL}{Th(V_D)^2}\right) \quad [23.1.11]$$

where:

| | |
|---|---|
| $C_L$ | concentration at distance L from origin assuming complete mixing of recharge with groundwater (mg/L) |
| $C_o$ | concentration at origin or at distance L = 0 (mg/L) |
| R | recharge mixing with groundwater (ft/yr) |
| W | width of area where recharge is mixing with groundwater (ft) |
| L | length of area where recharge is mixing with groundwater (ft) |
| Th | thickness of aquifer where groundwater flow is assumed to completely mix with recharge (ft) |
| $V_D$ | Darcy velocity of groundwater (ft/yr) |

### 23.1.2.6 Volatilization

Volatilization causes contaminants to transfer from the dissolved phase to the gaseous phase. In general, factors affecting the volatilization of contaminants from ground water into soil gas include the contaminant concentration, the change in contaminant concentration with depth, the Henry's Law constant and diffusion coefficient of the compound, mass transport coefficients for the contaminant in both water and soil gas, sorption, and the temperature of the water.[21]

The Henry's Law constant of a chemical determines the tendency of a contaminant to volatilize from ground water into the soil gas. Henry's Law states that the concentration of a contaminant in the gaseous phase is directly proportional to the compound's concentration in the liquid phase and is a constant characteristic of the compound:[11]

$$C_a = HC_l \qquad [23.1.12]$$

where:

H Henry's Law constant (atm m$^3$/mol)
$C_a$ concentration in air (atm)
$C_l$ concentration in water (mol/m$^3$)

Values of Henry's Law constants for selected chlorinated solvents are given in Table 23.1.2. As indicated in the table, values of H for chlorinated compounds also vary over several orders of magnitude. Chlorinated solvents have low Henry's Law constants, with the exception of vinyl chloride. Volatilization of chlorinated solvents compounds from ground water is a relatively slow process that generally can be neglected when modeling biodegradation.

**Table 23.1.2. Physical properties for chlorinated compounds**

| Constituent | CAS # | Molecular weight | | Diffusion coefficients | | | | log $K_{oc}$ (@20-25°C) | |
|---|---|---|---|---|---|---|---|---|---|
| | | | | in air | | in water | | | |
| | | $M_W$, g/mol | Ref | $D_{air}$, cm$^2$/s | Ref | $D_{wat}$, cm$^2$/s | Ref | Partition log(L/kg) | Ref |
| Bromodichloromethane | 75-27-4 | 163.8 | 22 | 2.98E-02 | 22 | 1.06E-05 | 22 | 1.85 | 22 |
| Carbon tetrachloride | 56-23-5 | 153.8 | 22 | 7.80E-02 | 22 | 8.80E-06 | 22 | 2.67 | 22 |
| Chlorobenzene | 108-90-7 | 112.6 | 22 | 7.30E-02 | 22 | 8.70E-06 | 22 | 2.46 | 22 |
| Chloroethane | 75-00-3 | 64.52 | 22 | 1.50E-01 | 22 | 1.18E-05 | 22 | 1.25 | 22 |
| Chloroform | 67-66-3 | 119.4 | 22 | 1.04E-01 | 22 | 1.00E-05 | 22 | 1.93 | 22 |
| Chloromethane | 74-87-3 | 51 | 23 | 1.28E-01 | 22 | 1.68E-04 | b | 1.40 | 28 |
| Chlorophenol, 2- | 95-57-8 | 128.6 | 22 | 5.01E-02 | 22 | 9.46E-06 | 22 | 2.11 | 22 |
| Dibromochloromethane | 124-48-1 | 208.29 | 22 | 1.99E-02 | 22 | 1.03E-05 | 22 | 2.05 | 22 |
| Dichlorobenzene, (1,2)(-o) | 95-50-1 | 147 | 22 | 6.90E-02 | 22 | 7.90E-06 | 22 | 3.32 | 22 |
| Dichlorobenzene, (1,4)(-p) | 106-46-7 | 147 | 22 | 6.90E-02 | 22 | 7.90E-06 | 22 | 3.33 | 22 |
| Dichlorodifluoromethane | 75-71-8 | 120.92 | 22 | 5.20E-02 | 22 | 1.05E-05 | 22 | 2.12 | 22 |

| Constituent | CAS # | Molecular weight | | Diffusion coefficients | | | | log $K_{oc}$ (@20-25°C) | |
|---|---|---|---|---|---|---|---|---|---|
| | | | | in air | | in water | | | |
| | | $M_W$, g/mol | Ref | $D_{air}$, $cm^2/s$ | Ref | $D_{wat}$, $cm^2/s$ | Ref | Partition log(L/kg) | Ref |
| Dichloroethane, 1,1- | 75-34-3 | 98.96 | 22 | 7.42E-02 | 22 | 1.05E-05 | 22 | 1.76 | 22 |
| Dichloroethane, 1,2- | 107-06-2 | 98.96 | 22 | 1.04E-01 | 22 | 9.90E-06 | 22 | 1.76 | 22 |
| Dichloroethene, cis-1,2- | 156-59-2 | 96.94 | 22 | 7.36E-02 | 22 | 1.13E-05 | 22 | 1.38 | c |
| Dichloroethene, 1,2-trans | 156-60-5 | 96.94 | 22 | 7.07E-02 | 22 | 1.19E-05 | 22 | 1.46 | 22 |
| Dichloromethane | 75-09-2 | 85 | 22 | 1.01E-01 | 22 | 1.17E-05 | 22 | 1.23 | 22 |
| Tetrachloroethane, 1,1,2,2- | 79-34-5 | 168 | 22 | 7.10E-02 | 22 | 7.90E-06 | 22 | 0.00 | 22 |
| Tetrachloroethene | 127-18-4 | 165.83 | 22 | 7.20E-02 | 22 | 8.20E-06 | 22 | 2.43 | 28 |
| Trichlorobenzene, 1,2,4- | 120-28-1 | 181.5 | 22 | 3.00E-02 | 22 | 8.23E-06 | 22 | 3.91 | 22 |
| Trichoroethane, 1,1,1- | 71-55-6 | 133.4 | 22 | 7.80E-02 | 22 | 8.80E-06 | 22 | 2.45 | 22 |
| Trichloroethane, 1,1,2- | 79-00-5 | 133.4 | 22 | 7.80E-02 | 22 | 8.80E-06 | 22 | 1.75 | 28 |
| Trichloroethene | 79-01-6 | 131.4 | 24 | 8.18E-02 | a | 1.05E-04 | b | 1.26 | d |
| Trichlorofluoromethane | 75-69-4 | 137.4 | 22 | 8.70E-02 | 22 | 9.70E-06 | 22 | 2.49 | 22 |
| Vinyl chloride | 75-01-4 | 62.5 | 22 | 1.06E-01 | 22 | 1.23E-05 | 22 | 0.39 | 26 |

[a]Calculated diffusivity using the method of Fuller, Schettler, and Giddings [from Reference 25]
[b]Calculated diffusivity using the method of Hayduk and Laudie and the reference 25
[c]Calculated using Kenaga and Goring $K_{ow}$/solubility regression equation from reference 25 and $K_{ow}$ data from reference 26, log (S, mg/L) = 0.922 log($K_{ow}$) + 4.184 d
[d]Back calculated from solubility [see note c, based on $K_{ow}$ from reference 26 and method from reference 27, log($K_{oc}$) = 0.00028 + 0.938 log ($K_{ow}$)]


### 23.1.2.7 Hydrolysis and dehydrohalogenation

Hydrolysis and dehydrohalogenation reactions are the most thoroughly studied abiotic attenuation mechanisms. In general, the rates of these reactions are often quite slow within the range of normal ground-water temperatures, with half-lives of days to centuries.[29,30] Hydrolysis is a substitution reaction in which a compound reacts with water, and a halogen substituent is replaced with a hydroxyl ($OH^-$) group resulting in the formation of alcohols and alkenes after:[31,32]

$$RX + HOH \rightarrow ROH + HX \qquad [23.1.13]$$

$$H_3C\text{-}CH_2X \rightarrow H_2C{=}CH_2 + HX \qquad [23.1.14]$$

The likelihood that a halogenated solvent will undergo hydrolysis depends in part on the number of halogen substituents. More halogen substituents on a compound will decrease the chance for hydrolysis reactions to occur,[29] and will therefore decrease the rate of the reaction. In addition, bromine substituents are more susceptible to hydrolysis than chlorine substituents;[29] for example, 1,2-dibromoethane is subject to significant hydrolysis reactions under natural conditions. McCarty[33] lists TCA (1,1,1-trichloroethane) as the only major chlorinated solvent that can be transformed chemically through hydrolysis (as well as elimination) leading to the formation of 1,1-DCE (l,1-dichlorethene) and acetic acid.

Locations of the halogen substituent on the carbon chain may also have some effect on the rate of reaction. The rate also may increase with increasing pH; however, a rate dependence upon pH is typically not observed below a pH of 11.[34,35] Rates of hydrolysis may also be increased by the presence of clays, which can act as catalysts.[29] Other factors that impact the level of hydrolysis include dissolved organic matter, and dissolved metal ions. Hydrolysis rates can generally be described using first-order kinetics, particularly in solutions in which water is the dominant nucleophile.[29] A listing of half-lives for abiotic hydrolysis and dehydrohalogenation of some chlorinated solvents is presented in Table 23.1.3. Note that no distinctions are made in the table as to which mechanism is operating; this is consistent with the references from which the table has been derived.[29,36]

**Table 23.1.3. Approximate half-lives of abiotic hydrolysis and dehydrohalogenation reactions involving chlorinated solvents**

| Compound | Half-Life (years) | Products |
|---|---|---|
| Chloromethane | no data | |
| Dichloromethane | 704[34] | |
| Chloroform | 3500,[34] 1800[37] | |
| Carbon tetrachloride | 41[37] | |
| Chloroethane | 0.12[29] | ethanol |
| 1,1-Dichloroethane | 61[37] | |
| 1,2-Dichloroethane | 72[37] | |
| 1,1,1-Trichloroethane | 1.7,[34] 1.1,[37] 2.5[38] | acetic acid, 1,1-DCE |
| 1,1,2-Trichloroethane | 140,[37] 170[34] | 1,1-DCE |
| 1,1,1,2-Tetrachloroethane | 47,[37] 380[34] | TCE |
| 1,1,2,2-Tetrachloroethane | 0.3,[29] 0.4,[37] 0.8[34] | 1,1,2-TCA, TCE |
| Tetrachloroethene | 0.7,[40]* 1.3E+06[37] | |
| Trichloroethene | 0.7,[40]* 1.3E+06[37] | |
| 1,1-Dichloroethene | 1.2E+10[37] | |
| 1,2-Dichloroethene | 2.1E+10[37] | |

*Butler and Barker[36] indicate that these values may reflect experimental difficulties and that the longer half-life [as calculated by Jeffers et al.[37]] should be used. [From T.H. Wiedemeier, H. S. Rifai, C. J. Newell and J.T. Wilson, **Natural Attenuation of Fuels and Chlorinated Solvents in the Subsurface**. Copyright © 1999 *John Wiley & Sons, Inc*. Reprinted by permission of John Wiley & Sons, Inc.]

One common chlorinated solvent for which abiotic transformations have been well studied is 1,1,1-TCA. 1,1,1-TCA may be abiotically transformed to acetic acid through a series of substitution reactions, including hydrolysis. In addition, 1,1,1-TCA may be reductively dehalogenated to form 1,1-DCA and then chloroethane (CA), which is then hydrolyzed to ethanol[38] or dehydrohalogenated to vinyl chloride.[37]

Dehydrohalogenation is an elimination reaction involving halogenated alkanes in which a halogen is removed from one carbon atom, followed by the subsequent removal of a hydrogen atom from an adjacent carbon atom. In this two-step reaction, an alkene is produced. Contrary to the patterns observed for hydrolysis, the likelihood that dehydrohalogenation will occur increases with the number of halogen substituents. It has been suggested that under normal environmental conditions, monohalogenated aliphatics apparently do not undergo dehydrohalogenation, and these reactions are apparently not likely to occur.[29,41] However, Jeffers et al.[37] report on the dehydrohalogenation of CA to VC. Polychlorinated alkanes have been observed to undergo dehydrohalogenation under normal conditions and extremely basic conditions.[29] As with hydrolysis, bromine substituents are more reactive with respect to dehydrohalogenation.

Dehydrohalogenation rates may also be approximated using pseudo-first-order kinetics. The rates will not only depend upon the number and types of halogen substituent, but also on the hydroxide ion concentration. Under normal pH conditions (i.e., near a pH of 7), interaction with water (acting as a weak base) may become more important.[29] Transformation rates for dehydrohalogenation reactions are presented in Table 23.1.3.

The organic compound 1,1,1-TCA is also known to undergo dehydrohalogenation.[38] In this case, TCA is transformed to 1,1-DCE, which is then reductively dehalogenated to VC. The VC is then either reductively dehalogenated to ethene or consumed as a substrate in an aerobic reaction and converted to $CO_2$. In a laboratory study, Vogel and McCarty[38] reported that the abiotic conversion of 1,1,1-TCA to 1,1-DCE has a rate constant of about 0.04 $year^{-1}$. Jeffers et a1.[37] reported that the tetrachloroethanes and pentachloroethanes degrade to TCE and PCE via dehydrohalogenation, respectively. Jeffers et al.[37] also report that CA may degrade to VC.

#### 23.1.2.8 Reduction reactions

Two abiotic reductive dechlorination reactions that may operate in the subsurface are hydrogenolysis and dihaloelimination. Hydrogenolysis is the simple replacement of a chlorine (or another halogen) by a hydrogen, while dihaloelimination is the removal of two chlorines (or other halogens) accompanied by the formation of a double carbon-carbon bond. While these reactions are thermodynamically possible under reducing conditions, they often do not take place in the absence of biological activity.[36,42-45] In general, microbes may produce reductants that facilitate such reactions in conjunction with minerals in the aquifer matrix. Moreover, the reducing conditions necessary to produce such reactions are most often created as a result of microbial activity. It is therefore not clear if some of these reactions are truly abiotic, or if because of their reliance on microbial activity to produce reducing conditions or reactants, they should be considered to be a form of cometabolism. In some cases, truly abiotic reductive dechlorination has been observed;[46,47] however, the conditions that favor such reactions may not occur naturally.

### 23.1.3 BIODEGRADATION OF CHLORINATED SOLVENTS

The biodegradation of organic chemicals can be grouped into two broad categories:

1) use of the organic compound as a primary growth substrate, and
2) cometabolism.

The use of chlorinated solvents as a primary growth substrate is probably the most important biological mechanism affecting them in the subsurface. Some chlorinated solvents are used as electron donors and some are used as electron acceptors when serving as primary growth substrates (meaning the mediating organism obtains energy for growth). When used as an electron donor, the chlorinated solvent is oxidized. Oxidation reactions can be aerobic or anaerobic. Conversely, when used as an electron acceptor, the chlorinated solvent is reduced via a reductive dechlorination process called halorespiration. It is important to note that not all chlorinated solvents can be degraded via all of these reactions. In fact, vinyl chloride is the only chlorinated solvent known to degrade via all of these pathways.

Chlorinated solvents can also be degraded via cometabolic pathways. During cometabolism, microorganisms gain carbon and energy for growth from metabolism of a primary substrate, and chlorinated solvents are degraded fortuitously by enzymes present in the metabolic pathway. Cometabolism reactions can be either oxidation or reduction reactions (under aerobic or anaerobic conditions), however based on data from numerous field sites, it does not appear that cometabolic oxidation will be a significant process in plumes of chlorinated solvents. Anaerobic reductive dechlorination can also occur via cometabolism. The process of cometabolic reductive dechlorination, however, is "sufficiently slow and incomplete that a successful natural attenuation strategy typically cannot completely rely upon it".[48]

The types of biodegradation reactions that have been observed for different chlorinated solvents are presented in Table 23.1.4. The remainder of this section will focus on describing the various mechanisms shown in Table 23.1.4.

**Table 23.1.4. Biological degradation processes for selected chlorinated solvents**

| Compound | Halorespiration | Direct aerobic oxidation | Direct anaerobic oxidation | Aerobic cometabolism | Anaerobic cometabolism |
|---|---|---|---|---|---|
| PCE | X | | | | X |
| TCE | X | | | X | X |
| DCE | X | X | X | X | X |
| Vinyl chloride | X | X | X | X | X |
| 1,1,1-TCA | X | | | X | X |
| 1,2-DCA | X | X | | X | X |
| Chloroethane | | X | | X | |
| Carbon tetrachloride | X | | | | X |
| Chloroform | X | | | X | X |
| Dichloromethane | | X | X | X | |

[From T.H. Wiedemeier, H. S. Rifai, C. J. Newell and J.T. Wilson, **Natural Attenuation of Fuels and Chlorinated Solvents in the Subsurface**. Copyright © 1999 *John Wiley & Sons, Inc*. Reprinted by permission of John Wiley & Sons, Inc.]

### 23.1.3.1 Halorespiration or reductive dechlorination using hydrogen

Reductive dechlorination is a reaction in which a chlorinated solvent acts as an electron acceptor and a chlorine atom on the molecule is replaced with a hydrogen atom. This results in the reduction of the chlorinated solvent. When this reaction is biological, and the organism is utilizing the substrate for energy and growth, the reaction is termed halorespiration. Only

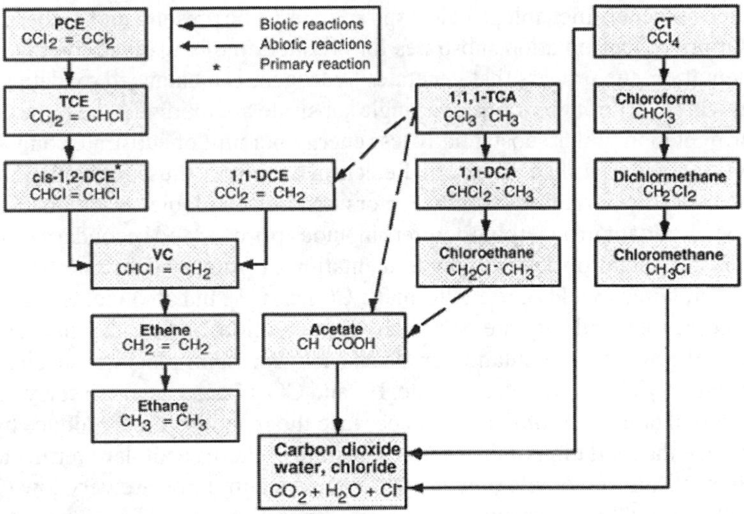


Figure 23.1.4. Abiotic and biological transformation pathways for selected chlorinated solvents. (From reference 1 after references 29, 53). [From T.H. Wiedemeier, H. S. Rifai, C. J. Newell and J.T. Wilson, **Natural Attenuation of Fuels and Chlorinated Solvents in the Subsurface**. Copyright © 1999 *John Wiley & Sons, Inc.* Reprinted by permission of John Wiley & Sons, Inc.]

recently have researchers demonstrated the existence of halorespiration.[49] Prior to this research, reductive dechlorination was thought to be strictly a cometabolic process. During halorespiration, hydrogen is used directly as an electron donor. The generalized reaction is given by:

$$H_2 + C\text{-}Cl \rightarrow C\text{-}H + H^+ + Cl^- \qquad [23.1.15]$$

where C-Cl represents a carbon-chloride bond in a chlorinated solvent. In this reaction, $H_2$ is the electron donor which is oxidized and the chlorinated solvent is the electron acceptor which is reduced. Although a few other electron donors (also fermentation products) besides hydrogen have been identified, hydrogen appears to be the most important electron donor for halorespiration. Only in the last four years have researchers begun to fully recognize the role of hydrogen as the electron donor in the reductive dechlorination of PCE and TCE.[48,50-52]

The hydrogen is produced in the subsurface by the fermentation of a wide variety of organic compounds including petroleum hydrocarbons and natural organic carbon. Because of its importance in the microbial metabolism of the halorespirators, the relative supply of hydrogen precursors compared to the amount of chlorinated solvent that must be degraded is an important consideration when evaluating natural attenuation. In general, reductive dechlorination of the ethenes occurs by sequential dechlorination from PCE to TCE to DCE to VC and finally to ethene. A summary of key biotic and abiotic reactions for the chlorinated ethenes, ethanes, and methanes first developed by Vogel[29] is shown in Figure 23.1.4.

For halorespiration to occur, Wiedemeier et al.[1] conclude that the following conditions must exist: "1) the subsurface environmental must be anaerobic and have a low oxidation-reduction potential (based on thermodynamic considerations, reductive dechlorination reactions will occur only after both oxygen and nitrate have been depleted from the aquifer);

2) chlorinated solvents amenable to halorespiration must be present; and 3) there must be an adequate supply of fermentation substrates for production of dissolved hydrogen."

Fermentation, the process that generates hydrogen, is a balanced oxidation-reduction reaction, in which different portions of a single substrate are oxidized and reduced, yielding energy. Fermentation yields substantially less energy per unit of substrate compared to oxidation reactions which utilize an external electron acceptor; thus, fermentation generally occurs when these external electron acceptors are not available. Bacterial fermentation which can be important in anaerobic aquifers includes primary and secondary fermentation.

Primary fermentation refers to the fermentation of primary substrates such as sugars, amino acids, and lipids yields acetate, formate, $CO_2$ and $H_2$, but also yields ethanol, lactate, succinate, propionate, and butyrate. Secondary fermentation, on the other hand, refers to the fermentation of primary fermentation products such as ethanol, lactate, succinate, propionate, and butyrate yielding acetate, formate, $H_2$, and $CO_2$. Bacteria which carry out these reactions are called obligate proton reducers because the reactions must produce hydrogen in order to balance the oxidation of the carbon substrates. These secondary fermentation reactions are energetically favorable only if hydrogen concentrations are very low ($10^{-2}$ to $10^{-4}$ atm or 8,000 nM to 80 nM dissolved hydrogen, depending on the fermentation substrate). Thus, these fermentation reactions occur only when the produced hydrogen is utilized by other bacteria, such as methanogens which convert $H_2$ and $CO_2$ into $CH_4$ and $H_2O$.

In the absence of external electron acceptors, the hydrogen produced by fermentation will be utilized by methanogens (methane producing bacteria). In this case, the ultimate end products of anaerobic metabolism of carbon substrates will be $CH_4$ (the most reduced form of carbon) and $CO_2$ (the most oxidized form of carbon). Methanogens will carry out the last step in this metabolism, the conversion of $H_2$ and $CO_2$ into $CH_4$. However, in the presence of external electron acceptors (halogenated organics, nitrate, sulfate, etc.), other products will be formed.[1]

There are a number of compounds besides the ones listed above that can be fermented to produce hydrogen. Sewell and Gibson[54] noted that petroleum hydrocarbons support reductive dechlorination. In this case, the reductive dechlorination is driven by the fermentation of biodegradable compounds such as the BTEX compounds in fuels. Metabolism of BTEX compounds to produce hydrogen likely requires the involvement of several strains of bacteria. Although the BTEX compounds are common fermentation substrates at chlorinated solvent sites, there are many other hydrocarbon substrates which are naturally fermented at sites and result in the generation of hydrogen such as acetone, sugars, and fatty acids from landfill leachate.

As hydrogen is produced by fermentative organisms, it is rapidly consumed by other bacteria. The utilization of $H_2$ by non-fermentors is known as interspecies hydrogen transfer and is required for fermentation reactions to proceed. Although $H_2$ is a waste product of fermentation, it is a highly reduced molecule which makes it an excellent, high-energy electron donor. Both organisms involved in interspecies hydrogen transfer benefit from the process. The hydrogen-utilizing bacteria gain a high energy electron donor, and, for the fermentors, the removal of hydrogen allows additional fermentation to remain energetically favorable.

A wide variety of bacteria can utilize hydrogen as an electron donor: denitrifiers, iron reducers, sulfate reducers, methanogens, and halorespirators. Thus, the production of hydrogen through fermentation does not, by itself, guarantee that hydrogen will be available for halorespiration. For dechlorination to occur, halorespirators must successfully compete against the other hydrogen utilizers for the available hydrogen. Smatlak et al.[51] suggest that the competition for hydrogen is controlled primarily by the Monod half-saturation constant $K_s(H_2)$, the concentration at which a specific strain of bacteria can utilize hydrogen at half the maximum utilization rate. Ballapragada et al.,[52] however, provide a more detailed discussion of halorespiration kinetics and point out that competition for hydrogen also depends on additional factors including the bacterial growth rate and maximum hydrogen utilization rate.

Smatlak et al.[51] have suggested that the steady-state concentration of hydrogen will be controlled by the rate of hydrogen production from fermentation. Both laboratory results and field observations have suggested, however, that the steady-state concentration of hydrogen is controlled by the type of bacteria utilizing the hydrogen and is almost completely independent of the rate of hydrogen production.[52,55-57] Under nitrate reducing conditions, steady-state $H_2$ concentrations were <0.05 nM; under iron reducing conditions, they were 0.2 to 0.8 nM; under sulfate reducing conditions, they were 1-4 nM, and under methanogenic conditions, they were 5-14 nM.[56,58,59] Finally, Carr and Hughes[57] show that dechlorination in a laboratory column is not impacted by competition for electron donor at high hydrogen concentrations. Thus, it is clear that an increased rate of hydrogen production will result in increased halorespiration without affecting the competition between bacteria for the available hydrogen.

#### *23.1.3.1.1 Stoichiometry of reductive dechlorination*

Under anaerobic conditions, Gossett and Zinder[48] showed that the reductive dehalogenation of the chlorinated ethenes occurs as a series of consecutive irreversible reactions mediated by the addition of 1 mole of hydrogen gas for every mole of chloride removed. Thus, the theoretical minimum hydrogen requirement for dechlorination can be calculated on a mass basis as shown below:[1]

1 mg $H_2$ will dechlorinate 21 mg of PCE to ethene
1 mg $H_2$ will dechlorinate 22 mg of TCE to ethene
1 mg $H_2$ will dechlorinate 24 mg of DCE to ethene
1 mg $H_2$ will dechlorinate 31 mg of VC to ethene

Complete fermentation of BTEX compounds is expected to yield 0.25 to 0.4 mg $H_2$ per mg BTEX. Therefore, for each mg of BTEX consumed, 4.5 to 7 mg of chloride could be released during reductive dechlorination. However, the utilization of hydrogen for dechlorination will never be completely efficient because of the competition for hydrogen in the subsurface discussed previously. One rule of thumb that has been proposed is the following: for reductive dechlorination to completely degrade a plume of dissolved chlorinated solvents, organic substrate concentrations greater than 25 to 100+ times that of the chlorinated solvent are required.[60]

#### *23.1.3.1.2 Chlorinated solvents that are amenable to halorespiration*

As shown in Table 23.1.4, all of the chlorinated ethenes (PCE, TCE, DCE, VC) and some of the chlorinated ethanes (TCA, 1,2-DCA) can be degraded via halorespiration; however, dichloromethane has not yet been shown to be degraded by this process. The oxidation state of a chlorinated solvent affects both the energy released by halorespiration and the rate at which the reaction occurs. In general, the more oxidized a compound is (more chlorine atoms on the organic molecule) the more amenable it is for reduction by halorespiration.

As with the ethenes, chlorinated ethanes will also undergo halorespiration. Dechlorination of 1,1,1-TCA has been described by Vogel and McCarty[38] and Cox et al.,[61] but understanding this pathway is complicated by the rapid hydrolysis reactions (e.g., half-life is 0.5-2.5 yrs) that can affect TCA.[30] Finally, halorespiration has been observed with highly chlorinated benzenes such as hexachlorobenzene, pentachlorobenzene, tetrachlorobenzene, and trichlorobenzene.[62-64] As discussed by Suflita and Townsend,[64] halorespiration of aromatic compounds has been observed in a variety of anaerobic habitats, including aquifer materials, marine and freshwater sediments, sewage sludges, and soil samples. However, isolation of specific microbes capable of these reactions has been difficult.

### 23.1.3.2 Oxidation of chlorinated solvents

In contrast to halorespiration, direct oxidation of some chlorinated solvents can occur biologically in groundwater systems. In this case, the chlorinated compound serves as the electron donor, and oxygen, sulfate, ferric iron or other compounds serve as the electron acceptor.

#### *23.1.3.2.1 Direct aerobic oxidation of chlorinated compounds*

Under direct aerobic oxidation conditions, the facilitating microorganism uses oxygen as an electron acceptor and obtains energy and organic carbon from the degradation of the chlorinated solvent. In general, the more-chlorinated aliphatic chlorinated solvents (e.g., PCE, TCE, and TCA) have not been shown to be susceptible to aerobic oxidation, while many of the progeny products (e.g., vinyl chloride, 1,2-DCA, and perhaps the isomers of DCE) are degraded via direct aerobic oxidation.

Hartmans et al.[65] and Hartmans and de Bont[66] show that vinyl chloride can be used as a primary substrate under aerobic conditions, with vinyl chloride being directly mineralized to carbon dioxide and water. Direct vinyl chloride oxidation has also been reported by Davis and Carpenter,[67] McCarty and Semprini,[53] and Bradley and Chapelle.[68] Aerobic oxidation is rapid relative to reductive dechlorination of dichloroethene and vinyl chloride. Although direct DCE oxidation has not been verified, a recent study has suggested that DCE isomers may be used as primary substrates.[68] Of the chlorinated ethanes, only 1,2-dichloroethane has been shown to be aerobically oxidized. Stucki et al.[69] and Janssen et al.[70] show that 1,2-DCA can be used as a primary substrate under aerobic conditions. In this case, the bacteria transform 1,2-DCA to chloroethanol, which is then mineralized to carbon dioxide. McCarty and Semprini[53] describe investigations in which 1,2-dichloroethane (DCA) was shown to serve as primary substrates under aerobic conditions.

Chlorobenzene and polychlorinated benzenes (up to and including tetrachlorobenzene) have been shown to biodegrade under aerobic conditions. Several studies have shown that bacteria are able to utilize chlorobenzene,[71] 1,4-DCB,[71-73] 1,3-DCB,[74] 1,2-DCB,[75] 1,2,4-TCB,[76,77] and 1,2,4,5-TeCB,[77] as primary growth substrates in aerobic systems. Nishino et al.[78] note that aerobic bacteria able to grow on chlorobenzene have been detected at a variety of chlorobenzene-contaminated sites but not at uncontaminated sites. Spain[79] suggests that this provides strong evidence that the bacteria are selected for their ability to derive carbon and energy from chlorobenzene degradation in situ. The pathways for all of these reactions are similar, bearing resemblance to benzene degradation pathways.[79,80]

### *23.1.3.2.2 Aerobic cometabolism of chlorinated compounds*

It has been reported that under aerobic conditions chlorinated ethenes, with the exception of PCE, are susceptible to cometabolic oxidation.[30,53,81,82] Vogel[30] further elaborates that the oxidation rate increases as the degree of chlorination decreases. Aerobic cometabolism of ethenes may be characterized by a loss of contaminant mass, the presence of intermediate degradation products (e.g., chlorinated oxides, aldehydes, ethanols, and epoxides), and the presence of other products such as chloride, carbon dioxide, carbon monoxide, and a variety of organic acids.[53,83] Cometabolism requires the presence of a suitable primary substrate such as toluene, phenol, or methane. For cometabolism to be effective, the primary substrate must be present at higher concentrations than the chlorinated compound, and the system must be aerobic. Because the introduction of high concentrations of oxidizable organic matter into an aquifer quickly drives the groundwater anaerobic, aerobic cometabolism typically must be engineered.

### *23.1.3.2.3 Anaerobic oxidation of chlorinated compounds*

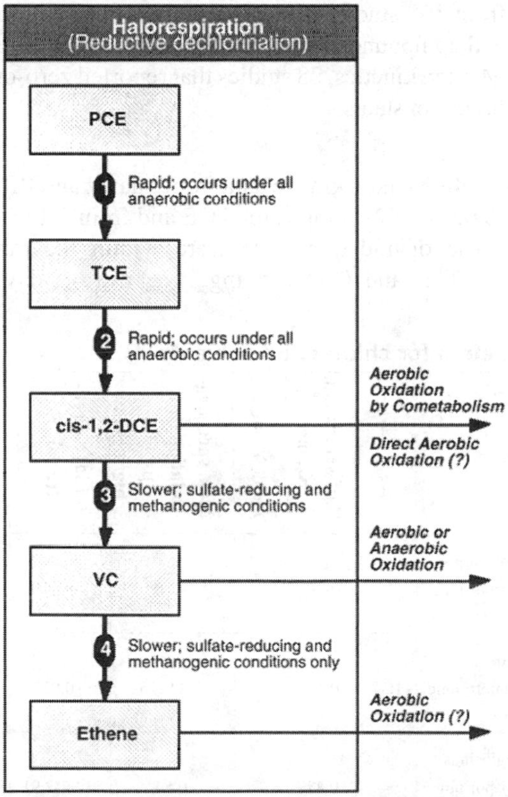


Figure 23.1.5. Reaction sequence and relative rates for halorespiration of chlorinated ethenes, with other reactions shown. [From T.H. Wiedemeier, H. S. Rifai, C. J. Newell and J.T. Wilson, **Natural Attenuation of Fuels and Chlorinated Solvents in the Subsurface**, reaction rates description from reference 88. Copyright © 1999 *John Wiley & Sons, Inc.* Reprinted by permission of John Wiley & Sons, Inc.]

Anaerobic oxidation occurs when anaerobic bacteria use the chlorinated solvent as an electron donor by utilizing an available electron acceptor such as ferric iron (Fe(III)). Bradley and Chapelle[84] show that vinyl chloride can be oxidized to carbon dioxide and water via Fe(III) reduction. In microcosms amended with Fe(III)-EDTA, reduction of vinyl chloride concentrations closely matched the production of carbon dioxide. Slight mineralization was also noted in unamended microcosms. The rate of this reaction apparently depends on the bioavailability of Fe(III). In a subsequent paper, Bradley and Chapelle[85] reported "significant" anaerobic mineralization of both DCE and VC in microcosms containing creek bed sediments. The sediments were taken from a stream where groundwater containing chlorinated ethenes continually discharges. Anaerobic mineralization was observed in both methanogenic and Fe(III) reducing conditions.

## 23.1.4 BIODEGRADATION RATES FOR CHLORINATED SOLVENTS

Overall, dechlorination is more rapid for highly chlorinated compounds than for compounds that are less chlorinated.[38,86,87] Figure 23.1.5 qualitatively shows the reaction rate and required conditions for halorespiration of PCE to ethene. PCE (four chlorines) degrades the fastest under all anaerobic environments, while VC (a single chlorine) will degrade only under sulfate-reducing and methanogenic conditions, with a relatively slow reaction rate.

At many chlorinated ethene sites, concentrations of cis-1,2-DCE are often higher than any of the parent chlorinated ethene compounds. The reason for the accumulation of 1,2-DCE may be due to either slower rates of DCE halorespiration, or the prevalence of organisms that reduce PCE as far as cis-1,2-DCE over ones that can reduce PCE all the way to ethene.[48] Although many researchers have commented that reductive dechlorination will result in the accumulation of VC (e.g., see 84, 89), at many field sites VC accumulation is much lower than cis-1,2-DCE. This may occur because the vinyl chloride in many chlorinated solvent plumes can migrate to zones that can support direct oxidation of VC oxidation, either aerobically and/or anaerobically.

Suarez and Rifai[90] analyzed data from 138 studies (field and laboratory) to estimate biodegradation coefficients for chlorinated compounds. Suarez and Rifai[90] found a total of thirteen studies that reported Michaelis-Menten kinetics, 28 studies that reported zero-order rates, and 97 studies that reported first-order constants.

### 23.1.4.1 Michaelis-Menten rates

The data in Table 23.1.5 present the Michaelis Menten kinetic data from Suarez and Rifai.[90] Half-saturation constants varied from 0.6 mg/L to 29.5 mg/L for TCE and from 0.17 mg/L to 28 mg/L for DCE. Maximum specific degradation rates were within the ranges 0.038-478.59 $mg_{compound}/mg_{protein}$-day for TCE, and 0-11,115 $mg_{compound}/mg_{protein}$-day for DCE.

**Table 23.1.5. Michaelis-Menten parameters for chlorinated solvents**

| Compound | Type of study | Redox environment | Culture | $\mu_{max}$, day$^{-1}$ | Half-saturation, $K_s$, mg/L | Yield, Y, mg/mg | Max. spec. degradation rate, $\mu_{max}/Y$, mg/mg-day | Initial concentration, $S_o$, mg/L | Ref |
|---|---|---|---|---|---|---|---|---|---|
| 1,1,1-TCA | Continuous reactor | Aerobic | Methylosinus trichosporium OB3b | | 28.46 | | 4.60 | >93.10 | 92 |
| 1,1-DCE | Growth reactor | Aerobic-Cometabolism (methane) | Mixed methanotrophic culture | 1.37 | 0.43 | | 0-11115 | 0.01 | 93 |
| | Continuous reactor | Aerobic | Methylosinus trichosporium OB3b | | 0.48 | | 0.84 | 1.94-2.91 | 92 |
| 1,2-DCA | Continuous reactor | Aerobic | Methylosinus trichosporium OB3b | | 7.62 | | 9.26 | 4.95-6.93 | 92 |

| Compound | Type of study | Redox environment | Culture | $\mu_{max}$, day$^{-1}$ | Half-saturation, $K_S$, mg/L | Yield, Y, mg/mg | Max. spec. degradation rate, $\mu_{max}/Y$, mg/mg-day | Initial concentration, $S_0$, mg/L | Ref |
|---|---|---|---|---|---|---|---|---|---|
| cis-1,2-DCE | Growth reactor | Aerobic-Cometabolism (phenol) | Filamentous phenol-oxidizers | | | | 0.27-1.50 | | 94 |
| | Continuous reactor | Aerobic | Methylosinus trichosporium OB3b | | 2.91 | | 25.40 | 12.60-25.20 | 92 |
| | Methanogenic fluidized bed reactor | Anaerobic | | | 28.00 | | | | 52 |
| trans-1,2-DCE | Continuous reactor | Aerobic | Methylosinus trichosporium OB3b | | 14.34 | | 46.19 | 8.72-14.74 | 92 |
| | Growth reactor | Aerobic-Cometabolism (methane) | Mixed methanotrophic culture | 0.68 | 0.17 | | 0.00-0.44 | 4.70 | 93 |
| PCE | Methanogenic fluidized bed reactor | Anaerobic | | | 12.00 | | | | 52 |
| | Biofilm reactor | | | | | | 0.00 | 0.99 | 95 |
| | Fed-batch reactor | Anaerobic | Methanogenic consortium | 0.47 | | | | | 96 |
| TCE | Growth reactor | Aerobic | | | 0.37 | | 0.53 | 14.70 | 97 |
| | Growth reactor | Aerobic-Cometabolism (formate) | | | 8.20 | | 7.60 | 10.10 | 97 |
| | Growth reactor | Aerobic-Cometabolism (methane) | Mixed methanotrophic culture | 1.07 | 0.13 | | 0.00-1.13 | 1.00 | 93 |
| | Methanogenic fluidized bed reactor | Anaerobic | | | 19.00 | | | | 52 |
| | Growth reactor | Aerobic-Cometabolism (phenol) | Filamentous phenol-oxidizers | | | | 0.10-0.25 | | 94 |

| Compound | Type of study | Redox environment | Culture | $\mu_{max}$, day$^{-1}$ | Half-saturation, $K_S$, mg/L | Yield, Y, mg/mg | Max. spec. degradation rate, $\mu_{max}/Y$, mg/mg-day | Initial concentration, $S_o$, mg/L | Ref |
|---|---|---|---|---|---|---|---|---|---|
| TCE | Microcosm | Aerobic-Cometabolism (toluene) | | 0.77-1.65 | | 0.52 | 1.50 | 0.66 | 98 |
| | Microcosm | Aerobic-Cometabolism (phenol) | | 0.88-1.43 | | 0.40 | 3.00 | 0.66 | 98 |
| | Growth reactor | Aerobic-Cometabolism (propane) | Propane-oxidizing culture | | 0.60 | | 0.04 | 3.00 | 99 |
| | Batch | Aerobic | Methylomonas methanica 68-1 | | 29.48 | 0.10 | 438.59 | | 100 |
| | Batch | | Methylosinus trichosporium OB3b | | 16.51 | 0.08 | 187.70 | | 100 |
| | Continuous reactor | Aerobic | Methylosinus trichosporium OB3b | | 19.00 | | 54.71 | 9.17-13.10 | 92 |
| | Continuous reactor | Aerobic-Cometabolism (methane) | Mixed methanotrophic culture | | | 0.01 | | | 101 |
| Vinyl chloride | Methanogenic fluidized bed reactor | Anaerobic | | | 23.00 | | | | 52 |
| | Small-Column Microcosm | Aerobic-Cometabolism (methane) | | | | 1.00-3.50 | | 1.00-17.00 | 102 |

[From M.P. Suarez and H.S. Rifai, *Bioremediation Journal,* **3**, 337-362. Copyright © 1999 *Battelle Memorial Institute*. Reprinted with permission.]

### 23.1.4.2 Zero-order rates

A summary of more than 40 studies reporting zero-order rates is included in Table 23.1.6. The reported zero-order rates ranged from 0 to 19.8 mg/L/day with mean values for anaerobic rates of 0.04, 2.14, 1.80, 1.74, and 0.11 mg/L/day for carbon tetrachloride, DCE, PCE, TCE, and vinyl chloride, respectively. TCE appeared to be reductively dechlorinated at the fastest rate coefficient, with a median equal to 0.76 mg/L/day. In contrast, vinyl chloride exhibited the slowest rate coefficient of reductive dechlorination with a median value of 0.01 mg/L/day.

**Table 23.1.6 Summary of zero-order decay rates for reductive dechlorination obtained from laboratory studies (mg/L-day).**

| | $CCl_4$ | cis -12-DCE | DCE (all other isomers) | PCE | TCE | Vinyl chloride |
|---|---|---|---|---|---|---|
| Number of rates[a] | 8 | 18 | 8 | 29 | 7 | 9 |
| minimum | 0.022 | 0.013 | 0.009 | 0.013 | 0.314 | 0.002 |
| 25th percentile | 0.024 | 0.183 | 0.023 | 0.288 | 0.511 | 0.006 |
| median | 0.029 | 0.511 | 0.250 | 0.577 | 0.760 | 0.011 |
| 75th percentile | 0.042 | 1.318 | 1.385 | 1.040 | 1.297 | 0.075 |
| 90th percentile | 0.049 | 3.348 | 2.021 | 2.801 | 3.798 | 0.379 |
| maximum | 0.054 | 16.958 | 3.470 | 19.800 | 7.490 | 0.495 |
| mean | 0.034 | 1.854 | 0.850 | 1.863 | 1.740 | 0.107 |
| standard deviation | 0.012 | 3.939 | 1.213 | 4.162 | 2.567 | 0.184 |
| geometric mean[b] | 0.000 | 15.513 | 1.471 | 17.323 | 6.590 | 0.034 |

[a]All the zero-order rates provided were calculated by the authors of the respective studies. [b]To calculate the geometric mean, values equal to zero were included as $10^{-10}$. [From M.P. Suarez and H.S. Rifai, *Bioremediation Journal,* **3**, 337-362. Copyright © 1999 *Battelle Memorial Institute*. Reprinted with permission]

### 23.1.4.3 First-order rate constants

Table 23.1.7 summarizes the first order decay coefficients from both field and laboratory studies. As can be seen in Table 23.1.7, first-order rate constants for chlorinated solvents varied from 0 to 1.03 $day^{-1}$ in the 90% of the cases, with mean values equal to 0.11, 0.02, 0.14, 0.05, 0.26, 0.17, and 0.23 $day^{-1}$ for carbon tetrachloride, DCA, DCE, PCE, TCA, TCE, and vinyl chloride, respectively. The range minimum-90th percentile for aerobic rates was 0-1.35 $day^{-1}$ while for anaerobic rates it was 0-1.11 $day^{-1}$. Field rates from aerobic/anaerobic studies ranged from 0 to 1.96 $day^{-1}$. The compound that showed the highest mean value under aerobic conditions was vinyl chloride (1.73 $day^{-1}$), while TCA exhibited the highest mean anaerobic rate coefficient (0.35 $day^{-1}$).

**Table 23.1.7. Summary of first-order decay rates for chlorinated solvents ($day^{-1}$).**

| | Number of rates | Number of reported rates | Number of calculated rates[b] | mean | standard deviation | 90th percentile | geometric mean[c] | range reported rates |
|---|---|---|---|---|---|---|---|---|
| **Carbon tetrachloride** | | | | | | | | |
| All studies | 13 | 0 | 13 | 0.108 | 0.134 | 0.216 | 0.054 | 0.0037-0.49 |
| Aerobic oxidation | | | | | | | | |
| In situ & laboratory | 1 | 0 | 1 | | | | | |
| In situ studies[a] | 1 | 0 | 1 | | | | | |
| Laboratory | | | | | | | | |

| | Number of rates | Number of reported rates | Number of calculated rates[b] | mean | standard deviation | 90th percentile | geometric mean[c] | range reported rates |
|---|---|---|---|---|---|---|---|---|
| Aerobic cometabolism | | | | | | | | |
| Field & laboratory | 1 | 0 | 1 | | | | | |
| In situ studies[a] | 1 | 0 | 1 | | | | | |
| Laboratory | | | | | | | | |
| Overall aerobic | 2 | 0 | 2 | 0.019 | | | | 0.016-0.022 |
| Aerobic/anaerobic (field studies) | | | | | | | | |
| Reductive dechlorination | | | | | | | | |
| Field & laboratory | 11 | 0 | 11 | 0.124 | 0.140 | 0.230 | 0.065 | 0.004-0.490 |
| Field/in situ studies[a] | 7 | 0 | 7 | 0.141 | 0.174 | 0.334 | 0.060 | 0.004-0.490 |
| Laboratory | 4 | 0 | 4 | 0.093 | | | 0.075 | 0.023-0.160 |
| Anaerobic oxidation | | | | | | | | |
| Field & laboratory | | | | | | | | |
| Field/in situ studies[a] | | | | | | | | |
| Laboratory | | | | | | | | |
| **DCA (all isomers)** | | | | | | | | |
| All studies | 25 | 16 | 9 | 0.017 | 0.036 | 0.046 | 0.001 | 0-0.131 |
| Aerobic oxidation | | | | | | | | |
| In situ & laboratory | 2 | 2 | 0 | | | | 0.000 | |
| In situ studies[a] | | | | | | | | |
| Laboratory | 2 | 2 | 0 | | | | 0.000 | |
| Aerobic cometabolism | | | | | | | | |
| Field & laboratory | 5 | 0 | 5 | 0.067 | 0.056 | 0.128 | 0.046 | 0.014-0.131 |
| In situ studies[a] | | | | | | | | |
| Laboratory | 5 | 0 | 5 | 0.067 | 0.056 | 0.128 | 0.046 | 0.014-0.131 |
| Overall aerobic | 7 | 2 | 5 | 0.048 | 0.056 | 0.126 | 0.000 | 0-0.131 |
| Aerobic/anaerobic (field studies) | | | | | | | | |
| Reductive dechlorination | | | | | | | | |
| Field & laboratory | 18 | 14 | 4 | 0.005 | 0.012 | 0.016 | 0.001 | 0-0.044 |
| Field/in situ studies[a] | 16 | 14 | 2 | 0.002 | 0.003 | 0.004 | 0.001 | 0-0.011 |
| Laboratory | 2 | 0 | 2 | 0.036 | | | 0.035 | 0.028-0.044 |
| Anaerobic oxidation | | | | | | | | |
| Field & laboratory | | | | | | | | |
| Field/in situ studies[a] | | | | | | | | |
| Laboratory | | | | | | | | |
| ***cis*-1,2-DCE** | | | | | | | | |
| All studies | 34 | 24 | 10 | 0.004 | 0.395 | 0.257 | 0.004 | 0-1.960 |

| | Number of rates | Number of reported rates | Number of calculated rates[b] | mean | standard deviation | 90[th] percentile | geometric mean[c] | range reported rates |
|---|---|---|---|---|---|---|---|---|
| Aerobic oxidation<br>In situ & laboratory<br>In situ studies[a]<br>Laboratory | | | | | | | | |
| Aerobic cometabolism<br>Field & laboratory<br>In situ studies[a]<br>Laboratory | <br>5<br>3<br>2 | <br>2<br>2<br>0 | <br>3<br>1<br>2 | <br>0.476<br>0.885<br>0.187 | <br>0.787<br>0.843<br>0.250 | <br>1.680<br>1.820<br>0.399 | <br>0.476<br>0.885<br>0.187 | <br>0.081-1.96<br>0.281-1.96<br>0.081-0.434 |
| Overall aerobic | | | | | | | | |
| Aerobic/anaerobic (field studies) | 4 | 4 | 0 | 0.000 | 0.003 | 0.006 | 0.000 | 0-0.008 |
| Reductive dechlorination<br>Field & laboratory<br>Field/in situ studies[a]<br>Laboratory | <br>25<br>17<br>8 | <br>18<br>13<br>5 | <br>7<br>4<br>3 | <br>0.004<br>0.002<br>0.014 | <br>0.048<br>0.031<br>0.069 | <br>0.069<br>0.013<br>0.117 | <br>0.004<br>0.002<br>0.014 | <br>0-0.200<br>0-0.130<br>0.001-0.200 |
| Anaerobic oxidation<br>Field & laboratory<br>Field/in situ studies[a]<br>Laboratory | | | | | | | | |
| **DCE (all other isomers)** | | | | | | | | |
| All studies | 27 | 14 | 13 | 0.149 | 0.302 | 0.666 | 0.003 | 0-1.150 |
| Aerobic oxidation<br>In situ & laboratory<br>In situ studies[a]<br>Laboratory | | | | | | | | |
| Aerobic cometabolism<br>Field & laboratory<br>In situ studies[a]<br>Laboratory | <br>8<br>4<br>4 | <br>2<br>0<br>2 | <br>6<br>4<br>2 | <br>0.458<br>0.720<br>0.196 | <br>0.416<br>0.316<br>0.347 | <br>0.845<br>1.012<br>0.521 | <br>0.002<br>0.670<br>0.000 | <br>0-1.150<br>0.390-1.150<br>0-0.714 |
| Overall aerobic | | | | | | | | |
| Aerobic/anaerobic (field studies) | | | | | | | | |
| Reductive dechlorination<br>Field & laboratory<br>Field/in situ studies[a]<br>Laboratory | <br>19<br>16<br>3 | <br>12<br>12<br>0 | <br>7<br>4<br>3 | <br>0.019<br>0.003<br>0.101 | <br>0.061<br>0.001<br>0.147 | <br>0.012<br>0.005<br>0.220 | <br>0.004<br>0.003<br>0.039 | <br>0.001-0.270<br>0.001-0.006<br>0.010-0.270 |

| | Number of rates | Number of reported rates | Number of calculated rates[b] | mean | standard deviation | 90th percentile | geometric mean[c] | range reported rates |
|---|---|---|---|---|---|---|---|---|
| Anaerobic oxidation | | | | | | | | |
| Field & laboratory | | | | | | | | |
| Field/in situ studies[a] | | | | | | | | |
| Laboratory | | | | | | | | |
| **PCE** | | | | | | | | |
| All studies | 50 | 31 | 19 | 0.051 | 0.084 | 0.153 | 0.000 | 0-0.410 |
| Aerobic oxidation | | | | | | | | |
| In situ & laboratory | 10 | 5 | 5 | 0.001 | 0.001 | 0.003 | 0.000 | 0-0.004 |
| In situ studies[a] | 3 | 3 | 0 | 0.000 | | | | 0.000 |
| Laboratory | 7 | 2 | 5 | 0.001 | 0.002 | 0.003 | 0.000 | 0-0.004 |
| Aerobic cometabolism | | | | | | | | |
| Field & laboratory | 3 | 1 | 2 | 0.025 | | | 0.000 | 0-0.054 |
| In situ studies[a] | | | | | | | | |
| Laboratory | 3 | 1 | 2 | 0.025 | | | 0.000 | 0-0.054 |
| Overall aerobic | 13 | 6 | 7 | 0.006 | 0.015 | 0.017 | 0.000 | 0-0.054 |
| Aerobic/anaerobic (field studies) | 1 | 1 | 0 | | | | | |
| Reductive dechlorination | | | | | | | | |
| Field & laboratory | 36 | 23 | 13 | 0.068 | 0.093 | 0.185 | 0.002 | 0-0.410 |
| Field/in situ studies[a] | 13 | 9 | 4 | 0.010 | 0.022 | 0.022 | 0.000 | 0-0.080 |
| Laboratory | 23 | 14 | 9 | 0.101 | 0.101 | 0.212 | 0.024 | 0-0.410 |
| Anaerobic oxidation | | | | | | | | |
| Field & laboratory | | | | | | | | |
| Field/in situ studies[a] | | | | | | | | |
| Laboratory | | | | | | | | |
| **TCA** | | | | | | | | |
| All studies | 47 | 27 | 20 | 0.261 | 0.502 | 1.026 | 0.000 | 0-2.330 |
| Aerobic oxidation | | | | | | | | |
| In situ & laboratory | 11 | 7 | 4 | 0.002 | 0.007 | 0.009 | 0.000 | 0-0.022 |
| In situ studies[a] | 2 | 2 | 0 | | | | | |
| Laboratory | 9 | 5 | 4 | 0.003 | 0.007 | 0.005 | 0.000 | 0-0.022 |
| Aerobic cometabolism | | | | | | | | |
| Field & laboratory | 5 | 1 | 4 | 0.247 | 0.522 | 0.723 | 0.001 | 0-1.180 |
| In situ studies[a] | | | | | | | | |
| Laboratory | 5 | 1 | 4 | 0.247 | 0.522 | 0.723 | 0.001 | 0.-1.180 |
| Overall aerobic | 16 | 8 | 8 | 0.079 | 0.294 | 0.030 | 0.000 | 0-1.180 |
| Aerobic/anaerobic (field studies) | | | | | | | | |

| | Number of rates | Number of reported rates | Number of calculated rates[b] | mean | standard deviation | 90[th] percentile | geometric mean[c] | range reported rates |
|---|---|---|---|---|---|---|---|---|
| Reductive dechlorination | | | | | | | | |
| Field & laboratory | 31 | 19 | 12 | 0.355 | 0.562 | 1.110 | 0.003 | 0-2.330 |
| Field/in situ studies[a] | 10 | 3 | 7 | 0.029 | 0.039 | 0.058 | 0.000 | 0-0.125 |
| Laboratory | 21 | 16 | 5 | 0.511 | 0.629 | 1.280 | 0.007 | 0-2.330 |
| Anaerobic oxidation | | | | | | | | |
| Field & laboratory | | | | | | | | |
| Field/in situ studies[a] | | | | | | | | |
| Laboratory | | | | | | | | |
| **TCE** | | | | | | | | |
| All studies | 86 | 52 | 34 | 0.173 | 0.475 | 0.636 | 0.001 | 0-3.130 |
| Aerobic oxidation | | | | | | | | |
| In situ & laboratory | 12 | 6 | 6 | 0.005 | 0.010 | 0.025 | 0.000 | 0-0.028 |
| In situ studies[a] | 2 | 2 | 0 | | | | | |
| Laboratory | 10 | 0 | 10 | 0.006 | 0.011 | 0.026 | 0.000 | 0-0.028 |
| Aerobic cometabolism | | | | | | | | |
| Field & laboratory | 17 | 7 | 10 | 0.586 | 0.566 | 1.418 | 0.309 | 0.024-1.650 |
| In situ studies[a] | 3 | 2 | 1 | 0.948 | | | 0.582 | 0.105-1.410 |
| Laboratory | 14 | 5 | 9 | 0.509 | 0.524 | 1.265 | 0.269 | 0.024-1.650 |
| Overall aerobic | 29 | 13 | 16 | 0.346 | 0.517 | 1.354 | 0.001 | 0-1.650 |
| Aerobic/anaerobic (field studies) | 1 | 1 | 0 | | | | | |
| Reductive dechlorination | | | | | | | | |
| Field & laboratory | 56 | 38 | 18 | 0.086 | 0.434 | 0.022 | 0.001 | 0-3.130 |
| Field/in situ studies[a] | 32 | 26 | 6 | 0.003 | 0.005 | 0.006 | 0.000 | 0-0.023 |
| Laboratory | 24 | 12 | 12 | 0.196 | 0.654 | 0.337 | 0.012 | 0-3.130 |
| Anaerobic oxidation | | | | | | | | |
| Field & laboratory | | | | | | | | |
| Field/in situ studies[a] | | | | | | | | |
| Laboratory | | | | | | | | |
| **Vinyl chloride** | | | | | | | | |
| All studies | 26 | 8 | 18 | 0.229 | 0.476 | 0.946 | 0.023 | 0-1.960 |
| Aerobic oxidation | | | | | | | | |
| In situ & laboratory | 4 | 0 | 4 | 0.087 | | | 0.080 | 0.043-0.125 |
| In situ studies[a] | | | | | | | | |
| Laboratory | 4 | 0 | 4 | 0.087 | | | 0.080 | 0.043-0.125 |
| Aerobic cometabolism | | | | | | | | |
| Field & laboratory | 4 | 0 | 4 | 1.023 | | | 0.552 | 0.055-1.960 |
| In situ studies[a] | 2 | 0 | 2 | 1.730 | | | 1.715 | 1.500-1.960 |
| Laboratory | 2 | 0 | 2 | 0.316 | | | 0.178 | 0.055-0.576 |

| | Number of rates | Number of reported rates | Number of calculated rates[b] | mean | standard deviation | 90th percentile | geometric mean[c] | range reported rates |
|---|---|---|---|---|---|---|---|---|
| Overall aerobic | 8 | 0 | 8 | 0.555 | 0.756 | 0.107 | 0.211 | 0.043-0.120 |
| Aerobic/anaerobic (field studies) | 3 | 2 | 1 | 0.004 | | | 0.002 | 0.001-0.009 |
| Reductive dechlorination | | | | | | | | |
| Field & laboratory | 8 | 5 | 3 | 0.153 | 0.228 | 0.499 | 0.007 | 0-0.520 |
| Field/in situ studies[a] | 4 | 4 | 0 | 0.003 | | | 0.001 | 0-0.007 |
| Laboratory | 4 | 1 | 3 | 0.303 | | | 0.036 | 0-0.520 |
| Anaerobic oxidation | | | | | | | | |
| Field & laboratory | 7 | 1 | 6 | 0.042 | 0.048 | 0.104 | 0.018 | 0.001-0.120 |
| Field/in situ studies[a] | 1 | 1 | 0 | | | | | |
| Laboratory | 6 | 0 | 6 | 0.049 | 0.048 | 0.107 | 0.028 | 0.008-0.120 |

[a]In situ studies include in situ microcosms and in situ columns
[b]When enough information was provided by the authors of a study, the authors of this paper calculated the rate coefficient assuming first-order kinetics
[c]To calculate the geometric mean, values equal to zero were included as $10^{-10}$
[From M.P. Suarez and H.S. Rifai, *Bioremediation Journal,* **3**, 337-362. Copyright © 1999 *Battelle Memorial Institute*. Reprinted with permission.]

The biodegradability under different electron acceptors for each one of the chlorinated solvents was also analyzed by Suarez and Rifai.[90] As summarized in Table 23.1.8, DCA presented very high potential for biodegradation via aerobic cometabolism and reductive dechlorination with none of the studies reporting recalcitrance. Median half-lives for this compound were 1,260 days and 15 days for reductive dechlorination and cometabolism, respectively. DCE exhibited high potential for aerobic cometabolism with 11% of the studies showing recalcitrance and a very short median half-life (1 day). None of the 44 studies on reductive dechlorination of DCE reported recalcitrance, which leads to the conclusion that DCE may undergo this process though with a relatively slow rate (median half-life equal to 234 days).

**Table 23.1.8. Biodegradability of chlorinated solvents**

| | All Studies | Process | | | |
|---|---|---|---|---|---|
| | | Aerobic oxidation | Cometabolism | Reductive dechlorination | Anaerobic oxidation |
| **Carbon tetrachloride** | | | | | |
| # rates | 13 | 1 | 1 | 11 | |
| # rates-recalcitrant | 0 | 0 | 0 | 0 | |
| half-life (days)[a] | 14 | NC | NC | 9 | |
| % rates recalcitrant | 0% | 0% | 0% | 0% | |
| potential for biodegradation[b] | almost always | NA | NA | almost always | |

| | All Studies | Process | | | |
|---|---|---|---|---|---|
| | | Aerobic oxidation | Cometabolism | Reductive dechlorination | Anaerobic oxidation |
| **DCA (all isomers)** | | | | | |
| # rates | 25 | 2 | 5 | 18 | |
| # rates-recalcitrant | 2 | 2 | 0 | 0 | |
| half-life (days)[a] | 990 | NC | 15 | 1260 | |
| % rates recalcitrant | 8% | 100% | 0% | 0% | |
| potential for biodegradation[b] | almost always | NA | almost always | almost always | |
| **DCE (all isomers)** | | | | | |
| # rates | 61 | | 13 | 44 | |
| # rates-recalcitrant | 3 | | 2 | 0 | |
| half-life (days)[a] | 173 | | 2 | 234 | |
| % rates recalcitrant | 5% | | 15% | 0% | |
| potential for biodegradation[b] | almost always | | frequently | almost always | |
| **PCE** | | | | | |
| # rates | 50 | 10 | 3 | 36 | |
| # rates-recalcitrant | 14 | 6 | 1 | 5 | |
| half-life (days)[a] | 80 | NC | 35 | 32 | |
| % rates recalcitrant | 28% | 60% | 35% | 14% | |
| potential for biodegradation[b] | sometimes | barely | NA | frequently | |
| **TCA** | | | | | |
| # rates | 47 | 11 | 5 | 31 | |
| # rates-recalcitrant | 14 | 8 | 1 | 5 | |
| half-life (days)[a] | 68 | NC | 53 | 24 | |
| % rates recalcitrant | 30% | 73% | 20% | 16% | |
| potential for biodegradation[b] | sometimes | barely | frequently | frequently | |
| **TCE** | | | | | |
| # rates | 85 | 11 | 17 | 56 | |
| # rates-recalcitrant | 12 | 6 | 0 | 5 | |
| half-life (days)[a] | 151 | NC | 3 | 201 | |
| % rates recalcitrant | 14% | 55% | 0% | 9% | |
| potential for biodegradation[b] | frequently | barely | almost always | almost always | |
| **Vinyl chloride** | | | | | |
| # rates | 27 | 4 | 5 | 15 | 7 |
| # rates-recalcitrant | 0 | 0 | 0 | 0 | 0 |
| half-life (days)[a] | 14 | 8 | 0.462 | 80 | 58 |
| % rates recalcitrant | 0 | 0% | 0% | 0% | 0% |
| potential for biodegradation[b] | almost always | almost always | almost always | almost always | almost always |

[a]Median value from the reported studies; [b]Quantitative estimation based on % occurrence of recalcitrance; NA Insufficient information; NC Not calculable ($\lambda$=0); Scale % recalcitrance - biodegradability: < 10% - Almost always, 10%-25% - Frequently, 25%-50% - Sometimes, 50%-75% - Barely, >75% - Almost never. [From M.P. Suarez and H.S. Rifai, *Bioremediation Journal,* **3**, 337-362. Copyright © 1999 Battelle Memorial Institute. Reprinted with permission.]

The process that exhibited the highest potential for biodegradation of PCE and TCA was reductive dechlorination with 86% and 84% of the analyzed studies showing biotransformation. Median half-lives for reductive dechlorination of PCE and TCA were 34 days and 24 days, respectively. With respect of TCE, none of the 17 studies reporting aerobic cometabolism (most of them laboratory studies) showed recalcitrance and the median half-life was very short (3 days). Reductive dechlorination also appeared to be a very good alternative for biotransformation of TCE with only 9% of 56 studies reporting recalcitrance and median half-life equal to 201 days. Finally, vinyl chloride exhibited very high potential for biodegradation under aerobic conditions with no studies showing recalcitrance and median half-lives of 8 days and 0.462 days for oxidation and cometabolism, respectively.

In addition to the data reported above by Suarez and Rifai,[90] a groundwater anaerobic biodegradation literature review was performed by Aronson and Howard.[102] Based on their review, Aronson and Howard[102] developed a range of "recommended values" for the anaerobic biodegradation first order decay rate coefficients. For many of the chlorinated solvents, the authors defined the low-end rate coefficient based on the lowest measured field value, and defined the high-end value as the mean rate coefficient for all the field/in-situ microcosm studies. Table 23.1.9 shows the resulting recommended ranges for first order anaerobic biodegradation rate coefficients for several chlorinated solvents along with the mean value of the field/in-situ microcosm studies (note that some minor discrepancies exist between the reported high-end rates and the mean value for the field/in-situ microcosm studies).

**Table 23.1.9. Mean and recommended first-order rate coefficients for selected chlorinated solvents presented by Aronson and Howard[102]**

| Compound | Mean of field/in-situ studies: 1st order rate coefficients, $day^{-1}$ | Mean of field/in-situ studies: 1st order rate half-lives, day | Mean of field/in-situ studies: number studies used for mean | Recommended 1st order rate coefficients: low-end 1st order rate coefficients, $day^{-1}$ | Recommended: low-end half-lives day | Recommended: high-end 1st order rate coefficients, $day^{-1}$ | Recommended: high-end half-lives day | Comments |
|---|---|---|---|---|---|---|---|---|
| PCE | 0.0029 | 239 | 16 | 0.00019 | 3,647 | 0.0033 | 210 | Lower limit was reported for a field study under nitrate-reducing conditions |
| TCE | 0.0025 | 277 | 47 | 0.00014 | 4,950 | 0.0025 | 277 | Lower limit was reported for a field study under unknown redox conditions |
| Vinyl chloride | 0.0079 | 88 | 19 | 0.00033 | 2,100 | 0.0072 | 96 | Lower limit was reported for a field study under methanogenic/sulfate-reducing conditions |
| 1,1,1-TCA | 0.016 | 43 | 15 | 0.0013 | 533 | 0.01 | 69 | Range not appropriate for nitrate-reducing conditions. Expect lower limit to be much less |
| 1,2-DCA | 0.0076 | 91 | 2 | 0.0042 | 165 | 0.011 | 63 | Range reported from a single field study under methanogenic conditions |

| Compound | Mean of field/in-situ studies | | | Recommended 1st order rate coefficients | | | | Comments |
|---|---|---|---|---|---|---|---|---|
| | 1st order rate | | number studies used for mean | low-end 1st order rate | | high-end 1st order rate | | |
| | coeffi-cients, day-1 | half-lives, day | | coeffi-cients, day-1 | half-lives day | coeffi-cients, day-1 | half-lives day | |
| Carbon tetrachloride | 0.37 | 1.9 | 9 | 0.0037 | 187 | 0.13 | 5 | Range not appropriate for nitrate-reducing conditions. Expect lower limits to be much less |
| Chloroform | 0.030 | 23 | 1 | 0.0004 | 1,733 | 0.03 | 23 | Only one field study available. Biodegradation under nitrate-reducing conditions expected |
| Dichloromethane | 0.0064 | 108 | 1 | 0.0064 | 108 | - | - | Rate constant reported from a single field study under methanogenic conditions |
| Trichlorofluoro-methane | - | - | - | 0.00016 | 4,331 | 0.0016 | 433 | All studies with very low concentrations of this compound |
| 2, 4-Dichlorophenol | 0.014 | 50 | 2 | 0.00055 | 1,260 | 0.027 | 26 | Range may not be appropriate for nitrate reducing conditions |

[From T.H. Wiedemeier, H. S. Rifai, C. J. Newell and J.T. Wilson, **Natural Attenuation of Fuels and Chlorinated Solvents in the Subsurface**. Copyright © 1999 *John Wiley & Sons, Inc.* Reprinted by permission of John Wiley & Sons, Inc.]

## 23.1.5 GEOCHEMICAL EVIDENCE OF NATURAL BIOREMEDIATION AT CHLORINATED SOLVENT SITES

### 23.1.5.1 Assessing reductive dechlorination at field sites

Assessing biological activity at a field site based on monitoring data can be difficult. However, there are a number of monitoring parameters that can be indicative of halorespiration. First, the presence of methane in the groundwater indicates that fermentation is occurring and that the potential for halorespiration exists. Second, the transformation of PCE and TCE has been studied intensely and many researchers report that of the three possible DCE isomers, 1,1-DCE is the least significant intermediate and that cis-1,2-DCE predominates over trans-1,2-DCE.[103-105] Third, because chlorinated ethenes are 55 to 85% chlorine by mass, the degradation of these compounds releases a large mass of chloride. Therefore, elevated chloride concentrations also indicate reductive dechlorination.

### 23.1.5.2 Plume classification schemes

Wiedemeier et al.[106] proposed a classification system for chlorinated solvent plumes based on the amount and origin of fermentation substrates that produce the hydrogen that drives halorespiration. Three types of groundwater environments and associated plume behavior, Type 1, Type 2, and Type 3, are described below. While the classification system can be used to represent entire plumes, it can also be used to define different zones within a chlorinated solvent plume.

#### *23.1.5.2.1 Type 1*

For highly chlorinated solvents to biodegrade, anaerobic conditions must prevail within the contaminant plume. Anaerobic conditions are typical at sites contaminated with fuel hydro-

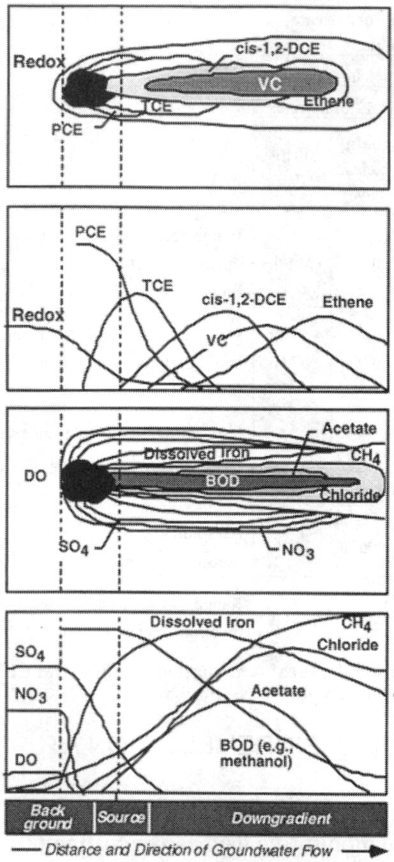


Figure 23.1.6. Conceptual model of Type 1 environment for chlorinated solvent plumes. [From T.H. Wiedemeier, H. S. Rifai, C. J. Newell and J.T. Wilson, **Natural Attenuation of Fuels and Chlorinated Solvents in the Subsurface**, after reference 88. Copyright © 1999 *John Wiley & Sons, Inc.* Reprinted by permission of John Wiley & Sons, Inc.]

carbons, landfill leachate, or other anthropogenic carbon because these organics exert a tremendous electron-acceptor demand on the system. This condition is referred to as a Type 1 environment. In a Type 1 environment, anthropogenic carbon is fermented to produce hydrogen which drives halorespiration.

The geochemistry of groundwater in a Type 1 environment is typified by strongly reducing conditions. This environment is characterized by very low concentrations of dissolved oxygen, nitrate, and sulfate and elevated concentrations of Fe(II) and methane in the source zone (Figure 23.1.6). The presence of methane is almost always observed and confirms that fermentation has been occurring at the site, generating hydrogen. If measured, hydrogen concentrations are typically greater than 1 nanomolar. Importantly, a Type 1 environment results in the rapid and extensive degradation of the more highlychlorinated solvents such as PCE, TCE, and DCE: PCE $\rightarrow$ TCE $\rightarrow$ DCE $\rightarrow$ VC $\rightarrow$ Ethene $\rightarrow$ Ethane

In this type of plume, cis-1,2-DCE and VC degrade more slowly than TCE; thus, they tend to accumulate and form longer plumes (Figure 23.1.6a). In Figure 23.1.6b, the PCE declines to zero and is replaced, in sequence, by a peak in TCE concentrations, followed by a peak in cis-1,2-DCE, VC, and ethene. Fermentation constituents (BOD and acetate) and inorganics are shown in Figure 23.1.6c and 23.1.6d. Figure 23.1.6d illustrates how the fermentation substrate (represented by BOD) extends beyond the source before being consumed. Both panels show long chloride and methane plumes extending far downgradient from the plume area, because chloride is conservative and methane cannot be biodegraded in an anaerobic environment. The acetate curve indicates where active primary fermentation is occurring; declining acetate concentrations are due to consumption by methanogens in the plume area.

#### *23.1.5.2.2 Type 2*

The classification system of Wiedemeier et al.[107] recognized that anaerobic conditions may also result from the fermentation of naturally-occurring organic material in the groundwater that flows through chlorinated solvent source zones. This Type 2 environment occurs in hydrogeologic settings that have inherently high organic carbon concentrations, such as coastal or stream/river deposits with high concentrations of organics, shallow aquifers with recharge zones in organic-rich environments (such as swamps), or zones impacted by natural oil seeps. A Type 2 environment generally results in slower biodegradation of the highly chlorinated solvents compared to a Type 1 environment. However, given sufficient organic loading, this environment can also result in rapid degradation of these compounds. A Type 2 environment typically will not occur in crystalline igneous and metamorphic rock (see discussion of likely hydrogeologic settings for Type 3 environments).

#### *23.1.5.2.3 Type 3*

A Type 3 environment is characterized by a well-oxygenated groundwater system with little or no organic matter. Concentrations of dissolved oxygen typically are greater than 1.0 mg/L. In such an environment, halorespiration will not occur and chlorinated solvents such as PCE, TCE, TCA, and CT will not biodegrade. In this environment, very long dissolved-phase plumes are likely to form. The most significant natural attenuation mechanisms for PCE and TCE will be advection, dispersion, and sorption. However, VC (and possibly DCE) can be rapidly oxidized under these conditions. A Type 3 environment is often found in crystalline igneous and metamorphic rock (fractured or unfractured) such as basalt, granite, schist, phyllite, glacial outwash deposits, eolian deposits, thick deposits of well-sorted, clean, beach sand with no associated peat or other organic carbon deposits, or any other type of deposit with inherently low organic carbon content if no anthropogenic carbon has been released.

Two conceptual models are provided for environments in which Type 3 behavior occurs. For sources with PCE and TCE, the major natural attenuation processes are dilution and dispersion alone (no biodegradation). As shown in 23.1.7, the PCE and TCE plumes extend from the source zone and concentrations are slowly reduced by abiotic processes. Chloride concentrations and oxidation-reduction potential will not change as groundwater passes through the source zone and forms the chlorinated ethene plume. If TCA is the solvent of interest, significant abiotic hydrolysis may occur, resulting in a more rapid decrease in TCA concentrations and an increase in chloride concentrations.

In Figure 23.1.7, a source releases VC and 1,2-DCA into the groundwater at a Type 3 site (an unlikely occurrence as more highly chlorinated solvents are typically released at sites). Because the VC and 1,2-DCA can be degraded aerobically, these constituents decline in concentration at a significant rate. Chloride is produced, and a depression in dissolved oxygen concentration similar to that occurring at fuel sites, is observed.

#### *23.1.5.2.4 Mixed environments*

As mentioned above, a single chlorinated solvent plume can exhibit different types of behavior in different portions of the plume. This can be beneficial for natural biodegradation of chlorinated solvent plumes. For natural attenuation, this may be the best scenario. PCE, TCE, and DCE are reductively dechlorinated with accumulation of VC near the source area (Type 1); then, VC is oxidized (Type 3) to carbon dioxide, either aerobically or via Fe(III) reduction further downgradient and does not accumulate. Vinyl chloride is removed from the system much faster under these conditions than under reducing conditions.

A less ideal variation of the mixed Type 1 and Type 3 environments is shown in the conceptual model in Figure 23.1.8. An extended TCE and 1,2-DCE plume results because insufficient fermentable carbon results in an anaerobic zone which is too short for complete biodegradation. Therefore, TCE extends well into the aerobic zone where no biodegradation occurs. A long DCE plume also extends into the aerobic zone, indicating in-

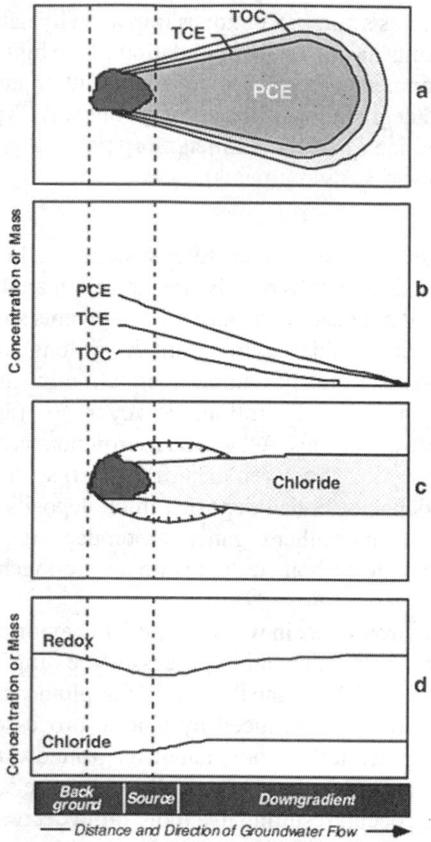


Figure 23.1.7. Conceptual model of Type 3 environment for chlorinated solvent plumes. [From T.H. Wiedemeier, H. S. Rifai, C. J. Newell and J.T. Wilson, **Natural Attenuation of Fuels and Chlorinated Solvents in the Subsurface**, after reference 88. Copyright © 1999 *John Wiley & Sons, Inc*. Reprinted by permission of John Wiley & Sons, Inc.]

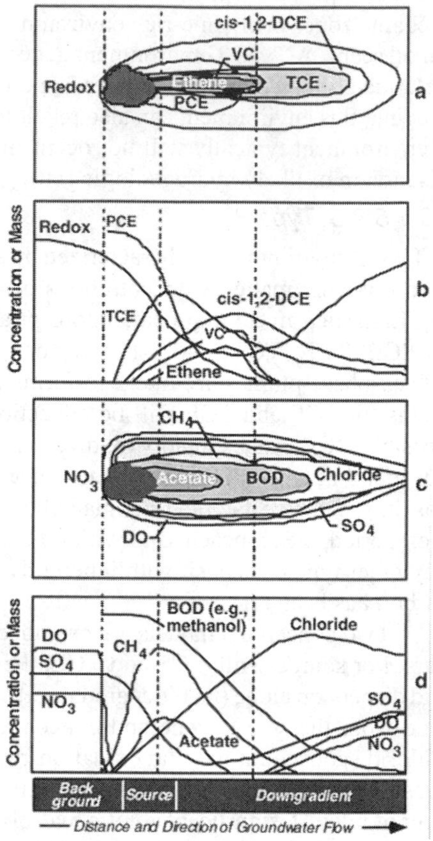


Figure 23.1.8. Conceptual model of mixed environments with Type 1 environment in source zone and Type 3 environment downgradient of source. [From T.H. Wiedemeier, H. S. Rifai, C. J. Newell and J.T. Wilson, **Natural Attenuation of Fuels and Chlorinated Solvents in the Subsurface**, after reference 88. Copyright © 1999 *John Wiley & Sons, Inc*. Reprinted by permission of John Wiley & Sons, Inc.]

significant direct aerobic biodegradation was assumed. While a long chloride plume will be observed, the short anaerobic zone means much less methane is produced, allowing dilution/dispersion to limit the extent of the methane plume.

## 23.1.6 CHLORINATED SOLVENT PLUMES - CASE STUDIES OF NATURAL ATTENUATION

### 23.1.6.1 Plume databases

Two different databases provided chlorinated solvent site data. The first database, the Hydrogeologic Database (HGDB),[2] provided information on plume length, plume width, plume thickness, and highest solvent concentration for 109 chlorinated solvent sites. The second database[1] condensed extensive site characterization data from 17 Air Force chlorinated solvent sites, with information on parent compounds vs. progeny products concentrations, competing electron acceptors, hydrogen, and metabolic by-products.

The data in the HGDB were broken into two groups: the chlorinated ethenes, where one or more of the chlorinated ethenes (PCE, TCE, DCE, or VC) was reported to be the major contaminant, and other chlorinated solvent sites, where all other chlorinated solvents besides the ethenes (e.g., TCA, DCA, chlorobenzene) were lumped together. As shown in Table 23.1.10, the median length of the 75 chlorinated ethene plumes was 1000 ft, with one site reporting a plume length of 13,200 ft. These median lengths are longer than those reported for fuel hydrocarbon plumes and this may be attributed to the competition for hydrogen during halorespiration.

**Table 23.1.10. Characteristics of chlorinated solvent plumes from HGDB database**

| | Plume length, ft | Plume width, ft | Vertical penetration, ft | Highest concentration, mg/L |
|---|---|---|---|---|
| Chlorinated ethenes (e.g., PCE, TCE, etc.) | | | | |
| Maximum | 13,200 | 4,950 | 500 | 28,000 |
| 75$^{th}$ percentile | 2,500 | 1,000 | 100 | 72 |
| Median | 1,000 | 500 | 40 | 8.467 |
| 25$^{th}$ percentile | 600 | 200 | 25 | 0.897 |
| Minimum | 50 | 15 | 5 | 0.001 |
| n | 75 | 75 | 78 | 81 |
| Other chlorinated solvents (e.g., TCA, DCA) | | | | |
| Maximum | 18,000 | 7,500 | 150 | 2,500 |
| 75$^{th}$ percentile | 2,725 | 1,000 | 51 | 13.250 |
| Median | 575 | 350 | 35 | 3.100 |
| 25$^{th}$ percentile | 290 | 188 | 24 | 0.449 |
| Minimum | 100 | 100 | 8 | 0.016 |
| n | 24 | 24 | 24 | 28 |

Note: Highest concentration for chlorinated ethenes (28,000 mg/L) was for TCE, which is above the solubility limit. The highest concentration for "other chlorinated solvents" (2500 mg/L) was for chloromethane and toluene. [From T.H. Wiedemeier, H. S. Rifai, C. J. Newell and J.T. Wilson, **Natural Attenuation of Fuels and Chlorinated Solvents in the Subsurface**, after reference 88. Copyright © 1999 *John Wiley & Sons, Inc.* Reprinted by permission of John Wiley & Sons, Inc.]

The other category, "other chlorinated solvent sites," had shorter plumes, with a median plume length of 575 ft compared to 1000 ft for the chlorinated ethene sites. Twelve of the 24 plumes were comprised of TCA, which is degraded biologically via halorespiration and other mechanisms and abiotically by hydrolysis (half life of 0.5 to 2.5 years). Despite the degradability of TCA, the TCA plumes had a median length of 925 ft. The shorter plumes in this database of 24 sites were reported to be comprised of either a general indicator, such as Total Organic Halogens, or individual compounds such as 1,1-dichloroethane, dichloromethane, or chlorobenzene. The median highest concentration at these "other chlorinated solvent sites" was 3.1 mg/L (see Table 23.1.10).

Data compiled from 17 Air Force sites using the AFCEE (Air Force Center for Environmental Excellence) natural attenuation protocol[107] showed a median plume length of 750 ft (based on 14 plumes). There were significant differences in plume length for different plume classes, with Type 1 plumes (sites with available man-made fermentation substrates such as BTEX) being shorter than Type 3 plumes (sites without available fermentation substrates). Twelve of the sites exhibiting Type 1 plumes had a median plume length of 625 ft, while the two sites with Type 3 plumes had lengths of 1100 and 5000 ft. Four mixed plumes (Type 1 in source zone, Type 3 in the downgradient part of the plume) had a median length of 2538 ft.

Site-specific biodegradation rate information was developed using several methods, including one developed by Buscheck and Alcantar,[108] one based on the use of conservative tracers (the trimethylbenzenes), and other methods such as model calibration. Rates varied significantly, with half-lives ranging from over 300 years for a Type 3 site located in Utah to 0.2 years for a Type 1 site (see Table 23.1.11). The median first order half-life for the 14 chlorinated plumes was 2.1 years.

**Table 23.1.11. Chlorinated solvent plume characteristics**

| No. | State | Type | Plume length, ft | Plume width, ft | Plume thickness, ft | Total chlor. solvents, mg/L | Seepage velocity, ft/yr | First-order biodegradation rate coeff. for solvents, $day^{-1}$ | Half-life, years | Method for calculating rate coefficient |
|---|---|---|---|---|---|---|---|---|---|---|
| 13 | UT | Type 3 | 5000 | 1400 | 40 | 4.953 | 60 | 0.000006 | 316.4 | Other |
| 9 | NY | Mixed | 4200* | 2050 | 60 | 774.721 | 139 | 0.001* | 1.9* | Other |
| 10 | NE | Mixed | 3500 | 1400 | 50 | 164.010 | 152 | 0.000001 | 1899 | Other |
| 8 | MA | Type 1 | 1800 | 1200 | 50 | 4.340 | 106 | 0.0005 | 3.8 | Other |
| 11 | FL | Mixed | 1575* | 400 | 15 | 1258.842 | 113 | 0.0009* | 2.1* | Other |
| 14 | AK | Type 3 | 1100 | 250 | 25 | 4.899 | 260 | 0.0065 | 0.3 | Other |
| 1 | SC | Type 1 | 750 | 550 | 5 | 328.208 | 1600 | - | - | - |
| 7 | MA | Type 1 | 750 | 250 | 50 | 50.566 | 20.8 | 0.0095 | 0.2 | Conserv. tracer |
| 12 | MS | Mixed | 750 | 550 | 5 | 0.472 | 1500 | 0.01 | 0.2 | Buscheck |
| 4 | NE | Type 1 | 650 | 450 | 30 | 47.909 | 6.7 | 0.0006 | 3.2 | Buscheck |
| 3 | FL | Type 1 | 600 | 350 | 20 | 0.429 | 36 | 0.0007 | 2.7 | Buscheck |
| 5 | WA | Type 1 | 550 | 300 | 10 | 3.006 | 32.9 | 0.001 | 1.9 | Buscheck |
| 2 | MI | Type 1 | 375 | 100 | 10 | 0.397 | 292 | - | - | - |
| 6 | OH | Type 1 | 100 | 60 | 10 | 15.736 | 25 | - | - | - |
| Maximum | | | 5,000 | 2,050 | 60 | 1,259 | 1,600 | 0.0095 | 316.4 | |
| 75th percentile | | | 1,744 | 1,038 | 48 | 136 | 233 | 0.00375 | 3.5 | |
| Median | | | 750.0 | 425.0 | 22.5 | 10.3 | 109.5 | 0.0009 | 2.1 | |
| 25th percentile | | | 613 | 263 | 10 | 3.3 | 34 | 0.00055 | 1.1 | |
| Minimum | | | 100 | 60 | 5 | 0.397 | 7 | 0.000006 | 0.2 | |
| n | | | 14 | 14 | 14 | 14 | 14 | 11 | 11 | |

*Plume discharges into stream; may not represent maximum potential plume length.
Mixed refers to Type 1 conditions in source zone, Type 3 conditions in downgradient part of plume. Median length Type 1 sites: 625 ft, mixed sites 2538 ft, Type 3: 3050 ft (two sites). [From T.H. Wiedemeier, H. S. Rifai, C. J. Newell and J.T. Wilson, **Natural Attenuation of Fuels and Chlorinated Solvents in the Subsurface**, after reference 88. Copyright © 1999 *John Wiley & Sons, Inc.* Reprinted by permission of John Wiley & Sons, Inc.]

### 23.1.6.2 Modeling chlorinated solvent plumes

Very few models exist (analytical or numerical) which are specifically designed for simulating the natural attenuation of chlorinated solvents in ground water. Ideally, a model for simulating natural attenuation of chlorinated solvents would be able to track the degradation of a parent compound through its daughter products and allow the user to specify differing decay rates for each step of the process. This may be referred to as a reactive transport model, in which transport of a solute may be tracked while it reacts, its properties change due to those reactions, and the rates of the reactions change as the solute properties change. Moreover, the model would also be able to track the reaction of those other compounds that react with or are consumed by the processes affecting the solute of interest (e.g., electron donors and acceptors).

Two models: BIOCHLOR and RT3D for the natural attenuation of chlorinated solvents have been presented recently in the general literature and will be briefly discussed in this section.

#### *23.1.6.2.1 BIOCHLOR natural attenuation model*

The BIOCHLOR Natural Attenuation Model[109] simulates chlorinated solvent natural attenuation using an Excel based interface. BIOCHLOR simulates the following reductive dechlorination process:

$$PCE \xrightarrow{k_1} TCE \xrightarrow{k_2} DCE \xrightarrow{k_3} VC \xrightarrow{k_4} ETH$$

The equations describing the sequential first order biodegradation reaction rates are shown below for each of the components:

$$r_{PCE} = -k_1 C_{PCE} \tag{23.1.16}$$

$$r_{TCE} = k_1 C_{PCE} - k_2 C_{TCE} \tag{23.1.17}$$

$$r_{DCE} = k_2 C_{TCE} - k_3 C_{DCE} \tag{23.1.18}$$

$$r_{VC} = k_3 C_{DCE} - k_4 C_{VC} \tag{23.1.19}$$

$$r_{ETH} = k_4 C_{ETH} \tag{23.1.20}$$

where:

| | |
|---|---|
| $k_1$, $k_2$, $k_3$, $k_4$ | the first order rate constants |
| $C_{PCE}$, $C_{TCE}$, $C_{DCE}$, $C_{VC}$ and $C_{ETH}$ | the aqueous concentration of PCE, TCE, DCE, vinyl chloride, and ethene, respectively. |

These equations assume no degradation of ethene.

To describe the transport and reaction of these compounds in the subsurface, one-dimensional advection, three-dimensional dispersion, linear adsorption, and sequential first order biodegradation are assumed as shown in the equations below. All equations, but the first, are coupled to another equation through the reaction term.

$$R_{PCE}\frac{dC_{PCE}}{dt} = -v\frac{dC_{PCE}}{dx} + D_x\frac{d^2C_{PCE}}{dx^2} + D_y\frac{d^2C_{PCE}}{dy^2} + D_z\frac{d^2C_{PCE}}{dz^2} - k_1C_{PCE} \tag{23.1.21}$$

$$R_{TCE}\frac{dC_{TCE}}{dt} = -v\frac{dC_{TCE}}{dx} + D_x\frac{d^2C_{TCE}}{dx^2} + D_y\frac{d^2C_{TCE}}{dy^2} + D_z\frac{d^2C_{TCE}}{dz^2} + k_1C_{PCE} - k_2C_{TCE} \quad [23.1.22]$$

$$R_{DCE}\frac{dC_{DCE}}{dt} = -v\frac{dC_{DCE}}{dx} + D_x\frac{d^2C_{DCE}}{dx^2} + D_y\frac{d^2C_{DCE}}{dy^2} + D_z\frac{d^2C_{DCE}}{dz^2} + k_2C_{TCE} - k_3C_{DCE} \quad [23.1.23]$$

$$R_{VC}\frac{dC_{VC}}{dt} = -v\frac{dC_{VC}}{dx} + D_x\frac{d^2C_{VC}}{dx^2} + D_y\frac{d^2C_{VC}}{dy^2} + D_z\frac{d^2C_{VC}}{dz^2} + k_3C_{DCE} - k_4C_{VC} \quad [23.1.24]$$

$$R_{ETH}\frac{dC_{ETH}}{dt} = -v\frac{dC_{ETH}}{dx} + D_x\frac{d^2C_{ETH}}{dx^2} + D_y\frac{d^2C_{ETH}}{dy^2} + D_z\frac{d^2C_{ETH}}{dz^2} + k_4C_{ETH} \quad [23.1.25]$$

where:

| | |
|---|---|
| $R_{PCE}$, $R_{TCE}$, $R_{DCE}$, $R_{VC}$, $R_{ETH}$ | retardation factors |
| V | seepage velocity |
| $D_x$, $D_y$, $D_z$ | dispersivities in the x, y, and z directions. |

BIOCHLOR uses a novel analytical solution to solve these coupled transport and reaction equations in an Excel spreadsheet. To uncouple these equations, BIOCHLOR employs transformation equations developed by Sun and Clement.[110] The uncoupled equations were solved using the Domenico model, and inverse transformations were used to generate concentration profiles. Details of the transformation are presented elsewhere.[110] Typically, source zone concentrations of cis-1,2-dichloroethythene (DCE) are high because biodegradation of PCE and TCE has been occurring since the solvent release.

BIOCHLOR also simulates different first-order decay rates in two different zones at a chlorinated solvent site. For example, BIOCHLOR is able to simulate a site with high dechlorination rates in a high-carbon area near the source that becomes a zone with low dechlorination rates downgradient when fermentation substrates have been depleted.

In addition to the model, a database of chlorinated solvent sites is currently being analyzed to develop empirical rules for predicting first-order coefficients that can be used in BIOCHLOR. For example, at sites with evidence of considerable halorespiration, the use of higher first order decay coefficients will be recommended. Indicators of high rates of halorespiration may include: i) high concentrations of fermentation substrates such as BTEX at the site, ii) high methane concentrations, which indicate high rates of fermentation, and iii) large ratios of progeny products to parent compounds, and iv) high concentrations of source zone chloride compared to background chloride concentrations.

The BIOCHLOR model was used to reproduce the movement of the Cape Canaveral plume from 1965 to 1998. The Cape Canaveral site (Figure 23.1.9) is located in Florida and exhibits a TCE plume which is approximately 1,200 ft long and 450 ft wide. TCE concentrations as high as 15.8 mg/L have been measured recently at the site. The site characteristics used in the BIOCHLOR model are listed in Table 23.1.12. The hydraulic conductivity assumed in the model was $1.8 \times 10^{-2}$ cm/sec and the hydraulic gradient was 0.0012. A porosity of 0.2 was assumed as well as the Xu and Eckstein model for longitudinal dispersivity.[8] The lateral dispersivity was assumed to be 10% of the longitudinal dispersivity and vertical dispersion was neglected.

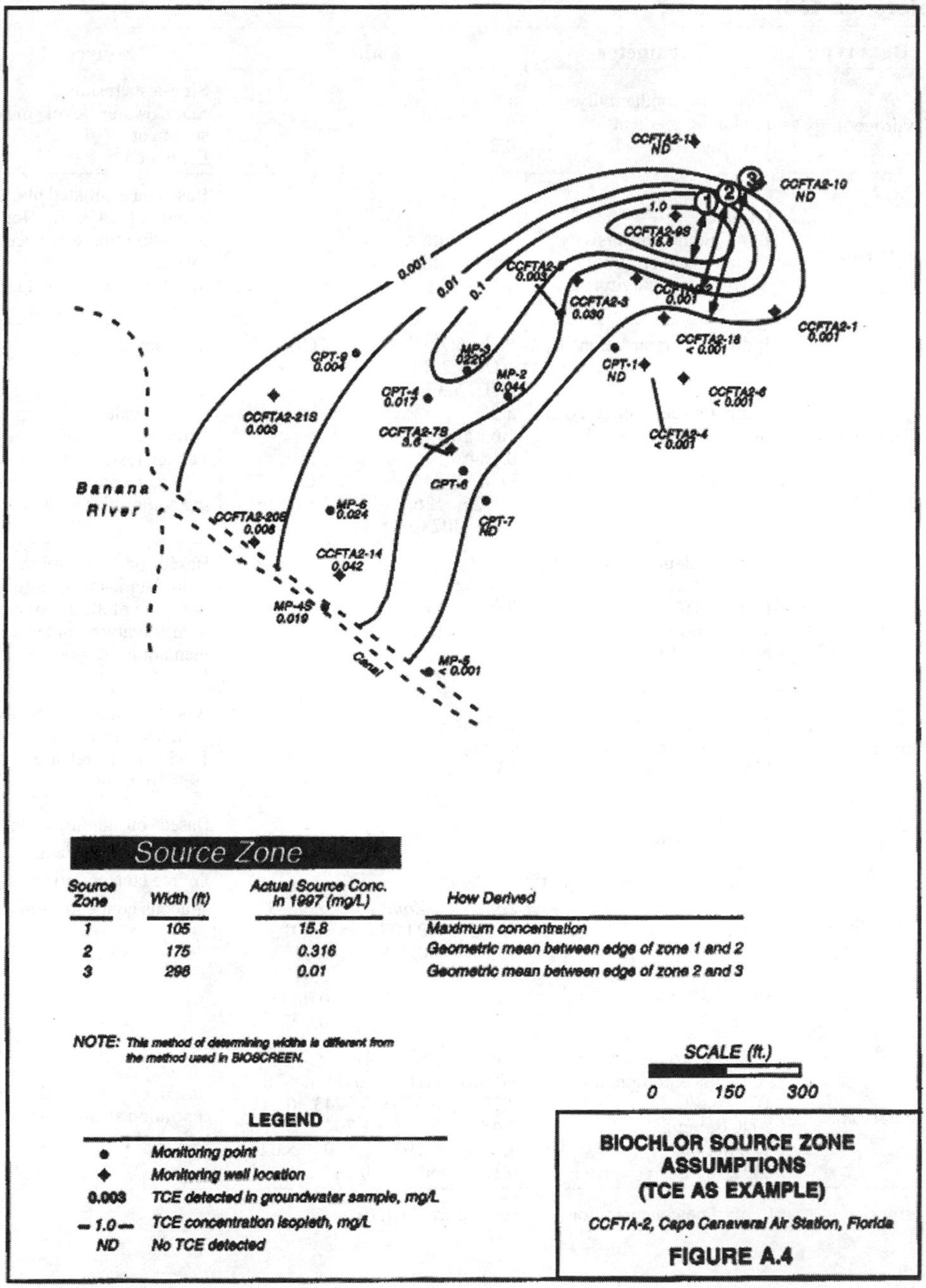


Figure 23.1.9. BIOCHLOR source zone assumptions (TCE as example), CCFTA-2, cape Canaveral Air Station, Florida [From reference 109].

**Table 23.1.12. BIOCHLOR example, Cape Canaveral Air Station, Florida. [From reference 109]**

| Data type | Parameter | Value | Source |
|---|---|---|---|
| Hydrogeology | Hydraulic conductivity:<br>Hydraulic gradient:<br>Porosity: | $1.8 \times 10^{-2}$ cm/s<br>0.0012 ft/ft<br>0.2 | Slug-tests results<br>Static water level measurement<br>Estimated |
| Dispersion | Original<br>Longitudinal dispersivity:<br>Transverse dispersivity:<br>Vertical dispersivity: | <br>varies with x<br>varies with x<br>0 ft | Based on estimated plume length of 1450 ft. Note: No calibration was necessary to match observed plume length |
| Adsorption | Individual retardation factors:<br><br>Common retardation factor:<br>Soil bulk density $\rho_b$:<br>$f_{oc}$:<br>$K_{oc}$: (L/kg) | PCE: 6.7 TCE: 2.8<br>c-DCE:2.8 VC: 5.6<br>ETH: 5.3<br>5.3<br>1.6 kg/L<br>0.184%<br>PCE: 398 TCE: 126<br>c-DCE: 126 VC: 316<br>ETH: 302 | Calculated<br><br><br>Median value<br>Estimated<br>Lab analysis<br>Literature correlation using solubilities at 20°C |
| Biodegradation | Biodegradation rate coefficients (1/year):<br>PCE → TCE<br>TCE → c-DCE<br>c-DCE → VC<br>C → ETH | <br>2.0<br>0.9<br>0.6<br>0.4 | Based on calibration to field data using a simulation time of 32 yr. Started with literature values and then adjusted model to fit field |
| General | Modeled area length<br>Modeled area width<br>Simulation time | 1085 ft<br>700 ft<br>33 yrs | Based on area of affected groundwater plume from 1965 (first release) to 1998 (present) |
| Source data | Source thickness<br>Source widths, ft<br><br>Source concentrations, mg/L<br><br>PCE<br>TCE<br>c-DCE<br>VC<br>ETH | 56 ft<br>*Zone 1* *Zone 2* *Zone 3*<br>105 175 298<br><br>*Zone 1* *Zone 2* *Zone 3*<br>0.056 0.007 0.001<br>15.8 0.318 0.01<br>98.5 1.0 0.01<br>3.080 0.089 0.009<br>0.030 0.013 0.003 | Based on geologic logs and monitoring data.<br>Source concentrations are aqueous concentrations |
| Actual data | Distance from source, ft<br>PCE concentration, mg/L<br>TCE, mg/L<br>c-DCE, mg/L<br>VC, mg/L<br>ETH, mg/L | *560* *650* *930* *1085*<br><0.001 ND <0.001 <0.001<br>0.22 0.0165 0.0243 0.019<br>3.48 0.776 1.200 0.556<br>3.080 0.797 2.520 5.024<br>0.188 ND 0.107 0.150 | Based on observed concentration at site near centerline of plume |
| Output | Centerline concentration | see Figure 23.1.10 | |

A median value for the retardation factor was used (R=5.3) since BIOCHLOR accepts only one value for this parameter. The site was modeled using one anaerobic zone with one set of rate coefficients as shown in Table 23.1.12. This is justified because the dissolved ox-

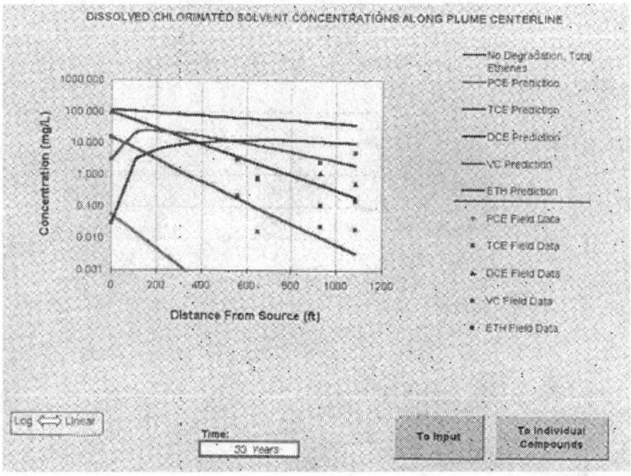


Figure 23.1.10. Centerline output. Cape Canaveral Air Force Base, Florida. [From references 109].

ygen readings at the site were less than 0.7 mg/L at all monitoring points. The rate coefficients were calculated by calibrating the model to the 1997 field data. The source zone was simulated as a spatially-variable source and the source concentrations ranged from 0.001 to 98.5 mg/L for the various compounds as shown in Table 23.1.12. The source thickness was estimated by using the deepest point in the aquifer where chlorinated solvents were detected.

Centerline concentrations for all five species (PCE, TCE, c DCE, VC and ETH) predicted by the model are shown in Figure 23.1.10. Figure 23.1.10 shows the centerline predictions for each chlorinated solvent and a no degradation curve for all of the chlorinated solvents as well as field data. The data in Figure 23.1.10 indicate that TCE concentrations discharging into the ocean will be less than 0.001 mg/L.

### 23.1.6.3 RT3D numerical model

RT3D (Reactive Transport in 3 Dimensions)[111] is a FORTRAN 90-based model for simulating 3D multi-species, reactive transport in groundwater. This model is based on the 1997 version of MT3D (DOD Version 1.5), but has several extended reaction capabilities. RT3D can accommodate multiple sorbed and aqueous phase species with any reaction framework that the user needs to define. RT3D can simulate different scenarios, since a variety of pre-programmed reaction packages are already provided and the user has the ability to specify their own reaction kinetic expressions. This allows, for example, natural attenuation processes or an active remediation to be evaluated. Simulations can be applied to modeling contaminants such as heavy metals, explosives, petroleum hydrocarbons, and/or chlorinated solvents.

RT3D's pre-programmed reaction packages include:

1 Two species instantaneous reaction (hydrocarbon and oxygen).
2 Instantaneous hydrocarbon biodegradation using multiple electron acceptors ($O_2$, $NO_3^-$, $Fe^{2+}$, $SO_4^{2-}$, $CH_4$).
3 Kinetically limited hydrocarbon biodegradation using multiple electron acceptors ($O_2$, $NO_3^-$, $Fe^{2+}$, $SO_4^{2-}$, $CH_4$).
4 Kinetically limited reaction with bacterial transport (hydrocarbon, oxygen, and bacteria).
5 Non-equilibrium sorption/desorption. Can also be used for non-aqueous phase liquid dissolution).
6 Reductive, anaerobic biodegradation of PCE, TCE, DCE, and VC.
7 Reductive, anaerobic biodegradation of PCE, TCE, DCE, and VC combined with aerobic biodegradation of DCE and VC.
8 Combination of #3 and #7.

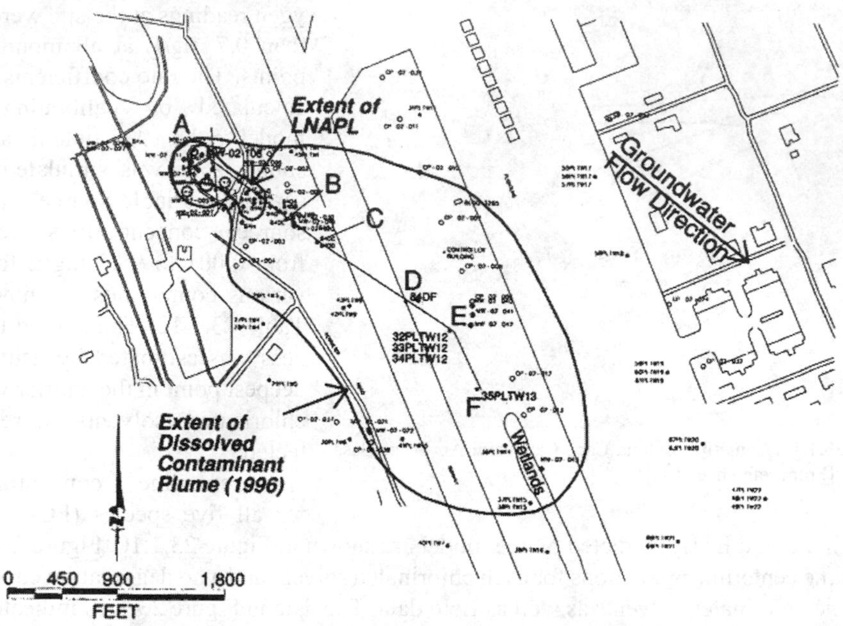


Figure 23.1.11. Site layout, Plattsburg Air Force Base, New York. [From T.H. Wiedemeier, H. S. Rifai, C. J. Newell and J.T. Wilson, **Natural Attenuation of Fuels and Chlorinated Solvents in the Subsurface**, after reference 88. Copyright © 1999 *John Wiley & Sons, Inc.* Reprinted by permission of John Wiley & Sons, Inc.]

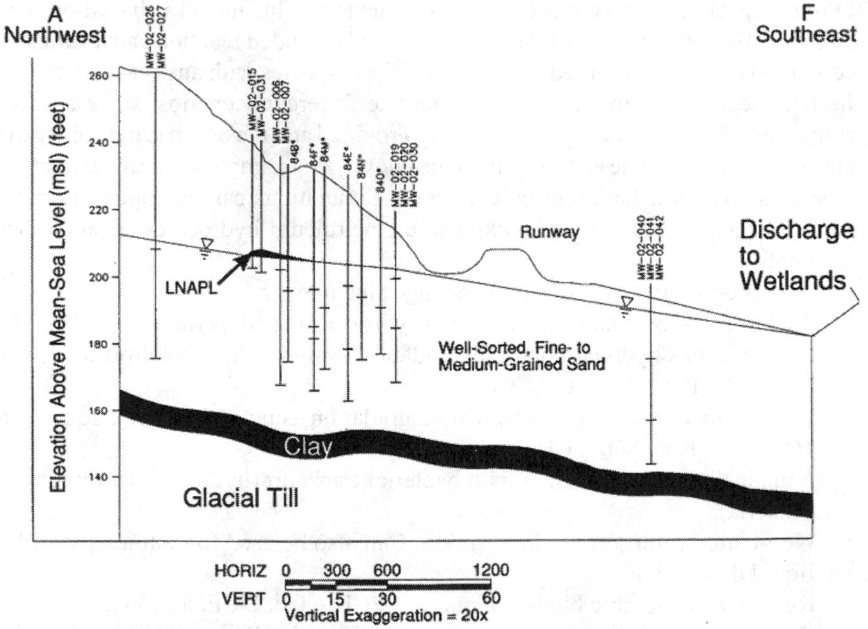


Figure 23.1.12. Hydrogeologic section, Plattsburg Air Force Base, New York. [From T.H. Wiedemeier, H. S. Rifai, C. J. Newell and J.T. Wilson, **Natural Attenuation of Fuels and Chlorinated Solvents in the Subsurface**, after reference 88. Copyright © 1999 *John Wiley & Sons, Inc.* Reprinted by permission of John Wiley & Sons, Inc.]

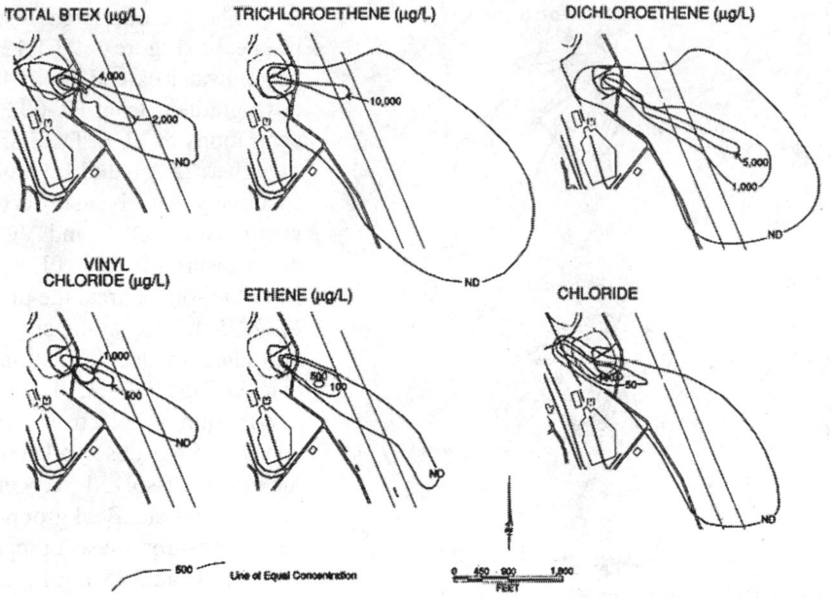


Figure 23.1.13. Chlorinated solvents and by-products, 1995, Plattsburg Air Force Base, New York. [From T.H. Wiedemeier, H. S. Rifai, C. J. Newell and J.T. Wilson, **Natural Attenuation of Fuels and Chlorinated Solvents in the Subsurface**, after reference 88. Copyright © 1999 *John Wiley & Sons, Inc*. Reprinted by permission of John Wiley & Sons, Inc.]

RT3D represents a remarkable breakthrough in the development and solving of optimization models of bioremediation design. It is a modular, three-dimensional simulator capable of predicting mufti-species (solutes and microbes), bio-reactive transport while understanding natural attenuation and active bioremediation processes.

#### 23.1.6.4 CS case study - The Plattsburgh Air Force Base

The Plattsburgh Air Force Base (AFB) in New York is a former fire training facility (site FT002). Activities at FT-002 (Figure 23.1.11) have caused contamination of shallow soils and groundwater with a mixture of chlorinated solvents and fuel hydrocarbons. Groundwater contaminants include TCE, cis-1,2-DCE, VC and BTEX. The site is underlain with 4 distinct stratigraphic units: sand, clay, till and carbonate bedrock. The depth to groundwater in the sand aquifer ranges from 45 ft below ground surface (BGS) on the west side of the site to zero on the east side of the runway (Figure 23.1.12). Groundwater flow is to the southeast and the average gradient is about 0.01 ft/ft.[1] Hydraulic conductivity values for the unconfined sand aquifer range from 0.059 to 90.7 ft/day. Wiedemeier et al.[1] estimated an average velocity of 142 ft/yr for the sand aquifer.

The extent of Light Non-Aqueous Phase (LNAPL) contamination at Plattsburgh is shown in Figure 23.1.13. The LNAPL is a mixture of jet fuel and waste solvents from which BTEX and TCE dissolve into the ground water (DCE and VC are not present in the waste mixture). The dissolved BTEX plume (Figure 23.1.13) extends approximately 2000 ft downgradient from the site and has a maximum width of about 500 ft. BTEX concentrations as high as 17 mg/L were measured in the source area. Historical data from FT-002 indicate that the dissolved BTEX plume has reached a quasi-steady state and is no longer expanding.

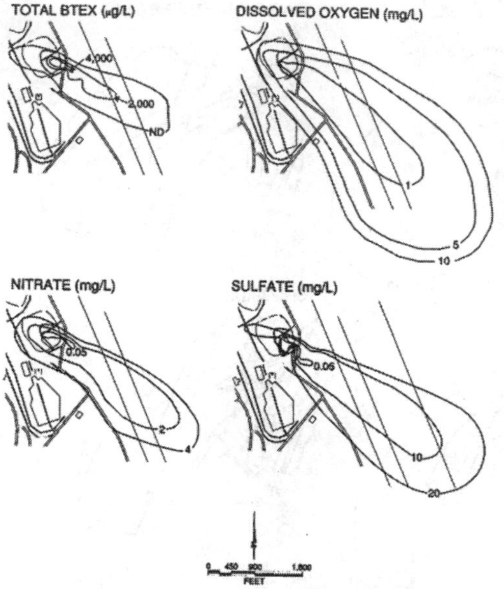


Figure 23.1.14. BTEX and electron acceptors, 1995, Plattsburg Air Force Base, New York. [From T.H. Wiedemeier, H. S. Rifai, C. J. Newell and J.T. Wilson, **Natural Attenuation of Fuels and Chlorinated Solvents in the Subsurface**, after reference 88. Copyright © 1999 *John Wiley & Sons, Inc*. Reprinted by permission of John Wiley & Sons, Inc.]

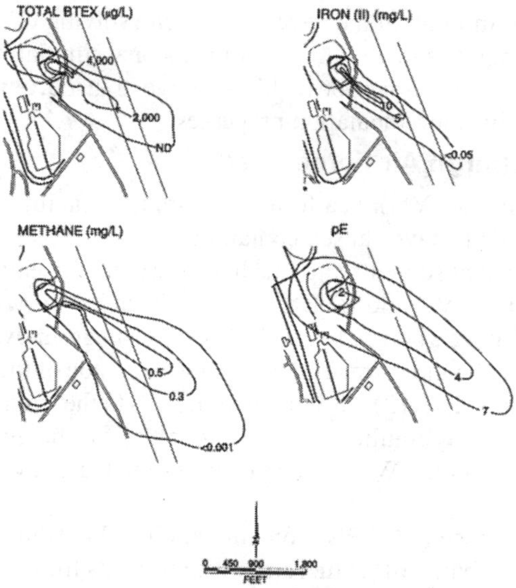


Figure 23.1.15. BTEX and metabolic by-products, 1995, Plattsburg Air Force Base, New York. [From T.H. Wiedemeier, H. S. Rifai, C. J. Newell and J.T. Wilson, **Natural Attenuation of Fuels and Chlorinated Solvents in the Subsurface**, after reference 88. Copyright © 1999 *John Wiley & Sons, Inc*. Reprinted by permission of John Wiley & Sons, Inc.]

The chlorinated solvent plumes (Figure 23.1.13) in groundwater extend about 4000 ft downgradient from FT-002. Concentrations of TCE, DCE and VC as high as 25, 51, and 1.5 mg/L, respectively, have been observed recently. Since DCE and VC were not measured in LNAPL samples from the source area, the presence of DCE and VC at the site can be attributed to dechlorination. The data in Figure 23.1.14 show the distribution of electron acceptor concentrations observed at the site including dissolved oxygen, nitrate and sulfate. Background concentrations for these compounds are 10, 10 and 25 mg/L, respectively and their absence within the contaminated zones is an indication of biodegradation of BTEX and chlorinated solvents at the site.

Figure 23.1.15, on the other hand, shows the distribution of metabolic by-products of the biodegradation reactions including ferrous iron and methane. The presence of these by-products is further evidence of biological activity in the aquifer. Elevated chloride and ethene concentrations as shown in Figure 23.1.13 suggest that TCE, DCE and VC are being biodegraded. Wiedemeier et al.[1] calculated apparent biodegradation constants for FT-002 using trimethylbenzene as a conservative tracer. Their results are shown in Table 23.1.13. The data in Table 23.1.13 indicate biodegradation rates of the chlorinated solvents at the site ranging between 0 and 1.27 per yr.

**Table 23.1.13. Approximate first-order biodegradation rate constants**

| Compound | Correction method | A-B 0-970 ft, $year^{-1}$ | B-C 970-1240 ft, $year^{-1}$ | C-E 1240-2560 ft, $year^{-1}$ |
|---|---|---|---|---|
| TCE | Chloride | 1.27 | 0.23 | -0.30 |
| | TMB | 1.20 | 0.52 | NA |
| | Average | 1.24 | 0.38 | -0.30 |
| DCE | Chloride | 0.06 | 0.60 | 0.07 |
| | TMB | 0.00 | 0.90 | NA |
| | Average | 0.03 | 0.75 | 0.07 |
| VC | Chloride | 0.00 | 0.14 | 0.47 |
| | TMB | 0.00 | 0.43 | NA |
| | Average | 0.00 | 0.29 | 0.47 |
| BTEX | Chloride | 0.13 | 0.30 | 0.39 |
| | TMB | 0.06 | 0.60 | NA |
| | Average | 0.10 | 0.45 | 0.39 |

[a]NA, not analyzed
[From T.H. Wiedemeier, H. S. Rifai, C. J. Newell and J.T. Wilson, **Natural Attenuation of Fuels and Chlorinated Solvents in the Subsurface**, after reference 88. Copyright © 1999 *John Wiley & Sons, Inc*. Reprinted by permission of John Wiley & Sons, Inc.]

Available geochemical data were analyzed by Wiedemeier et al.[1] and they concluded that the geochemistry of the ground water near the source area and for about 1500 ft downgradient is significantly different from the groundwater further downgradient from the source (between 1500 and 4000 ft downgradient). This led the authors to conclude that Plattsburgh exhibits Type 1 behavior near the source and Type 3 behavior within the leading edge of the plume (see Section 23.1.5.2).

In the area extending to 1500 ft downgradient from the source, BTEX and TCE are comingled in the ground water. This area is characterized by anaerobic conditions that are strongly reducing. BTEX is being used as a primary substrate and TCE is being reductively dechlorinated to cis-1,2-DCE and VC. Between 1500 and 2000 ft downgradient from the source, however, the majority of the BTEX has been biodegraded and the system exhibits Type 3 behavior. These conditions are not optimal for reductive dechlorination, and it is likely that VC is being oxidized via ferric reduction or aerobic respiration.

## REFERENCES


1 T. H. Wiedemeier, H. S. Rifai, C. J. Newell and J. T. Wilson, **Natural Attenuation of Fuels and Chlorinated Solvents in the Subsurface**, *John Wiley & Sons, Inc*, New York, NY, 1999.

2 C. J. Newell, L. P. Hopkins and P. B. Bedient, *Ground Water*, **28**, 703-714 (1990).

3 L. W. Gelhar, A. Montoglou, C. Welty and K. R. Rehfeldt, A Review of Field Scale Physical Solute Transport Processes in Saturated and Unsaturated Porous Media: Final Project Report, EPRI-EA-4190, Electric Power Research Institute, Palo Alto, CA, 1985.

4 L. W. Gelhar, L. Welty and K. R. Rehfeldt, *Water Resour. Res.*, **28**, 1955-1974 (1992).

5 P. Lallemand-Barres and P. Peaudecerf, *Etude Bibliographique Bulletin, Sec.*, **3/4**, 277-287 (1978).

6 J. F. Pickens and G. E. Grisak, *Water Resour. Res.*, **17**, 1191-1211 (1981).

7 K. Spitz and J. Moreno, **A Practical Guide to Groundwater and Solute Transport Modeling**, *Wiley, Inc.*, New York, New York, 1996.

8 M. Xu and Y. Eckstein, *Journal of Ground Water*, **33**, 905-908 (1995).

9 American Society for Testing and Materials (ASTM), Emergency Standard Guide for Risk-Based Corrective Action Applied at Petroleum Release Sites, ASTM E-1739, Philadelphia, 1995.

10 U.S. Environmental Protection Agency, Background Document for the Ground-Water Screening Procedure to Support 40 CFR Part 269: Land Disposal, EPA/530-SW-86-047, Washington, D.C., 1986.

11 W. J. Lyman, P. J. Reidy and B. Levy, **Mobility and Degradation of Organic Contaminants in Subsurface Environments**, Ed., C. K. Smoley, *CRC*, Boca Raton, FL, 1992.

12 J. Dragun, **The Soil Chemistry of Hazardous Materials**, Ed., *Hazardous Materials Control Research Institute*, Silver Spring, MD, 1988.

13 G. W. Bailey and J. L. White in **Residue Reviews**, F. A. Gunther and J. D. Gunther, Ed., *Springer Verlag*, New York, 1970, pp. 29-92.

14 S. W. Karickhoff, D. S. Brown and T. A. Scott, *Water Resour. Res.*, **13**, 241-248 (1979).

15 E. E. Kenaga and C. A. Goring, *ASTM Special Technical Publication*, **707**, ASTM, Philadelphia, 1980.

16 D. S. Brown and E. W. Flagg, *J. Environ. Qual.,* **10**, 382-386 (1981).

17 R. P. Shwarzenbach and J. Westall, *Environ. Sci. Tech.*, **G15**, 1360-1367 (1981).

18 J. J. Hassett, W. L. Banwart and R. A. Griffin in **Environment and Solid Wastes**, C. W. Francis and S. I. Auerbach, Ed., *Butterworth*, Boston, 1983, pp. 161-178.

19 C. T. Chiou, P. E. Porter and D. W. Schmedding, *Environ. Sci. Tech.*, **17**, 227-231 (1983).

20 R. A. Freeze and J. A. Cherry, *Groundwater*, *Prentice-Hall, Inc.*, Englewood Cliffs, New Jersey, 1979.

21 R. A. Larson and E. J. Weber, **Reaction Mechanisms in Environmental Organic Chemistry**, *Lewis Publishers*, Boca Raton, 1994.

22 U. S. E. P. Agency, Hazardous Waste Treatment, Storage, and Disposal Facilities (TSDF), EPA-450/3-87-026, USEPA, OAQPS, Air Emission Models, 1989.

23 K. Verschueren, **Handbook of Environmental Data on Organic Chemicals**, Second Ed., *Van Nostrand Reinhold Company Inc.*, New York, 1983.

24 NIOSH, Pocket Guide to Chemical Hazards, Ed., U.S. Dept. ofHealth & Human Services, Public Health Service, Centers for Disease Control, National Institute for Occupational Safety and Health, 1990.

25 W. J. Lyman, **Handbook of Chemical Property Estimation Methods**, *McGraw-Hill*, New York, 1982.

26

27 D. M. DiToro, *Chemosphere*, **14**, 1505-1538 (1985).

28 J. H. Montgomery, **Groundwater Chemicals Desk Reference**, 3rd Ed., *Lewis Publishers*, Chelsea, MI, 1990.

29 T. M. Vogel, C. S. Criddle and P. L. McCarty, *Environ. Sci. Tech.*, **21**, 722-736 (1987).

30 T. M. Vogel in **Handbook of Bioremediation**, R. D. Norris et al., Ed., *Lewis Publishers*, Boca Raton, Florida, 1994, pp. 201-225.

31 R. C. Knox, D. A. Sabatini and L. W. Canter, **Subsurface Transport and Fate Processes**, *Lewis Publishers*, Boca Raton, Florida, 1993.

32 R. L. Johnson, C. D. Palmer and W. Fish in **Fate and Transport of Contaminants in the Subsurface**, *U.S. EPA*, Cincinnati, OH and Ada, OK, 1989, pp. 41-56.

33 P. L. McCarty, Symposium on Natural Attenuation of Chlorinated Organics in Ground Water, Dallas, TX, September 11-13, 1996, U.S. EPA, EPA/540/R-96/509, 1996, pp. 5-9.

34 W. Mabey and T. Mill, *J. Phys. Chem. Ref. Data*, **7**, 383-415 (1978).

35 T. M. Vogel and M. Reinhard, *Environ. Sci. Tech.*, **20**, 992-997 (1986).

36 B. J. Butler and J. F. Barker in **Dense Chlorinated Solvents and other DNAPLs in Ground Water: History, Behavior, and Remediation**, J. F. Pankow and J. A. Cherry, Ed., *Waterloo Press*, Waterloo, Canada, 1996, pp.

37 P. M. Jeffers, L. M. Ward, L. M. Woytowitch and N. L. Wolfe, *Environ. Sci. Tech.*, **23**, 965-969 (1989).

38 T. Vogel and P. L. McCarty, *Environ. Sci. Tech.*, **21**, 1208-1213 (1987).

39 W. J. Cooper, M. Mehran, D. J. Riusech and J. A. Joens, *Environ. Sci. Tech.*, **21**, 1112-1114 (1987).

40 W. L. Dilling, N. B. Tfertiller and G. J. Kallos, *Environ. Sci. Tech.*, **9**, 833-838 (1975).

41 J. March, **Advanced Organic Chemistry**, 3rd ed. Ed., *Wiley*, New York, NY, 1985.

42 C. S. Criddle, P. L. McCarty, M. C. Elliott and J. F. Barker, *J. Contam. Hydrol.*, **1**, 133-142 (1986).

43 C. T. Jafvert and N. L. Wolfe, *Environ. Toxicol. Chem.*, **6**, 827-837 (1987).

44 M. Reinhard, G. P. Curtis and M. R. Kriegman, Abiotic Reductive Dechlorination of Carbon Tetrachloride and Hexachloroethane by Environmental Reductants: Project Summary, EPA/600/S2-90/040, U.S. EPA, Washington, DC., 1990.

45 D. W. Acton, Enhanced In Situ Biodegradation of Aromatic and Chlorinated Aliphatic Hydrocarbons in Anaerobic, Leachate-impacted Groundwaters, M. Sc. thesis University of Waterloo, Waterloo, Canada, 1990.

46 R. W. Gillham and S. F. O'Hannesin, *Ground Water*, **32**, 958-967 (1994).

47 T. C. Wang and C. K. Tan, *Bull. Environ. Contam. Toxicol.*, **45**, 149-156 (1990).

48 J. M. Gossett and S. H. Zinder, Symposium on Natural Attenuation of Chlorinated Organics in Groundwater, Dallas, TX, Sept. 11-13, 1996, U.S. EPA, EPA/540/R-96/509, 1996, pp.
49 C. Holliger and W. Schumacher, *Antonie Leeuwenhoek*, **66**, 239-246 (1994).
50 C. Holliger, G. Schraa, A. J. Stams and A. J. Zehnder, *Appl. Environ. Microbiol.*, 59, (1993).
51 C. R. Smatlak, J. M. Gossett and S. Zinder, *Environ. Sci. Tech.*, **30**, 2850-2858 (1996).
52 B. S. Ballapragada, H. D. Stensel, J. A. Puhakka and J. F. Ferguson, *Environ. Sci. Tech.*, **31**, 1728-1734 (1997).
53 P. L. McCarty and L. Semprini in **Handbook of Bioremediation**, R. D. N. e. al, Ed., *Lewis Publishers*, Boca Raton, Florida, 1994, pp. 87-116.
54 G. W. Sewell and S. A. Gibson, *Environ. Sci. Tech.*, **25**, 982-984 (1991).
55 D. E. Fennell, J. M. Gossett and S. H. Zindler, *Environ. Sci. Tech.*, **31**, 918-926 (1997).
56 F. H. Chapelle, P. B. McMahon, N. M. Dubrovsky, R. F. Fujii, E. T. Oaksford and D. A. Vroblesky, *Water Resour. Res.*, **31**, 359-371 (1995).
57 C. Carr and J. B. Hughes, *Environ. Sci. Tech.*, **30**, 1817-1824 (1998).
58 D. R. Lovley and S. Goodwin, *Geochim. Cosmochim. Acta*, **52**, 2993-3003 (1988).
59 D. R. Lovley, F. H. Chapelle and J. C. Woodward, *Environ. Sci. Tech.*, **28**, 1205-1210 (1994).
60 P. L. McCarty, (1997), personal communication.
61 E. Cox, E. Edwards, L. Lehmicke and D. Major in **Intrinsic Bioremediation**, R. E. Hinchee, J. T. Wilson and D. C. Downey, Ed., *Battelle Press*, Columbus, OH, 1995, pp. 223-231.
62 C. Holliger, G. Schraa, A. J. Stains and A. J. Zehnder, *Appl. Environ. Microbiol.*, **58**, (1992).
63 K. Ramanand, M. T. Balba and J. Duffy, *Appl. Environ. Microbiol.*, **59**, (1993).
64 J. M. Suflita and G. T. Townsend in **Microbial Transformation and Degradation of Toxic Organic Chemicals**, L. Y. Young and C. E. Cerniglia, Ed., *Wiley-Liss*, New York, 1995, pp. 654 pp.
65 S. Hartmans, J. A. M. de Bont, J. Tamper and K. C. A. M. Luyben, *Biotechnol. Lett.*, **7**, 383-388 (1985).
66 S. Hartmans and J. A. M. de Bont, *Appl. Environ. Microbiol.*, **58**, 1220-1226 (1992).
67 J. W. Davis and C. L. Carpenter, *Appl. Environ. Microbiol.*, **56**, 3878-3880 (1990).
68 P. M. Bradley and F. H. Chapelle, *Environ. Sci. Tech.*, **30**, 553-557 (1998).
69 G. Stucki, U. Krebser and T. Leisinger, *Experientia*, **39**, 1271-1273 (1983).
70 D. B. Janssen, A. Scheper, L. Dijkhuizen and B. Witholt, *Appl. Environ. Microbiol.*, **49**, 673-677 (1985).
71 W. Reineke and H. J. Knackmuss, *Eur. J. Appl. Microbiol. Biotechnol.*, **47**, 395-402 (1984).
72 G. Schraa, M. L. Boone, M. S. M. Jetten, A. R. W. van Neerven, P. J. Colberg and A. J. B. Zehnder, *Appl. Environ. Microbiol.*, **52**, 1374-1381 (1986).
73 J. P. Spain and S. F. Nishino, *Appl. Environ. Microbiol.*, **53**, 1010-1019 (1987).
74 J. A. M. de Bont, M. J. W. Vorage, S. Hartmans and W. J. J. van den Tweel, *Appl. Environ. Microbiol.*, **52**, 677-680 (1986).
75 B. E. Haigler, S. F. Nishino and J. C. Spain, *Appl. Environ. Microbiol.*, **54**, 294-301 (1988).
76 J. R. van der Meer, W. Roelofsen, G. Schraa and A. J. B. Zehnder, *FEMS Microbiol. Lett.*, **45**, 333-341 (1987).
77 P. Sander, R. M. Wittaich, P. Fortnagel, H. Wilkes and W. Francke, *Appl. Environ. Microbiol.*, **57**, 1430-1440 (1991).
78 S. F. Nishino, J. C. Spain and C. A. Pettigrew, *Environ. Toxicol. Chem.*, **13**, 871-877 (1994).
79 J. C. Spain, Symposium on Natural Attenuation of Chlorinated Organics in Groundwater, Dallas, TX, Sept. 11-13, 1996, U.S. EPA, EPA/540/R-96/509, Washington, D.C., 1996, pp.
80 F. Chapelle, **Groundwater Microbiology and Geochemistry**, *Wiley*, New York, 1993.
81 W. D. Murray and M. Richardson, *Crit. Rev. Environ. Sci. Technol.*, **23**, 195-217 (1993).
82 P. Adriaens and T. M. Vogel in **Microbial Transformation and Degradation of Toxic Organic Chemicals**, L. Y. Young and C. E. Cerniglia, Ed., *Wiley-Liss*, New York, 1995, pp. 654 pp.
83 R. E. Miller and F. P. Guengerich, *Biochemistry*, **21**, 1090-1097 (1982).
84 P. M. Bradley and F. H. Chapelle, *Environ. Sci. Tech.*, **30**, 2084-2086 (1996).
85 P. M. Bradley and F. H. Chapelle, *Environ. Sci. Tech.*, **31**, 2692-2696 (1997).
86 T. M. Vogel and P. L. McCarty, *Appl. Environ. Microbiol.*, **49**, 1080-1083 (1985).
87 E. J. Bouwer in **Handbook of Bioremediation**, R. D. N. e. al., Ed., *Lewis Publishers*, Boca Raton, FL, 1994, pp. 149-175.
88 Remediation Technologies Development Forum, Natural Attenuation of Chlorinated Solvents in Groundwater Seminar, Class notes Ed., RTDF, 1997.
89 J. W. Weaver, J. T. Wilson and D. H. Kampbell, EPA Project Summary, EPA/600/SV-95/001, U.S. EPA, Washington, D.C., 1995.
90 M. P. Suarez and H. S. Rifai, *Biorem. J.*, **3**, 337-362 (1999).

91 R. Oldenhuis, J. Y. Oedzes, J. van der Waarde and D. B. Janssen, *Appl. Environ. Microbiol.*, **57**, 7-14 (1991).
92 J. E. Anderson and P. L. McCarty, *Environ. Sci. Tech.*, **30**, 3517-3524 (1996).
93 A. R. Bielefeldt, H. D. Stensel and S. E. Strand in **Bioremediation of Chlorinated Solvents**, R. E. Hinchee, A. Leeson and L. Semprini, Ed., *Battelle Press*, Columbus, Ohio, 1994, pp. 237-244.
94 B. Z. Fathepure and J. M. Tiedje, *Environ. Sci. Tech.*, **28**, 746-752 (1994).
95 J. Gao, R. S. Skeen and B. S. Hooker in **Bioremediation of Chlorinated Solvents**, R. E. Hinchee, A. Leeson and L. Semprini, Ed., *Battelle Press*, Columbus, Ohio, 1995, pp. 53-59.
96 L. Alvarez-Cohen and P. L. McCarty, *Environ. Sci. Tech.*, **25**, 1381-1387 (1991).
97 U. Jenal-Wanner and P. L. McCarty, *Environ. Sci. Tech.*, **31**, 2915-2922 (1997).
98 J. E. Keenan, S. E. Strand and H. D. Stensel in **Bioremediation of Chlorinated and Polycyclic Aromatic Hydrocarbon Compounds**, R. E. Hinchee, A. Leeson, L. Semprini and S. K. Ong, Ed., *Lewis Publishers*, Boca Raton, Florida, 1994, pp. 1-13.
99 S.-C. Koh, J. P. Bowman and G. S. Sayler in **Bioremediation of Chlorinated and Polycyclic Aromatic Hydrocarbon Compounds**, R. E. Hinchee, A. Leeson, L. Semprini and S. K. Ong, Ed., *Lewis Publishers,* Boca Raton, Florida, 1994, pp. 327-332.
100 S. E. Strand, G. A. Walter and H. D. Stensel in **Bioremediation of Chlorinated Solvents**, R. E. Hinchee, A. Leeson and L. Semprini, Ed., *Battelle Press*, Columbus, Ohio, 1994, pp. 161-167.
101 M. E. Dolan and P. L. McCarty, *Environ. Sci. Tech.*, **29**, 1892-1897 (1995).
102 D. Aronson and P. H. Howard, Anaerobic Biodegradation of Organic Chemicals in Groundwater: A Summary of Field and Laboratory Studies, SRC TR-97-0223F, American Petroleum Institute, Washington, D.C., 1997.
103 F. Parsons, P. R. Wood and J. DeMarco, *J. Am. Water Works Assoc.*, **76**, 56-59 (1984).
104 G. Barrio-Lage, F. Z. Parsons, R. S. Nassar and P. A. Lorenzo, *Environ. Toxicol. Chem.*, **6**, 571578 (1987).
105 F. Parsons and G. Barrio-Lage, *J. Am. Water Works Assoc.*, **77**, 52-59 (1985).
106 T. H. Wiedemeier, M. A. Swanson, D. E. Montoux, J. T. Wilson, D. H. Kampbell, J. E. Hansen and P. Haas, Symposium on Natural Attenuation of Chlorinated Organics in Ground Water, Dallas, TX, September 11-13, 1996, EPA/540/R-96/509, 1996, pp. 35-59.
107 T. H. Wiedemeier, M. A. Swanson, D. E. Moutoux, E. K. Gordon, J. T. Wilson, B. H. Wilson, D. H. Kampbell, J. E. Hansen, P. Haas and F. H. Chapelle, Technical Protocol for Evaluating Natural Attenuation of Chlorinated Solvents in Groundwater, Air Force Center for Environmental Excellence, San Antonio, Texas, 1996.
108 T. E. Buscheck and C. M. Alcantar in **Intrinsic Remediation**, R. E. Hinchee, J. T. Wilson and D. C. Downey, Ed., *Battelle Press*, Columbus, Ohio, 1995, pp. 109-116.
109 C. E. Aziz, C. J. Newell, A. R. Gonzales, P. Haas, T. P. Clement and Y. Sun, BIOCHLOR Natural Attenuation Decision Support System User's Manual, prepared for the Air Force Center for Environmental Excellence, Brooks, AFB, San Antonio, 1999.
110 Y. Sun and T. P. Clement, *Transport in Porous Media J.*, (1998).
111 Y. Sun, J. N. Peterson, T. P. Clement and B. S. Hooker, Symposium on Natural Attenuation of Chlorinated Organics in Ground Water, Dallas, 1996, U. S. EPA, EPA/540/R-96/509, 1996, pp. 169.

# 23.2 REMEDIATION TECHNOLOGIES AND APPROACHES FOR MANAGING SITES IMPACTED BY HYDROCARBONS

Barry J. Spargo
**U.S. Naval Research Laboratory, Washington, DC, USA**
James G. Mueller
**URS/Dames & Moore, Chicago, IL, USA**

## 23.2.1 INTRODUCTION

Subsurface contamination of soils and aquifers by chlorinated hydrocarbons (CHC) and non-chlorinated hydrocarbons (HC) is likely the largest environmental issue in industrialized nations worldwide. Decades without controlled disposal practices, inadequate storage and distribution systems, and accidental releases have resulted in a large number of contaminated drinking water and aquifer systems. In addition, an untold number of ecosystems are subject to future contamination by impinging hydrocarbon plumes. The extent of potential contributors ranges from neighborhood facilities, such as laundries or gas stations, to major fuel refineries, industrial operations and chemical manufacturing facilities.

While characterization, control, and cleanup of these impacted areas may seem daunting, it is clear that not every impacted or potentially impacted area requires extensive remedial efforts. In fact, many impacts do not represent a significant risk to human health or the environment. In other areas, natural attenuation processes are effective in controlling the migration of the dissolved-phase plume (see Chapter 23.1). If, however, the presence of HCs elicits undesirable effects, then a number of strategies and developing remedial technologies can be used. This chapter will discuss these technologies and strategies and present a number of case studies documenting their effective implementation.

### 23.2.1.1 Understanding HC and CHC in the environment

As summarized by Mueller et al.,[1] hydrocarbons have been produced in the environment throughout geological time. Likewise CHCs are ubiquitous and are also of ancient ancestry.[2] It follows, therefore, that microorganisms have developed mechanisms for utilizing these compounds as growth substrates. Depending on the inherent recalcitrance of the HC, biodegradation mechanisms may be associated with various abiotic degradation processes. While catabolic interactions are often complex and not fully elucidated, a more thorough understanding of integrated processes allows today's scientists and engineers to design and implement more effective systems to mitigate situations where the concentration of HC in a given environment exceeds a desirable value.

### 23.2.1.2 Sources of HC in the environment

Hydrocarbons found in the environment are of diverse structure and are widely distributed in the biosphere, predominantly as surface waxes of leaves, plant oils, cuticles of insects, and the lipids of microorganisms.[3] Straight-chain HC, or alkanes, with carbon number maxima in the range of C17 to C21 are typically produced by aquatic algae. Conversely terrestrial plants typically produce alkanes with C25 to C33 maxima.[4] Plants also synthesize aromatic HC such as carotenoids, lignin, alkenoids, terpenes, and flavenoids.[5] Polycyclic aromatic hydrocarbons (PAH) are also of biogeochemical origin, formed whenever organic substances are exposed to high temperature via a process called pyrolysis. Here, the compounds formed are generally more stable than their precursors, usually alkylated benzene rings.[6] The alkyl groups can be of sufficient length to allow cyclization and then, with time, these cyclic moieties become aromatized. The temperature at which this process occurs determines the degree of alkyl substitution.

It follows, therefore, that fossil fuels such as coal and oil provide the largest source of mononuclear and polynuclear aromatic HC. Contemporary anthropogenic sources of HC in the environment thus originate from two primary sources: i) point source releases such as spills at industrial facilities which utilize large volume of fossil fuels, or ii) chronic lower-level inputs such as from atmospheric deposition.

#### 23.2.1.3 Sources of CHC in the environment

Chlorinated hydrocarbons are also abundant in nature.[2] For example, it is estimated that $5\times10^9$ Kg of chloromethane are produced annually, mainly by soil fungi.[7] However, CHCs of industrial origin perhaps represent an even greater contribution of CHC to the environment. These compounds include a variety of alkanes, alkenes, and aromatic compounds used principally as solvents and synthetic catalysts or intermediates. Reisch[8] estimated that approximately 18 billion pounds of 1,2-dichloromethane are produced in the United States annually. More information on the production and distribution of CHCs can be found herein (see Chapter 3).

While a majority of the CHC is used in a safe and conscientious manner, some material results in environmental contamination. This is often a result of accidental release, although improper disposal is also a common problem. Unfortunately, as a group, CHCs represent the most problematic of the environmental contaminants. This classification is based on their toxicity and environmental persistence. Thus effective means of remediating environments potentially impacted by CHCs is often necessary.

### 23.2.2 *IN SITU* BIOTREATMENT

When the degree of impact or nature of contamination exceeds safe or acceptable conditions then environmental remediation may be proposed. When the contamination is confined and physically accessible then relatively quick and simple remedial efforts can be implemented. For example, impacted soils can be excavated and disposed in a safe manner. Other related remedial efforts have been reviewed and discussed previously.[9-10]

However, many CHC impacts are not easy to remediate because the point of impact, volume of release, and/or the magnitude of the problem are often not known. Moreover, the physical nature of CHC is such that they often concentrate in areas as non-aqueous phases. In these situations, a variety of *in situ* remediation and source management strategies have been developed. The potential benefits of these *in situ* approaches are many, with the more important features being that they are: non-invasive, applicable to large areas of impact, and usually represent the most cost-efficient remedial alternative.

Various remedial approaches are presented below along with case studies summarizing their effective implementation.

#### 23.2.2.1 Microbial-enhanced natural attenuation/bioremediation

Remediation by monitored natural attenuation has been thoroughly reviewed in Chapter 23.1. In the environment, HCs are susceptible to a variety of physical, chemical, and microbiological transformation processes. Specifically, they can undergo biotransformation reac-

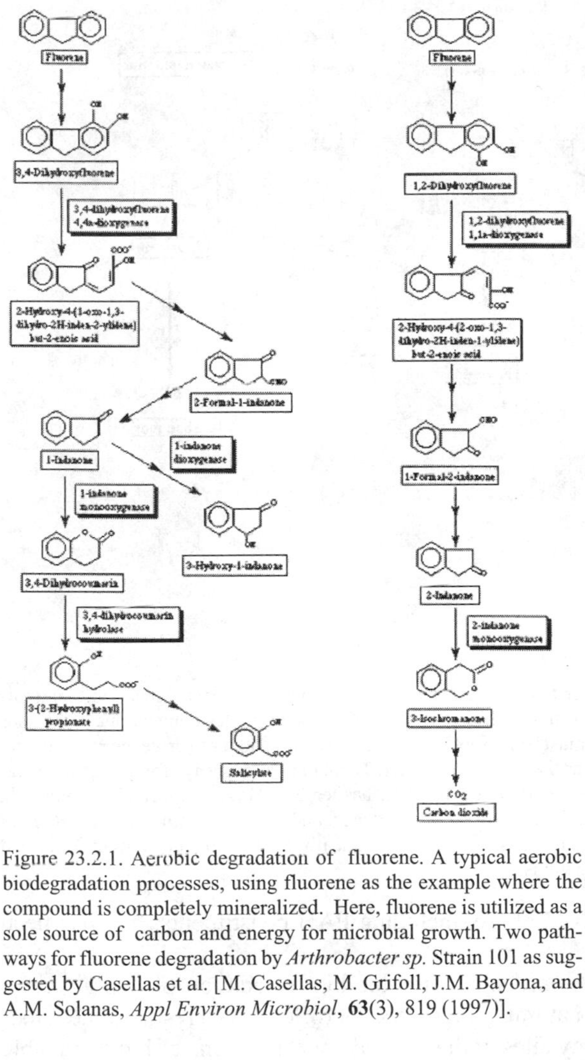


Figure 23.2.1. Aerobic degradation of fluorene. A typical aerobic biodegradation processes, using fluorene as the example where the compound is completely mineralized. Here, fluorene is utilized as a sole source of carbon and energy for microbial growth. Two pathways for fluorene degradation by *Arthrobacter sp.* Strain 101 as suggested by Casellas et al. [M. Casellas, M. Grifoll, J.M. Bayona, and A.M. Solanas, *Appl Environ Microbiol*, **63**(3), 819 (1997)].

tions under aerobic (presence of oxygen), hypoxic (low oxygen), and anaerobic (absence of oxygen) conditions. Examples of recognized biogeochemical reaction sequences are summarized in Figures 23.2.1 and 23.2.2. Some of these biological reactions are co-metabolic meaning that the microbes that catalyze them do not gain carbon or energy for growth and must therefore have a primary carbon source available to drive the processes.

Our ability to understand and capitalize on biological processes, such as biotransformation, in *in situ* and *ex situ* strategies, often result in a low cost, simple alternative to conventional treatment strategies. Several methods to better understand these processes, including direct measure of microbial activity,[11] transformation of tracer compounds into $CO_2$ or metabolic intermediates,[12] and microbial utilization of specific carbon sources through stable isotopes measurements[13] have been developed and applied to a number of hydrocarbon-impacted sites. Other indirect measurements used in the 1970's and 1980's such as plate counts have led investigators to believe that biodegradation was occurring, where direct measurements of biodegradation were not conducted to support those conclusions. Furthermore, the use of plate counts grossly underestimates the population of catabolically relevant biomass.[1] Biodegradation likely occurs in most systems, however the level of biodegradation may be insufficient to expect reasonable cleanup to target levels in the desired time frames.

The fact that HCs are amenable to aerobic biological treatment has been fully and convincingly established in the scientific literature (see also recent review[1]). In the absence of oxygen as an electron acceptor, microbially catalyzed reductive dehalogenation of CHC has been documented.[14-19] Recently, Yang and McCarthy[20] have demonstrated the reductive dechlorination of chlorinated ethenes at $H_2$ tension too low to sustain competitive growth of hydrogenotrophic methanogens. Anaerobic biotransformation of non-chlorinated HCs typ-

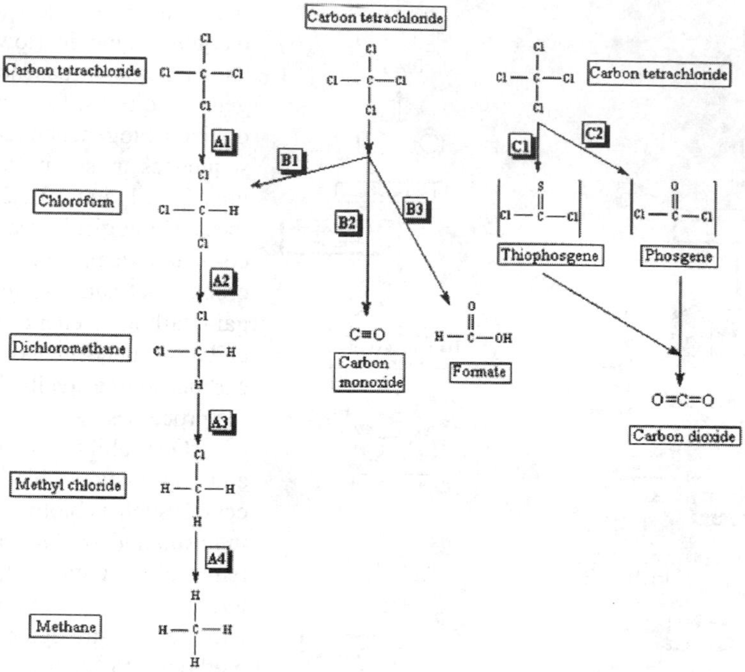


Figure 23.2.2. Anaerobic degradation of carbon tetrachloride. An example of anaerobic dehalogenation, using carbon tetrachloride as the model compound. In many cases, these reactions occur under cometabolic conditions meaning that an alternative growth substrate must be present to serve as an electron donor to drive the reduction reactions whereby carbon tetrachloride is used as the electron acceptor. Three known pathways for microbial degradation of carbon tetrachloride have been identified [U.E. Krone, R.K. Thauer, H.P. Hogenkamp, and K. Steinbach, *Biochemistry*, **30**(10), 2713 (1991); C.H. Lee, T.A. Lewis, A. Paszczynski, and R.L. Crawford, *Biochem Biophys Res Commun,* **261**(3), 562 (1999)]. These pathways are not enzymatically driven but rely on corrinoid and corrinoid-like molecules to catalyze these reactions.

ical of those prevalent in the dissolved phase, including PAH constituents, has also been demonstrated.[21-28]

It is therefore accepted that aerobic and anaerobic biodegradation of organic compounds occurs through the action of natural, indigenous microflora. As a result of these natural *in situ* microbial processes, many sites with elevated concentrations of biodegradable organics exhibit highly reducing and anaerobic conditions in areas containing elevated concentrations (i.e., suspected source areas). Moving outward laterally and down-gradient within the plume, the aquifers tend to become more oxidizing as a result of lower constituent levels, infiltration, and recharge with oxygenated water.

### *23.2.2.1.1 Case study - Cooper River Watershed, Charleston, SC, USA*

The Cooper River Watershed empties into the Charleston Harbor on the southern Atlantic coast of the United States. In the lower reaches, the Cooper River is a highly industrialized and urbanized watershed with storm sewer and surface run-off impact. The river supports industries such as a wood pulp processing plant, a former naval shipyard, and a chromium mining/processing facility. In addition a number of fossil fuel refineries, storage facilities,

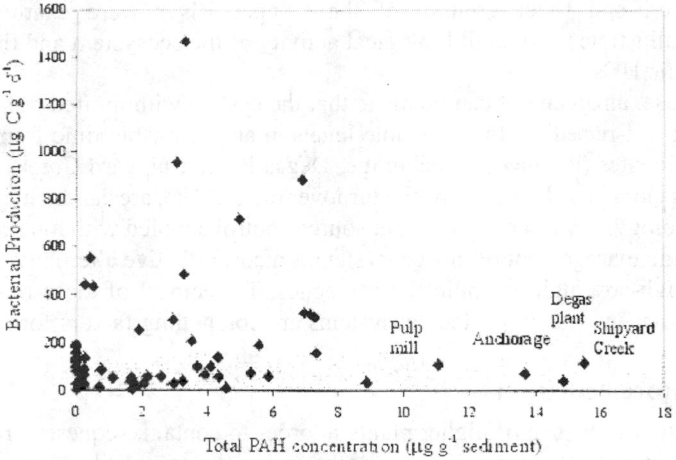


Figure 23.2.3. Bacterial productivity, a measure of bacterial activity, was compared with the level of contaminants at a number of site in the Cooper River watershed, South Carolina (USA). Where productivity is low and PAH concentration is high, the concentration of PAH or other mitigating factors suggest that the potential for PAH biodegradation may be limited in these systems compared to the other sampling points in this watershed. [Figure adapted from M. T. Montgomery, B. J. Spargo, and T.J. Boyd, Naval Research Laboratory, Washington, DC, NRL/MR/6115—98-8140, (1998)].

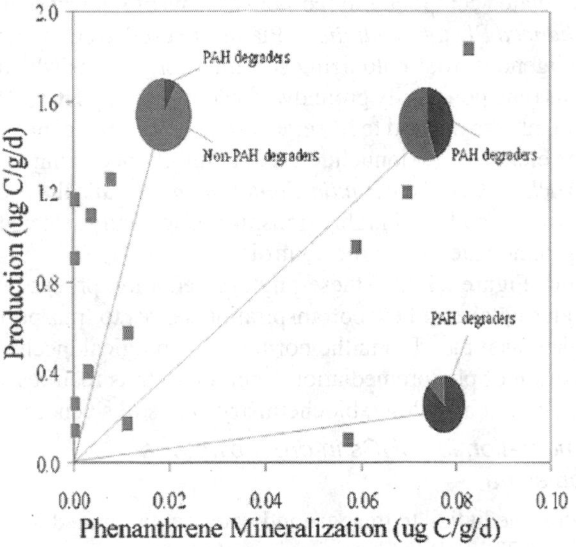


Figure 23.2.4. Microbial communities have the remarkable ability to adapt to utilize a number of carbon sources. Montgomery et al. suggest that the proportion of PAH degraders (shown as pie fraction) in a population can be reflected by their overall activity (measured by protein production) and their ability to mineralize the contaminant to $CO_2$. [Figure adapted from: Montgomery, M. T. , T. J. Boyd, J. K. Steele, D. M. Ward, D. C. Smith, B. J. Spargo, R. B. Coffin, J. W. Pohlman, M. Slenska, and J. G. Mueller, International Conference on Wetlands & Remediation, Salt Lake City, UT, November 16-17, 1999.]

and commercial shipping and recreational docks are scattered along the river. The physical characteristics of the Cooper River have been documented.[29]

In 1997 a study was initiated to examine the capacity of the Cooper River to "self-remediate" if source input was reduced or eliminated from the system. Specifically, pressure was placed on the Charleston Naval Shipyard to dredge the sediments in the area adjacent to the Navy property. A site study was conducted in this region and showed an elevated number of hydrocarbons, heavy metals and other regulated compounds. The study was expanded to include the larger watershed and to place the contaminant levels in the context reflective of regional inputs of contaminants. Concomitantly, the "bio-capacity" of the sediments, overlying boundary layer, and water column of the Cooper River were examined. Figures 23.2.3-23.2.4 illustrate the overall biological activity of the ecosystem and the capacity to degrade specific HCs.

In this case, an argument can be made that the system with limited non-point source HC input will self-remediate to acceptable levels in an reasonable time frame. However, several specific sites (Pulp Mill, Anchorage, Degas Plant, Shipyard Creek), where other contributing factors impede the microbial turnover rates of HC, are candidates for dredging or other technology. It was concluded that source control coupled with long term monitoring and strategic management of this ecosystem is a cost effective alternative to disruptive dredging or high-cost high technology approaches. The impact of these remediation approaches on adjacent, less-impacted ecosystems are compelling factors for avoiding their implementation.

### 23.2.2.2 Phytoremediation

Phytoremediation is the use of higher plants in order to contain, sequester, reduce, or degrade soil and groundwater contaminants for the eventual closure of hazardous waste sites. This rapidly emerging technology can be applied to a diverse range of environmental conditions and contains many potential advantages over conventional remediation technologies; such as substantially lower costs, improved safety, better aesthetics, and wider public acceptance.

There are at least three areas where phytoremediation *per se* can be utilized to treat soil or groundwater impacted by HCs and related compounds or co-constituents of interest such as heavy metals: 1) *Rhizosphere-Enhanced Phytoremediation*: Plants are used to stimulate the relevant catabolic activities of indigenous, root-colonizing soil microorganisms which results in enhanced remediation of soils (and potentially groundwater) impacted by HCs; 2) *Phytoextraction*: Specially selected plants are utilized to hyperaccumulate inorganic materials such as salts, heavy metals, trace elements, radionuclides, and naturally occurring radioactive materials (NORM), and; 3) *Plant-Based Hydraulic Containment*: Entails the use of the natural water uptake and transpiration ability of highly transpiring, specially selected trees or plants for either surficial or groundwater hydraulic control.

As summarized in the schematic (Figure 23.2.5), these phytoremediation processes often occur simultaneously. Here, water uptake and evapotranspiration serves to transport CHC and other solutes through the plant as it partakes in the normal physiological mechanisms of plant life. Successful application of phytoremediation technology thus requires a thorough understanding of agronomy, plant physiology, biochemistry, and soil sciences.

#### *23.2.2.2.1 Case study - phytoremediation for CHCs in groundwater at a chemical plant in Louisiana*

URS/Radian proposed the use of phytoremediation to treat groundwater contaminated with dissolved phase CHC at a chemical plant. The State of Louisiana regulatory agency recommended standard pump & treat technology. However, Radian was able to convince the State to use a new more cost effective remedy. Hybrid Poplar trees were planted to achieve hydraulic containment and phytoextraction of the entire dissolved plume. The shallow groundwater and long growing seasons were ideal for this remedial approach. Radian was eventually able to close the site and obtain a no further action letter from the State of Louisiana.

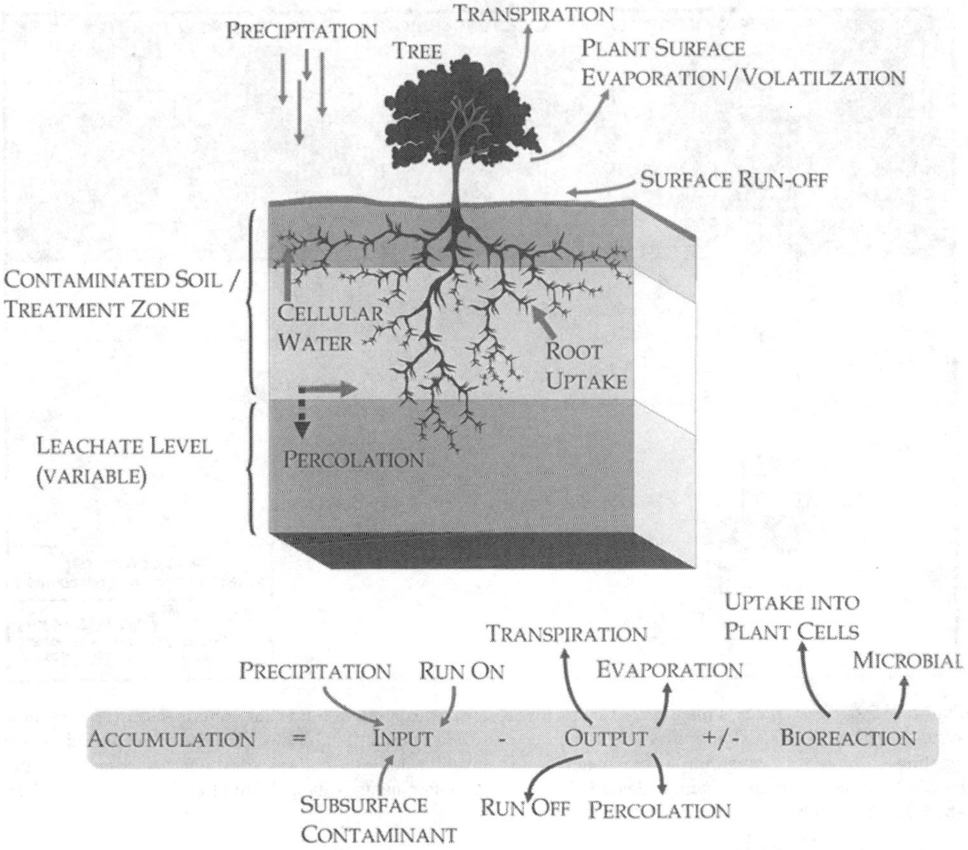


Figure 23.2.5. Phytoremediation: transport modes for HC and CHC in plant systems. Transformation of HC and CHC can be found in the root system with root-associated microbes and tissues of many plant species. HC and CHC are transported by normal plant physiological processes, such as water uptake and evaporation and transpiration. [Schematic courtesy M.A. Bucaro.]

## 23.2.3 *IN SITU* TREATMENT TECHNOLOGIES

### 23.2.3.1 Product recovery via GCW technology

Vertical groundwater circulation wells (GCW) create three-dimensional, *in situ* groundwater circulation cells to mobilize and transport dissolved phase constituents of interest from the aquifer to a central well for treatment via a number of biotic and/or abiotic processes. GCW technology relies on a positive- or negative-pressure stripping reactor in a specially adapted groundwater well. Often pressure is attained by an above-ground mounted blower and off-air treatment system, such as activated carbon. A generic GCW is shown in Figure 23.2.6. The basic principle of operation of one GCW technology depends on moving water within the well to a well screen area above the water table, where water cascades into the vadose zone surrounding the well. Water is drawn in at a screened area in the aquifer usually found below the contaminated zone creating vertical water circulation. Volatile organics are stripped from the dissolved phase by air-stripping and biodegradation is enhanced. For

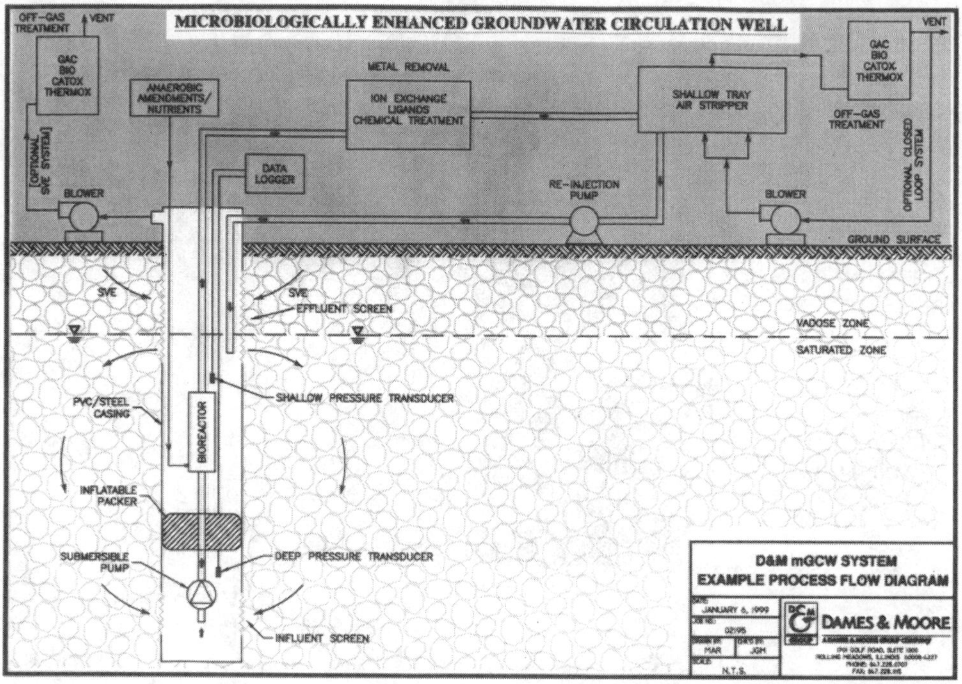


Figure 23.2.6. GCW process diagram. Effective hydrocarbon stripping in the water column is observed in these systems using a vacuum extraction. A circulation cell is created by directional flow of water in the vertical direction creating a capture zone extending several meters from the well. In addition a bioreactor (high surface area bacterial biofilm) can be used in the system to degrade low volatile contaminants [Adapted from Bernhartt et al., **U.S. Patent 5,910,245**, 1999]

more information on other GCW systems and their various modes of operation see Mueller et al.[30] and Allmon et al.[31]

Modifications of the technology have evolved as promising solutions for *in situ* remediation, source management, and/or accelerated recovery of phase-separated hydrocarbons. In each of these applications, however, success requires that the GCW zone of influence (ZOI) be validated. Accurate information on the direction and velocity of groundwater flow is equally important because it determines the rate of transport of contaminants being released or treated in the sub-surface. Toward this end, pressure transducers, conservative dye tracers, and *in situ* permeable flow sensors have been used to validate theoretical predictions of flow fields and the corresponding ZOI. The resulting information is then used to reassess the model and make operational changes to the system in order to meet clean-up goals and objectives.

#### *23.2.3.1.1 Case study - GCW recovery of creosote, Cabot/Kopper's Superfund Site, Gainesville, FL*

This site is a former pine tar and charcoal generation facility and a active wood treatment facility. Creosote used in the wood treating operation is the primary HC. The soil is 93% sand with some silt and clay. Remedial investigation results indicated that ground water in the shallow aquifer (10 to 23 ft below ground) had been impacted. Horizontal and vertical conductivities were $9\times10^{-3}$ cm/s and $9\times10^{-4}$ cm/s, respectively, with a horizontal gradient of 0.006. Initial total concentrations of PAHs in the soil exceeded 700 mg/kg. Total concentrations of PAHs in groundwater for all wells tested ranged from 5-50 mg/L. A single GCW well was installed at a depth of 3 to 8.5 m immediately down gradient from a lagoon area that had been identified as a source of creosote constituents. The system was started in February 1995 and continues to operate to date, having been taken over by the client in 1998. Samples taken after 18 months of operation indicated total PAH concentrations of 10-35 mg/L in up gradient wells. Concentrations in down gradient wells were measured at 0.04-2 mg/L indicating a marked reduction in HC in the aquifer as groundwater moved down gradient through the GCW circulation cell.

### 23.2.3.2 Surfactant enhanced product recovery

A possible remedy for CHC dense non-aqueous phase liquid (DNAPL) is to install an increased-efficiency pump-and-treat system based on introducing surfactants into the aquifer to increase the solubility of CHC and the rate at which it transfers into the water phase. In this type of system, groundwater is extracted, DNAPL is separated (if present), dissolved CHC is air-stripped or steam-stripped from the water, surfactant is added to the groundwater, and the surfactant-rich water is re-injected into the aquifer up-gradient of the suspected DNAPL deposit. As the surfactant-laden groundwater passes across the DNAPL zone it is capable of reaching a CHC saturation level that is many times the natural CHC solubility, thus removing DNAPL more efficiently.

Surfactant-aided product recovery differs significantly from surfactant flushing approaches in that the goal of the efforts is to increase the dissolution of DNAPL into the aqueous phase to expedite its removal. The goal is not to physically mobilize DNAPL through the addition of high concentrations of surface-active agents (i.e., surfactant flushing). In general, surfactant flushing *per se* is not advocated unless the geophysical properties of the aquifer are extremely well characterized, and the nature and source of CHC impact are well defined. Since these requisites are rarely met, the more aggressive use of surfactants is rarely considered.

#### *23.2.3.2.1 Case study - Surfactant-aided chlorinated HC DNAPL recovery, Hill Air Force Base, Ogden, Utah*

Hill Air Force Base at Ogden, Utah used large amounts of solvents as degreasing agents. From 1967 to 1975, unknown amounts of perchloroethene (PCE), trichloroethene (TCE), 1,1,1-trichloroethane (TCA), and dichloromethane (MC) were placed in shallow, unlined trenches as the means of disposal. In the mid-1980s, pools of CHC DNAPL were found at the base of the uppermost aquifer. These DNAPL pools were several feet thick in some areas and extended over 36 acres. As such, DNAPL was a significant and continuing source of off-site migration of dissolved-phase CHCs. As an interim remedial action, the Air Force designed and implemented conventional pump-and-treat technology to serve as a "Source Recovery System". In its first year of operation, the system recovered over 23,000 gallons of DNAPL. After several years of operation, DNAPL recovery decreased but thousands of gallons of product remained in the form of residual saturation (on the order of 20 percent of the pore volume). This residual material was not recoverable by normal pump-and-treat methods. In an effort to enhance the recovery of residual CHC DNAPL, URS/Dames & Moore initiated a surfactant-aided DNAPL recovery system. About 8 percent sodium dehexyl sulfosuccinate (an anionic surfactant), 4 percent isopropanol and 7,000 mg/L NaCl were combined to create an average DNAPL solubility of 620,000 mg/L (compared with a TCE solubility of 1,100 mg/L in natural groundwater). The solution was injected over a period of time, in an amount equal to 2.4 pore volumes in the test portion of the aquifer. DNAPL removal was 99 percent (estimated) and surfactant recovery at the extraction wells was 94 percent. Based on these results, a full-scale system was designed and implemented.

#### 23.2.3.3 Foam-enhanced product recovery

The use of foams to remove heavy immiscible fluids such as DNAPL from soil was developed by the petroleum industry for crude oil production. Subsequently, The Gas Research Institute developed the use of foams to release and mobilized DNAPL contaminants in the subsurface. Coupled with *in situ* or *ex situ* bioremediation, foam-enhanced product recovery can, potentially, transport CHC contaminants upward in the groundwater, thus reducing the potential for driving the contamination to previously non-impacted areas.

The use of foam for CHC DNAPL is currently viewed as experimental. The delivery of the foam, its sweep front, the foam stability, its ability to release CHC DNAPLs in the subsurface, and the resultant biodegradability of residuals can potentially be aided through the proper selection of foaming agents and nutrients. In theory, the technology can tailor the foam system to aerobic or anaerobic subsurface environments, depending on the selection of the carrier gas. This allows adequate biodegradation for the particular CHC in the foam-pollutant system. For example, DCE can be biodegraded aerobically, whereas PCE needs to be degraded anaerobically.

#### 23.2.3.4 Thermal desorption - Six Phase Heating

Six-Phase Heating$^{TM}$ (SPH) is a polyphase electrical technology that uses *in situ* resistive heating and steam stripping to achieve subsurface remediation. The technology was developed by Battelle's Pacific Northwest Laboratories for the U.S. Department of Energy to enhance the removal of volatile contaminants from low-permeability soils. The technology is also capable of enhancing the removal of DNAPLs from saturated zones.

SPH uses conventional utility transformers to convert three-phase electricity from standard power lines into six electrical phases. These electrical phases are then delivered throughout the treatment zone by steel pipe electrodes inserted vertically using standard drilling techniques. Because the SPH electrodes are electrically out of phase with each other, electricity flows from each electrode to the adjacent out-of-phase electrodes. *In situ* heating is caused by resistance of the subsurface to this current movement. In this manner, a volume of subsurface surrounded by electrodes is saturated with electrical current moving between the electrodes and heated. By increasing subsurface temperatures to the boiling point of water, SPH speeds the removal of contaminants such as CHCs via three primary mechanisms: increased volatilization, steam stripping, and enhanced residual mobility toward extraction wells via viscosity reduction.

Once subsurface soil and groundwater reach the boiling point of water, the *in situ* production of steam begins. Through preferential heating, SPH creates steam from within silt and clay stringers and lenses. As this steam moves towards the surface, it strips contaminants such as CHCs from both groundwater and soil matrix. Released steam can act as a carrier gas, sweeping CHC out of the subsurface and to extraction wells. However, it can also cause constituent migration and CHC displacement if the steam is allowed to condense prior to extraction.

#### *23.2.3.4.1 Case study - Six-Phase Heating removal of CHC at a manufacturing facility near Chicago, IL*

SPH has been employed to remove TCE and TCA from the subsurface at a former manufacturing facility near Chicago, Illinois.[32] Since 1991 combined steam injection with both ground water and soil vapor extraction had resulted in significant mass removal, but had left behind three hot spot areas after seven years of operation. These areas, which contained DNAPL in tight heterogeneous soil, were treated in less than four months by SPH.

Site lithology consisted of heterogeneous sandy silts to 18 ft below grade (bg) and a dense silty clay till from 18-55 ft bg. A shallow groundwater table was encountered at 7 ft bg and hydraulic conductivity through the remediation zone ranges from $10^{-4}$ - $10^{-5}$ cm/sec.

A network of 107 electrodes covering two-thirds of an acre was established. To treat beneath a warehouse, 85 of those electrodes were constructed directly through the floor of the building. Electrically conductive from 11-21 ft bg, the electrodes actively heated the depth interval from 5-24 ft bg. Once subsurface temperatures reach boiling, steam laden with chlorinated solvents was collected by a network of 37 soil vapor extraction wells screened to 5 ft bg.

SPH operations began on June 4, 1998. Within 60 days, temperatures throughout the entire 24,000 cubic yard treatment volume had reached the boiling point of water. With another 70 days of heating, separate phase DNAPL had been removed and TCE/TCA groundwater concentrations reduced to below the risk based target cleanup levels. Cleanup results are shown in Table 23.2.1.

**Table 23.2.1. Summary of groundwater cleanup results**

| Well | Compound | Jun. '98, μg/l | Oct. '98, μg/l | Reduction, % |
|---|---|---|---|---|
| B-3 | TCE | 58,000 | 790 | 98.6 |
| | TCA | 82,000 | non detect | >99.4 |
| Da2 | TCE | 370,000 | 8,800 | 97.6 |
| | TCA | 94,000 | 290 | 99.7 |
| F13 | TCE | 2,800 | 280 | 90.0 |
| | TCA | 150,000 | non detect | >99.9 |

In 100 days of heating, 23,000 cubic yards of DNAPL impacted subsurface were remediated to the site cleanup goals set by a State RBCA Tier III evaluation. Based upon these results, the site owner has elected to continue SPH to reach the lower cleanup goals to lessen long term liability. SPH preferentially heats subsurface zones with higher electrical conductivity. At this site, these zones included clay-rich soil lenses and areas with elevated chloride ion concentrations. DNAPL are typically trapped in silt and clay-rich stringers and lenses, while locations of elevated chloride ion concentrations, created from the biological dehalogenation of chlorinated solvents, also correspond to locations of elevated DNAPL concentrations. Thus, SPH targeted the specific subsurface locations of the DNAPL mass.

Calculations of costs included project permitting, preparation of work plans, installation and operations of the SPH, vapor extraction, air abatement, and condensate treatment systems, electrical use, waste disposal, and interim sampling and reporting. Final demobilization, sampling, and reporting were not in the costing calculations. As of 20 November 1998, remedial costs of SPH were estimated at $32 per cubic yard of treatment area. At this time 1,775 MW-hr of electrical energy were consumed, representing an electrical usage rate of $14,000 per month plus $40 per MW-hr for an electrical cost of $148,000 or $6.41 per cubic yard of treatment volume (personal communication, Greg Smith, URS/Radian).

#### 23.2.3.5 *In situ* steam enhanced extraction (Dynamic Underground Stripping)

Dynamic Underground Stripping (DUS), developed by Lawrence Livermore National Laboratory (LLNL) in Livermore, California and the College of Engineering at the University of California at Berkeley, is a combination of the following technologies: 1) Steam injection at the periphery of the contaminated area to heat permeable zone soils, vaporize volatile compounds bound to the soil, and drive the contaminants to centrally located vapor/groundwater extraction wells; 2) Electrical heating of less permeable clays and fine-grained sediments to vaporize contaminants and drive them into the steam zone; 3) Underground imaging, primarily Electrical Resistance Tomography (ERT) and temperature monitoring, to delineate the heated area and track the steam fronts to insure plume control and total cleanup; and 4) Vapor and steam extraction followed by treatment of effluent vapors, NAPL, and impacted groundwater before discharge.

DUS is potentially effective for material above and below the water table, and is also potentially suited for sites with interbedded sands and clay layers. DUS raises the temperature of the soil and groundwater leading to rapid removal of the contaminants due to the thermodynamic processes discussed above.

#### 23.2.3.6 *In situ* permeable reactive barriers (funnel and gate)

*In situ* permeable reactive barriers are used to convert CHC to less toxic and biodegradable intermediates using zero-valent metals such as iron. Permeable reactive barriers, primarily developed at the University of Waterloo, Groundwater Research Center, Canada and EnviroMetals Technologies, Inc. offer a unique cleanup option which does not require transport of contaminated materials (e.g., soil or groundwater) to the surface. Groundwater in the contaminate site can be directed to a permeable barrier region (usually through the use of non-permeable barriers) which is the reactive cell composed granular zero-valent iron. The thickness of the cell is based on the retention time (resident time of water within the cell, based on horizontal flow velocities), the ratio of granular iron to sand/pea gravel, and the types of contaminants. However, in the case study described below, 100% granular iron was used as an added safety factor to ensure complete transformation of the CHCs.

Degradation of CHC occurs through a reduction of iron. This is fundamentally an iron metal corrosion event, where elemental iron is converted to ferrous iron in the presence of water and hydroxyl ions. When dissolved oxygen or the oxygen tension of the surrounding groundwater is low, reactive hydrogen is produced, resulting in reductive dehalogenation of the CHC species, as shown:

$$Fe^{o} + X\text{-}Cl + H_2O \rightarrow X\text{-}H + Cl^{-} + OH^{-}$$

Using permeable reactive barriers, investigators have shown virtually complete dechlorination of CHCs, such as TCE to ethene or ethane (for review see[33]).

##### *23.2.3.6.1 Case study - CHC remediation using an in situ permeable reactive barrier at Naval Air Station Moffett Field, CA*

In late 1995, an *in situ* permeable reactive barrier demonstration at Naval Air Station Moffett Field near Mountain View, CA was constructed by URS/Dames & Moore. The pri-

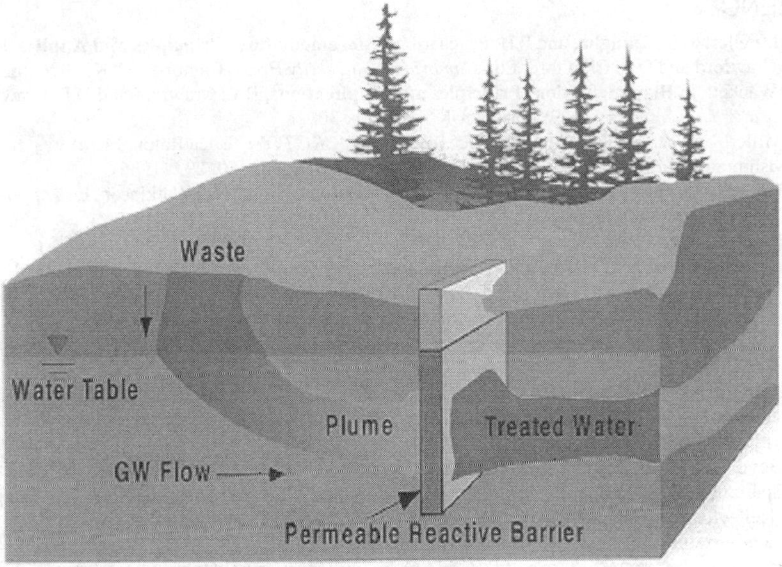


Figure 23.2.7. Schematic of a permeable reactive barrier. [From Permeable Reactive Barrier Technologies for Contaminant Remediation, EPA/600/R-98/125, 1998].

mary groundwater contaminants were TCE, PCE and *cis*-1,2-DCE (cDCE). The lithology of the site has been characterized as alluvial-fluvial clay, silt, sand and gravel, with an aquifer extending 5 to 60 ft below ground surface. The lithology is complex in this region and separates the aquifer into two zones with a discontinuous semi-confining aquitard. The mixed TCE, PCE, cDCE plume was ca. 10,000 ft by 5,000 ft extending along the direction of groundwater flow. Prior to installation of the permeable barrier system, TCE levels exceeded 5,000 μg/L, PCE 1,000 μg/L. A permeable reactive barrier was constructed with the dimensions 10 ft wide, 6 ft long, and 25 ft deep with a 2 ft pea gravel layer in front of and behind the cell. The cell was filled with granular zero-valent iron and a steel corrugate wall was constructed on each of the sides of the cell to form a funnel redirecting groundwater flow in that area through the iron cell. After nearly 4 years of operation, monitoring wells down gradient of the permeable barrier continue to shown non-detect for CHCs.

## 23.2.4 CONCLUSIONS

There are a number of technologies and strategies for managing hydrocarbon remediation. We have reviewed a number of the more promising, simple, and perhaps higher-technology solutions. However, a number of other demonstrated (pump and treat, air sparging) and emerging technologies do exist.[34] The basic themes of preventing migration passively, and removal of contaminant via degradation (biotic and abiotic) are seen in all the technologies. The proper use of particular technologies and strategies is very dependent on the extent and type of contamination, the site characteristics (hydrology, lithology, etc), the cleanup goals, and applicable regulations to name a few. Unfortunately, there are no hard and fast rules for the use of a particular technology; past experience suggests that the more that is known about site characteristics the greater the success of technology application.

## REFERENCES


1 J.G. Mueller, C.E. Cerniglia, and P.H. Pritchard. in **Bioremediation: Principles and Applications**, R.L. Crawford and D.L. Crawford, Ed., *Cambridge University Press*, Cambridge UK, 1996, pp. 125.

2 L.P. Wackett, in **Bioremediation: Principles and Applications**, R.L. Crawford and D.L. Crawford, Ed., *Cambridge University Press*, Cambridge UK, 1996, pp. 300.

3 F.J. Millero and M.L. Sohn, **Chemical Oceanography**, *CRC Press*, Boca Raton, FL, 1991.

4 M. Nishmura and E.W. Baker, *Geochemica et Cosmochimica Acta*, **50**, 299 (1986).

5 D.J. Hopper, in **Developments in Biodegradation of Hydrocarbons**, R.J. Watkinson, Ed., *Applied Science Publishers*, London., 1978, pp. 85.

6 M. Blumer, *Scientific American*, **234**, 35, (1976).

7 R.A. Rasmussen, M.A.R. Kahlil, and R.W. Dalluage, *J. of Geophysical Research*, **85**, 7350, (1980).

8 M.S. Reisch, *Chemical and Engineering News*, **72**(15), 12, (1994).

9 R.D. Norris, **Handbook of Bioremediation**, *Lewis Publishers*, Boca Raton, FL, 1994.

10 P.E. Flathman, E.J. Douglas and J.H. Exner, **Bioremediation: Field Experience**, *Lewis Publishers*, Boca Raton, FL, 1994.

11 M.T. Montgomery, T.J. Boyd, B.J. Spargo, J.G. Mueller, R.B. Coffin, and D.C. Smith, in **In Situ Bioremediation and Efficacy Monitoring**, B.J. Spargo, Ed., *Naval Research Laboratory*, Washington, DC, NRL/PU/6115—96-317, 1996, pp. 123.

12 T.J. Boyd, B.J. Spargo, and M.T. Mongtomery, in **In Situ Bioremediation and Efficacy Monitoring**, B.J. Spargo,Ed., *Naval Research Laboratory*, Washington, DC, NRL/PU/6115—96-317, 1996, pp. 113.

13 C.A. Kelley, B.A. Trust, and R.B. Coffin, *Envir. Sci. Technol.*, **31**, 2469 (1997).

14 D.F. Berry, A.J. Francis and J.-M. Bollag, *Microbiol. Rev.*, **51**, 43, (1987)

15 E. Arvin, B. Jensen, E.M. Godsy and D. Grbic-Galic, in: International Conference on Physiochemical and Biological Detoxification of Hazardous Wastes, Y.C. Wu (Ed.), 1988, p828.

16 M.D. Mikesell and S.A. Boyd, *Appl. Env. Microbiol.*, **52**, 861, (1986).

17 S.L Woods, J.F. Ferguson, and M.M. Benjamin, *Env. Sci. Technol.*, **23**, 62, (1989).

18 A.J. Frisbie and L. Nies, *Bioremediation Journal*, **1**, 65, (1997).

19 J. Dolfing and J.M. Tiedje, FEMS Microbiol. Ecol., **38**, 293, (1986).

20 Y.Yang and P.L. McCarthy, *Environ Sci. Technol.*, **32**, 3591, (1999).

21 E.M. Godsy and D.F. Goerlitz. U.S. Geological Survey Water-Supply Paper 2285, (1986).

22 J.R. Mihelcic, and R. G. Luthy, *Appl. Environ. Microbiol.*, **54**, 1182, (1988).

23 J.R. Mihelcic and R.G. Luthy, *Appl. Environ. Microbiol.*, **54**, 1188, (1988).

24 W.R. Mahaffey, UPRR, In Situ Treatment Process Development Program, Milestone II Report, Volume 3/ 4, (1988).

25 B. Al-Bashir, T. Cseh, R. Leduc, and R. Samson, *Appl. Microbiol. Biotechnology*, **34**, 414, (1990).

26 J. Thierren, G.B. Davis, C. Barber, B.M. Patterson, F. Pribac, T.R. Power and M. Lambert, *Groundwater*, **33**, 469, (1995).

27 A.A.M. Langenhoff, A.J.B. Zehnder and G. Schraa, *Biodegradation*, **7**, 267, (1996).

28 D. L. McNally, J.R. Mihelcic and D. R. Lueking, *Environ. Sci. Technol.*, **32**(17), 2633, (1998).

29 R.F. VanDolah, P.H. Wendt, and E.L. Wenner, South Carolina Coastal Council, NA87AA-D-CZ068. (1990)

30 J. Mueller, M. Ohr, B. Wardwell, and F. Lakhwala, *Soil & Groundwater Cleanup*, Oct/Nov, 8, (1999).

31 W.E. Allmon, L.G. Everett, A.T. Lightner, B. Alleman, T.J. Boyd, and B.J. Spargo, Naval Research Laboratory, Washington, DC, NRL/PU/6115—99-384, (1999).

32 B. Trowbridge, V. Jurka, G. Beyke and G. Smith, in Second International Symposium on Remediation of Chlorinated and Recalcitrant Compounds, Battelle Press, Columbus, OH, 2000.

33 U.S. EPA, In Situ Remediation Technology Status Report: Treatment Walla, EPA 542-K-94-004, Washington D.C., 1995.

34 S.K. Sikdar and R.L. Irvine, **Bioremediation: Principles and Practice**, *Battelle Press*, Columbus OH, 1997.

# 24

# PROTECTION

GEORGE WYPYCH
**ChemTec Laboratories, Inc., Toronto, Canada**

## 24.1 GLOVES

Many types of gloves are manufactured for different purposes. Proper gloves selection is important to ensure that the solvent or other chemical compound cannot penetrate through the glove where its subsequent evaporation would be prevented by the glove itself. Exposure to solvent entrapped between the glove and the skin is more severe than the exposure to solvent without glove.

The first requirement in the selection of a glove is that the materials from which gloves are produced are not dissolved or swollen by solvent. This alone is not sufficient. The other important requirement in glove selection is permeability of all of the materials in its structure. Permeability is measured according to ASTM Standard F 739 or European Standard EN 374. Permeability results are usually expressed in mg of permeate which penetrates 1 $m^2$ of material per minute. Good materials for gloves should have permeability below 1 mg $m^{-2}$ $min^{-1}$. Materials having permeability in the range from 1 to 10 mg $m^{-2}$ $min^{-1}$ may still be considered as suitable materials.

Relative to permeability, materials are rated by a five point scale of the permeation index number. The best choice of glove material has a permeation index number = 0. If the material has index 1 (permeability in the range from 1 to 10 mg $m^{-2}$ $min^{-1}$), the glove can be eventually accepted for harmful chemicals. Any higher index number makes justification for its use more difficult and in plain language such a glove should not be used with harmful chemicals.

Breakthrough time is another measure of glove fitness for purpose. This parameter measures the time in minutes during which a measurable amount of a particular chemical compound penetrated through a particular garment. Specialized monographs give data on breakthrough times for a large number of gloves and other protective clothing and numerous penetrants.[1,2] These sources[1,2] are the best collections of data which should be considered in glove selection. There is a relationship between permeation rate (and thus permeation index number) and breakthrough time. For practical purposes gloves are selected based on the assumption that they should resist penetrant breakthrough for more than eight hours. Such gloves are generally considered to have permeability index number = 0.

In addition to the properties of protective clothing materials, the toxicity of the penetrant should be considered. For less toxic solvents, materials with a permeation index number of 1 or even 2 may still be cautiously considered but will be less acceptable for more toxic penetrants.

Monitoring the quality of protection during use is not less important. Materials are not permanent. Frequent inspection of gloves is therefore very important. Any mechanical damage will compromise protection properties. The presence of cuts, cracks, or holes immediately renders the gloves useless. Gloves should be inspected for changes of color, texture, or hardening which are signs of material degradation.

It should be borne in mind that protective clothing is not universal for all chemical materials. For this reason, protection against one penetrant does not suggest that the glove has protecting qualities against the other materials. Mixtures of solvents behave in a different manner than do pure components. The temperature at which exposure occurs is also an important factor. Generally as temperature increases, permeability increases.

Table 24.1 gives general guidance regarding the applicability of different glove materials for different groups of solvents. Note that glove materials are not equal. They vary with the formulation used by a specific manufacturer and the thickness of the protective layer. For each selection, the manufacturer's specification should be consulted to estimate previously discussed parameters. The best practice requires that gloves are tested for particular conditions of their use. The other good option is to consult results of measurement included in the specialized monographs.[1,2]

**Table 24.1. General guidelines for selection of gloves for different groups of solvents**

| Solvent group | Suitable glove materials | Manufacturer |
|---|---|---|
| Acids | butyl<br>neoprene<br>Saranex-23 | North and many other<br>Ansell, Best, and many other<br>DuPont |
| Aldehydes | butyl<br>PE/EVAl/PE* | Best, North and many other<br>Safety4 |
| Amines | PE/EVAl/PE* | Safety4 |
| Esters | PE/EVAl/PE*<br>PVAl | Safety4<br>Edmont |
| Ethers | PE/EVAl/PE*<br>PVAl | Safety4<br>Edmont |
| Halogenated hydrocarbons | PE/EVAl/PE*<br>PVAl<br>Viton | Safety4<br>Edmont<br>North |
| Hydrocarbons, aliphatic | nitrile<br>PE/EVAl/PE*<br>Viton | Best, Mapa-Pioneer, North<br>Safety4<br>Best, North |
| Hydrocarbon, aromatic | PE/EVAl/PE*<br>PVAl<br>Viton | Safety4<br>Anselledmont, Comasec<br>North |

| Solvent group | Suitable glove materials | Manufacturer |
|---|---|---|
| Ketones | butyl<br>PE/EVAl/PE*<br>PVAl | Best, Comasec, Guardian, North<br>Safety4<br>Anselledmont, Edmont |
| Monohydric alcohols | butyl<br>nitrile<br>neoprene | Best, Guardian, North<br>Best, Comasec<br>Ansell, Best, Mapa-Pioneer |
| Phenols | butyl<br>neoprene<br>Syranex-23 | Best, Comasec, North<br>Anselledmont, Mapa-Pioneer<br>DuPont |
| Polyhydric alcohols | neoprene<br>nitrile rubber | many manufacturers |

*PE/EVAl/PE = polyethylene/ethylene vinyl alcohol/polyethylene

The above suggested selections are based on the test data included in the monograph.[1] Only the gloves which have a permeability index number = 0 were included. However, since solvents in each group frequently have diverse physical properties, in some specific cases the selection for that group may not be the most suitable for a particular member of the group. It should be also considered that some manufacturers have several grades of gloves in the same generic group of polymer therefore their specification should always be consulted.

## 24.2 SUIT MATERIALS

The selection of suit material is more complex because not only physical properties of the components but also the design of suit and structure of laminate used will determine its barrier properties. As a whole, these are usually proprietary and difficult to compare. The information in Table 24.1 can be used for general guidance but the industrial hygienist should make a full evaluation of the performance of the protective clothing in particular conditions and application.

## 24.3 RESPIRATORY PROTECTION

The selection of a respirator, general suggestions for its use, the adsorption capacity of filters, and breakthrough time are reviewed below.

The selection of respirator is based on the nature of respiratory hazards such as permissible exposure limits, established concentration immediately dangerous to life, actual concentration of hazardous substances, and amount of oxygen. When using filters, the ambient air must contain at least 20% oxygen at sea level. Filter protection should not be used for unknown contaminants and contaminants that are immediately dangerous to life and health.

Permissible exposure limits and actual concentration determine (among other factors) the breakthrough time as discussed below.

Only approved respirators should be used. Each country has a body which can approve respirators for use. In the USA, respirators are approved jointly by the National Institute for Occupational Safety and Health (US Department of Health and Human Services) and the Mine Safety and Health Administration (Department of Labor).

In addition to the the above parameters determining respirator selection, the following factors are also considered: nature of operation process, location of hazardous area (especially in respect to supply of respirable air), employee activity and intensity of effort required to perform the work (determines the amount of air which must be supplied to lungs), physical characteristics and limitations of various respirators.

The selection of respirator must be done with a full consideration of these many factors which require specialized knowledge going beyond this discussion. Further information on this subject can be found in the specialized monograph.[3]

Respirators can be generally divided into two main groups: air-purifying respirators and atmosphere-supplying respirators. The first group is more common and for this reason will be discussed in more detail below. Air-purifying respirators are still divided into non-powered and powered and these are further divided into gas/vapor and particle removing. Our main interest here is given to vapor-removing, non-powered respirators which are the most common in industry and laboratories in solvent applications.

Two additional subjects are integral part of good protection: respirator fit testing and training of users. Non-powered, vapor-removing respirators are also termed as negative pressure respirators. This means that respirator is under positive pressure during exhalation and under negative pressure during inhalation. If the respirator does not fit the user properly, some air from the surroundings will be drawn into the respirator during inhalation because of leaks. This will result in a diminishing efficiency of protection. For this reason, each user should be given help from the employer in selecting the most suitable type and size of respirator for the particular individual. In addition, the reasons for fitting should be clearly explained so that they are fully understood by the employee. It is only the employee who may later assure that the respirator is used properly and this will depend on his/her full understanding of the principles.

The first matter of importance is the rate of breathing under various conditions.The amount of air we breathe depends on our energy requirements or more precisely on the intensity with which we expand energy. If no work is performed, the average human being requires about 10 l of air per minute. The rate of breathing increases with the intensity of work (light work 20-30, heavy work 70-100 l/min). The use of a filter reduces air flow rate due to the pressure drop and many respirators cannot cope with high rates of flow which becomes the one important limitation of negative pressure respirators that must be evaluated during the selection process.

The pressure drop in the filter depends on its design. Chemical cartridges used for solvent adsorption are filled with activated carbon or activated charcoal which are amorphous forms of carbon characterized by their ability to rapidly absorb many gases and vapors. The carbon is obtained by pyrolysis of wood, nutshells, animal bones, petroleum products, and other carbonaceous materials. Activated carbon for respirators usually comes from coconut shells or coal after activation at temperature in the range from 800 to 900°C. It has a porous structure with a surface area of 10,000 $m^2/g$. On the one hand, maximizing surface contributes to increased capacity on the other it may lead to an increased pressure drop during breathing (depending on particles shapes and sizes). It is thus important to consider both capacity of the filter and its pressure drop.

The adsorption of vapor is a physical process which can be reversed. Desorption of vapor does not require a high energy and is equilibrium driven (equilibrium relative to the concentration of vapor in surrounding atmosphere and distribution of adsorbed vapor in activated carbon). Desorption during storage or non-use time can result in migration of previously adsorbed molecules. This phenomenon is very important for effective use of respirators. On one hand, it precludes the possibility of cartridge recycling. On the other hand, it imposes restriction on cartridge use. For good performance the mask should be worn 100% of the time. It is estimated that removal of respirator for 5 min in an 8-hour day causes a loss of 50% of the expected protection. It is also important to limit filter use to one shift even if it were not fully exhausted in that time.

The absorption capacity is calculated from the following equation:

$$W = \rho W_o \exp\left[-\frac{BT^2}{\beta}\left\{\log\frac{p_s}{p}\right\}^2\right] \qquad [24.1]$$

where:

| | |
|---|---|
| W | adsorption capacity per unit weight of carbon |
| $\rho$ | solvent density |
| $W_o$ | total volume of adsorption space |
| B | microporosity constant of carbon |
| T | temperature |
| $\beta$ | affinity coefficient of solvent vapor for carbon |
| $p_s$ | saturated vapor pressure of solvent at temperature T |
| p | equilibrium partial pressure of the solvent |

This equation shows that both solvent type and type of carbon affect performance. On the solvent side, its density, affinity, saturated vapor pressure and its actual concentration determine adsorption. On the carbon side, its porous structure and surface area available for adsorption determine the capacity of a particular filter. The data for the parameters of the equation can be found in chemical handbooks and therefore the equation can be used for predicting the adsorption capacity of a filter. The capacity of commercial filters is normally rated by experimental method in which amount of solvent adsorbed by filter is determined under conditions which specify concentration of solvent, rate of flow, and time. For example, a filter which contains 70 g of activated carbon was found to adsorb 20-25 g carbon tetrachloride present in concentration of 1000 ppm.

Under conditions of use it is important to predict how long a particular cartridge will last under real conditions. This is done by estimation of the breakthrough time from the following equation:

$$t = \frac{W\rho_c An}{QC_o}\left[z + \frac{1}{a_c\rho_c}\left(\frac{dG}{\eta}\right)^{0.41}\left(\frac{\eta}{\rho_a D}\right)^{0.67}\ln\left(\frac{C_b}{C_o}\right)\right] \qquad [24.2]$$

where:

| | |
|---|---|
| W | adsorption capacity per unit weight of carbon |
| $\rho_c$ | carbon density |
| A | cross-sectional area of adsorbent bed |
| n | number of cartridges tested |
| Q | flow rate |
| $C_o$ | concentration of solvent |
| z | V/A where V is carbon volume |
| $a_c$ | specific surface area |

| | |
|---|---|
| d | diameter of granule |
| G | mass velocity through the cartridge |
| $\eta$ | viscosity of air-vapor stream |
| $\rho_a$ | density of air-vapor stream |
| D | diffusion coefficient |
| $C_b$ | breakthrough concentration |

This equation gives information on the relevance of major parameters of performance of filters. Breakthrough time increases with the increase of the following parameters: adsorption capacity of carbon, its density, its volume, and the cross-sectional area of the adsorbent bed and size of the granules. It decreases with increased flow rate, concentration of solvent, and an increase in the value of the diffusion coefficient. The toxicity of solvent plays a role here as well because with the increase of solvent toxicity, the breakthrough concentration is decreased which decreases breakthrough time. Breakthrough time for 3M cartridges can be calculated using available software by providing the type of solvent, its concentration, and type of work (light, medium, heavy). In addition, relative humidity is accounted for in the calculations. The adsorption of vapors is not affected by humidity below 50% but decreases rapidly as the relative humidity increases above 50%. Adsorption temperature is also an essential factor.

The above discussion shows that, although there is generally one type of cartridge used for organic vapor, all cartridges are not the same in terms of performance. The conditions of use of these respirators determine if they can perform specific protective functions. Considering that one cartridge, typically disposed after one day work may contain 40 g of solvent which would otherwise be inhaled, the selection and use of respirators is not a trivial matter and should be given serious attention.

## REFERENCES

1 K Forsberg, L H Keith, **Chemical Protective Closing. Performance Index**. 2nd Ed., *John Wiley & Sons*, New York, 1999.
2 K Forsberg, S Z Mansdorf, **Quick Selection Guide to Chemical Protective Clothing**. 3rd Ed., *John Wiley & Sons*, 1997.
3 W H Revoir, C-T Bien, **Respiratory Protection Handbook**, *Lewis Publishers*, Boca Raton, 1997.

# 25

# NEW TRENDS BASED ON PATENT LITERATURE

GEORGE WYPYCH
**ChemTec Laboratories, Inc., Toronto, Canada**

Changes in production methods and the increased emphasis on environmental protection have produced continuous changes in solvent use and in the technology of solvent processing. The aim of this chapter is to analyze existing patent literature to illustrate the trends in

- the development of new solvents
- the use of solvents in various branches of industry
- the new methods of solvent processing with emphasis on solvent recycling.

## 25.1 NEW SOLVENTS

The developments and introduction of new solvents are discussed in Chapter 21 (supercritical fluids and ionic liquids) and Chapter 3 (terpenes). These three generic groups of materials are emerging technologies studied to address environmental concerns.

Cleaning solvents are used in many industrial applications. These solvents have typically been highly volatile, toxic, and flammable. Dichloromethane, methyl ethyl ketone, toluene, xylene are frequently used for cleaning. They have all negative features contributing to potential hazards such as high volatility, flammability, and toxicity, and they influence the natural balance of the environment. If they are to be replaced, their replacements must have equal performance – a difficult task considering their excellent cleaning properties. Two recent patents[1,2] suggest alternative solutions. 1-methyl-2-pyrrolidone is a suitable substitute for dichloromethane and methyl ethyl ketone in terms of its solvent abilities. It is less flammable and less volatile but its high cost prohibits its use in many applications. When it is blended with an alicyclic carbonate, a polar solvent, and terpene, a non-polar component, cost reaches an acceptable level and the blend can be used for cleaning paint equipment, reclamation of silkscreens, cleaning offset print rollers, printing surfaces, and as a general purpose cleaner.[1,2]

The removal of soiling and graffiti from exterior coatings requires the use of substantial quantities of solvents. Paint removers such as dichloromethane, benzene or toluene all of which which are hazardous solvents are commonly used. Several applications are needed before complete removal is achieved. Anti-graffiti coatings which are based on wax-containing compositions help to reduce solvent use. Such coating does not contain a high concentration of solvent and can be cleaned using water heated to temperatures from 120 to 190°C under elevated pressure. This cleans the graffiti but because the cleaning temperature must be higher than the melting temperature of wax, the coating is also removed and must be re-coated. An alternative method is based on the use of environmentally friendly 1-methyl-2-pyrrolidone which cleans graffiti without removing protective coating.[3]

Terpene, d-limonene, was selected for cleaning oil spills and waste petroleum products (e.g., tar or grease).[4] The combination of 1-methyl-2-pyrrolidone with d-limonene makes a good paint stripper.[5] It eliminates dichloromethane and other undesirable solvents.

Replacement of fluorine-containing solvents is yet another goal of recent inventions. A non-toxic solvent was developed for cleaning electrical parts and motors.[6] Such solvents must have high dielectric properties to prevent damage to equipment, electrocution, and/or fire. Chlorinated and fluorinated solvents which met these perfomance requirements can be replaced by a formulation according to the invention[6] which uses a mixture of aliphatic hydrocarbons and adipate, salicylate, and succinate esters. A similar composition is also proposed[7] as the solvents in the production of PVC foams.

Medical implants must have roughened surfaces or a porous coating to promote bone and tissue integration. During the manufacture of such implants, loosely adhering coating particulates and debris collect on the surface and must be removed. The conventional cleaning method involves surface blasting with particles of hard material such as alumina, glass beads, etc. This may produce cross-contamination which will adversely affect implant performance. Other methods include polishing with ice particles, dry ice particles, and chemical cleaning. An invention[8] which includes blasting of the surface with soluble solids (e.g., water soluble inorganic salts) which are then removed in ultrasonic water bath offers reduced contamination and eliminates the use of organic solvents.

A few general examples show that solvents can be optimized for better performance by mixing. Substantial amounts of solvents can be eliminated by the careful tailoring of the process and properties to the application. Post cleaning with solvent based materials can be minimized or eliminated.

## 25.2 ADHESIVES

When the impact of a particular adhesive technology is being assessed, all relevant factors must be considered. These include the method of adhesive production, solvents present in the adhesive itself and the cleaners and primers required for application. The current patent literature provides many examples which reduce pollution without compromising the quality of the joint. Regulations are primary drivers for this new technology with California regulations having the greatest impact. These regulations include the VOC limits imposed by the South Coast Air Quality Management District which reduced VOC limits from actual 750 to 850 g/l for some solvent contact adhesives to below 250 g/l. These limits depend on the technology and on the adhesive type.

A patent[9] provides a means to eliminate solvent from the production process of a pressure sensitive adhesive. The typical method of producing pressure sensitive adhesive involves mastication of the elastomer, dissolving the elastomer and additives in a hydrocarbon solvent, coating the solution onto a backing, and drying to remove solvent. The phenolic tackifying resin of the invention eliminates the use of solvent process.

But often solvents must be used in adhesives in order to lower their viscosity so that product can be applied without the use of extensive force and so that the adhesive can flow into small openings in the substrate surface to promote mechanical adhesion by anchoring. The easiest formulating technique is to add solvent. This is simple and there is no need to develop new polymeric material. This is particularly true of polyurethane based adhesives which require either a special blending techniques or manufacture of the polymer with low viscosity polyols to reduce its viscosity. However, formulation of polyurethanes without solvents is technically possible as shown in patents for an automotive adhesive[10] and a general purpose adhesive.[11] In these areas additional research will bring dividends but the current effort lags behind the needs.

In some adhesives, the total elimination of solvents is not possible. The best example is that of a solvent contact adhesive which is composed of solvent and a rheological additive. The solvent performs the essential function of swelling the adjacent surfaces of the polymeric material surfaces which are to be bonded. Until recently it seemed that coping with California regulations with such adhesives systems would not be possible. However, a recent patent[12] shows that not only can the amount of volatile solvent be reduced to fulfil California requirements but also the toxicity of the solvent can be reduced and its flash point increased without compromising the performance characteristics of the adhesive. In earlier concepts of contact adhesives very volatile solvents, such as tetrahydrofuran and methyl ethyl ketone were the ones most frequently used and this was based on the premiss that such solvents would penetrate the polymer rapidly and be removed from the material immediately after swelling occurred. This type of adhesive is particularly well suited when used in plastic piping systems where the assembled system must be tested before it is enclosed in partitions. This approach eliminated the use of very volatile solvents which evaporated to become air pollutants and, to provide adequate open time, were used in excess which generally led to dripping from the joint and consequent waste of adhesive. In the invention, 1-methyl-2-pyrrolidone is used in combination with dimethyl adipate and some aromatic hydrocarbon to produce an adhesive known for its ease of use, long term durability, application with no special equipment, low cost, fast cure, and low concentration of total solvent.

In other invention[13] concentration of solvent in a fumigation adhesive used for bonding polyethylene to polyethylene in agricultural applications was also reduced by the reformulation of the copolymer used in the adhesive.

For cleaning/priming operations there are also new developments. One trend is to apply the primer in a water based form.[14] In this invention, a water based primer was developed using water reducible polyurethanes. The primer is for general use but has been optimized for use with construction sealants where solvent based primers are most commonly used today. Another approach is to use environment friendly solvents such as terpenes for both cleaning and priming. But some attempts are misguided. In one case, it was proposed to replace the primer with an isocyanate, which does not evaporate readily therefore fulfils a requirement of the regulation, but exposes applicators to extremely toxic material. Attempts of this kind must be discouraged because regulations are not designed to cope with such a gross misuse of knowledge of chemistry.

One invention[15] consists of a multipurpose cleaner/primer for roofing membranes. These elastomeric sheets are normally powdered with talc to prevent interlayer sticking in the roll. Instead of using separate cleaning solvent followed by a primer application the product of invention was able to bond the talc and other contaminants to the membrane and provide a tacky surface to promote bonding to the roof substrate.

These examples show that there are many opportunities to reduce or eliminate solvents from adhesives. These usually require a more intensive R&D effort but lead to safer and higher performance products.

## 25.3 AEROSPACE

Information on inventions related to solvents used in the aerospace industry is sparse (13 patents in last 25 years compared with 935 for adhesives in the same period). The reason is the secretive nature of the industry with respect to new technology and the reliance on proprietary technology to maintain a technological advantage. There is certainly a move towards limiting the solvent use in aerospace industry as discussed in Chapter 14 but the primary objective is always the performance of the applied technology.

Aerospace materials must have high solvent resistance.[16-20] Patents[16-20] which demonstrate this also stress the need for high quality and uniformity in the materials used. For these reasons, solvents were used to extract low molecular weight components which could be detrimental in space where materials vaporize or degrade extremely rapidly. Also, solvent removal after processing is emphasized to eliminate odor and remove components which may cause degradation of material exposed to highly a degrading environment.

## 25.4 AGRICULTURE

Recent patents describe the use of solvents to improve the properties of agricultural chemicals. In one invention, a carrier was developed from an agglomerated composition of plant fibers and mineral filler. The purpose of the carrier is to absorb and hold a large quantity of pesticide until it is delivered to the application site. The pesticide must be in a form of low melting liquid. In order to reduce its melting point, solvents selected from aromatic hydrocarbons are used to dissolve pesticide.[21] In a water dispersible composition of insecticide, solvent is used to convert insecticide to a liquid form at room temperature.[22] Solvents proposed for this application are from a group of alkyl aromatic hydrocarbons, methyl esters of alkanoic acids, and ester mixtures derived from plant oils.

## 25.5 ASPHALT

In various asphalt processing methods, the rheological properties of the compounded asphalt are of primary importance. One invention[23] shows how, by blending different fractions of crude oil, the processing properties can be optimized for pavement applications. In another recent invention,[24] solvent-free polyurethane was compounded with asphalt. Such products have always contained substantial amounts of solvents to reduce viscosity to the level that allows the material to be cold applied with simple techniques. Using organoclay as compatibilizer for polyurethane and asphalt and a suitable plasticizer, elimination of solvent becomes possible. In addition to environmental safety, these products can be used safely on substrates which would otherwise be affected by solvents the formulation (e.g., polystyrene insulation boards). This is one example which shows that the benefits of solvent free systems go beyond the elimination of air pollutants.

## 25.6 AUTOMOTIVE APPLICATIONS

Cleaning products contain the majority of solvents used in automotive maintenance. New equipment has been designed for solvent recycling.[25,26] In one such design, solvent is dispensed from a holding reservoir to a wash basin. After the parts are washed, the spent solvent is directed to a distillation chamber and condensate is returned to the holding reservoir.[25] The equipment is designed in such a way as to prevent vapor from escaping to the atmosphere. In another development a biodegradable aqueous solution of detergent is used for cleaning.[26] The cleaning solution is periodically filtered to remove solid contamination then returned for reuse. Cleaning operations are conducted under elevated temperature to increase cleaning efficiency.

In automotive refinishing shops, various cleaning liquids are used most of which are based on organic solvents. In addition to the potential of pollution, such cleaning liquids are not suitable for removal of the water soluble dirt frequently found on car body parts (e.g., tree sap, bird droppings, etc.). A cleaning composition has been designed which can clean both solvent soluble and water soluble dirt.[27] It is composed of a water emulsion of hydrocarbon solvent, an alkyl ester, and a blend of surfactants.

Extensive efforts are known to be in progress which may lead to a reduction and even the elimination of solvents used in automotive painting operations. Each coating layer is applied by a different process which may be either a water based coating or a powder coating. An example of a water based coating for an automotive base coat can be found in a recent patent.[28] The resin is derived from propoxylated allyl alcohol containing both hydroxyl and carboxyl groups and can be cured by a variety of crosslinkers. This technology is also applicable to industrial appliance coatings and wood finishes.

## 25.7 COIL COATING

Coil coating is performed in an enclosed system. Generally less emphasis is placed on the elimination of solvents and more to the quality of the coating. There are technological barriers to the reduction of solvent use due to the stringent rheological requirements of the coatings.

A primer layer is applied after the metal surface has been cleaned and baked. Primer plays an important role in preventing metal corrosion, therefore several performance requirements must be met. The primer layer must have good flexibility to prevent mechanical fractures and loss of adhesion of the top coats. It must not have imperfections such as blisters, craters, pops, and voids any of those will conduct water to metal surface. The transport of water to the interface depends on the properties of the primer layer and its thickness. It is preferable that primer layers are thick and can be applied in one operation. This can be fulfilled if the primer is formulated to release solvent quickly during curing. After curing, the open structure of the film should harden to a uniform, non-permeable layer. In summary, the primer should form a flexible, uniform, and thick layer. This has been achieved by the application of a hydroxy-terminated polybutadiene component formulated with a large number of solvents from the group of petroleum distillates, ethers, ketones, and alcohols.[29] The permeability of the resin has a prevailing influence on primer properties.

There are on-going efforts to reduce the VOC content of coil coatings. High-solids coatings typically contain 50-70% solid resin. Coatings higher than 70 wt% solids are more difficult to design. The coatings termed "extra high solids" contain less than 20 wt% solvent. These extra high solids coatings were made possible by the design of the resin and the selection of solvents which, in combination, give low viscosity, good flow, and a rapid evaporation rate.[30] Solvents useful in this invention are xylenes, cyclohexanone, methyl ethyl ketone, 2-butoxyethanol, ethyl-3-ethoxypropionate, toluene, and n-butanol.

Like primers, top coatings have followed a similar development path. The goal is to obtain a thick coating in one pass. Several polymers are used for coil coatings, such as polyesters, polyester-polyurethane, PVC plastisols and organosols, polyvinylidene fluoride, silicone modified polyesters, epoxy resins, and acrylic resins. A typical film thickness with all resin systems except PVC plastisols is up to about 30 μm. PVC plastisol coatings are applied to give a film thickness in the range of 80-200 μm. Because of disposal issues, chlorinated polymers such as PVC are less desirable today and have been replaced by solvent based systems. It is possible to use hot melt coatings but the high cost of equipment limits their use. In one recent invention[31] a halogen-free, heat curing coating composition was developed based on a polyester solution. The solvent combination was such that one solvent would have a boiling point of at least 160°C and preferably 220°C to allow for curing process to be fast without causing a non-uniform film. High boiling solvents can be selected from the group of aromatic hydrocarbons, esters, glycol esters, and alcohols. The concentration of solvents in this coating depends on formulation. In pigmented formulations, solvents constitute 25-40 wt% whereas in unpigmented formulations 30-45 wt%.

Liquid, single component stoving enamels were developed based on polyurethanes in which polyisocyanate component was partially or completely blocked.[32] Solvents constituted 30-60 wt% and should be mixtures of two or more components. In still another development, a retroreflective coating composition was developed.[33] It is designed to reflect the incident light beam in the direction of light source. The major application is in road signs and automobile licence plates. These had not been previously produced by coil coating. In addition to the binder components the product contains more than 50 wt% reflective glass microspheres. The solvent content is over 30 wt% of the entire composition. Another invention deals with the role of a crosslinker.[34] The crosslinking mechanism plays an important role in the performance of coating. If the crosslinker acts too rapidly, it prevents the formation of an uniform film. But if the crosslinker acts too slowly the coater must be slowed to an inefficient rate (typically belt speeds are about 200 m/min).

## 25.8 COMPOSITES AND LAMINATES

The performance of composites and laminates depends heavily on interfacial adhesion between fibrous component and the binding layer. Traditional technologies combine layers of prefabricated textile or fibrous component which are adhered by layers of polymeric matrix. The most recent trend involves preparation of composites on a molecular level which increases the contact surface area between both components and thus performance of the fabricated materials. One such process makes a glass/polymer composites by mixing glass and polymer on molecular level. It employs a compatible solvent to disperse materials on molecular level. The preparation of the molecular phase composite involves four steps. The first is a selection of compatible solvents and solutes. The second involves mixing a solution of glass with polymer to form a homogeneous mixture at the molecular level. The third involves removing solvent while retaining a homogeneous molecular mixture. The fourth involves forming bonds between the glass and polymer. Polar solvents are used in this application with preference given to water. Consequently, polymers used are water soluble.

The traditional polymer/glass composites (and other combinations of fibrous material with polymer) are made either by dispersion of glass fibers in the polymer carrier matrix or by intercalation of the polymer phase through a glass carrier matrix. In these processes, it is important to limit the formation of air pockets and to increase adhesion because both affect the performance of composites. For this reason, the rheological properties of the polymer matrix must be carefully controlled through the use of appropriate solvents. In composites produced from cores, the strength of the bond between facing and core must be maximized. Improved bonding has been achieved by wetting the core with solvent. The solvent serves both as a wetting agent and the adhesive solvent.[36] Due to this wetting, compressive and shear modulus have been improved by a factor of 1.4. Depending on the adhesive and the core properties, an appropriate solvent was selected from a group including water, acetone, petroleum ether, xylene, methyl ethyl ketone, methyl isobutyl ketone, and some other solvents. In another invention,[37] a decorative laminate was produced from a fabric coated on both sides with PETG copolyester. The coating contains very little volatiles.[37]

It was found[38] that a selection of solvent has effect on film thickness which is related to the boiling point of solvent. Fast evaporating solvents such as n-butyl acetate and o-xylene gave thicker coatings. Thinner films were obtained when mesitylene or diglyme were used.

A solvent-free adhesive was used for laminating vinyl sheet to wood.[39] A new technology of recycling waste carpets and textiles was developed based on coating individual yarn fibers by an adhesive containing solvent.[40] The high surface contact area produced materials with excellent mechanical properties.

## 25.9 COSMETICS

Two groups of solvents are used in cosmetics industry. These are process solvents which are removed after the material is prepared and solvents which are retained in product. Preparing polyaminoacid particles for encapsulation of cosmetics requires a process solvent.[41] This particular process has several stages. First, an emulsion of aminoacids is prepared which is then polymerized and the organic solvent is removed. Hollow particles can be obtained by adding a poor solvent. Such solvents belong to one of the following groups: halogenated aliphatic hydrocarbons, halogenated aromatic hydrocarbons, esters, ethers, hydrocarbons. The preferred solvents are dichloromethane, chlorobenzene, and o-chlorobenzene.

Microsphere reservoirs for controlled release applications in medicine, cosmetics and fragrances are obtained evaporating solvent from an oil-in-water emulsion. The microcapsules have diameters of from 3 to 300 μm.[42] Many polymers are suitable for use in this invention which is based on the knowledge that the inclusion of plasticizer renders porous and spongy structure as opposed to the hollow core and relatively solid surface which results when no plasticizers are used.

Even smaller, nanometer size particles, can be prepared from cyclodextrin. These particles are useful as a carrier for pharmaceuticals and cosmetics. First, a solution of cyclodextrin in an organic solvent mixture is prepared, followed by the preparation of a water dispersion of surfactant. When the two components are mixed together a colloidal dispersion of microspheres is produced.[43] Particles sizes range from 90 to 900 nm. Typical solvents are methanol, ethanol, isopropanol, and acetone. Solvents are used in the purification of xanthan gum to lower ash and to obtain a product with no traces of solvents.[44] Lower alcohol is used as the solvent. Solvents have also been used to decolor fatty acid esters.[45] Crude oils extracted by pressing or with solvents cannot be used in cosmetic products. Refining processes include "chromatographic" decolorization which involves diluting the oil in an apolar solvent and absorbing impurities by contacting the resulting solution with a solid absorbent in a column. Food quality n-hexane is used as a solvent.

In some cases, solvents do remain in the final product. One such example occurs in the preparation of liquid vanillin composition used in food and cosmetics production.[46] The preparation of such a solution is complex. The solution must be pourable at room temperature, have high solids concentration (50-70%), be mechanically and chemically stable, be easy to dilute, be transparent, be stable to bacteria, and inexpensive. The solvents include water, ethanol, and propylene glycol. Polymeric liquid crystals are prepared by dispersing polysaccharide in water.[47] These liquid crystals are used for perfumes. Xanthan gum is also in use for thickening cosmetics.[48]

## 25.10 CLEANING

In this broad range of applications, recent work on household and electronic cleaners and metal degreasers have produced the most patents. Several patents have been selected to show progress in household cleaners. Soap scum removal and hard water deposit removal is done using a multi-component mixture containing hydrophobic cleaning solvents from the following group: dipropylene glycol mono-butyl ether, tripropylene glycol monobutyl ether, and ethylene glycol monohexyl ether.[49] In addition to the solvent, surfactants, polycarboxylic acids, hydrogen peroxide, a thickener, and water are used. A terpene-free microemulsion composition was developed based on combination of nonionic and anionic surfactants, methyl esters of $C_6$ to $C_{14}$ carboxylic acids, a co-surfactant and water.[50] This is VOC-free composition for cleaning hard surfaces such as metal, glass, ceramic, plastic, and linoleum.

An aerosol cleaning composition contains surfactant, a water-soluble or dispersible solvent and a chelating agent (potassium EDTA), a propellant, and water.[51] This composition is specifically designed for the hard surfaces found in a bathroom. The composition forms foam which collapses on the stained surface to deliver the cleaning agent. The novelty is in the use of the chelating agent which enhances soil and scum removal.

A furniture cleaning and polishing compound provides protection against water.[52] This is achieved by the use of a halofluoro polymer in combination with mineral oil, a hydrocarbon solvent, silicon oil, a detergent, and water.

A thickened aqueous cleaning compound was developed for cleaning hard surfaces, disinfection, and drain cleaning.[53] The thickened product increases the residence time and cleaning effectiveness especially on non-horizontal surfaces. A firearm cleaning agent has been designed to remove the fouling residue which contains carbon and metal residue.[54] This is achieved by a combination of butyl or amyl lactate which dissolves the carbon residue and detaches the metal residue, and an electrical neutralizing agent, namely a citrus distillate, which bonds with the positive ions of the metal residue.

There are other examples of efforts to eliminate or reduce solvents but the general trend is towards optimization of formulations to perform well in a specialized application.

A cleaning liquid for semiconductor devices comprises 1-5% $R_4NF$ (where R is a hydrogen atom or an alkyl group), 72-75% solvent, and water.[55] Solvents used in this application must be water soluble. Suitable solvents are dimethylformamide, dimethylacetamide, 1-methyl-2-pyrrolidone, and dimethylsulfoxide. Resist residues, which are formed during dry etching by the reaction of the photoresist with a reactive gas containing chlorine, must be removed because they may cause disconnections or abnormal wiring. In past, alkaline cleaning liquids were used which do not perform well under severe etching conditions.

A flux-removing cleaning composition is a water-based composition containing an alkaline salt component and a surfactant.[56,57] This formulation is designed to eliminate solvent based products.

A degreasing and cleaning compound for electronics has been developed based on a silicon containing cleaning agent (polyorganosiloxane), an isoparaffin ($C_3$-$C_{15}$), a surfactant, and a hydrophilic solvent (ethanol or acetone).[58] This formulation is intended to replace chlorinated solvents such as trichloroethane, trichloroethylene, tetrachloroethylene, and carbon tetrachloride which have been used up to now. In another invention,[59] a molecular level cleaning compound is proposed. It includes a specific solvent (n-propyl bromide), low boiling solvents (selected from the group containing nitromethane, 1,2-butylene oxide, 1,3-dioxolane, sec-butanol, ethanol, methanol), a saturated terpene, and a defluxing or ionic removing additive. The compound is designed for use in a vapor degreaser. It is suggested that this composition reduces the environmental impact because the solvents are non-flammable, non-corrosive, non-hazardous, and have very low ozone depletion potential. Superior cleaning capability of the compound is attributed to the strong solvent power of the mixture, bringing the material being cleaned into a contact with this fresh condensing solvent, and the solvent flashing action during its downward flow. In still another invention,[60] a degreasing and cleaning compound is based on non-polar hydrocarbons ($C_9$-$C_{12}$ hydrocarbons or dehydrogenated terpenes), an organic polar component (alcohol, amine, ketone, or ester), water and a surfactant. In this method, residual solvent is removed from the hot article by contacting it with an aqueous medium or low boiling solvent which are subsequently removed by evaporation.

An aqueous, solvent-free degreaser contains a blend of polyoxyalkylene block copolymers.[61] This compound is intended for cleaning oils and greases from a variety of surfaces but specifically printed electronic circuit boards and automotive parts. The formulation is designed to compete with cleaning products which contain solvents. It is formulated with nonionic polymeric surfactant dispersed in water. In another development,[62] a water-immiscible hydrocarbon is used in the first cleaning step. This is followed by treatment with an aqueous displacement solution which contains surfactant and a pH modifier. The displacement solution displaces the hydrophobic solvent residue from the surface of the substrate and prevents its redeposition. The invention is intended to replace ozone depleting CFCs and chlorine containing solvents.

These patents indicate that, although there are attempts to replace solvents from formulation or to use more environmentally friendly solvents, the main requirement remains the performance of the cleaner and this is best served by using solvents. The major push has been to formulate high performance, specialized materials and processes which are carried out in an enclosed environment to prevent pollution. It should be underlined that in these applications quality of cleaning is critical and any compromise in cleaning effectiveness would contribute to an even higher environmental impact because it would lead to production of non-performing products which would have to be discarded.

## 25.11 FIBERS

Solvents are used in the production of fibers and in their modification and recovery from wastes. Production of fibers with optoelectronic properties for optoelectronic modulators requires several steps.[63] In the first step, a metal hydrate or a hydrated metal compound (based on Pb, La, Zr, Ti) is dispersed in solvent (ethanol, n-propanol, isopropanol, n-butanol, isobutanol, 2-methoxyethanol, or 2-ethoxyethanol), the resulting dispersion is then heated to polymerize the material and stretched to gel the fiber. The final fiber formation is achieved by heating the gelatinized fiber.

Fibers containing intrinsically conductive polymer are manufactured from a composition containing an organic acid salt of an intrinsically conductive polymer, a matrix polymer, and a spinning solvent.[64] Polyaniline is the intrinsically conductive polymer. Polyaniline alone cannot be processed into fiber because of its low solubility (standard wet spinning methods require a polymer concentration of 15-20%). Many polymers can be used as the matrix polymer. For this application, polyacrylonitrile was selected. A variety of solvents can be used but dimethylacetamide was found to be the most useful.

There are many reasons for surface treatment of the fibers. In one process,[65] introduction of functional groups is described. The equipment designed for this process includes a cleaning vessel, a vacuum drier, and a plasma treatment vessel. The fiber is first treated with a solvent, which is subsequently removed in the vacuum drier to remove all residual solvent. The surface is then modified by a plasma treatment. Cleaning removes dirt from the natural fibers and impurities from man-made fibers. Water, hydrocarbons, and halogenated hydrocarbons are used in an enclosed system. Surface etching and cleaning techniques which were used in the past released solvents and other materials to environment, especially because the fibers were not sufficiently dried.

Fiber glass is modified by the application of a thermosetting resin in two pass process. Thus treated, these fibers meet the dust free requirement for use in microchip substrate.[66] Epoxy resin is applied from a solution in methyl ethyl ketone. In another invention,[67] cellulose fibers are treated with an aqueous solution to prevent fibrillation. Fibrillation gives a hairy appearance and occurs due to the mechanical abrasion of fibers when they are processed in a wet and swollen state.

Polyester can be recovered from contaminated polyester waste which contains non-polyester components.[68] The waste may be polyester blended with cotton or other fibers, polyester magnetic tapes, or coated polyester films and engineering resins. The waste is dissolved in dimethylterephthalate, methyl-p-toluate or dimethylisophthalate. The contaminants are filtered from solution and polyester is recovered by crystallization or used as a feedstock in a methanolysis process which forms dimethylterephthalate and alkylene glycol.

## 25.12 FURNITURE AND WOOD COATINGS

A water based wood copolymer coating was developed to accentuate the color and the natural grain structure of the original substrate.[69] This enables solvent based system such as nitrocellulose lacquers to be replaced by waterborne coating. In another invention,[70] a water based composition was developed for covering scratches, blemishes and other damage on finished wood articles such as furniture, wood trim and wooden flooring. The inventors claim that the product has performance characteristics similar to solvent based systems. In still another invention,[71] a repair kit was developed using soluble colorants. This kit contains solvent based product. An aqueous dispersion of polyurethane is proposed for coating wood substrates such as wood floorings, furniture, and marine surfaces.[72] This product is intended to replace solvent based system, although the emulsion does contain 7% of a polar coalescing solvent (selected from the group of 1-methyl-2-pyrrolidone, n-butyl acetate, N,N-dimethylformamide, toluene, methoxypropanol acetate, dimethyl sulfoxide, and others). There is also solvent based system (solvent selected from a group of ketones, tertiary alcohols, ethers, esters, hydrocarbons, chlorocarbons). Process solvents are later stripped in the manufacturing process.

There is an obvious move in the wood finishing industry to replace solvent based systems with water based systems. This is a necessary development given that finishing processes are often carried out in homes or in small plants or workshops which lack adequate ventilation.

## 25.13 PAPER

Solvents are relevant in pulping processes and waste paper processing. Solvent based coatings and inks are applied to paper and paper must be designed to accept such materials without absorbing an excessive amount of solvent.

The traditional pulping process uses various inorganic chemicals to separate cellulose from lignin and other components. A major problem associated with this process is a difficulty in recovering or destroying these inorganic chemicals before they become major environmental pollutants. Solvents may be part of the solution to this problem. A process was developed in which wood is digested in a single-phase mixture of alcohol, water, and a water-immiscible organic solvent such as a ketone.[73] After digestion and adjustment of the proportion between water and solvents, lignin is present in the organic solvent, cellulose in the solid pulp phase, and hemicelluloses and dissolved sugars are in the aqueous phase.

Acceptance of the solvent pulping method has been slow due to difficulties in solvent recovery. However, apparatus and technology have been developed which allow recovery of almost all the solvent.[74] In another development,[75] it was found that pulp bleaching can be performed more effectively in the presence of a solvent, either methanol or ethanol. Alcohol protects cellulose during the ozone bleaching stage better than does water. Better quality cellulose is the result.

Wax paper is recycled by immersing it in solvent, preferably n-hexane.[76] The wax is dissolved in the solvent and separated from the paper. Tipping paper used to produce cigarette filters is made with special grades of calcium carbonate which reduce the wear of cutting knifes and limit absorption of solvent from ink.[77] Paper stock was developed for release paper with a reduced tendency to absorb solvent from the coating. This allows thinner coats to be applied which retain good release performance.[78]

## 25.14 PRINTING

A rotogravure or flexographic printing ink has been developed which reduces wear on printing equipment.[79] Solvents used in these inks include aliphatic, aromatic, and naphthenic hydrocarbons, alcohols, glycol esters, ketones, esters, and water. With water, glycol esters are used as cosolvents.

A lithographic ink has been developed based on an ester-modified carboxyl-containing copolymer dissolved in a solvent system selected from a group which includes aliphatic and aromatic hydrocarbons.[80] The invention gives better pigment dispersion and color development. A printing ink for gravure printing is based on a polyurethane oligomer dissolved in organic solvents selected from aliphatic, aromatic, and naphthenic hydrocarbons, alcohols, ethers, and esters.[81] Here, again, the driving force behind the invention was to increase color strength. A novel method of screen printing was developed to apply surface layers of materials with properties such as phosphorescence, electrical conductivity, mechanical adhesion, dielectric properties, etc.[82] The choice of binder and solvent depends on the substrate involved. A metallic appearance on plastic products can be obtained by applying an ink coating.[83] Typical solvents include aromatic hydrocarbons, cyclohexanone, butyl acetate, diacetone alcohol, an glycol ether. The concentration of solvents is in the range from 75 to 90%. The method of processing is UV curing. The choice of solvent depends on the substrate. The method was developed as an alternative to hot foil stamping which has numerous limitations. In another invention,[84] a device for cleaning screen plates was developed. The process used a solvent and cleans with increased precision.

A water based ink was developed for writing, drawing and marking using fountain and ball pens.[85] An organic solvent based ink was developed for ink jet systems.[86] The ink contains 40 to 98% solvent. Typical solvents include ethanol, isopropanol, acetone and methyl ethyl ketone. In still another development,[87] a water based ink jet composition was developed.

These patents show that solvent replacement is not a primary activity in this field. There is substantial interest in developing new technology for printing industry but it is driven mostly by a desire to improve quality and to satisfy needs of emerging printing technologies.

## 25.15 STONE AND CONCRETE

A limited number of inventions were patented in this very broad and important field of construction technology. A solvent based coating for a containment structure is based on chlorosulfonated polyethylene.[88] The coating is intended for applications in trenches, tank interiors, containment lining, flooring systems and joint overlays. The aim is to improve adhesion of the coating to the substrates. Solvents are selected from acetone, xylene, methyl isobutyl ketone, diacetone alcohol, and several other ketones and esters. A water reducible coating for concrete sealing was developed based on an acrylic resin.[89] The coating is intended to compete with solvent based coatings. It contains butyl carbinol as the dispersing solvent. A waterproofing coating was developed based on a solution of styrene polymer.[90] Solvents are chosen from among dichloromethane, ethylene chloride, trichloroethane, chlorobenzene, acetone, ethyl acetate, propyl acetate, butyl acetate, benzene, toluene, xylene, ethyl benzene, and cyclohexanone.

Efflorescence of masonry, brick, concrete, and mortar can be eliminated by coating with a polyvinylalcohol solution in water.[91] A solvent addition during silica glass production prevents cracking.[92] The most effective solvents are methanol, ethanol, n-propanol, isopropanol, N,N-dimethylacetamide, N,N-dimethylformamide, methoxyethanol, and tetrafurfuryl alcohol. Oily stains on stonework can be cleaned with petroleum hydrocarbons.[93]

## 25.16 WAX

This section includes a discussion of formulated products which contain wax. A fluorinated copolymer was used in a wax composition for car body refinishing.[94] A mixture of ethyl acetate, toluene and heptane is used as a solvent. A cleaning and polishing composition for automobiles is also based on a fluorinated polymer and contains hydrocarbon solvent at up to 70%.[95] Universal auto lotion is a water/solvent emulsion of wax.[96] It contains isoparaffinic solvents. A car polish formulation contains 10-35% mineral spirits.[97] A semiconductor polishing composition contains alcohols, acid and water.[98]

## 25.17 SUMMARY

The patents evaluated show that the inventions address the needs of the industry. All industries seek to improve or optimize product performance. Elimination of solvents in these inventions is secondary. This does not mean that there is a lack of awareness of environmental impact. In some industries, such as, for example, the aerospace industry, quality is so important that the best solution must be found even if it is not the least polluting. The best opportunity for solvent replacement are in industries which are known to have limited capabilities to protect employees by the use of protective equipment, such as the furniture industry. There are industries which need to be more concerned about pollution but do not seem to have developed new products with this in mind (at least by the account of existing patents). This applies to the metal stamping and processing industries, industries supplying DIY markets, asphalt processing industries, concrete coatings, etc. There is little emphasis placed on pollution by these industries because they are made up of small, frequently, one-man operations and thus they are very difficult to assess. This also brings up the issue of the investment required to improve the environmental record. For example, automotive industry was able and did invest in environmental protection and the results are clearly beneficial.

It is interesting to ask the question "will solvents eventually disappear from the industrial environment?" The above discussion shows that many new technological processes are developed based on solvents. There are more new products containing solvents than there are inventions aimed at their elimination. So, in spite of a strong desire to eliminate solvents their production volume continues to increase slightly. One might ask if this approach is legally, politically and morally acceptable given that the general public see solvent elimination as a benefit. Or are the efforts being made simply directed at finding the least harmful means of delivering the product which the public demand. Above examples show that, although sometimes solvents are introduced in new processes which previously did not use them, the overall objective of having a lower impact on the environment seems to have been met. The paper industry example shows that the use of solvents in a well designed plant can be less harmful for the environment than traditional pulping. Technologies aimed at reprocessing wastes provide more insight. A large number of studies on polymer blending have shown that reprocessing of mixed plastics, usually conducted without solvents, results in products which are not useful. In order to make recycling work effectively (especially of products which are multi-component mixtures such as composites, laminates, coated fabrics, coextruded films, etc.) some means of material separation must be found and the use of solvents offers just such a possibility. This brings several questions: which solution is better? should we dispose of these materials in the soil and wait for 50-100 years until they are digested? or should we responsibly use solvent under controlled conditions which does not pollute water, air and soil and does not endanger workers and customers? Solvent use can meet public needs and modern technology can find ways to ensure that solvent emissions are not released to the external environment.

Chapter 14 has several examples that show that many solvent releases can be avoided by elimination of negligence and sloppiness. For example the majority of solvents emissions from the petroleum industry can be avoided if valve leaks are repaired. This does not require extensive research but adoption of simple engineering principles to rectify the situation. The patents discussed above teach us that solvents can be eliminated or their amount reduced if resins are reformulated to reduce their viscosity thus eliminating the need to add solvents. In many cases where products are reformulated, this option is completely neglected due to the fact that the formulator finds it easier to use solvents since they usually do not affect other process characteristics or material performance and provide the simplest tool to adjust application properties. More knowledge must be acquired and more research is required than is currently being expended in product development. Solvents are inexpensive and this tends to encourage their use to lower the cost of the product. If situation develops that allows manufacturers to be able to profit from solvent-reduced or solvent-free formulations many products will be reformulated accordingly. But this requires that the environmentally aware customers be willing to pay premium prices for products which are solvent-free.

The other option for future improvements, which is certainly being exercised as is shown in the above patents, is a rationalization of solvent composition. Here, regulations have played a progressive role in eliminating or reducing some very harmful materials. New inventions are still needed which aim to find more benign solvent mixtures or to replace resins which require aggressive or harmful solvents by ones which allow for more environmentally friendly compositions.

There are many technological processes which will continue to require solvents at the highest levels of technological development. In these instances, good engineering practices will take the place of solvent elimination. There is nothing wrong with using solvents in a responsible manner. Scientific and engineering evaluations of all options and an assessment of the total social costs and benefits will show the best ways to optimize products and processes.

## REFERENCES

1 A J Lucas, Z E Halar, **US Patent 5,665,690**, Inland Technology Inc., 1997.
2 A J Lucas, Z E Halar, **US Patent 5,449,474**, Inland Technology Inc., 1995.
3 D Perlman, R H Black, **US Patent 5,773,091**, Brandeis University, 1998.
4 W L Chandler, **US Patent 5,549,839**, 1996.
5 E Stevens, **US Patent 5,288,335**, Stevens Sciences Corp., 1994.
6 T Mancini, **US Patent 5,186,863**, US Polychemical Corporation, 1993.
7 T Mancini, **US Patent 5,158,706**, US Polychemical Corporation, 1992.
8 J A Davidson, A K Mishra, **US Patent 5,344,494**, Smith & Nephew, Inc., 1994.
9 D C Munson, A C Lottes, A Psellas, D A Brisson, **US Patent 5,914,157**, 3M, 1999.
10 H Khalil, W Majewski, G Nickel, G Wypych, J D van Heumen, **US Patent 5,288,797**, Tremco Ltd, 1994.
11 G Krawczyk, V Dreja, **US Patent 5,824,743**, Morton International GmbH, 1998.
12 C D Congelio, A M Olah, **US Patent 5,962,560**, BFGoodrich, 1999.
13 J L Troska, L M Kegley, W B Dances, **US Patent 5,932,648**, Shell Oil Company, 1999.
14 Y Tominaga, A Tsuchida, H Shiraki, **US Patent 5,866,657**, Takeda Chemical Industries Ltd., 1999.
15 R L Senderling, L E Gish, **US Patent 5,849,133**, Carlisle Companies Inc., 1998.
16 H R Lubowitz, C H Sheppard, **US Patent 5,446,120**, The Boeing Company, 1995.
17 R R Savin, **US Patent 5,098,938**, 1992.
18 R F Sutton, **US Patent 4,906,731**, DuPont, 1990.
19 C H Sheppad, H R Lubowitz, **US Patent 4,851,495**, The Boeing Company, 1989.
20 C E Sroog, **US Patent 4,687,611**, DuPont, 1987.
21 A D Lindsay, B A Omilinsky, **US Patent 5,843,203**, GranTec, Inc., 1998.
22 D L Miles, **US Patent 5,427,794**, Rhone-Poulenc, 1995.
23 R E Hayner, P K Doolin, J F Hoffman, R H Wombles, **US Patent 5,961,709**, Marathon Ashland Petroleum, 1999.
24 R J Janoski, **US Patent 5,421,876**, Tremco, Inc., 1995.

25 P G Mansur, **US Patent 5,549,128**, Mansur Industries Inc., 1996.
26 D Hartnell, **US Patent 5,464,033**, Major Industrial Technology, Inc., 1995.
27 J C Larson, G L Horton, **US Patent 5,230,821**, DuPont, 1993.
28 S-H Guo, **US Patent 5,959,035**, ARCO, 1999.
29 M T Keck, R J Lewarchik, J C Allman, **US Patent 5,688598**, Morton International Corporation, 1997.
30 T Kuo, **US Patent 5,922,474**, Eastman Chemical Company, 1999.
31 D. Keogler, M Schmitthenner, **US Patent 5,916,979**, Huels Aktiengesellschaft, 1999.
32 R Braunstein, F Schmitt, E Wolf, **US Patent 5,756,634**, Huels Aktiengesellschaft, 1998.
33 G L Crocker, R L Beam, **US Patent 5,736,602**, 1998.
34 W T Brown, **US Patent 5,936,043**, Rohm and Haas Company, 1999.
35 W D Samuels, G J Exarhos, **US Patent 5,422,384**, Battelle Memorial Institute, 19995.
36 K-H Holtz, J Luke, W Riederer, **US Patent 5,575,882**, Hoechst Aktiengesellschaft, 1996.
37 M D Eckart, R L Goodson, **US Patent 5,958,539**, Eastman Chemical Company, 1999.
38 D A Babb, W F Richey, K Clement, E R Peterson, A P Kennedy, Z Jezic, L D Bratton, E Lan, D J Perettie, **US Patent 5,730,922**, The Dow Chemical Company, 1998.
39 G R Magrum, **US Patent 5,837,089**, Ashland Inc., 1998.
40 A M Kotliar, S Michielsen, **US Patent 5,912,062**, Georgia Tech Research Corporation, 1999.
41 K Makino, S Fukuhara, K Kuroda, T Hayashi, **US Patent 5,852,109**, JSR Corporation, 1998.
42 R J R Lo, **US Patent 5,725,869**, Zeneca Limited, 1998.
43 M Skiba, D Wouessidjewe, A Coleman, H Fessi, J-P Devissaguet, D Duchene, F Puisieux, **US Patent 5,718,905**, Centre National de la Recherche Scientifique, 1998.
44 A Haze, K Ito, N Okutani, **US Patent 5,473,062**, Takeda Chemical Industries, Ltd., 1995.
45 P Gonus, H-J Wille, **US Patent 5,401,862**, Nestec S.A., 1995.
46 F Fournet, F Truchet, **US Patent 5,895,657**, Rhone-Poulenc Chimie, 1999.
47 M El-Nokaly, **US Patent 5,599,555**, The Procter & Gamble Company, 1997.
48 M H Yeh, **US Patent 5,552,462**, Rhone-Poulenc Inc., 1996.
49 A E Sherry, J L Flora, J M Knight, **US Patent 5,962,388**, The Procter & Gamble Company, 1999.
50 S F Gross, M J Barabash, J F Frederick, **US Patent 5,952,287**, Henkel Corporation, 1999.
51 M Ochomogo, T Brandtjen, S C Mills, J C Julian, M H Robbins, **US Patent 5,948,741**, The Clorox Company, 1999.
52 K J Flanagan, **US Patent 5,925,607**, Sara Lee Corporation, 1999.
53 C K Choy, **US Patent 5,705,467**, 1998.
54 P R Pomp, **US Patent 5,935,918**, 1999.
55 T Maruyama, R Hasemi, H Ikeda, T Aoyama, **US Patent 5,962,385**, Mitsubishi Gas Chemical Company, Inc., 1999.
56 F R Cala, R A Reynolds, **US Patent 5,958,144**, Church & Dwight, 1999.
57 F R Cala, R A Reynolds, **US Patent 5,932,021**, 1999.
58 M Inada, K Kabuki, Y Imajo, N Yagi, N Saitoh, **US Patent 5,888,312**, Toshiba Silicone Co. Ltd., 1999.
59 L A Clark, J L Priest, **US Patent 5,938,859**, Lawrence Industries Inc., 1999.
60 G J Ferber, G J Smith, **US Patent 5,300,154,** Bush Boake Allen Limited, 1994.
61 M C Welch, C O Kerobo, S M Gessner, S J Patterson, **US Patent 5,958,859**, BASF, 1999.
62 S B Awad, **US Patent 5,464,477**, Crest Ultrasonic Corporation, 1995.
63 K Kitaoka, **US Patent 5,911,944**, Minolta Co. Ltd., 1999.
64 P J Kinlen, B G Frushour, **US Patent 5,911,930**, Monsanto Company, 1999.
65 S Straemke, **US Patent 5,960,648**, 1999.
66 B K Appelt, W T Fotorny, R M Japp, K Papathomas, M D Poliks, **US Patent 5,928,970**, IBM, 1999.
67 C D Potter, **US Patent 5,882,356**, Courtaulds Fibers Limited, 1999.
68 W D Everhart, K M Makar, R G Rudolph, **US Patent 5,866,622**, DuPont, 1999.
69 A J Swartz, M E Curry-Nkansah, R P Lauer, M S Gebhard, **US Patent 5,922,410**, Rohm and Haas Company, 1999.
70 I J Barlow, **US Patent 5,849,838**, S C Johnson & Son, Inc., 1998.
71 R L Setzinger, **US Patent 5,590,785**, 1997.
72 K C Frisch, B H Edwards, A Sengupta, L W Holland, R G Hansen, I R Owen, **US Patent 5,554,686**, 3M, 1996.
73 S K Black, B R Hames, M D Myers, **US Patent 5,730,837**, Midwest Research Institute, 1998.
74 J H Lora, J P Maley, B F Greenwood, J R Phillips, D J Lebel, **US Patent 5,681,427**, Alcell Technologies Inc., 1997.

75 M Solinas, T H Murphy, A R P van Heiningen, Y Ni, **US Patent 5,685,953**, MacMillan Bloedel Limited, 1997.
76 R Vemula, **US Patent 5,891,303**, 1999.
77 L Snow, K Mahone, I Baccoli, **US Patent 5,830,318**, Schweitzer-Mauduit International Inc., 1998.
78 B Reinhardt, V Viehmeyer, M Hottentrager, **US Patent 5,807,781**, Kammerer GmbH, 1998.
79 J S Perz, S L Rotz, **US Patent 5,958,124**, The Lubrizol Corporation, 1999.
80 R H Boutier, B K McEuen, M F Heilman, **US Patent 5,948,843**, Elf Atochem North America, Inc., 1999.
81 P G Harris, P D Moore, **US Patent 5,886,091**, Milliken & Company, 1999.
82 D-A Chang, J-Y Lu, **US Patent 5,843,534**, Industrial Technology Research Institute, 1998.
83 B W Bechly, **US Patent 5,961,706**, Technigraph Corporation, 1999.
84 O Tani, **US Patent 5,901,405**, Tani Electronics Industry Co., Ltd., 1999.
85 R Fraas, **US Patent 5,961,703**, J S Staedtler GmbH & Co., 1999
86 S Engel, A Badewitz, **US Patent 5,935,310**, J S Staedtler GmbH & Co., 1999.
87 B A Lent, A M Loria, **US Patent 5,929,134**, Videojet Systems Interantional, Inc., 1999.
88 T G Priest, R P Chmiel, R L Iazzetti, E G Brugel, **US Patent 5,814,693**, Forty Ten LLC, 1998.
89 P A Smith, **US Patent 5,777,071**, ChemMasters, 1998.
90 J H Gaveske, **US Patent 5,736,197**, Poly-Wall International, Inc., 1998.
91 T Beckenhauer, **US Patent 5,681,385**, 1997.
92 K Takei, Y Machii, T Shimazaki, H Teresaki, H Banno, Y Honda, N Takane, **US Patent 5,871,558**, Hitachi Chemical Company, 1999.
93 J E Kerze, **US Patent 4,956,021**, 1990.
94 M R Wollner, **US Patent 5,962,074**, 3M, 1999.
95 P A Burke, K J Flanagan, A Mansur, **US Patent 5,782,962**, Sara Lee corporation, 1998.
96 R L Fausnight, D A Lupyan, **US Patent 5,700,312**, Blue Coral, Inc., 1997.
97 S Howard, R Frazer, **US Patent 5,288,314**, Crescent Manufacturing, 1994.
98 D D J Allman, W J Crosby, J A Mailo, **US Patent 5,861,055**, LSI Logic Corporation, 1999.

# ACKNOWLEDGMENTS

This following section contains acknowledgments included in the various sections of the book which were combined to form one section. For better identification, individual acknowledgments follow the reference to the title and authors of the book section.

Preface

*George Wypych*, ChemTec Laboratories, Inc., Toronto, Canada

I would like to thank Dr. Robert Fox and John Paterson who made all efforts that the language used in this book is simple to understand and the book is read with pleasure.

4.2 Polar solvation dynamics: Theory and simulations

*Abraham Nitzan*, School of Chemistry, the Sackler Faculty of Sciences, Tel Aviv University, Tel Aviv, 69978, Israel

This work was supported by Israel Science Foundation. I thank my E. Neria, R. Olender and P. Graf who collaborated with me on some of the works described in this report.

4.4 Methods for the measurement of solvent activity of polymer solutions

*Christian Wohlfarth*, Martin-Luther-University Halle-Wittenberg, Institute of Physical Chemistry, Geusaer Straße, D-06217 Merseburg, Germany

Thanks are given to G. Sadowski (TU Berlin) for providing Figure 4.4-7(b), B. A Wolf (Univ. Mainz) for providing Figure 4.4-13, and G. Maurer (Univ. Kaiserslautern) for providing Figure 4.4-6. Furthermore, I wish to thank M. D. Lechner (Univ. Osnabrück) and G. Sadowski for many helpful comments and discussions about this manuscript.

5.4 Mixed solvents, a way to change the polymer solubility

*Ligia Gargallo and Deodato Radic*, Facultad de Quimica, Pontificia Universidad Catolica de Chile, Casilla 306, Santiago 22, Chile

The authors wish to express their apprcciation to Mrs. Viviana Ulloa for her technical assistance in this work and to publishers and authors for permission to reproduce figures and tables from their publications as indicated specifically in the legends of the figures and tables.

6.1 Modern views on kinetics of swelling of crosslinked elastomers in solvents

*E. Ya. Dcnisyuk*, Institute of Continuous Media Mechanics; *V. V. Tereshatov*, Institute of Technical Chemistry, Ural Branch of Russian Academy of Sciences, Perm, Russia

This work was supported by a grant from Russian Fund of Fundamental Research (grant No 98-03-33333).

10.3 Solvent effects based on pure solvent scales
*Javier Catalán*, Departamento de Química Fisíca Aplicada, Universidad Autónoma de Madrid, Cantoblanco, E-28049, Madrid, Spain

The author wishes to thank all those who contributed to the development of our solvent scales (C. Díaz, P. Pérez, V. López, J.L. G de Paz, R. Martín-Villamil, J.G. González , J. Palomar, and F. García-Blanco) and also Spain's DGICyT (Project PB98-0063) for funding this work.

12.2 Chain conformations of polysaccharides in different solvents
*Ranieri Urbani and Attilio Cesàro,* Department of Biochemistry, Biophysics and Macromolecular Chemistry, University of Trieste, Italy

The paper has been prepared with financial support of University of Trieste and of Progetto Coordinato "Proprietà dinamiche di oligo e polisaccaridi", Grant CT97-02765.03 of the National Research Council of Italy (Rome). The authors wish also to thank dr. Paola Sist for patient technical assistance.

13.2 Solvent Effects on Free Radical Polymerization
*Michelle L. Coote and Thomas P. Davis*, Centre for Advanced Macromolecular Design, School of Chemical Engineering & Industrial Chemistry, The University of New South Wales, Sydney, Australia

We acknowledge the publishers Marcel Dekker for allowing us to reproduce sections of an earlier review, "A Mechanistic Perspective on Solvent Effects in Free Radical Polymerization".[128] MLC acknowledges the receipt of an Australian Postgraduate Award.

14.19.2 Recent advances in coalescing solvents for waterborne coatings
*David Randall*, Chemoxy International pcl, Cleveland, United Kingdom

I would like to acknowledge with much gratitude the help given by Mr R J Foster of Harco for his help in assembling the MFFT data for the presentation. I must also thank my colleagues at Chemoxy, Ms Carol White, who assembled much of the data used in this paper, and Miss Tracy McGough, who helped me produce the OHPs. Finally, I must acknowledge the assistance given by Mr T J P Thomas, who has acted as a consultant to Chemoxy International in this whole area.

I am indebted to Bob Foster at Harco, who kindly carried out some comparative formulations using Coasol, Di-isopropyl AGS and Di-isopropyl Adipate in comparison with a Monoester of Pentane Diol.

14.21.1 Use of solvents in the manufacture of drug substances (DS) and drug products (DP)
16.2 Residual solvents in pharmaceutical substances
Michel Bauer, International Analytical Department, Sanofi-Synthélabo, Toulouse, France; *Christine Barthélémy*, Laboratoire de Pharmacie Galénique et Biopharmacie, Faculté des Sciences Pharmaceutiques et Biologiques, Université de Lille 2, Lille, France

The authors thank Nick Anderson, Steve Byard, Juliette de Miras and Susan Richardson for their participation in the elaboration of this document.

15.2.2 A simple test to determine toxicity using bacteria
*James L. Botsford*, Department of Biology, New Mexico State University, Las Cruces, NM, USA
This work has been supported by the principal investigator's participation in several programs to assist ethnic minorities in the sciences. Many students have helped with this work.

20.3 Pregnancy outcome following maternal organic solvent exposure
*Kristen I. McMartin and Gideon Koren*, The Motherisk Program, Division of Clinical Pharmacology and Toxicology, Hospital for Sick Children, Toronto, Canada
Supported by grants from Imperial Oil Limited, Physician Services Incorporated, The Medical Research Council of Canada, and the CIBC Global Market Children's Miracle Foundation Chair in Child Health Research, The University of Toronto.

20.4 Industrial solvents and kidney disease
20.6 Chromosomal aberrations and sister chromatoid exchanges
20.7 Hepatotoxicity
*Nachman Brautbar*, University of Southern California, School of Medicine, Department of Medicine, Los Angeles, CA, USA
The author wishes to thank Ms. S. Loomis for her tireless work in transcribing this manuscript.

21.1 Supercritical solvents
*Aydin K. Sunol and Sermin G. Sunol*, Department of Chemical Engineering, University of South Florida, Tampa, FL, USA
Assistance of both Dr. John P. Kosky of MEI Corporation and Irmak E. Serifoglu with editing and typesetting are appreciated.

# Index

## A

*ab initio* 39,424,426,430,455,470,485,655,661,711
abiotic
  attenuation 1579
  degradation 1175
  process 1151
absorbance 343
absorption capacity 1635
acceptor number 566
acetaldehyde 1131
acetic acid 9,506,829
acetone 341-342,640,829,847,849,854,881,906,959,1149,1508
acetonitrile 2,13
acetophenone 819
acid/base
  character 580
  interaction 3,565,570,584,683,689
  properties 617
acidic groups 348
acidity 585,806,1053
acids 574
acoustic
  dissipation 364
  wave admittance analysis 342
acrylics 848
acrylonitrile 787
activated carbon 1509,1513,1543,1634
  service life 1531
activation
  energy 778,807
  theory 362
activity coefficient 177,282
  at infinite dilution 164
adhesion 14,339,570
adhesives 581,847,849,851,1638
  anaerobic 847
  chemically reactive 847
  polychloroprene 849
  pressure-sensitive 851
  UV-curable 847
adsorbent 1509
  inner surface area 1510
adsorption 570,1151
  capacity 1510
  enthalpy 1509
  fundamentals 1509
  heat 1526
  isotherm 1510
  kinetics 1509,1511
  preferential 274
adsorption-desorption 1160
adsorptive power 1510
aerosolization 354
aerospace 852,1640
AFM 569,706
agar 710
aggregate mass 689
  vesicular 690
aggregates 506,580,697
aggregation 565,570,578,689,697
agitation speed 798
agriculture 1640
air 1149
  contaminants, standard 925
  velocity 1526
airflow 353-354
airways 1082
Alagona approach 444
alcohol consumption 1396
alcohols 127,137,806,828,880
alkahest 7
alkalinity 1053
alkanes 127
N-alkylpyridinium 1468
allergic effects 1319
  potential 1320
Allerhand-Schleyer scale 588
p-allylanisole 826
allylation 811
altitude 1191
alumina 1513
aluminum stearate 693
amine 510
aminoacids 23
ammonia 16
amobarbital 286
amorphous
  phase 351
  structure 348
amphiphiles 691
amphotericity 577
n-amyl acetate 848-849
n-amyl ketone 849
analytical ultracentrifuge 186
anemia 1376

anhydrous form 282
aniline 9
  point 102,1056
anisole 819
annealing 254
anthraquinone 701
antifungal properties 881
antigens 1361
antiparallel orientation 12
Antoine equation 389
apolar 875,884
approximation, two-body 37
aprotic 762-763,798,811,813,829
aquifer 1572
Arrhenius description 573
asphalt 855
  processing 1640
association
  heteromolecular 505-507
  number 692,689
  self 350
asthma 1328
atomic surface tension 482
ATP 866
attenuation capacity 1163
autoignition temperature 1054
automotive 851
  finishing 1641
  industry 958
  maintenance 1641
  painting operations 1641
autoprotolysis 524
Axilrod-Teller term 453

## B

bacteria
  Gram-positive 862
  solvent-tolerant 859
Bader charges 444
barbituric acid 286
Barker-Henderson approach 749
barotropic effect 1425
barrier properties 693
basicity 585
bathochromic shift 590,592,602
battery electrolytes 1472
Beer law 353
Benard cells 412
benzene 1,39,105,127,342,697-698,812,817,
  829,848,882,952,957,1127,1133,1334
  exposure 1365
benzonitrile 794
benzyl alcohol 778,854
benzyl cyanide 815
BET 1510
binding energy 754
binodal 131,148,184,191-192,400,408
bioavailability 1130
biocatalysis 875
biodegradation 1151
  first-order 1153
  potential 1054
biofilm 1624
biological
  membrane 866,870
  monitoring 1080
biomarker 1375
bioreactor 1624
biosphere 856
biotic process 1151
biotransformation 859,1133,1618
biphenyls 287,301
bleaching 344
blends 184,694
blistering 956
blisters 1131
Blodgett-Langmuir film 496
blood 1081
blue shift 639
blush resistor 956
boiling
  point 1055
  temperature 15
Boltzmann factor 476,655
bond
  covalent 16
  energy 16
  intermolecular 345
  strength 849
Bondi approach 126
Bootstrap model 791
Born
  approach 482
  equation 769
  formula 677
  model 769
  radius 771
  term 652
  theory 133,738
Born-Kirkwood-Onsager expansion 665
Born-Oppenheimer approximation 422
Bose-Einstein statistics 429
boson 429
boundary
  condition 312
  surface 419
Brannon-Peppas equation 348
breakdown voltage 1059
breakthrough time 342,1631,1633,1635
breath monitoring 1079
breathing 1634
Brillouin theorem 659
Bristow-Watson method 246
bromine index 1055

bromobenzene 691-692,784
1-bromobutane 811
Brönsted equation 587
Brooker scale 590
Brownian dynamics 472
bubble
    collapse 360
    dynamics 356,366
    formation 354
    growth 360
    small 363
    surface 364
Buckhingam
    formula 682
    potential 459
burning, sustained 1069
butabarbital 286
1-butanol 545,694,854
2-butanol 694
t-butanol 694,1397
2-butanone 794,854
butyl acetate 848,881
butyl glycol 880
t-butyl methyl ether 2
2-butylhexanol 848
γ-butyrolactone 696

C

calorific value 1056
calorimeter 243,1056
calorimetric study 540
capillary
    condensation 1510,1524
    effect 379
    force 365
carbon dioxide 2,39,688,897
carbon tetrachloride 591,648,697,754,827,1133, 1382,1396
carbonyl
    frequency 698
    group 842
    stretching frequency 567
    stretching mode 349
carcinogen 1367
classification 1322
carcinogenic 2,1133
carcinogenicity 1132,1321
Car-Parrinello simulation 470
carrageenan 710
cartridge 1636
catalysis, triphase 830
catalyst 1467
    phase 826
    phase transfer 824
    separation 828
catalytic
    oxidation 1508
    reaction 2
catheters 956
cation radius 771
cavitation 371
    free energy 11,17
cavity 284,366
    boundary 498
    collapsing 366
    ellipsoidal 680
    field 682
    formation 292,648,663
    molecular 480
    radius 717,769
    shape 482,652,662,665
    surface 21,284,484
    surface area 284
    volume 648
cell
    membrane 860
    multiplication 1176
    physiology 866
    toluene-adapted 862
cellophane 1535
cellulose 127,694
    acetate butyrate 881
chain
    dimension 706,711
    flexibility 127,273
    rigid 127
    segment 345
    separation 128
    topology 709,711
    transfer 779
chair conformation 711
Chako formula 680
charge
    density 709
    distribution 132
    transfer 463,586,780,782
        complex 571
        state 641
Charnahan-Starling equation 762
chemical 344
    composition 419
    equilibrium 1
    potential 101,149,159,198,307,341,344-345, 575
    reaction rate 132
    reactivity 1,737,739,741,743,745,747,749, 751,753,755,757,759,761,763,765, 767,769,771,773,775
    resistance 339
    shift 591
chemisorption 1509
chlorinated hydrocarbons 827
chlorobenzene 506,794,810,812,830
1-chlorobutane 811

1-chlorooctane 826
chlorofluorocarbons 853,1189
chromatographic separation 2
chromophore 589,592
chromosomal changes 1375
chromosomes 1376
circulation cell 1625
clathrates 1130
Clausius-Clapeyron equation 244
Clausius-Mosotti
    relation 680
    theory 738
Clean Air Act 929
Clean Water Act 932
cleaners 1644
cleaning 853,894,957
    processes 1508
cleavage 702
climate forcing 1192
clothing 339
    materials 1631
    protective 342
cloud point 101,184,189-190,271,273,408,695
Cluff-Gladding method 334
coagulation bath 695
coal 952
    tar 855
coatings 386,565,963,1169,1555,1557
    environmental performance 965
    multilayer 35
    waterborne 352-353
coefficient
    hydrodynamic interaction 685
    mutual 340-341
    of diffusion 354
    of volumetric absorption 353
    self-diffusion 340-341,344
coffee 1441
cohesion energy 115,261
coil
    conformation 709
    dimension 273
    expansion 268
coil coating 880,1641
collisions 419
color 1056
colorimetric analysis 1056
combustion 1056
cometabolism 1582
compatibility 318
compatibilizer 693
complex formation 556
complexation 780,782,824
composites 1642
compressibility 126,200
    factor 244
    isothermal 648
Compton scattering 757
condensation 852
conductivity
    electrical 516,1126
    thermal 354
configuration 18
    interaction 431,658
conformation 18,25,707,710,788
conformational state 706
conformer 506,711
    energy 530
    equilibrium 530
    form 532
    stability 531
    transformation energy 531
conformic equilibrium 506
construction industry 1648
    products 855
contact angle 565
containers 693
continuum dielectric theory 133
continuum models 479
convective movement 1151
conversion 798
copolymer 686
copolymerization 779
    model 781
corrosion 339,852,1057
cosmetics 881,1319,1643
COSMO 481,484
cosolvency 267,793
    criterion 320
cosolvent 282,284-286,288,867,1175
    power 269
Coulomb
    contribution 444
    hole 430
    integral 448
    law 430,875
    term 459
Coulombic
    attraction 575
    interaction 439,451,574,727
Coupled-Cluster theory 431
coupling reactivity 816
covalence 463
cranial nerve 1328
cresol 957
    meta 590
crew-cut 690
critical
    behavior 191
    molar volume 340
    point 191-192,202
    region 214
    scattering, pulse-induced 184
    solution temperature 129

temperature 193
crosslink density 128
crosslinked elastomer 130
crosslinks 128,331,701
distance 345
cross-propagation reaction 786,790
cryoscopy 147,149,188
crystal
energy 282
lattice 1,1130
surface 1130
texture 1118
crystalline phase 350
crystallinity 351,694
degree 352
crystallization 348,567,694
crystals 694
curing time 343
cybotactic 585
cavity 585
region 584,608
cyclodextrin 294
cyclohexane 105,185,188,506,591,701,810, 826-827
cyclohexanone 698,704,813,956,1131,1165
cyclopentanone 956
cytochrome 1099,1112,1356,1379,1394


## D


Dalton law 364
Deborah number 339-340
Debye
longitudinal relaxation time 136
model 134
relaxation model 137
relaxation time 134
Debye-Hückel approximation 728
decaffeination 1441
n-decane 810,812,830
Dee-Walsh equation 203
defect, stress-related 386
defects 388
deformation 344
history 360
degreasing 1227,1233
dehydrobromination 826
dehydrogenase 1356
dehydrohalogenation 827,1572
de-icing fluids 853
DelPhi 481
demixing 176,184,189 - 190,206,273,695
density 505,516,521,1057
density-functional theory 574
deprotonation 33
dermal exposure 1079
dermatitis 881
des Cloizeaux Law 308
desorption 566
free energy 569
DFT 481
dialysis 956
dibromomethane 807
p-dichlorobenzene 699
1,2-dichloroethane 127,686,819,1133
dichloromethane 686,692,694-695,698,701,810, 812-813,815-817,819,827,826,829,956, 1396,1398
dichroism 592
dielectric constant 25,115,190,454,459,479,487, 652,738,754,763,781,829,875
continuum 584,649,663
response function 133
surface 494
continuum theory 674
enrichment 613
Diels-Alder 1470
diethyl ether 2
di-2-ethyl hexyl phthalate 686
diethylaniline 815
differential refractometer 158
diffusion 339
coefficient 38,310,351,388,1152,1157,1509
control 779
distance 346,348,1526
elastic 340
equation 307
gradient 339
molecular 1082
rate 344,350
time 340
translational 779
diffusive transport 340
diffusivity 341
dilute 282
dilution ratio 102,1058
dimeric interaction 425
2,3-dimethyl-butane 341
1,3-dimethyl-2-imidazolidinone 1136
N,N-dimethylacetamide 694,1135,1400
dimethyl-benzene 854
N,N-dimethylformamide 127,188,694-695,956, 1135,1170,1400
dimethylsulfoxide 302,507,694,709
Dimroth-Reichardt
scale 588
value 281
1,4-dioxane 686,691,694,709,956,1401
dip
coating 956
molding 956
diphenyl ether 2
diphenylhydantoin 286
dipolar moment 15

term 750
dipole moment 12,298,506,585,641,653,754, 763,807,875
dipole-dipole forces 584
interaction 12,138,528,530-531
permanent 767
Dirac function 662
dispersion 13
forces 454,649
disproportionation 783
dissociation 508
dissolution spontaneous 127
dissolving power 1422
distillation units 962
diurnal 1195
DMF 754
DNA 454
dodecane 826
domain
amorphous 350
crystalline 350
Dong-Winnick scale 589
donor
complex 785
number 566,738
donor-acceptor
approach 744
interaction 574,584
properties 527
reactions 571
Doppler broadening 644
double-liner system 1165
Drago
scale 591
equation 569
driveway sealer 855
driving force 339
drug acceptability 1130
degradation 1131
dry cleaners 1091
dry cleaning 883,1189
dryer 353 - 354,387
drying 339,565
conditions 386,1118
convection 353
defects 409
rate 396
technology 386
time 351
DTA 190
Dubois-Bienvenue scale 590
dye reduction 1100
dyes 1169
dynamic light scattering 689

**E**

ebulliometry 147,149,167
ecosystem 1617
ecotoxicity 1170
ecotoxicological effects 1185
effluents 1149
efflux pump 863
eigenfunction 427
eigenvalue 420
Einstein's equation 684
elastic linkage 361
elasticity potential 128
electron
affinity 574
cloud 469
transfer reaction 765
transport 1099
electron beam curing 880
electronegativity 754
equalization 444
electronic
nose 1222
state 462
transition 291,655
electronic industry 894,1644
electronic-vibrational spectra 639
electron-pair acceptors 3
donation 571
donors 3
electrophilic species 782
electrostatic
forces 649
interaction 12,453,541,651-652
potential 481
electrostatics 480
electrosynthesis 1473
elongational flow 360
embryotoxicity 1133,1334
Emergency Planing Act 933
emission 1149
limit value 1214
enantiomeric ratio 877
enantioselectivity 858,877
encephalopathy 1327
energy
consumption 388
minimum 18
of activation 341
potential 345
enhancement factor 1429
enthalpy 566-567,755
conformational 254
differential 195
evaporation 243,1509
exothermic 567
molar 345

of melting 346
entropic contribution 722
entropy 130
component 332
conformational 257
dilution 195
hydration 760
ideal 105
environmental fate 1149,1151,1153,1155,1157, 1159,1161,1170
laws 924
enzymatic
factors 1317
reaction 857,874
enzymes 857,872,1395
epoxy 847
equidensity 21
equilibrium
chemical process 528
constant 788
equipotential 21
ESR 2,700,790
ethanol 286,694,701,847,882
ethers 127
2-ethoxyhexyl acetate 848
ethyl acetate 342,694,816,827,881
ethyl benzene 341-342,854
N-ethyl carbazole 816
ethyl ether 810
ethyl formate 1131
ethylene glycol ether 1169
eutectic point 1465
eutectics 1466
EVA 347
evaporation
rate 353-354,956,1059
cooling 354
Ewald approach 487
excimer 268
exciplex 752,766 - 767
excitation energy 641
excited state 2
explosive limits 388,1060
exposure parameters 1315
extraction 1058,1068,1442
efficiency 940
process 2
temperature 943
extrusion 956

**F**

factor of chain expansion 685
fermentation 1584,1599
Fermi hole 430
Fermi-Dirac statistics 428
fermion 428
fetotoxicity 1338
fiber 695,1535,1645
diffraction 708
fibrosis 1394
Fick law 1084,1151
Fickian kinetics 340
film 693
casting 956
thickness 354
filters 1633
Fixman theory 728
flammability limits 1060
flanges 962
flash point 1060
flocculation 14
Flory equation 308
Flory matrix 723
Flory-Huggins
equation 365,389
interaction parameter 112,151,321,345,347, 569
theory 154,793,1125
Flory-Orwoll-Vrij model 199,206
Flory-Rehner
equation 128,321,331
theory 344
fluorescence 342,610,689
intensity 692
recovery 344
fluorescent radiation 701
fluorination 693
fluorine 693
fluorocarbon resin 880
foam-enhanced remediation 1626
folding rate 350
food 709,923
industry 1128
safety 934
force
field 451
repulsive 14
formaldehyde 1190
resin 881
formation heat 790
formic acid 591
fractionation 689
Franck-Condon
principle 291,752
state 644 - 645,654,658
free energy of mixing 345
free volume 125
freedom, degree 199-200,433,470,715
free-volume theory 339,353
freezing point 1061
frequency shift 699
Freundlich
equation 1152
isotherm 1510,1525,1575

Friedel-Crafts alkylation 1468
frontier orbital interaction 574
FTIR 2,342-343,396,565,570,698,756,759
fuel, sulfur 1056
fugacity 149,
    coefficient 170,195,207
furniture 1169,1205,1222-1223,1646
future trends 946,1637

## G

gas cleaning 1507
gas chromatography 1062,1064,1116,1135,
    1228,1562
gas tank 693
gasoline 1169
Gaussian chain 685
Gay-Berne potential 458
GC-IR 1131
GC-MS 1131
gel 333,344,707,710
    stable 194
    structure 698
gelation 697
genetic
    component 1361
geometry change 450
Gibbs
    energy of mixing 124
    free energy 101,148,152,191,705
Gibbs-Dugem-Margulis equation 157,553
glass transition temperature 176,254,340,
    348-349,703
glassy state 350
gloss 704
gloves 1631-1632
    surgical 956
glycerol 286
glycol ester 880
glycol ethers 1169
goniometer 689
grafting
    efficiency 703
green
    chemistry 1422
    strength 849
greenhouse gases 1189
ground transportation 950
groundwater 1150,1163,1571,1599
    well 1623
Grunwald-Winstein scale 587
Gutmann concept 738

## H

habit 1119
hair spray 881
halorespiration 1582
Hammond-Stokes relation 39
Hansen solubility 112
    parameters 106,264,695
hard sphere
    diameter 746
    potential 14,37
hard-core cell 201
hardness absolute 575
Hartree-Fock
    approximation 484
    description 430
    equation 665
    method 655
    procedure 423
    theory 661,666
hazard communication standard 926
hazardous wastes 1548
health and safety laws 924
heart 1415
heat
    capacity 254,1460
    conductivity 392
    dissipation 363
    flow 354
    of vaporization 387
    transport 1
Helmholtz free energy 493,566
hematopoietic effect 1365
hematoxicity 1365
Henry
    constant 173,177,206,213,1155
    law 148,389,1125,1150,1578
    diagram 112
hepatotoxicity 1337,1379,1385,1393
hepatoxins 1396
n-heptane 590,697,848,1126
herbicides 1102
hexamethylphosphoric triamide 2
n-hexane 2,127-128,323,590,697,812,828,
    848-849,1127
Hildebrand theory 124
Hildebrand-Scatchard
    equation 104
    hypothesis 345
hole 339-340
HOMO 571,573-574,789
homomorph 585-586,593,600
homopolymerization 777-778
homopropagation 778
homotactic interaction 722
Hook law 160
Hookean energy 732
hops 1441
HPLC 759
HSGC 154-155,167,178
Huggins equation 246
Hughes-Ingold theory 745

humidity 852
relative 353
Huyskens-Haulait-Pirson approach 103
hybridization 754
hydrate 282
hydrocarbons 102,941,1361,1617
hydrodynamic interaction 361
hydrodynamics 356
hydrogen
bond 15,683,709,760
bonding 111,573,584,613,703,740,1126
hydrolase 858
hydroperoxide 700
hydrophilicity 830
hydrophobic 866
effect 16,285,758,760
interaction 11,16,282
hydrophobicity 758,760,830,875
parameter 286
hydroxyl group 806
radicals 1559
hypernetted chain approximation 461,465
hypsochromic
sensitivity 607
shift 590,607,609
hypsochromism 588

I

IGC 148-149,154-155,164,166,177,565-566
immune system 1356,1367
incineration 880
infrared absorption 111
ingestion 1079
inhalation 1079,1316,1355
initiation 777
injection molding 956
ink 1647
inorganic chemicals 950
interaction
charge transfer 16
constant 566
dimeric 424
electrostatic 16
intermolecular 639
parameter 345
polymer-solvent 1125
potential 455
trimeric 424
interatomic force 470
interface 339,565
air-solvent 288
intermolecular
bonding 755
distance 12
forces 1
interaction 37,261
internal gradient 386
intersolute 282
effect 282,293
ion association 546
ion-dipole interaction 528,545
ionic
associate 507
liquid 2,1459
melts 616
radius 768
ion-ion interaction 528
ionization 816
iron and steel industry 951
irradiation 1131
isobutanol 351
isomerization 506,826
equilibrium 530
iosoctane 347
isopiestic measurement 158
isopleth diagram 1194
isopropanol 302,848,851,881,906,1397
isotactic 578
isotropic liquid 1465

J

Jackob number 376
joint filling 855
jump unit 339-340

K

Kapustinskii equation 1461
Kauri number 101
Kelvin equation 1510
ketones 880
kidney
disease 1355,1357,1359-1361,1655
function 1359
kinetic rate 807
Kirkwood
factor 754,762
theory 738
Kirkwood-Riseman-Zimm model 361
Kitaura-Morokuma decomposition scheme 429
knots 691
Knudsen method 245
Kohlemann-Noll theory 360
Kohn-Sham equation 471
Kosower scale 589

L

labeling 1062
laboratory hoods 962
laminar thickness 694
laminates 1642
land disposal 1149
Langevin dynamics 472
Langmuir
isotherm 1510,1575

method 245
laser 689
lattice-hole theory 203
LCST 198,268,372,380,384,1424
leachate concentration 1165
leather 1222
Leffler-Grunwald operator 285
Lennard-Jones
energy 749
interaction 138
potential 37,40,458,462,467,477
solvent 748
lethality 1176
leukemia 1365
Lewis
acidity 740
basicity 738
definition 571,617
donor 705
number 376,392
parameter 565
lifetime 654,1190
Lifshitz-van der Waals interaction 565
light absorption 2,680
limonene 864
linear
response 133
response theory 133
lipase 878
lipophilic contamination 898
liposomes 860
liquid
compressibility 356
constrained 495
crystals 485
general condition 357
motion 365
Newtonian 356
surface 493
system 419
lithology 1627
liver 1379,1393
function 1394
London
dispersion 469
equation 718
formula 447,672
Lorentz-Lorenz relation 680
LUMO 571,573-574,789
lungs 1082
Lux-Flood definition 573,618
lyophilization 874

**M**

macroscopic level 1
Manning theory 729
Marangoni number 413
Mark-Houwink equation 363,728
Markov chain theory 475
Martin correlation 362,390
mass transfer 387,816
material degradation 1632
Maxwell
element 361
equation 358
model 360
Mayo-Lewis model 780
mechanical
equilibrium 307
toughness 339
medical
applications 955
devices 955
melting 253
melts 200
membrane 689
cytoplasmatic 863
morphology 694
memory effect 357
Menschutkin reaction 1,9,30,742,745
menstruum 8
menstruum universale 7
Mercedes-Benz logo 761
mesomorphic solid 1465
mesophase 1465
mesopores 1513
metabolites 1317,1356
metal cation 450
metal casting 957
methane 208
methanol 188,286,341-342,547,694,826,829,957,1131
4-methyl-2-pentanone 854
N-methyl-2-pyrrolidone 398,686,694-695,698,778,841,847,956
2-methyl-pentane 341
methyl acetate 342,694,881,1131
methyl chloroform 1396
methyl ethyl ketone 127,129,687,842,847,849,854,881,952,956
methyl glycol acetate 881
methyl isobutyl ketone 703,881
1-methyl naphthalene 688
methylcyclopentane 129
1-methylimidizole 1460
micelles 690
microbial membrane 860
microcoulometry 1065
microdomains 350
microfiltration 694
microflora 1620
microorganisms 857,865
growth 1176
solvent resistant 865

survival 1176
micropores 1513
microscopy 689
microstructure 386
Millington equation 1152
mineral spirits 847-848,855,950
mining 1169
miscibility gap 268,273
mitochondria 1112
mixing
configurational entropy 124
entropy 124,292
free energy 104
model
classical 419
continuum 17,20,25,133
electrostatic 9
free volume 196
iceberg 16
lattice 198
lattice-fluid 196
supermolecule 18,22
molar
cohesive energy 103
volume 516,521
molecular
arrangement 698
design 36
dynamics 18,472
ensemble design 36
mass 345
mechanics 450
model 698
motion 254
orbital 423
orbital theory 571
orientation 752
packing 752
sieves 1513-1514
theory 361
toxicity 860
weight 343,351-352,689,703
molecule-molecule interaction 475
Møller-Plesset
perturbation 670
theory 431
treatment 671
molten salt 1459
Momany approach 444
monoclinic 1119
monomer
emission ratio 268
feed 781
monomer-monomer complex 785
Monte Carlo 17,19,37,208,472,476,662,723,760
morphological
feature 690
structure 689
morphology 339
of skin 339
Mulliken charges 444
mutagenicity 1132

## N

nail polish 881
remover 881
nanofibers 691
nanospheres 690
nanostructures 689
naphtha 848,854,881,950,957,1233
naphthalene 288,302
natural attenuation 1571,1573,1575,1577,1579,1581,1583,1585,1587,1589,1591,1593,1595,1597,1599,1601,1603,1605,1607,1609,1611,1613,1615
navy 1472
nephropathy 1359
nephrotoxicology 1356
Nernst layer 831
nervous system 1407
network 331
networks 710
neuropathy 1328
neuropsychological assessment 1329
neurotoxicity 1132
neutron scattering 185,698,756
small angle 147
Newton-Euler equation 37
Newtonian viscosity 361
nitrile rubber 848
nitrobenzene 641,819-820
nitrocellulose 848,881,1646
nitroethane 1400
nitromethane 754
nitroparaffins 1400
2-nitropropane 1400
NMR 2,342,351,388,705-706,713,759,788,793,1432
$^1$H-NMR 1135
shift 738
nondipolar 754
nonpolar 827
surface area 286
non-solvents 177
Norrish type II 701
nucleic acids 16
nucleophilic
agent 807
species 782
nucleophilicity 813
number of arms 689
Nusselt number 392

## O

octane 127,185
1-octanol 287-288
octupolar 754
odor 851,1219
  hedonic tone 1221
  in industrial plants 1222
  in materials 1222
  intensity 1220
  testing 1063
  unit 1220,1223
odorant concentration 1220
oils 923
Oldroyd equation 358,361
olfactometer 1223
Onsager
  cavity radius 654
  dipolar term 653
  model 651,667
  reaction field 654
  theory 655,715
Onsager-Bottcher theory 682
opalescence, critical 190
operation cost 1508
optical clarity 339
orbital radius 574
organic acids 16
organic chemical industry 962
orientational order 762
Ornstein-Zernike
  equation 464
  theory 490
Orofino-Flory theory 726
orthorhombic 1119
Osawa concept 729
OSHA standards 925
osmometry 147,149,154,168,195,267
  membrane 178,183
Ostvald-Freundlich equation 1491
oxidation degree 618
oxygen, singlet 702
ozone 1189,1192,1267
  depletion 1192
  precursor 1189
  production 1196
  tropospheric 1190

## P

pacemaker 956
packing density 746
Padé approximation 750
paints 354,581,709,1169,1235,1555,1557
  emissions 1559
  industry 1063
painting 950
paper 1647
Papirer's equation 566
partial pressure 1150
partition coefficient 288,291,860,875,1082,
  1149,1153
partitioning 2,867
patching 855
patent literature 1637
Pauli
  exclusion principle 427-430
  repulsion 649
Pauling crystal radius 770
PDAC 445
Pekar factor 766
Peklet number 363
PEL 925,1079
penetration 1422
perchloroethylene 914,956,1057
Percus-Yevick approximation 461,465
perfluorohexane 2
performance 1422
perichromism 2
permeability 693,860,1631
temperature effect 1632
permeation
  flux 694
  number 1631
permittivity 505-506,513,517-518,531,540,669
perturbation theory 436,461
pervaporation membrane 693
pesticides 1169
PET 688
pH 297,348,688,1064
pharmaceutical 709,1552
  industry 1468
  product 1119
  substances 282,1129,1135
pharmacopoeia 1129
phase
  boundary 344
  diagram 129-130
  homogenization 190
  separation 125,190,386
    thermally-induced 694
  toxicity 860
  transfer, catalyst 824
  transition 1420
phenol-formaldehyde resin 847
phenomenological model 288
phospholipids 860
photochemistry 1194
photodegradation 1151
photoefficiency 1559
photoionization 133
photolysis 701,1175,1192
photooxidation 700,1175
photoreactor 1563
photoresists 701,1546

photosensitizers 701
photostabilization 702
phytoremediation 1622
pickling 952,957
piezoelectric method 163
pinholes 956
Piola tensor 306
pipeline coatings 855
plasma cleaning 897
plasticizer 856
plasticizing action 701
plastisol 847
PLUME 1164
p-nitrophenol 699
Poisson equation 480
polar 750,754,760,807,824,842,897
bond 754
character 586
liquids 693
solvation 746
polarity 2,9,288,531,577,584-585,610,700,745,
803,813,815,875
effect 780,782
index 527
parameter 288
scale 584,588
polarity-polarizability scale 584
polarizability 298,584,655,740,749,785
polarizable continuum model 481
polarization 12,133,428,486
effect 459
orientational 669
poly(acrylic acid) 687
poly(amide-imide) 841
poly(ethylene glycol) 206
poly(ethylene oxide) 186,687,699
poly(vinyl methyl ether) 185
poly(vinylidene fluoride) 693
polyacrylonitrile 127,694
polyaddition 841,963
polyamide 694
polyamides 694,847
polyaniline 703,1126
polybutadiene 127
polybutylene 848
polycarbonate 694,700,820
polychloroprene 127,848-849
polycondensation 818,820,841,963
polycrystalline particles 1118
polydimethylsiloxane 185,206,704
polydispersity 154
polyelectrolyte 454
polyester 880-881
polyesters 694,847
polyether 818
polyetheretherketone 348-349,699
polyetherimide 694
polyethersulfone 694
polyethylene 128-129,693,701-702
polyethylene glycol 808,817,826
polyethylmethacrylate 103
polyimide 692,694,702,841,847
polyisobutylene 685,848
polyisobutylmethacrylate 103
polyisoprene 127
polymer
coil 271,779
drying 350
segment 339
semicrystalline 350
solubility 198
polymerization 842
free radical 777
polymers 253
polymethylmethacrylate 127,269,343,568,570,
578,686,1437
polymorph 755
polymorphic transition 1119
polymorphism 709
polyolefins 694
telechelic 694
polyphenylene 685
polyphenyleneoxide 694
polypropylene 693
polypyrrole 568
polysaccharides 706-707,709,711,713,715,717,
719,721,723,725,727,729,731,733,735,1654
polysiloxanes 273
polystyrene 127,129,186,188,269,342,344,686,
698
atactic 128
isotactic 128,697
syndiotactic 697-698
polysulfides 818,848,853
polysulfonate 818
polysulfone 694
polyurethanes 331,346,354,694,847-848,850,
855,956,1640
polyvinylacetate 185,351-352,848
polyvinylalcohol 127
polyvinylchloride 127,129,691-692,698,880
pore
diameter 1509,1524
size 694
system 1513
porins 862
porosity 1118
porous
matrices 1422
membrane 694
potential energy surface 422
potentiometric titration 619
powder coating 881
Poynting correction 171,1429

Prandtl number 392
Prausnitz theory 208
pregnancy 1333,1335,1337,1339,1341,1343, 1345,1347,1349,1351,1353,1655
pressure
  drop 1526
  partial 353
pressure-temperature relationship 1420
Prigogine
  approximation 199
  cell model 200
  theory 125
Prigogine-Flory-Patterson
  equation 202
  theory 199-200,207
primer 705,881
printing 1169,1647
probability function 430
process safety management 927
propagation 778
  heat 787,790
  pathway 785
  radical 793
  step 780,782
1-propanol 545,694
2-propanol 694
propyl acetate 342
propylene carbonate 591
propylene glycol ether 1169
protagonism 15
protection 1631
protective equipment 1080
proteins 16,754,866
protic 745,752,762,798,813,829
protolysis 524
proton
  influx 860
  leakage 860
  transfer 507
proton-donor function 510
pulmonary blood vessels 1082
pulping 1647
pump valves 962
pupil dilatation 1222
purity 1065
pycnometer 1057
pyridine 507

**Q**

QM/MM 478
quadrupolar 754
  forces 753
quantum
  description 419
  level 18
  number 430
  yield 702

**R**

radiant energy 353
radicals 700,782
  lifetime 701
  stability 790
radius of gyration 684-685,689,691
Raman 756
Raoult law 154,196,365,372,389,613,794
  activity coefficient 877
Rayleigh
  factor 182
  ratio 689
  time 363
reaction
  endothermic 1
  field 481,651-652
  field approach 487
  nucleophilic 9
  rate 798,807,811,815-816
reactive barriers 1628
reactivity 700
reactors 962
recrystallization 2,1130
recycling 880
red shift 752
refractive index 158,190,740,1066
regional scale 1193
regulations 1267,1269,1271,1273,1275,1277, 1279,1281,1283,1285,1287,1289,1291, 1293,1295,1297,1299,1301,1303,1305, 1307,1309
  in Europe 1311,1313
regulatory approaches 1199
Reichardt scale 740
reinforcing 339
relaxation
  enthalphy 254
  segmental 346
  spectrum 361
  time 133,340,358,361,654
relief devices 962
reproductive
  toxicity 1169
  toxicology 1353
reptation 344
repulsion 14
residence time 387
residual stress 339
resistance specific 1059
resistive heating 1626
resistivity 1058
resorcinol 699
Resource Conservation Act 932
respirator 1633
respirators 1371

respiratory
  physiology 1081
  protection 1633
respirometer 1099
retardation time 358
retention volume 566
retroreflective coating 881
rewind station 387
Reynolds number 366,370
rheological
  equation 356
  properties 683,704
Rider approach 122
rigid spheres 14
ring deformation 711
RISM 466
rod-like micelles 690
Röntgen model 755
rotamer 591
  energy 531
rotation 711
  barrier 532
  freedom 1477
Rouse theory 361
rubber 847
  elasticity 148
  natural 342
  nitrile 342
rubbery state 349
Runge-Kutta-Gill method 393

**S**

sample collectors 962
Sanchez-Lacombe
  equation 201
  theory 207
SANS 185-186,706
SAPT 440
SAXS 184-185,706
scaled-particle theory 461
Scatchard equation 263
Scatchard-Hildebrand theory 197
scattered intensity 689
scattering angle 689
Schlieren photography 186
Schrödinger equation 420,482,656-657,663,671,
  674,676
sealants 847-849,851,855
seals 962
second viral coefficient 689
sedimentation velocity 186
segment
  fraction 200
  size 200
  statistical 361
selectivity 694
self-consistent field 423
self-diffusion coefficient 38
self-remediation 1622
SEM 689
semiconductor 1546
shear modulus 347
Sherwood number 392
shoe industry 849
shrinkage 354
silica gel 1513
silicones 847-848
Simha-Somcynsky equation 203
similia similibus solvuntor 8
simulated-annealing 470
simulation 137
  numerical 132
skeletal structure 806
skin 1631
  absorption 1355
  layer thickness 697
  protective devices 1371
Slater determinant 669
small molecule 339
Small method 261
Smidsrød-Haug approach 729
Snyder's polarity 851
Soave-Redlich-Kwong eqaution 213
soil 1149,1157
  contamination 1617
solar cell 1476
solid matrix 339
solubility 16
  behavior 3
  demands 318
  equilibrium 282
  parameter 103,111,198,243,253,261,347
  relative 285
  supercritical pressure 1425
solute 684
  electrostatic field 133
  mass concentration 684
  polarization 11
  pure 282
  surface area 286
solute-solute interaction 282,292,301
solute-solvent interaction 455,481
solution 768
  dilute 362
  free energy 285
  inhomogeneous 148
  lattice 125
  processing 956
solvation 282,759-760,807
  continuum 482
  dynamics 133
  effect 283
  energy 133,505,528,639,657,667
  free energy 132

function 137
preferential 613
process 133
selective 538
shell 17,132,284,288,301,450
solvatochromic 288
analysis 609
dyes 660,705
shift 639,655,660,665,674,677,738
solvatochromism 2,590,593,601,610,751
negative 602
solvent
acidity 607
activity 146,194,207
adsorbed on crystal faces 1116
allergen 1319
basicity 600
binary 505,513
cavity 288,291
chemical potential 125
choice 872-873,875,877,879
cleaning 952
concentration 388
cost 657
crystalline 346
definition 1201,1234
degradation 700
diffusion 344,353
donor 787
effect 281
emission 1232
evaporation time 697
expansibility 747
explicit effect 780
exposure 1078
extraction 923
extraction process 935
flux 342
good 569,685
hydrophilic 1152
implicit effect 780
interaction 1130
ionizing power 587
loss 931
maximum contaminant level 1164
maximum permissible concentration 1164
metabolism 1317
mixed 505
mixture 267,281,352,683
motions 133,137
movement 339
neurotoxicity 1327
occluded 1116
oxidation 1559
partitioning 1067,1152
penetration 1631
polarity 3,593,610,702,705
polarity scale 588
poor 569
power 101
probe 585-586
quality 346
recovery 1507,1509,1511,1513,1515,1517,1519,1521,1523,1525,1527,1529,1531,1533,1535,1537,1539,1541,1543
reduction 847
removal 386
replacement 847
residual 349,1125,1129
retention 570
route of entry 1355
selection 1
separation 693
solubility 1150
state 345
symmetric 318
toxicity 1097,1132
traces 701
transport 3
transport phenomena 339
treated 1556
uptake 348
vapor phase 196
solvent activity
data 198
solvents 1150
applications 1507
binary 318
chlorinated 1572
classification 1235
cleaning 1056
critical properties 1420
diffusion 305
new 1637
oxygenated 694
polar 683
residual 1067
spent 1555
theta 685
use 847
solvent-solute interaction 282
solvent-solvent interaction 282,284,455
solvent-tolerant bacteria 859
solvolysis 297
rate 584,587
solvophobic effect 285
sorption isotherm 177
sorption-desorption hysteresis 177
spatial ordering 419
scale 1191
specific gravity 1057
specification 1068
spectral
band 639

broadening 643
line 639
shift 585,639
spectroscopy
dielectric 581
UV 689
spillages 962,1149
spin 430
spinodal 131,148,183-184,190-192,400,408
spin-spin interaction 530
spontaneous abortion 1343
Spriggs law 362
stability 700
stabilization 700
starlike 690
Staverman relation 203
steam-enhanced extraction 1628
stereochemical evidence 789
stereocomplex 578
stereoregularity 578
steric
properties 785
repulsion 448
Stockmayer
fluid 138,767
model 462
Stockmayer-Casassa relation 180
stoichiometric equilibrium 283
Stokes shift 592,612
storage tanks 962
strain, solvent-tolerant 862
strands 691
stress cracking 1059
tensor 356
stripping 853,950
reactor 1623
styrene 1385
styrene-butadiene copolymer 848
styrene-butadiene-styrene 848
subsurface contamination 1617
suit materials 1633
sulfur dioxide 853
supercritical
fluid 2,202
technology 1420
$CO_2$ 854
supermolecule approach 674
surface
crazing 700
distribution 569
energy 566
free energy 565
inhomogeneity 566
isopotential 653
layers 339
lubricity 956
modifier 565
solubility 1131
tension 14,284,288,301,352,364,567,570, 648,690,898,901-902,1460
surfactant-aided remediation 1625
surgical gloves 956
Svedberg equation 187
swell ratio 309
swelling 128,305,339,565
binary solvents 318
data 327
equilibrium 128,130,189,308,318,344-345, 348
extreme 320
kinetics 310,347
maximum 326
mode 312
pressure 195,346
stage 313
syndiotactic 578
synergistic effect 267
system free energy 132

## T

tableting 1552
tacticity 778
tautomeric equilibrium 506
tautomerism 1,33
Taylor diffusion coefficient 1432
tear strength 850
TEM 689
temperature 341,349,353,798,852
temperature-shift factor 362
teratogenic 1334
teratogenicity 1132,1336,1341
teratology 1337
termination 779
1,1,2,2-tetrachloroethane 1396,1398
tetrachloroethylene 348,1149,1334,1336,1388, 1396, 1399
tetrachloromethane 2,813
tetrahedric coordination 16
tetrahydrofuran 347,569,686,694,696,700-701, 754,956,1401
texture 1119
TGA 396,1134
theophylline 282
thermal
capacity 126
desorption 1626
expansion 190,198
oxidation 1508
thermochromic coefficient 752
thermodynamic 360
equilibrium 148,157-158,176
work of adhesion 565
thermolysis 702
thickness 343,353

change 350
time scale 339
time-temperature superposition 362
titanium dioxide 1559
TLV 1079
toluene 127,185,188,334,341-342,344,347,365, 686,704,794,816,826-827,829-830,847-849, 851,854,881,952,957,959,1328,1334,1336, 1385,1388
tomography 1628
torsional angle 711
Toth isotherm 1437
toxcicodynamics 1315
toxic residues 339
Toxic Substances Control Act 933
toxicokinetics 1315
transesterification 877
transfer reaction 779
transition state 9,786,807
tribological properties 339
1,1,1-trichloroethane 854,914,959,1138,1398
1,1,2-trichloroethane 695,1396
trichloroethylene 854,914,952,1328,1387,1396, 1399
triethylamine 30
trinitrotoluene 1400
Trouton
entropy 762
rule 762
tubes 956
TWA 925

## U

UCST 268-269,272,1424
Uhlig model 284
ultrafiltration 694
ultrasonic absorption 190
unimolecular heterolysis 1
unit cell 755
uptake
dermal 1316
routes 1315
UV 788
curing 880
radiation 853
UV/Vis 2

## V

van der Waals
dispersion 469
envelope 673
equation 467
equation of state 1423
forces 16,1477,1509
interaction 451,875
radius 663
repulsion 647
solid 653
sphere 662
surface 21,445,663,721
volume 206
van Oss-Good's expression 565
van't Hoff parameters 558
vapor
data 161
degreasing 1056, 1233
measurement 147,149,154
pressure 157,262,365,1069
solubility 1084
vaporization
energy 198
entropy 762,765
heat 15,261
vapor-liquid equilibrium 198
varnishes 344,583,1169,1235
velocity field 356
vents 962
vibronic transitions 589
N-vinylpyrrolidone 703
viscoelasticity 357
viscosity 351,505,515,517,683,844,1069,1432, 1460
intrinsic 268,684,687
measurement 689
Newtonian 684
relative 685
specific 684,692
VOC 854,959,1196,1201,1546
calculation 1069
calculation methods 1205-1209
definition 1202,1241,1267
emission 388,848
exempt 1070
Vogel-Tammann-Fulcher equation 1474
volume
excluded 273,689
fraction 308,345,347,350
free 274,340-341
hydrodynamic 272
molar 341-342,345
vorticity 360
Vrentas-Duda theory 353,388

## W

wash plant 1556
wastes 1163
water 15,137,1149
waterproofing 855
wave function 423
Wertheim pertubation theory 208,750
wettability 570
wetting 496,898,902
index 916
Wilke-Chang equation 38

wood coatings 1646
workplace
monitoring 1079
regulations 925
wound dressing 956

**X**

XPS 565,567,569
X-ray 708,756,770
small angle scattering 147
xylene 127,186,341,696,700-701,847-848,851,
854,881,952,959,1385

**Y**

yellowing 881
Yvon-Born-Green approximation 461

**Z**

Z scale 589
zeolites 495,1514
Zimm
plot 183
equation 689
zwitterion 1471